MOLECULAR DETECTION OF ANIMAL VIRAL PATHOGENS

MOLECULAR DETECTION OF ANIMAL VIRAL PATHOGENS

EDITED BY

DONGYOU LIU

CRC Press
Taylor & Francis Group
Boca Raton London New York

CRC Press is an imprint of the
Taylor & Francis Group, an **informa** business

CRC Press
Taylor & Francis Group
6000 Broken Sound Parkway NW, Suite 300
Boca Raton, FL 33487-2742

First issued in paperback 2022

© 2016 by Taylor & Francis Group, LLC
CRC Press is an imprint of Taylor & Francis Group, an Informa business

No claim to original U.S. Government works

Version Date: 20160419

ISBN 13: 978-1-03-240243-7 (pbk)
ISBN 13: 978-1-4987-0036-8 (hbk)

DOI: 10.1201/b19719

Library of Congress Cataloging-in-Publication Data

Names: Liu, Dongyou, editor.
Title: Molecular detection of animal viral pathogens / [edited by] Dongyou Liu.
Description: Boca Raton : CRC Press/Taylor & Francis, 2016. | Includes bibliographical references and index.
Identifiers: LCCN 2016003205 (print) | LCCN 2016003549 (ebook) | ISBN 9781498700368 (alk. paper) | ISBN 9781498700375
Subjects: | MESH: Viruses--pathogenicity | Animals | Viruses--isolation & purification | Molecular Diagnostic Techniques--methods | Virus Diseases--diagnosis
Classification: LCC QR364 (print) | LCC QR364 (ebook) | NLM QW 160 | DDC 579.2--dc23
LC record available at http://lccn.loc.gov/2016003205

Visit the Taylor & Francis Web site at
http://www.taylorandfrancis.com

and the CRC Press Web site at
http://www.crcpress.com

Dedication

This book is dedicated to a group of virologists who possess not only superior ingenuity, but also remarkable benevolence. Without their help, compilation of such an all-encompassing volume on major animal viral pathogens will be unimaginable.

Contents

Preface..xv
Editor xvii
Contributors xix

Chapter 1 Introductory Remarks 1

    Dongyou Liu

SECTION I Positive-Sense RNA Viruses

Chapter 2 *Betanodavirus* 9

    Dongyou Liu

Chapter 3 Taura Syndrome Virus 17

    *Parin Chaivisuthangkura, Akapon Vaniksampanna, Phongthana Pasookhush,
    Siwaporn Longyant, and Paisarn Sithigorngul*

Chapter 4 Bee Paralysis Virus 27

    Panuwan Chantawannakul

Chapter 5 *Iflavirus* (Deformed Wing Virus)... 37

    Jessica M. Fannon and Eugene V. Ryabov

Chapter 6 Avian Encephalomyelitis Virus.. 47

    Mazhar I. Khan, Zhixun Xie, Theodore Girshick, and Zhiqin Xie

Chapter 7 Duck Hepatitis Virus.. 53

    Dabing Zhang and Ning Liu

Chapter 8 Foot-and-Mouth Disease Virus... 61

    Dongyou Liu

Chapter 9 Porcine Encephalomyocarditis Virus.. 67

    Sandra Blaise-Boisseau, Aurore Romey, and Labib Bakkali Kassimi

Chapter 10 Porcine Teschovirus 79

    Cristina Cano-Gómez and Miguel Ángel Jiménez-Clavero

Chapter 11 Swine Vesicular Disease Virus 89

    *Jovita Fernández-Pinero, Giulia Pezzoni, Cristina Cano-Gómez, Paloma Fernández-Pacheco,
    Emiliana Brocchi, and Miguel Ángel Jiménez-Clavero*

Chapter 12   Avian Astroviruses 101

  Victoria Smyth

Chapter 13   Hepatitis E Virus 111

  Ilaria Di Bartolo, Fabio Ostanello, and Franco Maria Ruggeri

Chapter 14   Feline Calicivirus 121

  Rudi Weiblen, Luciane Teresinha Lovato, and Andréia Henzel

Chapter 15   Porcine Caliciviruses 129

  Dongyou Liu

Chapter 16   Vesicular Exanthema of Swine Virus 135

  Nick J. Knowles, Begoña Valdazo-González, Britta A. Wood, Katarzyna Bachanek-Bankowska, and Donald P. King

Chapter 17   Avian Leukosis Virus 145

  Yongxiu Yao and Venugopal Nair

Chapter 18   Bovine Leukemia Virus 157

  Silvina E. Gutiérrez and Agustina Forletti

Chapter 19   Caprine Arthritis–Encephalitis Virus and Visna–Maedi Virus 167

  Nuria Barquero, Ana Domenech, and Esperanza Gomez-Lucia

Chapter 20   Equine Infectious Anemia Virus 177

  Caroline Leroux and R. Frank Cook

Chapter 21   Feline Immunodeficiency Virus 191

  Dongyou Liu

Chapter 22   Feline Leukemia Virus 197

  Dongyou Liu

Chapter 23   Reticuloendotheliosis Viruses 205

  Dongyou Liu

Chapter 24   Border Disease Virus 211

  Jieyuan Jiang and Wenliang Li

Chapter 25   Bovine Viral Diarrhea Viruses 1 and 2 223

  Benjamin W. Newcomer

Chapter 26   Classical Swine Fever Virus 233

  Dongyou Liu

Chapter 27 Louping Ill Virus 241

Nicholas Johnson

Chapter 28 West Nile Virus 247

Dongyou Liu

Chapter 29 Eastern and Western Equine Encephalitis Viruses 255

Norma P. Tavakoli and Laura D. Kramer

Chapter 30 Venezuelan Equine Encephalitis Virus 269

María Belén Pisano, Marta Contigiani, and Viviana Ré

Chapter 31 Equine Arteritis Virus 277

Udeni B.R. Balasuriya

Chapter 32 Porcine Reproductive and Respiratory Syndrome Virus 287

Dongyou Liu

Chapter 33 Yellow Head Complex Viruses, including Gill-Associated Virus 295

James Munro

Chapter 34 Avian Infectious Bronchitis Virus 307

Vagner Ricardo Lunge, Aline Padilha de Fraga, and Nilo Ikuta

Chapter 35 Bovine Coronavirus 317

Dongyou Liu

Chapter 36 Feline Enteric Coronavirus and Feline Infectious Peritonitis Virus 323

Niels C. Pedersen

Chapter 37 Porcine Epidemic Diarrhea Virus 331

Dongyou Liu

Chapter 38 Toroviruses 339

Dongyou Liu

SECTION II Negative-Sense RNA Viruses

Chapter 39 Avian Bornavirus 347

Francesca Sidoti, Maria Lucia Mandola, Francesca Rizzo, Rossana Cavallo, and Cristina Costa

Chapter 40 Bovine Ephemeral Fever Virus 355

Hakan Bulut and Ahmet Kursat Azkur

Chapter 41 Rabies Virus 361

Dongyou Liu

Chapter 42 Vesicular Stomatitis Virus 369

Dongyou Liu and Sándor Belák

Chapter 43 Avian Influenza Virus 377

Erica Spackman

Chapter 44 Equine Influenza Virus 383

Thomas M. Chambers and Udeni B.R. Balasuriya

Chapter 45 Infectious Salmon Anemia Virus 393

Dongyou Liu

Chapter 46 Swine Influenza A Virus 399

Jianqiang Zhang, Phillip Gauger, and Karen Harmon

Chapter 47 Avian Metapneumovirus 407

José Francisco Rivera-Benitez, Jazmín De la Luz-Armendáriz, Alberto Bravo-Blas,
Luis Gómez-Núñez, and Humberto Ramírez-Mendoza

Chapter 48 Bovine Respiratory Syncytial Virus 413

Sara Hägglund and Jean François Valarcher

Chapter 49 Canine Distemper Virus 421

Rebecca P. Wilkes

Chapter 50 Hendra Virus 431

Dongyou Liu

Chapter 51 Menangle Virus 437

Timothy R. Bowden

Chapter 52 Newcastle Disease Virus 447

Dongyou Liu

Chapter 53 Nipah Virus 455

Supaporn Wacharapluesadee, Akanitt Jittmittraphap, Sangchai Yingsakmongkon, and
Thiravat Hemachudha

Chapter 54 Peste des Petits Ruminants Virus 467

Pam Dachung Luka

Chapter 55 Porcine Rubulavirus (Blue Eye Disease Virus)...477

José Iván Sánchez Betancourt and María Elena Trujillo Ortega

Chapter 56 Rinderpest Virus 485

Anke Brüning-Richardson

Chapter 57 Ebola Virus...495

Baochuan Lin and Anthony P. Malanoski

Chapter 58 Marburg Virus...507

Junping Yu and Hongping Wei

SECTION III   Negative- and Ambi-Sense RNA Viruses

Chapter 59 Akabane Virus 521

Akiko Uema, Yohsuke Ogawa, and Hiroomi Akashi

Chapter 60 Hantaviruses...533

Dongyou Liu

Chapter 61 Nairobi Sheep Disease Virus 543

Devendra T. Mourya and Pragya D. Yadav

Chapter 62 Rift Valley Fever Virus 553

William C. Wilson, Natasha N. Gaudreault, and Mohammad M. Hossain

Chapter 63 Schmallenberg Virus...563

Dongyou Liu

Chapter 64 Arenaviruses..569

Randal J. Schoepp and Aileen E. O'Hearn

Chapter 65 Picobirnaviruses..585

*Yashpal S. Malik, Naveen Kumar, Kuldeep Dhama, Krisztián Banyai,
Joana D'Arc Pereira Mascarenhas, and Raj Kumar Singh*

Chapter 66 Infectious Myonecrosis Virus 593

Maria Erivalda Farias de Aragão, Maria Verônyca Coelho Mello, and Maria de Lourdes Oliveira Otoch

Chapter 67 Infectious Bursal Disease Virus..601

Maja Velhner, Dejan Vidanović, and Ivan Dobrosavljević

Chapter 68 African Horse Sickness Virus...609

Dongyou Liu and Frank W. Austin

Chapter 69 Bluetongue Virus...619

    Fan Lee

Chapter 70 Epizootic Hemorrhagic Disease Virus 629

    Srivishnupriya Anbalagan

Chapter 71 Equine Encephalosis Virus 637

    Dongyou Liu

Chapter 72 Rotavirus 643

    Helen O'Shea, P.J. Collins, Lynda Gunn, Barbara A. Blacklaws, John McKillen, and
    Miren Iturriza Gomara

SECTION IV DNA Viruses

Chapter 73 Beak and Feather Disease Virus 659

    Ana Margarida Henriques, Teresa Fagulha, and Miguel Fevereiro

Chapter 74 Porcine Circovirus..671

    Yu-Liang Huang, Hui-Wen Chang, Victor Fei Pang, and Chian-Ren Jeng

Chapter 75 Aleutian Mink Disease Virus 679

    Trine Hammer Jensen and Åse Uttenthal

Chapter 76 Canine Parvovirus..687

    S. Parthiban

Chapter 77 Goose Parvovirus...693

    Yakun Luo and Shangjin Cui

Chapter 78 *Penaeus monodon* Hepandensovirus 699

    Dongyou Liu

Chapter 79 *Penaeus stylirostris* Penstyldensovirus...705

    Norma A. Macías-Rodríguez, Erika Camacho-Beltrán, Edgar Rodríguez-Negrete,
    Norma E. Leyva-López, and Jesús Méndez-Lozano

Chapter 80 Porcine Parvovirus..713

    Zhengyang Wang, Yu Zhu, Miaomiao Kong, and Shangjin Cui

Chapter 81 Polyomaviruses...719

    Wojciech Kozdruń

Chapter 82 Bovine Papillomaviruses...727

    M.A.R. Silva, R.C.P. Lima, M.N. Cordeiro, G. Borzacchiello, and A.C. Freitas

Chapter 83 Poultry Adenoviruses 735

Győző L. Kaján

Chapter 84 *Betaherpesvirinae* (*Suid herpesvirus* 2) 747

Timothy J. Mahony

Chapter 85 Iltovirus (Infectious Laryngotracheitis Virus) 753

Joseph J. Giambrone and Shan-Chia Ou

Chapter 86 Macavirus 765

Richa Sood, Kh. Victoria Chanu, and Sandeep Bhatia

Chapter 87 Marek's Disease Virus (*Gallid herpesvirus* 2) and Herpesvirus of Turkey (*Melagrid herpesvirus* 1) 773

Grzegorz Woźniakowski and Jowita Samanta Niczyporuk

Chapter 88 Cypriniviruses 783

Kenneth A. McColl, Agus Sunarto, and Lijuan Li

Chapter 89 *Ictalurid Herpesvirus* 1 797

Larry Hanson

Chapter 90 Abalone Herpesvirus 807

Mark St. J. Crane, Kenneth A. McColl, Jeff A. Cowley, Kevin Ellard, Keith W. Savin, Serge Corbeil, Nicholas J.G. Moody, Mark Fegan, and Simone Warner

Chapter 91 African Swine Fever Virus 817

Dongyou Liu

Chapter 92 Megalocytiviruses 825

Chun-Shun Wang and Chiu-Ming Wen

Chapter 93 Avipoxvirus 837

Bimalendu Mondal

Chapter 94 Capripoxviruses (Sheeppox, Goatpox and Lumpy Skin Disease Viruses) 845

E.H. Venter

Chapter 95 Myxoma Virus 855

June Liu and Peter Kerr

Chapter 96 Orthopoxvirus 869

Vinayagamurthy Balamurugan, Gnanavel Venkatesan, and Veerakyathappa Bhanuprakash

Chapter 97   Parapoxvirus... 881

   Graziele Oliveira, Galileu Costa, Felipe Assis, Ana Paula Franco-Luiz, Giliane Trindade,
   Erna Kroon, Filippo Turrini, and Jonatas Abrahão

Chapter 98   White Spot Syndrome Virus 891

   Dongyou Liu

SECTION V   Prions

Chapter 99   Bovine Spongiform Encephalopathy.. 901

   Akikazu Sakudo and Takashi Onodera

Chapter 100 Chronic Wasting Disease 913

   Akikazu Sakudo and Takashi Onodera

Index... 923

Preface

Being the smallest life form of all, viruses possess inadequate structural frameworks to lead an independent life of their own and rely on the living cells of other organisms (including bacteria, fungi, protozoa, helminths, insects, animals, humans, and plants) for replication and maintenance. The process of viruses settling inside their host organisms is commonly referred to as infection. As infectious agents, viruses are remarkably ingenious and highly efficient. While many viruses establish various symbiotic relationships with their hosts and play a beneficial role in augmenting genetic diversity, others may be extremely harmful and induce pathological changes in and even death to their hosts.

Animals constitute one of the largest and most diverse groups of living organisms on Earth. The significance of animals in the ecological balance of the environment and the well-being of human society is indisputable. Considering the devastating effects of pathogenic viruses on their animal hosts, it is crucial that animal-infecting viruses are rapidly and accurately detected and identified so that appropriate and timely control and prevention measures are implemented. In addition, as many viruses have the capacity to move from animals to humans, application of improved diagnostic techniques for animal-infecting viruses provides an effective early-warning system for any possible zoonotic infections that may exert a toll on human society.

With these issues in sight, this book aims to present an expert summary on the state-of-the-art diagnostic approaches for major animal viral pathogens. Each chapter contains a concise overview of the classification, morphology, epidemiology, clinical features, and diagnosis of one or a group of related viral pathogens; an outline of clinical sample collection and preparation procedures; a judicious selection of ready-to-use molecular detection protocols; and a discussion on further research requirements relating to improved diagnosis and control. With contributions from specialists in respective animal viral pathogen research, this book offers a trustworthy reference on molecular detection of major animal viral pathogens; an indispensable tool for medical, veterinary, and industrial laboratory scientists involved in virus determination; a convenient textbook for undergraduate and graduate students majoring in virology; and an essential guide for upcoming and experienced laboratory scientists wishing to acquire and polish their skills in molecular diagnosis of animal viral diseases.

Given the diverse nature of animal viral pathogens, a comprehensive book such as this is obviously beyond the capacity of an individual's undertaking. I was fortunate and honored to have a large group of virologists as the chapter contributors, whose detailed knowledge and technical insights on animal viral pathogen research have greatly enhanced its credibility. Additionally, the professionalism and dedication of CRC Press Executive Editor Chuck Crumly and Senior Project Coordinator Jill Jurgensen have increased its overall appeal. Finally, the understanding and support from my family—Liling Ma, Brenda, and Cathy—have helped maintain my focus during the compilation of this all-inclusive book.

Editor

Dongyou Liu, PhD, undertook veterinary science education at Hunan Agricultural University, China, and postgraduate training at the University of Melbourne, Victoria, Australia. Over the past two decades, he worked at several research and clinical laboratories in Australia and the United States of America, with focuses on molecular characterization and virulence determination of microbial pathogens, such as ovine footrot bacterium (*Dichelobacter nodosus*), dermatophyte fungi (*Trichophyton, Microsporum,* and *Epidermophyton*), and listeriae (*Listeria* spp.), and the development of nucleic acid–based quality assurance modules for security-sensitive and emerging viral pathogens. Dr. Liu is the primary author of more than 50 original research and review articles in various international journals, a contributor of 140 book chapters, and the editor of *Handbook of Listeria Monocytogenes* (2008), *Handbook of Nucleic Acid Purification* (2009), *Molecular Detection of Foodborne Pathogens* (2009), *Molecular Detection of Human Viral Pathogens* (2010), *Molecular Detection of Human Bacterial Pathogens* (2011), *Molecular Detection of Human Fungal Pathogens* (2011), Molecular Detection of Human Parasitic Pathogens (2012), and *Manual of Security Sensitive Microbes and Toxins* (2014), which were published by CRC Press. He is also a coeditor for *Molecular Medical Microbiology, Second Edition* (2014, Elsevier).

Contributors

Jonatas Abrahão
Department of Microbiology
Federal University of Minas Gerais
Belo Horizonte, Brazil

Hiroomi Akashi
Department of Veterinary Microbiology
University of Tokyo
Tokyo, Japan

Srivishnupriya Anbalagan
Newport Laboratories Inc.
Worthington, Minnesota

Felipe Assis
Department of Microbiology
Federal University of Minas Gerais
Belo Horizonte, Brazil

Frank W. Austin
Department of Microbiology
Federal University of Minas Gerais
Belo Horizonte, Brazil

Ahmet Kursat Azkur
Department of Virology
Kirikkale University
Kirikkale, Turkey

Katarzyna Bachanek-Bankowska
Vesicular Disease Reference Laboratory
The Pirbright Institute
Surrey, United Kingdom

Vinayagamurthy Balamurugan
National Institute of Veterinary Epidemiology and Disease
    Informatics
Indian Council of Agricultural Research
Bengaluru, Karnataka, India

Udeni B.R. Balasuriya
Department of Veterinary Science
University of Kentucky
Lexington, Kentucky

Krisztián Banyai
Centre for Agricultural Research
Institute for Veterinary Medical Research
Budapest, Hungary

Nuria Barquero
Departamento de Sanidad Animal
Universidad Complutense
Madrid, Spain

Sándor Belák
Department of Biomedical Sciences and Veterinary Public
    Health
Swedish University of Agricultural Sciences
Uppsala, Sweden

José Iván Sánchez Betancourt
Departamento de Medicina y Zootecnia de Cerdos
Universidad Nacional Autónoma de México
Distrito Federal, México

Veerakyathappa Bhanuprakash
Indian Veterinary Research Institute
Indian Council of Agricultural Research
Bengaluru, Karnataka, India

Sandeep Bhatia
Immunology Laboratory
National Institute of High Security Animal Diseases
Bhopal, Madhya Pradesh, India

Barbara A. Blacklaws
Department of Veterinary Medicine
University of Cambridge
Cambridge, United Kingdom

Sandra Blaise-Boisseau
Animal Health Laboratory
Paris-Est University, ANSES
Maisons-Alfort, France

G. Borzacchiello
University of Naples Federico II
Naples, Italy

Timothy R. Bowden
Australian Animal Health Laboratory
CSIRO Health and Biosecurity
Geelong, Victoria, Australia

Alberto Bravo-Blas
Centre for Immunobiology
University of Glasgow
Scotland, United Kingdom

Emiliana Brocchi
Istituto Zooprofilattico sperimentale della Lombardia e
    dell´Emilia Romagna
Brescia, Italy

Anke Brüning-Richardson
Leeds Institute of Cancer and Pathology
University of Leeds
Leeds, United Kingdom

Hakan Bulut
Department of Virology
Firat University
Elazig, Turkey

Erika Camacho-Beltrán
National Polytechnic Institute
Unidad Sinaloa
Guasave, Mexico

Cristina Cano-Gómez
Animal Health Research Centre
National Institute for Agricultural and Food Research and
    Technology
Madrid, Spain

Rossana Cavallo
Microbiology and Virology Unit
University Hospital "Città della Salute e della Scienza
    di Torino"
and
Department of Public Health and Pediatrics
University of Turin
Turin, Italy

Parin Chaivisuthangkura
Department of Biology
Srinakharinwirot University
Bangkok, Thailand

Thomas M. Chambers
Department of Veterinary Science
University of Kentucky
Lexington, Kentucky

Hui-Wen Chang
Graduate Institute of Molecular and Comparative Pathobiology
National Taiwan University
Taipei, Taiwan

Panuwan Chantawannakul
Department of Biology
Chiang Mai University
Chiang Mai, Thailand

Kh. Victoria Chanu
Immunology Laboratory
National Institute of High Security Animal Diseases,
Bhopal, Madhya Pradesh, India

P.J. Collins
Department of Biological Sciences
Cork Institute of Technology
Cork, Ireland

Marta Contigiani
Instituto de Virología "Dr. J.M. Vanella"
Universidad Nacional de Córdoba
Córdoba, Argentina

R. Frank Cook
Department of Veterinary Science
University of Kentucky
Lexington, Kentucky

Serge Corbeil
Australian Animal Health Laboratory
Commonwealth Scientific and Industrial Research Organisation
Geelong, Victoria, Australia

M.N. Cordeiro
Federal University of Pernambuco
Cidade Universitária
Recife, Brazil

Cristina Costa
Microbiology and Virology Unit
University Hospital "Città della Salute e della Scienza
    di Torino"
and
Department of Public Health and Pediatrics
University of Turin
Turin, Italy

Galileu Costa
Department of Microbiology
Federal University of Minas Gerais
Belo Horizonte, Brazil

Jeff A. Cowley
Queensland Bioscience Precinct
CSIRO Agriculture Flagship
St. Lucia, Queensland, Australia

Mark St. J. Crane
Australian Animal Health Laboratory
Commonwealth Scientific and Industrial Research
    Organisation
Geelong, Victoria, Australia

Shangjin Cui
Division of Pet Infectious Diseases
Chinese Academy of Agricultural Sciences
Beijing, People's Republic of China

Maria Erivalda Farias de Aragão
Department of Biology
University of State of Ceara
Fortaleza, Brazil

Jazmín De la Luz-Armendáriz
Department of Microbiology and Immunology
National Autonomous University of Mexico
Mexico City, Mexico

Kuldeep Dhama
Division of Pathology
Indian Veterinary Research Institute
Bareilly, Uttar Pradesh, India

Ilaria Di Bartolo
Department of Veterinary Public Health and Food Safety
Istituto Superiore di Sanità
Rome, Italy

Ivan Dobrosavljević
Veterinary Institute Požarevac
Požarevac, Serbia

Ana Domenech
Departamento de Sanidad Animal
Universidad Complutense
Madrid, Spain

Kevin Ellard
Department of Primary Industries Parks Water
    and Environment
New Town, Tasmania, Australia

Teresa Fagulha
Laboratório de Virologia
Instituto Nacional de Investigação Agrária e Veterinária
Oeiras, Portugal

Jessica M. Fannon
School of Life Sciences
University of Warwick
Coventry, United Kingdom

Mark Fegan
Department of Environment and Primary Industries
AgriBio Centre
Bundoora, Victoria, Australia

Paloma Fernández-Pacheco
Animal Health Research Centre
National Institute for Agricultural and Food Research and
    Technology
Madrid, Spain

Jovita Fernández-Pinero
Animal Health Research Centre
National Institute for Agricultural and Food Research and
    Technology
Madrid, Spain

Miguel Fevereiro
Laboratório de Virologia
Instituto Nacional de Investigação Agrária e Veterinária
Oeiras, Portugal

Agustina Forletti
Laboratorio Biológico de Tandil S.R.L.
Tandil, Argentina

Aline Padilha de Fraga
Molecular Diagnostics Laboratory
Universidade Luterana do Brasil
Canoas, Brazil

Ana Paula Franco-Luiz
Department of Microbiology
Federal University of Minas Gerais
Belo Horizonte, Brazil

A.C. Freitas
Federal University of Pernambuco
Cidade Universitária
Recife, Brazil

Natasha N. Gaudreault
School of Veterinary Medicine—
Diagnostic Medicine and Pathobiology
Kansas State University
Manhattan, Kansas

Phillip Gauger
Department of Veterinary Diagnostic and Production
    Animal Medicine
Iowa State University
Ames, Iowa

Joseph J. Giambrone
Department of Poultry Science
Auburn University
Auburn, Alabama

Theodore Girshick
Charles River Laboratory
Storrs, Connecticut

Miren Iturriza Gomara
Institute of Infection and Global Health
University of Liverpool
Liverpool, United Kingdom

Esperanza Gomez-Lucia
Departamento de Sanidad Animal
Universidad Complutense
Madrid, Spain

Luis Gómez-Núñez
Department of Virology
National Institute of Forestry, Agricultural and Livestock
    Research
Mexico City, Mexico

Lynda Gunn
Department of Biological Sciences
Cork Institute of Technology
Cork, Ireland

Silvina E. Gutiérrez
Facultad de Ciencias Veterinarias, Universidad Nacional del
    Centro de la provincia de Buenos Aires
Aires, Argentina

Sara Hägglund
Department of Clinical Sciences
Swedish University of Agricultural Sciences
Uppsala, Sweden

Larry Hanson
Department of Basic Sciences
Mississippi State University
Mississippi State, Mississippi

Karen Harmon
Department of Veterinary Diagnostic and Production
    Animal Medicine
Iowa State University
Ames, Iowa

Thiravat Hemachudha
World Health Organization Collaborating Centre
    for Research and Training on Viral Zoonoses
Chulalongkorn University
Bangkok, Thailand

Ana Margarida Henriques
Laboratório de Virologia
Instituto Nacional de Investigação Agrária e Veterinária
Oeiras, Portugal

Andréia Henzel
Laboratório de Microbiologia Molecular
Universidade Feevale
Feevale, Brazil

Mohammad M. Hossain
School of Veterinary Medicine—
Diagnostic Medicine and Pathobiology
Kansas State University
Manhattan, Kansas

Yu-Liang Huang
Division of Hog Cholera Research
Animal Health Research Institute
New Taipei City, Taiwan, Republic of China

Nilo Ikuta
Molecular Diagnostics Laboratory
Universidade Luterana do Brasil
Canoas, Brazil

Chian-Ren Jeng
Graduate Institute of Molecular and Comparative
    Pathobiology
National Taiwan University
Taipei, Taiwan, Republic of China

Trine Hammer Jensen
Department of Chemistry and Bioscience
Aalborg University/Aalborg Zoo
Aalborg, Denmark

Jieyuan Jiang
Key Laboratory of Veterinary Biological Engineering
    and Technology
Jiangsu Academy of Agricultural Sciences
Nanjing, People's Republic of China

and

Jiangsu Co-Innovation Center for Prevention and Control of
    Important Animal Infectious Diseases and Zoonoses
Yangzhou, People's Republic of China

Miguel Ángel Jiménez-Clavero
Center for Animal Health Research
National Institute for Agricultural and Food Research
    and Technology
Madrid, Spain

Akanitt Jittmittraphap
Department of Microbiology and Immunology
Mahidol University
and
World Health Organization Collaborating Centre
    for Research and Training on Viral Zoonoses
Chulalongkorn University
Bangkok, Thailand

Nicholas Johnson
Animal and Plant Health Agency
Surrey, United Kingdom

Győző L. Kaján
Institute for Veterinary Medical Research
Hungarian Academy of Sciences
Budapest, Hungary

Labib Bakkali Kassimi
Animal Health Laboratory
Paris-Est University, ANSES
Maisons-Alfort, France

Peter Kerr
School of Biological Sciences
University of Sydney
Sydney, New South Wales, Australia

Mazhar I. Khan
Department of Pathobiology and Veterinary Science
University of Connecticut
Storrs, Connecticut

Donald P. King
Vesicular Disease Reference Laboratory
The Pirbright Institute
Surrey, United Kingdom

Nick J. Knowles
Vesicular Disease Reference Laboratory
The Pirbright Institute
Surrey, United Kingdom

Miaomiao Kong
Division of Swine Infectious Diseases
Chinese Academy of Agricultural Sciences
Harbin, Heilongjiang, People's Republic of China

Wojciech Kozdruń
Department of Poultry Viral Diseases
National Veterinary Research Institute
Pulawy, Poland

Laura D. Kramer
Wadsworth Center
New York State Department of Health
and
Department of Biomedical Sciences
University at Albany
Albany, New York

Erna Kroon
Department of Microbiology
Federal University of Minas Gerais
Belo Horizonte, Brazil

Naveen Kumar
OIE Reference Laboratory for Avian Influenza
National Institute of High Security Animal Diseases
Bhopal, Madhya Pradesh, India

Fan Lee
Division of Epidemiology
Animal Health Research Institute
New Taipei City, Taiwan

Caroline Leroux
Retrovirus and Comparative Pathology
Université de Lyon
Lyon, France

Norma E. Leyva-López
National Polytechnic Institute
Unidad Sinaloa
Guasave, Mexico

Lijuan Li
College of Fisheries
Huazhong Agricultural University
Wuhan, People's Republic of China

Wenliang Li
Key Laboratory of Veterinary Biological Engineering
    and Technology
Jiangsu Academy of Agricultural Sciences
Nanjing, People's Republic of China

and

Jiangsu Co-Innovation Center for Prevention and Control
    of Important Animal Infectious Diseases and Zoonoses
Yangzhou, People's Republic of China

R.C.P. Lima
Federal University of Pernambuco
Cidade Universitária
Recife, Brazil

Baochuan Lin
Center for Bio/Molecular Science and Engineering
U.S. Naval Research Laboratory
Washington, DC

Dongyou Liu
RCPAQAP
Sydney, New South Wales, Australia

June Liu
Black Mountain Laboratories
CSIRO Biosecurity Flagship
Acton, Australian Capital Territory, Australia

Ning Liu
College of Veterinary Medicine
China Agricultural University
Beijing, People's Republic of China

Siwaporn Longyant
Department of Biology
Srinakharinwirot University
Bangkok, Thailand

Luciane Teresinha Lovato
Departamento de Microbiologia e Parasitologia
Universidade Federal de Santa Maria
and
Departamento de Medicina Veterinária Preventiva
Universidade Federal de Santa Maria
Santa Maria, Brazil

Pam Dachung Luka
Applied Biotechnology Division
National Veterinary Research Institute
Plateau State, Nigeria

Vagner Ricardo Lunge
Molecular Diagnostics Laboratory
Universidade Luterana do Brasil
Canoas, Brazil

Yakun Luo
Division of Swine Infectious Diseases
Chinese Academy of Agricultural Sciences
Harbin, Heilongjiang, People's Republic of China

Norma A. Macías-Rodríguez
National Polytechnic Institute
Unidad Sinaloa
Guasave, Mexico

Timothy J. Mahony
Centre for Animal Science
University of Queensland
St. Lucia, Queensland, Australia

Anthony P. Malanoski
Center for Bio/Molecular Science and Engineering
U.S. Naval Research Laboratory
Washington, DC

Yashpal S. Malik
Division of Biological Standardization
Indian Veterinary Research Institute
Bareilly, Uttar Pradesh, India

Maria Lucia Mandola
Laboratory of Avian Pathology and Molecular Virology
Istituto zooprofilattico sperimentale del Piemonte
Turin, Italy

Joana D'Arc Pereira Mascarenhas
Virology Section
Evandro Chagas Institute
Ananindeua, Brazil

Kenneth A. McColl
Australian Animal Health Laboratory
Commonwealth Scientific and Industrial Research
    Organisation
Geelong, Victoria, Australia

John McKillen
Veterinary Sciences Division
Agri-Foods and Biosciences Institute
Belfast, United Kingdom

Maria Verônyca Coelho Mello
Department of Nutrition
University of State of Ceara
Fortaleza, Brazil

Jesús Méndez-Lozano
National Polytechnic Institute
Unidad Sinaloa
Guasave, Mexico

Bimalendu Mondal
Eastern Regional Station
Indian Veterinary Research Institute
Kolkata, West Bengal, India

Nicholas J.G. Moody
Australian Animal Health Laboratory
Commonwealth Scientific and Industrial Research
    Organisation
Geelong, Victoria, Australia

Devendra T. Mourya
Maximum Containment Laboratory
National Institute of Virology
Pune, Maharashtra, India

James Munro
School of Animal and Veterinary Sciences
University of Adelaide
Adelaide, South Australia, Australia

Venugopal Nair
The Pirbright Institute
Surrey, United Kingdom

Benjamin W. Newcomer
Department of Pathobiology
Auburn University
Auburn, Alabama

Jowita Samanta Niczyporuk
Department of Poultry Viral Diseases
National Veterinary Research Institute
Pulawy, Poland

Yohsuke Ogawa
Department of Veterinary Microbiology
University of Tokyo
Tokyo, Japan

Aileen E. O'Hearn
Diagnostic Systems Division
U.S. Army Medical Research Institute of Infectious
    Diseases
Fort Detrick, Maryland

Graziele Oliveira
Department of Microbiology
Federal University of Minas Gerais
Belo Horizonte, Brazil

Takashi Onodera
Research Center for Food Safety
University of Tokyo
Tokyo, Japan

María Elena Trujillo Ortega
Departamento de Medicina y Zootecnia de Cerdos
Universidad Nacional Autónoma de México
Distrito Federal, México

Helen O'Shea
Department of Biological Sciences
Cork Institute of Technology
Cork, Ireland

Fabio Ostanello
Department of Veterinary Medical Sciences
University of Bologna
Bologna, Italy

Maria de Lourdes Oliveira Otoch
Department of Biology
University of State of Ceara
Fortaleza, Brazil

Shan-Chia Ou
Department of Poultry Science
Auburn University
Auburn, Alabama

Victor Fei Pang
Graduate Institute of Molecular and Comparative
    Pathobiology
National Taiwan University
Taipei, Taiwan, Republic of China

S. Parthiban
Department of Veterinary Microbiology
Veterinary College and Research Institute
Tirunelveli, Tamil Nadu, India

Phongthana Pasookhush
Department of Biology
Srinakharinwirot University
Bangkok, Thailand

Niels C. Pedersen
School of Veterinary Medicine
University of California, Davis
Davis, California

Giulia Pezzoni
Istituto Zooprofilattico sperimentale della Lombardia e
    dell´Emilia Romagna
Brescia, Italy

María Belén Pisano
Instituto de Virología "Dr. J.M. Vanella"
Universidad Nacional de Córdoba
Córdoba, Argentina

Humberto Ramírez-Mendoza
Department of Microbiology and Immunology
National Autonomous University of Mexico
Mexico City, Mexico

Viviana Ré
Instituto de Virología "Dr. J.M. Vanella"
Universidad Nacional de Córdoba
Córdoba, Argentina

José Francisco Rivera-Benitez
Department of Virology
National Institute of Forestry, Agricultural and Livestock
    Research
Mexico City, Mexico

Francesca Rizzo
Laboratory of Avian Pathology and Molecular Virology
Istituto Zooprofilattico Sperimentale del Piemonte
Turin, Italy

Edgar Rodríguez-Negrete
National Polytechnic Institute
Unidad Sinaloa
Guasave, Mexico

Aurore Romey
Animal Health Laboratory
Paris-Est University, ANSES
Maisons-Alfort, France

Franco Maria Ruggeri
Department of Veterinary Public Health and Food Safety
Istituto Superiore di Sanità
Rome, Italy

Eugene V. Ryabov
School of Life Sciences
University of Warwick
Coventry, United Kingdom

Akikazu Sakudo
Laboratory of Biometabolic Chemistry
University of the Ryukyus
Nishihara, Japan

Keith W. Savin
Department of Environment and Primary Industries
AgriBio Centre
Bundoora, Victoria, Australia

Randal J. Schoepp
Diagnostic Systems Division
U.S. Army Medical Research Institute of Infectious Diseases
Fort Detrick, Maryland

Francesca Sidoti
Microbiology and Virology Unit
University Hospital "Città della Salute e della Scienza
    di Torino"
and
Department of Public Health and Pediatrics
University of Turin
Turin, Italy

M.A.R. Silva
Federal Institute of Education Science and Technology of
    Paraiba
Paraíba, Brazil

Raj Kumar Singh
Veterinary Biotechnology
Indian Veterinary Research Institute
Bareilly, Uttar Pradesh, India

Paisarn Sithigorngul
Department of Biology
Srinakharinwirot University
Bangkok, Thailand

Victoria Smyth
Veterinary Sciences Division
Agri-Foods and Biosciences Institute
Belfast, United Kingdom

Richa Sood
Animal Bio-Containment Facility
National Institute of High Security Animal Diseases
Bhopal, Madhya Pradesh, India

Erica Spackman
Southeast Poultry Research Laboratory
U.S. Department of Agriculture
Athens, Georgia

Agus Sunarto
Fish Health Research Laboratory
Centre for Aquaculture Research and Development
Jakarta, Indonesia

and

Australian Animal Health Laboratory
Commonwealth Scientific and Industrial Research
Organisation
Geelong, Victoria, Australia

Norma P. Tavakoli
Wadsworth Center
New York State Department of Health
and
Department of Biomedical Sciences
University at Albany
Albany, New York

Giliane Trindade
Department of Microbiology
Federal University of Minas Gerais
Belo Horizonte, Brazil

Filippo Turrini
San Raffaele Scientific Institute
Milan, Italy

Akiko Uema
Department of Infection Control and Disease Prevention
University of Tokyo
Tokyo, Japan

Åse Uttenthal
National Veterinary Institute
Technical University of Denmark
Kalvehave, Denmark

Jean François Valarcher
Department of Clinical Sciences
Swedish University of Agricultural Sciences
Uppsala, Sweden

Begoña Valdazo-González
Division of Virology
The National Institute for Biological Standards and Control
Hertfordshire, United Kingdom

Akapon Vaniksampanna
Department of Biology
Srinakharinwirot University
Bangkok, Thailand

Maja Velhner
Scientific Veterinary Institute "Novi Sad"
Novi Sad, Serbia

Gnanavel Venkatesan
Division of Virology
Indian Veterinary Research Institute
Mukteswar, Uttarakhand, India

E.H. Venter
Department of Veterinary Tropical Diseases
University of Pretoria
Onderstepoort, South Africa

Dejan Vidanović
Veterinary Institute Kraljevo
Kraljevo, Serbia

Supaporn Wacharapluesadee
World Health Organization Collaborating Centre for
    Research and Training on Viral Zoonoses
Chulalongkorn University
Bangkok, Thailand

Chun-Shun Wang
Department of Life Sciences
National University of Kaohsiung
Kaohsiung, Taiwan, Republic of China

Zhengyang Wang
Division of Swine Infectious Diseases
Chinese Academy of Agricultural Sciences
Harbin, Heilongjiang, People's Republic of China

Simone Warner
Department of Environment and Primary Industries
AgriBio Centre
Bundoora, Victoria, Australia

Hongping Wei
Key Laboratory of Special Pathogens and Biosafety
Chinese Academy of Sciences
Wuhan, Hubei, People's Republic of China

Rudi Weiblen
Departamento de Medicina Veterinária Preventiva
Universidade Federal de Santa Maria
Santa Maria, Brazil

Chiu-Ming Wen
Department of Life Sciences
National University of Kaohsiung
Kaohsiung, Taiwan, Republic of China

Rebecca P. Wilkes
Veterinary Diagnostic and Investigational Laboratory
University of Georgia
Tifton, Georgia

William C. Wilson
Arthropod-Borne Animal Diseases Research Unit
U.S. Department of Agriculture
Manhattan, Kansas

Britta A. Wood
Vesicular Disease Reference Laboratory
The Pirbright Institute
Surrey, United Kingdom

Grzegorz Woźniakowski
Department of Poultry Viral Diseases
National Veterinary Research Institute
Pulawy, Poland

Zhiqin Xie
Department of Biotechnology
Guangxi Veterinary Research Institute
Nanning, Guangxi, People's Republic of China

Zhixun Xie
Institute Department of Biotechnology
Guangxi Veterinary Research
Nanning, Guangxi, People's Republic of China

Pragya D. Yadav
Maximum Containment Laboratory
National Institute of Virology
Pune, Maharashtra, India

Yongxiu Yao
The Pirbright Institute
Surrey, United Kingdom

Sangchai Yingsakmongkon
Department of Microbiology and Immunology
Kasetsart University
and
World Health Organization Collaborating Centre for
Research and Training on Viral Zoonoses
Chulalongkorn University
Bangkok, Thailand

Junping Yu
Key Laboratory of Special Pathogens and Biosafety
Chinese Academy of Sciences
Wuhan, Hubei, People's Republic of China

Dabing Zhang
College of Veterinary Medicine
China Agricultural University
Beijing, People's Republic of China

Jianqiang Zhang
Department of Veterinary Diagnostic and Production
    Animal Medicine
Iowa State University
Ames, Iowa

Yu Zhu
Division of Swine Infectious Diseases
Chinese Academy of Agricultural Sciences
Harbin, Heilongjiang, People's Republic of China

1 Introductory Remarks

Dongyou Liu

CONTENTS

1.1 Preamble 1
1.2 Virus Taxonomy 1
1.3 Virus Infections and Diseases 2
1.4 Virus Identification and Typing 2
1.5 Conclusion 5
References 5

1.1 PREAMBLE

As noncellular, submicroscopic agents that do not possess organelles for independent living and have to rely on cellular machinery of another organism for replication and maintenance, viruses are notably small. In fact, with a dimension of 20–400 nm (or 10^{-8}–10^{-6} mm), viruses are 10–100 times smaller than prokaryotes (10^{-7}–10^{-4} mm) and 1000 times smaller than eukaryotes (10^{-5}–10^{-3} mm). Being able to pass through conventional sterilizing filters (0.2 µm), viruses were initially referred to as filterable agents.

When not inside an infected cell or in the process of infecting a cell, viruses assume the form of independent particles (i.e., viral particles or virions). Structurally, viral particles (virions) consist of the genetic material (long molecules of either DNA or RNA that carry genetic information), a protein coat (also called the capsid or nucleocapsid, which surrounds and protects the genetic material), and in some cases an envelope [which is made up of host-derived cell membranes (phospholipids and proteins) and viral proteins, and which surrounds the protein coat when they are outside a cell]. Morphologically, viral particles (virions) may appear spherical, fusiform, rodlike, brick shaped, or pleomorphic. The envelope, when present, may be decorated with spikes, giving it a distinct appearance. The capsid (or nucleocapsid) may be icosahedral (isometric), helical, and complex in shape. The genetic material (or viral genome) may consist of either single-stranded (ss) or double-stranded (ds) ribonucleic acid (RNA) or deoxyribonucleic acid (DNA), in association with viral proteins (enzymes) that facilitate viral replication inside host cytoplasm (all RNA viruses) or nucleus (all DNA viruses apart from those belonging to the families *Asfarviridae* and *Poxviridae*).

Viruses are renowned for their ubiquity, abundance, and diversity. To date, >6000 distinct viruses have been identified from various ecological niches on Earth, including ocean, soil, air (aerosols), and inside bacteria, fungi, parasites, insects, animals, and plants. Besides ordinary viruses, there exists a category of viruses that have protein-coding deficiency and depend on other viruses for virion assembly, release, and subsequent infection of other cells (so-called viroids or subviral satellite). This is exemplified by hepatitis D virus, which is negative-sense, ssRNA virus considered as a subviral satellite of hepatitis B virus due to its reliance on the latter for the completion of its lifecycle. In addition, there is another category of viruses called prions, which is abbreviated from *pr*oteinaceous and *i*nfectious vir*ion*. Prion is noted for its ability to change the normal shape of a host protein into the prion shape, which converts even more proteins into prions. Finally, a large DNA virus called mimivirus (for microbe mimicking virus) was described in 1992 from amoeba cultures. Resembling gram-positive bacterium (with a diameter of about 750 nm), mimivirus shares some characteristics with iridovirus, asfarvirus, and phycodnavirus, all members of the nucelocytoplasmic large DNA virus (NCLDV) family.

1.2 VIRUS TAXONOMY

According to their nucleic acid compositions, viruses are divided into two major groups: (1) RNA viruses that include ribonucleic acids in their genomes and (2) DNA viruses that use deoxyribonucleic acids as genetic material. On the basis of the strandedness of their nucleic acids, viruses are separated into ss or ds RNA or DNA viruses. In view of the polarity of RNA strands, RNA viruses are differentiated into positive sense, negative sense, or ambisense. As the genomic RNA of positive-sense RNA viruses is identical to viral mRNA that can be directly translated into proteins by the host ribosomes, positive-sense RNA molecules can replicate without going through additional transcription. The resulting proteins then direct the replication of the genomic RNA. On the other hand, the genomic RNA of the negative-sense RNA viruses is complementary to viral mRNA and needs to be transcribed to positive-sense mRNA by a RNA-dependent RNA polymerase (RdRp) before replication. Therefore, purified RNA of a positive-sense virus may be infectious when transfected into cells, whereas purified RNA of a negative-sense virus is not immediately infectious [1].

The modern virus taxonomy reflects the morphological, physical, biological, and genetic (genome structure and sequence homology) properties of the viruses. The latest release

1

of International Committee on Taxonomy of Viruses classifies viruses into seven orders:

1. *Caudovirales* (*Myoviridae*, *Podoviridae* *Siphoviridae*),
2. *Herpesvirales* (*Alloherpesviridae*, *Herpesviridae*, *Malacoherpesviridae*),
3. *Ligamenvirales* (*Lipothrixviridae*, *Rudiviridae*),
4. *Mononegavirales* (*Bornaviridae*, *Filoviridae*, *Nyamiviridae*, *Paramyxoviridae*, *Rhabdoviridae*),
5. *Nidovirales* (*Arteriviridae*, *Coronaviridae*, *Mesoniviridae*, *Roniviridae*),
6. *Picornavirales* (*Dicistroviridae*, *Iflaviridae*, *Marnaviridae*, *Picornaviridae*, *Secoviridae*), and
7. *Tymovirales* (*Alphaflexiviridae*, *Betaflexiviridae*, *Gammaflexiviridae*, *Tymoviridae*).

There also are 78 families that have yet been assigned to an order (*Adenoviridae*, *Alphatetraviridae*, *Alvernaviridae*, *Amalgaviridae*, *Ampullaviridae*, *Anelloviridae*, *Arenaviridae*, *Ascoviridae*, *Asfarviridae*, *Astroviridae*, *Avsunviroidae*, *Baculoviridae*, *Barnaviridae*, *Benyviridae*, *Bicaudaviridae*, *Bidnaviridae*, *Birnaviridae*, *Bromoviridae*, *Bunyaviridae*, *Caliciviridae*, *Carmotetraviridae*, *Caulimoviridae*, *Chrysoviridae*, *Circoviridae*, *Clavaviridae*, *Closteroviridae*, *Corticoviridae*, *Cystoviridae*, *Endornaviridae*, *Flaviviridae*, *Geminiviridae*, *Globuloviridae*, *Guttaviridae*, *Hepadnaviridae*, *Hepeviridae*, *Hypoviridae*, *Hytrosaviridae*, *Inoviridae*, *Iridoviridae*, *Leviviridae*, *Luteoviridae*, *Marseilleviridae*, *Megabirnaviridae*, *Metaviridae*, *Microviridae*, *Mimiviridae*, *Nanoviridae*, *Narnaviridae*, *Nimaviridae*, *Nodaviridae*, *Nudiviridae*, *Ophioviridae*, *Orthomyxoviridae*, *Papillomaviridae*, *Partitiviridae*, *Parvoviridae*, *Permutotetraviridae*, *Phycodnaviridae*, *Picobirnaviridae*, *Plasmaviridae*, *Polydnaviridae*, *Polyomaviridae*, *Pospiviroidae*, *Potyviridae*, *Poxviridae*, *Pseudoviridae*, *Quadriviridae*, *Reoviridae*, *Retroviridae*, *Sphaerolipoviridae*, *Spiraviridae*, *Tectiviridae*, *Togaviridae*, *Tombusviridae*, *Totiviridae*, *Turriviridae*, *Virgaviridae*, and unassigned family). In turn, each family may contain one or more genera, each genus may include one or more species, and some species may have one or more isolates [2–4].

1.3 VIRUS INFECTIONS AND DISEASES

In general, the infection of host cells by a typical virus is a multistepped process: (1) after uptake by the host, virus utilizes surface proteins (enveloped) or capsid proteins (non-enveloped) to specifically bind to receptors on the host cellular membrane; (2) virus then enters into the cytoplasm through receptor-mediated endocytosis or membrane fusion; (3) inside the cytoplasm, virus uncoats or removes capsid by viral or host enzymes to release the viral genome; (4) viral polymerase transcribes from viral genome into viral mRNA and then translates into viral proteins; (5) viral proteins interact with new viral genome to form new virion particles; (6) newly assembled viruses bud off or release after cell lysis [1].

While many viruses reside in bacteria, fungi, parasites, insects, and plants, some find their way into animal hosts and cause a range of clinical symptoms (e.g., fever, inflammation, encephalitis, diarrhea), which are generally nonspecific. Tables 1.1 and 1.2 summarize major RNA and DNA virus families that are associated with clinical diseases in animals, although prions are not included in these tables. It is notable that a number of viruses are responsible for serious, and sometimes fatal diseases in animals, leading to significant economic losses and endangering the reliability of food supply. For instance, foot-and-mouth disease virus (see Chapter 8) and bluetongue virus (see Chapter 6) are associated with debilitating illnesses in livestock; canine parvovirus (see Chapter 76) causes deadly infection in pups; and deformed wing virus (see Chapter 5) induces stunted wings in honey bee. In addition, three RNA viruses [viral encephalopathy and retinopathy (see Chapter 2), yellow head virus (see Chapter 33), and gill associated virus (see Chapter 33)] and four DNA viruses [abalone viral ganglioneuritis (see Chapter 90), megalocytivirus (see Chapter 92), ostreid herpesvirus-1 microvariant (see Chapter 92), and white spot syndrome virus (see Chapter 98)] are implicated in serious contagious diseases in aquatic animals. Moreover, influenza viruses (see Chapters 41 and 46), rabies virus (see Chapter 43), hantaviruses (see Chapter 60), and arenaviruses (see Chapter 64) are examples of animal viral pathogens that pose health risks to human populations.

1.4 VIRUS IDENTIFICATION AND TYPING

Given the diversity of viruses and nonspecific signs they induce in animal hosts, an accurate diagnosis of viral infections and diseases requires the use of laboratory tests.

Traditionally, virus identification and characterization have relied heavily on phenotypic procedures. These include observing the morphological features of virions under electron microscopy; determining the viral stability by treatment at different temperatures, with various pH solutions, lipid solvents, and detergents; assessing the cytopathic effects of viruses *in vitro* using a variety of mammalian and insect cell lines; evaluating the viral pathogenicity *in vivo* using various animal models; analyzing the antigenicity and cross-reactivity of related viruses with serological tests; and ascertaining the number and sizes of viral segments by gel electrophoresis [5].

Since many of the phenotypic procedures lack the desired sensitivity, specificity, accuracy, intra- and interassay precision, as well as speed and cost-effectiveness, nucleic acid–based technologies such as gene probes, polymerase chain reaction (PCR), and nucleotide sequencing are increasingly adopted in clinical microbiological laboratories. In particular, due to their high sensitivity, specificity, and rapid testing turnaround, PCR and its derivatives (nested, multiplex, real time) have emerged as the method of choice for identification and typing of viruses and diagnosis of viral disease [6].

Nevertheless, due to the possibility that false-positive and false-negative results may occur in the nucleic acid amplification techniques such as PCR, it is critical to properly validate and standardize these test before routine application.

TABLE 1.1

Classification and Property of Animal-Infecting RNA Virus Families

Family	Strand/Sense	Genome Size (kb)	Structure (Segment)	Envelope	Virion Shape/Size (nm)	Nucleocapsid	Transmission Route	Examples
Nodaviridae	Single/positive	4.5	Linear (2)	−	Spherical/25–34	Icosahedral	Exposure, feeding	Betanodavirus
Dicistroviridae	Single/positive	7–10	Linear (1)	−	Spherical/30–32	Icosahedral	Close contact, feeding	Taura syndrome virus
Iflaviridae	Single/positive	8–10	Linear (1)	−	Spherical/26–32	Icosahedral	Feeding, mite vector	Deformed wing virus
Picornaviridae	Single/positive	7	Linear (1)	−	Spherical/28–30	Icosahedral	Feeding, droplet contact	Avian encephalitis virus; foot-and-mouth disease virus
Astroviridae	Single/positive	6–7	Linear (1)	−	Spherical/27–31	Icosahedral	Feeding	Avian astrovirus
Hepeviridae	Single/positive	7.2	Linear (1)	−	Spherical/27–34	Icosahedral	Feeding	Hepatitis E virus
Caliciviridae	Single/positive	8	Linear (1)	−	Spherical/35–39	Icosahedral	Feeding	Feline calicivirus; vesicular exanthema of swine virus
Retroviridae	Single/positive	7–11	Linear (1)	+	Spherical/80–100	Icosahedral	Close contact, mother's milk	Avian leukosis virus; feline immunodeficiency virus
Flaviviridae	Single/positive	10–12	Linear (1)	+	Spherical/45–60	Icosahedral	Insect bites; close contact, inhalation	Bovine viral diarrhea viruses; classical swine fever virus
Togaviridae	Single/positive	10–12	Linear (1)	+	Spherical/70	Icosahedral	Insect bite, droplet contact	Venezuelan equine encephalitis virus
Arteriviridae	Single/positive	15	Linear (1)	+	Spherical/45–60	Icosahedral	Close contact, feeding	Equine arteritis virus; porcine reproductive and respiratory syndrome virus
Roniviridae	Single/positive	26	Linear (1)	+	Rod shaped/160–200 × 34–63	Helical	Close contact, feeding	Yellow head virus
Coronaviridae	Single/positive	20–33	Linear (1)	+	Spherical/80–220	Helical	Droplet contact, inhalation	Avian infectious bronchitis virus; porcine epidemic diarrhea virus
Bornaviridae	Single/negative	9	Linear (1)	+	Spherical/90–150	Helical	Feeding, droplet contact	Avian bornavirus
Rhabdoviridae	Single/negative	11–15	Linear (1)	+	Bullet shaped/180 × 75	Helical	Insect bites, close contact, droplet contact	Bovine ephemeral fever virus; rabies virus
Orthomyxoviridae	Single/negative	12–15	Linear (7–8)	+	Pleomorphic/100	Helical	Droplet contact	Avian influenza virus; infectious salmon anemia virus
Paramyxoviridae	Single/negative	15–16	Linear (1)	+	Pleomorphic/150–300	Helical	Droplet contact, feeding	Canine distemper virus; Newcastle disease virus
Filoviridae	Single/negative	19	Linear (1)	+	Filamentous, pleomorphic/790–970 × 80	Helical	Direct contact	Ebola virus; Marburg virus
Bunyaviridae	Single/negative or ambisense	10–23	Linear (3)	+	Spherical, pleomorphic/80–120	Helical	Insect bite, inhalation	Akabane virus; Nairobi sheep disease virus
Arenaviridae	Single/ambisense	5–11	Circular (2)	+	Spherical/110–130	Helical	Close contact	Arenavirus
Picobirnaviridae	Double/ambisense	4–4.5	Linear (2)	−	Spherical/35–40	Icosahedral	Feeding, droplet contact	Picobirnavirus
Totiviridae	Double/ambisense	7.5	Linear (1)	−	Spherical/40	Icosahedral	Close contact, feeding	Infectious myonecrosis virus
Birnaviridae	Double/ambisense	6.1	Linear (2)	−	Spherical/60	Icosahedral	Feeding	Infectious bursal disease virus
Reoviridae	Double/ambisense	18–30	Linear (10–12)	−	Spherical/60–80	Icosahedral	Feeding; insect bite	Bluetongue virus; epizootic hemorrhagic disease virus

Sources: Knipe, D.M. and Howley, P.M. (eds.), *Fields Virology*, 5th edn., Lippincott Williams & Wilkins, Philadelphia, PA, 2007; King, A.M.Q. et al. (eds.), *Virus Taxonomy: Classification and Nomenclature of Viruses: Ninth Report of the International Committee on Taxonomy of Viruses*, Elsevier, San Diego, CA, 2012; http://viralzone.expasy.org/viralzone/all_by_species/19.html.

TABLE 1.2

Classification and Property of Animal-Infecting DNA Virus Families

Family	Strand	Genome Size (kb)	Structure (Segment)	Envelope	Virion Shape/Size (nm)	Nucleocapsid	Transmission Route	Examples
Circoviridae	Single	3	Circular	–	Spherical/17–24	Icosahedral	Direct contact	Beak and feather disease virus
Parvoviridae	Single	4–6	Linear (1)	–	Spherical/25	Icosahedral	Aerosol, direct contact	Porcine parvovirus
Polyomaviridae	Double	5	Circular (1)	–	Spherical/40	Icosahedral	Aerosol, direct contact	Polyomavirus
Papillomaviridae	Double	7–8	Circular (1)	–	Spherical/55	Icosahedral	Direct contact	Papillomavirus
Adenoviridae	Double	28–45	Linear (1)	–	Spherical/70–90	Icosahedral	Droplet contact, feeding	Poultry adenovirus
Herpesviridae	Double	120–240	Linear (1)	+	Spherical/150–200	Icosahedral	Direct contact, feeding	Infectious laryngotracheitis virus
Alloherpesviridae	Double	134–248	Linear (1)	+	Spherical/150–200	Icosahedral	Direct contact, feeding	Cyprinivirus; ictalurid herpesvirus
Malacoherpesviridae	Double	130–210	Linear (1)	+	Spherical/150–200	Icosahedral	Direct contact/exposure	Abalone herpesvirus
Asfarviridae	Double	130–190	Linear (1)	+	Spherical/175–215	Icosahedral	Direct contact, feeding, tick bites	African swine fever virus
Iridoviridae	Double	110–303	Linear (1)	+	Spherical/120–350	Icosahedral	Direct contact, feeding	Megalocytivirus
Poxviridae	Double	130–375	Linear (covalently closed)	+	Brick shaped or oval/250 × 360	Complex	Direct contact, insect bite	Fowl/turkey poxvirus; Lumpy skin disease virus

Sources: Knipe, D.M. and Howley, P.M. (eds.), *Fields Virology*, 5th edn., Lippincott Williams & Wilkins, Philadelphia, PA, 2007; King, A.M.Q. et al. (eds.), *Virus Taxonomy: Classification and Nomenclature of Viruses: Ninth Report of the International Committee on Taxonomy of Viruses*, Elsevier, San Diego, CA, 2012; http://viralzone.expasy.org/viralzone/all_by_species/19.html.

In addition, participation in external quality assurance programs provides an opportunity to assess the technical preparedness of the diagnostic laboratories involved in the molecular identification and typing of viruses.

1.5 CONCLUSION

With amazing diversity, versatility, and adaptability, viruses have found their way into virtually all ecological niches on Earth. Continuing population growths accompanied by habitat alteration and destruction, and increasing international trades and travels have contributed to the spread of viruses to where they were once absent. As a consequence, emerging and remerging viral pathogens are responsible for an-ever climbing number of epidemics in animals and humans worldwide. Owing to their time-consuming and variable nature, traditional phenotypic procedures are incapable of dealing this emergency. With exceptional sensitivity, specificity, detection limit, repeatability, and reproducibility, new generation genotypic procedures, especially PCR-based techniques, have risen to the challenge and are now widely adopted in routine diagnostic laboratories for detection and tying of viral pathogens of veterinary and medical significance. Nevertheless, certain phenotypic procedures such as virus isolation remain indispensable for advanced studies of viruses, including virus genome sequencing and characterization, analysis of immune responses, determination of pathogenicity in animal model, assessment of phenotypic antiviral susceptibility and vaccine preparations.

REFERENCES

1. Knipe, D.M. and Howley, P.M. (eds.), *Fields Virology*, 5th edn., 2007. Lippincott Williams & Wilkins, Philadelphia, PA.
2. King, A.M.Q., Adams, M.J., Carstens, E.B., and Lefkowitz, E.J. (eds.), *Virus Taxonomy: Classification and Nomenclature of Viruses: Ninth Report of the International Committee on Taxonomy of Viruses*, 2012. Elsevier, San Diego, CA.
3. Fauquet, C.M. and Fargette, D., International Committee on Taxonomy of Viruses and the 3,142 unassigned species. *Virol. J.*, 2, 64, 2005.
4. ICTV. International Committee on Taxonomy of Viruses. http://ictvonline.org/virustaxonomy.asp. Accessed on June 12, 2015.
5. Storch, G.A., Diagnostic virology. *Clin. Infect. Dis.*, 31, 739, 2000.
6. Vernet, G., Molecular diagnostics in virology. *J. Clin. Virol.*, 31, 239, 2004.
7. Swiss Institute of Bioinformatics. http://viralzone.expasy.org/viralzone/all_by_species/19.html. Accessed on June 12, 2015.

Section I

Positive-Sense RNA Viruses

2 Betanodavirus

Dongyou Liu

CONTENTS

2.1 Introduction 9
    2.1.1 Classification...9
    2.1.2 Morphology and Genome Structure 10
    2.1.3 Biology and Epidemiology 11
    2.1.4 Clinical Features and Pathogenesis 11
    2.1.5 Diagnosis 12
    2.1.6 Treatment and Prevention 12
2.2 Methods 12
    2.2.1 Sample Preparation...12
    2.2.2 Detection Procedures..13
        2.2.2.1 Conventional RT-PCR...13
        2.2.2.2 Real-Time RT-PCR 13
2.3 Conclusion 13
References..14

2.1 INTRODUCTION

Betanodavirus is the etiological agent of viral nervous necrosis (VNN), also known as vacuolating encephalopathy and retinopathy (VER) or encephalomyelitis, which is a devastating viral disease affecting marine and freshwater fish species worldwide. First observed in 1986 on an aquaculture farm of Japanese parrotfish (*Calotomus japonicus*) in Nagasaki, Japan, VNN was linked to loss of balance, necrosis in the brain and retina, and death of fish fry (of 6–25 mm length) during the summer (June–July) [1]. Almost simultaneously, VNN was discovered in Australia, Norway, and France, involving barramundi (Asian sea bass, *Lates calcarifer*), turbot (*Scophthalmus maximus*), European sea bass (*Dicentrarchus labrax*), red-spotted grouper (*Epinephelus akaara*), striped jack (*Pseudocaranx dentex*) [2–5], and other cultured warm-water and cold-water marine fish species (*n* = 30) throughout the world [1]. Given its capacity to induce high mortality (80%–100%) in juvenile fish, VNN exerts huge economic and ecological burdens on aquaculture industry and wild marine species [6]. In this chapter, we focus on betanodavirus in relation to its taxonomy, morphology, genome structure, epidemiology, clinical features, pathogenesis, diagnosis, and control, together with stepwise molecular protocols for its identification and diagnosis.

2.1.1 CLASSIFICATION

The family *Nodaviridae* (noda, derived from Nodamura, a Japanese village where Nodamura virus [NoV] was isolated) encompasses small (<5 kb), nonenveloped, isometric viruses with bipartite, positive-sense, single-stranded (ss) RNA genomes that are infective to insects (genus *Alphanodavirus*) and fish (genus *Betanodavirus*). Currently, the genus *Alphanodavirus* contains five recognized insect-infecting species (black beetle virus, Boolarra virus, flock house virus, NoV, and Pariacoto virus), while the genus *Betanodavirus* consists of four established fish-infecting species (barfin flounder nervous necrosis virus [BFNNV], red-spotted grouper nervous necrosis virus [RGNNV], striped jack nervous necrosis virus [SJNNV, type species], and tiger puffer nervous necrosis virus [TPNNV]). In addition, a large number of viruses remain unassigned in the family *Nodaviridae*, with >30 viruses being tentatively placed in the genus *Betanodavirus* alone [7–9].

Classification of betanodaviruses has been largely based on similarities detected in the variable region of the viral coat protein (CP) gene (located on RNA2 segment) [8]. Through comparative analysis of 78 full-length CP gene sequences of betanodaviruses originated from various geographical locations, six major clusters (groups I, II, III, IV, V, VI) are noted. Of these, groups II, III, IV, and VI are similar to the classic species, while groups I and V are single-isolate clusters that may be considered as new species. More specifically, group II (TPNNV) consists of two isolates from tiger puffer in Japan; group III (SJNNV) includes 10 isolates from striped jack, gilt-head sea bream, sole fish, and greasy grouper; group IV (BFNNV) has nine isolates in three subgroups (IVa, IVb, and IVc), with group IVa containing four isolates from Atlantic cod, group IVb having three isolates from barfin flounder, and group IVc having one isolate each from Atlantic cod and Malabar grouper;

and group VI (RGNNV) covers 55 isolates from 14 fish species with very different geographical locations in Asia, Europe, and Australia. Groups I and V are single-isolate clusters originated from European sea bass and turbot in colder regions (France and Norway, respectively) (Table 2.1) [10–13]. Indeed, despite its relatedness to group II (TPNNV) and group III (SJNNV) isolates based on comparative analysis of the partial RNA2 sequence, group I turbot nodavirus (AJ608266), originally isolated from turbot in Norway, was sufficiently different to represent possibly a fifth species in the genus *Betanodavirus* [14]. Similarly, placed in between the BFNNV and RGNNV clusters, *Dicentrarchus labrax* encephalitis virus (U39876) from sea bass in France may constitute another distinct species within the genus [8].

Due to the multiplicity of hosts, diversity of environments, and frequent movement of virus-infected fish through international trade, betanodaviruses with bipartite RNA genomes are constantly subjected to mutation (substitution) and reassortment pressures, which contribute to their genetic variability [15]. This is exemplified by the findings that group IV (BFNNV) can be subdivided into three subgroups (IVa, IVb, and IVc) on the basis of sequence differences in CP gene and that some RGNNV and SJNNV isolates from the Mediterranean possess RNA1 segment related to one particular species and RNA2 segment related to another species (i.e., RGNNV/SJNNV and SJNNV/RGNNV). Serologically, both BFNNV and

RGNNV belong to serotype C, despite being present mainly in cold-water species and warm-water species, respectively, while SJNNV is of serotype A and TPNNV of serotype B. This highlights the value of analyzing sequences from both RNA1 and RNA2 segments for precise determination of a given *Betanodavirus* isolate [8].

Interestingly, recent experimental study on the effect of two coexistent *Betanodavirus* viruses through coinfection and superinfection indicated that SJNNV replication is partially inhibited by the coexistence with RGNNV, whereas RGNNV replication is favored in coinfection or superinfection with SJNNV [16].

2.1.2 Morphology and Genome Structure

Members of the genus *Betanodavirus* are spherical, non-enveloped, and icosahedral viruses of about 25–34 nm in diameter. Being "nonenveloped," these viruses lack an outermost layer that is usually made up of lipids and proteins, and their genetic material is surrounded by icosahedral-shaped capsid (shell), which is constructed of 32 capsomers and shows a triangulation number = 3 ($T = 3$) icosahedral symmetry [17].

The genome of betanodaviruses is composed of two linear positive-sense, single-stranded RNA segments: RNA1 and RNA2 (both of which possess a 5′ terminal methylated cap and a non-polyadenylated 3′ terminal, i.e., without a poly(A) tail). The virion RNA is infectious and serves as both the genome and viral messenger RNA.

RNA1 (3103 nt in length with a G + C content of 49.6%) consists of an open reading frame (ORF) flanked by a 78 nt 5′-nontranslated regions (NTR) and a 77 nt 3′-NTR region. This ORF encodes a 982 aa protein with a calculated molecular mass of 110.74 kDa, which includes multiple functional domains: a mitochondrial targeting domain, a transmembrane domain, an RNA-dependent RNA polymerase (RdRp) domain, a self-interaction domain, and an RNA capping domain [6,8].

At the *N*-terminal region (the 3′ end) of RNA1 segment, a subgenomic RNA (sgRNA3 or RNA3, of 371 nt in length with a G + C content of 62%) exists. RNA3 contains an ORF that encodes a 75 aa peptide corresponding to a hypothetical B2 protein. This protein functions as an RNA silencing inhibitor of the host RNA interference system and facilitates intracellular viral RNA accumulation. RNA3 is present only in the nucleus of a virus-infected cell but not incorporated into the virion [6,18].

RNA2 (of 1433 nt in length with a G + C content of 53.24%) harbors an ORF flanked by 26 nt 5′-NTR and a 390 nt 3′-NTR. This ORF encodes a 338 aa viral capsid protein precursor α (of 42 kDa), which is autocleaved at a conserved Asn/Ala site during virus assembly into two mature proteins: a β-protein with a calculated molecular mass of 37.059 kDa and a γ-protein of 5 kDa [19]. The β-protein (CP) is the key component of the viral capsid and consists of a highly conserved region and a variable region [6].

TABLE 2.1
Phenotypic and Genotypic Characteristics of the Genus *Betanodavirus*

Species	Optimal Temperature for Replication	Serotype	Phylogroup
Striped jack nervous necrosis virus (SJNNV)	20°C–25°C	A	III
Tiger puffer nervous necrosis virus (TPNNV)	20°C	B	II
Barfin flounder nervous necrosis virus (BFNNV)	15°C–20°C	C	IV (IVa, IVb, IVc)
Red-spotted grouper nervous necrosis virus (RGNNV)	25°C–30°C	C	VI
Turbot nodavirus (TNV) (AJ608266/ Norway)	No growth on SSN-1 cells	Unknown	I
Dicentrarchus labrax encephalitis virus (U39876/France)	25°C	Unknown	V

Sources: Johansen, R. et al., *J. Fish Dis.*, 27(10), 591, 2004; Binesh, C.P. et al., *Arch. Virol.*, 158, 1589, 2013.

2.1.3 Biology and Epidemiology

Replication of betanodavirus involves a number of steps. These include (1) penetration into the host cell, (2) uncoating and release of the viral genomic RNA into the cytoplasm, (3) translation of RNA1 into RdRp protein, (4) synthesis of dsRNA genome from the genomic ssRNA(+), (5) transcription of dsRNA genome into viral mRNAs/new ssRNA(+) genomes, (6) expression of subgenomic RNA3, (7) translation of RNA2 into capsid protein α, (8) assembly of virus around genomic RNA1 and RNA2 in the cytoplasm, (9) cleavage of assembled capsid protein alpha into capsid protein β and γ, and (10) release of infectious particles [8].

Betanodavirus often enters the host via fin epithelial cells or skin epithelial cells. After passage through the fin skin into the fin tissue, the virus multiplies and spreads to the circulatory system. Large quantities of virus particles are distributed in primary tissues (brain and retina). Surviving fish may copious amounts of virus remaining in fin tissue, resulting in a carrier state. The release of the virus from fin epithelial cells into the seawater facilitates horizontal transmission to other fish [8,20–22].

Although the causative agent of VNN was first purified from diseased striped jack larvae and named striped jack nervous necrosis virus (SJNNV) in the later 1980s in Japan, Australia, Norway, and France [1], the earliest evidence related to the virus dates back to 1984 in Queensland, Australia, when clinical signs and histopathological lesions similar to VNN were observed in diseased larval barramundi (*Lates calcarifer*) [2].

A broad range of fish species, including marine and freshwater fish produced through aquaculture or living in the wild, are susceptible to betanodavirus infection [23,24]. To date, 39 fish species belonging to 22 families in 8 orders have been found to be prone to VNN. These include marine food and game fish species (red drum [*Sciaenops ocellatus*], cobia [*Rachycentron canadum*], sea bass [*Dicentrarchus labrax*], barramundi [*Lates calcarifer*], gilt-head sea bream [*Sparus aurata*], Pacific bluefin tuna [*Thunnus orientalis*], grouper [family *Serranidae*], halibut [*Hippoglossus hippoglossus*], and Japanese flounder [*Paralichthys oliveatus*]), marine ornamental fish species (convict surgeonfish [*Acanthurus triostegus*], lined surgeonfish [*Acanthurus lineatus*], narrowstripe cardinalfish [*Apogon exostigma*], threespot dascyllus [*Dascyllus trimaculatus*], Scopas tang [*Zebrasoma scopas*], blueband goby [*Valenciennea strigata*], tiger puffer [*Takifugu rubripes*], lemonpeel angelfish [*Centropyge flavissimus*], and orbicular batfish [*Platax orbicularis*]), freshwater species (tilapia [*Oreochromis niloticus*] and Chinese catfish [*Parasilurus asotus*]), and ornamental fish (guppy [*Poecilia reticulata*]) [25–34]. In laboratory studies, medaka (*Oryzias latipes*) and zebrafish (*Danio rerio*) are susceptible [35–37].

While fish at larvae or juvenile stages are often affected, significant mortality may also occur in older fish. In striped jack, VNN may occur between 1 and 20 days posthatching (8 mm total length), leading to an almost complete loss of larvae [1,38]. In European sea bass, mortality due to VNN is not seen until about 30 days posthatching. In seven-band grouper (*Epinephelus septemfasciatus*), VNN causes those of up to 1.8 kg body weight to exhibit belly-up swimming and inflation of the swim bladder, with lowered mortality in juvenile or older fish [39].

VNN has been reported on all continents apart from South America, including Asia (Japan, Korea, China, Taiwan, Vietnam, Thailand, Malaysia, Singapore, Indonesia, Brunei, Philippines, India), Middle East (Israel), Oceania (Australia, Tahiti), Europe (Croatia, Bosnia, Greece, Malta, Italy, France, Spain, Portugal, Tunisia, United Kingdom, Norway), and North America (United States, Canada) [6,40–44].

It appears that various betanodaviruses have adapted to tropical, subtropical, or cold-temperate environments. This is indicated by the preferred temperature ranges of 25°C–30°C, 20°C–25°C, 20°C, and 15°C–20°C, for RGNNV, SJNNV, TPNNV, and BFNNV, respectively (Table 2.1). Use of an established cell line (SSN-1) pinpoints that the optimal temperature for betanodavirus growth is 25°C and that the upper temperature limits for RGNNV are ~32°C [45].

Betanodaviruses can be spread horizontally (from infected fish to uninfected, naïve fish in the same water body) or acquired through exposure to virus-contaminated water [46]. They may also be transmitted through feeding of contaminated live foods, fish and mollusks, including copepod (*Tigriopus japonicus*), brine shrimp (*Artemia*), shrimp (*Acetesinte medius*), Japanese jack mackerel (*Trachurus japonicus*), and Japanese common squid (*Todarodes pacificus*) [47]. Furthermore, betanodaviruses may be transmitted vertically from brood stock to offspring, both within and on fertilized eggs [48–50].

2.1.4 Clinical Features and Pathogenesis

Betanodaviruses are neuropathogenic with the ability to cause damages throughout the central nervous system. Therefore, clinical signs of VNN-affected larvae and juvenile stages are largely neurological, ranging from anorexia, lethargy (clustering near the side of a pool), abnormal schooling and swimming behavior (vertical positioning and spinning), flexing of the body, muscle tremors, hyperinflation of the swim bladder (causing fish to float belly up at the surface), traumatic lesions (due to uncontrolled swimming/spinning), changes in skin pigmentation (either darkening or lightening), and scoliosis to death (up to 100% within 1 week of symptom onset).

Histopathologically, VNN is characterized by severely extended necrosis, vacuolation, and degeneration of the central nervous system (brain, spinal cord) and retina, but some fish in early larval stages may lack tissue vacuolation [42].

Betanodaviruses manage to evade the host's protective systems by interrupting and inhibiting host nonspecific (innate) and specific immune responses through production of the interferon-induced antiviral protein Mx in the infected fish [51,52]. This enables the virus to replicate and transmit progeny to other cells or to remain undetected by host immune surveillance in a latent condition [53]. In addition, betanodaviruses are adept in exploiting various environmental and procedural stressors (e.g., high density and high temperatures) that weaken the immune systems of fish and render them vulnerable to microbial infections.

2.1.5 DIAGNOSIS

Differential diagnosis of VNN includes various microbial infections (bacteria, virus, parasites), environmental factors (e.g., high ammonia levels in water), or toxins (e.g., pesticides), which may also contribute to changes in swimming behavior and pigmentation. A number of laboratory diagnostic tests can be used to confirm VNN, including histopathology, virus isolation, serology, and nucleic acid detection [34].

Histopathology involves preparation of thin, fixed-tissue sections on a glass slide, staining, and examination under microscope. Observation of vacuoles (holes) and necrosis (presence of dead tissue) in the brain, spinal cord, and retina of the eye by light microscopy provides presumptive diagnosis of VNN [34]. Electron microscopic examination reveals the virus in association with vacuolated cells and accumulating intracytoplasmically in paracrystalline arrays or as aggregates.

Virus isolation relies on cell culture to isolate and propagate the virus and to detect its cytopathic effects [34]. Several fish cell lines are now available for betanodavirus isolation: SSN-1 and E-11 derived from striped snakehead (*Ophicephalus striatus*) [54–56], GF-1 and other cell lines from groupers [57,58], and TV-1 and TF from turbot [59].

Serology checks for evidence of virus antigens or fish antibodies made against the virus. Commonly used serological tests include immunostaining methods, fluorescent antibody test, immunohistochemistry, and enzyme-linked immunoassay, as well as immunomagnetic reduction assay, which employs magnetic nanoparticles coated with dextran and antibodies [34,60].

Nucleic acid detection of betanodaviruses can be done by direct hybridization with a DNA probe or through amplification and detection by reverse transcription polymerase chain reaction (RT-PCR) or other techniques [61–64]. RT-PCR provides a rapid and convenient method for identification of betanodaviruses. Nested PCR involving second-round amplification is useful to diagnose asymptomatic fish (brood stock) [65–69]. In particular, the primer set R3 and F2 targeting the variable region (ca. 380 bases) of the SJNNV CP gene allows detection of most genotypic variants [65,66]. Recent development of real-time RT-PCR assay (rRT-PCR) enables highly sensitive detection and quantification of betanodaviruses [70–73]. Other nucleic acid detection approaches include loop-mediated isothermal amplification (LAMP), real-time LAMP, and microfluidic LAMP system [74–77]. Ideally, PCR detection is followed by another method, either histopathology with immunostaining or virus isolation in cell culture, for result validation [78].

2.1.6 TREATMENT AND PREVENTION

Currently, no effective treatments are available for VNN. Depopulation followed by disinfection is recommended. Good husbandry, close monitoring of fish behavior, and prompt identification of betanodavirus infection by laboratory tests are critical to limit the damages due to this disease. It is vital to isolate sick fish and handle them separately. Quarantine of new fish in a separate building before adding to existing farm populations is important.

To prevent horizontal transmission of infection, betanodaviruses can be inactivated with 50 ppm of chlorine, 50 mg/L sodium hypochlorite, calcium hypochlorite, benzalkonium chloride, and iodine for 10 min, 60% ethanol, 50% methanol, high alkalinities (pH 12 for 10 min), heat treatment (60°C for 30 min), ultraviolet (UV) radiation ($1.5–2.5 \times 10^5$ μ Ws/cm^2 at a UV intensity of 440 μ W/cm^2), acid–peroxygen, and ozone (at 0.1 μg/mL for 2.5 min). However, they are resistant to chloroform, ether, and formalin [79,80].

To prevent vertical transmission of infection from brood stock to offspring, washing fertilized eggs in ozone-treated seawater and disinfection of rearing water by ozone (e.g., 1 mg O_3/L for 1 min for striped trumpeter) are effective in controlling VNN in larvae production of striped jack, barfin flounder, Atlantic halibut, seven-band grouper, and striped trumpeter [22,81].

Improvement in spawning induction methods including brood stock feed and water supply is beneficial [81]. As betanodavirus can be spread by contaminated water, it is necessary to keep separate equipment (nets, siphon hoses) for each system, to disinfect systems or tanks between uses, to place footbaths and hand disinfectants (alcohol spray or handwashing stations) judiciously, and to use water from a clean source (e.g., deep wells) or water that has been processed with adequate UV sterilization or chlorinated and dechlorinated.

Vaccination of larvae and older grow-out fish with inactivated betanodavirus or viruslike particles expressed by baculovirus or *Escherichia coli*, synthetic peptides based on the virus capsid, recombinant proteins, and DNA vaccine increases the production of neutralizing antibodies against the homologous virus, activates genes associated with cellular and innate immunity, suppresses viral replication in the brains and kidneys, and reduces disease incidence [82–84]. Vaccination of brood stock reduces the risk of vertical transmission of NNV to offspring [50].

2.2 METHODS

2.2.1 SAMPLE PREPARATION

Samples are collected using aseptic technique from juvenile fish showing clinical signs of VNN. The preferred/primary samples for betanodavirus diagnosis are brain, eye (retina), and fin. Other samples (gill, head, kidney, heart, intestine, liver, spleen, muscle, and blood) may be used as alternatives.

The tissues are frozen in liquid nitrogen and homogenized in 10 volumes of L15 medium, centrifuged at 10,000 rpm for 20 min, and the supernatant is passed through a 0.22 μm filter and stored at −80°C until use.

Total RNA may be extracted from fish eyes/brain or cell culture supernatant using RNeasy Mini kit (Qiagen), NucleoSpin RNA virus kit (Macherey-Nagel), or TRIzol reagent (Invitrogen). For NucleoSpin RNA virus kit, 150 µL of cell culture supernatant or 150 µL of supernatant of organ pool (brain and eyes) ground up in cold phosphate buffered saline (PBS) (10%, w/v) is used. For TRIzol reagent (Invitrogen), 100 mg tissue is mixed with 1 mL TRIzol reagent and 200 µL of ice-cold chloroform. After isopropanol precipitation, the RNA pellet is resuspended in diethylpyrocarbonate-treated water. The RNA concentration and purity are determined at the absorbance of 260 nm (A_{260}) with a Nanodrop spectrophotometer (Thermo Fisher Scientific).

2.2.2 Detection Procedures

Among the numerous published RT-PCR methods for betanodavirus detection and identification, a majority focus on the RNA2 segment, which encodes the CP [34]. In the succeeding sections, two stepwise protocols are presented: one conventional PCR targeting RNA2 (which is valuable to laboratories without access to real-time PCR equipment) and one real-time RT-PCR targeting RNA1. Clearly, molecular procedures for VNN diagnosis are not limited to these two, and laboratories are advised to evaluate and adopt those most suited to their own circumstances.

2.2.2.1 Conventional RT-PCR

Kuo et al. [85] described the use of primers (203-F, 5′-GACGCGCTTCAAGCAACTC-3′, and 203-R, 5′-CGAACACTCCAGCGACACAGCA-3′) from the CP gene (of RNA2 segment) to amplify a 203 bp fragment from betanodaviruses.

RT is conducted with 2 µL of the extracted total RNA (TRIzol reagent) and Moloney murine leukemia virus reverse transcriptase (Promega) in accordance with the manufacturer's protocol.

PCR amplification may be performed with AmpliTaq Gold® PCR Master Mix (Applied Biosystems) in 25 µL volumes containing 2 µL of cDNA, 0.4 µM of each primer (203-F and 203-R), 1 µL of bovine serum albumin (BSA) 1%, and 2.5 mM $MgCl_2$.

PCR cycling conditions include 94°C for 4 min (initial denaturation), 45 cycles of 94°C for 30 s, 55°C for 30 s, and 72°C for 30 s and a final hold at 72°C for 5 min.

Amplification products are resolved on 1.5% agarose gels. Samples showing a 203 bp band are considered positive for betanodaviruses.

Note: If desired, the primer set (203-F and 203-R) can be applied in SYBR green-based real-time qPCR. PCR mixture (25 µL) is made up of 12.5 µL of 2× SYBR gren master mix (Applied Biosystems), 5 µM (each) forward and reverse primers, and 2 µL cDNA. Amplification and detection is performed in 96-well plate using GeneAmp 2700 thermocycler coupled with a GeneAmp 7000 sequence detection system (Applied Biosystems) in wells of a 96-well plate. The thermal profile consists of 1 cycle of 95°C for 2 min, 40 cycles of 95°C for 15 s, 60°C for 1 min, and 72°C for 20 s. The fluorescence of SYBR green against the internal passive reference dye ROX (ΔR_n) is measured at the end of each cycle. A sample is considered positive

when the ΔR_n value exceeds the threshold value. The threshold value is set at the midpoint of the ΔR_n versus the cycle number plot. For all the amplifications described in this study, the threshold value of ΔR_n is set as 0.25. The threshold cycle (C_T) is defined as the cycle at which a statistically significant increase in R_n is first detected. The copy number of the VNN sample is determined by normalizing the C_T values of the samples and then extrapolating the normalized C_T values to the standard curve of the corresponding virus. VNN-infected and healthy grouper had C_T values of 12–24 and 29–32, respectively.

2.2.2.2 Real-Time RT-PCR

Baud et al. [73] describe a generic real-time RT-PCR for detection and quantification of betanodaviruses in clinical specimens. The primers (oPVP154: 5′-TCCAAGCCGGTCCTAGTCAA-3′ and oPVP155: 5′-CACGAACGTKCGCATCTCGT-3′), each containing one base degeneration, amplify a 168 or 171 nt fragment from a conserved region at the 3′ end of the RNA1 segment, which is detected with the TaqMan tqPVP16 probe (Cy5-CGATCGATCAGCACCTSGTC-BHQ2). While the tqPVP16 probe and primer oPVP155 are located within the subgenomic RNA3, the primer PVP154 shares only 7 nt with this molecule.

An RT-PCR mixture (25 µL) is made up of 600 nM of each primer, 400 nM of probe, 5 µL of RNA, and 1× of Quantitect Probe RT-PCR master mix (Qiagen). Amplification and detection are conducted using CFX96 (Bio Rad) or RotorGene 6000 (Corbett), with the following parameters: 50°C for 30 min, and 95°C for 10 min, and then 40 cycles of denaturation/extension at 94°C for 15 s and at 60°C for 60 s.

Ct values are obtained starting with 10-fold dilutions of total RNA extracted from cell culture inoculated with an RGNNV strain.

Note: This RT-PCR enables detection of genotypes representative of the four recognized species of betanodaviruses, as well as five reassortants between RGNNV and SJNNV. Due to its rapid, sensitive, and specific features, the assay has potential for routine application in different geographic regions and host species. Sequencing analysis of the amplified product will yield genotyping data of the specimen.

2.3 CONCLUSION

Betanodaviruses are responsible for causing VNN or VER in marine finfish and some freshwater species worldwide, with mortalities of 80%–100%. While younger life stages are especially vulnerable, some older fish are also affected. As betanodaviruses are neuropathogenic (with a tendency to attack the brain, spinal cord, and eye), infected fish often display abnormal behavior (e.g., spinning or whirling, abnormal posture, muscle tremors) before succumbing to the disease. Since other diseases may also induce similar clinical signs, it is important to have VNN diagnosed early and appropriate control measures implemented promptly. Among the diagnostic methods available (light and electron microscopy, virus isolation, serology, and nucleic acid amplification), RT-PCR appears to be a standout, in terms of rapidity, sensitivity, and specificity [86]. Given the high frequency of mutations in betanodavirus populations, it is important to analyze the sequencing variations in outbreak isolates and update the primers and probes if necessary. In addition, if false-negative PCR results are suspected, use of another technique (e.g., cell culture, or another PCR targeting a different region of the genome) may be helpful.

REFERENCES

1. Mori K-I et al. Properties of a new virus belonging to nodaviridae found in larval striped jack (*Pseudocaranx dentex*) with nervous necrosis. *Virology.* 1992; 187:368–371.
2. Glazebrook JS, Heasman MP, and de Beer SW. Picorna-like viral particles associated with mass mortalities in larval barramundi, *Lates calcarifer* (Bloch). *J Fish Dis.* 1990; 13:245–249.
3. Bloch B, Gravningen K, and Larsen JL. Encephalomyelitis among turbot associated with a picornavirus-like agent. *Dis Aquat Org.* 1991; 10:65–70.
4. Breuil G, Bonami JR, Pepin JF, and Pichot Y. Viral infection (picorna-like virus) associated with mass mortalities in hatchery-reared sea-bass (*Dicentrarchus labrax*) larvae and juveniles. *Aquaculture.* 1991; 97: 109–116.
5. Munday BL, Langdon JS, Hyatt A, and Humphrey JD. Mass mortality associated with a viral-induced vacuolating encephalopathy and retinopathy of larval and juvenile barramundi, *Lates calcarifer* Bloch. *Aquaculture.* 1991; 103:197–211.
6. Shetty M et al. Betanodavirus of marine and freshwater fish: Distribution, genomic organization, diagnosis and control measures. *Indian J Virol.* 2012; 23(2):114–123.
7. Nishizawa T et al. Comparison of the coat protein genes of five fish nodaviruses, the causative agents of viral nervous necrosis in marine fish. *J Gen Virol.* 1995; 76(7):1563–1569.
8. Binesh CP, Greeshma C, and Jithendran KP. Genomic classification of betanodavirus by molecular phylogenetic analysis of the coat protein gene. *Arch Virol.* 2013; 158:1589–1594.
9. ICTV (International Committee on Taxonomy of Viruses). http://www.ictvdb.org/Ictv/index.htm. Accessed on June 12, 2015.
10. Thiery R et al. Genomic classification of new betanodavirus isolates by phylogenetic analysis of the coat protein gene suggests a low host-fish species specificity. *J Gen Virol.* 2004; 85:3079–3087.
11. Aspehaug V, Devold M, and Nylund A. The phylogenetic relationship of nervous necrosis virus from halibut (*Hippoglossus hippoglossus*). *Bull Eur Ass Fish Pathol.* 1999; 19:196–202.
12. Chi SC, Lo CF, Lin SC. Characterization of grouper nervous necrosis virus. *J Fish Dis.* 2001; 24:3–13.
13. Chi SC, Shieh JR, and Lin SC. Genetic and antigenic analysis of betanodaviruses isolated from aquatic organisms in Taiwan. *Dis Aquat Org.* 2003; 55:221–228.
14. Johansen R et al. Characterization of nodavirus and viral encephalopathy and retinopathy in farmed turbot, *Scophthalmus maximus* (L.). *J Fish Dis.* 2004; 27(10):591–601.
15. He M and Teng CB. Divergence and codon usage bias of Betanodavirus, a neurotropic pathogen in fish. *Mol Phylogenet Evol.* 2014; 83:137–142.
16. Lopez-Jimena B et al. A combined RT-PCR and dot-blot hybridization method reveals the coexistence of SJNNV and RGNNV betanodavirus genotypes in wild meagre (*Argyrosomus regius*). *J Appl Microbiol.* 2010;109(4):1361–1369.
17. Nakai T et al. Current knowledge of viral nervous necrosis (VNN) and its causative betanodaviruses. *Isr J Aquacult—Bamidgeh.* 2009; 61:198–207.

18. Iwamoto T et al. Characterization of striped jack nervous necrosis virus subgenomic RNA3 and biological activities of its encoded protein B2. *J Gen Virol.* 2005; 86:2807–2816.
19. Ransangan J and Manin BO. Genome analysis of Betanodavirus from cultured marine fish species in Malaysia. *Vet Microbiol.* 2012; 156(1–2):16–44.
20. Korsnes K, Karlsbakk E, Devold M, Nerland AH, and Nylund A. Tissue tropism of nervous necrosis virus (NNV) in Atlantic cod, *Gadus morhua* L., after intraperitoneal challenge with a virus isolate from diseased Atlantic halibut, *Hippoglossus hippoglossus* (L.). *J Fish Dis.* 2009; 32(8):655–665.
21. Nguyen HD, Mushiake K, Nakai T, and Muroga K. Tissue distribution of striped jack nervous necrosis virus (SJNNV) in adult striped jack. *Dis Aquat Organ.* 1997; 28:87–91.
22. Kuo HC et al. Nervous necrosis virus replicates following the embryo development and dual infection with iridovirus at juvenile stage in grouper. *PLoS One.* 2012; 7(4):e36183.
23. Bovo G et al. Viral encephalopathy and retinopathy outbreak in freshwater fish farmed in Italy. *Dis Aquat Organ.* 2011; 96(1):45–54.
24. Binesh CP, Renuka K, Malaichami N, and Greeshma C. First report of viral nervous necrosis-induced mass mortality in hatchery-reared larvae of clownfish, *Amphiprion sebae* Bleeker. *J Fish Dis.* 2013; 36(12):1017–1020.
25. Starkey WG et al. Nodavirus infection in Atlantic cod and Dover sole in the UK. *Vet Rec.* 2001; 149(6):179–181.
26. Tan C et al. Determination of the complete nucleotide sequences of RNA1 and RNA2 from greasy grouper (*Epinephelus tauvina*) nervous necrosis virus, Singapore strain. *J Gen Virol.* 2001; 82(3): 647–653.
27. Castric J et al. Sea bass *Sparus aurata*, an asymptomatic contagious fish host for nodavirus. *Dis Aquat Org.* 2001; 47:33–38.
28. Hegde A, Teh HC, Lam TJ, and Sin YM. Nodavirus infection in freshwater ornamental fish, guppy, *Poicelia reticulata*—Comparative characterization and pathogenicity studies. *Arch Virol.* 2003; 148:575–586.
29. Munday BL and Nakai T. Nodaviruses as pathogens in larval and juvenile marine finfish. *World J Microbiol Biotechnol.* 1997; 13:375–381.
30. Furusawa R, Okinaka Y, Uematsu K, and Nakai T. Screening of freshwater fish species for their susceptibility to a betanodavirus. *Dis Aquat Org.* 2007; 77:119–125.
31. Bigarré L et al. Outbreak of betanodavirus infection in tilapia, *Oreochromis niloticus* (L.), in fresh water. *J Fish Dis.* 2009;32(8):667–673.
32. David R et al. Molecular detection of betanodavirus from the farmed fish, *Platax orbicularis* (Forsskal) (Ephippidae), in French Polynesia. *J Fish Dis.* 2010; 33(5):451–454.
33. Liu XD et al. Infections of nervous necrosis virus in wild and cage-reared marine fish from South China Sea with unexpected wide host ranges. *J Fish Dis.* June 2015; 38(6):533–540. doi: 10.1111/jfd.12265.
34. OIE (Office International des Epizooties). *Manual of Diagnostic Tests for Aquatic Animals 2013.* Chapter 2.3.11: Viral encephalopathy and retinopathy. http://www.oie.int/fileadmin/Home/eng/Health_standards/aahm/current/2.3.11_VER.pdf. Accessed on June 12, 2015.
35. Furusawa R, Okinaka Y, and Nakai T. Betanodavirus infection in the freshwater model fish medaka (*Oryzias latipes*). *J Gen Virol.* 2006; 87:2333–2339.
36. Lu MW et al. Immunomagnetic reduction assay for nervous necrosis virus extracted from groupers. *J Virol Methods.* 2012; 181(1):68–72.

37. Binesh CP. Elevation of temperature and crowding trigger acute viral nervous necrosis in zebra fish, *Brachydanio rerio* (Hamilton-Buchanan), subclinically infected with betanodavirus. *J Fish Dis.* 2014; 37:279–282.

38. Arimoto M et al. Pathogenicity of the causative agent of viral nervous necrosis disease in striped jack, *Pseudocaranx dentex* (Bloch & Schneider). *J Fish Dis.* 1993; 16:461–469.

39. Fukuda Y, Nguyen HD, Furuhashi M, and Nakai T. Mass mortality of cultured seven-band grouper, *Epinephelus septemfasciatus*, associated with viral nervous necrosis. *Fish Pathol.* 1996; 31:165–170.

40. Curtis PA et al. Nodavirus infection of juvenile white seabass, *Atractoscion nobilis*, cultured in southern California: First record of viral nervous necrosis (VNN) in North America. *J Fish Dis.* 2001; 24:263–271.

41. Barke DE et al. First report of piscine nodavirus infecting wild winter flounder *Pleuronectes americanus* in Passamaquoddy Bay, New Brunswick, Canada. *Dis Aquat Org.* 2002; 49:99–105.

42. Munday BL, Kwang J, and Moody N. Betanodavirus infections of teleost fish: A review. *J Fish Dis.* 2002; 25:127–142.

43. Azad IS et al. Nodavirus infection causes mortalities in hatchery produced larvae of *Lates calcarifer*: First report from India. *Dis Aquat Org.* 2005; 63:113–118.

44. Cutrin JM et al. Emergence of pathogenic betanodaviruses belonging to the SJNNV genogroup in farmed fish species from the Iberian Peninsula. *J Fish Dis.* 2007; 30:225–232.

45. Hata N, Okinaka Y, Sakamoto T, Iwamoto T, and Nakai T. Upper temperature limits for the multiplication of betanodaviruses. *Fish Pathol.* 2007; 42:225–228.

46. Korsnes K, Karlsbakk E, Nylund A, and Nerland AH. Horizontal transmission of nervous necrosis virus between turbot Scophthalmus maximus and Atlantic cod Gadus morhua using cohabitation challenge. *Dis Aquat Org.* 2012; 99(1):13–21.

47. Gomez DK, Mori K, Okinaka Y, Nakai T, and Park SC. Trash fish can be a source of betanodaviruses for cultured marine fish. *Aquaculture.* 2010; 302:158–163.

48. Grotmol S, Bergh O, and Totland GK. Transmission of viral encephalopathy and retinopathy (VER) to yolk-sac larvae of the Atlantic halibut *Hippoglossus hippoglossus*: Occurrence of nodavirus in various organs and a possible route of infection. *Dis Aquat Org.* 1999;36:95–106.

49. Breuil G, Pepin JFP, Boscher S, and Thiery R. Experimental vertical transmission of nodavirus from broodfish to eggs and larvae of the sea bass, *Dicentrarchus labrax. J Fish Dis.* 2002; 25:697–702.

50. Kai YH, Su HM, Tai KT, and Chi SC. Vaccination of grouper broodfish (*Epinephelus tukula*) reduces the risk of vertical transmission by nervous necrosis virus. *Vaccine.* 2010; 28(4):996–1001.

51. Novel P et al. Two Mx genes identified in European sea bass (*Dicentrarchus labrax*) respond differently to VNNV infection. *Vet Immunol Immunopathol.* 2013; 153(3–4):240–248.

52. Hong JR. Betanodavirus: Mitochondrial disruption and necrotic cell death. *World J Virol.* 2013; 2(1):1–5.

53. Chen YM, Wang TY, and Chen TY. Immunity to betanodavirus infections of marine fish. *Dev Comp Immunol.* 2014; 43(2):174–183.

54. Frerichs GN, Rodger HD, and Peric Z. Cell culture isolation of piscine neuropathy nodavirus from juvenile sea bass, *Dicentrarchus labrax. J Gen Virol.* 1996; 77:2067–2071.

55. Iwamoto T, Mori K, Arimoto M, and Nakai T. High permissivity of the fish cell line SSN-1 for piscine nodaviruses. *Dis Aquat Org.* 1999; 39:37–47.

56. Iwamoto T, Nakai T, Mori K, Arimoto M, and Furusawa I. Cloning of the fish cell line SSN-1 for piscine nodaviruses. *Dis Aquat Org.* 2000; 43:81–89.

57. Chi SC, Hu WW, and Lo BL. Establishment and characterization of a continuous cell line (GF-1) derived from grouper, *Epinephelus coioides* (Hamilton): A cell line susceptible to grouper nervous necrosis virus (GNNV). *J Fish Dis.* 1999; 22:173–182.

58. Chi SC, Wu YC, and Cheng TM. Persistent infection of betanodavirus in a novel cell line derived from the brain tissue of barramundi *Lates calcarifer. Dis Aquat Org.* 2005; 65:91–98.

59. Aranguren R, Tafalla C, Novoa B, and Figueras A. Nodavirus replication in a turbot cell line. *J Fish Dis.* 2002; 25:361–366.

60. Arimoto M et al. 1992. Detection of striped jack nervous necrosis virus (SJNNV) by enzyme-linked immunosorbent assay (ELISA). *Fish Pathol.* 27:191–195.

61. Gomez DK et al. Detection of betanodaviruses in apparently healthy aquarium fishes and invertebrates. *J Vet Sci.* 2006; 7(4):369–374.

62. Gomez DK, Baeck GW, Kim JH, Choresca CH Jr, and Park SC. Molecular detection of betanodaviruses from apparently healthy wild marine invertebrates. *J Invertebr Pathol.* 2008; 97(3):197–202.

63. Suzuki K. Detection of DNAs homologous to betanodavirus genome RNAs in barfin flounder *Verasper moseri* and Japanese flounder *Paralichthys olivaceus. Dis Aquat Org.* 2006;72(3):225–239.

64. Toubanaki DK, Margaroni M, and Karagouni E. Nanoparticle-based lateral flow biosensor for visual detection of fish nervous necrosis virus amplification products. *Mol Cell Probes.* March 2015; 29(3):158–166.

65. Nishizawa T, Furuhashi M, Nagai T, Nakai T, and Muroga K. Genomic classification of fish nodaviruses by molecular phylogenetic analysis of the coat protein gene. *Appl Environ Microbiol.* 1997; 63(4):1633–1636.

66. Thiery R, Raymond JC, and Castric J. Natural outbreak of viral encephalopathy and retinopathy in juvenile sea bass, *Dicentrarchus labrax*: Study by nested reverse transcriptase-polymerase chain reaction. *Virus Res.* 1999; 63(1–2):11–17.

67. Grotmol S et al. Characterisation of the capsid protein gene from a nodavirus strain affecting the Atlantic halibut *Hippoglossus hippoglossus* and design of an optimal reverse-transcriptase polymerase chain reaction (RT-PCR) detection assay. *Dis Aquat Org.* 2000; 39:79–88.

68. Huang B et al. Detection of nodavirus in barramundi, *Lates calcarifer* (Bloch), using recombinant coat protein-based ELISA and RT-PCR. *J Fish Dis.* 2001; 24:135–141.

69. Gomez DK et al. PCR-based detection of betanodaviruses from cultured and wild marine fish with no clinical signs. *J Fish Dis.* 2004; 27:603–608.

70. Gomez DK, Baeck GW, Kim JH, Choresca CH Jr, and Park SC. Molecular detection of betanodavirus in wild marine fish populations in Korea. *J Vet Diagn Invest.* 2008; 20:38–44.

71. Hick P, Tweedie A, and Whittington RJ. Optimization of Betanodavirus culture and enumeration in striped snakehead fish cells. *J Vet Diagn Invest.* 2011; 23(3):465–475.

72. Mazelet L, Dietrich J, and Rolland JL. New RT-qPCR assay for viral nervous necrosis virus detection in sea bass, *Dicentrarchus labrax* (L.): Application and limits for hatcheries sanitary control. *Fish Shellfish Immunol.* 2011; 30(1):27–32.

73. Baud M et al. First generic one step real-time TaqMan RT-PCR targeting the RNA1 of betanodaviruses. *J Virol Methods.* 2015; 211:1–7.

74. Sung CH and Lu JK. Reverse transcription loop-mediated iso-thermal amplification for rapid and sensitive detection of nervous necrosis virus in groupers. *J Virol Methods.* 2009; 159(2):206–210.

75. Xu HD, Feng J, Guo ZX, Ou YJ, and Wang JY. Detection of red-spotted grouper nervous necrosis virus by loop-mediated iso-thermal amplification. *J Virol Methods.* 2010; 163(1):123–128.

76. Suebsing R, Oh MJ, and Kim JH. Development of a reverse transcription loop-mediated isothermal amplification assay for detecting nervous necrosis virus in olive flounder *Paralichthys olivaceus. J Microbiol Biotechnol.* 2012; 22(7):1021–1028.

77. Mekata T et al. Development of simple, rapid and sensitive detection assay for grouper nervous necrosis virus using real-time loop-mediated isothermal amplification. *J Fish Dis.* July 30, 2014. doi: 10.1111/jfd.12297.

78. Hick P and Whittington RJ. Optimisation and validation of a real-time reverse transcriptase-polymerase chain reaction assay for detection of betanodavirus. *J Virol Methods.* 2010; 163(2):368–367.

79. Arimoto M, Sato J, Maruyama K, Mimura G, and Furusawa I. Effect of chemical and physical treatments on the inactivation of striped jack nervous necrosis virus (SJNNV). *Aquaculture.* 1996; 143:15–22.

80. Frerichs GN, Tweedie A, Starkey WG, and Richards RH. Temperature, pH, and electrolyte sensitivity, and heat, UV and disinfectant inactivation of sea bass (*Dicentrarchus labrax*) neuropathy nodavirus. *Aquaculture.* 2000; 185:13–24.

81. Grotmol S and Totland GK. Surface disinfection of Atlantic halibut *Hippoglossus hippoglossus* eggs with ozonated sea-water inactivates nodavirus and increases survival of the lar-vae. *Dis Aquat Org.* 2000; 39:89–96.

82. Husgard S, Grotmol S, Hjeltnes B, Rodseth OM, and Biering E. Immune response to a recombinant capsid protein of striped jack nervous necrosis virus (SJNNV) in turbot *Scophthalmus maximus* and Atlantic halibut *Hippoglossus hippoglossus*, and evaluation of a vaccine against SJNNV. *Dis Aquat Org.* 2001; 45:33–44.

83. Thiery R et al. Induction of a protective immune response against viral nervous necrosis in the European sea bass *Dicentrarchus labrax* by using betanodavirus virus-like par-ticles. *J Virol.* 2006; 80(20):10201–10207.

84. Pakingking R Jr, Bautista NB, de Jesus-Ayson EG, and Reyes O. Protective immunity against viral nervous necrosis (VNN) in brown-marbled grouper (*Epinephelus fuscogutattus*) fol-lowing vaccination with inactivated betanodavirus. *Fish Shellfish Immunol.* 2010; 28(4):525–533.

85. Kuo HC et al. Real-time quantitative PCR assay for monitor-ing of nervous necrosis virus infection in grouper aquaculture. *J Clin Microbiol.* 2011; 49(3):1090–1096.

86. Haddad-Boubaker S et al. Molecular epidemiology of betanodaviruses isolated from sea bass and sea bream cultured along the Tunisian coasts. *Virus Genes.* 2013; 46(3):412–422.

3 Taura Syndrome Virus

Parin Chaivisuthangkura, Akapon Vaniksampanna, Phongthana Pasookhush, Siwaporn Longyant, and Paisarn Sithigorngul

CONTENTS

3.1 Introduction 17
    3.1.1 Classification... 17
    3.1.2 Genome Organization, Biology, and Epidemiology 17
    3.1.3 Clinical Features and Pathogenesis 19
    3.1.4 Diagnosis 19
        3.1.4.1 Conventional Techniques... 19
        3.1.4.2 Molecular Techniques.. 19
        3.1.4.3 Immunological Techniques.. 20
3.2 Methods 21
    3.2.1 Sample Preparation... 21
    3.2.2 Detection Procedures.. 22
        3.2.2.1 Gel-Based RT-PCR 22
        3.2.2.2 Quantitative RT-PCR 22
3.3 Conclusions... 23
References... 23

3.1 INTRODUCTION

Taura syndrome (TS) of peneid shrimp was described in 1992 in *Penaeus vannamei* harvested from shrimp farms near the Taura River in Ecuador [1]. TS later spread to many countries, including the United States, Taiwan, China, Thailand, and Indonesia [2–4]. The viral etiologic agent, Taura syndrome virus (TSV), was identified in 1994 through infectivity studies using specific pathogen-free *P. vannamei* as the host [5]. The principal hosts of TSV are white shrimp *P. vannamei* and blue shrimp *P. stylirostris*. However, other shrimp species such as *P. setiferus*, *P. monodon*, *P. chinensis*, *P. aztecus*, *P. duorarum*, *P. indicus*, and *Metapenaeus ensis* can be infected by the virus [6–12]. The disease often results in cumulative mortalities ranging from 40% to >90% of cultured populations of postlarval, juvenile, and subadult shrimp. Shrimp with TS are typically small juveniles weighing between 0.05 g and less than 5 g [6,8,13]. The horizontal transmission of TSV by cannibalism or contaminated water has been documented [5,6,8,14], but vertical transmission has not been experimentally verified. Currently, there are no vaccines or chemotherapy treatments available. Therefore, to reduce the risk of TSV infection during shrimp farming, screening of brood stock and spawned eggs/nauplii using polymerase chain reaction (PCR) has been applied. The development of specific pathogen-free shrimp stocks of *P. vannamei* and *P. stylirostris* has proven to be the most successful husbandry practice for the prevention and control of TS [15].

3.1.1 CLASSIFICATION

Characteristics of TSV virions include a buoyant density of approximately 1.338 g mL^{-1} in CsCl, icosahedral morphology, a size of 31–32 nm in diameter, and a positive sense single-stranded RNA (ssRNA) genome that is polyadenylated at the 3′ end [16]. Therefore, the virus was initially classified as a member of the family *Picornaviridae* [5,16]. However, genomic characterization and sequence comparison revealed that TSV is closely related to the cricket paralysis-like virus [17]. Presently, TSV has been classified by the International Committee on Taxonomy of Viruses into the novel genus *Aparavirus* in a new family *Dicistroviridae* in the order *Picornavirales* [18].

3.1.2 GENOME ORGANIZATION, BIOLOGY, AND EPIDEMIOLOGY

TSV is small, with a nonenveloped icosahedral shape. The ssRNA genome is 10,205 nucleotides in length, excluding the 3′-poly(A) tail, and contains two large open reading frames (ORFs) separated by an intergenic region (IGR) of 207 nucleotides [17]. The amino acid sequence of ORF1 consists of helicase, a protease, and an RNA-dependent RNA polymerase (RdRp). ORF2 contains the sequences of structural proteins, including three major capsid proteins, VP1 (CP2, 55 kDa), VP2 (40 kDa), and VP3 (24 kDa). ORF1 and ORF2 represent 92% of the TSV genome, and the remaining 8%

consists of an untranslated region (UTR). In the 3'-UTR, no putative polyadenylation signal (AAUAAA) was identified [17]. Sequence homology of the IGR between TSV and other members of the cricket paralysis-like viruses suggested that TSV ORF2 was translated by an internal ribosome entry site (IRES) [17]. Unlike other IRES-mediated translation initiations, the IGR-IRES-mediated initiation of translation in discistroviruses was shown to occur in the absence of base-pair interactions between the initiation codon and the anticodon triplet in initiator methionine tRNA (Met-tRNAi). Therefore, the IRES did not require an AUG initiation [19]. It has been shown that TSV translation from IGR-IRES started from a non-AUG start codon and did not involve initiator methionyl-tRNA[Met]. Translation from the TSV IGR-IRES initiated with alanyl-tRNA[Ala] even though there was an in-frame AUG codon positioned two codons upstream of the GCU-alanine codon of the TSV structural polyprotein [20].

In the absence of initiation factors, TSV IGR-IRES elements can recruit ribosomes and direct protein synthesis in a reconstituted system containing only purified ribosomal subunits, eukaryotic elongation factors 1A and 2, and aminoacylated tRNAs [20]. A computational-based structure search suggested that TSV had an internal stem-loop structure in the predicted IRES, and *in vitro* translation analysis showed that TSV had a structurally distinct IGR-IRES [21]. The IGR-IRESs contain three pseudoknots, one of which, called pseudoknot I (PKI), is essential for the function of IGR-IRES [22]. A cryomicroscopy structure analysis revealed that PKI of TSV IRES occupied the ribosomal decoding center at the aminoacyl (A) site in a manner similar to that of the tRNA anticodon/mRNA codon. Moreover, the structure indicated that TSV IRES initiated translation by an unprecedented mechanism. Specifically, the ORF of the IRES-driven mRNA was established by the placement of the tRNA–mRNA-like structure in the A site, whereas the 40S P site remained unoccupied during the initial step [23].

VP1 (equivalent to capsid protein 2; CP2) demonstrates greater variation in its amino acid sequence (3.5%) than VP2 (CP1) and VP3 (CP3) (both 0.8%) [24]. Therefore, the CP2 region has been used to establish the genetic relationship among TSV isolates. Currently, at least four genotypic variants have been identified according to the sequence of the VP1 structural protein. These variants are the Americas group, the Southeast Asian group, the Belize group, and the Venezuelan group [12,25,26]. In 2010, a newly emerged strain of TSV isolated from Colombian shrimp farms called CO 10 was analyzed. The phylogenetic analysis based on the CP2 amino acid sequence indicated that the CO 10 strain forms a new cluster and differs from the previous Colombia isolates. The results revealed the diversity of TSV CP2 in the Americas group leading to two new subclusters in South America (Venezuela and Colombia) and a new subcluster in Mexico in the Central America variant [27].

Infection of TSV occurs in tissues derived from the ectoderm and mesoderm. Infection in the cuticular epithelium was most prevalent in shrimp with acute infections, and the infection in lymphoid organs was observed in shrimp with chronic infections. Histological analysis revealed no signs of infection of TSV in the enteric organs [10,28–30].

In TSV vector studies, it has been demonstrated that the feces of seagulls (*Larus atricilla*) can contain infectious TSV particles for up to 6 h after consumption of viral-infected shrimp tissues. Therefore, this wild seabird may serve as a vector for transmission of TSV to aquatic environments [31]. In addition, the water boatman *Trichocorixa reticulata* (*Corixidae*), an aquatic insect collected from ponds with active TS, was shown to be TSV positive using *in situ* hybridization (ISH) assay in the gut [6,32].

A study on potential carriers of TSV was conducted in Thailand with five common native crustaceans found near shrimp ponds including two palaemonid shrimp species, *Palaemon styliferus* and *Macrobrachium lanchesteri*, and three crab species, *Sesarma mederi*, *Scylla serrata*, and *Uca vocans*. The results showed that *U. vocans*, *S. serrata*, and *S. mederi* gave strong reverse transcription polymerase chain reaction (RT-PCR) at 10, 15, and 50 days (d), respectively, after feeding with TSV-infecting *P. vannamei* carcasses. Also after feeding, a strong RT-PCR reaction was observed in *P. stylirostris* at 5 d postchallenge but no reaction occurred at 15 d. In the case of *M. lanchesteri*, a light RT-PCR reaction was observed at 2 d after feeding but no reaction occurred at 15 d. These results suggest that these crabs and palaemonid shrimp species can pose a risk as potential carriers for TSV transmission [33].

TS first appeared in Ecuador in 1992 [1] and may have caused up to US$100 million worth of damage [6]. TS has spread progressively to the United States, including the states of Hawaii, Texas, and Florida [34]. It was estimated that the outbreaks of TS in Southern and Central Texas led to US$10 million losses [6]. TS also caused an impact on shrimp farms in other countries in South and Central America, including Peru, Honduras, Guatemala, El Salvador, Brazil, Nicaragua, Belize, and Mexico [34]. TSV may have spread from South America to Asia due to importation of infected live *P. vannamei*, which became a primary shrimp species in Asia during the 1990s. The first TS outbreak in Asia occurred in Taiwan in late 1998 to early 1999, and shrimp production dropped abruptly to as low as 10% of the volume produced in early 1998. It was assumed that TSV was transferred to Taiwan by contaminated postlarvae and spawners from Ecuador and elsewhere [2]. In Thailand, the first TS outbreak was confirmed in 2003 after legal importation of *P. vannamei* in mid-2002. During that time, over 150,000 brood stock shrimp were officially imported from China and Taiwan [35]. In 2004, mass mortality of *P. vannamei* shrimp cultured in Korea occurred, and it was suggested that TSV was introduced to Korea via imported stocks of *P. vannamei*. Based on the partial nucleotide sequences of VP1, two Korean TSV isolates were 96%–99% similar to those of TSV isolates from the Americas, Taiwan, and Thailand [36].

During the mid-1990s, TSV caused devastation to the shrimp industries of Ecuador and Colombia. At that time, 20%–30% of shrimp survived, and various staff including technicians and researchers speculated that the survivors could be genetically resistant animals. Therefore, one of the major shrimp producers in Colombia initiated a program to select the survivors from infected ponds and used them as parents for the next generation in a simple mass selection process. Within two to three generations, the success of this simple scheme increased survival rates to previous levels [37].

In response to TSV outbreaks in the United States, the U.S. Marine Shrimp Farming Program initiated a selective breeding program to improve TSV resistance in *P. vannamei*, operated by the Oceanic Institute (OI). From 1995 to 1998, the OI (Waimanalo, Hawaii, United States) operated a breeding program based on a selection index weighted equally for growth and resistance to TSV. One line demonstrated an 18.4% increase in survival to TSV after one generation of selection compared with unselected control families [38]. The OI had been selecting *P. vannamei* for TSV resistance for over 15 generations, and several families per generation exhibited 100% survival after exposure [39]. Over the past decade, several shrimp breeding programs have distributed TSV-resistant brood stock to commercial shrimp hatcheries worldwide. Therefore, the availability of TSV-resistant shrimp is of great benefit to the shrimp farming industry, and TSV is no longer considered to be a major threat in many shrimp farming regions [40].

The first transcriptomic study in hemolymph and hemocytes of TSV-resistant and susceptible *P. vannamei* was recently conducted using high-throughput RNA-seq. The comparison of gene expression between resistant and susceptible shrimp revealed several differentially expressed genes involved in the immune response activity. These included (1) pathogen recognition through cues such as immune regulators, adhesive proteins, and signal transducers, (2) coagulation, (3) proPO pathway, (4) antioxidation, and (5) protease [41].

3.1.3 Clinical Features and Pathogenesis

TSV infection in *P. vannamei* exhibits gross pathology in three distinct phases including acute, transition, and chronic phases.

During the acute phase of disease outbreak in *P. vannamei*, onset of mortality is often sudden and massive with moribund shrimp approaching the pond edge and surface water. Moribund shrimp in the acute phase demonstrate symptoms such as a soft shell, pink to red coloration due to the expansion of red cuticular chromatophores especially in the tail fan, and visible necrosis of the epithelial tissue. Infected shrimps are lethargic, stop feeding despite having an empty gut, and typically survive less than 24 h. However, the acute phase in a shrimp farm pond may last for several days in an affected population. Death often occurs in the acute phase during molting [6,42,43].

In the transition phase, some affected shrimp may recover from the infection. The shrimp typically demonstrate randomly distributed variably sized melanized lesions in or under the cuticle where the acute phase necrosis occurs. Shrimp with these black lesions are at some risk of mortality during their succeeding molt. Shrimp that survive the next molting

process typically look normal and the lesions disappear from the cuticle. In the transition phase, death may also occur due either to osmotic failure as a consequence of widespread destruction of cuticular epithelium or to systematic infection from opportunistic bacteria [6,42,43].

In the chronic phase, shrimp may carry TSV throughout their life as a persistent infection. They may appear and behave normally but will show slightly less tolerance to low salinity stress than uninfected shrimp. The lymphoid organ of *P. vannamei* in this phase usually has spheroids (proliferative nodules of pale staining), vacuolated cells, and a lack of the central vessel that is typical for normal lymphoid organ tubules [6,42,43].

3.1.4 Diagnosis

3.1.4.1 Conventional Techniques

Histopathology changes in acute TSV-infected shrimp present as multifocal areas of necrosis in the cuticular epithelium and subcuticular connective tissue. Infection in the underlining muscle, heart, and lymphoid organ is occasionally observed. Pyknotic and karyorrhectic nuclei are scattered throughout the infected areas. Cytoplasmic remnants of the necrotic cells are abundant and present as spherical inclusion bodies that range from 1 to 20 µm in diameter and are often described as "peppered" or "buckshot" in appearance. In hematoxylin and eosin (H&E) staining, the inclusions show eosinophilic to pale basophilic remnants [6,42,43].

Shrimp that survive the acute infection may undergo the transition phase in which the typical acute-phase cuticular lesions decline in both abundance and severity. In the affected areas, infiltration and accumulation of hemocytes at the site of necrosis are conspicuous and become melanized, giving rise to the irregular black spots. Invasion of the *Vibrio* spp. may be observed on the corrosive cuticle. The lymphoid organ may appear normal with H&E staining [43,44].

In shrimp with chronic infection, the irregularly shaped and sized melanized lesions on the cuticle mostly disappear and display no gross signs of infection. The lymphoid organ in the chronic phase usually has spheroids, which are proliferative nodules of pale staining, vacuolated cells, and lack of a central vessel. When assayed via ISH with TSV probe or immunohistochemistry with monoclonal antibodies (MAbs), some cells gave positive signals for the virus [10,43,44].

3.1.4.2 Molecular Techniques

The early molecular assays consist of ISH and RT-PCR following the successful identification of TSV-specific gene targets [29,45,46]. The ISH method has been used for the routine histopathology diagnosis of TSV infection. However, false-negative results may be obtained if the shrimp tissues were preserved in Davidson's acetic acid–formaldehyde–alcohol solution for more than 24–48 h. The highly acidic pH (~3.5–4.0) of Davidson's fixative causes acid hydrolysis of TSV genomic RNA leading to false-negative probe results [28]. This problem can be avoided by using a neutral pH (~6.0–7.0) fixative called RNA-friendly fixative or through proper fixation (24–48 h) in Davidson's solution [28].

In the case of RT-PCR, Nunan et al. [46] designed a pair of primers called 9992F and 9195R based on a clone from TSV cDNA library. Using the one-step RT-PCR approach, the primers amplify a 231 base pair (bp) sequence of the TSV genome. Primer 9992F is located near the 3′ end of the IGR, and primer 9195R is located in the VP2 (CP1) gene of the TSV genome [17,46]. This RT-PCR assay allows detection of all known TSV genetic variants [43] and is recommended by the Office International des Epizooties (OIE) [15] for surveillance and diagnostic purposes.

An improved RT-PCR assay was developed by Navarro et al. [47] using newly designed primers, designated 7171F and 7511R, which amplify a 341 bp fragment. The detection limit of the improved RT-PCR assay was 20 copies of the TSV genome, 100 times more sensitive than the RT-PCR using the 9992F/9195R primers recommended by OIE. The primers 7171F/7511R were demonstrated to successfully detect TSV isolates from four phylogenetic lineages including Belize, the Americas, Southeast Asia, and Venezuela and displayed no cross-reactivity to infectious hypodermal and hematopoietic necrosis virus (IHHNV), white spot syndrome virus (WSSV), yellow-head virus (YHV), or infectious myonecrosis virus (IMNV) [47]. In addition to single RT-PCR methods, multiplex RT-PCR for detection of TSV and WSSV was developed and could be used to detect coinfection with WSSV and TSV in *P. vannamei* [48].

In addition to gel-based conventional RT-PCR, quantitative real-time RT-PCR assays were developed [49–51]. The real-time RT-PCR, using primers and a TaqMan probe designed from the ORF1 region of the TSV genomic sequence, demonstrated amplification with TSV isolates from Hawaii, Texas, Colombia, Mexico, Belize, Indonesia, and Thailand. No amplification was observed with RNA from healthy shrimp or an isolate of YHV. This real-time RT-PCR method has a detection limit of 100 copies and could be used to detect both acutely and chronically infected shrimp [49]. This TaqMan probe RT-PCR assay has been recommended by OIE [15]. The real-time multiplex PCR method was also developed for the detection of WSSV, IHHNV, and TSV using three sets of oligonucleotide primers and TaqMan probes specific for each virus. However, the assay has a sensitivity of 2000 copies of TSV, which is 20 times lower than that afforded by the single RT-PCR assay described earlier [50].

The SYBR Green RT-PCR was evaluated for TSV detection, and the results indicated that increasing the amplicon size from ~50 to ~70–100 bp enhanced its specificity and sensitivity [51].

The alternative nucleic acid amplification techniques, including nucleic acid–based amplification (NASBA) and loop-mediated isothermal amplification (LAMP), were also employed for specific detection of TSV [52–55]. NASBA coupled with dot-blotting procedure was approximately five times less sensitive than that of commercial nested RT-PCR kit IQ2000 TSV Detection and Prevention System (Farming IntelliGene Technology Corporation, Taiwan).

No false-positive reaction was obtained with nontarget shrimp RNA viruses including gill-associated virus (GAV), YHV, and IMNV [52].

The agarose gel–based RT loop-mediated isothermal amplification (RT-LAMP) demonstrated 10 times more sensitivity than the RT-PCR recommended by OIE but less sensitivity than the nested RT-PCR IQ2000 kit. No amplification was observed with nucleic acids from other shrimp pathogens including YHV and WSSV [53]. The colorimetric dot-blot hybridization (DBH) was used to detect RT-LAMP amplicons and exhibited the same sensitivity as the nested RT-PCR IQ2000 method. There was no cross-reaction with other shrimp viruses, including WSSV, IHHNV, IMNV, GAV, monodon baculovirus, and hepatopancreatic parvovirus. However, in some cases, tRNA samples or very low concentrations of the target templates could lead to the production of the aberrant products. Due to the utilization of hybridization probe specific to LAMP products, the DBH has the capacity to eliminate false-positive reactions [55].

Further development of RT-LAMP combined with lateral flow dipstick (LFD) for detection of LAMP amplicons was demonstrated [54]. This strip detected biotin-labeled LAMP products that had been hybridized with a fluorescein isothiocyanate (FITC)-labeled probe. Without relying on special equipment, the LFD is immersed into the appropriate buffer containing LAMP amplicons and the results are available within 10 min. This assay had a detection limit comparable to that of nested RT-PCR and exhibited high specificity to TSV with no cross-reactivity with WSSV, IHHNV, or YHV [54].

As the early RT-LAMP procedures included a step in which the reaction tube must be opened for further detection, the risk of contamination between samples could occur. Given the LAMP amplification produced magnesium pyrophosphate ($Mg_2P_2O_7$) by-products correlated with the amount of amplified LAMP amplicons, resulting in turbidity, a portable turbidimeter was designed and used to measure the turbidity without opening the reaction tubes. The results showed that this RT-LAMP coupled with turbidimeter is equally sensitive to agarose gel RT-LAMP and IQ2000-nested RT-PCR kit, with reduced contamination risk [56].

Recently, a hyperbranched rolling circle amplification (HRCA) assay combined with a strip test was developed for TSV detection. The sensitivity was approximately 10 copies, 100 times higher than that of RT-PCR. The HRCA test was TSV-specific and showed no cross-reaction to shrimp viruses including WSSV, YHV, and IHHNV [57].

3.1.4.3 Immunological Techniques

Chicken and mouse polyclonal TSV antisera and MAbs were produced using purified TSV as the antigen [58]. One of the MAbs designated 1A1, specific to VP1, was used to detect TSV via western blot dot and immunohistochemistry. However, while this MAb can react with TSV isolates from Taiwan, the United States, and Hawaii, it does not recognize TSV isolates from Mexico, Nicaragua, Belize, or Venezuela [25,26,59]. MAbs against recombinant structural capsid proteins, including VP1, VP2 and VP3, were also developed [60–62].

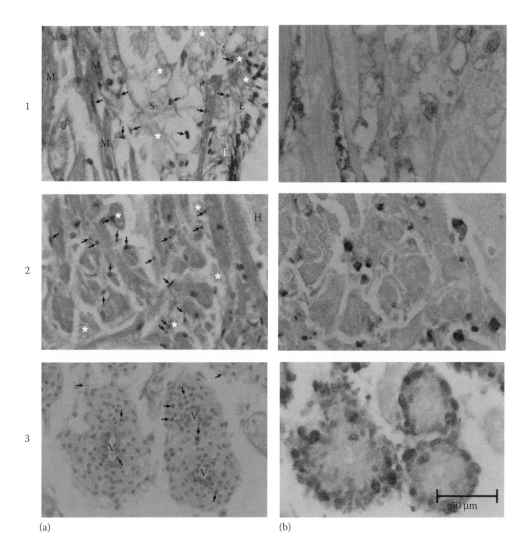

FIGURE 3.1   Histopathology and immunohistochemistry of tissues from naturally TSV-infected *P. vannamei* with asymptomatic infection in (1) cuticular epithelium, (2) heart, and (3) lymphoid organ. Two consecutive sections: (a) tissues were stained with hematoxylin and eosin (H&E) and (b) tissues were immunohistochemically stained with monoclonal antibody specific to VP2 of TSV and counterstained with eosin. Positive immunoreactivity appeared as brown staining. Intensely basophilic staining of pyknotic and karyorrhectic nuclei (arrow) and eosinophilic cytoplasmic remnants referred as "buckshot" appearance (*) are scattered in the lesion areas. *Note*: Lymphoid organ (3) demonstrated histologically normal in H&E staining; however, it demonstrated strong immunoreactivity of TSV infection. E, epithelium; S, subcuticular connective tissue; M, skeletal muscle; H, hemolymph; V, central vessel.

Through immunohistochemistry, these MAbs could be used to detect chronic and light TSV infection (Figure 3.1). It is likely that the combination of MAbs-specific to VP1, VP2, and VP3 may be useful for the detection of most TSV isolates from various geographic regions.

3.2 METHODS

3.2.1 SAMPLE PREPARATION

RT-PCR methods are recommended by the OIE for TSV monitoring and diagnostic purposes. Tissue samples such as hemolymph, pleopods, or whole small shrimp can be used for RT-PCR assays. Various commercial RNA extraction kits are available for TSV RNA isolation from fresh, frozen, and ethanol-preserved tissues.

In case of the High Pure RNA Tissue Kit (Roche, Penzberg, Germany), 400 µL of Lysis/Binding Buffer is added to the appropriate amount of tissue (max. 20–25 mg) in a nuclease-free 1.5 mL microcentrifuge tube. The tissues are disrupted and homogenized using a rotor–stator homogenizer, and the lysate is centrifuged for 2 min at maximum speed in a microcentrifuge. The supernatant is mixed with 200 µL of absolute ethanol. Next, the mixture is applied to the spin column combined with the collection tube (maximal volume 700 µL) and centrifuged at maximal speed ($13,000 \times g$) for 30 s, and then the flow-through is discarded. For each isolation, 10 µL of DNase I working solution is mixed with 90 µL of DNase incubation buffer, and the prepared solution is added to the spin column combined with the collection tube and incubated at 15°C–25°C for 15 min. The column is washed via the addition of 500 µL Wash Buffer I, centrifuged at $8000 \times g$ for 15 s, the flow-through is discarded, and the washing step is repeated with Wash Buffer II. Then, the column is washed by the addition of 300 μL Wash Buffer II and centrifuged at 13,000 × g for 2 min, and then the flow-through is discarded. Finally, the spin column is placed onto a new 1.5 mL microcentrifuge tube and 100 μL of elution buffer is added directly to the spin column and centrifuged at 8000 × g for 1 min. The eluted RNA can then be used for RT-PCRs or stored at −80°C for later analysis.

3.2.2 Detection Procedures

3.2.2.1 Gel-Based RT-PCR

Two oligonucleotide primers, designated 9992F and 9195R, are used in the RT-PCR, for specific amplification of a 231 bp product. The method outlined in the succeeding text for TSV detection is based on that described by Nunan et al. [46].

Primers	Product	Sequence	GC	Temperature
9992F	231 bp	5′-AAG TAG ACA GCC GCG CTT-3′	55%	69°C
9195R		5′-TCA ATG AGA GCT TGG TCC-3′	50%	63°C

The GeneAmp® EZ rTth RNA PCR kit (Applied Biosystems, Forster City, CA, USA) is used for the amplification reactions described in the succeeding text. The TSV positive and negative controls should be performed along with the no-template control.

3.2.2.1.1 Procedure

1. Prepare the RT-PCR reaction mix as follows (volumes listed are per specimen). (*Note*: Vortex and spin down all reagents before opening the tubes.)

Component	Volume (μL)
Diethyl pyrocarbonate (DEPC)-treated water	Variable
dNTPs (300 μM each)	Variable
9992F (0.46 μM)	Variable
9195R (0.46 μM)	Variable
Manganese acetate (2.5 mM)	Variable
5× EZ buffer	10
rTth DNA polymerase 2.5 U	1
RNA sample	10
Total volume	50

2. Add the RT-PCR reaction mix to a 0.2 mL PCR tube.
3. Spin the sample tubes in a microcentrifuge at 14,000 × g for 30 s at room temperature.
4. Place the sample tubes in a thermal cycler. RT may then proceed at 60°C for 30 min, followed by 94°C for 2 min. After the completion of RT, the samples are amplified for 40 cycles using the following conditions: denaturation at 94°C for 45 s, annealing/extension at 60°C for 45 s, and final extension at 60°C for 7 min.

5. Following the termination of the RT-PCR reaction, pulse spin the reaction tubes to pull down the condensation droplets at the inner wall of the tubes.
6. Analyze a 10 μL of the PCR products using 2.0% agarose gel, stained with ethidium bromide (0.5 g mL⁻¹), and electrophoresis in 0.5× TBE (Tris, boric acid, ethylenediaminetetraacetic acid). Then, visualize the sample under an ultraviolet light source.

3.2.2.2 Quantitative RT-PCR

The detection of TSV using quantitative RT-PCR is rapid, sensitive, and specific. The sensitivity of qRT-PCR is approximately 100 copies of the target sequence in the TSV genome [49]. The qRT-PCR method outlined in the succeeding text follows the procedure described by Tang et al. [49]. The designed primers and TaqMan probe are specific to the ORF1 region of the TSV genome (GenBank AF277675). The primers TSV1004F (5′-TTG GGC ACC AAA CGA CAT T-3′) and TSV1075R (5′-GGG AGC TTA AAC TGG ACA CAC TGT-3′) generate a 72 bp DNA fragment. The TaqMan probe TSV-P1 (5′-CAG CAC TGA CGC ACA ATA TTC GAG CAT C-3′) is labeled with fluorescent dyes 5-carboxylfluorescein (FAM) on the 5′ end and N,N,N',N'-tetramethyl-6-carboxyrhodamine (TAMRA) on the 3′ end (Applied Biosystems, Cat. no. 450025). The TSV-positive control and the no-template control should be included in each run.

3.2.2.2.1 Procedure

1. Extract RNA from shrimp tissue using the method described earlier.
2. Prepare the qRT-PCR reaction mix as follows using TaqMan One-Step RT-PCR master mixture (Applied Biosystems, part no. 4309169). (*Note*: Vortexing and spinning down the reaction mix before use is recommended.)

Component	Volume (μL)
2× Master mix without uracil-N-glycosylase (UNG)	12.50
40× MultiScribe and RNase Inhibitor Mix	0.63
TaqMan probe TSV-P1(0.1 μM)	Variable
TSV1004F (0.3 μM)	Variable
TSV1075R (0.3 μM)	Variable
RNA sample (5–50 ng)	Variable
DEPC-treated water	Variable
Total volume	25

3. Place the reaction tubes in the GeneAmp 5700 Sequence Detection System (Applied Biosystems); ABI PRISM 7000, 7300, and 7500 (newer models and brands are also suitable).
4. Perform the qRT-PCR cycling consisting of RT at 48°C for 30 min, initial denaturation at 95°C for

10 min, 40 cycles of denaturation at 95°C for 15 s, and annealing/extension at 60°C for 1 min.

5. Interpret the results from the threshold cycle obtained from real-time fluorescence measurements using a charge-coupled device camera. Samples will be interpreted as negative if the Ct (threshold cycle) value is 40 cycles or greater, whereas samples with Ct value lower than 40 cycles are interpreted as positive.

6. To confirm the results, the qRT-PCR product can be electrophoresed on a 4% agarose gel with ethidium bromide staining and exposed to UV light. The DNA fragment at 72 bp can be observed in the positive samples. (Optional)

3.3   CONCLUSIONS

TSV is a major viral pathogen in cultured white shrimp *P. vannamei*. First recognized in Ecuador in 1992 and later spread to many countries, TS disease usually results in mortalities between 40% and over 90% of infected shrimp. Based on genomic characterization and sequence comparison, TSV is classified to the novel genus *Aparavirus* in a new family *Dicistroviridae* (in the order *Picornavirales*). The ssRNA genome of TSV contains two ORFs separated by an IGR. It has been suggested that TSV ORF2 is translated by an IRES located within IGR. Currently, at least four genotypic lineages have been identified, including ones in Mexico, Southeast Asia, Belize/Nicaragua, and Venezuela/Aruba, according to the sequence of the VP1.

For molecular detection, OIE recommends the use of RT-PCR with a pair of primers called 9992F and 9195R, which can detect all known TSV genetic variants, and qRT-PCR using primers and TaqMan probes designed from the ORF1 region of the TSV genomic sequence for TSV surveillance and diagnostic purposes. However, several assays such as LAMP or monoclonal antibody-based techniques were developed for TSV detection with different sensitivities. The commercial nested RT-PCR kit is also available.

As no vaccine or chemotherapy treatment are presently available, breeding programs to improve TSV resistance in *P. vannamei* plays an important role in the control of TS. During the past several years, the shrimp breeding programs have distributed TSV-resistant brood stock to shrimp hatcheries worldwide, giving great benefit to the shrimp farming industry. As a consequence, in many shrimp farming regions, TSV is no longer considered a major threat.

REFERENCES

1. Jimenez, R., Sindrome de Taura (Resumen), in: Acuacultura del Ecuador. Camara Nacional de Acuacultura, Guayaquil, Ecuador, 1, 1992.
2. Tu, C. et al., Taura syndrome in Pacific white shrimp *Penaeus vannamei* cultured in Taiwan, *Dis. Aquat. Org.*, 38, 159, 1999.
3. Yu, C.I. and Song, Y.L., Outbreaks of Taura syndrome in Pacific white shrimp *Penaeus vannamei* cultured in Taiwan, *Fish Pathol.*, 35, 21, 2000.
4. Lien, T.W. et al., Genomic similarity of Taura syndrome virus (TSV) between Taiwan and western hemisphere isolates, *Fish Pathol.*, 37, 71, 2002.
5. Hasson, K.W. et al., Taura syndrome in *Penaeus vannamei*: Demonstration of a viral etiology, *Dis. Aquat. Org.*, 23, 115, 1995.
6. Brock, J.A., Special topic review: Taura syndrome, a disease important to shrimp farms in the Americas, *World J. Microbiol. Technol.*, 13, 415, 1997.
7. Chang, Y.S. et al., Genetic and phenotypic variations of isolates of shrimp Taura syndrome virus found in *Penaeus monodon* and *Metapenaeus ensis* in Taiwan, *J. Gen. Virol.*, 85, 2963, 2004.
8. Lightner, D.V., Epizootiology, distribution and the impact on international trade of two penaeid shrimp viruses in the Americas, *Rev. Sci. Tech. Office Int. Epiz.*, 15, 579, 1996.
9. Overstreet, R.M. et al., Susceptibility to Taura syndrome virus of some penaeid shrimp species native to the gulf of Mexico and southeastern United States, *J. Invert. Pathol.*, 69, 165, 1997.
10. Srisuvan, T., Tang, K.F.J., and Lightner, D.V., Experimental infection of *Penaeus monodon* with Taura syndrome virus (TSV), *Dis. Aquat. Org.*, 67, 1, 2006.
11. Stentiford, G.D., Bonami, J.R., and Alday-Sanz, V., A critical review of susceptibility of crustaceans to Taura syndrome, yellow head disease and white spot disease and implication of inclusion of these diseases in European legislation, *Aquaculture*, 291, 1, 2009.
12. Wertheim, J.O. et al., A quick fuse and the emergence of Taura syndrome virus, *Virology*, 390, 324, 2009.
13. Lotz, J.M., Effect of host size on virulence of Taura virus to the marine shrimp *Penaeus vannamei* (Crustacea: Penaeidae), *Dis. Aquat. Org.*, 30, 45, 1997.
14. White, B.L. et al., A laboratory challenge method for estimating Taura Syndrome virus resistance in selected lines of pacific white shrimp *Penaeus vannamei*, *J. World Aquacult. Soc.*, 33, 341, 2002.
15. OIE (Office International des Epizooties), *Manual of Diagnostic Tests for Aquatic Animal Diseases*, 7th edn., World Organization for Animal Health, Paris, France, 2014. http://www.oie.int/international-standard-setting/aquatic-manual/access-online/. Accessed on March 31, 2015.
16. Bonami, J.R. et al., Taura syndrome of marine penaeid shrimp: Characterization of the viral agent, *J. Gen. Virol.*, 78, 313, 1997.
17. Mari, J. et al., Shrimp Taura syndrome virus: Genomic characterization and similarity with members of the genus Cricket paralysis-like viruses, *J. Gen. Virol.*, 83, 917, 2002.
18. Chen, Y.P. et al., in: King, A.M.Q. et al. (eds.), Family *Dicistroviridae*. *Virus Taxonomy, Ninth Report of International Committee on Taxonomy of Viruses*, Elsevier, London, U.K., p. 840, 2012.
19. Sasaki, J. and Nakashima, N., Methionine-independent initiation of translation in the capsid protein of an insect RNA virus, *Proc. Natl. Acad. Sci. USA*, 97, 1512, 2000.
20. Cevallos, R.C. and Sarnow, P., Factor-independent assembly of elongation-competent ribosomes by an internal ribosome entry site located in an RNA virus that infects penaeid shrimp, *J. Virol.*, 79, 677, 2005.
21. Hatakeyama, Y. et al., Structural variant of the intergenic internal ribosome entry site elements in dicistroviruses and computational search for their counterparts, *RNA*, 10, 779, 2004.

22. Pfingsten, J.S., Costantino, D.A., and Kieft, J.S., Conservation and diversity among the three-dimensional folds of the Dicistroviridae intergenic region IRESes, *J. Mol. Biol.*, 370, 856, 2007.

23. Koha, C.S., Taura syndrome virus IRES initiates translation by binding its tRNA-mRNA–like structural element in the ribosomal decoding center, *Proc. Natl. Acad. Sci. USA*, 111, 9139, 2014.

24. Tang, K.F.J. and Lightner, D.V., Phylogenetic analysis of Taura syndrome virus isolates collected between 1993 and 2004 and virulence comparison between two isolates representing different genetic variants, *Virus Res.*, 112, 69, 2005.

25. Cote, I. et al., Taura syndrome virus from Venezuela is a new genetic variant, *Aquaculture*, 284, 62, 2008.

26. Erickson, H.S. et al., Taura syndrome virus from Belize represents a unique variant, *Dis. Aquat. Org.*, 64, 91, 2005.

27. Aranguren, L.F. et al., Characterization of a new strain of Taura syndrome virus (TSV) from Colombian shrimp farms and the implication in the selection of TSV resistant lines, *J. Invert. Pathol.*, 112, 68, 2013.

28. Hasson, K.W. et al., A new RNA-friendly fixative for the preservation of penaeid shrimp samples for virological detection using cDNA genomic probes, *J. Virol. Methods*, 66, 227, 1997.

29. Hasson, K.W. et al., The geographic distribution of Taura Syndrome Virus (TSV) in the Americas: Determination by histology and *in situ* hybridization using TSV-specific cDNA probes, *Aquaculture*, 171, 13, 1999.

30. Jimenez, R. et al., Periodic occurrence of epithelial viral necrosis outbreaks in *Penaeus vannamei* in Ecuador, *Dis. Aquat. Org.*, 42, 91, 2000.

31. Vanpatten, K.A., Nunan L.M., and Lightner, D.V., Seabirds as potential vectors of penaeid shrimp viruses and the development of a surrogate laboratory model utilizing domestic chickens, *Aquaculture*, 241, 31, 2004.

32. Lightner, D.V., Taura syndrome: An economically important viral disease impacting the shrimp farming industries of the Americas including the United States, *Proceedings of the 99th Annual Meeting US Animal Health Association*, Reno, NV, p. 36, 1995.

33. Kiatpathomchai, W. et al., Experimental infections reveal that common Thai crustaceans are potential carriers for spread of exotic Taura syndrome virus, *Dis. Aquat. Org.*, 79, 183, 2008.

34. Lightner, D.V. (ed.), *A Handbook of Shrimp Pathology and Diagnostic Procedures for Diseases of Cultured Penaeid Shrimp*, World Aquaculture Society, Baton Rouge, LA, 304pp., 1996.

35. Nielsen, L., Taura syndrome virus (TSV) in Thailand and its relationship to TSV in China and the Americas, *Dis. Aquat. Org.*, 63, 101, 2005.

36. Do, J.W. et al., Taura syndrome virus from *Penaeus vannamei* shrimp cultured in Korea, *Dis. Aquat. Org.*, 70, 171, 2006.

37. Cock, J. et al., Breeding for disease resistance of Penaeid shrimps, *Aquaculture*, 286, 1, 2009.

38. Argue, B.J. et al., Selective breeding of Pacific white shrimp *Litopenaeus vannamei* for growth and resistance to Taura syndrome virus, *Aquaculture*, 204, 447, 2002.

39. Moss, D.R. et al., Shrimp breeding for resistance to Taura syndrome virus, *Glob. Aquac. Advocate*, January/February, 40, 2011.

40. Moss, S.M. et al., The role of selective breeding and biosecurity in the prevention of disease in penaeid shrimp aquaculture, *J. Invert. Pathol.*, 110, 247, 2012.

41. Sookruksawong, S. et al., RNA-Seq analysis reveals genes associated with resistance to Taura syndrome virus (TSV) in the Pacific white shrimp *Litopenaeus vannamei*, *Dev. Comp. Immunol.*, 41, 523, 2013.

42. Flegel, T.W., Detection of major penaeid shrimp viruses in Asia, a historical perspective with emphasis on Thailand, *Aquaculture*, 258, 1, 2006.

43. Lightner, D.V., Virus diseases of farmed shrimp in the Western hemisphere (the Americas): A review, *J. Invert. Pathol.*, 106, 110, 2011.

44. Hasson, K.W. et al., Taura syndrome virus (TSV) lesion development and the disease cycle in the Pacific white shrimp *Penaeus vannamei*, *Dis. Aquat. Org.*, 36, 81, 1999.

45. Mari, J., Bonami, J.R., and Lightner, D.V., Taura syndrome of penaeid shrimp: Cloning of viral genome fragments and development of specific gene probes, *Dis. Aquat. Org.*, 33, 11, 1998.

46. Nunan, L.M., Poulos, B.T., and Lightner, D.V., Reverse transcription polymerase chain reaction (RT-PCR) used for the detection of Taura syndrome virus (TSV) in experimentally infected shrimp, *Dis. Aquat. Org.*, 34, 87, 1998.

47. Navarro, S.A. et al., An improved Taura syndrome virus (TSV) RT-PCR using newly designed primers, *Aquaculture*, 293, 290, 2009.

48. Tsai, J.M. et al., Simultaneous detection of white spot syndrome virus (WSSV) and Taura syndrome virus (TSV) by multiplex reverse transcription-polymerase chain reaction (RT-PCR) in Pacific white shrimp *Penaeus vannamei*, *Dis. Aquat. Org.*, 50, 9, 2002.

49. Tang, K.F.J., Wang, J., and Lightner, D.V., Quantitation of Taura Syndrome Virus by real-time RT-PCR with a TaqMan assay, *J. Virol. Methods*, 115, 109, 2004.

50. Xie, Z. et al., Development of a real-time multiplex PCR assay for detection of viral pathogens of penaeid shrimp, *Arch. Virol.*, 153, 2245, 2008.

51. Mouillesseaux, K.P., Klimpel, K.R., and Dhar, A.K., Improvement in the specificity and sensitivity of detection for the Taura syndrome virus and yellow head virus of penaeid shrimp by increasing the amplicon size in SYBR Green real-time RT-PCR, *J. Virol. Methods*, 111, 121, 2003.

52. Teng, P.H. et al., Rapid and sensitive detection of Taura syndrome virus using nucleic acid-based amplification, *Dis. Aquat. Org.*, 73, 13, 2006.

53. Kiatpathomchai, W. et al., Rapid and sensitive detection of Taura syndrome virus by reverse transcription loop-mediated isothermal amplification, *J. Virol. Methods*, 146, 125, 2007.

54. Kiatpathomchai, W. et al., Shrimp Taura syndrome virus detection by reverse transcription loop-mediated isothermal amplification combined with a lateral flow dipstick, *J. Virol. Methods*, 153, 214, 2008.

55. Teng, P.H. et al., Specific detection of reverse transcription-loop-mediated isothermal amplification amplicons for Taura syndrome virus by colorimetric dot-blot hybridization, *J. Virol. Methods*, 146, 317, 2007.

56. Sappat, A. et al., Detection of shrimp Taura syndrome virus by loop-mediated isothermal amplification using a designed portable multi-channel turbidimeter, *J. Virol. Methods*, 175, 141, 2011.

57. Zhao, Y. et al., Establishment of a sensitive and specific hyper-branched rolling circle amplification assay and test strip for TSV, *J. Virol. Methods*, 209, 41, 2014.

58. Poulos, B.T. et al., Production and use of antibodies for the detection of the Taura syndrome virus in penaeid shrimp, *Dis. Aquat. Org.*, 37, 99, 1999.

59. Erickson, H.S., Zarain-Herberg, M., and Lightner, D.V., Detection of Taura syndrome virus (TSV) strain differences using selected diagnostic methods: Diagnostic implications in penaeid shrimp, *Dis. Aquat. Org.*, 52, 1, 2002.

60. Longyant, S. et al., Specific monoclonal antibodies raised against Taura syndrome virus (TSV) capsid protein VP3 detect TSV in single and dual infections with white spot syndrome virus (WSSV), *Dis. Aquat. Org.*, 79, 75, 2008.

61. Chaivisuthangkura, P. et al., Improved sensitivity of Taura syndrome virus immunodetection with a monoclonal antibody against the recombinant VP2 capsid protein, *J. Virol. Methods*, 163, 433, 2010.

62. Hajimasalaeh, W. et al., Improved immunodetection of Taura syndrome virus using a monoclonal antibody specific for heterologously expressed VP1 capsid protein, *Arch. Virol.*, 158, 77, 2013.

4 Bee Paralysis Virus

Panuwan Chantawannakul

CONTENTS

4.1 Introduction 27
    4.1.1 Classification and Genome Organization 28
    4.1.2 Morphology, Biology, and Epidemiology 29
    4.1.3 Clinical Features and Pathogenesis 30
    4.1.4 Diagnosis 30
4.2 Methods 31
    4.2.1 Sample Preparation 31
        4.2.1.1 Sample Collection 31
        4.2.1.2 RNA Extraction Protocols 31
        4.2.1.3 Reverse Transcription for Conventional RT-PCR: cDNA Synthesis Protocol 31
    4.2.2 Detection Procedures 32
        4.2.2.1 Conventional RT-PCR Protocol 32
        4.2.2.2 Real-Time PCR Protocols 32
4.3 Conclusions 33
Acknowledgments 33
References 33

4.1 INTRODUCTION

Honeybees (*Apis* species) belong to the order *Hymenoptera*, which diverged from Diptera and Lepidoptera about 300 million years ago.[1] Apart from the consumable bee products (i.e., honey, royal jelly, propolis, bee wax, and pollen), honeybees play a crucial role as a key pollinator for plants and thus maintain natural ecosystems. Due to their importance to man, the complete genomes for the European honeybee (*Apis mellifera*), together with its pathogens and parasites,[2] were sequenced by 2006 to give insights into host–parasite relationships, transmission, and management of diseases to promote global bee populations.

Bee diseases are caused by microorganisms such as bacteria, fungi, protozoa, and viruses. Recently, bee viruses have garnered great interest as they could be found in seemingly healthy honeybee colonies; colonies may even contain multiple viruses that have potential to spread out by both horizontal and vertical transmission routes. It is believed that viruses are partly responsible for bee mortality especially when bee colony is weakened by parasitic bee mites known as *Varroa destructor*.[3,4] The unexplained incidence of rapid loss of adult bee populations or colony collapse disorder (CCD) has been a global concern since 2006.[5] Some beekeepers reported the losses of 50%–90% of bees[6] and many areas have suffered from the colony loss (including the United States, Canada, France, Sweden, and Germany[7,8]). A metagenomic survey recently detected Israeli acute paralysis virus (IAPV) in the United States, suggesting its connection to colony mortality.[9]

It is widely believed that the causes of bee losses are multifactorial, ranging from biological (e.g., virus, Nosema) to environmental factors (e.g., pesticides and nutrition). In view of the detrimental effects of bee viruses, it is necessary to better understand the relationship between honeybees and viruses for proper disease management and prevention of further colony losses.

Bee viruses are small and nonenveloped, and most of them have positive-sense single-stranded RNA. Based on their biophysical properties, they have been classified as picornavirus-like[10] along with other insect viruses, for example, Drosophila C virus,[11] cricket paralysis virus,[12] and aphid lethal paralysis virus.[13] The first bee virus discovered and reported was Sacbrood virus (SBV); it was described as a filterable agent that could cause the Sacbrood disease in honeybee after passing through fine unglazed porcelain filters.[14] Later, chronic bee paralysis virus (CBPV) and acute bee paralysis virus (ABPV) were found and isolated.[15–17] Since 1999, complete genomes of bee viruses have been accumulated, leading to their categorization on the basis of genomic structures. SBV,[18] deformed wing virus (DWV), Kakugo virus (KV),[19] and *Varroa destructor* virus (VDV-1)[20] are classified into the genus *Iflavirus*, which has a single large open reading frame (ORF) that encodes both structural and nonstructural proteins called monocistronic. The family *Dicistroviridae*, having the characteristic of two nonoverlapping ORFs (dicistronic), includes black queen cell virus,[21] ABPV,[22] Kashmir bee virus (KBV),[23] and IAPV.[24] CBPV has not yet been classified into any group due to lack of a full description of its genomic data. However, based on the deduced amino acid sequences of its conserved RNA-dependent RNA polymerase (RdRp) domains, CBPV appears to be phylogenically close to the *Nodaviridae* and *Tombusviridae* family clusters.[25]

Bee paralysis virus is aptly named based on clinical symptoms observed during infection. Infected bees become weak, tremble, and are unable to fly; therefore, they often crawl on the ground and may be seem *en masse* in front of hives. In severe case, the colonies collapse leaving the queen and a few workers on the neglected comb, which is reminiscent of CCD.[26] Paralysis in bees has been observed far back as Aristotle recorded hairless black bees having a dark and shiny appearance. The etiological agent, CBPV, was later confirmed and described.[17] Currently, two types of bee paralysis have been fully described: (1) acute bee paralysis, which is caused by ABPV and IAPV that are capable of killing the bees in a short time, and (2) chronic bee paralysis, of which CBPV is the causative agent. In this chapter, these bee paralysis viruses and characterization method are described.

4.1.1 CLASSIFICATION AND GENOME ORGANIZATION

RNA viruses contain only RNA as their genetic materials. Their classification is based on virus particle structure, genome, viral replication, host organism, and pathology. RNA virus has evolved to be a genetic element, protected by capsid proteins that can replicate in host cells. The RdRps, which are absent in cellular organisms, are universal and vital elements among RNA viruses. Thus, they are conserved and used as a standard tool to determine the evolutionary relationships among RNA viruses. The RdRps can recognize different

origins of replication at the 3′ termini of both plus and minus sense RNA strands to replicate viral genomes. RNA viruses lack proofreading machinery; therefore, coupled with a rapid, short replication time, a high mutation rate can be expected. RNA virus evolution can be more than a millionfold faster than its host. This suggests that the viruses facilitated independent and rapid evolution, resulting in better adapted host transmissibility. The RNA viruses are also known to have a quasi-species population structure.

In view of the genomic structure, ABPV is classified in the genus *Aparavirus* in the family *Dicistroviridae* along with KBV and IAPV.[27] They have a dicistronic genome organization, composed of two nonoverlapping ORFs that are spaced by an intergenic untranslated region (UTR). The 5′ proximal ORF encodes a nonstructural precursor that is autoproteolytically cleaved into an RNA helicase, cysteine protease with a chymotrypsin-like fold (Pro), and RdRp. The 3′ proximal ORF encodes capsid proteins (VP1-4) and has a poly-A tail (Figure 4.1). The intergenic UTR is known to contain an internal ribosome entry site (IRES) that initiates efficient translation.

Differences in the structure of intergenic internal ribosomal entry sites can be used to classify members of *Dicistroviridae*. *Aparaviruses* belong to Type II, which is based on the 5′ region sequence (UGAUCU and UGC Type I, UGGUUACCCAU and UAAGGCUU Type II). In addition, the structure of IRES elements in the two genera is also distinguished by the presence of an additional stem loop in the 3′ region of the intergenic region of IRES of *Aparavirus* but not in *Cripavirus*. The IRES site can facilitate production of the capsid protein. ABPV and IAPV have similar genomes; when excluding the poly-A tail, they are 9470 nt[22] and 9487 nt,[24] respectively. ABPV, IAPV, and KBV

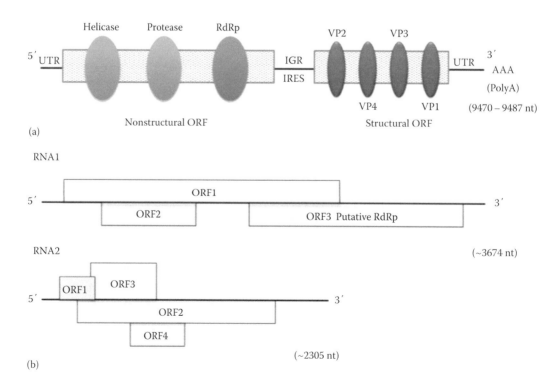

(a)

FIGURE 4.1 Genome organizations of (a) ABPV and IAPV and (b) CBPV.

are considered a closely related virus complex with common origins.[28] Most ORF sequences of IAPV share high similarities (more than 65%) with its closed relatives, KBV and ABPV; however, the homology of UTR sequences is weak among these three viruses.[24] The structural proteins among these three viruses also display high sequence identity. In contrast, antibodies of KBV, ABPV, and IAPV do not cross-react with proteins of these viruses; they are regarded as a unique viral entity.

CBPV is not yet classified into any genus or family, due to lack of full information about its genomic structure and protein function, despite sharing similarities with conserved RdRp sequence domains of the *Nodaviridae* and *Tombusviridae* families.[25] It is reported to be a positive single-stranded RNA virus with a characteristic segmented genome. The two RNAs were RNA1 (3674 nucleotides) and RNA2 (2305 nucleotides), which encode three and four putative overlapping ORFs, respectively.[29] The poly-A tails are not polyadenylated at the 3′ end as the ABPV genome, whereas the 5′ end has a cap structure that protects RNA from degradation by cellular exonuclease and ensure translation initiation by coupling with the cap-binding protein.

4.1.2 Morphology, Biology, and Epidemiology

ABPV and IAPV virions (size 30 nm) have icosahedral symmetry and do not have an envelope of distinct surface; conversely, CBPV virions (size 30–60 nm) are asymmetric.[30] These viruses are known to be host genus specific. Based on their genomes, ABPV and IAPV exhibit strong codon relatedness to the honeybee codon usage abundances, implying that viral proteins are more easily and highly expressed than if the codon usage abundances are unrelated like that of the fruit fly, red flour beetle, and mosquito.[31] This suggests that small host-specific viral genomes have the freedom to quickly optimize codon usage to successfully parasitize their preferred host. The optimization of codon usage toward highly abundant codons has previously been reported to occur in highly expressed genes[32]; optimization of honeybee viruses follows the same pattern. However, the viruses (ABP, IAPV, and KBV) belong to the family *Dicistroviridae*, which have more closely related codon usage than that of *Iflavirus* (KV, SBV, and VDV-1). This might be related to disease progression and virulence.[31] The main difference between ABPV, IAPV, and CBPV is that ABPV and IABPV cause sudden paralysis and death 1 day after infection, whereas CBPV displays lesser virulence and takes several days to kill the infected bees. CBPV also needs more viral titers to cause paralysis in the honeybee.[33]

When a virion attaches to the receptor of the host cell membrane, the RNA genome will be released into the host cell cytoplasm where replication of the viral genome occurs. The RNA also acts as mRNAs for the viral protein translation. By using the host translational machinery, the viral proteins are formed and the viral genome is replicated by using RdRp in the host cytoplasm. The newly formed positive strand RNA is assembled with the capsid proteins into progeny and ready to be released from the host cells. Interestingly, a segment

of IAPV can be integrated into the honeybee (*A. mellifera*) genome and vice versa.[24] The RNA segment thus acts as the recombination hotspot, and therefore, it explains the evidence for closely related codon usage in honeybee and virus. Interestingly, once the segment of IAPV is incorporated into the bee genome, the bee could be resistant to subsequent IAPV infection.[24] Information on the pathogenicity and cellular investigation and infection is not yet fully understood due to the lack of bee cell line cultures.

ABPV was first discovered during a laboratory study to characterize CBPV. ABPV is infective when injected into adult bees at similar doses to CBPV. The viruses are able to cause signs of disease such as weak appearance with trembling movement and inability to fly; they eventually kill both pupae and adult bees (ABPV 5–6 days, CBPV 7–8 days).

External stresses, including mite parasitism, starvation, or host nutritional status, which impair the host defense mechanism, can worsen the infected bees' condition causing clinical symptoms and morbidity. "Bee mite parasitic syndrome" is often used to refer when *V. destructor* affects the colony health by triggering severe viral infections in the honeybee colony. It is interesting to note that in areas where *V. destructor* is absent (e.g., Uganda), no bee samples could be detected having some known bee viruses including CBPV, ABPV, and IAPV.[34] Mite parasitism is known to weaken the infested bee colonies. There is evidence that *V. destructor* mite is associated with ABPV, IAPV, and KBV, and these viruses can cause bee mortality when the colony is infested with *V. destructor*.[4,35,36] It is known that *Varroa* can vector the viruses.[37,38] Nevertheless, CBPV has not yet been reported to associate with the mite as there is lack of evidence for CBPV presence in the parasitic mite.[25]

Both brood and adult bee can be infected with bee viruses. Horizontal and vertical transmissions have been reported as means of the spread of viral diseases in honeybees. Horizontal transmission could occur via contact, airborne, and fecal-oral route from food (honey, royal jelly, and pollen), which is contaminated with the virus from infected individuals, for example, salivary gland secretion. High population density within a colony leads to high contact and trophallactic rates. ABPV, IAPV, and CBPV were detected in colony foods such as pollen, honey, and royal jelly, but CBPV was found only in pollen.[38,39] The viruses can be found in bee digestive tract and feces[39] reaffirming the source of infection where the virus could cross the epithelial barrier of the digestive tract and invade different tissues. The viral load in the digestive tract is therefore much higher when compared to other tissues.[39] CBPV is concentrated in the head and also found in the mandibular and hypopharyngeal glands.[40,41] Under experimental conditions, following application of the virus on the surface of the cuticle of its freshly denuded hair, the virus readily infects the bee from close contact, possibly by invading through the epidermal cytoplasm. Bee viruses can be vertically transmitted through eggs (i.e., surface of eggs and transovarian route), sperm, and semen. The viruses can be detected in the ovaries of queen, eggs, and semen.[39] IAPV, ABPV, and CBPV genomic loads can be detected in high levels especially in summer time when the bee colonies are under high population density, thus possibly increasing the chances of transmission.

The prevalence of ABPV has been investigated in many countries worldwide.[42] Strain differences from clinical samples of different geographical origin are known and confirm the geographical segregation among viral strains.[43] This is also shown in CBPV strains in the United States.[44] ABPV is also commonly found in apparently healthy bees[45] and is the second most prevalent virus in Austria.[45] These surveys were done in European countries[45–47] and Thailand.[48] IAPV was first discovered in Israel.[24] Recently, it was found that the virus was present in five out of thirty five colonies (14%) in France by using reverse transcription polymerase chain reaction (RT-PCR). It is unlikely that there is a link between IAPV and colony loss during winter in France. However, the role of the virus in bee mortality is still unknown.[49] CBPV was found to attack adult bees and cause abnormal trembling of the body and wings. Other symptoms include a hairless, shiny, darkened appearance of the bees; these individuals are frequently attacked by guard bees when they return to their colonies. CBPV has a sporadic distribution including in the United States.[35,42,44] It was proposed that CBPV has a predominantly horizontal transmission.[25]

Bee viruses can cause simultaneous multiple infections in apiaries; for example, in Thailand, two viruses (DWV and ABPV) were found in 17% of samples, three viruses were found in 13%, and four viruses were found in 4%. The four viruses were DWV, ABPV, SBV, and KBV in samples collected from apiaries in Lampang province that were heavily infested with *V. destructor* mites.[48] Simultaneous multiple infections by bee viruses in honeybees are common in France, China, the United States, and Brazil.[4,50–52]

Apart from *A. mellifera* populations, another eight species of honeybee can be found in Asia, such as *Apis dorsata*, *A. florea*, *A. andreniformis*, and *A. cerana*. *A. cerana* is native to Asian countries where local beekeepers harvest their honey for consumption. By screening for all bee viruses in these bee species, IABPV and CBPV were present in *A. cerana cerana* in China[53] and IAPV was found in *A. cerana japonica*.[54] KBV and CBPV were found in the adult worker bees of *A. cerana* in South Korea at low levels (1.56% and 0.44%, respectively).[55] Honeybee viruses were also present in non-*Apis* hymenopteran pollinator such as bumblebees. ABPV was detected in five bumblebee species.[56] In containment greenhouse experiments, IAPV in honeybee could infect the bumblebees and vice versa within a week by sharing a food source (pollen).[57] This suggests that IAPV can be spread to the apiary environment.

4.1.3   CLINICAL FEATURES AND PATHOGENESIS

Most infected bees show no clear symptoms. The pathological effects of the infection mostly described are based mostly on the artificial injection/infection of the virus into honeybees under laboratory conditions. The natural tissue tropisms of the viruses are still not well documented. However, it is reported that environmental factors, for example, starvation and stress

arising from mite infestation, can hasten or aggravate the infection so that the bees show physical symptoms leading to mortality and eventually colony death. Environmental factors are considered to induce the changes from latent to lethal infection. Dicistrovirus prevalence and distribution in the honeybee populations is high; however, the symptoms or mortality are inconsistent. Virulence depends on the route of infection (e.g., infection of virus into the host hemocoel leads to rapid mortality). Varroa parasitism is associated with a lethal condition of the viruses. The virus particles can be injected during the mite feeding. Therefore, *V. destructor* are the vectors of viruses.[58] Infected bees may be asymptomatic or exhibit paralysis, uncoordinated movement, and death. APBV could cause bee mortality faster than CBPV.[17] Temperature is also important in the viral pathogenesis. ABPV-infected bees showed no symptoms at 35°C but died at 30°C.[59]

ABPV can be found in the brain tissue at high quantity in the dead bees.[60] Bees infected with IAPV also show shivering wings, paralysis, and death. IAPV can attack every stage and caste of honeybees and it replicates within all the bee tissues. However, it is concentrated in the gut and in the hypopharyngeal glands.[38] This explains the presence of IAPV in the food of bee colony. High viral load was also observed in the nerve tissues affecting the bee nervous system causing bees to exhibit paralysis, a typical nerve function impairment. The brood stage seems to be more susceptible to the viral infection than the adult bees.[38]

CBPV can damage the nervous system as it multiplies in the central nervous system, which contributes to the sensory processing, learning, memory storage, and control of motor patterns.[61–63] CBPV infects all developmental stages of honeybees. However, adult bees have high titers of virus than other brood stages, particularly in the old worker bees that function as guards at the entrance of the bee hive.[40] Interestingly, in the experimental viral injection of adult bees, the CBPV viral genome, without its capsid structure, is proved to be infectious. Naked CBPV RNAs of 10^9–10^{10} copies could replicate, produce viral particles in the host cells, and cause chronic paralysis symptom after 5 days of inoculation.[64]

Honeybees enlist different responses to defend viral infection. Adult bees tend to show a stronger response than the brood by altering gene expressions. Many studies focus on the immune response after infection. Immunity-related genes are triggered in adult bees more than that shown in brood after IAPV inoculation[38] so that the brood stage is more susceptible to infection than adult bees in which all defense mechanisms are fully developed.

4.1.4   DIAGNOSIS

The virus can be present in an "inapparent" infection of seemingly healthy bees or appears as an "overt" disease, and paralysis symptoms can be misinterpreted as other diseases and chemical intoxication. Therefore, appropriate diagnostic methods are necessary to confirm the cause of the disease. Based on the morphology or physical property of the bee viruses, they have buoyant densities in CsCl in the range of 1.33–1.42 g/mL with the sediment coefficients between 100S and 190S.[58] CBPV is isolated by separation on a 10%–40% (w/v) sucrose gradient.[29] The purified virus can be later observed under an electron microscope. However, the virions of honeybee viruses are similar in size and shape when observed under the electron microscope, and bee colonies or individual bee could be infected by multiple viral populations.[38] Therefore, it is difficult to observe in the field if the bees are infected or to purify each viral group for identification of the etiological agent. More specific and sensitive methods have been developed for bee virus detection. The conventional serological methods have been used for many years to detect or differentiate bee viruses. Enzyme-linked immunosorbent assay (ELISA) has proved to be useful to differentiate bee viruses.[65–67] However, KBV and ABPV sometimes show similarities in their serological reactions.[67] In addition, serological-based techniques require antibody production. With more information on viral genomes and also on the development of molecular techniques, ELISA and other immunological-based techniques have now been superseded by more specific and sensitive techniques, either conventional RT-PCR or real-time PCR. The molecular methods are widely applied nowadays for bee viral detection in honeybee or closely related insect samples.

4.2 METHODS

4.2.1 Sample Preparation

4.2.1.1 Sample Collection

Adult bees can be collected in a clean plastic bag. It is also recommended to take at least two types of samples (adult bees from the hive entrance and inside the hive) when studying the prevalence of bee viruses.[41] Honeybee eggs and brood can be collected in a clean plastic bag. Individual bee larvae and pupae can be collected in microcentrifuge tubes. The chemicals such as RNAlater® that prevent RNA degradation, or ethanol, can be added to this stage to preserve the samples. The samples should be kept on ice during transportation and frozen until RNA extraction.

4.2.1.2 RNA Extraction Protocols

Protocol A (Guanidine isothiocyanate-phenol-chloroform extraction)

1. Crush and homogenize the samples in the TRI_{ZOL}® reagent (1 mL/100 mg of tissue; *caution*: TRI_{ZOL} is toxic with skin and if swallowed).
2. Add 200 µL of chloroform into the sample (1 mL), shake vigorously, and incubate at room temperature for 3 min.
3. Centrifuge the solution at 12,000 × g for 10 min at 4°C. Transfer the aqueous phase (containing RNA) to a fresh tube.
4. Add 500 µL of isopropyl alcohol into a tube and incubate the sample at room temperature for 10 min.
5. Centrifuge at 12,000 × g for 10 min and discard the supernatant. The RNA precipitate forms a gel-like pellet on the bottom of the tube.

6. Wash the RNA precipitate by adding 1 mL of 75% ethanol.
7. Centrifuge at 7500 × g for 5 min at 4°C and pour off the supernatant and dry.
8. Dissolve RNA in 50 µL of RNase-free water and store the sample at −80°C.

Alternatively, RNA extraction for individual insect samples can be quickly done for the real-time PCR reaction as described Boonham et al.[68] This method is aimed at high-throughput testing:

Protocol B (Chelex resin method)

1. Individual insects are homogenized in a 0.5 mL microcentrifuge tube (using pellet grinders and matching tubes, Treff) with 50 µL diethylpyrocarbonate (DEPC)-treated water and stored on ice.
2. Chelex resin (Chelex 100, Biorad) (50 µL of a 50% w/v slurry) is added to each sample.
3. The samples are heated at 94°C for 5 min on a thermocycler.
4. The tubes are centrifuged for 5 min to pelletize debris and the supernatant is stored at −20°C for further PCR reactions/detection.

The RNA extraction also could be carried out following the conventional method or using RNA extraction kits that are commercially available.

4.2.1.3 Reverse Transcription for Conventional RT-PCR: cDNA Synthesis Protocol

The protocol is to synthesize cDNA from the RNA extract. The cDNA is prepared with random hexamer primers that can generate a bias-free cDNA copy of all RNA population, which is suitable for the virus survey or screening. The following is a standard RT protocol for cDNA synthesis:

1. Prepare the mixture on ice.

Component	Volume (µL)
5× first strand buffer	4
Moloney murine leukemia virus (M-MLV) reverse transcriptase (200 units)	1
A random hexamer (50 ng/µL)	1
Deoxynucleotide triphosphate (dNTP) mix (10 mM)	1
Dithiothreitol (DTT) (0.1 M)	2
DEPC-treated water	10
Sample RNA template (0.5 µg)	1

2. Mix gently by pipetting and spin the tube briefly to make sure that the contents are at the bottom of the tube.
3. Heat the mixture to 50°C for 50 min and terminate the reaction at 85°C for 5 min or 70°C for 15 min.[69]
4. Dilute the cDNA solution tenfold with nuclease-free water for further amplification reaction.

Alternatively, first-strand cDNA synthesis (Superscript® First-strand synthesis system for RT-PCR) (Invitrogen)

1. Prepare the RNA/primer mixture.

Component	Volume (µL)
RNA extract	1
Primer (50 µM oligo dT)	1
10 mM dNTP mix	1
DEPC-treated water	7

2. Incubate at 65°C for 5 min.
3. Place on ice for at least 1 min.
4. Prepare the cDNA synthesis mix.

Component	Volume (µL)
10× RT buffer	2
25 mM MgCl$_2$	4
0.1 M DTT	2
RNase OUT™ (40U/µL)	1
SuperScript™ III RT (200 U/µL)	1

5. Add 10 µL of the cDNA synthesis mix to each RNA/primer mixture.
6. Mix gently and incubate for 50 min at 50°C.
7. Terminate the reaction at 85°C for 5 min and chill on ice.
8. Collect the reactions by brief centrifugation and add 1 µL of RNase H to each tube.
9. Incubate at 37°C for 20 min and then the cDNA synthesis reaction can be used for further PCR reactions.

4.2.2 Detection Procedures

The difference of RT-PCR and real-time PCR is that the RT-PCR detects PCR products at an end point reaction but real-time PCR is used to detect the amplification reaction in real time. Real-time PCR is thus more sensitive and allows the quantification viral genome by Ct (threshold cycle values).

4.2.2.1 Conventional RT-PCR Protocol

Component	Volume (µL)
10× buffer	5
10 mM dNTP	1
50 mM MgCl$_2$	3
Primer F (10 µM)	2
Primer R (10 µM)	2
Taq polymerase (2.5 U)	0.5
cDNA	2
DEPC-treated water	34.5

Amplification is performed in a thermal cycler. The mixture is heated at 95°C for 2 min followed by 40 amplification cycles, each consisting of 30 s at 95°C, 45 s at 55°C, and 1 min at 72°C. Reactions are completed by a final extension step for 10 min at 72°C. The PCR products are viewed by 1.5% agarose gel electrophoresis. Fragment sizes are determined with reference to a 100 bp ladder. The primers used in the assay are shown in Table 4.1.

4.2.2.2 Real-Time PCR Protocols

Real-time PCR is conducted to quantify the viral numbers in the sample. There are two types of methods as follows:

4.2.2.2.1 SYBR® Green Assay

1. Amplification is performed in 10 µL reaction volumes consisting of the following:

Component	Volume (µL)
SsoFast EvaGreen supermix (SMX) (Biorad) (2 × real-time PCR mix contains dNTPs, Sso7d fusion polymerase, MgCl$_2$, EvaGreen dye, and stabilizers)	5
Primers (forward and reverse) 10 µM	1
Nuclease-free water	3
cDNA (template, diluted 1/10)	1

2. PCR reactions are carried out in a thermocycler, at 95°C for 30 s, 40 cycles of 95°C for 1 s, and 59°C for 5 s, followed by a melt-curve dissociation analysis.

4.2.2.2.2 TaqMan Assay

The master mix consists of the following:

Component	Volume (µL)
TaqMan buffer	2.5
MgCl$_2$	5.5
dNTP	2
Forward primer 300 nM	1
Reverse primer 300 nM	1
Probe 100 nM	0.5
AmpliTaq Gold (0.625 U)	0.125
MMLV (10 U)	0.125
Nuclease-free water	11.2

The reactions are set up in 96-well plates using TaqMan. For each reaction, 1 µL of the RNA extract is added to 24 µL of master mix in the appropriate well giving a final reaction volume of 25 µL. Plates are cycled using generic system conditions (48°C for 30 min, 95°C for 10 min, and 40 cycles of 60°C for 1 min plus 95°C for 15 s) within the 7900 Sequence Detection System (Applied Biosystems, Branchburg, New Jersey) using real-time data collection.

PCR products are then ligated into pGEM®-T easy vectors (Promega, USA) and transformed to *Escherichia coli* JM109 high-efficiency competent cells (Promega, USA) following the manufacturer's protocols. White bacterial

TABLE 4.1

Primers and Probes for the Detection of Acute Bee Paralysis Virus, Israeli Acute Bee Paralysis Virus, and Chronic Bee Paralysis Virus

Primer/Probe	Target/Size (bp)	Sequence (5'–3')	PCR	Reference/Gene ID
CBPV F	RdRp/455	AGTTGTCATGGTTAACAGGATACGAG	Conventional RT-PCR	AF375659[70]
CBPV R		TCTAATCTTAGCACGAAAGCCGAG		
CBPV F	RdRp/125	CAAAATCAACGAGCCAATCA	qPCR	AY004344.1[71]
CBPV R		AGTGTGAGGATCACCGGAAC		
CBPV 304F	RdRp/68	TCTGGCTCTGTCTTCGCAAA	TaqMan PCR	AF04230[37]
CBPV 371R		GATACCGTCGTCACCCTCATG		
CBPV 325T		TGCCCACCAATAGTTGGCAGTCTGC		
ABPV F	RdRp/452	TGAGAACACCTGTAATGTGG	Conventional RT-PCR	AF1506290[47]
ABPV R		ACCAGAGGGTTGACTGTGTG		
ABPV F	Capsid protein/124	ACCGACAAAGGGTATGATGC	qPCR	HM228893.1[71]
ABPV R		CTTGAGTTTGCGGTGTTCCT		
ABPV 95F	Capsid protein/65	TCCTATATCGACGACGAAAGACAA	TaqMan PCR	AF263733[37]
ABPV 159R		GCGCTTTAATTCCATCCAATTGA		
ABPV 121T		TTTCCCCGGACTTGAC		
IAPVF	Capsid protein/475	AGACACCAATCACGGACCTCAC	Conventional PCR	EF219380[24]
IAPVR		AGGATTTGTCTGTCTCCCAGTGCACAT		
IAPVF	RdRp/587	GCGGAGAATATAAGGCTCAG	qPCR	EF219380.1[71]
IAPVR		CTTGCAAGATAAGAAAGGGGG		

F, forward primer; R, reverse primer; T, probe. Probes consist of oligonucleotides with a 5'-reporter dye (FAM, 6-carboxyfluorescein) and a 3'-quencher (TAMRA, tetramethyl carboxyrhodamine).

colonies containing plasmids with inserts are selected and plasmid DNA is purified using the Wizard Plus SV Minipreps DNA purification system (Promega, USA). Purified plasmids are sequenced by DNA sequencing. The nucleotide sequences of amplicons are analyzed and compared with sequences published at Genbank, National Center for Biotechnology Information, NIH, in order to confirm specific amplification.

4.3 CONCLUSIONS

Honeybee virus paralysis diseases are known to cause bee death and colony collapse in honeybees worldwide. Due to the fact that the viruses can spread without showing any symptoms and simultaneous multiple infection commonly occurs, bee pathogen surveillance is therefore very important for efficient management schemes to prevent the spread of pathogens and diseases. This highlights the value of applying highly specific and sensitive detection/diagnosis methods. Given the current genomic knowledge of bee viruses, molecular diagnostic techniques are most efficient because of their high specificity and sensitivity. Furthermore, it is reliable and accessible to many laboratories compared to serological and other techniques. Nevertheless, RNA viruses are prone to mutate rapidly; therefore, the genetic variance of bee viruses should be studied further to correctly characterize the bee viruses. Host–parasite relationship is also an intriguing field of study, in particular when bee viruses can infect other insects outside their original host range.

ACKNOWLEDGMENTS

The author acknowledges the financial support of the Thailand Research Fund. Special thanks go to Dr. Geoffrey Williams and Dr. Alvin Yoshinaga for proofreading the manuscript.

REFERENCES

1. Grimaldi, D. and Engel, M.S. *Evolution of the Insects*, Cambridge University Press, Cambridge, U.K., 2005.
2. Weinstock, G.M. et al. Insights into social insects from the genome of the honeybee *Apis mellifera*. *Nature* 443, 931–949 (2006).
3. Ball, B.V. and Allen, M.F. The prevalence of pathogens in honey bee (*Apis mellifera*) colonies infested with the parasitic mite *Varroa jacobsoni*. *Annals of Applied Biology* 113, 237–244 (1988).
4. Chen, Y. et al. Multiple virus infections in the honey bee and genome divergence of honey bee viruses. *Journal of Invertebrate Pathology* 87, 84–93 (2004).
5. Stokstad, E. The case of the empty hives. *Science* 316, 970–972 (2007).
6. Frazier, M., van Engelsdorp, D., and Caron, D. Apiary news in Illinois colony collapse disorder in *Illinois State Beekeepers Association 1891*, Division of Natural Resources, Springfield, IL, 2007.
7. Underwood, R.M. and van Engelsdorp, D. Colony collapse disorder: Have we seen this before?. *Bee Culture* 35, 13–18 (2007).
8. Hamzelou, J. Where have all the bees gone? *The Lancet* 370, 639 (2007).
9. Cox-Foster, D.L. et al. A metagenomic survey of microbes in honey bee colony collapse disorder. *Science* 318, 283–287 (2007).

10. Moore, N.F., Reavy, B., and King, L.A. General characteristics, gene organization and expression of small RNA viruses of insects. *Journal of General Virology* 66, 647–659 (1985).

11. Johnson, K.N. and Christian, P.D. The novel genome organization of the insect picorna-like virus Drosophila C virus suggests this virus belongs to a previously undescribed virus family. *Journal of General Virology* 79, 191–203 (1998).

12. Wilson, J.E., Powell, M.J., Hoover, S.E., and Sarnow, P. Naturally occurring dicistronic cricket paralysis virus RNA is regulated by two internal ribosome entry sites. *Molecular and Cellular Biology* 20, 4990–4999 (2000).

13. Van Munster, M. et al. Sequence analysis and genomic organization of Aphid lethal paralysis virus: A new member of the family *Dicistroviridae*. *Journal of General Virology* 83, 3131–3138 (2002).

14. White, G.F. Sacbrood, a disease of bees, U.S. Department of Agriculture, Bureau of Entomology, Washington, DC, 1913.

15. Burnside, C.E. Preliminary observations on "paralysis" of honeybees. *Journal of Economic Entomology* 26, 162–168 (1933).

16. Burnside, C.E. The cause of paralysis of bees. *American Bee Journal* 85, 354–355 (1945).

17. Bailey, L., Gibbs, A.J., and Woods, R.D. Two viruses from adult honey bees (*Apis mellifera* Linnaeus). *Virology* 21, 390–395 (1963).

18. Ghosh, R.C., Ball, B.V., Willcocks, M.M., and Carter, M.J. The nucleotide sequence of sacbrood virus of the honey bee: An insect picorna-like virus. *Journal of General Virology* 80, 1541–1549 (1999).

19. Fujiyuki, T. et al. Novel insect picorna-like virus identified in the brains of aggressive worker honeybees. *Journal of Virology* 78, 1093–1100 (2004).

20. Ongus, J.R. et al. Complete sequence of a picorna-like virus of the genus *Iflavirus* replicating in the mite *Varroa destructor*. *Journal of General Virology* 85, 3747–3755 (2004).

21. Leat, N., Ball, B.V., Govan, V.A., and Davison, S. Analysis of the complete genome sequence of black queen-cell virus, a picorna-like virus of honey bees. *Journal of General Virology* 81, 2111–2119 (2000).

22. Govan, V.A., Leat, N., Allsopp, M., and Davison, S. Analysis of the complete genome sequence of acute bee paralysis virus shows that it belongs to the novel group of insect-infecting RNA viruses. *Virology* 277, 457–463 (2000).

23. de Miranda, J. et al. Complete nucleotide sequence of Kashmir bee virus and comparison with acute bee paralysis virus. *Journal of General Virology* 85, 2263–2270 (2004).

24. Maori, E. et al. Isolation and characterization of Israeli acute paralysis virus, a dicistrovirus affecting honeybees in Israel: Evidence for diversity due to intra- and inter-species recombination. *Journal of General Virology* 88, 3428–3438 (2007).

25. Ribière, M., Olivier, V., and Blanchard, P. Chronic bee paralysis: A disease and a virus like no other? *Journal of Invertebrate Pathology* 103, S120–S131 (2010).

26. Pettis, J.S., VanEngelsdorp, D., and Cox-Foster, D. Colony collapse disorder working group pathogen sub-group progress report. *American Bee Journal* 147, 595–597 (2007).

27. Mayo, M.A. Virology division news: Virus taxonomy—Houston 2002. *Archives of Virology* 147, 1071–1076 (2002).

28. de Miranda, J.R., Cordoni, G., and Budge, G. The Acute bee paralysis virus–Kashmir bee virus–Israeli acute paralysis virus complex. *Journal of Invertebrate Pathology* 103, S30–S47 (2010).

29. Olivier, V. et al. Molecular characterisation and phylogenetic analysis of Chronic bee paralysis virus, a honey bee virus. *Virus Research* 132, 59–68 (2008).

30. Bailey, L. and Ball, B.V. *Honey Bee Pathology*, 2nd edn., p. 193, Academic Press, London, U.K., 1991.

31. Chantawannakul, P. and Cutler, R.W. Convergent host–parasite codon usage between honeybee and bee associated viral genomes. *Journal of Invertebrate Pathology* 98, 206–210 (2008).

32. Cutler, R.W. and Chantawannakul, P. The effect of local nucleotides on synonymous codon usage in the honeybee (*Apis mellifera* L.) genome. *Journal of Molecular Evolution* 64, 637–645 (2007).

33. Bailey, L. Paralysis of the honey bee, *Apis mellifera* Linnaeus. *Journal of Invertebrate Pathology* 7, 132–140 (1965).

34. Kajobe, R. et al. First molecular detection of a viral pathogen in Ugandan honey bees. *Journal of Invertebrate Pathology* 104, 153–156 (2010).

35. Allen, M.F. and Ball, B.V. The incidence and world distribution of honey bee viruses. *Bee World* 77, 141–162 (1996).

36. Shen, M., Yang, X., Cox-Foster, D., and Cui, L. The role of varroa mites in infections of Kashmir bee virus (KBV) and deformed wing virus (DWV) in honey bees. *Virology* 342, 141–149 (2005).

37. Chantawannakul, P., Ward, L., Boonham, N., and Brown, M. A scientific note on the detection of honeybee viruses using real-time PCR (TaqMan) in Varroa mites collected from a Thai honeybee (*Apis mellifera*) apiary. *Journal of Invertebrate Pathology* 91, 69–73 (2006).

38. Chen, Y. et al. Israeli acute paralysis virus: Epidemiology, pathogenesis and implications for honey bee health. *PLOS Pathogens* 10, e1004261 (2014).

39. Chen, Y., Evans, J.D., and Feldlaufer, M. Horizontal and vertical transmission of viruses in the honey bee, *Apis mellifera*. *Journal of Invertebrate Pathology* 92, 152–159 (2006).

40. Blanchard, P. et al. Evaluation of a real-time two-step RT-PCR assay for quantitation of Chronic bee paralysis virus (CBPV) genome in experimentally-infected bee tissues and in life stages of a symptomatic colony. *Journal of Virological Methods* 141, 7–13 (2007).

41. Ball, B.V. Paralysis. In: Colin, M.E., Ball, B.V., and Kilani, M. (eds.), *Bee Disease Diagnosis*, pp. 81–89, Options Mediterannéennes, Zaragoza, Spain, 1999.

42. Ellis, J.D. and Munn, P.A. The worldwide health status of honey bees. *Bee World* 86, 88–101 (2005).

43. Bakonyi, T. et al. Phylogenetic analysis of acute bee paralysis virus strains. *Applied and Environmental Microbiology* 68, 6446–6450 (2002).

44. Chen, Y., Evans, J.D., and Pettis, J.S. The presence of chronic bee paralysis virus infection in honey bees (*Apis mellifera* L.) in the USA. *Journal of Apicultural Research* 50, 85–86 (2011).

45. Berényi, O., Bakonyi, T., Derakhshifar, I., Köglberger, H., and Nowotny, N. Occurrence of six honeybee viruses in diseased Austrian apiaries. *Applied and Environmental Microbiology* 72, 2414–2420 (2006).

46. Békési, L., Ball, B.V., Dobos-Kovács, M., Bakonyi, T., and Rusvai, M. Occurrence of acute bee paralysis virus of the honey bee (*Apis mellifera*) in a Hungarian apiary infested with the parasitic mite *Varroa jacobsoni*. *Acta Veterinaria Hungarica* 47, 319–324 (1999).

47. Tentcheva, D. et al. Prevalence and seasonal variations of six bee viruses in *Apis mellifera* L. and *Varroa destructor* mite populations in France. *Applied and Environmental Microbiology* 70, 7185–7191 (2004).

48. Sanpa, S. and Chantawannakul, P. Survey of six bee viruses using RT-PCR in Northern Thailand. *Journal of Invertebrate Pathology* 100, 116–119 (2009).

49. Blanchard, P. et al. First detection of Israeli acute paralysis virus (IAPV) in France, a dicistrovirus affecting honeybees (*Apis mellifera*). *Journal of Invertebrate Pathology* 99, 348–350 (2008).

50. Chen, Y., Pettis, J.S., and Feldlaufer, M.F. Detection of multiple viruses in queens of the honey bee *Apis mellifera* L. *Journal of Invertebrate Pathology* 90, 118–121 (2005).

51. Teixeira, E.W., Chen, Y., Message, D., Pettis, J., and Evans, J.D. Virus infections in Brazilian honey bees. *Journal of Invertebrate Pathology* 99, 117–119 (2008).

52. Grabensteiner, E., Bakonyi, T., Ritter, W., Pechhacker, H., and Nowotny, N. Development of a multiplex RT-PCR for the simultaneous detection of three viruses of the honeybee (*Apis mellifera* L.): *Acute bee paralysis virus, Black queen cell virus* and *Sacbrood virus. Journal of Invertebrate Pathology* 94, 222–225 (2007).

53. Ai, H., Yan, X., and Han, R. Complete genome sequence of a Chinese isolate of the Israel acute paralysis virus. *Sociobiology* 58, 49–66 (2012).

54. Kojima, Y. et al. Infestation of Japanese native honey bees by tracheal mite and virus from non-native European honey bees in Japan. *Microbial Ecology* 62, 895–906 (2011).

55. Choe, S.E. et al. Prevalence and distribution of six bee viruses in Korean *Apis cerana* populations. *Journal of Invertebrate Pathology* 109, 330–333 (2012).

56. Ribière, M., Olivier, V., Blanchard, P., Schurr, F., and Celle, O. The collapse of bee colonies: The CCD case (Colony collapse disorder) and the IAPV virus (Israeli acute paralysis virus). *Virologie* 12, 319–322 (2008).

57. Singh, R. et al. RNA viruses in hymenopteran pollinators: Evidence of inter-Taxa virus transmission via pollen and potential impact on non-*Apis* hymenopteran species. *PLoS One* 5, e14357 (2010).

58. Bailey, L. Viruses attacking the honey bee. *Advances in Virus Research* 20, 271–307 (1976).

59. Bailey, L. and Milne, R.G. The multiplication regions and interaction of acute and chronic bee paralysis viruses in adult honey bees. *Journal of General Virology* 4, 9–14 (1969).

60. Furgala, B. and Lee, P.E. Acute bee paralysis virus, a cytoplasmic insect virus. *Virology* 29, 346–348 (1966).

61. Heisenberg, M. Mushroom body memoir: From maps to models. *Nature Reviews Neuroscience* 4, 266–275 (2003).

62. Menzel, R. and Giurfa, M. Cognitive architecture of a mini-brain: The honeybee. *Trends in Cognitive Sciences* 5, 62–71 (2001).

63. Menzel, R. Searching for the memory trace in a mini-brain, the honeybee. *Learning & Memory* 8, 53–62 (2001).

64. Chevin, A., Schurr, F., Blanchard, P., Thiéry, R., and Ribière, M. Experimental infection of the honeybee (*Apis mellifera* L.) with the chronic bee paralysis virus (CBPV): Infectivity of naked CBPV RNAs. *Virus Research* 167, 173–178 (2012).

65. Allen, M.F., Ball, B.V., White, G.F., and Antoniw, J.F. The detection of acute paralysis virus in Varroa jacobsoni by the use of a simple indirect ELISA. *Journal of Apicultural Research* 25, 100–105 (1986).

66. Anderson, D.L. A comparison of serological techniques for detecting and identifying honeybee viruses. *Journal of Invertebrate Pathology* 44, 233–243 (1984).

67. Allen, M.F. and Ball, B.V. Characterisation and serological relationships of strains of Kashmir bee virus. *Annals of Applied Biology* 126, 471–484 (1995).

68. Boonham, N. et al. The detection of *Tomato spotted wilt virus* (TSWV) in individual thrips using real time fluorescent RT-PCR (TaqMan). *Journal of Virological Methods* 101, 37–48 (2002).

69. de Miranda, J.R. et al. Standard methods for virus research in *Apis mellifera. Journal of Apicultural Research* 52(2013).

70. Ribière, M. et al. Molecular diagnosis of chronic bee paralysis virus infection. *Apidologie* 33, 339–351 (2002).

71. Boncristiani, H. et al. Direct effect of acaricides on pathogen loads and gene expression levels in honey bees *Apis mellifera. Journal of Insect Physiology* 58, 613–620 (2012).

5 *Iflavirus* (Deformed Wing Virus)

Jessica M. Fannon and Eugene V. Ryabov

CONTENTS

5.1 Introduction...37
    5.1.1 Classification...37
    5.1.2 Epidemiology and Pathogenesis 39
    5.1.3 Diagnosis 40
        5.1.3.1 Conventional Techniques..40
        5.1.3.2 Molecular Techniques..40
5.2 Methods..40
    5.2.1 Sample Preparation 40
    5.2.2 Iflavirus Discovery 41
        5.2.2.1 Isolation of Iflavirus Particles from Honeybees..41
        5.2.2.2 Sequence-Independent Amplification of Viral cDNA 41
    5.2.3 Detection Procedures 42
        5.2.3.1 Isolation of Total RNA...42
        5.2.3.2 Reverse Transcription 43
        5.2.3.3 Nested PCR Detection of DWV, VDV1, and Their Recombinants 43
        5.2.3.4 Real-Time PCR Quantification of DWV...44
5.3 Conclusion and Future Perspective...44
References..45

5.1 INTRODUCTION

The genus *Iflavirus* includes diverse RNA viruses of arthropods, which have icosahedral nonenveloped virus particles about 26–32 nm in diameter (Figure 5.1a). Iflavirus virions contain a single copy of linear positive-sense 8–10 kb genomic RNA that possesses small genome-linked virus protein, Vpg, covalently linked to the 5′ terminus and poly(A) sequence at the 3′ terminus. Genomic RNAs of iflaviruses have extended 5′ untranslated region (5′UTR), reaching about 1 kb, which acts as an internal ribosome entry signal (IRES). The IRES is followed by a single open reading frame (ORF) coding for a large polyprotein, ranging in size from 2800 to 3200 amino acids (aa), with the structural proteins and the nonstructural proteins encoded by the 5′- and 3′-parts of the ORF, respectively (Figure 5.1) [1]. Iflavirus genomic RNA is infectious and serves as a template for the synthesis of viral polyprotein. The genus name is derived from the name of its type species, *infectious flacherie virus*, a pathogen of the silkworm *Bombyx mori* [2].

In recent years, more viruses with iflavirus characteristics have been discovered in a variety of arthropod species mostly as a result of transcriptome sequencing (Table 5.1, Figure 5.2). Despite high variation between the polyproteins of different iflaviruses, they share a number of common features. The short *N*-terminal leader protein (L) of unknown function, which is highly diverse between iflaviruses, precedes more conserved block of the structural proteins with three domains typical for the capsid proteins of icosahedral picornaviruses (VP1, VP3,

VP3) (Figure 5.1b) and the smallest structural protein VP4, which are arranged in the order VP2-VP4-VP1-VP3 in the polyprotein (Figure 5.1c). The structural proteins are proteolytically processed by the 3C protease domain encoded by the nonstructural block (Figure 5.1c), and precise processing sites have been experimentally determined for several iflaviruses, including deformed wing virus (DWV) [3] and *Lymantria dispar iflavirus* 1 [4] (Figure 5.1c). The C-terminal half of the iflaviral polyprotein contains domains with motifs typical for picornaviral 2C RNA helicase, a chymotrypsin-like 3C protease and 3D RNA-dependent RNA polymerase (Figure 5.1c) [5]. No experimental data are available on the procession of the nonstructural portion of iflavirus polyprotein, although it is suggested that the processing of the nonstructural proteins is facilitated by the 2C protease similar to the processing of the picornavirus polyproteins.

5.1.1 CLASSIFICATION

The genus *Iflavirus* covers a group of positive-sense ssRNA viruses in the family *Iflaviridae*, order *Picornavirales*. Currently, only six species are recognized in the genus *Iflavirus* according to the international committee on taxonomy of viruses (ICTV) report published in 2012 [1] (Table 5.1). However, at least 17 additional insect RNA viruses with fully sequenced genomes (or at least with fully sequenced protein-coding regions) may be considered as belonging to the genus *Iflavirus* according to

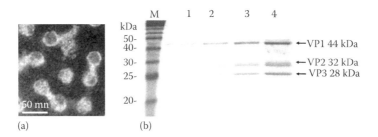

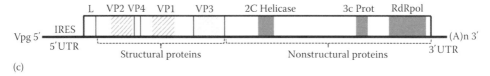

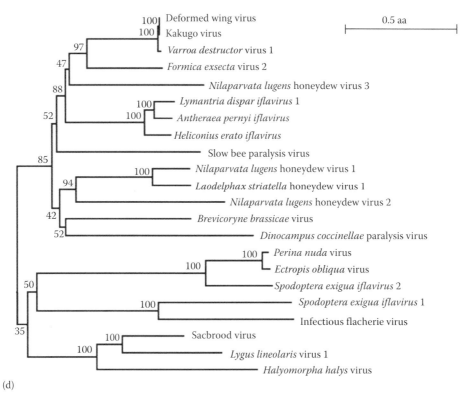

(d)

FIGURE 5.1 (a) Transmission electron photograph of deformed wing virus (DWV) particles, bar 50 nm. (b) sodium dodecyl sulfate poly-acrylamide gel electrophoresis (SDS-PAGE) of DWV virions showing three major structural proteins, VP1, VP2, and VP3. Lane M, molec-ular weight protein markers; lanes 1–4, dilutions of virus preparation. (c) Organization of DWV genomic RNA. The long box represents the single open reading frame with the conserved domains indicated. Vertical lines represent identified cleavage sites. Conserved domains are indicated as follows: boxes in the VP2 and VP1 proteins shown are picorna-like capsid drug-binding pocket domains, box in VP3 is a CrPV capsid-protein-like domain, and dark boxes in the nonstructural part are picorna-like 2C helicase, 3C protease, and RNA-dependent RNA polymerase (RdRp) domains. (d) Phylogram of the recognized and putative members of the genus *Iflavirus*. Full-length polyprotein sequences of the iflavirus isolates listed in Table 5.1 were used. The tree was produced using the neighbor-joining method and evaluated with bootstrap analysis, 1000 replicates, percentage of bootstrap support if each branch is indicated. Branch length indicates evolutionary distance; scale bar shows 0.5 amino acid substitutions per site.

their genome organization and their being significantly differ-ent from other iflaviruses (nucleotide [nt] identity less than 90%) (Table 5.1). Iflaviruses are genetically diverse, as phylo-genetic analysis showed that the viruses from different clades in Figure 5.2 have no regions of significant nt identity and only about 20%–30% aa identity. Among the evolutionarily close iflaviruses,

similarity can be much higher; the nt and aa identity between closely related honeybee viruses DWV and *Varroa destructor* virus 1 (VDV-1) are as high as 85% and 95%, respectively; and lepidopteran PnV and EoV have 82% nt and 87% aa identity.

In respect to the number of the *Iflavirus* group members, it should be noted that the real number of iflaviruses is likely to

TABLE 5.1

Members of the Genus *Iflavirus* with Full-Length Genomic RNA Sequences

Isolate Name (Acronym)	GenBank Accession Number	Host Common Name/Order	Genome	Detection[a]
Members of the genus *Iflavirus* recognized by the ICTV, 2012 report [1]				
Infectious flacherie virus (IFV)	NC_003781	Moths/*Lepidoptera*	[2]	
Perina nuda virus (PnV)	NC_003113	Moths/*Lepidoptera*	[20]	
Ectropis obliqua virus (EoV)	NC_005092	Moths/*Lepidoptera*	[21]	
Deformed wing virus—Italy (DWV)	NC_004830	Honeybee/*Hymenoptera* Mites/*Arachnida*	[3]	[6,10,19,22,23]
Deformed wing virus—Kakugo virus (KV)	NC_005876	Honeybee/*Hymenoptera*	[24]	[6,10]
Varroa destructor virus 1—Netherlands (VDV-1)	NC_006494	Honeybee/*Hymenoptera* Mites/*Arachnida*	[25]	[6,10,19]
Sacbrood virus—Rothamsted (SBV)	NC_002066	Honeybee/*Hymenoptera*	[8]	[11,19]
Putative members of the genus *Iflavirus*				
Formica exsecta virus 2 (Fex2)	KF500002	Ants/*Hymenoptera*	[26]	
Nilaparvata lugens honeydew virus 1 (NLHV-1)	AB766259	Planthopper/*Hemiptera*	[27]	
Nilaparvata lugens honeydew virus 2 (NLHV-2)	NC_021566	Planthopper/*Hemiptera*	[27]	
Nilaparvata lugens honeydew virus 2 (NLHV-2)	NC_021567	Planthopper/*Hemiptera*	[27]	
Lymantria dispar iflavirus 1 (LdIV1)	KJ629170	Moths/*Lepidoptera*	[4]	
Antheraea pernyi iflavirus—LnApIV (ApIV)	KF751885	Moths/*Lepidoptera*	[28]	
Heliconius erato iflavirus (HeIV)	KJ679438	Butterflies/*Lepidoptera*	[29]	
Slow bee paralysis virus (SBPV)	NC_014137	Honeybees/*Hymenoptera*	[30]	[19]
Laodelphax striatella honeydew virus 1 (LSHV1)	KF934491	Planthopper/*Hemiptera*		
Brevicoryne brassicae virus (BrBV)	NC_009530	Aphids/*Hemiptera*	[14]	
Dinocampus coccinellae paralysis virus—Quebec2013	KF843822	Parasitoid wasp/*Hymenoptera*		
Spodoptera exigua iflavirus 1 (SeIV-1)	NC_016405	Moths/*Lepidoptera*	[31]	
Spodoptera exigua iflavirus 2 (SeIV-2)	JN870848	Moths/*Lepidoptera*	[32]	
Lygus lineolaris virus 1 (LyLV-1)	JF720348	Plant bug/*Hemiptera*	[33]	
Halyomorpha halys virus—Beltsville	NC_022611	Stink bug/*Hemiptera*	[34]	

[a] References on detection, in addition to the genome organization reference.

be significantly higher than currently characterized and will continue to expand as more arthropod species are tested for the presence of viruses or subjected to transcriptome analysis. First, iflaviruses have currently been discovered only in economically important insect species, which have been specifically investigated for virus infections or subjected to comprehensive transcriptome analysis, such as several Lepidoptera species (moths and silkworm), as well as Lepidoptera cell lines, and Hymenoptera species (bees and parasitoid wasps). It is possible that the same species can act as a host to several *Iflavirus* species. For example, the honeybee *Apis mellifera*, one of the best investigated insect species, harbors a spectrum of iflaviruses, including DWV, closely related VDV-1 and DWV-VDV-1 recombinants [6], and more diverged slow bee paralysis virus (SBPV) and sacbrood virus (SBV), which, respectively, show only 26% and 19% aa identity with the DWV polyprotein. Second, iflaviruses have relatively limited host range. For example, DWV infects honeybees and non-*Apis* hymenopteran species, ranging from many solitary bees to bumblebees and wasps, but not Lepidoptera species [7]. Therefore, it is likely that unrelated arthropod groups harbor distinct *Iflavirus* species. Also, identification of iflaviruses is difficult because they often cause asymptomatic infections characterized by low level

of virus. This makes molecular characterization the only feasible way to discover novel iflaviruses.

Iflaviruses are highly diverse; phylogenetic analysis of the complete iflaviral polyproteins shows three major clades (Figure 5.1d). Notably, there is no relationship between the relatedness of the host species and the degree of similarity between viruses, for example, viruses infecting the same species may be evolutionarily distinct, for example, evolutionarily divergent SBV and DWV infect honeybees. This may reflect complex evolutionary history of iflaviruses, which may include host range shifts and recombination events.

5.1.2 EPIDEMIOLOGY AND PATHOGENESIS

Iflaviruses are transmitted vertically and horizontally. Members of this group may cause different degrees of symptom development, from asymptomatic to highly pronounced, corresponding to low or high levels of virus in the infected individuals. Usually, vertically transmitted iflaviruses do not accumulate to high levels and do not produce symptomatic infections. Iflaviruses that are transmitted horizontally often cause symptomatic infections with highly pronounced symptoms and insect death. Examples of such iflaviruses found in honeybees include orally transmitted SBV [8] and *V. destructor* mite–vectored DWV [9].

Deformed wing disease of honeybees is caused by pathogenic strains of DWV transmitted to honeybee pupae by *V. destructor* mite feeding on the hemolymph. This results in significant, thousandfold, increase of the levels of DWV, in particular, pathogenic strains of DWV-like viruses—recombinants with the VDV-1-derived structural genes and the DWV nonstructural genes [6,10]. Notably, depending on the route of transmission, the same virus, although not necessarily the same strains, could cause either symptomatic infection, characterized by high level of virus, or asymptomatic infection, low level of virus. Analysis of DWV infection in individual honeybees has shown that vertically transmitted genetically diverse strains of the virus usually accumulate to very low levels and infected bees remain asymptomatic, while horizontal transmission either by *V. destructor* mites or by artificial injection into honeybee hemolymph results in high levels of a limited number of DWV strains [10]. A similar picture emerged from the quantification of SBV in individual honeybees; asymptomatic bees were shown to harbor SBV, though at lower levels than those observed in individuals showing sacbrood disease symptoms [11].

5.1.3 Diagnosis

Members of the *Iflavirus* group are highly diverse viruses, making it impossible to devise a common method for iflavirus diagnosis. There are currently more than 20 characterized iflaviruses, and the number is likely to rise. Moreover, each *Iflavirus* species may contain a number of diverse isolates, which require modifications of the detection method, for example, use of different oligonucleotide primers for reverse transcription (RT)-PCR-based tests. Therefore, instead of listing methods for detection of all iflavirus strains, this chapter presents cost-effective, generic approaches that could be used (1) to discover novel iflaviruses or new strains of known iflaviruses in different arthropod species and (2) to devise methods for detection of particular iflaviruses or viral strains. The presented virus discovery methods specifically focus on characterization of genetic material associated with virus particles, rather than high-throughput transcriptome analysis, which also allows identification of novel RNA viruses. As an example, detailed methods describe discovery and detection of aphid iflaviruses and DWV-like viruses infecting honeybees. The same principles could be applied for detection of other iflaviruses.

5.1.3.1 Conventional Techniques

Electron microscopy is an essential tool for novel virus discovery. Visualization of virus particles in virus preparations is particularly important to confirm that the morphological characteristics of virus particles are typical for those of iflaviruses [4,6]. Nevertheless, the low concentration of virus particles in the asymptomatic individuals may make it difficult to detect virus particles in preparations from infected tissues. Immunological methods for iflavirus detection are not widely used for a number of reasons. It is often difficult to produce a sufficient amount of purified virus particles to raise

antibodies. Also, serological tests may not be sufficiently sensitive to detect asymptomatic iflavirus infections. For example, ELISAs have been developed for the detection of DWV, but it could reliably detect only high levels of DWV, with the virus in asymptomatic bees being below the detection threshold [12]. Furthermore, polyclonal antibody–based techniques could be used to differentiate different virus strains.

5.1.3.2 Molecular Techniques

Due to difficulties in using conventional techniques for identification purpose, most of the iflaviruses are detected by molecular approaches, such as RT-PCR and real-time RT-PCR. Identification of iflaviruses and characterization of their novel strains are essential steps in devising detection methods for iflaviruses in previously uncharacterized arthropod species. Apart from relatively expensive characterization of transcriptome (cloning of expressed sequence tags or high-throughput sequencing using the RNA-seq approach) that requires generation and sequencing of a high number of clones or reads and considerable bioinformatics effort, it is possible to use virus particles isolated from relatively small number of insects, or even individual insects, as a source of encapsidated viral RNA for sequence-independent amplification of the viral complimentary DNA (cDNA). The amplified fragments can be cloned into a plasmid vector and sequenced. Usually a library has 10%–50% of virus-derived clones, which provides initial sequence data for characterization of novel viruses and strains. Once information of sufficient stretches of viral RNA is obtained, primers for RT-PCR and real-time RT-PCR detection assays should be designed.

5.2 METHODS

In this section, we will cover molecular methods for the identification of novel iflaviruses and/or strains of previously characterized iflaviruses. We will present details of the methods used for detection, characterization, and quantification of DWV isolates in honeybees at different stages of development. DWV is a good example for the demonstration of the general principles of iflavirus detection because, depending on the inoculation route and the virus strain, DWV may accumulate to either low or high levels [6,10]. Molecular procedures, such as RT-PCR and real-time RT-PCR, are the methods of choice because they allow detection of DWV in cases of high-level and low-level infections.

5.2.1 Sample Preparation

Individual adult honeybees, larvae or pupae, as well as *V. destructor* mites should be frozen at −80°C as soon as possible following collection, while still alive, to preserve viral and host mRNA for analysis. It is possible to freeze insects in liquid nitrogen before storage at −80°C. The ideal strategy for collection of honeybee samples depends upon the required life stage of the bees to be studied, adult bees can be collected from frames or the hive entrance, or, if larvae, pupae, or eggs are required, then whole or partial frames can be removed for sampling. If the insects cannot be kept alive and immediate freezing is not possible, RNA could be preserved in tissues using RNAlater solution (Ambion). In this case, the larvae could be placed directly to the RNAlater solution, while the exoskeleton of pupae and adult bees should be broken open to allow fast diffusion of RNAlater solution into the tissues. Small arthropods, like aphids or *V. destructor* mites, could be collected while alive, paced individually or as groups into Eppendorf tubes, quickly frozen to −80°C, and stored frozen until extraction.

5.2.2 Iflavirus Discovery

5.2.2.1 Isolation of Iflavirus Particles from Honeybees

Different arthropod host species, or even different populations of the same host species, may be infected with iflaviruses significantly different from those previously described. Therefore, where possible, virus diversity should be characterized. The method given in the following for purifying honeybee viruses is suitable for nonenveloped viruses, including iflaviruses [6]. DWV-like viruses can be purified from honeybees of any developmental stage, but the greatest yield comes from symptomatic (deformed wings, stunted abdomen) adult bees or pupae heavily infested with *V. destructor*.

5.2.2.1.1 Procedure

1. Homogenize 5 g of insect tissue, frozen or live, with 35 mL of 100 mM sodium phosphate buffer, pH 7.4 (NaP buffer), supplemented with 0.05% of Tween 20 using a pestle and mortar.
2. Centrifuge for 12 min in a high-speed centrifuge (10,000 rpm, 4°C).
3. Carefully remove and retain the supernatant, leaving the fat layer (top) and the unhomogenized tissue (bottom).
4. Carefully layer the cleared supernatant over a 20% sucrose cushion (2.5 mL, 20% sucrose in NaP buffer) in a 30 mL ultracentrifuge tube, Beckman SW28 rotor. Top up with NaP buffer so that liquid reaches 0.5 cm from the top of the tube.
5. Centrifuge at 28,000 rpm, 18°C, for 3 h 30 min, pour off the supernatant, resuspend the pellet in 2 mL of NaP buffer, and leave overnight at 4°C to ensure complete resuspension of the virus particles.
6. Set up a discontinuous cesium chloride (CsCl) density gradient in a 35 mL ultracentrifuge tube, Beckman SW28 rotor, by layering from the bottom 5 mL aliquots of CsCl with the following densities: 1.6 g/cm^3, 1.5 g/cm^3, 1.4 g/cm^3, 1.3 g/cm^3, and 1.2 g/cm^3. The CsCl solutions are prepared using saturated CsCl (1.9 g/cm^3 in dH$_2$O) mixed with NaP buffer in varying proportions to achieve required density. Carefully layer the resuspended sucrose cushion pellet containing the virus particles over the CsCl gradient after being mixed with enough buffer to reach 0.5 cm from the top of the tube.
7. Centrifuge for 18 h, 28,000 rpm, at 4°C, using Beckman SW28 rotor.

8. Carefully collect separate 1.5 mL fractions, starting from the top.
9. Weigh each 1.5 mL fraction to determine density, collect fractions with a density of 1.30–1.40 g/cm^3, pool them together, and dilute 5–10-fold with NaP buffer.
10. Centrifuge the diluted fractions for 3 h, at 4°C, 28,000 rpm, using SW28 Beckman bucket rotor.
11. Immediately remove supernatant and resuspend the pellet with 200 μL 0.1 M sodium NaP buffer, pH 7.4. Note that pellet could be invisible. Store at −80°C.

Depending on the amount of the starting tissue material and the yield, virus preparations could be used to raise antisera and/or to isolate viral genomic RNA for further applications, including conventional cDNA synthesis or sequence-independent amplification of cDNA to viral RNA or next-generation sequencing [6,10] or sequence-independent amplification of viral cDNA described as follows.

5.2.2.2 Sequence-Independent Amplification of Viral cDNA

This method was adapted from the procedure for identification of human viruses described by [13] and was successfully used for identification of novel RNA viruses in aphids [14,15].

5.2.2.2.1 Procedure

1. Homogenize up to 50 adult aphids (total weight 15 mg) previously stored at −80°C, with 15 mL of 100 mM sodium phosphate buffer, pH 7.0.
2. Clarify spin the homogenate at 350 × *g* for 10 min in a benchtop Eppendorf centrifuge.
3. Filter the debris-free supernatant through a 0.80/0.22 μm filter (Millipore).
4. Centrifuge at 45,000 rpm in a Ti80 rotor (Beckman) at 4°C, for 2 h 30 min.
5. Completely remove supernatant and resuspend practically invisible pellet with 100 μL of 20 mM sodium phosphate buffer, pH 7.0.
6. In order to remove traces of DNA contamination, combine 15 μL of the virus suspension, 12 μL of RNA-free water, and 3 μL of the 10× transcription buffer and 1 μL of DNAseI from Ambion in vitro transcription kit (Cat. No. 1344), and incubate for 30 min at room temperature.
7. Extract RNA with RNeasy Plant mini kit (Qiagen) according to the manufacturer's instructions, and elute with 50 μL of RNase-free water.

The RNA preparation extracted from virus particles is reversely transcribed to produce the first cDNA strand; the second cDNA strand is produced using Klenow fragment (3′ → 5′ exo-). In both reactions, the tagged random hexanucleotide primer, "Tag-N6" (5′-GCCGGAGCTCTGCAGATATCNNNNNN-3′), is utilized.

First cDNA strand synthesis

Component	Volume (µL)
Column-purified encapsidated RNA	16
Oligonucleotide "Tag-N6," 10 µM	4

Incubate at 65°C for 5 min, then immediately chill in ice, and then add the following:

Component	Volume (µL)
5× first-strand buffer (Invitrogen)	8
Deoxynucleotide mix (5 mM each)	4
0.1 M DTT	2
RNaseOUT, Invitrogen (40 units/µL)	2
SuperScript II reverse transcriptase	2

Incubate at 25°C for 10 min and 42°C for 50 min.

For the second cDNA strand synthesis, which uses the same tagged hexanucleotide primer, incubate at 94°C for 3 min to denature, immediately chill in ice, add 5 units (1 µL) of Klenow fragment ($3' \rightarrow 5'$ exo-) (New England Biolabs), and incubate at 37°C for 1 h. Inactivate the enzyme by incubating at 75°C for 10 min.

To amplify the tagged cDNA fragments by PCR, set up 50 µL of the reaction by combining 5 µL of the tagged cDNA reaction, 5 µL of 10 µM primer 5′-GCCGGAGCTCTGCAGATATC-3′, and 2 µL of dNTP solution (5 mM each), 5 µL of Roche 10× PCR buffer, 32 µL of water, and 1 µL of Taq polymerase (Roche). After 5 min incubation at 94°C followed by 40 cycles (94°C for 1 min, 65°C for 1 min, and 72°C for 2 min), add a final 5 min incubation at 72°C. Separate the PCR products by agarose gel electrophoresis (1.2% gel on TAE buffer, 100 V for 30 min, stain with ethidium bromide). Extract the gel products ranging in size from 150 to 400 nt using QIAGEN gel extraction kit, and clone them into an AT cloning vector (e.g., pDrive vector, QIAGEN) to create a plasmid library. Sequence plasmids using Sanger dideoxy method, and carry out sequence similarity search using the BLAST program [16] to identify sequences coding for proteins with similarity to the viral proteins.

5.2.3 DETECTION PROCEDURES

5.2.3.1 Isolation of Total RNA

Samples must remain frozen at all times, prior to the addition of guanidinium thiocyanate-phenol-chloroform (TRI Reagent, Ambion) to prevent RNA degradation.

1. Collect clean pestles, mortars, and spatulas. Fill Dewar flask with liquid nitrogen and move collect frozen bees/pupae (from −80°C into Dewar flask).
2. Cool ceramic pestles and mortar with liquid nitrogen; rest the mortar inside to cool.
3. Top the liquid nitrogen, place the first sample into the mortar, and gently tap to break it up.

Keep covered in liquid nitrogen and grind the sample to a fine powder.

4. Allow the sample to "dry," and then immediately use a clean spatula to split evenly into two 1.5 mL microfuge tubes, and precool spatula and tubes in liquid nitrogen prior to use.
5. Place the tubes into the liquid nitrogen and move onto the next sample using a clean set of equipment.
6. Turn on the centrifuge to cool to 4°C. Take one aliquot of each sample to the 80°C for storage (to be kept as powder for protein or DNA analysis).
7. In the fume hood, add 1 mL TRI Reagent to each sample and mix well. Incubate for 5 min at room temperature.
8. Add 0.2 mL chloroform to each sample, shake for 15 s, and then incubate for 2 min at room temperature.
9. Centrifuge for 15 min, 12,000 rpm, at 4°C.
10. Transfer aqueous phase to fresh tube—this is the colorless liquid at the top of the tube. Dispose of (pink) phenol waste appropriately.
11. Add 0.5 mL isopropanol (propan-2-ol) and mix and incubate at −20°C for at least 1 h.
12. Centrifuge for 10 min, 12,000 rpm, at 4°C.
13. Remove the supernatant, leaving just the white pellet.
14. Add 1 mL of 75% ethanol, and vortex and centrifuge for 5 min, 12,000 rpm, at 4°C.
15. Carefully remove the supernatant, leaving just the white pellet, and air-dry pellet at room temperature for about 30 min.
16. Dissolve total RNA in 30 µL of RNAse-free water and store at −80°C.

Note: If sample has been stored in RNAlater solution (Ambion), then an additional step is required. The high salt concentration in the sample prevents RNA pellet formation. To avoid this difficulty, ensure all of the RNAlater solution is removed from the sample prior to homogenization, and, instead of pouring off all of the supernatant after the centrifugation with isopropanol (step 13), pipette off the top phase, and then precipitate the RNA contained within the bottom phase with the addition of 1 mL of 50% isopropanol. Recentrifuge and then proceed to the ethanol wash stage as per protocol.

Quantification of iflaviruses requires normalization to the levels of host mRNAs that are constitutively expressed at high levels, usually of ribosomal protein 49 (RP49) [6,10]. To ensure that transcript rather than genomic DNA is quantified and amplified, total RNA preparations should be treated with DNAseI (New England Biolabs). For DNAseI treatment, set up reaction with up to 5 µg of total RNA of 100 µL in 1× reaction buffer, add 2 units of DNAse I, and incubate at 37°C for 15 min, and then immediately run through an RNA purification column, "GeneJET RNA Purification Kit" (Thermo Scientific, Cat. No. K0731) or "RNeasy Plant Mini Kit" (QIAGEN, Cat. No. 74904), following the manufacturer's protocol.

RNA samples could be extracted directly from small insects like aphids, *V. destructor* mites, or fragments of arthropod tissues up to 25 mg by using "GeneJET RNA Purification Kit" or QIAGEN "RNeasy Plant Mini Kit," according to the manufacturer's instructions. Live aphids or mites could be placed in Eppendorf tubes and immediately homogenized with guanidine thiocyanate–containing lysis buffers using a mini pestle. When frozen material is used, the lysis buffer could be added before directly to the frozen sample in the Eppendorf tube and immediately homogenized using a mini pestle.

5.2.3.2 Reverse Transcription

Molecular methods are based on the detection of the viral cDNA, a copy of DNA complementary to viral RNA, which is produced by RNA-dependent DNA polymerases. cDNA is more stable than RNA and could be stored at −20°C; therefore, it is advisable to produce cDNA immediately after RNA cleanup. To ensure that the same cDNA preparation could be used with different primers for detection and quantification of different viruses and host transcripts, the following protocol uses a random hexamer oligonucleotide primer to produce fully representative cDNA copies of the RNA population:

Component	Volume (μL)
DNAseI-treated column-purified RNA	10
Random hexamer (N6) primer (100 μM)	1
Deoxynucleotide (dNTP) mix (10 mM each)	1

Heat mixture to 65°C for 5 min and quickly chill on ice. Collect the contents of the tube by brief centrifugation and add the following:

Component	Volume (μL)
5× first-strand buffer	4
0.1 M DTT	2
RNaseOUT, Invitrogen (40 units/μL)	1

Incubate at 25°C for 2 min and then add 1 μL (200 units) of SuperScript II RT (Invitrogen), and mix by pipetting gently up and down. Incubate the tube at 25°C for 10 min and then at 42°C for 50 min, and finally inactivate the reaction by heating at 70°C for 15 min. cDNA samples are diluted in 1:5 dH$_2$O and stored at −20°C.

5.2.3.3 Nested PCR Detection of DWV, VDV1, and Their Recombinants

Identification of the strains of DWV-like viruses, in particular highly virulent VDV-1-DWV recombinants, requires cloning and sequencing of cDNA fragments [6,10]. The levels of DWV in asymptomatic bees may not be sufficient for direct amplification of the cDNA fragments for further characterization. Nested PCR, which includes amplification of a fragment of cDNA in two stages, first with a pair of "outer primers" and then with the pair of "inner primers" located within the

reamplified region, allows amplification of low levels of template. In respect to DWV and VDV-1 infection, a highly characteristic region suitable for identification of the individual strains of DWV, VDV-1, and VDV-1-DWV recombinants is 1.3 kb fragments at the border between the structural genes (CP) and nonstructural (NS) gene blocks (corresponding to nucleotides 4926–6255 of the DWV genome; GenBank accession No. AJ489744). The outer and inner primer sequences are shown in Table 5.2.

Step 1—Using flanking primers "Nested-outer-universal-DWV-4700-For" and "Nested-outer-universal-DWV-6700-Rev" (Table 5.2).

For a 50 μL reaction volume mix

Component	Volume (μL)
GoTaq Green Master Mix, 2×	25
Forward primer (100 μM)	1
Reverse primer (100 μM)	1
cDNA template (diluted in 1:5 dH$_2$O)	5
dH$_2$O	18

Mix the components in a PCR tube and place into the thermal cycler for the following: 95°C, 2 min, 15 cycles (95°C, 30 s; 52°C, 1 min; 72°C, 2 min), 95°C, hold.

Step 2—Using DWV- or VDV-1-specific primer combinations. There are four possible inner primer combinations, allowing amplification of DWV, VDV-1, and VDV-1-DWV recombinants, including the nonrecombinant DWV ("Nested-Inner-DWV-4900-For" and "Nested-Inner-DWV-6500-Rev"); nonrecombinant VDV-1 ("Nested-Inner-VDV-4900-For" and "Nested-Inner-VDV-6300-Rev"), VDV-1 CP–DWV NS recombinants ("Nested-Inner-VDV-4900-For" and "Nested-Inner-DWV-6500-Rev"), and DWV CP–VDV-1 NS recombinants ("Nested-Inner-DWV-4900-For" and "Nested-Inner-VDV-4900-For") though the DWV CP-VDV-1 NS recombinant has never been detected.

Take 5 μL of the reaction in "Step 1" to serve as the template for each of these mixes in "Step 2":

For a 50 μL reaction volume mix

Component	Volume (μL)
GoTaq® Green Master Mix, 2×	25
Upstream primer (100 μM)	1
Upstream primer (100 μM)	1
Product from "Step 1" at 95°C	5
dH$_2$O	18

Mix the components in a PCR tube and place into the thermal cycler at 95°C for 2 min with 30 cycles (95°C for 30 s, 52°C for 1 min, 72°C for 2 min) and 72°C for 10 min.

The product of each reaction in "Step 2" can be analyzed by agarose gel electrophoresis (1% gel on TAE buffer, 100 V for 30 min, stain with ethidium bromide), to determine which strains of DWV-like viruses and which recombinants are present in the sample. Further information on virus identity for

TABLE 5.2

Oligonucleotide Primers for Detection of Deformed Wing Virus–Like Viruses

Primer ID	Primer Sequence (5′–3′)	Description
DWV-like-366	CTGTAGTTAAGCGGTTATTAGAA	qPCR-VDV-Struct-For
DWV-like-367	GGTGCTTCTGGAACAGCGGAA	qPCR-VDV-Struct-Rev
DWV-like-368	CTGTAGTCAAGCGGTTACTTGAG	qPCR-DWV-Struct-For
DWV-like-369	GGAGCTTCTGGAACGGCAGGT	qPCR-DWV-Struct-Rev
DWV-like-370	TTCATTAAAACCGCCAGGCTCT	qPCR-VDV-pol-For
DWV-like-371	CAAGTTCAGGTCTCATCCCTCT	qPCR-VDV-pol-Rev
DWV-like-372	TTCATTAAAGCCACCTGGAACA	qPCR-DWV-pol-For
DWV-like-373	CAAGTTCGGGACGCATTCCACG	qPCR-DWV-pol-Rev
Bee-actin-374	AGGAATGGAAGCTTGCGGTA	qPCR-Bee-actin-For
Bee-actin-375	AATTTTCATGGTGGATGGTGC	qPCR-Bee-actin-Rev
Varroa-actin-132	TGAAGGTAGTCTCATGGATAC	qPCR-Varroa-actin-For
Varroa-actin-130	TGAAGGTAGTCTCATGGATAC	qPCR-Varroa-actin-Rev
DWV-like-163	GTTTGTATGAGGTTATACTTCAAGGAG	qPCR-Universal-DWV-like-pol-R
DWV-like-164	GCCATGCAATCCTTCAGTACCAGC	qPCR-Universal-DWV-like-pol-R
DWV-like-155	CAGTAGCTTGGGCGATTGTTTCG	Nested-outer-Uni-DWV-4700-For
DWV-like-156	CGCGCTTAACACACGCAAATTATC	Nested-outer-Uni-DWV-6700-Rev
DWV-like-151	CTGTAGTCAAGCGGTTACTTGAG	Nested-Inner-DWV-4900-For
DWV-like-152	CTGTAGTTAAGCGGTTATTAGAA	Nested-Inner-VDV-4900-For
DWV-like-153	CTTGGAGCTTGAGGCTCTACA	Nested-Inner-DWV-6500-Rev
DWV-like-154	CTGAAGTACTAATCTCTGAG	Nested-Inner-VDV-6300-Rev

Sources: Moore, J. et al., *J. Gen. Virol.*, 92, 156, 2011; Ryabov, E.V. et al., *PLoS Pathog.*, 10, e1004230, 2014.

phylogenetic analysis can be obtained by cloning these PCR products into a plasmid vector (e.g., pGemT-Easy, Promega) and sequencing the plasmid using the Sanger dideoxy method.

5.2.3.4 Real-Time PCR Quantification of DWV

Real-time PCR (or qPCR) allows accurate and simple quantification of DWV-like viruses within individual bees. With qPCR, fluorescent dyes are used to label PCR products during thermal cycling, and the accumulation of fluorescent products is measured in real time. Here, we describe how the SYBR Green qPCR approach can be used to quantify viral load in individual bees, normalized to a constitutively expressed housekeeping gene (actin).

For a 20 µL reaction volume mix

Component	Volume (µL)
Brilliant III qPCR Master Mix, low ROX	10
dH$_2$O	7.6
Forward primer (100 µM)	0.2
Reverse primer (100 µM)	0.2
cDNA template	2

Reaction conditions: 95°C for 3 min, 50 cycles (95°C for 10 s, 60°C for 30 s).

For absolute quantification of viral copy number, serial dilutions of known amounts of template can be included on the qPCR plate and their ct scores used to plot a standard curve. Sequences of the oligonucleotide primers used for qRT-PCR quantification of the structural genes and nonstructural genes of DWV- and VDV-1-types, four pairs in total, as well as for the amplification of honeybee and *V. destructor* actin are shown in Table 5.2.

5.3 CONCLUSION AND FUTURE PERSPECTIVE

A growing body of evidence suggests that iflaviruses are widespread in a variety of arthropods, including Hymenoptera species (important pollinators like honeybee and bumblebees and parasitoid wasps) and Lepidoptera species, including pests and silkworm, as well as aphids. Although in the majority of cases iflaviruses are not present at high concentrations and do not cause obvious symptoms, they still have an effect on insect physiology and survival and affect host fitness. Also, rapid selection of pathogenic strains of iflaviruses may take place as a result of major changes in virus transmission. A well-studied example is the selection of highly virulent strains of DWV and/or DWV/VDV-1 recombinants that severely affect honeybee health following introduction of the mite *V. destructor* [10,17].

The development of tests for novel iflaviruses is important for insect cell culture–based research. Recently, it was shown that the widely used cell line that is derived from the gypsy moth *Lymantria dispar* is persistently infected with an LdIV1 [4]. Taking into account the impact of virus infections, including iflaviruses, on arthropod physiology, health, fecundity, and interspecies interactions, there is a need to expand screening for novel and characterized iflaviruses. Technical advances in high-throughput analysis of transcriptomes, in particular RNA-seq [18], have made possible to comprehensively characterize diversity of RNA viruses in any species using material extracted from individual organisms. Molecular tests, RT-PCR and real-time RT-PCR, or oligonucleotide array–based tests [19] similar to those that are already developed for the identified iflaviruses (Table 5.1) could be devised for the screening of particular iflavirus or even particular strain in a large number of individuals when required, when new sequencing data become available.

REFERENCES

1. Chen, Y.P., Nakashima, N., Christian, P.D., Bakonyi, T., Bonning, B.C., Valles, S.M., and Lightner, D.V. 2012. Family Iflaviridae. In: *Virus Taxonomy: Ninth Report of the International Committee on Taxonomy of Viruses*, pp. 846–849. A.M.Q. King, M.J. Adams, E.B. Carstens, and E.J. Lefkowitz (eds.). London, U.K.: Elsevier Academic.

2. Isawa, H., Asano, S., Sahara, K., Iizuka, T., and Bando, H. 1998. Analysis of genetic information of an insect picorna-like virus, infectious flacherie virus of silkworm: Evidence for evolutionary relationships among insect, mammalian and plant picorna(-like) viruses. *Arch. Virol.* 143, 127–143.

3. Lanzi, G., de Miranda, J.R., Boniotti, M.B., Cameron, C.E., Lavazza, A., Capucci, L., Camazine, S.M., and Rossi, C. 2006. Molecular and biological characterization of deformed wing virus of honeybees (*Apis mellifera* L.). *J. Virol.* 80, 4998–5009.

4. Carrillo-Tripp, J., Krueger, E.N., Harrison, R.L., Toth, A.L., Miller, W.A., and Bonning, B.C. 2014. *Lymantria dispar iflavirus* 1 (LdIV1), a new model to study iflaviral persistence in lepidopterans. *J. Gen. Virol.* 95, 2285–2296.

5. Koonin, E.V., Dolja, V.V., and Morris, T. J. 1993. Evolution and taxonomy of positive-strand RNA viruses: Implications of comparative analysis of amino acid sequences. *Crit. Rev. Biochem. Mol. Biol.* 28, 375–430.

6. Moore, J., Jironkin, A., Chandler, D., Burroughs, N., Evans, D.J., and Ryabov, E.V. 2011. Recombinants between Deformed wing virus and *Varroa destructor* virus-1 may prevail in *Varroa destructor*-infested honeybee colonies. *J. Gen. Virol.* 92, 156–161.

7. Singh, R., Levitt, A.L., Rajotte, E.G. et al. 2010. RNA viruses in hymenopteran pollinators: Evidence of inter-Taxa virus transmission via pollen and potential impact on non-Apis hymenopteran species. *PLoS One* 5, e14357.

8. Ghosh, R.C., Ball, B.V., Willcocks, M.M., and Carter, M.J. 1999. The nucleotide sequence of sacbrood virus of the honey bee: An insect picorna-like virus. *J. Gen. Virol.* 180, 1541–1549.

9. de Miranda, J.R. and Genersch, E. 2010. Deformed wing virus. *J. Invertebr. Pathol.* 103, S48–S61.

10. Ryabov, E.V., Wood, G.R., Fannon, J.M., Moore, J.D., Bull, J.C., Chandler, D., Mead, A., Burroughs, N., and Evans, D.J. 2014. A virulent strain of deformed wing virus (DWV) of honeybees (*Apis mellifera*) prevails after *Varroa destructor*-mediated, or in vitro, transmission. *PLoS Pathog.* 10, e1004230.

11. Blanchard, P., Guillot, S., Antùnez, K. et al. 2014. Development and validation of a real-time two-step RT-qPCR TaqMan assay for quantitation of Sacbrood virus (SBV) and its application to a field survey of symptomatic honey bee colonies. *J. Virol. Methods* 197, 7–13.

12. Martin, S.J., Ball, B.V., and Carreck, N.L. 2013. The role of deformed wing virus in the initial collapse of varroa infested honey bee colonies in the UK. *J. Apicult. Res.* 52, 251–258.

13. Allander, T., Tammi, M.T., Eriksson, M., Bjerkner, A., Tiveljung-Lindell, A., and Andersson, B. 2005. Cloning of a human parvovirus by molecular screening of respiratory tract samples. *Proc. Natl. Acad. Sci. USA* 102, 12891–12896.

14. Ryabov, E.V. 2007. A novel virus isolated from the aphid *Brevicoryne brassicae* with similarity to Hymenoptera picorna-like viruses. *J. Gen. Virol.* 88, 2590–2595.

15. Ryabov, E.V., Keane, G., Naish, N., Evered, C., and Winstanley, D. 2009. Densovirus induces winged morphs in asexual clones of the rosy apple aphid, *Dysaphis plantaginea*. *Proc. Natl. Acad. Sci. USA* 106, 8465–8470.

16. Altschul, S.F., Gish, W., Miller, W., Meyers, E.W., and Lipman, D.J. 1990. Basic local alignment search tool. *J. Mol. Biol.* 215, 403–410.

17. Martin, S.J., Highfield, A.C., Brettell, L., Villalobos, E.M., Budge, G.E., Powell, M., Nikaido, S., and Schroeder, D.C. 2012. Global honey bee viral landscape altered by a parasitic mite. *Science* 336, 1304–1306.

18. Liu, S., Vijayendran, D., and Bonning, B.C. 2011. Next generation sequencing technologies for insect virus discovery. *Viruses* 3, 1849–1869.

19. Glover, R.H., Adams, I.P., Budge, G., Wilkins, S., and Boonham, N. 2011. Detection of honey bee (*Apis mellifera*) viruses with an oligonucleotide microarray. *J. Invertebr. Pathol.* 107, 216–219.

20. Wu, C.Y., Lo, C.F., Huang, C.J., Yu, H.T., and Wang, C.H. 2002. The complete genome sequence of *Perina nuda* picorna-like virus, an insect-infecting RNA virus with a genome organization similar to that of the mammalian picornaviruses. *Virology* 294, 312–323.

21. Wang, X., Zhang, J., Lu, J., Yi, F., Liu, C., and Hu, Y. 2004. Sequence analysis and genomic organization of a new insect picorna-like virus, *Ectropis obliqua* picorna-like virus, isolated from *Ectropis obliqua*. *J. Gen. Virol.* 85, 1145–1151.

22. Genersch, E. 2005. Development of a rapid and sensitive RT-PCR method for the detection of deformed wing virus, a pathogen of the honeybee (*Apis mellifera*). *Vet. J.* 169, 121–123.

23. Yue, C. and Genersch, E. 2005. RT-PCR analysis of Deformed wing virus in honeybees (*Apis mellifera*) and mites (*Varroa destructor*). *J. Gen. Virol.* 86, 3419–3424.

24. Fujiyuki, T., Takeuchi, H., Ono, M., Ohka, S., Sasaki, T., Nomoto, A., and Kubo, T. 2004. Novel insect picorna-like virus identified in the brains of aggressive worker honeybees. *J. Virol.* 78, 1093–1100.

25. Ongus, J.R., Peters, D., Bonmatin, J.M., Bengsch, E., Vlak, J.M., and van Oers, M.M. 2004. Complete sequence of a picorna-like virus of the genus Iflavirus replicating in the mite *Varroa destructor*. *J. Gen. Virol.* 85, 3747–3755.

26. Johansson, H., Dhaygude, K., Lindstrom, S., Helantera, H., Sundstrom, L., and Trontti, K. 2013. A metatranscriptomic approach to the identification of microbiota associated with the ant *Formica exsecta*. *PLoS One* 8, e79777.

27. Murakami, R., Suetsugu, Y., Kobayashi, T., and Nakashima, N. 2013. The genome sequence and transmission of an iflavirus from the brown planthopper, *Nilaparvata lugens*. *Virus Res.* 176, 179–187.

28. Geng, P., Li, W., Lin, L., de Miranda, J.R., Emrich, S., An, L., and Terenius, O. 2014. Genetic characterization of a novel Iflavirus associated with vomiting disease in the Chinese oak silkmoth *Antheraea pernyi*. *PLoS One* 9, e92107.

29. Smith, G., Macias-Muñoz, A., and Briscoe, A.D. 2014. Genome sequence of a novel Iflavirus from mRNA sequencing of the butterfly *Heliconius erato. Genome Announc.* 2, e00398-14.

30. de Miranda, J.R., Dainat, B., Locke, B. et al. 2010. Genetic characterization of slow bee paralysis virus of the honeybee (*Apis mellifera* L.). *J. Gen. Virol.* 91, 2524–2530.

31. Millan-Leiva, A., Jakubowska, A.K., Ferre, J., and Herrero, S. 2012. Genome sequence of SeIV-1, a novel virus from the Iflaviridae family infective to *Spodoptera exigua. J. Invertebr. Pathol.* 109, 127–133.

32. Choi, J.Y., Kim, Y.-S., Wang, Y., Shin, S.W., Kim, I., Tao, X.Y., Liu, Q., Roh, J.Y., Kim, J.S., and Je, Y.H. 2012. Complete genome sequence of a novel picorna-like virus isolated from *Spodoptera exigua. J. Asia. Pac. Entomol.* 15, 259–263.

33. Perera, O.P., Snodgrass, G.L., Allen, K.C., Jackson, R.E., Becnel, J.J., O'Leary, P.F., and Luttrell, R.G. 2012. The complete genome sequence of a single-stranded RNA virus from the tarnished plant bug, *Lygus lineolaris* (Palisot de Beauvois). *J. Invertebr. Pathol.* 109, 11–19.

34. Sparks, M.E., Gundersen-Rindal, D.E., and Harrison, R.L. 2013. Complete genome sequence of a novel Iflavirus from the transcriptome of *Halyomorpha halys*, the brown marmorated stink bug. *Genome Announc.* 1, e00910-13.

6 Avian Encephalomyelitis Virus

Mazhar I. Khan, Zhixun Xie, Theodore Girshick, and Zhiqin Xie

CONTENTS

6.1 Introduction 47
    6.1.1 Classification, Morphology, and Chemical Composition 47
    6.1.2 Pathobiology and Epidemiology 47
    6.1.3 Clinical Features and Pathogenesis 48
    6.1.4 Diagnosis 48
        6.1.4.1 Conventional Techniques 48
        6.1.4.2 Molecular Techniques 49
6.2 Methods 49
    6.2.1 Sample Collection 49
    6.2.2 Detection Procedures 49
        6.2.2.1 Reverse Transcription Reaction 49
        6.2.2.2 PCR Detection of AEV 50
6.3 Conclusion 50
References 50

6.1 INTRODUCTION

Avian encephalomyelitis virus (AEV) causes central nervous system (CNS) signs (including ataxia, incoordination, paralysis, and rapid tremors) with high morbidity and some mortality in young chicks [1–7]. AEV was first described in 1932 in 2-week-old chicks showing tremors [5]. However, in adult laying birds, AEV infection causes no neurologic signs, apart from a slight reduction in egg production [3,7]. Outbreaks of AEV infections were observed in Connecticut, Maine, Massachusetts, and New Hampshire, which led to the naming of avian encephalomyelitis (AE) as "New England disease." Since then, AE has been identified in many parts of the world [7]. AEV has a limited host range; chickens, pheasant, Coturnix quail, and turkeys have been shown to be infected [7–9]. AEV is transmitted vertically to the progeny, but infection with AEV by the fecal–oral route also occurs [3,10–12]. In the mid-1950s, AE was successfully controlled by immunization [13].

6.1.1 CLASSIFICATION, MORPHOLOGY, AND CHEMICAL COMPOSITION

AEV is a member of the *Picornaviridae* family, which covers a large number of single-stranded RNA viruses organized in 21 genera [14]. Earlier studies [15] suggested that AEV belongs to the *Enterovirus* genus, but recent findings indicate that it shares high levels of protein homologies with the hepatitis A virus (HAV) [16]. Now, AEV has been tentatively classified as being a member of the *Hepatovirus* genus within the *Picornaviridae* family [17].

In addition to AEV, other notable viruses in the *Picornaviridae* family include poliovirus (human enterovirus C serotypes [PV-1, PV-2, PV-3]), human rhinovirus A, encephalomyocarditis virus, foot-and-mouth disease virus, HAV, human parechovirus (HPeV), equine rhinitis B virus, Aichi virus (AiV), and porcine teschovirus.

A purified virion particle of AEV has a diameter of 24–32 nm [18], is nonenveloped, and contains a single-stranded RNA genome of positive polarity [16,19]. It possesses a buoyant density of 1.31–1.33 g/mL and a sedimentation coefficient of 148 S [14–16,18]. The AEV genome is composed of 7032 nucleotides with an open reading frame of 6405 nts starting at nucleotide 495 [14,16]. AEV includes four structural proteins from the P1 region (VP1, VP2, VP3, and VP4) and seven nonstructural proteins from the P2 and P3 regions of the genome [16]. Shafren and Tannock [20] initially detected four virus-specific proteins with molecular weights of 43, 35, 33, and 14 kDa, respectively. One of the nonstructural protein-2As associated with AEV possesses a conserved motif shared with two other picornaviruses: HPeVs and AiV [21]. It has been shown that AEV is resistant to chloroform, acid, trypsin, and DNase and is protected against the effects of heat [22]. However, the AEV is susceptible to a single exposure to formaldehyde fumigation [23].

6.1.2 PATHOBIOLOGY AND EPIDEMIOLOGY

AEV is known to contain two distinct pathotypes: natural AEV field strains and embryo-adapted strains. (1) Natural AEV field strains are enterotropic. These strains infect chickens readily via the oral route and are shed in the feces. They are relatively nonpathogenic except in susceptible chicks infected by vertical transmission or by early horizontal transmission, in which case they cause neurologic signs. (2) Embryo-adapted strains are highly neurotropic and cause severe neurologic signs following intracerebral inoculation or parental routes such as intramuscular or subcutaneous inoculation. AEV-adapted Van Roekel (VR) strain can easily be replicated in chicken embryo brain cultures [20,24].

Both AEV pathotypes can replicate in embryos, but natural field strains do not cause obvious signs or gross lesions. On the other hand, adapted strains are pathogenic for embryos, causing muscular dystrophy and immobilization of skeletal muscles [25]. No differences in structural proteins have been observed between a field isolate and the embryo-adapted VR strain of AEV [26]. All chicken flocks eventually become infected with the AE virus, but the incidence of clinical disease is very low unless a breeder flock is not vaccinated and becomes infected with the AEV after the commencement of egg production. Turkey flocks apparently also experience high rates of natural infection based on serological surveys [8]. However, the rate of infection in pheasant and quail is not known [9].

6.1.3 CLINICAL FEATURES AND PATHOGENESIS

AE presents an interesting syndrome. In naturally occurring outbreaks, it usually makes its appearance when chicks are 1–2 weeks of age, although affected chicks have been observed at the same time of hatching. Affected chicks first show a slightly dull expression of the eyes, followed by a progressive ataxia from incoordination of the muscles, which may be detected readily by exercising the chicks [5]. As the ataxia grows more pronounced, chicks show an inclination to sit on their hocks. Some may refuse to move or may walk on their hocks and shanks. Fine tremors of the head and neck may become evident, the frequency and magnitude of which may vary. Exciting or disturbing the chicks may bring on the tremor, which may continue for variable period and recur at irregular intervals [4,5]. Ataxia usually progresses until the chick is incapable of moving around, and this stage is followed by initiation, prostration, and finally death [10]. Some chicks with definite signs of AE may survive and grow to maturity, and in some instances, signs may disappear completely. Survivors may later develop blindness from opacity giving a bluish discoloration to the lens [27]. There is marked age resistance to clinical signs in birds exposed after 2–3 weeks of age. Mature birds may experience a temporary drop in egg production (5%–10%) but do not develop neurologic signs. Virus may be propagated in the baby chick, chicken embryos from susceptible flocks, and a variety of cell culture systems. Several routes of inoculation in embryos have been used, but inoculation via the yolk sac at 5–7 days of embryonating egg is considered the method of choice [25]. Gross lesions are observed only with adapted strains.

Localization of the viral antigen using virus isolation, immunodiffusion, immunofluorescence, and enzyme-linked immunosorbent assay (ELISA) techniques has been reported by several investigators [3,20,28,29]. In young chicks exposed orally to field strains of AEV, primary infection of the alimentary tract is rapidly followed by a viremia and subsequent infection of the pancreas and other visceral organs (liver, spleen, heart, kidneys) and skeletal muscle and finally

the CNS. Alimentary tract infections involve muscular layers, and pancreatic infections are found in both the acinar and islet, persisting more in the latter. Viral antigen is relatively abundant in the CNS where the Purkinje neuron and the molecular layer of the cerebellum are apparently the favored sites of virus replication. Chicks with clinical signs at 10–30 days of age tend to have viral antigen mostly in the CNS and pancreas; less amounts of antigen have been seen in the heart and kidney; and only very small amounts have been seen in the liver and spleen. Avian encephalomyelitis virus studies of various experimental infections of adult chickens revealed that no viral antigen was found in the CNS; presumably, this lack of infection correlates with the absence of clinical disease in infected adults [29]. On the other hand, the pathogenesis of AEV in young chicks, naturally occurring or experimental infection, has been clearly defined [28]. It was described that chicks infected at day 1 of age generally died, whereas those infected at day 8 developed paresis but usually recovered, and infection at day 28 caused no clinical signs [30].

6.1.4 DIAGNOSIS

6.1.4.1 Conventional Techniques

Many methods have been developed for the diagnosis of AE, including virus isolation by intracerebral inoculation of 1-day-old chicks and yolk-sac inoculation of embryonated chicken eggs [6,25,31].

The brain is an excellent source of virus for isolation, although other tissues and organs induce the disease when injected into chicks [7,32]. Miyamae [33] found that in addition to the brain, the pancreas and duodenum were especially reliable sources of the virus. The need to titrate vaccine virus makes a sensitive method for virus detection very important. One system for virus assay is to inoculate embryos from a susceptible flock via the yolk sac at 5–7 days of age, allowing to hatch and observe chicks for signs of disease during the first 10 days [30,20]. When clinical signs appear, the brain, proventriculus, and pancreas should be examined for lesions as described in histopathology. Additionally or alternatively, the brain, pancreas, and duodenum of affected chicks can be examined for the specific viral antigen by immunofluorescence [34,35] or immunodiffusion tests [36,37].

Other methods for the diagnosis of AE include virus isolation by intracerebral inoculation of 1-day-old chicks and yolk-sac inoculation of embryonated chicken eggs [32,38]. Virus may be propagated in the baby chick, chicken embryos from susceptible flocks, and a variety of cell culture systems. Several routes of inoculation in embryos have been used, but inoculation via the yolk sac at 5–7 days of embryonating egg is considered the method of choice [32]. Various serologic methods have been developed, such as hemagglutination, complement fixation, indirect fluorescent-antibody technique, ELISA, virus neutralization, and agar-gel-precipitin tests [11,36,37,39–42].

However, isolation and serological methods are time-consuming and labor intensive. Furthermore, serological tests are often hampered by nonspecific reactions or cross-reactions.

6.1.4.2 Molecular Techniques

In recent years, the cDNA probe [43] and polymerase chain reaction (PCR) methods have been applied as a rapid diagnostic tool for the detection of avian viral and bacterial pathogens [44–48]. In this study, we have developed and optimized a reverse transcription-polymerase chain reaction (RT-PCR) assay using primers targeting to the VP2 gene to detect AEV [49].

6.2 METHODS

6.2.1 SAMPLE COLLECTION

Virus propagation: When clinical signs appear, brain, proventriculus, and pancreas tissues should be collected from affected chickens.

Preparation of clinical samples

1. Mince tissue specimens and add penicillin (10,000 IU/ mL) and streptomycin (100 µg/mL).
2. Incubate the mixture at 4°C for 4 h.
3. Centrifuge the mixture at $1000 \times g$ for 10 min.
4. Collect the supernatants.
5. Inoculate aliquots of 0.2 mL of supernatants into the allantoic cavity of 6-day-old embryonating specific-pathogen-free (SPF) eggs.
6. Collect the brains of infected embryos at 3 and 5 days postinoculation.
7. Resuspend tissues in phosphate-buffered saline (PBS) (10% w/v).
8. Disrupt brain tissues three times by freezing and thawing and store at −70°C.

Virus semipurification: Semipurified virus should be modified as described [38]:

1. The brains from six AEV-infected chick embryos can be suspended in 200 mL PBS.
2. Brain tissues should be treated with 0.5% (w/v) sodium dodecyl sulfate and disintegrated by shaking using glass beads for 30 min at 37°C.
3. It is followed by sonication for 20–30 s.
4. Virus suspensions are clarified by centrifugation at $1000 \times g$ for 30 min at 4°C.
5. The supernatants are further centrifuged at $80,000 \times g$ for 3 h at 4°C.
6. The resulting crude virus pellets should be resuspended in 1 mL of PBS.

These viral preparations will be used for extraction of RNA and stored at −20°C.

Negative pathogen controls for PCR assay: Infectious bronchitis virus (IBV), Newcastle disease virus (NDV), and adenovirus have been propagated in the allantoic cavity of 10-day-old SPF embryonated eggs [50]. Infectious laryngotracheitis virus (ILTV) is propagated on the chorioallantoic membrane in 10-day-old SPF embryonated eggs. Infectious bursal disease virus (IBDV) and reovirus grown in chicken embryo fibroblast monolayers using method described previously [47,48]. *Mycoplasma gallisepticum* (MG) is grown in Frey medium [51]. *Salmonella* sp. and *Escherichia coli* are grown in Luria–Bertani broth media [46].

Nucleic acid extraction: RNA extraction from AEV, IBV, NDV, IBDV, and reovirus is carried out using the Trizol LS manufacturer's protocol (Trizol LS, Life Technologies, Bethesda, MD). Deoxyribonucleic acid (DNA) from ILTV, adenovirus, MG, *Salmonella* sp., and *E. coli* are extracted using the phenol/chloroform/isoamyl alcohol solution (1:1:24 v/v) and purified (Gibco BRL, Grand Island, NY) using the method described by Pang et al. [50]. The concentrations of RNA and DNA determined by spectrophotometer using the Bio Mate 5 (Thermo Spectronic, Rochester, NY) and nucleic extractions are stored at −20°C.

Oligonucleotide primers: Pairs of primers that specifically amplify AEV are designated as

MK AE 1 (5′-CTT ATG CTG GCC CTG ATC GT-3′)

MK AE 2 (5′-TCC CAA ATC CAC AAA CCT AGC C-3′)

It is selected based on the published sequence data of AEV [14,16]. These primers flanked a 619 bp DNA sequence containing the VP2 gene. Primers are synthesized using Gibco (Invitrogen, Carlsbad, CA). Primers are aliquot into 50 µL volumes and stored at −20°C.

6.2.2 DETECTION PROCEDURES

The following RT-PCR protocol has been modified from the previous report [49].

6.2.2.1 Reverse Transcription Reaction

1. Prepare the RT mix as follows (volume indicated per specimen) (*Note*: vortex and spin down all microcentrifuge tubes of RT-PCR components before use):

Component	Volume (µL)
10× PCR buffer (500 mM KCl, 200 mM Tris_HCl [pH 8.4])	2.0
25 mM MgCl$_2$	4.0
Bovine serum albumin (BSA) (0.5 mg/mL)	2.5
Deoxynucleotide triphosphate (dntp) mix (10 mM)	2.0
RNase inhibitor (20 units/µL)	1.0
Downstream primer (MKAE2) 50 pmol	1.0
Moloney murine leukemia virus, reverse transcriptase (50 units/µL)	1.0
Diethylpyrocarbonate (DEPC)-treated water	6.5
Total	20.0

2. Add this 20 µL RT mix to each 0.5 mL tube followed by the addition of 5 µL (50 ng) of the individual RNA sample.
3. Spin the tube briefly to ensure that no reagent droplets remain on the inner wall of the tube.

4. Overlay the reaction mixture with one drop of mineral oil to prevent evaporation during the RT process.
5. Transfer the reaction tubes to a 42°C heat block for 30 min and then increase the temperature to 99°C for 5 min.
6. Rapidly chill the tubes on ice for 5 min.
7. The cDNA can be used directly in the PCR step or stored at −20°C.

6.2.2.2 PCR Detection of AEV

AEV is detected by PCR amplification of cDNA using primers MK AE 1/MK AE 2.

6.2.2.2.1 Procedure

1. Remove all aliquots of PCR reagents from the freezer. Vortex and spin down all reagents before opening the tubes.
2. Prepare the PCR reaction mix as follows:

Component	Volume (µL)
10× PCR buffer (500 mM KCl, 200 mM Tris_HCl [pH 8.4])	8.0
MgCl$_2$ 5 mM	8.0
BSA nuclease-free 0.5 mg/mL	2.5
Primer MK AE 1 60 pmol	1.5
AmpliTaq DNA polymerase 2.5 unit	6.0
DECP-distilled water	54.0
cDNA (RT reaction mix)	20.0
Total reaction	100.0

3. Turn on a thermocycler and preheat the block to 95°C.
4. Spin the sample tubes in a microcentrifuge at 10,000 × g for 30 s at room temperature.
5. Place the sample tubes in the thermal cycler for 35 cycles of 94°C for 1 min, 62°C for 1 min, 62°C for 1 min, and 1 cycle of 62°C for 10 min; keep at 4°C afterward.
6. Following amplification, pulse-spin the reaction tubes to pull down the condensation droplets at the inner wall of the tubes. The samples are now ready for electrophoresis or can be stored at −20°C.
7. Analyze the PCR products by agarose gel electrophoresis on 1.0% agarose gel in Tris/Borate/EDTA buffer at 80 V for 40 min [52].
8. Stain the gel with ethidium bromide and then visualize under ultraviolet light source.

Note: Negative control is included with clinical samples in order to confirm any possible contamination may have occurred during the PCR assay. The presence of 619 bp DNA product is indicative of AEV.

6.3 CONCLUSION

AEV is a single-stranded, positive-sense RNA virus that is capable of causing neurologic symptoms (e.g., ataxia, incoordination, paralysis, and rapid tremors) in young chicks, with high morbidity and moderate mortality. Additionally, this virus can reduce egg production in adult laying birds. Although outbreaks of AE were initially documented in Connecticut, Maine, Massachusetts, and New Hampshire, they have been subsequently shown to have a worldwide distribution. The implementation of immunization measures from the mid-1950s has largely put AE under control. Nevertheless, continuous vigilance and preparedness are necessary in order to minimize the damages from any future outbreaks of AE. Toward this goal, the availability of rapid, sensitive, and specific methods for the laboratory diagnosis of AEV infections is critical. Given the laborious and time-consuming nature of conventional techniques such as virus isolation and immunologic assays, the adoption of molecular techniques (e.g., RT-PCR) in routine diagnostic laboratories will enable prompt detection and identification of AEV and improve decision making in relation to the appropriate and effective control and prevention strategies against AE.

REFERENCES

1. Calnek, B. W., Taylor, P. J., and Sevoian, M. Studies on avian encephalomyelitis. IV. Epizootiology. Avian Dis. 4, 325–347 (1960).
2. Calnek, B. W., Taylor, P. J., and Sevoian, M. Studies on avian encephalomyelitis. V. Development and application of an oral vaccine. Avian Dis. 5, 297–312 (1961).
3. Calnek, B. W., Luginbuhl, R. E., and Helmboldt, C. F. Avian encephalomyelitis. In: Disease of Poultry, 10th edn., B. W. Calnek, H. J. Barnes, C. W. Beard, L. R. McDougald, and Y. M. Saif, eds. Iowa State University Press, Ames, IA, pp. 571–581, 1997.
4. Jones, E. E. An encephalomyelitis in chicken. Science. 76, 331–331 (1932).
5. Jones, E. E. Epidemic tremor, an encephalomyelitis affecting young chickens. J Exp Med. 59, 781–798 (1934).
6. McNulty, M. S., Connor, T. J., McNeilly, F., and McFerran, J. B. Biological characterization of avian enteroviruses and enterovirus-like viruses. Avian Pathol. 19, 75–78 (1990).
7. Tannock, G. A. and Shafren, D. R. Avian encephalomyelitis: A review. Avian Pathol. 23, 603–620 (1994).
8. Deshmukh, D. R., Larsen, C. T., Rude, T. A., and Pomeroy, B. S. Evaluation of live-virus vaccine against avian encephalomyelitis in turkey breeder hens. Am J Vet Res. 34, 863–867 (1973).
9. Hill, R. W. and Raymond, R. G. Apparent natural infection of Coturnix quail hens with the virus of avian encephalomyelitis. Avian Dis. 6, 226–227 (1962).
10. Olitsky, P. K. Experimental studies on the virus of infectious avian encephalomyelitis. J Exp Med. 70, 565–582 (1939).
11. Shafren, D. R., Tannock, G. A., and Groves, P. J. Antibody responses to avian encephalomyelitis virus when administered by different routes. Aust Vet. 69, 272–275 (1992).
12. Van Roekel, H., Bullis, K. L., and Clarke, M. K. Transmission of avian encephalomyelitis. J Am Vet Med Assoc. 99, 220 (1941).
13. Schaaf, K. Immunization for the control of avian encephalomyelitis. Avian Dis. 2, 279–289 (1958).
14. Todd, D., Weston, J. H., Mawhinney, K. A., and Laird, C. Characterization of the genome of avian encephalomyelitis virus with cloned cDNA fragments. Avian Dis. 43, 219–226 (1999).

15. Butterfield, W. K., Luginbuhl, R. E., and Helmboldt, C. F. Characterization of avian encephalomyelitis virus (an avian enterovirus). *Avian Dis.* 13, 363–378 (1969).

16. Marvil, P., Knowles, N. J., Mockett, P. A., Britton, P., Brown, T. D. K., and Cavanaugh, D. Avian encephalomyelitis virus is a picornavirus and is most closely related to hepatitis A virus. *J Gen Virol.* 80, 653–662 (1999).

17. International Committee on Taxonomy. In: Virus Taxonomy, Classification and Nomenclature of Viruses, M. H. V. Van Regenmortel, C. M. Fauquet, and D. H. L. Bishop, eds. Academic Press, San Diego, CA, 1162pp., 2000.

18. Gosting, L. H., Grinnell, B. W., and Matsumoto, M. Physicochemical characteristics of avian encephalomyelitis virus. *Vet Microbiol.* 5, 87–100 (1980).

19. Shafren, D. R. and Tannock, G. A. Further evidence that the nucleic acid of avian encephalomyelitis virus consists of RNA. *Avian Dis.* 36, 1031–1033 (1992).

20. Shafren, D. R. and Tannock, G. A. Pathogenesis of avian encephalomyelitis viruses. *J Gen Virol.* 72, 2713–2719 (1991).

21. Hishida, N., Odagiri, Y., Kotani, T., and Horiuchi, T. Morphological changes of neurons in experimental avian encephalomyelitis. *Jpn J Vet Sci.* 48, 169–172 (1986).

22. Bulow, V. V. Studies on the physio-chemical properties of the virus of avian encephalomyelitis (AE) with special reference to purification and preservation of virus suspension. *Zentralbl Veterinaermed [B].* 11, 674–686 (1964).

23. Ide, P. R. The sensitivity of some avian viruses to formaldehyde fumigation. *Can J Comp Med.* 43, 211–216 (1979).

24. Mancici, I. O. and Yates, V. J. Cultivation of avian encephalomyelitis virus in vitro. II. In chick embryo fibroblastic cell culture. *Avian Dis.* 12, 278–284 (1968).

25. Jungherr, E. L., Summner, F., and Luginbuhl, R. E. Pathology of egg-adapted avian encephalomyelitis. *Science.* 124, 80–81 (1956).

26. Butterfield, W. K., Helmboldt, C. M., and Luginbuhl, R. E. Studies on avian encephalomyelitis. IV. Early incidence and longevity of histopathologic lesions in chickens. *Avian Dis.* 13, 53–57 (1969).

27. Bridges, C. H. and Flowers, A. I. Isidocyclitis and cataracts associated with an encephalomyelitis in chickens. *J Am Vet Med Assoc.* 132, 79–84 (1958).

28. Braune, M. O. and Gentry, R. F. Avian encephalomyelitis virus. I. Pathogenesis in chicken embryos. *Avian Dis.* 15, 638–647 (1971).

29. Ikeda, S. and Matsuda, K. Susceptibility of chickens to avian encephalomyelitis virus. IV. Behavior of the virus in laying hens. *Natl Inst Anim Health Q Tokyo.* 16, 83–89 (1976).

30. Cheville, N. F. The influence of thymic and bursal lymphoid systems in the pathogenesis of avian encephalomyelitis. *Am J Pathol.* 58, 105–125 (1970).

31. Lukert, P. D. and Davis, R. B. An antigen used in the agar-gel precipitin reaction to detect avian encephalomyelitis virus antibodies. *Avian Dis.* 15, 935–938 (1971).

32. Wills, F. K. and Moulthrop, I. M. Propagation of avian encephalomyelitis virus in the chick embryo. *Southwest Vet.* 10, 39–42 (1956).

33. Miyamae, T. Invasion of avian encephalomyelitis virus from the gastrointestinal tract to the central nervous system in young chickens. *Am J Vet Res.* 44, 508–510 (1983).

34. Miyamae, T. Ecological survey by immunofluorescent method of virus in enzootics of avian encephalomyelitis. *Avian Dis.* 18, 369–377 (1974).

35. Miyamae, T. Immunofluorescent study on egg-adapted avian encephalomyelitis virus infection in chickens. *Am J Vet Res.* 38, 2009–2012 (1977).

36. Ikeda, S. Immunodiffusion tests in avian encephalomyelitis. I. Standardization of procedure and detection of antigens in infected chickens and embryos. *Natl Anim Health Q Tokyo* 17, 81–87 (1977).

37. Ikeda, S. Immunodiffusion tests in avian encephalomyelitis. II. Detection of precipitating antibody in infected chickens in comparison with neutralizing antibody. *Natl Inst Anim Health Q Tokyo.* 17, 88–94 (1977).

38. Tannock, G. A. and Shafren, D. R. A rapid procedure for the purification of avian encephalomyelitis virus. *Avian Dis.* 29, 312–321 (1985).

39. Halpin, F. B. A search for a hem-agglutinating property of the virus of infectious avian encephalomyelitis. *Avian Dis.* 10, 513–517 (1966).

40. Chol, W. P. and Miura, S. Indirect fluorescent antibody technique for the detection of avian encephalomyelitis antibody in chickens. *Avian Dis.* 16, 949–951 (1972).

41. Girshick, T. and Crary, Jr. C. K. Preparation of an agar-gel precipitating antigen for avian encephalomyelitis and its use in evaluating the antibody status of poultry. *Avian Dis.* 26, 798–804 (1982).

42. Sato, G., Watanabe, H., and Miura, S. An attempt to produce a complement-fixation antigen for the avian encephalomyelitis virus from infected chick embryo brains. *Avian Dis.* 13, 461–479 (1969).

43. Wei, I., Zhou, J., Wang, J., Shi, L., and Liu, J. Development of a non-radioactive digoxigenin cDNA probe for the detection of avian encephalomyelitis virus. *Avian Pathol.* 37, 187–191 (2008).

44. Mansy, M. S., Ashour, M. S. E., and Khan, M. I. Development of species-specific polymerase chain reaction (PCR) technique for *Proteus mirabilis. Mol Cell Probes.* 13, 133–140 (1999).

45. Nascimento, E. R., Yamamoto, R., and Khan, M. I. *Mycoplasma gallisepticum* vaccine strain-polymerase chain reaction. *Avian Dis.* 37, 203–211 (1993).

46. Nguyen, A. V., Khan, M. I., and Lu, Z. Amplification of *Salmonella* chromosomal DNA using polymerase chain reaction. *Avian Dis.* 38, 119–126 (1994).

47. Stram, Y., R. Meir, R., Molad, T., Blumenkranz, R., Malkinson, M., and Weisman, Y. Applications of the polymerase chain reaction to detect infectious bursal disease virus in naturally infected chickens. *Avian Dis.* 38, 879–884 (1994).

48. Xie, Z., Fadl, A. A., Girshick, T., and Khan, M. I. Amplification of avian reovirus RNA using the reverse transcriptase-polymerase chain reaction. *Avian Dis.* 41, 654–660 (1997).

49. Xie, Z., Khan, M. I., Girshick, T., and Xie, Z. Reverse transcriptase polymerase chain reaction to detect avian encephalomyelitis virus. *Avian Dis.* 49, 227–230 (2005).

50. Pang, Y. S., Khan, M. I., Wang, H., Xie, Z., and Girshick, T. Multiplex PCR and its application in experimentally infected SPF chickens with respiratory pathogens. *Avian Dis.* 46, 691–699 (2002).

51. Frey, M. I., Hanson, R. P., and Anderson, D. P. A medium for the isolation of avian mycoplasmas. *Am J Vet Res.* 29, 2163–2171 (1968).

52. Sambrook, J., Frisch, E. T., and Maniatis, T. *Molecular Cloning: A Laboratory Manual.* Cold Spring Harbor Laboratory Press, Cold Spring Harbor, NY, 1989.

7 Duck Hepatitis Virus

Dabing Zhang and Ning Liu

CONTENTS

7.1 Introduction 53
    7.1.1 Classification, Morphology, and Genome Organization 53
    7.1.2 Transmission, Clinical Features, and Pathogenesis... 55
    7.1.3 Diagnosis 55
        7.1.3.1 Conventional Techniques.. 56
        7.1.3.2 Molecular Techniques... 56
7.2 Methods 57
    7.2.1 Sample Collection and Preparation 57
    7.2.2 Detection Procedures... 57
        7.2.2.1 Reverse Transcription 57
        7.2.2.2 PCR Detection of DHAVs and DAstVs.. 57
        7.2.2.3 Sequence Analysis 58
7.3 Conclusion 58
References... 59

7.1 INTRODUCTION

Duck viral hepatitis (DVH) is an acute, rapidly spreading, highly fatal viral disease of young ducklings, 1–28 days of age [1]. The disease is characterized by clinical sign with opisthotonos and lesion of liver hemorrhages. All domestic ducklings are susceptible, with the exception of muscovies [2]. Ducklings under 3 weeks of age are highly susceptible. The disease may cause up to 90% mortality and is of economic importance to all duck-growing farms.

Previously, the causative agent of the disease was named duck hepatitis virus (DHV) and divided into three serotypes: DHV types 1 (DHV-1), 2 (DHV-2), and 3 (DHV-3) [1]. DHV-1 was first isolated by Levine and Fabricant from an outbreak of DVH in young white Pekin ducks on Long Island, New York, in 1949 [3], and later known to be worldwide in distribution [1]. This virus was classified as a picornavirus by Tauraso et al. [4]. DHV-2 and DHV-3 were originally recognized as separate serotypes because they induced DVH in ducklings vaccinated for DHV-1. DHV-2 was first described by Asplin in Norfolk, England, in 1965 [5]. This virus was shown to be distinct from DHV-1 by cross protection tests and therefore characterized as a new serotype of DHV. The disease caused by DHV-2 disappeared from commercial flocks by 1969 but reappeared in 1983/84 on three farms in Norfolk, England [6]. DHV-3-associated disease was first reported by Toth in 1969 and last observed in 1975 on a flock of white Pekin ducks on Long Island, New York [7,8]. The agent shared no common antigens with DHV-1 and DHV-2 in virus neutralization (VN) and fluorescent antibody (FA) tests and was characterized as the third type of DHV by Haider and Calnek [8]. It is generally considered that DHV-2 and DHV-3 have only been found in the United Kingdom and the United States, respectively.

Recent characterization of several DHV-2- and DHV-3-like viruses from ducks in China suggested that DHV-2 and DHV-3 may have been introduced into countries outside the United Kingdom and the United States [9,10].

In 2007, two new serotypes of DHV were reported by Tseng and Tsai in Taiwan and by Kim et al. in South Korea independently [11,12]. The Taiwanese new serotype was first noted by Lu et al. in 1989–1990 and isolated from a mule duckling in 1990 and white Roman goslings in 2004 [11,13]. The Korean new serotype of DHV was isolated from ducklings in the years of 2003 and 2004 [12]. Similar to DHV-2 and DHV-3, discoveries of the two new serotypes were also associated with DVH outbreaks in ducklings vaccinated for DHV-1. The two viruses were shown to be both serologically different from DHV-1 in VN tests [11,12], but they have never been compared with each other. The sequence difference between the two viruses in capsid-coding regions (approximately 30% divergence in nucleotide) suggests that they may belong to different serotypes [14].

7.1.1 CLASSIFICATION, MORPHOLOGY, AND GENOME ORGANIZATION

Although DHV-1 was first isolated 64 years ago and initial studies indicated that this virus had many characteristics similar to members of the genus *Enterovirus* [1,3,4], little was known about this virus in its molecular characteristics and precise taxonomical position in the family *Picornaviridae* for a long time. The molecular cloning of DHV-1 genome in 2006–2007 led to dramatic progress in understanding the molecular virology and taxonomy of the virus. Based on its genomic structure and sequence similarity

53

with previously known picornavirus genera, DHV-1 was proposed to be placed in a separate genus within the family *Picornaviridae* [15–17]. Meanwhile, the genome sequences of the two new DHV serotypes discovered in Taiwan and South Korea were also determined in 2007 [11,12]. Analyses of the complete genomes of both these viruses suggested that they would be classified in the same species as DHV-1 but as distinct genotypes (or serotypes) [11,12,14,18]. According to the *Ninth Report of International Committee for the Taxonomy of Viruses*, the three DHV types have been classified as members of the species *duck hepatitis A virus* (DHAV) in the genus *Avihepatovirus* in the family *Picornaviridae*. DHV-1 was renamed DHAV type 1 (DHAV-1), while the two newly discovered DHV serotypes were designated DHAV types 2 (DHAV-2) and 3 (DHAV-3), respectively [19].

Comparative sequence analysis of several DHAV strains indicated that the capsid sequences are highly conserved within the same type but highly divergent between different types. The nucleotide and amino acid identities within the same type are at least 90% and 92% in VP1, 94% and 96% in VP0, and 92% and 95% in VP3, whereas those between different types do not exceed 72% and 78% in VP1, 72% and 80% in VP0, and 74% and 83% in VP3, respectively [14]. In pairwise comparison of the 5′-untranslated region (UTR) sequences, DHAVs of the same type are also clearly distinguished from those of different types, and the limit of intratypic divergence appears to be about 7% nucleotide difference. In contrast, sequence variation among strains between types is considerably higher, ranging from 26% to 31%. Based on the sequence feature of 5′-UTR, an approach for molecular detection and typing of DHAV was reported by Fu et al. in 2008. This technique involved the amplification of an approximately 250 bp cDNA fragment from the 5′-UTR followed by nucleotide sequence determination and analysis of the amplified fragments. Classification based on the nucleotide sequence of the amplicon revealed that the sequence diverges by as much as 27%–36% between types. Within the same type, the partial 5′-UTR sequences are higher than 94% [18].

So far, data relating to molecular epidemiology of DHAV are limited. According to previous results obtained from serological tests, DHAV-1 is worldwide in distribution [1]; DHAV-2 has only been found in Taiwan, China [11]; and DHAV-3 has been reported in South Korean and mainland China [12,18]. Surveillance data of DHAV in China indicate that the DHAV-3 strains dominate as the cause of DVH outbreaks in recent years, although DHAV-1 is still circulating in China (GenBank EF093502.1).

The capsid protein precursor of DHAV-1 and DHAV-2 was predicted to be 731 amino acids (aa) in length, 2 aa shorter than that of DHAV-3. The subtle length difference between DHAV-3 and other DHAV types was found to be located at the VP1 region [11,14]. The DHAV virion was predicted to be composed of three structural capsid proteins (VP0, VP3, and VP1). The VP0 polypeptide does not appear to be proteolytically cleaved into VP4 and VP2 [11,12,15–17]. DHAV-1 has been estimated to be 20–40 nm in size [1], while the sizes of DHAV-2 and DHAV-3 have not been reported.

The genome of the DHAVs consists of positive-sense, single-stranded, polyadenylated RNA of 7689–7775 nucleotides (nt). Besides the 6 nt difference in VP1 mentioned earlier, the genomic length differences between the three types are mainly located at the UTRs. The polyadenylated genome contains a large open reading frame (ORF), encoding a putative polyprotein, which is preceded by a 5′-UTR and followed by 3′-UTR. The DHAV polyprotein appears to be cleaved into 10–12 mature products, forming its structural (VP0, VP3, and VP1) and nonstructural (2A, 2B, 2C, 3A, 3B, 3C, and 3D) proteins. A notable feature is the 2A protein, which contains three 2A motifs: (1) asparagine-proline-glycine-proline (NPGP) motif (related to the 2A of aphtho-, cardio-, erbo-, and teschoviruses), (2) avrRpt2-induced gene 1 like protein containing a GxxGxGKS nucleoside triphosphate binding motif, and (3) histidine and asparagine-cysteine box motif (similar to the 2A of parecho-, kobu-, and tremoviruses) [11,12,15–17]. It is therefore suggested by Tseng et al. in 2007 that DHAV may possess three putative 2A proteins [16]. However, it cannot be excluded that the 2A1 peptide sequence may be the carboxy-terminus of VP1, as is probably the case with the putative 2A1 of Ljungan virus (genus *Parechovirus*) [20]. It is also possible that 2A2 and 2A3 are a single multidomain protein [15,17]. The 5′-UTR was shown to possess a distinct hepacivirus/pestivirus-like internal ribosome entry site [21].

Originally thought to be a picornavirus, DHV-2 was characterized by Gough et al. in 1984/1985 as an astrovirus on the basis of morphology and cross protection tests [6]. A proposal was put forward that it should be renamed duck astrovirus (DAstV) [22]. The study by Todd et al., which showed that a 391 nt astrovirus-specific ORF1b was amplified from the DHV-2 M52 isolate, supported the view that DHV-2 belonged to an astrovirus [23]. DHV-3, originally regarded as a third serotype of picornavirus DHV [8], was identified by Todd et al. in 2009 as a second astrovirus in ducks because the 391 nt astrovirus-specific ORF1b sequence was also amplified from the DHV-3 X1222A Calnek isolate [23]. DHV-2 and DHV-3 are now classified as members of the genus *Avastrovirus* in the family *Astroviridae* [24]. Sequence comparison of the partial ORF1b region revealed that DHV-2 and DHV-3 share low nucleotide (64%) and amino acid (69%) identities with each other, suggesting that they may represent different avastrovirus species [23]. At present, the classification of avastroviruses is mainly based on genetic analysis of the complete capsid region at the amino acid level [24]. Due to the lack of the capsid sequences of DHV-2 and DHV-3, the taxonomic relationships of the two DAstVs to each other and to other avastroviruses remain to be resolved. The recent genomic characterization of a DHV-2-like isolate C-NGB and four DHV-3-like isolates represented by SL1 revealed that DHV-2 could be placed in species *Avastrovirus 2* together with turkey astrovirus 2, while DHV-3 may belong to an additional avastrovirus species [9,10,24].

The capsid protein of DAstVs is translated as one long precursor protein of approximately 81.7 kDa (DHV-2-like isolate C-NGB) and 80.7 kDa (DHV-3-like isolate SL1) [9,10], which is posttranslationally cleaved to form mature virion subunits in a mechanism that is not understood. Previous investigations have shown that DHV-2 is a small round, nonenveloped virus, 28–30 nm in diameter. The surface of the particles exhibits astrovirus-like morphology by Electron Microscope (EM) using negative staining [6]. However, DHV-3 appears to have a picornavirus-like morphology. Examination of duck kidney (DK) cells infected with DHV-3 by EM revealed crystalline arrays containing particles about 30 nm in diameter in the cytoplasm [8].

The genome of the DHV-2- and DHV-3-like astroviruses consists of positive-sense, single-stranded RNA. Excluding the poly(A) tail, the genomic lengths of the two DAstVs are 7722 and 7319 nt, respectively. The polyadenylated genome contains three ORFs, as well as a short 5′-UTR and a 3′-UTR. ORF1a and ORF1b encode the nonstructural proteins, which include several transmembrane helical motifs, a serine protease, a nuclear localization signal, and an RNA-dependent RNA polymerase motif. ORF2 encodes the capsid protein that is required for virion formation. By analogy to other astroviruses, a ribosomal frameshift signal can be observed in the overlap region of ORF1a and ORF1b of the DHV-2- and DHV-3-like astroviruses [9,10]. As a mechanism of virus evolution, recombination by avian astrovirus RNAs has been reported previously for turkey and chicken astroviruses [25–27]. Most recently, the DHV-2- and DHV-3-like isolates were shown to be closely related to different avastroviruses in different genomic regions, providing evidence that recombination events may have occurred during evolution of DAstVs [10]. Such phenomenon may complicate the classification of avastroviruses.

7.1.2 Transmission, Clinical Features, and Pathogenesis

As DHAV-1 is the most common virus reported in outbreaks worldwide, bulk of epizootiological data was obtained from studies on this particular virus. Transmission of the virus is usually by direct contact between birds or through fomites such as brooders, water, feed, and equipment. Egg transmission presumably does not take place [1]. However, a recent study in our laboratory revealed that DHAV-3 was easily detected from eggs produced by infected breeders, suggesting the possibility of egg transmission of DHAV-3 (GenBank KP995438.1).

The affected ducklings and recovered carriers are sources of infection. Recovered ducks can excrete DHAV-1 in feces for up to 8 weeks postinfection (PI) [1], which provides an extended opportunity for transmission of the virus to other birds. Although adult breeders on infected premises do not become clinical ill, they could be carriers. An epidemiological study in our laboratory indicated that a proportion of parent ducks may become infected and excrete DHAV-1, together with DHAV-3 and DHV-2-like astrovirus, throughout full laying period (GenBank KP995438.1). As a result, the ducklings may be infected by virus-contaminating eggs and brooders. Previous work by Asplin (1961) showed that DHAV-1 survived at least 10 weeks in uncleaned infected brooders and for longer than 37 days in moist feces stored in a cool shed [28]. Prolonged excretion after infection is likely to lead to high levels of viral contamination in brooders and farm environment.

The wild birds, rodents, and workers as mechanical carriers may also play a role in spreading the disease [1].

DHAV-1 usually affects ducklings under 4 weeks of age and often much younger. Morbidity may be 100% and mortality is variable. In some broods, less than 1-week-old mortality may reach 95%. In ducklings 1–3 weeks old, mortality may be 50% or less. In ducklings 4–5 weeks old, morbidity and mortality are low or negligible. Adult breeders on infected premises do not become clinically ill and continue in full production [1].

The clinical signs are characterized by lethargy and ataxia. Infected ducklings lose their balance, fall on their sides, and kick spasmodically prior to death. At death, the head is usually drawn back in the opisthotonos position. Onset and spread of the disease are very rapid. Death occurs within 1–2 h after signs are noted. Practically, all mortality in a flock will occur within 3–4 days, with most deaths on the second day. Gross pathological changes are found chiefly in the liver, which is enlarged and exhibits punctate and ecchymotic hemorrhages. The spleen is sometimes enlarged and mottled. In some cases, the kidney is swollen and renal blood vessels congested. Microscopic changes are mainly observed in the liver and are characterized by extensive hepatocyte necrosis and bile duct hyperplasia, together with varying degrees of inflammatory cell response and hemorrhage [1].

The oral and respiratory mutes can be the portal of virus entry. The target organ appears to be the liver. Survivors from the disease are immune to further infection [1].

DHAV-2 and DHAV-3 can also cause severe diseases. Mortality exceeding 70% in 1-week-old white Roman goslings and up to 80% in Pekin ducklings were recorded for DHAV-2 and DHAV-3, respectively [11,29]. Severity of the disease caused by DHV-2 was shown to be associated with age of ducklings, with losses up to 50% in 6–14-day-old ducklings and between 10% and 25% in 3–6-week-old ducklings [1,5,6]. The disease caused by DHAV-3 was observed to be less severe than DHAV-1 and mortality rarely exceeded 30% [1,7,8]. The clinical signs and pathological changes caused by these agents are similar to those of DHAV-1.

7.1.3 Diagnosis

The clinical and pathological observations are highly indicative of DVH. Therefore, diagnosis of the disease can usually be made based on the characteristic disease pattern in the flock and gross pathological changes. Initial diagnosis can be further confirmed by the recovery of virus from dead ducklings and the reproduction of the disease in susceptible ducklings [1]. As DVH may be caused by five different viruses, differential characterization of the causative agents is needed for definite diagnosis. However, cooccurrence of DHAV types 1–3, DHV-2, and DHV-3 disease outbreaks represents the main problem in differential diagnosis. Initial distinctions can be made from the responses of ducklings, embryonated eggs, and cell cultures to the isolated viruses [1]. More reliable methods of virus identification are serological tests or molecular techniques. These include VN tests, FA tests, reverse transcription polymerase chain reaction (RT-PCR), RT loop-mediated isothermal amplification

(RT-LAMP), real-time quantitative RT-PCR, and nucleotide sequence analysis. Among these methods, RT-PCR and nucleotide sequence analysis are widely used for the detection and typing of DHAVs and differential identification of DAstVs in recent years [9,10,14,18,23,30–33]. Using these techniques, diagnosis of the disease can be made more rapidly, accurately, and confidently. It is suggested that these techniques could be used to replace the traditional serological tests.

7.1.3.1 Conventional Techniques

DHAV-1 and DHAV-3 can readily be recovered by inoculating infective liver suspensions into the allantoic sac of embryonated duck eggs. The viruses can also be isolated by inoculation into embryonated chicken eggs. However, duck embryos are preferable to chicken embryos, since duck embryo mortality and lesions occur soon after inoculation and at earlier embryo passages [1]. This is especially true for isolation of DHAV-3. Following passages, the proportion of embryos that die with specific lesions tends to increase [1]. Isolation and propagation of DHAV-1 using primary duck embryo liver (DEL) cells, primary duck embryo kidney (DEK) cells, and day-old DK cells have also been described [16,34–36]. Tseng and Tsai reported that DHAV-2 can be isolated by inoculation onto day-old DK cells [11]. The presence of DHAVs can be further confirmed by reproduction of typical signs and lesions by inoculation of the isolate into ducklings between 1 and 7 days of age that are susceptible to DHAVs.

Serological tests including direct FA test and various VN tests have been described for DHAV-1. The direct FA technique performed on livers of natural cases or inoculated duck embryos was considered as a rapid and accurate diagnosis of DHAV-1 infection [1]. The VN tests include those performed in chicken embryos, duck embryos, and ducklings as well as microneutralization assay in DK and DEK cells and plaque reduction assay in DEL and DEK cells [1,16,34–39]. The VN tests have been used for virus identification, titration of serological response to vaccination, and epidemiologic surveys. However, they have not been used extensively for routine diagnosis of DHAV-1 infection because the clinical disease is too acute and the VN tests are time consuming. Restriction of the practical application of the VN tests is also due to the traditionally described DHV-2 and DHV-3 that have only been observed in the United Kingdom and the United States, respectively, in the past. Following emergence of new types of DHAV, cross neutralization tests may assume greater significance for typing of DHAV if antisera against the three DHAV types are available. The VN tests performed in day-old DK cells and ducklings and in chicken embryos have been employed for comparison of serological relationship of DHAV-2 and DHAV-3 to DHAV-1, respectively [11,12].

DHV-2 may be recovered by inoculation of liver suspensions into the amniotic cavity of embryonated chicken or duck eggs. But these are difficult and expensive processes because the embryos may respond erratically only after four passages and no deaths may be seen during earlier passages. Diagnosis of DHV-2 infection can also be made by inoculation of infective liver suspension into susceptible ducklings,

in which the response can be variable. A mortality rate of up to 20% may occur within a period of 2–4 days PI. VN tests in chicken embryos have been utilized for identification of DHV-2 isolate and detection of antibodies to DHV-2 in serum samples from survivors. But such tests were not routinely carried out because of the low and variable response of DHV-2 in embryonating chicken eggs [6]. Cross protection tests have been performed in ducklings; this technique was originally employed to distinguish DHV-2 from DHV-1 and to compare the antigenic relationship of DAstV isolated by Gough et al. to DHV-2 described by Asplin [5,6]. The most reliable diagnostic method for DHV-2 infection is EM examination of liver homogenates for the detection of astrovirus-like particles [1].

DHV-3 can be isolated by the inoculation of liver suspension onto the chorioallantoic membrane of embryonated duck eggs, which was considered as an effective method for diagnosis [1]. Alternatively, DHV-3 may be isolated in DK, DEL, or DEK cells. Although the virus does not cause cytopathic effect, the virus has been detected by a direct FA test [8]. The presence of DHV-3 can be confirmed by inoculation of infective liver suspensions into susceptible ducklings intramuscularly. The mortality rate may reach 20% with 60% morbidity. Intravenous inoculation is more effective [1].

7.1.3.2 Molecular Techniques

Following the successful cloning and sequencing of DHAV genome, various RT-PCR assays that were designed to amplify different genomic regions have been developed [18,30–33]. Using these assays, DHAV RNA, the most reliable marker for diagnosis of DHAV infections, can be readily produced from virus isolates and infective liver samples. Therefore, these molecular methods can be used for not only the identification of DHAV isolates but also the detection of DHAVs directly from clinical samples. Following nucleotide sequence determination and analysis of the amplified fragments, DHAV types can be identified. Type-specific RT-PCR assays have also been developed for rapid typing of DHAV. Considering that the 5′-UTR and capsid sequences are highly conserved within the same type but highly divergent between different types, the 5′-UTR and capsid regions are preferable to other genomic regions for molecular typing of DHAV [14,18]. Other molecular techniques such as RT-LAMP [40–42] and real-time quantitative RT-PCR [43,44], which is faster and more sensitive than standard RT-PCR, have recently been developed for rapid detection of DHAV-1 and DHAV-3.

In recent years, *Escherichia coli*–expressed DHAV VP1 proteins have been described. These VP1 proteins were subsequently used as antigens for development of indirect enzyme-linked immunosorbent assays (ELISA) that are suitable for investigations on serological epidemiology of DHAV infections [45,46]. Recent works in our laboratory demonstrated that VP1-based indirect ELISA has no type specificity, making it unsuitable for differential diagnosis of infections caused by different DHAV types. The VP1 proteins have also been employed to produce polyclonal and monoclonal immune sera that could then be utilized to establish ELISA-based diagnostic assays such as antigen-capture ELISA for detection of DHAV [47,48]. However, the limitation of ELISA-based diagnostic assays for DHAV detection is that only particular DHAV types for which specific monoclonal antibodies are available can be detected. Another issue that will need to be addressed is the ability of such tests to make differential diagnosis among different DHAV types.

For DAstVs, the molecular cloning and sequencing of a partial ORF1b region in 2009 led to a dramatic progress in understanding the classification of traditionally described DHV-2 and DHV-3 [23]. Using the partial ORF1b sequences as starting points, the DHV-2- and DHV-3-like astrovirus genomes have been successfully cloned and sequenced [9,10], which would be useful for development of molecular techniques for detection and characterization of DAstVs in both virus isolates and clinical liver samples. Specific RT-PCR assays have been developed for differential diagnosis between DHV-2- and DHV-3-like astroviruses [10]. An indirect ELISA for detecting antibodies of DHV-2-like astrovirus in duck sera has also developed [49].

7.2 METHODS

7.2.1 SAMPLE COLLECTION AND PREPARATION

Liver sample should be collected from dead ducklings at postmortem for the purpose of diagnosis. DHAV and DAstV can be detected in the samples stored at −20°C and −80°C for many years. The liver sample is homogenized in phosphate-buffered saline (20%, w/v). After centrifugation at $12,000 \times g$ at 4°C for 15 min, the supernatant is harvested and used for RNA extraction. The supernatant can also be stored at −20°C until use.

RNA can be extracted by using a commercial RNA extraction kit (e.g., TRIzol® Reagent, Life Techologies™). For TRIzol Reagent, RNA is extracted from supernatant of liver suspension following the manufacturer's instructions. Briefly, 250 µL of the supernatant is mixed with 750 µL of TRIzol Reagent in a 1.5 mL microcentrifuge tube, vortexed, and incubated at room temperature for 5 min. Then, 150 µL of chloroform is added, shaken vigorously for 15 s, and incubated at room temperature for 3 min. The mixture is centrifuged at $12,000 \times g$ for 15 min at 4°C, and the aqueous phase is removed into a new tube. Then, 500 µL of isopropanol is added into the tube. After incubation at room temperature for 10 min, the tube is centrifuged at $12,000 \times g$ for 10 min at 4°C. The resulting RNA pellet is washed with 1 mL of 75% ethanol, vortexed briefly, and centrifuged at $7500 \times g$ for 5 min at 4°C. Finally, the obtained RNA pellet is air dried for 5 min, resuspended in 50 µL of RNase-free water, and used as template. The genomic RNA can be stored at −80°C until RT is performed.

7.2.2 DETECTION PROCEDURES

Among the three types of DHAV, DHAV-2 has only been found in Taiwan, while DHAV-1 and DHAV-3 are cocirculating in South Korea and mainland China. Therefore, diagnosis and molecular epidemiological studies of DHAV conducted in mainland China

have mostly focused on the detection of DHAV-1 and DHAV-3 as well as searching for new DHAV types in recent years. Due to discoveries of DHV-2- and DHV-3-like astroviruses in mainland China, molecular detection of DAstVs have also been taken into consideration for the purpose of diagnosis of DVH and epidemiologic surveys of astroviruses in ducks.

7.2.2.1 Reverse Transcription

7.2.2.1.1 Procedure

1. Add 5 µL of extracted RNA and 20 pmol of reverse primer into a 0.2 mL RNase-/DNase-free microcentrifuge tube and then incubate at 70°C for 5 min and at 4°C for 5 min.
2. Prepare the RT mix (RT-mix) as follows. For each sample, 19 µL of RT-mix is prepared.

Component	Volume (µL)
M-MLV Reverse Transcriptase 5× buffer	5
deoxy-ribonucleoside triphosphate (dNTP) mixture (10 mM each)	5
Recombinant Ribonuclease Inhibitor (40 U/µL)	1
M-MLV Reverse Transcriptase (200 U/µL)	1
Diethyl pyrocarbonate-treated water	7
Total	19

3. Add 19 µL of the RT-mix to the 0.2 mL tube in the first step, mix gently by flicking the tube, and spin the tube in a microcentrifuge at top speed for 30 s.
4. Place the tube in the thermal cycler for incubation at 42°C for 60 min, followed by an enzyme inactivation step at 94°C for 5 min.
5. The cDNA can be used directly in the PCR step or stored at −20°C for future use.

Note: RT is performed for DHAV and DAstV independently.

7.2.2.2 PCR Detection of DHAVs and DAstVs

For differential diagnosis between DHAV and DAstV, PCR is performed by using primers DHAV-F5/DHAV-R5 and DAstV-F/DAstV-R, respectively.

7.2.2.2.1 Procedure

1. Remove all aliquots of PCR reagents from the freezer. Vortex and spin down all reagents before opening the tubes.
2. Prepare the PCR reaction mix as follows. For each sample, 40 µL of PCR reaction mix is prepared.

Component	Volume (µL)
Forward primer (20 pmol/µL)	2
Reverse primer (20 pmol/µL)	2
2× PCR SuperMix	25
ddH₂O	11
Total	40

Primer	Sequence (5′ → 3′)	Nucleotide Position	Expected Product (bp)	Specificity
DHAV-F5	GGAGGTGGTGCTGAAA	270–285[b]	250	DHAV
DHAV-R5	CCTCAGGAACTAGTCTGGA	519–501[b]		
DAstV-F[a]	GAYTGGACIMGITAYGAYGGIACIATICC	4528–4556[c]	438	DAstV
DAstV-R[a]	YTTIACCCACATICCRAA	4965–4948[c]		

[a] Primers reported by Todd et al. [23].
[b] Position in genome of DHAV-1 C80.
[c] Position in genome of DAstV C-NGB.

3. Add 10 µL of the RT product and 40 µL of the PCR reaction mix into a 0.2 mL PCR tube.
4. Turn on a thermocycler and preheat the block.
5. Spin the tube in a microcentrifuge at top speed for 30 s.
6. Place the sample tubes in the thermal cycler for amplification. PCR profile for DHAV comprises an initial denaturation at 94°C for 5 min, followed by 40 cycles of 94°C for 30 s, 50°C for 30 s, and 72°C for 30 s and a final extension at 72°C for 10 min. PCR profile for DAstV comprises an initial denaturation at 94°C for 5 min, followed by 45 cycles of 94°C for 1 min, 45°C for 1 min, and 72°C for 1.5 min and a final extension at 72°C for 10 min [23].
7. Analyze the PCR products by electrophoresis. For each sample, 8 µL of the PCR product is electrophoresed in a 1.5% agarose gel containing 0.5 mg/mL ethidium bromide in Tris-acetate-Ethylenediaminetetraacetic acid buffer at 130 V for 20 min. Visualize under ultraviolet (UV) light source or photograph under UV illumination with an imaging system.

Note: Negative and positive controls are concurrently included along with the test samples in order to determine whether the result is correct or wrong. The amplification of an approximately 250 bp product is indicative of DHAVs, while the amplification of a 438 bp product is indicative of DAstVs.

7.2.2.3   Sequence Analysis

Differential diagnosis among the causative agents of DVH can also be made by nucleotide sequence determination and analysis of the 250 and 438 bp fragments. Determination of the amplicons can be done by a sequencing company, using the purified PCR product as template in cycle sequencing reaction. Purification of the PCR products can be carried out by using a commercial gel extraction kit (e.g., QIAquick Gel Extraction Kit, Qiagen). The PCR products can also be sequenced after cloning. Briefly, the purified product is cloned into a vector (e.g., pGEM-T Easy Vector, Promega) and then transformed into competent *E. coli* (e.g., DH5α). After identification, plasmids extracted from the separated insert-positive clones are used for sequencing. Sequence identity and phylogenetic analyses should be performed to compare the sequence data with previously known sequences of the three DHAV types and the two DAstVs. For sequence analysis, the primer sequences should be removed from the sequences of the PCR products.

7.3   CONCLUSION

DVH is a severe disease of young ducklings because of the potential high mortality if not controlled. Up to now, members belonging to two virus families have been associated with the disease. Among the causative agents, DHAV-1 is known to be present in Northern America, Europe, and Asia, making this virus the most common cause of DVH. As newly discovered types of DHAV, DHAV-2 has only been reported in Taiwan, while DHAV-3 has been found in both South Korea and China. Although the traditionally described DHV-2 and DHV-3 are considered to be only present in the United Kingdom and the United States, respectively, recent characterization of several DHV-2-like and DHV-3-like astroviruses in China suggested that DHV-2 and DHV-3 were circulating outside the United Kingdom and the United States. Diagnosis of the disease can be readily made based on the clinical features and pathological changes. However, occurrence of similar disease outbreaks, caused by picornavirus DHAVs and astroviruses DHV-2 and DHV-3, offers the main problem in differential diagnosis. In the last century, characterization of DHAV-1, DHV-2, and DHV-3 strains relied mainly on the VN tests, EM examinations, and virus isolation, respectively.

Studies on molecular biology of DHAV began with the successful cloning of DHAV genome from several virus strains between 2006 and 2007. For DAstVs, the partial ORF1b sequences of DHV-2 and DHV-3 determined by Todd et al. in 2009 provide starting points from which the DHV-2- and DHV-3-like astrovirus genomes have been cloned and sequenced. The availability of the genomic sequences of DHAVs and DAstVs provides the possibility for development of new molecular techniques for detection and characterization of DHAV and DAstV strains both genetically and antigenically. The application of these sensitive methods for the detection of DHAV and DAstV in China contributes a greater recognition of molecular epidemiology of the viruses. More extensive surveillance for DHAVs and DAstVs in ducks of different species and in different geographic areas may enhance our understanding of molecular epidemiology and ecology of the viruses.

Traditional attenuated vaccine derived from serial passages in embryonated chicken or duck eggs has been reported for DHAV-1, DHAV-3, DHV-2, and DHV-3. All of them have been proven to be highly efficacious in controlling the disease caused by homologous virus. Among them, DHAV-1 vaccines derived from various strains of DHAV-1 have been used most commonly worldwide. It should be noted that such vaccines do not confer cross protection against the heterologous viruses. Therefore, a new vaccine consisting of a mixture of different strains of DHAV and/or DAstV should be developed in order to confer broad protection. Additionally, the implementation of different duck management procedures such as strict biosecurity and good sanitation may reduce the severity of the clinical effects.

REFERENCES

1. Woolcock, P.R., Duck hepatitis. In: Saif, Y.M., Barnes, H.J., Glisson, J.R., Fadly, A.M., McDougald, L.R., and Swayne, D.E. (Eds.), *Diseases of Poultry*, 11th edn. Iowa State Press, Ames, IA, pp. 343, 2003.
2. Sandhu, T.S., Ducks: Health management. *Encyclopedia Anim. Sci.*, 1, 1297, 2004.
3. Levine, P.P. and Fabricant, J., A hitherto-undescribed virus disease of ducks in North America. *Cornell Vet.*, 40, 71, 1950.
4. Tauraso, N.M., Coghill, G.E., and Klutch, M.J., Properties of the attenuated vaccine strain of duck hepatitis virus. *Avian Dis.*, 13, 321, 1969.
5. Asplin, F.D., Duck hepatitis: Vaccination against two serological types. *Vet. Rec.*, 77, 1529, 1965.
6. Gough, R.E. et al., An outbreak of duck hepatitis type II in commercial ducks. *Avian Pathol.*, 14, 227, 1985.
7. Toth, T.E., Studies of an agent causing mortality among ducklings immune to duck virus hepatitis. *Avian Dis.*, 13, 834, 1969.
8. Haider, S.A. and Calnek, B.W., In vitro isolation, propagation, and characterization of duck hepatitis virus type III. *Avian Dis.*, 23, 715, 1979.
9. Fu, Y. et al., Complete sequence of a duck astrovirus associated with fatal hepatitis in ducklings. *J. Gen. Virol.*, 90, 1104, 2009.
10. Liu, N. et al., Molecular characterization of a duck hepatitis virus 3-like astrovirus. *Vet. Microbiol.*, 170, 39, 2014.
11. Tseng, C.H. and Tsai, H.J., Molecular characterization of a new serotype of duck hepatitis virus. *Virus Res.*, 126, 19, 2007.
12. Kim, M.C. et al., Recent Korean isolates of duck hepatitis virus reveal the presence of a new geno- and serotype when compared to duck hepatitis virus type 1 type strains. *Arch. Virol.*, 152, 2059, 2007.
13. Lu, Y.S. et al., Infectious bill atrophy syndrome caused by parvovirus in a co-outbreak with duck viral hepatitis in ducklings in Taiwan. *Avian Dis.*, 37, 591, 1993.
14. Wang, L. et al., Classification of duck hepatitis virus into three genotypes based on molecular evolutionary analysis. *Virus Genes*, 37, 52, 2008.
15. Kim, M.C. et al., Molecular analysis of duck hepatitis virus type 1 reveals a novel lineage close to the genus Parechovirus in the family Picornaviridae. *J. Gen. Virol.*, 87, 3307, 2006.
16. Tseng, C.H., Knowles, N.J., and Tsai, H.J., Molecular analysis of duck hepatitis virus type 1 indicates that it should be assigned to a new genus. *Virus Res.*, 123, 190, 2007.
17. Ding, C. and Zhang, D., Molecular analysis of duck hepatitis virus type 1. *Virology*, 361, 9, 2007.
18. Fu, Y. et al., Molecular detection and typing of duck hepatitis A virus directly from clinical specimens. *Vet. Microbiol.*, 131, 247, 2008.
19. Knowles, N.J. et al., Picornaviridae. In: King, A.M.Q., Adams, M.J., Carstens, E.B., and Lefkowitz, E.J. (Eds.), *Virus Taxonomy: Classification and Nomenclature of Viruses: Ninth Report of the International Committee on Taxonomy of Viruses*. Elsevier, San Diego, CA, p. 855, 2011.
20. Johansson, E.S. et al., Molecular characterization of M1146, an American isolate of Ljungan virus (LV) reveals the presence of a new LV genotype. *J. Gen. Virol.*, 84, 837, 2003.
21. Pan, M. et al., Duck Hepatitis A virus possesses a distinct type IV internal ribosome entry site element of picornavirus. *J. Virol.*, 86, 1129, 2012.
22. Monroe, S.S. et al., Astroviridae. In: Fauquet, C.M., Mayo, A., Maniloff, J., Desselberger, U., and Ball, L.A. (Eds.), *Virus Taxonomy. Classification and Nomenclature of Viruses: Eighth Report of the International Committee on Taxonomy of Viruses*. Elsevier Academic Press, San Diego, CA, p. 859, 2005.
23. Todd, D. et al., Identification of chicken enterovirus-like viruses, duck hepatitis virus type 2 and duck hepatitis virus type 3 as astroviruses. *Avian Pathol.*, 38, 21, 2009.
24. Bosch, A. et al., Astroviridae. In: King, A.M.Q., Adams, M.J., Carstens, E.B., and Lefkowitz, E.J. (Eds.), *Virus Taxonomy. Classification and Nomenclature of Viruses: Ninth Report of the International Committee on the Taxonomy of Viruses*. Elsevier Academic Press, London, U.K., p. 953, 2011.
25. Pantin-Jackwood, M.J., Spackman, E., and Woolcock, P.R., Phylogenetic analysis of turkey astroviruses reveals evidence of recombination. *Virus Genes*, 32, 187, 2006.
26. Strain, E. et al., Genomic analysis of closely related astroviruses. *J. Virol.*, 82, 5099, 2008.
27. Todd, D. et al., Capsid protein sequence diversity of avian nephritis virus. *Avian Pathol.*, 40, 249, 2011.
28. Asplin, F.D., Notes on epidemiology and vaccination for virus hepatitis of ducks. *Off. Int. Epizoot. Bull.*, 56, 793, 1961.
29. Pan, M. et al., Recovery of duck hepatitis A virus 3 from a stable full-length infectious cDNA clone. *Virus Res.*, 160, 439, 2011.
30. Kim, M.C. et al., Development of one-step reverse transcriptase-polymerase chain reaction to detect duck hepatitis virus type 1. *Avian Dis.*, 51, 540, 2007.
31. Kim, M.C. et al., Differential diagnosis between type-specific duck hepatitis virus type 1 (DHV-1) and recent Korean DHV-1-like isolates using a multiplex polymerase chain reaction. *Avian Pathol.*, 37, 171, 2008.
32. Wen, X.J. et al., Detection, differentiation, and VP1 sequencing of duck hepatitis A virus type 1 and type 3 by a 1-step duplex reverse-transcription PCR assay. *Poult. Sci.*, 93, 2184, 2014.
33. Chen, L.L. et al., 2013. Improved duplex RT-PCR assay for differential diagnosis of mixed infection of duck hepatitis A virus type 1 and type 3 in ducklings. *J. Virol. Methods*, 192, 12, 2013.
34. Woolcock, P.R., An assay for duck hepatitis virus type I in duck embryo liver cells and a comparison with other assays. *Avian Pathol.*, 15, 75, 1986.
35. Woolcock, P.R., Chalmers, W.S.K., and Davis, D., A plaque assay for duck hepatitis virus. *Avian Pathol.*, 11, 607, 1982.
36. Kaleta, E.F., Duck viral hepatitis type 1 vaccination: Monitoring of the immune response with a microneutralisation test in Pekin duck embryo kidney cell cultures. *Avian Pathol.*, 17, 325, 1988.
37. Gough, R.E. and Spackman, D., Studies with inactivated duck virus hepatitis vaccines in breeder ducks. *Avian Pathol.*, 10, 471, 1981.
38. Hwang, J., Duck hepatitis virus-neutralization test in chicken embryos. *Am. J. Vet. Res.*, 30, 861, 1969.
39. Toth, T.E. and Norcross, N.L., Humoral immune response of the duck to duck hepatitis virus: Virus-neutralizing vs. virus-precipitating antibodies. *Avian Dis.*, 25, 17, 1981.
40. Li, C. et al., Rapid detection of duck hepatitis A virus genotype C using reverse transcription loop-mediated isothermal amplification. *J. Virol. Methods*, 196, 193, 2014.

41. Song, C. et al., 2012. Rapid detection of duck hepatitis virus type-1 by reverse transcription loop-mediated isothermal amplification. *J. Virol. Methods*, 182, 76, 2012.

42. Yang, L. et al., Development and application of a reverse transcription loop-mediated isothermal amplification method for rapid detection of Duck hepatitis A virus type 1. *Virus Genes*, 45, 585, 2012.

43. Huang, Q. et al., Development of a real-time quantitative PCR for detecting duck hepatitis a virus genotype C. *J. Clin. Microbiol.*, 50, 3318, 2012.

44. Yang, M. et al., 2008. Development and application of a one-step real-time TaqMan RT-PCR assay for detection of duck hepatitis virus type 1. *J. Virol. Methods*, 153, 55, 2008.

45. Liu, M. et al., Development and evaluation of a VP1-ELISA for detection of antibodies to duck hepatitis type 1 virus. *J. Virol. Methods*, 169, 66, 2010.

46. Ma, X. et al., Expression of VP1 gene and ELISA detection of antibodies against duck hepatitis virus. *Acta Microbiol. Sin.*, 48, 1110, 2008.

47. Li, C. et al., High yield expression of duck hepatitis A virus VP1 protein in *Escherichia coli*, and production and characterization of polyclonal antibody. *J. Virol. Methods*, 191, 69, 2013.

48. Zhang, T. et al., 2014. Characterization of monoclonal antibodies against duck hepatitis type 1 virus VP1 protein. *J. Virol. Methods*, 208C, 166, 2014.

49. Wang, X. et al., 2011. Expression of the C-terminal ORF2 protein of duck astrovirus for application in a serological test. *J. Virol. Methods*, 171, 8, 2011.

8 Foot-and-Mouth Disease Virus

Dongyou Liu

CONTENTS

8.1 Introduction 61
    8.1.1 Classification.. 61
    8.1.2 Morphology and Genome Organization 62
    8.1.3 Biology and Epidemiology 62
    8.1.4 Clinical Features and Pathogenesis 63
    8.1.5 Diagnosis 63
        8.1.5.1 Virus Isolation... 63
        8.1.5.2 Serological Assays 63
        8.1.5.3 Nucleic Acid Detection 64
    8.1.6 Treatment and Prevention 64
8.2 Methods 64
    8.2.1 Sample Preparation.. 64
    8.2.2 Detection Procedures... 65
        8.2.2.1 Conventional RT-PCR 65
        8.2.2.2 Real-Time RT-PCR 65
8.3 Conclusion 65
References.. 65

8.1 INTRODUCTION

Foot-and-mouth disease virus (FMDV), a single-stranded positive-sense RNA virus in the genus *Aphthovirus*, family *Picornaviridae*, is responsible for causing foot-and-mouth disease (FMD, or hoof-and-mouth disease) in cloven-hoofed animals, including cattle, water buffaloes, pigs, sheep, and goats, and about 70 wildlife species. Characterized by a transient high fever (for 2–3 days), formation of blisters inside the mouth and on the feet, and lameness, FMD causes enormous economical losses in animal production through culling, trade restriction, and prevention measures.

Despite its first description from cattle in Venice in 1514 and concerted eradication efforts during the past few decades, FMD is still found in more than 100 countries in Asia, Africa, and parts of South America. This may be attributable to its high contagiousness, widespread occurrence, broad host range, existence of carrier status, short duration of immunity and low cross-immunity (due to antigenic diversity and tendency to undergo mutational changes), poor surveillance and diagnostic facilities, and inadequate control programs. In this chapter, we present a brief overview on FMDV in relation to its classification, morphology, genome organization, epidemiology, clinical features, pathogenesis, diagnosis, treatment, and control, together with stepwise molecular protocols for its identification and typing.

8.1.1 CLASSIFICATION

FMDV is a single-stranded positive-sense RNA virus classified in the genus *Aphthovirus,* family *Picornaviridae*, order *Picornavirales*.

With 50 species identified to date, the family *Picornaviridae* is divided into 29 genera: *Aphthovirus*, *Aquamavirus*, *Avihepatovirus*, *Avisivirus*, *Cardiovirus*, *Cosavirus*, *Dicipivirus*, *Enterovirus*, *Erbovirus*, *Gallivirus*, *Hepatovirus*, *Hunnivirus*, *Kobuvirus*, *Kunsagivirus*, *Megrivirus*, *Mischivirus*, *Mosavirus*, *Oscivirus*, *Parechovirus*, *Pasivirus*, *Passerivirus*, *Rosavirus*, *Sakobuvirus*, *Salivirus*, *Sapelovirus*, *Senecavirus*, *Sicinivirus*, *Teschovirus*, and *Tremovirus* [1].

In turn, the genus *Aphthovirus* is separated into four species: bovine rhinitis A virus (formerly bovine rhinovirus types 1 and 3, with 2 serotypes), bovine rhinitis B virus (formerly bovine rhinovirus 2, with 1 serotype), equine rhinitis A virus (formerly equine rhinitis virus 1, with 1 serotype), and FMDV (the prototype species, with seven serotypes: O, A, C, Southern African Territories 1 [SAT1] , SAT2, SAT3, and Asia 1).

At the present, seven immunologically distinct serotypes (containing >65 subtypes) are recognized within the FMDV species. Of these, serotype O (for Oise) was first identified in France; serotype A (for Allemagne) in Germany; serotype C also in Germany; serotypes SAT1, SAT2, and SAT3 in South Africa; and serotype Asia 1 in Pakistan. Following successive vaccination and control campaigns, some FMDV serotypes have been eradicated in a number of countries. As a consequence, the presence of various FMDV serotypes is now largely confined to China, Southeast Asia (SEA), Africa, Middle East, the United Kingdom, and South America (Table 8.1) [2].

Within each serotype of FMDV, a spectrum of subtypes/ variants showing distinct antigenic, biological, and epidemiological characteristics has been identified using in vivo

TABLE 8.1

Geographic Distribution of Foot-and-Mouth Disease Virus Serotypes

Geographic Location	Presence of FMDV Serotype						
	O	A	C	SAT1	SAT2	SAT3	Asia I
China	Yes	Yes	No	No	No	No	Yes
Southeast Asia	Yes	Yes	No	No	No	No	Yes
Africa	Yes	No	No	Yes	Yes	Yes	No
Middle East	Yes	Yes	Yes	No	No	No	Yes
United Kingdom	Yes	No	No	No	No	No	No
South America	Yes	Yes	Yes	No	No	No	No

cross-protection tests and in vitro serological assays. The in vivo cross-protection tests differentiate FMDV subtypes/variants through the ability or inability of one subtype to confer the same level of immunity to another subtype of the same serotype in experimental animals. Due to their high cost and reliance on the use of animals, in vivo cross-protection tests have been largely superseded by serological assays (e.g., enzyme-linked immunosorbent assay [ELISA] and virus neutralization test [VNT]), which provide a rapid and economical way to determine antigenic relationships among various FMDV isolates. Furthermore, sequencing analysis of the viral protein 1 (VP1) coding gene in FMDV genome facilitates a precise discrimination of FMDV genotypes (topotypes) at the nucleic acid level.

On the basis of up to 15% difference in VP1 coding sequences, FMDV serotype O is separated into eight genetically and geographically distinct genotypes (topotypes): Middle East–South Asia (ME-SA), SEA, Cathay, Indonesia 1, Indonesia 2, East Africa, West Africa, and Europe–South America (Euro-SA). In addition, using the criteria of nucleotide sequence variations within the VP1 coding gene of >7.5% and other phylogenetic differences, four different lineages are identified within the ME-SA topotype, including the pandemic PanAsia lineage.

Similarly, FMDV serotype A can be distinguished into 26 genotypes (topotypes) on the basis of >15% difference in VP1 coding sequence. Of the three distinct genotypes (topotypes) from Asia, Europe–South America (Euro-SA), and Africa, topotype Asia is the most prevalent in the Middle East and South Asia and contains a number of lineages (e.g., A_{15}, A_{22}, A-IRN87, A-IRN96, A-IRN99, A-Iran05).

Compared to serotypes O and A, the FMDV serotype Asia 1 appears to have a relatively low genetic and antigenic diversity. Previous studies indicated that serotype Asia 1 isolates collected between 1952 and 1992 from 18 groups, while those collected over the last two decades in India belong to seven lineages. Additionally, the examination of serotype Asia 1 isolates from 2003 to 2007 led to the identification of six groups. Another serotype Asia 1 group, designated as group VII and later as Sindh-08, is found in the West Eurasian region [2].

8.1.2 MORPHOLOGY AND GENOME ORGANIZATION

FMDV is a small, spherical, nonenveloped particle with a smooth surface and a diameter of 25–30 nm. Its capsid is composed of 60 copies of the capsomers organized in a tightly packed icosahedral structure (assemble with a triangulation number of three). Each capsomer comprises four structural polypeptides known as VP1, VP2, VP3, and VP4, with VP2 and VP4 originating from the cleavage of precursor VP0. Exposed on the surface of the virion, VP1–VP3 (24 kDa) fold into an eight-stranded wedge-shaped β-barrel that constitutes the majority of the capsid structure, whereas VP4 (8.5 kDa) is located inside the capsid [1].

The capsid encapsidates a nonsegmented, single-stranded, positive-sense RNA of 8500 nt in length, which includes the 5′ untranslated region (5′ UTR, of 500–1200 nt in length), a long open reading frame (ORF, about 7000 nt in length, encoding a single polyprotein), the 3′ untranslated region (3′ UTR, of 30–650 nt in length), and a poly "A" tail.

The 5′ UTR is linked to a 24 or 25 residue proteins known as VPg (encoded by 3B) that serves as primer in replication. The 5′ UTR contains a short fragment called S-fragment (360 bases), a poly "C" tract, a series of RNA pseudoknot structures, a cis-acting replication element (cre) (also known as the 3B-uridylylation site [bus]), and the internal ribosome entry site (IRES). IRES serves as ribosomal entering site due to the absence of 5′ cap (7-methyguanosine), with the translation initiation starting at two AUG codons separated by 84 nt following the IRES.

The polyprotein consists of one structural and eight nonstructural proteins (NSPs) organized in the order L-1ABCD-2ABC-3ABCD, which result from cleavage by viral proteases. Of these, the 1A, 1B, 1C, and 1D proteins are the components (VP4, VP2, VP3, and VP1, respectively) of the capsid. VP2 and VP4 are cleavage products of the precursor VP0 (1AB) during virion maturation and after the capsid is formed. The L, 2A, and 3C proteins are proteinases that internally cleave the polyprotein. The 2B, 2C, and 3A proteins interfere with host cell function. The 3B protein is the VPg protein. The 3D protein is the RNA polymerase. It is noteworthy that among the viral structural proteins, VP1 plays a key role in virus attachment and entry, protective immunity, and serotype specificity. Analysis of the VP1 sequence offers clues to the evolutionary dynamics and phylogenetic relationships among FMDV isolates.

The 3′ UTR follows the ORF termination codon and contains a short stretch of RNA that folds into specific stem-loop structures followed by a poly "A" tract of variable length. This 3′ UTR also serves as some functions in genome replication and is specific for each picornavirus. Poly "A" is likely involved in FMDV translation and replication.

8.1.3 BIOLOGY AND EPIDEMIOLOGY

FMDV has the ability to hijack the host cellular machineries for its replication. After gaining entry to a susceptible host, FMDV particle binds to cell surface receptors, leading to the release of myristic acid, which is involved in the formation of a pore on the cell membrane to facilitate viral RNA injection. Within the cell, the RNA uncoats and a double-stranded RNA intermediate is generated from the positive-strand RNA genome by viral RDRP (RNA-dependent RNA polymerase). Initiated by IRES, the translation of viral proteins takes place in host cell ribosomes accompanied by a "shutoff" of cellular protein synthesis. The assembly of virus particles begins at 4–6 h after infection. Following death of the cell around 8 h postinfection, new virions are released.

FMDV is infective to >70 wild and domestic cloven-hoofed species, including cattle, water buffalo, sheep, goats, pigs, antelope, deer, bison, hedgehogs, mithun, yak, sambar, giraffes, elephants, and camels [3]. Llamas and alpacas may show mild symptoms, but do not pass it on to others of the same species. Mice, rats, and chickens may be infected under laboratory conditions, but not natural conditions.

FMDV is transmitted via contact with the infected host, body fluids, aerosols, fomites, fodder, motor vehicles, clothes and skin of animal handlers, standing water, uncooked food scraps, and feed supplements containing infected animal products. Cows can also catch FMD from the semen of infected bulls [4].

While cattle and sheep typically acquire the infection through respiration of aerosol containing the virus, swine are likely to be infected through ingestion or subcutaneous wounds. Some infected hosts (e.g., cattle and buffalo) may become asymptomatic and remain potentially contagious for up to 5 years, with the virus shedding in feces, urine, body fluids, and aerosols.

Among the seven serotypes, the most widely distributed viruses belong to serotypes O, A, and C, which are responsible for outbreaks in Europe, America, Asia, and Africa (Table 8.1) [5–7]. In particular, pandemic serotype O virus (the PanAsia strain) of the ME-SA topotype has been implicated in many recent FMD epidemics worldwide [8]. SAT1–SAT3 serotypes are mostly found in sub-Saharan Africa and maintained by African buffalo (*Syncerus caffer*). However, some limited outbreaks due to SAT1 viruses occurred in the Middle East and in Greece in the 1960s; and serotype SAT2 was associated with minor outbreaks in Yemen (1990), Kuwait and Saudi Arabia (2000), Northern African countries (Egypt and Libya), and Palestine and Bahrain in recent years [9–12]. Serotype Asia 1 is largely confined to Asia, despite its incursions into Greece in 1984 and 2000.

8.1.4 CLINICAL FEATURES AND PATHOGENESIS

With an incubation period ranging from 1 to 12 days, FMD is characterized by high fever (which declines rapidly after 2–3 days), appearance of blisters inside the mouth (leading to excessive secretion of stringy or foamy saliva and to drooling) and on the feet (with ruptured blisters causing pain and lameness), weight loss, swelling in the testicles of mature males, reduced milk production in cows, myocarditis (in calves), gastroenteritis (in pigs), and occasional death (about 2% in adults and 20% in newborns). Most infected animals eventually recover, but some remain asymptomatic carriers (with infectious virus

present in the esophagus and throat fluids), which represent a source of infection for other animals [13].

In cattle, both oral and pedal lesions are evident, and in swine, pedal lesions dominate. In addition, pigs do not act as asymptomatic carriers. In sheep, milder signs are mostly found, with 25% of infected sheep showing no lesions.

In human FMD cases (due mostly to laboratory accidents), symptoms range from malaise, fever, vomiting, red ulcerative lesions (surface-eroding damaged spots) of the oral tissues, to vesicular lesions (small blisters) of the skin.

As the respiratory tract is the chief route of virus entry in natural infection, virus multiplication initially occurs in the pharynx epithelium, generating primary vesicles (or "aphthae") that affect cells from the epithelial stratum spinosum. From there, the virus spreads to different organs and tissues, leading to the production of secondary vesicles in the mouth and feet. After 1 week, the disease modulates gradually as a result of emerging humoral response from the host.

8.1.5 DIAGNOSIS

FMD induces an acute, systemic vesicular disease in cloven-hoofed animals, which is clinically indistinguishable from vesicular stomatitis (VS), swine vesicular disease (SVD), and vesicular exanthema of swine (VES). Other diseases that may confuse with FMD include mucosal disease, malignant catarrhal fever, rinderpest, peste des petits ruminants, papular stomatitis, orf, blue tongue, and epizootic hemorrhagic disease. In addition, bacteria, chemicals, and mechanical trauma may be associated with stomatitis and pododermatitis. Therefore, use of laboratory methods is necessary for correct diagnosis of FMD [14,15].

Over the years, a variety of laboratory techniques have been developed for the identification and typing of FMDV, including virus isolation, serological assays, and nucleic acid detection.

8.1.5.1 Virus Isolation

Several stable cell lines (e.g., IBRS-2, MVPK-1 clone 7, LFBK cell line and baby hamster kidney 21 [BHK-21]) are suitable for FMDV isolation, although primary bovine thyroid cells represent a highly sensitive culture system for FMDV, facilitating detection of low amount of infectivity. It is notable that the use of BHK-21 cell line generally allows isolation of FMDV from only 10% to 30% of clinical materials. Another virus isolation technique involves inoculation of animals such as guinea pigs and cattle, which has been used previously for cross-immunity assessment of FMDV isolates [15].

8.1.5.2 Serological Assays

As faster, less expensive, and more precise alternatives to virus isolation, various serological assays have been developed for FMD diagnosis. These include complement fixation test (CFT), VNT, and ELISA. CFT was the test of choice for the diagnosis of FMD and virus typing until the 1970s, but its role has been largely taken over by more sensitive and easier to interpret tests such as ELISA. Considered as the *gold standard* for detection of antibodies to structural proteins of FMDV, VNT is a prescribed test for import/export certification of animals or animal products. Indirect ELISA detects FMDV antigen in infected cell culture fluid, mice carcass, and cattle tongue as well as antibodies in sera; sandwich ELISA using convalescent bovine immunoglobulin (Igs) as capture and anti-146S guinea pig sera as tracing sera offers a valuable means for detection and typing of FMDV directly from infected tissue culture and field materials. In comparison with conventional VNT, liquid-phase blocking ELISA using bovine convalescent sera is much more reproducible and less likely to be influenced by variations in tissue culture susceptibility. Therefore, it is no surprise that ELISA is considered by the World Organization for Animal Health (OIE) or Food and Agriculture Organization World Reference Laboratory for FMD as the preferred procedure for the detection of FMDV antigen and identification of viral serotypes. In a typical testing scheme, samples are first examined by ELISA, and ELISA-negative samples are inoculated into sensitive cell cultures followed by the confirmation of the virus serotype by ELISA [15,16].

Other types of serological assays include chromatographic strip, phage display, and biosensors. A monoclonal antibody-based chromatographic strip test for FMD diagnosis is as sensitive as the conventional antigen ELISA for the detection of FMDV in epithelial suspensions. Indeed, rapid chromatographic strip test and lateral flow device offer an added promise for pen-side diagnosis [17,18]. Phage display facilitates investigation on bovine antibody response to FMDV at a molecular level. An immunobiosensor using a piezoelectric crystal has been reported for FMD diagnosis and virus typing. The advantages of allosteric biosensor-based diagnosis are rapidity and potential for handling large number of samples and for automatic processing.

Another possible application of serological assays is in the differentiation of carrier animals from vaccinated animals. This is due to the fact that vaccinated animals do not produce antibodies to eight viral NSPs of FMDV, while infected animals do. Given that the 3D NSP may be occasionally present in the inactivated vaccine, the detection of antibodies to the 2C or the polyprotein 3ABC is a likely indicator of the previous infection (or carrier status), not a vaccinated host [19].

8.1.5.3 Nucleic Acid Detection

Given the time required for virus isolation (4–6 days) and serological assays (1 day + 4–6 days for virus isolation from initial ELISA-negative samples), nucleic acid–based detection methodologies have taken the spotlight. From early gene probe hybridization, nucleic acid amplification, to real-time platforms, we have witnessed the increasing sophistication of molecular tests and their evolution from pure research tools to workhorses in clinical diagnostic laboratories [20–23].

In particular, the development of reverse transcription polymerase chain reaction (RT-PCR) and its derivatives such as nested, multiplex, and real-time RT-PCR has opened new avenues for rapid, sensitive, and specific identification and typing

of FMDV [24–28]. For example, use of multiplex PCR enables simultaneous identification of viral vesicular diseases (FMD, VS, SVD, and VES) and for the differentiation of seven FMDV serotypes [29]. In addition, the availability of different probe chemistries (DNA-binding fluorophores, 5′ endonuclease, adjacent linear and hairpin oligoprobes, and self-fluorescing amplicons) permits convenient, real-time detection and quantification of FMDV, with a much reduced risk of cross contamination [30]. Real-time PCR assays incorporating universal primers and fluorescent-labeled probes from the conserved region within the 5′ UTR or conserved gene regions within the RNA-dependent RNA polymerase gene (3Dpol) have been recommended by the OIE for FMDV detection [15,31].

Other nucleic acid amplification methods, such as nucleic acid sequence-based amplification (NASBA) and RT loop-mediated isothermal amplification (RT-LAMP), have also been shown to aid the detection of FMDV [32,33]. Recent assessment of NASBA electrochemiluminescence and NASBA-enzyme-linked oligonucleotide capture supports their use as a rapid and sensitive diagnostic method for the detection and surveillance of FMDV, whereas RT-LAMP enables rapid, specific, and sensitive detection of FMDV in laboratory as well as field condition.

8.1.6 TREATMENT AND PREVENTION

No specific treatment is currently available for FMD. Although FMD is a disease of low mortality and high morbidity and most infected animals eventually recover with treatment, it exerts huge negative impacts on agriculture production and international trade, as well as economical burden associated with FMD control programs [34].

Therefore, apart from identification, quarantine, and culling of infected livestock, a most cost-effective way to prevent and control FMD is through vaccination. Many early vaccines have relied on inactivated FMDV, which is typically produced from live velogenic FMDV strain grown in BHK-21 cells, chemically inactivated (e.g., by aziridine or binary ethyleneimine), partially purified, and mixed with an oil or aluminum hydroxide/saponin adjuvant [35]. However, besides having inherent risk of introducing incompletely inactivated FMDV in naïve animals and causing outbreak, these inactivated vaccines suffer from the shortcomings of short shelf life, the need for adequate cold chain of formulated vaccines, and the difficulties of certain serotypes and subtypes to grow well in cell culture for vaccine production. Thus, alternative approaches are being taken to produce empty capsid vaccines, DNA vaccines, recombinant protein vaccines, and peptide vaccines, etc. [36].

8.2 METHODS

8.2.1 SAMPLE PREPARATION

The tissue of choice for the laboratory diagnosis of FMD is epithelium or vesicular fluid. At least 1 g of epithelial tissue is collected from an unruptured or recently ruptured vesicle located in the tongue, buccal mucosa, or feet, and placed in a transport medium (consisting of equal amounts of glycerol, 0.04 M phosphate buffer, pH 7.2–7.6, 1000 IU penicillin, 100 IU neomycin sulfate, 50 IU polymyxin B sulfate, 100 IU mycostatin). Tissue culture medium or phosphate-buffered saline in the range pH of 7.2–7.6 may be used in place of 0.04 M phosphate buffer. The sample is kept refrigerated or on ice until arrival in the laboratory. If epithelial tissue is unavailable (in advanced or convalescent cases), oesophageal–pharyngeal (OP) fluid may be collected by a probang (sputum) cup or swabbing the swab [15].

Upon receipt, a suspension of the epithelium (ES) is prepared in 0.04 M phosphate buffer (10% concentration w/v). Total RNA may be extracted from ES, OP fluid, saliva, or culture supernatant by Trizol or QIAamp Viral RNA Mini Kit (Qiagen).

8.2.2 DETECTION PROCEDURES

8.2.2.1 Conventional RT-PCR

Reid et al. [20] utilized an RT-PCR assay for FMD diagnosis with primers (F: 5′-GCCTG-GTCTT-TCCAG-GTCT-3′ and R: 5′-CCAGT-CCCCT-TCTCA-GATC-3′) from the 5′ UTR of FMDV genome.

8.2.2.1.1 Procedure

For RT, 2 μL of random hexamers (20 μg/mL) and 5 μL of nuclease-free water are added into a sterile 0.5 mL microcentrifuge tube. Added to the tube is a 5 μL of RNA purified by Trizol. After gentle mix by pipetting up and down, the tube is incubated at 70°C for 5 min and cooled at room temperature for 10 min. Each RT reaction mix is made up of 4 μL first strand buffer (5× conc.), 2 μL bovine serum albumin (acetylated, 1 mg/mL), 1 μL dNTPs (10 mM mixture each of dATP, dCTP, dGTP, and dTTP), 0.2 μL DTT (1 M), and 1 μL Moloney Murine Reverse Transcriptase (200 U/μL). A reaction mix of 8 μL is added to the 12 μL of random primer/RNA mix and then mixed by gently pipetting. The mix is incubated at 37°C for 45 min and kept on ice if the PCR amplification step is carried out shortly or stored at −20°C.

Each PCR mix is made up of 35 μL nuclease-free water, 5 μL PCR buffer (10× conc.), 1.5 μL MgCl$_2$ (50 mM), 1 μL dNTPs (10 mM each of dATP, dCTP, dGTP, and dTTP), 1 μL primer 1 (10 pmol/μL), 1 μL primer 2 (10 pmol/μL), and 0.5 μL Taq polymerase (5 units/μL). Then 45 μL of PCR mix and 5 μL of the RT product are added to a well of a PCR plate or to a 0.2 mL PCR tube. After spinning for 1 min, the PCR plate or tube is placed in a thermal cycler for PCR amplification with the following program: 1 cycle of 94°C for 5 min; 30 cycles of 94°C for 1 min, 55°C for 1 min, and 72°C for 2 min; and 1 cycle of 72°C for 7 min. After amplification, 20 μL aliquot of each PCR product is mixed with 4 μL of staining solution and loaded onto a 1.5% agarose gel. Upon electrophoresis, a positive result for FMDV is indicated by the presence of a 328 bp band.

8.2.2.2 Real-Time RT-PCR

Reid et al. [21] described primers (forward 5′-CACYT YAAGR TGACA YTGRT ACTGG TAC-3′ and reverse 5′-CAGAT YCCRA GTGWC ICITG TTA-3′) and TaqMan® probe (5′-CCTCG GGGTA CCTGA AGGGC ATCC-3′) from the 5′ UTR region for real-time PCR identification of FMDV.

8.2.2.2.1 Procedure

RT-PCR mix is prepared by using the Superscript III/Platinum Taq One-Step rRT-PCR Kit (Invitrogen). Each reaction (25 μL) is composed of 12.5 μL of 2× reaction mix, 0.5 μL of Superscript III/Platinum Taq Enzyme Mix (both supplied with the kit), 2 μL of each primer (20 pmol), 1.5 μL of probe (7.5 pmol), 1.5 μL of nuclease-free water (Promega), and 5 μL of purified nucleic acid. Amplification and detection are conducted in Mx4000 Multiplex Quantitative PCR System (Stratagene) with the following thermal cycling conditions: 60°C for 30 min and 95°C for 10 min, followed by 50 cycles of 95°C for 15 s and 60°C for 1 min. A sample with a value of CT32 is considered positive [31].

8.3 CONCLUSION

FMDV is the causative agent of a highly infectious disease in wild and domestic cloven-hoofed animals. Although clinical signs of FMD in adult animals are seemingly mild, and many infected animals often recover without treatment, the impacts of FMD lie in the loss of body weight and productivity in diseased or recovered livestock, trade restriction, and control cost. Therefore, it is crucial that FMD is correctly diagnosed and prevention measures are implemented. Toward this goal, various laboratory tests have been developed for the identification and typing of FMDV. Given the lengthy process and variability associated with virus isolation and serological assays, nucleic acid detection techniques (particularly RT-PCR) have been playing an increasingly prominent role in the routine detection and typing of FMDV.

REFERENCES

1. Knowles NJ et al. Picornaviridae. In: King AMQ et al. (eds.), *Virus Taxonomy: Classification and Nomenclature of Viruses: Ninth Report of the International Committee on Taxonomy of Viruses*, 2012, pp. 855–880. Elsevier, San Diego, CA.
2. Jamal SM, Belsham GJ. Foot-and-mouth disease: Past, present and future. *Vet Res*. 2013;44:116.
3. Bravo de Rueda C, de Jong MC, Eblé PL, Dekker A. Estimation of the transmission of foot-and-mouth disease virus from infected sheep to cattle. *Vet Res*. 2014;45:58.
4. Chase-Topping ME et al. Understanding foot-and-mouth disease virus transmission biology: Identification of the indicators of infectiousness. *Vet Res*. 2013;44:46.
5. Samuel AR, Knowles NJ. Foot-and-mouth disease type O viruses exhibit genetically and geographically distinct evolutionary lineages (topotypes). *J Gen Virol*. 2001;44:609–621.
6. Ryan E et al. Clinical and laboratory investigations of the outbreaks of foot-and-mouth disease in southern England in 2007. *Vet Rec*. 2008;163(5):139–147.

7. Banda F et al. Investigation of foot-and-mouth disease out-breaks in the Mable and Kazungula districts of Zambia. *Onderstepoort J Vet Res.* 2014;81(2):E1–E6.

8. Jamal SM et al. Evolutionary analysis of serotype A foot-and-mouth disease viruses circulating in Pakistan and Afghanistan during 2002–2009. *J Gen Virol.* 2011;44:2849–2864.

9. Habiela M et al. Molecular characterization of foot-and-mouth disease viruses collected from Sudan. *Transbound Emerg Dis.* 2010;57(5):305–314.

10. Elhaig MM, Elsheery MN. Molecular investigation of foot-and-mouth disease virus in domestic bovids from Gharbia, Egypt. *Trop Anim Health Prod.* 2014;46(8):1455–1456.

11. Knowles NJ et al. Outbreaks of foot-and-mouth disease in Libya and Saudi Arabia during 2013 due to an exotic O/ME-SA/Ind-2001 lineage virus. *Transbound Emerg Dis.* December 7, 2014. doi: 10.1111/tbed.12299.

12. Sallu RS et al. Molecular survey for foot-and-mouth disease virus in livestock in Tanzania, 2008–2013. *Onderstepoort J Vet Res.* 2014;81(2):E1–E6.

13. Jamal SM et al. Detection and genetic characterization of foot-and-mouth disease viruses in samples from clinically healthy animals in endemic settings. *Transbound Emerg Dis.* 2012;44:429–440.

14. Longjam N et al. A brief review on diagnosis of foot-and-mouth disease of livestock: Conventional to molecular tools. *Vet Med Int.* 2011;2011:905768.

15. OIE (Office International des Epizooties). *OIE Terrestrial Manual 2012.* Chapter 2.1.5: Foot and mouth disease. http://www.oie.int/fileadmin/Home/fr/Health_standards/tahm/2.01.05_FMD.pdf. Accessed on March 31, 2015.

16. Morioka K, Fukai K, Sakamoto K, Yoshida K, Kanno T. Evaluation of monoclonal antibody-based sandwich direct ELISA (MSD-ELISA) for antigen detection of foot-and-mouth disease virus using clinical samples. *PLoS One* 2014;9(4):e94.

17. Ferris NP et al. Development and laboratory validation of a lateral flow device for the detection of foot-and-mouth disease virus in clinical samples. *J Virol Methods.* 2009;155(1):10–17.

18. Fowler VL et al. Recovery of viral RNA and infectious foot-and-mouth disease virus from positive lateral-flow devices. *PLoS One.* 2014;9(10):e109322.

19. Hutber M, Pilipcinec E, Bires J. Foot and mouth disease virus. In: D. Liu (ed.), *Manual of Security Sensitive Microbes and Toxins,* 2014, pp. 655–663. CRC Press, Boca Raton, FL.

20. Reid SM, Ferris NP, Hutchings GH, Samuel AR, Knowles NJ. Primary diagnosis of foot-and-mouth disease by reverse transcription polymerase chain reaction. *J Virol Methods.* 2000;89(1–2):167–176.

21. Reid SM et al. Diagnosis of foot-and-mouth disease by real-time fluorogenic PCR assay. *Vet Rec.* 2001;149(20):621–623.

22. Pierce KE et al. Design and optimization of a novel reverse transcription linear-after-the-exponential PCR for the detection of foot-and-mouth disease virus. *J Appl Microbiol.* 2010;109(1):180–189.

23. Madi M, Mioulet V, King DP, Lomonossoff G, Montague N. Development of a non-infectious encapsidated positive control RNA for molecular assays to detect foot-and-mouth disease virus. *J Virol Methods.* 2015;220:27–34.

24. Reid SM et al. Utility of automated real-time RT-PCR for the detection of foot-and-mouth disease virus excreted in milk. *Vet Res.* 2006;37(1):121–132.

25. Reid SM et al. Performance of real-time reverse transcription polymerase chain reaction for the detection of foot-and-mouth disease virus during field outbreaks in the United Kingdom in 2007. *J Vet Diagn Invest.* 2009;21(3):321–330.

26. Reid SM et al. Pan-serotypic detection of foot-and-mouth disease virus by RT linear-after-the-exponential PCR. *Mol Cell Probes.* 2010;24(5):250–255.

27. Reid SM et al. Development of tailored real-time RT-PCR assays for the detection and differentiation of serotype O, A and Asia-1 foot-and-mouth disease virus lineages circulating in the Middle East. *J Virol Methods.* 2014;207:146–153.

28. McKillen J et al. Pan-serotypic detection of foot-and-mouth disease virus using a minor groove binder probe reverse transcription polymerase chain reaction assay. *J Virol Methods.* 2011;174(1–2):117–119.

29. Sharma GK et al. Comparison of stabilizers for development of a lyophilized multiplex reverse-transcription PCR mixture for rapid detection of foot and mouth disease virus serotypes. *Rev Sci Tech.* 2014;33(3):859–867.

30. Wernike K, Hoffmann B, Beer M. Simultaneous detection of five notifiable viral diseases of cattle by single-tube multiplex real-time RT-PCR. *J Virol Methods.* 2015;217:28–35.

31. Shaw AE et al. Implementation of a one-step real-time RT-PCR protocol for diagnosis of foot-and-mouth disease. *J Virol Methods* 2007;143:81–85.

32. Ding YZ et al. A reverse transcription loop-mediated isothermal amplification assay to rapidly diagnose foot-and-mouth disease virus C. *J Vet Sci.* 2014;15(3):423–426.

33. Ranjan R et al. Development and evaluation of a one step reverse transcription-loop mediated isothermal amplification assay (RT-LAMP) for rapid detection of foot and mouth disease virus in India. *Virusdisease.* 2014;25(3):358–364.

34. Knight-Jones TJ, Rushton J. The economic impacts of foot and mouth disease—What are they, how big are they and where do they occur? *Prev Vet Med.* 2013;112(3–4):161–173.

35. Ding YZ et al. An overview of control strategy and diagnostic technology for foot-and-mouth disease in China. *Virol J.* 2013;10:78.

36. Zhang L et al. Research in advance for FMD novel vaccines. *Virol J.* 2011;8:268.

9 Porcine Encephalomyocarditis Virus

Sandra Blaise-Boisseau, Aurore Romey, and Labib Bakkali Kassimi

CONTENTS

9.1 Introduction 67
    9.1.1 Classification 67
    9.1.2 Genomic Organization, Biology, and Epidemiology 67
        9.1.2.1 Genomic Organization 67
        9.1.2.2 Viral Proteins 69
        9.1.2.3 Three-Dimensional Structure and Assembly of the Capsid 69
        9.1.2.4 Infectious Cycle 69
        9.1.2.5 Epidemiology 71
    9.1.3 Clinical Features and Pathogenesis 72
        9.1.3.1 Swine 72
        9.1.3.2 Rodent 72
    9.1.4 Diagnosis 72
        9.1.4.1 Conventional Techniques 72
        9.1.4.2 Molecular Techniques 73
9.2 Methods 73
    9.2.1 Sample Preparation and Viral RNA Extraction 73
        9.2.1.1 Clinical Sample Preparation 73
        9.2.1.2 RNA Extraction 73
    9.2.2 Detection Procedures 73
        9.2.2.1 One-Step Conventional RT-PCR 73
        9.2.2.2 Real-Time RT-PCR 74
9.3 Conclusion 75
References 75

9.1 INTRODUCTION

Encephalomyocarditis virus (EMCV) was isolated for the first time in 1945 in Miami, Florida, from a captive chimpanzee and a gibbon that died suddenly of pulmonary edema and myocarditis [1]. Mice inoculated with a filtered edema fluid from the gibbon or the chimpanzee displayed paralysis of the posterior members and myocarditis followed by death in a week. The pathogenic agent was at that time given the name of EMCV. The virus had probably been transmitted from wild rats living in proximity to the monkeys, as nearly 50% of the captured rats had antibodies against EMCV. In 1948, in the Mengo district of Uganda, Dick et al. [2] isolated Mengo virus from a captive rhesus monkey that had developed a posterior member paralysis. In 1949, cross-neutralization studies showed that Mengo virus, EMCV, Columbia-SK (discovered in 1939), and MM (discovered in 1943) were antigenically indistinguishable but differed from Theiler's murine encephalomyelitis virus (TMEV) [3]. Following the Panama epizooty of 1958, Murnane et al. [4] described the isolation of EMCV from swine for the first time. Since then, the virus has been isolated in many countries, in swine and in wild animals.

9.1.1 CLASSIFICATION

EMCV belongs to the *Cardiovirus* genus in the *Picornaviridae* family, which covers a total of 26 genera [5,6]. The *Cardiovirus* genus is divided into three species: EMCV, Theilovirus, and Boone cardiovirus [7–9]. EMCV is further separated into two serotypes named EMCV-1 and EMCV-2, the latter of which was recently discovered in a wood mouse (*Apodemus sylvaticus*) in Germany [10]. EMCV-1 serotype includes Colombian-SK virus, MM virus, EMCV, Mengo virus, and Maus-Elberfeld (ME) virus. Theilovirus is made up of 15 genotypically distinct viruses. Boone cardiovirus is a newly described species from laboratory rats [9].

9.1.2 GENOMIC ORGANIZATION, BIOLOGY, AND EPIDEMIOLOGY

9.1.2.1 Genomic Organization

EMCV is a nonenveloped virus with an icosahedral capsid of approximately 30 nm in diameter [11] and a genome consisting of a positive-sense single-stranded RNA. The EMCV RNA genome is approximately 7840 bases long and composed

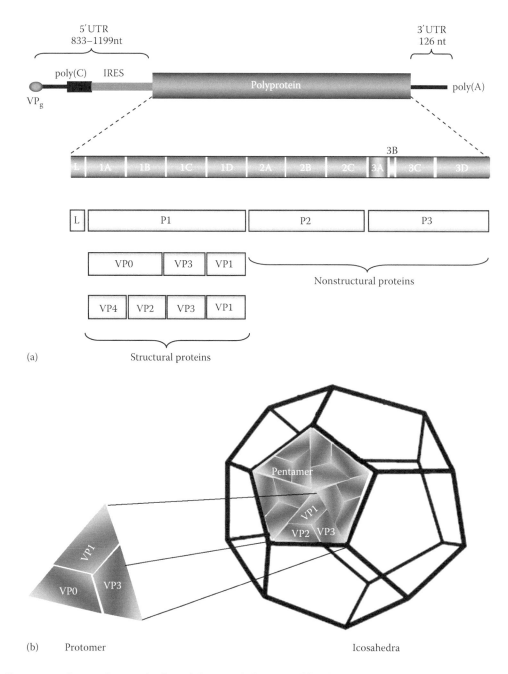

FIGURE 9.1 Structure and genomic organization of the encephalomyocarditis virus (EMCV). (a) Genome organization. The EMCV genome is a positive-sense single-stranded RNA shown with the viral protein (VP) VPg covalently linked to the 5′ end. The genomic RNA is composed of two untranslated regions with a poly(C) tract and internal ribosome entry site in the 5′ end, a poly(A) tail in the 3′ end and of a unique open reading frame that encodes for a single polyprotein. The polyprotein is processed to produce structural and nonstructural proteins. The P1 precursor is cleaved into capsid proteins (VP); P2 and P3 are cleaved to form the protease and the proteins that participate in viral RNA replication. (b) Diagram of the organization and assembly of EMCV capsid. The structural unit for the assembly of the capsid is the protomer made up of a single copy of VP1, VP3, and VP0 (precursor of VP2 and VP4). Protomers of 5 assemble into a pentamer, and then 12 pentamers form the icosahedra. During maturation, after genome encapsidation, VP0 is cleaved into VP2 and VP4. (VP4 is not visible in the schema because it is in the inner part of the capsid and interacts with the RNA).

of a unique coding region flanked by two untranslated regions (UTRs) of 833–1199 nucleotides (nt) and 126 nt in length at 5′ and 3′extremities (Figure 9.1a), respectively. The 5′ extremity is not capped but, rather, covalently linked to a viral protein (VP) of 20 amino acids (VPg) [8]. Starting approximately at 150 nt, there is a poly(C) tract, which can be of different length depending on the strain. Studies with Mengo virus in murine hosts suggest that the poly(C) length may play an important role in viral pathogenesis. However, LaRue et al. [12] reported that an EMCV strain containing a short poly(C) tract (7–10 nt) was pathogenic in mice, pigs, and cynomolgus macaques. Thus, the link between poly(C) length and pathogenesis is still controversial. Adjacent to this sequence, the RNA has a highly ordered structure made up of hairpin loops.

This domain is part of an internal ribosome entry site (IRES) [13]. Picornaviral IRES fall into four classes. IRES elements vary in size and have different complex secondary structures and distinct requirements for cellular proteins that underscore their functions [14,15]. EMCV IRES is a type II IRES. It allows ribosome binding and thus the initiation of translation of the unique open reading frame that encodes a single polyprotein of 2292 amino acids. A polyA tail of heterogeneous length (20–70 nt) is present at the 3′ end.

9.1.2.2 Viral Proteins

EMCV proteins and precursors get their names from their positions in the polyprotein (Figure 9.1a): L (leader), P1 (precursor of the capsid proteins 1A, 1B, 1C, and 1D named viral protein [VP]4, VP2, VP3, and VP1, respectively, when they are part of the virion), P2 (precursor of the nonstructural proteins 2A, 2B, and 2C), and P3 (precursor of nonstructural proteins 3A, 3B, 3C, and 3D) [8].

The L protein is only found in *Cardiovirus* and *Aphthovirus*. It is composed of a zinc finger domain at the N-termini and an acidic domain in the C-termini and can be phosphorylated on residues Thr47 and Tyr41 [8,16]. Study of deletions introduced into the L protein has established that the Mengo virus L protein is not necessary for the multiplication of this virus in BHK21 cells. However, a decrease in replication and plaque size was noticed when viruses mutated in this protein were cultured on L929 cells [17]. Those results could be explained by the role of the L protein as a transcription inhibitor of the interferon (IFN)-α4 and IFN-β genes (IFN-α4 and IFN-β genes are activated early during antiviral response) [18] and by the fact that BHK21 cells, in contrast with L929 cells, do not produce IFN. Indeed, following phosphorylation by the casein kinase II, the L protein inhibits iron–ferritin-mediated activation of NFκB and thus inhibits the synthesis of IFN-α and IFN-β [18]. The L protein may also be implicated in decreasing cellular protein synthesis and may interfere with the nucleocytoplasmic traffic through the binding of Ran-GTPase (Ras-related nuclear GTPase) [16].

The viral capsid is composed of four structural proteins: *VP1*, *VP2*, *VP3*, and *VP4*. Cleavage of the P2–P3 precursor generates mainly nonstructural proteins but also the *VPg* protein (3B), which is associated with the 5′ termini of the viral genome. While nearly all processing of the polyprotein is mediated by the *3C* protease, cleavage between the 2A and 2B proteins is protease independent. In fact, during the translation process, the C-ter region of the 2A protein adopts an unstable conformation that induces the cleavage between the two proteins specifically at the N-P-G-P sequence [8]. Moreover, the *2A* protein is possibly involved in the decrease in cellular mRNA translation in infected cells and localize to the nucleoli [19]. The 2A was also shown to be required for viral pathogenesis and inhibition of apoptosis during EMCV infection [20]. Recently, Petty et al. [21] demonstrated binding interactions between L and 2A protein; thus both of these "viral security" proteins should not be considered independent of one another. The role of the *2B* protein, however, is not

well established, but seems to be involved, along with the *2C*, in the formation of membranous vesicles during viral replication. On the other hand, it is well known that the 2B and 2C proteins and their precursor 2BC block the traffic of intracellular proteins from the endoplasmic reticulum to the Golgi and thus prevent the release of cellular proteins [8]. The EMCV "viroporin" 2B was shown to activate the NOD-like receptor family, pyrin domain containing 3 (NRLP3) inflammasome, a multiprotein complex containing the pattern recognition receptor NLRP3 that induces secretion of proinflammatory cytokines [22], and Hou et al. [23] demonstrated recently that the 2C protein along with the 3D protein are involved in EMCV-induced autophagy. Thus, these two nonstructural proteins are involved in host response to EMCV infection.

The 3AB precursor is associated to membranous vesicles by the *3A* protein. Its cleavage by the 3C protease liberates the 3B protein, which upon association with the RNA allows initiation of genome replication by the *3D* polymerase, which is the RNA-dependent RNA polymerase. EMCV replication is, however, regulated through the site-specific cleavage of the host poly(A)-binding protein by the 3C protease [24].

9.1.2.3 Three-Dimensional Structure and Assembly of the Capsid

Mengo virus has been crystallized and analyzed by x-ray diffraction [11,25]. Its viral capsid shows an organization similar to that of EMCV and other picornaviruses. The viral capsid is organized into an icosahedron with exactly 60 copies of each of the structural protein. The basic unit of the capsid is a protomer formed by the association of VP1, VP2, VP3, and VP4 (Figure 9.1b). The VP4 protein is located in the internal face of the capsid and interacts with the viral RNA. VP4 has a myristate domain in N-ter, necessary for protomers to assemble into pentamers.

The capsid is constituted of 12 pentamers; each one made up of 5 protomers. It is organized around symmetric axes of twofold, threefold, and fivefold symmetries. Five VP1 proteins are grouped around the fivefold symmetry axis, while VP2 and VP3 proteins alternate around the threefold symmetry axis.

9.1.2.4 Infectious Cycle

The viral cycle of EMCV is quite similar to that of the other picornaviruses (Figure 9.2). The first step is the attachment to a cellular receptor; VCAM1 (adhesion molecule of vascular cells) has been described as the EMCV receptor of murine endothelial cells. Mengo virus and EMCV are able to agglutinate human erythrocytes by virtue of their attachment to glycophorin A, the major sialoglycoprotein found at the erythrocyte surface. On nucleated human cells, the virus seems to attach via a sialoglycoprotein of 70 kDa, which is distinct from glycophorin A, but remains to be identified [8].

Interaction between the cellular receptor and viral capsid proteins induces conformational changes that affect contacts between pentamers and thus "prepare" the dissociation of the capsid. The exact mechanism of internalization

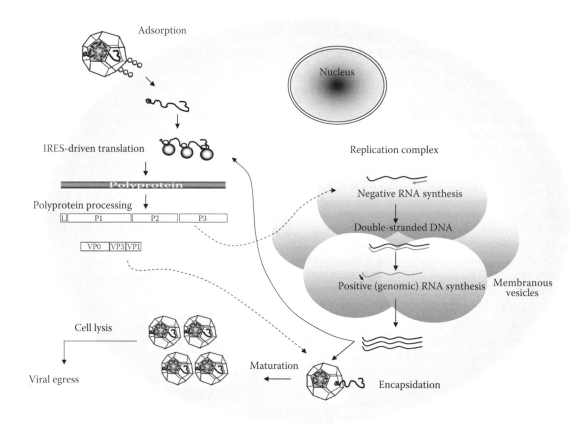

FIGURE 9.2 Encephalomyocarditis virus replication cycle. The virion binds to a cellular receptor. After uncoating, the genomic RNA is released by an unknown mechanism. Once in the cytoplasm, the viral protein (VP) VPg is detached to the 5′ end of the genome, the translation is initiated at the internal ribosome entry site, and the polyprotein is synthesized. The polyprotein is cleaved during and after the translation, leading to precursors or mature proteins. Some of those proteins go to the membranous vesicles where genome replication will occur. The positive genomic RNA is replicated into negative RNAs (linked to VPg in 5′ ends) that, in turn, serve as templates for the synthesis of positive RNAs. The new positive genomic RNAs will either serve for translation after removal of VPg or will be packaged into the procapsid. After RNA encapsidation, cleavage of the precursor VP0 into VP2 and VP4 allows maturation of the virion. Virions then egress by cell lysis.

and uncoating of the cardiovirus genome is poorly understood. Poliovirus and rhinovirus lose infectivity when cells are treated by chemical agents that increase the pH within endosomes, while *Cardiovirus* does not. This finding suggests that decapsidation of cardioviruses occurs at the cell membrane and may be due to the interaction between the virus and the cell receptor. Entry of the genomic RNA into the cytoplasm may depend on the myristoylated VP4 protein. Once the RNA is in the cytoplasm, the VPg protein is dissociated from the 5′ termini of the genome; the genome is then translated into VPs needed for the replication and the production of new viral particles [8].

The initiation of translation of EMCV is cap independent, thanks to its IRES that, with the help of some cellular factors (eukaryotic translation initiation factor [eIF]4G, eIF3, eIF4A), allows ribosome binding. EMCV uses different mechanisms to disrupt eIF4F complex formation, so to inhibit cap-dependent translation. For instance, during EMCV infection, the 4E-BP1 is dephosphorylated and sequesters eIF4E into an inactive eIF4E–4E-BP1 complex, thus inhibiting the

eIF4E–eIF4G interaction [26,27] required for the initiation of translation of cellular mRNA.

Polyprotein cleavage is carried out by the 3C protease of EMCV. However, the first cleavage occurs during elongation, prior to the synthesis of the 3C protease, between the 2A and 2B proteins. This could be due to a ribosome jump to the NPGP sequence, located between 2A and 2B, so as to prevent glycine and proline binding [28]. Once the 3C protease is translated, it is immediately active and starts to cleave the polyprotein in *cis*. When detached from the polyprotein, 3C cleaves newly synthesized polyproteins in *trans*. In this manner, maturation occurs and VPs are produced. It is believed that viral genome replication is initiated once VP concentration reaches a certain threshold.

Synthesis of a negative-strand RNA from the positive-strand genomic RNA is the first step in the genome replication of the virus. This negative-strand RNA will serve as a template to synthesize the genomic RNA. Replication takes place in the cytoplasm, within replication complexes that are essentially made up of RNA-dependent RNA polymerase

(3Dpol) linked to the 3C protease. Proteins 2B, 2C, and 3A induce the accumulation of membrane vesicles where, at the external part, units of replication associate. The 3AB precursor allows the anchorage of VPg and thus the initiation of synthesis. Regulatory mechanisms of EMCV RNA synthesis are not well known.

Encapsidation starts when the quantity of VPs reaches a certain threshold. Capsid precursor P1 is cleaved into VP0, VP1, and VP3 that spontaneously assemble into protomers and pentamers (see Section 9.1.2.3). Pentamers assemble with positive RNA to form virions. These virions become infectious when VP0 is cleaved into VP4 and VP2. Cells are lysed and thus virions are released. The length of the infectious cycle varies from 5 to 10 h, depending on several factors, such as viral strain, temperature, pH, host cell, and multiplicity of infection.

9.1.2.5 Epidemiology

9.1.2.5.1 *Host Range, Distribution, and Transmission*

EMCV can infect a wide range of animals, domestic as well as wild, including swine, boars, rodents, cattle, elephants, raccoons, marsupials, and tigers [29] and nonhuman primates such as bonobos [30], macaque [31], and orangutan [32]. EMCV within farms, zoological collection, and primate center cares is frequently fatal for a variety of animal species [33].

Of all domestic animals, pigs are the most sensitive to EMCV [34]; and swine infections have been responsible for severe economic losses on pig farms due to high mortality in piglets (myocarditis) and in sows (reproductive failure). Porcine EMCVs have been reported in many areas with swine industries including Europe and Asia [35]. In Europe, from 1986 to 1995, swine epizooties had been reported in Greece, Italy, and Belgium [36]. Serological investigations of swine herds and wild boars have been performed in France, Austria, the Netherlands, and Belgium and shown a seroprevalence ranged from 3% to 10% [36,37]. Interestingly, clinical EMCV outbreaks have never been reported in Sweden, although EMCV was detected by reverse transcription–polymerase chain reaction (RT-PCR) in 9% of rodents not in contact with farm animals as published recently by Backhans et al. [38].

Regarding Asia, epidemiological investigations were mostly conducted in China. Since its first detection in 2005 until 2012, six strains have been isolated from pigs of the Chinese mainland [39], and a porcine strain was recently isolated from northern China, showing a high pathogenicity in mice [40]. Serological survey conducted intensively in pig farms in China has shown a seroprevalence of neutralizing antibodies of 52%, with one farm in Beijing with 100% of tested animals giving positive results. Sows had the highest prevalence and could thus play an important role in the epidemiology of EMCV [41]. Serological investigations have also been conducted in Korean swine farms with a seroprevalence rate of 43.5% in pig herds compared to 9.1% in the total pig population [42].

Currently, two routes of infection are suggested for the introduction and spread of EMCV among domestic pigs (1) after reactivation of persistent EMCV infections or (2) by ingestion of feces or carcasses of infected rodents [43]. Rodents are indeed suspected to be the natural reservoir for the virus as they replicate and excrete virus until 29 days postinfection without any symptoms [44]. The presence of many infected rodents in contact with swine herds during EMCV epizooties suggests that they may play a key role in virus transmission [45,46]. Natural infections may be due to ingestion of contaminated food. In fact, horizontal transmission between swine during the short period of viremia seems limited, but should it occur might trigger the emergence of a major epizooty, depending on the viral strain and swine sensibility [47]. Wild boar may also play a critical role in EMCV epidemiology in domestic pigs [43].

9.1.2.5.2 *Zoonotic Potential*

Even if EMCV has often been described as a zoonotic agent, positive association with a human disease has not been established. Nevertheless, between 1940 and1950, childhood infections have been described in Germany and the Netherlands, associated with clinical signs like fever and encephalitis but no myocarditis. Different EMCV strains have been isolated after inoculation into rodents [48]. However, the appurtenance of these strains to the EMCV group had only been proved by serological tests without any virologic confirmation, and unfortunately those strains are no longer available for further characterization. Earlier, EMCV neutralizing antibodies had been discovered in seventeen soldiers that presented febrile symptoms for 3 days in Manila, and in three of four patients for whom several samples were available, a rise in antibody titers was demonstrated [49]. In 1948, following isolation of Mengo virus from a rhesus monkey, a researcher who was studying the virus and taking care of the infected animal developed a meningoencephalitis, from which he recovered. Later on, the virus was isolated from his blood [2]. In all of these cases, however, virus was isolated, but only from specimens obtained from nonsterile sites, thus precluding unequivocal attribution of the patient's symptoms to EMCV infection.

Since 1950 to 2009, no EMCV infections associated with clinical signs have been recorded. However, some serological studies made on healthy human populations revealed a prevalence of 2.3%–15%. In Austria, studies of prevalence in persons that work with animals revealed that 5% of employees were seropositive, as well as 15% of hunters [8]. More recently, a serological survey in Mexico among swine-specialist veterinarians revealed a prevalence of 27% (hemagglutination inhibition test) to 47% (viral neutralizing test) [50].

Brewer et al. [51] showed in 2001 that primary myocardial human cells are permissive to an EMCV strain isolated from aborted swine fetus. Hammoumi et al. [52] also reported that these cells as well as human spleen primary epithelial cells and human primary astrocytes are permissive to EMCV strains isolated from pig or rat.

In 2009, Oberste et al. [53] reported cases of human febrile illness in Peru, which were probably due to EMCV infection: in two febrile patients with nausea, headache, and dyspnea, EMCV was isolated from acute-phase sample, and subsequent molecular diagnostic was done. No other pathogens were detected from those patients using immunoglobulin M ELISA for flaviviruses, alphaviruses, bunyaviruses, arenaviruses, and rickettsia. Isolation of virus from an acute-phase serum sample strongly supports the role of EMCV in human infection and febrile illness. Same authors have reported that roughly one in five residents of Iquitos (Peru) had evidence for prior exposure to EMCV [54]. In addition, EMCV may be also a potential pathogen for human recipients in pig-to-human xenotransplantation [51,55].

9.1.2.5.3 Vaccine Strategies

Many attempts have been made to develop EMCV vaccines. In the past, adjuvanted vaccine was shown to be efficient in inducing high antibody titers in domestic and wild species. Moreover, a binary ethyleneimine-inactivated EMCV was licensed for use [56]. Recently, Chen et al. [57] have developed recombinant human adenovirus–based EMCV vaccines that can protect mice against a lethal challenge. This promising vaccine remains to be evaluated in swine. An alternative vaccine approach based on viruslike particles combined with an alum adjuvant was also proposed as a safe immunization strategy during sow gestation in order to prevent EMCV-induced reproductive failure in pig farms [58].

9.1.3 Clinical Features and Pathogenesis

Physiopathology of EMCV infection has been mainly studied in swine and rodents.

9.1.3.1 Swine

After an oral contamination, the initial target cells are monocytes, which may carry the infection from the tonsils to target organs. A few days postinfection, the virus can be isolated from the heart, brain, tonsils, salivary glands, spleen, liver, pancreas, lungs, kidney, and small intestine [59]. When piglets are infected "per os" the virus replicates in the intestine and can be detected as early as 24 h postinfection and for 3–5 days in the blood and feces. The high level of viruses in the spleen, lymph nodes, and mesenteric nodes indicates that viruses multiply in lymphoid tissues. On the other hand, it has been demonstrated that in infected piglets that survived EMCV infection, the virus can persist and be reactivated when animals are treated with dexamethasone [60]. Reactivated viruses became pathogenic for the animals, were excreted in feces, and were able to infect control negative piglets when housed together with treated piglets.

Usually, in pig and piglet, EMCV induces acute focal myocarditis with sudden death. Myocarditis is characterized by cardiac inflammation and cardiomyocyte necrosis. Other symptoms have been observed, such as anorexia, apathy, paralysis, or dyspnea [34,61]. Experimentally infected piglets showed high fever, followed by death within 2–11 days, or sometimes recovered with chronic myocarditis. Mortality in piglets before weaning can rise to 100% and decreases with aging [62]. Reproduction disorders including abortion, fetal death, or mummification have been described in infected females [63]. In dying piglets, during the acute-phase, cardiac disorders, including epicardial hemorrhages, are observed. Upon autopsy of experimentally infected piglets, hydropericardium, hydrothorax, lung edema, ascites, and lesions at the myocardium have often been described. The heart is often dilated and shows some focal areas of necrosis (with uneven grayish-white discoloration). In most cases, the virus is detected in the myocardium, even if myocardial lesions are small or absent. In piglets, histological analysis reveals myocarditis associated with scattered or localized accumulation of mononuclear cells, vascular congestion, edema, and myocardial fiber degeneration, with necrosis. In the brain, congestions are accompanied by meningitis, perivascular infiltration of mononuclear cells, and neuronal degeneration [64,65].

9.1.3.2 Rodent

In the rodent, EMCV infection can be asymptomatic [44], but mice can develop myocarditis, encephalitis, [66,67] member paralysis [12,68], or diabetes [69]. EMCV can also lead to reproduction disorders in pregnant mice [70] and testicular lesions in mice and hamster [71]. Genetic factors seem to influence the sensitivity of mice to EMCV infection [72]. It has been shown that males are more sensitive than females [73]. This difference might be due to the interferon immune response that happens earlier in females than in males after peritoneal injection [74].

9.1.4 Diagnosis

Until 1997, the diagnosis of EMCV infection was reliant on conventional methods like virus isolation from tissues and virus characterization by neutralization assay. This characterization is laborious and time consuming, and so research has been directed toward the detection of viral RNA by RT-PCR.

9.1.4.1 Conventional Techniques

The diagnosis of EMCV infection has traditionally been dependent on virus isolation and identification. Despite being technically straightforward, the virus is sometimes difficult to isolate, especially from pigs in the stage following development of circulating antibody. Furthermore, the virus could no longer be isolated from the blood or feces after 3 days of infection, when viremia disappears [75]. Thus, even if the virus persists for a long period in the heart or other organs, it is difficult to isolate owing to circulating antibodies [60]. Samples should be collected during the acute phase of infection and could be taken from tissue, such as the heart, as it is the main target organ in piglets, but also from the liver, kidney, or total blood.

Viral isolation is routinely performed on baby hamster kidney (BHK21) cell line or in mice. Briefly, homogenates of samples are inoculated into BHK21 cell monolayers. Then monolayers are examined daily for CPE for 72 h. Monolayers negative for CPE are passaged two more times to confirm negative virus isolation results. The viral isolation can also be performed by intracranial inoculation of mice with filtered crushed tissue or blood. Mice are examined daily for development of paralysis and then osculated for lesions in the brain and myocardium. Then, virus identification is performed by immunofluorescence on infected cell monolayers or by a viral neutralizing test.

Serological diagnosis relies on a hemagglutination inhibition or a seroneutralization test. Antibody titer superior to 1/16 is taken to mean that the animal has been previously exposed to the virus [76]. The seroneutralization test has a good specificity for antibodies directed against EMCV, because no cross-reactivity has been found with 62 human enterovirus serotypes, 11 porcine enterovirus serotypes, and at least 27 other viruses.

9.1.4.2 Molecular Techniques

As is described previously, the EMCV genome is a single-stranded RNA. Thus to detect the virus using molecular techniques, RT must be proceeded before PCR. RT-PCR methods allow the detection of viral RNA by primer-directed enzymatic amplification of specific target RNA sequences and avoid the time-consuming step of virus isolation. One-step RT-PCR methods have been further developed to minimize the risk of cross contamination encountered with two-step RT-PCR procedures. In addition, PCR products can be sequenced, allowing not only unambiguous identification of EMCV, but also epidemiologic analyses. Furthermore, PCR products can be detected directly by real-time RT-PCR by addition of specific probe, or by using SYBR Green mix, or even by one-step RT loop-mediated isothermal amplification (LAMP) method. Since 1997, several molecular methods have been described for rapid diagnosis of EMCV infections in pigs [77].

9.2 METHODS

9.2.1 SAMPLE PREPARATION AND VIRAL RNA EXTRACTION

In order to recover much viral RNA as possible, it is important to collect samples from suitable organs (i.e., where there might be virus at the highest concentration) and to apply an optimal RNA extraction method.

9.2.1.1 Clinical Sample Preparation

From live animals, blood, serum, or fecal samples should be used. Fecal extracts (10% w/v) are made in phosphate-buffered saline (pH 7.6) and clarified by centrifugation for 10 min at $4000 \times g$ to eliminate larger fecal debris. From death or euthanized animals, samples from the heart, spleen, liver, lungs, brain, tonsils, salivary glands, pancreas, kidney, and small intestine can be collected [59]. Supernatants of homogenized tissue suspensions (10% w/v) are used for virus detection. However, the most suitable organ to detect EMCV is the heart.

9.2.1.2 RNA Extraction

Three different RNA extraction methods can be used [78]:

1. *Guanidinium thiocyanate method*: Total RNAs are recovered from sample by an extraction method using commercially available mixtures of acid guanidinium thiocyanate and phenol–chloroform (TRIzol® LS Reagent, GIBCO-BRL, Life Technologies), according to the manufacturer's recommendations.

2. *Magnetic beads separation method*: This method relies on virus capture from sample on magnetic beads via monoclonal antibody specific for EMCV (immunomagnetic capture) and genomic RNA extraction from the captured particles by heating. The antibody used should be an anti-EMCV directed against epitopes present in a wide range of EMCV isolates. This method should be used to perform diagnosis from fecal samples, because it allows removal of PCR inhibitors found in feces.

3. *Guanidinium isothiocyanate (GITC)–silica methods*: RNA is extracted from samples using RNeasy Mini Kit (Qiagen) or QIAamp Viral RNA Mini Kit for purification of viral RNA (Qiagen Ref. 52906) according to the manufacturer's recommendations. Viral RNA is bound to the silica membrane, and contaminants are removed in two wash steps. Pure viral RNA is eluted in small volume of a low-salt buffer or RNAse-free water.

9.2.2 DETECTION PROCEDURES

The detection of EMCV viral nucleic acid from clinical samples was initially described by Vanderhallen and Koenen [77,79] who developed a two-step conventional RT-PCR using primers located in the 3D polymerase coding region. Oberste et al. [53] reported a similar two-step method for specific detection of EMCV in acute-phase serum of patients with febrile syndrome, after isolation of the virus on Vero-6 cell cultures. They used primer sets targeting the 5' UTR, the VP1, or the 3D coding region. Pérez and Díaz de Arce [80] proposed further an optimized protocol and using primers designed to amplify a 165 bp fragment from the 3D coding region with an analytical sensitivity of 2 (EMCV) 50% tissue culture infective dose ($TCID_{50}$) per 50 µL.

With the objective to increase rapidity, sensitivity, specificity, and simplicity and to remove RT-PCR inhibitors, procedures had evolved toward one-step conventional RT-PCR and recently real-time methods, described in the following sections.

9.2.2.1 One-Step Conventional RT-PCR

Kassimi et al. [78] adapted the two-step RT-PCR method developed by Vanderhallen and Koenen [77] using the same primers (P1 and P2) in a one-step RT-PCR assay. They described application of this method on EMCV-infected cells and on clinical samples according to the method used for RNA extraction.

9.2.2.1.1 Procedure

1. Prepare the one-step RT-PCR mixture (50 µL) using the one-step RT-PCR Kit (Qiagen). To amplify genome from virus captured on magnetic beads, prepare a first mix (30 µL) containing 400 µM each of deoxynucleotide triphosphates (dNTPs), 600 nM of each primer, and 10 U of RnaseOut. Add the mix

directly to the coated beads, heat at 100°C for 5 min, and chill on ice. Prepare a second mix (20 μL) containing 1× Qiagen one-step RT-PCR buffer and 2 μL of 1× Qiagen one-step RT-PCR enzyme mix. Add the second mix to the first mix just before the RT-PCR. To amplify RNA extracted by GITC–silica, prepare a first mix (8.5 μL) containing 10 U of RnaseOut, 0.4 mM each of dNTPs, and 600 nM of each primer. Add the mix to the 30 μL RNA eluted from the column, heat at 65°C for 5 min, and chill on ice. Prepare a second mix (10 μL) containing the one-step RT-PCR buffer (1×) and Qiagen one-step RT-PCR enzyme mix. Add the second mix to the first mix just before the RT-PCR.

2. Reverse-transcribe at 50°C for 30 min; conduct PCR amplification for one cycle at 95°C for 15 min and 35 cycles at 94°C for 30 s, 60°C for 30 s, and 72°C for 1 min and a final extension of 72°C for 10 min.

3. Analyze the PCR product by agarose gel electrophoresis, or use the product directly for sequencing.

Note: Further comparative studies [78] allowed to determine that for samples that do not inhibit the RT-PCR, the RNA extraction using GITC–silica combined with the one-step RT-PCR is the best choice. Immunocapture combined with the one-step RT-PCR is more suitable for samples containing substances that may inhibit RT-PCR.

9.2.2.2 Real-Time RT-PCR

9.2.2.2.1 TaqMan Real-Time RT-PCR

Yuan et al. [81] developed a TaqMan-based real-time RT-PCR assay targeting the EMCV 3D coding region and applied it to clinical samples (sera and tonsils [$n = 100$]). Primers (EMC-F and R) and probe (EMCV-P) were designed using the Primer Express Software (version 3.0; Applied Biosystem, United States). The probe was labeled with 5-carboxyfluorescein (FAM) at the 5′-end and *N,N,N′,N′*-tetramethyyl-6-carboxyrhodamine (TAMRA) at the 3′-end (EMCV-F: 5′-TCATTCGCCATTTCAACCCA-3′, EMCV-R: 5′-GAGATACAAACCCGCCCTAA-3′, EMCV-P: 5′-FAM-TCCCATCAGGTTGTGCAGGC GA-TAMRA-3′). The standard curve established with a recombinant plasmid containing the 3D target gene sequence had a wide dynamic range of 10^1–10^6 copies/μL with a linear correlation R^2 of 0.996 and a slope of −3.128 between the Ct value and the logarithm of the plasmid copy number. The detection limit of the TaqMan RT-PCR assay is 1.4×10^2 copies/μL. No cross-reaction was observed with other porcine viruses.

9.2.2.2.1.1 Procedure

1. Extract RNA from 150 μL of supernatant from virus-infected BHK-21 cells or tissue samples using the RNA extraction kit (Qiagen, United States) following the manufacturer's instruction.

2. Prepare real-time RT-PCR mixture (25 μL final) with 1 μL of extracted RNA, 12.5 μL of FastTaqMan

mixture (with ROX) (Cwbiotech, Beijing, China), 0.5 μL of Avian myeloblastosis virus (AMV) reverse transcriptase (Transgen, Beijing, China), 400 nM of each primer, and 200 nM of probe.

3. Carry out amplification and detection under the following conditions: RT step at 50°C for 20 min, PCR activation at 95°C for 3 min, and 40 cycles of amplification (5 s at 95°C and 40 s at 60°C). To analyze results, use the system software linked to the real-time thermocycler (Bio-Rad iQ5 was used to developed this real-time RT-PCR).

9.2.2.2.2 Real-Time RT-PCR

Carocci et al. [20] developed a real-time RT-PCR assay based on the use of two primers (TMF1 and TMR1) to amplify 90 bp located in the 3D region of the EMCV genome, by using a SYBR Green master mix (TMF1: 5′-GGGATCAGCTTTTACGGCTTT-3′, TMR1: 5′-TGCATCCGATAGAGAACTTAATGTCT-3′). The standard curve established with a recombinant plasmid containing EMCV genome had a linear correlation R^2 of 0.99 and a slope of −3.544 between the Ct value and the logarithm of the plasmid copy number. The detection limit of this real-time RT-PCR assay is 25 copies/5 μL. This method has been applied on infected cell culture and clinical samples.

9.2.2.2.2.1 Procedure

1. Extract RNA using RNeasy Mini Kit (Qiagen) as described by the manufacturer.

2. Prepare RT-PCR mixture with 2.8 μL of RNAse-free water, 10 μL of SYBR Green Quantitect RT-PCR master mix (QIAGEN Ref. 204243), 1 μL of each primer (TMF1 and TMR1 at 10 μM) at 0.5 μM final, 0.2 μL of RT enzyme (included in SYBR Green Quantitect RT-PCR master mix [QIAGEN Ref.204243]), and 5 μL of RNA.

3. Carry out amplification and detection under the following conditions: RT step at 50°C for 20 min, PCR activation at 95°C for 15 min, and 45 cycles of amplification (15 s at 94°C, 20 s at 55°C, and 10 s at 72°C) on LightCycler 480 (Roche).

4. Visualize target amplification by generating melting curve by incubating for 10 s at 95°C and for 1 min at 55°C and by raising the incubation temperature from 55°C to 95°C. Finally, let it cool down for 30 s at 40°C.

Alternative TaqMan protocol: Carocci et al. [20] designed a TaqMan probe annealing specifically between the TMF1 and TMR1 primers to an internal region of the PCR product (probe TM-FR1: 5′-FAM-CGATGCCAACGAGGACGCCC-MGB-3′). This method has been applied on RNA from infected cell culture or infected organ samples and remains to be validated.

9.2.2.2.2   Procedure

1. Extract RNA using the QIAamp Viral RNA Mini Kit (Qiagen) as described by the manufacturer.
2. Prepare RT-PCR mixture with 4.1 μL of RNAse-free water, 12.5 μL of AgPath buffer (Life Technologies, AGPATH-ID One-Step RT-PCR ref. 4387391), 1 μL of each primer (TMF1 and TMR1 at 10 μM) at 0.4 μM final, 1 μL of TM-FR1 probe (10 μM) at 0.16 μM final, 1 μL of enzyme (included in AGPATH-ID One-Step RT-PCR), and 5 μL of RNA.
3. Carry out amplification and detection under the following conditions: RT step at 45°C for 10 min, PCR activation at 95°C for 10 min, and 45 cycles of amplification (15 s at 95°C, 1 min at 60°C) on real-time thermocycler. Sigmoid amplification curves with Ct values under 40 are considered as positives results as "detection of EMCV genome."

Note: Even though RT-PCR allows rapid diagnosis of EMCV in samples that often need two passages on cell cultures to be isolated, attesting to the good sensitivity of the assay, a negative RT-PCR test should be verified by virus isolation and virus neutralization test.

9.2.2.2.3   One-Step RT-LAMP

Yuan et al. [82] developed a RT-LAMP assay to detect EMCV RNA. This technique relies on the use of 4 specific primers targeting the EMCV 3D coding region (forward outer [F3] 5′-GGCAGATCTTGGAGGAAGC-3′, backward outer [B3] 5′-AAGACTTGACGGGCAACG-3′, forward inner [F1c-F2] 5′-GCCCGAATCATCTCCTGTGACA-CCTTGGCATCCATTCTGCTG-3′, backward inner [B1c-B2]5′-AATGCCTTTGAGCCACAGGGTG-GGCGTAGTGCTGTT TTACGT-3′). The detection limit of RT-LAMP is 2.2×10^{-5} ng of cell culture–extracted RNA per μL. No cross-reactivity with classical swine fever virus (CSFV), bovine viral diarrhea virus (BVDV), porcine circovirus type 2, porcine epidemic diarrhea virus, and porcine reproductive and respiratory syndrome virus (PRRSV) was observed.

9.2.2.2.3.1   Procedure

1. Extract RNA from samples by using QIAamp viral RNA kit (Qiagen, Valencia, CA) according to the manufacturer's instructions.
2. Prepare one-step RT-LAMP mixture (25 μL final) with 0.5 μL of RNAsin inhibitor, 2.5 μL of 10× *Bst* DNA buffer, 2.5 μL of MgCl₂ (25 mM), 3 μL of dNTPs (2.5 mM), 0.5 μL of F3 primer (10 μM), 0.5 μL of B3 primer (10 μM), 2.5 μL of F1c-F2 primer (10 μM), 2.5 μL of B1c-B2 primer (10 μM), 1 μL of RNA template, 1 μL of AMV reverse transcriptase (200 U/μL), 1.5 μL of *Bst* DNA polymerase (8 U/μL), and 7 μL of diethylpyrocarbonate (DEPC) treated water.
3. Incubate for 50 min at 62°C.

4. Detection of amplification product can be done on 2% agarose gel electrophoresis: migration results are ladderlike bands. Alternatively, products can also be visualized readily in-tube under UV light with the addition of SYBR Green I.

Note: Due to high amplification of loop-DNA targets, potential risk cross contamination should be taken into account.

9.3 CONCLUSION

Since its first isolation in 1958 from fatal case of swine, EMCV has been recognized worldwide as the causal agent for myocarditis in young pigs or reproductive failure in sows, with severe economic consequence for swine industry. Although EMCV is often described as a zoonotic agent, no association with human disease has been clearly established. Nevertheless, given the recent progress in xenografts, especially pig-to-human xenotransplantation, the safety concerns about porcine EMCV need to be explored.

Traditionally, the diagnosis of EMCV infection has relied largely on virus isolation and virus neutralization assay. The availability of molecular methods provides the opportunity to further improve EMCV diagnosis in terms of manipulation and time required. In particular, the application of one-step conventional RT-PCR and real-time RT-PCR protocols has enhanced the accuracy and speed of EMCV identification. Additionally, the development of one-step RT-LAMP offers a rapid, accurate, and inexpensive means to detect EMCV under field conditions. Overall, the increased use of molecular tools makes it possible to better characterize and compare emerging pathogenic viral strains for clinical diagnosis as well as field surveillance.

REFERENCES

1. Helwig, F. C. and Schmidt, C. H., A filter-passing agent producing interstitial myocarditis in anthropoid apes and small animals, *Science*, 102, 31, 1945.
2. Dick, G. W. et al., Mengo encephalomyelitis; a hitherto unknown virus affecting mice, *Lancet*, 2, 286, 1948.
3. Dick, G. W., The relationship of Mengo encephalomyelitis, encephalomyocarditis, Columbia-SK and M.M. viruses, *J. Immunol.*, 62, 375, 1949.
4. Murnane, T. G. et al., Fatal disease of swine due to encephalomyocarditis virus, *Science*, 131, 498, 1960.
5. Adams, M. J. et al., Ratification vote on taxonomic proposals to the International Committee on Taxonomy of Viruses, *Arch. Virol.*, 158, 2023, 2013.
6. Knowles, N. J. et al., Picornaviridae. In: King, A. M. Q. et al. (eds.), *Virus Taxonomy: Classification and Nomenclature of Viruses: Ninth Report of the International Committee on Taxonomy of Viruses*, p. 855. Elsevier, San Diego, CA, 2012.
7. Bodewes, R. et al., Viral metagenomic analysis of feces of wild small carnivores, *Virol. J.*, 11, 89, 2014.
8. Carocci, M. and Bakkali-Kassimi, L., The encephalomyocarditis virus, *Virulence*, 3, 351, 2012.
9. Firth, C. et al., Detection of zoonotic pathogens and characterization of novel viruses carried by commensal Rattus norvegicus in New York City, *mBio*, 5, 2014.

10. Philipps, A. et al., Isolation and molecular characterization of a second serotype of the encephalomyocarditis virus, *Vet. Microbiol.*, 161, 49, 2012.

11. Luo, M. et al., The atomic structure of Mengo virus at 3.0 A resolution, *Science*, 235, 182, 1987.

12. LaRue, R. et al., A wild-type porcine encephalomyocarditis virus containing a short poly(C) tract is pathogenic to mice, pigs, and cynomolgus macaques, *J. Virol.*, 77, 9136, 2003.

13. Duke, G. M. et al., Sequence and structural elements that contribute to efficient encephalomyocarditis virus RNA translation, *J. Virol.*, 66, 1602, 1992.

14. Belsham, G. J., Divergent picornavirus IRES elements, *Virus Res.*, 139, 183, 2009.

15. Pan, M. et al., Duck Hepatitis A virus possesses a distinct type IV internal ribosome entry site element of picornavirus, *J. Virol.*, 86, 1129, 2012.

16. Basta, H. A. et al., Encephalomyocarditis virus leader is phosphorylated by CK2 and syk as a requirement for subsequent phosphorylation of cellular nucleoporins, *J. Virol.*, 88, 2219, 2014.

17. Zoll, J. et al., Mengovirus leader is involved in the inhibition of host cell protein synthesis, *J. Virol.*, 70, 4948, 1996.

18. Zoll, J. et al., The mengovirus leader protein suppresses alpha/beta interferon production by inhibition of the iron/ferritin-mediated activation of NF-kappa B, *J. Virol.*, 76, 9664, 2002.

19. Groppo, R. et al., Mutational analysis of the EMCV 2A protein identifies a nuclear localization signal and an eIF4E binding site, *Virology*, 410, 257, 2011.

20. Carocci, M. et al., Encephalomyocarditis virus 2A protein is required for viral pathogenesis and inhibition of apoptosis, *J. Virol.*, 85, 10741, 2011.

21. Petty, R. V. et al., Binding interactions between the encephalomyocarditis virus leader and protein 2A, *J. Virol.*, 88, 13503, 2014.

22. Ito, M. et al., Encephalomyocarditis virus viroporin 2B activates NLRP3 inflammasome, *PLoS Pathog.*, 8, e1002857, 2012.

23. Hou, L. et al., Nonstructural proteins 2C and 3D are involved in autophagy as induced by the encephalomyocarditis virus, *Virol. J.*, 11, 156, 2014.

24. Kobayashi, M. et al., Site-specific cleavage of the host poly(A) binding protein by the encephalomyocarditis virus 3C proteinase stimulates viral replication, *J. Virol.*, 86, 10686, 2012.

25. Luo, M. et al., Structure determination of Mengo virus, *Acta Crystallogr. B*, 45, 85, 1989.

26. Gingras, A. C. et al., Activation of the translational suppressor 4E-BP1 following infection with encephalomyocarditis virus and poliovirus, *P. Natl. Acad. Sci. USA*, 93, 5578, 1996.

27. Haghighat, A. et al., Repression of cap-dependent translation by 4E-binding protein 1: Competition with p220 for binding to eukaryotic initiation factor-4E, *EMBO J.*, 14, 5701, 1995.

28. Hahn, H. and Palmenberg, A. C., Deletion mapping of the encephalomyocarditis virus primary cleavage site, *J. Virol.*, 75, 7215, 2001.

29. Liu, H. et al., Isolation, molecular characterization, and phylogenetic analysis of encephalomyocarditis virus from South China tigers in China, *Infect. Genet. Evol.*, 19, 240, 2013.

30. Jones, P. et al., Encephalomyocarditis virus mortality in semi-wild bonobos (*Pan paniscus*), *J. Med. Primatol.*, 40, 157, 2011.

31. Masek-Hammerman, K. et al., Epizootic myocarditis associated with encephalomyocarditis virus in a group of rhesus macaques (*Macaca mulatta*), *Vet. Pathol.*, 49, 386, 2012.

32. Yeo, D. S. et al., A highly divergent Encephalomyocarditis virus isolated from nonhuman primates in Singapore, *Virol. J.*, 10, 248, 2013.

33. Canelli, E. et al., Encephalomyocarditis virus infection in an Italian zoo, *Virol. J.*, 7, 64, 2010.

34. Joo, H. S. Encephalomyocarditis virus. In: Straw, B. E., Leman, A. D., Mengeling, W. L., Mengeling, S., D'Allaire, S., and Taylor, D. J. (eds.), *Diseases of Swine*, Wolfe Publishing, London, p. 257, 1992.

35. Yuan, W. et al., Complete genome sequence of porcine encephalomyocarditis virus strain BD2, *Genome Announc.*, 1, 13, 2013.

36. Maurice, H. et al., The occurrence of encephalomyocarditis virus (EMCV) in European pigs from 1990 to 2001, *Epidemiol. Infect.*, 133, 547, 2005.

37. Bakkali Kassimi, L. et al., Serological survey of encephalomyocarditis virus infection in pigs in France, *Vet. Rec.*, 159, 511, 2006.

38. Backhans, A. et al., Occurrence of pathogens in wild rodents caught on Swedish pig and chicken farms, *Epidemiol. Infect.*, 141, 1885, 2013.

39. Lin, W. et al., Isolation, molecular characterization, and phylogenetic analysis of porcine encephalomyocarditis virus strain HB10 in China, *Infect. Genet. Evol.*, 12, 1324, 2012.

40. Yuan, W., Song, Q., Zhang, X., Zhang, L., and Sun, J. Isolation and molecular analysis of porcine encephalomyocarditis virus strain BD2 from northern China. *Infect. Genet. Evol.*, 21, 303, 2014.

41. Ge, X. et al., Seroprevalence of encephalomyocarditis virus in intensive pig farms in China, *Vet. Rec.*, 166, 145, 2010.

42. An, D. J. et al., Encephalomyocarditis in Korea: Serological survey in pigs and phylogenetic analysis of two historical isolates, *Vet. Microbiol.*, 137, 37, 2009.

43. Billinis, C., Encephalomyocarditis virus infection in wildlife species in Greece, *J. Wildl. Dis.*, 45, 522, 2009.

44. Psalla, D. et al., Pathogenesis of experimental encephalomyocarditis: A histopathological, immunohistochemical and virological study in rats, *J. Comp. Pathol.*, 134, 30, 2006.

45. Acland, H. M. and Littlejohns, I. R., Encephalomyocarditis virus infection of pigs. 1. An outbreak in New South Wales, *Aust. Vet. J.*, 51, 409, 1975.

46. Maurice, H. et al., Factors related to the incidence of clinical encephalomyocarditis virus (EMCV) infection on Belgian pig farms, *Prev. Vet. Med.*, 78, 24, 2007.

47. Kluivers, M. et al., Transmission of encephalomyocarditis virus in pigs estimated from field data in Belgium by means of R0, *Vet. Res.*, 37, 757, 2006.

48. Gajdusek, D. C., Encephalomyocarditis virus infection in childhood, *Pediatrics*, 16, 902, 1955.

49. Smadel, J. E. and Warren, J., The virus of encephalomyocarditis and its apparent causation of disease in man, *J. Clin. Invest.*, 26, 1197, 1947.

50. Rivera-Benitez, J. F. et al., Serological survey of veterinarians to assess the zoonotic potential of three emerging swine diseases in Mexico, *Zoonoses Public Health*, 61, 2014.

51. Brewer, L. A. et al., Porcine encephalomyocarditis virus persists in pig myocardium and infects human myocardial cells, *J. Virol.*, 75, 11621, 2001.

52. Hammoumi, S. et al., Encephalomyocarditis virus may use different pathways to initiate infection of primary human cardiomyocytes, *Arch. Virol.*, 157, 43, 2012.

53. Oberste, M. S. et al., Human febrile illness caused by encephalomyocarditis virus infection, Peru, *Emerg. Infect. Dis.*, 15, 640, 2009.

54. Czechowicz, J. et al., Prevalence and risk factors for encephalomyocarditis virus infection in Peru, *Vector Borne Zoonotic Dis.*, 11, 367, 2011.

55. Denis, P. et al., Genetic variability of encephalomyocarditis virus (EMCV) isolates, *Vet. Microbiol.*, 113, 1, 2006.

56. Huneke, R. B. et al., Antibody response in baboons (*Papio cynocephalus anubis*) to a commercially available encephalomyocarditis virus vaccine, *Lab. Anim. Sci.*, 48, 526, 1998.

57. Chen, Z. et al., Protective immune response in mice vaccinated with a recombinant adenovirus containing capsid precursor polypeptide P1, nonstructural protein 2A and 3C protease genes (P12A3C) of encephalomyocarditis virus, *Vaccine*, 26, 573, 2008.

58. Jeoung, H. Y. et al., A novel vaccine combined with an alum adjuvant for porcine encephalomyocarditis virus (EMCV)-induced reproductive failure in pregnant sows, *Res. Vet. Sci.*, 93, 1508, 2012.

59. Papaioannou, N. et al., Pathogenesis of encephalomyocarditis virus (EMCV) infection in piglets during the viraemia phase: A histopathological, immunohistochemical and virological study, *J. Comp. Pathol.*, 129, 161, 2003.

60. Billinis, C. et al., Persistence of encephalomyocarditis virus (EMCV) infection in piglets, *Vet. Microbiol.*, 70, 171, 1999.

61. Gaskin, J.M., Overview of encephalomyocarditis virus infection. *The Merck Manual*, 2013. available at http://www.merckvetmanual.com/mvm/generalized_conditions/encephalomyocarditis_virus_infection/overview_of_encephalomyocarditis_virus_infection.html. Accessed on March 31, 2015.

62. Littlejohns, I. R. and Acland, H. M., Encephalomyocarditis virus infection of pigs. 2. Experimental disease, *Aust. Vet. J.*, 51, 416, 1975.

63. Koenen, F. and Vanderhallen, H., Comparative study of the pathogenic properties of a Belgian and a Greek encephalomyocarditis virus (EMCV) isolate for sows in gestation, *Zbl. Vet. Med. B*, 44, 281, 1997.

64. Psychas, V. et al., Evaluation of ultrastructural changes associated with encephalomyocarditis virus in the myocardium of experimentally infected piglets, *Am. J. Vet. Res.*, 62, 1653, 2001.

65. Vlemmas, J. et al., Immunohistochemical detection of encephalomyocarditis virus (EMCV) antigen in the heart of experimentally infected piglets, *J. Comp. Pathol.*, 122, 235, 2000.

66. Petruccelli, M. A. et al., Cardiac and pancreatic lesions in guinea pigs infected with encephalomyocarditis (EMC) virus, *Histol. Histopathol.*, 6, 167, 1991.

67. Psalla, D. et al., Pathogenesis of experimental encephalomyocarditis: A histopathological, immunohistochemical and virological study in mice, *J. Comp. Pathol.*, 135, 142, 2006.

68. Kassimi, L. B. et al., Nucleotide sequence and construction of an infectious cDNA clone of an EMCV strain isolated from aborted swine fetus, *Virus Res.*, 83, 71, 2002.

69. Yoon, J. W. and Jun, H. S., Viruses cause type 1 diabetes in animals, *Ann. N.Y. Acad. Sci.*, 1079, 138, 2006.

70. Nakayama, Y. et al., Experimental encephalomyocarditis virus infection in pregnant mice, *Exp. Mol. Pathol.*, 77, 133, 2004.

71. Shigesato, M. et al., Early development of encephalomyocarditis (EMC) virus-induced orchitis in Syrian hamsters, *Vet. Pathol.*, 32, 184, 1995.

72. Kang, Y. and Yoon, J. W., A genetically determined host factor controlling susceptibility to encephalomyocarditis virus-induced diabetes in mice, *J. Gen. Virol.*, 74, 1207, 1993.

73. Scherr, G. H., Experimental studies on the pathogenesis of encephalomyocarditis virus for mice. I. The factor of sex of animal as affecting pathogenicity of inoculum and susceptibility to infection, *J. Bacteriol.*, 66, 105, 1953.

74. Pozzetto, B. and Gresser, I., Role of sex and early interferon production in the susceptibility of mice to encephalomyocarditis virus, *J. Gen. Virol.*, 66, 701, 1985.

75. Foni, E. et al., Experimental encephalomyocarditis virus infection in pigs, *Zbl. Vet. Med. B*, 40, 347, 1993.

76. Zimmerman, J. J. et al., Serologic diagnosis of encephalomyocarditis virus infection in swine by the microtiter serum neutralization test, *J. Vet. Diagn. Invest.*, 2, 347, 1990.

77. Vanderhallen, H. and Koenen, F., Rapid diagnosis of encephalomyocarditis virus infections in pigs using a reverse transcription-polymerase chain reaction, *J. Virol. Methods*, 66, 83, 1997.

78. Kassimi, L. B. et al., Detection of encephalomyocarditis virus in clinical samples by immunomagnetic separation and one-step RT-PCR, *J. Virol. Methods*, 101, 197, 2002.

79. Vanderhallen, H. and Koenen, F., Identification of encephalomyocarditis virus in clinical samples by reverse transcription-PCR followed by genetic typing using sequence analysis, *J. Clin. Microbiol.*, 36, 3463, 1998.

80. Pérez, L. J. and Díaz de Arce, H., A RT-PCR assay for the detection of encephalomyocarditis virus infections in pigs, *Braz. J. Microbiol.*, 40, 988, 2009.

81. Yuan, W. et al., Development of a TaqMan-based real-time reverse transcription polymerase chain reaction assay for the detection of encephalomyocarditis virus, *J. Virol. Methods*, 207, 60, 2014.

82. Yuan, W. et al., Rapid detection of encephalomyocarditis virus by one-step reverse transcription loop-mediated isothermal amplification method, *Virus Res.*, 189, 75, 2014.

10 Porcine Teschovirus

Cristina Cano-Gómez and Miguel Ángel Jiménez-Clavero

CONTENTS

10.1 Introduction 79
    10.1.1 Classification 79
    10.1.2 Morphology, Biology, and Epidemiology 80
    10.1.3 Clinical Features and Pathogenesis 82
    10.1.4 Diagnosis 82
        10.1.4.1 Conventional Techniques 82
        10.1.4.2 Molecular Techniques 83
10.2 Methods 84
    10.2.1 Sample Collection and Preparation 84
    10.2.2 Detection Procedures 84
        10.2.2.1 Conventional RT-PCR 84
        10.2.2.2 Real-Time RT-PCR 84
        10.2.2.3 Genotyping Based on VP1 and VP2 85
10.3 Conclusion and Future Perspectives 86
References 86

10.1 INTRODUCTION

Porcine teschoviruses (PTVs, genus *Teschovirus*, family *Picornaviridae*) are nonenveloped RNA viruses that infect swine populations. Complete genome sequence studies defined at least 13 distinct PTV serotypes (from PTV-1 to PTV-13),[1–3] although all of them belong to a single virus species, known as "porcine teschovirus."

Teschoviruses are transmitted by the fecal–oral route, and different studies indicate that PTVs are abundant and ubiquitous in healthy pig populations.[4–9] In fact, PTVs are generally nonpathogenic, and infected swine most often remain asymptomatic.[10] However, some virulent variants can cause a variety of clinical conditions, the most severe of which is a nonsuppurative viral encephalomyelitis known as "Teschen–Talfan disease."

Currently, PTV is considered an emerging virus threat for swine production in China and other Far Eastern countries.[11–15] Although most Teschen–Talfan disease outbreaks are caused by strains of PTV-1, no clear link has been found between pathogenicity and serotype so far.[11,13] Therefore, it is important to determine the serotype–genotype responsible for each outbreak so that pathogenic variants, recombinants, and newly emerging recombinant genotypes are clearly recognized and properly controlled. The laboratory methods are currently available for detecting and typing the known PTVs rapidly. However, new and sensitive techniques must be continuously updated in view of teschovirus' ability to evolve constantly.

10.1.1 CLASSIFICATION

PTV, the causal agent of Teschen–Talfan disease, is classified in the genus *Teschovirus* and family *Picornaviridae*.[16] Previously grouped within the genus *Enterovirus*, PTV is reclassified as belonging to the genus *Teschovirus* on the basis of (1) cytopathic effect (CPE), (2) physicochemical properties and replication in porcine cells, and (3) serological and molecular characteristics.[17–19]

Teschen is the name of the town in the Czech Republic where the disease was first documented in 1929.[20] In the 1930s–1950s, this swine disease was recognized as a particularly virulent, highly fatal, nonsuppurative form of encephalomyelitis that spread quickly throughout Europe and caused huge losses to the pig breeding industry.[21] Initially, it had numerous descriptive names as "Klobauk disease," "benign enzootic paresis," and "contagious porcine paralysis." Due to a gradual change in virulence, these viruses were associated with milder forms of the disease, first, in the United Kingdom, where it was called Talfan disease, and in Denmark, where it was called poliomyelitis suum.[10] At present, it is known as enterovirus encephalomyelitis or Teschen–Talfan disease. The last outbreaks of serotype 1 and other serotypes have been declared in Belarus, Canada, China, the United States, Haiti, Japan, Latvia, Madagascar, Moldova, Romania, Russia, Taiwan, Uganda, and Ukraine.[8,10–13,22–26] Enterovirus encephalomyelitis has not been reported in Western Europe since 1980 (Austria) and it is now considered rare. The disease did not meet the listing criteria notifiable and has been removed from the World Organisation for Animal Health (OIE) list in 2005.

There are at least 13 distinct serotypes described: PTV-1 to PTV-13.[1-3] The last described genotype PTV-13 has been found in asymptomatic wild boar by pyrosequencing using 454 GS FLX technology.[3] The PTV-13 sequence shows great variability, especially in the VP1 protein and P1 polyprotein (66%–74% and 74%–76% amino acid identity, respectively), as compared to other PTV sequences available. PTV speciation process seems to be still in progress and new serotypes are being identified.[27] The genetic distance of PTV serotypes within each group is quite similar, suggesting that the evolution of teschoviruses proceeded in two steps. Subsequently, three subgroups evolved, which then split into 12 or more serotypes. The first monophyletic clade includes genotypes PTV-1, PTV-3, PTV-10, and PTV-11. Phylogenetic relationships among some of the different PTV serotypes, particularly PTV-5, PTV-7, and PTV-9, could not be resolved, while monophyly of PTV serotypes 2, 4, 6, 8, and 12 could be established[1,2] (Figure 10.1). The estimated mean rate of teschovirus VP1 evolution under an uncorrelated relaxed-clock model ranged from 2.03×10^3 to 2.95×10^3 nucleotide substitutions per site per year (ns/s/y) with a mean rate of 2.46×10^3. The common ancestor (tMRCA) for the genus *Teschovirus* can be placed around 500 years (385–659 years); the clade including serotypes 1, 3, 10, and 11 may have originated around 200 years (180–273 years), while the second clade (serotypes 2, 4, 6, 8, and 12) seems to have originated earlier, around 400 years (335–526 years).[2]

10.1.2 MORPHOLOGY, BIOLOGY, AND EPIDEMIOLOGY

Virions are small nonenveloped particles of 25–30 nm in diameter, with a capsid of icosahedral symmetry that is made up of 60 units (protomers), each consisting of three surface coat proteins and a fourth internal capsid protein (Figure 10.2). The genetic diversity of these protomers seems to be responsible for the idiosyncrasies such as antigenicity, receptor recognition, buoyant density, and pH stability.

The genome of about 7.0–7.2 kb in length contains one long open reading frame encoding a single polyprotein flanked by 5′- and 3′-nontranslated region (NTR), followed by a poly(A) tail. Viral translation starts at the initiator codons, more likely at nt 432 (AUG). The polyprotein ranges from 2203 to 2207 amino acids.[1] The initiation of translation in picornavirus genome is considered to be strongly related to tertiary structure of the internal ribosome entry site contained in 5′ NTR.[18] The first excision produces polypeptides P1, P2, and P3, which will be subsequently processed into 12 products: leader protease (L) of 13.6 kDa, coat protein VP4 (internal capsid protein-1A) of 7.4 kDa, three surface coat protein (VP2-1B of 31.1 kDa, VP3-1C of 26.8 kDa, and VP1-1D of 28.9 kDa), four nonstructural proteins (2A of 2.1 kDa, 2B of 16.4 kDa, 2C of 36.8 kDa, and 3A of 10.8 kDa), a genome-linked protein (VPg-3B) of 2.8 kDa, and a protease (3C) of 22.7 kDa and polymerase (3D) of 51.2 kDa.

The highly conserved L protein has a protease active site like that of foot-and-mouth disease virus. Likewise, 2A protein possesses protease activity and diverges in its structure and function compared with other nonstructural proteins. VP4 is located on the internal side of the capsid associated with the RNA core, whereas VP1, VP2, and VP3 are exposed at the virion surface being the most divergent proteins. The epitopes responsible of serotype differentiation are located mostly in exposed areas within outer capsid proteins (VP1–VP3).[28] 3D region encodes a functionally important protein in the virus life cycle, the RNA polymerase. Viral genomic RNA has a viral protein (VPg) bound at its 5′ end instead of a methylated nucleotide cap structure. The shorter 3′ NTR is highly conserved and is important in the initiation of minus-strand synthesis.

Teschoviruses are highly abundant and prevalent in domestic pigs and wild boar, the only known natural hosts. Other animal species are not known to be susceptible to infection.[7,29,30] Suckling or weaned pigs are affected more commonly than older (growing pigs). There is no evidence that PTVs are zoonotic. Most strains are nonpathogenic but, occasionally, virulent forms trigger outbreaks of severe nonsuppurative encephalomyelitis. Transmission of Teschen–Talfan disease takes place between infected and susceptible pigs through direct contact or indirectly by contaminated food and water, as well as fomites. Less virulent strains are widespread in conventional swineherds and are thought to be maintained in weaned piglets, which are infected shortly after weaning, when maternal immunity disappears and litter pigs from multiple groups are mixed.[22,31] Infection occurs by the oral or intranasal route, and the virus is shed in feces, urine, and oral secretions in large quantities. The infective dose is very small and, in addition, these viruses are especially persistent, surviving more than 5 months in the environment.[32] Fomites (boots, clothing, or vehicles) can transport the virus long distances between farms without difficulty. PTVs are resistant to heat, pH ranging between 2 and 9, chloroform, lipid solvents, and common disinfectants. These properties make their spread even easier. However, the ionizing radiation, phenol, sodium hypochlorite, 70% ethanol, and iodophor acid disinfectants are able to inactivate this virus.[10]

PTVs have been identified as potential indicators of swine fecal contamination in surface waters as these viruses are excreted abundantly by their hosts.[33,34] Moreover, coinfections of PTVs with common swine pathogens such as enterovirus G, porcine adenovirus, porcine sapelovirus (PSV), porcine circovirus type II (PCV2), porcine reproductive and respiratory syndrome virus, and classical swine fever virus are rather frequent, reflecting the multi-infection status in the field.[5-9] Cocirculation of several serotypes in a geographic area[2] and coinfection of at least two different PTV serotypes in the same animals or herds[8] are common too. The long-lasting persistence observed for PTVs may promote the emergence of viable recombinants by cellular coinfection of viral variants, representing a key factor contributing to the existing viral diversity.[11] Nevertheless, mutation seems to be one of the driving forces in the evolution of the PTVs[2] generating genetic variation.

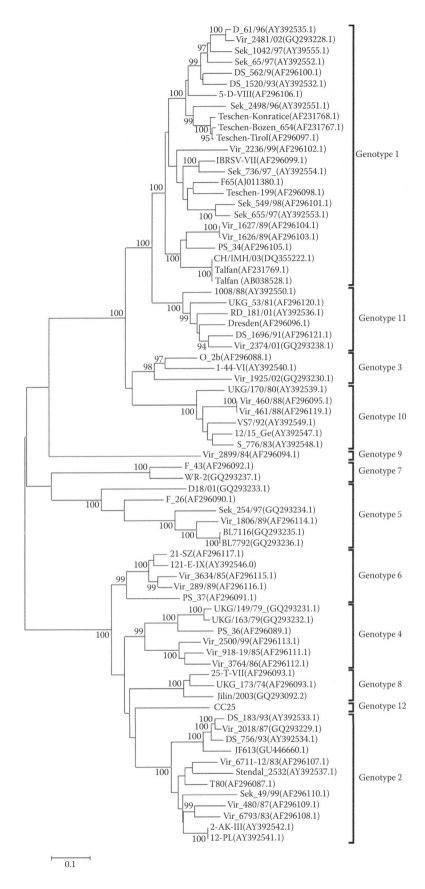

FIGURE 10.1 Phylogenetic analysis of the polyprotein porcine teschovirus (P1) using Bayesian analysis. (From Cano-Gómez, C., Epidemiological and molecular studies of porcine enteric viruses in Spain: Applications to diagnosis of Swine Vesicular Disease Virus. PhD thesis, Veterinary Medicine at the University Complutense of Madrid (UCM), Madrid, Spain 2013.)

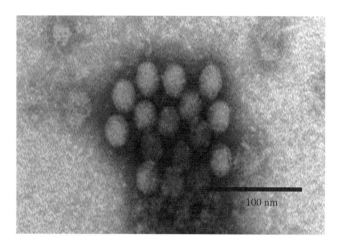

FIGURE 10.2 Electron microphotograph of isolate porcine teschovirus 1/Haiti/2009 (Genbank accession no. GQ914053). Spherical nonenveloped virus particles isolated from cultures of SK-6 cells showing cytopathic effect. (From Deng, M.Y. et al., *J. Vet. Diagn. Invest.*, 24(4), 671, 2012.)

10.1.3 Clinical Features and Pathogenesis

Although PTV infections are frequently asymptomatic, some strains have been associated with a wide variety of clinical conditions. The incubation period is variable, ranging between 5 and 40 days, but it depends on the virulence of the strain.[10,35] Mortality and morbidity are higher when animals are affected by a highly pathogenic strain (neurovirulent strains), up to 90% during the first days; however, milder strains cause variable mortality and morbidity. Currently, subclinical infections may be more common than clinically manifest diseases.

The disease is characterized by severe CNS disorder that courses with neurological symptoms. The clinical findings include fever and anorexia (not always), opisthotonos, nystagmus, locomotive disorders (hypersensitivity, tremors, clonic spasms, and ataxia), convulsions, paresis, and/or flaccid paralysis of the hindquarters that seldom progresses to total paralysis. Death occurs due to respiratory paralysis within 3–4 days after onset of nervous symptoms. Apart from neurological disease, teschoviruses may cause reproductive failure as stillbirths, mummification, embryonic death, and infertility (SMEDI syndrome),[36] respiratory disorders such as rhinitis and pneumonia,[37,38] enteral disease such as diarrhea,[39,40] myocarditis, and pericarditis,[41] and dermal lesions.[42]

Normally, the viral infection is limited to the gastrointestinal tract and associated lymphoid tissue (Peyer's patches) including the tonsils, where the virus initially replicates.[43] This is the case of piglets from immune sows, which are protected by maternal antibodies and will be replaced by own antibodies when developing active immunity. The virus can spread through the bloodstream (viremia) and reach the CNS, multiplying in the nerves in the absence of circulating antibodies. This situation can occur when the mother has not had contact with the virus, if the pig has not taken colostrum, or if the virus invades "naive" pigs for the first time, thereby affecting pigs of all ages.

No gross abnormalities are seen upon postmortem examinations. Sometimes, muscular atrophy and congestion of cerebrospinal meninges, and nasal mucosa can be perceived. Routine diagnostic pathological examination can confirm a multifocal nonsuppurative polioencephalomyelitis.[44] The degenerative pathological changes are observed in the gray matter of the diencephalon, cerebellum, medulla oblongata, and ventral horn of the spinal cord. The observed lesions are characterized by the presence of perivascular cuffing of the mononuclear cells (lymphocytes and plasma cells), meningitis, focal gliosis, satellitosis, neuronal necrosis, and neuronophagia.[10,12,24–26,45–49] These lesions, although not pathognomonic, support the diagnosis of Teschen–Talfan disease. The diagnosis can be confirmed by Immunohistochemistry (IHC) using specific antisera or monoclonal antibodies, allowing correlation of the pathological changes with the location of the agent.

Differential diagnosis includes Aujeszky's disease (pseudorabie), classical swine fever (CSFV-acute form), African swine fever, Japanese encephalitis, hemagglutinating encephalitis, edema disease (*escherichia coli* enterotoxemia), rabies, sapelovirus infection (PSV), bacterial meningoencephalitis (streptococcus suis), porcine reproductive and respiratory syndrome (highly virulent strains), salt poisoning, intoxication with lead or pesticides, toxins, and nutritional neuropathies.[10,50]

The health control measures include quarantines and movement controls, euthanasia of the animals affected, tracing of contacts, and ring vaccination. These measures should be implemented along with good husbandry practices (cleaning and disinfection) on the affected farms before restocking, in order to avoid the persistence of the virus in the environment. Humoral immune response is important in preventing PTV infection, and it has been suggested that immunosuppression derived from PCV2 coinfection may lead to clinical porcine teschoviral CNS disease. The clinical signs could be prevented by the use of inactivated or attenuated virus vaccines. Although active immunoprophylaxis was an important control measure in the past,[51,52] at present, commercially, there are neither vaccines nor treatment available.[10]

10.1.4 Diagnosis

For laboratory diagnosis of PTV infection, several methods are available. Diagnosis of the disease relies upon the observation of typical clinical signs plus histological lesions of the brain and spinal cord, together with identification of the virus in the CNS of affected pigs and detection of specific antibodies in the blood of convalescent animals. PTV infections are diagnosed by indirect immunofluorescence,[53,54] complement fixation,[55] virus-neutralization test (VNT) with standard antisera[56,57] and molecular detection and typing by reverse transcription polymerase chain reaction (RT-PCR), or a combination of RT-PCR and sequencing.[1,2,4,14,19,33,58–63]

10.1.4.1 Conventional Techniques

Generally, all PTV genotypes grow easily in cell culture. Monolayer cultures of primary porcine kidney or established cell lines derived from porcine tissue are suitable for isolation of PTV (e.g., PK-15 or IBRS-2).[64,65] Briefly, monolayers of cell at 90% of confluence growing in flask (25 cm²) are inoculated with 500–1000 μL of inoculum previously processed (essentially, homogenized clarified and sterile filtrated). Viral adsorption is allowed for 1 h and 30 min at 37°C. After that, the inoculum is removed and the infected cell monolayer is supplemented with Dulbecco's modified Eagle's medium (DMEM) with 2% of fetal bovine serum plus antibiotics (penicillin–streptomycin) to avoid bacterial contamination. Then, the infection is allowed to proceed by incubation at 37°C and 5% CO_2 until CPE develops in the cell monolayer. If the sample contains PTV, a characteristic CPE will be observed under the phase contrast microscope after 3–4 days. The CPE is characterized by small foci of rounded bright cells. After several passages, the virus grows faster and more efficiently, producing a complete CPE. Then, these infected cultures must be clarified by gentle centrifugation (600 × g for 10 min at 4°C), aliquoted, and preserved at −70°C.[66] In these conditions, the virus is stable for many years and is often referred as to virus isolate. The isolation must be confirmed by using specific antisera, monoclonal antibodies, VNT, IFA test, or RT-PCR. Once the virus is propagated in cell culture, the determination of the infectivity titer is recommended. This can be conveniently measured by infecting a cell line grown in an appropriate format (multiwall plates are often used for this purpose) with serial dilutions of the virus isolate and determining the highest dilution producing CPE in 50% of the inoculated cells. The 50% endpoint dilution, which is expressed as $TCID_{50}$/mL, can be calculated by using the statistical method described by Reed–Muench.[67] Once an isolate has been identified as PTV, piglet inoculation is the only certain means of determining that the given isolate is pathogenic.[10]

For serological diagnosis by seroneutralization, also known as VNT, the standard strains and monospecific hyperimmune sera for all PTV serotypes must be available at the laboratory. Alternatively, a method for the detection and titration of antibodies specific for PTV is enzyme-linked immunosorbent assay (ELISA).[68] It is performed in microtiter plates using PTV grown in cell cultures as antigen. Serotyping of PTV with hyperimmune sera is difficult due to the significant cross-reactivity existing between PTV serotype-specific antibodies.[14]

10.1.4.2 Molecular Techniques

RT-PCR techniques have substituted the traditional immunological tests and become the gold standard for diagnosis of PTV infections. Molecular methods for the specific detection of PTVs range from conventional (gel-based) RT-PCR,[14,58] nested RT-PCR,[19] real-time RT-PCR,[33,59,61,63] to reverse-transcription loop-mediated isothermal amplification (RT-LAMP),[62] The optimal strategy for molecular typing of PTVs should involve one or more of the surface proteins. Since neutralizing epitopes have been identified in both VP1 and VP2,[60] genotyping methods targeting partial sequences of both VP1[4] and VP2[60] have been designed. Indeed, these regions accumulate a huge divergence between genotype sequences (Figure 10.3). New approaches based on the analysis of the complete VP1 protein-coding gene or whole structural region (P1) of the viral genome have allowed molecular epidemiological studies demonstrating the genetic diversity of PTVs found in pig populations and establishing phylogenetic relationships among PTV viruses[1,2] (Figure 10.1).

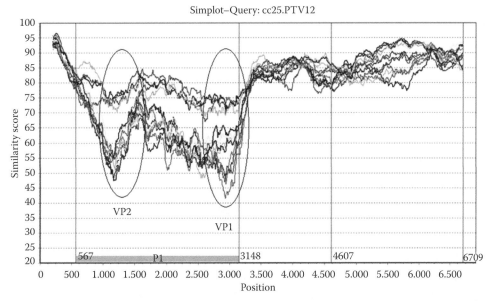

Window: 400 bp. step: 10 bp. gapstrip: on. Kimura (2-parameter). T/t: Variable

FIGURE 10.3 Nucleotide similarity of all described genotypes porcine teschovirus by Simplot program. (From Cano-Gómez, C., Epidemiological and molecular studies of porcine enteric viruses in Spain: Applications to diagnosis of Swine Vesicular Disease Virus. PhD thesis, Veterinary Medicine at the University Complutense of Madrid (UCM), Madrid, Spain 2013.)

10.2 METHODS

10.2.1 Sample Collection and Preparation

Brain and spinal cord samples should be collected from pigs slaughtered at an early clinical stage of the disease. When not processed immediately, the samples should be placed in a solution prepared with equal parts of phosphate-buffered isotonic saline solution (PBS), pH 7.4, and glycerol. Pieces of tissue are minced to prepare a 10% (w/v) suspension in PBS. The suspension is centrifuged at 800 g for 10 min and the supernatant fluid is used for inoculation of cell cultures.[10] Alternatively, the homogenization can be carried out in a mechanical homogenizer such as TissueLyser (Qiagen): for that 0.1 g of feces or tissues is diluted in 900 μL of sterile PBS 1× supplemented by penicillin–streptomycin for 4 min at 30 Hz (cycles/s) and clarified by centrifugation at $600 \times g$ for 10 min at 4°C. Also, homogenates can be sterilized by filtration using a filter with a pore diameter of 0.20 μm (Sartorius Stedim Biotech, Ministart®).[66] After that, the supernatant is ready for extraction of nucleic acids or for virus isolation.

For histological diagnosis, samples of cerebrum, cerebellum, diencephalon, medulla oblongata, and cervical and lumbar spinal cord should be collected. These tissues can be fixed in 10% neutral buffered formalin, paraffin embedded, sectioned at 4 μm, mounted on glass slides, and stained with hematoxylin and eosin following the process routinely in use for microscopic examination. For IHC, for example, dewaxed paraffin-embedded sections are subjected to a streptavidin–biotin complex peroxidase method using a Histfine Sab-PoTM Kit (Nichirei, Japan).[47] Heat denaturation is used for antigen retrieval.[45] Several primary antibodies like murine monoclonal anti-PTV1 (Talfan strain), rabbit anti-PTV1–7 (reference strains), and rabbit polyclonal antiserum against PTV-1 (VIR 2236/99 strain) have been applied.[45,47,66]

Paired serum samples are also helpful to assess seroconversions by either seroneutralization or ELISA. Also, stool samples collected from affected or nonaffected individuals can be valuable both in diagnostic strategy as well as in environmental investigations and epidemiological surveys.

Semiautomated workstation BioSprint 15 (Qiagen) can be used for obtaining viral RNA both from field samples and from experimental preparations previously processed (tissues, oral swabs, blood, feces, and cell culture isolates) following the manufacturer's instructions. The protocol may be modified by the addition of 1 μL of RNA carrier to increase the yield of purification. Likewise, viral RNA can be extracted using TRIzol reagent following the protocol described essentially by the manufacturers of TRIzol® LS Reagent Kit (Ambion). As positive control, an adequate dilution of the viral supernatants in culture medium (DMEM) to obtain a sharp band by electrophoresis and a threshold cycle (Ct) (in the case of real-time techniques) of 31 ± 2 should be prepared. Sterile deionized water (Milli-Q water) can be used as negative control.

10.2.2 Detection Procedures

10.2.2.1 Conventional RT-PCR

The following duplex RT-PCR protocol has been described by Palmquist (2002) and has been included as a detection method in OIE manual.[10] Using the same set of oligonucleotide primers allows to detect and differentiate simultaneously PTV and PSV on the basis of the size of the amplification product. The following protocol is an adaptation of this method using the commercial kit One-Step RT-PCR (Qiagen).

10.2.2.1.1 Procedure

1. Prepare the RT-PCR mix as follows: 5 μL of buffer 1× [Tris-Cl, KCl, $(NH_4)_2SO_4$, 2.5 mM $MgCl_2$], 0.2 mM dNTPs, 5 μL of Q solution 1×, 0.4 μM of each forward (1222-F: 5′-GTGGCGACAGGGTACAGAAGAG-3′) and reverse (1223-R: 5′-GGCCAGCCGCGACCC TGTCAG-3′) primers, 5 units of RNAse inhibitor (Applied Biosystems), and 1 μL of RT-PCR enzyme mix (Omniscript® and Sensiscript® Reverse Transcriptases, HotStartTaq® DNA polymerase). Then, add sterile water up to a volume of 23 μL of mix per reaction. Deliver 23 μL of the mix to each PCR tube. Tubes should be placed in a 96-well cooling block. Add 2 μL of the target RNA (sample) to the RT-PCR mix to complete the reaction final volume (25 μL). (*Note*: Vortex and spin down all microcentrifuge tubes of RT-PCR components before use and spin the tubes down briefly before putting them in thermocycler.)
2. RT and DNA amplification is carried in one step following the procedure described by the author: 30 min at 50°C, 15 min at 95°C, 40 cycles of 60 s at 94°C, 60 s at 56°C and 60 s at 72°C, and finally 10 min at 72°C.
3. Analyze the PCR products by electrophoresis through 3% agarose gel in TAE buffer at 150 V for 30 min.
4. Stain the gel with ethidium bromide or red gel and then visualize under ultraviolet light source. The presence of a 163 bp product is indicative of PTV, while the presence of a 180 bp product is indicative of PSV. Negative and positive controls are added in order to monitor the reaction procedure.

10.2.2.2 Real-Time RT-PCR

Real-time PCR methodology offers a number of advantages over traditional methods, such as its speed, sensitivity, and high throughput to analyze large numbers of samples without the need for gel electrophoresis at the end. The alignment of a conserved region (5′ noncoding region) of RNA sequences of all PTV prototype strains allowed to design a pair of oligonucleotide primers and a 5′ 6-FAM-3′ MGB labeled TaqMan probe for specific detection of all recognized PTV variants (i.e., serotypes) to date.[33,61]

This TaqMan-minor groove binder (MGB) real-time RT-PCR method is performed using QuantiTect Probe RT-PCR Kit (Qiagen), as described in the following protocol:

1. Prepare the RT-PCR mix as follows: 12.5 μL of a QuantiTect Probe RT-PCR master mix 1× (containing Tris-Cl, KCl, $(NH_4)_2SO_4$, 8 mM $MgCl_2$, dNTP mix, a reference dye (ROX) and HotStartTaq DNA polymerase), 0.4 μM of each forward (PTV-F1: 5′-CACCAGCGTGGAGTTCCTGTA-3′) and reverse (PTV-R1: 5′-AGCCGCGACCCTGTCA-3′) primers, 0.25 μM of the TaqMan-MGB probe (PTV-S1: 5′-FAM-TGCAGGACTGGACTTG-3′-MGB), and 0.25 μL of QuantiTect RT mix (Omniscript and Sensiscript Reverse Transcriptases), and add sterile water up to a volume of 15 μL of mix per reaction. Deliver 15 μL of mix to each optical tube. Tubes should be placed in a 96-well cooling block. Add 10 μL of target RNA (sample) to the RT-PCR mix to complete the reaction final volume (25 μL). (*Note*: Vortex and spin down all microcentrifuge tubes of RT-PCR components before use and spin the tubes down briefly before putting them in thermocycler.)

2. Perform RT-PCR reactions using the following conditions: 30 min at 50°C for RT, 15 min at 95°C for enzyme activation followed by 45 cycles of 15 s at 94°C, and 1 min at 60°C. Fluorescence level is read at the end of each cycle step (6-carboxyfluorescein [FAM] and ROX channels). Ct values below 38 will be considered positive.

10.2.2.3 Genotyping Based on VP1 and VP2

Molecular typing targeting the VP1 and VP2 regions has been recognized as the best choice because of good correlation with the classical serotype classification.[69] Specifically, the amplification of genomic fragments derived from the VP1- or VP2-coding regions proved to be broad in range and suitable for serotype assessment and therefore constitute a useful diagnostic tool for molecular diagnosis of PTV strains.[4,6,60,61] In the following section, two different ways of virus characterization are described: one based on the amplification and sequencing of the whole VP1-coding region and the other based on the amplification and sequencing of a fragment of the VP2-coding region.

10.2.2.3.1 Typing by VP1

Teschoviruses are molecularly typed by amplification with two degenerate primer pairs described by La Rosa et al.[4] The second pair (398-primers-ID: 1299-F 5′-GCA TCHAAYGARAAYCC-3′ and 1302-R 5′-CCAAAYC CAAARTCYTG-3′) is used in case of failure of the first one (397-primer-ID: 1298-F 5′-CCNGCNGARACAGGHTGTG-3′ and 1301-R 5′-TTCCANGTRAANGARGG-3′). Moreover, when a weak band is obtained, a second PCR reaction could

be attempted to improve the result. According to the protocol described by the authors, use 2 μL of viral RNA and 88 pmol of each primer. Thermocycling conditions for these reactions are as follows: RT at 42°C for 45 min; one cycle of template denaturation at 95°C for 2 min, primer annealing at 51°C for 1 min, and primer extension at 72°C for 1 min; and 35 cycles at 95°C for 1 min, 54°C for 1 min, and 72°C for 1 min, followed by one cycle of elongation for 10 min at 72°C. Bands of expected sizes of 431 bp (397-primer-ID) or 533 bp (398-primer-ID) are observed after electrophoresis on 2% agarose gel. Amplified cDNAs can be purified using High Pure PCR product Purification Kit (Roche), QIAquick Gel Extraction Kit (Qiagen), or ExoSAP-IT Kit (GE Healthcare) and then sequenced using the same primer sets as for RT-PCR assays. Sequences are edited using suitable assembler software such as SeqMan program (DNAstar, Lasergene) or Seqscape (Applied Biosystems). The consensus sequences of VP1 obtained are subjected to homology analysis using basic local alignment search tool (BLAST), searching for similarity as a first approach to PTV identification. Finally, multiple alignments can be carried out using Clustal or Muscle methods and phylogenetic analysis can be conducted by MEGA 5.05 or MrBayes 3.2à MEGA 5.05 or MrBayes 3.2 software programs in order to establish more accurate phylogenetic relationships between strains.

10.2.2.3.2 Typing by VP2

Kaku et al.[60] described a dominant antigenic site in the VP2 region of PTVs and designed a pair of degenerated primers to amplify about 320 bp by comparing the VP2 sequences of PTV prototype strains and field isolates. The single-step RT-PCR kit mentioned earlier (One-Step RT-PCR kit, Qiagen) is also used here for RT and cDNA amplification.

10.2.2.3.2.1 Procedure

1. Prepare the RT-PCR mix in the following way: 5 μL of buffer 1× [Tris-Cl, KCl, $(NH_4)_2SO_4$, 2.5 mM $MgCl_2$], 0.2 mM dNTPs, 5 μL of Q solution 1×, 0.6 μM of each forward (VP2-337F-puff: 5′-CACCARYTGCTTAARTGYKGTTGG-3′) and reverse (VP2-654R-puff: 5′-CACAGGGGTTGCT GAAGARTTTGT-3′) primers, 5 units of RNAse inhibitor (Applied Biosystems), and 1 μL of RT-PCR enzyme mix (Omniscript and Sensiscript Reverse Transcriptases, HotStartTaq DNA polymerase). Then, add sterile water up to a volume of 23 μL of mix per reaction. Add 2 μL of target RNA (sample) to the RT-PCR mix to complete the reaction final volume (25 μL).

2. Carry out RT and cDNA amplification in one step with the settings described by the author: 30 min at 50°C, 15 min at 95°C, 40 cycles of 30 s at 94°C, 60 s at 51°C and 60 s at 72°C, and finally 10 min at 72°C.

3. Analyze the PCR products by electrophoresis through 2% agarose gel in TAE buffer at 150 V for 30 min. Stain the gel with ethidium bromide or

red gel, and then visualize under ultraviolet light source. Amplicons of the expected size (~320 bp) are sequenced and analyzed in the same way as described for VP1.

10.3 CONCLUSION AND FUTURE PERSPECTIVES

PTVs are quite diverse, with some being pathogenic for domestic and wild suids, while certainly, most others are seemingly harmless. In addition, they are extremely ubiquitous, an endemic situation that is widely recognized, and cocirculation of different virus variants and even coinfections are frequently observed in swine populations. Indeed, the high potential of ssRNA viruses to undergo genetic changes through mutation and/or recombination mechanisms and their high replication rates is playing a role by continuously shaping their infectious behavior in their hosts.

From a diagnostic perspective, it is important to note that laboratory techniques are available for satisfactory detection and typing of the known range of PTVs. However, it is not unreasonable to assume that this range will be enlarged in the coming years, as more refined and sensitive techniques are being applied. For this reason, detection and typing techniques need to constantly take into account of all new variants of PTVs found and keep updated to cover its full range.

Similarly, sequence analysis and phylogenetic relation are powerful tools that play a critical role in the epidemiological investigation of PTVs. They facilitate comparisons between our strains and other strains circulating elsewhere so that the origin and possible pathogenicity of newly identified PTVs might be inferred. Nevertheless, current laboratory analyses are unable to assess the pathogenicity of a given PTV, without performing specific animal studies. Methods to correlate pathogenicity and possible genetic determinants of virulence are still lacking, and they are strongly needed in order to differentiate high and low pathogenicity PTV strains.

REFERENCES

1. Zëll, R., Dauber, M., Krumbholz, A. et al., Porcine teschoviruses comprise at least eleven distinct serotypes: Molecular and evolutionary aspects. *J Virol*, 75: 1620–1631, 2001.
2. Cano-Gómez, C., Palero, F., Buitrago, D. et al., Analyzing the genetic diversity of teschoviruses in Spanish pig populations using complete VP1 sequences. *Infect Genet Evol*, 11(8): 2144–2150, 2011.
3. Boros, A., Nemes, C., Pankovics, P. et al., Porcine teschovirus in wild boars in Hungary. *Arch Virol*, 157(8): 1573–1578, 2012.
4. La Rosa, G., Muscillo, M., Di Grazia, A. et al., Validation of rt-PCR assays for molecular characterization of porcine teschoviruses and enteroviruses. *J Vet Med B Infect Dis Vet Public Health*, 53: 257–265, 2006.
5. Buitrago, D., Cano-Gomez, C., Agüero, M. et al., A survey of porcine picornaviruses and adenoviruses in fecal samples in Spain. *J Vet Diagn Invest*, 22: 763–766, 2010.
6. Sozzi, E., Barbieri, I., Lavazza, A. et al., Molecular characterization and phylogenetic analysis of VP1 of porcine enteric picornaviruses isolates in Italy. *Transbound Emerg Dis*, 57: 434–442, 2010.
7. Prodelalova, J., The survey of porcine teschoviruses, sapeloviruses and enteroviruses B infecting domestic pigs and wild boars in the Czech Republic between 2005 and 2011. *Infect Genet Evol*, 12: 1447–1451, 2012.
8. Chiu, S., Hu, S., Chang, C. et al., The role of porcine teschovirus in causing diseases in endemically infected pigs. *Vet Microbiol*, 161(1–2): 88–95, 2012.
9. Donin, D.G., de Arruda Leme, R., Alfieri, A.F. et al., First report of Porcine teschovirus (PTV), Porcine sapelovirus (PSV) and Enterovirus G (EV-G) in pig herds of Brazil. *Trop Anim Health Prod*, 46(3): 523–528, 2014.
10. OIE (Office International des Epizooties), *Manual of Diagnostic Tests & Vaccines for Terrestrial Animals*. Chapter 2.8.10: Teschovirus encephalomyelitis (previously enterovirus encephalomyelitis or Teschen/Talfan disease). World Organisation for Animal Health, Paris, France, pp. 1146–1152, 2008.
11. Wang, B., Tian, Z.J., Gong, D.Q. et al., Isolation of serotype 2 porcine teschovirus in China: Evidence of natural recombination. *Vet Microbiol*, 146: 138–143, 2010.
12. Yamada, M., Nakamura, K., Kaku, Y. et al., Enterovirus encephalomyelitis in pigs in Japan caused by porcine teschovirus. *Vet Rec*, 155: 304–306, 2004.
13. Zhang, C.F., Cui, S.J., Hu, S. et al., Isolation and characterization of the first Chinese strain of porcine Teschovirus-8. *J Virol Methods*, 167(2): 208–213, 2010.
14. Liu, S., Zhao, Y., Hu, Q. et al., A multiplex RT-PCR for rapid and simultaneous detection of porcine teschovirus, classical swine fever virus and porcine reproductive and respiratory syndrome virus in clinical specimens. *J Virol Methods*, 172(1–2): 88–92, 2010
15. Lin, W., Cui, S., and Zell, R., Phylogeny and evolution of porcine teschovirus 8 isolated from pigs in China with reproductive failure. *Arch Virol*, 157: 1387–1391, 2012.
16. King, A.M.Q., Adams, M.J., Carstens, E.B., and Lefkowitz, E.J. (eds.), *Ninth Report of the International Committee on Taxonomy of Viruses. Virus Taxonomy: Classification and Nomenclature of Viruses*. Elsevier Academic Press, San Diego, CA, 2012.
17. Knowles, N.J., Buckley, L.S., and Pereira, H.G., Classification of porcine enteroviruses by antigenic analysis and cytopathic effects in tissue culture: Description of three new serotypes. *Arch Virol*, 62: 201–208, 1979.
18. Kaku, Y., Sarai, A., and Murakam, Y., Genetic reclassification of porcine enteroviruses. *J Gen Virol*, 82: 417–424, 2001.
19. Zëll, R., Krumbholz, A., Henke, A. et al., Detection of porcine enteroviruses by nRT-PCR: Differentiation of CPE groups I–III with specific primer sets. *J Virol Methods*, 88(2): 205–218, 2000.
20. Trefny, L., Massive illness of swine in Teschen area. *Zveroleki Obzori*, 23: 235–236, 1930
21. Klobouk, A., Encephalomyelitis enzootica suum (Teschen disease) spread outside of Teschen district. *Zvěrolékařské Rozpravy*, X: 146–149, 1936.
22. Pogranichniy, R.M., Janke, B.H., Gillespie, T.G. et al., A prolonged outbreak of polioencephalomyelitis due to infection with a group I porcine enterovirus. *J Vet Diagn Invest*, 15(2): 191–194, 2003.
23. Kouba, V., Teschen disease (Teschovirus encephalomyelitis) eradication in Czechoslovakia: A historical report. *Vet Med Czech*, 54: 550–560, 2009.

24. Bangari, D.S., Pogranichniy, R.M., Gillespie, T. et al., Genotyping of Porcine teschovirus from nervous tissue of pigs with and without polioencephalomyelitis in Indiana. *J Vet Diagn Invest*, 22(4): 594–597, 2010.

25. Salles, M.W., Scholes, S.F., Dauber, M. et al., Porcine teschovirus polioencephalomyelitis in western Canada. *J Vet Diagn Invest*, 23(2): 367–373, 2011.

26. Deng, M.Y., Millien, M., Jacques-Simon, R. et al., Diagnosis of Porcine teschovirus encephalomyelitis in the Republic of Haiti. *J Vet Diagn Invest*, 24(4): 671–678, 2012.

27. Sun, H., Gao, H., Chen, M. et al., New serotypes of porcine teschovirus identified in Shanghai, China. *Arch Virol*, 160(3):831-5, 2015.

28. Usherwood, E.J. and Nash, A.A., Lymphocyte recognition of picornaviruses. *J Gen Virol*, 76(Pt 3): 499–508, 1995.

29. Boros, A., Pankovics, P., and Reuter, G., Characterization of a novel porcine enterovirus in wild boars in Hungary. *Arch Virol*, 157: 981–986, 2012.

30. Cano-Gómez, C., García-Casado, M.A., Soriguer, R. et al., Teschoviruses and sapeloviruses in faecal samples from wild boar in Spain. *Vet Microbiol*, 165(1–2): 115–122, 2013.

31. Knowles, N.J., Porcine enteric picornaviruses. In: Straw, B.E., Zimmerman, J.J., D'Allaire, S., and Taylor, D.J. (eds.), *Diseases of Swine*, 9th edn. Blackwell, Oxford, U.K., pp 337–354, 2006.

32. Mahnel, H., Ottis, K., and Herlyn, M., Stability in drinking and surface water on nine virus species from different genera. *Zbl Bakt Hyg Abt OrigB*, 164: 64–84, 1977.

33. Jiménez-Clavero, M.A., Fernandez, C., Ortiz, J.A. et al., Teschoviruses as indicators of porcine fecal contamination of surface water. *Appl Environ Microbiol*, 69, 6311–6315, 2003.

34. Jones, T.H., Muehlhauser, V., and Thériault, G., Comparison of ZetaPlus 60S and nitrocellulose membrane filters for the simultaneous concentration of F-RNA coliphages, porcine teschovirus and porcine adenovirus from river water. *J Virol Methods*, 206: 5–11, 2014.

35. CFSPH, Teschovirus encephalomyelitis and porcine teschovirus infection, 2009. http://www.cfsph.iastate.edu./Factsheets/pdfs/enterovirus_encephalomyelitis.pdf (accessed on March 31, 2015).

36. Dunne, H.W., Gobble, J.L., Hokanson, J.F. et al., Porcine reproductive failure associated with a newly identified 'SMEDI' group of picornavirus. *Am J Vet Res*, 26: 1284–1297, 1965.

37. Meyer, R.C., Woods, G.T., and Simon, J., Pneumonitis in an enterovirus infection in swine. *J Comp Pathol*, 76: 397–405, 1966.

38. Liebke, H. and Schlenstedt, D., Eine Enterovirus (ECSO)-Infektion bei Schweinen mit nervösen Störungen und einer gleichzeitig vorhandenen Rhinitis. *Tierärztl Umscha*, 26(287–291): 324–330, 1971.

39. Izawa, H., Bankowski, R.A., and Howarth, J.A., Porcine enteroviruses. I. Properties of three isolates from swine with diarrhea and one from apparently normal swine. *Am J Vet Res*, 23: 1131–1141, 1962.

40. Edington, N., Christofini, G.J., and Betts, A.O., Pathogenicity of Talfan and Konratice strains of Teschen virus in gnotobiotic pigs. *J Comp Pathol*, 82: 393–399, 1972.

41. Long, J.F., Koestner, A., and Kasza, L., Infectivity of three porcine polioencephalomyelitis viruses for germfree and pathogen-free pigs. *Am J Vet Res*, 27: 274–279, 1966.

42. Knowles, N.J., The association of group III porcine enteroviruses with epithelial tissue. *Vet Rec*, 122: 441–442, 1988.

43. Long, J.F., Pathogenesis of porcine polioencephalomyelitis. In: Olsen, R.A., Krakowka, S., and Blakeslee, J.R. (eds.), *Comparative Pathology of Viral Diseases*, Vol. I. CRC Press, Boca Raton, FL, pp. 179–197, 1985.

44. Manuelidis, E.E., Sprinz, H., and Horstmann, D.M., Pathology of Teschen disease (virus encephalomyelitis of swine). *Am J Pathol*, 30: 567–597, 1954.

45. Yamada, M., Kaku, Y., Nakamura, K. et al., Immunohistochemical detection of porcine teschovirus antigen in the formalin-fixed paraffin-embedded specimens from pigs experimentally infected with porcine teschovirus. *J Vet Med A*, 54: 571–574, 2007.

46. Yamada, M., Kozakura, R., Kaku, Y. et al., Immunohistochemical distribution of viral antigens in pigs naturally infected with porcine teschovirus. *J Vet Med Sci*, 70: 305–308, 2008.

47. Yamada, M., Kozakura, R., Nakamura, K. et al., Pathological changes in pigs experimentally infected with porcine teschovirus. *J Comp Pathol*, 141(4): 223–228, 2009.

48. Yamada, M., Miyazaki, A., Yamamoto, Y. et al., Experimental teschovirus encephalomyelitis in gnotobiotic pigs. *J Comp Pathol*, 150(2–3): 276–286, 2014.

49. Takahashi, M., Seimiya, Y.M., Seki, Y. et al., A piglet with concurrent polioencephalomyelitis due to porcine teschovirus and postweaning multisystemic wasting syndrome. *J Vet Med Sci*, 70: 497–500, 2008.

50. Merck and Co., Inc., *Merk Veterinary Manual*, 5th edn. Whitehouse Station, NJ, 2000.

51. Klobouk, A., Active immunization against Teschen disease. *Zvěrolékařské Rozpravy, IX*, 14: 217–218, 1935.

52. Traub, E., Active immunization against Teschen disease using vaccines adsorbed on aluminium hydroxide. *Arch Tierheilkd*. 77: 52–66, 1942.

53. Romanenko, V.F., Pruss, O.G., Belyi, Y.U.A. et al., Immunofluorescent diagnosis of porcine encephalomyelitis. *Veterinariia*, 4: 69–72, 1982.

54. Auerbach, J., Prager, D., Neuhaus, S. et al., Grouping of porcine enteroviruses by indirect immunofluorescence and description of new serotypes. *Zbl Vet Med B*, 41: 277–282, 1994.

55. Knowles N.J. and Buckley, L.S., Differentiation of porcine enterovirus serotypes by complement fixation. *Res Vet Sci*, 29, 113–115, 1980.

56. Betts, A.O., Studies on enteroviruses of the pig. VI. The relationship of the T 80 strain of a swine polioencephalomyelitis virus to some other viruses as shown by neutralization tests in tissue cultures. *Res Vet Sci*, 1, 296–300, 1960.

57. Minor, P.D., Brown, F., Domingo, E. et al., Family picornaviridae. In: Murphey, F.A., Fauquet, C.M., Bishop, H.D.L., Ghabrial, S.H., Jarvis, A.W., Martelli, G.P., Mayo, M.A., and Summers, M.D. (eds.), *Virus Taxonomy: Classification and Nomenclature of Viruses*: *Sixth Report of the International Committee on Taxonomy of Viruses*. Elsevier, San Diego, CA; *Arch Virol*, 10: 329–336, 1995.

58. Palmquist, J.M., Munir, S., Taku, A. et al., Detection of porcine teschovirus and enterovirus type II by reverse transcription-polymerase chain reaction. *J Vet Diagn Invest*, 14(6): 476–480, 2002.

59. Krumbholz, A., Wurm, R., Scheck, O. et al., Detection of porcine teschoviruses and enteroviruses by LightCycler real-time PCR. *J Virol Methods*, 113: 51–63, 2003.

60. Kaku, Y., Murakami, Y., Sarai, A. et al., Antigenic properties of porcine teschovirus 1 (PTV-1) Talfan strain and molecular strategy for serotyping of PTVs. *Arch Virol*, 152(5): 929–940, 2007.

61. Cano-Gómez, C., Buitrago, D., Fernández-Pinero, J. et al., Evaluation of a fluorogenic real-time reverse transcription-polymerase chain reaction method for the specific detection of all known serotypes of porcine teschoviruses. *J Virol Methods*, 176(1–2): 131–134, 2011a.

62. Wang, B., Wang, Y., Tian, Z.J. et al., Development of a reverse transcription loop-mediated isothermal amplification assay for detection of Porcine teschovirus. *J Vet Diagn Invest*, 23(3): 516–518, 2011.

63. Zhang, C., Wang, Z., Hu, F. et al., The survey of porcine teschoviruses in field samples in China with a universal rapid probe real-time RT-PCR assay. *Trop Anim Health Prod*, 45(4): 1057–1061, 2013.

64. Madr, V., Propagation of the Teschen disease virus in cell cultures. *Veterinarstvi*, IX: 298–301, 1959.

65. Mayr, A. and Schwoebel, W., Propagation of the Teschen disease virus in porcine kidney cell cultures and properties of the cultured virus. 1.2.3. Part. *Zentralbl. Bakteriol*, 168: 329–359, 1957.

66. Cano-Gómez, C., Epidemiological and molecular studies of porcine enteric viruses in Spain: Applications to diagnosis of Swine Vesicular Disease Virus. PhD thesis, Veterinary Medicine at the University Complutense of Madrid (UCM), Madrid, Spain, 2013.

67. Reed, L.J. and Muench, H., A simple method of estimating fifty percent endpoints. *Am J Hyg*, 27: 493–497, 1938.

68. Hubschle, O.J.B., Rajoanarison, J., Koko, M. et al., ELISA for detection of Teschen virus antibodies in swine serum samples. *Dtsch Tierarztl. Wochenschr*, 90: 86–88, 1983.

69. Oberste, M.S., Nix, W.A., Maher, K. et al., Improved molecular identification of enteroviruses by RT-PCR and amplicon sequencing. *J Clin Virol*, 26, 375–377, 2003.

11 Swine Vesicular Disease Virus

Jovita Fernández-Pinero, Giulia Pezzoni, Cristina Cano-Gómez,
Paloma Fernández-Pacheco, Emiliana Brocchi, and Miguel Ángel Jiménez-Clavero

CONTENTS

11.1 Introduction 89
    11.1.1 Classification... 90
    11.1.2 Morphology and Biology... 90
        11.1.2.1 Virus: Structural and Genome Organization... 90
        11.1.2.2 Antigenic Structure... 90
        11.1.2.3 Virus–Cell Interactions... 91
        11.1.2.4 Genetic and Antigenic Variation 92
    11.1.3 Epidemiology... 92
        11.1.3.1 Geographic Range of Disease Occurrence 92
        11.1.3.2 Host Range... 92
        11.1.3.3 Disease Transmission and Virus Inactivation... 92
    11.1.4 Clinical Features, Pathogenesis, and Immunity 93
    11.1.5 Diagnosis 94
        11.1.5.1 Conventional Techniques 94
        11.1.5.2 Molecular Techniques 95
        11.1.5.3 On-Site Devices 95
11.2 Methods 96
    11.2.1 Sample Collection and Preparation 96
        11.2.1.1 Sample Collection.. 96
        11.2.1.2 Viral RNA Extraction.. 96
    11.2.2 Detection Procedures... 96
        11.2.2.1 Conventional RT-PCR Amplifying a 3D Gene Fragment 96
        11.2.2.2 Real-Time RT-PCR Amplifying a 5′ NCR Fragment... 97
        11.2.2.3 Molecular Characterization of SVDV Isolates 97
11.3 Conclusion and Future Perspectives 98
References.. 98

11.1 INTRODUCTION

Swine vesicular disease virus (SVDV) belongs to the *Enterovirus* genus within the family *Picornaviridae*. It causes an infectious and contagious disease in pigs called swine vesicular disease (SVD), characterized by the appearance of vesicles on the coronary bands of the feet, heels, skin, snout, tongue, lips, and teats, accompanied by fever. Its importance for animal health authorities relies in the fact that these signs are indistinguishable from those caused by other vesicular diseases of viral etiology affecting swine, including foot-and-mouth disease (FMD), vesicular stomatitis (VS), and vesicular exanthema of swine (VES), and therefore, differential diagnosis in the laboratory is essential to discern between them.

SVDV is closely related to coxsackievirus B5 (CV-B5),[1] a human virus, from which it is thought to derive from adaptation to pigs, possibly through changes in the receptor usage.[2] Although originally considered as zoonotic,[3] public health risk is negligible as its infection in humans has not been reported recently, and the late isolates of the virus have lost the ability to infect human cells,[2] thus decreasing its original zoonotic potential. So far, the disease, which was first described in Italy in 1966,[4] has only been reported in farmed pig populations in Europe and Far East Asia. In Europe, since 1994 and after a period of relatively high activity and broader geographic distribution, the disease was effectively cleared from most countries, except Italy, where SVD has been persistently reported causing occasional foci, and for this reason, intense surveillance and eradication plans are in place.[5] Remarkably, however, in the last 20 years, the disease has evolved to a milder form, usually asymptomatic. In Far East Asia, the disease is also likely to be present. The last reported case from this region occurred in Taiwan in 2000.[5]

SVD is highly contagious and spreads rapidly by direct contact with infected animals and by environmental contamination. This ability to spread, together with its clinical resemblance to FMD, made SVD one of the diseases listed by the World Organization for Animal Health (OIE) in the *OIE Terrestrial Animal Health Code*.[6] OIE-listed diseases are transmissible diseases that have the potential for very rapid spread, irrespective of national borders (often referred to as "transboundary diseases"), and are of obligatory declaration to the OIE due to their serious socioeconomic or public health consequences and impact in the international trade of animals and animal products. As the importance of SVD centers on its similarity to FMD, all biosecurity measures regarding SVDV handling, including diagnosis involving biologically active virus, must follow those in force for foot-and-mouth disease virus (FMDV) handling, that is, OIE containment group 4 (and equivalent biocontainment levels like BSL-3 and P3). This group includes "organisms that cause severe human or animal disease, may represent a high risk of spread in the community or animal population and for which there is usually no effective prophylaxis or treatment."[6] However, SVD is not a severe disease by itself, and as current diagnostic methods have evolved, the capability of differentiating SVD from FMD has improved, so the inclusion of this disease in the OIE list could be reevaluated in the future.

As SVD does not cause severe production losses, improved biosecurity plays a major role in the control and prophylaxis of the disease. The main sanitary measures in place are stamping out, restriction of pig movements (protection and surveillance zones), cleaning and disinfection, and restrictions on swill feeding and on importation of pig products from SVD-affected regions.[7] The often asymptomatic course of the infection presents a significant risk to livestock importation from countries where SVD status is not regularly assessed.

Vaccines against SVD are not available commercially, although experimental studies have been conducted to develop vaccines using different approaches.[8–12] In agreement with the status of SVD as an OIE-listed disease, where control is mainly achieved by stamping out and other sanitary measures, vaccination is disregarded, due to the general concern about the presence of specific SVDV antibodies in pigs, which would be raised after vaccination, and its generally mild course, discouraging investment in vaccine developments. Any attempt to develop a vaccine useful in the control of SVD has to consider the need for efficient methods to discriminate vaccinated from infected animals (differentiation of infected from vaccinated animals [DIVA] tests) accompanying the vaccine.

There is currently no treatment for SVD, although antiviral activity of pocket factor analogues in infections caused by enteroviruses (see Section 11.1.2.2) might open a valuable control strategy. Nonetheless, the low impact of this disease for swine health discourages investment in this type of approaches.

11.1.1 CLASSIFICATION

Taxonomically, SVDV (*Enterovirus* genus, *Picornaviridae* family) is considered a subspecies within the species CV-B5. Nucleotide sequence identity between both viruses is 75%–85%.[13] From their antigenic cross-reactivity, it was proposed as early as 1973 that SVDV arose in pigs by interspecies transmission of CV-B5 from humans.[14,15] Later, the analysis of nucleotide sequences of 45 isolates from outbreaks of Europe and Asia between 1966 and 1993 allowed estimating the date of the divergence of SVDV from CV-B5 between 1945 and 1965.[16]

11.1.2 MORPHOLOGY AND BIOLOGY

11.1.2.1 Virus: Structural and Genome Organization

The virion is a nonenveloped particle of 25–30 nm in diameter, with pseudospherical capsid of icosahedral T = 1 symmetry, which is composed of 60 protomers, each made up of one copy of each of the four structural proteins: VP1, VP2, VP3, and VP4. Underneath the capsid is a single-stranded positive sense RNA genome of approximately 7400 nucleotides,[17–19] whose organization is essentially identical to that of poliovirus 1, the type virus of the *Picornaviridae* family (Figure 11.1). Flanking a unique open reading frame encoding a precursor polyprotein of 2185 amino acid residues, there are two non-translated regions, one at the 5′ end encompassing a highly structured region, the internal ribosome entry site, where the ribosome assembles to initiate the translation of the polyprotein, and one at the 3′ end also with a highly structured segment probably involved in RNA replication, followed by a chain of polyadenines (poly-A) of variable length. The precursor polyprotein of SVDV is processed in the cytoplasm of the infected cell to yield 11 functional viral polypeptides. The first excision produces polypeptides P1, P2, and P3. After further processing, P1 gives rise to the structural proteins VP1–VP4, whereas P2 and P3 yield the nonstructural proteins (Figure 11.1). In the mature virion, proteins VP1, VP2, and VP3 (33, 32, and 29 kDa), respectively,[20] are exposed on the surface, forming a compact proteic shell. These three capsid proteins share a tertiary structure, highly conserved among picornaviruses, that consists of an eight stranded β-barrel constituting its central hydrophobic core. Conversely, VP4 (about 7.5 kDa) is facing the inner side of the capsid, interacting with the viral RNA molecule. Among the nonstructural polypeptides, there are two viral proteases (2A and 3C), one RNA polymerase (3D), the VPg, which is covalently bound to the 5′ end of the RNA molecule and participates in its replication, and other proteins whose functions are not well determined (2B, 2C, 3A, 3B).

11.1.2.2 Antigenic Structure

SVDV displays two types of antigenic sites: (1) those defined by monoclonal antibody-resistant mutant analyses, grouping in up to seven different neutralizing epitopes in the surface of the capsid,[21–23] and (2) those defined by *pepscan* analysis, essentially comprising internal epitopes not involved in neutralization,[23,24] with one exception: one antigenic site at the N-terminal region of the VP1 protein, which is internal, but emerges to the surface upon virus interaction with the cell receptor and is involved in neutralization.[25] The 3D structure of SVDV capsid has been solved,[26,27] and this fact has enabled the fine mapping of the antigenic sites. A singular feature is the existence of a sphingosine molecule on each VP1 subunit,

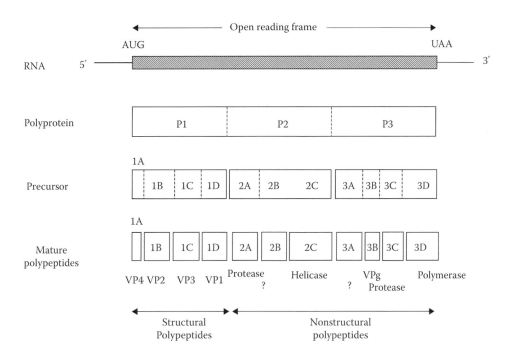

FIGURE 11.1 Genomic organization of SVDV and the posttranslational processing of polyprotein products. Discontinuous lines indicate the sites of proteolytic cleavage.

embedded in an internal cavity known as the "hydrophobic pocket" present in all enteroviruses studied, which contains lipids derived from cells known as "pocket factors," whose role is stabilizing the virus particle at the extracellular level.[28] Upon interaction with the receptor at the cell surface (see Section 11.1.2.3), these pocket factors are expelled, leading to unstable particles that, once internalized into the cell, undergo deep conformational changes, initiating virus uncoating and RNA release into the cytoplasm. Pocket factor analogues have been developed for use as antivirals,[28] but to date, no use in animal health is foresight for these compounds, at least in the near future.

11.1.2.3 Virus–Cell Interactions

All viruses belonging to the coxsackievirus B group, including SVDV, use a common surface molecule known as coxsackievirus–adenovirus receptor (CAR) as host cell receptor.[2,29] CAR is a 46 kDa transmembrane glycoprotein with two extracellular immunoglobulin-like domains, which also functions as an attachment molecule for adenovirus fiber proteins.[30] On the other hand, another surface molecule, DAF (decay accelerating factor, CD55), a 70 kDa protein involved in the control of homologous lysis by complement, has been postulated to act as a coreceptor used by some coxsackie B virus strains.[31–33] Evidence has been reported supporting also this role of DAF for SVDV but, interestingly, only earlier strains (It'66 and UK'72, isolated in Italy in 1966 and in the United Kingdom in 1972, respectively) are able to interact with DAF of human origin. By opposite, more recent strains (R1072, R1120, and SPA'93, isolated in Italy in the early 1990s and in Spain in 1993, respectively), have lost their ability to interact with human DAF and have not replaced this interaction with

the, otherwise different, porcine homologous version of the human DAF molecule.[2] This finding suggests that in its adaptation to swine from humans, SVDV has lost certain capacity to interact with human cells. In fact, recent SVDV isolates, though able to interact with CAR, have lost the capacity of infecting human HeLa cells, still present in earlier isolates.[2] This phenotypic change has been suggested to involve a lower tropism for human cells and, consequently, a reduced zoonotic potential for SVDV.[34]

In addition to binding to cell surface glycoproteins, the interaction of many viruses with the cell surface is mediated in many instances by sulfated glycosaminoglycans, particularly heparan sulfate proteoglycans.[35] Among the picornaviruses, FMDV-heparin/heparan sulfate interaction has been attributed to an adaptation to cell cultures.[36] In SVDV, interaction with heparan sulfate glycosaminoglycan has been shown to mediate the attachment to the host cell, indicating that this molecule could act as receptor or coreceptor for this virus.[37] The importance of this interaction in the process of cell attachment and entry remains to be determined.

The mechanism of cell entry of SVDV is not fully characterized. Similar to other picornaviruses, the capsid of SVDV undergoes a series of conformational changes upon interaction with its receptor, leading to a structurally and antigenically altered particle named "A particle." In this process, the N-terminus of VP1 becomes externalized and VP4 is lost. As noted earlier, this process uncovers a cryptic epitope that is involved in neutralization.[25] Those "A particles" are the most abundant form of the intracellular virus in the early stages of infection and seem to be a necessary intermediate preceding the uncoating and release of RNA into the cytoplasm in picornaviruses.

11.1.2.4 Genetic and Antigenic Variation

SVDV is considered a relatively recently emerged virus variant of CV-B5, which adapted to pigs,[1,13,16] providing a good example of the concept that adaptations and species barrier crossing can go either to or from humans.[38] Serologically, all isolates belong to a single serotype. However, in its short history, the virus has evolved into at least four recognized antigenic variants that can be distinguished by different MAb, correlating with four genetically distinct groups: group 1 comprises the first isolate (It'66); group 2 includes European and Asian SVD viruses isolated in the 1970s and early 1980s; group 3 is constituted by Italian isolates of the 1980s and 1990s; and group 4, isolates circulating in the late 1980s and early 1990s in a broad territory of Europe.[39] As mentioned earlier, in recent years, SVD has been almost exclusively reported in Italy, where an intensive surveillance program is in place. As a result of this surveillance, a number of SVDV isolates have been studied, all of which belong to antigenic and genetic group 4,[5] while no representative of antigenic/genetic groups 1–3 has been isolated in the last 20 years, a fact that might indicate a possible extinction of these three groups.

11.1.3 Epidemiology

11.1.3.1 Geographic Range of Disease Occurrence

The first occurrence of SVD was reported in Lombardy, Italy, in 1966, where it was clinically recognized as FMD, but physical and chemical analysis showed that the causative agent was an enterovirus originally termed "porcine enterovirus,"[4] though eventually it was renamed as "swine vesicular disease virus." The next outbreak was reported in Hong Kong in 1971, in an FMD vaccine trial.[40] Since then, SVD has been reported in different European and Asian countries (Table 11.1). More recently, besides Italy, where SVD reoccurs every year, outbreaks of SVD have occurred in Portugal in 2003–2004 (Leiria district) and in 2007 (Beja district).[41] Currently, the European Union countries have been declared disease-free, except Southern Italy, where SVD is endemic despite eradication campaigns in place in this country since 1995.[5,42,43] Africa, North and South America, and Oceania have remained free of the disease, but in various Asian countries, the disease is likely to be present,[42,44] the last case of SVD from Far East Asia being reported from Taiwan in 2000.[42]

11.1.3.2 Host Range

SVDV is considered to be highly contagious for pigs of all ages with variable morbidity, which can be as high as 80%–100% in individual pens, although it depends on a wide range of factors, including virulence of the viral strain involved, type of farm, and time elapsed between infection and detection, among others. Mortality is negligible. The susceptibility to natural infections has only been reported in swine (either wild boar or domestic pigs, including Euro-Asian pigs or American one-toed pigs).[45] Infections in nonswine species have been reported. Relatively high virus titers have been detected in the pharynx of sheep (but not in cattle) kept in

TABLE 11.1
Occurrence of SVD around the World

Country	Years of Report	Year of Most Recent Report
Europe		
Austria	1972–1976, 1978	1979
Belgium	1973, 1979, 1992	1993
Bulgaria	1971	1971
France	1973–1975, 1982	1983
Germany	1973–1977, 1979–1982	1985
Greece	1979	1979
Holland	1975, 1992	1994
Italy	1966, 1972–1984, 1988–1989, 1991–2010	2011
Malta	1975	1978
Poland	1972	1973
Portugal	1995, 2003–2004	2007
Romania	1972–1973	1987
Russia	1972	1976
Spain	1993	1993
Switzerland	1973	1975
The Netherlands	1975	1994
Ukraine	1972	1977
United Kingdom	1972–1977, 1979–1981	1982
Asia		
China	1971–1977, 1979–1981, 1984–1985, 1887–1989	1991
Japan	1973	1975
Macau	1989	1989
Taiwan	1997–1998	2000

Sources: Mowat, G.N. et al., Vet. Rec., 90, 618, 1972; Knowles, N.J. et al., Vet. Rec., 161, 71, 2007; Bellini, S. et al., Rev. Sci. Tech., 29, 639, 2010; EFSA, EFSA J., 10, 2631 (99pp.), 2012; Kanno, T. et al., J. Virol., 73, 2710, 1999.

close contact with SVD-infected pigs. Such sheep developed significant levels of neutralizing antibodies, indicating that they were infected, so the virus managed to replicate in their tissues, but without showing clinical signs. Despite this finding, sheep do not seem to play any role in the transmission of the disease.[46] Experimental infection has been induced in one-day-old mice being lethal to newborn mice.[4,47] Although SVDV is closely related to human CV-B5, and few cases of human illness have been reported caused by infections with earlier strains of SVDV, affecting people working in laboratories handling (and exposed to) high virus titers and/or infected pigs,[3] public health risk is not considered since, as mentioned earlier, the infection has not been reported in humans recently, and apparently, the virus has lost the ability to infect human cells, decreasing its original zoonotic potential.[2]

11.1.3.3 Disease Transmission and Virus Inactivation

The disease is transmitted mainly by direct contact with infected animals and their feces (fecal–oral route), fomites, and contaminated food. The virus resists fermentation and smoking processes used for conservation of certain food products, but is inactivated either after heat treatment (above 69°C) of pork products[48] or after treatment of cured ham for 1 year.[49] SVDV-contaminated environments have been shown to be as infectious as direct inoculation or contact with infected pigs. By opposite, airborne spread by aerosols from the lesions and skin is not an efficient way for SVDV transmission.[50,51] Nevertheless, inefficient disinfection of vehicles used in transport of pigs constitutes another important source of infection and disease dissemination from one farm to another.[43] A collective open drainage system or regular movements of infected pigs between pens increase the likelihood of spread within the farm, leading to the consideration of SVD as a "pen disease" rather than a "farm disease."[52]

The notable persistence and particular high stability of SVD virus particles play an essential role in the epidemiology of SVD.[53] The virus is resistant to most common alkaline or acidic disinfectants as well as to detergents and organic solvents such as ether and chloroform. Unaffected by desiccation and freezing, it can survive between 4 and 11 months over a pH range of 2.5–12 and temperature range of 12°C to –20°C. The inactivation of SVDV is achieved by heating at 56°C for 1 h, or by using disinfectants such as 1% sodium hydroxide (pH 12.4), 2% formaldehyde, and even 70% alcohol (allowing appropriate contact times).[54]

Once the infection is acquired, pigs secrete SVDV both orally and nasally, ceasing this secretion normally within 2 weeks. The virus is excreted in large amounts in feces starting up to 48 h before the onset of clinical signs and lasting usually around 10–20 days.[55] Nevertheless, occasionally excretion of the virus is observed during more time, even up to 70 days.[56] Furthermore, in certain circumstances (probably influenced by stress conditions), fecal excretion of the virus is resumed after cessation, and in this type of cases, the presence of the virus in feces has been observed up to 126 days after the initial infection.[57] Although the virus found in the tissues of infected pigs, particularly during the viremic phase, might bring on the contamination of pork products and introduce the disease agent into the food chain (particularly when the disease is not immediately recognized), in countries where this practice is not allowed, swill feeding is generally regarded as unimportant in the epidemiology of SVD.[43]

Generally, the highest viral shedding occurs from the vesicles within the first week after the infection,[51] and the virus does not persist in the tissues longer than 28 days following the infection. However, persistent infection of pigs has been observed experimentally using viral isolates of European origin, in which the virus remains in the tissues for longer than 3 months after the infection,[57,58] although it has proven difficult to reproduce these results due to the low frequency and lack of consistency observed for the phenomenon of SVDV persistency.[59] The epidemiological significance of persistent SVDV infection in pigs is uncertain. Experimental evidence indicates that in stress conditions like mixing of pigs from different pens, reactivation of latent virus infections for a short period of time is possible. However, to what extent this latency gives rise to epidemiologically relevant sources of infection remains unknown.

11.1.4 Clinical Features, Pathogenesis, and Immunity

This virus can enter the body through either broken skin lesions or mucous membranes of the digestive tract by ingestion of excreta or secretions from infected pigs.[60–62] The incubation period is usually 2–7 days, but it can extend longer if the dose of the virus is small. Viral replication starts in the primary site of infection, usually the gastrointestinal tract, spreading rapidly via the lymphatic system into the bloodstream at 1–2 dpi (days postinfection). This occurs evenly in pigs experimentally infected by direct inoculation of the virus as well as in pigs infected through contact with a contaminated environment.[51] The highest virus titers are found in the vesicles and tissues as lesions develop, decreasing as the antibody response rises, after the first week of clinical disease. The presence of circulating neutralizing antibodies, observed as early as 4 dpi, provides long-lasting protection against future reinfection. The tropism for epithelial tissues is well-known, but virus titers in the myocardium, brain, kidney, and spleen are higher than the titers in plasma or serum, being probably the sites of replication.[51,60,61] Similarly, the virus can be isolated from lymph nodes that may contain high viral titers[63] possibly due to either virus drainage from the tissues or local replication, or both.

Immunohistochemistry and in situ hybridization have been used to detect SVDV in epithelial and dermal cells and in infected tissues,[64] but more research must be implemented in order to know which cells are involved in SVDV replication, which would help to identify the mechanism behind the host tropism of the virus.[65]

As noted earlier, the development of neutralizing antibodies in the early stages of the infection is of paramount importance to impair virus dissemination throughout the body and limit the disease. IgM class antibodies dominate the earliest steps in the humoral immune response to SVDV. At 8–12 dpi, a switch to IgG occurs,[52,66] which provides a long-living immune protection to further SVDV infections. As only one serotype of SVDV is recognized, prior infection or vaccination with a given SVDV strain will be expected to protect against further infection with any other SVDV variant.

The importance of SVD relies on the fact that it is clinically indistinguishable from FMD, VS, and VES.[7] The differential diagnosis based on the observation of clinical signs is not feasible; thus, it is required to conduct specific laboratory investigations (see Section 11.1.5).

SVD courses most often as an acute, self-limited infection with the most perceptible clinical signs being the appearance of vesicles in the limbs, tongue, and snout. However, the infection may remain unnoticed since it is most frequently asymptomatic. Disease outcome may vary in appearance, and this variation poses some difficulties in the initial diagnosis. Affected pigs may show inappetence or even anorexia for 1 or 2 days, although this is not a consistent finding. Within 1–5 days after exposure, body temperature may rise to 41°C for 2–3 days; however, in experimental infection with European SVDV, pyrexia has not been observed.[53] Initially, blanching areas in the epithelium precede the formation of vesicles around the coronary bands and interdigital spaces and on the skin of the lower parts of the limbs, particularly at pressure points such as the knees. Vesicles may also appear on the snout, specifically on the dorsal surface, on the lips, tongue, and mammary glands, being less common in the oral cavity. Rupture of vesicles releases a straw-colored fluid (vesicular fluid) rich in infectious virus and reveals a shallow ulcer with a red hemorrhagic base. In few instances, punctiform ulcerative lesions have been found on the lower lips, but such lesions are not characteristic of the disease. Extension of the lesions on the coronary band may arise, and the hoof wall may separate from the underlying tissues, but the complete detachment of the hoof, as occurs with FMD, is uncommon.

Neurological signs have been observed, though rarely, in experimentally infected animals. They were reported as non-suppurative meningitis and panencephalomyelitis, principally affecting the midst and forebrain. These neurological signs may include trembling, unsteady gait, back arching, and chorea (rhythmic jerking).[59,67] Isolation from an aborted fetus suggests that abortion may occur, but there is no evidence that the pregnant sow is relevant for the perpetuation of the disease and dissemination of the infection. Experimentally, abortion is not typically seen and the virus is unlikely to infect or cross the placenta.[68] Other disease signs include lameness, salivation, and moderate general weakening. Normally, infected pigs recover after the acute phase of the disease within 2–3 weeks, healing all the skin lesions completely. Frequently, the only evidence of infection in the pig after careful examination is a dark horizontal line on the hooves where the growth was temporarily interrupted.

SVD may be subclinical, mild, or severe depending on the virulence of the strain and husbandry conditions. Regarding virulence, differences in pathogenicity between different strains have been assigned to specific regions and even single point mutations in the virus genome. Differences in pathogenicity between the virulent J1'73 and the avirulent H/3'76 strains were mapped in a region comprising nucleotides 2233–3368 of the viral RNA, corresponding to the C-terminus of VP3, the whole VP1, and the N-terminus of 2A viral protease.[44] Moreover, two different amino acid changes within this region, that is, glycine at the VP1–132 and proline at the 2A-20 positions, were identified as determinants of the high virulent phenotype. With regard to husbandry conditions, more severe lesions are seen when pigs are housed on concrete, particularly damp concrete, than on straw bedding or in grass. In addition, young pigs tend to show more severe clinical signs than older pigs. The only postmortem lesions of SVD are the vesicles that can be seen in live pigs. Nevertheless, the subclinical evolution of the disease observed during the last 20 years must be remarked. In fact, the infection is normally detected only by evidence of virus in feces of healthy pigs or by detection of antibodies.

11.1.5 Diagnosis

As SVD is clinically indistinguishable from FMD and other vesicular diseases of pigs (hence its status as an OIE-listed disease), the correct differential diagnosis is laboratory-based

and must be carried out using specific laboratory tests available. The standards for SVD diagnosis are described in the OIE *Manual of Diagnostic Test and Vaccines for Terrestrial Animals*.[7] Any vesicular condition in pigs may be caused by FMD; therefore, any sign of vesicular disease in pigs should be considered as FMD suspects until proven otherwise. Once suspicion of FMD is discarded, the diagnosis of SVD requires the facilities of a specialized laboratory, with appropriate biosecurity and biocontention measures. Countries lacking such facility should send samples for investigation to an OIE reference laboratory for SVD.[7]

The laboratory confirmation of SVD requires either the isolation of the virus or the detection of its genome or antigen, associated with the demonstration of specific antibodies.[53] A brief account of the current techniques and methods for the diagnosis of SVD is described in the following text, followed by a specific section describing the molecular methods implemented in the area in detail.

11.1.5.1 Conventional Techniques

In case of suspicion of the disease, samples of vesicular fluids and portions of vesicular lesion tissues are collected separately and aseptically in sterile containers. Fecal samples and serum from animals with or without lesions are collected for virus isolation, molecular analysis (usually, reverse transcription-polymerase chain reaction [RT-PCR]; see Section 11.1.5.2), and serological tests. Serum samples can be used for routine disease surveillance or export certification. Lymph nodes, thyroid and adrenal glands, kidneys, spleens, and heart tissues can be collected from slaughtered animals. Independent sets of sterile instruments should be used for sampling of each animal.

Virus neutralization test (VNT) and enzyme-linked immunosorbent assay (ELISA) are the techniques more commonly used for the detection of antibodies raised in response to SVDV infection. Various ELISA tests have been developed[66,69–71] including formats able to specifically differentiate IgM and IgG class immunoglobulins involved in SVDV binding.[66] Some of them are available commercially. While ELISA is the most appropriate technique for screening and has been shown to be efficient in large-scale serosurveillance programs,[63] VNT is currently the *gold standard* assay for confirmatory detection of antibodies to SVDV.[72] It is used to confirm the results after ELISA screening. Neutralizing antibodies can be detected as early as 3–4 days after experimental infection. The VNT is a laborious technique that requires 2–3 days to obtain the results. It is a quantitative test that is performed using IB-RS-2 cells or other susceptible porcine cells. The test measures the ability of serial dilutions of a given serum sample to neutralize (block) the infection of the cells by a known amount of infectious virus, previously titrated. It also needs inclusion of a reference antiserum of known titer, control cells, and infectious virus, which is used as antigen in the test.

Singleton reactors (SRs) are sera from individual pigs serologically positive for SVDV, but which have shown no clinical signs and for which there is neither a history of the disease on the holding nor contact with a known outbreak.[53] These SRs are very infrequent. In serological surveillance programs for SVD, about 1 in 220 tested samples yields positive by standard monoclonal antibody-competitive (MAC)-ELISA, and the gold standard VNT detects 1–3 SRs every 1000 sera tested. This low but consistent frequency of false-positive reactions is a common problem, provided that surveillance programs for SVD often analyze thousands of samples in some instances. The presence of an SR in a herd imposes restriction of movements to the farm in which it is found, leading to disruption of control measures and eradication programs and imposing quarantine restrictions to farms and bans for international trade.[53] The SR can be differentiated from true seropositive samples of infected pigs by resampling of the positive animals and their cohorts.[7] IgM class immunoglobulin is responsible for most SR reactions. The problem of SR can be reduced by applying a combination of VNT, MAC-ELISA, and isotype-specific ELISA.[73] The factors giving rise to SR sera are unknown; however, it seems likely that this condition is not due to a specific immune response to SVDV or other related viruses but to other responses resulting in a nonspecific IgM-mediated reaction.

Classical methods for identification of the virus rely on sandwich ELISA if sufficient antigen is present on the sample.[74] The antigen ELISA is the fastest and simplest test for SVDV detection in epithelium homogenates, where virus concentration is high enough to give a positive result in this test. However, the main drawback of this technique is its low sensitivity, making it unsuitable for the analysis of other types of samples. In particular, more sensitive tests are required for detection of SVDV in fecal samples, often the only available sample for virological diagnosis, since very frequently the infection courses without the development of vesicles/skin lesions. Virus isolation is the reference *gold standard* for virus detection and is used as a confirmatory method if a positive antigen ELISA or RT-PCR result is not associated with the detection of clinical signs of disease, the detection of seropositive pigs, or a direct epidemiological connection with a confirmed outbreak. If any inoculated culture subsequently develops a cytopathic effect (CPE), the demonstration of SVD antigen or genome in the culture supernatant by ELISA or by RT-PCR will suffice to make a positive diagnosis.[7]

The virus can be isolated from epithelium of skin lesions, fecal homogenates, and vesicular fluids. For this, the clarified homogenates are inoculated on monolayers of IB-RS-2 cells[75] or other susceptible porcine cells (SK6, PK-15), and these are examined daily for the development of CPE, for 2–3 days. The virus, if present in feces, is often found in small amounts, so its isolation from this type of sample often requires 2–3 successive blind passages and may be affected by other interfering enteric viruses that are highly prevalent in pig fecal samples.[76,77] If CPE is not observed after three blind passages, the sample is considered negative in virus isolation. The supernatant of the cell culture showing CPE is harvested and virus identification is performed by ELISA or RT-PCR.

11.1.5.2 Molecular Techniques

Several highly specific and sensitive RT-PCR techniques have been developed to detect SVDV RNA either in conventional and real-time formats, some of them enabling the simultaneous differentiation from FMDV and vesicular stomatitis virus (VSV).[55,64,78–83]

A recent development for multiplexing diagnosis is the nucleic acid–based multiplexed assay for detection of FMDV and ruling out for six other animal diseases that cause vesicular or ulcerative lesions in cattle, sheep, and swine (including SVD), based on a combination of magnetic bead-flow cytometry technology (Luminex) and RT-PCR.[84] Another approach in this regard uses solid-phase (chip) microarray technology combined with RT-PCR to address multiplexing of all the range of vesicular disease viruses.[85] This method enables the efficient detection and differentiation, including correct molecular serotyping of all known serotypes of FMDV, the two known serotypes of VSV and SVDV simultaneously.

Given that SVD has been occurring almost exclusively as a subclinical disease for a number of years, suitable samples for virus detection are feces, for which no differential diagnosis is required in relation to vesicular conditions. In fact, the unique molecular assay extensively validated in field conditions, particularly with feces samples, is the conventional one-step RT-PCR reported as reference test in the OIE manual.[7]

11.1.5.3 On-Site Devices

"On-site" or portable devices have also been incorporated to the range of SVDV diagnostic tools. These devices are sought aiming at enabling rapid virus detection and identification at the farm level, thus saving time from the first suspicion of the disease to its confirmation and the application of control measures.

A lateral flow device ("pen-side test") has been developed that showed a good performance in the detection of SVDV antigen in vesicular fluids and cell culture passage-derived supernatants.[86] Portable devices allowing "on-site" RT-PCR analysis have been developed and tested for its performance on detecting FMDV genome,[87] which could be suitable also for detection of other vesicular diseases, including SVDV, in the same setting, provided that there are RT-PCR techniques (specifically, fluorescence-based real-time formats are of particular utility in this regard) of enough sensitivity and specificity available. As explained in Section 11.2, there are a number of real-time RT-PCR techniques available for SVDV diagnostic, and their adaptation to portable devices should not be a problem. On the other hand, loop-mediated isothermal amplification (LAMP) assays provide some advantages in this regard with respect to classical or real-time RT-PCR methods, since they are faster and do not rely on expensive equipment such as portable thermal cyclers, providing robust performance in modestly equipped laboratories, as can be the case for field stations or mobile diagnostic units. A LAMP assay for the detection of SVDV has been recently developed,[88] although it needs to be improved in order to be able to detect all the circulating subgenomic lineages.

11.2 METHODS

11.2.1 SAMPLE COLLECTION AND PREPARATION

11.2.1.1 Sample Collection

Samples of vesicular fluids and portions of vesicular lesion tissues are the target samples for SVDV detection if clinical signs of vesicular disease are present. These should be collected separately and aseptically in sterile containers from individual pigs. Virus load in vesicular fluids and lesions are typically high (usually in the range of 10^8–10^9 $TCID_{50}$/mL), and proper measures to avoid any material contamination should be taken.

Fecal samples are the specimen of choice in case subclinical SVD is suspected, which can be collected from individual pigs or from the floor of premises suspected to contain, or to have contained, pigs infected with SVDV. Although the amount of virus present in feces is usually low (<10^4 $TCID_{50}$/mL), it may be excreted by this route for up to 3 months after infection,[7] being suitable for RT-PCR analysis.

The procedure for selecting tissues and organs for virus isolation as described in Section 11.1.5.1 is used for collection of samples for molecular analysis. Samples from the muscle of infected pigs and certain uncooked pig products may also be needed to undergo testing in order to assess the presence of SVDV (see Section 11.1.3.3) and should be processed in the same way.

Once under the adequate biosafety laboratory facilities, a suspension of each individual sample (vesicular lesions, feces, organs) is prepared by grinding the sample using a sterile pestle and mortar adding a volume of sterile phosphate buffered saline or tissue culture medium and antibiotics enough to get a 10% homogenate. The homogenate is vortexed and further clarified by centrifugation at $2000 \times g$ for 20–30 min, and the supernatant is collected for analysis, which should be previously filtered through a 0.45 µm filter. The treated samples are conveniently labeled, aliquoted, and stored refrigerated while diagnostic analysis is ongoing; the remaining material should be preserved and properly identified in an adequate freezer, preferably below −70°C.

11.2.1.2 Viral RNA Extraction

In the past, the viral RNA extraction was performed by the traditional organic procedure using phenol-chloroform and further alcoholic precipitation, which has been displaced by a variety of current commercial kits based on chaotropic salt lysis and silica nucleic acid affinity. Even more, the RNA extraction step can be achieved by manual or automated systems obtaining good quality RNA templates, which remove successfully the polymerase chain reaction (PCR) inhibitors present in the starting specimen.

Apart from the commercial kits mentioned earlier, the RNA can also be extracted by a virus immune-capture using an SVDV-specific monoclonal antibody (MAb) followed by guanidine thiocyanate lysis and ethanol precipitation.[89] This immune-capture technique has been shown to be effective with fecal samples enabling the immune-purification of the

virus and the overthrow of PCR inhibitory substances; the procedure is described in detail in the following.

An ELISA plate is coated with SVDV-specific MAb 5B7 at saturated concentration diluted in a carbonate–bicarbonate solution (200 µL/well) and incubated overnight at +4°C. After the plate wash, each sample (feces suspension) is distributed into three wells (200 µL/well, 600 µL of sample in total). After 1 h incubation at 37°C with slow shaking, the plate is washed manually to avoid cross-contamination. The viral lysis is performed by adding approximately 100 µL/well of lysis buffer (4M guanidine thiocyanate, 25 mM sodium citrate, pH7, 0.5% SarKosyl). After 3–5 min incubation, the sample from the three wells (300–350 µL) is collected and transferred in a single tube. RNA is then precipitated by a classical ethanol precipitation protocol and then resuspended in 20 µL of RNase-free water.

RNA from homogenates of vesicular fluids or tissues is extracted using the QIAamp viral RNA kit (Qiagen, Hilden, Germany) following the manufacturer instructions.

11.2.2 DETECTION PROCEDURES

Several molecular methods have been developed for SVDV detection (see Section 11.1.5.2), although RT-PCR technique is so far the most widely used for routine diagnostic purpose. Due to its high sensitivity and specificity, RT-PCR can be applied to any diagnostic specimen where SVDV presence is suspected. Several gel-based and real-time RT-PCR methods have been described for SVDV detection targeting different viral genome regions. Generic tests targeting conserved genome regions, such as 5′ NCR (noncoding region) or 3D gene, are preferred for diagnostic analysis to cover the range of viral genotypes. Methods amplifying variable genome regions, such as VP1 coding gene, are selected for further sequencing and molecular characterization of SVDV isolates. In this section, one conventional and one real-time RT-PCR assays for SVDV screening detection are described in detail.

11.2.2.1 Conventional RT-PCR Amplifying a 3D Gene Fragment

The described SVDV conventional RT-PCR method uses a primer set (SS4/SA2) designed in a highly conserved region within the viral 3D gene,[78] which ensures the detection of a wide range of SVDV isolates. This primer set amplifies a DNA fragment of 154 base pairs (bp), located from the nucleotide position 6875–7028 of the SVDV reference strain UKG 27/72 (GenBank accession no. X54521). The nucleotide sequences of the primers are described in Table 11.2.

In detail, the RT-PCR is performed using the commercial one-step RT-PCR kit (Qiagen, Hilden, Germany). RT-PCR mix is prepared in a final volume of 25 µL containing 2–5 µL of RNA sample, 1× one-step RT-PCR buffer, 1× Q solution (optional, but recommended if more specificity is needed), 0.2 mM dNTPs, 0.6 µM each forward and reverse primer (SS4/SA2), 0.2 U/µL RNAse inhibitor, 1 µL one-step RT-PCR enzyme mix. RT-amplification is then carried out in a single-tube one-step assay with the following incubation

TABLE 11.2

Nucleotide Sequences of the Primers and Probes Used for SVDV Molecular Detection and Characterization

PCR Target	SVDV Genome Region	Primer/Probe Name	Sequence (5′–3′)
SVDV conventional RT-PCR (Nunez et al. [78])	3D	SVDV-SS4 forward primer	TTCAGAATGATTGCATATGGGG
		SVDV-SA2 reverse primer	TCACGTTTGTCCAGGTTACY
SVDV[49]/β-actin real-time RT-PCR (Toussaint et al. [91])	5′ NCR	SVD-2B-IR252-F forward primer	CGAGAAACCTAGTACCACCATGAA
		SVD-2B-IR332-R reverse primer	CGGTGACTCATCGACCTGATC
		SVD-2B-IR289-P probe	6FAM-TCGCTCCGCACAACCCCAGTG-TAMRA
	Not applicable (internal control)	ACT-F1005-1029 forward primer	CAGCACAATGAAGATCAAGATCATC
		ACT-R1135-1114 reverse primer	CGGACTCATCGTACTCCTGCTT
		ACT-P1081-1105 probe	ROX-TCGCTGTCCACCTTCCAGCAGATGT-BHQ2
SVDV molecular characterization (Brocchi et al. [39])	1D (VP1)	GSVD-3 forward primer	ACACCCTTTATAAAACAGG
		NK44 reverse primer	CCACACACAGTTTTGCCAGTC
		GSVD-5 reverse primer	AACATGCTGTATGCGTTGCCTAT

program: 30 min at 55°C, 15 min at 95°C, 40 cycles at 94°C for 30 s, 57°C for 30 s and 72°C for 30 s, and a final extension step of 72°C for 7 min. Amplification products are then analyzed by electrophoresis on a 2% agarose gel containing 0.5 μg/mL ethidium bromide or other suitable DNA-intercalating agent such as GelRed™.

In a comparative study on positive fecal samples from many different outbreaks, the one-step RT-PCR showed the best performances with regard to the capability to reveal all the circulating genomic sublineages, with respect to the real-time RT-PCR targeting the 5′ NCR (described in the following section) and an RT-LAMP.[90]

11.2.2.2 Real-Time RT-PCR Amplifying a 5′ NCR Fragment

The real-time RT-PCR method described in this section for SVDV detection employs a primer set (SVD-2B-IR252-F/ SVD-2B-IR332-R) and a specific TaqMan probe (SVD-2B-IR289-P) directed to the highly conserved 5′ NCR of the viral genome, allowing as well the detection of a wide range of SVDV isolates.[55] The primers define a short DNA fragment of 81 bp, located from the nucleotide position 252–332 of the SVDV reference strain UKG 27/72 (GenBank accession no. X54521). TaqMan probe employed for amplified product detection is labeled with a reporter at 5′ end [6-carboxyfluorescein (FAM)] and a quencher at 3′ end (6-carboxytetramethylrhodamine).

Since feces contain a high level of PCR inhibitors and are the main target sample for the molecular diagnosis, the incorporation of an internal control in the PCR analysis is highly recommended to avoid false-negative results. This could be an endogenous gene such as the β-actin, which can be detected at the same time than SVDV in a duplex RT-PCR test, like the method described here. Specifically, a primer set and a TaqMan probe described for the amplification of a 136 bp DNA fragment of the β-actin gene[91] are added to the RT-PCR mix to carry out the simultaneous and differential detection of both SVDV and internal control in a single tube. To this end, the TaqMan probe for the internal control

is labeled with a different reporter dye, and in this case, ROX (6-carboxy-X-rhodamine) is selected for the 5′ end and a black hole quencher for the 3′ end.

Sequences of primers and probes are described in Table 11.2. In detail, the duplex RT-PCR is performed using the commercial *QuantiTect Multiplex RT-PCR NR kit (Qiagen, Hilden, Germany)*. RT-PCR mix is prepared in a final volume of 25 μL containing 5 μL of RNA sample, 1× QuantiTect Multiplex RT-PCR Master Mix, 0.9 μM SVD-2B-IR252-F forward primer, 0.3 μM SVD-2B-IR332-R reverse primer, 0.2 μM SVD-2B-IR289-P probe, 0.2 μM each β-actin forward and reverse primers (ACT-F1005-1029/ ACT-R1135-1114), 0.1 μM ACT-P1081-1105 probe, 0.2 U/μL RNAse inhibitor, and 0.25 μL QuantiTect Multiplex RT Mix. RT-amplification is then carried out in a single-tube one-step assay with the following incubation program: 30 min at 50°C, 15 min at 95°C, and 45 cycles at 94°C for 15 s, and 60°C for 60 s. Amplification products are detected by fluorescence reading at the end of each cycle in FAM (SVDV) and ROX (β-actin) channels.

11.2.2.3 Molecular Characterization of SVDV Isolates

All SVDV isolates are classified so far in a single serotype, though four genetic groups can be differentiated based on genome sequence variability.[39] Sequencing of the viral 1D gene, coding the major structural protein VP1, and further sequence comparison to other SVDV isolates allow establishing epidemiological relationships between viral isolates, inferring the most probable origin of a disease outbreak and tracing the evolution of an epidemic.

Amplification of the viral 1D gene for further sequencing can be achieved by conventional RT-PCR test using primers (GSVD-3/NK44) described in Table 11.1, defining a 980 bp DNA fragment. The amplified product can be directly sequenced by employing the same primers or using internal primers (GSVD-3/GSVD-5) to refine the sequencing to a shorter fragment of 613 bp.

Phylogenetic analyses based on sequences of a portion of the 3D gene (RNA-dependent RNA polymerase) show clusterization of SVDV isolates overlapping with that obtained with 1D sequences.[92]

The databases of 1D and 3D gene sequences of SVDV isolates are held at the OIE reference laboratories (Pirbright, UK; Brescia, Italy).[7]

11.3 CONCLUSION AND FUTURE PERSPECTIVES

SVDV is the causative agent of the disease of pigs whose clinical resemblance to FMD highlights the need for its accurate identification and discrimination. This fact imposes severe restrictions to the movements of pigs for which convincing evidence of being free from the virus (or of antibodies raised to it) is not provided. This poses a significant burden for the trade of these animals, and for this reason, the disease may cause severe economic losses for the producers, even though it is in fact a mild disease from the clinical point of view. Differentiation from FMD, though not possible clinically, is feasible if appropriate diagnostic tests are applied. Improvement of the diagnostic techniques is making differential diagnosis of vesicular diseases more and more affordable, feasible, and easy, and nowadays, portable devices allow a rapid and accurate differentiation of SVDV from FMDV infections on-site. As these tests become affordable and availability of competent laboratory services become more and more accessible, the constraint originally imposed to SVD due to its similarity to FMD will lose its sense. This, together with the fact that in recent years SVD is predominantly asymptomatic, makes necessary to rethink the measures currently in place for the control of SVD.

REFERENCES

1. Zhang, G., Wilsden, G., Knowles, N. J., and McCauley, J. W. Complete nucleotide sequence of a coxsackie B5 virus and its relationship to swine vesicular disease virus. *J Gen Virol* 74 (Pt 5), 845–853 (1993).
2. Jimenez-Clavero, M. A., Escribano-Romero, E., Ley, V., and Spiller, O. B. More recent swine vesicular disease virus isolates retain binding to coxsackie-adenovirus receptor, but have lost the ability to bind human decay-accelerating factor (CD55). *J Gen Virol* 86, 1369–1377 (2005).
3. Knowles, N. J. and Sellers, R. F. In *Handbook of Zoonoses, Section B Viral*, ed. Beran, G. W., pp. 437–444. CRC Press Inc., Boca Raton, FL, 1994.
4. Nardelli, L. et al. A foot and mouth disease syndrome in pigs caused by an enterovirus. *Nature* 219, 1275–1276 (1968).
5. Bellini, S., Santucci, U., Zanardi, G., Brocchi, E., and Marabelli, R. Swine vesicular disease surveillance and eradication activities in Italy. *Rev Sci Tech* 26, 585–593 (2007).
6. OIE. *Terrestrial Animal Health Code*. OIE World Organisation for Animal Health, Paris, France, 2011. http://www.oie.int/international-standard-setting/terrestrial-code/access-online/. Accessed on March 31, 2015.
7. OIE (Office International des Epizooties). *Manual of Diagnostic Tests Vaccines for Terrestrial Animals*, pp. 1–8. World Organisation for Animal Health, Paris, France, 2013. http://www.oie.int/fileadmin/Home/eng/Health_standards/tahm/2.08.09_SVD.pdf. Accessed on March 31, 2015.
8. Mowat, G. N., Prince, M. J., Spier, R. E., and Staple, R. F. Preliminary studies on the development of a swine vesicular disease vaccine. *Arch Gesamte Virusforsch* 44, 350–360 (1974).
9. Delagneau, J. F., Guerche, J., Adamowicz, P., and Prunet, P. Maladie vésiculeuse du porc: Propriétés physicochimiques et immunogènes de la souche France 1/73. *Ann Microbiol* 125, 559–574 (1974).
10. Gourreau, J. M., Dhennin, L., and Labie, J. Mise au point d'un vaccin a virus inactivé contre la maladie vesiculeuse du porc. *Rec Med Vet Ec Alfort* 151, 85–89 (1975).
11. McKercher, P. D. and Graves, J. H. A mixed vaccine for swine: An aid for control of foot-and-mouth and swine vesicular diseases. *Bol Cent Panameric Fiebre Aft* 23/24, 37–49 (1976).
12. Jimenez-Clavero, M. A., Escribano-Romero, E., Sanchez-Vizcaino, J. M., and Ley, V. Molecular cloning, expression and immunological analysis of the capsid precursor polypeptide (P1) from swine vesicular disease virus. *Virus Res* 57, 163–170 (1998).
13. Knowles, N. J. and McCauley, J. W. Coxsackievirus B5 and the relationship to swine vesicular disease virus. *Curr Top Microbiol Immunol* 223, 153–167 (1997).
14. Graves, J. H. Serological relationship of swine vesicular disease virus and Coxsackie B5 virus. *Nature* 245, 314–315 (1973).
15. Brown, F., Talbot, P., and Burrows, R. Antigenic differences between isolates of swine vesicular disease virus and their relationship to Coxsackie B5 virus. *Nature* 245, 315–316 (1973).
16. Zhang, G., Haydon, D. T., Knowles, N. J., and McCauley, J. W. Molecular evolution of swine vesicular disease virus. *J Gen Virol* 80 (Pt 3), 639–651 (1999).
17. Inoue, T., Suzuki, T., and Sekiguchi, K. The complete nucleotide sequence of swine vesicular disease virus. *J Gen Virol* 70 (Pt 4), 919–934 (1989).
18. Inoue, T., Yamaguchi, S., Kanno, T., Sugita, S., and Saeki, T. The complete nucleotide sequence of a pathogenic swine vesicular disease virus isolated in Japan (J1'73) and phylogenetic analysis. *Nucleic Acids Res* 21, 3896 (1993).
19. Seechurn, P., Knowles, N. J., and McCauley, J. W. The complete nucleotide sequence of a pathogenic swine vesicular disease virus. *Virus Res* 16, 255–274 (1990).
20. Tsuda, T., Tokui, T., and Onodera, T. Induction of neutralizing antibodies by structural proteins VP1 and VP2 of swine vesicular disease virus. *Nihon Juigaku Zasshi* 49, 129–132 (1987).
21. Kanno, T., Inoue, T., Wang, Y., Sarai, A., and Yamaguchi, S. Identification of the location of antigenic sites of swine vesicular disease virus with neutralization-resistant mutants. *J Gen Virol* 76 (Pt 12), 3099–3106 (1995).
22. Nijhar, S. K. et al. Identification of neutralizing epitopes on a European strain of swine vesicular disease virus. *J Gen Virol* 80 (Pt 2), 277–282 (1999).
23. Borrego, B., Carra, E., Garcia-Ranea, J. A., and Brocchi, E. Characterization of neutralization sites on the circulating variant of swine vesicular disease virus (SVDV): A new site is shared by SVDV and the related coxsackie B5 virus. *J Gen Virol* 83, 35–44 (2002).
24. Jimenez-Clavero, M. A., Douglas, A., Lavery, T., Garcia-Ranea, J. A., and Ley, V. Immune recognition of swine vesicular disease virus structural proteins: Novel antigenic regions that are not exposed in the capsid. *Virology* 270, 76–83 (2000).
25. Jimenez-Clavero, M. A., Escribano-Romero, E., Douglas, A. J., and Ley, V. The N-terminal region of the VP1 protein of swine vesicular disease virus contains a neutralization site that arises upon cell attachment and is involved in viral entry. *J Virol* 75, 1044–1047 (2001).

26. Verdaguer, N., Jimenez-Clavero, M. A., Fita, I., and Ley, V. Structure of swine vesicular disease virus: Mapping of changes occurring during adaptation of human coxsackie B5 virus to infect swine. *J Virol* 77, 9780–9789 (2003).

27. Fry, E. E. et al. Crystal structure of Swine vesicular disease virus and implications for host adaptation. *J Virol* 77, 5475–5486 (2003).

28. Tuthill, T. J., Groppelli, E., Hogle, J. M., and Rowlands, D. J. Picornaviruses. *Curr Top Microbiol Immunol* 343, 43–89 (2010).

29. Martino, T. A. et al. The coxsackie-adenovirus receptor (CAR) is used by reference strains and clinical isolates representing all six serotypes of coxsackievirus group B and by swine vesicular disease virus. *Virology* 271, 99–108 (2000).

30. Coyne, C. B. and Bergelson, J. M. CAR: A virus receptor within the tight junction. *Adv Drug Deliv Rev* 57, 869–882 (2005).

31. Bergelson, J. M. et al. Coxsackievirus B3 adapted to growth in RD cells binds to decay-accelerating factor (CD55). *J Virol* 69, 1903–1906 (1995).

32. Martino, T. A. et al. Cardiovirulent coxsackieviruses and the decay-accelerating factor (CD55) receptor. *Virology* 244, 302–314 (1998).

33. Shafren, D. R. et al. Coxsackieviruses B1, B3, and B5 use decay accelerating factor as a receptor for cell attachment. *J Virol* 69, 3873–3877 (1995).

34. Alexandersen, S. et al. Picornaviruses. In *Diseases of Swine*, eds. Zimmerman, J., Karriker, L., Ramirez, A., Schwartz, K., and Stevenson, G., pp. 587–620. John Wiley & Sons, Chichester, U.K., 2012.

35. Fears, C. Y. and Woods, A. The role of syndecans in disease and wound healing. *Matrix Biol* 25, 443–456 (2006).

36. Fry, E. E. et al. Structure of Foot-and-mouth disease virus serotype A10 61 alone and complexed with oligosaccharide receptor: Receptor conservation in the face of antigenic variation. *J Gen Virol* 86, 1909–1920 (2005).

37. Escribano-Romero, E., Jimenez-Clavero, M. A., Gomes, P., Garcia-Ranea, J. A., and Ley, V. Heparan sulphate mediates swine vesicular disease virus attachment to the host cell. *J Gen Virol* 85, 653–663 (2004).

38. Jiménez-Clavero, M. A. Animal viral diseases and global change: Bluetongue and West Nile fever as paradigms. *Front Genet* 3, 105 (2012).

39. Brocchi, E. et al. Molecular epidemiology of recent outbreaks of swine vesicular disease: Two genetically and antigenically distinct variants in Europe, 1987–94. *Epidemiol Infect* 118, 51–61 (1997).

40. Mowat, G. N., Darbyshire, J. H., and Huntley, J. F. Differentiation of a vesicular disease of pigs in Hong Kong from foot-and-mouth disease. *Vet Rec* 90, 618–621 (1972).

41. Knowles, N. J. et al. Reappearance of swine vesicular disease virus in Portugal. *Vet Rec* 161, 71 (2007).

42. Bellini, S., Alborali, L., Zanardi, G., Bonazza, V., and Brocchi, E. Swine vesicular disease in northern Italy: Diffusion through densely populated pig areas. *Rev Sci Tech* 29, 639–648 (2010).

43. EFSA. Scientific opinion on swine vesicular disease and vesicular stomatitis. *EFSA J* 10, 2631 (99pp.) (2012).

44. Kanno, T. et al. Mapping the genetic determinants of pathogenicity and plaque phenotype in swine vesicular disease virus. *J Virol* 73, 2710–2716 (1999).

45. Wilder, F. W. et al. Susceptibility of one-toed pig to certain diseases exotic to the United States. In *Proceedings of the 78th Annual Meeting of the US Animal Health Association*, Roanoke, VA, pp. 195–199, 1974.

46. Burrows, R., Mann, J. A., Goodridge, D., and Chapman, W. G. Swine vesicular disease: Attempts to transmit infection to cattle and sheep. *J Hyg* 73, 101–107 (1974).

47. Kadoi, K. The propagation of a strain of swine vesicular disease virus in one-day-old mice. *Nihon Juigaku Zasshi* 45, 821–823 (1983).

48. McKercher, P. D., Morgan, D. O., McVicar, J. W., and Shuot, N. J. Thermal processing to inactivate viruses in meat products. *Proc Annu Meet US Anim Health Assoc* 84, 320–328 (1980).

49. Mebus, C. et al. Survival of several porcine viruses in different Spanish dry-cured meat products. *Food Chem* 59, 555–559 (1997).

50. Herniman, K. A., Medhurst, P. M., Wilson, J. N., and Sellers, R. F. The action of heat, chemicals and disinfectants on swine vesicular disease virus. *Vet Rec* 93, 620–624 (1973).

51. Dekker, A., Moonen, P., de Boer-Luijtze, E. A., and Terpstra, C. Pathogenesis of swine vesicular disease after exposure of pigs to an infected environment. *Vet Microbiol* 45, 243–250 (1995).

52. Dekker, A., van Hemert-Kluitenberg, F., Baars, C., and Terpstra, C. Isotype specific ELISAs to detect antibodies against swine vesicular disease virus and their use in epidemiology. *Epidemiol Infect* 128, 277–284 (2002).

53. Lin, F. and Kitching, R. P. Swine vesicular disease: An overview. *Vet J* 160, 192–201 (2000).

54. Terpstra, C. Vesicular swine disease in The Netherlands. *Tijdschr Diergeneeskd* 117, 623–626 (1992).

55. Reid, S. M., Ferris, N. P., Hutchings, G. H., King, D. P., and Alexandersen, S. Evaluation of real-time assays for the detection of swine vesicular disease virus. *J Virol Methods* 116, 169–176 (2004).

56. Niedbalski, W. Application of different diagnostic methods for the detection of SVDV infection in pigs. *Bull Vet Inst Pulawy* 43, 11–18 (1999).

57. Lin, F., Mackay, D. K., and Knowles, N. J. The persistence of swine vesicular disease virus infection in pigs. *Epidemiol Infect* 121, 459–472 (1998).

58. Lahellec, M. and Gourreau, J. M. Maladie vésiculeuse du porc: Étude anatomo-pathologique. *Ann Rech Vet* 6, 179–186 (1975).

59. Lin, F., Mackay, D. K., Knowles, N. J., and Kitching, R. P. Persistent infection is a rare sequel following infection of pigs with swine vesicular disease virus. *Epidemiol Infect* 127, 135–145 (2001).

60. Chu, R. M., Moore, D. M., and Conroy, J. D. Experimental swine vesicular disease, pathology and immunofluorescence studies. *Can J Comp Med* 43, 29–38 (1979).

61. Lai, S. S., McKercher, P. D., Moore, D. M., and Gillespie, J. H. Pathogenesis of swine vesicular disease in pigs. *Am J Vet Res* 40, 463–468 (1979).

62. Mann, J. A. and Hutchings, G. H. Swine vesicular disease: Pathways of infection. *J Hyg* 84, 355–363 (1980).

63. Dekker, A. Swine vesicular disease, studies on pathogenesis, diagnosis, and epizootiology: A review. *Vet Q* 22, 189–192 (2000).

64. Mulder, W. A. et al. Detection of early infection of swine vesicular disease virus in porcine cells and skin sections. A comparison of immunohistochemistry and in-situ hybridization. *J Virol Methods* 68, 169–175 (1997).

65. Lin, F., Mackay, D. K., and Knowles, N. J. Detection of swine vesicular disease virus RNA by reverse transcription-polymerase chain reaction. *J Virol Methods* 65, 111–121 (1997).

66. Brocchi, E., Berlinzani, A., Gamba, D., and De Simone, F. Development of two novel monoclonal antibody-based ELISAs for the detection of antibodies and the identification of swine isotypes against swine vesicular disease virus. *J Virol Methods* 52, 155–167 (1995).

67. Lenghaus, C., Mann, J. A., Done, J. T., and Bradley, R. Neuropathology of experimental swine vesicular disease in pigs. *Res Vet Sci* 21, 19–27 (1976).

68. Watson, W. A. Swine vesicular disease in Great Britain. *Can Vet J* 22, 195–200 (1981).

69. Hamblin, C. and Crowther, J. R. A rapid enzyme-linked immunosorbent assay for the serological confirmation of swine vesicular disease. *Br Vet J* 138, 247–252 (1982).

70. Armstrong, R. M. and Barnett, I. T. An enzyme-linked immunosorbent assay (ELISA) for the detection and quantification of antibodies against swine vesicular disease virus (SVDV). *J Virol Methods* 25, 71–79 (1989).

71. Dekker, A., Moonen, P. L., and Terpstra, C. Validation of a screening liquid phase blocking ELISA for swine vesicular disease. *J Virol Methods* 51, 343–348 (1995).

72. Golding, S. M., Hedger, R. S., and Talbot, P. Radial immunodiffusion and serum-neutralisation techniques for the assay of antibodies to swine vesicular disease. *Res Vet Sci* 20, 142–147 (1976).

73. De Clercq, K. Reduction of singleton reactors against swine vesicular disease virus by a combination of virus neutralisation test, monoclonal antibody-based competitive ELISA and isotype specific ELISA. *J Virol Methods* 70, 7–18 (1998).

74. Ferris, N. P. and Dawson, M. Routine application of enzyme-linked immunosorbent assay in comparison with complement fixation for the diagnosis of foot-and-mouth and swine vesicular diseases. *Vet Microbiol* 16, 201–209 (1988).

75. De Castro, M. P. Behaviour of the foot and mouth disease virus in cell cultures: Susceptibility of the IB-RS-2 cell line. *Arq Inst Biol Sao Paulo* 31, 155–166 (1964).

76. Buitrago, D. et al. A survey of porcine picornaviruses and adenoviruses in fecal samples in Spain. *J Vet Diagn Invest* 22, 763–766 (2010).

77. Cano-Gomez, C. et al. Analyzing the genetic diversity of teschoviruses in Spanish pig populations using complete VP1 sequences. *Infect Genet Evol* 11, 2144–2150 (2011).

78. Nunez, J. I. et al. A RT-PCR assay for the differential diagnosis of vesicular viral diseases of swine. *J Virol Methods* 72, 227–235 (1998).

79. Fernandez, J. et al. Rapid and differential diagnosis of foot-and-mouth disease, swine vesicular disease, and vesicular stomatitis by a new multiplex RT-PCR assay. *J Virol Methods* 147, 301–311 (2008).

80. Rasmussen, T. B., Uttenthal, A., and Aguero, M. Detection of three porcine vesicular viruses using multiplex real-time primer-probe energy transfer. *J Virol Methods* 134, 176–182 (2006).

81. Hakhverdyan, M., Rasmussen, T. B., Thoren, P., Uttenthal, A., and Belak, S. Development of a real-time PCR assay based on primer-probe energy transfer for the detection of swine vesicular disease virus. *Arch Virol* 151, 2365–2376 (2006).

82. McMenamy, M. J. et al. Development of a minor groove binder assay for real-time one-step RT-PCR detection of swine vesicular disease virus. *J Virol Methods* 171, 219–224 (2011).

83. Benedetti D., Pezzoni, G., Grazioli, S., Bugnetti, M., and Brocchi, E. Detection of Swine Vesicular Disease Virus (SVDV) by a new 3D gene one-step real time PCR assay. In *16th International Symposium of the World Association of Veterinary Laboratory Diagnosticians WAVLD 2013, 10th OIE seminar, 32nd Symposium of AVID*, Berlin, Germany, pp. 143, 2013.

84. Lenhoff, R. J. et al. Multiplexed molecular assay for rapid exclusion of foot-and-mouth disease. *J Virol Methods* 153, 61–69 (2008).

85. Lung, O. et al. Multiplex RT-PCR detection and microarray typing of vesicular disease viruses. *J Virol Methods* 175, 236–245 (2011).

86. Ferris, N. P. et al. Development and laboratory evaluation of a lateral flow device for the detection of swine vesicular disease virus in clinical samples. *J Virol Methods* 163, 477–480 (2010).

87. Madi, M. et al. Rapid detection of foot-and-mouth disease virus using a field-portable nucleic acid extraction and real-time PCR amplification platform. *Vet J* 193, 67–72 (2012).

88. Blomstrom, A. L. et al. A one-step reverse transcriptase loop-mediated isothermal amplification assay for simple and rapid detection of swine vesicular disease virus. *J Virol Methods* 147, 188–193 (2008).

89. Fallacara, F., Pacciarini, M., Bugnetti, M., Berlinzani, A., and Brocchi, E. Detection of swine vesicular disease virus in faces samples by immune-PCR assay. In *Veterinary Virology in the New Millennium: Proceedings of the Fifth International Congress of the European Society for Veterinary Virology*, Brecia, Italy, pp. 173–174, 2000.

90. Benedetti, D., Pezzoni, G., Grazioli, S., Barbieri, I., and Brocchi, E. Comparative performance of three genome amplification assays for detection of swine vesicular disease virus in experimental and field samples. In *First Congress of the European Association of Veterinary Laboratory Diagnosticians*, Lelystad, the Netherlands, 2010.

91. Toussaint, J. F., Sailleau, C., Breard, E., Zientara, S., and De Clercq, K. Bluetongue virus detection by two real-time RT-qPCRs targeting two different genomic segments. *J Virol Methods* 140, 115–123 (2007).

92. Pezzoni, G., Grazioli, S., Tironi, D., Barbieri, I., and Brocchi., M. Validation of a loop mediated isothermal amplification assay for detection of swine vesicular disease virus in experimental and field samples. In *Fourth Annual Meeting EPIZONE "Bridges to the Future"*, Saint Malo, France, pp. 187, 2010.

12 Avian Astroviruses

Victoria Smyth

CONTENTS

12.1 Introduction 101
    12.1.1 Classification...101
    12.1.2 Structure, Biology, and Epidemiology...102
    12.1.3 Clinical Features and Pathogenesis..103
        12.1.3.1 Avian Nephritis Virus...104
        12.1.3.2 Chicken Astrovirus 104
        12.1.3.3 Turkey Astrovirus 105
        12.1.3.4 Duck Astrovirus...105
    12.1.4 Diagnosis 105
        12.1.4.1 Conventional Techniques...105
        12.1.4.2 Molecular Techniques..106
12.2 Methods 106
    12.2.1 Sample Preparation..106
    12.2.2 Detection Procedures...107
12.3 Conclusions and Future Perspectives..107
References..108

12.1 INTRODUCTION

Astroviruses are small, round, nonenveloped viruses typically 25–35 nm in diameter with a positive-sensed, single-stranded RNA genome ranging in length from approximately 6.2 kb (human) to 7.7 kb (duck).[1] They are generally classified according to the species they infect, although species crossover has been observed for some astroviruses, for example, astroviruses of chickens have been detected in turkeys.[2] Astroviruses have been discovered to infect many species including humans. The first observations of astroviruses in humans were in 1975 in the feces of children with diarrhea and gastroenteritis by Appleton and Higgens[3] and Madeley and Cosgrove, who named this group of viruses because of its characteristic star-like appearance.[4]

12.1.1 CLASSIFICATION

Astroviruses are classified in the family *Astroviridae*, with species infecting mammals forming the genus *Mamastrovirus* while those infecting birds forming the genus *Avastrovirus*. Currently, at least six species are identified in the genus *Mamastrovirus*, including bovine astrovirus (BAstV), feline astrovirus (FAstV), human astrovirus (HAstV), mink astrovirus (MAstV), ovine astrovirus (OAstV), and porcine astrovirus (PAstV). These viruses have been detected in the feces of cattle,[5,6] sheep,[7,8] deer,[9,10] pigs,[11,12] dogs,[13,14] mink,[15] sea lions,[16] rabbits,[17,18] rodents,[19,20] and bats.[21] For a recent review of human and mammalian astroviruses, see *Astrovirus Research, Essential Ideas, Everyday Impacts, Future Directions.*[22]

Classification of avian astroviruses is still somewhat in its infancy and, to date, only six different astroviruses are recognized as belonging to the *Avastrovirus* genus by the International Committee on Taxonomy of Viruses. The first report of disease caused by astroviruses was in ducklings in 1965;[23] however, it was not until the mid-1980s that it was recognized as an astrovirus by electron microscopy (EM).[24] This virus is referred to as duck hepatitis virus 2 (DHV-2) in older papers[25] but was reclassified as duck astrovirus serotype I (DAstV-1) more recently. A second avian astrovirus of ducks, antigenically and genetically distinct from DAstV-1, that also causes hepatitis in ducklings was reclassified as DAstV-2 after being previously known as DHV-3.[26]

There are two astrovirus species both associated with growth problems and kidney lesions in chickens. The first of these was isolated from a 1-week-old, normal broiler chick in 1976[27] and was subsequently termed avian nephritis virus[28] (ANV), but was originally described as a picornavirus and only later recognized as an astrovirus.[29] A second astrovirus species of chickens was discovered in 2004 and named chicken astrovirus[30] (CAstV), but earlier isolations of similar strains were described as enterovirus-like viruses.[26,31] Two astrovirus species were discovered in turkeys (TAstV): the first, turkey astrovirus serotype 1 (TAstV-1), in the UK in 1980[32] and the second, TAstV-2, was described in 2000.[33] In addition, avian astroviruses have been discovered in the feces of a number of wild birds including robins, shrikes, and great tits.[34]

Sections 12.1.4.2 and 12.2.2 will focus on members of the *Avastrovirus* genus, although much of the understanding of the generalized features of astroviruses relating to morphology and biology have been derived from studies into human astroviruses and are assumed to also extend to avian astroviruses with some corroborative evidence.

12.1.2 STRUCTURE, BIOLOGY, AND EPIDEMIOLOGY

Astroviruses are relatively simple viruses that share a generalized arrangement, in terms of both their component structures and their genomes. Astrovirus genome consists of only three open reading frames (ORF), the first two of which, ORF 1a and ORF 1b, encode nonstructural proteins including a protease (ORF 1a) and an RNA-dependent RNA polymerase (ORF 1b). The third ORF, ORF 2, encodes capsid protein, which is the most variable region of the genome, especially in the 3′ half.[35] There is a short 5′ untranslated region (UTR) prior to ORF 1a and a longer 3′ UTR after ORF 2. A polyadenylated tail at the extreme 3′ region completes the positive-sensed RNA genome, which, similar to messenger RNA, can be directly translated within the host cell.

While all astroviruses share this genetic arrangement, there are differences between species including overall genome length, the length of individual ORFs, and the frames in which the genomic or subgenomic transcripts are translated. In each arrangement exists a heptameric ribosomal frameshift signal (AAAAAAC) at the 3′ end of ORF 1a similar to those identified in retroviruses.[36] ORF 1b is in a different (+1) reading frame and the frame of ORF 2 determines which of three genomic arrangements the virus possesses: in ANV-like astroviruses, ORF 2 is in the same frame as ORF 1a; in CAstV-like genomes, ORF 2 is in the same frame as ORF 1b; and in DAstV-like astroviruses, ORF 2 is in a different reading frame to the preceding ORFs.[37]

Other notable sequences within the avian astrovirus genomes include a highly conserved 43-nucleotide-long s2m RNA stem-loop motif located at the 3′ end of the genome, which is also found in members of the *Mamastrovirus* genus, and picornavirus, calicivirus, and coronavirus families including SARS.[38] It appears to be one of the few mobile genetic elements found in viruses, and while its function is currently under investigation, the lack of the s2m motif in TAstV-2[33] and in certain mammalian astroviruses[20,21] would imply that it is not critical to astrovirus infection. There are short, conserved motifs in the 5′ UTR and in the region immediately upstream of ORF 2, which are possible promoter signals controlling transcription of the genomic and subgenomic messenger RNAs, respectively.

The morphology of astrovirus particles was originally observed by negative-stain EM of human astrovirus when a distinctive 5- or 6-pointed, star-shaped appearance was observed in about 10% of fecal preparations[3] (Figure 12.1). This characteristic morphology gave rise to the name astrovirus, from the Greek word *astron*, meaning stars. The star-like appendages are attributed to the capsid spikes that protrude from the capsid of the virion, which possesses a $T = 3$ icosahedral capsid symmetry comprised of 180 capsomers. The capsomers and spikes are formed from the translated product of ORF 2, VP90, which is matured posttranslationally by cleavage of VP90 into smaller subunits by cellular caspases[39] and the viral protease encoded in ORF 1a. Analysis of the capsid precursor protein across a range of human, avian, and animal astroviruses reveals a relatively conserved N terminal region from residue 1 to 415 followed by a hypervariable C terminal region.[35,40,41] The N terminus contains the protein region responsible for capsomer formation and ultimately capsid assembly, which requires a higher degree of stability out of functional necessity and therefore puts it under evolutionary restraint while the hypervariable C terminus, cleaved into smaller proteins, comprises the spikes, which interact with the hosts' immune systems; thus, genetic divergence to evade immune activity is advantageous.

Since some astroviruses can prove difficult to adapt to grow in cell culture, for instance, ANV, virus isolation is unsuitable and inconvenient as a diagnostic approach in most cases. Furthermore, since the characteristic star-like conformation is absent from many preparations of astroviruses, EM is also not a reliable tool for the diagnosis of avian astroviruses. Therefore, the majority of epidemiological data for avian astroviruses has been serological and, more recently, molecular in nature and predominantly relates to commercially farmed poultry especially chickens, ducks, and turkeys.

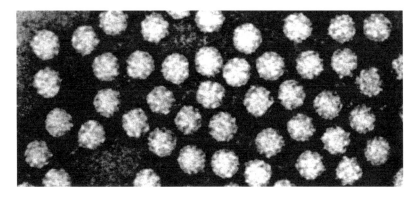

FIGURE 12.1   Negatively stained electron micrograph of turkey astrovirus displaying the characteristic star-like appearance. This morphology may not be present in most avian AstV preparations.

The presence of ANV in commercial poultry was identified much earlier than that of the distinct astrovirus species, CAstV. An indirect immunofluorescence (IIF) test for ANV serotype-1 (ANV-1) was developed and used to screen historical sera from Japanese broiler flocks finding an incidence of 12%–17% in 35%–62% of flocks from 1975 to 1978.[42] A serological survey of small numbers of historical sera (1975–1983) from 139 farms in Northern Ireland (NI) comprising broilers, broiler breeders, and layer chickens detected ANV in 135 farms, many with 100% detection rate.[43] Additionally, two-thirds of nine NI turkey flocks were also found to have seroconverted against ANV by IIF.[43]

A seroprevalence investigation of four generations of commercial chickens from great-grandparent (GGP) to broiler flocks in the United Kingdom for two strains of CAstV with low cross-reactivity by IIF, and which belong to separate CAstV genogroups, CAstV A and CAstV B, according to capsid–RNA sequence determination,[35] found the widespread presence of both strains (CAstV A 96%, CAstV B 93%) in 24–28 broiler breeder flocks aged 23–26 weeks from three organizations, and in 78% of the serum samples from 10 broiler flocks, 100% of which contained positive sera.[44] Antibodies to CAstV were also found in pooled serum samples from 25 parent flocks from eight countries spread across Northern, Western, and Eastern Europe.[44] This study also found that CAstV infections were less common in UK GGP (21%) and GP (28%) flocks than in the broiler and parent flocks.

Although serological tests such as IIF and ELISA continue to be useful tools in the detection and surveillance of avian astrovirus infections in poultry, there are limits to their usefulness as they may not detect different serotypes.[45] Since the late 1990s, there has been a move to the use of molecular tests such as reverse transcription polymerase chain reaction (RT-PCR) to diagnose the presence of these viruses in flocks based on specific genetic sequences of the viral genomes, which can be designed for broad detection across genotypes. Initially, the tests detected single species of astrovirus, for example, RT-PCR tests based on the conserved ORF 1b and variable ORF 2 of TAstV demonstrated the widespread presence of TAstV in flocks affected with poult enteritis and mortality[46] across the United States and also in turkey and guinea fowl flocks from northern Italy, which, when the PCR products were identified by DNA sequencing, were similar to TAstV-2.[47]

More recently, molecular tests have been designed to detect the presence of multiple enteric viruses including different astroviruses simultaneously (TAstV-1, TAstV-2, CAstV, ANV, and avian rotavirus)[2,48,49] and have been used to monitor chicken and turkey flocks in the United States, where astroviruses, especially TAstV-2, appear widespread.[2,50,51] While TAstVs are associated with diarrhea and enteritis in young poults, in common with other avian AstVs, they are also detected in apparently healthy flocks as well as those with reduced performance.[50] Although the application of molecular tests has improved the sensitivity of diagnosis of TAstV-2 compared to more traditional techniques such as EM,[50] there is little information available comparing prevalence of TAstV-2 in healthy versus affected flocks, but a recent study by Jindal et al. reports that enteric viruses including TAstV-2 have a lower prevalence in healthier flocks.[52]

In contrast to other avian AstVs, which are not typically associated with mortality, infections by the two known AstVs of ducks, DAstV-1 and DAstV-2, are reported to cause a fatal hepatitis in young ducklings. Infections by DAstV-1, previously known as duck hepatitis virus II, have resulted in losses of up to 50% in commercial duck farms in the United Kingdom[25] and China.[37] A limited serological survey of three commercial flocks in China reported DAstV-1 seroprevalence of 63% and 51% in two flocks that had experienced DAstV-1 infections, while a third flock that had not shown symptoms of a DAstV-1 infection had a seroprevalence of 28% with positive test values that were noticeably lower than those from the two diseased flocks.[53]

The application of specific molecular tests for TAstV-1, TAstV-2, CAstV, ANV, and DAstV (types 1 and 2) to organ samples from 22 Croatian duck flocks including duck embryos detected the presence of TAstV-2, ANV, or CAstV in 13 flocks but did not detect the presence of DAstV or TAstV-1 in these samples. Of these, ANV was the most predominant AstV detected in 77% of the AstV-positive flocks.[54] A second study by this group subsequently identified the presence of TAstV-1 in duck embryos from two flocks, and ANV was detected in both duck and goose embryos from flocks affected by stunting and prehatching mortality.[55] Epidemiological data is not widely available detailing the cross detections of various AstVs in different species of poultry flocks, which raises the question of how to develop a robust yet cost-effective testing strategy for the thorough detection of several AstVs simultaneously in order to provide adequate surveillance of commercially important birds.

12.1.3 CLINICAL FEATURES AND PATHOGENESIS

Avian astroviruses are enteric pathogens transmitted by fecal–oral route along with other viruses and bacteria, which comprise the intestinal virome and microbiome. Being non-enveloped, AstVs are more resistant to cleaning and may be persistent in poultry houses where darkling beetles can act as vectors for them.[56] For example, CAstV has been detected by RT-PCR in washings from the surface of darkling beetles and from internal tissues.[56] Astrovirus infections usually occur within the first weeks of life, and some can also be vertically transmitted from naïve in-lay parent birds, and the earlier they are contracted the worse the outcome, although this will depend not only on the particular AstV species but also the specific strain of infecting AstV that can vary substantially in pathogenicity. Other important factors include the load (dose) of infectious virus received and the presence of other enteric pathogens.

Generalized clinical features of astrovirus infections are the association with enteritis and, in some cases, diarrhea. However, a number of avian AstVs are also associated with diseases of other organs including nephritis and hepatitis. While some outbreaks may be associated with high mortality, in general, mortality is low for the majority of avian AstV infections. Morbidity, giving rise to reduced weight gain, increased culling, impaired development (requiring longer rearing times), uneven flock performance, and poorer feed conversion rates, is a common problem causing substantial economic losses to commercial producers. In some instances, impaired weight gain can exceed 50% and may affect a significant proportion of a flock. There have been difficulties in trying to experimentally determine the nature and extent of the problems caused by individual AstVs in birds due to the more complex etiology of enteric diseases and the multifactorial conditions that are present in wild-type infections. In addition, many avian AstVs, for example, ANV, can prove difficult to isolate in cell culture causing further problems.

12.1.3.1 Avian Nephritis Virus

Although ANV is highly contagious, it typically results in subclinical infections of young broiler chicks; however, its clinical effects can range from subclinical to growth retardation to mortality. Histological changes occur primarily in the kidney as a result of ANV infections, causing lesions in the kidneys and, to a lesser degree, in the intestines, where small numbers of shrunken, degenerate epithelial cells were found near the base of the villi in 3–6-day-old chicks 1 day after oral inoculation with ANV-1.[57] The detection of ANV in other organs including the liver, pancreas, and spleen has also been reported.

Isolates of the G-4260 (aka ANV-1) and three other strains of ANV resulted in interstitial nephritis when orally inoculated into 1-day-old specific-pathogen-free (SPF) chicks.[58] A third of these chicks inoculated with G4260 died with visceral urate deposits, and all chicks inoculated with G4260 were significantly lighter than control birds and yet a second strain, termed IR-N, which was classified as the same serotype as G-4260, resulted neither in mortality nor in significant weight inhibition, which demonstrates the diversity in pathogenicity exhibited by different ANV strains. The two other strains, which were identified as a different serotype (ANV-2) to G-4260 based on levels of sequence diversity suggestive of serotypic differences, caused limited mortality and reduced weight retardation.[58] Subsequently, sequencing of the ORF 2 of ANV strains from problem flocks in the United Kingdom, Germany, and Belgium has identified four putative new serotypes based on similar levels of sequence differences that exist between ANV-1 and ANV-2, but no clinical information is available for these new serotypes.[41]

Recently, de Wit and colleagues reported on the isolation of 10 new ANV strains from chicken and turkey flocks suffering from runting–stunting syndrome (RSS) and locomotory problems over an 18-year period from Germany, the Netherlands, and United Arab Emirates, which cluster relatively tightly on the basis of shared amino acid identity of 93%–99% with the group representative, isolate 19.[59] The group is tentatively named ANV-3 since the closest match to published avian AstVs was 88% with ANV-1 by amino acid sequence although antibodies to isolate 19 have little or no cross-reactivity with ANV-1 and vice versa.[59] These strains

were isolated from a variety of less common tissues including tracheas, cecal tonsils, tendons, and tibia of both turkeys and chickens, hock cartilage, joints, and synovial fluids suggesting that the tropism of this new ANV group may be wider than what has previously been observed for ANV-1 and ANV-2. Experimental inoculations of isolate 19 into 3-week-old SPF birds and the placing of noninoculated sentinel birds with the primary subjects resulted in limited diarrhea and some mortality, but the most common findings were interstitial nephritis and tubular degeneration although no symptoms were observed in the tendons.[59]

12.1.3.2 Chicken Astrovirus

CAstV was only recently recognized as a separate species from ANV when three similar isolates from Dutch broiler chicks with RSS were shown to be serologically distinct from ANV-1 and shared only 55% amino acid identity with ANV-1 in a region of the conserved nonstructural protein in ORF 1.[30] Subsequently, it has become apparent that, similar to other AstVs, there are many different strains of CAstV in circulation, again with varying degrees of pathogenicity and antigenic relatedness. These strains have been provisionally classified into two groups, CAstV A and CAstV B, according to a lower level of shared amino acid identity across ORF 2, where the hypervariable regions associated with antigenicity are located and which is supported by only a minor degree of cross-reactivity with the heterologous antisera.[35]

Similar clinical symptoms to those from ANV infections have been associated with CAstV infections of broiler flocks: diarrhea, enteritis, poor growth, and kidney disease. In addition, the involvement of CAstV in other symptoms such as poor feathering and clubbed down has been postulated but not conclusively demonstrated. In general, CAstV strains are not associated with mortality, but particular strains of CAstV, identified as a new subgroup of the CastV B group and designated Indian subgroup Bi, were implicated as the causative agent in outbreaks of visceral gout in week-old broiler flocks in India in 2011 and 2012 resulting in mortality approaching 40%.[60] Affected chicks presented with swollen kidneys, prominent ureters, visceral gout, and histopathological examinations revealed inflammation of the kidneys with necrosis and degeneration of epithelial cells in the proximal convoluted tubules.[60]

There is evidence also that some strains of CAstV may be vertically transmitted as in the case of the hatchability problem, "White Chicks."[61] This condition is characterized by weak, white chicks that fail to thrive and, since 2006, has led to increased embryo deaths annually and averaged a 29% reduction in hatchability in affected Finnish flocks over a brief period of about 2 weeks while the parent birds seroconverted against CAstV. Levels of CAstV RNA detected in the tissues of white chicks and dead embryos were high, while unaffected chicks were negative. The livers of white chicks were normal sized or smaller, greenish or mottled, with hypertrophic bile ducts. Inflammation was present in the heart, kidney, bursa, and perivascular tissue of the liver.[61] These strains have been identified also as a separate subgroup of CAstV B.[61]

12.1.3.3 Turkey Astrovirus

There are two serologically distinct species of TAstV, types 1 and 2, which are both reported to be widespread in the United States, with TAstV-2 the more prevalent.[50] TAstV-2 is associated with two major disease syndromes affecting turkey production: poult enteritis complex or syndrome (PEC or PES) and poult enteritis mortality syndrome (PEMS).[33,50,52] Symptoms, which are generally mild to moderate, usually appear between 1 and 3 weeks and can last up to 14 days[62] and include diarrhea, listlessness, nervousness, litter eating, and, in keeping with other avian AstVs, reduced weight gain. Subclinical TAstV infections may occur in healthy flocks.

Poults experimentally infected with TAstV-2 presented with swollen intestines, watery and persistent diarrhea, transiently reduced bursa size in the first 9 days postinfection, and substantially reduced growth.[63] In a further experimental inoculation of 2-day-old SPF poults with three strains of TAstV-2 that were deemed representative of the three most commonly circulating TAstV-2 groups in a 2005 survey of U.S. farms, clinical features also included diarrhea and growth depression.[64] Additional observations included distended intestines containing watery contents and undigested feed, and dilated ceca with foamy contents. In the first 6 days postinfection, mortality of poults ranged from 15% to 25% for the three isolates and was associated with weakness arising from anorexia resulting from the poults' refusal to eat and drink.[64] The main histopathological findings were microscopic intestinal lesions of mild intensity, consisting of shrunken, degenerate cells in the villi and crypt epithelium and crypt hyperplasia with increased crypt depth. Mild villous atrophy and mild lymphocytic infiltration were most apparent in the jejunum; but inflammation was also observed in the pancreas, liver, spleen, and kidneys.[64]

12.1.3.4 Duck Astrovirus

There are two known AstVs of ducks, DAstV-1 and DAstV-2, which were previously termed duck hepatitis virus 2 and 3, respectively, and which were originally thought to be picornaviruses but were later shown to be AstVs: DAstV-1 firstly by EM diagnosis based on virion morphology[24] and both DAstV-1 and DAstV-2 by RNA sequence analysis of ORF 1b.[26] Both viruses can cause a fatal hepatitis in young ducklings, which in UK cases involving DAstV-1 infections in the 1980s resulted in losses of up to 50% up to 2 weeks of age and mortality of between 10% and 25% for ducklings aged between 3 and 6 weeks.[25] The ducklings died less than 2 h after the clinical signs first appeared and, although in good condition, many had opisthotonos. Symptoms included polydipsia and loose droppings that were whitish due to extreme urate excretion. The livers contained multiple hemorrhages and extensive necrosis of the hepatocyte cytoplasm was observed, as was bile duct hyperplasia. Blood vessels stood out strongly against the pale and swollen kidneys.[25]

Similar levels of 50% mortality in young ducklings were experienced in China in 2008 during an outbreak of duck hepatitis caused by DAstV-1[37] whereas low mortality rates were reported for two Chinese duck flocks recently infected by DAstV-1 at 3 weeks (4.67%) and 6 weeks (0.63%)[53] indicating a range of pathogenicities for different strains of DAstV-1 as seen with other avian AstVs. Reports of DAstV-2 infections appear less common than DAstV-1 and are reportedly less severe; however, an outbreak of duck hepatitis caused by DAstV-2 in the 1960s in Long Island, USA, resulted in losses up to 30%.[65]

12.1.4 DIAGNOSIS

Astroviruses are one of a number of endemic viruses that can cause problems for the young of domestic poultry and are commonplace in commercial production units, although they are also present in subclinical infections. This complicates the process of diagnosis since the development of disease can be affected by a combination of factors, such as the susceptibility of breeds; the age of the birds when infected; the dose of virus they experience; the quality of their feed, which may promote or undermine their underlying gut health; and the combination of enteric pathogens they experience, including those that may be transmitted vertically as well as horizontally acquired infections. It may prove helpful to quantify specific AstV species and, where possible, identifying specific genotypes, and comparing these to known strains of clinical significance, is recommended.

12.1.4.1 Conventional Techniques

Astroviruses were first discovered in 1975 by the use of negative-contrast EM[3], and EM continued to be used to identify those avian AstVs that display the characteristic star-like morphology including TAstVs and DAstV-1[25,32] until the development of sensitive molecular tests from the late 1990s onward, which have largely replaced EM for routine diagnosis. There are difficulties with using EM directly on samples since not all AstVs are star-shaped and many preparations are indistinguishable from those containing other small, round viruses, and positive identification relies on the experience of trained operators. This process is time consuming, labor intensive, and relatively insensitive, so it is not suitable for testing large numbers of samples, although sensitivity can be enhanced by firstly propagating the virus in embryonated SPF eggs or, in some cases, primary cells, such as chick embryo liver cells, using purified sample homogenates.

The number and variety of conventional tests for diagnosis of specific astroviruses are relatively small, which is in part due to difficulties in producing convenient diagnostic tests through the limited availability of specific antibodies. An immunofluorescence (IF) test was developed and diagnosed the presence of DAstV-2 (DHV-III) virus in various tissues from experimentally infected ducklings,[66] and an indirect IF (IIF) test was used to examine the localization and intensity of the presence of ANV antigen in kidney and other tissues.[27,28] More recently, IIF tests were developed for detecting the two major groups of CAstV, Groups A and B, from chicken tissue samples.[35] However, these tests are labor intensive and required skilled operators to interpret the results, so more convenient tests have been developed, such as the enzyme-linked immunosorbent assay (ELISA), which can be automized to some degree and has the added benefit of more objective quantification through the analysis of optical densities by a colorimeter, which does not require subjective operator interpretation. An antigen-capture ELISA was described in 2005 for the detection of TAstV-2 antigen from fecal matter or gut contents,[48] but with limited availability of antibodies and a shift toward molecular detection of virus in recent years, most of the ELISAs developed have been produced to detect astrovirus antibodies in sera, which is a more convenient way to screen sera than the use of serum neutralization tests.

An ELISA for the detection of antibodies against ANV-1 was reported in 1991 using the G4260 strain grown in cell culture.[67] However, poor and inconsistent propagation of some avian AstVs in cell culture has led to an increase in the use of recombinant protein expression technologies to produce viral antigen. Recombinant protein technology has since been used to produce artificial viral antigen for two separate ELISAs for detection of CAstVs more recently[68,69] and also for DAstV[53] and for a Chinese-prevalent ANV strain that is genotypically different to both ANV-1 and ANV-2.[70] However, ELISAs based on single strains are unlikely to detect the full range of serotypic diversity exhibited by most AstVs.

12.1.4.2 Molecular Techniques

In keeping with diagnostic test development for other viruses, there has been a move toward the use of molecular techniques to provide rapid and sensitive results. One of the limitations when designing molecular tests is that they rely on available genetic sequence information, so in general, a test will only be as good at detecting diverse strains of a virus as the sequencing data that is available at that time. The first RT-PCR tests for TAstV were published in 2000, which detected the presence of TAstV RNA in samples from poults affected by PEMS.[46] The authors produced a pair of primers within the highly conserved ORF 1b region and a second pair in the more variable ORF 2 region, which is a useful strategy generally for avian AstVs depending on whether the primer pair is used to detect a wide range of strains, in which case a conserved region is better, or for amplification of very specific strains then a variable region would be recommended.

However, the level of sequence diversity is so extensive in avian AstVs, even within some relatively conserved nonstructural regions, that for broad coverage, it is necessary to compare the sequences of many strains and may also require the use of degenerate primers to detect as many different strains as possible. In primer regions, where different strains have different nucleotides at one or more sites, two or more individual primers can be produced, each containing one of the variant DNA nucleotides at a particular site, and these variant primers are then combined into one degenerate primer mix so that all variant permutations are covered. In degenerate primer nomenclature, these variant positions are given unique International Union of Pure and Applied Chemistry codes, for example, a cytosine or thymine at a single nucleotide position is denoted

as a "y." The high levels of sequence diversity exhibited by astroviruses complicate primer design and, in some instances, a compromise between specificity and sensitivity may be required as primer locations may, by necessity, be restricted to more conserved areas that may not be the most sensitive. Empirically, testing several pairs of primers within conserved areas in order to compare performance is therefore advisable.

12.2 METHODS

The molecular methods for detecting avian astroviruses covered in this section are all PCR based and comprise conventional RT-PCR and real-time, quantitative RT-PCR (RT-qPCR). The workflow consists of sample preparation, RNA purification, and application of avian AstV RT-(q)PCR tests to the purified RNA. The choice of sample and the format in which it is sampled, for example, tissues, swabs, and fecal matter, will depend not only on the specific virus and its known tropism but also on the age of the birds; in general, AstV infections of poultry may only be detected for 2–3 weeks in the tissues but will be excreted for longer in fecal matter. To screen flocks for the presence of avian AstVs, especially more expensive and longer-lived birds such as GPs, the use of cloacal swabs is recommended or boot swabs could also be used, which are noninvasive and require less sampling to achieve adequate flock coverage.

12.2.1 SAMPLE PREPARATION

Avian AstVs can be detected in a number of tissues, organs, and embryos in addition to intestinal contents and feces, all of which can be stored frozen at −20°C, or small amounts may be smeared onto FTA cards and stored in the short term at either +4°C, which is preferable, or at room temperature, which provide a convenient matrix for transportation. FTA cards are suitable for viral RNA preservation but appear to inactivate the virus making them unsuitable for virus isolation. In addition, the quality and quantity of recovered RNA may be less than that from fresh or frozen sample material, making FTA cards less suitable for large amplicons such as the capsid gene in ORF 2, although small amplicons can be successfully generated, which makes this substrate useful for real-time RT-qPCR tests. Cloacal swabs and fabric boot swabs have been successfully used for sampling of avian AstVs, although there is some debate as to whether wooden handled swabs contain substances that can inhibit the PCR process. Suitable tissues for sampling, in addition to feces, are described in Table 12.1.

Single or pooled swab tips are cut off and immersed in the minimal amount of room temperature phosphate buffered saline (PBS, pH7.2) needed to just cover them in a suitably sized container (0.5–5 mL depending on numbers and size of swab tips) for 15–30 min, followed by the addition of glass beads (5–10) and vortexing to produce a homogenate, which is then decanted for testing. Bootswabs, single or pooled in a small container, for example, beaker, are soaked in the minimal amount of PBS to cover them and incubated as for

TABLE 12.1
Tissue Types Suitable for Detection of Avian Astroviruses

	ANV and CAstV	DAstV	TAstV
Tissue	Intestines	Intestines	Intestines
	Kidney	Liver	Kidney
	Cloaca/vent swabs	Kidney	Thymus
	Liver	Spleen	Cecum
	Embryos	Embryos	Bursa
	Meconium		Spleen

swabs with occasional stirring. The homogenates may require centrifugation for 10–30 min at 2000 × *g* if the preparation still contains much solid matter after vortexing or stirring. Intestinal contents, meconium and feces, are made into a 10% homogenate in PBS (weight to volume), and then shaken vigorously or vortexed with glass beads and may require centrifugation as described to clarify. A small amount of tissue, ground in a mortar and pestle with a thin layer of sterile sand, is made into a 10% homogenate in PBS (weight to volume) and centrifuged as described earlier, which removes the sand and unground tissue. All supernatants are retained for use in testing.

12.2.2 Detection Procedures

Before molecular tests can be applied, the viral RNA must be purified from the clarified sample homogenate. Two of the most popular methods involve the use of proprietary RNA extraction kits that are either cartridge based or use the phenol–chloroform method. The advantage of the first method is that the RNA is conveniently eluted as a solution, which does not require further dissolving, but it is relatively expensive. The phenol–chloroform method is less expensive but requires dissolving a pellet of RNA, which, due to the minute amounts of RNA that are generally purified from these small viruses, is extremely small and practically invisible, making it difficult to dissolve by vigorous pipetting. Heating the RNA solution to 55°C for 10 min after the pipetting and then storing at −80°C for a few hours, preferably overnight, improves the RNA solubility. The amount of RNA obtained from avian AstVs is generally at subpicogram levels and therefore not readily quantifiable.

The choice of molecular tests for the detection of avian AstVs depends largely on the purpose for which the test is being run, for instance, a simple flock screening tool could make use of short amplicons in conserved regions to provide a broad detection range by either conventional RT-PCR or real-time RT-qPCR, whereas AstV phylogeny is probably most informative based on the highly variable ORF 2; the most reliably conserved regions within which to site primers are at the extreme 5′ and 3′ ends of the ORF, which produces a large amplicon greater than 2 kb.

Since the high degree of sequence variability typical of avian AstVs makes it difficult to design RT-PCR primers that can detect all strains from a particular species, a primer pair that can detect multiple strains from several species is all the more remarkable. Based on the relatively conserved ORF 1b, the degenerate primers AstVpol-For (5′-GAYTGGACIMGITAYGAYGGIACIATICC-3′) and AstVpol–Rev (5′-YTTIACCCACATICCRAA-3′) will amplify a fragment of approximately 434 bp, corresponding to nucleotides 3799–4233 in the genome of the G4260 strain of ANV-1 (GenBank accession no. AB033998) and allow detection of strains of ANV, CAstV, TAstV, and DAstV. This procedure can be used as a screening tool for unknown AstVs by conventional, gel-based PCR.[26] Furthermore, Tang and colleagues also designed a degenerate RT-PCR test (TAP-L1 and TAPG-R1) that generates a 597 bp fragment, corresponding to nucleotides 3550–4147 of the ANV-1 sequence, from TAstV-1, TAstV-2,[48] and ANV. But a similar drawback of both assays is their inability to distinguish which avian AstV has been detected without performing DNA sequencing on the amplicons.

A multiplex RT-PCR test has been designed that can detect TAstV-1 and TAstV-2 and ANV as well as avian rotavirus simultaneously in one reaction and differentiates on the basis of amplicon size;[49] therefore, sequencing of the amplicon is not necessary. A similar assay was designed to detect and distinguish ANV, CAstV, and avian rotavirus by the same authors.[49] A drawback of these tests is that they are relatively insensitive for all of the AstV species compared to the rotavirus, which, as the authors speculated, may be due to the use of degenerate primers for the AstVs but not for the rotavirus.[49] These multiplex assays were developed for use in flock screening where they suggest that lower sensitivity is less of an issue.[49]

In addition to the degenerate and multiplex assays, a number of conventional RT-PCR tests have been published for the detection of ANV,[71,72] CAstV,[73] DAstV,[55] and TAstV,[46,47] which are useful for screening of flocks. In addition, real-time RT-qPCR tests have been developed for both ANV and CAstV,[74] which can provide quantitative information regarding viral loads that may be helpful in determining the stage of infection and can shed light on whether infections are vertically or horizontally acquired. Another factor that can be helpful when investigating the clinical significance of an avian AstV infection is to determine which strain is present, and RT-PCR tests have been published that amplify all or part of the most variable region of the genome, ORF 2, including ANV,[41] CAstV,[35] DAstV,[53] and TAstV,[46] which facilitate comparisons with published strains associated with clinical significance.

12.3 CONCLUSIONS AND FUTURE PERSPECTIVES

Avian AstV infections are common in commercially farmed poultry where they are associated with PESs of turkeys, runting–stunting syndrome and kidney disease of chickens, hatchability problems in chicks, and a fatal hepatitis of ducklings. Equally, they can be detected in apparently healthy birds as subclinical infections, which suggests that other factors, such as the particular strain of AstV, the infecting virus load, or the presence of other pathogens, may be influencing the development of disease. Molecular tests are good for detecting virus but they may not detect disease. High AstV load may be necessary to correlate with clinical signs and so quantitative molecular tests, in particular, offer this facility. Avian AstVs display a high degree of genomic variability, especially in ORF 2, which codes for the immunogenic capsid protein, but even the more conserved areas are subject to substantial genomic instability, which makes designing comprehensive molecular tests that can detect multiple strains difficult. Molecular tests for avian AstVs are regularly being improved as the sequences from more strains become available through the public sequencing databases.

Recently designed molecular tests have recognized that other endemic viruses are often present as coinfections in poultry intestinal samples, such as avian reovirus, coronavirus, and rotavirus, and it is not uncommon for broiler chickens to be infected with both ANV and CAstV concomitantly.[74] Metagenomics is being used widely to research the communities of bacteria, known as microbiomes, that inhabit poultry guts, and a similar approach has started to investigate the communities of enteric viruses that populate the poultry gut. A recent metagenomic investigation of the intestinal tracts of turkeys from farms with a history of enteric disease by Day et al. has identified almost 20 genera of RNA enteric viruses in addition to bacteria, eukaryotes, and archaeotes.[75] Discovering which DNA viruses are also present and performing comparative metagenomics of the complete virome down to the strain level for birds with good and poor performance should start to reveal the underlying causes of these conditions and exclude benign strains or those resulting in subclinical infections. Improvements in the sequencing technologies, in particular achieving longer read lengths and the ability to tag individual sample preparations and multiplex them, combined with reductions in running costs and simplified bioinformatics analysis pipelines should turn comparative virome metagenomics into a routine diagnostic test in the not too distant future, which should be vastly more informative than current methods.

REFERENCES

1. Méndez, E. and Arias, C.F., Astrovirus, in Knipe, D.M. and Howley, P.M. (Eds.), *Fields Virology*, 6th edn. Lippincott Williams & Wilkins, Philadelphia, PA, p. 611, 2013.
2. Pantin-Jackwood, M.J., Spackman, E., and Woolcock, P.R., Molecular characterization and typing of chicken and turkey astroviruses circulating in the United States: Implications for diagnostics. *Avian Dis.*, 50, 397, 2006.
3. Appleton, H. and Higgens, P.G., Letter: Viruses and gastroenteritis in infants. *Lancet*, 305, 1297, 1975.
4. Madeley, C.R. and Cosgrove, B.P., 28 nm particles in faeces in infantile gastroenteritis. *Lancet*, 6, 451, 1975.
5. Woode, G.N. and Bridger, J.C., Isolation of small viruses resembling astroviruses and caliciviruses from acute enteritis of calves. *J. Med. Microbiol.*, 11, 441, 1978.
6. Tse, H. et al., Rediscovery and genomic characterization of bovine astroviruses. *J. Gen. Virol.*, 92, 1888, 2011.

7. Snodgrass, D.R. and Gray, E.W., Detection and transmission of 30 nm virus particles (astroviruses) in faeces of lambs with diarrhea. *Arch. Virol.*, 55, 287, 1977.
8. Jonassen, C.M., Jonassen, T., Sveen, T.Ø., Sveen, T.M., and Grinde, B., Complete genomic sequences of astroviruses from sheep and turkey: Comparison with related viruses. *Virus Res.*, 91, 195, 2003.
9. Tzipori, S., Menzies, J.D., and Gray, E.W., Detection of astrovirus in the faeces of red deer. *Vet. Rec.*, 108, 286, 1981.
10. Smits, S.L. et al., Identification and characterization of deer astroviruses. *J. Gen. Virol.*, 91, 2719, 2010.
11. Lan, D. et al., Molecular Characterization of a porcine astrovirus strain in China. *Arch. Virol.*, 156, 1869, 2011.
12. Xiao, C.-T., Halbur, P.G., and Opriessnig, T., Complete genome sequence of a newly identified porcine astrovirus genotype 3 strain US-MO123. *J. Virol.*, 86, 13126, 2012.
13. Williams Jr., F.P., Astrovirus-like, coronavirus-like, and parvovirus-like particles detected in the diarrheal stools of beagle pups. *Arch. Virol.*, 66, 215, 1980.
14. Martella, V. et al., Enteric disease in dogs naturally infected by a novel canine astrovirus. *J. Clin. Microbiol.*, 50, 1066, 2012.
15. Blomstrom, A.L. et al., Detection of a novel astrovirus in brain tissue of mink suffering from shaking mink syndrome by use of viral metagenomics. *J. Clin. Microbiol.*, 48, 4392, 2010.
16. Rivera, R. et al., Characterization of phylogenetically diverse astroviruses of marine mammals. *J. Gen. Virol.*, 91, 166, 2012.
17. Martella, V. et al., Astroviruses in rabbits. *Emerg. Infect. Dis.*, 17, 2287, 2011.
18. Stenglein, M.D. et al., Complete genome sequence of an astrovirus identified in a domestic rabbit (*Oryctolagus cuniculus*) with gastroenteritis. *Virol. J.*, 9, 216, 2012.
19. Phan, T.G. et al., The fecal viral flora of wild rodents. *PLoS Pathog.*, 7, e1002218, 2011.
20. Chu, D.K.W. et al., Detection of novel astroviruses in urban brown rats and previously known astroviruses in humans. *J. Gen. Virol.*, 91, 2457, 2010.
21. Chu, D.K.W. et al., Novel astroviruses in insectivorous bats. *J. Virol.*, 82, 9107, 2008.
22. Cattoli, G., Chu, D.K.W., and Peiris, M., Astrovirus infections in animal mammalian species, in Schultz-Cherry, S. (Ed.), *Astrovirus Research, Essential Ideas, Everyday Impacts, Future Directions*. Springer, New York, p. 135, 2013.
23. Asplin, F.D., Duck hepatitis: Vaccination against two serological types. *Vet. Rec.*, 77, 1529, 1965.
24. Gough, R.E., Collins, M.S., Borland, E., and Keymer, L.F., Astrovirus-like particles associated with hepatitis in ducklings. *Vet. Rec.*, 114, 279, 1984.
25. Gough, R.E., Borland, E.D., Keymer, L.F., and Stuart, J.C., An outbreak of duck hepatitis type II in commercial ducks. *Avian Pathol.*, 14, 227, 1985.
26. Todd, D. et al., Identification of chicken enterovirus-like viruses, duck hepatitis virus type 2 and duck hepatitis virus type 3 as astroviruses. *Avian Pathol.*, 28, 21, 2009.
27. Yamaguchi, S., Imada, T., and Kawamura, H., Characterization of a picornavirus isolated from broiler chicks. *Avian Dis.*, 23, 571, 1979.
28. Imada, T., Yamaguchi, S., and Kawamura, H., Pathogenicity for baby chicks of the G-4260 strain of the picornavirus "Avian Nephritis Virus". *Avian Dis.*, 23, 582, 1979.
29. Imada, T. et al., Avian nephritis virus (ANV) as a new member of the family *Astroviridae* and construction of infectious ANV cDNA. *J. Virol.*, 74, 8487, 2000.

30. Baxendale, W. and Metbatsion, T., The isolation and characterisation of astroviruses from chickens. *Avian Pathol.*, 33, 364, 2004.
31. McNeilly, F. et al., Studies on a new enterovirus-like virus isolated from chickens. *Avian Pathol.*, 23, 313, 1994.
32. McNulty, M.S., Curran, W.L., and McFerran, J.B., Detection of astroviruses in turkey faeces by direct electron microscopy. *Vet. Rec.*, 106, 561, 1980.
33. Koci, M.D., Seal, B.S., and Schultz-Cherry, S., Molecular characterization of an avian astrovirus. *J. Virol.*, 74, 6173, 2000.
34. Chu, D.K. et al., A novel group of avian astroviruses in wild aquatic birds. *J. Virol.*, 86, 13772, 2012.
35. Smyth, V.J. et al., Capsid protein sequence diversity of chicken astrovirus. *Avian Pathol.*, 39, 249, 2012.
36. Jiang, B. et al., RNA sequence of astrovirus: Distinctive genomic organization and a putative retrovirus-like ribosomal frameshifting signal that directs the viral replicase synthesis. *Proc. Natl. Acad. Sci. USA*, 90, 10539, 1993.
37. Fu, Y. et al., Complete sequence of a duck astrovirus associated with fatal hepatitis in ducklings. *J. Gen. Virol.*, 90, 1104, 2009.
38. Tengs, T. et al., A mobile genetic element with unknown function found in distantly related viruses. *Virol. J.*, 10, 132, 2013.
39. Mendéz, E., Salas-Ocampo, E., and Arias, C.F., Caspases mediate processing of the capsid precursor and cell release of human astroviruses. *J. Virol.*, 78, 8601, 2004.
40. Krishna, N.K., Identification of structural domains in astrovirus capsid biology. *Virol. Immunol.*, 18, 17, 2005.
41. Todd, D. et al., Capsid protein sequence diversity of avian nephritis virus. *Avian Pathol.*, 40, 249, 2011.
42. Imada, T. et al., Antibody survey against avian nephritis virus among chickens in Japan. *Natl. Inst. Anim. Health Q. (Tokyo)*, 20, 79, 1980.
43. Connor, T.J. et al., A survey of avian sera from Northern Ireland for antibody to avian nephritis virus. *Avian Pathol.*, 16, 15, 1987.
44. Todd, D. et al. A seroprevalence investigation of chicken astrovirus. *Avian Pathol.*, 38, 301, 2009.
45. Matsui, S.M. and Greenberg, H.B., Astroviruses, in Fields, B.N., Knipe, D.M., and Howley, P.M. (Eds.), *Fields Virology*, 3rd edn. Lippincott-Raven Publishers, Philadelphia, PA, p. 811, 1996.
46. Koci, M.D., Seal, B.S., and Schultz-Cherry, S., Development of an RT-PCR diagnostic test for an avian astrovirus. *J. Virol. Methods*, 90, 79, 2000.
47. Cattoli, G. et al., Co-Circulation of distinct genetic lineages of astroviruses in turkeys and guinea fowl. *Arch. Virol.*, 152, 595, 2007.
48. Tang, Y., Ismail, M.M., and Saif, Y.M., Development of antigen-capture enzyme-linked immunosorbent assay and RT-PCR for detection of turkey astroviruses. *Avian Dis.*, 49, 182, 2005.
49. Day, J.M., Spackman, E., and Pantin-Jackwood, M., A multiplex RT-PCR test for the differential identification of turkey astrovirus type-1, turkey astrovirus type-2, chicken astrovirus, avian nephritis virus and avian rotavirus. *Avian Dis.*, 51, 681, 2007.
50. Pantin-Jackwood, M.J. et al., Periodic monitoring of commercial turkeys for enteric viruses indicates continuous presence of astrovirus and rotavirus on the farm. *Avian Dis.*, 51, 674, 2007.
51. Pantin-Jackwood, M.J. et al. Enteric viruses detected by molecular methods in commercial chicken and turkey flocks in the United States between 2005 and 2006. *Avian Dis.*, 52, 235, 2008.
52. Jindal, N. et al., Detection and molecular characterization of enteric viruses from poult enteritis syndrome in turkeys. *Poult. Sci.*, 89, 217, 2010.
53. Wang, X. et al., Expression of the C-terminal ORF2 protein of duck astrovirus for application in a serological test. *J. Virol. Methods*, 171, 8, 2011.
54. Biđin, M. et al., Circulation and phylogenetic relationship of chicken and turkey-origin astroviruses detected in domestic ducks (Anas platyrhynchos domesticus). *Avian Pathol.*, 41, 555, 2012.
55. Biđin, M. et al., Astroviruses associated with stunting and pre-hatching mortality in duck and goose embryos. *Avian Pathol.*, 41, 91, 2012.
56. Rosenberger, J., Darkling beetles as vectors for bacterial and viral pathogens found in poultry litter. In *45th National Meeting on Poultry Health and Processing*, Ocean City, MD, 2010.
57. Smyth, J.A. et al., Studies on the pathogenicity of enterovirus-like viruses in chickens. *Avian Pathol.*, 36, 119, 2007.
58. Shirai, J. et al. Pathogenicity and antigenicity of avian nephritis isolates. *Avian Dis.*, 35, 49, 1991.
59. de Wit, J.J. et al., Detection and characterization of a new astrovirus in chicken and turkeys with enteric and locomotion disorders. *Avian Pathol.*, 40, 453, 2011.
60. Bulbule, N.R. et al., Role of chicken astrovirus as a causative agent of gout in commercial broilers in India. *Avian Pathol.*, 42, 464, 2013.
61. Smyth, V.J. et al., Chicken astrovirus detected in hatchability problems associated with "white chicks". *Vet. Rec.*, 173, 403, 2013.
62. Reynolds, D.L. and Saif, Y.M., Astrovirus: A cause of an enteric disease in turkey poults. *Avian Dis.*, 30, 728, 1986.
63. Koci, M.D. et al., Astrovirus induces diarrhea in the absence of inflammation and cell death. *J. Virol.*, 77, 11798, 2003.
64. Pantin-Jackwood, M.J., Spackman, E., and Day, J.M., Pathogenesis of type 2 turkey astroviruses with variant capsid genes in 2-day-old specific pathogen free poults. *Avian Pathol.*, 37, 193, 2008.
65. Toth, T.E., Studies of an agent causing mortality among ducklings immune to duck hepatitis virus. *Avian Dis.*, 13, 834, 1969.
66. Haider, S.A. and Calnek, B.W., In-vitro isolation, propagation and characterization of duck hepatitis virus type III. *Avian Dis.*, 23, 715, 1979.
67. Decaesstecker, M. and Meulemans, G., An ELISA for the detection of antibodies to avian nephritis virus and related entero-like viruses. *Avian Pathol.*, 20, 523, 1991.
68. Sellers, H. et al., A purified recombinant baculovirus expressed capsid protein of a new astrovirus provides partial protection to runting-stunting syndrome. *Vaccine*, 28, 1253, 2010.
69. Lee, A. et al., Chicken astrovirus capsid proteins produced by recombinant baculoviruses: Potential use for diagnosis and vaccination. *Avian Pathol.*, 42, 434, 2013.
70. Zhao, W. et al., Segmentation expression of capsid protein as an antigen for the detection of avian nephritis virus infection in chicken flocks. *J. Virol. Methods*, 179, 57, 2012.
71. Mandoki, M. et al., Phylogenetic diversity of avian nephritis virus in Hungarian chicken flocks. *Avian Pathol.*, 35, 224, 2006.
72. Todd, D. et al., Development and application of an RT-PCR test for detecting avian nephritis virus. *Avian Pathol.*, 39, 207, 2010.
73. Smyth, V.J. et al., Detection of chicken astrovirus by reverse transcriptase-polymerase chain reaction. *Avian Pathol.*, 38, 293, 2009.
74. Smyth, V.J. et al., Development and evaluation of real-time TaqMan® RT-PCR assays for the detection of avian nephritis virus and chicken astrovirus in chickens. *Avian Pathol.*, 39, 467, 2010.
75. Day, J.M., et al., Metagenomic analysis of the turkey gut RNA virus community. *Virol J.*, 7, 313, 2010.

13 Hepatitis E Virus

Ilaria Di Bartolo, Fabio Ostanello, and Franco Maria Ruggeri

CONTENTS

13.1 Introduction 111
    13.1.1 Classification...111
    13.1.2 Morphology and Genome Organization...112
    13.1.3 Clinical Features and Pathogenesis in Swine 112
    13.1.4 Diagnosis 113
        13.1.4.1 Conventional Techniques..113
        13.1.4.2 Molecular Techniques...114
13.2 Methods 115
    13.2.1 Sample Preparation...115
    13.2.2 Detection Procedures...115
        13.2.2.1 Conventional RT-PCR...116
        13.2.2.2 Real-Time RT-PCR 116
        13.2.2.3 Sequencing and Phylogenetic Analyses...117
13.3 Conclusion and Future Perspectives 117
References..118

13.1 INTRODUCTION

Hepatitis E is a viral disease that presents as acute hepatitis in humans. The etiological agent is hepatitis E virus (HEV), first identified in the early 1980s.[1] The disease is an important public health issue in developing countries where it is frequently epidemic. Industrialized countries were previously thought to be free from HEV, with a limited number of cases reported only in people who had traveled to endemic areas. However, more recent studies have documented an increasing number of sporadic cases in developed areas, among patients who had no history of travelling to countries endemic for hepatitis E. Furthermore, a high anti-HEV seroprevalence has been detected among healthy individuals in nonendemic countries.

Since the early 1990s, serological evidence of HEV infection and virus detection have been reported in many animal species in both developed and developing countries, suggesting that these host species may become infected with HEV-like viruses.[1] In 1997, a swine HEV strain was identified for the first time in the United States. This swine HEV strain correlated genetically to two human HEV strains isolated in the United States during the same period from patients who had not traveled to endemic areas.[2] Since then, swine HEV strains have been isolated across the globe. A strict genetic correlation between human and swine strains from the same geographic region has been observed frequently, and cross-species transmission of swine strains to humans and of human strains to nonhuman primates has been demonstrated.[3] Furthermore, several seroepidemiological studies have reported high antibody prevalence to HEV in people working in direct contact with swine or wild boar.[3]

The first direct evidence of a possible zoonotic transmission of HEV was provided in Japan in 2003, when cases of hepatitis E were associated to the ingestion of uncooked meat or organs from pigs, wild boar, or deer.[4] More recently, a study conducted in France confirmed that 13 human cases of hepatitis E were eventually linked to the consumption of raw figatellu pig liver sausages.[5] The disease is now recognized as an emerging zoonosis.

13.1.1 CLASSIFICATION

HEV is currently classified within the only genus *Hepevirus* in the family *Hepeviridae*,[1] although it was considered previously as a member of the family *Caliciviridae*. Subsequent genetic analyses demonstrated that the virus is distantly related to members of the family *Togaviridae*. Thus, a separate family (*Hepeviridae*) has been created for HEV to differentiate from caliciviruses. Given the genomic and amino acid diversity observed among HEV strains of different origins, a proposal for the establishment of additional genus within the family *Hepeviridae* has recently been presented to the International Committee on the Taxonomy of Viruses (ICTV) for approval.[6]

HEV strains detected in humans and other mammalian species constitute the major group within *Hepeviridae*. Four genotypes (G1–G4) of mammalian HEV are presently recognized, which primarily infect humans, domestic pigs, wild boar, deer, rabbit, and mongoose.[3] However, genetically distant HEV strains have been identified lately in moose, red fox, camel, rat, ferrets, wild boar, bat, mink, and cutthroat trout (*Oncorhynchus clarkii*).[7] Although avian HEV strains share

111

only 50%–60% nucleotide identity with mammalian HEV strains,[8] antibody cross-reaction with the capsid proteins of both groups of viruses was reported, suggesting the presence of common epitopes.[9] Nonetheless, avian HEV strains have never been associated with cases of infection in human beings.[10]

The proposed ICTV hierarchical system includes two genera in the family, named *Orthohepevirus* (all mammalian and avian HEV) and *Piscihepevirus* (for the cutthroat trout virus). The *Orthohepevirus* genus is divided into *Orthohepevirus A* (strains from humans, pig, wild boar, deer, mongoose, rabbit, and camel), *Orthohepevirus B* (strains from chicken), *Orthohepevirus C* (strains from rat, greater bandicoot, Asian musk shrew, ferret, and mink), and *Orthohepevirus D* (strains from bat).[6] The classification of genotypes and subgenotypes still remains included in the new taxonomic system. Despite the knowledge that different HEV genotypes occur, the virus seems otherwise to exist as a single serotype.[3]

In industrialized countries, autochthonous human cases are related to HEV strains belonging to genotypes 3 and 4,[1,11] which are still considered the only zoonotic genotypes. Genotype 3 was first identified in autochthonous human cases in the United States, when the two human strains US-1 and US-2 showed only 74%–75% nucleotide identity with the previously known genotypes 1 and 2 and were accordingly classified separately.[2] Since then, genotype 3 has been detected worldwide, associated to sporadic cases and small outbreaks in North America, Europe, Japan, and New Zealand. Consumption of raw pork meat or derivate products as liver sausage and exposure to pigs and wild boars are proposed as risk factors associated with HEV infection in industrialized countries.[12,13] Several papers confirmed the presence of genotype 3 HEV in raw pork liver sausage,[5,14,15] which in France has been reported to contain infectious HEV particles.[15]

HEV may contaminate surface waters and enter food production chains, in particular via shellfish culture areas and irrigation waters.[12] In fact, raw shellfish and vegetables cultivated in sewage-contaminated waters may be a significant source of HEV, and eating them shall be regarded as an additional risk factor for human infection.[16]

The other zoonotic genotype (genotype 4) is indigenous in Asia, where it was detected from both pigs and humans.[17] Genotype 4 is increasingly described as endemic in pigs in Asia and as cause of sporadic cases of hepatitis E in humans and of infection in the swine in China and Japan and recently in Europe.[17]

The recent and increasingly frequent detection of genotype 4 in Europe is of concern for public health and raises the question whether genotype 4 was somehow introduced into European domestic pigs and may be expected to spread further in European farms or whether it was transmitted through pig meat of Asian origin imported into Europe.[17]

13.1.2 Morphology and Genome Organization

HEV is a small (27–34 nm in diameter), icosahedral, non-enveloped, single-stranded, positive-sense RNA virus. The HEV genome is approximately 7.2 kb in length and presents a 7-methylguanosine cap followed by three overlapping open reading frames (ORFs) and a second noncoding region of approximately 65–74 nucleotides with a 3′ poly(A) tail. The genome length can vary slightly between animal strains although the genome organization seemed to be conserved in all these cases. The ORF1 (5073–5124 nt) codes for a nonstructural polyprotein of approximately 1690 amino acids, which is involved in viral genome replication and viral protein processing. The ORF3 (366–369 nucleotides) follows the ORF1 and overlaps the N-terminal portion of the ORF2 in a different reading frame, encoding a small phosphoprotein (pORF3). Recent studies have shown that pORF3 may be involved in virus release from infected cells.[18] The viral capsid protein encoded by ORF2 assembles into the complete virion. The virus capsid is made of 30 subunits containing homodimers of pORF2. Among the four major mammalian HEV genotypes, the sequence identity for the amino acid residues of the capsid protein is >85%, and most of the amino acid divergence is found in the N-terminal 111 residues. Expression of a truncated capsid protein lacking the first 111 amino acids and/or the C-terminal 59 amino acids using the baculovirus expression system in insect cells resulted in self-assembly of the capsid protein.[19]

13.1.3 Clinical Features and Pathogenesis in Swine

The pathogenesis of porcine HEV is largely unknown. As is the case for humans, also in the pig, it is not clear yet how the virus can reach the liver, once it has entered the host, and which its primary replication sites are. Although hepatocytes are the primary site for HEV detection,[2] little is known about viral detection and localization in the liver and extrahepatic tissues.[20] Using PCR or *in situ* hybridization in animals experimentally infected by the intravenous route, it was possible to detect viral RNA in several extrahepatic tissues such as lymph nodes, tonsils, spleens, stomachs, kidneys, salivary glands, lungs, and small and large intestines up to 20–27 days post-infection (pi), also in the absence of viremia. However, the negative-strand form of the virus genome, which is its replicative form, has been detected not only in the liver but also in the intestine principally and in the lymph nodes, between 7 and 27 days pi. Viremia can last for about 2 weeks, but the virus can be detected in the feces for a longer time, from 3 to 50 days pi. Seroconversion occurs about 2–3 weeks pi.[21,22] Tissues in which the virus first replicate and persists (from 3 to 27 days pi) are the liver, the small intestine, the colon, and the lymph nodes.[21] These observations, together with the fact that during experimental infections viral RNA can be detected in feces earlier than in the bile, and in quantities that are 10-fold higher, have led to argue that, after entering by the oral route and before inducing viremia, the virus replicates in the gastrointestinal tract.[22] Virological studies conducted on samples of serum and feces from swine herds have shown that HEV RNA can be primarily detected in pigs of 2–5 months of age, whereas animals younger than 2 months or older than 6–8 months are generally negative.[13] Considering these observations and given that maternal immunity is thought to last about 2 months, natural infection probably occurs as early as 1–3 months of age.[13,21,23]

In the swine, HEV is excreted in the feces as early as 7 days of postinfection, for about 3–4 weeks (with some differences between animal ages) with subsequent seroconversion and clearance of the infection by the immune system. Viremia follows after approximately 10–15 days since start of shedding and lasts for 1–2 weeks.[9,24] HEV infection seems therefore to be relatively short and self-limiting, resolving in a few weeks.[13,24] Pigs of any age are asymptomatic, whether infected with swine HEV naturally or experimentally.[22,23,25]

The HEV, at least the known genotype 3 strains, seems to be attenuated or at least not particularly virulent for domestic swine, while there is virtually no information available on its virulence in wild boar. The study of the disease in naturally or experimentally infected pigs has shown that HEV normally leads to subclinical infections with signs of hepatitis that can be detected only at histological analysis.[13,21,22] Some studies have highlighted a possible correlation between infection and liver damage,[25] and histopathological lesions have often been detected in the liver of experimentally and naturally infected animals.[21,22] These lesions include mild multifocal sinusoidal and periportal lymphoplasmocitary infiltrations and small areas of vacuolar degeneration and necrosis of hepatocites.[8,13,21] Hepatic inflammation and hepatocellular necrosis peaked in severity at 20 days pi.[21] In addition, slight subepithelial lymphohistiocytic cell infiltrations were observed in the gall bladder at 28 days pi.[24] An increasing trend in asparate aminotransferase (AST) and alanine aminotransferase (ALT) enzyme levels was also observed, which might suggest a slowly progressive development of liver damage occurring during HEV infection.[24]

In some cases, lymphoplasmocitary enteritis and multifocal interstitial lymphoplasmocitary nephritis have been detected. In experimentally infected animals, macroscopic lesions including mild to moderate enlargement of the mesenteric and hepatic lymph nodes have been detected sporadically.[21] In the ileum, a mild or moderate hyperplasia of Peyer's patches was observed. No lesions were observed in the spleen or pancreas.[24]

Passive maternal immunity has been shown to decline after 3–4 weeks of age, disappearing after 8–9 weeks. IgG seropositivity in piglets was associated with the serological status of the sows. However, it was uncommon to find seropositive animals below the age of 1–2 months, even if earlier results could have been biased by the low sensitivity of the serological tests used.[13]

The average time until antibody detection was found to be 2 weeks after the first HEV RNA detection in feces. Infected pigs develop an immune response against the virus that is characterized by an early IgM antibody response followed by an increase of IgGs approximately 1 week later. The IgM antibody titer decreases rapidly (1–2 weeks), whereas IgG concentration rises constantly for several weeks. The time determined before IgG development in pigs inoculated intravenously was 2-weeks of postinoculation at the earliest,[21,22] but times between 3 and 8 weeks of postinoculation were reported more often.[22]

In the field, seroconversion can be detected after the viremic phase at approximately 3–4 months of age (antibodies peak at 4 months), and animals show high serological titers until 5–6 months of age. At that time, the IgG titer starts to decline slowly.[11,13]

13.1.4 Diagnosis

For diagnosis of HEV infection, several methods are available. These include direct electron microscopy (EM), immune EM (IEM), enzyme immunoassay, immunofluorescence test, reverse transcription polymerase chain reaction (RT-PCR), real-time PCR, and nucleic acid sequence analysis. Among the different approaches, RT-PCR and nucleic acid sequence analysis are widely used for the detection and genotype identification of HEV.[26] These techniques have become the gold standard for diagnosis of HEV infections in the last decade.

13.1.4.1 Conventional Techniques

13.1.4.1.1 Direct Diagnosis

Detection of the virus has been performed originally by IEM, which provided the first evidence on the existence of HEV.[1] However, IEM has a low sensitivity compared to molecular detection methods and requires both expensive instruments and skilled personnel, which make it unsuitable for diagnostic routine.

Since the discovery of HEV, several attempts have been made to grow the virus in vitro, which at best led to only limited virus replication and low titers of progeny viruses. In general, the virus growth in cultured cells is very troublesome. Two main factors can hamper HEV growth in vitro: (1) the normally low viral load in test samples and (2) the virus integrity in the samples used for the inoculum. The first efficient cell culture systems for HEV that were capable of generating infectious HEV progenies in culture media were based on PLC/PRF/5 cells, derived from human hepatocellular carcinoma, and A549 cells, derived from human lung cancer. In the last few years, several papers reported attempts to cultivate different viral strains in vitro using additional cell lines of human or animal origin.[18] The experiments performed clearly demonstrated that both the initial day of appearance of HEV in the infected cell culture supernatant and its viral load (expressed in copy numbers, as determined by qPCR) were largely dependent on the original titer of the virus used for the inoculum. HEV could be first detected in the culture supernatant 8 days of postinoculation; however, it reached a concentration of 10^8 genome copies/mL only after several weeks (40–50 days).[18] The A549 and PLC/PRF/5 cell lines have been used in several studies, demonstrating efficient replication of both HEV genotypes 3 and 4 of both human and animal origin, including strains from swine and wild boar. Moreover, the virus progeny was infectious, as established by repeated passages in the same or other cell lines.[18,27] The PLC/PRF/5 cells used for monolayer cultures were also successfully used in a 3D growth system,[28] using special cell incubation conditions supported on microspheres. In addition, a recent study has further increased the number of cell lines permissive to

HEV growth, describing successful viral replication also in the human hepatoma-derived cell line HepaRG, and the porcine embryonic stem cell–derived cell line PICM-19.[29]

Some studies conducted using cell culture have demonstrated that pork liver sausage and swine liver sold as food can contain infectious HEV,[15,27] identifying the risk for foodborne HEV transmission. However, the use of *in vitro* HEV cultivation to assess the risk in food items is still too cumbersome for being presently considered suitable for application in the practice, calling for further studies in order to develop more reliable and handy protocols.

13.1.4.1.2 Indirect Diagnosis

Analyses of serum specimens during human acute infection and experimental infection in animals showed a classical serological pattern in the course of infection, which included the appearance of anti-HEV IgM followed by IgG. In humans, IgM antibodies are detectable since the early phases of illness after the incubation period up to 5 months. Anti-HEV IgG immunoglobulins appear shortly after the IgM and can persist up to 1–14 years.[30] Anti-HEV IgM can be detected in 13-week-old pigs and also in pigs at slaughter age, albeit less frequently. However, the detection of the long-lasting IgG can be conveniently used for monitoring HEV spreading both among pigs, in which the disease is normally asymptomatic, and the general human population, where the disease can also be asymptomatic, subclinical, or not promptly identified. HEV-specific antibodies remain detectable over a much longer period of time than does viral RNA, which allow performing a diagnosis during a longer detection window. This can be particularly helpful during outbreak investigations, taking into account the long incubation time of infection.

Serological diagnosis is generally performed using enzyme-linked immunosorbent assay (ELISA), western blot, or rapid immune chromatographic kits,[10] in which either recombinant HEV proteins or synthetic HEV peptides corresponding to antigen epitopes of the structural HEV proteins, such as pORF2 and pORF3, are used as target antigens.[24,31] Among the two peptides, recombinant pORF2 was proven to be a more sensitive and specific antigen than was pORF3. Given the existence of a single serotype including all known HEV genotypes infecting humans and pigs, a same antigen of any origin can in principle be used to test all HEV genotypes. However, as suggested in several studies, the use of antigens homologous to the infecting HEV genotype might have a better reactivity in antibody testing on either human or animal sera.[26] This apparent discrepancy does not have a clear explanation, although it could be linked to the existence of minor diversity in some protein epitopes between different genotypes that otherwise share common neutralization epitopes.[26]

The antibody detection methods used most frequently are based on ELISA test platforms. Presently, several commercial ELISA kits are available for detecting both IgM and IgG antibodies, using recombinant viral antigens attached to the solid phase surface, in an indirect or double-antigen sandwich ELISA format. There are numerous examples of application of recombinant pORF2, with or without pORF3, in diagnostic assays. Different antigens have been used, including *E. coli*–expressed HEV proteins and virus-like particles expressed in recombinant insect cells[19] or in plant cells systems.[26]

With some modifications, these antigens have been used for detection of IgG, IgM, and IgA antibodies.[32] IgM antibody detection, which has relevant clinical implications, can be hampered by competition with IgG immunoglobulin, generally present at higher titers. To circumvent these problems and to differentiate the detection of IgM and IgG responses, a possible approach is the use of secondary antibodies directed against either the γ- or the μ-chains of immunoglobulins, respectively.[30]

Animal strains of HEV, particularly those belonging to genotype 3, are correlated genetically and antigenically to human and, to a lesser extent, also to avian strains. A large part of the tests adopted for the detection of anti-HEV antibodies in animals are commercial human HEV immunoassays, where the secondary antibodies are replaced with antisera specific for the animal species concerned. Alternatively, in some studies, HEV-specific antibodies in animals have been detected by in-house assays developed using recombinant pORF2 or pORF3 proteins, originating from different genotypes.[33,34] The ample literature available supports that several epitopes, especially present in pORF2, cross-react between different species and strains. A species-independent double-antigen sandwich ELISA was consequently developed and used to detect antibodies against HEV in swine, wild boar, and deer, resulting both sensitive and specific for the target HEV.[35] Currently, immunoassays commercially available for the detection of specific anti-HEV antibodies during acute infection differ dramatically in their sensitivity and specificity, which may explain the discrepancies in the HEV seroprevalence reported in different studies. In addition, false reactivity for anti-HEV IgM with other hepatotropic viruses such as Epstein–Barr virus and cytomegalovirus has been reported.[36]

13.1.4.2 Molecular Techniques

In the last years, the molecular detection of HEV RNA has been largely improved by developing new primers and protocols and is now considered to be the gold standard for identification of viremia and virus shedding in both humans and animals. The recent advance of next-generation sequencing, by whole sequencing of RNA in biological samples, has enabled the discovery of HEV strains in novel animal hosts and reservoirs. Nevertheless, a major diagnostic difficulty in human infection is related to the short periods of both viremia (up to 2 weeks) and virus shedding in feces (up to 4 weeks),[37] which hamper diagnostic approaches based only on viral RNA detection.

Methods for RNA detection, mostly RT-PCR in conventional or real-time formats, are independent on the genotype and the type of specimen to be tested. Several RT-PCR protocols have been developed, which vary in the length and the position of the HEV genome segment being amplified, as well as in their ability to detect one or more genotypes. However, most primers normally in use enable the amplification of genomes of all genotypes, targeting conserved sequences.

Further analysis of the amplified PCR product for identification and characterization of the virus genotype is usually performed by nucleotide sequencing, by restriction with endonuclease enzymes, or by other detailed investigations following molecular cloning.

Recently, great improvement of diagnostic methods has been promoted by the implementation of several optimized real-time PCR protocols; this technique is not only able to quantify the virus load, but it also enhances significantly the rapidity, sensitivity, and specificity of the test.

The two real-time protocols most frequently used have been designed on the ORF2/ORF3 overlapping region and can be suitably used for detection of HEV genotypes 1–4.[38,39] However, a comparative study recently conducted by a WHO panel of laboratories expert in HEV detection showed some variability in the accuracy obtained with all these methods between different laboratories, in particular showing high interlaboratory variation in sensitivity. This study confirms the importance and need of multicentric standardization, as well as of suitable control materials to be used as international standards, for improving assays for quantitative determination of HEV RNA.[40]

13.2 METHODS

13.2.1 Sample Preparation

Nucleic acid extraction can be easily and efficiently performed using feces, serum, or a wide range of tissues. Several studies have proven that the highest sensitivity in HEV detection is achieved using swine samples such as bile, which cannot be obtained from live animals, followed by feces. The latter, together with sera, represents the type of sample used most commonly for monitoring virus circulation in pigs and for diagnosis in humans. Virus detection for risk assessment in food is generally performed in tissues such as liver, because the liver is used for producing sausage in many countries, but also in the intestine and lymph nodes.[20] Viral loads of 10^3–10^7 genome copies/g were estimated in positive liver, bile, and sausage samples.[14,41] Transmission of HEV through the consumption of contaminated food such as raw pork or wild boar meat or sausages has been described.[4,5] Nevertheless, as for hepatitis A, potential sources of contamination could be also in the crop field, including irrigation water, organic fertilizers, and human handling. Given the low infectious dose and complexity of the matrix (mainly for food and water) to be analyzed, methods for sample preparation should facilitate virus concentration, RNA recovery, and removal of inhibitors that might interfere with subsequent detection procedures.

In the last few years, several papers reported RNA extraction methods from food (i.e., sausages, pork chops) and water.[14,41–43] Most of the methods included the use of a process control, which is added to the samples before processing for monitoring the efficiency of the entire procedure. The viruses used as process control are genetically similar (at least in their genome organization) to the target virus and can be grown *in vitro*: those most commonly used are the feline calicivirus[44] or murine norovirus.[14] The process control used most commonly for hepatitis E is murine norovirus, which is co-concentrated and co-extracted with the target virus.

Different amounts of liver, sausage, and muscle have been used for HEV detection, ranging from 20 mg to 20 g.[41,43,45,46] Samples are chopped and homogenized in saline or lysis buffers and immediately processed for nucleic acid extraction. Fecal and bile samples should also be processed immediately after sampling or stored at −80°C. A 10% stool suspension is prepared in diethylpyrocarbonate (DEPC) water or saline buffer, and after vigorous shaking to release the virions and a centrifugation step to remove debris and bacteria, the supernatant is used for RNA extraction. The RNA virus extraction is obtained most commonly by commercial kits that are based on guanidine isothiocyanate (included in the lysis buffer) or using an organic solvent such as phenol–chloroform or chloroform–butanol[14,20,45] or solvents like Trizol. Virus concentration can also be carried out using 10% polyethylene glycol (PEG). Nucleic acid extraction is obtained by trapping RNA on magnetic silica or silica membranes. The eluted RNA is immediately processed for molecular detection or stored at −80°C.

As revealed by using the process control, the RNA recovery rate can vary remarkably depending on the matrix and the methods used, and the rate of recovery can range between 5.7% for dry pork liver sausages and 18% for *figatellu*[43] or other fresh liver sausages.[14] These recovery rates should be taken into account when performing a quantitative risk analysis.

13.2.2 Detection Procedures

For diagnosis of HEV infection in both animals and humans, analysis of RNA by nucleic acid amplification techniques is frequently used; this approach enables the identification of active infection during viremia (testing sera) and release of viruses (testing feces). Sequencing of amplified DNA is used to define the genotypes and subtypes and to determine the phylogenetic correlations among individual viral strains.

Several assays have been developed for the detection of HEV RNA, including conventional RT-PCR and nested RT-PCR protocols,[18,47] real-time RT-PCR,[39] and RT loop-mediated amplification.[48] Despite the high heterogeneity of the four genotypes infecting mammalians, all these assays are suitable for the detection of HEV genotypes 1–4. However, current methods of HEV nucleic acid detection have not been standardized, and different results have been observed in the performance of in-house tests,[49] thus confirming the necessity of control reagent panels to be used as internal controls for each specific genotype. The genotype 3 of HEV has been proposed as the best candidate for internal standard control, being the genotype most common worldwide.[40]

In addition to using positive controls, RT and PCR are also crucial steps to obtain an accurate detection of HEV RNA. Therefore, the detection of the process control should be first achieved to verify the efficiency of RNA extraction, and positive samples can then be further analyzed for the target virus. In addition, it is recommended to include a heterologous RNA or DNA as internal amplification control (IAC) in the samples to be analyzed. Several IACs have been designed, by including "probe" sequences not present in the HEV genome sequence.[50]

13.2.2.1 Conventional RT-PCR

The HEV genome is detected by RT followed by PCR and nested PCR or using a one-tube procedure, performing cDNA synthesis and the first PCR in a single step. This latter approach minimizes the handling of samples, reducing errors and contaminations.

Due to the low virus load, a nested PCR is often required in most samples, which is often decisive in case of low HEV RNA concentrations particularly in the swine. Moreover, due to the high genetic variability of HEV strains, detection by RT-PCR is performed using degenerate primers and analyzing different genomic regions.[49] Most protocols used for humans can be also applied satisfactorily in the case of animal testing, detecting strains belonging to genotypes 1–4. Most of the primers for universal use anneal in ORF2,[33,51] in the methyl transferase,[47] or in the conserved region of the RNA-dependent RNA polymerase[52] (Table 13.1). Recently, primers designed on genotypes 1–4 were found to detect suitably the novel HEV strains discovered in bats and wild rats.[53,54]

The cDNA synthesis is more frequently obtained using random hexamers, but gene-specific primer or oligo(dT) can also be used.

The RT step is carried out in reaction mixture containing buffer (composition depending on enzyme used), 0.01 M dithiothreitol (DTT), 1 mM dNTP, 50 pmol antisense primers or alternatively 0.5 µg of random primers, 20–40 U of RNase inhibitor, and 20–50 U of reverse transcriptase.

The mixture is then incubated at 37°C or 42°C for 1 h, depending on the enzymes used. The resulting cDNA is immediately used for the PCR.

The one-step reaction mixture includes 0.3–0.7 µM sense and antisense primers, 1× buffer (the buffer composition can vary depending on the enzymes used), 1.5–2 mM $MgCl_2$, 1 mM dNTPs, and enzyme mix (including both RT and DNA polymerase). RT is followed by PCR using thermal conditions optimized for the enzymes and primers used. For the second round nested PCR, inner primers (0.1–0–3 µM), 1× buffer (the buffer composition can vary depending on the enzymes used), 1.5–2 mM $MgCl_2$ or $MgSO_4$, 1 mM dNTPs, and 0.2–2 U Taq DNA polymerase are used with an amount of first RT-PCR template DNA corresponding to 10% of the final reaction volume.

13.2.2.2 Real-Time RT-PCR

Real-time RT-PCR is considered to be up to 100-fold more sensitive than conventional RT-PCR[55] and is frequently used for HEV detection in both humans and animals. Several assays based on TaqMan probe have been developed, mainly targeting the ORF1-2-3 regions, where the probe and primers are targeted. The different assays designed are able to detect all four mammalian genotypes.[39,55] A recent paper reports the validation of a multiplex real-time protocol for distinguishing between genotypes 3 and 4.[56] Despite the overall suitability of universal primers for any type of HEV, some differences in sensitivity have been shown investigating different subtypes of genotype 3 HEV by two common quantitative real-time RT-PCRs.[57] As for conventional PCR, real-time procedures can be performed in a one-step format or by synthesis of cDNA using random hexameric primers followed by qPCR. The protocol designed by

---

TABLE 13.1
Location of Target Regions Used for Polymerase Chain Reaction Detection of Hepatitis E Virus in Clinical, Food, and Environmental Matrices

Target Region	PCR	Primer		Sequence (5′ > 3′)	Product Length (bp)	Tm (°C)	Primer Position 5′–3′[a]	Genotype Specificity	Reference
Capsid	First PCR	HE044	Fw	CAAGGHTGGCGYTCKGTTGAGAC	505	55	5912–5934	Broad range (G1–G4)	[33]
		HE040	Rw	CCCTTRTCCTGCTGAGCRTTCTC			6395–6417		
	Nested	HE110-2	Fw	GYTCKGTTGAGCCTCYGGGGT	455	50	5922–5943		
		HE041	Rw	TTMACWGTCRGCTCGCCATTGGC			6356–6378		
Capsid	First PCR	3156	Fw	AATTATGCYCAGTAYCGRGTTG	730	52	5687–5708	Broad range (G1–G4)	[51]
		3157	Rw	CCCTTRTCYTGCTGMGCATTCTC			6395–6417		
	Nested	3158N	Fw	GTWATGCTYTGCATWCATGGCT	348	50	5971–5993		
		3159N	Rw	AGCCGACGAAATCAATTCTGTC			6297–6319		
Mtase	First PCR	HEVORF1s1	Fw	CTGGCATYACTACTGCYATTGAGC	418	50	55–79	G3/G4	[47]
		HEVORF1a1	Rw	CCATCRARRCAGTAAGTGCGGTC			450–473		
	Nested	HEVORF1s2	Fw	CTGCCYTKGCGAATGCTGTGG	287	50	103–124		
		HEVORF1a2	Rw	GGCAGWRTACCARCGCTGAACAT			367–390		
RdRp	First PCR	ESP 4213	Fw	CATGGTAAAGTGGGTCAGGGTAT	349	50	4213–4235	Broad range (G1–G4)	[52]
		EAP-4576	Rw	AGGGTGCCGGGCTCGCCGGA			4575–4595		
	Nested	ISP-4232	Fw	GTATTTCGGCCTGGAGTAAGAC	329	50	4232–4253		
		IAP-4561	Rw	TCACCGGAGTGYTTCTTCCAGAA			4560–4583		

[a] Primer positions are based on GenBank sequence accession no. M73218.

TABLE 13.2

Real-Time Polymerase Chain Reaction Detection of Hepatitis E Virus in Clinical, Food, and Environmental Samples

Target Region	Primer/Probe	Sequence (5′ > 3′)	Primer Position[a]	Genotype
ORF1/ORF3	HEV-F	GGTGGTTTCTGGGGTGAC	5260–5278	Broad range (G1–G4)
ORF2	HEV-R	AGGGGTTGGTTGGATGAA	5312–5330	
	HEV probe	5′-FAM-TGATTCTCAGCCCTTCGC–BGQ1-3′[b]	5285–5301	

Source: Enouf, V. et al., *J. Med. Virol.*, 78, 1076, 2006.

[a] Primer positions are based on GenBank sequence accession no. M73218.

[b] Black hole quencher or MGB can be used instead of TAMRA.

Jothikumar and coworkers[39] (Table 13.2) is frequently used, being cited in 157 papers, and has been applied to human hepatitis E diagnosis and detection of HEV in food and animals.[40,45]

According to the kit used, the thermal conditions, final concentrations of primers and probe, and reaction mixture components can vary.

Depending on the RNA elution volume, most protocols use 2–10 μL of RNA, 1× optimized buffer (including reverse transcriptase and DNA polymerase, dNTPs, $MgCl_2$), 100–250 nM Primer HEV-F, 250 nM Primer HEV-R, 100 nM-5 μM Probe HEV-P, and 1× ROX dye. The use of an IAC is suggested. The detection of IAC is obtained by performing a multiplex reaction including probes, labeled differently, for the detection of the target virus and the IAC. Detection of both IAC and target virus, or IAC alone, confirms that the real-time reaction succeeded, whereas the reaction is to be considered inhibited in the absence of IAC signal.[45]

13.2.2.3 Sequencing and Phylogenetic Analyses

To identify the HEV strains detected and to further characterize the phylogenetic correlations between strains, the DNA amplicons obtained can be subjected to sequencing. To this purpose, a suitable amount of amplified DNA (i.e., 3–15 ng for PCR products of 100–500 bp) of high purity is needed. There are two different approaches that can be followed, that is, the direct sequencing of PCR products after gel extraction or purification or the cloning of the PCR products using a commercial vector and the sequencing of resulting recombinant plasmids. The sequences obtained are compared with GenBank entries, and phylogenetic analyses are performed using one of several specific software. HEV is currently divided into four mammalian genotypes and three provisional genotypes detected in animals. The first proposal indicated that strains differing in nucleotide sequence by >20% in the ORF2 region should be classified into different genotypes. The subtype classification is controversial and recent studies have shown inconsistencies within this classification. However, the first scheme classification of 24 HEV subtypes, including 10 subtypes (3a–3j) of genotype 3 and 7 subtypes (4a–4g) of genotype 4, is still accepted.[58]

Problems that are currently encountered during phylogenetic comparisons are related to the use of different genomic regions for sequencing by different groups, often performed with short-stretch genomic fragments (300 bp), and using different phylogenetic methods. Classification into subtypes should be confirmed by good bootstrap values supporting the phylogenetic groupings, applying a model test for choosing the most suitable phylogenic methods to be applied in each case, and possibly analyzing longer stretches of the genome, including different coding regions.

13.3 CONCLUSION AND FUTURE PERSPECTIVES

Given the increasing number of epidemiological and virological reports available in the literature, HEV infection shall be considered an emerging zoonosis. Both in Europe and Asia, the swine appears to be the major animal reservoir for the virus and therefore represents a possible important threat for public health. Infection can be acquired by eating infected meat products, but this transmission mode can in principle be prevented or largely contained because HEV is heat inactivated during cooking. Nonetheless, the possibility of cross contamination between raw meat products and the risk of environmental virus spread through pig farming manure should also be taken into consideration because of the possible contamination of vegetables and drinking or bathing waters. Waterborne HEV may eventually end up in contaminating filtrating edible mollusks, thereby increasing public health risks further. HEV transmission to humans can also occur by direct contact with infected animals, highlighting a greater infection risk for farmers, workers, and veterinarians who are in contact with pigs. Also indirect contact with instruments and tools contaminated with infected animal feces cannot be disregarded. Although many veterinary aspects of infection are not known yet, the enzootic nature of swine HEV infection and HEV ability to cross the species barrier emphasizes zoonotic transmission risks, together with food and environmental safety issues. Concerning the virus, the information available on the genetics and evolutionary links between animal and human strains is limited, particularly concerning animal strains. The host range restriction of the infection is not fully known, and the human cases associated with the ingestion of uncooked meat from wild species suggest that the role of wild animals in the epidemiology of disease should be further assessed.

Despite the obvious health implications of this emerging zoonosis, information on the presence and circulation of HEV in pig herds or other animals is still insufficient, and both the natural history of infection in pigs and the economic impact of the disease on pig production also require further study. Evaluating the presence of HEV in other domestic and wild animal species is necessary, and genetic and epidemiological surveys of these viruses need to be conducted to assess the actual risks of zoonotic HEV transmission. Toward these aims, sensitive, specific, and inexpensive diagnostic tests represent a major need for the near future, and coordinated actions are desirable with large exchange of sequence and epidemiological data. Finally, the opportunity of vaccine strategies for control of disease in humans at risk should be evaluated, as well as a possible development of animal vaccines for containment of infection inside the food production and supply chain and in the environment.

REFERENCES

1. Emerson, S.U. and Purcell, R.H. Hepatitis E virus. *Rev Med Virol* 13, 145–154 (2003).
2. Meng, X.J. et al. A novel virus in swine is closely related to the human hepatitis E virus. *Proc Natl Acad Sci USA* 94, 9860–9865 (1997).
3. Meng, X.J. From barnyard to food table: The omnipresence of hepatitis E virus and risk for zoonotic infection and food safety. *Virus Res* 161, 23–30 (2011).
4. Tamada, Y. et al. Consumption of wild boar linked to cases of hepatitis E. *J Hepatol* 40, 869–70 (2004).
5. Colson, P. et al. Pig liver sausage as a source of hepatitis E virus transmission to humans. *J Infect Dis* 202, 825–834 (2010).
6. Smith, D.B. et al. Consensus proposals for classification of the family Hepeviridae. *J Gen Virol* 95, 2223–232 (2014).
7. Johne, R. et al. Hepeviridae: An expanding family of vertebrate viruses. *Infect Genet Evol* 27, 212–229 (2014).
8. Meng, X.J. Hepatitis E virus: Animal reservoirs and zoonotic risk. *Vet Microbiol* 140, 256–265 (2010).
9. Haqshenas, G., Shivaprasad, H.L., Woolcock, P.R., Read, D.H., and Meng, X.J. Genetic identification and characterization of a novel virus related to human hepatitis E virus from chickens with hepatitis-splenomegaly syndrome in the United States. *J Gen Virol* 82, 2449–2462 (2001).
10. Shata, M.T. et al. Protective role of humoral immune responses during an outbreak of hepatitis E in Egypt. *Trans R Soc Trop Med Hyg* 106, 613–618 (2012).
11. Scobie, L. and Dalton, H.R. Hepatitis E: Source and route of infection, clinical manifestations and new developments. *J Viral Hepat* 20, 1–11 (2013).
12. Van der Poel, W.H. Food and environmental routes of Hepatitis E virus transmission. *Curr Opin Virol* 4, 91–96 (2014).
13. Meng, X.J. et al. Prevalence of antibodies to hepatitis E virus in veterinarians working with swine and in normal blood donors in the United States and other countries. *J Clin Microbiol* 40, 117–122 (2002).
14. Di Bartolo, I., Angeloni, G., Ponterio, E., Ostanello, F., and Ruggeri, F.M. Detection of hepatitis E virus in pork liver sausages. *Int J Food Microbiol* 193, 29–33 (2015).
15. Berto, A. et al. Hepatitis E virus in pork liver sausage, France. *Emerg Infect Dis* 19, 264–266 (2013).
16. Maunula, L. et al. Tracing enteric viruses in the European berry fruit supply chain. *Int J Food Microbiol* 167, 177–185 (2013).
17. Colson, P. et al. Autochthonous infections with hepatitis E virus genotype 4, France. *Emerg Infect Dis* 18, 1361–1364 (2012).
18. Okamoto, H. Hepatitis E virus cell culture models. *Virus Res* 161, 65–77 (2011).
19. Li, T.C. et al. Empty virus-like particle-based enzyme-linked immunosorbent assay for antibodies to hepatitis E virus. *J Med Virol* 62, 327–333 (2000).
20. de Deus, N. et al. Detection of hepatitis E virus in liver, mesenteric lymph node, serum, bile and faeces of naturally infected pigs affected by different pathological conditions. *Vet Microbiol* 119, 105–114 (2007).
21. Halbur, P.G. et al. Comparative pathogenesis of infection of pigs with hepatitis E viruses recovered from a pig and a human. *J Clin Microbiol* 39, 918–923 (2001).
22. Meng, X.J. et al. Experimental infection of pigs with the newly identified swine hepatitis E virus (swine HEV), but not with human strains of HEV. *Arch Virol* 143, 1405–1415 (1998).
23. Kasorndorkbua, C. et al. Use of a swine bioassay and a RT-PCR assay to assess the risk of transmission of swine hepatitis E virus in pigs. *J Virol Methods* 101, 71–78 (2002).
24. Bouwknegt, M. et al. The course of hepatitis E virus infection in pigs after contact-infection and intravenous inoculation. *BMC Vet Res* 5, 7 (2009).
25. dos Santos, D.R. et al. Serological and molecular evidence of hepatitis E virus in swine in Brazil. *Vet J* 182, 474–480 (2009).
26. Aggarwal, R. Diagnosis of hepatitis E. *Nat Rev Gastroenterol Hepatol* 10, 24–33 (2013).
27. Takahashi, H. et al. A549 and PLC/PRF/5 cells can support the efficient propagation of swine and wild boar hepatitis E virus (HEV) strains: Demonstration of HEV infectivity of porcine liver sold as food. *Arch Virol* 157, 235–246 (2012).
28. Berto, A. et al. Replication of hepatitis E virus in three-dimensional cell culture. *J Virol Methods* 187, 327–332 (2013).
29. Rogee, S. et al. New models of hepatitis E virus replication in human and porcine hepatocyte cell lines. *J Gen Virol* 94, 549–558 (2013).
30. Khudyakov, Y. and Kamili, S. Serological diagnostics of hepatitis E virus infection. *Virus Res* 161, 84–92 (2011).
31. Ma, H. et al. Hepatitis E virus ORF3 antigens derived from genotype 1 and 4 viruses are detected with varying efficiencies by an anti-HEV enzyme immunoassay. *J Med Virol* 83, 827–832 (2011).
32. Takahashi, M. et al. Simultaneous detection of immunoglobulin A (IgA) and IgM antibodies against hepatitis E virus (HEV) is highly specific for diagnosis of acute HEV infection. *J Clin Microbiol* 43, 49–56 (2005).
33. Mizuo, H. et al. Polyphyletic strains of hepatitis E virus are responsible for sporadic cases of acute hepatitis in Japan. *J Clin Microbiol* 40, 3209–3218 (2002).
34. Ponterio, E. et al. Detection of serum antibodies to hepatitis E virus in domestic pigs in Italy using a recombinant swine HEV capsid protein. *BMC Vet Res* 10, 133 (2014).
35. Rutjes, S.A. et al. Seroprevalence and molecular detection of hepatitis E virus in wild boar and red deer in The Netherlands. *J Virol Methods* 168, 197–206 (2010).
36. Fogeda, M., de Ory, F., Avellon, A., and Echevarria, J.M. Differential diagnosis of hepatitis E virus, cytomegalovirus and Epstein-Barr virus infection in patients with suspected hepatitis E. *J Clin Virol* 45, 259–261 (2009).

37. Krawczynski, K., Meng, X.J., and Rybczynska, J. Pathogenetic elements of hepatitis E and animal models of HEV infection. *Virus Res* 161, 78–83 (2011).

38. Gyarmati, P. et al. Universal detection of hepatitis E virus by two real-time PCR assays: TaqMan and Primer-Probe Energy Transfer. *J Virol Methods* 146, 226–235 (2007).

39. Jothikumar, N., Cromeans, T.L., Robertson, B.H., Meng, X.J., and Hill, V.R. A broadly reactive one-step real-time RT-PCR assay for rapid and sensitive detection of hepatitis E virus. *J Virol Methods* 131, 65–71 (2006).

40. Baylis, S.A., Hanschmann, K.M., Blumel, J., and Nubling, C.M. Standardization of hepatitis E virus (HEV) nucleic acid amplification technique-based assays: An initial study to evaluate a panel of HEV strains and investigate laboratory performance. *J Clin Microbiol* 49, 1234–1239 (2011).

41. Leblanc, D., Poitras, E., Gagne, M.J., Ward, P., and Houde, A. Hepatitis E virus load in swine organs and tissues at slaughterhouse determined by real-time RT-PCR. *Int J Food Microbiol* 139, 206–209 (2010).

42. Martin-Latil, S., Hennechart-Collette, C., Guillier, L., and Perelle, S. Duplex RT-qPCR for the detection of hepatitis E virus in water, using a process control. *Int J Food Microbiol* 157, 167–173 (2012).

43. Martin-Latil, S., Hennechart-Collette, C., Guillier, L., and Perelle, S. Method for HEV detection in raw pig liver products and its implementation for naturally contaminated food. *Int J Food Microbiol* 176, 1–8 (2014).

44. Mattison, K. et al. The feline calicivirus as a sample process control for the detection of food and waterborne RNA viruses. *Int J Food Microbiol* 132, 73–77 (2009).

45. Di Bartolo, I. et al. Hepatitis E virus in pork production chain in Czech Republic, Italy, and Spain, 2010. *Emerg Infect Dis* 18, 1282–1289 (2012).

46. Pavio, N., Merbah, T., and Thebault, A. Frequent hepatitis E virus contamination in food containing raw pork liver, France. *Emerg Infect Dis* 20, 1925–1927 (2014).

47. Erker, J.C., Desai, S.M., and Mushahwar, I.K. Rapid detection of Hepatitis E virus RNA by reverse transcription-polymerase chain reaction using universal oligonucleotide primers. *J Virol Methods* 81, 109–113 (1999).

48. Zhang, L.Q. et al. Simple and rapid detection of swine hepatitis E virus by reverse transcription loop-mediated isothermal amplification. *Arch Virol* 157, 2383–2388 (2012).

49. La Rosa, G. et al. Molecular characterisation of human hepatitis E virus from Italy: Comparative analysis of five reverse transcription-PCR assays. *Virol J* 11, 72 (2014).

50. Diez-Valcarce, M., Kovac, K., Cook, N., Rodriguez-Lazaro, D., and Hernandez, M. Construction and analytical application of internal amplification controls (IAC) for detection of food supply chain-relevant viruses by real-time PCR-based assays. *Food Anal Methods* 4, 437–445 (2011).

51. Huang, F.F. et al. Detection by reverse transcription-PCR and genetic characterization of field isolates of swine hepatitis E virus from pigs in different geographic regions of the United States. *J Clin Microbiol* 40, 1326–1332 (2002).

52. Zhai, L., Dai, X., and Meng, J. Hepatitis E virus genotyping based on full-length genome and partial genomic regions. *Virus Res* 120, 57–69 (2006).

53. Drexler, J.F. et al. Bats worldwide carry hepatitis E virus-related viruses that form a putative novel genus within the family Hepeviridae. *J Virol* 86, 9134–9147 (2012).

54. Johne, R. et al. Detection of a novel hepatitis E-like virus in faeces of wild rats using a nested broad-spectrum RT-PCR. *J Gen Virol* 91, 750–758 (2010).

55. Enouf, V. et al. Validation of single real-time TaqMan PCR assay for the detection and quantitation of four major genotypes of hepatitis E virus in clinical specimens. *J Med Virol* 78, 1076–1082 (2006).

56. Zhang, X. et al. Validation of an internally controlled multiplex real time RT-PCR for detection and typing of HEV genotype 3 and 4. *J Virol Methods* 193, 432–438 (2013).

57. Abravanel, F. et al. Genotype 3 diversity and quantification of hepatitis E virus RNA. *J Clin Microbiol* 50, 897–902 (2012).

58. Lu, L., Li, C., and Hagedorn, C.H. Phylogenetic analysis of global hepatitis E virus sequences: Genetic diversity, subtypes and zoonosis. *Rev Med Virol* 16, 5–36 (2006).

14 Feline Calicivirus

Rudi Weiblen, Luciane Teresinha Lovato, and Andréia Henzel

CONTENTS

14.1 Introduction 121
   14.1.1 Classification.. 121
   14.1.2 Morphology, Genome Organization, Biology, and Epidemiology 122
   14.1.3 Clinical Features and Pathogenesis 123
   14.1.4 Diagnosis 123
      14.1.4.1 Conventional Techniques 123
      14.1.4.2 Molecular Techniques... 124
14.2 Methods 125
   14.2.1 Sample Collection... 125
   14.2.2 Detection Procedures.. 125
      14.2.2.1 Extraction of RNA 125
      14.2.2.2 Reverse Transcription 125
      14.2.2.3 PCR for FCV Detection... 125
      14.2.2.4 Multiplex PCR 125
      14.2.2.5 Real-Time RT-PCR 125
      14.2.2.6 Sequencing 126
14.3 Conclusion and Perspective 126
References... 127

14.1 INTRODUCTION

Respiratory diseases associated with infectious agents are a major problem in cats. Upper respiratory tract disease (URTD) has been the main cause of illness and health-associated euthanasia of cats in shelters in the United States. Bacteria like *Chlamydophila felis*, *Bordetella bronchiseptica*, and *Mycoplasma felis* are also frequently involved in URTD [12,37]. However, feline calicivirus (FCV) and felid herpesvirus type 1 (FeHV-1) are the main causes of respiratory disease in cats, and it is estimated that approximately 80% of URTD cases are due to one of the two viruses or both of them concurrently [12]. FCV infection has also been associated with oral ulcers and limping disease and to a severe disease-denominated virulent systemic disease (VSD), or FCV-VSD [31,36,37].

First described in New Zealand in 1957 [10], FCV is extensively distributed among the feline population worldwide. Wild and domestic members of the Felidae family are susceptible to FCV infection [36]. The prevalence observed in several countries among healthy domestic cats varied from 15% to 31%, while the prevalence among cats showing clinical signs of URTD was approximately 20%–53% [5]. Antibodies against FCV were also detected in wild feline such as puma, leopard, and jaguar [8]. A serologic survey performed in the southern region of Brazil demonstrated the presence of antibodies against FCV in 39.2% of 630 serum samples collected from domestic cats [17]. Furthermore, FCV was isolated alone or coinfecting with FeHV-1 in 34 (11.3%) of 302 samples from cats of the same region. An important observation is that 18 out of the 34 positive samples came from cats showing no clinical signs of disease, characterizing the carrier state [16].

Inactivated and attenuated vaccines against FCV are available, and most of them use a widely reactive FCV strain denominated F9 that originated from a virus isolated in 1970 [36,37].

14.1.1 CLASSIFICATION

FCV is a nonenveloped virus with a positive-sense RNA genome classified in the genus *Vesivirus*, family *Caliciviridae* [36,37], which consists of five related viral genera (*Lagovirus*, *Nebovirus*, *Norovirus*, *Sapovirus*, and *Vesivirus*). The term calicivirus is derived from the Greek word *calyx* (meaning cup or goblet), reflecting the fact that many calicivirus strains show visible cup-shaped depressions under electron microscopy. In turn, the genus *Vesivirus* comprises two recognized species, FCV, and vesicular exanthema of swine virus (VESV, the prototype species of the genus), in addition to the unclassified *Vesivirus*. Interestingly, VESV (see Chapter 16) causes a vesicular disease in pigs that is clinically indistinguishable from foot-and-mouth disease and swine vesicular disease (see Chapters 8 and 11).

FCV has a great variability but there is only one serotype [13,36]. Based on phylogenetic analysis of the FCV genome, no generally accepted genogroup or genotype classification can be put forward [36]. However, the presence of a genogroup was demonstrated among FCV strains isolated in a survey in Japan [38], and consistent clustering was observed among Australian strains when the whole capsid sequences were compared, but not when E-regions were analyzed separately [3]. It is also not yet possible to correlate any particular genome sequence with the different clinical manifestations caused by the viral infection in feline [13,36]. A number of epidemiological surveys performed in diverse geographic regions have shown that the most variable region at the genome sequence may vary its nucleotides from 20% to 40% among unrelated strains and 0% to 6% in epidemiologically related strains [36].

14.1.2 MORPHOLOGY, GENOME ORGANIZATION, BIOLOGY, AND EPIDEMIOLOGY

FCV has an icosahedral capsid about 27–40 nm [24,36,37]. The x-ray structure of a prototype of the *Vesivirus* genus, the San Miguel sea lion virus, has showed that the capsid of these viruses is formed by 180 copies of the VP1 protein and a few copies of VP2 [6]. The capsid protein of the *Caliciviridae* family are organized in 90 dimers forming 32 depressions, giving the viruses the characteristic arrangement suggesting the family denomination [6].

The FCV genome consists of a positive-sense RNA of about 7.7 kb, which is divided in three open reading frames (ORFs) [24,36,37]. According to strain FCV-F9, the part of the genome corresponding to the ORF1 maps approximately from nucleotide 20 to 5311, ORF2 from nucleotide 5314 to 7329, and ORF3 from 7326 to 7646 [25]. The 5′ end of the viral RNA genome is covalently linked to a VPg protein; the 3′ end of the genome is polyadenylated [18]. ORF1 is translated from the genomic RNA in a polyprotein precursor, while ORFs 2 and 3 are translated from subgenomic RNAs copied from the antigenomic RNA strand [18,24,40]. The subgenomic RNA copied from the antigenomic RNA is attached to VPg and is also polyadenylated [18]. ORF1 and ORF2 are cleaved by a 3C-like cysteine proteinase mapped to the C-terminal part of the ORF1 polyprotein [40]. The ORF3 codifies for VP2—a small capsid protein that is present in the capsid as 1–10 copies [31,36].

ORF1 is a very-well-conserved region localized at the 5′ end extending for about two-thirds of the genome [24,40]. The product of ORF1 is a 195 kDa polyprotein that is cleaved by a viral proteinase in five different sites resulting in six nonstructural proteins [40]. These proteins have been named according to their size or function as p5.6, p32, p39, p30, VPg, and Pro-Pol. Comparison of the FCV genome with the sequence of other members of the *Calicivirus* genus and functional studies have suggested the functions of some of the nonstructural proteins. Based on that evidence, p39 is an NTPase; Pro-Pol has proteinase and polymerase functions, respectively; and VPg, like *Picornavirus*, may play a role in the initiation of RNA translation by the cell machinery [24,40].

The main capsid protein is encoded by ORF2 that has been divided into six regions denominated A, B, C, D, E, and F [36].

The E-region is the most variable region of the genome, and it is further divided into one conserved region (conE) between two hypervariable regions (HVR) known as 5′ HVR_E and 3′ HVR_E [39]. The immunodominant epitopes eliciting neutralizing antibodies against the virus map to this region may also have a role in the viral evasion from the host immune response along with the persistent infection [21,34]. Conserved epitopes are identified within regions D and conE, but their role in antibody induction is unknown [34]. The product of ORF2 is a 73 kDa protein, further cleaved in the capsid protein of 60 kDa VP1, and a leader protein (LC) of 14.2 kDa, by an ORF1-codified proteinase [40].

The feline junctional adhesion molecule 1 (fJAM-A) was identified as the cell functional receptor bound by VP1 [27]. JAM-A is a member of the immunoglobulin superfamily apparently enrolled in apical tight junction assembly and leukocyte trafficking and platelet aggregation and expressed in epithelial cells, endothelial cells, leukocytes, platelets, and erythrocytes [23]. Further studies indicated that α-2,6-linked sialic acid may play a role as a component of the receptor for FCV on Crandell Rees feline kidney (CRFK) cells [41]. X-ray studies revealed that the VP1 is divided in two domains, S and P, and P is further divided in P1 and P2 subdomains [4]. Viruses with mutations in the P2 region were not able to bind to fJAM-A-soluble receptors, although these viruses bound to fJAM-A-expressing cells [27].

FCV has no other reservoir in nature besides feline [5]. Like other *Calicivirus*, FCV is very resistant in the environment at room temperature, and evidence showed that the virus maintains viability for more than 1 month at low temperature [36,37]. Then transmission of fomites from cages, water or feeding bowls, and others may happen. However, data from several studies have revealed that cats showing acute signs of the disease and carriers without any signs are the most important source of new infections [36]. Transmission occurs mainly through oral, nasal, and ocular secretions that carry a high load of the virus. After primary infection, the cat may become a carrier shedding the virus continually for months or years [36,37].

The prevalence of FCV is higher among cats living in places with dense populations compared to household cats. Epidemiological surveys showed that prevalence may vary from 50% to 90% among big populations of cats like those living in shelters, while a rate of 10% is commonly demonstrated in individual household cats [35]. The difference in prevalence may be attributed to reasons such as constant virus circulation, immune response pressure on the virus, and mutations occurring on the viral genome in carrier cats [36,37].

Immune response to the virus is directed against epitopes on the viral capsid major protein VP1 [34], and the antibody levels correlate well with protection against homologous but not heterologous viruses [20]. Following infection with FCV, neutralizing antibodies can be detected at 7 days postinfection [20], whereas it was also demonstrated that FCV infection may occur without detection of antibodies [28]. Existing vaccines protect against the disease but not against infection and the persistence of the virus. Due to the extensive antigenic variation, vaccines may not be protective against all the strains [36,37].

14.1.3 CLINICAL FEATURES AND PATHOGENESIS

FCV is mainly associated with acute oral disease and URTD, although a diverse range of clinical manifestations have already been described [31,36,37]. Cats exhibiting the acute oral disease usually display vesicles on the tongue, which further evolve to ulcers [36]. The oral disease may or may not be concomitant to mild respiratory disease resulting in oral and nasal discharges, and sometimes fever is detected [36]. In a number of cases, FCV infection is asymptomatic [36,37]. Infrequently, the infected cat may develop pneumonia [22]; however, severe and lethal respiratory disease occurs only among kittens [19]. A "limping syndrome" with lameness and fever characterized by the thickening of the synovial membrane and increased synovial fluid has also been observed [29].

In addition, since the 1990s, a severe systemic disease associated with high mortality has been described in different countries. This syndrome was earlier denominated as hemorrhagic-like fever and now known as FCV-associated virulent systemic disease or simply FCV-VSD [31,36,37]. The clinical signs of VSD are variable and may include severe respiratory tract disease; fever and ulcers on the oral cavity; cutaneous edema on the head, pinnae, and limbs; ulcerative lesions on the skin and paws; and jaundice [30,32,36,37]. Moreover, a systemic inflammatory response syndrome, intravascular coagulation, multiorgan failure [11], and necrosis of the liver, spleen, and pancreas have also been described [36]. The mortality rate is approximately 50%, but higher rates (67%) were also cited [11]. In contrast with other clinical manifestations of FCV infection, evidence suggests that adult cats are more susceptible, and the systemic disease is more severe in adults than kittens and young cats [19].

The most disturbing fact is that FCV vaccines currently in use apparently do not protect against FCV-VSD strains [36,37]. Another important finding about FCV-VSD isolated strains is that the virus apparently is more widely distributed on tissues than viruses isolated from other clinical manifestations [31]. The non-VSD strains are detected in skin epithelial cells, oral mucosa, and macrophages from joint fluid, while viruses isolated from VSD cases are detected in respiratory epithelial cells, pancreatic acinar cells, endothelial cells, and hepatocytes as well the tissues mentioned earlier [19,30,31].

FCV strains isolated from cats showing clinical manifestations of VSD are able to cause the same disease in experimentally infected cats. There is no consensus sequence that identifies the virus that causes VSD [31]. Research has been performed to find a molecular marker for FCV-VSD strains of the virus, but no consistent results have been demonstrated [31]. Evidence has suggested that mutations within the viral genome sequence may be the mechanism that generates the VSD strains of FCV; nonetheless, viruses isolated from different cases do not show sequence identity [1,31]. Then, mutation in different locations of the genome sequence could be responsible for the VSD phenotype [31], although host and immune factors may also play a role in this disease [36].

14.1.4 DIAGNOSIS

Clinical diagnosis of FCV infection as well as FCV-VSD syndrome is largely reliant on clinical signs, epidemiology data such as morbidity and mortality, and macroscopic and microscopic lesions that result from biopsies. Use of laboratorial methods for direct detection of the virus, viral antigen, and viral genome is crucial to increase the accuracy of FCV diagnosis. The available diagnostic methods for FCV range from conventional techniques such as viral isolation in cell culture and serological assays to more complex and currently used molecular techniques such as real-time polymerase chain reaction (qPCR).

FCV and its antigens are commonly detected by virus isolation (VI), electron microscopy (EM), immunofluorescence assay (IFA), immunohistochemistry (IHC), and histopathology. IHC and IFA have proven useful for the detection of viral antigens present in the tissues. The main serology assays performed in order to detect antibodies against FCV are enzyme-linked immunosorbent assay (ELISA) and virus neutralization (VN). In addition, molecular techniques such as the sequencing of VP1 can also be used in the diagnosis of FCV-VSD [37].

The current diagnostic techniques and their application are briefly discussed in the succeeding text; nevertheless, VI and conventional reverse transcription PCR (RT-PCR) represent the most common methods used for FCV diagnosis [37,42,44].

14.1.4.1 Conventional Techniques

VI is a valuable method for detecting FCV infection since it reveals the presence of replicating virus. FCV replicates readily and shows rapid growth in feline-established cell lines as well primary cell culture [16,37]. CRFK cell line [7] is the most widely used for culture isolation of FCV. Primary cell culture obtained from ovarian, pulmonary, and kidney embryonic or fetus tissues allows high replication rates of the virus. However, the procedure to obtain these tissues is laborious and contamination is frequent. Furthermore, such cell cultures do not tolerate many passages *in vitro*.

The virus can be isolated from nasal, conjunctival, oral, and pharyngeal swabs from cats showing clinical signs of the disease [16,36]. In addition, the lungs, tonsils, fragments of lesions from the oral cavity, liver, skin, paw pad, pancreas, nose, lips, and pinnae may be used for diagnosing the FCV-VSD syndrome [30].

VI from persistently infected cats showing no signs of disease may be more difficult, and in this case, pharyngeal swabs are recommended [21]. In our experience, FCV was isolated from 18 nasal swabs from cats without any clinical manifestation at the time of collection, and four cats with FeHV-1 coinfection were confirmed by PCR [16]. FCV isolation may also fail due to the small amount of virions in the sample, virus inactivation during shipment from the place of collection until arrival at the diagnostic laboratory, or the presence of antibodies in extracellular fluids that prevent virus replication *in vitro* [12,36,37]. Figure 14.1 shows FCV cytopathic effect on CRFK cells.

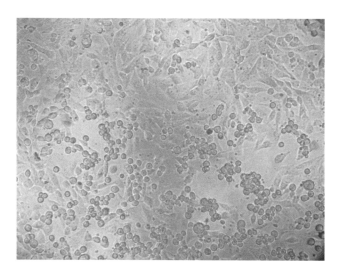

FIGURE 14.1 Cytopathic effect of feline calicivirus (FCV) in Crandell Rees feline kidney cells. FCV causes cellular rounding and detachment, lysis, and destruction of cell structure isolate SV65/90. (GenBank access number JX477806.1.)

EM offers a valuable approach for FCV detection although it often requires a high concentration of virus particles for a positive result. It implies that for successful results, the collection should be performed at the peak of viral replication, which normally corresponds to the period that cats are showing acute signs of the disease. EM may also be useful to confirm the presence of FCV after VI in cell culture. Figure 14.2 shows an electronic micrography of FCV isolate SV65/90 from a nasal swab of a cat aged 4 months from Southern Brazil.

IFA is usually performed to identify FCV after VI in cell culture. Polyclonal or monoclonal antibodies against the capsid protein may be utilized [30]. FCV1-43 is a primary monoclonal antibody against the capsid protein commonly used in IFA. This antibody is able to recognize the mature or the immature form of the capsid protein.

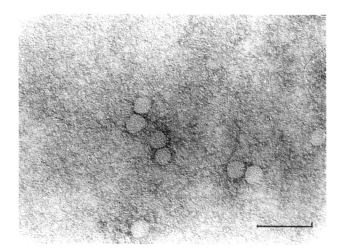

FIGURE 14.2 Electron microscopy of feline calicivirus (FCV) isolate FCV-SV65/90 (viral major capsid protein gene, partial cds_ GenBank access number JX477806.1). The bar equals to 90 nm.

IHC staining is a useful assay for postmortem specimens from cats with fatal URTD like severe pneumonia, FCV-VSD, or other conditions. IHC is widely applied for histopathology assays, but it is not a technique routinely performed in diagnostic laboratories. The use of the anti-FCV monoclonal antibody denominated FCV8-1C to perform IHC has been described [30].

Antibodies against FCV may be detected by VN or ELISA from serum, aqueous humor, and cerebrospinal fluid, although some cats survive a severe infection without developing detectable antibodies [37]. VN is widely used, because it is less expensive and easy to execute.

The antibody detection may have different objectives since it can be employed to support virus diagnosis and to evaluate the immune response regarding antibody titers from either vaccinated or infected cats. Serology may be used as an aid to viral diagnosis when paired serum samples are examined. However, difficulty in collecting sufficient amounts of blood from cats, the time required to obtain results, and low antibody titer in the convalescing are disadvantages associated with serological assays. Furthermore, the presence of antibodies may not correlate with active infection since the detection of antibodies fails to differentiate between the vaccinated and naturally infected cats [19,28,36].

The analysis of serology against FCV must be performed carefully either if it is used as an aid to the diagnosis of infection or to check protection against the virus by antibody titers. Levels of VN can be used to predict whether a cat is protected or not but must be interpreted properly, as false-negative results may be obtained if VN does not cross-react with the laboratory strains used in the test. Higher titers of antibodies may be observed with homologous strains than with heterologous strains [37]. When the strain used is not defined, it makes interpretation of the results difficult [9]. Consequently, the presence of specific antibodies is not a very useful resource to diagnose infection [12], and the absence of antibodies does not always correlate to unprotected host.

14.1.4.2 Molecular Techniques

Depending on clinical manifestations of FCV infection, RNA may be detected from nasal, oropharyngeal, or ocular swabs and tissue fragments obtained from biopsies or necropsy (mainly lung tissue and cutaneous scrapings) [37]. RT-PCR is widely applied for molecular diagnosis of FCV infection. qPCR has also been developed and it is another option when available. Genome sequencing is usually performed for research purposes, but it has also been applied for the diagnosis of FCV-VSD. Multiplex PCR to detect at the same time FCV and other feline infectious agents as herpesvirus type 1 (FeHV-1) and *Chlamydophila psittaci* may also be an option because these are agents commonly involved in feline respiratory disease [43].

The primers and viral strain used will determine the sensitivity of the techniques due to the genetic variability of the FCV genome [14,37,43,46]. It has been recommended to employ a large panel of strains for the optimization of molecular techniques in order to minimize false-negative results. The target gene as well as the primers (conserved or degenerated) used for amplification in molecular assays will depend on the purpose in each situation. The degenerate primers are useful to identify a wide range of FCV isolates, whereas conserved primers show better results in molecular analyses because they identify the correct or trusted genome sequence of the isolates. A similar approach may apply to the target gene used, since conserved genes are a good target for diagnosis/identification, while variable or highly variable regions (ORF2) are the target choice when studying the variability between isolates. An important point is that the primers should anchor in conserved regions at the start of the region to be amplified.

qPCR has higher sensitivity than conventional PCR and allows quantitation of the amount of virus in the sample. It is also less time consuming than conventional and nested PCR [1]. qPCR has been applied to the diagnosis of FCV as well as to FCV research. It may identify uniquely the virus strain aiding in molecular epidemiology and outbreak investigations. However, genetic markers associated with virulence, specifically hypervirulent strains, are not available [1,11].

Different qPCR protocols have been developed for FCV detection [1,14,47]. Helps et al. [14] were able to distinguish FCV isolates using SYBR Green I melting curve analysis. The assay was sensitive and allowed accurate quantitation of viral load [14]. qPCR using SYBR Green dye was further described by other research groups [45]. However, Abd-Eldaim et al. [2] found that applying a specific molecular probe in the TaqMan protocol increases the specificity of the assay and provides one more range of isolates detected. Both approaches will be presented in Section 14.2.

14.2 METHODS

14.2.1 Sample Collection

Swab samples (conjunctival, nasal, oral, and oropharyngeal) may be collected from cats with or without clinical signs of respiratory disease or oral lesions. Tissue fragments of lungs or oral lesions obtained from the autopsies are good samples to be analyzed by IFA or IHC.

14.2.2 Detection Procedures

14.2.2.1 Extraction of RNA

FCV RNA may be extracted from cell cultures inoculated with FCV or infected tissue fragments using TRIzol reagents (Invitrogen, Carlsbad, CA) following the manufacturer's protocol. After extraction, RNA is solubilized in 30 µL of ultrapure water with diethylpyrocarbonate and stored at −80°C until use. The QIAamp DNA Mini Kit (Qiagen, United States) has also been used to extract RNA from all samples.

14.2.2.2 Reverse Transcription

cDNA is synthesized in 20 µL of total solution containing the following: 2 µL of RNA (approximately 100 ng), 100 ng of random primers, 10× buffer from reverse transcriptase, 25 mM $MgCl_2$, 10 mM dNTPs, 0.1 mM DTT, 40 U RNaseOUT, and 200 U RT (SuperScript™ III RT, Invitrogen). The solution

is incubated at 65°C for 5 min, 25°C for 10 min, 42°C for 50 min, and 85°C for 5 min. cDNA is used as a template for the PCR for FCV identification.

14.2.2.3 PCR for FCV Detection

ORF2 (regions B–F), which encodes the major capsid protein [36], is a common target used for amplification of FCV cDNA. The following protocol was adapted by Henzel et al. [16] from the paper described by Ohe et al. [25]. The amplified product is a fragment of 955 base pairs (bp) from the capsid gene. The primer sequences are 8F (forward) 5′-CACSTTATGTCYGACACTGA-3′ (position 6142 B region) and 8R (reverse) 5′-CTRGADGTRTGCARRATTT-3′ (position 7097 F region), based on the FCV-F9 strain (GenBank access number M86379). The primers used are degenerate and the letters S, Y, R, and D refer to C/G, T/C, A/G, and A/C/T, respectively. The PCR conditions used are as follows: 94°C for 5 min for the initial denaturation, followed by 35 cycles of three steps of 94°C for 45 s, 48°C for 45 s, 72°C for 45 s, and a final extension of 7 min at 72°C. The resulting PCR products are electrophoresed in 1.5% agarose gel, stained with ethidium bromide, and visualized under UV light.

14.2.2.4 Multiplex PCR

The following multiplex PCR protocol was developed in 2001 by Sykes et al. [43] for simultaneous detection of FCV, FeHV-1, and *Chlamydia psittaci*. RT-PCR/PCR is performed in a single tube of 25 µL. For the RT step, final concentrations are 10 mM Tris–HCl (pH 8.3); 1.5 mM $MgCl_2$; 50 mM KCl; 0.01% gelatin; 5 mM DTT; 200 µM of each of dATP, dTTP, dGTP, and dCTP; 2.5 U RNase inhibitor (Roche Diagnostics, GmbH, Mannheim, Germany); and 50 U Moloney murine leukemia virus reverse transcriptase (Promega Corporation, Madison, WI). The extracted sample and reverse primer mixture are added; reactions are incubated at 42°C for 1 h and then heated to 100°C for 5 min. And the following are added to each reaction: a mixture of 0.5 µg of each primer, FCV capsid protein (U13992), CalcapF (5′-TTCGGCCTTTTGTGTTCC-3′ position 6398), and CalcapR (5′-TTGAGAATTGAACACATCAATAGATC-3′ position 7061); FeHV-1 thymidine kinase gene (M26660), HerpF (5′-GACGTGGTGAATTATCAGC-3′ position 510), and HerpR (5′-CAACTAGATTTCCACCAGGA-3′ position 797); and 1 µg of each of primers from *Chlamydia psittaci* target gene ompA (X56980), ChlaF (5′-ATGAAAAAACTCTTGAAATCGG-3′ position 364), and ChlaR (5′-CAAGATTTTCTAGACTTCATTTTGTT-3′) and 2.5 U of *Taq* polymerase (Promega). The following were the PCR conditions: 5 min denaturation at 95°C, followed by 40 cycles of 1 min at 91°C, 1 min at 56°C, and 1 min at 72°C. Water was the negative control, and a mixture of cell culture supernatants was the positive control [43].

14.2.2.5 Real-Time RT-PCR

The following protocol was extracted from Helps et al. [14]. qPCR is carried out using an iCycler IQ system (Bio-Rad). The primers were designed using MacVector 7.0 (Oxford Molecular) to amplify an 83 bp product from the ORF1 region of the FCV genome (nucleotides 2452–2534 from accession number M86379). The PCR product contains a 20 bp region between the forward and reverse primer binding sites that are variable in sequence. The RT-PCR reaction consists of 12.5 µL of 2× Platinum QRT-PCR Thermoscript (Invitrogen), 100 nM FCV-reverse primer 5′-CATATGCGGCTCTGATGGCTTGAAACTG-3′, 2.5 mL template RNA, 2.5 mL SYBR Green I (1 in 100,000 final dilution) (Sigma), and water to 22.5 mL. After an initial incubation at 50°C for 30 min and 70°C for 10 min to allow RT, 2.5 mL of 1 mM FCV-forward primer 5′-TAATTCGGTGTTTGATTTGGCCTGGGCT-3′ is added to each reaction. To inactivate the Thermoscript and activate the Platinum Taq, the reaction is incubated at 95°C for 5 min and then for 45 cycles of 95°C for 5 s and 60°C for 10 s. Fluorescence is detected at 530 nm at each annealing step (60°C). Immediately following PCR, a melting curve was performed by raising the incubation temperature from 80°C to 90°C in 0.2°C increments with a hold of 10 s at each increment. All reactions are run in triplicate [14].

The following real-time RT-PCR protocol, which used TaqMan, was extracted from Abd-Eldaim et al. [2]. A 120 bp fragment located at nucleotides 1–120 of the FCV genome (50 untranslated region; part of ORF1) is used as target. Primers and probe-specific sequences used are FCV-forward 5′-GTAAAAGAAATTTGAGACAAT-3′ (position in genome FCV-F9 1–21), FCV-reverse 5′-TACTGAAGWTCGCGYCT-3′ (position 120–104), and FCV-probe 26–49 5′-FAM-CAAACTCTGAGCTTCGTGCTTAAA-TAMRA-3′. Briefly, 5 µL of the synthesized cDNA is amplified in 25 µL total volume reactions using OmniMix HS (ready-to-use, lyophilized universal PCR reagent beads containing hot start Taq [Ta-KaRa Bio, Japan]), containing 200 µM dNTPs, 4 mM MgCl$_2$, 25 mM HEPES (pH 8.0), 200 nM probe, and 300 nM of each primer. The assay was carried out in the Cepheid SmartCycler II System (Cepheid, United States) with the following parameters: an initial activation step for the hot start Taq polymerase at 95°C for 2 min, then 45 three-step cycles consisting of denaturation at 95°C for 15 s, annealing at 56°C for 60 s, and extension at 72°C for 30 s. A threshold value of more than 40 is considered a negative result. DNase/RNase-free water is used as a negative control [2].

14.2.2.6 Sequencing

Sequencing is generally performed for research purposes, for the diagnosis of FCV-VSD, and, also, for the detection of *vaccine breakdown strains* (VBS).

The method described in the succeeding text was employed to study the variability of isolates in the HVR of the ORF2 gene. Region B–F of the ORF2 gene of FCV strains and isolates is initially amplified by PCR as described earlier. The primers were designed based on the FCV-F9 strain complete genome, including FCV_Capfor 5′-TTCGGCCGTTTGTCTTCC-3′ (position 6401–6419 [region B of ORF2]) and FCV_Caprev 5′-TTGTGAATTAAAGACATCAATAGACCT-3′ (position 7080–7053 [region F of ORF2]). A 679 bp product is amplified [15]. In order to optimize the PCR reaction and, also, as a

positive control, a commercial live attenuated vaccine can be used [16]. The negative control may consist of mock-infected CRFK cells. The PCR products may be purified with PureLink PCR kit (Invitrogen, Carlsbad, CA) and the purified product sequenced in a Megabace sequencer (Amersham Biosciences). The same primer sets that are used in the PCR assay can be used for the sequencing. The obtained sequence should be compared to published FCV sequences and the analysis conducted using evolutionary genetic analysis (MEGA4.0) [44].

The sequencing of the VP1 coding region has been used as an aid in the diagnosis of FCV-VSD [32], even though there are no sequencing markers identified so far. To amplify and sequence the E region of the FCV capsid gene, two specific primers are designed around the E region and their location on the sequence EU202915 was indicated (FCVCapseq4F [1127–1146] CAATC GACATTGGACTGAYA + FCVCapseq4R [1883–1902] GATA CACGGCAGADGARTCT). RT-PCRs are performed with the following conditions: 50°C for 30 min, 95°C for 15 min, 5 cycles of 94°C for 20 s/56°C for 20 s/72°C for 2 min, 5 cycles of 94°C for 20 s/54°C for 20 s/72°C for 2 min, 5 cycles 94°C for 20 s/52°C for 20 s/72°C for 2 min, 30 cycles of 94°C for 20 s/50°C for 20 s/72°C for 2 min, and 72°C for 10 min. PCR products obtained with this specific PCR are directly sequenced without cloning [32].

Sequencing may also be applied to detect isolates associated to the vaccine failures, VBS [26]. Sequencing of the 5′ HVR_E region of the capsid gene is often performed to differentiate field virus isolates from vaccine strains. Research executed applying the aforecited approach has found that a great part of the vaccine failure isolates are field viruses that differ from vaccine strains in 21.3%–38% [33].

14.3   CONCLUSION AND PERSPECTIVE

FCV is an important pathogen causing mainly respiratory and oral diseases in cats. FCV has also been eventually associated to a limping disease, jaundice, skin disorders, and, for the past few years, a virulent systemic disease (FCV-VSD). As an RNA virus, FCV shows high mutation rates due to the absence of the proofreading function of its polymerase. There is only one FCV serotype and genotype; yet mutation generates great genetic diversity, which may probably explain the wide variation of clinical manifestations, the poor efficacy of vaccines, and the difficulty to induce cross protection against different isolates, all of which may contribute to the viral persistence in the host.

The precise diagnosis of FCV will depend on the chosen technique and the sample collected; however, results should be interpreted carefully because in some situations, there is no correlation between the presence of the virus and clinical signs. It is due to the asymptomatic carrier state and also to the fact that cats vaccinated with live vaccines may occasionally shed the virus postvaccination. FCV laboratorial diagnosis is still performed mainly by VI and regular PCR. An idea to further facilitate the diagnosis could be the development of a fast method of FCV detection in the clinic. Rapid immunochromatographic tests, like the ones utilized in FIV and FeLV diagnosis, would help to avoid the introduction of FCV-positive cats to veterinary clinics and shelters.

The preference of cats as pets is growing worldwide, and only in Brazil, there is an estimated population of 23 million cats with the cat population growing at a rate of 8% a year. Brazil has the second largest cat population, only surpassed by the United States. Taking this into consideration, the future perspectives should involve studies with characterization of strains and reactive cross VN intending to generate outstanding vaccines that would prevent the disease in a larger percentage of the cat population. The new vaccines should be able to prevent a range of isolates including the novel isolates from the systemic syndrome (FCV-VSD). Besides, the research and development of antivirals to treat clinically ill cats would be another interesting issue to be explored.

REFERENCES

1. Abd-Eldaim, M., Potgieter, L., and Kennedy, M., Genetic analysis of feline caliciviruses associated with a hemorrhagic-like disease, *J. Vet. Diagn. Invest.*, 17, 420, 2005.
2. Abd-Eldaim, M. M. et al., Development and validation of a TaqMan real-time reverse transcription-PCR for rapid detection of feline calicivirus, *Arch. Virol.*, 154, 555, 2009.
3. Baulch-Brown, C., Love, D., and Meanger, J., Sequence variation within the capsid protein of Australian isolates of feline calicivirus, *Vet. Microbiol.*, 68, 107, 1999.
4. Bhella, D. et al., Structural insights into calicivirus attachment and uncoating, *J. Virol.*, 82, 8051, 2008.
5. Binns, S. H. et al., A study of feline upper respiratory tract disease with reference to prevalence and risk factors for infection with feline calicivirus and feline herpesvirus, *J. Feline Med. Surg.*, 2, 123, 2000.
6. Chen, R. et al., X-ray structure of a native calicivirus: Structural insights into antigenic diversity and host specificity, *Proc. Natl. Acad. Sci. USA*, 103, 8048, 2006.
7. Crandell, R. A., Fabricant, C. G., and Nelson-Rees, W. A., Development and characterization and viral susceptibility of a feline (*Felis catus*) renal cell line (CRFK), *In Vitro*, 9, 76, 1973.
8. Daniels, M. J. et al., Feline viruses in wildcats from Scotland, *J. Wildl. Dis.*, 35, 121, 1999.
9. Dawson, S. et al., A field trial to assess the effect of vaccination against feline herpesvirus, feline calicivirus and feline panleucopenia virus in 6-week-old kittens, *J. Feline Med. Surg.*, 3, 17, 2001.
10. Fastier, L. B., A new feline virus isolated in tissue culture, *Am. J. Vet. Res.*, 18, 382, 1957.
11. Foley, J. et al., Virulent systemic feline calicivirus infection: Local cytokine modulation and contribution of viral mutants, *J. Feline Med. Surg.*, 8, 55, 2006.
12. Gaskell, R., and Knowles, J., Feline respiratory disease update, *In Practice*, 11, 23, 1989.
13. Geissler, K. et al., Genetic and antigenic heterogeneity among feline calicivirus isolates from distinct disease manifestations, *Virus Res.*, 48, 193, 1997.
14. Helps, C. et al., Melting curve analysis of feline calicivirus isolates detected by real-time reverse transcription PCR, *J. Virol. Methods*, 106, 241, 2002.
15. Henzel, A. et al., Genetic and phylogenetic analyses of capsid protein gene in feline calicivirus isolates from Rio Grande do Sul in southern Brazil, *Virus Res.*, 163, 667, 2012.
16. Henzel, A. et al., Isolation and identification of feline calicivirus and feline herpesvirus in Southern Brazil, *Braz. J. Microbiol.*, 432, 560, 2012.
17. Henzel, A. et al., Serological survey of feline calicivirus and felid herpesvirus in Rio Grande do Sul, Brazil, *Acta Sci. Vet.*, 41, 1, 2013.
18. Herbert, T. P., Brierley, I., and Brown, T. D., Identification of a protein linked to the genomic and subgenomic mRNAs of feline calicivirus and its role in translation, *J. Gen. Virol.*, 78, 1033, 1997.
19. Hurley, K. F. and Sykes, E. S., Update on feline calicivirus: New trends, *Vet. Clin. North Am. Small Anim. Pract.*, 33, 759, 2003.
20. Kahn, D. E., Hoover, E. A., and Bittle, J. L., Induction of immunity to feline caliciviral disease, *Infect. Immun.*, 11, 1003, 1975.
21. Kreutz, L. C., Johnson, R. P., and Seal, B. S., Phenotypic and genotypic variation of feline calicivirus during persistent infection of cats, *Vet. Microbiol.*, 59, 229, 1998.
22. Love, D. N., Pathogenicity of a strain of feline calicivirus for domestic kittens, *Aust. Vet. J.*, 51, 541, 1975.
23. Mandell, K. J. et al., Junctional adhesion molecule 1 regulates epithelial cell morphology through effects on beta1 integrins and Rap1 activity, *J. Biol. Chem.*, 280, 11665, 2005.
24. Neill, J. D., Reardon, I. M., and Heinrikson, R. L., Nucleotide sequence and expression of the capsid protein gene of feline calicivirus, *J. Virol.*, 65, 5440, 1991.
25. Ohe, K. et al., Detection of feline calicivirus (FCV) from vaccinated cats and phylogenetic analysis of its capsid genes, *Vet. Res. Commun.*, 30, 293, 2006.
26. Ohe, K. et al., Genogrouping of vaccine breakdown strains (VBS) of feline calicivirus in Japan, *Vet. Sci. Commun.*, 31, 497, 2007.
27. Ossiboff, R. J. et al., Conformational changes in the capsid of a calicivirus upon interaction with its functional receptor, *J. Virol.*, 84, 5550, 2010.
28. Poulet, H. et al., Immunisation with a combination of two complementary feline calicivirus strains induces a broad cross-protection against heterologous challenges, *Vet. Microbiol.*, 106, 17, 2005.
29. Pedersen, N. C., Laliberte, L., and Ekman, S., A transient febrile "limping" syndrome of kittens caused by two different strains of feline calicivirus, *Feline Pract.*, 13, 26, 1983.
30. Pesavento, P. et al., Pathologic, immunohistochemical, and electron microscopic findings in naturally occurring virulent systemic feline calicivirus infection in cats, *Vet. Pathol.*, 41, 257, 2004.
31. Pesavento, P. A., Chang K. O., and Parker, J. S L., Molecular virology of feline calicivirus, *Vet. Clin. North Am. Small Anim. Pract.*, 38, 775, 2008.
32. Reynolds, B. S. et al., A nosocomial outbreak of feline calicivirus associated virulent systemic disease in France, *J. Feline Med. Surg.*, 11, 633, 2009.
33. Radford, A. D. et al., The use of sequence analysis of a feline calicivirus (FCV) hypervariable region in the epidemiological investigation of FCV related disease and vaccine failures, *Vaccine*, 15, 1451, 1997.
34. Radford, A. D. et al., The capsid gene of feline calicivirus contains linear B-cell epitopes in both variable and conserved regions, *J. Virol.*, 73, 8496, 1999.
35. Radford, A. D. et al., Molecular analysis of isolates of feline calicivirus from a population of cats in a rescue shelter, *Vet. Rec.*, 149, 477, 2001.
36. Radford, A. D. et al., Feline calicivirus, *Vet. Res.*, 38, 319, 2007.

37. Radford, A. D. et al., Feline calicivirus infection ABCD guidelines on prevention and management, *J. Feline Med. Surg.*, 11, 556, 2009.

38. Sato, Y. et al., Phylogenetic analysis of field isolates of feline calicivirus (FCV) in Japan by sequencing part of its capsid gene, *Vet. Res. Commun.*, 26, 205, 2002.

39. Seal, B. S., Ridpath, J. F., and Mengeling, W. L., Analysis of feline calicivirus capsid protein genes: Identification of variable antigenic determinant regions of the protein, *J. Gen. Virol.*, 74, 2519, 1993.

40. Sosnovtsev, S. V., Garfield, M., and Green, K. Y., Processing map and essential cleavage sites of the nonstructural polyprotein encoded by ORF1 of the feline calicivirus genome, *J. Virol.*, 76, 7060, 2002.

41. Stuart, A. D., and Brown, T. D., Alpha 2,6-linked sialic acid acts as a receptor for feline calicivirus, *J. Gen. Virol.*, 88, 177, 2007.

42. Sykes, J. E., Studdert, V. P., and Browning, G. F., Detection and strain differentiation of feline calicivirus in conjunctival swabs by RT-PCR of the hypervariable region of the capsid protein gene, *Arch Virol.*, 143, 1321, 1998.

43. Sykes, J. E. et al., Detection of feline calicivirus, feline herpesvirus 1 and *Chlamydia psittaci* mucosal swabs by multiplex RT-PCR/PCR, *Vet. Microbiol.*, 81, 95, 2001.

44. Tamura, K. et al., MEGA4: Molecular evolutionary genetics analysis (MEGA) software version 4.0, *Mol. Biol. Evol.*, 24, 1596, 2007.

45. Wilhelm, S. and Truyen, U., Real-time reverse transcription polymerase chain reaction assay to detect a broad range of feline calicivirus isolates, *J. Virol. Methods*, 133, 105, 2006.

15 Porcine Caliciviruses

Dongyou Liu

CONTENTS

15.1 Introduction 129
    15.1.1 Classification.. 129
    15.1.2 Morphology and Genome Organization... 130
    15.1.3 Replication and Epidemiology... 131
    15.1.4 Clinical Features and Pathogenesis 131
    15.1.5 Diagnosis 132
15.2 Methods 132
    15.2.1 Sample Preparation... 132
    15.2.2 Detection Procedures.. 132
        15.2.2.1 Conventional RT-PCR for Pan-Calicivirus Detection 132
        15.2.2.2 Real-Time RT-PCR for Detection of GII NoVs 132
        15.2.2.3 Real-Time RT-PCR for Detection of GIII SaVs... 132
15.3 Conclusion 132
References... 133

15.1 INTRODUCTION

Caliciviruses (family *Caliciviridae*) are small, positive-sense, single-stranded RNA viruses that are associated with a wide range of economically important diseases in animals and humans. The family *Caliciviridae* is complex, consisting of five recognized genera (*Vesivirus*, *Lagovirus*, *Norovirus*, *Sapovirus*, and *Nebovirus*) in addition to several proposed genera (e.g., *Bavovirus*, *Nacovirus*, *Recovirus*, *Secalivirus*, and *Valovirus*).

Caliciviruses affecting swine belong to the genera *Norovirus*, *Sapovirus*, and *Vesivirus*, as well as *Valovirus*, which is a yet to be approved genus. Of these, vesicular exanthema of swine virus (VESV) in the genus *Vesivirus* causes a vesicular disease in pigs with the formation of vesicles on the epithelium of the snout, lips, nostrils, and tongue. Norovirus genogroup II (genotypes 11, 18, and 19, or GII.11, GII.18, and GII.19) in the genus *Norovirus* and sapovirus genogroup III in the genus *Sapovirus* are detected in feces of pigs with diarrhea, and also associated with asymptomatic gastrointestinal infections in pigs. Furthermore, St-Valérien-like viruses (SVVs) in the tentative genus *Valovirus* are found in the feces of asymptomatic swine of all ages.

Caliciviruses affecting bovine are norovirus genogroup III in the genus *Norovirus* and Newbury-1 virus in the genus *Nebovirus*. Norovirus GIII includes two genotypes (GIII.1 and GIII.2), with genotype 2 (GIII.2) containing Newbury-2 virus, which was once thought to be related to Newbury-1 virus of the genus *Nebovirus*. Unlike caliciviruses affecting swine that sometimes produce asymptomatic infections in the gastrointestinal system, caliciviruses affecting bovine are

often responsible for inducing diarrhea, anorexia, xylose malabsorption, etc.

As VESV of the genus *Vesivirus* is discussed in Chapter 16, the focus of this chapter is caliciviruses that infect the gastrointestinal system of swine namely norovirus GII (GII.11, GII.18, and GII.19) and sapovirus GIII, which will be conveniently referred to as porcine caliciviruses here, while SVV of the tentative genus *Valovirus* will be mentioned briefly.

Although porcine caliciviruses are often implicated in mild or asymptomatic gastroenteritis in pigs, their importance in public health lies in their close genetic and antigenic relationships with human noroviruses (NoVs, particularly GII) that are predominant enteric pathogens of humans of all ages and human sapoviruses (SaVs) that are responsible for gastroenteritis in pediatric patients. The fact that human GII NoV-like RNA (GII.2, GII.3, GII.4, and GII.13) is occasionally detected in pigs highlights the risk of interspecies transmission. In addition, the cooccurrence of human NoV-like and porcine NoVs in pigs may facilitate the generation of novel recombinant strains with broader host ranges.

15.1.1 CLASSIFICATION

The family *Caliciviridae* (derived from the Greek word *calyx* meaning "cup" or "goblet," referring to cup-shaped depressions on virus surface) covers a group of nonenveloped, positive-sense, single-stranded RNA viruses, some of which are pathogenic to animals including humans, dolphins, reptiles, sheep, dogs, cattle, chickens, and amphibians. Currently, the family is composed of five genera: *Vesivirus*, *Lagovirus*, *Norovirus*, *Sapovirus*, and *Nebovirus*, although several tentative genera have been proposed but not approved (e.g., Atlantic salmon calicivirus, chicken calicivirus [Bavovirus], turkey calicivirus [Nacovirus], rhesus enteric calicivirus or Tulane virus [Recovirus], Secalivirus, and SVV [Valovirus]) [1–3].

Of the five recognized genera, the genus *Vesivirus* consists of seven species that infect dogs, cats, pigs, sea lion, walrus, and skunk (canine calicivirus, feline calicivirus [FCV], San Miguel sea lion virus (SMSV), skunk calicivirus, VESV, vesivirus Cro1, and walrus calicivirus). FCV causes respiratory disease in cats and is controlled by vaccination. VESV is responsible for causing a vesicular disease in pigs, which is characterized by the formation of vesicles on the epithelium of the snout, lips, nostrils, tongue, feet, and mammary glands. VESV only remains a concern among Californian pig farmers, given that the virus was eradicated in swine in the rest of the world in 1959. The source of VESV possibly comes from San Miguel Island sea lions, as an essentially similar virus (SMSV) was isolated from these sea lions in 1972, which causes typical signs of VESV in experimentally inoculated pigs.

The genus *Lagovirus* includes two rabbit-infecting species (European brown hare syndrome virus and rabbit hemorrhagic disease virus [RHDV]). RHDV is associated with a fatal liver disease in adult European rabbits (*Oryctolagus cuniculus*) worldwide, which contributes to significant economic losses in the rabbit meat and fur industry and has negative impacts on wild rabbit populations and its dependant predators.

The genus *Norovirus* contains a single species, Norwalk virus, which is further separated into five genogroups (GI, GII, GIII, GIV, and GV) on the basis of 45%–61% differences in capsid genes, with a tentative new genogroup (GVI) that includes human and canine NoVs being proposed recently. In addition, 31 distinct NoV clusters or genotypes are identified within the genogroups (8 GI, 19 GII, 2 GIII, 1 GIV, and 1 GV genotypes) on the basis of 14%–44% differences in capsid genes. NoV strains are determined within a genotype by 0%–14% differences in capsid genes. NoV nomenclature usually contains place names where they were first identified and also references to their genogroup and genotype. For example, the Farmington Hills strain (which was linked to gastroenteritis cases in Farmington Hills, MI) is referred to as a GII.4 (genogroup II, genotype 4) strain. While NoVs infecting humans belong to GI, GII, and GIV, those infecting animals belong to GII (swine), GIII (bovine, consisting of genotypes 1 and 2; genotype 2 contains Newbury-2 virus, which was once thought to be related to Newbury-1 virus of the genus *Nebovirus*), GIV (lions and dogs), and GV (mice) [4]. In particular, GII NoVs account for most NoV outbreaks in humans and are also associated with infections in finisher pigs. More specifically, porcine NoVs belong to GII genotype 11 (GII.11), GII.18, and GII.19; human NoVs comprise the other GII genotypes (1–10, 12–17). The close genetic and antigenic relationships between human and porcine GII NoVs and the occasional detection of human GII NoV-like RNA (GII.2, GII.3, GII.4, and GII.13) in pigs raise serious concerns

of possible interspecies transmission [5]. Recent studies provided evidence on the cooccurrence of human NoV-like and porcine NoVs in commercially reared pigs, which could lead to novel recombinant strains with broader host ranges [6].

The genus *Sapovirus* includes a single species, Sapporo virus (discovered in 1977 in Sapporo, Japan), which is further divided into seven genogroups (GI–GVII), with another tentative genogroup (GVIII) being proposed [7–10]. SaVs have been identified in bats, California sea lions, dogs, pigs, and mink as well as humans, with GI, GII, GIV, and GV SaVs implicated in mild gastroenteritis in young children and GIII SaVs occurring in all age groups of pigs. Recent phylogenetic analysis of complete capsid sequences from 95 human and animal SaV strains available in GenBank suggested the existence of 14 genogroups. Specifically, GV includes human, porcine, and sea lion SaVs. Human SaVs belong to GI, GII, GIV, and GV. Porcine SaVs belong to GIII, GVI, GVII, GVIII, GIX?, GX?, and GXI? in addition to GV. GV and GVIII porcine SaVs appear to be more closely related to human SaVs than to other porcine strains, with GVIII SaVs becoming more genetically variable in the United States. Mink, dog, and bat SaVs belong to GXII?, GXIII?, and GXIV?, respectively. Previously unassigned porcine strains Po/F2-4, Po/F8-9, Po/DO19, and Po/2014P2 are classified into GVII, which forms at least three distinct genetic clusters in the capsid phylogenetic tree, including the prototypical GVII SaVs (Po/K7, Po/F2-4, and Po/DO19, respectively). In addition, Po/F16-7 is classified into GIX? (19), Po/K8 into GX?, and Po/2053P4 into GXI? [5].

The genus *Nebovirus* contains a single species, Newbury-1 virus, which is infective to cattle, resulting in fecal color change, increased fecal output, anorexia, and xylose malabsorption [11,12].

The proposed genus *Valovirus* covers a novel group of swine caliciviruses known as the SVV. These viruses were first identified in the feces of asymptomatic finisher swine in the province of Quebec, Canada, and showed an overall prevalence of 23.8% (range 2.6%–80.0%) in finisher pigs in North Carolina [13], as well as 10% in pig serum samples (including 3.3% [2/60]) in the 3–6 months age group, 8.4% (18/214) in the 7–10 months age group, and 12.6% (43/340) in the >10 months age group in Italy [14]. The pathogenic potential of SVV in pigs remains to be determined.

15.1.2 MORPHOLOGY AND GENOME ORGANIZATION

Caliciviruses are small, nonenveloped viruses of 27–40 nm in diameter. The capsid is composed of 180 capsid proteins and appears hexagonal/spherical with T = 3 icosahedral symmetry.

Beneath the capsid is a nonsegmented, polyadenylated, positive-sense, single-stranded RNA genome of 7.5–8.5 kb, whose noncapped 5′-terminus is linked to a VPg protein, followed by three open reading frames (ORF1–ORF3) and a 3′-terminus with a poly(A) tail [15].

ORF1 encodes a large polyprotein that is cleaved post- and cotranslationally by the virus-encoded cysteine proteinase (3C or NS6), yielding the nonstructural proteins

(Hel, VPg, Pro, and POL). The VPg protein (typically 13–15 kDa in size) is involved in the initiation of protein translation from viral RNA.

ORF2 encodes the major capsid protein viral protein 1 (VP1), which is the key determinant of calicivirus antigenicity. The VP1 coding gene sequences have been used extensively for the identification of calicivirus genogroups and genotypes. For example, norovirus genogroups are defined by the differences of 45%–61%, genotypes by the differences of 14%–44%, and strains by the differences of 0%–14% in capsid genes.

ORF3 encodes the minor capsid protein VP2. A shorter, subgenomic RNA (sgRNA) is also produced from ORF3 during replication, which is 3′ coterminal to the genomic RNA.

Although most caliciviruses have three ORFs, mouse norovirus appears to be an exception, with a fourth overlapping ORF, which produces a virulence factor 1 within the VP1 coding region (ORF2). Furthermore, RHDV and some SaVs (GII and GIII) possess another capsid-coding region in frame with ORF1, so that VP1 can be synthesized as part of the large polyprotein from ORF1 and also as a single protein from OFR2. In addition, FCV genome encodes a leader capsid peptide at the 5′ end of VP1.

Analysis of the complete genome of a porcine calicivirus strain (Ah-1) prevalent in pig groups in Anhui Province, China, demonstrated the presence of a 9-nucleotide (nt) 5′ untranslated region (UTR), two partially overlapping ORFs, and a 70-nt 3′ UTR. ORF1 is positioned between nt 10 and 6775 and encodes 2255 aa protein. ORF2 is located between nt 6771 and 7286 and encodes 172 aa protein [16,17].

15.1.3 Replication and Epidemiology

In general, replication of caliciviruses is a multistepped process. The initial attachment of the virus to the host receptors mediates its entry via endocytosis. Inside the cell, the virus uncoats and releases the viral RNA into the cytoplasm. Initiated by VPg, translation of ORF1 polyprotein from the viral RNA and subsequent processing yield replication proteins, which facilitate synthesis of dsRNA genome from the genomic ssRNA(+). Transcription and replication from dsRNA genome generate viral mRNAs/new ssRNA(+) genomes. In addition, translation of sgRNA gives rise to the capsid protein and VP2, which are utilized for assembly of new virus particles, which are then released after cell lysis.

The RNA of calicivirus virion is infectious and serves as both the genome and viral messenger RNA. Transmission of caliciviruses is mostly through the fecal–oral route upon contamination of shell fish, leafy greens, soft red fruits, water, or fomites. Transmission by close contact or the respiratory route (aerosolized vomitus particles) is also possible. Caliciviruses have a worldwide distribution and infect a diversity of animals (including humans, cattle, pigs, cats, chickens, mink, reptiles, dolphins, and amphibians) [18,19].

NoVs and SaVs of the family *Caliciviridae* are emerging enteric pathogens in humans and animals. The prototype norovirus (Norwalk virus) was first identified as the cause of a gastroenteritis outbreak among schoolchildren in Norwalk, Ohio, in 1968, and subsequently characterized in 1972. Since then, a large number of NoVs belonging to different genotypes have been shown to be the culprit for acute gastroenteritis in humans of all ages. Due to their high infectivity, persistence in the environment, resistance to common disinfectants, and difficulty in controlling their transmission through routine sanitary measures, NoVs are the major foodborne and waterborne pathogens, causing extended outbreaks in human populations [20,21]. It is estimated that NoVs are associated with nearly 23 million infections per year in the United States alone, with 50,000 hospitalizations and 300 deaths. In developing countries, NoVs are responsible for over 1 million hospitalizations and 200,000 deaths in young children annually [20,21].

The first evidence of norovirus in pigs was obtained in 1980 [22,23]. The investigation of NoVs in three pig farms in North Carolina revealed a prevalence of 20% in finisher pigs [24]. Other studies showed a prevalence of 25% of porcine NoVs in Canada, 2%–6% in Europe, 9% in New Zealand, 8% in Brazil, and <1%–15% in Asia [25–31]. Recent identification of humanlike NoVs (GII.2, GII.4) in pigs that also harbor other swine NoVs implies the possibility of novel recombinant NoVs emerging in the future that show higher pathogenic potential and broader host specificity [32,33].

SaVs were first observed by direct electron microscopy in stools of diarrheic children in 1976. In 1980, a porcine SaV (Cowden strain, previously known as porcine enteric calicivirus [PEC]) was found from the fecal samples of diarrheic piglets by electron microscopy in the United States [23]. Since then, porcine SaVs have been described in South Korea, Japan, China, Czech, Hungary, Germany, Belgium, Italy, Canada, Brazil, Venezuela, etc. [26,34–52]. Originated from diarrheic piglets, the porcine sapovirus strains Cowden and LL14/02/ US, representatives of sapovirus genogroup GIII, have the capacity to induce enteric diseases and lesions in experimentally infected pigs [53–55].

15.1.4 Clinical Features and Pathogenesis

Porcine caliciviruses (NoVs and SaVs) are implicated in gastroenteritis in pigs, as they were first identified from diarrheic piglets. More often, infection with porcine caliciviruses may be asymptomatic, as documented in a number of recent reports.

In cattle, infection with norovirus GIII.2 led to persisting diarrhea and prolonged fecal shedding, but with no significant intestinal lesions in Gn calves. On the other hand, norovirus GIII.1 induced remarkably severe atrophic jejunitis in conventional calves [56].

In humans, calicivirus infections commonly cause acute gastroenteritis (inflammation of the stomach and intestines), with vomiting and diarrhea being the main clinical symptoms. Other common symptoms include nausea, abdominal cramps, low-grade fever, and malaise. Compared to human norovirus infections, SaV infections have a shorter duration of viral shedding and are less associated with projectile vomiting.

Considering that porcine caliciviruses often cause persistent, asymptomatic infections, it suggests the host's inability to effectively clear the invading viral pathogens. One possible mechanism may be the ability of porcine caliciviruses to constantly mutate in areas predicted to be important for immune recognition during chronic infection (e.g., the hypervariable P2 capsid domain), which helps the virus to evade host immune surveillance functions and partially explains the repeated susceptibility to calicivirus infections. Alternatively, porcine caliciviruses may use its products to block or shut off some cellular signaling pathways in the host that are key to the normal immune responses. Indeed, there is evidence that the functionality of type I interferon signaling pathways is critical to prevent serious and even lethal infections.

15.1.5 Diagnosis

Diagnosis of porcine calicivirus infections is possible through electron microscopy, virus isolation, serological assays, and nucleic acid detection. Use of electron microscopy has been fundamental to the initial discovery of caliciviruses. The subsequent application of stable porcine kidney cell line (LLC-PK cells) in the presence of the intestinal content fluid filtrate from uninfected gnotobiotic pigs, or bile acids, has enabled better characterization of PEC [57]. More recently, the development of nucleic acid detection methods such as RT-PCR has made it possible to detect and subtype porcine caliciviruses from pigs with or without overt clinical symptoms [58–61].

15.2 METHODS

15.2.1 Sample Preparation

Fecal suspensions are prepared in minimal essential medium at 10% (wt/vol) (Invitrogen). The RNA is then extracted from 200 µL of the supernatants using a QIAamp Viral RNA Mini Kit (Qiagen). The RNA is eluted into 50 µL RNase-free water. Viral RNA may be also extracted from stool specimens using the TRIzol method (Life Technologies).

15.2.2 Detection Procedures

15.2.2.1 Conventional RT-PCR for Pan-Calicivirus Detection

Jiang et al. [62] designed a pair of primers (P290, 5′-GATTACTCCAAGTGGGACTCCAC-3′; P289, 5′-TGACA ATGTAATCATCACCATA-3′) from the conserved region of calicivirus RNA polymerase gene for amplification of a specific 319 or 331 bp fragment in NoVs or SaVs, respectively. RT-PCR may be performed using the PrimeScript® One-Step RT-PCR Kit Ver.2 (Dye Plus) (TAKARA) with the following cycling programs: 45°C for 30 min and 94°C for 5 min, followed by 40 cycles of 94°C for 1 min, 48°C for 1 min and 20 s, and 72°C for 1 min, plus a final extension of 72°C for 10 min. The amplified products are separated by 2% agarose gel electrophoresis in the presence of SYBR safe DNA gel stain (Invitrogen) and are visualized under UV light. The presence of 319 or 311 bp is indicative of NoVs or SaVs, respectively [63].

15.2.2.2 Real-Time RT-PCR for Detection of GII NoVs

Wang et al. [64] described a one-step TaqMan RT-qPCR for detection of GII NoVs (consisting of both human- and animal-infecting NoVs) using primers COG2F (5′-CAR GAR BCN ATG TTY AGR TGG ATG AG-3′) and COG2R (5′-TCG ACG CCA TCT TCA TTC ACA-3′) and the probe RING2 (5′,6-FAM–TGG GAG GGC GAT CGC AAT CT–Black Hole Quencher) [65]. The 20 µL RT-PCR mixture is composed of 2 µL of sample RNA, 400 µM of deoxynucleoside triphosphates (dNTPs), 200 nM of each primer, and 100 nM of the probe by using a Qiagen One-Step RT-PCR kit (Qiagen). The assay is performed with the following cycling conditions: 1 cycle at 50°C for 30 min, 1 cycle at 95°C for 15 min, and 45 cycles at 95°C for 15 s and 57.5°C for 60 s in an Eppendorf Mastercycler RealPlex instrument (Eppendorf). The detection limit of this assay is 10 genomic equivalents (GE) per reaction (20 µL).

15.2.2.3 Real-Time RT-PCR for Detection of GIII SaVs

Wang et al. [64] developed a one-step TaqMan RT-qPCR for the detection of porcine genogroup III (GIII) SaVs. The forward primer (5′-CCA GAA GTG TTC GTG ATG GAG-3′; nt 5125–5145), reverse primer (5′-GCC CRG CTG GYT GGA CTG-3′; nt 5230–5213), and probe (5′,6-carboxyfluorescein [5′,6-FAM]–TGCGAGCAACCCAGAGGGCACTCA–Iowa Black fluorescence Quencher; nt 5169–5192) target the beginning of the major capsid VP1 gene in the SaV Cowden strain.

RT-PCR mixture (20 µL) is made up of 2 µL of sample RNA, 400 µM of dNTPs, 200 nM of each primer, and 100 nM of the probe by using a Qiagen One-Step RT-PCR Kit (Qiagen). The assay is carried out with the following cycling conditions: 1 cycle at 50°C for 30 min, 1 cycle at 95°C for 15 min, and 45 cycles at 95°C for 15 s and 57.5°C for 60 s in an Eppendorf Mastercycler RealPlex instrument (Eppendorf).

The assay detects ≥10 GE per reaction (20 µL) and is nonreactive with other genogroups of porcine SaVs or porcine NoVs.

15.3 CONCLUSION

The family *Caliciviridae* encompasses a diverse and complex group of positive-sense, single-stranded RNA viruses that are infective to a range of animals as well as humans. In humans, caliciviruses are responsible for diarrhea, which can be fatal in young children if loss of body fluids is not rectified and replenished. In swine, caliciviruses tend to cause mild or asymptomatic gastrointestinal infections. Nevertheless, considering the close genetic and antigenic relationships between human and porcine caliciviruses and the coexistence of these viruses in swine, there is a real danger that novel recombinant viruses with potent pathogenicity and trans-species capability may emerge in the future. Therefore, it is important to have sufficient preparedness to deal with these events if they do occur. This preparedness should include the ability to promptly and accurately identify and type porcine and human caliciviruses. Toward this end, various nucleic acid–based diagnostic techniques such as RT-PCR have been developed and have shown promise for direct detection and quantification of caliciviruses from swine and human specimens.

REFERENCES

1. Berke T, Matson DO. Reclassification of the Caliciviridae into distinct genera and exclusion of hepatitis E virus from the family on the basis of comparative phylogenetic analysis. *Arch Virol*. 2000;145(7):1421–1436.
2. Green KY. Caliciviridae: The noroviruses. In: Knipe DM, Howley PM, eds., *Fields Virology*, 5th edn. Lippincott Williams & Wilkins, Philadelphia, PA, 2007, pp. 949–979.
3. L'Homme Y et al. Genomic characterization of swine caliciviruses representing a new genus of Caliciviridae. *Virus Genes*. 2009;39:66–75.
4. Jung K, Scheuer KA, Zhang Z, Wang Q, Saif LJ. Pathogenesis of GIII.2 bovine norovirus, CV186-OH/00/US strain in gnotobiotic calves. *Vet Microbiol*. 2014;168(1):202–207.
5. Scheuer KA et al. Prevalence of porcine noroviruses, molecular characterization of emerging porcine sapoviruses from finisher swine in the United States, and unified classification scheme for sapoviruses. *J Clin Microbiol*. 2013;51(7):2344–2353.
6. Zheng DP et al. Norovirus classification and proposed strain nomenclature. *Virology*. 2006;346(2):312–323.
7. Farkas T et al. Genetic diversity among sapoviruses. *Arch Virol*. 2004;149:1309–1323.
8. Hansman GS et al. Intergenogroup recombination in sapoviruses. *Emerg Infect Dis*. 2005;11:1916–1920.
9. Jeong C et al. Genetic diversity of porcine sapoviruses. *Vet Microbiol*. 2007;122:246–257.
10. Barry AF, Alfieri AF, Alfieri AA. High genetic diversity in RdRp gene of Brazilian porcine sapovirus strains. *Vet Microbiol*. 2008;131(1–2):185–191.
11. Oliver SL, Asobayire E, Dastjerdi AM, Bridger JC. Genomic characterization of the unclassified bovine enteric virus Newbury agent-1 (Newbury1) endorses a new genus in the family Caliciviridae. *Virology* 2006;350(1):240–250.
12. Kaplon J, Guenau E, Asdrubal P, Pothier P, Ambert-Balay K. Possible novel nebovirus genotype in cattle, France. *Emerg Infect Dis*. June 2011;17(6):1120–1123. http://dx.doi.org/10.3201/eid1706.100038. Accessed on March 31, 2015.
13. Wang Q et al. Characterization and prevalence of a new porcine Calicivirus in Swine, United States. *Emerg Infect Dis*. 2011;17(6):1103–1106.
14. Di Martino B et al. Seroprevalence of St-Valerien-like caliciviruses in Italian swine. *J Gen Virol*. 2012;93(1):102–105.
15. Alhatlani B, Vashist S, Goodfellow I. Functions of the 5′ and 3′ ends of calicivirus genomes. *Virus Res*. February 9, 2015;206:134–143.
16. Yang S et al. Molecular characterization and phylogenetic analysis of the complete genome of a porcine sapovirus from Chinese swine. *Virol J*. 2009;6:216.
17. Fan K, Wang R. Complete genome of a porcine calicivirus strain in Anhui province, China, is significantly shorter than that of the other Chinese strain. *J Virol*. 2012;86(24):13823.
18. Guo M, Evermann JF, Saif LJ. Detection and molecular characterization of cultivable caliciviruses from clinically normal mink and enteric caliciviruses associated with diarrhea in mink. *Arch Virol*. 2001;146:479–493.
19. Halaihel N et al. Enteric calicivirus and rotavirus infections in domestic pigs. *Epidemiol Infect*. 2010;138(4):542–548.
20. Richards GP, Watson MA, Meade GK, Hovan GL, Kingsley DH. Resilience of norovirus GII.4 to freezing and thawing: Implications for virus infectivity. *Food Environ Virol*. 2012;4(4):192–197.
21. Karst SM, Zhu S, Goodfellow IG. The molecular pathology of noroviruses. *J Pathol*. 2015;235(2):206–216.
22. Bridger JC. Detection by electron microscopy of caliciviruses, astroviruses and rotavirus-like particles in the faeces of piglets with diarrhoea. *Vet Rec*. 1980;107:532–533.
23. Saif LJ, Bohl EH, Theil KW, Cross RF, House JA. Rotavirus-like, calicivirus-like, and 23-nm virus-like particles associated with diarrhea in young pigs. *J Clin Microbiol*. 1980;12:105–111.
24. Wang QH, Souza M, Funk JA, Zhang W, Saif LJ. Prevalence of noroviruses and sapoviruses in swine of various ages determined by reverse transcription-PCR and microwell hybridization assays. *J Clin Microbiol*. 2006;44:2057–2062.
25. Farkas T et al. Seroprevalence of noroviruses in swine. *J Clin Microbiol*. 2005;43:657–661.
26. Keum HO et al. Porcine noroviruses and sapoviruses on Korean swine farms. *Arch Virol*. 2009;154(11):1765–1774.
27. Reuter G et al. Incidence, diversity, and molecular epidemiology of sapoviruses in swine across Europe. *J Clin Microbiol*. 2010;48(2):363–368.
28. Song YJ et al. Identification of genetic diversity of porcine Norovirus and Sapovirus in Korea. *Virus Genes*. 2011;42(3):394–401.
29. Chao DY, Wei JY, Chang WF, Wang J, Wang LC. Detection of multiple genotypes of calicivirus infection in asymptomatic swine in Taiwan. *Zoonoses Public Health*. 2012;59(6):434–444.
30. Wilhelm B et al. Survey of Canadian retail pork chops and pork livers for detection of hepatitis E virus, norovirus, and rotavirus using real time RT-PCR. *Int J Food Microbiol*. 2014;185:33–40.
31. Silva PF et al. High frequency of porcine norovirus infection in finisher units of Brazilian pig-production systems. *Trop Anim Health Prod*. 2015;47(1):237–241.
32. Wang QH et al. Genetic diversity and recombination of porcine sapoviruses. *J Clin Microbiol*. 2005;43:5963–5972.
33. Wang QH, Han MG, Cheetham S, Souza M, Funk JA, Saif LJ. Porcine noroviruses related to human noroviruses. *Emerg Infect Dis*. 2005;11:1874–1881.
34. Kim HJ, Cho HS, Cho KO, Park NY. Detection and molecular characterization of porcine enteric calicivirus in Korea, genetically related to sapoviruses. *J Vet Med B Infect Dis Vet Public Health*. 2006;53(4):155–159.
35. Martínez MA et al. Molecular detection of porcine enteric caliciviruses in Venezuelan farms. *Vet Microbiol*. 2006;116(1–3):77–84.
36. Yin Y et al. Genetic analysis of calicivirus genomes detected in intestinal contents of piglets in Japan. *Arch Virol*. 2006;151:1749–1759.
37. Reuter G, Bíró H, Szűcs Gy. Enteric caliciviruses in domestic pigs in Hungary. *Arch Virol*. 2007;152:611–614.
38. Martella V et al. Genetic heterogeneity of porcine enteric caliciviruses identified from diarrhoeic piglets. *Virus Genes*. 2008;36(2):365–373.
39. Martella V et al. Identification of a porcine calicivirus related genetically to human sapoviruses. *J Clin Microbiol*. 2008;46:1907–1913.
40. Mauroy A et al. Noroviruses and sapoviruses in pigs in Belgium. *Arch Virol*. 2008;153(10):1927–1931.
41. Zhang W et al. The first Chinese porcine sapovirus strain that contributed to an outbreak of gastroenteritis in piglets. *J Virol*. 2008;82:8239–8240.
42. Collins PJ, Martella V, Buonavoglia C, O'Shea H. Detection and characterization of porcine sapoviruses from asymptomatic animals in Irish farms. *Vet Microbiol*. 2009;139(1–2):176–182.

43. Shen Q et al. Molecular detection and prevalence of porcine caliciviruses in eastern China from 2008 to 2009. *Arch Virol.* 2009;154(10):1625–1630.

44. Shen Q et al. Prevalence of hepatitis E virus and porcine caliciviruses in pig farms of Guizhou province, China. *Hepat Mon.* 2011;11(6):459–463.

45. Cunha JB, de Mendonça MC, Miagostovich MP, Leite JP. Genetic diversity of porcine enteric caliciviruses in pigs raised in Rio de Janeiro State, Brazil. *Arch Virol.* 2010;155(8):1301–1305.

46. Liu GH et al. RT-PCR test for detecting porcine sapovirus in weanling piglets in Hunan Province, China. *Trop Anim Health Prod.* 2012;44(7):1335–1339.

47. Dufkova L, Scigalkova I, Moutelikova R, Malenovska H, Prodelalova J. Genetic diversity of porcine sapoviruses, kobuviruses, and astroviruses in asymptomatic pigs: An emerging new sapovirus GIII genotype. *Arch Virol.* 2013;158(3):549–558.

48. das Merces Hernandez J et al. Genetic diversity of porcine sapoviruses in pigs from the Amazon region of Brazil. *Arch Virol.* 2014;159(5):927–933.

49. Di Bartolo I et al. Detection and characterization of porcine caliciviruses in Italy. *Arch Virol.* 2014;159(9):2479–2484.

50. Liu W, Yang B, Wang E, Liu J, Lan X. Complete sequence and phylogenetic analysis of a porcine sapovirus strain isolated from western China. *Virus Genes* 2014;49(1):100–105.

51. Liu ZK, Li JY, Pan H. Seroprevalence and molecular detection of porcine sapovirus in symptomatic suckling piglets in Guangdong Province, China. *Trop Anim Health Prod.* 2014;46(3):583–587.

52. Machnowska P, Ellerbroek L, Johne R. Detection and characterization of potentially zoonotic viruses in faeces of pigs at slaughter in Germany. *Vet Microbiol.* 2014;168(1):60–68.

53. Guo M et al. Molecular characterization of a porcine enteric calicivirus genetically related to Sapporo-like human caliciviruses. *J Virol.* 1999;73(11):9625–9631.

54. Guo M et al. Comparative pathogenesis of tissue culture-adapted and wild-type Cowden porcine enteric calicivirus (PEC) in gnotobiotic pigs and induction of diarrhea by intravenous inoculation of wild-type PEC. *J Virol.* 2001;75(19):9239–9251.

55. Flynn WT, Saif LJ, Moorhead PD. Pathogenesis of porcine enteric calicivirus-like virus in four-day-old gnotobiotic pigs. *Am J Vet Res.* 1988;49:819–825.

56. Mijovski JZ, Poljsak-Prijatelj M, Steyer A, Barlic-Maganja D, Koren S. Detection and molecular characterisation of noroviruses and sapoviruses in asymptomatic swine and cattle in Slovenian farms. *Infect Genet Evol.* 2010;10(3):413–420.

57. V, Kim Y, Chang KO. The crucial role of bile acids in the entry of porcine enteric calicivirus. *Virology.* 2014;456–457:268–278.

58. Le Guyader F et al. Evaluation of a degenerate primer for the PCR detection of human caliciviruses. *Arch Virol.* 1996;141:2225–2235.

59. Vinjé J et al. Molecular detection and epidemiology of "Sapporo-like viruses." *J Clin Microbiol.* 2000;38:530–536.

60. Mauroy A, Van der Poel WH, der Honing RH, Thys C, Thiry E. Development and application of a SYBR green RT-PCR for first line screening and quantification of porcine sapovirus infection. *BMC Vet Res.* 2012;8:193.

61. Stals A et al. Molecular detection and genotyping of noroviruses. *Food Environ Virol.* 2012;4(4):153–167.

62. Jiang X et al. Design and evaluation of a primer pair that detects both Norwalk- and Sapporo-like caliciviruses by RT-PCR. *J Virol Methods.* 1999;83:145–154.

63. Sato G et al. Characterization of St-Valerien-like virus genome detected in Japan. *J Vet Med Sci.* 2014;76(7):1045–1050.

64. Wang Q, Zhang Z, Saif LJ. Stability of and attachment to lettuce by a culturable porcine sapovirus surrogate for human caliciviruses. *Appl Environ Microbiol.* 2012;78(11):3932–3940.

65. Kageyama T et al. Broadly reactive and highly sensitive assay for Norwalk-like viruses based on real-time quantitative reverse transcription-PCR. *J Clin Microbiol.* 2003;41(4):1548–1557.

16 Vesicular Exanthema of Swine Virus

Nick J. Knowles, Begoña Valdazo-González, Britta A. Wood,
Katarzyna Bachanek-Bankowska, and Donald P. King

CONTENTS

16.1 Introduction 135
    16.1.1 Classification, Morphology, and Genome Organization 136
    16.1.2 Transmission, Clinical Features, and Pathogenesis... 137
        16.1.2.1 Transmission 137
        16.1.2.2 Clinical Features.. 137
        16.1.2.3 Pathogenesis.. 138
    16.1.3 Diagnosis 139
16.2 Methods 140
    16.2.1 Sample Collection and Preparation 140
    16.2.2 Detection Procedures... 140
        16.2.2.1 Real-Time RT-PCR 140
        16.2.2.2 Sequence Analysis 140
16.3 Future Prospective 141
References.. 141

16.1 INTRODUCTION

Vesicular exanthema of swine (VES) was first recognized on five pig farms in California in April 1932 as a clinical syndrome that was indistinguishable from foot-and-mouth disease (FMD) [1]. Although experimental inoculation of material from these cases failed to reproduce vesicular lesions in either cattle or guinea pigs, the causative agent was presumed to be an atypical FMD virus. However, four novel immunologically distinct viruses were subsequently isolated, which were named VES viruses (VESV) A, B, C, and D, and one of these (VESV type B or 1934B) from 1934 is still available (at the Pirbright Institute, UK). Similar outbreaks (of up to 184 cases/year) were observed every year between 1933 and 1951 (except in 1937–1938). Typically, these outbreaks affected mainly (but not exclusively) garbage-fed pigs in California. Between 1940 and 1942, three immunologically distinct viruses were recovered, but these were subsequently lost or destroyed. A number of pig-passaged tissue samples of another VESV serotype, 101–143 (which was isolated at the Pirbright Institute from a sample collected from pigs in San Francisco stockyards in 1943), are still available, but it has not yet been possible to recover viable virus from these samples. On June 16, 1952, VES appeared at a biological manufacturing plant in Grand Island, Nebraska. The source of this infection was traced to Cheyenne, Wyoming, where pigs had been fed garbage from transcontinental trains that had originated in California. Unfortunately, before the disease was detected in Grand Island, VES-infected pigs had been shipped to Omaha stockyards and sold on. In a little over a month (by July 29), outbreaks had occurred in 19 states, and by September 1953, a total of 42 states and the District of Columbia had

reported cases of VES. Outbreaks continued in California until 1955 and in New Jersey until 1956 [2]. In 1948, a new collection of VESV isolates was started and new serotypes were discovered almost every year. The viruses were named alphabetically and with the year that the serotype was first isolated as a suffix, namely, A48, B51, C52, D53, and E55. Due to the retrospective nature of this analysis, some of the isolates were named out of chronological sequence (see later text). Subsequently, four unique serotypes were detected on a single farm in San Mateo County, California, in December 1954 (H54) and May (I55), September (F55), and October 1955 (G55). It was speculated that pigs on this farm were persistently infected and that new serotypes evolved by mutation [2]. The VES outbreaks that occurred outside California were initially only caused by B51. However, in 1954 and 1956, two new serotypes (designated K54 and J56) appeared on two farms in Secaucus, New Jersey [3]. As these serotypes were not found elsewhere in the United States, it was postulated that these viruses had evolved from the B51 serotype [4]. On October 22, 1959, the U.S. Secretary of Agriculture announced that VES had been eradicated and subsequently classified as a reportable foreign animal disease [2].

In 1972, viruses were isolated from rectal swabs of California sea lions (*Zalophus californianus*) that had recently aborted and from the nose of an emaciated northern seal (*Callorhinus ursinus*) pup, all of which were collected from animals on San Miguel Island, off the coast of southern California [5]. The agent was named San Miguel sea lion virus 1 (SMSV-1) and was soon recognized as being closely related to VESV. Between 1972 and 1991, many more SMSV serotypes were isolated from a range of marine mammals; and there are now 16 SMSV serotypes (1–2 and 4–17; type 3 was found to be a mixture of types 1 and 2). Interestingly, a candidate enterovirus isolated from a California gray whale (*Eschrichtius robustus*) was subsequently reclassified as VESV-A48 [6,7]. In 1978 and 1979, a number of serotypically distinct SMSV-like viruses were isolated from various animals at the San Diego Zoo, including cetaceans [8], primates [9–11], and reptiles [12]. In 1981, another serotypically distinct SMSV was found in cattle in Oregon [13]. SMSV-5 and a serotypically novel virus (provisionally named McAll) have been isolated from cases of vesicular disease in humans [14]. Other VESV/SMSV-related viruses have been isolated from skunks [15] and rabbits [16].

16.1.1 CLASSIFICATION, MORPHOLOGY, AND GENOME ORGANIZATION

Classification: VESV belong to the species *vesicular exanthema of swine virus* in the genus *Vesivirus*, family *Caliciviridae* [17]. The family consists of four other genera, *Lagovirus*, *Nebovirus*, *Norovirus*, and *Sapovirus* [17]. Five unclassified caliciviruses (Atlantic salmon calicivirus, "Bavovirus," "Nacovirus," "Recovirus," and "Valovirus") may represent additional new genera. The *Vesivirus* genus contains one other recognized species, *Feline calicivirus*, and a number of unclassified viruses (including mink and canine caliciviruses), which may form at least two new vesivirus species (Figure 16.1). In this chapter, VESV and viruses related to the *vesicular exanthema of swine virus* species will be referred to as the VESV/SMSV complex. There have been more than 30 VESV/SMSV serotypes described; however, not all have been serologically compared with each other and the lack of general availability of certain prototype viruses (e.g., from SMSV-15 to SMSV-17) prevents a proper serological or genetic

classification. Recent determination of the genome sequence of one of the atypical SMSVs, serotype 8 (KM244552), shows that it does not belong to the *vesicular exanthema of swine virus* species (Figure 16.1) [18]. Failure to detect SMSV-12 by VESV/SMSV-specific western blot assay [15] and reverse transcription polymerase chain reaction (RT-PCR) targeted to highly conserved genome regions [19,20] also suggests that this virus may not belong to this virus species.

Morphology and genome organization: VES virions are nonenveloped and measure 27–40 nm in diameter by negative-stain electron microscopy (NS-EM) and 35–40 nm by cryo-electron microscopy. Most, but not all, caliciviruses have 32 distinctive cup-shaped depressions (L. calix, cup) on their surface, generating the characteristic "Star of David" appearance by NS-EM. The capsid is composed of 90 dimers of the major structural protein VP1 arranged on a T = 3 icosahedral lattice. The x-ray crystal structure of SMSV-4 has been determined to 3.2 Å resolution [21].

VESV/SMSVs possess a single-stranded positive-sense RNA genome of c. 8300 nucleotides. The 3′ end of the genome is polyadenylated and at the 5′ end, a small virus-encoded protein (VPg) is covalently attached to the RNA via a tyrosine at residue 24 [22] (Figure 16.2). VPg is involved in both genome replication and translation, the latter via an interaction with elongation initiation factor 4E [23]. The genome has a very short 5′ untranslated region (19 nt) and encodes three open reading frames (ORFs) (Figure 16.2); ORF1 encodes the nonstructural proteins, ORF2 encodes the major capsid polypeptide (VP1), and ORF3 encodes a small basic protein (VP2) [17]. ORF1 and ORF2 of the VESV/SMSV complex are separated by 5 nt (CCACT/C). A 3′ coterminal subgenomic RNA of 2.2–2.4 kb is

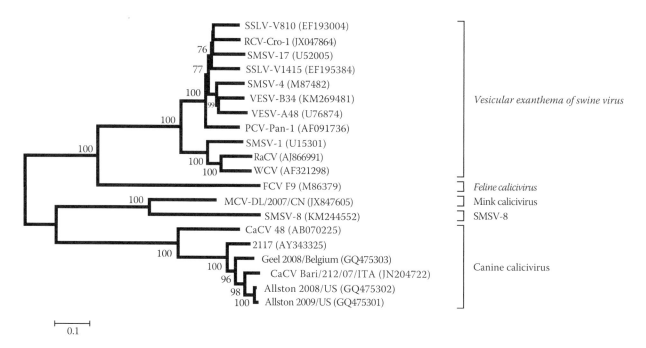

FIGURE 16.1 Genetic relationships between vesicular exanthema of swine virus, San Miguel sea lion virus, and related vesiviruses. Neighbor-joining tree (JTT amino acid substitution model) showing the relationship of the VP1 capsids. Accepted species of *Vesivirus* genus are indicated in italics to distinguish them from unclassified viruses. Tree calculated using MEGA 6.02.

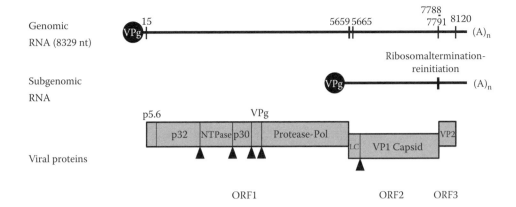

FIGURE 16.2 Genome organization of a typical vesicular exanthema of swine virus showing genomic and subgenomic RNA. Location of the three open reading frames and autoprotease cleavage sites (▲) are shown. Genome numbering in the figure is based upon VESV B1-34 (GenBank KM269481).

synthesized intracellularly and is apparently VPg linked. This also encodes the VP1 capsid polypeptide allowing an abundant expression of capsid proteins.

16.1.2 Transmission, Clinical Features, and Pathogenesis

16.1.2.1 Transmission

It has been suggested that the primary reservoir host of VESV/SMSV is the opaleye perch (*Girella nigricans*); however, other types of fish or poikilotherms from the Pacific Rim of North America and marine mammals (e.g., pinnipeds and cetaceans) might also serve as reservoirs [5,8,24–26]. Parasites such as the sea lion liver fluke (*Zalophotrema* sp.) and the fur seal lungworm (*Parafilaroides* sp.) may act as mechanical vectors [25,27]. VESV/SMSV is viable in seawater at 15°C for more than 14 days, suggesting that transmission may occur via contaminated water [28]. Although controversial, antibody surveys suggest that VESV/SMSV may be endemic in swine [26], cattle [29], and horses [30] in the United States. A carrier state has been suggested in pigs [4] and cattle [31], and is supported by the observation that primate calicivirus (PCV) Pan-1 (a member of the VESV/SMSV complex; see Figure 16.3) was isolated from the throat of a pygmy chimpanzee (*Pan paniscus*) after 6 months of infection [9].

With pigs, transmission occurs typically via direct contact with infected animals and their products and, less frequently, indirectly via contaminated feed, water, urine, and feces. In experimental settings, VESV/SMSV has been successfully transmitted by the application of virus suspensions to scarified skin of the snout or by intradermal, intranasal, and oral routes in pigs [1,32,33] and pinnipeds [33–35]. Additionally, the pig body louse (*Haematopinus suis*) may act as a mechanical vector, acquiring the infection from viremic pigs and transmitting to susceptible pigs [2]. The virus can remain intact/infectious in meat (even when decomposed) at a temperature of 7°C for 4 weeks and at −70°C for 18 weeks, and is not readily inactivated after 10 min at 80°C [1]. Therefore, with most cases of VES in pigs, the initial infection occurred by ingestion

of virus-contaminated raw garbage containing waste from dead/slaughtered pigs, marine mammals, or fish. The incubation period under both experimental and field conditions is usually 24–72 h, but it can be as long as 12 days. Once infected, pigs remain contagious for approximately 5 days [32]. Horizontal transmission occurs readily, as vesicular fluid and vesicle epithelium contain large quantities of virus, which can be shed into the environment. Heavily contaminated premises may remain infectious for several months. Infection with VESV/SMSV by direct contact has also been reported in pinnipeds [35], cattle [13], and reptiles [12].

16.1.2.2 Clinical Features

The clinical signs of VESV/SMSV infections range from asymptomatic (e.g., opaleye perch and swine) to varying degrees of vesicular disease severity (e.g., swine, pinnipeds, and humans) depending on the host and the virus strain. Both natural and experimental VESV/SMSV infections have been associated with abortion (swine, pinnipeds, and cattle), pneumonia (swine, pinnipeds, cattle), diarrhea (swine and cattle), myocarditis (swine and pinnipeds), encephalitis (swine, pinnipeds, and primates), and hepatitis (swine and humans) [5,26,32,36]. The vesicular lesions caused by VESV/SMSV are indistinguishable from other vesicular diseases affecting swine, namely, FMD, vesicular stomatitis, and swine vesicular disease. These three diseases are highly infectious and require notification to the World Organization for Animal Health. In general, experimental infections of pigs with VESV and SMSVs are indistinguishable from each other, although some SMS viruses appear to be more pathogenic for pigs than some of the original VESV serotypes [32]. Disease in pigs is characterized by fever (40.5°C–42°C), the formation of vesicles in the oral cavity, and on the snout and feet (soles, coronary bands, and interdigital areas). Occasionally, vesicles also appear in the mammary glands of the sows. Vesicles are usually flat, blanched areas of 5–30 mm in diameter and raised 10–20 mm above the surrounding unaffected epithelium [1]. Vesicles that contain a clear fluid and a high quantity of virus rupture easily

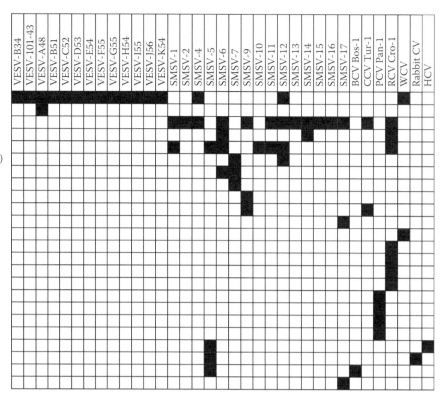

FIGURE 16.3 Broad species distribution of vesicular exanthema of swine viruses/San Miguel sea lion virus and related vesiviruses. Note: Shaded boxes in figure denote host species from which VESV/SMSVs have been isolated. (Data compiled from various publications [see listed references] and from the Laboratory for Calicivirus Studies list of Calicivirus, Oregon State University, Oregon, USA. [Vesiviruses cell culture adapted] isolated and in repository, http://www.kindplanet.org/smith/isolations.htm, Accessed on March 31, 2015.)

leaving raw, bleeding ulcers that are subsequently covered with a fibrinous membrane. Secondary vesicles may appear adjacent to the ruptured vesicles. Oral lesions cause excessive salivation, tongue protrusion, dysphagia, anorexia, and subsequent weight loss due to reduced food intake. Nasal lesions may cause airway obstruction and associated signs. Lesions in the feet cause swelling, inflammation, and lameness, which can be exacerbated by secondary bacterial infections. In severe cases, inflammation of the legs and joints may also occur. The vesicular forms of VESV/SMSV can appear concurrently with diarrhea in infected animals, as well as abortion, stillbirth, cessation of milk production, mortality in suckling piglets, and weakness or death in growing pigs. Morbidity ranges from 30% to 90%, but mortality in adult animals is low. Lesions normally heal in 5–7 days but can be complicated by secondary bacterial infections causing recovery to take up to 90 days. The hooves of previously infected pigs can have a dark horizontal line between the old and the healed wall, which is a characteristic mark of VES and other vesicular diseases.

Natural infections of pinnipeds with VESV/SMSV are associated with clinical signs, including vesicles (fluid-filled blisters) up to 25 mm in diameter on the hairless parts of the dorsal and ventral aspects of flippers and, less frequently, around the mouth and on the hairy regions of the body [33–35]. Vesicles may rupture leaving erosion or may regress without rupturing. Abortion and premature births, and possibly encephalitis and pneumonitis, occur in pinnipeds [5]. In humans, SMSV-5 caused systemic illness, including vesicular lesions on all four extremities [14].

16.1.2.3 Pathogenesis

Pathogenesis of the disease after intradermal inoculation of pigs can be divided into two phases: formation of "primary" lesions at the site of inoculation (12–48 h postinfection) accompanied by fever, followed by formation of "secondary" lesions (48–72 h later), at different locations to the primary lesions. These secondary vesicles may occur as a result of local spread of the virus from the primary lesions, or an effect of systemized viremia. Subcutaneous, intramuscular, and intravenous inoculations are followed by the development of vesicles at any susceptible site within 24–96 h. Vesicles consist of an accumulation of cell debris and fluid, as a consequence of viral replication in the cells of the epidermis, and spread to adjacent cells [1]. Inflammation occurs in subcutaneous connective tissues. The virus has been isolated from the draining lymph nodes of cutaneous lesions (such as mandibular, prescapular lymph nodes), as well as superficial inguinal lymph nodes [32,33]. Pigs remain infectious for about 5 days and the length of the viremia and the generation of neutralizing antibodies depend on the particular strain of the virus and the host [32–35]. In pigs, the viremia can be difficult to detect as it is intermittent, but never more than 4 days in length [32]. However, in pinnipeds, viremia may last up to 11 days [35]. Neutralizing antibodies in pigs appear 3–5 days after infection and can be maintained at high titers anywhere from 90 days to 30 months [2]; in pinnipeds, neutralizing antibodies may last for more than 75 days [35].

16.1.3 DIAGNOSIS

Conventional techniques: As clinical differentiation of VESV from other notifiable vesicular diseases of pigs is impossible, numerous methods have been developed for the detection of VESV, including cell culture virus isolation [37], indirect immunofluorescence [38], indirect antigen ELISA [39], complement fixation [37,40,41], agar diffusion [37,41], cross-immunity challenges [3], and virus neutralization [3,37]. Before the advent of molecular assays such as RT-PCR [19,42,43] and real-time RT-PCR (rRT-PCR [20]), diagnosis was primarily based on the detection of VESV-specific antibodies. The most commonly used assays have been complement fixation [37,40,41] and virus neutralization [3,37]. However, as these assays can be time consuming (e.g., 4 days to obtain results by virus neutralization) and are dependent upon antiserum and cell culture availability, additional diagnostic methods were developed. These additional methods included VESV antigen detection by indirect immunofluorescence [38] and indirect antigen ELISA [39]. Although these assays did not eliminate the need for antiserum and cell culture, diagnostic results could be obtained rapidly, and as demonstrated previously [39], the antigen ELISA was more sensitive than complement fixation. An advantage of a number of these conventional assays (e.g., complement fixation, virus neutralization, and antigen ELISA) is the capability to determine the serotype of the VESV [3,39,40].

Molecular techniques: The development of conventional (agarose gel–based) RT-PCR assays for the detection of VESV/SMSV has enhanced the ability to diagnose VES; however, to ensure diagnostic sensitivity, these assays should be used in conjunction with virus isolation. An advantage of the conventional RT-PCR method is the ability to sequence the DNA amplicon. As the RT-PCR assays developed for VESV/SMSV are directed at both hypervariable and highly conserved regions of the genome, these assays provide tools to both detect and study the genetic relatedness of these viruses using phylogenetic reconstruction methods [19,42].

Increased diagnostic sensitivity has been achieved with the development of rRT-PCR assays for VESV/SMSV [20,44]. With this approach VES/SMS viruses can be detected in clinical samples without virus isolation. In addition, the VESV/SMSV-specific rRT-PCR assay can be performed in parallel with other assays used for the detection of FMDV, VSV, and SVDV, and therefore it can be applied as a differential diagnostic tool for these notifiable vesicular diseases [20]. Other methods for differential diagnosis of vesicular diseases, including a conventional multiplex RT-PCR assay [43] and microarrays [43,45], have been described but these methods are not currently used for routine diagnostics. Of these methods, rRT-PCR followed by DNA sequencing (as outlined later in Section 16.2.2) provides the most effective approach for the detection and characterization of VES/SMS viruses.

The use of next-generation sequencing is being used increasingly in molecular epidemiology studies of viruses as new approaches for obtaining full genome sequences are being developed. Recently, a new Illumina protocol for whole genome sequencing of positive-stranded RNA viruses, including caliciviruses, was described (Figure 16.4) [46]. In this protocol, the RNA in genomic DNA-depleted samples is transcribed using nonspecific primers and second-strand DNA is synthesized prior to library preparation. With this

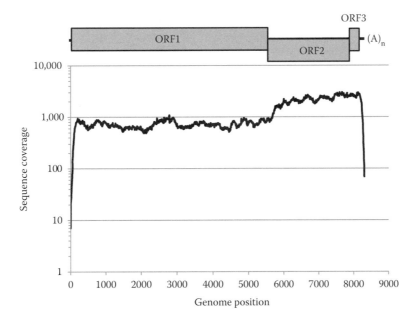

FIGURE 16.4 Sequence coverage of a complete vesicular exanthema of swine virus genome (VESV B1-34; GenBank KM269481) generated by a next-generation sequencing (Illumina) protocol. The genome is annotated to highlight the three different vesicular exanthema of swine virus open reading frames.

approach, viral genome sequences can be obtained from a limited quantity of starting material and without the requirement for genome specific amplification.

16.2 METHODS

16.2.1 SAMPLE COLLECTION AND PREPARATION

Vesicular epithelium is the preferred sample type for VES diagnosis; however, virus can be detectable at low levels in the blood of acutely infected pigs [47]. Epithelium samples should be collected into phosphate-buffered glycerine (pH 7.4) and stored at −20°C or below until processed. Tissue suspensions are prepared by grinding a small portion of vesicular epithelium in a mortar and pestle with sterile sand and phosphate-buffered solution (approximately 10% w/v; [39]). Suspensions are clarified by centrifugation and stored at −80°C. Viral antigens and/ or genome may be detectable in these original suspensions. The original suspension can also be used to inoculate cells for virus propagation in susceptible cell culture systems. Porcine kidney cells such as PK-15 [37,38] and IB-RS-2 [39] are permissive to VESV/SMSV infection and develop cytopathic effect (CPE). Cell culture supernatant samples are clarified by centrifugation and stored at 4°C for immediate use or −80°C for long-term storage (also at −20°C in 50% glycerol).

16.2.2 DETECTION PROCEDURES

16.2.2.1 Real-Time RT-PCR

Nucleic acid can be extracted from the original tissue suspension or from cell culture–propagated virus. Several methods have been used for the extraction of VESV RNA, including manual extraction with TRIzol (Gibco; [19]) or QIAzol (Qiagen; [43]), spin columns (RNeasy mini kits, Qiagen; [45]), and automated extraction with a MagNA Pure LC robotic machine (Roche; [20,48]). The following protocol is based on the one-step RT-PCR protocol for the diagnosis of FMDV and related vesicular diseases used at the Pirbright Institute (based on [49,50]).

16.2.2.1.1 Procedure

1. Dilute the primers and probe to the working concentrations. The primers and probe were developed for the universal detection of marine caliciviruses, including VESV, and target a conserved region of the RNA polymerase gene [20].

Oligo Name	Sequence (5′–3′)	Working Concentration (pmol/µL)
Forward primer	GAY GAC GGT GTY TAC ATY GTY C	10
Reverse primer	GGG AYI GGC GTT ATY TCA GCR T	10
Probe	CTG AAR CCG ACY CGG ACC GAC A	5

2. Prepare the one-step RT-PCR master mix in the following. Note: Gently mix all of the components with a pipette and spin down before use.

Component	Volume per Reaction (µL)
2× reaction mix	12.5
Nuclease-free water	1.5
Forward primer	2
Reverse primer	2
TaqMan probe	1.5
Superscript III RT/Platinum Taq mix	0.5
Total	20

3. Add 20 µL of master mix per well of the PCR plate (or PCR tube).
4. Add 5 µL of extracted/sample RNA per well and then seal the plate.
5. Spin the plate for 1 min.
6. Load the PCR plate in the real-time PCR machine and use the following program: 30 min at 60°C, followed by 10 min at 95°C, and then 50 cycles of 95°C for 15 s, and 60°C for 1 min.

Note: The cycling conditions listed earlier can also be used for FMDV[5′ UTR] [51], FMDV[3D] [52], and SVDV[2B-IR] [49] one-step RT-PCR assays. As such, RNA samples can be tested simultaneously for FMDV, SVDV, and VESV on a single PCR plate in three or four separate one-step RT-PCR assays.

16.2.2.2 Sequence Analysis

To investigate genetic relatedness of VESV/SMSV, RT-PCR amplicons of different parts of the genome can be sequenced and compared by phylogenetic analysis. A number of different conventional RT-PCR assays are suitable for the generation of amplicons for sequencing using the Sanger approach, including those listed in the following:

Targeted Region	Primer Name	Product Size (bp)	Nucleotide Variability	Sequence (5′–3′)	Reference
RNA helicase-like coding region	Hel1	357	Conserved	GTCCCAGTATTCGGATTTGTCTGCG	[42]
	Hel2			AGCGGGTAGTTCAGTCAAGTTCACC	
RNA-dependent RNA polymerase coding region	Pol1	419	Conserved	GCCTTCTGGTATGCCACTAACATCC	[42]
	Pol3			GGTGGAACGGTCCAATTTTCAGTG	
N-terminal capsid coding region	1F	768	Intermediate	GTGAGGTGTTTGAGAATTAG	[19]
	1R			ACATCAATTCCGCCAGACCA	
Capsid protein precursor gene—A region	A left	520	Intermediate	CCACTATGGCTACTACTCACACGC	[53]
	A right			TGTTGAACGACAGTGCCTTG	
Capsid protein precursor gene—E region	E left	666	Variable	TGATTCCACAACAACTGGTTGGTC	[53]
	E right			TGAGATGTCATCAGAATCTGGCTG	

Prior to the sequencing procedure, the concentration of RT-PCR products must be quantified and unused primers and nucleotides must be removed from the reaction. This can be achieved with a commercial post-PCR purification kit, such as QIAquick PCR purification kit (Qiagen) or illustra GFX PCR DNA and gel band purification kits (GE Healthcare Life Sciences) used according to manufacturer's instructions. The concentration of the PCR product can be estimated using electrophoresis by running the sample in parallel to a standard DNA, or by using a spectrophotometer (e.g., NanoDrop, Thermo Scientific). Purified PCR products of known concentrations can be used as a template in cycle sequencing reactions, such as the protocol detailed in the following (BigDye Terminator v3.1 Cycle Sequencing Kit; Applied Biosystems, Foster City, CA).

16.2.2.2.1 Procedure

1. For each sequencing reaction, mix the following reagents:

Component	Volume per Reaction (μL)
Terminator ready reaction mix (v3.1)	0.5
5× sequencing buffer	2
Purified PCR product (~100 ng)	1
Sequencing primer (10 μM) (refer to the previous table)	1
Nuclease-free water	5.5
Total	10

2. Centrifuge at $14,000 \times g$ for 30 s.
3. Incubate the sample in a thermocycler for 1 min at 96°C followed by 25 cycles of 96°C for 10 s, 50°C for 5 s, and 60°C for 4 min.
4. Purify the product of sequencing amplification by ethanol-EDTA precipitation, wash with 70% ethanol, and dry the pellet.
5. Resuspend the pellet in 20 μL formamide and analyze the nucleotide sequence of the DNA product using an automated DNA sequencer, such as ABI 3100 (Applied Biosystems, Foster City, CA).

After obtaining good-quality sequence data, compare the sequence to other publicly available sequences in the National Center for Biotechnology Information, GenBank, European Molecular Biology Laboratory, or DNA Data Bank of Japan databases. MEGA software [54] can be used to perform phylogenetic comparisons.

16.3 FUTURE PROSPECTIVE

In view of the widespread nature of the VESV/SMSV complex in the aquatic environment in the Pacific Ocean, the possibility that vesicular disease might reoccur in domesticated species (e.g., pigs) in the United States or elsewhere is real. Therefore, the maintenance of diagnostic testing laboratories and the development of new approaches that can be used to detect and characterize these viruses remain an important priority.

REFERENCES

1. Madin, S. H. and Traum, J., Vesicular exanthema of swine. *Bacteriol. Rev.* 19, 6–21, 1955.
2. Bankowski, R. A., Vesicular exanthema. *Adv. Vet. Sci.* 10, 23–64, 1965.
3. Holbrook, A. A., Geleta, J. N., and Hopkins, S. R., Two new immunological types of vesicular exanthema virus. *Proc. U.S. Livest. Saint. Assoc.* 63, 332, 1959.
4. Bankowski, R. A., Perkins, A. G., Stuart, E. E., and Kummer, M., Recovery of new immunological types of vesicular exanthema virus. *Proc. 60th Annu. Meet. U.S. Livest. Sanit. Assoc.* 302–320, 1956.
5. Smith, A. W., Akers, T. G., Madin, S. H., and Vedros, N. A., San Miguel sea lion virus isolation, preliminary characterization and relationship to vesicular exanthema of swine virus. *Nature* 244, 108–110, 1973.
6. Watkins, H. M. S, Worthington, G. R. L., and Latham, A. B., Isolation of enterovirus from the California gray whale (*Eschrichtius gibbosus*). *Bacteriol. Proc. Am. Soc. Microbiol.* 69, 180, 1969.
7. Smith, A. W., Skilling, D. E., and House, J. A., Re-isolation and reclassification of a whale enterovirus as a calicivirus serotype A48. Unpublished data from the Laboratory for Calicivirus Studies, Oregon State University, Corvallis, and the Plum Island National Animal Disease Laboratory, Plum Island, New York, 1998.
8. Smith, A. W., Skilling, D. E., and Ridgway, S., Calicivirus-induced vesicular disease in cetaceans and probable interspecies transmission. *J. Am. Vet. Med. Assoc.* 183, 1223–1225, 1983.
9. Smith, A. W., Skilling, D. E., Ensley, P. K., Benirschke, K., and Lester, T. L., Calicivirus isolation and persistence in a pygmy chimpanzee (*Pan paniscus*). *Science* 221, 79–81, 1983.
10. Smith, A. W., Skilling, D. E., Anderson, M. P., and Benirschke, K., Isolation of primate calicivirus Pan paniscus type 1 from a douc langur (*Pygathrix nemaeus* L.). *J. Wildl. Dis.* 21, 426–428, 1985.
11. Smith, A. W., Skilling, D. E., and Benirschke, K., Calicivirus isolation from three species of primates: An incidental finding. *Am. J. Vet. Res.* 46, 2197–2199, 1985.
12. Smith, A. W., Anderson, M. P., Skilling, D. E., Barlough, J. E., and Ensley, P. K., First isolation of calicivirus from reptiles and amphibians. *Am. J. Vet. Res.* 47, 1718–1721, 1986.
13. Smith, A. W., Mattson, D. E., Skilling, D. E., and Schmitz, J. A., Isolation and partial characterization of a calicivirus from calves. *Am. J. Vet. Res.* 44, 851–855, 1983.
14. Smith, A. W. et al., In vitro isolation and characterization of a calicivirus causing a vesicular disease of the hands and feet. *Clin. Infect. Dis.* 26, 434–439, 1998.
15. Seal, B. S., Lutze-Wallace, C., Kreutz, L. C., Sapp, T., Dulac, G. C., and Neill, J. D., Isolation of caliciviruses from skunks that are antigenically and genotypically related to San Miguel sea lion virus. *Virus Res.* 37, 1–12, 1995.
16. Martín-Alonso, J. M. et al., Isolation and characterization of a new vesivirus from rabbits. *Virology* 337, 373–383, 2005.
17. Clarke, I. N. et al., *Caliciviridae*. In: King, A. M. Q., Adams, M. J., Carstens, E. B., and Lefkowitz, E. J. (eds.), *Virus Taxonomy: Classification and Nomenclature of Viruses: Ninth Report of the International Committee on Taxonomy of Viruses.* San Diego, CA: Elsevier, pp. 977–986, 2012.

18. Neill, J. D., The complete genome sequence of the San Miguel sea lion virus-8 reveals that it is not a member of the vesicular exanthema of swine virus/San Miguel sea lion virus species of the *Caliciviridae*. *Genome Announc.* 2(6), e01286-14, 2014.

19. Reid, S. M., Ansell, D. M., Ferris, N. P., Hutchings, G. H., Knowles, N. J., and Smith, A. W., Development of a reverse transcription polymerase chain reaction procedure for the detection of marine caliciviruses with potential application for nucleotide sequencing. *J. Virol. Methods* 82, 99, 1999.

20. Reid, S. M. et al., Development of a real-time reverse transcription polymerase chain reaction assay for detection of marine caliciviruses (genus *Vesivirus*). *J. Virol. Methods* 140, 166, 2007.

21. Chen, R., Neill, J. D., Estes, M. K., and Prasad, B. V., X-ray structure of a native calicivirus: Structural insights into antigenic diversity and host specificity. *Proc. Natl. Acad. Sci. USA* 103, 8048–8053, 2006.

22. Mitra, T., Sosnovtsev, S. V., and Green, K. Y., Mutagenesis of tyrosine 24 in the VPg protein is lethal for feline calicivirus. *J. Virol.* 78, 4931–4935, 2004.

23. Goodfellow, I. et al., Calicivirus translation initiation requires an interaction between VPg and eIF 4 E. *EMBO Rep.* 6, 968–972, 2005.

24. Smith, A. W. and Akers, T. G., Vesicular exanthema of swine. *J. Am. Vet. Med. Assoc.* 169, 700–703, 1976.

25. Smith, A. W., Skilling, D. E., Dardiri, A. H., and Latham, A. B., Calicivirus pathogenic for swine: A new serotype isolated from opaleye *Girella nigricans*, an ocean fish. *Science* 209, 940–941, 1980.

26. Smith, A. W. and Skilling, D. E., Can the oceans make you sick? In *Proceedings of the 10th Biennial Conference of the International Institute of Fisheries Economics and Trade.* Corvallis, OR ,2000. http://oregonstate.edu/dept/IIFET/2000/papers/skilling.pdf. Accessed November 22, 2014.

27. Smith, A. W., Skilling, D. E., and Brown, R. J., Preliminary investigation of a possible lungworm (*Parafilaroides decorus*), fish (*Girella nigricans*), and marine mammal (*Callorhinus ursinus*) cycle for San Miguel sea lion virus type 5. *Am. J. Vet. Res.* 41, 1846–1850, 1980.

28. Smith, A. W. and Boyt, P. M., Caliciviruses of ocean origin. *J. Zoo Wildl. Med.* 21, 3–23, 1990.

29. Kurth, A., Evermann, J. F., Skilling, D. E., Matson, D. O., and Smith, A. W., Prevalence of vesivirus in a laboratory-based set of serum samples obtained from dairy and beef cattle. *Am. J. Vet. Res.* 67, 114–119, 2006.

30. Kurth, A., Skilling, D. E., and Smith, A. W., Serologic evidence of vesivirus-specific antibodies associated with abortion in horses. *Am. J. Vet. Res.* 67, 1033–1039, 2006.

31. Barlough, J. E., Berry, E. S., Smith, A. W., and Skilling, D. E., Prevalence and distribution of serum neutralizing antibodies to Tillamook (bovine) calicivirus in selected populations of marine mammals. *J. Wildl. Dis.* 23, 45–51, 1987.

32. Gelberg, H. B. and Lewis, R. M., The pathogenesis of vesicular exanthema of swine virus and San Miguel sea lion virus in swine. *Vet. Pathol.* 19, 424–443, 1982.

33. Gelberg, H. B., Dieterich, R. A., and Lewis, R. M., Vesicular exanthema of swine and San Miguel sea lion virus: Experimental and field studies in otarid seals, feeding trials in swine. *Vet. Pathol.* 19, 413–423, 1982.

34. Gelberg, H. B., Mebus, C. A., and Lewis, R. M., Experimental vesicular exanthema of swine virus and San Miguel sea lion virus infection in phocid seals. *Vet. Pathol.* 19, 406–412, 1982.

35. Van Bonn, W., Jensen, E. D., House, C., House, J. A., Burrage, T., and Gregg, D. A., Epizootic vesicular disease in captive California sea lions. *J. Wildl. Dis.* 36, 500–507, 2000.

36. Smith, A. W., Skilling, D. E., Matson, D. O., Kroeker, A. D., Stein, D. A., Berke, T., and Iversen, P. L., Detection of vesicular exanthema of swine-like calicivirus in tissues from a naturally infected spontaneously aborted bovine fetus. *J. Am. Vet. Med. Assoc.* 220, 455–458, 2002.

37. Edwards, J. F., Yedloutschnig, R. J., Dardiri, A. H., and Callis, J. J., Vesicular exanthema of swine: Isolation and serotyping of field samples. *Can. J. Vet. Res.* 51, 358, 1987.

38. Wilder, F. W., Detection of swine caliciviruses by indirect immunofluorescence. *Can. J. Comp. Med.* 44, 87, 1980.

39. Ferris, N. P. and Oxtoby, J. M., An enzyme-linked immunosorbent assay for the detection of marine caliciviruses. *Vet. Microbiol.* 42, 229, 1994.

40. Bankowski, R. A., Wichmann, R., and Kummer, A., A complement-fixation test for identification and differentiation of immunological types of the virus of vesicular exanthema of swine. *Am. J. Vet. Res.* 14, 145, 1953.

41. Barber, T. L., Moulton, W. M., and Stone, S. S., The identification and typing of vesicular exanthema by complement fixation and agar diffusion tests. *Proc. 62th Annu. Meet. U.S. Livest. Sanit. Assoc.* 1–7, 1958.

42. Neill, J. D. and Seal, B. S., Development of PCR primers for specific amplification of two distinct regions of the genomes of San Miguel sea-lion and vesicular exanthema of swine viruses. *Mol. Cell. Probes* 9, 33, 1995.

43. Lung, O., Fisher, M., Beeston, A., Hughes, K. B., Clavijo, A., Goolia, M., Pasick, J., Mauro, W., and Deregt, D., Multiplex RT-PCR detection and microarray typing of vesicular disease viruses. *J. Virol. Methods* 175, 236, 2011.

44. McClenahan S. D., Bok, K., Neill, J. D., Smith, A. W., Rhodes, C. R., Sosnovtsev, S. V., Green, K. Y. and Romero, C. H., A capsid gene-based real-time reverse transcription polymerase chain reaction assay for the detection of marine vesiviruses in the Caliciviridae. *J. Virol. Methods* 161, 12–18, 2009.

45. Jack, P. M. J. et al., Microarray-based detection of viruses causing vesicular or vesicular-like lesions in livestock animals. *Vet. Microbiol.* 133, 145, 2009.

46. Logan G. et al., A universal protocol to generate consensus level genome sequences for foot-and-mouth disease virus and other positive-sense polyadenylated RNA viruses using the Illumina MiSeq. *BMC Genomics* 15, 828, 2014.

47. Smith, A. W. and Madin, S. H., Vesicular exanthema. In: Leman, A. D., Straw, B., Glock, R. D., Mengeling, W. L., Penney, R. H. C., and Scholl, E. (eds.), *Diseases of Swine*, 6th edn. Ames, IA: Iowa State University Press, pp. 358, 1986.

48. Hindson, B. J. et al., Diagnostic evaluation of multiplexed reverse transcription-PCR microarray assay for detection of foot-and-mouth and look-alike disease viruses. *J. Clin. Microbiol.* 46, 1081, 2008.

49. Reid, S. M., Ferris, N. P., Hutchings, G. H., King, D. P., and Alexandersen, S., Evaluation of real-time reserve transcription polymerase chain reaction assays for the detection of swine vesicular disease virus. *J. Virol. Methods* 116, 169, 2004.

50. Reid, S. M. et al., Performance of real-time reverse transcription polymerase chain reaction for the detection of foot-and-mouth disease virus during field outbreaks in the United Kingdom in 2007. *J. Vet. Diagn. Invest.* 21, 321–30, 2009.

51. Reid, S. M., Ferris, N. P., Hutchings, G. H., Zhang, Z., Belsham, G. J., and Alexandersen, S., Detection of all seven serotypes of foot-and-mouth disease virus by real-time, fluorogenic reverse transcription polymerase chain reaction assay. *J. Virol. Methods* 105, 67, 2002.

52. Callahan, J. D. et al., Use of a portable real-time reverse transcriptase-polymerase chain reaction assay for rapid detection of foot-and-mouth disease virus. *J. Am. Vet. Med. Assoc.* 220, 1636, 2002.

53. Neill, J. D., Meyer, R. F., and Seal, B. S., The capsid protein of vesicular exanthema of swine virus serotype A48: Relationship to the capsid protein of other animal caliciviruses. *Virus Res.* 54, 39–50, 1998.

54. Tamura, K., Stecher, G., Peterson, D., Filipski, A., and Kumar, S., MEGA6: Molecular evolutionary genetics analysis version 6.0. *Mol. Biol. Evol.* 30, 2725–2729, 2013.

17 Avian Leukosis Virus

Yongxiu Yao and Venugopal Nair

CONTENTS

17.1 Introduction 145
    17.1.1 Etiology.. 145
        17.1.1.1 Classification... 145
        17.1.1.2 Virus Structure and Replication 146
        17.1.1.3 Avian Leukosis Virus Envelope Subgroups.. 146
        17.1.1.4 Pathogenicity... 147
    17.1.2 Transmission and Epidemiology... 147
    17.1.3 Clinical Features and Pathogenesis 147
    17.1.4 Diagnosis 148
        17.1.4.1 Conventional Techniques 148
        17.1.4.2 Molecular Techniques 149
17.2 Methods 150
    17.2.1 Sample Preparation... 150
    17.2.2 Detection Procedures.. 150
        17.2.2.1 Conventional PCR Detection of ALV... 150
        17.2.2.2 Multiplex PCR Detection of ALV... 150
        17.2.2.3 Quantitative PCR Detection of ALV 152
        17.2.2.4 LAMP Detection of ALV 153
17.3 Conclusion 154
References... 154

17.1 INTRODUCTION

The leukosis/sarcoma (L/S) group of diseases encompasses a variety of transmissible benign and malignant neoplasms of chickens caused by members of the family *Retroviridae* [1]. The term "leukosis" embraces several different leukemia-like proliferative diseases of the cells of the hemopoietic system due to avian leukosis virus (ALV), with an undertone that a leukemic blood picture is not always present [2,3]. The components of the hemopoietic system that may undergo neoplastic change include the lymphopoietic (lymphocytic) system, the erythropoietic (red cell) system, and the myelopoietic (myelocytic) system. Presenting signs of the leukosis are mostly nonspecific (loss of appetite, weakness, emaciation, diarrhea, pale wattles). Lymphoid leukosis (LL) has been the most common form of L/S group of diseases seen in field flocks, although more recently, myeloid leukosis has become increasingly prevalent.

ALV infection is commonly encountered in chicken flocks and is known to be of significant economic importance. Losses from ALV-induced diseases are attributed to two sources. First, the tumor mortality that usually accounts for 1%–2%, with occasional losses as high as 20% in some flocks. Second, subclinical infection by ALV, to which most flocks are subjected, produces a depressive effect on a number of key performance traits, including egg production and quality [4–6]. Economic losses due to ALV tumor mortality and reduced productivity were estimated to be in millions of U.S. dollars each year [7]. Currently, the virus is causing huge economic losses to the poultry industry in countries such as China [8,9].

17.1.1 ETIOLOGY

17.1.1.1 Classification

Viruses of the avian leukosis/sarcoma group (ALSV) belong to the genus *Alpharetrovirus*, subfamily *Orthoretrovirinae*, family *Retroviridae* (which contains another subfamily *Spumaretrovirinae*). Whereas the subfamily *Orthoretrovirinae* includes six genera (*Alpharetrovirus*, *Betaretrovirus*, *Deltaretrovirus*, *Epsilonretrovirus*, *Gammaretrovirus*, and *Lentivirus*), the subfamily *Spumaretrovirinae* consists of a single genus *Spumavirus*. In turn, the genus *Alpharetrovirus* is divided into nine species: *avian carcinoma Mill Hill virus 2*, *ALV* (the type species), *avian myeloblastosis virus*, *avian myelocytomatosis virus 29*, *avian sarcoma virus CT10*, *Fujinami sarcoma virus*, *Rous sarcoma virus* (RSV), *UR2 sarcoma virus*, and *Y73 sarcoma virus* [10]. In common with other members of this RNA virus family, ALVs are characterized by the unique possession of the enzyme reverse transcriptase that drives the generation of the DNA provirus that is integrated into the host genome during viral replication.

17.1.1.2 Virus Structure and Replication

Structurally, ALVs are simple viruses (Figure 17.1). From the 5′ end to the 3′ end of the genome are located four main genes, *gag/pro-pol-env*, which encode, respectively, the proteins of the virion group-specific (gs) antigens and protease, the enzyme reverse transcriptase, and the envelope glycoprotein. These structural genes are flanked by genomic sequences associated with the regulation of viral replication, which in the DNA provirus form the viral long terminal repeats (LTRs) that carry promoter and enhancer sequences. Some laboratory strains and field isolates of ALV also possess one (or rarely two) viral oncogene(s) inserted within the genome. Such viruses have acquired the oncogene by transduction of a cellular oncogene during oncogenesis. These viruses are usually genetically defective, with deletions within the genome, and require the presence of "helper viruses" (coinfecting nondefective ALVs) to enable replication. As with other retroviruses, replication of ALSV is characterized by the formation, under the direction of reverse transcriptase, of a DNA provirus that becomes linearly integrated into the host cell

genome. Subsequently, the proviral genes are transcribed into viral RNAs, which are translated to produce precursor and mature proteins that constitute the virion.

17.1.1.3 Avian Leukosis Virus Envelope Subgroups

ALVs can be classified as endogenous (subgroup E) or exogenous viruses according to their mode of transmission. Exogenous ALVs from chickens can be classified into five subgroups (A, B, C, D, and J), on the basis of differences in the viral envelopes that affect antigenicity, as determined by induction of virus neutralizing antibodies, host range, and the ability to interfere with infection by other ALVs of the same or differing subgroups. The F, G, H, and I subgroups of ALV are found in ring-necked and golden pheasants, Hungarian partridge, and Gambel's quail, respectively. Viruses of subgroups A, B, C, D, and J are oncogenic, causing mainly lymphoid leukosis (subgroups A and B) or myeloid leukosis (subgroup J). Subgroup E viruses are not oncogenic. Viruses within a subgroup usually cross-neutralize to varying extents, although antigenic variants of subgroup J ALV exist that do not cross-neutralize. A partial cross-neutralization occurs between subgroup B and D viruses, but otherwise viruses in the different subgroups do not cross-neutralize each other.

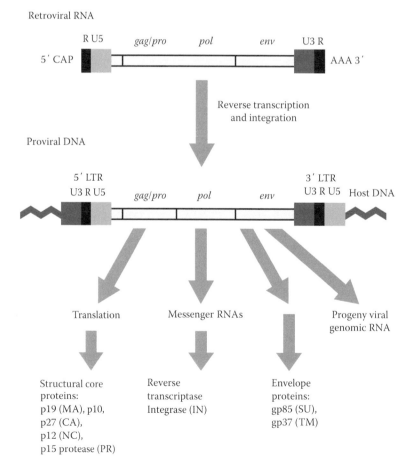

FIGURE 17.1 Key features of the viral RNA and proviral DNA forms of the genome of avian leukosis virus. CAP, 5′ end structure; AAA, polyadenylation of 3′ end; R, repeat sequence; U5, unique 5′ end sequence; U3, unique 3′ end sequence; LTR, long terminal repeat; MA, matrix protein; CA, capsid protein; NC, nucleocapsid protein; PR, protease; SU, surface protein; TM, transmembrane.

Strains of ALSV can also be placed into two major classes in respect of rapidity of induction of tumors: *acutely transforming viruses* and *slowly transforming viruses*.

1. *Acutely transforming viruses* are those that carry viral oncogenes in their genome. These viruses can induce neoplastic transformation, *in vivo* or *in vitro*, within a few days or weeks. Depending on the oncogene possessed by the virus, acutely transforming ALVs induce different types of neoplasms, for example, v-myc, myeloid leukosis (myelocytoma); v-myb, myeloid leukosis (myeloblastosis); *v-erbB*, erythroid leukosis; and v-src, sarcoma [11–15]. Such ALVs are termed "acutely transforming," and neoplastic cells are induced within a few days after infection.

2. *Slowly transforming viruses* do not carry viral oncogenes. They induce tumors by activation of a cellular proto-oncogene to bring about neoplastic transformation and development of tumors over many weeks or months [12,15–18].

17.1.1.4 Pathogenicity

Strains of ALV often produce more than one type of neoplasm, and the oncogenic spectrum of each strain tends to be characteristic but often overlaps with responses to other strains. Viral factors including the origin and dose and host factors such as route of inoculation, age, genotype, and sex will influence the incidence and types of neoplasms induced.

17.1.2 TRANSMISSION AND EPIDEMIOLOGY

Exogenous ALVs are almost ubiquitous in commercial chickens on a worldwide basis, although many primary egg-type and meat-type breeding companies institute ALV eradication schemes. Chickens are the natural hosts for all viruses of L/S group [19]; these viruses have not been isolated from other avian species except pheasants, partridges, and quail.

Exogenous ALVs are transmitted in two ways: vertically from infected hens to their offspring through the egg and horizontally from bird to bird by direct or indirect contact [7,9,20]. Although usually only a small percentage of chicks are infected vertically, this route of transmission is important in transmitting the infection from one generation to the next and in providing a source of contact infection to other chicks. Most chickens become infected by close contact with congenitally infected birds. Sources of virus from infected birds include feces, saliva, and desquamated skin. The period of survival of ALV outside the body is relatively short (a few hours), and consequently ALV is not highly contagious. Although vertical transmission is important in the maintenance of the infection, horizontal infection may also be necessary to maintain a rate of vertical transmission sufficient to prevent the infection from dying out [21]. The infection does not spread readily from infected birds to birds in indirect contact (in separate pens or cages), probably because of the relatively short life of the virus outside the birds. However, contact exposure at hatch was shown to be an effective method of spread of ALV-J among broiler breeder chickens [22–24] and was prevented by small group rearing [25].

Four classes of ALV infection statuses are recognized in mature chickens: (1) no viremia, no antibody (V−A−); (2) no viremia, with antibody (V−A+); (3) with viremia, with antibody (V+A+); and (4) with viremia, no antibody (V+A−) reviewed by [7,9,20,26]. Birds in an infection-free flock and genetically resistant birds in a susceptible flock fall into the category V−A−. Genetically susceptible birds in an infected flock fall into one of the other three categories. Most are V−A+, and a minority, usually less than 10%, is V+A−. Most V+A− hens transmit ALV to a varying but relatively high proportion of their progeny [9,20]. A small proportion of V−A+ hens transmit the virus congenitally and do so more intermittently; the tendency for congenital transmission of ALV in this category was found to be more frequent in hens with low antibody titer [27]. Congenitally infected embryos develop immunologic tolerance to the virus and after hatching make up the V+A− class, with high levels of virus in the blood and tissues and an absence of antibodies. By 22 weeks of age, up to 25% of meat-type chickens exposed to ALV-J at hatch were found to be V+A−, although this could be affected by a number of factors including the virus strain [28].

Birds from infected flocks, and products, such as eggs and meat, could spread infection to other flocks and locations. As ALV is vertically transmitted, infection may spread internationally in hatching eggs and day-old chicks, and spread in semen is possible. Importers need to guard against the introduction of ALV by requiring health certificates from the exporter relating to the specific disease and infection status of source flocks. Introduction of a new ALV into susceptible poultry populations could have serious consequences because of the absence of immunity and of effective control measures in breeding and production populations other than at the primary breeding level.

17.1.3 CLINICAL FEATURES AND PATHOGENESIS

Outward signs of the leukotic diseases are mostly nonspecific. They include loss of appetite, weakness, diarrhea, dehydration, and emaciation. In lymphoid leukosis especially, there may be abdominal enlargement. The comb may be pale, shriveled, or occasionally cyanotic. In erythroblastosis and myeloblastosis, hemorrhage from feather follicles also may occur. After clinical signs develop, the course is usually rapid, and birds die within a few weeks. Other affected birds may die without showing obvious signs.

Fully developed LL occurs in chickens of about 4 months of age and older. Grossly visible tumors almost invariably involve the liver, spleen, and bursa of Fabricius. Other organs often grossly involved include the kidney, lung, gonad, heart, bone marrow, and mesentery. The tumor cells are B cells, expressing immunoglobulin M (IgM) and other B-cell markers, which originate in the bursa of Fabricius but subsequently develop metastasis to other visceral organs.

Natural cases of erythroblastosis (erythroid leukosis) usually occur in birds between 3 and 6 months of age. The liver and kidney are moderately swollen, and the spleen often is greatly enlarged. The enlarged organs are usually cherry red to dark mahogany and are soft and friable. The bone marrow is bright red and liquid. Affected birds are often anemic, with muscle hemorrhages and occasionally abdominal hemorrhage from a ruptured liver. The disease is an intravascular erythroblastic leukemia.

Myeloblastosis (myeloblastic myeloid leukosis) was predominately a sporadic disease of adult chickens. The liver is greatly enlarged and firm with diffuse grayish tumor infiltrates, which give a mottled or granular ("Morocco leather") appearance. The spleen and kidneys are also diffusely infiltrated and moderately enlarged. The bone marrow is replaced by a solid, yellowish-gray tumor cell infiltration. A severe leukemia exists, with myeloblasts comprising up to 75% of peripheral blood cells and forming a thick buffy coat and usually an anemia and thrombocytopenia.

Tumors of myelocytomatosis (myelocytic myeloid leukosis) are distinctive and can be recognized on gross examination with some degree of certainty. Characteristically, they occur on the surface of bones in association with the periosteum and near cartilage, although any tissue or organ can be affected. Myelocytomas often develop at the costochondral junctions of the ribs, on the inner sternum, on the pelvis, and on the cartilaginous bones of the mandible and nares. Flat bones of the skull are also commonly affected. Tumors may also be seen in the oral cavity, on the trachea, and in and around the eye [29]. The tumors are usually nodular and multiple, with a soft, friable consistency and of creamy color. In the disease caused by subgroup J ALV, myelocytomatous infiltration often causes enlargement of the liver and spleen and other organs, in addition to skeletal tumors [30] and myelocytic leukemia [31].

Hemangioma is found in the skin or in visceral organs in chickens of various ages. They appear as blood-filled cystic masses (blood blisters) or more solid tumors and consist of distended blood-filled spaces lined by endothelium or as more cellular, proliferative, lesions [32]. They are often multiple and may rupture, causing fatal hemorrhage [33]. More recently, many workers have reported the incidence of hemangiomas in layer chickens infected with ALV-J in China [34–36].

Renal tumors may cause paralysis due to pressure on the sciatic nerve. Sarcomas and other connective tissue tumors may be seen in the skin and musculature. When advanced, these various other tumors may be accompanied by the nonspecific signs given previously. Benign tumors may follow a long course, malignant tumors a rapid one.

In osteopetrosis, the long bones of the limbs are commonly affected. Uniform or irregular thickening of the diaphyseal or metaphyseal regions can be detected by inspection or palpation. The affected areas are often unusually warm. Birds with advanced lesions have characteristic "bootlike" shanks. Affected birds usually are pale and stunted and walk with a stilted gait or limp. In recent years, ALV has also been shown to be associated with the so-called fowl glioma [37], associated with cerebellar hypoplasia and myocarditis [38–42].

17.1.4 Diagnosis

17.1.4.1 Conventional Techniques

17.1.4.1.1 Virological Tests

As ALV is widespread among chickens, virus isolation and the demonstration of antigen or antibody have limited or no value in diagnosing field cases of lymphomas. However, assays for the detection of ALV are very useful in the identification and classification of new isolates, safety testing of vaccines, and testing pathogen-free and other breeder flocks for freedom from virus infection. Virus isolation is generally the ideal detection method (the so-called gold standard) since the technique can detect all ALVs and is the starting point for various other virus studies. Samples most commonly used for detection of ALV include blood, plasma, serum, meconium, cloacal and vaginal swabs, oral washings, egg albumen, embryos, and tumors [7,43,44]. Virus also can be isolated from albumen of newly laid eggs or the 10-day-old embryo of eggs laid by hens that are transmitting virus vertically, from feather pulp, and from semen. All ALSVs are very thermolabile and can be preserved for long periods only at temperatures below −60°C. Thus, materials used for biological assays for infectious virus should be collected and placed on melting ice or stored at −70°C until assayed. In contrast, samples for the detection of ALV gs antigens by direct assays can be stored at −20°C.

Considering that most strains of ALV produce no visible morphologic changes in cell culture, assays for the presence of ALV are based on the following: (1) detection of specific proteins or glycoproteins coded for by one or more of the three major genes of ALV—gag, pol, and env genes—(2) detection of specific proviral DNA or viral RNA sequences of ALV by the polymerase chain reaction (PCR) and reverse transcription (RT)-PCR, respectively. Several biological, molecular, and serological assays can be used for the detection of endogenous and exogenous ALVs.

As most ALVs produce no visible morphologic changes in culture, detection of ALV p27 forms the basis of several diagnostic tests for virus. Indirect biologic assays such as complement fixation (CF) for avian leukosis, ELISA for ALV, phenotypic mixing (PM), resistance-inducing factor (RIF), and nonproducer (NP) cell activation tests are used for the detection of ALVs. Of all such assays, ELISA is the most commonly used test. But, p27 is shared by both exogenous and endogenous viruses; hence, the assay cannot be used to differentiate between these two groups of viruses. All these biological assays require the use of chicken embryo fibroblasts (CEFs) with specific host range. CEFs that are resistant to infection with endogenous ALV (C/E) are desirable to use in tests for detection and isolation of exogenous ALV. Other cells, such as those resistant to subgroup A (C/A) and resistant to subgroup J ALV (C/J) [45], can also be used to confirm the subgroup of isolated ALV. Testing samples on CEFs that are susceptible to all subgroups of ALV (C/O) and those that are resistant to subgroup E (C/E) can be used in differentiating exogenous and endogenous ALV. If a positive test is obtained from using C/O but not C/E CEFs, the sample is positive for endogenous ALV. Positive tests using both C/E and C/O indicate the presence of exogenous ALV. Recently, a flow cytometry method using a highly specific alloantibody termed R2 has been described for the detection of endogenous ALV envelope in chicken plasma [46,47]. It should be noted that some tests such as CF and ELISA and possibly NP, PM, R(-)Q cell, and fluorescent-labelled antibody (FA) can be suitable for all leukosis and sarcoma viruses. The RIF test can be performed only on ALVs that are not rapidly cytopathogenic. Other tests are specific for certain virus strains. Rapid transformation of fibroblast cultures is produced only by certain RSV and of hematopoietic cell cultures only by defective ALV.

Direct [48] and indirect [49] immunofluorescence assay as well as flow cytometry [45,50] have been used to detect viral antigen in CEF cultures. Flow cytometry has also been shown to be a very useful tool in identifying the subgroup of ALV strains contaminating commercial Marek's disease vaccines [51–53]. Assays for RT activities have been used for the detection of oncogenic RNA viruses including all ALSVs [54]. Detection of this enzyme, either directly using a correct template for detecting enzymatic activity [55,56] or indirectly using radioimmunoassay [57], is an indication of the presence of the virus. Test for adenosine triphosphatase activity specifically for avian myeloblastosis virus has also been reported.

17.1.4.1.2   Characterization of ALV Subgroups

ALV isolates belonging to different subgroups can be characterized using the following methods:

1. Viral interference assays, which test the ability of an isolate to interfere with the formation of transformed foci in C/E CEF cultures by RSV of known subgroup.
2. Virus neutralization assays, in which an isolate is placed in a subgroup according to susceptibility to neutralization by chicken antisera with known subgroup neutralizing activity. The isolate or the RSV pseudotype is exposed to antiserum and then examined for growth or focus formation, respectively, in C/E CEF cultures. An RSV pseudotype is an acutely transforming RSV that can be created by coating a replication-defective RSV with the viral envelope of an ALV and hence endowing the virus with the subgroup characteristics of the ALV, in this case of the isolate. Fluorescent antibody staining of infected cultures using antisera against different subgroups is also used. These methods are usually satisfactory, although in the case of subgroup J isolates, in which *env* gene mutations and consequent antigenic changes are frequent, subgroup J antisera may fail to neutralize new variants. However, mouse monoclonal antibodies have been developed that appear in fluorescent antibody tests to detect a wider range of subgroup J isolates.
3. Host range assays, in which an isolate, or the RSV pseudotype, is placed in a subgroup according to the ability to grow in, or transform, respectively, CEFs of varying ALV subgroup susceptibility phenotypes.

For example, a subgroup A ALV would be identified if the virus grew in C/E CEFs but did not grow in C/AE CEFs (resistant to subgroups A and E). This method relies on the availability of CEF phenotypes that exclude certain ALV subgroups. Naturally occurring CEFs that exclude subgroup J have never been detected, although workers in the United States have developed a C/J CEF line that stably expresses J *env* gene [18].
4. PCR assays specific for various ALV subgroups have been developed [16,37,38], as discussed later.

17.1.4.1.3   Serological Assays

Detection of antibodies against ALV is used in flock surveillance to detect the presence or absence of infection by exogenous ALV and to identify particular classes of birds in epidemiological studies and ALV eradication programs. Plasma, serum, and egg yolk are suitable samples for the detection of antibodies to ALSVs. Virus neutralization and antibody ELISA can be used as described in the following:

1. Virus neutralization tests. Antibody is detected by its ability to neutralize infectivity of ALVs of known subgroup or of the RSV pseudotypes. Usually, a 1:5 dilution of heat-inactivated (56°C for 30 min) serum is mixed with an equal quantity of a standard preparation of RSV of a known pseudotype. After incubation, the residual virus is quantitated by any one of many procedures, the cell culture assay being the most commonly used. A microneutralization test to assay for residual virus can be used for the detection of ALV antibody [44]. The test can be conducted in 96-well plates, and the neutralization of the virus is determined by ELISA on culture fluids [58]. Positive ELISA indicates no antibody, whereas a negative ELISA indicates neutralization of ALV and the presence of antibody. Although virus and antibody within subgroups usually cross-neutralize, antibodies against variant viruses may fail to neutralize a representative subgroup virus. This has been observed particularly with subgroup J, due to virus mutation, and is a limitation to the use of virus neutralization (VN) tests. These tests are slow requiring 7–10 days and are technically demanding.
2. Antibody ELISA tests. Viral antigens may be used in ELISA tests to detect sub-gs ALV antibodies. Tests for antibodies to subgroups A, B, and J are commercially available. The subgroup J antibody ELISA uses recombinant *env* antigen produced in baculovirus system and appears to detect antibodies to all variant viruses studied. The antibody ELISA tests are rapid (requiring one day), specific, and suitable for large-scale testing.

17.1.4.2   Molecular Techniques

A more effective control of ALV infections mainly depends on the early detection and removal of infected birds to reduce contact with infected birds and the incidence of horizontal spread. Therefore, achieving rapid detection of infection is imperative in the effective control of the spread of ALVs. PCR-based methods had been developed to provide a rapid tool for the detection and identification of ALV's proviral DNA and viral RNA including subgroup E viruses. RT-PCR has also been used to detect several subgroups of ALV [59,60]. A specific PCR for ALV subgroup A can be used to detect proviral DNA and viral RNA in various tissues from ALV-infected chickens [61]. Reverse transcriptase–nested PCR (RT-nested PCR) test that amplifies a fragment of the LTR of exogenous ALV subgroups A, B, C, D, and J, but not endogenous retroviral sequences, has been described [62]. Several primers specific for the detection of the most commonly isolated ALVs, particularly subgroup A [63], and the new subgroup ALV-J [64–66] have been developed. Other primers specific for endogenous, subgroup E ALV can also be used to detect cell culture infected with endogenous ALV-E, but not those infected with exogenous ALV of subgroups A, B, C, D, and J [44]. Multiplex PCR is a useful technique for the rapid differential diagnosis of avian viruses and the detection of multiple infections of avian viruses under field conditions. Recently, a sensitive and specific multiplex PCR method for the detection of ALV-A, ALV-B, and ALV-J has been developed [67]. This novel method allows for three very common subgroups of exogenous ALVs (ALV-A, ALV-B, and ALV-J) to be detected and differentiated in one reaction. Real-time RT-PCR using the H5/H7 primer pair has been shown to be highly reproducible [68,69]. The rapid detection and quantification of ALV-J with proviral DNA [69] as well as ALV-A and ALV-B (duplex real-time RT-PCR) [60] using TaqMan-based real-time PCR method have also been reported.

As further modification of PCR tests, loop-mediated isothermal amplification (LAMP) method for ALV subgroup A and J has been developed [70,71]. The LAMP method was developed by Notomi et al. [72]. This novel technique generally requires isothermal conditions and four different primers for DNA amplification [72,73] and has been applied to the detection of several pathogens. The LAMP reaction requires 30–60 min and can be performed at a single temperature ranging from 60°C to 65°C. LAMP does not require the DNA denaturation, annealing, and extension PCR cycles [72,73]. In addition, the results can be ascertained easily by the naked eye [73,74].

17.2 METHODS

17.2.1 SAMPLE PREPARATION

The most commonly used samples for proviral DNA extraction include cultured cells and clinical samples such as cloacal swabs, albumen, feather shaft, blood, tissue samples, and tumors. For the extraction of DNA from cultured cells, 1×10^6 infected cells were pelleted by centrifugation at $1000 \times g$ for 10 min. The resulting pellet was resuspended in 200 μL of lysis buffer (4 M guanidine hydrochloride, 25 mM sodium citrate, 1% sarkosyl) and extracted with 300 μL

phenol/chloroform/iso-amyl alcohol (25:24:1) three times. The DNA was precipitated with isopropanol in the presence of 2 M ammonium acetate and centrifuged for 10 min at $10,000 \times g$ in a microcentrifuge. The pellet was washed with 70% ethanol and dried at room temperature. Subsequently, the DNA was resuspended in nuclease-free water and stored at −20°C. For the extraction of DNA from blood, 25 μL of whole blood was lysed in addition to 200 μL of lysis buffer, and 25 μL of this was diluted to 200 μL of water and then phenol extracted as for cultured cells. DNA from the feather shaft was obtained by incubating 5–10 mm portions in lysis buffer at 37°C overnight, followed by the extraction of DNA. Tissue samples, homogenized in phosphate-buffered saline containing 1000 units/mL of penicillin and streptomycin, were centrifuged at $6000 \times g$ for 5 min at 4°C, and an aliquot of the supernatant was subjected to proviral DNA extraction as described.

The total RNA from the virus stocks, serum samples, feather shaft, chicken egg albumen, and cell lysate can be extracted using a commercial reagent such as TRIzol according to the manufacturer's recommendations. After resuspension in diethyl pyrocarbonate–treated water, the RNA sample was transcribed with reverse transcriptase according to the manufacturer's protocol. The cDNA was treated with RNase H and the concentration was measured. Both the RNA and cDNA samples were stored at −70°C until use.

17.2.2 DETECTION PROCEDURES

17.2.2.1 Conventional PCR Detection of ALV

The PCR is the most common DNA-based test used for detection and identification of ALV including subgroup E viruses. Most sequences used for developing primers are located in the *pol*, *env*, and LTR regions. Several pairs of primers were designed and tested for the detection of ALV subgroups by PCR [53,66,75,76]. The sequences of the primers, the target subgroups, and the expected size of PCR products are listed in Table 17.1. Of those, the forward primers H5 and all were conserved across several ALV subgroups so that they can be used in combination with several sub-gs reverse primers as detailed in the table. The PCR conditions can be referenced to the original publications accordingly.

17.2.2.2 Multiplex PCR Detection of ALV

In the multiplex PCR method for the detection of the major ALV subgroups A, B, and J [67], a common forward primer, PF, and three downstream primers, AR, BR, and JR were designed to amplify 715 bp for subgroup A, 515 bp for subgroup B, and 422 bp for subgroup J simultaneously in one reaction (Table 17.1). The common forward primer PF, based on the 3′ region of the pol gene, was conserved across ALV-A, ALV-B, and ALV-J. The downstream primers were chosen from the *env* gene, which allows for discrimination of the three subgroups.

The DNA samples or cDNAs from RT reaction prepared earlier were used as templates in PCR reactions with different

TABLE 17.1

Target Avian Leukosis Virus Subgroups and Oligonucleotide Primers Used in Polymerase Chain Reaction/Reverse Transcription–Polymerase Chain Reaction

Target Subgroup	Primer/Orientation	Sequence (5′–3′)[a]	Product Size (bp)	Reference
A	H5/forward	GGATGAGGTGACTAAGAAAG	694	[75]
	envA/reverse	AGAGAAAGAGGGGCGTCTAAGGAGA		
	A/reverse[b]	CCCATTTGCCTCCTCTCCTTGT	1300	[53]
	PA1/forward	CTACAGCTGTTAGGTTCCCAGT	229	[76]
	PA2/reverse	GTCACCACTGTCGCCTATCCG		
	PA10/forward	GGCTTCAGGCCAAAAGGGGT	232	[76]
	PA20/reverse	GTGCATTGCCACAGCGGTACTG		
B	envB/reverse[c]	ATGGACCAATTCTGACTCATT	846	[83][e]
	PB1/forward	GGCTTTACCCCATACGATAG	259	[76]
	PB2/reverse	ACACATCCTGACAGATGGACCA		
C	envC/reverse[c]	GAGGCCAGTACCTCCCACG	859	[83][e]
	C/reverse[b]	CCCATATACCTCCTTTTCCTCTG	1400	[53]
	PC1/forward	TATTTCGCCCCAAGGGCCAC	238	[76]
	PC2/reverse	CCACGTCTCCACAGCGGTAAGT		
D	envD/reverse[c]	ATCCATACGCACCACAGTATTCG	797	[83][e]
	PD1/forward	GGCTTCACCCCATACGGCAG	258	[76]
	PD2/reverse	CCATACGTCCTCACAGATAGAATA		
E	E/reverse[b]	GGCCCCACCCGTAGACACCACTT	1250	[53]
	PE1/forward	GGCTTCGCCCCACACTCCAA	265	[76]
	PE2/reverse	GCACATCTCCACAGGTGTAAAT		
	H3/forward	AACAACACCGATTTAGCCAGC	360	[83][e]
	H8/reverse	TGGTGAATCCACAATATCTACGAC		
J	H7/reverse[c]	CGAACCAAAGGTAACACACG	545	[66]
	H7b/reverse[c]	GAACCAAAGGTAACACGT	544	[75]
	J/forward	CTTGCTGCCATCGAGAGGTTACT	2300	[53]
	J/reverse	AGTTGTCAGGGAATCGAC		
B, D	BD/reverse[b]	AGCCGGACTATCGTATGGGGTAA	1100	[53]
A, B, C, D, E	AD1/reverse[c]	GGGAGGTGGCTGACTGTGT	295–326[d]	[66]
	PU1/forward	CTRCARCTGYTAGGYTCCCAGT	229[d]	[76]
	PU2/reverse	GYCAYCACTGTCGCCTRTCCG		
	All/forward	CGAGAGTGGCTCGCGAGATGG	2400[d]	[53]
	All/reverse	ACACTACATTTCCCCCTCCCTAT		
A, B, C, D, E, J	AJ1/reverse[c]	ATGAACGGCCCATTCCCCTATTCC	1672[d]	[83][e]
A, B, J (multiplex PCR)	PF/forward	CGGAGAAGACACCCTTGCT		[67]
	AR/reverse	GCATTGCCACAGCGGTACTG	715	
	BR/reverse	GTAGACACCAGCCGGACTATC	515	
	JR/reverse	CGAACCAAAGGTAACACACG	422	

[a] R = A/G, Y = T/C.

[b] In combination with all/forward primer.

[c] In combination with H5/forward primer.

[d] Size can vary with different subgroups.

[e] Primers used in Nair's lab.

pair/group of primers accordingly. PCR amplification was carried out as described, and the products were analyzed by 1.0% agarose gel electrophoresis. To further confirm the result and gain the detailed sequencing information, the PCR products from the gel were excised and purified using commercial DNA gel extraction kit. The purified DNA can be cloned into TA cloning vector for sequencing or sequenced directly using the same primer for PCR. The specificity and the divergence of the sequences obtained with respect to the corresponding sequences of ALV subgroup were determined using BLAST [77]. For DNA alignment and phylogenetic analysis, the nucleotide sequences were aligned with reference sequences from GenBank using the ClustalW program, version 1.8 [78]. A neighbor-joining tree can be drawn using the MEGA program, version 3.1 [79], with confidence levels assessed using 1000 bootstrap replications.

17.2.2.3 Quantitative PCR Detection of ALV

For the detection and quantitation of ALV subgroup A and B, a duplex quantitative real-time RT-PCR was developed [59]. The assay detected as few as 56 gp85 cDNA copies and was 100-fold more sensitive than a conventional RT-PCR. A highly conserved region within gp85 was used to design primers and the TaqMan probe using Beacon Designer software version 7.0 (Palo Alto, CA). β-*actin* was used as a reference gene. The sequences of the primers and probes were listed in Table 17.2. For construction of standard curves, two conventional RT-PCR products of *gp85* and β-*actin* genes with the same primers used for TaqMan were cloned, and the plasmid DNAs were prepared as described [60]. The duplex quantitative RT-PCR (qRT-PCR) was performed in a 25 μL reaction system containing 2.5 μL of 10× buffer, 2 μL of 2.5 mmol/L deoxynucleoside triphosphates (dNTPs), 1 μL of 10 μmol/L R-F primer, 1 μL of 10 μmol/L R-R primer, 0.5 μL of 5′-6-carboxy-fluorescein (FAM)-labeled gp85 probe, 1 μL of 10 μmol/L β-R primer, 1 μL of 10 μmol/L β-F primer, 0.5 μL of 5′-6-carboxy-X-rhodamine (ROX)-labeled β-actin probe, 0.5 μL of 5 U Hotstart ExTaq polymerase, 1 μL of template DNA, and 14 μL of distilled water. The amplification was run using a Rotor Gene 3000 Real-time Thermal Cycler (Corbett Research, Australia) with the following protocol: 94°C for 3 min followed by 40 cycles of 94°C for 15 s and 61.3°C for 50 s. The serially diluted recombinant plasmids, gp85 and β-actin, together with the cDNA of samples for the detection, were tested simultaneously. The copy numbers of virus and housekeeping genes were extrapolated from the standard curves for *gp85* and β-*actin* genes for each run.

For ALV-J quantification, Kim et al. developed and modified a quantitative competitive RT-PCR (QC-RT-PCR) method and a real-time RT-PCR method [68,80]. QC-RT-PCR method is based on coamplification of ALV-J genomic RNA and a known amount of a synthesized RNA competitor. The real-time RT-PCR method used the H5/H7 primer pair. These methods are very specific, sensitive, easy to perform,

highly reproducible, and rapid compared with conventional method with 50% tissue culture infective dose (TCID$_{50}$) [68]. To improve the sensitivity, a TaqMan-based real-time PCR method for the rapid detection and quantification of ALV-J with proviral DNA (Table 17.2) was developed [69]. The forward primer ALV-JNF was designed against the 3′ region of the *pol* gene, which is conserved among several subgroups but not in the EAV family, according to the ALV-J sequence (HPRS-103; GenBank accession number Z46390.1). The reverse primer ALV-JNR was designed from a well-conserved region of the *gp85* gene of ALV-J, which could be distinguished from other subgroups. The TaqMan probe was a 23-bp oligonucleotide located downstream of ALV-JNF. Real-time PCRs were performed with a LightCycler 480 real-time thermocycler (Roche Instrument Center, Switzerland). To obtain optimal specific fluorescent signals, the annealing temperature and primer, probe, and Mg^{2+} concentration were optimized. The reaction was performed in a 25 μL system containing 1 μL of cDNA, 2.5 μL of 10× *Ex Taq* buffer, 2 μL of dNTP (2.5 mM), 1 μL of MgCl$_2$ (75 mM), 1 μL of ALV-JNF (10 pM), 1 μL of ALV-JNR (10 pM), 0.5 μL of probe (10 pM), 1 unit of *Ex Taq* HS (TaKaRa, China), and the appropriate volume of double-distilled water (ddH$_2$O). The real-time PCR was carried out with a predenaturation step at 95°C for 5 min and amplification for 40 cycles at a melting temperature of 95°C for 10 s and an annealing/elongation temperature of 65°C for 40 s. Fluorescent signals were collected during the elongation step. A 214 bp fragment of ALV-J was amplified with the primer pair ALV-JNF/ALV-JNR and then cloned into the pMD-18T vector (TaKaRa) to obtain the recombinant plasmid pMD-JNS. The concentration of the plasmid was determined with a UV spectrophotometer, and the plasmid copy number was calculated using the following formula: number of copies = (concentration in ng × 6.022 × 10^{23})/(genome length × 10^9 × 650). Serial dilutions from 1 × 10^1 to 1 × 10^{10} copies/μL of the plasmid standard DNA were used to produce a standard curve.

TABLE 17.2
Primers and Probes Used in Real-Time Polymerase Chain Reaction

Target	Primers/Probes	Sequence (5′–3′)	Product Size (bp)	Reference
ALV-A+B	R-F	GACGCCCCTCTTTCTCTAACTC		
	R-R	ATATGCACCGCAGTACTCACTC	87	[60]
	R[a]	TGCCTATCCGCTGTCACCACTGTAA		
β-actin	β-F	CCAGCCATGTATGTAGCCATCC		
	β-R	CACCATCACCAGAGTCCATCAC	95	[60]
	β[b]	CTGTGCTGTCCCTGTATGCCTCTGG		
ALV-J	ALV-JNF	TTGCAGGCATTTCTGACTGG		
	ALV-JNR	ACACGTTTCCTGGTTGTTGC	214	[69]
	Probe[c]	ACACGTTTCCTGGTTGTTGC		

[a] The probe was labeled with FAM and 3′-Black Hole Quencher 2 (BHQ2).

[b] The probe was labeled with ROX and 3′-6-carboxytetramethyl-rhodamine (TAMRA).

[c] The probe was labeled with FAM and 3′-BHQ1.

17.2.2.4 LAMP Detection of ALV

LAMP is a type of *Bst* polymerase–dependent nucleic acid amplification technology based on isothermal strand displacement amplification. Compared to conventional PCR using two primers, this technology has high specificity since it requires 4 (or 6) primers to identify 6 (or 8) specific domains. LAMP assay has been developed for simple, rapid, and sensitive detection of ALV-A and ALV-J [70,71].

The online software Primer Explorer V4 (https://primer-explorer.jp) was used to design the LAMP primers (including a pair of outer primers, F3 and B3; a pair of inner primers, forward inner primer [FIP] and back inner primer [BIP]; and a pair of loop primers, loop forward [LF] and loop back [LB]) within the gp85 as shown in Table 17.3 for ALV-A detection. A Loopamp DNA Amplification Kit (Eiken Chemical Co., Ltd., Tokyo, Japan) was used for the LAMP reaction. Each 25 µL reaction system contained the following: 12.5 µL of the reaction mix, 1 µL of *Bst* DNA polymerase, 1.6 µM of each inner primer (FIP and BIP), 0.8 µM of each loop primer (LF and LB), 0.2 µM of each outer primer (F3 and B3), 1 µL of fluorescent detection reagent (FDR, Eiken Chemical Co., Ltd., Tokyo, Japan), and 1 µL of DNA as template. Incubation at 65°C for 45 min was selected as the optimal reaction condition after optimization of different temperatures (61°C, 63°C, 65°C, and 67°C) and different time periods (45, 60, 90, 120 min). A Loopamp Realtime Turbidimeter LA-320C (Eiken Chemical Co., Ltd., Tokyo, Japan) was used to measure the turbidity and determine the reaction temperature and time. Using this method, 20 copies of viral nucleic acid sequence were detected within 45 min, which was 100 times more sensitive than conventional PCR. The fluorescent dye fluorescent detection reagent was added during the preparation of the reaction solution, this enabled the results to be assessed directly based on color changes (orange for negative and green for positive) without opening the lid of the reaction tube, which reduces the possibility of contamination.

Primers specific for ALV-J LAMP detection were designed using Primer Explorer V3 software (https://primerexplorer.jp) based on the *pol* and gp85 regions of the prototype strain of ALV-J, HPRS-103, which effectively avoid amplification of other subgroups and the EAV-HP family. The four primers consisted of one pair each of forward and reverse outer primers (F3 and B3) and forward and reverse inner primers (FIP and BIP) (Table 17.3). The 25 µL LAMP reaction was carried out with a set of specific primers containing 1.0 µM (each) FIP and BIP, 0.25 µM (each) F3 and B3, 0.8 mM each dNTP, 0.4 M betaine (Sigma-Aldrich, Inc., MO), 20 mM Tris-HCl, 10 mM KCl, 10 mM $(NH_4)_2SO_4$, 6 mM $MgSO_4$, 0.1% Tween 20, 8 U of Bst DNA polymerase (New England Biolabs), and 1 µL DNA template. The mixture was incubated at 63°C for 45 min in a water bath and terminated by incubation at 84°C for 5 min. The LAMP products were detected by several methods: 2% agarose gel electrophoresis with UV light transillumination, observation of turbidity, and naked-eye determination of color change after addition of SYBR Green I dye (2 µL per reaction tube) under normal light conditions (Cambrex BioSciences, Inc., ME). After the addition of SYBR Green I dye to the terminated reaction, the color indicating the positive reaction changed to yellowish green, whereas indicators of negative reactions remained reddish orange. The addition of SYBR Green I dye to the completed LAMP reaction mixture was advised for easy handling. The detection limit of the LAMP assay was 5 copies (approximately 0.02 fg) per reaction during a 45 min reaction, whereas that of conventional PCR was 100 copies per tube. Thus, the sensitivity of the LAMP assay was at least 20-fold higher than that of conventional PCR. Detection of positive samples by the use of the LAMP assay (26/49) demonstrated the higher sensitivity of this assay in comparison to conventional PCR (21/49) and virus isolation (19/46).

TABLE 17.3

Primers and Sequences for Avian Leukosis Virus Loop-Mediated Isothermal Amplification Detection

Target	Primer	Sequence (5′–3′)[a]	Position (5′–3′)[b]	Reference
ALV-A	F3	TAGACAGGAAGCCACGCG	673–690	[70]
	B3	CACTGGGTTTCAGGATGACC	878–897	
	FIP	GCCTATCCGCTGTCACCACTG	F1c: 738–758	
		TCTCCTTAGACGCCCCTC	F2: 694–711	
	BIP	ACTGCGGTGCATATGGCTACAG	B1c: 783–804	
		CATTGCCACAGCGGTACT	B2: 843–860	
	LF	ACGGTTTCGAGGAGTTAGAGAAA	712–734	
	LB	TGGAACATATATAACTGCTCACAGG	809–833	
ALV-J	F3	GGRAAGGTGAGCAAGAAGG	5323–5341	[71]
	B3	YGTCTTATTTGCCCAGGTGA	5513–5532	
	FIP	GTAACCTCTCGAYGGCAGC	F1c: 5397–5415	
		ACTCTAAGAAGAAGCCGCCA	F2: 5342–5361	
	BIP	ATTTCYGTTGTCCCAGGGGTG	B1c: 5452–5472	
		CCCACACGTTTCCTGGTTG	B2: 5494–5512	

[a] R = A/G, Y = T/C.

[b] Positions are shown relative to RAV-1 (M19113) for ALV-A and HPRS-103 (Z46390) for ALV-J.

17.3 CONCLUSION

ALVs are still considered ubiquitous in commercial chickens, notwithstanding the eradication programs instituted by many primary breeding companies. With few exceptions, infections occur in all chicken flocks. No commercial vaccine is available for the protection of chickens from infection with ALV. Recombinant ALVs expressing subgroup A and J envelope glycoproteins have been produced that could have potential as vaccines to protect against horizontal transmission. Congenitally infected chicks are immunologically tolerant and, thus, cannot be immunized, even if a suitable vaccine was available. These chickens constitute major source of ALV transmission and are the most likely to develop neoplasms.

Control of AL/SV infection depends mainly on early detection and removal of virus-shedding birds to reduce spread of congenital and contact infection in other birds. There are a number of biological, molecular, and serological assays that can be used for the isolation and identification of ALV. These assays are based on the detection of proteins coded for by one or more of the three major genes of ALV: gag, pol, and env. Biological assays for the isolation and identification of ALV including virus neutralization tests require the use of CEFs with specific host range and take more than a week to obtain results. Biological and molecular assays, but not direct assays based on the detection of gs p27, can be used to differentiate between endogenous and exogenous ALV. It has been suggested that the unusually high rate of horizontal transmission and the high frequency of molecular and antigenic variation of ALV-J may interfere with the success of eradication programs of ALV-J in broiler breeder flocks.

In recent years, there have been several reports of ALV-J outbreaks in parent and commercial layer flocks as well as some local breeds in China. Meanwhile, strains of ALV-A and ALV-B have been isolated in China, especially from the native chicken breeds, which suggest the widespread and complex nature of ALV infection in China. In addition, the coinfection of ALV-A and ALV-J as well as ALV-A and ALV-B detected provides a potential opportunity for recombination between different ALV subgroups [75,81]. Moreover, three strains (JS11C1, JS11C2, JS11C3), suggested to be a new subgroup K of ALV (but not confirmed) that is different from all six known subgroups, have been isolated from Chinese native chicken breed "luhua" in 2012 [82]. Thus, it is essential to have a continuous epidemiological surveillance in place to detect new variants potentially escaping the current detection methods.

REFERENCES

1. C. M. Fauquet, M. A. Mayo, J. Maniloff, U. Desselberger, and L. A. Ball, *Virus Taxonomy: VIIIth Report of the International Committee on Taxonomy of Viruses*. Elsevier-Academic Press, Burlington, MA, 2005.
2. G. F. de Boer (ed.), *Avian Leukosis*. Martinus Nijhoff Publishing Group, Boston, MA, 1987.
3. L. N. Payne and A. M. Fadly, Leukosis/Sarcoma group. In *Diseases of Poultry*, 10th edn., B. W. Calnek, H. J. Barnes, C. W. Beard, L. R. McDougald, and Y. M. Saif (eds.). Iowa State University Press, Ames, IA, 1997, pp. 416–466.
4. J. S. Gavora, Influences of avian leukosis virus infection on production and mortality and the role of genetic selection in the control of lymphoid leukosis. In *Avian Leukosis*, G. F. de Boer (ed.). Martinus Nijhoff, Boston, MA, 1987, pp. 241–260.
5. J. Gavora, J. Spencer, and J. Chambers, Performance of meat-type chickens test-positive and -negative for lymphoid leukosis virus infection. *Avian Pathol* 11, 29–38 (1982).
6. J. S. Gavora, J. L. Spencer, R. S. Gowe, and D. L. Harris, Lymphoid leukosis virus infection: Effects on production and mortality and consequences in selection for high egg production. *Poult Sci* 59, 2165–2178 (1980).
7. A. M. Fadly and V. Nair, Leukosis/Sarcoma group. In *Diseases of Poultry*, Y. M. Saif, A. M. Fadly, J. R. Glisson, L. R. McDougald, L. K. Nolan, and D. E. Swayne (eds.). Blackwell Publishing Ltd, Ames, IA, 2008, Chapter 15, pp. 514–568.
8. Y. L. Gao et al., Avian leukosis virus subgroup J in layer chickens, China. *Emerg Infect Dis* 16, 1637–1638 (2010).
9. L. N. Payne and V. Nair, The long view: 40 years of avian leukosis research. *Avian Pathol* 41, 1–9 (2012).
10. International Committee on Taxonomy of Viruses: Virus Taxonomy 2014 Release. Available at http://ictvonline.org/virusTaxonomy.asp; accessed March 30, 2015.
11. P. Enrietto and M. Hayman, Structure and virus-associated oncogenes of avian sarcoma and leukemia viruses. In *Avian Leukosis*, G. F. De Boer (ed.). Martinus Nijhoff, Boston, MA, 1987, pp. 29–46.
12. P. J. Enrietto and J. A. Wyke, The pathogenesis of oncogenic avian retroviruses. *Adv Cancer Res* 39, 269–314 (1983).
13. T. Graf and H. Beug, Avian leukemia viruses: Interaction with their target cells in vivo and in vitro. *Biochim Biophys Acta* 516, 269–299 (1978).
14. C. Moscovici and L. Gazzolo, Virus-cell interaction of avian sarcoma and defective leukemia viruses. In *Avian Leukosis*, G. F. De Boer (ed.). Martinus Nijhoff, Boston, MA, 1987, pp. 153–170.
15. H. J. Kung and J. L. Liu, Retroviral oncogenesis. In *Viral Pathogenesis*, N. Nathanson (ed.). Lippincott-Raven Publishers, Philadelphia, PA, 1997, pp. 235–266.
16. J. M. Coffin, S. H. Hughes, and H. E. Varmus (eds.). *Retroviruses*. Cold Spring Harbor Laboratory Press, Cold Spring Harbor, NY, 1997.
17. V. Nair, Retrovirus-induced oncogenesis and safety of retroviral vectors. *Curr Opin Mol Ther* 10, 431–438 (2008).
18. H. Fan and C. Johnson, Insertional oncogenesis by non-acute retroviruses: Implications for gene therapy. *Viruses* 3, 398–422 (2011).
19. L. N. Payne, Epizootiology of avian leukosis virus infection. In *Avian Leukosis*, G. F. De Boer (ed.). Martinus Nijhoff, Boston, MA, 1987, pp. 47–76.
20. L. N. Payne, Biology of avian retroviruses. In *The Retroviridae*, J. Levy (ed.). Plenum Press, New York, 1992, vol. 1, pp. 299–404.
21. L. N. Payne and N. Bumstead, Theoretical considerations on the relative importance of vertical and horizontal transmission for the maintenance of infection by exogenous avian lymphoid leukosis virus. *Avian Pathol* 11, 547–553 (1982).
22. R. L. Witter, Determinants of early transmission of ALV-J in commercial broiler breeder chickens. In *Proceedings, International, Symposium on ALV-J and Other Avian Retroviruses*, E. F. Kaleta, L. N. Payne, and U. Heffels-Redmann (eds.). World Veterinary Poultry Association, Rauischholzhausen, Germany, 2000, pp. 216–225.

23. R. L. Witter, L. D. Bacon, H. D. Hunt, R. E. Silva, and A. M. Fadly, Avian leukosis virus subgroup J infection profiles in broiler breeder chickens: Association with virus transmission to progeny. *Avian Dis* 44, 913–931 (2000).

24. A. M. Fadly and E. J. Smith, Isolation and some characteristics of a subgroup J-like avian leukosis virus associated with myeloid leukosis in meat-type chickens in the United States. *Avian Dis* 43, 391–400 (1999).

25. R. L. Witter and A. M. Fadly, Reduction of horizontal transmission of avian leukosis virus subgroup J in broiler breeder chickens hatched and reared in small groups. *Avian Pathol* 30, 641–654 (2001).

26. A. M. Fadly, Avian retroviruses. *Vet Clin North Am Food Anim Pract* 13, 71–85 (1997).

27. K. Tsukamoto, M. Hasebe, S. Kakita, Y. Taniguchi, H. Hihara, and Y. Kono, Sporadic congenital transmission of avian leukosis virus in hens discharging the virus into the oviducts. *J Vet Med Sci* 54, 99–103 (1992).

28. A. R. Pandiri, W. M. Reed, J. K. Mays, and A. M. Fadly, Influence of strain, dose of virus, and age at inoculation on subgroup J avian leukosis virus persistence, antibody response, and oncogenicity in commercial meat-type chickens. *Avian Dis* 51, 725–732 (2007).

29. C. R. Pope, E. M. Odor, and M. Salem, Unusual eye lesions associated with ALV-J virus. In *Proceedings, 48th Western Poultry Disease Conference*, 1999, pp. 96–97.

30. S. M. Williams, S. D. Fitzgerald, W. M. Reed, L. F. Lee, and A. M. Fadly, Tissue tropism and bursal transformation ability of subgroup J avian leukosis virus in White Leghorn chickens. *Avian Dis* 48, 921–927 (2004).

31. L. N. Payne, S. R. Brown, N. Bumstead, K. Howes, J. A. Frazier, and M. E. Thouless, A novel subgroup of exogenous avian leukosis virus in chickens. *J Gen Virol* 72 (Pt 4), 801–807 (1991).

32. J. G. Campbell, *Tumours of the Fowl*. William Heinemann Medical Books, London, U.K., 1969.

33. D. Soffer, N. Resnick-Roguel, A. Eldor, and M. Kotler, Multifocal vascular tumors in fowl induced by a newly isolated retrovirus. *Cancer Res* 50, 4787–4793 (1990).

34. W. Pan et al., Novel sequences of subgroup J avian leukosis viruses associated with hemangioma in Chinese layer hens. *Virol J* 8, 552 (2011).

35. H. Lai et al., Isolation and characterization of emerging subgroup J avian leukosis virus associated with hemangioma in egg-type chickens. *Vet Microbiol* 151, 275–283 (2011).

36. H. N. Zhang et al., An ALV-J isolate is responsible for spontaneous haemangiomas in layer chickens in China. *Avian Pathol* 40, 261–267 (2011).

37. K. Ochiai, K. Ohashi, T. Mukai, T. Kimura, T. Umemura, and C. Itakura, Evidence of neoplastic nature and viral aetiology of so-called fowl glioma. *Vet Rec* 145, 79–81 (1999).

38. T. Toyoda et al., Cerebellar hypoplasia associated with an avian leukosis virus inducing fowl glioma. *Vet Pathol* 43, 294–301 (2006).

39. N. Iwata, K. Ochiai, K. Hayashi, K. Ohashi, and T. Umemura, Nonsuppurative myocarditis associated with so-called fowl glioma. *J Vet Med Sci* 64, 395–399 (2002).

40. H. Hatai et al., Nested polymerase chain reaction for detection of the avian leukosis virus causing so-called fowl glioma. *Avian Pathol* 34, 473–479 (2005).

41. H. Hatai et al., Prevalence of fowl glioma-inducing virus in chickens of zoological gardens in Japan and nucleotide variation in the env gene. *J Vet Med Sci* 70, 469–474 (2008).

42. S. Nakamura, K. Ochiai, H. Hatai, A. Ochi, Y. Sunden, and T. Umemura, Pathogenicity of avian leukosis viruses related to fowl glioma-inducing virus. *Avian Pathol* 40, 499–505 (2011).

43. A. M. Fadly, Isolation and identification of avian leukosis viruses: A review. *Avian Pathol* 29, 529–535 (2000).

44. A. M. Fadly and R. L. Witter, Oncornaviruses: Leukosis/sarcoma and reticuloendotheliosis. In *A Laboratory Manual for the Isolation and Identification of Avian Pathogens*, 4th edn., J. R. Glisson, D. J. Jackwood, J. E. Pearson, W. M. Reed, and D. E. Swayne (eds.). American Association of Avian Pathologists, Kennet Square, PA, 1998, pp. 185–196.

45. H. D. Hunt, L. F. Lee, D. Foster, R. F. Silva, and A. M. Fadly, A genetically engineered cell line resistant to subgroup J avian leukosis virus infection (C/J). *Virology* 264, 205–210 (1999).

46. L. D. Bacon, Detection of endogenous avian leukosis virus envelope in chicken plasma using R2 antiserum. *Avian Pathol* 29, 153–164 (2000).

47. L. D. Bacon, E. J. Smith, A. M. Fadly, and L. B. Crittenden, Development of an alloantiserum (R2) that detects susceptibility of chickens to subgroup E endogenous avian leukosis virus. *Avian Pathol* 25, 551–568 (1996).

48. G. Kelloff and P. K. Vogt, Localization of avian tumor virus group-specific antigen in cell and virus. *Virology* 29, 377–384 (1966).

49. F. E. Payne, J. J. Solomon, and H. G. Purchase, Immunofluorescent studies of group-specific antigen of the avian sarcoma-leukosis viruses. *Proc Natl Acad Sci USA* 55, 341–349 (1966).

50. H. Hunt, B. Lupiani, and A. M. Fadly, Recombination between ALV-J and endogenous subgroup E viruses. In *Proceedings, International Symposium on ALV-J and Other Avian Retroviruses*, E. F. Kaleta, L. N. Payne, and U. Heffels-Redmann (eds.). Justus Liebig-Gesellschaft zu Giessen, Rauischholzhausen, Germany, 2000, pp. 50–60.

51. A. Fadly, R. Silva, H. Hunt, A. Pandiri, and C. Davis, Isolation and characterization of an adventitious avian leukosis virus isolated from commercial Marek's disease vaccines. *Avian Dis* 50, 380–385 (2006).

52. T. Barbosa, G. Zavala, and S. Cheng, Molecular characterization of three recombinant isolates of avian leukosis virus obtained from contaminated Marek's disease vaccines. *Avian Dis* 52, 245–252 (2008).

53. R. F. Silva, A. M. Fadly, and S. P. Taylor, Development of a polymerase chain reaction to differentiate avian leukosis virus (ALV) subgroups: Detection of an ALV contaminant in commercial Marek's disease vaccines. *Avian Dis* 51, 663–667 (2007).

54. H. M. Temin, The cellular and molecular biology of RNA tumor viruses, especially avian leukosis-sarcoma viruses, and their relatives. *Adv Cancer Res* 19, 47–104 (1974).

55. A. Tereba and K. G. Murti, A very sensitive biochemical assay for detecting and quantitating avian oncornaviruses. *Virology* 80, 166–176 (1977).

56. G. Kelloff, M. Hatanaka, and R. V. Gilden, Assay of C-type virus infectivity by measurement of RNA-dependent DNA polymerase activity. *Virology* 48, 266–269 (1972).

57. A. Panet, D. Baltimore, and T. Hanafusa, Quantitation of avian RNA tumor virus reverse transcriptase by radioimmunoassay. *J Virol* 16, 146–152 (1975).

58. A. M. Fadly, T. F. Davison, L. N. Payne, and K. Howes, Avian leukosis virus infection and shedding in brown leghorn chickens treated with corticosterone or exposed to various stressors. *Avian Pathol* 18, 283–298 (1989).

59. D. Hauptli, L. Bruckner, and H. P. Ottiger, Use of reverse transcriptase polymerase chain reaction for detection of vaccine contamination by avian leukosis virus. *J Virol Methods* 66, 71–81 (1997).

60. G. Zhou et al., A duplex real-time reverse transcription polymerase chain reaction for the detection and quantitation of avian leukosis virus subgroups A and B. *J Virol Methods* 173, 275–279 (2011).

61. P. A. M. Van Woensel, A. V. Blaaderen, R. J. M. Mooman, and G. F. D. Boer, Detection of proviral DNA and viral RNA in various tissues early after avian leukosis virus infection. *Leukemia* 6 (Suppl. 3), 135S–137S (1992).

62. M. Garcia et al., Development and application of reverse transcriptase nested polymerase chain reaction test for the detection of exogenous avian leukosis virus. *Avian Dis* 47, 41–53 (2003).

63. B. Lupiani, H. Hunt, R. Silva, and A. Fadly, Identification and characterization of recombinant subgroup J avian leukosis viruses (ALV) expressing subgroup A ALV envelope. *Virology* 276, 37–43 (2000).

64. R. F. Silva, A. M. Fadly, and H. D. Hunt, Hypervariability in the envelope genes of subgroup J avian leukosis viruses obtained from different farms in the United States. *Virology* 272, 106–111 (2000).

65. E. J. Smith, S. M. Williams, and A. M. Fadly, Detection of avian leukosis virus subgroup J using the polymerase chain reaction. *Avian Dis* 42, 375–380 (1998).

66. L. M. Smith et al., Development and application of polymerase chain reaction (PCR) tests for the detection of subgroup J avian leukosis virus. *Virus Res* 54, 87–98 (1998).

67. Q. Gao et al., Development and application of a multiplex PCR method for rapid differential detection of subgroup A, B, and J avian leukosis viruses. *J Clin Microbiol* 52, 37–44 (2014).

68. Y. Kim, S. M. Gharaibeh, N. L. Stedman, and T. P. Brown, Comparison and verification of quantitative competitive reverse transcription polymerase chain reaction (QC-RT-PCR) and real time RT-PCR for avian leukosis virus subgroup J. *J Virol Methods* 102, 1–8 (2002).

69. L. Qin et al., Development and application of real-time PCR for detection of subgroup J avian leukosis virus. *J Clin Microbiol* 51, 149–154 (2013).

70. Y. Wang et al., Development of loop-mediated isothermal amplification for rapid detection of avian leukosis virus subgroup A. *J Virol Methods* 173, 31–36 (2011).

71. X. Zhang et al., Development of a loop-mediated isothermal amplification assay for rapid detection of subgroup J avian leukosis virus. *J Clin Microbiol* 48, 2116–2121 (2010).

72. T. Notomi et al., Loop-mediated isothermal amplification of DNA. *Nucleic Acids Res* 28, E63 (2000).

73. Y. Mori, K. Nagamine, N. Tomita, and T. Notomi, Detection of loop-mediated isothermal amplification reaction by turbidity derived from magnesium pyrophosphate formation. *Biochem Biophys Res Commun* 289, 150–154 (2001).

74. B. R. Bista et al., Development of a loop-mediated isothermal amplification assay for rapid detection of BK virus. *J Clin Microbiol* 45, 1581–1587 (2007).

75. S. P. Fenton, M. R. Reddy, and T. J. Bagust, Single and concurrent avian leukosis virus infections with avian leukosis virus-J and avian leukosis virus-A in Australian meat-type chickens. *Avian Pathol* 34, 48–54 (2005).

76. T. D. Pham, J. L. Spencer, and E. S. Johnson, Detection of avian leukosis virus in albumen of chicken eggs using reverse transcription polymerase chain reaction. *J Virol Methods* 78, 1–11 (1999).

77. Basic Local Alignment Search Tool. Available at http://blast.ncbi.nlm.nih.gov/blast/Blast.cgi. Accessed on March 31, 2015.

78. S. Kumar, K. Tamura, and M. Nei, MEGA3: Integrated software for Molecular Evolutionary Genetics Analysis and sequence alignment. *Brief Bioinform* 5, 150–163 (2004).

79. J. D. Thompson, T. J. Gibson, F. Plewniak, F. Jeanmougin, and D. G. Higgins, The CLUSTAL_X windows interface: Flexible strategies for multiple sequence alignment aided by quality analysis tools. *Nucleic Acids Res* 25, 4876–4882 (1997).

80. Y. Kim and T. P. Brown, Development of quantitative competitive-reverse transcriptase-polymerase chain reaction for detection and quantitation of avian leukosis virus subgroup J. *J Vet Diagn Invest* 16, 191–196 (2004).

81. B. Xu, W. Dong, C. Yu, Z. He, Y. Lv, Y. Sun, X. Feng, N. Li, L. F. Lee, and M. Li, Occurrence of avian leukosis virus subgroup J in commercial layer flocks in China. *Avian Pathol* 33, 13–17 (2004).

82. X. Wang, P. Zhao, and Z. Z. Cui, Identification of a new subgroup of avian leukosis virus isolated from Chinese indigenous chicken breeds. *Bing Du Xue Bao (Chin J Virol)* 28, 609–614 (2012).

83. V. Nair, *Molecular Diagnostic Manual*. Viral Oncogenesis Group, The Pirbright Institute, Berkshire, U.K., 2015.

18 Bovine Leukemia Virus

Silvina E. Gutiérrez and Agustina Forletti

CONTENTS

18.1 Introduction 157
    18.1.1 Classification, Morphology, and Genome Organization 157
    18.1.2 Biology, Transmission, and Epidemiology 158
    18.1.3 Pathogenesis and Clinical Features 159
    18.1.4 Diagnosis 160
        18.1.4.1 Conventional Techniques 160
        18.1.4.2 Molecular Techniques 161
18.2 Methods 161
    18.2.1 Sample Collection and Preparation 161
    18.2.2 Detection Procedures 161
        18.2.2.1 Nested PCR for the Detection of BLV *env* Sequences 161
        18.2.2.2 Real-Time PCR for the Detection of BLV *pol* Sequences 162
18.3 Conclusions and Future Perspectives.. 163
References... 164

18.1 INTRODUCTION

Bovine leukosis (lymphosarcoma, leukemia, or lymphoma) is one of the most frequent neoplastic diseases of cattle. Two types of bovine leukosis are recognized on the basis of their epidemiology: enzootic bovine leukosis and sporadic bovine leukosis. The enzootic form, which is caused by bovine leukemia virus (BLV), is the most frequent. The sporadic bovine leukosis, which includes the juvenile, thymic, and cutaneous forms of the disease, is not associated with BLV or any other infectious agents [1].

First described in 1871, bovine leukemia only attracted significant attention in the early twentieth century in several European countries, notably Denmark and Germany, where clusters of herds with a high incidence of a similar disease suggested a viral etiology. However, the viral etiology of this disease was not described until 1969 [2].

BLV is the type species of the genus *Deltaretrovirus* in the *Retroviridae* family. This genus also covers human T-cell leukemia viruses (HTLV-1 to HTLV-3) and simian T-cell leukemia viruses (STLV-1 to STLV-3), some of which are associated with proliferative or neurologic diseases of human and nonhuman primates [2,3]. Deltaretroviruses share many particular molecular and biological properties: they infect and transform lymphocytes *in vivo* and *in vitro*, cause disease only in a low percentage of infected hosts and after a long latency period, and do not induce viremia or detectable antigens *in vivo*; and in addition to the structural classical retroviral genes (*gag–pol–env*), they harbor open reading frames (ORFs) that codify accessory and regulatory proteins. Given these similarities, BLV infection in animals is regarded as a useful model for the study of the HTLVs and their associated diseases [3–5].

The majority of infected cattle remain asymptomatic throughout their lives, and only a small fraction (usually less than 5%) of the infected individuals develop malignant lymphoma. Persistent lymphocytosis (PL), a benign proliferation of lymphoid cells, is developed in approximately 30% of BLV-infected cattle [1].

Infection by BLV is widespread in all continents, except in Europe, where most of the member states of the European Economic Community are officially free of BLV [5]. BLV infection constitutes an important economic problem in the dairy cattle industry, due not only to the mortality of cattle suffering from lymphosarcoma but also to the regulatory restriction in the international trade of cattle and their by-products.

Bovine lymphosarcoma is an invariably fatal disease, for which no effective treatment currently exists. Despite advances in research on experimental vaccines, there is as yet no vaccine commercially available for the control of enzootic bovine leukosis.

18.1.1 CLASSIFICATION, MORPHOLOGY, AND GENOME ORGANIZATION

BLV is a complex exogenous oncogenic retrovirus with transacting activity that principally infects B lymphocytes. According to the present classification of viruses, BLV is the type species of the genus *Deltaretrovirus* in the family *Retroviridae*. This genus also includes viruses from human and nonhuman primates (collectively grouped in the species primate T-lymphotropic viruses 1, 2, and 3) that share biological and molecular characteristics with BLV [6].

Morphologically, BLV virions are type C particles with a diameter between 60 and 125 nm and have a three-layered structure: an inner core composed of genome–nucleoprotein complex with helical symmetry, surrounded by an icosahedral capsid composed primarily of the p24 protein (CA p24) codified by the *gag* gene. The CA p24 protein appears to be a major target for the host's humoral immune response. The external envelope, derived from the cell membrane, contains the transmembrane and surface glycoproteins (TM gp30 and SU gp51), both codified by the virus. While TM gp30 is poorly immunogenic, it is a key factor during the fusion between the virion envelope and new target lymphocytes [7]. The SU gp51 induces an important humoral immune response, a property used for diagnostics. Specific antibodies directed to three conformational epitopes (designated F, G, and H) located in the amino terminal moiety of the SU gp51 molecule exhibit neutralizing activity. This region of the SU gp51 also contains the putative receptor-binding domain for the recognition of the specific receptor in the target cells. The cellular receptor for BLV still remains elusive [8].

The virus particle is diploid, that is, it contains two copies of positive ssRNA. Genomic RNA is 8420 bp long and presents a 5′ CAP and a polyadenylated tail in its 3′ end [9]. The structural genes (*gag*, *pol*, and *env*) are flanked by the long terminal repeats (LTRs) in the proviral form of the virus. The 5′ LTR contains several *cis*-acting elements involved in the regulation of BLV transcription. BLV is a complex retrovirus that, apart from encoding *gag*, *pol*, and *env* genes, present in all retroviruses, encodes several other regulatory and accessory proteins in ORFs. The regulator of expression (p18 REX) and the transcriptional activator of viral expression (p34 Tax) proteins are both codified by a double-spliced mRNA. Other accessory proteins are R3 and G4, of 5.5 and 11.6 kDa, respectively, codified by separate mRNAs. These accessory proteins are dispensable for infectivity *in vivo* but essential for viral propagation and pathogenicity [10]. The analysis of five fully sequenced isolates of BLV from different regions of the world showed that they are >95% homologous at the nucleotide level [11]. The phylogenetic and similarity analyses performed on a conserved region of the *env* gene allowed the classification of BLV isolates into seven genotypes [12].

18.1.2 Biology, Transmission, and Epidemiology

The major target of the virus is the B lymphocyte [13,14], although evidence of BLV infection in other cell types such as monocytes and macrophages has been reported both *in vivo* and *in vitro* [15,16]. In cattle, the majority of infected B cells express the CD5 marker [17–19]. Soon after infection of a target cell, the RNA genome is copied into DNA by the virus-encoded reverse transcriptase. The provirus then integrates into the host genomic DNA. The integrated provirus is by far the most commonly detectable viral form observed in persistently infected animals. It is believed that once the specific immune response is developed, the replicative cycle is blocked or abolished and the virus replicates only by mitotic expansion of provirus-carrying cells [20].

BLV integrates randomly in the cellular genome and therefore insertional mutagenesis is not the mechanism of transformation of infected cells [8]. It also lacks a known oncogene, as in rapidly transforming retroviruses. An intriguing fact is that infection of cattle proceeds without any evidence of viral protein expression. In effect, BLV virions or viral proteins cannot be detected in the peripheral blood by any currently available method [21]. Moreover, evidence for the presence of viral transcripts in peripheral blood cells is controversial [22–25]. It is believed that the virus is repressed at the transcriptional level *in vivo* and that the virus can be easily derepressed upon *in vitro* culture for a few hours [1]. Repression of the viral expression may be due, at least in part, to a suboptimal enhancer sequence located in the proviral 5′ LTR. An elusive plasma factor related to fibronectin could also be responsible for the repressed state in which BLV is maintained *in vivo* [26–29]. It seems that the repression of the viral expression is a mechanism that BLV has evolved to evade the immune surveillance and hence to persist in its host and in nature.

BLV naturally infects cattle, zebus, buffalos, and capybaras and can be experimentally transmitted to sheep, goats, rabbits, chickens, pigs, and rats (reviewed in [30]). Humans do not seem to be susceptible to BLV infection; however, human cells have been infected with BLV *in vitro* [31]. Although sheep are not natural hosts for BLV, they are considered a very good animal model for studying BLV pathogenesis, as they are sensitive to BLV infection and develop the disease much earlier and with higher frequency than cattle (almost all infected sheep develop lymphosarcoma after being experimentally infected) [10].

BLV has worldwide distribution. Most Western European countries have controlled the infection by following the implementation of strict control measures. On the other hand, in the Americas, where the animal health sanitary authorities have not implemented compulsory control or eradication programs, prevalence of BLV infection ranges from 25% to 50% at the individual level for most countries, affecting 80%–90% of the herds (reviewed in [5]).

Since infectious, cell-free BLV is rarely produced *in vivo* and is very unstable, most susceptible cattle are thought to become infected by exposure to infected lymphocytes and not by a cell-free virus. Thus, infected lymphocytes from blood, milk, or any other secretion or excretion may potentially transmit the infection upon contact with susceptible cattle. *In utero* vertical transmission occurs by transplacental infection after immune competence of the fetus is established (i.e., the third month of gestation). This way of transmission is infrequent under natural conditions, occurring in 3%–8% of calves born to infected cows and most frequently in cows having PL [32,33]. Colostrum contains infected lymphocytes, but it is not a common way of transmission, likely due to the protective role of the antibodies it contains. Transmission of BLV to calves by feeding bulk milk from infected cows has been documented, being the susceptibility of calves dependent on the presence of specific antibodies obtained from the dam's colostrum and the age of the calf [33,34]. In natural conditions, iatrogenic procedures that involve the contact of infected blood like dehorning, castration, venipuncture, and application of ear tags are risk factors highly associated with BLV transmission, when practiced without proper disinfection of material [33].

18.1.3 Pathogenesis and Clinical Features

After a variable period of 2–8 weeks upon infection, cattle develop humoral and cellular immune responses to capsid and envelope proteins that are inefficient to clear the infection. Once established, the immune responses efficiently abolish the viral replicative cycle, and the virus replicates only by mitotic division of provirus-carrying cells. Hence, the virus persists indefinitely throughout life [20].

BLV infection is characterized by the "iceberg principle" typical of many viral diseases. The infection is asymptomatic in the majority (approximately 70%) of infected cattle, while one-third of the infected cattle develop a permanent and relatively stable increase in the number of B lymphocytes in the peripheral blood termed PL. PL is the result of the accumulation of untransformed B lymphocytes and thus is considered a benign condition. Cattle with clinical manifestations of disease, usually less than 5% of infected cattle, are represented by the tip of the iceberg.

It is believed that the disease (lymphoma or lymphosarcoma) and the subclinical stage of PL are results from a complex interplay between the virus and the animal. From the beginning, it was observed that lymphosarcoma and PL appeared more frequently in familial clusters. Epidemiological data indicate that both conditions are under the control of host genetic factors [1]. The molecular basis for the genetic resistance and susceptibility of cattle to the development of PL was mapped to the major histocompatibility complex (MHC) class II BoLA DRB3.2 gene [35,36]. Among the asymptomatic BLV-infected cattle, some are particularly resistant to the *in vivo* spread of BLV, maintaining a very low proviral load in the peripheral blood [37]. This condition is also strongly associated with the MHC class II BoLA DRB3.2 gene, suggesting that the cellular immune response is important in controlling the *in vivo* spread of BLV [38].

Polymorphism in the promoter region of the tumor necrosis factor-α (TNF-α) gene, which is located within the class III region of the MHC, has also been associated with the progression of the disease in BLV-infected cattle. A single nucleotide polymorphism at position −824 in the TNF-α gene, which determines low transcriptional activity, has been associated with lymphosarcoma induced by BLV and high proviral load in BLV-infected cattle [39]. In the early phase of infection, high levels of TNF-α may contribute to the elimination of BLV [40].

The clinical form of the disease results from the accumulation of transformed lymphocytes in one or more organs after a long latency period of 1–8 years. This form is rarely seen in animals under 2 years of age and is most common in the 4–8-year age group. The pathologic condition is a multicentric lymphosarcoma. Lesions can be localized in almost any organ, but the abomasum, heart, visceral and peripheral lymph nodes, spleen, uterus, and kidneys are most frequently affected. Lesions can be observed as white firm tumor masses or as a diffuse tissue infiltrate in any organ. The latter pattern results in an enlarged pale organ. There may be an accompanying lymphocytosis and some degree of anemia. The presence of large numbers of immature or abnormal lymphocytes is a more reliable indication of the presence of the disease. The clinical signs and the duration of the illness are variable, depending on the localization and speed of grow of tumors. In most cases, the course is subacute to chronic, initiated by a marked loss of weight and appetite and weakness. The beginning of the disease often concurs with parturition and the consequent lactation is compromised. Most frequent signs are decreased milk production, external and internal lymphadenopathy, and posterior paresis. Other signs that may be present are fever, bilateral or unilateral exophthalmos, and diarrhea. In 5%–10% of the cases, the course is peracute, and the affected animals suddenly die without any previous evidence of disease. This clinical presentation may be due to the involvement of adrenal glands or an internal hemorrhage caused by a spleen rupture or breakage of an abomasal ulcer [41].

The mechanisms by which BLV induces PL and cellular transformation are not fully understood. Tumors are mostly monoclonal for BLV integration within a single animal, but the virus does not seem to have preferential integration sites among tumors from different animals. Most often, the CD5 marker is expressed on tumor cells [30]. BLV does not carry any known oncogene, but a number of experimental observations indicate that Tax protein can act as an oncogene [4].

PL is the result of a disruption of the target cell homeostasis, which is a complex balance between proliferation and apoptosis rates. As a consequence of impaired cell proliferation and decreased cell death, the B-cell turnover is reduced in PL cattle [42]. This situation finally leads to the accumulation of infected cells in the blood.

The role of TNF-α in the progression of BLV-induced lymphocytosis is related to an augmented expression of the cytokine and imbalance of TNF-α receptors. TNF-α is augmented in peripheral blood mononuclear cells (PBMCs) from BLV-infected cattle and induces proliferative responses and inhibition of apoptosis by acting on TNF-α type II receptors, whose expression is elevated in BLV-infected cattle and PL cattle [43,44]. T-cell-derived IL-2 may also be implicated in the development or maintenance of PL [45]. An indirect role of p34 Tax in this process has been proposed on the basis of its interaction with the cellular protein tristetraprolin, which is a posttranscriptional repressor of TNF-α expression. p34 Tax, by interacting with tristetraprolin, could inhibit its repressor activity, resulting in an augmented expression of TNF-α [46].

The imbalance in B-cell homeostasis may contribute to the accumulation of mutations that finally lead to the tumoral transformation. Mutations of p53 tumor suppressor gene have been found in approximately half of BLV-induced tumors in cattle [47–49]. Inactivation of p53 by mutation appears to be one of the critical events in the transition from PL to the tumoral phase [10]. An increased relation in B-cell leukemia/lymphoma-2 (Bcl-2)/bax expression has also been implicated in disease progression [50].

The p34 Tax protein is a key contributor on the oncogenic potential of BLV. It is believed that Tax plays a role in the initiation of tumorigenesis, but it does not seem to be essential for the maintenance of the transformed phenotype [30]. Apart from its action on the transcriptional activation of viral expression, Tax has been shown to immortalize primary rat embryo fibroblasts and to cooperate with the Ha-ras oncogene in cellular transformation [51]. This property is also present in the BLV G4 protein [52]. The mechanism by which Tax induces transformation is not fully understood. The cytokine-independent growth induced by Tax in ovine B cells correlates with increased Bcl-2 levels and signaling through NF-κB [53]. Tax also inhibits the base excision DNA repair of oxidative damage, thus allowing the accumulation of mutations in cellular DNA [54].

To unravel the functional role of Tax, insights into the Tax-responsive genes have been carried out by microarray-based gene expression analysis. Studies using the ovine model, human cDNA microarrays and Tax-transfected human cells showed that Tax deregulates a broad network of interrelated pathways rather than a single B-lineage-specific regulatory process. Differentially expressed genes include genes related to apoptosis, DNA transcription, and repair; proto-oncogenes; cell cycle regulators; transcription factors, stress response, and immune response; and small Rho GTPases/GTPase-binding proteins [55,56]. Using the ovine model, genes known to be associated with human neoplasia, especially B-cell malignancies, were extensively represented [56]. Genes related to the immune response, mainly of the interferon family of antiviral factors, were preferentially downregulated by Tax [55].

18.1.4 DIAGNOSIS

Due to the absence of viral expression and viremia in BLV-infected animals, direct identification of BLV or its antigens in peripheral blood leucocytes requires *in vitro* culture of these cells. This methodology involves separation of mononuclear cells in a Ficoll-Paque gradient and subsequent culture for at least 24 h. For this reason, the molecular methods, mainly the PCR, have been widely used to identify proviral sequences in PBMCs from infected cattle and are currently regarded as the most useful and rapid approach for direct detection of BLV. The PCR combined with restriction fragment length polymorphism (RFLP) and/or sequencing is used to study the genomic variability of the virus. Finally, as virtually all infected cattle develop antibodies to viral antigens, serological tests are commonly used for the diagnosis of BLV infection in cattle over 6–9 months of age.

18.1.4.1 Conventional Techniques

Methods for the direct detection of BLV, as stated earlier, require the isolation and short-term (24–48 h) culture of PBMCs and subsequent identification of the virus or its antigens by electron microscopy, immunological methods, or syncytia infectivity assay. The observation of budding particles by electron microscopy in short-term cultures of lymphocytes from cattle with PL or lymphosarcoma was decisive in the discovery of BLV and most useful in the early studies on BLV infection. This technique is not specific and, due to the cost and expensive equipment required, is no more used for diagnostic purposes. Other techniques used for the identification of BLV in the cultured PBMCs (reviewed in [1]) are the syncytia infectivity assay [57,58], the radioimmunoassay, immunoperoxidase assay [59,60], and ELISA. From these, only the detection of BLV antigen by ELISA has, nowadays, practical application for assessing the cattle with higher risk of transmitting the infection [61,62]. The selective culling of cattle on the basis of antigen expression in PBMCs cultures, as determined by ELISA, could aid in the control of BLV infection in heavily infected herds [63].

The sheep bioassay has proven to be the most sensitive test to detect infective BLV, however, due to the time required and cost, is restricted to the research area. In the sheep bioassay, blood or any other fluid such as milk, colostrum, and urine is inoculated in lambs, and infection is confirmed by seroconversion (detection of anti-gp51 antibodies) and reisolation of the virus from PBMCs. This test has been used to determine the level of infectivity of different body fluids and to demonstrate protection in vaccination trials [64–66]. This method is also used to confirm infection when conflicting results from serological and PCR methods arise [67,68].

Serological methods for the detection of specific antibodies in serum or milk are among the most practical, economical, and widely used approaches to the diagnosis of BLV infection in cattle older than 6–9 months of age. Serological tests developed early after the discovery of BLV including the agar gel immunodiffusion (AGID) test and the radioimmunoprecipitation (RIP) tests using gp51 and p24 as antigens were followed by several ELISAs in more recent years. The AGID test was rapidly adopted and widely used because of its simplicity and specificity. Extensive comparative studies have shown that the RIP test using gp51 as antigen was the most sensitive test for the identification of infected cattle, followed by RIP using p24 as antigen [69,70]. False-negative AGID results are frequent in samples from cows in the periparturient period or in those cattle that develop low titers of antibodies. Several indirect and blocking ELISAs have been developed and are commercially available in many countries [68,71,72]. Antibodies against the envelope gp51 usually reach higher titers compared to the p24 antigen; hence, ELISA tests using gp51 as antigen are preferred. Due to the relatively expensive equipment and hazards regarding the use of radioactive isotopes, the RIP has been gradually replaced by the gp51-ELISAs. The ELISA has the additional advantage that can be used not only in serum or plasma but also in individual or pooled milk. As antibody titers in milk are significantly reduced compared to serum, the sensitivity of the ELISA to be used in pooled milk samples is critical. Herds with infection rates as low as 2.5%–5% gave positive result in the pooled milk ELISA [68,73]. Although most of the ELISAs are significantly more sensitive compared to the AGID, both the ELISA and AGID are prescribed for international trade by the World Organization for Animal Health [74]. Anti-BLV antibodies passively acquired by calves upon ingestion of colostrum persist until the age of 6–9 months. During this period the only way to detect the infection is by a direct method, such as PCR.

18.1.4.2 Molecular Techniques

The detection of the integrated BLV provirus by PCR is generally used as a complement to serology for confirmatory testing. The main applications of PCR are (1) diagnosing infection in calves below 6–9 months of age due to the interference of colostral antibodies with serological methods, (2) early diagnosing before specific antibodies are developed, (3) clarifying inconsistent or doubtful results obtained by serological tests, (4) differentiating between clinical cases of sporadic and enzootic lymphomas, and (5) confirming the etiology of tumor cases from material obtained at slaughterhouses. PCR has also been used as an alternative to the sheep bioassay to ensure that cattle used in the production of whole blood vaccines or cell lines are free from BLV.

Many variants of the PCR method including simple amplification [75–77], nested PCR [68,78], ELISA-PCR [79,80], *in situ* PCR [81], and loop-mediated isothermal amplification [82] have been developed to detect or semiquantify the BLV provirus. It is well established that, for diagnostic purposes, in order to achieve an acceptable level of sensitivity, it is necessary to carry out the PCR in a nested format. Variable efficiencies in the detection of the provirus by PCR have been observed, depending on the target gene and primers used. The best results have been obtained with primers targeting the *pol* and *env* regions [78,83–85], due to high sequence conservation between strains. Over the last decade, real-time PCRs (qPCRs) using TaqMan chemistry have been developed to quantify BLV provirus [83–86].

The PCR is highly sensitive for the diagnosis of BLV infection when compared to ELISA, mainly because of the detection by PCR of infected cattle at the initial stage of infection, before seroconversion [75,78,87]. However, the possibility of BLV infection should not be excluded by a negative PCR result in PBMCs alone. Negative PCR results in serologically positive cattle have been reported [68,78,88–90]. This situation may be due, at least in some cases, to an extremely low proviral load in the peripheral blood in genetically resistant infected cattle [38]. Conversely, cattle carrying the provirus without a detectable serological response of antibodies have also been reported; however, there is no agreement if this situation is associated or not with a particular genotype of the virus [91–93].

BLV-infected cattle are more consistently detected by PCR from the blood compared to milk leukocytes [84,94]. Lew et al. [83] further showed that the sensitivity and specificity of the PCR were improved when the DNA was extracted from purified PBMCs (compared to whole blood) by using a commercial kit (compared to a conventional DNA extraction method). The extraction of DNA from PBMCs by commercial column-based methods is specially recommended for real-time PCR.

18.2 METHODS

18.2.1 Sample Collection and Preparation

The peripheral blood is obtained by jugular or coccygeal venipuncture in individual syringes either with sodium heparin (5 U/mL) or with sodium citrate (4% final concentration). PBMCs may be obtained using a ficoll–diatrizoate gradient (Ficoll-Paque from GE Healthcare). An alternative and economical procedure is used to obtain total leucocytes from the blood, with similar results for PCR. The buffy coat layer obtained from 10 mL of each blood sample, or 3 mL of whole blood, is mixed for 1 min with 11 mL of cold ammonium chloride buffer (150 mM NH_4Cl, 8 mM Na_2CO_3, 6 mM EDTA) to lyse the erythrocytes. After centrifugation ($1000 \times g$ for 7 min at 4°C), the leukocytes are washed once with phosphate-buffered saline and may be stored as a pellet at −20°C or used immediately for DNA extraction.

DNA extraction for conventional PCR can be performed by a phenol–chloroform procedure, followed by ethanol precipitation, as described in [95]. For real-time PCR, extraction of DNA by the use of high-quality commercial kits (illustra blood genomicPrep Mini Spin Kit, GE Healthcare; DNeasy Blood and Tissue Kit, Qiagen) is recommended. When using the illustra blood genomicPrep Mini Spin Kit from GE Healthcare, we obtain the peripheral blood leukocytes from 3 mL of blood as explained earlier and follow the instructions of the manufacturer (protocol for 1 mL of blood with the following modifications: volume of Proteinase K and lysis buffer and time of incubation at the stage of lysis are doubled). Quantification of DNA can be carried out by measuring the absorbance at 260 nm. Purified genomic DNA is stored at −20°C.

18.2.2 Detection Procedures

The nested PCR method described by Fechner et al. [78] has been reproduced in many laboratories and is one of the procedures recommended by the World Organization for Animal Health for the international trade of cattle. However, nested PCR is laborious and time consuming, and there is always the risk of contamination by amplicon carryover. The introduction in recent years of equipment and reagents that enable to follow the reaction in real time has favored the application of the real-time PCR for the detection and quantification of pathogens. A qPCR employing SYBR Green® chemistry, which allows the detection and quantification BLV *pol* sequences, is also described.

18.2.2.1 Nested PCR for the Detection of BLV *env* Sequences [78]

The method described is based on primer sequences matching the *env* gene of BLV, according to the sequence available from GenBank accession no. K02120 (Table 18.1). This region is highly conserved among BLV isolates. In order to avoid contamination with amplicons, which is a serious problem in nested PCR, special precautions should be adopted. DNA isolation and preparation of amplification mixtures should be carried out in a separate room and with different micropipettes from that used in postamplification procedures. When possible, amplification mixtures are prepared under a biological cabinet that is previously irradiated with UV light.

18.2.2.1.1 Procedure

1. Dilute DNA samples to 25 ng/µL, so as to have 500 ng in 20 µL.
2. Prepare the reaction mix for the first round as follows: Before pipetting, vortex and spin down all the reagents. Volumes are indicated for a 50 µL reaction, but they can be scaled down to a 20 µL reaction. One negative control (DNA from a BLV-free bovine), one positive control, and no template control (double-distilled water) should be added in each experiment. Calculate the final volume of the reaction mix by multiplying the indicated volumes by the total number of samples including controls, plus one. External primers used for the first round of PCR are env-1F and env-2R (Table 18.2).
3. Add 30 µL of the PCR mixture and 20 µL of each DNA sample or control to each tube.
4. Spin the sample tubes in a microcentrifuge at 14000 × g for 30 s at room temperature.
5. Place the tubes in the thermal cycler. Set the program for the first round of PCR in the thermal cycler: initial incubation at 94°C for 2 min; 40 cycles of 95°C for 30 s, 62°C for 30 s, and 72°C for 1 min; and final extension at 72°C for 4 min.

6. Prepare the reaction mix for the second round of PCR. The PCR mixture is prepared with the same components and quantities indicated for the first round, except that the volume of distilled water is augmented to 34.75 µL per reaction. Internal primers used for the second round are env-3F and env-4R.
7. Add 47 µL of the PCR mixture and 3 µL of the product of amplification from the first round of PCR to each tube. Spin the tubes and place them in the thermal cycler.
8. Set the program for the second round of PCR in the thermal cycler: initial incubation at 94°C for 2 min; 40 cycles of 95°C for 30 s, 70°C for 30 s, and 72°C for 1 min; and final extension at 72°C for 4 min.
9. Spin down the PCR tubes and load 10 µL of each PCR product with 10 µL of 2× loading buffer in the wells of a 1.5% agarose gel. Load 10 µL of a 100 bp ladder to check the size of the amplification products.
10. Analyze the PCR products by electrophoresis in tris–borate–EDTA buffer. Run the gel at 100 V for approximately 30 min to 1 h or until the dye has reached the limit of the gel.
11. Stain the gel by placing it in a container with ethidium bromide solution at 0.5 µg/mL in agitation for 15–20 min.
12. Visualize the amplification products under UV illumination.
13. Interpretation of results:
    Controls: Check the absence of amplification in the nontemplate and negative controls. An amplification product of the expected size (444 bp) should be observed in the positive control lane.

Samples giving an amplification product of the expected size are considered positive for BLV infection.

Note: Further confirmation of the specificity of the amplification products can be performed by sequencing or by RFLP analysis [91].

18.2.2.2 Real-Time PCR for the Detection of BLV pol Sequences

The protocol described is used to amplify a 59 bp fragment of the pol gene of BLV using primers pol-F and pol-R. The amplification is monitored by incorporation of SYBR Green dye; subsequent melting curve analysis allows checking for specificity. This protocol has been set up in an ABI 7500 Real-Time PCR System (Applied Biosystems) using the 7500 Software v.2.0.6, but should work well in any other similar equipment (Table 18.3).

TABLE 18.1
Sequence and Position of Primers for BLV env Nested PCR

Primer	Sequence (5′–3′)	Nucleotide Position	Amplicon Length (bp)
env-1F	TCTGTGCCAAGTCTCCCAGATA	5032–5053	598
env-2R	AACAACAACCTCTGGGAAGGG	5629–5608	
env-3F	CCCACAAGGGCGGCGCCGGTTT	5099–5121	444
env-4R	GCGAGGCCGGGTCCAGAGCTGG	5542–5521	

TABLE 18.2
Composition of the PCR Mixture

Component	Volume (µL)	Final Concentration
PCR buffer (10×)	5	1×
dNTPs (100 mM)	1.5	300 µM
MgCl₂ (25 mM)	3	1.5 mM
Primer F (20 pmol/µL)	1.25	0.2 µM
Primer R (20 pmol/µL)	1.25	0.2 µM
Taq DNA polymerase (5 U/µL)	0.25	1.25 U/50 µL reaction
Double-distilled water	17.75	
Total	30	

TABLE 18.3
Sequence of Primers for BLV pol Real-Time PCR

Primer	Sequence (5′–3′)	Nucleotide Position
pol-F	CACCATTCACCCCCACTTG	4618–4636
pol-R	TCAGAGCCCTTGGGTGTTTC	4657–4676

18.2.2.2.1 Preparation of DNA from Samples and Standard Curve

DNA from samples should be adjusted to a concentration of 10 ng/μL.

To prepare the standard (calibration) curve used to quantify BLV copy number, DNA is extracted from the fetal lamb kidney (FLK) cell line [96], which harbors 4 proviral copies per cell. DNA extracted from Vero cell line is used as a source of BLV-negative DNA to compensate the DNA input. A dilution containing 10,000 BLV copies in 3 μL is prepared by mixing 500 ng of DNA extracted from FLK cells with 500 ng of DNA from Vero cells in a final volume of 100 μL of UltraPure DNase/RNase-Free Distilled Water. Four serial 10-fold dilutions are prepared in DNA from Vero cells at a concentration of 10 ng/μL, so as to obtain standards with 1000, 100, 10, and 1 BLV proviral copies in 3 μL.

18.2.2.2.1.1 Procedure

1. Calculate the amount of reactions to be done. All reactions should be carried out in triplicate. For quantification of copy number, the whole standard curve (from 10,000 to 1 BLV copies) should be run. A negative control (DNA from BLV-free cattle) and a nontemplate control are always included.
2. Thaw all the solutions and reagents, spin them down and mix by pipetting up and down, and then store on ice.
3. Prepare the PCR mixture in a 2 mL tube on ice by adding the following components. Volumes are indicated for a 20 μL reaction. Multiply the indicated volumes by the number of reactions to be run, plus one (Table 18.4).
4. Mix the solution carefully by pipetting up and down. Add 17 μL of the mixture in each of the PCR optical tubes or wells of the optical microplate.
5. Add 3 μL of template DNA (10 ng/μL). UltraPure water is added in nontemplate control tubes/wells. Mix carefully by pipetting up and down.
6. Seal tubes with optical tube caps or microplates with optical adhesive foil. If possible, centrifuge microplates or tubes at $1000 \times g$ for 1 min.
7. Program the real-time PCR equipment following the operator's manual of the supplier with the following cycling conditions: initial incubation at 50°C for 2 min and 95°C for 10 min, followed by 40 cycles of 95°C

for 15 s and 60°C for 1 min. Melting curves are performed after amplification, in order to test specificity.
8. Place the tubes or microplate in the instrument and start the reaction.

18.2.2.2.2 Analysis and Interpretation of Results

Examine the amplification plot (ΔRn vs. cycle) and correct baseline and threshold values. Examine the standard curve (C_t vs. log concentration of template); the slope of the regression line should be near −3.32, which indicates 100% amplification efficiency. Efficiency values between 90% and 110% are acceptable. A value of $R^2 > 0.99$ is desirable. It may be necessary to omit any well from the analysis that differs significantly from the average of associated replicate wells to fit these requirements. Examine melting curves of amplified products; a peak at 76.8°C ± 0.1°C is indicative of specific amplification.

Following this methodology, mean Cq values of 24.3, 27.9, 31.2, and 33.8 are obtained for the standards with 10,000, 1,000, 100, and 10 proviral copies. Samples giving a Cq value >36 are considered negative by this qPCR. Calculation of proviral copy number in positive samples is carried out by interpolation within the calibration curve. Samples giving Cq values that fall out of the dynamic range (i.e., below the Cq obtained in the standard with 10,000 proviral copies) should be diluted in BLV-negative DNA at a concentration of 10 ng/μL and retested to determine their copy number.

18.3 CONCLUSIONS AND FUTURE PERSPECTIVES

BLV infection is an important veterinary problem in those countries that have not implemented control or eradication programs, like the United States, Canada, and majority of the countries in South America. In this situation, the virus has spread, and nearly 80% of the dairy herds are infected, with high infection rates in most cases.

Molecular and serological methods developed to diagnose BLV infection have been instrumental for the identification and characterization of infected cattle capable of maintaining very low levels of proviral load [37]. These cattle, which carry the BoLA DRB3.2*0902 marker, are believed to be naturally resistant to the spread of the virus and not capable of transmitting BLV under normal dairy farm conditions [38]. The study of the basis of this natural resistance to BLV spread is of paramount importance not only to fully understand the virus–host interplay but also to delineate preventive or therapeutic strategies to control this pathogen.

Given the wide spread of the virus in many countries where the infection has not been early controlled, the classical control program (test and eliminate) is not effective, due to an unfavorable cost/benefit relationship. A control and eradication program based on the marker-assisted selection of resistant cattle has been recently proposed. It appears that this strategy, which is at the experimental level, would be economically feasible for the control of BLV infection in heavily infected herds [97].

TABLE 18.4
Real-Time PCR Mixture

Component	Volume (μL)	Final Concentration
SYBR Green Master Mix (Rox) (2×)	10	1×
Primer pol-F (10 μM)	0.6	0.3 μM
Primer pol-R (10 μM)	0.6	0.3 μM
UltraPure DNase-Free Distilled Water	5.8	
Total	17	

A potential therapeutic treatment has been proposed, based on the fact that inhibition of histone deacetylases induces the expression in latently infected cells. This approach is also founded on the premise that the lack of viral expression protects infected cells from immune clearance and that once expression is induced, the immune system should eliminate infected cells. This therapy has proven to be efficient in tumor regression in the sheep model, but its efficacy in infected cows is presently unknown [98].

REFERENCES

1. Ferrer, J.F., Bovine lymphosarcoma. *Adv Vet Sci Comp Med*, 1980. 24: 1–68.
2. Maclachlan, N.J. and E.J. Dubovi, Chapter 14: Retroviridae, in *Fenner's Veterinary Virology*, A. Press, ed., 2011, Elsevier Science, Amsterdam, the Netherlands, pp. 243–274.
3. Barbeau, B. et al., *Conference Highlights of the 16th International Conference on Human Retrovirology: HTLV and Related Retroviruses*, Montreal, Quebec, Canada, June 26–30, 2013; *Retrovirology*, 11: 19.
4. Aida, Y. et al., Mechanisms of pathogenesis induced by bovine leukemia virus as a model for human T-cell leukemia virus. *Front Microbiol*, 2013. 4: 328.
5. Rodriguez, S.M. et al., Preventive and therapeutic strategies for bovine leukemia virus: Lessons for HTLV. *Viruses*, 2011. 3(7): 1210–1248.
6. ICTV (International Committee on Taxonomy of Viruses). 2013 Release, EC45, Edinburg. July 2013 (Available from: http://www.ictvonline.org/virusTaxonomy.asp. Accessed on September 8, 2014.).
7. Voneche, V. et al., Fusogenic segments of bovine leukemia virus and simian immunodeficiency virus are interchangeable and mediate fusion by means of oblique insertion in the lipid bilayer of their target cells. *Proc Natl Acad Sci USA*, 1992. 89(9): 3810–3814.
8. Gillet, N. et al., Mechanisms of leukemogenesis induced by bovine leukemia virus: Prospects for novel anti-retroviral therapies in human. *Retrovirology*, 2007. 4: 18.
9. Vogt, V.M., Retroviral virions and genomes, in *Retroviruses*, J.M. Coffin, Hughes, S.H., and Varmus, H.E., eds., 1997, Cold Spring Harbor Laboratory Press, New York, pp. 27–69.
10. Willems, L. et al., Genetic determinants of bovine leukemia virus pathogenesis. *AIDS Res Hum Retroviruses*, 2000. 16(16): 1787–1795.
11. Dube, S. et al., The complete genomic sequence of an in vivo low replicating BLV strain. *Virol J*, 2009. 6: 120.
12. Rodriguez, S.M. et al., Bovine leukemia virus can be classified into seven genotypes: Evidence for the existence of two novel clades. *J Gen Virol*, 2009. 90(Pt 11): 2788–2797.
13. Paul, P.S. et al., Evidence for the replication of bovine leukemia virus in the B lymphocytes. *Am J Vet Res*, 1977. 38(6): 873–876.
14. Mirsky, M.L. et al., The prevalence of proviral bovine leukemia virus in peripheral blood mononuclear cells at two subclinical stages of infection. *J Virol*, 1996. 70(4): 2178–2183.
15. Domenech, A. et al., In vitro infection of cells of the monocytic/macrophage lineage with bovine leukaemia virus. *J Gen Virol*, 2000. 81(Pt 1): 109–118.
16. Schwartz, I. et al., In vivo leukocyte tropism of bovine leukemia virus in sheep and cattle. *J Virol*, 1994. 68(7): 4589–4596.
17. Mirsky, M.L., Y. Da, and H.A. Lewin, Detection of bovine leukemia virus proviral DNA in individual cells. *PCR Methods Appl*, 1993. 2(4): 333–340.
18. Meirom, R., J. Brenner, and Z. Trainin, BLV-infected lymphocytes exhibit two patterns of expression as determined by Ig and CD5 markers. *Vet Immunol Immunopathol*, 1993. 36(2): 179–186.
19. Depelchin, A. et al., Bovine leukemia virus (BLV)-infected B-cells express a marker similar to the CD5 T cell marker. *Immunol Lett*, 1989. 20(1): 69–76.
20. Florins, A. et al., Cell dynamics and immune response to BLV infection: A unifying model. *Front Biosci*, 2007. 12: 1520–1531.
21. Kettmann, R. et al., Genomic integration of bovine leukemia provirus and lack of viral RNA expression in the target cells of cattle with different responses to BLV infection. *Leuk Res*, 1980. 4(6): 509–519.
22. Haas, L., T. Divers, and J.W. Casey, Bovine leukemia virus gene expression in vivo. *J Virol*, 1992. 66(10): 6223–6225.
23. Jensen, W.A., J. Rovnak, and G.L. Cockerell, In vivo transcription of the bovine leukemia virus tax/rex region in normal and neoplastic lymphocytes of cattle and sheep. *J Virol*, 1991. 65(5): 2484–2490.
24. Rovnak, J. and J.W. Casey, Assessment of bovine leukemia virus transcripts in vivo. *J Virol*, 1999. 73(10): 8890–8897.
25. Alexandersen, S. et al., Identification of alternatively spliced mRNAs encoding potential new regulatory proteins in cattle infected with bovine leukemia virus. *J Virol*, 1993. 67(1): 39–52.
26. Gupta, P. and J.F. Ferrer, Expression of bovine leukemia virus genome is blocked by a nonimmunoglobulin protein in plasma from infected cattle. *Science*, 1982. 215(4531): 405–407.
27. Gupta, P., S.V. Kashmiri, and J.F. Ferrer, Transcriptional control of the bovine leukemia virus genome: Role and characterization of a non-immunoglobulin plasma protein from bovine leukemia virus-infected cattle. *J Virol*, 1984. 50(1): 267–270.
28. van den Heuvel, M.J., B.J. Jefferson, and R.M. Jacobs, Purified bovine plasma blocking factor decreases Bovine leukemia virus p24 expression while increasing protein synthesis and transcriptional activity of peripheral blood mononuclear cells in short-term culture. *Can J Vet Res*, 2005. 69(3): 186–192.
29. van den Heuvel, M.J., B.J. Jefferson, and R.M. Jacobs, Isolation of a bovine plasma fibronectin-containing complex which inhibits the expression of bovine leukemia virus p24. *J Virol*, 2005. 79(13): 8164–8170.
30. Schwartz, I. and D. Levy, Pathobiology of bovine leukemia virus. *Vet Res*, 1994. 25(6): 521–536.
31. Derse, D. and L. Martarano, Construction of a recombinant bovine leukemia virus vector for analysis of virus infectivity. *J Virol*, 1990. 64(1): 401–405.
32. Lassauzet, M.L. et al., Factors associated with in utero or periparturient transmission of bovine leukemia virus in calves on a California dairy. *Can J Vet Res*, 1991. 55(3): 264–268.
33. Hopkins, S.G. and R.F. DiGiacomo, Natural transmission of bovine leukemia virus in dairy and beef cattle. *Vet Clin North Am Food Anim Pract*, 1997. 13(1): 107–128.
34. Dimmock, C.K., Y.S. Chung, and A.R. MacKenzie, Factors affecting the natural transmission of bovine leukaemia virus infection in Queensland dairy herds. *Aust Vet J*, 1991. 68(7): 230–233.
35. Xu, A. et al., Polymorphism in BoLA-DRB3 exon 2 correlates with resistance to persistent lymphocytosis caused by bovine leukemia virus. *J Immunol*, 1993. 151(12): 6977–6985.

36. Zanotti, M. et al., Association of BoLA class II haplotypes with subclinical progression of bovine leukaemia virus infection in Holstein-Friesian cattle. *Anim Genet*, 1996. 27(5): 337–341.

37. Juliarena, M.A., S.E. Gutierrez, and C. Ceriani, Determination of proviral load in bovine leukemia virus-infected cattle with and without lymphocytosis. *Am J Vet Res*, 2007. 68(11): 1220–1225.

38. Juliarena, M.A. et al., Association of BLV infection profiles with alleles of the BoLA-DRB3.2 gene. *Anim Genet*, 2008. 39(4): 432–438.

39. Konnai, S. et al., Tumor necrosis factor-alpha genetic polymorphism may contribute to progression of bovine leukemia virus-infection. *Microbes Infect*, 2006. 8(8): 2163–2171.

40. Kabeya, H. et al., Up-regulation of tumor necrosis factor alpha mRNA is associated with bovine-leukemia virus (BLV) elimination in the early phase of infection. *Vet Immunol Immunopathol*, 1999. 68(2–4): 255–265.

41. Radostits, O. et al., Chapter 21: Disease associated with viruses and chlamydia I, in *Veterinary Medicine*, O.M. Radostits, and Done, S.H., eds., 2007, Elsevier, New York, pp. 1209–1221.

42. Debacq, C. et al., Reduced cell turnover in bovine leukemia virus-infected, persistently lymphocytotic cattle. *J Virol*, 2003. 77(24): 13073–1383.

43. Konnai, S. et al., Tumor necrosis factor-alpha up-regulation in spontaneously proliferating cells derived from bovine leukemia virus-infected cattle. *Arch Virol*, 2006. 151(2): 347–360.

44. Konnai, S. et al., Imbalance of tumor necrosis factor receptors during progression in bovine leukemia virus infection. *Virology*, 2005. 339(2): 239–248.

45. Trueblood, E.S. et al., B-lymphocyte proliferation during bovine leukemia virus-induced persistent lymphocytosis is enhanced by T-lymphocyte-derived interleukin-2. *J Virol*, 1998. 72(4): 3169–3177.

46. Twizere, J.C. et al., Interaction of retroviral Tax oncoproteins with tristetraprolin and regulation of tumor necrosis factor-alpha expression. *J Natl Cancer Inst*, 2003. 95(24): 1846–1859.

47. Dequiedt, F. et al., Mutations in the p53 tumor-suppressor gene are frequently associated with bovine leukemia virus-induced leukemogenesis in cattle but not in sheep. *Virology*, 1995. 209(2): 676–683.

48. Ishiguro, N. et al., p53 mutation as a potential cellular factor for tumor development in enzootic bovine leukosis. *Vet Immunol Immunopathol*, 1997. 55(4): 351–358.

49. Zhuang, W. et al., Point mutation of p53 tumor suppressor gene in bovine leukemia virus-induced lymphosarcoma. *Leukemia*, 1997. 11(Suppl. 3): 344–346.

50. Reyes, R.A. and G.L. Cockerell, Increased ratio of bcl-2/bax expression is associated with bovine leukemia virus-induced leukemogenesis in cattle. *Virology*, 1998. 242(1): 184–192.

51. Willems, L. et al., Cooperation between bovine leukaemia virus transactivator protein and Ha-ras oncogene product in cellular transformation. *EMBO J*, 1990. 9(5): 1577–1581.

52. Kerkhofs, P. et al., In vitro and in vivo oncogenic potential of bovine leukemia virus G4 protein. *J Virol*, 1998. 72(3): 2554–2559.

53. Szynal, M. et al., Disruption of B-cell homeostatic control mediated by the BLV-Tax oncoprotein: Association with the upregulation of Bcl-2 and signaling through NF-kappaB. *Oncogene*, 2003. 22(29): 4531–4542.

54. Philpott, S.M. and G.C. Buehring, Defective DNA repair in cells with human T-cell leukemia/bovine leukemia viruses: Role of tax gene. *J Natl Cancer Inst*, 1999. 91(11): 933–942.

55. Arainga, M., E. Takeda, and Y. Aida, Identification of bovine leukemia virus tax function associated with host cell transcription, signaling, stress response and immune response pathway by microarray-based gene expression analysis. *BMC Genomics*, 2012. 13: 121.

56. Klener, P. et al., Insights into gene expression changes impacting B-cell transformation: Cross-species microarray analysis of bovine leukemia virus tax-responsive genes in ovine B cells. *J Virol*, 2006. 80(4): 1922–1938.

57. Ferrer, J.F. and C.A. Diglio, Development of an in vitro infectivity assay for the C-type bovine leukemia virus. *Cancer Res*, 1976. 36(3): 1068–1073.

58. Ferrer, J.F., C. Cabradilla, and P. Gupta, Use of a feline cell line in the syncytia infectivity assay for the detection of bovine leukemia virus infection in cattle. *Am J Vet Res*, 1981. 42(1): 9–14.

59. Esteban, E.N., R.M. Thorn, and J.F. Ferrer, An amplified immunoperoxidase assay to detect bovine leukemia virus expression: Development and comparison with other assays. *Cancer Res*, 1985. 45(7): 3231–3235.

60. Jerabek, L., P. Gupta, and J.F. Ferrer, An infectivity assay for bovine leukemia virus using the immunoperoxidase technique. *Cancer Res*, 1979. 39(10): 3952–3954.

61. Cowley, J.A. et al., Infectivity of bovine leukaemia virus infected cattle: An ELISA for detecting antigens expressed in in vitro cultured lymphocytes. *Vet Microbiol*, 1992. 30(2–3): 137–150.

62. Miller, L.D. et al., Blood from bovine leukemia virus-infected cattle: Antigen production correlated with infectivity. *Am J Vet Res*, 1985. 46(4): 808–810.

63. Molloy, J.B. et al., Control of bovine leukaemia virus transmission by selective culling of infected cattle on the basis of viral antigen expression in lymphocyte cultures. *Vet Microbiol*, 1994. 39(3–4): 323–333.

64. Miller, J.M., M.J. Van der Maaten, and M.J. Schmerr, Vaccination of cattle with binary ethylenimine-treated bovine leukemia virus. *Am J Vet Res*, 1983. 44(1): 64–67.

65. Miller, J.M. and M.J. Van der Maaten, Infectivity tests of secretions and excretions from cattle infected with bovine leukemia virus. *J Natl Cancer Inst*, 1979. 62(2): 425–428.

66. Kanno, T. et al., Effect of freezing treatment on colostrum to prevent the transmission of bovine leukemia virus. *J Vet Med Sci*, 2013. 76(2): 255–257.

67. Eaves, F.W. et al., A field evaluation of the polymerase chain reaction procedure for the detection of bovine leukaemia virus proviral DNA in cattle. *Vet Microbiol*, 1994. 39(3–4): 313–321.

68. Gutierrez, S.E. et al., Development and evaluation of a highly sensitive and specific blocking enzyme-linked immunosorbent assay and polymerase chain reaction assay for diagnosis of bovine leukemia virus infection in cattle. *Am J Vet Res*, 2001. 62(10): 1571–1577.

69. Miller, J.M., M.J. Schmerr, and M.J. Van Der Maaten, Comparison of four serologic tests for the detection of antibodies to bovine leukemia virus. *Am J Vet Res*, 1981. 42(1): 5–8.

70. Gupta, P. and J.F. Ferrer, Comparison of various serological and direct methods for the diagnosis of BLV infection in cattle. *Int J Cancer*, 1981. 28(2): 179–184.

71. Have, P. and R. Hoff-Jorgensen, Demonstration of antibodies against bovine leukemia virus (BLV) by blocking ELISA using bovine polyclonal anti-BLV immunoglobulin. *Vet Microbiol*, 1991. 27(3–4): 221–229.

72. Portetelle, D., M. Mammerickx, and A. Burny, Use of two monoclonal antibodies in an ELISA test for the detection of antibodies to bovine leukaemia virus envelope protein gp51. *J Virol Methods*, 1989. 23(2): 211–222.

73. Klintevall, K. et al., Evaluation of an indirect ELISA for the detection of antibodies to bovine leukaemia virus in milk and serum. *J Virol Methods*, 1991. 33(3): 319–333.

74. OIE (Office International des Epizooties), *OIE Terrestrial Manual 2012.* Enzootic bovine leukosis. World Organisation for Animal Health, Paris, France, pp. 1–11.

75. Beier, D., P. Blankenstein, and H. Fechner, Possibilities and limitations for use of the polymerase chain reaction (PCR) in the diagnosis of bovine leukemia virus (BLV) infection in cattle. *Dtsch Tierarztl Wochenschr*, 1998. 105(11): 408–412.

76. Beier, D. et al., Identification of different BLV provirus isolates by PCR, RFLPA and DNA sequencing. *Berl Munch Tierarztl Wochenschr*, 2001. 114(7–8): 252–256.

77. Dube, S. et al., Degenerate and specific PCR assays for the detection of bovine leukaemia virus and primate T cell leukaemia/lymphoma virus pol DNA and RNA: Phylogenetic comparisons of amplified sequences from cattle and primates from around the world. *J Gen Virol*, 1997. 78(Pt 6): 1389–1398.

78. Fechner, H. et al., Evaluation of polymerase chain reaction (PCR) application in diagnosis of bovine leukaemia virus (BLV) infection in naturally infected cattle. *J Vet Med B*, 1996. 43(10): 621–630.

79. Naif, H.M. et al., Early detection of bovine leukemia virus by using an enzyme-linked assay for polymerase chain reaction-amplified proviral DNA in experimentally infected cattle. *J Clin Microbiol*, 1992. 30(3): 675–679.

80. Rola, M. and J. Kuzmak, The detection of bovine leukemia virus proviral DNA by PCR-ELISA. *J Virol Methods*, 2002. 99(1–2): 33–40.

81. Duncan, R.B., Jr., W.K. Scarratt, and G.C. Buehring, Detection of bovine leukemia virus by in situ polymerase chain reaction in tissues from a heifer diagnosed with sporadic thymic lymphosarcoma. *J Vet Diagn Invest*, 2005. 17(2): 190–194.

82. Komiyama, C. et al., Development of loop-mediated isothermal amplification method for diagnosis of bovine leukemia virus infection. *J Virol Methods*, 2009. 157(2): 175–179.

83. Lew, A.E. et al., Sensitive and specific detection of proviral bovine leukemia virus by 5′ Taq nuclease PCR using a 3′ minor groove binder fluorogenic probe. *J Virol Methods*, 2004. 115(2): 167–175.

84. Kuckleburg, C.J. et al., Detection of bovine leukemia virus in blood and milk by nested and real-time polymerase chain reactions. *J Vet Diagn Invest*, 2003. 15(1): 72–76.

85. Rola-Luszczak, M. et al., Development of an improved real time PCR for the detection of bovine leukaemia provirus nucleic acid and its use in the clarification of inconclusive serological test results. *J Virol Methods*, 2013. 189(2): 258–264.

86. Heenemann, K. et al., Development of a Bovine leukemia virus polymerase gene-based real-time polymerase chain reaction and comparison with an envelope gene-based assay. *J Vet Diagn Invest*, 2012. 24(4): 649–655.

87. Klintevall, K. et al., Bovine leukaemia virus: Rapid detection of proviral DNA by nested PCR in blood and organs of experimentally infected calves. *Vet Microbiol*, 1994. 42(2–3): 191–204.

88. Jacobs, R.M. et al., Proviral detection and serology in bovine leukemia virus-exposed normal cattle and cattle with lymphoma. *Can J Vet Res*, 1992. 56(4): 339–348.

89. Murtaugh, M.P. et al., Detection of bovine leukemia virus in cattle by the polymerase chain reaction. *J Virol Methods*, 1991. 33(1–2): 73–85.

90. Reichel, M.P. et al., Evaluation of alternative methods for the detection of bovine leukaemia virus in cattle. *NZ Vet J*, 1998. 46(4): 140–146.

91. Fechner, H. et al., Provirus variants of the bovine leukemia virus and their relation to the serological status of naturally infected cattle. *Virology*, 1997. 237(2): 261–269.

92. Licursi, M. et al., Genetic heterogeneity among bovine leukemia virus genotypes and its relation to humoral responses in hosts. *Virus Res*, 2002. 86(1–2): 101–110.

93. Monti, G., R. Schrijver, and D. Beier, Genetic diversity and spread of bovine leukaemia virus isolates in Argentine dairy cattle. *Arch Virol*, 2005. 150(3): 443–458.

94. Martin, D. et al., Comparative study of PCR as a direct assay and ELISA and AGID as indirect assays for the detection of bovine leukaemia virus. *J Vet Med B Infect Dis Vet Public Health*, 2001. 48(2): 97–106.

95. Maniatis, T.F., Fritsch, E., and Sambrook, J., *Molecular Cloning: A Laboratory Manual*, 1982, Cold Spring Harbor Laboratory Press, Cold Spring Harbor, NY.

96. Van Der Maaten, M.J. and J.M. Miller, Replication of bovine leukemia virus in monolayer cell cultures. *Bibl Haematol*, 1975(43): 360–362.

97. Esteban, E.N. et al., Bovine leukemia virus (BLV), proposed control and eradication programs by marker assisted breeding of genetically resistant cattle, in *Animal Genetics*, L.J. Rechi, ed., 2009, Nova Science Publishers, Inc., Hauppauge, NY, pp. 107–130.

98. Achachi, A. et al., Valproate activates bovine leukemia virus gene expression, triggers apoptosis, and induces leukemia/lymphoma regression in vivo. *Proc Natl Acad Sci USA*, 2005. 102(29): 10309–10314.

19 Caprine Arthritis–Encephalitis Virus and Visna–Maedi Virus

Nuria Barquero, Ana Domenech, and Esperanza Gomez-Lucia

CONTENTS

19.1 Introduction 167
    19.1.1 Classification 167
    19.1.2 Morphology, Biology, and Epidemiology 167
    19.1.3 Pathogenesis and Clinical Features 168
        19.1.3.1 Pathogenesis 168
        19.1.3.2 Clinical forms 169
    19.1.4 Diagnosis 169
        19.1.4.1 Conventional Diagnosis 169
        19.1.4.2 Molecular Techniques 170
19.2 Methods 171
    19.2.1 Sample Collection and Preparation 171
        19.2.1.1 Milk Samples 171
        19.2.1.2 Blood Samples 171
        19.2.1.3 DNA Extraction 172
    19.2.2 PCR Detection 172
        19.2.2.1 Procedure 172
    19.2.3 SRLV Cloning, Sequence Analysis, and Genotype Identification 173
        19.2.3.1 Procedure 173
19.3 Conclusion 173
References 173

19.1 INTRODUCTION

Visna–maedi virus (VMV, also known as maedi–visna virus [MVV] or ovine lentivirus) and caprine arthritis–encephalitis virus (CAEV) are retroviruses that affect small ruminants (sheep and goats). They are grouped together as small ruminant lentiviruses (SRLVs), due to their genomic and antigenic similarities [1].

VMV was first isolated by Sigurdardottir and Thormar [2]. The infection by this virus causes a chronic and generally subclinical disease called visna–maedi (VM). Clinical symptoms can present in the form of respiratory, nervous, joint, and/or mammary signs. Maedi refers to the respiratory signs (interstitial pneumonia) and visna alludes to the nervous form (progressive inflammatory disease of the central nervous system).

CAEV was first isolated by Crawford and colleagues [3] from goats showing joint and nervous clinical signs, from which the disease gets its name (caprine arthritis–encephalitis [CAE]). The clinical form associated with the infection is usually arthritis; however, any of the other clinical forms can be present in affected animals.

19.1.1 CLASSIFICATION

VMV and CAEV belong to the genus *Lentivirus*, subfamily *Orthoretrovirinae*, family *Retroviridae*. SRLVs have been classified according to their phylogenetic relationships into five groups: A, B, C, D, and E [4]. The genotype A includes VMV-like strains, genotype B comprises CAEV-like strains, genotype C contains SRLV isolated in Norwegian sheep and goats, genotype D is found in Spanish and Swiss sheep, and finally genotype E is present in Italian goats [5].

19.1.2 MORPHOLOGY, BIOLOGY, AND EPIDEMIOLOGY

SRLV particle is composed of an envelope and a core (Figure 19.1), with a diameter of 90–120 nm. The envelope is formed by a bilayer of phospholipids derived from the host cell and by glycoproteins codified by the gene *env* of the viral genome. These glycoproteins are surface protein (SU) or gp135SU and transmembrane protein (TM) or gp46TM [1]. The SU (gp135SU) is a very glycosylated protein that is attached through noncovalent sites to the hydrophobic end of the TM [6]. Four immunodominant epitopes have been identified in TM, three of which are associated with clinical arthritis [7], and gp46TM has been used successfully in enzyme-linked immunosorbent assay (ELISA) for the detection of antibodies. This protein is responsible for the fusion between the viral envelope and the lipid membrane of the host cell [8]. The core contains the viral RNA and is formed by three proteins codified by the gene *gag*. The capsid (CA)

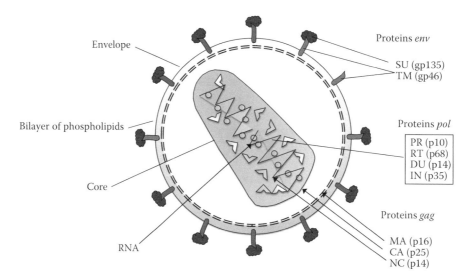

Envelope

Proteins *env*

SU (gp135)
TM (gp46)

Bilayer of phospholipids

Proteins *pol*

PR (p10)
RT (p68)
DU (p14)
IN (p35)

Core

Proteins *gag*

RNA

MA (p16)
CA (p25)
NC (p14)

FIGURE 19.1   Structure of a typical lentivirus.

protein (or p25CA) induces the production of antibodies in the host; thus, it is used in ELISA methods. The other two proteins are matrix protein (MA) or p16MA and nucleocapsid protein (NC) or p14NC. Inside the capsid, there are the reverse transcriptase, dUTPase, integrase, and protease, necessary for the transcription and viral integration (Figure 19.1).

The genomes of VMV and CAEV consist of around 9,000 to 10,000 nucleotides (nt). They are constituted by two independent linear strands of positive-sense RNA and contain three structural genes (*gag*, *pol*, and *env*), which synthesize viral proteins and regulatory accessory genes (*vif*, *rev* and *nef*, and *vpr-like* in some strains). Accessory proteins regulate viral replication. SRLV replicates by producing a DNA intermediate called "provirus" that integrates into the host cell genome. The proviral DNA is flanked by long terminal repeat (LTR) sequences that contain promoters that initiate the transcription of the DNA [1].

SRLVs may infect goats and sheep; both viruses were considered previously to be species specific, but some studies have shown that they can be transmitted between goats and sheep [9]. SRLV infections are widespread worldwide with the exception of Iceland [10]. Considering each virus separately, VMV is spread across the world except in Iceland, Australia, and New Zealand [11]. Seroprevalence is variable between and within countries. Seroprevalence in the United States and Canada was estimated to be 28% of the sheep and 61% of the flocks [12]. Studies in Spain have reported that between 12% and 66% of the animals and 30% and 100% of the flocks are infected, but prevalence varies according to the region [13]. Seroprevalence of CAEV varies between countries from 5.2% to 42% in the goats and from 4% to 73% in the flocks. Studies showed that 31% of goats in the United States [14], 42% in Switzerland [15], 12.5% in Syria [16], 1.9% in Turkey [17], 8% in Brazil [18], and 5.2% in the Sultanate of Oman [19] are infected. In addition, 73% of the flocks in the United States [14], 35% in Brazil [18], and 4% in Italy [20] are affected. SRLV prevalence is higher in developed countries, possibly related to the practice of feeding lambs with a pool of colostrum/milk, which facilitates the spread of the infection [21], and to the intensive farming [22].

Transmission may occur among animals of the same flock or from mother to offspring [23] through body fluids, mainly lung secretions (aerosol droplet infection) [24], and shedding of virus particles, infected macrophages, and epithelial cells in the colostrum and milk [25]. However, it is not clear whether the SRLVs are transmitted between animals as free virions or within the infected cells [1]. Other routes such as venereal and transplacental transmission have not been fully demonstrated [23,26].

Currently, no effective treatment or vaccine has been developed against SRLV infections. Therefore, the most effective means of controlling the infection is an early diagnosis and preventing viral transmission through effective prophylactic measures [1].

19.1.3   Pathogenesis and Clinical Features

19.1.3.1   Pathogenesis

Although SRLV may infect a variety of cells, the major targets are those of the monocyte/macrophage lineage, where viral expression is closely dependent on the differentiation of monocyte to macrophage, and dendritic cells. Other additional cell targets can act as viral reservoirs, including epithelial cells of the mammary gland, which may be correlated with the transmission of the infection through milk [1].

Once the virus has entered the body, there is a short phase of postinfection viremia. In this stage, transitory immunopathological alterations take place. The virus infects macrophages and dendritic cells of the respiratory or intestinal mucosa, depending on the site of entry. The dendritic cells migrate to the lymph nodes where the virus is transferred to the macrophages. When these cells leave the lymph node, they spread the infection to other organs [1]. It is possible that the infected macrophages enter the bone marrow where they can infect myeloid cells or stromal cells. This would cause a continuous production of infected cells, which translates into the lifelong chronic infection [1]. The replication in monocytes and macrophages does not take place until these cells differentiate in the target organs [27] and is thought to depend on specific transcription factors that bind sequences in the LTR, thus triggering replication. This mechanism allows continual SRLV replication in the tissues and the ensuing immune response causes a chronic inflammation that originates the pathological changes observed in target organs of SRLV-infected animals [1].

The main alteration in affected tissues is the infiltration by mononuclear cells (lymphocytes, macrophages, and plasma cells) [1].

The final stage of the pathogenesis begins when the clinical disease starts. Depending on the clinical form, the animal can die in a short time or remain chronically infected. In spite of the high prevalence in some flocks, clinical signs usually are hardly noticeable, as there are many factors that influence the pathogenesis, mostly the viral strain, the age and breed of the animal, the exposure route, the secondary infections, and the management conditions.

19.1.3.2 Clinical forms

SRLV-infected sheep and goats develop a slow progressive inflammatory disease affecting the lungs, nervous system, joints, and/or mammary glands. In general, the process caused by SRLV is subclinical; however, a small percentage of the animals can develop one or more clinical forms associated to the location of the pathological alteration. Progressive wasting and, in some cases, cachexia are signs that may also accompany the main disease. Usually, the respiratory and mammary forms are predominant in VM, whereas the nervous and joint forms are more important in CAE. The disease does not progress uniformly in all individuals and may depend on the infecting virus strain, the host's genetics, and the breed, as some breeds are more likely to develop some clinical signs than others [28].

Respiratory disease in infected animals is characterized by lymphocytic interstitial pneumonia and fibrosis of the lungs. Lungs are enlarged and heavy, up to three times the normal size, uniformly swollen, with rounded edges, rubbery texture, grayish-yellow coloration, and dry surface. The lungs can show well-demarcated pale-gray areas of consolidation. At the histological level, the lesion is characterized by chronic interstitial pneumonia with the infiltration of lymphocytes, monocytes, macrophages, and plasma cells into the interalveolar septum, which contributes to their thickening, and hyperplasia of the perialveolar smooth muscle. The first stage of the disease usually passes unnoticed, but the affected animals are delayed if the flock is moved. As the lesions progress, the clinical signs are dyspnea, abdominal breathing, extension of the neck, dilation of the nostrils, and breathing through the open mouth. Nasal discharge is only present concomitant to the secondary bacterial infections.

The mammary form is characterized by subclinical, indurative, chronic, diffuse, bilateral, nonsuppurative, and nonpainful mastitis, with enlarged mammary lymph nodes. Histologically, it is a chronic interstitial mastitis, characterized by the infiltration of mononuclear cells around the acinus and milk ducts.

Chronic progressive arthritis has a higher incidence in goats than in sheep. The most affected joints are the carpus and/or tarsus. However, other joints, such as the atlantooccipital and coxofemoral joints, may be affected as well [14]. The first clinical sign is arthritis with joint enlargement, with edema and congestion of the synovial membrane and joint capsule, which causes the thickening of both structures [14]. The arthritis is characterized by mononuclear cell infiltration, with villous hypertrophy, angiogenesis, and necrosis of the synovial membrane [1].

The nervous disease is more frequent in goats than in sheep, where it is rarely seen. The disease is characterized by meningoencephalitis, astrocytosis, microgliosis, and focal secondary demyelination in the brain and spinal cord [1]. Affected animals may present lack of coordination, hind limb paresis, and ataxia.

19.1.4 Diagnosis

As no gold standard diagnostic test exists for SRLV infection, a combination of clinical signs, pathological lesions, and laboratorial testing are relied upon for SRLV diagnosis. Due to the fact that SRLV infections may be asymptomatic and that clinical signs are not specific to SRLV, the antibody and virus detection are indicated for early diagnosis [11]. Laboratory diagnosis can be done using either indirect techniques, which detect antibodies, or direct or etiological techniques, which detect the virus itself or its nucleic acid. The World Organisation for Animal Health (OIE) has recommended the use of either agar gel immunodiffusion (AGID) or ELISA for the detection of specific antibodies [11,29]. Virus detection can be achieved using molecular biology techniques such as polymerase chain reaction (PCR), reverse transcription PCR (RT-PCR), and real-time PCR (rtPCR) or quantitative PCR (qPCR) for provirus detection.

19.1.4.1 Conventional Diagnosis

19.1.4.1.1 Clinical and Epidemiological Diagnosis

It is based on the presence of the different clinical forms (mammary, arthritis, pulmonary, and nervous forms) and epidemiological features of SRLV infection, such as age of presentation and chronic nature of the processes.

SRLV-induced respiratory disease must be differentiated from ovine pulmonary adenomatosis ("jaagsiekte"), parasitic lung infections, and caseous lymphadenitis.

The mammary form must be differentiated from contagious agalactia produced by *Mycoplasma agalactiae*. Differential diagnosis in neurological cases includes listeriosis, scrapie, rabies, louping ill, parasitic infections of the central nervous system, and brain tumors. The differential diagnosis for animals with arthritis may include traumatic arthritis and infectious arthritis caused by *Mycoplasma* species [30].

19.1.4.1.2 Pathology Diagnosis

It is only reliable in animals with pulmonary disease, and it is based on microscopic observation of lesions.

19.1.4.1.3 Serological Diagnosis

Serological diagnosis is the most commonly approach to detect SRLV infection. Some of the diagnosis techniques used are AGID, radioimmunoprecipitation assay (RIPA), western blot (WB), and ELISA. Being the first serological tests developed for the diagnosis of SRLV, RIPA and AGID are usually regarded as reference standards for confirmation [31]. WB and RIPA are mainly used to confirm inconclusive results.

19.1.4.1.4 AGID

AGID has a specificity of 100% but low sensitivity, and it is often related to a subjective interpretation. Other disadvantages are the inapplicability for the determination of antibodies in milk, lack of automation, and subjective interpretation of the results [10]. The test is run with whole virus, which needs to be concentrated from culture supernatants and treated with detergents [32] to inactivate it and to expose internal epitopes. The most widely used test contains CA (p25 for VMV or p28 for CAEV) and gp135SU, which results in two precipitation bands. Due to the antigenic and genomic similarities between VMV and CAEV, the test reagents may be used for both viruses.

19.1.4.1.5 RIPA

RIPA requires, as in the case of AGID, that the peptide and protein components of tissues are solubilized and denatured with detergents. Antibodies are added to cellular lysates or tissue homogenates to form antigen–antibody complexes. These complexes are precipitated by a solid-phase immune sorbent, usually formed by a second antibody bound to Sepharose (an insoluble particulate resin), or by protein A (surface protein of *Staphylococcus aureus* with great affinity for the Fc fragment of the Ig) bound to Sepharose. The complexes are pelleted by centrifugation and washed and detected by polyacrylamide gel electrophoresis with sodium dodecyl sulfate (SDS-PAGE) with radiolabeled-specific antibodies [33].

19.1.4.1.6 Western Blot

WB involves the separation of viral proteins by SDS-PAGE before being transferred to a nitrocellulose membrane. When they are exposed to the sheep serum, specific antibodies will remain attached to the nitrocellulose membrane in the bands where viral proteins have migrated [33] and are detected using enzyme-labeled antisheep immunoglobulin antibodies. In general, it is considered as a good reference test with very high specificity and sensitivity [31,32] and is used to confirm inconclusive results.

19.1.4.1.7 ELISA

ELISA has been widely used for the detection of SRLV, and more than 30 publications describing different ELISA techniques are available [32]. ELISA used for the detection of SRLV can be classified into two categories: indirect ELISA that uses whole virus or recombinant proteins (or synthetic peptides) as antigens, and competitive ELISA that uses monoclonal antibodies (p25CA, p16MA, p14NC, and gp46-50TM) [32].

ELISA, for the detection of SRLV infections, is relatively easy to apply, and its cost is quite reasonable [10]. Some shortcomings associated with ELISA as well as with other serological methods are the possibly prolonged time between infection and a detectable humoral immune response. In addition, antibody titers fluctuate throughout the animal's life, which may also make the detection difficult. Moreover, young animals infected at birth that have been fed with milk or colostrum containing maternal antibodies for at least 2 or 3 months may stay seronegative until they are between 6 and 12 months of age [10]. Also, it has a low sensitivity in detecting antibodies in the early stages of the infection with non-genotype A isolates [34]. For this reason, it is recommended to develop a specific ELISA according to the predominant strain in each geographical area [11].

19.1.4.1.8 Coculture and Virus Isolation

Different laboratories have applied direct diagnostic methods such as viral isolation, a very laborious and expensive method whose main disadvantages are the lack of permissive cell lines and limited viral production in cultured cells. Although this method is definitive in positive samples, it is not suitable for large-scale studies and has the possibility of false negatives, which hamper its routine use. It is performed by coculturing infected fluids or cells [11]. A sample suspicious to be infected by SRLV is cultured in vitro with a cell line that allows viral replication (usually fibroblasts or goat synovial membrane cells), and the presence of virus is confirmed by the detection of cytopathic effect (syncytium formation) or by reverse transcriptase activity in cell supernatants [10].

19.1.4.1.9 Other Techniques for the Detection of Viral Proteins or Viral Genomes

Immunohistochemistry: It is based on the detection of viral antigen in tissues using labeled antibodies [25]. Its disadvantages are that it is expensive and cumbersome and that it needs to be done postmortem.

In situ *hybridization*: It consists on the detection of viral nucleic acids in suspicious tissues using probes labeled radioactively or with fluorochromes. It is expensive and cumbersome, and it is not used routinely.

19.1.4.2 Molecular Techniques

Molecular techniques represent an alternative to overcome the shortages of serological diagnosis. As in other retroviruses, the RNA genome of SRLV is transcribed into DNA by the reverse transcriptase, and it integrates in the genome of the host cell in the form of provirus. This proviral DNA may be detected in many tissues and fluids by PCR. Some of the samples from which proviral DNA may be extracted include the lung, bronchial alveolar fluid, colostrum, milk, peripheral blood mononuclear cells, mammary gland and udder, brain, carpal synovial membrane, heart, kidney, liver, spleen, lymph nodes, semen, testes, uterus, oviduct, cumulus cells, and bone marrow [31]. Viral RNA may be detected in exudates that contain free viral particles after it is transcribed in vitro into DNA, using RT-PCR. However, SRLV RNA has not been directly detected in the tissues cited earlier [31].

19.1.4.2.1 PCR

Several PCR techniques have been developed for the diagnosis of SRLVs. As discussed in the introduction, SRLVs have high mutability, and thus it is difficult to find a standard PCR technique. The efficiency of PCR depends on the specificity of the primers designed. Target sequences for PCR primers include LTR, *gag*, *pol*, and *env* regions. In addition to conventional one-step PCR (cPCR), nested PCR (nPCR) and seminested PCR (snPCR) (involving two or more amplification rounds) have been developed in order to increase the sensitivity [10]. Comparison of some PCR techniques are summarized in Table 19.1.

19.1.4.2.2 Reverse Transcription PCR

Unlike PCR that detects integrated DNA, RT-PCR parts from RNA that is transcribed to double-stranded DNA by reverse transcriptase. From this point on, the reactions are like PCR. Its advantage is that it amplifies the virus itself, but the levels of free extracellular SRLV particles are usually low. For this reason, the amplification from DNA may be more effective.

19.1.4.2.3 Real-Time or Quantitative PCR

rtPCR or qPCR is increasingly used in routine laboratories for the detection and quantification of viral nucleic acids in different cells or tissues, including diagnosis for SRLV, for which some techniques have been developed in the last few years [35–37 among others]. Even though qPCR allows the quantification of the DNA target and limits the risks of amplicon contamination [30], its efficacy is limited due the high mutability of the SRLV. For this reason, its use in routine diagnosis is not widespread (Table 19.1) [10].

19.2 METHODS

19.2.1 Sample Collection and Preparation

Sample preparation varies according to the type of sample and detection method. In general, the samples used more often are blood and milk. However, tissues from affected organs are used sometimes to confirm results.

19.2.1.1 Milk Samples

Milk is taken aseptically from the udder in sterile containers after disinfecting the end of the nipple with 70% alcohol and discarding the first foremilk. The samples are kept at 4°C during the transport to the laboratory. Upon arrival to the laboratory, DNA is extracted immediately from 1 mL. The procedure is easier if milk samples are kept refrigerated till just before extraction. Several 1 mL aliquots can be stored in Eppendorf tubes at −20°C. When needed, they can be thawed in a 37°C water bath.

19.2.1.1.1 Procedure

1. Add 0.5 mL of cold PBS to 1 mL milk sample.
2. Centrifuge at 4000 rpm for 10 min.
3. Separate the thick cream layer from the walls of the Eppendorf tube with a sterile needle or toothpick and discard this layer.
4. Discard the underlying supernatant.
5. Repeat washing with PBS three times.
6. Count the number of cells.

19.2.1.2 Blood Samples

Blood samples are collected by venipuncture from the jugular vein into the vacuum lithium heparin tubes. Similar to

TABLE 19.1
Comparison between Different Polymerase Chain Reaction Techniques

PCR	Primers Location into the Viral Genome	Comparison between Techniques	Reference
cPCR	*gag* and *pol*	*pol*-PCR is more sensitive than *gag*-PCR.	[38]
snPCR	*pol* and LTR	LTR-PCR is more sensitive than *pol*-PCR.	[39]
nPCR	*gag* and *pol*	*gag*-PCR is more sensitive than *pol*-PCR.	[40]
snPCR	*gag*	*gag*-PCR is less sensitive than AGID.	[41]
snPCR	*pol*	High sensitivity, by using degenerate primers.	[42]
nPCR	*gag*	*gag*-PCR is more sensitive than AGID in seronegative animals.	[43]
nPCR	*gag* and LTR	*gag*-PCR is more sensitive than LTR-PCR.	[44]
cPCR	*env* and *pol*	*env*-PCR is more sensitive than *gag*-PCR.	[45]
cPCR	LTR	LTR-PCR has a sensitivity of 98% with reference to AGID and ELISA.	[46]
cPCR	LTR, *gag*, and *env*	LTR-PCR is more sensitive than *gag*-PCR and *env*-PCR.	[47]
cPCR	*gag*	*gag*-PCR is less sensitive than AGID.	[48]
cPCR	*gag*	*gag*-PCR is more sensitive than the ELISA and WB.	[49]
nPCR	*gag*	*gag*-PCR increases its sensitivity when used along with hybridization.	[50]
snPCR	*pol* and LTR	*pol*-PCR is more sensitive than LTR-PCR.	[51]

Source: Adapted from Ramirez, H. et al., *Viruses*, 5(4), 1175, 2013.

cPCR, conventional one-step PCR; nPCR, nested PCR; snPCR, seminested PCR.

milk, blood samples are kept at 4°C during the transport to the laboratory.

1. Treat sample with a lysis buffer (10 mM KHCO₃, 460 mM NH₄Cl, 0.1 mM EDTA·2H₂O) to lyse red blood cells.
2. Once lysed (obvious change in the color and density of the blood), centrifuge the samples at 5000 rpm for 5 min.
3. Discard the supernatant.
4. Wash the pellet with PBS and centrifuge at 10,000 rpm for 3–5 min. This process is repeated until a clean pellet is obtained (minimum of three washings).

19.2.1.3 DNA Extraction

1. Add 0.5 mL of extraction buffer (200 mM Tris-HCl, pH 7.5, 250 mM NaCl, 25 mM EDTA, and 0.5% SDS) to the blood or milk sample pellet. Vortex vigorously until the pellet is resuspended and incubate on ice for 10–15 min.
2. Centrifuge 8 min at 4000 rpm.
3. Decant the supernatant into another Eppendorf tube and add 0.25 mL of phenol and 0.25 mL of chloroform/isoamyl alcohol (24:1). Invert several times.
4. Centrifuge at 14,000 rpm for 3 min.
5. Transfer very carefully the upper aqueous layer (above the interface) into another Eppendorf tube. Make sure that no material from the interface is transferred.
6. Add 0.5 mL of chloroform/isoamyl alcohol and invert repeatedly.
7. Centrifuge at 14,000 rpm for 3 min.
8. Transfer carefully the upper aqueous layer to a clean Eppendorf tube. Add 500 μL of cold isopropanol to precipitate DNA.
9. Invert repeatedly. If the amount of DNA is high, a cotton-like substance will be seen in the tube.
10. Centrifuge at 14,000 rpm for 3 min.
11. Discard the supernatant. Add 1 mL of cold 96%–100% ethanol. Centrifuge as above.
12. Discard the supernatant. Let the pellet dry completely to eliminate all ethanol remains that may hamper the PCR.
13. Resuspend the pellet in 40 μL of double-distilled water (ddH₂O).

Alternatively, QIAamp® DNA Blood Mini Kit (Qiagen) can be used for DNA extraction.

19.2.2 PCR DETECTION

In this chapter, we describe a nested LTR-PCR technique that has been used in several studies throughout the years. It was first developed in 2000 by Ryan and colleagues and is still in use (Table 19.2) [52,53].

19.2.2.1 Procedure

1. Thaw all the PCR reagents, vortex them, and spin briefly.
2. Prepare the PCR reaction mix as follows (nPCR is performed in reaction mix volumes of 22 μL):

	First PCR
Diethylpyrocarbonate (DEPC)-treated water	16 μL
PCR buffer 10× containing MgCl₂ (2 mM)	2.5 μL
Deoxynucleoside triphosphate (dNTP) mix at a final concentration of 0.225 mM each	0.5 μL
Primer LTR External forward (EFW) at a final concentration of 1 mM	1 μL
Primer LTR External reverse (ERV) at a final concentration of 1 mM	1 μL
Taq DNA polymerase at 0.04 U/μL	1 μL

3. Add 3 μL of DNA to the mix.
4. Always include a positive and a negative control; positive controls could be DNA extracted from a sample of a positive animal or from a tissue culture infected in vitro. A negative control could be prepared by adding water instead of DNA to the PCR mix or could be DNA extracted from a negative animal.
5. Preheat the block of the thermocycler to 94°C and set the conditions to the following: initial denaturation step at 94°C followed by 30 cycles of 94°C, 55°C (hybridization), and 72°C (extension), consecutively, each for 30 s, followed by a final extension step of 10 min at 72°C.

TABLE 19.2
Primers Used for the Long Terminal Repeat Polymerase Chain Reaction [52]

Primer	Sequence (5′–3′)	Nucleotide Position	Expected Product (bp)
LTR EFW	ACTGTCAGGRCAGAGAACARATGCC	8914–8938	
LTR ERV	CTCTCTTACCTTACTTCAGG	328–309	
LTR IFW	AAGTCATGTAKCAGCTGATGCTT	9049–9071	203
LTR IRV	TTGCACGGAATTAGTAACG	129–111	

R = A or G; K = G or T.

6. Spin the samples at 14,000 rpm for 30 s at room temperature.

7. Prepare the PCR reaction mix for the second round as follows:

	2nd PCR
DEPC-treated water	16 µL
PCR buffer 10× containing MgCl$_2$ (2 mM)	2.5 µL
dNTP mix at a final concentration of 0.225 mM each	0.5 µL
Primer LTR internal forward (IFW) at a final concentration of 1 mM	1 µL
Primer LTR internal reverse (IRV) at a final concentration of 1 mM	1 µL
Taq DNA polymerase at 0.04 U/µL	1 µL

8. Add 1 µL of the first reaction and 2 µL of DEPC water.

9. Spin the sample tubes at 14,000 rpm for 30 s at a room temperature.

10. Repeat PCR as done earlier except that the extension should be at 50°C.

11. After the amplification, spin down briefly the PCR tubes to pull down the condensation droplets at the inner wall of the tubes.

12. The PCR products may now be analyzed or stored at −20°C.

13. Resolve the PCR products using 2% agarose gel electrophoresis in tris-acetic acid-EDTA (TAE) buffer at 100 V for 30 min.

14. Stain the gel with ethidium bromide (or similar nonmutagenic intercalating dyes) and visualize the results using an ultraviolet light source.

19.2.3 SRLV CLONING, SEQUENCE ANALYSIS, AND GENOTYPE IDENTIFICATION

Following diagnosis and confirmation of a positive sample, sequence analysis can be carried out in order to identify the specific genotype that is causing the infection.

19.2.3.1 Procedure

1. Perform a PCR technique with primers selected from the LTR, *env*, *pol*, or *gag* region, following a procedure similar to the one described earlier.

2. PCR amplicons are determined using 1.5% agarose gels electrophoresis, and bands of the expected size are cut out and purified with QIAquick Gel Extraction Kit (Qiagen).

3. DNA bands are cloned into the vector pGEMT-easy® (Promega) following the manufacturer's instructions.

4. Transformed colonies (*Escherichia coli* XL1-Blue) are grown overnight at 37°C in LB medium (1% Bacto Tryptone, 0.5% yeast extract, and 1% NaCl) with 100 µg/mL ampicillin.

5. Screening of positive clones is made by restriction enzyme digestion (EcoRI).

6. Purified plasmid DNA from bacterial culture is performed using the Quantum Prep® Plasmid Miniprep Kit (Bio-Rad) according to the manufacturer's instructions.

7. Sequencing of purified plasmid DNA is done using an ABI PRISM 310 Genetic Analyzer (Applied Biosystems) or similar sequencer.

8. Partial sequences from LTR, *env*, *pol*, and *gag* regions are aligned with the ClustalW2 and PHYLIP programs (or similar) and phylogenetic trees produced by the neighbor-joining method with Kimura's correction, using 1000 bootstrap confidence.

19.3 CONCLUSION

SRLV infections are prevalent worldwide, affecting the production and animal welfare of sheep and goats. For that reason accurate diagnosis is important. However, diagnosis is difficult because there is not a *gold standard* diagnostic test. Several studies have shown the efficacy of combining ELISA and PCR techniques to detect SRLV infection. This combination of diagnostic tests increases the number of animals detected and therefore the effectiveness of monitoring and control programs of the disease.

REFERENCES

1. Blacklaws, B.A., Small ruminant Lentiviruses: Immunopathogenesis of visna-maedi and caprine arthritis and encephalitis virus, *Comp Immunol Microbiol Infect Dis*, 35(3): 259–269, 2012.

2. Sigurdardottir, B. and Thormar, H., Isolation of a viral agent from the lungs of sheep affected with Maedi, *J Infect Dis*, 114: 55–60, 1964.

3. Crawford, T.B., Adams, D.S., Sande, R.D., Gorham, J.R., and Henson, J.B., The connective tissue component of the caprine arthritis-encephalitis syndrome, *Am J Pathol*, 100: 443–454, 1980.

4. Reina, R. et al., Molecular characterization and phylogenetic study of maedi visna and caprine arthritis encephalitis viral sequences in sheep and goats from Spain, *Virus Res*, 121(2): 189–198, 2006.

5. Glaria, I. et al., Visna/maedi virus genetic characterization and serological diagnosis of infection in sheep from a neurological outbreak, *Vet Microbiol*, 155(2–4): 137–146, 2012.

6. Narayan, O., The lentiviruses of sheep and goats. In *The Retroviridae*, Vol. 2, J.A. Levy (ed.). Plenum Press, New York, pp. 229–255, 1993.

7. Bertoni, G. et al., Antibody reactivity to the immunodominant epitopes of the caprine arthritis-encephalitis virus Gp38 transmembrane protein associates with the development of arthritis, *J Virol*, 68(11): 7139–7147, 1994.

8. Crane, S.E., Kanda, P., and Clements, J.E., Identification of the fusion domain in the visna virus transmembrane protein, *Virology*, 185: 488–492, 1991.

9. Pisoni, G., Quasso, A., and Moroni, P., Phylogenetic analysis of small-ruminant lentivirus subtype B1 in mixed flocks: Evidence for natural transmission from goats to sheep, *Virology*, 339(2): 147–152, 2005.

10. Ramirez, H., Reina, R., Amorena, B., de Andres, D., and Martinez, H.A., Small ruminant lentiviruses: Genetic variability, tropism and diagnosis, *Viruses*, 5(4): 1175–1207, 2013.

11. Reina, R. et al., Prevention strategies against small ruminant lentiviruses: An update, *Vet J*, 182(1): 31–37, 2009.

12. Arsenault, J., Dubreuil, P., Girard, C., Simard, C., and Belanger, D., Maedi-visna impact on productivity in Quebec sheep flocks (Canada), *Prev Vet Med*, 59: 125–137, 2003.

13. Barquero, N. et al., Diagnostic performance of PCR and ELISA on blood and milk samples and serological survey for small ruminant lentiviruses in central Spain, *Vet Rec*, 168(1): 20, 2010.

14. Rowe, J.D. and East, N.E., Risk factors for transmission and methods for control of caprine arthritis-encephalitis virus infection, *Vet Clin North Am Food Anim Pract*, 13(1): 35–53, 1997.

15. Krieg, A. and Peterhans, E., Caprine arthritis-encephalitis in Switzerland: Epidemiologic and clinical studies, *Schweiz Arch Tierheilkd*, 132(7): 345–352, 1990.

16. Giangaspero, M., Vanopdenbosch, E., and Nishikawa, H., Lentiviral arthritis and encephalitis in goats in North-West Syria, *Rev Elev Med Vet Pays Trop*, 45(3–4): 241, 1992.

17. Burgu, I., Akca, Y., Alkan, F., Ozkul, A., Karaoglu, T., and Cabalar, M., Antibody prevalence of caprine arthritis encephalitis virus (CAEV) in goats in Turkey, *Dtsch Tierarztl Wochenschr*, 101(10): 390–391,1994.

18. Bandeira, D.A., de Castro, R.S., Azevedo, E.O., de Souza Seixas Melo, L., and de Melo C.B., Seroprevalence of caprine arthritis-encephalitis virus in goats in the Cariri Region, Paraiba State, Brazil, *Vet J*, 180(3): 399–401, 2009.

19. Tageldin, M.H., Johnson, E.H., Al-Busaidi, R.M., Al-Habsi, K.R., and Al-Habsi, S.S., Serological evidence of caprine arthritis-encephalitis virus (CAEV) infection in indigenous goats in the Sultanate of Oman, *Trop Anim Health Prod*, 44(1): 1–3, 2012.

20. Gufler, H. and Baumgartner, W., Overview of herd and CAEV status in dwarf goats in South Tyrol, Italy, *Vet Q*, 29(2): 68–70, 2007.

21. De la Concha-Bermejillo, A., Maedi-visna and ovine progressive pneumonia, *Vet Clin North Am Food Anim Pract*, 13: 13–33, 1997.

22. Leginagoikoa, I. et al., Extensive rearing hinders maedi-visna virus (MVV) infection in sheep, *Vet Res*, 37(6): 767–778, 2006.

23. Blacklaws, B.A. et al., Transmission of small ruminant lentiviruses, *Vet Microbiol*, 101(3): 199–208, 2004.

24. Narayan, O., Sheffer, D., Clements, J.E., and Tennekoon, G., Restricted replication of lentiviruses. Visna viruses induce a unique interferon during interaction between lymphocytes and infected macrophages, *J Exp Med*, 162: 1954–1969, 1985.

25. Bolea, R. et al., Maedi-visna virus infection of ovine mammary epithelial cells, *Vet Res*, 37(1): 133–144, 2006.

26. Peterson, K., Brinkhof, J., Houwers, D.J., Colenbrander, B., and Gadella, B.M., Presence of pro-lentiviral DNA in male sexual organs and ejaculates of small ruminants, *Theriogenology*, 69(4): 433–442, 2008.

27. Narayan, O., Role of macrophages in the immunopathogenesis of visna-maedi of sheep, *Prog Brain Res*, 59: 233–235, 1983.

28. Larruskain, A. and Jugo B.M., Retroviral infections in sheep and goats: Small ruminant lentiviruses and host interaction, *Viruses*, 5(8): 2043–2061, 2013.

29. OIE (Office International des Epizooties). 2008. *OIE Terrestrial Manual 2008*. Chapter 2.7.3/4: Caprine arthritis/encephalitis & Maedi-Visna. http://www.oie.int/fileadmin/Home/eng/Health_standards/tahm/2.07.03-04_CAE_MV.pdf. Accessed November 26, 2015.

30. Brinkhof, J.M., Houwers, D.J., Moll, L., Dercksen, D., and van Maanen, C., Diagnostic performance of ELISA and PCR in identifying SRLV-infected sheep and goats using serum, plasma and milk samples and in early detection of infection in dairy flocks through bulk milk testing, *Vet Microbiol*, 142: 193–198, 2010.

31. Herrmann-Hoesing, L.M., Diagnostic assays used to control small ruminant lentiviruses, *J Vet Diagn Invest*, 22: 843–855, 2010.

32. de Andres, D. et al., Diagnostic tests for small ruminant lentiviruses, *Vet Microbiol*, 107(1–2): 49–62, 2005.

33. Knowles, D.P., Jr., Laboratory diagnostic tests for retrovirus infections of small ruminants, *Vet Clin North Am Food Anim Pract*, 13(1): 1–11, 1997.

34. Lacerenza, D. et al., Antibody response in sheep experimentally infected with different small ruminant lentivirus genotypes, *Vet Immunol Immunopathol*, 112(3–4): 264–271, 2006.

35. Brinkhof, J.M., van Maanen, C., Wigger, R., Peterson, K., and Houwers, D.J., Specific detection of small ruminant lentiviral nucleic acid sequences located in the proviral long terminal repeat and leader-gag regions using real-time polymerase chain reaction, *J Virol Methods*, 147: 338–344, 2008.

36. De Regge, N. and Cay, B., Development, validation and evaluation of added diagnostic value of a Q(rt)-PCR for the detection of genotype A strains of small ruminant lentiviruses, *J Virol Methods*, 194(1–2): 250–257, 2013.

37. Kuhar, U., Barlic-Maganja, D., and Grom, J., Development and validation of TaqMan probe based real time PCR assays for the specific detection of genotype A and B small ruminant lentivirus strains, *BMC Vet Res*, 9: 172, 2013.

38. Haase, A.T., Retzel, E.F., and Staskus, K.A., Amplification and detection of lentiviral DNA inside cells, *Proc Natl Acad Sci USA*, 87: 4971–4975, 1990.

39. Leroux, C., Chastang, J., Greenland, T., and Mornex, J.F., Genomic heterogeneity of small ruminant lentiviruses: Existence of heterogeneous populations in sheep and of the same lentiviral genotypes in sheep and goats, *Arch Virol*, 142(6): 1125–1137, 1997.

40. Alvarez, V. et al., PCR detection of colostrum-associated maedi-visna virus (MVV) infection and relationship with ELISA-antibody status in lambs, *Res Vet Sci*, 80(2): 226–234, 2006.

41. Barlough, J. et al., Double-nested polymerase chain reaction for detection of caprine arthritis-encephalitis virus proviral DNA in blood, milk, and tissues of infected goats, *J Virol Methods*, 50(1–3): 101–113, 1994.

42. Celer, V., Jr., Celer, V., Nejedla, E., Bertoni, G., Peterhans, E., and Zanoni, R.G., The detection of proviral DNA by semi-nested polymerase chain reaction and phylogenetic analysis of Czech Maedi-Visna isolates based on gag gene sequences, *J Vet Med B Infect Dis Vet Public Health*, 47(3): 203–215, 2000.

43. Eltahir, Y.M. et al., Development of a semi-nested PCR using degenerate primers for the generic detection of small ruminant lentivirus proviral DNA, *J Virol Methods*, 135(2): 240–246, 2006.

44. Zanoni, R.G., Cordano, P., Nauta, I.M., and Peterhans, E., PCR for the detection of lentiviruses from small ruminants, *Schweiz Arch Tierheilkd*, 138(2): 93–98, 1996.

45. Extramiana, A.B., Gonzalez, L., Cortabarria, N., Garcia, M., and Juste, R.A, Evaluation of a PCR technique for the detection of MVV proviral DNA in blood milk and tissue samples in naturally infected sheep, *Small Rum Res*, 44: 109–118, 2002.

46. Ali Al Ahmad, M.Z. et al., Detection of viral genomes of caprine arthritis-encephalitis virus (CAEV) in semen and in genital tract tissues of male goat, *Theriogenology*, 69(4): 473–480, 2008.

47. Zanoni, R.G., Nauta, I.M., Kuhnert, P., Pauli, U., Pohl, B., and Peterhans, E., Genomic heterogeneity of small ruminant lentiviruses detected by PCR, *Vet Microbiol*, 33(1–4): 341–351, 1992.

48. Johnson, L.K., Meyer, A.L., and Zink, M.C., Detection of ovine lentivirus in seronegative sheep by in situ hybridization, PCR, and cocultivation with susceptible cells, *Clin Immunol Immunopathol*, 65: 254–260, 1992.

49. Wagter, L.H., Jansen, A., Bleumink-Pluym, N.M., Lenstra, J.A., and Houwers, D.J., PCR detection of lentiviral gag segment DNA in the white blood cells of sheep and goats, *Vet Res Commun*, 22(5): 355–362, 1998.

50. Karanikolaou, K., Angelopoulou, K., Papanastasopoulou, M., Koumpati-Artopiou, M., Papadopoulos, O., and Koptopoulos, G., Detection of small ruminant lentiviruses by PCR and serology tests in field samples of animals from Greece, *Small Rum Res*, 58: 181–187, 2005.

51. Barquero, N. et al., Comparison of two PCR and one ELISA techniques for the detection of small ruminant lentiviruses (SRLVs) in milk of sheep and goats, *Res Vet Sci*, 94(3): 817–819, 2013.

52. Ryan, S., Tiley, L., McConnell, I., and Blacklaws, B., Infection of dendritic cells by the maedi-visna lentivirus, *J Virol*, 74(21): 10096–10103, 2000.

53. Ramirez, H.R. et al., Study of compartmentalization in the visna clinical form of small ruminant lentivirus infection in sheep, *BMC Vet Res*, 8: 8, 2012.

20 Equine Infectious Anemia Virus

Caroline Leroux and R. Frank Cook

CONTENTS

20.1 Introduction 177
    20.1.1 Classification, Morphology, and Genome Organization 178
    20.1.2 Epidemiology 179
    20.1.3 Transmission 180
    20.1.4 Clinical Features and Pathogenesis 181
    20.1.5 Diagnosis 182
        20.1.5.1 Conventional Techniques 182
        20.1.5.2 Molecular Techniques 183
20.2 Methods 183
    20.2.1 Sample Collection and Preparation 183
    20.2.2 Detection Procedures 184
20.3 Conclusion and Future Perspectives 185
References 187

20.1 INTRODUCTION

Equine infectious anemia (EIA) has an almost worldwide distribution and is considered to be a significant threat to the equine industry as evidenced by the fact that it is one of the 11 notifiable equine-specific diseases listed by the World Organisation for Animal Health. Disease signs in horses consistent with EIA were first described in 1843[1] with a more comprehensive account of each clinical phase coupled with the demonstration that it is a blood-borne infection published by Vallée and Carré.[2] Moreover, these authors established that EIA was caused by a "filterable agent" making it one of the first diseases recognized as having a viral etiology (equine infectious anemia virus [EIAV]).

Diagnosis is complicated by the fact that clinical signs are generally nonspecific and vary considerably between individual animals and possibly even between equid species.[3] Although equids remain infected for life, most are able to control viral replication to such an extent that they eventually become free of overt disease signs and are therefore classified as "inapparent carriers." This common feature of EIA complicates detection and contributes to inadvertent spread of the disease. At present, there are no effective commercially available vaccines against EIA and so control is solely dependent on the identification and removal or quarantine of infected animals.

The first practical diagnostic assay, the agar gel immunodiffusion (AGID) or "Coggins test," for the disease was developed in the 1970s,[4,5] but despite its age and the fact that it is an indirect, serological procedure, it remains the only officially recognized test in many countries throughout the world. With the exception of Italy and Romania that have instigated national surveillance programs for all equids 6 months or older, EIA testing in most countries is restricted to selected equine populations. These generally include breeding stock, animals that participate in sporting events, animals for sale at public auction, and animals that are being transported across state lines (e.g., in the United States) or national boundaries unless restricted by free trade agreements such as within the European Union. In cases where EIA control programs have been effectively instigated, the incidence of disease in tested equine populations is usually low. When routine testing in the United States began in 1972, approximately 100,000 samples were evaluated with 4% identified as positive. In 2012, there were 1,443,959 EIA tests and only 36 (0.0025%) positive cases were discovered. However, despite the apparent success of current EIA control programs, there is considerable room for improvement.

Recent studies have highlighted the fact that while the AGID test has an enviable reputation for specificity, it is prone to delivering false-negative results. For example, in comparative experiments conducted as part of the Italian National Surveillance Program, it was demonstrated that when serum samples were screened by enzyme-linked immunosorbent assay (ELISA) instead of AGID, the number of EIA positive cases identified increased by 17%.[6,7] However, while more sensitive serological assays are available for EIA diagnosis, including ELISA and immunoblot, they all suffer from the disadvantage of a delay between the time of exposure to EIAV and the development of detectable humoral responses. Unfortunately, while equine cells of the monocyte/macrophage lineage are believed to comprise the predominant host cell type for EIAV, *ex vivo* cultures of these cells are too insensitive and too variable in terms of viability for routine use in conventional virus isolation techniques.[8] Although field strains of EIAV have been adapted to replicate in equine or canine fibroblastic cell types, this is a time-consuming, inefficient process that invariably produces a significant number of mutagenic changes in the viral genome resulting in an attenuated phenotype *in vivo*.[9] Therefore, direct diagnosis of EIAV infections must at present rely on the detection of viral structural components with the most commonly selected target for investigators being nucleic acid.

20.1.1 CLASSIFICATION, MORPHOLOGY, AND GENOME ORGANIZATION

EIAV has been described as the "country cousin" of human immunodeficiency virus (HIV) because like its famous relative, it is classified in the genus *Lentivirus*, subfamily *Orthoretrovirinae*, family *Retroviridae*.[10] As its name suggests, lentiviruses (*lente-*, Latin for "slow") involves a long incubation period. These viruses both incorporate their

genetic information into the DNA of the host cell and have the unique ability to infect nondividing cells, making them one of the most efficient gene delivery vectors. At present, five serogroups (i.e., bovine, equine, feline, ovine/caprine, and primate lentivirus groups) are recognized within the genus *Lentivirus*, reflecting the vertebrate hosts with which they are associated. Besides EIAV, other notable lentiviruses include HIV-1, HIV-2, feline immunodeficiency virus, and visna–maedi virus.

Electron microscopy of EIAV preparations reveals a mixture of 115 nm diameter oval and circular particles in which the typical lentiviral conical core is encased by a proteinaceous matrix that in turn is bounded by a lipid membrane containing numerous 6–8 nm projections (Figure 20.1a).[11,12] Within the core are two copies of a single-stranded genomic RNA molecule that after infection undergo a reverse transcription (RT) reaction by virally encoded reverse transcriptase and RNase H enzymes to produce double-stranded proviral DNA. As with all retroviruses, this proviral DNA

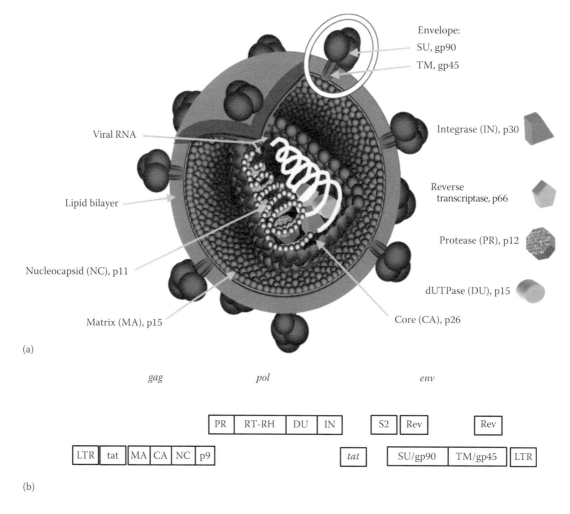

FIGURE 20.1 (a) Model of an equine infectious anemia virus (EIAV) particle showing the location and identity of the structural proteins. Although envelope glycoproteins (SU/TM) are depicted as comprising a trimeric structure as in HIV-1,[97,98] this has not been verified for EIAV. Antigens used in approved serological diagnostic tests, albeit not in all countries, are labeled. Both TM and p26 are included as antigens in one EIA-ELISA test approved for use in the United States. However, because these antigens are combined, this test does not discriminate between antibodies produced against TM and p26. (b) EIAV proviral genome organization showing the relative positions of the long terminal repeats, the three major genes (*gag, pol,* and *env*), and the additional open reading frames encoding (Tat, Rev, and S2).

may become integrated into host cell chromatin via the action of the EIAV integrase protein such that it is transcribed as if it were a normal cellular gene.

All complete EIAV proviral DNA molecules sequenced to date are approximately 8200 base pairs in length giving EIAV the honor of having the shortest genome of any extant lentivirus discovered to date with, as expected, the prototypical retroviral genome organization (Figure 20.1b) in which the major genes *gag*, *pol*, and *env* are bounded by long terminal repeats (LTR).[13–17] However, lentiviruses are classified as complex retroviruses because they contain small open reading frames (ORF) encoding so-called ancillary proteins in addition to the three major genes. The number of ORFs and the functions of the ancillary proteins they encode differ between lentiviruses with only Rev (essential for export of unspliced and singly spliced viral RNA transcripts from the host cell nucleus) being common to all members of the genus. Although all lentiviruses contain an ORF designated *tat*, the function(s) of the protein it encodes may differ. In most cases, this ORF encodes a transactivator protein (Tat) that is essential for transcription of the viral genome; however, in the small ruminant lentiviruses, the encoded protein promotes cell cycle arrest and therefore may be analogous to the Vpr protein of HIV-1.[18,19] Polypeptides encoded by the other ORFs appear to be involved in pathogenesis and/or combating retroviral restriction factor (RRF) host cell defense systems.

In addition to being the smallest living lentivirus in terms of genome size, EIAV also possesses the simplest genomic organization containing just three additional ORFs encoding Tat, the second exon of Rev, and S2 (Figure 20.1b), with the latter being a unique feature within lentiviruses. In contrast, HIV-1, which is one of the most genetically complex lentiviruses, has six ORFs coding for accessory or regulatory proteins. Furthermore, EIAV is the only surviving lentivirus that does not possess an additional ORF encoding a Vif ortholog. This ancillary lentiviral protein promotes degradation of host retroviral defense proteins in the apolipoprotein-β editing complex 3 family or APOβEC3–cytidine deaminase proteins. These molecules are incorporated into progeny retroviral particles where they can introduce potentially lethal mutations in the viral genome by converting cytidine to uracil. The absence of a Vif ortholog in EIAV is somewhat unexpected since horses possess more APOβEC3 genes than any other nonprimate species.[20] However, despite possessing just 84% of the coding capacity of HIV, EIAV is, from the virus point of view, a highly successful pathogen, and just like its more famous relatives, it establishes persistent infections by resisting all host-specified defense mechanisms.

20.1.2 Epidemiology

At more than 400 million years old, members of the family *Retroviridae* are ancient and perhaps not surprisingly have an enormous host range that includes all known vertebrates.[21] In contrast, the current host range of lentiviruses is limited to equids, bovids, wild and domestic small ruminants, wild and domestic *Felidae*, African nonhuman primates,

and man, suggesting the genus arose only relatively recently in evolutionary terms. Indeed, it has been calculated based on nucleotide substitution rates that lentiviruses emerged around 1 million years ago.[22] However, discoveries such as the observation that similar defective endogenous lentiviral proviruses are present within the genomes of European rabbits and hares despite these being separate leproid species for 12 million years[23,24] suggest that calculations based on nucleotide substitution rates underestimates the age of lentiviruses by at least an order of magnitude. Nonetheless, compared to other members of the *Retroviridae*, lentiviruses are probably "recent arrivals," and this may account for the fact that within the *Perissodactyla*, only the *Equidae* act as hosts for this retroviral genus. The assumption being that early members of this family were exposed to a lentiviral infection sometime after divergence from the rhinoceros and tapir (55–40 million years ago) but as EIAV infects all equids, before the speciation of modern equines (2–4 million years ago).

With the exception of Iceland that operates a strict "closed horse herd" policy, EIA occurs worldwide with outbreaks being recently reported in Asia, Europe, North America, and South America.[14,25–31] On the basis of discussions with veterinarians and animal health-care professionals from different countries, it appears that the incidence of EIA in routinely tested equid populations such as performance horses is relatively low (2%–0.0025%). However, in equids not subjected to regular testing, prevalence may be much higher such as >30% reported in a survey of farms located in the Pantanal region in Brazil.[31] In the United States, new cases of EIA are almost always found in animals offered for sale for the first time at public auction after living their lives on a single farm.

Unfortunately, there is a relative paucity of information concerning the molecular epidemiology of EIAV, and to date, only four complete genomic sequences from field isolates have been published: $EIAV_{LIA}$ from China, $EIAV_{IRE}$ from Ireland, $EIAV_{MIY}$ from Japan, and $EIAV_{WY}$ from the United States. These share ≥80% nucleotide sequence identity and show no evidence of recombination with one another, and phylogenetic analysis suggests that each comprises a separate monophyletic group or clade.[16,17] Therefore, it appears as if all four have evolved independently since diverging from a common ancestor. These observations suggest that the global molecular epidemiology of EIAV is likely to be extremely complex, an assertion supported by the type of analysis shown in Figure 20.2 where 19 *gag* sequences encoding the matrix protein (p15) are found to comprise 13 monophyletic groups. Based on this level of complexity, it might be anticipated that the design of universally successful molecular diagnostic assays and vaccines will be difficult in the case of this virus. However, we should keep in mind that the current phylogenetic reconstructions for EIAV are all based on a very limited set of sequences, and that the picture may change considerably as more sequences from geographically distinct origins become available.

Differences in the extent of nucleotide or amino acid sequence identity between EIAV field isolates are not distributed evenly throughout the genome. In some cases such as Gag

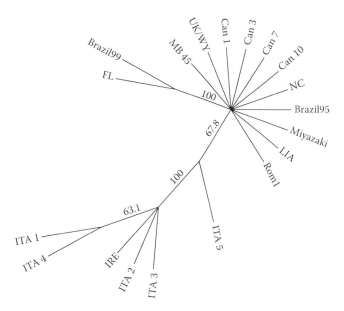

FIGURE 20.2 Phylogenetic analysis of equine infectious ane-mia virus (EIAV) isolates based on nucleotide sequences encod-ing Gag p15 (matrix protein). A phylogenetic tree of aligned EIAV *gag* p15 sequences was constructed by the neighbor-joining method with bootstrap values determined over 1000 iterations. Branch lengths are proportional to the distance existing between sequences. The *gag* p15 sequences included in the analysis are from Canada (Can 1, 3, 7, 10 [GenBank accession numbers EF418582, EF418583, EF418584, EF418585]), China (Liaoning [LIA] [GenBank accession number AF327877]), Ireland (IRE [GenBank accession number JX480631]), Italy (ITA 1, 2, 3, 4, 5 [Genbank accession numbers EU240733, EU375543, EU375544, EU741609, GQ265785]), Japan (Miyazaki [GenBank acces-sion number JX003263]), Romania (Rom 1 [GenBank accession number GQ229581]), and the United States (Wyoming UK/WY [GenBank accession number AF033820]). In addition, sequences were included from Florida (FL), North Carolina (NC), Brazil (Brazil 95, 99),[29] and Argentina (MB 45).

p26, Gag p11, and the *pol* gene products, it is relatively high (79%–89% amino acid identity), whereas in Gag p9 and S2 amino acid, identity is <50%. The fact that there is a high level of amino acid sequence identity between isolates in Gag p26 is fortunate because this antigen forms the basis of almost all commercially available serological assays for EIA including the AGID (Coggins) test. However, while there are certainly genetic differences between geographically distinct EIAV iso-lates, almost all of the structural/functional motifs within viral proteins that were identified mainly using laboratory-generated derivatives of the EIAV$_{WY}$ strain are either preserved or con-tain highly conservative amino acid substitutions.[16,26,29,32] For example, despite less than 50% amino acid sequence iden-tity, all Gag p9 molecules analyzed to date contain a "late domain" or YPDL motif consisting of tyrosine (Y), proline (P), aspartic acid (D), and leucine (L)[14,16,26,29,33] that have been shown to be essential for the release of progeny virions from the host cell.[34,35] Analysis of the immunologically important surface unit (SU) envelope glycoprotein (gp90) demonstrates that amino acid substitutions between strains are distributed throughout the molecule with the exception of the amino

terminus including the six amino acid signal peptide and the relative positions of 16 cysteine residues suggesting that disul-phide bridges are essential to the structural and functional integrity of this viral glycoprotein.[16,32] As with all retroviruses studied to date, the EIAV reverse transcriptase lacks "proof-reading" ability in terms of 3′–5′ exonuclease activity and is error prone with a tendency to produce A:C mismatches.[36] Furthermore, initial viral replication occurs in the cytoplasm where host cell-proofreading activities are absent. This combi-nation creates an environment where there is a high probability that mutations will be introduced during the production of pro-viral DNA molecules from the genomic viral RNA, resulting in the rapid generation of a population of related but geneti-cally diverse viruses ("quasispecies"). Compared with homog-enous viral populations, such quasispecies are better equipped to deal with multiple selective pressures such as that exerted by the immune system. Consequently, within infected equids, the genetic composition of the EIAV population changes with time, although this is limited by the requirement to preserve the structural and functional properties of many viral pro-teins along with some secondary structural motifs within viral nucleic acids. In general terms, it has been shown that during the course of infection, *gag* and *pol* remain relatively conserved, while extensive genetic variation occurs in *env* and in ORF 3 encoding the second exon of Rev.[37–40] Somewhat surprisingly, Gag p9 and S2, the two proteins with the lowest amino acid sequence identity between EIAV isolates and there-fore presumably possessing considerable potential for change, are extensively conserved over time within individual infected hosts.[41] An important factor contributing to the initiation and maintenance of a persistent infection is that the SU envelope glycoprotein of EIAV can undergo major "antigenic drift"[42,43] meaning that each febrile episode in an infected animal is associated with a different antigenic variant or "immunologi-cal escape mutant." This conclusion is supported by the fact that the nucleotide sequences of SU molecules differs within each clinical episode.[39,40,44] However, unlike the relatively uniform distribution observed between EIAV strains, nonsyn-onymous nucleotide substitutions occurring during the course of an infection are not randomly distributed but are generally restricted to eight hypervariable domains.[39]

20.1.3 Transmission

EIAV is generally considered as a blood-borne disease although viral nucleic acids have been detected in swabs taken from nasal passages, the buccal cavity, and genitalia.[28] Under natu-ral conditions, it appears to be transmitted predominantly by blood-feeding insects comprising members of the *Tabanidae* (horseflies), *Chrysops* (deer flies), and *Stomoxys calcitrans* (stable flies).[45–50] Mosquitoes are unlikely to play a major role in EIAV transmission because lentiviruses, unlike some virus types, cannot migrate from the insect gut to the salivary gland in order to be injected into a new host and therefore, once ingested by mosquitoes, are probably destroyed by digestive enzymes. Importantly, EIAV does not replicate in insect tis-sues and so transmission is purely mechanical with efficiency determined in part by the volume of the blood meal that is retained on hematophagous insect mouthparts. Consequently, size of the inoculum is an important factor for efficient transmission although even in the case of large horseflies such as *Tabanus fuscicostatus*, the volume of blood retained post-feeding is very limited and estimated at only 10 ± 5 nL.[51] As these flies generally need just a single blood meal to complete their life cycle, transmission of EIAV can only occur if feeding is interrupted, for example, by defensive reactions of the equid in response to a painful bite, so that the fly travels to a second host in order to feed to repletion. The primary factor determining whether a fly will return to the original or seek a new host is distance. It has been demonstrated that 99% of horseflies will return to their original target if no other equid is positioned less than ~45 m (50 yards) away.[50] In those jurisdictions, particularly in the United States, where euthanasia is not mandatory, it is required that EIAV-infected equids are quarantined at a minimum of four times this distance (~183 m) from all uninfected equines.

As a result of low residual blood meal volumes, the innate behavioral characteristics of insect vectors and the fact that EIAV probably does not survive significantly longer than 30 min on fly mouthparts,[47] natural insect-mediated transmission appears to be a relatively inefficient process only having a high probability of occurrence when susceptible animals are in close proximity and when the donor equid experiences clinical episodes associated with high levels of blood-associated virus (see Section 20.1.4). In contrast, iatrogenic transmission is highly effective with man potentially having the greatest influence on the dissemination of EIAV worldwide. Recent examples include the high-profile 2006 outbreaks of EIA in Ireland and Italy that originated from the administration of contaminated horse blood products to thoroughbred foals.[27] Furthermore, reuse of veterinary equipment without sterilization is commonplace in many countries. It is worthwhile remembering that residual blood volumes in hypodermic syringes are likely to be 1,000–10,000-fold higher than found on horsefly mouthparts and that EIAV remains infectious in syringe needles for at least 96 h.[52]

In addition, to insects and man, EIAV has been reported to be transmitted *in utero* although this only seems to occur if the mare experiences a febrile episode during gestation suggesting that high tissue-associated viral loads are required before the placental barrier is crossed.[53] The fact that healthy EIA-free foals can be born and until weaned remain in close proximity, even in areas with high insect-vector populations, demonstrates that inapparent carrier mares, in which viral loads are low, do not present a major risk to their offspring.

Although a blood-borne etiology for EIAV has been extensively documented, the recent finding of viral p26 major capsid antigen expression in alveolar and bronchiolar epithelial cells of the lung parenchyma coupled with an associated interstitial disease[54] suggests that airborne transmission could also occur with this virus. This is certainly not an unreasonable hypothesis as this is the major route for transmission of small ruminant lentiviruses in domestic sheep and goats.[55] Therefore, secretion into the lumen of the equine lung could

result in the exhalation of EIAV in droplets. However, further studies are required to determine if and under what circumstances airborne transmission of EIAV can occur although it is interesting to note that nose-to-nose contact as a behavioral form of greeting frequently occurs between equids in the same social group.

20.1.4 CLINICAL FEATURES AND PATHOGENESIS

From the beginning of the early twentieth century,[2] EIA has consistently been described in the literature as comprising of three clinical phases. The first is designated the acute stage, it lasts 1–3 days and is usually characterized by fever and thrombocytopenia. The second or chronic phase consists of multiple sequential disease episodes in which fever and thrombocytopenia may be accompanied by one or more of the following: anemia, edema, depressed neurological reactions, petechial hemorrhaging on mucous membranes, and cachexia. Eventually after 12–24 months, the frequency of these episodes gradually diminishes, and the equid enters a prolonged third or inapparent carrier stage where there are no overt clinical signs. Although this classical pattern of disease progression is frequently observed in experimental infections,[44,56,57] the situation may be very different in the field where clinical signs can be extremely variable ranging from unresolvable high fever resulting in death to cases where there are either no disease signs or they are so mild as to go unnoticed by owners. In addition to potential differences in pathogenicity between virus strains, man may play a significant role in determining clinical outcome because exposure to high inoculum levels of EIAV, as is likely to occur during iatrogenic transmission, can significantly enhance disease signs.[58]

In common with other lentiviruses, the host cell tropism of EIAV includes cells of the monocyte/macrophage lineage although progeny virus production only occurs following differentiation into mature tissue macrophages or dendritic cells.[59] However, in contrast to HIV, but in common with small ruminant lentiviruses, EIAV cannot infect $CD4^+$ T-helper lymphocytes, and so it does not induce long-term immunodeficiency in equids. Furthermore, some EIAV strains may infect endothelial cells,[60] and as outlined earlier, viral p26 antigen is expressed in lung alveolar and bronchiolar epithelial cells of persistently infected horses[54] although the significance of this relatively new discovery in terms of progeny virus production and transmission remains to be determined.

During the acute and chronic stages of EIA, induction of clinical disease signs is dependent on tissue-associated viral loads. When these reach a threshold level that experimentally equates with plasma-associated EIAV RNA burdens of 5×10^7 to 1×10^8 copies/mL,[61] they trigger release of proinflammatory cytokines that include tumor necrosis factor-α (TNFα), interleukin 1-α and 1-β (IL-1α, IL-1β), IL-6, and transforming growth factor-β (TGFβ). IL-6 and TNFα induce febrile responses by activating the arachidonic pathway to increase production of prostaglandin E2 (PGE2), while TNFα/TGFβ contributes to thrombocytopenia by suppressing equine megakaryocyte growth and TNFα promotes anemia by downregulating erythropoiesis.[62–65] Furthermore, in some species, TNFα induces severe thrombocytopenia by stimulating release of platelet agonists including thrombin, plasmin, and serotonin.[66] At later stages, adaptive immune responses may also contribute to pathogenesis by immune-mediated destruction of antibody-coated platelets, phagocytosis of complement C3-coated erythrocytes resulting in the presence of hemosiderin granules in macrophages, and thickened glomerular tufts within the kidney caused by excess levels of C3.[67–69] Induction of oxidative stress produced by changes in glutathione peroxidase and uric acid levels might also play a role in the pathogenesis of EIAV infections by escalating inflammatory responses while simultaneously decreasing immune cell proliferation.[70]

As demonstrated by experiments in foals with severe combined immunodeficiency, resolution of clinical signs following exposure to EIAV is dependent on adaptive immune responses.[71,72] In experimentally infected, immune competent horses or ponies, humoral responses against EIAV are detectable at 14–28 days postinfection (pi) using sensitive immunoblot or ELISA assays. However, antibodies with viral-neutralizing activity are usually not observed until 38–87 days pi and may not reach maximal levels until 90–148 days pi.[42,56] Therefore, as virus-specific cytotoxic T-lymphocytes are detectable by 14 days pi,[73] it is generally believed that cell-mediated and not humoral immune responses are responsible for the initial control of EIAV replication. Early immune responses are "strain specific," and as discussed earlier, each febrile episode is associated with an antigenically distinct virus population derived from the original infecting virus.[42,43] Therefore, the equid immune system is forced into a cycle of "catch-up" where it must respond to the emergence of each new dominant variant in the viral population. Fortunately, this cycle does not continue indefinitely and is eventually broken (enabling infected equids to become inapparent carriers) by maturation and broadening of the immune response from strain specific to strain cross-reactive.[42,56,74,75] The requirement for active adaptive immune responses for maintenance of the inapparent carrier state is evidenced by the increase of plasma-associated virus load and recrudescence of disease after corticosteroid-induced immunosuppression.[76–78]

20.1.5 DIAGNOSIS

The variation in disease signs that are observed between individual cases precludes diagnosis of EIA based solely on clinical profile. Furthermore, there is no universally accepted virus isolation or viral component detection techniques. Therefore, identification of EIAV-infected equids is currently dependent on indirect, serological methods.

20.1.5.1 Conventional Techniques

In most countries, the AGID ("Coggins") test is the "gold standard" for determining the serological status of equids to EIAV, and only those animals producing positive reactions in this assay are considered, on an official basis, to be seropositive for this virus. However, as discussed earlier, AGID test can

produce a significant number of false-negative reactions,[6,7] a sensitivity issue in part caused by the relatively large amounts of antigen/antibody complexes required to form visible lines of precipitation. Furthermore, in the case of weak reactors, interpretation of AGID results can be highly subjective creating a situation where a serum sample is classified as "positive" in one laboratory but "negative" in another.[6,7] Although some nations such as the United States and more recently the European Union permit alternative serological testing procedures based on the ELISA format (both competitive and noncompetitive), samples producing positive reactions usually have to be confirmed by AGID before an official diagnosis is given. Almost all commercially produced AGID and ELISA test kits for the serological diagnosis of EIA employ recombinant forms of the viral p26 major core protein derived from the EIAV$_{WY}$ strain as the detection antigen. Recombinant forms of p26 have also been used in several published noncommercial EIA-ELISAs along with an immunochromatographic lateral flow test.[7,79–82] In addition to the p26 protein, ELISA and fluorescence polarization assays have been developed using recombinant antigens or synthetic peptides based on the SU or Transmembrane (TM) envelope glycoproteins. These antigens have less than 65% amino acid sequence identity between EIAV strains, and none of these assays have as yet been officially adopted as approved diagnostic tests.[30,83–86] However, one of the most sensitive serological diagnostic procedures discovered to date is an immunoblot test based on gradient purified whole viral antigen derived from EIAV$_{PV}$, a fibroblast-adapted variant of the EIAV$_{WY}$ strain.[87] Although EIAV$_{WY}$-like viruses share less than 80% nucleotide sequence identity with the other full-length EIAV genomes sequenced to date,[32] the EIAV$_{PV}$ antigen has successfully detected virus-specific antibodies in field cases of EIA in Europe, Asia, and North and South America.[7,27,30,33,88] At the present time, the EIAV-immunoblot test is not officially approved, although in the United States Department of Agriculture EIA Uniform Methods and Rules, it is described as a "supplemental laboratory test" that "may be used to resolve equivocal results on official laboratory tests." However, currently, this test can only be performed at the National Veterinary Services Laboratory, Ames, Iowa, and the University of Kentucky. Samples are considered positive in the EIAV-immunoblot test if they react with two or more of the three major viral antigens SU, TM, or p26.

On the basis of experiences gained during the National Surveillance Program, it is probable that Italy will adopt a three-tiered strategy for EIAV diagnosis. In this, all samples are subjected to initial screening using an ELISA-based test[7] with those found positive for EIA confirmed by AGID. In the few cases where the results do not agree, additional immunoblot tests are performed using the EIAV$_{PV}$ antigen. Therefore, in contrast to an AGID only system, screening is accomplished using a potentially more sensitive technique that, particularly when combined with an ELISA reader, produces less subjective results. Moreover, the increased tendency that EIA-ELISA tests have for producing "false-positive" results is counteracted by the use of AGID and/or EIA immunoblot.

20.1.5.2 Molecular Techniques

The major disadvantage of indirect serological techniques is the delay between exposure to a pathogen and the development of detectable humoral responses, referred to as the "serological window." Although most experimentally EIAV-infected equids become seropositive (even in AGID tests) by 45 days pi, delays as long as 157 days have been occasionally reported in field cases.[27] Importantly, tissue-associated viral loads can reach their highest levels in the days immediately following the initial exposure; therefore, in recently infected equids, there is a high risk for transmission before they become detectable in any of the currently approved diagnostic tests. Indeed during the 2006 EIA outbreak in Italy, a farm was declared "free" of the disease by AGID testing following a 90-day quarantine period only to have new cases appear a short time later.[26] Consequently, there is a significant requirement for the development of direct methods of detection particularly in cases where recent exposure is suspected. To meet this need in the absence of reliable, sensitive virus isolation techniques, researchers have focused on the development of methods to detect EIAV nucleic acids either in the form of proviral DNA or viral RNA. To date, these methods have all been based on the polymerase chain reaction (PCR) involving either direct amplification with proviral DNA or, in the case of viral RNA, following production of complimentary DNA (cDNA) by RT. However, with EIAV, DNA amplification techniques such as PCR face two major hurdles. The first is these techniques depend on the presence of extensive nucleotide sequence identity between the oligonucleotide primers and the target DNA with mismatches resulting in decreased sensitivity or even complete inhibition of the amplification process. As described earlier, EIAV sequences are highly variable between geographically distinct isolates,[16,29,32] and in individually infected animals, some regions of the viral genome undergo significant mutational changes over time.[37,39] Consequently, before successful PCR-based assays can be developed, it is essential to identify highly conserved regions within the EIAV genome, a process hindered by the current lack of published nucleotide sequence information for this virus particularly in regions outside of the *gag* gene. The second major hurdle facing the development of direct molecular diagnostic assays for EIAV is that tissue- and blood-associated viral loads in inapparent carriers are often at extremely low levels[89,90] making detection difficult even for highly sensitive techniques such as PCR. Nonetheless, a number of PCR-based assays have been described for EIAV.[25,26,28,29,33,91–94] Unfortunately, the earlier assays were developed when almost all published sequences were derived from variants of the EIAV$_{WY}$ strain and so may be of limited use for diagnosis in the field.[91,92] The TaqMan-style assay described by Cook et al.[92] was designed in and around sequences (slippery sequence and pseudoknot) responsible for the ribosomal frame-shifting event that is required for the translation of the *pol* gene[95] in the belief that these will be highly conserved between isolates. However, it has since been discovered that although the slippery knot and pseudoknot motifs are conserved, the surrounding sequences encoding Gag p9 vary extensively between different EIAV isolates.[28,29] Therefore, this real-time RT-PCR assay is only effective for detection of viral RNA from EIAV$_{WY}$-like strains. Although the oligonucleotide primers described by Nagarajan and Simard[25] successfully detected proviral DNA from seropositive horses in Canada, resulting in them becoming recommended reagents for detection of EIAV viral nucleic acids by the OIE, they failed because of a significant number of mismatches (Figure 20.3) to amplify proviral sequences from EIA-seropositive horses in Europe[26,28] and Japan.[33] Recently, however, several PCR-based assays have been reported that have successfully detected viral-specific nucleic acids from seropositive field cases of EIA.[7,26,28,29,33,93,94] It should be noted that some of these published assays[29,93] were designed to amplify entire EIAV *gag* gene sequences and therefore should perhaps be considered as analytical rather than diagnostic assays. In addition, to amplifying viral nucleic acids from EIA field cases, some recently described PCR-based assays have been reported to correlate well with conventional serological diagnostic tests even with samples collected from clinically inapparent equids.[26,33]

20.2 METHODS

20.2.1 Sample Collection and Preparation

Febrile episodes of EIA are correlated with high viral loads in most tissues including blood[89] and therefore provide the best conditions for detection of viral nucleic acids. Under such favorable circumstances, the only problem facing the application of PCR-based techniques is the extent of nucleotide sequence identity between the oligonucleotide primers and the target sequences within the viral genome. However, tissue-/blood-associated viral loads decline by at least 1000-fold coupled with the resolution of overt clinical signs,[61,89] thereby adding the limits of detection to the difficulties that must be overcome before molecular-based methods can be used for routine EIAV diagnosis. In experimentally infected inapparent carrier horses, the organs with the highest levels of EIAV nucleic acids appear to be the liver and spleen.[89] This also seems to be the case in naturally infected equids as shown by the detection of EIAV proviral sequences by a single-stage PCR in spleen DNA samples collected from all 22 seropositive horses, whereas less than 20% of these animals were PCR positive using DNA isolated from peripheral blood mononuclear cells (PBMC) samples.[94] Unfortunately, difficulties associated with performing routine biopsies on these organs, especially if large numbers of animals are involved, limits their usefulness for diagnosis. Consequently, PBMCs as a source of proviral DNA and plasma for viral RNA are currently viewed as the most practical sample materials for detection of EIAV infection. Although some authors have reported that PCR-based amplification of EIAV sequences is accomplished more reliably from PBMC DNA,[26,33] viral RNA has been successfully

```
Consensus   G TAA – T TGGGCGC TA AGT T TGG   GACAGCAAGA T T TAT TAGGGG
            ══════════ N & S P1 ══════════   ══════════ N & S P2 ══════════

 1. PA        . . . . – . . . . . . . . . . T . . . . C . . . .   . . . . . . . . . . . . . . . . . . . . . . .
 2. IRE       . C T – G . . . . . . . . . . . . . . . . . . . .   . . . . . . . T . . . . . . . . . . . . . . .
 3. WY        . . . . – . . . . . . . . . . . . . C . A . . . .   . . . . . . . . . G . . . . . . . . A . .
 4. Brazil 77 . . . . – . . . . . . T . . . . . . . . . . . . .   . . G . . . . . . . . . . . . . . . A . .
 5. Brazil 95 . . . . – . . . . . . T . . . . . . . . . . . . .   . . G . . . . . . . . . . . . . . . A . .
 6. Brazil 99 . . . AC . . . . . . . . . . . C . . . . . . . .   . . . . . . . . . . . C . . . . . . A . .
 7. NC        . . . . – . . . . . . . . . . . . . . . . . . . .   . . . . . . . . . . . . . . . . . . . . . .
 8. LIA       . . . . – . . . . . . . . . G . A . . . . . . . .   . . . . . . . . . G . . C . . . . . . . .
 9. FL        . . . AC . . . . . . . . . . . C . . . . . . . .   . . . . . . . . . . . . . . . . . . A . .
10. Can1      . . . . – . . . . . . . . . . . . . C . A . . . .   . . . . . . . . . . . . . . . . . . A . .
11. Can3      . . . . – . . . . . . T . . . . . . . . . . . . .   . . . . . . . . . . . . . . . . . . A . .
12. Can7      . . . . – C . . . . . . . . . . . . A . . . . . .   . . . . . . . . . . . . . . . . . . A . .
13. Can10     . . . . – . . . . . . . . . . . . . C . A . . . .   . . . . . . . . . . . . . . . . . . . A
14. MIY       . . . T – . . . . . . . . . . . . . C . A . . . .   . . . . . . . . . G . . . . . . . . A . .
15. ITA1      . C T – G . . . . . . . . . . . . . . . . . . . .   . . . . . . . T . . . . . . . . . . . . . .
16. ITA2      . C T – G . . . . . . . . . . . . . . . . . . . .   . . . . . . . T . . . . . . . . . . . . . .
17. ITA3      . C T – G . . . . . . . . . . . . . . . . . . . .   . . . . . . . T . . . . . . . . . . . . . .
18. ITA4      . C T – G . . . . . . . . . . . . . . . . . . . .   . . . . . . . T . . . . . . . . . . . . . .
19. ITA5      . . . . . . . . . . . . . . . . . . . . . . . . .   . . . . . . . T . G . . . . . . . . . . . .
20. ROM1      . . . . – C . . . . . . . T . . C . . . . . . . .   . . . . . . . . . G . . C . . . . . . . .


Consensus   TGGG TAAA TAC TA TA CAGC AA   T TGC CAAA TGC TC CAC TGG
            ══════════ N & S P4 ══════════   ══════════ N & S P5 ══════════

 1. PA        . . . . . . . . . . . . C . . T . . . . .   . . . . . G . . . . . C . . . . T .
 2. IRE       . . . . . . . . . . . . . . . A . . G . .   . . . . . . . . . . . . . . A . . . .
 3. WY        . . . . . G . . . . . C . . . . A C . . .   . A . . G . . . . . . . . . . . . . .
 4. Brazil 77 . . . . . . . . . . . . . . . . . A C . .   . A . A . . . C C . A . . TA . . .
 5. Brazil 95 . . . . . . . . . . . . . . . . . A C . .   . A . A . . . C C . A . . TA . . .
 6. Brazil 99 . . . . . G . C . . . . . . . A C . . . .   . . . . . G . . . . . . . C . . T .
 7. NC        . . . . . . . . . . . C . . . . A C . . .   . . . . . . . . . . . G . . . . A
 8. LIA       . . . . . . . . . . . . . . . . . . . A .   . . . . . C . . . . T . A .
 9. FL        . . . . . G . C . . . . . . . A C . . . .   . . . . . . . . . . . . . C . T .
10. Can1      . . . . . G . . . . . . . . . A C . . . .   . A . G . . . . . . . . . . . . .
11. Can3      . . . . . G . . . . . C . . . GC . . . .   . A . G . . . . . . . . . . . . .
12. Can7      . . . . . . . . . . . C . . . A C . . . .   . A . G . . . . . . . . . . . . .
13. Can10     . . . . . . . . . . . C . . AAC T C . . .   . . . . . C . . G . . T .
14. MIY       . . . . . G . C . . . . . . . A C . . . .   . A . G . . . . . G . . . . . .
15. ITA1      . . . . . . . . . . . . . . . A . . G . .   . . . . . . . . . . . . A . . . .
16. ITA2      . . . . . . . . . . . . . . . A . . G . .   . . . . . . . . . . . . A . . . .
17. ITA3      . . . . . . . . . . . . . . . A . . G . .   . . . . . . . . . . . . A . . . .
18. ITA4      . . . . . . . . . . . . . . . A . . G . .   . . . . . . . . . . . . A . . . .
19. ITA5      . . . . . . . . . . . . . . . A . . G . .   . . . . . . . . . . . . A . . . .
20. ROM1      . . . . . . . . . . . . . . . A . . G . .   . . . . . . . . . . . . A . . . .
```

FIGURE 20.3 Variation between different viral isolates in binding regions of oligonucleotide primers currently recommended by the World Organisation for Animal Health. Binding regions for the current World Organisation for Animal Health–recommended oligonucleotide primers N and SP1, N and SP2, N and SP4, and N and SP5[25] are shown for American, Asian, and European equine infectious anemia virus isolates (GenBank accession numbers AB008196, AF016316, AF327877, EF418582, EF418583, EF418584, EF418585, EU24073, EU375543, EU375544, EU741609, GQ265785, GQ229581, GQ855755, JX480631, and JX003263[29]). A dot indicates nucleotide sequence identity, whereas a dash indicates a deletion.

detected by nested RT-PCR in plasma samples obtained from asymptomatic, naturally EIAV-infected Romanian horses. This suggests the need for further studies to compare sample source materials, nucleic acid extraction techniques, and different PCR-based methods.

At present, there are no standardized protocols for nucleic acid isolation and purification prior to analysis for EIAV nucleotide sequences, and several methods have been described. DNA from PBMC has been isolated directly from whole blood using a GenElute Blood Genomic DNA Kit (Sigma, St. Louis, MO)[26] or DNeasy Blood and Tissue Kit (Qiagen, Germantown, MD).[33]. Alternatively, some researchers have attempted to enrich PBMC prior to nucleic acid extraction by pretreating whole blood samples for 10 min with red cell lysis buffer (154.98 mM ammonium chloride [NH$_4$Cl], 9.99 mM potassium hydrogen carbonate [KHCO$_3$], 1.27 mM ethylene diamine tetra-acetic acid [EDTA]).[94] Similarly, a number of commercially available methods have been described for

extracting EIAV RNA from plasma samples collected by jugular venipuncture into evacuated glass tubes containing EDTA. These include the High Pure Viral RNA Kit (Roche Molecular Biochemicals, Indianapolis, IN) and the QIAamp Viral RNA Mini Kit (Qiagen).[7,28,29]

20.2.2 DETECTION PROCEDURES

The oligonucleotide primer sequences (and probe sequences in the case of real-time TaqMan assays) along with the published thermocycling conditions of recently described PCR-based assays for EIAV are shown in Table 20.1. Assay conditions will vary depending on the source of the PCR reagents. Furthermore, subsequent analysis demonstrates that none of the oligonucleotide primer binding sequences described to date are completely conserved between viral isolates although there appears to be higher levels of nucleotide sequence identity within the LTR R/U5 and 5′ untranslated

TABLE 20.1

Polymerase Chain Reaction–Based Assays Designed to Amplify Equine Infectious Anemia Virus Sequences in Naturally Infected Equids

Primer/Probe Sequence (5′–3′)	Location[a]	Assay Conditions[b]	Detection	Reference
First round	*gag*	30 cycles (20″ 94°C, 30″ 66°C, 30″ 68°C)	547 bp	[26]
F: GACATGGAGCAAAGCGCTCA				
R: CTGCCCAGGCACCACATCTA				
Second round		30 cycles (20″ 94°C, 30″ 64°C, 30″ 68°C)	313 bp	
F: TGTGGGCGCTAAGTTTGGTG				
R: TTTCTGTTTCCAGCCCCATC				
First round	*5′ UTR*	35 cycles (30″ 94°C, 30″ 52°C, 40″ 72°C)	246 bp	[33]
F: GACAGTTGGGCACTCAGATT				
R: CAGGAACACCTCCAGAAGAC				
Second round		35 cycles (30″ 94°C, 30″ 58°C, 40″ 72°C)	198–203 bp	
F: ATTCTGCGGTCTGAGTCCCT				
R: TAAGTTCTCCTCTGCTGTCC				
F: TGCGGTCTGAGTCCCTTCTC	*LTR/5′ UTR*	35 cycles (30″ 94°C, 30″ 60°C; 60″ 72°C)	285 bp	[96]
R: ACTTCTTGAGCGCYTTGCTC				
RT with random hexamers	*gag*	30′ 95°C, 5′ 25°C, 10′ 48°C	TaqMan	[28]
F: CAATGCAGAAATGCGCAAAA		45 cycles (15″ 95°C, 60″ 60°C)	FAM probe	
R: GCTGACCCTTCTGCTGTATGG				
Pr: CCTCCCCTGAGCCCC				
RT with reverse PCR primer (R)	*LTR/5′ UTR*	5′ 50°C, 30″95°C	TaqMan	[7]
F: GGCGCCCGAACAGGGACC		35 cycles (15″ 95°C, 30″ 52°C, 60″ 60°C)	FAM probe	
R: TGGCCAGGAACACCTCCAGAAGAC				
Pr: TGAACCTGGCTGATCGTAGGA				

LTR, long terminal repeat; 5′ UTR, 5′ untranslated region; F, forward primer; R, reverse primer; Pr, probe.

[a] Location of oligonucleotide sequences within the EIAV genome.

[b] Assay conditions as published; these will vary depending on the source of the PCR reagents.

region than within Gag protein coding sequences (Figure 20.4). However, this conclusion is based on a very small number of published sequences and must be confirmed by additional studies.

The fact that there does not appear to be a "universal" PCR-based system for EIAV suggests that researchers attempting to detect nucleic acid sequences from field cases of suspected exposure to this virus should not limit themselves to a single published assay. Furthermore, they might consider testing a range of annealing temperatures in each assay using a temperature gradient and/or design "degenerate" primers based on published sequences to overcome potential mismatches. However, equids as with all mammals possess significant amounts of retroviral-derived material in their chromosomal DNA. Therefore, particularly when using cellular DNA as a template in PCR-based assays, it is possible that lowering annealing temperatures and/or using degenerate primers will result in false-positive amplification of sequences that are not of EIAV origin. Although these can usually be excluded on the basis of size, modified PCR assays should be tested using samples obtained from EIAV seronegative animals and all DNA products sequenced to determine their authenticity.

20.3   CONCLUSION AND FUTURE PERSPECTIVES

EIAV possesses the smallest, least complex genome of any extant lentivirus, and yet it is a highly successful pathogen (from a viral perspective) establishing persistent infections by resisting or evading host-defensive responses. Furthermore, with its widespread distribution, this virus poses a significant threat to most of the world's equine populations. The lack of pathognomonic signs or lesions coupled with the fact that most EIAV-infected equids eventually become inapparent carriers enhances the risks of inadvertent transmission and complicates diagnosis by, with currently available technology, necessitating the use of laboratory-based testing procedures.

Conventional direct diagnostic techniques such as virus isolation in equine macrophage cultures derived from bone marrow biopsies or PBMC are not practical for routine use and therefore confirming exposure to this virus is at present dependent on indirect serological methods with most countries reliant exclusively on the AGID or "Coggins" test. This 40-year-old assay has a well-deserved reputation for specificity, but recent studies have confirmed that it is prone to producing false-negative results.[6,7] Although the probability of detecting EIAV-infected equids is improved by adoption of

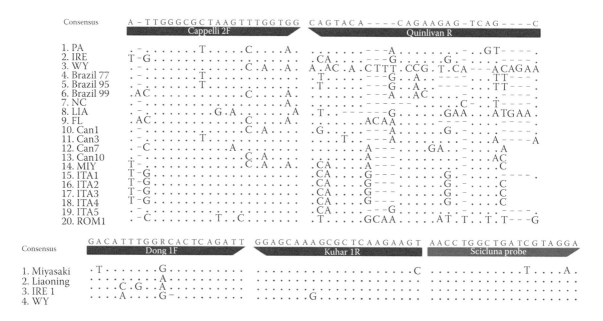

FIGURE 20.4 Examples of variation within primer or TaqMan probe binding regions of recently described PCR-based assays for equine infectious anemia virus. Primer binding regions for PCR-based assays described by Cappelli et al.[26] and Quinlivan et al.[28] are within the viral *gag* gene (comparison sequences: GenBank accession numbers AB008196, AF016316, AF327877, EF418582, EF418583, EF418584, EF418585, EU24073, EU375543, EU375544, EU741609, GQ265785, GQ229581, GQ855755, JX480631, and JX003263[29]) whereas those by Dong et al.,[33] Scicluna et al.,[7] and Kuhar et al.[94] are located within the U3 region of the 5′ long terminal repeat and/or the 5′ untranslated region for which less comparative nucleotide sequence information is available (GenBank accession numbers AF327877, AF016316, JX480631, JX003263). A dot indicates nucleotide sequence identity, whereas a dash indicates a deletion.

a three-tier diagnostic system in which all serum samples are initially screened by less subjective, potentially more sensitive EIA-ELISA with confirmation by AGID followed by immunoblot testing if required,[6,7] there is an urgent need for direct techniques that can detect infected equids shortly after exposure to EIAV without having to wait for production of virus-specific antibodies.

To fulfill this requirement, a number of PCR-based tests for detection of EIAV nucleic acids have been described although these have yet to be validated against viral strains from different geographical regions. In addition, alignment studies using published EIAV sequences suggest that none of the primer binding sites (or probe binding sites in the case of TaqMan-type real-time PCR assays), described to date, including those currently recommended by the OIE, are located in regions of the viral genome that show complete conservation between isolates. However, three of the PCR assays described to date have primer (and TaqMan probe in the case of[7]) binding sites within the LTR U5 domain and/or the 5′ untranslated region,[7,33,94] areas of the EIAV genome that appear to contain the highest levels of nucleotide sequence identity between strains. Unfortunately, this conclusion is based on limited experimental information but, if confirmed by additional studies, suggests that one or more of these assays, perhaps with modifications such as the use of degenerate primers (or probes) to compensate for the relatively small amounts of nucleotide variation present in these regions of the viral genome, could form the basis of a viable direct diagnostic test for EIAV. Alternatively, these regions of the viral genome could be used to design new amplification primers and detection probes.

While correct primer (and probe) design is obviously important for the detection of EIAV nucleic acids using either conventional PCR-based techniques or some of the newer approaches such as isothermal amplification, efficient sample material extraction/purification to ensure maximum recovery is also a critical parameter especially in the case of inapparent carriers where viral loads are expected to be low. In some cases, it might even be worth investigating if detection rates can be improved by sequence-independent amplification of viral RNA.[96] Finally, some researchers have reported that for most inapparent carriers, EIAV sequences can only be amplified from DNA extracted from PBMC and not from plasma RNA preparations suggesting that during this stage of the disease proviral, DNA in PBMC is more abundant than plasma-associated viral RNA.[26,33] However, this finding is not universal and so it must be determined if these discrepant results reflect differences in horse/viral populations or methodologies.

In view of the extent of genetic variation between geographically distinct EIAV strains, development of broadly reactive nucleic acid detection techniques for this virus will almost certainly require international, multicenter collaborative efforts. The laboratories involved will need to carefully evaluate different viral nucleic acid extraction and amplification techniques using a broad range of samples. Furthermore, they will need to determine if proviral DNA, viral RNA, or both should be the preferred starting material.

REFERENCES

1. Lignee, M. Memoire et observations sur ure maladie de sarg, connue sous le nom d'anhemie hydrohemie: Cachexie aqueuse du cheval. *Rec Med Vet* 20, 30 (1843).

2. Vallée, H. and Carré, H. Sur la natur infectieuse de l'anenie du cheval. *C R Acad Sci* 139, 331–333 (1904).

3. Cook, S.J., Cook, R.F., Montelaro, R.C., and Issel, C.J. Differential responses of Equus caballus and Equus asinus to infection with two pathogenic strains of equine infectious anemia virus. *Vet Microbiol* 79, 93–109 (2001).

4. Coggins, L., Norcross, N.L., and Nusbaum, S.R. Diagnosis of equine infectious anemia by immunodiffusion test. *Am J Vet Res* 33, 11–18 (1972).

5. Pearson, J.E. Protocol for the immunodiffusion (Coggins) test for equine infectious anemia. *U-S-Livest-Sanit-Assoc-Proc* 75, 649–659 (1971).

6. Issel, C.J. et al. Challenges and proposed solutions for more accurate serological diagnosis of equine infectious anaemia. *Vet Rec* 172, 210 (2013).

7. Scicluna, M.T. et al. Is a diagnostic system based exclusively on agar gel immunodiffusion adequate for controlling the spread of equine infectious anaemia? *Vet Microbiol* 165, 123–134 (2013).

8. Hines, R. and Maury, W. DH82 cells: A macrophage cell line for the replication and study of equine infectious anemia virus. *J Virol Methods* 95, 47–56 (2001).

9. Carpenter, S. and Chesebro, B. Change in host cell tropism associated with in vitro replication of equine infectious anemia virus. *J Virol* 63, 2492–2496 (1989).

10. Leroux, C., Cadore, J.L., and Montelaro, R.C. Equine Infectious Anemia Virus (EIAV): What has HIV's country cousin got to tell us? *Vet Res* 35, 485–512 (2004).

11. Matheka, H.D., Coggins, L., Shively, J.N., and Norcross, N.L. Purification and characterization of equine infectious anemia virus. *Arch Virol* 51, 107–114 (1976).

12. Weiland, F., Matheka, H.D., Coggins, L. and Hartner, D. Electron microscopic studies on equine infectious anemia Virus (EIAV). *Arch Virol* 55, 335–340 (1977).

13. Cook, R.F., Berger, S.L., Cook, S.J., Ghabrial, N.N., Leroux, C., Montelaor, R.C., and Issel, C.J. Induction of disease in horses and ponies by molecularly cloned equine infectious anemia virus (EIAV). *Proceedings of the Eighth International Conference on Equine Infectious Diseases*, R&W Publications (Newmarket) Limited, Great Britain, U.K., p. 397 (1999).

14. Dong, J.B. et al. Identification of a novel equine infectious anemia virus field strain isolated from feral horses in southern Japan. *J Gen Virol* 94, 360–365 (2013).

15. Perry, S.T. et al. The surface envelope protein gene region of equine infectious anemia virus is not an important determinant of tropism in vitro. *J Virol* 66, 4085–4097 (1992).

16. Quinlivan, M., Cook, F., Kenna, R., Callinan, J.J., and Cullinane, A. Genetic characterization by composite sequence analysis of a new pathogenic field strain of equine infectious anemia virus from the 2006 outbreak in Ireland. *J Gen Virol* 94, 612–622 (2013).

17. Tu, Y.B. et al. Long terminal repeats are not the sole determinants of virulence for equine infectious anemia virus. *Arch Virol* 152, 209–218 (2007).

18. Villet, S. et al. Maedi-visna virus and caprine arthritis encephalitis virus genomes encode a Vpr-like but no Tat protein. *J Virol* 77, 9632–9638 (2003).

19. Villet, S. et al. Lack of trans-activation function for Maedi Visna virus and Caprine arthritis encephalitis virus Tat proteins. *Virology* 307, 317–327 (2003).

20. Bogerd, H.P., Tallmadge, R.L., Oaks, J.L., Carpenter, S., and Cullen, B.R. Equine infectious anemia virus resists the anti-retroviral activity of equine APOBEC3 proteins through a packaging-independent mechanism. *J Virol* 82, 11889–11901 (2008).

21. Han, G.Z. and Worobey, M. An endogenous foamy-like viral element in the coelacanth genome. *PLoS Pathog* 8, e1002790 (2012).

22. Wertheim, J.O. and Worobey, M. Dating the age of the SIV lineages that gave rise to HIV-1 and HIV-2. *PLoS Comput Biol* 5, e1000377 (2009).

23. Katzourakis, A., Tristem, M., Pybus, O.G., and Gifford, R.J. Discovery and analysis of the first endogenous lentivirus. *Proc Natl Acad Sci USA* 104, 6261–6265 (2007).

24. Keckesova, Z., Ylinen, L.M., Towers, G.J., Gifford, R.J., and Katzourakis, A. Identification of a RELIK orthologue in the European hare (*Lepus europaeus*) reveals a minimum age of 12 million years for the lagomorph lentiviruses. *Virology* 384, 7–11 (2009).

25. Nagarajan, M.M. and Simard, C. Detection of horses infected naturally with equine infectious anemia virus by nested polymerase chain reaction. *J Virol Methods* 94, 97–109 (2001).

26. Cappelli, K. et al. Molecular detection, epidemiology, and genetic characterization of novel European field isolates of equine infectious anemia virus. *J Clin Microbiol* 49, 27–33 (2011).

27. Cullinane, A. et al. Diagnosis of equine infectious anaemia during the 2006 outbreak in Ireland. *Vet Rec* 161, 647–652 (2007).

28. Quinlivan, M., Cook, R.F., and Cullinane, A. Real-time quantitative RT-PCR and PCR assays for a novel European field isolate of equine infectious anaemia virus based on sequence determination of the gag gene. *Vet Rec* 160, 611–618 (2007).

29. Capomaccio, S. et al. Detection, molecular characterization and phylogenetic analysis of full-length equine infectious anemia (EIAV) gag genes isolated from Shackleford Banks wild horses. *Vet Microbiol* 157, 320–332 (2012).

30. Reis, J.K. et al. Recombinant envelope protein (rgp90) ELISA for equine infectious anemia virus provides comparable results to the agar gel immunodiffusion. *J Virol Methods* 180, 62–67 (2012).

31. Borges, A.M. et al. Prevalence and risk factors for Equine Infectious Anemia in Pocone municipality, northern Brazilian Pantanal. *Res Vet Sci* 95, 76–81 (2013).

32. Dong, J. et al. Comparative analysis of LTR and structural genes in an equine infectious anemia virus strain isolated from a feral horse in Japan. *Arch Virol* 159, 3413–3420 (2014).

33. Dong, J.B. et al. Development of a nested PCR assay to detect equine infectious anemia proviral DNA from peripheral blood of naturally infected horses. *Arch Virol* 157, 2105–2111 (2012).

34. Puffer, B.A., Parent, L.J., Wills, J.W., and Montelaro, R.C. Equine infectious anemia virus utilizes a YXXL motif within the late assembly domain of the GAP p9 protein. *J Virol* 71, 6541–6546 (1997).

35. Jin, S., Chen, C., and Montelaro, R.C. Equine infectious anemia virus Gag p9 function in early steps of virus infection and provirus production. *J Virol* 79, 8793–8801 (2005).

36. Bakhanashvili, M. and Hizi, A. Fidelity of DNA synthesis exhibited in vitro by the reverse transcriptase of the lentivirus equine infectious anemia virus. *Biochemistry* 32, 7559–7567 (1993).

37. Belshan, M., Harris, M.E., Shoemaker, A.E., Hope, T.J., and Carpenter, S. Biological characterization of Rev variation in equine infectious anemia virus. *J Virol* 72, 4421–4426 (1998).

38. Salinovich, O. et al. Rapid emergence of novel antigenic and genetic variants of equine infectious anemia virus during persistent infection. *J Virol* 57, 71–80 (1986).

39. Leroux, C., Issel, C.J., and Montelaro, R.C. Novel and dynamic evolution of equine infectious anemia virus genomic quasispecies associated with sequential disease cycles in an experimentally infected pony. *J Virol* 71, 9627–9639 (1997).

40. Zheng, Y.H. et al. Insertions, duplications and substitutions in restricted gp90 regions of equine infectious anaemia virus during febrile episodes in an experimentally infected horse. *J Gen Virol* 78, 807–820 (1997).

41. Li, F. et al. The S2 gene of equine infectious anemia virus is a highly conserved determinant of viral replication and virulence properties in experimentally infected ponies. *J Virol* 74, 573–579 (2000).

42. Rwambo, P.M. et al. Equine infectious anemia virus (EIAV) humoral responses of recipient ponies and antigenic variation during persistent infection. *Arch Virol* 111, 199–212 (1990).

43. Kono, Y. Antigenic variation of equine infectious anemia virus as detected by virus neutralization. *Arch Virol* 98, 91–97 (1988).

44. Leroux, C., Craigo, J.K., Issel, C.J., and Montelaro, R.C. Equine infectious anemia virus genomic evolution in progressor and nonprogressor ponies. *J Virol* 75, 4570–4583 (2001).

45. Stein, C.D., Lotze, J.C., and Mott, L.O. Transmission of equine infectious anemia by the stable fly, *Stomoxys calcitrans*, the horsefly, *Tabanus sulcifrons* (Macquart), and by injection of minute amounts of virus. *Am J Vet Res* April, 183–193 (1942).

46. Hawkins, J.A., Adams, W.V., Cook, L., Wilson, B.H., and Roth, E.E. Role of horse fly (*Tabanus fuscicostatus* Hine) and stable fly (*Stomoxys calcitrans* L.) in transmission of equine infectious anemia to ponies in Louisiana. *Am J Vet Res* 34, 1583–1586 (1973).

47. Hawkins, J.A., Adams, W.V., Wilson, B.H., Issel, C.J., and Roth, E.E. Transmission of equine infectious anemia virus by *Tabanus fuscicostatus*. *J Am Vet Med Assoc* 168, 63–64 (1976).

48. Kemen, M.J., McClain, D.S., and Matthysse, J.G. Role of horse flies in transmission of equine infectious anemia from carrier ponies. *J Am Vet Med Assoc* 172, 360–362 (1978).

49. Foil, L., Adams, W.V.J., Issel, C.J., and Pierce, R. Tabanid (Diptera) populations associated with an equine infectious anemia outbreak in an inapparently infected herd of horses [*Hybomitra lasiophthalma, Tabanus lineola*, virus]. *J Med Entomol* 21, 28–30 (1984).

50. Issel, C.J. and Foil, L.D. Studies on equine infectious anemia virus transmission by insects. *J Am Vet Med Assoc* 184, 293–297 (1984).

51. Foil, L.D., Adams, W.V., McManus, J.M., and Issel, C.J. Bloodmeal residues on mouthparts of *Tabanus fuscicostatus* (Diptera: Tabanidae) and the potential for mechanical transmission of pathogens. *J Med Entomol* 24, 613–616 (1987).

52. Williams, D.L., Issel, C.J., Steelman, C.D., Adams, W.V.J., and Benton, C.V. Studies with equine infectious anemia virus: Transmission attempts by mosquitoes and survival of virus on vector mouthparts and hypodermic needles, and in mosquito tissue culture. *Am J Vet Res* 42, 1469–1473 (1981).

53. Kemen, M.J. and Coggins, L. Equine infectious anemia: Transmission from infected mares to foals. *J Am Vet Med Assoc* 161, 496–499 (1972).

54. Bolfa, P. et al. Interstitial lung disease associated with Equine Infectious Anemia Virus infection in horses. *Vet Res* 44, 113 (2013).

55. Peterhans, E. et al. Routes of transmission and consequences of small ruminant lentiviruses (SRLVs) infection and eradication schemes. *Vet Res* 35, 257–274 (2004).

56. Hammond, S.A., Cook, S.J., Lichtenstein, D.L., Issel, C.J., and Montelaro, R.C. Maturation of the cellular and humoral immune responses to persistent infection in horses by equine infectious anemia virus is a complex and lengthy process. *J Virol* 71, 3840–3852 (1997).

57. Hammond, S.A. et al. Immune responses and viral replication in long-term inapparent carrier ponies inoculated with equine infectious anemia virus. *J Virol* 74, 5968–5981 (2000).

58. Kemeny, L.J., Mott, L.O., and Pearson, J.E. Titration of equine infectious anemia virus: Effect of dosage on incubation time and clinical signs. *Cornell Vet* 61, 687–695 (1971).

59. Maury, W. Monocyte maturation controls expression of equine infectious anemia virus. *J Virol* 68, 6270–6279 (1994).

60. Maury, W., Oaks, J.L., and Bradley, S. Equine endothelial cells support productive infection of equine infectious anemia virus. *J Virol* 72, 9291–9297 (1998).

61. Cook, R.F. et al. Enhancement of equine infectious anemia virus virulence by identification and removal of suboptimal nucleotides. *Virology* 313, 588–603 (2003).

62. Costa, L.R., Santos, I.K., Issel, C.J., and Montelaro, R.C. Tumor necrosis factor-alpha production and disease severity after immunization with enriched major core protein (p26) and/or infection with equine infectious anemia virus. *Vet Immunol Immunopathol* 57, 33–47 (1997).

63. Sellon, D.C., Russell, K.E., Monroe, V.L., and Walker, K.M. Increased interleukin-6 activity in the serum of ponies acutely infected with equine infectious anaemia virus. *Res Vet Sci* 66, 77–80 (1999).

64. Tornquist, S.J. and Crawford, T.B. Suppression of megakaryocyte colony growth by plasma from foals infected with equine infectious anemia virus. *Blood* 90, 2357–2363 (1997).

65. Tornquist, S.J., Oaks, J.L., and Crawford, T.B. Elevation of cytokines associated with the thrombocytopenia of equine infectious anaemia. *J Gen Virol* 78, 2541–2548 (1997).

66. Tacchini-Cottier, F., Vesin, C., Redard, M., Buurman, W., and Piguet, P.F. Role of TNFR1 and TNFR2 in TNF-induced platelet consumption in mice. *J Immunol* 160, 6182–6186 (1998).

67. Perryman, L.E., McGuire, T.C., Banks, K.L., and Henson, J.B. Decreased C3 levels in a chronic virus infection: Equine infectious anemia. *J Immunol* 106, 1074–1078 (1971).

68. Sentsui, H. and Kono, Y. Phagocytosis of horse erythrocytes treated with equine infectious anemia virus by cultivated horse leukocytes. *Arch Virol* 95, 67–78 (1987).

69. Henson, J.B. and McGuire, T.C. Immunopathology of equine infectious anemia. *Am J Clin Pathol* 56, 306–313 (1971).

70. Bolfa, P.F. et al. Oxidant-antioxidant imbalance in horses infected with equine infectious anaemia virus. *Vet J* 192, 449–454 (2012).

71. Perryman, L.E., O'Rourke, K.I., and McGuire, T.C. Immune responses are required to terminate viremia in equine infectious anemia lentivirus infection. *J Virol* 62, 3073–3076 (1988).

72. Mealey, R., Fraser, D., Oaks, J., Cantor, G., and Mcguire, T. Immune reconstitution prevents continuous equine infectious anemia virus replication in an Arabian foal with severe combined immunodeficiency: Lessons for control of lentiviruses. *Clin Immunol* 101, 237–247 (2001).

73. Mealey, R.H. et al. Early detection of dominant Env-specific and subdominant Gag-specific CD8+ lymphocytes in equine infectious anemia virus-infected horses using major histocompatibility complex class I/peptide tetrameric complexes. *Virology* 339, 110–126 (2005).

74. Taylor, S.D., Leib, S.R., Carpenter, S., and Mealey, R.H. Selection of a rare neutralization-resistant variant following passive transfer of convalescent immune plasma in equine infectious anemia virus-challenged SCID horses. *J Virol* 84, 6536–6548 (2010).

75. Taylor, S.D. et al. Protective effects of broadly neutralizing immunoglobulin against homologous and heterologous equine infectious anemia virus infection in horses with severe combined immunodeficiency. *J Virol* 85, 6814–6818 (2011).

76. Kono, Y., Hirasawa, K., Fukunaga, Y., and Taniguchi, T. Recrudescence of equine infectious anemia by treatment with immunosuppressive drugs. *Natl Inst Anim Health Q* 16, 8–15 (1976).

77. Tumas, D.B., Hines, M.T., Perryman, L.E., Davis, W.C., and McGuire, T.C. Corticosteroid immunosuppression and monoclonal antibody-mediated CD5+ T lymphocyte depletion in normal and equine infectious anaemia virus-carrier horses. *J Gen Virol* 75, 959–968 (1994).

78. Craigo, J.K. et al. Immune suppression of challenged vaccinates as a rigorous assessment of sterile protection by lentiviral vaccines. *Vaccine* 25, 834–845 (2007).

79. Kong, X.G. et al. Evaluation of equine infectious anemia virus core proteins produced in a baculovirus expression system in agar gel immunodiffusion test and enzyme-linked immunosorbent assay. *J Vet Med Sci* 60, 1361–1362 (1998).

80. Kong, X.G. et al. Application of equine infectious anemia virus core proteins produced in a baculovirus expression system to serological diagnosis. *Microbiol Immunol* 41, 975–980 (1997).

81. Alvarez, I., Gutierrez, G., Barrandeguy, M., and Trono, K. Immunochromatographic lateral flow test for detection of antibodies to Equine infectious anemia virus. *J Virol Methods* 167, 152–157 (2010).

82. Singha, H., Goyal, S.K., Malik, P., Khurana, S.K., and Singh, R.K. Development, evaluation, and laboratory validation of immunoassays for the diagnosis of equine infectious anemia (EIA) using recombinant protein produced from a synthetic p26 gene of EIA virus. *Ind J Virol* 24, 349–356 (2013).

83. Tencza, S.B. et al. Development of a fluorescence polarization-based diagnostic assay for equine infectious anemia virus. *J Clin Microbiol* 38, 1854–1859 (2000).

84. Ball, J.M., Henry, N.L., Montelaro, R.C., and Newman, M.J. A versatile synthetic peptide-based ELISA for identifying antibody epitopes. *J Immunol Methods* 171, 37–44 (1994).

85. Soutullo Adriana, V.V., Mario, R., Roberto, P., and Georgina, T. Design and validation of an ELISA for equine infectious anemia (EIA) diagnosis using synthetic peptides. *Vet Microbiol* 79, 111–121 (2001)

86. Thomas, L.M. et al. A soluble recombinant fusion protein of the transmembrane envelope protein of equine infectious anaemia virus for ELISA. *Vet Microbiol* 31, 127–137 (1992).

87. Rwambo, P.M., Issel, C.J., Hussain, K.A., and Montelaro, R.C. In vitro isolation of a neutralization escape mutant of equine infectious anemia virus (EIAV). *Arch Virol* 111, 275–280 (1990).

88. McConnico, R.S. et al. Predictive methods to define infection with equine infectious anemia virus in foals out of reactor mares. *J Equine Vet Sci* 20, 387–392 (2000).

89. Harrold, S.M. et al. Tissue sites of persistent infection and active replication of equine infectious anemia virus during acute disease and asymptomatic infection in experimentally infected equids. *J Virol* 74, 3112–3121 (2000).

90. Issel, C.J. and Adams, W.V.J. Detection of equine infectious anemia virus in a horse with an equivocal agar gel immunodiffusion test reaction. *J Am Vet Med Assoc* 180, 276–278 (1982).

91. Langemeier, J.L. et al. Detection of equine infectious anemia viral RNA in plasma samples from recently infected and long-term inapparent carrier animals by PCR. *J Clin Microbiol* 34, 1481–1487 (1996).

92. Cook, R.F, Cook, S.J., Li, F., Montelaro, R.C., and Issel, C.J. Development of a multiplex real-time reverse transcriptase-polymerase chain reaction for equine infectious anemia virus (EIAV). *J Virol Methods* 105, 171–179 (2002).

93. Boldbaatar, B. et al. Amplification of complete gag gene sequences from geographically distinct equine infectious anemia virus isolates. *J Virol Methods* 189, 41–46 (2013).

94. Kuhar, U., Zavrsnik, J., Toplak, I., and Malovrh, T. Detection and molecular characterisation of equine infectious anaemia virus from field outbreaks in Slovenia. *Equine Vet J* 46, 386–391 (2014).

95. Chen, C. and Montelaro, R.C. Characterization of RNA elements that regulate gag-pol ribosomal frameshifting in equine infectious anemia virus. *J Virol* 77, 10280–10287 (2003).

96. Malboeuf, C.M. et al. Complete viral RNA genome sequencing of ultra-low copy samples by sequence-independent amplification. *Nucleic Acids Res* 41, e13 (2013).

97. Pinter, A. et al. Oligomeric structure of gp41, the transmembrane protein of human immunodeficiency virus type 1. *J Virol* 63, 2674–2679 (1989).

98. Tran, E.E. et al. Structural mechanism of trimeric HIV-1 envelope glycoprotein activation. *PLoS Pathog* 8, e1002797 (2012).

21 Feline Immunodeficiency Virus

Dongyou Liu

CONTENTS

21.1 Introduction 191
    21.1.1 Classification..191
    21.1.2 Morphology and Genome Organization 192
    21.1.3 Replication and Epidemiology..192
    21.1.4 Clinical Features and Pathogenesis 193
    21.1.5 Diagnosis 194
    21.1.6 Treatment and Prevention...194
21.2 Methods 194
    21.2.1 Sample Preparation...194
    21.2.2 Detection Procedures 195
        21.2.2.1 Procedure 195
21.3 Conclusion 195
References...195

21.1 INTRODUCTION

Feline immunodeficiency virus (FIV) is a retrovirus that is responsible for an immune disorder in cats throughout the world. Clinically, FIV infection is characterized by an acute phase with fever, leucopenia, gingivitis, and generalized lymphadenopathy, followed by a latent phase without obvious symptoms, and a last, symptomatic phase with CD4+ lymphocyte depletion and increased susceptibility to opportunistic infections and neoplasia. Given the occurrence of two other retroviruses in cats, that is, feline leukemia virus (FeLV) and feline foamy virus (FeFV), that induce similar clinical symptoms to FIV, it is important that FIV infection is accurately diagnosed for implementation of appropriate disease management and prevention. The purpose of this chapter is to present a brief overview on FIV relating to its classification, morphology, genome organization, epidemiology, clinical features, pathogenesis, treatment, and prevention, together with a streamline molecular protocol for FIV identification.

21.1.1 CLASSIFICATION

FIV is a member of the genus *Lentivirus*, family *Retroviridae*, which covers a group of retroviruses with the unique property of transcribing their RNA into DNA after entering a cell. Subsequent integration of the retroviral DNA into the chromosomal DNA of the host cell allows its propagation in situ. Of the two subfamilies within the *Retroviridae* family, the subfamily *Orthoretrovirinae* is divided into six genera (*Alpharetrovirus*, *Betaretrovirus*, *Deltaretrovirus*, *Epsilonretrovirus*, *Gammaretrovirus*, and *Lentivirus*), whereas the subfamily *Spumaretrovirinae* is made up of a single genus (*Spumavirus*). The genus *Lentivirus* consists

of 10 recognized species that are organized into five groups: bovine lentivirus group (bovine immunodeficiency virus [BIV] and Jembrana disease virus), equine lentivirus group (equine infectious anemia virus [EIAV]), feline lentivirus group (FIV and puma lentivirus), ovine/caprine lentivirus group (caprine arthritis–encephalitis virus [CAEV] and maedi–visna virus [MVV]), and primate lentivirus group (human immunodeficiency virus 1 [HIV-1, the type species], human immunodeficiency virus 2 [HIV-2], and simian immunodeficiency virus [SIV]). It is notable that members of the genus *Lentivirus* (lenti-, Latin for "slow") cause immunodeficiency and/or multiorgan diseases in respective hosts, which are characterized by a long incubation period before onset of clinical symptoms, disease progression despite a strong host immune response, and a predictably deadly outcome.

Based on their cellular tropism and disease manifestations, members of the genus *Lentivirus* may be separated into two groups. The first group includes HIV-1, HIV-2, SIV, FIV, and BIV, which infect lymphocytes and cells of the monocyte/macrophage lineage, leading to immunodeficiency syndrome. The second group contains EIAV, MVV, and CAEV, which replicate predominantly in macrophages, leading to multiorgan disease.

Phylogenetic analysis of the envelope (*env*), group-specific antigen (*gag*), and polymerase (*pol*) genes has enabled further identification of five subtypes (A, B, C, D, E) in addition to a putative subtype F within the FIV species [1]. Of these, subtypes A and B commonly occur in the United States and Canada; subtypes B, C, and D are present in Asia; subtypes A, B, C, and D are found in Europe; subtype A is identified in Australia and Africa; and subtypes B and E are detected in South Africa [2].

21.1.2 MORPHOLOGY AND GENOME ORGANIZATION

Similar to other lentiviruses, FIV is an enveloped, spherical to pleomorphic particle of 80–100 nm in diameter. The virion contains a host cell–acquired membrane (the lipid bilayer) from which viral glycoproteins protrude as spike-like projections (8 nm). Mature capsid is composed of 1572 capsid proteins.

The genome of FIV is dimeric, comprising two identical, linear, positive-sense, single-stranded RNA of about 9400 nt each. Typical of genome structures found in other retroviruses, the FIV genome is organized in the order 5′-LTR-gag(MA-CA-NC)-pol(PR-RT-DU-IN)-vif-orfA-rev-env-rev-LTR-3′ [3].

The long terminal repeats located at the 5′ and 3′ ends are about 600 nt in length, which include the U3, R, and U5 regions. The 5′ end is capped and contains a primer binding site. The 3′ end has a polypurine tract and poly-A tail.

Of the open reading frames contained in the FIV genome, the *gag*, *pol* (polymerase), and *env* genes are found in all retroviruses, the *vif* (viral infectivity factor) and *rev* (regulator of virion expression) genes are lentivirus-specific accessory genes, and the *orf-A* gene is a FIV-specific accessory gene (which is thought to have functional overlap with the *vpr*, *vpu*, and *nef* genes in HIV-1).

The Gag polyprotein encoded by *gag* is cleaved into matrix (MA, also known as p15), capsid (CA, p24), and nucleocapsid (NC, p13) proteins. Cleavage between CA and NC generates a 9 aa peptide, and cleavage at the C-terminus of NC produces a 2 kDa fragment (p2). Structurally, the purified FIV CA protein p24 is comprised mainly of α-helices, appearing as monomeric at low concentration and dimeric at high concentration. During virus assembly, the CA protein p24 is packaged into a viral core (the protein shell of a virus, protecting the viral genome) and the MA protein forms a shell underneath the lipid bilayer [3].

The Pol polyprotein is translated from *pol* via ribosomal frame-shifting mechanism. Cleavage of Pol by the viral protease produces protease (PR, p55, and p61 dimers), reverse transcriptase, deoxyuridine triphosphatase (dUTPase or DU), and integrase (IN). FIV dUTPase reduces levels of dUTP by converting it to dUMP, a precursor for dTTP synthesis, and thus limits uracil misincorporation during reverse transcription (RT). In dividing cells, cellular dUTPase levels are higher and dUTP levels are low, but in nondividing cells, the opposite is true. FIV IN contains three domains: N and C terminal domains (for 3′ end processing and joining) and a catalytic domain in the core [3].

The Env polyprotein (150 kDa) encoded by *env* is processed into a 130 kDa molecule with the removal of a leader peptide (L, of 174 residues). The 130 kDa Env precursor is further cleaved into surface unit (SU, gp95) and transmembrane (TM, gp38) glycoproteins. While FIV SU mediates binding to the cell surface receptors, TM promotes fusion of the viral and cellular membranes during virus entry. Being heavily glycosylated, the SU and TM glycoproteins may mask potential neutralizing epitopes that are recognized by the B cell and increase the virus resistance to host's neutralizing antibodies [4].

The Vif protein promotes viral replication by binding host APOBEC3 (A3) cytidine deaminases and inducing their degradation via proteasome. Without A3 degradation, the virus is prone to G-to-A hypermutation and yields reduced amount of RT products, leading to low or no viral replication [5].

The Rev protein binds to a highly structured *cis*-acting RNA element known as the Rev responsive element (RRE) and helps (1) shuttling of viral structural protein mRNAs from nucleus to cytoplasm, (2) interacting with the RRE to suppress mRNA splicing, (3) increasing stability and translation level of the incompletely spliced viral mRNAs, and (4) enhancing viral RNA encapsidation, facilitating structural protein expression and genomic RNA production [6].

The OrfA protein (77 aa) encoded by *orfA* (also known as *orf2*) is a multifunctional protein involved in transactivation of viral protein expression, virion formation and infectivity, cell cycle arrest, and downregulation of cell surface expression of the primary FIV receptor CD134 [5].

21.1.3 REPLICATION AND EPIDEMIOLOGY

Like other lentiviruses, FIV replicates in two ways: lytic and latent replications. Lytic replication occurs in the nucleus and begins with virus attachment via the SU glycoprotein to host receptor CD134. This initial binding changes the shape of the SU protein, which facilitates interaction between SU and the chemokine receptor CXCR4. This interaction then results in the fusion of viral and cellular membranes (which is mediated by TM glycoprotein) and subsequent viral nucleocapsid entry into the host cell cytoplasm [7]. Inside the cell, the virus uncoats partially, and a linear dsDNA molecule is synthesized from ssRNA (+) genome by the reverse transcriptase. The viral dsDNA enters the nucleus and is covalently and randomly integrated into the cell's genome by the viral integrase (provirus). Transcription of provirus by Pol II yields viral spliced and unspliced RNAs, and further translation of spliced viral RNAs produces Rev and other proteins. Nuclear export of the incompletely spliced RNA is mediated by Rev, and translation of unspliced viral RNAs generates Env, Gag, and Gag–Pol polyproteins. New virion is assembled at the host cellular membrane with packaging of the viral RNA genome. After budding through the plasma membrane, the virions are released. Proteolytic processing of the precursor polyproteins by viral protease ushers in maturation of the virions [3].

Latent replication takes place with a provirus integrated in the host chromosome. Driven by the promotor elements in the 5′ LTR, provirus transcribes into the unspliced full length mRNA, which serves as genomic RNA for packaging into virions or as a template for translation of gag and gag (pr) pol (1 ribosomal frameshift) polyproteins. The incompletely spliced mRNA encodes Env and the accessory protein Vif. Rev is then cleaved into SU and TM envelope proteins, which escort unspliced and incompletely spliced RNAs out of the nucleus of infected cells, and are subsequently incorporated into virions [8].

First isolated in 1986 from domestic cats (*Felis catus*) with immunodeficiency-like syndromes in a Northern California cattery, FIV has been shown to be widely distributed in the world, infecting many of the 37 *Felidae* species, such as European wild cat, African wild cat, desert cat, black-footed cat, jungle cat, pallas cat, puma, cheetah, bobcat, ocelot, Geoffroy's cat, lion, leopard, and tiger. [9–19].

The primary mode of FIV transmission is deep bite wounds and scratches, with blood-tainted saliva from the infected cat entering the other cat's bloodstream. Other means of transmission include blood transfusion and artificial insemination [20]. Occasionally, FIV may be vertically transmitted from pregnant females to their offspring in utero, usually during passage through the birth canal or when the newborn kittens ingest infected milk. This is especially so if the queen is undergoing an acute infection.

Risk factors for FIV infection are being of the male sex, adulthood, and outdoor access. Males are more likely to be infected than females due to the bite wounds from fights over territory.

Cats with FIV are persistently infected in spite of their ability to mount antibody and cell-mediated immune responses.

21.1.4 Clinical Features and Pathogenesis

The initial stage (or acute phase) of FIV infection in domestic cats is accompanied by mild symptoms, such as fever, lethargy, poor coat condition, anorexia, pyrexia, enteritis, stomatitis, dermatitis, conjunctivitis, respiratory tract disease, and peripheral lymphadenopathy.

Lasting several days to a few weeks, this initial stage is followed by the latent asymptomatic stage, in which the infected cats may appear healthy for months or years. Factors that influence the length of the asymptomatic stage include the virulence of FIV subtype (A–E), exposure to other pathogens, and the age of the cat [21].

With the increasing depletion of CD4+ T cells and inversion of CD4/CD8 ratio, the infection progresses to the last, symptomatic phase (commonly known as the feline acquired immune deficiency syndrome, or FAIDS). Impairment of the cat's immune functions renders it vulnerable to secondary infections and neoplasia, with cats becoming infected with bacteria, viruses, protozoa, and fungi that usually do not affect healthy animals. The emergence of secondary infections and heightened susceptibility to neoplasia are responsible for many of the symptoms associated with FIV. These include gingivitis (inflammation of the gums), stomatitis (mouth inflammation), and chronic rhinitis; hyperthyroidism and diabetes mellitus; persistent diarrhea; uveitis and chorioretinitis; progressive weight loss and severe wasting; lymphomas (mostly B-cell lymphomas), cutaneous squamous cell carcinoma, anemia, leukemia, fibrosarcoma, and mast cell tumors; hypergammaglobulinemia; abortion of kittens; psychotic behavior, twitching movements of the face and tongue, compulsive roaming, dementia, loss of bladder and rectal control, disturbed sleep patterns, nystagmus, ataxia, seizures, intention tremors, and other neuropathological disorders; and death.

Histologically, chronic ulcero-proliferative gingivostomatitis is evidenced by invasion of plasma cells and lymphocytes in the mucosa accompanied by variable degrees of neutrophilic and eosinophilic inflammation. Pathological findings in the caudate nucleus, midbrain, and rostral brain stem include the presence of perivascular infiltrates of mononuclear cells, diffuse gliosis, glial nodules, and white matter pallor.

With proper care, FIV-infected cats can live many years and commonly die at an old age from causes unrelated to FIV infection.

Although the primary cellular targets of FIV are monocytes and monocyte-derived macrophages, the Env protein produced by FIV preferably binds CD134 (OX40), which is largely present in activated CD4+ T cells. Shortly after entry to the host via bite wound, the virus travels to the local lymph nodes, where it reproduces in T lymphocytes. The virus then spreads to other lymph nodes in the body, and its tropism extends further to both CD8+ T cells and B cells, leading to a generalized but usually temporary enlargement of the lymph nodes, along with fever [22].

By replicating in lymphocytes, FIV causes reduction in CD4+ T cells through decreased production secondary to bone marrow or thymic infection, lysis of infected cells (cytopathic effects), destruction of virus-infected cells by the immune system, or death by apoptosis [23,24]. This alters the ratio of CD4 and CD8 T cells [25]. As CD4+ cells have critical roles in promoting and maintaining both humoral and cell-mediated immunity, their reduction weakens host immunosurveillance mechanisms against microbial pathogens and neoplasia, leading to secondary infections and tumor development [26,27]. In addition to immunosuppression, FIV-infected cats may also develop immune-mediated diseases such as hypergammaglobulinemia and hyperproteinemia due to an overactive immune response. Hypergammaglobulinemia reflects polyclonal B-cell stimulation with an excessive antibody production against chronic/persistent infection. Instead of neutralizing the invading pathogens, these overproduced antibodies form antigen–antibody complexes, which deposit in narrow capillary beds, leading to glomerulonephritis, polyarthritis, uveitis, and vasculitis.

Abnormal neurologic function and virus-induced CNS lesions result from neuronal damages by FIV, such as neuronal apoptosis, effects on the neuron supportive functions of astrocytes (the most common cell type of the brain), toxic products released from infected microglia, and cytokines produced in response to viral infection. One of the most important functions of astrocytes is the regulation of extracellular glutamate, a major excitatory neurotransmitter that accumulates as a consequence of neuronal activity. FIV infection of feline astrocytes may inhibit their glutamate-scavenging ability and result in excessive accumulation of extracellular glutamate that contributes to increased susceptibility to oxidative injury, alterations in mitochondrial membrane potential (and disruption of feline astrocytes by FIV infection in the energy-producing capacities of the cell), neuronal toxicity, and death. Occasionally, neurologic signs may be also attributable to opportunistic toxoplasmosis and cryptococcosis.

21.1.5 Diagnosis

Felidae species are susceptible to a number of viruses, including three from the family *Retroviridae*, that is, FIV of the genus *Lentivirus*, FeLV of the genus *Gammaretrovirus*, and FeFV of the genus *Spumavirus*. Differential diagnosis should include these viruses. Although both FIV and FeLV may produce similar clinical symptoms in cats, FIV is associated mainly with B-cell lymphoma, while FeLV causes mainly T-cell lymphoma. In addition, FIV sometime induces immune-mediated diseases such as hyperglobulinemia, while FeLV rarely does so [28,29].

Therefore, it is critical that laboratory methods (e.g., virus isolation, serological assays, and nucleic acid detection) are applied for accurate differentiation of FIV from FeLV and other viral pathogens [30,31].

Considered the gold standard for FIV infection, virus isolation from peripheral blood lymphocytes is a reliable and definitive diagnostic test, but its routine use is hampered by the high cost and lengthy time required [32].

Serological assays detect the presence of antibody (typically against FIV p24 and p15) in the blood of infected cats using an ELISA (e.g., SNAP® FIV/FeLV Combo and PetChek FIV antibody test; IDEXX Laboratories, Westbrook, Maine). Most cats will show detectable antibodies to FIV within 60 days of exposure. Western blot and immunofluorescent antibody assays may be employed as confirmatory tests in seropositive cats with no history of FIV vaccination [33].

Nucleic acid detection procedures such as PCR identify parts of FIV's genetic material (RNA or DNA) and provide a valuable means of determining a cat's true infection status, especially for FIV antibody–positive cats that have an unknown vaccination history or that have been vaccinated against FIV but still suspected to be infected [34–36]. Real-time PCR assay for FIV (e.g., FIV RealPCR™, IDEXX Laboratories) permits quantification of viral DNA in peripheral blood leukocytes [37,38]. Nevertheless, a negative PCR for FIV RNA may possibly reflect a level of viral nucleic acid below the limit of detection or a strain of FIV that is not detected by the test. For this reason, PCR targeting FIV proviral DNA integrated into the genome of infected cells is preferred. Indeed, the development of a *gag*-targeted dual-emission fluorescence resonance energy transfer (FRET) PCR has been shown to reliably and sensitively detect and differentiate FIV subtypes A, B1, C, D, and B2/E (which have identical target sequence) [39].

21.1.6 Treatment and Prevention

FIV is a cat lentivirus with structural, genomic, and pathogenic parallels to HIV. Approaches that have been explored for the development of antiretroviral drugs against HIV infection will enhance the design of FIV-specific treatment. With the goal to interrupt the key steps in the virus replication cycle, anti-HIV drugs have focused on the following areas: (1) inhibition of virus attachment to host cell surface receptors and coreceptors to stop virus entry into susceptible cells; (2) inhibition of fusion of the virus membrane with the cell membrane; (3) blockade of RT of viral genomic RNA; (4) interruption of nuclear translocation and viral DNA integration into host genomes; (5) prevention of viral transcript processing and nuclear export; and (6) inhibition of virion assembly and maturation. This has led to the commercial availability or clinical trials of several classes of anti-HIV drugs (e.g., fusion or entry inhibitors, nucleoside reverse transcriptase inhibitors (NRTI), non-NRTI, NRTI, viral IN blockers, and PR inhibitors). Nonetheless, the effectiveness of these anti-HIV drugs against FIV is yet to be assessed [40–42].

Therefore, current management of FIV infection in cats centers on proactive actions such as (1) neutering/spraying of asymptomatic FIV-infected cats to avoid fighting and virus transmission, (2) housing infected cats in individual cages and proper sterilizing of needles and surgical instruments used on FIV-positive cats, (3) adopting only infection-free cats into a household with uninfected cats, (4) feeding cats with nutritionally complete and balanced diets to enhance natural immunity, (5) avoiding uncooked food (raw meat and eggs and unpasteurized dairy products) that may contain foodborne microbial pathogens, and (6) vaccinating cats against possible infection [40,43].

The first commercial vaccine against FIV (Fel-O-Vax FIV®, Fort Dodge Animal Health, Fort Dodge, IA) has been marketed since 2002 in the United States and several other countries. Based on formalin-inactivated whole viruses of FIV subtype A (Petaluma) and subtype D (Shizuoka), the vaccine has been shown to protect against FIV subtype B, but only partially against FIV subtype A challenge [44,45].

Furthermore, Lymphocyte T-Cell Immunomodulator (manufactured and distributed exclusively by T-Cyte Therapeutics, Inc) is a strongly cationic glycoprotein (50 kDa) with the capacity to regulate CD-4 lymphocyte production and function. By increasing lymphocyte numbers and interleukin-2 production in animals, it aids in the treatment of cats infected with FeLV and/or FIV and the associated symptoms of lymphocytopenia, opportunistic infection, anemia, granulocytopenia, or thrombocytopenia [43].

21.2 METHODS

21.2.1 Sample Preparation

Whole blood samples are collected by either jugular or saphenous venipuncture into EDTA tubes before transportation in insulated containers packed with ice bricks.

Total nucleic acid may be extracted by glass fiber matrix binding and elution with the High Pure PCR template preparation kit (Roche Molecular Biochemicals). For each sample, 400 μL EDTA whole blood is mixed with an equal volume of binding buffer (6 M guanidine-HCl, 10 mM urea, 20% [vol/vol] Triton X-100, 10 mM Tris-HCl, pH 4.4), and nucleic acid is eluted in 40 μL elution buffer [39].

21.2.2 Detection Procedures

Wang et al. [39] developed a real-time PCR for detection and quantification of FIV serotype. While the upstream and downstream primers hybridize at the 5' end of the *gag* gene, the most conserved region of the FIV genome, the probes (F1-ABCE, F1-D, F2-ABCE, and F3-D) hybridize to the polymorphic regions of the *gag* amplification target and allow maximum differentiation of FIV subtypes (Table 21.1).

21.2.2.1 Procedure

PCR buffer (1×) is composed of 20 mM Tris-HCl, pH 8.4, 50 mM KCl, 4.5 mM MgCl$_2$, 0.05% Nonidet™ P-40, 0.05% Tween®, 20, 0.03% acetylated bovine serum albumin, and 200 µM each dATP, dCTP, dGTP, and 600 µM dUTP. Master mixtures are prepared freshly from 10× PCR buffer, 5× oligonucleotide mixture, 50× nucleotide mixture, and enzymes. Each 20-µL reaction contains 15 µL of 1.33× master mixture, 5 µL of sample nucleic acids, 1.5 U Platinum® *Taq* DNA polymerase (Invitrogen), 0.2 U heat-labile uracil-N-glycosylase (Roche Applied Science), and 0.0213 U ThermoScript™ reverse transcriptase (Invitrogen). Primers are used at 0.8 µM, Lightcycler Red 640 probes are used at a concentration of 0.2 µM, and the carboxyfluorescein probes are used at 0.1 µM. For convenient pipetting, ThermoScript reverse transcriptase is used at a 1:140 dilution in storage buffer.

PCR is conducted on a LightCycler 1.5 real-time PCR platform with software version 3.53 (Roche Applied Science). Thermal cycling is preceded by a 20 min RT step at 55°C, followed by a 5 min incubation at 95°C. Thermal cycling consists of a step-down protocol of 6 cycles for 12 s at 64 or 60°C, for 11 s at 72°C, and for 0 s at 95°C; 9 cycles for 12 s at 62 or 58°C, for 11 s at 72°C, and for 0 s at 95°C; 3 cycles for 12 s at 60 or 56°C, for 11 s at 72°C, and for 0 s at 95°C; and 40 cycles for 8 s at 54 or 50°C. Fluorescence acquisition is undertaken for 11 s at 72°C and for 0 s at 95°C.

After completion of the PCR, the melting temperature (T_m) of probe hybridization to the targets is determined by melting curve analysis as the peak of the second derivative of the fluorescence released during a temperature increase from 35°C to 75°C. In channel F2/F1, subtypes A and C can be discriminated by their higher T_ms (58°C–60°C) from those of the B subtype (T_m = 54°C) and the D subtype (no melting peak). Fluorescence F3/F1 demonstrates distinct T_m differences between FIV-A (45°C), FIV-B1 (49°C), FIVB2/E (52°C), FIV-C (55°C), and FIV-D (68°C) subtypes.

Note: This *gag* gene-based dual-emission FRET real-time PCR identifies FIV genotypes by differential melting temperatures of partially mismatched FRET probes at two fluorescence emission wavelengths. In addition, the assay allows conclusive discrimination between FIV antibody-positive cats from natural infection and cats with antibody positivity by vaccination.

21.3 CONCLUSION

Being one of the most common infectious agents of cats, FIV is associated with a severe immune disease that resembles acquired immune deficiency syndrome (AIDS) in humans. Conventional approaches for diagnosis of FIV infection have depended on virus isolation and serological assays. In view of the high cost involved in cell culture and the difficulty of using serological methods to differentiate naturally infected cats from vaccinated cats, molecular techniques (especially PCR) have been increasingly utilized in routine diagnostic laboratories for detection and quantification of FIV RNA and proviral DNA, the latter of which is integrated into the genome of infected cells.

REFERENCES

1. Marçola TG et al. Identification of a novel subtype of feline immunodeficiency virus in a population of naturally infected felines in the Brazilian Federal District. *Virus Genes.* 2013;46(3):546–550.
2. Bachmann MH et al. Genetic diversity of feline immunodeficiency virus: Dual infection, recombination, and distinct evolutionary rates among envelope sequence clades. *J Virol.* 1997;71:4241–4253.
3. Kenyon JC, Lever AM. The molecular biology of feline immunodeficiency virus (FIV). *Viruses.* 2011;3(11):2192–2213.
4. Pancino G et al. Structure and variations of feline immunodeficiency virus envelope glycoproteins. *Virology.* 1993;192:659–662.
5. Troyer RM, Thompson J, Elder JH, VandeWoude S. Accessory genes confer a high replication rate to virulent feline immunodeficiency virus. *J Virol.* 2013;87(14):7940–7951.
6. Na H, Huisman W, Ellestad KK, Phillips TR, Power C. Domain- and nucleotide-specific Rev response element regulation of feline immunodeficiency virus production. *Virology.* 2010;404(2):246–260.
7. Willett BJ, Hosie MJ. The virus-receptor interaction in the replication of feline immunodeficiency virus (FIV). *Curr Opin Virol.* 2013;3(6):670–675.
8. McDonnel SJ, Sparger EE, Murphy BG. Feline immunodeficiency virus latency. *Retrovirology.* 2013;10:69.

TABLE 21.1

Oligonucleotide Primers and Probes for Real-Time PCR Detection of FIV

Primer/Probe	Sequence (5'→3')ᵃ
Upstream primer	ATGGGGAAYGGACAGGGGCGAGA
Downstream primer	TCTGGTATRTCACCAGGTTCTCGTCCTGTA
F1-ABCE	(6-FAM)-TACTCTTSSCCCCT ACTCCTACAGCA-(6-FAM)
F1-D	(6-FAM)-TACTTCGTTGTCCCG TACCTACAGCA-(phosphate)
F2-ABCE	(LC Red 640)-CATTACTACATCTYT TWATGGCCAYTTTCCA-(phosphate)
F3-D	CCATCTAAAATTTCCTTCCCCGA ACTTC-(Cy5.5)

ᵃ Y represents C or T, R represents A or G, S represents C or G, and W represents A or T.

9. Pedersen NC, Ho EW, Brown ML, Yamamoto JK. Isolation of a T-lymphotropic virus from domestic cats with an immunodeficiency-like syndrome. *Science*. 1987; 235(4790):790.

10. Pecoraro MR et al. Genetic diversity of Argentine isolates of feline immunodeficiency virus. *J Gen Virol*. 1996;77:2031–2035.

11. Steinrigl A, Klein D. Phylogenetic analysis of feline immunodeficiency virus in Central Europe: A prerequisite for vaccination and molecular diagnostics. *J Gen Virol*. 2003;84:1301–1307.

12. Weaver EA, Collisson EW, Slater M, Zhu G. Phylogenetic analyses of Texas isolates indicate an evolving subtype of the clade B feline immunodeficiency viruses. *J Virol*. 2004;78(4):2158–2163.

13. Gleich SE, Krieger S, Hartmann K. Prevalence of feline immunodeficiency virus and feline leukaemia virus among client-owned cats and risk factors for infection in Germany. *J Feline Med Surg*. 2009;11(12):985–992.

14. Ravi M, Wobeser GA, Taylor SM, Jackson ML. Naturally acquired feline immunodeficiency virus (FIV) infection in cats from western Canada: Prevalence, disease associations, and survival analysis. *Can Vet J*. 2010;51(3):271–276.

15. Kelly PJ et al. Identification of feline immunodeficiency virus subtype-B on St. Kitts, West Indies by quantitative PCR. *J Infect Dev Ctries*. 2011;5(6):480–483.

16. Oğuzoğlu TC et al. First molecular characterization of feline immunodeficiency virus in Turkey. *Arch Virol*. 2010;155(11):1877–1881.

17. Teixeira BM, Hagiwara MK, Cruz JC, Hosie MJ. Feline immunodeficiency virus in South America. *Viruses*. 2012;4(3):383–396.

18. Chang-Fung-Martel J, Gummow B, Burgess G, Fenton E, Squires R. A door-to-door prevalence study of feline immunodeficiency virus in an Australian suburb. *J Feline Med Surg*. 2013;15(12):1070–1078.

19. Mora M, Npolitano C, Ortega R, Poulin E, Pizarro-Lucero J. Feline immunodeficiency virus and feline leukemia virus infection in free-ranging guignas (*Leopardus guigna*) and sympatric domestic cats in human perturbed landscapes on Chiloé Islanda, Chile. *J Wildl Dis*. 2015;51(1):199–208.

20. Jordan HL et al. Transmission of feline immunodeficiency virus in domestic cats via artificial insemination. *J Virol*. 1996;70:8224–8228.

21. Stickney AL, Dunowska M, Cave NJ. Sequence variation of the feline immunodeficiency virus genome and its clinical relevance. *Vet Rec*. 2013;172(23):607–614.

22. Scott VL et al. Immunomodulator expression in trophoblasts from the feline immunodeficiency virus (FIV)-infected cat. *Virol J*. 2011;8:336.

23. Carreño AD, Mergia A, Novak J, Gengozian N, Johnson CM. Loss of naïve (CD45RA+) CD4+ lymphocytes during pediatric infection with feline immunodeficiency virus. *Vet Immunol Immunopathol*. 2008;121(1–2):161–168.

24. Achleitner A, Clark ME, Bienzle D. T-regulatory cells infected with feline immunodeficiency virus up-regulate programmed death-1 (PD-1). *Vet Immunol Immunopathol*. 2011;143(3–4):307–313.

25. Miller MM, Thompson EM, Suter SE, Fogle JE. CD8+ clonality is associated with prolonged acute plasma viremia and altered mRNA cytokine profiles during the course of feline immunodeficiency virus infection. *Vet Immunol Immunopathol*. 2013;152(3–4):200–208.

26. Magden E et al. Acute virulent infection with feline immunodeficiency virus (FIV) results in lymphomagenesis via an indirect mechanism. *Virology*. 2013;436(2):284–294.

27. Zielonka J, Münk C. Cellular restriction factors of feline immunodeficiency virus. *Viruses*. 2011;3(10):1986–2005.

28. Sand C, Englert T, Egberink H, Lutz H, Hartmann K. Evaluation of a new in-clinic test system to detect feline immunodeficiency virus and feline leukemia virus infection. *Vet Clin Pathol*. 2010;39(2):210–214.

29. Hartmann K. Clinical aspects of feline retroviruses: A review. *Viruses*. 2012;4(11):2684–2710.

30. Ammersbach M, Bienzle D. Methods for assessing feline immunodeficiency virus infection, infectivity and purification. *Vet Immunol Immunopathol*. 2011;143(3–4):202–214.

31. Litster A, Lin JM, Nichols J, Weng HY. Diagnostic utility of CD4%:CD8 low% T-lymphocyte ratio to differentiate feline immunodeficiency virus (FIV)-infected from FIV-vaccinated cats. *Vet Microbiol*. 2014;170(3–4):197–205.

32. Hartmann K et al. Quality of different in-clinic test systems for feline immunodeficiency virus and feline leukaemia virus infection. *J Feline Med Surg*. 2007;9:439–445.

33. Hosie MJ, Pajek D, Samman A, Willett BJ. Feline immunodeficiency virus (FIV) neutralization: A review. *Viruses*. 2011;3(10):1870–1890.

34. Crawford PC, Slater MR, Levy JK. Accuracy of polymerase chain reaction assays for diagnosis of feline immunodeficiency virus infection in cats. *J Am Vet Med Assoc*. 2005;226(9):1503–1507.

35. Ertl R, Birzele F, Hildebrandt T, Klein D. Viral transcriptome analysis of feline immunodeficiency virus infected cells using second generation sequencing technology. *Vet Immunol Immunopathol*. 2011;143(3–4):314–324.

36. Ammersbach M, Little S, Bienzle D. Preliminary evaluation of a quantitative polymerase chain reaction assay for diagnosis of feline immunodeficiency virus infection. *J Feline Med Surg*. 2013;15(8):725–729.

37. Morton JM, McCoy RJ, Kann RK, Gardner IA, Meers J. Validation of real-time polymerase chain reaction tests for diagnosing feline immunodeficiency virus infection in domestic cats using Bayesian latent class models. *Prev Vet Med*. 2012;104(1–2):136–148.

38. Kann RK et al. Association between feline immunodeficiency virus (FIV) plasma viral RNA load, concentration of acute phrotase peins and disease severity. *Vet J*. 2014;201(2):181–183.

39. Wang C et al. Dual-emission fluorescence resonance energy transfer (FRET) real-time PCR differentiates feline immunodeficiency virus subtypes and discriminates infected from vaccinated cats. *J Clin Microbiol*. 2010;48(5):1667–1672.

40. Hosie MJ et al. Feline immunodeficiency. ABCD guidelines on prevention and management. *J Feline Med Surg*. 2009;11(7):575–584.

41. Fogle JE, Tompkins WA, Campbell B, Sumner D, Tompkins MB. Fozivudine tidoxil as single-agent therapy decreases plasma and cell-associated viremia during acute feline immunodeficiency virus infection. *J Vet Intern Med*. 2011;25(3):413–418.

42. Mohammadi H, Bienzle D. Pharmacological inhibition of feline immunodeficiency virus (FIV). *Viruses*. 2012;4(5):708–724.

43. Levy J et al. American Association of Feline Practitioners' feline retrovirus management guidelines. *J Feline Med Surg*. 2008;10(3):300–316.

44. Yamamoto JK, Pu R, Sato E, Hohdatsu T. Feline immunodeficiency virus pathogenesis and development of a dual-subtype feline-immunodeficiency-virus vaccine. *AIDS*. 2007;21(5):547–563.

45. Little S et al. Feline leukemia virus and feline immunodeficiency virus in Canada: recommendations for testing and management. *Can Vet J*. 2011;52(8):849–855.

22 Feline Leukemia Virus

Dongyou Liu

CONTENTS

22.1 Introduction 197
    22.1.1 Classification 197
    22.1.2 Morphology and Genome Organization 198
    22.1.3 Replication and Epidemiology 198
    22.1.4 Clinical Features and Pathogenesis 199
        22.1.4.1 Clinical Features 199
        22.1.4.2 Pathogenesis 199
    22.1.5 Diagnosis 200
        22.1.5.1 Virus Isolation 200
        22.1.5.2 Serological Assays 200
        22.1.5.3 Nucleic Acid Detection 201
    22.1.6 Treatment and Prevention 201
22.2 Methods 201
    22.2.1 Sample Preparation 201
    22.2.2 Detection Procedures 201
        22.2.2.1 Real-Time PCR Detection of FeLV DNA 201
        22.2.2.2 Real-Time RT-PCR Detection of FeLV RNA 201
22.3 Conclusion 202
References 202

22.1 INTRODUCTION

Feline leukemia virus (FeLV) is a single-stranded, positive-sense RNA of the family *Retroviridae* that is associated with proliferative, degenerative, and malignant diseases of myeloid, erythroid, and lymphoid origins in cats worldwide. Despite having a relatively simple genome, FeLV is capable of producing variants of enhanced pathogenic potential through genetic recombinations with endogenous FeLV (enFeLV)-related sequences in the feline genome. By suppressing the host immune system and predisposing cats to secondary infections, FeLV is responsible for up to 85% of mortality in persistently infected felines within 3 years of diagnosis.

22.1.1 CLASSIFICATION

FeLV is classified in the genus *Gammaretrovirus*, subfamily *Orthoretrovirinae*, family *Retroviridae*, which includes another subfamily *Spumaretrovirinae*. Members of the family *Retroviridae* are characterized by their unique ability to transcribe their RNA into DNA upon entry into host cell and integrate their DNA (provirus) into host chromosomal DNA as an alternative means of propagation and also a potential source of oncogenic development. Taxonomically, the subfamily *Orthoretrovirinae* is made up of six genera (*Alpharetrovirus*,

Betaretrovirus, *Deltaretrovirus*, *Epsilonretrovirus*, *Gammaretrovirus*, and *Lentivirus*), while the subfamily *Spumaretrovirinae* is currently composed of a single genus (*Spumavirus*).

Within the genus *Gammaretrovirus*, four major groups are recognized: *mammalian virus group* (FeLV, gibbon ape leukemia virus, guinea pig type C oncovirus, porcine type C oncovirus, and murine leukemia virus), *replication-defective virus group* (Finkel–Biskis–Jinkins murine sarcoma virus, Gardner–Arnstein feline sarcoma virus, Hardy–Zuckerman feline sarcoma virus, Harvey murine sarcoma virus, Kirsten murine sarcoma virus, Moloney murine sarcoma virus, Snyder–Theilen feline sarcoma virus, and Woolly monkey sarcoma virus), *reptilian virus group* (Viper retrovirus), and *avian (reticuloendotheliosis) virus group* (chick syncytial virus, duck infectious anemia virus, spleen necrosis virus, nondefective REV-A, and defective REV-T) [1].

Furthermore, phylogenetic analyses of FeLV *env* and *gag* gene sequences led to the identification of at least three subgroups within the species: FeLV-A, FeLV-B, and FeLV-C. FeLV-A subgroup includes naturally occurring, horizontally transmissible viruses that spread cat-to-cat in nature. Once considered as of low pathogenicity, FeLV-A is clearly involved in neoplastic diseases (typically thymic lymphoma of T-cell origin, after a protracted asymptomatic phase). FeLV-B and FeLV-C subgroups are thought not to be horizontally transmissible in nature but arise *de novo* through recombination between FeLV-A and enFeLV-related elements in the feline genome. Specifically, FeLV-B arose through recombination in the *env* region between FeLV-A and enFeLV sequences present in the feline genome, and FeLV-C arose through deletion and mutation of the FeLV-A *env* gene. Whereas FeLV-B is commonly found in tissues from lymphoma relative to asymptomatic infected cats or other disease conditions, FeLV-C is associated with pure red cell aplasia (nonregenerative anemia). There is evidence that FeLV-A is able to transmit efficiently between cats, but FeLV-B and FeLV-C replication *in vivo* often requires the presence (assistance) of FeLV-A, despite their full replication competency *in vitro*. Indeed, FeLV-B and FeLV-C viruses are often identified alongside a concurrent FeLV-A infection [2]. In addition to FeLV-B and FeLV-C, other subgroups such as FeLV-AC, FeLV-T, and FeLV-D have been occasionally described in the literature. These subgroups may be also derived from recombination events taking place between FeLV-A and enFeLV sequences [3–6].

Recent examination of nonrecombinant full-length (1.9 kb) *env* genes from FeLV isolated in Japan and other parts of the world allows the definition of three genotypes (I–III) in line with three well-known subgroups. Genotype I is further divided into seven clades (GI/Clade1, GI/Clade2, GI/Clade3, GI/Clade4, GI/Clade5, GI/Clade6, and GI/Clade7). Genotype II appears to form a single clade with limited number of strains (n = 7) analyzed. Notably, Genotypes I and II strains came exclusively from Japan. Group III is separated into two clades: GIII/Clade1 and GIII/Clade2, with strains isolated in Europe, South America, and the United States, including known FeLV strains such as FeLV-A clone 945, FeLV-A 61E, FeLV-T 61C, FeLV-A Glasgow-1, FeLV-A Richard (pFRA), FeLV-AC FY981, FeLVs derived from Iberian lynx, Brazilian FeLVs, and the exogenous FeLV components of the FeLV-C/Sarma and FeLV-B GA strains. However, GM35 sequences from Gunma Prefecture did not fit in any of the major genotypes [7]. These findings suggest that FeLV has undergone extensive evolutionary diversification within Japan, and the genetic distinctness of Japanese FeLV strains may indicate a long history of geographical isolation from the rest of the world [7].

22.1.2 MORPHOLOGY AND GENOME ORGANIZATION

FeLV is an enveloped, spherical particle of 80–100 nm in diameter. The virion possesses densely dispersed surface projections, which appear as small distinctive glycoprotein (gp70) spikes evenly covering the surface of the virion.

The genome of FeLV is dimeric, containing two monopartite linear ssRNAs(+) of about 8.3 kb each, which is organized in the order of 5′-LTR-gag(MA-CA-NC)-pol(PR-RT-IN)-env(SU-TM)-LTR-3′.

The long terminal repeats (LTRs) located at the 5′ and 3′ ends measure about 600 nt in length and include the U3, R, and U5 regions. The 5′ end is capped and contains a primer binding site; the 3′ end consists of a polypurine tract and a poly-A tail. The U3 possesses the transcriptional promoter and potent enhancer sequences that can act on adjacent viral genes as well as cellular genes at the sites of proviral integration. If host genes with oncogenic potential are activated by the LTR, malignancy may arise [8].

The *gag* (group-specific antigen) gene encodes a 65 kDa Gag polyprotein, which plays a part in budding and is processed by the viral protease during virion maturation outside the cell into four subunits: a 15 kDa matrix protein (MA, p15), a 12 kDa RNA-binding phosphoprotein (pp12), a 30 kDa capsid protein (CA, p39), and a 10 kDa nucleocapsid protein (NC, p10). Matrix protein p15 targets Gag and gag–pol polyproteins to the plasma membrane and also mediates nuclear localization of the preintegration complex [9]. Capsid protein p30 forms the spherical core of the virion that encapsulates the genomic RNA–nucleocapsid complex. Nucleocapsid protein p10 is involved in the packaging and encapsidation of two copies of the genome and binds to conserved elements within the packaging signal, located near the 5′ end of the genome.

The *pol* (polymerase) gene encodes a polyprotein that is cleaved into three proteins: protease (p14), reverse transcriptase (p80), and integrase (p46). These enzymes are involved in posttranslational modification of viral proteins and synthesis of viral RNA and DNA during viral assembly.

The *env* (envelope) gene (1.9 kb) encodes an 85 kDa polyprotein (Env), which is cleaved into two functional units: surface unit (SU, gp70) and transmembrane (TM, p15E) glycoproteins [10]. Residing on the particle surface and anchored to the TM protein, the SU protein is ideally situated to attach to the host cell surface and to interact directly with the receptor. FeLV SU protein consists of an amino-terminal receptor-binding domain (RBD) followed by a proline-rich region (PRR). Within RBD, two variable regions (designated VRA and VRB) exist, with VRA being the primary determinant of receptor binding and VRB being a secondary determinant required for efficient infection. PRR mediates conformational changes that are necessary for virus entry. In infected cell, gp70 may block viral receptors to prevent further infection by the same subgroup [11].

22.1.3 REPLICATION AND EPIDEMIOLOGY

Replication of FeLV involves several steps. The virus utilizes the SU glycoprotein to attach to host receptors (i.e., thiamine transporter FeTHTR1 for FeLV-A, phosphate transporters FePiT1 or FePiT2 for FeLV-B, heme transporter FLVCR for FeLV-C, and FePiT1 plus cofactor FELIX for FeLV-T), and the TM glycoprotein to mediate fusion with cell membrane [12,13]. Following internalization, the virus uncoats partially and uses the reverse transcriptase to yield linear dsDNA molecule from ssRNA(+) genome. Entering into the nucleus when the nuclear membrane is disassembled at mitosis, viral dsDNA is then integrated covalently and randomly into the cell's genome by the viral integrase (forming so-called provirus). Driven by in the promoter elements in the 5′ LTR, provirus is transcribed by Pol II into unspliced mRNA that will serve as genomic RNA to be packaged into virions or as a template for translation of gag and gag–pro–pol (1 ribosomal readthrough of the gag gene termination codon). The unspliced mRNA is transported out of the nucleus and translated into the Env, Gag, and Gag–Pol polyproteins. These proteins and unspliced mRNA are packaged into new virions at the host cellular membrane. After budding through the plasma membrane, the new virions are released. Subsequent proteolytic processing of the precursors (polyproteins) by viral protease results in mature virions [1].

Since its initial discovery in domestic cats in 1964, FeLV has been shown to be widely distributed [14–16]. Apart from domestic cats, a number of wild and captive cats (including lynx, cheetahs, and lions) are also susceptible to FeLV infection [17–19]. However, FeLV appears to be species specific and does not infect either humans or dogs. Depending on various geographical and risk factors (e.g., young age, poor health/vaccination status, and high population density), the prevalence of FeLV in domestic cats ranges from 2.9% to 18.9% in Asia (e.g., 2.9% in Japan, 6% in Taiwan and Thailand, 14.7% in Singapore, and up to 18.9% in Malaysia), 3.4% to 5.3% in North America (e.g., 3.4% in Canada and up to 5.3% in Raleigh, FL), and 4.6% in Egypt.

FeLV is shed in very high quantities in saliva and nasal secretions, urine, feces, milk, and blood, but only survives a few hours outside a cat's body. Its main route of transmission is via saliva during aggressive behaviors (fighting/biting) or exchange of bodily fluids (saliva, nasal secretions, and semen) through mating, mutual grooming, blood transfusion, or shared litter box or food dish. Vertical transmission from infected mother cat to kittens may occasionally occur *in utero*, via saliva during grooming, or via milk feeding [1].

22.1.4 Clinical Features and Pathogenesis

22.1.4.1 Clinical Features

Named after a tumor that first attracted the attention of veterinary scientist, FeLV induces a variety of clinical symptoms, including tumors, anemia, immunosuppression, hematologic disorders, neuropathy, reproductive disorders, and fading kitten syndrome [1].

During the early stages of FeLV infection, cats may exhibit no signs. With the cat's health progressively deteriorating over time (weeks, months, or even years), various clinical symptoms may appear, including loss of appetite; lethargy; poor coat condition; litter box avoidance; progressive weight loss; severe wasting; enlarged lymph nodes (lymphadenopathy); fever; pale gums and other mucus membranes; inflamed gums (gingivitis) and mouth (stomatitis); bladder, skin, or upper respiratory infections; breathing difficulty; abscesses; diarrhea and vomiting; jaundice (yellow color in the mouth and whites of eyes); seizures, behavior changes, and other neurological disorders; anisocoria (uneven pupils) or other eye problems; liver disease; aplastic anemia (pancytopenia); abortion of kittens or other reproductive failures (in females); and neoplasia (thymic, alimentary, and multicentric tumors) [20,21].

Interestingly, of 8642 FeLV-infected cats presented to veterinary teaching hospitals in North America, stomatitis (and coinfection with feline immunodeficiency virus [FIV] infection, feline infectious peritonitis [FIP], upper respiratory infection, and hemotropic mycoplasmosis) was observed in 15% of the cases, followed by anemia (11%), lymphoma (6%), leukopenia or thrombocytopenia (5%), and leukemia or myeloproliferative diseases (4%) [22].

At the nucleic level, FeLV proviruses from lymphomas typically show 2–3 tandem direct repeats of enhancer elements in the LTR, whereas FeLV LTRs derived from nonneoplastic disease or weakly pathogenic strains contain only a single copy of the enhancer or other repeated elements such as the upstream region of the enhancer in MDS and AML and the 21-bp triplication in non-T-cell disease [23].

22.1.4.2 Pathogenesis

FeLV infection in cats is a recurrent illness interspersed with periods of relative health. The manifestation of clinical signs is associated with the progression of infection stages, that is, abortive, regressive, progressive, and local infections [24]. There is evidence that changes in FeLV LTR and SU gene are particularly relevant to its pathogenesis, contributing to distinctive clinical features and oncogenic mechanisms.

22.1.4.2.1 Abortive Infection

Upon entry into the host, FeLV initially replicates in the local lymphoid tissue in the oropharyngeal area. However, viral replication in the case of low dose infection is largely terminated by an effective humoral and cell-mediated immune response from immunocompetent/healthy cats ("regressor cats"). These asymptomatic cats have high levels of neutralizing antibodies in the blood, but neither FeLV antigen nor viral RNA or proviral DNA is detectable. Nevertheless, the virus may still be found later in tissue samples of these cats.

22.1.4.2.2 Regressive Infection

Following an effective immune response, virus replication is contained prior to or shortly after bone marrow infection. Some FeLVs manage to spread through infected mononuclear cells (lymphocytes and monocytes). Infected cats have detectable free antigen in plasma by enzyme-linked immunosorbent assay (ELISA) and shed virus in saliva. However, viremia is terminated within weeks or months (so-called transient viremia). Bone marrow cells in cats with residual viremia may still be infected, with hematopoietic precursor cells developing into granulocytes and platelets that carry the virus to the entire body. Although the viremia may be eliminated afterward, proviral DNA integrated in cellular chromosomal DNA (e.g., bone marrow stem cells) will be maintained in the host cells. This condition is called "latent infection" (now considered a part of the regressive infection). Antigen-negative, provirus-positive cats may be considered FeLV carriers, and following reactivation, they can act as an infection source.

22.1.4.2.3 Progressive Infection

Without being contained early in the infection, FeLV replicates extensively in the lymphoid tissues and then in the bone marrow and mucosal and glandular epithelial tissues, resulting to "persistent viremia," a condition now known as progressive infection. Cats with progressive infection manifest various clinical signs/diseases, and up to 85% of cats die within 3.5 years. Progressive infection can be distinguished from regressive infection by the presence of higher viral load in the peripheral blood for 16 weeks and longer after infection.

22.1.4.2.4 Focal infection

Rarely occurring in the field, focal infection (or "atypical infection") is observed in 10% of experimental infections. It is characterized by a persistent atypical local viral replication (e.g., in mammary glands, bladder, eyes), leading to intermittent or low grade production of antigen and weakly positive or discordant results in antigen tests [1].

It is noteworthy that neonatal kittens are most vulnerable to FeLV infection, with marked thymic atrophy appearing after infection ("fading kitten syndrome"), leading to severe immunosuppression, wasting, and early death (due mostly to leukemias, lymphomas, and nonregenerative anemias). Mature cats acquire a progressive resistance toward FeLV and tend to have abortive or regressive infections, or progressive infections that are accompanied by milder signs and a more protracted period of health.

FeLV has the capacity to induce both cytoproliferative (e.g., lymphoma or myeloproliferative disorder) and cytosuppressive (e.g., immunodeficiency or myelosuppression) disease. Its ability to cause malignancy is attributable to the fact that the FeLV DNA (provirus) inserted/integrated into host genome near a cellular oncogene [e.g., c-myc, flvi-1, flvi-2 (contains bmi-1), fit-1, pim-1, and flit-1] may activate its overexpression, resulting in uncontrolled proliferation of these cells (clone). In addition, FeLV may incorporate cellular oncogene sequences to form a recombinant virus (e.g., FeLV-B, FeSV), which may induce neoplasms once inside a new cell [25,26]. For example, FeSV, a recombinant virus between the FeLV-A genome and cellular oncogenes (such as fes, fms, or fgr), is associated with fibrosarcomas in cats in the presence of FeLV-A as a helper virus that supplies proteins (such as Env) [27]. Due to FeSV, fibrosarcomas often grow rapidly and produce multiple cutaneous or subcutaneous nodules that are locally invasive and metastasize to the lung and other sites [28,29].

The ability of FeLV to cause thymic atrophy and depletion of lymph node paracortical zones is likely responsible for immunosuppression, leading to decreased response to T-cell mitogens, prolonged allograft reaction, reduced immunoglobulin production, depressed neutrophil function, and complement depletion, with increased vulnerability to secondary infectious diseases and decreased tumor surveillance functions.

The mechanisms for FeLV to induce hematopoietic disorders may include interruption or inactivation of cellular genes in the infected cells, alteration in expression of neighboring genes, and induction of antigen expression on the cell surface, leading to an immune-mediated destruction of the cell.

The neurologic signs observed in FeLV-infected cats are largely due to lymphoma and lymphocytic infiltrations in the brain or spinal cord leading to compression or FeLV-induced neurotoxicity, as a polypeptide of FeLV-C envelope has been associated with dose-dependent alterations in intracellular calcium ion concentration, neuronal survival, and neurite outgrowth.

22.1.5 Diagnosis

Cats infected with FeLV often show anemia, immunodeficiency, leukemia, lymphoma, and other symptoms that are similar to those caused by FIV and feline foamy virus. In addition, FIV and feline corona virus causing FIP may co-occur with FeLV in cats [30,31]. Several features may be exploited to distinguish between FeLV and FIV, including FeLV's tendency to cause symptomatic illness instead of FIV's usual asymptomatic infection and FeLV's association with T-cell lymphoma in comparison with FIV's linkage to B-cell lymphoma. However, laboratory methods based on the detection of virus or viral antigen in the plasma, serum, or whole blood are indispensable for accurate identification of FeLV and for differential diagnosis of diseases caused by FeLV, FIV, and other viral pathogens.

22.1.5.1 Virus Isolation

Virus isolation is a useful technique for the production of a large quantity of viruses of high purity for subsequent confirmation of FeLV by electron microscopy, serological, and/or molecular assays. Typically, Vero cells are grown in Dulbecco's Modified Eagle's Medium (DMEM; Invitrogen) supplemented with 10% fetal bovine serum (FBS; Invitrogen) at 37°C in the presence of 5% CO_2. Before infection with the virus, the cells are washed with phosphate-buffered saline and inoculated with viral specimen for 6 h in DMEM containing 5% FBS at 37°C in 5% CO_2. After 4 days, the culture media is changed and incubated until 80%–90% of the cells are floating or lightly attached to the T-75 flask (typically 10 days postinfection). The viruses are harvested, purified by density gradient ultracentrifugation (3% sucrose), and used in serological or molecular assays.

22.1.5.2 Serological Assays

A number of serological assays have been developed for the diagnosis of FeLV infection. These include indirect immunofluorescence assay (IFA), ELISA, virus neutralization test, passive hemagglutination test, complement fixation test, lateral flow rapid test, and Western blot [32]. IFA detects FeLV structural (gag) antigens in the cytoplasm of infected leukocytes and platelets during the progressive phase of the infection (secondary viremia, usually 6–8 weeks after initial infection), and cats with IFA-positive results generally have a poor long-term prognosis. ELISA detects FeLV extracellular p27 capsid antigen in plasma or sera, which is present in progressive infection (with persistent antigenemia) but variable during early viremia and with latent infections [33].

22.1.5.3 Nucleic Acid Detection

The use of molecular techniques such as PCR to detect FeLV RNA and provirus DNA provides a rapid and reliable approach for diagnosing FeLV infection, as exogenous viral RNA and endogenous provirus DNA represent superior predictors of progressive and latent infections, respectively [34–39].

Recent development of a real-time PCR targeting a 74 bp sequence within the FeLV U3 LTR portion permits specific detection and quantification of exogenous FeLV provirus sequences present in blood leukocytes [40]. In addition, the application of a multiplex reverse transcription PCR (RT-PCR) assay permits the differential diagnosis of FeLV vaccine and wild strains [41].

A combination of qPCR detecting FeLV DNA (in the blood, bone marrow, saliva, and tissues) and ELISA detecting p27 capsid antigen enables discrimination of cats with detectable nucleic acids and undetectable antigenemia (latent infections) from cats with both detectable nucleic acids and antigenemia (active infections).

Interestingly, cats with progressive infection become persistently positive for the provirus and viral RNA and have high viral loads, whereas cats with regressive infection have lower provirus and viral RNA loads than cats with progressive infection. Only rarely, the provirus becomes undetectable with regressive infection. Moreover, cats with regressive infections are negative for FeLV antigen using ELISA and IFA and also negative in blood culture; however, FeLV proviral DNA is detected in the blood by PCR.

22.1.6 Treatment and Prevention

As no specific cure is available for virus infection, lymphocyte T-cell immunomodulator (manufactured by T-Cyte Therapeutics, Inc.) and Interferon-ω (omega) (manufactured by Virbac) may be employed as a treatment aid for FeLV infection. Secondary infections and neoplasia may be treated as the situation demands [42].

Keeping healthy cats indoors and away from FeLV-infected cats represents an important way to control the spread of the disease. Other useful measures involve housing infected cats separately, adopting only infection-free cats into households with uninfected cats, etc.

Vaccination offers another effective tool to prevent FeLV infection. Currently, four FeLV vaccines (e.g., ATCvet, Purevax FeLV) are commercially available in the United States. Purevax FeLV (manufactured by Merial) is a recombinant live virus originated from a bird host, which does not replicate in mammals [42].

22.2 METHODS

22.2.1 Sample Preparation

Samples are collected from cats that are sedated with ketamine hydrochloride (11 mg/kg). Blood, buffy coat cell pellets, plasma, and tissues (thymus, tonsil, retropharyngeal lymph nodes, bone marrow, spleen, and mesenteric lymph nodes) are collected and stored at −80°C.

Peripheral blood mononuclear cells (PBMCs) are isolated from the blood by Ficoll-Hypaque (Histopaque-1077; Sigma Diagnostics) density gradient centrifugation, separated into 1×10^6 PBMC/mL aliquots, and stored at −80°C. DNA is purified from PBMC using QIAamp DNA Blood Mini Kit (Qiagen) and eluted in 100 μL of elution buffer, and DNA concentrations are determined spectrophotometrically [43].

DNA is extracted and RNA digested from tissues using a QIAamp® DNA Mini Kit and RNase A (Qiagen), respectively, eluted in 100 μL of elution buffer, and quantified spectrophotometrically.

Viral RNA is extracted from plasma using High Pure Viral RNA Purification Kit (Roche), while genomic DNA is isolated from whole blood using Qiagen DNA extraction kit (Qiagen) [44].

22.2.2 Detection Procedures

22.2.2.1 Real-Time PCR Detection of FeLV DNA

Torres et al. [43] described a real-time PCR assay to amplify exogenous and not enFeLV sequences within the U3 region of FeLV-A/61E LTR. The forward primer (5′ AGTTCGACCTTCCGCCTCAT 3′, nt 241–260) and reverse primer (5′ AGAAAGCGCGCGTACAGAAG 3′, nt 308–289) cover a 68 bp fragment. The corresponding probe (5′ TAAACTAACCAATCCCCATGCCTCTCGC 3′, nt 262–289) is labeled with the reporter dye, FAM (6-carboxyfluorescein), at the 5′ end and the quencher dye, TAMRA (6-carboxytetramethylrhodamine; Applied Biosystems) or Black Hole Quencher-1 (BioSource International, Inc., Camarillo, CA), at the 3′ end. Both probes are blocked at the 3′ end to prevent extension, and they give similar performance in the assay.

22.2.2.1.1 Procedure

The 25 μL reaction mixture is composed of 400 nM of each primer, 80 nM of fluorogenic probe, 12.5 μL of TaqMan Universal PCR Master Mix (Applied Biosystems), 3.5 μL of PCR-grade H_2O, and 5 μL of sample or plasmid standard DNA. The master mix is supplied at a 2× concentration and contained AmpliTaq Gold DNA Polymerase, AmpErase Uracil N-glycosylase (UNG), dNTPs with dUTP, and optimized buffer components.

Amplification and detection are performed in triplicate using an iCycler iQ Real-time PCR Detection System (Bio-Rad Laboratories, Inc.). A template control (no DNA, PCR-grade H_2O only) and a negative control (FeLV-naïve SPF cat DNA) are included in every reaction plate. Thermal cycling conditions consist of 2 min at 50°C to allow enzymatic activity of UNG and 10 min at 95°C to reduce UNG activity, to activate AmpliTaq Gold DNA Polymerase, and to denature the template DNA, followed by 40 cycles of 15 s at 95°C for denaturation and 60 s at 60°C for annealing/extension.

22.2.2.2 Real-Time RT-PCR Detection of FeLV RNA

Torres et al. [44] utilized the primer–probe set mentioned earlier to detect FeLV RNA.

22.2.2.2.1 Procedure

The 25 μL one-tube reaction mixture is made up of 400 nM of each primer, 80 nM of fluorogenic probe, 12.5 μL of TaqMan® One-Step RT-PCR Master Mix (Applied Biosystems), 0.625 μL of MultiScribe™ Reverse Transcriptase and RNase Inhibitor Mix (Applied Biosystems), 2.875 μL of PCR-grade H_2O, and 5 μL of sample or RNA standard. The master mix is supplied at a 2× concentration and contained AmpliTaq® Gold DNA Polymerase, dNTPs with dUTP, and optimized buffer components.

Amplification and detection are conducted in triplicate using an iCycler iQ™ Real-time PCR Detection System (Bio-Rad Laboratories, Inc.). Every reaction plate contained a negative control (FeLV-naïve SPF cat RNA), a template control (no RNA, PCR-grade H_2O), and an extraction control (extracted PCR-grade H_2O). Thermal cycling conditions include 30 min at 48°C for the RT reaction and 10 min at 95°C to activate AmpliTaq® Gold DNA Polymerase and to denature the template cDNA, followed by 40 cycles of 15 s at 95°C for denaturation and 60 s at 60°C for annealing/extension.

Note: The threshold cycle (C_T) value of each nucleic acid sample is compared with the standard curve of the coamplified standard template to determine the viral loads of the samples. End-point dilutions of the standard templates yield lower detection limits of 5 copies/DNA qPCR reaction and 10 copies/RNA qPCR reaction (1150 RNA copies/mL plasma).

The RNA qPCR enables detection of viral RNA in the absence of antigenemia, and the DNA qPCR appears to be more closely correlated to the p27 ELISA [45].

22.3 CONCLUSION

Transmitted by saliva (bites), FeLV is a deadly feline pathogen with widespread distribution. Having the ability to recombinate with endogenous proviral sequence in the cat genome, FeLV infection causes a complex spectrum of clinical diseases, including anemia, immunodeficiency, leukemia, and lymphoma [46,47]. As clinical presentations of FeLV infection resemble to those caused by FIV and other microbial pathogens, laboratory methods are required for correct diagnosis. While ELISA offers a convenient technique for detection of FeLV antigens and antibodies, PCR facilitates identification of both exogenous and enFeLV RNA and DNA. Therefore, a combined application of ELISA and PCR enables rapid and reliable diagnosis of FeLV infection and contributes to improved control and prevention of this important feline pathogen.

REFERENCES

1. Willett BJ, Hosie MJ. Feline leukaemia virus: Half a century since its discovery. *Vet J.* 2013;195(1):16–23.
2. Kawamura M et al. Genetic diversity in the feline leukemia virus gag gene. *Virus Res.* 2015;204:74–781.
3. Phipps AJ, Chen H, Hayes KA, Roy-Burman P, Mathes LE. Differential pathogenicity of two feline leukemia virus subgroup A molecular clones, pFRA and pF6A. *J Virol.* 2000;74:5796–5801.
4. Levy LS. Advances in understanding molecular determinants in FeLV pathology. *Vet Immunol Immunopathol.* 2008;123(1–2):14–22.
5. Polani S, Roca AL, Rosensteel BB, Kolokotronis SO, Bar-Gal GK. Evolutionary dynamics of endogenous feline leukemia virus proliferation among species of the domestic cat lineage. *Virology.* 2010;405(2):397–407.
6. Stewart H, Adema KW, McMonagle EL, Hosie MJ, Willett BJ. Identification of novel subgroup A variants with enhanced receptor binding and replicative capacity in primary isolates of anaerogenic strains of feline leukaemia virus. *Retrovirology.* 2012;9:48.
7. Watanabe S et al. Phylogenetic and structural diversity in the feline leukemia virus env gene. *PLoS One.* 2013;8(4):e61009.
8. Chandhasin C, Lobelle-Rich PA, Levy LS. Feline leukemia virus LTR variation and disease association in a geographic and temporal cluster. *J Gen Virol.* 2004;85:2937–2942.
9. Johnson C, Levy LS. Matrix attachment regions as targets for retroviral integration. *Virology J (Biomed Central).* 2005;2:68.
10. Gwynn SR, Hankenson FC, Lauring AS, Rohn JL, Overbaugh J. Feline leukemia virus envelope sequences that affect T-cell tropism and syncytium formation are not part of known receptor-binding domains. *J Virol.* 2000;74:5754–5761.
11. Chandhasin C, Coan PN, Levy LS. Subtle mutational changes in the SU protein of a natural feline leukemia virus subgroup A isolate alter disease spectrum. *J Virol.* 2005;79:1351–1360.
12. Lauring AS, Anderson MM, Overbaugh J. Specificity in receptor usage by T-cell-tropic feline leukemia viruses: Implications for the in vivo tropism of immunodeficiency-inducing variants. *J Virol.* 2001;75:8888–8898.
13. Shalev Z et al. Identification of a feline leukemia virus variant that can use THTR1, FLVCR1, and FLVCR2 for infection. *J Virol.* 2009;83:6706–6716.
14. Gleich SE, Krieger S, Hartmann K. Prevalence of feline immunodeficiency virus and feline leukaemia virus among client-owned cats and risk factors for infection in Germany. *J Feline Med Surg.* 2009;11(12):985.
15. Little S et al. Feline leukemia virus and feline immunodeficiency virus in Canada: Recommendations for testing and management. *Can Vet J.* 2011;52(8):849.
16. Englert T, Lutz H, Sauter-Louis C, Hartmann K. Survey of the feline leukemia virus infection status of cats in Southern Germany. *J Feline Med Surg.* 2012;14(6):392–398.
17. Meli ML et al. Feline leukemia virus infection: A threat for the survival of the critically endangered Iberian lynx (*Lynx pardinus*). *Vet Immunol Immunopathol.* 2010;134(1–2):61–67.
18. O'Brien SJ et al. Emerging viruses in the Felidae: Shifting paradigms. *Viruses.* 2012;4(2):236–257.
19. Mora M, Napolitano C, Ortega R, Poulin E, Pizarro-Lucero J. Feline immunodeficiency virus and feline leukemia virus infection in free-ranging guignas (*Leopardus guigna*) and sympatric domestic cats in human perturbed landscapes on Chiloé Island, Chile. *J Wildl Dis.* 2015;51(1):199–208.
20. Hartmann K. Clinical aspects of feline immunodeficiency and feline leukemia virus infection. *Vet Immunol Immunopathol.* 2011;143(3–4):190–201.
21. Hartmann K. Clinical aspects of feline retroviruses: A review. *Viruses.* 2012;4(11):2684–2710.
22. Stützer B et al. Incidence of persistent viraemia and latent feline leukaemia virus infection in cats with lymphoma. *J Feline Med Surg.* 2011;13(2):81–87.

23. Helfer-Hungerbuehler AK et al. Dominance of highly divergent feline leukemia virus A progeny variants in a cat with recurrent viremia and fatal lymphoma. *Retrovirology.* 2010;7:14.

24. Bolin LL, Levy LS. Viral determinants of FeLV infection and pathogenesis: Lessons learned from analysis of a natural cohort. *Viruses.* 2011;3(9):1681–1698.

25. Athas GB, Choi B, Prabhu S, Lobelle-Rich PA, Levy LS. Genetic determinants of feline leukemia virus-induced multicentric lymphomas. *Virology.* 1995;214:431–438.

26. Athas GB, Lobelle-Rich P, Levy LS. Function of a unique sequence motif in the long terminal repeat of feline leukemia virus isolated from an unusual set of naturally occurring tumors. *J Virol.* 1995;69:3324–3332.

27. Fujino Y, Ohno K, Tsujimoto H. Molecular pathogenesis of feline leukemia virus-induced malignancies: Insertional mutagenesis. *Vet Immunol Immunopathol.* 2008;123(1–2):138–143.

28. Bande F et al. Prevalence and risk factors of feline leukaemia virus and feline immunodeficiency virus in peninsular Malaysia. *BMC Vet Res.* 2012;8:33.

29. Bande F, Arshad SS, Hassan L, Zakaria Z. Molecular detection, phylogenetic analysis, and identification of transcription motifs in feline leukemia virus from naturally infected cats in malaysia. *Vet Med Int.* 2014;2014:760961.

30. Wolf-Jäckel GA et al. Quantification of the humoral immune response and hemoplasma blood and tissue loads in cats coinfected with 'Candidatus Mycoplasma haemominutum' and feline leukemia virus. *Microb Pathog.* 2012;53(2):74–80.

31. Najafi H et al. Molecular and clinical study on prevalence of feline herpesvirus type 1 and calicivirus in correlation with feline leukemia and immunodeficiency viruses. *Vet Res Forum.* 2014;5(4):255–261.

32. Kim WS et al. Development and clinical evaluation of a rapid diagnostic kit for feline leukemia virus infection. *J Vet Sci.* 2014;15(1):91–97.

33. Boenzli E et al. Detection of antibodies to the feline leukemia virus (FeLV) transmembrane protein p15E: An alternative approach for serological FeLV detection based on antibodies to p15E. *J Clin Microbiol.* 2014;52(6):2046–2052.

34. Hofmann-Lehmann R et al. Feline leukemia provirus load during the course of experimental infection and in naturally infected cats. *J Gen Virol.* 2001;82:1589–1596.

35. Hofmann-Lehmann R et al. How molecular methods change our views of FeLV infection and vaccination. *Vet Immunol Immunopathol.* 2008;123(1–2):119–123.

36. Cattori V et al. Real-time PCR investigation of feline leukemia virus proviral and viral RNA loads in leukocyte subsets. *Vet Immunol Immunopathol.* 2008;123(1–2):124–128.

37. Nakamura M et al. Differential diagnosis of feline leukemia virus subgroups using pseudotype viruses expressing green fluorescent protein. *J Vet Med Sci.* 2010;72(6):787–790.

38. Sand C, Englert T, Egberink H, Lutz H, Hartmann K. Evaluation of a new in-clinic test system to detect feline immunodeficiency virus and feline leukemia virus infection. *Vet Clin Pathol.* 2010;39(2):210–214.

39. Helfer-Hungerbuehler AK et al. Long-term follow up of feline leukemia virus infection and characterization of viral RNA loads using molecular methods in tissues of cats with different infection outcomes. *Virus Res.* 2015;197:137–150.

40. Tandon R, Cattori V, Willi B, Lutz H, Hofmann-Lehmann R. Quantification of endogenous and exogenous feline leukemia virus sequences by real-time PCR assays. *Vet Immunol Immunopathol.* 2008;123:129–133.

41. Ho CF et al. Development of a multiplex amplification refractory mutation system reverse transcription polymerase chain reaction assay for the differential diagnosis of feline leukemia virus vaccine and wild strains. *J Vet Diagn Invest.* 2014;26(4):496–506.

42. Lutz H et al. Feline leukaemia: ABCD guidelines on prevention and management. *J Feline Med Surg.* 2009;11(7):565–574.

43. Torres AN, Mathiason CK, Hoover EA. Re-examination of feline leukemia virus: Host relationships using real-time PCR. *Virology.* 2005;332:272–283.

44. Torres AN, O'Halloran KP, Larson LJ, Schultz RD, Hoover EA. Development and application of a quantitative real-time PCR assay to detect feline leukemia virus RNA. *J Vet Immunol Immunopathol.* 2008;123:81–89.

45. Torres AN, O'Halloran KP, Larson LJ, Schultz RD, Hoover EA. Feline leukemia virus immunity induced by whole inactivated virus vaccination. *Vet Immunol Immunopathol.* 2010;134(1–2):122–131.

46. Chandhasin C et al. Unique long terminal repeat and surface glycoprotein gene sequences of feline leukemia virus as determinants of disease outcome. *J Virol.* 2005;79:5278–5287.

47. Krunic M et al. Decreased expression of endogenous feline leukemia virus in cat lymphomas: A case control study. *BMC Vet Res.* 2015;11(1):90.

23 Reticuloendotheliosis Viruses

Dongyou Liu

CONTENTS

23.1 Introduction 205
    23.1.1 Classification..205
    23.1.2 Morphology and Genome Organization..206
    23.1.3 Biology and Epidemiology 206
    23.1.4 Clinical Features and Pathogenesis 206
    23.1.5 Diagnosis 207
    23.1.6 Treatment and Prevention 207
23.2 Methods 207
    23.2.1 Sample Preparation..207
    23.2.2 Detection Procedures...208
        23.2.2.1 Conventional PCR..208
        23.2.2.2 Real-Time PCR 208
23.3 Conclusion 209
References...209

23.1 INTRODUCTION

Reticuloendotheliosis is a neoplastic disease of poultry caused by a group of reticuloendotheliosis viruses (REVs) belonging to the genus *Gammaretrovirus*, family *Retroviridae*. Affecting a variety of avian species such as chicken, turkey, duck, goose, pheasant, and peafowl, reticuloendotheliosis is characterized by immunosuppression, runting–stunting syndrome, chronic lymphomas, and chronic neoplasm, which resemble both Marek's disease and avian leukosis. Interestingly, the etiologic agent for Marek's disease is Marek's disease virus (MDV) of the genus *Alphaherpesvirus*, family *Herpesviridae*, and that for avian leukosis is avian leukosis virus (ALV) of the genus *Alpharetrovirus*, family *Retroviridae*. These avian oncogenic viruses have a huge economic impact on poultry industry worldwide, and their overlapping disease profiles make differential diagnosis on clinical ground difficult, if not impossible. As the topics on MDV and ALV are covered in Chapters 17 and 87, the main focus here is REV.

23.1.1 CLASSIFICATION

REVs are classified taxonomically in the genus *Gammaretrovirus*, subfamily *Orthoretrovirinae*, family *Retroviridae*. The family *Retroviridae* is noted for its unique ability to transcribe viral RNA into DNA inside the host cell and to integrate its DNA (provirus) into host chromosomal DNA for propagation. Historically, classification of retroviruses was based on the morphological features detected by electron microscopy, which led to the identification of four groups (A–D). Type A retroviruses ("intracisternal particles")

are nonenveloped, immature particles that are found only inside cells (possibly representing endogenous retrovirus-like genetic elements). Type B retroviruses are enveloped, extracellular particles with a condensed, acentric core and prominent envelope spikes. Type C retroviruses are similar to type B, but have a central core and barely visible spikes. Type D retroviruses are slightly larger (up to 120 nm in diameter), with less prominent spikes. However, this morphology-based classification scheme has been largely superseded by sequence-based phylogenetic analyses.

Currently, two subfamilies (*Orthoretrovirinae* and *Spumaretrovirinae*) are recognized within the family *Retroviridae*. Whereas the subfamily *Orthoretrovirinae* includes six genera (*Alpharetrovirus*, *Betaretrovirus*, *Deltaretrovirus*, *Epsilonretrovirus*, *Gammaretrovirus*, and *Lentivirus*), the subfamily *Spumaretrovirinae* consists of a single genus (*Spumavirus*).

In turn, the genus *Gammaretrovirus* is separated into four major groups: *mammalian virus group* (feline leukemia virus, gibbon ape leukemia virus, guinea pig type C oncovirus, porcine type C oncovirus, and murine leukemia virus), *replication defective virus group* (Finkel–Biskis–Jinkins murine sarcoma virus, Gardner–Arnstein feline sarcoma virus, Hardy–Zuckerman feline sarcoma virus, Harvey murine sarcoma virus, Kirsten murine sarcoma virus, Moloney murine sarcoma virus, Snyder–Theilen feline sarcoma virus, and woolly monkey sarcoma virus), *reptilian virus group* (viper retrovirus), and *avian (reticuloendotheliosis) virus group* (chick syncytial virus [CSV], duck infectious anemia virus [DIAV], spleen necrosis virus [SNV], nondefective REV-A, and defective REV-T).

As type C retroviruses, *avian (reticuloendotheliosis) virus group* (REVs) comprises a number of strains, including CSV, DIAV, SNV, nondefective REV-A, and defective REV-T. Apart from defective REV-T, all known REV strains belong to a single serotype, with limited variation in genetic sequences. Nonetheless, on the basis of neutralization tests and differential reactivity with monoclonal antibodies, three REV subtypes have been noted. In addition, REV strains can be further defined by their ability to replicate in cell culture. While most field isolates are nondefective for replication in cell cultures and possess no viral oncogene, a laboratory strain (defective REV-T) is known to be a defective replication in cell cultures and possesses a viral oncogene (*v-rel*). This oncogene appears to be the source of acute reticulum cell neoplasia in experimental chicks, which gave the name reticuloendotheliosis to the virus group. The recent isolation of 10 additional REV strains from wild birds in Northeast China further highlights the complexity of the *reticuloendotheliosis virus group* [1].

23.1.2 Morphology and Genome Organization

Like other gammaretroviruses, REVs possess enveloped, spherical virions of about 100 nm in diameter. The envelope is dotted with small projections (spikes) consisting of a virus-encoded glycoprotein (surface unit [SU]), which evenly cover the surface of the virion. The outer envelope glycoprotein (SU) is linked by disulfide bonds to the transmembrane (TM) glycoprotein, which holds the SU protein in the envelope. Underneath the envelope is the amorphous matrix (MA) protein followed by the icosahedral capsid (CA). Inside the CA is the conical, electron-dense core, which is made up of RNA genome, nucleocapsid (NC) protein, reverse transcriptase, and integrase (IN) [2].

The genome of REV comprises two identical molecules of linear, single-stranded, positive-sense RNA of 8284 nt, which is organized in the order of 5′-LTR-gag-pol-*env*-LTR-3′. These two molecules are physically linked as a dimer by hydrogen bonds [3]. The detailed nucleotide locations of REV genes and noncoding sequences are as follows: long terminal repeat (LTR), 1–543; primer binding site (PBS), 544–933; *gag*, 934–2433; *pol*, 2434–6015; *env*, 5952–7712; polypurine tract (PPT), 7713–7741; and LTR, 7742–8284 [4].

The gag precursor protein is 499 amino acids long and is cleaved into 4 structural proteins, matrix (MA, p12) extending from amino acids 2 to 114, p18 (RNA binding phosphoprotein) from amino acids 115 to 199, capsid (CA, p30) from amino acids 200 to 442, and NC (p10) from amino acids 443 to 494. The CA p30 (30 kDa) protein encapsulates the genomic RNA-NC complex and is the major REV group-specific antigen [5–10].

The pol precursor protein is 1193 amino acids long, and is cleaved into two proteins: reverse transcriptase/ribonuclease H and IN. During replicative cycle, reverse transcriptase (with associated RNase H activity) reverse transcribes the RNA genome into viral DNA, which is then integrated into the host chromosome as the provirus by IN.

The *env* precursor is 587 amino acids long with a signal peptide of 36 amino acids located in the NH2 terminal region, and is cleaved by cellular furin on amino acid 398 to produce SU (gp90) and TM (gp20) glycoproteins. The mature envelope protein (Env) consists of a trimer of SU–TM heterodimers attached by a labile interchain disulfide bond. Associated with the extravirion surface through its binding to TM, SU binds to a specific receptor on the surface of the host target cell. This interaction triggers the refolding of TM, facilitating its fusion with host cell plasma membrane, and subsequent entry of the NC into the cytoplasm.

23.1.3 Biology and Epidemiology

Originated from a retrovirus that circulated in ancestral mammals, REVs have undergone recombination events to assume their current identity. The extremely low genetic diversity among all avian REVs isolates and sequences points to their very emergence from a single founder in birds [11].

REVs are mostly transmitted horizontally through fecal–oral route. By integrating their DNA (provirus) into the nuclear DNA of the host cell, REVs find an effective way to propagate and maintain in poultry host. Occasionally, REV infection of germ cells permits their integration into host germ line, which may be vertically inherited as host alleles (so-called endogenous retroviruses [ERVs]). Indeed, vertical transmission of REV has been documented in chickens and turkeys. The virus has been transmitted accidentally through the use of contaminated vaccines [12–14]. The fact that partial or complete genomic insertions of REV are observed in the genome of other avian viruses—fowlpox virus (FPV) and MDV—suggests an alternative means of REV maintenance [11,12,15–26].

Since the isolation of the prototypic strain from a turkey in 1957, REV has been shown to infect a range of avian species, including domestic poultry (e.g., chickens, turkeys, ducks, geese, and quail) and game birds and waterfowl, including peafowl, pheasants, partridges, prairie chickens, and mallards [27–30].

Serologic surveys have shown seropositive prevalence rates of 3.3%–40% in chicken flocks, which tend to increase with the age of the birds. Although infection is widespread, REV outbreaks occur only sporadically, are relatively rare, and are frequently associated with REV-contaminated vaccines, such as FPV vaccines [31]. Chicken flocks that seroconvert after 10 weeks of age are usually without clinical disease or viral shedding to progeny.

23.1.4 Clinical Features and Pathogenesis

The nondefective strains of REV induce three distinct syndromes in poultry: nonneoplastic runting, chronic lymphoid neoplasms (e.g., B and T lymphomas), and acute reticulum cell neoplasia (reticuloendotheliosis).

Clinical manifestations of the runting syndrome include weight loss, paleness, occasional paralysis, and abnormal feathering. Typically, the runting syndrome occurs 4–10 weeks after the administration of contaminated vaccines to

1-day-old chicks. The abnormal feather lesion ("Nakanuke"), with the barbules being compressed to the shaft over a small part of its length, may be of diagnostic value [32].

Acute and chronic neoplasia may involve lymphoid organs, the liver, the spleen, the intestine, and the heart, as well as the reticulum cell. Regardless of type or host species, the tumors are usually composed of uniform, large, lymphoreticular cells. In chickens, bursa-related B-cell tumor may appear similar to lymphoid leukosis caused by ALV, whereas nonbursal (T-cell) lymphomas with shorter latent periods and lesions may resemble superficially those of Marek's disease.

Other clinical signs due to REV infection include anemia, immunosuppression, bursal and thymic atrophy, and enlarged nerves. In turkeys, prominent lesions include enlarged livers and nodular lesions on the intestines.

Gross examination of the liver with tumor from REV-infected chicken reveals the massively enlarged, hemorrhagic liver. Multifocal irregular pale areas are present on the cut and uncut surfaces. The hepatic parenchyma contains large neoplastic lumps or multicentric and expansive nodular lesions, suggestive of degeneration, cell lysis, and massive infiltration of neoplastic lymphoid cells. Histopathological examination of tumor samples using hematoxylin and eosin stain (H&E) shows the characteristic REV reticular cells proliferation and infiltration, and blood vessel dilation and congestion.

23.1.5 Diagnosis

MDV, ALV, and REV represent important avian oncogenic viruses with significant economic impact on poultry industry. MDV causes skin and visceral tumors in addition to immunosuppression, neurological symptoms, and ocular lesions. ALV is associated with lymphoid leukosis, primarily in the bursa of Fabricius and visceral organs, as well as late-onset myelocytomatosis, resulting from infection with ALV subgroup J (ALV-J). Although not yet as ubiquitous as ALVs and MDVs, REV is responsible for inducing bursal and T-cell lymphomas in chickens and turkeys, of which the T-cell lymphomas of chickens are not easily be distinguished from Marek's disease and the chronic B-cell lymphomas are largely indistinguishable from those of lymphoid leukosis. Furthermore, the REV-induced runting syndrome may be confused with immunosuppressive syndromes caused by other viral agents. This emphasizes the need to use laboratory methods to differentiate REV from ALV and MDV.

Traditionally, laboratory differentiation of avian oncogenic viruses is based on virus isolation (using DF1 cells or chicken embryo fibroblast cells) and histopathological examination of tumor tissues [33]. Liver histopathology may reveal typical pathognomonic lesions of REV, with severe congestion in the central and portal veins, hepatic sinusoids associated with degeneration and necrosis in the hepatocytes, and massive number of pleomorphic lymphoid cells and reticular cell infiltration replacing the hepatic parenchyma.

Due to the time and specialized skills required for conducting these classic tests, serology, immunohistochemistry, and, more recently, nucleic acid amplification have been developed for rapid and accurate detection of REV in avian blood and tissue samples.

The commonly used serological assays include agar gel precipitin test, indirect fluorescent antibody test, enzyme-linked immunosorbent assay (ELISA), and serum neutralization test. ELISA is a sensitive and reliable method for detecting REV antigens (e.g., p30 group-specific antigen) and antibodies, and commercial ELISA kits for antibody detection (e.g., FlockChek) are available.

Nucleic acid detection procedures such as in situ hybridization, PCR, and microarrays allow detection of REV RNA and proviral DNA from clinical specimens [34,35]. With excellent speed, sensitivity, specificity, and reproducibility, PCR represents the method of choice for the diagnosis of avian oncogenic viruses including REV. The use of PCR and its derivatives (e.g., nested PCR, multiplex PCR, real-time PCR, reverse transcription polymerase chain reaction (RT-PCR)) has facilitated precise discrimination of replication-defective and replication-competent REV provirus, and REV field and vaccine strains [36–39].

23.1.6 Treatment and Prevention

No effective treatment is currently available for REV infection in poultry. Prevention measures include rapid diagnosis, removal of potential transmitter hens, rearing progeny under isolated conditions, and vaccination of naïve flocks. Several studies have clearly demonstrated that the gp90 protein of REV is immunogenically potent and capable of inducing sustained high levels of neutralizing antibodies against REV. Vaccination with gp90 protein (in recombinant format and/or DNA vaccine construct) protects chicken from viremia after REV challenge [40–43].

23.2 METHODS

23.2.1 Sample Preparation

Blood samples are collected aseptically from the infected chickens. The liver and spleen are excised from poultry postmortem.

Genomic DNA may be extracted by phenol–chloroform. Briefly, 10 µL of whole blood is added to 500 µL of 1× STET buffer (100 mM NaCl, 10 mM Tris–HCl [pH 8.0], 1 mM ethylenediaminetetraacetic acid [pH 8.0], and 5% Triton X-100) and 20 µL of proteinase K (20 mg/mL) and incubated for 15 min at 56°C. Then, 500 µL of buffered phenol–chloroform–isoamyl alcohol 25:24:1 is added, and the sample is vortex mixed and then centrifuged for 5 min at 12,000 × g. The aqueous phase is collected and transferred to a new tube, and then 500 µL of chloroform–isoamyl alcohol (24:1) is added, vortexed, and centrifuged for 5 min at 12,000 × g. The aqueous phase is transferred to a new tube, and the DNA is precipitated by adding 100% ethanol (1:2, v:v) and 3 M sodium acetate (pH 5.2 [1:0.1, v:v]), mixed by inverting several times, and incubated at −20°C for at least 2 h. The DNA is pelleted by centrifugation at 12,000 × g for 5 min, air-dried, and resuspended in 90 µL of nuclease-free water [36].

Alternatively, genomic DNA may be extracted from blood sample, tissue, or cell culture using Qiagen DNA Kit (Omega). DNA is eluted into 50 μL elution buffer and stored at −80°C.

23.2.2 DETECTION PROCEDURES

23.2.2.1 Conventional PCR

23.2.2.1.1 Protocol of Jiang et al. [1]

Jiang et al. [1] utilized primers RF (5′-GCCTTAGCCGCCATTGTA-3′) and RL (5′-CCAGCCTACACCACGAACA-3′) designed from LTR for specific PCR amplification of a 383 bp fragment from REV proviral DNA in wild birds.

PCR mixture is prepared using PrimeSTAR HS DNA polymerase (Takara Biotechnology) with the proviral genomic DNA as template. PCR conditions consist of an initial denaturation cycle for 5 min at 95°C, followed by 35 cycles of denaturation for 30 s at 94°C, annealing for 30 s at 53°C, and an extension for 1 min 20 s at 72°C, with a final extension of 10 min at 72°C. The products are detected by agarose gel electrophoresis.

Note: A second pair of primers (gp90U: 5′-CAGGAATTCATGGACTGTCTCACC-3′ and gp90L: 5′-AGAGTCGACCTTATGACGCCTAGC-3′) from REV gp90 gene may be utilized for result verification if desired.

23.2.2.1.2 Protocol of El-Sebelgy et al. [39]

El-Sebelgy et al. [39] employed primers LTRF2 (5′-GCGCTGGCTCGCTAACTG-3′) and LTRR2 (5′-TTCGATCTCGTGTTTGTTCGTGATT-3′) to amplify a 200 bp amplicon from REV LTR.

PCR mixture (50 μL) is composed of 25 μL Master Mix (Taq PCR Master Mix, Promega), 1 μL forward primer, 1 μL reverse primer, 22 μL nuclease-free water, and 1 μL extracted DNA. PCR amplification is conducted in Applied Biosystems GeneAmp PCR System 9700, with the following program: initial denaturation cycle at 94°C for 5 min, 40 cycles of denaturation at 94°C for 30 s, annealing at 58.5°C for 30 s, extension at 72°C for 45 s, and a final extension cycle at 72°C for 5 min. The PCR product is separated by electrophoresis of 5 μL product in 1.5% agarose in 1× TAE, and ethidium bromide is added to a concentration of 0.5 μg/mL for nucleic acid visualization.

23.2.2.2 Real-Time PCR

23.2.2.2.1 Protocol of Sun et al. [36]

Sun et al. [36] developed a duplex real-time PCR for sensitive and specific detection of REV in whole blood samples. The primers and probes come from the envelope protein gene (env) and the LTR region of REV (Table 23.1).

PCR mixture is made up of 10 pmol of each env primer, 20 pmol of each LTR primer, 3 pmol of env probe, 6 pmol of LTR probe, and 3 μL of template DNA in a 25 μL final reaction volume. A no template control and a PCR amplification control (DNA from an REV-positive sample) are included as negative and positive controls, respectively. The PCR amplification and fluorescence detection are carried out in a real-time

TABLE 23.1
Identity of Primer and Probes for Polymerase Chain Reaction Detection of REV

Primer/Probe	Sequence (5′–3′)	Product (bp)
env-F	TCACTCTCGATGGAAATTGCAG	96
env-R	CCAGTCCTATTGTCTGCTTCCC	
env-probe	6FAM-TAGATGTCAACTGCTATGCA-MGBNFQ	
LTR-F	AGGCTCATAAACCATAAAAGGAAATGT	119
LTR-R	CCTTTACAACCATTGGCTCAGTATG	
LTR-probe	VIC-ACAAACACGAGATCGAACTA-MGBNFQ	

PCR thermocycler, with initial incubations at 50°C for 2 min and 95°C for 10 min, followed by 40 cycles at 95°C for 15 s and 60°C for 60 s. Fluorescence is collected during the second step (annealing/extension) of the PCR process.

23.2.2.2.2 Protocol of Li et al. [41]

Li et al. [41] described a real-time PCR for REV detection and quantification targeting a 146 bp fragment of REV gp90 gene with primers (gp90F and gp90R) and probe (gp90P) labeled fluorescently with FAM at 5′ end and with BHQ1 at 3′ end (Table 23.2). The inclusion of primers (actinF and actinR) and probe (actinP) from chicken β-actin gene provides an internal control to monitor the efficiency of DNA extraction and PCR amplification (Table 23.2).

Real-time PCR mixture (25 μL) is made up of 2.5 μL of 10× Ex Taq Buffer, 2 μL of dNTP (2.5 mM), 3 μL of MgCl₂ (25 mM), 1 μL of each primer (10 pM), 0.5 μL of probe (10 pM), 1 U of Ex Taq HS (TaKaRa), 1 μL of DNA, and appropriate amount of ddH₂O.

The mixture is subjected to predenaturation at 95°C for 5 min, followed by 40 cycles of denaturation at 95°C for 10 s, annealing at 55°C for 30 s, and elongation at 72°C for 20 s, using LightCycler®480 (Roche Diagnostics).

The fluorescent signals are measured during the elongation step. Negative control is set up by substituting the DNA template with ddH₂O. In each run, a series of dilutions of the

TABLE 23.2
Identity of Polymerase Chain Reaction Primer and Probe Sequences

Target Gene	Primer/Probe	Sequence (5′–3′)
gp90	gp90F	AAGAATCTGTGCGTGAAAG
	gp90R	TAAGGACCTGGTGAGTAGC
	gp90P	FAM-CCCCGACCAAGAGGAGTAGAT-BHQ-1
β-Actin	actinF	CTGGCATTGCTGACAGGAT
	actinP	GCCTCCAATCCAGACAGAGT
	actinP	FAM-AGAAGGAGATCACAGCCCTGGCACC-BHQ-1

plasmid standard is also included along with DNA samples. The Cp value is determined using the second derivative method of the LightCycler software, version 1.5.0 (Roche Diagnostics).

Note: This TaqMan real-time PCR method enables rapid, efficient, and reliable detection and quantitation of REV in the blood. With a minimum detection limit of 10 proviral DNA copies, this assay is 100 times more sensitive than conventional PCR for REV detection.

23.3 CONCLUSION

REVs are responsible for nonneoplastic runting, chronic lymphoid neoplasms, and acute reticulum cell neoplasia in both domestic and wild poultry species. Given the similarities of clinical symptoms caused by REV to those by MDV and ALV, it is important that these oncogenic viruses are correctly and rapidly identified. While conventional laboratory methods (e.g., virus isolation, histopathological examination, and serological assays) have proven valuable for the diagnosis of REV infection, their obvious shortcomings (as exemplified by their time-consuming and variable nature) render them pale by comparison to the new generation molecular techniques. Indeed, application of real-time PCR allows not only rapid detection but also accurate quantification of REV RNA and proviral DNA from blood and tissue specimens. This has greatly enhanced our capacity in dealing with this important poultry pathogen in terms of rapid response to potential REV outbreaks and implementation of appropriate control and prevention measures.

REFERENCES

1. Jiang L et al. Molecular characterization and phylogenetic analysis of the reticuloendotheliosis virus isolated from wild birds in Northeast China. *Vet Microbiol.* 2013;166(1–2):68–75.
2. Barbosa T, Zavala G, Cheng S, Villegas P. Full genome sequence and some biological properties of reticuloendotheliosis virus strain APC-566 isolated from endangered Attwater's prairie chickens. *Virus Res.* 2007;124(1–2):68–77.
3. Darlix JL, Gabus C, Allain B. Analytical study of avian reticuloendotheliosis virus dimeric RNA generated in vivo and in vitro. *J Virol.* 1992;66(12):7245–7252.
4. Jiang T et al. Complete genomic sequence of a Muscovy duck-origin reticuloendotheliosis virus from China. *J Virol.* 2012;86(23):13140–13141.
5. Wang Y, Cui Z, Jiang S. Sequence analysis for the complete proviral genome of reticuloendotheliosis virus Chinese strain HA9901. *Sci China C: Life Sci.* 2006;49(2):149–157.
6. Lin CY, Chen CL, Wang CC, Wang CH. Isolation, identification, and complete genome sequence of an avian reticuloendotheliosis virus isolated from geese. *Vet Microbiol.* 2009;136(3–4):246–249.
7. Liu Q, Zhao J, Su J, Pu J, Zhang G, Liu J. Full genome sequences of two reticuloendotheliosis viruses contaminating commercial vaccines. *Avian Dis.* 2009;53(3):341–346.
8. Li J et al. Complete genome sequence of reticuloendotheliosis virus strain MD-2, isolated from a contaminated turkey herpesvirus vaccine. *Genome Announc.* 2013a;1(5):e00785-13.
9. Li J et al. Isolation, identification, and whole genome sequencing of reticuloendotheliosis virus from a vaccine against Marek's disease. *Poult Sci.* 2015;94(4):643–649.
10. Bao KY et al. Isolation and full-genome sequence of two reticuloendotheliosis virus strains from mixed infections with Marek's disease virus in China. *Virus Genes.* 2015;50(3):418–424.
11. Niewiadomska AM, Gifford RJ. The extraordinary evolutionary history of the reticuloendotheliosis viruses. *PLoS Biol.* 2013;11(8):e1001642.
12. Singh P, Schnitzlein WM, Tripathy DN. Reticuloendotheliosis virus sequences within the genomes of field strains of fowlpox virus display variability. *J Virol.* 2003;77(10):5855–5862.
13. Fadly A, Garcia MC. Detection of reticuloendotheliosis virus in live virus vaccines of poultry. *Dev Biol (Basel).* 2006;126:301–305; discussion 327.
14. Tadese T, Fitzgerald S, Reed WM. Detection and differentiation of re-emerging fowlpox virus (FWPV) strains carrying integrated reticuloendotheliosis virus (FWPV-REV) by real-time PCR. *Vet Microbiol.* 2008;127(1–2):39–49.
15. Reimann I, Werner O. Use of the polymerase chain reaction for the detection of reticuloendotheliosis virus in Marek's disease vaccines and chicken tissues. *Zentralbl Veterinarmed B.* 1996;43(2):75–84.
16. Takagi M et al. Detection of contamination of vaccines with the reticuloendotheliosis virus by reverse transcriptase polymerase chain reaction (RT-PCR). *Virus Res.* 1996;40(2):113–121.
17. Hertig C, Coupar BE, Gould AR, Boyle DB. Field and vaccine strains of fowlpox virus carry integrated sequences from the avian retrovirus, reticuloendotheliosis virus. *Virology.* 1997;235(2):367–376.
18. Moore KM, Davis JR, Sato T, Yasuda A. Reticuloendotheliosis virus (REV) long terminal repeats incorporated in the genomes of commercial fowl poxvirus vaccines and pigeon poxviruses without indication of the presence of infectious REV. *Avian Dis.* 2000;44(4):827–841.
19. Kim TJ, Tripathy DN. Reticuloendotheliosis virus integration in the fowl poxvirus genome: Not a recent event. *Avian Dis.* 2001;45(3):663–669.
20. García M, Narang N, Reed WM, Fadly AM. Molecular characterization of reticuloendotheliosis virus insertions in the genome of field and vaccine strains of fowl poxvirus. *Avian Dis.* 2003;47(2):343–354.
21. Zhang Z, Cui Z. Isolation of recombinant field strains of Marek's disease virus integrated with reticuloendotheliosis virus genome fragments. *Sci China C: Life Sci.* 2005;48(1):81–88.
22. Cui Z, Zhuang G, Xu X, Sun A, Su S. Molecular and biological characterization of a Marek's disease virus field strain with reticuloendotheliosis virus LTR insert. *Virus Genes.* 2010;40(2):236–243.
23. Sun AJ et al. Functional evaluation of the role of reticuloendotheliosis virus long terminal repeat (LTR) integrated into the genome of a field strain of Marek's disease virus. *Virology.* 2010;397(2):270–276.
24. Biswas SK, Jana C, Chand K, Rehman W, Mondal B. Detection of fowl poxvirus integrated with reticuloendotheliosis virus sequences from an outbreak in backyard chickens in India. *Vet Ital.* 2011;47(2):147–153.
25. Su S et al. Complete genome sequence of a recombinant Marek's disease virus field strain with one reticuloendotheliosis virus long terminal repeat insert. *J Virol.* 2012;86(24):13818–13819.
26. Su S et al. Sequence analysis of the whole genome of a recombinant Marek's disease virus strain, GX0101, with a reticuloendotheliosis virus LTR insert. *Arch Virol.* 2013;158(9):2007–2014.

27. Bohls RL et al. Phylogenetic analyses indicate little variation among reticuloendotheliosis viruses infecting avian species, including the endangered Attwater's prairie chicken. *Virus Res*. 2006;119(2):187–194.

28. Cheng Z et al. Occurrence of reticuloendotheliosis in Chinese partridge. *J Vet Med Sci*. 2007;69(12):1295–1298.

29. Mays JK, Silva RF, Lee LF, Fadly AM. Characterization of reticuloendotheliosis virus isolates obtained from broiler breeders, turkeys, and prairie chickens located in various geographical regions in the United States. *Avian Pathol*. 2010;39(5):383–389.

30. Jiang L et al. First isolation of reticuloendotheliosis virus from mallards in China. *Arch Virol*. 2014;159(8):2051–2057.

31. Awad AM, Abd El-Hamid HS, Abou Rawash AA, Ibrahim HH. Detection of reticuloendotheliosis virus as a contaminant of fowl pox vaccines. *Poult Sci*. 2010;89(11):2389–2395.

32. Filardo EJ, Lee MF, Humphries EH. Structural genes, not the LTRs, are the primary determinants of reticuloendotheliosis virus A-induced runting and bursal atrophy. *Virology*. 1994;202(1):116–128.

33. Davidson I, Alphandary R, Novoseler M, Malkinson M. Replication of non-defective reticuloendotheliosis viruses in the avian embryo assayed by PCR and immunofluorescence. *Avian Pathol*. 1997;26(3):579–593.

34. Deng X et al. Development of a loop-mediated isothermal amplification method for rapid detection of reticuloendotheliosis virus. *J Virol Methods*. 2010;168(1–2):82–86.

35. Wang LC, Huang D, Pu CE, Wang CH. Avian oncogenic virus differential diagnosis in chickens using oligonucleotide microarray. *J Virol Methods*. 2014;210C:45–50.

36. Sun F et al. A duplex real-time polymerase chain reaction assay for the simultaneous detection of long terminal repeat regions and envelope protein gene sequences of Reticuloendotheliosis virus in avian blood samples. *J Vet Diagn Invest*. 2011;23(5):937–941.

37. Li K et al. Development of TaqMan real-time PCR assay for detection and quantitation of reticuloendotheliosis virus. *J Virol Methods*. 2012;179(2):402–408.

38. Cao W et al. Use of polymerase chain reaction in detection of Marek's disease and reticuloendotheliosis viruses in formalin-fixed, paraffin-embedded tumorous tissues. *Avian Dis*. 2013;57(4):785–789.

39. El-Sebelgy MM et al. Molecular detection and characterization of reticuloendotheliosis virus in broiler breeder chickens with visceral tumors in Egypt. *Int J Vet Sci Med*. 2014;2(1):21–26.

40. Li K et al. Recombinant gp90 protein expressed in Pichia pastoris induces a protective immune response against reticuloendotheliosis virus in chickens. *Vaccine*. 2012;30(13):2273–2281.

41. Li K et al. Protection of chickens against reticuloendotheliosis virus infection by DNA vaccination. *Vet Microbiol*. 2013;166(1–2):59–67.

42. Li K et al. Enhancement of humoral and cellular immunity in chickens against reticuloendotheliosis virus by DNA prime-protein boost vaccination. *Vaccine*. 2013;31(15):1944–1949.

43. Drechsler Y et al. A DNA vaccine expressing ENV and GAG offers partial protection against reticuloendotheliosis virus in the prairie chicken (Tympanicus cupido). *J Zoo Wildl Med*. 2013;44(2):251–261.

24 Border Disease Virus

Jieyuan Jiang and Wenliang Li

CONTENTS

24.1 Introduction 211
   24.1.1 Classification, Genome Organization, and Morphology 211
      24.1.1.1 Classification 211
      24.1.1.2 Genome Organization 212
      24.1.1.3 Morphology 213
   24.1.2 Epidemiology, Transmission, and Pathogenesis 213
      24.1.2.1 Epidemiology 213
      24.1.2.2 Transmission 214
      24.1.2.3 Pathogenesis 214
   24.1.3 Clinical Symptoms and Pathologic Changes 214
      24.1.3.1 Acute Infection 214
      24.1.3.2 Fetal Infection 214
   24.1.4 Diagnosis 215
      24.1.4.1 Conventional Techniques 215
      24.1.4.2 Molecular Techniques 215
24.2 Methods 216
   24.2.1 Virus Isolation 216
      24.2.1.1 Sample Preparation 216
      24.2.1.2 Isolation Procedure 216
      24.2.1.3 Immunofluorescence or Immunoperoxidase Assay 217
   24.2.2 RT-PCR 217
      24.2.2.1 Sample Preparation 217
      24.2.2.2 Detection Procedure 217
24.3 Conclusion and Future Perspectives 219
References 219

24.1 INTRODUCTION

Border disease (BD) is a major viral disease of sheep and goats caused by border disease virus (BDV). The disease is first recognized in the border region of England and Wales [1], and the clinical signs of infected sheep herds mainly include barren ewes, abortions, stillbirths, and the birth of small weak lambs. Affected lambs can show tremor, abnormal body figure, and hairy fleeces (also called "hairy shaker" or "fuzzy"), so the disease has also been referred to as "hairy shaker disease" [2]. The disease is widely distributed in the world. Prevalence rates in sheep vary from 5% to 50% between countries and from region to region within countries [3].

24.1.1 CLASSIFICATION, GENOME ORGANIZATION, AND MORPHOLOGY

24.1.1.1 Classification

BDV is a helical, enveloped, positive-sense single-stranded RNA (+ssRNA) virus, and belongs to the genus *Pestivirus*

within the family *Flaviviridae*. In addition to BDV, the *Pestivirus* genus also includes three related species responsible for livestock diseases: bovine viral diarrhea viruses-1 (BVDV-1), BVDV-2, and classical swine fever (hog cholera) virus (CSFV) [4]. These viruses share physical, chemical, and biological characteristics and also partially share common antigens. These viruses have been confirmed to infect across species. The antibodies against pestivirus have been detected in more than 40 ruminant species including cattle, sheep, and goats, and cross-species reactions have also been observed [5]. BDV and BDV infections are now full appreciated because of the development of molecular biological techniques.

Historically, the viruses in the genus *Pestivirus* have been named and classified based on host species origins. BDV hosts are sheep and goats, BVDV-1 and BVDV-2 are bovine related, and CSFV infects pigs [6]. Since it is later discovered that pestivirus could easily infect cross-species, this earlier classification scheme has serious limitations and falls short in practical application. This is further evident with more unclassified pestiviruses being discovered recently, including "Hobi-like" pestiviruses of bovine origin, mainly detected from contaminated fetal calf serum [7–10], the pestivirus pronghorn from antelope [11], and Bungowannah virus from pigs [12]. These additional pestiviruses have been postulated to form novel groups/species.

Genotyping and phylogenetic analysis of pestiviruses is widely used for pestivirus grouping. Typing is crucial for classifying novel viruses and revealing their evolutionary history. Based on phylogenetic studies of the sequences of the 5′ untranslated region (UTR) and/or N^{pro} available in GenBank [13,14], BDV isolates are divided into seven phylogenetic groups. BDV-1 is present in sheep from the United States [15], the United Kingdom [16], Australia [17], and New Zealand [18]. BDV-2, BDV-3, and BDV-4 are now in Europe [8,19–21] and BDV-5 and BDV-6 in France [22]. Isolates from Turkey form BDV-7 [23]. The four BDV isolates from goats and sheep in China belong to the BDV-3 group [24,25].

Some pestivirus strains previously considered as BDV have been reclassified into the other pestivirus species based on the phylogenetic types; one pestivirus isolate from goats showing the BD-type signs in Korea is identified as the BVDV-2 genotype [26]. Several pestiviruses, isolated from lambs and kids with BD-like syndromes in Italy during the 1990s and characterized as BDV [27], have been recognized as BVDV-1 or BVDV-2 strains [28]. In addition, five BVDV positive-diagnosed UK cattle between 2006 and 2008 are classified as BDV positive, and two of them belong to the BDV-1a group, and the other three to the BDV-1b group [13]. The suspected BVDV-1 isolate FNK2012-1 from swine in Japan is clustered into BDV genotype 1 [29].

One valuable genotyping method is based on pestivirus 5′ UTR sequences and their secondary structure variation, analyzed with the palindromic nucleotide substitutions. Of the 536 pestivirus strains examined by using this method, 131 strains are clustered as BDV species and further classified into at least 8 genotypes [30]. Based on this knowledge, one software tool for the numerical taxonomy of the genus *Pestivirus* has been developed. Analysis of 543 pestivirus isolates led to the discrimination of 9 species including BVDV-1, BVDV-2, BVDV-3, CSFV, Pronghorn, Giraffe, Bungowannah, BDV-1, and BDV-2. Interestingly, while 131 strains typed as BDV-1 come from sheep, Pyrenean chamois, cattle, pig, reindeer, and wisent, 5 BDV-2 strains are from sheep and goats [31].

Antigenic typing is also used to cluster BDV strains. Eighteen BDV strains including two atypical porcine pestiviruses (lower reaction with the monoclonal antibodies [mAbs] specific for CSFV or BVDV) can be divided into three groups with mAbs against the BDV proteins E2, E^{rns}, or NS2/3 [14]. Four Italian BDV isolates are recognized and grouped with the mAbs known to cross-react with referenced BDV-1 and BDV-5 strains. However, all four isolates also show varied reaction with a distinct set of mAbs against E2, E^{rns}, and NS2/3 for the antigenic characterization of BDV, BVDV, and CSFV [32].

24.1.1.2   Genome Organization

The complete genome sequences of genotypes BDV-1, BDV-2, BDV-3, BDV-4, BDV-5, and BDV-7 have been determined [25,33–37]. The full sequences of the reported BDV strains comprise a single-stranded positive-oriented RNA genome (+ssRNA) being a length of ~12.3 kb, which is similar to the other species of the genus *Pestivirus*. The 3′ terminus of the genome is not polyadenylated but terminates with a short poly(C) tract. There is an internal ribosome entry site at the 5′ end to control translation initiation. The genome of the first sequenced BDV strain BD31 isolated from a lamb with hairy shaker syndrome is 12,268 nucleotide (nt) long and has a single open reading frame (ORF) of 11,688 nt beginning at nucleotide 357 [38]. BDV strain Aveyron [37] is 12,284 nt and contains one 11,700 nt large ORF with a 5′ UTR being 370 nt long and a 3′ UTR of 214 nt. The single ORF encodes one viral polyprotein of 3,899 amino acids (aa) length. The Japanese BDV-1 strain FNK2012-1 comprises 12,327 nt containing a large ORF of 11,685 nt [29]. The genome of BDV strain H2121 isolated from Pyrenean chamois and typed as BDV-4 is 12,305 nt long with the ORF encoding 3,899 aa with 76 nt long 5′ UTR and 229 nt long 3′ UTR [36]. In the polyprotein, the mature viral proteins are arranged in the following order (from the N to C terminus): N^{pro}, C, E^{rns} (E0), E1, E2, p7, NS2-3, (NS2), (NS3), NS4A, NS4B, NS5A, and NS5B. The polyprotein is cleaved by viral and cellular proteases into four structural proteins: the nucleocapsid protein (C) and three envelope glycoproteins E^{rns}, E1, and E2, plus at least 7–8 nonstructural proteins (N^{pro}, p7, NS2, NS3, NS4A, NS4B, NS5A, and NS5B) [37,39,40]. Of the glycoproteins, E2 is the major immunodominant envelope protein and induces major neutralization antibody response in animals.

Two phenotypes of BDV strains: cytopathogenic (cp) and non-cp (ncp) have been identified by their effects on cell cultures [41–43]. There are insertions of (cp) Cumnock and (cp) Moredun comprising 387 and 333 nt, respectively; which are from host cellular components (Jiv-coding insertions) at an identical genomic region encoding the C-terminal part of NS2 [44]. The BDV-4 isolate Chamois-1 has four additional aa (AEVG) insertion in the N-terminal region of the gene encoding NS2 and grows without cytopathic effect (CPE) capability [36]. BDV-5 strain Aveyron also contains an unknown original insertion of four aa (KAPD) in the C-terminal region of NS2 and has CPE function [37]. The two viruses produce significantly larger amounts of NS3 protein than other BDV strains based on Western blot analysis [36,37].

Comparison of complete coding sequences of BDV strains shows that the different strains of BDV may share more than 75% nucleotide sequence identities except for some BDV-7 strains, 70% to <75% with CSFV strains, 67%–70% with BVDV-1, and less than 67% with BVDV-2 viruses [25,37]. However, pestiviral proteins have higher identities than the corresponding nucleotide sequences. The BDV isolates show >85% identity in the entire polyprotein sequences, which is notably higher than that observed in CSFV strains and other pestiviruses [36]. The viral proteins of pestiviruses also demonstrate varied levels of homology. While NS3, NS4A, NS4B, and E1 are the most conserved pestiviral proteins, E2, p7, and NS5A appear to be less conserved. The protein NS2 varies due to insertions of the host protein factors [34,36,45].

24.1.1.3 Morphology

BDV particles are very similar to the other viruses in the genus *Pestivirus*. The mature BDV virion is roughly circular and about 40–46 nm in diameter with a 20–25 nm core [46]. The virions contain three virus-encoded membrane proteins (E^{rns}, E1, and E2) in addition to the capsid proteins (C) binding with viral genome RNA. The buoyant density is about 1.115 g/mL in sucrose gradient [41]. The virus is inactivated by heating to 60°C for 90 min and sensitive to ether treatment [47]. BDV isolates exist in two kinds of biological phenotypes according to their effects on cell cultures. Most BDV strains do not produce CPE. However, a few cp BDVs are isolated from sheep [41,42], but one of them has been confirmed to be undistinguishable from BVDV *in vitro* [42]. The BDV strains can grow in many ovine and bovine primary cells and cell lines, and now BDV isolation is usually performed on these cell lines. The BDV Japan FNK2012-1 isolated from pigs also grows in porcine primary cells and cell lines; however, it grows better in bovine and ovine cells [29].

24.1.2 Epidemiology, Transmission, and Pathogenesis

24.1.2.1 Epidemiology

BD in sheep occurs in Europe, North America, Australia, and New Zealand [3]. BDV infections have also been found in Asian countries including India, Turkey [23], Japan [48], and China [24]. Antibody prevalence rates among adult sheep vary from 5% to 50% between countries and the regions within countries. As in Northern Ireland, 5.3% of 918 ewe sera from 30.4% of 92 flocks in 10 regions are positive for BDV antibody. The geographical variations are very significant, with flock prevalence ranging from 0 in the Enniskillen region to 70% in the Coleraine. The seropositive proportion is not influenced by infections of BVDV-1 on the farms [49]. Presence of BDV is positively detected in 8 of the 25 commercial flock farms in Northern Spain (32%) [50]; 17.6% (29/165) of samples from sheep are seropositive for BDV in Japan. And the mean seropositive prevalence in individual herds range from 5.0% to 42.9% in 9 sampling groups with no statistical difference in the prevalence among sheep from 3 to 6 years old [48].

Similar clinical and pathological characteristics are observed in goats experimentally and naturally infected with BDV as are found in sheep. Compared to sheep BDV infections, the BDV infections of goats are not diagnosed much but recognized increasingly. Despite the relatively few clinical reports, serologic surveys of BD on goats without clinical BDs in Canada in 1984 show that about 16% of goats are positive in specific antibody reaction and 2.6%–45.5% variation between areas [51]. The specific antibodies to pestivirus including BVDV-1 and BDV are 11.5% of individual prevalence and 31.3% of flock prevalence in 549 goats in 80 flocks from four regions of Austria in 2006 [52]. BDV seropositivity is 63.6% of the surveyed goats in Eastern and Southeastern Anatolia of Turkey [53]. Sera from 34 of 126 dairy goat farms in Netherlands are 27% positive for BDV with a herd prevalence of 32% (ranging from 1% to 100%) [54]. In Switzerland, 5059 sera from 382 sheep herds and 503 sera from 54 goat flocks are about 9% and 6% positive for BDV, respectively [55]. It should be noted that previous serosurveys have great difficulty in distinguishing BVDV infection from BDV infection. Three previous BDV isolates from goats are reclassified as BVDV-1 by 5′ UTR and N^{pro} gene sequencing and typing [6]; and the BD-like diseases reported in Korea were confirmed to be BVDV-2 virus infections [26]. Only a few cases of goat BDV infections have truly been identified. The experimental infections of BDV in pregnant goats are performed in 1975 [56] and in 1987 [57], respectively. The first case in goats naturally infected with BDV occurs in 1982 [58]. BD outbreaks in five herds of goats are caused by contaminated vaccines in Norway [59]. Detection of BDV in goats with diarrhea is first confirmed and further classified as BDV-3 infection in China in 2013 [24]. One BDV-3 isolate is obtained from 2 Italian goat herds, and positive antibodies are detected in 61/67 and in 38/169 goats, respectively [45].

Since 2001, disease outbreaks associated with BDV-4 infection have been observed in chamois. The infections result in a significant reduction of the chamois population in the French and Spanish Central Pyrenees [60–63]. Two epidemiological pathways of BDV chamois infection were noted in the Val d'Aran and Pallars Sobirà (VAPS) and Cerdanya, Alt Urgell, Berguedà and Solsonès (CAUBS) areas of Spain. The endemic infections in the VAPS area have resulted in frequent BD outbreaks and significant decrease of chamois numbers; however, lower BDV prevalence allows chamois population to quickly recover in CAUBS [64].

Bovines are infected with BDV in nature [13,65], and most of these infections have been confused with BVDV infections. Only by use of sequencing and genotyping methods can BDV infections be precisely determine in samples from infected cattle [66]. Australian V/TOB isolate of bovine origin, different from other BVDV isolates in antigenicity, is genotyped as one BDV [6]. Cattle showing BVDV signs are also diagnosed with BDV infection between 2006 and 2008 in the United Kingdom [13]. Persistent infection (PI) of BDV in cattle is detected in one bull that grows slower and is smaller in size than the healthy animals at the same age in New Zealand [67]. However, BDV infects calves when the previously seronegative calves mingled with PI BDV sheep in fields [68,69]. Furthermore, direct contact between bovines and PI sheep more easily allows infection by BDV than direct oral inoculation of virus culture [70].

Apart from the high seroprevalence of pestivirus infections, rare cases of BDV infections have been recorded in other farm animals and wild ruminants. Pigs are infected with BDV in nature and experimentally. BDV-infected pigs show no clinical signs; the viremia is usually detected from 2 to 10 days postinfection (dpi), and antibodies are induced from 14 dpi and continue to increase subsequently. The first BDV isolated from pigs with hemorrhagic lesions was genotyped by N^{pro} gene homology [71]. Recently, two new BDV isolates have been isolated: one is from a BDV persistently infected pig in Japan [72] and another in Spain [73].

In Japan, the BDV isolated pig farm has a >58.5% seropositivity rate of the surveyed pigs against CSFV E2 group antigen without CSFV infection history [72]. In addition, one BDV isolate has also been characterized in captive zoo reindeer and bison in 2003 [19].

24.1.2.2 Transmission

BDV is usually very sensitive to environmental conditions outside its hosts and cannot survive for a long period. There are at least two transmission pathways for BDV in susceptible sheep flocks [74,75]: vertical and horizontal transmission. Direct contact (feeding or oral–nasal route) easily accelerates horizontal transmission due to excretion of viruses from infected animals. Vertical transmission is via placenta and more crucial for animals, which may cause PI in newborns during the defined pregnant period when fetuses with immature immune systems are infected with the viruses [76]. The diseased and PI sheep are the sources of disease, and PI animals are more dangerous [77,78]. PI lambs are the most potent source of infections because the animals are viremic and constantly excrete viruses [45]. BDV infection usually results in presentation of PI lambs in sheep flocks. Recently, other species of PI BDV animals are also detected such as PI cattle [67], PI pigs [72], and PI goats [45], but the exact models of virus transmission are not clear yet.

BDV cross-species transmissions occur exactly as for the other pestiviruses. Sheep BD outbreaks have been caused by transmission of virus from cattle to sheep [68,79]. Calves can seroconvert and induce antibody response against BDV when kept together with infected sheep [69]. BDV can naturally infect pigs reared in the same facility with experimentally infected animals, and the infections induce the pigs to produce protective antibody response against wild CSFV challenge [80]. Wild boar (*Sus scrofa*) in Europe may be infected with BDV and show positive neutralizing antibodies [81].

Pestiviruses, including BDV contaminants in modified live virus (MLV) vaccines, may be one important transmission mode for BDV in animals if MLV vaccines are produced in the cells and sera derived from virus-infected ovine, goats, and other animals. The Ch1Es cell line of caprine origin, contaminated with BDV, is detected by reverse transcription polymerase chain reaction (RT-PCR) in 1995 [82]. Outbreaks of BD in sheep and goats have been associated with the use of such vaccines [83,84].

24.1.2.3 Pathogenesis

Adult, immunocompetent small ruminants can produce strong immune responses against BDV intrusion and do not show any signs of acute infection. When pregnant animals are infected with BDV, the viruses can infect the embryo or fetus via the placenta. The outcome of fetal BD infection depends largely on the gestation period of infection. In early gestation, the virus infection easily kills the embryo, resulting in the fetus being resorbed or aborted. In most cases, persistent viremic BD lambs happen, which are mainly infected prenatally before maturation of immune system between about 60 and 85 days of gestation in sheep (80 and 100 days in goats) [85]. The fetuses are immunotolerant and the virus may grow and exist widespread in

all organs. The animals, if they survive, are usually persistently infected for life. When the infection happens in the later stages of gestation and the fetuses have a mature immune system, the infections usually damage the fetal central nervous system and result in hypomyelination, degeneration of oligodendroglial cells by direct damage, or indirect functions such as changing hormone secretion and the inflammation of the nervous system [86]. The infection in later stages of gestation also affects fetal development, and the bones and muscles may be damaged [87,88]. In addition, BDV infections in pregnant goats result virtually exclusively in abortions and malformations in fetuses and neonates. Therefore, PI goats are rarely found [23].

PI sheep have significantly more CD8+ T lymphocytes, and the total numbers of CD4+ cells are not significantly changed. The proportions of lymphocyte subpopulations are altered with a very high ratio of the CD8+/CD4+. Peripheral lymphocytes and the lymphocytes expressing class I MHC antigen are significantly reduced [89,90].

Evasion of the innate immune system is crucial for the establishment of BDV PI in affected lambs. Activating caspase-3-related apoptosis happens in BDV-infected sheep. Many Terminal deoxynucleotidyl transferase dUTP nick end labeling (TUNEL) positive and activated caspase-3 elements are found in affected glial cells, endothelial cells, and neurons of BD animals [91]. Activated caspase-3 in neuronal cells commonly plays major roles in neuroprotection and neurodamage by mediating sublethal challenges for apoptotic disintegration of the nuclear matrix [92]. The strong activation and distribution of caspase-9, similar to activated caspase-3, in the BD animals suggests that the intrinsic pathway of apoptosis may play important roles in the occurrence and progression of the disease [91].

24.1.3 Clinical Symptoms and Pathologic Changes

24.1.3.1 Acute Infection

Normally, healthy newborn and adult sheep and goats infected with BDV only appear with mild or without clinical symptoms [93]. Some infected animals have slight fever and a mild leukopenia without diagnostic characteristics. The animals have a short viremia detectable between 2 and 10 dpi and produce specific positive antibody response after 14 days.

Occasionally, during the outbreaks of BDV in new regions, sick animals show high fever, anorexia, conjunctivitis, nasal discharge, dyspnea, profound and prolonged leukopenia, and sometimes with serious diarrhea. Morbidity and mortality are high in these situations [24,94–97]. Recently, BDV-infected European chamois were shown to have depression, weakness, and movement difficulties. The animals have abnormal behavior; absence of flight reaction; asymmetric patches on the head, neck, and trunk; and different degrees of alopecia with skin hyperpigmentation. Chronic wasting is also a common feature in all affected chamois [60,63].

24.1.3.2 Fetal Infection

The BDV infected fetuses show different clinical BD symptoms and fetal death may occur at any stage of pregnancy, and more commonly during early gestation. The mortality of BD lambs and kids is usually high during birth and the early periods of life. At the later stage of gestation, abortion of larger fetuses, stillbirths, and the premature births of small, weak babies are often seen. Infection in goats is less common with abortion being the main presenting sign [2,87,88].

The live affected lambs and kids may be normal or PI. The affected young lambs have small stature, domed heads, are weak, and present paralysis. The diseased animals have difficulty standing up and walking, with ataxia or not, and show uncoordinated movements and posterior wobbling with involuntary muscular tremors [23,50,59,98]. One of the most conspicuous characteristics is abnormal fleece, consisting of long and straight birth coats as "hairy shaker" in many sick lambs. The affected lambs may also have short legs. With careful nursing, a proportion of BD lambs can survive and become normal. The hair changes disappear in 9–12 weeks and the nervous symptoms gradually recover by 20 weeks of age [2,23]. The animals appear to be tolerant of the virus and usually remain PI for life. Typically, there is no or little inflammatory reaction. However, the BD animals often grow slowly under normal field conditions [2,96].

The infected maternal animals manifest subclinical or mild symptoms except for characteristic features such as placentitis and high rates of abortion, stillbirth, or the premature births of small and weak lambs. The birth rates of babies have been low, generally without lambs at lambing time, or with signs of BD in young animals present in the flocks, that is normally the first sign of BD.

No gross lesions are observed at necropsy of the infected lambs and kids. Spinal cord cross-sectional areas are significantly reduced at all four levels and in both grey and white matter [99]. Macroscopical changes are observed. Fetal anomalies include cerebellar hypoplasia, hydranencephaly, porencephaly, and arthrogryposis. Cotyledons are small or immature, with areas of pinpoint grayish necrosis [50,78,86,91].

In the mucosal cases of BDV infections, the tonsils and lymph nodes are usually edematous, enlarged, and hemorrhagic. A few round, hyperemic ulcers can be found on the hard palate and pinpoint shallow erosions on the soft palate. The esophagus and parts of the large intestines show numerous oval and linear mucosal ulcers [24,91,95,100]. In BD chamois, the most consistent microscopic changes are in the skin and the brain. Skin lesions include severe alopecia with follicular atrophy, hyperplasia, and melanosis with evident orthokeratotic hyperkeratosis. The brains consistently appear with edema, diffuse moderate spongiosis, and neuronal degeneration and death [62].

24.1.4 Diagnosis

A variety of techniques may be employed for BDV diagnosis, especially in PI animals. These include virus isolation, virus neutralization test (VNT), immunohistochemistry, indirect or direct immunofluorescence assay (IFA), enzyme-linked immunosorbent assay (ELISA), agar gel immunodiffusion test, RT-PCR, nested RT-PCR (nRT-PCR), and real-time RT-PCR (qRT-PCR) [49,72,98,101–104]. ELISA and VNT are frequently used for serological investigation, while virus isolation, nucleotide acid detection assays (RT-PCR, nRT-PCR, and qRT-PCR), and IFA are always used for agent examination. Furthermore, nucleotide sequencing plus phylogenetic analysis of RT-PCR products is frequently used for identification and genotyping of clinical BDV strains [16,23,105].

24.1.4.1 Conventional Techniques

Serological testing is widely used to determine the prevalence of BDV infection in a flock, region, or country, especially in previously BDV-free areas. Antibody positive results usually imply the introduction and prevalence of field BDV. VNT is commonly used but is work intensive and time consuming, and thus not suitable for detection of large numbers of samples. Both cytopathic and noncytopathic strains of BDV can be used for the VNT, but cytopathic BDV (e.g., Moredun, 137/4) is recommended because of the simplicity of operation. Furthermore, because of the cross-reactivity between antibodies against different pestivirus members, a differential VNT is recommended in which sera are titrated out against viruses from each of the four *Pestivirus* groups, that is, BDV, BVDV-1, BVDV-2, and CSFV.

ELISA is the most valuable screening tool and is suitable for large-scale epidemiological investigation. The mAb-capture ELISA has been developed [106]. mAb VPM22 bound to 96-well microplates is used to capture antigen from detergent-solubilized BDV-infected cells. Diluted test sera are added for BDV-specific antibody detection. The results have good qualitative correlation with the VNT. Recently, a blocking ELISA applied to bulk-tank milk samples has been developed for the detection of BDV positive dairy sheep flocks [107]. ELISA kits that specifically detect antibodies targeting p80/125 protein, common to all BVDV and BDV strains (e.g., Synbiotics, Pourquier, IDEXX), can be used [64,102]. BDV-specific commercial ELISA kits are also available now (e.g., Svanova) [49]. In clinical investigation, ELISA kits for pestivirus and BDV-specific ELISA/VNT can be combined and used for more efficient and precise screening. The less sensitive agar gel immunodiffusion assay may also be used.

Viral antigen detection is important to identify the PI animals. The most sensitive and reliable way to confirm BDV infection is to isolate infectious virus from clinical samples. Commercial ELISA kits detecting pestivirus p125/p80 antigen (e.g., Synbiotics) can be used to detect antigen in PI sheep. In 1990, a mAb-capture ELISA is developed to detect specific antigen in leukocytes of sheep persistently infected with BDV. Blind trial is conducted to compare the specificity of the ELISA with virus isolation, and there is total agreement between the two tests [108]. However, the ELISA is usually not sufficiently sensitive to detect acutely viremic animals.

24.1.4.2 Molecular Techniques

Following the development of molecular biology and sequencing of pestivirus genomes, nucleic acid amplification methods (RT-PCR, nRT-PCR, and qRT-PCR) are increasingly used for diagnosing BDV infection. Pan-pestivirus primers (primers 324/326) are valuable for detecting all species of pestivirus and thus can be used for preliminary test, but the positive

TABLE 24.1

Reported Primers Could Be Used for RT-PCR, nRT-PCR, or qRT-PCR Detection

Primers	Sequences (5′–3′)	Target	Size (bp)	Reference
324	ATGCCCWTAGTAGGACTAGCA	5′ UTR	288	[109]
326	TCAACTCCATGTGCCATGTAC			
PBD1	TCGTGGTGAGATCCCTGAG	5′ UTR	225	[104]
PBD2	GCAGAGATTTTTTATACTAGCCTATRC			
320F	GCCTGATAGGGTGYWGCAGAG	Npro-C	740	[13]
1040R	TTYCCTTTCTTCTTYACCTGGTA			
BD1	TCTCTGCTGTACATGGCACATG	Npro-C	738	[16]
BD2	TTGTTRTGGTACARRCCGTC			
BDV87F	CCGTGTTAACCATACACGTAGTAGGA	5′ UTR	155	[103]
BDV237	GCCCTCGTCCACGTAGCA			
BDV136T (probe)	CTCAGGGATCTCACCACGA			

results need further sequence analysis to exclude BVDV existence [109]. Accompanied with the genome sequencing of several BDV strains and comparing them with those of other pestiviruses, BDV-specific primers have been designed and validated for RT-PCR [13,16,104]. The specific primers can be used alone in RT-PCR or following the use of pan-pestivirus primers in nRT-PCR (Table 24.1). The nRT-PCR method not only increases the specificity and sensitivity but also the possibility of sample contamination. An nRT-PCR is also developed to distinguish three ruminant pestiviruses, BVDV-1, BVDV-2, and BDV. The consensus PCR product is subjected to second PCR using type-specific primers and is able to differentiate the three ruminant pestiviruses [110]. A closed one-tube RT-PCR assay has also been developed, which is very sensitive and less prone to giving false-positive results compared to nRT-PCR.

qRT-PCR, which uses BDV-specific primers, or plus specific fluorescent probes, has advantages in high specificity, sensitivity, and prevention of contamination. A one-step qRT-PCR has been designed to detect and type BDV, BVDV-1, and BVDV-2 in ovine samples [103]. This qRT-PCR is validated on clinical samples and shows more sensitive than virus isolation and nRT-PCR, which also reduces the potential cross-contamination of samples. In addition, the results of qRT-PCR virus typing agree completely with sequencing. Substitution of different fluorescent probes results in specific tests for BDV or BVDV-2 [111]. A platform included a TaqMan qRT-PCR for the universal detection of pestiviruses, and a microarray assay that can use the amplicons produced in the qRT-PCR to identify the specific pestivirus has been reported. This assay combines the advantages of qRT-PCR with the multiplexing capability of microarray technology [112] and will be an efficient screening tool.

24.2 METHODS

24.2.1 Virus Isolation

24.2.1.1 Sample Preparation

To isolate infectious virus from viremic animals, sterile serum can be used directly. However, it is necessary to wash leukocytes from anticoagulated whole blood samples repeatedly before cocultivating them with susceptible cells in order to increase sensitivity. Tissues (spleen, thyroid, thymus, kidney, brain, lymph nodes, and gut lesions) collected from dead animals should be homogenized in phosphate-buffered saline (PBS) or culture medium, centrifuged at 8000 rpm for 10 min to remove debris, and the supernatants would be passed through 0.45 μm filters before inoculating onto cell monolayers.

Semen can also be examined for BDV but must be diluted at least 10-fold.

24.2.1.2 Isolation Procedure

A series of primary ovine cell cultures (e.g., kidney, testes, and lung) and semicontinuous cell lines derived from fetal lamb muscle, whole embryo [83], or sheep choroid plexus can also be used. Several bovine cell cultures (e.g., testicular, embryonic tracheal, or turbinate cells) and cell lines (e.g., Madin–Darby bovine kidney) may also be used. However, they are less sensitive for the primary isolation and growth of some BDV strains. It is necessary that the cells and fetal bovine serum used for virus isolation are pestivirus-free (no antipestivirus activity and no pestivirus).

A practical sensitive isolation procedure is as follows, and it should be optimized for maximum sensitivity when different field strains are isolated.

1. Subconfluent cell monolayers cultured in flask or plate are washed three times with Hanks' balanced salt solution and then inoculated with prepared samples to adsorb for 2 h at 37°C.
2. Cultures are washed with Hanks' balanced salt solution or medium, and an appropriate volume of maintenance medium is added.
3. Cultures are incubated for 5–7 days at 37°C and examined for CPE daily.
4. Cultures are frozen and thawed and passaged more times as before.
5. After three to five passages, cultures could be tested for virus presence by RT-PCR (24.2.2), immunofluorescence, or immunoperoxidase methods (24.2.1.3).

24.2.1.3 Immunofluorescence or Immunoperoxidase Assay

1. Cells are seeded onto plastic plates or cover slips, and virus is inoculated. Controls of negative cells and standard BDV-inoculated cells must be included.
2. Three to five days later, the culture medium is removed, and the cells are washed gently with PBS, air dried, and fixed with acetone or ethanol (previously cooled to −20°C) for 20 min.
3. The acetone or ethanol is removed, and the cells are gently washed two or more times with PBS.
4. Antibodies (e.g., BDV antiserum, BDV mAb) diluted in PBS with 1% Tween 20 (PBST) are added. The plates or slips are incubated at 37°C for 30–60 min in a humid atmosphere.
5. Primary antibodies are removed, and the cells are washed three times with PBS + Tween20(PBST).
6. Fluorescence- or HRP-conjugated secondary antibody diluted in PBST is added and incubated at 37°C for 30–60 min in a humid atmosphere.
7. Secondary antibodies are removed, and the cells are washed three times with PBST.
8. For immunofluorescence assay, positive cells are determined under fluorescence microscopy.
9. For immunoperoxidase assay, substrate (e.g., 3-amino-9-ethyl carbazole [AEC]) is added for color development, and positive cells (brown) are examined under the microscope. AEC stock solution: AEC (0.1 g) dissolved in dimethylformamide (15 mL). AEC working solution: add 0.3 mL stock solution to 4.7 mL membrane-filtered acetate buffer (0.05 M, pH 5.0), and then add 5 µL 30% H_2O_2.

24.2.2 RT-PCR

24.2.2.1 Sample Preparation

RT-PCR is used widely for diagnosing BDV infection.

1. Serum can be used directly.
2. Anticoagulated whole blood samples should be washed repeatedly to separate the leukocytes.
3. Tissues samples should be homogenized in PBS and centrifuged at 8000 rpm for 10 min to remove debris.

All samples would be stored at −70°C until RNA extraction.

24.2.2.2 Detection Procedure

24.2.2.2.1 RNA Extraction

Total RNA is extracted by the conventional guanidinium isothiocyanate–phenol–chloroform extraction method (e.g., TRIzol, Invitrogen) or other commercially available kit (e.g., QIAamp Viral RNA Mini kit, Qiagen).

For QIAamp Viral RNA Mini Kit, 140 µL samples are mixed with 560 µL of AVL viral lysis buffer. The mixture is incubated at room temperature for 10 min, and 560 µL of ethanol is added. The mixture is applied onto the spin column, centrifuged at $6000 \times g$ for 1 min, and 500 µL of AW1 buffer is added. The column is centrifuged at $6000 \times g$ for 1 min to remove unbound materials and washed with 500 µL of AW2 buffer. Then, the column is centrifuged at full speed for 3 min and placed into a new 1.5 mL microcentrifuge tube. Next, 60 µL of AVE buffer is added and the column is incubated at room temperature for 1 min. Then, the column is centrifuged at $6000 \times g$ for 1 min to elute viral RNA. The obtained RNA is used as a template for RT-PCR amplification. The RNA can be stored at −70°C until RT-PCR assay is performed.

24.2.2.2.2 Reverse Transcription

For two-step RT-PCR, either random hexamers or specific primers may be used in RT stage (*Note*: Oligo(dT) is not suitable). Many commercial RT kits are available, and RT may be performed according to the following basic protocol:

1. Prepare the RT mix in 0.2 mL tubes as follows (Table 24.2), according to the number of samples.
2. Aliquot 15 µL of RT mix to each 0.2 mL tube and 5.0 µL of individual RNA sample.
3. Spin the tubes briefly to ensure that no reagent droplets remain on the inner wall of the tubes.
4. Transfer the reaction tubes to a 42°C water bath for 30–60 min, and then increase temperature to 85°C for 5 min to inactivate the reverse transcriptase.
5. Place the tubes on ice for 5 min.
6. The cDNA can be used directly or stored at −20°C for future use.

24.2.2.2.3 PCR Amplification

1. Prepare the PCR reaction mix as follows (Table 24.3), according to the number of samples you detected.
2. Add 22.0 µL of PCR mix and 3.0 µL of cDNA into a 0.2 mL PCR tube. Positive and negative controls must be included.
3. Spin the sample tubes in a microcentrifuge.
4. Place the sample tubes in the thermal cycler, and the reaction procedure is denaturation at 94°C for 5 min; 35 cycles of denaturation at 94°C for 30 s, annealing at 54°C for 30 s, and extension at 72°C

TABLE 24.2

Reverse Transcription Volumes Indicated per Specimen

Component	Volume (µL)
RNase-free water	3.5
2× reaction buffer	10.0
Reverse transcriptase	0.5
Random primer/specific reverse primer	0.5
RNase inhibitor	0.5
Total	15.0

Note: The volume of each component could be changed according to the instructions of the kits you choose and the RNA concentration.

TABLE 24.3

Polymerase Chain Reaction Volumes Indicated per Specimen

Component	Volume (μL)
RNase-free water	17.25
10× reaction buffer (containing MgCl₂)	2.5
dNTPs mix (2.5 mM)	1
Forward primer (20 pmol/μL)	0.5
Reverse primer (20 pmol/μL)	0.5
Taq DNA polymerase	0.25
Total	22.0

Note: The RT-PCR can also be done in a single PCR tube as one-step RT-PCR format using commercially available kits.

TABLE 24.4

Real-Time Reverse Transcription Polymerase Chain Reaction Volume as Each Sample

Component	Volume (μL)
RNase-free water	8.0
2× reaction buffer	12.5
Forward primer (20 pmol/μL)	0.5
Reverse primer (20 pmol/μL)	0.5
Probe	0.5
Dye	0.5
Enzyme mix	0.5
Total	23.0

for 45 s; and is terminated with a final extension of 10 min at 72°C.

5. Analyze the PCR products by agarose gel electrophoresis through 1.5% agarose gel and further sequencing.

24.2.2.2.4 nRT-PCR

nRT-PCR is commonly used to increase the specificity and sensitivity of clinical detection. The reaction could be done as in the following procedure:

1. Using pan-pestivirus primers (324/326) (Table 24.1) in the primary RT-PCR for about 25–30 cycles as RT-PCR method (Section 24.2.2.2.3).
2. Using BDV-specific primers (PBD1/PBD2) (Table 24.1) in a secondary PCR with the diluted products (at least 1:10 dilution) of primary RT-PCR as template and with more volume of RNAse-free water added to keep the final volume of 25 μL; for the method, refer to Section 24.2.2.2.3.
3. The products are checked on 1.5% agarose gel electrophoresis and further sequencing.

24.2.2.2.5 qRT-PCR

Since the technique is very sensitive and also easy to be contaminated during operation, one-step qRT-PCR is commonly recommended. The primers and probe [103] listed in Table 24.1 might be used.

1. In one-step qRT-PCR, the commercially available kit can be used according to the manufacturer's instructions. Prepare the reaction mix as follows (Table 24.4), according to the number of samples you detected. The reactions are done in triplicate or duplicate.
2. Add 23.0 μL of reaction mix and 2.0 μL of RNA into a qRT-PCR tube. Positive, negative, and blank controls must be included.
3. After a short spin, put on the machine for qRT-PCR operation: one cycle at 45°C for 5 min (RT step); one

cycle at 95°C for 2 min; followed by 40 cycles of 95°C for 15 s and 60°C for 30 s.

4. Data are collected and summed as described in the machine instructions. Samples are usually considered positive if the cycle threshold value is 35 cycles or less. It is better to determine positivity by setting up the positive curve at the same reactions.

24.2.2.2.6 Sequencing

For final BDV detection using the gene RNA amplification, gene sequencing must be performed to distinguish between various pestivirus species. There are usually two procedures for sequencing the final PCR products. Direct sequencing of the amplified DNA is fast but may lead to some mistakes because of the short PCR products. Another method is to clone the PCR products into a PCR plasmid before sequencing. Here, PCR plasmid cloning is briefly introduced as follows:

1. The PCR-amplified DNAs are purified with DNA purification methods.
2. Purified DNA is ligated with PCR plasmid in 10 μL volume at 4°C overnight (TA cloning requires only a short time).
3. Cloned plasmids are transformed into competent *Escherichia coli* cells (such as TOPO-10 or DH5α) by "hot-shocking" technique.
4. Transformed cells are grown on lysogeny broth (LB) plates containing penicillin.
5. Five to six colonies are chosen for PCR checking, and three positive colonies are used for sequencing (the specific primers for BDV or common plasmid primers can be used).

Sequencing of BDV gene amplified products can be done by companies or by the lab themselves. The sequencing methods are performed according to standard procedures by companies or as detailed in other descriptions for other viruses in this volume.

The sequenced data of virus genomic amplifications are *Blast* with the information in *GenBank* and determined whether BDV or other pestiviruses, and also being which genotype basically.

24.3 CONCLUSION AND FUTURE PERSPECTIVES

BDV has worldwide distribution and has become endemic in small ruminant populations in rearing areas and can cross-infect other animal species including cattle and pigs. BDV belongs to the genus Pestivirus, similar to BVDV and CSFV. Viruses within the genus have relatively high levels of homology in their genomes and share cross-reactive immunogenicity in serological assays. Nearly all BDV isolates are ncp in cell culture with a very few cp BDV isolates reported. There are no defined serotypes but BDV strains from sheep exhibit considerable antigenic diversity. Until now, BDV isolates have been divided into at least seven genotypes and more may be identified later. Clinical signs mainly are maternal productive syndromes and infected lambs showing "hairy shaker disease" signs. The acute mucosal disease cases have occasionally occurred in new BDV outbreaks in sheep and goats. Vertical transmission plays an important role in the epidemiology of the disease. Infection of fetuses can result in the birth of PI lambs, which are the most potent source of infection. There are different serological methods that have been developed and detail introduced in World Organization for Animal Health (OIE) protocols [3]. Great achievements in molecular biological techniques have recently been made based on the BDV genome sequences. These methods, including RT-PCR, nRT-PCR, and qRT-PCR combining with sequencing, are more accurate, rapid, and sensitive for BDV detection, and also help to identify out BDV infections in cattle and pigs and differentiate BDV from other pestivirus species. Sequencing plus phylogenetic analysis has been one of the most used typing methods for BDV isolates around the world.

With the recent rapid development of sheep and goat industries in the world, especially in developing countries, it is urgently important to pay more attention to the prevention and control of BDV infections. Until now, there have not been any systematic procedures for BDV infection control. The BD control programs should ideally consist of comprehensive strategies and measures. It is very important to maintain absence of PI animals within herds and to cut off the infectious source for breeding or trading purposes. Establishing accurate commercial BDV detection kits might be considered. The ELISA based on specific mAbs for BDV and the automatic viral RNA detection methods are useful for large numbers of sample screenings. There is no available commercial vaccines for BD. Inactivated vaccines have been developed for use in sheep; but they require regular boosters to maintain strong protection against field challenge [113,114]. The BDV E2 gene, encoding the major immunological antigen, has larger sequence diversity and significantly effects immune results. Effective vaccines must be considered, which either consist of viral strains exhibiting a wide degree of cross-protectively or incorporate several viral strains. Effective BDV vaccines must also be effective at stopping transplacental spread of the virus. Treatment strategies may be another consideration for prevention and control of BDV infections [101]. The two important diseases of pestivirus (CSFV and BVDV) have been successfully controlled or eradicated in several countries. Effective technique can be used for BDV prevention, and also will be practical and achieved. In the near future, it may be possible to develop comprehensive strategies to improve current control, intervention, and treatment methods for BDV infections; this will greatly benefit the small ruminant industry.

REFERENCES

1. Hughes, L. E., Kershaw, G. F., and Shaw, I. G., Border disease: An undescribed disease of sheep. *Vet. Rec.*, 71, 313, 1967.
2. Nettleton, P. F. et al., Border disease of sheep and goats. *Vet. Res.*, 29, 327,1998.
3. Nettleton, P. F. and Willoughby, K., Border disease. In: *OIE Manual of Diagnostic Tests and Vaccines for Terrestrial Animals.* OIE, Paris, France, Chapter 2.7.1, p. 963, 2010.
4. Paton, D. J., Pestivirus diversity. *J. Comp. Pathol.*, 112, 215, 1995.
5. Hamblin, C. and Hedger, R. S., The prevalence of antibodies to bovine viral diarrhoea/mucosal disease virus in African wildlife. *Comp. Immunol. Microbiol. Infect. Dis.*, 2, 295, 1979.
6. Becher, P. et al., Phylogenetic analysis of pestiviruses from domestic and wild ruminants. *J. Gen. Virol.*, 78, 1357, 1997.
7. Schirrmeier, H. et al., Genetic and antigenic characterization of an atypical pestivirus isolate, a putative member of a novel pestivirus species. *J. Gen. Virol.*, 85, 3647, 2004.
8. Stalder, H. P. et al., Genetic heterogeneity of pestiviruses of ruminants in Switzerland. *Prev. Vet. Med.*, 72, 37, 2005.
9. Stahl, K. et al., Natural infection of cattle with an atypical 'HoBi'-like pestivirus—Implications for BVD control and for the safety of biological products. *Vet. Res.*, 38, 517, 2007.
10. Mao, L. et al., Genome sequence of a novel Hobi-like pestivirus in China. *J. Virol.*, 86, 12444, 2012.
11. Vilcek, S. et al., Characterization of a novel pestivirus originating from a pronghorn antelope. *Virus Res.*, 108, 187, 2005.
12. Kirkland, P. D. et al., Identification of a novel virus in pigs—Bungowannah virus: A possible new species of pestivirus. *Virus Res.*, 129, 26, 2007.
13. Strong, R. et al., Antigenic and genetic characterisation of border disease viruses isolated from UK cattle. *Vet. Microbiol.*, 141, 208, 2010.
14. Giammarioli, M. et al., Genetic and antigenic typing of border disease virus (BDV) isolates from Italy reveals the existence of a novel BDV group. *Vet. Microbiol.*, 147, 231, 2011.
15. Sullivan, D. G., Chang, G. J., and Akkina, R. K., Genetic characterization of ruminant pestiviruses: Sequence analysis of viral genotypes isolated from sheep. *Virus Res.*, 47, 19, 1997.
16. Vilcek, S. et al., Molecular characterization of ovine pestiviruses. *J. Gen. Virol.*, 78, 725, 1997.
17. Becher, P. et al., Molecular characterization of border disease virus, a pestivirus from sheep. *Virology*, 198, 542, 1994.
18. Vilcek, S. et al., Genetic typing of pestiviruses from New Zealand. *N. Z. Vet. J.*, 46, 35, 1998.
19. Becher, P. et al., Genetic and antigenic characterization of novel pestivirus genotypes: Implications for classification. *Virology*, 311, 96, 2003.
20. Krametter-Froetscher, R. et al., Influence of communal alpine pasturing on the spread of pestiviruses among sheep and goats in Austria: First identification of border disease virus in Austria. *Zoonos. Public Health*, 54, 209, 2007.

21. Valdazo-Gonzalez, B., Alvarez-Martinez, M., and Sandvik, T., Genetic and antigenic typing of border disease virus isolates in sheep from the Iberian Peninsula. *Vet. J.*, 174, 316, 2007.

22. Dubois, E. et al., Genetic characterization of ovine pestiviruses isolated in France, between 1985 and 2006. *Vet. Microbiol.*, 130, 69, 2008.

23. Oguzoglu, T. C. et al., Border disease virus (BDV) infections of small ruminants in Turkey: A new BDV subgroup? *Vet. Microbiol.*, 135, 374, 2009.

24. Li, W. et al., Detection of border disease virus (BDV) in goat herds suffering diarrhea in eastern China. *Virol. J.*, 10, 1, 2013.

25. Liu, X. et al., Genome Sequence of Border Disease Virus Strain JSLS12-01, Isolated from Sheep in China. *Genome Announc.*, 1, e00502-13, 2013.

26. Kim, I. J. et al., Identification of bovine viral diarrhea virus type 2 in Korean native goat (Capra hircus). *Virus Res.*, 121,103, 2006.

27. Buonavoglia, C. et al., Persistent pestivirus infection in sheep in Apulia (southern Italy). *New Microbiol.*, 17, 163, 1994.

28. Pratelli, A. et al., Genomic characterization of pestiviruses isolated from lambs and kids in southern Italy. *J. Virol. Methods*, 94, 81, 2001.

29. Nagai, M. et al., Molecular, biological, and antigenic characterization of a Border disease virus isolated from a pig during classical swine fever surveillance in Japan. *J. Vet. Diagn. Invest.*, 26, 547, 2014.

30. Giangaspero, M., Genetic variation of Border disease virus species strains. *Vet. Ital.*, 47, 415, 2011.

31. Giangasperoa, M., Apicellab, C., and Harasawa, R., Numerical taxonomy of the genus Pestivirus: New software for genotyping based on the palindromic nucleotide substitutions method. *J. Virol. Methods*, 192, 59, 2013.

32. Paton, D. J., Sands, J. J., and Edwards, S., Border disease virus: Delineation by monoclonal antibodies. *Arch. Virol.*, 135, 241, 1994.

33. Avalos-Ramirez, R. et al., Evidence for the presence of two novel pestivirus species. *Virology*, 286, 456, 2001.

34. Becher, P., Orlich, M., and Thiel, H. J., Complete genomic sequence of border disease virus, a pestivirus from sheep. *J. Virol.*, 72, 5165, 1998.

35. Rasmussen, T. B. et al., Generation of recombinant pestiviruses using a full-genome amplification strategy. *Vet. Microbiol.*, 142, 13, 2010.

36. Vilcek, S. et al., Complete genomic sequence of a border disease virus isolated from Pyrenean chamois. *Virus Res.*, 152, 164, 2010.

37. Vilcek, S. et al., Molecular characterization of border disease virus strain Aveyron. *Vet. Microbiol.*, 171, 87, 2014.

38. Ridpath, J. F. and Bolin, S. R., Comparison of the complete genomic sequence of the border disease virus, BD31, to other pestiviruses. *Virus Res.*, 50, 237, 1997.

39. Wiskerchen, M. and Collett, M. S., Pestivirus gene expression: Protein p80 of bovine viral diarrhea virus is a proteinase involved in polyprotein processing. *Virology*, 184, 341, 1991.

40. Xu, J. et al., Bovine viral diarrhea virus NS3 serine proteinase: Polyprotein cleavage sites, cofactor requirements, and molecular model of an enzyme essential for pestivirus replication. *J. Virol.*, 71, 5312, 1997.

41. Vantsis, J. T. et al., Experiments in Border disease. VIII. Propagation and properties of a cytopathic virus. *J. Comp. Pathol.*, 86, 111, 1976.

42. Laude, H. and Gelfi, J., Properties of Border disease virus as studied in a sheep cell line. *Arch. Virol.*, 62, 341, 1979.

43. Nettleton, P. F. et al., The production and survival of lambs persistently infected with border disease virus. *Comp. Immunol. Microbiol. Infect. Dis.*, 15, 179, 1992.

44. Becher. P. et al., Cytopathogenicity of border disease virus is correlated with integration of cellular sequences into the viral genome. *J. Virol.*, 70, 2992, 1996.

45. Rosamilia, A. et al., Detection of border disease virus (BDV) genotype 3 in Italian goat herds. *Vet. J.*, 199, 446, 2014.

46. Gray, E. W. and Nettleton, P. F., The ultrastructure of cell cultures infected with border disease and bovine virus diarrhoea viruses. *J. Gen. Virol.*, 68, 2339, 1987.

47. Gardiner, A. C. et al., Experiments in border disease. V. Preliminary investigations on the nature of the agent. *J. Comp. Pathol.*, 82, 159, 1972.

48. Giangaspero, M. et al., Epidemiological survey of Border disease virus among sheep from northern districts of Japan. *J. Vet. Med. Sci.*, 73, 1629, 2011.

49. Graham, D. A. et al., Pestiviral infections in sheep and pigs in Northern Ireland. *Vet. Rec.*, 148, 69, 2001.

50. Garcia-Perez, A. L., Clinical and laboratorial findings in pregnant ewes and their progeny infected with Border disease virus (BDV-4 genotype). *Res. Vet. Sci.*, 86, 345, 2009.

51. Lamontagne, L. and Roy, R., Presence of antibodies to bovine diarrhea-mucosal disease virus (border disease) in sheep and goat flocks in Quebec. *Can. J. Comp. Med.*, 48, 225, 1984.

52. Krametter-Froetscher, R. et al., Prevalence of antibodies to pestiviruses in goats in Austria. *J. Vet. Med. B Infect. Dis. Vet. Public Health*, 53, 48, 2006.

53. Ataseven, V. S. et al., Seropositivity of agents causing abortion in local goat breeds in Eastern and South-eastern Anatolia. *Turkey Revue Med. Vet.*, 157, 545, 2006.

54. Orsel, K. et al., Seroprevalence of antibodies against pestiviruses in small ruminants in The Netherlands. *Tijdschr. Diergeneeskd.*, 134, 380, 2009.

55. Danuser, R. et al., Seroprevalence and characterization of pestivirus infections in small ruminants and new world camelids in Switzerland. *Schweiz. Arch. Tierheilkd.*, 151, 109, 2009.

56. Barlow, R. M. et al., Experiments in Border disease. VII. The disease in goats. *J. Comp. Pathol.*, 85, 291, 1975.

57. Loken, T., Experimentally-induced border disease in goats. *J. Comp. Pathol.*, 97, 85, 1987.

58. Loken, T., Bjerkas, I., and Hyllseth, B., Border disease in goats in Norway. *Res. Vet. Sci.*, 33, 130, 1982.

59. Loken, T., Krogsrud, J., and Bjerkas, I., Outbreaks of border disease in goats induced by a pestivirus-contaminated orf vaccine, with virus transmission to sheep and cattle. *J. Comp. Pathol.*, 104, 195, 1991

60. Arnal, M. et al., A novel pestivirus associated with deaths in Pyrenean chamois (*Rupicapra pyrenaica* pyrenaica). *J. Gen. Virol.*, 85, 3653, 2004.

61. Hurtado, A. et al., Molecular identification of a new pestivirus associated with increased mortality in the Pyrenean chamois (*Rupicapra pyrenaica* pyrenaica) in Spain. *J. Wildl. Dis.*, 40, 796, 2004.

62. Marco, I. et al., Severe outbreak of disease in the southern chamois (*Rupicapra pyrenaica*) associated with border disease virus infection. *Vet. Microbiol.*, 120, 33, 2007.

63. Marco, I. et al., Border disease virus among chamois, Spain. *Emerg. Infect. Dis.*, 15, 448, 2009.

64. Fernandez-Sirera, L. et al., Two different epidemiological scenarios of border disease in the populations of Pyrenean chamois (*Rupicapra p.* pyrenaica) after the first disease outbreaks. *PLoS One*, 7, 1, 2012.

65. Cranwell, M. P. et al., Detection of Border disease virus in cattle. *Vet. Rec.*, 161, 211, 2007.

66. Hornberg, A. et al., Genetic diversity of pestivirus isolates in cattle from Western Austria. *Vet. Microbiol.*, 135, 205, 2009.

67. McFadden, A. M. et al., The first case of a bull persistently infected with Border disease virus in New Zealand. *N. Z. Vet. J.*, 60, 290, 2012.

68. Braun, U. et al., Infection of cattle with Border disease virus by sheep on communal alpine pastures. *Schweiz. Arch. Tierheilkd.*, 155, 123, 2013.

69. Krametter-Froetscher, R. et al., Transmission of border disease virus from sheep to calves a possible risk factor for the Austrian BVD eradication programme in cattle? *Wien. Tierärztl. Monatsschr.*, 95, 200, 2008.

70. Braun, U. et al., Sheep persistently infected with Border disease readily transmit virus to calves seronegative to BVD virus. *Vet. Microbiol.*, 168, 98, 2014.

71. Roehe, P. M., Woodward, M. J., and Edwards, S., Characterisation of p20 gene sequences from a border disease-like pestivirus isolated from pigs. *Vet. Microbiol.*, 33, 231, 1992.

72. Kawanishi, N. et al., First isolation of border disease virus in Japan is from a pig farm with no ruminants. *Vet. Microbiol.*, 171, 210, 2014.

73. Rosell, R. et al., Identification of a porcine pestivirus as a border disease virus from naturally infected pigs in Spain. *Vet. Rec.*, 174, 18, 2014.

74. Gardiner, A. C. and Barlow, R. M., Vertical transmission of Border disease infection. *J. Comp. Pathol.*, 91, 467, 1981.

75. Shaw, I. G., Border disease of sheep (hypomyelinogenesis). *Agriculture*, 78, 373, 1971.

76. Plant, J. W. et al., Pathology in the ovine foetus caused by an ovine pestivirus. *Aust. Vet. J.*, 60, 137, 1983.

77. Barlow, R. M. et al., Mechanisms of natural transmission of Border disease. *J. Comp. Pathol.*, 90, 57, 1980.

78. Potts, B. J. et al., Viral persistence and abnormalities of the central nervous system after congenital infection of sheep with border disease virus. *J. Infect. Dis.*, 151, 337, 1985.

79. Carlsson, U., Border disease in sheep caused by transmission of virus from cattle persistently infected with bovine virus diarrhoea virus. *Vet. Rec.*, 128, 145, 1991.

80. Leforban, Y., Vannier, P., and Cariolet, R., Protection of piglets born from ruminant pestivirus experimentally infected sows, and their contacts, to the challenge with hog cholera virus. *Ann. Rech. Vet.*, 23, 73, 1992.

81. Roic, B. et al., Serum antibodies directed against classical swine fever virus and other pestiviruses in wild boar (Sus scrofa) in the Republic of Croatia. *Dtsch. Tierarztl. Wochenschr.*, 114, 145, 2007.

82. Harasawa. R. and Mizusawa, H., Demonstration and genotyping of pestivirus RNA from mammalian cell lines. *Microbiol. Immunol.*, 39, 979, 1995.

83. Thabti, F. et al., Experimental model of Border Disease Virus infection in lambs: Comparative pathogenicity of pestiviruses isolated in France and Tunisia. *Vet. Res.*, 33, 35, 2002.

84. Nettleton, P. F. and Entrican, G., Ruminant pestiviruses. *Br. Vet. J.*, 151, 615, 1995.

85. Oguzoglu, T. C., A review of border disease virus infection in ruminants: Molecular characterization, pathogenesis, diagnosis and control. *Ani. Heal. Prod. Hyg.*, 1, 1, 2012.

86. Zakarian, B., Barlow, R. M., and Rennie, J. C., Periarteritis in experimental Border disease of sheep. I. The occurrence and distribution of the lesion. *J. Comp. Pathol.*, 85, 453, 1975.

87. Barlow, R. M., Experiments in border disease. IV. Pathological changes in ewes. *J. Comp. Pathol.*, 82, 151, 1972.

88. Barlow, R. M. et al., Experiments in Border disease. II. Some aspects of the disease in the foetus. *J. Comp. Pathol.*, 80, 635, 1970.

89. Burrells, C. et al., Lymphocyte subpopulations in the blood of sheep persistently infected with border disease virus. *Clin. Exp. Immunol.*, 76, 446, 1989.

90. Woldehiwet, Z. and Sharma, R., Alterations in lymphocyte subpopulations in peripheral blood of sheep persistently infected with border disease virus. *Vet. Microbiol.*, 22, 153, 1990.

91. Toplu, N. et al., Neuropathologic study of border disease virus in naturally infected fetal and neonatal small ruminants and its association with apoptosis. *Vet. Pathol.*, 48, 576, 2011.

92. McLaughlin, B. et al., Caspase 3 activation is essential for neuroprotection in preconditioning. *Proc. Natl. Acad. Sci. USA*, 100, 715, 2003.

93. Valdazo-Gonzalez, B., Alvarez, M., and Sandvik, T., Prevalence of border disease virus in Spanish lambs. *Vet. Microbiol.*, 128, 269, 2008.

94. Roeder, P. L., Jeffrey, M., and Drew, T. W., Variable nature of border disease on a single farm: The infection status of affected sheep. *Res. Vet. Sci.*, 43, 28, 1987.

95. Monies, R. J., Paton, D. J., and Vilcek, S., Mucosal disease-like lesions in sheep infected with Border disease virus. *Vet. Rec.*, 155, 765, 2004.

96. Gonzalez, J. M. et al., Natural border disease virus infection in feedlot lambs. *Vet. Rec.*, 174, 69, 2014.

97. Gardiner, A. C., Nettleton, P. F., and Barlow, R. M., Virology and immunology of a spontaneous and experimental mucosal disease-like syndrome in sheep recovered from clinical border disease. *J. Comp. Pathol.*, 93, 463, 1983.

98. Toplu, N., Oguzoglu, T. C., and Albayrak, H., Dual infection of fetal and neonatal small ruminants with border disease virus and peste des petits ruminants virus (PPRV): Neuronal tropism of PPRV as a novel finding. *J. Comp. Pathol.*, 146, 289, 2012.

99. Done, J. T. et al., Border disease of sheep: Spinal cord morphometry. *J. Comp. Pathol.*, 95, 325, 1985.

100. Kessell, A., Finnie, J., and Windsor, P., Neurological diseases of ruminant livestock in Australia. IV: Viral infections. *Aust. Vet. J.*, 89, 331, 2011.

101. Newcomer, B. W. and Givens, M. D., Approved and experimental countermeasures against pestiviral diseases: Bovine viral diarrhea, classical swine fever and border disease. *Antivir. Res.*, 100, 133, 2013.

102. Marco, I. et al., Serologic and virologic investigations into pestivirus infection in wild and domestic ruminants in the Pyrenees (NE Spain). *Res. Vet. Sci.*, 87, 149, 2009.

103. Willoughby, K. et al., Development of a real time RT-PCR to detect and type ovine pestiviruses. *J. Virol. Methods*, 132, 187, 2006.

104. Vilbek, S. and Paton, D. J., A RT-PCR assay for the rapid recognition of border disease virus. *Vet. Res.*, 31, 437, 2000.

105. Thabti, F. et al., Detection of a novel border disease virus subgroup in Tunisian sheep. *Arch. Virol.*, 150, 215, 2005.

106. Fenton, A. et al., A monoclonal antibody capture ELISA to detect antibody to border disease virus in sheep serum. *Vet. Microbiol.*, 28, 327, 1991.

107. Corbiere, F. et al., Short communication: Performance of a blocking antibody ELISA bulk-tank milk test for detection of dairy sheep flocks exposed to border disease virus. *J. Dairy Sci.*, 95, 6542, 2012.

108. Fenton, A. et al., An ELISA for detecting pestivirus antigen in the blood of sheep persistently infected with border disease virus. *J. Virol. Methods*, 27, 253, 1990.

109. Vilcek, S. et al., Pestiviruses isolated from pigs, cattle and sheep can be allocated into at least three genogroups using polymerase chain reaction and restriction endonuclease analysis. *Arch. Virol.*, 136, 309, 1994.
110. Fulton, R. W. et al., Nested reverse transcriptase-polymerase chain reaction (RT-PCR) for typing ruminant pestiviruses: Bovine viral diarrhea viruses and border disease virus. *Can. J. Vet. Res.*, 63, 276, 1999.
111. McGoldrick, A. et al., Closed one-tube reverse transcription nested polymerase chain reaction for the detection of pestiviral RNA with fluorescent probes. *J. Virol. Methods*, 79, 85, 1999.
112. LeBlanc, N. et al., A novel combination of TaqMan RT-PCR and a suspension microarray assay for the detection and species identification of pestiviruses. *Vet. Microbiol.*, 142, 81, 2010.
113. Vantsis, J. T. et al., Immunization against Border disease. *J. Comp. Pathol.*, 90, 349, 1980.
114. Brun, A. et al. Evaluation of the potency of an inactivated vaccine against border disease pestivirus infection in sheep. In: *Proceedings of the Second Symposium on Pestiviruses*, ed. Edwards, S. Fondation Marcel Merieux, Annecy, France, p. 257, 1993.

25 Bovine Viral Diarrhea Viruses 1 and 2

Benjamin W. Newcomer

CONTENTS

25.1 Introduction 223
    25.1.1 Classification...223
    25.1.2 Morphology and Genome Organization...225
    25.1.3 Clinical Features and Transmission 226
        25.1.3.1 Transient Infection 226
        25.1.3.2 Reproductive Tract Infection 226
        25.1.3.3 Persistent Infection...227
        25.1.3.4 Transmission 227
    25.1.4 Diagnosis 227
        25.1.4.1 Conventional Techniques..228
        25.1.4.2 Molecular Techniques...229
25.2 Methods 229
    25.2.1 Sample Preparation...229
        25.2.1.1 Serum...229
        25.2.1.2 White Blood Cells..229
        25.2.1.3 Nasal Swab Samples...229
        25.2.1.4 Semen...230
        25.2.1.5 Tissue Sample 230
        25.2.1.6 RNA Extraction 230
    25.2.2 Detection Procedures..230
        25.2.2.1 Standard RT-PCR 230
        25.2.2.2 Quantitative RT-PCR 231
25.3 Conclusions and Future Perspectives..231
References...231

25.1 INTRODUCTION

First described as "an apparently new transmissible disease of cattle" in 1946 [1], bovine viral diarrhea virus (BVDV) has become one of the most widespread viral pathogens of cattle. Since 2007, bovine viral diarrhea has been listed by the International Office of Epizootics as a reportable disease of cattle. Despite its moniker, the effects of BVDV infection are manifested not only in the gastrointestinal tract but also in the respiratory, reproductive, cardiovascular, lymphatic, immune, integumentary, or central nervous systems [2] of cattle, as well as several domestic and feral mammalian species of the order *Artiodactyla* [3]. The virus is responsible for a wide variety of clinical syndromes, but infection may also be subclinical or asymptomatic, emphasizing the need for accurate and efficient diagnostic tests. This is especially true in the case of persistent infection, where animals may appear phenotypically normal but consistently shed high amounts of live virus, playing a significant role in the epidemiology of the virus. The clinical and economic importance of BVDV as a livestock pathogen demands fast and accurate diagnostic tests at both the herd and individual animal levels.

25.1.1 CLASSIFICATION

BVDV is the prototypical virus in the *Pestivirus* genus of the *Flaviviridae* family. In 1993, the virus was established as the causative agent of severe morbidity and mortality in adult cattle and veal calves in Canada [4]. Subsequent genomic characterization of virus isolates from the outbreak revealed large differences in sequence homology to known BVDV strains [5], leading to the division of BVDV into two distinct genotypes, BVDV1 and BVDV2. The two viruses were first separated by their phylogenetic differences [6]. Although antigenic differences in the two species have also been demonstrated, the two species remain serologically cross-reactive.

Both species can be further divided into subgenotype groupings through phylogenetic analysis of conserved regions of the genome although the clinical significance of such groupings is unknown [7,8]. Phylogenetic analysis allows the characterization of BVDV1 into at least 11 groups (Figure 25.1); subtypes 1a and 1b are the predominant strains in North America [7,9]. Classification of BVDV2 isolates into subgenotypes is less common due to the smaller number of described isolates [10]. Two subgenotypes of BVDV2 have been reported in North America with type 2a more commonly isolated than type 2b [11].

Viral isolates from either BVDV genotype can be characterized by their activity in cultured epithelial cell lines. Infection with cytopathic (CP) isolates causes vacuolation and cell death in cultured cell monolayers, whereas infections with noncytopathic (NCP) stains do not. Biotype is not an accurate predictor of *in vivo* pathogenicity [12]. In nature, NCP strains predominate. The CP strains are believed to arise from NCP strains by mutagenesis, and the CP strain and its NCP counterpart are known as a viral pair. This occurs most commonly as a result of a recombination event where extra pieces of genetic code are inserted into an NCP genome. However, not all biotype changes are associated with a recombination event as CP viruses without insertions have been identified [13]. Consequently, CP strains are more variable in length than NCP strains [14,15]. Biotype is not an indicator of virulence as animals infected with either CP or NCP strains exhibit a wide range of clinical syndromes.

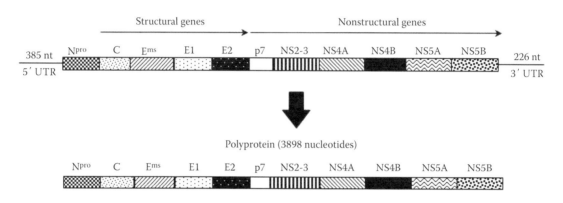

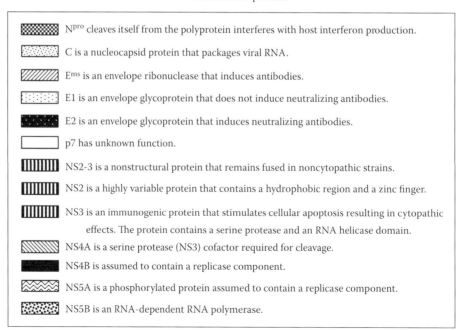

FIGURE 25.1 Genomic structure of bovine viral diarrhea virus. nt, nucleotides; NTR, nontranslated region; RNA, ribonucleic acid. (Reprinted from Newcomer, B.W. and Givens, M.D., *Antiviral Res.*, 100, 133, 2013. With permission.)

25.1.2 Morphology and Genome Organization

The BVDV virion is typical of the pestiviruses and measures 40–60 nm in diameter with genomic ribonucleic acid (RNA) and the nucleocapsid protein contained in the central core of an icosahedral capsid [16]. Two glycoproteins are anchored in the membranous envelope with a third peptide loosely attached by undefined interactions. The BVDV genome consists of a single strand of positive-sense RNA. Unlike most eukaryotic messenger RNA, the BVDV genome lacks a 5′ methylguanosine cap, instead possessing a 5′ untranslated region (UTR) of approximately 380 nucleotides [17]. Likewise, a 229–273 untranslated nucleotide sequence on the 3′ end of the genome

(3′ UTR) takes the place of a 3′ polyadenine tail. The genome of approximately 12.5 kb encodes a lone open reading frame that is translated into a single polyprotein of approximately 3900 amino acids. CP strains of the virus commonly contain insertions, duplications, and other rearrangements resulting in larger genomic RNAs [16]; however, nucleotide deletions have resulted in genomes smaller than the type strain [15,18]. Both viral and host endoproteinases are involved in co- and posttranslational processing of the polypeptide product into at least 11 mature viral proteins (Figure 25.2). The individual proteins in order from 5′ to 3′ are the amino-terminal autoprotease (Npro), the nucleocapsid or core protein (C), the

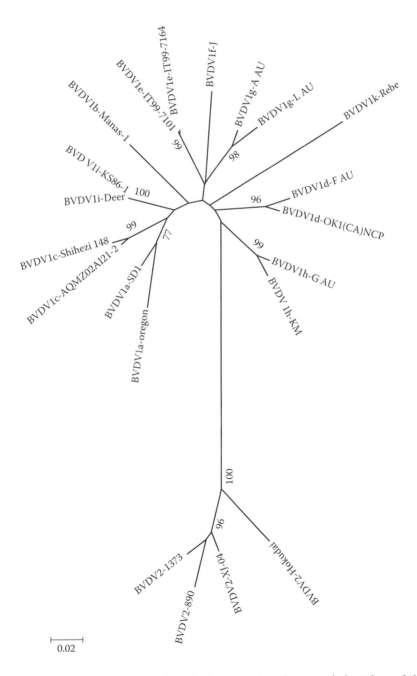

FIGURE 25.2 Diversity of selected strains of bovine viral diarrhea virus based on genetic homology of the 5′ untranslated region. (Reprinted from Newcomer, B.W. and Givens, M.D., *Antiviral Res.*, 100, 133, 2013. With permission.)

ribonuclease-soluble envelope glycoprotein (E^{rns}), the primary envelope glycoprotein (E1), the secondary envelope glycoprotein (E2), protein p7, the fused second and third nonstructural proteins (NS2-3), the amino- and carboxy-terminal portions of the fourth nonstructural protein (NS4A, NS4B), and the amino- and carboxy-terminal portions of the fifth nonstructural protein (NS5A, NS5B) [18].

The BVDV capsid containing the single-stranded RNA genome is enveloped in a host-derived lipid membrane and is comprised of the four virion-encoded structural proteins, core (C), E^{rns}, E1, and E2. The nucleocapsid, or core, protein is comprised of 102 amino acids and functions in the processing and packaging of genomic RNA into the nascent virion [18]. The E^{rns} envelope glycoprotein is made up of 227 amino acids. The amino-terminus, created when the nucleocapsid protein is cleaved from the polypeptide chain, contains a signal sequence for translocation into the endoplasmic reticulum. Infected cattle generate high levels of antibody to the E^{rns} protein, which forms the basis of some commercially available diagnostic tests for BVDV. However, these antibodies are somewhat limited in their ability to neutralize the virus. The E1 protein is not highly immunogenic and functions primarily in viral cell entry in conjunction with E2 [19]. As the immunodominant protein, E2 is extremely antigenic, and exposure to the protein through vaccination or natural infection results in the stimulation of neutralizing antibody production in exposed cattle [18]. In addition to being variable in length, the amino acid sequence contains three hypervariable regions believed to reflect immunologic selective pressure. The high antigenicity of the protein has led to the study of E2 subunit vaccines as a potential alternative to conventional vaccines that would allow differentiation of vaccinated animals from those naturally infected; to date, this potential has yet to be realized commercially.

The first nonstructural protein to be translated is the N^{pro} autoprotease. In addition to its role as a protease, the protein is also involved in the suppression of host immune defenses via interference with production of type I interferon through the activity of interferon regulatory factor 3 [20]. All proteins necessary for viral replication are nonstructural [18]. The role of protein p7 is not fully understood, but the protein likely functions in late-stage assembly and viral release, much like the analogous protein of hepatitis C virus. The remaining nonstructural proteins (NS2-3, NS4, and NS5) are each subject to posttranslational processing that yields two distinct peptides. High levels of neutralizing antibodies to NS2-3 are seen after natural infection with BVDV or following vaccination with a modified-live vaccine [18]. Cleavage of the protein yields the second (NS2) and third (NS3) nonstructural proteins, and constitutive expression of NS3 is a hallmark of CP strains as the cleavage rate of the protein falls dramatically shortly after infection with NCP strains. The NS3 protein includes the protease and helicase domains of NS2-3 and is one of the more widely studied proteins of BVDV. The NS4 protein is cleaved by the NS3 protease to yield NS4A, a cofactor to the serine protease activity of NS3, and NS4B of unknown function. The NS5 protein is incompletely cleaved to yield NS5A and NS5B, which functions as the RNA-dependent RNA polymerase [18].

25.1.3 CLINICAL FEATURES AND TRANSMISSION

Clinical disease caused by BVDV infection can take several forms and is a source of significant economic losses in cattle worldwide. The clinical effects of BVDV are manifested in the respiratory, gastrointestinal, reproductive, cardiovascular, lymphatic, immunologic, integumentary, or central nervous systems of the affected cattle throughout the world. However, reproductive losses and respiratory disease associated with BVDV infection are thought to have the largest negative economic impact for producers. Economic losses are substantial in both the beef and dairy industries. Clinical manifestations of BVDV infection are often broadly discussed in the categories of (1) transient infection, (2) infection of the reproductive tract, and (3) persistent infection.

25.1.3.1 Transient Infection

Transient infection occurs when the virus infects seronegative, immunocompetent cattle and may result in any of several different clinical syndromes that are influenced by viral strain and several host factors (e.g., age, immune status, gestational status). Clinical signs due to acute BVDV infection include depression, inappetence, fever, decreased milk production, oculonasal discharge, and oral ulcerations [2]. Despite its name, BVDV infection resulting in diarrhea is poorly characterized and inconsistently seen in transient infections, particularly of adult cattle. Immunosuppression may lead to secondary bacterial infections, particularly of the respiratory tract, and qualities that enhance the pathogenesis of other viral or bacterial pathogens. Consequently, BVDV has been implicated along with other pathogens (e.g., *Pasteurella multocida*, *Mannheimia haemolytica*, parainfluenza-3 virus, bovine respiratory syncytial virus) in the pathogenesis of the bovine respiratory disease complex [21].

In the mid-1990s, a syndrome caused by acute BVDV infection marked by severe thrombocytopenia and hemorrhage was described [5]. Mortality rates in both mature cattle and heifers were higher than commonly seen in previous outbreaks of BVDV, reaching 9% and 53%, respectively [4]. The viral isolates responsible for the outbreak were classified as BVDV2 due to differences in sequence homology from previous BVDV isolates, although hemorrhagic diatheses can result from infection with strains of either viral genotype. Due to the highly varied nature of BVDV-associated disease, clinical diagnoses of BVDV infection should be confirmed with laboratory diagnostic testing as both false-positive and false-negative diagnoses are common when the diagnosis is based solely on clinical signs.

25.1.3.2 Reproductive Tract Infection

Transient infection of susceptible, pregnant females as BVDV may result in fetal infection as the virus is able to readily cross the placenta and infect the fetus. The effect of infection on the fetus is largely determined by the stage of gestation when the dam is infected. Abortion or early embryonic death occurs commonly in earlier stages of gestation, due, at least in part, to endometrial inflammation resulting from the viral infection.

Persistent infection may occur if the dam is infected between 45 and 125 days of gestation and is discussed in detail in the following section. Transplacental infection between 100 and 150 days of gestation commonly results in congenital defects, particularly central nervous system malformations due to the ability of BVDV to cross the blood–brain barrier. Cerebellar hypoplasia is the most notable sign seen, but other congenital defects include growth retardation, hypomyelinogenesis, hydranencephaly, hydrocephalus, microencephaly, microphthalmia, ocular cataracts, retinal degeneration, optic neuritis, alopecia, hypotrichosis, brachygnathism, thymic aplasia, and pulmonary hypoplasia [22]. Infections that occur in the last trimester are generally neutralized by the fetus's immune response but may result in abortion.

In 1998, a new face was added to the multifaceted spectrum of clinical BVDV syndromes when a seropositive, nonviremic bull at an AI center was found to have a unique, localized persistent testicular infection (PTI) [23]. Since then, only one other report exists of a similarly infected bull, also found at a bull stud [24]. Bulls with PTI are nonviremic but consistently shed virus in the semen that is detectable using routine virus isolation techniques. The virus is confined to the testicular tissue of affected bulls where it is associated with both Sertoli and germinal cells [24]. Although the prevalence of bulls with PTI is thought to be very low, their presence may go undetected as the infection does not result in clinical signs. Consequently, unless these bulls are identified through appropriate diagnostic testing, they can propagate the spread of BVDV through the presence of infectious virus in their semen. The pathophysiology of PTI in bulls is currently not fully understood and remains a topic of further research.

25.1.3.3 Persistent Infection

Infection of the fetus with an NCP strain of the virus between the end of the embryonic stage of gestation and before the development of fetal immunocompetence (45–125 days of gestation) may result in persistent infection [2]. Persistently infected (PI) animals consistently shed high levels of live virus and are thus important in the epidemiologic aspects of BVDV propagation. Immunotolerance characterizes persistent infection; PI animals do not mount an immune response to the infecting NCP strain although seropositivity to heterologous viral strains may be seen. Cows that are PI will invariably give birth to PI calves. At birth, many PI animals are weak and ill thrifty and most die before 1 year of age. However, others may not show signs of disease and serve as important reservoirs of the virus for transmission to susceptible cohorts. The identification and removal of PI animals is one of the cornerstones of BVDV management, and consequently, diagnostic tests must be able to differentiate transient and persistent infections to be the most useful in making decisions regarding BVDV control at the herd and individual animal levels.

Mucosal disease (MD) is a syndrome unique to PI animals that are superinfected with the CP strain of BVDV homologous to the NCP infecting strain. Such strains may arise by mutation of the original infecting strain or may come from external sources such as modified-live vaccines or herdmates [25]. Development of MD is almost always fatal although recovery of isolated cases has been described [26]. Death may be preceded by widespread ulceration of the upper gastrointestinal tract and hemorrhagic lesions in the abomasum and elsewhere [2]. Multiple forms of MD have been described, including acute MD, chronic MD, and delayed onset MD [26].

25.1.3.4 Transmission

Animals are usually infected through the respiratory or gastrointestinal mucosa; subsequent primary viremia results in widespread dissemination of the virus. Although typically much shorter, viremia may last up to 15 days [27] and is often accompanied by pyrexia and leukopenia. Leukopenia in affected cattle is characterized by lymphopenia and potentially neutrophilia [28]. Tropism of BVDV is incompletely understood, but the virus is most reliably isolated from cells of hematopoietic or lymphoid tissue (e.g., white blood cells, spleen) but can also be found in tissues from multiple organ systems. The lymphotropism exhibited by BVDV affects cells of both the innate and adaptive immune systems [2]. The PI animal serves as the principal viral reservoir and the major source of infection to other cattle through direct contact with bodily secretions including saliva, urine, tears, feces, milk, and semen. Mechanical and arthropod vectors are also capable of viral transmission. Airborne transmission in the absence of direct contact is not a major route of transmission and occurs only over very short distances. Venereal transmission is possible from bulls experiencing transient, testicular, or persistent infection and calves born to PI dams are consistently PI themselves. The virus is capable of infecting several wildlife species of the order *Artiodactyla*, and cross-species transmission to cattle has been documented in specific experimental situations although the risk of interspecies transmission remains largely uncharacterized.

25.1.4 DIAGNOSIS

A definitive diagnosis of BVDV infection is only made through laboratory testing due to the diverse clinical and subclinical syndromes displayed by infected animals. Several tests are available, and selection of the appropriate test will be dictated by several factors, including the management system of the affected farm, financial constraints, and availability of tests at a given laboratory [2]. Since not all available tests are appropriate for each clinical situation, care should be made when selecting a test in order to reach a successful solution quickly and efficiently [29]. Most BVDV testing focuses on the detection and identification of PI animals in order to remove the said animals from the herd to limit virus shedding and spread; not all tests are appropriate for identification of PI animals. Essentially all diagnostic tests for BVDV fall into one of four main categories: virus isolation, antigen detection, serology, and molecular techniques.

25.1.4.1 Conventional Techniques

25.1.4.1.1 Virus Isolation

Historically, isolation of live virus from tissues or secretions of infected animals is considered the "gold standard" of diagnostic test for BVDV. While the test is still widely used in research, in many diagnostic settings, it has been replaced by tests that are more economical and have a shorter turnaround time. The best application of the test is to confirm infection in an individual animal showing clinical signs of transient infection. Many cells from several different species will support the growth of BVDV in culture but cell lines used in BVDV isolation assays are generally limited to bovine turbinate, bovine testicle, and Madin–Darby bovine kidney cells [29]. Infection with CP strains of BVDV results in characteristic cytopathic effects (CPE) in cultured monolayers; cells display vacuolization and apoptosis within 48 h of infection (Figure 25.3). However, most field strains of BVDV are of the NCP biotype and do not result in observable CPE. Consequently, alternative

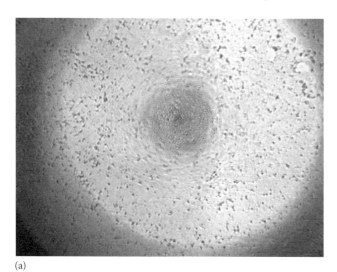

(a)

(b)

FIGURE 25.3 Microscope photographs (100×) of Madin–Darby bovine kidney cell cultures 72 h after infection with a noncytopathic (a) or cytopathic (b) strain of BVDV.

visualization techniques such as immunofluorescence or immunostaining must be employed to detect infected cells.

White blood cells extracted from whole blood samples are the preferred sample when collecting samples from the live animal for detection of BVDV by virus isolation [29]. Serum samples are also commonly used although false-negative results may be obtained as a result of the presence of serum-neutralizing antibodies in the serum. In one study of 23 separate diagnostic laboratories using seven sample–test combinations, virus isolation performed on serum resulted in the lowest consistency in detecting positive samples and the lowest level of agreement between laboratories [30]. Nasal swab samples may also be subjected to virus isolation assays as live virus is often shed in the nasal secretions of infected cattle. Tissue from the lymphoid organs such as the thymus, spleen, Peyer's patches, and mesenteric lymph nodes is the preferred sample to be collected from a dead animal or aborted fetus for submission for virus isolation [29]. A single virus isolation test is not capable of differentiating persistent and transient infections. Isolation of virus from samples obtained from an animal at least 3 weeks apart is indicative of persistent infection.

25.1.4.1.2 Antigen Detection

Direct antigen detection is generally less sensitive and less reliable than virus isolation techniques; however, antigen detection assays are generally quicker and more economical to perform than virus isolation assays [29]. Antigen detection tests are generally used as screening tests to detect PI animals although additional testing may be warranted in certain circumstances due to the occurrence of false positives. Commonly used antigen detection methods include antigen capture ELISAs (ACE) and immunohistochemistry (IHC) techniques. IHC techniques have been the traditional test for the detection of PI animals and are still widely used although the more recent development of economical and user-friendly ACE kits are being used more routinely. Most commercial IHC testing is performed on fresh or formalin-fixed tissue samples commonly taken from the ear and referred to as "ear notches." While IHC tests are primarily carried out by trained personnel at diagnostic laboratories, commercially available ACE kits have been used cow-side for the identification of PI animals. Several studies have demonstrated a high level of sensitivity for the commercial ACE kits when used as a screening test to detect PI animals [30,31]. Commercial ACE kits rely on monoclonal antibodies targeting the E^{rns} glycoprotein of BVDV; consequently, differing strains of BVDV may not be detected by the test [32].

25.1.4.1.3 Serology

Serologic tests can be used to demonstrate previous exposure to the virus, but care must be taken when interpreting the results as serologic tests are unable to differentiate antibodies produced in response to natural exposure versus those produced after vaccination or those present as a result of passive transfer through the ingestion of maternal colostrum [2]. However, in the right circumstances, serologic testing may prove beneficial to assess compliance with a vaccine protocol or to assess the herd's exposure status to BVDV [29]. In unvaccinated herds or in regions where the virus has been eradicated, testing of seronegative sentinel animals may be useful to determine herd exposure. The serum neutralization assay is the most commonly used serologic test for BVDV infection and can be used to differentiate exposure to BVDV1 and BVDV2. However, comparison of results between different laboratories is complicated as results will differ to some degree depending on the similarity of the infecting and reference strains. Additionally, anti-BVDV antibodies may be detected by various ELISA tests although these are no longer routinely used in most laboratories. Serologic assays are of little value in detecting PI animals as such individuals do not mount an antibody response to the infecting strain of the virus.

25.1.4.2 Molecular Techniques

Although virus isolation is regarded as the historical "gold standard" of BVDV diagnostic assays, the technique has largely been supplanted by the use of reverse transcription polymerase chain reaction (RT-PCR). The advantages of RT-PCR relative to virus isolation include high sensitivity, rapid turnaround time, low relative cost, obviation of the need for dedicated cell culture laboratory space and equipment, and the ability to distinguish the various genotypes and subgenotypes of BVDV. Additionally, RT-PCR is not adversely affected by the presence of neutralizing antibodies in serum samples. Commercially available kits with simple viral RNA extraction steps have encouraged the acceptance of RT-PCR as the primary herd screening assay used by many diagnostic laboratories. Additionally, RT-PCR can be used to detect BVDV in a wide variety of samples including serum, whole blood, milk, tissues, nasal swabs, semen, follicular fluid, and embryos [33]. The RT-PCR assay should be validated for each sample type as poor RNA extraction will adversely affect the sensitivity of the assay [29]. Detection of viral nucleic acid does not necessarily mean that infectious virus is present in the sample; thus, clinicians must take care to interpret assay results in conjunction with the clinical scenario.

The high sensitivity of the RT-PCR assay has made the testing of pooled samples an effective and economical means of screening whole herds or other groups of animals. The assay is sensitive enough that a single viremic sample in a pool of 100 negative sera has been detected by RT-PCR with the sensitivity of the assay estimated to exceed that of virus isolation by 10^1–10^4-fold [34]. However, sensitivity of the test decreases as the number of samples included in the pooled aliquot increases leading to controversy as to the optimal number of samples to be included in each pooled test. Pooling of milk samples for BVDV detection by RT-PCR is also done commonly due to the ease of sample collection and economic benefit of batch testing combined with the superior test sensitivity of RT-PCR. However, when using pooled samples, it is important to remember which samples are not included in the assay. For example, when testing for BVDV infection on a dairy using bulk tank milk samples, the results will evidently not reflect the infection status of any nonlactating animals

(e.g., bulls, dry cows, heifers) on the farm. Recently, a diagnostic test using RT-PCR has been described for nonlactating animals that do not require individual sampling [35]. By swabbing consumption surfaces and assaying the swabs by RT-PCR, the investigators were able to determine if PI animals were present in a group of cattle. While promising, this test is not yet available on a commercial basis.

Several molecular diagnostic test protocols have been described for the detection of BVDV in laboratory and clinical samples. Quantitative RT-PCR (RT-qPCR) and standard RT-PCR protocols for the detection of BVDV have been well described and the choice of specific protocol will be determined by the equipment available, goal of the diagnostic test, and operator preference. Sequencing of the amplicon following standard RT-PCR allows for genetic sequencing and identification of the isolate (sub)genotype, while RT-qPCR provides a quantitative assessment of viral load in the sample. Multiplex PCRs are performed commonly in diagnostic laboratories for the detection of multiple pathogens, including BVDV, in a single clinical sample.

25.2 METHODS

25.2.1 Sample Preparation

25.2.1.1 Serum

Serum samples should be collected in sterile blood collection tubes without anticoagulants. The sample is left to sit at room temperature for <2 h to allow for clot formation, and serum is collected by centrifugation at $200 \times g$ (1100 rpm) for 20 min. Samples that are assayed immediately may be aliquoted into sterile tubes and stored at −80°C until the time of testing.

25.2.1.2 White Blood Cells

Whole blood samples are collected in sterile blood collection tubes containing EDTA. The samples are centrifuged at $700 \times g$ (2020 rpm) at 4°C for 30 min. The buffy coat band containing the white blood cells is extracted using a sterile Pasteur pipette and placed in a 15 mL centrifuge tube containing 10 mL of 0.15 M NH_4Cl. After mixing, the tube is centrifuged at $700 \times g$ (2020 rpm) at 4°C for 10 min and the supernatant is poured off. To the remaining sample, 10 mL of minimum essential media (MEM) is added and mixed before centrifugation at $700 \times g$ (2020 rpm) at 4°C for 10 min a final time. The supernatant is removed and the white blood cell pellet is resuspended in 0.5 mL of MEM. The sample may then be passaged or submitted directly to RNA extraction. White blood cell samples should not be frozen before RNA extraction.

25.2.1.3 Nasal Swab Samples

For nasal swab samples, one nasal swab is placed in a sterile tube containing 3 mL of MEM + 0.1 gentamicin and allowed to sit for at least 1 h at 4°C before processing. Samples not to be immediately processed may be allowed to sit in the solution for longer periods (up to at least 48–72 h) at 4°C. Immediately before RNA extraction, the tube is agitated gently and the nasal swab removed; RNA is extracted from the media.

25.2.1.4　Semen

Raw semen samples are partially extended (1:8) in egg yolk–citrate extender by adding 250 μL of raw semen to 1750 μL of the extender [36]. The extended semen sample is then processed through a Sephacryl S-400 column according to the manufacturer's instructions before RNA extraction is initiated.

25.2.1.5　Tissue Sample

Various tissues can be assayed for the presence of BVDV viral nucleic acid using RT-PCR including tissue samples obtained at necropsy, testicular biopsy samples, and ear notch samples. For a tissue sample that measures approximately 1 cm × 1 cm, add the tissue and 3 mL of MEM to a stomach bag and stomach at high speed for 2 min. Additional MEM may be added for larger samples. If necessary, the resulting contents may be centrifuged gently to remove gross particulate matter. The supernatant may then be stored at −80°C or submitted directly for RNA extraction.

25.2.1.6　RNA Extraction

An efficient and simple method of RNA extraction that is validated for each sample type to be tested is a requirement for achieving highly sensitive results by RT-PCR. Several kits are commercially available for RNA extraction (e.g., QIAamp Viral RNA Mini Kit, Qiagen, Valencia, CA). When testing multiple sample types, validation of the chosen extraction kit for each sample type is advised to minimize the opportunity for false-negative test results.

In our laboratory, the High Pure RNA Isolation Kit (Roche, Indianapolis, IN) is used for the extraction of RNA from buffy coat samples while the QIAamp Viral RNA Mini Kit has been used effectively on infected uterine tubal cells, embryo sonicate fluid, follicular fluid, cell culture supernatant, serum, semen, and nasal swab and tissue samples. RNA extraction is performed according to the manufacturer's instructions for each kit. Briefly, for the QIAamp Viral RNA Mini Kit, 140 μL of the sample supernatant is added to 560 μL of the viral lysis buffer. Incubation at room temperature for 10 min is followed by the addition of 560 μL of absolute alcohol to the sample. Then, 630 μL of the solution is added to the spin column and centrifuged (all centrifugation is performed at 25,000 × g) for 1 min. The sample is then successively washed by adding 500 μL of AW1 and AW2 buffer with a centrifugation step after each wash lasting 1 and 3 min, respectively. Finally, the RNA is eluted by the addition of 60 μL of AVE buffer and centrifuged for 1 min after incubation at room temperature for 1 min. The RNA may be used immediately for RT-PCR or stored at −80°C until RT-PCR will be performed.

25.2.2　Detection Procedures

25.2.2.1　Standard RT-PCR

The following protocol is a nested RT-PCR (RT-nPCR) assay performed in a single closed-tube reaction modified from a previously published report [37].

1. The following mixture is aliquoted into the cap of a 200 μL tube and allowed to dry for 2 h at room temperature:

Order	Reagent	Concentration	Volume (μL)
1	Trehalose	22% w/v	5.0
2	Deoxynucleotide mix (dNTP)	10 mM	1.0
3	Taq polymerase	5 U/μL	0.25
4	BVDV 180 primer 5′-CCTGAGTACAGGGDAGT CGTCA-3′	50 μM	0.4
5	Hepatitis C virus (HCV) 368 primer 5′-CCATGTGCCATGTACAG-3′	50 μM	0.4

2. The RT-PCR master mix is then prepared as follows (volumes shown are per PCR):

Order	Reagent	Concentration	Volume (μL)
1	RNAse-free water		25.5
2	MgCl$_2$	25 mM	8.0
3	10× PCR buffer		5.0
4	dNTPs	10 mM	2.0
5	BVDV 100 primer 5′-GGCTAGCCATGCCCTTAG-3′	5 μM	1.0
6	HCV 368 primer 5′-CCATGTGCCATGTACAG-3′	5 μM	1.0
7	Triton X-100	10% stock	1.0
8	Dithiotreitol (DTT)	100 mM	0.25
9	RNAsin	40 U/μL	0.25
10	Taq polymerase	5 U/μL	0.5
11	M-MLV reverse transcriptase (MMLV RT)	200 U/μL	0.5
	Total		45

3. Briefly, 45 μL of the master mix is added to a thin-walled PCR tube in a frozen (−20°C) rack and overlaid with 2 drops (50 μL) of mineral oil.
4. Then, 5 μL of extracted RNA is added to the master mix through the oil layer for a total.
5. The tube is then subjected to the following cycle parameters: 37°C for 45 min, 95°C for 5 min, and then 20 cycles at 94°C for 1 min, 55°C for 1 min, and 72°C for 1 min, followed by a final elongation cycle of 72°C for 10 min.
6. The tube is inverted several times and then centrifuged at 14,000 × g for 12 s.
7. Then, 30 cycles are run at 94°C for 1 min, 55°C for 1 min, and 72°C for 45 s before a final elongation step of 72°C for 10 min.
8. The amplicons are maintained at 4°C until visualization of the PCR products is achieved by agarose gel electrophoresis.

25.2.2.2 Quantitative RT-PCR

The following protocol is currently used in our laboratory and uses the iTaq Universal SYBR Green One-Step Kit in a 96-well-plate format.

1. The RT-qPCR master mix is prepared as follows:

Reagent	Concentration	Volume (µL)
One-step SYBR Green reaction mix	1×	10
iScript reverse transcriptase		0.25
RNAse-free water		4.15
BVDV forward primer 5′-TAGCCATGCCCTTAGTAGGAC-3′	0.3 µM	0.3
BVDV reverse primer 5′-GACGACTACCCTGTACTCAGG-3′	0.3 µM	0.3
Total		15

2. In each well of the plate, 5.0 µL of extracted RNA template is added to the 15 µL of the master mix.
3. The thermocycler is programmed for the following cycle parameters: the plate is held at 50°C for 10 min and then 95°C for 1 min followed by 39 cycles at 95°C for 10 s and 56.5°C for 30 s.
4. The melting-curve analysis cycle is then performed at 65°C–95°C in 0.5°C increments with each step lasting 2–5 s.

25.3 CONCLUSIONS AND FUTURE PERSPECTIVES

BVDV is one of the primary viral pathogens of cattle and continues to cause significant economic losses to cattle producers worldwide. Despite causing a wide variety of clinical syndromes, the lack of pathognomonic symptoms following infection dictates the need for sensitive and specific diagnostic tools that are practical, scalable, and economical. In several countries and geographic regions of the world, voluntary or mandatory BVDV eradication programs have been implemented. Diagnostic sensitivity and specificity is important to the success of such programs as well as for surveillance activities following eradication. In areas where BVDV infection is endemic, diagnostic accuracy will continue to play an important role in control of the virus and viral transmission. Particularly important in the epidemiology of BVDV infections is the presence of PI animals. Thus, diagnostic tools must not only be able to determine the presence or absence of viral infection from clinical samples but, when infection is detected, to be able to classify the infection as a transient or persistent infection.

With the first description of the genomic sequence of the virus in 1988 [38], the era of molecular diagnostics was initiated. Since that time, advances in diagnostic detection have led to the widespread use of RT-PCR as the principal diagnostic assay for BVDV infection in many laboratories. Compared to virus isolation, the historical gold standard for the diagnosis of BVDV infection, RT-PCR has high sensitivity, rapid turnaround time, low relative cost, and no need for dedicated cell culture laboratory space and equipment. However, virus isolation and other conventional diagnostic tools are not without their benefits and should not be overlooked or forgotten by the diagnostician dealing with BVDV.

Current diagnostics are unable to readily differentiate infected from vaccinated animals. With a continued effort in many areas to eradicate the disease, future labors will focus on the development of assays to distinguish between field and vaccinal exposure. Additionally, the involvement of BVDV in multiagent infections, particularly in bovine respiratory disease, supports the continued development and use of multiplex RT-PCR assays to determine the presence or absence of multiple pathogens from a single clinical sample. The impact of BVDV infection on cattle health in populations worldwide underscores the continued need for accurate and economical diagnostic assays to safeguard cattle health and well-being.

REFERENCES

1. Olafson, P., Maccallum, A. D., and Fox, F. H., An apparently new transmissible disease of cattle, *Cornell Vet.*, 36, 205, 1946.
2. Walz, P. H. et al., Control of bovine viral diarrhea virus in ruminants, *J. Vet. Intern. Med.*, 24, 476, 2010.
3. Passler, T. and Walz, P. H., Bovine viral diarrhea virus infections in heterologous species, *Anim. Health Res. Rev.*, 11, 191, 2010.
4. Carman, S. et al., Severe acute bovine viral diarrhea in Ontario, 1993–1995, *J. Vet. Diagn. Invest.*, 10, 27, 1998.
5. Pellerin, C. et al., Identification of a new group of bovine viral diarrhea virus strains associated with severe outbreaks and high mortalities, *Virology*, 203, 260, 1994.
6. Ridpath, J. F., Bolin, S. R., and Dubovi, E. J., Segregation of bovine viral diarrhea virus into genotypes, *Virology*, 205, 66, 1994.
7. Vilcek, S. et al., Bovine viral diarrhoea virus genotype 1 can be separated into at least eleven genetic groups, *Arch. Virol.*, 146, 99, 2001.
8. Flores, E. F. et al., Phylogenetic analysis of Brazilian bovine viral diarrhea virus type 2 (BVDV-2) isolates: Evidence for a subgenotype within BVDV-2, *Virus Res.*, 87, 51, 2002.
9. Fulton, R. W. et al., Bovine viral diarrhoea virus (BVDV) subgenotypes in diagnostic laboratory accessions: Distribution of BVDV1a, 1b, and 2a subgenotypes, *Vet. Microbiol.*, 111, 35, 2005.
10. Vilcek, S. et al., Genetic diversity of BVDV: Consequences for classification and molecular epidemiology, *Prev. Vet. Med.*, 72, 31, 2005.
11. Ridpath, J. F., Practical significance of heterogeneity among BVDV strains: Impact of biotype and genotype on U.S. control programs, *Prev. Vet. Med.*, 72, 17, 2005.
12. Bezek, D. M., Grohn, Y. T., and Dubovi, E. J., Effect of acute infection with noncytopathic or cytopathic bovine viral diarrhea virus isolates on bovine platelets, *Am. J. Vet. Res.*, 55, 1115, 1994.
13. Quadros, V. L. et al., A search for RNA insertions and NS3 gene duplication in the genome of cytopathic isolates of bovine viral diarrhea virus, *Braz. J. Med. Biol. Res.*, 39, 935, 2006.

14. Meyers, G. et al., Rearrangement of viral sequences in cytopathogenic pestiviruses, *Virology*, 191, 368, 1992.

15. Tautz, N. et al., Pathogenesis of mucosal disease—A cytopathogenic pestivirus generated by an internal deletion, *J. Virol.*, 68, 3289, 1994.

16. Lindenbach, B. D. and Rice, C. M., Flaviviridae: The viruses and their replication, Chapter 32 in *Fields Virology*, eds. D. M. Knipe and P. M. Howley, Lippincott Williams & Wilkins, Philadelphia, PA, 2001.

17. Brock, K. V., Deng, R., and Riblet, S. M., Nucleotide sequencing of 5′ and 3′ termini of bovine viral diarrhea virus by RNA ligation and PCR, *J. Virol. Methods*, 38, 39, 1992.

18. Donis, R. O., Molecular biology of bovine viral diarrhea virus and its interactions with the host, *Vet. Clin. North Am. Food Anim. Pract.*, 11, 393, 1995.

19. Ronecker, S. et al., Formation of bovine viral diarrhea virus E1–E2 heterodimers is essential for virus entry and depends on charged residues in the transmembrane domains, *J. Gen. Virol.*, 89, 2114, 2008.

20. Hilton, L. et al., The NPro product of bovine viral diarrhea virus inhibits DNA binding by interferon regulatory factor 3 and targets it for proteasomal degradation, *J. Virol.*, 80, 11723, 2006.

21. Ridpath, J., The contribution of infections with bovine viral diarrhea viruses to bovine respiratory disease, *Vet. Clin. North Am. Food Anim. Pract.*, 26, 335, 2010.

22. Grooms, D. L., Reproductive consequences of infection with bovine viral diarrhea virus, *Vet. Clin. North Am. Food Anim. Pract.*, 20, 5, 2004.

23. Voges, H. et al., Persistent bovine pestivirus infection localized in the testes of an immuno-competent, non-viraemic bull, *Vet. Microbiol.*, 61, 165, 1998.

24. Newcomer, B. W. et al., Laboratory diagnosis and transmissibility of bovine viral diarrhea virus from a bull with a persistent testicular infection, *Vet. Microbiol.*, 170, 246, 2014.

25. Brownlie, J., Pathogenesis of mucosal disease and molecular aspects of bovine virus diarrhoea virus, *Vet. Microbiol.*, 23, 371, 1990.

26. Bolin, S. R., The pathogenesis of mucosal disease, *Vet. Clin. North Am. Food Anim. Pract.*, 11, 489, 1995.

27. Duffell, S. J., and Harkness, J. W., Bovine virus diarrhoea-mucosal disease infection in cattle, *Vet. Rec.*, 117, 240, 1985.

28. Newcomer, B. W. et al., Effect of treatment with a cationic antiviral compound on acute infection with bovine viral diarrhea virus, *Can. J. Vet. Res.*, 77, 170, 2013.

29. Saliki, J. T. and Dubovi, E. J., Laboratory diagnosis of bovine viral diarrhea virus infections, *Vet. Clin. North Am. Food Anim. Pract.*, 20, 69, 2004.

30. Edmondson, M. A. et al., Comparison of tests for detection of bovine viral diarrhea virus in diagnostic samples, *J. Vet. Diagn. Invest.*, 19, 376, 2007.

31. Kuhne, S. et al., Detection of bovine viral diarrhoea virus infected cattle—testing tissue samples derived from ear tagging using an Erns capture ELISA, *J. Vet. Med. B: Infect. Dis. Vet. Public Health*, 52, 272, 2005.

32. Gripshover, E. M. et al., Variation in E(rns) viral glycoprotein associated with failure of immunohistochemistry and commercial antigen capture ELISA to detect a field strain of bovine viral diarrhea virus, *Vet. Microbiol.*, 125, 11, 2007.

33. Lanyon, S. R. et al., Bovine viral diarrhoea: Pathogenesis and diagnosis, *Vet. J.*, 199, 201, 2014.

34. Weinstock, D., Bhudevi, B., and Castro, A. E., Single-tube single-enzyme reverse transcriptase PCR assay for detection of bovine viral diarrhea virus in pooled bovine serum, *J. Clin. Microbiol.*, 39, 343, 2001.

35. Givens, M. D. et al., Detection of cattle persistently infected with bovine viral diarrhea virus using a non-invasive, novel testing method, *Proceedings of the Fifth United States BVDV Symposium*, San Diego, CA, p. 105, 2011.

36. Givens, M. D. et al., Detection of bovine viral diarrhea virus in semen obtained after inoculation of seronegative postpubertal bulls, *Am. J. Vet. Res*, 64, 428, 2003.

37. McGoldrick, A. et al., Closed one-tube reverse transcription nested polymerase chain reaction for the detection of pestiviral RNA with fluorescent probes, *J. Virol. Methods*, 79, 85, 1999.

38. Collett, M. S. et al., Molecular-cloning and nucleotide-sequence of the pestivirus bovine viral diarrhea virus, *Virology*, 165, 191, 1988.

39. Newcomer, B. W. and Givens, M. D., Approved and experimental countermeasures against pestiviral diseases: Bovine viral diarrhea, classical swine fever and border disease, *Antiviral Res.*, 100, 133, 2013.

26 Classical Swine Fever Virus

Dongyou Liu

CONTENTS

26.1 Introduction 233
    26.1.1 Classification 233
    26.1.2 Morphology and Genome Organization 234
    26.1.3 Replication and Epidemiology 234
    26.1.4 Clinical Features and Pathogenesis 235
        26.1.4.1 Acute Disease 235
        26.1.4.2 Subacute and Chronic Disease 235
        26.1.4.3 Congenital Disease 235
    26.1.5 Diagnosis 236
    26.1.6 Treatment and Prevention 236
26.2 Methods 237
    26.2.1 Sample Preparation 237
    26.2.2 Detection Procedures 237
        26.2.2.1 Conventional RT-PCR 237
        26.2.2.2 Real-Time RT-PCR 237
26.3 Conclusion 237
References 238

26.1 INTRODUCTION

First recognized in Tennessee, United States, in 1810, classical swine fever (CSF, alternatively known as hog cholera, pig plague, or swine fever) is a highly contagious, fatal disease of pigs, wild boar, and collared peccaries. The etiological agent for CSF is a single-stranded and positive-sense RNA virus in the genus *Pestivirus* of the family *Flaviviridae* with the namesake of classical swine fever virus (CSFV). Typical clinical symptoms include fever, skin lesions, convulsions, and death (usually within 15 days, especially in young animals), which are indistinguishable from those of African swine fever (see Chapter 91). Once widely spread in many parts of the world, CSF has been successfully eradicated from Canada, the United States, Australia, and New Zealand, but still occurs in Asia, South America, Eastern Europe, and Russia, with significant economic impacts through production losses, trade bans, surveillance, and prevention costs.

26.1.1 CLASSIFICATION

CSFV is a member of the genus *Pestivirus*, family *Flaviviridae*. The family *Flaviviridae* (*flavus* means "yellow" in Latin, derived from yellow fever virus, a type virus of the family) consists of four genera (*Flavivirus*, *Hepacivirus*, *Pegivirus*, *Pestivirus*). In turn, the genus *Pestivirus* (*pestis* means "plague" in Latin) comprises four recognized species (bovine viral diarrhea virus genotype 1, bovine viral diarrhea virus genotype 2, border disease virus (BDV), and CSFV) together with five proposed species (antelope pestivirus, bovine viral diarrhea virus genotype 3 [atypical bovine pestivirus], Bungowannah pestivirus, giraffe pestivirus, Hobi pestivirus, and Tunisian sheep virus)] [1–6].

Composed of a single serotype, CSFV is further divided into three major groups (genotypes) and 10 subgroups (subtypes) (i.e., subgroup 1.1, 1.2, 1.3; subgroup 2.1, 2.2, 2.3; subgroup 3.1, 3.2, 3.3, 3.4) based on phylogenetic analysis of the polymerase gene NS5B, the envelope glycoprotein gene E2, and the 5′ nontranslated region (NTR) [7,8]. Additionally, distinct strains that circulate in certain countries/regions may be identified within subgroups as clades (e.g., clade 2.1a, 2.1b, and 2.1c). In general, genotype 1 encompasses the highly virulent, historical strains that were isolated from various parts of the world and still occur in South and Central America as well as the Caribbean Sea; all live-attenuated vaccine strains also belong to genotype 1. Genotype 2 includes the strains that are largely found in Europe and that have caused epidemics since the 1980s. Genotype 3 contains the strains that are mainly distributed in Asia (e.g., Taiwan, Korea, Japan, and Thailand) [9]. Genotypes 2 and 3 strains are often of moderate to low virulence and account for mild CSF that may be overlooked or dismissed as other viral disease at the early stages, contributing to the delay in implementation of necessary control measures and also the increase in disease prevalence [10]. There is evidence that some CSFV strains may switch from the historical group 1 or 3 to the more recently prevalent group 2 in Europe and Asia, further complicating the diagnostic and preventive efforts [11,12].

Differing from other flaviviruses by the presence of the N^{pro} autoprotease, members of the genus *Pestivirus* are infective to nonhuman mammals, including various bovine (cattle, sheep, and goats), and swine species, resulting in diseases of varying severity (from subclinical to overt clinical syndromes such as fever, diarrhea, hemorrhagic syndrome, abortion, and death). It is notable that while most *Pestivirus* species are capable of infecting animals out of their natural host, CSFV is the only species that is restricted to domestic pigs and wild boars.

26.1.2 Morphology and Genome Organization

CSFV possesses an enveloped, spherical virion of 40–60 nm in diameter. Formed during maturation and assembly of the virion by budding through membranes of infected cells, the lipid envelope is covered by dimers with a T = 3-like organization.

The genome of CSFV consists of a monopartite, linear, single-stranded, positive-sense RNA of 12.3 kb in length, which is made up of a large open reading frame (ORF) flanked by 5′ and 3′ NTRs. The detailed arrangement of CSFV RNA is as follows: 5′-NTR-Npro-C-E^{rns}-E1-E2-p7-NS2-3-NS4A-NS4G-NS5A-NS5B-NTR-3′. Translation of the ORF results in a 3898-amino-acid polyprotein, which is further processed by viral Npro, NS2, and NS3 proteinases and by host signal peptidase to generate four structural (C, E^{rns}, E1, and E2) and eight nonstructural (N^{pro}, p7, NS2, NS3, NS4A, NS4B, NS5A, and NS5B) proteins (Table 26.1) [13–15].

Unlike some other flaviviruses that carry a methylated nucleotide cap at their 5′ UTR, CSFV has a noncapped 5′ UTR (375 nt), which includes an internal ribosome entry site (located between 60 and 375 nt) with binding sites for NS3 and NS5A and is involved in the initiation of cap-independent translation as well as RNA replication. The 3′ UTR (232 nt) in CSFV is uracil rich, contains two NS5A binding sites (3′ UTRSL-1 and 3′ UTRSL-2), but lacks a poly A tail (an indication for the absence of posttranscriptional modifications).

In addition to mature structural proteins and nonstructural proteins, disulfide-linked heterodimers (E^{rns}-E2 and E1-E2) and homodimers (E2-E2 and E^{rns}-E^{rns}) may be observed in CSFV virions. Furthermore, two precursor proteins E2-P7 (60 kDa, 443 aa) and NS2-3 (120 kDa, 1140 aa) are produced from the polyprotein, which may play accessory roles in virus replication and which are present in infected cells but not in pelleted virions [16].

26.1.3 Replication and Epidemiology

Upon entry into the host, CSFV utilizes E(rns) and E2 envelope glycoproteins for attaching to host cellular receptors (heparan sulfate [HS] and laminin receptor [LamR], which interacts

TABLE 26.1
Characteristics of CSFV Structural and Nonstructural Proteins

Gene	Protein	Size	Function
Npro	Npro (N-terminal autoprotease)	23 kDa, 168 aa	Cysteine protease for cotranslational cleavage from the nascent downstream nucleocapsid protein C; antagonist of host IFN-α/β induction pathway
C	C (capsid or core protein)	14.3 kDa, 99 aa	Enhancement of CSFV RNA replication; regulation of SUMOylation pathway for immune evasion
E^{rns}	Erns (envelope protein RNase secreted) or E0 glycoprotein	44 kDa, 227 aa	Attachment to laminin receptor on cell surface; endonuclease/ribonuclease; inhibition of RNA and protein synthesis in host cells; induction of apoptosis of lymphocytes; inhibition of early-stage immunization of host
E1	E1 (envelope) glycoprotein	33 kDa, 195 aa	Involvement in virus attachment and invasion
E2	E2 (envelope) glycoprotein	55 kDa, 373 aa	Type I transmembrane protein; involvement in virus attachment and invasion; primary neutralizing antigen
p7	P7 (nonstructural protein)	7 kDa, 70 aa (E2-P7, 60 kDa, 443 aa)	Viroporin; involved in infectious virus production (budding process and virus packaging), but not associated with virus particles
NS2	NS2 (nonstructural protein)	54 kDa, 457 aa (NS2–3, 120 kDa, 1140 aa)	Autoprotease; involvement in Cyclin degradation and synthesis; activation of NF-κB; upregulation of IL-8; antagonization of type I IFN response
NS3	NS3 (nonstructural protein)	75 kDa, 683 aa (NS2–3)	Multifunctional protein with serine protease, RNA helicase, nucleoside triphosphatase (NTPase) activity; involvement in viral transcription and translation
NS4A	NS4A (nonstructural protein)	11 kDa, 64 aa	Cofactor of NS3 serine protease; involvement in RNA replication and formation of infectious virion
NS4B	NS4B (nonstructural protein)	30 kDa, 347 aa	NTPase; essential role in RNA replication
NS5A	NS5A (nonstructural protein)	56 kDa, 497 aa	Replicase component; stimulation of virus replication and inhibition of virus translation (i.e., switch from translation to replication); induction of host cell autophagy and reactive oxygen species production
NS5B	NS5B (nonstructural protein)	78 kDa, 718 aa	RNA-dependent RNA polymerase (RdRp); initiation of translation; replicase (production of negative-strand RNA from positive-strand RNA); nucleotidyl transferase

Sources: Lamp, B. et al., *J. Virol.*, 85(7), 3607, 2011; Zhang, H. et al., *Isr. J. Vet. Med.*, 66(3), 89, 2011.

with E[rns] protein) [17]. This leads to the fusion between the viral envelope and the cell plasma membrane and the subsequent endocytosis mediated by clathrin. Inside the cytoplasm, the viral nucleocapsid is removed, and the positive-sense genomic ssRNA is translated (via cap-independent mechanism) into a polyprotein, which is cleaved into four structural and eight nonstructural proteins. Synthesis of a complementary RNA(−) from the genomic ssRNA(+) and formation of a dsRNA genome occur at the surface of endoplasmic reticulum. Transcription/replication of the dsRNA genome results in viral mRNAs/progeny ssRNA(+) genomes, which are incorporated into new virus particles. The new virion buds at the endoplasmic reticulum, migrates to the Golgi apparatus, and exits the cell via the secretory pathway. While NS3, NS4A, NS4B, NS5A, and NS5B are required for RNA replication, all mature proteins (except Npro) and the precursor NS2–3 play critical roles in the life cycle of CSFV [18–20].

CSFV initially infects the host via the oronasal route, and its first replication takes places in the tonsils, where it infects the crypt epithelial cells, macrophages, lymphocytes, and endothelial cells. It then spreads to the regional lymph nodes and disseminates systemically via the bloodstream to the spleen, kidneys, distal ileum, brain, etc. Congenitally, infected piglets are inapparent carriers, which are persistently viremic, and shed the virus for 6–12 months before dying. With a positive polarity, CSFV RNA is infectious and serves as both transcriptional and translational templates.

CSFV survives for 3 days at 50°C and 7–15 days at 37°C, for 3–4 days in decomposing organs and 15 days in decomposing blood and bone marrow, for up to 4 weeks in winter in pens, for 17 to >180 days during salt curing and smoking, for months in refrigerated meat, and for years in frozen meat. The virus is sensitive to drying and ultraviolet light and readily inactivated by heating at 65.5°C for 30 min or at 71°C for 1 min. In addition, it is rapidly killed by pH at pH < 3.0 or pH > 11.0, and is susceptible to ether, chloroform, β-propiolactone (0.4%), chlorine-based disinfectants, cresol (5%), sodium hydroxide (2%), formalin (1%), sodium carbonate (4% anhydrous or 10% crystalline with 0.1% detergent), ionic and nonionic detergents, and strong iodophors (1%) in phosphoric acid [21].

Secretions and excretions (oronasal and lachrymal discharges, urine, feces, and semen), blood, and tissues of sick or dead animals represent the main sources of infection. Ingestion, contact with the conjunctiva or mucous membranes, skin abrasions, genital transmission, artificial insemination, percutaneous blood, and transplacenta are typical routes of transmission. The virus may be spread by farm visitors, veterinarians, and pig traders; by indirect contact through premises, implements, vehicles, clothes, instruments, and needles; by wind (up to 1 km); and by ingestion of insufficiently cooked waste food. The farming practice of keeping semi-feral pigs in Southeast Europe may be another risk factor [22].

CSFV infection appears to be limited to domestic pigs, feral pigs, and wild boars. As wild boar populations may harbor the virus, domestic pigs in the nearby area are at a high risk. CSF has been previously found in all continents. Following adoption of compulsory vaccination or slaughter and eradication policies, a number of countries have declared CSF-free, including Canada (1963), the United States (1978), the United Kingdom, Western Europe, Australia, and New Zealand as well as Japan (2007). Currently, CSF is present in much of Asia, Central and South America, parts of Eastern Europe (e.g., Latvia and Russia), and South Africa [23,24]. In China, CSFV subgroups/subgenotypes 2.1 (particularly clades 2.1b and 2.1c), 2.2, 2.3, and 1.1 have been identified [25,26].

26.1.4 Clinical Features and Pathogenesis

CSF is a highly contagious hemorrhagic viral disease of pigs, with a diverse range of clinical signs. The disease severity may be influenced by the virulence of CSFV strains, age, and immune status of susceptible hosts. In general, strains of high virulence are responsible for acute (i.e., rapid), serious disease, while strains of intermediate or low virulence induce subacute or chronic (i.e., long-lasting) diseases. Younger piglets tend to be more severely affected than older pigs. The incubation period of CSF varies from 2 to 14 days, but clinical signs may not be apparent until after 2 to 3 weeks [27].

26.1.4.1 Acute Disease

Appearing in a small number of growing pigs, the initial nonspecific signs include high fever (40°C–42°C), anorexia, lethargy, depression, sleepiness, reluctance to get up or to eat, and walk/stand with heads down and tails limp. In the few days that follow, increasing numbers of pigs become affected. Clinical signs may change from slight constipation to a yellow-grey diarrhea and conjunctivitis (inflammation of the eye surface) with thin to thick discharges, resulting in some of the eyelids being completely closed and adhered. Younger piglets may shiver and huddle together. Further disease progression may render the affected pigs very thin and weak, with a staggering walk (due to damage in the spinal nerves), a drunken walk (due to partial paralysis of the hind end), and a tendency to be in a sitting or lying position. As a result of hemorrhages, the color of affected pigs' skins may turn purple, first on the ears and tail, then the snout, lower legs, belly, and back, and affected pigs may go into convulsions and die in 5–20 days [28].

26.1.4.2 Subacute and Chronic Disease

The growing pigs with subacute disease may show similar but less severe signs to those with acute disease. These include dullness, capricious appetite, pyrexia, diarrhea for up to 1 month, ruffled appearance, growth retardation, apparent recovery with eventual relapse, and death within about 3 months. Some older animals may show transient pyrexia and inappetence and recovery with lifelong immunity.

26.1.4.3 Congenital Disease

This form of disease tends to occur in piglets in the sow's uterus through transplacental route. This often involves sows with inadequate vaccination (e.g., with original attenuated virus vaccines) or infection by CSFV strain of low virulence. Piglets may be stillborn, mummified, deformed, aborted, or weak. Some piglets may be born apparently healthy (if the virus crosses the placenta before the piglets' immune systems have developed) and grow on to be persistent carriers without clinical signs. After several weeks or months of age, the piglets may develop typical but milder clinical signs (tremor, weakness, runting, and poor growth) without high temperatures, and succumb to the disease in 12 months [28].

In the field, CSF is often manifested as subclinical infections or coinfections with other viruses and/or bacteria.

Postmortem examination may reveal many small hemorrhages throughout the body and larger hemorrhages in the lymph nodes, the lungs, and the skin. The tonsils may show inflammation and necrotic foci. The kidney surfaces are covered with bloody spots, like mottled ducks' eggs. The spleen may have dark raised areas of dead tissue (multifocal infarction of the margin), which may be pathognomonic. The lungs may show severe pneumonia, hemorrhage, and pleurisy due to secondary bacterial infection. The stomach and gut may contain brightly colored liquid. The inner lining of the large intestine near its junction with the small intestine is raised (the so-called button ulcers, which are often seen in subacute/chronic form of the disease and are unique to CSF). The bladder may also display multiple red spots. In congenital form of the disease, lesions may include central dysmyelinogenesis, cerebellar hypoplasia, microencephaly, pulmonary hypoplasia, hydrops, and other malformations [21].

CSFV proteins Npro, Core, Erns, E1, E2, P7, and NS4B have been shown to be related to virulence [29,30]. The systemic hemorrhage seen in CSF is a result of CSFV-induced necrosis of endothelial cells with subsequent vasculitis and hemorrhaging, together with severe anemia, thrombocytopenia, and disturbance of fibrinogen synthesis. The immunosuppression results from downregulation and necrosis of bone marrow hematopoietic cells and lymphocytes [31].

26.1.5 DIAGNOSIS

Differential diagnoses for CSF include other febrile and/or hemorrhagic diseases of pigs, such as African swine fever and other diseases producing septicemias (e.g., salmonellosis, streptococcosis, pasteurellosis, actinobacillosis, erysipelas, eperythrozoonosis, and *Haemophilus parasuis*), hemorrhage (porcine dermatitis and nephropathy syndrome, hemolytic disease of the newborn, coumarin poisoning, and thrombocytopenic purpura), runting (postweaning multisystemic wasting syndrome, enterotoxicosis, swine dysentery, and campylobacteriosis), abortion (pseudorabies virus, encephalomyocarditis virus infection, porcine reproductive and respiratory syndrome, and parvovirus), and nervous signs (viral encephalomyelitis and salt poisoning).

Pigs suspected with CSFV infection should be confirmed by laboratory tests (virus culture, serology, and molecular techniques). Suitable samples include the tonsils, spleen, lymph nodes, kidney, and blood [32].

Virus isolation is often conducted on porcine kidney cells (i.e., PK 15 and SK6 cells) with clarified organ suspension or with whole blood, buffy coat, plasma, or serum. As CSFV

does not produce cytopathic effects, it needs to be confirmed by immunofluorescence or immunoperoxidase tests.

Serology is routinely used for diagnosis and surveillance of CSF as CSFV antigens are detectable 2–15 days postinfection and antibodies are detected in blood 2–3 weeks after infection and as long as animals survive. The commonly used serological assays include virus neutralization tests (VNT), direct fluorescent antibody test (FAT), indirect hemagglutination test, and enzyme-linked immunosorbent assays (ELISA).

VNT is useful for CSF diagnosis. However, because VNT relied on cultured virus for detection, it is time consuming and not routinely applied. FAT provides a rapid way to detect CSFV antigen, although it may cross-react with BVDV of cattle and BDV of sheep, either of which may sometimes infect pigs. Due to its relatively low sensitivity, antigen capture ELISA based on Erns glycoprotein is mainly used for herd checking. ELISA employing E2 glycoprotein is widely utilized to screen for antibodies during and after outbreaks, to monitor CSFV infections in wild boar, and to test coverage of immunization of wild boars after vaccination.

Molecular techniques such as PCR provide the most sensitive method for detection of CSFV both in early infection and for a longer period in recovering pigs [33–41]. Targets for diagnosis of CSF by reverse transcription polymerase chain reaction (RT-PCR) include the 5′ NTR (as exemplified by pestivirus-specific primers 324/326, which amplify a 288 bp fragment), the E2 glycoprotein gene, the NS5B nonstructural protein gene, and the Npro gene [42,43]. In addition, the use of SYBR green and TaqMan probes in real-time RT-PCR allows convenient detection of amplification products and enables improved differentiation of infected from vaccinated animals [44–50].

26.1.6 TREATMENT AND PREVENTION

No specific treatment is available for CSF, and antibiotic therapy is helpful in relieving symptoms due to secondary bacterial infections.

In case of CSF outbreak, it is important to slaughter all pigs on affected farms, safely dispose carcasses and bedding, disinfect farm equipment and surrounding areas, control pig movements, and trace possible source of infection [51].

Prevention of CSV is achieved by adopting two strategies: nonvaccination stamping out policy and systematic prophylactic vaccination program [52].

Nonvaccination stamping out policy has been successfully applied in Europe that led to the eradication of CSF in most EU countries. However, given the high cost associated with nonvaccination stamping out policy, many other countries utilize systematic prophylactic vaccination program to control and prevent CSF.

Live attenuated vaccines containing the C strain (Chinese strain) is widely used to control the disease in endemic areas. These vaccines are usually produced in primary bovine testicle cells or continuous swine testicle cells and tested for quality and efficacy in rabbits. In comparison with the subunit E2 vaccine that only induces production of an antibody against

CSFV, the C-strain vaccine upregulates the antibody level and also activates T-cell responses [53,54]. Pigs often develop protective immunity 1 week to 10 days after vaccination, and the immunity lasts for 2–3 years [55]. Suckling piglets are protected by colostrum from vaccinated sows for about 6–8 weeks, and vaccination may be carried out after 8 weeks [56].

26.2 METHODS

26.2.1 SAMPLE PREPARATION

Whole blood is collected in 5 mL plastic EDTA tubes, and whole blood for serum is collected in 10 mL plastic blood serum tubes.

RNA may be extracted from whole blood or different fractions of whole blood (WBC, RBC, serum, or plasma) using commercial kits such as QIAamp® Viral RNA Mini kit. Briefly, up to 200 μL sample is added with 20 μL of proteinase K and 200 μL of the kit-supplied AVL lysis buffer supplemented with 28 μg of the kit-supplied carrier RNA. The content is mixed and incubated at 56°C for 15 min to complete cellular lysis. Next, 250 μL of 96%–100% ethanol is added to the crude lysate, and the entire content applied onto the kit-supplied spin column and centrifuged. In subsequent steps, the spin column with bound RNA is washed twice with 500 μL of high-salt wash buffer AV1, once with 500 μL of low-salt wash buffer AV2, and finally with 500 μL of 96%–100% ethanol. The washed RNA is eluted from the column with 50 μL of the kit-supplied elution buffer AVE. An additional washing step between the low- and high-salt washes with a customized high-salt wash buffer consisting of 2 M sodium chloride plus 2 mM EDTA may be also employed to remove PCR inhibitors [57].

About 0.2–0.5 g of tissue samples (tonsil, pharyngeal/mesenteric lymph nodes, spleen, kidney, or distal ileum) are homogenized in 1.5 mL PBS using an Omni-TH Tissue Homogenizer (Omni International), and RNA is extracted from homogenate using the QIAamp Viral RNA Mini kit (Qiagen).

26.2.2 DETECTION PROCEDURES

26.2.2.1 Conventional RT-PCR

Das et al. [57] described the use of primers (forward: 5′-TGCCCAAGACACACCTTAACC-3′ and reverse: 5′-GGCCTCTGCAGCGCCCTAT-3′) from the highly conserved 5′-untranslated region (UTR) of the CSFV genome for diagnosis of CSF.

RT-PCR mixture (25 μL) is made up of 0.2 μM forward primer, 0.4 μM reverse primer, 1 μL of RT-PCR enzyme mix, 12.5 μL of the kit-supplied 2× buffer, 2.5 μL of template RNA, and nuclease-free water (up to 25 μL).

Thermocycling parameters consist of 1 cycle of RT at 55°C for 30 min, 1 cycle of heat denaturation at 95°C for 2 min, 35 cycles of heat denaturation at 95°C for 5 s, annealing at 60°C for 30 s, and extension at 72°C for 15 s, followed by a final extension of amplification at 72°C for 2 min.

The amplified product is separated by agarose gel, stained and visualized under UV light.

TABLE 26.2
Identity of RT-PCR Primers and Probes

Target	Primer/Probe	Sequence (5′–3′)
CSFV	CSF100-F	ATG CCC AYA GTA GGA CTA GCA
	CSF-Probe 1	FAM-TGG CGA GCT CCC TGG GTG GTC TAA GT-BHQ1
	CSF192-R	CTA CTG ACG ACT GTC CTG TAC
IC	EGFP1-F	GAC CAC TAC CAG CAG AAC AC
	EGFP1-HEX	HEX-AGC ACC CAG TCC GCC CTG AGC A-BHQ1
	EGFP2-R	GAA CTC CAG CAG GAC CAT G

26.2.2.2 Real-Time RT-PCR

Haines et al. [58] employed primers (CSF 100-F and CSF 192-R) and probe (CSF-Probe 1) from the 5′ UTR of CSFV for diagnosis of CSF (Table 26.2). The inclusion of internal control primers (EGFP1-F and EGFP2-R) and probe (EGFP1-HEX) allows monitoring of RNA extraction and amplification efficiency (Table 26.2).

RT-PCR mixture (25 μL) consists of 8.4 μL RNA extract, 12.5 μL 2× reaction mix, 5 mM MgSO₄, 5 U RNAsin, 50 nM ROX, and 0.5 U Superscript III reverse transcriptase/Platinum Taq mix, together with 600 nM of CSF100-F and CSF 192-R primers, 200 nM of CSF-Probe, 1, 200 nM of EGFP1-F and EGFP2-R, and 100 nM EGFP-HEX probe.

Amplification is conducted in Mx3005P QPCR System (Agilent Technologies) with the following conditions: RT at 50°C for 15 min, incubation at 95°C for 2 min, then 50 cycles of denaturation at 95°C for 15 s, and annealing and extension at 60°C for 1 min. After amplification, a threshold cycle (C_T) value is assigned to each sample.

Note: This one-step, real-time RT-PCR detects 4.9 copies of CSFV and is suitable for routine diagnosis of CSF.

26.3 CONCLUSION

CSF is a highly contagious viral disease of domestic and wild swine, presenting with acute hemorrhagic fever and respiratory, gastrointestinal, and neurological symptoms, which lead to high mortality and significant loss of productivity. Given the clinical similarities between CSF and other diseases such as African swine fever, it is essential to use laboratory tests to correctly diagnose CSF. While virus isolation and serological assays have contributed greatly to the identification of CSFV in the past, their slow turnover and low sensitivity are increasingly limiting their continued application, especially in a time when frequent international trades and population movements have made it feasible for once-isolated, sporadic cases of infection to rapidly spread to other parts of the world as pandemics. With the capacity to specifically amplify from a single copy of gene fragment, RT-PCR circumvents the shortcomings of virus isolation and serological tests and enables much improved CSFV identification, subtyping, and differentiation of infected from vaccinated animals.

REFERENCES

1. Liu L, Xia H, Wahlberg N, Belák S, Baule C. Phylogeny, classification and evolutionary insights into pestiviruses. *Virology*. 2009;385(2):351–357.
2. Giangaspero M. Genetic variation of Border disease virus species strains. *Vet Ital*. 2011;47(4):415–435.
3. Bauermann FV, Flores EF, Ridpath JF. Antigenic relationships between Bovine viral diarrhea virus 1 and 2 and HoBi virus: Possible impacts on diagnosis and control. *J Vet Diagn Invest*. 2012;24(2):253–261.
4. Bauermann FV, Ridpath JF, Weiblen R, Flores EF. HoBi-like viruses: An emerging group of pestiviruses. *J Vet Diagn Invest*. 2013;25(1):6–15.
5. Neill JD et al. Complete genome sequence of pronghorn virus, a pestivirus. *Genome Announc*. 2014;2(3):e00575-14.
6. Richter M, Reimann I, Schirrmeier H, Kirkland PD, Beer M. The viral envelope is not sufficient to transfer the unique broad cell tropism of Bungowannah virus to a related pestivirus. *J Gen Virol*. 2014;95(10):2216–2222.
7. Paton DJ et al. Genetic typing of classical swine fever virus. *Vet Microbiol*. 2000;73(2–3):137–157.
8. Postel A, Moennig V, Becher P. Classical swine fever in Europe—The current situation. *Berl Munch Tierarztl Wochenschr*. 2013;126(11–12):468–475.
9. Patil SS et al. Genetic typing of recent classical swine fever isolates from India. *Vet Microbiol*. 2010;141(3–4):367–373.
10. Floegel-Niesmann G, Bunzenthal C, Fischer S, Moennig V. Virulence of recent and former classical swine fever virus isolates evaluated by their clinical and pathological signs. *J Vet Med B: Infect Dis Vet Public Health*. 2003;50(5):214–220.
11. Lowings P, Ibata G, Needham J, Paton D. Classical swine fever virus diversity and evolution. *J Gen Virol*. 1996;77 (Pt 6):1311–1321.
12. Sarma DK et al. Phylogenetic analysis of recent classical swine fever virus (CSFV) isolates from Assam, India. *Comp Immunol Microbiol Infect Dis*. 2011;34(1):11–15.
13. Lamp B et al. Biosynthesis of classical swine fever virus nonstructural proteins. *J Virol*. 2011;85(7):3607–3620.
14. Lamp B, Riedel C, Wentz E, Tortorici MA, Rümenapf T. Autocatalytic cleavage within classical swine fever virus NS3 leads to a functional separation of protease and helicase. *J Virol*. 2013;87(21):11872–11883.
15. Zhang H, Cao HW, Wu ZJ, Cui YD. A review of molecular characterization of classical swine fever virus. *Isr J Vet Med*. 2011;66(3):89–95.
16. Elbers K et al. Processing in the pestivirus E2-NS2 region: Identification of proteins p7 and E2p7. *J Virol*. 1996;70(6):4131–4135.
17. Chen J et al. The laminin receptor is a cellular attachment receptor for classical swine fever virus. *J Virol*. 2015;89(9):4894–4906.
18. Chen Y et al. Classical swine fever virus NS5A regulates viral RNA replication through binding to NS5B and 3′ UTR. *Virology*. 2012;432(2):376–388.
19. Sheng C et al. Classical swine fever virus NS5A protein interacts with 3′-untranslated region and regulates viral RNA synthesis. *Virus Res*. 2012;163(2):636–643.
20. Sheng C et al. Classical swine fever virus NS5B protein suppresses the inhibitory effect of NS5A on viral translation by binding to NS5A. *J Gen Virol*. 2012;93(5):939–950.
21. OIE (Office International des Epizooties). *Terrestrial Manual 2014*. Chapter 2.8.3: Classic swine fever. http://www.oie.int/fileadmin/Home/eng/Health_standards/tahm/2.08.03_CSF.pdf. Accessed on June 12, 2015.
22. Ji W, Guo Z, Ding NZ, He CQ. Studying classical swine fever virus: Making the best of a bad virus. *Virus Res*. 2015;197:35–47.
23. Vlasova A et al. Molecular epidemiology of classical swine fever in the Russian Federation. *J Vet Med B Infect Dis Vet Public Health*. 2003;50(8):363–367.
24. Song JY et al. Prevalence of classical swine fever virus in domestic pigs in South Korea: 1999–2011. *Transbound Emerg Dis*. 2013;60(6):546–551.
25. Sun SQ et al. Genetic typing of classical swine fever virus isolates from China. *Transbound Emerg Dis*. 2013;60(4):370–375.
26. Luo Y, Li S, Sun Y, Qiu HJ. Classical swine fever in China: A minireview. *Vet Microbiol*. 2014;172(1–2):1–6.
27. Moennig V, Floegel-Niesmann G, Greiser-Wilke I. Clinical signs and epidemiology of classical swine fever: A review of new knowledge. *Vet J*. 2003;165(1):11–20.
28. The Pig Site. http://www.thepigsite.com/pighealth/article/447/classical-swine-fever-csf-hog-cholera-hc/, accessed on June 1, 2015.
29. Gladue DP et al. Classical swine fever virus p7 protein is a viroporin involved in virulence in swine. *J Virol*. 2012;86(12):6778–6791.
30. Leifer I et al. Differentiation of C-strain "Riems" or CP7_E2alf vaccinated animals from animals infected by classical swine fever virus field strains using real-time RT-PCR. *J Virol Methods*. 2009;158(1–2):114–122.
31. Feng L et al. In vitro infection with classical swine fever virus inhibits the transcription of immune response genes. *Virol J*. 2012;9:175.
32. Bouma A et al. Evaluation of diagnostic tests for the detection of classical swine fever in the field without a gold standard. *J Vet Diagn Invest*. 2001;13(5):383–388.
33. Biagetti M, Greiser-Wilke I, Rutili D. Molecular epidemiology of classical swine fever in Italy. *Vet Microbiol*. 2001;83(3):205–215.
34. Jemersić L et al. Genetic typing of recent classical swine fever virus isolates from Croatia. *Vet Microbiol*. 2003;96(1):25–33.
35. Blacksell SD et al. Genetic typing of classical swine fever viruses from Lao PDR by analysis of the 5′ non-coding region. *Virus Genes*. 2005;31(3):349–355.
36. Po F, Le Dimna M, Le Potier MF. Five years' experience of classical swine fever polymerase chain reaction ring trials in France. *Rev Sci Tech*. 2011;30(3):797–807.
37. Podgórska K, Kamieniecka K, Stadejek T, Pejsak Z. Comparison of PCR methods for detection of classical swine fever virus and other pestiviruses. *Pol J Vet Sci*. 2012;15(4):615–620.
38. Liu JK, Wei CH, Yang XY, Dai AL, Li XH. Multiplex PCR for the simultaneous detection of porcine reproductive and respiratory syndrome virus, classical swine fever virus, and porcine circovirus in pigs. *Mol Cell Probes*. 2013;27(3–4):149–152.
39. Chowdry VK et al. Development of a loop-mediated isothermal amplification assay combined with a lateral flow dipstick for rapid and simple detection of classical swine fever virus in the field. *J Virol Methods*. 2014;197:14–18.
40. Kim YK et al. The CSFV DNAChip: A novel diagnostic assay for classical swine fever virus. *J Virol Methods*. 2014;204:44–48.
41. Lung O et al. Insulated isothermal reverse transcriptase PCR (iiRT-PCR) for rapid and sensitive detection of classical swine fever virus. *Transbound Emerg Dis*. January 27, 2015. doi: 10.1111/tbed.12318. [Epub ahead of print].
42. Desai GS, Sharma A, Kataria RS, Barman NN, Tiwari AK. 5′-UTR-based phylogenetic analysis of classical swine fever virus isolates from India. *Acta Virol*. 2010;54(1):79–82.

43. Hoffmann B et al. Classical swine fever virus detection: Results of a real-time reverse transcription polymerase chain reaction ring trial conducted in the framework of the European network of excellence for epizootic disease diagnosis and control. *J Vet Diagn Invest*. 2011;23(5):999–1004.

44. Kaden V et al. Diagnostic procedures after completion of oral immunisation against classical swine fever in wild boar. *Rev Sci Tech*. 2006;25(3):989–997.

45. Leifer I, Ruggli N, Blome S. Approaches to define the viral genetic basis of classical swine fever virus virulence. *Virology*. 2013;438(2):51–55.

46. Lin YC et al. Application of real-time quantitative polymerase chain reaction to monitoring infection of classic swine fever virus and determining optimal harvest time in large-scale production. *Vaccine*. 2013;31(47):5565–5567.

47. Wernike K, Beer M, Hoffmann B. Rapid detection of foot-and-mouth disease virus, influenza A virus and classical swine fever virus by high-speed real-time RT-PCR. *J Virol Methods*. 2013;193(1):50–54.

48. Dias NL et al. Validation of a real time PCR for classical Swine Fever diagnosis. *Vet Med Int*. 2014;2014:171235.

49. Everett HE, Crudgington BS, Sosan-Soulé O, Crooke HR. Differential detection of classical swine fever virus challenge strains in C-strain vaccinated pigs. *BMC Vet Res*. 2014;10(1):281.

50. Widén F et al. Comparison of two real-time RT-PCR assays for differentiation of C-strain vaccinated from classical swine fever infected pigs and wild boars. *Res Vet Sci*. 2014;97(2):455–457.

51. Penrith ML, Vosloo W, Mather C. Classical swine fever (hog cholera): Review of aspects relevant to control. *Transbound Emerg Dis*. 2011;58(3):187–196.

52. Newcomer BW, Givens MD. Approved and experimental countermeasures against pestiviral diseases: Bovine viral diarrhea, classical swine fever and border disease. *Antiviral Res*. 2013;100(1):133–150.

53. Ganges L et al. Recent advances in the development of recombinant vaccines against classical swine fever virus: Cellular responses also play a role in protection. *Vet J*. 2008;177(2):169–177.

54. Huang YL, Deng MC, Wang FI, Huang CC, Chang CY. The challenges of classical swine fever control: Modified live and E2 subunit vaccines. *Virus Res*. 2014;179:1–11.

55. Chander V, Nandi S, Ravishankar C, Upmanyu V, Verma R. Classical swine fever in pigs: Recent developments and future perspectives. *Anim Health Res Rev*. 2014;15(1):87–101.

56. Blome S, Meindl-Böhmer A, Loeffen W, Thuer B, Moennig V. Assessment of classical swine fever diagnostics and vaccine performance. *Rev Sci Tech*. 2006;25(3):1025–1038.

57. Das A, Beckham TR, McIntosh MT. Comparison of methods for improved RNA extraction from blood for early detection of classical swine fever virus by real-time reverse transcription polymerase chain reaction. *J Vet Diagn Invest*. 2011;23(4):727–735.

58. Haines FJ, Hofmann MA, King DP, Drew TW, Crooke HR. Development and validation of a multiplex, real-time RT PCR assay for the simultaneous detection of classical and African swine fever viruses. *PLoS One*. 2013;8(7):e71019.

27 Louping Ill Virus

Nicholas Johnson

CONTENTS

27.1 Introduction 241
    27.1.1 Classification 241
    27.1.2 Biology and Epidemiology 241
    27.1.3 Clinical Features and Pathogenesis 242
    27.1.4 Diagnosis 242
27.2 Methods 243
    27.2.1 Sample Preparation 243
    27.2.2 Reverse Transcription Loop-Mediated Amplification (RT-LAMP) 244
27.3 Conclusions and Future Perspectives 244
Acknowledgments 245
References 245

27.1 INTRODUCTION

Louping ill (louping, derived from an old Scottish word "loup," which means to spring into the air, reflecting the key effect of the disease in sheep) is a tick-transmitted viral disease of primarily sheep and occasionally other hosts. The causal agent louping ill virus (or LIV) is a single-stranded positive-sense RNA virus belonging to the *Flaviviridae* family. Affecting the central nervous system (CNS), louping ill (also known as ovine encephalomyelitis, infectious encephalomyelitis of sheep, or trembling ill) is characterized by a biphasic fever, muscular tremors, ataxia (particularly of the hindlimbs), depression, coma, and death. The disease is prevalent in the rough hill grazing land of the British Isles where the sheep tick, *Ixodes ricinus*, is present. However, similar diseases of sheep caused by related viruses have been recently reported in Norway, Spain, Turkey, and Bulgaria.

27.1.1 CLASSIFICATION

LIV is classified within the order *Mononegavirales*, family *Flaviviridae*, and genus *Flavivirus* [1]. Viruses within this genus are transmitted by vectors, particularly arthropods. The most prominent members are transmitted by mosquitoes or ticks. LIV is grouped with the tick-borne viruses and is transmitted by the common sheep tick, *I. ricinus*. It is closely related to tick-borne encephalitis virus (TBEV), which as its name suggests is a collection of viruses transmitted by hard ticks and causes encephalitis [2]. The TBEV complex of viruses has been considered by some authors as forming a single virus species consisting of a number of types [3]. These viruses are found along a global cline coincident with temperate latitudes of the Northern Hemisphere [4].

However, the varying phenotypic properties of the virus and distinct genetic differences suggest that the designation of viruses within the group as species or subspecies is unresolved. It is certainly true that LIV shares many properties with TBEV including a conserved polyprotein 3414 amino acids in length (for detail, see Section 27.1.2), the same tick host, and the causative agent of a febrile illness that can develop into a neurological disease. However, certain features of LIV mark it apart from TBEV. First, and most relevant to this volume, is its virulence for domestic livestock, especially sheep (*Ovis aries*) [5,6] and red grouse (*Lagopus scoticus*), a feature not shared by TBEV. A number of viruses within the TBEV complex share this virulence for sheep and goats including Spanish sheep encephalitis virus (SSEV), Greek goat encephalitis virus, and Turkish sheep encephalitis virus. There have also been reports of encephalitic disease in sheep from Bulgaria. Second, LIV is found almost exclusively in the British Isles, whereas TBEV has never been isolated in these islands, although it is reported throughout continental Europe and Asia. Finally, TBEV causes significant human morbidity and mortality while the zoonotic potential of LIV is limited with most evidence being restricted to serological studies in humans and laboratory-acquired infections [7,8].

27.1.2 BIOLOGY AND EPIDEMIOLOGY

LIV is exclusively transmitted by life stages of the tick *I. ricinus*. There are three stages to the tick life cycle. The juvenile ticks (larva and nymph) feed on small mammals although it is not certain which species are critical to persistence of LIV, and a recent report has suggested that they may not play a role in LIV transmission as they are not abundant at upland sites [9]. The hare (*Lepus timidus*) is susceptible to LIV infection; although considered a nonviremic host, it is still capable of transmitting the virus [10]. The adult tick feeds on large ungulates such as deer and livestock. The distribution of LIV within the British Isles shows a strong association with areas associated with upland farming of sheep (Figure 27.1). Moorland is particularly favorable for ticks as the thick layer of vegetation provides a humid environment that enables all stages of the tick to avoid desiccation. During spring and autumn, juvenile and adult ticks emerge and climb vegetation in a behavior described as questing. When a suitable mammal passes, the tick clambers onto the host and inserts its mouth parts into the skin. The tick will remain on the host, feeding for a number of days until it becomes fully engorged. If the tick is infected with LIV, this is transmitted to the host. Infection in mammals results in a viremia of short duration and variable level dependent on the host [11]. The highest levels of viremia are observed in sheep. Infection during cofeeding is believed to be more important for infection of ticks, a process where an uninfected tick becomes infected when it feeds in close proximity to an infected tick [12]. From the perspective of diagnosis, blood samples are rarely useful for detection of virus in livestock as the viremia has usually declined to undetectable levels when neurological signs develop and disease is suspected. However, serological testing is useful in confirming recent infection.

Until recent times, LIV was considered exclusive to the British Isles. However, an LIV isolate has been reported in Norway, and recent serological studies suggest that it is cocirculating with TBEV in parts of Southern Norway [13]. A further report of a virus isolated in goats from Spain [14]

concluded that this virus was distinct from SSEV and might be LIV, suggesting that the distribution of LIV and closely related viruses that cause encephalomyelitis in ovines is more extensive than previously thought.

27.1.3 CLINICAL FEATURES AND PATHOGENESIS

Sheep are particularly susceptible to disease following infection with LIV [15]. The early manifestation of infection is pyrexia that is often not recognized in livestock. This often subsides but is then followed by the onset of neurological signs. These include nonspecific behaviors such as panting and nibbling, followed by more overt signs such as circling, unprovoked leaping, and ataxia. There is no effective treatment for infection and, during the later stages, animals become recumbent and some die. Those that survive exhibit sequelae such as torticollis and limb paralysis. Other domestic animals can be infected with LIV including cattle [16], horses [17], and South American camelids [18].

Infection follows the bite of an infected tick. Local replication at the bite site leads to viremia that is associated with fever. In sheep, neuroinvasion is associated with disease signs. The mechanism of entry of LIV into the CNS has not been studied but once within the CNS, extensive virus replication occurs. LIV readily replicates in neuronal cells, and widespread lesions are observed in the brain of infected sheep. Experimental infections have indicated a short incubation period from infection to death of only 9 days [5].

27.1.4 DIAGNOSIS

The primary serological test for LIV infection is the hemagglutination inhibition (HAI) test. This was originally developed by Clark and Casals [19] for investigating a range of arthropod-borne viruses and subsequently adapted by Williams and Thorburn [20] to measure LIV seroprevalence. The method has the benefit that it is not species specific and can be applied to a range of domestic and wildlife species [9]. An alternative to measuring seroprevalance is the plaque reduction neutralization test. This method takes longer to perform, typically 3–5 days, and involves culture of both virus and cells [21]. The serum sample is mixed with a fixed quantity of virus and added to a cell monolayer, then incubated with a gel overlay that restricts the spread of nonneutralized virus. In the presence of virus-neutralizing antibody, the monolayer remains intact, whereas if neutralizing antibody is absent, discrete areas of the monolayer or plaques form.

Until the 1990s, detection of virus itself was limited to immunohistochemical detection in fixed brain tissue or isolation from infected tissue in a cell line such as Vero or porcine kidney cells. The development of a cytopathic effect in the form of cell lysis indicates the presence of a virus. However, confirmation that LIV is the causative agent required further analysis including the molecular detection techniques described in later text.

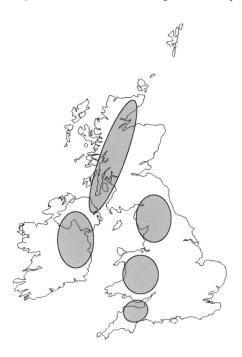

FIGURE 27.1 Map of the British Isles showing main areas where LIV is detected. Shaded area indicates distribution of louping ill disease in the British Isles. These areas correspond closely to upland sheep grazing moors.

The development of virus-specific detection techniques such as reverse transcription polymerase chain reaction (RT-PCR) has greatly improved the ability to detect and differentiate viruses. RT-PCR has been used to amplify fragments of the LIV genome for the purposes of complete genome sequencing [22,23] and subsequent phylogenetic analysis [24]. Recently, RT-PCR has been used to detect LIV diagnostically. The first RT-PCR assays used for virus detection were developed in the 1990s [25]. This has been updated with a TaqMan real-time assay that rapidly detects virus in mammalian and avian samples, including blood and neurological tissue [26]. An alternative RT-PCR approach is to use degenerate primers, where alternative bases are included at a particular position, capable of amplifying all members of the *Flavivirus* genus. Such universal primer sets have been reported in numerous papers, mainly amplifying conserved region of the NS5 gene. Two methods have been applied to the detection of LIV [27,28]. Complete methodologies can be found in published papers but critical details of each RT-PCR, including primer sequences and amplification conditions, are provided in Table 27.1.

An alternative to RT-PCR is RT loop-mediated amplification, a method for amplifying short specific fragments of RNA that has the benefit that it requires a single temperature. This method has been used extensively to detect both human and animal viruses [29] based on isothermal amplification of target cDNA through binding of a set of 4 primers and the

strand-displacement polymerase activity of the enzyme *Bst1*. Reverse transcriptase is included in the reaction to enable RT of the virus genome. Both RT and amplification occur at a single temperature, 65°C. A series of concatenated products are generated that when separated on an agarose gel gives a distinctive "ladder." Six primer sets were designed based on a 200 bp sequence within the LIV NS5 gene. Each was assessed using a small panel of archived LIV isolates. Assays 3 and 4 amplified laboratory isolates from Scotland and Northern England but not from Wales. A further primer set was designed (designated LIV6, see Table 27.2), which amplified isolates from all three regions. A protocol for this technique is provided in Section 27.2.2.

27.2 METHODS

27.2.1 SAMPLE PREPARATION

The neurological signs characteristic of louping ill reflect infection of the CNS, and virus can be detected throughout the brain and spinal cord [30]. Viral RNA can be extracted using a range of methods, including TRIzol (Invitrogen). Using the TRIzol protocol, a small quantity of brain or spinal cord is homogenized in TRIzol solution and centrifuged with added chloroform. This leads to effective partitioning of RNA in an upper aqueous phase and allows sufficient recovery of total RNA from a sample of ovine spinal cord to sequence a complete LIV genome [23].

TABLE 27.1
PCR Methodologies for Identification of Louping III Virus

Reference	Primers and Probe (5′–3′)	Conditions
Marriott et al. [26]	Forward: GCTGTCAAGATGGATGTGTACA Reverse: ACTTGTTTCCCTCAATGCTGAA MGB probe: FAM-CTGGAGTGCTGCTGAA-MGB	RT: 48°C for 30 min PCR: 50 cycles of 95°C for 15 s and 60°C for 60 s
Moureau et al. [27]	Forward: TGYRTBTAYAACATGATGGG Reverse: GTGTCCCAICCNGCNGTR	RT: 50°C for 30 min PCR: 40 cycles of 94°C for 15 s, 50°C for 30 s, and 72°C for 45 s
Johnson et al. [28]	Flavi-For: GCMATHTGGTWCATGTGG Flavi-Rev: GTRTCCAKCCDGCNGTRTC	RT: 50°C for 30 min PCR: 40 cycles of 94°C for 30 s, 60°C for 30 s, and 72°C for 30 s
Degenerate symbols: Y	C+T	
R	A+G	
B	G+T+C	
I	Inosine	
N	A+C+G+T	
M	A+C	
W	A+T	
K	G+T	
D	G+A+T	

TABLE 27.2

LIV Primers for RT-LAMP

Assay	Primers	Sequence (5′–3′)
LIV6	LIV6F3 (outer primer)	GGCCATCTGGTACATGTGG
	LIV6B3 (outer primer)	GCGGTGTCATCATCTGCGTAG
	LIV6FIP (inner primer)	AGTGGAGGAGGAGTTGAGGGGAA
		TTTTGGCCTCCATTCAAGGTTCAT
	LIV6BIP (inner primer)	GGACTCTCTGGAAGCCCAGTGATT
		TTGCCGCTTCCTGGAGTTTG

27.2.2 Reverse Transcription Loop-Mediated Amplification (RT-LAMP)

Five μL of extracted RNA are added to LAMP master-mix (GeneSys Ltd) with additional ThermoScript Reverse Transcriptase (Invitrogen) in a total volume of 25 μL. A complete list of the reaction mix is provided in the following:

Isothermal master mix		15 μL
ThermoScript RT		1 μL
Inner primers	LIV6F3 (5 picomoles)	1 μL
	LIV6B3 (5 picomoles)	1 μL
Outer primers	LIV6FIP (50 picomoles)	1 μL
	LIV6BIP (50 picomoles)	1 μL
RNA template		5 μL

The mixture is incubated at 65°C for 60 min in a water bath, heat block, or standard thermal cycler, and the amplification products are detected on an agarose gel. Figure 27.2a shows the result of separate amplification products on an agarose gel stained with DNA-binding dyes (ethidium bromide or SYBR safe, Invitrogen). RT-LAMP positive samples give a distinctive ladder appearance representing the concatenated products of variable length generated by the loop primers during the amplification process.

Alternatively, addition of the intercalating dye, SYBR green, into the master mix enables RT-LAMP products to be detected in real time in a manner similar to real-time RT-PCR (Figure 27.2b). In this case, LIV was detected in RNA extracted from the spinal cord of a ewe with neurological disease. Initial amplification was observed after 10 min and the reaction reached completion by 20 min.

27.3 CONCLUSIONS AND FUTURE PERSPECTIVES

LIV is a neglected disease due to its relative rarity, limited distribution, reduced market prices for sheep, low zoonotic potential, and the availability of an effective vaccine. However, the disease persists and continues to kill hundreds of sheep and grouse every year [15]. Serology, in the form of the HAI test, is still used routinely to confirm the presence

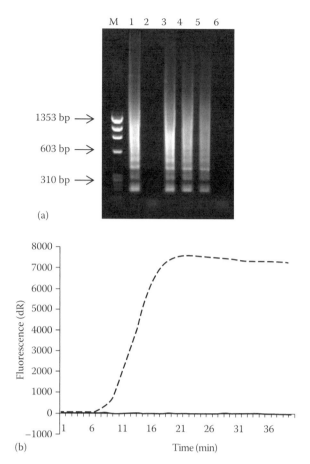

(a)

(b)

FIGURE 27.2 Detection of LIV RT-LAMP amplification products. (a) Detection of LIV in RNA samples extracted from sheep brain by agarose gel electrophoresis. M, DNA molecular weight markers (φx Promega, size of molecular markers are indicated in base pairs—bp); 1, LIV strain LI/31; 2, no template control; 3, LIV isolate Penrith 1 2009; 4, LIV isolate Penrith 2 2009; 5, LIV isolate Aberystwyth 2009; 6, no template control. (b) Detection of LIV RNA extracted from the spinal cord of a ewe (dashed line) using real-time detection. A no-template control has been included for comparison (solid line).

of LIV in sheep flocks and has recently been used to suggest the presence of an LIV-related virus in Northern Spain [14]. Immunohistochemistry can also be used to confirm encephalitis and, through the use of specific antibodies, to confirm infection. This and virus isolation in tissue culture are useful but relatively slow, expensive techniques. Methods such as RT-PCR and RT-LAMP have been developed to confirm diagnosis in fallen livestock, particularly sheep, more rapidly. Other livestock species, horses and camelids, are also susceptible to infection in areas where ticks are abundant. A diagnosis of LIV should be considered in these species, especially if there is evidence of neurological disease. RT-PCR has been the method of choice in recent years due to its rapid turnaround time and application for a range of samples (brain, blood). Alternative methods such as RT-LAMP remain at the experimental stage but could in the future be adapted for pen-side methods that could be applied at the farm level.

LIV cases within England, Wales, and Scotland are reported every year to the APHA, and there is no evidence that this will decline despite the availability of a vaccine and tick management strategies [31]. Recent seroprevalence studies indicate that the virus also persists in Ireland [32]. This virus will continue to cause disease in livestock for the foreseeable future. Although neglected, LIV continues to present a number of challenges. Its distribution had appeared restricted to the British Isles although there is evidence for its presence in tick populations of Southern Norway [13]. TBEV is present throughout continental Europe but absent from the United Kingdom. The reason for this sharp separation is not clear. Sheep movements are believed to be responsible for the dispersal of LIV to different regions of the British Isles [24]. Such a method may also explain the introduction of the virus from the continent. The recent report of a closely related virus detected in goats in Northern Spain [14] that is distinct from the previously described SSEV [33] may suggest a possible common ancestry for these viruses.

A further question is the varied virulence between LIV and TBEV for ovines and humans, respectively. Despite close examination of the genomes of both viruses and comparative studies in sheep, there is as yet no strong explanation for the differences in pathogenesis between each virus. There have been reported cases of human infection with LIV although these have been rare in recent years. It is possible that this is due to underdiagnosis of human encephalitis. If this is true, then access to rapid detection methods, such as those described in this chapter that can be adapted for both human and veterinary use, will be of great benefit in diagnosing cases of neurological disease.

ACKNOWLEDGMENTS

This work was supported by Defra, UK grant SV3045, and European Union FP7 project Anticipating the Global Onset of Novel Epidemics (ANTIGONE), project number 278976.

REFERENCES

1. Heinze, F.X. et al., Family flaviviridae. In: van Regenmortel, M.H.V. et al. (eds.), *Virus Taxonomy*. Academic Press, San Diego, CA, p. 859, 2000.
2. Gritsun, T.S., Nuttall, P.A., and Gould, E.A., Tick-borne flaviviruses. *Adv. Virus Res.*, 61, 317, 2003.
3. Grard, G. et al., Genetic characterization of tick-borne flaviviruses: New insights into evolution, pathogenic determinants and taxonomy. *Virology*, 361, 80, 2007.
4. Heinze, D.M., Gould, E.A., and Forrester, N.L., Revisting the clinical concept of evolution and dispersal for the tick-borne flaviviruses by using phylogenetic and biogeographic analyses. *J. Virol.*, 16, 8663, 2012.
5. Sheahan, B.J., Moore, M., and Atkins, G.J., The pathogenicity of louping ill virus for mice and lambs. *J. Comp. Pathol.*, 126, 137, 2002.
6. Mansfield, K.L. et al., Innate and adaptive immune responses in sheep to infection with tick-borne flaviviruses (in preparation).
7. Davidson, M.M., Williams, H., and Macleod, J.A.J., Louping ill in man: A forgotten disease. *J. Infect.*, 23, 241, 1991.
8. Mansfield, K.L. et al., Tick-borne encephalitis virus—A review of an emerging zoonosis. *J. Gen. Virol.*, 90, 1781, 2009.
9. Gilbert, L. et al., Role of small mammals in the persistence of louping-ill virus: Field survey and tick co-feeding studies. *Med. Vet. Entomol.*, 14, 277, 2000.
10. Norman, R. et al., The role of non-viraemic transmission on the persistence and dynamics of a tick borne virus—Louping ill in red grouse (*Lagopus lagopus scoticus*) and mountain hares (*Lepus timidus*). *J. Math. Biol.*, 48, 119, 2004.
11. Reid, H., Epidemiology of louping-ill. In: Mayo, M.A. and Harrap, K.A. (eds.), *Vectors in Virus Biology*. Academic Press, London, U.K., pp. 161–178, 2004.
12. Labuda, M. et al., Amplification of tick-borne encephalitis virus infection during co-feeding or ticks. *Med. Vet. Entomol.*, 7, 339, 1993.
13. Ytrehus, N. et al., Tick-borne encephalitis virus and louping-ill virus may co-circulate in Southern Norway. *Vector Borne Zoonotic Dis.*, 13, 762, 2013.
14. Balsiero, A. et al., Louping ill in goats, Spain, 2011. *Emerg. Infect. Dis.*, 18, 976, 2012.
15. Jeffries, C.L. et al., Louping ill virus: An endemic tick-borne disease of Great Britain. *J. Gen. Virol.*, 95, 1005, 2014.
16. Twomey, D.F. et al., Louping ill on Dartmoor. *Vet. Rec.*, 149, 687, 2001.
17. Hyde, J., Nettleton, P., Marriot, L., and Willoughby, K., Louping ill in horses. *Vet. Rec.*, 160, 532, 2007.
18. Cranwell, M.P. et al., Louping ill in alpaca. *Vet. Rec.*, 162, 28, 2008.
19. Clarke, D.H. and Casals, J. Techniques for hemagglutination and hemagglutination-inhibition with arthropod-borne viruses. *Am. J. Trop. Med. Hyg.*, 7, 561, 1958.
20. Williams, H. and Thorburn, H. The serological response of sheep to infection with louping-ill virus. *J. Hyg.*, 59, 437, 1961.
21. Mansfield, K.L. et al., Flavivirus-induced antibody cross-reactivity. *J. Gen. Virol.*, 92, 2821, 2011.
22. Gritsun, T.S. et al., Complete sequence of two tick-borne flaviviruses isolated from Siberia and the UK: Analysis and significance of the 5′ and 3′-UTRs. *Virus Res.*, 49, 27, 1997.
23. Marston, D. et al., Louping ill virus genome sequence derived from the spinal cord of an infected lamb. *Genome Announc.*, 1, e00454-3e13, 2013.
24. McGuire, K. et al., Tracing the origins of louping ill virus by molecular phylogenetic analysis. *J. Gen. Virol.*, 79, 981, 1998.
25. Johnson, N. et al., Rapid molecular detection methods for arboviruses of livestock of importance to northern Europe. *J. Biomed. Biotechnol.*, 2012, 719402, 2012.
26. Marriott, L. et al., Detection of Louping ill virus in clinical specimens from mammals and birds using TaqMan RT-PCR. *J. Virol. Methods*, 137, 21, 2006.
27. Moureau, G., Temmam, S., Gonzalez, J.P., Charrel, R.N., Grard, G., and De Lamballerie, X. A real-time RT-PCR method for the universal detection and identification of flaviviruses. *Vector Borne Zoonotic Dis.*, 7, 467, 2007.
28. Johnson, N. et al., Assessment of a novel real-time pan-flavivirus RT-polymerase chain reaction. *Vector Borne Zoonotic Dis.*, 10, 667, 2010.
29. Mori, Y. and Notomi, T. Loop-mediated isothermal amplification (LAMP): A rapid, accurate, and cost-effective diagnostic method for infectious diseases. *J. Infect. Chemother.*, 15, 62, 2009.

30. Doherty, P.C. and Reid, H.W. Louping-ill encephalomyetlitis in the sheep. II. Distribution of virus and lesions in nervous tissue. *J. Comp. Pathol.*, 81, 531, 1971.

31. Laurenson, M.K. et al., Prevalence, spatial distribution and the effect of control measures on louping-ill virus in the Forest of Bowland, Lancashire. *Epidemiol. Infect.*, 135, 963, 2007.

32. Barret, D. et al., Seroprevalence of louping ill virus (LIV) antibodies in sheep submitted for post-mortem examinations in the North West of Ireland in 2011. *Ir. Vet. J.*, 65, 20, 2012.

33. Marin, M.S. et al., The virus causing encephalomyelitis in sheep in Spain: A new member of the tick-borne encephalitis group. *Res. Vet. Sci.*, 58, 11, 1995.

28 West Nile Virus

Dongyou Liu

CONTENTS

28.1 Introduction 247
   28.1.1 Classification.. 247
   28.1.2 Morphology and Genome Structure 248
   28.1.3 Biology and Epidemiology 248
   28.1.4 Clinical Features and Pathogenesis 249
   28.1.5 Diagnosis 249
      28.1.5.1 Virus Isolation.. 249
      28.1.5.2 Serological Assays 249
      28.1.5.3 Nucleic Acid Detection 250
   28.1.6 Treatment and Prevention 250
28.2 Methods 250
   28.2.1 Sample Preparation.. 250
   28.2.2 Detection Procedures... 251
      28.2.2.1 Standard RT-PCR 251
      28.2.2.2 Real-Time RT-PCR 251
      28.2.2.3 Phylogenetic Analysis 252
28.3 Conclusion 252
References.. 252

28.1 INTRODUCTION

West Nile virus (WNV) is a zoonotic pathogen that affects birds, horses, and humans, as well as other vertebrates worldwide. Maintained in a mosquito–bird–mosquito transmission cycle, WNV infection in birds may lead to death within 24 h of symptom onset. In humans and horses, WNV is responsible for causing fever, meningitis, and encephalitis, with long-term sequelae, or death a possibility. Due to its nonspecific clinical presentations, it is vital that specific and sensitive laboratory methods are available and applied for differential diagnosis of West Nile disease. This will help prevent the further spread of this emerging disease.

28.1.1 CLASSIFICATION

WNV is a member of the genus *Flavivirus*, family *Flaviviridae*, which covers monopartite, linear, single-stranded, positive-sense RNA viruses of 9.6–12.3 kb in length. Within the family *Flaviviridae*, four genera are recognized: *Flavivirus*, *Hepacivirus*, *Pegivirus*, and *Pestivirus*. Consisting of >60 viruses, the genus *Flavivirus* is organized into tick-borne viruses, mosquito-borne viruses, and viruses with no known arthropod vector [1].

The tick-borne viruses are further separated into two groups: (1) *mammalian tick-borne virus group* (Absettarov virus, Alkhurma virus, deer tick virus, Gadgets Gully virus, Kadam virus, Karshi virus, Kyasanur forest disease virus, Langat virus, louping ill virus, Omsk hemorrhagic fever virus, Powassan virus, Royal Farm virus, Sokuluk virus, tick-borne encephalitis virus, Turkish sheep encephalitis virus) and (2) *seabird tick-borne virus group* (Kama virus, Meaban virus, Saumarez Reef virus, Tyuleniy virus).

The mosquito-borne viruses are divided into eight groups: (1) *Aroa virus group* (Aroa virus and Bussuquara virus), (2) *dengue virus group* (dengue virus and Kedougou virus), (3) *Japanese encephalitis virus group* (Alfuy virus, Bussuquara virus, Cacipacore virus, Japanese encephalitis virus [JEV], Koutango virus [KOUV], Murray Valley encephalitis virus, Rocio virus, St. Louis encephalitis virus, Usutu virus, WNV, and Yaounde virus), (4) *Kokobera virus group* (Kokobera virus), (5) *Ntaya virus group* (Bagaza virus, Baiyangdian virus, duck egg drop syndrome virus, Ilheus virus, Israel turkey meningoencephalomyelitis virus, Jiangsu virus, Ntaya virus, and Tembusu virus), (6) *Spondweni virus group* (Zika virus), (7) *yellow fever virus group* (Banzi virus, Bouboui virus, Edge Hill virus, Jugra virus, Saboya virus, Sepik virus, Uganda S virus, Wesselsbron virus, and yellow fever virus), and (8) *viruses without known vertebrate host* (Aedes flavivirus, Barkedji virus, Calbertado virus, cell fusing agent virus, Chaoyang virus, Culex flavivirus, Culex theileri flavivirus, Donggang virus, Ilomantsi virus, Kamiti river virus, Lammi virus, Marisma mosquito virus, Nakiwogo virus, Nhumirim virus, Nounane virus, Spanish Culex flavivirus, Spanish Ochlerotatus flavivirus, and Quang Binh virus).

Viruses with no known arthropod vector are differentiated into three groups: (1) *Entebbe virus group* (Entebbe bat virus and Yokose virus), (2) *Modoc virus group* (Apoi virus, Cowbone Ridge virus, Jutiapa virus, Modoc virus, Sal Vieja virus, San Perlita virus), and (3) *Rio Bravo virus group* (Bukalasa bat virus, Carey Island virus, Dakar bat virus, Montana myotis leukoencephalitis virus, Phnom Penh bat virus, and Rio Bravo virus).

Composed of a single serotype, WNV is classified into five distinct lineages [2,3]. Lineage 1 consists of two clades: 1a and 1b, with clade 1a including isolates from Africa, Europe, the Middle East, Asia, and the Americas; and clade 1b being represented by the Australian Kunjin virus [4]. Lineage 2 isolates are detected primarily in sub-Saharan Africa and Madagascar, with recent incursions into Europe (Greece, Hungary, and Italy) and Russia [5]. Lineage 3 contains a single isolate from the Czech Republic border region near Rabensburg, Austria [6–8]. Lineage 4 is present in the Caucasus region of Russia. Lineage 5 (formerly clade 1c) is identified in India only [9,10]. Other isolates that may form putative lineages of WNV include KOUV from Senegal, a group of isolates from Spain, and a variant Kunjin virus from Sarawak, Malaysia [11]. It is noteworthy that only lineages 1, 2, and 5 have been associated with significant outbreaks in humans [12].

28.1.2 Morphology and Genome Structure

WNV possesses a spherical virion of about 50 nm in diameter, which is covered by an envelope made up of two viral glycoproteins (envelope protein E and membrane protein M), with the exodomains of envelope (E) protein dimers showing a head-to-tail position on the outer surface of the virion membrane. Beneath the envelope is an icosahedral core (made up of capsid protein C) of 30 nm in diameter that encases a linear ssRNA(+) of approximately 11,000 nt in length [1,13,14].

The genome of WNV contains a single, long open reading frame (ORF) flanked by two untranslated regions (UTR), one each at the 5′ and 3′ ends. The 5′ UTR (of 96 nt in length) carries a methylated nucleotide cap, while the 3′ UTR (of 337–649 nt in length) lacks a polyadenylation tail, but terminates with a conserved CU_{OH} instead. The ORF (of 10,301 nt in length) encodes a polyprotein of 3,430 aa (consisting of C-prM-E-NS1-NS2a-NS2b-NS3-NS4a-NS4b-NS5), which is processed co- and posttranslationally by the viral serine protease complex (NS2B-NS3) and various cellular proteases into 10 mature proteins, including 3 structural proteins (capsid [C], premembrane [prM, which later matures into membrane M], and envelope [E]) and 7 nonstructural proteins (NS1, NS2a, NS2b, NS3, NS4a, NS4b, and NS5) [13,14].

C (123 aa) is a key component of viral capsid. prM (167 aa) acts as a chaperone for envelope protein E during intracellular virion assembly and is cleaved by host furin to release pr peptide and M in the last step of virion assembly. After cleavage of prM, E (417 aa) dissociates from M and homodimerizes. E is involved in binding to host cell surface receptor for virus internalization through clathrin-mediated endocytosis,

as well as in membrane fusion between virion and host endosomes [15].

NS1 (352 aa) is a glycoprotein containing three conserved *N*-linked glycosylation sites and multiple conserved cysteines that form disulfide bonds essential for virus viability. NS1 monomers are soluble and hydrophilic while NS1 homodimers associate with membranes. In addition, NS1 may be also secreted by mammalian cells as a soluble hexamer [14].

NS3 (619 aa) comprises a serine protease (175 aa) at its *N*-terminal, which is active only when complexed with NS2B. The NS3-NS2B protease complex cleaves the viral polyprotein at multiple sites. The *C*-terminal domain of NS3 has been shown to possess RNA-stimulated nucleoside triphosphatase, ATPase/helicase, and 5′-triphosphatase activities [14].

NS5 (905 aa) contains an *S*-adenosyl methionine methyltransferase (MTase) domain at its *N*-terminal region, which has both N7 and 2′-*O* MTase activities and also acts as a guanylyltransferase. In addition, NS5 contains conserved sequence motif characteristic of all RNA-dependent RNA polymerases at its *C*-terminal region [14].

On the other hand, NS2a (231 aa), NS2b (131 aa), NS4a (149 aa), and NS4b (256 aa) are small, hydrophobic proteins without obvious enzymatic activities. These proteins facilitate the assembly and/or anchoring of viral replication complexes on the endoplasmic reticulum membrane [14].

28.1.3 Biology and Epidemiology

WNV was first isolated in 1937 from the blood of a febrile patient in the West Nile district of Uganda (currently Nile Province). Subsequently, outbreaks due to this virus were recorded in Israel (1951–1952, 1957 and 1962, 1998–2000), France (1962–1965, 2000), South Africa (1974 and 1983–1984), Algeria (1994 and 1997), Morocco (1996), Romania (1996), Tunisia (1997), and Russia (1999). In 1999, WNV was detected in New York, United States, and soon spread to large parts of America including southern Canada, Mexico, Cayman Islands, Guadeloupe Islands, El Salvador, Cuba, Colombia, and Argentina [12,14–18].

WNV is maintained in nature in an enzootic transmission cycle between wild birds (reservoir) and mosquitoes (vector). Wild birds acquire the virus after being bitten by an infected mosquito belonging to several ornithophilic species of the *Culex* genus (e.g., *Culex pipiens*, *Cx. nigripalpus*, *Cx. quinquefasciatus*, and *Cx. restuans*). Crows and magpies, house sparrows, house finches, and other passerines often develop viremia 2–7 days after infection. WNV may persist in the skin for an undetermined period of time, allowing mosquito vector (e.g., *Aedes vexan* and *Ochlerotatus* spp.) that feed both on birds and mammals to transmit the virus. Significant mortalities due to WNV have been observed in domestic geese, migrating white storks, lappet-faced vulture, and white-eyed gull. In addition, birds may acquire the virus via mosquito-independent transmission such as ingestion of infected mosquitoes, mice, and birds, as well as contaminated water.

WNV can be transported over long distances in migratory birds. Resting in wetland areas, migrating birds may come in contact with the local mosquito species, possibly initiating a local amplification cycle. In temperate regions, the virus may overwinter in female mosquitoes as well as in birds, so there is no need for continuous reintroductions [10].

In the enzootic areas of Europe and North America, WNV infection builds up in wild birds in the spring and early summer, with peak mortality in birds appearing from the mid-summer to early fall. One to a few weeks after bird mortality begins, human and horse cases may emerge. The overwintering of the virus in mosquitoes may be another contributing factor for seasonal reemergence of the WNV. The virus can be transmitted vertically from infected female mosquitoes to a small percentage of their eggs, helping virus survival through the winter. However, in the tropics, WNV infection in birds may persist year-round, and recently infected birds migrating north may reintroduce the virus the following spring.

Apart from wild and domestic birds, WNV is infective to humans, horses, rodents, reptilians, and amphibians. In humans, besides mosquito bites, WNV may also be transmitted via transplanted organs, blood transfusions, skin wounds, breastfeeding, or during pregnancy (from mother to child) [19]. In general, humans, horses, and other mammals do not develop a sufficient viremia to allow mosquito infection, and they are considered dead-end hosts for WNV. On the other hand, some infected reptiles and amphibians may develop adequate blood concentrations of virus for mosquito transmission to occur [20].

28.1.4 Clinical Features and Pathogenesis

WNV infection in birds typically lasts a few hours to a few days, with clinical manifestations ranging from nonspecific signs, ocular disease (e.g., anisocoria and impaired vision), lethargy and ruffled feathers, regurgitation of feedstuffs, decreased appetite, complete anorexia, enteritis, decrease in body weight, polyuria and biliverdinuria, external hemorrhage from the mouth or cloaca, dull mentation, unusual posture, inability to hold head upright, torticollis, opisthotonus, seizures, to death (within 24 h of the symptom onset in highly susceptible species, such as corvids). Some parrots may display persistent neurological signs after recovery (e.g., ataxia, tremors, abnormal head posture, circling, and convulsions) [19,21–23].

In horses, the majority of the infections are asymptomatic, and only about 10% of infected horses suffer from a severe neuroinvasive disease that is characterized by encephalitis, paralysis of the limbs, facial tremors, and death (in one-third of the cases). Gross visceral lesions are nonspecific, and the most significant histologic lesions are nonsuppurative meningoencephalomyelitis, particularly in the brainstem and spinal cord (including occasional meningitis, focal spinal ventral horn poliomalacia, dorsal and lateral horn poliomyelitis, leukomyelitis, asymmetrical ventral motor spinal neuritis, and frequent olfactory region involvement). The most significant histologic lesions in this case were nonsuppurative [24–26].

In humans, with an incubation period of 2–14 days, WNV infections may be either asymptomatic (80%) or evolve into West Nile fever (20%) or severe West Nile disease (<1%). The high number of nonsymptomatic cases may increase the risk of WNV transmission through blood donation or organ transplants.

West Nile fever, the mild form of the disease, lasts 3–6 days and may show signs such as fever, malaise, anorexia, nausea, vomiting, eye pain, headache, body aches, myalgia, occasional skin rash (on the trunk of the body), and swollen lymph glands. In a small number of cases (<1%), severe West Nile disease (also called West Nile encephalitis or meningitis or West Nile poliomyelitis) may present with encephalitis, meningitis, and acute flaccid paralysis. Encephalitis due to WNV may show depression, altered levels of consciousness, lethargy, changes in personality, and fever. Typical signs of meningitis consist of headache, fever or hypothermia, rigidity, Kernig (the leg cannot be fully extended in a sitting position) or Brudzinski signs (flexion of the neck leads to flexion of the hip and knee), stupor, disorientation, photophobia, phonophobia, tremors, convulsions, muscle weakness, coma, and paralysis. Fatality rates in patients showing the neurological manifestations may approach 4%–18%, and neurological deficits may linger in this group of patients [19].

Microscopic examination of brains from affected humans reveals the activation of resident microglia; perivascular and parenchymal accumulation of macrophages, CD4+ and CD8+ T cells and B cells; formation of microglial nodules; reactive proliferation of astrocytes; and eventual neuronal cell death [12,27].

28.1.5 Diagnosis

Given the emergence and reemergence of WNV in various parts of the world, and the nonspecific signs induced, it is important that laboratory methods are developed and applied for its identification and surveillance. Virus isolation, serological assays, and nucleic acid detection are excellent examples of laboratory methods that have proven valuable for diagnosis of WNV infection in mosquito, cerebrospinal fluid (CSF), serum, blood, and other specimens.

28.1.5.1 Virus Isolation

WNV can be grown on Vero, RK-13, CHO, ATC-15, and AP61 cells. Typically, aliquots of clarified supernatant of test tissues or serum are inoculated into confluent monolayers and observed daily for evidence of cytopathic effects (CPEs). If CPEs are not evident, serological (indirect fluorescence assay [IFA] with WNV-specific MAbs) or molecular (reverse transcription polymerase chain reaction [RT-PCR]) tests may be performed to confirm the identity. Although virus isolation takes 6 days and requires level 3 biosafety containment and samples of good quality (infectious virus-containing samples), it remains a valuable method for WNV detection in mosquito pools, and vertebrate and avian samples.

28.1.5.2 Serological Assays

Serological assays offer a rapid approach for the diagnosis of WNV infection. Antigens for these assays may be prepared from WNV-infected cell culture (e.g., Vero/C6/CHO/ATC-15 cells) supernatants, WNV-infected suckling mouse brains, or recombinant proteins (e.g., E, DIII of E protein, NS1 and NS3 and NS5, and flavivirus-like particles from prM and E proteins).

Plaque reduction neutralization test (PRNT) at the 90% plaque reduction level offers a confirmation and a titration method of WNV-specific Abs (from serum or CSF). The demonstration of a fourfold increase in Abs titers comparing acute and convalescent sera using PRNT is a clear indication of WNV infection. PRNT is conducted in a biosafety level 3 laboratory. Briefly, heat-inactivated CSF or serum samples are tested at 1:100 final dilution. Equal volume of serum and medium containing 100 plaque-forming units of WNV are incubated for 75 min at 37°C before inoculation onto confluent monolayers of Vero E6 cells grown in 25 cm² flasks. After the inoculum is adsorbed for 1 h at 37°C, cells are overlayed with agarose-containing medium and incubated for 72 h at 37°C. Then, a second agarose overlay containing 0.003% neutral red dye is applied to each flask for plaque visualization. After a further overnight incubation at 37°C, the number of virus plaques per flask is assessed. Endpoint titers are determined as the greatest dilution in which >90% neutralization of the challenge virus is achieved. Samples with reciprocal 90% neutralization titers of >10 were considered positive.

Immunohistochemistry (IHC) may be performed on variable tissues and animal tissues, and remains useful for specific confirmation of WNV infection in fatal WN encephalitis cases or in bird disease. The recently improved IFA affords a cost-effective and sensitive detection of both IgM and IgG anti-WNV Abs.

Enzyme-linked immunosorbent assay (ELISA) is currently the mostly widely applied primary screening method for WNV diagnosis. Three types of ELISA are in common use: IgG, MAC-ELISA, and epitope-blocking ELISA. The IgM antibody-capture ELISA (MAC-ELISA) detects early antibodies (IgM), permitting diagnosis of acute infections (sera sampled 8–45 days after infection) from serum or CSF and differentiation between old and recent infections. ELISA and IFA protocols (incorporating urea treatment of antigen–antibody complexes) for determination of IgG avidity help differentiate between recent and past infections. However, IgG ELISA is less specific for arboviral antigen than IgM. Epitope-blocking ELISA is an antibody competition assay that is species independent and has been tested with success in multiple avian species and domestic animals. ELISA has the advantages of being rapid, reproducible, and less expensive than PRNT, IHC, indirect hemagglutination assay (IHA), and IFA. However, it tends to show cross-reaction with other flaviviruses, especially those from the same serocomplex. Thus, confirmation by PRNT is necessary for ELISA-positive sera.

28.1.5.3 Nucleic Acid Detection

Nucleic acid–based techniques such as RT-PCR represent a powerful alternative for WNV detection [28–31]. RT-PCR enables in vitro amplification of WNV RNA from minute amounts of the starting material, in a variety of formats, including (1) standard RT-PCR, (2) nested RT-PCR, (3) multiplex RT-PCR, and (4) real-time RT-PCR [32–37]. To further verify the identity of WNV, the amplified products from the standard or nested RT-PCR protocol can be sequenced. Other nucleic acid detection methods include nucleic acid sequence–based amplification, loop-mediated isothermal amplification, and branched DNA.

RT-PCR has been utilized for the detection of WNV in serum, CSF, fresh tissues, and formalin-fixed human tissues. Briese et al. [38] utilized two primer–probe sets from the NS3 and NS5 protein sequences for real-time RT-PCR detection of 50–100 WNV molecules. Lanciotti and Kerst [39] reported a real-time RT-PCR assay targeting the E gene with a sensitivity of 0.1 PFU of viral RNA. As viremia begins within a few days after the infection and usually precedes the clinical signs, detection of viral RNA provides a clear indicator of recent infection. Nevertheless, viral RNA detection has limited value in late diagnosis since viremia is short lived [40–48].

28.1.6 Treatment and Prevention

Currently, there is a lack of specific treatment for WNV infections in humans and animals. Administration of neutralizing antibody may help alleviate the clinical symptoms of WNV encephalitis in WNV-infected patients, while use of antibiotics reduces the damages caused by secondary bacterial infection.

A number of vaccine candidates with varied efficacy are being developed to prevent WNV infection. These include DNA-based vaccines, live chimeric/recombinant vaccine constructs, live attenuated virus, and inactivated or subunit vaccines. DNA vaccines expressing the prM and Env proteins from WNV or the domain III (DIII) region of the Env protein have been tested in both mice and horses. DNA vaccine constructs encoding for single-round infectious particles or a full-length cDNA copy of the attenuated Kunjin strain of WNV are also being evaluated [49,50].

In the absence of approved vaccines for WNV, vector management remains an important option to prevent and control outbreaks of the disease. Clearly, this requires a strong organizational backbone and coordination between the different stakeholders, skilled technicians and operators, appropriate equipment, and sufficient financial resources [51].

28.2 METHODS

28.2.1 Sample Preparation

WNV RNA may be extracted from blood, tissues, and mosquitoes using either the traditional phenol–chloroform protocol or commercial kits based on a silica gel–based membrane or other principles.

For RNA extraction from mosquitoes using Qiagen RNeasy Kit, pools of 10–50 individual mosquitoes are homogenized in diluent containing 20% fetal bovine serum,

50 μg of streptomycin per mL, 50 U of penicillin, and 2.5 μg of amphotericin B per mL in phosphate-buffered saline in a Spex CertiPrep (Metuchen, NJ) 8000-D mixer mill for 3 min. Alternatively, 50 mg (3 by 3 by 6 mm) of vertebrate tissues is homogenized in 700 μL of RNeasy lysis buffer in a Spex CertiPrep 8000-D mixer mill for 3 min. RNA is extracted from 350 μL of homogenized samples. The extracted RNA is eluted in a total volume of 50 μL of RNase-free water.

28.2.2 Detection Procedures

28.2.2.1 Standard RT-PCR

Lanciotti et al. [52] designed a pair of primers (forward: 5′-TTGTGTTGGCTCTCTTGGCGTTCTT-3′, nt 233–257 and reverse: 5′-CAGCCGACAGCACTGGACATTCATA-3′, nt 616–640) from the C-terminal portion of the C gene and the N-terminal part of the prM gene for amplification of a 408-bp fragment from WNV RNA.

28.2.2.1.1 Procedure

1. Standard RT-PCR mixture (50 μL) is composed of 1× reaction buffer, 0.4 mM dNTPs, 0.6 μM primers (forward and reverse), 1× Q solution, and 1 μL of reverse transcriptase-DNA *Taq* polymerase enzyme mix (One-Step RT-PCR Kit, Qiagen), 5 μL of RNA eluate extracted with RNeasy (or 20 μL of RNA eluate extracted with ABI Prism 6700 workstation).

2. The mixture is incubated at 50°C for 30 min to synthesize the first-strand cDNA, at 95°C for 15 min to inactivate the reverse transcriptase and to activate DNA *Taq* polymerase, 35 cycles of 94°C for 45 s, 56°C for 45 s, and 72°C for 1 min for PCR amplification, and a final elongation at 72°C for 10 min.

3. The RT-PCR products (20 μL) is analyzed on a 1.5% agarose gel with TAE buffer and stained with ethidium bromide (0.5 μg/mL).

Note: The sensitivity of this standard RT-PCR can be increased by approximately 10-fold with a nested PCR using internal primer set (forward: 5′-CAGTGCTGGATCGATGGAGAGG-3′, nt 287–308 and reverse: 5′-CCGCCGATTGATAGCACTGGT-3′, nt 370–390) [42]. Briefly, nested PCR mixture (25 μL) is made up of 1× reaction buffer, 0.2 mM dNTPs, 0.3 μM inner primers, 1 μL of first round RT-PCR template (diluted 10,000-fold),

and 0.625 U of *Taq* DNA polymerase (the *Taq* PCR Core Kit, Qiagen). The mixture is subjected to an initial 94°C for 3 min; 22 cycles of 94°C for 45 s, 58°C for 45 s, and 72°C for 1 min; and a final 72°C for 10 min. The resulting product (15 μL) is analyzed on a 1.5% agarose gel.

28.2.2.2 Real-Time RT-PCR

Jiménez-Clavero et al. [44] described a real-time RT-PCR protocol based on a 5′-Taq nuclease-3′ minor groove binder DNA probe (TaqMan MGB) for the detection of WNV lineages 1 and 2. The primers and probe set are directed to a highly conserved sequence within the 3′ NC region of the WNV genome. A second TaqMan MGB probe detects WNV lineage 2 isolates whose genomes differ in two nucleotide positions from lineage 1 genomes (Table 28.1).

28.2.2.2.1 Procedure

1. WNV is grown in Vero cells or baby hamster kidney-21 cells. Viruses are titrated by a standard limiting dilution assay. Viral RNA is extracted from the clarified supernatants of virus cultures, or 0.1 mL suspensions of the lyophilized infected plasma using High Pure Viral Nucleic Acid Extraction Kit.

2. The TaqMan MGB-RRT-PCR mixture (25 μL) is composed of 2 μL of isolated RNA, 12.5 μL of 2× QuantiTect Probe RT-PCR Master Mix, 0.625 μL of QuantiTect RT Mix, 0.4 μM of WNV-specific primers (WN-LCV-F1 and WN-LCV-R1), and 0.25 μM of the fluorogenic TaqMan probes (WN-LCV-S1 and WN-LCV-S2) and is RNase free.

3. The mixture is subjected to an RT step at 50°C for 30 min, a hot start at 95°C for 15 min, and 45 cycles of 95°C for 15 s and 60°C for 1 min using Smart Cycler II equipment and software.

Note: The assay detects WNV isolates belonging to lineage 1 with clade 1a and clade 1b (Kunjin) as well as lineage 2 (B956) with a sensitivity of 0.01–0.001 pfu/tube. Performed in 96-well format, this assay is suitable for the large-scale surveillance of areas where both WNV lineages 1 and 2 exist or potentially spread. By contrast, the TaqMan RRT-PCR of Lanciotti and Kerst [39] using 3′ NC primers and probe set and tetramethylrhodamine as a quencher detects only lineage 1 clade 1a isolates, and not Kunjin (clade 1b) and B956 (lineage 2).

TABLE 28.1
TaqMan Minor Groove Binder Real-Time RT-PCR Primers and Probes

Primer/Probe	Sequence (5′–3′)	Nucleotide Positions	Specificity
WN-LCV-F1	GTGATCCATGTAAGCCCTCAGAA	10,597–10,619	
WN-LCV-R1	GTCTGACATTGGGCTTTGAAGTTA	10,649–10,672	
WN-LCV-S1	FAM-AGGACCCCACATGTT-MGB	10,633–10,647	WNV lineage 1
WN-LCV-S2	FAM-AGGACCCCACGTGCT-MGB	10,633–10,647	WNV lineage 2

28.2.2.3 Phylogenetic Analysis

Bakonyi et al. [53] employed a universal JEV group–specific primer pair from the NS5 and 3′ UTR (forward primer: 5′-GARTGGATGACVACRGAAGACATGCT-3′ and reverse primer: 5′-GGGGTCTCCTCTAACCTCTAGTCCTT-3′) for RT-PCR amplification followed by sequencing for phylogenetic analysis of WNV isolates.

28.2.2.3.1 Procedure

1. Brain specimens from diseased goose and goshawk are homogenized in ceramic mortars by using sterile quartz sand, and the homogenates are suspended in RNase-free distilled water. Samples are stored at −80°C until nucleic acid extraction.
2. RT-PCR mixture (25 µL) is composed of 5 µL of 5× buffer (final MgCl$_2$ concentration 2.5 mmol/L), 0.4 mmol/L of each dNTP, 10 U RNasin RNase Inhibitor (Promega), 20 pmol of the genomic and reverse primers, 1 µL enzyme mix (containing Omniscript and Sensiscript RTs and HotStarTaq DNA polymerase) (One-Step RT-PCR Kit, Qiagen), and 2.5 µL template RNA.
3. RT is carried out at 50°C for 30 min, followed by denaturation at 95°C for 15 min. The cDNA is then amplified in 40 cycles of 94°C for 40 s, 57°C for 50 s, and 72°C for 1 min. The reaction is completed by a final extension at 72°C for 7 min.
4. After RT-PCR, 10 µL of the amplicon is electrophoresed in a 1.2% TAE agarose gel at 5 V/cm for 80 min. The gel is stained with ethidium bromide and bands are visualized under UV light and photographed. Product sizes are determined with reference to a 100 bp DNA ladder (Promega).
5. The PCR products of the expected sizes are excised from the gel, and the DNA is extracted by using the QIAquick Gel Extraction Kit (Qiagen). Fluorescence-based direct sequencing is performed in both directions on PCR products using the ABI Prism Big Dye Terminator Cycle Sequencing Ready Reaction Kit (Perkin-Elmer) and an ABI Prism 310 Genetic Analyzer (Perkin-Elmer). Nucleotide sequences are identified by basic local alignment search tool (BLAST) search against gene bank databases.

28.3 CONCLUSION

WNV is a flavivirus with the capacity to induce significant mortality in birds and horses as well as long-term sequelae and occasional death in humans. Diagnosis of West Nile disease has previously relied on virus isolation and serological assays. Due to the lengthy time and specialized skills required for conducting these tests, nucleic acid detection methods, especially RT-PCR, have been developed as rapid, specific, and sensitive alternatives for improved identification and typing of WNV. Being amenable to automation, RT-PCR provides a complete solution for nucleic acid extraction, amplification, and product detection, thus dramatically increasing the throughput and versatility of diagnosis, especially for the confirmation of a recent infection. Surveillance and preventive measures are an ongoing need to reduce the public health impact of WNV in areas with favorable transmission conditions.

REFERENCES

1. Chambers TJ, Hahn CS, Galler R, Rice CM. Flavivirus genome organization, expression, and replication. *Annu Rev Microbiol.* 1990;44:649–688.
2. Poidinger M, Hall RA, Mackenzie JS. Molecular characterization of the Japanese encephalitis serocomplex of the flavivirus genus. *Virology.* 1996;218(2):417–421.
3. Charrel RN et al. Evolutionary relationship between Old World West Nile virus strains evidence for viral gene flow between Africa, the Middle East, and Europe. *Virology.* 2003;315(2):381–388.
4. Mackenzie JS et al. Arboviruses causing human disease in the Australasian zoogeographic region. *Arch Virol.* 1994;136(3–4):447–467.
5. Hernández-Triana LM et al. Emergence of west nile virus lineage 2 in europe: A review on the introduction and spread of a mosquito-borne disease. *Front Public Health.* 2014;2:271.
6. Bakonyi T, Hubálek Z, Rudolf I, Nowotny N. Novel flavivirus or new lineage of West Nile virus, central Europe. *Emerg Infect Dis.* 2005;11(2):225–231.
7. Hubálek Z et al. Mosquito (Diptera: Culicidae) surveillance for arboviruses in an area endemic for West Nile (Lineage Rabensburg) and Tahyna viruses in Central Europe. *J Med Entomol.* 2010;47(3):466–472.
8. Aliota MT et al. Characterization of Rabensburg virus, a flavivirus closely related to West Nile virus of the Japanese encephalitis antigenic group. *PLoS One.* 2012;7(6):e39387.
9. Gaunt MW et al. Phylogenetic relationships of flaviviruses correlate with their epidemiology, disease association and biogeography. *J Gen Virol.* 2001;82(8):1867–1876.
10. Chancey C, Grinev A, Volkova E, Rios M. The global ecology and epidemiology of West Nile Virus. *Biomed Res Int.* 2015;2015:376230.
11. McMullen AR. Molecular evolution of lineage 2 West Nile virus. *J Gen Virol.* 2013;94(2):318–325.
12. Donadieu E et al. Differential virulence and pathogenesis of West Nile viruses. *Viruses.* 2013;5(11):2856–2880.
13. Lanciotti RS et al. Complete genome sequences and phylogenetic analysis of West Nile virus strains isolated from the United States, Europe, and the Middle East. *Virology.* 2002;298(1):96–105.
14. Brinton MA. Replication cycle and molecular biology of the West Nile virus. *Viruses.* 2013;6(1):13–53.
15. Koo QY et al. Conservation and variability of West Nile virus proteins. *PLoS One.* 2009;4(4):e5352.
16. Lanciotti RS et al. Origin of the West Nile virus responsible for an outbreak of encephalitis in the northeastern United States. *Science.* 1999;286(5448):2333–2337.
17. Zeller HG, Schuffenecker I. West Nile virus: An overview of its spread in Europe and the Mediterranean basin in contrast to its spread in the Americas. *Eur J Clin Microbiol Infect Dis.* 2004;23(3):147–156.
18. Roehrig JT. West Nile virus in the United States—A historical perspective. *Viruses.* 2013;5(12):3088–3108.
19. Campbell GL, Marfin AA, Lanciotti RS, Gubler DJ. West Nile virus. *Lancet Infect Dis.* 2002;2(9):519–529.

20. Suthar MS, Pulendran B. Systems analysis of West Nile virus infection. *Curr Opin Virol.* 2014;6:70–75.

21. Reisen WK et al. Chronic infections of West Nile virus detected in California dead birds. *Vector Borne Zoonotic Dis.* 2013;13(6):401–405.

22. Ziegler U et al. Pathogenesis of West Nile virus lineage 1 and 2 in experimentally infected large falcons. *Vet Microbiol.* 2013;161(3–4):263–273.

23. Pérez-Ramírez E, Llorente F, Jiménez-Clavero MÁ. Experimental infections of wild birds with West Nile virus. *Viruses.* 2014;6(2):752–781.

24. Ibarra-Juarez L et al. Detection of West Nile virus-specific antibodies and nucleic acid in horses and mosquitoes, respectively, in Nuevo Leon State, northern Mexico, 2006–2007. *Med Vet Entomol.* 2012;26(3):351–354.

25. Williams JH, van Niekerk S, Human S, van Wilpe E, Venter M. Pathology of fatal lineage 1 and 2 West Nile virus infections in horses in South Africa. *J S Afr Vet Assoc.* 2014;85(1):1105.

26. Niczyporuk JS et al. Occurrence of West Nile virus antibodies in wild birds, horses, and humans in Poland. *Biomed Res Int.* 2015;2015:234181.

27. Suen WW, Prow NA, Hall RA, Bielefeldt-Ohmann H. Mechanism of West Nile virus neuroinvasion: A critical appraisal. *Viruses.* 2014;6(7):2796–2825.

28. Niedrig M, Linke S, Zeller H, Drosten C. First international proficiency study on West Nile virus molecular detection. *Clin Chem.* 2006;52(10):1851–1854.

29. Kesavaraju B et al. Evaluation of a rapid analyte measurement platform for West Nile virus detection based on United States mosquito control programs. *Am J Trop Med Hyg.* 2012;87(2):359–363.

30. Zink SD, Jones SA, Maffei JG, Kramer LD. Quadraplex qRT-PCR assay for the simultaneous detection of Eastern equine encephalitis virus and West Nile virus. *Diagn Microbiol Infect Dis.* 2013;77(2):129–132.

31. Worwa G et al. Allele-specific qRT-PCR demonstrates superior detection of single nucleotide polymorphisms as genetic markers for West Nile virus compared to Luminex® and quantitative sequencing. *J Virol Methods.* 2014;195:76–85.

32. Gaunt MW, Gould EA. Rapid subgroup identification of the flaviviruses using degenerate primer E-gene RT-PCR and site specific restriction enzyme analysis. *J Virol Methods.* 2005;128(1–2):113–127.

33. Barros SC et al. Simultaneous detection of West Nile and Japanese encephalitis virus RNA by duplex TaqMan RT-PCR. *J Virol Methods.* 2013;193(2):554–557.

34. Del Amo J et al. A novel quantitative multiplex real-time RT-PCR for the simultaneous detection and differentiation of West Nile virus lineages 1 and 2, and of Usutu virus. *J Virol Methods.* 2013;189(2):321–327.

35. Lim SM, Koraka P, Osterhaus AD, Martina BE. Development of a strand-specific real-time qRT-PCR for the accurate detection and quantitation of West Nile virus RNA. *J Virol Methods.* 2013;194(1–2):146–153.

36. Faggioni G et al. Rapid molecular detection and genotyping of West Nile Virus lineages 1 and 2 by real time PCR and melting curve analysis. *J Virol Methods.* 2014;207:54–59.

37. Kumar JS, Saxena D, Parida M. Development and comparative evaluation of SYBR Green I-based one-step real-time RT-PCR assay for detection and quantification of West Nile virus in human patients. *Mol Cell Probes.* 2014;28(5–6):221–227.

38. Briese T, Glass WG, Lipkin WI. Detection of West Nile virus sequences in cerebrospinal fluid. *Lancet.* 2000;355(9215):1614–1615.

39. Lanciotti RS, Kerst AJ. Nucleic acid sequence-based amplification assays for rapid detection of West Nile and St. Louis encephalitis viruses. *J Clin Microbiol.* 2001;39(12):4506–4513.

40. Kuno G, Chang GJ, Tsuchiya KR, Karabatsos N, Cropp CB. Phylogeny of the genus Flavivirus. *J Virol.* 1998;72(1):73–83.

41. Johnson DJ, Ostlund EN, Pedersen DD, Schmitt BJ. Detection of North American West Nile virus in animal tissue by a reverse transcription-nested polymerase chain reaction assay. *Emerg Infect Dis.* 2001;7(4):739–741.

42. Shi P-Y et al. High-throughput detection of West Nile virus RNA. *J Clin Microbiol.* 2001;39(4):1264–1271.

43. Eisler DL, McNabb A, Jorgensen DR, Isaac-Renton JL. Use of an internal positive control in a multiplex reverse transcription-PCR to detect West Nile virus RNA in mosquito pools. *J Clin Microbiol.* 2004;42(2):841–843.

44. Jiménez-Clavero MA, Agüero M, Rojo G, Gómez-Tejedor C. A new fluorogenic real-time RT-PCR assay for detection of lineage 1 and lineage 2 West Nile viruses. *J Vet Diagn Invest.* 2006;18(5):459–462.

45. Amanna IJ, Slifka MK. Current trends in West Nile virus vaccine development. *Expert Rev Vaccines.* 2014;13(5):589–608.

45. Linke S, Ellerbrok H, Niedrig M, Nitsche A, Pauli G. Detection of West Nile virus lineages 1 and 2 by real-time PCR. *J Virol Methods.* 2007;146(1–2):355–358.

46. Tang Y, Anne Hapip C, Liu B, Fang CT. Highly sensitive TaqMan RT-PCR assay for detection and quantification of both lineages of West Nile virus RNA. *J Clin Virol.* 2006;36(3):177–182.

47. Zaayman D, Human S, Venter M. A highly sensitive method for the detection and genotyping of West Nile virus by real-time PCR. *J Virol Methods.* 2009;157(2):155–160.

48. Schuh AJ, Tesh RB, Barrett AD. Genetic characterization of Japanese encephalitis virus genotype II strains isolated from 1951 to 1978. *J Gen Virol.* 2011;92(3):516–527.

50. Ulbert S, Magnusson SE. Technologies for the development of West Nile virus vaccines. *Future Microbiol.* 2014;9(10):1221–1232.

51. Bellini R, Zeller H, Van Bortel W. A review of the vector management methods to prevent and control outbreaks of West Nile virus infection and the challenge for Europe. *Parasit Vectors.* 2014;7:323.

52. Lanciotti RS et al. Rapid detection of West Nile virus from human clinical specimens, field-collected mosquitoes, and avian samples by a TaqMan reverse transcriptase-PCR assay. *J Clin Microbiol.* 2000;38(11):4066–4071.

53. Bakonyi T et al. Lineage 1 and 2 strains of encephalitic West Nile virus, central Europe. *Emerg Infect Dis.* 2006;12(4):618–623.

29 Eastern and Western Equine Encephalitis Viruses

Norma P. Tavakoli and Laura D. Kramer

CONTENTS

29.1 Introduction 255
    29.1.1 Classification, Morphology, and Genome Organization 255
        29.1.1.1 Classification.. 255
        29.1.1.2 Morphology and Genome Organization... 256
    29.1.2 Biology and Epidemiology 256
    29.1.3 Clinical Features and Pathogenesis 258
    29.1.4 Diagnosis 259
        29.1.4.1 Conventional Techniques 260
        29.1.4.2 Molecular Techniques.. 260
29.2 Methods 261
    29.2.1 Sample Preparation.. 261
        29.2.1.1 Animal Specimens... 261
        29.2.1.2 Mosquito Specimen Collection for Surveillance 261
        29.2.1.3 Sample Processing... 261
        29.2.1.4 Nucleic Acid Extraction 262
    29.2.2 Detection Procedures.. 262
        29.2.2.1 Real-Time RT-PCR for Detection of EEEV 262
        29.2.2.2 Real-Time RT-PCR for Detection of WEEV 263
        29.2.2.3 Standard RT-PCR 263
29.3 Conclusion and Future Perspectives 264
Acknowledgments.. 265
References.. 265

29.1 INTRODUCTION

29.1.1 CLASSIFICATION, MORPHOLOGY, AND GENOME ORGANIZATION

29.1.1.1 Classification

The family *Togaviridae* is comprised of four genera: *Alphavirus* (26 species), *Rubivirus* (1 species), *Pestivirus* (3 species), and *Arterivirus* (1 species) [1,2]. The alphaviruses include at least seven antigenic complexes, two of which will be reviewed in this chapter: Western equine encephalitis virus (WEEV) and Eastern equine encephalitis virus (EEEV), and five others—the Middelburg virus, Ndumu virus, Semliki Forest virus, Venezuelan equine encephalitis virus (VEEV), and Barmah Forest virus [3]. EEEV is the sole species in the EEEV antigenic complex. North and South American antigenic varieties can be distinguished serologically [4], geographically, epidemiologically, ecologically, genetically, and by level of pathogenicity. In view of clear biologic and genetic differences, South American EEEV was recently reclassified as a separate species named Madariaga virus (MADV) [5]. North American EEEV (Lineage I) is found mainly in the Eastern United States, Canada, and the Caribbean Islands [6], and the isolates form a highly conserved lineage. In addition to MADV, there are old reports of EEEV isolations made in the Philippines, Thailand, the Asiatic region of Russia, and the former Czechoslovakia [7].

The WEEV complex forms a monophyletic group and includes the prototype virus, WEEV, as well as Buggy Creek virus, Fort Morgan virus (FMV), and Highlands J virus (HJV) in North America, Aura virus in South America, and Sindbis virus (SINV) and Whataroa viruses in the Old World. The Aura virus is most distant by neutralization test [8]. HJV is the sole representative of this complex on the East Coast of the United States, while WEEV complex viruses have not been reported in tropical regions of Central America. Genetic variations among selected WEEV strains isolated in California since 1938 were analyzed by sequencing the E2 protein, leading to the identification of four major lineages [9]. Apparently, WEEV arose by a single recombination event within the E3 gene between a member of the EEEV lineage and the SINV lineage approximately 1300–1900 years ago (Figure 29.1) [10]. The envelope glycoproteins and a portion of the 3′ nontranslated region were derived from a SINV-like ancestor, while the capsid and nonstructural proteins were derived from the EEEV ancestor. Aura virus, a New World member in the WEEV complex, SINV, and Whataroa virus are not recombinants.

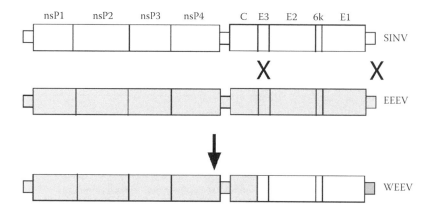

FIGURE 29.1 Schematic representation of the recombination event that produced the western equine encephalitis virus (WEEV). The crossover points by which WEEV was produced are indicated. SINV, sindbis virus; EEEV, eastern equine encephalitis virus. (Adapted from Hahn, C.S. et al., *Proc. Natl. Acad. Sci. USA*, 85, 5997, 1988. With permission.)

WEEV strains can be divided into high-virulence and low-virulence pathotypes following intranasal [11] and subcutaneous [12,13] infection of mice, and these differences correlate with amino acid differences concentrated in the structural genes. It has been hypothesized that epizootics may arise by mutation of avirulent strains circulating enzootically [14].

29.1.1.2 Morphology and Genome Organization

All alphaviruses are similar on a molecular level, both structurally and functionally. The virion is roughly spherical, enveloped, 65–70 nm in diameter, and has icosahedral symmetry. The virion has an outer glycoprotein shell surrounding a host-derived lipid bilayer and a core consisting of the viral RNA, complexed with the capsid protein. The genomic RNA is single-stranded, positive-sense, nonsegmented, capped on the 5′ end, and polyadenylated on the 3′ end. The length of the complete genome is approximately 11.7 kb, not including the cap and tail, the latter regions averaging approximately 70 nucleotides. The RNA is infectious and can initiate replication in a susceptible cell [15]. When released into the cytoplasm of the host cell, the 49S RNA serves as messenger for translation of three of the four nonstructural proteins and template for transcription of replicative intermediate RNA. The genome contains two open reading frames. The nonstructural proteins (nsP1, nsP2, nsP3, and nsP4) are translated directly from the 5′ two-thirds of the genomic RNA, and the structural proteins at the 3′ end by a subgenomic mRNA, 26S RNA. A junction region is situated between the structural and nonstructural regions [16]. There are four nonstructural proteins: nsP1, nsP2, nsP3, and nsP4; the structural proteins include the capsid, E3, E2, and E1. The latter two proteins are glycoproteins that form a functional dimer on the envelope of the virus [17] and a heterodimer in the infected cell [18]. E2 contains the major neutralization epitope [19], while E1 contains the major fusion activity [20]. Replication takes place in the cytoplasm of the infected cell.

29.1.2 BIOLOGY AND EPIDEMIOLOGY

Passerine birds are the predominant vertebrate hosts for WEEV and EEEV, leading to widespread geographic

distribution; but otherwise, WEEV and EEEV differ antigenically and ecologically. The first report of an epizootic of equine encephalomyelitis dates back to 1831 when more than 75 horses died in Massachusetts [21]. During a major equine outbreak of the disease in coastal areas of Delaware, Maryland, New Jersey, and Virginia in 1933, the virus was isolated from horses for the first time [22].

Studies in the 1930s showed that mosquito species including *Aedes*, *Culex*, and *Coquillettidia* were capable of transmitting EEEV from one vertebrate to another [23–25]. An outbreak of human disease in 1938 resulted in 30 cases of fatal encephalitis in children living in the Northeastern United States, where a concurrent outbreak was occurring in horses. During this outbreak, EEEV was isolated from the central nervous system (CNS) of humans [26,27]. In 1949, for the first time, the virus was isolated from a naturally infected mosquito (*Coquillettidia perturbans*) in Georgia and shortly thereafter in Louisiana from *Culiseta melanura*, the most common mosquito vector of EEEV [28,29]. In the 1930s outbreaks, there was a suspicion that birds were involved in the transmission cycle, and finally the virus was isolated for the first time from a wild bird in 1950 [30]. Since then it has been shown that many birds are susceptible to EEEV infection [31] and outbreaks of encephalitis have been recorded in pigeons (*Columba livia*) and exotic birds such as pheasants, Pekin ducks (*Anas platyrhynchos domestica*), and chukars (*Alectoris chukar*) [32–35], among others.

In North America, EEEV is found mainly in the eastern half of the United States with most transmissions occurring near freshwater hardwood swamps in states along the Atlantic seaboard and the Gulf Coast, and in the Great Lakes region [36,37]. Between 1964 and 2007, approximately 254 EEEV infections were confirmed in the United States [38]. The greatest numbers of these cases occurred in Florida, Massachusetts, Georgia, and New Jersey. The virus also has been isolated from mosquitoes in other eastern and central states [39–42] (see Figure 29.2) and in South America.

In most northern states, cases of EEE in horses and humans occur in the period between July and October [43], and in Florida, they occur year-round, but mainly between

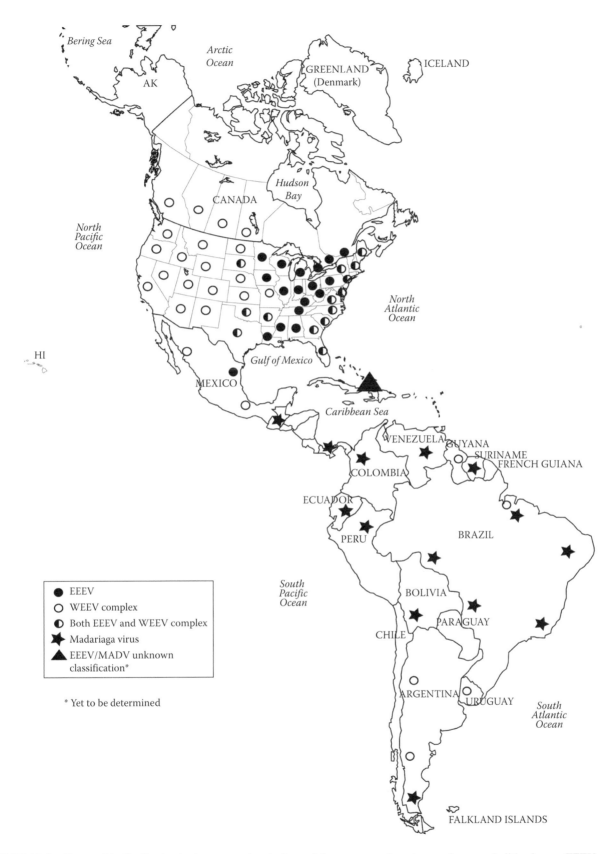

FIGURE 29.2 Geographic distribution in the western hemisphere of the eastern and western equine encephalitis viruses. EEEV, eastern equine encephalitis virus; WEEV, western equine encephalitis virus; MADV, Madariaga virus.

May and August. Generally, wild birds, pheasants, or horses are infected prior to human infections. Epizootics occur every 5–10 years when populations of epizootic and enzootic mosquito vectors increase [36,44,45].

EEEV is rural in distribution and virus activity is generally observed in forests and wetlands. In North America, the main transmission cycle involves *Cs. melanura* as the principal endemic vector [46–48] and possibly the overwintering host, and passerines, wading birds, and starlings as primary amplifying hosts [47,49,50]. Humans, equines, and other small mammals and domestic fowl are dead-end hosts because, in general, viremia in these hosts is not high enough to infect mosquitoes. However, in South America, MADV, which cycles between small mammals and mosquitoes, does mount a high enough viremia to infect mosquitoes [4]. This virus has been isolated primarily from *Culex* mosquitoes, and even though in certain areas mosquito surveillance suggests enzootic circulation and the seroprevalence is reported as being fairly high, human disease is very rare [51–53]. It has been hypothesized that the paucity of human disease due to MADV in South America is a result of South American strains not being highly pathogenic [54].

The first epizootic of WEE was noted in the United States in 1912 when 25,000 horses died from encephalitic disease in the Central Plains [55]. At this time, intermittent activity had been observed in the plains and agricultural regions of Western and Central United States and across South-Central Canada from approximately Lake Superior to the Rocky Mountains, as well as British Columbia [56]. The causative agent, WEEV, was first isolated in 1930 from a horse (*Equus caballus*) in Merced County, California, during an epizootic that infected approximately 6000 animals with a 50% case fatality rate [57]. Most WEEV activity is found within the range of *Culex tarsalis*, but the virus also can be isolated in Central and South America [58] (Figure 29.2).

The first human isolate was made in 1938 from the brain of a fatal case in a child [59]. The incidence of WEE in humans and equids peaked during the mid-twentieth century. Major human epidemics occurred between the Mississippi River delta and the Rocky Mountains, where the June isotherm is higher than 21.1°C. In the subsequent decades, the number of human cases declined to fewer than one to two annually, with no cases detected since 1994. Using a mouse model, there is no evidence of change in virulence of the virus suggesting that ecological factors affecting the exposure of humans and horses are more likely the cause for the decline in cases [12]. Greater adoption of personal protective measures or increased use of air-conditioning may have contributed to the drop in cases [60]. Development of vector control programs and widespread use of an equine vaccine are responsible as well for decreased infection rates in equines and humans.

WEEV is maintained in an enzootic cycle generally between *Culex tarsalis* mosquitoes and songbirds, especially finches and house sparrows, in agricultural regions of North America, but the transmission cycle has not been well characterized in South America. A late season enzootic cycle between *Aedes melanimon* and black-tailed jackrabbits has

been observed [61]. Equines and humans are dead-end hosts in that they do not mount sufficiently high viremias to infect mosquitoes. It is not definitively known how the virus overwinters, but genetic analyses of isolates from California over time and space suggest the virus most likely is maintained in separate geographic areas of the state through local persistence in enzootic foci [9].

HJV, an east coast member of the WEEV complex in the United States, has a natural cycle and geographic distribution similar to that of EEEV and is transmitted by *Cs. melanura* mosquitoes to songbirds in freshwater swamps. However, unlike EEEV and WEEV, HJV does not generally cause disease in horses and is not known to have been a significant causative agent of disease in humans. However, in the past, some cases of HJV in the Eastern United States may have been mistakenly identified as WEEV [62]. FMV, transmitted between cliff swallows (*Petrochelidon pyrrhonota*) and house sparrows (*Passer domesticus*) that occupy nests in swallow colonies, and cimicid bugs (Hemiptera, *Cimicidae*, *Oeciacus vicarius*), has been detected only in western states of the United States, specifically CO, SD, WA, and western TX, but is most closely related to HJV serologically [63]. Both these viruses, when unpassaged, are attenuated in infant mice, and mammals in general. The Buggy Creek virus has the same ecological niche as FMV and is closely related genetically [64]. It appears from phylogenetic analysis that they may be strains of the same virus [65].

29.1.3 Clinical Features and Pathogenesis

As their name suggests, WEEV and EEEV cause encephalitis in humans and horses although their virulence and incidence vary. The case fatality rate of WEEV in horses ranges from 3% to 50%, while that of EEEV is estimated at 70% to 90% [66]. Horses infected with EEEV or WEEV may or may not show symptoms. In symptomatic cases, infection with either virus begins with fever, dullness, lethargy, and loss of appetite followed by various degrees of excitability and rapidly progresses to drowsiness, apparent blindness, nystagmus, head pressing, circling, ataxia, paresis, depression, seizures, and coma in fatal cases [36]. Neurological sequelae are common in surviving horses infected with EEEV [46].

Emus (*Dromaius novaehollandiae*) [67], ostriches (*Struthio camelus*), whooping cranes (*Grus americana*) [68], and swine (*Sus domesticus*) can also be infected with EEEV/WEEV and develop clinical disease. However, the infections in these species may present with prominent abdominal visceral lesions such as hemorrhagic enterocolitis and splenic and hepatic necrosis resulting in hepatitis and/or gastrointestinal tract disease, rather than encephalitis. Pheasants (*Phasianus colchicus*) and emus are particularly susceptible to EEEV infections [69,70]. In emus, EEEV causes multiple organ hemorrhage, hepatocellular necrosis, lymphoid necrosis, and necrotizing vasculitis in the spleen, as well as necrosis of endothelial cells [67,69]. Necrosis of the liver, lung, spleen, intestines, kidney, lung, adrenal gland, and gonads is observed in infected whooping cranes [68]. Experimentally infected starlings are more susceptible to EEE-related deaths with a mortality corresponding to peak viremia [49]. Mortality in whooping cranes, house sparrows (*Passer domesticus*), cowbirds (*Molothrus ater*), grackles (*Quiscalus quiscula*), and red-winged blackbirds (*Agelaius phoeniceus*) is relatively high, whereas the bobwhite quail (*Colinus virginianu*) and white-throated sparrows (*Zonotrichia albicollis*) have a high survival rate [68,71]. Songbirds are the enzootic vertebrate hosts of EEEV in North America where significant mortality is observed among penned exotic game birds [47,72]. Among birds, younger cases have a greater cerebral tissue sensitivity to the virus, are more viremic, and are therefore more susceptible to CNS disease.

Natural EEEV infection has been reported in cows, sheep, dogs, and white-tailed deer [41,73–75]. In sheep, front limb incoordination progressed to paralysis of the front and hind limbs within 2 days. The clinical signs in sheep are similar to that of horses except that the reported sheep remained alert until euthanized whereas horses are generally dull and somnolent [74]. In dogs, it appears that the disease first manifests with pyrexia, anorexia, and diarrhea and rapidly progresses to clinical signs including recumbency, nystagmus, depression, and seizures [75]. Multiple EEEV-infected white-tailed deer displayed various combinations of clinical signs including head tilt, ataxia, confusion, circling, blindness, prostration, dyspnea, and emaciation [41]. The differences in pathologic manifestations of virus infection in different species include genetic differences in susceptibility to infection and immunologic response to viral infection, exposure to different viral strains that vary in pathogenicity, naiveness to viral infection, and differences in exposure to infected vectors due to variations in weather, habitat, and ecology.

Although not extensively studied, there are some recent investigations of EEE and WEE in animals. The pathogenesis of EEE and WEE, similar to other alphavirus encephalitis cases, appears to be characterized by three phases: an initial peripheral phase of virus replication and spread, the neuroinvasion phase, and a terminal CNS phase in which the virus spreads within the brain, infects primarily the neurons, and results in an often fatal neurodegenerative state [76]. Neuroinvasiveness and encephalitis of EEEV have been studied in murine and hamster models [77]. It appears that the virus invades the CNS of infected animals via infected leukocytes or the vascular route by passive transfer across the blood–brain barrier [77]. Young mice infected with EEEV develop a biphasic disease course with initial viral replication in peripheral tissues followed by viremia, widespread infection of neurons, CNS invasion, and encephalitis characterized by seizures [77]. In mouse brain, there is widespread neuronal necrosis, rarefaction of the adjacent neutrophil in the gray matter, and some white matter tracts. Outside the brain, EEEV also targets fibroblasts, skeletal muscle, and myocytes, osteoblasts, ovarian stromal cells, and keratinocytes, among other cells. In the hamster model, neurons are the main viral target in the brain and the first antigen positive neuronal cells are in the basal ganglia and brainstem with the virus subsequently infecting periventricular and perivascular neurons

and the hippocampus [78]. In horses, both the cerebellum and brainstem are affected. The inflammatory cell infiltrate consists of neutrophils, lymphocytes, macrophages, and plasma cells scattered throughout the gray and white matter or arranged in perivascular cuffs [79]. Additionally, multifocal hemorrhage, vasculitis, and meningitis have been identified [79]. Nonsuppurative encephalitis and vasculitis have been noted in birds. In sheep, lesions were observed in both grey and white matter and consisted of multifocal glial nodules, neuronal necrosis, neuronophagia, perivascular cuffing, and multifocal lymphocytic meningitis [79]. In dogs infected with EEEV, the gray matter of the cerebral cortex and midbrain shows random distribution of perivascular cuffs of lymphocytes, plasma cells, and a few histiocytes and neutrophils [75].

Experimental infection of cynomolgus macaques (*Macaca fascicularis*) with EEEV resulted in elevated serum levels of blood urea nitrogen, sodium, and alkaline phosphatase, and prominent leukocytosis. The leukocytes were primarily granulocytes [80]. In addition, the onset and severity of neurological signs were similar to what is seen in human EEEV cases and also in cynomolgus macaques experimentally infected with WEEV [80,81].

In mouse models of WEE, virulent viral strains caused histological lesions in the brain characterized by laminar or multifocal neuronal necrosis, edema, mild infiltrates of lymphocytes in the meninges and subependymal areas, and occasional fibrin thrombi [13]. Less virulent strains caused limited neuronal necrosis and exhibited glial nodules and perivascular cuffs [13]. In adult hamsters infected with WEEV, lethal disease was characterized by tachypnea, conjunctivitis, incoordination, and seizures [82]. Encephalitis affecting all parts of the brain, especially the olfactory regions, was accompanied by high titers of virus in the brain. Neuronal necrosis, infiltration of meninges by lymphocytes, astrocytosis, microgliosis, hemorrhage, and spongy degeneration was observed in brain lesions [82]. Neurologic disease caused by WEEV has been reported in turkeys, emus, and an opossum [83–85]. In emus, clinical signs varied from mild to severe and included anorexia, lethargy with sternal recumbency, ataxia, muscle tremors, head tilt, unnatural positioning of the head on the back, acute onset of paralysis, and lateral recumbency with paddling [84].

29.1.4 Diagnosis

The diagnosis of EEE and WEE is challenging because the clinical signs of the diseases are nonspecific and any of a panel of viral and nonviral agents could be responsible [86]. However, accurate diagnosis is important because although there is no specific treatment available, steps can be taken (e.g., vaccination and vector control) to prevent further spread of disease. Furthermore, surveillance in animals is helpful in preventing spread to humans as frequently animal infections foreworn infections in humans.

The diagnosis of disease in animals is generally based on clinical history, serological testing, virus isolation, molecular testing, and identification of characteristic histopathological changes in the brains of infected animals.

29.1.4.1 Conventional Techniques

The identification of alphaviruses including EEEV and WEEV has traditionally been achieved by inoculation of cell culture or suckling mouse brain and identification of the specific isolate using immunofluorescence assays.

29.1.4.1.1 Virus Isolation

Virus isolation is generally performed on Vero cells since the cell line is easy to culture and most arboviruses can be amplified on these cells. Virus-positive samples are identified further by indirect immunofluorescence assay and/or reverse transcription-PCR (RT-PCR) [87]. EEEV is a select agent and also requires BSL-3 containment. Many diagnostic laboratories are therefore not equipped to isolate EEEV. In animals as well as humans, cells in several organs are infected with EEEV. However, the CNS tissue remains the target tissue for virus isolation and immunohistochemical evaluation.

29.1.4.1.2 Enzyme Immunoassay

The enzyme immunoassay (EIA) procedure is a simple, rapid, and accurate assay for the detection of EEEV antigen in mosquito pools and blood and brain specimens from infected birds [88,89]. For EEEV detection, the protocol involves coating plates with polyclonal mouse EEE-specific capture antibody, addition of blocking solution followed by addition of the test specimen. If there are EEE-specific antigens present in the specimen, they will bind to the polyclonal capture antibody. The addition of polyclonal rabbit EEE-specific detection antibody and goat anti-rabbit IgG conjugated to enzyme followed by a substrate will complete the reaction. During the first 24 h after an infection, virus isolation is more sensitive than an EIA. However, 24 h or longer after infection, the EIA is more sensitive in identifying infected birds [89].

The IgM antibody capture ELISA (MAC-ELISA) test was developed to specifically detect IgM antibody, which is produced early in an immune response [90] and is therefore a valuable method for detecting acute infections. MAC-ELISA assays have been developed for the detection of EEEV and WEEV [91], and the versatility of the assay lends itself to combining multiple assays [92].

Interpretation of serological results based on one serum specimen is difficult without a supporting clinical history or if an animal has been vaccinated.

29.1.4.1.3 Immunohistochemical Assay

Virus antigen can be detected in fresh or formalin-fixed paraffin-embedded tissue using immunohistochemical staining. The assay requires antibody to EEEV antigen, serum-blocking solution, biotinylated secondary antibody, enzyme conjugate, and substrate/chromogen mixture and slides can be examined via light microscopy [79].

29.1.4.1.4 Plaque Reduction Neutralization Test

The plaque reduction neutralization test (PRNT) is considered the gold standard procedure for the identification of arboviral antibody, and protocols are well established [93,94]. The test can help to distinguish false-positive results arising from ELISAs and other serologic assays. The disadvantage of performing PRNTs is that they have a long turnaround time, they are labor intensive, and a live virus is needed for the procedure, and this is a problem for performing testing for EEEV because, as previously discussed, this virus has to be handled in a BSL-3 containment facility with select agent security. Some of the latter issues can be overcome by using attenuated chimeric viruses with equivalent antigenic makeup in a relatively benign backbone such as that of SINV [95].

29.1.4.2 Molecular Techniques

The detection of viral nucleic acid in serum or cerebrospinal fluid has been increasingly used in recent decades for alphavirus diagnosis. The detection of nucleic acid is only effective during the viremic phase of the disease, which generally lasts 3–5 days. As the immune response develops and the virus is cleared, it becomes difficult to detect an infection by molecular methods, and therefore after the viremic phase, a negative result using a molecular test is not an indication of absence of a particular infection.

Various molecular methods including PCR and real-time PCR have been used for the detection of EEE and WEE viruses. The advantages of using molecular methods such as PCR are their superior speed, sensitivity, and specificity. The disadvantages are that virus will only be detected during the viremic stage, as discussed previously, and that depending on the primers used, different strains of the viruses may not be detected. For this reason, primers and probes should be designed in conserved regions of the genome. Even so, viral genomes, especially those of RNA viruses, mutate, thus making assay design challenging.

One strategy for ruling in or ruling out alphavirus infection is to perform a genus-specific RT-PCR assay followed by species-specific RT-PCR or sequencing to determine the species. Pfeffer et al. [96] developed an alphavirus genus-specific seminested RT-PCR assay, which targets the NSP1 gene. An alternative generic nested PCR assay targeted the NSP4 gene and was followed by sequencing to detect and identify members of the *Alphavirus* genus [97]. Of late, a generic alphavirus diagnostic assay has been developed that includes more recently identified alphavirus strains and targeted the NSP4 region of the genome [98]. The limit of detection for the assay was 50 gene copies for EEEV and 5 gene copies for WEEV [98].

In the 1990s, several PCR assays for the detection of EEE and WEE viruses were designed, but they were unable to distinguish between closely related heterologous alphaviruses [99–101]. A more sensitive and specific RT-PCR assay was developed that targeted the E2 gene of EEEV and WEEV [102]. The EEE RT-PCR assay was a nested PCR assay and detected approximately 30 RNA molecules, and the WEE assay was a seminested PCR assay that detected approximately 20 RNA molecules [102].

Real-time RT-PCR assays have been developed for the detection of alphaviruses. Real-time RT-PCR assays are rapid and have improved sensitivity compared to traditional methods such as conventional PCR, IFA, ELISA, and virus isolation. In addition, multiplexing is possible as different target sequences can be detected simultaneously using several fluorophores within the same reaction. Despite the possibility for multiplexing, in general, real-time RT-PCR diagnostic assays for WEEV and EEEV are not multiplexed together as the two viruses occur in geographically distinct areas, which do not overlap. One of the exceptions is a duplex real-time PCR assay targeting the NSP3 sequence of WEEV and the E3 sequence (encoding an envelope protein) of EEEV, which was developed in China [103]. The assay had a sensitivity of one copy/reaction for each virus, which was 10 times more sensitive than virus isolation [103]. No cross-reaction was observed against other encephalitic viruses including the Japanese encephalitis virus, tick-borne encephalitis virus, Chikungunya virus, dengue virus (DENV), yellow fever virus, and Saint Louis encephalitis virus (SLEV) [103].

The emergence of the West Nile virus (WNV) in the United States in 1999 led to the development of a multiplex nested RT-PCR assay to detect EEEV and WNV in horse tissue and tissues of other mammalian and bird species [104]. The primers targeted the highly conserved E2 region of EEEV and resulted in the amplification of a 565 base pair (bp) followed by a 140 bp region, which could be detected by gel electrophoresis [104]. The assay was 100-fold more sensitive than cell culture and did not cross-react with the related alphaviruses WEEV or VEEV [104].

In 2003, a comparison of three molecular methods showed that nucleic acid sequence-based amplification (NASBA), standard RT-PCR, and TaqMan real-time PCR were all useful in detecting WEEV and EEEV in mosquito and vertebrate tissue samples [105]. All three methods were shown to be reproducible and specific with the TaqMan assay being the most sensitive, followed by the NASBA assay [105]. It should be noted that sensitivity was determined using plaque forming unit (pfu) values rather than copy number and is therefore not entirely accurate. With a turnaround time of 4 h, both the TaqMan and the NASBA assays are rapid, especially compared to traditional methods of virus isolation and immunofluorescence, which can take approximately 1 week to complete [105]. In 2011, the EEEV TaqMan assay failed to detect four isolates of EEEV in Connecticut due to mutations in the probe binding sites, indicating that a single nucleotide substitution within the probe binding site may lead to TaqMan assay failure [106]. In order to improve the reliability of molecular detection of rapidly evolving viruses, it is therefore advantageous to develop multitarget TaqMan assays and create backup targets for virus detection. In addition, assays should be continually monitored and redesigned if mutations in viral targets arise too frequently [106].

A multilocus RT-PCR assay for the detection of New and Old World alphaviruses was reported in 2007 [107]. The resultant amplicons were analyzed by electrospray ionization mass spectrometry and base compositions of the amplicons were used to assign the viral subtype [107]. This assay is an accurate, rapid, high-throughput assay that can be used for surveillance purposes and has the added advantage of being able to detect previously uncharacterized alphaviruses.

Multiplex assays to detect EEEV with other arboviruses have also been developed. For example, a duplex assay to detect EEEV with SLEV, which targeted the conserved E1 gene of EEEV, was developed [108]. This assay was specific and reproducible and had a sensitivity of 5 gene copies [108]. The control used for the assay was a short RNA transcript that allowed the assay to be performed in a BSL-2 laboratory overcoming the requirement to handle genomic EEEV RNA in a BSL-3 environment by nonselect agent approved personnel [108]. A quadruplex real-time PCR assay targeting the envelope and nonstructural protein of both EEEV and WNV is regularly used in NYS to perform surveillance for these two viruses in mosquito and mammalian specimens, and is also specific and reproducible [109].

Commercially available kits such as the BioinGentech Veterinary PCR kits (Concepcion, Chile) are available for the detection of both EEEV and WEEV RNA in animals. These kits include reagents for the isolation of RNA as well as standard RT-PCR reagents and controls. They are reported as being reliable and accurate, and with a turnaround time of 3 h, they are rapid and therefore convenient for performing diagnostic testing in horses.

29.2 METHODS

29.2.1 SAMPLE PREPARATION

29.2.1.1 Animal Specimens

Animal samples (tissue and serum) for nucleic acid testing must be collected and either shipped immediately on a cold pack to the laboratory for testing, or frozen and then shipped frozen to the laboratory. Samples must not be kept at room temperature for long periods of time nor be repeatedly freeze-thawed; either treatment will compromise the test results.

29.2.1.2 Mosquito Specimen Collection for Surveillance

CDC light traps baited with dry ice, and gravid traps, are most commonly used to collect adult *Cx. tarsalis* for WEEV, and predominantly *Cs. melanura* for EEEV detection, during the transmission season. Trapped mosquitoes must be transported to the laboratory, where they should be immobilized by chilling for subsequent identification, or anesthetized with Trimethylamine™ and sorted according to genus, species, and sex. Identified mosquitoes should be grouped into pools of 50 or fewer (pools of same species and sex) and placed into 2 mL safe-lock microfuge tubes, each containing one 4.5 mm diameter zinc-plated steel BB, and shipped to the testing laboratory on dry ice [110].

29.2.1.3 Sample Processing

The methods used for processing of either animal or surveillance specimens for analysis are based on specimen type and assay. Except for serum samples, the initial step usually involves homogenization of material. For RT-PCR assays, RNA is extracted, ideally by homogenization in the presence of a lysis buffer that creates a highly denaturing environment, which will inactivate RNases released from the tissue. However, if the specimen cannot be divided and is to be used for virus isolation as well as RNA purification, it must be homogenized in a medium that will not inactivate the virus. For example, when brain tissue is to be tested by both cell culture and RT-PCR, separate excisions can be processed for the two assays. Mosquito samples, however, cannot be split easily and are therefore generally first homogenized in phosphate-buffered saline (PBS) or culture medium, with aliquots of the homogenate subsequently taken for RNA purification and cell culture assay. Tissue can be homogenized by mixer mill (Retsch, MM 301; Retsch, Inc., Newtown, PA), placing the sample (20 mg or less) in a 2 mL safe-lock microfuge tube (Eppendorf, Westbury, NY) containing two stainless steel beads (5mm; Qiagen, Valencia, CA), and up to 1 mL of PBS, culture medium, or lysis buffer. The tubes are placed in prechilled 24-well adapter racks (TissueLyser adapter set; Qiagen) homogenized for 30 s at 24 cycles/s, and then placed on ice for 5 min. After clarification by centrifuge, the homogenate is ready for virus isolation or purification of viral RNA.

29.2.1.4 Nucleic Acid Extraction

Multiple methods for RNA extraction are in use depending on the preference of the individual laboratory. TRIzol LS (Life Technologies, Gaithersburg, MD) has been used for many years and is a reliable method for extraction of RNA. In addition, many kits are available, including the QIAamp Viral RNA mini kit (Qiagen), MagMAX viral RNA Isolation kit (Life Technologies), MasterPure RNA Extraction kit (Epicentre,, Madison, WI), E.Z.N.A. Viral RNA kit (Omega Bio-Tek, Norcross, GA), High Pure Viral Nucleic Acid Kit (Roche Diagnostics, Indianapolis, IN), and ZR Viral RNA kit (Zymo Research, Irvine, CA). RNA extraction can also be performed with fully automated instruments including m2000 (Abbott Laboratories, Abbott Park, IL), MagNA Pure LC 2.0 (Roche Diagnostics), BioRobot M48 (Qiagen), MagMax Express Magnetic Particle Processor (Life Technologies), and NucliSENS easyMAG (bioMerieux, Durham, NC). Automation of the extraction procedure reduces technician hands-on time and errors and can minimize contamination issues. Automation is obviously suitable for high-throughput processing.

At least one negative extraction control should be included in each set of samples to be extracted for detection of potential cross-contamination. An internal control must also be used for the extraction process to identify inefficient nucleic extraction and possible PCR inhibition. Some specimen types such as blood and serum are inhibitory to PCR; for these matrices, a negative result in the viral assay does not necessarily denote absence of viral RNA. In such cases, inhibition of the internal control PCR will alert the technician. Additionally, occurrence of a Ct value outside the acceptable range for the internal control will draw attention to errors that could have arisen in the nucleic acid extraction process.

We have developed a method based on spiking a lysed sample before the extraction process with a known amount of transcript RNA from a portion of the green fluorescent protein (GFP) gene [108]. Once the nucleic acid is extracted, a real-time RT-PCR assay is performed to detect the GFP RNA [108]. Test results are valid if the Ct value of the GFP real-time RT-PCR assay falls within the predefined range.

For extraction of nucleic acid from tissue samples, including animal tissue and mosquito homogenate, we use the MagMax RNA Viral Isolation kit (Life Technologies) on a Tecan Evo 150 liquid handler (Tecan, Morrisville, NC). Following homogenization and centrifugation, RNA is extracted from an aliquot of the homogenate according to the manufacturer's instructions in a final elution volume of 50 µL. To minimize RNA degradation, samples should be kept chilled during the various stages of the procedure.

29.2.2 Detection Procedures

29.2.2.1 Real-Time RT-PCR for Detection of EEEV

The assay described here was published in 2013 and targets conserved regions of E1 and NS1 of EEEV [109]. As published, the assay can also include primers and probes to detect WEEV but the absence of these components will not negatively impact the assay. The assay is performed using two sets of primers and probes targeting EEEV (Table 29.1). The E1 probe was a minor groove binder labeled with 6-carboxyfluorescein (6-FAM) at the 5′ end, and the NS1 probe is labeled with Cyanine 5 and a black hole quencher (BHQ) at the 3′ end. To eliminate the possibility that the presence of enzyme inhibitors in the sample contributes to negative PCR results, a separate real-time RT-PCR for GFP should be performed with the primers and probes (GFP forward, GFP reverse, and GFP probe) listed in Table 29.1 [111]. The GFP probe is labeled with the reporter 6-FAM at the 5′ end and the quencher 6-carboxy-tetramethyl-rhodamine [TAMRA] at the 3′ end.

29.2.2.1.1 Procedure

1. Prepare the reaction mix using 10 µL mastermix from Quanta Biosciences (Gaithersburg, MD) qScript XLT One-Step qRT-PCR Toughmix kit, 5 µL template, 0.3 µL of 100 µmol/L each primer and 0.06 µL of 25 µmol/L probe, and deionized water to a final volume of 25 µL [109].
2. Incubate the mix at 50°C for 5 min, followed by 30 s at 95°C then 45 cycles of 95°C for 10 s, and 60°C for 1 min on an ABI 7500 real-time PCR machine (Life Technologies).
3. Prepare a separate RT-PCR mix consisting of universal buffer (Life Technologies), the forward and reverse primers at 900 nM each, and 250 nM probe for detection of GFP [111].
4. Incubate the mix at 48°C for 30 min, followed by 95°C for 10 min, 45 cycles of 95°C for 15 s, and 60°C for 1 min on an ABI 7500 or 7900 instrument (Life Technologies).

TABLE 29.1

Primers and Probes Used in Real-Time RT-PCR Assays for EEEV and WEEV and for Internal Control Detection

Primer/Probe	Sequence (5' → 3')	Nucleotide Start	Reference
EEE-E1-F	CCCTAGTTCGATGTACTTCCG	11,365	Zink et al. [109]
EEE-E1-R	GCATTATGCACTGCCCTTAG	11,476	Zink et al. [109]
EEE-E1-FAM	6-FAM-CCGCCGATGCAGTG-MGB	11,387	Zink et al. [109]
EEV-NS1-F	ATGAAGAGCGCAGAAGACCC	273	Zink et al. [109]
EEV-NS1-R	GCGTCGACATTACTGTTAGC	393	Zink et al. [109]
EEV-NS1-CY5	CY5-CAGACTCTACCGCTACGCAGACAAG-BHQ	296	Zink et al. [109]
WEE 10,248	CTGAAAGTCGGCCTGCGTAT	10,248	Lambert et al. [105]
WEE 10,314c	CGCCATTGACGAACGTATCC	10,314	Lambert et al. [105]
WEE probe	6-FAM-ATACGGCAATACCACCGCGCACC-BHQ1	10,271	Lambert et al. [105]
GFP forward	CACCCTCTCCACTGACAGAAAT	549	Tavakoli et al. [111]
GFP reverse	TTTCACTGGAGTTGTCCCAATTC	470	Tavakoli et al. [111]
GFP probe	6-FAM-TGTGCCCATTAACATCACCATCTAATTCAACA-TAMRA	525	Tavakoli et al. [111]

EEEV, eastern equine encephalitis virus; GFP, green fluorescent protein; 6-FAM, 6-carboxyfluorescein; MGB, minor groove binder (Applied Biosystems); BHQ1, black hole quencher; TAMRA, 6-carboxy-tetramethyl-rhodamine. EEEV sequence is from GenBank accession number KJ659366 (EEEV/Culicidae/USA/2708/2007). WEEV sequence is from GenBank accession number AF214040 (strain 71V_1658). GFP sequence is from GenBank accession number EU341596 (cloning vector pGFPm-T).

Note: EEEV is a select agent and its RNA is infectious; therefore, work with culture or RNA must be done in a select agent–approved BSL-3 laboratory. For construction of a positive control for the assay suitable for use in a BSL-2 laboratory where diagnostic assays are performed on patient or animal specimens, the target sequence for the real-time PCR assay can be cloned into the PCR-Blunt II-TOPO plasmid (Life Technologies) [108]. The plasmid can be used to transcribe a control RNA transcript containing the real-time PCR target sequence in the RiboMax large-scale RNA production system—T7 (Promega, Madison, WI) [108]. The purified transcript should be quantified by measurement of the absorbance at A260 nm. A negative RT-PCR control in which water is substituted for sample should be included in each PCR sample set, for the detection of potential PCR cross-contamination.

The EEEV real-time assay described here is a rapid, accurate method for performing high-throughput vector surveillance and has the added advantage of targeting two independent regions of the EEEV genome, thus minimizing false-negative results due to strain variations [109].

29.2.2.2 Real-Time RT-PCR for Detection of WEEV

Lambert et al. [105] described a TaqMan real-time RT-PCR assay for the detection of WEEV; they used a primer/probe set that targets the E1 region of WEEV (Table 29.1). The WEEV probe was labeled at the 5' end with the 6-FAM reporter dye and at the 3' end with BHQ1 quencher molecule.

29.2.2.2.1 Procedure

1. Prepare RT-PCR mix (50 µL) using TaqMan One-Step RT-PCR master mix (Life Technologies) with 1 µM of each primer, 200 nM of probe, and 5 µL of extracted RNA [105]. The negative control consists of water in place of extracted RNA.

2. Incubate the mix at 48°C for 30 min for RT, 95°C for 10 min, and 45 cycles of 95°C for 15 s and 60°C for 1 min on an ABI 7700 instrument in the original method (in our laboratory an ABI 7500 instrument is used).

Note: WEE standards, prepared from RNA extracted from WEE stock that was amplified in and titered on Vero cells, were divided into aliquots, stored at −80°C, and used in each assay at sequential tenfold dilutions equivalent to 1000–0.01 pfu. The WEEV real-time RT-PCR assay detected <0.1 pfu of virus and did not cross-react with any of the related alphaviral or unrelated arboviral RNA tested [105].

29.2.2.3 Standard RT-PCR

Laboratories that do not have the capability of performing real-time RT-PCR can conduct standard RT-PCR, although the assays may not be as sensitive [105]. Standard RT-PCR assays specifically for the detection of WEEV and EEEV have previously been reported and should be performed as documented [102,105,112]. Alternatively, group-specific RT-PCR assays for the detection of members of the *Alphavirus* genus, followed by sequence analysis for identification of members of the group, can be performed [96,97].

In the following, we present two nested PCR assays for the detection of EEEV and WEEV that were reported originally by Linssen et al. [102]. The primers for these assays were designed from the sequences of the structural polyprotein coding regions. For EEEV detection, the RT-PCR primers are EEE-4 and cEEE-7 and the nested PCR primers are EEE-5 and cEEE-6 (Table 29.2). For WEEV detection, the RT-PCR primers are WEE-1 and cWEE-3, and the nested PCR primers are WEE-2 and cWEE-3 (Table 29.2).

TABLE 29.2
Primers for Nested PCR Detection of EEEV and WEEV

Primer	Sequence (5′ → 3′)	Nucleotide Position	Expected Product (bp)	Specificity
EEE-4	CTAGTTGAGCACAAACACCGCA	9377 (E2)	464	
cEEE-7	CACTTGCAAGGTGTCGTCTGCCCTC	9817 (E2)		
EEE-5	AAGTGATGCAAATCCAACTCGAC	9457 (E2)	262	EEEV
cEEE-6	GGAGCCACACGGATGTGACACAA	9697 (E2)		
WEE-1	GTTCTGCCCGTATTGCAGACACTCA	1157 (E2)	354	
cWEE-3	CCTCCTGATCTTTTTCTCCACG	1490 (E2)		
WEE-2	GTCTTTCGACCACGACCATG	1316 (E2)	195	WEEV
cWEE-3	As above	As above		

Source: Adapted from Linssen, B. et al., *J. Clin. Microbiol.*, 38, 1527, 2000. With permission.

29.2.2.3.1 Procedure

1. Prepare the RT-PCR mix (100 μL) containing 5 mM dithiothreitol, 10 mM Tris-HCl (pH 8.3), 50 mM KCl, 0.01% gelatin, 2 mM MgCl₂, 200 μM dNTP, 2.5 U of TaqExtender PCR additive (Agilent Technologies, Santa Clara, CA), 2 U of RAV-2 reverse transcriptase (GE Healthcare Bio-sciences, Pittsburgh, PA), 2 U of AmpliTaq DNA polymerase (Life technologies), 0.1 μM each of EEE-4, and cEEE-7 primers (for EEEV detection); or 0.2 μM each of WEE-1 and cWEE-3 primers (for WEEV detection) and 5 μL of RNA template (extracted by Qiagen kit).

2. For EEEV detection, incubate the RT-PCR mix at 50°C for 30 min, 1 cycle of 94°C for 90 s, 64°C for 90 s, and 72°C for 90 s; 35 cycles of 94°C for 20 s, 64°C for 30 s, and 72°C for 20 s; and the final step, 72°C for 5 min. For WEEV detection, incubate the RT-PCR mix at 50°C for 30 min; 1 cycle of 94°C for 90 s, 68°C for 60 s, and 72°C for 90 s; 35 cycles of 94°C for 20 s, 68°C for 30 s, and 72°C for 17 s; and a final 72°C for 5 min.

3. Prepare the nested PCR (100 μL) containing 10 mM Tris-HCl (pH 8.3), 50 mM KCl, 0.01% gelatin, 5 mM dithiothreitol, 2 mM MgCl₂, 200 μM dNTP, 1.5 U of AmpliTaq DNA polymerase, 0.3 μM each of EEE-5 and cEEE-6 primers (for EEEV detection), or 0.3 μM each of WEE-2, cWEE-3 primers (for WEEV detection), and 2 μL of cDNA template from the previous RT-PCR.

4. For EEEV detection, carry out the nested PCR with 1 cycle of 94°C for 90 s; 25 cycles of 94°C for 20 s, 65°C for 35 s, and 72°C for 17 s; and a final 72°C for 4 min. For WEEV detection, carry out the nested PCR with 1 cycle of 94°C for 90 s; 25 cycles of 94°C for 20 s, 63°C for 35 s, and 72°C for 15 s; and a final 72°C for 4 min.

5. Electrophorese the nested PCR products on 2% agarose gel, and visualize with ethidium bromide stain.

Note: The expected nested PCR products for EEEV and WEEV are 262 and 195 bp, respectively. The sensitivity of the nested PCR assay is 30 (EEEV) or 20 (WEEV) RNA molecules [102].

29.3 CONCLUSION AND FUTURE PERSPECTIVES

EEE and WEE, although rare, are important arbovirus diseases affecting humans and animals in the United States. In recent years, cases of WEE have diminished. However, EEE, which has a greater mortality rate, continues to cause severe disease accompanied by long-term brain damage and death in humans and horses in the Atlantic and Gulf coast states. Many states in the United States have disease surveillance programs, which serve as early warning systems in the detection of mosquito-borne viruses that can infect animals and humans. Mosquito testing, baited traps, and wild bird and sentinel chicken testing all aid in the early detection of mosquito-transmitted diseases such as EEE and WEE. The control of mosquito populations is critical in protecting humans from infectious bites.

Several vaccines are under development for EEE and WEE although there are none currently available for humans. Equine vaccines are in use in some countries and help reduce infections in horses. Inactivated-virus vaccines derived from a North American strain of EEEV [113] and WEEV [114] are both available for veterinary use. Safety is a concern because inactivated-virus vaccines have the potential to cause disease if the virus used in the manufacture of the vaccine is not completely inactivated. Individuals such as laboratory workers, who are at high-risk of infection at work, can be immunized against both viruses, but the vaccines are not available for the general public [115]. Efforts are being made to develop recombinant vaccines, which are safer to use and which induce rapid and effective humoral and cellular immunity [116]. In addition, the potential use of recombinant, chimeric WEEV and EEEV as the basis for vaccines is under investigation [117].

There is an urgent need to develop rapid diagnostic methods for the detection of arboviruses in human and other animal hosts for testing of vector mosquito populations. Due to the small sample volume in some cases, multiplexed testing would be advantageous. Multiplexing would reduce hands-on time, the number of analytical instruments required, and reagent costs. The types of technology that allow multiplexing include multiplex real-time RT-PCR, microarrays, and xMAP technology [118,119]. More recently, massively parallel DNA-sequencing platforms, also termed next-generation sequencing technologies, are being used more frequently in diagnostic laboratories for pathogen discovery and diagnostic applications [120,121]. Whole genome sequencing and phylogenetic analysis have been used to detect HJV in the brain of a Mississippi sandhill crane (*Grus canadensis*) [122]. The feasibility of using pyrosequencing to detect DENV in *Aedes aegypti* mosquito pools was investigated and the method was found to be sufficiently sensitive to perform arbovirus surveillance [123]. Similarly, massively parallel sequencing using the Personal Genome Machine (PGM; Life Technologies) has demonstrated the potential for this technology in assessing arbovirus ecology and evaluating novel control strategies [124]. No doubt with the availability of these advanced molecular detection methodologies, the routine diagnostic testing for arboviruses such as EEEV and WEEV will be changing dramatically in the years to come.

ACKNOWLEDGMENTS

We thank Elizabeth Kauffman and Mary Franke of the Wadsworth Center for their invaluable assistance with the references and formatting of this chapter.

REFERENCES

1. Matthews, R. E. F., Classification and nomenclature of viruses; fourth report of the International Committee on Taxonomy of Virus, *Intervirology*, 17, 1, 1982.
2. Westaway, E. G. et al., Togaviridae, *Intervirology*, 24, 125, 1985.
3. Calisher, C. H. et al., Reevaluation of the western equine encephalitis antigenic complex of alphaviruses (family *Togaviridae*) as determined by neutralization tests, *Am. J. Trop. Med. Hyg.*, 38, 447, 1988.
4. Casals, J., Antigenic variants of Eastern equine encephalitis virus, *J. Exp. Med.*, 119, 547, 1964.
5. Arrigo, N. C. et al., Evolutionary patterns of eastern equine encephalitis virus in North versus South America suggest ecological differences and taxonomic revision, *J. Virol.*, 84, 1014, 2010.
6. Weaver, S. C. et al., Evolution of alphaviruses in the eastern equine encephalomyelitis complex, *J. Virol.*, 68, 158, 1994.
7. von Sprockhoff, H. and Ising, E., On the presence of viruses of the American equine encephalomyelitis in Central Europe. Review, *Arch. Gesamte Virusforsch.*, 34, 371, 1971.
8. Calisher, C. H. and Karabatsos, N., Arbovirus serogroups definition and geographic distribution, in *The Arboviruses: Epidemiology and Ecology*, ed. T. P. Monath (CRC Press, Boca Raton, FL, 1988), p. 19.
9. Kramer, L. D. and Fallah, H. M., Genetic variation among isolates of western equine encephalomyelitis virus from California, *Am. J. Trop. Med. Hyg.*, 60, 708, 1999.
10. Hahn, C. S. et al., Western equine encephalitis virus is a recombinant virus, *Proc. Natl. Acad. Sci. USA*, 85, 5997, 1988.
11. Nagata, L. P. et al., Infectivity variation and genetic diversity among strains of Western equine encephalitis virus, *J. Gen. Virol.*, 87, 2353, 2006.
12. Forrester, N. L. et al., Western Equine Encephalitis submergence: Lack of evidence for a decline in virus virulence, *Virology*, 380, 170, 2008.
13. Logue, C. H. et al., Virulence variation among isolates of Western equine encephalitis virus in an outbred mouse model, *J. Gen. Virol.*, 90(Pt 8), 1848–1858, 2009.
14. Bianchi, T. I. et al., Western equine encephalomyelitis: Virulence markers and their epidemiologic significance, *Am. J. Trop. Med. Hyg.*, 49, 322, 1993.
15. Strauss, E. G. and Strauss, J. H., Structure and replication of the alphavirus genome, in *The Togaviruses and Flaviviruses*, eds. S. Schlesinger and M. Schlesinger (Plenum Press, New York, 1986), p. 35.
16. Ou, J. H. et al., Sequence studies of several alphavirus genomic RNAs in the region containing the start of the subgenomic RNA, *Proc. Natl. Acad. Sci. USA*, 79, 5235, 1982.
17. Bracha, M. and Schlesinger, M. J., Inhibition of Sindbis virus replication by zinc ions, *Virology*, 72, 272, 1976.
18. Rice, C. M. and Strauss, J. H., Association of sindbis virion glycoproteins and their precursors, *J. Mol. Biol.*, 154, 325, 1982.
19. Dalrymple, J. M. et al., Antigenic characterization of two sindbis envelope glycoproteins separated by isoelectric focusing, *Virology*, 69, 93, 1976.
20. Garoff, H. et al., Nucleotide sequence of cDNA coding for Semliki Forest virus membrane glycoproteins, *Nature*, 288, 236, 1980.
21. Hanson, R. P., An epizootic of equine encephalomyelitis that occurred in Massachusetts in 1831, *Am. J. Trop. Med. Hyg.*, 6, 858, 1957.
22. TenBroeck, C. and Merrill, M. H., Transmission of eastern equine encephalomyelitis, *Proc. Soc. Exp. Biol. Med.*, 31, 217, 1933.
23. Merrill, M. H. et al., Mosquito transmission of equine encephalomyelitis, *Science*, 80, 251, 1934.
24. TenBroeck, C. and Merrill, M. H., Transmission of equine encephalomyelitis by mosquitoes, *Am. J. Pathol.*, 11, 847, 1935.
25. Davis, W. A., A study of birds and mosquitoes as hosts for the virus of Eastern equine encephalomyelitis, *Am. J. Hyg.*, 32, 45, 1940.
26. Webster, L. T. and Wright, F. H., Recovery of Eastern equine encephalomyelitis virus from brain tissue of human cases of encephalitis in Massachusetts, *Science*, 88, 305, 1938.
27. Fothergill, L. D. et al., Human encephalitis caused by the virus of the Eastern variety of equine encephalomyelitis, *N. Engl. J. Med.*, 219, 411, 1938.
28. Howitt, B. F. et al., Recovery of the virus of Eastern equine encephalomyelitis from mosquitoes (*Mansonia perturbans*) collected in Georgia, *Science*, 110, 141, 1949.
29. Chamberlain, R. W. et al., Recovery of virus of Eastern equine encephalomyelitis from a mosquito, *Culiseta melanura* (Coquillett), *Proc. Soc. Exp. Biol. Med.*, 77, 396, 1951.
30. Kissling, R. E. et al., Recovery of virus of Eastern equine encephalomyelitis from blood of a purple grackle, *Proc. Soc. Exp. Biol. Med.*, 77, 398, 1951.
31. Kissling, R. E. et al., Studies on the North American arthropod-borne encephalitides. III. Eastern equine encephalitis in wild birds, *Am. J. Hyg.*, 62, 233, 1955.

32. Tyzzer, E. E. et al., The occurrence in nature of equine encephalomyelitis in the ring-necked pheasant, *Science*, 88, 505, 1938.

33. Beaudette, F. R. and Black, J. J., Equine encephalomyelitis in New Jersey pheasants in 1945 and 1946, *J. Am. Vet. Med. Assoc.*, 112, 140, 1948.

34. Moulthrop, I. M. and Gordy, B. A., Eastern viral encephalomyelitis in chukar (*Alectoris graeca*), *Avian Dis.*, 4, 247, 1960.

35. Dougherty, E., III and Price, J. I., Eastern encephalitis in white Peking ducklings on Long Island, *Avian Dis.*, 4, 247, 1960.

36. Morris, C. D., Eastern equine encephalomyelitis, in *The Arboviruses: Epidemiology and Ecology*, ed. T. P. Monath (CRC Press, Inc, Boca Raton, FL, 1988), p. 1.

37. Griffin, D. E., Alphaviruses, in *Fields Virology*, eds. D. M. Knipe and P. M. Howley (Lippincott, Williams & Wilkins, Philadelphia, PA, 2001), p. 917.

38. CDC, Confirmed and probable eastern equine encephalitis cases, human, United States, 1964–2007.

39. Ortiz, D. I. et al., Isolation of EEE virus from *Ochlerotatus taeniorhynchus* and *Culiseta melanura* in Coastal South Carolina, *J. Am. Mosq. Control Assoc.*, 19, 33, 2003.

40. Beckwith, W. H. et al., Isolation of eastern equine encephalitis virus and West Nile virus from crows during increased arbovirus surveillance in Connecticut, 2000, *Am. J. Trop. Med. Hyg.*, 66, 422, 2002.

41. Schmitt, S. M. et al., An outbreak of eastern equine encephalitis virus in free-ranging white-tailed deer in Michigan, *J. Wildl. Dis.*, 43, 635, 2007.

42. Cupp, E. W. et al., Transmission of eastern equine encephalomyelitis in central Alabama, *Am. J. Trop. Med. Hyg.*, 68, 495, 2003.

43. Hachiya, M. et al., Human eastern equine encephalitis in Massachusetts: Predictive indicators from mosquitoes collected at 10 long-term trap sites, 1979–2004, *Am. J. Trop. Med. Hyg.*, 76, 285, 2007.

44. Grady, G. F. et al., Eastern equine encephalitis in Massachusetts, 1957–1976. A prospective study centered upon analyses of mosquitoes, *Am. J. Epidemiol.*, 107, 170, 1978.

45. Letson, G. W. et al., Eastern equine encephalitis (EEE): A description of the 1989 outbreak, recent epidemiologic trends, and the association of rainfall with EEE occurrence, *Am. J. Trop. Med. Hyg.*, 49, 677, 1993.

46. Scott, T. W. and Weaver, S. C., Eastern equine encephalomyelitis virus: Epidemiology and evolution of mosquito transmission, *Adv. Virus Res.*, 37, 277, 1989.

47. Dalrymple, J. M. et al., Ecology of arboviruses in a Maryland freshwater swamp. 3. Vertebrate hosts, *Am. J. Epidemiol.*, 96, 129, 1972.

48. LeDuc, J. W. et al., Ecology of arboviruses in a Maryland freshwater swamp. II. Blood feeding patterns of potential mosquito vectors, *Am. J. Epidemiol.*, 96, 123, 1972.

49. Komar, N. et al., Eastern equine encephalitis virus in birds: Relative competence of European starlings (*Sturnus vulgaris*), *Am. J. Trop. Med. Hyg.*, 60, 387, 1999.

50. McLean, R. G. et al., Experimental infection of wading birds with eastern equine encephalitis virus, *J. Wildl. Dis.*, 31, 502, 1995.

51. Turell, M. J. et al., Isolation of viruses from mosquitoes (Diptera: Culicidae) collected in the Amazon Basin region of Peru, *J. Med. Entomol.*, 42, 891, 2005.

52. Scherer, W. F. et al., Serologic surveys for the determination of antibodies against the Eastern, Western, California and St. Louis encephalitis and dengue 3 arboviruses in Central America, 1961–1975, *Bol. Oficina Sanit. Panam.*, 87, 210, 1979.

53. Dietz, W. H., Jr. et al., Eastern equine encephalomyelitis in Panama: The epidemiology of the 1973 epizootic, *Am. J. Trop. Med. Hyg.*, 29, 133, 1980.

54. Aguilar, P. V. et al., Endemic eastern equine encephalitis in the Amazon region of Peru, *Am. J. Trop. Med. Hyg.*, 76, 293, 2007.

55. Sabattini, M. S. et al., Arbovirus investigations in Argentina, 1977–1980. I. Historical aspects and description of study sites, *Am. J. Trop. Med. Hyg.*, 34, 937, 1985.

56. Reisen, W. K. and Monath, T. P., Western equine encephalitis, in *The Arboviruses: Epidemiology and Ecology*, ed. T. P. Monath (CRC Press, Boca Raton, FL, 1988), p. 89.

57. Meyer, K. F. et al., The etiology of epizootic encephalomyelitis of horses in the San Joaquin Valley, 1930, *Science*, 74, 227, 1931.

58. Johnston, R. E. and Peters, C. J., Alphaviruses, in *Fields Virology*, eds. B. N. Fields et al. (Lippincott, Williams & Wilkins, Philadelphia, PA, 1996), p. 843.

59. Howitt, B., Recovery of the virus of equine encephalomyelitis from the brain of a child, *Science*, 88, 455, 1938.

60. Gahlinger, P. M. et al., Air conditioning and television as protective factors in arboviral encephalitis risk, *Am. J. Trop. Med. Hyg.*, 35, 601, 1986.

61. Calisher, C. H., Medically important arboviruses of the United States and Canada, *Clin. Microbiol. Rev.*, 7, 89, 1994.

62. Karabatsos, N. et al., Identification of Highlands J virus from a Florida horse, *Am. J. Trop. Med. Hyg.*, 39, 603, 1988.

63. Calisher, C. H. et al., Characterization of Fort Morgan virus, an alphavirus of the western equine encephalitis virus complex in an unusual ecosystem, *Am. J. Trop. Med. Hyg.*, 29, 1428, 1980.

64. Padhi, A. et al., Phylogeographical structure and evolutionary history of two Buggy Creek virus lineages in the western Great Plains of North America, *J. Gen. Virol.*, 89, 2122, 2008.

65. Pfeffer, M. et al., Phylogenetic analysis of Buggy Creek virus: Evidence for multiple clades in the Western Great Plains, United States of America, *Appl. Environ. Microbiol.*, 72, 6886, 2006.

66. Zacks, M. A. and Paessler, S., Encephalitic alphaviruses, *Vet. Microbiol.*, 140, 281, 2010.

67. Tully, T. N., Jr. et al., Eastern equine encephalitis in a flock of emus (*Dromaius novaehollandiae*), *Avian Dis.*, 36, 808, 1992.

68. Dein, F. J. et al., Mortality of captive whooping cranes caused by eastern equine encephalitis virus, *J. Am. Vet. Med. Assoc.*, 189, 1006, 1986.

69. Veazey, R. S. et al., Pathology of eastern equine encephalitis in emus (*Dromaius novaehollandiae*), *Vet. Pathol.*, 31, 109, 1994.

70. Whiteman, C. E.; Bickford, A. A. *Avian Disease Manual*, 3rd edn. (Kendal/Hunt, Dubuque, IA, 1989).

71. Williams, J. E. et al., Wild birds as eastern (EEE) and western (WEE) equine encephalitis sentinels, *J. Wildl. Dis.*, 7, 188, 1971.

72. Weinack, O. M. et al., Pheasant susceptibility at different ages to Eastern encephalitis virus from various sources in Massachusetts, *Avian Dis.*, 22, 378, 1978.

73. McGee, E. D. et al., Eastern equine encephalomyelitis in an adult cow, *Vet. Pathol.*, 29, 361, 1992.

74. Bauer, R. W. et al., Naturally occurring eastern equine encephalitis in a Hampshire wether, *J. Vet. Diagn. Invest.*, 17, 281, 2005.

75. Farrar, M. D. et al., Eastern equine encephalitis in dogs, *J. Vet. Diagn. Invest.*, 17, 614, 2005.

76. Steele, K. E. and Twenhafel, N. A., Review paper: Pathology of animal models of alphavirus encephalitis, *Vet. Pathol.*, 47, 790, 2010.

77. Vogel, P. et al., Early events in the pathogenesis of eastern equine encephalitis virus in mice, *Am. J. Pathol.*, 166, 159, 2005.

78. Paessler, S. et al., The hamster as an animal model for eastern equine encephalitis—And its use in studies of virus entrance into the brain, *J. Infect. Dis.*, 189, 2072, 2004.

79. Patterson, J. S. et al., Immunohistochemical diagnosis of eastern equine encephalomyelitis, *J. Vet. Diagn. Invest.*, 8, 156, 1996.

80. Reed, D. S. et al., Severe encephalitis in cynomolgus macaques exposed to aerosolized Eastern equine encephalitis virus, *J. Infect. Dis.*, 196, 441, 2007.

81. Reed, D. S. et al., Aerosol exposure to western equine encephalitis virus causes fever and encephalitis in cynomolgus macaques, *J. Infect. Dis.*, 192, 1173, 2005.

82. Zlotnik, I. et al., The pathogenesis of western equine encephalitis virus (W.E.E.) in adult hamsters with special reference to the long and short term effects on the C.N.S. of the attenuated clone 15 variant, *Br. J. Exp. Pathol.*, 53, 59, 1972.

83. Woodring, F. R., Naturally occurring infection with equine encephalomyelitis virus in turkeys, *J. Am. Vet. Med. Assoc.*, 130, 511, 1957.

84. Ayers, J. R. et al., An epizootic attributable to western equine encephalitis virus infection in emus in Texas, *J. Am. Vet. Med. Assoc.*, 205, 600, 1994.

85. Emmons, R. W. and Lennette, E. H., Isolation of western equine encephalitis virus from an opossum, *Science*, 163, 945, 1969.

86. Kennedy, J. S. et al., Induction of human T cell-mediated immune responses after primary and secondary smallpox vaccination, *J. Infect. Dis.*, 190, 1286, 2004.

87. Kauffman, E. B. et al., Virus detection protocols for West Nile virus in vertebrate and mosquito specimens, *J. Clin. Microbiol.*, 41, 3661, 2003.

88. Hildreth, S. W. and Beaty, B. J., Application of enzyme immunoassays (EIA) for the detection of La Crosse viral antigens in mosquitoes, *Prog. Clin. Biol. Res.*, 123, 303, 1983.

89. Scott, T. W. et al., A prospective field evaluation of an enzyme immunoassay: Detection of eastern equine encephalomyelitis virus antigen in pools of *Culiseta melanura*, *J. Am. Vet. Med. Assoc.*, 3, 412, 1987.

90. Goldsby RA et al., Antibodies: Structure and function, in *Immunology*, 5th edn., eds. Goldsby, R. A. et al. (W.H. Freeman and Company, New York, 2003), p. 76.

91. Calisher, C. H. et al., Specificity of immunoglobulin M and G antibody responses in humans infected with eastern and western equine encephalitis viruses: Application to rapid serodiagnosis, *J. Clin. Microbiol.*, 23, 369, 1986.

92. Martin, D. A. et al., Standardization of immunoglobulin M capture enzyme-linked immunosorbent assays for routine diagnosis of arboviral infections, *J. Clin. Microbiol.*, 38, 1823, 2000.

93. Beaty, B. J. et al., Arboviruses, in *Diagnostic Procedures for Viral, Rickettsial and Chlamydial Infections*, eds. N. J. Schmidt and R. W. Emmons (American Public Health Association, Washington, DC, 1989), p. 797.

94. Calisher, C. H. et al., Relevance of detection of immunoglobulin M antibody response in birds used for arbovirus surveillance, *J. Clin. Microbiol.*, 24, 770, 1986.

95. Johnson, B. W. et al., Use of sindbis/eastern equine encephalitis chimeric viruses in plaque reduction neutralization tests for arboviral disease diagnostics, *Clin. Vaccine Immunol.*, 18, 1486, 2011.

96. Pfeffer, M. et al., Genus-specific detection of alphaviruses by a semi-nested reverse transcription-polymerase chain reaction, *Am. J. Trop. Med. Hyg.*, 57, 709, 1997.

97. Sanchez-Seco, M. P. et al., A generic nested-RT-PCR followed by sequencing for detection and identification of members of the alphavirus genus, *J. Virol. Methods*, 95, 153, 2001.

98. Grywna, K. et al., Detection of all species of the genus Alphavirus by reverse transcription-PCR with diagnostic sensitivity, *J. Clin. Microbiol.*, 48, 3386, 2010.

99. Armstrong, P. et al., Sensitive and specific colorimetric dot assay to detect eastern equine encephalitis viral RNA in mosquitoes (Diptera: Culicidae) after polymerase chain reaction amplification, *J. Med. Entomol.*, 32, 42, 1995.

100. Vodkin, M. H. et al., A rapid diagnostic assay for eastern equine encephalomyelitis viral RNA, *Am. J. Trop. Med. Hyg.*, 49, 772, 1993.

101. Linssen, B., Kinney, R., Kaaden, O. R., and Pfeffer, M., *Equine Infectious Diseases VIII* (R&W, Newmarket, Ontario, Canada, 1999).

102. Linssen, B. et al., Development of reverse transcription-PCR assays specific for detection of equine encephalitis viruses, *J. Clin. Microbiol.*, 38, 1527, 2000.

103. Kang, X. et al., A duplex real-time reverse transcriptase polymerase chain reaction assay for detecting western equine and eastern equine encephalitis viruses, *Virol. J.*, 7, 284, 2010.

104. Johnson, A. J. et al., Detection of anti-West Nile virus immunoglobulin M in chicken serum by an enzyme-linked immunosorbent assay, *J. Clin. Microbiol.*, 41, 2002, 2003.

105. Lambert, A. J. et al., Detection of North American eastern and western equine encephalitis viruses by nucleic acid amplification assays, *J. Clin. Microbiol.*, 41, 379, 2003.

106. Armstrong, P. M. et al., Development of a multi-target TaqMan assay to detect eastern equine encephalitis virus variants in mosquitoes, *Vector Borne Zoonotic Dis.*, 12, 872, 2012.

107. Eshoo, M. W. et al., Direct broad-range detection of alphaviruses in mosquito extracts, *Virology*, 368, 286, 2007.

108. Hull, R. et al., A duplex real-time reverse transcriptase polymerase chain reaction assay for the detection of St. Louis encephalitis and eastern equine encephalitis viruses, *Diagn. Microbiol. Infect. Dis.*, 62, 272, 2008.

109. Zink, S. D. et al., Quadruplex qRT-PCR assay for the simultaneous detection of Eastern equine encephalitis virus and West Nile virus, *Diagn. Microbiol. Infect. Dis.*, 77, 129, 2013.

110. Kauffman, E. B. et al., West Nile virus laboratory surveillance program: Cost and time analysis, *Ann. N. Y. Acad. Sci.*, 951, 351, 2001.

111. Tavakoli, N. P. et al., Detection and typing of human herpesvirus 6 by molecular methods in specimens from patients diagnosed with encephalitis or meningitis, *J. Clin. Microbiol.*, 45, 3972, 2007.

112. O'Guinn, M. L. et al., Field detection of eastern equine encephalitis virus in the Amazon Basin region of Peru using reverse transcription-polymerase chain reaction adapted for field identification of arthropod-borne pathogens, *Am. J. Trop. Med. Hyg.*, 70, 164, 2004.

113. Maire, L. F., III et al., An inactivated eastern equine encephalomyelitis vaccine propagated in chick-embryo cell culture. I. Production and testing, *Am. J. Trop. Med. Hyg.*, 19, 119, 1970.

114. Barber, T. L. et al., Efficacy of trivalent inactivated encephalomyelitis virus vaccine in horses, *Am. J. Vet. Res.*, 39, 621, 1978.

115. Bartelloni, P. J. et al., An inactivated eastern equine encephalomyelitis vaccine propagated in chick-embryo cell culture. II. Clinical and serologic responses in man, *Am. J. Trop. Med. Hyg.*, 19, 123, 1970.

116. Wu, J. Q. et al., Complete protection of mice against a lethal dose challenge of western equine encephalitis virus after immunization with an adenovirus-vectored vaccine, *Vaccine*, 25, 4368, 2007.

117. Schoepp, R. J. et al., Recombinant chimeric western and eastern equine encephalitis viruses as potential vaccine candidates, *Virology*, 302, 299, 2002.

118. Palacios, G. et al., Panmicrobial oligonucleotide array for diagnosis of infectious diseases, *Emerg. Infect. Dis.*, 13, 73, 2007.

119. Thiemann, T. C. et al., Development of a high-throughput microsphere-based molecular assay to identify 15 common bloodmeal hosts of Culex mosquitoes, *Mol. Ecol. Resour.*, 12, 238, 2012.

120. Barzon, L. et al., Next-generation sequencing technologies in diagnostic virology, *J. Clin. Virol.*, 58, 346, 2013.

121. Capobianchi, M. R. et al., Next-generation sequencing technology in clinical virology, *Clin. Microbiol. Infect.*, 19, 15, 2013.

122. Ip, H. S. et al., Identification and characterization of Highlands J virus from a Mississippi sandhill crane using unbiased next-generation sequencing, *J. Virol. Methods*, 206, 42, 2014.

123. Bishop-Lilly, K. A. et al., Arbovirus detection in insect vectors by rapid, high-throughput pyrosequencing, *PLoS Negl. Trop. Dis.*, 4, e878, 2010.

124. Hall-Mendelin, S. et al., Detection of arboviruses and other micro-organisms in experimentally infected mosquitoes using massively parallel sequencing, *PLoS One*, 8, e58026, 2013.

30 Venezuelan Equine Encephalitis Virus

María Belén Pisano, Marta Contigiani, and Viviana Ré

CONTENTS

30.1 Introduction 269
    30.1.1 Classification 269
    30.1.2 Virion Structure and Genome Organization 270
    30.1.3 Pathogenesis 270
    30.1.4 Epidemiology and Distribution 271
        30.1.4.1 Epidemic/Epizootic Viruses 271
        30.1.4.2 Enzootic Viruses 272
        30.1.4.3 Origin of Epidemics 272
    30.1.5 Clinical Features 272
    30.1.6 Diagnosis 273
        30.1.6.1 Conventional Techniques 273
        30.1.6.2 Molecular Techniques 273
30.2 Methods 273
    30.2.1 Sample Collection and Preparation 273
    30.2.2 Detection Procedures 274
        30.2.2.1 Reverse Transcription 274
        30.2.2.2 Nested-PCR Detection 274
30.3 Conclusions 275
References 275

30.1 INTRODUCTION

In recent decades, there has been a worldwide resurgence of viral pathogens transmitted by arthropods (arboviruses), particularly those transmitted by mosquitoes. Diseases caused by many of them (e.g., dengue, Venezuelan equine encephalitis virus, Chikungunya) constitute a major health problem of high impact in temperate and subtropical regions of the world [1,2]. The emergence of these viruses is usually associated with variants that have recently evolved, although the current knowledge shows that most are zoonotic arboviruses with specific ecological niches, residing naturally in wild animal species (hosts). Access to other host populations, such as domestic animals and humans, is a consequence of multiple factors, including environmental and ecological changes (relating to the structure of the soil, wildlife, animal migration, vector distribution, human migration, and trafficking of pets) as well as viral RNA genome plasticity.

Venezuelan equine encephalitis viruses (VEEV) are zoonotic pathogens of medical and veterinary importance circulating in the United States. Since its first isolation and identification in the 1930s [3], outbreaks of VEEV in human and equine populations have been periodically recognized in Venezuela, Colombia, Ecuador, Mexico, Trinidad, Panama, Guatemala, Peru, and the United States [2,4]. In humans, VEEV infections range from asymptomatic to encephalitis and death. In many

cases, VEEV is clinically indistinguishable from dengue fever and other arboviral diseases (with symptoms like headache, chills, tachycardia, fever, myalgia, arthralgia, retro-orbital pain, rash, nausea, and vomiting), and confirmatory diagnosis requires the use of specialized laboratory tests [2,5]. In horses, depending on VEEV subtype and strain, infection may produce (1) unapparent infection followed by seroconversion; (2) systemic disease characterized by tachycardia, fever, depression, and anorexia; or (3) encephalitis [6–8]. Recent implementation of surveillance systems has allowed the detection of new human cases in countries and areas with unknown VEEV activity, showing endemic circulation of these viruses [2]. As part of viral control, molecular techniques represent a valuable tool. The potential emergence of epizootic viruses from enzootic progenitors further highlights the need to strengthen the control strategies against the disease [2].

30.1.1 CLASSIFICATION

VEEV form a complex of the same name within the *Alphavirus* genus of the *Togaviridae* family [9]. Due to their shared requirement for arthropods as transmission vectors, the genus *Alphavirus* and several genera of the families *Flaviviridae* and *Bunyaviridae* are collectively known as arthropod-borne viruses or "arboviruses." Members of the genus *Alphavirus* are often classified as either New World alphaviruses or Old World alphaviruses, depending on the geographic location from which they were originally isolated [10]. New World alphaviruses include Eastern equine encephalitis virus (EEEV), VEEV, Western equine encephalitis virus (WEEV), and Mayaro virus (MAYV), among others; and they generally cause encephalitis in humans and other mammals, with the exception of MAYV, which causes febrile illness, arthralgia, and rash. Old World alphaviruses such as Chikungunya virus, O'Nyong-Nyong virus, Ross River virus, Semliki Forest virus, and Sindbis virus can cause fever, rash, and arthralgia syndrome that is rarely fatal [11,12].

VEEV complex was originally formed by six serological subtypes (I–VI). VEEV was first isolated in 1938 during an outbreak of equine encephalitis in the Guajira region, Venezuela, from the brain of a sick horse. In the following years, VEE-related viruses were isolated and identified in many locations in South America, Central America, the Caribbean islands, and the southern regions of the United States [8]. Using a short incubation hemagglutination inhibition (HI) test, isolates of these viruses were distinguished into subtypes I–IV [13], which together form the VEEV complex. Subsequently, Cabassou virus (CABV) and Rio Negro virus (RNV) were isolated and showed to fall within the VEEV antigenic complex, forming subtypes V and VI, respectively [14,15]. Thus, the VEEV complex contains six serological subtypes, which are currently considered as viral species (Table 30.1).

30.1.2 Virion Structure and Genome Organization

Similar to other alphaviruses, VEEV is spherical and enveloped with an icosahedral symmetry of approximately 60–70 nm diameter. The virus possesses a positive-sense RNA genome, contained within a capsid formed by the protein C [8]. The nucleocapsid (RNA + capsid) is surrounded by a lipid envelope derived from the host cell membrane as well as the viral glycoproteins E1 and E2 [8,12]. In total, alphaviruses have five structural proteins, C, E3, E2, 6K, and E1, which are part of the viral particle structure, and four nonstructural proteins (nsP), nsP1–4, involved in RNA replication and production of subgenomic RNA from which structural proteins are synthesized [10].

The genome is composed of a single-stranded positive-sense RNA molecule of approximately 11,000 bp. It is organized with the nonstructural protein genes at the 5' end (two-thirds of the genome) and the structural protein genes at the 3' end (one-third left of the genome) [8,9] (Figure 30.1). Regions of glycoproteins E3-E2-E1 have been extensively used to compare divergence and construct phylogenetic trees for the genus *Alphavirus* since they are the most divergent [8,16]. In addition, the region encoding glycoprotein E2 may contain major determinants of equine virulence and amplification potential, which determine the strain phenotype [2].

30.1.3 Pathogenesis

Transmission to vertebrate hosts (including humans) occurs mainly through the bites of infected mosquitoes. During feeding, infected mosquitoes deposit virus-containing saliva extravascularly [17]. After subcutaneous inoculation, the virus infects skeletal muscle at the local site or Langerhans cells of the skin [18,19]. Langerhans cells, as well as dendritic cells, transport the virus to the regional lymph nodes, where replication takes place. The viral particles reach plasma circulation through which they infect other tissues [8]. Viremia is usually

TABLE 30.1

Members of the VEEV Complex: Origin and Current Distribution

Serologic Classification (Subtype)	Virus/Prototype Strain	Origin	Source	Current Distribution
IAB	VEEV—Trinidad donkey	Trinidad, 1943	Donkey	North, Central, and north of South America
IC	VEEV—P676	Venezuela, 1963	Mosquito	North of South America
ID	VEEV—3880	Panama, 1961	Human	South and Central America
IE	VEEV—Mena II, 68U201	Guatemala, 1968	Hamster	Central America and Mexico
IF	Mosso das Pedras virus—78V3531	Brazil, 1978	Mosquito	Brazil
II	Everglades virus (EVEV)—Fe3-7c	Florida, 1963	Mosquito	Florida (United States)
IIIA	Mucambo virus (MUCV)—BeAn8	Brazil, 1954	Monkey	North of South America and Trinidad
IIIB	Tonate virus—CaAn410d	French Guiana, 1973	Birds	French Guiana
	Bijou Bridge virus—Bijou Bridge	Colorado, 1974	Mosquito	West of the United States (Colorado)
IIIC	MUCV—71D1252	Peru, 1971	Mosquito	Peru
IIID	MUCV—V407660	Peru, 1998	Rodent, mosquito, and human	Peru
IV	Pixuna virus (PIXV)—BeAr35645	Brazil, 1961	Mosquito	Brazil
V	Cabassou virus (CABV)—CaAr508	French Guiana, 1968	Mosquito	French Guiana
VI	Rio Negro virus (RNV)—AG80-663	Argentina, 1980	Mosquito	Argentina

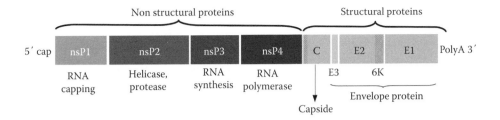

FIGURE 30.1 VEEV (and all alphaviruses) genome.

accompanied by interferon production, proinflammatory cyto-kines, and fever, and ends with the appearance of the humoral immune response approximately 5 days postinfection [8].

In humans, target tissues include the central nervous system (CNS) (producing encephalitis) [8]. The specific mechanism by which VEEV enters the CNS is not entirely clear. Recent studies demonstrate that VEEVs may enter through olfac-tory neurons and start replication in the brain, which induces opening the blood–brain barrier, allowing a second input of virus from the periphery to the CNS [20]. Neuroinvasiveness varies according to the virus and strain [21]. Once in the CNS, the virus spreads from cell to cell or through the cerebrospi-nal fluid (CSF). For most neurotropic alphaviruses, including VEEV, the target cell within the CNS is the neuron, and the damage can be severe and irreversible. VEEV can also cause persistent infection of the CNS including microglia and oligo-dendroglial cells, causing demyelization [8].

30.1.4 EPIDEMIOLOGY AND DISTRIBUTION

VEEV subtypes are divided into two epidemiological groups: epidemic/epizootic viruses and enzootic viruses.

30.1.4.1 Epidemic/Epizootic Viruses

Subtypes IAB and IC belong to the group of epidemic/epizootic viruses, which emerge periodically causing out-breaks in humans and equines [8,16], with high morbidity and mortality rates [22]. Transmission cycle for these subtypes involves equines, which act as highly efficient amplification hosts, developing high titers of the virus in the blood, and mosquito vectors [6]; humans are end hosts (Figure 30.2). The virulence in equines and the induction of viremia are the most important attributes of the epizootic phenotype. When horses become infected, humans living in close association with these animals in agricultural settings may also acquire the infection, resulting in epidemic disease [6]. Although infection by epidemic/epizootic strains has been observed in humans, sheep, dogs, bats, rodents, and some birds, there have never been any recorded major epidemics in the absence of horses [23]. Humans develop similar viral titers to equines, but do not usually act as amplifying hosts due to a limited exposure to mosquito bites [23]. Nevertheless, transmission by humans cannot be discounted completely. Infected people show high titers of virus with both epizootic and enzootic strains. Mosquito vectors with peridomestic urban habits,

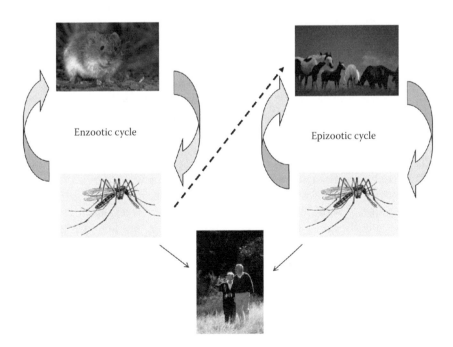

FIGURE 30.2 Transmission cycles of VEE complex viruses. The dotted line symbolizes the emergence of epidemic/epizootic strains from mutations of enzootic strains. For both cycles, humans are end hosts.

such as *Stegomyia (Aedes) aegypti* and *Ae. albopictus* can transmit the virus after ingestion of oral doses of the virus similar to titers of viremia developed in human infections. Therefore, the occurrence of human–mosquito–human transmission is possible [24,25].

The epidemic/epizootic strains are opportunistic in the use of mosquito vectors during outbreaks. Field studies have indicated that more than one mosquito species can be involved in viral transmission during an outbreak [26]. Although susceptibility to infection is required for biological transmission of viruses and some species are almost completely refractory, ecological and behavioral traits such as longevity, food preference (host), survival, and population size are probably more important than the differences in susceptibility of vectorial capacity. Some species that appear to be moderately susceptible to viral infection have been implicated as major vectors for outbreaks of VEEV [23].

It has been described that the main species of mosquitoes involved in the transmission of these viruses would be *Aedes* spp. and *Psorophora* spp. [27]. *Ochlerotatus (Oc.) sollicitans* and *Oc. taeniorhynchus* also exhibit high infection rates, depending on the U.S. region [26]. *Oc. taeniorhynchus* is probably the most important epizootic vector in South America [23].

30.1.4.2 Enzootic Viruses

Subtypes ID, IE, IF, and II–VI have not been originally associated with epidemics or epizootics. They are called enzootic because they carry out their cycle in wild habitats, involving mosquito vectors and rodent reservoir hosts (Figure 30.2) [2]). These viruses are avirulent and unable to amplify in equines, but most of them can cause illness in humans. Only subtype IE has been related to an outbreak in horses in Mexico [28]. Interestingly, epizootic IE strains cause encephalitis in horses but with low titers of virus [29], and are unable to amplify in these animals [2]. For this reason, equines may represent an end host for these strains.

Enzootic strains are involved in active transmission cycles in tropical and subtropical areas of the United States: Everglades virus (EVEV) in Florida, VEEV ID in Central and South America, VEEV IE in Central America, and VEEV IF (Mosso das Pedras Virus), Mucambo virus (MUCV), Pixuna virus (PIXV), CABV, and RNV in South America [2,8]. In endemic areas, mosquito isolates mainly belong to *Culex (Melanoconion)* spp., which live in swampy areas and breed near aquatic plants [8]. These mosquitoes feed on a variety of rodents, birds, and other vertebrates. However, isolations have also been obtained from the genera *Aedes, Stegomyia, Mansonia, Psorophora, Haemagogus, Sabethes, Deinocerites,* and *Anopheles* [2,5]. Wild birds are susceptible to infection, but mammals (mainly rodents) are the most probable reservoirs as shown by viral isolations, levels of viremia, serology, and disease resistance [23,27,30–32]. Enzootic transmission cycles for VEEV ID, VEEV IE, EVEV, MUCV, and Tonate virus (Bijou Bridge virus) have been described; all of them (with the exception of Bijou Bridge) are maintained in cycles involving rodents and mosquitoes in the subgenus *Culex (Melanoconion)* [2]. Bijou Bridge virus is transmitted by the bedbug *Oeciacus vicarious* to the birds of western North America [33].

30.1.4.3 Origin of Epidemics

Venezuelan equine encephalitis epidemics or epizootics have occurred in intervals of approximately 10–20 years in livestock in many places of South America, when epizootic mosquito populations increase [34]. For several years, the main enigma regarding VEEV epidemiology was to detect the source of epidemic/epizootic virus and to uncover the mechanisms of persistence between outbreaks. To explain this, several hypotheses have been postulated: (1) the emergence of epizootic subpopulations within enzootic VEEV populations [35], (2) the initiation of outbreaks by the administration of improperly inactivated vaccines [16], (3) the emergence of subtypes IAB and IC from cryptic transmission cycles [6], (4) the maintenance of epizootic strains in latent infections in horses or other animals [23], and (5) the periodic emergence of epizootic IAB, IC, and IE strains from the evolution (mutations) of enzootic VEEV progenitors [16]. The last hypothesis has been supported by extensive phylogenetic analysis [2,9,16] and is the most accepted so far. It has been documented that remarkable similarities between epizootic IC and enzootic ID viruses exist. A total of 15 amino acid differences were identified between strains of both viruses isolated in Venezuela, and two of them were located within the E2 glycoprotein region, which is suspected to be the major determinant of equine virulence and amplification potential [2,6]. Some studies have reported that a single amino acid substitution in the position 213 of the glycoprotein E2 in ID Venezuelan strains (enzootic) resulted in a change in the equine viremia phenotype, generating high-titer viremia in horses, as occurred with epizootic strains [2,36].

30.1.5 Clinical Features

In horses, VEEV infection can result in (1) unapparent infection followed by seroconversion; (2) systemic disease characterized by tachycardia, fever, depression, and anorexia; or (3) encephalitis [6].

Enzootic strains can infect equines, but these infections are asymptomatic and with low titer of the virus, and may confer immunity in the animals against epizootic strains [37,38].

Epizootic strains produce illness in horses. The virus is found in nasal secretions, eyes, throat, urine, and milk and is transmitted by the respiratory route, as well as by mosquito bites [8]. The incubation period is short (1–5 days), then fever appears, along with viremia. After the disappearance of fever, some cases progress with neurological symptoms: hypersensitivity to noise and touch, incoordination, circular movement, disorientation, involuntary muscle movements, depression, drowsiness, paralysis, convulsions, and sometimes death [7,8]. Equine infections by these strains are associated to leukopenia and high-titered and prolonged viremia (sometimes $>10^8$ infectious particles/mL of blood, for 4–5 days) [8], which allows one horse to be the source of the virus for thousands of mosquitoes in only 1 day, facilitating the virus spread [7]. Encephalitis and death correlate to the magnitude of the viremia and the ability of the virus to replicate outside the CNS and then spread to the brain [23].

In animals that recover, antibodies (Abs) appear approximately 7 days after infection [39].

In humans, infection occur mainly by mosquito bites, although it is also possible for an infection to occur through the respiratory tract, as has been demonstrated by the occurrence of infection in laboratory workers [7,8,40]. The incubation period is short (1–5 days). Most infections are asymptomatic or mild, and the individuals go through a short discomfort, with spontaneous recovery. Clinical manifestations that can occur are similar to those of dengue or influenza, with sudden onset of severe headache, chills, tachycardia, fever, myalgia, arthralgia, retro-orbital pain, nausea, and vomiting [5,40] accompanied by leukopenia [23]. Some cases may have a biphasic febrile course: after few days of fever, there may be signs of CNS involvement, ranging from somnolence, encephalitis, photophobia, disorientation, convulsions, paralysis, coma, to death [5,34]. Neurologic disease occurs in 4%–14% of the cases [16] and many of them may remain as sequelae [8].

30.1.6 Diagnosis

For the diagnosis of human and/or animal cases with clinical symptoms compatible with VEE, it is essential to know the epidemiological situation in the region. This is because in some parts of the United States, there is cocirculation of VEEVs and other arboviruses producing similar clinical manifestations. VEE can be misdiagnosed as diseases caused by dengue or Oropouche, group C bunyavirus, and/or Guaroa viruses, since they are clinically indistinguishable [2]. The differential diagnostic of neurological cases should be performed including other alphaviruses, such as WEEV and EEEV, as well as some neurotropic flaviviruses, like the Saint Louis encephalitis virus or the West Nile virus.

Laboratory diagnosis of VEEV is based on viral isolation, detection of the viral genome, and antigens, and/or detection of specific Abs (IgG or IgM).

30.1.6.1 Conventional Techniques

Since its first isolation in the 1930s, VEEV has been diagnosed through virus isolation or detection of specific immune response. Virus isolation is a very sensitive and specific technique, which can be performed using cell culture or suckling mice, and is the "gold standard" technique for virus detection [41,42]. Successful attempts of viral isolation from clinical specimens depend on the care taken for the collection, handling, and shipping of the sample to the laboratory. For virus isolation, the most adequate human samples are CSF and blood from the early days of symptom onset (viral load is very short) or autopsy materials (from the CNS). For equines, the sample of choice is blood and/or the brains of the fatal cases. Further identification of the isolations is performed using PCR, immunofluorescence (IF), and ELISA, among others. Although virus isolation presents many advantages, it is time consuming and requires sterile handling of tissue cultures under biosafety level 2 or 3 conditions (depending on the virus), or the use of experimental animals [41].

Serological diagnosis can be carried out through detection of IgG or IgM Abs. In the case of specific IgM detection, ELISA and indirect IF assays can be utilized using anti-IgM conjugates for both humans and horses from serum or CSF samples. Positive IgM results must be confirmed by seroconversion using plaque reduction neutralization test (PRNT), which detects total Abs and defines viral subtypes. For diagnosis or confirmation of infection using PRNT, it is necessary to use two paired samples, obtained at different times. The first sample must be obtained during the first days of the disease (this sample can be also used for isolation or amplification of the genome if it was taken under sterile conditions and maintained refrigerated); the other sample must be taken after the third week of the onset of symptoms. Seroconversion is defined by a titer difference of fourfold or greater between these samples or a negative result for the first sample and a positive result for the second one.

HI is another conventional technique for VEEV Ab detection based on the bind of VEEV specific Abs to viral particles, which inhibits agglutination of avian erythrocytes. This is a simple and accessible technique that can be used for quantification of specific Abs [8,43] but requires obtaining avian red blood cells periodically.

30.1.6.2 Molecular Techniques

Molecular techniques such as reverse transcription polymerase chain reaction (RT-PCR) and RT-nested PCR offer a valuable tool for rapid, sensitive, and reliable detection of VEEVs and can be used for both diagnostic application and epidemiological surveillance. Some of these techniques allow genus-specific amplification, followed by sequencing of the amplified fragments for specific identification of all alphaviruses [44]. These assays are useful in geographical regions with cocirculation of many alphaviruses that produce similar symptoms. Other techniques target several genome regions of VEEV complex, permitting specific discrimination of various subtypes [41,42]. Molecular techniques are particularly useful as a surveillance method in endemic and nonendemic areas, utilizing other biological samples such as mosquito homogenates or mammalian serum samples (rodent). These techniques provide a means to conduct phylogenetic analyses based on the amplification of the PE2 region, whose notable sequence variations help reveal details about the genetic variability of the strains [16,45].

With a higher sensitivity, nested PCR is clearly advantageous over single-round PCR. This is important for VEEV detection from clinical samples due to short viremia, as well as from mosquito samples in which the number of virus present is very low.

30.2 METHODS

30.2.1 Sample Collection and Preparation

For diagnosis, it is important to consider the time in which the sample is taken since the period of viremia (in humans and equines) is very short (3–5 days), so blood must be taken up to 5 days after the onset of the symptoms.

After collection, the blood must be separated into serum, as this is what will be analyzed. It is also possible to analyze brain samples (biopsy or autopsy) if the infected equine dies. These samples must be refrigerated as soon as possible upon arrival at the laboratory to prevent the disintegration of the virus and the fall of viral titer. In cases when the sample could not be processed within 24 h, it must be frozen and stored at −80°C.

For brain sample, it is necessary to cut the sample in 2 or 3 parts, triturate in a cold mortar, homogenize with a sterile minimal essential medium supplemented with 10% fetal bovine serum and 1% gentamicin, and clarify by centrifugation at 10,000 rpm for 30 min. The supernatant is poured into individual screw-cap vials and frozen at −80°C until viral RNA detection.

For mosquito samples, pool the collected mosquitoes by species, date and place them in lots of 1–50 unengorged individuals, and store at −80°C until processing. Then, each mosquito pool is processed the same as described for brain sample, and the supernatant is used for virus detection.

Several methods are used for VEEV RNA extraction from clinical or field specimens. These include conventional RNA separation with the guanidine isothiocyanate method, as well as the use of commercial extraction kits, which are highly sensitive and reproducible. Commercial kits used for viral RNA extraction are mainly based on columns containing silica gel, which retain the nucleic acid. The RNA binds to the silica membrane, and contaminants are efficiently washed away using different wash buffers. High-quality RNA is eluted in a special RNase-free buffer, ready for direct use or safe storage. The purified RNA is generally free of protein, nucleases, and other contaminants and inhibitors.

Trizol® reagent (Invitrogen BRL, Life Technologies, Rockville, MD) is a complete, ready-to-use solution of phenol and guanidine isothiocyanate for isolation of total RNA. Using this procedure, 150 μL of the sample is mixed with 700 μL of Trizol reagent, 1 μL of glycogen, and 200 μL of chloroform. The mixture is vortexed for 2 min, incubated for 20 min at room temperature, and centrifuged at 13,000 rpm for 20 min. Total RNA is precipitated by isopropyl alcohol and ethanol, air-dried, and dissolved in 20 μL of sterile water. Once extracted, RNA should be stored at −80°C until RT-PCR.

30.2.2 Detection Procedures

30.2.2.1 Reverse Transcription

The passage of genome RNA to DNA copy (cDNA) can be performed using two kinds of primers. It is possible to use either reverse oligonucleotide, which is going to be used in the PCR amplification, or random hexamer primers, which are a mixture of six nucleotides primers of all possible combinations, with the capacity to generate cDNA of multiple lengths. The use of random primers has the advantage of converting any RNA that is present in the sample into cDNA, which can be then amplified by PCR. This is very useful for the

diagnosis of VEE, which shares clinical symptoms with other arboviral diseases; a single RT reaction will provide templates for several specific PCR amplifications (for each arbovirus).

30.2.2.1.1 Procedure

Prepare the RT-mix as shown in Table 30.2. The volume of RT-mix for one sample is 10 μL, to which 10 μL of the extracted sample is added to obtain a final volume of cDNA, after the retrotranscription reaction of 20 μL.

First, the 10 μL of RNA is incubated at 65°C for 10 min, then 10 μL of RT-mix is added, and the mixture is incubated at 45°C for 60 min.

30.2.2.2 Nested-PCR Detection

All VEEV complex members are amplified by nested PCR using primers VEE45, VEE176, VEE83, and VEE138 (Table 30.3) [42].

30.2.2.2.1 Procedure

For the first amplification (PCR I), 5 μL of cDNA is added to 45.25 μL PCR I mix (50.25 μL final volume), prepared as shown in Table 30.4. The mix is subjected to an initial denaturation step at 94°C for 2 min, followed by 40 cycles of denaturation at 94°C for 30 s, primer annealing at 64°C for 1 min, extension at 72°C for 30 s, and a final extension at 72°C for 5 min.

For nested PCR (PCR II), 2 μL of each PCR I product is transferred to 48.25 μL nested PCR mixture (50.25 μL final volume), prepared as shown in Table 30.4. The second PCR is carried out with an initial denaturation at 94°C for 2 min, followed by 40 cycles of 94°C for 30 s, 61°C for 40 s, and 72°C for 30 s. A final extension at 72°C for 5 min is performed.

PCR products, of 80 bp (10 μL), are analyzed by electrophoresis using a 2% agarose gel containing 0.5 μg/mL of SYBR® Safe (Life Technologies) in TBE buffer gels and visualized under UV light. A molecular weight marker of 50 bp is included on each gel.

A negative control is concurrently included along with the test samples in order to monitor any possible contamination that might occur during the PCR procedure. Positive control is also included to test that reaction conditions are right.

TABLE 30.2
Components and Volumes of the RT-Mix

Component	Volume (μL)
Buffer 5×	4
MgCl$_2$ 25 mM	2.4
dNTPs 10 mM	1
Random hexamer primers 10 μM	1
Reverse transcriptase	1
RNase out	0.5
Water	0.1
Total	10

TABLE 30.3

Primers Used for VEEV Amplification

Oligonucleotide	Sequence (5′–3′)	Position
VEE45 (+1)	ATGGAGAARGTTCACGTTGAYATCG	45–70 (nsP1 region)
VEE176 (−1)	YTCGATYARYTTNGANGCYARATGC	176–201 (nsP1 region)
VEE83 (+2)	ARGAYAGYCCNTTCCTYMGAGC	83–105 (nsP1 region)
VEE138 (−2)	CRTTAGCATGGTCRTTRTCNGTNAC	138–163 (nsP1 region)

Source: Pisano, M.B. et al., *J. Virol. Methods*, 186, 203, 2012.

TABLE 30.4

Components and Volumes of PCR Mix for the Amplification of VEEV

Component	Mix PCR I Volume (μL)	Mix PCR II (Nested PCR) Volume (μL)
Water	30	33
Buffer 5× GoTaq	10	10
dNTPs 10 mM	1	1
VEE45 (PCR I)/VEE83 (PCR II) 10 μM	2	2
VEE176 (PCR I)/VEE138 (PCR II) 10 μM	2	2
GoTaq DNA polymerase	0.25	0.25
Total	45.25	48.25

It is important to note that this technique is not useful to identify VEEV subtype so that a positive result only indicates the presence of VEEV genome in the sample.

30.3 CONCLUSIONS

VEEVs cause human and equine diseases throughout the United States, with symptoms ranging from mild illness to encephalitis and death, in addition to a high proportion of subclinical cases. Clinically, VEE is indistinguishable from dengue fever and other arboviral diseases, and confirmatory diagnosis is required with the use of specialized laboratory tests. The disease burden of endemic VEE in developing countries remains largely unknown [2]. The potential emergence of epizootic viruses from enzootic progenitors highlights the need to strengthen surveillance activities, as well as employ fast and sensitive techniques for viral detection.

Diagnosis of acute VEEV infection can be done by detection of viral antigens or nucleic acids or by virus isolation. These techniques are only successful if the blood or CSF is collected during the viremic phase of infection, which lasts for 3–5 days. Detection of RNA by RT-nested PCR is a fast, sensitive, and specific alternative for diagnosis of VEE acute infection and can also be useful as a surveillance method.

Real-time PCR is another alternative for VEEV detection, using either SYBR Green reagent or specific probes. This technique has the advantages of requiring less processing time than a conventional PCR, and being highly sensitive and reproducible. Furthermore, it can be used for viral quantification, which is essential for determining the amount of virus in infected hosts and mosquitoes, with the goal to evaluate virus transmission.

REFERENCES

1. LeBeaud, A. D., Why arboviruses can be neglected tropical diseases, *PLOS Negl. Trop. Dis.*, 2, e247, 2008.
2. Aguilar, P. V. et al., Endemic Venezuelan equine encephalitis in the Americas: Hidden under the dengue umbrella, *Future Virol.*, 6, 721, 2011.
3. Beck, C. E. and Wyckoff, R. W., Venezuelan equine encephalomyelitis, *Science*, 88, 530, 1938.
4. Aguilar, P. V. et al., Endemic Venezuelan equine encephalitis in northern Peru, *Emerg. Infect. Dis.*, 10, 880, 2004.
5. Epidemiologic Alert, Vector Borne Equine Encephalitis, Pan American Health Organization (PAHO), July 20, 2010. http://new.paho.org/hq/index.php?option=com_content&task=view&id=1239&Itemid=1091&lang=es. Accessed on March 31, 2015.
6. Greene, I. et al., Envelope glycoprotein mutations mediate equine amplification and virulence of epizootic Venezuelan equine encephalitis virus, *J. Virol.*, 79, 9128, 2005.
7. Contigiani, M., Togaviridae, in: *Microbiología Biomédica*, p. 896, Basualdo, J., Coto, C., and de Torres, R. (eds.), Atlante, Buenos Aires, AA, 2005.
8. Griffin, D., Alphaviruses, in: *Fields Virology*, p. 1023, Fields, B. and Knipe, D. (eds.), Lippincott Williams & Wilkins, Philadelphia, PA, 2007.
9. Powers, A. M. et al., Evolutionary relationships and systematics of the alphaviruses, *J. Virol.*, 75, 10118, 2001.

10. Strauss, J. H. and Strauss, E. G., The alphaviruses: Gene expression, replication and evolution, *Microbiol. Rev.*, 58, 491, 1994.

11. Ryman, K. D. and Klimstra, W. B., Host responses to alphavirus infection, *Immunol. Rev.*, 225, 27, 2008.

12. Leung, J. Y., Ng, M. M., and Chu, J. J., Replication of alphaviruses: A review on the entry process of alphaviruses into cells, *Adv. Virol.*, 2011, 249640, 2011.

13. Young, N. A. and Johnson, K. M., Antigenic variants of Venezuelan equine encephalitis virus: Their geographic distribution and epidemiologic significance, *Am. J. Epidemiol.*, 89, 286, 1969.

14. Digoutte, J. and Girault, G., The protective properties in mice of Tonate virus and two strains of Cabassou virus against neurovirulent everglades Venezuelan encephalitis virus, *Ann. Microbiol. (Paris)*, 127B, 429, 1976.

15. Calisher, C. H., Arbovirus investigations in Argentina, 1977–1980. III. Identification and characterization of viruses isolated, including new subtypes of western and Venezuelan equine encephalitis viruses and four new bunyaviruses (Las Maloyas, Resistencia, Barranqueras and Antequera), *Am. J. Trop. Med. Hyg.*, 34, 956, 1985.

16. Powers, A. M. et al., Repeted emergence of epidemic/epizootic Venezuelan Equine Encephalitis from a single genotype of enzootic subtype ID virus, *J. Virol.*, 71, 6697, 1997.

17. Turell, M. J., Tammariello, R. F., and Spielman, A., Nonvascular delivery of St. Louis encephalitis and Venezuelan equine encephalitis viruses by infected mosquitoes (Dipteria: Culicidae) feeding on a vertebrate host, *J. Med. Entomol.*, 32, 563, 1995.

18. Grimley, P. M. and Friedman, R. M., Arboviral infection of voluntary striated muscles, *J. Infect. Dis.*, 122, 42, 1970.

19. Liu, C. et al., A comparative study of the pathogenesis of western equine and eastern equine encephalomyelitis viral infections in mice by intracerebral and subcutaneous inoculations, *J. Infect. Dis.*, 122, 53, 1970.

20. Schafer, A. et al., The role of the blood-brain barrier during Venezuelan equine encephalitis virus infection, *J. Virol.*, 85, 10682, 2011.

21. Dubuisson, J. et al., Genetic determinants of Sindbis virus neuroinvasiveness, *J. Virol.*, 71, 2636, 1997.

22. Wang, E. et al., A novel, rapid assay for detection and differentiation of serotype-specific antibodies to Venezuelan Equine Encephalitis complex alphaviruses, *Am. J. Trop. Med. Hyg.*, 72, 805, 2005.

23. Weaver, S. C. et al., Venezuelan Equine encephalitis, *Annu. Rev. Entomol.*, 49, 141, 2004.

24. Harrington, L. C., Edman, J. D., and Scott, T. W., Why do female *Aedes aegypti* (Dipteria: Culicidae) feed preferentially and frequently on human blood?, *J Med. Entomol.*, 38, 411, 2001.

25. Weaver, S. C. and Reisen, W. K., Present and future arboviral threats, *Antiviral Res.*, 85, 328, 2010.

26. Sudia, W. D. et al., Epidemic Venezuelan equine encephalitis in North America in 1971: Vector studies, *Am. J. Epidemiol.*, 101, 17, 1975.

27. Carrara, A. S. et al., Venezuelan equine encephalitis virus infection of spiny rats, *Emerg. Infect. Dis.*, 11, 663, 2005.

28. Brault, A. C. et al., Venezuelan Equine Encephalitis emergence: Enhanced vector infection from a single aminoacid substitution in the envelope glycoprotein, *Proc. Natl. Acad. Sci. USA*, 101, 11344, 2004.

29. Gonzalez-Salazar, D. et al., Equine amplification and virulence of subtype IE Venezuelan equine encephalitis viruses isolated during the 1993 and 1996 Mexican epizootics, *Emerg. Infect. Dis.*, 9, 161, 2003.

30. Barrera, R. et al., Contrasting sylvatic foci of Venezuelan equine encephalitis virus in northern South America, *Am. J. Trop. Med. Hyg.*, 67, 324, 2002.

31. Estrada-Franco, J. G. et al., Venezuelan equine encephalitis virus, southern Mexico, *Emerg. Infect. Dis.*, 10, 2113, 2004.

32. Grayson, M. A. and Galindo, P., Epidemiologic studies of Venezuelan equine encephalitis virus in Almirante, Panama, *Am. J. Epidemiol.*, 88, 80, 1968.

33. Monath, T. P. et al., Recovery of Tonate virus ('Bijou Bridge' strain), a member of the Venezuelan equine encephalomyelitis virus complex, from cliff swallow nest bugs (*Oeciacus vicarius*) and nestling birds in North America, *Am. J. Trop. Med. Hyg.*, 29, 969, 1980.

34. Rivas, F. et al., Epidemic Venezuelan equine encephalitis in La Guajira, Colombia, 1995, *J. Infect. Dis.*, 175, 828, 1997.

35. Stanick, D. R., Wiebe, M. E., and Scherer, W. F., Markers of Venezuelan encephalitis virus which distinguish enzootic strains of subtype ID from those of IE, *Am. J. Epidemiol.*, 122, 234, 1985.

36. Anishchenko, M. et al., Venezuelan Encephalitis emergence mediated by a phylogenetically predicted viral mutation, *PNAS*, 103, 4994, 2006.

37. Fillis, C. and Calisher, C., Neutralizing antibody responses of humans and mice to vaccination with Venezuelan encephalitis (TC-83) virus, *J. Clin. Microbiol.*, 10, 544, 1979.

38. Henderson, B. E. et al., Experimental infection of horses with three strains of Venezuelan equine encephalitis virus. I. Clinical and virological studies, *Am. J. Epidemiol.*, 93, 194, 1971.

39. Kissling, R. E. et al., Venezuelan equine encephalomyelitis in horses, *Am. J. Hyg.*, 63, 274, 1956.

40. Ehrenkranz, N. J. and Ventura, A. K., Venezuelan equine encephalitis virus infection in man, *Ann. Rev. Med.*, 25, 9, 1974.

41. Linssen, B. et al., Development of reverse transcription-PCR assay specific for detection of equine encephalitis viruses, *J. Clin. Microb.*, 38, 1527, 2000.

42. Pisano, M. B. et al., Specific detection of all members of the Venezuelan equine encephalitis complex: Development of a RT-Nested PCR, *J. Virol. Methods*, 186, 203, 2012.

43. Clarke, D. H. and Casals, J., Techniques for hemagglutination and hemagglutination-inhibition with arthropod borne viruses, *Am. J. Trop. Med. Hyg.*, 1959, 561, 1959.

44. Sánchez Seco, M. P. et al., A generic nested-RT-PCR followed by sequencing for detection and identification of members of the Alphavirus genus, *J. Virol. Methods*, 95, 153, 2001.

45. Pisano, M. B. et al., Genetic and evolutionary characterization of Venezuelan equine encephalitis virus isolates from Argentina, *Inf. Gen. Evol.*, 26, 72, 2014.

31 Equine Arteritis Virus

Udeni B.R. Balasuriya

CONTENTS

31.1 Introduction 277
    31.1.1 Classification 277
    31.1.2 Morphology, Biology, and Epidemiology 277
    31.1.3 Clinical Features and Pathogenesis 279
    31.1.4 Diagnosis 280
        31.1.4.1 Conventional Techniques 280
        31.1.4.2 Molecular Techniques 280
31.2 Methods 281
    31.2.1 Sample Collection and Preparation 281
        31.2.1.1 Sample Collection 281
        31.2.1.2 Processing of Clinical Samples for Virus Isolation 281
        31.2.1.3 Processing of Clinical Samples for Molecular Testing 281
    31.2.2 Detection Procedures 282
        31.2.2.1 Standard RT-PCR and RT-nPCR 282
        31.2.2.2 Real-Time RT-PCR 282
        31.2.2.3 Insulated Isothermal RT-PCR 282
31.3 Conclusion and Future Perspectives 282
Acknowledgment 283
References 283

31.1 INTRODUCTION

31.1.1 CLASSIFICATION

Equine arteritis virus (EAV), the causative agent of equine viral arteritis (EVA), was first isolated from the lung of an aborted fetus following an extensive outbreak of respiratory disease among horses on a Standardbred breeding farm in Bucyrus, Ohio, in 1953.[1,2] EAV is the prototype virus in the family *Arteriviridae* (genus *Arterivirus*), order *Nidovirales*, which also includes porcine reproductive and respiratory syndrome virus, simian hemorrhagic fever virus, lactate dehydrogenase–elevating virus of mice, and recently identified wobbly possum disease virus of free-ranging Australian brushtail possums (*Trichosurus vulpecula*) in New Zealand.[3,4] All members of the order *Nidovirales* are enveloped viruses with linear, positive-sense, single-stranded ribonucleic acid (RNA) genomes. Furthermore, the genome organization and replication strategy of EAV strikingly resemble those of corona- and toroviruses, including the production of a 3′-coterminal nested set of subgenomic mRNAs (sg mRNAs), and key domains of the arterivirus replicase are homologs of domains found in corona- and torovirus replicases.[5] However, EAV and other arteriviruses differ considerably in their genetic complexity and virion architecture from the other nidoviruses.

31.1.2 MORPHOLOGY, BIOLOGY, AND EPIDEMIOLOGY

The EAV virion is an enveloped, spherical, 50–65 nm particle with an isometric core that contains a single-stranded, positive-sense RNA molecule.[5] The EAV genome length varies between 12,704 and 12,731 bp among different virus strains and includes a 5′ leader sequence (224 nucleotides) and at least ten open reading frames (ORFs).[6–11] The two most 5′-proximal ORFs (1a and 1b) occupy approximately three-fourths of the genome and encode two replicase polyproteins (pp1a and pp1ab). These two precursor proteins are extensively processed after translation into at least 13 nsps (nsp1–12, including nsp7 α/β) by three viral proteases (nsp1, nsp2, and nsp4).[5,12–13] The remaining ORFs are located in the 3′-proximal quarter of the genome and encode for the structural proteins of the virus, which include seven envelope proteins (E, GP2, GP3, GP4, ORF5a protein, GP5, and M) and a nucleocapsid protein (N), which are encoded by ORFs 2a, 2b, 3–4, 5a, 5b, and 6–7 that are located at the 3′-proximal quarter of the genome.[5,14,15] These structural proteins are expressed from six viral sg mRNAs that form a 3′-coterminal nested set and contain a common leader sequence encoded by the 5′-end of the genome.[16] The major envelope proteins of EAV are GP5 (glycosylated) and M (unglycosylated) and they form a disulfide-linked heterodimer in the mature virus particles.

It has been shown that the GP5 is the principal target of neutralizing antibodies to EAV and four major neutralization sites (A, B, C, and D) are located in the amino terminal ectodomain of the protein. It has been demonstrated that GP5–M heterodimer formation is critical for the expression of the neutralization epitopes in their authentic form.[17–20] The M protein likely acts as an essential scaffold on which the GP5 protein folds to form the epitopes that induce neutralizing antibodies in horses. In addition, the EAV envelope contains a heterotrimer of three minor membrane glycoproteins (GP2, GP3, and GP4) and two unglycosylated envelope proteins (E and ORF5a protein).[14,15] It has been shown that the ectodomains of GP5 and M are not the major determinants of cellular tropism, but the minor envelope proteins are the critical proteins in mediating cellular tropism of EAV.[21–23] EAV is readily inactivated by lipid solvents (e.g., ether and chloroform) and by common disinfectants and detergents. The virus survives in cryopreserved semen samples and embryos for many years. EAV remains infectious for 75 days at 4°C, between 2 and 3 days at 37°C, and 20–30 min at 56°C. The virus is also highly stable and remains infectious for decades in organ samples and tissue culture fluid when stored frozen (−70°C to −80°C).

The advent of recombinant DNA technology and establishment of reverse genetics systems (e.g., infectious cDNA clone) has significantly advanced the knowledge on the molecular biology of EAV.[21] Specifically, these studies have lead to the identification of major viral determinants of virulence, neutralization, persistence, and protective host immunity.[24–27] Recently, the advances in EAV molecular biology and the disease EVA have been summarized in several review articles.[15,21,28]

EAV can infect horses, donkeys, mules, and zebras.[15] Based on serological surveys, EAV infection has been common among horses in North and South America, Europe, Australia, Africa, and Asia.[29,30] Other countries such as Iceland and Japan are apparently free of the virus.[15] Recent studies have shown that New Zealand is also free of EAV.[31] However, the seroprevalence of EAV infection of horses varies between countries and among horse breeds within a country. EAV infection is considered endemic in Standardbred but not in Thoroughbred horses in the United States, with 77.5%–84.3% of all Standardbreds but only up to 5.4% of Thoroughbreds being seropositive to the virus.[29,32–37] The seroprevalence of EAV infection of Warmblood stallions is also very high in a number of European countries, with some 55%–93% of Austrian Warmblood stallions being seropositive to EAV.[38] Similarly, there is high seroprevalence among mares and stallions of Hucul horses in Poland, 53.2% and 68.2%, respectively.[39] Seroprevalence to EAV increases with age, indicating that horses may be repeatedly exposed to the virus as they age.

It has been reported that 30%–70% of stallions that are exposed to EAV can become persistently infected carriers and continue to shed virus in their semen for variable time periods.[15] The duration of the EAV carrier state in the stallion can be short (2 months), intermediate (7 months), or long term (many years or sometimes lifelong). Persistently infected carrier stallions function as the natural reservoir of EAV and disseminate the virus to susceptible mares at breeding (Figure 31.1). Some 85%–100% of seronegative mares bred to long-term carrier stallions become infected with the virus and seroconvert within 28 days. Mares that become infected

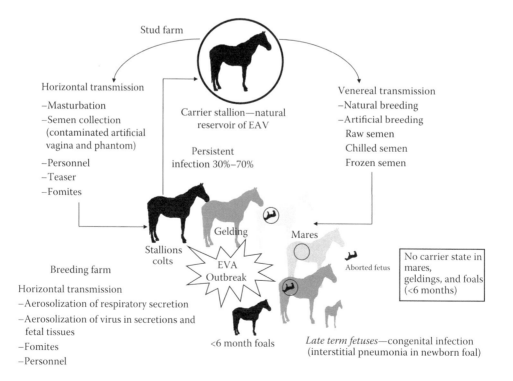

FIGURE 31.1 Transmission of equine arteritis virus.

following natural or artificial breeding then can readily transmit the virus by nasal aerosol to susceptible cohorts in close proximity. Thus, the two principal modes of EAV transmission are horizontal transmission by aerosolization of infectious respiratory tract secretions from acutely infected horses and venereal transmission during natural or artificial insemination with infective semen from persistently infected stallions. However, direct and close contact is necessary for aerosol transmission of EAV between horses.[36,40] EAV can also be transmitted by aerosol from urine and other body secretions of acutely infected horses, aborted fetuses and their membranes, and the masturbates of acutely or chronically infected stallions.[38,41–46] Recently, it has been demonstrated that the virus can also be transmitted during embryo transfer from mares inseminated with infective semen.[15,28] EAV can also be transmitted through indirect contact with fomites or personnel. Congenital infection results from vertical transmission (transplacental transmission) of the virus when pregnant mares are infected late in gestation.

There is only one known serotype of EAV and field isolates from North America and Europe are genetically closely related and differ only by 15% at the nucleotide level. Phylogenetic analyses based on ORF5 sequence cluster EAV field strains into two distinct clades, reflecting their North American (NA) and European (EU) origin.[15,47–49] Recent phylogenetic analysis shows that EAV isolates in the EU group could be further divided into two subgroups, EU subgroup-1 (EU-1) and EU subgroup-2 (EU-2). Data from individual EU and NA isolates indicate movement of viruses between the two continents due to the transport of carrier stallions and/or their semen. The major nucleotide and amino acid differences between NA and EU EAV strains occur in the nsp2-encoding part of ORF1a, as well as in ORFs 3 and 5, which encode the GP3 and GP5 envelope proteins, respectively.[50] EAV tends to evolve during persistent infection in the reproductive tract of the carrier stallions that leads to the emergence of novel genetic and antigenic variants that could initiate new disease outbreaks.

31.1.3 Clinical Features and Pathogenesis

The vast majority of EAV infections are subclinical (or inapparent) in nature, and therefore, it is underdiagnosed and underreported in countries where the disease is endemic.[15,28] The clinical signs displayed by EAV-infected horses depend on a variety of factors including the genetics, age, and physical condition of the horse, challenge dose and route of infection, strain of virus, and environmental conditions. With the sole and notable exception of the experimentally derived and highly horse-adapted EAV VBS, other strains and field isolates of EAV very rarely cause fatal infection in adult horses.[49,51] The incubation period of 3–14 days (usually 6–8 days following venereal exposure) is followed by pyrexia of up to 41°C that may persist for 2–9 days. Acutely infected horses may develop a wide range of clinical signs including pyrexia, depression, anorexia, dependent edema (prepuce, scrotum, ventral trunk, and limbs), stiffness of gait, conjunctivitis,

lacrimation and swelling around the eyes (periorbital and supraorbital edema), respiratory distress, urticaria, and leukopenia. The virus causes abortion in pregnant mares, and abortion rates during natural outbreaks of EVA can vary from 10% to 60% of infected mares. EAV-induced abortions can occur at any time between 3 and 10 months of gestation. The aborted fetuses are usually partially autolyzed at the time of expulsion. Aborted fetuses may exhibit interlobular pulmonary edema, pleural and pericardial effusion, and petechial and ecchymotic hemorrhages on the serosal and mucosal surfaces of the small intestine. EAV infection can cause a severe fulminating interstitial pneumonia and a progressive pneumoenteric syndrome in neonatal foals.

With the exception of persistently infected stallions, EAV is cleared from the tissues of infected horses by 28 days after the exposure. A significant number of acutely infected stallions (10%–70%) become persistently infected and shed the virus in semen; however, in contrast, there is no evidence of any analogous persistent infection in mares, geldings, or foals (<6 months of age). The virus persists in the ampulla and in the accessory sex glands of the male reproductive tract, and the establishment and maintenance of the carrier state in stallions are testosterone dependent.[29] Stallions may undergo a period of temporary subfertility associated with decreased libido and poor semen quality parameters (e.g., reduced sperm motility and concentration and an increased percentage of morphologically abnormal sperm in ejaculates) during the acute phase of EAV infection. The effect on semen quality is exerted by scrotal edema and fever during acute infection of the stallions and could be observed between 9 and 76 days postinfection (DPI).[29,52,53] However, semen quality is apparently normal in persistently infected stallions, despite high-titer virus shedding in the semen, which suggests that the virus seems to exert little to no direct effect after semen quality returned to normal.[53] Venereal infection of mares by persistently infected carrier stallions does not appear to result in subsequent fertility problems.[29]

Initial multiplication of EAV takes place in the upper respiratory tract epithelial cells and alveolar macrophages in the lung following respiratory exposure, and the virus soon appears in the regional lymph nodes, especially the bronchial nodes. Within 3 days, the virus is present in the blood (viremia lasts for 3–19 DPI) and in virtually all organs and tissues where it replicates in macrophages and endothelial cells. The pathogenesis of EVA is not clearly defined. *In vitro* and *in vivo* studies have demonstrated increased transcription of genes encoding proinflammatory mediators (IL-1β, Il-6, IL-8, and TNF-α) following EAV infection suggesting that these cytokine mediators are critical in determining the outcome of the infection and severity of the disease.[24,54] Furthermore, virulent and avirulent strains of EAV induced different quantities of TNF-α and other proinflammatory cytokines from both infected endothelial cells and macrophages. The clinical manifestations of EVA reflect endothelial cell injury in small arteries and consequently increased vascular permeability.[15] Recently, it has been shown that EAV can infect CD3+ T cells from some horses *in vitro* but not from others.[55]

Further studies have shown that based on this *in vitro* susceptibility of CD3+ T cells to EAV infection, the horses could be divided into susceptible and resistant groups and the stallions with CD3+ susceptible phenotype may represent those at higher risk of becoming carriers compared to those that lack this phenotype.[28] Subsequently, a genome-wide association study identified a common, genetically dominant haplotype associated with the *in vitro* CD3+ susceptible phenotype, as well as EAV carrier state in the region of equine chromosome 11. Experimental inoculation with EAV into horses with *in vitro* CD3+ susceptibility or resistance showed a significant difference between the two groups of horses in terms of proinflammatory and immunomodulatory cytokine mRNA expression and evidence of increased clinical signs in horses possessing the *in vitro* CD3+ T cell–resistant phenotype.[24]

The characteristic histological feature of EVA is a severe necrotizing panvasculitis of small blood vessels.[56,57] Affected arteries show foci of intimal, subintimal, and medial necrosis with edema and infiltration of lymphocytes and neutrophils. Prominent vascular lesions are also seen in the placenta, brain, liver, and spleen of aborted fetuses. The lungs of affected neonatal foals have severe interstitial pneumonia. The characteristic vascular lesions of EVA do not appear to be the result of immune-mediated injury. Vascular injury in EVA is likely the result of direct virus-mediated injury to the lining (endothelium) and walls (media) of affected vessels. Increased vascular permeability and leukocyte infiltration resulting from generation of chemotactic factors lead to hemorrhage and edema around these vessels. However, the relative roles and importance of direct virus-induced injury versus virus-induced proinflammatory mediators have not yet been defined.

31.1.4 DIAGNOSIS

Clinical signs of EVA resemble many other respiratory and reproductive diseases of horses, and therefore, diagnosis of EVA has to be further confirmed by laboratory diagnostic testing of appropriate clinical specimens. It is also important to identify EAV carrier stallions prior to breeding and/or international movement of their germplasm. During outbreaks, EAV can spread rapidly to susceptible horses and can result in very high morbidity. Therefore, rapid and accurate diagnosis of EVA is critical for implementation of prevention and control measures to avoid the spread of EAV and to reduce the economic impact of the disease. Various laboratory testing procedures based on both conventional and contemporary techniques have been described for the diagnosis of EVA in horses.[15]

31.1.4.1 Conventional Techniques

The confirmation of diagnosis of EVA is currently based on virus detection and/or serological demonstration of elevated neutralizing antibody titers (fourfold or greater) in paired serum samples taken at a 14–28-day interval. EAV can be isolated from nasal swabs or anticoagulated blood collected from horses with signs of EVA or the tissues (including placenta) of

aborted equine fetuses. Putative carrier stallions are first identified by serology as they are always seropositive, and persistent infection is confirmed by virus isolation from semen in cell culture or by test-breeding using seronegative mares (and monitoring these for seroconversion to EAV after breeding).

Virus neutralization test (VNT) is the principal serological assay used to detect evidence of EAV infection by most laboratories around the world, and it continues to be the current World Organisation for Animal Health (Office International des Epizooties [OIE]) prescribed standard test for EVA (gold standard).[58] Several laboratories have developed and evaluated enzyme-linked immunosorbent assays (ELISAs) to detect antibodies to EAV using whole virus, synthetic peptides, or recombinant viral proteins (e.g., GP5, M, and N) as antigens. However, none of these ELISAs or recently described microsphere immunoassay (Luminex) has yet been shown to be of equivalent sensitivity and specificity to the VNT.[59] Interestingly, a recently developed commercial cELISA assay (VMRD, Pullman, WA) appears to have very high sensitivity and specificity compared to the VNT.[60,61,62] This new cELISA requires further validation before it can be used as a prescribed test for screening horses for international movement.

Virus isolation (VI) is currently the OIE-approved gold standard for the detection of EAV in semen and is the prescribed test for international trade. Isolates of EAV from clinical specimens, including semen, are often confirmed in a one-way neutralization assay using polyclonal equine sera raised against the prototype Bucyrus strain of EAV or by immunofluorescent assay using monoclonal antibodies to EAV.[58,59] These methods, although of proven reliability, are time-consuming, expensive, and cumbersome. The availability of modern nucleic acid–based assays has revolutionized diagnostic testing of clinical specimens for many infectious disease agents.[63–72] Compared to traditional VI, these assays are frequently more sensitive, less expensive, and less time-consuming.[64–66,71]

Histopathological examination utilizing indirect immunohistochemical staining is also useful for diagnosis of abortion in particular.[70] An avidin–biotin complex (ABC) for immunoperoxidase staining using monoclonal antibodies to individual EAV proteins has been successfully used to detect viral antigens in formalin-fixed, paraffin-embedded samples, as well as in frozen tissue sections.[56,57,74] Tissue samples should be fixed and saved in 10% neutral buffered formalin and submitted for histopathologic and immunohistochemical examination at an appropriate laboratory.

31.1.4.2 Molecular Techniques

Sensitive standard reverse transcription polymerase chain reaction (RT-PCR), RT-nested PCR (RT-nPCR), and real-time RT-PCR (rRT-PCR) assays for the detection of EAV nucleic acid in tissue culture supernatants and clinical specimens have been developed,[63,67,72,75–78] and these assays are being increasingly used for routine diagnostic purposes. These assays target different genes (ORFs 1b, 3, 4, 5, 6, and 7), and their sensitivity and specificity vary considerably. The sensitivity of RT-PCR-based assays is significantly increased by using either RT-nPCR that incorporate two primer pairs specific for ORF 1b or real-time

TaqMan RT-PCR that uses primers and a probe specific for a highly conserved region of ORF7.[80,81] RT-PCR-based assays have several potential advantages over the current VI procedure including fast and reliable means of virus detection.

31.2 METHODS

31.2.1 SAMPLE COLLECTION AND PREPARATION

31.2.1.1 Sample Collection

The most appropriate specimens for VI from horses with clinical signs include nasopharyngeal swabs or washings, conjunctival swabs, and citrated or ethylenediaminetetraacetic acid (EDTA) blood samples for separation of buffy coat cells. Heparinized blood is not suitable for virus isolation because of the inhibitory effect of heparin on the isolation of EAV in cell culture.[82] The sperm-rich fraction of the ejaculate is optimal for virus isolation from equine semen samples.[29] Placenta, fetal fluids, lung, spleen, and lymphoid tissues should be collected for virus isolation to confirm cases of EAV-induced abortion. A wide variety of organs and lymph nodes associated with the alimentary and respiratory tracts should be collected for virus isolation in suspected cases of "pneumoenteric" forms of EVA in young foals.

Specimens should be collected as soon as possible after the onset of clinical signs or suspected EAV infection. Nasopharyngeal and conjunctival swabs should be immediately placed in transport medium (any cell culture medium or balanced salt solution containing 2%–5% fetal bovine or calf serum) and either refrigerated or, preferably, frozen at −20°C or lower.[29] All other specimens for virus isolation should be packed in dry ice and dispatched by overnight delivery to an appropriate laboratory, except for blood samples that should be refrigerated.

For serologic diagnosis of acute EAV infection, acute and convalescent blood samples should be collected in tubes without anticoagulant. Serum is separated by centrifugation and either refrigerated or, preferably, frozen at −20°C or lower. Tissue samples should be fixed and saved in 10% neutral buffered formalin and submitted for histopathologic and immunohistochemical examination at an appropriate laboratory.

31.2.1.2 Processing of Clinical Samples for Virus Isolation

Nasal swabs and lacrimal swabs in virus transport medium are centrifuged at $500 \times g$ for 10 min at 4°C to eliminate cellular debris and then filtered through 0.45 μm syringe filters. The filtrate is then used for inoculation of cell monolayers of susceptible cell lines or can be stored at −80°C until further use. Blood samples collected into EDTA anticoagulant tubes are centrifuged at $500 \times g$ for 10 min. Plasma and buffy coat cells are aspirated and placed in 15 mL conical centrifuge tubes. Buffy coat cells are pelleted by centrifugation at $1500 \times g$ for 10 min at 4°C. The plasma is aspirated and discarded, and the cell pellet is washed in Eagle's Minimum Essential Medium (EMEM; Mediatech, Herndon, VA) twice and resuspended in 2 mL of EMEM. The cell suspensions are

aliquoted and frozen at −80°C until VI is performed. Tissue samples should be ground aseptically in EMEM to make 10% tissue homogenates for VI.

Cell culture isolation of EAV is usually done in the rabbit kidney-13 (RK-13) continuous cell line, and development of a cytopathic effect (CPE) in inoculated cells indicates the presence of virus. The identity of the virus isolate should always be confirmed by immunofluorescence or immunoperoxidase staining or by microneutralization assay with EAV-specific antiserum or monoclonal antibodies.

Isolation of EAV from equine semen samples is attempted in both high (passage level 399–409 [KY]) and low passage RK-13 (passage level 194–204; ATCC CCL-37, Manassas, VA) cell lines according to the OIE-described protocol.[58] It has been shown that the high passage RK-13 cell line is superior for the primary isolation of EAV particularly from semen as compared to the low passage RK-13 cell line. Briefly, semen samples are sonicated for 45 s (3 × 15 s), and sperm and cellular debris are sedimented by centrifugation ($230 \times g$, 10 min) at 4°C. Serial 10-fold dilutions (10^{-1}–10^{-3}) of the seminal plasma of each sample are made in supplemented EMEM, and 1 mL of each dilution is inoculated into each of 2 × 25 cm² flasks containing confluent monolayers of RK-13 cells. Inoculated flasks are incubated at 37°C for 1 h before being overlaid with supplemented EMEM containing 0.75% CMC. Flasks are incubated at 37°C and checked for the appearance of CPE on days 3 and 4 post inoculation. If there is no detectable CPE, a second blind passage is performed on day 4. The RK-13 cell monolayers are fixed and stained with a 0.2% crystal violet solution containing 2% formaldehyde on day 5 post inoculation in the case of the first passage and on day 4 post inoculation in the case of the second passage in cell culture.[83]

31.2.1.3 Processing of Clinical Samples for Molecular Testing

Any of the clinical samples including semen samples can be tested for viral RNA by molecular diagnostic assays. A prerequisite for the performance of molecular diagnostic assays is an efficient method to isolate and purify viral RNA from clinical specimens from horses.[84–86] Viral RNA is extremely susceptible to degradation by endogenous RNases (ribonucleases) that are present in all living cells. Therefore, the key to successful isolation of high-quality RNA for laboratory diagnosis is to ensure that neither endogenous nor exogenous RNases are introduced prior to or during the nucleic acid extraction procedure. The extracted RNA should be aliquoted into multiple tubes and stored at −20°C (short-term) or −80°C (long-term) to avoid degradation prior to testing.

Currently, there are many methods to isolate and purify viral RNA from clinical specimens, but only a very few studies have been published that describe the comparison of various RNA extraction methods of samples from animal origin including equine.[81,87,88] Most of the commercial reagent kits for RNA extraction are based on fluid phase, solid phase, or magnetic separation techniques.[89–91] The traditional RNA extraction method using acid guanidinium thiocyanate–phenol–chloroform technique developed by Chomczynski and Sacchi in 1979 has been replaced by commercial reagent kits that use glass fiber filters, silica gel spin columns, and magnetic beads to trap RNA present in the clinical specimens.[89–92] Furthermore, the new commercial reagent kits have increased the yield, purity, and quality of extracted viral RNA by removing contaminating genomic DNA, other RNAs, and RT-PCR enzyme inhibitors present in the clinical specimens. However, these kits vary significantly in technological principles used for nucleic acid extraction. There are various pros and cons to each technological principle, and therefore, the quality and purity of RNA extracted from various reagent kits can vary significantly.

Several automated nucleic acid isolation kits using magnetic beads and magnetic particle processors are commercially available.[93] Unlike automated liquid handling systems that move reagents into and out of a single well of a multiwell plate to perform the different steps of an RNA isolation procedure, the magnetic particle processors use permanent magnetic rods to collect magnetic beads from the solution and release them into other wells containing reagents for the subsequent steps of the protocol. The effectiveness of bead collection and transfer leads to superior washing and elution efficiency, as well as rapid processing of a large number of samples (e.g., 24, 48, or 96 samples in a batch). Additionally, the magnetic bead–based method can also be used for manual RNA purification. Previous studies have demonstrated that viral nucleic acid extraction using the magnetic bead extraction method increases sensitivity of the rRT-PCR assay compared to the column-based extraction method for the detection of EAV nucleic acid.[81]

31.2.2 DETECTION PROCEDURES

31.2.2.1 Standard RT-PCR and RT-nPCR

As indicated earlier, standard RT-PCR (two-step and one-step) and RT-nPCR assays for the detection of EAV in clinical samples have been previously described in the literature. RT-nPCR assay is more sensitive than the standard RT-PCR but prone to give false-positive results due to sample carryover and cross-contamination.[67]

31.2.2.2 Real-Time RT-PCR

The rRT-PCR has the following important advantages over the standard two-step RT-PCR: (1) eliminating the possibility of cross-contamination between samples with previously amplified products since the sample tube is never opened and (2) reducing the chance of false-positive reactions because the rRT-PCR product is detected with a sequence-specific probe. Therefore, standard RT-PCR and RT-nPCR assays have now been replaced by rRT-PCR assays for diagnosis of EAV. It has been demonstrated that at least one of the rRT-PCR assays described in literature has equal to or higher sensitivity than VI for the detection of EAV nucleic acid in semen samples.[81]

31.2.2.3 Insulated Isothermal RT-PCR

Recently, GeneReach (GeneReach USA, Lexington, MA) has developed a TaqMan® probe-based insulated isothermal RT-PCR (iiRT-PCR) assay for the detection of EVA nucleic acid in semen and tissue samples.[94,95] The assay is based on iiRT-PCR for qualitative detection of EAV RNA in clinical samples. Fluorogenic probe hydrolysis chemistry is used to generate a fluorescent signal when a specific RNA sequence of EAV is amplified. This assay is specially designed to be used on a compatible iiRT-PCR instrument, POCKIT™ Nucleic Acid Analyzer (GeneReach USA, Lexington, MA).[92,93] The POCKIT Xpress™ Portable PCR Platform for on-site testing is suitable to be used in veterinary clinics, racetracks, breeding facilities, and diagnostic laboratories.[94]

31.3 CONCLUSION AND FUTURE PERSPECTIVES

EAV infection continues to pose a great threat to the equine industry in the United States and around the world. Over the years, the importation of carrier stallions and virus-contaminated semen into the United States unquestionably has been responsible for the introduction of new strains of EAV and precipitated outbreaks of EVA.[97–100] In 2006, a multistate occurrence of EVA was confirmed in Quarter horses for the first time, beginning in New Mexico and spreading to at least 10 other states (Utah, Alabama, Oklahoma, Kansas, Montana, Colorado, Texas, California, and Idaho).[9] This was the most extensive recorded occurrence of EVA in the United States and was primarily the result of interstate shipment of infective semen (fresh, cooled, or frozen) from virus-shedding stallions. Embryo transfer from mares bred to carrier stallions has increased the spread of EAV around the world. Therefore, the identification of persistently infected stallions and the institution of biosecurity and management practices to prevent the introduction of EAV-infected horses into new premises during acute outbreaks can prevent further outbreaks of EVA. Very clearly, rRT-PCR has significant advantages over VI in terms of reproducibility between laboratories, ease and speed of completion, and cost; thus, it is logical that the most accurate rRT-PCR assay would replace VI as the prescribed test for international trade. Furthermore, validation of the recently described cELISA could help to overcome the limitations associated with the VNT, which is the current "gold standard" for detection of serum antibodies to EAV.

There has been significant recent progress in understanding the molecular biology of EAV and mechanism of virus replication in cells. The application of cutting-edge molecular biology techniques along with development of reverse genetic systems to manipulate the viral genome has facilitated these advancements. However, neither the location of EAV persistence in the male reproductive tract nor the mechanisms of persistence have been examined in depth. Ongoing studies using contemporary molecular and genomic techniques will facilitate the identification of the cellular factors and pathways associated with the assembly of EAV particles in infected cells and the establishment of the carrier state in stallions. Such studies are prerequisite to the logical design and development of antiviral drugs to clear the virus from the reproductive tract of carrier stallions. Recent studies indicate that the existing vaccines may not induce complete protection against all field strains of EAV. Future development of efficacious vaccines that will engender broad heterotypic immunity to all strains of EAV will require ongoing comparative sequence analysis of the ORF5 of field isolates of the virus from around the world. Logically, future vaccines should be developed utilizing recombinant DNA technology, such as the use of infectious clones as a stable genetic background to develop marker (differentiating infected from vaccinated animals [DIVA]) vaccines that distinguish naturally infected horses from vaccinated ones.

ACKNOWLEDGMENT

This work is partially supported by the USDA-National Institute of Food and Agriculture Research Initiative competitive grant numbers 2013-68004-20360.

REFERENCES

1. Doll, E.R., Bryans, J.T., McCollum, W.H., and Crowe, M.E. Isolation of a filterable agent causing arteritis of horses and abortion by mares; its differentiation from the equine abortion (influenza) virus. *Cornell Vet* **47**, 3–41 (1957).
2. Doll, E.R., Knappenberger, R.E., and Bryans, J.T. An outbreak of abortion caused by the equine arteritis virus. *Cornell Vet* **47**, 69–75 (1957).
3. Dunowska, M., Biggs, P.J., Zheng, T., and Perrott, M.R. Identification of a novel nidovirus associated with a neurological disease of the Australian brushtail possum (*Trichosurus vulpecula*). *Vet Microbiol* **156**, 418–424 (2012).
4. Snijder, E.J., Kikkert, M., and Fang, Y. Arterivirus molecular biology and pathogenesis. *J Gen Virol* **94**, 2141–2163 (2013).
5. Snijder, E.J., and Spann, W. J. M. Arteriviruses. In *Fields Virology* (eds. Knipe, D.M., Howley, Peter M., Griffin, Diane E., Lamb, Robert A., Martin, Malcolm A., Roizman, B., and Straus, S. E.) pp. 1337–1355 (Lippincott Williams & Wilkins, Philadelphia, PA, 2007).
6. van Dinten, L.C., den Boon, J.A., Wassenaar, A.L., Spaan, W.J., and Snijder, E.J. An infectious arterivirus cDNA clone: Identification of a replicase point mutation that abolishes discontinuous mRNA transcription. *Proc Natl Acad Sci USA* **94**, 991–996 (1997).
7. den Boon, J.A. et al. Equine arteritis virus is not a togavirus but belongs to the coronavirus-like superfamily. *J Virol* **65**, 2910–2920 (1991).
8. Miszczak, F. et al. Emergence of novel equine arteritis virus (EAV) variants during persistent infection in the stallion: Origin of the 2007 French EAV outbreak was linked to an EAV strain present in the semen of a persistently infected carrier stallion. *Virology* **423**, 165–174 (2012).
9. Zhang, J. et al. Molecular epidemiology and genetic characterization of equine arteritis virus isolates associated with the 2006–2007 multi-state disease occurrence in the USA. *J Gen Virol* **91**, 2286–2301 (2010).
10. Balasuriya, U.B. et al. Development and characterization of an infectious cDNA clone of the virulent Bucyrus strain of Equine arteritis virus. *J Gen Virol* **88**, 918–924 (2007).
11. Balasuriya, U.B. et al. Genetic characterization of equine arteritis virus during persistent infection of stallions. *J Gen Virol* **85**, 379–390 (2004).
12. Ziebuhr, J., Snijder, E.J., and Gorbalenya, A.E. Virus-encoded proteinases and proteolytic processing in the Nidovirales. *J Gen Virol* **81**, 853–879 (2000).
13. van Dinten, L.C., Rensen, S., Gorbalenya, A.E., and Snijder, E.J. Proteolytic processing of the open reading frame 1b-encoded part of arterivirus replicase is mediated by nsp4 serine protease and is essential for virus replication. *J Virol* **73**, 2027–2037 (1999).
14. Firth, A.E. et al. Discovery of a small arterivirus gene that overlaps the GP5 coding sequence and is important for virus production. *J Gen Virol* **92**, 1097–1106 (2011).
15. Balasuriya, U.B., Go, Y.Y., and MacLachlan, N.J. Equine arteritis virus. *Vet Microbiol* **167**, 93–122 (2013).
16. den Boon, J.A., Kleijnen, M.F., Spaan, W.J., and Snijder, E.J. Equine arteritis virus subgenomic mRNA synthesis: Analysis of leader-body junctions and replicative-form RNAs. *J Virol* **70**, 4291–4298 (1996).
17. Balasuriya, U.B. et al. Expression of the two major envelope proteins of equine arteritis virus as a heterodimer is necessary for induction of neutralizing antibodies in mice immunized with recombinant Venezuelan equine encephalitis virus replicon particles. *J Virol* **74**, 10623–10630 (2000).
18. Balasuriya, U.B. et al. Alphavirus replicon particles expressing the two major envelope proteins of equine arteritis virus induce high level protection against challenge with virulent virus in vaccinated horses. *Vaccine* **20**, 1609–1617 (2002).
19. Ma, Y., Jiang, Y.B., Xiao, S.B., Fang, L.R., and Chen, H.C. Co-expressed GP5 and M proteins of porcine reproductive and respiratory syndrome virus can form heterodimers. *Wei Sheng Wu Xue Bao* **46**, 639–643 (2006).
20. Jiang, Y. et al. Immunogenicity and protective efficacy of recombinant pseudorabies virus expressing the two major membrane-associated proteins of porcine reproductive and respiratory syndrome virus. *Vaccine* **25(3)**, 547–560.
21. Balasuriya, U.B., Zhang, J., Go, Y.Y., and MacLachlan, N.J. Experiences with infectious cDNA clones of equine arteritis virus: Lessons learned and insights gained. *Virology* **462–463**, 388–403 (2014).
22. Lu, Z. et al. Chimeric viruses containing the N-terminal ectodomains of GP5 and M proteins of porcine reproductive and respiratory syndrome virus do not change the cellular tropism of equine arteritis virus. *Virology* **432**, 99–109 (2012).
23. Go, Y.Y. et al. Complex interactions between the major and minor envelope proteins of equine arteritis virus determine its tropism for equine CD3+ T lymphocytes and CD14+ monocytes. *J Virol* **84**, 4898–4911 (2010).
24. Go, Y.Y. et al. Assessment of correlation between in vitro CD3+ T cell susceptibility to EAV infection and clinical outcome following experimental infection. *Vet Microbiol* **157**, 220–225 (2012).
25. Zhang, J. et al. Development and characterization of an infectious cDNA clone of the modified live virus vaccine strain of equine arteritis virus. *Clin Vaccine Immunol* **19**, 1312–1321 (2012).
26. Balasuriya, U.B. et al. Development and characterization of an infectious cDNA clone of the virulent Bucyrus strain of Equine arteritis virus. *J Gen Virol* **88**, 918–924 (2007).
27. Balasuriya, U.B. et al. Equine arteritis virus derived from an infectious cDNA clone is attenuated and genetically stable in infected stallions. *Virology* **260**, 201–208 (1999).
28. Balasuriya, U.B. Equine viral arteritis. *Vet Clin North Am Equine Pract* **30**, 543–560 (2014).
29. Timoney, P.J. and McCollum, W.H. Equine viral arteritis. *Vet Clin North Am Equine Pract* **9**, 295–309 (1993).
30. Echeverria, M.G., Pecoraro, M.R., Galosi, C.M., Etcheverrigaray, M.E., and Nosetto, E.O. The first isolation of equine arteritis virus in Argentina. *Rev Sci Tech* **22**, 1029–1033 (2003).

31. McFadden, A.M. et al. Evidence for absence of equine arteritis virus in the horse population of New Zealand. *N Z Vet J* **61(5)**, 300–304 (2013).

32. McCollum, W.H. and Bryans, J.T. Serological identification of infection by equine arteritis virus in horses of several countries. In *Proceedings of the Third Interantional Conference of Equine Infectious Diseases* (eds. Bryans, J.T. and Gerber, H.) pp. 256–263 (S. Karger, Basel, Germany, 1973).

33. McCue, P.M., Hietala, S.K., Spensely, M.S., Mihalyi, J., and Hughes, J.P. Prevalence of equine viral arteritis in California horses. *California Vet* **45**, 24–26 (1991).

34. McKenzie, J. Survey of stallions for equine arteritis virus. *Surveillance* **16**, 17–18 (1996).

35. Moraillon, A. and Moraillon, R. Results of an epidemiological investigation on viral arteritis in France and some other European and African countries. *Ann Rech Vet* **9**, 43–54 (1978).

36. Timoney, P.J. Equine viral arteritis: Epidemiology and control. *J Equine Vet Sci* **8**, 54–59 (1988).

37. Hullinger, P.J., Gardner, I.A., Hietala, S.K., Ferraro, G.L., and MacLachlan, N.J. Seroprevalence of antibodies against equine arteritis virus in horses residing in the United States and imported horses. *J Am Vet Med Assoc* **219**, 946–949 (2001).

38. Burki, F., Hofer, A., and Nowotny, N. Objective data plead to suspend import-bans for seroreactors against equine arteritis virus except for breeder stallions. *J Appl Anim Res* **1**, 31–42 (1992).

39. Rola, J., Larska, M., Rola, J.G., Belak, S., and Autorino, G.L. Epizotiology and phylogeny of equine arteritis virus in hucul horses. *Vet Microbiol* **148**, 402–407 (2011).

40. Collins, J.K., Kari, S., Ralston, S.L., Bennet, D.G., Traub-Dargatz, J.L., and McKinnon, A.O. Equine viral arteritis in a veterinary teaching hospital. *Prev Vet Med* **4**, 389–397 (1987).

41. Guthrie, A.J. et al. Lateral transmission of equine arteritis virus among Lipizzaner stallions in South Africa. *Equine Vet J* **35**, 596–600 (2003).

42. McCollum, W.H., Timoney, P.J., and Tengelsen, L.A. Clinical, virological and serological responses of donkeys to intranasal inoculation with the KY-84 strain of equine arteritis virus. *J Comp Pathol* **112**, 207–211 (1995).

43. McCollum, W.H., Prickett, M.E., and Bryans, J.T. Temporal distribution of equine arteritis virus in respiratory mucosa, tissues and body fluids of horses infected by inhalation. *Res Vet Sci* **12**, 459–464 (1971).

44. McCollum, W.H. Pathologic features of horses given avirulent equine arteritis virus intramuscularly. *Am J Vet Res* **42**, 1218–1220 (1981).

45. Glaser, A.L., de Vries, A.A., Rottier, P.J., Horzinek, M.C., and Colenbrander, B. Equine arteritis virus: Clinical symptoms and prevention. *Tijdschr Diergeneeskd* **122**, 2–7 (1997).

46. Glaser, A.L., de Vries, A.A., Rottier, P.J., Horzinek, M.C., and Colenbrander, B. Equine arteritis virus: A review of clinical features and management aspects. *Vet Q* **18**, 95–99 (1996).

47. Balasuriya, U.B., Timoney, P.J., McCollum, W.H., and MacLachlan, N.J. Phylogenetic analysis of open reading frame 5 of field isolates of equine arteritis virus and identification of conserved and nonconserved regions in the GL envelope glycoprotein. *Virology* **214**, 690–697 (1995).

48. Stadejek, T. et al. Genetic diversity of equine arteritis virus. *J Gen Virol* **80(3)**, 691–699 (1999).

49. McCollum, W.H. and Timoney, P.J. Experimental observations on the virulence of isolates of equine arteritis virus. In *Proceedings of the Eighth International Conference Equine Infectious Diseases* (eds. Wernery, U., Wade, J.F., Mumford, J.A., and Kaaden, O.-R.) pp. 558–559 (R & W Publications (Newmarket) Limited, Dubai, United Arab Emirates, 1998).

50. Balasuriya, U.B. and MacLachlan, N.J. The immune response to equine arteritis virus: Potential lessons for other arteriviruses. *Vet Immunol Immunopathol* **102**, 107–129 (2004).

51. Pronost, S. et al. Description of the first recorded major occurrence of equine viral arteritis in France. *Equine Vet J* **42**, 713–720 (2010).

52. Neu, S.M., Timoney, P.J., and Lowry, S.R. Changes in semen quality following experimental equine arteritis virus infection in the stallion. *Theriogenology* **37**, 407–431 (1992).

53. Campos, J.R., Breheny, P., Araujo, R.R., Troedsson, M.H.T., Squires, E.L., Timoney, P. J., and Balasuriya, U.B.R. Semen quality of stallions challenged with the Kentucky 84 strain of EAV. *Theriogenology* **82(8)**, 1068–1079 (2014).

54. Moore, B.D. et al. Virulent and avirulent strains of equine arteritis virus induce different quantities of TNF-alpha and other proinflammatory cytokines in alveolar and blood-derived equine macrophages. *Virology* **314**, 662–670 (2003).

55. Go, Y.Y. et al. Genome-wide association study among four horse breeds identifies a common haplotype associated with in vitro CD3+ T cell susceptibility/resistance to equine arteritis virus infection. *J Virol* **85**, 13174–13184 (2011).

56. Del Piero, F. Equine viral arteritis. *Vet Pathol* **37**, 287–296 (2000).

57. MacLachlan, N.J. et al. Fatal experimental equine arteritis virus infection of a pregnant mare: Immunohistochemical staining of viral antigens. *J Vet Diagn Invest* **8**, 367–374 (1996).

58. OIE (Office International des Epizooties). *OIE Manual of Diagnostic Tests and Vaccines for Terrestrial Animals* (World Organisation for Animal Health, Paris, France, 2004).

59. Go, Y.Y., Wong, S. J., Branscum, A., Demarset, V. L., Shuck, K. M., Vickers, M. L., Zhang, J., McCollum, W. H., Timoney, P. J., and Balasuriya, U. B. R. Development of a Fluorescent Microsphere Immunoassay for Detection of Antibodies Specific to Equine Arteritis Virus and Comparison with the Virus Neutralization Test. *Clin Vaccine Immunol* 15, 12 (2007).

60. Chung, C., Wilson, C., Timoney, P., Balasuriya, U., Adams, E., Adams, D. S., Evermann, J. F., Clavijo, A., Shuck, K., Rodgers, S., Lee, S. S. & McGuire, T. C. Validation of an Improved Competitive Enzyme-Linked Immunosorbent Assay to Detect Equine Arteritis Virus Antibody. *J Vet Diagn Invest* **25**, 727–735 (2013).

61. Chung, C.J. et al. Enhanced sensitivity of an antibody competitive blocking enzyme-linked immunosorbent assay using Equine arteritis virus purified by anion-exchange membrane chromatography. *J Vet Diagn Invest.* 27, 728–738 (2015).

62. Pfahl, K. et al. Further evaluation and validation of a commercially available competitive ELISA (cELISA) for the detection of antibodies specific to equine arteritis virus (EAV). *Vet. Rec.* 178, 95 (2016).

63. Balasuriya, U.B. et al. Detection of equine arteritis virus by real-time TaqMan reverse transcription-PCR assay. *J Virol Methods* **101**, 21–28 (2002).

64. Belak, S. The molecular diagnosis of porcine viral diseases: A review. *Acta Vet Hung* **53**, 113–124 (2005).

65. Belak, S. Molecular diagnosis of viral diseases, present trends and future aspects: A view from the OIE Collaborating Centre for the Application of Polymerase Chain Reaction Methods for Diagnosis of Viral Diseases in Veterinary Medicine. *Vaccine* **25**, 5444–5452 (2007).

66. Belak, S. and Thoren, P. Molecular diagnosis of animal diseases: Some experiences over the past decade. *Expert Rev Mol Diagn* **1**, 434–443 (2001).

67. Gilbert, S.A., Timoney, P.J., McCollum, W.H., and Deregt, D. Detection of equine arteritis virus in the semen of carrier stallions by using a sensitive nested PCR assay. *J Clin Microbiol* **35**, 2181–2183 (1997).

68. Hussy, D., Stauber, N., Leutenegger, C.M., Rieder, S., and Ackermann, M. Quantitative fluorogenic PCR assay for measuring ovine herpesvirus 2 replication in sheep. *Clin Diagn Lab Immunol* **8**, 123–128 (2001).

69. Leutenegger, C.M. et al. Real-time TaqMan PCR as a specific and more sensitive alternative to the branched-chain DNA assay for quantitation of simian immunodeficiency virus RNA. *AIDS Res Hum Retroviruses* **17**, 243–251 (2001).

70. McKillen, J. et al. Molecular beacon real-time PCR detection of swine viruses. *J Virol Methods* **140**, 155–165 (2007).

71. Pusterla, N., Madigan, J.E., and Leutenegger, C.M. Real-time polymerase chain reaction: A novel molecular diagnostic tool for equine infectious diseases. *J Vet Intern Med* **20**, 3–12 (2006).

72. Westcott, D.G. et al. Use of an internal standard in a closed one-tube RT-PCR for the detection of equine arteritis virus RNA with fluorescent probes. *Vet Res* **34**, 165–176 (2003).

73. Del Piero, F. Diagnosis of equine arteritis virus infection in two horses by using monoclonal antibody immunoperoxidase histochemistry on skin biopsies. *Vet Pathol* **37**, 486–487 (2000).

74. Lopez, J.W., del Piero, F., Glaser, A., and Finazzi, M. Immunoperoxidase histochemistry as a diagnostic tool for detection of equine arteritis virus antigen in formalin fixed tissues. *Equine Vet J* **28**, 77–79 (1996).

75. Chirnside, E.D. and Spaan, W.J. Reverse transcription and cDNA amplification by the polymerase chain reaction of equine arteritis virus (EAV). *J Virol Methods* **30**, 133–140 (1990).

76. St-Laurent, G., Morin, G., and Archambault, D. Detection of equine arteritis virus following amplification of structural and nonstructural viral genes by reverse transcription-PCR. *J Clin Microbiol* **32**, 658–665 (1994).

77. Sekiguchi, K. et al. Detection of equine arteritis virus (EAV) by polymerase chain reaction (PCR) and differentiation of EAV strains by restriction enzyme analysis of PCR products. *Arch Virol* **140**, 1483–1491 (1995).

78. Ramina, A. et al. Detection of equine arteritis virus in semen by reverse transcriptase polymerase chain reaction-ELISA. *Comp Immunol Microbiol Infect Dis* **22**, 187–197 (1999).

79. Starick, E. Rapid and sensitive detection of equine arteritis virus in semen and tissue samples by reverse transcription-polymerase chain reaction, dot blot hybridisation and nested polymerase chain reaction. *Acta Virol* **42**, 333–339 (1998).

80. Lu, Z. et al. Comparison of two real-time reverse transcription polymerase chain reaction assays for the detection of Equine arteritis virus nucleic acid in equine semen and tissue culture fluid. *J Vet Diagn Invest* **20**, 147–155 (2008).

81. Miszczak, F. et al. Evaluation of two magnetic-bead-based viral nucleic acid purification kits and three real-time reverse transcription-PCR reagent systems in two TaqMan assays for equine arteritis virus detection in semen. *J Clin Microbiol* **49**, 3694–3696 (2011).

82. Asagoe, T. et al. Effect of heparin on infection of cells by equine arteritis virus. *J Vet Med Sci* **59**, 727–728 (1997).

83. McCollum, W.H., Doll, E.R., Wilson, J.C., and Cheatham, J. Isolation and propagation of equine arteritis virus in monolayer cell cultures of rabbit kidney. *Cornell Vet* **52**, 452–458 (1962).

84. Murphy, J. and Bustin, S.A. Reliability of real-time reverse-transcription PCR in clinical diagnostics: Gold standard or substandard? *Expert Rev Mol Diagn* **9**, 187–197 (2009).

85. Bustin, S.A. and Mueller, R. Real-time reverse transcription PCR (qRT-PCR) and its potential use in clinical diagnosis. *Clin Sci (Lond)* **109**, 365–379 (2005).

86. Bustin, S.A., Benes, V., Nolan, T., and Pfaffl, M.W. Quantitative real-time RT-PCR—a perspective. *J Mol Endocrinol* **34**, 597–601 (2005).

87. Deng, M.Y., Wang, H., Ward, G.B., Beckham, T.R., and McKenna, T.S. Comparison of six RNA extraction methods for the detection of classical swine fever virus by real-time and conventional reverse transcription-PCR. *J Vet Diagn Invest* **17**, 574–578 (2005).

88. Knepp, J.H., Geahr, M.A., Forman, M.S., and Valsamakis, A. Comparison of automated and manual nucleic acid extraction methods for detection of enterovirus RNA. *J Clin Microbiol* **41**, 3532–3536 (2003).

89. Zahringer, H. Old and new ways to RNA. *Lab Times*, p. 10 (2012).

90. Fang, X., Willis, R.C., Burrell, A., Evans, K., Hoang, Q., Xu, W., and Bounpheng, M. Automation of nucleic acid isolation on KingFisher magnetic particles. *JALA* **12**, 7 (2007).

91. Berensmeier, S. Magnetic particles for the separation and purification of nucleic acids. *Appl Microbiol Biotechnol* **73**, 495–504 (2006).

92. Chomczynski, P. and Sacchi, N. Single-step method of RNA isolation by acid guanidinium thiocyanate-phenol-chloroform extraction. *Anal Biochem* **162**, 156–159 (1987).

93. Carossino, M., Lee, P. Y. A., Nam, B., Skillman, A., Shuck, K. M. Timoney, P. J., Tsai, Y. L., Li-Juan Ma, L. J., Chang, H. F. G, Wang, H. T. T., and Balasuriya, U. B. R. Development and evaluation of a reverse transcription-insulated isothermal polymerase chain reaction (RT-iiPCR) assay for detection of equine arteritis virus in equine semen and tissue samples using the POCKITTM system. *J. Virol. Methods (submitted)*.

94. Balasuriya, U.B. Type A influenza virus detection from horses by real-time RT-PCR and insulated isothermal RT-PCR. *Methods Mol Biol* **1161**, 393–402 (2014).

95. Tsai, Y.L. et al. Development of TaqMan probe-based insulated isothermal PCR (iiPCR) for sensitive and specific on-site pathogen detection. *PLoS One* **7**, e45278 (2012).

96. Chang, H.F. et al. A thermally baffled device for highly stabilized convective PCR. *Biotechnol J* **7**, 662–666 (2012).

97. Balasuriya, U.B., Maclachlan, N.J., De Vries, A.A., Rossitto, P.V., and Rottier, P.J. Identification of a neutralization site in the major envelope glycoprotein (GL) of equine arteritis virus. *Virology* **207**, 518–527 (1995).

98. Balasuriya, U.B. et al. Serologic and molecular characterization of an abortigenic strain of equine arteritis virus isolated from infective frozen semen and an aborted equine fetus. *J Am Vet Med Assoc* **213**, 1586–1289, 1570 (1998).

99. McCollum, W.H., Timoney, P.J. Lee Jr, J.W., Habacker, P.L., Balasuriya, U.B.R., and MacLachlan, N.J. Features of an outbreak of equine viral arteritis on a breeding farm associated with abortion and fatal interstitial pneumonia in neonatal foals. In *Proceeding of the Eight International Conference (Equine Infectious Diseases VIII)* (eds. Wernery, U., Wade, J.F., Mumford, J.A., and Kaaden, O.-R.) pp. 559–560 (R & W Publications (Newmarket) Limited, Dubai, United Arab Emirates, 1998).

100. Balasuriya, U.B. et al. Genetic stability of equine arteritis virus during horizontal and vertical transmission in an outbreak of equine viral arteritis. *J Gen Virol* **80(8)**, 1949–1958 (1999).

32 Porcine Reproductive and Respiratory Syndrome Virus

Dongyou Liu

CONTENTS

32.1 Introduction 287
    32.1.1 Classification..287
    32.1.2 Morphology and Genome Organization...288
    32.1.3 Biology and Epidemiology 289
    32.1.4 Clinical Features and Pathogenesis 289
        32.1.4.1 Sows..289
        32.1.4.2 Piglets...289
        32.1.4.3 Boars...289
        32.1.4.4 Weaners and Growers 289
    32.1.5 Diagnosis 290
    32.1.6 Treatment and Prevention 290
32.2 Methods 291
    32.2.1 Sample Preparation...291
    32.2.2 Detection Procedures..291
        32.2.2.1 Conventional RT-PCR..291
        32.2.2.2 Real-Time RT-PCR 291
32.3 Conclusion 292
References...292

32.1 INTRODUCTION

Porcine reproductive and respiratory syndrome (PRRS) is an economically important disease of swine resulting from the infection with a single-stranded, positive-sense RNA virus called porcine reproductive and respiratory syndrome virus (PRRSV). First recognized in the United States in 1987 and subsequently detected in Europe and other parts of the world, PRRS is characterized by late-term reproductive failure in sows (e.g., infertility, abortions, stillbirths, or the birth of weak piglets) and respiratory distress in piglets. With the capacity to induce severe clinical disease and maintain a life-long subclinical infection in swine population, PRRSV is a serious threat to the swine industry worldwide. In the United States, the estimated cost associated with PRRS amounts to >600 million U.S. dollars per year.

32.1.1 CLASSIFICATION

PRRSV is a single-stranded RNA virus classified in the genus *Arterivirus*, family *Arteriviridae*, order *Nidovirales*.

The order *Nidovirales* (nidus means nest in Latin, reflecting to the fact that members of this order produce a 3′-coterminal nested set of subgenomic mRNA during infection) covers RNA viruses that possess seven conserved domains (5′-TM2–3CLpro-TM3-RdRp-Zm-HEL1-NendoU-3′). Of these, TM2 and TM3 are transmembrane domains; 3CLpro is a 3C-like protease (a catalytic His-Cys dyad); RdRp is the RNA polymerase; Zm is a Zn cluster–binding domain fused with a helicase (HEL1); and NendoU is an uridylate-specific endonuclease. While the first three are encoded in open reading frame (ORF)1a, the remaining four are encoded in ORF1b.

Depending on the size of the genome, the order *Nidovirales* is conveniently arranged into two divisions: the large nidoviruses with genomes of 26.3–31.7 kb are found in the families *Coronaviridae* and *Roniviridae*, while the small nidoviruses with genomes of 12.7–15.7 kb are included in the families *Arteriviridae* and *Mesoniviridae* [1].

Currently, the family *Arteriviridae* consists of a single genus *Arterivirus*, in which five species are recognized: equine arteritis virus (EAV, the type species; see Chapter 31), lactate dehydrogenase elevating virus (LDV), PRRSV, simian hemorrhagic fever virus (SHFV), and wobbly possum disease virus (WPDV) [2]. Interestingly, EAV and PRRSV are associated with respiratory and reproductive disease in equids and swine, respectively. LDV is generally nonpathogenic and occurs in field and the laboratory mice. SHFV is found in colobus monkey (*Procolobus rufomitratus tephrosceles*) and in African red-tailed (guenon) monkeys (*Cercopithecus ascanius*). WPDV causes neurologic disease among free-ranging Australian brushtail possums (*Trichosurus vulpecula*) in New Zealand.

PRRSV is further distinguished into genotype 1 (European genotype or PRRSV-1) and genotype 2 (Northern American genotype or PRRSV-2). While PRRSV-1 mainly occurs in Europe, PRRSV-2 is present in North and South America as well as Asia, including a highly virulent type 2 PRRSV in Southeast Asia that harbors a discontinuous 30-amino-acid deletion in its nonstructural protein (nsP) 2 [3–5]. Although both genotypes 1 and 2 are infective to cells of the monocyte/macrophage lineage, they demonstrate differences at the protein and nucleic acid levels. For example, PRRSV-1 and PRRSV-2 share 75.2%–81.6% identity in the structural M protein, 73.2%–75.0% in nsP9, 50.5%–54.3% in nsP1, and 24.4%–28.0% in nsP2.

Recent examinations of ORF5 and ORF7 gene sequences and nucleocapsid protein from East European PRRSV isolates revealed the existence of three subtypes within PRRSV-1, that is, a pan-European subtype 1 and East European subtypes 2 and 3, whose nucleocapsid proteins are of 128, 125, and 124 amino acids (aa) in size, respectively [6]. Additionally, phylogenetic analysis of ORF5 gene sequence enabled the identification of 12 clades (A–L) within PRRSV-1 subtype 1 and 9 lineages within PRRSV-2 [7,8].

32.1.2 MORPHOLOGY AND GENOME ORGANIZATION

Like other arteriviruses, PRRSV possesses an enveloped, pleomorphic but roughly spherical virion of 45–60 nm in diameter. The virion displays surface patterns with no obvious spikes. Underneath the envelope is an icosahedral core (shell) measuring 25–35 nm in diameter, which encapsidates a monopartite, single-stranded RNA(+) of 15.1–15.4 kb in length.

The genome of PRRSV consists of a methyl-capped 5′ untranslated region (UTR; 156–224 nt long), 10 ORFs (ORF1a, ORF1b, ORF2a, ORF2b, ORF3, ORF4, ORF5, ORF5a, ORF6, ORF7), and a 3′ UTR (59–117 nt long) with a polyadenylated [poly(A)] tail [5].

Located at the 5′ end and accounting for 75% of the viral genome, ORF1a and ORF1b encode two long polyproteins (pp) (1727–2502 aa for pp1a and 3175–3959 aa for pp1ab), which, after enzymatic cleavage by four viral proteases (encoded in the nsP1α, nsP1β, nsP2, and nsP4 regions), produce 14 nsPs, including nsP1α, nsP1β, nsP2, and nsP3–6, nsP7α, nsP7β, nsP8–12, and two additional proteins, nsP2TF and nsP2N, which result from ribosomal frameshifting (Table 32.1) [9].

ORF2–ORF7 are located at the 3′ end and encode eight structural proteins designated as glycoprotein (GP) 2, E, GP3, GP4, GP5, ORF5a, M, and N (through an expression from a 3′-coterminal nested set of subgenomic RNAs) (Table 32.1). It is of interest that ORF5 (encoding the major

---

TABLE 32.1
Porcine Reproductive and Respiratory Syndrome Virus Proteins and Functions

ORF	Protein	Molecular Mass (kDa)	Function
1a	nsP1α	19	Papain-like cysteine protease
	nsP1β	25	Papain-like cysteine protease
	nsP2	114	Cysteine protease
	nsP3	29	Transmembrane protein
	nsP4	21	Serine protease
	nsP5	NA	Transmembrane protein
	nsP6	NA	Modulation of nsP7 function
	nsP7α	16	RNA synthesis and protein translation
	nsP7β	14	RNA synthesis and protein translation
	nsP8	5	Modulation of nsP7 function
1b	nsP9	85	RNA-dependent RNA polymerase
	nsP10	49	Helicase
	nsP11	25	Endoribonuclease
	nsP12	21	Genome transcription and replication
2a	GP2	30	Minor envelope protein, interacting with CD163
2b	E	10	Minor envelope protein, possibly forming oligomeric ion channel
3	GP3	45	Minor envelope protein
4	GP4	35	Minor envelope protein, interacting with CD163
5	GP5	26	Major envelope protein, interacting with sialoadhesin
6	M	19	Major envelope protein, interacting with heparan sulfate
7	N	15	Nucleocapsid

Source: Adapted from Charerntantanakul, W., *World J. Virol.*, 1(1), 23, 2012.

nsP, nonstructural protein; GP, glycoprotein; ORF, open reading frames; NA, not available.

envelope glycoprotein) exhibits marked genetic variation and represents an ideal target for phylogenetic assessment of PRRSV isolates.

32.1.3 Biology and Epidemiology

The primary target cells of PRRSV infection in vivo are porcine pulmonary alveolar macrophages. PRRSV relies on the GP5 to attach to heparan sulfate on the cell surface and utilizes the GP2, GP3, and GP4 trimer to interact with host CD163 (cysteine-rich scavenger receptor, or SRCR). As the primary and core receptor for PRRSV, CD163 determines the susceptibility of cells to the virus. Sialoadhesin (CD169/ Siglec-1) is an immunoglobulin-like lectin (carbohydrate-binding protein) that carries sialic acids and mediates endocytosis. Other cellular molecules with potential receptor role in PRRSV binding are vimentin, CD151, and dendritic cell–specific intercellular adhesion molecule-3-grabbing nonintegrin (CD209) [10].

Upon entry into the cell via clathrin-mediated endocytosis, the viral membrane fuses with the endomal membrane, and the viral ssRNA is released into the cytoplasm. After translation from the viral RNA, replicase pp1a and pp1ab interact with nsPs to form replication and transcription complex, which generates a full-length genome and a minus RNA strand. The synthesis of sgmRNA from the minus RNA strand facilitates the expression of structural proteins. Subsequently, the new genomes are assembled into the nucleocapsids at the host endoplasmic reticulum and then released via exocytosis [11].

In experimentally infected sows, PRRSV is detectable in the tonsils, heart, uterus, kidneys, and lymph nodes. In piglets from intranasally inoculated sows, PRRSV replicates primarily in lymphoid tissues and then spreads to the lungs, heart, liver, spleen, and kidneys. In boars, PRRSV persists in lymphoid tissues (such as the tonsils), although some infected macrophages may bring the virus to the tissues of the male reproductive tract and semen. Once contaminated semen appears in the uterus, PRRSV infection may begin from the endometrial tissues and regional lymph nodes following hematogenic or lymphogenic dissemination throughout the sow organism.

PRRS occurs only in pigs, and its horizontal transmission is made possible by close contacts between infected and naïve animals, through nasal secretions, saliva, feces, urine, and semen. Other factors that facilitate the virus spread include movement of carrier pigs, aerosols, droplets, dust, and contaminated boots, clothing, equipment, vehicles, as well as infected mallard duck.

Following its first report in the United States in 1987, type 1 (European genotype) and type 2 (North American genotype) PRRSV were isolated in Europe and North America in the early 1990s. Of the three subtypes within PRRSV-1 (European genotype), subtype 1 (Lelystad virus-like) is distributed throughout Europe and the world, while subtypes 2 and 3 are mainly present in Eastern Europe. Interestingly, PRRSV-1 isolates exhibit length differences in the nucleocapsid protein (ranging in size from 124, 128, to 132 aa) that are not observed in PRRSV-2. In addition to PRRSV-2 that occurs in North and South America and Asia, a type 2 highly pathogenic PRRSV (HP-PRRSV) strain was identified in swine farms in China in 2006 [3]. HP-PRRSV is characterized by a unique discontinuous deletion of 30 aa in the nsP2 and is responsible for causing high fever and severe morbidity and mortality in pigs of all ages. Subsequently, HP-PRRSV was also detected in other Asian countries such as Vietnam [3].

32.1.4 Clinical Features and Pathogenesis

PRRSV infection may result in a range of clinical presentations that are influenced by the host age and immune status, as well as the virulence of PRRSV strains involved. For example, PRRSV is commonly found in grower pigs in comparison with sows, possibly due to the fact that sows tend to be vaccinated while grower pigs are not.

32.1.4.1 Sows

Clinical signs of PRRS in dry sows during the first month of infection include inappetence (spreading over 7–14 days, affecting 10%–15% of sows), fever (39°C–40°C), abortions (often late term, affecting 1%–6% of sows), transient discoloration (bluing) of the ears (blue ear disease, affecting 2% of sows), early farrowings (10%–15% over the first 4 weeks), prolonged anestrus and delayed returns to heat postweaning, and coughing and respiratory signs [12].

Clinical signs of PRRS in farrowing sows in the first month of infection include inappetence over the farrowing period, reluctance to drink, no milk (agalactia) and mastitis, early farrowings, discoloration of the skin (including cyanosis or bluing of the vulva and ears) and pressure sores, lethargy, respiratory signs (coughing and pneumonia), mummified piglets (10%–15% death in the last 3–4 weeks of pregnancy), stillbirth (up to 30%), and very weak piglets at birth [12].

32.1.4.2 Piglets

Clinical signs of PRRS in piglets include diarrhea, increased respiratory infections (e.g., Glasser's disease), and decreased survival rate.

32.1.4.3 Boars

Clinical signs of PRRS in boars include inappetence, fever, lethargy, loss of libido, reduced fertility, poor litter sizes, and lowered sperm output.

32.1.4.4 Weaners and Growers

Clinical signs of PRRS in growing herd include slight inappetence, mild coughing, hairy wasting pigs, or no symptoms. In untreated herd, the disease may exacerbate, with acute extensive consolidating pneumonia, formation of multiple abscesses, loss of condition, diarrhea, pale skin, mild coughing, sneezing, discharges from the eyes, increased respiratory rates, and mortality (12%–15%) [12].

In 4–12-week-old pigs born from infected sows, clinical signs include inappetence, malabsorption, wasting, coughing, pneumonia, and mortality (12% or higher) in this postweaning period. Secondary bacterial infections in pigs from 12 to 16 weeks of age may lead to abscesses in the lungs and throughout the body, lameness with abscesses, and poor, stunted growth [12].

In grower pigs, lesions are mainly observed in cranioventral lung lobes that are characterized by consolidation, discoloration, and failure to collapse when the thorax is opened. Microscopically, the alveolar walls are thickened due to the infiltration of macrophages, lymphocytes, and neutrophils. Tonsillitis in the grower pigs is characterized by erosion of the crypt epithelia and dilated crypt lumina filled with macrophages, neutrophils, lymphocytes, and cryptal epithelial cells [12].

Histological examination of placenta from infected sows reveals local separation between the uterine epithelium and trophoblast, complete degradation of the fetal placental mesenchyme, and appearance of apoptotic (and necrotic) cells, which may be initiated by PRRSV damage of susceptible fetal placental macrophages and bystander cells. The deterioration of placental functions may contribute to fetal death and clinical representations of congenital PRRSV infection [13,14].

PRRSV nsPs have the ability to inhibit the production of type 1 interferons (particularly IFN-α and IFN-β), induce abnormal B cell proliferation and repertoire development (often lymphopenia and thymic atrophy), and suppress innate immunity, contributing to its survival in the host [15–19]. By damaging the mucociliary transport system, PRRSV impairs the function of porcine alveolar macrophages, induces apoptosis of immune cells, and thus increases susceptibility to secondary bacterial infections [20–23].

32.1.5 Diagnosis

Associated with severe weight loss (wasting), failure to thrive, and pneumonia, PRRS resemble to the postweaning multisystemic wasting syndrome caused by porcine circovirus type 2 (PCV2), which is a nonenveloped single-stranded DNA virus of the genus *Circovirus*, family *Circoviridae* [24]. Therefore, differential diagnosis for PRRS includes PCV2 infection, in addition to classical swine fever (caused by a positive-sense single-stranded RNA virus of the genus *Pestivirus*, family *Flaviviridae*), African swine fever (caused by a large enveloped, double-stranded DNA virus of the genus *Asfivirus*, family *Asfarviridae*), and Aujeszky's disease (pseudorabies, due to Suid herpesvirus 1, a member of the genus *Varicellovirus*, subfamily *Alphaherpesvirinae*, family *Herpesviridae*) [25–29].

Traditionally, diagnosis of PRRS is based on clinical signs, postmortem examinations, and detection of the virus in the herd by virus isolation and serological assays. Recent development of molecular techniques such as reverse-transcription polymerase chain reaction (RT-PCR) offers a rapid, sensitive, and specific alternative for identification and typing of PRRSV [30,31].

PRRSV is difficult to isolate using subcultured cells. Indeed, the initial isolation of PRRSV was achieved with primary porcine pulmonary macrophages [32]. Years later, Marc-145 cells and other cell lines are found to accommodate PRRSV growth in vitro [33].

The current serological assays for PRRSV detection include immunoperoxidase monolayer assay and enzyme-linked immunosorbent assay (ELISA). The commercially available ELISA (HerdChek® 2XR/3XR PRRS ELISA, IDEXX Laboratories, Inc.) provides a reliable approach for antibody detection with a rapid turnaround time.

Use of nucleic acid amplification technologies such as RT-PCR facilitates improved identification and genotyping of PRRSV [34–46]. Indeed, real time RT-PCR assays have been reported for specific determination of PRRSV-1 (European genotype), PRRSV-2 (North American genotype), and HP-PRRSV [47–50]. Several real-time RT-PCR assays are commercially available (e.g., VetMax, Applied Biosystems; VetAlert, Tetracore; and AcuPig, AnDiaTec) [51].

Using real-time RT-PCR, PRRSV may be detected in boar serum, blood swabs, and oral fluids as early as 24–48 h postinfection and in semen as early as 48–120 h postinfection. Sensitivity-wise, serum appears to be numerically superior to blood swabs > semen > oral fluid/frothy saliva for early detection of PRRSV infection in boars [52,53].

32.1.6 Treatment and Prevention

PRRSV is responsible for reproductive failure in sows, preweaning mortality in piglets, and reduced performance in growing pigs. Control and elimination of PRRS at the farm level are dependent on adopting several management procedures, including pig flow, gilt acclimation, and vaccination. The management of pig flow (or herd closure) is based on the delayed introduction of the new herd until the resident virus dies out. Vaccination of pigs with modified live virus vaccines (MLV vaccines or attenuated vaccines) and killed virus vaccines (KV vaccines or inactivated vaccines) provides full protection against homologous PRRSV challenges. Currently, MLV and KV vaccines for both European and North American PRRSV strains are available commercially (e.g., RespPRRS/Ingelvac PRRS MLV, Boehringer Ingelheim Animal Health) [9].

In general, attenuated vaccines are prepared by in vitro cell culture passaging of virulent virus until an attenuated phenotype appears, whereas inactivated vaccines are produced by chemical or physical inactivation of virulent viruses. With the capacity to induce virus neutralizing antibodies, attenuated vaccines prevent virus replication in target cells, viremia, and clinical symptoms. On the other hand, without inducing virus neutralizing antibodies, inactivated vaccines cannot prevent PRRSV infections. Nevertheless, MLV vaccines have an unwanted tendency to spread within the vaccinated herds as well as to neighboring nonvaccinated herds, and revert or recombinate genetically, leading to abortion and high mortality in pregnant sows. KV vaccines are safer but less efficacious in conferring protection [54].

To eliminate the possible side effects of MLV vaccines and improve the efficacy of KV vaccines, plasmid DNA and viral vector vaccines that contain PRRSV structural proteins (GP3, M, and/or GP5) are being developed [55]. This may also help overcome the drawback of current PRRSV vaccines relating to the difficulty to distinguish vaccinated animals from pigs that have recovered from a natural infection. However, the ability of the new generation vaccines to protect pigs against PRRS remains to be ascertained [9].

32.2 METHODS

32.2.1 SAMPLE PREPARATION

Blood is collected from the anterior vena cava of pigs and placed in 5 mL centrifuge tubes without coagulant or anticoagulant. After centrifugation at $1000 \times g$ for 10 min, serum is stored at $-80°C$ until use.

Frothy saliva (from the submandibular salivary gland of aroused boars), oral fluid, and semen are collected from boars. Semen is centrifuged and semen supernatant and cell fractions are tested separately.

Tissues (from the tonsils, lungs, liver, spleens, and lymph nodes) are collected from pigs suspected of PRRSV infection. The tissues are first homogenized, 20% (w/v), in 10% (w/v) sterile phosphate-buffered saline and filtered through 0.22 mm membrane filters.

Viral RNA is extracted from serum, semen, or tissue filtrate by using QIAamp Viral RNA Mini Kit (Qiagen). Internal control RNA may be similarly processed. The extracted RNA is eluted in a 50 μL kit elution buffer.

32.2.2 DETECTION PROCEDURES

32.2.2.1 Conventional RT-PCR

Xiao et al. [56] utilized primers nsP2-F (5'-AACA CCCAGGCGACTTCA-3') and nsP2-R (5'-GCATGT CAACCCTATCCCAC-3') from the nsP2 gene in RT-PCR for differential diagnosis of PRRSV and HP-PRRSV.

RT-PCR mixture (10 μL) is composed of 0.5 μL cDNA, 5.0 μL 2 × PCR reaction mix (TaKaRa), 0.4 μL nsP2-F (10 μM) primer, 0.4 μL nsP2-R (10 μM) primer, 0.1 μL Taq DNA polymerase (2.5 U/μL, TaKaRa), and 3.6 μL H$_2$O. amplification parameters that include the following: 95°C for 3 min; 30 cycles of 94°C for 30 s, 60°C for 30 s, and 72°C for 1 min; and a final extension at 72°C for 5 min. The resulting products are detected by 1.5% agarose gel electrophoresis in 1 × TAE.

Note: The presence of an 874 bp product is indicative of PRRSV, while the presence of a 787 bp product is indicative of HP-PRRSV.

32.2.2.2 Real-Time RT-PCR

Wernike et al. [57] developed a multiplex real-time quantitative reverse transcription polymerase chain reaction (RT-qPCR) for specific identification of PRRSV genotype 1 (European type 1), genotype 2 (North American type 2), and highly pathogenic North American strain (HP-PRRSV) (Table 32.2).

RT-PCR mixture (25 μL) is made up of 8.75 μL RNase-free water, 5.0 μL RNA UltraSense™ 5× Reaction Mix (RNA UltraSense One-Step Quantitative RT-PCR System, Invitrogen), 1.25 μL RNA UltraSense Enzyme Mix, 1.0 μL type 1 PRRSV–specific FAM-labeled primer–probe mix (20 pmol/μL PRRSV-EU-2.1F, 20 pmol/μL PRRSV-EU-2.1R, 6.25 pmol/μL PRRSV-EU-2.1FAM), 1.0 μL type 2 PRRSV–specific Texas Red-labeled primer–probe mix (10 pmol/μL PRRSV-US-1dF, 10 pmol/μL PRRSV-US-1R, 2.5 pmol/μL PRRSV-US-1TEX), 1.0 μL HP-PRRSV-specific Cy5-labeled primer–probe mix (10 pmol/μL PRRSV-HP-1F, 10 pmol/μL PRRSV-HP-1R, 2.5 pmol/μL PRRSV-HP-1Cy5aS), 2 μL IC-specific HEX-labeled primer–probe mix (2.5 pmol/μL EGFP-11-F, 2.5 pmol/μL EGFP-10R, 1.25 pmol/μL EGFP-HEX) (Table 32.2), and 5 μL RNA template.

Amplification and detection are carried out in a Bio-Rad CFX96 Real-Time Detection System (Bio-Rad) using the following thermal profile: RT at 50°C for 15 min, initial Taq activation at 95°C for 2 min; 45 cycles of denaturation at 95°C for 15 s, annealing at 56°C for 20 s, and extension at 72°C for 30 s.

Note: The multiplex PRRSV RT-qPCR allows rapid detection and differentiation of both genotypes 1 and 2 PRRSV as well

TABLE 32.2
Identity of Porcine Reproductive and Respiratory Syndrome Virus and Internal Control Primers and Probes

Name	Sequence 5′-3′	Genome Position
PRRSV-US-1dF	ATRATGRGCTGGCATTC	15257–15273
PRRSV-US-1R	ACACGGTCGCCCTAATTG	15353–15370
PRRSV-US-1TEX	ACACGGTCGCCCTAATTG	15307–15330
PRRSV-HP-1F	CCGCGTAGAACTGTGACAAC	2914–2933
PRRSV-HP-1R	TCCAGGATGCCCATGTTCTG	3035–3016
PRRSV-HP-1Cy5aS	Cy5-ACGCACCAGGATGAGCCTCTGGAT-BHQ3	2941–2964
PRRSV-EU-2.1F	GCACCACCTCACCCRRAC	14792–14809
PRRSV-EU-2.1R	CAGTTCCTGCRCCYTGAT	14851–14868
PRRSV-EU-2.1FAM	FAM-CCTCTGYYTGCAATCGATCCAGAC-BHQ1	14819–14842
EGFP-11-F	CAGCCACAACGTCTATATCATG	537–558
EGFP-10-R	CTTGTACAGCTCGTCCATGC	813–794
EGFP-HEX	HEX-AGCACCCAGTCCGCCCTGAGCA-BHQ1	703–724

as HP-PRRSV, with an analytical sensitivity of <200 copies per μL for PRRSV type 1 assay and 20 copies per μL for PRRSV type 2 and HP-PRRSV. The inclusion of EGF primers and probes as an internal control ensures the authenticity of test results.

32.3 CONCLUSION

PRRS is an economically significant disease of swine that is characterized by the appearance of severe clinical symptoms (e.g., embryonic death, late-term abortions, early farrowing, mummified fetuses, and weak-born piglets) and maintenance of a life-long subclinical infection, resulting in enormous production losses and increased control and prevention costs. The ability to rapidly and accurately identify PRRSV genotypes and subtypes is crucial for successful management of this disease. Although virus isolation and serological assays remain useful for diagnosis of PRRS, molecular techniques especially RT-PCR have become an indispensable tool for detecting and genotyping PRRSV and for selecting the most appropriate type-specific vaccines to control the disease.

Owing to the obvious shortcomings of current commercial PRRSV vaccines (MLV and KV vaccines), continuous efforts are necessary to develop a universal vaccine that provides broad protection against circulating PRRSV strains, without the risk of vaccine-introduced clinical signs. Clearly, this will require detailed knowledge as to how PRRSV suppresses and evades host innate and adaptive immune responses, what molecules are linked to such immune suppression and evasion, which PRRSV epitopes are able to confer broad protection, and how PRRSV nsPs and structural proteins influence virus replication, virulence, immunity, and protection [9,17].

REFERENCES

1. Nga PT et al. Discovery of the first insect nidovirus, a missing evolutionary link in the emergence of the largest RNA virus genomes. *PLoS Pathog.* 2011;7(9):e1002215.
2. Archambault D et al. Animal arterivirus infections. *BioMed Res Int.* 2014;2014:303841.
3. Wang FX et al. Isolation and sequence analysis of highly pathogenic porcine reproductive and respiratory syndrome virus from swine herds in the Jilin Province of China. *Indian J Virol.* 2013;24(1):90–92.
4. Kappes MA, Faaberg KS. PRRSV structure, replication and recombination: Origin of phenotype and genotype diversity. *Virology.* 2015;479–480:475–486.
5. Lu ZH et al. Complete genome sequence of a pathogenic genotype 1 subtype 3 porcine reproductive and respiratory syndrome virus (strain SU1-Bel) from pig primary tissue. *Genome Announc.* 2015;3(3):e00340-15.
6. Stadejek T et al. Definition of subtypes in the European genotype of porcine reproductive and respiratory syndrome virus: Nucleocapsid characteristics and geographical distribution in Europe. *Arch Virol.* 2008;153:1479–1488.
7. Shi M et al. Molecular epidemiology of PRRSV: A phylogenetic perspective. *Virus Res.* 2010;154(1–2):7–17.

8. Shi M et al. Phylogeny-based evolutionary, demographical, and geographical dissection of North American type 2 porcine reproductive and respiratory syndrome viruses. *J Virol.* 2010;84(17):8700–8711.
9. Charerntantanakul W. Porcine reproductive and respiratory syndrome virus vaccines: Immunogenicity, efficacy and safety aspects. *World J Virol.* 2012;1(1):23–30.
10. Zhang Q, Yoo D. PRRS virus receptors and their role for pathogenesis. *Vet Microbiol.* 2015;177(3–4):229–241.
11. Yun SI, Lee YM. Overview: Replication of porcine reproductive and respiratory syndrome virus. *J Microbiol.* 2013;51(6):711–723.
12. The Pig Site. Porcine reproductive & respiratory syndrome (PRRS). http://www.thepigsite.com/diseaseinfo/97/porcine-reproductive-respiratory-syndrome-prrs/, accessed June 1, 2015.
13. Karniychuk UU, Nauwynck HJ. Pathogenesis and prevention of placental and transplacental porcine reproductive and respiratory syndrome virus infection. *Vet Res.* 2013;44:95.
14. Salguero FJ et al. Host-pathogen interactions during porcine reproductive and respiratory syndrome virus 1 infection of piglets. *Virus Res.* 2015;202:135–143.
15. Rodríguez-Gómez IM, Gómez-Laguna J, Carrasco L. Impact of PRRSV on activation and viability of antigen presenting cells. *World J Virol.* 2013;2(4):146–151.
16. Stukelj M, Toplak I, Svete AN. Blood antioxidant enzymes (SOD, GPX), biochemical and haematological parameters in pigs naturally infected with porcine reproductive and respiratory syndrome virus. *Pol J Vet Sci.* 2013;16(2):369–376.
17. Huang C, Zhang Q, Feng WH. Regulation and evasion of antiviral immune responses by porcine reproductive and respiratory syndrome virus. *Virus Res.* 2014;202:101–111.
18. Wang R, Zhang YJ. Antagonizing interferon-mediated immune response by porcine reproductive and respiratory syndrome virus. *Biomed Res Int.* 2014;2014:315470.
19. Rascón-Castelo E, Burgara-Estrella A, Mateu E, Hernández J. Immunological features of the non-structural proteins of porcine reproductive and respiratory syndrome virus. *Viruses.* 2015;7(3):873–886.
20. Wang YX et al. Identification of immunodominant T-cell epitopes in membrane protein of highly pathogenic porcine reproductive and respiratory syndrome virus. *Virus Res.* 2011;158(1–2):108–115.
21. Snijder EJ, Kikkert M, Fang Y. Arterivirus molecular biology and pathogenesis. *J Gen Virol.* 2013;94(10):2141–2163.
22. Butler JE et al. Porcine reproductive and respiratory syndrome (PRRS): An immune dysregulatory pandemic. *Immunol Res.* 2014;59(1–3):81–108.
23. García-Nicolás O et al. Comparative analysis of cytokine transcript profiles within mediastinal lymph node compartments of pigs after infection with porcine reproductive and respiratory syndrome genotype 1 strains differing in pathogenicity. *Vet Res.* 2015;46(1):34.
24. Chang CY et al. The application of a duplex reverse transcription real-time PCR for the surveillance of porcine reproductive and respiratory syndrome virus and porcine circovirus type 2. *J Virol Methods.* 2014;201:13–19.
25. Ogawa H et al. Multiplex PCR and multiplex RT-PCR for inclusive detection of major swine DNA and RNA viruses in pigs with multiple infections. *J Virol Methods.* 2009;160(1–2):210–214.
26. Xu XG et al. Development of multiplex PCR for simultaneous detection of six swine DNA and RNA viruses. *J Virol Methods.* 2012;183(1):69–74.

27. Liu JK, Wei CH, Yang XY, Dai AL, Li XH. Multiplex PCR for the simultaneous detection of porcine reproductive and respiratory syndrome virus, classical swine fever virus, and porcine circovirus in pigs. *Mol Cell Probes.* 2013;27(3–4):149–152.

28. Wernike K, Hoffmann B, Beer M. Single-tube multiplexed molecular detection of endemic porcine viruses in combination with background screening for transboundary diseases. *J Clin Microbiol.* 2013;51(3):938–944.

29. Rao P et al. Development of an EvaGreen-based multiplex real-time PCR assay with melting curve analysis for simultaneous detection and differentiation of six viral pathogens of porcine reproductive and respiratory disorder. *J Virol Methods.* 2014;208:56–62.

30. OIE (Office International des Epizooties). *Terrestrial Manual 2010.* Chapter 2.8.7: Porcine reproductive and respiratory syndrome. http://www.oie.int/fileadmin/Home/eng/Health_standards/tahm/2.08.07_PRRS.pdf. Accessed on June 12, 2015.

31. Benson JE et al. A comparison of virus isolation, immunohistochemistry, fetal serology, and reverse-transcription polymerase chain reaction assay for the identification of porcine reproductive and respiratory syndrome virus transplacental infection in the fetus. *J Vet Diagn Invest.* 2002;14(1):8–14.

32. Valícek L et al. Isolation and identification of porcine reproductive and respiratory syndrome virus in cell cultures. *Vet Med (Praha).* 1997;42(10):281–287.

33. Ding Z et al. Proteomic alteration of Marc-145 cells and PAMs after infection by porcine reproductive and respiratory syndrome virus. *Vet Immunol Immunopathol.* 2012;145(1–2):206–213.

34. Spagnuolo-Weaver M et al. The reverse transcription polymerase chain reaction for the diagnosis of porcine reproductive and respiratory syndrome: Comparison with virus isolation and serology. *Vet Microbiol.* 1998;62(3):207–215.

35. Umthun AR, Mengeling WL. Restriction fragment length polymorphism analysis of strains of porcine reproductive and respiratory syndrome virus by use of a nested-set reverse transcriptase-polymerase chain reaction. *Am J Vet Res.* 1999;60(7):802–806.

36. Dee SA et al. Identification of genetically diverse sequences (ORF 5) of porcine reproductive and respiratory syndrome virus in a swine herd. *Can J Vet Res.* 2001;65(4):254–260.

37. O'Sullivan TL, Friendship RM, Pearl DL, McEwen B, Dewey CE. Identifying an outbreak of a novel swine disease using test requests for porcine reproductive and respiratory syndrome as a syndromic surveillance tool. *BMC Vet Res.* 2012;8:192.

38. Toplak I et al. Identification of a genetically diverse sequence of porcine reproductive and respiratory syndrome virus in Slovenia and the impact on the sensitivity of four molecular tests. *J Virol Methods.* 2012;179(1):51–56.

39. Yang K et al. A one-step RT-PCR assay to detect and discriminate porcine reproductive and respiratory syndrome viruses in clinical specimens. *Gene.* 2013;531(2):199–204.

40. Drigo M et al. Validation and comparison of different end point and real time RT-PCR assays for detection and genotyping of porcine reproductive and respiratory syndrome virus. *J Virol Methods.* 2014;201:79–85.

41. Errington J, Jones RM, Sawyer J. Use of tissue swabbing as an alternative to tissue dissection and lysis prior to nucleic acid extraction and real-time polymerase chain reaction detection of Bovine viral diarrhea virus and Porcine reproductive and respiratory syndrome virus. *J Vet Diagn Invest.* 2014;26(3):418–422.

42. Gou H et al. Rapid and sensitive detection of type II porcine reproductive and respiratory syndrome virus by reverse transcription loop-mediated isothermal amplification combined with a vertical flow visualization strip. *J Virol Methods.* 2014;209:86–94.

43. Lu ZH, Archibald AL, Ait-Ali T. Beyond the whole genome consensus: Unravelling of PRRSV phylogenomics using next generation sequencing technologies. *Virus Res.* 2014;194:167–174.

44. Steinrigl A et al. Comparative evaluation of serum, FTA filter-dried blood and oral fluid as sample material for PRRSV diagnostics by RT-qPCR in a small-scale experimental study. *Berl Munch Tierarztl Wochenschr.* 2014;127(5–6):216–221.

45. Trang NT et al. Detection of porcine reproductive and respiratory syndrome virus in oral fluid from naturally infected pigs in a breeding herd. *J Vet Sci.* 2014;15(3):361–367.

46. Trang NT et al. Enhanced detection of porcine reproductive and respiratory syndrome virus in fixed tissues by in situ hybridization following tyramide signal amplification. *J Vet Diagn Invest.* 2015;27(3):326–331.

47. Lurchachaiwong W et al. Rapid detection and strain identification of porcine reproductive and respiratory syndrome virus (PRRSV) by real-time RT-PCR. *Lett Appl Microbiol.* 2008;46(1):55–60.

48. Xiao XL et al. Rapid detection of a highly virulent Chinese-type isolate of Porcine Reproductive and Respiratory Syndrome Virus by real-time reverse transcriptase PCR. *J Virol Methods.* 2008;149(1):49–55.

49. Kang K et al. A direct real-time polymerase chain reaction assay for rapid high-throughput detection of highly pathogenic North American porcine reproductive and respiratory syndrome virus in China without RNA purification. *J Anim Sci Biotechnol.* 2014;5(1):45.

50. Wu H, Rao P, Jiang Y, Opriessnig T, Yang Z. A sensitive multiplex real-time PCR panel for rapid diagnosis of viruses associated with porcine respiratory and reproductive disorders. *Mol Cell Probes.* 2014;28(5–6):264–270.

51. Gerber PF et al. Comparison of commercial real-time reverse transcription-PCR assays for reliable, early, and rapid detection of heterologous strains of porcine reproductive and respiratory syndrome virus in experimentally infected or noninfected boars by use of different sample types. *J Clin Microbiol.* 2013;51(2):547–556.

52. Christopher-Hennings J, Nelson EA. PCR analysis for the identification of porcine reproductive and respiratory syndrome virus in boar semen. *Methods Mol Biol.* 1998;92:81–88.

53. Pepin BJ et al. Comparison of specimens for detection of porcine reproductive and respiratory syndrome virus infection in boar studs. *Transbound Emerg Dis.* 2015;62(3):295–304.

54. Martelli P et al. Efficacy of a modified live porcine reproductive and respiratory syndrome virus (PRRSV) vaccine in pigs naturally exposed to a heterologous European (Italian cluster) field strain: Clinical protection and cell-mediated immunity. *Vaccine.* 2009;27(28):3788–3799.

55. Renukaradhya GJ, Meng XJ, Calvert JG, Roof M, Lager KM. Inactivated and subunit vaccines against porcine reproductive and respiratory syndrome: Current status and future direction. *Vaccine.* 2015;33(27):3065–3072.

56. Xiao S et al. Simultaneous detection and differentiation of highly virulent and classical Chinese-type isolation of PRRSV by real-time RT-PCR. *J Immunol Res.* 2014;2014:809656.

57. Wernike K et al. Detection and typing of highly pathogenic porcine reproductive and respiratory syndrome virus by multiplex real-time RT-PCR. *PLoS One.* 2012;7:e38251.

33 Yellow Head Complex Viruses, including Gill-Associated Virus

James Munro

CONTENTS

33.1 Introduction 295
   33.1.1 Classification and Genome Organization 296
   33.1.2 Morphology, Biology, and Epidemiology 296
   33.1.3 Clinical Features and Pathogenesis 298
      33.1.3.1 Clinical Signs 298
      33.1.3.2 Pathogenesis 298
   33.1.4 Diagnosis 300
      33.1.4.1 Conventional Techniques 300
      33.1.4.2 Molecular Techniques 301
33.2 Methods 301
   33.2.1 Sample Preparation 301
   33.2.2 Detection Procedures 301
      33.2.2.1 RT-PCR for Specific Detection of Genotype 1 in Diseased Shrimp 301
      33.2.2.2 Nested RT-PCR for Differential Detection of Genotype 1 and 2 in Healthy or Diseased Shrimp 302
      33.2.2.3 RT-nPCR for Detection of All Currently Known Genotypes in the Yellow Head Complex 302
      33.2.2.4 Qualitative RT-PCR for Detection of All Currently Known Genotypes in the Yellow Head Complex 303
33.3 Conclusions and Future Perspectives... 303
References... 303

33.1 INTRODUCTION

With a constantly increasing requirement for shrimp products, accompanied by an ever-advancing technology, penaeid culture has progressed from extensive subsistence farming to a major industry, with a worldwide production of over 4 million tons annually, valued at approximately US$19 billion in 2012.[1] The importance of shrimp disease has increased proportionally along with the growth of this new industry.

During the period of 1970s–1990s, shrimp aquaculture exhibited an astonishing expansion of approximately 24% per annum.[2] Since then, however, diseases such as yellow head disease (YHD) have had a devastating impact on the industry. An economic impact assessment from Fegan and Clifford[3] reported that in 1994, just over US$2 billion was lost due to disease, reducing the industry growth to approximately 8% per annum, the lowest in a 30-year period. More recently, Lightner[4] estimated that the costs, in 2001, of White Spot Syndrome Virus, Taura Syndrome Virus, Yellow Head Virus (YHV), and Infectious Hypodermal and Hematopoietic Necrosis Virus exceeded US$ 7 billion, and Briggs et al.[5] estimated that mortality due to shrimp viruses probably exceeded US$1 billion per year.

Recently, shrimp aquaculture has again shown rapid growth. The production period between 2000 and 2008 had an average annual increase of close to 15%, yielding approximately 3 million tons of shrimp products in 2008. This was reportedly due somewhat to the adoption of the white leg shrimp (*Litopenaeus vannamei*) within the Asian continent, after successful introduction from Latin America.[2] Disease outbreaks continue to cause major losses within the shrimp industry. Shrimp diseases have traditionally been a localized problem, but with the expansion and globalization of the industry, diseases that have been restricted to one region are now rapidly spreading cross geographic borders, placing a heavy reliance on sensitive detection methods to uphold biosecurity in the receiving region or country.

This chapter will focus on the yellow head complex nidoviruses, including gill-associated virus (GAV) with respect to the present literature, with a focus on the molecular detection of the closely related viruses.

YHV and GAV are considered to be genotypes within the yellow head complex, which includes four other closely related genotypes[6] with recombination between genotypes also being evident.[7] YHV emerged in 1990 in Thailand and its first description was published by Limsuwan in 1993.[8] The virus was reported to cause severe disease with 100% mortality within 3 days after signs of infection. While the disease-causing etiology was unknown, the syndrome was dubbed yellow head disease based on the yellow coloration of the dorsal cephalothorax due to the enlarged hepatopancreas and the generally pale appearance of affected shrimp. The etiological agent of the syndrome was identified as being viral in 1993.[8]

GAV is the junior synonym of lymphoid organ virus (LOV), which was reported in Australia in 1995[9] and was initially considered to be a nonpathogenic variant of GAV within wild and farmed black tiger shrimp, *Penaeus monodon*.[10] GAV was reported in 1997 as a pathogenic variant of LOV.[10] In 2000, a sequence analysis indicated that LOV had a 98.9% nucleotide identity to GAV, indicating they are the same virus presenting as either a chronic or an acute infection.[11]

33.1.1 Classification and Genome Organization

The yellow head viral complex consists of six recognized genetic lineages (Figure 33.1), which are distributed across most of the natural geographic range of penaeid shrimp. Of these, only genotype 1, which is the causative agent of YHD resulting in high mortalities, and genotype 2 (GAV), which is regarded as mildly pathogenic, have been associated with causing disease in penaeid shrimp.[8,10,12,13] Genotypes 3–6 were detected exclusively as low-level infections in apparently healthy shrimp.[6] The homologous genetic recombination of these genotypes has also been reported, both naturally and experimentally. Natural recombination was detected between YHV genotypes 2, 3, and 5, while experimental genetic recombination was demonstrated between pathogenic genotypes 1 and 2.[7] The high level of genetic exchange among genotypes indicates the expanding genetic diversity and has wider implications on detection methods, virulence, and international trade. Currently, YHD is considered to be an infection with YHV genotype 1 and is the only listed agent for YHD.[14] Trade barriers, disease definition, and obligatory reporting of viral detection have not yet been challenged with genome sequence containing a partial sequence of genotype 1, resulting in a mixed classification.

The collective groups of viruses in the yellow head complex are enveloped, rod-shaped, positive-sense RNA viruses that are classified as a single species in the genus *Okavirus*, family *Roniviridae*, order *Nidovirales*.[15] The 26,662 nucleotide (nt) genome of YHV (genotype 1) contains four long open reading frames (ORFs). ORF1a encodes a large polyprotein (pp1a) containing cis-acting papain-like protease (PLP) and 3C-like protease (3CLP) domains.[16] ORF1b overlaps ORF1a and is expressed only as a result of a (−1) ribosomal frame shift at a "slippery" sequence upstream of a predicted pseudoknot structure in the overlap region. The extended polyprotein (pp1ab) contains "SDD" RNA-dependent RNA polymerase, helicase, metal ion binding, exonuclease, uridylate-specific endoribonuclease, and ribose- 2′-O-methyl transferase domains of the replication complex.[16,17] ORF2 encodes the nucleoprotein (p20), which is the only polypeptide component of the helical nucleocapsid.[18,19] ORF3 encodes a polyprotein (pp3) that is processed at signal peptidase type 1 cleavage sites to generate the two virion envelope glycoproteins, gp116 and gp64, and an N-terminal triple-membrane-spanning fragment of unknown function.[20] The 26,235 nt YHV genotype 2 genome is slightly smaller than the YHV genotype 1, primarily due to significant deletions in intergenic regions (IGRs), but shares ~79% overall nucleotide sequence identity with YHV genotype 2.[16,21] The genome organization of genotype 2 is similar to genotype 1 and all identified functional domains are preserved, but YHV genotype 2 contains an additional small ORF near the 3′-terminus (ORF4, 83 aa) that may be expressed at low levels in infected cells.[11,21–23] Like other nidoviruses, the YHV complex transcribes a nested set of 3′-coterminal, polyadenylated, genomic, and subgenomic mRNAs.[16,21] Partial genome sequence analysis of YHV genotypes 3, 4, and 5 has indicated that they share a similar genome organization and transcription strategy and are more closely related in sequence to genotype 2 than to genotype 1.[6]

33.1.2 Morphology, Biology, and Epidemiology

YHV virions are rod-shaped, enveloped particles containing helical nucleocapsids that mature by the process of budding through intracytoplasmic membranes.[8,10] The nucleocapsids exhibit striations with a periodicity of approximately 7 nm and are often observed in association with the distended endoplasmic reticulum.[10]

The virions vary from 160 to 200 by 34–63 nm in size and are often packed densely into vesicles, resembling paracrystalline arrays. Free virions are also observed in intercellular spaces probably via release from disintegrating cells.[8,24] Within all stages of virally infected cells, virogenic stroma and filamentous viral nucleocapsids, 116–435 by 16–18 nm in size, are often observed scattered randomly within the cytoplasm.[24] The nucleocapsid of the YHV complex becomes enveloped by passage through the endoplasmic reticulum or the virions have occasionally been observed invading the interstitial spaces of the lymphoid organ and gain their envelope by passage through the plasma membrane.[24]

YHD outbreaks have been reported only in the black tiger prawn (*P. monodon*) and the white Pacific shrimp (*P. vannamei*).[8,25] Stentiford et al.[26] has produced a detailed review of hosts susceptible to YHV infection, according to scientific literature. About 18 potential host species were identified from the scientific literature. Stentiford et al.[26] used criteria to ascertain the validity of the reports, which included whether the exposure method was invasive or noninvasive, and scientific proof of the replication, viability, pathology, and location of the virus. From this review, Stentiford et al.[26] concluded that there are scientific data available to support susceptibility of *P. monodon*, *P. merguiensis*, *P. vannamei*, *P. setiferus*, *P. aztecus*, *P. duorarum*, *Metapenaeus brevicornis*, *M. affinis*, and *Palaemon styliferus*, and there are scientific data suggesting susceptibility of *P. esculentus*, *P. japonicus*, *P. stylirostris*, *M. ensis*, and *M. bennettae* with uncertainty of virus identification. Information on *P. esculentus*, *M. ensis*, *M. bennettae*, and *Macrobrachium lanchesteri* was considered insufficient to scientifically assess susceptibility with regard to criteria, and the scientific data suggesting susceptibility of *M. lanchesteri*, *M. sintangense*, and *Palaemon serrifer* were essentially experimental and invasive. Additional studies that surveyed 16 crab species collected from the vicinity of shrimp farms in

FIGURE 33.1 Neighbor-joining phylogenetic trees constructed from a ClustalX multiple alignments of (a) a ~ 671 nt sequence in the ORF1b gene and (b) a ~ 1.25 kb sequence in the ORF3 gene amplified by RT-nested PCR from 28 yellow head complex viruses and the reference strains of YHV (THA-98-Ref) and GAV (AUS-96-Ref). Bootstrap values shown at branch points indicate branching frequency in 1000 replicates. The clustering of isolates into six genetic lineages is indicated and the 10 isolates that clustered differently in the two analyses are shaded. (From Wijegoonawardane, P.K. et al., *Virology*, 390, 79, 2009.)

Thailand detected no evidence of either natural infection or experimental susceptibility[27] and, to date, infections by genotypes 2–6 in the YHV complex have been detected only in *P. monodon*.[6]

33.1.3 CLINICAL FEATURES AND PATHOGENESIS

33.1.3.1 Clinical Signs

Initially, infection of YHV (genotype 1) in *P. monodon* of cultured populations and experimental trials affected juvenile to subadult shrimp and often induced 100% mortality in shrimp within ponds within 3 days from the onset of disease.[8] However, natural disease outbreaks from YHV (genotype 2) have been reported in *P. monodon* up to 40 g,[10] showing genotypic variation in virulence.

The initial gross sign of infection is a rapid increase in feeding at an abnormally high rate for several days, followed by a sudden decline in appetite;[8] however, this is not seen in all cases. Within 1 day of the shrimp ceasing to feed, they begin either slowly or erratically swimming near the edge of the pond, and mortalities soon follow. Dead shrimp are found at the edge of the pond and scattered evenly over the entire bottom of the pond. The disease is often characterized by a pale to yellowish coloration of the cephalothorax and gills due to the underlying yellow hepatopancreas showing through the translucent carapace of the prawn and also a generally pale or bleached appearance of affected shrimp.[8,28] It has also been reported that YHV (genotype 2) from Australia causes the color of the prawn to change to a degree of pink to red, with primarily the appendages, tail fan, and mouth parts being most noticeable; however, this characteristic can often be attributed to general stress and is not necessarily caused only by YHV infection. Spann et al.[10] also reported that the gills changed from the normal clear/yellow to pink, and the shrimp exhibited fouling of the gills, shell, and tail rot.

33.1.3.2 Pathogenesis

Clinical signs of YHD occur in *P. monodon* within 7–10 days of exposure. YHV replicates in the cytoplasm of infected cells in which long filamentous prenucleocapsids are abundant, and virions bud into cytoplasmic vesicles in densely packed paracrystalline arrays for egress at the cytoplasmic membrane.[8] YHV infection can be transmitted horizontally by injection, by ingestion of infected tissue, by immersion in sea water containing tissue extracts filtered to be free of bacteria, or by cohabitation of naive shrimp with infected shrimp.[29,30] Infection of shrimp has also been established by injection of extracts of shrimp homogenate (*Acetes* sp.) collected from infected ponds.[31] Vertical transmission of genotype 2 has been shown to occur from both male and female parents, possibly by surface contamination or infection of tissue surrounding fertilized eggs.[21] The dynamics of how YHV infection spreads within aquaculture ponds is poorly understood. However, the rapid accumulation of mortalities during disease outbreaks suggests that horizontal transmission occurs very effectively. Low-level mortality on farms, throughout the production period, has also been associated with YHV (genotype 2) infection (Table 33.1), indicating that mortality attributed to YHV is often multifactorial.[13]

YHV (genotype 2) persists as a chronic infection for at least 50 days in *P. esculentus* that survive experimental challenge.[32] The high prevalence of subclinical or chronic infection often found in healthy *P. monodon* infected with YHV genotypes 2–6 from postlarval stages onward suggests that these infections can persist for life.[6,33] There is also evidence that genotype 1 can persist in survivors of experimental infection.[27,34]

The modes of infection of YHV have been placed into two general groups: horizontal transmission and vertical transmission. Horizontal transmission can occur when YHV-free susceptible shrimp feed on infected carcasses, experience bath exposure to membrane-filtered tissue extracts, either by

TABLE 33.1

Average Prevalence and Loading of GAV in Three Shrimp Farms Growing *Penaeus monodon* at 1-Month Poststocking and 1-Week Preharvest, Indicating the Percentage Increase between the Sampling Periods and Percentage of Daily Increase (*n* = 1800)

	Disease Outbreak, Low Production (*n* = 300[a])	Disease Outbreak, Emergency Harvest (*n* = 220[a])	No Disease Outbreak, Low Production (*n* = 200[a])	No Disease Outbreak, High Production (*n* = 180[a])
Prevalence (%)				
Poststocking	80.67	75.50	59.38	64.44
Preharvest	96.33	94.55	93.00	73.89
% Increase	16.26	20.14	36.16	12.78
Loading				
Poststocking	1.07	0.91	0.87	0.83
Preharvest	1.74	1.51	1.46	0.99
% Increase	38.39	40.07	40.03	16.31
Survival				
% Increase	27.00	52.04	34.45	57.28

Source: Munro, J. et al., *J. Fish Dis.*, 34, 13, 2011.

[a] Sample number for each sample period (i.e., either poststocking or preharvest).

cohabitation with infected shrimp or by direct experimental injection of the viral inoculum.[33] Experimental infections with YHV (genotype 2) indicate that larger (~20 g) *P. japonicus* are less susceptible to disease than smaller (~6–13 g) shrimp of the same species.[35] However, experimental infection of *P. monodon* indicated an inverse relationship with shrimp size and mortality in the same study. This may have been due to the initial viral load of the chronically infected *P. monodon* compared to the uninfected *P. japonicus*, where the larger *P. monodon* may have had an elevated initial viral load compared to the smaller *P. monodon*. However, this was not discussed in the paper.

P. monodon are susceptible to YHV (genotype 1) infection beyond postlarvae (PL) 15.[36] However, this research was conducted under controlled environmental conditions and did not give a clear indication of susceptibility of postlarvae shrimp at stage PL15 or earlier, under suboptimal conditions. An experiment to determine the susceptibility of postlarvae *P. monodon* to YHV (genotype 2) by ingestion showed that PL20 shrimp died 7–10 days postinfection while PL_{15} survived the exposure;[33] these findings support the earlier research conducted by Khongpradit et al.[36] The available data suggests that the disease was associated with viral loading. Walker et al.[33] reported that YHV from a diseased *P. monodon* caused mortalities after ingestion or immersion exposure while extracts from YHV-infected healthy *P. monodon* caused infection without mortality. However, after the viral concentration of the healthy shrimp had an equivalent titer to the diseased shrimp, YHV extracts from chronically infected, healthy *P. monodon* also induced disease, indicating that the disease is associated with viral loading. However, in these studies, there was no referral to the possibility of multiple viral infections.

To determine whether vertical transmission of YHV contributes to the high prevalence of chronic infections in wild and farmed *P. monodon* in eastern Australia, Cowley et al.[21] tested gonads and lymphoid organs for signs of YHV (genotype 2) from healthy male and female *P. monodon* broodstock and in fertilized eggs in addition to nauplii spawned from wild-fertilized females, using reverse transcription (RT)-nPCR as the detection method. The results indicated that the level of YHV in wild *P. monodon* was generally low. However, high levels of YHV were detected in moribund male broodstock reared in captivity for more than 12 months. The RT-nPCR-nPCR products from spermatophores in these shrimp were also significantly greater than those amplified from the lymphoid organ, which had previously been identified as the primary site of YHV replication in chronically infected *P. monodon*.[9] It was also reported that one in three spermatophores examined by TEM had mature YHV virions in the seminal fluid but not in the sperm cells. The RT-nPCR for YHV in eggs were positive; however, nauplii and protozoa were generally negative. This implies that YHV is associated with the egg surface and the majority of the virus is lost when the nauplii hatch and that the infection levels in the protozoa remain low. Cowley et al.[21] reported that this could be due to the lack of development of the lymphoid organ in larval and early postlarval life stages, which is likely to limit potential infection levels. RT-nPCR has detected YHV in PL_5 to

PL_{15} shrimp from hatcheries and experimental spawning's of *P. monodon*,[33] suggesting that viral replication in postlarvae is occurring at sufficient levels to be detected.[33] Cowley et al.[21] reported that the identification of lymphoid organ spheroid bodies and YHV particles in approximately 1.2 g juvenile *P. monodon* grown from hatchery stocks (PL_6 and PL_{20}), suggesting that at least some of the postlarvae were infected with YHV. However, as individual postlarvae were not grown in isolation, it strongly raises the distinct possibility that some juvenile infections occurred during the course of the grow-out through cannibalism or water-borne transmission. Clearly, this horizontal transmission could promote translocation of YHV and the potential infection of wild *P. monodon* in the vicinity of farms via water or through the escape of infected farmed shrimp. The authors concluded that the high prevalence of chronic YHV infection in *P. monodon* broodstock from Northeastern Australia and farmed shrimp produced from these broodstock promotes the idea that this is perpetuated primarily by vertical transmission both in the wild and in hatcheries.[21] Unfortunately, however, the paper failed to comment on the likely survival of the YHV-infected progeny. The study only tested eggs and nauplii for the presence of YHV (genotype 2). It is only a speculation that these postlarvae survive through adulthood.

The infection prevalence of yellow head complex viruses in healthy *P. monodon* (as detected by RT-nPCR) can be high (50%–100%) in farmed and wild populations in Australia, Asia, and East Africa as well as in *L. vannamei* farmed in Mexico.[6,13,33,37,38] The prevalence of individual genotypes varies according to the geographical origin of the shrimp. In contrast, except in situations of disease outbreaks in aquaculture ponds, the prevalence of genotype 1 is more commonly low (<1%) in healthy wild or farmed *P. monodon*. The use of detection methods less sensitive than PCR (e.g., histology, immunoblot, dot blot, *in situ* hybridization, etc.) is likely, in most cases, to result in the real infection prevalence among populations of shrimp being underestimated.

Tissue distribution of YHV: A study by Lu et al.[39] reported that YHV particles were detected in the gill, lymphoid organ, head soft tissue, heart, midgut, hepatopancreas, abdominal muscle, eyestalk, and nerve cord of an experimentally infected *Penaeus vannamei*. Lu et al.[39] reported that the lymphoid organ, gill, and head muscle had a 50% tissue culture infectious dose assay ($TCID_{50}$) titer (mL^{-1}) of 10^6, while the midgut, abdominal muscle, and heart had a $TCID_{50}$ titer (mL^{-1}) of 10^5, and the nerve cord, hepatopancreas, and eyestalk had a $TCID_{50}$ titer (mL^{-1}) of 10^4. These findings suggest that the lymphoid organ, gill, and head muscle contained the highest number of infectious virions compared with the other tested tissue/organs. Cowley et al.[21] tested gonads from *P. monodon* for signs of YHV (genotype 2) and reported that the RT-nPCR products had a greater intensity from the spermatophores than those amplified from the lymphoid organ. These results, combined, indicate that the viral infection is systemic. Virions with a morphological appearance similar to YHV have also been reported in the optic nerve fibers and in the nerve cord of *P. monodon*.[12,40]

33.1.4 Diagnosis

For diagnosis of YHV infection, several methods are available with varying sensitivity and specificity. These include histological and electron microscopy methods, antibody-based assays, and molecular techniques that involves traditional PCR, quantitative PCR, nucleic acid sequence analysis, and *in situ* hybridization. The PCR methods are considered the most sensitive and specific assays available and are regarded as the gold standard for the diagnosis of YHV, replacing the traditional transmission electron microscopy technique to visualize virions, which has a lower sensitivity and is unable to distinguish between specific genotypes. While the actual choice of assay is dependent on geographic location, cost, and the specificity and sensitivity required, all detection methods require a confirmation assay to confirm a positive result.

There are regular publications of detection methods for YHV, some of which will be discussed in this chapter. The World Organisation for Animal Health (OIE) publishes detection methods that are regarded as diagnostic procedures of choice when infection presence requires reporting. The molecular diagnostic methods discussed in the following will include the recommendations from the OIE as these specific techniques are regularly reviewed.

33.1.4.1 Conventional Techniques

33.1.4.1.1 Histopathology

Spann et al.[10] reported that diseased shrimp display disorganization and loss of a normal, defined tubule structure in the lymphoid organ, which could be seen on histological sections stained with hematoxylin and eosin (H&E). The gills of diseased shrimp displayed structural damage such as fusion of gill filament tips, general necrosis, and loss of cuticle from primary and secondary lamellae. However, this detection method and signs of disease only indicate a generic disease and, as mentioned previously, these symptoms are also characteristic of penaeid shrimp infected with other viruses.

Flegel et al.[41] reported that YHV can be diagnosed histologically in moribund shrimp by the presence of intensely basophilic inclusions in shrimp tissues and that these inclusions can be seen best with H&E staining of sectioned stomach and gill tissue. However, the cytoplasmic, virus-associated inclusions stain deeply basophilic in the same manner as pyknotic nuclei, and it can be difficult to differentiate between the two without using an electron microscope, suggesting that there is no clear way to characterize and diagnose YHV infection using standard histological examination with H&E-stained preparations alone.[8] The lymphoid organ is distinctly abnormal in YHV infections of acute levels, showing nuclear abnormalities, cytoplasmic abnormalities, and necrotic cells, which are also a characteristic of penaeid shrimp infected with other viruses.[8]

33.1.4.1.2 Electron Microscopy

Transmission electron microscopy, which gives an image of the virus, is the traditional gold standard test for the visualization and confirmation of viral infection, such from YHV.

However, this method is not useful for detecting early stage infections, for on-farm applications or when high throughput is required. It is time consuming and requires expensive equipment, and therefore, it is not applicable for on-farm use or for screening hatchery broodstock.

33.1.4.1.3 Serology

Polyclonal antisera have been produced for the detection of YHV.[42] However, the assay (Western blot analysis) was not highly sensitive and was not applicable to on-farm field examinations.[43] Sithigorngul et al.[43,44] produced monoclonal antibodies (MAbs) specific to YHV-enveloped protein. Most of the YHV-specific MAbs were specific to the 67 kDa protein and only a few were specific to the 135 and 22 kDa protein. Several of these antibodies against YHV did bind to hemolymph and tissues from uninfected shrimp.[43] This would be expected to some degree; as the virus matures and protrudes through intracytoplasmic membranes, host material is incorporated into the viral envelope. Alternatively, the reported uninfected shrimp may have had an undetected low level of infection, resulting in elevated binding. Munro and Owens[45] developed chicken polyclonal antibodies (PAbs) and mice MAbs toward YHV (genotype 2) from Australia. The PAbs reacted with 63, 110, and 170 kDa proteins of YHV (genotype 2) while the MAbs reacted with the 20 kDa protein of YHV. These antibodies were used to develop an enzyme-linked immunosorbent assay (ELISA), which was capable of quantitative detection of YHV (genotype 2) in individual shrimp. However, the ability of the ELISA to detect other genotypes was not determined.

YHV has been reported to hemagglutinate chicken erythrocytes.[42] Hemagglutination (HA) activity was determined via a qualitative and quantitative detection method yielding an HA end-point titer of 1:256, and the virus was not eluted after 24 h, suggesting that the reaction was stable and that the virus lacked receptor-destroying enzymes. Munro and Owens[46] used this HA activity of YHV to develop a low-cost detection method, while they demonstrated that YHV-negative shrimp caused negligible HA. The method of Munro and Owens[46] was developed using genotype 2 of the yellow head complex, and it is only assumed that all the genotypes cause the same level of HA.

As previously mentioned, the decision on which detection method to use is dependent on many factors. Munro and Owens[47] compared the sensitivity, specificity, positive predictive value, negative predictive value, accuracy, and cost-effectiveness between RT-n PCR designed by Cowley et al.[11] and an ELISA designed by Munro and Owens,[45] with a subsequent comparison between the ELISA and an HA assay (Munro and Owens[46]) toward YHV (genotype 2).

Using the RT-nPCR as the gold standard, the ELISA had an accuracy of 91.7% when using a cut-off optical density >0.75 as a positive result. When compared with the ELISA, the HA had an accuracy of 73% when using an HA titer cut-off greater than 16 as a positive result. These results indicate that alternative tests for GAV (ELISA and HA) can be used to explore multiple questions about the disease status of *P. monodon* stocks in a cost-effective manner albeit less sensitive and specific.

33.1.4.2 Molecular Techniques

33.1.4.2.1 In Situ Hybridization

In situ hybridization has been developed[48] for the detection of YHV infection in *P. vannamei* using a cDNA fragment labeled with digoxigenin that resulted in a highly sensitive test. This same probe was subsequently used to detect YHV from Australia (genotype 2).[49] Spann et al.[32] have also reported the development of an *in situ* hybridization probe for the detection of YHV in *P. monodon* and *P. esculentus* within Australia.

33.1.4.2.2 Polymerase Chain Reaction

Wongteerasupaya et al.[50] developed an RT-PCR for the detection of YHV (genotype 1) in Thailand. This test was specific and sensitive for the selected region of the YHV genome, with other nucleic acid templates (WSSV and HPV) commonly present in this region, giving no amplification signal. The RT-PCR was able to amplify as little as 0.01 pg of YHV-RNA and showed evidence of infection in *P. monodon* at 6–12 h after experimental exposure to the virus. However, it is more likely that the RT-PCR detected the YHV genome that was injected into the prawn and circulated in the hemolymph, than it was to be detecting infection at such an early time after injection. Cowley et al.[11] developed an RT-nPCR for the detection of YHV within Australia (genotype 2). The specific genome amplification with the one-step PCR is able to detect 0.01 pg cDNA, and while using the two-step PCR technique, 0.01 fg cDNA was detectable. The YHV genome could be detected within 6 h of experimental infection of *P. japonicus*. However, again it is probable that the test was detecting genome that was injected into the shrimp and does not necessarily indicate viral replication within cells. Gill biopsies can be used as a tissue sample[11] that offer a nonlethal sampling method; additionally, it has been reported that dried hemolymph can be used for the detection of YHV with RT-PCR,[51] resulting in the ability to sample broodstock. A multiplex RT-nPCR has been developed for the differentiation between genotype 1 and 2.[22] It can be used to differentiate the two genotypes in diseased shrimp or for screening for asymptomatic carriers. This test will not detect all six known genotypes, and genotype 3 may generate a PCR product indistinguishable in size from that generated with genotype 2.

The third assay is a multiplex RT-nested PCR protocol described by Wijegoonawardane et al.[52] This multiplex RT-nPCR is capable of detecting the six genotypes of the yellow head complex of viruses but will not discriminate between genotypes. Assignment and differentiation of genotype can be achieved by nucleotide sequence analysis of the RT-PCR product. Finally, Wijegoonawardane et al.[53] have reported a SYBR Green real-time quantitative RT-PCR for the detection of the six yellow head genotypes. The RT-nPCR has a sensitivity of approximately 1.25 RNA copies while the quantitative RT-PCR can detect down to approximately 2.5 RNA copies. All PCR products should be sequenced as a confirmation step when reporting the presence of positive samples.

33.2 METHODS

33.2.1 Sample Preparation

For juvenile or adult shrimp, gill tissue is easily accessible and may be taken with minimal damage to the individual and used to prepare total RNA. Alternatively, lymphoid organ, hemolymph, or pleopods may be used; however, viral loading may vary between organs and standardization must be prepared if quantitative methods are being used. Fresh tissue is preferred. Tissue preserved in 95% analytical grade ethanol or RNAlater, or stored frozen at −70°C, is also suitable for total RNA preparation. Homogenize 10–20 mg of tissue or dispense 50 μL hemolymph in 500 μL Trizol reagent and extract total RNA according to the manufacturer's instructions. Resuspend RNA in 25 μL of molecular grade water, heat at 55°C for 10 min, cool by semi-immersion in an ice slurry, and use immediately or store at −70°C until required. The quality and quantity of isolated RNA should be determined by testing against UV absorbance at 260 and 280 nm, with an optimal ratio of 1.8:1–2:1. RNA yield will vary with the type and freshness of tissues as well as the quality of the preservative used and the length of time tissue has been preserved. However, RNA yields from fresh tissues would be expected to vary from 0.2 to 2.0 μg/μL and about half these amounts from alcohol-preserved tissues.

Postlarval shrimp can be sampled by pooling individuals with positive selection of weak individuals increasing the likelihood of detection. Assumed infection prevalence must be taken into account when deciding on numbers to be pooled and processed for RNA extraction using Trizol.

Within each test group, a negative control of either molecular grade water or YHV-free RNA sample and a positive control of YHV-positive RNA should be used.

33.2.2 Detection Procedures

33.2.2.1 RT-PCR for Specific Detection of Genotype 1 in Diseased Shrimp

To synthesize cDNA, mix 2 μL RNA in 20 μL PCR buffer (10 mM Tris/HCl pH 8.3, 50 mM KCl) containing 2.5 U of M-MLV (Moloney murine leukemia virus) reverse transcriptase, 1.0 U ribonuclease inhibitor, 0.75 μM antisense primer 144R, 1 mM of dNTP stock mixture, and 5 mM MgCl₂, and incubate at 42°C for 15 min. Incubate the mixture at 100°C for 5 min and cool the mixture to 5°C. Add PCR mixture (10 mM Tris/HCl pH 8.3, 50 mM KCl) containing 2.5 U *Taq* DNA polymerase, 2 mM MgCl₂, and 0.75 μM of primer 10F to give a final volume of 100 μL. Conduct PCR amplification for 40 cycles at 94°C for 30 s, 58°C for 30 s, 72°C for 30 s, and finishing at 72°C for 10 min. Aliquot a portion of the product to a 2% agarose/TAE (Tris-acetate-EDTA [ethylenediamine tetra-acetic acid]) gel containing 0.5 μg/mL ethidium bromide and following electrophoresis, visualize the 135 bp DNA band expected for YHV with a suitable DNA ladder using a UV transilluminator.

The sensitivity of the PCR is approximately 0.01 pg of purified YHV RNA ($\approx 10^3$ genomes).

PCR primer sequences: 10F: 5'-CCG-CTA-ATT-TCA-AAA-ACT-ACG-3'

144R: 5'-AAG-GTG-TTA-TGT-CGA-GGA-AGT-3'

33.2.2.2 Nested RT-PCR for Differential Detection of Genotype 1 and 2 in Healthy or Diseased Shrimp

For cDNA synthesis, 2 µL RNA (ideally 1.0 µg total RNA), 0.7 µL of 50 pmol/µL primer GY5, and molecular grade water are added to total 6 µL; the mixture is then incubated at 70°C for 10 min and chilled on ice. Add 2 µL Superscript III buffer × 5 (250 mM Tris/HCl pH 8.3, 375 mM KCl, 15 mM MgCl$_2$), 1 µL 100 mM DTT, and 0.5 µL 10 mM dNTP stock and mix gently. Preheat to 42°C for 2 min, add 0.5 µL 200 U/µL reverse transcriptase, and incubate at 42°C for 1 h. Heat the reaction at 70°C for 10 min, chill on ice, and spin briefly in a microcentrifuge to collect the contents of the tube.

For the first PCR step, prepare a 50 µL reaction mixture containing 1 × *Taq* buffer (10 mM Tris/HCl pH 8.3, 50 mM KCl, 0.1% Triton X-100), 1.5 mM MgCl$_2$, 35 pmol of each primer GY1 and GY4, 200 µM of stock dNTP, and 2.5 U *Taq* polymerase in a 0.5 mL thin-walled tube. Overlay the reaction mixture with 50 µL liquid paraffin, heat at 85°C for 2 min, and then add 1 µL of prepared cDNA. Conduct PCR amplification using 35 cycles at 95°C for 30 s, 66°C for 30 s, and 72°C for 45 s, followed by final extension at 72°C for 7 min. For the second (nested) PCR step, prepare a 50 µL reaction mixture containing 2 µL of the first step PCR product, 1 × *Taq* buffer, 1.5 mM MgCl$_2$, 35 pmol of each primer GY2, Y3, and G6, 200 µM of stock dNTP, and 2.5 U *Taq* polymerase in a 0.5 mL thin-walled tube. Conduct PCR using amplification conditions as described earlier. Visualize the final PCR product using the same procedures as previously described.

If the viral load is sufficiently high, a 794 bp DNA will be amplified from either genotype 1 or 2 in the first PCR step. In the second PCR step, a 277 bp product indicates detection of genotype 1 and a 406 bp product indicates detection of genotype 2. The presence of both 406 and 277 bp products indicates a dual infection with both genotype 1 and 2. The detection sensitivity of the second-step PCR is ~1000-fold greater than the first-step PCR, and genotype 1 or 2 can be detected to a limit of 10 fg lymphoid organ total RNA.

The sequences of RT-PCR primers generic for genotype 1 and 2 (GY) or specific for genotype 1 (Y) or genotype 2 (G) are as follows:

GY1: 5'-GAC-ATC-ACT-CCA-GAC-AAC-ATC-TG-3'
GY2: 5'-CAT-CTG-TCC-AGA-AGG-CGT-CTA-TGA-3'
GY4: 5'-GTG-AAG-TCC-ATG-TGT-GTG-AGA-CG-3'
GY5: 5'-GAG-CTG-GAA-TTC-AGT-GAG-AGA-ACA-3'
Y3: 5'-ACG-CTC-TGT-GAC-AAG-CAT-GAA-GTT-3'
G6: 5'-GTA-GTA-GAG-ACG-AGT-GAC-ACC-TAT-3'

33.2.2.3 RT-nPCR for Detection of All Currently Known Genotypes in the Yellow Head Complex

For cDNA synthesis, mix 2 µL RNA (ideally 1.0 µg total RNA), 50 ng random hexamer primers, and 1.0 µL 10 mM dNTP and make up to a total volume of 14 µL in sterile molecular grade water, incubate at 65°C for 5 min, and chill on ice. Add 4.0 µL Superscript III buffer × 5, 1.0 µL 100 mM DTT, 1.0 µL 40 U/µL RNaseOUT™, and 1.0 µL 200 U/µL reverse transcriptase and mix gently. Incubate at 25°C for 5 min and then at 42°C for 55 min, stop the reaction by heating at 70°C for 15 min, chill on ice, and spin briefly in a microcentrifuge to collect the contents of the tube. For the first PCR step, add 1 µL cDNA to a total 25 µL reaction mixture containing 1 × *Taq* buffer (10 mM Tris/HCl, pH 9.0, 50 mM KCl, 0.1% Triton X-100), 1.5 µL 25 mM MgCl$_2$, 0.35 µL primer mix containing 25 pmol/µL of each primer pool YC-F1ab and YC-R1ab, 0.5 µL 10 mM dNTP mix and 0.25 µL 5 U/µL *Taq* DNA polymerase. Conduct PCR amplification by denaturation at 95°C for 1 min followed by 35 cycles at 95°C for 30 s, 60°C for 30 s, and 72°C for 40 s, followed by a final extension at 72°C for 7 min. For the second PCR step, use 1 µL of the first PCR product in the reaction mixture as prepared previously but substituting primer pools YC-F2ab and YC-R2ab. Conduct PCR amplification using denaturation at 95°C for 1 min followed by 35 cycles at 95°C for 30 s, 60°C for 30 s, and 72°C for 30 s, followed by a final extension at 72°C for 7 min. Visualize the final PCR product using the same procedures as previously described.

If the viral load is sufficiently high, a 358 bp DNA is amplified in the first PCR step. The second (nested) PCR step amplifies a 146 bp product. The detection of these products indicates detection of at least one of the six genotypes in the yellow head complex. Further assignment of genotype is possible by nucleotide sequence analysis of the PCR product followed by comparison with sequences of the known genotypes by multiple sequence alignment and phylogenetic analysis. The detection sensitivity limits of the first PCR step and nested PCR step are 2500 and 2.5 RNA templates, respectively.

PCR primer sequences are as follows (each primer comprises a pool of equal quantities of two related oligonucleotide sequences):

YC-F1ab pool: 5'-ATC-GTC-GTC-AGC-TAC-CGC-AAT-ACT-GC-3'
5'-ATC-GTC-GTC-AGY-TAY-CGT-AAC-ACC-GC-3'
YC-R1ab pool: 5'-TCT-TCR-CGT-GTG-AAC-ACY-TTC-TTR-GC-3'
5'-TCT-GCG-TGG-GTG-AAC-ACC-TTC-TTG-GC-3'
YC-F2ab pool: 5'-CGC-TTC-CAA-TGT-ATC-TGY-ATG-CAC-CA-3'
5'-CGC-TTY-CAR-TGT-ATC-TGC-ATG-CAC-CA-3'
YC-R2ab pool: 5'-RTC-DGT-GTA-CAT-GTT-TGA-GAG-TTT-GTT-3'
5'-GTC-AGT-GTA-CAT-ATT-GGA-GAG-TTT-RTT-3'
Mixed base codes: R(AG), Y(CT), M(AC), K(GT), S(GC), W(AT), H(ACT), B(GCT), V(AGC), D(AGT), N(AGCT).

33.2.2.4 Qualitative RT-PCR for Detection of All Currently Known Genotypes in the Yellow Head Complex

cDNA synthesis is achieved by using the same method as for the multiplex nRT-PCR (see previous texts). The primers are eightfold degenerate and target the same ORF1b gene sequences as the nested RT-PCR primers used earlier.

The qRT-PCR is performed by diluting 2 μL of cDNA into 8 μL of molecular grade water and adding this to an 18 μL reaction mixture containing 7 μL of 2× SYBR Green Master mix and 5 pmol/μL each primers, YHc-F2 and YHc-R2. Dispense 5 μL aliquots as 3 replicates into a 384-well real-time PCR plate. PCR cycling conditions are 50°C for 2 min and 95°C for 10 min and 40 cycles of 95°C for 15 s and 60°C for 1 min, followed by incubation at 95°C for 15 s, 60°C for 15 s, and 95°C for 15 s to generate a DNA dissociation curve.

PCR primer sequences are as follows:

YHc-F2: 5′-CGCTTYCARTGTATCTGYATGCACCA-3′
YHc-R2: 5′-RTCAGTGTACATGTTKGAGAGTTTRTT-3′

33.3 CONCLUSIONS AND FUTURE PERSPECTIVES

YHV was discovered as the etiological agent of shrimp mortalities in Thailand over 10 years ago. Since its discovery, YHV is now present in most shrimp-producing countries and is a major concern to the shrimp industry causing mortalities in endemic areas and placing biosecurity limitations on YHV naïve or specific genome naive shrimp farming areas. The relatively poor understanding of the mechanisms of viral infection and replication[54] and the limited understanding of the shrimp immune system limit the control of the virus.

The risk of homologous recombination within the YHV complex stands to be a future problem not only for prescribed detection methods that rely on a specific molecular sequence but also for further trade and biosecurity measures. Currently, border biosecurity refers to specific genotypes within the complex, which does not address the risk of a hybridized genotype that could potentially harbor highly virulent genes while presenting molecularly as a nonpathogenic genotype.

The molecular detection of YHV is extremely specific and sensitive with the capabilities of high-throughput automated screening of samples. The application of the available assays has contributed to driving our current knowledge and therapeutics for the industry, ultimately being a major contributor to controlling the health and economic impact of the virus. While the available molecular assays are extremely beneficial to the shrimp industry, the high cost involved in molecular testing can be a limitation for use in developing countries, where the virus is often prevalent. When combined with the limited availability of alternative, low-cost detection methods, the viral and disease spread will continue until effective treatment is developed. Treatment methods for the virus, and shrimp viruses in general, are still within the research realm. Some methods such as RNA interference,[55–59] which is driven by our molecular understanding of the viral and host complex, show promising results if effective delivery methods for application at a farm level and effective sequence targets can be optimized.

REFERENCES

1. FAO Yearbook. Fishery and Aquaculture Statistics 2012. (Food and Agricultural Organization, Rome, Italy, 2014).
2. Food and Agricultural Organization. *World Aquaculture 2010*, Vol. 500/1 105 (Food and Agricultural Organization, Rome, Italy, 2011).
3. Fegan, D.F. and Clifford, H.C. Health management for viral diseases in shrimp farms. In *The New Wave* (eds. Browdy, C.L.J. and Jory, D.E.), pp. 168–198 (World Aquaculture Society, Baton Rouge, LA, 2001).
4. Lightner, D.V. The penaeid shrimp viral pandemics due to IHHNV, WSSV, TSV and YHV: History in the Americas and current status. In *Proceedings of the 32nd Joint UJNR Aquaculture Panel Symposium*, pp. 17–20 (Davis and Santa Barbara, CA, 2003).
5. Briggs, M., Funge-Smith, S., Subasinghe, R., and Phillips, M. *Introductions and Movement of Penaeus vannamei and Penaeus stylirostris in Asia and the Pacific*, Vol. 476 (Food and Agricultural Organization, Rome, Italy, 2005).
6. Wijegoonawardane, P.K. et al. Genetic diversity in the yellow head nidovirus complex. *Virology* 380, 213–225 (2008).
7. Wijegoonawardane, P.K. et al. Homologous genetic recombination in the yellow head complex of nidoviruses infecting *Penaeus monodon* shrimp. *Virology* 390, 79–88 (2009).
8. Chantanachookin, C.B. et al. Histology and ultra-structure reveal a new granulosis-like virus in *Penaeus monodon* affected by yellow-head disease. *Dis Aquat Organ* 17, 145–157 (1993).
9. Spann, K.M., Vickers, J.E., and Lester, R.J. Lymphoid organ virus of *Penaeus monodon* from Australia. *Dis Aquat Organ* 23, 127–134 (1995).
10. Spann, K.M., Cowley, J.A., Walker, P.J., and Lester, R.J.G. A yellow-head-like virus from *Penaeus monodon* cultured in Australia. *Dis Aquat Organ* 31, 169–179 (1997).
11. Cowley, J.A., Dimmock, C.M., Spann, K.M., and Walker, P.J. Detection of Australian gill-associated virus (GAV) and lymphoid organ virus (LOV) of *Penaeus monodon* by RT-nested PCR. *Dis Aquat Organ* 39, 159–167 (2000).
12. Callinan, R.B. and Jiang, L. Fatal, virus-associated peripheral neuropathy and retinopathy in farmed *Penaeus monodon* in eastern Australia. II. Outbreak descriptions. *Dis Aquat Organ* 53, 195–202 (2003).
13. Munro, J., Callinan, R., and Owens, L. Gill-associated virus and its association with decreased production of *Penaeus monodon* in Australian prawn farms. *J Fish Dis* 34, 13–20 (2011).
14. OIE (Office International des Epizooties). *Manual of Diagnostic Tests for Aquatic Animals. Yellowhead disease*, pp. 204–217 (World Organization for Animal Health, Paris, France, 2009).
15. Cowley, J.A., Walker, P.J., Flegel, T.W., Lightner, D.V., Bonami, J.R., Snijder, E.J., and Degroot, R.J. Family Roniviridae. In *Virus Taxonomy, IXth Report of the International Committee on Taxonomy of Viruses* (eds. King, A., Adams, M., Carstens, E., and Lefkowitz, E.J.), pp. 797–801 (Elsevier, London, U.K., 2012).
16. Sittidilokratna, N., Dangtip, S., Cowley, J.A., and Walker, P.J. RNA transcription analysis and completion of the genome sequence of yellow head nidovirus. *Virus Res* 136, 157–165 (2008).

17. Sittidilokratna, N. et al. Complete ORF1b-gene sequence indicates yellow head virus is an invertebrate nidovirus. *Dis Aquat Organ* 50, 87–93 (2002).

18. Sittidilokratna, N., Phetchampai, N., Boonsaeng, V., and Walker, P.J. Structural and antigenic analysis of the yellow head virus nucleocapsid protein p20. *Virus Res* 116, 21–29 (2006).

19. Soowannayan, C. et al. Detection and differentiation of yellow head complex viruses using monoclonal antibodies. *Dis Aquat Organ* 57, 193–200 (2003).

20. Jitrapakdee, S. et al. Identification and analysis of gp116 and gp64 structural glycoproteins of yellow head nidovirus of *Penaeus monodon* shrimp. *J Gen Virol* 84, 863–873 (2003).

21. Cowley, J.A., Hall, M.R., Cadogan, L.C., Spann, K.M., and Walker, P.J. Vertical transmission of gill-associated virus (GAV) in the black tiger prawn *Penaeus monodon*. *Dis Aquat Organ* 50, 95–104 (2002).

22. Cowley, J.A. et al. Multiplex RT-nested PCR differentiation of gill-associated virus (Australia) from yellow head virus (Thailand) of *Penaeus monodon*. *J Virol Methods* 117, 49–59 (2004).

23. Cowley, J.A. and Walker, P.J. Molecular biology and pathogenesis of roniviruses. In *Nidoviruses* (eds. Perlman, S., Gallagher, T., and Snijder, E.J.), pp. 361–377 (ASM Press, Washington, DC, 2008).

24. Spann, K.M. and Lester, R.J.G. Viral diseases of penaeid shrimp with particular reference to four viruses recently found in shrimp from Queensland. *World J Microbiol Biotechnol* 13, 419–426 (1997).

25. Senapin, S. et al. Impact of yellow head virus outbreaks in the whiteleg shrimp, *Penaeus vannamei* (Boone), in Thailand. *J Fish Dis* 33, 421–430 (2010).

26. Stentiford, G.D., Bonami, J.R., and Alday-Sanz, V. A critical review of susceptibility of crustaceans to Taura syndrome, Yellowhead disease and White Spot Disease and implications of inclusion of these diseases in European legislation. *Aquaculture* 291, 1–17 (2009).

27. Longyant, S.S., S. Chaivisuthangkura, P., Rukpratanporn, S., Sithigorngul, W., and Sithigorngul, P. Experimental infection of some penaeid shrimps and crabs by yellow head virus (YHV). *Aquaculture* 257, 83–91 (2006).

28. Cowley, J.A. et al. Yellow head virus from Thailand and gill-associated virus from Australia are closely related but distinct prawn viruses. *Dis Aquat Organ* 36, 153–157 (1999).

29. Flegel, T.W., Sriurairatana, S., Wongterrasupaya, C., Boonsaeng, V., Panyim, S., and Withyachumnarnkul, B. Progress in characterisation and control of yellow head virus of *Penaeus monodon*. In *Swimming through Troubled Water* (eds. Browdy, C.L.H. and Hopkins, J.S.), pp. 76–83 (World Aquaculture Society, Baton Rouge, LA, 1995).

30. Lightner, D.V. (ed.) *Handbook of Pathology and Diagnostic Procedures for Diseases of Penaeid Shrimp* (World Aquaculture Society, Baton Rouge, LA, 1996).

31. Flegel, T.W., Fegan, D.F., and Sriurairatana, S. Environmental control of infectious shrimp diseases in Thailand. In *Asian Fisheries Society* (eds. Shariff, M., Subasinghe, R.P., and Arthur, J.R.), pp. 65–79 (Asian Fisheries Society, Manila, Philippines, 1995).

32. Spann, K.M., McCulloch, R.J., Cowley, J.A., East, I.J., and Walker, P.J. Detection of gill-associated virus (GAV) by in situ hybridization during acute and chronic infections of *Penaeus monodon* and *P. esculentus*. *Dis Aquat Organ* 56, 1–10 (2003).

33. Walker, P.J., Cowley, J.A., Spann, K.M., Hodgson, R.A.J., Hall, M., and Withyachumnarnkul, B. Yellow head complex viruses: Transmission cycles and topographical distribution in the Asia-Pacific region. In *The New Wave* (eds. Browdy, C.L.J. and Jory, D.E.), pp. 227–237 (World Aquaculture Society, Baton Rouge, LA, 2001).

34. Longyant, S. et al. Differences in susceptibility of palaemonid shrimp species to yellow head virus (YHV) infection. *Dis Aquat Organ* 64, 5–12 (2005).

35. Spann, K.M., Donaldson, R.A., Cowley, J.A., and Walker, P.J. Differences in the susceptibility of some penaeid prawn species to gill-associated virus (GAV) infection. *Dis Aquat Organ* 42, 221–225 (2000).

36. Khongpradit, R., Kasornchandra, J., and Boonyaratalin, S. Susceptibility of the postlarval stages of black tiger shrimp (*Penaeus monodon*) to yellow-head baculovirus (YBV). In *Diseases in Asian Aquaculture II* (eds. Shariff, M., Subasinghe, R.P., and Arthur, J.R.), p. 6 (Asian Fisheries Society, Manila, Philippines, 1995).

37. Castro-Longoria, R., Quintero-Arredondo, N., Grijalva-Chon, J.M., and Ramos-Paredes, J. Detection of the yellow-head virus (YHV) in wild blue shrimp, *Penaeus stylirostris*, from the Gulf of California and its experimental transmission to the Pacific white shrimp, *Penaeus vannamei*. *J Fish Dis* 31, 953–956 (2008).

38. Sánchez-Barajas, M., Liñán-Cabello, M.A., and Mena-Herrera, A. Detection of yellow-head disease in intensive freshwater production systems of *Litopenaeus vannamei*. *Aquacult Int* 17, 101–112 (2009).

39. Lu, Y., Tapay, L.M., Loh, P.C., Brock, J.A., and Gose, R.B. Distribution of yellow-head virus in selected tissues and organs of penaeid shrimp *Penaeus vannamei*. *Dis Aquat Organ* 23, 67–70 (1995).

40. Smith, P.T. Diseases of the eye of farmed shrimp *Penaeus monodon*. *Dis Aquat Organ* 43, 159–173 (2000).

41. Flegel, T.W., Booyaratpalin, S., and Withyachumnarnkul, B. Progress in research on yellow head virus and white spot virus in Thailand. In *Disease in Asian Aquaculture III* (eds. Flegel, T.W. and MacRae, I.H.), pp. 285–295 (Asian Fisheries Society, Manila, Philippines, 1997).

42. Nadala, E.C., Jr., Tapay, L.M., Cao, S., and Loh, P.C. Yellowhead virus: A rhabdovirus-like pathogen of penaeid shrimp. *Dis Aquat Organ* 31, 141–146 (1997).

43. Sithigorngul, P. et al. Monoclonal antibodies specific to yellow-head virus (YHV) of *Penaeus monodon*. *Dis Aquat Organ* 49, 71–76 (2002).

44. Sithigorngul, P. et al. Development of a monoclonal antibody specific to yellow head virus (YHV) from *Penaeus monodon*. *Dis Aquat Organ* 42, 27–34 (2000).

45. Munro, J. and Owens, L. Production of polyclonal and monoclonal antibodies against gill-associated virus and the development of an ELISA. *Aquaculture* 262, 173–182 (2007).

46. Munro, J. and Owens, L. Haemagglutination as a low-cost detection method for gill-associated virus and by inference, yellowhead virus in *Penaeus monodon* Fabricius, 1798. *Aquacult Res* 36, 1369–1373 (2005).

47. Munro, J. and Owens, L. Sensitivity and specificity of current diagnostic tests for gill-associated virus in *Penaeus monodon*. *J Fish Dis* 29, 649–655 (2006).

48. Tang, K.F. and Lightner, D.V. A yellow head virus gene probe: Nucleotide sequence and application for in situ hybridization. *Dis Aquat Organ* 35, 165–173 (1999).

49. Tang, K.F., Spann, K.M., Owens, L., and Lightner, D.V. In-situ detection of Australian gill associated virus with a yellow head virus gene probe. *Aquaculture* 205, 1–5 (2002).

50. Wongteerasupaya, C. et al. Detection of yellow-head virus (YHV) of *Penaeus monodon* by RT-PCR amplification. *Dis Aquat Organ* 31, 181–186 (1997).

51. Kiatpathomchai, W., Jitrapakdee, S., Panyim, S., and Boonsaeng, V. RT-PCR detection of yellow head virus (YHV) infection in *Penaeus monodon* using dried haemolymph spots. *J Virol Methods* 119, 1–5 (2004).

52. Wijegoonawardane, P.K., Cowley, J.A., and Walker, P.J. Consensus RT-nested PCR detection of yellow head complex genotypes in penaeid shrimp. *J Virol Methods* 153, 168–175 (2008).

53. Wijegoonawardane, P.K., Cowley, J.A., and Walker, P.J. A consensus real-time RT-PCR for detection of all genotypic variants of yellow head virus of penaeid shrimp. *J Virol Methods* 167, 5–9 (2010).

54. Posiri, P., Kondo, H., Hirono, I., Panyim, S., and Ongvarrasopone, C. Successful yellow head virus infection of *Penaeus monodon* requires clathrin heavy chain. *Aquaculture* 435, 480–487 (2015).

55. La Fauce, K. and Owens, L. RNA interference with special reference to combating viruses of crustacea. *Ind J Virol* 23, 226–243 (2012).

56. Yodmuang, S., Tirasophon, W., Roshorm, Y., Chinnirunvong, W., and Panyim, S. YHV-protease dsRNA inhibits YHV replication in *Penaeus monodon* and prevents mortality. *Biochem Biophys Res Commun* 341, 351–356 (2006).

57. Tirasophon, W., Roshorm, Y., and Panyim, S. Silencing of yellow head virus replication in penaeid shrimp cells by dsRNA. *Biochem Biophys Res Commun* 334, 102–107 (2005).

58. Saksmerprome, V., Charoonnart, P., Gangnonngiw, W., and Withyachumnarnkul, B. A novel and inexpensive application of RNAi technology to protect shrimp from viral disease. *J Virol Methods* 162, 213–217 (2009).

59. Assavalapsakul, W., Chinnirunvong, W., and Panyim, S. Application of YHV-protease dsRNA for protection and therapeutic treatment against yellow head virus infection in *Litopenaeus vannamei*. *Dis Aquat Organ* 84, 167–171 (2009).

34 Avian Infectious Bronchitis Virus

Vagner Ricardo Lunge, Aline Padilha de Fraga, and Nilo Ikuta

CONTENTS

34.1 Introduction 307
    34.1.1 Classification, Morphology, and Genome Organization 307
    34.1.2 Biology..309
    34.1.3 Epidemiology..309
    34.1.4 Transmission, Pathogenesis, and Clinical Features..310
    34.1.5 Diagnosis 311
        34.1.5.1 Conventional Techniques 311
        34.1.5.2 Molecular Techniques 311
34.2 Methods 311
    34.2.1 Sample Collection and Preparation 312
    34.2.2 Detection Procedures..312
34.3 Conclusion 313
References..313

34.1 INTRODUCTION

Viral respiratory diseases are important causes of economic losses to the poultry industry worldwide. They are responsible for impaired growth of the flocks, reduced egg production, and quality, mortality, and slaughter downgrading. In addition, they may be associated with intercurrent bacterial infections, contributing to increased economic losses (including costs for diagnoses, vaccines, antimicrobials, and treatment). Two viral respiratory diseases (the very virulent Newcastle and the highly pathogenic avian influenza) have the potential to devastate poultry industry, but biosecurity efforts have kept the world's main poultry-producing regions free of these pathogens. Three other viral respiratory diseases (infectious laryngotracheitis, swollen head syndrome, and infectious bronchitis [IB]) are highly prevalent and frequent in industrial poultry flocks worldwide. Of these, IB has been considered the most economically important [1].

IB is an acute and highly contagious disease that affects only domestic fowl (*Gallus gallus*). The disease occurs in birds of all ages and any type of chicken (broiler, breeder, layers). Young birds are particularly susceptible and mortality is usually high when opportunistic pathogens complicate the disease. IB virus (IBV) is prevalent in chicken flocks, and it has also been found in peafowl (*Galliforme*). Furthermore, IBV-like viruses have been isolated from pigeon (*Columbiforme*), turkey (*Galliforme*), teal, duck, and goose (*Anseriforme*) [2].

IBV initially infects the upper respiratory tract (mainly trachea) and may later spread to the kidney, oviduct, and other organs and tissues. Clinical signs include cough, nasal discharge, and watery eyes in birds with respiratory manifestation.

When young pullets are affected, damage to the reproductive tract can result in layers and breeders failing to come into production. Reduced egg quality can also be observed in IBV-infected layers and breeders. Decreased feed conversion efficiency and increased carcass condemnations are common in broiler, reducing the overall performance of the flocks. Morbidity is typically 100%, and mortality is variable according to the IBV strain and characteristics of the flock (age, type of chicken, breed, immune status, etc.). In flocks with nephritis, mortality can be higher than 50%. All these losses determine the economic impact of this disease in the different levels of the poultry industry chain [2,3].

34.1.1 CLASSIFICATION, MORPHOLOGY, AND GENOME ORGANIZATION

Avian IBV is a member of the genus *Gammacoronavirus* (formerly group 3), subfamily *Coronavirinae*, family *Coronaviridae*, order *Nidovirales*. The genus *Gammacoronavirus* currently consists of two species: avian coronavirus (type species) and Beluga whale coronavirus SW1. IBV and other previously existing coronaviruses of birds, such as turkey, duck, goose, pheasant, and pigeon, represent strains of avian coronavirus [4].

IBV is round to pleomorphic, with a mean diameter of 120 nm. It possesses an envelope and club-shaped surface projections (the glycosylated spike protein) of about 20 nm in length. Virions form by budding at internal cell membranes and their characteristic appearance under electron microscopy is a fringe of bulbous surface projections forming an image reminiscent of the solar corona. This morphology is created by the viral spike (S) peplomers, proteins of the viral surface. After spontaneous disruption of IBV particles, the core structure could be observed as strands of only 1–2 nm in diameter [2].

In addition to the S protein, there are three other proteins that contribute to the overall IBV structure: membrane (M), envelope (E), and nucleocapsid (N). S protein is a dimer or trimer in coronaviruses and has two basic functions: (1) to attach the virus to receptor molecules on host cells and (2) to activate fusion of the viral envelope with cellular membranes, releasing the viral RNA into the host cell's cytoplasm to be replicated [5]. IBV has the S protein in a cleaved form, with amino-terminal S1 (535 amino acids) and carboxy-terminal S2 (625 amino acids) subunits. The bulbous head of the S protein is formed largely by the S1 subunit, and it is anchored in the membrane by the carboxy-terminal portion of S2. S proteins interact with the transmembrane regions of a large copy number of a small integral M glycoprotein and with low amounts of the very small, membrane-associated, and nonglycosylated E protein. Both of these proteins are required for virus particle formation [2,5]. The genome is packaged into a helical nucleocapsid with the N phosphoprotein, creating the ribonucleoprotein [2].

The single-stranded, positive-sense RNA genome contains a 5′ cap structure and a 3′ poly-A tail and has about 27.6 kb in length in classical IBV (Figure 34.1). It includes several open reading frames (ORFs) that encode both the structural and nonstructural viral proteins. The whole genome is predominantly arranged in the order 5′ untranslated region (UTR)–replicase gene–structural protein genes (S, E, M, and N)–3′ UTR. Both UTR have around 500 nucleotides essential to viral RNA transcription and replication [6]. There are further nonstructural accessory protein genes interspersed between the structural protein genes that can vary in number among the IBV groups [2,7].

The replicase segment is also called Gene 1 and encodes the RNA-dependent RNA polymerase (RdRp), which is responsible for replicating the IBV genome. Gene 1 has a length of 20 kb and consists of two independent ORFs (ORF1a and ORF1b). These ORFs are translated into two polyproteins that are further cleaved into 15 products, among which the first two N-terminal ones are separated by a papain-like proteinase (PL^{pro}) and the remaining of the C-terminal products by a 3C-like proteinase (3CL^{pro}) [8]. The RNA genome also encodes the four structural proteins S (Gene 2), E (Gene 3c), M (Gene 4), and N (Gene 6), as well as the small nonstructural accessory proteins 3a, 3b, 5a, and 5b [2]. S is the most highly variable gene in the genome, mainly the S1 fragment, and responsible for the high diversity of IBV serotypes/genotypes. It also plays a key role in determining the species and tissue–cell tropism of all avian coronavirus [2,9–11].

The complex and large positive-sense RNA genome can have different sizes and organization in some variant IBV as well as in avian coronavirus that infects other bird species (turkey, pigeon, duck, etc.). For example, genomes that lack either all or most of the genes coding for nonstructural proteins at the 3′ end were observed in Australian types of IBV subgroup 2. These viruses had a unique ORF called X1 instead of the Genes 3 and 5, presenting two possible organizations: 5′ UTR–replicase gene–S–X1–E–M–N–3′ UTR or 5′ UTR–replicase gene–S–X1–E–M–5b–N–3′ UTR [12].

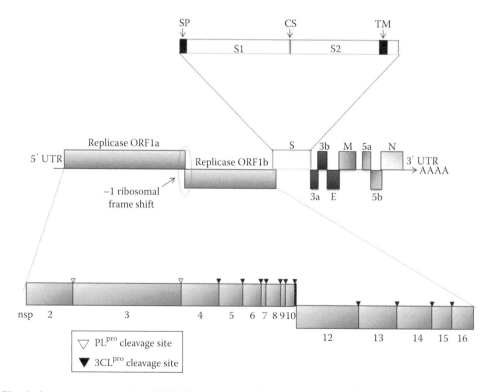

FIGURE 34.1 Classical genome organization of IBV. The genome is 27,620 nt long, excluding poly (A) tract. Middle: 10 genes and ORFs. Ribosomal frame shift is indicated. Top: details of spike protein, signal peptide (SP), spike protein cleavage (CS), transmembrane domain of spike protein (TM). Bottom: putative domains of ORF1a/1b polyprotein: nonstructural protein (nsp).

A recent study demonstrated that Korean variant strains are defective in segments of the Genes 3a and 3b [13]. On the other hand, turkey coronavirus (TCoV) has a genome very similar to IBV (>80%), and the main difference is in the S1 sequence (<40% of similarity) [7,14]. The quite common emergence of IBV variants as well as other avian coronaviruses in the field seems to be a result of mutation events in the S gene [13,15–18].

34.1.2 Biology

Initially, IBV replicates in the avian respiratory tract, but it has also tropism to epithelial cells from the kidney, oviduct, and gut. Viral entry relies on the interplay between the virion and the host cell, and it is initiated by the interaction of the S1 viral protein with specific receptor domains on the cell surface. Experimental data demonstrate that tropism to different cells (trachea, kidney, oviduct, cecal tonsils, etc.) is also determined by the S protein [10,19–21]. After initial binding, the coronavirus fuses the envelope with the host cell membrane to deliver the nucleocapsid into the target cell. The fusion involves large conformational changes of the S protein [5].

In order to infect cells, IBV enters by endocytosis. Inhibitory drugs of the clathrin-mediated pathway (such as chlorpromazine) abolished IBV infection. IBV virions harbor cleaved S glycoproteins between the S1 and S2 domains, proteolytically matured during infection in the host. This processing is carried out by the cellular proprotein convertase furin. IBV Beaudette strain has the peculiar feature to contain a second furin cleavage site in the S2 domain of the spike. Infection was inhibited by the presence of a furin inhibitor or amino acid substitutions in the S1–S2 cleavage site in Beaudette strain. Mutants containing a minimal cleavage site are still infectious, but they are dependent on serine proteases for productive infection [22].

After entrance, IBV replicates in the cytoplasm. Translation of the plus-stranded genome is initiated at the Gene 1 to produce the RdRp. The replicase–transcriptase proteins are encoded in the ORF1a and ORF1b and are synthesized initially as two large polyproteins: pp1a and pp1ab. The synthesis of pp1ab involves programmed ribosomal frame shifting during translation of ORF1a, resulting in the C-terminal extension of a relatively small fraction of ORF1a-encoded polypeptide (pp1a) with the ORF1b-encoded polypeptide. During or after synthesis, these polyproteins are co- and posttranslationally processed by the autoproteinases (PL^{pro} and 3CL^{pro}) that reside in pp1a. Among the 15 nonstructural proteins (nsp) produced, nsp2–nsp11 are encoded in ORF1a, and nsp12–nsp16 are encoded in ORF1b (Figure 34.1) [23,24]. RdRp replicates the genome through a continuous process that starts with the synthesis of an intermediate full-length minus-strand RNA, which then serves as the template for the asymmetric synthesis of progeny plus-strand genomes. A process of discontinuous transcription during negative-strand synthesis, regulated by short AU-rich sequences known as transcription regulatory sequences (TRSs), leads to the production of a nested set of subgenomic mRNAs (sgmRNAs) sequences

and expression of the structural and accessory proteins. Three of the sgmRNAs (from the Genes 2, 4, and 6) are responsible for production of the S, M, and N proteins, respectively. Two sgmRNAs (3 and 5) encode three (3a, 3b, and E) and two (5a and 5b) nonstructural proteins. However, the precise mechanisms of sgmRNA synthesis are yet not fully understood, and probably, more transcripts are produced [23–25].

With the synthesis of structural proteins, N binds genomic RNA and the M and S proteins are integrated into the membrane of the endoplasmic reticulum (ER). Assembled nucleocapsids with helical twisted RNAs bud into the ER lumen and are encased with the membrane. The viral progeny is transported by Golgi vesicles to the cell membranes, and the viral release is by exocytose into the extracellular space. New virus starts to appear 3–4 h after infection, with maximum output per cell being reached within 12 h at 37°C [3,23].

Rapid replication, high mutation rate, and genomic recombination result in extensive IBV genetic diversity [26]. The precise mechanism behind the emergence of this high diversity is still largely unknown, but it is well known that RdRp has low fidelity and limited ability to correct mistakes, although it has an associated exoribonuclease. High mutation rates are attributed to this low proofreading capability that generates around 10^{-3}–10^{-5} substitutions per site per year [5,7]. Point mutations, insertions, and deletions are observed along the genome of different IBV strains, especially in the gene encoding the S glycoprotein. In the most conserved parts of the genome, synonymous replacements also accumulate significantly [25]. Further, avian coronaviruses exchange genetic material (recombination) when different viral genotypes cocirculate in the same host [5]. Recent studies have also shown that recombination is the main mechanism behind the emergence of the novel *Gammacoronavirus* of turkey (TCoV) [14,27]. It is suggested that TCoV is a recombination between the classical IBV genome with the S fragment of another coronavirus from a still unknown source [14]. Similar studies have demonstrated that these two phenomenons (mutation and recombination) are also associated to the emergence of new IBV variants in the field worldwide [12,15,16,18,28].

34.1.3 Epidemiology

IBV is ubiquitous in most poultry-producing regions of the world and able to spread very rapidly in nonprotected birds. The virus is highly infectious, and it spreads by aerosol as well as by mechanical means. It is shed via both the respiratory tract and the feces and persists in the flocks for several weeks or months. Furthermore, one or more serotype/genotype can circulate in a specific geographic region [3].

An overview of the IBV history is essential to understand the current epidemiological situation. IB was first reported in the United States as a respiratory disease of young birds by Schalk and Hawn in 1931. Viral etiology was established some years after by Beach and Schalm in 1936. The first vaccine against IBV, strain M41 of the serotype Massachusetts (Mass), was developed in 1941. Other clinical manifestations, such as decline in egg production and more typical clinical respiratory signs, were demonstrated in 1950. A different serotype named Connecticut (Conn) was isolated in 1956 and considered the first IBV variant strain [29,30]. In the following decade, H52 and H120 vaccines (both of Mass serotype) were developed by serial passages in eggs and have been widely used in poultry flocks until today [31].

Injuries in kidney tissue were associated with IBV in the 1960s [32]. Other variant serotypes (Florida, Clark 333, and Arkansas) were described in the United States in the following years. Arkansas was one of the most important IBV variant strains found in the United States, and it is still associated to economic losses [26]. The emergence of viral field variants required the development of new vaccines (against variant serotypes), since Mass did not provide adequate immune protection [30].

The involvement of the enteric system (mainly the intestine) with IBV infection was demonstrated only in the 1970s, with several reports of viral excretion in feces but no apparent effect on the gut function [29].

Several variants (such as D274 and D1466) were reported in Europe in the 1980s. The control of these variants was performed with the use of attenuated live and inactivated vaccines against these specific strains. With the development of new diagnostic tools using molecular biology techniques, other variants (such as California, DE072, Vic S, N2/75, QX, and BR) were reported worldwide [29]. Despite this great diversity of variants, a small number of them (Mass, 4/91, QX-like) spread globally and are currently found in different regions of the world [33–36]. The main example of a globally distributed strain is the Mass, probably due to the intensive vaccination of the poultry industrial flocks with strains of this serotype. Other strains, as for example 4/91 and QX, disseminated naturally and are now present in important poultry-producing regions and backyard chickens [37,38].

34.1.4 Transmission, Pathogenesis, and Clinical Features

The main mode of IBV transmission is believed to be airborne, but the frequency of the air spread between flocks is unknown. Infection can also occur with the ingestion of water and contaminated food with fecal material, since IBV can survive for days in the environment, especially at low temperatures [39]. There are reports of survival for up to 20 weeks in the feces, a potential source of infection for birds [40]. It is also suggested that other species of birds can act as vectors, since IBV was demonstrated to infect more species [3,41].

IBV is highly contagious and has a very short incubation period up to the colonization of the trachea and appearance of the first respiratory signs (24–48 h after infection). Most IBV strains are isolated from the trachea with high concentrations during the first 3–5 days postinfection (PI) [42]. The viral titer drops rapidly in the second week and virus isolation is not always possible [43]. The progression of tracheal lesions can be classified into three stages: (1) degeneration, with deciliation and shedding of the epithelium

(first 2 days PI); (2) hyperplasy, with the formation of new epithelial cells (3–10 days); and (3) regeneration, with the repair of the injured tissue (10–20 days) [44]. When secondary pathogens like avian mycoplasmas, *Escherichia coli*, *Ornithobacterium rhinotracheale*, and/or *Bordetella* complicate the disease, pneumonia, airsacculitis, and peritonitis can also occur [3,7].

After respiratory tract infection, IBV spreads to various tissues. The urinary (mainly kidneys), reproductive, and digestive (mainly cecal tonsil) systems are the preferential places for viral replication [2,29]. In the kidney, viral replication occurs in the proximal convoluted tubules and distal and collecting tubules. IBV can cause granular degeneration, vacuolization, and desquamation of the tubular epithelium [3]. The injury of the epithelial cells may result in changes in the transport of liquids and acute renal failure [45]. Poultry infected with nephropathogenic strains (e.g., T strain, Holte, Gray, SAIBK, and B1648) may show exclusively these renal damages, without any respiratory lesion [3].

In the digestive tract, the virus replicates in different tissues, such as esophageal, proventricular, duodenal, jejunal, cecal tonsil, rectal, and cloacal [2]. In the intestine, IBV replication has been reported in histiocyte-like and lymphoid cells of the cecal tonsils [45]. IBV was detected in the digestive tract and cloacal swabs from chickens infected with an IBV enterotropic variant strain (named G) for a period of up to 28 days. This strain has also been associated with the presence of lesions in the intestine and beyond the trachea, kidney, and reproductive tract [46,47].

Viral excretion occurs intermittently during the IBV persistence in chicken bodies. In a classic experiment, it was demonstrated that the onset of lay is preceded or coincided with a period of IBV excretion [48]. The intermittent excretion is the main way to spread among birds of the same flock by horizontal transmission (with direct and indirect contact) and even to new flocks [3,49].

Clinical presentation in the field includes respiratory, renal, reproductive, and digestive typical signs. The main affected tissue and/or organ depends on the tropism of the infecting strain, but respiratory signs are quite common for all strains and include cough, sneezing, depression, catarrhal inflammation in the airways and sinuses with nasal discharge, and watery eyes. The consequence is a decrease in food consumption and weight loss. Adult birds usually do not show respiratory signs or they are discrete [45].

Renal manifestation is observed in some cases and can be confused with other diseases. Signs of this manifestation are depression, ruffled feathers, wet droppings, watery stools, and increased water consumption. Mortality can be high (>50%) when renal injury is accompanied by urolithiasis [3].

Reproductive signs are mainly observed in layer and breeder flocks. IBV infection of 1-day-old chicks and even pullets can cause permanent damage to developing organs of the reproductive system. In adult hens, a decrease in egg production, and reduced internal quality (low albumen protein synthesis) and shell quality (soft-shelled or rough-shelled) are observed [45]. The severity of the production decline may vary with the period of lay and infecting virus strain [3].

Chickens with the digestive form of the disease may have watery stools and/or severe diarrhea. Some lesions can be found along the gastrointestinal tract and cecal tonsils [2,3]. However, enteric IBV infection is usually asymptomatic in chickens. Interestingly, the main clinical signs of turkeys infected with TCoV are enteric. Birds exhibit depression, anorexia, decreased water consumption, watery diarrhea, dehydration, hypothermia, and weight loss [50].

34.1.5 DIAGNOSIS

Diagnosis of IB is initially based on the clinical history of the flock and gross lesions of the affected birds. Infected chickens have serous, catarrhal, or caseous exudates in the trachea, nasal passages, sinuses, and air sacs. Pneumonia signs may be observed around the large bronchi. Swollen and pale kidneys occur in the nephropathogenic infections, while oviducts and ovaries with reduced length and weight are eventually present in layers and breeders [3]. However, all of these clinical signs and lesions are also seen with other diseases, and confirmation of the IBV infection must be carried out by laboratory testing.

34.1.5.1 Conventional Techniques

The most classical methodology is virus isolation by inoculation into embryonating chicken eggs. IBV isolates usually replicate well in 10–11-day-old embryonating chicken eggs following inoculation of the allantoic cavity [51]. Field isolates often require adaptation via several (three or more) passages to achieve high viral titers. Characteristic IBV-infected chicken embryos (stunting, dwarfing, curling of the embryo) and increasing mortality rates are observed with more passages. A common lesion of the IBV-infected embryo is the presence of urates in the mesonephros of the embryonic kidney [3].

IBV can also be isolated in an organ (mainly trachea) or cell cultures. Twenty-day-old embryos' tracheal rings are usually prepared and maintained in roller tubes. After IBV infection, ciliostasis occurs within 3–4 days; and it is easily observed by microscopy. However, some strains with affinity to other tissues cannot cause visible ciliostasis [52]. Further, IBV can also be propagated in chicken kidney cell cultures. This cell type forms syncytia, which quickly round up and detach from the culture surface and has been largely used with several IBV isolates. However, some strains (e.g., Beaudette) cannot form syncytia and the cells remain attached to the culture surface. These strains should be maintained in other cell type, such as Vero [53].

In any system, virus isolation is generally a sensitive and useful procedure for the diagnosis of viral infection when clinical specimens are of good quality and have been collected ideally within 3–5 days PI. The main biological samples for virus isolation are trachea (especially with the first week of infection) and cloacal swabs or cecal tonsils collected during postmortem examination (mainly when more than 1 week has elapsed since the start of the infection). IBV is preferably isolated from a specific-pathogen-free (SPF) source (embryonating eggs or cells). The whole process is lengthy and laborious because field isolates usually require more passages [3,42].

Observations in embryonating eggs, organ, and cell cultures are not in themselves enough to detect IBV. Virus isolates must be tested with serological or molecular assays to confirm IBV presence and determine the identity of a field strain. Virus neutralization (VN) and enzyme-linked immunosorbent assay (ELISA) are considered type specific, although some cross-reactions between serotypes can occur. VN is based on the reaction of the IBV isolate with serotype-specific antibodies raised in chickens. Chicken embryos are most often employed, but antibodies can be measured using tracheal organ cultures or cell culture systems [54]. Two strains (A and B) are considered to be of the same serotype when two-way heterologous reactions (antiserum A with virus B, and antiserum B with virus A) differ less than 20-fold from the homologous reaction titer (antiserum A with virus A, antiserum B with virus B) in both directions [42]. ELISA is also widely used, and there are several commercial kits available. IBV antigen can be captured by anti-IBV antibodies that are coated on a microtiter plate. Subsequent incubations with anti-IBV conjugate and substrate will show positive optical densities when IBV is present in the sample. Samples have been previously inoculated in eggs since high amounts of virus are needed for antigen detection [42]. However, ELISA has many advantages for routine virus testing, being cost effective, robust, easy to use, and scalable to testing large numbers of samples [55].

34.1.5.2 Molecular Techniques

Molecular nucleic acid amplification techniques have been increasingly used for IBV detection and genotyping [3]. The main technique, reverse transcription polymerase chain reaction (RT-PCR), has been applied worldwide to detect IBV directly in clinical samples, after passages in embryonating eggs or in organ and cell cultures. Further, other molecular assays such as restriction fragment length polymorphism (RFLP) and nucleotide sequencing are able to genotype or even indirectly serotype the IBV sample, determining the identity of the strain. There is a good agreement between data represented by the VN serotype and S1 nucleotide sequence [3,56].

IBV molecular detection methods have highly conserved genomic regions as targets, such as the nucleocapsid gene [57,58] and 5′ UTR [59]. On the other hand, only S1 gene is currently used as a target in the genotyping assays [3,60]. Previous studies also developed RT-PCR genotype-specific assays, targeted on S1 gene, to detect only specific variant strains [61–63].

34.2 METHODS

Molecular methods to detect and genotype IBV require an adequate sample collection, transport, and storage as well as an efficient RNA extraction. In general, these first steps are laborious and time-consuming but essential to obtain a conclusive result in the analysis. Afterward, the main molecular biology technique used is RT-PCR either to detect or to genotype IBV. IBV can be detected by traditional RT-PCR and gel electrophoresis [64,65] or more recently by real-time RT-PCR procedures [57,59]. Some methods include a nested amplification step to improve the sensitivity and specificity of the assay. In contrast, the main methods to genotype IBV require an additional nucleotide sequence characterization step: RFLP or sequencing [66,67].

34.2.1   Sample Collection and Preparation

The main clinical specimen for this analysis is traditionally the trachea, considered the first replication site in the beginning of the infection, especially within the first week of infection. However, the cloaca and cecal tonsils are also recommended because IBV can replicate in the digestive tract in later stages of the infection. Other organs and tissues (lung, kidney, and oviduct) should also be considered to detect IBV according to the clinical history of the disease [3,42]. The samples for analysis could be either trachea/cloacal swabs or any organ/tissue collected during postmortem examination. Procedures for sample collection and processing have been previously described in detail [42]. The recommendations should be implemented on the farm. In addition, specimens should be transported to the laboratory from the collection site with a coolant to maintain a refrigerated temperature of approximately 2°C–8°C or preferably frozen [54,68].

Clinical samples are usually prepared in pools (set of organs collected from three to five birds) prior to the RNA extraction. Each pool should be composed of organs from a specific physiological system (lung, trachea, or tracheal swabs for respiratory; cecal tonsils, intestine, and cloacal swabs for digestive; kidney, ovary, and oviduct for urinary/reproductive), but it is also common to mix organs from different systems. Afterward, the extraction and purification of nucleic acids are prerequisites for nucleic acid–based molecular diagnostics. The main steps in all nucleic acid extractions are as follows: cell lysis, inactivation of cellular nucleases, and extraction of the RNA [69]. The main methods to extract RNA include the use of phenol and chloroform organic solvents, silica beads, and magnetism.

Phenol–chloroform alkaline extraction involves the lysis of the biological sample by a detergent or chaotropic substance with or without protein-degrading enzymes, followed by several processing steps applying organic solvents such as phenol and/or chloroform or ethanol and nucleic acid precipitation [70]. Silica particle extraction is a solid-phase method for isolating nucleic acid from biological samples. Silica beads are used to bind to nucleic acids in the presence of a chaotropic substance. This is one of the most widespread method for isolating nucleic acids from biological samples and is known as simple, rapid, and reliable [71]. Finally, magnetic separation of nucleic acids is an emerging technology that uses magnetism for the efficient separation of nucleic acids in solution. Magnetic carriers with immobilized affinity ligands or prepared from a biopolymer exhibiting affinity to the target

nucleic acid are used to isolate nucleic acid [72]. Despite the method used, extracted RNA should be stored between –20°C and –80°C until tested. It is advised to keep RNA at –80°C for long-term storage [54].

34.2.2   Detection Procedures

Molecular assays used to detect IBV RNA include RT-PCR and its variations (nested RT-PCR and real-time RT-PCR). In traditional RT-PCR, IBV RNA is first reverse transcribed to a complementary DNA (cDNA) by a reverse transcriptase. Subsequently, the cDNA is submitted to PCR. Amplification products are then submitted to electrophoresis in polyacrylamide or agarose gel and stained with silver nitrate or ethidium bromide, respectively [73,74].

An additional amplification step ("nested" to the first amplicon) can improve the sensitivity [75]. However, nested PCR should be used carefully in routine diagnosis because of the high risk of contamination ("carryover"), leading to false-positive results. Another very important topic to the development of a molecular technique for IBV is the target RNA region. Genome fragments commonly used in detection tests are the highly conserved N gene, 5′ UTR, and 3′ UTR [57–59]. However, these assays may be unable to distinguish between IBV and other avian coronaviruses.

Several studies describe RT-PCR for IBV detection, but two methods have been widely used [57,59]. The assay described by Callison et al. is a real-time RT-PCR that uses a primer pair and a TaqMan® dual-labeled probe to amplify a 143-bp fragment of the 5′ UTR of IBV genome [59]. Meir et al. chose a primer pair and a TaqMan probe to amplify a 130-bp fragment of nucleocapsid gene [57] (Table 34.1).

Genotyping techniques are usually performed after an RT-PCR of the S1 gene. RFLP includes restriction digestion with selected enzymes followed by gel electrophoresis. The banding patterns are observed after gel staining and then compared with expected patterns of known genotypes [3,66].

However, nucleotide sequencing of the hypervariable fragment of the S1 gene is the most useful technique for IBV genotyping. RT-PCR product cycle sequencing may be used diagnostically to identify previously characterized strains and also unrecognized field isolates and variants [3,33]. One important step in all sequencing methods is the careful preparation of the template with adequate concentration and purity of the RT-PCR products. Prior to sequencing reaction, the unincorporated nucleotide and primers remaining after the PCR amplification must be removed using commercial kits [76]. S1 gene sequences should then be comparatively analyzed with reference data to determine the relatedness among the IBV strains [3].

S1 genotype-specific RT-PCR may be used to identify IBV field isolates as well as vaccine strains in avian flock samples. S1 gene primers specific for several genotypes have been described in previous studies [61–63,67,77]. More recently, real-time PCR assays have been developed for the detection of Mass, Conn, Arkansas, and Delaware genotypes, and specific primers and probes are described in Table 34.1 [63].

TABLE 34.1

Sequence of Primers and Probes Used for Detection and Genotyping of IBV

Primer/Probe	Sequence (Primers and Probes)	Target Region	Reference
IBV5′ GU391	5′-GCTTTTGAGCCTAGCGTT-3′	5′ UTR[a]	[59]
IBV5′ GL533	5′-GCCATGTTGTCACTGTCTATTG-3′	5′ UTR[a]	[59]
IBV5′ G probe	5′-FAM CACCACCAGAACCTGTCACCTC-BHQ1-3′	5′ UTR[a]	[59]
AIBV-fr	5′-ATGCTCAACCTTGTCCCTAGCA-3′	N[a]	[57]
AIBV-as	5′-TCAA-ACTGCGGATCA-TCACGT-3′	N[a]	[57]
Probe AIBV-TM	5′-FAM-TTGGAAGTAGAGTGACGCCCAAACTTCA-BHQ1-3′	N[a]	[57]
Ark-F9	5′-GGTGAAGTCACTGTTTCTA-3′	S1[b]	[63]
Ark-R9	5′-AGCACTCTGGTAGTAATAC-3′	S1[b]	[63]
Ark-P	5′-TET-TRTATGACAACGAATC-MGBNFQ-3′	S1[b]	[63]
Mass-F9	5′-CGTKTACTACTAYCAAAGTGC-3′	S1[b]	[63]
Mass-R9	5′-CCATGAATARTACCAACARTACAC-3′	S1[b]	[63]
Mass-P	5′-FAM-AGCCTGCATTATTARAT-MGBNFQ-3′	S1[b]	[63]
Conn-F9	5′-ATGCRGTAGTTAATACTTC-3′	S1[b]	[63]
Conn-R9	5′-CGWCATAGCTATAGARGAA-3′	S1[b]	[63]
Conn-P	5′-CY5-ACCAATAATACCAACAATACACTCTCTTAA-BHQ-2-3′	S1[b]	[63]
Del/GA-F9	5′-AGGCGTTTGTACTGYATA-3′	S1[b]	[63]
Del/GA-R9	5′-GCCATGCCTTAAAATTTG-3′	S1[b]	[63]
Del/GA-P	5′-TET-TTKTGACACACTGTGGT-MGBNFQ-3′	S1[b]	[63]

[a] IBV detection.
[b] IBV genotyping.

34.3 CONCLUSION

IBV has been controlled by the extensive use of vaccines, but this virus remains as a major economic problem. The constant emergence of variant strains has challenged vaccination strategies. The advent of molecular techniques greatly improved the detection of this important pathogen in industrial poultry flocks. Further, the availability of molecular techniques to amplify and sequence viruses provides tools for the characterization of IBV classical and variant strains. In previous years, the application of these methods in veterinary laboratories contributed to improve the health status of domestic fowl flocks and to determine the vaccination strategies for this virus in the main chicken-producing regions of the world. The continuous use of these methodologies will certainly help in the detection of new outbreaks as well as allow the surveillance of the emergence of new genetic variants, displaying different biological and antigenic properties, which may require a revision of the current vaccination program in a specific region or country.

REFERENCES

1. Jones, R. C. Viral respiratory diseases (ILT, aMPV infections, IB): Are they ever under control? *Br Poult Sci.* 51(1), 1–11, 2010.
2. Cavanagh, D. Coronavirus avian infectious bronchitis virus. *Vet Res.* 38, 281–297, 2007.
3. Cavanagh, D., Gelb Jr., J. Infectious bronchitis. In: Saif, Y. M. (Ed.), *Diseases of Poultry*, 12th edn. Ames, IA: Blackwell, 2008.
4. International Committee on Taxonomy of Viruses. 2013. http://ictvonline.org/virusTaxonomy.asp?taxnode_id=20130566. Accessed on October 30, 2014.
5. Belouzard, S., Millet, J. K., Licitra, B. N., Whittaker, G. R. Mechanisms of Coronavirus cell entry mediated by the viral spike Protein. *Viruses* 4, 1011–1033, 2012.
6. De Vries, A. A. F., Horzinek, M. C., Rottier, P. J. M., De Groot, R. J. The genome organization of the Nidovirales: Similarities and differences between Arteri-, Toro-, and coronaviruses. *Sem Virol.* 8, 33–47, 1997.
7. Jackwood, M. W., Hall, D., Handel, D. Molecular evolution and emergence of avian gammacoronaviruses. *Infect Genet Evol.* 12, 1305–1311, 2012.
8. Hagemeijer, M. C., Rottier, P. J. M., Haan, A. M. Biogenesis and dynamics of the Coronavirus replicative structures. *Viruses* 4, 3245–3269, 2012.
9. Hodgson, T., Britton, P., Cavanagh, D. Neither the RNA nor the proteins of open reading frames 3a and 3b of the coronavirus infectious bronchitis virus are essential for replication. *J Virol.* 80, 296–305, 2006.
10. Hodgson, T., Casais, R., Dove, B., Britton, P., Cavanagh, D. Recombinant infectious bronchitis Coronavirus Beaudette with the spike protein gene of the pathogenic M41 strain remains attenuated but induces protective immunity. *J Virol.* 78(24), 13804–13811, 2004.
11. Toro, H., Van Santen, V. L., Jackwood, M. W. Genetic diversity and selection regulates evolution of infectious bronchitis virus. *Avian Dis.* 56(3), 449–455, 2012.
12. Mardani, K., Noormohammadi, A. H., Hooper, P., Ignjatovic, J., Browningi, G. F. Infectious bronchitis viruses with a novel genomic organization. *J Virol.* 82(4), 2013–2024, 2008.

13. Mo, M., Hong, S., Kwon, H., Kim, I., Song, C., Kim, J. Genetic diversity of spike, 3a, 3b and E genes of infectious bronchitis viruses and emergence of new recombinants in Korea. *Viruses* 5, 550–567, 2013.

14. Jackwood, M. W. et al. Emergence of a group 3 coronavirus through recombination. *Virology* 398, 98–108, 2010.

15. Kusters, J. G., Jager, E. J., Niesters, H. G. M., van der Zeijst, B. A. M. Sequence evidence for RNA recombination in field isolates of avian coronavirus infectious bronchitis virus. *Vaccine* 8, 605–608, 1990.

16. Lee, C. W., Jackwood, M. W. Evidence of genetic diversity generated by recombination among avian coronavirus IBV. *Arch Virol.* 145, 2135–2148, 2000.

17. Liu, X. et al. Comparative analysis of four Massachusetts type infectious bronchitis coronavirus genomes reveals a novel Massachusetts type strain and evidence of natural recombination in the genome. *Infect Genet Evol.* 14, 29–38, 2013.

18. Wang, L., Xu, Y., Collisson, E. W. Experimental confirmation of recombination upstream of the S1 hypervariable region of infectious bronchitis virus. *Virus Res.* 49, 139–145, 1997.

19. Ballesteros, M. L., Sánchez, C. M., Enjuanes, L. Two amino acid changes at the N-terminus of transmissible gastroenteritis Coronavirus spike protein result in the loss of enteric tropism. *Virology* 227, 378–388, 1997.

20. Sánchez, C. M. et al. Targeted recombination demonstrates that the spike gene of transmissible gastroenteritis Coronavirus is a determinant of its enteric tropism and virulence. *J Virol.* 73(9), 7607–7618, 1999.

21. Casais, R., Dove, B., Cavanagh, D., Britton, P. Recombinant avian infectious bronchitis virus expressing a heterologous spike gene demonstrates that the spike protein is a determinant of cell tropism. *J Virol.* 77(16), 9084–9089, 2003.

22. Yamada, Y., Liu, D. X. Proteolytic activation of the spike protein at a novel RRRR/S motif is implicated in furin-dependent entry, syncytium formation, and infectivity of Coronavirus infectious bronchitis virus in cultured cells. *J Virol.* 83(17), 8744–8758, 2009.

23. Pasternak, A. O., Spaan, W. J. M., Snijder, E. J. Nidovirus transcription: How to make sense…? *J Gen Virol.* 87, 1403–1421, 2006.

24. Sawicki, S. G., Sawicki, D. L., Siddell, S. G. A contemporary view of Coronavirus transcription. *J Virol.* 81(1), 20–29, 2007.

25. Gorbalenya, A. E., Enjuanes, L., Ziebuhr, J., Snijder, E. J. Nidovirales: Evolving the largest RNA virus genome. *Virus Res.* 117, 17–37, 2006.

26. Jackwood, M. W. Review of infectious bronchitis virus around the world. *Avian Dis.* 56(4), 634–641, 2012.

27. Lin, T. L., Loa, C. C., Wu, C. C. Complete sequences of 3 end coding region for structural protein genes of turkey coronavirus. *Virus Res.* 106, 61–70, 2004.

28. Mardani, K., Noormohammadi, A. H., Ignjatovic, J., Browning, G. F. Naturally occurring recombination between distant strains of infectious bronchitis virus. *Arch Virol.* 155(10), 1581–1586, 2010.

29. Cook, J. K. A., Jackwood, M., Jones, R. C. The long view: 40 years of infectious bronchitis research. *Avian Pathol.* 41(3), 239–250, 2012.

30. De Wit, J. J., Cook, J. K. A., Van Der Heijden, H. M. J. Infectious bronchitis virus variants: A review of the history, current situation and control measures. *Avian Pathol.* 40(3), 223–235, 2011.

31. Bijlenga, G., Cook, J. K. A., Gelb Jr., J., De Wit, J. J. Development and use of the H strain of avian infectious bronchitis virus from the Netherlands as a vaccine: A review. *Avian Pathol.* 33(6), 550–557, 2004.

32. Cumming, R. B. The etiology of "uraemia" of chickens. *Aust Vet J.* 38(11), 554, 1962.

33. Fraga A. P. et al. Emergence of a new genotype of avian infectious bronchitis virus in Brazil. *Avian Dis.* 57, 225–232, 2013.

34. Ignjatovic, J., Gould, G., Sapats, S. Isolation of a variant infectious bronchitis virus in Australia that further illustrates diversity among emerging strains. *Arch Virol.* 151, 1567–1585, 2006.

35. Jackwood, M. W., Hilt, D. A., Lee, C., Kwon, H. M., Callison, S. A. Data from 11 years of molecular typing infectious bronchitis virus field isolates. *Avian Dis.* 49:4, 614–618, 2005.

36. Ma, H. et al. Genetic diversity of avian infectious bronchitis Coronavirus in recent years in China. *Avian Dis.* 56(1), 15–28, 2012.

37. Abro, S. H., Renström, L. H. M., Ullman, K., Belák, S., Baule, C. Characterization and analysis of the full-length genome of a strain of the European QX-like genotype of infectious bronchitis virus. *Arch Virol.* 157, 1211–1215, 2012.

38. Worthington, K. J., Currie, R. J. W., Jones, R. C. A reverse transcriptase-polymerase chain reaction survey of infectious bronchitis virus genotypes in Western Europe from 2002 to 2006. *Avian Pathol.* 37(3), 247–257, 2008.

39. Brandão, P. E., Lovato, L. T., Slhessarenko, R. D. Coronaviridae. In: Flores, E. F. (Ed.), *Virologia Veterinária*, 2nd edn. Santa Maria, Spain: Editora da UFSM, 2012.

40. Ignjatović, J., Sapats, S. Avian infectious bronchitis virus. *Rev Sci Tech Off Int Epiz.* 19(2), 493–508, 2000.

41. Cavanagh, D. Coronaviruses in poultry and other birds. *Avian Pathol.* 34(6), 439–448, 2005.

42. De Wit, J. J. Technical review: Detection of infectious bronchitis virus. *Avian Pathol.* 29(2), 71–93, 2000.

43. MacLachlan, N. J., Edward, J. D. Coronaviridae. In: Fenner's, F. (Ed.), *Veterinary Virology*, 4th edn. San Diego, CA: Elsevier, 2011.

44. Nakamura, K., Cook, J. K. A., Otsuki, K., Huggins, M. B., Frazier, J. A. Comparative study of respiratory lesions in two chicken lines of different susceptibility infected with infectious bronchitis virus: Histology, ultrastructure and immunohistochemistry. *Avian Pathol.* 20(2), 241–257, 1991.

45. Dhinakar Raj, G., Jones R. C. Infections bronchits virus: Immunopathogenesis of infection in the chicken. *Avian Pathol.* 26, 677–706, 1997.

46. El-Houadfi, M. D., Jones, R. C. Isolation of avian infectious bronchitis viruses in Morocco including an enterotropic variant. *Vet Rec.* 116(16), 445, 1985.

47. Balestrin, E., Fraga, A. P., Ikuta, N., Canal, C. W., Fonseca, A. S. K., Lunge, V. R. Infectious bronchitis virus: Dissemination and main circulating genotypes in different avian physiological systems from Brazilian poultry commercial flocks. *Poult Sci.* 93(8), 1922–1929, 2014.

48. Jones, L. C., Ambali, A. G. Re-excretion of an enterotropic infectious bronchitis virus by hens at point of lay after experimental infection at day old. *Vet Rec.* 120, 617–618, 1987.

49. Di Fábio, J., Buitrago, L. Y. V. Bronquite Infecciosa das Galinhas. In: Berchieri Júnior, A., Silva, E. N., Di Fábio, J., Sesti, L., Zuanaze, M. A. F. (Eds.), *Doenças das Aves*, 2nd edn. Campinas, Brazil: FACTA, 2009.

50. Guy, J. S. Turkey coronavirus enteritis. In: Saif, Y. M. (Ed.), *Diseases of Poultry*, 12th edn. Ames, IA: Blackwell, 2008.

51. Darbyshire, J. H., Cook, J. K. A., Peters, R. W. Comparative growth kinetic studies on avain infectious bronchitis virus in different systems. *J Comp Pathol*. 85, 623–630, 1975.

52. Cook, J. K. A., Darbyshire, J. H., Peters, R. W. The use of chicken tracheal organ cultures tor the isolation and assay of avian infectious bronchitis virus. *Arch Virol*. 50, 109–118, 1976.

53. Otsuki, K., Yamamoto, H., Tsubokura, M. Studies on avian infectious bronchitis virus (IBV). *Arch Virol*. 60(1), 25–32, 1979.

54. OIE (Office International des Epizooties). *Terrestrial Manual*. Chapter 2.3.2: Avian infectious bronchitis virus, pp. 1–15, 2013. http://www.oie.int/fileadmin/Home/eng/Health_standards/tahm/2.03.02_AIB.pdf. Accessed on October 9, 2014.

55. Boonham, N., Kreuze, J., Winter, S., van der Vlugt, R., Bergervoet, J., Tomlinson, J., Mumford, R. Methods in virus diagnostics: From ELISA to next generation sequencing. *Virus Res*. 186, 20–31, 2014.

56. Ladman, B. S., Loupos, A. B., Gelb Jr., J. Infectious bronchitis virus S1 gene sequence comparison is a better predictor of challenge of immunity in chickens than serotyping by virus neutralization. *Avian Pathol*. 35(2), 127–133, 2006.

57. Meir, R., Maharat, O., Farnushi, Y., Simanov, L. Development of a real-time TaqMan® RT-PCR assay for the detection of infectious bronchitis virus in chickens, and comparison of RT-PCR and virus isolation. *J Virol Methods* 163(2), 190–194, 2010.

58. Zwaagstra, K. A., Van Der Zeijst, B. A. M., Kusters, J. G. Rapid detection and identification of avian infectious bronchitis virus. *J Clin Microbiol*. 30(1), 79–84, 1992.

59. Callison, S. A. et al. Development and evaluation of a real-time TaqMan RT-PCR assay for the detection of infectious bronchitis virus from infected chickens. *J Virol Methods* 138, 60–65, 2006.

60. Villarreal, L. Y. B. Diagnosis of infectious bronchitis: An overview of concepts and tools. *Braz J Poult Sci*. 12(2), 111–114, 2010.

61. Acevedo, A. M. et al. A duplex SYBR Green I-based real-time RT-PCR assay for the simultaneous detection and differentiation of Massachusetts and non-Massachusetts serotypes of infectious bronchitis virus. *Mol Cell Probes* 27(5–6), 184–192, 2013.

62. Chen, H. W., Wang, C. H. A multiplex reverse transcriptase-PCR assay for the genotyping of avian infectious bronchitis viruses. *Avian Dis*. 54(1), 104–108, 2010.

63. Roh, H., Jordan, B. J., Hilt, D. A., Jackwood, M. W. Detection of infectious bronchitis virus with the use of real-time quantitative reverse transcriptase–PCR and correlation with virus detection in embryonated eggs. *Avian Dis*. 58(3), 398–3403, 2014.

64. Adzhar, A., Shaw, K., Britton, P., Cavanagh, D. Universal oligonucleotides for the detection of infectious bronchitis virus by the polymerase chain reaction. *Avian Pathol*. 25(4), 817–836, 1996.

65. Falconi, E., D'Amore, E., Trani, L. D., Sili, A., Tollis, M. Rapid diagnoses of avian infectious bronchitis virus by the polymerase chain reaction. *J Virol Methods* 64, 125–130, 1997.

66. Kwon, H. M., Jackwood, M. W., Gelb, J. J. Differentiation of infectious bronchitis virus serotypes using polymerase chain reaction and restriction fragment length polymorphism analysis. *Avian Dis*. 37, 194–202, 1993.

67. Keeler Jr., C. L., Reed, K. L., Nix, W. A., Gelb Jr., J. Serotype identification of avian infectious bronchitis virus by RT-PCR of the peplomer (S-1) gene. *Avian Dis*. 42, 275–284, 1998.

68. Butot, S., Zuber, S., Baert, L. Sample preparation prior to molecular amplification: Complexities and opportunities. *Curr Opin Virol*. 4, 66–70, 2014.

69. Rahman, M. M., Elaissari, A. Nucleic acid sample preparation for in vitro molecular diagnosis: From conventional techniques to biotechnology. *Drug Discov Today* 17(21/22), 1199–1207, 2012.

70. Chomczynski, P., Sacchi, N. The single-step method of RNA isolation by acid guanidinium thiocyanate–phenol–chloroform extraction: Twenty-something years on. *Nat Protoc*. 1(2), 581–585, 2006.

71. Boom, R., Sol, C. J., Salimans, M. M., Jansen, C. L., Wertheim-Van Dillen, P. M., Van Der Noordaa, J. Rapid and simple method for purification of nucleic acids. *J Clin Microbiol*. 28, 495–503, 1990.

72. Berensmeier, S. Magnetic particles for the separation and purification of nucleic acids. *Appl Microbiol Biotechnol*. 73, 495–504, 2006.

73. Jin, L. T., Choi, J. K. Usefulness of visible dyes for the staining of protein or DNA in electrophoresis. *Electrophoresis* 25(15), 2429–2438, 2004.

74. Stellwagen, N. C. Electrophoresis of DNA in agarose gels, polyacrylamide gels and in free solution. *Electrophoresis* 30(Suppl. 1), 188–195, 2009.

75. Cavanagh, D., Mawditt, K., Britton, P., Naylor, C. J. Longitudinal studies of infectious bronchitis virus and avian pneumovirus in broilers using type-specific polymerase chain reactions. *Avian Pathol*. 28, 593–605, 1999.

76. Linnarsson, S. Recent advances in DNA sequencing methods—General principles of sample preparation. *Exp Cell Res*. 316, 1339–1343, 2010.

77. Handberg, K. J., Nielsen, O. L., Pedersen, M. W., Jorgensen, P. H. Detection and strain differentiation of infectious bronchitis virus in tracheal tissues from experimentally infected chickens by reverse transcription polymerase chain reaction. Comparison with an immunohistochemical technique. *Avian Pathol*. 28(4), 327–335, 1999.

35 Bovine Coronavirus

Dongyou Liu

CONTENTS

35.1 Introduction 317
    35.1.1 Classification..317
    35.1.2 Morphology and Genome Organization...318
    35.1.3 Biology and Epidemiology 318
    35.1.4 Clinical Features and Pathogenesis 319
        35.1.4.1 CD (calf scours) 319
    35.1.5 Diagnosis 319
    35.1.6 Treatment and Prevention 320
35.2 Methods 320
    35.2.1 Sample Preparation 320
    35.2.2 Detection Procedures.. 320
        35.2.2.1 Conventional RT-PCR... 320
        35.2.2.2 Real-Time RT-PCR 321
35.3 Conclusion 321
References.. 321

35.1 INTRODUCTION

Bovine coronavirus (BCoV) is a single-stranded, positive-sense RNA virus implicated in three distinct clinical syndromes in cattle: (1) calf diarrhea (CD), (2) winter dysentery (WD) with hemorrhagic diarrhea in adults, and (3) bovine respiratory disease complex (BRDC) or shipping fever of feedlot cattle. Given its common occurrence in dairy and beef cattle farms worldwide and its negative impact on milk production in dairy herds and weight gain in calves and adult cattle, considerable efforts have been made in recent years toward the elucidation of BCoV's genetic characteristics, epidemiology, transmission, clinical features, pathogenesis, diagnosis, treatment, and prevention.

35.1.1 CLASSIFICATION

Coronaviruses are enveloped, single-stranded, positive-sense RNA viruses classified in the family *Coronaviridae*, order *Nidovirales*. Of the two subfamilies within the family *Coronaviridae*, the subfamily *Coronavirinae* is separated into four genera, *Alphacoronavirus* and *Betacoronavirus* naturally occurring in bats and *Deltacoronavirus* and *Gammacoronavirus* residing primarily in birds and pigs, while the subfamily *Torovirinae* is divided into two genera, *Bafinivirus* and *Torovirus*, being present in fish and mammals (causing gastroenteritis in mammals such as cattle, horses, and pigs, but rarely humans), respectively [1].

Currently, the genus *Alphacoronavirus* consists of at least 16 species: transmissible gastroenteritis virus, porcine respiratory CoV, canine CoV, feline infectious peritonitis virus, mink CoV, rhinolophus bat CoV HKU2, miniopterus bat CoV 1A, miniopterus bat CoV 1B, miniopterus bat CoV HKU8, human CoV 229E (HCoV-229E), human CoV NL63 (HCoV-NL63), porcine epidemic diarrhea virus (see Chapter 37), scotophilus bat CoV 512, hipposideros bat CoV HKU10, rousettus bat CoV HKU10, and mystacina tuberculata bat CoV [2].

The genus *Betacoronavirus* is divided into four lineages (A–D). *Lineage A* includes rabbit CoV HKU14, human CoV-OC43 (HCoV-OC43), porcine hemagglutinating encephalomyelitis virus (PHEV), dromedary camel coronavirus HKU23, canine respiratory CoV, bovine CoV, giraffe CoV, sable antelope CoV, equine CoV (ECoV), human CoV-HKU1 (HCoV-HKU1), murine hepatitis virus, rat CoV, and China Rattus coronavirus HKU24. *Lineage B* contains severe acute respiratory syndrome (SARS)-related palm civet CoV, SARS-associated human CoV, SARS-related Chinese ferret badger CoV, and SARS-related rhinolophus bat CoV HKU3. *Lineage C* comprises Middle East respiratory syndrome CoV (MERS-CoV), pipistrellus bat CoV HKU5, and tylonycteris bat CoV HKU4. *Lineage D* consists of rousettus bat coronavirus HKU9 [2–5].

The genus *Deltacoronavirus* contains wigeon CoV HKU20, common moorhen CoV HKU21, night heron CoV HKU19, bulbul CoV HKU11 (type species), munia CoV HKU13, magpie–robin CoV HKU18, thrush CoV HKU12, white-eye CoV HKU16, porcine CoV HKU15, and sparrow CoV HKU17 [2].

The genus *Gammacoronavirus* consists of infectious bronchitis virus (IBV), partridge CoV (IBV-partridge), turkey CoV, peafowl CoV (IBV-peafowl), beluga whale CoV SW1, and bottlenose dolphin CoV HKU22 [2].

The genus *Bafinivirus* contains a single species white bream virus and the genus *Torovirus* consists of equine torovirus (type species; formerly Berne virus), human torovirus, porcine torovirus, and bovine torovirus (BoTV; formerly Breda virus, BRV). Of particular relevance, BoTV is a kidney-shaped virus associated with diarrhea in calves worldwide (see Chapter 38) [1].

Interestingly, although coronaviruses in the subfamily *Coronavirinae* do not usually produce clinical symptoms in their natural hosts (bats and birds), accidental transmission of these viruses to humans and other animals may result in respiratory, enteric, hepatic, or neurologic diseases of variable severity. Coronaviruses that have been shown to cause diseases in humans include HCoV-229E (alphacoronavirus 1b), HCoV-NL63 (alphacoronavirus 1b), HCoV-OC43 (betacoronavirus 2a), HCoV-HKU1 (betacoronavirus 2a), SARS-HCoV (betacoronavirus 2b), and MERS-CoV (betacoronavirus 2c) [2].

Belonging to Lineage A, genus *Betacoronavirus*, subfamily *Coronavirinae*, and family *Coronaviridae*, BCoV consists of a single serotype that encompasses a diversity of strains/isolates (e.g., BCoV Bubalus, BCoV cow, BCoV alpaca, bovine enteric coronavirus, bovine respiratory coronavirus, human enteric coronavirus 4408, sambar deer coronavirus, waterbuck coronavirus, and white-tailed deer coronavirus) that demonstrate genetic and antigenic proximities to HCoV-OC43, PHEV, and ECoV. Besides inducing diarrhea in newborn calves, WD with hemorrhagic diarrhea in adult cattle, and respiratory tract infections in calves and feedlot cattle, BCoV has the capacity to cause transboundary infection/disease in buffalos, lamas, alpacas, deer, and giraffes [6].

Phylogenetic examination of BoCV isolates reveals the existence of two genomic clades (1 and 2). While reference enteric BoCV strains and a vaccine strain belong to clade 1, BCoV from respiratory tract, nasal swab, and bronchoalveolar washing fluids belongs to clade 2. In addition, the respiratory isolates from Oklahoma are further separated into three subclades, 2a, 2b, and 2c [7,8].

35.1.2 MORPHOLOGY AND GENOME ORGANIZATION

BCoV possesses an enveloped, pleomorphic/spherical virion of 65–210 nm in diameter, with a double layer of short (hemagglutinin) and long (spike) surface projections, the latter of which appear club-shaped and measure about 20 nm in length. The lipoprotein envelope carries four structural proteins: membrane (M) glycoprotein, spike (S) glycoprotein, hemagglutinin–esterase (HE) glycoprotein, and small envelope (E) protein. Beneath the envelope is a flexible and helical nucleocapsid formed by nucleoprotein (N) in association with the viral genome.

The genome of BCoV is a linear, nonsegmented, single-stranded, positive-sense RNA of 31,028 nt and includes 13 open reading frames (ORFs) flanked by 5′- (nt 1–210) and 3′- (nt 30,740–31,028) untranslated regions (UTRs). Its 5′-UTR is capped, and its 3′-UTR contains a polyadenylated tail. Some overlappings are observed in ORF1a (nt 211–13,362, including stop codon) and ORF1b (13,332–21,494); ORF4 (S, 23,641–27,732)

and ORF5 (N_s "4.9 kDa", 27,722–27,811); ORF7 (N_s "12.7 kDa", 28,106–28,435) and ORF8 (E, 28,422–28,676); and ORF10 (N, 29,393–30,739) and ORF11 (I, 29,454–30,077). Intergenic sequences are identified in ORF1b and ORF2 (N_s 32 kDa, 21,504–22,340); ORF2 and ORF3 (HE, 22,352–23,626); ORF3 and ORF4; ORF5 and ORF6 (N_s "4.8 kDa", 27,889–28,026); ORF6 and ORF7; ORF8 and ORF9 (M, 28,691–29,383); and ORF9 and ORF10] [9].

Comparison of genome sequences of two field isolates of BCoV, representing respiratory (BCoV-R) and enteric (BCoV-E) strains, respectively, revealed differences in 107 out of 31,028 positions, scattered throughout the genome except the 5′-UTR. Differences in 25 positions are nonsynonymous, leading to 24 amino acid changes in all proteins except pp1b. The remaining 82 nucleotide differences do not cause amino acid changes, although they might modulate phenotypic properties by affecting the RNA structure and/or RNA interaction(s). Six replicase mutations are identified within or immediately downstream of the predicted largest pp1a-derived protein, p195/p210. It is possible that single amino acid changes within p195/p210 as well as within the S glycoprotein may have contributed to the different phenotypes of the BCoV isolates [9].

Of the five structural proteins (nucleocapsid [N], transmembrane, small envelope [E], hemagglutinin–esterase [HE], and spike [S] proteins) encoded by the BCoV genome, the N protein (50 kDa) is highly conserved and offers an ideal target for molecular diagnostic application; the surface HE glycoprotein (120–140 kDa) acts a receptor destroying enzyme (esterase) to reverse hemagglutination; and the outer surface S glycoprotein (190 kDa, 1363 aa) forms large club (petal)-shaped spikes on the surface of the virion. Cleavage of the S protein by an intracellular protease between aa 768 and 769 results in a variable S1 N-terminal domain (subunit) and a more conserved S2 C-terminal domain (subunit). The S1 subunit is involved in virus binding to host–cell receptors, induction of neutralizing antibody expression, and hemagglutinin activity. With a highly variable/mutable sequence, the S1 is associated with changes in antigenicity, tissue tropism, viral pathogenicity, and host range [10]. Indeed, a single amino acid change (A528V) within the BCoV S1 subunit has been shown to confer resistance to VN [11]. The S2 is the transmembrane subunit with a critical role in the mediation of fusion between viral and cellular membranes. Interestingly, no consistent differences are observed in the full-length S gene between BCoV strains causing respiratory and enteric diseases [12].

35.1.3 BIOLOGY AND EPIDEMIOLOGY

BCoV has a predilection for intestinal and pulmonary epithelial cells. First recognized as enteric pathogens (BCoV-E) in association with CD and WD in adult cattle, BCoV was also identified as respiratory pathogens (BCoV-R) in association with BRDC or shipping fever of feedlot cattle. Although BCoV strains causing enteric (BCoV-E) or respiratory (BCoV-R) infection may show distinct biological properties and host cell tropism, they are closely related antigenically and genetically [13–17]. In fact, some BCoV strain may cause simultaneous enteric and respiratory tract infections.

Transmission of BCoV is mainly through fecal–oral and respiratory (aerosol) routes. As in the case of BRDC, BCoV may enter the host via aerosol and undergo initial and extensive replication in the nasal mucosa. Following ingestion of mucus secretions containing large quantities of infectious virus with protective coating, cattle with respiratory infection may develop gastrointestinal symptoms such as diarrhea. Clinically or subclinically infected calves and young adult cattle function as reservoirs for BCoV in the herd, with sporadical shedding of the virus in feces and nasal secretions [18].

BCoV has a widespread presence in both dairy and beef cattle. WD has been described in Europe, North America, and East Asia, with a peak incidence in winter, while CD and shipping fever are found in various regions of the world [19–21]. Although BCoV outbreaks typically occur in autumn and winter, severe cases of infection in adult cattle may also emerge in warmer seasons. Economic losses due to BCoV infection are attributable to dramatic reduction in milk yield in dairy herds and loss of body condition in both calves and adults, with death being a not uncommon outcome for affected calves.

Unlike many other coronaviruses that have restricted host ranges, BCoV has the ability to cause transboundary infections. This is evidenced by the findings that bovine-like CoV is present in sambar deer (*Cervus unicolor*), white-tailed deer (*Odocoileus virginianus*), waterbuck (*Kobus ellipsiprymnus*), elk (*Cervus elephus*), caribou (*Rangifer tarandus*), giraffe, water buffalo calves, alpacas, and dogs [22–27], and that bovine-like CoV is implicated in deer and waterbuck with bloody diarrhea resembling WD in cattle. Furthermore, isolation of a BCoV-like human enteric coronavirus-4408/US/94 from a child with acute diarrhea highlights the public health impact of BCoV.

As an enveloped virus, BCoV is sensitive to detergents and lipid solvents such as ether and chloroform. It is also inactivated by conventional disinfectants, formalin, and heat [28].

35.1.4 Clinical Features and Pathogenesis

With an incubation of 3–8 days, BCoV is implicated in enteric disease in newly born calves (CD) and adult cattle (WD) as well as respiratory tract infections of growing calves and shipping fever pneumonia in feedlot cattle. Although BCoV infection rarely causes death, its morbidity can be as high as 100%, contributing to sudden and dramatic reduction in milk production and loss of body weight.

35.1.4.1 CD (calf scours)

BCoV infection in calves between the ages of 1 and 3 weeks often produces gastrointestinal signs such as profuse diarrhea (watery yellow, gray, or greenish containing blood or mucus), dehydration, depression, weakness, weight loss, anorexia, convulsions, and sometimes death. Notable lesions include colon villous atrophy. Some calves (2–6 months of age) may develop respiratory infection with a serous to purulent nasal discharge, coughing, rhinitis, pneumonia, fever, and inappetence,

often with concurrent diarrhea. Sporadic shedding of virus in nasal secretions and feces may begin 5 days after infection. Secondary bacterial infections with other common respiratory viruses (e.g., bovine rotavirus A), bacteria, and mycoplasms may exacerbate clinical signs [2,29].

WD with hemorrhagic diarrhea in adults: BCoV infection in adults is normally subclinical, although it may cause WD outbreaks in housed cattle over the winter months, with a sudden onset of semiliquid, often hemorrhagic diarrhea, leading to severe anemia, marked drop in milk production in dairy herds, and death. Histopathologically, loss of surface epithelial cells and necrosis of crypt epithelial cells in the large intestine containing detectable virus particles are noted [30–32].

BRDC or shipping fever: BCoV infection in young adult feedlot cattle (6–10 months of age) may develop fever, coughing, dyspnea, rhinitis, bronchopneumonia, anorexia, diarrhea, and weight loss, accompanied by necrotizing lung lesions (e.g., interstitial emphysema, bronchiolitis, and alveolitis). Death of infected cattle may occur as a result of necrotizing pneumonia. Shedding of respiratory BCoV in nasal secretions and feces begins 4–10 days after infection, with nasal shedding consistently preceding fecal shedding. Interestingly, cattle with relatively high respiratory BCoV antibody ELISA titers or neutralizing antibodies in serum are less likely to shed respiratory BCoV [33,34].

The pathogenesis of BCoV-induced diarrhea lies in its ability to destroy surface epithelial cells of the intestinal villi in both the small and large intestines, with remarkable loss of absorptive capacity, increase in the gut fluid volume, and osmotic imbalance. This contributes to rapid loss of water and electrolytes, hypoglycemia, lactic acidosis, hypervolemia, acute shock, cardiac failure, and death. The severity of the disease is influenced by the host age, nutritional and immune status, virus load, stresses, and copresence of other pathogenic organisms. For example, the onset of BRDC may be linked to coinfection with bovine respiratory syncytial virus, parainfluenza-3 virus, bovine herpesvirus, and bovine viral diarrhea virus that have the capacity to sabotage host's immune mechanisms. In addition, long-distance shipping with cattle from multiple feedlots may create physical stresses that compromise host's immune defense against BCoV and other pathogens, and also increase exposure of cattle to high concentrations of new pathogens not previously encountered. Chemotherapy (e.g., corticosteroids and dexamethasone) may reduce the numbers of cells (CD4 and CD8 T cells) and levels of cytokines that directly impact on host immune fitness.

35.1.5 Diagnosis

BCoV is associated with CD, WD in adults, and BRDC (shipping fever). Diagnosis of BCoV infection requires differentiation from other diseases with similar symptoms. For example, bovine rotavirus A (BRV), *Escherichia coli*, *Salmonella enterica*, *Giardia intestinalis*, and *Cryptosporidium parvum* are known to cause diarrhea in calves, while *Mannheimia haemolytica*, *Pasteurella multocida*, *Histophilus somni*,

Arcanobacterium pyogenes, and *Mycoplasma* are often involved in bovine respiratory disease [35,36].

Although a presumptive diagnosis can be made on the basis of history and clinical signs, definitive diagnosis of BCoV infection relies on laboratory confirmation of the presence of virus particles/antigens and antiviral antibodies in nasal secretions, feces, and tissues.

Virus isolation using the G clone of human rectal tumor (HRT)-18 cells, or Vero cells, provides a useful way to diagnose BCoV infection [37].

Electron microscopy allows direct observation of BCoV particles in feces and other samples. Typically, 5 g of feces is resuspended in 25 mL of phosphate buffered saline (PBS) and centrifuged at $5000 \times g$ for 20 min. The clarified supernatant (12 mL) is centrifuged again at $100,000 \times g$ for 1 h, and the resultant pellet is resuspended in 1 mL of water. Then, 5 µL of sample is mixed with 5 µL of 2% phosphotungstic acid (pH 6.9), and 1 µL of the mixture is applied to the center of a 300-mesh grid. The sample is side blotted with a piece of torn filter paper to remove the majority of the sample, thus leaving a fine layer to air dry. Samples are examined under a microscope at 30,000× magnification [10].

Serological assays such as direct fluorescent antibody test, virus neutralization (VN), hemagglutination inhibition, and enzyme-linked immunosorbent assay (ELISA) enable detection of viral antigens and antiviral antibodies in nasal secretions, feces, serum, and frozen or paraffin-embedded tissues (trachea, lung, ileum, and colon) [10].

Nucleic acid amplification techniques such as PCR allow sensitive detection of viral nucleic acids directly from clinical specimens [38–46]. In particular, real-time reverse transcription (RT)-PCR targeting conserved regions of the BCoV genome (polymerase or N protein) facilitates rapid detection of divergent strains [47,48]. In addition, sequence analysis of the S gene facilitates subtyping of BCoV strains causing intestinal and respiratory infections [49–53].

35.1.6 Treatment and Prevention

Treatments of animals with BCoV infections are largely aimed at alleviation of clinical symptoms, including intravenous fluid therapy and provision of adequate colostrum intake in newborn calves. In addition, isolation of calves with diarrhea and ventilation of housing are helpful in reducing disease incidence.

Control of BCoV-related diarrhea is possible by vaccinating calves on farm prior to shipping to auction barns or feedlots with a live vaccine (ATCvet code QI02). Although no vaccines are currently available to prevent BCoV-associated pneumonia in young calves or in the BRDC of feedlot cattle, there is evidence that intranasal (IN) vaccination of feedlot calves with a modified live BCoV calf vaccine on entry to a feedlot reduces the risk of treatment for BRDC in calves. In addition, IN administration of a BCoV CD strain induces cross-protection against field exposure to a respiratory BCoV [54,55]. Future research on the combined use of strains representing CD, WD, and respiratory isolates may open a way for a single broad-spectrum BCoV vaccine against BCoV infections [10].

35.2 METHODS

35.2.1 Sample Preparation

Samples (feces, fecal swabs, and nasal swabs) are collected from suspected cases of gastrointestinal and/or respiratory tract infections. Swabs are placed immediately after collection in 2 mL minimum essential media (MEM) containing 1000 U penicillin and 1 mg streptomycin. After transportation on dry ice, samples are mixed by pulse-vortexing for 15 s, and swabs are discarded. Sample suspensions are diluted 1:10 in MEM and centrifuged at 5000 rpm for 15 min at 4°C. The clarified supernatants are transferred to sterile vials, and aliquots of 140 µL are separated for RNA extraction using QIAamp Viral RNA Extraction Kit (Qiagen).

Fecal suspensions (approximately 0.1 g of feces or 100 µL in 1 mL 1× PBS, pH 7.5–8.0) are mixed for 1 min at 15 Hz on a tissue lyser. Fecal suspension of 50 µL is mixed with 400 µL of lysis solution, 1 µL of carrier RNA (1 µg/µL), and 1 µL of exogenous internal positive control (XIPC) RNA (at 10,000 copies/µL) for 5 min at 15 Hz and centrifuged at $20,000 \times g$ for 3 min to generate the clarified lysate. The clarified lysate (350 µL) is transferred to a 96-well, deep-well plate containing 20 µL of magnetic bead mix (10 µL of lysis/binding enhancer and 10 µL of RNA binding beads), and 200 µL of isopropanol is added. The plate is loaded onto an automated magnetic particle processor for nucleic acid purification: lysis/binding for 5 min, one 2 min wash 1, one 2 min wash 2, 1 min dry step, and a 3 min heated elution step at 70°C. Purified nucleic acid is then denatured at 95°C for 5 min in a thermal cycler to denature the double-stranded BRV RNA for RT-PCR.

Alternatively, feces are diluted in nuclease-free water (Promega) at a ratio of 3:1 (v/v), and suspensions are centrifuged at $5000 \times g$ for 10 min at 4°C. RNA is extracted from 250 µL of supernatant recovered using the TRIzol Reagent (Invitrogen), and the RNA pellet is diluted in 30 µL of nuclease-free water (Promega).

Tissues from the upper respiratory tract (nasal, pharyngeal tissues, trachea) and lung, and tissues from the distal small intestine and colon may be collected from suspect necropsied disease cases. RNA is extracted from these tissues by using Qiagen RNeasy Kit.

35.2.2 Detection Procedures

35.2.2.1 Conventional RT-PCR

Erles et al. [56] reported a gel-based RT-PCR for the detection of BCoV RNA in clinical samples with primers from the spike-protein gene (Sp1: 5′-CTTATAAGT GCCCCCAAACTAAA-3′, nt 25,277–25,300 and Sp2: 5′-CC TACTGTGAGATCACATGTTTG-3′, nt 25,876–25,898), which generate a specific product of 622 bp.

RT-PCR is conducted with SuperScript™ One-Step RT-PCR Kit (Invitrogen) using the following cycling program: RT at 50°C for 30 min, inactivation of Superscript II RT at 94°C for 2 min, and 45 cycles of 94°C for 30 s, 55°C for 30 s, and 68°C for 30 s, with a final extension at 68°C

TABLE 35.1

Identity of Primers and Probe for Real-Time RT-PCR Detection of BCoV

Primer/Probe	Sequence (5′–3′)
BCVF[c]	GCTACATCAATCTCAAGGACATTGGT
BCVR[c]	CTTAAAACAACTAGTCCCACATCCTGT
BCVPb[c]	FAM-TAAAAGCCATACATACCAAGGCCA-BHQ1

for 10 min. PCR products are separated by electrophoresis in 1.5% agarose gels and visualized under UV light after ethidium bromide staining [56,57].

35.2.2.2   Real-Time RT-PCR

Cho et al. [58] utilized primers BCVF and BCVR from the spike (S) gene sequence for specific detection of BCoV. The TaqMan probe BCVPb is labeled with 6-carboxyfluorescein (FAM) at the 5′ end and with a nonfluorescent quencher 1 (NFQ1) at the 3′ end (Table 35.1).

RT-PCR mixture (25 µL) is prepared with AgPath-ID™ Multiplex RT-PCR Kit (Applied Biosystems), containing 5 µL extracted template, 500 nM each of BCVF and BCVR, and 150 nM BCVPb.

Amplification and detection are carried out on an automated real-time PCR system (Smart Cycler® II, Cepheid, Inc.) with the following cycling parameters: 50°C for 30 min, 95°C for 15 min, and 35 cycles of 94°C for 15 s and 60°C for 60 s.

Samples with a Ct of 35 cycles or less are considered positive.

35.3   CONCLUSION

With a specific tropism for intestinal and pulmonary epithelial cells, BcoV is associated with CD, WD in adult cattle, and respiratory distresses in cattle of all ages. Economical losses due to BCoV outbreaks are not limited by animal mortalities (which can be as high as 80% in complicated cases), as dramatic reduction in milk yield in dairy herds and poor growth in calves and cattle together with cost of veterinary care and medications can add a significant burden on the affected farms. Due to the fact that a number of other microbial pathogens cause similar clinical signs, laboratory identification of BCoV antigen, antibodies, and nucleic acids is essential for correct diagnosis. In particular, the application of real-time RT-PCR has made it possible to detect and identify BCoV with unprecedented sensitivity, specificity, and speed [59].

REFERENCES

1. Aita T et al. Characterization of epidemic diarrhea outbreaks associated with bovine torovirus in adult cows. *Arch Virol.* 2012;157(3):423–431.
2. Woo PC et al. Novel betacoronavirus in dromedaries of the Middle East, 2013. *Emerg Infect Dis.* 2014;20(4):560–572.
3. Vijgen L et al. Evolutionary history of the closely related group 2 coronaviruses: Porcine hemagglutinating encephalomyelitis virus, bovine coronavirus, and human coronavirus OC43. *J Virol.* 2006;80:7270–7274.
4. Zhang J et al. Genomic characterization of equine coronavirus. *Virology.* 2007;369(1):92–104.
5. Lau SK et al. Discovery of a novel coronavirus, China Rattus coronavirus HKU24, from Norway rats supports the murine origin of Betacoronavirus 1 and has implications for the ancestor of Betacoronavirus lineage A. *J Virol.* 2015;89(6):3076–3092.
6. Barros IN et al. A multigene approach for comparing genealogy of Betacoronavirus from cattle and horses. *Sci World J.* 2013;2013:349702.
7. Martínez N et al. Molecular and phylogenetic analysis of bovine coronavirus based on the spike glycoprotein gene. *Infect Genet Evol.* 2012;12(8):1870–1878.
8. Fulton RW, Ridpath JF, Burge LJ. Bovine coronaviruses from the respiratory tract: Antigenic and genetic diversity. *Vaccine.* 2013;31(6):886–892.
9. Chouljenko VN, Lin XQ, Storz J, Kousoulas KG, Gorbalenya AE. Comparison of genomic and predicted amino acid sequences of respiratory and enteric bovine coronaviruses isolated from the same animal with fatal shipping pneumonia. *J Gen Virol.* 2001;82(12):2927–233.
10. Saif LJ. Bovine respiratory coronavirus. *Vet Clin North Am Food Anim Pract.* 2010;26(2):349–364.
11. Yoo D, Deregt D. A single amino acid change within antigenic domain II of the spike protein of bovine coronavirus confers resistance to virus neutralization. *Clin Diagn Lab Immunol.* 2001;8:297–302.
12. Bidokhti MR et al. Tracing the transmission of bovine coronavirus infections in cattle herds based on S gene diversity. *Vet J.* 2012;193(2):386–390.
13. Hasoksuz M, Lathrop SL, Gadfield KL, Saif LJ. Isolation of bovine respiratory coronaviruses from feedlot cattle and comparison of their biological and antigenic properties with bovine enteric coronaviruses. *Am J Vet Res.* 1999;60:1227–1233.
14. Gelinas AM, Sasseville AM, Dea S. Identification of specific variations within the HE, S1, and ORF4 genes of bovine coronaviruses associated with enteric and respiratory diseases in dairy cattle. *Adv Exp Medicine Biol.* 2001;494:63–67.
15. Lin XQ, O'Reilly KL, Storz J. Antibody responses of cattle with respiratory coronavirus infections during pathogenesis of shipping fever pneumonia are lower with antigens of enteric strains than with those of a respiratory strain. *Clin Diagn Lab Immunol.* 2002;9:1010–1013.
16. Zhang X et al. Quasispecies of bovine enteric and respiratory coronaviruses based on complete genome sequences and genetic changes after tissue culture adaptation. *Virology.* 2007;363(1):1–10.
17. Fulton RW et al. Enteric disease in postweaned beef calves associated with Bovine coronavirus clade 2. *J Vet Diagn Invest.* 2015;27(1):97–101.
18. Thomas CJ et al. Transmission of bovine coronavirus and serologic responses in feedlot calves under field conditions. *Am J Vet Res.* 2006;67:1412–1420.
19. Park SJ et al. Detection and molecular characterization of calf diarrhoea bovine coronaviruses circulating in South Korea during 2004–2005. *Zoonoses Public Health.* 2007;54(6–7):223–230.
20. Kanno T, Kamiyoshi T, Ishihara R, Hatama S, Uchida I. Phylogenetic studies of bovine coronaviruses isolated in Japan. *J Vet Med Sci.* 2009;71(1):83–86.
21. Stipp DT et al. Frequency of BCoV detection by a semi-nested PCR assay in faeces of calves from Brazilian cattle herds. *Trop Anim Health Prod.* 2009;41(7):1563–1567.

22. Hasoksuz M et al. Biologic, antigenic, and full-length genomic characterization of a bovine-like coronavirus isolated from a giraffe. *J Virol.* 2007;81:4981–4990.

23. Hasoksuz M, Vlasova A, Saif LJ. Detection of group 2a coronaviruses with emphasis on bovine and wild ruminant strains. Virus isolation and detection of antibody, antigen, and nucleic acid. *Methods Mol Biol.* 2008;454:43–59.

24. Kaneshima T et al. The infectivity and pathogenicity of a group 2 bovine coronavirus in pups. *J Vet Med Sci.* 2007;69(3):301–303.

25. Alekseev KP et al. Bovine-like coronaviruses isolated from four species of captive wild ruminants are homologous to bovine coronaviruses, based on complete genomic sequences. *J Virol.* 2008;82:12422–12431.

26. Decaro N et al. Characterisation of bubaline coronavirus strains associated with gastroenteritis in water buffalo (*Bubalus bubalis*) calves. *Vet Microbiol.* 2010;145(3–4):245–251.

27. Chung JY, Kim HR, Bae YC, Lee OS, Oem JK. Detection and characterization of bovine-like coronaviruses from four species of zoo ruminants. *Vet Microbiol.* 2011;148(2–4):396–401.

28. Mullis L, Saif LJ, Zhang Y, Zhang X, Azevedo MS. Stability of bovine coronavirus on lettuce surfaces under household refrigeration conditions. *Food Microbiol.* 2012;30(1):180–186.

29. Tsunemitsu H, Saif LJ. Antigenic and biological comparisons of bovine coronaviruses derived from neonatal calf diarrhea and winter dysentery of adult cattle. *Arch Virol.* 1995;140:1303–1311.

30. Traven M et al. Experimental reproduction of winter dysentery in lactating cows using BCV—Comparison with BCV infection in milk-fed calves. *Vet Microbiol.* 2001;81:127–151.

31. Jeong JH et al. Detection and isolation of winter dysentery bovine coronavirus circulated in Korea during 2002–2004. *J Vet Med Sci.* 2005;67(2):187–189.

32. Natsuaki S et al. Fatal winter dysentery with severe anemia in an adult cow. *J Vet Med Sci.* 2007;69(9):957–960.

33. Lathrop SL, Wittum TE, Loerch SC, Perino LJ, Saif LJ. Antibody titers against bovine coronavirus and shedding of the virus via the respiratory tract in feedlot cattle. *Am J Vet Res.* 2000;61:1057–1061.

34. Hick PM et al. Coronavirus infection in intensively managed cattle with respiratory disease. *Aust Vet J.* 2012;90(10):381–386.

35. Storz J et al. Isolation of respiratory bovine coronavirus, other cytocidal viruses, and *Pasteurella* spp. from cattle involved in two natural outbreaks of shipping fever. *J Am Vet Med Assoc.* 2000;216:1599–1604.

36. Fulton RW et al. Lung pathology and infectious agents in fatal feedlot pneumonias and relationship with mortality, disease onset, and treatments. *J Vet Diagn Invest.* 2009;21(4):464–477.

37. Hansa A et al. Isolation of bovine coronavirus (BCoV) in vero cell line and its confirmation by direct FAT and RT-PCR. *Pak J Biol Sci.* 2013;16(21):1342–1347.

38. Tsunemitsu H, Smith DR, Saif LJ. Experimental inoculation of adult dairy cows with bovine coronavirus and detection of coronavirus in feces by RT-PCR. *Arch Virol.* 1999;144:167–175.

39. Cho KO et al. Cross-protection studies between respiratory and calf diarrhea and winter dysentery coronavirus strains in calves and RT-PCR and nested PCR for their detection. *Arch Virol.* 2001;146:2401–2419.

40. Loa CC et al. Differential detection of turkey coronavirus, infectious bronchitis virus, and bovine coronavirus by a multiplex polymerase chain reaction. *J Virol Methods.* 2006;131(1):86–91.

41. Takiuchi E, Stipp DT, Alfieri AF, Alfieri AA. Improved detection of bovine coronavirus N gene in faeces of calves infected naturally by a semi-nested PCR assay and an internal control. *J Virol Methods.* 2006;131(2):148–154.

42. Asano KM et al. Multiplex semi-nested RT-PCR with exogenous internal control for simultaneous detection of bovine coronavirus and group A rotavirus. *J Virol Methods.* 2010;169(2):375–379.

43. Zhu W, Dong J, Haga T, Goto Y, Sueyoshi M. Rapid and sensitive detection of bovine coronavirus and group a bovine rotavirus from fecal samples by using one-step duplex RT-PCR assay. *J Vet Med Sci.* 2011;73(4):531–534.

44. Fukuda M et al. Development and application of one-step multiplex reverse transcription PCR for simultaneous detection of five diarrheal viruses in adult cattle. *Arch Virol.* 2012;157(6):1063–1069.

45. Schroeder ME et al. Development and performance evaluation of calf diarrhea pathogen nucleic acid purification and detection workflow. *J Vet Diagn Invest.* 2012;24(5):945–953.

46. Liang X, Chigerwe M, Hietala SK, Crossley BM. Evaluation of Fast Technology Analysis (FTA) Cards as an improved method for specimen collection and shipment targeting viruses associated with Bovine Respiratory Disease Complex. *J Virol Methods.* 2014;202:69–72.

47. Amer HM, Almajhdi FN. Development of a SYBR Green I based real-time RT-PCR assay for detection and quantification of bovine coronavirus. *Mol Cell Probes.* 2011;25(2–3):101–107.

48. Amer HM et al. A new approach for diagnosis of bovine coronavirus using a reverse transcription recombinase polymerase amplification assay. *J Virol Methods.* 2013;193(2):337–340.

49. Hasoksuz M, Sreevatsan S, Cho KO, Hoet AE, Saif LJ. Molecular analysis of the S1 subunit of the spike glycoprotein of respiratory and enteric bovine coronavirus isolates. *Virus Res.* 2002;84:101–109.

50. Jeong JH, Kim GY, Yoon SS. Molecular analysis of S gene of spike glycoprotein of winter dysentery bovine coronavirus circulating in Korea during 2002–2003. *Virus Res.* 2005;108:207–212.

51. Liu L et al. Molecular epidemiology of bovine coronavirus on the basis of comparative analyses of the S gene. *J Clin Microbiol.* 2006;44(3):957–960. Erratum in: *J Clin Microbiol.* 2006;44(5):1924.

52. Kanno T, Hatama S, Ishihara R, Uchida I. Molecular analysis of the S glycoprotein gene of bovine coronaviruses isolated in Japan from 1999 to 2006. *J Gen Virol.* 2007;88:1218–1224.

53. Takiuchi E, Alfieri AF, Alfieri AA. Molecular analysis of the bovine coronavirus S1 gene by direct sequencing of diarrheic fecal specimens. *Braz J Med Biol Res.* 2008;41(4):277–282.

54. Han MG, Cheon DS, Zhang X, Saif LJ. Cross-protection against a human enteric coronavirus and a virulent bovine enteric coronavirus in gnotobiotic calves. *J Virol.* 2006;80(24):12350–12356.

55. Plummer PJ et al. Effect of intranasal vaccination against bovine enteric coronavirus on the occurrence of respiratory tract disease in a commercial backgrounding feedlot. *J Am Vet Med Assoc.* 2004;225:726–731.

56. Erles K, Toomey C, Brooks HW, Brownlie J. Detection of a group 2 coronavirus in dogs with canine infectious respiratory disease. *Virology.* 2003;310(2):216–223.

57. Decaro N et al. Detection of bovine coronavirus using a TaqMan-based real-time RT-PCR assay. *J Virol Methods.* 2008;151(2):167–171.

58. Cho YI, Kim WI, Liu S, Kinyon JM, Yoon KJ. Development of a panel of multiplex real-time polymerase chain reaction assays for simultaneous detection of major agents causing calf diarrhea in feces. *J Vet Diagn Invest.* 2010;22(4):509–517.

59. Cho KO, Hoet AE, Loerch SC, Wittum TE, Saif LJ. Evaluation of concurrent shedding of bovine coronavirus via the respiratory tract and enteric route in feedlot cattle. *Am J Vet Res.* 2001;62:1436–1441.

36 Feline Enteric Coronavirus and Feline Infectious Peritonitis Virus

Niels C. Pedersen

CONTENTS

36.1 Introduction 323
   36.1.1 Classification, Morphology, and Genome Organization 324
   36.1.2 Transmission, Clinical Features, and Pathogenesis 324
      36.1.2.1 FECV 324
      36.1.2.2 FIPV 325
36.2 Methods 325
   36.2.1 Sample Collection and Preparation 325
   36.2.2 Detection Procedures 326
      36.2.2.1 Serologic Testing 326
      36.2.2.2 RT-PCR 326
      36.2.2.3 Immunohistochemistry 327
      36.2.2.4 Sequence Analysis and Genotype Comparison 327
36.3 Conclusion 328
References 328

36.1 INTRODUCTION

Feline infectious peritonitis (FIP) was recognized as a new and emerging disease entity in the late 1950s [1]. It has been subsequently identified worldwide and in virtually all species of *Felidae*, with domestic cats being the major target host [2]. The disease occurs primarily in an enzootic form [2] and epizootics are noteworthy because of their rarity [3]. The incidence of FIP was reported at 1:300 cats seen in North American veterinary institutions [4] and from <1% to >5% of kittens and younger cats from shelters, catteries, and other such high-density environments [2]. Once clinical signs appear, the mortality is extremely high; and there is currently no effective way to prevent or treat the disease.

The viral etiology of FIP was not established until 1966 when viral particles were observed in diseased tissues of experimentally infected cats [5]. The feline infectious peritonitis virus (FIPV) was linked to a coronavirus in 1970 [6]. The causative agent was subsequently found to be closely related to transmissible gastroenteritis virus of pigs, canine coronavirus (CCoV), and human coronavirus 229E [7]. Genetic comparisons and phylogenetic analysis of feline coronavirus (FCoV), CCoV, and porcine coronavirus were not carried out until almost two decades later [8,9].

There was initial confusion about the exact coronavirus responsible for FIP. Coronavirus seropositivity was found to be widespread among both diseased and healthy cats in the same environments, although cats with FIP tended to have higher antibody titers than healthy cats [10]. Since the incidence of FIP was only a fraction of the incidence of seropositivity, it was initially hypothesized that FIP was an infrequent clinical manifestation of a common and largely innocuous coronavirus infection. However, cell culture–propagated FIPV was subsequently found to cause FIP in virtually all experimentally infected cats [11]. In contrast, fecal extracts from healthy seropositive cats in environments where FIP occurred caused a transient mild to unapparent enteritis and not FIP [12]. The fecal virus from healthy cats was then given the name feline enteric coronavirus (FECV), while the virus from FIP diseased tissues retained the name feline infectious peritonitis virus (FIPV). It was subsequently postulated that FIPVs evolved by internal mutation from FECVs, a theory that was ultimately proven correct [13]. The precise incidence of this transformation is unknown, but one study using cats infected with feline immunodeficiency virus suggests that FIPV mutants may arise in as high as 20% of FECV infections [14]. No particular strain of FECV is more likely to mutate to become an FIPV than another [13,15].

FECVs have a highly specific tropism for the mature apical epithelium of the villi in the lower intestine, while FIPVs have lost their tropism for the intestine and have gained a specific tropism for peritoneal macrophages [13]. Three specific types of mutations have been currently associated with the FECV-to-FIPV transformation: (1) the accumulation of mutations in the 3′ portion of the accessory 3c gene that leads either to truncations of the 3c gene product or to significant amino changes [15]; (2) the single nucleotide polymorphism (SNP) mutations in the S1/S2 cleavage site of the structural spike (S) gene that can have a negative, positive, or neutral effect on the rate of furin cleavage [16]; and (3) an SNP mutation in the fusion domain of the S gene [17]. The first two mutations are highly variable and unique to each FIPV and absent from the infecting FECV strain, while the third mutations in the fusion region of the S gene are common to all FIPVs and absent from all FECVs. It is yet to be determined how these various mutations affect enterocyte vs. macrophage tropism and host/virus interactions.

FECV and FIPV, although different in cell tropism and disease potential, are virtually identical and do not meet the definition for distinct species or even strains. Therefore, they are designated as biotypes or, in some instances, pathotypes. The biotype designations of FECV and FIPV are analogous to that of mouse hepatitis virus (MHV), and there are interesting analogies between the cat and mouse viruses in a biotype-associated disease. MHV, biotype enteric, is a common enteric and largely inconsequential pathogen of wild mice but a bane for experimental mouse colonies [18]. In contrast, MHV, biotype systemic, causes a range of systemic diseases not seen in nature but characteristic of each laboratory where they have been cell cultured and animal passaged for decades.

36.1.1 Classification, Morphology, and Genome Organization

FECV and FIPV are positive-sense, single-stranded RNA viruses (Group IV) belonging to the family *Coronaviridae*, subfamily *Alphacoronavirus*. They are about 120 nm in diameter with an envelope displaying numerous large petal or spikelike progressions. The genomic structure and replication strategy of FIPV is similar to other coronaviruses [19]. The genome of FCoVs consists of >29,000 nucleotides and 11 open reading frames (ORFs) encoding structural, nonstructural, and accessory genes. Two-thirds of the 5' end of the genome is made up of ORF1ab, which is translated into polyproteins pp1a and pp1ab and processed into the viral polymerase and other nonstructural proteins involved in proteolytic cleavage and RNA synthesis. One-third of the 3' end of the genome encodes for structural (S, M, M, N) and accessory (3abc, 7ab) genes. Structural and accessory genes are expressed as genomic and subgenomic RNAs, and each is translated to yield only a protein encoded by the 5'-most ORF.

FIPVs are virtually identical to the FECVs from which they arise, a concept that is sometimes difficult to articulate. This has led many researchers to use the collective nomenclature of FCoV for FIPVs and FECVs. However, this often leads to greater confusion, as it is often difficult to ascertain which biotype is being researched [2]. Moreover, the use of the term FCoV does not take precedence into account. FIPV was the name originally given to the causative agent of FIP found in diseased tissues or effusions, while FECV was the name originally given to the enteric form of the virus isolated from feces of groups of cats that suffer periodic loses to FIP. If the term FCoV is to be used, it should be prefaced to indicate the biotype, that is, FCoV biotype enteric or biotype FIP.

FECVs, and their FIPV biotypes, exist as a myriad of strains with 70%–98%+ relatedness [2]. Although there is considerable genetic differences between FECVs and FIPVs based on geography and population isolation, strains in cohoused groups of cats tend to remain the same between cats over time [20]. These facts make it possible to trace many isolates back to their geographic or colony origins.

In addition to the two biotypes, all FECVs (and, therefore, all FIPVs) occur in two serotypes, 1 and 2 (types 1 and 2 or types I and II) [21]. These serotypes are determined by the makeup of the 3' region of the S and adjacent 3a genes [15,22]. Serotype 1 FECVs and FIPVs have characteristic catlike S genes, while serotype 2 viruses arise by recombination with CCoV. The exact extent of the recombination varies from one serotype 2 isolate to another [23]. Serotype 1 viruses make up from 70% to 90%+ of isolates around the world [24,25], with serotype 2 strains being somewhat more common in Asia than North America and Europe [26].

Serotypes 1 and 2 FECVs, and, to a lesser extent, serotype 1 FIPVs, have been extremely difficult to culture *in vitro* [27]. Serotype 2 FIPVs appear to be more easily cultivated and on a wider range of cat cell lines. FECV appears to be highly tropic for mature intestinal epithelium, and only recently have intestinal cell lines been developed that appear to support viral growth [28]. Likewise, macrophage cultures that can be readily infected and can replicate serotype 1 FIPVs have not been available. An occasional serotype 1 isolate of FIPV has been adapted to grow on continuous macrophage-like cells [11].

36.1.2 Transmission, Clinical Features, and Pathogenesis

36.1.2.1 FECV

FECV is the most common in the large and diverse group of enteropathogens detected at similar prevalence in feces from diarrheic and healthy cats [29]. The major site for virus replication in healthy shedders is in enterocytes of the colon and cecum [30]. FECV is enzootic in cat populations throughout the world and is most prevalent in dense multicat environments where kittens are present, such as shelters, catteries, and large urban/suburban feral cat colonies. The infection is less common among rural cats where population densities are low, and opportunity for cat-to-cat transmission through the fecal–oral route is lower. It is absent from the isolated Falkland Islands [31]. The virus is present in very high levels in feces, especially during the acute stage of infection, and to a greater level in kittens than adults [32,33].

FECV transmission is by the fecal–oral route and facilitated by the shared use of litter pans and other communal defecation sites. The virus can survive for up to 2 weeks in dried feces and fomites and can be easily transferred on clothes and body by people [12]. Kittens are usually not infected before 9 weeks of age, suggesting some degree of maternal immunity [32]. Clinical signs of infection are usually inapparent or mild, consisting of a transient diarrhea lasting for only several days. Therefore, there is only a slightly greater prevalence of fecal FECV shedding between diarrheic and healthy cats [29]. FECV-infected cats will continue to shed virus for weeks, months, and sometimes indefinitely following initial infection [32]. A weak immunity is eventually achieved and virus shedding ceases. However, the loss of infection is also associated with a loss of immunity and secondary infections are not unusual.

FECV is not an important enteropathogen of cats in its own right [29]. Rather, it is significant for two other reasons: (1) FECVs mutate within the cat's body to become FIPVs, and FIP is a highly fatal disease that is one of the leading infectious causes of death in cats, and (2) the extreme genotypic and phenotypic similarities of FECVs and FIPVs confound many diagnostic procedures for FIPV infection [27].

36.1.2.2 FIPV

FIPVs reflect the strain and clade diversity of FECVs within the same environments [2,34]. FIP is not considered to be an infectious disease in the classic sense, because infectious virus is not shed in the feces [2,13]. Therefore, cat-to-cat transmission of FIPVs does not occur to any degree. FIPVs are highly macrophage associated and are usually infectious only when diseased tissues or fluids are directly inoculated into susceptible cats. There are rare exceptions, however, where horizontal transmission has been documented [3].

The name "feline infectious peritonitis" is derived from the most common clinical manifestation of the disease, a diffuse subacute to chronic inflammation of the omentum, serosal surfaces, and mesenteries of the abdomen, and less commonly the pleura and pericardium [1,2] (Figure 36.1). This inflammation leads to an outpouring of a characteristic exudate that is yellowish in color, mucinous in character, high in proteins, often containing fibrinous tags, and with macrophages, neutrophils, and lymphocytes being the prominent

cell types. The same virus was later found to be the cause of a second form of FIP characterized by larger, focal, and more typically granulomatous lesions resembling tumors [2,9]. The two forms of FIP are referred to as wet/effusive/nonparenchymatous and dry/noneffusive/parenchymatous. Noneffusive FIP appears to evolve from an initial bout of effusive disease, and terminally, some cats with noneffusive FIP may retrogress into a more effusive disease and *vice versa*.

The classical lesion of the effusive form of FIP is the pyogranuloma, an accumulation of macrophages and neutrophils, appearing grossly as numerous small whitish plaques surrounding small venules in target tissues [2] (Figure 36.1). Venous congestion, microhemorrhage, edema, and inflammatory effusion are prominent features of effusive FIP. The lesions of noneffusive FIP are much less diffuse and tend to target selective areas of the abdomen (kidney, mesenteric lymph node), chest (pericardium), eyes (uveal tract), brain (ependyma and meninges), and spinal cord (meninges). Granulomatous lesions contain small foci of infected macrophages that appear to be remnants of surface pyogranulomas that have followed blood vessels deeper into the parenchyma. The mass of the lesion consists of a dense lymphocytic/plasmacytic infiltrate that extends some distance into the surrounding parenchyma. Inflammatory effusions are either absent or comparatively minor.

The consequence of FECV-to-FIPV mutations depends on the nature of the host's immune response. Ample evidence suggests that only cell-mediated immunity is effective in eliminating the virus once it enters the peritoneal macrophages [13]. Humoral immunity appears to actually be harmful in that it enhances the uptake of virus by macrophages and participates along with the viral antigen and complement in an Arthus-type vasculitis [2,13]. Most cats, in nature, are apparently able to mount a protective immune response. Those that fail to do so will develop FIP, which has an extremely high mortality. The form of disease following FIPV infection is also dependent on the balance between humoral and cellular immunity. Effusive FIP indicates a failure of cellular immunity, thus allowing the virus to replicate to a high level and spread rapidly from macrophage to macrophage within the peritoneal cavity and to distant sites. Noneffusive FIP is a product of a partially effective cellular immune response. Cellular immunity is sufficient to localize the virus to specific tissues but insufficient to eliminate it from the body.

36.2 METHODS

36.2.1 Sample Collection and Preparation

Proper sample collection is essential for successful identification of FECV or FIPV infections. First and foremost, samples for the detection of FECV or FIPV by reverse transcription PCR (RT-PCR) or RT-quantitative PCR (RT-qPCR) must contain sufficient levels of viral RNA. This is not a problem with FECV, because FECV shedding is extremely high and even very small amounts of feces will yield a positive result. However, attention should be paid to eliminating fecal

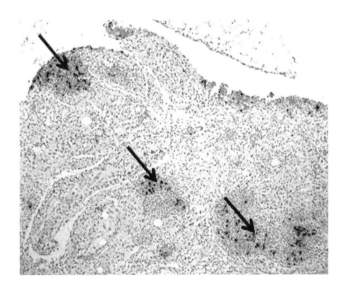

FIGURE 36.1 Photomicrograph of inflamed omentum from a cat with the effusive form of FIP. Arrows indicate pyogranulomas that contain macrophages staining positive by immunoperoxidase for FIPV antigen.

inhibitors of PCR [35]. The greater, and more significant, problem lies with diagnosing FIPV infection. FIPV is highly macrophage associated (Figure 36.1), and infected macrophages are found only within diseased tissues. There is not a significant viremia at any stage of the infection, making blood an unsuitable tissue to sample. The levels of FIPV are high in diseased tissues and abdominal or pleural exudates from cats with effusive FIP, while virus levels are much lower in diseased tissues of cats with noneffusive FIP [2]. Abdominal or thoracic exudate from cats with effusive FIP is usually rich with virus, but over 90% of the virus is cell associated. Therefore, testing should be concentrated on the cellular fraction and not cell-free supernatant. Biopsies of diseased tissues, such as would be taken in cats with noneffusive FIP in organs such as the kidney or mesenteric lymph nodes, should be tested by immunohistochemistry to make sure that the diseased tissue was present in the biopsy. Samples taken at necropsy should contain mainly diseased tissues and not normally appearing tissues from a grossly diseased organ. This puts the onus on the veterinarian or veterinary pathologist to provide the proper type of sample.

36.2.2 Detection Procedures

The diagnosis of FECV and FIPV infections, and the pitfalls of indirect and direct diagnostic procedures, has been amply reviewed [27]. The definitive diagnosis of FIPV as a cause of disease requires identification of the virus within typical lesions or effusions by an RT-PCR-based test or immunohistochemistry. The question is whether it is necessary to confirm the FIP biotype of this virus. It can be argued that this is not necessary, because it would be extremely unlikely to find the FECV biotype in a typical FIP lesion.

36.2.2.1 Serologic Testing

Misinformation still exists over the use and interpretation of FCoV antibody titers in serum or plasma [2]. The main problem with antibody tests is that both FECVs and FIPVs, being virtually identical to each other, evoke the same antibody responses. Titers also tend to be high among healthy cats in the same environments that foster FIP. FCoV antibody titers, if accurately done, are nonetheless of some value. Although many healthy FECV-exposed cats have titers by indirect immunofluorescence assay from 1:100 to 1:400 [32], as do many cats with FIP, the likelihood of a titer being associated with FIP increases with its magnitude. Few healthy cats have titers of 1:1600, and titers of 1:3200 and above are highly indicative of FIP [27]. Healthy cats with titers below 1:100 infrequently shed FECV in their feces, while cats with titers of 1:400 are usually positive shedders [32].

36.2.2.2 RT-PCR

Tests based on RT-PCR have been used to help diagnose FIP for almost two decades [36]. Nested PCR is a method used to amplify very small amounts of FCoV cDNA and, therefore, increase sensitivity. A nested PCR was reportedly over 90% sensitive and specific in detecting FIPV in ascites from cats

with effusive FIP [37]. However, nested PCRs are extremely sensitive to contamination of the laboratory with PCR products, causing false-positive reactions. The problem of laboratory contamination with PCR products is greatly reduced with RT-qPCR.

A detailed description of a diagnostic RT-qPCR for identifying FECV/FIPV genomic RNA was described by Gut and colleagues [38]. RT-qPCR was reportedly 10–100 times more sensitive in detecting FIPV mRNA than nested RT-PCR [30]. Primers used by Gut and colleagues [38] were from the accessory 7b gene, which is more conserved than the 3c, S, and M genes and less conserved than the 7a, 3a, and 3b genes [39]. As a result of its 5′ location in the genome, it is also the most plentiful subgenomic mRNA. However, it is difficult to design a single primer pair that is truly universal, so several sets of primers that are based on the researches of known FECV and FIPV genomes have been used in combination. It is also common to incorporate a recombinant FCoV plasmid (pT7-StFCoV) produced in E. coli with a test sample to assure the integrity of the test procedure and to quantify viral genomic copies [38]. Commercial tests often include a standard negative control and a negative extraction control. PCR product carryover as a source of false-positive reactions can be virtually eliminated by incorporating the AmpErase UNG system (Applied Biosystems) [40] to ensure that PCR products cannot be amplified in subsequent PCR reactions. The laboratory should also be routinely monitored for random positive PCR signals due to cDNA contamination. Test procedures should also take into account PCR inhibitors, especially in fecal samples [35].

Differentiating FECVs from FIPVs is a problem with any PCR-based test, because FECV and FIPV sequences are virtually identical [41]. However, it was assumed at the time that FECV replicates only in the intestine, while FIPV replicates only in macrophages within extraintestinal tissues. Based on this premise, Simons and colleagues [42] developed primers to the highly conserved M gene that would only detect subgenomic mRNA. This test was found to identify 46% of FIPV in the blood of cats suspected of having FIP, but only 5% of healthy cats. Positivity increased to 93% in cats with confirmed FIP, while none of the non-FIP cats tested positive. However, a second study using the same procedure found that 54% of healthy cats, especially in the 6–12-month-old range also tested positive [43]. This was explained subsequently by experimental infection studies of FECV by Kipar et al. [30] and Vogel et al. [33], who identified replicating forms of FECV in blood monocyte/macrophages.

Different methodologies have been used to increase the sensitivity and spectrum of FECV and FIPV detection. Hornyák et al. [44] developed an RT-qPCR using the principle of primer-probe energy transfer that combined the detection of the M gene subgenomic messenger, genome copy quantitation, and a broad range of genetic variants. They claimed that this procedure was 10–50,000 times more sensitive in detecting subgenomic mRNA in tissues of cats with FIP than in the feces of healthy FECV shedders. Although such procedure may increase the sensitivity and limit the detection to only replicating virus, it would still not differentiate FIPVs from FECVs.

Present attempts to differentiate FIPVs from FECVs have centered on FIP-associated mutations. Mutations in the 3c and S1/S2 furin cleavage peptide are unique to each FIPV and are therefore not good candidates. The most FIPV-specific mutations are two alternative single-nucleotide changes within the fusion protein region of the S protein [17]. One or the other of these mutations occurs in over 95% of FIPVs detected in diseased tissues.

It is generally assumed that current RT-qPCR is highly accurate in detecting FCoV RNA in feces and diseased tissues/effusions. Although a positive result is highly specific, the sensitivity of RT-qPCR is dependent on obtaining samples with sufficient levels of coronavirus RNA. Insufficient viral genomic RNA would lead to false-negative tests. This is usually not a problem with the detection of FECV in feces, which contain high levels of virus. It can be a significant problem with inadequate sampling of tissues or effusions of cats with FIP. FIPV is highly macrophage bound, so effusions must include the cellular fraction. Blood and nondiseased tissues contain extremely low levels of virus and are therefore not reliable substrates for testing. Finally, no test with this complexity and sampling limitations is expected to be 100% sensitive and specific. Indeed, PCR-based testing was found to be only 80%–90% accurate in confirming the presence of FIPV in diseased tissues in one field study [45].

36.2.2.3 Immunohistochemistry

Immunostaining of diseased tissues (immunohistochemistry) of fluids by either immunofluorescence or immunoperoxidase can be as reliable a method as RT-PCR, but the accuracy is also limited by the quality of the reagents used, the presence of infected macrophages within the tissues sampled, and the expertise of the microscopist in discerning true-positive staining within macrophages in tissue lesions or from effusions. Staining is either by immunofluorescence or immunoperoxidase, with the former being more sensitive. However, immunofluorescence requires frozen sections either fresh or in glycerol/resin protection, while immunoperoxidase can be done on formalin-fixed tissues. Formalin fixation and wax embedding should be done as rapidly as possible following tissue collection. Although immunohistochemistry is considered to be an accurate way to make a definitive diagnosis of FIP [46], there are cases of FIP that will test negative

depending on the quality of the tissues, the presence of adequate lesions within the material examined, and the quality of reagents and test performance. Nonspecific-positive staining of macrophages may also be a problem. Litster et al. [47] compared results from direct immunofluorescence on antemortem feline effusions with necropsy results in 17 cats with abdominal or thoracic effusions. Histopathologic examination of tissues collected at necropsy confirmed FIP in 10/17 cases and ruled out FIP in 7/17 cases. Antemortem direct immunofluorescence testing was positive in all 10 cases confirmed as FIP at necropsy. In the seven cats where FIP was ruled out at necropsy, direct immunofluorescence was negative in five cases and positive in the remaining two cases. The calculated sensitivity of immunohistochemistry testing was 100% and the specificity was 71.4%.

Immunohistochemistry should be used much more often with effusions or fluids from cats suspected of having FIP. Effusions often contain numerous virus-filled macrophages that can be concentrated onto slides. This technique has also been used successfully to detect FIPV-infected macrophages within the cerebrospinal fluid of a cat with neurologic disease [48]. Immunoperoxidase was used to diagnose FIPV in macrophages in the skin of two cats and atypical skin lesions (multiple papular lesions) [49,50].

36.2.2.4 Sequence Analysis and Genotype Comparison

All known FCoVs belong to a single species with numerous genotypes and at least two serotypes and two biotypes. Genotypic differences are a result of the inherent mutability of RNA viruses [13]. The S, 3a–c, E, M, N, and 7a/b structural and accessory genes are the most common regions for genotyping [39] (Figure 36.2). The large ORF1ab is highly conserved between isolates and more difficult to amplify. The S gene, while less conserved, is very large and, therefore, difficult to amplify without creating progressive primer sets. Therefore, it is common to sequence fragments B (3a–c, E, M) or C (N, 7a, 7b). The 3c is the most polymorphic, followed by the S and M genes [39]. The 3a, 3b, 7a, and 7b accessory genes are the most conserved. The 7b gene is often used for diagnostic RT-qPCR, because 7b subgenomic mRNA is transcribed at the highest concentration and the 7b sequence is the most conserved between isolates.

Genotypic differences vary according to the relative geographic isolation of a given cat population. FECV isolates

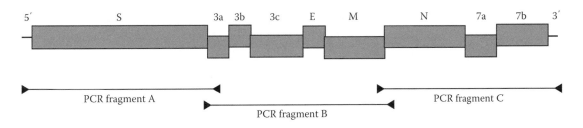

FIGURE 36.2 Structural and accessory genes of FECV/FIPV. Fragment B is commonly used for genotyping. The 3′ region of S or the 5′ region of 3a can be used for identifying serotype 1 and 2 strains.

within isolated groups of cats tend to remain virtually identical over time [20]. As expected, FIPV isolates usually differ by less than 1%–2% from FECVs in the same environments [14,15]. Strains differing by just a few SNPs can be frequently found within the same infected cat [39]. FECV and FIPV isolates from different geographic regions of the Western United States can differ by as much as 6%–16% [39]. Variation up to 30% or so can occur between more disparate regions.

Serotype 2 FECVs and FIPVs occur as a result of recombination between FCoV and CCoV. This recombination always involves the 3′ region of S and the 5′ region of 3a (Figure 36.2) [39]. The segments that recombine vary in size and genotype, thus making each serotype 2 isolate unique. Although sequencing of the large S gene is often used to identify CCoV sequence, sequencing of the smaller 3a gene is simpler and just as accurate [39]. Serotype 1 and 2 isolates can also be differentiated by virus neutralization, as the recombination involves the virus neutralization epitopes of either FCoV or CCoV.

Biotype conversion has been one of the most intriguing and often debated topics related to FIP pathogenesis. The concept of FECV-to-FIPV internal mutation was first confirmed from animal studies conducted by Poland et al. [14], but based on sequencing of FIPV/FECV pairs subsequently published by Vennema et al. [51]. The original mutation was found in deletions, insertions, or premature stop codons that involved the accessory 3c gene. Subsequent studies confirmed the presence of functional 3c gene mutations in FIPVs and the role of an intact 3c gene in intestinal tropism [15]. However, further studies indicated that only about two-thirds to three-fourths of FIPVs had mutations in 3c leading to a truncated protein [39,52]. The conclusion of one study was that 3c gene mutations played no role in FIP because functional mutations were not universal in all FIPVs [52]. However, the second study found that the 3′ region of nonfunctionally mutated 3c gene was also highly mutated in FIPVs due to the accumulation of nonsynonymous SNPs causing major amino acid substitutions. This led to the conclusion that different types of mutations in the 3′ region of 3c could be involved in FIPVs. Based on the conclusion that 3c mutations were not important for the FIP biotype, Chang and colleagues [17] sequenced whole genomes of 11 FECV/FIPV pairs searching for other possible mutations. Other than functional mutations in about two-thirds of 3c genes, they found only two mutations that were consistently found in FIPVs and not in their parental FECVs. These two mutations were SNPs in the fusion domain of the S gene, the first at position 23,531 and the second two residues downstream. The first mutation was present in 108/118 FIPVs sequenced and none in 183 FECVs, while the second mutation was present in 5/118 FIPVs and none in FECVs. The first mutation caused a methionine-to-leucine change, while the second mutation caused a serine-to-alanine substitution. Chang and colleagues [17] postulated that these particular mutations, which were present in at least 95% of all FIPVs, might affect virus/cell fusion and were solely responsible for the change in cell tropism and virulence of FECVs. However, a recent study by Porter et al. [53] demonstrated the presence of the methionine-to-leucine-associated SNP in tissues of healthy cats, leading them to conclude that this mutation is associated with spread of the virus from the intestine but not in the pathogenesis of FIP disease. Studies conducted at around this same period identified an additional highly mutable region in the S1/S2 furin cleavage peptide [16]. SNP mutations in this short region were unique to each FIPV and caused a number of amino acid changes, some enhancing furin cleavage, some inhibiting furin cleavage, and some without effect. This suggests that furin cleavage peptide mutations, like mutations in the fusion domain, are related to changes in cell tropism and not to the interaction of infected virus with the host immune system.

36.3 CONCLUSION

The diagnosis of FIP is tantamount to a death sentence and owners of affected cats often require a definitive diagnosis before conceding to euthanasia. A diagnosis of FIP can be made with a very high degree of certainty based on the age, breed, and origin of the cat; clinical signs; and identifying certain abnormalities in complete blood counts, blood proteins (albumin, globulin, A:G ratio), urine, blood chemistry, analysis of fluid, histopathology on biopsies, and serology [27]. However, classical changes are not always present in these various tests and a diagnosis is frequently based on indirect rather than direct evidence. A definitive diagnosis requires the identification of viral proteins or RNA in diseased tissues or effusions [27]. These substances are found only within macrophages within diseased tissues and rarely in the blood. Moreover, they are not always found at levels that are measureable, especially in cases of noneffusive FIP where there may be a paucity of virus relative to the degree of damaged tissue. Tests such as RT-qPCR and immunochemistry remain the mainstay in making a definitive diagnosis of FIP, and even these tests and published reports indicate that they may only confirm a diagnosis in around 80% of cases. This is not a problem with the tests, but rather with the type and condition of samples that are submitted. Samples to be tested must contain adequate numbers of infected macrophages and be properly stored, transported, and tested.

REFERENCES

1. Holzworth, J.E., Some important disorders of cats. *Cornell Vet.* 53, 157, 1963.
2. Pedersen, N.C., A review of feline infectious peritonitis virus infection: 1963–2008. *J. Feline Med. Surg.* 11, 225, 2009.
3. Wang, Y.T. et al., An outbreak of feline infectious peritonitis in a Taiwanese shelter: Epidemiologic and molecular evidence for horizontal transmission of a novel type II feline coronavirus. *Vet. Res.* 44, 57, 2013.
4. Rohrbach, B.W. et al., Epidemiology of feline infectious peritonitis among cats examined at veterinary medical teaching hospitals. *J. Am. Vet. Med. Assoc.* 218, 1111, 2001.
5. Wolfe, L.G. and Griesemer, R.A., Feline infectious peritonitis. *Pathologica Veterinaria* 3, 255, 1966.

6. Ward, J., Morphogenesis of a virus in cats with experimental feline infectious peritonitis. *Virology* 41, 191, 1970.

7. Pedersen, N.C., Ward, J., and Mengeling, W.L., Antigenic relationship of the feline infectious peritonitis virus to coronaviruses of other species. *Arch. Virol.* 58, 45, 1978.

8. Motokawa, K. et al., Comparison of the amino acid sequence and phylogenetic analysis of the peplomer, integral membrane and nucleocapsid proteins of feline, canine and porcine coronaviruses. *Microbiol. Immunol.* 40, 425, 1996.

9. Montali, R.J. and Strandberg, J.D., Extraperitoneal lesions in feline infectious peritonitis. *Vet. Pathol.* 9, 109, 1972.

10. Pedersen, N.C., Serologic studies of naturally occurring feline infectious peritonitis. *Am. J. Vet. Res.* 37, 1447, 1976.

11. Pedersen, N.C., Boyle, J.F., and Floyd, K., Infection studies in kittens utilizing feline infectious peritonitis virus propagated in cell culture. *Am. J. Vet. Res.* 42, 363, 1981.

12. Pedersen, N.C. et al., An enteric coronavirus infection of cats resembling transmissible gastroenteritis of swine, and its relationship to feline infectious peritonitis. *Am. J. Vet. Res.* 42, 368, 1981.

13. Pedersen, N.C., An update on feline infectious peritonitis: Virology and immunopathogenesis. *Vet. J.* 201, 123, 2014.

14. Poland, A.M. et al., Two related strains of feline infectious peritonitis virus isolated from immunocompromised cats infected with a feline enteric coronavirus. *J. Clin. Microbiol.* 34, 3180, 1996.

15. Pedersen, N.C. et al., Feline infectious peritonitis: Role of the feline coronavirus 3c gene in intestinal tropism and pathogenicity based upon isolates from resident and adopted shelter cats. *Virus Res.* 165, 17, 2012.

16. Licitra, B.N. et al., Mutation in spike protein cleavage site and pathogenesis of feline coronavirus. *Emerg. Infect. Dis.* 19, 1066, 2013.

17. Chang, H.W. et al., Spike protein fusion peptide and feline coronavirus virulence. *Emerg. Infect. Dis.* 18, 1089, 2012.

18. Compton, S.R. et al., Pathogenesis of enterotropic mouse hepatitis virus in immunocompetent and immunodeficient mice. *Comp. Med.* 54, 681, 2004.

19. Hagemeijer, M.C. et al., Biogenesis and dynamics of the coronavirus replicative structures. *Viruses* 4, 3245, 2012.

20. Addie, D.D. et al., Persistence and transmission of natural type I feline coronavirus infection. *J. Gen. Virol.* 84, 2735, 2003.

21. Pedersen, N.C. et al., Pathogenic differences between various feline coronavirus isolates. Coronaviruses; molecular biology and pathogenesis. *Adv. Exp. Med. Biol.* 173, 365, 1984.

22. Herrewegh, A.A. et al., Feline coronavirus type II strains 79–1683 and 79–1146 originate from a double recombination between feline coronavirus type I and canine coronavirus. *J. Virol.* 72, 4508, 1998.

23. Lin, C.N. et al., Full genome analysis of a novel type II feline coronavirus NTU156. *Virus Genes* 46, 316, 2013.

24. Benetka, V. et al., Prevalence of feline coronavirus types I and II in cats with histopathologically verified feline infectious peritonitis. *Vet. Microbiol.* 99, 31, 2004.

25. Kummrow, M. et al., Feline coronavirus serotypes 1 and 2: Seroprevalence and association with disease in Switzerland. *Clin. Diagn. Lab. Immunol.* 12, 1209, 2005.

26. Hohdatsu, T. et al., The prevalence of types I and II feline coronavirus infections in cats. *J. Vet. Med. Sci.* 54, 557, 1992.

27. Pedersen, N.C., An update on feline infectious peritonitis: Diagnostics and therapeutics. *Vet. J.* 201, 133, 2014.

28. Desmarets, L.M. et al., Establishment of feline intestinal epithelial cell cultures for the propagation and study of feline enteric coronaviruses. *Vet. Res.* 44, 71, 2013.

29. Sabshin, S.J. et al., Enteropathogens identified in cats entering a Florida animal shelter with normal feces or diarrhea. *J. Am. Vet. Med. Assoc.* 241, 331, 2012.

30. Kipar, A. et al., Sites of feline coronavirus persistence in healthy cats. *J. Gen. Virol.* 91, 1698, 2010.

31. Addie, D.D. et al., Quarantine protects Falkland Islands (Malvinas) cats from feline coronavirus infection. *J. Feline Med. Surg.* 14, 171, 2012.

32. Pedersen, N.C., Allen, C.E., and Lyons, L.A., Pathogenesis of feline enteric coronavirus infection. *J. Feline Med. Surg.* 10, 529, 2008.

33. Vogel, L. et al., Pathogenic characteristics of persistent feline enteric coronavirus infection in cats. *Vet. Res.* 41, 7, 2010.

34. Harley, R.D. et al., Phylogenetic analysis of feline coronavirus strains in an epizootic outbreak of feline infectious peritonitis. *J. Vet. Int. Med.* 27, 445, 2013.

35. Monteiro, L. et al., Polysaccharides as PCR inhibitors in feces: *Helicobacter pylori* model. *J. Clin. Microbiol.* 35, 995, 1997.

36. Li, X. and Scott, F.W., Detection of feline coronavirus in cell cultures and in fresh and fixed tissues using polymerase chain reaction. *Vet. Microbiol.* 42, 65, 1994.

37. Gamble, D.A. et al., Development of a nested PCR assay for detection of feline infectious peritonitis virus in clinical specimens. *J. Clin. Microbiol.* 35, 673, 1997.

38. Gut, M. et al., One-tube fluorogenic reverse transcription-polymerase chain reaction for the quantitation of feline coronaviruses. *J. Virol. Methods* 77, 37, 1999.

39. Pedersen, N.C. et al., Significance of coronavirus mutants in feces and diseased tissues of cats suffering from feline infectious peritonitis. *Viruses* 1, 166, 2009.

40. Pang, J. et al., Use of modified nucleotides and uracil-DNA glycosylase (UNG) for the control of contamination in the PCR-based amplification of RNA. *Mol. Cell. Probes* 6, 251, 1992.

41. Herrewegh, A.A. et al., Detection of feline coronavirus RNA in feces, tissues, and body fluids of naturally infected cats by reverse transcriptase PCR. *J. Clin. Microbiol.* 33, 684, 1995.

42. Simons, A.F. et al., A mRNA PCR for the diagnosis of feline infectious peritonitis. *J. Virol. Methods* 124, 111, 2005.

43. Can-Sahna, K. et al., The detection of feline coronaviruses in blood samples from cats by mRNA RT-PCR. *J. Feline Med. Surg.* 9, 369, 2007.

44. Hornyák, A. et al., Subgenomic mRNA of feline coronavirus by real time polymerase chain reaction based on primer-probe energy transfer (P-sg-QPCR). *J. Virol. Methods* 181, 155, 2012.

45. Sharif, S. et al., Diagnostic methods for feline coronavirus: A review. *Vet. Med. Int.* 2010, 7, pii: 809480, 2010.

46. Giori, L. et al., Performances of different diagnostic tests for feline infectious peritonitis in challenging clinical cases. *J. Small Anim. Pract.* 52, 152, 2011.

47. Litster, A.L., Pogranichniy, R., and Lin, T.L., Diagnostic utility of a direct immunofluorescence test to detect feline coronavirus antigen in macrophages in effusive feline infectious peritonitis. *Vet. J.* 198, 362, 2013.

48. Ives, E.J., Vanhaesebrouck, A.E., and Cia, F., Immunocytochemical demonstration of feline infectious peritonitis virus within cerebrospinal fluid macrophages. *J. Feline Med. Surg.* 15, 1149, 2013.

49. Bauer, B.S. et al., Positive immunostaining for feline infectious peritonitis (FIP) in a Sphinx cat with cutaneous lesions and bilateral panuveitis *Vet. Ophthol.* 16(Suppl. 1), 160, 2013.

50. Declercq, J. et al., Papular cutaneous lesions in a cat associated with feline infectious peritonitis. *Vet. Dermatol.* 19, 255, 2008.

51. Vennema, H. et al., Feline infectious peritonitis viruses arise by mutation from endemic feline enteric coronaviruses. *Virology* 243, 150, 1998.

52. Chang, H.W. et al., Feline infectious peritonitis: Insights into feline coronavirus pathobiogenesis and epidemiology based on genetic analysis of the viral 3c gene. *J. Gen. Virol.* 91, 415, 2010.

53. Porter, E. et al., Amino acid changes in the spike protein of feline coronavirus correlate with systemic spread of virus from the intestine and not with feline infectious peritonitis. *Vet. Res.* 45, 49, 2014.

37 Porcine Epidemic Diarrhea Virus

Dongyou Liu

CONTENTS

37.1 Introduction 331
    37.1.1 Classification..331
    37.1.2 Morphology and Genome Organization..332
    37.1.3 Biology and Epidemiology 332
    37.1.4 Clinical Features and Pathogenesis 333
    37.1.5 Diagnosis 333
        37.1.5.1 Virus Isolation...333
        37.1.5.2 Electron Microscopy..334
        37.1.5.3 Immunohistochemistry...334
        37.1.5.4 Indirect Fluorescent Antibody Test 334
        37.1.5.5 ELISA 334
        37.1.5.6 Nucleic Acid Detection..334
    37.1.6 Treatment and Prevention 334
37.2 Methods 335
    37.2.1 Sample Preparation 335
    37.2.2 Detection Procedures..335
        37.2.2.1 Conventional RT-PCR..335
        37.2.2.2 Real-Time RT-PCR 335
37.3 Conclusion 335
References...336

37.1 INTRODUCTION

Porcine epidemic diarrhea (PED), also known as PED syndrome, is a serious disease of pigs due to a coronavirus named porcine epidemic diarrhea virus (PEDV). Clinically, the disease is characterized by acute diarrhea, vomiting, dehydration, and high mortality, especially in suckling piglets. First identified in Europe in the 1970s, PED has been reported in Asia and North America. In view of its serious disease outcome, widespread distribution, and clinical resemblance to diarrheal illness associated with transmissible gastroenteritis coronavirus (TGEV), rotavirus, and other pathogens, prompt and accurate diagnosis is vital to reduce the economical impact of PED on swine farming industry worldwide.

37.1.1 CLASSIFICATION

PEDV is a member of the genus *Alphacoronavirus*, subfamily *Coronavirinae*, family *Coronaviridae*. The family *Coronaviridae* (order *Nidovirales*) covers an expanding group of enveloped, single-stranded, positive-sense RNA viruses that are organized into two subfamilies. The subfamily *Coronavirinae* consists of four genera: *Alphacoronavirus* and *Betacoronavirus* naturally occurring in bats, and *Deltacoronavirus* and *Gammacoronavirus* residing primarily in birds and pigs; the subfamily *Torovirinae* includes

two genera: *Bafinivirus* and *Torovirus*, which are present in fish and mammals (causing gastroenteritis in mammals such as cattle, horses, and pigs, but rarely humans), respectively.

Currently, the genus *Alphacoronavirus* contains at least 16 species: transmissible gastroenteritis virus (TGEV), porcine respiratory coronavirus (PRCV), canine coronavirus, feline infectious peritonitis virus, mink coronavirus, *Rhinolophus* bat coronavirus HKU2, *Miniopterus* bat coronavirus 1A, *Miniopterus* bat coronavirus 1B, *Miniopterus* bat coronavirus HKU8, human coronavirus 229E, human coronavirus NL63, PEDV, *Scotophilus* bat coronavirus 512, *Hipposideros* bat coronavirus HKU10, *Rousettus* bat coronavirus HKU10, and *Mystacina tuberculata* bat coronavirus [1].

Phylogenetic analysis of the full-length genomic sequences of PEDV classifies PEDV strains into two distinct genogroups, with the genogroup 1 (G1) consisting of PEDV field strains isolated before 2010 and the derived vaccine strains while the genogroup 2 (G2) including PEDV strains isolated since 2011 in China and the United States (U.S. PEDV) [2,3].

G1 can be further separated into at least three clusters (subgroups 1a, 1b, and R). The subgroup 1a includes strains with the same SL2 and SL4 sequences at the 5′-untranslated region (UTR) (e.g., the prototype CV777 strain as well as strains LZC and SM98). The subgroup 1b consists of strains with the 8-aa deletion in nsp3 and the large open reading frame 3

(ORF3) deletion at the C-terminus (e.g., the DR13-attenuated vaccine strain as well as strains JS2008 and SD-M). The subgroup R comprises strains that may represent recombinants of other genogroups (e.g., the virulent DR13 strain from South Korea and the oldest Chinese PEDV strain, CH/S) [2].

G2 is distinguished into two subgroups (2a and 2b). The subgroup 2a contains strains that share several unique nucleotides (e.g., the U.S. strains MN, IA2, CO/13, IA2013, and IN2013 and a Chinese strain, AH2012), which may be further divided into two sublineages: MN-IA2 and IA1-CO/13-IA2013-IN2013. The subgroup 2b includes strains mainly from southeastern China [2,4,5].

The classification of PEDV strains into two distinct genogroups (G1 and G2) based on whole genome sequences is supported by data generated from comparison of the spike (S), ORF3, envelope (E), membrane (M), and nucleocapsid (N) gene sequences of PEDV strains [6].

Besides PEDV, other swine-infecting coronaviruses include TGEV and PRCV (both belonging to the genus *Alphacoronavirus*) as well as porcine deltacoronavirus (PdCV) (belonging to the genus *Deltacoronavirus*).

TGEV was identified in 1946 as the cause of a life-threatening acute enteric disease of pigs in the United States. Similar to PEDV, TGEV replicates in the differentiated enterocytes covering the villi of the porcine small intestine and induces profuse watery diarrhea, vomiting, emesis, and dehydration in suckling piglets, with morbidity rates of 80%–100% and mortality rates of 50%–90% [7].

First noted in the United States in 1984, PRCV is a naturally occurring deletion mutant of TGEV, with a 224-aa deletion at positions 21–244 in the S gene. This alters viral tropism from intestinal to respiratory epithelia. By replicating almost exclusively in the respiratory tract of swine, PRCV infection closely resembles other porcine viral pneumonias.

PdCV emerged in early 2014 as the culprit for fatal cases of PEDV-negative diarrhea in Ohio piglets and quickly spread to >20 states in the United States and Canada. Typical of TGEV and PEDV infections, clinical presentation of PdCV-related illness includes vomiting, diarrhea, and dehydration, with high rates of morbidity and mortality [8].

37.1.2 Morphology and Genome Organization

Similar to other coronaviruses, PEDV has an enveloped, pleomorphic/spherical virion of 80–120 nm in diameter, with a double layer of short (hemagglutinin) and long (S, club-shaped, 20 nm long) surface projections. The lipoprotein envelope is composed of four structural proteins: M glycoprotein, S glycoprotein, hemagglutinin–esterase (HE) glycoprotein, and small E protein. Underneath the envelope is a flexible and helical nucleocapsid that is formed by nucleoprotein (N) in association with the viral genome.

The genome of PEDV is composed of a linear, positive-sense, single-stranded RNA. Analysis of the genome sequences of a virulent PEDV strain (CH/ZMDZY/11) from a suckling piglet with severe diarrhea in China revealed a genome of 28,038 nucleotides (nt) in length, excluding the poly(A) tail, with the

typical gene order 5′ UTR-ORF1a/1b–S–ORF3–E–M–N-3′ UTR. The 5′ and 3′ ends of the genome contain 292 and 334 nt UTRs, respectively, which flank seven ORFs within the genome. The ORF1a/1b gene (nt 293–20,637) encodes two large nonstructural precursor polyproteins: replicase 1a (nt 293–12,601) and replicase 1b (nt 12,601–20,637). The S gene consists of 4161 nt (nt 20,634–24,794), ORF3 of 675 nt (nt 24,794–25,468), the E gene of 231 nt (nt 25,449–25,679), the M gene of 681 nt (nt 25,687–26,368), and the N gene of 1326 nt (nt 26,379–27,704) [9].

Interestingly, the complete genome sequence of CH/ZMDZY/11 exhibits varied nucleotide sequence identities with those of other Chinese PEDV strains: 99.1% with BJ-2011-1 and GD-B; 98.9% with CH-FIND-3-2011; 98.8% with AJ1102 and LC; 98.2% with GD-A, CH-GD-1, and ZJCZ4; and 97.3% with CH–S. In addition, the S gene of CH/ZMDZY/11 displays 92.5%–99.1% amino acid homology to those of other reported Chinese strains. These genetic variations might have contributed to antigenic changes in the strains under investigation, and thus decreased efficiency of the CV777-based vaccines against these strains [9–11].

The S glycoprotein (1365 aa, 180–220 kDa, encoded by ORF2; also known as E2, peplomer protein) consists of S1 domain (718 aa) at the N-terminal and S2 domain (649 aa) at the C-terminal. S1 domain interacts with cell receptors and induces conformational changes in the S protein unmasking the fusion peptide of S2 domain to facilitate fusion of viral and host cell membranes [12–14].

The M protein (226 aa, 27–32 kDa, encoded by ORF5; also known as E1 glycoprotein, matrix glycoprotein, M glycoprotein) is a component of the viral envelope (which also includes E and S proteins) that interacts with E protein in the budding compartment of the host cell, located between the endoplasmic reticulum (ER) and the Golgi complex. Through its interaction with N protein, the M protein participates in RNA packaging into the virus. Besides its involvement in the virus assembly process, the M protein induces antibodies that neutralize virus in the presence of complement [15].

The nucleoprotein N (441 aa, 58 kDa; also known as Nucleocapsid protein) is a major structural component of virion that associates with genomic RNA to form a long, flexible, helical nucleocapsid. Localized in the ER, the PEDV N protein induces cell cycle prolongation at the S-phase, ER stress, and upregulation interleukin-8 expression. The N protein may be also involved in viral transcription and translation and/or replication [16,17].

The envelope small membrane protein E (76 aa, 7 kDa encoded by ORF4; also known as sM protein) is localized in the ER, with small quantities localized in the nucleus. Through its interaction with the M protein, the PEDV E protein induces ER stress, activates NF-κB, and thus upregulates IL-8 and Bcl-2 expression [18].

37.1.3 Biology and Epidemiology

PEDV targets the epithelial cells of small and large intestines of pigs for replication and is not known to infect hosts other than pigs. Therefore, it is not considered a zoonotic pathogen and poses no risk to human health or to food safety.

As PEDV is discharged in feces, its transmission is mainly through oral–fecal route, although aerosols may play an accessory role in its transmission [19]. Feces-contaminated feed, service vehicles, equipment, and personnel represent additional risk factors for disease spread [20].

The virus can survive at least 28 days in slurry at 4°C, 7 days in feces-contaminated dry feed at 25°C, up to 14 days at 25°C in wet feed, and at least 28 days in wet feed mixture at 25°C but loses infectivity above 60°C. In addition, it is stable at pH 6.5–7.5 at 37°C but susceptible to formalin (1%), anhydrous sodium carbonate (4%), lipid solvents, iodophors in phosphoric acid (1%), and sodium hydroxide (2%) [21,22].

First emerged in the United Kingdom in 1971 as the cause of a swine disease resembling transmissible gastroenteritis, PEDV prototype strain CV777 was identified in Belgium in 1978. In the subsequent years, PEDV was reported in other European countries (including Hungary, Italy, Germany, France, Switzerland, and the Czech Republic), Asia, and Americas [23–25]. Currently, PEDV causes gastroenteritis in pigs of all ages with high morbidity but low mortality in Europe and is responsible for high mortality in suckling piglets in Asia, especially China, South Korea, Japan, Thailand, the Philippines, and Vietnam [26–32]. In 2010, massive outbreaks of PED in China resulted in near 100% illness among infected swine herds and 80%–100% mortality among infected suckling piglets (<10 days of age) [33,34]. In April 2013, a highly virulent PEDV variant was behind explosive epidemics on swine farms in the United States, leading to mortality rates of up to 95% among suckling pigs [35,36]. By December 2014, PEDV has been reported in 41 U.S. states and Canada [37]. In addition, a less virulent PEDV variant (strain OH851) that causes mild disease in sows instead of younger animals was also identified in the United States [38,39]. In Canada, PEDV was first noted in January 2014 on a pig farm in Ontario Province and has since appeared on farms in Manitoba, Prince Edward Island, and Quebec Provinces [40]. Furthermore, PEDV was detected in swine samples from Mexico in 2013.

37.1.4 Clinical Features and Pathogenesis

With an incubation period of 1–4 days, PED is a highly contagious, acute enteric disease of swine characterized by vomiting, watery diarrhea (which may appear as early as 2 days postinfection and subside by 10 days postinfection), dehydration, and death, particularly in neonates. The severity of PED is dependent on age and immunological status of the pigs, previous exposure and presence of secondary infection, etc. [41].

In adult swine, the primary signs are watery diarrhea, pyrexia, anorexia, and lethargy, with a mortality rate of <5%.

In suckling piglets, clinical signs may range from watery diarrhea, dehydration, metabolic acidosis, to death (as high as 90% in pigs less than 7 days of age but very low in pigs older than 2 weeks).

In weanlings/feeders, clinical signs include watery diarrhea, vomiting, high morbidity (almost all animals are sick), and reduced weight gain, with a mortality rate of <10% [42].

PEDV is detectable in feces by PCR from 1 day to 24 days postinfection. Microscopic lesions consist of severe diffuse atrophic enteritis with significantly reduced villous length (this thinning of the intestines is notably seen in the small intestines).

Being enteropathogenic, PEDV enters the host via oral ingestion and replicates in the villous epithelial cells of the entire small and large intestines, especially the jejunum and ileum [43,44]. This results in degeneration of enterocytes and shortening of the villi, which are typical of acute, severe atrophic enteritis, with profound watery diarrhea, viremia, vomiting, extensive dehydration, and death as main clinical manifestations.

A 204-aa deletion at positions 713–916 in the C-terminus of S1 and N-terminus of S2 in the PEDV variant destroys four N-linked glycosylation sites at positions 728, 745, 783, and 875, as well as two neutralizing epitopes, SS2 (753–760) and SS6 (769–776), leading to conformational change of S protein and subsequent alteration in antigenicity/immunogenicity [45,46].

37.1.5 Diagnosis

PEDV induces acute diarrhea, dehydration, and other nonspecific symptoms in pigs, which are clinically indistinguishable from those caused by TGEV, rotavirus, bacteria, (e.g., *Clostridium*, *E. coli*, *Salmonella*, *Brachyspira*, *Lawsonia intracellularis*, etc.) or parasites (*Isospora suis*, *Cryptosporidium*, *nematodes*, etc.). Therefore, laboratory confirmatory tests are necessary to make a final and definitive diagnosis.

Laboratory diagnostic methods for PED are essentially reliant on the detection of virus particles, antigens (by virus isolation, electron microscopy, immunohistochemistry, antigen enzyme-linked immunosorbent assays [ELISA], nucleic acid detection, etc.), or antibodies (ELISA, immunofluorescence, serum neutralization, etc.).

37.1.5.1 Virus Isolation

PEDV can be isolated by using Vero (African green monkey kidney cell line ATCC® CCL-81™), Vero 76 (ATCC® CRL-1586™), MARC-145 (M145), ST (pig testis cell line), IPEC-J2, or MK-DIEC (duck intestinal epithelial cell) cells [47].

Prior to inoculation, intestines from infected pigs are homogenized in phosphate-buffered saline, and the debris is removed by centrifugation at $10,000 \times g$ for 10 min followed by filtration through a 0.2 μm filter. Alternatively, intestinal swabs and intestinal tissue emulsions are treated with a 100× antibiotic cocktail (10× final concentration) containing 10,000 IU/mL penicillin G, 10 mg/mL streptomycin, 10 mg/mL kanamycin, 5000 U/mL nystatin, and 1500 U/mL polymyxin B sulfate at 1:10 for 30 min at room temperature. The 10× antibiotic cocktail treated swabs and emulsions are then clarified by centrifugation.

Vero cells (or other cells) are cultured and maintained in minimum essential medium supplemented with 10% fetal bovine serum, 2 mM l-glutamine, 0.05 mg/mL gentamicin, 10 unit/mL penicillin, 10 µg/mL streptomycin, and 0.25 µg/mL amphotericin. Confluent Vero cells in 6-well plates are washed twice with the postinoculation medium and inoculated with 300 µL of sample and 100 µL of postinoculation medium. The postinoculation medium is MEM supplemented with tryptose phosphate broth (0.3%), yeast extract (0.02%), and trypsin 250 (5 µg/mL). After 2 h incubation at 37°C with 5% CO_2, 3.6 mL postinoculation medium is added to each well. Inoculated cells (passage 0 [P0]) are incubated at 37°C with 5% CO_2. When a 70% cytopathic effect (CPE) develops, the plates are subjected to freeze-thaw once. The mixtures are centrifuged at $3000 \times g$ for 10 min at 4°C. The supernatants are harvested for further propagation or saved at −80°C. If no CPE was observed at 7 days postinoculation, the plates are frozen and thawed once and the supernatants are inoculated on new Vero cells for a second passage. Inoculated cells at each passage may be tested by an immunofluorescence assay (IFA). If CPE and IFA staining are negative after four passages, the virus isolation result is considered negative.

37.1.5.2 Electron Microscopy

The small intestine homogenate is centrifuged at $4,200 \times g$ for 10 min, and the supernatants are subjected to ultracentrifugation at $30,000 \times g$ for 30 min to pellet the virus particles, which are then negatively stained with 2% phosphotungstic acid (pH 7.0) and examined with electron microscope.

Vero cells infected with PEDV are trypsinized at 24 h postinfection and centrifuged at $800 \times g$ for 5 min. The cell pellets are resuspended in 0.01 M PBS (pH 7.2–7.4) and centrifuged again at $800 \times g$ for 5 min. The cell pellets are fixed with 2.5% glutaraldehyde-0.1 M sodium cacodylate. The cell pellets are postfixed in 1% osmium tetroxide for 90 min. The samples are dehydrated in an ascending ethanol series followed by propylene oxide and embedded in Eponate 12 resin (Ted Pella Inc). Ultrathin sections are stained with uranyl acetate and lead citrate and examined with electron microscope.

37.1.5.3 Immunohistochemistry

Five micrometer sections are cut from embedded tissues (jejunum, ileum, colon), air-dried overnight, and placed into a 60°C oven for 1 h. The deparaffinized and rehydrated sections are quenched for 10 min in aqueous 3% hydrogen peroxide and rinsed in MilliQ water. Epitopes are retrieved using Dako Target Retrieval Solution in Biocare Medical Decloaking Chamber. Once slides are cooled, they are placed into Tris-buffered saline plus Tween (TBST; MediMabs) for 5 min. The slides are incubated with mouse monoclonal antibody directed against the PEDV S protein at a dilution of 1:1500 for overnight incubation at 4°C. Slides are rinsed with TBST and incubated for 30 min with an Envision + antimouse kit (horseradish peroxidase labeled) (Dako) and a TBST rinse. Diaminobenzidine (DAB; Dako) is used as the substrate chromogen, and the slides are counterstained with Gill's hematoxylin [48,49].

37.1.5.4 Indirect Fluorescent Antibody Test

PEDV-infected and mock-infected cells grown on glass cover slips in 24-well tissue culture plates are fixed with 3.7% polyoxymethylene, washed three times with 0.1 M PBS, and permeabilized for 5 min with 0.1% Triton X-100. The cells are incubated with a 1% solution of bovine serum albumin (BSA) (30 min, room temperature, reverse transcription [RT]), then stained with fluorescein isothiocyanate (FITC)-conjugated TGEV polyclonal antibody (VMRD) and monoclonal mouse anti-PEDV M protein, respectively (overnight, 4°C in dark conditions). After washing five times with 0.1 M PBS, a Cy3-conjugated sheep antimouse IgG (A0521, Beyotime) is added to the cells (1:400, 1 h, RT). The negative control slices are treated in an identical manner except the primary antibodies that are omitted. Cells are washed 3 times with 0.01 M PBS, and images are acquired using a fluorescence microscope (Zeiss).

Alternatively, infected Vero monolayers are fixed in cold ethyl alcohol, and polyclonal rabbit anti-PEDV nucleoprotein (NP) antiserum is added at 1:500. Cells are rinsed and then incubated with FITC-labeled goat antirabbit IgG (Jackson Immunoresearch) at a dilution of 1:50, and then read using a fluorescent microscope.

37.1.5.5 ELISA

PEDV S protein antibodies are detected with a complex-trapping-blocking ELISA using monoclonal antibodies and PEDV grown on Vero cells as antigen. Samples with a blocking percentage >50% are considered positive, samples with <40% blocking are considered negative, and samples between 40% and 50% blocking as dubious or suspicious. Samples with initial results in the dubious/suspicious range are retested in the ELISA and also tested by an IFA using PEDV BR1/87-infected Vero cells fixed with ice-cold methanol [50–52].

37.1.5.6 Nucleic Acid Detection

Nucleic acid amplification techniques such as RT polymerase chain reaction (RT-PCR) and its derivatives offer a rapid and sensitive approach for PEDV identification and phylogenetic analysis [53–59]. The common targets for RT-PCR include the S and N genes [60–63]. Exploiting the fact that classically attenuated cell culture passaged PEDV shows mutations in ORF3, which changes to restriction fragment length polymorphism (RFLP) patterns, an RT-PCR-based RFLP assay was developed to detect PEDV in clinical specimens and differentiate between field PEDV isolates and a vaccine strain. Furthermore, genotyping and phylogenetic analysis on the S gene enabled differentiation of pandemic (PX) and classical (CX) groups of PEDV.

37.1.6 Treatment and Prevention

Other than symptomatic relief of diarrheal and control of secondary infections, no specific treatment is available for PED. The use of electrolyte and energy supplements may prevent dehydration in young pigs. Colostrum from immune sows can protect neonates against infection. Unless secondary infections occur, most growing pigs recover in 7–10 days without treatment. Adequate quarantine, isolation of cases, disinfection of transportation and equipment vehicles, appropriate disposal of dead pigs, and slurry are helpful to prevent the disease spread in the herd [41].

Modified live vaccines (MLV) have proven valuable for the control of PEDV [64]. The cell culture attenuated, subgroup I strain 83P-5 with 14 amino acid changes in the immunodominant S protein is available as an attenuated live PEDV vaccine. The J-vac (Nisseiken) attenuated live vaccine derived from the 83P-5 strain can be also utilized. In addition, the live attenuated, subgroup II DR13 vaccine strain with 13 mutations in the S protein was developed. MLV is able to stimulate a more robust and protective immune response than inactivated virus vaccines [65]. However, due to viral mutations and accompanying antigenic changes, long-term protection is often lacking. Furthermore, attenuated, live vaccine may have a risk for reversion to virulence [66]. For these reasons, continuing effort is required to develop safer vaccines based on protective viral protein subunits [41,67].

37.2 METHODS

37.2.1 Sample Preparation

Total RNA is extracted from 500 µL of the clarified supernatant of the 10% tissue emulsion using Qiagen RNeasy® Mini Kit and eluted in 50 µL of nuclease-free water. Alternatively, total RNA is purified by Trizol reagent (Invitrogen). Briefly, 100 µL of unclarified emulsion is added to 900 µL of Trizol reagent and vortexed thoroughly. After adding 200 µL of chloroform, the mixture is vortexed thoroughly for ~30 s and incubated for 10 min. The aqueous and organic phases are separated by centrifugation, and the aqueous phase is transferred to an equal volume of isopropanol. The RNA precipitate is then pelleted, dried, and dissolved in 20 µL of nuclease-free water.

37.2.2 Detection Procedures

37.2.2.1 Conventional RT-PCR

Song et al. [68] described a multiplex RT-PCR for simultaneous detection of PEDV and TGEV. While primers P1 and P2 generate a specific fragment of 651 bp from the PEDV S gene, primers T1 and T2 produce a specific fragment of 859 bp from the TGEV S gene (Table 37.1).

TABLE 37.1

Primers for Identification of Porcine Epidemic Diarrhea Virus and Transmissible Gastroenteritis Coronavirus

Primer	Sequence (5′–3′)	Position on the S Gene	Product (bp)
P1	TTCTGAGTCACGAACAGCCA	1466–1485	651
P2	CATATGCAGCCTGCTCTGAA	2097–2116	
T1	GTGGTTTTGGTYRTAAATGC	16–35	859
T2	CACTAACCAACGTGGARCTA	855–874	

For RT, 10 µL of extracted RNA and 1 µL (1 µg/µL) of random primer (hexadeoxyribonucleotide mixture) are mixed. The mixture is heated at 95°C for 3 min and immediately placed on ice. The remaining reagents, which consisted of 10 µL of 53 first strand buffer (50 mM Tris-HCl, 75 mM KCl, 3 mM $MgCl_2$), 10 mM DL-Dithiothreitol, 0.3 µM of each deoxyribonucleotide triphosphate (dNTP), and 100 units of M-MLV reverse transcriptase in a final volume of 50 µL, are added. The mixture is incubated at 37°C for 60 min, and the reaction is stopped by heating to 95°C for 2–3 min. The cDNA is either stored at 22°C or amplified immediately.

For PCR amplification, 2 µL of cDNA is mixed with a reaction mixture containing 2.5 µL of 10× Taq DNA polymerase buffer, 3 mM of $MgCl_2$, 2.0 µL of dNTPs (2.5 µM/µL), 0.5 µL of each primer (10 pmol each), and 1 µL of Taq DNA polymerase. MilliQ water is added to make up a total volume of 25 µL.

Amplification is performed at 94°C for 5 min; 30 cycles of 94°C for 30 s, 53°C for 60 s, 72°C for 60 s; and a final extension at 72°C for 5 min.

The RT-PCR products are analyzed by electrophoresis in 1.5% agarose gel containing ethidium bromide. The presence of a 651 bp band is indicative of PEDV, while the presence of an 859 bp band is indicative of TGEV.

Note: This multiplex RT-PCR provides a rapid, sensitive, and cost-effective tool for the identification of PEDV and TGEV. If desired, SuperScript II RT/Platinum Taq One-Step RT-PCR Kit (Invitrogen) may be utilized for RT and amplification in a single step [69].

37.2.2.2 Real-Time RT-PCR

Yu et al. [70] utilized primers PED-NF (5′-CGCAAAG ACTGAACCCACTAATTT-3′) and PED-NR (5′-TTGCC TCTGTTGTTACTTGGAGAT-3′) and probe PED-Cy5 (Cy5-TGTTGCCATTGCCACGACTCCTGC-BHQ3) from the N gene for quantitative detection of PEDV by RT-PCR [71].

RT-PCR mixture (20 µL) is made up of 0.8 µL of ThermoScript™ Plus/Platinum® Taq Enzyme Mix (Invitrogen), 10 µL of 2× ThermoScript Reaction Mix (a final concentration of 3 mM $MgCl_2$), 0.5 µL of both PEDV forward and reverse primers, 0.5 µL of PEDV-Cy5 probe, 2 µL of RNA, and 5.7 µL of water.

Amplification is conducted in an ABI7500 (Applied Bio systems) under the following conditions: initial reverse transcription at 58°C for 30 min, followed by initial denaturation at 95°C for 5 min, 40 cycles of denaturation at 95°C for 30 s, and annealing and extension at 60°C for 1 min.

The intensities of the fluorescent dyes in each reaction are read automatically during PCR cycling and optical data are analyzed with 7500 software v2.0.6.

37.3 CONCLUSION

PEDV is implicated in acute, severe atrophic enteritis of swine that presents diarrhea, vomiting, dehydration, and death as main symptoms. With a widespread distribution and a propensity to form mutational variants that escape the reign of host immune systems, PEDV is responsible for devastating economic losses in pig farming industry throughout Europe, Asia, and North America. Owing to the clinical similarity between PED and diseases caused by other microbial pathogens, PEDV has proven to be a challenge for laboratories engaged in its diagnosis and relevant authorities in its control. Fortunately, with serological assays as a backup and through its intrinsic sensitivity, specificity, and speed, RT-PCR provides an assurance on our technical capability to reliably diagnose and track PEDV [72–74]. This will greatly enhance our decision-making as to the most appropriate management and intervention strategies against this important pathogen in the event of an outbreak.

REFERENCES

1. Woo PC. et al. Novel betacoronavirus in dromedaries of the Middle East, 2013. *Emerg Infect Dis.* 2014;20(4):560–572.
2. Huang YW. et al. Origin, evolution, and genotyping of emergent porcine epidemic diarrhea virus strains in the United States. *MBio.* 2013;4(5):e00737-13.
3. Sun M. et al. Genomic and epidemiological characteristics provide new insights into the phylogeographical and spatiotemporal spread of porcine epidemic diarrhea virus in Asia. *J Clin Microbiol.* 2015;53(5):1484–1492.
4. Gao Y. et al. Phylogenetic analysis of porcine epidemic diarrhea virus field strains prevailing recently in China. *Arch Virol.* 2013;158(3):711–715.
5. Song D. et al. Molecular characterization and phylogenetic analysis of porcine epidemic diarrhea viruses associated with outbreaks of severe diarrhea in piglets in Jiangxi, China 2013. *PLoS One.* 2015;10(3):e0120310.
6. Vlasova AN. et al. Distinct characteristics and complex evolution of PEDV strains, North America, May 2013–February 2014. *Emerg Infect Dis.* 2014;20(10):1620–1628.
7. Lin CM. et al. Antigenic relationships among porcine epidemic diarrhea virus and transmissible gastroenteritis virus strains. *J Virol.* 2015;89(6):3332–3342.
8. Ma Y. et al. Origin, evolution, and virulence of porcine deltacoronaviruses in the United States. *MBio.* 2015;6(2):e00064.
9. Wang X-M. et al. Complete genome sequence of a variant porcine epidemic diarrhea virus strain isolated in central China. *Genome Announc.* 2013;1(1):e00243-12.
10. Zhou YJ. et al. Complete genome sequence of a virulent porcine epidemic diarrhea virus strain. *J Virol.* 2012;86(24):13862.
11. Song D. et al. Full-length genome sequence of a variant porcine epidemic diarrhea virus strain, CH/GDZQ/2014, responsible for a severe outbreak of diarrhea in piglets in Guangdong, China, 2014. *Genome Announc.* 2014;2(6):e01239-14.
12. Sun DB. et al. Spike protein region (aa 636789) of porcine epidemic diarrhea virus is essential for induction of neutralizing antibodies. *Acta Virol.* 2007;51(3):149–156.
13. Sato T. et al. Mutations in the spike gene of porcine epidemic diarrhea virus associated with growth adaptation in vitro and attenuation of virulence in vivo. *Virus Genes.* 2011;43:72–78.
14. Lawrence PK, Bumgardner E, Bey RF, Stine D, Bumgarner RE. Genome sequences of porcine epidemic diarrhea virus: In vivo and in vitro phenotypes. *Genome Announc.* 2014;2(3):e00503-14.
15. Marthaler D. et al. Complete genome sequence of porcine epidemic diarrhea virus strain USA/Colorado/2013 from the United States. *Genome Announc.* 2013;1:e00555.
16. Xu X. et al. Porcine epidemic diarrhea virus E protein causes endoplasmic reticulum stress and up-regulates interleukin-8 expression. *Virol J.* 2013;10:26.
17. Pan X. et al. Monoclonal antibody to N protein of porcine epidemic diarrhea virus. *Monoclon Antib Immunodiagn Immunother.* 2015;34(1):51–54.
18. Xu X. et al. Porcine epidemic diarrhea virus N protein prolongs S-phase cell cycle, induces endoplasmic reticulum stress, and up-regulates interleukin-8 expression. *Vet Microbiol.* 2013a;164(3–4):212–221.
19. Alonso C. et al. Evidence of infectivity of airborne porcine epidemic diarrhea virus and detection of airborne viral RNA at long distances from infected herds. *Vet Res.* 2014;45:73.
20. Pasick J. et al. Investigation into the role of potentially contaminated feed as a source of the first-detected outbreaks of porcine epidemic diarrhea in Canada. *Transbound Emerg Dis.* 2014;61(5):397–410.
21. OIE (Office International des Epizooties). Infection with porcine epidemic diarrhea virus: Aetiology, epidemiology, diagnosis, prevention and control. OIE Technical Factsheet. http://www.oie.int/doc/ged/D13924.PDF. Accessed on June 12, 2015.
22. Gerber PF. et al. The spray-drying process is sufficient to inactivate infectious porcine epidemic diarrhea virus in plasma. *Vet Microbiol.* 2014;174(1–2):86–92.
23. Martelli P. et al. Epidemic of diarrhoea caused by porcine epidemic diarrhoea virus in Italy. *Vet Rec.* 2008;162(10):307–310.
24. Hanke D. et al. Comparison of porcine epidemic diarrhea viruses from Germany and the United States, 2014. *Emerg Infect Dis.* 2015;21(3):493–496.
25. Lin CN. et al. Isolation and characterization of porcine epidemic diarrhea viruses associated with the 2013 disease outbreak among swine in the United States. *J Clin Microbiol.* 2014;52(1):234–243.
26. Puranaveja S. et al. Chinese-like strain of porcine epidemic diarrhea virus, Thailand. *Emerg Infect Dis.* 2009;15:1112–1115.
27. Tian Y. et al. Molecular characterization and phylogenetic analysis of new variants of the porcine epidemic diarrhea virus in Gansu, China in 2012. *Viruses.* 2013;2013(5):1991–2004.
28. Kim SH. et al. Genetic characterization of porcine epidemic diarrhea virus in Korea from 1998 to 2013. *Arch Virol.* 2015;160(4):1055–1064.
29. Kim YK. et al. A novel strain of porcine epidemic diarrhea virus in Vietnamese pigs. *Arch Virol.* 2015;160(6):1573–1577.
30. Wang J. et al. Porcine epidemic diarrhea virus variants with high pathogenicity, China. *Emerg Infect Dis.* 2013;19(12):2048–2049.
31. Lin CN. et al. US-like strain of porcine epidemic diarrhea virus outbreaks in Taiwan, 2013–2014. *J Vet Med Sci.* 2014;76(9):1297–1299.
32. Zhao PD. et al. Genetic variation analyses of porcine epidemic diarrhea virus isolated in mid-eastern China from 2011 to 2013. *Can J Vet Res.* 2015;79(1):8–15.
33. Pan Y. et al. Isolation and characterization of a variant porcine epidemic diarrhea virus in China. *Virol J.* 2012;9:195.
34. Sun RQ. et al. Outbreak of porcine epidemic diarrhea in suckling piglets, China. *Emerg Infect Dis.* 2012;18(1):161–163.
35. Stevenson GW. et al. Emergence of porcine epidemic diarrhea virus in the United States: Clinical signs, lesions, and viral genomic sequences. *J Vet Diagn Invest.* 2013;25(5):649–654.
36. Chen Q. et al. Isolation and characterization of porcine epidemic diarrhea viruses associated with the 2013 disease outbreak among swine in the United States. *J Clin Microbiol.* 2014;52:234–243.
37. Marthaler D, Bruner L, Collins J, Rossow K. Third strain of porcine epidemic diarrhea virus, United States. *Emerg Infect Dis.* 2014;20(12):2162–2163.

38. Wang L, Byrum B, Zhang Y. New variant of porcine epidemic diarrhea virus, United States, 2014. *Emerg Infect Dis.* 2014;20:917–919.

39. Bowman AS, Krogwold RA, Price T, Davis M, Moeller SJ. Investigating the introduction of porcine epidemic diarrhea virus into an Ohio swine operation. *BMC Vet Res.* 2015;11(1):38.

40. Ojkic D. et al. The first case of porcine epidemic diarrhea in Canada. *Can Vet J.* 2015;56(2):149–152.

41. Jung K, Saif LJ. Porcine epidemic diarrhea virus infection: Etiology, epidemiology, pathogenesis and immunoprophylaxis. *Vet J.* 2015;204:134–143.

42. Alvarez J, Sarradell J, Morrison R, Perez A. Impact of porcine epidemic diarrhea on performance of growing pigs. *PLoS One.* 2015;10(3):e0120532.

43. Cong Y. et al. Porcine aminopeptidase N mediated polarized infection by porcine epidemic diarrhea virus in target cells. *Virology.* 2015;478:1–8.

44. Huan CC. et al. Porcine epidemic diarrhea virus uses cell-surface heparan sulfate as an attachment factor. *Arch Virol.* 2015;160(7):1621–1628.

45. Madson DM. et al. Pathogenesis of porcine epidemic diarrhea virus isolate (US/Iowa/18984/2013) in 3-week-old weaned pigs. *Vet Microbiol.* 2014;174(1–2):60–68.

46. Park S, Kim S, Song D, Park B. Novel porcine epidemic diarrhea virus variant with large genomic deletion, South Korea. *Emerg Infect Dis.* 2014;20(12):2089–2092.

47. Khatri M. Porcine epidemic diarrhea virus replication in duck intestinal cell line. *Emerg Infect Dis.* 2015;21(3):549–550.

48. Guscetti F. et al. Immunohistochemical detection of porcine epidemic diarrhea virus compared to other methods. *Clin Diagn Lab Immunol.* 1998;5(3):412–414.

49. Kim O, Chae C. Comparison of reverse transcription polymerase chain reaction, immunohistochemistry, and in situ hybridization for the detection of porcine epidemic diarrhea virus in pigs. *Can J Vet Res.* 2002;66(2):112–116.

50. Sozzi E. et al. Comparison of enzyme-linked immunosorbent assay and RT-PCR for the detection of porcine epidemic diarrhoea virus. *Res Vet Sci.* 2010;88(1):166–168.

51. Paudel S, Park JE, Jang H, Shin HJ. Comparison of serum neutralization and enzyme-linked immunosorbent assay on sera from porcine epidemic diarrhea virus vaccinated pigs. *Vet Q.* 2014;34(4):218–223.

52. Wang Z. et al. Development of an antigen capture enzyme-linked immunosorbent assay for virus detection based on porcine epidemic diarrhea virus monoclonal antibodies. *Viral Immunol.* 2015;28(3):184–189.

53. Ogawa H. et al. Multiplex PCR and multiplex RT-PCR for inclusive detection of major swine DNA and RNA viruses in pigs with multiple infections. *J Virol Methods.* 2009;160(1–2):210–214.

54. Ben Salem AN. et al. Multiplex nested RT-PCR for the detection of porcine enteric viruses. *J Virol Methods.* 2010;165(2):283–293.

55. Ren X, Li P. Development of reverse transcription loop-mediated isothermal amplification for rapid detection of porcine epidemic diarrhea virus. *Virus Genes.* 2011;42(2):229–235.

56. Wang L, Zhang Y, Byrum B. Development and evaluation of a duplex real-time RT-PCR for detection and differentiation of virulent and variant strains of porcine epidemic diarrhea viruses from the United States. *J Virol Methods.* 2014;207:154–157.

57. Zhao J. et al. A multiplex RT-PCR assay for rapid and differential diagnosis of four porcine diarrhea associated viruses in field samples from pig farms in East China from 2010 to 2012. *J Virol Methods.* 2013;194(1–2):107–112.

58. Zhao PD. et al. Development of a multiplex TaqMan probe-based real-time PCR for discrimination of variant and classical porcine epidemic diarrhea virus. *J Virol Methods.* 2014;206:150–155.

59. Gou H. et al. Rapid and sensitive detection of porcine epidemic diarrhea virus by reverse transcription loop-mediated isothermal amplification combined with a vertical flow visualization strip. *Mol Cell Probes.* 2015;29(1):48–53.

60. Chen JF. et al. Molecular characterization and phylogenetic analysis of membrane protein genes of porcine epidemic diarrhea virus isolates in China. *Virus Genes.* 2008;36(2):355–364.

61. Chen J. et al. Molecular epidemiology of porcine epidemic diarrhea virus in China. *Arch Virol.* 2010;155(9):1471–1476.

62. Chen J. et al. Detection and molecular diversity of spike gene of porcine epidemic diarrhea virus in China. *Viruses.* 2013;22;5(10):2601–2613.

63. Chen J. et al. Genetic variation of nucleocapsid genes of porcine epidemic diarrhea virus field strains in China. *Arch Virol.* 2013;158(6):1397–1401.

64. Goede D. et al. Previous infection of sows with a "mild" strain of porcine epidemic diarrhea virus confers protection against infection with a "severe" strain. *Vet Microbiol.* 2015;176(1–2):161–164.

65. Song D, Park B. Porcine epidemic diarrhoea virus: A comprehensive review of molecular epidemiology, diagnosis, and vaccines. *Virus Genes.* 2012;44:167–175.

66. Tian PF. et al. Evidence of recombinant strains of porcine epidemic diarrhea virus, United States, 2013. *Emerg Infect Dis.* 2014;20(10):1735–1738.

67. Meng F. et al. Evaluation on the efficacy of immunogenicity of recombinant DNA plasmids expressing spike genes from porcine transmissible gastroenteritis virus and porcine epidemic diarrhea virus. *PLoS One.* 2013;8(3):e57468.

68. Song DS. et al. Multiplex reverse transcription-PCR for rapid differential detection of porcine epidemic diarrhea virus, transmissible gastroenteritis virus, and porcine group A rotavirus. *J Vet Diagn Invest.* 2006;18(3):278–281.

69. Kim SH. et al. Multiplex real-time RT-PCR for the simultaneous detection and quantification of transmissible gastroenteritis virus and porcine epidemic diarrhea virus. *J Virol Methods.* 2007;146(1–2):172–177.

70. Yu X. et al. Development of a real-time reverse transcription loop-mediated isothermal amplification method for the rapid detection of porcine epidemic diarrhea virus. *Virol J.* 2015;12(1):76.

71. Bridgen A, Duarte M, Tobler K, Laude H, Ackermann M. Sequence determination of the nucleocapsid protein gene of the porcine epidemic diarrhoea virus confirms that this virus is a coronavirus related to human coronavirus 229E and porcine transmissible gastroenteritis virus. *J Gen Virol.* 1993;74:1795–1804.

72. Li Z. et al. Sequence and phylogenetic analysis of nucleocapsid genes of porcine epidemic diarrhea virus (PEDV) strains in China. *Arch Virol.* 2013;158(6):1267–1273.

73. Li ZL. et al. Molecular characterization and phylogenetic analysis of porcine epidemic diarrhea virus (PEDV) field strains in south China. *Virus Genes.* 2012;45(1):181–185.

74. Yuan W, Li Y, Li P, Song Q, Li L, Sun J. Development of a nanoparticle-assisted PCR assay for detection of porcine epidemic diarrhea virus. *J Virol Methods.* 2015;220:18–20.

38 Toroviruses

Dongyou Liu

CONTENTS

38.1 Introduction 339
   38.1.1 Classification 339
   38.1.2 Morphology and Genome Organization 340
   38.1.3 Biology and Epidemiology 340
   38.1.4 Clinical Features and Pathogenesis 340
   38.1.5 Diagnosis 341
      38.1.5.1 Virus Isolation 341
      38.1.5.2 Electron Microscopy 341
      38.1.5.3 Serological Assays 341
      38.1.5.4 Nucleic Acid Detection 341
   38.1.6 Treatment and Prevention 341
38.2 Methods 342
   38.2.1 Sample Preparation 342
   38.2.2 Detection Procedures 342
      38.2.2.1 Conventional RT-PCR 342
      38.2.2.2 Real-Time RT-PCR 342
38.3 Conclusion 343
References 343

38.1 INTRODUCTION

Toroviruses are a distinct group of positive single-stranded RNA viruses belonging to the genus *Torovirus*, family *Coronaviridae*. Of the four torovirus species recognized to date, equine torovirus (EToV), bovine torovirus (BToV), and porcine torovirus (PToV) are associated with enteric infections and diarrhea in horses, cattle, and pigs, respectively, while human torovirus (HToV) is implicated in nosocomial gastroenteritis, especially in individuals with suppressed immune functions. Although toroviruses generally cause subclinical infection or mild diarrheal disease in adults, they can be deadly when naïve young animals and children are involved. Given their widespread distribution and their ability to cross species boundary, toroviruses certainly deserve more attention in the areas of rapid diagnostic methodologies and effective control measures than they are currently receiving.

38.1.1 CLASSIFICATION

Toroviruses are classified in the genus *Torovirus*, subfamily *Torovirinae*, family *Coronaviridae*, order *Nidovirales*.

The family *Coronaviridae* covers an expanding group of enveloped, nonsegmented, single-stranded, positive-sense RNA viruses that are separated into two subfamilies: *Coronavirinae* and *Torovirinae*. The subfamily *Coronavirinae* consists of four genera, including bat-infecting *Alphacoronavirus* and

Betacoronavirus and bird-/pig-infecting *Deltacoronavirus* and *Gammacoronavirus*; the subfamily *Torovirinae* contains two genera: fish-infecting *Bafinivirus* and mammal-infecting *Torovirus* [1–3].

Currently, the genus *Torovirus* is divided into four species: EToV (type species; formerly Berne virus, or BEV), BToV (formerly Breda virus, or BRV), PToV, and HToV [2].

EToV was identified from a diarrheic fecal sample of a horse with diarrhea in Berne, Switzerland, in 1972, and thus named Berne virus (BEV). It is also the first torovirus successfully cultured in vitro [4].

BToV was isolated from a case of neonatal calf diarrhea in Breda, Iowa, in 1979 [5–7]. Morphologically and antigenically related to EToV, BToV (also known as Breda virus, or BRV) was shown to cause calf diarrhea in experimentally infected gnotobiotic calves and under field conditions. Various BToV strains have been propagated in the human rectal adenocarcinoma cell line (HRT-18) and other cells. Besides its link to neonatal calf diarrhea, BToV-like particles have been described in children with gastrointestinal and respiratory symptoms.

PToV was identified and characterized in the feces of swine in the Netherlands in 1998 [8]. Analysis of the amino acid sequence of PToV HE protein led to the separation of two HE lineages, represented by strains Markelo and P4, respectively, which share amino acid sequence homology of 80% and are antigenically distinct [9].

HToV was identified from fecal samples of humans with enteric disease in the 1980s [10–13]. A recent study showed that persons infected with torovirus are frequently immuno-compromised and nosocomially infected and tend to have bloody diarrhea [14].

38.1.2 Morphology and Genome Organization

Morphologically, toroviruses are spherical, oval, elongated, kidney shaped, enveloped particles of 100–140 nm (average 120 nm) in diameter, covered with two sets of surface projections. The longer projections (made of S protein) are drumstick or petal shaped, extending for approximately 19 nm in length, while the short projections (made of HE protein) are club shaped, measuring about 6 nm in length. Underneath the envelope is an elongated, tubular (or torus) nucleocapsid of helical symmetry (from which the virus gets its name) that encases a monopartite, linear, positive-sense, single-stranded RNA of approximately 28,000 nucleotides (e.g., BToV Breda-1 strain is 28,475 nt, PToV SH1 strain is 28,301 nt, and PToV-NPL/2013 strain is 28,305 nt in length) [15–17].

Besides a capped 5′ untranslated region (UTR) and a poly-adenylated 3′ UTR, the torovirus genome encompasses two large, overlapping open reading frames, ORF1a and ORF1b, coding for nonstructural proteins with replicase/transcriptase activity, as well as four small ORFs, ORFs 2–5, coding for spike (S), membrane (M), hemagglutinin–esterase (HE), and nucleocapsid (N) structural proteins, respectively. In culture-adapted EToV (BEV) and some BToV strains, the HE gene is partially deleted although the corresponding mRNA is produced.

Analysis of the complete genome sequence of PToV-NPL/2013 strain revealed a 28,305 nt genome, which shares a 92% identity to that of PToV-SH1 strain. The 5′ UTR and 3′ UTR are 792 and 196 nt in length, respectively. The poly-protein-encoding ORF (ORF1ab) is located between nt 794 and 20,910 and encodes a polyprotein of 6,706 aa. ORF1a is located between nt 794 and 14,047 and encodes a replicase of 4,418 aa. Genes encoding S, M, HE, and N proteins are 4722, 702, 1284, and 492 nt in length, respectively, and predicted proteins share 94%, 99%, 92%, and 96% identity to those of PToV-SH1, respectively [16,17].

The S protein is a major glycoprotein forming the large spikes that protrude from the viral particles. Upon entry to the host, the S protein binds to host cell receptors, induces virus-mediated cell fusion, and elicits neutralizing antibodies.

The HE protein is a class I membrane glycoprotein (65 kDa) belonging to the receptor destroying enzyme protein family. It forms homodimers, which make up the smaller spikes, and has the ability to bind sialic acids and catalyze the disruption of that binding by means of its acetyl-esterase activity. Torovirus HE protein is thought to function as a viral coreceptor, an enzyme that digests mucus layers to allow virus access to the target cells in the respiratory and/or enteric tracts, and a molecule that influences host/cell tropism. Interestingly, culture-adapted EToV and BToV strains harbor deletions or mutations in the HE gene, which are acquired during in vitro growth, leading to a malfunctioned HE protein. The fact that the HE protein is intact in a vast majority of PToV and BToV field strains highlights its indispensible role during in vivo infection [9].

38.1.3 Biology and Epidemiology

Similar to other coronaviruses, torovirus employs the viral S protein (also HE protein if present) to attach to host receptors and mediate its entry via endocytosis into the host cell. Subsequent fusion of virus membrane with the endosomal membrane facilitates the release of ssRNA(+) genome into the cytoplasm, from which the replicase polyprotein is translated and proteolytically cleaved into the viral polymerase (RdRp). Transcription of genomic ssRNA(+) into ssRNA(−) enables formation of a dsRNA genome, whose transcription/replication generates viral mRNAs/new ssRNA(+) genomes. Following synthesis of structural proteins from subgenomic mRNAs, virus assembly takes place at membranes of the endoplasmic reticulum and/or the Golgi complex. New virions are then released after budding or cell lysis.

Although EToV is the first torovirus identified and represents the type species of the genus, BToV and PToV have attracted much attention lately due to their common occurrence and association with diarrheal diseases in cattle and pigs.

Since its first observation from a case of neonatal calf diarrhea in Iowa, United States, in 1979, BToV has been detected in the Netherlands, Switzerland, Germany, Belgium, Hungry, United Kingdom, France, Spain, Italy, Japan, South Korea, India, Turkey, New Zealand, South Africa, and Brazil [18–25].

BToV is sensitive to lipid solvent (10% chloroform) and heat at 56°C for 30 min, but is somewhat resistant to acid (pH 3.0 at 22°C for 3 h) and 5-iodo-2′-deoxyuridine (IUdR). In addition, BToV does not pass through 50-nm-pore-size (or less) filter.

After initial identification in the feces of swine in the Netherlands in 1998, PToV has been reported in many other countries, including Hungary, Italy, Spain, United Kingdom, United States, Canada, South Korea, China, and South Africa [18,26–28].

In addition to horses, cattle, pigs, and humans, toroviruses have been detected in sheep, goats, lagomorphs, rodents, and domestic cats. In cats, torovirus infection is associated with diarrhea and protrusion of nictitating membranes. This highlights the promiscuous nature of toroviruses and the need to clarify the roles of these additional animal hosts in the epidemiology of toroviruses.

38.1.4 Clinical Features and Pathogenesis

Toroviruses are implicated in enteric diseases in both animals and humans, with BToV and PToV being recognized as infectious gastrointestinal agent in cattle and as a predominant cause of acute enteric infections in piglets.

Clinical signs of BToV infection in cattle include diarrhea, dehydration, pyrexia, lethargy, depression, and anorexia, together with mucoid feces, trembling, paralysis, and sudden death in calves [29,30]. Experimental infection of gnotobiotic calves or colostrum-deprived calves results in mild to moderate diarrhea, depression, and anorexia. Histopathologically, BToV infection is evident in villous and crypt enterocytes of the mid-jejunum, ileum, colon, and cecum, leading to severe vacuolar degeneration, necrosis and exfoliation of enterocytes, villous atrophy, crypt hyperplasia, and, in some cases, fused villi [31].

Although most pigs with PToV infection appear to be subclinical, they shed virus in feces that represent a constant source of infection to susceptible, naïve piglets. This was demonstrated by PToV-related epidemic outbreaks of diarrhea in piglets that occurred in China in 2011. With high morbidity and mortality, these outbreaks resulted in devastating losses to pig farming industry [28].

EToV (BEV) was shown to cause cytopathic effect (CPE) (e.g., cell rounding and detachment) in E. Derm cells 24 h postinfection (hpi) (at 5 pfu/cell), with some cells displaying characteristic signs of apoptosis, including disorganized nucleus, chromatin condensation, membrane blebbing, altered mitochondria, and cell disassembly in vesicles. At 48 hpi, extensive CPE causes many cells to float in the medium [32].

38.1.5 DIAGNOSIS

A number of microbial pathogens may induce diarrhea in cattle and pigs. For example, in pigs, potential diarrhea-causing viruses include PToV, porcine epidemic diarrhea coronavirus, porcine kobuvirus, porcine group A–C rotaviruses, transmissible gastroenteritis coronavirus, astroviruses, mammalian orthoreovirus, porcine sapovirus, and porcine norovirus [28]. Therefore, definitive diagnosis of torovirus infections are dependent on the detection of viral particles, antigens, antibodies, and nucleic acids using various laboratory procedures such as virus isolation, electron microscopy, serological assays, and nucleic acid detection.

38.1.5.1 Virus Isolation

EToV (BEV) has been shown to grow in E. Derm and embryonic mule skin cells, while BToV is able to replicate in human rectal adenocarcinoma HRT-18 cells [33]. Typically, confluent HRT-18 cell monolayers in 24-well plates are washed with Eagle minimal essential medium (EMEM) and inoculated with 0.1 mL of the intestinal-content suspensions. After adsorption for 60 min at 37°C, the cells are washed with EMEM and added 0.5 mL of EMEM. The cells are incubated for 7 days at 37°C and examined for CPEs. After incubation, the cells and supernatant were frozen and thawed once to harvest cell lysates, and subsequent passages are carried out in the same manner with 0.1 mL of cell lysates.

38.1.5.2 Electron Microscopy

Fecal specimens are diluted with an equal volume of 1% (wt/vol) ammonium acetate and clarified by centrifugation at $9000 \times g$ for 15 min at 4°C. The supernatant is transferred to a new tube and centrifuged at $12,000 \times g$ for 15 min at 4°C.

Each sample is applied to a 400-mesh grid precoated with polyvinyl formal and carbon. The grids are stained for 1 min with 2% phosphotungstic acid (pH 7.0) and examined by negative-contrast electron microscope with a Philips EM 300 microscope, at a magnification of ×50,000.

For infected cell culture supernatants, partial purification is carried out by ultracentrifugation through a 20% (wt/wt) sucrose cushion. The cells are negatively stained with 2% ammonium molybdate and examined under electron microscope. For ultrastructural observation, infected cells are fixed with 2% glutaraldehyde, postfixed with 1% osmium tetroxide, dehydrated in graded ethanol solutions, and embedded in an Epon mixture. Ultrathin sections are stained with a lead citrate-uranyl acetate solution and observed under electron microscope.

38.1.5.3 Serological Assays

A variety of serological procedures are useful for detection of torovirus antigens and antibodies. These include indirect fluorescent antibody test, hemagglutination inhibition, serum neutralization, and enzyme-linked immunosorbent assay (ELISA). Serum neutralization (plaque reduction) test allows detection of serum neutralizing antibodies against torovirus. The presence of specific IgG antibodies against PToV in the sera may be tested with an indirect ELISA using the highly conserved, recombinant viral N protein as antigen [34].

38.1.5.4 Nucleic Acid Detection

Recent development of nucleic acid amplification techniques such as reverse transcription-polymerase chain reaction (RT-PCR) has enabled rapid, sensitive, and precise detection of toroviruses [35–38]. In particular, the use of SYBR Green chemistry and/or the TaqMan probes in real-time RT-PCR allows quantitative detection of toroviruses.

Comparative phylogenetic analysis of the genes encoding the structural proteins S, M, HE, and N of BToV and PToV from European swine and cattle herds showed that all newly characterized BToV variants might have arisen from a genetic exchange, during which the 3′ end of the HE gene, the N gene, and the 3′ UTR of a BToV Breda virus-like parent had been swapped for those of PToV. In addition, some BToV and PToV variants with chimeric HE genes may have resulted from recombination events involving unknown toroviruses [39,40].

38.1.6 TREATMENT AND PREVENTION

No specific treatment is currently available for torovirus infection. Symptomatic treatments such as fluid therapy (to prevent dehydration), colostrum containing antibodies (to strengthen the immune response), and antibiotics (to control secondary bacterial infections) may be applied to animals suffering from torovirus disease.

Other control and prevention measures include isolation of affected animals from apparently healthy animals, wearing gloves and boots when handling affected animals, composting of fecal matters, proper disposal of bedding material, and disinfection of floor and walls [41].

38.2 METHODS

38.2.1 SAMPLE PREPARATION

Samples (feces, rectal swabs, intestinal contents from the ileum and rectum, blood) are collected from animals. Fecal samples are diluted 1:1 in sterile phosphate buffered saline (PBS), homogenized by vortexing, and centrifuged at $3000 \times g$ for 10 min at 4°C. The supernatant is aliquoted and kept at −80°C. Rectal swabs are embedded in 900 µL of sterile PBS solution and stored at −80°C. Prior to RNA isolation, tubes containing rectal swab samples embedded in PBS are vortexed, and 250 µL of the fecal suspension is transferred to a new tube. Fecal debris is discarded by centrifugation at 9000 rpm, 10 min at 4°C. Intestinal contents are diluted 1:10 in 0.01 M PBS (pH 7.4) and clarified by low-speed centrifugation at $3000 \times g$ for 10 min.

Total RNA is extracted from 200 µL of supernatant or clarified rectal swab homogenate using TRIzol reagent (Invitrogen) or High Pure RNA isolation kit (Roche Applied Science), resuspended in 50 µL of RNase-free water, and stored at −80°C until analysis.

38.2.2 DETECTION PROCEDURES

38.2.2.1 Conventional RT-PCR

38.2.2.1.1 BToV Detection

Hoet et al. [42] reported an RT-PCR for detection of BToV with forward primer (5′-GTG TTA AGT TTG TGC AAA AAT G-3′, nt 37–57) and reverse primer (5′-TGC ATG AAC TCT ATA TGG TGT-3′, nt 758–777). The predicted RT-PCR product is 741 bp from the 5′ end region of the spike gene.

RT-PCR is performed by pretreating 4 µL of the extracted RNA with 1 µL of dimethyl sulfoxide for 10 min at 70°C–75°C. Then, 5 µL of the pretreated RNA sample is mixed with the RT-PCR mixture (5 µL of 10× buffer, 5 mL MgCl₂ [25 mM], 1 mL dNTPs [10 mM], 1 µL of forward primer [200 ng/µL], 1 µL reverse primer [200 ng/µL], 0.5 µL of Avian myeloblastosis virus (AMV)-RT [10 U/µL], 0.5 mL of RNasin [40 U/mL], 0.5 µL of Taq DNA polymerase [5 U/µL], and 31.5 µL of dH₂O, in a total of 50 µL). The mixture is subjected to 1 RT phase of 90 min at 42°C, an initial denaturation step of 5 min at 94°C, and 30 cycles of 1 min at 94°C, 2 min at 55°C, and 2 min at 72°C. The final extension step is 10 min at 72°C. The RT-PCR products are visualized on 1.5% agarose gels stained with ethidium bromide, and their size is determined by comparison against DNA molecular weight markers.

38.2.2.1.2 PToV Detection

Zhou et al. [43] developed an RT-PCR for PToV detection with forward primer (5′-ACCCCTGCCTGAGGTTTCYTT-3′) and reverse primer (5′-AGCACGACGTTGTCTRCGTGT-3′), which amplifies a 451 bp fragment from the conserved region of the S gene.

Reverse transcription is conducted with TaKaRa RT system. PCR amplification is carried out with 200 µM of each dNTP,

10 pmol of each primer, 1.0 U Taq DNA polymerase (Promega), and 1.5 mM MgCl₂, in a total volume of 40 µL.

PCR cycling parameters include denaturation at 94°C for 2 min, 30 cycles of amplification (94°C for 30 s, 57°C for 30 s, and 72°C for 30 s), and a final extension of 72°C for 7 min.

The PCR product is resolved using 1% agarose gel electrophoresis, stained with ethidium bromide, and visualized under ultraviolet light (Bio-Rad gel imaging system, Hercules).

38.2.2.2 Real-Time RT-PCR

38.2.2.2.1 BToV and PToV Detection

Hosmillo et al. [44] described a SYBR Green real-time RT-PCR assay with a universal primer pair (F: 5′-TTACTGGYTATTGGGCMYT-3′, nt 26029–26047; R: 5′-AAAGGRGTGCAGTGWAGCTT-3′ nt 26196–26215) for the detection and quantitation of both BToV and PToV in bovine and porcine stool samples. These primers come from the membrane (M) gene of bovine and porcine torovirus strains and generate a 187 bp product from BToV and PToV. Subsequent melting curve analysis permits differentiation between BToV and PToV.

RT-PCR mixture (25 µL) is composed of 5 µL of RNA template, 12.5 µL of SensiMix one-step mixture (Quantace), 1 µL each of 0.5 µM forward and reverse primers (final concentration of each primer: 20 nM), 0.5 µL of 50× SYBR Green solution (final concentration: 1×), 0.5 µL of RNase inhibitor (final concentration: 10 units), 0.5 µL of MgCl₂ (final concentration: 4.0 mM), and 4 µL of RNase-free water.

Amplification and detection are carried out in Corbett Research Rotor-Gene 6000 series Real-Time Amplification system (Corbett Research), with the following conditions: RT at 42°C for 30 min, activation of the hot-start DNA polymerase at 95°C for 10 min, and 45 cycles of 94°C for 15 s, 50°C for 45 s, and 72°C for 20 s.

The melting curve analysis is performed after 45 reaction cycles; the temperature ramp is programmed from 72°C to 95°C in increments of 1°C, waiting 5 s before each acquisition. Samples are considered positive if both an exponential increase of fluorescence and a BToV- or PToV-specific melting peak are observed.

38.2.2.2.2 PToV Detection

Pignatelli et al. [45] employed forward primer rtNII5′ (5′-CCCTGGTTTTAGACCTATGTTTCA-3′) and reverse primer rtNII3′ (5′-GCAGCACTCTTAGCAATCT-TCACTA-3′) flanking a 279 bp conserved region of PToV N gene for PCR detection of PToV.

RT reaction is carried out with SuperScript III reverse transcriptase (Invitrogen). Briefly, 9 µL of viral RNA purified from field samples is mixed with 0.5 mM of dNTPs mix (Roche Diagnostic) and 0.5 µM of random hexamers mix (Roche Diagnostic S), incubated for 10 min at 65°C and kept on ice for 1 min. Then, a mixture containing 4 µL of 5× reaction buffer, 1 µL of 0.1 M DTT, 2 µL of 50 mM MgCl₂, and 1 µL of SuperScript III reverse transcriptase (1 U/µL) are added to the mixture. The reaction mix is incubated for 5 min at 25°C, followed by 45 min at 50°C, and finally 15 min at 75°C to inactivate reverse transcriptase. The cDNA preparations are aliquoted and stored at −20°C until use.

Real-time PCR is conducted with Power PCR SYBR Green kit (Applied Biosystems). The PCR mixture (20 µL) is made up of 2 µL of RT reaction, 10 µL of Power PCR SYBR Green mastermix (including hot-start DNA polymerase, reaction buffer, dNTP mix, and SYBR Green dye; Applied Biosystems), 0.5 µM of each primer, and 12.8 µL of PCR grade water.

PCR amplification is performed on an ABI 7000 sequence detector (Applied Biosystems) using universal thermal conditions: 10 min at 95°C and 40 cycles of 15 s at 95°C and 1 min at 60°C. A denaturation step is added at the end of amplification reaction for Tm analysis.

The results are analyzed using SDS v1.2 software (Applied Biosystems). The baseline, threshold, Ct (Ct indicates the cycle in which a target sequence is first detected), and Tm values of amplified products are automatically determined.

38.3 CONCLUSION

Toroviruses are causal agents of enteric infections and diarrhea in various mammal hosts including cattle, pig, horses, and humans. Due to the fact that many other viruses, bacteria, and parasites may induce similar clinical symptoms in these hosts, use of laboratory methods is essential for definitive diagnosis of torovirus infection. With the rapid advances in nucleic acid detection technologies, especially the development of real-time RT-PCR, precise and speedy identification of toroviruses has never been so straightforward. In comparison with the increasing sophistication of diagnostic methodologies, however, our understanding of torovirus pathogenesis and our available tools for torovirus control are far from adequate. Unquestionably, the ability of toroviruses to undergo constant mutations in response to changing external environments and the absence of a reliable in vitro culture system that does not lead to mutated toroviruses may have hampered the progress of our research efforts. Plugging such knowledge gaps will help us keep an upper hand over these opportunistic pathogens.

REFERENCES

1. González JM, Gomez-Puertas P, Cavanagh D, Gorbalenya AE, Enjuanes L. A comparative sequence analysis to revise the current taxonomy of the family Coronaviridae. *Arch Virol.* 2003;148(11):2207–2235.
2. De Groot RJ. et al. Family Coronaviridae. In: King AMQ, Lefkowitz E, Adams MJ, Carstens EB. (eds.), *Virus Taxonomy*, Elsevier, Oxford, U.K., 2011; pp. 806–828.
3. Woo PC. et al. Novel betacoronavirus in dromedaries of the Middle East, 2013. *Emerg Infect Dis.* 2014;20(4):560–572.
4. Weiss M, Steck F, Horzinek M. Berne virus: A new equine RNA-virus. I. Virus structure and seroepidemiology. *Zentral Bakteriol Mikrobiol Hyg.* 1984;258:537.
5. Woode GN, Reed DE, Runnels PL, Herrig MA, Hill HT. Studies with an unclassified virus isolated from diarrheic calves. *Vet Microbiol.* 1982;7:221–240.
6. Duckmanton L, Carman S, Nagy E, Petric M. Detection of bovine torovirus in fecal specimens of calves with diarrhea from Ontario farms. *J Clin Microbiol.* 1998;36(5):1266–1270.
7. Duckmanton LM, Tellier R, Liu P, Petric M. Bovine torovirus: Sequencing of the structural genes and expression of the nucleocapsid protein of Breda virus. *Virus Res.* 1998;58(1–2):83–96.
8. Kroneman A. et al. Identification and characterization of a porcine torovirus. *J Virol.* 1998;72(5):3507–3511.
9. Pignatelli J, Alonso-Padilla J, Rodríguez D. Lineage specific antigenic differences in porcine torovirus hemagglutinin-esterase (PToV-HE) protein. *Vet Res.* 2013;44:126.
10. Beards GM, Green BJ, Flewett TH. Preliminary characterisation of torovirus-like particles of humans: Comparison with Berne virus of horses and Breda virus of calves. *J Med Virol.* 1986;20:67–78.
11. Duckmanton L, Luan B, Devenish J, Tellier R, Petric M. Characterization of torovirus from human fecal specimens. *Virology.* 1997;239(1):158–168.
12. Jamieson FB. et al. Human torovirus: A new nosocomial gastrointestinal pathogen. *J Infect Dis.* 1998;178:1263–1269.
13. Lodha A, de Silva N, Petric M, Moore AM. Human torovirus: A new virus associated with neonatal necrotizing enterocolitis. *Acta Paediatr.* 2005;94(8):1085–1088.
14. Gubbay J. et al. The role of torovirus in nosocomial viral gastroenteritis at a large tertiary pediatric centre. *Can J Infect Dis Med Microbiol.* 2012;23(2):78–81.
15. Draker R, Roper RL, Petric M, Tellier R. The complete sequence of the bovine torovirus genome. *Virus Res.* 2006;115(1):56–68.
16. Anbalagan S, Peterson J, Wassman B, Elston J, Schwartz K. Genome sequence of torovirus identified from a pig with porcine epidemic diarrhea virus from the United States. *Genome Announc.* 2014;2(6):e01291-14.
17. Sun H. et al. Molecular characterization and phylogenetic analysis of the genome of porcine torovirus. *Arch Virol.* 2014;159(4):773–778.
18. Matiz K. et al. Torovirus detection in faecal specimens of calves and pigs in Hungary: Short communication. *Acta Vet Hung.* 2002;50(3):293–296.
19. Ito T, Okada N, Fukuyama S. Epidemiological analysis of bovine torovirus in Japan. *Virus Res.* 2007;126(1–2):32–37.
20. Kirisawa R, Takeyama A, Koiwa M, Iwai H. Detection of bovine torovirus in fecal specimens of calves with diarrhea in Japan. *J Vet Med Sci.* 2007;69(5):471–476.
21. Park SI. et al. Molecular epidemiology of bovine noroviruses in South Korea. *Vet Microbiol.* 2007;124(1–2):125–133.
22. Park SJ. et al. Molecular epidemiology of bovine toroviruses circulating in South Korea. *Vet Microbiol.* 2008;126(4):364–371.
23. Alonso-Padilla J. et al. Seroprevalence of porcine torovirus (PToV) in Spanish farms. *BMC Res Notes.* 2012;5:675.
24. Nogueira JS. et al. First detection and molecular diversity of Brazilian bovine torovirus (BToV) strains from young and adult cattle. *Res Vet Sci.* 2013;95(2):799–801.
25. Gülaçtı I, Işıdan H, Sözdutmaz I. Detection of bovine torovirus in fecal specimens from calves with diarrhea in Turkey. *Arch Virol.* 2014;159(7):1623–1627.
26. Pignatelli J. et al. Molecular characterization of a new PToV strain: Evolutionary implications. *Virus Res.* 2009;143(1):33–43.
27. Shin DJ. et al. Detection and molecular characterization of porcine toroviruses in Korea. *Arch Virol.* 2010;155(3):417–422.
28. Zhou Y, Chen L, Zhu L, Xu Z. Molecular detection of porcine torovirus in piglets with diarrhea in southwest China. *Sci World J.* 2013;2013:984282.

29. Hoet AE. et al. Enteric and nasal shedding of bovine torovirus (Breda virus) in feedlot cattle. *Am J Vet Res.* 2002;63(3):342–348.

30. Hoet AE. et al. Association of enteric shedding of bovine torovirus (Breda virus) and other enteropathogens with diarrhea in veal calves. *Am J Vet Res.* 2003;64(4):485–490.

31. Aita T. et al. Characterization of epidemic diarrhea outbreaks associated with bovine torovirus in adult cows. *Arch Virol.* 2012;157(3):423–431.

32. Maestre AM, Garzón A, Rodríguez D. Equine torovirus (BEV) induces caspase-mediated apoptosis in infected cells. *PLoS One.* 2011;6(6):e20972.

33. Kuwabara M, Wada K, Maeda Y, Miyazaki A, Tsunemitsu H. First isolation of cytopathogenic bovine torovirus in cell culture from a calf with diarrhea. *Clin Vaccine Immunol.* 2007;14(8):998–1004.

34. Pignatelli J, Grau-Roma L, Jiménez M, Segalés J, Rodríguez D. Longitudinal serological and virological study on porcine torovirus (PToV) in piglets from Spanish farms. *Vet Microbiol.* 2010;146(3–4):260–268.

35. Hoet AE. et al. Detection of bovine torovirus and other enteric pathogens in feces from diarrhea cases in cattle. *J Vet Diagn Invest.* 2003;15(3):205–212.

36. Ito T. et al. Genetic and antigenic characterization of newly isolated bovine toroviruses from Japanese cattle. *J Clin Microbiol.* 2010;48(5):1795–1800.

37. Zlateva KT. et al. Design and validation of consensus-degenerate hybrid oligonucleotide primers for broad and sensitive detection of corona- and toroviruses. *J Virol Methods.* 2011;177(2):174–183.

38. Fukuda M. et al. Development and application of one-step multiplex reverse transcription PCR for simultaneous detection of five diarrheal viruses in adult cattle. *Arch Virol.* 2012;157(6):1063–1069.

39. Smits SL. et al. Phylogenetic and evolutionary relationships among torovirus field variants: Evidence for multiple intertypic recombination events. *J Virol.* 2003;77(17):9567–9577.

40. Cong Y. et al. Evolution and homologous recombination of the hemagglutinin-esterase gene sequences from porcine torovirus. *Virus Genes.* 2013;47:66–74.

41. Dhama K, Pawaiya RVS, Chakraborty S, Tiwari R, Verma AK. Toroviruses affecting animals and humans: A review. *Asian J Anim Vet Adv.* 2014;9:190–201.

42. Hoet AE, Chang KO, Saif LJ. Comparison of ELISA and RT-PCR versus immune electron microscopy for detection of bovine torovirus (Breda virus) in calf fecal specimens. *J Vet Diagn Invest.* 2003;15(2):100–106.

43. Zhou L. et al. Molecular epidemiology of Porcine torovirus (PToV) in Sichuan Province, China: 2011–2013. *Virol J.* 2014;11:106.

44. Hosmillo MD. et al. Development of universal SYBR Green real-time RT-PCR for the rapid detection and quantitation of bovine and porcine toroviruses. *J Virol Methods.* 2010;168(1–2):212–217.

45. Pignatelli J, Jiménez M, Grau-Roma L, Rodríguez D. Detection of porcine torovirus by real time RT-PCR in piglets from a Spanish farm. *J Virol Methods.* 2010;163(2):398–404.

Section II

Negative-Sense RNA Viruses

39 Avian Bornavirus

Francesca Sidoti, Maria Lucia Mandola, Francesca Rizzo,
Rossana Cavallo, and Cristina Costa

CONTENTS

39.1 Introduction 347
    39.1.1 Classification 347
    39.1.2 Morphology, Genome Structure, and Epidemiology 347
    39.1.3 Clinical Features and Pathogenesis 348
    39.1.4 Diagnosis 349
        39.1.4.1 Conventional Techniques: Indirect/Direct Diagnosis 349
        39.1.4.2 Molecular Techniques 351
39.2 Methods 351
    39.2.1 Sample Preparation 351
    39.2.2 Detection Procedures 351
        39.2.2.1 Detection of Avian Bornavirus in Psittacine Birds by RT-PCR 351
        39.2.2.2 Detection of Avian Bornavirus in Canary Birds 352
        39.2.2.3 Detection of Avian Bornavirus in Wild Canada Geese and Swans by Real-Time RT-PCR 353
39.3 Future Perspectives 353
References 353

39.1 INTRODUCTION

39.1.1 CLASSIFICATION

Proventricular dilatation disease (PDD) is a common progressive and often fatal neurologic disease of captive psittacine birds worldwide, affecting more than 70 species, and has been reported in several captive and free-ranging birds representing at least five additional orders, including toucans, honey creepers, weaver finches, water fowl, raptors, and passerines. PDD has long been considered an infectious disease, with multiple viruses having been involved as possible etiological agents. In 2008, two independent research groups in Israel and the United States, using pyrosequencing of cDNA from the brain tissue of parrots with PDD, identified two strains of a novel, nonsegmented, negative-sense, single-stranded RNA viruses, which were classified as distinct genotypes of a new genus of the *Bornaviridae* family denominated as avian bornavirus (ABV).[1,2] ABV RNA was detected by real-time polymerase chain reaction (PCR) in the brain, proventriculus, and adrenal gland of three parrots that died from PDD but not in four unaffected birds.[2] Kistler and colleagues identified by microarray a bornavirus hybridization signature in five of eight PDD cases and none of the controls. High-throughput pyrosequencing in combination with conventional PCR cloning and sequencing allowed to define a complete viral genome sequence. Gray et al.[3] succeeded in culturing ABV by inoculating primary mallard embryo fibroblasts with a fresh brain suspension from the brain of psittacines diagnosed with PDD. Based on

ABV isolates, seven distinct genotypes were identified, each sharing only approximately 65% nucleotide sequence identity with previously known members of the *Bornaviridae* family, all of which originated from mammalian hosts, and approximately 85%–95% with other ABV genotypes. The presence of different genotypes can be explained by the wide diversity of species affected by PDD.

In short, ABVs are classified as distinct serotypes/genotypes within the species Borna disease virus (BDV), genus *Bornavirus*, family *Bornaviridae*, order *Mononegavirales* on the basis of several unique features of the genome and mechanism of replication (in particular, nuclear replication rather than cytoplasmic replication). As the only species in the *Bornaviridae* family, BDV is a neurotropic virus responsible for an encephalitic disease in horses, sheep, and occasionally other domesticated mammalians as well as birds. There is also correlative evidence suggesting a possible link between BDV infection and neuropsychiatric disorders (e.g., bipolar disorder) in humans.

39.1.2 MORPHOLOGY, GENOME STRUCTURE, AND EPIDEMIOLOGY

ABVs are spherical and enveloped particles with a diameter ranging from 40 to 190 nm. While particles of 90–150 nm with electron-dense core presumably represent infectious virions, smaller particles are proposed to be defective. Morphologically, ABV particles are characterized by spikes with lengths of about 7 nm, observable by electronic microscope, that represent viral glycoproteins.[4]

The genome of ABVs is a linear, negative-stranded, nonsegmented RNA comprising approximately 8900 nt that encodes 6 viral proteins. In particular, the first monocistronic mRNA (1.2 kb) encodes for the viral structural protein corresponding to the nucleoprotein N (p40). The second transcription unit is bicistronic and encodes the X protein (p10) and the phosphoprotein P (p23). The third tricistronic mRNA (2.8 or 7.1 kb) encodes matrix protein M (p16), glycoprotein G (p57), and L-polymerase (p190). Expression of these viral proteins is regulated by alternative splicing. In contrast to the other members of the *Mononegavirales* order, ABV replicates and transcribes in the cell nucleus.[5] The glycoprotein G plays an essential role in initiation of infection. In particular, this structural protein has been implicated in binding to one or more still unidentified cellular surface receptors.[6] Subsequently, ABV requires a low pH to initiate fusion and is therefore internalized by endocytic compartments. Clathrin-mediated endocytosis represents the model for virus entry; after binding to the cell surface and endocytosis, the low pH of the endosome activates fusion of the viral membrane with that of the endosome by leading the release of the ribonucleoprotein (RNP) complex (genomic RNA packaged by the nucleoprotein N and assembled with the phosphoprotein P) into the cytosol.[7] The release of the RNP complex is also mediated by the matrix protein M; in particular, this protein is associated with infectious particles and it functions in early events of viral infection by causing the release of RNP complex into the cytoplasm.[8] Subsequently, viral RNP complex is imported into the nucleus where viral RNA synthesis occurs. The X protein mediates nuclear shuttling of viral gene products such as RNP particles, and it is also involved in the regulation of the viral polymerase.[9] Nuclear replication and transcription of the ABV genome are catalyzed by the same viral polymerase complex; the genomic RNA packaged by the nucleoprotein N constitutes the ribonucleocapsid that serves as template for the associated polymerase complex components L and P. The six major open reading frames (ORFs) (N, X, P, M, G, and L) are expressed from three 5′-capped and 3′-polyadenylated RNAs. Replication of negative-strand RNA genome is facilitated by the synthesis of a full-length positive-strand copy of the viral genome (antigenome) that serves as a template for new negative-strand progeny genomes.

The recent discovery of sequences distantly related to L, M, and N genes in the ABV genome belonging to the same phylogenetic group indicates the presence of several genotypes globally distributed that infect a wide range of vertebrate species. Initially, particularly in 2008, ABVs were identified in parrots and related species (order *Psittaciformes*) suffering from PDD.[2] To date, 7 ABV genotypes (ABV-1 to ABV-7), located to 2 different phylogenetic groups, have been detected in about 30 psittacine species from four different continents. ABV-2 and ABV-4 represent the predominant genotypes in Europe, whereas ABV-6 sequences were detected in birds from Austria and Switzerland.[1] Additional ABV genotypes were discovered in 2 other host orders, passerines (*Passeriformes*) (genotypes ABV-C1 to ABV-C3 and

genotype ABV-EF) and waterfowl (*Anseriformes*) (genotype ABV-CG). In particular, genotypes ABV-C1, ABV-C2, ABV-C3 were described in common canaries, whereas genotype ABV-EF was isolated from estrildid finches.[10–12] ABV-CG was isolated from geese and swans and is widely distributed across North America. Recently, a new ABV genotype has been characterized and isolated from the brains of wild ducks.[13] This new genotype, provisionally designated as ABV genotype MALL, was detected in 12 of 83 mallards, and 1 of 8 wood ducks collected at a single location in central Oklahoma. It has 72% nucleotide identity and 83% amino acid identity with the ABV-CG genotype previously shown to be present in geese and swans. ABV strains of the same genotype usually exhibit at least 95% nucleotide identity with each other. For genotypes belonging to the same phylogenetic group, sequence identity is approximately 80%–85%, whereas viruses from different phylogenetic groups are only 65%–75% identical. So far, only infections with a single genotype of ABV have been attributed to PDD cases in psittacine birds. However, a recent study described cases of double infection with two different ABV genotypes indicating the need for a more systematic search for mixed infections. Two cases showed a mixed infection with ABV-2 and ABV-4 genotypes, and one case showed a mixed infection with ABV-2 and ABV-6.[14] Issues that remain to be resolved are whether psittacine birds constitute, also, the natural reservoir of ABVs or whether other species, in which disease probably does not develop upon infection, will be identified in the future. Many important questions regarding the biology and epidemiology of ABV remain to be answered.

39.1.3 Clinical Features and Pathogenesis

PDD (synonyms: proventricular dilatation syndrome, macaw wasting/fading syndrome, neuropathic gastric dilatation of psittaciformes, psittacine encephalomyelitis, myenteric ganglioneuritis, and infiltrative splanchnic neuropathy) is a progressive fatal inflammatory disease that affects mainly, but not exclusively, psittacine birds (order: *Psittaciformes*). PDD was first recognized in the 1970s in imported macaws (*Ara* sp.) in Europe and North America and subsequently has been reported in Australia, Middle East, South America, and South Africa.[15] PDD has been termed after the predominant feature of the disease in parrots, consisting in a dilatation of the proventriculus by ingested food resulting from defects in intestinal motility. It seems that intestinal abnormalities derive from virus-induced damage to the enteric nervous system, in that birds are unable to empty their digestive tract or digest the food, with consequent weight loss, crop stasis, proventricular and intestinal dilatation, vomiting/regurgitation, maldigestion with passing of undigested seeds in their feces, starvation, and eventually death. As relevant damage in the central nervous system is also present, neurological signs are typical, with encephalitis and myelitis resulting in depression, seizures, ataxia, blindness, and tremors. Both gastrointestinal and neurological signs can be observed in affected birds. At physical examination, birds suffering from PDD can be dehydrated and mildly to severely emaciated, with atrophied pectoral muscles. The proventriculus may or may not be dilated in all affected birds, but in nearly 70% of the cases it appears distended with seeds and thin walled. Sometimes, the proventriculus may rupture with spillage of food into the coelomic cavity, with consequent peritonitis. Also duodenum may appear distended, as well as the adrenal glands. Occasionally, signs are almost unapparent and birds die suddenly. Definitive diagnosis is based on histopathological examination on crop biopsy or necropsy findings, with pathognomonic lesions being the occurrence of a lymphoplasmacytic infiltration in the ganglia and myenteric plexus of the gastrointestinal tract or central nervous system. Microscopic lesions can be found in several organs including the gastrointestinal tract; central, peripheral, and autonomic nervous system; heart; adrenal glands; and occasionally in the nerves and ganglia of various visceral organs. Infiltrate consists of a few to large number of lymphocytes mixed with some plasma cells occurring in the proventriculus and/or gizzard, duodenum, and other parts of the intestine. In the proventriculus, it can be observed an attenuation of glands and fibrosis of the mucosa. In other parts of the gastrointestinal tract and other organs, lesions tend to be less consistent.

The incubation period of PDD is extremely variable, from a minimum of 11 days reported in one study to approximately 1 month or more, up to years in some cases.[16] Clinical presentation is not peculiar and none of the signs should be considered pathognomonic; moreover, the severity of both gastrointestinal and neurologic symptoms may vary. Other differential diagnoses should always be considered. Among laboratory abnormalities, aplastic anemia is the most commonly seen and is likely related to malabsorption. Some birds present leukocytosis. Moreover, catabolic condition may account for reduction in total protein and albumin levels, increase in muscular enzymes such as lactate dehydrogenase, creatine kinase, and aspartate aminotransferase. No other significant abnormalities are present, as well as no fecal or cytologic findings are specific for PDD. However, these investigations are usually performed for the differential diagnosis process. Diagnostic imaging is useful for making the diagnosis, although in the absence of specific signs, usually a moderately to markedly distended proventriculus is found with ingesta and gas. The gold standard for diagnosing PDD is the histologic examination on a bioptic specimen from a relevant anatomic site of the proventriculus or the crop. This is a simpler and less invasive approach; the sensitivity of crop biopsy for PDD may vary and can be increased by proper selection of the biopsy site and utilization of multiple biopsy sections. The recent discovery of ABV has had a great impact on diagnostic process of PDD, with the development of specific molecular and serologic assays for its detection.

39.1.4 Diagnosis

Although PDD clinical symptoms are not pathognomonic and vary greatly between patients, a marked weight loss, vomiting/regurgitation, and the presence of undigested food in feces should be considered as an alert for the gastrointestinal PDD form, mostly if they are concomitant. The central nervous system form has a wider range of nonspecific clinical symptoms

from slight to deep neurologic disorders such as seizures and ataxia, with the most severe cases reaching status epilepticus.[15] In the clinical diagnosis, a differential diagnosis has always to be considered for the nonspecific nature of symptoms; and it should be noted that mixed gastrointestinal and central nervous system PDD forms may commonly affect birds.[17]

Fecal and crop cytology are necessary in differential diagnosis for ruling in or out avian yeasts and helminth, causing gastrointestinal signs similar to PDD. Diagnostic imaging techniques such as survey radiography, contrast radiography, contrast fluoroscopy, and ultrasonography are helpful diagnostic tools, but not appropriate for PDD confirmation or exclusion. It has to be noticed that while the presence of classical clinical signs represents a strong indication for PDD, in the absence of clinical signs, freedom from disease and/or infection could not be assumed. Although this variety of clinical techniques has been used for tentative antemortem diagnosis, histopathologic examination of tissues remains the gold standard for a definitive PDD diagnosis. Therefore, considering the uneven distribution of lesions in biopsy material, false-negative results may occur.[18]

The significant genetic variation within the *Bornaviridae* family and, particularly, the higher genetic variability of strains recognized within *Avian Bornavirus* genus, represent a challenge for veterinary laboratory diagnosticians. Since 1980, different diagnostic assays have been developed for both indirect and direct diagnosis of ABV by employing different type of samples from dead or live birds belonging to different avian species. It has to be taken into account that ABV RNA detection is not indicative of PDD. Moreover, seropositive birds may shed the virus in their urofeces and past exposure to one ABV strain does not provide protection against a different one. Thereby, diagnostic tools should be used carefully, due to the fragmentary knowledge on ABV and the inherent limitations of the techniques.

39.1.4.1 Conventional Techniques: Indirect/Direct Diagnosis

As concerns indirect diagnosis, the detection of specific anti-ABV antibodies is performed by serological tests. Over the last two decades, indirect fluorescent antibody tests (IFAT), enzyme-linked immunosorbent assays (ELISA), and Western blot have been the most used tests.[19,20] Direct diagnosis is mainly based on identification and characterization of viral antigens by using IFAT, immunohistochemistry (IHC), in situ hybridization, and virus isolation (VI) on dead birds and crop biopsies. As a consequence of bornavirus infection, animals develop a detectable antibody response that seems to correlate positively with the onset of PDD.[21] Serology provides a much superior diagnostic procedure compared to crop biopsy although apparently healthy normal birds may possess detectable antibodies to ABV.[22] However, Heffels-Redmann and colleagues demonstrated that virus shedding and antibody production coincided in only one-fifth of the positive psittacine birds.[23] Therefore, they recommended to use swabs from crop and cloaca along with serum for a reliable diagnosis of

ABV infections in live birds. Besides, the same authors in a later study outlined that birds with high ABV RNA load in crop and cloaca together with high anti-ABV antibodies demonstrate a high risk of PDD onset, indicating that humoral antibodies do not protect against the disease. To date, as outlined by Gancz and colleagues, the definitive diagnosis of PDD continues to be based on histology by using serology, PCR, and IHC as confirmatory tests.[15,24]

39.1.4.1.1   Serology

As indication of a previous exposure to the virus, serological tests are used to detect the presence of specific antibodies to ABV in the bloodstream. The two major ABV immunogenic proteins are the nucleoprotein N and the phosphoprotein P.[15] Detection of bornavirus-specific antibodies is routinely performed by serological assays, such as IFAT, Western blot, or ELISA assays, even though serologic tests appear to be of limited usefulness in disease diagnosis. IFAT and Western blot assays are the most sensitive and specific tests, with 90%–95% sensitivity and specificity.[21,25]

Considering that ABV strains do not replicate equally well in all cell types, numerous ABV/BDV persistently infected cells lines of avian and mammalian origin have been tested as target cells for ABV antibody detection by IFAT.[3,26–28] Recently, the effect of antigenic variability on the sensitivity of antibody detection was assessed by IFAT on sera from naturally or experimentally infected animals by using various cell lines infected with different ABV/BDV genotypes. The principal outcome of the study is that antibody titers are highest when the homologous target virus is used. Furthermore, a good cross-reactivity between closely related genotypes belonging to a common phylogenetic group exists, but detection sensitivity is reduced when target viruses are not closely related. It is therefore suggested that antibodies detected with any target bornavirus should be considered as bornavirus-reactive rather than genotype-specific.[20]

A homemade ELISA assay using ABV recombinant P40, P24, and P16 proteins produced in *Escherichia coli* demonstrated a considerable variation in the nature and titer of antibodies in ABV naturally infected psittacine birds. Hence, variability in the presence of anti-ABV P24 phosphoprotein and P16 matrix protein antibodies makes these antibodies less useful for the diagnosis of ABV infection. Conversely, the constant identification of anti-P40 protein antibodies allows to identify past/present infection in many birds.[29] Recently, an indirect ELISA for the detection of anti-ABV antibodies in captive canary birds has been set up.[28] Plates were coated with recombinant ABV N protein and serum antibodies were detected by rabbit anti-canary IgG serum. As evidenced in the study, seroconversion did not correlate with virus tissue distribution or viral shedding, in accordance with previous observations from naturally infected psittacines.[24]

The assessment of the specificity and sensitivity of serologic tests is difficult due to different assays, poor understanding of disease epidemiology, absence of gold standard to identify infection in live birds, difficulty in identifying a true control group of uninfected birds, variability among species, and relatively poor correlation between viral shedding, pathology, and presence of antibodies.

39.1.4.1.2   Virus Isolation

Viral propagation on cell lines remains a very sensitive tool for the detection of new ABV genotypes, even though time consuming. Therefore VI, rarely used for routine diagnosis, finds application in research as a source of virus for genome sequencing, in vivo experimental inoculation, development of new molecular techniques, and serologic assays.[12,27]

Quail cell line CEC-32 and primary duck embryo fibroblasts have been the most commonly cell lines used for ABV isolation.[27,28] In particular, CEC and QM7 quail cells have shown best results for ABV isolation from canary birds, while, as concerns psittacine, significant results have been obtained by using exclusively quail cell line CEC-32.

Bornaviruses establish persistent infections in cell culture, without causing an evident cytopathic effect. Hence, the virus identification is usually performed by immunofluorescence staining for viral antigen. Bornavirus-specific antigens could be detected by using IFATs with polyclonal anti-ABV-2 or anti-BDV-1 rabbit sera on organ samples of various psittacine birds.[10,12,26–28,30]

In particular, in a recent study a panel of different polyclonal rabbit sera and monoclonal mouse antibodies for antigen was systematically tested for the detection of 4 bornavirus genotypes (ABV-2, ABV-C2, ABV-EF, and BDV-1) from different phylogenetic groups.[20] Polyclonal rabbit sera directed against N, X, P, and M proteins showed good cross-reactivity with all 4 genotypes tested. Although these recent data confirm previous evidences of a high degree of cross-reactivity between different ABV and BDV genotypes, more distantly related virus strains that are able to escape detection by such antibodies may exist.[10,11,26,30] Nevertheless, VI might complement ABV diagnosis when the presence of a new or uncommon virus genotype is suspected.[27]

Swabs or blood samples are commonly collected from live birds. In particular, swabs, placed in transport medium containing antibiotics, are considered a good sample to investigate the shedding of infectious virus.[27] Any prolonged storage of infected materials should be at −70°C or below to minimize loss of virus titer, avoiding repeated freezing and thawing.

39.1.4.1.3   Immunohistochemistry

Histopathological diagnosis is still considered the gold standard test for PDD diagnosis; nevertheless, histopathology might also show false-negative results, due to the not constant distribution of PDD lesions.[18] Anyway, a recent study demonstrated an excellent statistical association between histopathologic diagnosis of the disease and results obtained by ABV-specific reverse transcription polymerase chain reaction (RT-PCR) and IHC.[18] IHC has been applied directly on paraffin-embedded organ samples to the identified cell and organ distribution of viral antigens in psittacine birds, using polyclonal antisera raised against purified recombinant ABV nucleoprotein N and ABV phosphoprotein P.[18,30,31] In live birds, IHC on crop biopsy is a simple and less invasive diagnostic approach, although its sensitivity is controversial, with a prevalence of ganglioneuritis in crops of patients with PDD ranging from 22% to 76%.[15,16,32] It has to be noted that proper selection of the site and the setting up of multiple sections have been suggested to increase the sensitivity of the method.

The brain, spinal cord, adrenal gland, pancreas, and kidney are considered the best samples for IHC, in the presence of postmortem diagnosis, due to the clear ABV antigen localization and to the minimal unspecific background staining. Unspecific background staining might be also influenced by samples autolysis, making the use of fresh samples essential for IHC test.[23] ABV was also detected in multiple other organs, including the anterior gastrointestinal tract (crop, proventriculus, ventriculus, and duodenum) as well as the heart, testes, ovary, and thyroid. The broad tissue distribution of viral antigens has been confirmed also in canary birds.[10,15,28]

39.1.4.2 Molecular Techniques

Specific molecular techniques for ABV RNA identification represent an efficient tool to verify the infection of a bird in suspected cases of PDD.[33] Nonetheless, the selection of samples for PCR testing is crucial to obtain a reliable result. However, due to the limited knowledge gained on this virus, molecular tools should be used with caution. At the moment, RT-PCR is routinely used for the detection of RNA ABV. RT-PCR on feather calami, a relatively noninvasive sample collection, may be a reliable method for the diagnosis of a current ABV infection in live birds. Although single feathers may be sufficient for the detection of viral RNA, processing feathers in pool is advisable for a meaningful result.[29]

The presence of PCR inhibitors should be taken into account when performing PCR on feather and fecal materials. In these cases, the use of inhibitors removal procedures, such as bovine serum albumin added to PCR mix or simply sample dilution, would help to avoid false negative.[34] As noticed by some authors,[27] RT-PCR assays may fail to detect distantly related or yet unknown genotypes, so classical viral isolation and serology could represent valid alternatives. Moreover, given that birds shed the virus intermittently in their droppings, diagnosis requires multiple PCR tests to confirm results.[35] However, RT-PCR for the identification of ABV infections represents, actually, the better assay to use for screening of psittacine birds at risk for PDD, especially in early infections.[18] Moreover, gene sequencing and comparison with the available ABV sequences present in public databases allow the characterization of genotype and the phylogenetic reconstruction.

39.2 METHODS

39.2.1 Sample Preparation

Crop biopsy and cloacal, choanal, pharyngeal, and crop swabs, as well as urofeces, white blood cells, and feather calami are recommended samples for ABV PCR testing in case of intravitam diagnosis.[23] Fresh urofecal droppings or cloacal swab collection are simple methods to evaluate ABV infection, because birds excrete the greatest amount of virus

in the urine.[36] Nonetheless, shedding appears to be intermittent in healthy infected subjects, consequently the test of a single dropping from a single bird is of limited usefulness. A pool of multiple droppings from single or different birds over a week of collection period may improve sensitivity.[35] Feather sampling is relatively noninvasive approach and gives the advantage that ABV RNA remains stable for at least 4 weeks when the feather is stored at room temperature.[29] Based on results from Delnatte and colleagues, bone marrow is to be considered as an accessible antemortem sample for the diagnosis of ABV infection.[37] In the same study, a tissue ranking obtained by threshold cycle (Ct) values of the real-time PCR has been applied to determine the viral load of each organ suggesting proventriculus, kidney, colon, cerebrum, and cerebellum as the best samples for the diagnosis by RT-PCR from dead birds. On the other hand, ABV RNA was reported in the nervous system, in the gastrointestinal tract, and in nearly all other tissues in PDD-affected dead psittacine birds.[1–3,18,21,26,31,38] For a meaningful diagnostic result, collection of samples should be made from freshly dead or euthanized birds and organs transported to the laboratory at 4°C within 48 h. In case of long-range delivery, tissue samples should be transported in an RNA stabilization reagent or at least frozen.[23]

39.2.2 Detection Procedures

Due to the marked genetic variability of ABV, several primer sets have been designed able to identify nucleic acid of different avian orders. Degenerated primer sets targeting ABV/BDV consensus sequences of the N (Ncon), M (Mcon), or L (Lcon) gene have been widely and successfully applied for the detection of 7 ABV genotypes isolated in psittacine birds.[1,26,27,31] RT-PCR assays targeting the highly expressed N gene and the conserved M gene regions demonstrate a comparable sensitivity, higher than L gene assays. Specific primers set, targeting consensus sequences of the N (Ccon) and P (Fcon) genes of ABV genotypes isolated from canary birds, have been used in combination with the Ncon and Mcon pairs to identify ABV strains in canary birds belonging to the *Passeriformes* order.[28] This study highlighted remarkable variations in sensitivity between different primer sets used in the detection of ABV strains from canary bird.

Two real-time RT-PCR protocols designed on the P and M genes have been applied successfully on different psittacine and anserine bird tissues.[2,37] In particular, real-time RT-PCR targeting M gene appeared more sensitive than conventional M gene RT-PCR and IHC in tissues with low viral load, reducing also the number of false-negative results.[37,38]

39.2.2.1 Detection of Avian Bornavirus in Psittacine Birds by RT-PCR

Weissenbock and colleagues described two RT-PCR assays for conserved genes of ABV and nucleotide sequence analysis used to examine paraffin wax-embedded or frozen tissue samples of 31 psittacine birds with PDD.[30]

Name	Sequence (5′–3′)	Nucleotide Position	Product Size (bp)
N632-F	CATGAGGCTATWGATTGGATTA	632–653	389
N1020-R	TAGCCNGCCMKTGTWGGRTTYT	1020–999	
M1908-F	CAAGGTAATYGTYCCTGGATGG	1908–1929	352
M2259-R	ACCAATGTTCCGAAGMCGAWAY	2259–2238	

39.2.2.1.1 Procedure

1. Design bornavirus-specific universal oligonucleotide primer pairs targeting putative N protein and M protein gene fragments. Align BDV-specific nucleic acid sequences deposited in GenBank, including representative sequences of the 5 ABV genotypes described by Kistler and colleagues, and search for conserved genomic regions.[1]

2. Extract viral RNA from 10 μm sections of paraffin wax-embedded psittacine tissue samples. Process samples in pools of 3–5 sections of each block. Remove paraffin from the tissue section by incubation with 1mL xylene for 20 min at 37°C, followed by pelleting tissues by centrifugation at 16,000 × g for 5 min at room temperature. Remove xylene and resuspend the pellets in 1 mL RNase-free ethanol for 5 min at room temperature. Samples were centrifuged again at 16,000 × g for 5 min at room temperature, and repeat the ethanol treatment. After centrifugation, remove the ethanol, and let the pellets air-dry. After 10 min, resuspend the tissue samples in 250 μL ATL tissue lysis buffer (QIAGEN, Hilden, Germany) and 25 μL Proteinase K (QIAGEN) was added. Samples were digested with proteinase for 16 h at 55°C, followed by an enzyme inactivation step for 8 min at 95°C. Use 140 μL of tissue lysates to extract viral RNA using the QIAamp Viral RNA Mini Kit (QIAGEN) according to the manufacturer's recommendations.

3. Carry on one-step RT and amplification using a One-Step RT-PCR Kit (QIAGEN) according to the manufacturer's instructions and with a final primers concentration of 0.8 μmol/L. Perform amplification reaction in a GeneAmp PCR System 2700 Thermocycler (Applied Biosystems, Foster City, CA) with the following thermal profile: 30 min at 50°C, 15 min at 95°C, 45 cycles × (30 s at 94°C, 30 s at 50°C, and 30 s at 72°C), and 7 min at 72°C. Use RNA extracts from psittacine organs without indication of PDD as negative controls.

4. Run PCR products by electrophoresis in 1.5% Tris-acetate–EDTA agarose gel and stained with ethidium bromide.

5. For sequencing, purify PCR products with the Quantum Prep PCR Kleen Spin Columns (Bio-Rad, Hercules, CA) according to the manufacturer's protocol. Perform a fluorescence-based direct sequencing of amplicons in both directions by using the ABI PRISM Big Dye Terminator Cycle Sequencing Ready Reaction Kit (Applied Biosystems).

6. Identify nucleotide sequences by the Basic Local Alignment Search Tool performing the alignment with the Align Plus program, version 4.1 (Scientific and educational Software, Cary, NC). Create multiple alignments for phylogenetic analyses using the Clustal X program and conduct phylogenetic analyses by the neighbor-joining algorithm. Perform bootstrap resampling analyses of phylogenetic trees on 1000 replicates. Draw phylogenetic trees with the help of the TreeView 1.6.6 software.

Note: Due to the nature of sample material (paraffin wax–embedded tissues), primers for the amplification of relatively short PCR products were designed to reduce the chance of false-negative reactions due to the RNA fragmentation effect of the formaldehyde fixation.

39.2.2.2 Detection of Avian Bornavirus in Canary Birds

Rubbenstroth and colleagues applied four different RT-PCR protocols to investigate the presence of ABV in captive canary birds and to analyze the genetic variability.[28] In the initial detection of ABV RNA from canary swabs and organs, the widely used RT-PCR assays Ncon and Mcon were applied.[1] Following the detection of two ABV strains, two new PCR assays specific for passerine birds were designed: the Ccon primer pair targeting the N gene consensus sequence of two canary isolates and the Fcon targeting the P gene consensus sequence of same isolates plus an ABV sequence derived from a Bengalese finch.

Name	Sequence (5′–3′)	Nucleotide Position	Product Size (bp)
Ncon F	CCHCATGAGGCTATWGATTGGA TTAACG	629–656	426
Ncon R	GCMCGGTAGCCNGCCATTGTDGG	1026–1004	
Mcon F	GGRCAAGGTAATYGTYCCTGGA TGGCC	1905–1931	359
Mcon R	CCAACACCAATGTTCCGAAGMCG	2264–2242	
Ccon_F	GGTGTGGTGATTGGKTCTTC	490–509	219
Ccon_R	SGGYGAYTCAAAGTCTGTAG	708–689	
Fcon_F	CCGCAGACAGYACGTCGC	1323–1340	167
Fcon_R	TCATCATTCGAYARCTGCTCCCTTCC	1489–1465	

39.2.2.2.1 Procedure

1. Extract RNA from swab samples using the QIAamp Viral RNA Mini Kit (Qiagen, Hilden, Germany) and organ samples using phenol–chloroform extraction with Trifast (Peqlab, Erlangen, Germany).

2. Perform RT of approximately 2 μg RNA per sample using the Revertaid Reverse Transcription Kit (Fermentas, St. Leon-Rot, Germany).

3. Carry out the initial detection of ABV RNA from canary samples by two RT-PCRs with Ncon and Mcon degenerated primer sets. Prepare PCR mix with MgCl$_2$ 3 mM, dNTPs 0.2 mM each, primers 0.2 μM each, and 0.625 units of Taq Polymerase (Fermentas) in a 25 μL reaction. Set up the cycling condition for both assays with a denaturation step (2 min at 94°C), followed by 40 cycles of denaturation (94°C, 20 s), primer annealing (30 s, 48.5°C), and elongation (72°C, 25 s) with a final elongation step (72°C, 5 min).

4. Design Ccon and Fcon primers, specific for the N and P genes of canary birds, using complete genome sequence of canary isolates obtained in the study and a putative ABV sequence originating from a Bengalese finch (GenBank DC290659).

5. Apply four PCR assays to screen samples derived from canary birds for the presence of ABV. Perform the Ncon and Mcon assays as described in procedure 4 and for the Ccon and Fcon methods prepare the PCR mix with MgCl$_2$ 2 mM, dNTPs 0.2 mM each, primers 0.2 μM each, and 0.625 units of Taq Polymerase (Fermentas) in a 25 μL reaction. Set up cycling conditions respecting time and temperatures indicated for each of the following primer set: denaturation step (2 min at 94°C), followed by 40 cycles of denaturation (94°C, 20 s), primer annealing (30 s, 48.5°C for Fcon, 50.5°C for Ccon), elongation (72°C, 15 s for Ccon and Fcon), and a final elongation step (72°C, 5 min).

6. Perform the Sanger sequencing of PCR products of the Ncon PCR and a phylogenetic analysis together with ABV and BDV sequences derived from GenBank, including two ABV sequences from canary birds.

Note: Ccon and Ncon appeared to be the most sensitive assays for the detection of ABV strains from canary birds and related finches.

39.2.2.3 Detection of Avian Bornavirus in Wild Canada Geese and Swans by Real-Time RT-PCR

Delnatte and colleagues described a duplex real-time RT-PCR assay targeting both psittacine ABV (ABV_M set) and geese ABV (ABVG_M set) matrix gene sequences to identify ABV in frozen and fixed paraffin-embedded tissues from free-ranging waterfowls.[37]

Name	Sequence (5′–3′)
ABV_M_F_120201	CAAGGTAATYGTYCCTGGATGGCC
ABV_M_R_120201	TCACTGAAAGAAANGGTATRTTGAT
ABV_M_Pr_120201	TAATGTTGGARATAGACTTTGTTGG
ABVG_M_F_111029	CGAGGGAGAAGAGACTGGTTGATT
ABVG_R_111029	ACTGCCAAAGAGTTGAGCGT
ABVG_M_Pr_111029	ATGTGGAACCTGCTGGTCACTCA

39.2.2.3.1 Procedure

1. Perform RNA extraction on FFPE sections of tissues using the RNeasy FFPE Kit (Qiagen Inc., Mississauga, Ontario, Canada) according to the manufacturer's instructions.

2. Set up the duplex amplification mixture in a 25 μL final volume using the AgPath-ID One-Step RT-PCR Kit (Applied Biosystems, Inc.), according to the manufacturer's recommendations.

3. Carry out amplification reactions in a LightCycler 480 Real-Time PCR System (Roche, Laval, Quebec, Canada) with the following thermal profile: 10 min at 45°C, 10 min at 95°C, 45 cycles × (5 s at 94°C, 60 s at 60°C).

Note: In the interpretation of the results, a Ct value <37.00 was considered positive, Ct from 37.00 to 39.99 inconclusive, and Ct ≥ 40.00 negative.

39.3 FUTURE PERSPECTIVES

A huge collection of data has been accumulated through research all over the world from the first description of ABV in 2008 in psittacine birds. Despite all this considerable work, several questions about avian bornaviruses are still looking for an answer. In the next future, research efforts should be dedicated at elucidating the factors that induce the development of disease in infected subjects, since many aspects in the pathogenesis of PDD remain unclear.[37]

Moreover, the great genetic heterogeneity as well as the supposed existence of novel genotypes in other avian and nonavian species poses a major challenge for the near future in the diagnosis of ABV infections.[27] Robust and reliable tools for the molecular characterization of ABV genotypes are necessary and would be of great value to understand the antigenic and evolutionary relationships within the virus family.

REFERENCES

1. Kistler, A.L. et al. Recovery of divergent avian bornaviruses from cases of proventricular dilatation disease: Identification of a candidate etiologic agent. *Virol J* 5, 88 (2008).
2. Honkavuori, K.S. et al. Novel borna virus in psittacine birds with proventricular dilatation disease. *Emerg Infect Dis* 14, 1883–1886 (2008).
3. Gray, P. et al. Use of avian bornavirus isolates to induce proventricular dilatation disease in conures. *Emerg Infect Dis* 16, 473–479 (2010).
4. Kohno, T. et al. Fine structure and morphogenesis of Borna disease virus. *J Virol* 73, 760–766 (1999).
5. Briese, T., Hatalski, C.G., Kliche, S., Park, Y.S., and Lipkin, W.I. Enzyme-linked immunosorbent assay for detecting antibodies to Borna disease virus-specific proteins. *J Clin Microbiol* 33, 348–351 (1995).
6. Schneider, P.A., Hatalski, C.G., Lewis, A.J., and Lipkin, W.I. Biochemical and functional analysis of the Borna disease virus G protein. *J Virol* 71, 331–336 (1997).

7. Clemente, R. and de la Torre, J.C. Cell entry of Borna disease virus follows a clathrin-mediated endocytosis pathway that requires Rab5 and microtubules. *J Virol* 83, 10406–10416 (2009).

8. Gonzalez-Dunia, D., Cubitt, B., and de la Torre, J.C. Mechanism of Borna disease virus entry into cells. *J Virol* 72, 783–788 (1998).

9. Wolff, T., Unterstab, G., Heins, G., Richt, J.A., and Kann, M. Characterization of an unusual importin alpha binding motif in the borna disease virus p10 protein that directs nuclear import. *J Biol Chem* 277, 12151–12157 (2002).

10. Weissenbock, H., Sekulin, K., Bakonyi, T., Hogler, S., and Nowotny, N. Novel avian bornavirus in a nonpsittacine species (Canary; Serinus canaria) with enteric ganglioneuritis and encephalitis. *J Virol* 83, 11367–11371 (2009).

11. Payne, S. et al. Detection and characterization of a distinct bornavirus lineage from healthy Canada geese (*Branta canadensis*). *J Virol* 85, 12053–12056 (2011).

12. Rubbenstroth, D. et al. Discovery of a new avian bornavirus genotype in estrildid finches (Estrildidae) in Germany. *Vet Microbiol* 168, 318–323 (2014).

13. Guo, J. et al. Characterization of a new genotype of avian bornavirus from wild ducks. *Virol J* 11, 197 (2014).

14. Nedorost, N. et al. Identification of mixed infections with different genotypes of avian bornaviruses in psittacine birds with proventricular dilatation disease. *Avian Dis* 56, 414–417 (2012).

15. Gancz, A.Y., Clubb, S., and Shivaprasad, H.L. Advanced diagnostic approaches and current management of proventricular dilatation disease. *Vet Clin North Am Exot Anim Pract* 13, 471–494 (2010).

16. Gregory, C.R. et al. Proventricular dilatation disease: A viral epornitic. *Proceedings of the Annual Conference of the Association Avian Veterinarians*, Reno, NV, pp. 43–52 (1997).

17. Shivaprasad, H.L. et al. Spectrum of lesions (pathology) of proventricular dilatation syndrome. *Proceedings of the Annual Conference of the Association Avian Veterinarians*, Philadelphia, PA, pp. 505–506 (1995).

18. Raghav, R. et al. Avian bornavirus is present in many tissues of psittacine birds with histopathologic evidence of proventricular dilatation disease. *J Vet Diagn Invest* 22, 495–508 (2010).

19. Briese, T., de la Torre, J.C., Lewis, A., Ludwig, H., and Lipkin, W.I. Borna disease virus, a negative-strand RNA virus, transcribes in the nucleus of infected cells. *Proc Natl Acad Sci USA* 89, 11486–11489 (1992).

20. Zimmermann, V., Rinder, M., Kaspers, B., Staeheli, P., and Rubbenstroth, D. Impact of antigenic diversity on laboratory diagnosis of Avian bornavirus infections in birds. *J Vet Diagn Invest* 26, 769–777 (2014).

21. Villanueva, I. et al. The diagnosis of proventricular dilatation disease: Use of a Western blot assay to detect antibodies against avian Borna virus. *Vet Microbiol* 143, 196–201 (2010).

22. De Kloet, S.R. and Dorrestein, G.M. Presence of avian bornavirus RNA and anti-avian bornavirus antibodies in apparently healthy macaws. *Avian Dis* 53, 568–573 (2009).

23. Heffels-Redmann, U. et al. Occurrence of avian bornavirus infection in captive psittacines in various European countries and its association with proventricular dilatation disease. *Avian Pathol* 40, 419–426 (2011).

24. Heffels-Redmann, U. et al. Follow-up investigations on different courses of natural avian bornavirus infections in psittacines. *Avian Dis* 56, 153–159 (2012).

25. Lierz, M. et al. Experimental infection of cockatiels with different avian bornavirus genotypes. *Proceedings of the Annual Conference of the Association Avian Veterinarians*, Louisville, KY (2012).

26. Rinder, M. et al. Broad tissue and cell tropism of avian bornavirus in parrots with proventricular dilatation disease. *J Virol* 83, 5401–5407 (2009).

27. Rubbenstroth, D., Rinder, M., Kaspers, B., and Staeheli, P. Efficient isolation of avian bornaviruses (ABV) from naturally infected psittacine birds and identification of a new ABV genotype from a salmon-crested cockatoo (*Cacatua moluccensis*). *Vet Microbiol* 161, 36–42 (2012).

28. Rubbenstroth, D. et al. Avian bornaviruses are widely distributed in canary birds (*Serinus canaria* f. domestica). *Vet Microbiol* 165, 287–295 (2013).

29. de Kloet, A.H., Kerski, A., and de Kloet, S.R. Diagnosis of Avian bornavirus infection in psittaciformes by serum antibody detection and reverse transcription polymerase chain reaction assay using feather calami. *J Vet Diagn Invest* 23, 421–429 (2011).

30. Weissenbock, H. et al. Avian bornaviruses in psittacine birds from Europe and Australia with proventricular dilatation disease. *Emerg Infect Dis* 15, 1453–1459 (2009).

31. Kistler, A.L., Smith, J.M., Greninger, A.L., Derisi, J.L., and Ganem, D. Analysis of naturally occurring avian bornavirus infection and transmission during an outbreak of proventricular dilatation disease among captive psittacine birds. *J Virol* 84, 2176–2179 (2010).

32. Berhane, Y. et al. Peripheral neuritis in psittacine birds with proventricular dilatation disease. *Avian Pathol* 30, 563–570 (2001).

33. Hoppes, S., Gray, P.L., Payne, S., Shivaprasad, H.L., and Tizard, I. The isolation, pathogenesis, diagnosis, transmission, and control of avian bornavirus and proventricular dilatation disease. *Vet Clin North Am Exot Anim Pract* 13, 495–508 (2010).

34. Das, A., Spackman, E., Pantin-Jackwood, M.J., and Suarez, D.L. Removal of real-time reverse transcription polymerase chain reaction (RT-PCR) inhibitors associated with cloacal swab samples and tissues for improved diagnosis of Avian influenza virus by RT-PCR. *J Vet Diagn Invest* 21, 771–778 (2009).

35. Hoppes, S.M., Tizard, I., and Shivaprasad, H.L. Avian bornavirus and proventricular dilatation disease: Diagnostics, pathology, prevalence, and control. *Vet Clin North Am Exot Anim Pract* 16, 339–355 (2013).

36. Heatley, J.J. and Villalobos, A.R. Avian bornavirus in the urine of infected birds. *Vet Med Res Rep* 3, 19–23 (2012).

37. Delnatte, P. et al. Pathology and diagnosis of avian bornavirus infection in wild Canada geese (*Branta canadensis*), trumpeter swans (*Cygnus buccinator*) and mute swans (*Cygnus olor*) in Canada: A retrospective study. *Avian Pathol* 42, 114–128 (2013).

38. Lierz, M. et al. Anatomical distribution of avian bornavirus in parrots, its occurrence in clinically healthy birds and ABV-antibody detection. *Avian Pathol* 38, 491–496 (2009).

40 Bovine Ephemeral Fever Virus

Hakan Bulut and Ahmet Kursat Azkur

CONTENTS

40.1 Introduction 355
    40.1.1 Classification.. 355
    40.1.2 Morphology, Biology, and Epidemiology 355
    40.1.3 Clinical Features and Pathogenesis 356
    40.1.4 Diagnosis 357
        40.1.4.1 Conventional Techniques... 357
        40.1.4.2 Molecular Techniques.. 357
40.2 Methods 358
    40.2.1 Sample Preparation.. 358
    40.2.2 Detection Procedures... 358
        40.2.2.1 One-Step RT-PCR.. 358
        40.2.2.2 Sequencing and Phylogenetic Analysis... 358
40.3 Future Perspectives.. 358
References.. 359

40.1 INTRODUCTION

Bovine ephemeral fever (BEF; also known as 3-day sickness) is a vector-borne, viral disease of cattle and water buffaloes that leads to dramatic loss in milk production. The disease in cattle is characterized by acute febrile reaction, stiffness, lameness, and spontaneous recovery within 3 days [1]. The mortality rate was reported as 1%–3% in the past outbreaks, but about 20% was noted in some herd of the recent outbreaks [2]. The most important vector of the causal agent bovine ephemeral fever virus (BEFV) is *Culicoides* flies. Although the disease has been reported in the tropical and subtropical regions of Australia, Africa, and Asia, considering that *Culicoides* flies are transported to long distances especially by wind, BEFV may spread to wider geographies or continents in recent years [1–3]. Given the increased number of epidemics and high mortality rates observed in recent outbreaks of BEF, with devastating economic consequences, it is critical that rapid, sensitive, and specific diagnostic methods are available and applied for improved detection and monitoring of BEFV.

40.1.1 CLASSIFICATION

BEFV is classified as a member of the genus *Ephemerovirus* in the family *Rhabdoviridae*. Besides BEFV, the genus *Ephemerovirus* also includes Berrimah virus, Adelaide River virus, Kotonkan virus, Malakal virus, Kimberley virus (KIMV), and Obodhiang virus. As Berrimah, Kimberley, and Malakal viruses are morphologically similar to BEFV and react with BEFV serologically, they are collectively called

BEFV serogroup. Some of the more recent phylogenetic studies have shown that BEFV serogroup viruses may have evolved as a geographic variant of BEFV due to high percentage of the genome and amino acid sequence similarity. Additionally, BEFV serogroup viruses showed cross-reactions with rabies-related viruses of the *Lyssavirus* genus. For this reason, it was suggested by some researchers that BEFV serogroup viruses and rabies-related viruses may combine to form a large complex new group. As BEFV genome and amino acid organization demonstrate significant differences from other rhabdoviruses, it is accepted as the most realistic approach that BEFV should remain a distinct genus within the family *Rhabdoviridae* [4,5].

40.1.2 MORPHOLOGY, BIOLOGY, AND EPIDEMIOLOGY

BEFV virions are enveloped, bullet- or cone-shaped, with a negative-sense single-stranded RNA of approximately 14.9 kb in size. The BEFV genome contains 10 long open reading frames, arranged in the order $3'$-N-P-M-G-G_{NS}-$\alpha 1$-$\alpha 2$-β-γ-L-$5'$, encoding the five common rhabdovirus structural proteins—nucleoprotein (N), phosphoprotein (P), matrix (M), glycoprotein (G), and large (L)—and several additional proteins of unknown functions [5–7].

The N protein (52 kDa) is a phosphorylated protein associated with nucleocapsid structure. N protein plays a canonical role in the transcription and translation complex and is also involved in the virus assembly. BEFV is similar to KIMV, both having a group-reactive antigen region in the N protein [6].

G protein is an 81 kDa transmembrane glycoprotein that contains type-specific, neutralizing antigenic sites. It can be removed from the virion by treatment with nonionic detergents. G protein is considered as the main discriminative antigen of BEFV with four distinct antigenic sites (G_1, G_2, G_3, and G_4) on its surface. Although BEFV G_1 region is a linear antigenic structure that can react with only anti-BEFV antibody, G_2, G_3, and G_4 regions are able to react with BEFV and BEFV-related viruses. BEFV expresses a 90 kDa nonstructural glycoprotein (GNS), which also has the characteristics of a rhabdovirus glycoprotein identified in BEFV-infected cells, including putative signals and transmembrane domains. The role of GNS is still obscure. However, antibodies against this protein do not have any neutralization activity [7–10].

The other small BEFV accessory genes ($\alpha1$, $\alpha2$, β, and γ) encode proteins of unknown functions that have not previously been detected in infected cells or in virions. All other ephemeroviruses (Berrimah virus, KIMV, Malakal virus, Adelaide River virus, Obodhiang virus, and Kotonkan virus) encode $\alpha1$ protein in the long G–L intergenic region, directly downstream of the GNS gene. BEFV $\alpha1$ proteins play a role as a viroporin and promote release of viral particles from cells [7,11].

BEFV entry is through a clathrin-mediated and dynamin 2-dependent endocytic pathway. BEFV enters host cells after receptor-mediated endocytosis and depends on acidic cellular compartments for productive infection [5–7]. BEFV-induced apoptosis is reported in BHK-21, Vero, HeLa, NS-1, and MDCK cell lines. Indeed, replication of BEFV has higher efficiency and apoptosis induction is greater in Vero cell line [12–14]. Viral gene expression is required for apoptosis induction in BEFV-infected cells. BEFV induces apoptosis through Fas signaling and mitochondrial pathway and causes caspase-3, caspase-8, and caspase-9 activation [13]. Further investigations revealed that Src-dependent JNK signaling pathway has a key role on apoptosis triggered by BEFV [14].

Although the virus has one serotype, in recent years, virological studies and field observations showed that isolated virus from several outbreaks might have different phenotypes (virulence, transmission, etc.) or antigenic variations. Mutations on BEFV genome, especially the G gene, result in different phenotypes or antigenic variants. The nucleotide sequences of G gene showed three different clusters or genotypes. For this reason, it is suggested that different phenotypes isolated may have been responsible for various morbidity and mortality rates observed in the same country in different years [2,15–18].

The first scientific account of ephemeral fever was documented under the title "Epizootic dengue fever of cattle." It is then estimated that the infection spread in East African countries and others [1]. BEFV is reported in tropical regions of Asia and Australia in the following years. BEF endemic or epidemic mainly occurs in summer or rainy seasons in some regions of Africa, Asia, and Australia [1–3,15–18]. As *Culicoides* midges are abundant in the emerging region of BEFV, they form an important link in the BEFV epidemiology. Although countries with outbreaks or high prevalence of BEFV employ a wide variety of measures to reduce *Culicoides* biting attacks on cattle or other animals, this control approach

is neither effective nor practical. During the past few years, a number of articles investigating how the virus is spread over continents and vast distances have been published. The findings revealed that the mosquitoes can be dispersed overnight up to distances of 650 km compared to only 50 km during the day [2,16,19]. The epizootic swept 1000 km westward into the Kimberley region of Western Australia and eastward into Queensland and then progressively moved 3000 km southward on a wide front through eastern states to reach Victoria in February 1937 [17].

Vaccination is recommended in countries where BEFV is prevalent and outbreaks occur often [1,3,8]. BEFV-free countries with neighbors where the disease is present and has the appropriate environmental conditions for insect vectors should be carefully monitored for the epidemic. BEFV-free countries should take the necessary precautions and plan for emergency vaccination. Attenuated and inactivated vaccines have been used in Japan, Australia, and South African countries [1,3,20,21]. Although natural infection provides long-lasting protective antibodies against BEFV for years, attenuated vaccine may protect cattle for approximately 1 year. It is clear that the attenuation may reduce the immunogenicity of the virus. Additionally, attenuated vaccines may cause clinical disease because attenuated BEFV vaccines, like other RNA virus vaccines, have a much higher mutation rate. Thus, there is a possibility of the vaccine reverting to a more virulent strain in the cattle. Furthermore, the vaccines may be contaminated with other viruses, mainly pestiviruses. However, inactive vaccines that include a high amount of viruses should be used to achieve the desired level of protective immune response, at least twice [21]. Although BEFV has one serotype, there are many genotypes and different strains [17,22]. The preferred vaccine strains that provide optimal immune responses are those isolated from BEF outbreak seen in the country. As animals that recover from natural infection develop lifelong immune protection, they do not show any clinical symptoms following new outbreaks [21]. In the new outbreaks, young animals (born after the previous epidemic) living in a herd that has naturally infected are affected severely. Therefore, young animals should be vaccinated.

40.1.3 Clinical Features and Pathogenesis

While BEF-affected cattle and water buffalo show clinical symptoms, other ruminants have subclinical symptoms [1,3]. Clinical symptoms of the infection range from asymptomatic infection to death. It is thought that BEFV may cause distinct clinical symptoms based on animal factors such as age, reinfection, and genetic background. However, compared to host (animal) factors, virus-related factors may be less significant. Cattle with BEFV infection develop sudden onset of fever (41°C–42°C), usually biphasic or triphasic. During the first fever, milk production in lactating cows often drops dramatically. Animals may have tachypnea, depression, anorexia, ruminal atony, serous or mucoid discharges from the nose and eyes, waves of shivering, joint pain, stiffness, and shifting lameness. Many animals become recumbent for 8 h to several days. During the second fever, the symptoms are more severe. Level of milk production will never go back as it was before the disease. One of the major clinical signs is paralyses in the limbs. Recumbence may be seen progressing from sternal to lateral. These paralyses are temporary and the affected animals may promptly recover following the fever stage. Symptoms in severe cases can include significantly increased salivation, dyspnea, progressive loss of reflexes, and finally death. Other complications may include pneumonia, mastitis, temporary infertility in later pregnancy, and abortion in the cow. Clinical symptoms and death rate are lower in young cattle than in the older ones, with the former having higher average daily weight gain and higher milk production. Mild and uncomplicated cases recover in a short period of time and recovery rate is 95%–97%. Although there are some preliminary data about which genotype and phenotype of the virus exerts a notable influence on the severity of the disease and clinical observations, further studies are required to support this conclusion [1–3,23].

The pathogenesis of the disease is complex and remains obscure [1,2]. Injury of the endothelial lining of small blood vessel is central to expression of BEFV, but there is no evidence that the virus causes widespread tissue destruction. The primary effect of the inflammation is an increased permeability of small blood vessels, which has been demonstrated in BEFV infection. The first main signs of BEF are depression and cessation of milk secretion. Animals have fever, depression, anorexia, tachycardia, permeability of blood vessels, joint swelling, subcutaneous edema, ocular and nasal discharges, and neutrophilia with many immature forms that represents one of the most important hematological changes. Hypocalcemia is caused by the development of immature neutrophilia. Hematological and biochemical changes caused by BEFV are similar to periparturient hypocalcemia symptoms. These changes may lead to subcutaneous emphysema in a small percentage of cases especially in hot weather. And finally, another sign of BEFV-affected animals is the prolonged paralysis that remains after the fever terminates. Significant hematological and biochemical changes have been reported in cattle suffering from BEFV disease. In short, the clinical disease of BEF is accompanied by a marked neutrophilia, lymphopenia, increased plasma fibrinogen levels, and decreased serum calcium levels. The clinical signs of BEFV infection can be ameliorated without preventing viremia by treatment with anti-inflammatory and with calcium influx due to these parameters. Polyserositis and synovitis are recorded in lymph nodes; focal necrosis and edema of the lungs are also seen in certain muscle groups as a result of pathophysiological effects [1,3,24].

40.1.4 Diagnosis

40.1.4.1 Conventional Techniques

Although BEF demonstrates characteristic epidemiological and clinical signs, laboratory diagnosis should be mandatory for definitive diagnosis [1,3]. Virus isolation in cell culture and suckling mice are used as a gold standard. For these purposes, whole blood with EDTA should be collected from sick animals that have fever. It is known that the gold standard workflow for BEFV identification is time consuming and labor intensive. As an alternative diagnostic approach, sera from cattle should be taken again 2 weeks later and increased virus titer might be determined in this second sample by microneutralization test. The microneutralization test is very sensitive but expensive and technically demanding. Therefore, application of ELISA for BEFV diagnosis is more popular in recent years than neutralization test. Among numerous ELISA systems available, a blocking ELISA is preferable. The most important problem in the serological diagnosis of BEFV is KIMV because of their close antigenic relationships. KIMV causes subclinical infection in cattle and induces low titer neutralization antibody that does not provide any protection against BEFV. This situation creates difficulty for the evaluation of serological testing. To overcome this problem, G1-ELISA can be used to detect BEFV only. Given that sera positive to KIMV did not cross-react in the blocking ELISA or virus neutralization test, it is possible that sera from animals with KIMV infection would not test positive with indirect ELISA as recently reported [18,25].

40.1.4.2 Molecular Techniques

BEFV may spread very fast following the infection focus to impede infect broad geographical region. Therefore, at the beginning of the BEF-suspected outbreak, it is mandatory to use a reliable and rapid diagnostic method. Due to the fact that virus isolation and microneutralization are time consuming, molecular diagnostic methods are increasingly used in recent years [1–3]. Reverse transcription-polymerase chain reaction (RT-PCR) is the most preferable method for diagnosis of BEFV [1–3,15–17,22]. RT and PCR stages may be carried out separately but also performed in a single tube known as one-step. One-step RT-PCR is a commonly used system because it is rapid, inexpensive, and non-labor intensive. To help choose a particular molecular test for adoption, it is desirable to consider various elements of primer design requirements into one process such that users can simply input the template and obtain the desired target-specific primers. Primers specific for BEFV G region are used commonly for diagnosis [2,15–17,22]. Some of the positive RT-PCR products are chosen for sequencing confirmation of RT-PCR. Evolutionary analysis conducted on nucleotide sequences of the G gene is important to find genotypes and to better understand the molecular epidemiology of BEFV. Phylogenetic analysis of the BEFV G gene sequences from field isolates or clinical samples showed that BEFV isolates are grouped into distinct genotypes or clusters. To date, these genotypes are mainly based on the comparison of the different geographical regions. These genotype analyses are sometimes differently interpreted because each researcher chooses different parts and various lengths of the G gene for phylogenetic analysis. Based on phylogenetic analysis of the BEFV G gene sequences, a genogroup or cluster (genotype I-3 virus strains) includes Far East Asian countries such as Taiwan, China, and Japan. Virus strains isolated from Turkey, Egypt, and Israel are assigned in a different cluster [2,15,16,22]. Australian strains are placed in an independent cluster [17]. As a result, phylogenetic studies showed that, although BEFV is serologically monotypic, it displays significant antigenic variations among its genotypes.

Real-time PCR and the novel RT-loop-mediated isothermal amplification (RT-LAMP) methods are later developments in the molecular diagnosis and genotype determination of BEFV [26,27]. RT-LAMP appears to be more sensitive and allows earlier detection of BEFV in clinical samples compared with conventional RT-PCR and the virus isolation method [27]. However, the use of RT-LAMP method is not as widespread as conventional RT-PCR.

40.2 METHODS

40.2.1 SAMPLE PREPARATION

Whole blood samples containing EDTA as anticoagulant that have been taken from the acute phase of the disease should be used for RT-PCR. It is suggested that blood samples should be taken from the recumbent cattle or those with high fever. Samples for virus isolation must be transported to the laboratory immediately on ice and may be stored at 4°C in the laboratory for a few days if needed. If a portion of the samples is stored at −70°C, these samples may be used for virus isolation in future. Although both conventional procedures and commercial kits may be used for BEFV RNA isolation, commercial extraction kits are generally preferred because these kits provide larger quantity and higher quality of RNA than conventional isolation procedures.

Before RNA extraction, 2 mL of blood cells should be suspended in 4 volumes of sterile 0.2% NaCl to lyse erythrocytes, and 7.2% NaCl should be added to reconstitute and washed with PBS. While many commercial kits (e.g., QIAamp Viral RNA Mini Kit, Qiagen, and ZR Viral RNA Mini Kit, Zymo Research Corp.) are available for RNA isolation from blood samples with EDTA, the following protocol is based on ZR Viral RNA Mini Kit. Briefly, the pelleted cell is mixed with 400 μL of RNA lysis buffer. The mixture is incubated at room temperature for 15 min. Then, the lysate is transferred to Zymo-Spin IC Column in a collection tube and centrifuged at 16,000 × g for 2 min. The flow-through is discarded and the Zymo-Spin IC Column is placed back into the collection tube. Next, 500 μL of viral RNA wash buffer is added to the column. Centrifugation is repeated at 16,000 × g for 2 min and the column is transferred into the emptied RNase-free collection tube. Then, 25 μL of DNase/RNase-free water is added to the column and incubated at room temperature for 1 min. The RNA is eluted from column with centrifugation at 16,000 × g for 30 s. The RNA samples can be used immediately or stored at −70°C.

40.2.2 DETECTION PROCEDURES

In this section, a one-step RT-PCR protocol is presented for the detection of BEFV from clinical blood samples. Subsequent phylogenic analysis reveals additional details on the genotypes of BEFV strains.

40.2.2.1 One-Step RT-PCR

The following one-step RT-PCR assay utilizes the QIAGEN OneStep RT-PCR Kit (Qiagen, Valencia, CA). The assay is carried out with 50 μL total volume containing 10 μL of 5× RT-PCR buffer, 400 μM of each dNTPs, 2 μL of RT-PCR enzyme, 400 nM of each forward (bef19F; 5′-ATTTACAATGTTCCGGTGAA-3′) and reverse (bef523R 5′-GGTATCCATGTTCCGGTTAT-3′) primers, 1 μL of 1.5 mM MgCl$_2$, 3 μL of template RNA, and 30 μL of RNase-free water. Amplification is performed with the conditions involving an RT step at 53°C for 30 min followed by denaturation at 95°C for 15 min, 35 cycles of denaturation at 94°C for 1 min, annealing at 52°C for 1 min and extension at 72°C for 1 min, and a final step of extension at 72°C for 10 min. The PCR products are electrophoresed on 1.5% agarose gel. With the RT-PCR procedure, the nucleotide sequences (520 bp) of the partial gene region of G gene of BEFV are amplified. But, for amplification of the full G gene of BEFV, BEF-GF, 5′-ATGTTCAAGGTCCTMATAATTACC-3′, and BEF-GR, 5′-TTAATGATCAAAGAACCTATC-3′, primers can be used from RT-PCR.

40.2.2.2 Sequencing and Phylogenetic Analysis

The RT-PCR products may be sequenced for confirmation of amplification and phylogenetic analysis. For this purpose, positive RT-PCR products are purified using gel purification kits (e.g., QIAquick Gel Extraction Kit, Qiagen, Valencia, CA). Then, 2 μL of purified product is visualized using agarose gel before sequencing and DNA concentration and purity of DNA are measured by the spectrometer (Thermo Scientific NanoDrop 2000 Spectrophotometer). The product is sequenced in both directions with the forward and reverse primers by using the ABI 310 Genetic Analysis System (Iontec Co., Istanbul, Turkey). The gene sequences are aligned using ClustalW program and then compared to the other sequences of BEFV obtained from the GenBank database. The position of the sequences in the phylogenetic tree may be checked by a neighbor-joining method from the PHYLIP software. In addition to nucleotide sequence analysis, converting gene sequences into amino acid sequences is suggested for the determination of amino acid variations in the antigenic region that induces neutralizing antibodies.

40.3 FUTURE PERSPECTIVES

Vaccine provides an important advantage fighting against BEFV because of one serotype. Additionally, current epidemiological studies that determine virus circulation in countries or continents are a useful tool. Given the possibility that new BEFV serotypes may emerge due to high mutation rate and that BEFV vector, *Culicoides* species, may find new life zone due to changing ecological balance, BEF will spread to countries or regions more than expected. Therefore, continuing efforts in BEFV vaccine developments are warranted to provide sustained, lifelong immunity to animal hosts against existing as well as newly emerged BEFV serotypes.

While the application of molecular techniques has uncovered valuable insights on the molecular epidemiology of BEFV, there are still controversial issues and some problems that remain unsolved. It is known that appropriate vectors and susceptible hosts are crucial elements for a vector-borne disease to emerge. One unsolved issue concerning BEF is that BEFV cases may disappear after an outbreak in a region but continue for many years in other regions. In addition, it is not fully elucidated whether there is any relationship between distinct mortality and morbidity rates according to the different epidemic years, virus genotypic, and phenotypic variations. To address these questions, we need further investigations on molecular evolution and epidemiology of BEFV, contributing to our fight against BEF.

Given our incomplete knowledge about the functions of BEFV proteins especially accessory proteins to date, it is essential that further studies are conducted in these areas. The full elucidation of the functions of the BEFV protein will not only improve our understanding of rhabdovirus biology but also open potential new avenues for effective control of BEF in the future.

REFERENCES

1. Nandi, S. and Negi, B.S. Bovine ephemeral fever: A review. *Comp. Immun. Microbiol. Infect. Dis.*, 1999; 22, 81–91.
2. Tonbak, S., Berber, E., Yoruk, M.D. et al. A large-scale outbreak of bovine ephemeral fever in Turkey, 2012. *J. Vet. Med. Sci.*, 2013; 75, 1511–1514.
3. MacLachlan, N.J. and Dubovi, E. 2011. Rhabdoviridae. In *Fenner's Veterinary Virology*, 4th edn., N.J. Maclachlan, E.J. Dubovi, and F. Fenner (eds.), pp. 338–340. Academic Press, Amsterdam, the Netherlands.
4. Wang, Y., Cowley, J.A., and Walker, P.J. Adelaide River virus nucleoprotein gene: Analysis of phylogenetic relationships of ephemeroviruses and other rhabdoviruses. *J. Gen. Virol.*, 1995; 76, 995–999.
5. Blasdell, K.R., Voysey, R., Bulach, D.M. et al. Malakal virus from Africa and Kimberley virus from Australia are geographic variants of a widely distributed ephemerovirus. *Virology*, 2012; 433, 236–244.
6. Walker, P.J., Wang, Y., Cowley, J.A. et al. Structural and antigenic analysis of the nucleoprotein of bovine ephemeral fever rhabdovirus. *J. Gen. Virol.*, 1994; 75, 1889–1899.
7. Walker, P.J., Dietzgen, R.G., Joubert, D.A. et al. Rhabdovirus accessory genes. *Virus Res.*, 2011; 162, 110–125.
8. Hertig, C., Pye, A.D., Hyatt, A.D. et al. Vaccinia virus-expressed bovine ephemeral fever virus G but not G(NS) glycoprotein induces neutralizing antibodies and protects against experimental infection. *J. Gen. Virol.*, 1996; 77, 631–640.
9. Walker, P.J., Byrne, K.A., Cybinski, D.H., Doolan, D.L., and Wang, Y.H. Proteins of bovine ephemeral fever virus. *J. Gen. Virol.*, 1991; 72, 67–74.
10. Uren, M.F., Walker, P.J., Zakrzewski, H. et al. Effective vaccination of cattle using the virion G protein of bovine ephemeral fever virus as an antigen. *Vaccine*, 1994; 1, 845–850.
11. Joubert, D.A., Blasdell, K.R., Audsley, M.D. et al. Bovine ephemeral fever rhabdovirus α1 protein has viroporin-like properties and binds importin β1 and importin 7. *J. Virol.*, 2014; 88, 1591–1603.
12. Chang, C.J., Shih, W.L., Yu, F.L., Liao, M.H., and Liu, H.J. Apoptosis induced by bovine ephemeral fever virus. *J. Virol. Methods*, 2004; 122, 165–170.
13. Lin, C.H., Shih, W.L., Lin, F.L. et al. Bovine ephemeral fever virus-induced apoptosis requires virus gene expression and activation of Fas and mitochondrial signaling pathway. *Apoptosis*, 2009; 14, 864–877.
14. Chen, C.Y., Chang, C.Y., Liu, H.J. et al. Apoptosis induction in BEFV-infected Vero and MDBK cells through Src-dependent JNK activation regulates caspase-3 and mitochondria pathways. *Vet. Rec.*, 2010; 41, 15.
15. Kato, T., Aizawa, M., Takayoshi, K. et al. Phylogenetic relationships of the G gene sequence of bovine ephemeral fever virus isolated in Japan, Taiwan and Australia. *Vet. Microbiol.*, 2009; 137, 217–223.
16. Aziz-Boaron, O., Klausner, Z., Hasoksuz, M. et al. Circulation of bovine ephemeral fever in the Middle East-strong evidence for transmission by winds and animal transport. *Vet. Microbiol.*, 2012; 158, 300–307.
17. Trinidad, L., Blasdell, K.R., Joubert, D.A. et al. Evolution of bovine ephemeral fever virus in the Australian episystem. *J. Virol.*, 2014; 88, 1525–1535.
18. Zheng, F.Y., Lin, G.Z., Qiu, C.Q. et al. Development and application of G1-ELISA for detection of antibodies against bovine ephemeral fever virus. *Res. Vet. Sci.*, 2009; 87, 211–212.
19. Finlaison, D.S., Read, A.J., and Kirkland, P.D. An epizootic of bovine ephemeral fever in New South Wales in 2008 associated with long-distance dispersal of vectors. *Aust. Vet. J.*, 2010; 88, 301–306.
20. Tzipori, S. and Spradbrow, P.B. A cell culture vaccine against bovine ephemeral fever. *Aust. Vet. J.*, 1978; 54, 323–328.
21. Aziz-Boaron, O., Gleser, D., Yadin, H. et al. The protective effectiveness of an inactivated bovine ephemeral fever virus vaccine. *Vet. Microbiol.*, 2014; 173, 1–8.
22. Ting, L.J., Lee, M.S., Lee, S.H. et al. Relationships of bovine ephemeral fever epizootics to population immunity and virus variation. *Vet. Microbiol.*, 2014; 173, 241–248.
23. Young, P.L. and Spradbrow, P.B. Clinical response of cattle to experimental infection with bovine ephemeral fever virus. *Vet. Rec.*, 1990; 126, 86–88.
24. St George, T.D., Cybinski, D.H., Murphy, G.M. et al. Serological and biochemical factors in bovine ephemeral fever. *Aust. J. Biol. Sci.*, 1984; 37, 341–349.
25. Zheng, F.Y., Lin, G.Z., Qiu, C.Q. et al. Serological detection of bovine ephemeral fever virus using an indirect ELISA based on antigenic site G1 expressed in Pichia pastoris. *Vet. J.*, 2010; 185, 211–215.
26. Stram, Y., Kuznetzova, L., Levin, A. et al. A real-time RT-quantitative(q) PCR for the detection of bovine ephemeral fever virus. *J. Virol. Methods*, 2005; 130, 1–6.
27. Zheng, F., Lin, G., Zhou, J. et al. A reverse-transcription, loop-mediated isothermal amplification assay for detection of bovine ephemeral fever virus in the blood of infected cattle. *J. Virol. Methods*, 2011; 171, 306–309.

41 Rabies Virus

Dongyou Liu

CONTENTS

41.1 Introduction 361
    41.1.1 Classification.. 361
    41.1.2 Morphology and Genome Structure 362
    41.1.3 Biology and Epidemiology 362
    41.1.4 Clinical Features and Pathogenesis 363
    41.1.5 Diagnosis 363
        41.1.5.1 Virus Isolation... 363
        41.1.5.2 Histopathology 363
        41.1.5.3 Serological Tests 363
        41.1.5.4 Nucleic Acid Detection 364
    41.1.6 Treatment and Prevention 364
41.2 Methods 364
    41.2.1 Sample Preparation... 364
    41.2.2 Detection Procedures.. 365
        41.2.2.1 RT-PCR for Rabies Virus Detection.. 365
        41.2.2.2 RT-PCR for Pan-Lyssavirus Detection 365
41.3 Conclusion 365
References... 366

41.1 INTRODUCTION

Rabies virus is the causative agent of a zoonotic disease called rabies (a term derived from Latin word rabies, meaning "fury" or "madness"), which has been known since antiquity. The disease is easily recognizable when a dog infected with rabies virus loses its usual temperament and attacks/bites people without provocation. Considered a scourge for its prevalence in the nineteenth century, rabies has created irrational fear among people due to the absence of efficacious treatment. Following the discovery of the causative agent and the development of postexposure prophylaxis (PEP), this once ferocious disease has been largely brought under control. As a consequence, rabies has become a neglected disease in modern times, in spite of its stubbornly high prevalence and its association with around 50,000 deaths each year worldwide. In this chapter, we redraw readers' attention to rabies virus, with a concise overview on its classification, morphology, genome organization, epidemiology, clinical features, pathogenesis, diagnosis, treatment, and prevention. This is followed by the presentation of streamlined molecular protocols for its detection and identification, and a discussion of future research needs that will help us keeping an upper hand over this deadly pathogen.

41.1.1 CLASSIFICATION

Rabies virus (RABV) is negative-sense, single-stranded (ss) RNA virus classified in the genus *Lyssavirus*, family *Rhabdoviridae*, order *Mononegavirales*. Consisting of over 200 viruses, the family *Rhabdoviridae* is separated into 11 cleared defined genera (*Cytorhabdovirus*, *Ephemerovirus*, *Lyssavirus*, *Novirhabdovirus*, *Nucleorhabdovirus*, *Perhabdovirus*, *Sigmavirus*, *Sprivivirus*, *Tibrovirus*, *Tupavirus*, and *Vesiculovirus*) as well as nine proposed genera (*Almendravirus*, *Bahiavirus*, *Bracorhabdovirus*, *Curiovirus*, *Dichorhabdovirus*, *Hapavirus*, *Ledantevirus*, *Sawgravirus*, and *Sripuvirus*).

In turn, the genus *Lyssavirus* (lyssa, derived from Greek word "lud" or "violent") is divided into 12 recognized species (RABV [type species], Lagos bat virus [LBV], Mokola virus [MOKV], Duvenhage virus [DUVV], European bat lyssavirus types 1 and 2 [EBLV-1 and EBLV-2], Australian bat lyssavirus [ABLV], Aravan virus [ARAV], Khujand virus [KHUV], Irkut virus [IRKV], West Caucasian bat virus [WCBV], and Shimoni bat virus [SHIBV]) as well as two recently discovered species (Bokeloh bat lyssavirus [BBLV] and Ikoma virus [IKOV]) [1].

Based on phylogenetic analyses, lyssaviruses have been differentiated into seven genotypes: RABV (genotype 1), LBV (genotype 2), MOKV (genotype 3), DUVV (genotype 4), EBLV type 1 (EBLV-1; genotype 5), EBLV type 2 (EBLV-2; genotype 6), and ABLV (genotype 7). The five recently identified lyssaviruses (ARAV, KHUV, IRKV, WCBV, and SHIBV) may represent distinct genotypes in accordance with current approaches for genotype definition (amount of nucleotide identity of the N gene and bootstrap support of joining to certain phylogenetic groups). Nevertheless, experimental data suggest that KHUV demonstrates relatedness to genotype 6; ARAV shows moderate similarity to genotypes 4, 5, and to a lesser extent 6; IRKV may be related to genotypes 4 and 5; WCBV has limited relationship to genotypes 2 and 3; and SHIBV is most similar to genotype 2 [1].

Through serological examination, members of the genus *Lyssavirus* have been distinguished into 2 phylogroups, with phylogroup 1 covering RABV, DUVV, EBLV-1, EBLV-2, ABLV, ARAV, KHUV, and IRKV and phylogroup 2 containing LBV, MOKV, and SHIBV. WCBV and IKOV are non-cross-reactive serologically with any of the two phylogroups, with WCBV possibly forming a separate phylogroup 3, while IKOV being part of phylogroup 3 or constituting a fourth phylogroup [1].

Furthermore, the classic RABV may be divided into the following two major phylogenetic groups: one associated with carnivores (typically dogs) in an "urban" cycle, and another associated with bats, raccoons, and skunks in a rural or "sylvatic" cycle. The former is distributed widely and contains six clusters (clades): the Africa 2, Africa 3, Arctic-related, Asian, cosmopolitan, and Indian subcontinent clades. The other lyssavirus species are largely maintained in bats with a more restricted distribution: DUVV, LBV, MOKV, SHIBV, and IKOV are present in Africa; EBLV-1, EBLV-2, and BBLV in Europe; ABLV in Australia; and ARAV, KHUV, IRKV, and WCBV in Asia [1,2].

41.1.2 Morphology and Genome Structure

As its family name (*Rhabdoviridae*, rhabdo is Greek for rod) suggests, lyssavirus is an enveloped, bullet-shaped (or elongated, rod-like) virus of about 75 nm in diameter and about 180 nm in length. The lipid envelope is decorated with spikes of 9–11 nm (made of glycoprotein [G] trimers), which interact with the matrix protein M. The matrix protein M forms a protective coating for the nucleocapsid underneath, which is composed of the nucleoprotein (N), large protein (L), and phosphoprotein (P). Coiling symmetrically to form a cylindrical structure with a striated appearance, the nucleocapsid encapsidates a linear, nonsegmented, negative-sense, ssRNA genome of 11,924 nucleotides [3–5].

The genome of RABV comprises five protein-coding genes arranged in the order of 3′ N-P-M-G-L 5′, encoding nucleoprotein (N), phosphoprotein (P), matrix protein (M), glycoprotein (G), and large protein (L), respectively [3–5]. Of these, the glycoprotein (G) is the only viral protein exposed on the virion surface and is involved in viral attachment and cell entry. Not surprisingly, as the primary surface antigen, the trimeric G protein is the main target of host neutralizing antibodies. The nucleoprotein (N), the phosphoprotein (P), and the large protein (L) interact with each other to form the nucleocapsid (also referred to as ribonucleoprotein complex, for which the matrix protein (M) provides a protective coating). The phosphoprotein (P) and the large protein (L) function as viral RNA polymerase, which is involved in the

transcription from the original negative-strand RNA into five mRNA strands and a positive strand of RNA [6]. After sufficient proteins (P, L, N, G, and M) are translated from five mRNA strands, new negative strands of RNA are synthesized from the template of the positive-strand RNA. These negative strands form complexes with N, P, L, and M proteins and then interact with the G protein to generate new viruses [1].

41.1.3 Biology and Epidemiology

RABV infection usually begins with the transdermal inoculation of infected saliva in the bite of a rabid dog or bat. On rare occasions, rabies can be transmitted via routes other than bites (e.g., aerosols, corneal, and organ transplants as in the case of human rabies). The virus initially enters into muscle cells close to the biting site, with the assistance of its G protein (trimeric spikes) that binds to acetylcholine receptors (p75NR) on the cell surface [7]. After replication in muscle cells without disturbing the host's immune system, the virus utilizes its P protein to interact with dynein, a protein present in the cytoplasm of nerve cells, facilitating its travel to the central nervous system (CNS). After undergoing further replication in motor neurons of the brain, the virus migrates centrifugally to the peripheral and autonomic nervous systems and then the salivary glands (as well as nasal and oral cavities, adrenal glands, kidney, cardiac muscle, and hair follicles), where the virus is ready to be transmitted to the next host [8].

Many warm-blooded species are susceptible to RABV. These include humans, monkeys, raccoons, cattle, groundhogs, weasels, bears, wolves, coyotes, dogs, cats, foxes, mongooses, skunks, and bats [9–12]. While birds infected with RABV are often asymptomatic, squirrels, hamsters, guinea pigs, gerbils, chipmunks, rats, mice, rabbits, and hares are apparently refractory to RABV.

The main reservoir hosts and transmission vectors for RABV are carnivores (order Carnivora) and bats (order Chiroptera), although other animals (e.g., kudu antelope in Namibia and nonhuman primates in Brazil) may be potential reservoir hosts [13–16].

Dogs (and to a lesser extent cats) are involved in the urban cycle of rabies infection and represent the primary RABV vector in Asia and in parts of the United States and Africa [17–24]. Human mortality from urban canine rabies is estimated to be 55,000 deaths per year, with 56% of the deaths occurring in Asia (especially India and Vietnam) and 44% in Africa. The majority of the western European countries are free of urban canine rabies (RABV) [25].

Bats appear to be the original reservoir for lyssaviruses. Bats (red fox, raccoon dog, raccoon, and skunks) are associated with the sylvatic (wildlife) cycle of rabies infection, which remains widespread throughout parts of Europe, Africa, and North America [26]. In Europe, fox rabies is common, while the raccoon dog is often implicated in the epidemiology of rabies in the Baltic countries. In addition, rabies cases in Europe are mostly attributable to Lyssavirus genotypes 1 (RABV, classical rabies), 5, and 6 (EBLV types 1 and 2) [27].

41.1.4 CLINICAL FEATURES AND PATHOGENESIS

RABV infections in mammals involves three stages: prodromal, excitement (furious), and paralytic (dumb), although in atypical cases, not all stages are necessarily observed [1].

The first (prodromal) stage lasts 1–3 days and is characterized by behavioral changes and nonspecific signs typical of a viral encephalitis. In addition, due to viral replication in dorsal root ganglia and ganglionitis, there is neuropathic pain at the site of infection (bite wound).

The second (excitative) stage (often known as "furious rabies") lasts 3–4 days and is characterized by the tendency of the affected animal to be hyperreactive to external stimuli and bite at anything near.

The third (paralytic) stage is caused by damage to motor neurons and characterized by rear limb paralysis, drooling, and swallowing difficulty (paralysis of facial and throat muscles). Death is usually an outcome of respiratory arrest.

Interestingly, different animal species tend to exhibit varying times of virus excretion before the onset of clinical signs (cats <24 h, dogs <13 days, and foxes <29 days postinfection).

Postmortem examination of nonspecific lesions in the cervical spinal cord, hypothalamus, and pons may reveal ganglioneuritis in nerve centers, histolymphocytic cuffs, and gliosis. The only specific lesions may be oval-shaped intracytoplasmic eosinophil inclusions (Negri bodies) of 4–5 μm in size located in Ammon's horn, which correlate to the aggregation of developing RABV particles and which have provided a basis for diagnosing rabies for many years prior to the availability of specific and sensitive diagnostic methodologies.

In humans, although the period between infection and the first flu-like symptoms (e.g., fever) is typically 2–12 weeks, incubation periods as short as 4 days and longer than 6 years have nevertheless been observed. Clearly, a deep bite near the head of the victim may entail a short incubation period. With progression toward acute encephalitis, clinical signs may expand to hypersalivation, paresthesia (itching) around biting site, progressive cranial nerve paralysis, leg weakness, anxiety, insomnia, confusion, agitation, abnormal behavior, paranoia, terror, hydrophobiasis, and hallucinations, progressing to delirium, coma, and death (which almost always occurs 2–10 days after first symptoms) [28].

As the historic name for rabies, hydrophobia ("fear of water," which is commonly associated with furious rabies that affects 80% of infected people) refers to a set of symptoms (including difficulty swallowing, panicking when presented with liquids to drink, and thirst due to increased saliva production) that appear in the later stages of an infection. As the multiplication and assimilation of rabies are concentrated in the salivary glands, there is an urge to further transmit the virus through biting. The remaining 20% may experience a paralytic form of rabies marked by muscle weakness, loss of sensation, and paralysis, without usual fear of water [1].

41.1.5 DIAGNOSIS

Animals infected with rabies generally show nonspecific symptoms such as frenzy, extreme tremors, salivation, and paresis that are indistinguishable to encephalitic conditions caused by canine distemper virus or acute trauma. Other differential diagnoses include herpes simplex virus type one, varicella zoster virus, echoviruses, polioviruses, human enteroviruses 68–71, West Nile virus, Nipah virus, transmissible spongiform encephalopathies, tetanus, listeriosis, poisoning, and Aujeszky's disease. Paralytic rabies may be sometimes mistaken for Guillain–Barré syndrome. Secondary infections such as malaria may also complicate rabies diagnosis.

For many years, histopathological detection of Negri bodies in the CNS has provided a basis for diagnosing of rabies. However, this postmortem diagnostic approach is largely irrelevant or unsuitable for guiding PEP. Therefore, alternative laboratory methods (e.g., virus isolation, serological tests, and nucleic acid detection) have been developed and applied for confirmation of rabies [29–32].

41.1.5.1 Virus Isolation

Virus isolation involves the inoculation of brain homogenates suspected to harbor RABV onto neuroblastoma cells (whose cell membranes contain acetylcholine receptors and neurotransmitter synthetic enzymes that render them susceptible to wild RABV infection). Positive results are often available within 2–4 days.

Another useful way to isolate and detect RABV is the mouse inoculation test (MIT), which involves the intracerebral inoculation of mice with a clarified supernatant of a homogenate of brain material (cortex, Ammon's horn, cerebellum, medulla oblongata) and observation for up to 28 days. Clinical signs suggestive of rabies may appear as early as 6–8 days.

41.1.5.2 Histopathology

Observation of specific histologic changes (e.g., Negri bodies) in CNS tissues has been commonly applied for rabies diagnosis before sensitive and specific diagnostic tests became available. This procedure is no longer routinely performed in clinical diagnostic laboratories, due to its poor reliability, particularly for decomposed material.

41.1.5.3 Serological Tests

One of the most widely applied serological tests for rabies diagnosis is the fluorescent antibody test (FAT). This technique utilizes fluorescently labeled antirabies antibodies to detect RABV nucleocapsid protein in smears prepared from the hippocampus, cerebellum, medulla oblongata, or spinal cord tissue as well as cell monolayers. The FAT takes a few hours to complete and has a sensitivity of 95%–99%. For the antemortem diagnosis of rabies using nonneural material (e.g., saliva, cerebrospinal fluid, skin biopsies) or decomposed tissues, the performance of FAT may be less optimal, although in tissue sections of skin biopsies, especially in the nerve endings surrounding the hair follicles, viral antigens may be detected with the FAT [1].

Other serological tests for RABV include rapid immunodiagnostic assay, histochemical test, enzyme-linked immunosorbent assay (ELISA), rapid rabies enzyme immunodiagnosis, and rapid fluorescent focus inhibition test.

41.1.5.4 Nucleic Acid Detection

Methods for detecting rabies nucleic acids range from in situ hybridization, reverse transcription polymerase chain reaction (RT-PCR), nucleic acid sequence-based amplification, to microarray [33–35].

Among these, RT-PCR (and its various derivatives, such as standard, nested, multiplex, and real time) is widely adopted in clinical laboratories for qualitative and quantitative detection of rabies from a variety of samples (saliva, urine, and cerebrospinal fluid, as well as decomposed or paraffin-fixed archival specimens) [36–42]. Many of these assays target the highly conserved N and L genes of RABV and lyssaviruses. To date, a large number of RT-PCR procedures for RABV detection have been reported [36–42]. Further advances in real-time RT-PCR technologies allow instant result readout and eliminate potential cross-contamination between samples [43]. Real-time RT-PCR for ARAV, KHUV, IRKV, and ABLV are also available [44,45]. Recently, a one-step quantitative real-time PCR assay utilizing a single primer–probe set for detecting four African lyssaviruses (RABV, MOKV, LBV, and DUVV) was described [46]. However, species-specific assays for individual African lyssaviruses (MOKV, LBV, DUVV, SHIBV, and IKOV) and WCBV have not been published. Fortunately, the use of pan-lyssavirus real-time RT-PCR will help cover this limitation (particularly, a recent study indicates that the sensitivity of RT-PCR is similar to that of virus isolation concerning SHIBV and WCBV) [1].

41.1.6 Treatment and Prevention

Treatments for rabies are largely based on the use of PEP. In humans, following exposure to an animal with suspected rabies, the biting/scratch wound should be carefully cleaned and washed for 5–15 min, with running water, soap/detergent, then povidone-iodine, alcohol, or other antiseptic; and a vaccine produced on cell culture (e.g., a vaccine prepared through the chemical inactivation of rabies-infected tissue culture human diploid cells HDCV) administered. Injection of rabies immunoglobulins (RIGs, especially immunoglobulins of human origin and fragments F(ab′)$_2$ of equine origin) is advisable in case of high-risk exposure (e.g., transdermal bites/scratches, licks on broken skin, contamination of mucous membrane with saliva from licks, and exposure of bites/scratches and wound or mucous membrane to bat saliva). For humans, the Centers for Disease Control and Prevention in the United States recommends one dose of human RIG (<20 units per kg body weight, injected around the bite, as well as by deep intramuscular injection away from vaccination site) and four doses of rabies vaccine (at days 0, 3, 7, and 14) at the deltoid area or lateral thigh in infants. PEP may be stopped if laboratory

examination rules out the suspected animal of rabies. In general, PEP is not effective after onset of neurological signs (generally after 10 days of infection) [47–49].

The control measures for rabies should center on notification, isolation, disinfection, and immunization. Vaccination of dogs and improved access to PEP represent effective approach for the prevention of rabies. Several vaccine preparations (e.g., SAG2 [RABIGEN® SAG2], Street Alabama Dufferin B19 [SAD B19, and SAD Bern]) have proven to be safe and potent for animals and are used in Europe. SAG2 vaccine is a modified live attenuated RABV vaccine, selected from the SAD Bern strain. This attenuated strain does not spread in vivo or induce a persistent infection [50–53]. Recombinant vaccines based on vaccinia virus and human adenovirus vector, both expressing the rabies glycoprotein of the ERA strain (V-RG and ONRAB, respectively), are in use in the United States and Canada to prevent outbreaks of rabies in wildlife.

The annual World Rabies Day on September 28 provides the general public with information about rabies, its prevention, and elimination.

41.2 METHODS

41.2.1 Sample Preparation

Tissues from the thalamus, pons, and medulla are recommended for rabies diagnosis in animals. The hippocampus (Ammon's horn), cerebellum, and cerebrum have been shown to give negative results in up to 11% of positive brains, while the thalamus gives 100% positive results. Therefore, a pool of brain tissues including the brain stem may be collected for testing. Brain tissue collected via transorbital or transforamen magnum route is also usable. If large numbers of animal carcasses are involved, the brain material may be collected by passing a plastic straw through the occipital foramen, the so-called "straw technique." The use of glycerine preservation (temperature: +4°C or −20°C) or dried smears of brain tissue on filter paper (temperature: +30°C) may facilitate safe transportation of infected material.

If brain tissue is not available, other types of specimens (e.g., saliva, spinal fluid, tears, oral and fecal swabs, skin biopsies from wound site or nape of neck, and corneal impression) may be tested [54].

For virus isolation, 1 g of brain sample may be homogenized with sand (Sigma) in 10 mL of alpha-minimum essential medium (α-MEM; Gibco BRL), or using the MagnaNA lyser tissue homogenizer tubes containing 1.4 mm (diameter) ceramic beads (Roche Applied Science) and 1.0 mL of MEM-10 as a diluent. The homogenized sample is centrifuged at $8000 \times g$ for 5 min, and the supernatant filtered through a 0.45 μm membrane (Millipore). Briefly, 100 μL of filtrate is inoculated onto a monolayer of murine NG108–15 neuroblastoma cells (ATCC HB-12317) in a 24-well plate. The cells are maintained in α-MEM supplemented with antibiotics (100 IU/mL penicillin, 10 μg/mL streptomycin, and 0.25 μg/mL amphotericin B) and 10% heat-inactivated fetal bovine serum (Gibco BRL) at 37°C in a 5% CO_2 incubator for 1 h. Afterward, the inoculation

TABLE 41.1

Primer and Probe Mix for Lyssavirus- and Rabies Virus–Specific RT-PCR

Assay	Primer/Probe	Sequence (5′–3′)	Position	Concentration (µM)
R13 MP	JW12	ATG TAA CAC CYC TAC AAT G	55–73	10
	N165–146	GCA GGG TAY TTR TAC TCA TA	165–146	10
	LysGT1-FAM	FAM-ACA AGA TTG TAT TCA AAG TCA ATA ATC AG-BHQ1	81–109	2.5
	LysGT5-HEX	HEX-AA CAR GGT TGT TTT YAA GGT CCA TAA-BHQ1	80–105	2.5
	LysGT6-Cy5	Cy5-ACA RAA TTG TCT TCA ARG TCC ATA ATC AG-BHQ3	81–109	2.5
	BBLV-1TEX	TEX-CTC TGA CAA GAT TGT CTT CAA AGT C-BHQ2	76–101	2.5
R14 MP	RV-N-196-F	GAT CCT GAT GAY GTA TGT TCC TA	266–288	10
	RV-N-283-R	RGA TTC CGT AGC TRG TCC A	353–335	10
	RabGT1-B-FAM	FAM-CAG CAA TGC AGT TYT TTG AGG GGA C-BHQ1	297–321	2.5

medium is removed and replaced with fresh α-MEM. Cells are observed for cytopathic effects for 5 days.

For FAT, frozen thin sections of brain tissue (e.g., Ammon's horn tissue) on slides are fixed in 80% chilled acetone (−20°C) for 20 min. After three washes with phosphate buffer saline (PBS; pH 7.2), the slides are incubated with a monoclonal antibody (Jeno Biotech) against rabies for 45 min at 37°C, and then stained with fluorescein isothiocyanate–conjugated goat anti-mouse IgA, IgG, and IgM (KPL). After rinsing with PBS, the slides are air dried and mounting buffered glycerin (Southern Biotechnology Associate, Birmingham, AL) is applied. The slides are examined under cover slips at magnification of 400 using a fluorescent microscope (Nikon). Positive and negative controls are run together with the test samples. The slide showing specific fluorescence is confirmed as positive.

Viral RNA may be extracted from brain samples using the Qiagen RNeasy Mini Kit (for brain and cell culture medium) or Qiagen QIAamp Viral RNA Mini Kit (for serum and CSF). Briefly, brain samples (400 mg) are first homogenized in 1 mL of PBS with a Potter homogenator. Then, 85 µL of the brain homogenate is mixed with 265 µL of lysis buffer (RLT) and used for RNA extraction. For RNA extractions from infected cell culture supernatants, a volume of 150 µL is homogenized in 200 µL of RLT buffer, as starting material for the RNA extraction. Starting from serum or CSF, a volume of 140 µL is used for RNA extraction with the Qiagen QIAamp Viral RNA Mini Kit. The extracted RNA is eluted in 50 µL of RNase- and DNase-free water.

41.2.2 DETECTION PROCEDURES

41.2.2.1 RT-PCR for Rabies Virus Detection

Yang et al. [29] utilized primer sets (RVNDF: 5′-GRAA TTGGGCTTTGACTGGA-3′ and RVNDR: 5′-AAAGGG GCTGTCTCGAAAAT-3′) for amplification of 181 bp product from the N gene for specific detection of RABV.

RT-PCR mixture is composed of 5 µL of denatured RNA, 1 µL of each primer (50 pmol), 5 µL of 5 × buffer (12.5 mM $MgCl_2$), 1 µL of dNTP mix, 1 µL of an enzyme mix (reverse transcriptase and Taq DNA polymerase), and 11 µL of distilled water (Qiagen).

The cycling program includes cDNA synthesis at 42°C for 30 min, 45 cycles of 95°C for 15 s, 55°C for 15 s, and 72°C for 15 s, and a final extension at 72°C for 5 min.

RT-PCR products are separated and visualized on 1.8% agarose gels containing ethidium bromide. Samples showing a 181 bp band are considered positive.

41.2.2.2 RT-PCR for Pan-Lyssavirus Detection

Fischer et al. [43] described the use of pan-lyssavirus and RABV-specific real-time RT-PCR for the detection of lyssaviruses and RABV, with primers and probes designed from the N gene (Table 41.1).

RT-PCR mixture is prepared by using AgPath-ID™ One-Step RT-PCR Kit (Applied Biosystems), with each 12.5 µL reaction containing 2.75 µL (R13 MP) or 1.25 µL (R14 MP) RNase-free water, 6.25 µL 2 × RT-PCR buffer, 0.5 µL 25 × RT-PCR enzyme mix, 2.5 µL RNA template or RNase-free water for the no template control, and 0.5 µL of primer–probe mixes (Table 41.1).

Amplification and detection are performed in duplicates in Bio-Rad 96-well PCR plates using a CFX96 quantitative PCR system (Bio-Rad Laboratories), with the following thermal program: 1 cycle of 45°C for 10 min and 95°C for 10 min, followed by 45 cycles of 95°C for 15 s, 56°C for 20 s, and 72°C for 30 s.

For each RT-qPCR, a quantification cycle number (C_q) is determined according to the PCR cycle number at which the fluorescence of the reaction crosses a value that is statistically higher than the background. Finally, mean C_q values are calculated from duplicates. A cutoff >42 is defined for negative results.

41.3 CONCLUSION

As a zoonotic pathogen with a worldwide distribution, RABV is maintained in two lifecycles: while urban rabies involves dogs and cats as the primary transmission vector, sylvatic rabies utilizes bats and other wildlife species as the transmission vector. Rabies usually begins following a bite from an infected animal and quickly develops clinical signs typical of encephalitis. In humans, rabies is fatal if PEP is not administered before severe symptoms emerge (about 10 days after infection). Therefore, the ability to rapidly and accurately detect and diagnose rabies plays a crucial role in the decision to implement PEP (which is costly). For this reason, WHO and OIE recommend the FAT, the rabies tissue culture infection test, and the MIT for the diagnosis of rabies. With the recent development in nucleic acid amplification technologies such as RT-PCR, rapid, sensitive, and specific detection and typing of rabies are no longer unreachable [55–57].

REFERENCES

1. McElhinney LM, Leech S, Marston DA, Fooks AR. Rabies virus. In: Liu D (ed.), *Molecular Detection of Human Viral Pathogens*. 2011; pp. 425–435, CRC Press, Boca Raton, FL.
2. Rupprecht CE, Hanlon CA, Hemachudha T. Rabies re-examined. *Lancet Infect Dis*. 2002;2:327–343.
3. Zhang G, Fu ZF. Complete genome sequence of a street rabies virus from Mexico. *J Virol*. 2012;86(19):10892–10893.
4. Zhou M et al. Complete genome sequence of a street rabies virus isolated from a dog in Nigeria. *Genome Announc*. 2013;1(1):e00214-12.
5. Yu F et al. Comparison of complete genome sequences of dog rabies viruses isolated from China and Mexico reveals key amino acid changes that may be associated with virus replication and virulence. *Arch Virol*. 2014;159(7):1593–1601
6. Wiltzer L et al. Interaction of rabies virus P-protein with STAT proteins is critical to lethal rabies disease. *J Infect Dis*. 2014;209(11):1744–1753.
7. Lafon M. Rabies virus receptors. *J Neurovirol*. 2005;11:82–87.
8. Shi N et al. Alterations in microRNA expression profile in rabies virus-infected mouse neurons. *Acta Virol*. 2014;58(2):120–127.
9. Carnieli P, Castilho JG, Fahl WDO, Véras NMC, Timenetsky MDCST. Genetic characterization of Rabies virus isolated from cattle between 1997 and 2002 in an epizootic area in the state of São Paulo, Brazil. *Virus Res*. 2009;144:215–224.
10. Aréchiga-Ceballos N et al. New rabies virus variant found during an epizootic in white-nosed coatis from the Yucatan Peninsula. *Epidemiol Infect*. 2010;138:1586–1589.
11. Condori-Condori RE, Streicker DG, Cabezas-Sanchez C, Velasco-Villa A. Enzootic and epizootic rabies associated with vampire bats Peru. *Emerg Infect Dis*. 2013;19:1463–1469.
12. Araujo DB et al. Antibodies to rabies virus in terrestrial wild mammals in native rainforest on the north coast of São Paulo State, Brazil. *J Wildl Dis*. 2014;50(3):469–477.
13. Badrane H, Tordo N. Host switching in Lyssavirus history from the Chiroptera to the Carnivora orders. *J Virol*. 2001;75:8096–8104.
14. Borucki MK et al. Ultra-deep sequencing of intra-host rabies virus populations during cross-species transmission. *PLoS Negl Trop Dis*. 2013;7:e2555.
15. Scott TP et al. Complete genome and molecular epidemiological data infer the maintenance of rabies among kudu (*Tragelaphus strepsiceros*) in Namibia. *PLoS One*. 2013;8:e58739.
16. Mollentze N, Biek R, Streicker DG. The role of viral evolution in rabies host shifts and emergence. *Curr Opin Virol*. 2014;8:68–72.

17. Horton DL et al. Rabies in Iraq: Trends in human cases 2001–2010 and characterisation of animal rabies strains from Baghdad. *PLoS Negl Trop Dis*. 2013;7(2):e2075.
18. Ellison JA et al. Bat rabies in Guatemala. *PLoS Negl Trop Dis*. 2014;8(7):e3070.
19. Stahl JP et al. Update on human rabies in a dog- and fox-rabies-free country. *Med Mal Infect*. 2014;44(7):292–301.
20. Tang HB et al. Re-emergence of rabies in the Guangxi province of Southern China. *PLoS Negl Trop Dis*. 2014;8(10):e3114.
21. Wang L, Tang Q, Liang G. Rabies and rabies virus in wildlife in mainland China, 1990–2013. *Int J Infect Dis*. 2014;25:122–129.
22. Yin JF et al. Identification of animal rabies in Inner Mongolia and analysis of the etiologic characteristics. *Biomed Environ Sci*. 2014;27(1):35–44.
23. Zieger U et al. The phylogeography of rabies in Grenada, West Indies, and implications for control. *PLoS Negl Trop Dis*. 2014;8(10):e3251.
24. Dibia IN et al. Phylogeography of the current Rabies viruses in Indonesia. *J Vet Sci*. 2015. [Epub ahead of print]
25. Knobel DL et al. Re-evaluating the burden of rabies in Africa and Asia. *Bull World Health Organ*. 2005;83:360–368.
26. Tasioudi KE et al. Recurrence of animal rabies, Greece, 2012. *Emerg Infect Dis*. 2014;20(2):326–328.
27. Schatz J et al. Lyssavirus distribution in naturally infected bats from Germany. *Vet Microbiol*. 2014;169(1–2):33–41.
28. Pathak S et al. Diagnosis, management and post-mortem findings of a human case of rabies imported into the United Kingdom from India: A case report. *Virol J*. 2014;11:63.
29. Yang DK et al. Comparison of four diagnostic methods for detecting rabies viruses circulating in Korea. *J Vet Sci*. 2012;13(1):43–48.
30. Banyard AC, Horton DL, Freuling C, Müller T, Fooks AR. Control and prevention of canine rabies: The need for building laboratory-based surveillance capacity. *Antiviral Res*. 2013;98(3):357–364.
31. Silva SR et al. Biotechnology advances: A perspective on the diagnosis and research of rabies virus. *Biologicals*. 2013;41(4):217–223.
32. Morgan SM, Pouliott CE, Rudd RJ, Davis AD. Antigen detection, rabies virus isolation, and Q-PCR in the quantification of viral load in a natural infection of the North American beaver (*Castor canadensis*). *J Wildl Dis*. 2015;51(1):287–289.
33. Fischer M, Hoffmann B, Freuling CM, Müller T, Beer M. Perspectives on molecular detection methods of lyssaviruses. *Berl Munch Tierarztl Wochenschr*. 2012;125(5–6):264–271.
34. Suin V et al. A two-step lyssavirus real-time polymerase chain reaction using degenerate primers with superior sensitivity to the fluorescent antigen test. *Biomed Res Int*. 2014;2014:256175.
35. Wang L et al. A SYBR-green I quantitative real-time reverse transcription-PCR assay for rabies viruses with different virulence. *Virol Sin*. 2014;29(2):131–132.
36. Heaton PR et al. A hemi-nested PCR assay for the detection of six genotypes of rabies and rabies-related viruses. *J Clin Microbiol*. 1997;35:2762–2766.
37. Coertse J et al. A case study of rabies diagnosis from formalin-fixed brain material. *J S Afr Vet Assoc*. 2011;82(4):250–253.
38. Biswal M, Ratho RK, Mishra B. Role of reverse transcriptase polymerase chain reaction for the diagnosis of human rabies. *Indian J Med Res*. 2012;135(6):837–842.
39. David D. Role of the RT-PCR method in ante-mortem and post-mortem rabies diagnosis. *Indian J Med Res*. 2012;135(6):809–811.

40. Beltran FJ, Dohmen FG, Del Pietro H, Cisterna DM. Diagnosis and molecular typing of rabies virus in samples stored in inadequate conditions. *J Infect Dev Ctries.* 2014;8(8):1016–1021.

41. Conrardy C et al. Molecular detection of adenoviruses, rhabdoviruses, and paramyxoviruses in bats from Kenya. *Am J Trop Med Hyg.* 2014;91(2):258–266.

42. McElhinney LM, Marston DA, Brookes SM, Fooks AR. Effects of carcase decomposition on rabies virus infectivity and detection. *J Virol Methods.* 2014;207:110–113.

43. Fischer M et al. Molecular double-check strategy for the identification and characterization of European Lyssaviruses. *J Virol Methods.* 2014;203:23–32.

44. Foord AJ et al. Molecular diagnosis of lyssaviruses and sequence comparison of Australian bat lyssavirus samples. *Aust Vet J.* 2006;84(7):225–230.

45. Hughes GJ. Experimental infection of big brown bats (*Eptesicus fuscus*) with Eurasian bat lyssaviruses Aravan, Khujand, and Irkut virus. *Arch Virol.* 2006;151(10):2021–2035.

46. Coertse J, Weyer J, Nel LH, Markotter W. Improved PCR methods for detection of African rabies and rabies-related lyssaviruses. *J Clin Microbiol.* 2010;48(11):3949–3955.

47. Hemachudha T et al. Human rabies: Neuropathogenesis, diagnosis, and management. *Lancet Neurol.* 2013; 12(5):498–513.

48. Weant KA, Baker SN. Review of human rabies prophylaxis and treatment. *Crit Care Nurs Clin North Am.* 2013;25(2):225–242.

49. Kim YR. Prophylaxis of human hydrophobia in South Korea. *Infect Chemother.* 2014;46(3):143–148.

50. Cheong Y et al. Strategic model of national rabies control in Korea. *Clin Exp Vaccine Res.* 2014;3(1):78–90.

51. Driver C. Rabies: Risk, prognosis and prevention. *Nurs Times.* 2014;110(14):16–18.

52. Fooks AR et al. Current status of rabies and prospects for elimination. *Lancet.* 2014;384(9951):1389–1399.

53. Mähl P et al. Twenty year experience of the oral rabies vaccine SAG2 in wildlife: A global review. *Vet Res.* 2014;45(1):77.

54. Wacharapluesadee S et al. Detection of rabies viral RNA by TaqMan real-time RT-PCR using non-neural specimens from dogs infected with rabies virus. *J Virol Methods.* 2012;184(1–2):109–112.

55. Mani RS, Madhusudana SN. Laboratory diagnosis of human rabies: Recent advances. *Sci World J.* 2013;2013:569712.

56. Fischer M et al. A step forward in molecular diagnostics of lyssaviruses—Results of a ring trial among European laboratories. *PLoS One.* 2013;8(3):e58372.

57. Aravindhbabu RP, Manoharan S, Ramadass P. Diagnostic evaluation of RT-PCR-ELISA for the detection of rabies virus. *Virus Dis.* 2014;25(1):120–124.

42 Vesicular Stomatitis Virus

Dongyou Liu and Sándor Belák

CONTENTS

42.1 Introduction 369
    42.1.1 Classification 369
    42.1.2 Morphology and Genome Organization 370
    42.1.3 Replication and Epidemiology 370
    42.1.4 Clinical Features and Pathogenesis 371
    42.1.5 Diagnosis 372
        42.1.5.1 Virus Isolation 372
        42.1.5.2 Serological Detection of Viral Antigens and Antibodies 372
        42.1.5.3 RNA Detection 372
    42.1.6 Treatment and Prevention 372
42.2 Methods 373
    42.2.1 Sample Preparation 373
        42.2.1.1 Sample Collection 373
        42.2.1.2 Virus Isolation 373
        42.2.1.3 RNA Isolation 373
    42.2.2 Detection Procedures 373
        42.2.2.1 Multiplex RT-PCR for the Detection of Vesicular Disease Viruses 373
        42.2.2.2 Real-Time RT-PCR for the Detection of VSV-IN and VSV-NJ 373
        42.2.2.3 RT-PCR and Enzyme Digestion for the Detection of VSV-IN and VSV-NJ 374
42.3 Conclusion 374
References 374

42.1 INTRODUCTION

Vesicular stomatitis (VS; also known as vesicular stomatitis virus disease, vesicular stomatitis fever, and Indiana fever) is an arthropod-borne, zoonotic vesicular disease caused by vesicular stomatitis viruses (VSV) in the genus *Vesiculovirus*, family *Rhabdoviridae*. Clinically, VS is characterized by the formation of vesicles, erosions, and ulcers on the tongue, oral tissues, feet, and teats, along with pain, anorexia, and secondary mastitis, leading to substantial loss of productivity in all affected species (primarily horses and cattle and occasionally swine, sheep, goats, llamas, and alpacas) and death in swine. With the exception of its appearance in horses, VS resembles other vesicular diseases in animals, that is, foot-and-mouth disease (FMD), swine vesicular disease (SVD), and vesicular exanthema of swine (VES), which all display vesicular lesions on the mouth and/or feet, and some (e.g., FMD) of which have significant impact on international trade. Therefore, differentiation of VS from other vesicular diseases is vital to contain the spread of these diseases and reduce their drain on agriculture industry.

42.1.1 CLASSIFICATION

VSV are negative-sense single-stranded (ss) RNA viruses classified in the genus *Vesiculovirus*, family *Rhabdoviridae*, order *Mononegavirales*. Encompassing >200 different viruses that are present in a broad range of hosts (including insects, fish, mammals, and plants), the family *Rhabdoviridae* (*rhabdo* is Greek for "rod") is organized into 11 genera (*Cytorhabdovirus*, *Ephemerovirus*, *Lyssavirus*, *Novirhabdovirus*, *Nucleorhabdovirus*, *Perhabdovirus*, *Sigmavirus*, *Sprivivirus*, *Tibrovirus*, *Tupavirus*, and *Vesiculovirus*) together with nine proposed genera (*Almendravirus*, *Bahiavirus*, *Bracorhabdovirus*, *Curiovirus*, *Dichorhabdovirus*, *Hapavirus*, *Ledantevirus*, *Sawgravirus,* and *Sripuvirus*) [1].

Interestingly, while members of the genus *Lyssavirus* (e.g., rabies virus) replicate exclusively in mammals, viruses belonging to the genera *Vesiculovirus* and *Ephemerovirus* (e.g., bovine ephemeral fever virus) occur in both vertebrate and invertebrate hosts. The agents of the genus *Novirhabdovirus* are present in fish and other aquatic animals, including invertebrates, and those of the genera *Cytorhabdovirus* and *Nucleorhabdovirus* are plant pathogens.

Currently, the genus *Vesiculovirus* consists of nine recognized species: Carajas virus, Chandipura virus (CHPV), Cocal virus (COCV; formerly the Indiana 2 subtype of VSV), Isfahan virus (ISFV), Maraba virus, Piry virus (PIRYV), vesicular stomatitis Alagoas virus (VSV-AV; formerly the Indiana 3 subtype of VSV), vesicular stomatitis Indiana virus (VSV-IN, the type species; formerly the Indiana 1 subtype of VSV, often still referred to as VSV), and vesicular stomatitis New Jersey virus (VSV-NJ). An additional 19 viruses are also tentatively assigned to this genus.

It is notable that Chandipura, Piry, and Isfahan viruses cause acute febrile illness and meningoencephalitis in humans, with CHPV and ISFV being identified in Asia and PIRYV in South America. VSV-IN, VSV-NJ, VSV-AV, and COCV are responsible for VS in animals (particularly horses, cattle, and pigs) and are collectively known as VSV [2,3]. VSV are endemic in North, Central, and South America, with VSV-IN and VSV-NJ being of importance to animal health [4]. Except for COCV, VSV may occasionally infect humans [5].

42.1.2 Morphology and Genome Organization

Similar to other members of the genus *Vesicuovirus*, VSV are enveloped, bullet-shaped (or elongated, rodlike) particles of approximately 75 nm in diameter and 185 nm in length. The envelope is made up of glycoprotein (G) and matrix (M) protein, with a trimer of the G protein forming spikes of 9–11 nm on surface and M protein lying underneath. Within the envelope is the helical nucleocapsid that is coiled symmetrically into a cylindrical structure with a striated appearance [6]. Composed of the nucleoprotein (N) and large (L) protein, the nucleocapsid wraps around a linear, nonsegmented, negative-sense, single-stranded RNA of 11,161 nucleotides (as exemplified by the prototype VSV-IN), protecting the RNA from ribonuclease digestion and maintaining the RNA in a configuration suitable for transcription [7].

Structurally, the VSV genome is organized in the order 3′ N-P-M-G-L 5′, encoding (via five monocistronic mRNAs) five multifunctional proteins: nucleocapsid (N) protein, phosphoprotein (P, a component of the viral RNA polymerase), M protein, G protein, and L protein (a component of the viral RNA polymerase). There is evidence that the gene transcripts are synthesized both in sequential order and quantity according to the same order of the genes in the genome (i.e., N > P > M > G > L) [8].

The nucleoprotein (N) is a 52 kDa molecule that forms a complex with the P protein, thus preventing the concentration-dependent aggregation (oligomerization) of N and keeping the N protein in a monomeric, encapsidation-competent form. The N protein encapsidates nascent genome RNA (through binding to the ribose-phosphate backbone of the RNA) into an RNase resistant form (which functions as the active template for transcription and replication) to protect from cellular RNAse activity in the absence of polynucleotide synthesis.

Binding to the nascent leader RNA in its unphosphorylated form, the P protein promotes read-through of the transcription termination signals and initiates nucleocapsid assembly on the nascent RNA chain. Together with L protein, P protein forms viral RNA-dependent RNA polymerase required for replication of the mRNA, while L protein maintains catalytic functions for RNA polymerization, capping, and Poly-A polymerase.

M protein is a 229 amino acid protein of 25 kDa in size that forms a layer between the G protein containing outer membrane and the nucleocapsid core (consisting of nucleoprotein [N], polymerase [L], P protein, and RNA genome), condensing the nucleocapsid core into the "skeletons" seen in mature virions. The N terminus of M possesses several positively charged amino acid residues that interact directly with negatively charged membranes.

G protein is a 67 kDa (495 amino acid) molecule with N-linked glycan. Spiking out of the membrane, G protein mediates viral attachment to an LDL receptor (LDLR) or an LDLR family member present on the host cell and contains major antigenic determinant [9]. Indeed, the VSV G protein has been shown to be a determinant of higher virulence of VSV-NJ than of VSV-IN in swine [10].

42.1.3 Replication and Epidemiology

In general, the replication of vesiculoviruses involves three distinct phases: uncoating, transcription/replication, and viral assembly. First, G protein binds to LDL receptor on the surface of host cells and mediates the viral entry via endocytosis and fusion of the viral envelope with the endosomal membrane. The virus then enters the cell cytoplasm through partially clathrin-coated vesicles, which recruit components of the actin machinery to induce its own uptake. Once inside the cytoplasm, the L + P polymerase complex (i.e., the viral RNA-dependent RNA polymerase) attaches to the encapsidated genome at the leader region and initiates sequential transcription of the open reading frames (ORFs) in the virus genome to generate five monocistronic mRNAs, with the intergenic sequence directing termination and reinitiation of transcription by the polymerase between genes. The mRNA is capped and polyadenylated by the L protein during synthesis, and progeny vRNA is made from a positive-sense intermediate. Virions are then assembled around the nucleoprotein core and subsequently bud from cytoplasmic membranes and the outer membrane of the cell, acquiring the M + G proteins during the process, which are responsible for the characteristic bullet-shaped morphology of the virus [11].

VSV mainly infect insects, horses, cattle, swine, sheep, goats, llamas, alpacas, donkeys, and mules. Other susceptible animals include deer, pronghorn antelope, South American camelids, raccoons, lynx, bobcats, bears, coyotes, foxes, dogs, deer, rabbits, skunk, armadillo, opossum, rodents, bats, turkeys, and ducks, which may function as potential reservoir host for VSV. In addition, guinea pigs, hamsters, mice, ferrets, and chickens have been infected experimentally. Humans are also susceptible [12].

The most important vectors for VSV transmission are sand flies (*Lutzomyia* sp. especially *Lutzomyia shannoni*). Other potential biological vectors include blackfly (*Simulium*), horsefly (*Tabanus*), deerfly (*Chrysops*), stablefly (*Stomoxys*), biting midges (*Culicoides*), and mosquitoes (*Aedes* and *Culex*). Flies in the genus *Musca* and eye gnats (*Chloropidae*) may act as mechanical vectors for VSV. In addition, cattle that ingest experimentally infected grasshoppers (*Melanoplus sanguinipes*) develop clinical signs [13].

Natural VSV infections involve cytolytic infections of mammalian hosts and subsequent transmission by insects. Within insects, infections are noncytolytic persistent. Once introduced into a herd, VS may spread from animal to animal by direct contact with broken skin or mucous membranes, vesicular material, saliva and nasal secretions, and exposure to contaminated fomites including food, water, and milking machines. Animals may acquire infection by aerosols, but most species do not develop skin lesions via this route. In addition, VSV do not seem to cross the placenta or cause fetal seroconversion [14].

VSV in saliva can remain viable for 3–4 days on fomites but do not survive for long periods in the environment except in cool, dark places. The virus is inactivated by sunlight, formaldehyde, ether, 1% sodium hypochlorite, 70% ethanol, 2% glutaraldehyde, 2% sodium carbonate, 4% sodium hydroxide, 2% iodophore, 1% cresylic acid, 2.5% phenol, 0.4% HCl, 2% sodium orthophenylphenate, and 4% sodium hypochlorite [15]. Insect vectors may acquire the virus by feeding on lesion material or contaminated secretions, as domesticated or wild animals generally have insufficient viremia for insect vectors to become infected. Humans may acquire the infection by contact with vesicular fluid and saliva from infected animals, or through insect bites [16].

VS was first described in 1916 on horses showing the clinical signs typical of VSV infection in Denver, Colorado stockyards. In the same year, the disease was also reported in Illinois, Iowa, Kansas, Missouri Montana, Nebraska, South Dakota, Utah, and Wyoming. In 1925, the causal agent for VS was detected from healthy cattle arrived in Indiana from Kansas City and named vesicular stomatitis Indiana strain. In 1926, investigation of a VS outbreak in cattle in New Jersey led to the identification of vesicular stomatitis New Jersey strain, which appears distinct from vesicular stomatitis Indiana strain. In the subsequent years, sporadic VS cases have been described throughout the United States, with the exception of New England [4]. The most recent VS outbreak in the United States occurred in 2012, when two horses on an equine property in Otero County, New Mexico, had clinical signs consistent with VS and positive VS virus serology. The causal agent was identified as VSV-NJ, and a total of 51 horses on 36 premises in 2 states (including 34 in New Mexico and 2 in Colorado) were confirmed positive. Phylogenetic analysis of the virus indicated its close relationship to viruses detected in the state of Veracruz, Mexico, in 2000 [17].

At the moment, VS is endemic in Mexico, Central America, and South America (particularly Brazil), with seasonal outbreaks emerging at the end of the rainy season or early in the dry season. Sporadic VS may appear in other parts of the Western Hemisphere, both north and south of the endemic area. While VSV-NJ and VSV-IN are present in North, Central, and South America, VSV-AV (Indiana–3) and cocal virus (Indiana–2) are mainly found in parts of South America such as Argentina and Southern Brazil. Besides the Western Hemisphere, VS has been recorded in France and South Africa.

42.1.4 CLINICAL FEATURES AND PATHOGENESIS

With an incubation period of 2–8 days, the initial symptoms of VS are fever (24–48 h after infection) and excessive salivation (due to rupture of the oral vesicles). This is followed by the appearance of vesicles (blisters, of varied sizes), papules, erosions, and ulcers (which result from rupture of vesicles) around the mouth and on the feet, udder, and prepuce. Due to the pain associated with these lesions, animals may show inappetance, anorexia, lethargy, and pyrexia and refuse to drink and walk. Other clinical signs may include sloughing off in the epithelium of the tongue or muzzle, swollen nostrils and muzzle, catarrhal nasal discharge, bleeding from ulcers, and fetid odor in the mouth. Although most (80%–95%) of the animals (especially those of under a year of age) in a herd may have subclinical infections, the morbidity rate in the symptomatic animals (particularly pigs) ranges from 5% to 90%. If no secondary bacterial infections or other complications develop, affected animals often recover in approximately 2–3 weeks [8].

In humans, infections with Indiana and New Jersey VSV serotypes may show biphasic high fever and other flulike symptoms such as severe malaise, headaches, myalgia, arthralgia, retrosternal pain, eye aches, and nausea. Occasionally, vesicles may form on the oral mucosa, lips, nose, and hands, with a rare occurrence of cervical lymphadenopathy. In human cases of Alagoas virus infections reported in Brazil, flulike symptoms resolve within 3–4 days of appearance. Due to its zoonotic nature, human VS is only transmitted from animals, not from other humans.

VSV have been shown to induce intercellular edema in the Malphigian layer of the epithelium, leading to the separation of the epithelial cells by vacuolar cavities. Subsequent infiltration of inflammatory cells, such as granulocytes and monocytes, contributes to cellular lysis and necrosis. As the necrotic, edematous mucosa breaks free from underlying tissue, cellular exudates fill the cavity and form the vesicles. The whole process of vesicle formation (involving intercellular edema, cellular necrosis, and inflammatory cell infiltration) takes about 48 h upon infection and vesicles may begin to disappear with seepage of edematous fluid afterward [8].

VSV M protein has the capacity to interrupt cellular transcription pathways and to block mRNA export from the nucleus, both of which contribute to its aversion of the host type I interferon response and its disease process in mammalian hosts [18].

42.1.5 Diagnosis

With the exception of horses, VS in most animal species is indistinguishable on the basis of clinical manifestations from other vesicular diseases including FMD, vesicular exanthema, and SVD. Other differential diagnoses consist of footrot, rinderpest, infectious bovine rhinopneumonitis, bovine viral diarrhea, malignant catarrhal fever, epizootic hemorrhagic disease, bluetongue, contagious ecthyma, lip and leg ulceration, and chemical or thermal burns. Therefore, laboratory tests are required for accurate diagnosis of vesicular stomatitis. The current recommendations from World Organization for Animal Health (Office International des Epizooties) for the laboratory diagnosis of VS include virus isolation, serological detection of viral antigens and antibodies, and RNA detection [15].

42.1.5.1 Virus Isolation

African green monkey kidney (Vero), baby hamster kidney (BHK-21), and IB-RS-2 cells are useful for VSV isolation and detection, as VSV, FMDV, and SVDV cause different cytopathic effects on these cells. Therefore, inoculation of all three cell lines with the test sample enhances the diagnosis. In addition, embryonated eggs, or infant mice may be employed for VSV cultivation.

42.1.5.2 Serological Detection of Viral Antigens and Antibodies

Although electron microscopy can help differentiate vesicle-causing viruses, detection of viral antigens and antibodies provides a cost-effective approach for confirmation of the virus identity. The widely used serological assays include serum neutralization test (SNT), complement fixation test (CFT), and enzyme-linked immunosorbent assays (ELISA). SNT, CFT, and competitive ELISA (cELISA) have been recognized as prescribed tests by the OIE for international trade purposes. In addition, immunochromatographic (ICT) assay based on colloidal gold label has proven useful for identification of VSV in comparison with reverse transcription (RT)-PCR (which employs forward primer: 5′-CGGATCCGTCAAAATGCCCAAGAGTCACA-3′ and reverse primer: 5′-CCTCGAGTGGGTTTTTAGGAGCA AGATAGC-3′) [19].

42.1.5.3 RNA Detection

VSV RNA may be amplified and detected by (RT-PCR on a variety of formats (heminested, multiplex, and real-time assays) [20–28]. Recently, a multiplex RT-PCR with an accompanying microarray assay was reported for simultaneous detection and typing of FMDV and VSV and for the detection of SVDV and VESV [28].

Other useful molecular technologies for detection and identification of VSV RNA and related pathogens include isothermal amplification, nucleic acid hybridization-based procedures (e.g., macroarray, microarray, and proximity ligation assay), metagenomic analysis, next generation sequencing, microfluidic analysis systems, and nanotechnology [29–32]. The recent description of padlock probes and microarrays highlights the power of these alternative technologies for simultaneous detection and differentiation of FMDV, SVDV, and VSV and for uncovering serotype details in cases of VSV infection. In contrast to other array methods, the padlock probe/microarray assay requires conserved target sequences only for the padlock probes and thus is sensitive to genomic variations only during the probe hybridization. Furthermore, the intramolecular hybridization and ligation of padlock probes makes it easy to include additional probes in the pool without affecting the protocol, in order to cater for new emerging strains that are not recognized by the current padlock probe set. Therefore, this multiplex detection platform based on padlock probes and microarrays provides an effective screening tool for viruses causing similar vesicular symptoms and facilitates rapid counteractions, especially in case of FMD outbreaks [33].

42.1.6 Treatment and Prevention

As no specific treatment or cure is currently available for vesicular stomatitis, symptomatic treatments, such as cleansing the lesions with a mild antiseptic solution to eliminate secondary bacterial infections and provision of softened feed, will help reduce the impact of the infection.

Good sanitation, disinfection, and quarantine practice will facilitate the containment of the disease. Feed and water troughs should be cleaned regularly. Milking equipment should be disinfected between uses using 1% sodium hypochlorite, 70% ethanol, 2% glutaraldehyde, 2% sodium carbonate, 4% sodium hydroxide, iodophore disinfectants, formaldehyde, or chlorine dioxide. VSV may be also inactivated by UV light, lipid solvents, or heat. Uninfected livestock should be kept away by stabling from any infected/symptomatic animals during outbreaks to prevent the spread of VSV between animals by direct contact. Unless the animals are going directly to slaughter, no movement of animals from an infected property should be allowed for at least 21 days after all lesions are healed. Insect breeding areas should be sprayed with insecticides, and treated eartags are used on animals [15].

Human infection with VSV often results from handling affected animals, contaminated fomites, tissues, blood, or virus cultures. To prevent VSV infection in humans, protective clothing and gloves should be used when handling infected animals, and biological safety precautions should be taken in the laboratory.

It is of interest to note that while only a limited number of vaccine preparations are available for use in animals, but not in humans, against VSV infection, many vaccine candidates based on recombinant VSV have been developed for treatment/control of cancers and certain viral diseases (e.g., diseases caused by filoviruses, HCV, HIV, etc.) [34,35]. Given that most cancer cells have impaired antiviral responses related to type I interferon pathways, they are more susceptible to VSV than normal cells. In addition, with an easily manipulated genome, relative independence of a receptor or cell cycle, cytoplasmic replication without risk of host cell transformation, and lack of preexisting immunity in humans, VSV offer a promising candidate for clinical use [35].

42.2 METHODS

42.2.1 SAMPLE PREPARATION

42.2.1.1 Sample Collection

As VS is a zoonotic disease, samples (serum, blister fluid, the epithelium covering unruptured vesicles, epithelial flaps from freshly ruptured vesicles, swabs of the ruptured vesicles, esophageal/pharyngeal fluid collected with a probang cup from cattle, or throat swabs from pigs) should be collected and handled with all appropriate precautions, and sent frozen with dry ice within 2 days of collection to authorized laboratories to prevent the spread of the disease [36].

42.2.1.2 Virus Isolation

Clinical specimens are inoculated onto African green monkey kidney (Vero), baby hamster kidney (BHK-21), and IB-RS-2 cells maintained in minimum essential medium supplemented with 10% fetal bovine serum (FBS) or the sand fly cell line maintained in Grace's insect cell culture medium supplemented with 15% FBS. The inoculated cultures are checked under an inverted microscope regularly for cytopathic effects and stained after 48 h postinfection with amido black. Tissue culture fluids from cultures showing cytopathic effects are stained with 1% sodium phosphotungstic acid pH 6·0 and examined with a transmission electron microscope.

42.2.1.3 RNA Isolation

Total RNA is extracted from 100 to 200 µL amounts of samples (cell cultures, serum, EDTA-blood, nasal and pharyngeal swabs, and feces or tissue homogenates 10% in PBS) with a commercial reagent (e.g., Tripure Isolation Reagent or High Pure PCR Template Preparation Kit, Roche Diagnostics; QIAamp viral RNA minikit, Qiagen). RNA is resuspended in 10–50 µL of Milli-Q water. Nasal and pharyngeal swabs are solubilized in 2 mL of PBS for 2 h prior to RNA extraction. Interestingly, a recent study by Arshed et al. [27] showed that a guanidine-thiocyanate-based RNA extraction (Qiagen RNeasy Mini Kit, Qiagen, Valencia, CA) followed by column-based purification is superior to triphasic extraction (Tri-reagent method) for preparation of VSV-IN and VSV-NJ RNA from cell lysate and spiked tissue samples as assessed by one-step RT-PCR.

42.2.2 DETECTION PROCEDURES

42.2.2.1 Multiplex RT-PCR for the Detection of Vesicular Disease Viruses

Lung et al. [28] described a multiplex RT-PCR for detection of vesicular disease viruses with species-specific primers obtained from gene regions of FMDV, VSV, SVDV, and VESV (Table 42.1). The multiplex RT-PCR is highly specific as it does not react with genomic nucleic acids of eight nontarget livestock viruses tested: exotic malignant catarrhal fever virus, rinderpest virus, bluetongue virus, bovine herpesvirus 1, bovine viral diarrhea virus type I and type II, orf virus, and porcine reproductive and respiratory syndrome virus. In addition, nucleic acids from oral swabs collected from healthy cattle, sheep, and pigs are not recognized by the multiplex RT-PCR.

42.2.2.2 Real-Time RT-PCR for the Detection of VSV-IN and VSV-NJ

Rasmussen et al. [23] developed a quantitative RT-PCR assay for the detection and differentiation of the two main VSV serotypes based on primer–probe energy transfer (PriProET) detection system, which combines probe-based real-time monitoring of PCR amplification with confirmation of probe hybridization from the melting temperature (T_m) curve. A pair of primers (one of which was labeled with the donor fluorophore 6-carboxyfluorescein [FAM]) is designed from the most conserved regions of the L gene (encoding the RNA-dependent RNA polymerase) of the VSV genome, and two serotype-specific fluorescent probes are labeled with a reporter fluorophore, Texas Red (a serotype Ind-1-specific probe), or cyanine 5 (Cy5; a serotype NJ-specific probe), facilitating differentiation on the basis of the corresponding emission spectra (Texas Red, 610 nm;

TABLE 42.1
Primers Used for Multiplex RT-PCR

Virus	Primers (5′–3′)	Nucleotide Position	Amplicon (bp)
FMDV	GGT CTT GCC CAG TAC TAC RCM CAG TA	1C gene (nt 2837–2862)[a]	1190
	GCG GAC ACC AGA CGG TTR AAG TC	2B gene (nt 4004–4026)[a]	
VSV	GGA GAT AAW TGG CAT RAA CTT CC	L gene (nt 6054–6076)[b]	1064
	GAC CAT CTY TTK GTY TCT ARY CCT C	L gene (nt 7093–7117)[b]	
SVDV	CAG CGG CAC TCC TCA GAC ACT AC	3C gene (nt 5765–5787)[c]	791
	GAG TTT CAG GCA CGT AAA CCA CAC	3D gene (nt 6432–6555)[c]	
VESV	CGA CTC GAT GGA CCT GTT CAC ATA CG	Polymerase gene (nt 5101–5126)[d]	649
	CGT AGA GGT CGG TTA GGT CCT TTC TG	Polymerase gene (nt 5724–5749)[d]	

[a] FMDV reference strain A24 Cruzeiro/Brazil/55 (GenBank accession no. AY593768).

[b] VSV Indiana reference strain 98COE (GenBank accession no. AF473864).

[c] SVDV reference strain Itl1/97 (GenBank accession no. EU151456).

[d] VESV reference strain (GenBank accession no. NC_002551).

TABLE 42.2

PCR Primers and Probes for Detection of VSV

VSV Serotype	Primer or Probes	Sequence (5'–3')[a]	Nucleotide Position[b]
VSV	Forward	FAM-TAAATGAPGATGAKACPATGCAATC	7017–7041
VSV	Reverse	ACKCAIGTPACPCGPGACCATCT	7131–7109
Ind-1	Ind-1-specific probe	CGGTATTTTTCCATAATTCAAGTAATCTGCT-Texas Red	7072–7042
New Jersey	NJ-specific probe	GGAATTTTCCCATAGTTCAAATAGTCTGCT-Cy5	7071–7042

[a] K (dK), P (dP), or I (inosine).

[b] Nucleotide positions are based on the VSV Ind-1 sequence (GenBank accession no. J02428).

Cy5, 660 nm) (Table 42.2). This RT-PCR method provides a sensitive, fast, and robust tool for the simultaneous detection, differentiation, and quantification of the two major VSV serotypes, Ind-1 and NJ.

42.2.2.3 RT-PCR and Enzyme Digestion for the Detection of VSV-IN and VSV-NJ

VSV specific primers (VSV-1/VSV-2) targeting the highly conserved L gene region enable amplification of a 110 bp fragment from VSV-IN and VSV-NJ (Table 42.3). Subsequent digestion of the PCR product with BsmA I restriction endonuclease generates two fragments of 86 and 24 bp for VSV-IN and two fragments of 102 and 8 bp for VSV-NJ [24].

42.2.2.3.1 Procedure

1. Prepare RT-PCR mixture (25 μL) containing 2 μL of RNA sample, 1× Buffer (Tris-Cl, KCl, [NH$_4$]$_2$SO$_4$), 2.5 mM MgCl$_2$, 1× Q solution, 0.2 mM dNTPs, 0.7 μM of each primer VSV-1/VSV-2, 5 U RNase Inhibitor (Applied Biosystems), and 1 μL Enzyme Mix (Omniscript and Sensiscript Reverse Transcriptases, HotStartTaq DNA polymerase).

2. Conduct RT amplification in a one step with the following conditions: 30 min at 50°C, 15 min at 95°C, 40 cycles at 94°C in 30 s, 60°C in 30 s, 72°C in 30 s, and finally 7 min at 72°C.

3. Digest 10 μL of VSV amplification products with 5 U of BsmA I in a volume of 20 μL for at least 1 h 30 min at 55°C.

4. Analyze the restriction pattern on a 3% agarose gel containing 0.5 μg/mL of ethidium bromide.

TABLE 42.3

PCR Primers from VSV L Gene for Detection of VSV-IN and VSV-NJ

Primer	Sequence (5'–3')	Nucleotide Position	Amplicon (bp)
VSV-1	AATGACGATGAGACYATGCAATC	nt 7019–7041	110
VSV-2	CAAGTCACYCGTGACCATCT	nt 7128–7109	

42.3 CONCLUSION

VSV are causative agents of an acute vesicular disease in animals (particularly cattle, horses, and pigs), with clinical manifestations ranging from vesicles, erosions, and ulcers on the tongue, oral tissues, feet, and teats. Although it is a nonlethal disease with geographical distribution largely limited to the West hemisphere (i.e., North, Central, and South America), VS is important in its resemblance to other viral vesicular diseases of livestock, FMDV in cattle, and SVDV and VESV in pigs. However, VSV are the only one of these vesicular disease viruses that infect horses. Indeed, inoculation of a horse's tongue was used historically to differentiate VSV from the other vesicular disease viruses. In addition, VSV is a zoonotic agent that may occasionally infect humans, while FMDV, SVDV, and VESV are in general, considered as noninfective to human hosts.

Laboratory differentiation of VS from FMD, SVD, and VES has traditionally relied on virus isolation and serological detection of viral antigens and antibodies. As these procedures are time consuming and pose significant health risk to laboratory personnel, nucleic acid amplification techniques such as RT-PCR have increasingly been adapted in clinical laboratories for improved identification and typing of VSV.

Considering the notable absence of effective vaccines, the increasing intercontinental travel, the insatiable demand for susceptible animals (cattle, pigs, and horses) as food source, and the remarkable plasticity of the VSV genome, there is a potential for VS outbreaks to occur more frequently in the future. Therefore, in addition to have an accurate and sensitive diagnostic tool at our disposal, further investigations on the pathogenic mechanisms of VSV will aid the development of novel control measures against vesicular stomatitis.

REFERENCES

1. Lyles DS, Rupprecht CE. Rhabdoviridae. In: Knipe DM, Howley PM (eds.), *Fields Virology*, 5th edn., 2007; pp. 1363–1408, Lippincott Williams & Wilkins, Philadelphia, PA.

2. Travassos da Rosa AP et al. Carajas and Maraba viruses, two new vesiculoviruses isolated from phlebotomine sand flies in Brazil. *Am J Trop Med Hyg.* 1984;33(5):999–1006.

3. Tesh RB et al. Natural infection of humans, animals, and phlebotomine sand flies with the Alagoas serotype of vesicular stomatitis virus in Colombia. *Am J Trop Med Hyg.* 1987;36(3):653–661.

4. Rodriguez LL. Emergence and re-emergence of vesicular stomatitis in the United States. *Virus Res.* 2002;85:211–219.

5. Lichty BD, Power AT, Stojdl DF, Bell JC. Vesicular stomatitis virus: Re-inventing the bullet. *Trends Mol Med.* 2004;10(5):210–216.

6. Luo M. The nucleocapsid of vesicular stomatitis virus. *Sci China Life Sci.* 2012;55(4):291–300.

7. Rodriguez LL et al. Full-length genome analysis of natural isolates of vesicular stomatitis virus (Indiana 1 serotype) from North, Central and South America. *J Gen Virol.* 2002;83:2475–2483.

8. Salmon MD, McCluskey BJ. Vesicular stomatitis virus. In: Liu D (ed.), *Manual of Security Sensitive Microbes and Toxins*, 2014; pp. 737–745, CRC Press, Boca Raton, FL.

9. Albertini AA, Baquero E, Ferlin A, Gaudin Y. Molecular and cellular aspects of rhabdovirus entry. *Viruses.* 2012;4(1): 117–139.

10. Martinez I et al. Vesicular stomatitis virus glycoprotein is a determinant of pathogenesis in swine, a natural host. *J Virol.* 2003;77(14):8039–8047.

11. Hastie E, Cataldi M, Marriott I, Grdzelishvili VZ. Understanding and altering cell tropism of vesicular stomatitis virus. *Virus Res.* 2013;176(1–2):16–32.

12. McCluskey BJ, Mumford EL. Vesicular stomatitis and other vesicular, erosive, and ulcerative diseases of horses. *Vet Clin North Am Equine Pract.* 2000;16(3):457–469.

13. Stallknecht DE et al. Potential for contact and mechanical vector transmission of vesicular stomatitis virus New Jersey in pigs. *Am J Vet Res.* 1999;60:43–48.

14. Letchworth GJ, Rodriguez LL, Barrera JDC. Vesicular stomatitis. *Vet J.* 1999;157(3):239–260.

15. OIE. Vesicular stomatitis. Aetiology, epidemiology, diagnosis, prevention and control. http://www.oie.int/fileadmin/Home/eng/Animal_Health_in_the_World/docs/pdf/Disease_cards/VESICULAR_STOMATITIS.pdf; accessed May 11, 2015.

16. Schmitt B. Vesicular stomatitis. *Vet Clin North Am Food Anim Pract.* 2002;18(3):453–459.

17. McCluskey BJ, Pelzel-McCluskey AM, Creekmore L, Schiltz J. Vesicular stomatitis outbreak in the southwestern United States, 2012. *J Vet Diagn Invest.* 2013;25(5):608–613.

18. Faul EJ, Lyles DS, Schnell MJ. Interferon response and viral evasion by members of the family rhabdoviridae. *Viruses.* 2009;1(3):832–851.

19. Sun C et al. Development of a convenient immunochromatographic strip for the diagnosis of vesicular stomatitis virus serotype Indiana infections. *J Virol Methods.* 2013;188(1–2):57–63.

20. Höfner MC, Carpenter WC, Ferris NP, Kitching RP, Ariza Botero F. A hemi-nested PCR assay for the detection and identification of vesicular stomatitis virus nucleic acid. *J Virol Methods.* 1994;50(1–3):11–20.

21. Rodriguez LL, Letchworth GJ, Spiropoulou CF, Nichol ST. Rapid detection of vesicular stomatitis virus New Jersey serotype in clinical samples by using polymerase chain reaction. *J Clin Microbiol.* 1993;31(8):2016–2020.

22. Magnuson RJ et al. A single-tube multiplex reverse transcription-polymerase chain reaction for detection and differentiation of vesicular stomatitis Indiana 1 and New Jersey viruses in insects. *J Vet Diagn Invest.* 2003;15(6):561–567.

23. Rasmussen TB, Uttenthal A, Fernández J, Storgaard T. Quantitative multiplex assay for simultaneous detection and identification of Indiana and New Jersey serotypes of vesicular stomatitis virus. *J Clin Microbiol.* 2005;43(1):356–362.

24. Fernández J et al. Rapid and differential diagnosis of foot-and-mouth disease, swine vesicular disease, and vesicular stomatitis by a new multiplex RT-PCR assay. *J Virol Methods.* 2008;147(2):301–311.

25. Wilson WC et al. Field evaluation of a multiplex real-time reverse transcription polymerase chain reaction assay for detection of Vesicular stomatitis virus. *J Vet Diagn Invest.* 2009;21(2):179–186.

26. Hole K, Velazquez-Salinas L, Clavijo A. Improvement and optimization of a multiplex real-time reverse transcription polymerase chain reaction assay for the detection and typing of Vesicular stomatitis virus. *J Vet Diagn Invest.* 2010;22(3): 428–433. Erratum in: *J Vet Diagn Invest.* 2010;22(6):1018.

27. Arshed MJ et al. Comparison of RNA extraction methods to augment the sensitivity for the differentiation of vesicular stomatitis virus Indiana1 and New Jersey. *J Clin Lab Anal.* 2011;25(2):95–99.

28. Lung O et al. Multiplex RT-PCR detection and microarray typing of vesicular disease viruses. *J Virol Methods.* 2011;175(2):236–245.

29. Belák S, Karlsson OE, Leijon M, Granberg F. High-throughput sequencing in veterinary infection biology and diagnostics. *Rev Sci Tech.* 2013;32(3):893–915.

30. Granberg F, Karlsson OE, Leijon M, Liu L, Belák S. Molecular approaches to recognize relevant and emerging infectious diseases in animals. *Methods Mol Biol.* 2015;1247:109–124.

31. Granberg F, Karlsson OE, Belák S. Metagenomic approaches to disclose disease-associated pathogens: Detection of viral pathogens in honeybees. *Methods Mol Biol.* 2015;1247:491–511.

32. Van Borm S et al. Next-generation sequencing in veterinary medicine: How can the massive amount of information arising from high-throughput technologies improve diagnosis, control, and management of infectious diseases? *Methods Mol Biol.* 2015;1247:415–436.

33. Baner J et al. Microarray-based molecular detection of foot-and-mouth disease, vesicular stomatitis and swine vesicular disease viruses, using padlock probes. *J Microbiol Methods.* 2007;143:200–206.

34. Geisbert TW, Feldmann H. Recombinant vesicular stomatitis virus-based vaccines against Ebola and Marburg virus infections. *J Infect Dis.* 2011;204(Suppl. 3):S1075–S1081.

35. Hastie E, Grdzelishvili VZ. Vesicular stomatitis virus as a flexible platform for oncolytic virotherapy against cancer. *J Gen Virol.* 2012;93(12):2529–2545.

36. OIE (Office International des Epizooties). *OIE Terrestrial Manual 2010.* Chapter 2.1.19: Vesicular stomatitis. http://www.oie.int/fileadmin/Home/eng/Health_standards/tahm/2.01.19_VESICULAR_STOMITIS.pdf; accessed May 11, 2015.

43 Avian Influenza Virus

Erica Spackman

CONTENTS

43.1 Introduction 377
   43.1.1 Classification... 377
   43.1.2 Epidemiology... 378
   43.1.3 Clinical Features.. 378
   43.1.4 Diagnosis 378
      43.1.4.1 Virus Isolation... 378
      43.1.4.2 Clinical Sample Collection 378
      43.1.4.3 Molecular Detection Assays 379
43.2 Methods 379
   43.2.1 Sample Processing.. 379
   43.2.2 Conventional RT-PCR 379
   43.2.3 Real-Time RT-PCR.. 380
43.3 Conclusions and Future Needs 381
Disclaimer.. 381
References.. 381

43.1 INTRODUCTION

Avian influenza virus (AIV) is type A influenza, which is adapted to an avian host. Although avian influenza has been isolated from numerous avian species, the primary natural hosts for the virus are dabbling ducks, shorebirds, and gulls.[1] The virus can be found worldwide in these species and in other wild bird species. In poultry (chickens, turkeys, and more rarely domestic ducks), AIV can cause substantial production losses. Therefore, it is among the most important diseases of animal production. AIV also gains notoriety because, although very uncommon, it can be transmitted to people so there is always awareness that the virus could constitute a public health threat.

Given the importance of AIV for agriculture, some strains (i.e., H5 and H7) are reportable to the World Organization for Animal Health (OIE). This adds a regulatory framework to the detection of the virus and necessitates the standardization of tests at the national and international level. In the past decade, molecular tests for AIV detection have been extensively validated by numerous animal health and veterinary diagnostic organizations (e.g., OIE, OFFLU, United States Department of Agriculture (USDA)). As AIV is a rapidly evolving virus, the validated tests are continually monitored for performance, and updates are made for new virus variants as appropriate. The current tests will be presented here. Finally, since AIV is a reportable disease, laboratories unfamiliar with working with AIV should check with local veterinary authorities on specific regulatory requirements.

43.1.1 CLASSIFICATION

Type A influenza, including AIV, is a member of the *Orthomyxoviridae* family and harbors a negative-sense RNA genome comprised of eight segments. Reassortment among the genomic segments and the high replication error rate typical of RNA viruses result in a high level of population diversity, which allows for rapid evolution of the virus, and may complicate detection. Therefore, among AIVs, there is tremendous biological and genetic diversity.

Of the eight genome segments, the segment encoding the two matrix proteins (M1 and M2), which are structural, and the segment encoding the nucleoprotein (NP) are the most conserved and, therefore, are the gene segments most frequently targeted for diagnostic tests. The shortest segment encodes two "nonstructural" genes (NS1 and NS2). The longest three segments each encode the three polymerase complex genes (PA, PB1, and PB2). Finally, the coat proteins, which determine the serological subtypes, are the hemagglutinin (HA), which binds the host cell receptor, and the neuraminidase (NA), which cleaves newly budded virus from the host cell. Identification of the HA and NA subtypes is diagnostically important but is difficult due to their high variability.

There are 16 HA subtypes and 9 NA subtypes of AIV. Of the 144 possible combinations, most have been isolated from the natural host species of the virus, where the greatest diversity of AIV is found. The subtypes vary in prevalence; H3, H4, and H6 are quite common while H14 and H15 are very rare. Importantly, the HA and NA subtype only denotes the serological reactivity of a strain; there can be tremendous biological, host range, virulence, and genetic variation within a given subtype. For example, within the H7 subtype, sequences can vary close to 30% among lineages.[2–4] In addition, because of the segmented genome of AIV, the other six segments from two isolates of the same HA and NA subtype may be from different lineages. The practical benefit of this variation is that it provides valuable clues to molecular epidemiological studies.

43.1.2 EPIDEMIOLOGY

The natural host species of AIV are members of the Anatidae (dabbling ducks), Scolopacidae (shorebirds), and Laridae (gulls) families.[1] Some species from within these families have higher infection prevalence, for example, mallards (*Anas platyrhynchos*) tend to have higher prevalence among the Anatidae. AIV is essentially endemic in these populations and typically does not cause any clinical disease. Experimental infections and rare isolations from wild birds of other families demonstrate that other wild avian species may be infected, but data are limited or missing for a vast number of avian species. Domestic species (chickens and turkeys) are not natural host species; AIV strains have different levels of adaptation for different avian species, for example, the infectious dose for chickens can vary as much as 5 $\log_{10}$ among isolates.[5]

AIV ecology and epidemiology are best described in domestic poultry species, especially chickens and turkeys. In domestic birds, AIV is initially introduced from wild birds, although the specific exposure route is not always apparent unless the birds have outside access or their water source is untreated surface water.[6] AIV is highly contagious and host-adapted strains spread rapidly among susceptible birds. Movement of people (contaminated equipment, vaccination crews, vehicle movement) is likely the most important means of AIV spread. However, spread by wind up to about 1 km has been documented and birds on farms near roads on which infected birds were transported during outbreaks have become infected.[7] Therefore, biosecurity and increased surveillance are crucial elements of AIV control in domestic poultry.

43.1.3 CLINICAL FEATURES

AIV is classified as low pathogenic (LP) or highly pathogenic (HP) in chickens and turkeys based on specific tests prescribed by the OIE. There is an *in vivo* test with a mortality end point and a molecular test that is based on the protein sequence at a known marker for virulence in chickens (the HA proteolytic cleavage site).[8] Only isolates of the H5 and H7 subtypes have ever been classified as HP so these two subtypes, even when LP, are frequently more highly controlled than other subtypes. It is important to understand that these criteria only apply to gallinaceous birds and are regulatory, not necessarily biological. The term "high pathogenicity" tends to be used as a relative term for other species which creates confusion.

Disease from AIV can vary widely depending on the infected species and virus strain. In natural host species (dabbling ducks), infection typically does not cause disease at all, even when ducks are exposed to forms that are HP for chickens.[9] The rare exceptions are some strains of the H5N1 HPAIV from Asia, which can cause severe disease and death. Clinical signs in young Pekin ducks are similar to HPAIV in chickens.[10]

In chickens and turkeys, HPAIV is characterized by rapid (can be as early as 24–48 h postexposure) and high mortality (as high as 100%).[9] Prior to death, birds will become severely lethargic and neurological signs may be present. Subcutaneous hemorrhages may be observed on the shanks and/or in the wattles and combs. Periorbital swelling and green diarrhea are also common. In contrast, infection with the LP form of the virus may be subclinical. Disease from the LP form is essentially respiratory; the birds present with congestion, periorbital swelling, and lethargy. Birds that lay eggs will often stop laying.

43.1.4 DIAGNOSIS

43.1.4.1 Virus Isolation

Although the focus of this chapter is on molecular methods, virus isolation will be mentioned briefly. Virus isolation will always need to be conducted on some types of samples (e.g., index cases), both as a confirmatory test and to provide an isolate for characterization. Virus isolation for AIV is most commonly conducted with embryonated chickens eggs (ECE), which are free of maternal antibody to AIV. Virus isolation is highly sensitive but is nonspecific (numerous agents will replicate in ECE); it is expensive, requires a source of ECE, and demands more biosecurity and biosafety controls than molecular methods.[11] Positive samples may yield a result in a couple of days, but it can take up to 2 weeks to confirm a negative.

43.1.4.2 Clinical Sample Collection

The optimal specimens from avian species are oropharyngeal swabs (including swabbing the choanal cleft) or tracheal swabs and cloacal swabs. Oral/tracheal swabs are preferred from chickens and turkeys, and cloacal swabs are preferred from waterfowl. However, collecting both is optimal if resources allow. Lung, kidney, and spleen are among the best tissues for virus detection, but tissues are not necessarily better than swabs. Several recent reports have successfully utilized feather pulp for the detection of the highly pathogenic form of the virus.[12–14] Environmental swabs may be collected from any surface; however, in most cases, it is advised to perform virus isolation with environmental samples if they are being used to confirm cleaning and disinfection because molecular methods can detect inactivated virus.

Proper sample collection and transport are crucial to achieving the best accuracy of any diagnostic test. For AIV, the specifics of sample collection and transport have been empirically established. Also, most AIV clinical specimens will be tested by more than one type of assay (e.g., real-time reverse transcription polymerase chain reaction [RT-PCR] [rRT-PCR] and virus isolation); therefore, the sample needs to be in a form that will work for most tests, and there needs to be in a sufficient volume for subsequent tests without diluting the sample.

Oral and cloacal swabs should be collected in a well-buffered, salt-balanced media, which contains protein such as brain–heart infusion broth.[15,16] Media that lack protein do not preserve live virus well. The media volume depends on whether swabs will be pooled; 3 mL is recommended for 1–5 swabs per vial and 5.5 mL is recommended for 6–11 swabs per vial. Specimens should never be transported dry as this will reduce detection by both virus isolation and molecular methods. Flocked nylon swabs have been shown to be optimal, but urethane foam and Dacron wound tip swabs are satisfactory.[16] Swabs with wooden shafts and calcium alginate swabs should be avoided as their use can inhibit RT-PCR and virus isolation. Finally, specimens should be maintained at refrigeration temperatures at all times, and freeze-thaw cycles and warm temperatures should be avoided.

43.1.4.3 Molecular Detection Assays

rRT-PCR is the most widely used format, and well-validated tests are available.[17,18] Conventional RT-PCR and isothermal nucleic acid detection assays have been reported,[17,19–23] but few have been adequately validated. When determining which test to use, it is best to start with validated tests and to consider what the goals are for testing. Most AIV tests will target a conserved gene therefore will have specificity for influenza A. Type-specific tests are much more reliable than subtype tests, although one should consider that false negatives still are a possibility due to sequence variation in the virus. Subtyping tests are best when used to confirm the presence of a specific virus lineage and are not suitable for screening samples for novel AIVs since they are prone to false negatives and even false positives because of the variability of the HA and NA genes. Gene sequencing is the most reliable method for subtype identification. It is also important to note that it is not possible to reliably identify the pathotype of an AIV isolate by RT-PCR.

Commercial kits for RT-PCR are widely available and one-step procedures are recommended. The procedures described here are presented with specific commercial kits; however, one may optimize most any kit for use with these tests. Note that for real-time RT-PCR using hydrolysis probes, the enzyme must have 5′ exonuclease activity.

43.2 METHODS

43.2.1 Sample Processing

Samples should be preserved on ice throughout processing to preserve the live virus if virus isolation will be attempted with the specimen after testing by RT-PCR. The RNA extraction methods utilized are another crucial element of the successful testing.[24] The sample type will dictate which RNA extraction method is used. Commercial kits and reagents are used frequently and provide good results. Oropharyngeal and tracheal swab samples may be extracted successfully with most any method as these samples are relatively clean. Cloacal and

environmental swabs should be extracted with kits and methods that are designed for use with samples containing fecal material.[24] A 10% w/v homogenate of tissues should be prepared in a viral transport media, then the RNA may be extracted using a suitable method for tissues.

43.2.2 Conventional RT-PCR

Although rRT-PCR is more common than conventional RT-PCR for AIV detection, conventional assays are used when real-time PCR equipment is not available. The assay presented here targets the NS gene segment, the shortest influenza gene segment. The primers (Table 43.1) are directed to the conserved ends of the gene segment[25] and will detect most influenza A viruses. This procedure was optimized with the Qiagen OneStep RT-PCR kit (Qiagen Inc., Valencia, CA), and the volumes correspond to this kit. Positive controls and template-free negative controls (nuclease-free water in place of template) should be included in each set of reactions run.

1. Label the RT-PCR reaction tubes and place them on ice or in a cooling rack.
2. Prepare the master mix in a 1.5 mL microfuge tube as shown in Table 43.2. Prepare 10% extra to compensate for pipettor error.
3. Mix by vortexing and pulse spin the master mix.
4. Add 22 μL of reaction mix with primers to each tube.
5. Add 3 μL of template RNA to the appropriate tube.
6. Close the tubes and centrifuge for 3–5 s before loading the thermalcycler.
7. Run the thermalcycler with the conditions in Table 43.3.
8. Analyze the results by running the reaction on a 1% agarose gel. The expected size is approximately 890 bp.

TABLE 43.1

Primer Sequences for Conventional and Real-Time Reverse Transcription Polymerase Chain Reaction for Avian Influenza Virus

Assay	Component	Name	Sequence
Conventional	Forward primer	NS F1	5′-AGC AAA AGC AGG GTG ACA A-3′
	Reverse primer	NS R890	5′-AGT AGA AAC AAG GGT GTT-3′
Real-time	Forward primer	M F25	5′-AGA TGA GTC TTC TAA CCG AGG TCG-3′
	Probe	M F64	5′-[FAM]-TCA GGC CCC CTC AAA GCC GA-[BHQ]-3′
	Reverse primer 1	M R124-2002	5′-TGC AAA AAC ATC TTC AAG TCT CTG-3′
	Reverse primer 2	M R124-2009	3′-TGC AAA GAC ACT TTC AG TCT CTG-3′

TABLE 43.2

Reverse Transcription Polymerase Chain Reaction Conditions for Conventional RT-PCR for the Influenza NS Gene[25] Using the Qiagen OneStep RT-PCR Kit

Component	Volume (µL)	Final Concentration
Nuclease-free water	13	
dNTPs (kit supplied)	1	320 µM ea. dNTP
5× buffer (kit supplied)	5	1×
Forward primer (NS F1)	1	20 pmol
Reverse primer (NS R890)	1	20 pmol
Enzyme (kit supplied)	1	
Master mix total	22	
Template	3	
Final volume	25	

TABLE 43.3

Cycling Conditions for Conventional Reverse Transcription Polymerase Chain Reaction for the AIV NS Gene

Phase	Number of Cycles	Time (min)	Temperature (°C)
RT	1 cycle	30	50
		15	95
PCR	35 cycles	1	94
		1	53
		1	72
Final extension	1 cycle	10	72
Soak	1 cycle	Infinity	4

Note: The RT phase is specific for the Qiagen OneStep kit.

43.2.3 REAL-TIME RT-PCR

The procedure described in the following is based on an rRT-PCR test that targets the M gene of influenza A[18], which is the basis of the U.S. Department of Agriculture AIV test, is referenced in the OIE manual and has been updated periodically to ensure optimal performance.[26] This test has been widely used with avian and mammalian specimens and extensive validation data have been generated that have demonstrated reliable performance with the detection of influenza A virus. As with any molecular test for an RNA virus, false negatives are still a possibility due to unpredictable sequence variation in novel isolates; therefore, results should be confirmed with virus isolation if necessary.

The procedure here is described with the AB7500 FAST real-time PCR instrument (Life technologies, Grand Island, NY) and Ambion AgPath-ID™ kit (Life technologies, Grand Island, NY). With optimization, other instruments and reagents may be used successfully. Positive controls and template-free negative controls (nuclease-free water in place of template) should be included in each set of reactions run.

1. Place a 96-well reaction plate in the manufacturer supplied rack, and prepare a key for the template positions in the plate.
2. Prepare the master mix in a 1.5 mL microfuge tube as shown in Table 43.4. Prepare 10% extra to compensate for pipettor error. Note that there are two reverse primers in the reaction; this is to provide the broadest reactivity.
3. Mix by vortexing and pulse spin the master mix.
4. Add 17 µL of reaction mix with primers and probes to the appropriate wells of the reaction plate.
5. Add 8 µL of template RNA to the appropriate well.
6. Seal the plate with the plate sealer provided by the manufacturer and centrifuge (pulse spin for 30 s) in a centrifuge equipped with cups for microtiter plates before loading the thermal cycler.
7. Run the thermalcycler with the conditions shown in Table 43.5.
8. Analyze the results: usually the automatic analysis setting will be accurate. First, ensure that the negative and positive controls are valid. Then examine each sample individually. With clinical samples, cycle threshold values over 35 are typically considered "suspect" and should be confirmed or additional samples should be collected from a flock if possible. With experimental samples, a standard curve may be used to establish an end point (be sure to prepare the standard curve with the same isolate, which will be in the samples being tested).

TABLE 43.4

Real-Time Reverse Transcription Polymerase Chain Reaction Master Mix Volumes per Reaction for the AI Matrix Assay Using the Ambion AgPath-ID™

	Volume per Reaction (µL)	Final Concentration
AgPath kit 2× buffer	12.5	1×
Nuclease-free water	0.33	
AgPath kit enzyme mix	1.0	
Forward primer M F25 (20 µmol/µL stock)	0.25	20 pmol
Reverse primer 1 M R124 2009 (20 µmol/µL stock)	0.25	20 pmol
Reverse primer 2 M R124 2002 (20 µmol/µL stock)	0.25	20 pmol
Probe M F64 (20 µmol/µL stock)	0.25	3 pmol
AgPath kit detection enhancer	1.67	
RNase inhibitor	0.5	13 units
MM per reaction	17	
Template	8	
Total	25 µL	

Note that the ROX™ passive reference dye, dNTP's, and additional buffer components are included in the kit supplied 2× buffer.

TABLE 43.5

Amplification Parameters for the Ambion AgPath-ID™ Reverse Transcription Polymerase Chain Reaction Reagents and the ABI 7500 Fast Instrumentation for Avian Influenza Virus (M Gene)

Probe/ Primer Set		Step	Time	Temperature (°C)
Reverse transcription	1 Cycle	RT	10 min	45
		Heat activation of *Taq*	10 min	94
PCR	45 Cycles	Denaturation	1 s	94
		Annealing (acquire fluorescence)	30 s	60

Note that the RT step is specific for the enzyme supplied with this kit.

43.3 CONCLUSIONS AND FUTURE NEEDS

Detection of AIV by type-specific molecular methods is fast and accurate; however, a reliable method of rapidly identifying subtypes, which is robust to sequence changes, would be highly valuable. Microarray-based tests have been developed with this goal, but their performance: time, skills and equipment needed, cost, and poor sensitivity, have been barriers to validation and adoption.[27] RT-PCR, which utilizes mass spectroscopy to identify the product, is among the current up-and-coming technologies, and if validated and the cost were to be reduced to a reasonable level, it could provide an improvement to current methods.[28,29]

Finally, the technologies of viral diagnostics moves rapidly and influenza is a proof-of-concept agent for many new methods. New technologies will often be first validated for public health use but can often be validated for veterinary use for avian influenza if the fit for purpose for veterinary diagnosis is met.

DISCLAIMER

Mention of trade names or commercial products in this chapter is solely for the purpose of providing specific information and does not imply recommendation or endorsement by the author or U.S. Department of Agriculture.

REFERENCES

1. Stallknecht, D.E. and Brown, J.D. Ecology of avian influenza in wild birds. In *Avian Influenza* (ed. Swayne, D.E.), pp. 43–58 (Blackwell, Ames, IA, 2008).
2. Suarez, D.L. et al. Recombination resulting in virulence shift in avian influenza outbreak in, Chile. *Emerg Infect Dis* **10**, 693–699 (2004).
3. Pasick, J. et al. Intersegmental recombination between the haemagglutinin and matrix genes was responsible for the emergence of a highly pathogenic H7N3 avian influenza virus in British Columbia. *J Gen Virol* **86**, 727–731 (2005).
4. Kageyama, T. et al. Genetic analysis of novel avian A(H7N9) influenza viruses isolated from patients in China, February to April 2013. *Euro Surveill* **18**, 1–15 (2013).
5. Swayne, D.E. and Slemons, R.D. Using mean infectious dose of high- and low-pathogenicity avian influenza viruses originating from wild duck and poultry as one measure of infectivity and adaptation to poultry. *Avian Dis* **52**, 455–460 (2008).
6. Hinshaw, V.S., Webster, R.G., and Rodriguez, R.J. Influenza A viruses: Combinations of hemagglutinin and neuraminidase subtypes isolated from animals and other sources: Brief review. *Arch Virol* **62**, 281–290 (1979).
7. Bowes, V.A. et al. Virus characterization, clinical presentation, and pathology associated with H7N3 avian influenza in British Columbia broiler breeder chickens in 2004. *Avian Dis* **48**, 928–934 (2004).
8. OIE (Office International des Epizooties). *Manual of Diagnostic Tests and Vaccines for Terrestrial Animals*. Avian influenza. www.oie.int (World Organization for Animal Health, Paris, France, 2012).
9. Swayne, D.E., Suarez, D.L., and Sims, L.D. Influenza. In *Diseases of Poultry* (ed. Swayne, D.), pp. 181–218 (Blackwell, Ames, IA, 2013).
10. Pantin-Jackwood, M.J. and Swayne, D.E. Pathobiology of Asian highly pathogenic avian influenza H5N1 virus infections in ducks. *Avian Dis* **51**, 250–259 (2007).
11. Spackman, E. and Killian, M. Avian influenza virus isolation, propagation and titration in embryonated chicken eggs. In *Animal Influenza Virus*, Vol. 1161 (ed. Spackman, E.), pp. 125–140 (Springer, 2014).
12. Busquets, N. et al. Persistence of highly pathogenic avian influenza virus (H7N1) in infected chickens: Feather as a suitable sample for diagnosis. *J Gen Virol* **91**, 2307–2313 (2010).
13. Yamamoto, Y. et al. Detecting avian influenza virus (H5N1) in domestic duck feathers. *Emerg Infect Dis* **14**, 1671–1672 (2008).
14. Parker, C.D. et al. Outbreak of Eurasian lineage H5N1 highly pathogenic avian influenza in turkeys in Great Britain in November 2007. *Vet Rec* **175**, 282 (2014).
15. Erickson, G.A., Brugh, M., and Beard, C.W. Newcastle disease and avian influenza virus stability under simulated shipping conditions. In *Twenty-First Annual Meeting of American Association of Veterinary Laboratory Diagnosticians*, pp. 309–318 (Buffalo, NY, 1978).
16. Spackman, E., Pedersen, J.C., McKinley, E.T., and Gelb, J., Jr. Optimal specimen collection and transport methods for the detection of avian influenza virus and Newcastle disease virus. *BMC Vet Res* **9**, 35 (2013).
17. Slomka, M.J. et al. Identification of sensitive and specific avian influenza polymerase chain reaction methods through blind ring trials organized in the European Union. *Avian Dis* **51**, 227–234 (2007).
18. Spackman, E. et al. Development of a real-time reverse transcriptase PCR assay for type A influenza virus and the avian H5 and H7 hemagglutinin subtypes. *J Clin Microbiol* **40**, 3256–3260 (2002).
19. Bao, H. et al. Development of a reverse transcription loop-mediated isothermal amplification method for the rapid detection of avian influenza virus subtype H7. *J Virol Methods* **179**, 33–37 (2012).
20. Moore, C. et al. Development and evaluation of a real-time nucleic acid sequence based amplification assay for rapid detection of influenza A. *J Med Virol* **74**, 619–628 (2004).
21. Starick, E., Romer-Oberdorfer, A., and Werner, O. Type- and subtype-specific RT-PCR assays for avian influenza A viruses (AIV). *J Vet Med B Infect Dis Vet Public Health* **47**, 295–301 (2000).

22. Hoffmann, E., Stech, J., Guan, Y., Webster, R.G., and Perez, D.R. Universal primer set for the full-length amplification of all influenza A viruses. *Arch Virol* **146**, 2275–2289 (2001).

23. Fouchier, R.A. et al. Detection of influenza A viruses from different species by PCR amplification of conserved sequences in the matrix gene. *J Clin Microbiol* **38**, 4096–4101 (2000).

24. Spackman, E. and Lee, S.A. Avian influenza virus RNA extraction. *Methods Mol Biol* **1161**, 93–104 (2014).

25. Suarez, D.L. and Perdue, M.L. Multiple alignment comparison of the non-structural genes of influenza A viruses. *Virus Res* **54**, 59–69 (1998).

26. Slomka, M.J. et al. Real time reverse transcription (RRT)-polymerase chain reaction (PCR) methods for detection of pandemic (H1N1) 2009 influenza virus and European swine influenza A virus infections in pigs. *Influenza Other Respir Viruses* **4**, 277–293 (2010).

27. Brown, I.H. Advances in molecular diagnostics for avian influenza. *Dev Biol (Basel)* **124**, 93–97 (2006).

28. Ecker, D.J. et al. Rapid identification and strain-typing of respiratory pathogens for epidemic surveillance. *Proc Natl Acad Sci USA* **102**, 8012–8017 (2005).

29. Patel, D.A. et al. Development and evaluation of a PCR and mass spectroscopy (PCR-MS)-based method for quantitative, type-specific detection of human papillomavirus. *J Virol Methods* **160**, 78–84 (2009).

44 Equine Influenza Virus

Thomas M. Chambers and Udeni B.R. Balasuriya

CONTENTS

44.1 Introduction 383
   44.1.1 Classification 383
   44.1.2 Morphology, Biology, and Epidemiology 383
   44.1.3 Clinical Features and Pathogenesis 384
   44.1.4 Diagnosis 385
      44.1.4.1 Conventional Techniques 385
      44.1.4.2 Molecular Techniques 386
44.2 Methods 386
   44.2.1 Sample Collection and Preparation 386
      44.2.1.1 Nasopharyngeal Swabs 386
      44.2.1.2 Serum Samples 386
   44.2.2 Detection Procedures 386
      44.2.2.1 Virus Isolation 386
      44.2.2.2 Rapid Antigen Detection Kits 387
      44.2.2.3 Retrospective Serology 387
      44.2.2.4 Nucleic Acid Tests 389
44.3 Conclusion and Future Perspectives 390
References 390

44.1 INTRODUCTION

44.1.1 CLASSIFICATION

Equine influenza virus (EIV) that causes equine influenza (EI) is the most important equine respiratory viral pathogen and has the greatest economic impact internationally, because it has a very high morbidity rate during outbreaks leading to disruption of major equestrian events.[1–3] EIV can also infect other equids such as donkeys, mules, and zebras. EIV is a member of the family Orthomyxoviridae and belongs to the genus *Influenzavirus A* (influenza A type).[4] The first strain of EIV isolated in 1956 was of H7N7 configuration (subtype 1) and designated influenza virus A/equine/Prague/56, and it caused epidemics during the 1960s and 1970s.[5,6] The last confirmed outbreak caused by an H7N7 subtype in horses was recorded in 1979[5,7] and therefore, the H7N7 subtype is thought to be extinct or possibly still circulating at a very low level in nature.[8–11] A second EIV subtype (subtype 2), H3N8, emerged in 1963 and was designated as influenza virus A/equine/Miami/63.[7,12] This subtype has been associated with all confirmed outbreaks of EI since 1980. Extensive antigenic drift has been detected in this virus over the years (see later).[13–19]

44.1.2 MORPHOLOGY, BIOLOGY, AND EPIDEMIOLOGY

EIV is pleomorphic and has spherical or filamentous virions with a diameter of 80–120 nm. It is an enveloped virus with a segmented, single-stranded, negative-sense RNA genome. The entire EIV RNA genome is approximately 13.5 kb and contains 8 RNA segments that encode for 11 viral proteins: PB2 (encoded by segment 1), PB1 and PB1-F2 (segment 2), PA (segment 3), HA (segment 4), NP (segment 5), NA (segment 6), M1 and M2 (segment 7), and NS1 and NEP (segment 8). The viral proteins include three surface glycoproteins (hemagglutinin [H], neuraminidase [N], and ion channel protein [M2]) embedded in the host-cell-derived lipid envelope, five structural proteins (M1 matrix protein, nucleoprotein [NP], and polymerase complex proteins [PA, PB1, and PB2]), and three nonstructural proteins (NS1, NS2 [nuclear export protein—NEP], and PB1-F2). The RNA gene segments are closely associated with the NP and are surrounded by the M1 protein which is very closely associated with the lipid envelope containing the three surface glycoproteins. Both the HA and NA are major surface glycoproteins and horses mount a strong humoral antibody response to these two proteins following infection; the major neutralization epitopes of EIV are located in the HA protein. The HA protein has an important role in determining the binding to the host cell receptors that have α-2-6-linked or α-2-3-linked sialic acid moieties (α2-3 sialic acid in horses), as well as a cleavage site that must be cleaved by host cell proteases. HA protein also causes hemagglutination of red blood cells. The NA glycoprotein plays an important role in budding and release of progeny virus from the host cell surface by cleaving the glycosidic bonds of the mannosaccharide, neuraminic acid.

Both HA and NA genes (segments 4 and 6, respectively) play a very important role in the evolution of EIV and other type A influenzaviruses.[19,20] Point mutations in these two genes lead to antigenic drift whereas reassortment of HA and NA with other subtypes leads to antigenic shift (emergence of new subtypes). However, the antigenic drift in EIV H3N8 strains has been reported to be less than that reported for the human influenza A viruses. Since its emergence in 1963, EIV H3N8 subtype (represented by A/equine/Miami/63) evolved as a single lineage for two decades.[20] However, in the mid-1980s, the H3N8 subtype evolved into the Eurasian lineage and the American lineage due to natural selection of variants within equine hosts and reassortment between strains during outbreaks.[21,22] The viruses belonging to the American lineage (represented by A/equine/Newmarket/1/93 and A/equine/Kentucky/94) were predominantly isolated from horses on the North American continent, whereas the viruses belonging to the Eurasian lineage (represented by A/equine/Newmarket/2/93) were mainly isolated from horses in Europe and Asia. Recent sequence and phylogenetic analysis has shown that currently circulating H3N8 strains primarily belong to American lineage and over time, the American lineage viruses have further evolved into a Florida lineage with at least two distinct clades (Florida clade 1 and Florida clade 2).[23] The EIV strains belonging to the Florida sublineage have been spread across Europe, North America, South America, Japan, and Australia in the early part of the twenty-first century. A majority of the EIV H3N8 strains isolated in Europe since 2003 belong to the Florida sublineage. Two distinct clades of Florida sublineage have been identified; Florida clade 1 viruses isolated in North America in 2003 are distinct from Florida clade 2 viruses which spread to Europe in 2003. However, H3N8 strains belonging to both Florida clade 1 and 2 were isolated from horses in the United Kingdom from 2007 to 2010.[19,21,22] EIV H3N8 strains belonging to Florida clade 1 caused outbreaks in South Africa in 2003 and subsequently in Japan and Australia in 2007. EIV H3N8 strains belonging to Florida clade 2 were associated with the major outbreaks in China, Mongolia, and India from 2007 to 2009. Since 2005, the Eurasian lineage has been unimportant, although it may still circulate at low levels.

Interspecies transmission of EIV H3N8 has been reported both in the United States and United Kingdom.[4] A significant outbreak of canine influenza (CI) was reported in greyhound dogs in Florida in early 2004, which was shown to be due to cross-species transmission of H3N8 EIV.[58] Since the initial CI outbreak in Florida, evidence of virus activity has been reported in 30 states in the United States, reportedly affecting tens of thousands of dogs.[13] Sequence analysis of canine influenza virus (CIV) from the original outbreak (A/canine/FL/43/04) revealed that all eight gene segments of this virus shared ≥96% nucleotide sequence homology with that of the H3N8 EIV subtype, suggesting that CIV resulted from direct interspecies transmission of EIV to dogs without genetic reassortment.[13,58] The sequence changes in the hemagglutinin gene (H3 HA) are indicative of adaptive evolution of the virus in its new host.[25] Phylogenetic analysis of

four greyhound isolates and two pet dog isolates revealed that these viruses were closely related to EIV strains recovered from horses since 2000, belonging to the Florida sublineage of the American lineage of H3N8 EIV.[24,58] The isolation of four closely related H3N8 EIV subtype strains from dogs that died in different locations over a 2-year period of time, together with serological evidence of widespread infection among greyhounds in different states, strongly confirmed sustained circulation of CIV involving dog-to-dog transmission of the virus.[2,13,24] Retrospective studies have also shown that transmission of EIV H3N8 from horses to foxhounds has occurred in the United Kingdom. Similarly, transmission of EIV to dogs has been also reported in the recent extensive outbreak of EI in Australia. Recent studies have shown that avian influenza viruses are capable of replicating in explanted equine tracheal epithelial cultures, but these viruses failed to establish productive infections in experimentally inoculated ponies.[27] EIV of the H3N8 subtype was isolated from pigs in China[28]; however, experimental infection of pigs with H3N8 EIV failed to demonstrate replication competence in that species.

Currently, equine H3N8 influenza virus continues to be the most important equine respiratory pathogen of horses in many countries around the world. EI is considered enzootic in the United States, United Kingdom, and many other European countries.[29] New Zealand and Iceland are the only countries that have remained continuously free of EI.[30] The epidemiology of EI is influenced by viral factors (e.g., emergence of variant strains), host factors (e.g., immune status of the horse), and environmental factors such as management practices and movement of horses. A major factor in the spread of EI in the last three decades has been the increase in long-distance international movement of horses by air. Horses incubating the disease, either clinically or subclinically infected, can introduce the infection into susceptible horses if they are not quarantined for an adequate amount of time. Such quarantine failures have led to major EI outbreaks in the United Kingdom, South Africa, Hong Kong, Dubai, Japan, and most recently Australia. Some of the horses vaccinated against EIV may be partially protected and such horses may not show any signs of clinical infection but continue to shed virus. Such horses may play an important role in maintenance and perpetuation of the virus in the equine population.

44.1.3 CLINICAL FEATURES AND PATHOGENESIS

Clinical presentation of EI disease was described by Landolt et al.[31] and is similar in its main aspects to typical influenza in humans.[32] The initial signs are pyrexia of up to 41°C and frequent dry hacking cough. Serous nasal discharge soon appears that changes to mucopurulent discharge within 1–2 days. Respiratory rate and heart rate become elevated and auscultation of the lungs often reveals an inspiratory wheeze. Horses may become lethargic and anorexic; donkeys less so—perhaps elevating their risk as they can be sicker than they appear. These signs vary in severity depending on the immune status of the horse from vaccination or prior exposure, and EI-diseased horses can be sources of contagion without themselves being ill based on casual inspection. Mild uncomplicated cases often seem to resolve within a few days, although virus shedding may persist a few days beyond the cessation of overt disease signs, and cough may persist for weeks. However EI in horses, as with influenza in humans, predisposes the host to secondary infections owing to the loss of muco-ciliary escalator function, a key physical defense mechanism of the respiratory tract. Virus-mediated destruction by apoptosis of the ciliated respiratory epithelium denudes the trachea and bronchial tree of ciliated cells and ciliary fluid flow,[33,34] permitting bacterial pathogens, including *Streptococcus zooepidemicus*, to transit and colonize in the lungs resulting in bacterial bronchopneumonia that can be life threatening if untreated.[35,36] A secondary spike of pyrexia at 2–5 days after the initial spike is characteristic of secondary infections. Other complications of EI infection that have been observed include lower limb edema, myositis, and myocarditis.[32,37] Influenza-associated encephalopathy has been observed although rarely.[38] Given adequate veterinary care, EI is rarely fatal in otherwise healthy horses, even seronegative ones. However, in third world countries large outbreaks with significant mortality have been reported.[39] Neonatal foals are also at risk of lethal disease if they fail to ingest colostrum containing EI-specific maternal antibodies.[40,41]

44.1.4 Diagnosis

EI is the most important respiratory virus infection of horses and can disrupt major equestrian events and cause significant economic losses to the equine industry worldwide. Influenza H3N8 virus spreads rapidly in susceptible horses and can result in very high morbidity within 24–48 h after exposure to the virus. Therefore, rapid and accurate diagnosis of EI is critical for implementation of prevention and control measures to avoid the spread of EIV and to reduce the economic impact of the disease. Various laboratory testing procedures have been described for the diagnosis of EI in horses (Table 44.1). Traditionally, the gold standard laboratory test for the diagnosis of EI was attempted virus isolation (VI) from nasal swabs/washings in embryonated hens' eggs.[42] Following isolation, the virus is subtyped by means of the hemagglutination-inhibition (HI) test using sera specific for the H3N8 or H7N7 subtypes. These methods are time consuming and cumbersome. In the past decade, antigen detection immunoassays such as the Directigen™ Flu A test kit (Becton-Dickinson, Sparks, MD) and nucleic acid amplification-based assays (standard reverse transcription PCR [RT-PCR] or real-time RT-PCR [rRT-PCR]) were developed and evaluated by various groups.[43–53] The antigen detection immunoassay kits are designed to detect the NP of both influenza A and B viruses.[7,43] Of the commercially available antigen detection immunoassays, the Directigen Flu A test has been used to detect EIV in nasal swabs for a considerable time by certain laboratories.[43,49] While this assay has been found to be useful as an initial screening test to confirm a diagnosis of EI during an outbreak, its limited sensitivity does not make

TABLE 44.1
Laboratory Techniques Used for Diagnosis of EIV

Laboratory Techniques

Detection of virus
1. Virus isolation (VI) in embryonated hens' eggs (10–11-day-old embryos inoculated via amnion or allantois) or cell culture (e.g., MDCK cells)
2. Electron microscopy (EM)

Detection of viral antigens
1. Immunofluorescence assay (IFA)
2. Antigen detection immune assays (rapid diagnostic assays)

Detection of viral nucleic acid using contemporary molecular biology methods
1. Standard PCR-based assays
2. Quantitative real-time PCR assays
3. Insulated isothermal RT-PCR (iiRT-PCR)
4. Loop-mediated isothermal amplification assay (LAMP)

Indirect demonstration of EIV infection (Serology)
1. Hemagglutination-inhibition (HI) assay
2. Single radial hemolysis (SRH) assay

it an ideal method for the diagnosis of EIV infection on an individual animal basis.[50]

44.1.4.1 Conventional Techniques

The longstanding method for diagnosis of influenza has been VI on culture of samples from respiratory secretions. In the horse, respiratory secretions of diagnostic utility include nasal or nasopharyngeal swab samples or transtracheal washes. Of these, nasopharyngeal swabs present the best combination of convenience and diagnostic sensitivity. Nasal swabs using, for example, applicator sticks (15 cm) are more convenient and more readily tolerated by the horse compared to nasopharyngeal swabs, which should be inserted 20–25 cm up the nasal meatus; but the amount of virus per sample and the duration of virus detectability postinfection are much superior with nasopharyngeal swabs.[54] The enhanced sensitivity is valuable since horses may not be sampled until after the peak of virus shedding has already passed. It is highly desirable that swab samples be collected as soon as possible after the first presentation of clinical signs.

In the laboratory, diagnostic specimens are usually cultured in 11-day-old embryonated eggs or alternatively in Madin-Darby canine kidney (MDCK) cells. Virus growth may be observed by various means such as hemagglutination assay or immunofluorescence assay using EIV-specific monoclonal antibodies. Once detected, the isolates are available for characterization of antigenic or genetic similarity to known EIV strains, an advantage not afforded by other detection methods. EIV readily mutates to adapt to replicate in embryonated hens' eggs or continuous cell lines,[55,56] but it is sometimes seen that virus replication to detectable levels requires one or more blind passages.

As a diagnostic tool, virus culture is disadvantageous because it requires a virological laboratory and availability of embryonated eggs of the right age, or MDCKs ready for use.

For field use a number of rapid influenza detection kits (IDK) have been developed. These are typically self-contained antigen-capture ELISA devices that most often detect the influenza NP protein and can produce a result within a few minutes. While designed for detection of human influenza infection, the NP of human influenza viruses and EIV are sufficiently antigenically similar that some of these kits also detect EIV. Their sensitivity, however, is less by some orders of magnitude compared to virus culture or alternatively to RT-PCR, so effectively a positive test result is useful but a negative result is not. A laboratory NP-ELISA has been developed, which has superior sensitivity albeit less convenience.[57]

Postmortem diagnosis of EI infection by immunofluorescence or immunohistochemistry analysis of lung tissue is useful for confirmation of EI as an etiologic agent of a disease outbreak causing fatalities, such as the first identification of CI.[58] Electron microscopy for EIV is rarely done today thanks to the availability of more convenient and agent-specific methods.

Retrospective diagnosis can also be done by using serological techniques, generally single radial hemolysis or HI assays, to detect EIV-specific antibodies. Paired acute and convalescent serum samples are required. This is of little use in the face of an ongoing outbreak.

44.1.4.2 Molecular Techniques

Following the introduction of EI into Australia in 2007, an rRT-PCR developed to detect the avian influenza virus matrix gene was used as the molecular diagnostic method of choice for EI.[51,59] Recently, an insulated isothermal RT-PCR (iiRT-PCR) for the detection of EIV H3N8 subtype has been described.[60]

44.2 METHODS

44.2.1 Sample Collection and Preparation

44.2.1.1 Nasopharyngeal Swabs

Nasopharyngeal swabs and transport medium for use by equine practitioners are available from vendors or obtained from the Maxwell H. Gluck Equine Research Center. Using care not to contaminate the gauze end of the swab, it is inserted into the nasal meatus angling inward and down to avoid the blind pouch of the nostril, then slid as far as possible up the meatus (about 25 cm in an adult horse) and, once fully inserted, rotated a few times to enhance contact with mucosal fluids. Upon removal the gauze end is inserted into transport medium (the formulation in the OIE Manual of Diagnostic Tests and Vaccines for Terrestrial Animals[61] is phosphate-buffered saline with either 40% glycerol or 2% tryptose phosphate broth, and antibiotic solution including penicillin [200 units], streptomycin [200 units], and fungizone [5 mg]). Sufficient transport medium must be used to completely immerse the gauze. Once collected, these samples must be kept cold (4°C) during transport to the laboratory. If they are not to be processed for testing within 1–2 days, they should be stored at −70°C. Processing is as described in the OIE Manual.[61] In brief, as much as possible of the liquid fraction

of the sample is collected. It may be clarified of particulates by low-speed centrifugation if necessary, and further antibiotics may be added if the samples appear to be heavily contaminated. The authors do not recommend filtration to remove contaminants, as virus may stick to the filter. For injection into embryonated eggs, the sample may be used undiluted or with minimal dilution. For culture on MDCK cells, dilution to reduce the glycerol concentration below 5% is advisable.

44.2.1.2 Serum Samples

Equine sera for retrospective serology are obtained by standard blood collection from the jugular vein into sterile blood collection tubes without any clotting inhibitors (e.g., heparin, citrate, or EDTA). The whole blood is allowed to clot (this may be speeded by incubation at 37°C), then centrifuged at $1000 \times g$ to pellet the clotted cells, and the clear serum fraction is removed with a pipette and stored at 4°C (for short periods) or −20°C (for long-term storage). Sodium azide (0.02%) may be added to inhibit growth of contaminants (but consider whether azide may interfere with subsequent assays such as ELISA).

44.2.2 Detection Procedures

44.2.2.1 Virus Isolation

Methods for VI have been published in the OIE Manual[61] and reviewed in Chambers and Reedy.[62] By either culture method, if virus is not detected upon first passage a sterile aliquot should be obtained for one or more subsequent blind passages. Virus isolates have to be further confirmed by hemagglutination assay or immunofluorescence assay using EIV-specific monoclonal antibodies.

44.2.2.1.1 Virus Isolation in Embryonated Eggs

Embryonated hens' eggs (such as SPAFAS eggs) are allowed to develop in a humidified, rotating egg incubator until 10–11 days of age. The injection site is determined by candling to identify the line of the air sac and a place along the circumference that is relatively free of major blood vessels. The surface of the egg is topically disinfected, for example, with 70% ethyl alcohol, and a small injection hole is drilled through the hard shell but not so deeply as to penetrate the inner shell membrane. Using a tuberculin syringe, 0.1–0.2 mL of the processed diagnostic sample is injected straight down into the allantoic cavity or alternatively the amniotic cavity. For diagnostic purposes, no dilution or minimal dilution of the sample is desirable (in contrast to amplification of high-titer virus stocks where a dilution of >1:100 is used to prevent production of defective interfering particles). Following inoculation, the injection site is resealed with paraffin wax or equivalent. Eggs are incubated at 35°C–37°C for 2–3 days, then chilled (e.g., overnight in a refrigerator). To harvest a small sample for virus analysis, the wax seal is melted, a sterile Pasteur pipet is inserted into the allantoic cavity, and a small amount of fluid is withdrawn taking care not to collect egg yolk. This sample can be analyzed without dilution by hemagglutination assay using 1% chicken erythrocytes in U-bottom or V-bottom wells. If detectable virus growth in eggs has occurred, the allantoic fluid should yield a positive HA test result. The allantoic fluid may also be tested by RT-PCR or rapid IDK if desired.

44.2.2.1.2 Virus Isolation in MDCK Cell Culture

For virus culture in MDCK cells, the cells (ATCC CCL34) are grown to near confluency in tissue culture tubes or small flasks and then washed with PBS or with serum-free tissue culture medium. The inoculum, diluted with serum-free tissue culture medium containing antibiotics, is added and allowed to adsorb for 1 h. After washing and replacement of medium, exogenous trypsin protease activity must be supplied, and this is typically by addition of tosyl-phenylethyl-chloromethyl ketone (TPCK)-treated trypsin (1–2 µg/mL final concentration). It is sometimes useful also to pretreat the inoculum with TPCK-trypsin before inoculation of cells. Following infection, cells are incubated at 37°C and examined daily for cytopathic effects. If such effects are observed, the supernatant medium can be harvested and infection confirmed by, for example, HA or RT-PCR tests; supernatants should be so tested after 4–5 days regardless of appearance of cytopathic effect.

44.2.2.2 Rapid Antigen Detection Kits

These may be membrane or lateral-flow rapid ELISA modules and directions for use are manufacturer specific. The swab sample should be tested directly without any preprocessing or dilution other than as directed by the manufacturer. The test modules include positive controls, and these should be carefully observed to support a valid interpretation. As mentioned, these kits were not designed to detect EIV; where they do, it is a side feature and not every rapid IDK is effective or equally sensitive for the purpose of EIV detection. Positive results on EIV detection have been published for the Directigen Flu A and Flu A+B kits (Becton-Dickinson), Espline (Fujirebio), and Quick S-influ A•B (Denka Seiken),[43,49,63] and the author's lab has used the BinaxNOW Influenza A&B test (Inverness Medical) with success.

44.2.2.3 Retrospective Serology

Methods have been published in the OIE Manual[61] in detail.

44.2.2.3.1 Single Radial Hemolysis Test

The Single Radial Hemolysis (SRH) test measures the size of the zone of hemolysis created when EIV-specific antibodies diffuse radially through a matrix of complement-sensitized erythrocytes conjugated with virus. Paired acute and convalescent sera should be tested side by side, together with a subtype-specific positive control serum and if possible a known-negative serum. Sera should be heat inactivated by incubation at 56°C for 30 min.

44.2.2.3.1.1 Procedure

1. Obtain sheep whole blood in Alsever's solution and separate the red blood cells (RBCs) by low-speed centrifugation. Wash the RBCs 3× in saline/HEPES solution (0.85% NaCl, 0.05 M HEPES, and 0.02% sodium azide as preservative; adjust to pH 6.5). Resuspend the RBCs to make an 8% solution (v/v packed cells) in saline/HEPES, allowing a minimum of 1 mL of 8% RBCs for each 6 cm × 11 cm immunodiffusion plate.

2. Add a stock solution of clarified virus to the RBCs. The optimum ratio of virus to RBC volume is that which yields the largest and clearest zone of hemolysis and must be predetermined by titration for each virus strain; it typically lies within a range of 50–1500 HAU per mL of RBCs. Also, as negative control, prepare RBCs mixed with PBS-A only or mixed with a non-cross-reactive influenza virus strain such as influenza A/PR/8/34 (H1N1). Mix gently and incubate at 4°C for 10 min (PBS-A, aka PBS-London: 10 g NaCl, 0.25 g KCl, 1.45 g Na_2PO_4, 0.25 g KH_2PO_4, and 0.2 g Na azide per liter in distilled water).

3. Slowly stir in ½ volume of a freshly prepared 1:400 dilution (in normal saline) of $CrCl_3$ stock solution (6 g/10 mL distilled water). Incubate at room temperature for 5 min with occasional mixing and then centrifuge at $1500 \times g$ for 5 min to pellet the RBCs. Discard the supernatant and resuspend the RBCs in saline/HEPES with 0.2% (w/v) bovine serum albumin. Repeat the centrifugation step and resuspend the RBCs to 8% (v/v) in PBS-A.

4. Add 0.9 mL of virus-sensitized 8% RBCs to 7.8 mL of 1% agarose that has been melted and tempered to 42°C. Mix gently, then immediately add 0.3 mL of undiluted guinea pig complement. Mix again and pour into 6 cm × 11 cm immunodiffusion plates (10 cm diameter Petri dishes made be used but will require more RBC-agarose mixture). Allow to set with the lid off for 5 min and then store the prepared plates at 4°C until use.

5. To perform the test, punch 3 mm wells in agarose about 12–15 mm apart, including a well on each plate for positive control serum. Add 10 µL of test or control serum to a well. Paired acute and convalescent sera should be tested side by side, in duplicate. Incubate at 34°C for 20 h in a humidified chamber. To read the result, illuminate the plate from beneath on a light box and measure the average diameter of each zone of hemolysis. Calculate the area of each zone of hemolysis in mm^2, subtracting the area of the well (expected to be about 7 mm^2).

6. An increase in area of hemolysis of 25 mm^2, or 50%, whichever is smaller, is considered a significant increase in antibody titer between paired acute and convalescent serum samples and is expected to represent a twofold increase in antibody titer. Positive and negative sera should give the expected results for the test to be valid. If sera also produce zones of hemolysis on the negative control plate (RBCs conjugated with an irrelevant virus or with no virus), the sera should be preadsorbed with fresh RBCs and retested.

44.2.2.3.2   Hemagglutination-Inhibition Assay

Methods have been published in the OIE Manual in detail.[61] The widely used HI assay is a quantal assay, meaning that the titers are expressed in successive stepwise multiples of (usually) 2 instead of continuous distributions as in SRH; thus, the difference between a titer of 512 and one of 1024 is a difference of only one well, which is within the accepted range of experimental error. A fourfold difference in titer is the minimum considered a significant difference. For data analysis, this quantal nature of the assay must be taken into account: geometric means (based on $\log_2$-transformed data) are used instead of arithmetic means.

44.2.2.3.2.1   Procedure

1. Removal of nonspecific inhibitors of hemagglutination
   Equine sera contain a natural inhibitor of hemagglutination, α-2 macroglobulin, which must be inactivated before the serum can be tested. There are several methods to do this, the most common being kaolin, receptor-destroying enzyme (RDE), and trypsin-periodate treatment. The kaolin method is adequate for HI against equine H3 HA but can yield false-positive HI reactions against equine H7 HA.[64]

2. Serum pretreatment with kaolin
   Mix 2 volumes of serum with 10 volumes of 10% kaolin (v/v in PBS) for 20 min followed by centrifugation at $1200 \times g$ for 20 min. To the supernatant, add 1 volume of 5% chicken RBCs and 7 volumes of PBS. Centrifuge at $400 \times g$ for 20 min. The supernatant, a 1:10 dilution of kaolin-treated serum, is ready for use.

3. Serum pretreatment with RDE
   Prepare an RDE working solution of 10–100 units/mL in calcium/saline solution (1.0 g $CaCl_2 \cdot 2H_2O$, 9.0 g NaCl, 1.203 g H_3BO_3, and 0.052 g $Na_2B_4O_7 \cdot 10H_2O$ per liter in H_2O). Mix 0.1 mL of serum with 0.4 mL of freshly prepared RDE working solution. Mix and incubate 12–18 h at 37°C. Add 0.3 mL of 2.5% (w/v in H_2O) sodium citrate solution. Mix, securely seal the tube, and incubate at 56°C for 30 min. Chill on ice and add 0.2 mL PBS. The serum, now at 1:10 dilution, is ready for use.

4. Serum pretreatment with trypsin-periodate
   Mix 0.1 mL of serum with 0.1 mL of 0.4% trypsin working solution (1.6 mL of 25 g porcine trypsin/L mixed with 8.4 mL of 0.1M sodium phosphate buffer, pH 7.2). Incubate at 56°C for 30 min and then cool to room temperature. Add 0.3 mL of 0.01 M $NaIO_4$ in PBS, mix and incubate 15 min at RT. Add 0.3 mL of 1% glycerol (v/v in normal saline) and incubate for 15 min at RT. The treated serum can then be tested as is (1:8 dilution) or 0.2 mL of normal saline may be added to achieve a 1:10 dilution.

5. Antigen pretreatment with Tween-80/ether
   Pretreatment of the virus antigen with Tween-80/ether is often done because it increases the sensitivity of the test to low levels of serum antibodies. Standardization of the resulting artificial antigen is, however, a problem and batch-to-batch variation can affect the reproducibility of the test result. For this antigen pretreatment, in a glass flask mix 39.5 mL of virus-containing allantoic fluid with 0.5 mL of 10% Tween-80 in PBS. Mix gently for 5 min at RT, chill on ice, transfer to a fume hood, and add 20 mL of fresh diethyl ether. Mix well for 15 min at 4°C, then let stand 1 h to allow the aqueous and organic layers to separate. Remove the aqueous layer into a new glass bottle and allow to stand overnight in the fume hood so that residual ether will evaporate. Aliquot and store the treated antigen at −70°C until use.

6. Performing the HI assay
   In a microtiter plate with U- or V-shaped wells, add 25 μL of PBS to all wells of a row (or column) except the first, for each serum sample to be tested, including a positive control serum. A negative control well contains 25 μL of PBS only. To the first well add 50 μL of the serum sample, then perform successive twofold serial dilutions by transferring 25 μL to the next well, mix and repeat down the row to the end, discarding the last 25 μL. Thus, if starting from a serum treated to 1:10 dilution by kaolin, RDE, and trypsin-periodate methods, the dilution series in reciprocal will be 10, 20, 40, 80, 160, and so on. To each well add 25 μL of virus antigen that has been diluted and carefully titrated to a concentration of 4 HAU/25 μL (or 8 HAU/50 μL). Mix gently and allow to stand for 30 min at room temperature. Add 50 μL of 0.5% chicken RBCs (v/v packed cells in PBS, freshly prepared) to each well, mix gently and incubate at room temperature for an additional 30 min, before reading the result. If insufficient hemagglutination-inhibiting antibodies are present at a given serum dilution, virus-mediated hemagglutination will occur and the RBCs will have settled into a fuzzy film across the bottom of the well, resembling the negative control well. Where antibodies have completely inhibited hemagglutination, RBCs will roll to the bottom of the well forming a compact button; and if the plate is tilted this button should form a tear-drop shape. The HI titer is the reciprocal of the greatest serum dilution that completely inhibits hemagglutination. If no inhibition is observed even in the first well, the titer is recorded as "<10," not as zero (assuming the starting dilution was 1:10). For purposes of computing means and variances, the authors' own arbitrary convention is to set "<10" as equal to 5.

44.2.2.4 Nucleic Acid Tests

44.2.2.4.1 Viral Nucleic Acid Extraction

The use of molecular diagnostic assays for the detection of viral agents in clinical specimens requires stringent methods to ensure the stability of the nucleic acids (RNA).[65–67] Sample collection and purification of RNA mark the initial step of every molecular diagnostic assay and therefore are the most important determinants of the reproducibility and biological relevance of subsequent RT-PCR results. Thus, it is prudent to start with proper sample collection, transport, and storage, followed by adherence to rigorous laboratory protocols when extracting nucleic acids and properly storing purified nucleic acids. Naked RNA is extremely susceptible to degradation by endogenous RNases (ribonucleases) that are present in all living cells. Therefore, the key to the successful isolation of high-quality RNA for laboratory diagnosis is to ensure that neither endogenous nor exogenous RNases are introduced prior to or during the nucleic acid extraction procedure. It is recommended to store the extracted RNA at −20°C (short term) or −80°C (long term) until further use. Care should be exercised every time the sample is taken out of storage for analysis. Repeated freeze–thaw cycles will degrade the RNA and therefore, it is recommended that the eluted RNA be aliquoted into multiple tubes for future use.

A prerequisite for the performance of molecular diagnostic assays is an efficient method to isolate and purify viral RNA from clinical specimens such as nasal swabs or nasal secretions from horses. Currently, there are many methods to isolate and purify viral RNA from clinical specimens, but there are only a very few studies that describe comparison of various RNA extraction methods of samples from animal origin including equine.[68–70] Most of the commercial reagent kits for RNA extraction are based on fluid phase, solid phase, or magnetic separation techniques.[71–73] The traditional RNA extraction method using acid guanidinium thiocyanate–phenol–chloroform technique developed by Chomczynski and Sacchi has been replaced by commercial reagent kits that use glass fiber filters, silica-gel spin columns, and magnetic beads to trap RNA present in the clinical specimens.[71–74] Furthermore, the new commercial reagent kits have increased the yield, purity, and quality of extracted viral RNA by removing contaminating genomic DNA, other RNAs, and RT-PCR enzyme inhibitors present in the clinical specimens. However, these kits vary widely in technological principles used for nucleic acid extraction. There are various pros and cons to each technological principle and therefore, the quality and purity of RNA extracted from various reagent kits can vary significantly.

Several automated nucleic acid isolation kits using magnetic beads and magnetic particle processors are commercially available.[75] Unlike automated liquid handling systems that move reagents into and out of a single well of a multi-well plate to perform the different steps of an RNA isolation procedure, the magnetic particle processors use permanent magnetic rods to collect magnetic beads from the solution and release them into other wells containing reagents for the subsequent steps of the protocol. The effectiveness of bead collection and transfer leads to superior washing and elution efficiency, as well as rapid processing of a large number of samples (e.g., 24, 48, or 96 samples in a batch). Additionally, the magnetic bead-based method can also be used for manual RNA purification.

44.2.2.4.2 Standard RT-PCR

Two-step and one-step standard RT-PCR assays for the detection of EIV in clinical samples have been previously described in the literature.[44,45,47] These include universal influenza A assays or assays that are specific for EIV. Quinlivan et al. described a nested RT-PCR assay targeting the nucleoprotein gene and a standard RT-PCR targeting the matrix gene of EIV.[50] However, most of these RT-PCR assays have now been replaced by rRT-PCR assays for routine diagnosis of EIV.

44.2.2.4.3 Real-Time RT-PCR

In the past decade, nucleic acid amplification-based assays (standard RT-PCR or rRT-PCR) were developed and evaluated by various groups.[43–51] However, such assays were not widely used for the routine diagnosis of this disease. This changed following the introduction of EI into Australia in 2007, when an rRT-PCR developed to detect the avian influenza virus matrix gene was used as the molecular diagnostic method of choice for EI.[51,52,59] A panel of rRT-PCR assays capable of detecting a wide range of EIV strains comprising both subtypes of EIV (H7N7 and H3N8) was developed.[76] These rRT-PCR assays use TaqMan® minor groove binding (MGB™) probes targeting the NP, M, H3, and H7 HA genes of the virus and they provide a fast and reliable means of virus detection and disease surveillance. Of this panel of assays, the NP-specific rRT-PCR assay has been routinely used for the detection of EIV clinical samples.

44.2.2.4.4 Insulated Isothermal RT-PCR

Recently, GeneReach (GeneReach USA, Lexington, MA) has developed a TaqMan probe-based iiRT-PCR assay for the detection of EIV H3N8 subtype, and this assay has a very high specificity and equal or higher sensitivity as compared to the NP-specific rRT-PCR.[53,60,77] The assay is based on iiRT-PCR for qualitative detection of influenza A virus subtype H3N8 in clinical samples. Fluorogenic probe hydrolysis chemistry is used to generate a fluorescent signal when a specific RNA sequence of influenza A virus subtype H3N8 is amplified. This assay is specially designed to be used on a compatible iiRT-PCR instrument, POCKIT™ Nucleic Acid Analyzer (GeneReach USA, Lexington, MA).[77,78] The POCKIT Xpress™ Portable PCR Platform for onsite testing is suitable for use in veterinary clinics, racetracks, breeding facilities, and diagnostic laboratories.[53]

44.2.2.4.5 Loop-Mediated Isothermal Amplification Assay

RT loop-mediated isothermal amplification (RT-LAMP) assay was developed to detect EIV H3N8 subtype. The RT-LAMP primer set was designed to target the HA gene of this subtype. It was reported this assay was 10 times more sensitive than an RT-PCR assay and highly specific for EIV.

44.3 CONCLUSION AND FUTURE PERSPECTIVES

Timely diagnosis of EI depends critically upon prompt collection of appropriate specimens for testing. Once samples are in hand, there are various methods of greater or lesser utility available to perform that testing. The nasopharyngeal swab is the best sample and obtaining that swab is the weak link in the chain, as horses may be uncooperative. Horses are large and strong enough to be potentially dangerous and equine practitioners understandably may be reluctant to force the issue. Thus, a sampling method better tolerated by horses would be an important advance. Perhaps, with the development of highly sensitive field-capable diagnostic technologies, specimens such as saliva or visible discharges from the nares may be demonstrated to contain sufficient influenza virus for reliable diagnosis.

A further disincentive is that there is no disease-specific treatment for EI infection, other than management of clinical signs and control of secondary infections—which practitioners are likely to do with or without a laboratory diagnosis. Some anti-influenza drugs used in humans are indeed effective against EI[79–81]; but the prospect of viruses developing resistance to these drugs leads to their restriction to human use. Possibly once a drug becomes obsolete for human use, it may start a second life as a veterinary drug and be licensed for use in horses.

The benefit of timely diagnosis is both tactical and strategic disease control. Movements of affected horses can be curtailed, and at-risk horses can be freshly vaccinated. Characterization of virus isolates responsible for outbreaks is vital because influenza vaccines require periodic updating to match newly emerging variant strains. Improvement in diagnostic technologies, including sampling technologies, coupled with improvement in vaccine technologies that facilitate rapid updating, promises great benefits for control of influenza in horses.

REFERENCES

1. Cullinane, A. and Newton, J.R. Equine influenza—A global perspective. *Vet Microbiol* 167, 205–214 (2013).
2. Elton, D. and Bryant, N. Facing the threat of equine influenza. *Equine Vet J* 43, 250–258 (2011).
3. Landolt, G.A. Equine influenza virus. *Vet Clin North Am Equine Pract* 30, 507–522 (2014).
4. Daly, J.M., MacRae, S., Newton, J.R., Wattrang, E., and Elton, D.M. Equine influenza: A review of an unpredictable virus. *Vet J* 189, 7–14 (2011).
5. Webster, R.G. Are equine 1 influenza viruses still present in horses? *Equine Vet J* 25, 537–538 (1993).
6. Sovinova, O., Tumova, B., Pouska, F., and Nemec, J. Isolation of a virus causing respiratory disease in horses. *Acta Virol* 2, 52–61 (1958).
7. van Maanen, C. and Cullinane, A. Equine influenza virus infections: An update. *Vet Q* 24, 79–94 (2002).
8. Wright, P.F., Neumann, G., and Kawaoka, Y. Orthomyxoviruses. In: *Fields Virology*, Vol. 2 (eds. Knipe, D.M. and Howley, P.M.), pp. 1691–1740 (Lippincott Williams & Wilkins, Philadelphia, PA, 2007).
9. Ismail, T.M., Sami, A.M., Youssef, H.M., and Abou Zaid, A.A. An outbreak of equine influenza type 1 in Egypt in 1989. *Vet Med J Giza* 38, 195–206 (1990).
10. Madic, J., Martinovic, S., Naglic, T., Hajsig, D., and Cvetnic, S. Serological evidence for the presence of A/equine-1 influenza virus in unvaccinated horses in Croatia. *Vet Rec* 138, 68 (1996).
11. Singh, G. Characterization of A/eq-1 virus isolated during the equine influenza epidemic in India. *Acta Virol* 38, 25–26 (1994).
12. Waddell, G.H., Teigland, M.B., and Sigel, M.M. A new influenza virus associated with equine respiratory disease. *J Am Vet Med Assoc* 143, 587–590 (1963).
13. Dubovi, E.J. and Njaa, B.L. Canine influenza. *Vet Clin Small Anim Pract* 38, 827–835 (2008).
14. Daniels, R.S., Skehel, J.J., and Wiley, D.C. Amino acid sequences of haemagglutinins of influenza viruses of the H3 subtype isolated from horses. *J Gen Virol* 66, 457–464 (1985).
15. Klingeborn, B., Rockborn, G., and Dinter, Z. Significant antigenic drift within the influenza equi 2 subtype in Sweden. *Vet Rec* 106, 363–364 (1980).
16. Oxburgh, L., Berg, M., Klingeborn, B., Emmoth, E., and Linne, T. Evolution of H3N8 equine influenza virus from 1963 to 1991. *Virus Res* 34, 153–165 (1994).
17. Ozaki, H. et al. Antigenic variation among equine H 3 N 8 influenza virus hemagglutinins. *Jpn J Vet Res* 48, 177–186 (2001).
18. van Oirschot, J.T., Masurel, N., Huffels, A.D., and Anker, W.J. Equine influenza in the Netherlands during the winter of 1978–1979; antigenic drift of the A-equi 2 virus. *Vet Q* 3, 80–84 (1981).
19. Bryant, N.A. et al. Antigenic and genetic variations in European and North American equine influenza virus strains (H3N8) isolated from 2006 to 2007. *Vet Microbiol* 138, 41–52 (2009).
20. Lewis, N.S. et al. Antigenic and genetic evolution of equine influenza A (H3N8) virus from 1968 to 2007. *J Virol* 85, 12742–12749 (2011).
21. Woodward, A.L. et al. Development of a surveillance scheme for equine influenza in the UK and characterisation of viruses isolated in Europe, Dubai and the USA from 2010–2012. *Vet Microbiol* 169, 113–127 (2014).
22. Bryant, N.A. et al. Isolation and characterisation of equine influenza viruses (H3N8) from Europe and North America from 2008 to 2009. *Vet Microbiol* 147, 19–27 (2011).
23. Murcia, P.R. et al. Intra- and interhost evolutionary dynamics of equine influenza virus. *J Virol* 84, 6943–6954 (2010).
24. Payungporn, S. et al. Influenza A virus (H3N8) in dogs with respiratory disease. *Emerg Infect Dis* 14, 902–908 (2008).
25. Pecoraro, H.L., Bennett, S., Spindel, M.E., and Landolt, G.A. Evolution of the hemagglutinin gene of H3N8 canine influenza virus in dogs. *Virus Genes* 49, 393–399 (2014).
26. Hoelzer, K. et al. Intrahost evolutionary dynamics of canine influenza virus in naive and partially immune dogs. *J Virol* 84, 5329–5335 (2010).
27. Chambers, T.M., Balasuriya, U.B., Reedy, S.E., and Tiwari, A. Replication of avian influenza viruses in equine tracheal epithelium but not in horses. *Influenza Other Respir Viruses* 7(Suppl. 4), 90–93 (2013).

28. Tu, J. et al. Isolation and molecular characterization of equine H3N8 influenza viruses from pigs in China. *Arch Virol* 154, 887–890 (2009).

29. Wilson, W.D. Equine influenza. *Vet Clin North Am Equine Pract* 9, 257–282 (1993).

30. Cullinane, A., Elton, D., and Mumford, J. Equine influenza—Surveillance and control. *Influenza Other Respir Viruses* 4, 339–344 (2010).

31. Landolt, G.A., Townsend, H.G.G., and Lunn, D.P. Equine influenza infection. In: *Equine Infectious Diseases* (eds. Sellon, D.C. and Long, M.T.), pp. 141–151 (Saunders, St. Louis, MO, 2014).

32. Gerber, H. Clinical features, sequelae, and epidemiology of equine influenza. In: *Equine Infectious Diseases II* (eds. Bryans, J.T. and Gerber, H.), pp. 63–80 (Karger, Basel, Switzerland, 1970).

33. Lin, C. et al. The involvement of a stress-activated pathway in equine influenza virus-mediated apoptosis. *Virology* 287, 202–213 (2001).

34. O'Niell, F.D., Issel, C.J., and Henk, W.G. Electron microscopy of equine respiratory viruses in organ cultures of equine fetal respiratory tract epithelium. *Am J Vet Res* 45, 1953–1960 (1984).

35. Sarasola, P., Taylor, D.J., Love, S., and McKellar, Q.A. Secondary bacterial infections following an outbreak of equine influenza. *Vet Rec* 131, 441–442 (1992).

36. Anzai, T., Walker, J.A., Blair, M.B., Chambers, T.M., and Timoney, J.F. Comparison of the phenotypes of *Streptococcus zooepidemicus* isolated from tonsils of healthy horses and specimens obtained from foals and donkeys with pneumonia. *Am J Vet Res* 61, 162–166 (2000).

37. Willoughby, R. et al. The effects of equine rhinovirus, influenza virus and herpesvirus infection on tracheal clearance rate in horses. *Can J Vet Res* 56, 115–121 (1992).

38. Daly, J.M. et al. Investigation of equine influenza cases exhibiting neurological disease: Coincidence or association? *J Comp Pathol* 134, 231–235 (2006).

39. Yondon, M. et al. Isolation and characterization of H3N8 equine influenza A virus associated with the 2011 epizootic in Mongolia. *Influenza Other Respir Viruses* 7, 659–665 (2013).

40. Gilkerson, J.R. Equine influenza in Australia: A clinical overview. *Aust Vet J* 89(Suppl. 1), 11–13 (2011).

41. Peek, S.F. et al. Acute respiratory distress syndrome and fatal interstitial pneumonia associated with equine influenza in a neonatal foal. *J Vet Intern Med* 18, 132–134 (2004).

42. OIE (Office International des Epizooties). Equine influenza. In: *Manual of Diagnostic Tests and Vaccines for Terrestrial Animals* (eds. V. Caporale et al.), pp. 546–557 (OIE, Paris, France, 2000).

43. Chambers, T.M., Shortridge, K.F., Li, P.H., Powell, D.G., and Watkins, K.L. Rapid diagnosis of equine influenza by the Directigen FLU-A enzyme immunoassay. *Vet Rec* 135, 275–279 (1994).

44. Donofrio, J.C., Coonrod, J.D., and Chambers, T.M. Diagnosis of equine influenza by the polymerase chain reaction. *J Vet Diagn Invest* 6, 39–43 (1994).

45. Fouchier, R.A. et al. Detection of influenza A viruses from different species by PCR amplification of conserved sequences in the matrix gene. *J Clin Microbiol* 38, 4096–4101 (2000).

46. Munch, M., Nielsen, L.P., Handberg, K.J., and Jorgensen, P.H. Detection and subtyping (H5 and H7) of avian type A influenza virus by reverse transcription-PCR and PCR-ELISA. *Arch Virol* 146, 87–97 (2001).

47. Oxburgh, L. and Hagstrom, A. A PCR based method for the identification of equine influenza virus from clinical samples. *Vet Microbiol* 67, 161–174 (1999).

48. Quinlivan, M., Dempsey, E., Ryan, F., Arkins, S., and Cullinane, A. Real-time reverse transcription PCR for detection and quantitative analysis of equine influenza virus. *J Clin Microbiol* 43, 5055–5057 (2005).

49. Yamanaka, T., Tsujimura, K., Kondo, T., and Matsumura, T. Evaluation of antigen detection kits for diagnosis of equine influenza. *J Vet Med Sci* 70, 189–192 (2008).

50. Quinlivan, M. et al. Comparison of sensitivities of virus isolation, antigen detection, and nucleic acid amplification for detection of equine influenza virus. *J Clin Microbiol* 42, 759–763 (2004).

51. Foord, A.J. et al. Real-time RT-PCR for detection of equine influenza and evaluation using samples from horses infected with A/equine/Sydney/2007 (H3N8). *Vet Microbiol* 137, 1–9 (2008).

52. Donofrio, J.C., Coonrod, J.D., Davidson, J.N., and Betts, R.F. Detection of influenza A and B in respiratory secretions with the polymerase chain reaction. *PCR Methods Appl* 1, 263–268 (1992).

53. Balasuriya, U.B. Type A influenza virus detection from horses by real-time RT-PCR and insulated isothermal RT-PCR. *Methods Mol Biol* 1161, 393–402 (2014).

54. Paillot, R. et al. Duration of equine influenza virus shedding and infectivity in immunised horses after experimental infection with EIV A/eq2/Richmond/1/07. *Vet Microbiol* 166, 22–34 (2013).

55. Ilobi, C.P. et al. Antigenic and molecular characterization of host cell-mediated variants of equine H3N8 influenza viruses. *J Gen Virol* 75, 669–673 (1994).

56. Ilobi, C.P. et al. Direct sequencing of the HA gene of clinical equine H3N8 influenza virus and comparison with laboratory derived viruses. *Arch Virol* 143, 891–901 (1998).

57. Cook, R.F., Sinclair, R., and Mumford, J.A. Detection of influenza nucleoprotein antigen in nasal secretions from horses infected with A/equine influenza (H3N8) viruses. *J Virol Methods* 20, 1–12 (1988).

58. Crawford, P.C. et al. Transmission of equine influenza virus to dogs. *Science* 310, 482–485 (2005).

59. Spackman, E. et al. Development of a real-time reverse transcriptase PCR assay for type A influenza virus and the avian H5 and H7 hemagglutinin subtypes. *J Clin Microbiol* 40, 3256–3260 (2002).

60. Balasuriya, U.B. et al. Rapid detection of equine influenza virus H3N8 subtype by insulated isothermal RT-PCR (iiRT-PCR) assay using the POCKIT Nucleic Acid Analyzer. *J Virol Methods* 207, 66–72 (2014).

61. OIE (Office International des Epizooties). Equine influenza. In: *Manual of Diagnostic Tests and Vaccines for Terrestrial Animals* (eds. V. Caporale et al.), Vol. 2, pp. 865–878 (World Organisation for Animal Health (OIE), Paris, France, 2012).

62. Chambers, T.M. and Reedy, S.E. Equine influenza culture methods. *Methods Mol Biol* 1161, 403–410 (2014).

63. Galvin, P. et al. The evaluation of three diagnostic tests for the detection of equine influenza nucleoprotein in nasal swabs. *Influenza Other Respir Viruses* 8, 376–383 (2014).

64. Boliar, S., Stanislawek, W., and Chambers, T.M. Inability of kaolin treatment to remove nonspecific inhibitors from equine serum for the hemagglutination inhibition test against equine H7N7 influenza virus. *J Vet Diagn Invest* 18, 264–267 (2006).

65. Murphy, J. and Bustin, S.A. Reliability of real-time reverse-transcription PCR in clinical diagnostics: Gold standard or substandard? *Expert Rev Mol Diagn* 9, 187–197 (2009).

66. Bustin, S.A. and Mueller, R. Real-time reverse transcription PCR (qRT-PCR) and its potential use in clinical diagnosis. *Clin Sci (Lond)* 109, 365–379 (2005).

67. Bustin, S.A., Benes, V., Nolan, T., and Pfaffl, M.W. Quantitative real-time RT-PCR—A perspective. *J Mol Endocrinol* 34, 597–601 (2005).

68. Miszczak, F. et al. Evaluation of two magnetic-bead-based viral nucleic acid purification kits and three real-time reverse transcription-PCR reagent systems in two TaqMan assays for equine arteritis virus detection in semen. *J Clin Microbiol* 49, 3694–3696 (2011).

69. Deng, M.Y., Wang, H., Ward, G.B., Beckham, T.R., and McKenna, T.S. Comparison of six RNA extraction methods for the detection of classical swine fever virus by real-time and conventional reverse transcription-PCR. *J Vet Diagn Invest* 17, 574–578 (2005).

70. Knepp, J.H., Geahr, M.A., Forman, M.S., and Valsamakis, A. Comparison of automated and manual nucleic acid extraction methods for detection of enterovirus RNA. *J Clin Microbiol* 41, 3532–3536 (2003).

71. Zahringer, H. Old and new ways to RNA. *Lab Times*, p. 10 (2012).

72. Fang, X., Willis, R.C., Burrell, A., Evans, K., Hoang, Q., Xu, W., and Bounpheng, M. Automation of nucleic acid isolation on KingFisher magnetic particles. *JALA* 12, 7 (2007).

73. Berensmeier, S. Magnetic particles for the separation and purification of nucleic acids. *Appl Microbiol Biotechnol* 73, 495–504 (2006).

74. Chomczynski, P. and Sacchi, N. Single-step method of RNA isolation by acid guanidinium thiocyanate-phenol-chloroform extraction. *Anal Biochem* 162, 156–159 (1987).

75. Balasuriya, U.B. RNA extraction from equine samples for equine influenza virus. *Methods Mol Biol* 1161, 379–392 (2014).

76. Lu, Z. et al. Development and evaluation of one-step TaqMan real-time reverse transcription-PCR assays targeting nucleo-protein, matrix, and hemagglutinin genes of equine influenza virus. *J Clin Microbiol* 47, 3907–3913 (2009).

77. Tsai, Y.L. et al. Development of TaqMan probe-based insulated isothermal PCR (iiPCR) for sensitive and specific on-site pathogen detection. *PLoS One* 7, e45278 (2012).

78. Chang, H.F. et al. A thermally baffled device for highly stabilized convective PCR. *Biotechnol J* 7, 662–666 (2012).

79. Yamanaka, T., Tsujimura, K., Kondo, T., Hobo, S., and Matsumura, T. Efficacy of oseltamivir phosphate to horses inoculated with equine influenza A virus. *J Vet Med Sci* 68, 923–928 (2006).

80. Yamanaka, T. et al. Efficacy of a single intravenous dose of the neuraminidase inhibitor peramivir in the treatment of equine influenza. *Vet J* 193, 358–362 (2012).

81. Rees, W.A. et al. Pharmacokinetics and therapeutic efficacy of rimantadine in horses experimentally infected with influenza virus A2. *Am J Vet Res* 60, 888–894 (1999).

45 Infectious Salmon Anemia Virus

Dongyou Liu

CONTENTS

45.1 Introduction 393
    45.1.1 Classification 393
    45.1.2 Morphology and Genome Organization 394
    45.1.3 Biology and Epidemiology 394
    45.1.4 Clinical Features and Pathogenesis 395
    45.1.5 Diagnosis 395
        45.1.5.1 Virus Isolation 396
        45.1.5.2 Electron Microscopy 396
        45.1.5.3 Serological Assays 396
        45.1.5.4 Nucleic Acid Detection 396
    45.1.6 Treatment and Prevention 396
45.2 Methods 396
    45.2.1 Sample Preparation 396
    45.2.2 Detection Procedures 396
        45.2.2.1 Conventional RT-PCR 396
        45.2.2.2 Real-Time RT-PCR 397
45.3 Conclusion 397
References 397

45.1 INTRODUCTION

Infectious salmon anemia virus (ISAV) is a member of the family *Orthomyxoviridae*, responsible for causing infectious salmon anemia (ISA) in farmed Atlantic salmon (*Salmo salar*). Characterized by lethargy, anemia, hemorrhage of the internal organs, and death, ISA syndrome has been described in Norway, Scotland, Canada, the United States, the Faroe Islands, and Chile. Posing a serious threat to the economic sustainability of aquaculture industries worldwide, ISA is listed as a reportable disease by the OIE.

45.1.1 CLASSIFICATION

ISAV belongs to the genus *Isavirus*, family *Orthomyxoviridae*. The family *Orthomyxoviridae* (orthos, Greek for "straight"; myxa, Greek for "mucus") classifies a group of enveloped, segmented, negative-sense, single-stranded RNA viruses into six genera: *Influenza virus A*, *Influenza virus B*, *Influenza virus C*, *Isavirus*, *Thogotovirus*, and *Quaranjavirus*. Of these, *Influenza virus A*, *Influenza virus B*, and *Influenza virus C* contain one species (or type) each, with influenza A infecting birds, humans, pigs, horses; influenza B infecting humans and seal; and influenza C infecting humans, pigs, and dogs. *Isavirus* includes a single species that causes disease in Atlantic salmon. *Thogotovirus* consists of two species Thogoto virus (THOV, the type species) and Dhori virus (DHOV), the latter

of which includes Batken virus and Bourbon virus, as well as tentative viruses Araguari virus, Aransas Bay virus, Jos virus, and Upolu virus as subtype/strains; these viruses infect ticks, mosquitoes, and mammals, including humans. *Quaranjavirus* (named from two of the constituent viruses *Quaran*fil virus and *Johnston Atoll* virus) comprises six species (Cygnet River virus, Johnston Atoll virus, Lake Chad virus, Quaranfil virus, Tyulek virus, and Wellfleet Bay virus) that are found in arthropods, birds, and occasionally humans (Quaranfil virus only); *Quaranjavirus* differs from other genera of the family in having insect (tick) vectors and 6 monocistronic genome segments, with distantly related tick-borne viruses containing 6 or 7 segments being classified in *Thogotovirus*.

The genus *Isavirus* consists of a single species, that is, ISAV, which is distinguished into two major genotypes (groups), genotype I (European) and genotype II (North American), on the basis of sequence variations within a small, highly polymorphic region (HPR) located in the 5′-end of the hemagglutinin-esterase (HE) gene (segment 6) in addition to changes in segments 2 and 8 sequences. Indeed, the presence of gaps rather than single nucleotide mutations in the HPR, together with single amino acid substitution or a sequence insertion near the putative cleavage site of the fusion (F) protein, has been implicated in ISAV virulence [1,2].

ISAV genotype I (European group) covers viruses originated from Norway, Scotland, and Chile, and genotype II (North American group) consists of isolates from Canada and the United States. Typically, the homology shared between European and North American groups in the hemagglutinin nucleotide sequence is 79%, whereas that shared within each group is >90% [2,3].

In addition, focusing on the extracellular region of hemagglutinin, the European isolates may be further separated into three subgroups, with G1 containing isolates from Norway between 1996 in 2000, G2 comprising isolates from Norway, one from Scotland and one from the Faroe Islands, and G3 including isolates from Norway between 1987 and 2000 and Scotland [1].

As the name implies, ISAV induces severe anemia in sea-farmed Atlantic salmon (*Salmo salar*) as well as subclinical infections in sea-run Brown trout (*Salmo trutta*) and Rainbow trout (*Onchorhyncus mykiss*), which may act as important carriers and reservoirs of the virus.

45.1.2 Morphology and Genome Organization

ISAV is an enveloped, pleiomorphic (usually rounded but occasionally filamentous) particle of 100–130 nm in diameter, covered by surface projections of approximately 20 nm.

Encapsidated by nucleoprotein (N), the ISAV genome (13.5 kb in total) consists of eight linear, negative-sense, single-stranded RNA segments of 890–2341 nt, each of which contains the 3′ and 5′ noncoding sequences, and which together encode 11 proteins (Table 45.1) [1,2].

The HE protein is a glycoprotein involved in receptor-binding and receptor-destroying activities. The interaction between HE and host cell receptor [5-*N*-acetyl-4-*O*-acetyl neuraminic acid (Neu4,5Ac$_2$) residue] facilitates the fusion of the membranes and the entrance of the ribonucleoprotein complex. In pathogenic ISAV from diseased fish with clinical disease and pathological signs, a HPR containing gaps rather than single-nucleotide substitutions is identified before the transmembrane domain in the HE gene. Of the 28 different HPRs described to date, 26 are found in European group, 2 in

North American group, and 28 in ISAV isolates from Chile, with HPR7b being the most prevalent. On the other hand, in nonpathogenic ISAV from both apparently healthy wild and farmed Atlantic salmon, the HPR in the HE gene contains all intact motifs (so-called HPR0). It is suggested that pathogenic ISAV arises from nonpathogenic ISAV through the deletions/alterations in HRP0 of the HE gene [3–5].

Segment 7 is capable of generating these proteins via alternative splicing mechanism. Located in the +1 reading frame, ORF-1 of segment 7 encodes a polypeptide (NS1) of 301 aa with an antagonist activity to signaling by type I interferon (Table 45.1). Retaining the first 66 nucleotides of ORF-1, ORF-2 of segment 7 eliminates an intron of 526 bases to continue with reading frame +2, and produces a polypeptide of 159 aa, which operates as nuclear export protein through exporting the ribonucleoprotein complex from the nucleus to the cytoplasm. Retaining the first 22 amino acids of ORF-1 (via splicing of an intron of 257) and inclusion of 81 amino acids from the +3 reading frame (ORF3), a protein (NS3) of 81 aa with unknown function is generated [3,6].

Similarly, segment 8 consists of two possible overlapping ORFs, with ORF-1 producing a 24 kDa protein (M1) in vitro, but a 22 kDa protein in vivo, that is related to the viral matrix (membrane), and ORF-2 generating a structural protein (M2) of 27 kDa, with RNA binding and type I interferon antagonistic properties [3].

The 3′ and 5′ untranslated regions (UTR) are highly conserved. The 5′-UTR may be involved in the process of replication and transcription, acting jointly with the 3′-UTR [7,8].

45.1.3 Biology and Epidemiology

The primary target cells for ISAV are endothelial cells lining blood vessels, indicating that virus replication may occur in any organ. Virus replication may also occur in leukocytes and sinusoidal macrophages in kidney tissue.

TABLE 45.1
ISAV Genome Segments and Encoded Proteins

Genome Segment	Segment Size (kb)	ORF (nt)	Encoded Protein	Protein Size (aa)	Protein Mass (kDa)
1	2.3–2.4	2166	PB2 (viral polymerase B2)	722	79.5
2	2.3–2.4	2127	PB1 (viral polymerase B1)	709	80.5
3	2.2	1851	N (nucleoprotein)	616	66
4	1.9–2.0	1737	PA (viral polymerase A)	579	65.3
5	1.6–1.7	1332	F (fusion protein)	444	50
6	1.5	1167	HE (hemagglutinin-esterase protein)	389	42
7	1.3	ORF1, 903	NS1 (nonstructural protein 1 with type I interferon antagonistic property)	301	35
		ORF2, 477	NEP (nuclear export protein)	159	18
		ORF3, 243	NS3 (putative nonstructural protein 3 with unknown function)	81	10.6
8	1.0	ORF1, 591	M1 (membrane protein 1)	197	22–24
		ORF2, 726	M2 (membrane protein 2, a RNA-binding structural protein with type I interferon antagonistic property)	242	26.4

Sources: Kibenge, F.S. et al., *Anim. Health Res. Rev.*, 5(1), 65, 2004; Cottet, L. et al., *Virus Res.*, 155(1), 10, 2011; Aamelfot, M. et al., *J. Fish Dis.*, 37(4), 291, 2014.

Upon entry into the host via the gills, ISAV utilizes its HE protein to bind to host glycoprotein receptors containing 4-*O*-acetylated sialic acid residues, which also functions as a substrate for the receptor-destroying enzyme. This binding facilitates clathrin-mediated endocytosis of the virus, and subsequent fusion of virus membrane with the vesicle (endosome) membrane releases encapsidated RNA segments into cytoplasm. Migrating to the nucleus, genome segments are transcribed into genomic RNA of both negative and positive polarity and mRNA by viral polymerase (PB1, PB2, and PA). The viral transcripts then move out of the nucleus for translation. The translated proteins are transported from the cytoplasm to the nucleus, and ribonucleoprotein complex forms between genomic RNA, nucleoprotein, and RNA polymerase complex. In association with nuclear export protein (NEP), the ribonucleoprotein complex moves to the cytoplasm, and interacts with the plasma membrane region where virus assembly and budding of progeny virions take place [1].

ISA is primarily a disease of Atlantic salmon (*Salmo salar*). While it does not seem to occur at the embryo stage, ISA can affect the fish at any life stage after hatching. First coming to notice in Norway in 1984, when up to 80% of Atlantic salmon stock in some hatcheries died of a previously undocumented disease, the etiological agent for ISA was characterized as ISAV 10 years later using salmon head kidney (SHK-1) cell line. In 1996, a deadly disease tentatively named "hemorrhagic kidney syndrome" occurred in Atlantic salmon farm in Canada. Subsequent study confirmed that the culprit for both ISA and hemorrhagic kidney syndrome is ISAV. Since then, ISA has been reported in farmed Atlantic salmon in Scotland (1998), Faroe Islands (2000), USA (2001), and Chile (2007) [3,9,10].

Although Atlantic salmon does not appear to suffer life-long infection, ISAV may persist in the population at the farm level by continuous infection of new individuals that may or may not develop clinical signs. Besides farmed Atlantic salmon, ISAV has been shown to cause subclinical infection in feral Atlantic salmon, brown trout, sea trout (*Salmo trutta*), cod (*Gadus morhua*), Coho salmon (*Oncorhynchus kisutch*), and pollock (*Pollachius virens*). In experimental infection, rainbow trout (*Oncorhynchus mykiss*), cod (*Gadus morhua*), Arctic char (*Salvelinus alpinus*), and Atlantic herring (*Clupea harengus*) are also susceptible to ISAV. These asymptomatically infected fish may shed ISAV through natural excretions/secretions and act as reservoir for this virus [11,12].

ISAV is largely transmitted by contact with infected fish, their secretions, or contaminated water. Contact with contaminated equipment or people who have handled infected fish may also facilitate its transmission. Another important factor is geographical proximity to infected marine sites or slaughterhouses/processing plants, which release unprocessed, contaminated water. ISA outbreaks tend to occur in late spring (rising water temperatures) and late autumn/early winter (falling water temperatures) [13].

The optimal temperature for replication of ISAV is between 10°C and 15°C. ISAV survives well at 15°C for 10 days or 4°C for 14 days, but lose infecting ability at 56°C for 30 min.

In addition, the virus retains its infectivity in sea water for 20 h and in blood and liver for 6 days at 6°C. It is sensitive to UV irradiation and ozone, and inactivated by pH 4 or pH 12 for 24 h, or chlorine (100 mg/mL for 15 min).

45.1.4 CLINICAL FEATURES AND PATHOGENESIS

ISA is a multisystemic disease with high mortality rates in farmed Atlantic salmon (*Salmo salar*). Depending on infective dose, virus strain, temperature, age, and immune status of the host, ISAV infection often induces a variety of clinical signs, ranging from loss of appetite, lethargy, swimming slowly and gasping at the surface, pale gills, exophthalmia, scale pocket edema, swollen abdomen, bleeding along the belly and sides, blood in the anterior eye chamber, to death (average 30%, but as high as 100%). Some fish may maintain a normal appetite without displaying any external signs of illness and then suddenly die [3].

Postmortem examination often reveals dark and swollen kidney, liver, and spleen; pinpoint bleeding in the fatty tissue surrounding organs; pale heart; bloody fluid in the abdominal cavity and around the heart; swim bladder edema; intestinal congestion; circulatory system malfunction; and ascites. Histological findings include hemorrhages and/or hepatocyte necrosis in the liver, erythrocyte accumulation in spleen stroma, and diffuse interstitial hemorrhage with tubular necrosis [14,15].

At the genetic level, ISAV pathogenicity is associated with the existence of a high polymorphism region (HPR) in segment 6 and insertion in segment 5 sequence [16]. As ISAV targets vascular endothelial cells and macrophages, the underlying genetic variations in ISAV high and low pathogenic strains influence the expression of both innate and adaptive immune response relevant genes and outcome of the disease [17,18]. For example, ISAV high pathogenic strain induces hemagglutination that results in virus uptake and productive infection of Atlantic salmon erythrocytes whereas low pathogenic strain does not [17]. In addition, ISAV low pathogenic strain tends to cause the highest fold changes of most immune-relevant genes (including interferon-inducible protein Gig1, Mx1 protein, interferon-induced protein with tetratricopeptide repeats 5, Radical S-adenosyl methionine domain-containing protein [viperin], and several genes involved in the ISGylation pathway) in comparison with high pathogenic strain [18]. This suggests that ISAV high pathogenic strain is capable of evading and suppressing host immune responses for its own advantage.

45.1.5 DIAGNOSIS

Diagnosis of ISA on the basis of clinical manifestations is difficult as several other diseases may also induce anemia, such as erythrocytic inclusion body syndrome, winter ulcer, and septicemia due to *Moritella viscosa*.

Laboratory methods that have been used for ISV diagnosis include virus culture, electron microscopy, serological assays, and nucleic acid detection, with the goal to detect virus particles, antigens, antibodies, and nucleic acids [19–21].

45.1.5.1 Virus Isolation

Atlantic salmon kidney (ASK) and Atlantic SHK-1 are recommended for in vitro isolation and growth of ISAV. Other cell lines such as TO (Atlantic salmon macrophage/dendritic-like cell), epithelioma papulosum cyprinid, and CHSE-214 (chinook salmon embryo) cells may support ISAV growth but do not always produce distinct cytopathic effect [18]. ASK cells are generally maintained at 20°C in Leibovitz's L-15 cell culture medium supplemented with fetal bovine serum (5% or 10%), L-glutamine (4 mM), and gentamicin (50 μg/mL) and 2-mercaptoethanol (40 μM) (optional) [22,23].

45.1.5.2 Electron Microscopy

Virus particles may be observed in endothelial cells and leukocytes from tissue preparations using electron microscopy.

45.1.5.3 Serological Assays

Indirect fluorescent antibody test using monoclonal antibodies (MAbs) against ISAV HE on kidney smears (imprints) or on frozen tissue sections of kidney, heart, and liver provides a useful way to detect ISAV. Immunohistochemistry based on polyclonal antibody against ISAV nucleoprotein allows detection of ISAV on paraffin sections from formalin-fixed tissue (of mid-kidney and heart). Hemagglutination assay may be applied to determine the level of hemagglutination that is induced by the highly pathogenic but not the low pathogenic ISAV strains. Enzyme-linked immunosorbent assays with either purified virus or lysates from ISAV-infected cell cultures permits detection of ISAV-specific antibodies [24,25].

45.1.5.4 Nucleic Acid Detection

Reverse transcription polymerase chain reaction (RT-PCR) and real-time RT-PCR enable detection of both European and North-American HPR-deleted ISAV, as well as HPR0 ISAV, with primer sets derived from genomic segment 8 and segment 7 [26–31]. Samples with positive PCR results by using segment 7 or 8 primer sets may be confirmed by sequencing the HPR of segment 6. This will help determine the ISAV HPR variant present (HPR-deleted or HPR0 or both) [32,33].

Given its high sensitivity and rapid turnaround time, RT-PCR is increasingly adopted in clinical laboratories for routine laboratory diagnosis of ISAV [34,35]. Using RT-PCR, ISAV is detected in the cephalic and middle kidney 8 days postinfection, and in the liver, heart, and gills 13 days postinfection [36].

45.1.6 TREATMENT AND PREVENTION

Although no treatment is available for infected fish, recent research indicated that the broad-spectrum antiviral drug ribavirin (1-β-d-ribofuranosyl-1,2,4-triazole-3-carboxamide) is effective in inhibiting ISAV replication both in vitro and in vivo.

The current control and prevention approaches for ISA rely on early detection, eradication, and vaccination [37].

For example, implementation of husbandry practices has proven valuable to reduce the incidence of ISA. These include daily removal of dead/moribund fish, compulsory slaughter ("all in/all out"), disinfection of offal and wastewater in infected farms, strict movement controls on suspect farms, and surveillance of farms in the vicinity of an outbreak.

Despite being only partially effective, commercial vaccines have been employed for protection of Atlantic salmon against ISA in North America from 1999, the Faroe Islands from 2005, Norway from 2009, and Chile from 2010 [3,38,39]. In an attempt to further improve the efficacy of existing anti-ISAV vaccines, inactivated ISAV antigen was used to immunize against ISAV challenge, leading to a relative percent survival as high as 86 [40]. Moreover, intramuscular injection of plasmid expressing the type I interferons has been shown to confer strong immunity to Atlantic salmon against ISAV infection [41].

45.2 METHODS

45.2.1 SAMPLE PREPARATION

Blood is collected from live fish in heparin or EDTA tubes. Heart (including valves and bulbus arteriosus), mid-kidney, liver, pancreas/intestine, spleen, and gills are collected from dead fish.

Fish tissue is weighed and macerated to a 10% suspension (w/v) in PBS with 10× antibiotics. The homogenates are then diluted 1:1 in sterile PBS and clarified by centrifugation at 3000 rpm for 15 min, and the supernatants are used for RNA extraction.

Total RNA is extracted from 375 μL of infected fish tissue supernatant, purified virus preparation, or virus-infected tissue culture suspensions using 1.25 mL of Trizol reagent (Invitrogen). The extracted RNA is eluted in 20–50 μL of nuclease-free water and treated with DNAse I (Roche). RNA is quantitated by UV spectrophotometry.

45.2.2 DETECTION PROCEDURES

45.2.2.1 Conventional RT-PCR

Munir and Kibenge [42] described a one-tube RT-PCR for ISAV with primers F5 (5′-GAA GAG TCA GGA TGC CAA GAC G-3′)/R5 (5′-GAA GTC GAT GAT CTG CAG CGA-3′) for amplification of 211 bp product from ISAV RNA segment 8.

RT-PCR mixture (50 μL) is made up of 0.8 mM deoxynucleotide triphosphates, 0.8 μM of each primer, 5 mM dithiothreitol solution, 0.25 μL RNase inhibitor (40 U/μL), 2 μL template (>0.632 ng/μL), and 1.0 μL enzyme mix (AMV and Expand High Fidelity PCR-System, Titan One Tube RT–PCR System Kit, Roche Molecular Biochemicals) and 23.25 μL of PCR-grade water.

Amplification is carried out in a PTC-200 DNA Engine Peltier thermal cycler (MJ Research Inc.) with the following thermal profile: a single cycle of RT for 30 min at 55°C; a predenaturation step for 2 min at 94°C; 40 amplification cycles each consisting of denaturation for 30 s at 94°C, annealing for 45 s at 61°C, and extension for 90 s at 72°C; and a final extension step of 10 min at 72°C.

The amplified PCR products are resolved by 1% agarose gel electrophoresis in 0.5× TBE buffer and visualized by staining with ethidium bromide and photographed under 304 nm UV light.

45.2.2.2 Real-Time RT-PCR

McBeath et al. [43] utilized primers 404F_ISA8 (5′ tgg gca atg gtg tat ggt atg a-3′) and RA3 (583R)-ISA8 (5′ gaa gtc gat gaa ctg cag cga-3′) and FAM labeled probe (491_ISA8) (6-FAM cag gat gca gat gta tgc-MGB quencher) from segment 8 [44] for quantitative RT-PCR detection of ISAV.

RT-PCR mixture (20 µL) is composed of 8 µL of RNA and 12 µL of master mix containing 9.28 µL LC 480 RNA Master hydrolysis probe (one-step RT-PCR kit, Roche), 1.88 µL of activator $Mn(OAc)_2$ (50 mM), 1 µL of enhancer (20×), 1.13 µL of ISAV Segment 8 Forward primer and Reverse primer (20 µM each), and 1.04 µL of ISAV segment 8 probe (6 µM).

Amplification and detection are performed in 96-well plate using the LC 480 instrument (Roche) with absolute quantitation methods. The cycling conditions are 1 cycle of RT for 3 min at 63°C, denaturation at 95°C for 3 s, and 45 cycles of denaturation at 95°C for 15 s, annealing at 60°C for 1 min and amplification and detection at 72°C for 1 s.

For generation of the standard curve, the in vitro transcribed RNA standards are run in 5 replicates. The standard curve is constructed automatically with LC software version 4.0 (Roche) using the Ct values obtained when the serial 10-fold dilutions of the in vitro transcribed RNA samples with known numbers of RNA transcripts are used as templates. The standard curve obtained is then used as an external standard curve in all subsequent TaqMan® qRT-PCR assays on LC480.

For calculating ISAV RNA copy number equivalents per mL of unknown sample, the ISAV RNA copy number equivalents/20 µL RT-PCR reaction is multiplied by a factor of 20/8 × 1000/375 based on the use of 8 µL of the total 20 µL RNA eluted from 375 µL of virus lysate used for RT-PCR reaction.

45.3 CONCLUSION

ISAV is an orthomyxovirus implicated in ISA, which is a fatal disease of farmed Atlantic salmon, with symptoms ranging from severe anemia, exophthalmia, petechial hemorrhages in the visceral fat and skin, hemorrhagic hepatic necrosis and kidney tubular necrosis, interstitial hemorrhages, to death. Although ISA currently affects aquaculture in Norway, Scotland, Canada, the United States, the Faroe Islands, and Chile, it has potential to be transmitted to other Atlantic salmon farming regions in the world [45]. Considering its high sensitivity, specificity, and speed, RT-PCR has become a method of choice for routine identification and quantification of ISAV from clinical specimens and also for differentiation of high and low pathogenic ISAV strains [46].

REFERENCES

1. Cottet L et al. Infectious salmon anemia virus—Genetics and pathogenesis. *Virus Res.* 2011;155(1):10–19.
2. Aamelfot M, Dale OB, Falk K. Infectious salmon anaemia—Pathogenesis and tropism. *J Fish Dis.* 2014;37(4):291–307.
3. Kibenge FS, Munir K, Kibenge MJ, Joseph T, Moneke E. Infectious salmon anemia virus: Causative agent, pathogenesis and immunity. *Anim Health Res Rev.* 2004;5(1):65–78.
4. Christiansen DH, Østergaard PS, Snow M, Dale OB, Falk K. A low-pathogenic variant of infectious salmon anemia virus (ISAV-HPR0) is highly prevalent and causes a non-clinical transient infection in farmed Atlantic salmon (*Salmo salar* L.) in the Faroe Islands. *J Gen Virol.* 2011;92(4):909–918.
5. Godoy MG et al. Genetic analysis and comparative virulence of infectious salmon anemia virus (ISAV) types HPR7a and HPR7b from recent field outbreaks in Chile. *Virol J.* 2014;11(1):204.
6. Kibenge FS, Xu H, Kibenge MJ, Qian B, Joseph T. Characterization of gene expression on genomic segment 7 of infectious salmon anaemia virus. *Virol J.* 2007;4:34.
7. Kulshreshtha V et al. Identification of the 3′ and 5′ terminal sequences of the 8 RNA genome segments of European and North American genotypes of infectious salmon anemia virus (an orthomyxovirus) and evidence for quasispecies based on the non-coding sequences of transcripts. *Virol J.* 2010;7:338.
8. Fourrier M, Heuser S, Munro E, Snow M. Characterization and comparison of the full 3′ and 5′ untranslated genomic regions of diverse isolates of infectious salmon anaemia virus by using a rapid and universal method. *J Virol Methods.* 2011;174(1–2):136–143.
9. Godoy MG et al. First detection, isolation and molecular characterization of infectious salmon anaemia virus associated with clinical disease in farmed Atlantic salmon (*Salmo salar*) in Chile. *BMC Vet Res.* 2008;4:28.
10. Kibenge FS et al. Infectious salmon anaemia virus (ISAV) isolated from the ISA disease outbreaks in Chile diverged from ISAV isolates from Norway around 1996 and was disseminated around 2005, based on surface glycoprotein gene sequences. *Virol J.* 2009;6:88.
11. Grove S, Hjortaas MJ, Reitan LJ, Dannevig BH. Infectious salmon anaemia virus (ISAV) in experimentally challenged Atlantic cod (*Gadus morhua*). *Arch Virol.* 2007;152(10):1829–1837.
12. Skår CK, Mortensen S. Fate of infectious salmon anaemia virus (ISAV) in experimentally challenged blue mussels *Mytilus edulis. Dis Aquat Organ.* 2007;74(1):1–6.
13. Nylund A et al. Transmission of infectious salmon anaemia virus (ISAV) in farmed populations of Atlantic salmon (*Salmo salar*). *Arch Virol.* 2007;152(1):151–179.
14. Jørgensen SM, Hetland DL, Press CM, Grimholt U, Gjøen T. Effect of early infectious salmon anaemia virus (ISAV) infection on expression of MHC pathway genes and type I and II interferon in Atlantic salmon (*Salmo salar* L.) tissues. *Fish Shellfish Immunol.* 2007;23(3):576–588.
15. Hetland DL, Dale OB, Skjødt K, Press CM, Falk K. Depletion of CD8 alpha cells from tissues of Atlantic salmon during the early stages of infection with high or low virulent strains of infectious salmon anaemia virus (ISAV). *Dev Comp Immunol.* 2011;35(8):817–826.
16. Svingerud T, Holand JK, Robertsen B. Infectious salmon anemia virus (ISAV) replication is transiently inhibited by Atlantic salmon type I interferon in cell culture. *Virus Res.* 2013;177(2):163–170.
17. Workenhe ST et al. Infectious salmon anaemia virus replication and induction of alpha interferon in Atlantic salmon erythrocytes. *Virol J.* 2008;5:36.

18. Workenhe ST, Hori TS, Rise ML, Kibenge MJ, Kibenge FS. Infectious salmon anaemia virus (ISAV) isolates induce distinct gene expression responses in the Atlantic salmon (*Salmo salar*) macrophage/dendritic-like cell line TO, assessed using genomic techniques. *Mol Immunol*. 2009;46(15):2955–2974.

19. McClure CA, Hammell KL, Stryhn H, Dohoo IR, Hawkins LJ. Application of surveillance data in evaluation of diagnostic tests for infectious salmon anemia. *Dis Aquat Organ*. 2005;63(2–3):119–127.

20. Nérette P, Dohoo I, Hammell L. Estimation of specificity and sensitivity of three diagnostic tests for infectious salmon anaemia virus in the absence of a gold standard. *J Fish Dis*. 2005;28(2):89–99.

21. Nérette P et al. Estimation of the repeatability and reproducibility of three diagnostic tests for infectious salmon anaemia virus. *J Fish Dis*. 2005;28(2):101–110.

22. Abayneh T, Toft N, Mikalsen AB, Brun E, Sandberg M. Evaluation of histopathology, real-time PCR and virus isolation for diagnosis of infectious salmon anaemia in Norwegian salmon using latent class analysis. *J Fish Dis*. 2010;33(6):529–532.

23. Molloy SD, Thomas E, Hoyt K, Bouchard DA. Enhanced detection of infectious salmon anaemia virus using a low-speed centrifugation technique in three fish cell lines. *J Fish Dis*. 2013;36(1):35–44.

24. Gustafson L et al. Estimating diagnostic test accuracy for infectious salmon anaemia virus in Maine, USA. *J Fish Dis*. 2008;31(2):117–125.

25. Nérette P, Stryhn H, Dohoo I, Hammell L. Using pseudo-gold standards and latent-class analysis in combination to evaluate the accuracy of three diagnostic tests. *Prev Vet Med*. 2008;85(3–4):207–225.

26. Starkey WG et al. Detection of infectious salmon anaemia virus by real-time nucleic acid sequence based amplification. *Dis Aquat Organ*. 2006;72(2):107–113.

27. Caraguel C, Stryhn H, Gagné N, Dohoo I, Hammell L. Traditional descriptive analysis and novel visual representation of diagnostic repeatability and reproducibility: Application to an infectious salmon anaemia virus RT-PCR assay. *Prev Vet Med*. 2009;92(1–2):9–19.

28. Caraguel C, Stryhn H, Gagné N, Dohoo I, Hammell L. A modelling approach to predict the variation of repeatability and reproducibility of a RT-PCR assay for infectious salmon anaemia virus across infection prevalences and infection stages. *Prev Vet Med*. 2012;103(1):63–73.

29. Caraguel C, Stryhn H, Gagné N, Dohoo I, Hammell L. Use of a third class in latent class modelling for the diagnostic evaluation of five infectious salmon anaemia virus detection tests. *Prev Vet Med*. 2012;104(1–2):165–173.

30. Snow M, McKay P, Matejusova I. Development of a widely applicable positive control strategy to support detection of infectious salmon anaemia virus (ISAV) using TaqMan real-time PCR. *J Fish Dis*. 2009;32(2):151–156.

31. Carmona M, Sepúlveda D, Cárdenas C, Nilo L, Marshall SH. Denaturing gradient gel electrophoresis (DGGE) as a powerful novel alternative for differentiation of epizootic ISA virus variants. *PLoS One*. 2012;7(5):e37353.

32. Giray C, Opitz HM, MacLean S, Bouchard D. Comparison of lethal versus non-lethal sample sources for the detection of infectious salmon anemia virus (ISAV). *Dis Aquat Organ*. 2005;66(3):181–185.

33. McBeath AJ, Bain N, Snow M. Surveillance for infectious salmon anaemia virus HPR0 in marine Atlantic salmon farms across Scotland. *Dis Aquat Organ*. 2009;87(3):161–169.

34. Workenhe ST, Kibenge MJ, Iwamoto T, Kibenge FS. Absolute quantitation of infectious salmon anaemia virus using different real-time reverse transcription PCR chemistries. *J Virol Methods*. 2008;154(1–2):128–134.

35. Godoy MG et al. TaqMan real-time RT-PCR detection of infectious salmon anaemia virus (ISAV) from formalin-fixed paraffin-embedded Atlantic salmon *Salmo salar* tissues. *Dis Aquat Organ*. 2010;90(1):25–30.

36. Graham DA, Brown A, Savage P, Frost P. Detection of salmon pancreas disease virus in the faeces and mucus of Atlantic salmon, *Salmo salar* L., by real-time RT-PCR and cell culture following experimental challenge. *J Fish Dis*. 2012;35(12):949–951.

37. LeBlanc F, Laflamme M, Gagné N. Genetic markers of the immune response of Atlantic salmon (*Salmo salar*) to infectious salmon anemia virus (ISAV). *Fish Shellfish Immunol*. 2010;29(2):217–232.

38. Gomez-Casado E, Estepa A, Coll JM. A comparative review on European-farmed finfish RNA viruses and their vaccines. *Vaccine*. 2011;29(15):2657–2671.

39. Collet B. Innate immune responses of salmonid fish to viral infections. *Dev Comp Immunol*. 2014;43(2):160–173.

40. Lauscher A et al. Immune responses in Atlantic salmon (*Salmo salar*) following protective vaccination against infectious salmon anemia (ISA) and subsequent ISA virus infection. *Vaccine*. 2011;29(37):6392–6401.

41. Chang CJ, Robertsen C, Sun B, Robertsen B. Protection of Atlantic salmon against virus infection by intramuscular injection of IFNc expression plasmid. *Vaccine*. 2014;32(36):4695–4702.

42. Munir K, Kibenge FS. Detection of infectious salmon anaemia virus by real-time RT-PCR. *J Virol Methods*. 2004;117(1):37–47.

43. McBeath A, Bain N, Fourrier M, Collet B, Snow M. A strand specific real-time RT-PCR method for the targeted detection of the three species (vRNA, cRNA and mRNA) of infectious salmon anaemia virus (ISAV) replicative RNA. *J Virol Methods*. 2013;187(1):65–71.

44. Snow M et al. Development, application and validation of a TaqMan real-time RT-PCR assay for the detection of infectious salmon anaemia virus (ISAV) in Atlantic salmon (*Salmo salar*). *Dev Biol (Basel)*. 2006;126:133–145; discussion 325–326.

45. Crane M, Hyatt A. Viruses of fish: An overview of significant pathogens. *Viruses*. 2011;3(11):2025–2046.

46. Lyngstad TM et al. Low virulent infectious salmon anaemia virus (ISAV-HPR0) is prevalent and geographically structured in Norwegian salmon farming. *Dis Aquat Organ*. 2012;101(3):197–206.

46 Swine Influenza A Virus

Jianqiang Zhang, Phillip Gauger, and Karen Harmon

CONTENTS

46.1 Introduction 399
    46.1.1 Classification, Morphology, and Genome Organization 399
    46.1.2 Epidemiology and Ecology... 400
        46.1.2.1 North American Swine IAVs.. 400
        46.1.2.2 European Swine IAVs.. 400
        46.1.2.3 Asian Swine IAVs 401
        46.1.2.4 Swine IAVs in South America and Africa... 401
        46.1.2.5 Role of Swine in IAV Ecology... 401
    46.1.3 Clinical Features and Pathogenesis 401
    46.1.4 Diagnosis 402
        46.1.4.1 Conventional Techniques... 402
        46.1.4.2 Molecular Techniques.. 402
46.2 Methods 403
    46.2.1 Sample Collection and Preparation 403
    46.2.2 Detection Procedures.. 403
        46.2.2.1 rRT-PCR for Swine IAV Detection.. 403
        46.2.2.2 rRT-PCR for Swine IAV Subtyping... 404
46.3 Conclusion and Future Perspectives 404
References.. 405

46.1 INTRODUCTION

Influenza A virus (IAV) belongs to the family *Orthomyxoviridae*, and the genome is composed of eight single-stranded, negative-sense RNA segments. IAV can infect a wide range of host species such as birds (wild and domestic), bats, and mammals (including humans, pigs, horses, dogs, and cats) [1]. The epidemiology and ecology of IAVs are complicated due to the multihost and segmented genome. In swine, IAVs are one of the major pathogens that cause acute respiratory disease and result in substantial economic burden to swine producers. In addition to swine IAVs, human and avian IAVs can also infect swine. Swine have been thought to serve as an intermediate "mixing vessel" for the generation of diversified and novel influenza viruses [2]. There have been numerous reports of zoonotic human infections with swine IAVs and reverse zoonotic swine infections with human IAVs [3–7]. Therefore, swine hosts and swine IAVs are important components of the "one health" concept. Detection and characterization of swine IAVs are not only necessary for surveillance, control, and prevention of swine influenza but also critical for public health purposes. In this chapter, swine IAV classification, genome organization, epidemiology, clinical features and pathogenesis, and diagnosis are described.

46.1.1 CLASSIFICATION, MORPHOLOGY, AND GENOME ORGANIZATION

Influenza viruses belong to the family *Orthomyxoviridae* that includes six genera: *Influenzavirus A*, *Influenzavirus B*, *Influenzavirus C*, *Thogotovirus*, *Isavirus*, and *Quaranfilvirus* [8]. Influenza viruses are classified by type, and IAVs are further classified into subtypes. IAVs are named according to the following convention [8]: genus (type), species of origin (omitted if human), location of isolation, isolate number, year of isolation, and the subtype based on hemagglutinin (HA or H) and neuraminidase (NA or N), for example, A/swine/Iowa/A01477800/2014(H1N1). Currently, 18 HA subtypes (H1–H18) and 11 NA subtypes (N1–N11) have been identified for IAVs including two recently described subtypes of bat IAVs (H17N10 and H18N11) [9,10].

The genome of IAV possesses eight segments of negative-sense, single-stranded RNA: polymerase basic 2 (PB2), polymerase basic 1 (PB1), polymerase acid (PA), hemagglutinin (HA), nucleoprotein (NP), neuraminidase (NA), matrix (M), and nonstructural (NS) segments. Each of these segments encode one protein except the PB1, M, and NS segments where the PB1 segment encodes PB1 protein, PB1-F2 protein, and PB1-N40 protein; the M segment encodes M1 protein and M2 protein; the NS segment encodes NS1 protein and nuclear export protein (NEP)/NS2 protein [8]. IAVs are enveloped viruses. The envelope is a lipid membrane derived from the host cell and harbors the HA, NA, and M2 proteins with abundances in the decreased order of HA, NA, and M2. In the virion, HA protein is present as a trimer, whereas NA and M2 proteins are present as tetramers [8]. The characteristic morphology of IAV particles is their distinctive spikelike projections on the virion surface formed by the HA and NA proteins [11]. The M1 protein lies beneath the lipid envelope of the virion, and M1 is the most abundant structural protein in the virion. The NEP/NS2 protein is also enclosed inside the virion. The core of the virus particle is the ribonucleoprotein complex that consists of eight RNA segments coated with multiple copies of the NP and bound by the polymerase proteins PB2, PB1, and PA [8]. The IAV virion particles are roughly spherical but could be pleomorphic and have a diameter of approximately 80–120 nm [11].

There are two main mechanisms for molecular evolution of IAVs: antigenic drift (natural mutation process due to the lack of viral RNA polymerase proofreading activity) and antigenic shift (reassortment process by which two or more different influenza viruses infect one cell or one host and exchange their genomic segments). Continuous antigenic drift and shift have resulted in growing diversity of IAVs. Understanding IAV epidemiology and evolution requires not only the type and subtype information but also genotype information. Genotyping involves determining the genetic sequences of each RNA segment of IAVs and performing phylogenetic analyses to define the evolutionary lineages. In recent years, genotyping has been increasingly used to understand the origins and evolutionary relationships of IAVs worldwide. Therefore, influenza viruses probably should be classified by type, subtype, and genotype (phylogenetic cluster).

46.1.2 Epidemiology and Ecology

IAVs are present throughout much of the world particularly where large numbers of pigs are raised. However, the viruses circulating in different geographic regions vary in their subtypes and genotypes. Here, a brief summary is provided with an emphasis on North American swine IAVs. Please refer to several published reviews for details [11–17].

46.1.2.1 North American Swine IAVs

In North America, swine influenza was first documented in pigs coincident with the 1918 human Spanish flu pandemic. The virus was isolated in 1930 and was referred to as classical swine H1N1 (cH1N1) [18]. The cH1N1 IAVs remained genetically and antigenically conserved in U.S. swine population until 1998 when a novel triple-reassortant H3N2 virus (HA, NA, and PB1 segments from human seasonal H3N2; PB2 and PA segments from North American avian IAV; NP, M, and NS segments from swine cH1N1) emerged and established itself in U.S. swine [19,20]. The unique composition of the internal genes (human PB1, avian PB2 and PA, and swine NP, M, and NS) of the triple-reassortant H3N2 virus was referred to as the triple-reassortant internal gene (TRIG) cassette [21].

Subsequently, the triple-reassortant H3N2 viruses reassorted with cH1N1, resulting in the emergence and spread of reassortant H1N2 and H1N1 IAVs that contained the TRIG cassette with the HA gene from cH1N1 [22,23]. From 2003 to 2005, two new swine H1N1 and H1N2 viruses emerged and spread in U.S. swine. These viruses contained either the HA or both HA and NA genes derived from human seasonal IAVs, but their internal gene constellation was similar to the TRIG genes of contemporary triple-reassortant viruses [24]. Further studies indicated that these swine IAVs emerging in 2003–2005 were likely from at least two separate introductions of human seasonal viruses [25] and the cross-species transmission from human to swine was estimated to occur during August 2002–March 2003 [26]. In 2009, pandemic H1N1 IAV (H1N1pdm09) emerged in humans and subsequently spread to swine. The H1N1pdm09 virus possesses PB2 and PA genes of North American avian virus origin, PB1 gene of human H3N2 virus origin, HA (H1), NP and NS genes of classical swine cH1N1 origin, NA (N1), and M genes of Eurasian avian virus origin [27]. This particular constellation of segments had not been previously detected in swine or other species prior to its emergence in humans in 2009 [27]. The H1N1pdm09 virus has further reassorted with contemporary North American swine IAVs and brought additional diversity to the genetic complexity of IAV [28–31].

Due to increasing diversity of swine IAVs, a phylogenetic cluster classification system was established to help characterize the viruses [24]. Over the years, the HA gene of the classical swine H1N1 viruses have evolved into H1α, H1β, and H1γ clusters; the HA genes of two swine H1 viruses introduced in 2002–2003 from the human seasonal IAVs formed H1δ1 and H1δ2 clusters; the HA gene of the H1N1pdm09 formed the H1pdm09 cluster [13]. The HA genes of swine H3 viruses have evolved into H3I, II, III, and IV clusters with H3IV cluster being the most common [32]. The H3IV cluster viruses have further diversified into subclusters A–F [33]. Therefore, the IAVs currently circulating in the North American swine populations include at least 10 phylogenetically distinct HA clusters: H1α, H1β, H1γ, H1δ1, H1δ2, H1pdm09, H3I, H3II, H3III, and H3IV (H3IV A–F) [32,34]. The NA genes of swine IAVs in North America include classical swine lineage N1, human origin N2 lineage introduced in 1998, human origin N2 lineage introduced in 2002–2003, human origin N1 lineage introduced in 2002–2003, and N1pdm09 lineage although the prevalence of these NA lineages are variable [26,32].

IAVs with other subtypes (e.g., H2N3, H4N6, and H3N1) were also detected in the North American swine populations [35–38], but none of these viruses have become established in swine populations. Contemporary IAVs currently circulating in North American swine populations are still mainly H1N1, H1N2, and H3N2 subtypes in spite of the increasing diversity of these viruses.

46.1.2.2 European Swine IAVs

IAVs circulating in the European swine populations are predominantly H1N1, H1N2, and H3N2 subtypes. However, the genetic lineages of these swine IAVs are distinct from the

North American swine IAVs [11]. The classical swine H1N1 virus that previously circulated in European swine populations was replaced with an antigenically and genetically distinct European avian-like H1N1 virus around 1979 [14]. Subsequently, humanlike H3N2 and humanlike H1N2 viruses were introduced into and established infections in European swine [11–14]. Emergence of H1N1pdm09 and reassortment with endemic swine IAVs in Europe further diversified European swine IAV ecology [39–41]. Some unusual subtypes such as H3N1 and H1N7 viruses were detected in European swine but apparently failed to persist and spread in pigs [42,43].

46.1.2.3 Asian Swine IAVs

The epidemiology of swine IAVs in Asia is more complex than that in North America and Europe. Some swine IAVs from North America (e.g., classical swine H1N1 and triple-reassortant H3N2 and H1N2 viruses) and Europe (e.g., European avian-like H1N1 and European H3N2 reassortant viruses) and their reassortant descendants had been detected in Asia [15,16]. Some humanlike H3N2 viruses local to Asian countries were detected in Asian swine as well [15,16]. The H1N1pdm09 and its reassortants were also detected in Asia [44–47]. In addition, some avian-like IAVs such as H9N2, H5N1, H5N2, H4N8, and H6N6 and equine H3N8 IAV have been sporadically detected in pigs from some Asian countries [15,16,48–50]. Concurrent circulation of so many genetically diverse IAVs in swine has resulted in very complex swine IAV ecology in Asia.

46.1.2.4 Swine IAVs in South America and Africa

IAVs have also been detected in South American swine populations and African swine populations, as reviewed by Vincent et al. [12].

46.1.2.5 Role of Swine in IAV Ecology

Wild aquatic birds are the natural reservoir for IAV H1-H16 and N1-N9 subtypes [51]. H17-H18 and N10-N11 subtypes of IAV have recently been detected in bats [9,10]. However, it appears that only certain subtypes of IAV readily established their infections in mammalian species. For example, the IAV subtypes in global swine populations are predominantly H1N1, H1N2, and H3N2 although other subtypes can be sporadically detected. Swine tissues are found to express both α-2,6-linked sialic acid (receptor preferred by human/swine IAVs) and α-2,3-linked sialic acid (receptor preferred by avian IAVs), and swine were thought as a "mixing vessel" for genetic reassortment between avian and mammalian IAVs [2]. However, it has been argued whether swine are necessary intermediate hosts for avian-to-human infections because human and some land-based poultry were found to also express both types of sialic acids [11]. Nevertheless, swine play an important role in the ecology of IAV, and this is especially true at the human–swine interface. Zoonotic transmission of swine-adapted IAVs to humans has been documented. Searching PubMed database in April 2006 identified 50 cases of zoonotic swine IAV infections though the actual number could be higher [52]. From 2005 to 2014, there were a total of 372 cases of human infections with swine origin variant influenza viruses in the United States, including 351 H3N2v, 16 H1N1v, and 5 H1N2v [53]. Among which, an H3N2 variant (H3N2v) containing seven gene segments from the triple-reassortant swine H3N2 and one gene segment (M gene) from the H1N1pdm09 has caused 343 human infections between 2011 and 2014 in the United States [54]. Interestingly, reverse zoonotic transmission of IAVs from human to swine have occurred more frequently than the opposite [7]. Globally, there were at least 49 human-to-swine transmissions of the H1N1pdm09 during 2009–2011 and 23 separate introductions of human seasonal IAVs into swine during 1990–2011 [4].

46.1.3 Clinical Features and Pathogenesis

The transmission routes of IAV between pigs are thought to be through direct contact via nasal or oral secretions and through inhalation of aerosolized virus in droplets [11]. The clinical outcomes of swine IAV infection could be dependent on multiple factors including virus load, age of pigs, immune status, housing conditions, and concurrent infections. Typical clinical influenza in pigs is characterized as an acute, high-morbidity, low-mortality respiratory disease in uncomplicated infections [11]. At the first 1–2 days postinfection, pigs develop high fever (>40.5°C, 105°F) followed by lethargy, anorexia, and nasal discharge. Conjunctivitis, tachypnea, and dyspnea could occur. By days 3 and 4, pigs may develop a harsh deep barking cough considered as the hallmark clinical sign of swine influenza [55]. Pigs of all ages are susceptible to IAV infection, but clinical disease is often more mild in nursing pigs. In the absence of concurrent infections with other pathogens, pigs usually recover in 7–10 days but not all pigs in a population will get infected and recover simultaneously [55].

IAV infects the epithelial cells lining the surface of the respiratory tract of pigs (nasal mucosa, trachea, bronchi, bronchioles, and lung) with a preference for the lower instead of the upper respiratory tract [11,55]. Virus can be detected from the nasal or oropharyngeal swabs, bronchoalveolar lavage fluid, and respiratory tract tissues including lungs [11]. Virus shedding in nasal secretions begins by 1–3 days postinfection, and the duration of the shedding lasts approximately 4–7 days [55]. Swine IAV infection is generally limited to the respiratory tract and is occasionally detected from extra-respiratory sites. Low level and transient influenza viremia was only observed in one study [56]. In another study, virus was detected in intestines and spleen by reverse transcription polymerase chain reaction (RT-PCR) but not by virus isolation and in brainstem by both RT-PCR and virus isolation [57].

The most common gross lesions found in swine IAV-infected pigs are bronchopneumonia predominantly in the cranioventral lung lobes, but the consolidation level varies [11,55]. In cases involving concurrent bacterial bronchopneumonia, lesions could be more extensive [55].

The hallmark microscopic lesion of swine influenza is necrotizing bronchitis and bronchiolitis with variable severity of interstitial pneumonia [55,58]. Within the first 24 h postinfection (hpi), neutrophil infiltration and accumulation start to occur in the lumens of alveolar capillaries. Extensive necrosis and sloughing of epithelial cells into airway lumens accompanied by more evident neutrophil accumulation occur at 24–48 hpi, causing obstruction of the airways. From 48 to 72 hpi, some airways are in active necrosis accompanied by peribronchiolar and perivascular infiltration of lymphocytes. By 72–96 hpi and beyond, airways are in repair. During the recovery phase, the epithelial cells lining the airways proliferate and result in an irregular hyperplastic appearance; lymphocytic cuffs also become more prominent. Over the following days, the epithelial hyperplasia resolves and peribronchiolar lymphocytic cuffing may remain.

46.1.4 Diagnosis

Clinical signs of swine IAV infection are not pathognomonic and could be similar to symptoms caused by other respiratory pathogens. A definitive diagnosis needs swine IAV-specific laboratory tests. Numerous virological and serological assays are available for swine IAV detection and characterization, but in this chapter, only virological techniques are discussed. Here, we refer to the RT-PCR and sequencing as "molecular techniques" and all other virological methods as "conventional techniques." But it is noteworthy that some methods categorized under "conventional techniques" are still commonly used for swine IAV detection and diagnosis.

46.1.4.1 Conventional Techniques

Electron microscopy (EM) can directly visualize virus particles in examined specimens if sufficient amounts of virus are present. However, EM is not suitable for routine use due to the relatively lower analytical sensitivity and time-consuming procedures. The direct or indirect immunofluorescence assay (IFA) test using different swine IAV antibodies can detect viral antigen in fresh or frozen tissue sections. Immunohistochemistry (IHC) to detect viral antigen on formalin-fixed tissue sections is still frequently performed to confirm the histopathological observations. IHC can also facilitate comparison of virus distribution and cellular localization to determine whether viral antigen distribution coincides with lesions in the tissues. The NP is conserved among IAVs, and anti-NP antibodies can be used in IFA and IHC assays to detect multiple subtypes of IAVs. In contrast, the HA protein is subtype specific, and HA-specific antibodies can be used to detect specific subtypes of influenza virus. A number of commercial antigen capture enzyme-linked immunosorbent assay (ELISA) kits developed for detection of human and avian IAVs can be used to detect swine influenza virus antigen from clinical samples. These kits detect HA, NA, or NP of IAV [59]. The antigen ELISAs are a rapid assay and can be conducted in the field, but the sensitivity may not be satisfactory [59,60]. Swine influenza virus isolation can be conducted in embryonated chicken eggs and/or in a number of primary

cells or continuous cell lines (the most commonly used cell line is Madin–Darby canine kidney) [61]. Virus isolation is an important technique to obtain a live and infectious virus for antigenic characterization, pathogenesis investigation, and vaccine production. Proper sample storage to maintain viable virus is critical for virus isolation success. IAV hemagglutinates chicken or turkey erythrocytes. Hemagglutination (HA) test is not often directly used on clinical samples but frequently used to test egg fluids or cell culture fluids for confirming virus isolation results.

46.1.4.2 Molecular Techniques

Since PCR technique became available, conventional gel-based PCR and RT-PCR have been widely used as a tool for diagnosis and research. Subsequent innovation and introduction of real-time PCR (rPCR) and real-time RT-PCR (rRT-PCR) provided a more powerful tool for diagnosis. Currently, rRT-PCR assays are the method of choice in many laboratories for the detection and subtyping of IAV in swine. Compared to other methods such as IFA, IHC, and virus isolation for swine IAV detection, rRT-PCR has a number of advantages: rapid turnaround time between sample receipt and obtaining results, high sensitivity, high specificity, high throughput, and capability for genomic quantitation. In addition, rRT-PCR can be used to test various specimen types, which can be problematic for other detection methods. Following detection and subtyping of IAV using rRT-PCR, sequencing technique can be used to further genetically characterize the virus.

46.1.4.2.1 rRT-PCR for Swine IAV Detection

rRT-PCR has been widely employed as a diagnostic and/or surveillance test for the detection of swine IAV. To detect a broad range of IAVs, swine IAV-screening rRT-PCR assays with primers and probes targeting the conserved M and/or NP gene are generally used. An avian influenza rRT-PCR targeting the M gene [62] was adapted for the detection of IAV in swine samples and has been validated by the USDA and is routinely used for swine IAV surveillance in the United States. Commercial swine influenza virus test kits are also available, some of which contain an internal control that provides an additional quality assurance measure and helps troubleshoot problems encountered during the rRT-PCR testing [63].

46.1.4.2.2 rRT-PCR for Swine IAV Subtype Identification

Traditionally, determination of the IAV HA and NA subtypes was performed on the cultured virus using hemagglutination inhibition test and neuraminidase inhibition test. However, both assays are time consuming and require standardized reference antigens and antisera for HA and NA, which are often difficult to obtain. Currently, RT-PCR is routinely used for IAV subtyping; in addition, subtyping RT-PCR can be conducted directly on clinical samples and not necessarily on the cultured viruses. A number of RT-PCR assays have been described [64–68]. In the United States, a multiplex swine IAV HA (H1 and H3) and NA (N1 and N2) subtyping PCR assay commercially available from Life Technologies™ has been used by the USDA National Animal Health

Laboratory Network laboratories for surveillance testing [63]. Contemporary IAVs circulating in the swine populations are predominantly H1N1, H1N2, and H3N2 subtypes although other subtypes can be sporadically detected. If any samples are strongly positive by the swine IAV screening PCR but negative for H1, H3, N1, and N2, the HA and NA genes from those samples can be sequenced to determine if any uncommon subtypes are present.

46.1.4.2.3 Determination of Swine IAV Genome Sequences

Detection of swine IAV by screening PCR followed by subtyping PCR is commonly performed for swine IAV diagnosis and surveillance. However, as mentioned in Section 46.1.3, continuous antigenic drift and shift have resulted in growing diversity of IAVs, and the type and subtype information is not sufficient to characterize the detected IAVs. Sequencing followed by comparative sequence analysis and phylogenetic analysis has been frequently conducted in recent years to further characterize IAVs and to enhance our understanding of the epidemiology and evolutionary relationships of the viruses. Many laboratories are still using the traditional Sanger methods to determine genetic sequences. However, more and more laboratories have started to employ next-generation sequencing technologies to obtain more extensive sequence data.

46.2 METHODS

46.2.1 SAMPLE COLLECTION AND PREPARATION

Antemortem specimens such as nasal swabs and oral fluids and postmortem specimens such as nasal turbinates, tonsil, trachea, lung, and bronchoalveolar lavage can be used for swine IAV detection by RT-PCR. The most common specimens submitted to diagnostic laboratory for swine IAV testing are nasal swabs, oral fluids, and lung tissues. It is noteworthy that oral fluids are becoming more common for IAV prognostic profiling [60,69–71]. Oral fluids are collected by hanging absorbent cotton ropes above or on the sides of pens. Pigs interact with and chew on the rope. Usually within 30 min, the rope will be soaked with saliva and other oral secretions. Normally, oral fluids can be successfully collected from pigs older than 3 weeks of age but may be difficult to collect from suckling pigs [72]. Due to short periods of virus shedding, nasal swabs and oral fluids should be obtained from acutely ill animals.

During collection, samples should be kept on ice if at all possible. Following collection, samples should be transported to the laboratory as soon as possible. Samples can generally be stored at 4°C for up to 48 h. However, if testing cannot be initiated within 48 h, samples should be frozen either at −20°C (preferably in a non-frost-free freezer) for a month or less or at −70°C or below for long-term storage.

Numerous commercial kits for viral RNA extraction are available. Although manual RNA extraction is still being used in some laboratories, automated extraction of nucleic acid from specimens using 24-well, 96-well, or 384-well format is used in many laboratories for high-throughput extraction. The nucleic acid extraction procedures vary by specimen type. For example, in our laboratory, we use one procedure to extract viral RNA from nasal swabs, lung tissue homogenates, tracheal tissue homogenates, bronchoalveolar lavage, cell culture fluids, or embryonated chicken egg fluids and another procedure to extract viral RNA from oral fluids. Both extraction procedures are magnetic bead-based methods using the commercial 5× MagMAX™ Pathogen RNA/DNA Kit (4462359, Life Technologies). For nucleic acid extraction from nasal swab, lung, and other samples, the standard protocol from the manufacturer is followed. Briefly, 50 μL samples are mixed with 130 μL Lysis/Binding Solution and 20 μL Bead Mix for 5 min. RNA Binding Beads are captured and supernatants discarded. Two washes each with 150 μL Wash Solution 1 and 150 μL Wash Solution 2 follow. Beads are dried by shaking for 2 min. Viral RNA is eluted into 90 μL elution buffer. Compared to the nucleic acid extraction protocol from nasal swab and lung samples, the extraction protocol from oral fluids mainly differs in two aspects: (1) Sample preparation before extraction: Oral fluids are first vortexed, and 300 μL vortexed oral fluids are added to 450 μL Lysis/Binding Solution and shaken for 5 min. Lysate is clarified by centrifugation at $2500 \times g$ for 5 min. Clarified lysate (600 μL) is used for nucleic acid extraction. (2) A 300 μL of Wash Solution 1 and 450 μL Wash Solution 2 are used for each respective wash step. Elution volume remains at 90 μL. The RNA sample should be tested by rRT-PCR as soon as possible after extraction. RNA samples can be stored at −70°C or colder for longer term storage.

46.2.2 DETECTION PROCEDURES

In this section, the USDA-validated IAV M gene–based rRT-PCR is described for swine IAV detection, and a multiplex swine IAV HA (H1 and H3) and NA (N1 and N2) subtyping PCR assay from Life Technologies is described for swine IAV subtyping. Previously described sequencing approaches using the Sanger method [73] can be followed for determining the whole genome sequences of eight segments of IAV. In recent years, next-generation sequencing technology using different platforms such as Illumina® MiSeq® and Life Technologies Ion Torrent™ has been employed to determine the whole genome sequences of IAVs. It is beyond the scope of this chapter to describe the detailed sequencing procedures and sequence analysis methods. Thus, swine IAV whole genome sequencing and analysis procedures are not provided here.

46.2.2.1 rRT-PCR for Swine IAV Detection

The primers and probe of the USDA-validated swine IAV M gene–based rRT-PCR assay are M+25 forward primer (M+25F) 5′-AGATGAGTCTTCTAACCGAGGTCG-3′, M-124 reverse primer (M-124R) 5′-TGCAAAAACATCTTCAAGTCTCTG-3′, M-124 modified reverse primer (M-124R Mod) 5′-TGCAAA GACACTTTCCAGTCTCTG-3′, and M+64 probe (M+64P) 5′-FAM-TCAGGCCCCCTCAAAGCCGA-BHQ1-3′.

TABLE 46.1

USDA-Validated Influenza A Virus Real-Time RT-PCR (M Gene) Reaction Mix Volumes Using the AgPath-ID One-Step RT-PCR Kit (Life Technologies)

Component	Volume per Reaction (µL)	Final Concentration
AgPath-ID 2× master mix	12.5	1×
AgPath-ID 25× enzyme mix	1	
AgPath-ID detection enhancer	1.5	
M25F (20 µM)	0.5	0.4 µM
M124R (20 µM)	0.5	0.4 µM
M124R mod (20 µM)	0.5	0.4 µM
M64P (6 µM)	0.5	0.12 µM
RNA template	8	
Total volume	25	

The AgPath-ID™ One-step RT-PCR reagents from Life Technologies are used.

1. A 25 µL RT-PCR reaction is prepared as shown in Table 46.1 and is added to a 96-well reaction plate for the AB 7500 Fast. It is important to include a positive amplification control, a positive RNA extraction control, a negative RNA extraction control, and a negative amplification control (e.g., nuclease-free water).
2. Seal the plate tightly and centrifuge at $2000 \times g$ for approximately 15 s.
3. Place the sealed reaction plate into the AB 7500 Fast instrument (Life Technologies) and run the rRT-PCR with the following conditions: 1 cycle of 45°C for 10 min, 1 cycle of 95°C for 10 min, and 45 cycles of 95°C for 15 s and 60°C for 45 s.
4. Analysis of results. Set the threshold to 0.2 and use the "autobaseline setting." Examine the positive and negative control to ensure that the reaction results are correct. The positive amplification control is expected to have a Ct value of 25, and the Ct value should be between 21 and 29 for the run to be valid. The positive extraction control should be positive. The negative extraction control and negative amplification control should each be clearly negative. Evaluate the raw data from each sample individually using the instrument analysis software. This swine IAV matrix gene rRT-PCR assay is run 45 cycles, and any Ct value less than 45 is reported.

46.2.2.2 rRT-PCR for Swine IAV Subtyping

The Swine Influenza Virus Subtyping RNA Reagents (primers and probes) (Life Technologies) and Path-ID™ Multiplex One-Step RT-PCR Kit (Life Technologies) include 2× Multiplex RT-PCR Buffer, 10× Multiplex RT-PCR Enzyme Mix, 25× H1H3 Primer Probe Mix, and 25× N1N2 Primer Probe Mix.

TABLE 46.2

Swine Influenza Virus Subtyping One-Step RT-PCR (Life Technologies) Reaction Mix Volumes

Component	Component	Volume per Reaction (µL)
H1H3 subtyping	2× multiplex RT-PCR buffer	12.5
	10× multiplex RT-PCR enzyme mix	2.5
	25× H1H3 primer probe mix	1
	Nuclease-free water	1
	RNA template	8
	Total volume	25
N1N2 subtyping	2× multiplex RT-PCR buffer	12.5
	10× multiplex RT-PCR enzyme mix	2.5
	25× N1N2 primer probe mix	1
	Nuclease-free water	1
	RNA template	8
	Total volume	25

1. RT-PCR reaction preparations are shown in Table 46.2. Set up two reactions for each sample in 96-well reaction plate for AB 7500: one reaction is for HA (H1 and H3) subtyping, and the other is for NA (N1 and N2) subtyping. It is important to include a positive amplification control (H1, H3, or N1, N2) and a negative amplification control (e.g., nuclease-free water).
2. Seal the plate tightly and centrifuge at $2000 \times g$ for approximately 15 s.
3. Place the sealed reaction plate into the AB 7500 Fast instrument (Life Technologies) and run the rRT-PCR with the following conditions: 1 cycle of 48°C for 10 min, 1 cycle of 95°C for 10 min, and 40 cycles of 95°C for 15 s and 60°C for 45 s.
4. Analysis of results. Set the threshold to 0.2 and use the "autobaseline setting." Examine the positive and negative controls to ensure that the reaction results are correct. The positive H1 control should have a Ct 25–29 in the VIC channel (H1 probe), and the positive H3 control should have a Ct 25–29 in the FAM channel (H3 probe). Similarly, the N1 and N2 controls should have a Ct 25–29 in the VIC and FAM channels (N1 and N2 probes), respectively. The negative amplification control should be negative for H1, H3, N1, and N2. Evaluate the raw data from each sample individually using the instrument analysis software. If the sample has a Ct < 40 in a specific assay, the sample is positive for that subtype.

46.3 CONCLUSION AND FUTURE PERSPECTIVES

Swine influenza is a highly contagious, acute respiratory disease in pigs and can cause significant economic losses for pork producers. However, the multihost nature and interspecies transmission of IAV ensure that IAV in swine is not only a viral pathogen for pigs but also has its unique niche in view of the whole IAV ecology. It was formerly believed that swine are intermediate hosts for avian IAV transmission to humans. However, recent advancement in IAV receptor studies suggests that swine are not more suitable hosts than humans for stable transmission of avian IAVs. Nonetheless, there is no doubt swine can receive and harbor a large number of IAVs with different lineages and then via reassortment generate novel viruses that can further transmit to other species including humans. In fact, numerous human-to-swine and swine-to-human transmissions of IAV have demonstrated the importance of the human–swine interface in influenza virus ecology.

Currently, there are no interventions that can effectively prevent influenza viruses from evolving through antigenic shift and antigenic drift. Efforts to better monitor IAV evolution through enhanced surveillance in both swine and human populations are warranted. Molecular techniques provide efficient tools to rapidly detect and characterize IAVs. Screening RT-PCR assays that can identify almost all IAVs are used for detection of IAV, and then subtyping RT-PCR and genetic sequencing can further characterize the detected viruses. In addition, antigenic evaluation of viruses should be performed in conjunction with genetic analysis. Together, these approaches can assist in understanding the epidemiology and evolutionary relationships of IAVs worldwide. The acquired information can be further used to update vaccine formulations to include contemporary circulating virus strains or to develop more effective vaccines for prevention and control of swine influenza.

REFERENCES

1. Yoon, S.W., R.J. Webby, and R.G. Webster, Evolution and ecology of influenza A viruses. *Curr Top Microbiol Immunol*, 2014. 385: 359–375.
2. Ma, W., R.E. Kahn, and J.A. Richt, The pig as a mixing vessel for influenza viruses: Human and veterinary implications. *J Mol Genet Med*, 2008. 3(1): 158–166.
3. Lindstrom, S. et al., Human infections with novel reassortant influenza A(H3N2)v viruses, United States, 2011. *Emerg Infect Dis*, 2012. 18(5): 834–837.
4. Nelson, M.I. et al., Global transmission of influenza viruses from humans to swine. *J Gen Virol*, 2012. 93(10): 2195–2203.
5. Nelson, M.I. et al., Evolution of novel reassortant A/H3N2 influenza viruses in North American swine and humans, 2009–2011. *J Virol*, 2012. 86(16): 8872–8878.
6. Nelson, M.I. et al., Introductions and evolution of human-origin seasonal influenza a viruses in multinational swine populations. *J Virol*, 2014. 88(17): 10110–10119.
7. Nelson, M.I. and A.L. Vincent, Reverse zoonosis of influenza to swine: New perspectives on the human-animal interface. *Trends Microbiol*, 2015. 23(3): 142–153.
8. Shaw, M.L. and P. Palese, Orthomyxoviridae, in *Fields Virology*, D.M. Knipe and Howley P.M. (eds.), 2013, Wolters Kluwer/Lippincott Williams & Wilkins, Philadelphia, PA, pp. 1151–1185.
9. Wu, Y. et al., Bat-derived influenza-like viruses H17N10 and H18N11. *Trends Microbiol*, 2014. 22(4): 183–191.
10. Tong, S. et al., New world bats harbor diverse influenza A viruses. *PLoS Pathog*, 2013. 9(10): e1003657.
11. Van Reeth, K., I.H. Brown, and C.W. Olsen, Influenza virus, in *Diseases of Swine*, J. Zimmerman (ed.) 2012, Wiley-Blackwell, Chichester, U.K., pp. 557–571.
12. Vincent, A. et al., Review of influenza A virus in swine worldwide: A call for increased surveillance and research. *Zoonoses Public Health*, 2014. 61(1): 4–17.
13. Vincent, A.L., K.M. Lager, and T.K. Anderson, A brief introduction to influenza A virus in swine. *Methods Mol Biol*, 2014. 1161: 243–258.
14. Brown, I.H., History and epidemiology of Swine influenza in Europe. *Curr Top Microbiol Immunol*, 2013. 370: 133–146.
15. Choi, Y.K., P.N. Pascua, and M.S. Song, Swine influenza viruses: An Asian perspective. *Curr Top Microbiol Immunol*, 2013. 370: 147–172.
16. Zhu, H. et al., History of Swine influenza viruses in Asia. *Curr Top Microbiol Immunol*, 2013. 370: 57–68.
17. Lorusso, A. et al., Contemporary epidemiology of North American lineage triple reassortant influenza A viruses in pigs. *Curr Top Microbiol Immunol*, 2013. 370: 113–132.
18. Shope, R.E., Swine influenza: III. Filtration experiments and etiology. *J Exp Med*, 1931. 54(3): 373–385.
19. Zhou, N.N. et al., Genetic reassortment of avian, swine, and human influenza A viruses in American pigs. *J Virol*, 1999. 73(10): 8851–8856.
20. Webby, R.J. et al., Evolution of swine H3N2 influenza viruses in the United States. *J Virol*, 2000. 74(18): 8243–8251.
21. Vincent, A.L. et al., Swine influenza viruses a North American perspective. *Adv Virus Res*, 2008. 72: 127–154.
22. Karasin, A.I. et al., Genetic characterization of H1N2 influenza A viruses isolated from pigs throughout the United States. *J Clin Microbiol*, 2002. 40(3): 1073–1079.
23. Webby, R.J. et al., Multiple lineages of antigenically and genetically diverse influenza A virus co-circulate in the United States swine population. *Virus Res*, 2004. 103(1–2): 67–73.
24. Vincent, A.L. et al., Characterization of a newly emerged genetic cluster of H1N1 and H1N2 swine influenza virus in the United States. *Virus Genes*, 2009. 39(2): 176–185.
25. Lorusso, A. et al., Genetic and antigenic characterization of H1 influenza viruses from United States swine from 2008. *J Gen Virol*, 2011. 92(4): 919–930.
26. Nelson, M.I. et al., Spatial dynamics of human-origin H1 influenza A virus in North American swine. *PLoS Pathog*, 2011. 7(6): e1002077.
27. Smith, G.J. et al., Origins and evolutionary genomics of the 2009 swine-origin H1N1 influenza A epidemic. *Nature*, 2009. 459(7250): 1122–1125.
28. Ducatez, M.F. et al., Multiple reassortment between pandemic (H1N1) 2009 and endemic influenza viruses in pigs, United States. *Emerg Infect Dis*, 2011. 17(9): 1624–1629.
29. Kitikoon, P. et al., Pathogenicity and transmission in pigs of the novel A(H3N2)v influenza virus isolated from humans and characterization of swine H3N2 viruses isolated in 2010–2011. *J Virol*, 2012. 86(12): 6804–6814.
30. Liu, Q. et al., Emergence of novel reassortant H3N2 swine influenza viruses with the 2009 pandemic H1N1 genes in the United States. *Arch Virol*, 2012. 157(3): 555–562.
31. Tremblay, D. et al., Emergence of a new swine H3N2 and pandemic (H1N1) 2009 influenza A virus reassortant in two Canadian animal populations, mink and swine. *J Clin Microbiol*, 2011. 49(12): 4386–4390.

32. Anderson, T.K. et al., Population dynamics of cocirculating swine influenza A viruses in the United States from 2009 to 2012. *Influenza Other Respir Viruses*, 2013. 7(Suppl 4): 42–51.

33. Kitikoon, P. et al., Genotype patterns of contemporary reassorted H3N2 virus in US swine. *J Gen Virol*, 2013. 94(6): 1236–1241.

34. Rajao, D.S. et al., Pathogenesis and vaccination of influenza a virus in swine. *Curr Top Microbiol Immunol*, 2014. 385: 307–326.

35. Ma, W. et al., Identification of H2N3 influenza A viruses from swine in the United States. *Proc Natl Acad Sci USA*, 2007. 104(52): 20949–20954.

36. Karasin, A.I. et al., Isolation and characterization of H4N6 avian influenza viruses from pigs with pneumonia in Canada. *J Virol*, 2000. 74(19): 9322–9327.

37. Karasin, A.I. et al., H4N6 influenza virus isolated from pigs in Ontario. *Can Vet J*, 2000. 41(12): 938–939.

38. Ma, W. et al., Isolation and genetic characterization of new reassortant H3N1 swine influenza virus from pigs in the midwestern United States. *J Virol*, 2006. 80(10): 5092–5096.

39. Howard, W.A. et al., Reassortant Pandemic (H1N1) 2009 virus in pigs, United Kingdom. *Emerg Infect Dis*, 2011. 17(6): 1049–1052.

40. Moreno, A. et al., Novel H1N2 swine influenza reassortant strain in pigs derived from the pandemic H1N1/2009 virus. *Vet Microbiol*, 2011. 149(3–4): 472–477.

41. Starick, E. et al., Reassorted pandemic (H1N1) 2009 influenza A virus discovered from pigs in Germany. *J Gen Virol*, 2011. 92(5): 1184–1188.

42. Moreno, A. et al., Novel swine influenza virus subtype H3N1 in Italy. *Vet Microbiol*, 2009. 138(3–4): 361–367.

43. Brown, I.H. et al., Isolation of an influenza A virus of unusual subtype (H1N7) from pigs in England, and the subsequent experimental transmission from pig to pig. *Vet Microbiol*, 1994. 39(1–2): 125–134.

44. Song, M.S. et al., Evidence of human-to-swine transmission of the pandemic (H1N1) 2009 influenza virus in South Korea. *J Clin Microbiol*, 2010. 48(9): 3204–3211.

45. Vijaykrishna, D. et al., Reassortment of pandemic H1N1/2009 influenza A virus in swine. *Science*, 2010. 328(5985): 1529.

46. Lam, T.T. et al., Reassortment events among swine influenza A viruses in China: Implications for the origin of the 2009 influenza pandemic. *J Virol*, 2011. 85(19): 10279–10285.

47. Zhu, H. et al., Novel reassortment of Eurasian avian-like and pandemic/2009 influenza viruses in swine: Infectious potential for humans. *J Virol*, 2011. 85(20): 10432–10439.

48. Su, S. et al., Complete genome sequence of an avian-like H4N8 swine influenza virus discovered in southern China. *J Virol*, 2012. 86(17): 9542.

49. Zhang, G. et al., Identification of an H6N6 swine influenza virus in southern China. *Infect Genet Evol*, 2011. 11(5): 1174–1177.

50. Tu, J. et al., Isolation and molecular characterization of equine H3N8 influenza viruses from pigs in China. *Arch Virol*, 2009. 154(5): 887–890.

51. Brown, J.D., R. Poulson, and D.E. Stallknecht, Wild bird surveillance for avian influenza virus. *Methods Mol Biol*, 2014. 1161: 69–81.

52. Myers, K.P., C.W. Olsen, and G.C. Gray, Cases of swine influenza in humans: A review of the literature. *Clin Infect Dis*, 2007. 44(8): 1084–1088.

53. Centers for Disease Control and Prevention, Reported Infections with Variant Influenza Viruses in the United States since 2005. Available at http://www.cdc.gov/flu/swineflu/variant-cases-us.htm; accessed on March 30, 2015.

54. Centers for Disease Control and Prevention, Case Count: Detected U.S. Human Infections with H3N2v by State since August 2011. Available at http://www.cdc.gov/flu/swineflu/h3n2v-case-count.htm; accessed on March 30, 2015.

55. Janke, B.H., Clinicopathological features of Swine influenza. *Curr Top Microbiol Immunol*, 2013. 370: 69–83.

56. Brown, I.H. et al., Pathogenicity of a swine influenza H1N1 virus antigenically distinguishable from classical and European strains. *Vet Rec*, 1993. 132(24): 598–602.

57. De Vleeschauwer, A. et al., Comparative pathogenesis of an avian H5N2 and a swine H1N1 influenza virus in pigs. *PLoS One*, 2009. 4(8): e6662.

58. Janke, B.H., Influenza A virus infections in swine: Pathogenesis and diagnosis. *Vet Pathol*, 2014. 51(2): 410–426.

59. Detmer, S. et al., Diagnostics and surveillance for Swine influenza. *Curr Top Microbiol Immunol*, 2013. 370: 85–112.

60. Goodell, C.K. et al., Probability of detecting influenza A virus subtypes H1N1 and H3N2 in individual pig nasal swabs and pen-based oral fluid specimens over time. *Vet Microbiol*, 2013. 166(3–4): 450–460.

61. Zhang, J. and P.C. Gauger, Isolation of swine influenza virus in cell cultures and embryonated chicken eggs. *Methods Mol Biol*, 2014. 1161: 265–276.

62. Spackman, E. et al., Development of a real-time reverse transcriptase PCR assay for type A influenza virus and the avian H5 and H7 hemagglutinin subtypes. *J Clin Microbiol*, 2002. 40(9): 3256–3260.

63. Zhang, J. and K.M. Harmon, RNA extraction from swine samples and detection of influenza A virus in swine by real-time RT-PCR. *Methods Mol Biol*, 2014. 1161: 277–293.

64. Stockton, J. et al., Multiplex PCR for typing and subtyping influenza and respiratory syncytial viruses. *J Clin Microbiol*, 1998. 36(10): 2990–2995.

65. Li, J., S. Chen, and D.H. Evans, Typing and subtyping influenza virus using DNA microarrays and multiplex reverse transcriptase PCR. *J Clin Microbiol*, 2001. 39(2): 696–704.

66. Fereidouni, S.R. et al., Rapid molecular subtyping by reverse transcription polymerase chain reaction of the neuraminidase gene of avian influenza A viruses. *Vet Microbiol*, 2009. 135(3–4): 253–260.

67. He, J. et al., Rapid multiplex reverse transcription-PCR typing of influenza A and B virus, and subtyping of influenza A virus into H1, 2, 3, 5, 7, 9, N1 (human), N1 (animal), N2, and N7, including typing of novel swine origin influenza A (H1N1) virus, during the 2009 outbreak in Milwaukee, Wisconsin. *J Clin Microbiol*, 2009. 47(9): 2772–2778.

68. Yang, Y. et al., Simultaneous typing and HA/NA subtyping of influenza A and B viruses including the pandemic influenza A/H1N1 2009 by multiplex real-time RT-PCR. *J Virol Methods*, 2010. 167(1): 37–44.

69. Detmer, S.E. et al., Detection of Influenza A virus in porcine oral fluid samples. *J Vet Diagn Invest*, 2011. 23(2): 241–247.

70. Romagosa, A. et al., Sensitivity of oral fluids for detecting influenza A virus in populations of vaccinated and non-vaccinated pigs. *Influenza Other Respi Viruses*, 2012. 6(2): 110–118.

71. Ramirez, A. et al., Efficient surveillance of pig populations using oral fluids. *Prev Vet Med*, 2012. 104(3–4): 292–300.

72. Culhane, M.R. and S.E. Detmer, Sample types, collection, and transport for influenza A viruses of swine. *Methods Mol Biol*, 2014. 1161: 259–263.

73. Hoffmann, E. et al., Universal primer set for the full-length amplification of all influenza A viruses. *Arch Virol*, 2001. 146(12): 2275–2289.

47 Avian Metapneumovirus

José Francisco Rivera-Benitez, Jazmín De la Luz-Armendáriz,
Alberto Bravo-Blas, Luis Gómez-Núñez, and Humberto Ramírez-Mendoza

CONTENTS

47.1 Introduction 407
    47.1.1 Classification, Genome Structure, and Function 407
    47.1.2 Epidemiology, Clinical Features, and Pathogenesis 408
    47.1.3 Diagnosis 409
        47.1.3.1 Conventional Techniques 409
        47.1.3.2 Molecular Techniques 409
47.2 Methods 410
    47.2.1 Sample Collection and Preparation 410
    47.2.2 Detection Procedures..410
        47.2.2.1 cDNA Synthesis..410
        47.2.2.2 Duplex Nested Polymerase Chain Reaction Detection of Avian Metapneumovirus A and B Subtypes 410
47.3 Conclusion and Future Perspective...411
References..411

47.1 INTRODUCTION

Avian metapneumovirus (aMPV), previously referred to as turkey rhinotracheitis,[1] is an emerging viral pathogen in poultry, responsible for a highly contagious disease particularly in chickens and turkeys.[2] Although aMPV infection often involves the respiratory tract, it can also affect the oviduct, contributing to reduced egg production.[3] Birds of any age are susceptible to aMPV infection, and secondary infection with bacteria such as *Escherichia coli*, *Bordetella*, and *Pasteurella* can complicate the situation, resulting in a condition known as swollen head syndrome (SHS), with adverse impact on infected animals.[3–5]

aMPV was first described in South Africa in 1978, but the disease rapidly spread to other regions in Europe and Asia[2,4] and more recently in the Americas (United States, Brazil, and Mexico).[6–10] The only countries reported as aMPV free are Australia and Canada.[11] aMPV belongs to the family *Paramyxoviridae* and is further divided into four subtypes. Subtypes A and B were identified in South Africa and Europe, subtype C in Asia and the United States, and subtype D was detected in France.[2,10,12–16] In Latin America, only subtypes A and B have been described, mainly in Brazil and Mexico.[6–8,17]

47.1.1 CLASSIFICATION, GENOME STRUCTURE, AND FUNCTION

aMPV is a member of the genus *Metapneumovirus*, subfamily *Pneumovirinae*, family *Paramyxoviridae*.[1] Apart from the subfamily *Pneumovirinae*, the family *Paramyxoviridae* contains a second subfamily *Paramyxovirinae*. While the subfamily *Pneumovirinae* comprises two genera (*Metapneumovirus* and *Pneumovirus*), the subfamily *Paramyxovirinae* encompasses seven genera (*Aquaparamyxovirus*, *Avulavirus*, *Ferlavirus*, *Henipavirus*, *Morbillivirus*, *Respirovirus*, and *Rubulavirus*).

Besides aMPV (which was isolated for the first time in South Africa in the late 1970s), the genus *Metapneumovirus* also includes *human metapneumovirus* (hMPV), which was described in the Netherlands in 2001. Whereas hMPV is separated into subtypes A and B,[18] aMPV is shown to contain four subtypes: A, B, C, and D.

aMPV is a pleomorphic, enveloped, unsegmented, single-stranded, negative-sense RNA virus of approximately 14,000 nucleotides, with helical symmetry and projections on its surface. The genome of aMPV encodes eight structural proteins: nucleoprotein (N), phosphoprotein (P), two matrix proteins (M1 and M2), fusion protein (F), small hydrophobic protein (SH), surface glycoprotein (G), and RNA-dependent polymerase (L).[1,2] The division of aMPV into subtypes (A–D) is based on the variations of the surface glycoprotein sequences as well as the antigenic differences among viral strains.[10,12,14,15]

As the best characterized structural proteins of aMPV, glycoprotein G is associated with viral coupling and pathogenesis.[19] Both the aMPV and hMPV differ from the genus *Pneumovirus* by their lack of genes required for the synthesis of nonstructural proteins NS1 and NS2 and structural protein M2, by the specific order of genes that code for other structural proteins in the viral genome, and by the order in which their protein-coding genes are located within the viral genome.[20] Part of the structural proteins is the polymerase complex, which includes protein N, which represents a key element of the nucleocapsid and an essential component of the polymerase complex. Protein P is an important part of the replication and transcription complexes. Analysis of the interaction between protein P and protein N has shown that the carboxy-terminal domain of protein P contains most of the necessary elements to couple with protein N. Protein L is considered as the main component of the polymerase complex due to its role in the synthesis of viral RNA. Thus, proteins P, N, and L together with the viral RNA genome form the ribonucleoprotein of aMPV. In addition, gene M2 is composed of two open reading frames that code for proteins M2-1 (184–186 amino acids) and M2-2 (71–73 amino acids). It has been suggested that M2 is also linked to the nucleocapsid and that it is involved in the regulation of viral replication and transcription. Due to this feature, M2 has also been referred to as transcriptional elongation factor.[20] Protein M surrounds the nucleocapsid, and this facilitates its coupling to the virion's lipid envelope. Results from *in vitro* experiments suggest that protein M from hMPV influences the immune response, by promoting activation of macrophages and dendritic cells when shed as a soluble molecule. Another member of the *Paramyxoviridae* family, human respiratory syncytial virus (of the genus *Pneumovirus*), also has a protein M that can be shed during cell death or as an active process. It has also been proposed that protein M could have a role as part of the viral assembly, just like in other paramyxoviruses.[20–23]

aMPV contains three surface proteins. The first one, protein G, is a type II, highly glycosylated membrane protein, which is the most variable among MPVs and interacts with its receptors located on the target cell surface. It has been shown that protein G of hMPV has an important role as inhibitory factor for the immune response of the host by blockage IFN-α, IFN-β, and inducible chemokine production. Protein G was first thought to be the most important surface protein; however, studies have shown that some protein G–deficient, subtype C aMPV mutants are able to infect cells and replicate efficiently *in vitro*. Interestingly, the same aMPV mutant produced a milder disease in infected turkeys, as shown by lower clinical score and antibodies when compared with the wild-type virus. The results from these studies lead to the idea that *in vitro* replication of aMPV is protein G independent and that the presence of protein F is enough for viral coupling and fusion.[24,25]

The second surface protein, F, is a type I membrane protein, synthesized as an inactive precursor (F0), giving rise to subunits F1 and F2 after its cleavage by proteases in the host cell. Protein F promotes a pH-independent fusion between the viral envelope and the cell surface. It also induces the development of syncytia in infected cells and determines the viral host-specific tropism. This was experimentally shown by the inability of hMPV to infect turkey poults, even though aMPV can infect and replicate in mammal cell culture.[24,26]

Finally, protein SH is an integral type II membrane protein, which first allocates in the endoplasmic reticulum (ER), before being transported to the cell surface via ER–Golgi apparatus. It has been found that replication of hMPV can be SH independent, but it is involved in the modulation of immune responses by affecting nuclear factor kappa-light-chain-enhancer

of activated B cells (NF-κB)-dependent gene transcription.[20,24,27–29] It is important to mention that the nature of the cell receptor for aMPV is yet to be elucidated, although it has been suggested that one or more glycosaminoglycans or heparin-like molecules could be involved.[20]

47.1.2 Epidemiology, Clinical Features, and Pathogenesis

Initial studies on glycoprotein G using molecular sequence analysis as well as neutralization with monoclonal antibodies suggested that two subtypes (A and B) of aMPV existed within the same serotype. These subtypes are the most frequent in Europe, Israel, Japan, and Brazil.[22,30] Later, in 1996, a third subtype (C) was reported in the United States, whose amino acid sequence was very similar to that of hMPV.[27] First emerged from the Colorado outbreak of turkey in 1996, aMPV subtype C was implicated in a subsequent, much bigger outbreak in Minnesota, the largest turkey producer in that country. The cost of aMPV subtype C to poultry industry in the United States has been estimated at US$15 million per year, from 1997 to 2002.[2,22,24,31] It was suggested that migratory birds could be involved in the dissemination of the virus because aMPV was detected in nasal turbinates of sparrows, geese, swallows, and starlings captured in the central region of the United States, showing a nucleotide identity of 90%–95% and an amino acid identity of 97%–99% when compared with the virus isolated from turkeys.[32–34] Moreover, aMPV outbreaks in the United States showed a seasonal pattern, with 80% of them happening during spring (April to May) and fall (October to December).[22,35] A retrospective study reported two aMPV isolates in France with low nucleotide identity to subtypes A, B, and C, therefore they were classified as part of subtype D.[12,15]

Subtypes A, B, and D mainly affect turkeys and chickens, although subtype A has also been detected in pheasants and guinea fowls.[2,36] Subtype C was first isolated from turkeys albeit reports of similar isolates from duck, sparrows, geese, swallows, and gulls also exist.[2,32,34] Migratory birds are a key in aMPV transmission due to the vast areas that they cover, although local wild birds have also been implicated in the viral dissemination, due to the source of food and shelter that a poultry farm represents. Turkeys are the main natural host for aMPV, followed by chickens and hens. Infection can occur from very early age in turkeys and chickens, although birds of any age can become infected. aMPV is transmitted by direct contact with infected birds, which usually disseminate the virus around 2–3 days after the first exposure and the viral shedding continues for up to 1 week. Airborne transmission is the least relevant form of transmission, and no vertical transmission has been demonstrated up to date.[2,3,5,20]

The severity of the disease is highly influenced by secondary bacterial infection together with poor husbandry standards. The pathogens involved in the complication of aMPV infection are *Bordetella avium*, *Pausterella*, *E. coli*, *Mycoplasma gallisepticum*, *Chlamydophila*, and *Ornithobacterium rhinotracheale*. The onset of clinical signs and viral shedding usually begins as soon as 2–4 h postinfection. Different strains of aMPV show tropism for turkeys and chickens; however, other species such as sparrows, gulls, pheasants, wild ducks, and geese can get a subclinical form of infection by aMPV.

Clinical signs of the disease caused by aMPV infection are highly variable depending on species, age, environmental conditions, and secondary infections. However, in general terms, viral replication is limited to the trachea and lungs in both turkeys and chickens. In turkey poults, viral replication is limited to the upper respiratory tract, together with the onset of viremia. The clinical signs shown can be snicking, sneezing, crepitations, ocular and nasal discharge, conjunctivitis, foaming, swollen infraorbital sinuses, submandibular edema, and depression. When aMPV infection is not complicated by a secondary infection, recovery is rapid and can happen after approximately 14 days.[2,3,5,20] However, the risk of airsacculitis, pericarditis, pneumonia, perihepatitis, and increased morbidity and mortality increases significantly when a secondary bacterial infection complicates the initial aMPV infection. While morbidity can reach 100%, mortality of 80% of turkey poults is likely and only 0.5% of adults. In older animals, clinical signs include coughing and shaking head. Egg-laying hens can develop uterine prolapse due to the coughing, and egg production can drop down to 30%, affecting the quality of the egg shells and increase the risk of peritonitis. Mortality of laying hens ranges from 0% to 50%.

Clinical signs in broiler chickens are less characteristic than in turkeys and are often mistaken for infections of the upper respiratory system caused by different viral agents; however, just like in turkeys, a considerably more severe illness develops when a secondary agent, such as infectious bronchitis virus, influenza, *Mycoplasma*, or *E. coli*, complicates the disease. Evidence strongly suggests that aMPV is one of the main causes for SHS in broiler chickens. SHS is characterized by respiratory illness, depression, swollen infraorbital sinuses, unilateral or bilateral periorbital, and facial swelling, which in some cases can involve the whole head. These clinical signs can be followed by cerebral disorientation, torticollis, and opisthotonos. Even though normally the mortality does not exceed 1%–2%, morbidity can reach 10% and egg production is frequently affected. Similarly to turkeys, chicken mortality rate will increase in the presence of a secondary infection.

On histological analysis, macroscopic lesions in turkeys include loss of tracheal cilia, together with watery-to-mucous or purulent exudate. The reproductive tract can show regression of ovary and oviduct, abnormalities of the yolk and albumin, as well as peritonitis. Macroscopic lesions in chickens are less evident and only found in the head, although in cases of SHS, yellow jellylike or purulent edema and swollen periorbital sinuses can be observed. Histological analysis can show severe tracheal and pulmonary inflammation, loss of epithelial architecture, loss of cilia, and inflammatory cell infiltrates, sometimes combined with hemorrhage, lymphocytic infiltrates, and tracheal hyperplasia.[2–5,11,13,20,22,24,37,38]

47.1.3 DIAGNOSIS

Several diseases can display similar signs to aMPV infection, such as Newcastle disease, avian paramyxovirus infection, infectious bronchitis, laryngotracheitis, and avian influenza. For the diagnosis of aMPV from live birds, it is recommended to collect blood samples or oropharyngeal swabs for antigen or antigen-specific antibody detection. aMPV infection is diagnosed by the presence of viral particles, viral nucleic acids from infected tissues, or detection of anti-aMPV antibodies from recovering birds. Sample collection should be done promptly, aiming for the acute phase of the disease, and samples should be collected also from birds without clinical signs, as long as they are part of the same flock. Postmortem examination samples from turbinates, trachea, and lungs should be collected, as well as swabs from infraorbital area and nasal sinuses.[2,3,11]

47.1.3.1 Conventional Techniques

Serology is the most common method for diagnosis of aMPV infection, particularly in nonvaccinated flocks, because aMPV can be difficult to isolate and identify. The most widely used method is enzyme-linked immunosorbent assay (ELISA) because it is a rapid test and its cost is lower than reverse transcription polymerase chain reaction (RT-PCR) or viral isolation.[2] When this test is used from chicken samples, the response is weaker than from turkeys.

The best method for viral isolation is the primary culture of tracheal tissue or embryonated eggs, followed by subsequent passages in cell culture.[2,11] Serial aMPV passages in tracheal and pulmonary primary cultures with further adaptation to Vero cell cultures have proved to be an adequate and sensitive method for aMPV isolation.[6] aMPV isolation using chicken or turkey embryonated eggs is done via the yolk sac. If during the first passage there is no evident infection, the yolk sac is processed and passed onto a second embryonated egg, and then the infection is evident usually after 7–9 days. Usually, several passages are required for aMPV to cause embryonic death; therefore, this method is considered as slow, expensive, and labor intensive.[2,11]

Neutralization assays with monoclonal anti–glycoprotein G antibodies are used to differentiate between subtypes A and B, whereas neutralization assays based on the use polyclonal antisera have shown that subtypes A and B belong within the same serotype.[13] aMPV subtype C is not neutralized by monoclonal antibodies that differentiates between subtypes A and B, and on top of that, it is poorly neutralized by polyclonal antibodies.[15]

47.1.3.2 Molecular Techniques

RT-PCR and real-time RT-PCR[14,39–41] are the most widely used assays for molecular detection of aMPV infection. Among the advantages of these methods are their high sensitivity and their speed. However, one important disadvantage is that viral RNA can be detected only in the initial phase of the disease; thus, 2–10 days postinfection, clinical signs have already appeared and viral detection will be difficult even from the affected birds. Therefore, it is recommended to make sure that the samples are collected from birds without clinical signs but that are not in the recovery phase either. Usually, the RT-PCR is designed to amplify RNA fragments from the F, M, N, or G genes of aMPV.[11]

Moreover, due to the high similarities between subtypes and the inability of the first RT-PCR to discriminate subtype A from B, a nested RT-PCR is required to amplify a subtype-specific fragment of the gene G.[14,41] Once the target fragment has been amplified, it can be purified and used for sequencing analysis in order to compare different genotypes among subtypes A, B, C, or D.[6]

47.2 METHODS

47.2.1 Sample Collection and Preparation

The diagnosis protocol presented here is aimed to detect aMPV from respiratory tissue samples. This method is based on the amplification of a gene fraction coding for protein G and should be used to discriminate between aMPV subtypes A or B. It is important that tracheal tissue samples are collected at the earliest stages of infection. It is recommended to do postmortem analysis from 10 birds with clinical signs of the disease and 10 birds without signs but from the same flock. In case that the affected flock includes birds of different ages, samples representing each age group should be collected. Make sure that the tissue samples are collected individually with scissors and forceps, place the samples in cryotubes, and store in liquid nitrogen (this protocol is useful when viral isolation is required). Samples can also be stored in containers with dry ice or cooling packs, in case that only molecular analysis is required and the transit to the laboratory is shorter than 24 h. Once in the laboratory, use liquid nitrogen to macerate 0.2 g of sample with mortar and pestle. It is possible to pool tissues from up to five animals per sample (preferentially if the birds presented similar characteristics, age, or clinical signs). Once macerated, tissue samples should be rehydrated using minimum essential media or phosphate buffered saline (PBS) (pH7.4) at a 1:10 w/v ratio. Next, centrifuge the tissue suspension for 20 min at $2000 \times g$ at 4°C and use the supernatant for total RNA extraction.

RNA extraction is based on cell lysis (guanidinium thiocyanate buffer), capture in a silica-gel membrane, RNAse inactivation, RNA washes with ethanol, and elution with diethylpyrocarbonate (DEPC) water. According to the RNeasy Mini kit (Qiagen), vortex 200 μL of 10% (w/v) tissue supernatant with 600 μL RLT buffer. Add 800 μL of 70% ethanol and mix well by pipetting. Transfer the sample to a column and centrifuge for 30 s at $8000 \times g$ at 4°C. Next, samples are washed three times, first using 700 μL of RW1 buffer, followed by a second wash with 500 μL of RPE buffer, both washing steps include centrifugation for 30 s at $8000 \times g$ at 4°C, and for the third wash, add 500 μL of RPE buffer to the column and centrifuge for 2 min at $8000 \times g$ at 4°C. Finally, add 30–50 μL of DEPC water to the column for the elution step. This is followed by centrifugation for 1 min at $8000 \times g$ at 4°C. The resulting RNA should be kept at −80°C until required. RNA quantification should be performed before starting the RT-PCR, using a

spectrophotometer with a wavelength of 260 nm with the criterion of inclusion for the RNA being >1.8 OD_{260}/OD_{280} ratio.

47.2.2 Detection Procedures

As mentioned earlier, duplex nested RT-PCR allows discrimination between two of the most important aMPV subtypes for commercial aviculture (chickens and hens), and it is usually the first method of choice. Once the aMPV subtype has been identified, the following step is to sequence the amplified fragments for molecular epidemiology analysis. Alternatively, a subtype-specific quantitative RT-PCR can be used to determine the viral load from each sample. The following protocol has been modified from previous reports.[14,41]

47.2.2.1 cDNA Synthesis

47.2.2.1.1 Procedure

1. Add each of the components from the succeeding list into a sterile 0.2 mL PCR tube.

Component	Volume/Sample (μL)
Up to 500 ng total RNA	n
Primer G6– (10 pmol/μL)	4
5′-CTGACAAATTGGTCCTGATT-3′	
10 mM deoxynucleotide (dNTP) mix	1
DEPC-treated water	Up to 10

2. Mix gently by pipetting and briefly centrifuge.
3. Incubate the PCR reaction tube at 65°C for 5 min and immediately place on ice for 1 min.
4. Prepare a master mix reaction for cDNA synthesis adding each of the components in the order given.

Component	Volume/Sample (μL)
10× RT buffer	2
25 mM MgCl₂	4
0.1 M dithiothreitol (DTT)	2
RNase OUT™ (40 U/μL)	1
SuperScript® III RT (200 U/μL)	1

5. Incubate at 50°C for 50 min, then at 85°C for 5 min, and immediately place on ice.
6. Add 1 μL RNase H and incubate at 37°C for 20 min.
7. The cDNA obtained can be immediately used in PCR or stored at −20°C until required.

47.2.2.2 Duplex Nested Polymerase Chain Reaction Detection of Avian Metapneumovirus A and B Subtypes

Amplification of gene G from aMPV is done in two steps. The first step is to amplify the outer fragment (444 bp). Do not forget to include positive and negative controls (no template control) for each step of the PCR.

1. For the amplification of the outer fragment, add each of the following components into a sterile 0.2 mL PCR tube:

Component	Volume/Sample (µL)
5× Colorless GoTaq PCR buffer (Promega)	5
10 mM dNTP mix	2
Primer G6– (10 pmol/µL)	1.25
Primer G1+ (10 pmol/µL)	1.25
5'-GGGACAAGTATC T/C C/A T/G AT-3'	
GoTaq DNA polymerase	0.5
DEPC-treated water	10
cDNA	5

2. Place the mix tube in a thermocycler, using the following program: 3 min at 94°C; 30 cycles of 1 min at 94°C, 1.5 min at 50°C, 2 min at 72°C; and a final 10 min at 72°C.

Once the first amplification is completed, the second round of PCR will amplify the inner fragments (nested). The amplified fragments of gene G will be of 268 bp for aMPV subtype A and 361 bp for aMPV subtype B.

Prepare a master mix reaction for nested PCR adding each of the components in the order given.

Component	Volume/Sample (µL)
5× Colorless GoTaq PCR buffer (Promega)	5
10 mM dNTP mix	2
Primer G5– (10 pmol/µL)	1.25
5'-CAAAGA A/G CCAATAAGCCCA-3'	
Primer G8+A (10 pmol/µL)	1.25
5'-CACTCACTGTTAGCGTCATA-3'	
Primer G9+B (10 pmol/µL)	1.25
5'-TAGTCCTCAAGCAAGTCCTC-3'	
GoTaq DNA polymerase	0.5
DEPC-treated water	8.75
DNA from the first PCR	5

The program required for the thermocycler is the same as that used for the first PCR.

The amplified products should be analyzed using electrophoresis in 2% agarose gel in tris-acetate (TAE) buffer for 45 min at 80 V. Next, the gel should be stained using Midori Green and visualized using an UV transilluminator. Presence of a band size of approximate 268 bp indicates the presence of aMPV subtype A. Presence of a band size of approximate 361 bp indicates the presence of aMPV subtype B. The presence of two bands in the gel suggests a coinfection by both aMPV subtypes.

47.3 CONCLUSION AND FUTURE PERSPECTIVE

aMPV is an endemic agent in many zones with intensive poultry farming, affecting the most important regions for poultry meat type and egg production (United States, China, Brazil, and the European Union), therefore its economic impact is major. Measures to prevent aMPV infection include vaccination with attenuated and inactivated virus; however, it has been recently observed that some vaccine strains—derived mainly from subtypes A and B—have reverted their virulence, which has raised concerns about poor standards in the viral attenuation process.[42–44] Currently, molecular diagnosis of aMPV allows the detection of vaccine-derived virulent strains. This is done thanks to the partial sequencing of the G gene, which facilitates the detection of changes in nucleotide 6358, thus making possible to discriminate a vaccine derived from a wild-type virus.[44] In the current era of mass gene sequencing, it would be of great benefit if more studies focused on the complete analysis of the aMPV genome. This could facilitate the detection of new variants of the virus and assess any potentially detrimental effect of such variants on poultry industry. Similarly, the analysis of the whole aMPV genome will provide useful information for accurate selection of new vaccine candidates. This information would enhance the current control measures and possibly will lead to the eradication of this important avian disease.

REFERENCES

1. Wang, L.-F. et al. Paramyxoviridae. In *Virus Taxonomy: Classification and Nomenclature of Viruses Ninth Report of the International Committee on Taxonomy of Viruses* (eds. King, A.M.Q., Adams, M.J., Carstens, E.B., and Lefkowitz, E.J.), pp. 640–653 (Elsevier Academic Press, San Diego, CA, 2011).
2. Cook, J.K. Avian rhinotracheitis. *Rev Sci Tech* 19, 602–613 (2000).
3. Cook, J.K. Avian pneumovirus infections of turkeys and chickens. *Vet J* 160, 118–125 (2000).
4. Buys, S.B., du Preez, J.H., and Els, H.J. Swollen head syndrome in chickens: A preliminary report on the isolation of a possible aetiological agent. *J S Afr Vet Assoc* 60, 221–222 (1989).
5. Cook, J.K. et al. Avian pneumovirus infection of laying hens: Experimental studies. *Avian Pathol* 29, 545–556 (2000).
6. Rivera-Benitez, J.F., Martinez-Bautista, R., Rios-Cambre, F., and Ramirez-Mendoza, H. Molecular detection and isolation of avian metapneumovirus in Mexico. *Avian Pathol* 43, 217–223 (2014).
7. Chacon, J.L., Brandao, P.E., Buim, M., Villarreal, L., and Ferreira, A.J. Detection by reverse transcriptase-polymerase chain reaction and molecular characterization of subtype B avian metapneumovirus isolated in Brazil. *Avian Pathol* 36, 383–387 (2007).
8. Dani, M.A., Durigon, E.L., and Arns, C.W. Molecular characterization of Brazilian avian pneumovirus isolates: Comparison between immunochemiluminescent Southern blot and nested PCR. *J Virol Methods* 79, 237–241 (1999).
9. Govindarajan, D., Buchholz, U.J., and Samal, S.K. Recovery of avian metapneumovirus subgroup C from cDNA: Cross-recognition of avian and human metapneumovirus support proteins. *J Virol* 80, 5790–5797 (2006).
10. Seal, B.S. Matrix protein gene nucleotide and predicted amino acid sequence demonstrate that the first US avian pneumovirus isolate is distinct from European strains. *Virus Res* 58, 45–52 (1998).

11. Pedersen, J. and Gough, R. Turkey rhinotracheitis (avian metapneumovirus). In: *OIE Manual of Diagnostic Tests and Vaccines for Terrestrial Animals*, OIE, 590-8, 2008.

12. Bayon-Auboyer, M.H., Arnauld, C., Toquin, D., and Eterradossi, N. Nucleotide sequences of the F, L and G protein genes of two non-A/non-B avian pneumoviruses (APV) reveal a novel APV subgroup. *J Gen Virol* 81, 2723–2733 (2000).

13. Cook, J.K. and Cavanagh, D. Detection and differentiation of avian pneumoviruses (metapneumoviruses). *Avian Pathol* 31, 117–132 (2002).

14. Juhasz, K. and Easton, A.J. Extensive sequence variation in the attachment (G) protein gene of avian pneumovirus: Evidence for two distinct subgroups. *J Gen Virol* 75, 2873–2880 (1994).

15. Toquin, D., Bayon-Auboyer, M.H., Senne, D.A., and Eterradossi, N. Lack of antigenic relationship between French and recent North American non-A/non-B turkey rhinotracheitis viruses. *Avian Dis* 44, 977–982 (2000).

16. Wei, L. et al. Avian metapneumovirus subgroup C infection in chickens, China. *Emerg Infect Dis* 19, 1092 (2013).

17. D'Arce, R.C. et al. Subtyping of new Brazilian avian metapneumovirus isolates from chickens and turkeys by reverse transcriptase-nested-polymerase chain reaction. *Avian Pathol* 34, 133–136 (2005).

18. Feuillet, F., Lina, B., Rosa-Calatrava, M., and Boivin, G. Ten years of human metapneumovirus research. *J Clin Virol* 53, 97–105 (2012).

19. Cecchinato, M. et al. Avian metapneumovirus (AMPV) attachment protein involvement in probable virus evolution concurrent with mass live vaccine introduction. *Vet Microbiol* 146, 24–34 (2010).

20. Easton, A.J., Domachowske, J.B., and Rosenberg, H.F. Animal pneumoviruses: Molecular genetics and pathogenesis. *Clin Microbiol Rev* 17, 390–412 (2004).

21. Munir, S. and Kapur, V. Regulation of host cell transcriptional physiology by the avian pneumovirus provides key insights into host-pathogen interactions. *J Virol* 77, 4899–4910 (2003).

22. Njenga, M.K., Lwamba, H.M., and Seal, B.S. Metapneumoviruses in birds and humans. *Virus Res* 91, 163–169 (2003).

23. Bagnaud-Baule, A. et al. The human metapneumovirus matrix protein stimulates the inflammatory immune response in vitro. *PLoS One* 6, e17818 (2011).

24. de Graaf, M. et al. Fusion protein is the main determinant of metapneumovirus host tropism. *J Gen Virol* 90, 1408–1416 (2009).

25. Bao, X. et al. Human metapneumovirus glycoprotein G inhibits innate immune responses. *PLoS Pathog* 4, e1000077 (2008).

26. Schowalter, R.M., Smith, S.E., and Dutch, R.E. Characterization of human metapneumovirus F protein-promoted membrane fusion: Critical roles for proteolytic processing and low pH. *J Virol* 80, 10931–10941 (2006).

27. Alvarez, R., Lwamba, H.M., Kapczynski, D.R., Njenga, M.K., and Seal, B.S. Nucleotide and predicted amino acid sequence-based analysis of the avian metapneumovirus type C cell attachment glycoprotein gene: Phylogenetic analysis and molecular epidemiology of US pneumoviruses. *J Clin Microbiol* 41, 1730–1735 (2003).

28. Dar, A.M., Munir, S., Goyal, S.M., and Kapur, V. Sequence analysis of the matrix (M2) protein gene of avian pneumovirus recovered from turkey flocks in the United States. *J Clin Microbiol* 41, 2748–2751 (2003).

29. Deng, Q. et al. Topology and cellular localization of the small hydrophobic protein of avian metapneumovirus. *Virus Res* 160, 102–107 (2011).

30. Banet-Noach, C., Simanov, L., and Perk, S. Characterization of Israeli avian metapneumovirus strains in turkeys and chickens. *Avian Pathol* 34, 220–226 (2005).

31. Velayudhan, B.T. et al. Emergence of a virulent type C avian metapneumovirus in turkeys in Minnesota. *Avian Dis* 49, 520–526 (2005).

32. Song, M.-S. et al. Genetic characterization of avian metapneumovirus subtype C isolated from pheasants in a live bird market. *Virus Res* 128, 18–25 (2007).

33. Turpin, E., Stallknecht, D., Slemons, R., Zsak, L., and Swayne, D. Evidence of avian metapneumovirus subtype C infection of wild birds in Georgia, South Carolina, Arkansas and Ohio, USA. *Avian Pathol* 37, 343–351 (2008).

34. Bennett, R. et al. Evidence of avian pneumovirus spread beyond Minnesota among wild and domestic birds in central North America. *Avian Dis* 48, 902–908 (2004).

35. Shin, H.-J., Njenga, M.K., McComb, B., Halvorson, D.A., and Nagaraja, K.V. Avian pneumovirus (APV) RNA from wild and sentinel birds in the United States has genetic homology with RNA from APV isolates from domestic turkeys. *J Clin Microbiol* 38, 4282–4284 (2000).

36. Catelli, E., Terregino, C., De Marco, M., Delogu, M., and Guberti, V. Serological evidence of avian pneumovirus infection in reared and free-living pheasants. *Vet Rec* 149, 56–58 (2001).

37. Aung, Y., Liman, M., and Rautenschlein, S. Experimental infections of broilers with avian Metapneumovirus subtype A and B. *World's Poultry Sci J* 62, 134 (2006).

38. Sugiyama, M. et al. Drop of egg production in chickens by experimental infection with an avian metapneumovirus strain PLE8T1 derived from swollen head syndrome and the application to evaluate vaccine. *J Vet Med Sci* 68, 783 (2006).

39. Guionie, O. et al. Laboratory evaluation of a quantitative real-time reverse transcription PCR assay for the detection and identification of the four subgroups of avian metapneumovirus. *J Virol Methods* 139, 150–158 (2007).

40. Cecchinato, M. et al. Development of a real-time RT-PCR assay for the simultaneous identification, quantitation and differentiation of avian metapneumovirus subtypes A and B. *Avian Pathol* 42, 283–289 (2013).

41. Naylor, C., Shaw, K., Britton, P., and Cavanagh, D. Appearance of type B avian Pneumovirus in great Britain. *Avian Pathol* 26, 327–338 (1997).

42. Catelli, E., Cecchinato, M., Savage, C.E., Jones, R.C., and Naylor, C.J. Demonstration of loss of attenuation and extended field persistence of a live avian metapneumovirus vaccine. *Vaccine* 24, 6476–6482 (2006).

43. Cecchinato, M. et al. Reversion to virulence of a subtype B avian metapneumovirus vaccine: Is it time for regulators to require availability of vaccine progenitors? *Vaccine* 32, 4660–4664 (2014).

44. Lupini, C., Cecchinato, M., Ricchizzi, E., Naylor, C.J., and Catelli, E. A turkey rhinotracheitis outbreak caused by the environmental spread of a vaccine-derived avian metapneumovirus. *Avian Pathol* 40, 525–530 (2011).

48 Bovine Respiratory Syncytial Virus

Sara Hägglund and Jean François Valarcher

CONTENTS

48.1 Introduction 413
    48.1.1 Classification, Morphology, and Genome Organization 413
    48.1.2 Biology and Epidemiology 413
    48.1.3 Clinical Features and Pathogenesis 414
    48.1.4 Diagnosis 414
        48.1.4.1 Conventional Techniques..414
        48.1.4.2 Molecular Techniques..416
48.2 Methods 416
    48.2.1 Sample Collection and Preparation 416
    48.2.2 Detection Procedure 417
48.3 Conclusion and Future Perspectives 417
References...418

48.1 INTRODUCTION

Bovine respiratory syncytial virus (BRSV) is a major pathogen involved in bovine respiratory disease (BRD), which is one of the disease complexes responsible for severe economic losses in cattle production [1]. Pure BRSV infections alone may cause acute and fatal illness, but coinfections with other respiratory viruses are often diagnosed simultaneously [2] and viral infections commonly precede bacterial colonization of the bovine lung [3]. The prevalence of BRSV is considered high worldwide, even based on direct detection by conventional techniques with moderate sensitivity, or indirect detection by serology, which is hampered by the inhibitory effect of preexisting immunity on humoral responses to infections [4]. However, novel, high-throughput molecular techniques with improved sensitivity provide new means to understand the real direct and indirect impact of this infection on agricultural profits, animal welfare, and the use of antibiotics.

48.1.1 CLASSIFICATION, MORPHOLOGY, AND GENOME ORGANIZATION

BRSV is an enveloped RNA virus classified in the genus *Pneumovirus*, family *Paramyxoviridae*, order *Mononegavirales* [5]. Virions are spherical to filamentous, with a diameter of 100 nm to 1 μm for spherical particles, and a width of 70–190 nm and a length of up to 2 μm for filamentous particles [6]. Its genomic RNA is single stranded, nonsegmented, negative sense, and approximately 15,000 nucleotides long and contains 10 genes that code for 11 proteins. The fusion glycoprotein (F), the membrane glycoprotein (G), and the small hydrophobic protein (SH) are incorporated in the host cell membrane–derived virus envelope, whereas the matrix protein (M) is present under this envelope. Within the virion, a helical nucleocapsid consisting of nucleoproteins (N) encapsidates a virus RNA and mediates interaction with the polymerase consisting of the large protein (L), the phosphoprotein (P), and the polymerase cofactor M2-1. Other proteins are M2-2, which mediates a switch from transcription to RNA replication, and the nonstructural proteins NS1 and NS2, which inhibit viral RNA transcription and replication, as well as host cell immune responses to virus infection through inhibition of and resistance to type 1 interferons [4].

48.1.2 BIOLOGY AND EPIDEMIOLOGY

Virus assembly and spread occur by budding from the cell membrane, by cell fusion resulting in formation of multinucleated syncytia, and by lysis of cells [7]. The membrane glycoproteins F and G have essential functions in virus attachment and in cell membrane fusion of host cells. Virus–cell fusion likely occurs through an interaction between F and a cell surface receptor called nucleolin [8], and following virus entry into the cell, viral RNA transcription as well as replication takes place in the cytoplasm of the cell. The major target cells for replication are epithelial cells in the respiratory tract and pneumocytes [9,10], though replicating viruses have also been detected in B-lymphocytes within lymph nodes [11].

BRSV exists as a single serotype, and antibody cross reactivity occurs between BRSV, human RSV (HRSV), caprine RSV, and ovine RS virus (RSV) [12]. According to reactions with panels of monoclonal antibodies, BRSV can be divided into antigenic subgroups (A, AB, B, and untyped) [13–15] and sequence data on F, G, and N reveal 5, 6, and 5 distinct phylogenetic subgroups, respectively, with isolates from different time periods and geographical origin [16]. The average amino acid homology between BRSV strains is 83.3% and 99% for the G and F protein, respectively [16], as compared with 53% and 89% for G and F between HRSV strains, respectively, of different subtypes (A and B) [17]. Bovine RSV and HRSV share 52% and 78% amino acid identity in G and F, respectively [16], and although these viruses do not seem to cross the species barrier naturally, probably due to differences in the F and NS proteins [4], infections have been reproduced experimentally with HRSV in cattle [18]. BRSV is most closely related to the caprine RSV [19], and high seroprevalence to ovine RSV has been reported in cattle, indicating transmission between different ruminants [20].

Clinical disease caused by BRSV is mainly diagnosed in autumn and winter in temperate climate zones [21]. It is unclear which factors favor disease during these periods and whether the virus continues to circulate between animals and herds when it is less frequently detected. It has been suggested that BRSV persistently infect animals between outbreaks, based on repeated outbreaks in herds, and antibody titer increases in the absence of reinfections [22]. Furthermore, viral messenger RNA has been detected in lymph nodes of animals 71 days after infection [11]. Although attempts to detect shedding of reactivated virus following provocation (with cortisol or other infections) have failed, BRSV-specific antibody titers increased also in serum of provoked animals, which suggested low-level virus activation [23]. However, this could presumably also be a result of polyclonal stimulation of BRSV-specific memory B cells through bystander T-cell activation or toll-like receptor engagement [24]. Indeed, sequence data from recurrent outbreaks in the same herds in different years have revealed considerable nucleotide variation, implying that new viral strains were introduced or that latent strains evolve dramatically [25] and consensus sequences of BRSV strains obtained in a herd during one outbreak are identical, even if obtained 2–3 weeks apart [16,25,26].

48.1.3 Clinical Features and Pathogenesis

Infections with BRSV cause clinical signs ranging from transient fever to acute respiratory distress with abdominal dyspnea [22,27]. In animal groups containing many naïve individuals, the morbidity is often close to 100%, whereas the mortality is usually low, but may reach 20%–30% [28,29]. Disease is generally observed in calves between 1 and 6 months of age in areas where the infection is endemic, probably because most older animals are immune and clinically protected [22], although these may shed virus following reinfection [30–32]. Furthermore, high levels of MDA provide at least partial protection in young calves [32–34].

When 13 groups of 1–15-month-old calves with serologically confirmed BRSV infection were investigated for clinical signs of disease, animals in 3 groups showed only mild clinical signs of BRD (increased coughing and nasal discharge), whereas 47% of animals in 10 groups showed abdominal dyspnea, 50% showed abnormal breathing sounds on auscultation, 59% had cough, and 62% had nasal discharge [35]. Pyrexia, lacrimation, tachypnea, and abdominal dyspnea are common clinical signs, whereas subcutaneous emphysema, anorexia, and reluctance to lie down are more occasionally observed—the latter is associated with poor prognosis. Death is generally preceded by strong depression, severe abdominal dyspnea, bradypnea, and cyanosis (unpublished observations including outbreaks described in [16]).

The clinical disease is partly caused by virus destruction of cells in the respiratory tract and partly by detrimental effects of inflammatory responses. Direct immunopathological mechanisms have also been suggested to be involved in the pathogenesis of BRSV, such as excessive immunoglobulin (Ig) complement binding and mast cell activation [36,37]. Bronchiolar obstruction by mucus and cell debris results in atelectasis and emphysema. In parallel, the mucociliary damage and virus-induced alterations of pulmonary immune responses lead to the predisposition of secondary bacterial or mycoplasma infections that can prolong and aggravate the disease [38]. Consequently, necropsy in late stages of respiratory disease often reveal bacterial infections when BRSV is no longer detectable, which illustrates the importance of early diagnosis in BRSV outbreaks [35,39].

48.1.4 Diagnosis

48.1.4.1 Conventional Techniques

48.1.4.1.1 Indirect Laboratory Diagnosis Based on Detection of BRSV-Specific Antibodies

BRSV-specific IgG antibody detection in serum or milk is commonly used for diagnosis of BRSV infection at a herd level, as well as for serosurveillance [40–42]. Since BRSV is endemic in most cattle populations and the seroprevalence consequently is high, the detection of increased antibody titers in paired samples from the same individual is necessary to diagnose a recent infection. When interpreting such data, it should be taken in consideration that animals with high levels of BRSV-specific IgG (actively or passively derived) at the time of infection commonly do not show an increase in antibody titer. In fact, although a weak IgM response is sometimes observed, both mucosal and serum IgG_1, IgG_2, and IgA responses may be completely suppressed [32,43,44]. For serological diagnosis of a recent BRSV infection at a herd level, paired samples for IgG analysis should therefore preferably be obtained from 6- to 7-month-old calves, which have lost their MDA but have likely not been repeatedly infected while being seronegative, which would result in persisting high titers of BRSV-specific IgG. If other animal age groups are targeted, results should be interpreted cautiously. Samples are preferably collected from three to five individuals during the first week after the onset of clinical signs in the herd and again 3–4 weeks later.

The kinetics of BRSV-specific antibodies can be used to date an infection. IgM and IgA antibodies are detected in serum from experimentally infected, previously naïve, colostrum-deprived calves, from day 8 to 14 and day 12 to 29, respectively, after infection [32]. Serum IgG becomes apparent from day 13 to 17 (IgG_1) or day 25 to 90 (IgG_2) and are detectable for at least 4 months [32,43,45]. Following BRSV infection of calves that have previously been effectively primed with BRSV (by infection or vaccination), strong and rapid memory responses are observed already from day 6 to 8 in serum (IgA, IgG$_1$, and IgG$_2$, no IgM) and from day 4 to 11 in respiratory secretions (IgA) [32,44,46,47]. At least in some adult individuals, IgG antibodies may persist in serum and milk for more than 3 years after natural infection, without new reinfections [48]. The duration of humoral responses will depend on the magnitude of the initial antibody response (in turn influenced by preexisting antibodies, such as MDA) and the antibody half-life (26.6 days for IgG$_1$ [43]), as well as the possible induction of long-lived, antibody-producing plasma cells in the bone marrow and the persistence of antigen in lymph nodes [24].

The antibody kinetics will also depend on the sensitivity and specificity of the diagnostic assay utilized in different laboratories. Indirect antibody ELISAs, based on conventional or molecular techniques, have high sensitivity and specificity in general [49]. When the BRSV antigen in these assays derives from an infected cell lysate (or similarly a lysate of cells infected with a vector virus that expresses a BRSV protein [50]), samples should also be analyzed for antibody binding to antigens based on uninfected cell lysate so that a corrected optical density value ($COD = OD_{\text{BRSV-infected cell lysate}} - OD_{\text{uninfected cell lysate}}$) can be calculated to achieve specific results. Unspecific binding can be circumvented by molecular methods, for example, by using a highly purified RSV protein for coating [51]. However, in both the classic and molecular assays, the sensitivity may be impaired by antibody competition in which the dominating antibody isotype inhibits the detection of other isotypes, specific for the same epitope but present in minority [52]. This does not occur in BRSV capture ELISAs [43], but in such assays the sensitivity may decrease when analyzing samples containing much antibodies of the targeted antibody isotype, of other specificities.

Both the indirect and capture ELISAs detect a repertoire of antibodies with different BRSV epitope specificity, some of which are involved in the protection against BRSV disease. The contribution of different antibodies directed to various viral proteins or epitopes to protection against disease may vary considerably. The classic way to measure the protective efficacy of the total antibody response is to determine the titer of BRSV neutralizing (VN) antibodies, which are induced by both BRSV F and G [50]. However, the correlation between VN titers and the degree of protection against infection was not always consistent, probably because cytotoxic T lymphocytes and mucosal IgA are also important in controlling the viral infection [44,47,53]. Partly because of difficulties to standardize VN assays between laboratories, higher throughput blocking or competitive ELISAs (based on an RSV-neutralizing monoclonal antibody, palivizumab), are complementing, or even largely replacing, the classic HRSV-neutralizing assay in human medicine [54], and an equivalent assay was developed for BRSV [55]. While not being able to distinguish between antibody isotypes, this assay allows detection of BRSV-specific antibodies in samples from any animal species without changing assay components. However, unspecific antibody binding can block the particular epitope for which the assay is specific, by steric hindrance, which decreases specificity.

48.1.4.1.2 Direct Laboratory Diagnosis of BRSV

In contrast to indirect diagnosis, the conventional direct laboratory diagnosis of BRSV is disadvantaged by the short duration of shedding of high titers of the virus. The conventional direct detection of the virus is applicable only during a few days after the onset of clinical signs, whereas the most sensitive molecular techniques permit virus detection during a 2-week period [39,56,57].

Commonly used conventional techniques for direct detection of BRSV are isolation of the virus in tissue culture and immune-based detection of antigens in histological sections or body secretions (immunohistochemistry, antigen ELISAs, or pen-side antigen tests). Virus isolation, which is more used in research than in routine diagnostics today, is time consuming, cannot be automatized, requires expertise for result interpretation, and is dependent on virus viability. The previous extensive use of this technique for diagnosis of BRSV, which is a fragile virus, might have resulted in an underestimation of the frequency of infections in the past [58]. BRSV is preferably primary propagated on bovine embryonic turbinate [57], lung [59], or kidney cells [60] and is most easily isolated from infected respiratory cells obtained by nasopharyngeal brushing, transtracheal aspiration, or bronchoalveolar lavage (BAL) after removal of mucus (by filtering through a gauze) and centrifugation (at $200 \times g$ for 10 min) of BAL and aspirate. The cell pellet rather than the supernatant is used for inoculation of tissue culture cells, after cell lysis by freezing in cell culture medium and thawing, since the virus is commonly present in high quantities in cells, or is attached to them, and since respiratory secretions may have a toxic effect on cells *in vitro*. If the sample, nevertheless, appears toxic to the cultured cells, the sample can be diluted or several virus passages can be performed. At first inoculation or after passage, multinucleated giant cells (syncytia) are usually observed in light microscope 3–5 days post infection and are thereafter lysed. Three passages are generally required to declare a sample negative. In addition to using immune staining of cells (e.g., by immunofluorescense), or staining for morphologic inspection of cytopathic effects (such as syncytia or lysed cells), live cultured cells can be imaged during the course of an infection by using dual-labeled hairpin oligonucleotide probes (molecular beacons) specific for BRSV and confocal microscopy [61]. Virus isolation always implies a risk of amplification of coinfections that are better gained by the culture conditions than the actual pathogen.

Immunohistochemistry and antigen ELISAs are widely used for BRSV antigen detection in tissue samples and body fluids [10], but the sensitivity and the specificity of these assays are not so high [62] or depend on the experience of the examiner. Several rapid pen-side antigen test kits for HRSV have been used for BRSV diagnosis in the field [57,63]. The BRSV-detection limit for three of these tests was reported to be ~156 TCID$_{50}$, resulting in a sensitivity higher than that of a BRSV antigen ELISA but lower than that of a BRSV reverse transcription (RT)-PCR [63]. This is in agreement with previous findings for a similar test (88 PFU) [57]. However, the test accuracy for BRSV may vary considerably between trademarks [64].

48.1.4.2 Molecular Techniques

The introduction of RT-PCR for BRSV RNA detection during the 1990s resulted in dramatic diagnostic improvements [39,57,65]. The real-time RT-PCR, also called RT-quantitative PCR (RT-qPCR), is now probably the most commonly used direct diagnostic technique in routine laboratory diagnosis of this virus. In contrast to the classic, nested, and gel-based RT-PCR, which is labor intensive and consequently expensive, the RT-qPCR is rapid and user friendly, as well as easily standardized, automated, and scaled up to simultaneous analysis of up to 96 samples. The RT and PCR process is often made sequentially in the same reaction tube and only one round of amplification is performed, which decrease labor time and risks for sample cross-contamination. Moreover, no gel-based post-PCR processing is needed.

Several RT-qPCR assays were published for different target genes of BRSV such as F, G, and N [62,66–68]. Some of these assays are commercially available [68], and some are multiplexed for the detection of multiple respiratory viruses [69]. Since the RT-qPCR is based on a more objective detection of amplicons rather than the classic PCR, it can be used to quantify viral RNA in the original sample [46,66]. A range of different systems is available to detect amplicons in real time (e.g., TaqMan®, molecular beacon, CYBR Green, and Primer-Probe energy transfer), all of which generate a fluorogenic signal that can be measured in each cycle and quantified in "real time" by a computer. The TaqMan system is used in a high-speed droplet-real-time PCR, which enables the detection of BRSV in a few minutes, with maintained sensitivity compared to RT-qPCR [70].

48.2 METHODS

48.2.1 SAMPLE COLLECTION AND PREPARATION

For optimal diagnosis of a BRSV outbreak, one or several (preferably three to five) animals with acute clinical signs of BRSV (e.g., fever, tachypnea, and serous nasal discharge for less than 3 days) are selected for sampling. When using direct detection techniques with moderate to high sensitivity, it is possible to make a herd diagnosis by pooling samples from several animals to lower laboratory costs.

Nasal swabs are easily collected from cattle (see Figure 48.1) and are minimally invasive, which limits stress of sick animals and allows repeated samplings on consecutive days for research purposes. When using RT-qPCR analysis on this type of sample, the diagnostic window is at least 4 days (between days 3 and 7 after infection [44]), but can reach 12 days (between days 7 and 19 after infection [56]). Swabs are immediately placed in a commercial virus transport medium (e.g., Virocult®) or 1 mL

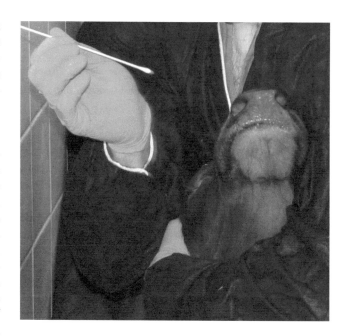

FIGURE 48.1 Collection of nasal swab sample from calf for detection of BRSV by RT-qPCR. To avoid stress, which may aggravate hypoxia caused by respiratory disease, the calf is gently fixed by placing its hindquarters into a corner of the pen and by holding the base of its right ear with the sampler's left hand. The calf nozzle is directed upward and the sampler's right wrist is placed on the labial commissure of the calf while taking the sample so that any movement can easily be followed. A sterile, preferably flexible, nasal cotton swab is inserted approximately 5 cm into the nostril, swabbed against the nasal mucosa and then immediately placed in a virus transport medium.

Dulbecco's modified Eagle medium (Lonza, Belgium), supplemented with buffer and antibiotics as described in [46] in addition to 2% fetal calf serum, and stored at +4°C for analysis the same day or at −75°C for later RNA extraction and analysis by RT-qPCR. Though not suitable for animals with severe disease, transtracheal aspiration or BAL (obtained through a fiber-optic bronchoscope after sedation) can be performed with sterile saline as described earlier [71,72], and infected cells are recovered and stored at −75°C (in RNase inactivating solution or lysis buffer for PCR analysis, or in cell culture medium for virus isolation). Alternatively, postmortem samples can be collected by swabbing lung tissue from a densified lobe or by performing BAL through the trachea, with up to 500 mL PBS or cell culture medium. Besides virus isolation or PCR, antibody staining may be performed on such respiratory secretions after cytospin, on frozen cryosections of fresh lungs, or on sections of tissue fixed in 10% buffered formalin and after pretreatment, as described previously [11].

BRSV RNA may be extracted in different ways prior to RT-qPCR. The classic way, which is effective and which minimize any cross-contamination of samples, is to use phenol/guanidinium isothiocyanate and chloroform for cell solubilization, protein denaturation, and separation of RNA from proteins and DNA. Briefly, for each specimen, 250 μL sample material (e.g., nasal swab transport medium or lung tissue homogenate in equal amounts of PBS [w/v]) is mixed with

750 μL lysis buffer (e.g., TRIzol Reagent, Invitrogen/Life Technologies) and incubated for 5 min at room temperature (RT°C). Thereafter, 200 μL chloroform is added and then the sample is mixed, incubated for 2 min at RT°C, and centrifuged at $13,900 \times g$ and 4°C for 15 min. The aqueous phase is then transferred to a new tube containing 500 μL isopropanol and incubated overnight at −20°C. After centrifugation at $9,070 \times g$ and 4°C for 15 min, the isopropanol is removed and the RNA pellet washed with −20°C 1 mL of 80% ethanol. After an additional centrifugation at $13,900 \times g$ and 4°C for 5 min, the ethanol is next removed and the RNA pellet is left to air-dry at RT°C for at least 1 h. The RNA is thereafter dissolved in 30 μL of diethylpyrocarbonate (DEPC)-treated water, preferably stored at −75°C and handled on wet ice. Another, quicker alternative is to use an RNA extraction kit, according to the manufacturer's instructions. It may then be suggested to discard fluids after centrifugation by pipetting, instead of pouring, and to open tubes with tube openers (e.g., USASI1400-1008, VWR, one tool per sample) instead of fingers. Extraction robots have also been used with success to study the kinetics of virus shedding after experimental challenge.

48.2.2 DETECTION PROCEDURE

The RT-qPCR protocol detailed in the following has been modified from previous publications [44,46,47,62]:

1. Dilute primers and probes (designed as detailed in Table 48.1) in DEPC-treated water into stocks (100 μM for primers and probes) and working solutions (10 μM for primers and 5 μM for probes). Aliquot and store at −20°C.
2. Store Moloney murine leukemia virus reverse transcriptase enzyme at −75°C. Do not thaw this enzyme, but pipette it slowly when viscous and then immediately put it back at −75°C. Use a kit (e.g., iScript™ One-Step RT-PCR Kit for Probes, Bio-Rad Laboratories, or equivalent) and proceed as described in Table 48.2, or add separate ingredients to a reaction mix, as detailed previously [62].
3. Vortex all reagents, including RNA, and rapidly spin them in a microcentrifuge before pipetting into

TABLE 48.1
TaqMan RT-qPCR System Specific for the BRSV F Gene, Generating Products of 85 Base Pairs

Type	Sequence (5′–3′)
Forward primer	AAGGGTCAAACATCTGCTTAACTAG
Reverse primer	TCTGCCTGWGGGAAAAAAG
TaqMan® probe[a]	AGAGCCTGCATTRTCACAATACCACCCA

Source: Hakhverdyan, M. et al., *J. Virol. Methods*, 123, 195, 2005.
[a] The complementary probe is labeled with FAM at the 5′ end and with TAMRA at the 3′ end.

TABLE 48.2
TaqMan BRSV RT-qPCR Protocol

Component	Volume (μL)[a]	Final Concentration
DEPC-treated water	5.5	
iScript reaction mix[b]	12.5	0.5×
RNA	5.0	
Forward primer[c]	0.75	300 nM
Reverse primer[c]	0.75	300 nM
TaqMan® probe[d]	0.5	100 nM
iScript MMLV reverse transcriptase[a]	0.5	0.5×
Total	25	

[a] Volume per specimen.
[b] Reagent included in iScript™ One-Step RT-PCR Kit for Probes, Bio-Rad Laboratories.
[c] Working solution at 10 μM.
[d] Working solution at 5 μM.

TABLE 48.3
Thermodynamic Profile for BRSV RT-qPCR

Time	Temperature (°C)	Action	Cycle Repetition
10 min	50	Complementary DNA synthesis	×1
5 min	95	Reverse transcriptase inactivation	×1
15 s	95	DNA amplification	×45
30 s	60	Primer and probe annealing[a]	

[a] Data collection.

qPCR reaction tubes kept on wet ice. Protect the probe from light at all steps. When pipetting, avoid adding air bubbles in the reaction mix. Include several negative (in which RNA is replaced with DEPC-treated water) and positive (containing quantified plasmids or RNA extracted from different dilutions of BRSV infected cell lysate) control samples in each run. Use lids to close reaction tubes before analysis.

4. Place the reaction tubes in a thermal cycler, programmed as described in Table 48.3.

48.3 CONCLUSION AND FUTURE PERSPECTIVES

A range of sensitive techniques are now available for BRSV diagnosis, *in vivo* or *postmortem*, and the key to successful detection remains the selection of animals to be sampled. An increasingly available and successively cheaper novel technology brings new perspectives: high-throughput sequencing and new generation vaccines will probably influence the way of controlling and understanding BRSV. Deep sequencing will enable virus tracing during epidemics to identify routes of transmission between herds and consequently to develop appropriate biosecurity measures.

Research on virus persistence will clarify whether some animals reexcrete BRSV following a primary infection, and research on virulence factors might come to explain the varying degree of disease severity in different BRSV outbreaks. Furthermore, high-throughput sequencing, proteomics, and the development of multiplex, broad diagnostic tests will aid in the detection of unknown respiratory pathogens and coinfections involved in BRD, as well as in the mapping of protective immune effectors. Such results will guide the development of new medical measures of pathogen control.

Currently, novel differentiation of infected from vaccinated animals (DIVA) vaccines, in which one or several virus proteins are omitted and used for serological diagnosis, are under development. These can potentially be used to protect naïve animals (e.g., in control programs currently discussed in Scandinavian countries) and allow serological monitoring of virus spread, vaccine safety, and efficacy (e.g., duration of protection) under field conditions. DIVA ELISAs will have to be developed and evaluated for the detection of antibodies directed against omitted virus proteins in vaccine preparations used to immunize animals. These tests will probably be used to indicate unsatisfactory virological protection at the population level, due to lack of vaccine efficacy, short duration of vaccine-induced protection, or antigenic drift due to evolution of circulating BRSV strains.

REFERENCES

1. Nicholas, R.A. and Ayling, R.D., Mycoplasma bovis: Disease, diagnosis, and control. *Res Vet Sci* 74, 105, 2003.
2. Valarcher, J.F. and Hägglund, S., Viral respiratory diseases in cattle. In *World Buiatric Congress*, H. Navetat and F. Schelcher (eds.), pp. 384–397 Nice, France: WBC, October 15–19, 2006.
3. Babiuk, L.A., Lawman, M.J., and Ohmann, H.B., Viral-bacterial synergistic interaction in respiratory disease. *Adv Virus Res* 35, 219, 1988.
4. Valarcher, J.F. and Taylor, G., Bovine respiratory syncytial virus infection. *Vet Res* 38, 153, 2007.
5. International Committee on Taxonomy of Viruses. http://ictvonline.org/virusTaxonomy.asp (accessed on January 8, 2015).
6. Liljeroos, L., Krzyzaniak, M.A., Helenius, A. et al., Architecture of respiratory syncytial virus revealed by electron cryotomography. *Proc Natl Acad Sci USA* 110, 11133, 2013.
7. Kingsbury, D.W., Paramyxoviridae: The viruses and their replication. In *Virology*, 2nd edn., B.N. Fields and D.M. Knipe (eds.), 945pp. New York: Ravens Press, 1990.
8. Mastrangelo, P. and Hegele, R.G., The RSV fusion receptor: Not what everyone expected it to be. *Microbes Infect* 14, 1205, 2012.
9. Viuff, B., Tjornehoj, K., Larsen, L.E. et al., Replication and clearance of respiratory syncytial virus: Apoptosis is an important pathway of virus clearance after experimental infection with bovine respiratory syncytial virus. *Am J Pathol* 161, 2195, 2002.
10. Viuff, B., Uttenthal, A., Tegtmeier, C. et al., Sites of replication of bovine respiratory syncytial virus in naturally infected calves as determined by in situ hybridization. *Vet Pathol* 33, 383, 1996.

11. Valarcher, J.F., Bourhy, H., Lavenu, A. et al., Persistent infection of B lymphocytes by bovine respiratory syncytial virus. *Virology* 291, 55, 2001.
12. Stott, E.J. and Taylor, G., Respiratory syncytial virus. Brief review. *Arch Virol* 84, 1, 1985.
13. Furze, J., Wertz, G., Lerch, R. et al., Antigenic heterogeneity of the attachment protein of bovine respiratory syncytial virus. *J Gen Virol* 75, 363, 1994.
14. Grubbs, S.T., Kania, S.A., and Potgieter, L.N., Validation of synthetic peptide enzyme immunoassays in differentiating two subgroups of ruminant respiratory syncytial virus. *J Vet Diagn Invest* 13, 123, 2001.
15. Langedijk, J.P., Meloen, R.H., Taylor, G. et al., Antigenic structure of the central conserved region of protein G of bovine respiratory syncytial virus. *J Virol* 71, 4055, 1997.
16. Valarcher, J.F., Schelcher, F., and Bourhy, H., Evolution of bovine respiratory syncytial virus. *J Virol* 74, 10714, 2000.
17. McIntosh, K. and Chanock, R.M., Respiratory syncytial virus. In *Virology*, 2nd edn., B.N. Fields and D.M. Knipe (eds.), 1045pp. New York: Ravens Press, 1990.
18. Thomas, L.H., Stott, E.J., Collins, A.P. et al., Infection of gnotobiotic calves with a bovine and human isolate of respiratory syncytial virus. Modification of the response by dexamethasone. *Arch Virol* 79, 67, 1984.
19. Trudel, M., Nadon, F., Simard, C. et al., Comparison of caprine, human and bovine strains of respiratory syncytial virus. *Arch Virol* 107, 141, 1989.
20. Grubbs, S.T., Kania, S.A., and Potgieter, L.N., Prevalence of ovine and bovine respiratory syncytial virus infections in cattle determined with a synthetic peptide-based immunoassay. *J Vet Diagn Invest* 13, 128, 2001.
21. Stott, E.J., Thomas, L.H., Collins, A.P. et al., A survey of virus infections of the respiratory tract of cattle and their association with disease. *J Hyg* 85, 257, 1980.
22. Van der Poel, W.H., Kramps, J.A., Middel, W.G. et al., Dynamics of bovine respiratory syncytial virus infections: A longitudinal epidemiological study in dairy herds. *Arch Virol* 133, 309, 1993.
23. Van der Poel, W.H., Langedijk, J.P., Kramps, J.A. et al., Serological indication for persistence of bovine respiratory syncytial virus in cattle and attempts to detect the virus. *Arch Virol* 142, 1681, 1997.
24. Amanna, I.J. and Slifka, M.K., Mechanisms that determine plasma cell lifespan and the duration of humoral immunity. *Immunol Rev* 236, 125, 2010.
25. Larsen, L.E., Tjornehoj, K., and Viuff, B., Extensive sequence divergence among bovine respiratory syncytial viruses isolated during recurrent outbreaks in closed herds. *J Clin Microbiol* 38, 4222, 2000.
26. Larsen, L.E., Uttenthal, A., Arctander, P. et al., Serological and genetic characterisation of bovine respiratory syncytial virus (BRSV) indicates that Danish isolates belong to the intermediate subgroup: No evidence of a selective effect on the variability of G protein nucleotide sequence by prior cell culture adaption and passages in cell culture or calves. *Vet Microbiol* 62, 265, 1998.
27. Larsen, L.E., Bovine respiratory syncytial virus (BRSV): A review. *Acta Vet Scand* 41, 1, 2000.
28. Bryson, D.G., McFerran, J.B., Ball, H.J. et al., Observations on outbreaks of respiratory disease in housed calves–(1) Epidemiological, clinical and microbiological findings. *Vet Rec* 103, 485, 1978.

29. Schreiber, P., Matheise, J.P., Dessy, F. et al., High mortality rate associated with bovine respiratory syncytial virus (BRSV) infection in Belgian white blue calves previously vaccinated with an inactivated BRSV vaccine. *J Vet Med B, Infect Dis Vet Public Health* 47, 535, 2000.

30. Baker, J.C., Ames, T.R., and Markham, R.J., Seroepizootiologic study of bovine respiratory syncytial virus in a dairy herd. *Am J Vet Res* 47, 240, 1986.

31. Taylor, G., Bruce, C., Barbet, A.F. et al., DNA vaccination against respiratory syncytial virus in young calves. *Vaccine* 23, 1242, 2005.

32. Kimman, T.G., Westenbrink, F., Schreuder, B.E. et al., Local and systemic antibody response to bovine respiratory syncytial virus infection and reinfection in calves with and without maternal antibodies. *J Clin Microbiol* 25, 1097, 1987.

33. Kimman, T.G., Zimmer, G.M., Westenbrink, F. et al., Epidemiological study of bovine respiratory syncytial virus infections in calves: Influence of maternal antibodies on the outcome of disease. *Vet Rec* 123, 104, 1988.

34. Belknap, E.B., Baker, J.C., Patterson, J.S. et al., The role of passive immunity in bovine respiratory syncytial virus-infected calves. *J Infect Dis* 163, 470, 1991.

35. Verhoeff, J. and van Nieuwstadt, A.P., BRS virus, PI3 virus and BHV1 infections of young stock on self-contained dairy farms: Epidemiological and clinical findings. *Vet Rec* 114, 288, 1984.

36. Jolly, S., Detilleux, J., and Desmecht, D., Extensive mast cell degranulation in bovine respiratory syncytial virus-associated paroxystic respiratory distress syndrome. *Vet Immunol Immunopathol* 97, 125, 2004.

37. Kimman, T.G., Terpstra, G.K., Daha, M.R. et al., Pathogenesis of naturally acquired bovine respiratory syncytial virus infection in calves: Evidence for the involvement of complement and mast cell mediators. *Am J Vet Res* 50, 694, 1989.

38. Bryson, D.G., McFerran, J.B., Ball, H.J. et al., Observations on outbreaks of respiratory disease in housed calves--(2) Pathological and microbiological findings. *Vet Rec* 103, 503, 1978.

39. Larsen, L.E., Tjornehoj, K., Viuff, B. et al., Diagnosis of enzootic pneumonia in Danish cattle: Reverse transcription-polymerase chain reaction assay for detection of bovine respiratory syncytial virus in naturally and experimentally infected cattle. *J Vet Diagn Invest* 11, 416, 1999.

40. Elvander, M., Severe respiratory disease in dairy cows caused by infection with bovine respiratory syncytial virus. *Vet Rec* 138, 101, 1996.

41. Hägglund, S., Svensson, C., Emanuelson, U. et al., Dynamics of virus infections involved in the bovine respiratory disease complex in Swedish dairy herds. *Vet J* 172, 320, 2006.

42. Ohlson, A., Blanco-Penedo, I., and Fall, N., Comparison of Bovine coronavirus-specific and Bovine respiratory syncytial virus-specific antibodies in serum versus milk samples detected by enzyme-linked immunosorbent assay. *J Vet Diagn Invest* 26, 113, 2014.

43. Uttenthal, A., Larsen, L.E., Philipsen, J.S. et al., Antibody dynamics in BRSV-infected Danish dairy herds as determined by isotype-specific immunoglobulins. *Vet Microbiol* 76, 329, 2000.

44. Blodörn, K., Hägglund, S., Fix, J. et al., Vaccine safety and efficacy evaluation of a recombinant bovine respiratory syncytial virus (BRSV) with deletion of the SH gene and subunit vaccines based on recombinant human RSV proteins: N-nanorings, P and M2-1, in calves with maternal antibodies. *PLoS One* 9, e100392, 2014.

45. Schrijver, R.S., Langedijk, J.P., van der Poel, W.H. et al., Antibody responses against the G and F proteins of bovine respiratory syncytial virus after experimental and natural infections. *Clin Diagn Lab Immunol* 3, 500, 1996.

46. Hägglund, S., Hu, K., Vargmar, K. et al., Bovine respiratory syncytial virus ISCOMs-Immunity, protection and safety in young conventional calves. *Vaccine* 29, 8719, 2011.

47. Hägglund, S., Hu, K.F., Larsen, L.E. et al., Bovine respiratory syncytial virus ISCOMs—Protection in the presence of maternal antibodies. *Vaccine* 23, 646, 2004.

48. Klem, T.B., Tollersrud, T., Osteras, O. et al., Association between the level of antibodies in bulk tank milk and bovine respiratory syncytial virus exposure in the herd. *Vet Rec* 175, 47, 2014.

49. Elvander, M., Edwards, S., Naslund, K. et al., Evaluation and application of an indirect ELISA for the detection of antibodies to bovine respiratory syncytial virus in milk, bulk milk, and serum. *J Vet Diagn Invest* 7, 177, 1995.

50. Taylor, G., Thomas, L.H., Furze, J.M. et al., Recombinant vaccinia viruses expressing the F, G or N, but not the M2, protein of bovine respiratory syncytial virus (BRSV) induce resistance to BRSV challenge in the calf and protect against the development of pneumonic lesions. *J Gen Virol* 78(Pt 12), 3195, 1997.

51. Riffault, S., Meyer, G., Deplanche, M. et al., A new subunit vaccine based on nucleoprotein nanoparticles confers partial clinical and virological protection in calves against bovine respiratory syncytial virus. *Vaccine* 28, 3722, 2010.

52. Graham, D.A., Foster, J.C., Mawhinney, K.A. et al., Detection of IgM responses to bovine respiratory syncytial virus by indirect ELISA following experimental infection and reinfection of calves: Abolition of false positive and false negative results by pre-treatment of sera with protein-G agarose. *Vet Immunol Immunopathol* 71, 41, 1999.

53. Taylor, G., Rijsewijk, F.A., Thomas, L.H. et al., Resistance to bovine respiratory syncytial virus (BRSV) induced in calves by a recombinant bovine herpesvirus-1 expressing the attachment glycoprotein of BRSV. *J Gen Virol* 79 1759, 1998.

54. Glenn, G.M., Smith, G., Fries, L. et al., Safety and immunogenicity of a Sf9 insect cell-derived respiratory syncytial virus fusion protein nanoparticle vaccine. *Vaccine* 31, 524, 2013.

55. Hägglund, S., Hu, K., Blodörn, K. et al., Characterization of an experimental vaccine for bovine respiratory syncytial virus. *Clin Vaccine Immunol* 21, 997, 2014.

56. Timsit, E., Le Drean, E., Maingourd, C. et al., Detection by real-time RT-PCR of a bovine respiratory syncytial virus vaccine in calves vaccinated intranasally. *Vet Rec* 165, 230, 2009.

57. Valarcher, J.F., Bourhy, H., Gelfi, J. et al., Evaluation of a nested reverse transcription-PCR assay based on the nucleoprotein gene for diagnosis of spontaneous and experimental bovine respiratory syncytial virus infections. *J Clin Microbiol* 37, 1858, 1999.

58. Baker, J.C., Werdin, R.E., Ames, T.R. et al., Study on the etiologic role of bovine respiratory syncytial virus in pneumonia of dairy calves. *J Am Vet Med Assoc* 189, 66, 1986.

59. Tjornehoj, K., Uttenthal, A., Viuff, B. et al., An experimental infection model for reproduction of calf pneumonia with bovine respiratory syncytial virus (BRSV) based on one combined exposure of calves. *Res Vet Sci* 74, 55, 2003.

60. Taylor, G., Thomas, L.H., Wyld, S.G. et al., Role of T-lymphocyte subsets in recovery from respiratory syncytial virus infection in calves. *J Virol* 69, 6658, 1995.

61. Santangelo, P., Nitin, N., LaConte, L. et al., Live-cell characterization and analysis of a clinical isolate of bovine respiratory syncytial virus, using molecular beacons. *J Virol* 80, 682, 2006.

62. Hakhverdyan, M., Hagglund, S., Larsen, L.E. et al., Evaluation of a single-tube fluorogenic RT-PCR assay for detection of bovine respiratory syncytial virus in clinical samples. *J Virol Methods* 123, 195, 2005.

63. Socha, W. and Rola, J., Use of rapid human respiratory syncytial virus strip tests for detection of bovine respiratory syncytial virus in experimentally vaccinated calves. *Pol J Vet Sci* 15, 629, 2012.

64. Urban-Chmiel, R., Wernicki, A., Grooms, D.L. et al., Rapid detection of bovine respiratory syncytial virus in Poland using a human patient-side diagnostic assay. *Transbound Emerg Dis* 62, 407, 2015.

65. Vilcek, S., Elvander, M., Ballagi-Pordany, A. et al., Development of nested PCR assays for detection of bovine respiratory syncytial virus in clinical samples. *J Clin Microbiol* 32, 2225, 1994.

66. Boxus, M., Letellier, C., and Kerkhofs, P., Real time RT-PCR for the detection and quantitation of bovine respiratory syncytial virus. *J Virol Methods* 125, 125, 2005.

67. Socha, W. and Rola, J., Comparison of four RT-PCR assays for detection of bovine respiratory syncytial virus. *Pol J Vet Sci* 14, 449, 2011.

68. Timsit, E., Maingourd, C., Le Drean, E. et al., Evaluation of a commercial real-time reverse transcription polymerase chain reaction kit for the diagnosis of bovine respiratory syncytial virus infection. *J Vet Diagn Invest* 22, 238, 2010.

69. Thonur, L., Maley, M., Gilray, J. et al., One-step multiplex real time RT-PCR for the detection of bovine respiratory syncytial virus, bovine herpesvirus 1 and bovine parainfluenza virus 3. *BMC Vet Res* 8, 37, 2012.

70. Uehara, M., Matsuda, K., Sugano, M. et al., A new high-speed droplet-real-time polymerase chain reaction method can detect bovine respiratory syncytial virus in less than 10 min. *J Vet Med Sci* 76, 477, 2014.

71. Espinasse, J., Alzieu, J.P., Papageorgiou, C. et al., Use of transtracheal aspiration to identify pathogens in pneumonic calves. *Vet Rec* 129, 339, 1991.

72. Pringle, J.K., Viel, L., Shewen, P.E. et al., Bronchoalveolar lavage of cranial and caudal lung regions in selected normal calves: Cellular, microbiological, immunoglobulin, serological and histological variables. *Can J Vet Res* 52, 239, 1988.

49 Canine Distemper Virus

Rebecca P. Wilkes

CONTENTS

49.1 Introduction 421
    49.1.1 Classification 421
    49.1.2 Genome Organization, Biology, and Epidemiology 421
    49.1.3 Clinical Features and Pathogenesis 422
    49.1.4 Diagnosis 423
        49.1.4.1 Conventional Techniques 423
        49.1.4.2 Molecular Techniques 425
49.2 Methods 425
    49.2.1 Sample Preparation 425
    49.2.2 Detection Procedures 426
        49.2.2.1 Methods for Detection 426
        49.2.2.2 Methods for Genetic Lineage Typing 427
        49.2.2.3 Methods to Distinguish Vaccine Shedding from True Infection 427
49.3 Future Perspectives 427
References 428

49.1 INTRODUCTION

49.1.1 CLASSIFICATION

Canine distemper virus (CDV) is a member of the order *Mononegavirales*, family *Paramyxoviridae*, subfamily *Paramyxovirinae*, and genus *Morbillivirus*. It is closely related to viruses responsible for measles in humans, rinderpest (now extinct) in cattle, peste des petits ruminants in small ruminants, phocine distemper in seals, and cetacean morbilliviruses in dolphins, porpoises, and whales.[1–3]

49.1.2 GENOME ORGANIZATION, BIOLOGY, AND EPIDEMIOLOGY

CDV is an enveloped virus with a single-stranded, negative-sense RNA genome, 15,690 nucleotides long. The genomic RNA is enclosed by a helical nucleocapsid composed of the nucleocapsid (N) protein and is replicated by the viral polymerase (L) protein and its cofactor phosphoprotein (P). The P gene also encodes two nonstructural proteins (C and V). The N, P, and L proteins together with the viral RNA constitute the ribonucleoprotein (RNP) complex, which directs the synthesis of capped and polyadenylated mRNAs.[4] The viral envelope contains two glycoproteins, the fusion (F) and hemagglutinin (H) proteins, and a membrane-associated protein (matrix [M]), which links the envelope with the RNP. The H protein facilitates viral attachment to receptors on the host cell and is therefore the major determinant of viral tropism.[4] Neutralizing antibodies produced against H gene epitopes are important for protective immunity by preventing viral attachment to host cells. The F protein executes the fusion of the host cell plasma membrane and the viral envelope, resulting in entry of the viral RNP into the cytoplasm. The F protein also promotes fusion between host cell membranes[4,5]; however, experiments with recombinant viruses suggest that it is the H protein, not the F protein, that is actually the main determinant of fusion efficiency and, therefore, the cytopathogenicity determinant at least in vitro.[4]

Of all the proteins encoded by CDV, the H protein has the most variability.[4] In fact, compared to other morbilliviruses, the H protein of CDV has the greatest variability. This results in a very broad host range, which is not seen with other morbilliviruses. CDV infects several mammalian species in the order *Carnivora*, including members of the families *Canidae* (dog, fox, coyote, wolf), *Mustelidae* (ferret, badger, mink, weasel, otter, fisher), *Mephitidae* (skunk), *Procyonidae* (raccoon), Felidae (lion, tiger), *Ailuridae* (red panda), *Hyaenidae* (hyena), *Phocidae* (seal), *Ursidae* (bear), and *Viverridae* (palm civet, genets).[1,5–7] The virus also causes disease in noncarnivores, including collared peccaries (*Tayassu tajacu*)[8] and macaques (*Macaca*).[9,10]

Like other enveloped viruses, CDV is quickly inactivated in the environment, and routine disinfectants and cleaning readily abolish virus infectivity. Transmission mainly occurs by direct contact or via exposure to infectious aerosols. The virus is shed primarily by oronasal secretion, but the virus can be detected at high titers from all secretions and excretions, including urine.[1,5]

Transmission of CDV between different species has resulted in devastating epidemics. Domestic dogs have been implicated in the epidemic that led to the death of 20% of the Serengeti and Mara lions in the 1990s.[11] Terrestrial carnivores are thought to be the source of the CDV strain that killed thousands of Caspian seals (*Pusa caspica*) between 1997 and 2001,[12] and raccoons have been implicated in the transmission of CDV to animals in zoo collections and conservation parks.[7] CDV is considered a disease of global importance in both common and endangered carnivores.[13] CDV contributed to the endangerment of the North American black-footed ferret[14] and is an emerging disease in the endangered wild Amur tigers in the Russian Far East.[13]

Temporal fluctuations in disease prevalence have been observed, with increased frequency during the cold season. Dogs of any age are susceptible, but an age-related susceptibility to infection is evident because many adult dogs are protected by vaccine immunization. Disease tends to be more common in dogs 3–6 months old, correlating with the decline in maternally derived immunity during this period.[5] The infectious disease presents worldwide distribution and maintains a high incidence and high levels of lethality, with no specific treatment. Though intensive vaccination programs in developed countries have controlled the virus, there have been reports of increased activity of CDV worldwide.[6,15–20]

The use of molecular techniques facilitates in-depth study of CDV epidemiology and the dynamics of circulation of various CDV strains around the world in susceptible animals. The genetic diversity of the H gene allows identification of distinct genotypes, which cluster according to geographic patterns, irrespective of the species of origin.[5] These lineages have been defined based on the regions where they have been detected and include Asia-1, Asia-2, Europe, European Wildlife, Arctic-like, and Africa.[15,17–19,21] There have only been two (North) American lineages clearly defined, including America-1, which includes strains used in vaccine production in the 1950s and still in use today, and America-2. However, we recently sequenced CDV positive samples from the United States from 2010 to 2014, and based on our phylogenetic evaluation with previously identified strains from the United States, there appear to be at least four different "North American" genotypes currently circulating in the United States, including a new genotype of CDV we discovered circulating in the Southeast United States.[20] Recent reports suggest multiple genotypes are also circulating in South America, including two newly defined "South American" genotypes, designated South America-2[22] and South America-3.[19] The European lineage is also circulating in South America (including Brazil and Uruguay), resulting in a proposal to change the lineage name from Europe to Europe/South America-1.[19,22]

Some CDV strains seem to be more virulent or are associated with different tropism, but this relies on individual variations among the various strains rather than on peculiar properties inherent to a given CDV lineage.[5] Another research suggests that differences in neurovirulence are related to disease duration rather than strain variability.[23] Despite the presence of multiple genotypes, CDV is currently considered

a single serotype, as defined by polyclonal antisera.[2,5] The greatest genetic and antigenic diversity is between the vaccine strains (America-1 lineage) and the other CDV lineages (approximately 10% amino acid variability for H protein). Sera raised against field CDV isolates may have neutralizing titers up to 10-fold higher against the homologous virus than against vaccine strains. Although it is unlikely that such antigenic variation may affect the protection induced by vaccine immunization, there have been several reports of vaccine failures in multiple countries, including Japan,[24] Mexico,[25] Argentina,[26] the United States,[16] and Spain.[27] Many have questioned the efficacy of the vaccines against currently circulating field strains. It is possible that critical amino acid substitutions in key epitopes of the H protein may allow escape, particularly from the limited antibody repertoire of maternal origin of young unvaccinated pups, increasing the risk for infection by field CDV strains.[5] However, there has been no research published to date that definitively proves that the currently available vaccines are not protective.

49.1.3 CLINICAL FEATURES AND PATHOGENESIS

CDV infection produces a systemic disease. The incubation period may range from 1 to 4 weeks or more. Following aerosol infection, the virus replicates in macrophages and lymphocytes of the upper respiratory tract, with subsequent infection of local lymphoid tissues, including the tonsils. This leads to the dissemination of the virus by the lymphatics and blood to distant hematopoietic tissues and the onset of lymphopenia. This initial viremia results in clinical signs that may include fever, lethargy, anorexia, ocular and nasal discharge, and tonsillitis. The virus then infects all the lymphoid tissues including lymph nodes, spleen, thymus, mucosa-associated lymphoid tissues (MALTs), and bone marrow, resulting in severe immunosuppression. A robust antiviral immune response, particularly production of detectable antibodies to the viral glycoproteins,[28] may enable the infected animal to eliminate the virus at this point in the infection, but in animals with a weak response, a secondary viremia follows and is associated with a high fever and infection of epithelial cells of multiple organs throughout the body. CDV can be detected in cells of the respiratory system, gastrointestinal and urinary tract, endocrine system, lymphoid tissues, central nervous system, and vasculature including thrombocytes and different lymphoid cell subsets, as well as keratinocytes of the skin. Morbilliviruses are highly cell associated, but massive amounts of virus are released as cells succumb to infection; this is particularly true of lung or bladder epithelia.[6,29,30] Clinical signs relate to areas of virus replication and can include gastrointestinal signs (vomiting and diarrhea) and/or respiratory signs that may be complicated by secondary bacterial infection (purulent nasal discharge, coughing, dyspnea, pneumonia). Secondary bacterial infections are common as a result of the immunosuppression caused by the virus. Dogs may also display ocular disease, enamel hypoplasia, or cutaneous lesions, such as a pustular dermatitis of the inner thighs, ventral abdomen, and inner surface of the ear pinnae or hyperkeratosis of the foot pads and nose. The initial clinical signs disappear, but the virus persists for extended periods in the uvea, neurons, or urothelium and in some skin areas (foot pads).[5]

Disease may progress to neurologic signs. Some dogs develop acute demyelinating lesions with CNS involvement occurring parallel to systemic disease and succumb to encephalitis early (within 2–4 weeks of infection). Some have subacute to chronic demyelinating encephalitis, with CNS signs appearing weeks to months following systemic signs, or even occurring without any previous history of systemic signs.[28] Once in the CNS, CDV can invade multiple cells, resulting in acute or chronic demyelination. Perivascular cuffing with lymphocytes, plasma cells, and monocytes in the demyelinated areas may be present in chronic cases. Both white and gray matter may be affected. A rare outcome of CDV infection is chronic encephalomyelitis of older dogs, termed "old dog encephalitis,"[5] but this manifestation has not been well defined. Most infected dogs have CNS involvement, but only about 30% of dogs (those with low or no antibody response) exhibit neurological signs.[23,27] Typical signs include myoclonus of the temporal and forelimb muscles and chewing-gum movements of the mouth or may include circling, head tilt, nystagmus, paresis or paralysis, seizures, and dementia/behavioral changes. CNS disease is typically progressive (i.e., cerebellar or vestibular signs progressing to tetraparesis or paralysis) and many dogs die. Some dogs may recover but likely display lifelong persistent residual signs such as myoclonus.[5,27,30,31] There have also been recent reports of unusual acute neurological signs as a result of CDV infection in vaccinated adult dogs. Signs include acute or peracute onset of cerebellar and/or vestibular signs and tetraparesis/plegia in the absence of systemic signs, with progression to severe generalized seizures that were unresponsive to anticonvulsant therapy.[27] In a separate study, demyelinating lesions were a constant finding and chronic encephalomyelitis was predominant in mature dogs with distemper encephalomyelitis that presented exclusively with neurological disease.[31] CDV should always be considered as an important differential diagnosis in dogs with progressive and multifocal neurological disease, even when the typical CDV systemic signs are not present, particularly in areas where the virus is endemic.[31]

49.1.4 Diagnosis

49.1.4.1 Conventional Techniques

A practical diagnosis of distemper is based on multiple systemic clinical signs (respiratory, gastrointestinal, neurological, or dermatological); however, dogs may lack the classical clinical presentation and/or signs may be confounded with other respiratory or enteric diseases. Therefore, laboratory confirmation is required for suspected cases.[31,32] Some antemortem clinical information is useful, such as the presence of lymphopenia and viral inclusions in white blood cells. Predominant lymphocytic pleocytosis in the CSF (>5 cells/μL) may suggest the diagnosis of CDV infection, but not all dogs infected with CDV will have changes in CSF, and none of these clinical findings are definitive.[33,34] Conclusive diagnosis can be difficult due to the unpredictable and variable course of distemper, for example, length of viremia, organ manifestation, and a lack of or delayed humoral and cellular immune responses.[35] There are multiple methods for detection in addition to molecular methods, and understanding what the tests detect and how they can be used are important for appropriate application for CDV diagnosis. Summaries of serological tests (Table 49.1) and virus detection methods (Table 49.2) are provided.

TABLE 49.1
Serological Tests for Canine Distemper Virus

Test	Sample Type	Limitations	Significance	Comments
IgM titer	Serum	Low or absent very early in infection	Signifies recent exposure	Does not differentiate between recent vaccine exposure and infection
Serum neutralization (SN)	Serum	Low or absent early in infection	Measures functional antibodies, including maternally derived antibodies	Results may vary depending on strain used in testing; single test does not differentiate vaccine response from infection
CSF IgG	CSF and serum recommended	Testing CSF alone may misdiagnose CDV	Measures local production of IgG in CNS as a result of CDV encephalitis	Must insure intact BBB and no blood contamination of CSF sample
Indirect immunofluorescence antibody assay (IFA)	Serum	May be negative early in infection; subjectivity in assay variability between labs	Measures all CDV specific antibodies, not just neutralizing ones	May be more sensitive than SN, but as with SN, single test cannot differentiate vaccine response from infection
ELISA (TiterCHEK)	Serum	Commercially available test used mainly for detection of protective titers from vaccination	Promoted for detection of protective antibody titer; may be negative in acute infection	Protective titers not well defined in the literature

TABLE 49.2

Detection Methods for Canine Distemper Virus

Test	Sample Type	Limitations	Significance
Virus isolation	Whole blood (EDTA), tissues, and swabs	Vero/SLAM cells have made efficient isolation possible, but it is limited to research or some diagnostic labs; requires special sample handling to maintain virus viability	Isolated virus is viable; if virus detected in antemortem sample, suggests animal is contagious
Direct immunofluorescence antibody assay (IFA)	Slide made from swabs and tissues	Poor sensitivity, particularly with mucosal samples taken 3 weeks or more postinfection	Positive result is conclusive, but many false negatives occur; some subjectivity with test evaluation and variation between labs
Immunochromatography	Conjunctival swab	Commercially available; not as sensitive as molecular detection	Can be used patient side; false-negative results may occur
Immunohistochemistry	Formalin-fixed paraffin-embedded tissues	Mainly for postmortem testing	Confirms pathological lesions are associated with CDV
Molecular detection of CDV RNA	Whole blood (EDTA), swabs, tissues, urine, and feces	Limited to research and diagnostic labs but is the most sensitive method for virus detection	Detects RNA; virus does not have to be viable for detection

Detection of CDV IgM antibodies is reliable for the identification of recent CDV infection or exposure. It is important to note that an IgM response will also be detected within a week after vaccination and persists for at least a month, whereas IgM titers persist for at least 3 months following natural infection.[36]

The determination of CDV neutralizing antibodies (serum neutralization [SN] assay) in serum or CSF may be helpful in some animals, but the results are variable and depend on the stage of the disease. Titers will be low or absent during the acute phase of the infection. In addition, a positive titer may be the result of a vaccine-induced immune response or the presence of maternally derived antibodies. The neutralization assay measures only the functional antibodies that bind to CDV hemagglutinin or fusion protein and interfere with virus uptake[37]; however, neutralizing activity against the America-1 lineage Onderstepoort strain, which is commonly used in these assays, may not correspond to neutralizing activity against field isolates and, therefore, may not accurately reflect the titer or protection.[35]

Additional serological tests to detect CDV-specific antibodies include ELISA and indirect immunofluorescence antibody assay (IFA).[38] IFA and whole virus ELISA measure not only functional antibodies but all CDV-specific antibodies. The ELISA method has been shown to be more sensitive than the SN assay, but as with the SN assay, a single test cannot discriminate a vaccine titer from a titer resulting from infection.[5,38] Serologic methods, including a commercially available ELISA, are routinely used and marketed to detect "protective" antibody titers in previously vaccinated dogs.[39]

Detection of immunoglobulin G (IgG) in CSF is an additional serological method that is used to diagnose canine distemper encephalitis (CDE). Local antibody responses to CDV occur in the CNS in cases of persistent infection. Transfer of blood proteins, including IgG antibodies, is limited by the blood–brain barrier (BBB) to maintain homeostasis.

Therefore, the amount of IgG in CSF is 1/256 to 1/2048 of that in serum in subjects with a normal BBB. CDV antibody titer in serum and CSF should be simultaneously examined for an accurate diagnosis of CDE by confirming local antibody production in the CNS. However, this ratio can be misleading if there is a nonspecific collapse of the BBB or an excess blood mixed with the CSF during sampling, which can result in misdiagnosis. Therefore, it is recommended that a reference antibody test (canine parvovirus or adenovirus) also be performed on the CSF sample to facilitate a more accurate diagnosis.[40] An increase in CDV IgG in the CSF, in the absence of IgG titers to parvovirus or adenovirus in the CSF, confirms a diagnosis of CDV encephalitis.

Other methods used to diagnose CDV involve detection of the virus itself. These methods include virus isolation, CDV antigen detection by an IFA, immunochromatography or immunohistochemistry, and molecular methods to detect CDV RNA. Virus isolation can best be made from buffy coat cells during the early phase of infection or from suspensions from macerated tissues (postmortem). This method involves cocultivation with mitogen-stimulated canine blood lymphocytes or the use of a marmoset lymphoid cell line, B95a.[1] Historically, this was an inefficient method because field strains do not replicate well in vitro and virus adaptation to cell culture is very time-consuming. However, with the engineering of Vero cells that express the canine signaling lymphocyte activation molecule (SLAM), which is the lymphotropic receptor (expressed on activated T and B cells, monocytes, and dendritic cells) used for viral attachment, efficient isolation of field strains is now possible.[41] CDV replication in cells usually induces formation of giant cells (syncytia) with intracytoplasmic and intranuclear eosinophilic inclusion bodies, both in vitro and in vivo (Figure 49.1).[5] Samples must be handled properly to maintain virus viability for isolation (kept cool and moist), and this can be difficult with an enveloped virus like CDV.

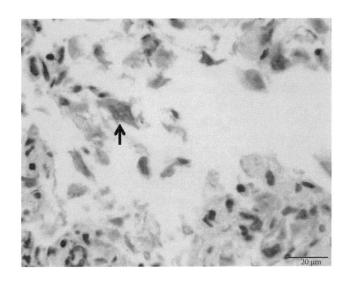

FIGURE 49.1 Raccoon, lung, H&E: multinucleated syncytial cell with eosinophilic intracytoplasmic inclusion body within alveoli (arrow). (Photo courtesy of Dr. Jenny Payette Pope, Department of Biomedical and Diagnostic Sciences, College of Veterinary Medicine, University of Tennessee, Knoxville, TN.)

IFA on conjunctival, nasal, and vaginal smears that are incubated with polyclonal or monoclonal anti-CDV antibodies marked with fluorescein has been widely used for virus detection. However, this method lacks sensitivity, particularly in subacute or chronic forms of the disease. For antemortem diagnosis, this method works best within the first 3 weeks after infection, when the virus is still present in epithelial cells associated with the mucous membranes.[5,38]

Immunohistochemistry represents a highly sensitive and specific method for the detection of CDV antigen in tissue obtained postmortem. In chronic CDV encephalomyelitis, viral antigen may disappear from the inflammatory demyelinating lesions due to the antiviral immune responses, and false-negative results have been reported.[31] This method is suitable only within limits for the diagnosis of distemper in living animals (i.e., from biopsy samples).[35]

In 2008, an immunochromatography rapid test was developed.[42] This commercially available test (Anigen Rapid CDV Ag Test Kit, Animal Genetics, Suwon, Korea) is fast and user-friendly but is not as sensitive for CDV diagnosis as molecular assays.[34]

49.1.4.2 Molecular Techniques

Use of a rapid detection method is desirable to properly manage CDV infection because of the contagious nature and the high mortality rates associated with this disease. The assay should not only be rapid, but it should be sensitive and specific as well to detect even small amounts of virus early in infection.[32] The sensitivity, specificity, and rapidity of molecular methods compared with conventional methods make molecular techniques preferable for CDV diagnosis.[32]

Molecular techniques are very sensitive, but sensitivity varies with sample source, viral nucleic acid extraction method, molecular platform (i.e., reverse transcription polymerase chain reaction [RT-PCR] vs. nested RT-PCR vs. real-time

RT-PCR), and choice of primers. Additionally, endogenous RNases and lack of accessibility of partially degraded RNA may influence the sensitivity of RT-PCR.[35]

Multiple assay platforms have been developed for CDV RNA detection. Several RT-PCR protocols were originally developed.[26,35,43] Nested RT-PCR protocols followed and were developed to increase sensitivity because detection of CDV with the RT-PCR was not satisfactory during the first and end stages of the infection[16,38,44,45]; however, nested PCR is inherently prone to contamination and time-consuming. More recently, real-time RT-PCR protocols have been published.[32,34] Real-time assays are faster and less prone to contamination, while maintaining the sensitivity or even increasing the sensitivity obtainable by nested protocols. Another major advantage of real-time RT-PCR is the ability to quantitate the viral load in clinical specimens, whereas conventional RT-PCR allows only qualitative analysis. This makes the real-time assay useful for pathogenesis studies.[32] Various ways these molecular assays are used for CDV diagnosis are described in the following section.

49.2 METHODS

49.2.1 Sample Preparation

Suitable samples for the molecular detection of CDV RNA include serum, whole blood, CSF, urine, saliva, nasal samples, rectal swabs, and conjunctival swabs.[22,32,35,43] Some studies suggest that all these sample types are useful, irrespective of the clinical form of the disease.[35,43] Other studies suggest that this is not actually the case and relate stage of infection to variation in the best sample type.[34,45] Conjunctival swabs have been determined to be the best sample for diagnosis of acute CDV in experimentally infected dogs, compared to blood, CSF, nasal irrigation solution, and urine,[45] whereas urine samples have been found to be more sensitive than serum or leukocytes and at least as sensitive, if not more sensitive, as CSF for detection of CDV RNA in dogs with neurological signs.[31,45] In a recent report of CDV-induced acute onset of neurological signs in adult dogs, RT-PCR on CSF was negative in two out of four cases.[27] Lack of amplification of CDV in the CSF does not exclude distemper, because CDV is not always present in the CSF, even when distemper encephalitis is confirmed by immunohistochemistry.[35] Urine is much easier to collect than CSF and has been demonstrated to be a suitable sample type in our laboratory as well.

We suggest the most logical approach to best sample selection involves determination of the stage of infection. We recommend whole blood during febrile periods (presence of viremia), nasal and conjunctival (for conjunctivitis) swabs for acute infection in dogs displaying respiratory signs, and rectal swabs or feces for dogs showing gastrointestinal signs. Urine is our specimen of choice for dogs with neurologic signs. It is important to note that of these sample types, urine has been associated the most with PCR inhibition (second only to formalin-fixed paraffin-embedded tissue)[46] and an extraction procedure approved for urine should be used to ensure removal of inhibitors. We routinely use the QIAamp

Viral RNA Mini Kit (Qiagen). Extraction protocols specific for swabs are also available. Some protocols are marketed for the extraction of DNA from swabs, but we have found that one of these protocols works very well for the extraction of CDV RNA (buccal swab protocol, QIAamp DNA Mini Kit, Qiagen). Two swabs can be processed together in the same tube. Tissues useful for postmortem detection include lymphatic (i.e., lymph nodes, spleen, tonsil), renal pelvis, bladder, lung, and brain tissues. Minced tissues can be mixed 1:1 with phosphate-buffered saline and macerated. Supernatant can then be used for extraction with a protocol for blood. Additionally, extraction protocols for small pieces of tissue also exist. There are numerous extraction kits, protocols, and methods. The most important thing to do is to ensure that whatever method is used, it consistently produces efficient quality and quantities of RNA for amplification.

49.2.2 Detection Procedures

49.2.2.1 Methods for Detection

The preferable molecular method for rapid and sensitive detection of CDV is real-time RT-PCR, which targets a conserved region of the genome. Lineages, while identified most commonly in areas where they are originally detected, can also be detected in distant areas[47] as a result of trade and global movement of animals.[48] Therefore, the use of a conserved region of the genome is necessary for adequate detection of different strains from around the world. The N gene has been used for primer production for most published assays because it has areas of high conservation[49]; however, considering that CDV is an RNA virus (lacks proofreading during replication), silent mutations do commonly occur. These changes are not limited only to highly variable regions such as the H gene but occur in the N gene as well, including the middle region that is considered the most conserved area.[49]

We routinely use a published real-time RT-PCR assay that detects the N gene in our lab.[32] Primers and probe for this assay are included in Table 49.3. Briefly, the SuperScript III Platinum One-Step qRT-PCR Kit (Invitrogen, Life Technologies, Grand Island, NY, USA) is used with 300 nM of each primer and 200 nM of probe in a 25 μL total volume

reaction with 5 μL RNA. Amplification is carried out with an initial RT step of 42°C for 30 min, followed by 95°C for 2 min and 45 cycles of 95°C for 15 s, 48°C for 1 min, and 60°C for 1 min. This assay successfully detected several strains that are currently circulating in the United States,[20] and the assay was originally validated with clinical samples from dogs in Italy, but the samples were not genotyped.[32] It is important to note that this assay failed to detect several Brazilian strains, which was thought to be the result of point mutations that were detected in the primer binding regions.[34]

Mutations in primer and probe binding regions may not necessarily inhibit detection (depending on where the mutations occur), but they will certainly affect the threshold cycle (Ct) value of a real-time RT-PCR assay. The Ct value is routinely used to correlate the significance of the positive result to clinical signs and also to distinguish modified live virus (MLV) vaccine shedding from true infection (see methods to determine vaccine shedding).

While real-time RT-PCR may be a preferred method, it does require expensive equipment and highly trained personnel, limiting the assay to diagnostic and research laboratories. Recently, we evaluated an RT-insulated isothermal (ii) PCR-based method to detect CDV. This platform is marketed for affordable, point-of-care molecular detection of CDV. The method uses a convective PCR platform, which uses natural thermal convection to drive PCR by providing a constant heat source at the bottom of the reaction tube. A temperature gradient is created through which the liquid is cycled, driving the PCR. This method provides natural flexibility in the annealing temperature and lessens the effect of base–pair mismatches on primer binding. RT-iiPCR is a useful technique for RNA viruses that commonly contain sequence variation between strains. However, due to concerns associated with specificity, this particular platform also incorporates a TaqMan-based probe in the reaction to increase test specificity. This specific assay was found to be just as sensitive and specific as a published real-time RT-PCR method for the detection of U.S. strains, while potentially having the added benefit of detecting a wider range of geographic lineages.[50] Additional testing with more lineages would confirm this.

An additional molecular method that has the benefit of use without the need for specialized equipment is RT loop-mediated

TABLE 49.3
Primer and Probe Sequences for Molecular Assays

Primer/Probe	Source	Sequence (5′-3′)	Location[a]	Application
CDV-F	Elia et al. [32]	AGCTAGTTTCATCTTAACTATCAAATT	905–931	Detection
CDV-R		TTAACTCTCCAGAAAACTCATGC	966–987	
CDV-Pb		FAM-ACCCAAGAGCCGGATACATAGTTTCAATGC-BHQ1	934–963	
CDV vaccine Fw	Wilkes et al. [20]	ATAATGATGTTATCATCAGYGATGAT	4390–4415	Differentiation from vaccine
CDV vaccine Rev		CTTGGTCCGATAATGATCAACC	4683–4662	strains (except Vanguard
CDV probe AM1		FAM-CTTAGTAGCAYTGCCCAAGATCCCTTGATC-BHQ1	4471–4500	vaccine) and genotyping (if combined with sequencing)

[a] Based on genome sequence for Onderstepoort strain.

isothermal amplification (RT-LAMP). RT-LAMP is an autocycling strand displacement DNA synthesis carried out using a reverse transcriptase and a DNA polymerase (Bst DNA polymerase) with a high strand displacement activity. A set of specific primers that can recognize a total of six distinct sequences in the target cDNA are used; thus, the technique amplifies the target cDNA sequence with a high selectivity and the reaction can be performed with just a water bath. The method described for CDV in the literature was developed with primers based on the N gene sequence from the America-1 lineage Onderstepoort strain.[51] The assay was tested with clinical whole blood samples from dogs in Korea, but it is unknown how well the assay works with all the various genotypes.

49.2.2.2 Methods for Genetic Lineage Typing

While the N gene has been the target of choice for CDV detection, the variable H gene is commonly used to genotype CDV strains.[15,48,52–54] Another area that is useful for genotyping is the M-F intergenic region, which also contains high variability between strains.[20,55] Several different molecular methods have been published for genetic lineage typing for epidemiological studies, including RT-PCRs with sequencing,[15,54] heminested protocols with genotype specific probes,[48] and real-time RT-PCR with sequencing.[20] For our method,[20] 5 μL RNA samples are run in 25 μL total volume reactions using a commercially available master mix for real-time RT-PCR (SuperScript III Platinum One-Step qRT-PCR Kit, Invitrogen, Life Technologies, Grand Island, NY, USA). The optimized mix contains 200 μM deoxyribonucleotide triphosphates, 3 mM $MgCl_2$, 200 nM of the 6-carboxyfluorescein (FAM)-labeled probe, and 300 nM of each primer (Table 49.3). One unit of RNase inhibitor (RNaseOUT, Invitrogen, Life Technologies, Grand Island, NY, USA) is also added to the mix. Samples are amplified in a real-time PCR machine with an RT step at 50°C for 30 min and activation step for the Hot Start Taq Polymerase at 95°C for 10 s, followed by 45 cycles of denaturation at 95°C for 5 s, annealing at 48°C for 1 min, and elongation at 60°C for 1 min. The probe detects America-1 vaccine strains (additional methods described in Section 49.2.2.3); wild-type strains can be detected by running the sample on a gel. The real-time PCR products can be sequenced for genotyping.[20]

49.2.2.3 Methods to Distinguish Vaccine Shedding from True Infection

Many of the methods used for genotyping can also be used for differentiation of America-1 lineage vaccine strains from field strains. All the currently used MLV vaccines contain an America-1 strain except one,[52] which contains a Rockborn-like strain[56] and resembles the more contemporary field strains. False-positive results may occur with the molecular detection methods by amplification of MLV vaccine shedding.[20,38,43] Vaccine virus may be detected from day 2 post-vaccination up to day 14 in multiple sample types, including PBMCs, ocular discharge, nasal discharge, saliva, feces, and urine.[44] There is a recombinant pox-vectored vaccine containing H and F genes

from an America-1 lineage strain that is available for CDV that is not shed by the vaccinate,[57] but it is more costly than MLV vaccines and not as widely used.[20] Differentiation tests are designed to take advantage of the sequence differences between the vaccine strains and wild-type strains.[58] Most of the assays used for differentiation are gel-based and use restriction fragment length polymorphism (RFLP).[26,53,59] Another assay uses separate primers for differentiation.[54] Most of these assays are a single RT-PCR and lack the necessary sensitivity for differentiation during acute stages of infection. This is particularly problematic for certain situations, such as outbreaks in shelter populations that are routinely vaccinated on intake.[20,60] A commercial laboratory has developed a quantitative RT-PCR (CDV Quant RealPCR™ Test, IDEXX) for differentiation. This quantitative information allows discrimination of vaccine interference from infection with a wild-type strain of distemper virus as the amount of virus present during infection is typically exponentially more than would be detected because of recent vaccination. This method cannot distinguish between the two when virus shedding is low (less than 1 million particles), which may occur during acute infection. Therefore, we developed a real-time RT-PCR with a probe that detects the America-1 lineage vaccines.[20] One of the other real-time PCR assays also has a method to differentiate between vaccine and field strains by using RFLP.[34] Differentiation of the Rockborn-like vaccine remains a problem and though one published assay has a way of distinguishing this vaccine from the Asia-1 lineage, the method will not work for all lineages.[55] Differentiation of this vaccine strain from field strains by sequencing works when all other methods fail.[20]

49.3 FUTURE PERSPECTIVES

Despite historically effective vaccination programs, CDV continues to be a worldwide problem for domestic dogs and wild carnivores. Several researchers have suggested the need for an updated vaccine for CDV. Vaccines strains have not changed since their development in the 1950s. However, there is currently no research to definitively substantiate the need for a new vaccine or to demonstrate what new strain or strains would be best if a new vaccine were made. Studies to determine how genetic differences influence antigenicity are warranted. It is important that any new vaccine developed be easily differentiated from field strains. Continual monitoring of field strains is needed to maintain assays with the ability to detect a broad range of genotypes. This is particularly important as global travel increases, with moving of animals and CDV strains into new areas.

Epidemiological evaluation has been made easier with the use of whole genome sequencing,[61] but full genome sequences are still needed for several genotypes. This molecular platform will likely provide data to produce improved primers and probes targeting different regions of the virus for CDV detection. As prices for whole genome sequencing continue to decrease, this method will potentially be used not only for research purposes but for diagnostics in the veterinary community as well.

REFERENCES

1. Appel, M.J. and Summers, B.A. Pathogenicity of morbilliviruses for terrestrial carnivores. *Vet Microbiol* **44**, 187–191 (1995).

2. Harder, T.C. and Osterhaus, A.D. Canine distemper virus—A morbillivirus in search of new hosts? *Trends Microbiol* **5**, 120–124 (1997).

3. Barrett, T. Morbillivirus infections, with special emphasis on morbilliviruses of carnivores. *Vet Microbiol* **69**, 3–13 (1999).

4. von Messling, V., Zimmer, G., Herrler, G., Haas, L., and Cattaneo, R. The hemagglutinin of canine distemper virus determines tropism and cytopathogenicity. *J Virol* **75**, 6418–6427 (2001).

5. Martella, V., Elia, G., and Buonavoglia, C. Canine distemper virus. *Vet Clin North Am Small Anim Pract* **38**, 787–797, vii–viii (2008).

6. Beineke, A., Puff, C., Seehusen, F., and Baumgartner, W. Pathogenesis and immunopathology of systemic and nervous canine distemper. *Vet Immunol Immunopathol* **127**, 1–18 (2009).

7. Kapil, S. and Yeary, T.J. Canine distemper spillover in domestic dogs from urban wildlife. *Vet Clin North Am Small Anim Pract* **41**, 1069–1086 (2011).

8. Appel, M.J. et al. Canine distemper virus infection and encephalitis in javelinas (collared peccaries). *Arch Virol* **119**, 147–152 (1991).

9. Sun, Z. et al. Natural infection with canine distemper virus in hand-feeding Rhesus monkeys in China. *Vet Microbiol* **141**, 374–378 (2010).

10. Sakai, K. et al. Lethal canine distemper virus outbreak in cynomolgus monkeys in Japan in 2008. *J Virol* **87**, 1105–1114 (2013).

11. Roelke-Parker, M.E. et al. A canine distemper virus epidemic in Serengeti lions (*Panthera leo*). *Nature* **379**, 441–445 (1996).

12. Wilson, S.C., Eybatov, T.M., Amano, M., Jepson, P.D., and Goodman, S.J. The role of canine distemper virus and persistent organic pollutants in mortality patterns of Caspian seals (*Pusa caspica*). *PLoS One* **9**, e99265 (2014).

13. Seimon, T.A. et al. Canine distemper virus: An emerging disease in wild endangered Amur tigers (*Panthera tigris altaica*). *MBio* **4, e00410** (2013).

14. Wisely, S.M., Buskirk, S.W., Fleming, M.A., McDonald, D.B., and Ostrander, E.A. Genetic diversity and fitness in black-footed ferrets before and during a bottleneck. *J Hered* **93**, 231–237 (2002).

15. Mochizuki, M., Hashimoto, M., Hagiwara, S., Yoshida, Y., and Ishiguro, S. Genotypes of canine distemper virus determined by analysis of the hemagglutinin genes of recent isolates from dogs in Japan. *J Clin Microbiol* **37**, 2936–2942 (1999).

16. Kapil, S. et al. Canine distemper virus strains circulating among North American dogs. *Clin Vaccine Immunol* **15**, 707–712 (2008).

17. Gamiz, C. et al. Identification of a new genotype of canine distemper virus circulating in America. *Vet Res Commun* **35**, 381–390 (2011).

18. Budaszewski Rda, F. et al. Genotyping of canine distemper virus strains circulating in Brazil from 2008 to 2012. *Virus Res* **180**, 76–83 (2014).

19. Espinal, M.A., Diaz, F.J., and Ruiz-Saenz, J. Phylogenetic evidence of a new canine distemper virus lineage among domestic dogs in Colombia, South America. *Vet Microbiol* **172**, 168–176 (2014).

20. Wilkes, R.P., Sanchez, E., Riley, M.C., and Kennedy, M.A. Real-time reverse transcription polymerase chain reaction method for detection of Canine distemper virus modified live vaccine shedding for differentiation from infection with wild-type strains. *J Vet Diagn Invest* **26**, 27–34 (2014).

21. Woma, T.Y., van Vuuren, M., Bosman, A.M., Quan, M., and Oosthuizen, M. Phylogenetic analysis of the haemagglutinin gene of current wild-type canine distemper viruses from South Africa: Lineage Africa. *Vet Microbiol* **143**, 126–132 (2010).

22. Panzera, Y. et al. Evidence of two co-circulating genetic lineages of canine distemper virus in South America. *Virus Res* **163**, 401–404 (2012).

23. Bonami, F., Rudd, P.A., and von Messling, V. Disease duration determines canine distemper virus neurovirulence. *J Virol* **81**, 12066–12070 (2007).

24. Lan, N.T. et al. Comparative analyses of canine distemper viral isolates from clinical cases of canine distemper in vaccinated dogs. *Vet Microbiol* **115**, 32–42 (2006).

25. Simon-Martinez, J., Ulloa-Arvizu, R., Soriano, V.E., and Fajardo, R. Identification of a genetic variant of canine distemper virus from clinical cases in two vaccinated dogs in Mexico. *Vet J* **175**, 423–426 (2008).

26. Calderon, M.G. et al. Detection by RT-PCR and genetic characterization of canine distemper virus from vaccinated and non-vaccinated dogs in Argentina. *Vet Microbiol* **125**, 341–349 (2007).

27. Galan, A. et al. Uncommon acute neurologic presentation of canine distemper in 4 adult dogs. *Can Vet J* **55**, 373–378 (2014).

28. Rima, B.K., Duffy, N., Mitchell, W.J., Summers, B.A., and Appel, M.J. Correlation between humoral immune responses and presence of virus in the CNS in dogs experimentally infected with canine distemper virus. *Arch Virol* **121**, 1–8 (1991).

29. von Messling, V., Milosevic, D., and Cattaneo, R. Tropism illuminated: Lymphocyte-based pathways blazed by lethal morbillivirus through the host immune system. *Proc Natl Acad Sci USA* **101**, 14216–14221 (2004).

30. Carvalho, O.V. et al. Immunopathogenic and neurological mechanisms of canine distemper virus. *Adv Virol* **2012**, 163860 (2012).

31. Amude, A.M., Alfieri, A.A., and Alfieri, A.F. Antemortem diagnosis of CDV infection by RT-PCR in distemper dogs with neurological deficits without the typical clinical presentation. *Vet Res Commun* **30**, 679–687 (2006).

32. Elia, G. et al. Detection of canine distemper virus in dogs by real-time RT-PCR. *J Virol Methods* **136**, 171–176 (2006).

33. Amude, A.M., Alfieri, A.A., and Alfieri, A.F. Clinicopathological findings in dogs with distemper encephalomyelitis presented without characteristic signs of the disease. *Res Vet Sci* **82**, 416–422 (2007).

34. Fischer, C.D. et al. Detection and differentiation of field and vaccine strains of canine distemper virus using reverse transcription followed by nested real time PCR (RT-nqPCR) and RFLP analysis. *J Virol Methods* **194**, 39–45 (2013).

35. Frisk, A.L., Konig, M., Moritz, A., and Baumgartner, W. Detection of canine distemper virus nucleoprotein RNA by reverse transcription-PCR using serum, whole blood, and cerebrospinal fluid from dogs with distemper. *J Clin Microbiol* **37**, 3634–3643 (1999).

36. Blixenkrone-Moller, M. et al. Studies on manifestations of canine distemper virus infection in an urban dog population. *Vet Microbiol* **37**, 163–173 (1993).

37. Perrone, D., Bender, S., and Niewiesk, S. A comparison of the immune responses of dogs exposed to canine distemper virus (CDV)—Differences between vaccinated and wild-type virus exposed dogs. *Can J Vet Res* **74**, 214–217 (2010).

38. Jozwik, A. and Frymus, T. Comparison of the immunofluorescence assay with RT-PCR and nested PCR in the diagnosis of canine distemper. *Vet Res Commun* **29**, 347–359 (2005).

39. Gray, L.K., Crawford, P.C., Levy, J.K., and Dubovi, E.J. Comparison of two assays for detection of antibodies against canine parvovirus and canine distemper virus in dogs admitted to a Florida animal shelter. *J Am Vet Med Assoc* **240**, 1084–1087 (2012).

40. Soma, T. et al. Canine distemper virus antibody test alone increases misdiagnosis of distemper encephalitis. *Vet Rec* **173**, 477 (2013).

41. Seki, F., Ono, N., Yamaguchi, R., and Yanagi, Y. Efficient isolation of wild strains of canine distemper virus in Vero cells expressing canine SLAM (CD150) and their adaptability to marmoset B95a cells. *J Virol* **77**, 9943–9950 (2003).

42. An, D.J., Kim, T.Y., Song, D.S., Kang, B.K., and Park, B.K. An immunochromatography assay for rapid antemortem diagnosis of dogs suspected to have canine distemper. *J Virol Methods* **147**, 244–249 (2008).

43. Shin, Y.J. et al. Comparison of one-step RT-PCR and a nested PCR for the detection of canine distemper virus in clinical samples. *Aust Vet J* **82**, 83–86 (2004).

44. Kim, Y.H., Cho, K.W., Youn, H.Y., Yoo, H.S., and Han, H.R. Detection of canine distemper virus (CDV) through one step RT-PCR combined with nested PCR. *J Vet Sci* **2**, 59–63 (2001).

45. Saito, T.B. et al. Detection of canine distemper virus by reverse transcriptase-polymerase chain reaction in the urine of dogs with clinical signs of distemper encephalitis. *Res Vet Sci* **80**, 116–119 (2006).

46. Buckwalter, S.P. et al. Inhibition controls for qualitative real-time PCR assays: Are they necessary for all specimen matrices? *J Clin Microbiol* **52**, 2139–2143 (2014).

47. Martella, V. et al. Heterogeneity within the hemagglutinin genes of canine distemper virus (CDV) strains detected in Italy. *Vet Microbiol* **116**, 301–309 (2006).

48. Martella, V. et al. Genotyping canine distemper virus (CDV) by a hemi-nested multiplex PCR provides a rapid approach for investigation of CDV outbreaks. *Vet Microbiol* **122**, 32–42 (2007).

49. Yoshida, E. et al. Molecular analysis of the nucleocapsid protein of recent isolates of canine distemper virus in Japan. *Vet Microbiol* **59**, 237–244 (1998).

50. Wilkes, R.P. et al. Rapid and sensitive detection of canine distemper virus by one-tube reverse transcription-insulated isothermal polymerase chain reaction. *BMC Vet Res* **10**, 213 (2014).

51. Cho, H.S. and Park, N.Y. Detection of canine distemper virus in blood samples by reverse transcription loop-mediated isothermal amplification. *J Vet Med B Infect Dis Vet Public Health* **52**, 410–413 (2005).

52. Demeter, Z., Palade, E.A., Hornyak, A., and Rusvai, M. Controversial results of the genetic analysis of a canine distemper vaccine strain. *Vet Microbiol* **142**, 420–426 (2010).

53. Di Francesco, C.E. et al. Detection by hemi-nested reverse transcription polymerase chain reaction and genetic characterization of wild type strains of Canine distemper virus in suspected infected dogs. *J Vet Diagn Invest* **24**, 107–115 (2012).

54. Yi, L. et al. Development of a combined canine distemper virus specific RT-PCR protocol for the differentiation of infected and vaccinated animals (DIVA) and genetic characterization of the hemagglutinin gene of seven Chinese strains demonstrated in dogs. *J Virol Methods* **179**, 281–287 (2012).

55. Chulakasian, S. et al. Multiplex Amplification Refractory Mutation System Polymerase Chain Reaction (ARMS-PCR) for diagnosis of natural infection with canine distemper virus. *Virol J* **7**, 122 (2010).

56. Martella, V. et al. Lights and shades on an historical vaccine canine distemper virus, the Rockborn strain. *Vaccine* **29**, 1222–1227 (2011).

57. Paoletti, E. Applications of pox virus vectors to vaccination: An update. *Proc Natl Acad Sci USA* **93**, 11349–11353 (1996).

58. Si, W., Zhou, S., Wang, Z., and Cui, S.J. A multiplex reverse transcription-nested polymerase chain reaction for detection and differentiation of wild-type and vaccine strains of canine distemper virus. *Virol J* **7**, 86 (2010).

59. Wang, F. et al. Differentiation of canine distemper virus isolates in fur animals from various vaccine strains by reverse transcription-polymerase chain reaction-restriction fragment length polymorphism according to phylogenetic relations in china. *Virol J* **8**, 85 (2011).

60. Lechner, E.S. et al. Prevalence of protective antibody titers for canine distemper virus and canine parvovirus in dogs entering a Florida animal shelter. *J Am Vet Med Assoc* **236**, 1317–1321 (2010).

61. Marcacci, M. et al. Whole genome sequence analysis of the arctic-lineage strain responsible for distemper in Italian wolves and dogs through a fast and robust next generation sequencing protocol. *J Virol Methods* **202**, 64–68 (2014).

50 Hendra Virus

Dongyou Liu

CONTENTS

50.1 Introduction 431
    50.1.1 Classification...431
    50.1.2 Morphology and Genome Organization 432
    50.1.3 Biology and Epidemiology 433
    50.1.4 Clinical Features and Pathogenesis 433
    50.1.5 Diagnosis 434
    50.1.6 Treatment and Prevention 434
50.2 Methods 435
    50.2.1 Sample Preparation... 435
    50.2.2 Detection Procedures.. 435
        50.2.2.1 Conventional RT-PCR.. 435
        50.2.2.2 Real-Time RT-PCR 435
50.3 Conclusion 435
References...435

50.1 INTRODUCTION

Hendra virus (HeV) is a zoonotic pathogen that was first identified in Hendra, a suburb of Brisbane, Queensland, Australia, in 1994. Commonly present in fruit bats (flying foxes) without overt clinical symptoms, HeV is capable of causing acute, rapidly progressive, and fatal disease in horses and humans. Since its initial description, HeV has continued to emerge in Queensland and northern New South Wales, Australia. By June 2014, a total of 50 sporadic HeV outbreaks have been recorded involving horses as the intermediate host, resulting in the death or euthanasia of 83 horses (with a mortality rate of 75%) and 7 human cases (including four casualties) after exposure to infected horses.

50.1.1 CLASSIFICATION

HeV is classified in the genus *Henipavirus*, subfamily *Paramyxovirinae*, family *Paramyxoviridae*, order *Mononegavirales*.

The order *Mononegavirales* covers viruses with linear, monopartite, single-stranded, negative-sense RNA genomes, which are organized into five families: *Bornaviridae* (containing spherical virions of 70–130 nm in diameter), *Filoviridae* (containing filamentous virions of 790–970 nm in length), *Nyamiviridae* (containing spherical virions of 100–130 nm in diameter), *Paramyxoviridae* (containing spherical virions of about 150 nm in diameter), and *Rhabdoviridae* (containing bullet-shaped virions of 180 nm in length and 75 nm in width). Some of the best-known or most deadly animal and human pathogens are found within the order *Mononegavirales*,

including Ebolavirus, Marburgvirus (*Filoviridae*), measles virus, mumps virus, Newcastle disease virus, rinderpest virus, *Rubulavirus* (*Paramyxoviridae*), rabies virus, and vesicular stomatitis virus (*Rhabdoviridae*).

The family *Paramyxoviridae* is divided into two subfamilies: *Paramyxovirinae* and *Pneumovirinae*. Whereas the subfamily *Paramyxovirinae* consists of eight genera (*Aquaparamyxovirus*, *Avulavirus*, *Ferlavirus*, *Henipavirus*, *Morbillivirus*, *Respirovirus*, *Rubulavirus*, and type species Tupaia paramyxovirus (TPMV)-like viruses), the subfamily *Pneumovirinae* comprises of two genera (*Pneumovirus* and *Metapneumovirus*).

Representing a relatively new addition to the family *Paramyxoviridae*, the genus *Henipavirus* contains three recognized species, HeV (the type species), Nipah virus (NiV), and Cedar virus (CedPV), as well as a few tentative viruses [1].

HeV (originally referred to as equine morbillivirus) was discovered in September 1994 in Hendra, a suburb of Brisbane, Queensland, Australia, as the causal agent of a zoonotic disease that led to the deaths of 13 horses and a trainer at a training complex [2,3]. By June 2014, a total of 50 HeV outbreaks have occurred in Queensland and neighboring New South Wales, Australia, resulting in 83 horse deaths and 4 human casualties. Bats of the order *Chiroptera* appear to act as the natural host/reservoir for HeV, which may spread from fruit bats to horses and then from horses to humans in occasional "spillover" events [4,5].

NiV was found in April 1999 in the village of Sungai Nipah, Malaysia, as the cause of a neurological and respiratory disease of pigs and humans, involving 257 human cases (including 105 human deaths) and the culling of one million pigs.

FIGURE 50.1 Schematic presentation of Hendra virus genome. The 3′ leader (55 nucleotide [nt]) and 5′ trailer (33 nt) regions of *Henipavirus* genome serve as promoters for both transcription and replication. The coding sequence of each gene is flanked by untranslated regions that contain a gene-start sequence (12 nt) and a gene-end sequence (12 nt). Immediately following the gene-end sequence is a trinucleotide non-coding intergenic region (GAA) that precedes the gene-start sequence of the following gene. Besides the P, the *Henipavirus* P gene encodes at least three nonstructural proteins (C, V, and W). While the V and W proteins are generated by RNA editing, the C protein is encoded by a second open reading frame (ORF) that initiates 23 nt downstream of the translational initiation site for the P ORF. An additional accessible ORF within the P gene is identified as the small basic protein.

Since then, NiV outbreaks have been also reported in Bangladesh and India. It appears that NiV naturally occurs in bats of the order *Chiroptera*, domestic animals (pigs, cattle, and goats), and humans may acquire the virus after ingestion of date palm sap or other foods that are contaminated with bat saliva or urine. In addition, NiV transmissions from domestic animals to humans or from humans to humans are also possible [1].

CedPV was first observed in pteropid urine in 2009 in Queensland, Australia, while undertaking research on HeV. Despite its similarity to both HeV and NiV, CedPV does not induce disease in laboratory animals that are usually susceptible to paramyxoviruses. To date, CedPV has been largely detected in bats of the order *Chiroptera* [6].

Three tentative henipaviruses (BatPV/Eid.hel/GH10/2008, BatPV/Eid.hel/GH45/2008, and BatPV/Eid.hel/GH48/2008) were identified from feces of African straw-colored fruit bats (*Eidolon helvum*) roosting in zoological garden in the center of Kumasi, Ghana. Preliminary data suggest a limited risk of exposure to virus from bat feces, as virus RNA concentrations are low based on quantitative reverse transcription (RT)-PCR assessment [7]. In addition, molecular or serological evidence points out the possible existence of HeV-like RNA or proteins in bat bush meat from Congo, fruit bats *Pteropus rufus*, *Eidolon dupreanum*, *Rousettus madagascariensis*, and *Rousettus leschenatli* and *Myotis* spp. microbats in China [8].

50.1.2 MORPHOLOGY AND GENOME ORGANIZATION

Morphologically, HeV is an enveloped, pleomorphic (spherical or filamentous) particle of approximately 150–200 nm in diameter, which is covered by surface projections (spikes) of 10–18 nm. The spikes are composed of attachment glycoprotein (G) and fusion (F) protein. Beneath the envelope is the matrix (M) protein, which encases the nucleocapsid core (also referred to as the ribonucleoprotein [RNP] complex) consisting of the nucleocapsid (N) protein–encapsidated viral genome and the viral RNA-dependent RNA polymerase (RdRp) complex formed by the large (L) protein and phosphoprotein (P).

The genome of HeV is a nonsegmented, negative-sense, single-stranded RNA of 18,234 nucleotides (nt) in length, which contains six open reading frames (ORFs) in the order of 3′-N-P-M-F-G-L-5′. In all, the HeV genome encodes six major structural proteins (N protein, P, M protein, F protein, G, and L protein or RNA polymerase) and three nonstructural

proteins (C, V, and W) as well as a small basic (SB) protein (Figure 50.1 and Table 50.1) [9–11].

The N protein (also known as nucleocapsid) homomultimerizes and binds to the P (via its carboxy terminal domain) and viral genomic RNA (via its amino terminal domain), forming the nucleocapsid core. This helps protects the viral RNA genome and position the polymerase on the template.

The P is produced from the unedited P gene and serves as a scaffold between the L protein and the encapsidated genome. RNA editing of the P gene consists of an insertion of one or two guanine nucleotides, with the +1G edited RNA giving rise to the V protein, and the +2G edited RNA giving rise to the W protein. The C protein is produced from leaky scanning of the P mRNA, via a second ORF that initiates 23 nt downstream of the translational initiation site for the P ORF, while an additional accessible ORF between the C and V proteins in the P gene generates the SB protein. The P and V proteins are localized in the cytoplasm, whereas the W protein is present in the nucleus. Functionally, the V and W proteins inhibit host immune response to infection by blocking the type I interferon (IFN) signaling pathway (through interaction with host STAT1 and STAT2 and thus inhibition of their

TABLE 50.1

Hendra Virus Genes and Encoded Proteins

Gene	mRNA (nt)	5′-UTR (nt)	3′-UTR (nt)	Protein	Protein (aa)	Protein Mass (kDa)
N	2224	57	568	Nucleocapsid	532	58
P	2698	105	469	Phosphoprotein	707	78
V				V	457	50
W				W	448	49
C				C	166	19
M	1359	100	200	Matrix	352	40
F	2331	272	418	Fusion	546	60
G	2564	233	516	Glycoprotein	604	67
L	6955	153	67	RNA polymerase	2244	257

Sources: Smith, I.L. et al., Hendra virus, in: Liu, D. (ed.), *Molecular Detection of Human Viral Pathogens*, Taylor & Francis/CRC Press, Boca Raton, FL, 2010, pp. 471–478; Uniprot, http://www.uniprot.org/uniprot/; accessed on June 12, 2015; Wang, L.F. et al., *J. Virol.*, 74(21), 9972, 2000.

Note: aa, amino acid; kDa, kilodalton; nt, nucleotide; UTR, untranslated region.

phosphorylation and subsequent nuclear translocation) as well as by efficiently blocking the type II IFN signaling pathway. The C protein is also a weak antagonist of IFN and a regulator of viral RNA synthesis [12].

The M protein recruits cellular host factors for viral budding and provides rigidity and structure to the virion through interactions with the cytoplasmic tail of the F protein, the RNP complex, and the inner surface of the virion envelope. During bud formation, the M protein associates at the inner side of the plasma membrane of infected cells.

The F protein is a type I transmembrane protein that forms trimers on the cell surface and mediates the fusion of virus and host cell membranes during infection. The F protein is synthesized as an inactive precursor, F0, which is subsequently cleaved at amino acid K109 into the active subunits F1 and F2 by the host cellular protease cathepsin L after endocytosis of the virion. Both the F and G proteins are required for fusion [13].

The attachment G (also known as glycoprotein G) is a type II membrane protein that lacks hemagglutination and neuraminidase activities, forms tetramers upon cell surface expression, and binds to the protein-based receptor molecule EphrinB2 on neurons, smooth muscle, and endothelial cells surrounding small arteries [14]. The binding of glycoprotein G to the receptor induces a conformational change that allows the F protein to trigger virion/cell membrane fusion [15].

The L protein (also known as RNA-directed RNA polymerase L, large structural protein, replicase, transcriptase) displays RdRp, mRNA guanylyltransferase, mRNA (guanine-N(7)-)-methyltransferase, and poly(A) synthetase activities that are crucial for viral RNA synthesis (including initiation, elongation, and termination of both mRNA transcription and genome replication). The L protein interacts with the P to form the viral RdRp complex, which associates with the N protein–encapsidated genome to form the RNP complex [12].

50.1.3 Biology and Epidemiology

Upon entry into the host, HeV attaches to host cell surface receptors (such as EphrinB2) through glycoprotein G. This triggers F protein to undergo a conformational cascade, leading to membrane fusion between the virion and host cell and subsequent release of the nucleocapsid into the cytoplasm. The viral RdRp binds the encapsidated genome at the 3′ leader region and sequentially transcribes each gene by recognizing start and stop signals flanking viral genes, producing positive (+)-sense antigenomes and then negative (−)-sense genomes and capped and polyadenylated mRNA. Subsequent translation of viral genomes into nucleoprotein and other viral proteins facilitates the formation of newly formed RNP complex, which interacts with the M protein under the plasma membrane and initiates budding to release the virion.

HeV naturally occurs in fruit bats (or flying foxes) of *Pteropus* spp., which serve as the reservoir host and show no recognizable clinical symptoms despite having detectable virus in the blood, urine, feces, saliva, placental material, aborted fetuses, and birthing fluids. Of the four flying fox

species found in Australia, that is, the spectacled flying fox (*Pteropus conspicillatus*), the little red flying fox (*Pteropus scapulatus*), the black flying fox (*Pteropus alecto*), and the gray-headed flying fox (*Pteropus poliocephalus*), all have neutralizing antibodies against HeV. Transmission among flying foxes likely takes place within the roost, where uninfected animals acquire the virus through exposure to urine or feces matter excreted by other infected animals [16].

As the spillover host for HeV, horses may become infected through contact with pasture or feed contaminated with flying fox urine, feces, and plant materials masticated by bats. Horse-to-horse transmission is possible when uninfected horses have direct contact with infectious bodily fluids such as blood, urine, saliva, or nasal discharge from an infected horse or surfaces or equipment contaminated with infectious material [17].

Human infection with HeV mostly results from exposure to mucous membranes and nonintact skin to nasal and respiratory secretions, blood, and urine from infected horses. Participation in autopsy of infected animals or direct close contact with infected horses creates opportunity for exposure via droplets, cuts, or abrasions. Human-to-human or bat-to-human transmission has yet to be described.

Besides horses and humans, HeV appears to be infective to a number of mammals including African green monkeys, pigs, cats, guinea pigs, hamsters, and ferrets. However, experimentally infected dogs, chickens, rabbits, rats, and mice are asymptomatic, in spite of the fact that HeV-infected dogs, rabbits, and rats may mount a neutralizing antibody response [18].

HeV is sensitive to desiccation, temperature, and pH, but can persist in the environment (such as nasal secretions) for several days under optimum conditions.

50.1.4 Clinical Features and Pathogenesis

With an incubation period (i.e., the time from exposure to the appearance of the first clinical signs) of 5–16 days, HeV infection may result in a range of clinical symptoms in infected hosts.

In horses, HeV tends to induce severe respiratory (e.g., increased respiratory and heart rate and labored breathing) and neurological (e.g., ataxia, head tilt, facial nerve paralysis, and encephalitis) signs along with high fever, facial swelling, depression, weakness, anorexia, and nasal discharge. Histological examination reveals interstitial pneumonia with alveolar edema associated with hemorrhage, alveolar thrombosis, and necrosis, in addition to necrosis of the walls of small blood vessels. Virus is present in the lung, spleen, kidney, brain, lymph nodes, blood, saliva, and urine. There is evidence that HeV may appear in nasopharyngeal secretions at least 2 days before the onset of clinical signs [19].

In pigs, oronasal or intranasal inoculation of HeV leads to respiratory symptoms and mild neurological disease. The virus is found in nasal, oral, and ocular swabs as well as in feces [20].

In cats, subcutaneous, oral, or intranasal administration of HeV results in acute febrile disease with pulmonary signs (e.g., severe dyspnea, open-mouth breathing, pulmonary edema, necrosis of alveolar walls, and vascular lesions), similar to those observed in horses and humans. Virus is found in the lung, brain, spleen, kidney, rectum, trachea, bladder, heart muscle, liver lymph nodes, pleural fluid, urine, and blood, although the presence of virus in the brain is not accompanied by neurological signs. Cat-to-cat transmission may occur when housed in the same cage, and the cat's urine is infective to horses. However, no horse-to-cat transmission has been reported [21].

In pteropid bats, HeV infection is asymptomatic although specific antibodies are detectable. In some animals, histological lesions are present and immunostaining for HeV in endothelial cells of some animals is positive [16].

In African green monkeys (*Chlorocebus aethiops*), intratracheal inoculation of HeV results in a severe systemic disease, with resemblance to that observed in humans. Clinical signs range from decreased activity, depression, severe respiratory and neurological disease, generalized vasculitis, to death (nearly 100% in 8.5 days). Postmortem examination reveals pulmonary edema, meningeal hemorrhage and edema, endothelial syncytia, and vasculitis. Virus is detectable in the kidneys, heart, brain, gall bladder, stomach, sex organs, skeletal muscles, plasma, and whole blood [22]. Overall, the observed clinical signs observed in African green monkeys resemble those observed in human cases [23].

In humans, HeV infection begins with influenza-like illness (e.g., headache, high fever, and drowsiness) and progresses to interstitial pneumonia, encephalitis, seizures, segmental myoclonus, hypertension, tachycardia, areflexia, hypotonia, convulsion, coma, and death [36]. Histopathological examination reveals disseminated endothelial-cell involvement with vasculitis (in the brain, lung, heart, and kidneys), arterial thrombosis, ischemia, small necrotic or vacuolar plaques (in the cerebrum and cerebellum), and mild meningeal and brain parenchymal inflammation. Virus is detectable in CNS tissues (e.g., neurons, ependyma, and blood vessels), type II pneumocytes, pulmonary blood vessels, glomeruli and renal tubules, and blood vessels [24].

HeV demonstrates a predominantly respiratory or neurological tropism, with the ability to spread to other organs by the hematogenous route and form syncytia in vascular epithelial tissues. Assisted by its V and W as well as C proteins, HeV limits the induction of type I IFN in mammalian host cells and evades host immune surveillance, so that the virus can replicate inside the host cell and spread to other cells unhindered [25,26].

50.1.5 Diagnosis

Considering that HeV initially induces flulike symptoms followed by respiratory and neurological signs at the late stage, differential diagnosis should include a large number of viral and bacterial infections. Under the circumstance, use of laboratory methods (e.g., virus isolation, electron microscopy, immunohistochemistry, serological assays, and nucleic acid detection) is essential for achieving a precise and timely diagnosis [27].

Virus isolation: HeV is a Biosafety Level 4 (BSL 4) organism and a category C priority agent in the National Institute of Allergy and Infectious Diseases (NIAID) Biodefense Research Agenda. Isolation of HeV therefore requires the highest level of containment (BSL 4). Vero E6 (ATCC C1008) or Vero (ATCC CCL81) (African green monkey kidney), RK13 (rabbit kidney), MDBK (bovine kidney), LLC-MK2 (monkey kidney), BHK (baby hamster kidney), Hep-2 and HeLa cells, and embryonated chicken eggs may be utilized for HeV isolation from nasal swabs and throat swabs and the blood, urine, lung, liver, spleen, kidney, and lymph nodes of infected horses, from serum, urine, or nasopharyngeal aspirates (NPA), or from the kidney, spleen, lung, and brain of infected humans. Typically, 200–500 μL of serum, urine, or NPA (diluted in media) or tissue (kidney, spleen, lung, and brain) suspension (10% w/v, prepared in media) is clarified and filtered through a 0.2 μm filter and placed onto cell monolayers of Vero E6 (ATCC C1008) or Vero cells (ATCC CCL81) and incubated for 1 h at 37°C. Additional media is furbished and incubated at 37°C for 5 days. On Vero monolayer, HeV often produces large syncytia containing multiple nuclei. Two more passages are undertaken to ensure the success of isolation [27].

Electron microscopy: HeV particles can be observed by negative contrast electron microscopy. In addition, the location of HeV within tissues and cells can be pinpointed by immunoelectron microscopy.

Immunohistochemistry: Immunohistochemical detection of viral antigen in formalin-fixed tissues is useful for retrospective diagnosis of HeV.

Serological assays: Several serological assays have been used for detection of HeV antigens and antibodies [28,29]. These include neutralization test, indirect immunofluorescence assays (IFA), and enzyme-linked immunoassay (ELISA). Demonstration of a fourfold rise in antibody titer confirms a recent infection. In particular, ELISA allows efficient screening of large numbers of samples, especially during epidemiological and surveillance studies, while IFA is useful for testing small numbers of samples. Neutralization test is considered as the reference standard for definitive or confirmatory diagnosis.

Nucleic acid detection: A variety of molecular protocols has been developed for rapid and sensitive detection and quantitation of HeV RNA. These include conventional RT-PCR and real-time RT-PCR based on SYBR Green dye or TaqMan probe [30]. Additionally, loop-mediated isothermal amplification-lateral flow dipstick (LAMP–LFD) procedure offers a valuable approach for HeV identification [16].

50.1.6 Treatment and Prevention

Currently, no antiviral drugs are approved for animal or human infections with HeV. A recombinant human neutralizing monoclonal antibody targeting the ephrin receptor binding site has shown promise for HeV treatment and may be administered to humans following high-risk exposures to HeV. In addition, poly(I)-poly ($C_{12}U$)—a strong inducer of IFN—heptad peptides corresponding to the C-terminal heptad repeat of the F protein, and RNA interference technology are being evaluated for the treatment of HeV infection [1].

Vaccination with a recombinant G surface antigen of HeV has resulted in the production of high levels of protective antibodies against HeV challenge in animals [31]. This vaccine (trade name Equivac HeV®) is available in Australia since November 2012 for horses.

Furthermore, HeV vaccine based on the soluble G-glycoproteins has been developed by NIAID-supported researchers. This vaccine is capable of inducing strong neutralizing antibodies in animals [32].

50.2 METHODS

50.2.1 SAMPLE PREPARATION

Blood, nasal swabs, and urine from live animals and the brain, kidney, lung, and spleen from dead animals are suitable for HeV diagnosis. Fresh samples are sent chilled and not frozen to the laboratory.

QIAamp viral RNA mini extraction kit (Qiagen) is used to extract viral RNA from serum, nasopharyngeal aspirate (NPA), and cerebrospinal fluid (CSF), while QIAshredder and RNeasy mini kit (Qiagen) are employed for RNA extraction from tissues.

50.2.2 DETECTION PROCEDURES

50.2.2.1 Conventional RT-PCR

Halpin et al. [33] described a RT-PCR with primers (forward, 5′ GGC TAC AAC GAG AAA TTT GTG 3′; reverse, 5′ TTC TAG CAT TGT CCT TGG GAT 3′) for specific amplification of a 200 bp fragment from the M protein gene of HeV.

For RT, 2.5 μL of template RNA is annealed with 10 pmol of reverse-sense primer in a total volume of 5 μL at 65°C for 10 min. The mixture is then chilled on ice. RT reagents are added to give a total volume of 20.5 μL containing synthesis buffer (250 mM Tris–HCl, 200 mM KCl, 2.5 mM MgCl$_2$, 2.5% Tween 20, v/v, pH 8.3), 100 mM dithiothreitol, 1 mM (each) deoxynucleoside triphosphates (dATP, dCTP, dGTP, dTTP), 20 U of RNase inhibitor and 50 U of expand reverse transcriptase (expand RT preamplification system, Boehringer Mannheim). This mixture is incubated at 42°C for 60 min and then chilled on ice.

PCR mixture (50 μL) is made up of synthesis buffer, 2.5 mM of MgCl$_2$, 10 pmol of each primer, 2 μL of the reverse-transcribed product, and 1.5 U *Tth* of DNA polymerase (Biotech International).

Amplification is performed in DNA Thermal Cycler 480 (Perkin Elmer) with the following programs: initial 94°C for 2 min and 30 cycles of denaturation (94°C, 1 min), primer annealing (45°C, 1 min), and extension (72°C, 1 min).

PCR product is separated by electrophoresis in 1.2% agarose gel prepared in 0.5× Tris/Borate/EDTA (TBE) buffer containing 0.5 μg/mL of ethidium bromide at 80 V for 60 min and visualized on a UV transilluminator.

50.2.2.2 Real-Time RT-PCR

Smith et al. [34] developed a TaqMan-based assay for real-time RT-PCR detection of HeV, with primers (forward, 5′-CTTCGACAAAGACGGAACCAA-3′, nt 5755-5775; reverse, 5′-CCAGCTCGTCGGACAAAATT-3′, nt 5823-5804) and probe [5′-(FAM) TGGCATCTTTCATGCTCCATCTCGG (TAMRA)-3′, nt 5778–5802] from a conserved region of the M gene.

RT-PCR mixture (50 μL) is composed of 1× TaqMan buffer A (50 mM KCl, 10 mM Tris–HCl, 0.01 mM EDTA, 60 nM ROX passive reference [pH 8.3], 5.5 mM MgCl$_2$, 300 μM of each of dATP, dCTP and dGTP, 600 μM dUTP), 150 nM of probe and 300 nM of each of forward and reverse primers, 12.5 U Multiscribe RT, 20 U RNase inhibitor and 1.25 U Taq Gold polymerase (Applied Biosystems), and 5 μL extracted RNA.

Amplification and detection are performed on Applied Biosystems 7700 machine with the following cycling parameters: 48°C for 30 min and 40 cycles of 95°C for 15 s and 60°C for 1 min.

50.3 CONCLUSION

HeV is an emerging zoonotic pathogen with the ability to evade host immune surveillance and replicate within and spread between the host cells undisturbed [35]. Not surprisingly, the outcome of HeV infection in horses and humans as well as other mammals is predictably grim [36]. Due to the noncharacteristic clinical symptoms caused by HeV, it is important to use laboratory means for its diagnosis. Negating the shortcomings of traditional procedures, RT-PCR has proven to be a powerful tool in the identification and quantification of HeV from various specimens. Given the mobility of its reservoir host (fruit bats), the abundance of its intermediate host (horses), and the breadth of its potential hosts, HeV is poised to extend its current geographic and species boundaries and pose increasing risk to human population and animal husbandry. Continuing efforts in anti-HeV drug and vaccine research and development together with refinement in diagnostic methodologies will be crucial to stop the spread of HeV outbreaks.

REFERENCES

1. Aguilar HC, Lee B. Emerging paramyxoviruses: Molecular mechanisms and antiviral strategies. *Expert Rev Mol Med.* 2011;13:e6.
2. Murray K et al. A morbillivirus that caused fatal disease in horses and humans. *Science.* 1995;268:94–97.
3. Selvey LA et al. Infection of humans and horses by a newly described morbillivirus. *Med J Aust.* 1995;162:642–645.
4. Plowright RK et al. Urban habituation, ecological connectivity and epidemic dampening: The emergence of Hendra virus from flying foxes (*Pteropus* spp.). *Proc Biol Sci.* 2011;278:3703–3712.
5. Mahalingam S et al. Hendra virus: An emerging paramyxovirus in Australia. *Lancet Infect Dis.* 2012;12:799–807.

6. Marsh GA et al. Cedar virus: A novel Henipavirus isolated from Australian bats. *PLoS Pathog*. 2012;8(8):e1002836.

7. Drexler JF et al. Henipavirus RNA in African bats. *PLoS One*. 2009;4(7):e6367.

8. Hayman DTS et al. Evidence of henipavirus infection in West African fruit bats. *PLoS One*. 2008;3(7):e2739.

9. Wang LF et al. The exceptionally large genome of Hendra virus: Support for creation of a new genus within the family *Paramyxoviridae*. *J Virol*. 2000;74:9972–9979.

10. Marsh GA et al. Genome sequence conservation of Hendra virus isolates during spillover to horses, Australia. *Emerg Infect Dis*. 2010;16:1767–1769.

11. Smith IL, Smith GA, Broos A. Hendra virus. In: Liu D. (ed.), *Molecular Detection of Human Viral Pathogens*. Taylor & Francis/CRC Press, Boca Raton, FL, 2010, pp. 471–478.

12. Uniprot. http://www.uniprot.org/uniprot/; accessed on June 12, 2015.

13. Popa A, Pager CT, Dutch RE. C-terminal tyrosine residues modulate the fusion activity of the Hendra virus fusion protein. *Biochemistry*. 2011;50:945–952.

14. Bonaparte MI et al. Ephrin-B2 ligand is a functional receptor for Hendra virus and Nipah virus. *Proc Natl Acad Sci USA*. 2005;102:10652–10657.

15. Bowden TA et al. Dimeric architecture of the Hendra virus attachment glycoprotein: Evidence for a conserved mode of assembly. *J Virol*. 2010;84:6208–6217.

16. Field H et al. Hendra virus infection dynamics in Australian fruit bats. *PLoS One*. 2011;6(12):e28678.

17. Marsh GA et al. Experimental infection of horses with Hendra Virus/Australia/Horse/2008/Redlands. *Emerg Infect Dis*. 2011;17:2232–2238.

18. Rockx B, Winegar R, Freiberg AN. Recent progress in henipavirus research: Molecular biology, genetic diversity, animal models. *Antiviral Res*. 2012;95(2):135–149.

19. Ball MC, Dewberry TD, Freeman PG, Kemsley PD, Poe I. Clinical review of Hendra virus infection in 11 horses in New South Wales, Australia. *Aust Vet J*. 2014;92(6):213–218.

20. Li M, Embury-Hyatt C, Weingartl HM. Experimental inoculation study indicates swine as a potential host for Hendra virus. *Vet Res*. 2010;41:33.

21. Weingartl HM, Berhane Y, Czub M. Animal models of henipavirus infection: A review. *Vet J*. 2009;181:211–220.

22. Williamson MM, Torres-Velez FJ. Henipavirus: A review of laboratory animal pathology. *Vet Pathol*. 2010;47:871–880.

23. Rockx B et al. A novel model of lethal Hendra virus infection in African green monkeys and the effectiveness of ribavirin treatment. *J Virol*. 2010;84:9831–9839.

24. Playford EG et al. Human Hendra virus encephalitis associated with equine outbreak, Australia, 2008. *Emerg Infect Dis*. 2010;16:219–223.

25. Shaw ML. Henipaviruses employ a multifaceted approach to evade the antiviral interferon response. *Viruses*. 2009;1:1190–1203.

26. Virtue ER, Marsh GA, Wang LF. Interferon signaling remains functional during henipavirus infection of human cell lines. *J Virol*. 2011;85:4031–4034.

27. Wang LF, Daniels P. Diagnosis of henipavirus infection: Current capabilities and future directions. *Curr Top Microbiol Immunol*. 2012;359:179–196.

28. Kaku Y et al. Antigen capture ELISA system for henipaviruses using polyclonal antibodies obtained by DNA immunization. *Arch Virol*. 2012;157(8):1605.

29. McNabb L et al. Henipavirus microsphere immuno-assays for detection of antibodies against Hendra virus. *J Virol Methods*. 2014;200:22–28.

30. Feldman K et al. Design and evaluation of consensus PCR assays for henipaviruses. *J Virol Methods*. 2009;161:52–57.

31. Pallister J et al. A recombinant Hendra virus G glycoprotein-based subunit vaccine protects ferrets from lethal Hendra virus challenge. *Vaccine*. 2011;29:5623–5630.

32. Broder CC et al. A treatment for and vaccine against the deadly Hendra and Nipah viruses *Antiviral Res*. 2013;100:8–13.

33. Halpin K et al. Isolation of Hendra virus from pteropid bats: A natural reservoir of Hendra virus. *J Gen Virol*. 2000;81:1927–1932.

34. Smith IL, Halpin K, Warrilow D, Smith GA. Development of a fluorogenic RT-PCR assay (TaqMan) for the detection of Hendra virus. *J Virol Methods*. 2001;98:33–40.

35. Rodriguez JJ, Wang LF, Horvath CM. Hendra virus V protein inhibits interferon signaling by preventing STAT1 and STAT2 nuclear accumulation. *J Virol*. 2003;77:11842–11845.

36. Wong KT et al. Human Hendra virus infection causes acute and relapsing encephalitis. *Neuropathol Appl Neurobiol*. 2009;35:296–305.

51 Menangle Virus

Timothy R. Bowden

CONTENTS

51.1 Introduction 437
    51.1.1 Classification and Morphology..437
    51.1.2 Biology and Epidemiology 438
    51.1.3 Clinical Features and Pathogenesis 440
        51.1.3.1 Pigs...440
        51.1.3.2 Humans 441
        51.1.3.3 Fruit Bats...441
    51.1.4 Diagnosis 441
        51.1.4.1 Conventional Techniques 441
        51.1.4.2 Molecular Techniques 442
51.2 Methods 442
    51.2.1 Sample Preparation...442
    51.2.2 Detection Procedures...443
        51.2.2.1 Procedure 443
51.3 Conclusions and Future Perspectives...444
Acknowledgment 444
References..445

51.1 INTRODUCTION

Menangle virus, a recent addition to the family *Paramyxoviridae*, subfamily *Paramyxovirinae*, was isolated in 1997 from stillborn piglets at a large commercial piggery in New South Wales, Australia, during the investigation of an outbreak of severe reproductive disease, which persisted from April to September of that year.[1] The index piggery, which housed 2600 sows in four separate breeding units, was located approximately 60 km southwest of Sydney on a property that was adjacent to the Nepean River.[2,3] The disease was characterized by a reduction in both the farrowing rate and the number of live piglet births per litter, occasional abortions, and an increase in the proportion of mummified and stillborn piglets, some of which had deformities.[1,2,4] Although Menangle virus was only ever isolated from affected stillborn piglets, subsequent seroepidemiological investigations led to the implementation of a successful eradication program,[3] the identification of a likely natural host (fruit bats, also known as flying foxes, in the genus *Pteropus*),[1] and the unexpected realization that the virus had infected two piggery workers, causing severe influenza-like illness and a rash.[5]

51.1.1 CLASSIFICATION AND MORPHOLOGY

Paramyxoviruses are a diverse group of enveloped viruses that infect vertebrates, primarily mammals and birds, as well as reptiles and fish. The family *Paramyxoviridae* is divided into two subfamilies, *Paramyxovirinae* and *Pneumovirinae*, the member species of which can be distinguished on the basis of ultrastructure, genome organization, sequence relatedness of the encoded proteins, antigenic cross-reactivity, and biological properties of the attachment proteins (presence or absence of hemagglutinating and neuraminidase activities).[6,7] *Pneumovirinae* contains two genera, *Pneumovirus* and *Metapneumovirus*, whereas *Paramyxovirinae* is separated into seven genera, *Rubulavirus*, *Avulavirus*, *Respirovirus*, *Aquaparamyxovirus*, *Ferlavirus*, *Henipavirus*, and *Morbillivirus*, the type species of which are mumps virus, Newcastle disease virus, Sendai virus, Atlantic salmon paramyxovirus, Fer de Lance paramyxovirus, Hendra virus, and measles virus, respectively.

Following its original isolation and propagation in cell culture, Menangle virus was shown to be a member of the subfamily *Paramyxovirinae* by virtue of its ultrastructure as determined by electron microscopy (Figure 51.1).[1] Virions in this subfamily are typically 150–350 nm in diameter and are usually spherical in shape, although pleomorphic and filamentous forms are often observed.[6,7] Within each virion is the ribonucleoprotein core, a single strand of nonsegmented, negative-sense RNA, which is tightly bound along its entire length by nucleocapsid (N) proteins such that the genome is rendered insensitive to attack by nucleases.[8–10] The ribonucleoprotein complex, or nucleocapsid, is 19 ± 4 nm in diameter and has a helical symmetry with a pitch of 5.8 ± 0.4 nm,[1] thereby resulting in its characteristic "herringbone" appearance. Also associated with the nucleocapsid is the viral RNA-dependent RNA polymerase, which consists of two protein subunits: L and P. Surrounding the core is a lipid envelope containing two surface glycoproteins, which are visualized by electron

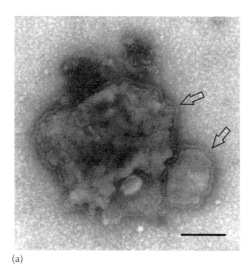
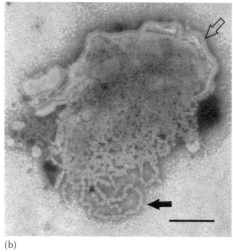

(a)   (b)

FIGURE 51.1 Transmission electron micrographs of Menangle virus negatively stained with 2% phosphotungstic acid (pH 6–8). (a) An intact virion with single fringe of surface projections (open arrows) extending from the viral envelope; bar, 100 nm. (b) A disrupted virion, viral envelope with surface projections (open arrow) and ribonucleoprotein complex or nucleocapsid (solid arrow); bar, 100 nm. (Adapted from Philbey, A.W. et al., *Emerg. Infect. Dis.*, 4, 269, 1998; Electron micrographs kindly provided by the AAHL Biosecurity Microscopy Facility, CSIRO, Australia.)

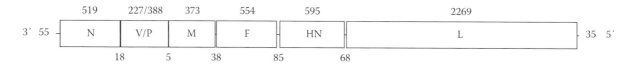

FIGURE 51.2 Genome organization of Menangle virus. The full-length genome (15,516 nucleotides) comprises six genes, each of which is drawn to scale. Numbers above each gene indicate the length, in amino acids, of the encoded proteins. Numbers below the gene boundaries indicate the length, in nucleotides, of the intergenic regions. Numbers immediately to the left and right of the genome indicate the length, in nucleotides, of the extragenic leader and trailer sequences, respectively. Note that unedited V/P gene transcripts encode the V protein, whereas insertion of two nontemplated G residues at the editing site (not shown) results in synthesis of the P gene mRNA. Note also that the negative-sense genome is depicted rather than the complementary antigenome, which is in the same sense as the viral mRNA molecules. GenBank accession number AF326114.

microscopy as an external fringe of projections 17 ± 4 nm long.[1] One glycoprotein, HN for rubulaviruses, avulaviruses, respiroviruses, aquaparamyxoviruses, and ferlaviruses, H for morbilliviruses, and G for henipaviruses, mediates attachment of the virion to the cellular receptor; the other glycoprotein, F, is responsible for mediating fusion between the virion envelope and the cell plasma membrane, a process that delivers the viral genome into the cell cytoplasm where viral transcription and replication occur.[7] Additionally, located on the inner surface of the viral envelope is a nonglycosylated matrix (M) protein, which is thought to play a central role in coordinating virion assembly and budding.[7]

Characterization and analysis of the complete genome sequence (15,516 nucleotides), and the encoded proteins, subsequently established that Menangle virus should be classified as a new member of the genus *Rubulavirus*.[11,12]

51.1.2 BIOLOGY AND EPIDEMIOLOGY

As for most other members of the genus *Rubulavirus*, there are six transcriptional units, in the order 3'-N-V/P-M-F-HN-L-5', which are separated by intergenic regions of variable length

and sequence (Figure 51.2). Although believed to transcribe each gene sequentially from the 3' end of the genome, the viral polymerase does not always reinitiate RNA synthesis at every gene junction following release of the newly synthesized mRNA. Instead, the polymerase detaches from the genome template, at which point it must reenter the genome at its 3' terminus for further transcription to take place. Consequently, an mRNA gradient is produced such that genes located closest to the 3' terminus of the genome are expressed at greater levels than their downstream neighbors.[7,13,14]

In contrast to the N, M, F, HN, and L genes, all of which encode single proteins using individual open reading frames, the V/P gene encodes at least two different proteins from two or more overlapping reading frames using a mechanism known as RNA editing, during which the polymerase adds one or more nontemplated G residues, cotranscriptionally, to a percentage of the newly synthesized mRNAs.[7] As such, two or more different transcripts are synthesized, which code for proteins that possess common N-termini but different C-termini, resulting from a translational shift downstream of the editing site. For Menangle virus, as for other rubulaviruses[7] and ferlaviruses,[15] faithful gene transcripts encode the V protein, whereas the predominant editing event (the insertion of two nontemplated G residues at the editing site) yields the P mRNA, which codes for the P protein.[11] Significantly, many of the *Paramyxovirinae* accessory gene products, which are generated using a variety of distinct mechanisms, have important roles such as regulating viral RNA synthesis, mediating virulence *in vivo* and/or counteracting the induction of an interferon-mediated antiviral state by host cells following infection.[7]

Several novel findings were also apparent, the most significant of which would appear to be the limited sequence homology of the deduced Menangle virus HN protein when compared to attachment proteins of other *Paramyxovirinae*.[11] Although the predicted structure of Menangle virus HN was most similar to other rubulavirus HN proteins, it lacked the majority of the amino acids that are considered critical determinants of both sialic acid binding and hydrolysis. This unexpected finding is unique among all known rubulavirus, avulavirus, respirovirus, aquaparamyxovirus, and ferlavirus HN proteins, with the exception of that from the closely related Tioman virus,[16] another rubulavirus that was later isolated from fruit bats in Malaysia during the search for the natural reservoir of Nipah virus.[17] This lack of conservation in functional amino acids suggests that the Menangle and Tioman virus attachment proteins, in contrast to those from all other known members of the rubulavirus, avulavirus, respirovirus, aquaparamyxovirus, and ferlavirus genera, are unlikely to use sialic acid as a cellular receptor, a prediction that in part would explain the observation made at the time of its initial isolation that Menangle virus is nonhemadsorbing and nonhemagglutinating using erythrocytes from various species.[1] The uniqueness of the Menangle and Tioman virus HN proteins is further highlighted by the apparent marked divergence in their evolutionary development in comparison to the N, P, M, F, and L proteins, the significance of which is not yet understood.[11,12,16]

The isolation of Menangle virus in cell culture provided a means of detecting Menangle virus–specific neutralizing antibodies in sera, thus enabling seropositive animals to be identified. Sera from pigs that had been collected during and prior to 1996, as well as samples collected as late as February 1997, before the onset of clinical disease, were all seronegative for antibodies against the new virus. However, 58 of 59 samples collected from three of the four breeding units in late May, immediately preceding the onset of disease in these units, had high neutralizing antibody titers (256–4096) against Menangle virus.[3] Subsequent epidemiological investigations using archival sera collected during separate national surveys for unrelated infectious disease agents, trace forward and trace back piggeries, and pigs with and without reproductive disease elsewhere in Australia, revealed that infection with Menangle virus was confined to the index piggery and two associated piggeries, also within New South Wales, which were located several hundred kilometers away at Young and Trunkey Creek.[3] Neither of these piggeries was a breeding establishment. Instead, each was used for rearing batches of 12–14-week-old pigs, originating from the index piggery, until slaughter at 24 weeks of age.

Although the reproductive parameters had returned to normal by mid-September 1997, two separate cross-sectional studies, the first conducted during August and September 1997, the other in March 1998, subsequently revealed that the virus was being maintained in the affected piggeries by infection of successive batches of growing pigs as they lost their protective maternal antibody at about 12–14 weeks of age.[3] Whereas colostral antibodies had declined to almost undetectable levels by this age, 95% of pigs subsequently developed neutralizing antibody titers ≥128 by the time of slaughter, at 24–26 weeks of age, or by the time they entered the breeding herd as replacement gilts at 28 weeks of age.[3] Significantly, this immunity appeared to be protective because no further reproductive disease was seen after September 1997 and sows, which had produced affected litters during the outbreak, farrowed normal litters subsequently.[2,3]

Menangle virus was only ever isolated from the brain, heart, and lungs of affected stillborn piglets.[1,2] Thus, it was not determined how the virus was spread between pigs. However, serological monitoring of various groups of sentinel pigs provided circumstantial evidence regarding likely modes of transmission. Sentinel pigs placed into the grower and grower–finisher areas of one of the units during September and October 1997, respectively, when infection was known to be active, took up to 1 month to seroconvert, despite the fact that the sentinel pigs in the grower–finisher area were allowed free access to walkways, between pens, to maximize their exposure to other pigs.[3] This suggested that transmission between pigs was relatively slow and required close contact, and thus probably resulted from viral excretion in feces or urine rather than in respiratory aerosols.[3] Furthermore, 180 gilts and sows, which had been introduced into the breeding herd in November 1997, were seronegative to Menangle virus when tested in February 1999. This finding demonstrated that Menangle virus infection was not active in the breeding herd after the period of reproductive disease and that persistent shedding did not occur.[3] Finally, a large group of 8-week-old pigs that was moved into an uncleaned building, which, until 3 days earlier, was occupied by pigs known to be infected with Menangle virus, was seronegative at slaughter age suggesting that survival of the virus in the environment was not prolonged, and that maternally derived antibodies were protective in weaned pigs less than 8 weeks of age.[3] The realization that active Menangle virus infection was confined to the grower–finisher age group enabled a strategy for virus eradication to be developed. Implementation of this program, together with depopulation of the piggery at Young, resulted in successful elimination of the virus from the remaining two piggeries by February 1999.[3]

Determining the probable source of the outbreak was facilitated by the fact that fruit bats had previously been identified as the likely natural host of two newly described zoonotic viruses: Hendra virus,[18] a paramyxovirus that had caused the death of two humans in Queensland within the 3 years preceding the discovery of Menangle virus,[19–24] and Australian bat lyssavirus, a rhabdovirus that, although first identified in a sick fruit bat,[25–27] had subsequently caused the death of a bat carer.[28,29] A large breeding colony of 20,000–30,000 fruit bats, which consisted primarily of grey-headed fruit bats (*Pteropus poliocephalus*), but also little red fruit bats (*Pteropus scapulatus*), roosted annually, from October to April, within 150–200 m of the nearest breeding unit, and had done so for at least 30 years.[1,3] Serum samples were initially collected from this colony, as well as from fruit bats elsewhere in New South Wales and Queensland. Forty-two of 125 samples were positive in a virus neutralization test with titers against Menangle virus ranging from 16 to 256.[1] Considering that antibodies were not detected in a range of wild and domestic species found within the vicinity of the piggery, including rodents, birds, cattle, sheep, cats, and a dog, fruit bats were implicated as the likely natural host of Menangle virus.[1] More comprehensive serological surveys have since demonstrated that antibodies to Menangle virus are widespread in at least three of the four pteropid species endemic to Australia.[3,30] However, despite finding paramyxovirus-like particles in fruit bat feces collected from the colony nearby the index piggery, attempts to isolate Menangle virus from these samples were unsuccessful.[3,30] Nevertheless, following the discovery of Menangle virus, two other paramyxoviruses were identified in Malaysia, Nipah and Tioman viruses, both of which were isolated from fruit bat urine.[17,31] Using similar sampling procedures and newly established primary cell lines of bat origin, Menangle virus was subsequently isolated from fruit bat (*Pteropus alecto*) urine samples that were collected in South East Queensland.[32] It would therefore seem probable that transmission of Menangle virus from fruit bats to pigs occurred indirectly, perhaps through ingestion of fruit bat excreta, and this may have been associated with the movement of pigs between buildings using uncovered walkways, a management practice that had only been implemented 2 years before the outbreak.[3]

Prior to the isolation of Menangle virus from pigs, only two paramyxoviruses were considered to be zoonotic: Newcastle disease virus, which in humans can cause conjunctivitis,[33,34] and Hendra virus, which was known to have infected three people, two of whom subsequently died.[19–24] Considering that the relationship between Hendra and Menangle viruses had not been clearly established, but that both were considered to have "jumped" from fruit bats to infect other species (horses and humans in the case of Hendra virus), a serological survey was conducted to assess the zoonotic potential of the new virus. A total of 251 people, all of whom were considered to have been exposed to potentially infected pigs, were tested. This included all staff at the index (n = 33) and grower (n = 5) piggeries, abattoir workers (n = 142), researchers and animal handlers (n = 41), veterinarians and pathology laboratory workers (n = 24), and others (n = 6).[5] Two piggery staff, one from the index piggery and the other from one of the two associated grower piggeries, were shown to be seropositive to Menangle virus with neutralizing antibody titers of 128 and 512, respectively.[5] Further investigations revealed that both workers had contracted very similar illnesses in early June 1997, during the period when reproductive disease in pigs was present. The first worker had regular prolonged contact with farrowing pigs, which resulted in splashing of amniotic fluid and blood to his face, and he also frequently received minor wounds to his hands and forearms.[5] Although the second worker had no exposure to birthing pigs, he performed necropsies on pigs without wearing protective eye wear or gloves and reported that exposure to pig secretions, including feces and urine, was common.[5] Neither worker had any contact with bats. Extensive testing of both men in September 1997 failed to identify an alternative cause for the illnesses, which were therefore considered to have resulted from infection with Menangle virus. Although the mode of transmission from pigs to humans was not established, the available evidence suggested that Menangle virus did not readily transmit to people, and that infection probably required parenteral or permucosal exposure to infectious materials.[5]

Experimental infection of weaned pigs, by intranasal inoculation, has since demonstrated that Menangle virus is shed in nasal and oral secretions, feces, and urine, typically for periods of less than 1 week, following an incubation period of only a few days.[35] Although the stability of Menangle virus in these samples was not evaluated, these findings suggested that transmission is possible not only by direct contact between pigs, by exposure to infectious nasal and oral secretions, but also by indirect means, following contamination of the environment, food, and water with virus. Furthermore, the presence of infectious virus in nasal and oral secretions indicates that spread by aerosol droplet might also play a role in transmission. Nevertheless, the low concentrations of virus shed in each sample type would appear to correlate with the slow rate of spread observed in the index piggery, which led to the hypothesis that Menangle virus was more likely to be spread by fecal or urinary excretion than by respiratory aerosols.[2,3]

51.1.3 CLINICAL FEATURES AND PATHOGENESIS

51.1.3.1 Pigs

In naive pigs, infection with Menangle virus causes a marked decline in reproductive performance, characterized by a reduction in both the farrowing rate and the number of live piglet births per litter, occasional abortions, and an increase in the proportion of mummified and stillborn piglets, some of which have deformities; there are otherwise no clinical signs attributable to infection with the virus in postnatal, growing, or adult pigs of any age.[1–4]

The first affected litter, detected during the week commencing April 21, 1997, consisted of one stillborn piglet and seven mummified fetuses.[2] The proportion of affected litters (those containing fewer than six live-born piglets) in this unit increased significantly in the following weeks and peaked at 64.3% 5 weeks later, at which time only 17 of 45 mated sows farrowed.[2] During the period of disease in this unit, which lasted for 15 weeks, the farrowing rate decreased from a pre-outbreak average of 80.2% to 63.2%, and the number of piglets born alive per litter decreased from an average of 9.6 ± 2.9 to 8.3 ± 3.9.[2] The disease occurred sequentially in the remaining three breeding units, commencing 7, 8, and 11 weeks after the disease was initially recognized in the first affected unit, and continued for periods of 12, 12, and 11 weeks, respectively. The reproductive performance in all four units had returned to normal by mid-September 1997, 7 months after the virus was considered to have first entered the breeding herd.[2]

The composition of affected litters varied markedly throughout the Menangle virus outbreak. Some litters contained only one or two mummified or stillborn fetuses, suggesting that infection had occurred early during gestation, with subsequent death and resorption of the embryos.[2] In contrast, others contained a mixture of mummified and stillborn piglets in various stages of development, as well as apparently normal fetuses. Like porcine parvovirus,[36] this suggested that transplacental infection of fetuses was somewhat variable and that subsequent spread of the virus from infected to neighboring fetuses, *in utero*, was possible. The size of mummified fetuses, and the stage of development at which pathology was induced, also indicated that infection prior to day 70 of gestation was required for fetuses to be adversely affected,[2] which is consistent with other infectious causes of reproductive failure in swine.

Pathologic changes in affected stillborn piglets were characterized by severe degeneration, or even absence, of the brain and spinal cord (particularly the cerebral hemispheres, cerebellum, and brain stem), arthrogryposis, brachygnathia, and kyphosis,[1,2,4] although the pattern of central nervous system lesions observed within affected litters appeared to change over time.[4] Occasionally, marked pulmonary hypoplasia was observed, and in 41% of stillborn piglets examined, an excessive amount of straw-colored fluid, sometimes fibrinous, was present in body cavities.[1,4] Histologically, pathology was most marked in the central nervous system and was characterized by extensive degeneration and necrosis of grey and white matter, in association with infiltrates of macrophages and lymphocytes.[1,4] In some lesions within the brain and spinal cord, intranuclear and intracytoplasmic inclusion bodies were present in neurons and, occasionally, nonsuppurative myocarditis and leptomeningitis were also observed.[1,4]

Following experimental challenge of weaned pigs intranasally, a viremia of short duration (lasting only a matter of days) and low titer was detected.[35] Menangle virus was shed at low to moderate concentrations in nasal and oral secretions, feces, and urine, typically for periods of less than 1 week. Infectious virus and viral RNA were detected in a wide range of solid tissues, including tonsil, mandibular lymph node, jejunum, ileum, and mesenteric lymph node, indicating that secondary lymphoid organs and gastrointestinal tissues were major sites of viral replication and dissemination. This was subsequently confirmed by abundant immunolabeling of viral nucleocapsid protein in lymphoid tissues and throughout the intestinal epithelium, particularly within jejunum, which would explain the presence of virus in feces.[35]

51.1.3.2 Humans

Only two people are suspected to have been infected with Menangle virus, both of whom experienced a severe influenza-like illness with a rash. In each case, the diagnosis was made retrospectively after demonstrating that the men had high neutralizing antibody titers to the virus, with no other likely cause identified. The first patient experienced sudden onset of malaise and chills, followed by drenching sweats and fever, and was confined to bed with severe headaches and myalgia for 10 days.[5] By the fourth day, he had developed a spotty red rash and, on examination, exhibited upper abdominal tenderness, lymphadenopathy, and a rubelliform rash. He was absent from work for 14 days, tired easily on his return, and reported weight loss of 10 kg (~22 lb) due to this illness. The second patient experienced fever, chills, rigors, drenching sweats, marked malaise, back pain, severe frontal headache of 4–5-day duration, and photophobia.[5] A spotty, red, nonpruritic rash developed on his torso 4 days after the illness started and persisted for 7 days. He had essentially recovered after 10 days, having lost 3 kg (~6.6 lb).[5] Neither patient had cough, vomiting, or diarrhea.

51.1.3.3 Fruit Bats

Infection of pteropid bats with Menangle virus has not been associated with disease,[30] and this is consistent with the perceived role of bats as reservoir hosts of not only this and other paramyxoviruses that have emerged in the Australasian region since 1994, including Hendra, Nipah, Tioman, and Cedar viruses, but also many other viruses from a diverse spectrum of viral families.[32,37–41]

51.1.4 Diagnosis

Suspicion or confirmation of Menangle virus infection must be reported to relevant government authorities. In the United States, Menangle virus is listed as a U.S. Department of Agriculture Select Agent on the National Select Agent Register. Although the recommended sample types for confirming disease in humans will need to be guided by future characterization of the infection in people, should this opportunity arise, multiple laboratory tests have been developed and used, both during and since the outbreak, to increase understanding of the epidemiology and pathogenesis of Menangle virus, principally in pigs.

51.1.4.1 Conventional Techniques

51.1.4.1.1 Virus isolation

Menangle virus will grow in a wide range of cell types, including cells of porcine, human, and pteropid bat origin, and possesses neither hemadsorbing nor hemagglutinating activities using erythrocytes from several species.[1,32,35] The cytopathic effect is similar to that caused by other paramyxoviruses and is characterized by vacuolation of cells, formation of syncytia, and, eventually, lysis of the cell monolayer. During the outbreak, however, virus was only isolated from lung (n = 9), brain (n = 9), and heart (n = 5) of 10 affected, stillborn piglets, after 3–5 blind passages in baby hamster kidney (BHK$_{21}$) cells.[1,2] Considering that 170 tissues (including lung, brain, heart, kidney, and spleen from 57 affected piglets) were tested, virus was isolated from fewer than 1 in 7 samples. It has since been determined that virus is more likely to be isolated from piglets that exhibit gross or histological pathology of the central nervous system.[4]

51.1.4.1.2 Serology

Detection of Menangle virus–specific antibodies is undertaken using the virus neutralization test.[2,35] This test requires the use of live virus and cell culture, and routinely takes 5–7 days to complete. Neutralizing antibody titers >16 are considered positive in humans and pigs,[2,3] whereas titers ≥8 are considered positive in fruit bats.[3,30] The development of a prototype enzyme-linked immunosorbent assay, based on recombinant Menangle virus nucleocapsid proteins, offers the future prospect of a rapid and convenient serological test with no requirement for infectious reagents.[42]

51.1.4.1.3 Immunohistochemistry

In formalin-fixed pig tissues, demonstration of viral antigen in lesions can assist with diagnosis, either at the time of an outbreak investigation or retrospectively.[43] Following experimental infection of weaned pigs with Menangle virus, by intranasal inoculation, viral antigen was readily detected in lymphoid tissues and intestinal mucosa, and in limited amounts in other tissues, using a rabbit hyperimmune serum raised against purified recombinant viral nucleocapsid protein.[35]

51.1.4.1.4 Electron microscopy

Negative contrast[1] or ultrathin section[44] electron microscopical analysis of virus propagated in cell culture reveals the presence of pleomorphic enveloped virions and nucleocapsids, the ultrastructures of which are consistent with viruses in the family *Paramyxoviridae*, subfamily *Paramyxovirinae*.

51.1.4.2 Molecular Techniques

51.1.4.2.1 Real-Time Reverse Transcription Polymerase Chain Reaction

During the Menangle virus outbreak, molecular tests such as the reverse transcription polymerase chain reaction (RT-PCR) assay were not available to assist with disease investigation and control activities. Several attempts to amplify segments of the Menangle virus genome using primers based on conserved regions of other viruses in the subfamily *Paramyxovirinae* were unsuccessful. It was not until later, when a PCR-based cDNA subtraction strategy was used to isolate, clone, and sequence viral cDNAs from Menangle virus–infected cells, that the viral genome sequence was determined.[11,12] This led to the development of a multiplexed real-time RT-PCR TaqMan assay that, to date, has only been evaluated using clinical specimens (blood and solid tissues) collected from experimentally infected pigs.[35] Significantly, viral RNA was detectable in blood for a substantially longer period of time than infectious virus (at least 2 weeks), suggesting that blood might be a useful diagnostic indicator for confirming recent infection of pigs with Menangle virus. The assay clearly had both the specificity to distinguish Menangle virus from closely related viral genomes, including Tioman virus, with which it shares 72% nucleotide identity in the region corresponding to the target amplicon, and the analytical sensitivity to detect very low amounts of template in a wide range of cells and tissues. Simultaneous detection of 18S rRNA as an internal standard also provided a simple and effective means of excluding false-negative results.

In a diagnostic sense, several significant advantages of real-time PCR technologies are readily apparent, compared to conventional PCR: there is no requirement for any post-PCR processing, such as agarose gel electrophoresis or DNA sequencing, to confirm the identity of amplicons, and this greatly enhances the efficiency of sample testing; the sensitivity of the system is at least as good as conventional nested PCR tests; detection is performed in a closed-tube system, which greatly reduces the risk of carry-over contamination, either by previously amplified products or by samples containing high endogenous concentrations of target sequences; the methodologies are readily automated to increase sample throughput, if required; finally, because all TaqMan assays conform to universal cycling conditions, multiple assays targeting different pathogens can be run simultaneously in the same 96- or 384-well plate. Other improvements include hot start modifications to increase fidelity[45] and the inclusion of uracil N-glycosylase[46] for the prevention of carry-over contamination. Nevertheless, separation of work space into designated areas exclusively for reagent preparation, sample processing (nucleic acid extraction), PCR amplification, and post-PCR sample analysis, should this be required, is always recommended.[47]

51.2 METHODS

In experimentally infected pigs, viral RNA was detected in blood (for up to 21 days following challenge) and in solid tissues, including jejunum, ileum, nasal mucosa, lung, duodenum, colon, cecum, rectum, and spleen, all of which could conceivably be used to confirm recent infection with Menangle virus.[35] In a disease investigation scenario, however, brain, lung, and heart from affected stillborn piglets would be the samples of choice, since these are the tissues from which infectious virus was isolated during the outbreak in 1997.[1,2] Prior to the development of neutralizing antibodies in naive pigs, it is also likely that infectious virus would be present in nasal and oral secretions, as well as in feces and urine,[35] although the kinetics of viral RNA shedding in these samples has not been evaluated. The following two sections are therefore restricted to summarizing the extraction of Menangle virus RNA from blood and solid tissues, and its subsequent detection by real-time RT-PCR, since these are the sample categories that are likely to be of most relevance in the context of an outbreak investigation in pigs.

51.2.1 Sample Preparation

For whole blood, total RNA is isolated using the S.N.A.P. Total RNA Isolation Kit (Invitrogen) as recommended by the manufacturer. Briefly, 450 μL of Binding Buffer (7 M guanidine HCl, 2% Triton X-100) containing 2 mg/mL Proteinase K (Qiagen) is added to 150 μL of blood and vortexed vigorously for 1 min. Samples are subsequently incubated at 37°C for 20 min in a water bath and centrifuged for 3 min at 13,000 rpm (13,400–16,100 × g) in a benchtop microcentrifuge, after which the supernatants are transferred to new 2 mL polypropylene tubes with O-ring caps (Sarstedt). To each supernatant is added 300 μL of isopropanol, which is mixed by pipetting. Each sample is passed several times through a 21 gauge needle attached to a sterile 1 mL plastic syringe, transferred to a S.N.A.P. Total RNA column and centrifuged for 1 min at 13,000 rpm (13,400–16,100 × g). Columns are washed once with 600 μL of Super Wash Solution (3.5 M guanidine HCl, 0.67% Triton X-100, 0.33% [v/v] isopropanol), twice with RNA Wash Solution (100 mM NaCl, 76% [v/v] ethanol), and are dried by centrifugation at 13,000 rpm (13,400–16,100 × g) for 2 min. Nucleic acid is eluted in 135 μL of RNase-free water, to which is added 15 μL of 10 × DNase buffer and 1 μL (2 U) of RNase-free DNase. After incubation at 37°C for 10 min, Binding Buffer (450 μL) and isopropanol (300 μL) are added to each reaction, which is mixed and transferred to a new column. Columns are centrifuged for 1 min at 13,000 rpm (13,400–16,100 × g), washed twice with RNA Wash Solution, and centrifuged for an additional 2 min to dry the resin. RNA is then eluted in 125 μL of RNAse-free water and stored at −80°C before use.

For fresh or RNAlater (Ambion) stabilized tissues, total RNA is extracted using the RNeasy Mini Kit (Qiagen) following the instructions of the manufacturer, with minor modifications as follows. Routinely, 10–20 mg of each tissue is cut into cubes of edge length 1–2 mm using a separate sterile pair of forceps, disposable scalpel, and tissue culture dish (as cutting surface) per sample. Cut tissues are transferred to 2 mL polypropylene tubes with O-ring caps (Sarstedt), to which have been added sterile 1.0 mm zirconia/silica beads (Biospec Products) to the 250 μL gradation, and 600 μL of lysis Buffer RLT. Samples are cooled on ice and homogenized in a Mini-BeadBeater-8 (Biospec Products) for two 1 min periods, which are separated by an additional incubation on ice. Samples are centrifuged for 3 min at 13,000 rpm (13,400–16,100 × g) in a benchtop microcentrifuge, after which supernatants are transferred to new 2 mL polypropylene tubes with O-ring caps (Sarstedt). To each lysate is added 600 μL of 70% (v/v) ethanol, which is mixed by pipetting. Successive aliquots (600 μL) of each sample are applied to an RNeasy column and centrifuged at 13,000 rpm (13,400–16,100 × g) for 1 min. Columns are washed once with 700 μL of Buffer RW1, twice with 500 μL of Buffer RPE, and are dried by centrifugation for 1 min at 13,000 rpm (13,400–16,100 × g). RNA is then eluted from each column with two successive 30 μL aliquots of RNase-free water, quantified by spectrophotometry, and stored at −80°C prior to use.

51.2.2 DETECTION PROCEDURES

A two-step real-time RT-PCR assay has been developed for the quantitation of Menangle virus RNA in clinical samples from pigs.[35] The particular system used was the TaqMan assay (Applied Biosystems), a probe-based technology that provides a means of detecting only specific PCR products. The probe contains a fluorescent reporter dye, 6-carboxyfluorescein (FAM), at its 5′ end and a quencher dye,

6-carboxytetramethylrhodamine (TAMRA), at its 3′ end. The virus-specific primers and probe were designed to amplify and detect a short cDNA target of 66 nucleotides in length (corresponding to nucleotides 478–543 of the viral antigenome, within the N gene of Menangle virus). Their sequences are as follows: MP80F (5′-CGGATTTGAGCCTGGTACGT-3′), MP81R (5′-ACCTCTCCATTTGTCATCGGA-3′), and MT01-FAM (5′-FAM-TTCTCGCATTTGCCCTTAGCCGG-TAMRA-3′). Simultaneous detection and quantitation of an endogenous control (18S rRNA), in a multiplexed format, provides a means of excluding the presence of RT or PCR inhibitors in every sample tested, which provides a convenient and simple method of excluding false-negative results. The 18S rRNA primers and probe, the sequences of which are proprietary, are supplied in the TaqMan Ribosomal RNA Control Reagents Kit (Applied Biosystems).

51.2.2.1 Procedure

Synthesis of cDNA is performed using the TaqMan Reverse Transcription Reagents Kit (Applied Biosystems) according to the manufacturer's recommendations. Routinely, RT reactions are prepared in thin-walled 0.2 mL polypropylene capped tubes (Axygen Scientific Inc.) in 10 μL volumes: each reaction contains 1 × RT buffer, 5.5 mM MgCl$_2$, 500 μM of each dNTP, 2.5 μM of random hexamers, 0.4 U/μL of RNase inhibitor, 3.125 U/μL of MultiScribe Reverse Transcriptase, and either 2 μL (if extracted from whole blood) or 100 ng (if extracted from solid tissue) of RNA template.[35] Reaction mixtures without enzyme are also included as negative RT controls. Reaction tubes are capped, gently mixed, briefly centrifuged, and transferred to a GeneAmp PCR System 9700 Thermal Cycler (Applied Biosystems). RT is performed using the following cycling conditions: 25°C for 10 min (to maximize primer-RNA template binding), 37°C for 60 min (cDNA synthesis), and 95°C for 5 min (to inactivate the reverse transcriptase enzyme). Following RT, cDNA samples are generally stored at −20°C prior to use as template for real-time PCR.

Typically, real-time PCR assays are set up in polypropylene optical 96-well reaction plates (Applied Biosystems) in 25 μL volumes. Each reaction is set up in triplicate and comprises 12.5 μL of TaqMan Universal PCR Master Mix (Applied Biosystems) containing an internal reference dye (ROX) to normalize for non-PCR-related fluctuations in fluorescence occurring from well to well, 300 nM of each Menangle virus primer, 250 nM of Menangle virus probe, 50 nM of each 18S rRNA primer, 200 nM of 18S rRNA probe, 2 μL (if derived from whole blood) or 1 μL (if derived from solid tissue) of cDNA template, and nuclease-free water to 25 μL.[35] Plates are sealed with optical adhesive covers and transferred to an ABI Prism 7700 Sequence Detection System (Applied Biosystems) for amplification and detection of products using the following cycling conditions: 2 min at 50°C (to activate AmpErase uracil N-glycosylase), and 10 min at 95°C (to activate AmpliTaq Gold DNA Polymerase), and 50 cycles of 15 s at 95°C and 1 min at 60°C. Results are expressed in terms of the threshold cycle (C_T), the fractional cycle number at which the change in the reporter dye fluorescence (ΔR_n) passes an arbitrary, fixed threshold set in the log (exponential) phase of amplification. Fixed threshold ΔR_n values are set at 0.25 for the Menangle virus–specific products and at 0.08 for the 18S rRNA-specific products (although these settings will need to be established for the specific real-time PCR platform in use).

Note: Using random hexamer primers to initiate cDNA synthesis enables viral genome, antigenome, and N mRNA (the most abundant viral transcript) to be amplified and detected simultaneously, thereby maximizing test sensitivity. For the Menangle virus assay, the amplification efficiency (98%) is constant for template copy numbers that vary by at least five orders of magnitude (corresponding to upper and lower C_T values of approximately 36 and 18, respectively). Notably, however, despite the N genes of the pig and bat isolates of Menangle virus sharing 95% nucleotide identity, two mutations in the 3' terminal region of the reverse primer would prevent the bat variant from being detected by this assay. Development of additional virus-specific real-time RT-PCR assays will therefore be required to ensure that all currently known pig[11,12] and bat[32,41] isolates of Menangle virus are detectable.

51.3 CONCLUSIONS AND FUTURE PERSPECTIVES

Menangle virus is one of nine novel paramyxoviruses of fruit bat origin to be identified in the Australasian region within the last 2 decades. However, while it has caused disease in people, it is currently not in the same category as Hendra (Chapter 50) and Nipah (Chapter 53) viruses, which have caused serious illness and multiple fatalities in humans. Isolated in 1997 from stillborn piglets at a large commercial piggery in New South Wales, Australia, Menangle virus is a newly described cause of serious reproductive disease in pigs and may also cause severe influenza-like illness, with a rash, in humans. Although successfully eradicated from the affected piggeries within 2 years of entering the breeding herd of the index farm, many of the details relating to its pathogenesis and epidemiology, not only in pigs but also particularly in fruit bats and humans, remain unknown. Menangle virus probably remains endemic in fruit bats, the likely natural host of the virus, and therefore a continuing risk of reinfection exists.

Fortunately, it would appear that spillover of Menangle virus from the reservoir host to pigs is a rare event, with only one such occurrence documented to date.[1–5] In addition, it would seem that transmission to humans likely requires intense occupational exposure to recently infected pigs, including contamination of cuts and abrasions or mucosal surfaces, such as conjunctivae, with infectious materials.[5] Despite the lack of knowledge regarding the ecology of the virus, basic controls that restrict direct and indirect exposure of fruit bats to pigs should prevent future disease outbreaks in intensively managed piggeries that have effective biosecurity measures in place. An understanding of how and when the virus is shed from the natural host will help to define the risk factors associated with spillover to pigs, including extensively farmed and feral species, and

real-time RT-PCR would be an ideal tool to facilitate screening of such samples prior to confirming their infectivity using cell culture. Increased awareness and recognition of Menangle virus as a potential cause of serious reproductive disease in swine, and influenza-like illness, with a rash, in humans, should also ensure its prompt consideration in the differential diagnosis of similar disease episodes in the future.

Also intriguing is the existence, in Malaysia, of Tioman virus, which cross-reacts with Menangle virus–specific antisera[17,48] and is closely related genetically.[11,12,16,17] Considering that antibodies to Tioman virus, or a Tioman-like virus, have since been detected in fruit bats in Madagascar,[49] novel rubulaviruses have been identified in, and isolated from, fruit bats in China,[50] Ghana,[51,52] and Australia,[41] and an estimated 66 new paramyxoviruses have been discovered in a worldwide surveillance study targeting 119 bat and rodent species,[39] it would appear that related viruses with an unknown propensity to cause disease in domestic animal species or humans are far more abundant and widespread than previously thought. Although Tioman virus has not been associated with disease in pigs or humans, a recent study has established that weaned pigs are susceptible to experimental infection.[53] Whether the virus is pathogenic in pregnant pigs, however, is not known. Furthermore, low neutralizing antibody titers to Tioman virus, or a closely related virus, have been detected in sera from 3 of 169 villagers on the island from which it was originally isolated, suggesting possible prior infection by this or a similar virus.[54] In the absence of any identifiable link to pigs, it was postulated that human infection might have occurred through consumption of fruit that had been partially eaten by bats, a potential risk factor common to 32 of the surveyed villagers, including two of the three who had detectable neutralizing antibodies to the virus.[54] It would not seem unreasonable to presume, therefore, that direct transmission of Menangle virus from fruit bats to people, such as wildlife carers, might be possible without prior amplification in pigs.

Many fundamental questions remain to be answered, such as how Menangle virus is maintained in fruit bats and over what geographic range it should be considered endemic, as well as which cellular receptor is used for viral attachment and how it relates to pathogenesis in each susceptible host species. Obtaining answers to complex questions such as these will be challenging and will require a multidisciplinary approach. However, until such time as Menangle virus infection in humans is better characterized, the true zoonotic potential of this recently emerged paramyxovirus will remain unclear.

ACKNOWLEDGMENT

The author would like to acknowledge that parts of this chapter, including the figures, are adapted with permission from the texts published previously in *Molecular Detection of Human Viral Pathogens*, 2010, edited by Dongyou Liu, CRC Press, Boca Raton, FL (ISBN: 9781439812365) and in *Manual of Security Sensitive Microbes and Toxins*, 2014, edited by Dongyou Liu, CRC Press, Boca Raton, FL (ISBN: 9781466553965).

REFERENCES

1. Philbey, A. W. et al., An apparently new virus (family *Paramyxoviridae*) infectious for pigs, humans, and fruit bats, *Emerg. Infect. Dis.*, 4, 269, 1998.
2. Love, R. J. et al., Reproductive disease and congenital malformations caused by Menangle virus in pigs, *Aust. Vet. J.*, 79, 192, 2001.
3. Kirkland, P. D. et al., Epidemiology and control of Menangle virus in pigs, *Aust. Vet. J.*, 79, 199, 2001.
4. Philbey, A. W. et al., Skeletal and neurological malformations in pigs congenitally infected with Menangle virus, *Aust. Vet. J.*, 85, 134, 2007.
5. Chant, K. et al., Probable human infection with a newly described virus in the family *Paramyxoviridae*, *Emerg. Infect. Dis.*, 4, 273, 1998.
6. Wang, L. F. et al., Family *Paramyxoviridae*, in *Virus Taxonomy: Classification and Nomenclature of Viruses. Ninth Report of the International Committee on Taxonomy of Viruses*, eds. A. M. Q. King et al. (Elsevier/Academic Press, San Diego, CA, 2012), 672.
7. Lamb, R. A. and Parks, G. D., *Paramyxoviridae*, in *Fields Virology*, eds. D. M. Knipe et al. (Lippincott Williams & Wilkins, Philadelphia, PA, 2013), 957.
8. Kingsbury, D. W. and Darlington, R. W., Isolation and properties of Newcastle disease virus nucleocapsid, *J. Virol.*, 2, 248, 1968.
9. Compans, R. W. and Choppin, P. W., The nucleic acid of the parainfluenza virus SV5, *Virology*, 35, 289, 1968.
10. Heggeness, M. H., Scheid, A., and Choppin, P. W., Conformation of the helical nucleocapsids of paramyxoviruses and vesicular stomatitis virus: Reversible coiling and uncoiling induced by changes in salt concentration, *Proc. Natl. Acad. Sci. U.S.A.*, 77, 2631, 1980.
11. Bowden, T. R. et al., Molecular characterization of Menangle virus, a novel paramyxovirus which infects pigs, fruit bats, and humans, *Virology*, 283, 358, 2001.
12. Bowden, T. R. and Boyle, D. B., Completion of the full-length genome sequence of Menangle virus: Characterisation of the polymerase gene and genomic 5′ trailer region, *Arch. Virol.*, 150, 2125, 2005.
13. Sedlmeier, R. and Neubert, W. J., The replicative complex of paramyxoviruses: Structure and function, *Adv. Virus Res.*, 50, 101, 1998.
14. Curran, J. and Kolakofsky, D., Replication of paramyxoviruses, *Adv. Virus Res.*, 54, 403, 1999.
15. Kurath, G. et al., Complete genome sequence of Fer-de-Lance virus reveals a novel gene in reptilian paramyxoviruses, *J. Virol.*, 78, 2045, 2004.
16. Chua, K. B. et al., Full length genome sequence of Tioman virus, a novel paramyxovirus in the genus *Rubulavirus* isolated from fruit bats in Malaysia, *Arch. Virol.*, 147, 1323, 2002.
17. Chua, K. B. et al., Tioman virus, a novel paramyxovirus isolated from fruit bats in Malaysia, *Virology*, 283, 215, 2001.
18. Young, P. L. et al., Serologic evidence for the presence in *Pteropus* bats of a paramyxovirus related to equine morbillivirus, *Emerg. Infect. Dis.*, 2, 239, 1996.
19. Murray, K. et al., A novel morbillivirus pneumonia of horses and its transmission to humans, *Emerg. Infect. Dis.*, 1, 31, 1995.
20. Murray, K. et al., A morbillivirus that caused fatal disease in horses and humans, *Science*, 268, 94, 1995.
21. Selvey, L. A. et al., Infection of humans and horses by a newly described morbillivirus, *Med. J. Aust.*, 162, 642, 1995.
22. Rogers, R. J. et al., Investigation of a second focus of equine morbillivirus infection in coastal Queensland, *Aust. Vet. J.*, 74, 243, 1996.

23. Hooper, P. T. et al., The retrospective diagnosis of a second outbreak of equine morbillivirus infection, *Aust. Vet. J.*, 74, 244, 1996.
24. O'Sullivan, J. D. et al., Fatal encephalitis due to novel paramyxovirus transmitted from horses, *Lancet*, 349, 93, 1997.
25. Crerar, S. et al., Human health aspects of a possible lyssavirus in a black flying fox, *Commun. Dis. Intell.*, 20, 325, 1996.
26. Fraser, G. C. et al., Encephalitis caused by a lyssavirus in fruit bats in Australia, *Emerg. Infect. Dis.*, 2, 327, 1996.
27. Hooper, P. T. et al., A new lyssavirus—The first endemic rabies-related virus recognized in Australia, *Bull. Inst. Pasteur.*, 95, 209, 1997.
28. Allworth, A., Murray, K., and Morgan, J., A human case of encephalitis due to a lyssavirus recently identified in fruit bats, *Commun. Dis. Intell.*, 20, 504, 1996.
29. Samaratunga, H., Searle, J. W., and Hudson, N., Non-rabies Lyssavirus human encephalitis from fruit bats: Australian bat Lyssavirus (pteropid Lyssavirus) infection, *Neuropathol. Appl. Neurobiol.*, 24, 331, 1998.
30. Philbey, A. W. et al., Infection with Menangle virus in flying foxes (*Pteropus* spp.) in Australia, *Aust. Vet. J.*, 86, 449, 2008.
31. Chua, K. B. et al., Isolation of Nipah virus from Malaysian Island flying-foxes, *Microbes Infect.*, 4, 145, 2002.
32. Barr, J. A. et al., Evidence of bat origin for Menangle virus, a zoonotic paramyxovirus first isolated from diseased pigs, *J. Gen. Virol.*, 93, 2590, 2012.
33. Trott, D. G. and Pilsworth, R., Outbreaks of conjunctivitis due to the Newcastle disease virus among workers in chicken-broiler factories, *Br. Med. J.*, 5477, 1514, 1965.
34. Mustaffa-Babjee, A., Ibrahim, A. L., and Khim, T. S., A case of human infection with Newcastle disease virus, *Southeast Asian J. Trop. Med. Public Health*, 7, 622, 1976.
35. Bowden, T. R. et al., Menangle virus, a pteropid bat paramyxovirus infectious for pigs and humans, exhibits tropism for secondary lymphoid organs and intestinal epithelium in weaned pigs, *J. Gen. Virol.*, 93, 1007, 2012.
36. Mengeling, W. L., Porcine parvovirus, in *Diseases of Swine*, eds. B. E. Straw et al. (Blackwell Publishing, Ames, IA, 2006), 373.
37. Calisher, C. H. et al., Bats: Important reservoir hosts of emerging viruses, *Clin. Microbiol. Rev.*, 19, 531, 2006.
38. Marsh, G. A. et al., Cedar virus: A novel henipavirus isolated from Australian bats, *PLoS Pathog.*, 8, e1002836, 2012.
39. Drexler, J. F. et al., Bats host major mammalian paramyxoviruses, *Nat. Commun.*, 3, 796, 2012.
40. Wong, S. et al., Bats as a continuing source of emerging infections in humans, *Rev. Med. Virol.*, 17, 67, 2007.
41. Barr, J. et al., Isolation of multiple novel paramyxoviruses from pteropid bat urine, *J. Gen. Virol.*, 96, 24, 2015.
42. Juozapaitis, M. et al., Generation of Menangle virus nucleocapsid-like particles in yeast *Saccharomyces cerevisiae*, *J. Biotechnol.*, 130, 441, 2007.
43. Hooper, P. T. et al., Immunohistochemistry in the identification of a number of new diseases in Australia, *Vet. Microbiol.*, 68, 89, 1999.
44. Yaiw, K. C. et al., Viral morphogenesis and morphological changes in human neuronal cells following Tioman and Menangle virus infection, *Arch. Virol.*, 153, 865, 2008.
45. Birch, D. E. et al., Simplified hot start PCR, *Nature*, 381, 445, 1996.
46. Pang, J., Modlin, J., and Yolken, R., Use of modified nucleotides and uracil-DNA glycosylase (UNG) for the control of contamination in the PCR-based amplification of RNA, *Mol. Cell Probes*, 6, 251, 1992.

47. Storch, G. A. and Wang, D., Diagnostic virology, in *Fields Virology*, eds. D. M. Knipe et al. (Lippincott Williams & Wilkins, Philadelphia, PA, 2013), 414.

48. Petraityte, R. et al., Generation of Tioman virus nucleocapsid-like particles in yeast *Saccharomyces cerevisiae*, *Virus Res.*, 145, 92, 2009.

49. Iehle, C. et al., Henipavirus and Tioman virus antibodies in pteropodid bats, Madagascar, *Emerg. Infect. Dis.*, 13, 159, 2007.

50. Lau, S. K. et al., Identification and complete genome analysis of three novel paramyxoviruses, Tuhoko virus 1, 2 and 3, in fruit bats from China, *Virology*, 404, 106, 2010.

51. Baker, K. S. et al., Co-circulation of diverse paramyxoviruses in an urban African fruit bat population, *J. Gen. Virol.*, 93, 850, 2012.

52. Baker, K. S. et al., Novel, potentially zoonotic paramyxoviruses from the African straw-colored fruit bat Eidolon helvum, *J. Virol.*, 87, 1348, 2013.

53. Yaiw, K. C. et al., Tioman virus, a paramyxovirus of bat origin, causes mild disease in pigs and has a predilection for lymphoid tissues, *J. Virol.*, 82, 565, 2008.

54. Yaiw, K. C. et al., Serological evidence of possible human infection with Tioman virus, a newly described paramyxovirus of bat origin, *J. Infect. Dis.*, 196, 884, 2007.

52 Newcastle Disease Virus

Dongyou Liu

CONTENTS

52.1 Introduction 447
    52.1.1 Classification.. 447
    52.1.2 Morphology and Genome Organization.. 448
    52.1.3 Biology and Epidemiology 449
    52.1.4 Clinical Features and Pathogenesis 449
    52.1.5 Diagnosis 450
        52.1.5.1 Virus Isolation... 450
        52.1.5.2 Bioassays.. 450
        52.1.5.3 Serological Assays 450
        52.1.5.4 Nucleic Acid Detection 450
    52.1.6 Treatment and Prevention 450
52.2 Methods 451
    52.2.1 Sample Preparation.. 451
    52.2.2 Detection Procedures... 451
        52.2.2.1 Conventional RT-PCR... 451
        52.2.2.2 Real-Time RT-PCR 451
52.3 Conclusion 452
References... 452

52.1 INTRODUCTION

Newcastle disease (ND) is a highly contagious and often fatal disease of birds resulting from infection with the virulent strains of Newcastle disease virus (NDV). Since its first descriptions in the 1920s, ND has been responsible for drops in egg production and loss of stocks (up to 100%) in poultry farming industry through successive outbreaks. Given the non-specific respiratory, gastrointestinal, and nervous symptoms it induces, the use of laboratory techniques such as reverse transcription polymerase chain reaction (RT-PCR) is crucial for rapid and accurate identification, virulence determination, and epidemiological tracking of this deadly pathogen.

52.1.1 CLASSIFICATION

NDV is a member of the genus *Avulavirus*, subfamily *Paramyxovirinae*, family *Paramyxoviridae*, order *Mononegavirales*.

Of the two subfamilies within the family *Paramyxoviridae*, the subfamily *Paramyxovirinae* comprises eight genera (*Aquaparamyxovirus*, *Avulavirus*, *Ferlavirus*, *Henipavirus*, *Morbillivirus*, *Respirovirus*, *Rubulavirus*, and *TPMV-like viruses*), whereas the subfamily *Pneumovirinae* includes two genera (*Pneumovirus* and *Metapneumovirus*).

The genus *Avulavirus* covers viruses that once belonged to the genus *Rubulavirus* and currently consists of 13 recognized species, namely, avian paramyxovirus serotype 1 (APMV-1, commonly known as NDV), APMV-2, APMV-3, APMV-4, APMV-5, APMV-6, APMV-7, APMV-8, APMV-9, APMV-10, APMV-11, APMV-12, and APMV-13, on the basis of hemagglutination inhibition (HI), neuraminidase inhibition, and phylogenetic assays (Table 52.1) [1–7].

Consisting of a single serotype, NDV, or APMV-1, can be separated into two classes (I and II) based on genome lengths and fusion (F) gene phylogenetic reconstruction.

Class I includes viruses that possess genomes of 15,198 nt, usually occur in waterfowl and shorebirds, and are of low virulence for chickens (except for one strain).

Class II encompasses pathogenic strains that have genomes of 15,186 nt (or 15,192 nt) and that are distinguished into 16 genotypes (I–XVI), with isolates from 1930 to 1960 belonging to I, II, III, IV, IX and those appearing after 1960 belonging to V, VI, VII, VIII, X, XI, XII, XIII, XIV, XV, XVI [8,9]. In addition, class II genotypes VI and VII are further subdivided into eight (a–h) and five (a–e) subgenotypes, respectively (Table 52.2). It is notable that of the 16 genotypes within class II, genotypes V–VIII harbor genomes of 15,192 nt, while others possess genomes of 15,186 nt. The increase in genome sizes (from 15,186 to 15,192 nt) results from insertion of 6 nucleotides in the 5′ noncoding region of the nucleoprotein (N) gene [10].

In accordance to their virulence in poultry, APMV-1 isolates can be differentiated into three pathotypes: lentogenic,

TABLE 52.1

Characteristics of Avian Paramyxovirus Serotypes 1–13

APMV	Genome (nt)	Host	Disease	F Protein Cleavage Site[a]
APMV-1 (velogenic)	15,186 (or 15,192)	>250 avian species; chickens most severely affected	High mortality	GRRQKR/F
APMV-1 (lentogenic)	15,198	Waterfowl, shorebirds	Asymptomatic or subclinical	GGRQGR/L
APMV-2	14,904	Turkey, chickens, psittacines, rails, passerines	Mild respiratory disease and drop in egg production; infertility in turkey	DKRAGR/F
APMV-3	16,272	Turkey, psittacines, passerines, flamingos	Encephalitis and high mortality in caged birds; respiratory disease in turkey and stunted growth in young chickens	PRPSGR/L
				ARRRGR/L
APMV-4	15,054	Ducks, geese, chickens	Mild interstitial pneumonia and catarrhal tracheitis	ADIQPR/F
APMV-5	17,262	Budgerigars, lorikeets	Depression, dyspnea, diarrhea, torticollis, and acute enteritis with high mortality in immature budgerigars	GKRKKR/F
APMV-6	16,236	Ducks, geese, turkey	Mild respiratory disease and drop in egg production in turkey; avirulent in chickens	PAREPR/L
APMV-7	15,480	Pigeons, doves, turkey	Mild multifocal nodular lymphocytic air sacculitis and decreased egg production in turkey	TLPSSR/F
APMV-8	15,342	Ducks, geese	Unknown	GYRQTR/L
APMV-9	15,438	Ducks	Unknown	RIREGR/I
APMV-10	15,226	Rockhopper penguins	Unknown	DKPSQR/I
APMV-11	17,412	Common snipe	Unknown	DSGTKR/F
APMV-12	15,312	Wigeon	Unknown	RGREPR/L
APMV-13	Incomplete	Geese	Unknown	QVRENR/L

[a] The amino acid sequence of the cleavage site of F protein precursor F0 to F1–F2 subunits is a primary determinant of avian paramyxovirus infectivity. NDV virulent strains contain at least three basic residues (lysine and arginine [K and R] between positions 113 and 116 at the C terminus of the F2 protein [cleavage site motif ^{111}GRRQKR/F^{117}] and phenylalanine [F] at the N terminus of the F1 protein [residue 117]).

mesogenic, or velogenic. The lentogenic pathotype is the least virulent and is associated with subclinical (mild respiratory or enteric) infections in both domestic poultry and wild bird populations. The mesogenic pathotype is moderately virulent and induces moderate respiratory and occasional nervous signs with low to moderate mortality. The velogenic pathotype is the most virulent and causes nervous signs or extensive hemorrhagic lesions in the gastrointestinal tract with high mortality (up to 100%). The velogenic pathotype may be further classified as neurotropic or viscerotropic depending on its predilection site [10].

52.1.2 MORPHOLOGY AND GENOME ORGANIZATION

NDV is an enveloped, pleomorphic (usually spherical, but sometimes filamentous) particle of about 200–300 nm in diameter. Derived from the plasma membrane of the host cell, the lipid bilayer envelope is covered with spike-like surface projections of 8 nm in length, which are made up of hemagglutinin–neuraminidase (HN) protein tetramers and F protein trimers. Beneath the envelope is the nonglycosylated matrix (M) protein, which encases the nucleocapsid core formed by the nucleocapsid (N) protein that encapsidates the

viral genomic RNA. The phosphoprotein (P) and large (L) protein remain bound to the nucleocapsid core, forming the herringbone-like ribonucleoprotein complex.

Being a monopartite, negative-sense, single-stranded RNA of 15,186 nt (seen in isolates before 1960), 15,192 nt (seen in isolates from China), and 15,198 nt (seen in avirulent strain from Germany), which follows the "rule of six," the genome of NDV includes 55 nt leader at its 3′ end and 114 nt trailer at its 5′ end, flanking six open reading frames (ORF) in the order of 3′-N-P-M-F-HN-L-5′ [11–13].

The six ORFs (N, P, M, F, HN, L) are separated from each other by intergenic sequences, with each possessing its own gene start and gene end signal sequences, facilitating the production of the N protein, M protein, phosphoprotein (P), F protein, HN protein, and L-polymerase protein (Table 52.3). In addition, insertion of one G residue in the P ORF at consensus sequence "^{394}AAAAGGG^{401}" during RNA editing produces accessory protein V, which is present only in the virus-infected cells and is involved in anti-IFN activity (Table 52.3). Occasionally, insertion of two G residues at the RNA editing site in the P ORF generates W mRNA, which is present in NDV infected cells, but W protein has not been detected.

TABLE 52.2
Characteristics of Newcastle Disease Virus Classes and Genotypes

Class	Genotype	Subgenotype	Pathogenicity
I (between 1930 and 1960)			Avirulent
II (from 1960)	I		Mostly avirulent (including vaccine strain VG/GA)
	II		Avirulent (including vaccine strains B1, Clone30, LaSota, and MET95)
	III		Virulent
	IV		Virulent
	V		Virulent
	VI	a–h	Virulent
	VII	a–e	Virulent
	VIII		Virulent
	IX		Virulent
	X		Avirulent
	XI		Virulent
	XII		Virulent
	XIII		Virulent
	XIV		Virulent
	XV		Virulent
	XVI		Virulent

TABLE 52.3
Newcastle Disease Virus Genes and Encoded Proteins

Gene	Protein	Protein (aa)	Protein Mass (kDa)
N	Nucleocapsid (coat protein)	489	55
P	Phosphoprotein	395	42
	V	239	26
	W	179	19
M	Matrix protein	364	40
F	Fusion protein (glycosylated envelope protein)	553	59
H	Hemagglutinin–neuraminidase attachment protein	616	68
L	RNA polymerase	2204	248

52.1.3 Biology and Epidemiology

NDV has a tropism for avian respiratory epithelium cells. Upon entry to the respiratory tract, the virus attaches to host cell surface receptors through HN glycoprotein. This leads to fusion between virus and plasma membranes and release of ribonucleocapsid into the cytoplasm [14]. The viral RNA-dependent RNA polymerase sequentially transcribes each gene in the negative-sense RNA genome into viral mRNA, positive-sense antigenome, and then negative-sense genome. Following translation from mRNA and viral genome, the

nucleoprotein encapsidates neosynthesized viral genomes to form the ribonucleocapsid. Then, the ribonucleocapsid binds to the M protein and the newly assembled virions bud at the plasma membrane or release after cell lysis.

NDV is an avian virus that is infective to over 250 species of birds of all age groups. As natural reservoir for the virus, wild birds tend to have asymptomatic infection. In susceptible hosts (such as domestic poultry), NDV infection causes severe illness, with high concentration of the virus in birds' bodily discharges. NDV is rarely isolated from naturally infected nonavian and nonhuman hosts except for one relating to cattle in 1952 and another relating to sheep in 2012 [15]. However, NDV is potentially pathogenic to humans and may cause conjunctivitis if exposed to the eyes.

NDV is relatively stable in nature and survives for several weeks in a warm and humid environment on birds' feathers, manure, and other materials, and indefinitely in frozen material. It also tolerates a wide range of pH (2–11). The virus is sensitive to heat at 56°C and is inactivated rapidly by dehydration and by the ultraviolet rays in sunlight. In addition, it is sensitive to detergents, lipid solvents, formaldehyde, and oxidizing agents.

NDV is spread primarily through direct contact between healthy birds and the bodily discharges of infected birds (including droppings and secretions from the nose, mouth, and eyes). Virus-bearing material picked up on shoes and clothing from infected flocks may spread the infection to healthy flocks [16].

Since its first descriptions in Java, Indonesia, in 1926 and Newcastle-upon-Tyne, England, in 1927, NDV has been responsible for causing devastating diseases in domestic poultry worldwide. Currently, NDV is endemic in Africa, Asia, and Americas. Between the years 2011 and 2014, NDV outbreaks have been reported in Israel (involving little owls and African penguins); Qatar; Cameroon, Central African Republic, Côte d'Ivoire, and Nigeria; China, Vietnam, Cambodia, Malaysia, and Indonesia; Dominican Republic; and Minnesota, Massachusetts, Maine, New Hampshire, and Maryland (involving cormorants and gulls) [17–23].

52.1.4 Clinical Features and Pathogenesis

With an incubation period of 2–15 days, ND is a highly contagious disease that affects a broad range of domestic and wild bird species and results in a spectrum of clinical diseases (from undetectable to rapidly fatal), the severity of which is dependent on the age and immune status of the host, pathogenic potential of NDV strains, and concurrent diseases [10].

Clinically, ND is characterized by sneezing, gasping, coughing, rales, anorexia, ruffled feathers, depression, listlessness, hyperthermia, hypothermia, greenish/watery diarrhea, edema of the eyes and neck, rough- or thin-shelled eggs, and reduced egg production. In some cases, neurological signs (e.g., circling, backward progression, convulsive somersaulting, head tremor, torticollis, ataxia, wing and leg paralysis) may be present. Death is a usual outcome, with sudden death occurring in acute cases [24].

Postmortem lesions include petechiae in the proventriculus and on the submucosae of the gizzard, and severe enteritis in the duodenum [25].

Differential diagnoses consist of highly pathogenic avian influenza, infectious laryngotracheitis, salmonellosis, spirochaetosis, mycoplasmosis, and hemorrhagic diseases.

NDV has the ability to undergo nucleotide change to enhance its pathogenic potential. One example relates to the F protein, which plays a critical role in virus entry into the cytoplasm. Highly pathogenic NDV strains have risen from changes in the glycosylation site or the cytoplasmic domain of F protein, leading to increased viral replication and pathogenesis in chickens. In addition, NDV virulent strains have a tendency to acquire at least three basic residues (lysine and arginine [K and R] between positions 113 and 116 at the C terminus of the F2 protein [cleavage site motif ^{111}GRRQKR/F^{117}] and phenylalanine [F] at the N terminus of the F1 protein [residue 117]). This renders the precursor protein F0 more susceptible to cleavage (into mature F1 and F2 subunits) by the ubiquitous proteases of the host, resulting in increased virulence. On the other hand, NDV strains of low virulence have fewer basic amino acid residues at the C terminus of the F2 protein and leucine (L) at position 117 (cleavage site motif ^{111}GGRQGR/L^{117}), which requires trypsin-like proteases to cleave the precursor [26].

Another example relates to the HN protein, whose receptor recognition and neuraminidase activities determine and also contribute to virulence of NDV. The HN protein from NDV virulent strains are less likely to have substitution of residues F220, S222, and L224 in the membrane proximal part of dimer interface of the HN globular domain, which weakens its association with receptors as well as its interaction with the F protein. Similarly, the HN protein of NDV virulent strains rarely has substitution of amino acid residue (L94) in the region intervening two heptad repeats in HN stalk structure, which affects fusion by hindering interaction with F protein [10].

Due to these nucleotide and amino acid changes, NDV virulent strains demonstrate enhanced ability to suppress or evade the host immune functions that are at harm's way to the invading virus [27].

52.1.5 Diagnosis

Laboratory methods for diagnosis of NDV infection rely on virus isolation, bioassays, serological tests, and nucleic acid detection.

52.1.5.1 Virus Isolation

NDV is often isolated in embryonated chicken eggs or African green monkey kidney (Vero) cell line. For virus isolation in embryonated chicken eggs, 0.2 mL of filtered sample is inoculated into 9–11-day-old specific pathogen free (SPF) embryonated chicken eggs via allantoic route, and the eggs are incubated at 37°C until death of embryos or a maximum period of 120 h, whichever is earlier. The embryos are candled every day, the dead embryos are chilled at 4°C overnight, and the harvested allantoic fluids are tested for

hemagglutination (HA) activity and by bioassays. The allantoic fluids, which are negative for HA activity, are further passaged at least once more in embryonated chicken eggs. The presence of NDV in the allantoic fluids may be confirmed by HI test employing LaSota strain, specific serum, and RT-PCR.

For virus isolation on African green monkey kidney (Vero) cell line, the filtered sample is inoculated on cell monolayer in 25 cm^2 tissue culture flasks and incubated at 37°C for 1 h. Maintenance media (Eagle MEM) without fetal calf serum and with antibiotics (penicillin 10,000 IU/mL, streptomycin 100 μg/mL) is added. The flasks are incubated at 37°C and are observed daily for the appearance of cytopathic effects (CPE) for up to 7 days postinoculation. When 100% CPE is observed, the flasks are frozen at −20°C. The virus is harvested after two freeze–thaw cycles. At least three passages are carried out to confirm the isolation.

52.1.5.2 Bioassays

The pathogenicity of NDV isolates may be determined on the basis of the intracerebral pathogenicity index (ICPI) in dayold chicks, the intravenous pathogenicity index in 6-week-old chickens, or the mean death time (MDT) in chickens.

52.1.5.3 Serological Assays

HI and enzyme-linked immunosorbent assay (ELISA) including indirect, sandwich, and blocking or competitive ELISA are useful for screening and confirmation of NDV infection.

52.1.5.4 Nucleic Acid Detection

The virulence of NDV mesogenic or velogenic isolates is related to the basic amino acids at the cleavage site of the F protein, as lentogenic NDV isolates harbor fewer basic amino acids at that particular location. RT-PCR amplification and sequencing analysis of the gene segment underlying the F protein cleavage site provides a rapid and accurate approach for virulence determination of NDV strains [28,29]. The use of specific probes in real-time PCR makes pathotyping possible without sequencing analysis. In addition, various RT-PCR procedures have been developed for sensitive and specific identification and phylogenetic examination of NDV strains directly from clinical specimens [30–38].

ND is a list A disease by the Office Internationale des Epizooties (OIE). For NDV pathotyping, the OIE recommends the determination of ICPI with a value of 0.7 or higher and demonstration of amino acid motif in the F gene cleavage site after sequencing. According to the OIE guidelines, the presence of one basic amino acid at residues 116 and 115 of the F protein cleavage site and a phenylalanine at residue 117 along with a basic amino acid (R) at amino acid position 113 are considered markers of a virulent pathotype, and such isolates are notifiable to OIE [39].

52.1.6 Treatment and Prevention

As no specific treatment is available for NDV infection, the implementation of quarantine and sanitary measures and the use of prophylactic vaccines will help reduce the occurrence or recurrence of ND outbreaks. These include immediate quarantine of any animals showing symptoms of ND, thorough cleaning and disinfection of facility, and vaccination of new birds before introduction to a flock [39].

Vaccines based on both live and killed NDV strains (including lentogenic vaccine strains Hitchner-B1, LaSota, V4, NDW, I2, and mesogenic vaccine strains Roakin, Mukteswar, and Komarov) are commercially marketed. Thermostable NDV vaccines have the advantages of easy preservation and transportation. Typically, broiler chickens are vaccinated via intraocular route on day one and boosted through drinking water on day 14. Layers are vaccinated first with live vaccine and then with inactivated vaccine via intramuscular or subcutaneous route [39,40].

NDV demonstrates a number of distinct features that make it a valuable vector for delivering antimicrobial and anticancer molecules in both animals and humans. These include its high growth efficiency in embryonated chicken eggs, cell culture, and the respiratory tract of avian and nonavian species; its predilection for the respiratory tract, which is ideal for delivering protective antigens from respiratory pathogens; its ability to elicit both humoral and cellular immune response; its relatively simple genome for easy manipulation; its tendency to replicate in host cytoplasm without integration into host genome; and its inherent ability to grow in IFN-deficient cells such as tumor cells.

Incorporating human influenza virus hemagglutinin (HA) protein, HIV and SIV Gag protein, HIV glycoproteins, human respiratory syncytial virus F glycoprotein, human parainfluenza virus 3 HN protein, or SARS-CoV spike glycoprotein, recombinant NDV strains have shown promise to confer protective immunity to respective pathogens in human recipients. In addition, recombinant NDV expressing VP2 or HA has been shown to protect chickens against infectious bursal disease, or H5N1 and H7N7 influenza virus, while that expressing gN and gG glycoproteins, gD glycoprotein, or G and F proteins protects mice against Rift Valley fever virus, bovine herpesvirus, or Nipah virus infection.

Furthermore, both NDV lytic (mesogenic and velogenic PV701, 73-T and MTH-68) and nonlytic (lentogenic Ulster) strains have proven to be potent anticancer agents. Recombinant NDV expressing immunostimulatory molecules (e.g., complement regulatory protein, IFNs, IL-2, GM-CSF, IgG antibody, and TNF-α) demonstrates enhanced oncolytic property.

52.2 METHODS

52.2.1 SAMPLE PREPARATION

The trachea, liver, kidney, and brain as well as tracheal and cloacal swabs are collected from dead birds while blood samples are collected from alive but sick birds.

The mucosal layer of the trachea is scratched by a scalpel blade, and tissues of 0.5 cm^3 is removed from each liver and kidney. The specimen is then crushed in a sterile mortar with 1 mL sterile phosphate buffered saline (PBS) buffer. Brain tissue is triturated and homogenized in PBS with antibiotics (penicillin 1000 IU/mL and streptomycin 1000 µg/mL) making up a 10% (w/v) solution. The 10% suspension is clarified by centrifugation at $2500 \times g$ for 15 min. The supernatant is collected and stored at −20°C until further use. Part of this suspension is used for RNA isolation for RT-PCR, and part is inoculated onto Vero cell culture after filtration through 0.22 µm filter for virus isolation. The samples may be also passaged in 10-day-old SPF embryonated eggs. The allantoic fluids are used for the assessment of pathogenicity (ICPI, MDT), a detection efficacy for validated real-time PCR and for sequence analysis [41].

Total RNA is extracted using RNeasy mini kit (Qiagen) or TRIzol reagent (Life Technologies) and stored at −70°C until use.

52.2.2 DETECTION PROCEDURES

52.2.2.1 Conventional RT-PCR

Mase and Kanehira [42] described an RT-PCR-restriction enzyme analysis (REA) method for identification and differentiation of virulent and avirulent NDV isolates, with primers NDV-F2 (5′-TGGAGCCAAAACCGCGCACCTGCGG-3′; nt 4235–4258) and NDV-R2 (5′-GGAGGATGTTGGCAGCAT-3′; nt 5000–4983), which covers the cleavage site of F protein and the region to construct a phylogenetic tree for genotyping the virus.

RT is performed with 9 mers random primers and Super-Script™ III (Invitrogen). PCR amplification is conducted with resulting cDNA and Takara Ex Taq PCR kit (Takara), using the following cycling parameters: 35 cycles of 94°C for 30 s, 50°C for 30 s, and 72°C for 30 s.

The expected PCR product is about 750 bp. Restriction enzyme Hin1 I digestion of this product from avirulent NDV strains yields several fragments, but not from virulent NDV strains, with the exception of the JP/Miyadera/51 strain (genotype II). However, the PCR product from the JP/Miyadera/51 strain is digested into 450 and 316 bp fragments with restriction enzyme Apa I, while those from other strains are not digested.

Note: This RT-PCR-REA method allows rapid differentiation of avirulent and virulent NDV strains. Further details on the cleavage site of F protein may be obtained by sequencing the 750 bp product.

52.2.2.2 Real-Time RT-PCR

Guo et al. [9] utilized degenerate primers (forward: 5′-GAGCTAATGAACATTCTTTC-3′; reverse: 5′-AAYAGGC GRACCACATCTG-3′; where R = A/G and Y = C/T) and probe (5′-FAM-TCATACACTATTATRGCRTCATTCTT-BHQ1-3′) from the L gene sequence for quantitative detection of NDV.

RT-PCR mixture (20 µL) is made up of 0.4 µL (10 µM) of primers and probe, 10 µL of 2× One-Step RT-PCR Buffer III (One-Step PrimeScript RT-PCR Kit, TaKaRa), 0.4 µL of TaKaRa Ex Taq HS (5 U/µL), 0.4 µL of PrimeScript RT Enzyme Mix II, 6 µL of RNase free dH$_2$O, and 2 µL of template RNA sample.

Amplification and detection are carried out in LightCycler 480 II Real-time PCR system (Roche), with the following parameters: 42°C for 5 min, 95°C for 10 s, 40 cycles of 95°C for 5 s, and 52°C for 30 s.

Note: To verify the results, RT-PCR products may be separated by electrophoresis on a 2.0% agarose gel and visualized using ethidium bromide.

52.3 CONCLUSION

Despite a history that spans for 90 odd years and our continuing efforts to put it under control during this time, ND has been far from being defeated or eradicated. This may be attributable to the fact that the etiologic agent of ND—NDV—naturally occurs in wild birds, which help spread and replenish the virus during seasonal migrations. Given the potential of NDV to cause serious economical damages, to the domestic poultry industry, it is vital that rapid, sensitive, and specific diagnostic methods are utilized for its identification, pathotyping, and monitoring. This will ensure the implementation of appropriate control and prevention measures including vaccination that will help limit the spread of this severe and often fatal disease of birds.

REFERENCES

1. Miller PJ et al. Evidence for a new avian paramyxovirus serotype 10 detected in rockhopper penguins from the Falkland Islands. *J Virol.* 2010;84(21):11496–11504.
2. Briand FX, Henry A, Massin P, Jestin V. Complete genome sequence of a novel avian paramyxovirus. *J Virol.* 2012;86(14):7710.
3. Briand FX, Henry A, Brown P, Massin P, Jestin V. Complete genome sequence of a newcastle disease virus strain belonging to a recently identified genotype. *Genome Announc.* 2013;1(1):e00100–e00112.
4. Nayak B et al. Avian paramyxovirus serotypes 2–9 (APMV-2–9) vary in the ability to induce protective immunity in chickens against challenge with virulent Newcastle disease virus (APMV-1). *Vaccine.* 2012;30(12):2220–2227.
5. Khattar SK et al. Evaluation of the replication, pathogenicity, and immunogenicity of avian paramyxovirus (APMV) serotypes 2, 3, 4, 5, 7, and 9 in rhesus macaques. *PLoS One.* 2013;8(10):e75456.
6. Terregino C et al. Antigenic and genetic analyses of isolate APMV/wigeon/Italy/3920-1/2005 indicate that it represents a new avian paramyxovirus (APMV-12). *Arch Virol.* 2013;158(11):2233–2243.
7. Yamamoto E, Ito H, Tomioka Y, Ito T. Characterization of novel avian paramyxovirus strain APMV/Shimane67 isolated from migratory wild geese in Japan. *J Vet Med Sci.* 2015;77(9):1079–1085.
8. Duan X et al. Characterization of genotype IX Newcastle disease virus strains isolated from wild birds in the northern Qinling Mountains, China. *Virus Genes.* 2014;48(1):48–55.
9. Guo H et al. A comparative study of pigeons and chickens experimentally infected with PPMV-1 to determine antigenic relationships between PPMV-1 and NDV strains. *Vet Microbiol.* 2014;168(1):88–97.
10. Ganar K, Das M, Sinha S, Kumar S. Newcastle disease virus: Current status and our understanding. *Virus Res.* 2014;184:71–81.
11. Courtney SC et al. Complete genome sequencing of a novel newcastle disease virus isolate circulating in layer chickens in the Dominican Republic. *J Virol.* 2012;86(17):9550.
12. Kim SH et al. Complete genome sequence of a novel Newcastle disease virus strain isolated from a chicken in West Africa. *J Virol.* 2012;86(20):11394–11395.
13. Xie Z et al. Complete genome sequence analysis of a Newcastle disease virus isolated from a wild egret. *J Virol.* 2012;86(24):13854–13855.
14. Jardetzky TS, Lamb RA. Activation of paramyxovirus membrane fusion and virus entry. *Curr Opin Virol.* 2014;5:24–33.
15. Sharma B et al. Isolation of Newcastle disease virus from a non-avian host (sheep) and its implications. *Arch Virol.* 2012;157(8):1565–1567.
16. Ramey AM et al. Genetic diversity and mutation of avian paramyxovirus serotype 1 (Newcastle disease virus) in wild birds and evidence for intercontinental spread. *Arch Virol.* 2013;158(12):2495–2503.
17. Munir M, Zohari S, Berg M. Newcastle disease virus in pakistan: Genetic characterization and implication in molecular diagnosis. *Indian J Virol.* 2012;23(3):368–373.
18. Courtney SC et al. Highly divergent virulent isolates of Newcastle disease virus from the Dominican Republic are members of a new genotype that may have evolved unnoticed for over 2 decades. *J Clin Microbiol.* 2013;51(2):508–517.
19. Kumar A et al. Detection and molecular characterization of Newcastle disease virus in peafowl (*Pavo cristatus*) in Haryana State, India. *Indian J Virol.* 2013;24(3):380–385.
20. Samuel A et al. Phylogenetic and pathotypic characterization of Newcastle disease viruses circulating in west Africa and efficacy of a current vaccine. *J Clin Microbiol.* 2013;51(3):771–781.
21. Hao H et al. Genomic characterisation of two virulent Newcastle disease viruses isolated from crested ibis (*Nipponia nippon*) in China. *Gene.* 2014;553(2):84–89.
22. Marks FS et al. Targeted survey of Newcastle disease virus in backyard poultry flocks located in wintering site for migratory birds from Southern Brazil. *Prev Vet Med.* 2014;116(1–2):197–202.
23. Haroun M, Mohran KA, Hassan MM, Abdulla NM. Molecular pathotyping and phylogenesis of the first Newcastle disease virus strain isolated from backyard chickens in Qatar. *Trop Anim Health Prod.* 2015;47(1):13–19.
24. Śmietanka K et al. Experimental infection of different species of birds with pigeon paramyxovirus type 1 virus—evaluation of clinical outcomes, viral shedding, and distribution in tissues. *Avian Dis.* 2014;58(4):523–530.
25. Lu A et al. Evaluation of histopathological changes, viral load and immune function of domestic geese infected with Newcastle disease virus. *Avian Pathol.* 2014;43(4):325–332.
26. Mehrabanpour MJ, Khoobyar S, Rahimian A, Nazari MB, Keshtkar MR. Phylogenetic characterization of the fusion genes of the Newcastle disease viruses isolated in Fars province poultry farms during 2009–2011. *Vet Res Forum.* 2014;5(3):187–191.
27. Kapczynski DR, Afonso CL, Miller PJ. Immune responses of poultry to Newcastle disease virus. *Dev Comp Immunol.* 2013;41(3):447–453.
28. De Battisti C et al. Rapid pathotyping of Newcastle disease virus by pyrosequencing. *J Virol Methods.* 2013;188(1–2):13–20.
29. Desingu PA et al. A rapid method of accurate detection and differentiation of Newcastle disease virus pathotypes by demonstrating multiple bands in degenerate primer based nested RT-PCR. *J Virol Methods.* 2015;212:47–52.

30. Wise MG et al. Development of a real-time reverse-transcription PCR for detection of Newcastle disease virus RNA in clinical samples. *J Clin Microbiol.* 2004;42:329–338.

31. Mia Kim L, Suarez DL, Afonso CL. Detection of a broad range of class I and II Newcastle disease viruses using a multiplex real-time reverse transcription polymerase chain reaction assay. *J Vet Diagn Invest.* 2008;20(4):414–425.

32. Abdel-Glil MY, Mor SK, Sharafeldin TA, Porter RE, Goyal SM. Detection and characterization of Newcastle disease virus in formalin-fixed, paraffin-embedded tissues from commercial broilers in Egypt. *Avian Dis.* 2014;58(1):118–123.

33. Miller PJ, Torchetti MK. Newcastle disease virus detection and differentiation from avian influenza. *Methods Mol Biol.* 2014;1161:235–239.

34. Nanthakumar T, Kataria RS, Tiwari AK, Butchaiah G., Kataria JM. Pathotyping of Newcastle disease viruses by RT-PCR and restriction enzyme analysis. *Vet Res Commun.* 2000;24:275–286.

35. Qiu X et al. Development of strand-specific real-time RT-PCR to distinguish viral RNAs during Newcastle disease virus infection. *Sci World J.* 2014;2014:934851.

36. Rabalski L, Smietanka K, Minta Z, Szewczyk B. Detection of Newcastle disease virus minor genetic variants by modified single-stranded conformational polymorphism analysis. *Biomed Res Int.* 2014;2014:632347.

37. Saba Shirvan A, Mardani K. Molecular detection of infectious bronchitis and Newcastle disease viruses in broiler chickens with respiratory signs using Duplex RT-PCR. *Vet Res Forum.* 2014;5(4):319–323.

38. Xie Z et al. Simultaneous typing of nine avian respiratory pathogens using a novel GeXP analyzer-based multiplex PCR assay. *J Virol Methods.* 2014;207:188–195.

39. OIE (Office International des Epizooties). *OIE Terrestrial Manual 2012.* Chapter 2.3.14: Newcastle disease. http://www.oie.int/fileadmin/Home/fr/Health_standards/tahm/2.03.14_NEWCASTLE_DIS.pdf. Accessed on June 12, 2015.

40. Tsunekuni R, Hikono H, Saito T. Evaluation of avian paramyxovirus serotypes 2 to 10 as vaccine vectors in chickens previously immunized against Newcastle disease virus. *Vet Immunol Immunopathol.* 2014;160(3–4):184–191.

41. Liu L et al. Assessment of preparation of samples under the field conditions and a portable real-time RT-PCR assay for the rapid on-site detection of Newcastle disease virus. *Transbound Emerg Dis.* September 11, 2014.

42. Mase M, Kanehira K. Simple differentiation of avirulent and virulent strains of avian paramyxovirus serotype-1 (Newcastle disease virus) by PCR and restriction endonuclease analysis in Japan. *J Vet Med Sci.* 2012;74(12):1661–1664.

53 Nipah Virus

Supaporn Wacharapluesadee, Akanitt Jittmittraphap,
Sangchai Yingsakmongkon, and Thiravat Hemachudha

CONTENTS

53.1 Introduction 455
    53.1.1 Classification, Morphology, and Genome Organization 456
    53.1.2 Epidemiology 456
        53.1.2.1 Epidemiology of Nipah Virus Outbreak in Malaysia 456
        53.1.2.2 Epidemiology of Nipah Virus Outbreak in Bangladesh 456
        53.1.2.3 Infection in Bats 457
    53.1.3 Clinical Features and Pathogenesis 457
        53.1.3.1 In Humans 457
        53.1.3.2 In Pigs 458
        53.1.3.3 In Bats 458
        53.1.3.4 In Other Animals 458
    53.1.4 Diagnosis 458
        53.1.4.1 Conventional Techniques 458
        53.1.4.2 Molecular Techniques 460
53.2 Methods 461
    53.2.1 Sample Collection and Preparation 461
        53.2.1.1 Sample Collection 461
        53.2.1.2 Sample Preparation 461
    53.2.2 Detection Procedures 462
        53.2.2.1 One-Step RT-PCR QIAGEN 462
        53.2.2.2 Nested PCR 462
        53.2.2.3 Heminested PCR and Sequencing 463
53.3 Conclusion 464
Acknowledgments 464
References 464

53.1 INTRODUCTION

Nipah virus (NiV) is a bat-borne paramyxovirus that causes disease in both humans and animals. The first recorded outbreak of NiV infection occurred in Malaysia from September 1998 to May 1999 [1]. The disease was named after Kampung Sungai Nipah (Nipah River Village), where the first viral isolate was obtained. Retrospective investigations suggested that NiV was responsible for sporadic diseases in pigs in Peninsular Malaysia since late 1996 but was not recognized as the clinical signs, morbidity, and mortality were not markedly different from those of several endemic pig diseases [2].

NiV is a highly pathogenic paramyxovirus and biosafety level (BSL)-4 containment is required to work with live virus in laboratories. The virus causes severe febrile encephalitis and respiratory distress syndrome with high mortality rate in humans. During the outbreaks in Malaysia and Singapore, most human infections occurred in workers in pig slaughterhouse who had direct contact with NiV-infected pigs. A relatively low mortality rate was noted in pigs. Approximately 1.1 million pigs were culled for a successful outbreak control [3]. Bats of the genus *Pteropus* were identified as the natural reservoir hosts of NiV.

NiV encephalitis was subsequently described in India in 2001 [4], and several NiV outbreaks were also reported in Bangladesh from 2001 to present. Drinking raw date palm sap contaminated with NiV from bat urine or saliva has been identified as a major route of transmission of NiV to humans in Bangladesh. There has been no intermediate animal host as in the outbreak in Malaysia [5]. Human-to-human transmission among family members, community care givers [6], and care workers has been documented in Bangladesh [7]. In 2014, there have been reports of 17 persons in the Philippines associated with NiV infection [8]. They had acute encephalitis, meningitis, or influenza-like illness. Almost all had participated in horse slaughtering and/or horse meat consumption.

53.1.1 CLASSIFICATION, MORPHOLOGY, AND GENOME ORGANIZATION

Classification: NiV is a single-stranded, negative-sense RNA virus belonging to the *Henipavirus* genus in the *Paramyxovirinae* subfamily, *Paramyxoviridae* family [9]. The *Paramyxoviridae* family is divided into two subfamilies: the *Paramyxovirinae* and the *Pneumovirinae* [10]. In turn, the subfamily *Paramyxovirinae* is separated into five genera: *Rubulavirus*, *Respirovirus*, *Avulavirus*, *Morbillivirus*, and *Henipavirus*, the latter of which has been added recently. Hendra virus (HeV), initially discovered in 1994, is the prototype of the genus *Henipavirus*, while NiV is the second species, which was first identified in Malaysia in 1999. Recently, a new paramyxovirus, Cedar virus, was isolated from Australian bats that share significant features with the known henipaviruses and has been classified as the third species within this genus [11].

Morphology: NiV shares common morphological characteristics with other members of the family *Paramyxoviridae*, including a herringbone-like nucleocapsid structure and nucleocapsid accumulation along plasma membranes during budding. Its RNA genome consists of helical nucleocapsids encased in an envelope forming relatively spherical, pleomorphic virus particle. However, some unique characteristics exist, including unusual cytoplasmic ringlike reticular inclusions in the proximity to the membranes of the endoplasmic reticulum and long cytoplasmic tubules in the periphery of NiV-infected cells. Nipah virions on average are larger in diameter than typical paramyxoviruses (150–400 nm) [1,12]. Extracellular virus particles are pleomorphic, with an average diameter of 500 nm.

Genome organization: The genome of NiV isolates from Malaysia is 18,246 nucleotides (nt) in size [13], 12 nt longer than the HeV. NiV isolates from Bangladesh contain 18,252 nt [14], one of the largest genomes among paramyxoviruses, with the exception of the rodent-borne J-virus and Beilong virus [15]. The genome size of Cedar virus, the new member of *Henipavirus* is 18,162 nt [11]. The *Henipavirus* genome consists of six genes encoded from 3′ to 5′, the nucleocapsid (N), phosphoprotein (P), matrix (M), fusion (F), attachment glycoprotein (G), and large (L) polymerase, respectively. The NiV N protein consists of 532 amino acids (aa) and is responsible for encapsidating the viral genome and its typical herringbone-like structures seen in electron micrographs [16]. The NiV M protein, 352 aa long, interacts with the cytoplasmic tail of the F protein, the ribonucleoprotein complex, and the inner surface of the virion envelope, to provide rigidity and structure to the virion [16]. The NiV F protein is a 546 aa long type I transmembrane protein, which forms trimers on the cell surface. It is responsible for mediating the fusion of virus and host cell membranes during infection [9]. The NiV G protein is a 604 aa long type II membrane protein that forms tetramers upon cell surface expression. It serves primarily in the cellular receptor binding [9]. The L gene of NiV is 6955 nt in length and contains an open reading frame (ORF) that is predicted to encode an L protein consisting of 2244 aa. It is the largest and the least abundant viral protein. There are many different enzymatic functions attributed to it, including initiation, elongation, and termination of both mRNA transcription (capping, methylation, and polyadenylation of mRNA) and genome replication [16]. The NiV P is 709 aa long and is the only essential gene product encoded by the P gene for genome replication [17]. In addition to coding for the P, the P gene of NiV encodes three other nonstructural proteins (C, V, and W). The V and W proteins are generated by RNA editing, while the C protein is encoded by a second ORF downstream of the translational initiation site for the P ORF [18].

Nucleotide homology in the ORFs of the various genes of HeV and NiV exceeds 70%, with more than 80% amino acid identity in most genes, but less than 49% nucleotide similarity compared with other members of the *Paramyxovirinae* family [18]. Sequence analyses of the N, P, M, F, and G genes of NiV isolated from human and pigs in Malaysia were nearly identical, while NiV isolated from *Pteropus hypomelanus* shortly after the outbreak had 5–6 nt differences when compared to the human isolates [19].

53.1.2 EPIDEMIOLOGY

53.1.2.1 Epidemiology of Nipah Virus Outbreak in Malaysia

The first incidence of human NiV infection was recognized in a large outbreak in Peninsular Malaysia from September 1998 to May 1999 [1]. The initial human cases were identified among pig farmers in the city of Ipoh, northwestern part of Peninsular Malaysia, in late September 1998. By December 1998, larger clusters of similar cases were reported 300 km south of Ipoh. A total of 283 cases were reported with 109 (39%) fatalities [20]. Between March 10 and 19, 1999, there was a subsequent outbreak in one of Singapore's abattoirs due to infected pigs imported from Malaysia. Eleven workers developed NiV-associated encephalitis or pneumonia, with one fatality [21]. The identical sequences of NiV isolates from humans and pigs, combined with the observation that humans with NiV infection had close contact with pigs' secretions and excretions, suggested that NiV was transmitted from infected pigs to humans through direct contact with pigs or fresh pig products [22]. There was no report of person-to-person transmission of NiV in Malaysia or Singapore.

Since 1999, NiV infection in human and animal has not been detected in Malaysia or Singapore but has been recurring in India and Bangladesh since 2001. The alarming one was an outbreak in the Philippines in 2014 [8].

53.1.2.2 Epidemiology of Nipah Virus Outbreak in Bangladesh

Nipah encephalitis/acute respiratory distress syndrome with very high case fatalities in humans have been recognized almost annually in Bangladesh since 2001. More than 70% of the NiV-infected patients have died [23]. As of January 2014,

304 human cases with NiV infection, leading to 232 (76%) fatalities, in Bangladesh have been reported, [24]. Drinking raw date palm sap contaminated with bat urine or saliva containing NiV has been demonstrated as a mode of transmission of NiV to humans in Bangladesh. There were no intermediate animal hosts [5]. Other mode of transmission is via human to human, usually nosocomial, or corpse to human [7]. A small number of NiV-infected cases in Bangladesh had history of contact with sick cows [25]. The retrospective study suggested that NiV patients were more likely than noninfected patients to have had contact with an ill cow (OR 7.9, CI 2.2–27.7). However, there was no definite proof due to unavailability of cow specimens for testing.

53.1.2.3 Infection in Bats

Bats of the genus *Pteropus* have been suggested as the potential natural reservoir hosts for NiV [19]. Table 53.1 shows a summary of testing for NiV in bats from outbreak and non-outbreak countries. Successful virus isolation from bats has been reported from the urine of *Pteropus vampyrus* [38] and *P. hypomelanus* [19] in Malaysia and *P. lylei* in Cambodia [39]. NiV infection in non-*Pteropus* bat was evident by the response of antibody (Table 53.1). Viral RNA was detected in bats from India, Thailand, Indonesia, and East Timor. The seasonal prevalence of NiV in *P. lylei* was demonstrated by the presence of viral RNA from bat urine [40] whose sequences were similar to those found in several outbreaks in

Bangladesh during the early part of the year. Nested reverse transcription polymerase chain reaction (RT-PCR) method has applied to detect NiV RNA in the urine, saliva, and serum of bats [41–43].

53.1.3 CLINICAL FEATURES AND PATHOGENESIS

53.1.3.1 In Humans

NiV outbreak in Malaysia caused severe, rapidly progressive encephalitis with a mortality rate of nearly 40%. The incubation period in humans ranged from 4 days to 2 months, where more than 90% of the cases had an incubation period of 2 weeks or less [20]. The main presenting features were fever, headache, dizziness, and vomiting. Distinctive clinical signs included segmental myoclonus, areflexia and hypotonia, hypertension, and tachycardia, suggesting brain stem and upper cervical cord involvement [44]. A few patients developed atypical pneumonia. Two unique features of encephalitis were observed in the cases in Malaysia as acute encephalitis with vasculitis and late onset or relapse with multifocal cortical gray matter involvement [45].

On the other hand, the case-fatality rate was higher in Bangladesh and India than in Malaysia. The clinical characteristics of NiV infection in Bangladesh other than encephalitis include a severe respiratory component, which was not reported in the outbreaks in Malaysia [46]. Other common clinical features include fever, altered mental status, headache,

TABLE 53.1
Summary of Bat Species Positive for Nipah Virus

Country	Bat Species	Testing Method	Year of Study (References)
Outbreak countries			
Malaysia	*Pteropus hypomelanus, Eonycteris spelaea, Cynopterus brachyotis, Scotophilus kuhlii, Pteropus vampyrus*	Antibodies	1999 [26]
	P. hypomelanus	Virus isolation	1999–2000 [19]
	P. vampyrus	Antibodies, virus isolation	2004–2005 [38]
	P. hypomelanus (antibodies only), *P. vampyrus*	Antibodies and RNA	2004 [27]
Bangladesh	*Pteropus giganteus*	Antibodies	2003, 2007 [25,28]
Nonoutbreak countries			
Africa	*Eidolon helvum*	Antibodies	1992, 2007–2011 [29,30]
Cambodia	*Pteropus lylei*	Antibodies, virus isolation	2000–2001 [39]
China	*Rousettus leschenaultia, Hipposideros armiger, H. pomona, Miniopterus* spp., *Myotis daubentonii, M. ricketti, Rhinolophus affinis, R. sinicus*	Antibodies	2004–2007 [31]
East Timor	*P. vampyrus, R. amplexicaudatus*	RNA	2004–2005 [32]
Ghana	*E. helvum, Epomophorus gambianus, Hypsignathus monstrosus*	Antibodies	2007 [76]
India	*P. giganteus*	Antibodies	2003 [33]
	P. giganteus	RNA, antibodies	2009–2010 [34]
Indonesia	*P. vampyrus*	RNA, antibodies	2009 [79]
Madagascar	*Eidolon dupreanum, Pteropus rufus*	Antibodies	2003–2005 [35]
Papua New Guinea	*Dobsonia magna, Pteropus alecto, P. conspicillatus*	Antibodies	1996–1997, 1999, 2008 [78,36]
Thailand	*P. hypomelanus, P. vampyrus, P. lylei, Hipposideros larvatus*	Antibodies	2002–2005 [41]
	P. lylei, H. larvatus	RNA	2002–2005 [41]
Vietnam	*Rousettus leschenaultia, Cynopterus sphinx*	Antibodies	2007–2008 [37]

cough, respiratory symptoms, vomiting, and convulsions. The median incubation period is 9 days (range 6–11 days). Cases with late-onset or relapsed encephalitis were not mentioned in Bangladesh and India.

Clinical features of patients in the Philippines include acute encephalitis, meningitis, and influenza-like illness where mortality was found only in the case of encephalitis (82%) [8]. There was also human-to-human transmission. One serum sample from patient was positive by real-time PCR for NiV, and one cerebrospinal fluid (CSF) sample from another of these patients was NiV positive by next-generation sequencing (single-sequence read with 71 bp of the P gene of NiV).

53.1.3.2 In Pigs

NiV infection in pigs was reported only from Malaysia during 1998–1999. The spread of NiV infection among pigs is easily spread through transport of pigs from farm to farm. The incubation period was estimated to be 7–14 days [47]. The clinical patterns of naturally infected pigs varied according to the age of the pigs. Sows were noted to primarily present the neurological symptoms, while porkers predominantly exhibited respiratory symptoms. However, many infected pigs appeared to be asymptomatic.

Infection in suckling pigs (piglets) presented with higher mortality rate (approximately 40%). The major symptoms include open-mouth breathing, leg weakness with muscle tremors, and neurological twitches.

Pigs from 4 weeks to 6 months of age (weaners and porkers) usually presented acute febrile illness (>39.9°C) with respiratory signs ranging from rapid and labored respiration to a harsh nonproductive cough (loud barking cough). The mortality rate is less than 5%, but the morbidity rate approaches 100%. In less severe cases, open-mouth breathing was presented, while hemoptysis was found in severe cases. One or more of neurological signs usually accompanied the respiratory signs, for example, trembling and neurological twitches, muscle spasms and myoclonus, rear leg weakness and varying degrees of spastic paresis or lameness, uncoordinated gait when driven and hurried, and generalized pain especially in the hindquarters.

Similar symptoms were presented in boars and sows. Sudden death could be observed. However, the usual presentations were acute febrile illness (>39.9°C), with labored breathing, increased salivation (drooling or frothy), nasal discharge (serous or mucopurulent or bloody), and possible early abortion for sows (first trimester). Some or all of the following neurological signs, such as agitation and head pressing, tetanus-like spasm and seizures, nystagmus, champing of the mouth, and apparent pharyngeal muscle paralysis, were also present.

53.1.3.3 In Bats

Fruit bats of the genus *Pteropus* are the natural hosts for NiV, but infection in bats causes no apparent disease. The longitudinal study of captive *P. vampyrus* bats strongly suggests that NiV circulates in wild bat populations and their antibodies can be maintained for long periods [48]. It was also found that maternal antibodies against NiV were transmitted passively.

This study proposes that NiV was recrudesced and then transmitted horizontally between bats during the study period (1 year).

Experimental subcutaneous NiV infection in gray-headed fruit bats (*Pteropus poliocephalus*) resulted in the production of neutralizing antibodies in serum and histological changes in different tissues. However, these bats developed subclinical infection that was characterized by the transient presence of the virus within selected viscera, in episodic viral excretion, and in seroconversion, despite the inoculation of rather high viral doses [49]. The virus isolation was successful from the urine, kidney, and uterus.

53.1.3.4 In Other Animals

Other animals found to be infected with NiV during the initial outbreak in Malaysia were found to be naturally infected with NiV, such as cats, dogs, and horses [1,50]. Serological evidence of dogs in the areas in proximity with the farms during the outbreak in Malaysia revealed that dogs were susceptible to NiV infection [51]. Pulmonary vasculitis, alveolar edema, and inflammation were demonstrated in two naturally infected dogs [52]. A number of species of animals have been employed as animal models for NiV infection including fruit bats, pigs, horses, cats, ferrets, squirrel monkeys, African green monkeys, mice, hamsters, and guinea pigs [53]. Cats are susceptible to NiV infection under natural or experimental conditions. Severe pulmonary inflammation and bronchial epithelium involvement may be prominent. Vasculopathy was observed in many organs, except the brain parenchyma, but meninges were involved [52].

During 2014, *Henipavirus* infection caused severe illness and death in horses in southern Philippines [8]. Ten horses died with neurological symptoms (9/10), but no samples were available for testing. Four cats and one dog that ate horse meat died. Four serum samples from dogs were positive for neutralizing antibodies against NiV.

53.1.4 Diagnosis

A major challenge for NiV diagnosis is that the virus is listed as a BSL-4 agent. The first-class equipment for advanced diagnostic platforms are lacking in many developing countries identified as hot spots for NiV outbreaks. The diagnostic capability (exploiting both phenotypic and genotypic characteristics of NiV) has continually been developed and refined, but no commercial diagnostic kit for NiV is currently available.

53.1.4.1 Conventional Techniques

Conventional techniques useful for NiV detection range from virus isolation, electron microscopy (EM) examination, immunohistochemistry, to serological tests.

Virus isolation and EM are important for identification and characterization. NiV grows well in African green monkey kidney (Vero) cells, and the cytopathic effect (CPE) is usually seen within 3–5 days. CPE is not specific because many other viruses can cause similar syncytia formation;

thus, another test to identify the virus is required. Isolation of NiV requires laboratories with BSL-4 containment and has a very slow turnaround time in inoculation, with low viral titer sample. NiV was isolated from two bat species in Malaysia, *P. hypomelanus* (from urine and a partially eaten fruit swab) [19] and *P. vampyrus* (from urine) [38] and from the urine of *P. lylei* in Cambodia [39]. NiV from pigs was isolated after infecting Vero cells with lung tissue samples from NiV-infected pigs that died from acute respiratory distress syndrome and encephalitis [54]. It is recommended that tissue samples from the brains, lungs, kidneys, and spleens of suspected NiV-infected animals should be submitted for viral isolation [55]; otherwise, they can be tested by molecular assays. Samples should be transported at a temperature of 4°C to the designated laboratory within 48 h or should be sent frozen on dry ice, or nitrogen vapors should be used if shipping is over 48 h [56]. However, in the absence of BSL-4 containment, primary virus isolation from suspected samples can be conducted under strict BSL-3 conditions [57]. When CPE with syncytia is observed, the cell culture and manipulation of potentially infectious specimens should be stopped. Fixed cells should be confirmed using proper techniques: immunofluorescence assay or RT-PCR assay in supernatant fluids. In NiV-positive cases, the specimen should be transferred to an appropriate laboratory for further identification and analysis. Recently, different pteropid bat cell lines have been developed at Australian Animal Health Laboratory as a useful additional tool for virus isolation [58].

Immunohistochemistry proved to be one of the most useful tests in NiV investigation during the initial outbreak in Malaysia [59]. This technique is safe and can be performed using formalin-fixed tissue specimens. There is a wide range of tissues for NiV antigen detection. During the outbreak in Malaysia, the presence of NiV antigens were demonstrated by immunohistochemistry in syncytial cells in the lungs, in epithelial cells lining the upper airway passages, and in intraluminal necrotic debris in the upper and lower airways of infected pigs [1]. NiV antigen was also detected in the kidneys by focally staining the renal tubular epithelial cells. Ideal specimens for immunohistochemistry testing should include samples from various regions of the brain, lungs, mediastinal lymph nodes, spleen, and kidneys. In pregnant animals, the uterus, placenta, and fetal tissues should be included [55].

Initially, the immunostaining for NiV antigens was performed with rabbit HeV antiserum and later with convalescent pig and cat anti-NiV antisera [55]. At present, it has been replaced by hyperimmune mouse ascetic fluids [57] or rabbit polyclonal antibodies against NiV [60]. Monoclonal antibodies (MAbs) against NiV antigen were developed, and their abilities to detect NiV antigen in formalin-fixed tissues of naturally infected pig were tested [61]. The results were consistent with the previous assays using rabbit polyclonal antibody.

Several serological tests have been developed including serum neutralization test (SNT), enzyme-linked immunosorbent assay (ELISA), Luminex-based binding and inhibition antibody tests, and SNT based on pseudotyped viruses. Detection of antibodies by immunofluorescence assay, Western blotting, and immunoprecipitation assays is possible, but these methods are not routinely used [57].

SNT is the gold standard for serology and the confirmation test for ELISA results. It uses live virus to measure antibody function and requires BSL-4 containment [55]. A rapid immune plaque assay has been developed to detect neutralizing NiV antibodies [62]. The turnaround time is reduced from up to 3 days in the traditional microneutralization assay to 1 day in the immune plaque assay. Recently, several systems of SNT assays using pseudotyped viruses expressing F and G protein of NiV as targeted antigens have been developed, and the assay can be performed at BSL-2 laboratory [63]. The NiV recombinant vesicular stomatitis virus systems with luciferase [64] and green fluorescent protein (GFP) reporters [65] have been developed. Recently, the pseudovirus system using nonfluorescent reporter has been developed, where the GFP reporter gene was replaced by secreted alkaline phosphatase (SEAP). SEAP activity in culture supernatant can then be measured on an ELISA plate reader [66].

ELISA is convenient, rapid, and widely used as it can be safely performed at BSL-2 laboratory. Test serum is either inactivated by γ-irradiation or heated at 56°C. ELISA procedures for the detection of NiV-specific immunoglobulin (Ig) G and IgM have been developed at U.S. Centers for Disease Control and Prevention (CDC) [57]. They can be used for the detection of antibody in humans, livestock, and wildlife animals without a nonspecific reaction. NiV–Vero E6, which is prepared in a BSL-4 laboratory and inactivated by γ-irradiation, is used as an antigen, while a mock-infected Vero E6 is used as a control [57]. It has been found that CDC ELISA and SNT yield a relative sensitivity of 87% and specificity of 99% in bat sera testing [39]. The recombinant NiV G antigens expressed from baculoviruses or *Escherichia coli* [67,68], and NiV N antigen expressed from *E. coli* can be produced in a BSL-2 laboratory [69]. Results from the initial tests are promising. A MAb-based solid-phase blocking ELISA for the detection of NiV antibodies has been developed as a screening test for NiV infection and evaluated using samples from bats and pigs [70].

Anti-N and anti-P MAbs were used to develop two antigen capture ELISAs, one for virus detection and the other for differentiation between NiV and HeV. The lower limit of detection of the capture assay with both MAbs was 400 pfu [71]. The purified polyclonal antibodies derived from NiV G-encoding DNA-immunized rabbits were also developed for the detection of NiV antigen by ELISA [72]. The lowest detection limit against live Bangladesh strain viruses was 2.5×10^4 $TCID_{50}$ (50% tissue culture infectious doses). The applications of both capture ELISAs on natural biological specimens remain to be evaluated. In future, this system may provide a high-throughput format with simple and affordable methodology that could serve as an alternative rapid diagnostic assay.

Without the need for a BSL-4 facility, ELISA remains to be the most affordable serological test for the diagnosis of NiV infection, but it requires inactivated NiV–Vero E6 cell slurry antigen. During the past decade, novel serological technologies have been developed, increasing the surveillance capacity especially in developing countries. Examples of new serological technologies include the new format of NiV antigens, which have been expressed by recombinant DNA technology, multiplex detection of multiple pathogen antibodies using microsphere technology, and the NiV–pseudovirus for the SNT where the test can be performed in a low-containment-level laboratory.

Luminex-based multiplexed microsphere assays have been developed specifically for the detection of *Henipavirus* (HeV and NiV) antibodies [73]. Two different test formats were developed, a total antibody-binding and a restricted receptor-blocking format designed as a surrogate for virus neutralization. The target antigen for both assays is recombinant-expressed surface G protein of HeV and NiV that are coated on two Luminex beads with distinctive fluorescent spectra. For binding assays, the bound antibodies were detected by biotinylated protein A/G, followed by streptavidin–phycoerythrin. On the other hand, in the blocking format, the cellular receptor, a recombinant ephrin-B2 labeled with biotin, will bind to the G protein on the beads in the absence of *Henipavirus*-specific antibodies. This supports the binding of streptavidin–phycoerythrin molecule, resulting in a positive binding signal. The Luminex blocking assay may be a substitute for SNT that can be performed in the absence of live virus at a standard laboratory without any high biocontainment need and requires very small amounts of sera while providing valuable data quickly (within 2–3 h) [73]. In addition, Luminex assays have significantly enhanced the capability and capacity to confirm HeV infection in humans [74], animal models [75], and international serosurveillance for henipaviruses in bats and other species including fruit bats and domestic pigs in West Africa [76,77], *Pteropid* bats in Papua New Guinea [78], and *P. vampyrus* bats in Indonesia [79].

53.1.4.2 Molecular Techniques

By targeting the nucleic acids, molecular techniques play an increasingly important role in the diagnosis of viral infection since they have high sensitivity and specificity and do not require a high-containment-level laboratory. The commonly applied molecular techniques include conventional PCR, real-time PCR, sequencing, and viral multiplex detection by microsphere suspension array system. Together with ELISA, PCR is indispensable during NiV surveillances in bats, the natural reservoir hosts for NiV. The conventional PCR system can be found in most laboratories in developing countries, and the positive control RNA or DNA can easily be constructed by direct synthesis or using in vitro transcription technology [80].

Genome sequence of NiV clearly places it in the *Paramyxoviridae* family and is closely related to HeV. NiV RNA was initially amplified using RT-PCR assay with primers designed to amplify 141 bp of the P gene, which has conserved sequences across many members of the *Paramyxoviridae* family [1]. The PCR primers used to amplify complete N, P, and M genes of NiV were designed from the initial short PCR products of these genes and sequence data from HeV in a stepwise fashion [1]. These three genes from NiV share 70% to 78% nucleotide homology with HeV [1]. NiV N-specific PCR

used to amplify the 228 bp amplicon was developed to study NiV infection in tissues and serum specimens from infected humans and pigs during the outbreaks in Malaysia and Singapore and specimens from infected cats and dogs from Malaysia [1]. The sequences derived from isolates of human cases and pigs from both countries were identical.

Two conventional RT-PCR assays were developed during the recent NiV outbreaks in India and Bangladesh. The primer set NVNBF-4 (5′-GGAGTTATCAATCTAAGTTAG-3′) and NVNBR-4 (5′-CATAGAGATGAGTGTAAAAGC-3′) amplifies 159 bp of the N gene of NiV. The primer set NVBMFC1 (5′-CAATGGAGCCAGACATCAAGAG-3′) and NVBMFR2 (5′-CGGAGAGTAGGAGTTCTAGAAG-3′) amplifies 320 bp of the M gene [4,6]. NiV RNA in urine samples from 5 of 6 patients was detected by RT-PCR of N gene, while RNA from the M gene was detected in 3 of these 5 samples [4]. Sequence analysis of partial N and M gene sequences (159 and 320 bp, respectively) can be used to confirm NiV infection and differentiate between those in Bangladesh in 2004 and Malaysia in 1999 [4].

Another conventional PCR method, the duplex-nested RT-PCR, which was initially developed for surveillance study in bats [43], can be useful for NiV RNA detection in animals and humans. It has been shown to detect and differentiate NiV infection between that from Malaysia and that from Bangladesh. This method is based on the nested RT-PCR at N gene, yielding a high sensitivity; however, it requires standardization and stringent quality control to avoid false-positive result from cross contamination or false-negative result from PCR inhibitors. Using a positive control RNA construct of different sizes and nucleotide sequences can help differentiating the positive control from wild-type NiV [80]. Further, coamplification of extrinsic RNA with wild-type RNA, by adding a minimum amount of extrinsic RNA to the PCR reagents before starting the reaction, could monitor the PCR inhibitor contamination in the biological specimens [43]. The PCR product should be sequenced for confirmation. Sizes of first- and second-round PCR products are 398 and 227 bp, respectively. Two heminested PCRs can be performed from the same set of primers, and 357 bp sequences of NiV RNA can be obtained for further bioinformatics analysis. Details of the method can be found in the next section. Real-time RT-PCR methods for the detection and quantitation of NiV have been developed for the screening of a population of known negative pig samples [80], assessment of NiV infection in a feline model under the vaccine study [81], quantitation of NiV replication kinetics in Vero cells [82], and detection and characterization of NiV in cell culture and in biological samples [83]. All four PCR methods target the N gene for amplification, but at different positions. The array consensus PCR assays for the detection of henipaviruses at the N, P, M, and L genes by using the SYBR Green and TaqMan real-time PCRs and a conventional heminested PCR have been developed [84]. It has been found that N gene SYBR Green assay is the most useful assay for the investigation of unknown henipaviruses. The detection procedures of these five methods have been published [85].

Rather than serological assay, microsphere array system can be applied for simultaneous fluorescent identification of multiple separate nucleotide targets in a single reaction. Microsphere suspension array system (Luminex MagPlex-TAG) targets multiple sites within the N and P genes for the simultaneous detection and differentiation between HeV and NiV [86]. The analytical sensitivity of the microsphere array assays coordinated with the standard qPCR assay. The assays offer advantages over qPCR as it can multiplex; however, it requires more turnaround time. Two PCRs are required before hybridization followed by fluorescence detection using the Bio-Plex instrument.

53.2 METHODS

53.2.1 Sample Collection and Preparation

53.2.1.1 Sample Collection

Human antemortem specimens that can be used for NiV testing are CSF, saliva, urine, and nasal and pharyngeal secretions [3]. The specimens should be collected during the acute, febrile stages of the illness and frozen on dry ice or liquid nitrogen for further diagnosis by virus isolation or molecular techniques. NiV was isolated and identified using molecular techniques from porcine lung tissue, canine brain tissue, and feline brain tissue during the first outbreak in Malaysia [1]. Excretions and secretions such as urine, saliva, and pharyngeal and respiratory secretions of infected pigs were the source of transmission between pigs in the same farm during the outbreak in Malaysia and can be used for molecular testing [3]. In experimentally infected pigs, NiV could be detected in CSF, pharyngeal swabs and nasal washes, brain tissue, olfactory bulb, trigeminal ganglion, turbinate, trachea, submandibular lymph node, and bronchiolar lymph node using real-time RT-PCR [87].

Formalin-fixed tissue and paraffin-embedded blocks are suitable for immunohistochemistry assay and can be shipped at room temperature. Specimens for serology should be preserved and transported at a temperature of −20°C or below [57]. NiV isolation was successfully performed from bat urine and a swab from a partially eaten fruit [19]. NiV RNA has been detected in urine, blood, saliva, liver and bladder from naturally infected bats using PCR assay [34,41,42,79].

53.2.1.2 Sample Preparation

Excretions and secretions such as urine, saliva, and pharyngeal and respiratory secretions of pigs, porcine lung tissue, canine brain tissue, and feline brain tissue are sources of specimen for detection of NiV by molecular assays [1]. All types of specimens should be transported to the laboratory within 24 h and kept cool at 4°C during transportation. For long-term storage, specimens should be frozen at −80°C. Lysis buffer containing a chaotropic salt (guanidine isothiocyanate) can be used to collect the specimen for field sampling in order to preserve the nucleic acid and kill the virus at site [40].

Samples for NiV RNA testing can be collected directly from captured bats (saliva, urine, or blood) or from the

environment using plastic sheets placed directly under bat roosts to collect their urine [40,41]. Oropharyngeal or urine swab is placed in a cryovial filled with 1 mL of NucliSENS Lysis Buffer (bioMérieux). Every 4–6 drops of urine swab collected from plastic sheet should be put immediately into 9 mL of NucliSENS Lysis Buffer [40]. Two cotton swabs containing approximately 1.2 mL of bat urine are pooled in each lysis buffer tube. The total nucleic acid is then extracted directly from these specimens using the Boom technology method according to the manufacturer's instructions (bioMérieux).

A dried blood spot (DBS) is an alternative type for sample collection for NiV surveillance in resource-limited settings. A blood sample of 50 μL is dropped onto the Flinders Technology Associates filter paper cards [FTA® cards] (Whatman), air-dried at room temperature, and kept in plastic bag containing a desiccator for further analysis or long-term storage (at −80°C). DBS filter paper containing 50 μL of bat blood is a source for NiV RNA PCR detection. NiV RNA extracted from DBS showed concordant results with RNA extracted from bat saliva [56].

Several commercially available methods can be used for extraction of viral RNA from biological specimens. Boom technology (bioMérieux) is a protocol for the extraction of RNA and DNA and is capable of simultaneously processing various sample types (serum, plasma, whole blood, urine, nasal washes, nasal swab, throat swab, bronchoalveolar lavages, sputum, feces, DBS, cells, or tissue specimens) and different volumes, using the same reagent kit. For specimens in the liquid form, place the secretion sample in the lysis buffer at an optimal ratio of 1 mL of specimen to 2 mL lysis buffer, except whole blood samples, where 0.1–2 mL ratio is allowed. The lysate is incubated at room temperature for 10 min before proceeding with the standard extraction steps according to the manufacturer's instructions (bioMérieux). Preextraction step is required for the extraction of nucleic acid from the DBS paper. Two blood spots (2 spots of 50 μL/extraction) are cut from the filter paper, using a pair of scissors, and added into a 2 mL NucliSENS Lysis Buffer tube (bioMérieux). Tubes are incubated horizontally on a roller mixer for 30 min at room temperature and centrifuged at $1500 \times g$ for 15 s to spin down the fluid. The supernatant lysate is then transferred for nucleic acid extraction process.

Preextraction step is required to remove lipid component from fatty tissues, such as brain and adipose tissue. Recently, high yields and high quality of total RNA without phenol contamination kit has been introduced (RNeasy Lipid Tissue Mini Kit, Qiagen). The kit combines organic extraction and chaotropic disruption that contribute to the high lysis efficiency of QIAzol Lysis Reagent. Tissue sample (10–100 mg) is homogenized in QIAzol Lysis Reagent using a machine (TissueRuptor or TissueLyser, Qiagen) or manually gridded using a Micropestle (Eppendorf). After adding chloroform, the homogenate is separated into aqueous and organic phases by centrifugation. The upper, aqueous phase (RNA component) is extracted, and an equal volume of 70% ethanol is added to provide proper binding conditions. The sample is then applied to the RNeasy Mini spin column, as per the manufacturer's instructions. Briefly, 700 µL of the mixture is transferred to an RNeasy Mini spin column, placed in a 2 mL collection tube, and centrifuged at $8000 \times g$ for 15 s, after which 700 µL of RW1 buffer is added to wash the membrane. The column is centrifuged again at $8000 \times g$ for 15 s and washed by adding 500 µL Buffer RPE twice. To elute the RNA, 30–50 µL of RNase-free water is added directly to the spin column membrane and centrifuged at $8000 \times g$ for 1 min. Up to 100 µg of total RNA (mini kit) binds to the membrane, while other contaminants are efficiently washed away.

53.2.2 Detection Procedures

The molecular methods for the detection of NiV have been previously summarized; additional information not covered here can be found from a similar review [85]. The following RT-PCR protocol for the screening and sequencing characterization of NiV has been modified from previous report [43].

53.2.2.1 One-Step RT-PCR QIAGEN

53.2.2.1.1 Procedure

1. Thaw primer solutions, dNTP mix, 5× OneStep RT-PCR Buffer, and RNase-free water, and place them on ice.
2. Prepare a master mix according to Table 53.2.
3. Mix the master mix thoroughly, and dispense 43 µL into PCR tubes.
4. Add 2 µL of internal control (IC) RNA (2000 molecules of Kanamycin Positive Control RNA; Promega) to the individual PCR tubes.
5. Add 5 µL of individual RNA sample to each PCR tube.
6. Transfer the reaction tubes to thermal cycler with a heated lid.
7. Program the thermal cycler for RT and PCR as follows: 30 min at 50°C for RT followed by 15 min at 95°C for HotStarTaq DNA polymerase activation and reverse transcriptase inhibition and 30 repetitive cycles of 1 min of denaturation at 94°C, 1 min of annealing at 55°C, and 1 min of elongation at 72°C. Elongation is extended for 10 additional min in the last cycle.
8. Start the RT-PCR program while PCR tubes are still on ice. Wait until the thermal cycler has reached 50°C. Then place the PCR tubes in the thermal cycler.
9. Following amplification, pulse-spin the reaction tubes. Proceed to the next step of nested PCR or store at –20°C.

53.2.2.2 Nested PCR

53.2.2.2.1 Procedure

1. Prepare a master mix according to Table 53.3.
2. Mix the master mix thoroughly, and dispense 49 µL into PCR tubes.
3. Add 1 µL of individual first-round PCR sample to each PCR tube.
4. Transfer the reaction tubes to thermal cycler with a heated lid.

TABLE 53.2
Reaction Components for One-Step Reverse Transcription Polymerase Chain Reaction (First-Round Polymerase Chain Reaction)

Component (Volumes Indicated per Specimen)	Volume (µL)	Final Concentration
Master mix		
RNase-free water	23.75	—
5× OneStep RT-PCR Buffer containing 12.5 mM MgCl₂	10.0	1×
25 mM MgCl₂	1.0	0.5 mM
dNTP Mix (containing 10 mM of each dNTP)	2.0	400 µM of each dNTP
20 µM CONINT1F (IC forward primer) (5′-CTGGCCTGTTGAACAAGTCT-3′)	0.5	0.2 µM
20 µM CONINT1R (IC reverse primer) (5′-GATCTGATCCTTCAACTCAGC-3′)	0.5	0.2 µM
20 µM NP1F (NiV forward primer) (5′-CTTGAGCCTATGTATTTCAGAC-3′)	1.5	0.6 µM
20 µM NP1R (NiV reverse primer) (5′-GCTTTTGCAGCCAGTCTTG-3′)	1.5	0.6 µM
OneStep RT-PCR Enzyme Mix	2.0	—
RNase inhibitor (40 U/µL)	0.25	0.2 U
Template (add at steps 4–5)		
Internal control RNA (1000 molecule/µL)	2.0	—
Template RNA or total nucleic acid of specimens	5.0	—
Total volume	50.0	—

TABLE 53.3
Reaction Components for Nested PCR (Second-Round PCR)

Component (Volumes Indicated per Specimen)	Volume (µL)	Final Concentration
Master mix		
RNase-free water	29.5	—
5× Buffer (Promega)	10.0	1×
25 mM MgCl$_2$	4.0	0.5 mM
dNTP Mix (containing 10 mM of each dNTP)	1.0	400 µM of each dNTP
20 µM 1UPS (IC forward primer) (5′ GCCATTCTCACCGGATTCAGTCGTC 3′)	0.5	0.2 µM
20 µM 1DS (IC reverse primer) (5′-AGCCGCCGT CCCGTCAAG TCAG-3′)	0.5	0.2 µM
20 µM NP2F (nested NiV forward primer) (5′-CTGCTGCAGTTCAGGAAACATCAG-3′)	1.5	0.6 µM
20 µM NP2R (nested NiV reverse primer) (5′-ACCGGATGTGCTCACAGAACTG-3′)	1.5	0.6 µM
5 U/µL Taq DNA polymerase	0.5	2.5 U
Template (add at step 3)		
First-round PCR product	1.0	—
Total volume	50.0	—

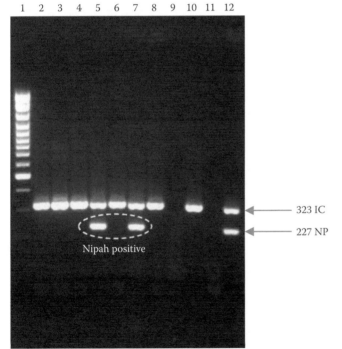

FIGURE 53.1 Duplex-nested polymerase chain reaction (PCR) detection of Nipah virus (NiV) RNA. The upper band (323 bp) is the internal control band. The lower band (227 bp) is the NiV-specific band. Lane 1 is the 100 bp DNA ladder markers (Fermentas), starting with 100 bp marker. NiV-positive samples are lanes 5 and 7. Lane 9 is a sample with PCR inhibitor. Lanes 10 and 12 are the negative and positive controls, respectively.

5. Program the thermal profile in the following order: 5 min denaturation at 94°C and 35 repetitive cycles of 30 s of denaturation at 94°C, 30 s of annealing at 55°C, and 1 min of elongation at 72°C. Elongation is extended for 5 additional min in the last cycle.

6. Analyze the PCR products (5–10 µL) by agarose gel electrophoresis through 2% agarose gel containing 0.5 µg/mL of ethidium bromide in Tris-borate-EDTA buffer for 45 min and visualized under UV light.

7. The presence of a 227 bp product is indicative of NiV positive, while the product size of IC is 323 bp (Figure 53.1).

53.2.2.3 Heminested PCR and Sequencing

The heminested PCR is prepared from the first-round positive PCR samples in the same manner as nested PCR, but without IC primers. The heminested primer pairs were NP1F/NP2R and NP1R/NP2F, which result in 342 bp and 283 bp PCR products, respectively. The 357 bp nt sequences at position 1197–1553 (according to GenBank accession no. NC 002728) can be obtained from these two PCR products.

53.2.2.3.1 Procedure

1. Prepare master mix as Table 53.3 of two separate PCRs from different 2 primer sets, NP1F/NP2R and NP1R/NP2F, but replace the IC primers from the reaction with 1 µL water.

2. Add 1 µL of product from the first round of PCR into each reaction tube.

3. Process all steps as stated in Section 53.2.2.2 (nested PCR). The resulting products of NP1F/NP2R and NP1R/NP2F are 342 bp and 283 bp, respectively (Figure 53.2).

4. Purify the PCR product by using a commercial spin column kit.

5. Add the primer NP2R to the sequencing reaction of NP1F/NP2R PCR product.

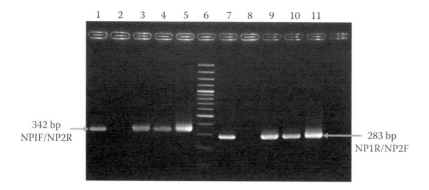

FIGURE 53.2   Heminested polymerase chain reaction detection of Nipah virus (NiV) RNA without internal control RNA. The left band (342 bp) is the NiV-specific band from the NP1F/NP2R primers. The right band (283) is the NiV-specific band from the NP1R/NP2F primers. Each pair of NiV-positive samples is shown in lanes 3 and 9, 4 and 10, and 5 and 11, respectively. Lane 6 is the 100 bp DNA ladder markers (Fermentas). Lanes 1 and 7 are the positive controls, while lanes 2 and 8 are the negative controls, respectively.

6. Add the primer NP2F to the sequencing reaction of NP1R/NP2F PCR product.
7. Perform the sequencing reaction according to the manufacturer's protocol (Applied Biosystems, Foster City, CA).
8. Analyze nucleotide sequence of the DNA product using an automated ABI PRISM 3730 sequencer.
9. Edit sequences using BioEdit program and assemble sequences from 2 PCR products; the nucleotide sequences are located at positions 1197–1553 (according to GenBank accession no. NC 002728).
10. Compare the 357 bp sequences with those of the reference strains available in databases such as the National Center for Biotechnology Information GenBank database.

53.3   CONCLUSION

NiV outbreaks have occurred in Malaysia, India, and Bangladesh and in the Philippines. The first outbreak in Malaysia (September 1998–May 1999) nearly collapsed the billion-dollar pig-farming industry. NiV continues to cause annual outbreaks of fatal human encephalitis/respiratory distress syndrome in Bangladesh due to spillover from its bat reservoir host. Recently, contact with NiV-infected horses and consumption of uncooked horse meat resulted in the outbreak in the Philippines. Currently, it lacks approved therapeutics and vaccines. Many countries are at risk of NiV infection outbreak due to the wide home range of NiV natural hosts, *Pteropus* bats.

The application of molecular methods allows the diagnosis in suspected human and animal cases. In addition, molecular methods have been used as a tool for NiV surveillance in bats in several countries such as Malaysia, India, Thailand, Indonesia, and East Timor. Although antibody screening assay is a powerful tool for NiV surveillance in many countries, it cannot characterize strain of viruses, and thus its use is limited compared to the PCR method.

There are many well-established PCR protocols with relatively low-cost methods. Transferring of these molecular protocols to the national or regional diagnostic laboratories in developing countries should enable a faster response and reduce the impact of future outbreaks. The surveillance of NiV infection should be conducted in natural hosts (bats), targeted livestock animals (pigs), and encephalitis/respiratory syndrome patients of unknown cause from at-risk countries. In order to have a complete understanding of the ecology and transmission of NiV, a comprehensive analysis of bats across their migratory routes in Africa, Southeast Asia, and Australia should be conducted. Knowledge of NiV molecular diversity and the presence of cross-species dissemination in bats and other animals will help in providing a clear assessment of the zoonotic potential of the bat-borne NiV.

ACKNOWLEDGMENTS

This work was supported by research grants from the Ratchadaphiseksomphot Endowment Fund of Chulalongkorn University (RES560530148-HR), Health and Biomedical Science Research Program by the National Research Council of Thailand and Health System Research Institute (P-13-50154), and the Research Chair Grant, the National Science and Technology Development Agency, Thailand. We gratefully acknowledge Siriporn Khai for checking the English usage throughout this chapter.

REFERENCES

1. Chua, K. B. et al. Nipah virus: A recently emergent deadly paramyxovirus. *Science* 288, 1432, 2000.
2. Aziz, J. et al. Nipah virus infections of animals in Malaysia. Paper *Presented at the Abstracts of the XI International Congress of Virology*, Sydney, New South Wales, Australia, 1999.
3. Chua, K. B. Epidemiology, surveillance and control of Nipah virus infections in Malaysia. *Malays. J. Pathol.* 32, 69, 2010.
4. Chadha, M. S. et al. Nipah virus-associated encephalitis outbreak, Siliguri, India. *Emerg. Infect. Dis.* 12, 235, 2006.

5. Luby, S. P. et al. Foodborne transmission of Nipah virus, Bangladesh. *Emerg. Infect. Dis.* 12, 1888, 2006.
6. Gurley, E. S. et al. Person-to-person transmission of Nipah virus in a Bangladeshi community. *Emerg. Infect. Dis.* 13, 1031, 2007.
7. Sazzad, H. M. et al. Nipah virus infection outbreak with nosocomial and corpse-to-human transmission, Bangladesh. *Emerg. Infect. Dis.* 19, 210, 2013.
8. Ching, P. K. G. et al. Outbreak of Henipavirus infection, Philippines, 2014. *Emerg. Infect. Dis.*, 21, 328, 2015.
9. Wang, L. et al. Molecular biology of Hendra and Nipah viruses. *Microbes Infect.* 3, 279, 2001.
10. Mayo, M. A. A summary of taxonomic changes recently approved by ICTV. *Arch. Virol.* 147, 1655, 2002.
11. Marsh, G. A. et al. Cedar virus: A novel Henipavirus isolated from Australian bats. *PLoS Pathog.* 8, e1002836, 2012.
12. Goldsmith, C. S. et al. Elucidation of Nipah virus morphogenesis and replication using ultrastructural and molecular approaches. *Virus Res.* 92, 89, 2003.
13. Chan, Y. P. et al. Complete nucleotide sequences of Nipah virus isolates from Malaysia. *J. Gen. Virol.* 82, 2151, 2001.
14. Harcourt, B. H. et al. Genetic characterization of Nipah virus, Bangladesh, 2004. *Emerg. Infect. Dis.* 11, 1594, 2005.
15. Jack, P. J. et al. The complete genome sequence of J virus reveals a unique genome structure in the family Paramyxoviridae. *J. Virol.* 79, 10690, 2005.
16. Lamb, R. A. and Kolakofsky, D. Paramyxoviridae: The viruses and their replication. In *Fields Virology*, eds. Fields, B. N., Knipe, D. M., and Howley, P. M. Philadelphia, PA: Lippincott Williams & Wilkins, 2007.
17. Halpin, K. et al. Nipah virus conforms to the rule of six in a minigenome replication assay. *J. Gen. Virol.* 85, 701, 2004.
18. Harcourt, B. H. et al. Molecular characterization of Nipah virus, a newly emergent paramyxovirus. *Virology* 271, 334, 2000.
19. Chua, K. B. et al. Isolation of Nipah virus from Malaysian Island flying-foxes. *Microbes Infect.* 4, 145, 2002.
20. Chua, K. B. Nipah virus outbreak in Malaysia. *J. Clin. Virol.* 26, 265, 2003.
21. Paton, N. I. et al. Outbreak of Nipah-virus infection among abattoir workers in Singapore. *Lancet* 354, 1253, 1999.
22. Chua, K. B. et al. Fatal encephalitis due to Nipah virus among pig-farmers in Malaysia. *Lancet* 354, 1257, 1999.
23. Luby, S. P. and Gurley, E. S. Epidemiology of henipavirus disease in humans. *Curr. Top. Microbiol. Immunol.* 359, 25, 2012.
24. Simons, R. R. et al. Potential for introduction of bat-borne zoonotic viruses into the EU: A review. *Viruses* 6, 2084, 2014.
25. Hsu, V. P. et al. Nipah virus encephalitis reemergence, Bangladesh. *Emerg. Infect. Dis.* 10, 2082, 2004.
26. Yob, J. M. et al. Nipah virus infection in bats (order Chiroptera) in peninsular Malaysia. *Emerg. Infect. Dis.* 7, 439, 2001.
27. Shirai, J. et al. Nipah virus survey of flying foxes in Malaysia. *Jarq Jpn. Agric. Res. Q* 41, 69, 2007.
28. Homaira, N. et al. Nipah virus outbreak with person-to-person transmission in a district of Bangladesh, 2007. *Epidemiol. Infect.* 138, 1630, 2010.
29. Peel, A. J. et al. Use of cross-reactive serological assays for detecting novel pathogens in wildlife: Assessing an appropriate cutoff for henipavirus assays in African bats. *J. Virol. Methods* 193, 295, 2013.
30. Peel, A. J. et al. Continent-wide panmixia of an African fruit bat facilitates transmission of potentially zoonotic viruses. *Nat. Commun.* 4, 2770, 2013.
31. Li, Y. et al. Antibodies to Nipah or Nipah-like viruses in bats, China. *Emerg. Infect. Dis.* 14, 1974, 2008.
32. Breed, A. C. et al. The distribution of henipaviruses in Southeast Asia and Australasia: Is Wallace's line a barrier to Nipah virus? *PLoS One* 8, e61316, 2013.
33. Epstein, J. H. et al. Henipavirus infection in fruit bats (*Pteropus giganteus*), India. *Emerg. Infect. Dis.* 14, 1309, 2008.
34. Yadav, P. D. et al. Detection of Nipah virus RNA in fruit bat (*Pteropus giganteus*) from India. *Am. J. Trop. Med. Hyg.* 87, 576, 2012.
35. Iehle, C. et al. Henipavirus and Tioman virus antibodies in pteropodid bats, Madagascar. *Emerg. Infect. Dis.* 13, 159, 2007.
36. Field, H. et al. Henipaviruses and fruit bats, Papua New Guinea. *Emerg. Infect. Dis.* 19, 670, 2013.
37. Hasebe, F. et al. Serologic evidence of Nipah virus infection in bats, Vietnam. *Emerg. Infect. Dis.* 18, 536, 2012.
38. Rahman, S. A. et al. Characterization of Nipah virus from naturally infected Pteropus vampyrus bats, Malaysia. *Emerg. Infect. Dis.* 16, 1990, 2010.
39. Reynes, J. M. et al. Nipah virus in Lyle's flying foxes, Cambodia. *Emerg. Infect. Dis.* 11, 1042, 2005.
40. Wacharapluesadee, S. et al. A longitudinal study of the prevalence of Nipah virus in Pteropus lylei bats in Thailand: Evidence for seasonal preference in disease transmission. *Vector Borne Zoonotic Dis.* 10, 183, 2010.
41. Wacharapluesadee, S. et al. Bat Nipah virus, Thailand. *Emerg. Infect. Dis.* 11, 1949, 2005.
42. Wacharapluesadee, S. et al. Drinking bat blood may be hazardous to your health. *Clin. Infect. Dis.* 43, 269, 2006.
43. Wacharapluesadee, S. and Hemachudha, T. Duplex nested RT-PCR for detection of Nipah virus RNA from urine specimens of bats. *J. Virol. Methods* 141, 97, 2007.
44. Goh, K. J. et al. Clinical features of Nipah virus encephalitis among pig farmers in Malaysia. *New. Engl. J. Med.* 342, 1229, 2000.
45. Tan, C. T. et al. Relapsed and late-onset Nipah encephalitis. *Ann. Neurol.* 51, 703, 2002.
46. Hossain, M. J. et al. Clinical presentation of Nipah virus infection in Bangladesh. *Clin. Infect. Dis.* 46, 977, 2008.
47. Mohd Nor, M. N. et al. Nipah virus infection of pigs in peninsular Malaysia. *Rev. Sci. Tech.* 19, 160, 2000.
48. Sohayati, A. R. et al. Evidence for Nipah virus recrudescence and serological patterns of captive Pteropus vampyrus. *Epidemiol. Infect.* 139, 1570, 2011.
49. Middleton, D. J. et al. Experimental Nipah virus infection in pteropid bats (*Pteropus poliocephalus*). *J. Comp. Pathol.* 136, 266, 2007.
50. Hooper, P. et al. Comparative pathology of the diseases caused by Hendra and Nipah viruses. *Microbes Infect.* 3, 315, 2001.
51. Field, H. et al. The natural history of Hendra and Nipah viruses. *Microbes Infect.* 3, 307, 2001.
52. Wong, K. T. and Ong, K. C. Pathology of acute henipavirus infection in humans and animals. *Pathol. Res. Int.* 2011, 567248, 2011.
53. Dhondt, K. P. and Horvat, B. Henipavirus infections: Lessons from animal models. *Pathogens* 2, 264, 2013.
54. AbuBakar, S. et al. Isolation and molecular identification of Nipah virus from pigs. *Emerg. Infect. Dis.* 10, 2228, 2004.
55. Daniels, P. et al. Laboratory diagnosis of Nipah and Hendra virus infections. *Microbes Infect.* 3, 289, 2001.
56. Daniels, P. et al. Hendra and Nipah virus diseases. In *OIE Manual of Diagnostic Tests and Vaccines for Terrestrial Animals Seventh Edition,* ed. The OIE Biological Standards Commission, pp. 1225–1238. Paris: World Organisation for Animal Health (OIE), 2012.

57. Rollin, P. E. et al. Hendra and Nipah viruses. In *Manual of Clinical Microbiology*, ed. J. Versalovic, pp. 1479–1487. Washington, DC: ASM Press, 2012.

58. Crameri, G. et al. Establishment, immortalisation and characterisation of pteropid bat cell lines. *PLoS One* 4, e8266, 2009.

59. Wong, K. T. et al. Nipah virus infection: Pathology and pathogenesis of an emerging paramyxoviral zoonosis. *Am. J. Pathol.* 161, 2153, 2002.

60. Middleton, D. J. et al. Experimental Nipah virus infection in pigs and cats. *J. Comp. Pathol.* 126, 124, 2002.

61. Tanimura, N. et al. Monoclonal antibody-based immunohistochemical diagnosis of Malaysian Nipah virus infection in pigs. *J. Comp. Pathol.* 131, 199, 2004.

62. Crameri, G. et al. A rapid immune plaque assay for the detection of Hendra and Nipah viruses and anti-virus antibodies. *J. Virol. Methods* 99, 41, 2002.

63. Wang, L. F. and Daniels, P. Diagnosis of henipavirus infection: Current capabilities and future directions. *Curr. Top. Microbiol. Immunol.* 359, 179, 2012.

64. Tamin, A. et al. Development of a neutralization assay for Nipah virus using pseudotype particles. *J. Virol. Methods* 160, 1, 2009.

65. Kaku, Y. et al. A neutralization test for specific detection of Nipah virus antibodies using pseudotyped vesicular stomatitis virus expressing green fluorescent protein. *J. Virol. Methods* 160, 7, 2009.

66. Kaku, Y. et al. Second generation of pseudotype-based serum neutralization assay for Nipah virus antibodies: Sensitive and high-throughput analysis utilizing secreted alkaline phosphatase. *J. Virol. Methods* 179, 226, 2012.

67. Eshaghi, M. et al. Nipah virus glycoprotein: Production in baculovirus and application in diagnosis. *Virus Res.* 106, 71, 2004.

68. Eshaghi, M. et al. Purification of the extra-cellular domain of Nipah virus glycoprotein produced in *Escherichia coli* and possible application in diagnosis. *J. Biotechnol.* 116, 221, 2005.

69. Tan, W. S. et al. Solubility, immunogenicity and physical properties of the nucleocapsid protein of Nipah virus produced in *Escherichia coli*. *J. Med. Virol.* 73, 105, 2004.

70. Kashiwazaki, Y. et al. A solid-phase blocking ELISA for detection of antibodies to Nipah virus. *J. Virol. Methods* 121, 259, 2004.

71. Chiang, C. F. et al. Use of monoclonal antibodies against Hendra and Nipah viruses in an antigen capture ELISA. *Virol. J.* 7, 115, 2010.

72. Kaku, Y. et al. Antigen capture ELISA system for henipaviruses using polyclonal antibodies obtained by DNA immunization. *Arch. Virol.* 157, 1605, 2012.

73. Bossart, K. N. et al. Neutralization assays for differential henipavirus serology using Bio-Plex protein array systems. *J. Virol. Methods* 142, 29, 2007.

74. Playford, E. G. et al. Human Hendra virus encephalitis associated with equine outbreak, Australia, 2008. *Emerg. Infect. Dis.* 16, 219, 2010.

75. Dups, J. et al. A new model for Hendra virus encephalitis in the mouse. *PLoS One* 7, e40308, 2012.

76. Hayman, D. T. et al. Evidence of henipavirus infection in West African fruit bats. *PLoS One* 3, e2739, 2008.

77. Hayman, D. T. et al. Antibodies to henipavirus or henipa-like viruses in domestic pigs in Ghana, West Africa. *PLoS One* 6, e25256, 2011.

78. Breed, A. C. et al. Prevalence of henipavirus and rubulavirus antibodies in pteropid bats, Papua New Guinea. *Emerg. Infect. Dis.* 16, 1997, 2010.

79. Sendow, I. et al. Nipah virus in the fruit bat Pteropus vampyrus in Sumatra, Indonesia. *PLoS One* 8, e69544, 2013.

80. Chen, J. M. et al. A stable and differentiable RNA positive control for reverse transcription-polymerase chain reaction. *Biotechnol. Lett.* 28, 1787, 2006.

81. Mungall, B. A. et al. Feline model of acute Nipah virus infection and protection with a soluble glycoprotein-based subunit vaccine. *J. Virol.* 80, 12293, 2006.

82. Chang, L. Y. et al. Quantitative estimation of Nipah virus replication kinetics in vitro. *Virol. J.* 3, 47, 2006.

83. Guillaume, V. et al. Specific detection of Nipah virus using real-time RT-PCR (TaqMan). *J. Virol. Methods* 120, 229, 2004.

84. Feldman, K. S. et al. Design and evaluation of consensus PCR assays for henipaviruses. *J. Virol. Methods* 161, 52, 2009.

85. Rota, P. A. et al. Nipah virus. In *Molecular Detection of Human Viral Pathogens*, ed. L. Dongyou, pp. 543–555. Boca Raton, FL: CRC Press, 2011.

86. Foord, A. J. et al. Microsphere suspension array assays for detection and differentiation of Hendra and Nipah viruses. *BioMed Res. Int.* 2013, 289295, 2013.

87. Weingartl, H. M. et al. Recombinant Nipah virus vaccines protect pigs against challenge. *J. Virol.* 80, 7929, 2006.

54 Peste des Petits Ruminants Virus

Pam Dachung Luka

CONTENTS

54.1 Introduction 467
    54.1.1 Classification..468
    54.1.2 Virus Structure, Genome Organization, and Geographical Distribution................................468
        54.1.2.1 Virus Structure and Genome Organization 468
        54.1.2.2 Geographical Distribution 469
    54.1.3 Epidemiology..469
        54.1.3.1 Transmission 469
        54.1.3.2 Host Range and Pathogenicity 469
        54.1.3.3 Pattern of the Disease 469
    54.1.4 Clinical Features and Pathogenesis 470
        54.1.4.1 Clinical Features and Pathogenesis 470
        54.1.4.2 Pathology 470
        54.1.4.3 Postmortem Findings...470
        54.1.4.4 Immunity 471
    54.1.5 Diagnosis 471
        54.1.5.1 Conventional Techniques...471
        54.1.5.2 Molecular Techniques 471
54.2 Methods 472
    54.2.1 Sample Preparation...472
    54.2.2 Detection Procedure 472
        54.2.2.1 RT-PCR Amplification...472
        54.2.2.2 Sequencing Analysis and Phylogeny 472
54.3 Conclusion and Future Perspectives 473
References..474

54.1 INTRODUCTION

Peste des petits ruminants (PPR), caused by PPR virus (PPRV), is an acute and highly contagious disease of both domestic and wild small ruminants related to the eradicated rinderpest (RP) disease [1]. The disease is characterized by high fever, ocular and nasal discharges, pneumonia, necrosis and ulceration of the mucous membrane, and inflammation of the gastrointestinal tract leading to severe diarrhea and death [2,3]. PPR is now widely spread from the North, West, Central, and East Africa; Arabian Peninsula; Middle East; and Asian subcontinent [4].

First described in West Africa in 1942 [5] and later recognized to be endemic in West and Central Africa [6], PPR spreads into the Middle East and Asian subcontinent within a large area encompassing the developing world where subsistence small ruminants farming serves as a major source of food and trade. The high morbidity and mortality of PPR exerted a detrimental impact on food production and animal welfare in endemic areas [7].

PPRV belongs to the genus *Morbillivirus* and family *Paramyxoviridae*. It is genetically related to RP virus (RPV) of cattle, measles virus (MV) of humans, canine distemper virus (CDV), and the morbilliviruses of aquatic mammals [8]. All morbilliviruses are related serologically, and sequence data also reveal a high degree of homology at the nucleic acid level. The genome contains six tandemly arranged transcription units encoding six structural proteins: the nucleocapsid (N), phosphoprotein (P), matrix (M), surface glycoproteins F and H, and the polymerase or large (L) protein, as well as two nonstructural proteins C and V, respectively [9–11].

Clinically, PPR cannot be differentiated in sheep and goats from RPV or other similar diseases. Therefore, a laboratory confirmation is important for accurate diagnosis. Conventional serological tests like agar gel immunodiffusion (AGID) and counterimmunoelectrophoresis (CIEP) have been used in diagnosis but are not sensitive enough for reliability. Specific diagnosis can be made by virus neutralization test (VNT), competitive enzyme-linked immunosorbent assays (c-ELISA), and immunocapture ELISA [12], but the reverse transcription polymerase chain reaction (RT-PCR) is a more sensitive test that has been developed and widely used for the detection of PPRV. Forsyth and Barrett [13] developed an RT-PCR using a set of primers to amplify the conserved region of fusion protein (F) gene, while Couacy-Hymann et al. [14] designed the nucleoprotein (N) gene primers. These techniques have become popular for specific diagnosis and molecular epidemiology of PPRV. Other techniques such as the real-time RT-PCR has also been developed to detect, quantify, and differentiate lineages [15,16]. Using molecular epidemiology tools, PPRV has been placed into four groups or lineages circulating in the world.

54.1.1 CLASSIFICATION

The classification of PPRV has been achieved by sequencing the F and N gene segments [13,14] into four distinct lineages. Out of the four phylogenetic lineages, three are located in Africa. The fourth lineage is the only one present in the Indian subcontinent but also coexists in the Middle East and some parts of Africa.

It is still not clear whether the apparent geographical spread of the disease in the last 50 years is real or reflects increased awareness, wider availability of diagnostic tools, or even a change in the virulence of the virus. It seems most likely that a combination of factors has been responsible for the current distribution of the disease [4]. It is also known that confusion of PPR with pneumonic pasteurellosis and other pneumonic diseases of small ruminants has precluded and delayed its recognition in some countries [17].

54.1.2 VIRUS STRUCTURE, GENOME ORGANIZATION, AND GEOGRAPHICAL DISTRIBUTION

PPR is caused by a virus that was assumed for a long time to be a variant of RP adapted to small ruminants. However, virus cross neutralization studies and electron microscopy revealed that it was a morbillivirus that had the physicochemical characteristic of a distinct virus but biologically and antigenically related to RPV. It has also been demonstrated to be an immunologically distinct virus with a separate epizootiology in areas where both viruses were enzootic [18]. The development of specific nucleic acid probes for hybridization studies and nucleic acid sequencing has proved that PPRV is quite distinct from RPV [19].

54.1.2.1 Virus Structure and Genome Organization

When viewed through electron microscope, morbilliviruses display the typical structure of *Paramyxoviridae*: a pleomorphic particle with a lipid envelope that encloses a helical nucleocapsid [2,20]. The nucleocapsids have a characteristic herringbone appearance. The virus is equally very fragile and cannot survive for a very long time outside its host. Its half-life has been estimated to be 22 min at 6°C and 3.3 h at 37°C. Morbilliviruses have been described as linear, nonsegmented, single-stranded, negative-sense RNA with genomes approximately 15–16 kb in size [21] and organized into six

contiguous, nonoverlapping transcription units corresponding to it six structural viral proteins in the order of 3′-N-P-M-F-H-L-5′ in the genome sense [11,22]. The virus has a variable diameter between 150 and 700 nm and mean of 500 nm [23]. This size variation is linked to the number of nucleocapsids incorporated into the virus particles. In the case of MV, this number might be more than 30 [24]. The full-length genome sequences of MV of humans [25], RPV [10], CDV of dogs and some wild carnivores [26], PPRV [9], and the dolphin morbilliviruses [27] have been generated and used in reverse genetics for the determination of gene function [28,29].

The nucleocapsid (N) protein is the major and most abundant viral protein both in the virion and in infected cells where it is directly associates with the RNA genome to form the typical herringbone structure of morbillivirus nucleocapsids [30]. It is located at the 3′ end of the genome adjacent to a 52–56 bases of the leader sequence in the genomic RNA [31,32]. Several roles have been proposed for the leader RNA, among which are its role as a binding site for the polymerase enzyme to initiate replication to generate positive strands and also in downregulating host cellular transcription [21,32,33].

Phosphoprotein (P), which is next to the N protein, is an essential subunit of the viral RNA-dependent RNA polymerase [34]. It encodes the nonstructural P, C, and V proteins. The P protein of morbilliviruses interacts with both the N and L proteins to form the viral polymerase. There is a great similarity between the N termini of V and P, which has been shown to play an important role in RPV replication [35]. C-minus mutants have been demonstrated to express growth defects in vitro, which were also evident by the level of mRNA transcripts generated. Contrarily, the V-minus mutants behave different from the C-minus mutants with an altered cytopathic effect (CPE) and increased genome/antigenome RNA production. The C and V proteins of paramyxoviruses also act as interferon antagonists, modifying the cellular immune response to infection [36,37].

The matrix (M) proteins are the most conserved protein within morbilliviruses and interact with surface glycoproteins (H and F) within the lipid envelope as well as the viral RNA. This was reflected in the conservation of the PPRV M protein when compared to that of other morbilliviruses with ranging identities from 91% between CDV and PDV to 76% between RPV and CDV, which PPRV and RPV share 84% identities [38]. This high conservation is probably due to the pivotal role the M protein plays in virion morphogenesis. The matrix protein is likely to be stable to variations giving its antigenic similarities within the family. M interacts with both the nucleocapsid and the cytoplasmic tails of the F and H glycoproteins [39].

The fusion (F) protein and hemagglutinin (H) protein are two surface glycoproteins produced by morbilliviruses that are buried into the viral envelope and project outward as spikes. Penetration of the host cell is made possible by the F protein, whereby it mediates the fusion of the viral and cellular membranes. Cooperation of the H protein is required to promote the activity of the F protein [40,41]. One of the unique features of the morbillivirus F mRNA is the fact that it has a long 5′-untranslated region. When translated, F mRNA, which is named F0, undergoes cleaving by endoproteases into the active F protein [42]. Virus assembly can proceed without the F0 cleavage, but infectivity and pathogenesis are dependent on it for all paramyxoviruses [43].

The H protein is responsible for the attachment of the virus to the cell receptors [34,44], the main being the signaling lymphocyte activation molecule (which is also called CD 150) [45]. This biological activity of the H protein is one of the classification criteria for *Paramyxoviridae*. The H proteins are highly variable in their activity within members of the family [46], but for its activity to be elicited, the cooperation of F is required [47]. Near the N-terminus is a long hydrophobic domain, which acts as a signal peptide but does not require cleaving like the F protein. Structurally, the N-terminal is on the cytoplasmic side of the membrane, while the C-terminus projects outward. This particular arrangement classifies H as a type II glycoprotein, which is characterized by a disulfide-liked homodimer [48,49]. For a mature H protein to emerge at the cell surface, it has to go through glycosylation, folding, and oligomerization through the rough endoplasmic reticulum and Golgi apparatus [48,50,51].

The large (L) protein is the major component of viral transcriptases and replicase with several functions. The functions performed are polymerase activity, methylation, capping, and polyadenylation activities [34].

54.1.2.2 Geographical Distribution

PPR is known to be present in a broad belt of sub-Saharan Africa, Arabia, the Middle East, and Asia subcontinent. Major outbreaks in Turkey, Morocco, Algeria, and Tibet China in recent years have indicated a marked rise in the global incidence of PPR [4]. Although the virus was first isolated in Nigeria [52], serological evidences suggest its presence in many more countries. PPRV isolates have been grouped into four distinct lineages based on the partial sequence analysis of either nucleoprotein or fusion (F) protein genes. Genetic relationship between PPR isolates from different geographical regions has been studied by sequence comparison of the F and N protein genes. Using the two gene targets, four lineages were revealed [4,53]. Lineage 1 and 2 viruses have been found exclusively to West Africa from isolates of the 1970s to the 1980s. More so, viruses of the lineage 3 were found in East Africa, Arabia, and India; however, it was never found in India again since the first report in 1992 [54]. Lineage 4 has been distinct to Asia even though it has been reported in other parts of Africa [4,55].

54.1.3 Epidemiology

54.1.3.1 Transmission

PPRV transmission requires close contact between infected animals in the febrile stage and susceptible ones [56]. Discharges from the eyes, nose, and mouth, as well as loose feces, are a source of infection because they contain virus and when an infected animal coughs or sneezes, infective droplets are released into the environment [57,58]. Animals in close contact inhale the droplets and are likely to become infected. Besides, there has been no report of a carrier state for PPRV. Therefore, movement of infected animals and contact with susceptible herds usually is the major source in spread especially when infected animal cough and sneeze or share common drinking or grazing site [57]. Trades in small ruminants at markets where animals from different sources are brought into close contact with one another are at risk of infection. In addition, the extensive or semi-intensive farming system where animals move and interact with other herds freely increases the chances for PPR transmission.

54.1.3.2 Host Range and Pathogenicity

PPR is mainly a disease of goats and sheep where the virus exhibits different levels of virulence. The virulence varies from strain to strain with sheep generally undergoing a mild form of the disease, while goats are severely affected [17] but sometimes the reverse occurs with certain field strains [59]. Breed may also affect the outcome of PPRV infection and its epidemiology where certain breeds such as the West African dwarf, Iogoon, and Kindi are highly susceptible, while the Wall breeds only developed mild form of the disease [17,59–61].

Field outbreaks have also been reported from both zoological gardens, and the wild with wild animals such as gazelle, antelope, Nubian ibex (*Capra nubiana*), gemsbok (*Oryx gazella*), Laristan sheep (*Ovis orientalis laristani*), Nigale (Tragelaphinae) deer, and Asian lion (*Panthera leo persica*) have been confirmed with the virus [7,62,63].

Cattle are not affected with PPR, whether RP vaccinated or not, even if they are in contact with affected goats and sheep. Cattle and pigs are known to be a dead-end host, and all attempt to induce clinical disease in adult cattle experimentally failed; they undergo a silent or subclinical infection that protect them against subsequent challenge with virulent strain of RP [2,59]. Antibodies have been detected in camels and wild animals have been reported to be infected with the disease whether experimentally or under field condition [64]

54.1.3.3 Pattern of the Disease

Generally, morbidity is common, particularly in fully susceptible goat populations. Milder forms of the disease may occur in sheep and partially immune goat populations. There are considerable differences in the epidemiological pattern of the disease in the different ecological systems and geographical areas. Morbidity can be up to 80%–90% with accompanied mortality between 50% and 80% [17]. In arid and semiarid regions, PPR is rarely fatal but usually occurs as a subclinical or inapparent infection opening the door for other infections such as pasteurellosis [17]. At the same time, the disease can also be exacerbated and confused with other diseases such as pasteurellosis, contagious ecthyma, contagious caprine pleuropneumonia, bluetongue, heartwater, coccidiosis, mineral poisoning, and foot-and-mouth disease [64].

Serological data revealed that antibodies occur in all age groups from 4 to 24 months indicating a constant circulation of the virus [65–67]. Therefore, an increase in incidence is a reflection of an increase in the number of susceptible young goats recruited into the flocks. Yearlings have been proved to be highly susceptible than the adults and unweaned animals compared to adults [68].

54.1.4 CLINICAL FEATURES AND PATHOGENESIS

54.1.4.1 Clinical Features and Pathogenesis

When an infection is introduced into a herd, the virus is incubated for 3–4 days within the lymph nodes draining the oropharynx before being disseminated to other parts of the animal via the circulatory system leading to pneumonia. The animal eventually begins to exhibit clinical signs that can be categories into three forms: peracute, acute, and mild forms. The acute form is the most common; it starts with a sudden rise in body temperature from 39.5°C to 41°C and may remain high for 5–8 days. These may return to normal prior to recovery or drops below normal before death. There is hyperpnea that may be characterized by rocking movements of the chest and abdominal walls as the animal breaths. Severely affected animals may show labored and noisy breathing characterized by extension of the head and neck, dilation of the nostrils, and protrusion of the tongue with painful coughs. A clear watery discharge starts to issue from the eyes, nose, and mouth, later becoming thick and yellow as a result of secondary bacterial infection. The appearance of a serous to mucopurulent nasal discharge may form crust and occlude the nostril and sneezing. There could be ocular discharge resulting in matting of the eyelids. The discharges wet the chin and the hair below the eye and tend to dry, causing matting together of the eyelids, obstruction of the nose, and difficulty in breathing [57,69].

One to two days after fever has set in, the mucous membranes of the mouth and eyes become reddened. The epithelial necrosis causes small pinpoint grayish areas on the gums, dental pad, palate, lips, inner aspects of the cheeks, and upper surface of the tongue. These areas increase in number and size and join together. The lining of the mouth changes in appearance. It becomes pale and coated with dying cells, and in some cases, the normal membrane may be completely obscured by a thick cheesy material. Underneath the dead surface cells, there are shallow erosions. Gentle rubbing across the gum and palate with a finger may yield a foul-smelling material containing shreds of epithelial tissue [56].

Following the onset of fever, diarrhea appears about 2–3 days after which may not be evident in either early or mild infections. The feces are initially soft and then watery, and foul-smelling and may contain blood streaks and pieces of dead gut tissues. Where diarrhea is not an obvious sign, inserting a cotton wool swab into the rectum may reveal evidence of soft feces, which may be stained with blood. Such victims may eventually become dehydrated with sunken eyeballs, and death often follows within 7–10 days from the onset of clinical reaction. Other animals may recover after a protracted convalescence [57].

54.1.4.2 Pathology

PPRV has affinity for lymphoid and epithelial tissues where it induces severe lesions in organs and systems rich in them [6]. The respiratory system is the most likely for viral entry. After viral entry through the respiratory tract, the virus first becomes localized and replicates in the pharyngeal and mandibular lymph nodes as well as tonsil. The virus buds and becomes available for systemic spread 2–3 days after infection to organs such as spleen, bone marrow, and mucosa of the gastrointestinal tract and respiratory system, and 1–2 days later, the first clinical signs begin to appear [6].

54.1.4.3 Postmortem Findings

Following the death of an affected animal, the carcass is usually emaciated, with the hindquarters soiled with feces and sunken eyeballs. The eyes and nose contain dried-up discharges with swollen lips with erosions and sometimes scabs or nodules in late cases. The nasal orifices are reddened and congested with clear or creamy yellow exudates and erosions. These erosions may be dry on the gums, on soft and hard palates, on tongue and cheeks, and into the esophagus. Grossly, the lung appears dark red or purple with some firm to touch areas mainly on the anterior and cardiac lobes, which is an indication of pneumonia. Associated lymph nodes become swollen and soft when touched in which the abomasum is congested and hemorrhagic [70].

The pathology caused by PPR may vary, but necrotic and ulcerative lesions within the oral cavity and gastrointestinal tract are predominant [71]. While oral erosions are a constant feature, there is rarely any lesion within the rumen and reticulum and omasum. Lesions within the small intestine are generally moderate. Comparatively, the large intestine is usually more severely affected, with congestion around the ileocecal valve, at the cecocolic junction, and in the rectum. The posterior part of the colon and the rectum are characterized by discontinuous streaks of congestion "zebra stripes" on the crests of the mucosal folds.

Erosion and petechiae may be obvious on the nasal mucosa, turbinates, larynx, and trachea of the respiratory systems sometimes accompanied with bronchopneumonia. Histopathologically, PPRV causes epithelial necrosis of the mucosa of the alimentary and respiratory tracts marked by the presence of eosinophilic intracytoplasmic and intranuclear inclusion bodies. Multinucleated giant cells (syncytia) can be observed in all affected epithelia as well as in the lymph nodes [69]. In the spleen, tonsil, and lymph nodes, the virus causes necrosis of lymphocytes evidenced by pyknotic nuclei and karyorrhexis. Brown and others [69] using immunohistochemical methods detected viral antigen in the cytoplasm and nuclei of tracheal, bronchial and bronchioepithelial cell, type II pneumocytes, syncytial cells, and alveolar macrophages. Small intestines are congested with lining hemorrhages and erosions. Large intestines (cecum, colon, and rectum) have small red hemorrhages along the folds of the lining, joining together as time passes and becoming darker, even green/black in stale carcasses.

54.1.4.4 Immunity

The surface glycoproteins hemagglutinin (H) and fusion protein (F) of morbilliviruses are highly immunogenic and confer protective immunity. PPRV is antigenically closely related to RPV, and antibodies against PPRV are both cross neutralizing and cross protective [18].

54.1.5 Diagnosis

For accurate and timely diagnosis of PPR to be possible, the appropriate samples need to be collected for the right control strategy to be implemented. Where PPR is suspected, samples can be collected from either live or dead animals for diagnosis. From live animals, ocular and nasal swabs and debris from oral lesions should be collected for diagnosis, while whole blood (in heparin or EDTA) should be taken for virus isolation especially at the acute phase of the disease. Biopsy samples of lymph nodes or spleen may also be useful. Samples for virus isolation should be collected during the acute stage of the disease, when clinical signs are present, whenever possible, these samples should be taken from animals with high fever and before the onset of diarrhea. At postmortem, samples can be collected from lymph nodes (particularly the mesenteric and mediastinal nodes), lungs, spleen, tonsils, and affected sections of the intestinal tract (e.g., ileum and large intestine). These samples should be taken from euthanized or freshly dead animals. Samples for virus isolation should be transported on ice. Similar samples should be collected in formalin for histopathology.

54.1.5.1 Conventional Techniques

Clinically, PPR suspected cases can be confirmed using a number of conventional laboratory tests. These conventional diagnostic tests include AGID test, CIEP, ELISA, VNT, and isolation of the virus in cell culture, which is time consuming and labor intensive [72]. But with the advent of molecular biology, RT-PCR techniques enable rapid and specific diagnosis of PPR possible [13–16].

AGID test: AGID is a very simple and inexpensive test that is widely used for the confirmation of PPR antigen [73]. The test can be ran within 2–4 h using standard positive or negative hyperimmune serum, respectively, to determine either antigen or antibodies in a suspected sample. AGID has a specificity of up to 92% and can be used on the field.

CIEP: This technique has been described as the most rapid test for viral antigen. It was described to utilize a glass slide, electrophoresis tank/buffer, and electric current to run through. Agarose is melted in 0.025 M barbitone acetate solution and poured onto the horizontal slide. Pair of wells is created on the agar, and the antigen and antiserum are poured site by site, and 10–12 mA current is allowed to pass through for 30–60 min in a 0.1 M barbitone acetate running buffer. The run is viewed under light for the formation of precipitation lines between a positive serum and tissue, respectively [73].

ELISA: Two ELISAs have been developed for the diagnosis of PPR: immunocapture and competitive ELISAs.

The immunocapture (sandwich) ELISA is suitable for routine diagnosis of field samples such as ocular and nasal swabs [74]. The test utilizes monoclonal antibodies developed against the nucleocapsid protein in tissues or secretions of infected animals with high levels of sensitivity [72,75] and a quick result within 2 h that differentiates between PPR and RP.

The c-ELISA was developed to capture the hemagglutinin protein of PPR either in a blocking ELISA as developed by Saliki et al. [76] or in a c-ELISA as described by Anderson and McKay [77], all for the purpose of differentiating PPRV from RPV.

VNT: The VNT is a gold standard test for the diagnosis of PPRV and RPV. It is very reliable, sensitive, and specific but time consuming, laborious, and expensive [78].

Virus isolation: In addition to the rapid diagnostic techniques, virus isolation of field samples should be carried out to further understand the biology of the virus. Primary lamb kidney or African green monkey kidney (Vero) cell tissue cultures can be used to achieve this objective. Monolayer cultures are inoculated with suspect material (swab material, buffy coat, or 10% tissue suspensions) and examined daily for evidence of CPE. The CPE produced by PPRV can develop within 5 days and consists of cell rounding and aggregation culminating in syncytia formation in lamb kidney cells. In Vero cells, it is sometimes difficult to see the syncytia. However, in stained, infected Vero cells, small syncytia are always seen. Syncytia are recognized by a circular arrangement of nuclei giving a "clockface" appearance. Coverslip cultures may give a CPE earlier than day 5. There is also the presence of intracytoplasmic and intranuclear inclusions bodies [52]. After 5–6 days postinoculation, blind passages should always be carried out because CPE may take time to appear.

54.1.5.2 Molecular Techniques

RT-PCR: RT-PCR is applied to detect the PPRV genome in blood and ocular or nasal swab tissue samples. Small fragments of viral RNA are amplified by RT-PCR even when no infectious virus is detected by virus isolation. The RT-PCR provides a sensitive, specific, and rapid alternative to virus isolation for the detection of PPRV for appropriate control strategies to be implemented. However, the sensitivity of RT-PCR assays makes them prone to cross contamination, thereby giving a false result. The detection of genomic RNA by RT-PCR is the most sensitive technique for detecting the presence of the agent in persistently infected animals and is particularly useful if samples submitted are unsuitable for virus isolation and antigen detection because they have undergone degradation. PPRV can be detected by RT-PCR from a very early stage of infection in tissues and swabs.

Real-time RT-PCR: The real-time RT-PCR approach is widely preferred. The aim is to detect amplification of target product by fluorescence signals from target-specific "oligonucleotide probes," which has several advantages over gel-based conventional RT-PCR methods such as increased speed, sensitivity, reduced chances of cross contamination, and provision of a quantitative result. A good number of real-time RT-PCR assays have been described for the detection, quantification, and lineage differentiation of PPRV [15,16]. The assay was found to be both sensitive and specific, with a higher analytical sensitivity.

Loop-mediated isothermal amplification (*LAMP*): LAMP is a faster and more sensitive method for amplifying viral RNA without having to use the thermal cycler. It follows the principle of strand displacement reaction and stem–loop structure allowing the amplification of the target with high specificity, selectivity, and rapidity under the isothermal conditions. Li et al. [79] described assay using six set of primers designed from the matrix gene to amplify PPRV RNA by incubated at 63°C for 60 min in a heating block or water bath. The sensitivity of LAMP was reported to be similar to real-time RT-PCR and 10-fold higher than conventional RT-PCR.

54.2 METHODS

54.2.1 Sample Preparation

Collection of samples: Both ocular and nasal swabs should be collected from the conjunctival and/or nasal discharges of live animals. Where there are buccal lesions, mucosal tissues should also be collected. During the peracute or early stage of the disease, whole blood is also to be collected in anticoagulant for virus isolation and PCR. Blood can equally be collected into sample bottles without anticoagulant for serum, which can be used for serum diagnosis. At necropsy, mesenteric and bronchial nodes, lungs, spleen, and intestinal mucosae should also be collected aseptically, chilled on ice, and transported under refrigeration to the laboratory. Fragments of organs collected for histopathology are to be placed in 10% formalin. During or after the outbreak, blood can be collected for serological diagnosis and monitoring [78].

Sample processing: Blood samples collected without anticoagulant are allowed to clot, and the serum is decanted into sterile tubes and kept at −20°C for antibody detection by c-ELISA, while blood collected in anticoagulant tubes are to be spun at 5000 rpm for 5 min and the plasma decanted and the buffy coat is collected and kept at −20°C for antigen detection. The ocular and nasal swabs, collected into a viral transport medium, are to be vortexed and centrifuged, and the supernatant is decanted into sterilized tubes and kept at −70°C for RNA extraction. From the tissues collected, 1 g is weighed and homogenized, and 9 mL of PBS is added and centrifuged in a refrigerated centrifuge at 10,000 rpm for 5 min to make 10% tissue suspension. The supernatant is to be decanted into a sterile tube and kept at 4°C for RNA extraction [78] and the pellet discarded into a disinfectant.

RNA extraction: The quality of RNA is important to the outcome or success of amplification. Over time, the procedure of isolating the RNA has evolved from the conventional method to the use of commercial kits. One of the kits used is QIAamp Viral RNA Mini Kit, Qiagen. Briefly, 560 μL

of prepared Buffer AVL (viral lysis buffer) containing carrier RNA is mixed with 140 μL of the sample incubated for 10 min at room temperature. A 560 μL of ethanol (96%–100%) is added and mixed, and 630 μL of the solution is applied to QIAamp Mini spin column and spun at 6000 g for 1 min. After applying all the solution, the QIAamp Mini spin column is carefully opened and 500 μL of wash buffer 1 (AW1) was added and centrifuged and the flow through is discarded, and then wash buffer 2 (AW2) is applied and spun at high speed for 3 min and the flow through discarded. The QIAamp Mini spin column is placed in a 1.5 mL microcentrifuge and 60 μL of the elution buffer (AVE) was added and incubated for 1 min at room temperature and spun at $6000 \times g$ for 1 min to elute the RNA and kept at −20°C for RT-PCR.

54.2.2 Detection Procedure

54.2.2.1 RT-PCR Amplification

Five microliters extracted RNA is reverse transcribed to cDNA using OmniScript™ Reverse Transcriptase Kit (Qiagen, Germany) using the following conditions: 1 μL of random hexamer, 0.125 mM each dNTPs, 10 units of ribonuclease inhibitor (MBI Fermentas), 2 units of OmniScript Reverse Transcriptase, and the buffer provided by the manufacturer. The reaction is carried out at 37°C for 60 min and stopped by incubation at 95°C for 10 min (Table 54.1).

Three microliters of the reverse transcribed (cDNA) product is used as template for the PCR. The reaction is carried out in 200 μL thin-walled PCR tubes, in a final volume of 25 μL containing the following reagents: 12.5 μL of DreamTaq PCR Master Mix (2×) containing ready-to-use DreamTaq DNA polymerase, buffer, $MgCl_2$, and dNTPs (MBI Fermentas), 1 μL of forward and reverse primers (10 pmoles of each primer) designed by Forsyth and Barrett [13] and Couacy-Hymann et al. [14], NP3/NP4, and 7.5 μL of nuclease-free water (Tables 54.2 and 54.3). Amplification using those primers (Table 54.1) was carried under the following conditions: initial denaturation step 95°C for 5 min followed by denaturation at 94°C for 30 min, annealing at 50°C at 60 s, extension at 72°C for 2 min for 35 cycles, and final extension at 72°C for 25 min to amplify a 372 bp product. The reaction conditions for NP3/NP4 primers targeting 351 bp product of the nucleoprotein gene are as follows: initial denaturation at 95°C at 5 min, followed by 35 cycles of denaturation at 94°C for 30 s, annealing at 55°C for 30 s and extension for 72°C for 1 min, and a final extension for 72°C for 25 min. The PCR products are analyzed on 1.5% agarose gel electrophoresis stain with ethidium bromide or gel red and viewed under ultraviolet light source.

54.2.2.2 Sequencing Analysis and Phylogeny

After amplifying a positive PPRV sample, characterization can be achieved by direct sequencing of the partial N or F protein genes segments. The products can be cleaned up using a commercial kit such as Wizard SV Gel and PCR Clean-Up System (Promega) or QIAquick PCR Purification

TABLE 54.1
Types of Assays and Details of Primers for Molecular Diagnosis

Assay Type	Primer	Primer Sequence	Target Gene Position	Expected Band Size	Reference
RT-PCR	F1 (F)	5'-ATC ACA GTG TTA AAG CCT GTA GAG G-3'	F gene 777–801	372 bp	Forsyth and Barett [13]
	F2 (R)	5'-GAG ACT GAG TTT GTG ACC TAC AAG C-3'	1124–1148		
RT-PCR	NP3 (F)	5'-TCT CGG AAA TCG CCT CAC AGA CTG-3'	N gene 1232–1255	351 bp	Couacy-Hymann et al. [14]
	NP4(R)	5'-CCT CCT CCT GGT CCT CCA GAA TCT-3'	1583–1560		
RT-PCR	N1 (F)	5' GAT GGT CAG AAG ATC TGC A 3'	N gene 1208–1226	463 bp	Kerur et al. [80]
	N2 (R)	5' CTT GTC GTT GTA GAC CTG A 3'	1670–1652		
Multiplex RT-PCR	Fr2	5' ACAGGCGCAGGTTTCATTCTT 3'	N gene	337 bp	Balamurugan et al. [81]
			1270–1290		
	Re1	5' GCTGAGGATATCCTTGTCGTTGTA	1584–1606		
	MF-Morb	5' CTTGATACTCCCCAGAGATTC 3'	M gene 477–497	191 bp	
	MR-PPR3	5'TTCTCCCATGAGCCGACTATGT 3'	646–667		
Real-time PCR	F	5'CACAGCAGAGGAAGCCAAACT	1213–1233	94 bp	Bao et al. [16]
	P	FAM 5' CTCGGAAATCGCC TCGCAGGCT 3' TAMRA	1237–1258		
			1327–1307		
	R	5' TGTTTTGTGCTGGAGGAAGGA			

TABLE 54.2
Reaction Mixture for Reverse Transcriptase

Sl. No.	Component	Volume/Reaction (μL)	Final Concentration
1	RNase-free water	10.25	—
2	10× buffer RT	2.00	1×
3	dNTP Mix (5 mM each dNTP)	1.00	0.125 mM each dNTP
4	Random hexamer	1.00	100 ng
5	Ribonuclease inhibitor	0.25	10 units
6	OmniScript Reverse Transcriptase	0.50	2 units
	Total	15.00	—
7	Template RNA	5.00	
	Grand total	20.00	

TABLE 54.3
Reaction Mixture for Polymerase Chain Reaction

Sl. No.	Component	Volume (μL)	Final Concentration
1	DNase–RNase-free water	7.50	—
2	2× PCR master mix	12.50	1×
3	Forward primer (10 pmol/μL)	1.00	10 pmol
4	Reverse primer (10 pmol/μL)	1.00	10 pmol
	Total	22.00	—
5	cDNA template	3.00	—
	Grand total	25.00	—

Kit (Qiagen). The purified product can be checked on an agarose gel before using it as a template for sequencing. Sequencing is performed using the BigDye Terminator v3.1 Cycle Sequencing Kit (Applied Biosystems, Foster City, CA) based on the manufacturer's protocol.

Sequences generated should be compared with other PPRV published or referenced sequences within either National Center for Biotechnology Information, GenBank databases, European Molecular Biology Laboratory, or DNA Data Bank of Japan. Phylogenetic analysis of quality sequences can be carried out using Molecular Evolutionary Genetics Analysis version 4 [82].

54.3 CONCLUSION AND FUTURE PERSPECTIVES

PPRV is a very important pathogen of sheep and goats that has continued to spread rapidly with high morbidity and mortality in spite of the availability of a protective vaccine. This spread has brought about attendant losses of livestock, thereby threatening PPR-free neighboring states. Since the virus has been identified to have a separate epizootiology from RP, it has been categorized as List A disease by the Office of International Des Epizooties. It has also received the attention of the international community toward a possible global eradication.

However, several factors and drivers have been implicated in the continuous spread and persistence of the disease to include population growth, trade, globalization, and inadequate control strategy to combat the disease. Although the virus is a single strain with four lineages, inadequate control measures have allowed the virus to spread and persist within susceptible populations.

The advent of molecular techniques has made diagnosis highly sensitive, specific, and rapid. It also gives room for a thorough understanding of the molecular biology of the PPRV. However, there are still a lot of gabs in the understanding of the virus and its epidemiology in areas such as virus–host interaction, epidemiological factors, evolutionary mechanisms, and clinical outcome.

For a future global eradication campaigns to be successful, there is the need for coordination at national and regional levels among various stakeholders in the industry. However, there is the need to evaluate the effectiveness of the vaccination, and a new one should be developed to take care of the handling and storage factors. All this will enable the formulation of a comprehensive PPR control.

REFERENCES

1. Furley, C.W., W.P. Taylor, and T.U. Obi, An outbreak of peste des petits ruminants in a zoological collection. *Vet Rec*, 1987. 121:443–447.
2. Gibbs, P.J.E. et al., Classification of the peste des petits ruminants virus as the fourth member of the genus Morbillivirus. *Intervirology*, 1979. 11:268–274.
3. Abubakar, M., Q. Ali, and H.A. Khan, Prevalence and mortality rate of peste des petits ruminant (PPR): Possible association with abortion in goats. *Trop Anim Health Prod*, 2008. 40:317–321.
4. Banyard, A. et al., Global distribution of peste des petits ruminants virus and prospects for improved diagnosis and control. *J Gen Virol*, 2010. 91:2885–2897.
5. Gargadennec, L.L.A., La peste des petits ruminants. *Bulletin des Services Zoo Techniques et des Epizzoties de l'Afrique Occidentale Francaise*, 1942. 5:16–21.
6. Scott, G.R., Rinderpest and peste des petits ruminants. In Gibbs, E.P.J. (Ed.), *Virus Diseases of Food Animals*, Vol. II: Disease Monographs. Academic Press, New York, pp. 401–425, 1981.
7. Abu-Elzein, E.M.E. et al., Isolation of peste des petits ruminants from goats in Saudi Arabia. *Vet Rec*, 1990. 127(12):309–310.
8. Barrett, T. et al., The molecular-biology of rinderpest and peste-des-petits ruminants. *Ann Med Vet*, 1993. 137:77–85.
9. Bailey, D. et al., Full genome sequence of peste des petits ruminants virus, a member of the Morbillivirus genus. *Virus Res*, 2005. 110:119–124.
10. Baron, M.D. and T. Barrett, Sequencing and analysis of the nucleocapsid (N) and polymerase (L) genes and the extragenic domains of the vaccine strain of rinderpest virus. *J Gen Virol*, 1995. 76:593–603.
11. Diallo, A., Morbillivirus group: Genome organisation and proteins. *Vet Microbiol*, 1990. 23(1–4):155–163.
12. Libeau, G. et al., Rapid differential diagnosis of RP and PPR using an immunocapture ELISA. *Vet Rec*, 1994. 134:300–304.
13. Forsyth, M.A. and T. Barrett, Evaluation of polymerase chain reaction for the detection and characterisation of rinderpest and peste des petits ruminants viruses for epidemiological studies. *Virus Res*, 1995. 39:151–163.
14. Couacy-Hymann, E. et al., Rapid and sensitive detection of peste des petits ruminants virus by a polymerase chain reaction assay. *J Virol Methods*, 2002. 100:17–25.
15. Kwiatek, O. et al., Quantitative one-step real time RT-PCR for the fast detection of the four genotypes of PPRV. *J Virol Methods*, 2010. 165(2):168–177.
16. Bao, J. et al., Development of one-step real time RT-PCR assay for the detection and quantification of Peste des petits ruminants virus. *J Virol Methods*, 2008. 148(1–2):232–236.
17. Lefevre, P.C. and A. Diallo, Peste des petits ruminants virus. *Rev Sci Tech Off Int Epiz*, 1990. 9:951–965.
18. Taylor, W.P., Protection of goats against PPR with attenuated RP virus. *Res Vet Sci*, 1979a. 27:321–324.
19. Diallo, A. et al., Differentiation of rinderpest and peste des petits ruminants viruses using specific cDNA clones. *J Virol Methods*, 1989. 23(2):127–136.
20. Meng, X., Y. Dou, and X. Cai, Ultrastructural features of PPRV infection in Vero cells. *Virol Sin*, 2014. 29(5):311–313.
21. Norrby, E. and M.N. Oxman, Measles virus, In Fields, B.N. et al. (Eds.), *Virology*, 2nd edn., Vol. I., Raven Press, New York, pp. 1013–1044, 1990.
22. Barrett, T. et al., The molecular biology of the morbilliviruses. In Kingsbury, D.W. (Ed.), *The Paramyxoviruses*. Plenum Press, New York, pp. 83–102, 1991.
23. Durojaiye, O.A., W.P. Taylor, and C. Smale, The ultrastructure of peste des petits ruminants virus. *Zentralblatt für Veterinärmedizin*, 1985. 32:460–465.
24. Roger, F. et al., Investigation of a new pathological condition of camels in Ethiopia. *J Camel Prac Res*, 2000. 7:163–166.
25. Cattaneo, R. et al., Measles virus editing provides an additional cysteine-rich protein. *Cell*, 1989. 56:759–764.
26. Barrett, T. et al., The nucleotide sequence of the gene encoding the F protein of canine distemper virus: A comparison of the deduced amino acid sequence with other paramyxoviruses. *Virus Res*, 1987. 8:373–386.
27. Rima, B.K., A. Collin, and P.J.A. Earle, *Completion of the Sequence of a Cetacean Morbillivirus and Comparative Analysis of the Complete Genome Sequences Four Morbilliviruses*. National Center for Biotechnology Information (NCBI) NIH, Bethesda, MD, 2003.
28. Nagai, Y., Paramyxovirus replication and pathogenesis: Reverse genetics transforms understanding. *Rev Med Virol*, 1999. 9:83–99.
29. Neumann, G., M. Whitt, and Y. Kawaoka, A decade after the generation of a negative-sense RNA virus from cloned cDNA what have we learned? *J Gen Virol*, 2002. 83:2635–2662.
30. Diallo, A. et al., Comparison of proteins induced in cells infected with RP and PPR viruses. *J Gen Virol*, 1987. 68:2033–2038.
31. Crowley, J.C. et al., Sequence variability and function of measles 3′ and 5′ ends and intercistronic regions. *Virology*, 1988. 164:498–506.
32. Ray, J., L. Whitton, and R.S. Fujinami, Rapid accumulation of measles virus leader RNA in the nucleus of infected HeLa cells and human lymphoid cells. *J Virol*, 1991. 65:7041–7045.
33. Walpita, P., An internal element of the measles virus antigenome promoter modulates replication efficiency. *Virus Res*, 2004. 100:199–211.
34. Lamb, R.A. and D. Kolakofsky, Paramyxoviridae: The viruses and their replication. In Knipe, D.M. and Howley, P.M. (Eds.), *Fields Virology*, 3rd edn. Lippincott Williams & Wilkins, Philadelphia, PA, pp. 1305–1340, 2001.

35. Baron, M.D. and T. Barrett, Rinderpest viruses lacking the C and V proteins show specific defects in growth and transcription of viral RNAs. *J Virol*, 2000. 74:2603–2611.

36. Gotoh, B. et al., Paramyxovirus accessory proteins as interferon antagonists. *Microbiol Immunol*, 2001. 45:787–800.

37. Horvath, C.M., Silencing STATs: Lessons from paramyxovirus interferon evasion. *Cytokine Growth Factor Rev*, 2004. 15:117–127.

38. Haffar, A. et al., The matrix protein gene sequence analysis reveals close relationship between PPR virus and dolphin morbillivirus. *Virus Res*, 1999. 64:69–75.

39. Coronel, E.C. et al., Nucleocapsid incorporation into parainfluenza virus is regulated by specific interaction with matrix protein. *J Virol*, 2001. 75:1117–1123.

40. Moll, M., H.D. Klenk, and A. Maisner, Importance of the cytoplasmic tails of the measles virus glycoproteins for fusogenic activity and the generation of recombinant measles viruses. *J Virol*, 2002. 76:7174–7186.

41. Plemper, R.K. et al., Strength of envelope protein interaction modulates cytopathicity of measles virus. *J Virol*, 2002. 76:5051–5061.

42. Meyer, G. and A. Diallo, The nucleotide sequence of the fusion protein gene of the PPR virus: The long untranslated region in the 5-end of the F-protein gene of morbilliviruses seems to be specific to each virus. *Virus Res*, 1995. 37:23–38.

43. Watanabe, M. et al., Delayed activation of altered fusion glycoprotein in a chronic measles virus variant that causes subacute sclerosing panencephalitis. *J Neurovirol*, 1995. 1(2):177–188. Corrected and republished in: *J Neurovirol* (1): 412–423.

44. Choppin, P. and A. Scheid, The role of viral glycoproteins in adsorption, penetration and pathogenicity of viruses. *Rev Infect Dis*, 1980. 2:40–61.

45. Tatsuo, H., O. Nobuyuki, and Y. Yusuke, Morbilliviruses use signaling lymphocyte activation molecules (CD150) as cellular receptors. *J Virol*, 2001. 75:5842–5850.

46. Blixenkrone-Müller, M. et al., Comparative analysis of the attachment protein (H) of dolphin morbillivirus. *Virus Res*, 1996. 40:47–55.

47. Das, S.C., M.D. Baron, and T. Barrett, Recovery and characterization of a chimeric rinderpest virus with the glycoproteins of peste-des-petits-ruminants virus: Homologous F and H proteins are required for virus viability. *J Virol*, 2000. 74:9039–9047.

48. Plemper, R.K., A.L. Hammond, and R. Cattaneo, Characterization of a region of the measles virus hemagglutinin sufficient for its dimerization. *J Virol*, 2000. 74:6485–6493.

49. Vongpunsawad, S. et al., Selectively receptor blind measles viruses: Identification of residues necessary for SLAM- or CD46-induced fusion and their localization on a new hemagglutinin structural model. *J Virol*, 2004. 78:302–313.

50. Hu, A.H. et al., Role of N-linked oligosaccharide chains in the processing and antigenicity of measles virus hemagglutinin protein. *J Gen Virol*, 1994. 75:1043–1052.

51. Blain, F., P. Liston, and D.J. Briedis, The carboxy-terminal 18 amino acids of the measles virus hemagglutinin are essential for its biological function. *Biochem Biophys Commun*, 1995. 214:1232–1238.

52. Taylor, W.P. and A. Abegunde, The isolation of peste des petits ruminants virus from Nigerian sheep and goats. *Res Vet Sci*, 1979. 26:94–96.

53. Dhar, B.P. et al., Recent epidemiology of peste de petits ruminants virus. *Vet Micro*, 2002. 88:153–159.

54. Nanda, Y.P. et al., The isolation of peste des petits ruminants virus from Northern India. *Vet Micro*, 1996. 51:207–216.

55. Luka, P.D. et al., Molecular characterization of Peste des petits ruminants virus from North eastern region of Karamoja, Uganda. *Arch Virol*, 2012. 157(1):29–35.

56. Braide, V.B., Peste des petits ruminants. *World Anim Rev*, 1981. 39:25–28.

57. Bundza, A. et al., Experimental PPR (goat plague) in goats and sheep. *Can J Vet Res*, 1988. 52:46–52.

58. Ezeibe, M.C.O. et al., Persistent detection of peste de petits ruminants antigen in the faeces of recovered goats. *Trop Anim Health Prod*, 2008. 40:517–519.

59. Taylor, W.P., The distribution and epidemiology of PPR. *Prev Vet Med*, 1984. 2:157–166.

60. El-Hag, B. and T.W. Taylor, Isolation of PPR virus from Sudan. *Res Vet Sci*, 1984. 36:1–4.

61. Diop, M., J. Sarr, and G. Libeau, Evaluation of novel diagnostic tools for peste des petits ruminants virus in naturally infected goat herds. *Epidemiol Infect*, 2005. 133:711–717.

62. Abubakar, M. et al., Peste des petits ruminants (PPR): Disease appraisal with global and Pakistan perspective. *S Rum Res*, 2011. 96:1–10.

63. Balamurugan, V. et al., Peste des petits ruminants virus detected in tissues from an Asiatic lion (*Panthera leo persica*) belongs to Asian lineage IV. *J Vet Sci*, 2012. 13(2):203–206.

64. Couacy-Hymann, E. et al., Surveillance of wildlife as a tool for monitoring rinderpest and peste des petits ruminants in West Africa. *Rev Sci Tech*, 2005. 24:869–877.

65. Taylor, W.P., Serological studies with the virus of peste des petits ruminants in Nigeria. *Res Vet Sci*, 1979. 26:236–242.

66. Luka, P.D. et al., Seroprevalence of Peste des petits ruminants antibodies in sheep and goats after vaccination in Karamoja, Uganda: Implication on control. *Int J Anim Vet Adv*, 2011. 3(1):18–22.

67. Waret-Szkuta, A. et al., Peste des Petits Ruminants (PPR) in Ethiopia: Analysis of a national serological survey. *BMC Vet Res*, 2008. 4:34.

68. Taylor, W.P., S. Abusaidy, and T. Barret, The epidemiology of PPR in the sultanate of Oman. *Vet Micro*, 1990. 22:341–352.

69. Brown, C.C., J.C. Mariner, and H.J. Olander, An immunohistochemical study of the pneumonia caused by PPR virus. *Vet Pathol*, 1991. 28:166–170.

70. Roeder, P.L. and T.U. Obi, Recognizing peste des petites ruminants: A field manual. *FAO Anim Health Man*, 1999. (5):28.

71. Roeder, P.L., G. Abraham, and T. Barrett, PPR in Ethiopian goats. *Trop Anim Health Prod*, 1994. 26:69–73.

72. Libeau, G. et al., Rapid differential diagnosis of RP and PPR using an immunocapture ELISA. *Vet Rec*, 1994. 134:300–304.

73. Obi, T.U., The detection of PPR virus antigen by agar gel precipitation test and counter-immunoelectrophoresis. *J Hyg*, 1984. 93:579–586.

74. Diallo, A. et al., Recent Developments in the diagnosis of rinderpest and peste de petits ruminants. *Vet Microbiol*, 1995. 44:307–317.

75. Saliki, J.T. et al., Comparison of monoclonal antibodies-based sandwich ELISA and virus isolation for detection of PPR virus in goats tissue and secretion. *J Clin Microbiol*, 1994. 32:1349–1353.

76. Saliki, J. et al., Monoclonal antibody based blocking ELISA for specific detection and titration of PPR virus antibody in caprine and ovine sera. *J Clin Microbiol*, 1993. 31:1075–1082.

77. Anderson, J. and J.A. McKay, The detection of antibodies against peste des petits ruminants virus in cattle, sheep and goats and the possible implications to rinderpest control programmes. *Epidemiol Infect*, 1994. 1(12):225–231.

78. OIE (Office International des Epizooties), *Manual, Manual of Diagnostic Tests and Vaccines for Terrestrial Animals*, 5th edn., Chapter 2.1.10, pp. 211–220, 2004.

79. Li, L. et al., Rapid detection of peste des petits ruminants virus by a reverse transcriptase loop-mediated isothermal assay. *J Virol Methods*, 2010. 170:37–41.

80. Kerur, N., Jhala, M.K., and Joshi, C.G. Genetic characterization of Indian peste des petits ruminants virus (PPRV) by sequencing and phylogenetic analysis of fusion protein and nucleoprotein gene segments. *Res Vet Sci*, 2008. 85(1):176–183. Epub September 11, 2007.

81. Balamurugan, V. et al. One-step multiplex RT-PCR assay for the detection of peste des petits ruminants virus in clinical samples. *Vet Res Commun*, 2006. 30(6):655–666.

82. Tamura, K. et al., Molecular evolutionary genetics analysis (MEGA) software version 4.0. *Mol Biol Evol*, 2007. 24(8):1596–1599.

55 Porcine Rubulavirus (Blue Eye Disease Virus)

José Iván Sánchez Betancourt and María Elena Trujillo Ortega

CONTENTS

55.1 Introduction 477
    55.1.1 Classification..477
    55.1.2 Morphology, Genome Organization, Biology, and Epidemiology 477
        55.1.2.1 Morphology and Genome Organization 477
        55.1.2.2 Biology...478
        55.1.2.3 Epidemiology 479
    55.1.3 Clinical Features and Pathogenesis 479
        55.1.3.1 Clinical Features...479
        55.1.3.2 Pathogenesis...479
    55.1.4 Diagnosis 480
        55.1.4.1 Conventional Techniques...480
        55.1.4.2 Molecular Techniques..480
55.2 Methods 480
    55.2.1 Sample Preparation..480
    55.2.2 Detection Procedures...480
        55.2.2.1 Hemagglutination Inhibition..480
        55.2.2.2 Seroneutralization..481
        55.2.2.3 Viral Isolation 481
        55.2.2.4 Titration of Isolated Virus...481
        55.2.2.5 RT-PCR 481
55.3 Conclusion 482
References..482

55.1 INTRODUCTION

The blue eye disease, affecting pigs of all ages, is caused by a virus of the *Rubulavirus* genus, *Paramyxoviridae* family [1]. The first cases were reported in 1980, in hog farms at La Piedad, Michoacán, Mexico. Clinical signs, mostly of nervous nature, were observed in piglets younger than 30 days. New outbursts were reported later on the same year in the nearby states of Jalisco and Guanajuato and in 1983 in Mexico City, Nuevo León, Hidalgo, Tlaxcala, Tamaulipas, Puebla, Campeche, and Querétaro. Since its emergence, the disease has been difficult to control, and by 1992, it had spread to 16 states in Mexico [2]. According to a seroprevalence survey of porcine rubulavirus (PRV) performed in 2004, a prevalence of 10%–30% was found in 18 Mexican states [3].

55.1.1 CLASSIFICATION

Based on its biologic and structural characteristics including the presence of hemagglutinating and neuraminidase activities, the blue eye virus is classified in the species *Porcine rubulavirus* (*Rubulavirus porcino*), genus *Rubulavirus*, subfamily *Paramyxovirinae*, family *Paramyxoviridae*, order *Mononegavirales* [4–6]. Other species within the genus *Rubulavirus* include *Achimota virus 1*, *Achimota virus 2*, *Mapuera virus*, *Menangle virus*, *Mumps virus*, *Parainfluenza type 2*, *Parainfluenza type 4*, *Parainfluenza type 5*, *Simian virus 41*, *Tioman virus*, *Tuhokovirus virus 1*, *Tuhokovirus virus 2*, and *Tuhokovirus virus 3*.

The *Paramyxoviridae* family comprises enveloped viruses that harbor a single-stranded RNA genome divided into six or seven genes. These genes code for an equal or greater number of proteins, including viral RNA polymerase. Paramyxoviruses are highly variable, showing one mutation in 10,000 bases on each replication [6–8].

The blue eye disease virus does not cross-react with antigens from parainfluenza 1, 2, 3, and 4 viruses, neither with Newcastle disease, measles, respiratory syncytial, or mumps viruses.

55.1.2 MORPHOLOGY, GENOME ORGANIZATION, BIOLOGY, AND EPIDEMIOLOGY

55.1.2.1 Morphology and Genome Organization

Besides sharing a tissular tropism and several pathological traits, PRV demonstrates significant structural and genetic similarity to human mumps virus [9].

Under electron microscope the virus appears as pleomorphic virions, most of which are spherical in shape, 180–300 nm in diameter. Its lipoprotein membrane shows 8–12 nm protrusions associated with HN and F glycoproteins, responsible for the hemagglutinating, hemolytic, neuraminidase, and syncytium-forming activities seen in PRV [4,9].

The genome of PRV consists of six genes, in the order of 3'-N-P/V-M-F-HN-L-5'. These genes encode ten proteins with structural, regulating, and enzymatic functions. The N, M, HN, and L genes code for one protein each; the F gene codes for a polypeptide called F_0, split by a cell protease into the F_1 and F_2 proteins; the P gene codes for several proteins, depending on the mRNA editing region [10–14].

Among the structural proteins, the HN is a multifunctional glycoprotein, consisting of 576 amino acids and having a molecular weight of 66 kDa. It is responsible for receptor recognition and for the adhesion to target cells during infection due to a specific recognition of the N-acetylneuraminic acid residues linked by α-2,3-glycosidic bonds to galactose (NeuAcα-2,3 Gal). Being the determinant in the susceptibility of hosts to PRV infection and the tissular tropism of the virus, the expression of NeuAcα-2,3 Gal is higher in the central nervous system of neonate pigs and also in the reproductive tract of adult hogs [15–17]. The HN protein exhibits hemagglutinating and neuraminidase activities. It has two functional domains: one is the ability of agglutinating erythrocytes and the other is the ability of deleting sialic acid groups. It agglutinates erythrocytes of a wide variety of animal species, including hogs, sheep, cows, dogs, rabbits, mice, rats, hamsters, guinea pigs, chickens, and humans with blood types A, B, and O due to a specific recognition of the NeuAcα-2,3 Gal expressed on the surface of all erythrocytes [5,17–19]. The neuraminidase or sialidase activity of HN allows for hydrolyzing the sialic acid residues of cell receptors; this activity has been proposed as a virulence factor [18,20]. This protein is regarded as immunodominant in PRV infections. Recent studies have reported changes in the nucleotide sequence of the gene coding for the HN protein, resulting in increased virulence and dissimilar clinical signs for each isolate [21,22].

The nucleoprotein (NP), an integral part of the nucleocapsid, is composed of 576 amino acids, with a molecular weight of 68 kDa. During replication/transcription, each NP molecule is linked to six high-affinity nucleotides, thus preventing the presence of uncoated PRV genome within the host cell and protecting the viral RNA against degradation in the cell [20,23–25].

The P protein, made up of 404 amino acids, has a molecular weight of 52 kDa. It is associated to the L protein and to NP. Additionally, it forms a complex with the L protein exhibiting polymerase activity, responsible for transcribing and replicating the viral genome [26].

The L protein, the largest protein in the virion, is made up of 2251 amino acids and has a molecular weight of 200 kDa. As the catalytic component of RNA polymerase, it plays a role in the initiation, elongation, methylation, edition, and polyadenylation processes. Its functions include catalytic activities in the synthesis of genomic and messenger RNA for

transcribing and replicating viral particles without entering the host cell nucleus, since these events take place in the cytoplasm [14,20,23,27].

The matrix (M) protein is made up of 369 amino acids—with a high proportion of basic residues—and has a molecular weight of 41.7 kDa. It is located in the inner face of the viral envelope. Due to its net positive charge, it shows a great affinity for NP. This feature is crucial, since it allows the M protein to participate in the viral particle assembly with lower energy expenditure. It also plays a role in maintaining the virion structure and in the assembly of new virions [10,13,20,28].

The F glycoprotein is made up of 541 amino acids and has a molecular weight of 59 kDa. A highly hydrophobic domain in this protein takes part in the fusion of the viral envelope to the membrane of the host cell, allowing the virus to enter the cytoplasm and be transmitted cell-to-cell without being exposed to the extracellular medium [10,20,28]. This trait gives rise to the multinucleated giant cells (syncytia) visible in tissues and cell cultures infected by paramyxoviruses [29,30].

The nonstructural proteins V (249 amino acids and 26.1 kDa), I (174 amino acids and 18.1 kDa), P (404 amino acids and 42 kDa), and C (126 amino acids and 14.7 kDa) are all coded by the viral genome after editing the P gene. Apparently, these proteins are involved in transcriptional regulation and evasion of antiviral responses [4,20,29].

55.1.2.2 Biology

The replication cycle of rubulavirus comprises several phases: recognition, adsorption to cell surface, membrane fusion, cell penetration, uncoating, transcription, genome replication, virion assembly, and virion release.

A viral particle is adsorbed to the host cell membrane through the HN protein, which recognizes a specific receptor—usually glycosylated. Then, a conformational change in the HN protein activates the F protein, exposing a highly hydrophobic domain through which the cell membrane and the viral envelope are fused. This event takes place on the cell surface and at a neutral pH, and therefore it does not depend on lysosomal enzymes, as it is the case with other viruses.

The integration of the viral envelope to the cell membrane system results in the nucleocapsid being released into the cytoplasm. Since the virion generates its own polymerase, a complete copy in positive sense of the viral genome (which will serve as template for the new genomes) is synthesized once the viral RNA is released. In parallel, the synthesis of messenger RNA coding for viral proteins is started.

Translational products are directed to the assembly site. The NP, P, and L proteins are bound to the newly synthesized RNA, and the M protein accumulates in the inner region of the cell membrane. The HN and F proteins, synthesized in the endoplasmic reticulum, are glycosylated in the Golgi apparatus and then expressed in the outer face of the membrane, in sites with close contact to the M protein.

The affinity of NP to M protein and of the latter to glycoproteins is critical in the assembly of virions, which are subsequently released by exocytosis [28]. Since rubulavirus replication occurs entirely in the cytoplasm, nucleocapsid agglomerations are produced and visible as large cytoplasmic inclusions within infected cells [26,27,31].

Viral genetic material could also pass directly from an infected cell to a neighboring cell, with no viral particle release. This is due to the fusion of cell membranes caused by HN and F proteins, giving rise to giant multinucleated cells (syncytia), the signature of paramyxovirus infections [28].

PRV is sensitive to lipid solvents, such as ether and chloroform, and also to formaldehyde and β-propiolactone. The virus is inactivated at 56°C after 4 h, while viral hemagglutinating activity is lost only after 48–60 h at 56°C [5]. PRV has been cultured in a wide array of cell lines such as PK-15, Vero, MDBK, ST, Pt, BT, Bt, BHK-21, GMK, CK, and BEK and in the primary cultures of cat kidney, bovine embryo, equine dermis, human fetal cells, porcine choroid plexus, bovine choroid plexus, and mink lung, causing a cytopathic effect (CPE) in all of them; it can also be replicated in chicken embryo [4,30,32–34].

55.1.2.3 Epidemiology

Hog is the only animal known to be clinically affected by PRV under natural conditions. PRV has been experimentally inoculated to mice, rats, chicken embryos, rabbits, cats, and dogs. None of these hosts showed infection-related signs. However, antibodies against the virus have been found in experimentally infected rats, rabbits, dogs, cats, and peccaries [4,35–37].

Blue eye disease has only been described in Mexico. Currently, the disease has set in and is occasionally reported in the states of Michoacán, Jalisco, Guanajuato, Tlaxcala, México, Hidalgo, Mexico City, and Querétaro [2,38,39].

55.1.3 Clinical Features and Pathogenesis

55.1.3.1 Clinical Features

In 1980, virtually all affected piglets were younger than 15 days. Clinical features included fever, bristly hair, and arched back, sometimes with constipation or diarrhea. Later, nervous signs like ataxia, weakness, rigidity (mainly in the hind legs), muscular tremor, and abnormal postures (such as the sitting dog posture) were observed [40,41].

After 1983, severe outbreaks of encephalitis and respiratory signs were observed, with mortality rates up to 30% in 15–45 kg pigs [34]. Reproductive failure became apparent on the same year with an increase in the rates of repeat breeding, stillborn pigs, and mummified fetuses, along with a small increase in abortions [42].

In 1988, reports of orchitis, epididymitis, and testicular atrophy with severe semen alterations were linked to the blue eye paramyxovirus. On that year, the morbidity rate for piglets born during an outbreak was 20%–50%, and mortality was 89%–90% [43].

In 1998, severe cases of the disease were reported in association to the porcine reproductive and respiratory syndrome virus (VPRRS). In these cases, reproductive parameters were further altered with respect to PRV and VPRRS alone. Pregnant sows exhibited anorexia and corneal opacity

in some cases. Reproductive failure affected several parameters for 2–11 months, 4 months as an average. The number of piglets born alive decreased, while the number of mummified fetuses and piglet mortality increased, and therefore the number of weaned piglets dropped dramatically. A marked increase in the number of stillborn and mummified was also observed, and consequently a drop in the number of born alive and then in the total number of born piglets followed [40,44].

Some cases of animals showing respiratory signs such as dyspnea, cough, and sneeze were reported as associated to the blue eye disease, but they were found related to bacterial infective processes instead. Nevertheless, viral isolates were obtained from lungs and other organs from these animals [45].

55.1.3.2 Pathogenesis

The virus is transmitted by direct contact and by inhalation of virus-contaminated aerosol particles expelled by carrier animals through nasal discharges [2,26,40]. PRV spreads horizontally, from infected animals to susceptible animals; there is also evidence of vertical transmission since the virus is capable of crossing the placental barrier [46–48].

Under natural conditions, PRV enters the body through the oronasal cavity. When virus-contaminated particles are large, replication occurs first at the oronasal cavity, mainly at the nasopharyngeal mucosa and the associated lymphatic tissue (tonsils). When particles are small, the virus enters the lung with the inhaled air, settling on the upper airways, where it replicates profusely; the virus is then carried through the blood by erythrocytes and leucocytes, starting the systemic infection, propagation, and replication in a number of immunoprivileged sites in lymphatic and reproductive organs [1,49,50].

PRV may enter the central nervous system (CNS) through the olfactory nerve endings in the nasal mucosa, and then it can spread to the hippocampus, brain stem, and cerebellum [20,49,50].

Experimental PRV infections have been produced by intranasal, intratracheal, intraocular, intracranial, intraperitoneal, and intramuscular inoculation [51,52].

PRV distribution varies according to the age of infected pigs. In neonates, the main site for viral replication is tissues of the CNS (olfactory bulb, piriform cortex lobe, hippocampus, thalamus, brain stem, and cerebellum) and the respiratory tract; in weaned pigs, the infection occurs in the respiratory tract and in the CNS [4,53].

In adult boars, the virus replicates mainly in tissues from the respiratory and reproductive tract (testicle and epididymis). In experimentally infected pregnant sows, PRV was isolated mainly from the respiratory tract, tonsil, and ovaries [34,54].

The tissue distribution of PRV is determined by the expression of cell receptors, specifically of N-acetylneuraminic acid α-2,3 galactose (NeuAcα-2,3 Gal); this molecule is very abundant in the nervous, respiratory, and lymphatic systems of neonate piglets and in the respiratory and reproductive systems of adult pigs [17,49,51].

Studies reporting virulence changes in four isolates (PAC-6, PAC-7, PAC-8, and PAC-9) allowed subdividing PRV strains into three groups: group 1 was formed by LPMV and PAC-4; group 2 was formed by PAC-2, PAC-3, and CI-IV; group 3 was formed by isolates PAC-6, PAC-7, PAC-8, and PAC-9. These changes resulted from sequence alterations of the gene coding for the HN protein, which confer virulence and cause nervous signs in adult animals and in the fattening line [54]. The humoral response against PRV starts with the presence of specific antibodies in titers of 4.6 $\log_2$ since days 9 and 10 postinfection (PI); maximum titers have been observed at weeks 7 and 9 PI, with values of 7.6 $\log_2$.

Studies on antibody levels in long-term experimental infections have detected hemagglutination-inhibiting antibodies since week 1 PI in a titer of 2.85 $\log_2$, reaching a peak at week 9 PI in a titer of 5.0 $\log_2$ and maintaining constant levels up to 19 weeks. These antibodies are directed against proteins HN, NP, and M [55–59].

55.1.4 DIAGNOSIS

55.1.4.1 Conventional Techniques

Serological tests have proven valuable for PRV diagnosis. Hemagglutination inhibition (HI) is the most used test to assess PRV spread in Mexico, while seroneutralization (SN), ELISA, and immunoperoxidase assays are also applicable [57].

Several studies have been undertaken to establish the most reliable serological test for PRV diagnosis. In one study involving 46 animals after an acute outbreak, the sensitivity of SN, ELISA, and HI for PRV detection was 89.1%, 89.1%, and 84.7%, respectively. In 35 animals immunized with an experimental vaccine, SN sensitivity was 94.2%, ELISA sensitivity was 91.4%, and HI sensitivity was 80%. In 94 vaccinated animals at a farm during an outbreak, ELISA sensitivity was 91.5% and HI sensitivity was 98.9%. Thus, SN and ELISA represent the most sensitive assays, while HI can be used as a herd-testing tool. Since HI detects a small percentage of false positives, it is not appropriate for individual testing [57].

PRV diagnosis is also achievable by virus detection using viral isolation, indirect immunofluorescence, or electron microscopic observation in infected tissues. The most used cell lines for PRV isolation are Vero (African green monkey kidney), PK-15 (porcine kidney) [60], ST (swine testicle), and BHK-21 (baby hamster kidney). As a characteristic CPE in these cells, PRV promotes syncytium formation. Additionally, intracytoplasmic inclusion bodies have been observed in PK-15 cells [61].

55.1.4.2 Molecular Techniques

A reverse transcription polymerase chain reaction (RT-PCR) targeting the HN gene has been used for some years to detect PRV. Several modifications to the technique have been proposed [26,62,63]. A PCR assay amplifying the strains with genetic variations reported by Sánchez-Betancourt et al. [54] will be described in the Methods section, as the primers therein used are designed for conserved regions of the variants reported so far.

In 2013, a real-time RT-PCR (qRT-PCR) assay was first reported [64]. This test is more sensitive than and as specific

as gel-based RT-PCR assay. In addition, this qRT-PCR has the capability of quantifying viral RNA, thus allowing stage-wise evaluation of a clinical infection. This technique should be considered when studies on viral infection dynamics with different viral isolates are to be performed. Since some of these strains are genetically and antigenically characterized, the qRT-PCR may also be used to establish strain virulence [62].

55.2 METHODS

55.2.1 SAMPLE PREPARATION

Animals showing clinical signs suggestive of this disease can be diagnosed by obtaining a blood sample to perform HI or SN tests, as described in later text. It is advisable to obtain the highest possible amount of serum from blood samples; for this, an anticoagulant-free blood collection tube (e.g., Vacutainer®) should be kept horizontally immediately after collecting the sample. Once serum is obtained, complement proteins should be inactivated by incubating the tube in a water bath at 56°C for 30 min.

For antigen detection, viral isolation or RT-PCR can be performed (see Sections 55.2.2.3 and 55.2.2.4). The recommended swine samples are nasal swabs, tonsils, lung, and a brain sample in those cases with nervous signs. In any case, the sample should be taken within 1 h after the death of the suspect pig.

55.2.2 DETECTION PROCEDURES

55.2.2.1 Hemagglutination Inhibition

To adsorb and eliminate nonspecific hemagglutination inhibitors, sera are treated as follows: 200 µL of serum is placed in a V-bottom 96-well culture plate, 100 µL of kaolin suspension and 100 µL of 5% bovine erythrocytes are added, and the plate is incubated at 4°C for 24 h.

Adsorbed sera are placed in U-bottom 96-well microplates and 50 µL of PBS (phosphate buffered saline, pH 7.2) is added to all wells. Then, 50 µL of the serum under investigation (supernatant) are placed in line A; double serial dilutions are made from lines A through H, and 50 µL is discarded at the end. It is advisable to reserve the last two rows as controls (positive and negative).

Then, 50 µL of viral antigen with 8 UHA is added to each well, from lines A through H. Plates are incubated for 30 min at room temperature, and 50 µL of 0.5% bovine erythrocytes is added to each well. Finally, plates are incubated at room temperature for 30–60 min.

As a positive control, one row per plate is added with a serum for which the titer of hemagglutination-inhibiter antibodies is known, along with PBS, viral antigen, and erythrocytes. In this case, a sample is regarded as positive when a button is formed by erythrocyte sedimentation.

As a negative control, a serum, known as not having antibodies against PRV, is added, along with PBS, viral antigen, and erythrocytes. Since no antibodies are present in this case, the antigen is bound to erythrocytes and hemagglutination is observed.

In HI tests, a serum is regarded as positive when its titer is equal to or greater than 1:16 [19].

55.2.2.2 Seroneutralization

Serum samples were inactivated at 56°C for 30 min. A sterile, flat bottom, 96-well microplate is used to make serum dilutions.

To all wells, 50 μL of serum-free medium is added from columns 1 through 10. Then, 50 μL of the serum under investigation is added to line A, and the plate is homogenized; then, 50 μL is passed to line B and so on until line H, and 50 μL is discarded at the end. Dilutions start at 1:2 up to 1:256. Positive and negative control sera are placed in the last columns.

At a dose of 300 DICC$_{50}$%, 50 μL of antigen (previously titered for virus) is added and is incubated for 60 min at room temperature. Then, 100 μL from each well line is transferred to a plate containing previously prepared Vero cells.

Plates are incubated for 72 h, and 50 μL of supernatant is transferred to a U-bottom plate. Then, 50 μL of 0.5% bovine erythrocytes is added to each well to read for agglutinating activity. The CPE caused by the virus is read after 120 h.

Positive result: The serum has antibodies, and when these antibodies bind to the antigen, no CPE is observed.

Negative result: Since no antibody is present, the virus causes CPE. A result is regarded as positive when it is higher than 1:32 [19].

55.2.2.3 Viral Isolation

Tissue samples should be macerated and filtered through a 0.22 μ filter under sterile conditions. Then, 1 mL of the suspect sample is placed on a monolayer of Vero cells (African green monkey kidney) with 24 h of growth in 25 cm^2 flasks, supplemented with Eagle's minimum essential medium (MEM) and 5% fetal bovine serum (FBS). Flasks are incubated at 37°C. Cultures are maintained with MEM, with no FBS. Flasks are checked daily, and the presence or absence of CPE is observed at days three and six postinoculation [64].

To observe the presence or absence of PRV-caused agglutination, 50 μL of cell culture supernatant are obtained and successively compared with 50 μL of 0.5% erythrocytes from bird, bovine, and guinea pig. Agglutination is reported as positive and classified as low, moderate, or high according to its intensity.

55.2.2.4 Titration of Isolated Virus

In all wells of a U-bottom 96-well microplate, 50 μL of PBS (phosphate buffer saline, pH 7.2) is placed. Then, 50 μL of clarified supernatant are placed in row 1, using two columns for each virus. Double serial dilutions are made, starting with 1:2 up to 1:4096. A positive and a negative control should be included in all cases, being the negative an erythrocyte sample and the positive a known virus.

Once dilutions are made, 50 μL of 0.5% bovine erythrocytes is added to all wells, from rows 1 through 12. The hemagglutination final titer is regarded as the wall prior to the

one showing red cell sedimentation. Virus replication can be achieved until the third blind pass.

55.2.2.5 RT-PCR

55.2.2.5.1 *Extraction of Viral RNA*

If the suspect sample consists of tissue or nasal swab, it will be ground with liquid nitrogen in a mortar; 200 μL of MEM or PBS will be added, and the sample will be kept frozen until processed.

If the sample was taken from cell cultures, cells will be centrifuged at 4500 rpm for 15 min at 4°C. The cell button is then resuspended in 200 μL of sterile PBS, and 5 mL of glycogen is added, leaving it to stand for 4 min at 4°C.

Both sample types are processed as follows: 1 mL of TRIzol® reagent is added and the sample is incubated at 4°C for 10 min. Then, 200 μL of chloroform is added; the tube is vortexed for 15 s, incubated at 4°C for 10 min, and centrifuged at 12,000 rpm for 15 min at 4°C.

The phase containing RNA is separated in Eppendorf® tubes, and one volume of isopropanol is added. The tube is incubated for 15 min at 4°C and centrifuged at 12,000 rpm for 10 min at 4°C. Isopropanol is decanted, and the button (RNA) is washed with 200 μL of 70% ethanol. The tube is centrifuged at 12,000 rpm for 10 min at 4°C; ethanol is decanted and the button is left to dry. The button is resuspended in 50 μL of DEPC water (diethyl pyrocarbonate 0.1%).

55.2.2.5.2 *Reverse Transcription*

RNA from the previous step is used to obtain complementary DNA as follows: 7 μL of RNA, 1 μL of dNTPs, and 1.25 μL of the HNISBR1 primer are added to 5.8 μL of DEPC water. The tube is placed in a thermocycler at 65°C for 5 min and left to stand at 4°C for 5 min. Then, 5 μL of buffer (5×), 1 μL of DTT, and 1 μL of RNAseOut (40 unit/μL) are added, and the tube is incubated at 42°C for 2 min. Then, 1 μL of reverse transcriptase (200 units/μL) is added; the tube is incubated at 42°C for 50 min and finally at 70°C for 15 min.

55.2.2.5.3 *PCR*

Conditions for PCR are as follows: one initial denaturation cycle at 94°C for 4 min, 30 denaturation cycles at 94°C for 30 s, annealing at 54°C for 30 s, extension at 72°C for 1 min, and one final extension cycle at 72°C for 3 min. Amplifying reactions will include 4 ng/μL of template DNA, 0.5 pM of both primers, 0.2 mM of dNTPs, 0.15 mg/mL of BSA, Triton® 0.1%, buffer C 1X (1.5 mM MgCl$_2$), 0.125 U/μL of Taq polymerase, and H$_2$O in an amount enough to 20 μL.

55.2.2.5.4 *Electrophoresis*

PCR products are evaluated by electrophoresis on 2% agarose gel and ethidium bromide staining. BstEII-digested λ bacteriophage DNA (λ/BstEII 100 ng/μL) can be used as a molecular weight marker or any other marker containing DNA with 1000–2000 bp. PCR products are then viewed under ultraviolet light.

55.3 CONCLUSION

A number of vaccines have been developed to prevent the disease [65–70], with differing antigen concentrations and differing vaccination and evaluation schemes. However, the performed studies have consistently reported a viral evolution, evidenced in the genetic variants found; on the other hand, the reported viral clusters also indicate an antigenic change among the analyzed virus, as serological tests have shown [63,70].

The genetic and antigenic variants of PRV found to date help us to answer many questions about how to control the disease and about the failure of the marketed vaccines to prevent it. Also, it is now clear that the protection conferred by commercial vaccines will depend on the antigen they contain, since the deficient capacity of antibodies to protect against heterologous viruses, as commonly detected when evaluating commercial vaccines, is proved.

It is advisable to use the HI test as a diagnostic screening and the SN test to assess the protector capacity of postvaccine antigens. RT-PCR or qRT-PCR will be required for antigen detection, along with viral isolation, which remains as the gold standard to demonstrate the presence of viral antigen.

REFERENCES

1. Ramírez-Mendoza H, Hernández JP, Reyes LJ, Zenteno E, Moreno LJ, Kennedy S. Lesions in the reproductive tract of boars experimentally infected with porcine rubulavirus. *J. Comp. Pathol.* 1997; 117: 237.
2. Fuentes RMJ, Gay GM, Herradora LMA, Retana RA. Evaluación de una vacuna experimental contra ojo azul en cerdos mediante las pruebas de inmunogenicidad, inocuidad, potencia y medición de la inmunidad pasiva en lechones. *Vet. Méx.* 1994; 25: 3.
3. Milián SF, García CL, Anaya EA. Prevalencia de enfermedades de importancia económica de los cerdos en México. *Memorias de la XL Reunión Nacional de Investigación Pecuaria.* 2004; 3.
4. Moreno-López J, Correa-Girón P, Martínez A, Ericsson A. Characterization of a Paramyxovirus isolated from the brain of a piglet in Mexico. *Arch. Virol.* 1986; 91: 221–231.
5. Stephano HA, Gay GM, Ramirez TC. Encephalomyelitis, reproductive failure and corneal opacity (blue eye) in pigs, associated with a paramyxovirus infection. *Vet. Rec.* 1988; 122: 6.
6. Rima B et al. Family *Paramyxoviridae.* In: Murphy FA, Fauquet CM, Bishop DH, Ghabrial SA, Jarvis AW, Matelli GP, Mayo MA, Summers MD (eds.), *Virus Taxonomy, Classification and Nomenclature of Viruses.* New York: Springer-Verlag, 1995.
7. Pringle CR. Virus taxonomy. The universal system of virus taxonomy, updated to include the new proposals ratified by the International Committee on Taxonomy of Viruses during. *Arch. Virol.* 1998; 144: 421–429.
8. Drake JW. Rates of spontaneous mutation among RNA viruses. *Proc. Natl. Acad. Sci. USA* 1993; 90: 4171–4175.
9. Reyes-Leyva J. et al. Mecanismos moleculares de patogenicidad viral: Estudios con el rubulavirus porcino. In: *Mensaje Bioquímico XXVI*, pp. 99–127, Mexico: Facultad de Medicina, UNAM, 2002.
10. Berg M, Sundqvist A, Moreno-López J, Linné T. Identification of the porcine paramyxovirus LPM matrix gene: Comparative sequence analysis with other paramyxoviruses. *J. Gen. Virol.* 1991; 72; 1045–1050.
11. Linné T, Berg M, Bergvall AC, Hjertner B, Moreno- López, J. The molecular biology of the porcine paramyxovirus LPMV. *Vet. Microbiol.* 1992; 33: 263–273.
12. Sundqvist A, Berg M, Moreno-López J, Linné T. The haemagglutinin-neuraminidase glycoprotein of the porcine paramyxovirus LPMV: Comparison with other paramyxoviruses revealed the closest relationship to simian virus 5 and mumps virus. *Arch. Virol.* 1992; 122: 331–340.
13. Wang LF et al. Full-length genome sequence and genetic relationship of two paramyxoviruses isolated from bats and pigs in the Americas. *Arch. Virol.* 2007; 152: 1259–1271.
14. Svenda M, Berg M, Moreno- López J, Linné, T. Analysis of the large (L) protein gene of the porcine Rubulavirus LPMV: Identification of possible functional domains. *Virus Res.* 1997; 48: 57–70.
15. Reyes- Leyva J et al. NeuAcα2,3 Gal-Glycoconjugate expression determines cell susceptibility to the porcine Rubulavirus LPMV. *Comp. Biochem. Physiol.* 1997; 118: 327–332.
16. Zenteno-Cuevas R, Hernandez J, Espinosa B, Reyes-Leyva J, Zenteno, E. Secondary structure prediction of the hemagglutinin-neuraminidase from a porcine Rubulavirus. *Arch. Virol.* 1998; 143: 333–352.
17. Vallejo V, Reyes-Leyva J, Hernandez J, Ramirez P, Delanoy P, Zenteno E. Differential expression of sialic acid on porcine organs during maturation process. *Comp. Biochem. Physiol.* 2000; 126: 415–424.
18. Reyes-Leyva J, Hernandez-Jáuregui P, Montaño LF, Zenteno E. The porcine paramyxovirus LPM specifically recognizes sialyl (alpha 2,3) lactose-containing structures. *Arch. Virol.* 1993; 133: 195–200.
19. Ramírez MH, Carreón NR, Mercado GC, Rodríguez TJ. Hemoaglutinación e inhibición de la hemoaglutinación del paramixovirus porcino a través de la modificación de algunas variables que participan en la prueba. *Vet. Méx.* 1996; 27: 257–259.
20. Santos-López G, Hernández J, Borraz-Argüello MT, Ramírez-Mendoza H, Vallejo V, Reyes-Leyva J. Estructura, función e implicaciones patológicas de las proteínas del *Rubulavirus porcino. Arch. Med. Vet.* 2004; 36: 119–136.
21. Hernández J, Reyes-Leyva J, Zenteno R, Ramirez H, Hernández-Jáuregui P, Zenteno E. Immunity to porcine rubulavirus infection in adult swine. *Vet. lmmunol. Immunopathol.* 1998; 64: 367.
22. Sánchez-Betancourt JI et al. Molecular characterization of the hemagglutinin-neuraminidase gene of porcine Rubulavirus isolates associated with neurological disorders in fattening and adult pigs. *Res. Vet. Sci.* 2008; 85: 359–367.
23. Sundqvist A, Berg M, Hernandez-Jauregui P, Linné T, Moreno-López J. The structural proteins of a porcine paramyxovirus (LPMV). *J. Gen. Virol.* 1990; 71: 609–613.
24. Kolakofsky D, Pelet T, Garcin D, Hausmann S, Curran J, Roux L. Paramyxovirus RNA synthesis and the requirement for hexamer genome length: The rule of six revisited. *J. Virol.* 1998; 72: 891–899.
25. Calain P, Roux L. The rule of six, a basic feature for efficient replication of Sendai virus defective interfering RNA. *J. Virol.* 1993; 67: 4822–4830.
26. Berg M, Hjertner B, Moreno- López J, Linné T. The P gene of the porcine paramyxovirus LPMV encodes three possible polypeptides P, V and C: The P protein mRNA is edited. *J. Gen. Virol.* 1992; 73: 1195–1200.

27. Lamb RA, Kolakofsky D. Paramyxoviridae: The viruses and their replication. In: Fields BN, Knipe DM, Howlet PM (eds.), *Fields Virology*, 3rd edn. New York: Lippincot-Raven Publishers, 1996.

28. Berg M, Bergvall AC, Svenda M, Sundqvist A, Moreno-López J, Linné T. Analysis of the fusion protein gene of the porcine rubulavirus LPMV: Comparative analysis of the paramyxovirus F proteins. *J. Gen. Virol.* 1997; 14: 55–61.

29. Morrison TG, Pother A. Structure, function and intracellular processing of glycoproteins of *Paramyxoviridae*. In: Kingsbury DW (ed.), *The Paramyxoviruses*. New York: Plenum Press, 1991.

30. Hernández-Jáuregui P, Yacoub A, Kennedy S, Curran B, Téllez C, Svenda M, Ljung L, Moreno-López J. Uptake of porcine rubulavirus (LPMV) by PKI5 cells. *Arch. Med. Res.* 2001; 32: 400–409.

31. Martínez LA, Correa GP, Fajardo MR, Garibay SM. Aislamiento y estudio de un virus porcino parecido a los paramixovirus. In: Correa PY, Morilla A (eds.), *Encuentro sobre enfermedades infecciosas del Cerdo*, pp. 15–21. México, 1985.

32. Ramírez NR, Martínez LA, Correa GP, Colinas TA. Un brote de paramixovirosis encefalitis en cerdos de una granja del Estado de México. Mem. II Congreso ALVEC, XII Convención AMVEC, pp. 64–67, III Encuentro UNPC. Acapulco, Guerrero, México, 1987.

33. Stephano HA, Gay, GM. El síndrome del Ojo Azul en granjas engordadoras. In: *Memorias del XIX Congreso de la Asociación Mexicana Veterinarios Especialistas en Cerdos*, pp. 71–74. México: AMVEC, 1985.

34. Arellanes AE, Fuentes RM, Carreón NR, Ramírez MH. Inoculación experimental del paramixovirus del ojo azul en el gato doméstico (*Felis catus*). *Rev. Vet. Méx.* 1994; 25: 239–241.

35. Cuetero RS, Ramírez MH, Carreón NR, Campuzano GJ. Inoculación experimental del paramixovirus del ojo azul en ratas de laboratorio (cepa Wistar) vía intramuscular. *Vet. Méx.* 1995; 26: 231–236.

36. Campos M. E. Síndrome del ojo azul o cerdos zarcos. Mem. XVII convención AMVEC-Ixtapa. Zihuatanejo, Guerrero, México, 1981.

37. Morilla A, Gonzalez-Vega D, Estrada E. Diosdado F. Seroepidemiology of blue eye disease. In: Morilla A, Yoon K, Zimmerman J (eds.), *Trends in Emerging Viral Infections of Swine*. Ames, IA: Iowa State University Press, 2002.

38. Kirkland PD, Stephano A. Paramyxoviruses: Rubulavirus, Menangle and Nipah virus infections. In: Straw B, Zimmerman J, D'Allaire S, Taylor D (eds.), *Diseases of Swine*, 9th edn. New York: Blackwell Publishing, 2006.

39. Stephano HA. La enfermedad del ojo azul. Signos clínicos y lesiones. In: *Memorias del Simposio Internacional sobre Enfermedades Emergentes del Cerdo*, pp. 1–10. México, 2000.

40. Taylor DJ. Rubulavirus infection and "Blue Eye". In: Zimmerman J, Karriker L, Ramirez A (eds.), *Pig Diseases*, 7th edn., pp. 54–55. Glasgow, U.K.: Wiley-Blackwell Publishing, 1999.

41. Stephano HA, Gay GM. Experimental studies of a new viral syndrome in pigs called "Blue Eye" characterized by encephalitis and corneal opacity. In: *Proceedings of the 8th International Pig Veterinary Society Congress*, 71pp., Ghent, Belgium, 1984.

42. Campos HR, Carbajal SM. Trastornos reproductivos de los sementales de una granja porcina de ciclo completo ante un brote de ojo azul. In: *Memorias del XIV Congreso de la Asociación de Veterinarios Especialistas en Cerdos*, pp. 62–64. México: AMVEC, 1989.

43. Sánchez BJI, Trujillo OME, Doporto DJM, Becerra AF. Identificación de las afecciones productivas en hembras reproductoras después de un brote de Ojo Azul. In: XXXIX Congreso Nacional de la Asociación Mexicana de Médicos Veterinarios Especialistas en Cerdos Mexico: AMVEC, 2004.

44. Sánchez BJI, Doporto DJM, Trujillo OM, Carreón NR, Becerra FA. Aislamiento del paramixovirus porcino a partir de animales de engorda con cuadro respiratorio. Memorias XXXVII Congreso AMVEC 2002, pp. 110–111. Puerto Vallarta, Jalisco, Mexico.

45. Martínez LA, Correa GP, Colinas TA, Galina PL. Opacidad corneal bilateral congénita en lechones de cerdas expuestas al paramixovirus porcino de La Piedad, Michoacán. Memorias del XIV Congreso Nacional de la Asociación Mexicana de Veterinarios Especialistas en Cerdos. Morelia (Michoacán) México (DF): Asociación Mexicana de Veterinarios Especialistas en Cerdos, AC, 1989.

46. Hernández-Jáuregui P, Ramírez MH, Mercado GC, Moreno-López J, Kennedy S. Experimental porcine rubulavirus (La Piedad-Michoacán virus) infection in pregnant gilts. *J. Comp. Pathol.* 2004; 130: 1–6.

47. Stephano, H.A., Gay, G.M. Experimental studies on a new syndrome in pigs called Blue Eye characterized by encephalitis and corneal opacity. In: *Proceedings of the 8th Congress of the International Pig Veterinary Society*, 71pp., Ghent, Belgium, 1984.

48. Allan GM et al. A sequential study of experimental porcine paramyxovirus (LPMV) infection of pigs: Immunostaining of cryostat sections and virus isolation. *J. Vet. Diagn. Invest.* 1996; 8: 405.

49. Reyes-Leyva J, García-Morales O, Santos-López G, Vallejo V, Ramírez-Mendoza H, Hernández J. Detección de viremia en la infección experimental por Rubulavirus porcino. *Arch. Med. Vet.* 2004; 36: 41.

50. Ramírez-Herrera MA, Mendoza-Magaña ML, Dueñas SH. Experimental infection of swine and cat central nervous systems by the pig paramyxovirus of the blue eye disease. *Zentralbl. Veterinarmed.* [B] 1997; 44: 461.

51. Ramírez-Herrera MA, Mendoza-Magaña ML, Dueñas-Jiménez JM, Mora-Galindo J, Dueñas-Jiménez SH. Electrophysiological and morphological alterations in peripheral nerves by the pig paramyxovirus of blue eye disease in neonatal pigs. *J. Vet. Med. B* 2001; 48: 477–487

52. McNeilly F et al. Comparative study on the use of virus and antibody detection techniques for the diagnosis of La Piedad Michoacán paramixovirus (LPMV) infection in pigs. *J. Vet. Diagn. Invest.* 1997; 9: 3–9.

53. Rivera-Benitez JF. Persistencia del *Rubulavirus porcino* en semen de verracos infectados experimentalmente (tesis de Maestría). México (DF): FMVZ-UNAM, 2009.

54. Sánchez-Betancourt BJI, et al. Molecular characterization of the hemagglutinin-neuraminidase gene of porcine rubulavirus isolates associated with neurological disorders in fattening and adult pigs. *Res. Vet. Sci.* 2008; 85: 359.

55. Fujinami RS, Oldstone MBA. Alterations in expression of measles virus polypeptides by antibody: Molecular events in antibody induced antigenic modulation. *J. Immunol.* 1980; 125: 78–85.

56. Hernández J, Reyes-Leyva J, Ramírez H, Valenzuela O, Zenteno E. Características de la respuesta inmune de cerdos infectados con el *Rubulavirus porcino*. *Vet. Méx.* 2004; 35: 65–74.

57. Morilla GA, Diosdado VF, González VD, Ojeda ZP, Mercado PM, Campomanes CA, Hernández JP, Moreno LJ. Estudio comparativo entre las pruebas de inmunoperoxidasa, ELISA e inhibición de la hemoaglutinación para el diagnóstico serológico de la enfermedad de ojo azul en cerdos. Memorias del Simposio internacional de enfermedades emergentes del cerdo. Academia Veterinaria Mexicana, A.C. Irapuato (Guanajuato) México, 2000.

58. Hernández J, Garfias Y, Reyes-Leyva J, Chávez R, Lascurain R, Vargas J, Zenteno E. Peanut and Amaranthus leucocarpus lectins discriminate between memory and naive/quiescent porcine lymphocytes. *Vet. Immunol. Immunopathol.* 2002; 84: 71–82.

59. Hernández LJ, Ramírez MH, Zenteno CR, Monroy BJ, Reyes LJR, Zenteno E. Neumonitis inducida por el rubulavirus porcino. *Rev. Instituto Nacional de Enfermedades Respiratorias* 1997; 10: 250–255.

60. Wiman AC et al. Porcine rubulavirus LPMV RNA persists in the central nervous system of pigs after recovery from acute infection. *J. Neurovirol.* 1998; 4: 545–552.

61. Cuevas JS, Rodríguez-Ropón A, Kennedy S, Moreno-López J, Berg M, Hernández-Jáuregui P. Investigation of T-cell responses and viral mRNA persistence in lymph nodes of pigs infected with porcine rubulavirus. *Vet. Immunol. Immunopathol.* 2009; 127: 148–152.

62. Rivera-Benitez JF, García-Contreras A, Reyes-Leyva J, Hernández J, Sánchez-Betancourt JI, Ramírez-Mendoza H. Efficacy of quantitative RT-PCR for detection of the nucleoprotein gene from different porcine rubulavirus strains. *Arch. Virol.* 2013; 158: 1849–1856

63. Sánchez Betancourt JI, Trujillo ME, Mendoza ES, Reyes-Leyva J, Alonso MR. Genetic and antigenic changes in porcine rubulavirus. *Can. J. Vet. Res.* 2012; 76: 36–37.

64. Colinas TA, Martinez LA, Correa GP, Fajardo MR. Curva de crecimiento del paramixovirus porcino de La Piedad Michoacán en la línea celular Pk-15. Mem. del II Congreso ALVEC; XII convención AMVEC, pp. 68–71. Acapulco, Guerrero, México, 1987.

65. Correa-Girón P, Martínez LA, Solís HM, Pérez SJ, Coba AMA, Aguirre F. Antigenicidad de diferentes dosis de la vacuna vs la enfermedad del ojo azul (EOA) del INIFAP en cerdos con PRRS. In: *Memorias de la XXXVII Reunión Nacional de Investigación Pecuaria,* 220pp., México, 2001.

66. Correa-Girón P, Martínez LA, Pérez SJ, Coba AMA, Solís HM. Vaccinación against blue eye disease. In: Morilla A, Yoon K, Zimmerman J (eds.), *Trends Emerging Viral Infections of Swine,* 1st edn., pp. 65–69. Ames, IA: Iowa State Press, 2002.

67. Iglesias G, Rincón F, Vazquez B, Tapia J, Domínguez F. Evaluación de una vacuna para la prevención de la enfermedad causada por el paramixovirus porcino. In: *First International Symposium upon Pig Paramixovirus,* pp. 43–44. México, 1994.

68. Martínez LA et al. Evaluación en condiciones de campo de tres lotes diferentes de la vacuna del INIFAP contra la EOA: I) en cerdas reproductoras. In: *XXXVI Congreso AMVEC,* 70pp., Mexico, 2001.

69. Zenteno-Cuevas, R. Purificación y predicción de determinantes antigénicos y de estructura secundaria en la hemaglutinina-neuraminidasa del paramixovirus porcino de La Piedad Michoacán. Tesis de Maestría, Facultad de Medicina, UNAM, México, 1997.

70. Escobar-López C, Rivera Benites F, Castillo H. Ramírez-Mendoza H, Trujillo Ortega ME, Sánchez Betancourt JI. Identification of antigenic variants of the porcine Rubulavirus in sera of field swine and their seroprevalence. *Transbound. Emerg. Dis.* 2012; 59: 416–420.

56 Rinderpest Virus

Anke Brüning-Richardson

CONTENTS

56.1 Introduction 485
   56.1.1 Classification and Genome Organization 486
   56.1.2 Morphology, Biology, and Epidemiology 486
   56.1.3 Clinical Features 487
   56.1.4 Diagnosis 487
      56.1.4.1 Conventional Technique 487
      56.1.4.2 Molecular Techniques 488
56.2 Methods 490
   56.2.1 Sample Preparation 490
   56.2.2 Detection Procedures 490
      56.2.2.1 AGID Test 490
      56.2.2.2 IC-ELISA 490
      56.2.2.3 Rapid Chromatographic Strip Test 490
      56.2.2.4 PCR 491
      56.2.2.5 Competitive ELISA 491
56.3 Conclusions and Future Perspectives 493
References 493

56.1 INTRODUCTION

One of the most devastating veterinary diseases in terms of economic and animal losses, rinderpest was declared globally eradicated in 2011,[1] an incredible feat considering the worldwide distribution of the disease, only matched by the global eradication of smallpox 31 years earlier.[2] Rinderpest was an ancient disease known to Romans and Egyptians alike, which accompanied humankind into the twenty-first century.[3] It has been estimated to have caused losses totaling $2 billion in, for example, Nigeria alone and brought huge suffering to communities depending on farming or meat production for their likelihood sweeping from Eastern Europe to Africa, where its devastating effects were especially felt.[4,5] The first major epidemic in Africa in the 1890s is believed to have killed more than 80% of the African livestock, and as a consequence, one-third of Ethiopians and two-thirds of the Maasai people starved to death.[6] The second African epidemic and outbreaks in many Asian regions in the 1980s led to the loss of an estimated 100 million animals.[6] The disease was devastating due to the mortality rates of up to 90% among susceptible animals.[7] Ultimately, the threat of rinderpest was so great that it led to the establishment of the first veterinary schools in Europe and the introduction of control measures such as slaughter policies and animal isolation when it became clear that close animal contact was the mode of virus transmission.[7–11] In the twentieth century, a concerted effort to bring the disease under control led to the announcement of the Global Rinderpest Eradication Program (GREP) by the UN's Food and Agricultural Organization (FAO), which aimed at an internationally coordinated collaboration to eradicate rinderpest worldwide.[12] The success of this program was, according to Professor John Anderson, "the development of the right technology for field use in Africa and Asia, successful transfer of that technology along with technical backup, and the provision of standardized diagnostic kits that everyone could use." In addition, the development of a live attenuated vaccine that gave lifelong immunity, rinderpest education programs, relevant training of staff involved in the eradication program, and the collaboration of people at national and international levels all contributed to the eradication.

Rinderpest is caused by one of the members of the morbilliviruses, a group of viruses that includes measles virus (MV) and distemper virus (DV). As such, the causative agent, the rinderpest virus (RPV), is an encapsidated virus characterized by the possession of a negative single-stranded RNA with a total genome of 15,200–15,900 nucleotides. It was prevalent in most parts of the world with the exception of the New World with several strains displaying different levels of virulence. The virus was spread by close animal contact through particularly ocular and nasal secretions and conquered new environments by animal movement and trade.

56.1.1 Classification and Genome Organization

According to the Baltimore classification, RPV belongs to group V and is a single-stranded, negative-sense RNA virus. The viruses in this group fall within the order of *Mononegavirales* and are part of the family *Paramyxoviridae* and the genus *Morbillivirus*. Other members of the genus *Morbillivirus* include peste des petits ruminants virus (PPRV), canine DV (CDV), human MV, and viruses that affect marine mammals including seals, porpoises, and dolphins[13,14] as well as Atlantic pinniped and cetacean species.[15,16] All morbilliviruses share similar physicochemical properties that are characterized by similar cytopathic effects in cell culture and common antigens.

The negative-stranded, nonsegmented RNA encodes eight viral proteins: nucleoprotein (N), phosphoprotein (P), matrix (M)-protein, fusion (F)-protein, hemagglutinin (H)-protein, polymerase (L), and two nonstructural proteins (C and V).[17] Gene expression is initiated when the RNA polymerase binds the encapsidated genome at the leader region followed by a sequential transcription of each gene after the recognition of start and stop signals of flanking viral genes. Capping and polyadenylation of mRNAs is regulated by the L-protein during synthesis. RNA editing generates the V-protein and the C-protein by leaky scanning of the P-protein mRNA.[17]

56.1.2 Morphology, Biology, and Epidemiology

The morphology of RPV was first described by Plowright et al.[18] in a study utilizing electron microscopy. Virus particles are measured to be between 1200 and 3000 Å in diameter with an internal helical component of about 175 Å. They are shown to be pleomorphic in appearance, and all possess an outer membrane and appear variable in shape and size, which are either spherical or filamentous. The hemagglutinin–neuraminidase and F-glycoproteins are located as spikes on the surface of the virus. The nucleocapsid appears hollow, coiled, and rod shaped with the herringbone pattern of subunits.[2] Over the years, more details have emerged, and we now know that there are clear functions for each protein. The H-glycoprotein is involved in the adsorption of the virus to the susceptible cells, whereas the F-protein plays a role in cell fusion. Made up of two polypeptides F1 and F2 via a disulfide bond, the F-protein is believed to contribute to the cytopathic effect of the virus. The L-, P-, and N-polypeptides are all parts of the nucleocapsid with the P-protein believed to be the viral RNA polymerase and the N-protein with a function in virus replication. The M-protein, or matrix protein, interacts strongly with the nucleocapsid and associates with the internal part of the virus envelope.[17] The replication of the RPV begins in the host cell cytoplasm after the attachment of the virus to the host cell surface receptors via the H-glycoprotein. Once fusion with the membrane has taken place, the ribonucleocapsid is released into the cytoplasm. Sequential gene transcription occurs, followed by the capping of viral mRNA and polyadenylation. Replication is completed with the encapsidation of newly synthesized viral genomes

with nucleoprotein. The viruses exit the host cell after the interaction of the ribonucleocapsid with the M-proteins while still located close to the plasma membrane.[19]

Natural hosts for RPV are members of even-toed ungulates belonging to the order *Artiodactyla* including cattle, buffaloes and other bovines, goats, sheep, and pigs, but the virus can also affect wildlife such as giraffe, eland, kudu, various antelopes, and wildebeest. Although only one serotype of the virus exists, there are nevertheless differences in virulence with highly virulent strains such as RPV-Saudi and mild strains including RPV-Egypt.[20] The phylogenetic relationship of the RPVs revealed that there are three main lineages: the *African lineage 1* (Egypt/84, RBS/Rb-72, Nigeria/Buffalo/83, RBK/W.Pokot/86, RBK/Kiambu/88, RBS/Wakobu/92/1 and RBK/Marsabit/87/1), the *African lineage 2* (RBT/1, Nigeria/Sokoto/64, Nigeria/Sokoto/83, RGK/1 and RbuffaloK/95), and the *Asian lineage* (KAG, Sri Lanka/87, India Bison/89, India/Mukteshwar, Saudi/81 and Pakistan/84).[21] The disease is passed on to a susceptible animal through close contact with an infected animal, and its virus-laden nasal and ocular secretions. Spread through contaminated bedding, animal food and water are less common due to the instability of the virus particles outside its animal host. The virus is especially sensitive to high humidity (50%–60%), light, and high and low pHs. As an enveloped virus, it is also efficiently inactivated by lipid solvents.[22]

The disease is believed to have originated from Mongolia and then spread throughout Europe during the sixteenth century, leading to the devastating losses among the naïve cattle populations.[7] Rinderpest was introduced into Africa in the nineteenth century, resulting in the first great panzootic wave that hit Africa. Before the disease burnt itself out, it had affected the fauna and flora of Africa, leading to livestock losses in millions. A second epizootic occurred during the 1980s, culminating in the demise of again millions of animals.[7,23] Rinderpest also affected many regions in Asia.[24,25] Outbreaks were reported in China and Korea, which spread through military conflicts and eventually subsided possibly due to the interruption of host–host interactions.[26] Rinderpest was introduced from Korea to Japan in the seventeenth century, resulting in the loss of over 500,000 animals and continued to affect livestock into the twentieth century.[27,28] India and Pakistan were badly affected incessantly until the completion of GREP. Middle Eastern regions including Turkey, Kuwait, Saudi Arabia, Bahrain, Qatar, and the United Arab Emirates have been ceaselessly exposed to rinderpest invasion.[7] Only the New World appears to have been spared the devastation caused by the disease.[29–31]

The main driver in the spread of the disease is the import of the infected livestock into regions with naïve cattle populations, as once recovered from the disease, the animal requires lifelong immunity. This mechanism of disease establishment is believed to have been introduced with the invasion of Europe by the Mongol hordes, who brought infected yaks that had acquired innate resistance to the disease and was also responsible for the introduction of the virus into Africa through European soldiers and infected livestock taken as food supply.[5,32] Thus, the changing status of the disease between an enzootic state and the epizootic state depends on the circulatory activity of the virus. Animals continuously exposed to the virus acquires an inherited innate resistance, resulting in a slow spread of the disease, reduced clinical response to the disease, and increased survival rates. Mainly young animals that has lost protection from maternal antibodies in the colostrum after weaning are susceptible in enzootic regions.[22] When newly introduced to a naïve host population, the virus is able to spread rapidly, causing a high number of losses among the whole population and resulting in epizootics as seen in the African continent.

56.1.3 CLINICAL FEATURES

Historically, rinderpest clinical signs manifested themselves in the three Ds: discharge, diarrhea, and death.[22] The disease is characterized by an incubation period of 3–15 days, which is followed by a prodromal fever, anorexia, and the affected animal showing signs of depression. Mucopurolent discharge from the eye and nose precedes necrotic lesions on the gum and buccal mucosa. The final stages of disease are characterized by shooting diarrhea, which can be watery and bloody and may contain mucous. At this stage the animal may show extreme distress caused by abdominal pain and thirst. Its death is mainly due to severe dehydration.[7,22] Most recent cases of rinderpest manifested themselves in less severe clinical signs with slight transient fever caused by less virulent strains of rinderpest, which made a precise diagnosis difficult. The peracute, acute, or subacute status of infection is dependent on strain virulence and individual susceptibility of the animal. Upon the postmortem examination, the disease presents itself as pathological changes especially throughout the gastrointestinal and upper respiratory tract with pronounced necrotic regions and erosion. Congestion and hemorrhage are also seen leading to the characteristic zebra striping in the lower GI tract. Lesions are often found on the gums and associated regions of the oral tract. Other observations are the enlargement of lymph nodes with associated edema and the appearance of necrotic foci in Peyer's patches. Overall, the carcass appears dehydrated, emaciated, and soiled.[22]

56.1.4 DIAGNOSIS

56.1.4.1 Conventional Technique

The diagnosis of rinderpest is either a direct diagnosis revealing the presence of the RPV or viral antigens or indirect diagnosis detecting the presence of antibodies to the virus. Once the causative agent of a rinderpest-like disease is suspected, various diagnostic techniques are available to confirm the presence of RPV. Classically, direct detection of RPV includes virus isolation, agar gel immunodiffusion (AGID), counterimmunoelectrophoresis, passive hemagglutination, immunofluorescence (IF), immunoperoxidase staining, and most recently immunocapture ELISA (IC-ELISA) and a chromatographic strip test. Here, virus isolation, ELISA, and chromatographic strip test technology will be discussed in

more detail, as these are the more routinely used techniques for a large-scale testing or penside diagnosis during GREP. For information on the other techniques, there are several in-depth publications available.[7,22] For the detection of antibody, the virus neutralization assays and ELISA technologies were the most commonly developed assays.

Isolation of RPV involves the inoculation of a tissue suspension from a sample of a rinderpest suspect animal into the cell layers of susceptible cell lines such as bovine kidney or Vero cells, allowing the virus to infect the chosen recipient cell line. Virus infection is noted by microscopic examination of the recipient cells and manifested itself in early cytopathic changes, which eventually result in rinderpest-induced syncytia in the cell cultures. Once syncytia are observed, the addition of rinderpest immune serum and inhibition of cytopathic effects will confirm the presence of RPV and rinderpest infection. However, the fact that PPR infections can also produce similar cytopathic effects and are also inhibited by rinderpest immune serum activity due to cross-reactivity and the need for a sophisticated tissue culture laboratory setup with associated materials and equipment made this a cumbersome technique.

In contrast, AGID proves a simplistic and easy-to-use technology. Based on the interaction between viral antigen and antibodies, a soluble antigen (suspect sample) is placed in a clear medium, which precipitates out after the reaction with the rinderpest-specific antibody. The technique is improved by adding the sample in wells cut into agar contained within a Petri dish or on a microslide. Controls including a standard antigen and serum are also incorporated in the test, which is deemed positive for rinderpest if a precipitation line occurs between the sample and the standard serum in the agar.

AGID is easy to prepare and has been routinely used until it is replaced by an IC-ELISA for large-scale screening and, in the later stages of GREP, the chromatographic strip test for diagnosis at penside.[33,34] The IC-ELISA is an ELISA based on monoclonal antibodies (Mabs) specific for RPV N-protein. Here, suspect tissue is added to an ELISA plate precoated with an RPV-trapping antibody. Any antigen in the sample is trapped by the RPV-specific antibody on the plate, which is then detected after the addition of a biotinylated RPV-detecting secondary antibody. RPV's presence is visualized by color development, which is read and recorded using an ELISA reader. The advantages of the ELISA over the AGID are sensitivity as well as the ability to test several samples at once and reproducibility. It still requires access to laboratory equipment and an associated ELISA reader.

In the later stages of GREP, a rapid diagnostic chromatographic strip test was developed, which allows the fast diagnosis of rinderpest in the field or at penside.[34,35] This test relies like the other tests on the interaction of a rinderpest-specific antibody to antigen contained within a suspect sample. In the original test, an RPV H-protein-specific antibody labeled with blue colloidal dye is bound in a dehydrated state on a nitrocellulose strip encased in a plastic device. Upon rehydration by the addition of the sample, usually lachrymal fluid collected with a cotton swab, any antigen present is trapped by the dye-labeled antibody and travels along the nitrocellulose strip by capillary action. This complex is then detected as a blue line in a result window, which contains another antibody line reacting with the rinderpest antibody. A control is also incorporated to confirm that the test has worked, so that a two-line result indicates the presence of RPV and a one-line result the absence of the virus. The test is easy to use, fast (in most cases a result will be obtained within 5 min), and robust enough for application in the field under even adverse conditions. It enables a rapid response to a suspect rinderpest outbreak. This technology is especially useful in the final stages of GREP when a differential diagnosis has to be made in the field. It led to the detection of remaining rinderpest foci in Africa and Pakistan.[36,37]

Indirect RPV detection relies on the detection of the presence of antibodies in RPV. The most commonly used technology is an assay in ELISA format. This format is ideal as antibody detection during the earlier stages of GREP, which is important for measuring the impact of mass vaccination on cattle populations and immunity status. Here, seroconversion of vaccinated animals is a valuable indicator of herd immunity, which needs to reach at least 80% in order to eradicate rinderpest. In this ELISA format initially, ELISA plates are coated with rinderpest antigen, and sera from suspect animals are added to the plates and then reacted with secondary antibodies to the species of animal to be tested. As for the IC-ELISA, colorimetric results are obtained with the ELISA. This test is fast (3 h) and highly sensitive; however, the need to produce secondary antibodies to a range of host species renders it cumbersome.[38] For this reason, a new format is adopted (competitive ELISA [C-ELISA]), which is based on rinderpest-specific Mabs.[39] In this assay, again, the first step is coating the ELISA plates with RPV antigen. The test serum is then added to the plate after which there is a further step with the addition of the rinderpest-specific Mab to the RPV H-protein. Any antibody present in the test serum competes with the mouse-derived Mab ELISA kit for antigenic sites in the samples leading to a reduced reaction between Mab and antigen. This competition event results in a colorimetric readout with reduced color development if antibody is present in the test sera. The percentage inhibition (PI) can then be calculated. The test is produced in a kit format by the World Reference Laboratory in Pirbright and is available through the FAO.[39] Reproducibility of results is consistently good, and associated training courses in this technique ensure that C-ELISA users are confident to carry out this test with high success rates.

56.1.4.2 Molecular Techniques

With the advent and advancement of detecting single copies of DNA by means of the polymerase chain reaction (PCR),[40,41] it becomes possible to detect RPV at the molecular level. Preceding PCR technology attempts have already made use of molecular probes such as single-stranded nucleic acid hybridization probes complementary to a single strand in the sample with an initially radioactive label, which is later replaced by a nonradioactive label.[42] For the rinderpest-specific assay,

these are cDNA clones corresponding to the RPV N gene. Drawbacks of this technique are low sensitivity and the need for pure target RNA. However, the development of a PCR for the detection of RPV is an integral driver in the advancement of GREP. For the first time, whole virus genomes can be sequenced, which is especially important with regard to epidemiological studies to establish global geographical locations and movements of viral strains as well as to locate the origin of new outbreaks of rinderpest.

A rinderpest-specific reverse transcription (RT)-PCR was developed by Forsyth and Barrett.[43] In this assay, suspect sample is homogenized and then subjected to the phenol/chloroform extraction to purify the RNA. The RNA is used to make a complementary DNA (cDNA) by RT using random hexanucleotide primers. The generated cDNA is then amplified with the help of rinderpest-specific amplification primers, which are based on the virus F-protein gene of the virus. A universal primer set is generally used, which is based on sequences in the P gene of the morbilliviruses, allowing the amplification of copy DNA from all known morbilliviruses. This reaction is included to ensure that any morbillivirus is identified in the first reaction, which then can be used to make a precise diagnosis using the rinderpest-specific primer sets. The generated data can then also be utilized for sequencing allowing the identification of the RPV strain in question for epidemiological studies. Although these PCRs are routinely carried out during GREP at the World Reference Laboratory at Pirbright, other studies have aimed to improve upon the original assay.

Most recently, there have been publications on modified PCR assays for the detection of RPV. Carrillo et al.[44] reported a real-time RT-PCR system, which is highly specific recognizing virus strains belonging to the three lineages and also sensitive. Primer sets are designed to target the viral gene L10 as this probe detects all RPV tested and does not show cross-reactivity with PPRV or other viruses that cause diseases clinically similar to rinderpest. The assay detects RPV in samples from dead and live animals, and the authors suggested that the format, although lab based for the time being, can also be adapted for field use.

Differential diagnosis of RPV infections to distinguish RP infections from other viral infections are becoming increasingly important after considering that rinderpest is often concomitant with other diseases of veterinary importance with clinical signs similar to rinderpest. To address this, Yeh et al.[45] developed a specific and sensitive one-step multiplex RT-PCR assay for the simultaneous and differential detection of Rift Valley fever virus, bluetongue virus, RPV, and PPRV. This assay is based on novel dual-priming oligonucleotides that contain two separate priming regions joined by a polydeoxyinosine linker. Incorporation of the linker blocks extension of nonspecifically primed templates and allows the early diagnosis of the four viruses, when trialed using whole blood or oral mucosal tissue, as samples. Main events in rinderpest eradication and concomitant developments in diagnostic technologies are summarized in Figure 56.1.

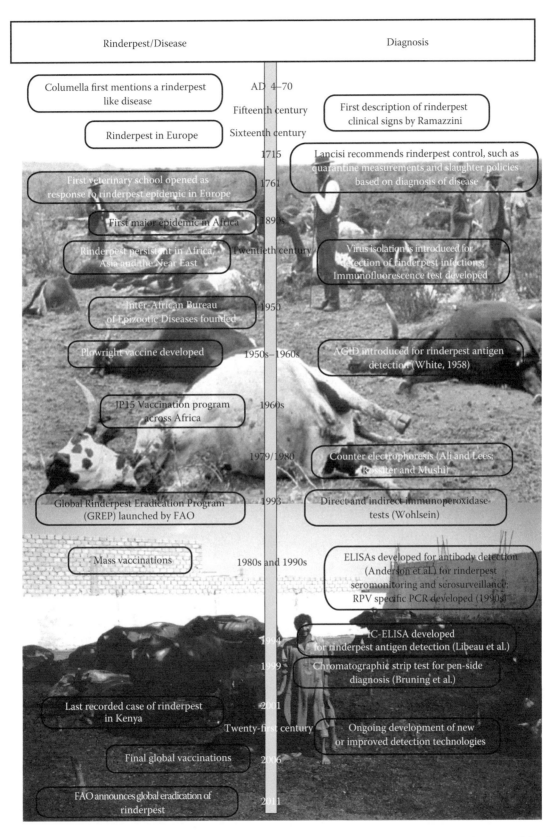

Rinderpest/Disease		Diagnosis
Columella first mentions a rinderpest like disease	AD 4–70	
	Fifteenth century	First description of rinderpest clinical signs by Ramazzini
Rinderpest in Europe	Sixteenth century	
	1715	Lancisi recommends rinderpest control, such as quarantine measurements and slaughter policies based on diagnosis of disease
First veterinary school opened as response to rinderpest epidemic in Europe	1761	
First major epidemic in Africa	1890s	
Rinderpest persistent in Africa, Asia and the Near East	Twentieth century	Virus isolation is introduced for detection of rinderpest infections; Immunofluorescence test developed
Inter-African Bureau of Epizootic Diseases founded	1950	
Plowright vaccine developed	1950s–1960s	AGID introduced for rinderpest antigen detection (White, 1958)
JP15 Vaccination program across Africa	1960s	
	1979/1980	Counter electrophoresis (Ali and Lees; Rossiter and Mushi)
Global Rinderpest Eradication Program (GREP) launched by FAO	1993	Direct and indirect immunoperoxidase tests (Wohlsein)
Mass vaccinations	1980s and 1990s	ELISAs developed for antibody detection (Anderson et al.) for rinderpest seromonitoring and serosurveillance; RPV specific PCR developed (1990s)
	1994	IC-ELISA developed for rinderpest antigen detection (Libeau et al.)
	1999	Chromatographic strip test for pen-side diagnosis (Bruning et al.)
Last recorded case of rinderpest in Kenya	2001	
	Twenty-first century	Ongoing development of new or improved detection technologies
Final global vaccinations	2006	
FAO announces global eradication of rinderpest	2011	

FIGURE 56.1 Simplified timeline of major events in the history of rinderpest disease and concomitant development of the diagnosis of rinderpest.

56.2 METHODS

56.2.1 Sample Preparation

Sample preparation is an important initial step for any of the diagnostic techniques discussed in this chapter. As RPV is particularly susceptible to humidity, light, and low and high pHs and is shed in great numbers only during the prodromal stage of infection, it is advantageous to sample efficiently and swiftly during the acute phase of the disease. Animals at the diarrheic stage of disease or recovering animals are less likely to yield enough detectable virus. For samples from carcasses, time is of essence due to autolysis and putrefaction. To maximize virus retrieval for analysis, it is advisable to sample as many animals as possible from an affected herd. Samples are taken from different sites (e.g., from the nose, eyes, and gums), placed into a cold box, and transported to the nearest reference laboratory for analysis. Lymph node and spleen biopsies, gum debris, and ocular and nasal discharges appear to harbor most viral particles. When using animal tissues, these are prepared for analysis by placing into a tissue grinder and working into a tissue suspension. Nasal and ocular secretions are collected with cotton swabs, which are then immersed in the appropriate extraction buffer to release the virus. Whole blood also contains virus. As RPV is associated with the leukocyte fraction, peripheral blood mononuclear cells (PBMCs) have to be isolated from the whole blood. To do this, Ficoll reagents or sequential centrifugation of whole blood are used to separate the buffy coat from the whole blood sample. All samples prepared in this way are then used in the direct or indirect conventional diagnostic techniques described previously. Sample preparation for PCR technologies requires the extraction of RNA from whole samples by phenol/chloroform extraction, but due to toxicity issues, this is replaced by commercially available RNA extraction kits based on column separation.

56.2.2 Detection Procedures

The procedures for the most commonly detection methods are described here and are the condensed version of protocols as they appear in the accompanying protocols of commercial kits or guidelines by the FAO.[22,46] An exhaustive literature exists on the lesser used technologies and can be found elsewhere.[7,22]

56.2.2.1 AGID Test

1. Prepare fresh samples (lymph node, spleen) as tissue suspensions or extract lachrymal/nasal secretion into phosphate-buffered saline (PBS buffer). Gum scrapings can also be used, which are added directly to the agar plate wells.
2. Prepare agar (4 g agar/400 mL distilled water) as previously described[22] and pour into Petri dishes or onto microslide to a depth of 4 and 3 mm, respectively.
3. Allow agar to solidify overnight. Use a template for well patterns to cut out wells in a hexagonal pattern in the solidified agar and remove agar plugs with dissecting needle.

4. Mark out the control antigen and sample position on the plate or slide.
5. Add the control antigen into the appropriate well and add the test antigen to the appropriate well using the Pasteur pipette.
6. Add prepared hyperimmune serum to the designated well.
7. Place plate/slide into a humidified chamber at room temperature less than 40°C. Remove suspect antigen preparation within 30 min to improve speed and accuracy of reaction.
8. Check plates at 2, 12, 24, and 36 h intervals. Record the lines of white precipitate indicating the presence of antigen in the sample. Ensure that the positive control lines are clearly visible.

Requirements for this technique are reagents such as agar, Petri dish/microslides, humidifying chamber, pastettes, control rinderpest antigen, and hyperimmune serum.

56.2.2.2 IC-ELISA[33]

1. Prepare samples as for the AGID test.
2. Add the trapping antibody to ELISA plate as per instruction and layout.
3. Incubate at 37°C with agitation.
4. Wash the plate 3× with a washing buffer and remove excess fluid.
5. Then add 50 μL of the sample in the supplied blocking buffer in duplicate, add 50 μL of reference antigen as per instruction, add negative controls (buffer only) to the wells as instructed, add 25 μL biotinylated RPV- and PPRV-specific Mabs to the assigned wells as instructed, and add 25 μL of supplied streptavidin–peroxidase conjugate in blocking buffer.
6. Incubate for 60 min at 37°C with agitation.
7. Wash the plate as in step 4.
8. Add 100 μL of substrate/chromogen as instructed and allow color development for 10 min.
9. Stop the color reaction by adding 100 μL of 1 M H_2SO_4.
10. Read the plate on ELISA reader at 492 nm.
11. Interpretation of results: A sample with an optical density (OD) value greater than twice the mean OD value of the blank controls is considered to be positive.

Requirements: Basic laboratory setup, washing facilities, buffer-making facilities, pipettes, shaking incubator, and ELISA reader.

56.2.2.3 Rapid Chromatographic Strip Test[34,35]

1. Collect lachrymal fluid with cotton swab; make sure to take a sample from both eyes.
2. Prepare lachrymal swab by immersing in the supplied extraction buffer (200 μL) and squeeze the swab gently to obtain as much fluid as possible.

3. Remove strip test from protective pouch. Lay on a flat surface.
4. Add 200 μL of lachrymal fluid in the extraction buffer to the sample window (blue pad at the bottom of the device).
5. Allow to react for 5–20 min.
6. Record the appearance of blue lines in the result window. Interpretation of test: Two blue lines mean positive, while one blue line means negative.

Requirements: All parts of the test are included in the test. It is advisable to keep samples on ice or at 4°C until tested so a cold box is desirable.

56.2.2.4 PCR[43]

Extract RNA from tissue or PBMCs using phenol/chloroform or RNA extraction kit according to the manufacturer's instructions.

1. RT of RNA
   For tissue-extracted RNA, dissolve RNA pellet in RNAse-free water at 1 mg/mL. For PBMC-derived RNA, allow ¼—1/10 of RNA/120 mL whole blood. Use 0.75 mL microcentrifuge tubes for thermocyclers.
   RNA—RT reaction:
   5 μL RNA solution
   2 μL random hexanucleotide primers (50 ng/μL)
   3 μL pure water to give 10 μL reaction mix
2. Incubate for 5 min at 70°C, and quickly spin to collect reaction mix in the bottom of the tube.
3. Prepare RT reaction mix: 4 μL of 5× RT buffer, 2 μL DTT (0.1 M), 2 μL BSA (acetylated, 0.1 mg/mL), 1 μL MoMuLV reverse transcriptase (200 units), and 1 μL deoxynucleotide triphosphates (dNTPs) (10 mM each). Then, add 10 μL of this to each RNA sample.
4. Mix carefully and centrifuge briefly for a few seconds to spin reaction to the bottom of the tube.
5. Incubate at room temperature (25°C) for 5 min and then at 37°C for 30 min; the RT products can be stored at −20°C until use.
6. For PCR, remove 5 μL of RT reaction and add 5 μL 10× PCR buffer, 36.5 μL pure water, 0.5 μL Taq polymerase (5 U/μL), 1 μL dNTPs (10 mM each), 1 μL forward primer, and 1 μL reverse primer.
7. Place the tubes in thermocycler and use appropriate cycling program such as the following:
   Step 1: 94°C 5 min
   Step 2: 50°C 1 min
   Step 3: 72°C 2 min
   Step 4: 94°C 1 min
   Step 5: Go to step 2 30 times
   Step 6: 50°C 1 min
   Step 7: 72°C 10 min
   Step 8: 4°C Hold

Primer sequences:
Universal primer set 1 for P gene conserved across the *Morbillivirus* genus:
UPP1 5′ ATGTTTATGATCACAGCGGT 3′
UPP2 5′ ATGGGTTGCACCACTTGCT 3′
RPV-specific primer set 2 for the virus F gene conserved in each morbillivirus but different among RPV strains:
RPVF3 5′ GGGACAGTGCTTCAGCCTATTA AGG 3′
RPVF4 5′ CAGCCCTAGCTTCTGACCCAC GATA 3′
Nested PCR set 3 for RPV set within the amplified F gene fragment to confirm origin of DNA:
RPVF3a 5′ GCTCTGAACGCTATTACTAAG 3′
RPVF4a 5′ CTGCTTGTCGTATTTCCTCAA 3′

8. Analyze RT-PCR products by running samples on a 1.5% agarose gel; determine fragment size by comparing to reference ladder on gel, with the following expected product sizes: universal primers 429 bp, F-specific primer 372 bp, and nested PCR 235 bp.

Requirements: Standard RNA extraction kits or reagents, access to PCR-associated reagents, RNAse-free water, sterile tissue culture hood for reaction setup, pipettes, sterile and filtered pipette tips and plastics, thermocycler, running tanks for agarose gels, running buffer, agarose, and ethidium bromide.

Examples of the main detection techniques described here are illustrated in Figure 56.2.

56.2.2.5 Competitive ELISA[39]

1. Prepare serum from blood sample of suspect animal using standard techniques.
2. Prepare antigen as recommended by manufacturers. Add 50 μL of antigen to each well in ELISA plate. Incubate at 37°C for 1 h with agitation.
3. Wash the plate three times by immersing wells with PBS and expelling PBS.
4. Add 40 μL of blocking buffer to wells, add a further 10 μL to Mab control wells and 60 μL to conjugate control wells, add 10 μL test serum to appropriate wells, add 10 μL of strong positive control serum to appropriate wells, add 10 μL of weak positive control serum to appropriate wells, add 10 μL of negative serum to appropriate wells, and add 50 μL of Mab prepared in blocking buffer to all appropriate wells.
5. Incubate for 1 h at 37°C with agitation.
6. Wash the plate three times as under step 3.
7. Add 50 μL of the anti-mouse conjugate prepared in the blocking buffer to all wells at the predetermined concentration.
8. Incubate for 1 h at 37°C with agitation.
9. Wash the plate as before under step 3.

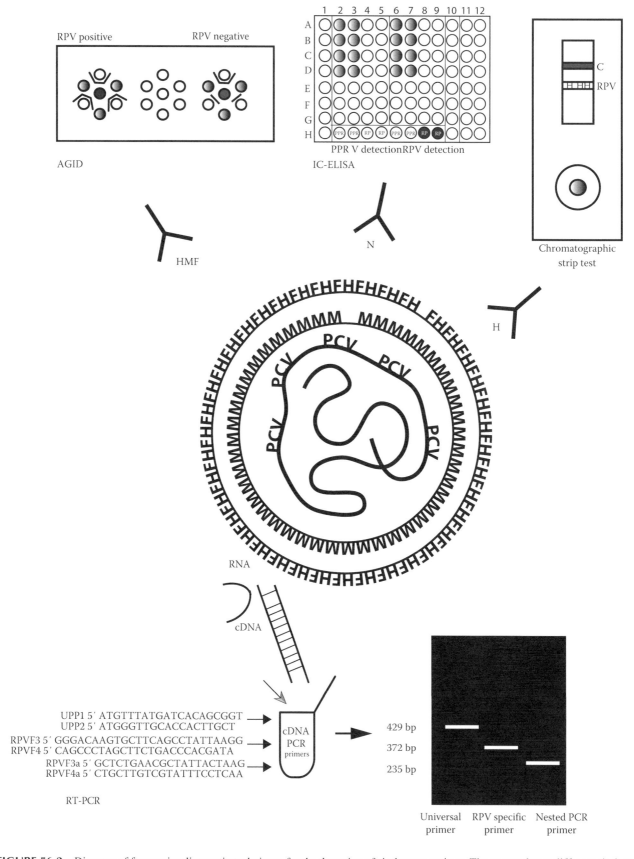

FIGURE 56.2 Diagram of four major diagnostic techniques for the detection of rinderpest antigen. The assays detect different rinderpest proteins as indicated by capital letters or RPV cDNA generated from RNA by RT. Dark shading indicates a control antigen and graduated shading indicates the suspect rinderpest sample. Y symbolizes antibodies used in the assays.

10. Make up OPD solution as per kit instruction and add hydrogen peroxidase solution just before use. Add 50 μL of substrate/chromogen/well; allow it to incubate for 10 min for color development.
11. Stop the reaction by adding 50 μL of 1 M H_2SO_4.
12. Read the plate on ELISA plate reader at 492 nm.

Interpretation of results is achieved by calculating the mean of the Mab control wells and determining the PI as follows:

PI = 100− [(Mean OD of test wells/Mean OD of Control Mab wells) × 100]

Inhibition values greater than 50% are RPV positive.

The kit also includes internal controls to ensure that the results are valid and not affected by external factors such as the quality of the water to rehydrate the supplied kit reagent with.

Requirements: Basic laboratory setup, water systems, shaking incubator and the like, pipettes, buffer-making facilities, and ELISA plate reader.

56.3 CONCLUSIONS AND FUTURE PERSPECTIVES

The success of GREP was multifactorial. The exploitation of virus biology has made eradication an achievable target as there is only one known serotype of the virus and infection conferred lifelong immunity if the animal has not succumbed to the disease, unlike diseases such as foot-and-mouse disease where several serotypes exist and vaccination only protects against infection with the virus in circulation at the time of infection.[47–50] When it became clear that a vaccine could protect animals from infection and that vaccination gave lifelong immunity to rinderpest, the possibility for global eradication of the virus became an attractive option. Through the collaboration of people at national and international levels, from herd farmer to clinical director of reference laboratories, an efficient global eradication program was conceived, which relied heavily on the mass vaccination of vulnerable animals. However, the rinderpest eradication success story was also made possible by the development of—in parallel—powerful diagnostic techniques, which reliably detected rinderpest infections or antibodies to rinderpest in vaccinated populations or in animals suspected to be infected with the virus. These started off as isolation of virus in cell cultures, detection of virus by visualization of antigen/antibody interactions in agar plugs, and other tests such as the IF test leading to more sophisticated technologies including ELISAs. For the first time, the mass screen and detection of present and past infections was possible, allowing the assessment of RPV infection status and the impact of vaccination programs on herd immunity. The development of a simple diagnostic chromatographic strip test facilitated the rapid diagnosis of the remaining rinderpest foci in the field leading to a swift implementation of emergency vaccinations. In addition, advances in the field of molecular biology resulted in the development of highly sensitive molecular probes for the detection of even low copy numbers of RPV RNA in samples supporting the diagnostic side of the rinderpest eradication program in addition to allowing the epidemiological study of virus origin, movement, and evolution in terms of virus virulence.

Even with the completion of GREP, new technologies are being developed or are being improved, perhaps reassuringly so bearing in mind that rinderpest viral stocks still exist in laboratories around the world and the threat of accidental or intended re-release of the virus into the environment remains.[51] Not only does the scientific interest in rinderpest and its biology remain, but it also extends beyond rinderpest related research into other fields. For example, the possibility to eradicate the close relative of RPV–PPRV has become a possibility through the lessons learned from GREP. Powerful diagnostic techniques similar to those developed for rinderpest eradication already exist for the fight against PPRV,[48,52–54] and the infrastructure left behind by GREP is still in place to be adopted by a similar eradication program for PPR.

REFERENCES

1. OIE, Global rinderpest eradication, 2011. http://www.oie.int/for-the-media/rinderpest/, accessed on March 31, 2015.
2. FAO, Emergencies preparedness, 2012. http://www.who.int/csr/disease/smallpox/en/, accessed on March 31, 2015.
3. Spinage, A.A., *Cattle Plague: A History*. Kluwer Academic/Plenum Publishers, New York, 2003.
4. Pankhurst, R., The great Ethiopian famine of 1888–1892. Anew assessment. *J Hist Med Allied Sci*, 21, 95, 1996.
5. Kjekshus, H., *Ecology Control and Economic Development in East African History: The Case of Tanganyika, 1850–1950*. Berkeley, CA: University of California Press, 1977.
6. Anderson, J., Lessons learned from global rinderpest eradication, 2010. http://www.foodsecurity.ac.uk/blog/index.php/2010/10/lessons-from-rinderpest-eradication/, accessed on March 31, 2015.
7. Barrett, T., Pastoret, P.P., and Taylor, W.P., *Rinderpest and peste de petits ruminants—Virus Plagues of Large and Small Ruminants*. Amsterdam, the Netherlands: Academic Press, 2006.
8. Leclainche, E., *Historie de la medicine veterinaire*. Toulouse, France: Office du Livre, 1936.
9. Lancisi, G.M., *Dissertatio historica de bovilla peste, ex campaniae finibus anno MDCCXIII Latio importata*. Rome, Italy: Ex Typographia Joannis Mariae Salvioni, 1715.
10. Wilkinson, L., Rinderpest and mainstream infectious disease concepts in the eighteenth century. *Med His*, 28, 129, 1984.
11. Wilkinson, L., *Animals and Disease: An Introduction to the History of Comparative Medicine*. Cambridge, U.K.: Cambridge University Press, 1992.
12. Roeder, P. and Rich, K., Conquering the cattle plague. In D.J. Spielman and R. Pandya-Lorch (eds.), *Millions Fed: Proven Successes in Agricultural Development*. Washington, DC: IFPRI, 2009.
13. Cosby, S.L. et al., Characterisation of a seal morbillivirus. *Nature*, 336, 115, 1988.
14. Domingo, M. et al., Morbillivirus infection in dolphins. *Nature*, 348, 21, 1990.
15. Duignan, P.J. et al., Morbillivirus infection in pilot whales (Globicephala melas and G. macrorhynchus) from the western Atlantic. *Mar Mamm Sci*, 11, 150, 1995.
16. Duignan, P.J. et al., Morbillivirus infection in cetaceans of the western Atlantic. *Vet Microbiol*, 44, 241, 1995.

17. Diallo, A., Morbillivirus group: Genome organisation and proteins. *Vet Microbiol*, 23, 155, 1990.

18. Plowright, W., Cruickshank, J.G., and Waterson, A.P., The morphology of rinderpest virus. *Virology*, 17, 118, 1962.

19. Viralzone, Rinderpest virus (RPV). http://viralzone.expasy.org/all_by_species/86.html, accessed on March 31, 2015.

20. Taylor, W. P., Epidemiology and control of rinderpest. *Rev Sci Tech*, 5, 407, 1986.

21. Barrett, T., Morbilliviruses into the twenty-first century. In: *Proceedings of the FAO Technical Consultation on the Global Rinderpest Eradication Programme*. Rome, Italy: FAO, 1996.

22. Anderson, J., Barrett, T., and Scott, G.R., *Manual of the Diagnosis of Rinderpest*. Rome, Italy: FAO, 1996.

23. African Union—Interafrican Bureau for Animal Resources (AU-IBAR), The Eradication of Rinderpest from Africa: A Great Milestone. African Union—Interafrican Bureau for Animal Resources (AU-IBAR), Kenya, 2011.

24. Matin, M.A. and Rafi, M.A., Present status of Rinderpest diseases in Pakistan. *J Vet Med B*, 53, 26, 2006.

25. Beregoy, N.Y., The Fight against Cattle Plague in Russia, 1830–1902: A brief overview of methodical approaches. *Stud Hist Biol*, 4, 1, 2012.

26. Kim, D.J. and Yoo, H.S., Globalism of outbreak and prevalence of Cattle Plague (Rinderpest) around Byengjahoran (1636–1638). *Uisahak*, 22, 41, 2013.

27. Kishi, H., A historical study on outbreaks of rinderpest during the Edo era in Japan. *Yamaguchi J Vet Med*, 3, 33, 1976.

28. Kishi, H., A historical study on the diagnosis of rinderpest at the beginning of Meiji period. *Jpn J Vet Hist*, 11, 1, 1977.

29. Bureau of Government Laboratories, *A Preliminary Report on Rinderpest of Cattle and Carabaos in the Philippine Islands* (Vol. 4, No. 1). Manila, Philippines: Bureau of Government Laboratories, Bureau of Printing, 1904.

30. NAHIS, Information sheet rinderpest, 2012. http://nahis.animalhealthaustralia.com.au/pmwiki/pmwiki.php?n=Factsheet.782?skin=factsheet; accessed on June 30, 2015.

31. Animal and Plant Health Inspection Service, Consideration of the Status of Rinderpest in South America. https://www.aphis.usda.gov/newsroom/2014/01/pdf/status_of_rinderpest_in_sa.pdf, 2011. Accessed on March 31, 2015.

32. Mack, M., The great African cattle plague epidemic of the 1890s. *Trop Anim Health Prod*, 2, 210, 1970.

33. Libeau, G., Diallo, A., Colas, F., and Guerre, L., Rapid differential diagnosis of rinderpest and peste des petits ruminants using an immunocapture ELISA. *Vet Rec*, 134, 300, 1994.

34. Brüning, A., Bellamy, K., Talbot, D., and Anderson, J., A rapid chromatographic strip test for the pen-side diagnosis of rinderpest virus. *J Virol Methods*, 81, 143, 1999.

35. Brüning-Richardson, A., Akerblom, L., Klingeborn, B., and Anderson, J., Improvement and development of rapid chromatographic strip-tests for the diagnosis of rinderpest and peste des petits ruminants viruses. *J Virol Methods*, 174, 42, 2011.

36. Wambura, P.N. et al., Diagnosis of rinderpest in Tanzania by a rapid chromatographic strip-test. *Trop Anim Health Prod*, 32, 141, 2000.

37. Hussain, M., Iqbal, M., Taylor, W.P., and Roeder, P.L., Pen-side test for the diagnosis of rinderpest in Pakistan. *Vet Rec*, 149, 300, 2001.

38. Anderson, J., Rowe, L.W., and Taylor, W.P., Use of an enzyme-linked immunosorbent assay for the detection of IgG antibodies to rinderpest virus in epidemiological surveys. *Res Vet Sci*, 34, 77, 1983.

39. Anderson, J., McKay, J.A., and Butcher, R.N., The use of monoclonal antibodies in competitive ELISA for the detection of antibodies to rinderpest and peste des petits ruminants. In: The seromonitoring of Rinderpest throughout Africa. Phase One. Proceedings of Final Research Co-Ordination Meeting. Joint FAO/IAEA (Food and Agriculture Organisation of the United Nations/International Atomic Agency) Division, Vienna, Austria, 43, 1991.

40. Mullis, K.B., Ehrlich, H.A., Arnheim, N., Horn, G.T., Saiki, R.K., and Scharf, S.J., One of the first polymerase chain reaction (PCR) patents. US4683195 A. Application number US 06/828, 144, 1986.

41. Mullis, K.B., Target amplification for DNA analysis by the polymerase chain reaction. *Ann Biol Clin*, 48, 579, 1990.

42. Diallo, A., Barrett, T., Barbron, M., Subbarao, S.M., and Taylor, W.P., Differentiation of rinderpest and peste des petits ruminants viruses using specific cDNA clones. *J Virol Methods*, 23, 127, 1989.

43. Forsyth, M.A. and Barrett, T., Evaluation of polymerase chain reaction for the detection and characterisation of rinderpest and peste des petits ruminants viruses for epidemiological studies. *Virus Res*, 39, 151, 1995.

44. Carrillo, C. et al., Specific detection of Rinderpest virus by real-time reverse transcription-PCR in preclinical and clinical samples from experimentally infected cattle. *J Clin Microbiol*, 48, 4094, 2010.

45. Yeh, Y. et al., Simultaneous detection of rift valley fever, bluetongue, Rinderpest, and Peste des Petits ruminants viruses by a single-tube multiplex reverse transcriptase-PCR assay using a dual-priming oligonucleotide system. *J Clin Microbiol*, 49, 1389, 2011.

46. FAO, Global Rinderpest Eradication Programme (GREP), 2014. http://www.fao.org/ag/againfo/programmes/en/grep/home.html, accessed on March 31, 2015.

47. Arzt, J., White, W.R., Thomsen, B.V., and Brown, C.C., Agricultural Diseases on the move early in the third millennium. *Vet Pathol*, 47, 15, 2010.

48. Anderson, J. et al., Rinderpest eradicated: What next? *Vet Rec*, 169, 10, 2011.

49. Butler, D., Officials act to secure cattle-plague virus. *Nature*, 488, 15, 2012.

50. Klepac, P., Metcalf, C.J.E., McLean, A.R., and Hampson, K., Towards the endgame and beyond: Complexities and challenges for the elimination of infectious diseases. *Philos Trans R Soc Lond B Biol Sci*, 368, 1, 2013.

51. Fournié, G. et al., The risk of rinderpest re-introduction in post-eradication era. *Prev Vet Med*, 113, 175, 2014.

52. Anderson, J. and McKay, J.A., The detection of antibodies against peste des petits ruminants virus in cattle, sheep and goats and the possible implications to rinderpest control programmes. *Epidemiol Infect*, 112, 225, 1994.

53. Baron, M.A., Parida, S., and Oura, C.A.L., Peste des petits ruminants: A suitable candidate for eradication? *Vet Rec*, 169, 16, 2011.

54. Mariner, J.C. et al., Rinderpest eradication: Appropriate technology and social innovations. *Science*, 337, 1309, 2012.

57 Ebola Virus

Baochuan Lin and Anthony P. Malanoski

CONTENTS

57.1 Introduction 495
    57.1.1 Classification...495
    57.1.2 Morphology, Genome Organization, Biology, and Epidemiology 496
        57.1.2.1 Morphology and Genome Organization 496
        57.1.2.2 Biology..498
        57.1.2.3 Epidemiology 499
    57.1.3 Clinical Features and Pathogenesis 499
    57.1.4 Diagnosis 500
57.2 Methods 501
    57.2.1 Sample Preparation..501
    57.2.2 Detection Procedures...502
57.3 Conclusion and Future Perspective...502
References...503

57.1 INTRODUCTION

Viral hemorrhagic fever (VHF) is a general term for a severe illness characterized by fever and bleeding disorder caused by five families of RNA viruses, including *Arenaviridae* (Lassa fever, Junin, and Machupo), *Bunyaviridae* (Crimean-Congo hemorrhagic fever, Rift Valley fever, Hantaan hemorrhagic fevers), *Filoviridae* (Ebola and Marburg), *Flaviviridae* (yellow fever, dengue, Omsk hemorrhagic fever, Kyasanur forest disease), and *Rhabdoviridae* (Bas-Congo virus) [1–5]. Epidemic hemorrhagic fever caused by Ebola and Marburg viruses represents a major public health issue in Sub-Saharan Africa with the highest fatality rates of the VHFs and lack of any validated prophylaxis or treatment options. The current ongoing (largest known) outbreak of Ebola hemorrhagic fever (EHF) in West Africa (2014), including Guinea, Liberia, and Sierra Leone, brought the spotlight on Ebola virus and highlights the virulent nature of the pathogen [6].

Ebola viruses cause a severe, often fatal disease in humans and nonhuman primates (such as monkeys, gorillas, and chimpanzees). Since its discovery in 1976 in what is now the Democratic Republic of the Congo (DRC) near the Ebola River, Ebola viruses have caused small sporadic outbreaks except for the most recent one (Table 57.1) [3,7]. Before 2007, there were four identified species of Ebola viruses, *Zaire ebolavirus*, *Sudan ebolavirus*, *Taï Forest (Côte d'Ivoire) ebolavirus*, and *Reston ebolavirus*. Among these, *Z. ebolavirus* and *S. ebolavirus* have caused larger human outbreaks in Africa with high mortality rates (≈50%–90% and 40%–65%, respectively) [8]. *R. ebolavirus* was first reported in 1989 in association with an outbreak of VHF among monkeys imported from the Philippines to Reston, Virginia, United States. Subsequent outbreaks of *R. ebolavirus* were also associated with monkeys from the Philippines in 1990, 1992, and 1996 [9–13]. Although infection with this virus can be fatal in monkeys, the 13 infections confirmed in humans were asymptomatic [3,7]. Recently, *R. ebolavirus* was isolated from pigs coinfected with porcine reproductive and respiratory syndrome virus that were experiencing a severe respiratory disease syndrome in the Philippines in 2008. Phylogenetic analysis of these viruses suggested that *R. ebolavirus* has been circulating since, and possibly before, the initial identification of this virus in monkeys in 1989. These findings have raised concerns about the potential disease emergence in humans and other livestock animals in the future [14]. A fifth species of Ebola virus, *Bundibugyo ebolavirus*, was identified as a cause of EHF outbreak in Uganda from late 2007 to early 2008 with 25% fatality rate and a subsequent outbreak in the DRC in 2012 with 36.5% mortality rate [7,15–17].

57.1.1 CLASSIFICATION

Ebola viruses and its lesser known cousin, Marburg virus, constitute the *Filoviridae* family in the order of *Mononegavirales*. Based on genetic diversity and antigenic characterization, there are five designated specie, *Z. ebolavirus*, *S. ebolavirus*, *Taï Forest ebolavirus* (also known as *Ivory coast ebolavirus* or *Cote d'Ivoire ebolavirus*), *B. ebolavirus*, and *R. ebolavirus* [3,18]. The *Z. ebolavirus* is most pathogenic for human followed by *S. ebolavirus* and *B. ebolavirus*, while *Taï Forest ebolavirus* and *R. ebolavirus* cause lethal infections in nonhuman primates, but have not yet been associated with fatal human cases (Table 57.1). Genomic surveillance of current

TABLE 57.1
List of Ebola Outbreaks

Ebola Subtype	Year	Country	Reported Cases[a]	Mortality Rate[b]
Zaire	1976	DRC	318	88%
	1977	DRC	1	100%
	1994	Gabon	52	60%
	1995	DRC	315	81%
	1996 (Jan–Apr)	Gabon	37	57%
	Jul 1996–Jan 1997	Gabon	60	74%
	Oct 2001–Mar 2002	Gabon	65	82%
		RC	57	75%
	Dec 2002–Apr 2003	RC	143	89%
	2003 (Nov–Dec)	RC	35	83%
	2007	DRC	264	71%
	Dec 2008–Feb 2009	DRC	32	47%
	Mar 2014–present	Guinea, Liberia, Sierra Leone	745	Unknown
Sudan	1976	Sudan, England	284	53%
			1	0%
	1979	Sudan	34	65%
	2000–2001	Urganda	425	53%
	2004	Sudan	17	41%
	May 2011	Urganda	1	100%
	2012 (Jun–Oct)	Urganda	11	36.4%
	Nov 2012–Jan 2013	Urganda	6	50%
Taï Forest	1994	Ivory Coast	1	0%
Bundibugyo	Dec 2007–Jan 2008	Urganda	149	25%
	2012 (Jun–Nov)	DRC	36	36.1%

Sources: Leroy, E.M. et al., *Clin. Microbiol. Infect.*, 17, 964, 2011; CDC, Chronology of Ebola Hemorrhagic Fever Outbreaks, Known cases and outbreaks of Ebola Hemorrhagic Fever, in Chronological Order, available from: http://www.cdc.gov/vhf/ebola/resources/outbreak-table.html.

Note: DRC, Democratic Republic of Congo; RC, Republic of Congo.

[a] Number of human cases.

[b] Mortality rate is calculated based on reported number of deaths among reported number of human cases.

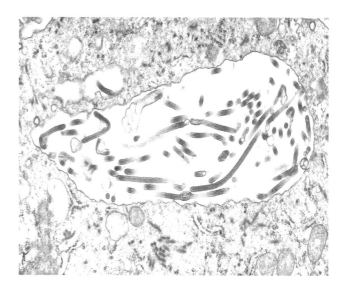

FIGURE 57.1 Ebola virus. Transmission electron micrograph of tissue samples infected with Ebola viruses showing the presence of numbers of Ebola virions. (Image downloaded from public health image library [ID #13218] of Centers for Disease Control and Prevention, Dekalb County, GA.)

outbreak strains identified *Z. ebolavirus* as the causative agent and suggested an independent zoonotic transmission event since all strains diverged from a common ancestor followed by human-to-human transmission [6,19].

57.1.2 MORPHOLOGY, GENOME ORGANIZATION, BIOLOGY, AND EPIDEMIOLOGY

57.1.2.1 Morphology and Genome Organization

Ebola virus has a unique morphology: it appears as filamentous particles about 1.4 μm in length and 80 nm in diameter and forms twisted filaments (hence the name

filovirus) (Figure 57.1). The filamentous, enveloped virus contains linear, nonsegmented, single-stranded negative-sense RNA. Along the central axis of the filamentous virion resides a nucleocapsid (NC), a right-handed double-layered helix with an outer diameter of 41 nm and a hollow inner channel 16 nm in diameter [20,21].

The single-stranded, negative-sense RNA of the virus is ~18.9 kb long and encodes seven structural proteins and two nonstructural proteins. The genomes of all sequenced Ebola viruses share a conserved structure with a 3′ leader sequence (~55 bases long), followed by in sequence nucleoprotein (NP), P protein (VP35), matrix protein (VP40), glycoprotein (GP), second NP (VP30), envelop protein (VP24), RNA-dependent RNA polymerase (L), and finally trailer sequences (Figure 57.2) [18,22–25]. The genomes of Ebola viruses contain long 5′ trailer sequences and a conserved transcriptional start and stop (polyadenylation) signal, containing the sequence 3′-UAAUU, which are unique characteristics of *filovirus* genomes. Another distinct feature of the genomes of *filovirus* is the positioning of some of the genes that overlap one another without separating intergenic regions. For Ebola viruses, these overlaps, which are limited to the conserved transcriptional signals, were found between the VP35 and VP40, the GP and VP30, and the VP24 and L [23]. A final specific feature of Ebola virus is the organization of the GP gene that encodes two ORFs expressed through transcriptional editing [26]. It should be noted that the genome of *R. ebolavirus*, the less virulent strain in humans, showed similar sequences of proteins and 3′ leader to other Ebola viruses, but did not share any similarity in noncoding and intergenic regions as well as a much shorter 5′ trailer sequence (Figure 57.2) [25,27].

Five of the nine proteins encoded by Ebola viruses form a ribonucleoprotein (RNP) complex, including NP, VP35, VP30,

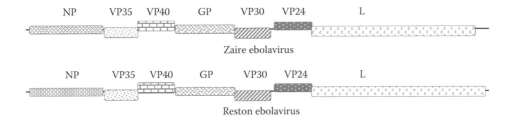

FIGURE 57.2 Genome structures of Ebola viruses indicating the proteins encoded by the viruses. NP, NP, VP35, VP4, GP, sGP, VP30, VP24, and L protein. The genome structures were constructed based on the sequences deposited in GenBank (*Zaire ebolavirus*, NC_002549; *Sudan ebolavirus*, NC_006432; *Cote d'Ivoire ebolavirus*, NC_014372; *Bundibugyo ebolavirus*, NC_014373; *Reston ebolavirus*, AB050936). The genome structure of *Z. ebolavirus* was used as a representative for *S. ebolavirus*, *Cote d'Ivoire ebolavirus*, and *B. ebolavirus*.

VP24, and L protein, which is responsible for the transcription and replication of the viral genome [20,21,28–30]. NP functions in RNA encapsidation and consists of 739 amino acids (~104 kDa). The N-terminal 450 amino acids are required for the NP–NP interaction and formation of NP tube structure, while the region from amino acids 1 to 600 are essential for the formation of NC-like structure and for viral genome replication [31]. VP35 serves as RNA-dependent RNA polymerase cofactor and consists of 340 amino acids (~35 kDa) with a highly conserved C-terminal basic amino acid motif that displays remarkable similarity to the N-terminal RNA-binding domain of the NS1 protein of influenza A virus. Also similar to NS1 protein of influenza A virus, VP35 functions as antagonist of type 1 interferon (IFN) by preventing the activation of interferon regulatory factor (IRF)-3 and inhibiting IRF-3-dependent gene expression [32].

VP30 acts as an Ebola virus–specific transcription activator and is 288 amino acids in length (~30 kDa). It consists of a zinc-finger motif, two serine clusters (amino acids 29–31 and 42–46) each containing three serine residues, a stretch of hydrophobic amino acids (94–112), and C-terminal dimeric domain (amino acids 142–272). The functions of VP30 depend on homooligomerization through hydrophobic amino acids and C-terminal dimeric domain that leads to hexamerization as well as phosphorylation at the two serine clusters. Evidence suggests that the balance between hexamer and dimer forms of VP30 is fundamental to the regulation of transcription and replication activity. The phosphorylation of VP30 modulates the composition of the viral polymerase complex by dissociating from the VP35-L complex, thereby regulating transcription and replication activities of the polymerase [33–36].

VP24 is believed to be a secondary matrix protein, consists of 251 amino acids (~24 kDa) with a hydrophobic region (amino acids 90–130), and features commonly associated with a viral matrix protein that are crucial in virion assembly and budding. Similar to the matrix protein (VP40), the N-terminal region of VP24 is critical for self-oligomerization, preferentially tetramers. Evidences also suggested that VP24 possesses membrane binding capability potentially through a consensus motif of a functional L-domain (YxxL) (amino acids 172–175). However, the precise locations of these domains have yet to be defined [37]. VP24 is also essential for the formation of functional RNP complex by ensuring the

correct assembly of a fully functional NC with NP and VP35; however, VP24 is not required for viral transcription and replication. Further evidence indicated that VP24 inhibits transcription and replication of the viral genome by hampering the functions of VP30 and L via binding to NP, and the association of VP24 with the RNP complex may serve as a converting signal from a transcription and replication-competent form to one that is ready for viral assembly [38,39]. In addition, VP24 functions as IFN antagonist by competing with tyrosine-phosphorylated signal transduction and transcription 1 (STAT1) protein to bind to karyopherin α1, blocking the nuclear accumulation of tyrosine-phosphorylated STAT1 and leading to the inhibition of IFN signaling and evade the antiviral effects of IFNs [40,41].

L (RNA-dependent RNA polymerase) protein, the largest and least abundant protein, is 2212 amino acids in length with a molecular weight of ~252 kDa. The L protein of Ebola virus, like most L proteins of nonsegmented negative-strand RNA viruses, consists of an RNA-binding element (amino acids 553–571), a putative RNA template recognition and/or phosphodiester bond formation domain (amino acids 738–744), and an ATP and/or purine ribonucleotide triphosphate–binding domain (amino acids 1815–1841), and highly conserved twin cysteine residues (amino acids 1351–1352) that potentially stabilize the secondary structure of the proteins for the proper conformation of the putative active sites [42].

The matrix protein (VP40, ~40 kDa) of Ebola virus, functionally similar to Gag protein of retroviruses and the M proteins of rhabdoviruses and paramyxoviruses, is the most abundant protein in mature filovirus virions and the key building block for virion maturation as well as budding. It is believed that VP40 consists of three functional domains, the M, I, and L, based on the working model of Gag protein. The M domain mediates the localization and binding to the plasma membrane, followed by self-interaction or oligomerization adjudicated by the I domain, and then the L domain, key C-terminal sequences of VP40, directed the interactions with host proteins Tsg101 and Nedd4 that lead to budding or "pinching off" from the cell surface. Evidence suggests that the N-terminal region of Ebola VP40 mediated the membrane binding and formation of hexamers; however, the precise locations of M and I domains have yet to be defined [43,44].

One distinct feature of Ebola viruses is the production of soluble and transmembrane glycoproteins (sGP and GP). The GP gene of Ebola virus encodes, through transcriptional editing events, four GPs, secreted GP (sGP), GP1, GP2, and small sGP (ssGP). The primary transcript (70%) is a single polypeptide at ~110 kDa, cleaved by furin to generate sGP that forms disulfide-linked homodimers and secreted Δ-peptide [45,46]. About 25% of transcripts generated experience transcriptional stutter resulting in another single polypeptide, the virion-attached or envelope spike GP. This polypeptide is cleaved by furin into GP1 and GP2 subunits that form disulfide-linked heterodimer and then assembled to trimers on the virus surface. The GP1 subunit, similar to the HA1 subunit of influenza virus, contains a receptor-binding site and mediates cellular attachment. The GP2 subunit contains a helical heptad-repeat region, transmembrane anchor, and a cytoplasmic tail. It is responsible for anchoring the GP into the viral membrane and the fusion of the viral and cellular membranes. Transcription stutter also results in the production of ssGP that forms disulfide-linked homodimers (5%) [45–48]. In addition to sGP, Ebola virus–infected cells also secrete shed GP that is produced by proteolytic cleavage of virion-attached GP by tumor necrosis factor (TNF)-α– converting enzyme that removes transmembrane anchor and releases the GP1-truncated GP2 heterodimers in the blood of infected animals. The functions of sGPs and shed GP is likely to facilitate immune evasion, acting as "antibody sinks" or decoys for host immunity, by competing with virion-attached GP for antibody binding that inhibits the neutralizing activity of Ebola virus antibodies [46,49].

57.1.2.2 Biology

Productive infection of Ebola viruses begins with the unique entry pathway into cells, which is complex and highly orchestrated requiring the binding of GP1 with adherence factors, including C-type lectins, β1 integrins, and Tyro3 (TAM) family tyrosine kinase receptors, as well as cell receptors, such as T-cell immunoglobulin (Ig) mucin domain-1 and various yet unidentified surface receptors [50]. Upon binding to the receptor, the virus is internalized into endosomes mainly through macropinocytosis, a mechanism that allows the uptake of large particle, where GP1 is cleaved to exposed receptor-binding site by cathepsin L and B and then interacts with endosomal/lysosomal membrane protein Niemann–Pick C1 (NPC1). The GP1/NPC1 interaction is important for *filovirus* entry events, viral escape, which occur in late endosomes and/or in lysosomes. Once the GP1/NPC1 interaction occurs, GP2 initiates viral/cellular membrane fusion via mechanisms similar to other enveloped viruses, releasing the NC into the cytoplasm and allowing the transcription and replication of the virions [51–54].

The newly entered encapsulated RNAs act as template for initial transcription of polyadenylated monocistronic mRNAs in a 3′–5′ direction via a polymerase complex consisting of NP, L, VP35, and nonphosphorylated VP30 proteins, which is critical to initiate transcription. Once initiated, the polymerase complex produces capped and polyadenylated mRNAs (Figure 57.3) [33]. Transcription is regulated by conserved start (RNA secondary structure) and stop signals at the viral gene border. Translation of these mRNAs is managed by cellular translation machinery at cellular ribosomes that leads

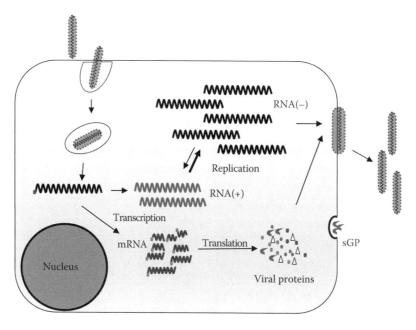

FIGURE 57.3 Life cycle of Ebola virus. Upon binding to the receptor, the virus is internalized into endosomes and release the nucleocapsid into the cytoplasm. The encapsulated RNAs act as template for transcription of mRNAs that were translated by cellular translation machinery and led to viral protein buildup. Phosphorylation of VP30 induces changes in the activity of the polymerase complex, triggering a transition between transcription and replication. In replication mode, copies of the viral genome are synthesized and serve as templates for replication of negative-strand viral RNAs that were assembled at the plasma membrane and bud from the cell surface. (Adapted from Takada, A., *Front. Microbiol.*, 3, 34, 2012.)

to viral protein buildup. Phosphorylation of VP30 by a still unidentified cellular kinase induces changes in the activity of the polymerase complex, triggering a transition between transcription and replication [18]. In replication mode, full-length positive-sense copies of the viral genome are synthesized and serve as templates for replication of negative-strand viral RNA. At the plasma membrane, NP-encapsidated full-length viral RNAs and the other viral structural proteins are assembled with VP40 and GP and incorporated into enveloped virus particles that bud from the cell surface (Figure 57.3) [20,30]. During replication, the phosphorylated VP30 dissociates from the polymerase complex, due to the impaired interaction with VP35 and possibly the viral RNA. The released phosphorylated VP30 remains bound to the NC via a strengthened interaction with NP. If VP30 becomes dephosphorylated by the cellular phosphatases PP1 or PP2A, the nonphosphorylated VP30 is again recruited to the polymerase complex to switch on the transcription mode [33].

57.1.2.3 Epidemiology

The first EHF outbreaks were described in 1976 in two neighboring locations, southern Sudan and northern Zaire (now the Democratic Republic of the Congo, DRC) [8]. The causative agents of these two epidemics were identified as two different strains of Ebola virus, S. and Z. ebolavirus. A small outbreak caused by S. ebolavirus was reported in 1979, and no further outbreak was further reported until 1994, which was followed by annual epidemics from 1995 to 1997, again from 2000 to 2004 and from 2007 to 2008 (Table 57.1). One case of EHF caused by a third strain of Ebola virus, Taï Forest (Côte d'Ivoire) ebolavirus, was reported in 1994. From late 2007 to early 2008, a new strain of Ebola virus, B. ebolavirus, was identified as a cause of EHF outbreak in Uganda and a subsequent epidemic in the DRC in 2012. With the current outbreak still raging in Guinea, Liberia, and Sierra Leone, EHF remains epidemic in equatorial Africa and a health threat for African populations [3,7,8].

It is suggested that fruit bats are the natural reservoir of filoviruses based on the successful isolation of Marburg virus from Rousettus aegyptiacus [55]. Experimental inoculation of Z. ebolavirus in 24 species of plants and 19 species of vertebrates and invertebrates indicated that fruit and insectivorous bats, Epomophorus wahlbergi, Tadarida condylura and Tadarida pumila, support the replication and circulation of high titers of Ebola virus without becoming ill [56]. Further studies detected seroprevalence of Ebola virus–specific IgG in serum from three different bat species, Hypsignathus monstrosus, Epomops franqueti, and Myonycteris torquata, which support the idea that bats are the reservoir for Ebola viruses; however, no viral RNAs were detected in these fruit bats [57,58].

Besides the fact that bats are probable natural reservoir, EHF is known to regularly infect primates living in Africa and the Philippines [59]. Epidemiologic studies show that almost all human EHF outbreaks in Gabon and in the Republic of Congo resulted from introductions of the virus from different infected animal sources, that is, gorilla, chimpanzee, or duiker. These studies also indicated that increased animal mortality always preceded the human outbreak. The index cases (mainly hunters) were all infected when handling dead animals, and subsequent transmission occurred by direct person-to-person contact [59,60]. While R. ebolavirus does not cause human illness, highly fatal infections in cynomolgus macaques (Macaca fascicularis) have occurred in the United States and Italy between 1990 and 1992 among monkeys imported from the Philippines [61–63]. Until recently, it was not known to infect domesticated animals, R. ebolavirus was identified in pigs in the Philippines in 2009 suggesting that R. ebolavirus has been circulating possibly before its initial discovery in monkeys that has raised concerns about the potential disease emergence in humans and other livestock animals in the future [14].

EHF is a classic zoonotic disease and is not persistent in the human population. Large outbreaks are caused by person-to-person transmission after the introduction of the virus into humans from a zoonotic reservoir. Human-to-human transmission occurs through direct physical contact with an infected person or contact through body fluids. The most important risk factor was direct repeated contact with a sick person's body fluids, and the risk was higher when the exposure took place during the late stage of the disease. Transmission through heavily contaminated fomites might also be a risk for transmission [64–66]. The examination of the presence of Ebola virus in various body fluids during the acute phase of illness found that Ebola viruses existed in a wide variety of bodily fluids, including saliva, breast milk, semen, stool, and tears [67].

57.1.3 Clinical Features and Pathogenesis

Illness can occur 2–21 days after infection, but most typically within 7–14 days, beginning with a sudden onset of malaise and fever, accompanied by headache, vomiting, and diarrhea. Three to four days after the onset of the initial symptoms, patients typically experienced progressively severe sore throat, abdominal pain, joint/muscle pain, and anorexia/weight loss. Hemorrhagic symptoms, such as internal (mainly the gastrointestinal tract) and subcutaneous bleeding, vomiting of blood, and reddening of the eyes, were experienced by 30%–50% of patients. Some patients also suffered from difficulty in swallowing/breathing, skin rash, red eyes, chest pain, and hiccups. In severe and fatal forms, patients suffered multiorgan dysfunction, including hepatic damage and renal failure, which leads to shock and death [16,68–70].

The characteristics of EHF are immune suppression and a systemic inflammatory response that result in the impairment of the vascular, coagulation, and immune systems that leads to multiple organ failure [8]. EHF generally occurs in resource-limited regions, where little data can be obtained from patients, and most of our current knowledge of pathogenesis comes from the studies in macaques. These studies indicated that the major early targets of Ebola viruses' infection are macrophages and dendritic cells (DCs). Macrophages initiate antiviral defense by employing innate immune mechanisms resulting the release of proinflammatory cytokines, while DCs initiate adaptive immune response by presenting antigens to naïve T cells. Ebola virus infection partially impairs the functions of both macrophages and DC; however, both cells are still able to initiate an inflammation response that further attracts target cells into the infection site. The systemic spread of viruses occurs via release of newly made virions from infected macrophages and DCs that infect resident macrophages and DCs in tissues, while natural killer cells and T lymphocytes remain uninfected. This causes a further spike in proinflammatory mediators and vasoactive substances and results in multifocal tissue necrosis. The host immune response by macrophages and DC causes disseminated intravascular coagulation (DIC), which is a prominent characteristic of Ebola virus infection. DIC is similar to septic shock caused by endotoxin and other bacterial products and is a syndrome characterized by coagulation abnormalities including systemic intravascular activation of coagulation leading to widespread deposition of fibrin in the circulation, which contributes to the multiple organ failure and high mortality rates characteristic of Ebola virus infections [4,22,71,72].

A few studies have documented the serologic and cytokine profiles of Ebola virus infections that indicated abnormal innate immune response and suppression of adaptive immunity with characteristic low levels of proinflammatory cytokines, high levels of inflammatory cytokines, immunosuppressor, and antigen titers. The study on patients infected with Z. ebolavirus serological profiles indicated increased levels of numerous proinflammatory cytokines, such as interleukin (IL)-1β, IL-1RA, IL-6, IL-8, IL-15, IL-16, and TNF-α and chemokines and growth factors, including macrophage inflammatory protein (MIP)-1α, MIP-1β, monocyte chemotactic protein (MCP)-1, macrophage colony-stimulating factor, macrophage migration inhibitory factor, IFN-inducible protein 10, growth-regulated oncogene α, and eotaxin. In addition, very low levels of T lymphocytes' cytokines, that is, IL-2, IL-3, IL-4, IL-5, IL-9, and IL-13, along with significantly decreased CD4 and CD8 peripheral lymphocytes, were found in nonsurviving Z. ebolavirus–infected patients [73,74]. Similarly, patients infected with S. ebolavirus showed increased levels of IL-1β, IP-10, regulated-on-activation normal T cell expressed and secreted (RANTES), IFN-α, IL-6, IL-8, IL-10, and MIP-1β. It was implicated that higher levels of IL-6, IL-8, IL-10, and MIP-1β were observed in fatal S. ebolavirus infection, while surviving patients have higher level of IFN-α [75]. These results suggesting that cytokines produced by macrophages and other infected or activated cells, such as endothelial cells, may play a significant role in the hemorrhage and vascular shock seen in VHFs.

Similar results were obtained in the study of nonhuman primates with Ebola virus infection. Nonhuman primates with the highest antigen titers also had high levels of circulating inflammatory cytokines, such as TNF-α, IFN-γ, IL-2, IL-1β, and IL-6 that can induce vascular permeability

changes *in vitro* and have been associated to some degree with vascular shock. In addition, two cytokines, MCP-1 and RANTES, can affect vascular permeability by inducing the release of superoxide radicals, and histamine also appeared to be higher in the same nonhuman primates. RANTES may also affect platelet function through the release of arachidonic acid [76,77]. However, a recent study on serological samples from patients with *B. ebolavirus* infection showed low levels of proinflammatory cytokines, such as IL-1α, IL-1β, IL-6, and TNF-α, as well as high levels of immunosuppressor cytokines like IFN-α2, IFN-γ, and IL-10 and viral antigen titers, which suggested that pathogenesis mechanism of *B. ebolavirus* is different from other *ebolavirus* infection [78].

57.1.4 Diagnosis

Clinical diagnosis of EHF is possible when patients present with severe and acute febrile diseases and with a history of travel to an endemic area if other prominent causes, in areas endemic for Ebola virus, such as malaria, typhoid fever, and shigellosis, can be ruled out. However, differential diagnosis of EHF is still difficult due to the fact that a wide range of causes in area endemic for Ebola viruses can result in similar clinical features. In addition to the aforementioned diseases, patients with meningococcal septicemia, plague, leptospirosis, anthrax, relapsing fever, typhus, murine typhus, yellow fever, Chikungunya fever, and fulminant viral hepatitis can all present similar symptoms [18]. Laboratory diagnosis is needed to confirm Ebola virus infection. Laboratory diagnosis is based on two principles, measurement of host-specific immune response or the presence of viral particles or particle components [8].

Conventional techniques: Due to its highly virulent nature, conventional virus isolation using tissue or organ culture and virus plaque assay are too dangerous for routine detection of the presence of Ebola virus in clinical samples. Virus isolation should only be attempted in high-level containment environment, that is, biosafety level 4 laboratory. Instead, serologic testing, such as antigen capture enzyme-linked immunosorbent assays (ELISA), is the most rapid means for diagnosing infection. The most common antibodies used in ELISA are against GP proteins as they are the antigenic determinants of Ebola virus [79–82]. ELISA using antibodies against NP protein also proved to be sensitive and specific [83–87]. Furthermore, ELISA detecting Ig antibodies, against Ebola virus, and IgM or IgG ELISA can be used to determine the seroprevalence of Ebola virus infection [88–90]. Other conventional techniques, including immunohistochemistry and indirect immunomicroscopy, were also used to identify viral particles in serum and tissue samples [12,63,91–94]

Molecular techniques: Molecular techniques, such as reverse transcription polymerase chain reaction (RT-PCR), nested RT-PCR, real-time RT-PCR, and RT loop-mediated isothermal amplification, were also developed for Ebola virus infection detection. For these molecular assays, L gene seems to be the main target, followed by NP and GP genes (Table 57.2) [63,94–103]. Although these molecular assays

TABLE 57.2

List of Primers and Probes Used for Ebola Virus Detection

Primer Name	Sequence (5′ → 3′)	Gene Target	References
Primer 1	ATCGGAATTTTCTTTCTCATTGAAAGA	L Zaire	[95,97]
Primer 2	ATGTGGTGGATTATAATAATCACTGACATGCAT		
Filo-A	ATCGGAATTTTTCTTTCTCATT	L Zaire	[63,102]
Filo-B	ATGTGGTGGGTTATAATAATCACTGACATG		
GAB-1	GAATGTAGGTAGAACCTTCGG	L Zaire	[102]
GAB-2	GCATATAACACTGTGGGATTG		
F3	AATAACGAAAGGAGTCCCTA	L Zaire	[100]
B3	CTGACAGGATATTGATACAACA		
FIP	GTCACACATGCTGCATTGTGTTTTCTATATTTAGCCTCTCTCTCCCT		
BIP	GCCGCAATGAATTTAACGCAATTTTAATATGAGCCCAGACCTT		
EBO-GP1	AATGGGCTGAAAATTGCTACAATC	GP Zaire	[63]
EBO-GP2	TTTTTTTAGTTTCCCAGAAGGCCCACT		
EBO-1	TGGGTAATYATCCTYTTCCA	GP Zaire	[102]
EBO-2	ACGACACCTTCAGCRAAAGT		
EBO-3	GTTTGTCGKGACAAACTGTC		
EBO-4	TGGAARGCWAAGTCWCCGG		
GPZAI1201F	ACAGTCAAGGAAGGGAAG	GP Zaire	[98]
GPZAI1485R	GTTTTGGGGACTTGTTGT		
EBOGP-1DF	TGGGCTGAAAAYTGCTACAATC	GP Zaire/Sudan	[103]
EBOGP-1DR	CTTTGTGMACATASCGGCAC		
EBOGP-1DZPrb	FAM-CTACCAGCAGCGCCAGACGG-TAMRA		
EBOGP-1DSPrb	VIC-TTACCCCCACCGCCGGATG-TAMRA		
RES-NP1	GTATTTGGAAGGTCATGGATTC	NP Reston	[63]
RES-NP2	CAAGAAATTAGTCCTCATCAATC		
ZAI-NP1	GGACCGCCAAGGTAAAAAATGA	NP Zaire	
ZAI-NP2	GCATATTGTTGGAGTTGCTTCTCAGC		
SudZaiNP1(+)	GAGACAACGGAAGCTAATGC	NP Sudan	[101]
SudZaiNP1(−)	AACGGAAGATCACCATCATG		
SudZaiNP2(+)	GGTCAGTTTCTATCCTTTGC		
SudZaiNP2(−)	CATGTGTCCAACTGATTGCC		
Forward-1	GAAAGAGCGGCTGGCCAAA	NP Sudan	[101]
Reverse-1	AACGATCTCCAACCTTGATCTTT		
Probe	FAM-TGACCGAAGCCATCACGACTGCAT-QSY7		
Forward-2	TGGAAAAAAACATTAAGAGAACACTTGC	NP Zaire	[94,99]
Reverse-2	AGGAGAGAAACTGACCGGCAT		
TaqMan probe	FAM-CATGCCGGAAGAGGAGACAACTGAAGC-BHQ1		

provide higher detection sensitivity, common practice for outbreak investigation is to use both ELISA and RT-PCR to analyze clinical samples to confirm Ebola virus infections.

57.2 METHODS

Ebola virus is known to shed in a wide variety of bodily fluids; thus, various samples can be used for detecting Ebola virus infection, including tissue samples (liver, lung, and spleen), serum, blood, urine, saliva, breast milk, semen, sweat, and tear with serum as the most commonly used specimen. Due to its high infectivity, it is recommended that an inactivation treatment of specimens suspected of Ebola virus infection be applied using gamma irradiation from a cobalt-60 source before performing ELISA or molecular assays [18,67,104].

57.2.1 SAMPLE PREPARATION

For ELISA, serum samples are used directly [105]. For molecular assays, samples can be purified with commercially available RNA extraction kits, such as Qiagen RNeasy or RNeasy Mini Kits and QIAamp Viral RNA kit [95,97]. Total RNA can also be extracted using the acid guanidine–phenol extraction procedure. Briefly, specimens are homogenized in 5 volumes of 4 M guanidine thiocyanate, 25 mM sodium citrate, 0.5% sodium sarkosyl, 100 mM b-mercaptoethanol, and 200 mM sodium acetate (pH 4.0) and then extracted with equal volume of phenol and 0.3 volume of chloroform–isoamyl alcohol (49:1) by incubating on ice for 20 min. After incubation, samples are centrifuged at $14,000 \times g$ for 15 min at 4°C, and the aqueous phase containing RNA can be recovered by isopropanol precipitation and then resuspended in diethyl pyrocarbonate–treated water [98].

57.2.2 DETECTION PROCEDURES

The two-step RT-PCR methods targeting L gene were developed by Sanchez et al. [63]. The RNA was first transcribed to cDNA in a 30 μL reaction volume containing 10 μL template, 2.5 μL each of primers EBO-GP1 and Filo-A (100 ng/mL, Table 57.2), 1 μL of RNase inhibitor, 1× RT buffer, 1 μM dNTPs, and 2 μL of murine leukemia virus RT, and the reaction was carried out at 42°C for at least 30 min. PCR was then performed in a 100 μL volume containing 15 μL cDNA, 1x PCR buffer, 2 mM MgCl₂, 200 mM dNTPs, 3 μL each of primers Filo-A and Filo-B (Table 57.2), and 0.5 U Taq DNA polymerase. The reaction was carried out for 3 cycles at 94°C for 30 s, 37°C for 30 s, and 72°C for 2 min followed by 30 cycles at 94°C for 30 s, 45°C for 30 s, and 72°C for 1 min [63]. This method is modified by Morvan et al. [102] using Filo-A and Filo-B primers (50 ng/μL each) for RT-PCR followed by nested PCR. The RT reaction was carried out in a 20 μL volume at 42°C for 60 min, and the PCR was carried out in a 50 μL volume with 5 μL cDNA for 3 cycles at 94°C for 1 m, 37°C for 30 s, and 74°C for 2 min followed by 35 cycles at 94°C for 1 min, 42°C for 30 s, and 74°C for 2 min. For nested PCR, Gabon subtype-specific primers (GAB-1 and GAB-2) were designed to amplify the L gene; the PCRs were carried out in a 50 μL volume containing 1 μL of RT-PCR product and 100 ng/μL each of primers for initial denaturation at 94°C for 1 min followed by 40 cycles at 94°C for 40 s, 55°C for 40 s, and 72°C for 1 min and then a final extension at 72°C for 2 min [102]. It was further modified by Leroy et al. [97] using 1 mM random hexamer in RT reaction that was carried out at 42°C for 90 min followed by heat inactivation of reverse transcriptase at 65°C for 5 min. PCR was then performed in a 50 μL volume containing 4 μL cDNA, 1× PCR buffer, 0.5 U Taq DNA polymerase, 200 mM each of dNTPs, and 400 nM each of primers 1 and 2 (Table 57.2). The reaction was carried out for 40 cycles at 94°C for 30 s, 55°C for 30 s, and 72°C for 90 s [97].

The RT-LAMP (loop mediated isothermal amplification) reaction by Kurosaki et al. [100] was performed with a Loopamp RNA Amplification Kit (Eiken Chemical Co. Ltd., Tokyo, Japan) in a 25 μL reaction mixture containing 1.6 μM each of primers FIP and BIP and 200 nM each of the outer primers F3 and B3 (Table 57.2) and 2.0 μL of RNA sample, and the reaction was incubated at 63°C for 60 min and then heated at 80°C for 2 min to terminate the reaction. For real-time RT-LAMP, the reaction mixtures were observed by spectrophotometric analysis using a real-time turbidimeter (LA-200, Teramecs, Kyoto, Japan). The time of positivity observed through real-time RT-LAMP assay was determined as the time at which the turbidity reached the threshold value fixed at 0.1, which doubles the average turbidity value of negative controls of several replicates [100].

RT-PCR for GP gene was developed by Sanchez et al. [63] using same PCR conditions as L gene with EBO-GP1 and EBO-GP2 primers (Table 57.2). This method is modified by Morvan et al. [102] using EBO-1 and EBO-2 primers at 100 ng/μL each for RT-PCR followed by nested PCR. The RT-PCR was carried out in a 100 μL volume at 41°C for 60 min and 94°C for 1 min followed by 40 cycles at 94°C for 40 s, 38°C for 40 s,

and 74°C for 1 min with a final extension at 74°C for 2 min. For nested PCR, the PCRs were carried out in a 10 μL volume containing 3 μL of RT-PCR product and 100 ng/μL each of EBO-3 and EBO-4 primers (Table 57.2) for initial denaturation at 94°C for 1 min followed by 40 cycles at 94°C for 40 s, 43°C for 40 s, and 72°C for 1 min and then a final extension at 72°C for 2 min [102]. Rodriguez et al. [98] also developed a one-step RT-PCR using primers GPZA1201F and GPZAI1485R (Table 57.2) with cycling condition at 50°C for 30 min followed by 35 cycles at 94°C for 15 s and at 50°C for 60 s with a final extension at 60°C for 10 min. The real-time RT-PCR for GP genes was developed by Gibb et al. [103] using TaqMan EZ RT-PCR Kit (Life Technologies, New York, United States) with 0.5 μM each of EBOGP-1DF and EBOGP-1DR primers and 0.2 μM each of EBOGP-1DS and EBOGP-1DZ probes. The PCR was performed in a 50 μL reaction volume with cycling condition at 55°C for 45 min and 94°C for 1 min followed by 40 cycles at 94°C for 15 s and 60°C for 30 s [103].

A one-step RT-PCR for NP gene was developed by Sanchez et al. [63] using EZ rTth RNA PCR Kit (Roche, Indianapolis, IN) with RES-NP1 and RES-NP2 or ZAI-NP1 and ZAI-NP2 (Table 57.2) in a 50 μL reaction volume with 1–5 μL RNA template. The reaction was carried at 50°C for 30 min, followed by 35 cycles at 94°C for 15 s and 50°C for 30 s [63]. Nested PCR followed by RT-PCR for NP gene was developed by Towner et al. [101] using Access RT-PCR kits (Promega, Madison, WI) and 5 μL of RNA for RT-PCR in a 50 μL reaction volume containing SudZaiNP1(+) and SudZaiNP1(−) primers (Table 57.2). The reaction was carried out at 50°C for 30 min and 94°C for 2 min, followed by 38 cycles at 94°C for 30 s, 50°C for 30 s, and 68°C for 1 min. Nested PCR was performed using SudZaiNP2(+) and SudZaiNP2(−) primers (Table 57.2) with 38 cycles at 94°C for 30 s, 50°C for 30 s, and 72°C for 1 min [101]. In addition, Towner et al. [101] also developed a real-time RT-PCR assay for NP gene. The RNA was first transcribed to cDNA in a 10 μL RT reaction volume with 1 μM forward-1 primer (Table 58.2) and 1 μL RNA. The reaction was carried out at 55°C for 10 min, followed by heat inactivation at 95°C for 30 min. The real-time PCRs were performed in a 25 μL volume containing 5 μL of cDNA, 1× TaqMan Universal Master Mix (Life Technologies, New York), 1 μM each of forward-1 and reverse-1 primers, 200 nM of fluorogenic probe. The reaction was carried out at 50°C for 2 min and 95°C for 10 min, followed by 40 cycles at 95°C for 15 s and 60°C for 1 min [101]. The same group revised the real-time RT-PCR assay in 2007 using random hexamer to transcribe RNA into cDNA and then using cDNA in 50 μL PCR with 900 nM each of forward-2 and reverse-2 primers and 200 nM TaqMan probe (Table 57.2) utilizing the same PCR cycling conditions [94].

57.3 CONCLUSION AND FUTURE PERSPECTIVE

Ebola viruses represent a serious health threat to human and animal populations, and the historic frequency of outbreaks along with the increasing size of outbreaks suggests that the threat will only become greater in the future. The continuing growth of human populations encroaching on wildlife habitats where bats (potential natural reservoir) and transmission vector animals, that is, primates, live ensures that the risks will increase. While only the *R. ebolavirus*, which so far has not proven fatal in humans, has been shown to have been transmitted to domesticated animals, such as pigs, this raises concern that a more lethal strain might make the same transition or that the *R. ebolavirus* might mutate in the domesticated animals to a more lethal version. This represents a serious concern as it would greatly increase the contact points and transmission vectors for the disease.

Detection methods for these viruses are unlikely to undergo modification unless a serious weakness is observed with the current tests. This has less to do with whether new methods might prove beneficial or not and more to do with the serious risks involved in handling the live pathogen. The risk is too great to warrant the testing required to validate new methods unless the payoff for the work is great enough.

The future research efforts on treatment and prevention of Ebola virus infection will be not only important, but also profitable. The identification of the proteins involved in the entry and initial transcription presents the possibility of study for chemicals that will interfere with one or the other mechanism. The recognition that many of the most severe consequences of the disease are due to runaway or over responses of the host immune system also presents many avenues of research for methods to mitigate this response and increase the survivability of the disease once it occurs.

REFERENCES

1. Schattner, M. et al., Pathogenic mechanisms involved in the hematological alterations of arenavirus-induced hemorrhagic fevers, *Viruses*, 5, 340–351, 2013.
2. Walter, C.T. and J.N. Barr, Recent advances in the molecular and cellular biology of bunyaviruses, *J Gen Virol*, 92, 2467–2484, 2011.
3. Leroy, E.M. et al., Ebola and Marburg haemorrhagic fever viruses: Major scientific advances, but a relatively minor public health threat for Africa, *Clin Microbiol Infect*, 17, 964–976, 2011.
4. Bray, M., Pathogenesis of viral hemorrhagic fever, *Curr Opin Immunol*, 17, 399–403, 2005.
5. Grard, G. et al., A novel rhabdovirus associated with acute hemorrhagic fever in central Africa, *PLoS Pathog*, 8, e1002924, 2012.
6. Baize, S. et al., Emergence of Zaire Ebola virus disease in Guinea—Preliminary report, *N Engl J Med*, 371, 1418–1425, 2014.
7. Martinez, R., Chronology of Ebola Virus Disease outbreaks, 1976–2014. Available at: http://healthintelligence.drupalgardens.com/content/chronology-ebola-virus-disease-outbreaks-1976-2014; last accessed March 31, 2015.
8. Feldmann, H. and T.W. Geisbert, Ebola haemorrhagic fever, *Lancet*, 377, 849–862, 2011.
9. Rollin, P.E. et al., Ebola (subtype Reston) virus among quarantined nonhuman primates recently imported from the Philippines to the United States, *J Infect Dis*, 179(Suppl 1), S108–S114, 1999.
10. Jahrling, P.B. et al., Preliminary report: Isolation of Ebola virus from monkeys imported to USA, *Lancet*, 335, 502–505, 1990.
11. CDC, Update: Ebola-related filovirus infection in nonhuman primates and interim guidelines for handling nonhuman primates during transit and quarantine, *MMWR Morb Mortal Wkly Rep*, 39, 22–24, 29–30, 1990.
12. Geisbert, T.W. et al., Association of Ebola-related Reston virus particles and antigen with tissue lesions of monkeys imported to the United States, *J Comp Pathol*, 106, 137–152, 1992.
13. Hayes, C.G. et al., Outbreak of fatal illness among captive macaques in the Philippines caused by an Ebola-related filovirus, *Am J Trop Med Hyg*, 46, 664–671, 1992.
14. Barrette, R.W. et al., Discovery of swine as a host for the Reston ebolavirus, *Science*, 325, 204–206, 2009.
15. MacNeil, A. et al., Filovirus outbreak detection and surveillance: Lessons from Bundibugyo, *J Infect Dis*, 204(Suppl 3), S761–S767, 2011.
16. MacNeil, A. et al., Proportion of deaths and clinical features in Bundibugyo Ebola virus infection, Uganda, *Emerg Infect Dis*, 16, 1969–1972, 2010.
17. Albarino, C.G. et al., Genomic analysis of filoviruses associated with four viral hemorrhagic fever outbreaks in Uganda and the Democratic Republic of the Congo in 2012, *Virology*, 442, 97–100, 2013.
18. Sanchez, A. et al., Filoviridae: Marburg and Ebola Viruses, in *Fields Virology*, D.M. Knipe and P.M. Howley, Eds., pp. 1279–1304. Lippincott Williams & Wilkins: Philadelphia, PA, 2001.
19. Gire, S.K. et al., Genomic surveillance elucidates Ebola virus origin and transmission during the 2014 outbreak, *Science*, 345, 1369–1372, 2014.
20. Noda, T. et al., Assembly and budding of Ebolavirus, *PLoS Pathog*, 2, e99, 2006.
21. Beniac, D.R. et al., The organisation of Ebola virus reveals a capacity for extensive, modular polyploidy, *PLoS One*, 7, e29608, 2012.
22. Hoenen, T. et al., Ebola virus: Unravelling pathogenesis to combat a deadly disease, *Trends Mol Med*, 12, 206–215, 2006.
23. Sanchez, A. et al., Sequence analysis of the Ebola virus genome: Organization, genetic elements, and comparison with the genome of Marburg virus, *Virus Res*, 29, 215–240, 1993.
24. Towner, J.S. et al., Newly discovered ebola virus associated with hemorrhagic fever outbreak in Uganda, *PLoS Pathog*, 4, e1000212, 2008.
25. Ikegami, T. et al., Genome structure of Ebola virus subtype Reston: Differences among Ebola subtypes. Brief report, *Arch Virol*, 146, 2021–2027, 2001.
26. Sanchez, A. et al., The virion glycoproteins of Ebola viruses are encoded in two reading frames and are expressed through transcriptional editing, *Proc Natl Acad Sci USA*, 93, 3602–3607, 1996.
27. Groseth, A. et al., Molecular characterization of an isolate from the 1989/90 epizootic of Ebola virus Reston among macaques imported into the United States, *Virus Res*, 87, 155–163, 2002.
28. Huang, Y. et al., The assembly of Ebola virus nucleocapsid requires virion-associated proteins 35 and 24 and posttranslational modification of nucleoprotein, *Mol Cell*, 10, 307–316, 2002.
29. Groseth, A. et al., The Ebola virus ribonucleoprotein complex: A novel VP30-L interaction identified, *Virus Res*, 140, 8–14, 2009.
30. Bharat, T.A. et al., Structural dissection of Ebola virus and its assembly determinants using cryo-electron tomography, *Proc Natl Acad Sci USA*, 109, 4275–4280, 2012.

31. Watanabe, S. et al., Functional mapping of the nucleoprotein of Ebola virus, *J Virol*, 80, 3743–3751, 2006.

32. Hartman, A.L. et al., A C-terminal basic amino acid motif of Zaire ebolavirus VP35 is essential for type I interferon antagonism and displays high identity with the RNA-binding domain of another interferon antagonist, the NS1 protein of influenza A virus, *Virology*, 328, 177–184, 2004.

33. Biedenkopf, N. et al., Phosphorylation of Ebola virus VP30 influences the composition of the viral nucleocapsid complex: Impact on viral transcription and replication, *J Biol Chem*, 288, 11165–11174, 2013.

34. Martinez, M.J. et al., Role of Ebola virus VP30 in transcription reinitiation, *J Virol*, 82, 12569–12573, 2008.

35. Hartlieb, B. et al., Oligomerization of Ebola virus VP30 is essential for viral transcription and can be inhibited by a synthetic peptide, *J Biol Chem*, 278, 41830–41836, 2003.

36. Hartlieb, B. et al., Crystal structure of the C-terminal domain of Ebola virus VP30 reveals a role in transcription and nucleocapsid association, *Proc Natl Acad Sci USA*, 104, 624–629, 2007.

37. Han, Z. et al., Biochemical and functional characterization of the Ebola virus VP24 protein: Implications for a role in virus assembly and budding, *J Virol*, 77, 1793–1800, 2003.

38. Noda, T. et al., Regions in Ebola virus VP24 that are important for nucleocapsid formation, *J Infect Dis*, 196(Suppl 2), S247–S250, 2007.

39. Watanabe, S. et al., Ebola virus (EBOV) VP24 inhibits transcription and replication of the EBOV genome, *J Infect Dis*, 196(Suppl 2), S284–S290, 2007.

40. Reid, S.P. et al., Ebola virus VP24 binds karyopherin alpha1 and blocks STAT1 nuclear accumulation, *J Virol*, 80, 5156–5167, 2006.

41. Reid, S.P. et al., Ebola virus VP24 proteins inhibit the interaction of NPI-1 subfamily karyopherin alpha proteins with activated STAT1, *J Virol*, 81, 13469–13477, 2007.

42. Volchkov, V.E. et al., Characterization of the L gene and 5′ trailer region of Ebola virus, *J Gen Virol*, 80(Pt 2), 355–362, 1999.

43. Harty, R.N., No exit: Targeting the budding process to inhibit filovirus replication, *Antiviral Res*, 81, 189–197, 2009.

44. Harty, R.N. et al., A PPxY motif within the VP40 protein of Ebola virus interacts physically and functionally with a ubiquitin ligase: Implications for filovirus budding, *Proc Natl Acad Sci USA*, 97, 13871–13876, 2000.

45. Barrientos, L.G. et al., Disulfide bond assignment of the Ebola virus secreted glycoprotein SGP, *Biochem Biophys Res Commun*, 323, 696–702, 2004.

46. Cook, J.D. and J.E. Lee, The secret life of viral entry glycoproteins: Moonlighting in immune evasion, *PLoS Pathog*, 9, e1003258, 2013.

47. Brindley, M.A. et al., Ebola virus glycoprotein 1: Identification of residues important for binding and postbinding events, *J Virol*, 81, 7702–7709, 2007.

48. Manicassamy, B. et al., Comprehensive analysis of ebola virus GP1 in viral entry, *J Virol*, 79, 4793–4805, 2005.

49. Dolnik, O. et al., Ectodomain shedding of the glycoprotein GP of Ebola virus, *EMBO J*, 23, 2175–2184, 2004.

50. Takada, A., Filovirus tropism: Cellular molecules for viral entry, *Front Microbiol*, 3, 34, 2012.

51. Hunt, C.L. et al., Filovirus entry: A novelty in the viral fusion world, *Viruses*, 4, 258–275, 2012.

52. Carette, J.E. et al., Ebola virus entry requires the cholesterol transporter Niemann-Pick C1, *Nature*, 477, 340–343, 2011.

53. Cote, M. et al., Small molecule inhibitors reveal Niemann-Pick C1 is essential for Ebola virus infection, *Nature*, 477, 344–348, 2011.

54. Miller, E.H. et al., Ebola virus entry requires the host-programmed recognition of an intracellular receptor, *EMBO J*, 31, 1947–1960, 2012.

55. Towner, J.S. et al., Isolation of genetically diverse Marburg viruses from Egyptian fruit bats, *PLoS Pathog*, 5, e1000536, 2009.

56. Swanepoel, R. et al., Experimental inoculation of plants and animals with Ebola virus, *Emerg Infect Dis*, 2, 321–325, 1996.

57. Leroy, E.M. et al., Fruit bats as reservoirs of Ebola virus, *Nature*, 438, 575–576, 2005.

58. Pourrut, X. et al., Spatial and temporal patterns of Zaire ebolavirus antibody prevalence in the possible reservoir bat species, *J Infect Dis*, 196(Suppl 2), S176–S183, 2007.

59. Leroy, E.M. et al., Multiple Ebola virus transmission events and rapid decline of central African wildlife, *Science*, 303, 387–390, 2004.

60. Formenty, P. et al., Ebola virus outbreak among wild chimpanzees living in a rain forest of Cote d'Ivoire, *J Infect Dis*, 179(Suppl 1), S120–S126, 1999.

61. Miranda, M.E. et al., Epidemiology of Ebola (subtype Reston) virus in the Philippines, 1996, *J Infect Dis*, 179(Suppl 1), S115–S119, 1999.

62. Miranda, M.E. et al., Chronological and spatial analysis of the 1996 Ebola Reston virus outbreak in a monkey breeding facility in the Philippines, *Exp Anim*, 51, 173–179, 2002.

63. Sanchez, A. et al., Detection and molecular characterization of Ebola viruses causing disease in human and nonhuman primates, *J Infect Dis*, 179(Suppl 1), S164–S169, 1999.

64. Francesconi, P. et al., Ebola hemorrhagic fever transmission and risk factors of contacts, Uganda, *Emerg Infect Dis*, 9, 1430–1437, 2003.

65. Dowell, S.F. et al., Transmission of Ebola hemorrhagic fever: A study of risk factors in family members, Kikwit, Democratic Republic of the Congo, 1995. Commission de Lutte contre les Epidemies a Kikwit, *J Infect Dis*, 179(Suppl 1), S87–S91, 1999.

66. MacNeil, A. and P.E. Rollin, Ebola and Marburg hemorrhagic fevers: Neglected tropical diseases? *PLoS Negl Trop Dis*, 6, e1546, 2012.

67. Bausch, D.G. et al., Assessment of the risk of Ebola virus transmission from bodily fluids and fomites, *J Infect Dis*, 196(Suppl 2), S142–S147, 2007.

68. Bennett, D. and D. Brown, Ebola virus, *BMJ*, 310, 1344–1345, 1995.

69. Dixon, M.G. and I.J. Schafer, Ebola viral disease outbreak—West Africa, 2014, *MMWR Morb Mortal Wkly Rep*, 63, 548–551, 2014.

70. International Commission, Ebola haemorrhagic fever in Zaire, 1976, *Bull World Health Organ*, 56, 271–293, 1978.

71. Bray, M. and T.W. Geisbert, Ebola virus: The role of macrophages and dendritic cells in the pathogenesis of Ebola hemorrhagic fever, *Int J Biochem Cell Biol*, 37, 1560–1566, 2005.

72. Geisbert, T.W. et al., Pathogenesis of Ebola hemorrhagic fever in primate models: Evidence that hemorrhage is not a direct effect of virus-induced cytolysis of endothelial cells, *Am J Pathol*, 163, 2371–2382, 2003.

73. Wauquier, N. et al., Human fatal zaire ebola virus infection is associated with an aberrant innate immunity and with massive lymphocyte apoptosis, *PLoS Negl Trop Dis*, 4(10), e837, 2010.

74. Baize, S. et al., Inflammatory responses in Ebola virus-infected patients, *Clin Exp Immunol*, 128, 163–168, 2002.

75. Hutchinson, K.L. and P.E. Rollin, Cytokine and chemokine expression in humans infected with Sudan Ebola virus, *J Infect Dis*, 196(Suppl 2), S357–S363, 2007.

76. Hutchinson, K.L. et al., Multiplex analysis of cytokines in the blood of cynomolgus macaques naturally infected with Ebola virus (Reston serotype), *J Med Virol*, 65, 561–566, 2001.

77. Villinger, F. et al., Markedly elevated levels of interferon (IFN)-gamma, IFN-alpha, interleukin (IL)-2, IL-10, and tumor necrosis factor-alpha associated with fatal Ebola virus infection, *J Infect Dis*, 179(Suppl 1), S188–S191, 1999.

78. Gupta, M. et al., Serology and cytokine profiles in patients infected with the newly discovered Bundibugyo ebolavirus, *Virology*, 423, 119–124, 2012.

79. Shahhosseini, S. et al., Production and characterization of monoclonal antibodies against different epitopes of Ebola virus antigens, *J Virol Methods*, 143, 29–37, 2007.

80. Ksiazek, T.G. et al., Enzyme immunosorbent assay for Ebola virus antigens in tissues of infected primates, *J Clin Microbiol*, 30, 947–950, 1992.

81. Ksiazek, T.G. et al., ELISA for the detection of antibodies to Ebola viruses, *J Infect Dis*, 179(Suppl 1), S192–S198, 1999.

82. Ksiazek, T.G. et al., Clinical virology of Ebola hemorrhagic fever (EHF): Virus, virus antigen, and IgG and IgM antibody findings among EHF patients in Kikwit, Democratic Republic of the Congo, 1995, *J Infect Dis*, 179(Suppl 1), S177–S187, 1999.

83. Groen, J. et al., Serological reactivity of baculovirus-expressed Ebola virus VP35 and nucleoproteins, *Microbes Infect*, 5, 379–385, 2003.

84. Ikegami, T. et al., Antigen capture enzyme-linked immunosorbent assay for specific detection of Reston Ebola virus nucleoprotein, *Clin Diagn Lab Immunol*, 10, 552–557, 2003.

85. Ikegami, T. et al., Immunoglobulin G enzyme-linked immunosorbent assay using truncated nucleoproteins of Reston Ebola virus, *Epidemiol Infect*, 130, 533–539, 2003.

86. Ikegami, T. et al., Development of an immunofluorescence method for the detection of antibodies to Ebola virus subtype Reston by the use of recombinant nucleoprotein-expressing HeLa cells, *Microbiol Immunol*, 46, 633–638, 2002.

87. Niikura, M. et al., Analysis of linear B-cell epitopes of the nucleoprotein of ebola virus that distinguish ebola virus subtypes, *Clin Diagn Lab Immunol*, 10, 83–87, 2003.

88. Gonzalez, J.P. et al., Ebola and Marburg virus antibody prevalence in selected populations of the Central African Republic, *Microbes Infect*, 2, 39–44, 2000.

89. Heffernan, R.T. et al., Low seroprevalence of IgG antibodies to Ebola virus in an epidemic zone: Ogooue-Ivindo region, Northeastern Gabon, 1997, *J Infect Dis*, 191, 964–968, 2005.

90. Leroy, E.M. et al., A serological survey of Ebola virus infection in central African nonhuman primates, *J Infect Dis*, 190, 1895–1899, 2004.

91. Geisbert, T.W. and P.B. Jahrling, Use of immunoelectron microscopy to show Ebola virus during the 1989 United States epizootic, *J Clin Pathol*, 43, 813–816, 1990.

92. Geisbert, T.W. et al., Rapid identification of Ebola virus and related filoviruses in fluid specimens using indirect immunoelectron microscopy, *J Clin Pathol*, 44, 521–522, 1991.

93. Zaki, S.R. et al., A novel immunohistochemical assay for the detection of Ebola virus in skin: Implications for diagnosis, spread, and surveillance of Ebola hemorrhagic fever. Commission de Lutte contre les Epidemies a Kikwit, *J Infect Dis*, 179(Suppl 1), S36–S47, 1999.

94. Towner, J.S. et al., High-throughput molecular detection of hemorrhagic fever virus threats with applications for outbreak settings, *J Infect Dis*, 196(Suppl 2), S205–S212, 2007.

95. Formenty, P. et al., Detection of Ebola virus in oral fluid specimens during outbreaks of Ebola virus hemorrhagic fever in the Republic of Congo, *Clin Infect Dis*, 42, 1521–1526, 2006.

96. Gonzalez, J.P. et al., Ebola virus circulation in Africa: A balance between clinical expression and epidemiological silence, *Bull Soc Pathol Exot*, 98, 210–217, 2005.

97. Leroy, E.M. et al., Diagnosis of Ebola haemorrhagic fever by RT-PCR in an epidemic setting, *J Med Virol*, 60, 463–467, 2000.

98. Rodriguez, L.L. et al., Persistence and genetic stability of Ebola virus during the outbreak in Kikwit, Democratic Republic of the Congo, 1995, *J Infect Dis*, 179(Suppl 1), S170–S176, 1999.

99. Stephens, K.W. et al., Cross-platform evaluation of commercial real-time reverse transcription PCR master mix kits using a quantitative 5′ nuclease assay for Ebola virus, *Mol Cell Probes*, 24, 370–375, 2010.

100. Kurosaki, Y. et al., Rapid and simple detection of Ebola virus by reverse transcription-loop-mediated isothermal amplification, *J Virol Methods*, 141, 78–83, 2007.

101. Towner, J.S. et al., Rapid diagnosis of Ebola hemorrhagic fever by reverse transcription-PCR in an outbreak setting and assessment of patient viral load as a predictor of outcome, *J Virol*, 78, 4330–4341, 2004.

102. Morvan, J.M. et al., Identification of Ebola virus sequences present as RNA or DNA in organs of terrestrial small mammals of the Central African Republic, *Microbes Infect*, 1, 1193–1201, 1999.

103. Gibb, T.R. et al., Development and evaluation of a fluorogenic 5′ nuclease assay to detect and differentiate between Ebola virus subtypes Zaire and Sudan, *J Clin Microbiol*, 39, 4125–4130, 2001.

104. Elliott, L.H. et al., Inactivation of Lassa, Marburg, and Ebola viruses by gamma irradiation, *J Clin Microbiol*, 16, 704–708, 1982.

105. Breman, J.G. et al., A search for Ebola virus in animals in the Democratic Republic of the Congo and Cameroon: Ecologic, virologic, and serologic surveys, 1979–1980. Ebola Virus Study Teams, *J Infect Dis*, 179(Suppl 1), S139–S147, 1999.

58 Marburg Virus

Junping Yu and Hongping Wei

CONTENTS

58.1 Introduction 507
    58.1.1 Classification 507
    58.1.2 Morphology, Biology, and Epidemiology 507
        58.1.2.1 Morphology 507
        58.1.2.2 Biology 508
        58.1.2.3 Epidemiology 508
    58.1.3 Transmission, Clinical Features, and Pathogenesis 509
        58.1.3.1 Transmission 509
        58.1.3.2 Clinical Features 509
        58.1.3.3 Pathogenesis 509
    58.1.4 Diagnosis 511
        58.1.4.1 Conventional Techniques 511
        58.1.4.2 Molecular Techniques 511
58.2 Methods 513
    58.2.1 Sample Collection and Preparation 513
    58.2.2 Detection Procedures 514
        58.2.2.1 RNA Extraction 514
        58.2.2.2 One-Step Real-Time RT-PCR 514
        58.2.2.3 TaqMan™ Real-Time RT-PCR 515
        58.2.2.4 Antigen Detection ELISA 515
58.3 Conclusion and Future Perspective 515
References 516

58.1 INTRODUCTION

Marburg virus (MARV) is the causal agent of MARV disease (MVD) (formerly known as Marburg hemorrhagic fever) in human and nonhuman primates, which was first identified in 1967 during epidemics in Marburg and Frankfurt in Germany and Belgrade in the former Yugoslavia from importation of infected monkeys from Uganda [1,2]. Similar to the Ebola virus (EBOV) disease, MVD is a severe and highly fatal disease. Both MARV and EBOV are among the most virulent pathogens known to infect humans. Both diseases are rare, but have a capacity to cause dramatic outbreaks with high fatality. Despite significant progress in drug development for MVD [3], there are no approved vaccines or postexposure treatment drugs available yet. The predominant treatment is general supportive therapy.

MARV is transmitted by direct contact with blood, body fluids, and tissues of infected persons. Transmission of MARV may also occur by handling ill or dead infected wild animals (monkeys, fruit bats, etc.). Due to the fact that a number of viruses, such as Lassa virus, Hantavirus, and Yellow fever virus, could cause fever and bleeding disorders similar to MVD, laboratory diagnosis is absolutely necessary to identify MARV infections.

58.1.1 CLASSIFICATION

MARV belongs to the *Filoviridae* family. The genus *Marburgvirus* contains one species, *Marburg Marburgvirus*, formerly referred to as Lake Victoria marburgvirus [2,4,5], which has two lineages: Ravn virus (RAVV) (including isolates from 1987 outbreak, one isolate from the Democratic Republic of the Congo [DRC] outbreak in 1998–2000, and one human and several bat isolates from infections that took place in Uganda in 2007) and MARV (other isolates) [6]. Phylogenetic analysis of genomic sequence data divides the known members of the species into at least five different lineages, four of which are very close with the nucleotide sequences difference within 7%, while the fifth has a divergency of up to 21% [7–9].

58.1.2 MORPHOLOGY, BIOLOGY, AND EPIDEMIOLOGY

58.1.2.1 Morphology

Filoviruses have the shape of filament, so early electron microscopy (EM) of blood and tissues proposed that the causative agent of the disease might be related to rhabdoviruses or spiral-shaped bacterium *Leptospira*, both of which resemble filoviruses to some extent [8,10]. Later, the distinct nature of filoviruses was recognized and the pathogen was identified as the newly discovered MARV [10]. Marburg virions are pleomorphic, with the appearance of rod- or ring-like, crook- or six-shaped, or branched structures [10]. Cryo-EM showed the morphological forms of purified MARV released from infected Vero cells: filamentous 30%, six-shaped 37%, and round 33% [11]. The MRAV filamentous particles have a mean length of 892 nm and a mean diameter of 91 nm, both of which are longer than the conventional EM (diameter 80 nm and length 795–828 nm [10,12,13]), ascribing to the less compression and shrinkage effects of cryo-EM [11]. MARV virions are similar to EBOV, since both belong to *Filoviridae*. However, although MARV genomes are slightly longer than EBOV genomes, morphologically MARV virions are apparently shorter than EBOV virions (EBOV with a length of around 1000 nm) [10].

58.1.2.2 Biology
Like EBOV, MARV is a single-stranded nonsegmented negative-sense (NNS) RNA virus. The genome of MARV is about 19.1 kb in size, containing seven genes in a linear order of *NP-VP35-VP40-GP-VP30-VP24-L* (Figure 58.1) [10,14]. Each MARV viral gene possesses a highly conserved transcription initiation site, a transcription termination site, an unusually long 3′ and 5′ untranslated regions, and an open reading frame [8,14]. Most of the genes are separated by short intergenic regions of 4–9 nucleotides (nts). There is only one long intergenic region of 97 nts located between the glycoprotein (GP) and the viral protein (VP) 30 gene. The transcription stop site of the *VP30* gene and the transcription start site of the *VP24* gene overlap, sharing a highly conserved sequence of 5 nts [15]. Gene overlap is found among all filoviruses and is unique among members of the order *Mononegavirales* (Figure 58.1) [15].

MARV particles are composed of seven structural proteins, five of which, nucleoprotein (NP), VP35 and VP40, GP, and RNA-dependent RNA polymerase (L), are counterparts of proteins found in all NNS RNA viruses [8]. The GP, the only surface protein, has the structural features of a type I transmembrane GP, mediating receptor binding and fusion during virus entry [16,17]. The nucleocapsid is surrounded by a layer of matrix protein VP40, which plays a key role in the budding of viral particles. The VP24, which is unique to the *Filoviridae* family, is a minor matrix protein, with a structural similarity to the NP and involvement in nucleocapsid formation and assembly. The four remaining proteins, NP, VP35, VP30, and L, associated with the viral RNA genome, form the MARV nucleocapsid [8,15]. The NP is not only a major structural component of the nucleocapsid but also playing a key role in viral replication and transcription. The VP35 can mask doubled-stranded RNA (dsRNA) to counter innate immunity through binding to dsRNA [18,19]. Once the nucleocapsid is released into the cytoplasm of infected cell, transcription and replication of the viral RNA genome start [8]. Replication of MARV begins with the synthesis of a positive-sense reverse complement of the genome, the antigenome. The antigenome, in turn, serves as a template for the synthesis of new genomes. The minimal protein requirement for replication is NP, VP35, and L. During transcription, the NNS genome is transcribed into seven monocistronic mRNA species. These mRNAs are capped and polyadenylated by the viral RNA-dependent RNA polymerase. NP, VP35, and L are sufficient to support the transcription of MARV minigenomes [8,15].

58.1.2.3 Epidemiology
MARV first emerged in August 1967 in Marburg and Frankfurt in Germany and Belgrade in Yugoslavia (now Serbia) among laboratory workers who handled African green monkeys (*Chlorocebus aethiops*) imported from Uganda. There were

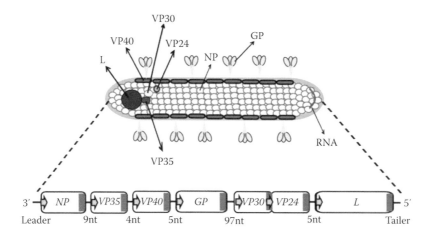

FIGURE 58.1 Marburg virus structure and genome organization. Top portion, schematic of Marburg virion. The RNA genome is encapsidated by nucleoprotein, viral protein (VP) 35, VP30, and L. VP40 and VP24 are matrix proteins. Glycoprotein trimers are inserted into the viral membrane. Bottom portion, structure of the MARV genome with transcription signals. The genes are illustrated as boxes and nontranscribed regions as black lines (leader, tailer, and intergenic regions). Transcription start signals are described as light gray arrows and stop signals as dark bars. A gene overlap appears between *VP30* and *VP24*. The lengths of the intergenic regions are shown in Table 58.1.

31 MVD cases (25 primary, 6 secondary infections) and 7 deaths. After 3 months, the unknown deadly infectious agent was isolated, characterized, and identified as MARV [1,20]. MVD had not been heard of for 8 years. In 1975, a young Australian who had traveled throughout Zimbabwe died in Johannesburg in South Africa [21]. The infection spread to his traveling companion and later also to a nurse. The latter two infected persons recovered and MARV was determined as the infectious agent [21]. During the period of 1975–1985, only sporadic outbreaks that affected small numbers of individuals were noted (Table 58.1). As the fatality rates of MVD were lower than those seen in the devastating outbreaks of EBOV (up to 90%), MARV had been thought to be less threatening until the outbreaks in the DRC in 1998–2000 and then in Angola in 2004–2005. A total number of 406 cases and the high fatality rates (83% in DRC and 90% in Angola) revealed that MARV is as threatening as EBOV [7,22].

MARV is a risk for travelers to sub-Saharan Africa, who, with their return home, may increase the chance of importing and spreading MVD to other countries. This is reflected by the two recent cases in 2008 of a Dutch and an American tourist who both presumably got infected during a visit to the Python Cave in Uganda (Table 58.1). In the nearby Kitaka mine in the Kamwenge District of Uganda, a gold mining activity resulted in four cases of MARV infection in June–September 2007 [23]. In the two more recent outbreaks in 2012 and 2014, while no epidemiologic data were available, sequence analysis of viral genomes in 2012 showed high similarity (~99.3%) to MARV isolates from *Rousettus aegyptiacus* bats [24]. So far, no genome information was obtained from the case confirmed on October 5, 2014. In addition, two laboratory infections have been reported and one of which had a fatal outcome [10,25]. Work with MARV is restricted to high-containment biosafety level 4 (BSL-4) laboratories owing to the lack of an approved vaccine or treatment, the high infectivity and fatality, and the potential of aerosol transmission [26]. Since MARV poses a severe threat to public health and safety, it has been classified as Select Agent by the Centers for Disease Control and Prevention, as Risk Group 4 Agent by the World Health Organization, and as Category A Priority Pathogen by the National Institute of Allergy and Infectious Diseases (Table 58.1).

58.1.3 Transmission, Clinical Features, and Pathogenesis

58.1.3.1 Transmission

Primary MARV infections are essentially acquired via exposure to an infected animal, either a reservoir host (several bat species) or a spillover host such as nonhuman primates. African green monkeys (*Cercopithecus aethiops*) imported from Uganda were the source of infection for humans during the first Marburg outbreak. Experimental inoculations in pigs with different EBOVs have showed that pigs are susceptible to filovirus infection and shed the virus. Therefore, pigs should be considered as a potential amplifier host during MVD outbreaks.

After transmission to humans, MARV can spread between individuals through direct contact with blood or other body fluids (e.g., saliva, tears, semen, sweat, urine, faces, and breast milk) from infected patients. Typical risks of exposure include administration of medical care to infected individuals as well as handling of corpses without the use of proper protection [31]. MARV disappears from blood and most tissues after the acute stage of the disease. However, it may persist for some time in certain anatomic sites, such as the testes or the anterior chamber of the eye [32].

58.1.3.2 Clinical Features

Much of what we know about MVD clinical data was obtained from the three main outbreaks: the first outbreak in 1967, the 1998–2000 DRC outbreak, and the 2004–2005 Angola outbreak. The incubation period in humans ranges from 2 to 21 days (with an average of 5–9 days) [8,33]. MVD course is divided largely into three distinct phases: a generalization phase (days 1–4), an early organ phase (days 5–13), and a late organ or convalescence phase (day 13 onward), depending on the outcome of infection [8]. The generalization phase starts abruptly with influenza-like symptoms including high fever (~40°C), severe headache, chills, myalgia, and malaise. Severe watery diarrhea, abdominal pain and cramping, rashes on face, nausea, and vomiting are also symptoms in this phase [32]. In the early organ phase, patients may have the symptoms of prostration, dyspnea, exanthema, and abnormal vascular permeability (including conjunctival injection and edema). Severe hemorrhagic manifestations appear in the later part of the early organ phase, such as petechiae, mucosal bleeding, uncontrolled leakage from venipuncture sites, visceral hemorrhagic effusions, melena, bloody diarrhea, hematemesis, and ecchymoses [33]. In fatal cases, death occurs most often between 8 and 9 days after symptom onset, usually preceded by severe blood loss and shock. In the late organ phase, the patients get into a critical state showing convulsions, severe metabolic disturbances, diffuse coagulopathy, multiorgan failure, and shock. Survivors do not normally have severe hemorrhagic manifestations and may not even reach the late organ phase [33].

58.1.3.3 Pathogenesis

The pathogenesis of MARV infections in humans is poorly understood. Several animal models have been developed for MARV infections using nonhuman primates and rodents, such as cynomolgus macaques and guinea pigs, which have contributed to understanding the pathogenesis of MVD [34]. MARV initially targets monocytes, macrophages, and dendritic cells in the lymphoid tissues as well as Kupffer cells and cells lining the sinusoids in the liver, progresses to infection of parenchymal cells in the liver and adrenal gland and high endothelial venules in lymphoid tissues, and finally spreads to endothelial cells in a variety of tissues [35]. The virus or viral antigen is detected in the liver, lymph nodes, spleen, adrenal grand, kidney, and blood in infected cynomolgus macaques [35]. Under experimental conditions, aerosol transmission of MARV is possible in macaque models, although such a transmission route has not been described in human outbreaks [36].

TABLE 58.1
Known Cases and Outbreaks of Marburg Virus Disease (in Chronological Order)

Year(s)	City/Country	Apparent or Suspected Origin/Mode of Infection	Reported Number (%) of Deaths among Cases	Isolate Designation (Abbreviation)	Reference
1967	Marburg and Frankfurt in Germany; Belgrade in Yugoslavia (now Serbia)	Uganda/handling of infected African green monkey tissues	7/31 (23%)	MARV Ci67, Flak (MARV Flak), Hartz (MARV Hartz), MARV "L," Porton (MARV Porton), Poppinga (MARV Pop), Ratayczak (MARV Rat), Voege (MARV Voe)	[20]
1975	Johannesburg in South Africa	Zimbabwe/unknown–most likely: sleeping in rooms inhabited by bats or visit to Sinoia caves (now Chinhoyi caves)	1/3 (33%)	Cruikshank (MARV Cru), Hogan (MARV Hogan), Ozolin (MARV Ozo)	[21]
1980	Nairobi in Kenya	Kenya/working close to Kitum Cave, Mount Elgon National Park	1/2 (50%)	Musoke (MARV Mus)	[27]
1987	Nairobi in Kenya	Kenya/visit to Kitum Cave/Mount Elgon National Park	1/1 (100%)	Ravn (RAVV Ravn), R1 (RAVV R1)	[28]
1988	Koltsovo in Soviet Union (now Russia)	Soviet Union (now Russia)/laboratory infection: needlestick injury	1/1 (100%)	"Variant U" (MARV "U")	[10]
1990	Koltsovo in Soviet Union (now Russia)	Soviet Union (now Russia)/laboratory infection: unspecified violation of safety requirements	1/1 (100%)	—	[25]
1998–2000	Durba, Watsa in Democratic Republic of Congo	Durba, DRC/gold mining in Goroumbwa cave	128/154 (83%)	MARV 01DRC99, MARV 02DRC99, MARV 03DRC99, MARV 04DRC99, MARV 05DRC99, MARV 06DRC99, MARV 08DRC. MARV 10DRC99, MARV 11DRC99, MARV 12DRC00, MARV 13DRC00, MARV 14DRC00, MARV 15 DRC00, MARV 16DRC00, MARV17 DRC00, MARV 18DRC00, MARV 19DRC00, MARV 20DRC00, MARV 21DRC00, MARV 22DRC00, MARV 22DRC00, MARV 23DRC00, MARV 24DRC00, MARV 25DRC00, MARV 26DRC00, MARV 27DRC00, MARV 28DRC00, MARV 29DRC00, MARV 30DRC00, MARV 31DRC00, MARV 32DRC00, MARV 33DRC00, MARV 34DRC00, MARV DRC 5/99 Aru, MARV DRC 5/99 Dra, RAVV 09DRC99	[22]
2004–2005	Uíge Province, Angola	Uíge Province, Angola/unknown	227/252 (90%)	MARV Angola	[7]
2007	Kamwenge, Uganda	Lead and gold mine in Kamwenge District, Uganda/gold mining in Kitaka Cave	1/4 (25%)	MARV-01Uga 2007, RAVV-02Uga 2007	[23]
2008	Colorado City, USA ex Uganda	Uganda/visit of Python Cave in Maramagambo Forest	0/1 (0)	—	[29]
2008	Leiden, Netherlands ex Uganda	Uganda/visit of Python Cave in Maramagambo Forest	1/1 (100%)	MARV Leiden	[30]
2012	Uganda	Kabale, Ibanda, Mbarara, and Kampala Districts of Uganda/unknown	4/15 (27%)	MARV Ibanda-423, MARV Kabale-422	[24]
2014	Uganda	Kampala District of Uganda/unknown	1/1 (until Nov. 2014)	—	Ministry of Health (MoH) of Uganda

Sources: Modified from Brauburger, K. et al., *Viruses-Basel*, 4, 1878, 2012; CDC, USA, http://www.cdc.gov/vhf/marburg/resources/outbreak-table.html.

58.1.4 DIAGNOSIS

Diagnostic tests for MARV need to have desired sensitivity, specificity, and reliability because any misdiagnosis will create havoc to the society. Therefore, MVD diagnosis should not rely on a single diagnostic method alone. Many of the signs and symptoms of MVD are similar to those of other more frequent infectious diseases, such as malaria or typhoid fever, making diagnosis of MVD difficult. This is especially true if only a single case is involved. However, if a person has the early symptoms of MVD and there is reason to suspect that MVD should be considered, the patient should be isolated and public health professionals notified. Samples from the patient can then be collected and tested to confirm infection. A BSL-3 laboratory with class II biosafety cabinet should be used to process clinical specimens and detect MARV, which general hospitals and health departments usually do not have. Therefore, clinical assessment is critical for initial diagnosis of a MARV infection.

MARV infections can be diagnosed definitively only in laboratories, by a number of different tests:

- Enzyme-linked immunosorbent assay (ELISA), IgM, and antigen detection
- Serum neutralization test
- Reverse transcription polymerase chain reaction (RT-PCR) assay
- Virus isolation by cell culture

Samples from suspected MVD patients should be handled minimally and with the utmost caution.

58.1.4.1 Conventional Techniques

Conventional diagnostic techniques include virus isolation, [37,38] followed by EM [13,39], serum neutralization test, and immunohistochemistry (IHC) [40].

Virus isolation and EM play pivotal roles to identify and diagnose MVD in the early years [41]. MARV grows well in cell lines MA-104, Vero E6, Vero 76, Hela-229, SW-1, and DBS-FRhL-2. Cell Vero E6 is used commonly for MARV isolation. However, the virus isolation can be laborious, time consuming, and costly, because high-level containment is warranted and a BSL-4 laboratory is required owning to the extreme hazard associated with handling this virus [37,42]. EM-based techniques can reveal the virions visually and have played a critical role in the discovery and recognition of MARV in 1967 [43]. As a simple and fast method, EM has proven valuable to assist MARV detection and provide insight into the pathogenesis of the agent. However, EM-based MARV diagnostic methods demand sophisticated instruments and require high virus titers (10^5–10^6/mL). Furthermore, some viruses with similar morphologies might complicate the EM observation.

After mixing a testing serum sample with neutralization antibodies specific to MARV, the neutralization test could be used to verify if the serum could stop the sample from causing cytopathic effects to Vero cells. If neutralization happens, it will confirm the presence of MARV in the sample. Similar to virus isolation, a BSL-4 laboratory is required for the serum neutralization test.

IHC is used to detect MARV in formalin-fixed biopsy/autopsy materials. An antibody specific to MARV is used to react with MARV first. Then an enzyme or fluorescent-labeled second antibody is used to react with the antibody attached to the MARV virions. Finally, the fluorescence or the signal generated by the enzyme is measured to test MARV. The significant advantage of IHC is that no need of high-level biocontainment is required during the test [40,44], but IHC usually needs experienced personnel to interpret the results.

58.1.4.2 Molecular Techniques

Molecular diagnostic methods for MARV mainly fall into two types: RT-PCR and ELISA.

As an important and widely used molecular method for detecting virus, RT-PCR has become increasingly popular, due to its speed, simplicity, sensitivity, and high-throughput capacity. MARV nucleic acid can be detected in the blood from day 3 up to 7–16 days after the onset of symptoms. RT-PCR is proven effective for detecting MARV RNA in body fluids and tissues of infected individuals and is capable of generating results in less than 2 h [44]. As summarized in Table 58.2, many nucleic acid amplification techniques, such as conventional RT-PCR [45,46], quantitative real-time RT-PCR (qRT-PCR) [46–54], and RT loop-mediated isothermal amplification methods (RT-LAMPs) as well as RT isothermal recombinase polymerase amplification assays (RT-RPAs) [55,56], have been developed for the detection of MARV RNA in clinical specimens. The fact that MARV is inactivated by a chaotropic agent guanidinium isothiocyanate, a common commercial component in RNA extraction buffers, facilitates safe handling of clinical materials [57]. In recent years, further improvements have been made on PCR-based techniques for MARV detection. Ogawa et al. [45] reported pan-MARV or pan-filovirus RT-PCR assays to detect all known MARV strains or even all known filovirus species using consensus PCR primer sets specific for the viral NP gene. Kurosaki et al. [55] described an RT-LAMP technique to effectively detect MARV, which is suitable for use as field diagnostics and is highly applicable to areas where MARV is endemic. Euler et al. [56] designed a panel of very rapid and highly sensitive isothermal real-time RPAs for the detection of category A bioterrorism agents including MARV and EBOV. This syndromic panel has the potential to be implemented onto microfluidic platforms (Table 58.2).

ELISA-based methods consist of two main varieties: antigen detection and antibody detection [37]. In deadly MARV infections with high mortality, patients usually die before specific antibody response is mounted. Fatal infections of MARV often end with high viremia and no evidence of an immune response. The presence of high titers of infectious MARV in the blood and tissues of patients at the early stage of illness suggests the appropriateness of antigen detection for diagnosis of the illness at this stage. Either hyperimmune serum or

TABLE 58.2

Reverse Transcription Polymerase Chain Reaction Assays Developed for Marburg Virus Detection (in Chronological Order)

Assay Method	Primer and Probe	Target Gene	Reference
Real-time RT-PCR (TaqMan™)	Forward: F(1179–1203) Reverse: R(1310–1331) P(1209–1236)	MARV NP (153 bp)	[54]
RT-RPA	Forward: MAR1 RPA FP Reverse: MAR1 RPA RP Probe: MAR1 RPA P	MARV NP (FJ750959, 1121–1256 bp)	[56]
	Forward: MAR2 RPA FP Reverse: MAR2 RPA RP Probe: MAR2 RPA P	MARV (FJ750953, 1121–1260 bp)	
One-step RT-PCR	Forward: FiloNP-Fm Reverse: FiloNP-Rm	MARV NP (594 bp)	[45]
	Forward: Filo-A Reverse: Filo-B	Filovirus L (419 bp)	
Real-time RT-PCR (one-step)	Forward: MARVLF Reverse: MARVLR	MARV L (AY358025; 13321–133517 bp)	[48]
	Forward: MARVGPF Reverse: MARVGPR	MARV GP (AY358025; 6131–6355 bp)	
	Forward: MARVNPF Reverse: MARVNPR	MARV NP (AY358025; 967–1146 bp)	
Real-time RT-PCR (TaqMan™)	Forward: F391 Reverse: R455-3 Probe: p429A	Marburg Musoke-MGB NP (65 bp)	[53]
	Forward: F391 Reverse: R455 Probe: p413S	Marburg Ci67-MGB NP (65 bp)	
	Forward: F350 Reverse: R415 Probe: p371S	Marburg Ravn-MGB NP (66 bp)	
	Forward: F4573T Reverse: R4638G Probe: p4601CT	Marburg Angola-MGB VP40 (66 bp)	
	Forward: F1788 Reverse: R1864 Probe: p1815S	Marburg Ravn-TM NP (77 bp)	
	Forward: F985 Reverse: R1064R Probe: p1035A	Marburg Angola-TM NP (80 bp)	
RT-LAMP	Outer primers: Mus-F3, Mus-B3 Forward inner primer: Mus-FIP Reverse inner primer: Mus-BIP Forward loop primer: Mus-LF Reverse loop primer: Mus-LB	MARV Musoke NP (AY358025; 967–1146 bp)	[55]
	Outer primers: Rav-F3, Rav-B3 Forward inner primer: Rav-FIP Reverse inner primer: Rav-BIP Forward loop primer: Mus-LF Reverse loop primer: Rav-LB	MARV Ravn NP (AY358025; 967–1146 bp)	
Real-time RT-PCR (one-step)	Forward: FiloA Reverse: Filo B Probe: FAMMBG	MARV L (419 bp)	[50]

(Continued)

TABLE 58.2 (Continued)
Reverse Transcription Polymerase Chain Reaction Assays Developed for Marburg Virus Detection (in Chronological Order)

Assay Method	Primer and Probe	Target Gene	Reference
One-step RT-PCR	Forward: Greene-Filo-U12683-A	Filovirus L (640 bp)	[49]
	Reverse: Greene-Filo-L13294-A		
	Forward: Greene-Filo-U12683-B		
	Reverse: Greene-Filo-L13294-B		
	Forward: Greene-Filo-U12683-C		
	Reverse: Greene-Filo-L13294-C		
	Forward: Greene-Filo-U12683-D		
	Reverse: Greene-Filo-L13294-D		
Real-time RT-PCR (two steps)	Forward	MARVAngola-05: VP40	[7]
	Reverse		
	Probe		
Real-time RT-PCR (TaqMan™)	Forward: MN FP	MARV NP (77 bp)	[52]
	Reverse: MN RP		
	Probe: MN P		
Real-time RT-PCR (one-step)	Forward: FiloA	Filovirus L (419 bp)	[47]
	Reverse: Filo B		
Real-time RT-PCR (one-step and TaqMan™)	Forward: MBGGP3 forward	MARV GP (143 bp)	[46]
	Reverse: MBGGP3 reverse		
	Probe: MBGGP3Prb		

virus protein-specific (e.g., NP) antibodies can be utilized to capture MARV antigen [37]. Generation of monoclonal antibodies to MARV protein (e.g., NP, GP) by immunizing mice or rabbits with native or recombinant proteins has increased the utility of antigen detection ELISA for diagnosis of MVD [58–61].

On the other hand, the host immune response can be effectively evaluated by using antibody detection ELISA. This method is particularly suitable for the diagnosis of MVD in patients who survive but not in those who succumb to the infection, given the high fatality and weak antibody response at an early stage of MVD. Direct IgM and IgG ELISAs, as well as IgM-capture ELISAs, are commonly used for the detection of virus-specific antibodies. MARV-specific IgM antibodies can appear as early as 2 days post-onset of symptoms and disappear from 30 to 168 days after infection, while IgG antibodies can persist for many years [33]. Accordingly, IgM-capture ELISAs are more frequently used for the diagnosis of acute illness, while IgG ELISAs are primarily used to identify individuals who have recovered from MARV infection or for conducting epidemiological (serosurveys) and epizoological studies. Since manipulation of live MARV is limited to a BSL-4, usually recombinant proteins of MARV are used as antigens to capture or detect the MARV antibodies. The usefulness of recombinant MARV proteins for detecting specific antibodies in IgG- and IgM-ELISAs has been confirmed. The carboxy-terminal half of MARV NP (341–695 amino acid residues), expressed as glutathione S-transferase-tagged recombinant proteins in an *Escherichia coli* system, was used to develop highly efficient

and specific IgG-ELISA for MARV antibody detection [62]. His-tagged secreted form of MARV GP was also used as antigens for the detection of MARV-specific antibodies. The GP-based ELISA was evaluated by testing antisera collected from mice immunized with viruslike particles as well as from humans and nonhuman primates infected with MARV [63]. This GP-based ELISA allows detection of MARV-specific IgM antibodies in specimens collected from patients during the acute phase of infection, indicating its value for diagnosis as well as ecological and serosurvey studies [64]. Recently, a protein microarray comprised of NP, GP, and VP40 from all Ebola and MARV species was developed for detection of filovirus antibodies in sera. It not only permits detection of IgG responses to specific filovirus antigens resulting from vaccination or viral challenge but also provides a useful approach to measure the relative levels of other antibody isotypes (IgM) [65].

58.2 METHODS

58.2.1 Sample Collection and Preparation [47,48]

As blood and tissues from patients at the early stage of illness contain high titers of MARV particles, they represent an ideal source of specimen for MVD diagnosis. However, other specimens such as saliva (oral swab) and urine (less reliable), as well as breast milk, may serve as alternative source materials if blood is not available. Suspected MARV blood samples are collected with personal protective equipment (PPE), including surgical mask (N95), cap, goggles, gown or apron, and two pairs of gloves and boots. Blood is collected using vacuumed blood tubes with clot activator and polymer gel for serum separation (3 mL in each tube). Then, the tubes are sealed by parafilm and put inside a plastic bag or a centrifuge tube to seal again. The surfaces of the tubes and the bags are disinfected with 1% hypochlorite solution before shipping to a laboratory for MARV detection.

Infectious specimens should be handled in the field laboratory by personnel wearing Tyvek suits and high-efficiency particulate arrestance (HEPA) filter-equipped air-purifying respirators in a class II biosafety cabinet inside a BSL-3 or BSL-4 laboratory.

58.2.2 Detection Procedures

RT-PCR and ELISA have been widely applied for rapid and sensitive detection of MARV. The conserved fragments of the NP, GP, L, or VP40 genes can be targeted for RT-PCR and antibodies to recombinant NP or GP can be used for ELISA. Two real-time RT-PCR-based methods (one-step and TaqMan™ real-time RT-PCR methods) and an antigen detection ELISA method are presented in the succeeding text.

58.2.2.1 RNA Extraction

Before RT-PCR, MARV RNA extraction should be completed in a BSL-3 laboratory with a class II biosafety cabinet. QIAamp Viral RNA Mini Kit is used for MARV RNA extraction from serum samples. (If other RNA extraction kits are used, please refer to the corresponding protocols.)

1. Pipette 560 μL of the Buffer AVL (to inactivate RNases and lyse viruses in the samples) into a 1.5 mL microcentrifuge tube.
2. Add 140 μL of serum to the aforementioned tube with Buffer AVL. Mix by pulse vortexing for 15 s.
3. Incubate at room temperature (15°C–25°C) for 10 min.
4. Briefly centrifuge the tube to remove drops from the inside of the lid.
5. Add 560 μL of ethanol (96%–100%) to the sample, and mix by pulse vortexing for 15 s. After mixing, briefly centrifuge the tube.
6. Carefully apply 630 μL of the solution from step 3 to the QIAamp Mini column (in a 2 mL collection tube) without wetting the rim. Close the cap, and centrifuge at 6000 × g (8000 rpm) for 1 min. Place the QIAamp Mini column into a clean 2 mL collection tube, and discard the tube containing the filtrate.
7. Carefully open the QIAamp Mini column, and repeat step 6.
8. Carefully open the QIAamp Mini column, and add 500 μL of Buffer AW1. Close the cap, and centrifuge at 6000 × g (8000 rpm) for 1 min. Place the QIAamp Mini column in a clean 2 mL collection tube (provided), and discard the tube containing the filtrate.
9. Carefully open the QIAamp Mini column, and add 500 μL of Buffer AW2. Close the cap and centrifuge at full speed (20,000 × g; 14,000 rpm) for 3 min.

Continue directly to step 11 or to eliminate any chance of possible Buffer AW2 carryover, perform step 10, and then continue to step 11.

10. Recommended: Place the QIAamp Mini column in a new 2 mL collection tube (not provided), and discard the old collection tube with the filtrate. Centrifuge at full speed for 1 min.
11. Place the QIAamp Mini column in a clean 1.5 mL microcentrifuge tube (not provided). Discard the old collection tube containing the filtrate. Carefully open the QIAamp Mini column and add 60 μL of Buffer AVE equilibrated to room temperature. Close the cap, and incubate at room temperature for 1 min. Centrifuge at 6000 × g (8000 rpm) for 1 min.

The extracts are resuspended in 60 μL of AVE buffer, aliquoted, and stored at −70°C before RT-PCR amplification is carried out. All work should be performed with PPE as outlined earlier. All waste materials should be treated with disinfectant (e.g., 1% hypochlorite solution) for at least 10 min and incinerated on the same day.

58.2.2.2 One-Step Real-Time RT-PCR

Please refer to Table 58.2 for the primers and probes used for MARV real-time RT-PCR detection. The following procedure for one-step real-time RT-PCR has been described by previous reports [47,48]. SuperScript® III One-Step RT-PCR System with Platinum® Taq DNA Polymerase (Life Technologies, Karlsruhe, Germany) is used in SYBR Green Real-Time RT-PCR assays.

58.2.2.2.1 Procedure

1. Add the following to a 0.2 mL, nuclease-free, thin-walled PCR tube on ice. For multiple reactions, prepare a master mix to minimize reagent loss and enable accurate pipetting: 12.5 μL of 2× reaction mix provided with the kit (including 0.4 mM of each deoxy-ribonucleoside triphosphate (dNTP), 3.2 mM MgSO₄), 0.5 μL of 10 μM each primer, 2 μL of RNA, 1 μL of SuperScript® III RT/Platinum® Taq Mix, 1 μL of SYBR Green stock solution (Roche Molecular Biochemicals, Mannheim, Germany) diluted 1:5000 in dimethyl sulfoxide, and RNase-free water to 25 μL total volume.
2. Gently mix and make sure that all the components are at the bottom of the amplification tube. Centrifuge briefly if needed. Depending on the thermal cycler used, overlay with silicone oil if necessary.
3. RT-PCR detection involved the following steps: RT at 50°C for 20 min; initial denaturation at 94°C for 2 min; 40 cycles with 94°C for 15 s (denature), 60°C for 30 s (anneal), and 68°C for 1 min; and final extension of 68°C for 5 min. Collect the data and analyze the results.
4. Since SYBR Green intercalates nonspecifically into PCR products, a melting curve analysis is performed following PCR to identify the correct

amplification products by its specific melting temperature. Melting curve analysis included the following steps: 95°C for 5 s and 65°C for 15 s, with heating temperature of 95°C at a rate of 0.1°C/s with continuous reading of fluorescence. Samples were considered positive if they produced melting point–confirmed amplification products in two assays. Amplification products were later confirmed by sequencing.

Note: Negative control is concurrently included along with the test samples in order to monitor any possible contamination that might be occurred during the PCR procedure. The presence of a corresponding product is indicative of positive serum. All work should be performed with PPE as outlined earlier. All waste materials should be treated with disinfectant (1% hypochlorite solution suggested) for at least 10 min and incinerated on the same day.

58.2.2.3   TaqMan™ Real-Time RT-PCR

The primers and probes used for MARV real-time RT-PCR detection are shown in Table 58.2. The following procedure for one-step real-time RT-PCR has been reported previously [51,53,54]. One-step TaqMan™ real-time qRT-PCR is performed using AgPath-ID™ One-Step RT-PCR Kit (Applied Biosystems) according to the manufacturer's instructions. (If other RT-PCR kits are used, please refer to the corresponding protocols.)

58.2.2.3.1   Procedure

1. Remove all aliquots of PCR reagents (diethylpyrocarbonate [DEPC]-treated water, 5× Colorless GoTaq PCR buffer, 2.5 mM dNTP mix, primers, and GoTaq DNA polymerase) from the freezer. Vortex and spin down all reagents before opening the tubes.
2. Prepare the RT-PCR reaction mix as follows: 12.5 μL of 2 × RT-PCR buffer, 1 μL of 10 μM each primer, 1 μL 3 μM TaqMan™ probe, 5 μL of RNA, 1 μL of enzyme mix, and RNase-free water to 25 μL total volume.
3. Real-time RT-PCR cycling is performed on Bio-Rad CFX96 or ABI7500 system as follows: 50°C for 30 min, 95°C for 10 min, and then 40 cycles of 15 s at 95°C and 45 s at 60°C.

The threshold cycle (CT) value for a positive sample is adjustable according to a liner range of a typical standard curve for each virus detection assay, which was also a strict standard for positive determination. Negative control should be set along with the test samples in order to monitor any possible contamination that might occur during the test.

58.2.2.4   Antigen Detection ELISA

MARV NP and GP have been tested as the antigens to detect MARV in plasma or serum samples by ELISA. The following procedure is based on that described by previous reports [38,61].

58.2.2.4.1   Procedure

1. Before detection, inactivate the MARV samples at 60°C for 1 h in a BSL-3 laboratory with class II biosafety cabinet, and leave the samples on ice.
2. Coat the purified monoclonal antibody onto microwell immunoplates at >100 ng/well in 100 μL of phosphate-buffered saline solution (PBS) at 4°C overnight, followed by blocking with PBS containing 0.05% Tween-20 and 5% nonfat milk (PBST-M) for 1 h at room temperature.
3. Wash the plates with PBST, and add 100 μL of serially diluted MARV samples and incubate the plates at 37°C for 1 h.
4. Wash the plates with PBST, and add 100 μL of rabbit polyclonal antibody against MARV antigen (diluted 1:500 with PBST-M) to each well.
5. After 1 h incubation at 37°C, wash the plates with PBST. Then, add the horseradish peroxidase–conjugated goat antirabbit IgG. The plates are incubated for 1 h at RT.
6. After another extensive wash with PBST, add 100 μL of ABTS substrate solution [4 mM 2,2′-azino-di-(3-ethyl-benzthiazoline-6-sulphonic acid)], and add solution 2.5 mM hydrogen superoxygenphosphate (pH 4.2)] (Roche Diagnostics, Mannheim, Germany) to each well. Measure the optical density (OD) at a wavelength of 405 nm with a reference wavelength 490 nm after 30 min of incubation at 37°C.

Note: As a negative control, mock antigen-inoculated wells are tested. The adjusted OD values (OD_{405}nm) are calculated by subtracting the OD of the negative control well from the corresponding OD values.

58.3   CONCLUSION AND FUTURE PERSPECTIVE

Given the seriousness of MARV infection, it is important that sensitive, specific, and reliable tests are available and used at national and international reference laboratories for MVD diagnosis. While virus isolation is simple, its application is hampered by its time-consuming nature, and its requirement for high-level biocontainment and skilled operators. EM enables visualization of viral particles, but its usefulness is limited by its low sensitivity. Additionally, virus isolation and EM both involve handling live viruses in the highest laboratory BSL-4, restricting their use to well-resourced reference laboratories. ELISA and RT-PCR are sensitive and can detect viruses after inactivation of the viral particles. Therefore, these methods could be performed in a BSL-3 laboratory with class II biosafety cabinet and are thus more widely applied than virus isolation and EM for MARV detection.

Although the current methods are effective for detection of MARV, virus mutations should be monitored closely. RT-PCR and ELISA have shown their values for early detection of the viruses due to their high sensitivity and fast speed, but they might be ineffective dealing with mutated viruses.

Therefore, when new outbreaks emerge, the current molecular methods and reagents may need validation again to confirm their capability to detect mutated viruses. Traditional methods such as virus isolation and electron microscopy are less prone to miss mutated viruses, and they are invaluable for assisting the identification of filoviruses during disease outbreaks.

To date, MVD outbreaks have occurred mainly in underdeveloped Africa or originated from underdeveloped Africa (Table 58.1), such as Uganda and Congo. Due to limited power supply and refrigeration in these areas, field detection of MARV can be problematic using the current molecular techniques. Therefore, there is an urgent need to develop MARV test kits that do not depend on expensive instruments, have minimum sample handling and technical requirements, and generate the results within a short timeframe (preferably <30 min). Such devices would greatly aid the control of MARV infections in the future.

REFERENCES

1. Kissling, R. E., Robinson, R. Q., Murphy, F. A., Whitfield, S. G. Agent of disease contracted from green monkeys. *Science* 1968, 160, 888.
2. Kuhn, J. H. et al. Proposal for a revised taxonomy of the family Filoviridae: Classification, names of taxa and viruses, and virus abbreviations. *Arch Virol* 2010, 155, 2083.
3. Ursic-Bedoya, R. et al. Protection against lethal Marburg virus infection mediated by lipid encapsulated small interfering RNA. *J Infect Dis* 2014, 209, 562.
4. Adams, M. J., Carstens, E. B. Ratification vote on taxonomic proposals to the International Committee on Taxonomy of Viruses (2012). *Arch Virol* 2012, 157, 1411.
5. Kuhn, J. H. et al. Filoviridae. In *Virus Taxonomy: Classification and Nomenclature of Viruses: Ninth Report of the International Committee on Taxonomy of Viruses*; King, A.M.Q., Adams, M.J., Carstens, E.B., Lefkowitz, E.J. Eds. Elsevier Academic Press: Amsterdam, the Netherlands 2011, vol. 9, 7pp.
6. Peterson, A. T., Holder, M. T. Phylogenetic assessment of filoviruses: How many lineages of Marburg virus? *Ecol Evol* 2012, 2, 1826.
7. Towner, J. S. et al. Marburgvirus genomics and association with a large hemorrhagic fever outbreak in Angola. *J Virol* 2006, 80, 6497.
8. Brauburger, K., Hume, A. J., Muhlberger, E., Olejnik, J. Forty-five years of Marburg virus research. *Viruses-Basel* 2012, 4, 1878.
9. Towner, J. S. et al. Isolation of genetically diverse Marburg viruses from Egyptian Fruit Bats. *PloS Pathog* 2009, 5, e1000536.
10. Kuhn, J. H. Filoviruses. A compendium of 40 years of epidemiological, clinical, and laboratory studies. *Arch Virol Suppl* 2008, 20, 13.
11. Bharat, T. A. M. et al. Cryo-electron tomography of Marburg virus particles and their morphogenesis within infected cells. *PloS Biol* 2011, 9, e1001196.
12. Welsch, S. et al. Electron tomography reveals the steps in filovirus budding. *PloS Pathog* 2010, 6, e1000875.
13. Geisbert, T. W., Jahrling, P. B. Differentiation of filoviruses by electron microscopy. *Virus Res* 1995, 39, 129.
14. Feldmann, H. et al. Marburg virus, a filovirus—Messenger-RNAs, gene order, and regulatory elements of the replication cycle. *Virus Res* 1992, 24, 1.
15. Muhlberger, E. Filovirus replication and transcription. *Future Virol* 2007, 2, 205.
16. Dolnik, O. et al. Interaction with Tsg101 is necessary for the efficient transport and release of nucleocapsids in Marburg virus-infected cells. *PloS Pathog* 2014, 10, e1004463.
17. Will, C. et al. Marburg virus gene 4 encodes the virion membrane-protein, a type-I transmembrane glycoprotein. *J Virol* 1993, 67, 1203.
18. Ramanan, P. et al. Structural basis for Marburg virus Vp35-mediated immune evasion mechanisms. *Proc Natl Acad Sci USA* 2012, 109, 20661.
19. Bale, S. et al. Marburg virus Vp35 can both fully coat the backbone and cap the ends of dsRNA for interferon antagonism. *PloS Pathog* 2012, 8, e1002916.
20. Siegert, R., Shu, H. L., Slenczka, W. Identification and isolation of Marburg virus. *Deut Med Wochenschr* 1968, 93, 604.
21. Gear, J. S. S. et al. Outbreak of Marburg virus disease in Johannesburg. *Br Med J* 1975, 4, 489.
22. Bausch, D. G. et al. Marburg hemorrhagic fever associated with multiple genetic lineages of virus. *N Engl J Med* 2006, 355, 909.
23. Adjemian, J. et al. Outbreak of Marburg hemorrhagic fever among miners in Kamwenge and Ibanda districts, Uganda, 2007. *J Infect Dis* 2011, 204, S796.
24. Albarino, C. G. et al. Genomic analysis of filoviruses associated with four viral hemorrhagic fever outbreaks in Uganda and the Democratic Republic of the Congo in 2012. *Virology* 2013, 442, 97.
25. Nikiforov, V. V. et al. A case of a laboratory infection with Marburg fever. *Zh Mikrobiol Epidemiol Immunobiol* 1994, 3, 104.
26. USDHHS. *Biosafety in Microbiological and Biomedical Laboratories*, 5th edn. Department of Health and Human Services: Washington, DC, 2009.
27. Smith, D. H. et al. Marburg-virus disease in Kenya. *Lancet* 1982, 1, 816.
28. Johnson, E. D. et al. Characterization of a new Marburg virus isolated from a 1987 fatal case in Kenya. *Arch Virol Suppl* 1996, 11, 101.
29. Centers for Disease Control and Prevention (CDC). Imported case of Marburg hemorrhagic fever—Colorado, 2008. *MMWR Morb Mortal Wkly Rep* 2009, 58, 1377.
30. Timen, A. et al. Response to imported case of Marburg hemorrhagic fever, the Netherlands. *Emerg Infect Dis* 2009, 15, 1171.
31. Bausch, D. G. et al. Risk factors for Marburg hemorrhagic fever, Democratic Republic of the Congo. *Emerg Infect Dis* 2003, 9, 1531.
32. Kortepeter, M. G., Bausch, D. G., Bray, M. Basic clinical and laboratory features of filoviral hemorrhagic fever. *J Infect Dis* 2011, 204, S810.
33. Mehedi, M., Groseth, A., Feldmann, H., Ebihara, H. Clinical aspects of Marburg hemorrhagic fever. *Future Virol* 2011, 6, 1091.
34. Nakayama, E., Saijo, M. Animal models for Ebola and Marburg virus infections. *Front Microbiol* 2013, 4, 267.
35. Hensley, L. E. et al. Pathogenesis of Marburg hemorrhagic fever in cynomolgus macaques. *J Infect Dis* 2011, 204, S1021.
36. Alves, D. A. et al. Aerosol exposure to the Angola strain of Marburg virus causes lethal viral hemorrhagic fever in cynomolgus macaques. *Vet Pathol* 2010, 47, 831.

37. Saijo, M. et al. Laboratory diagnostic systems for Ebola and Marburg hemorrhagic fevers developed with recombinant proteins. *Clin Vaccine Immunol* 2006, 13, 444.

38. Saijo, M. et al. Characterization of monoclonal antibodies to Marburg virus nucleoprotein (Np) that can be used for Np-capture enzyme-linked immunosorbent assay. *J Med Virol* 2005, 76, 111.

39. Geisbert, T. W., Rhoderick, J. B., Jahrling, P. B. Rapid identification of Ebola virus and related filoviruses in fluid specimens using indirect immunoelectron microscopy. *J Clin Pathol* 1991, 44, 521.

40. Geisbert, T. W., Jaax, N. K. Marburg hemorrhagic fever: Report of a case studied by immunohistochemistry and electron microscopy. *Ultrastruct Pathol* 1998, 22, 3.

41. Siegert, R., Shu, H. L., Slenczka, W. Isolation, identification and diagnosis of Marburg-virus infection harvested in Monkeys. *Klin Wochenschr* 1969, 47, 1010.

42. Wang, Y. P., Zhang, X. E., Wei, H. P. Laboratory detection and diagnosis of filoviruses. *Virol Sin* 2011, 26, 73.

43. Peters, D. Electron microscopy of Marburg-virus. *Klin Wochenschr* 1969, 47, 1010.

44. Grolla, A., Lucht, A., Dick, D., Strong, J. E., Feldmann, H. Laboratory diagnosis of Ebola and Marburg hemorrhagic fever. *Bull Soc Pathol Exot* 2005, 98, 205.

45. Ogawa, H. et al. Detection of all known filovirus species by reverse transcription-polymerase chain reaction using a primer set specific for the viral nucleoprotein gene. *J Virol Methods* 2011, 171, 310.

46. Gibb, T. R., Norwood, D. A., Woollen, N., Henchal, E. A. Development and evaluation of a fluorogenic 5′-nuclease assay to identify Marburg virus. *Mol Cell Probes* 2001, 15, 259.

47. Drosten, C. et al. Rapid detection and quantification of RNA of Ebola and Marburg viruses, Lassa virus, Crimean-Congo hemorrhagic fever virus, Rift Valley fever virus, dengue virus, and yellow fever virus by real-time reverse transcription-PCR. *J Clin Microbiol* 2002, 40, 2323.

48. Grolla, A. et al. The use of a mobile laboratory unit in support of patient management and epidemiological surveillance during the 2005 Marburg outbreak in Angola. *PloS Negl Trop Dis* 2011, 5, e1183.

49. Zhai, J. et al. Rapid molecular strategy for filovirus detection and characterization. *J Clin Microbiol* 2007, 45, 224.

50. Panning, M. et al. Diagnostic reverse-transcription polymerase chain reaction kit for filoviruses based on the strain collections of all European biosafety level 4 laboratories. *J Infect Dis* 2007, 196(Suppl. 2), S199.

51. Huang, Y. et al. Rapid detection of filoviruses by real-time TaqMan polymerase chain reaction assays. *Virol Sin* 2012, 27, 273.

52. Weidmann, M., Muhlberger, E., Hufert, F. T. Rapid detection protocol for filoviruses. *J Clin Virol* 2004, 30, 94.

53. Trombley, A. R. et al. Comprehensive panel of real-time TaqMan polymerase chain reaction assays for detection and absolute quantification of filoviruses, arenaviruses, and New World hantaviruses. *Am J Trop Med Hyg* 2010, 82, 954.

54. Pang, Z. et al. Comprehensive multiplex one-step real-time TaqMan qRT-PCR assays for detection and quantification of hemorrhagic fever viruses. *PloS One* 2014, 9, e95635.

55. Kurosaki, Y., Grolla, A., Fukuma, A., Feldmann, H., Yasuda, J. Development and evaluation of a simple assay for Marburg virus detection using a reverse transcription-loop-mediated isothermal amplification method. *J Clin Microbiol* 2010, 48, 2330.

56. Euler, M. et al. Development of a panel of recombinase polymerase amplification assays for detection of biothreat agents. *J Clin Microbiol* 2013, 51, 1110.

57. Blow, J. A., Dohm, D. J., Negley, D. L., Mores, C. N. Virus inactivation by nucleic acid extraction reagents. *J Virol Methods* 2004, 119, 195.

58. Sherwood, L. J., Osborn, L. E., Carrion, R., Patterson, J. L., Hayhurst, A. Rapid assembly of sensitive antigen-capture assays for Marburg virus, using in vitro selection of llama single-domain antibodies, at biosafety level 4. *J Infect Dis* 2007, 196, S213.

59. Zhang, J. B. et al. Production and characterization of monoclonal antibodies to nucleoprotein of Marburg virus. *Hybridoma* 2008, 27, 423.

60. Vladyko, A. S., Chepurnov, A. A., Bystrova, S. I., Lemeshko, N. N., Lukashevich, I. S. Marburg virus—Detection of antigen by enzyme-linked-immunosorbent-assay. *Voprosy Virusologii* 1991, 36, 419.

61. Saijo, M. et al. Marburgvirus nucleoprotein-capture enzyme-linked immunosorbent assay using monoclonal antibodies to recombinant nucleoprotein: Detection of authentic Marburgvirus. *Jpn J Infect Dis* 2006, 59, 323.

62. Saijo, M. et al. Enzyme-linked immunosorbent assays for detection of antibodies to Ebola and Marburg viruses using recombinant nucleoproteins. *J Clin Microbiol* 2001, 39, 1.

63. Nakayama, E. et al. Enzyme-linked immunosorbent assay for detection of filovirus species-specific antibodies. *Clin Vaccine Immunol* 2010, 17, 1723.

64. van der Groen, G., Kurata, T., Mets, C. Modifications to indirect immunofluorescence tests on Lassa, Marburg, and Ebola material. *Lancet* 1983, 1, 654.

65. Kamata, T., Natesan, M., Warfield, K., Aman, M. J., Ulrich, R. G. Determination of specific antibody responses to the six species of Ebola and Marburg viruses by multiplexed protein microarrays. *Clin Vaccine Immunol* 2014, 21, 1605.

Section III

Negative- and Ambi-Sense RNA Viruses

59 Akabane Virus

Akiko Uema, Yohsuke Ogawa, and Hiroomi Akashi

CONTENTS

59.1 Introduction 521
    59.1.1 Classification... 521
    59.1.2 Morphology, Biology, and Epidemiology 523
    59.1.3 Clinical Features and Pathogenesis 524
    59.1.4 Diagnosis 525
        59.1.4.1 Conventional Techniques... 525
        59.1.4.2 Molecular Techniques... 526
59.2 Methods 526
    59.2.1 Sample Preparation: Extraction of Viral RNA 526
        59.2.1.1 Animal Tissue... 526
        59.2.1.2 Cell Cultures... 526
    59.2.2 Detection Procedures.. 526
        59.2.2.1 RT-PCR and Nested PCR 526
        59.2.2.2 Cloning and Sequencing 527
        59.2.2.3 Quantitative RT-PCR 528
    59.2.3 Rescue of Virus by Reverse Genetics System 528
        59.2.3.1 Plasmid Construction.. 528
        59.2.3.2 Transfection 529
59.3 Conclusion and Perspectives.. 529
References.. 529

59.1 INTRODUCTION

Akabane virus (AKAV) causes reproductive and congenital problems in cattle, sheep, and goats, resulting in severe economic losses. AKAV is transmitted primarily by biting midges (*Culicoides* species). Although pregnant animals infected with AKAV exhibit no clinical signs, *in utero* fetal exposure to the virus results in abortion, premature birth, stillbirth, and congenital deformities such as arthrogryposis and hydranencephaly (AH) syndrome.[1]

Arthropod-borne diseases like Akabane disease, AH abnormalities of the calves, were first reported in Australia in 1951–1955,[2] but viral isolation was not attempted in that case. In 1959, the JaGAr39 strain, a reference strain of AKAV, was first isolated from mosquitoes (*Aedes vexans* and *Culex tritaeniorhynchus*) in Akabane village of the Gunma Prefecture, Japan, by the survey of *Aedes* arboviruses.[3,4] The first impact of the virus was reported in 1974 in Australia, in 1972–1975 in Japan, and in 1969–1970 in Israel,[1,5,6] and outbreaks of the disease have occurred in Japan, Israel, Turkey, and Australia.[6–8] In Japan, more than 42,000 congenital abnormalities, abortions, stillbirths, and premature births were caused in cattle.[1] Furthermore, AKAV has been detected in many tropical and subtropical regions,[9–13] and consequently, AKAV is widely distributed throughout Australia, Southeast Asia, East Asia, the Middle East, and Africa.

Vaccination has reduced the prevalence of Akabane disease; however, cases still occur in Japan and Korea.[14,15] Recently, antigenic and pathogenic variants of AKAV have been isolated.[16–19] For example, the Iriki strain caused nonsuppurative encephalitis and neurological symptoms to a calf in southern Japan in 1984.[20] Since the variants such as the Iriki strain showed low cross-reactivity with antiserum against the reference strain of AKAV by neutralization tests, those may be the reason for disease continuation in areas where vaccines are administered. Therefore, further investigation of both epidemiology and pathogenesis is required to elucidate the etiological agent. Besides, molecular studies will facilitate novel strategies for the control of Akabane disease in the future.

59.1.1 CLASSIFICATION

AKAV belongs to the genus *Orthobunyavirus* of the *Bunyaviridae* family. Based on serology, AKAV is a member of the Simbu serogroup, which consists of 25 viruses distributed globally. The viruses belonging to the serogroup cause disease in human and ruminants and are transmitted by biting midges and mosquitoes.[21] Although this group has been isolated from all continents except Europe,[21] Schmallenberg virus (SBV), which is likely a reassortant virus of Sathuperi virus and Shamonda virus[22] or an ancestor of Shamonda virus,[23] was discovered in a vast region of Europe in 2011.[24–26] In the Simbu serogroup, Aino virus (AINOV) and SBV that affect Japan and Australia[27,28] and European countries, respectively, cause abortion and congenital abnormalities in cattle, sheep, and goats as well as AKAV, while *Oropouche virus* causes dengue-like febrile illness in human populations in South America.[29]

Currently, AKAV is divided into four clusters based on the S RNA segment sequences reflecting geographical origin such as Japan, Australia, Taiwan, Israel, and Kenya. Two clusters comprised the Japanese, Taiwanese, and Israeli isolates; hence, Asian strains seemed to have evolved in a common gene pool. The Australian and Kenyan isolates were each placed in the independent third (genotypes III) and fourth (genotypes IV) clusters, respectively. Phylogenetic analysis revealed that the Japanese field isolates could be divided into two major clusters (genotypes I and II), and genotype II was subdivided into two branches (IIa and IIb) (Figure 59.1). These data suggest that AKAV have evolved in multiple lineages, and the various genotypes might be introduced continuously from an exotic source. The geographic divergence is considered to drive AKAV variation.[17,30] In order to examine to what extent variation of the viral genome is reflected in the antigenic properties, monoclonal antibodies (MAbs) to AKAV were produced by immunization with OBE-1 strain and tested serologically.[16] There was no distinct correlation between the

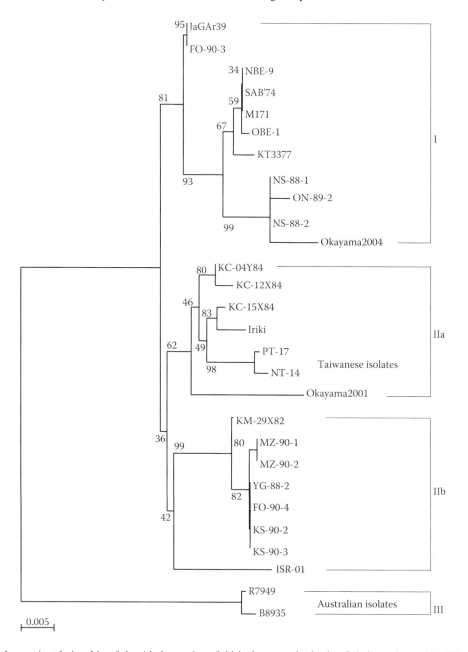

FIGURE 59.1 Phylogenetic relationship of the Akabane virus field isolates on the basis of their nucleocapsid (N) open reading frame (ORF) nucleotide sequences. Phylogenetic analyses of the N ORFs (between nucleotide positions −34 and −735, total 702 nucleotides) show that four groups are extant. Numbers indicate bootstrap percentages out of 1000 replicates. The scale bar represents 0.005% substitution per base.

TABLE 59.1

Enzyme-Linked Immunosorbent Assay Tests with Akabane Virus Monoclonal Antibodies

Protein Specificity	Gc													N		
Isolates / Mab clone	IB3	4F9	ID3	4F7	2B12	3A5	4A10	4D4	5G4	5E5	2F1	9G10	9G12	2A7	5F11	5E8
JaGAr39	<10	400	25,600	12,800	12,800	6,400	3,200	3,200	12,800	1,600	1,600	3,200	160	51,200	12,800	6,400
B8935	<10	<10	12,800	12,800	1,600	1,600	800	<10	<10	1,600	1,600	<10	<10	51,200	6,400	12,800
R7949	<10	<10	51,200	25,600	12,800	6,400	6,400	10	10	6,400	12,800	12,800	320	51,200	12,800	12,800
OBE-1	12,800	51,200	51,200	12,800	25,600	6,400	25,600	12,800	25,600	6,400	6,400	6,400	320	102,400	25,600	12,800
NBE-9	<10	<10	12,800	6,400	6,400	1,600	1,600	1,600	6,400	1,600	400	1,600	80	25,600	12,800	6,400
SAB'74	<10	<10	12,800	6,400	6,400	1,600	12,800	<10	<10	1,600	400	1,600	160	51,200	12,800	12,800
KT3377	<10	800	25,600	12,800	400	10	40	1,600	6,400	<10	6,400	3,200	160	102,400	12,800	6,400
M171	<10	400	25,600	12,800	12,800	6,400	12,800	6,400	25,600	3,200	6,400	12,800	320	102,400	25,600	12,800
KM-29X82	<10	400	25,600	6,400	6,400	3,200	10	800	3,200	<10	1,600	1,600	40	51,200	12,800	3,200
KC-12X84	<10	<10	51,200	25,600	12,800	6,400	3,200	6,400	12,800	<10	6,400	12,800	320	102,400	25,600	12,800
KC-15X84	<10	<10	51,200	25,600	25,600	25,600	<10	6,400	400	6,400	6,400	12,800	320	102,400	25,600	12,800
KC-04Y84	<10	<10	25,600	25,600	25,600	12,800	3,200	3,200	12,800	<10	6,400	6,400	320	102,400	25,600	12,800
Iriki	<10	<10	51,200	12,800	25,600	51,200	<10	1,600	400	<10	25,600	12,800	160	102,400	12,800	6,400
NS-88-1	<10	1,600	51,200	6,400	12,800	6,400	80	1,600	6,400	160	25,600	3,200	80	51,200	51,200	3,200
NS-88-2	<10	1,600	12,800	6,400	12,800	3,200	40	800	1,600	40	20	400	20	25,600	12,800	3,200
YG-88-2	<10	20	6,400	6,400	10	10	40	1,600	3,200	20	800	200	80	51,200	25,600	6,400
ON-89-2	<10	3,200	51,200	51,200	102,400	25,600	1,600	6,400	12,800	25,600	25,600	6,400	320	51,200	25,600	<10
MZ-90-1	<10	20	12,800	12,800	25,600	12,800	160	800	6,400	20	12,800	1,600	160	102,400	25,600	6,400
MZ-90-2	<10	20	10	<10	12,800	3,200	400	160	6,400	40	160	1,600	160	51,200	12,800	6,400
FO-90-3	<10	400	51,200	51,200	102,400	25,600	800	3,200	12,800	1,600	25,600	3,200	160	51,200	12,800	6,400
FO-90-4	<10	800	25,600	25,600	102,400	25,600	1,600	3,200	25,600	1,600	25,600	6,400	640	102,400	51,200	12,800
KS-90-2	<10	10	<10	<10	102,400	6,400	200	400	800	200	6,400	800	160	102,400	12,800	6,400
KS-90-3	<10	160	51,200	51,200	102,400	25,600	800	1,600	12,800	20	12,800	3,200	160	102,400	25,600	6,400

phylogenic relationship and their reactivity to MAbs that recognized Gc and nucleocapsid (N) protein (Table 59.1). For example, only the OBE-1 strain responded with MAb 1B3 and showed high response to MAb 4F9 although NBE-9 and SAB′74 strain, which are phylogenetically close to the OBE-1 strain, showed no response to the MAbs. This suggests that the epitope is absent from those viruses, and therefore there is antigenic variation besides genomic diversity.

59.1.2 MORPHOLOGY, BIOLOGY, AND EPIDEMIOLOGY

AKAV is an enveloped virus and has a tripartite negative-sense, single-stranded RNA genome. The viral particle is roughly spherical, sometimes pleomorphic, and has a diameter of 70–130 nm (typically 90–100 nm). The virion surface consists of an envelope with a spikelike projection.[31] The envelope is a host cell–derived lipid layer, while spikes are comprised two viral glycoproteins (Gn and Gc). In the *Orthobunyavirus* genus, the virion has three genome segments: L, M, and S.[32] Each genome is encapsulated with N proteins and an RNA-dependent RNA polymerase (RdRp), referred to as the ribonucleoprotein (RNP) complex in a helical configuration. The large (L) segment encodes the large (L) protein, which acts as the RdRp. The medium (M) segment encodes a precursor polyprotein containing NH_2-Gn-NSm-Gc-COOH.

Afterward, the polyprotein is cleaved into two glycoproteins, Gn and Gc, and a nonstructural protein (NSm). Gn and Gc are responsible for viral neutralization and attachment to cell receptors,[33] and analysis of AKAV antigenic properties using MAbs showed that Gc possessed hemagglutination activity (HA).[16] The function of NSm remains unknown, but it may be responsible for apoptosis signaling[34] and virus assembly.[35,36] The small (S) segment encodes N and a nonstructural protein (NSs) in an overlapping open reading frame (ORF). N plays a role in RNP formation, and NSs is responsible for interferon antagonism and regulation of host protein synthesis.[37,38] Additionally, the untranslated region (UTR) of each segment is suggested to play an essential role in RNA regulation and genome synthesis.[39]

Orthobunyaviruses enter cells by attaching to a cell surface receptor and penetrating through the endocytic pathway[40]; however, the complete details of viral entry remain unclear.

Most of the bunyaviruses utilize clathrin-dependent endocytosis,[40–43] although some viruses enter cells via another pathway such as caveolae-dependent endocytosis for Rift Valley phlebovirus to mammalian cell lines[44] and clathrin-independent endocytosis for Uukuniemi phlebovirus and Andes hantavirus.[45,46] Recently, it is suggested that AKAV enters bovine-derived cell lines in a manner of a nonclathrin, noncaveolae endocytic pathway that is dependent on dynamin, while AKAV infection in nonbovine-derived cell lines is dependent on clathrin endocytosis.[47] After viral uptake into the cytosol, virus particles fuse with the endosome to gain acidity and release viral genome into cytoplasm. Subsequently, primary transcription and translation of viral proteins begin. The new virion assembles and buds into the Golgi compartment, followed by exocytosis from cells.[32,36,48]

The epidemiology of AKAV is consistent with the classification as arbovirus, that is, its appearance is seasonal, being restricted to period when insects are active, and it has a limited geographical distribution. The outbreaks of Akabane disease are associated with transmission by arthropod vectors in summer and autumn.[1,2] Since infected midges cannot survive during the winter, the outbreak gradually subsides.[49–51] In Australia, AKAV is transmitted by *Culicoides brevitarsis*,[52–54] which is a common biting pest of domestic ruminants. Due to suitable environmental conditions, the habitats of *C. brevitarsis* are widespread throughout the year in midnorthern and northern to southern coastal plains of New South Wales, Australia, and AKAV is an endemic disease in these areas.[55–57] AKAV-neutralizing antibodies have been found in other host species such as horse and camel, which are not common hosts of *C. brevitarsis* in the endemic areas, and therefore, it is suggested that these animals are accidental hosts of *C. brevitarsis* or that this virus has other vectors.[57] In the southern parts of Japan, *C. oxystoma* is the principle vector of AKAV.[58,59] On the other hand, *C. brevitarsis* has also been detected in the area, leading to concerns of arbovirus outbreaks involving AKAV and other orthobunyaviruses.[60] In Oman, *C. imicola* was reported as an AKAV vector.[61] In Israel, the definitive vector remains unknown. Furthermore, there is a report that AKAV can replicate in *C. variipennis* and *C. nubeculosus*, which are widespread in America and Europe, respectively. In the report, *C. variipennis* was susceptible to oral infection, and *C. nubeculosus* was susceptible to infection only when intrathoracic inoculation *in vitro*,[62] while there have been no reports of virus isolation from these midges in field surveys of these regions.

AKAV also have been isolated from mosquitoes, such as *A. vexans* and *C. tritaeniorhynchus* in Japan[3] or *Anopheles funestus* in Kenya.[63] However, experimental attempt to infect via the *Culex* and *Aedes* spp. have been unsuccessful although AKAV was isolated from the cephalothoraxes of *Aedes albopictus* fed blood meal in the laboratory.[64] Recently, AKAV was isolated from nonblood-fed *C. tritaeniorhynchus*, *Culex vishnui*, *Anopheles vagus*, and *Ochlerotatus* species in northern Vietnam.[9] Furthermore, AKAV was propagated and maintained in *Aedes albopictus* cell culture as a persistent infection.[65] Thus, it remains possible to AKAV transmission by mosquitoes.

59.1.3 Clinical Features and Pathogenesis

AKAV-associated congenital abnormalities have been reported in cattle, sheep, and goats.[1,53,66,67] In Taiwan, AKAV has been isolated in pigs, which showed convulsion and diarrhea.[68] Moreover, horse, donkey, camel, and wildlife animals

have been found to be seropositive for AKAV, while pathogenic lesions have not been reported.[12,69]

The host immune system against AKAV infection has been suggested to be both humoral and cellular immunity, which decrease or eliminate AKAV from blood and infected tissues. Thus, animals with a mature immune system have the potential to resist AKAV infection,[70,71] whereas animals with an immature immune system, predominately the fetus, are susceptible to virus infection and exhibit pathogenic lesions.[72]

The well-known severe sign is delivery of fetus with congenital deformities, AH syndrome. Infected pregnant cattle, sheep, and goats show dystocia, abortion, stillbirth, and premature birth, while delayed labor can be observed with only pregnant goats.[73] Infected fetuses are generally dead during birth. The live newborns show neurological problems including the inability to stand, blindness, lack of a sucking reflex, nystagmus, and death within 3 days due to an inability to feed.[74,75] The fetuses with the nonpurulent encephalomyelitis and myositis are usually born with mild to moderate neurological and muscular problems such as ataxia, muscle atrophy, paralysis, and blindness.[75]

AKAV infection in newborn and adults generally results transient viremia without any clinical signs.[70,76] However, infection with the highly pathogenic Iriki strain, which was isolated from the cerebellum specimen of a calf, resulted in nonpurulent encephalitis.[20] Iriki-like strains also caused encephalomyelitis in young and adult animals.[11,77] Various clinical signs have been observed in those animals, including lassitude, anorexia, loss of milk yield, tremor, stretching of the legs, and ataxia. Since there is no typical therapy and symptomatic animals are suggestive of a poor prognosis, the animals should be euthanized.

Histopathological and immunohistochemical studies have shown that nervous system damage by AKAV is induced directly and indirectly. The direct effects are caused by virus infection into neurons, resulting in cytolysis,[78] and the indirect effects are caused by cytotoxic cytokine responses, resulting in inflammatory reactions such as vascular damage.[79] AKAV causes inflammation in all parts of the central nervous system (CNS), the cerebrum, cerebellum, brainstem, and spinal cord, and these inflammatory lesions typically develop into severe encephalitis such as necrotizing encephalitis.

Gross appearance of AH syndrome shows disorders in two major organs, the CNS and muscular systems. The lesions in the brain present hypoplasia in both sides of the cerebrum, such as hydranencephaly, microencephaly, and porencephaly, where the cranial cavity is replaced by a clear cerebrospinal fluid. These are produced by the circulating cerebrospinal fluid in necrosis areas. The midbrain, cerebellum, medulla oblongata, and spinal cord are reduced in size.[70,76] The cystic cavity and small subcortical cysts in brain tissue can also be observed.[80] Histopathological lesions in the CNS show nerve cell degeneration. "Perivascular cuffs" are prominent in perivascular layer, that is, cells including plasma cells, histiocytes, and lymphocytes infiltrate into perivascular layer.[79] The ventral horn of the spinal cord shows loss of motor neurons, gliosis, and perivascular cuffs.[81] Skeletal deformities present scoliosis, kyphosis, torticollis, and brachygnathism.[70,79] Arthrogryposis is produced by the decrease in the number of ventral horn neurons that control muscle fiber. Skeletal muscle is atrophied and edematous, predominately in the limbs. Joints are fixed in flexion.[79] In case of polymyositis in skeletal muscle, the muscle system shows myotubule and myofiber degeneration and infiltration of inflammatory cells in skeletal muscle tissue.[82] Markedly, degenerative myotubules are broken into tissue fragments, and some disappear. Abnormal myotubules are swollen and their nuclei are pyknotic. Myofibers undergo degeneration, disruption, and fragmentation. Some of the fragmented fibers lose striation. With regard to encephalitis in newborns and adults, no macroscopic lesions are identified in the CNS. Histopathological signs of nonpurulent encephalomyelitis comprise perivascular cuffs and nerve cell degeneration in the midbrain, cerebellopontine, and medulla oblongata and rarely in the cerebrum and cerebellum.[83]

AKAV infection is initiated by intravenous route in the dermis by AKAV-infected arthropod vectors. After virus invasion, viremia typically occurs in 1–9 days, whereas the main cells responsible for viral multiplication remain unclear. AKAV-neutralizing antibody is detected 1–3 days after viremia and can be persistently detected for more than a year. During this period, the virus spreads to the brain, spinal cord, and skeletal muscle by a hematogenous route and also to the placenta, ovary, cornea, uvea, and lens fiber.[84,85] Transplacental vertical transmission is the route of AKAV infection into fetuses.[86] The pathological changes are closely related to the age at which the fetus is infected.

AKAV pathogenesis is divided into the following three phases: the nonpurulent encephalomyelitis phase, polymyositis phase, and AH phase. The nonpurulent encephalomyelitis phase is the primary pathological change in AKAV infectious stage. The fetus exposed to the virus after the middle stage of gestation produces only the pathological change.[70,77,80,87] Moreover, multifocal encephalitis but not the necrotizing one is produced during this stage. The polymyositis phase is the second pathological change. There is evidence that this lesion is produced by AKAV infection directly and indirectly only during the myotubule and myofiber development phases in muscular of the fetus, which correspond to first to middle stage of gestation, and results in abnormal myotubules and inflammation. Therefore, infection of AKAV at the late stages of gestation will not result in polymyositis.[82] The AH phase is the final stage of Akabane disease lesion development. These are sequential changes of nonpurulent encephalomyelitis and polymyositis, respectively.[79] The AH lesions are produced in the fetus exposed to AKAV during the early stage, that is, 62–96 days in cattle[76] and 26–69 days in goats and sheep[70,80,88] to middle stages of gestation.

59.1.4 DIAGNOSIS

59.1.4.1 Conventional Techniques

Since Akabane disease spreads rapidly via arthropod-borne means, the diagnosis must be rapid and reliable in order to prevent transmission to noninfected animals. A field diagnosis is based on clinical manifestation and detection of arthropod vector activity. Postmortem changes are examined by autopsy or necropsy as well as histological examination. However, since other orthobunyaviruses including AINOV and Peaton virus (PEAV) have been associated with bovine congenital abnormalities in Japan, it is difficult to differentiate with clinical features alone. Serological analysis is a common and convenient method for AKAV detection due to its accuracy and high sensitivity. The analysis is generally used for epidemiological surveys and immunity levels in vaccinated animals. The neutralization test is the most commonly used for serological diagnosis of AKAV infection.[70,89–91] Detection of neutralizing antibody in precolostral or fetal serum is available. Since cattle, sheep, and goats cannot transfer maternal antibodies to the fetus through the placenta and newborn animals acquire antibodies from the dam via the colostrum, the presence of anti-AKAV antibodies in precolostral or fetal serum suggests that the fetus had been infected with the virus *in utero*. Additionally, the hemagglutination inhibition test,[92] complement fixation test (CFT),[93] hemolysis inhibition test,[94] and enzyme-linked immunosorbent assay (ELISA)[95,96] have also been used as serological diagnosis of AKAV infection, whereas the method such as CFT and ELISA detect cross-reactive antibodies against viruses of the same serogroup. Moreover, because these tests detect immunoglobulin (Ig) G, the antibody detection period after virus exposure can be prolonged. Whether observed animals were recently infected should be evaluated with paired serum samples taken at 1-week interval[1,74] or IgM (a recent infection indicator) assay simultaneously with IgG.[96]

Viral isolation using cell culture is a general technique that can be applied to virus identification and vaccine preparation. Generally, defibrinated blood from sentinel cattle and CNS and skeletal muscle tissues from aborted fetuses are used for virus isolation. The cell types for standard AKAV isolation involve HmLu-1 cells[66] derived from hamster lung, BHK-21 cells from hamster kidney,[19] and Vero cells from African green monkey kidney,[71,84] due to their sensitivity based on the cytopathic effect (CPE) and plaque formation.[97] Moreover, experimental animals such as suckling mice, hamsters,[4] or chicken embryos[98] are commonly used for AKAV isolation. However, this technique is not appropriate in routine diagnosis work because it takes a time to detect and fresh materials are required. Abnormally delivered animals cannot be also used due to the presence of antibody. In some cases, virus isolation is not successful but other tests show positive for AKAV.[18,81,99]

The establishment of anti-AKAV antibodies is the key factor for immunological tests including immunohistochemistry[18,77,83,87] and immunofluorescence[78] assays. Immunohistochemical detection of virus antigen has become an important tool for pathological studies and diagnosis of infectious diseases. Although brain and skeletal muscle tissues are typically subjected to histopathological examination, it is important to verify AKAV specificity of the antibody because of the cross-reactivity among Simbu serogroup viruses.

59.1.4.2 Molecular Techniques

With the advancements in molecular biology technologies, amplification of AKAV genetic materials has become one of the most valuable tools for the confirmation of AKAV infection. Reverse transcription polymerase chain reaction (RT-PCR), as well as nested RT-PCR,[100] has been proposed as the new standard method for AKAV detection due to the rapidity and sensitivity. Moreover, multiplex RT-PCR[101,102] and quantitative real-time RT-PCR[102] facilitate rapid identification of viruses to the serotype level with improved accuracy.

59.2 METHODS

59.2.1 Sample Preparation: Extraction of Viral RNA

59.2.1.1 Animal Tissue

Tissue samples collected from infected animals should be placed on dry ice or flash frozen in liquid nitrogen as soon as possible after autopsy. The samples can be stored at −80°C for many years. Viral RNA is extracted using TRIzol Reagent (Life Technologies), according to the manufacturer's instructions. Briefly, 50–100 mg of tissue sample is added with 1 mL of TRIzol Reagent and homogenized. The homogenized sample can be stored at −80°C for at least 1 month or proceeded sequentially to the following phase separation. After incubation at room temperature for 5 min to permit complete dissociation of the nucleoprotein complex, the homogenate is added with 0.2 mL of chloroform and incubated for 2–3 min at room temperature, and then, the mixture is centrifuged at $12,000 \times g$ for 15 min. The mixture separates into a lower phenol–chloroform phase, an interphase, and an upper aqueous phase. The RNA remains exclusively in the aqueous phase. The aqueous phase is added with 0.5 mL of isopropanol and incubated at room temperature for 10 min and then centrifuged at $12,000 \times g$ for 10 min to precipitate the RNA. The supernatant is removed and RNA pellet is washed with 1 mL of 75% ethanol. The RNA in ethanol is centrifuged at $7,500 \times g$ for 5 min and the wash is discarded. The RNA pellet is dried for 5–10 min but not completely. The RNA pellet is added with 20–50 μL of RNase-free water or 0.5% sodium dodecyl sulfate (SDS) solution and incubated at 55°C–60°C for 10–15 min in order to resuspend completely. The RNA solution can be stored at −80°C until the following assay is performed.

Note: For liquid samples such as blood plasma or cerebrospinal fluid, TRIzol LS Reagent is used instead of TRIzol Reagent. A sample of 250 μL is added with 750 μL of TRIzol LS Reagent and homogenized by pipetting several times.

59.2.1.2 Cell Cultures

AKAV genomes are commonly isolated using cell culture techniques. Viral RNA is extracted from the supernatant of cells infected by viral solution or infected animal tissue homogenates in cold phosphate-buffered saline, using QIAamp Viral RNA Mini Kit (Qiagen), according to the

manufacturer's instructions. Briefly, 140 μL of cell culture supernatant is mixed to 560 μL of AVL buffer containing carrier RNA. The mixture is incubated at room temperature for 10 min and 560 μL of ethanol is added. Then, the mixture is applied onto the spin column and centrifuge at $6000 \times g$ for 1 min, and 500 μL of AW1 buffer is added. The column is centrifuged at $6000 \times g$ for 1 min to remove unbound materials and washed by the addition of AW2 buffer. Then, the column is centrifuged at full speed for 3 min and placed into a new 1.5 mL microcentrifuge tube. Finally, 60 μL of AVE buffer is added directly onto the column to elute AKAV genomic RNA. After incubation at room temperature for 1 min, the column is centrifuged at $6000 \times g$ for 1 min. The obtained RNA can be stored at −80°C until the following assay is performed. The RNA is used as a template for RT-PCR, nested PCR, or quantitative real-time RT-PCR amplification.

59.2.2 Detection Procedures

59.2.2.1 RT-PCR and Nested PCR

Since AKAV, AINOV, PEAV, and some other Simbu serogroup orthobunyaviruses have been isolated from bovine and insect specimens in Japan and AINOV causes similar congenital deformities such as AH syndrome, the differential diagnosis using molecular techniques is needed. Based on the high degree of sequence conservation among orthobunyavirus S RNA species, the following specific primer sets for AKAV and AINOV were developed to detect and differentiate these viruses by RT- and nested PCR.

Primers	Sequence (5′–3′)	Nucleotide Positions
First round for Aino virus		
AISF132	CCC AAC TCA ATT TCG ATA CC	132-151
AISR780	TTT GGA ACA CCA TAC TGG GG	780-761
First round for Akabane virus		
AKSF19	TAA CTA CGC ATT GCA ATG GC	19-38
AKSR740	TAA GCT TAG ATC TGG ATA CC	740-721
Nested for Aino virus		
AISF313	CCA TCG TCT CTC AGG ATA TC	313-332
AISR657	ACA GCA TTG AAG GCT GCA CG	657-638
Nested for Akabane virus		
AKSF177	GAA GGC CAA GAT GGT CTT AC	177-196
AKSR407	GGC ATC ACA ATT GTG GCA GC	407-388

These primer sets have been examined for their capacity to amplify PCR products using seven Simbu serogroup viruses (AKAV, AIVNOV, PEAV, Douglas virus, Tinaroo virus [(TINV)], *Thimiri virus*, and Facey's Paddock virus) isolated in Japan and Australia. RT- and nested PCR products with AKAV-specific primers are obtained for AKAV and TINV (Figure 59.2a). These results are expected on the basis of the

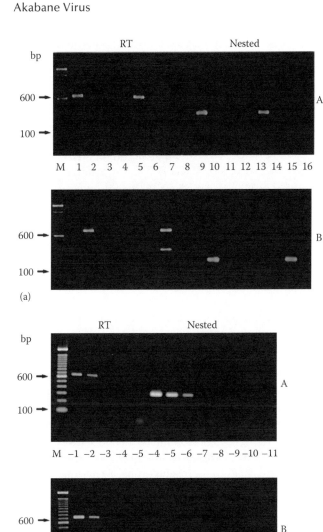

(a)

(b)

FIGURE 59.2 Agarose gel electrophoresis of the (A) Aino- and (B) Akabane-specific reverse transcription polymerase chain reaction (RT-PCR) and nested PCR products. (a) M, molecular weight marker; 1, 9 Aino virus; 2, 10 Akabane virus; 3, 11 Douglas virus; 4, 12 Facey's Paddock virus; 5, 13 Peaton virus; 6, 14 Thimiri virus; 7, 15 Tinaroo virus; 8, 16 cell culture fluid control. Numbers 1–8 show the result of RT-PCR; 9–16 show nested PCR. (b) Tenfold serial dilutions of RNA from 10^5 plaque-forming unit of each virus were amplified using the specific primers by RT-PCR and nested PCR. The figure shows from 10^{-1} to 10^{-5} dilutions for RT-PCR and from 10^{-4} to 10^{-11} for nested PCR.

their S RNA sequence similarity.[103] The nested PCR product of AKAV is digested only by *HphI*, and the TINV product can be cut only with *BstEII*. Similarly, PEAV yields RT-PCR and nested PCR products with AINOV-specific primers (Figure 59.2a). This nested PCR product of PEAV fails to be digested by three enzymes (*AvaII*, *EcoRI*, and *HaeII*), whereas that of AINOV is digested.

To determine the sensitivity of RT- and nested PCR, template RNA dilution was performed. RNAs extracted from approximately 100 plaque-forming units (PFU) of AINOV and 10 PFU of AKAV were detected by RT-PCR, whereas 10^{-3} PFU of AINOV and 10^{-5} PFU of AKAV were detected by nested PCR (Figure 59.2b). These results show that the nested PCR followed by restriction fragment length polymorphism analysis provides a simple, rapid, and reliable means for diagnosing AKAV and AINOV infection. Thus, these methods are currently being used in the field.

59.2.2.2 Cloning and Sequencing

A phylogenic analysis by N or M ORFs is carried out to characterize AKAV genotypes, and the identification of complete sequences of three genomes can clarify the presence of a mutant or unidentified strain. The following is the protocol of complete genome sequencing, which has been modified from previous reports[19,104]:

59.2.2.2.1 Procedure

AKAcon primer (sequence in the following table) used for cDNA synthesis is designed using the 3'- and 5'-end complementary sequences of each segment of the OBE-1 strain. The RT mix is present as follows. First-strand cDNA is synthesized using the SuperScript III Reverse Transcriptase (Life Technologies), according to the manufacturer's instructions.

Note: After the reaction, 1 μL of RNase H (Life Technologies) may be added and incubated at 37°C for 20 min to remove RNA complementary to the cDNA.

Component	Volume (μL)
RNA	7
AKAcon primer (1 pM)	2
dNTP (deoxyribonucleotide triphosphate) mix (2.5 mM each)	4
5× first-strand buffer	4
0.1 MDTT (M dithiothreitol)	1
RNaseOut	1
SuperScript III Reverse Transcriptase	1
Total	20

Subsequently, to prepare the PCR reaction mix of LA Taq (Takara Bio), 3 μL of cDNA is added to 22 μL of PCR mixture according to the manufacturer's instructions. Full-length S (856 bp: OBE-1), M (4309 bp: OBE-1), and L (6868 bp: OBE-1) RNA segments are amplified by PCR with the primer pairs, SF and SR, MF and MR, and LF and LR, respectively. The amplicon is ligated into the TOPO TA Cloning Kit (Life Technologies). Three clones are sequenced using the BigDye Terminator v3.1 Cycle Sequencing Kit (Applied Biosystems), according to the manufacturer's protocol. To confirm the 3'- and 5'-UTR sequences of the viral-sense RNA (vRNA), PCR is performed with AKAcon and S400-381 for the 3'-end of the S vRNA, S459-478 for the 5'-end of the S vRNA, M303-284 for the 3'-end of the M vRNA, M4059-4080 for the 5'-end of the M vRNA, L-4R for the 3′-end of the L vRNA, or L5′RACE for the 5′-end of the L vRNA. The sequences of the UTRs are confirmed by direct sequences of the PCR products. The primer sequences are present as follows:

RNA Segment	Primer	Sequence (5′–3′)	Nucleotide Positions (OBE-1)
S, M, L	AKAcon	ggggctctagaAGTAGTG	
S	SF	AGTAGTGAACTCCACTA TTAACTACGC	1-28
	SR	AGTAGTGTTCTCCACTAA TTAACTATA	858-832
M	MF	AGTAGTGAACTACCAC AACAAAATG	1-25
	MR	AGTAGTGTTCTACCAC AACAAATAATTAT	4309-4281
L	LF	AGTAGTGTACCCCTAAAT AC AACATACA	1-28
	LR	AGTAGTGTGCCCCTAAAT GCAATAATAT	6868-6841
S	S400-381	CAATTGTGGCAGCTGCC TCT	400-381
	S459-478	GGGATTTGCCCCTGGTG CTG	459-478
M	M303-284	TCTTGAACTGGATTGCA CTC	303-284
	M4059-4080	AACCTATATGGAATGAGT TAAG	4059-4080
L	L-4R	TCTCATAATCCAGGCCC AAT	400-381
	L5′RACE	CCCTGGATATCGATAAT GAA	6643-6662

59.2.2.3   Quantitative RT-PCR

Quantitative determination of viral genome provides information about viremia, and a quantitative PCR (qPCR) assay using AKAV-specific primers is useful to distinguish from infections of AINOV.[102]

59.2.2.3.1   Procedure: RT

1. cDNA for qPCR is synthesized briefly using the PrimeScript RT Reagent Kit (Takara Bio). Prepare the RT mix to each 0.5 mL tube as follows:

Component	Volume (μL)
RNA	2
5× PrimeScript Buffer (for real time)	2
PrimeScript RT Enzyme Mix I	0.5
Oligo dt primer (50 μM)	0.5
Random 6 mers (100 μM)	0.5
RNase-free water	4.5
Total	10

2. Transfer the reaction tubes to 37°C heat box for 15 min (RT reaction), and increase temperature to 85°C for 5 s (inactivity of enzyme), and then chill the tubes.

3. The cDNA can be used directly in the following qPCR step or stored at −20°C for future use.

59.2.2.3.2   Procedure: qPCR

1. Prepare the qPCR mix to 1.5 mL tubes as follows (volumes indicated per specimen) and place on ice. For qPCR, ak151f (5′-GCTAGAGTCTTCTT CCTCAACCAGAA-3′) and ak231r (5′-AAAAGTA AGATCGACACTTGGTTGTG-3′) primers[102] are used to amplify the AKAV cDNA (81 bp).

Component	Volume (μL)
Power SYBR Green PCR Master Mix	12.5
ak151f primer (10 μM)	0.5
ak231r primer (10 μM)	0.5
Sterilized water	9.5
Total	23

2. Put the mix into each well of 96-well plate, and add 2 μL of individual cDNA sample.

3. Spin the plate briefly to pull down the droplets at the inner wall of the wells.

4. Place the sample plate in a thermocycler with a real-time system. Perform the reaction under the following cycling conditions, 95°C for 10 min and 45 cycles of 95°C for 15 s and 60°C for 30 s.

5. Finally, analyze the results using a real-time PCR software.

Note: Especially, in case of animal tissues, an endogenous control gene such as GAPDH or 18S rRNA should be tested, using primers for those control genes.

For the purpose of absolute qPCR, the addition of external standard such as a plasmid coding AKAV target sequence is required in every reaction to determine the absolute amount of the target gene from each sample. The external standard should contain sequences that are the same as the target sequence.

59.2.3   Rescue of Virus by Reverse Genetics System

Reveres genetics systems for bunyaviruses have been developed based on the methodology originally established for the prototype of the genus *Orthobunyavirus*, *Bunyamwera virus* (BUNV).[105] The rescue systems using bacteriophage T7 RNA polymerase have been reported for BUNV, *La Crosse virus*, and *Rift Valley fever virus*,[105–107] and the system using RNA polymerase I has been established for AKAV.[108] The following is the protocol of virus rescue using RNA polymerase I–based reverse genetics, which has been modified from the previous report.

59.2.3.1   Plasmid Construction

The plasmids expressing the L and N protein of AKAV should be prepared for the supporting proteins. The pcDNA/*myc*-His A vector is used for cloning of the L or N coding region using restriction enzyme sites *EcoRI* and *XhoI,* resulting in pcDNA/L and pcDNA/S, respectively. By change of ATG start codon to ACG of NSs, the mutated plasmid, pcDNA/SΔNSs, which is a mutant with a knockout of the nonstructural protein NSs and encodes N protein alone, is constructed. pcDNA/SΔNSs is recommended instead of pcDNA/S, since the NSs protein suggests to be a negative regulator of gene activity.[104]

The full-length L, M, and S segment should be inserted into the plasmid containing the murine RNA polymerase I promoter and terminator, such as pRF42 vector (Ramon Flick, University of Texas Medical Branch, and Ralf F. Pettersson, Karolinska Institutet, Sweden), and result in the pRF42/L, pRF42/M, and pRF42/S plasmids can exhibit RNA polymerase I–driven transcription of the respective genomic RNA.

59.2.3.2 Transfection
Procedure

The following manipulation should be conducted in a sterile environment:

1. Approximately 18–24 h before transfection, plate HmLu-1 cells in a 2 mL growth medium per well in a 6-well plate. Incubate the cells at 37°C overnight. Subconfluent cells should be used for transfection.
2. Prepare the plasmid mix as follows (volumes indicated per well):

	Component	6 Wells	12 Wells
Plasmid	pRF42/S	0.5 µg	0.25 µg
	pRF42/M	1.5 µg	0.75 µg
	RF42/L	2 µg	1 µg
	pcDNA/SΔNS	0.1 µg	0.05 µg
	pcDNA/L	0.4 µg	0.2 µg
	Total	4.5 µg	2.25 µg
	OPTI-MEM	100 µL	50 µL
	transIT-LT1	9 µL	4.5 µL

3. Add the plasmid mix into 100 µL of OPTI-MEM (Life Technologies) in a sterile tube, and mix gently.
4. Drop 9 µL transIT-LT1 (Mirus), warmed at room temperature before using, into the OPTI-MEM including the DNA mixture without contact with the sides of the tube, and mix gently.
5. Incubate at room temperature for 15–30 min.
6. Add the transIT-LT1 reagent: DNA complex dropwise to different areas of one well, and rock the culture vessel gently in order to evenly distribute the complex. Incubate the transfected cells at 37°C for approximately 2–5 days, and observe the occurrence of CPE.

Note: If the efficiency of the rescue of the virus is poor, two or more wells should be transfected with each aforementioned volume.

59.3 CONCLUSION AND PERSPECTIVES

Since there is no practical treatment for infected animals, Akabane disease is required to protect by vaccination and vector control. Inactivated vaccines have been used in Japan, South Korea, and Australia,[15,109,110] and live-attenuated vaccines have been used only in Japan and South Korea.[15,111] In case of inactivated vaccines developed in Japan, yearly booster immunizations were recommended for cows in terms of the safety and effectiveness.[109] The inactivated vaccine available to Australia is administrated by intramuscular injection on two occasions, 4 weeks apart, to heifers and cows prior to mating.[110] Recently, formalin-inactivated trivalent vaccines for AKAV, AINOV, and Kasba virus have been developed in Japan and South Korea. This vaccine exhibits improved safety and immunogenicity in pregnant cows.[15] The attenuated live vaccine has been established in Japan. The vaccine strain TS-C2[111] is a temperature-sensitive mutant derived from the OBE-1 strain, which was isolated from a naturally infected bovine fetus in 1974. The safety and immunogenicity of the strain have been showed in pregnant cows, but only the immunogenicity in pregnant ewes because the inoculation of the strain into pregnant ewes exhibited viremia.[111,112] Therefore, this vaccine should be used for only cows.

Although these vaccines have decreased the incidence of Akabane disease, sporadic cases still occur in Japan, South Korea, and Australia.[14] The extensive genetic and antigenic variants may be responsible for vaccine failure. While vaccine development is a long-term process, molecular biological approaches contribute to application to an emerging virus. Reverse genetics technology yielded a recombinant AKAV from cloned cDNA,[108] and the manipulation can be exploited in molecular analysis and vaccine development of AKAV.

Vector control is one of the most reliable methods of arbovirus disease prevention. This process includes the elimination of larvae and adult arthropods (biting midges and mosquitoes) using insecticides or the avoidance of exposure to arthropods by screening of animal sheds.[113] However, these vector control measures may be impractical and difficult to apply in the field due to a temporary effect and a negative impact on the environment.

REFERENCES

1. Inaba, Y., Kurogi, H., and Omori, T. Letter: Akabane disease: Epizootic abortion, premature birth, stillbirth and congenital arthrogryposis-hydranencephaly in cattle, sheep and goats caused by Akabane virus. *Aust Vet J* 51, 584–585 (1975).
2. Blood, D.C. Arthrogryposis and hydranencephaly in newborn calves. *Aust Vet J* 32, 125–131 (1956).
3. Matsuyama, T. et al. Isolation of arbor viruses from mosquitoes collected at live-stock pens in Gumma Prefecture in 1959. *Jpn J Med Sci Biol* 13, 191–198 (1960).
4. Oya, A., Okuno, T., Ogata, T., Kobayashii, and Matsuyama, T. Akabane, a new arbor virus isolated in Japan. *Jpn J Med Sci Biol* 14, 101–108 (1961).

5. Hartley, W.J., Wanner, R.A., Della-Porta, A.J., and Snowdon, W.A. Serological evidence for the association of Akabane virus with epizootic bovine congenital arthrogryposis and hydranencephaly syndromes in New South Wales. *Aust Vet J* 51, 103–104 (1975).

6. Shimshony, S. An epizootic of Akabane disease in bovines, ovines and caprines in Israel, 1969–70: Epidemiological assessment. *Acta Morphol Acad Sci Hung* 28, 197–199 (1980).

7. Inaba, Y. Akabane disease: An epizootic congenital arthro-gryposis-hydranencephaly syndrome in cattle, sheep and goats, caused by Akabane virus. *Jpn Agric Res Q* 13, 123–133 (1979).

8. Yonguç, A.D., Taylor, W.P., Csontos, L., and Worrall, E. Bluetongue in western Turkey. *Vet Rec* 111, 144–146 (1982).

9. Bryant, J.E. et al. Isolation of arboviruses from mosqui-toes collected in northern Vietnam. *Am J Trop Med Hyg* 73, 470–473 (2005).

10. Chang, C.W., Liao, Y.K., Su, V., Farh, L., and Shiuan, D. Nucleotide sequencing of S-RNA segment and sequence analysis of the nucleocapsid protein gene of the newly iso-lated Akabane virus PT-17 strain. *Biochem Mol Biol Int* 45, 979–987 (1998).

11. Liao, Y.K., Lu, Y.S., Goto, Y., and Inaba, Y. The isolation of Akabane virus (Iriki strain) from calves in Taiwan. *J Basic Microbiol* 36, 33–39 (1996).

12. Al-Busaidy, S., Hamblin, C., and Taylor, W.P. Neutralising antibodies to Akabane virus in free-living wild animals in Africa. *Trop Anim Health Prod* 19, 197–202 (1987).

13. Taylor, W.P. and Mellor, P.S. The distribution of Akabane virus in the Middle East. *Epidemiol Infect* 113, 175–185 (1994).

14. Inaba, Y. and Matsumoto, M. Akabane virus. in *Virus Infections of Ruminants*, Vol. 3 (eds. Dinter, Z. and Morein, B.), pp. 467–480 (Elsevier Science Publishers, Amsterdam, the Netherlands, 1990).

15. Kim, Y.H. et al. Development of inactivated trivalent vac-cine for the teratogenic Aino, Akabane and Chuzan viruses. *Biologicals* 39, 152–157 (2011).

16. Akashi, H. and Inaba, Y. Antigenic diversity of Akabane virus detected by monoclonal antibodies. *Virus Res* 47, 187–196 (1997).

17. Yamakawa, M., Yanase, T., Kato, T., and Tsuda, T. Chronological and geographical variations in the small RNA segment of the teratogenic Akabane virus. *Virus Res* 121, 84–92 (2006).

18. Kamata, H. et al. Encephalomyelitis of cattle caused by Akabane virus in southern Japan in 2006. *J Comp Pathol* 140, 187–193 (2009).

19. Ogawa, Y. et al. Comparison of Akabane virus isolated from sentinel cattle in Japan. *Vet Microbiol* 124, 16–24 (2007).

20. Miyazato, S. et al. Encephalitis of cattle caused by Iriki iso-late, a new strain belonging to Akabane virus. *Jpn J Vet Sci* 51, 128–136 (1989).

21. Calisher, C.E. History, classification, and taxonomy of viruses in the family *Bunyaviridae*. In *The Bunyaviridae* (ed. Elliott, R.M.), pp. 1–17 (Plenum Press, New York, 1996).

22. Yanase, T. et al. Genetic reassortment between Sathuperi and Shamonda viruses of the genus Orthobunyavirus in nature: Implications for their genetic relationship to Schmallenberg virus. *Arch Virol* 157, 1611–1616 (2012).

23. Goller, K.V., Hoper, D., Schirrmeier, H., Mettenleiter, T.C., and Beer, M. Schmallenberg virus as possible ancestor of shamonda virus. *Emerg Infect Dis* 18, 1644–1646 (2012).

24. Kupferschmidt, K. Infectious disease. Scientists rush to find clues on new animal virus. *Science* 335, 1028–1029 (2012).

25. Garigliany, M.M. et al. Schmallenberg virus in calf born at term with porencephaly, Belgium. *Emerg Infect Dis* 18, 1005–1006 (2012).

26. Bilk, S. et al. Organ distribution of Schmallenberg virus RNA in malformed newborns. *Vet Microbiol* 159, 236–238 (2012).

27. Cybinski, D.H. and St George, T.D. A survey of antibody to Aino virus in cattle and other species in Australia. *Aust Vet J* 54, 371–373 (1978).

28. Miura, Y. et al. Serological comparison of Aino and Samford viruses in Simbu group of bunyaviruses. *Microbiol Immunol* 22, 651–654 (1978).

29. Tesh, R.B. The emerging epidemiology of Venezuelan hemor-rhagic fever and Oropouche fever in tropical South America. *Ann N Y Acad Sci* 740, 129–137 (1994).

30. Akashi, H., Kaku, Y., Kong, X.G., and Pang, H. Sequence determination and phylogenetic analysis of the Akabane bun-yavirus S RNA genome segment. *J Gen Virol* 78, 2847–2851 (1997).

31. Ito, Y. et al. Electron microscopy of Akabane virus. *Acta Virol* 23, 198–202 (1979).

32. Schmaljohn, C.S. and Hooper, J.W. *Bunyaviridae*: The viruses and their replication. In *Fields Virology*, 4th edn., Vol. 2 (eds. Knipe, D.M. and Howley, J.W.), pp. 1581–1602 (Lippincott Williams & Wilkins, Philadelphia, PA, 2001).

33. Ludwig, G.V., Israel, B.A., Christensen, B.M., Yuill, T.M., and Schultz, K.T. Role of La Crosse virus glycoproteins in attach-ment of virus to host cells. *Virology* 181, 564–571 (1991)

34. Acrani, G.O. et al. Apoptosis induced by Oropouche virus infection in HeLa cells is dependent on virus protein expres-sion. *Virus Res* 149, 56–63 (2010).

35. Shi, X., Kohl, A., Léonard, V.H., Li, P., McLees, A., and Elliott, R.M. Requirement of the N-terminal region of orthobunyavirus nonstructural protein NSm for virus assem-bly and morphogenesis. *J Virol* 80, 8089–8099 (2006).

36. Fontana, J., Lopez-Montero, N., Elliott, R.M., Fernandez, J.J., and Risco, C. The unique architecture of Bunyamwera virus factories around the Golgi complex. *Cell Microbiol* 10, 2012–2028 (2008).

37. Bouloy, M. et al. Genetic evidence for an interferon-antago-nistic function of rift valley fever virus nonstructural protein NSs. *J Virol* 75, 1371–1377 (2001).

38. Weber, F. et al. Bunyamwera bunyavirus nonstructural protein NSs counteracts the induction of alpha/beta interferon. *J Virol* 76, 7949–7955 (2002).

39. Lowen, A.C. and Elliott, R.M. Mutational analyses of the non-conserved sequences in the Bunyamwera Orthobunyavirus S segment untranslated regions. *J Virol* 79, 12861–12870 (2005).

40. Hollidge, B.S. et al. Orthobunyavirus entry into neurons and other mammalian cells occurs via Clathrin-mediated endocy-tosis and requires trafficking into early endosomes. *J Virol* 86, 7988–8001 (2012).

41. Jin, M., Park, J., Lee, S., Park, B., Shin, J., Song, K.J., Ahn, T.I., Hwang, S.Y., Ahn, B.Y., and Ahn, K. Hantaan virus enters cells by clathrin-dependent receptor-mediated endocytosis. *Virology* 294, 60–69 (2002).

42. Santos, R.I., Rodrigues, A.H., Silva, M.L., Mortara, R.A., Rossi, M.A., Jamur, M.C., Oliver, C., and Arruda, E. Oropouche virus entry into HeLa cells involves clathrin and requires endosomal acidification. *Virus Res* 138, 139–143 (2008).

43. Simon, M., Johansson, C., and Mirazimi, A. Crimean-Congo hemorrhagic fever virus entry and replication is clathrin-, pH- and cholesterol-dependent. *J Gen Virol* 90, 210–215 (2009).

44. Harmon, B., Schudel, B.R., Maar, D., Kozina, C., Ikegami, T., Tseng, C.T., and Negrete, O.A. Rift Valley fever virus strain MP-12 enters mammalian host cells via caveola-mediated endocytosis. *J Virol* 86, 12954–12970 (2012).

45. Lozach, P.Y., Mancini, R., Bitto, D., Meier, R., Oestereich, L., Overby, A.K., Pettersson, R.F., and Helenius, A. Entry of bunyaviruses into mammalian cells. *Cell Host Microbe* 7, 488–499 (2010).

46. Ramanathan, H.N. and Jonsson, C.B. New and Old World hantaviruses differentially utilize host cytoskeletal components during their life cycles. *Virology* 374, 138–150 (2008).

47. Bangphoomi, N., Takenaka-Uema, A., Sugi, T., Kato, K., Akashi, H., and Horimoto, T. Akabane virus utilizes alternative endocytic pathways to entry into mammalian cell lines. *J Vet Med Sci* 76, 1471–1478 (2014).

48. Elliott, R.M. Emerging viruses: The *Bunyaviridae*. *Mol Med* 3, 572–577 (1997).

49. Murray, M.D. Akabane epizootics in New South Wales: Evidence for long-distance dispersal of the biting midge *Culicoides brevitarsis*. *Aust Vet J* 64, 305–308 (1987).

50. Bishop, A.L., Barchia, I.M., and Spohr, L.J. Models for the dispersal in Australia of the arbovirus vector, *Culicoides brevitarsis* Kieffer (Diptera: Ceratopogonidae). *Prev Vet Med* 47, 243–254 (2000).

51. Bishop, A.L., Spohr, L.J., and Barchia, I.M. Factors affecting the spread of *Culicoides brevitarsis* at the southern limit of distribution in eastern Australia. *Vet Ital* 40, 316–319 (2004).

52. Doherty, R.L. Arboviruses of Australia. *Aust Vet J* 48, 172–180 (1972).

53. Della-Porta, A.J., Murray, M.D., and Cybinski, D.H. Congenital bovine epizootic arthrogryposis and hydranencephaly in Australia. Distribution of antibodies to Akabane virus in Australian Cattle after the 1974 epizootic. *Aust Vet J* 52, 496–501 (1976).

54. St George, T.D., Standfast, H.A., and Cybinski, D.H. Isolations of akabane virus from sentinel cattle and *Culicoides brevitarsis*. *Aust Vet J* 54, 558–561 (1978).

55. Murray, M.D. The seasonal abundance of female biting-midges, *Culicoides brevitarsis* Kieffer (Diptera, Ceratopogonidae), in Coastal South-Eastern Australia. *Aust J Zool* 39, 333–342 (1991).

56. Jagoe, S., Kirkland, P.D., and Harper, P.A. An outbreak of Akabane virus-induced abnormalities in calves after agistment in an endemic region. *Aust Vet J* 70, 56–58 (1993).

57. Cybinski, D.H., St George, T.D., and Paull, N.I. Antibodies to Akabane virus in Australia. *Aust Vet J* 54, 1–3 (1978).

58. Kurogi, H., Akiba, K., Inaba, Y., and Matumoto, M. Isolation of Akabane virus from the biting midge *Culicoides oxystoma* in Japan. *Vet Microbiol* 15, 243–248 (1987).

59. Yanase, T. et al. Isolation of bovine arboviruses from *Culicoides* biting midges (Diptera: Ceratopogonidae) in southern Japan: 1985–2002. *J Med Entomol* 42, 63–67 (2005).

60. Yanase, T. et al. Detection of *Culicoides brevitarsis* activity in Kyushu. *J Vet Med Sci* 73, 1649–1652 (2011).

61. Al-Busaidy, S.M. and Mellor, P.S. Isolation and identification of arboviruses from the Sultanate of Oman. *Epidemiol Infect* 106, 403–413 (1991).

62. Jennings, M. and Mellor, P.S. Culicoides: Biological vectors of Akabane virus. *Vet Microbiol* 21, 125–131 (1989).

63. Metselaar, D. and Robin, Y. Akabane virus isolated in Kenya. *Vet Rec* 99, 86 (1976).

64. Yamashita, N., Sakaguchi, M., Saito, H., and Chiba, S. Vectorial role of aedine mosquitoes in Akabane virus transmission (in Japanese). *J Jpn Vet Med Assoc* 52, 763–767 (1999).

65. Han, H.D. Propagation and persistent infection of Akabane virus in cultured mosquito cells. *J Vet Med Sci* 43, 689–697 (1981).

66. Kurogi, H. et al. Epizootic congenital arthrogryposis-hydranencephaly syndrome in cattle: Isolation of Akabane virus from affected fetuses. *Arch Virol* 51, 67–74 (1976).

67. Parsonson, I.M., Della-Porta, A.B., Snowdon, W.A., and Murray, M.D. Letter: Congenital abnormalities in foetal lambs after inoculation of pregnant ewes with Akabane virus. *Aust Vet J* 51, 585–586 (1975).

68. Huang, C.C., Huang, T.S., Deng, M.C., Jong, M.H., and Lin, S.Y. Natural infections of pigs with akabane virus. *Vet Microbiol* 94, 1–11 (2003).

69. Yang, D.K. et al. Serosurveillance for Japanese encephalitis, Akabane, and Aino viruses for Thoroughbred horses in Korea. *J Vet Sci* 9, 381–385 (2008).

70. Parsonson, I.M., Della-Porta, A.J., and Snowdon, W.A. Congenital abnormalities in newborn lambs after infection of pregnant sheep with Akabane virus. *Infect Immun* 15, 254–262 (1977).

71. Della-Porta, A.J. et al. Akabane disease: Isolation of the virus from naturally infected ovine foetuses. *Aust Vet J* 53, 51–52 (1977).

72. McClure, S. et al. Maturation of immunological reactivity in the fetal lamb infected with Akabane virus. *J Comp Pathol* 99, 133–143 (1988).

73. Hashiguchi, Y., Nanba, K., and Kumagai, T. Congenital abnormalities in newborn lambs following Akabane virus infection in pregnant ewes. *Natl Inst Anim Health Q (Tokyo)* 19, 1–11 (1979).

74. Haughey, K.G., Hartley, W.J., Della-Porta, A.J., and Murray, M.D. Akabane disease in sheep. *Aust Vet J* 65, 136–140 (1988).

75. Coverdale, O.R., Cybinski, D.H., and St George, T.D. Congenital abnormalities in calves associated with Akabane virus and Aino virus. *Aust Vet J* 54, 151–152 (1978).

76. Kurogi, H., Inaba, Y., Takahashi, E., Sato, K., and Satoda, K. Congenital abnormalities in newborn calves after inoculation of pregnant cows with Akabane virus. *Infect Immun* 17, 338–343 (1977).

77. Lee, J.K. et al. Encephalomyelitis associated with akabane virus infection in adult cows. *Vet Pathol* 39, 269–273 (2002).

78. Kitani, H., Yamakawa, M., and Ikeda, H. Preferential infection of neuronal and astroglia cells by Akabane virus in primary cultures of fetal bovine brain. *Vet Microbiol* 73, 269–279 (2000).

79. Konno, S., Moriwaki, M., and Nakagawa, M. Akabane disease in cattle: Congenital abnormalities caused by viral infection. Spontaneous disease. *Vet Pathol* 19, 246–266 (1982).

80. Narita, M., Inui, S., and Hashiguchi, Y. The pathogenesis of congenital encephalopathies in sheep experimentally induced by Akabane virus. *J Comp Pathol* 89, 229–240 (1979).

81. Hartley, W.J., De Saram, W.G., Della-Porta, A.J., Snowdon, W.A., and Shepherd, N.C. Pathology of congenital bovine epizootic arthrogryposis and hydranencephaly and its relationship to Akabane virus. *Aust Vet J* 53, 319–325 (1977).

82. Konno, S. and Nakagawa, M. Akabane disease in cattle: Congenital abnormalities caused by viral infection. Experimental disease. *Vet Pathol* 19, 267–279 (1982).

83. Kono, R. et al. Bovine epizootic encephalomyelitis caused by Akabane virus in southern Japan. *BMC Vet Res* 4, 20 (2008).

84. Parsonson, I.M., Della-Porta, A.J., Snowdon, W.A., and O'Halloran, M.L. The consequences of infection of cattle with Akabane virus at the time of insemination. *J Comp Pathol* 91, 611–619 (1981).

85. Ushigusa, T., Uchida, K., Murakami, T., Yamaguchi, R., and Tateyama, S. A pathologic study on ocular disorders in calves in southern Kyushu, Japan. *J Vet Med Sci* 62, 147–152 (2000).

86. Kirkland, P.D., Barry, R.D., Harper, P.A., and Zelski, R.Z. The development of Akabane virus-induced congenital abnormalities in cattle. *Vet Rec* 122, 582–586 (1988).

87. Uchida, K. et al. Detection of Akabane viral antigens in spontaneous lymphohistiocytic encephalomyelitis in cattle. *J Vet Diagn Invest* 12, 518–524 (2000).

88. Kurogi, H., Inaba, Y., Takahashi, E., Sato, K., and Goto, Y. Experimental infection of pregnant goats with Akabane virus. *Natl Inst Anim Health Q (Tokyo)* 17, 1–9 (1977).

89. Miura, Y., Hayashi, S., Ishihara, T., Inaba, Y., and Omori, T. Neutralizing antibody against Akabane virus in precolostral sera from calves with congenital arthrogryposis-hydranencephaly syndrome. *Arch Gesamte Virusforsch* 46, 377–380 (1974).

90. Sellers, R.F. and Herniman, K.A. Neutralising antibodies to Akabane virus in ruminants in Cyprus. *Trop Anim Health Prod* 13, 57–60 (1981).

91. Mohamed, M.E., Mellor, P.S., and Taylor, W.P. Akabane virus: Serological survey of antibodies in livestock in the Sudan. *Rev Elev Med Vet Pays Trop* 49, 285–288 (1996).

92. Goto, Y. et al. Hemagglutination-inhibition test applied to the study of akabane virus infection in domestic animals. *Vet Microbiol* 3, 89–99 (1978).

93. Sato, K. et al. Appearance of slow-reacting and complement-requiring neutralizing antibody in cattle infected with Akabane virus. *Vet Microbiol* 14, 183–189 (1987).

94. Goto, Y. et al. Hemolytic activity of Akabane virus. *Vet Microbiol* 4, 261–278 (1979).

95. Blacksell, S.D., Lunt, R.A., and White, J.R. Rapid identification of Australian bunyavirus isolates belonging to the Simbu serogroup using indirect ELISA formats. *J Virol Methods* 66, 123–133 (1997).

96. Ungar-Waron, H., Gluckman, A., and Trainin, Z. ELISA test for the serodiagnosis of Akabane virus infection in cattle. *Trop Anim Health Prod* 21, 205–210 (1989).

97. Kurogi, H., Inaba, Y., Takahashi, E., Sato, K., and Satoda, K. Development of Akabane virus and its immunogen in HmLu-1 cell culture. *Natl Inst Anim Health Q (Tokyo)* 17, 27–28 (1977).

98. Miah, A.H. and Spradbrow, P.B. The growth of akabane virus in chicken embryos. *Res Vet Sci* 25, 253–254 (1978).

99. Oem, J.K. et al. Bovine epizootic encephalomyelitis caused by Akabane virus infection in Korea. *J Comp Pathol* 47, 101–105 (2012).

100. Akashi, H., Onuma, S., Nagano, H., Ohta, M., and Fukutomi, T. Detection and differentiation of Aino and Akabane Simbu serogroup bunyaviruses by nested polymerase chain reaction. *Arch Virol* 144, 2101–2109 (1999).

101. Ohashi, S., Yoshida, K., Yanase, T., Kato, T., and Tsuda, T. Simultaneous detection of bovine arboviruses using single-tube multiplex reverse transcription-polymerase chain reaction. *J Virol Methods* 120, 79–85 (2004).

102. Stram, Y. et al. Detection and quantitation of akabane and aino viruses by multiplex real-time reverse-transcriptase PCR. *J Virol Methods* 116, 147–154 (2004).

103. Akashi, H., Kaku, Y., Kong, X., and Pang, H. Antigenic and genetic comparisons of Japanese and Australian Simbu serogroup viruses: Evidence for the recovery of natural virus reassortants. *Virus Res.* 50, 205–213 (1997).

104. Ogawa, Y., Kato, K., Tohya, Y., and Akashi, H. Sequence determination and functional analysis of the Akabane virus (family *Bunyaviridae*) L RNA segment. *Arch Virol* 152, 971–979 (2007).

105. Bridgen, A. and Elliott, R.M. Rescue of a segmented negative-strand RNA virus entirely from cloned complementary DNAs. *Proc Natl Acad Sci USA* 93, 15400–15404 (1996).

106. Blakqori, G. and Weber, F. Efficient cDNA-based rescue of La Crosse bunyaviruses expressing or lacking the nonstructural protein NSs. *J Virol* 79, 10420–10428 (2005).

107. Ikegami, T., Won, S., Peters, C.J., and Makino, S. Rescue of infectious rift valley fever virus entirely from cDNA, analysis of virus lacking the NSs gene, and expression of a foreign gene. *J Virol* 80, 2933–2940 (2006).

108. Ogawa, Y., Sugiura, K., Kato, K., Tohya, Y., and Akashi, H. Rescue of Akabane virus (family *Bunyaviridae*) entirely from cloned cDNAs by using RNA polymerase I. *J Gen Virol* 88, 3385–390 (2007).

109. Kurogi, H. et al. Development of inactivated vaccine for Akabane disease. *Natl Inst Anim Health Q (Tokyo)* 18, 97–108 (1978).

110. Charles, J.A. Akabane virus. *Vet Clin North Am Food Anim Pract* 10, 525–546 (1994).

111. Kurogi, H. et al. An attenuated strain of Akabane virus: A candidate for live virus vaccine. *Natl Inst Anim Health Q (Tokyo)* 19, 12–22 (1979).

112. Hashiguchi, Y., Murakami, Y., and Nanba, K. Responses of pregnant ewes inoculated with Akabane disease live virus vaccine. *Natl Inst Anim Health Q (Tokyo)* 21, 113–120 (1981).

113. Carpenter, S., Mellor, P.S., and Torr, S.J. Control techniques for Culicoides biting midges and their application in the U.K. and northwestern Palaearctic. *Med Vet Entomol* 22, 175–187 (2008).

60 Hantaviruses

Dongyou Liu

CONTENTS

60.1 Introduction 533
   60.1.1 Classification 533
   60.1.2 Morphology and Genome Organization 535
   60.1.3 Biology and Epidemiology 536
   60.1.4 Clinical Features and Pathogenesis 536
   60.1.5 Diagnosis 537
      60.1.5.1 Phenotypic Techniques 537
      60.1.5.2 Genotypic Techniques 537
   60.1.6 Treatment and Prevention 537
60.2 Methods 538
   60.2.1 Sample Preparation 538
   60.2.2 Detection Procedures 538
      60.2.2.1 Conventional RT-PCR 538
      60.2.2.2 Real-Time RT-PCR 538
60.3 Conclusion 539
References 539

60.1 INTRODUCTION

Hantaviruses are a group of segmented, negative-sense, single-stranded RNA viruses that commonly circulate in rodents without inducing overt clinical signs. However, accidental hantavirus infections in humans through inhalation of aerosolized rodent excretes may result in two serious clinical diseases namely hemorrhagic fever with renal syndrome (HFRS) and hantavirus pulmonary syndrome (HPS) or hantavirus cardiopulmonary syndrome (HCPS), depending on hantavirus species involved. Hantaviruses causing HFRS primarily occur in Europe and Asia (Old World) and employ rodents of the Murinae and Arvicolinae subfamilies for maintenance, while those causing HPS/HCPS are mostly found in the Americas (New World) and use rodents of the Sigmodontinae subfamily as reservoir. In order to implement effective control and prevention measures against HFRS and HPS/HCPS in human populations, it is vital to have the responsible agents correctly identified.

60.1.1 CLASSIFICATION

Hantaviruses are classified in the genus *Hantavirus*, family *Bunyaviridae*, which covers about 350 enveloped, tripartite, negative-sense, single-stranded RNA viruses, including 97 recognized species, 81 tentative species, and a large number of isolates with undefined taxonomic status. The family *Bunyaviridae* is divided into five genera (*Hantavirus*, *Nairovirus*, *Orthobunyavirus*, *Phlebovirus*, and *Tospovirus*), of which the genera *Orthobunyavirus*

and *Phlebovirus* utilize mosquitoes and mammals for their life cycles, with humans becoming infected via mosquito bites; the genus *Nairovirus* circulates between ticks and mammals, with humans acquiring the viruses through tick bites; the genus *Hantavirus* is maintained by cyclical transmission between persistently infected small mammals (rodents and occasionally insectivorous shrews and bats), with incidental human infections arising primarily from inhalation of aerosolized rodent excreta; and the genus *Tospovirus* involves trips (commonly known as thunderflies and corn lice) and plants in its life cycle.

Consisting of more than 80 antigenically and genetically related viruses, the genus *Hantavirus* currently includes 24 recognized species (Andes virus [ANDV], Bayou virus [BAYV], Black Creek Canal virus [BCCV], Caño Delgadito virus [CADV], Dobrava-Belgrade virus [DOBV], El Moro Canyon virus [ELMCV], Hantaan virus [HTNV], Isla Vista virus [ISLAV], Khabarovsk virus [KHAV], Laguna Negra virus [LANV], Muleshoe virus [MULV], New York virus [NYV], Prospect Hill virus [PHV], Puumala virus [PUUV], Rio Mamore virus [RIOMV], Rio Segundo virus [RIOSV], Saaremaa virus [SAAV], Sangassou virus [SANGV], Seoul virus [SEOV], Sin Nombre virus [SNV], Thailand virus [THAIV], Thottapalayam virus [TMPV], Topografov virus [TOPV], and Tula virus [TULV]) as well as a number of less well known or unassigned viruses (Table 60.1). The criteria for the demarcation of hantavirus species reflect the recently updated rule of a 10% difference in small (S) segment similarity and a 12% difference in medium (M) segment similarity based on complete amino acid sequences [1–4].

TABLE 60.1
Classification and Characteristics of the Genus *Hantavirus*

Species	Isolate	Abbreviation	Reservoir Host	Place of First Isolation	Human Disease
			Order Rodentia		
			Family Muridae		
			Subfamily Murinae		
Hantaan virus	Hantaan virus	HTNV	*Apodemus agrarius*	Korea	Severe HFRS
	Amur virus	AMRV	*Apodemus peninsulae*	Russia	HFRS
	Dabieshan virus	DBSV	*Niviventer confucianus*	China	
Seoul virus		SEOV	*Rattus norvegicus*	Korea	Moderate HFRS
Thailand virus		THAIV	*Bandicota indica*	Thailand	
Dobrava-Belgrade virus		DOBV	*Apodemus flavicollis*	Slovenia	Severe HFRS
Saaremaa virus		SAAV	*Apodemus agrarius*	Finland	Mild HFRS
Sangassou virus		SANGV	*Hylomyscus simus*	Guinea	
			Order Rodentia		
			Family Muridae		
			Subfamily Arvicolinae		
Puumala	Puumala virus	PUUV	*Myodes glareolus*	Sweden	HFRS
	Hokkaido virus	HOKV	*Myodes rutilus*	Japan, Siberia	
			Myodes rufocanus		
	Muju virus	MUJV	*Myodes regulus*	Korea	
Prospect Hill virus	Prospect Hill virus	PHV	*Microtus pennsylvanicus*	Maryland, United States	
	Bloodland Lake virus	BLLV	*Microtus ochrogaster*	Missouri, United States	
Isla Vista virus		ISLAV	*Microtus californicus*	California, United States	
Tula virus		TULV	*Microtus arvalis*	Russia	HFRS
			Microtus rossiaemeridionalis		
Khabarovsk virus		KHAV	*Microtus fortis*	Eastern Russia	
Topografov virus		TOPV	*Lemmus sibiricus*	Siberia	
			Order Rodentia		
			Family Muridae		
			Subfamily Sigmodontinae		
Sin Nombre virus	Sin Nombre virus	SNY	*Peromyscus maniculatus*	New Mexico, United States	HPS/HCPS
	Blue River virus	BRV	*Peromyscus leucopus*	Indiana, United States	
	Monongahela virus	MGLV	*Peromyscus maniculatus*	Pennsylvania, United States	
New York virus		NYV	*Peromyscus leucopus*	New York, United States	HPS/HCPS
Black Creek Canal virus		BCCV	*Sigmodon hispidus*	Florida, United States	HPS/HCPS
Bayou virus		BAYV	*Oryzomys palustris*	Louisiana, United States	HPS/HCPS
Muleshoe virus		MULV	*Sigmodon hispidus*	Texas, United States	HPS/HCPS
El Moro Canyon virus		ELMCV	*Reithrodontomys megalotis*	New Mexico, United States	
Rio Segundo virus		RIOSV	*Reithrodontomys mexicanus*	Costa Rica	
Cano Delgadito virus		CADV	*Sigmodon alstoni*	Venezuela	
Andes virus	Andes virus	ANDV	*Oligoryzomys longicaudatus*	Argentina	HPS/HCPS
	Pergamino virus	PERV	*Akodon azarae*	Argentina	
	Maciel virus	MACV	*Bolomys obscurus*	Argentina	
	Araraquara virus	ARQV	*Bolomys lasiurus*	Brazil	HPS/HCPS
	Oran virus	ORNV	*Oligoryzomys longicaudatus*	Argentina	HPS/HCPS
	Bermejo virus	BMJV	*Oligoryzomys chacoensis*	Argentina	HPS/HCPS
	Lechiguanas virus	LECV	*Oligoryzomys flavescens*	Argentina	HPS/HCPS
Laguna Negra virus		LANV	*Calomys laucha*	Paraguay	HPS/HCPS
Rio Mamore virus		RIOMV	*Oligoryzomys microtis*	Bolivia	
			Order Insectivora		
			Family Soricidae		
Thottapalayam virus		TMPV	*Suncus murinus*	India, Southeast Asia	

(Continued)

TABLE 60.1 (*Continued*)
Classification and Characteristics of the Genus *Hantavirus*

Species	Isolate	Abbreviation	Reservoir Host	Place of First Isolation	Human Disease
Tanganya virus (unassigned)		TGNV	*Crocidura theresae*	Guinea	
Azagny virus (unassigned)		AZGV	*Crocidura obscurior*	Guinea	
Imjin virus (unassigned)		MJNV	*Crocidura lasiura*	Korea	
			Order Chiroptera		
			Family Hipposideridae		
Xuan Son virus (unassigned)			*Hipposideros pomona*	Vietnam	
			Order Chiroptera		
			Family Nycteridae		
Magboi virus (unassigned)		MGBV	*Nycteris hispida*	Sierra Leone	
			Order Chiroptera		
			Family Vespertilionidae		
Mouyassué virus (unassigned)			*Neoromicia nanus*	Côte d'Ivoire	

Sources: Klein, S.L. and Calisher, C.H., *Curr. Top. Microbiol. Immunol.*, 315, 217, 2007; Firth, C. et al., *J. Virol.*, 86, 13756, 2012; International Committee on Taxonomy of Viruses (ICTV), http://www.ictvonline.org/virusTaxonomy.asp/.
HFRS, hemorrhagic fever with renal syndrome; HPS, hantavirus pulmonary syndrome; HCPS, hantavirus cardiopulmonary syndrome.

According to the types of rodent hosts involved in transmission, hantaviruses are separated into *Murinae group* (e.g., HTNV, SEOV, Amur virus [AMRV], DOBV, and SAAV, as they are transmitted by Murinae rodents [Old World mice and rats]), *Arvicolinae group* (e.g., PUUV and TULV, as they are transmitted by Arvicolinae rodents [voles]), and *Sigmodontinae group* (e.g., SNV, NYV, BCCV, BAYV, ANDV, and LANV, as they are transmitted by Sigmodontinae rodents [New World mice and rats]) (Table 60.1). For convenience, hantaviruses belonging to the Murinae and Arvicolinae groups are referred to as the Old World hantaviruses, whereas those belonging to the Sigmodontinae group are referred to as the New World hantaviruses.

Interestingly, DOBV, SAAV, HTNV, AMRV, SEOV, PUUV, and TULV (so-called Old World hantaviruses) are linked to HFRS in Europe and Asia, responsible for 150,000–200,000 cases per year and mortality rates of up to 12%. On the other hand, ANDV, BAYV, BCCV, LANV, NYV, and SNV (so-called New World hantaviruses) are implicated in HPS or HCPS in the Americas, involving >200 cases per year in South America and mortality rate of around 40% [5,6].

Phylogenetic analyses have shown that HTNV contains nine close-related clades (designated subtypes HTN 1–9); and SEOV from different geographic locations forms five clades (designated subtypes SEO 1–5) [7].

PUUV is differentiated into eight lineages in Eurasia: (1) the Central European lineage, (2) the Alpe-Adrian lineage, (3) the Latvian lineage, (4) the Danish lineage, (5) the South Scandinavian lineage, (6) the North Scandinavian lineage, (7) the Finland lineage, and (8) the Russian lineage [8].

ANDV consists of three clades: *Andes clade* includes 12 viruses, which are further differentiated into three well-supported groups: (1) Castelo dos Sonhos virus (CASV), CASV-2, and Tunari virus; (2) Pergamino virus, Maciel virus, Araraquara virus, and Paranoa virus; and (3) Oran virus, Bermejo virus (BMJV), BMJV-Neembucu virus, Lechiguanas

virus, and Andes Central Plata virus. These viruses are found in Argentina, Bolivia, Brazil, Chile, Paraguay, and Uruguay and carried by rodents of genera *Akodon*, *Necromys*, and *Oligoryzomys*. *Laguna Negra clade* contains LANV (causing HPS in Paraguay, Bolivia, and Brazil and carried by *Calomys laucha*) and LANV-2 (causing HPS in Brazil); *Rio Mamore clade* consists of Alto Paraguay virus (Paraguay), Anajatuba virus (ANJV; Brazil), Maripa virus (MARV; French Guiana), Rio Mearim virus (Brazil), Rio Mamore virus (RIOMV; Bolivia and Peru), and RIOMV-3 and RIOMV-4 (Brazil). Of these, only ANJV and MARV are associated with HPS. Within the Rio Mamore clade, only RIOMV is currently recognized as a species by the International Committee on Taxonomy of Viruses. Other viruses within this clade may be unique genotypes of RIOMV. The reservoir of RIOMV in Peru, Bolivia, and Amazonas, Brazil (RIOMV-3), is *Oligoryzomys microtis*, while the host of RIOMV-4 is a related species similar to *Oligoryzomys* sp. RR-2010a [9–11].

60.1.2 Morphology and Genome Organization

Hantaviruses are enveloped, cytoplasmic viruses with a spherical to pleomorphic appearance and a diameter of 80–120 nm. Viewed by electron microscopy, negatively stained hantaviruses show a grid-like structure composed of surface projections (spikes) of 8 nm, which are surrounded by a prominent fringe embedded in a lipid bilayer envelope of 5 nm thickness. These spikes are made up of glycoprotein Gn and Gc heterodimers (encoded by the M segment).

Inside the envelope exist three filamentous, helical nucleocapsids, measuring 200–3000 nm in length and 2–2.5 nm in width. These nucleocapsids are made up of multicopies of nucleocapsid protein N encapsidating each of three segments of antisense (and sometimes ambisense), single-stranded RNA, designated S (small), M (medium), and L (large), amounting to about 12,000 nucleotides (nt) in total.

The S segment (1696–2083 nt in length) encodes the nucleocapsid (N) protein and possible a NSs protein in some isolates. The M segment (3613–3707 nt in length) encodes a polyprotein that is cotranslationally cleaved to yield the envelope glycoproteins Gn (formerly G1) and Gc (formerly G2). The L segment (6530–6550 nt in length) encodes the L protein, which is a RNA-dependent RNA polymerase with replicase and transcriptase activities.

Each of the three segments possesses short noncoding regions (NCR) at the 5′ and 3′ ends. The NCR at the 5′ end is 37–51 nt for all three segments, and the NCR at the 3′ end is 370–730 nt in S segment, 168–229 nt in M segment, and 38–43 nt in L segment. A consensus 3′-terminal nucleotide sequence (AUCAUCAUC) is found in the 3′ NCR, which is complementary to the 5′ terminal sequence. The 5′ and 3′ NCRs form panhandle structures and play a role in replication and encapsidation facilitated by binding with the viral nucleocapsid (N) protein [12].

60.1.3 Biology and Epidemiology

Hantaviruses commonly occur in rodents of subfamilies Arvicolinae, Murinae, and Sigmodontinae in the family Muridae (order Rodentia), which are widely distributed throughout the world [13–20]. In Europe and Asia (the Old World), hantaviruses are present in rodents of the Murinae and Arvicolinae subfamilies, while in the Americas (the New World), hantaviruses exist in rodents of the Sigmodontinae subfamily. Each hantavirus has one or possibly a few specific rodent hosts (Table 60.1). In addition, a hantavirus with unknown human pathogenic potential, Thottapalayam virus, appears in insectivore Asian house shrew (*Suncus murinus* of family Soricidae, order Insectivora), and several unassigned hantaviruses are found in insectivorous shrews and bats as well (Table 60.1) [21–28].

As reservoir hosts for hantaviruses, rodents may become infected for weeks to years without displaying overt clinical diseases, with viruses being concentrated and secreted via saliva, urine, and feces. Under experimental conditions, hantaviruses are generally eliminated during the first 2 months of the life of a rodent, but the elimination of viruses in the wild can take up about 8 months, as shown in bank voles.

Hantaviruses can survive for longer periods if protected by organic material. However, they are susceptible to drying in the environment and are vulnerable to disinfectants (e.g., 1% sodium hypochlorite, 2% glutaraldehyde, 70% ethanol, and detergents). Further, hantaviruses are susceptible to acid (pH 5) and heat (60°C for over 30 min).

Transmission of hantaviruses within rodents is facilitated by aerosols (particularly aerosolized urine), virus-containing saliva and feces, or bites during fighting. Humans generally acquire hantaviruses through inhalation of aerosolized excreta (urine, feces, and saliva) from infected rodents. From the respiratory tract, hantaviruses quickly move to the pulmonary epithelium and endothelium and then disseminate to other tissues and cell types (such as lungs).

Continuing expansion in agricultural or mining activities that disrupt predator–prey relationships has contributed to dramatic rise in rodent populations, and frequent leisure travel and excursions increase human exposure to hantaviruses. Together, these activities are reasons behind the re/emergence of hantavirus outbreaks in human populations [29,30].

Hantavirus infection tends to peak in spring and fall, and occupations at high risk of hantavirus infection include rodent control workers, farmers, forestry workers, hunters, military personnel, field biologists, and laboratory technicians during close contact with infected rodents in disease-endemic areas. In addition, camping or staying in rodent-infested cabins may also increase the risk. A few minutes of exposure to aerosolized virus is sufficient for the infection to establish. Furthermore, broken skin, the conjunctiva and other mucous membranes, rodent bites, and possibly ingestion may be other alternative route of hantavirus entry [31].

60.1.4 Clinical Features and Pathogenesis

In rodents, hantaviruses cause chronic, persistent, and sometimes lifelong infections with relatively few adverse signs, although viruses may be discharged through saliva, urine, and feces over a long period of time and are typically detected in lung and kidney tissue as well as the heart, spleen, liver, and nervous system during the acute phase of infection [32].

However, studies on wild rodents seropositive for the SNV and NYV revealed a high prevalence of pulmonary edema and periportal hepatitis, suggestive of an intermediate form of HPS. In addition, ANDV-infected hamsters showed the sudden and rapid onset of severe respiratory distress similar to HPS observed in humans [33]. In experimentally infected infant rats with SEOV, fatal meningoencephalitis was observed [34].

Overall, an ostensive proinflammatory response against hantaviruses is generally absent in rodent reservoir host. This may be attributable to the fact that (1) hantaviruses tend to cause a nonlytic infection, which does not attract as much attention from host immune surveillance network as in a lytic infection; (2) hantaviruses have the ability to replicate in immune cells such as macrophages, monocytes, and T lymphocytes; and (3) hantaviruses downregulate the cyclic phases of viral replication—the existence of a subtle balance between an efficient immune response to limit viral replication and a strong uncontrollable response that results in fatal illness [35].

In humans, hantaviruses are responsible for two major clinical syndromes: HFRS due to the Old World hantaviruses (such as HTNV, AMRV, SEOV, DOBV, SAAV, PUUV, and TULV), and HPS or HCPS due to the New World hantaviruses (such as ANDV, BAYV, BCCV, LANV, NYV, and SNV) [36].

HFRS, formerly known as Korean hemorrhagic fever, epidemic hemorrhagic fever, or nephropathia epidemica (a mild form of HFRS often caused by PUUV), was first noticed in Korea in 1950s and characterized in 1978. With an incubation period of 1–6 weeks, HFRS demonstrates several stages of progression (febrile, hypotensive/proteinuric, oliguric, diuretic, and convalescent). The initial clinical signs of HFRS are sudden and may include fever, chills, prostration, headache, backache, and blurred vision, along with gastrointestinal signs (e.g., nausea, vomiting, and abdominal pain) that mimic appendicitis. Additionally, some patients may have a flushed face and conjunctivae, or a petechial rash on the palate or trunk, and photophobia (the febrile or prodromal stage). Later symptoms may include low blood pressure (hypotension), acute shock, vascular leakage, disseminated intravascular coagulation, and acute kidney failure, leading to severe fluid overload (the proteinuric stage) and death from acute shock. In general, HTNV, DOBV, and AMRV cause most severe symptoms (with mortality rates of 10%–15% for HTNV and 7%–12% for DOBV), SEOV results in more moderate disease (with mortality rate of 1%–5%), and PUUV induces typically mild disease (with mortality rate of 0.1%–0.4%). Complete recovery takes weeks or months [31,37].

HPS or HCPS came to prominence in 1993 when a novel hantavirus named Sin Nombre virus (SNV) was described from an outbreak of a respiratory distress syndrome with high case-fatality rate in Southwestern United States. This was followed by the identification of several other HCPS-causing hantaviruses (i.e., ANDV, BCCV, LANV, Juquitiba virus, and Araraquara virus) in North and South America in subsequent years. Since then, HPS/HCPS has been recognized as an emergent disease in the Americas and a significant public health problem in South America [31,38].

The clinical symptoms of HPS/HCPS range from a mild course disease without sequelae to fulminant respiratory distress with high lethality (up to 50% for SNV). At the early (prodrome) phase, the signs of HPS/HCPS may resemble those of other infectious diseases such as influenza and may include fever, myalgia, respiratory abnormality, and headache. However, a lack of rhinorrhea and sore throat may help differentiate HCPS from illness caused by influenza. Often, the patients develop acute noncardiac pulmonary edema and hypotension within 2–15 days, with bilateral infiltrates and occasionally pleural effusions. Marked conjunctival injections, facial flushing, and petechiae are also noted. Laboratory results such as thrombocytopenia, hemoconcentration, neutrophilic leukocytosis, and circulating immunoblasts in patients in living risk regions are additional indicator for HPS/HCPS. Patients often show signs of cardiopulmonary involvement by the time they are hospitalized due to the rapid disease progression; nearly 20% of the individuals may succumb to an acute respiratory distress syndrome before clinical and laboratory investigations are completed. Patients surviving the acute phase of the disease tend to recover normally within 5–7 days without any noticeable sequelae [31,39].

The clinical variability of hantavirus infections may be related to distinct factors, such as viral strain, size of inoculums inhaled, or genetic markers of the host. Hantaviruses appear to target endothelial cells, and viral entry is mediated by β_3-integrins [40]. Hantavirus infection of endothelial cells in the kidneys and lungs leads to increased vascular permeability and subsequent dysfunction mediated by immune responses underscores the distinct clinical manifestations of HFRS and HPS/HCPS [41,42]. As an immunopathologic disease with no evident cytopathic effects in hantavirus-infected human cells, patients with HPS/HCPS often show elevated levels of plasmatic IFN-γ, IL-2, IL-4, IL-6 and TNF-α, and presence of T lymphocyte (CD8+) and other cells producing proinflammatory cytokines in lung tissues [43,44].

60.1.5 Diagnosis

Hantaviruses are responsible for causing HFRS in Eurasia and hantavirus pulmonary syndrome (HPS)/hantavirus cardiopulmonary syndrome (HCPS) in the Americas. As clinical symptoms of these diseases are similar to those caused by influenza and other viral infections, laboratory methods (based on phenotypic and genotypic principles) are required for their correct diagnosis [37,45].

60.1.5.1 Phenotypic Techniques

Hantaviruses may be grown in Vero E6 cells (African green monkey kidney cell line) for initial identification by electron microscopy [46]. Subsequent examination by plaque-reduction neutralization tests provides a definitive diagnosis of hantavirus infection.

In case that in vitro recovery of relevant virus is unsuccessful, serological confirmation of specific IgM in acute phase sera or a rise in IgG titer is diagnostic. The commonly used serological tests include immunofluorescent antibody test, enzyme-linked immunosorbent assays, Western blot (immunoblotting), virus neutralization, and immunochromatographic tests. However, serological tests are only useful after seroconversion, and IgG response may take weeks to develop. In addition, these tests have limited specificity due to the high similarity of hantavirus proteins [47,48].

60.1.5.2 Genotypic Techniques

As an alternative to serological tests, nucleic acid–based procedures offer a more specific and sensitive approach for hantavirus detection, differentiation, and quantification in an early state of infection, before the onset of clinical symptoms [49]. Typically, total RNA is purified from whole blood or serum during the acute phase of infection or from autopsy tissue samples. Primers derived from the conserved regions of hantaviruses are applied in reverse transcription polymerase chain reaction (RT-PCR). The incorporation of SYBR Green dye or fluorescent probes into RT-PCR enables real-time quantification of hantavirus RNA from clinical specimens [48,50–54].

60.1.6 Treatment and Prevention

For patients with HFRS and HPS/HCPS, several treatment/prophylaxis approaches may be attempted. If used early, antiviral ribavirin (1-β-dribofuranosyl-1,2,4-triazole-3-carboxamide) may help reduce renal insufficiency. Administration of human neutralizing antibodies during the acute phase is another beneficial treatment and/or prophylaxis for hantaviral infections. In addition, provision of supportive treatment enhances natural recovery. Patients with suspected hantavirus infection may be given oxygen and mechanical ventilation support to relieve breathing difficulty during the acute pulmonary stage [55].

Prevention of hantavirus infection in humans should focus on (1) limiting exposure to rodent saliva, excreta, and bites;

(2) control of rats and mice in areas frequented by humans; (3) disposing of rodent nests, sealing any cracks and holes in homes where mice or rats may get in; and (4) setting up traps, laying down rodenticides or using natural predators such as cats at home [55].

When handling infected wild mice and rats, protective apparels (e.g., gloves, goggles, rubber boots or disposable shoe covers, coveralls or gown, and respirator masks equipped with N-100 filters) should be worn. Decontamination of infected areas is accomplished by soaking with a 10% (v/v) solution of household bleach [55].

A number of vaccines against hantavirus infections in humans are being developed or in early stages of clinical trials. For example, a vaccinia virus–vectored vaccine containing the M and S genes of HTNV was shown to elicit neutralizing antibodies. A vesicular stomatitis virus expressing hantavirus glycoproteins protects mice against HTNV infection. DNA vaccines containing HTNV, SEOV, PUUV, or ANDV M gene sequences stimulate high-titer neutralizing antibodies against hantavirus challenges [56–58].

60.2 METHODS

60.2.1 Sample Preparation

Considering the dangers posed by hantavirus-containing aerosols, tissues, and serum specimens, laboratory testing of hantaviruses needs extra caution, and all handlings should be preferably in biosafety cabinets with biosafety level 3 (BSL-3) or 4 (BSL-4) rating.

Total RNA is extracted from sera tissues and other specimens by using the TRIzol reagent (Invitrogen) or column-based kits (e.g., QIAamp Viral Mini Kit). The extracted RNA may be quantitated using spectrophotometer and stored at −80°C until used.

60.2.2 Detection Procedures

60.2.2.1 Conventional RT-PCR

Klempa et al. [21] developed a nested RT-PCR assay to detect currently known and possible novel members of the genus *Hantavirus*, with degenerated primers from the highly conserved L segment. Specifically, HAN-L-F1 and HAN-L-R1 are used for the first-round PCR, while primers HAN-L-F2 and HAN-L-R2 are used for the second-round PCR (Table 60.2).

Total RNA is extracted from rodent blood (preserved in liquid nitrogen) using Blood RNA Kit (Peqlab) and reverse transcribed with random hexamers.

The 412 nt sequence of the first PCR product is determined by amplification, cloning, and sequencing of overlapping fragments generated by two seminested PCR.

Note: To further verify the results, additional fragments of 837 nt and 694 nt are amplified from the S and M segments, respectively, by PCR and sequenced. The resulting sequences are subjected to maximum likelihood and neighbor-joining phylogenetic analyses.

TABLE 60.2
Primers for Nested Polymerase Chain Reaction Detection of Hantaviruses

PCR	Primer	Sequence (5′–3′)	Product (bp)
First round	HAN-L-F1	ATGTAYGTBAGTGCWGATGC′	412
	HAN-L-R1	AACCADTCWGTYCCRTCATC	
Second round	HAN-L-F2	TGCWGATGCHACIAARTGGTC	
	HAN-L-R2	GCRTCRTCWGARTGRTGDGCAA	

60.2.2.2 Real-Time RT-PCR

60.2.2.2.1 Pan-Hantavirus Detection

Mohamed et al. [59] described a SYBR Green–based RT-PCR for sensitive detection (<100 copies) of hantavirus RNA of different clades, with six degenerated PanHanta primers from highly conserved regions (2940 and 3481 nt) of the L segment (Table 60.3).

At nonconserved positions within the proposed primer sequences, where three or four different nucleotides are found, deoxyinosine are inserted to reduce the degeneracy of the PCR primers.

RT-PCR mixture (25 μL) is made up of 1 μL cDNA template, 25 μM of each primer, and 12.5 μL of KAPA master mix (KAPA SYBR® FAST qPCR Kit, KAPA Biosystems).

Amplification and detection are carried out in ABI Prism 7900HT Sequence Detection System 2.0 (Applied Biosystems), with the following cycling conditions: an initial denaturation step at 95°C for 3 min, 35 cycles of a denaturation step at 95°C for 15 s, and an annealing temperature of 56°C for 15 s.

60.2.2.2.2 Specific Detection of DOBV, HTNV, PUUV, and SEOV

Aitichou et al. [60] reported a one-step real-time RT-PCR for specific identification of DOBV, HTNV, PUUV, and SEOV, with primers from the S segment, which recognize all four viruses, and virus-specific TaqMan probes containing the fluorescent dyes fluorescein aminohexyl (FAM) at the 5′ terminus and tetramethyrhodamine (TAMRA) at the 3′ terminus (Table 60.4).

TABLE 60.3
Identity of Pan-Hantavirus Primers from Large Segment

Primer	Sequence (5′–3′)	T_m
PanHanta-F1	ATGTATGTIAGTGCWGATGC	54.3–56.4
PanHanta-R1	ACCAITCWGWICCATCAYC	53.0–59.5
PanHanta-F2	TGCWGATGCIACRAAATGGTC	57.5–61.2
PanHanta-R2	GCATCATCWGARTGATGIGCAA	58.4–62.1
PH-F	TYTTTGARTTTGCICATCAYTCW GATGATGC	65.6–70.9
PH-R	TCATGYAIATTRAACATICTYTTCCACA	59.9–67.2

T_m, melting temperature; I or dI, deoxyinosine.

TABLE 60.4
Identity of Primers and TaqMan Probes

Primer/Probe	Sequences (5′–3′)	Position
HANTAV1U	GWG-GVC-ARA-CAG-CWG-AYT	374-391
HANTAV1L	TCC-WGG-TGT-AAD-YTC-HTC-WGC	604-624
DOP1P	FAM-AAG CAT TGT GAT CTA CCT GAC ATC A-TAMRA	395-419
HTN1P	FAM-AGC ATC ATC GTC TAT CTT ACA TCC-TAMRA	397-420
PUU1P	FAM-TTC ACA ATT CCT ATC ATT TTG AAG GC-TAMRA	427-452
SEO1P	FAM-CCA TAA TTG TCT ATC TGA CAT CA-TAMRA	404-426

RT-PCR mixture (30 µL) is made up of 0.2 mM dNTP, 4 mM MgSO$_4$, 0.6 µL of RT/Platinum Taq mix (SuperScript™ III One-Step RT-PCR System with Platinum® Taq High Fidelity, Invitrogen), 0.4 µM of each primer, 0.25 µM of TaqMan probe, and various amounts of viral RNA. Separate RT-PCR mixture is prepared for each virus.

The RT and PCR amplification are carried out in a single tube on Smart Cycler (Cepheid) with 1× 50°C for 30 min, 1× 95°C for 2 min, and 45 cycles of 95°C for 15 s, 52°C for 30 s, and 72°C for 1 min.

Note: The RT-PCR assays have detection limits of 25, 25, 25, and 12.5 plaque-forming units for DOBV, HTNV, PUUV, and SEOV, respectively. The DOBV, HTNV, and SEOV assays demonstrate a specificity of 100%, while the PUUV assay shows a specificity of 98%. Therefore, these assays offer rapid, sensitive, and specific approaches for identification and differentiation of four Old World hantaviruses.

60.2.2.2.3 Specific Detection of SNY

Botten et al. [61] reported a real-time RT-PCR targeting the S segment for specific identification of SNV. Use of sense primer SNV S-179f (5′-GCAGACGGGC AGCTGTG-3′) and antisense primer SNV S-245r (5′-AGATCAGCCAGTTCCCGCT-3′) allows amplification of a 66 bp fragment from the SNV S segment, which is detected with a dual-labeled (FAM, TAMRA) fluorescent probe (positive-sense probe starting at nt 198, 5′-TGCATTGGAGACCAAACTCGGAGAACTT-3′).

RT-PCR mixture (50 µL) is made up of 45 µL of master mix (One-Step RT-PCR Master Mix Reagent Kit, Applied Biosystems) containing 0.2 µM of each primers and probe and 5 µL of extracted RNA.

Amplification and detection involve RT (50°C for 30 min), initial denaturation (95°C for 10 min), and 40 cycles of 95°C for 10 s, 50°C for 10 s, and 72°C for 30 s. Data acquisition occurs at the end of the annealing stage (50°C for 10 s) of each amplification cycle.

60.2.2.2.4 Specific Detection of ANDV

Rowe and Pekosz [46] employed a real-time RT-PCR for quantitative detection of ANDV. The primers (forward, 5′-GGAAAACATCACAGCACACGAA-3′, nt 66–87, and reverse, 5′-CTGCCTTCTCGGCATCCTT-3′, nt 118–136) and probe (5′-FAM-AACAGCTCGTGACTGCTCGGCAAAA-TAMRA-3′, nt 89–113) are derived from the ANDV S segment, with the probe containing FAM/TAMRA labels (FAM is 6-carboxyfluorescein; TAMRA is 6-carboxytetramethylrhodamine).

RT-PCR mix is prepared with the ABI TaqMan EZ RT-PCR kit according to the manufacturer's protocol (Applied Biosystems).

Amplification and detection are carried out on ABI 7000 Real-Time PCR System (Applied Biosystems) with the following conditions: RT at 60°C for 30 min, denaturation at 95°C for 2 min, and 40 cycles of denaturation at 95°C for 30 s and annealing and extension at 60°C for 1 min.

60.3 CONCLUSION

Commonly present in rodents of the Murinae, Arvicolinae, and Sigmodontinae subfamilies, hantaviruses are implicated in two clinical diseases in humans: hemorrhagic fever with renal syndrome (HFRS) in Europe and Asia and hantavirus pulmonary syndrome (HPS) or hantavirus cardiopulmonary syndrome (HCPS) in the Americas. Given the existence of a large number of hantavirus species and isolates and the severity of HFRS and HPS/HCPS, laboratory methods are indispensible for diagnosis of hantavirus infections.

Although phenotypic techniques such as virus isolation and serological assays are useful for hantavirus identification, they generally have low sensitivity, poor specificity, and slow result availability. The recent development and application of nucleic acid detection technologies such as RT-PCR have made rapid, sensitive, and specific detection and differentiation of hantaviruses a reality. It is envisaged that these high-performing molecular diagnostic methods will play a vital role in the epidemiological monitoring of hantavirus outbreaks and contribute to the improved control and prevention strategies against these deadly pathogens.

REFERENCES

1. Maes P et al. A proposal for new criteria for the classification of hantaviruses, based on S and M segment protein sequences. *Infect Genet Evol.* 2009;9(5):813–820.
2. Klempa B et al. Sangassou virus, the first hantavirus isolate from Africa, displays genetic and functional properties distinct from those of other murinae-associated hantaviruses. *J Virol.* 2012;86(7):3819–3827.
3. Briese T, Calisher CH, Higgs S. Viruses of the family Bunyaviridae: Are all available isolates reassortants? *Virology.* 2013;446(1–2):207–216.
4. International Committee on Taxonomy of Viruses (ICTV). http://www.ictvonline.org/virusTaxonomy.asp/. Accessed on March 31, 2015.
5. Bi Z, Formenty PB, Roth CE. Hantavirus infection: A review and global update. *J Infect Dev Ctries.* 2008;2(1):3–23.
6. Krüger DH, Figueiredo LT, Song JW, Klempa B. Hantaviruses—Globally emerging pathogens. *J Clin Virol.* 2015;64:128–136.

7. Wang H et al. Genetic diversity of hantaviruses isolated in China and characterization of novel hantaviruses isolated from *Niviventer confucianus* and *Rattus rattus*. *Virology*. 2000;278(2):332–345.

8. Ali HS et al. First molecular evidence for Puumala hantavirus in Poland. *Viruses*. 2014;6(1):340–353.

9. Medina RA et al. Ecology, genetic diversity, and phylogeographic structure of Andes virus in humans and rodents in Chile. *J Virol*. 2009;83:2446–2459.

10. Firth C et al. Diversity and distribution of hantaviruses in South America. *J Virol*. 2012;86(24):13756–13766.

11. de Oliveira RC et al. Hantavirus reservoirs: Current status with an emphasis on data from Brazil. *Viruses*. 2014;6(5):1929–1973.

12. Yoshimatsu K, Arikawa J. Antigenic properties of N protein of hantavirus. *Viruses*. 2014;6(8):3097–3109.

13. Cross RW et al. Old World hantaviruses in rodents in New Orleans, Louisiana. *Am J Trop Med Hyg*. 2014;90(5):897–901.

14. Dupinay T et al. Detection and genetic characterization of Seoul virus from commensal brown rats in France. *Virol J*. 2014;11:32.

15. Esteve-Gassent MD et al. Pathogenic landscape of transboundary zoonotic diseases in the Mexico-US border along the Rio Grande. *Front Public Health*. 2014;2:177.

16. Montoya-Ruiz C, Diaz FJ, Rodas JD. Recent evidence of hantavirus circulation in the American tropic. *Viruses*. 2014;6(3):1274–1293.

17. Yanagihara R, Gu SH, Arai S, Kang HJ, Song JW. Hantaviruses: Rediscovery and new beginnings. *Virus Res*. 2014;187:6–14.

18. Yu XJ, Tesh RB. The role of mites in the transmission and maintenance of Hantaan virus (Hantavirus: Bunyaviridae). *J Infect Dis*. 2014;210(11):1693–1699.

19. Klein TA et al. Hantaan virus surveillance targeting small mammals at nightmare range, a high elevation military training area, Gyeonggi Province, Republic of Korea. *PLoS One*. 2015;10(4):e0118483.

20. Verner-Carlsson J et al. First evidence of Seoul hantavirus in the wild rat population in the Netherlands. *Infect Ecol Epidemiol*. 2015;5:27215.

21. Klempa B et al. Hantavirus in African wood mouse, Guinea. *Emerg Infect Dis*. 2006;12:838.

22. Klempa B et al. Novel hantavirus sequences in shrew, Guinea. *Emerg Infect Dis*. 2007;13:520–522.

23. Song JW et al. Characterization of Imjin virus, a newly isolated hantavirus from the Ussuri white-toothed shrew (*Crocidura lasiura*). *J Virol*. 2009;83(12):6184–6191.

24. Kang HJ, Kadjo B, Dubey S, Jacquet F, Yanagihara R. Molecular evolution of Azagny virus, a newfound hantavirus harbored by the West African pygmy shrew (*Crocidura obscurior*) in Cote d'Ivoire. *Virol J*. 2011;8:373.

25. Sumibcay L et al. Divergent lineage of a novel hantavirus in the banana pipistrelle (*Neoromicia nanus*) in Côte d'Ivoire. *Virol J*. 2012;9:34.

26. Weiss S et al. Hantavirus in bat, Sierra Leone. *Emerg Infect Dis*. 2012;18:159–161.

27. Arai S et al. Novel bat-borne hantavirus, Vietnam. *Emerg Infect Dis*. 2013;19(7):1159–1161.

28. Gu SH et al. Molecular phylogeny of hantaviruses harbored by insectivorous bats in Côte d'Ivoire and Vietnam. *Viruses*. 2014;6(5):1897–1910.

29. Essbauer S et al. A new Puumala hantavirus subtype in rodents associated with an outbreak of Nephropathia epidemica in South-East Germany in 2004. *Epidemiol Infect*. 2006;134(6):1333–1344.

30. Holmes EC, Zhang YZ. The evolution and emergence of hantaviruses. *Curr Opin Virol*. 2015;10:27–33.

31. Manigold T, Vial P. Human hantavirus infections: Epidemiology, clinical features, pathogenesis and immunology. *Swiss Med Wkly*. 2014;144:w13937.

32. Klein SL, Calisher CH. Emergence and persistence of hantaviruses. *Curr Top Microbiol Immunol*. 2007;315:217–252.

33. Safronetz D, Ebihara H, Feldmann H, Hooper JW. The Syrian hamster model of hantavirus pulmonary syndrome. *Antiviral Res*. 2012;95(3):282–292.

34. Badole SL, Yadav PD, Patil DR, Mourya DT. Animal models for some important RNA viruses of public health concern in SEARO countries: Viral hemorrhagic fever. *J Vector Borne Dis*. 2015;52(1):1–10.

35. Schountz T, Prescott J. Hantavirus immunology of rodent reservoirs: Current status and future directions. *Viruses*. 2014;6(3):1317–1335.

36. Cruz CD et al. Novel strain of Andes virus associated with fatal human infection, central Bolivia. *Emerg Infect Dis*. 2012;18(5):750–757.

37. Sargianou M et al. Hantavirus infections for the clinician: From case presentation to diagnosis and treatment. *Crit Rev Microbiol*. 2012;38(4):317–329.

38. Dvorscak L, Czuchlewski DR. Successful triage of suspected hantavirus cardiopulmonary syndrome by peripheral blood smear review: A decade of experience in an endemic region. *Am J Clin Pathol*. 2014;142(2):196–201.

39. Pinto Junior VL, Hamidad AM, Albuquerque Filho Dde O, dos Santos VM. Twenty years of hantavirus pulmonary syndrome in Brazil: A review of epidemiological and clinical aspects. *J Infect Dev Ctries*. 2014;8(2):137–142.

40. Cifuentes-Muñoz N, Salazar-Quiroz N, Tischler ND. Hantavirus Gn and Gc envelope glycoproteins: Key structural units for virus cell entry and virus assembly. *Viruses*. 2014;6(4):1801–1822.

41. Schönrich G, Krüger DH, Raftery MJ. Hantavirus-induced disruption of the endothelial barrier: Neutrophils are on the payroll. *Front Microbiol*. 2015;6:222.

42. Hepojoki J, Vaheri A, Strandin T. The fundamental role of endothelial cells in hantavirus pathogenesis. *Front Microbiol*. 2014;5:727.

43. Terajima M, Ennis FA. T cells and pathogenesis of hantavirus cardiopulmonary syndrome and hemorrhagic fever with renal syndrome. *Viruses*. 2011;3(7):1059–1073.

44. Charbonnel N et al. Immunogenetic factors affecting susceptibility of humans and rodents to hantaviruses and the clinical course of hantaviral disease in humans. *Viruses*. 2014;6(5):2214–2241.

45. Vaheri A, Vapalahti O, Plyusnin A. How to diagnose hantavirus infections and detect them in rodents and insectivores. *Rev Med Virol*. 2008;18(4):277–288.

46. Rowe RK, Pekosz A. Bidirectional virus secretion and nonciliated cell tropism following Andes virus infection of primary airway epithelial cell cultures. *J Virol*. 2006;80(3):1087–1097.

47. Latus J et al. Detection of Puumala hantavirus antigen in human intestine during acute hantavirus infection. *PLoS One*. 2014;9(5):e98397.

48. Oldal M et al. Identification of hantavirus infection by Western blot assay and TaqMan PCR in patients hospitalized with acute kidney injury. *Diagn Microbiol Infect Dis.* 2014;79(2):166–170.

49. Hu D et al. Development of reverse transcription loop-mediated isothermal amplification assays to detect Hantaan virus and Seoul virus. *J Virol Methods.* 2015;221:68–73.

50. Trombley AR et al. Comprehensive panel of real-time TaqMan polymerase chain reaction assays for detection and absolute quantification of filoviruses, arenaviruses, and New World hantaviruses. *Am J Trop Med Hyg.* 2010;82(5):954–960.

51. Jiang W et al. Quantification of Hantaan virus with a SYBR green I-based one-step qRT-PCR assay. *PLoS One.* 2013;8(11):e81525. Erratum in: *PLoS One.* 2013;8(12).

52. Jiang W et al. Development of a SYBR Green I based one-step real-time PCR assay for the detection of Hantaan virus. *J Virol Methods.* 2014;196:145–151.

53. Li Z et al. A two-tube multiplex real-time RT-PCR assay for the detection of four hemorrhagic fever viruses: Severe fever with thrombocytopenia syndrome virus, Hantaan virus, Seoul virus, and dengue virus. *Arch Virol.* 2013;158(9):1857–1863.

54. Mossbrugger I, Felder E, Gramsamer B, Wölfel R. EvaGreen based real-time RT-PCR assay for broad-range detection of hantaviruses in the field. *J Clin Virol.* 2013;58(1):334–335.

55. Maes P, Clement J, Gavrilovskaya I, Van Ranst M. Hantaviruses: Immunology, treatment, and prevention. *Viral Immunol.* 2004;17(4):481–497.

56. Maes P, Clement J, Van Ranst M. Recent approaches in hantavirus vaccine development. *Expert Rev Vaccines.* 2009;8(1):67–76.

57. Schmaljohn C. Vaccines for hantaviruses. *Vaccine.* 2009;27(Suppl. 4):D61–D64.

58. Krüger DH, Schönrich G, Klempa B. Human pathogenic hantaviruses and prevention of infection. *Hum Vaccin.* 2011;7(6):685–693.

59. Mohamed N et al Development and evaluation of a broad reacting SYBR-green based quantitative real-time PCR for the detection of different hantaviruses. *J Clin Virol.* 2013;56(4):280–285.

60. Aitichou M et al. Identification of Dobrava, Hantaan, Seoul, and Puumala viruses by one-step real-time RT-PCR. *J Virol Methods.* 2005;124(1–2):21–26.

61. Botten J et al. Shedding and intercage transmission of Sin Nombre hantavirus in the deer mouse (*Peromyscus maniculatus*) model. *J Virol.* 2002;76(15):7587–7594.

61 Nairobi Sheep Disease Virus

Devendra T. Mourya and Pragya D. Yadav

CONTENTS

61.1 Introduction 543
    61.1.1 Classification and History...544
    61.1.2 Morphology, Genomic Organization, Biology, and Epidemiology 544
        61.1.2.1 Morphology...544
        61.1.2.2 Genomic Organization..545
        61.1.2.3 Virus Replication 545
        61.1.2.4 Phylogenetic Analysis 545
        61.1.2.5 Epidemiology 546
    61.1.3 Pathogenesis and Clinical Features 547
        61.1.3.1 Pathogenesis..547
        61.1.3.2 Clinical Features in Animals...548
        61.1.3.3 Clinical Features in Humans 548
    61.1.4 Diagnosis 548
    61.1.5 Prevention and Control 549
        61.1.5.1 Management Practices 549
61.2 Methods 550
    61.2.1 Sample Preparation..550
        61.2.1.1 Virus Isolation from Ticks 550
    61.2.2 Detection Procedures..551
        61.2.2.1 Virus Neutralization Test..551
        61.2.2.2 Reverse Transcription Polymerase Chain Reaction..551
61.3 Conclusion 551
References...551

61.1 INTRODUCTION

Arthropod-borne viruses (arboviruses) constitute one of the largest groups of vertebrate viruses of which a majority have the potential to cause devastating epidemics of public health importance. The last few decades witnessed a spurt in the number and frequency of arboviral epidemics impeding socioeconomic development of many tropical countries, majority with a weaker health system. During these outbreaks, death rates were exceptionally high, and the survivors were left in a state of debilitating chronic arthralgia or with neurological complications for the rest of their lives. Introduction of the West Nile virus to the United States in 1999; explosive outbreaks of the Chikungunya virus in Indian Ocean islands, India, and Southeast Asian countries; outbreaks of the tick-borne encephalitis virus in Scandinavian countries; and geographic expansion of the Crimean–Congo hemorrhagic fever virus (CCHFV) in Europe, Africa, and Asia, including India, are some of the major arboviral outbreaks that caused high mortality and morbidity in recent years. Dengue, caused by four different serotypes of dengue virus, continued to be a major global threat with approximately 390 million cases and one million deaths annually. Global warming in association with rapid increase in population, urbanization, and advancements

made in human transport and the changing agricultural practices are thought to be the major reasons for the spurt in arboviral outbreaks/epidemics. The changing environmental conditions favored the growth of vector populations, which showed a rapid increase in recent years thereby enhancing the transmission of arboviral infections. Recent decades have shown heavy migration of people across the nations and continents in search of opportunities either exposing themselves or exposing the naïve population to infections with an exotic agent. Despite considerable national and international efforts, arboviral infections continued to be a major public health issue, mainly due to the nonavailability of prophylactics and therapeutics.

Recent decades also witnessed a boom in the incidence of zoonotic viruses causing human outbreaks probably either due to the increased association with animals as part of animal husbandry or due to the explosive increase in vectors. Nairobi sheep disease virus (NSDV), an important member of the genus *Nairovirus* in the family *Bunyaviridae*, is highly pathogenic to sheep and goats in Africa causing acute hemorrhagic gastroenteritis with 70%–90% mortality in susceptible populations. The virus is viscerotropic in sheep and produces ecchymosis in the intestines and hyperplasia of the mesenteric nodes. Virus transmission is mainly through the bite of infected ticks.

Although the virus is deadly to sheep and goat, it is not a major health problem to humans so far as it causes only self-limiting febrile illness. Hybridization studies using RNA probes demonstrated that NSDV is closely related to Hazara virus (HAZV), a member of the CCHFV serogroup, which also includes the highly pathogenic agent CCHFV. NSDV exists in India as Ganjam virus (GANV), which is a milder virus that causes febrile illness and lumbar paralysis in native sheep and goats. This chapter is focused on NSDV biology with its association with other antigenically related viruses.

61.1.1 CLASSIFICATION AND HISTORY

NSDV is the prototype virus of the genus *Nairovirus*, family *Bunyaviridae*. The genus *Nairovirus* comprises seven serogroups consisting of 34 predominantly tick-borne viruses, majority of which are associated with severe human and livestock diseases [1,2]. CCHFV, Dugbe virus (DUGV), Kupe virus, and GANV are some of the other economically important viruses in the genus. CCHFV is the most important human pathogen among nairoviruses, with a widespread geographical distribution covering Africa, Asia, the Middle East, and Eastern Europe with a case fatality rate ranging from 15% to 70%. This disease affects those people who work in close association with livestock such as agricultural workers, slaughterhouse workers, and veterinarians. Man acquires infection from direct contact with blood or other infected tissues from livestock or through the bite of infected tick. High mortality usually results when the transmission is through blood or respiratory secretions, which are most often intrafamilial or nosocomial in nature. A variant of NSDV is prevalent in the Indian subcontinent as GANV, causing febrile infections in humans and mild infections in sheep and goats [3].

Acute hemorrhagic gastroenteritis in sheep and goats with very high morbidity and mortality rates occurred in large scale in Nairobi, Kenya, and has led to the first isolation of NSDV in 1910 [2]. Inoculation of the blood of a sheep infected with NSDV in susceptible sheep induced acute gastroenteritis in the latter confirming NSDV as the causative agent of the disease [4,5]. The ixodid tick, *Rhipicephalus appendiculatus*, was identified as the transmitting agent of the virus subsequently [6]. Initially, it was thought that NSDV is only endemic in East Africa; however, recent data have shown that it is also prevalent in India and Sri Lanka [3]. Recent studies at the genetic and serologic levels have demonstrated GANV as an Asian variant of NSDV as the two viruses differed only by 10% and 3% at nucleotide and amino acid levels, respectively [3,7]. GANV was first isolated from *Haemaphysalis intermedia* ticks collected from goats suffering from lumbar paralysis in Orissa, India, during 1954–1955 and named after the place of isolation [8]. Subsequent studies have yielded several isolations mainly from *Haemaphysalis* ticks followed by a few from mosquitoes, sheep, and man [9].

61.1.2 MORPHOLOGY, GENOMIC ORGANIZATION, BIOLOGY, AND EPIDEMIOLOGY

61.1.2.1 Morphology

NSDV virions are mostly spherical (elongated forms can also been seen) having a diameter in the range of 80–120 nm (Figure 61.1). It has a lipid bilayer envelope approximately 5–7 nm thick on which glycoprotein spikes or peplomers are embedded. Peplomers are 5–10 nm long projections from the surface of the envelope and are thought to be consisting of heterodimers of two viral glycoproteins, Gn and Gc. Electron microscopy and biochemical studies suggest that there may be 270–1400 GP peplomers per virion [10]. Being enveloped, the virions are very sensitive to heat, acid pH levels, lipid solvents, detergents, and disinfectants. The viral envelope gets removed after treatment with lipid solvents or nonionic detergents and results in the loss of infectivity [11].

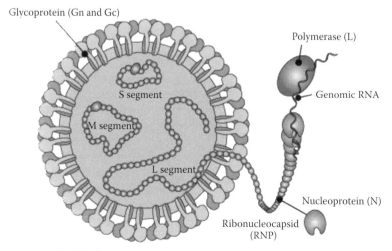

Enveloped, spherical. Diameter from 80 to 120 nm.

FIGURE 61.1 Structure of Nairobi sheep disease virus. (Courtesy of ExPASY Bioinformatics Resource Portal, SIB Swiss Institute of Bioinformatics, Lausanne, Switzerland.)

61.1.2.2 Genomic Organization

NSDV virion envelope comprises circular and helical nucleocapsid segments, which form the genomic structure. Panhandles, that is, noncovalent bonds between palindromic sequences on the 3′ and 5′ ends of each RNA genome segment, form the nucleocapsid circles [12]. The three RNA segments and the terminal sequences are identical for all the members of the genus *Nairovirus*. The genome consists of three segments of negative-sense, single-stranded RNA, namely, the small (S), medium (M), and large (L) segments (Figure 61.2). The S segment encodes the viral nucleocapsid (N) protein; the M segment encodes precursors to both Gn and Gc (preGn and preGc), which are generated by cotranslational cleavage. Posttranslational cleavage of preGn yields Gn and two other amino-terminal products of unknown function: a mucin-rich product (hypervariable mucin-like domain that is highly O-linked glycosylated) and an additional glycoprotein (GP38). Although a cleavage site at the carboxyl terminus of preGn is not yet identified, evidence of a polypeptide removal exists. The L segment encodes a large viral protein, namely the RNA–dependent RNA polymerase (transcriptase) protein [12]. The carboxyl-terminal half of this protein contains most of the polymerase motifs. The amino-terminal part is largely of unknown function. The L segment, being extremely long (>12 kb) and encoding a single protein of 450 kDa, is unusual among bunyaviruses [13]. Genomic segments can reassort between closely related viruses.

61.1.2.3 Virus Replication

After receptor-mediated endocytosis, virus replication takes place in the cytoplasm of the host cells. The first step after penetration of the host cell and uncoating is the activation of the virion RNA polymerase (transcriptase) and transcription of viral mRNAs from each of the three virion RNAs as genome of the single-stranded, negative-sense RNA viruses cannot be translated directly. The earlier procedure utilizes a cap-snatching mechanism (cleaving 5′-methylated caps from host mRNAs and adding these to viral mRNAs to prime transcription) mediated by the RNA polymerase

as it also has endonuclease activity [14]. Translation of L, M, and S mRNAs takes place in the endoplasmic reticulum with cotranslational cleavage of M segment polyprotein and dimerization of Gn and Gc. Membrane-associated RNA replication of the virion RNA occurs after primary viral mRNA transcription. Structural proteins for virion synthesis are formed after a second round of transcription and translation. Morphogenesis requires localization of the N protein in budding compartments with transport of dimerized Gn and Gc to the Golgi where glycosylation takes place, and virions acquire envelopes by budding into the Golgi network in the final stages of assembly. They are then transported in the secretory vesicles through the cytoplasm and released by exocytosis at the cell surface [15]. In some particular cell types, viruses have been observed to bud directly from the host cell plasma membrane.

61.1.2.4 Phylogenetic Analysis

Molecular characterization of NSDV and GANV with respect to S, M, and L segments and its phylogenetic comparison with other members in the *Nairovirus* genus revealed antigenic relationship between the two [3]. Based on the sequence information available in GenBank, it appears that the S segment of GANV and NSDV is considerably shorter (1589 nt) than other characterized nairoviruses (1716, 1672, and 1677 nt for DUGV, CCHFV, and HAZV, respectively). The complete M segment generally encodes two structural glycoproteins for all nairoviruses, while an additional encoded nonstructural protein, NSm, has been described for members of the *Orthobunyavirus*, *Phlebovirus*, and *Tospovirus* genera. Mature glycoproteins encoded by the M segment are generally processed by cotranslational cleavage [16]. However, in DUGV, two surface glycoproteins have been reported: a 73 kDa G1 and a 35 kDa G2 protein [17], while in HAZV, a member of the CCHF serogroup has 3 surface glycoproteins [18]. Analysis of the glycoprotein precursor at amino acid sequence demonstrated the absence of RSKR, which is present in all CCHFV strains but not in DUGV or GANV.

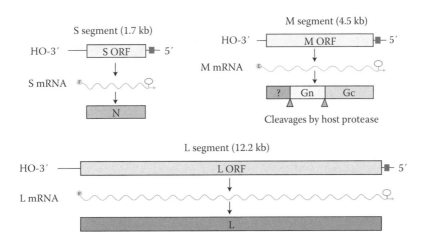

FIGURE 61.2 Genome organization of nairoviruses. (Courtesy of ExPASY Bioinformatics Resource Portal, SIB Swiss Institute of Bioinformatics, Lausanne, Switzerland.)

Phylogenetics of complete S, M, and L segment of Nairobi sheep disease virus and Ganjam virus: Amplification and sequencing of complete S segment of 1590 bp for eight strains of GANV predicted that open reading frame (ORF) encodes an N protein of 482 amino acids in length. The amplified products of M segment of GANV (strain G 619, 779, 6233) and NSDV (strain 708) were sequenced by different primers, and sequences were aligned for phylogenetic study. The complete nucleotide sequences of 5083 and 5063 nt of M gene of GANV and NSDV contained one long ORF of 1624 and 1627 amino acids and is predicted to encode a polyprotein of 180 and 181 kDa, respectively. The N-terminal 24 and 23 amino acids are likely to function as a signal peptide for the two viruses. Similar to CCHFV, topology predictions suggest that GANV and NSDV M polyprotein has the ability to traverse the membrane five times. Although the N-terminal with mucin-like domain characteristic of CCHFV is absent in GANV and NSDV, possibility of O-glycosylation at the N-terminal exists due to the presence of serines and threonines. Unlike CCHFV M genome, RSKR was not available in GANV and NSDV, which is conserved in all CCHFV strains. In CCHFV amino acid motif RRLL, in DUGV amino acid motif RRKL, and in GANV and NSDV amino acid motif RRLM has been found at same position. In CCHF amino acid motif RKPL, in DUGV amino acid motif RKLL, and in NSDV and GANV amino acid motif RRLL has been found at same position. In CCHFV M (IbAr10200) segment, motifs RRLL519 and RKPL1040 were the cleavage sites of Gn (mediated by SKI-1 protease) and Gc (mediated by unknown protease), respectively. Interestingly, in GANV and NSDV, potential SKI-1 or SKI-1-like cleavage sites are located at aa 449 and 453 (RRLM) and aa 972 and 976 (RRLL), respectively, raising the possibility for SKI-1-like proteases in the processing of glycoproteins. However, RKPL found at aa 137 in GANV is not represented in NSDV. A long fragment of amino acid is deleted in CCHFV in comparison to GANV, NSDV, and DUGV from position aa 781–872. An RSxR247 motif responsible for furin/PC-mediated cleavage in CCHFV M segment is absent in GANV and NSDV. Five N-linked glycosylation sites (NxT/S) are present in both viruses, and actual utilization will depend on their topological location and the presence of prolines in the motif. A highly conserved CPYC motif characteristically present in thioredoxin superfamily of proteins is found in two locations in both NSDV and GANV as well as in CCHFV. A hydropathy plot of the M-segment protein shows at least six highly hydrophobic potential membrane–spanning regions. N-terminal mucin-like domain, which is characteristic of CCHFV (aa 247), is comparatively smaller in GANV (aa 95) and DUGV (aa 47). Similarly, 14% of mucin-like domain of CCHF is O-glycosylated but in GANV it is only 10% according to the prediction. L genome of GANV (G 619 and 779) and NSDV is 12080 and 12064 nt, respectively, which is the second largest genome in Nairovirus group after DUGV. The ORF of GANV and NSDV starts from 42nd and 43rd positions to 12,015th and 12,016th positions, which encodes 3991 amino acids of 453 kDa for both viruses. GANV polymerase gene is having one ribosome-binding factor A situated at 3813–3835 positions, which is not found in NSDV and other nairoviruses.

OTU-like cysteine protease and helicase are located at 29–158 and 35–157 positions, which are similar to that of CCHFV and DUGV. Catalytic sequence in OTU region of GANV differs from CCHFV as GDGNCFYHSIAX100 HFD is found in the latter, while ADGNCFYHSIAX HFD is found in GANV and NSDV. Zinc finger motif is present at same positions 604–627 in both viruses. RNA-dependent RNA polymerase is present at positions 2082–2823, while RNA-directed RNA polymerase is at positions 2378–2593 in both viruses. Sequence comparison showed that DUGV is the biggest genome in the genus Nairovirus. KTHIG is absent in NSDV, which is present in all CCHFV [7].

61.1.2.5 Epidemiology

R. appendiculatus (brown ear tick) is the principal vector of NSDV (Figure 61.3). The virus is maintained in the tick life cycle through transovarial and transstadial routes and can persist up to 2 years in adults. Transmission to goats and sheep occurs through the bite of larvae, nymphs, or adults. Transovarial transmission of NSDV was also demonstrated in R. pulchellus in Somalia [19]. Amblyomma variegatum (African bont tick) also plays an important role in virus transmission as it has been held responsible for an outbreak of Nairobi sheep disease (NSD) in Kenya [20]. However, NSDV was isolated only from R. appendiculatus out of eight species of ticks representing three genera (Amblyomma, Hyalomma, and Rhipicephalus) in Kenya [2,4]. A closer relationship exists between the presence of NSDV antibody in sheep and goats and infestations with R. appendiculatus rather than with A. variegatum [21]. In India, H. intermedia is the main vector of GANV, even though the virus has been isolated from H. wellingtoni and Rhipicephalus haemaphysaloides [22,23]. Perera et al. [24] reported isolation of a virus belonging to the NSD antigenic group from H. intermedia in Sri Lanka. NSDV infection should be suspected when there is movement of animals from endemic areas to nonendemic areas and vice versa.

Outbreaks also follow due to incursions of ticks, particularly following heavy rains [25]. Vertebrates develop high viremia, sufficient for virus transmission by ticks [26]. Domestic animals, namely, cattle, sheep, goats, and horses, as well as wild animals such as antelopes and bison are hosts for ticks. Virus spread from the pastoral communities, which exist near wildlife parks, and occasional mixing of wildlife and livestock allows transfer of ticks and possibly viruses between these animal groups. Livestock marketing practices allow movement of animals across the borders in the region, which allows the tick-borne virus to move between countries [27]. Poor husbandry and grazing practices put pressure on land resources resulting in continuous movement of herds of animals in search of pasture causing transmission of infected ticks. When susceptible herds were moved from northern districts to Nairobi area, which is endemic to NSDV, large epizootics of NSDV have been reported [2]. Maternal antibodies protect young animals in endemic areas while they acquire immune protection through active immunity when exposed to the virus. Apart from goats and sheep, no known vertebrate host is infected with the virus even in endemic areas. Human

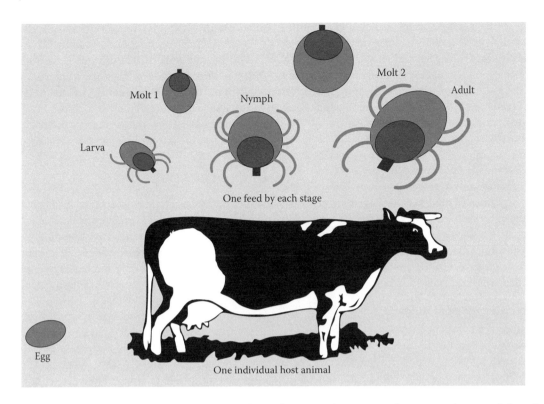

FIGURE 61.3 One host life cycle of *Rhipicephalus appendiculatus* (diagrammatic representation as cows do not get infected with Nairobi sheep disease virus/Ganjam virus). (From Walker, A.R., A one-host life-cycle of a hard tick such as Rhipicephalus (Boophilus) microplus feeding on one individual cow, author's own work, November 28, 2011. Ticks of Domestic Animals. Wikipedia. https://en.wikipedia.org/wiki/Ticks_of_domestic_animals. Last modified July 29, 2015.)

infections are common among pastoral communities in Nairobi as livestock are frequently maintained in enclosures close to human habitation or sometimes inside the houses due to security reasons. The close association increases the potential for virus transmission between animals and humans.

61.1.3 Pathogenesis and Clinical Features

61.1.3.1 Pathogenesis

There have been no detailed studies on the nature of the pathogenesis in NSDV/GANV infections as it was either ignored or misdiagnosed as peste des petits ruminants or Rift Valley fever virus. NSDV infection often results in severe disease in sheep and goats, which are implicated as being important vertebrate hosts [2]. In experimental NSDV infections of sheep, goats, and suckling mice, the virus is found replicating to high titers in the lung, liver, spleen, and other organs of the reticuloendothelial system [28]. Also, in India, GANV has not been taken seriously due to the mild nature of infection in native animals. However, the virus was found highly pathogenic to exotic and crossbred breeds inducing mortality [7,9]. Nevertheless, serology on NSDV/GANV cases has shown increase in IL-6, IL-10, and Tumour Necrosis Factor α (TNF-α) in clinically severe cases. Similarly, it was found that monocyte-derived dendritic cells infected with NSDV/GANV also release IL-6, IL-10, and TNF-α. Tarif et al. showed that pathogenicity of NSDV/GANV is associated with increase in all the aforementioned cytokines and IL-12 as well as due to a decrease in IL-4,

all concordant with a Th1, proinflammatory response. Vascular endothelium appears to be the primary cell target. Virus replication in the endothelial cells and release of the aforementioned cytokines results in edema and necrosis of the capillary walls of the mucosal surfaces of the intestine, gall bladder, and female genital tract, leading to congestion, hemorrhage, and catarrhal inflammation. The highly pathogenic isolate leads to more prolonged pyrexia and profound leukopenia that has not previously been reported for NSDV infections, although it is a common clinical sign of viral hemorrhagic fever and may be caused by the large-scale apoptosis of leukocytes [29].

Although differences are observed among various breeds and strains of sheep and goats in the susceptibility to both viruses, clinical symptoms are almost identical. NSDV infection is characterized by acute hemorrhagic gastroenteritis [5]. Clinical signs begin with marked pyrexia and an increase in temperature to 40°C–41°C. This is followed by hyperventilation accompanied by a prominent clinical depression, anorexia, and a disinclination to move. Leukopenia is very prominent during this period of hyperthermia [28]. Diarrhea usually develops within 36–56 h postonset of the febrile reaction, which is profuse, watery, and fetid. Feces become hemorrhagic and mucoid (dysentery) followed by a temperature decline accompanied by colicky pains and tenesmus. Animals stand with lowered head and show conjunctivitis-like symptoms with mucopurulent or serosanguinous nasal discharge. Breathing becomes rapid and painful. Lymph nodes become enlarged and can be palpable. In hyperacute infections, a sudden rise in body temperature occurs and abruptly declines on the third to sixth day followed by collapse and death, which is associated with severe diarrhea and dehydration. In less acute cases, the course of the disease is slower and death may occur up to 11 days after the onset of clinical signs or survive. It is the native breeds of sheep that show more severe disease than in Merinos with a mortality rate ranging from 40% to 90% [25,30]. East African hair sheep and Persian fat-tailed sheep are highly susceptible to NSD with case fatality rates exceeding 75%. Nonnative to Africa, Romney and Corriedale sheep show a low mortality rate ranging from 30% to 40%. Although mortality rates as high as 90% have been reported in some indigenous breeds of goats, NSDV is less virulent to goats than sheep. Contradictory to NSDV, GANV causes mild infections such as febrile illness and lumbar paralysis to native breeds of goat and sheep and causes high morbidity and mortality in exotic and crossbred sheep and goats [9].

61.1.3.2 Clinical Features in Animals

1. Gross lesions: Pathological lesions due to NSDV in sheep and goats can be misleading as many other febrile diseases of sheep reported from the NSDV endemic areas induce similar types of lesions. Generalized lesions include lymphadenitis with petechial and ecchymotic hemorrhages on the serous surfaces of the alimentary tract, spleen, heart, and other organs [5]. However, in infected animals, hemorrhages on the mucosa of the abomasum, especially along the folds in the ileocaecal valve and most commonly in the colon and rectum, where zebra striping (congestion or hemorrhages appearing as longitudinal striations in the mucosa) may be seen are characteristic of NSDV. Zebra striping of the rectum is more conspicuous than that of the colon. On occasions, enlarged and hemorrhagic gall bladder is also seen. Inflammatory lesions with hemorrhage may also be seen in the female genital tract in case of abortion.

2. Microscopic lesions: Most of the microscopic lesions are seen in the vascular endothelium, which appears to be the primary cell target as virus replication mainly takes place in the endothelial cells. Release of proinflammatory and other cytokines results in edema and necrosis of the capillary walls of the intestine, gall bladder, and female genital tract; leading to congestion, hemorrhage, and catarrhal inflammation. Other common histopathological/microscopic lesions observed are myocardial degeneration, nephritis, and necrosis of the gall bladder.

61.1.3.3 Clinical Features in Humans

Although highly pathogenic to sheep and goats, NSDV and GANV are apparently rare zoonotic agents that cause infections in humans. Natural infections are mostly subclinical or febrile illness with influenza-like symptoms. The clinical signs may include fever, headache, back and abdominal pain, joint pain, nausea, and vomiting as well as conjunctival infection. Laboratory infection has been associated with fever and

joint pain in Africa [31], while laboratory-acquired infections with GANV have been reported in India [32].

Specific antibodies to these viruses have been detected in the general population, laboratory workers, and agricultural workers in Uganda, India, and Sri Lanka [8,33]. Even though NSDV and GANV are classified as Biosafety Level 3 agents, the diseases caused by these viruses are not considered to be of public health concern [34].

61.1.4 Diagnosis

NSDV infection should be suspected in sheep or goats, which exhibit hemorrhagic gastroenteritis and nasal discharge in or near an enzootic area. Tick infestation is also supportive. Diagnosis is particularly likely when native sheep/goats do not show clinical signs and only the recently imported sheep or goat shows clinical symptoms. Abortion is a common sequel to NSDV infection. A tentative diagnosis based on postmortem lesions can be sometimes difficult because many animals dying from NSDV do not show these lesions.

Therefore, the use of laboratory techniques is vital for confirmation of NSDV and GANV infections. Virus isolation in either infant mouse or invertebrate cell cultures can be carried out as these systems are susceptible to both NSDV and GANV.

Viral antigens and antibodies in infected or recovered animals can be detected by immunodiffusion, complement fixation (CF), indirect fluorescent antibody test (indirect IFA), hemagglutination, enzyme-linked immunosorbent assay (ELISA), neutralization test (NT), etc. Monoclonal antibodies have been developed against the antigens of NSDV strain I-34 and are being evaluated for their application as diagnostic reagents (OIE).

In particular, direct fluorescent antibody test is used for the identification of virus in tissue culture when no cytopathic effect (CPE) is evident [15]. Hyperimmune mouse ascitic fluids and immune serum prepared in rabbit or sheep conjugated with a fluorescent stain can be used for detection of antigen directly. While the method is simple, nonspecific fluorescence due to other members of the family may occur, which is a major drawback. Further confirmation tests will be required.

Indirect IFA is one of the most suitable tests recommended for detection of antigen/antibodies for the members of *Nairovirus* serogroup. Some drawbacks are associated with this test, including low specificity and erroneous visual reading of results. Since IFA is an indirect test where most of the nonspecific reactions can be avoided, it is one of the methods for screening large number of samples with high specificity. However, cross-reactions with closely associated viruses, namely, DUGV and CCHFV, could be observed [2,8].

ELISA is one of the best diagnostic tests used for detection of antigen and antibodies. With the standardization of techniques using partially purified antigen, diagnosis is specific and rapid [35].

Virus NT is one of the earliest techniques used for detecting antibody prevalence in human and animal sera. It is a fundamental observation in virology that serum from an animal infected with a virus can neutralize the infectivity of the virus when allowed to react together. Hence, the test can be done virtually with every virus even if the titer of antibodies is low. The loss of infectivity is brought about by interference of the bound antibody with any one of the steps leading to the release of the viral genome into the host cells. The presence of unneutralized virus can be detected by CPE.

In recent years, a range of molecular techniques have been developed and applied for detection of NSDV and GANV nucleic acids. Molecular procedures such as single-step reverse transcription polymerase chain reaction (RT-PCR) and real-time RT-PCR have been standardized for rapid and specific diagnosis of both NSDV and GANV [27,36] (Table 61.1). RNA probes developed from S and M genome segments of DUGV have the potential to be applied as diagnostic tools for NSDV and GANV [17,37].

61.1.5 PREVENTION AND CONTROL

61.1.5.1 Management Practices

It has been shown by epidemiological investigations that outbreaks of NSDV do not occur in a state of enzootic stability. The disease spreads as a result of animal movements from endemic areas to disease-free areas and vice versa. The disease can be controlled by proper management of animals through strict vigilance. Since ticks are the vector of the disease, deticking is important and carried out with the supervision of veterinarians. Although it is largely a disease of sheep and goat, its potential to affect humans represents a serious health concern. People should be made aware of the transmission mechanism of the virus by posters/literature in the endemic areas. Professionals like slaughter house workers, sheep shearers, veterinarians, and others who are directly involved with livestock industry are at risk of contacting the infection. They should be educated to use personal protective

TABLE 61.1
Summary of Polymerase Chain Reaction Primers for Identification of Nairobi Sheep Disease Virus/Ganjam Virus

Target Gene	Primer	Sequence (5′–3′)	Nucleotide Position	Size (bp)	References
A. RT-PCR primers					
S	NSD12f	GAA TGG TCG AAC GTG GAC	311-328	869	Marczinke and Nichol [3]
	NSD16r	TGC TGT CAG GAC ACC AGG	1197-1180		Sang et al. [27]
	Ganj1f	GGG ATG ATC CAA CGT GGC	310-327	962	Marczinke and Nichol [3]
	Ganj4r	AGT ATG TTC TTG ATA TGA CCT	1292-1272		
	Nairo forward	TCTCAAAGAAACACGTGCCGC	1-21	400	Lambert and Lanciotti [37]
	Nairo reverse	GTCCTTCCTCCACTTGWGRG CAGCCTGCTGGTA	463-431		
L	6942+	ATGATTGCIAAYAGIAAYTTYAA		443	Honig et al. [36]
	7385-	ACAGCARTGIATIGGC			
Partial S, M, L	GANVVS-F1	TCAGGATATGAGGTTAGCAT	101-120	396	Yadav et al. [7]
	GANVVS-R1	TAATCATGTCGGATAGCATGT	574-554		
	GANV VL-F1	TCTGACTGGTAGAACAACTGGT	13-34	425	
	GANV VL-R1	TGTGTTGTCTCCTGATATACAGATT	437-418		
	GANVVM-F1	GCACTAGCAGGTTAATCTCAA	20-39	701	
	GANVVM-710R	GAGCTACAATTGCTGTATCTTC	721-700		
Complete S, M, L	Nairo1F	TCTCAAAGACAc/aACa/gTGCCGC	1-21	1590	Yadav et al. [7]
	Nairo11R	TCTCAAAGAt/gATCGTTGCCg/aCAC	1590-1568		
	CCHF-L1F	TCT CAA AGA TAT CAA TCC CCC C	1-22	7793	
	CCHF-L1R	TTG GCA CTA TCT TTC ATTTGA C	7772-7752	6700	
	CCHF-L2F	GAA GAG CTA TAT GAC ATA AGG C	6136-6157		
	CCHF-L2R	TCT CAA AGA AAT CGT TCC CCC CAC	12081-12058	5083 (GANV)	
	NAIROTERM-1 RXN	TCTCAAAGAIATASTKGCGGCAC		5063(NSD)	
B. Real-time RT-PCR primers					
S	NSDV/GANV F1	TGACCATGCAGAACCAGATYG	40-58	147	Tarif et al. [29]
	NSDV/GANV R1A	GAAACAAGCCTCATGCTAACCT	205-184		
	NSDV/GANV (F2)	GGAGAATGGCAAAGAGGTTGT	854-874	299	
	NSDV/GANV(R2)	GTAAATCCGATTGGCAGTGAAG	1174-1153		
	NSDV/GANV F3b (RT primer)	GTCTTTGAACTYTGACCA	28-45	177	
	NSDV/GANV(F1A)	TGA CCA TGC AGA ACC AGA TYG	40-58		
	NSDV/GANV(R1A)	GAA ACA AGC CTC ATGCTA ACC T	205-184		

equipment while handling infected animals. Similarly, the use of tick repellents should also be considered while handling livestock as the immature stages and adult ticks are also infective and can transmit the virus by bite. Dead sheep should be buried or incinerated.

Vaccination: No licensed vaccine is available commercially to date either for humans or animals though several studies have been initiated. An attenuated vaccine has been developed by serial passaging; NSDV 35 passages in adult mice were found to cause severe reactions in some breeds of sheep and hence discontinued. A similar vaccine formulation by further mouse brain passages was developed in Entebbe but was found not applicable for vaccination due to its adverse effects. A tissue culture–based vaccine has been found to induce immune response and protection following two inoculations given at an interval of 14 days. However, this vaccine is not produced commercially [2,4].

Treatment: There is no specific treatment available for NSDV. Symptomatic treatment should be practiced against pyrexia, diarrhea, hemorrhages, depression, and other symptoms. Supportive treatment of patients is by means of fluid replacement therapy. Availability of a good-quality feed and protection from climatic adversities may reduce the mortality rate.

61.2 METHODS

61.2.1 Sample Preparation

61.2.1.1 Virus Isolation from Ticks

Virus isolation from ticks can be carried out as described by Joshi et al. [23]. In brief, ticks were allowed to digest blood, pooled according to species, sex, and locality, and triturated in chilled mortars and pestles in 1.5–2 mL minimum essential medium to fine suspension after washing the ticks thoroughly in water followed by surface sterilization with 70% ethanol. The suspension is centrifuged in a refrigerated centrifuge at $6000 \times g$; the supernatant is collected, Millipore filtered, aliquoted, and inoculated into the cell culture system or infant mice. High-speed homogenizers generate linear motion to disrupt samples thoroughly in just seconds. Specialized matrix beads can also be used for homogenization of ticks. An aliquot of the suspension can be used for RNA extraction using Qiagen (Valencia, CA, USA) RNA extraction kit as per manufacturer's instructions.

Virus isolation from animals: Samples such as blood, plasma mesenteric lymph nodes, and spleen tissue from febrile or dead animals can be used for virus isolation. Serum or plasma can be used directly as inoculums. Lymph nodes or spleen should be homogenized to make an approximate 10% (w/v) in growth medium or Bovine albumin in phosphate saline (BAPS) and inoculated in infant mice or cell culture for virus isolation as described earlier. Human serum can be used as inoculum for virus isolation.

Virus isolation in cell culture: Baby hamster kidney (BHK-21-C13) cells, Vero cell line, or any other susceptible cell line can be used for virus isolation [38]. Sample prepared as described earlier or mouse brain filtrate can be used as inoculum. Confluent monolayer cultures of BHK-21-C13 or Vero prepared in 24-well tissue culture plates (NUNC/Corning) were inoculated with the suspension at the rate of 100 μL per well, adsorbed for 1 h at 37°C with rocking at every 10 min and fed with maintenance medium [39]. Cultures were incubated in a 37°C incubator stationary and were observed daily for CPE. Cells were harvested when the cells show 50%–75% CPE. The cultures will undergo freeze–thaw cycle thrice followed by centrifugation in a refrigerated centrifuge as described earlier, and the supernatant will be aliquoted and stored at −80°C as stock virus. One aliquot will be used for confirmation of NSDV using molecular tools or immunofluorescent antibody technique (IFA) or any other confirmation technique (CF, indirect IFA, hemagglutination assay, ELISA, NT, etc.).

Virus isolation in animals: Suspected sample prepared as discussed earlier is inoculated in 2-day-old infant mice through the intracerebral route as described by Joshi et al. [23]. The mice are observed daily for sickness, and when acute sickness is observed, brains are harvested. Virus suspension is prepared in BAPS and confirmed by further passaging in mice followed by other diagnostic techniques such as RT-PCR and IFA.

Samples for agar gel immunodiffusion (AGID) test: AGID can be a valuable primary diagnostic tool for detection of NSDV antigen in tissues. This test has been used successfully with crude antigens prepared either from spleen and mesenteric lymph nodes of infected sheep or infected mouse brain. In brief, approximately 0.5– 1 g of the aforementioned tissue is homogenized to give 10%–20% suspensions in phosphate-buffered saline (PBS). The suspension is centrifuged at 1000 g for 10–15 min and the supernatant is used for the test. Hyperimmune serum can be prepared in sheep, mice, or rabbits using isolated virus or tissues such as spleen or brain from an NSDV-infected sheep.

Samples for CF test: The infected tissue culture fluids can be used as antigens for CF tests for virus identification. However, with the standardization of ELISAs and molecular tools, CF test is rarely used as it is complicated by anticomplementary activity of sheep sera [40].

Samples for ELISA: Tissue culture or mouse brain antigen can be used for ELISA for virus identification. Uninfected cells can be used for the preparation of negative control. Positive antigen and negative controls are adsorbed directly onto the ELISA plates, and the test is carried out with NSD immune as well as normal serum [35].

Antigen preparation for IFA: Monolayer cultures of the Vero or BHK cells grown on cover slips either in Leighton tubes or in 24-well plates are infected with characterized strain of NSDV and incubated for 2–3 days at 37°C. At the commencement of infective foci, the cover slips are removed from the growth medium, washed once with PBS, and fixed in chilled acetone for 10 min at room temperature. The cover slips are air-dried and stored at −20°C until use for IFA.

Uninfected cover slips grown with Vero and BHK cells are also fixed and stored to be used as negative control in IFA.

For serodiagnosis of NSDV infection, paired acute and convalescent serum samples should be collected.

61.2.2 DETECTION PROCEDURES

61.2.2.1 Virus Neutralization Test

Davies et al. described an NT protocol for NSDV detection [2,8]. In brief, serial twofold dilutions of heat-inactivated test serum are placed in 96-well tissue culture plates in quadruplicate and incubated with 2-log $TCID_{50}$ of the virus for 1 h at 37°C. After incubation, the serum–virus mixture is transferred onto susceptible cells (in this case, BHK cells) grown in 96-well plates as designed and incubated for a period of 4–5 days. The cultures are observed for CPE (if not neutralized) and scored under an inverted microscope, stained with Amido Black, and the antibody titer is determined. The disadvantage of this test is that it gives equivocal results with closely related viruses and hence unreliable in this disease.

61.2.2.2 Reverse Transcription Polymerase Chain Reaction

1. RT-PCR for detection of the family *Bunyaviridae*
   Lambert and Lanciotti developed an RT-PCR based on S segment, which is not specific for NSDV [41]. They evaluated numerous alternatives prior to the selection of primers and assay reaction, and then reaction conditions were selected based on the ability to detect the largest number of bunyaviruses with the highest relative sensitivity (Table 61.1).
2. RT-PCR for detection of NSDV
   S segment: Marczinke and Nichol [3] used nucleotide sequence data obtained from their initial PCR products to design additional primers specific for NSDV and GANV. This strategy resulted in the successful determination of the S RNA segment sequence of both viruses. Sang et al. [27] also reported RT-PCR-based diagnosis depending on S as the target gene segment, which encodes nucleoprotein regions (Table 61.1).
   L segment: Honig et al. [36] focused primer designing efforts on the highly conserved polymerase domains located within the L segment of *Nairovirus* genome. Primer targeting for the specific diagnosis of NSDV/GANV needs further testing and refinement (Table 61.1).
   S, M, and L gene segments: Full-length S, M, and L gene segments may also be amplified and sequenced to prove the relationship of NSDV to GANV [7] (Table 61.1).
3. Real-time RT-PCR for detection of NSDV/GANV
   NSDV-specific real-time RT-PCR has been described and is claimed to be more sensitive than other RT-PCRs [29]. They used primer sets from

regions that were conserved in NSDV/GANV and not present in other nairoviruses, making it specific for NSDV/GANV. However, it has not yet been fully validated for diagnostic use (Table 61.1).

61.3 CONCLUSION

NSDV is an important disease of sheep and goats in northeast Africa with a mortality rate ranging from 50% to 90% impeding the economies of several countries. GANV, an Asian variant of NSDV, is widely prevalent in India although it causes mild infections in animals. Although GANV is less pathogenic to native animals, it was detrimental to crossbred and exotic varieties of sheep and goats causing huge economic loss in sheep/goat farmers in Tamil Nadu and Andhra Pradesh in India. While both viruses are not found to be pathogenic to humans, necessary preparedness to develop diagnostics and vaccines against them are warranted as they are closely related to the dreaded virus CCHFV, which has the potential to cause large outbreaks with case fatality rates exceeding 50%. Although many conventional techniques are available for the diagnosis of NSDV, such as CF, IFA, and AGID, tools for rapid diagnosis of the virus are the need of the hour. Although RT-PCR and real-time RT-PCR have been developed, more efforts are required to standardize these sensitive and specific procedures. There is also an urgent need for efficient vaccines for animals and humans for protecting humans and livestock in the endemic and nonendemic areas. A comprehensive approach from governments, scientists, and the local people is needed for the development of effective prophylactics/therapeutics to combat against diseases like NSDV.

REFERENCES

1. Nichol, S.T., Bunyaviruses. In: *Fields Virology*, pp. 1603–1634, eds. D.M. Knipe, P.M. Howley, D.E. Griffin, R.A. Lamb, M.A. Martin, B. Roizman, and Strauss, S.E. Philadelphia, PA: Lippincott Williams & Wilkins, 2001.
2. Davies, F.G., Nairobi sheep disease in Kenya, The isolation of virus from sheep and goats, ticks and possible maintenance hosts, *J. Hyg.*, 81, 259–265, 1978.
3. Marczinke, B.I. and Nichol, S.T., Nairobi sheep disease virus, an important tick-borne pathogen of sheep and goats in Africa, is also present in Asia, *Virology*, 303, 146–151, 2002.
4. Davies, F.G., Nairobi sheep disease Kenya, The isolation of virus from goats, wild ruminants and rodents within Kenya, *J. Hyg. Camb.*, 81, 251, 1978b.
5. Montgomery, R., On a tick-borne gastro-enteritis of sheep and goats occurring in British East Africa, *J. Comp. Pathol.*, 30, 28–57, 1917.
6. Daubney, R. and Hudson, J.R., Nairobi sheep disease, *Parasitology*, 23, 507–524, 1931.
7. Yadav, P.D., Vincent, M.J., Khristova, M., Kale, C., Nichol, S.T., Mishra A.C., and Mourya, D.T., Genomic analysis reveals Nairobi sheep disease virus to be highly diverse and present in both Africa, and in India in the form of the GANV variant, *Infect. Genet. Evolut.*, 11, 1111–1120, 2011.

8. Dandawate, C.N. and Shah, K.V., GANV: A new arbovirus isolated from ticks *Haemaphysalis intermedia* Warburton and Nuttal, 1909 in Orissa, India, *Ind. J. Med. Res.*, 57, 799–804, 1969.

9. Sudeep, A.B., Jadi, R.S., and Mishra, A.C., GANV. *Ind. J. Med. Res.*, 130, 514–519, 2009.

10. Obijeski, J.F. et al., Structural proteins of La Crosse virus, *J. Virol.*, 19, 985–997, 1976.

11. Obijeski, J.F. and Murphy, F.A., *Bunyaviridae*: Recent biochemical developments, *J. Gen. Virol.*, 37, 1–14, 1977.

12. Schmaljohn, C.S. and Nichol, S.T., *Bunyaviridae*. In: *Fields Virology*, pp. 1741–1790, eds. D.M. Knipe and Howley, P.M. Philadelphia, PA: Lippincott Williams & Wilkins, 2007.

13. Frias-Staheli, N., Medina, R.A., and Bridgen, A., Nairovirus molecular biology and interaction with host cells. In: *Bunyaviridae: Molecular and Cellular Biology*, eds. A. Plyusnin and Elliott, R.M. Caister, U.K.: Academic Press, 2011.

14. Murphy, F.A., Gibbs, E.P.J., Horzinek, M.C., and Studdert, M.J., Laboratory diagnosis of viral diseases. In: *Veterinary Virology*, 3rd edn., pp. 193–224, eds. F.A. Murphy, E.P. Gibbs, M.J. Studdert, and Horzinek, M.C. San Diego, CA: Academic Press, 1999.

15. Quinn, P.J., Markey, B.K., Carter, M.E., Donnelly, W.J., and Leonard, F.C., *Veterinary Microbiology and Microbial Disease*. Oxford, U.K.: Blackwell Sciences, 2002.

16. Schmaljohn, C.S. and Hooper, J.W., *Bunyaviridae*: The viruses and their replication. In: *Fields Virology*, pp. 1581–1602, eds. D.M. Knipe and Howley, P. Philadelphia, PA: Lippincott Williams & Wilkins, 2001.

17. Marriott, A.C., El-Ghorr, A.A., and Nuttall, P.A., Dugbe *Nairovirus* M RNA: Nucleotide sequence and coding strategy, *Virology*, 190, 606–615, 1992.

18. Foulke, R.S., Rosato, R.R., and French, G.R., Structural polypeptides of Hazara virus, *J. Gen. Virol.*, 53, 169–172, 1981.

19. Edelsten, R.M., The distribution and prevalence of Nairobi sheep disease and other tick-borne infections of sheep and goats in northern Somalia, *Trop. Anim. Health Prod.*, 7, 29–34, 1975.

20. Daubney, R. and Hudson, J.R., Nairobi sheep disease, *Parasite*, 23, 507, 1934.

21. Howarth, J.A. and Terpstra, C., The propagation of Nairobi sheep virus in goat testes, goat kidneys and hamster kidneys, *J. Comp. Pathol.*, 75, 437–441, 1965.

22. Rajagopalan, P.K., Sreenivasan, M.A., and Paul, S.D., Isolation of GANV from bird tick *Haemaphysalis wellingtoni Nuttall* and *Warburton*, *Ind. J. Med. Res.*, 58, 1195–1196, 1970.

23. Joshi, M.V., Geevarghese, G., Joshi, G.D., Ghodke, Y.S., Mourya, D.T., and Mishra, A.C., Isolation of GANV from ticks collected off domestic animals around Pune, Maharashtra, India. *J. Med. Entomol.*, 42, 204–206, 2005.

24. Perera, L.P., Peiris, J.S.M., Weilgana, D.J., Calisher, C.H., and Shope, R.E., Nairobi Sheep disease virus isolated from *Haemaphysalis intermedia* ticks collected in Sri Lanka, *Ann. Trop. Med. Parasitol.*, 90, 91–93, 1996.

25. Davies, F.G., Nairobi sheep disease, *Parasitology*, 39, 95–98, 1997.

26. Davies, F.G. and Mwakima, F., Qualitative studies of the transmission of Nairobi sheep disease virus by *Rhipicephalus appendiculatus* (*Ixodoidea, ixodidae*), *J. Comp. Pathol.*, 92, 15–20, 1982.

27. Sang, R. et al., Tickborne arbovirus surveillance in market livestock, Nairobi, Kenya, *Emerg. Infect. Dis.*, 12, 1074–1080, 2006.

28. Mugera, G.M. and Chema, S., Nairobi sheep disease: A study of its pathogenesis in sheep, goats and suckling mice, *Bull. Epiz. Dis. Afr.*, 15, 337–354, 1967.

29. Tarif, B.A., Lasecka, L., Holzer, B., and Baron, M.D., GANV/Nairobi sheep disease virus induces a pro-inflammatory response in infected sheep, *Vet. Res.*, 43, 2–12, 2012.

30. OIE (Office International des Epizooties), *Manual of Diagnostic Tests and Vaccines for Terrestrial Animals*. Paris, France: OIE, 2008, Bunyaviral diseases of animals. http://www.oie.int/eng/normes/mmanual/2014/pdf/2.09.01_bunyaviral_diseases. pdf. Accessed on December 20, 2014.

31. Terpstra, C., Nairobi sheep disease. In: *Infectious Diseases of Livestock with Special Reference to Southern Africa*, pp. 718–722, eds. J.A.W. Coetzer, G.R. Thomson, and Tustin, R.C. Cape Town, South Africa: Oxford University Press, 1994.

32. Rao, C.V. et al., Laboratory infections with GANV, *Ind. J. Med. Res.*, 74, 319–324, 1981.

33. Zeller, H. and Bouloy, M., Infections by viruses of the families *Bunyaviridae* and *Filoviridae*, *Rev. Sci. Tech. Off. Int. Epiz.*, 19, 79–91, 2000.

34. The Center of Food Security and Public Health, Iowa State University, Nairobi sheep disease, http://www.cfsph.iastate.edu/Factsheets/pdfs/nairobi_sheep_disease.pdf, 2009. Accessed on December 21, 2014.

35. Munz, E., Reimann, M., Fritz, T., and Meier, K., An enzyme-linked immunosorbent assay (ELISA) for the detection of antibodies to Nairobi sheep disease virus in comparison with an indirect immunofluorescent and haemagglutination test. II. Results observed with sera of experimentally infected rabbits and sheep and with African sheep sera, *Zentrabl. Veterinarmed.*, 31, 537–549, 1984.

36. Honig, J.E., Osborne, J.C., and Nichol, S.T., The high genetic variation of viruses of the genus *Nairovirus* reflects the diversity of their predominant tick hosts, *Virology*, 318, 10–16, 2004.

37. Ward, V.K., Marriott, A.C., Polyzoni, T., El Ghorr, A.A., Antoniadis, A., and Nuttall, P.A., Expression of the nucleocapsid protein of Dugbe virus and antigenic cross-reactions with other nairoviruses, *Virus Res.*, 24, 223–229, 1992.

38. Shepherd, N.C., Gee, C.D., Jessep, T., Timmins, G., Carroll, S.N., and Bonner, R.B. Congenital bovine epizootic arthrogryposis and hydranencephaly, *Aust. Vet. J.*, 54, 171–177, 1978.

39. Macpherson, I. and Stoker, M.G.P., Polyoma transformation of hamster cell clones, an investigation of genetic factors affecting cell competence, *Virology*, 16, 147–151, 1962.

40. Kilelu, E.S. et al., Diagnosis of Nairobi sheep Disease virus in sheep using complement fixation test. *Kenya Vet.* 77(93), 77, 1983.

41. Lambert, A.J. and Lanciotti, R.S., Consensus amplification and novel multiplex sequencing method for S segment species identification of 47 viruses of the *Orthobunyavirus*, *Phlebovirus*, and *Nairovirus* genera of the family *Bunyaviridae*, *J. Clin. Microbiol.*, 47, 2398–2404, 2009.

42. Walker, A.R., A one-host life-cycle of a hard tick such as Rhipicephalus (Boophilus) microplus feeding on one individual cow, author's own work, November 28, 2011.

62 Rift Valley Fever Virus

William C. Wilson, Natasha N. Gaudreault, and Mohammad M. Hossain

CONTENTS

62.1 Introduction 553
   62.1.1 Classification, Morphology, and Genome Organization 553
   62.1.2 Epidemiology 554
   62.1.3 Clinical Features and Pathogenesis 554
   62.1.4 Diagnosis 555
      62.1.4.1 Conventional Techniques 555
      62.1.4.2 Molecular Techniques 555
62.2 Methods 557
   62.2.1 Sample Preparation 557
   62.2.2 Detection Procedures 557
62.3 Conclusion 559
References 559

62.1 INTRODUCTION

Rift Valley fever virus (RVFV) is a mosquito-transmitted virus (or arbovirus) that is endemic in sub-Saharan Africa. In the last decade, Rift Valley fever (RVF) outbreaks have resulted in the loss of human and animal life [1], with significant economic impact [2]. The disease in livestock is primarily associated with sheep, goats, and cattle, leading to abortions and hemorrhagic disease. Camel deaths were also reported in the 2010 RVF outbreak in Mauritania, which has raised concerns about the effects of the virus on these important domestic animals [3]. Severe disease outbreaks that involve human cases and mortality can be predicted by high rainfall and other measurable environmental conditions that contribute to the increased mosquito populations [4]. Similarly, prevalence of the disease in domestic and wild animals is associated with ideal environmental conditions [5]. Climate change and human activities (trade, irrigation, etc.) have increased concerns of arboviral disease outbreaks and movement of arboviruses into new ecosystems [6]. The potential of RVFV 1 day being introduced into new ecosystems such as Europe [7] or North America [8] has emphasized the need for the development and application of highly specific and sensitive diagnostic assays.

62.1.1 CLASSIFICATION, MORPHOLOGY, AND GENOME ORGANIZATION

Bunyaviridae is a large family of viruses with at least five genera: *Bunyavirus*, *Phlebovirus*, *Nairovirus*, *Hantavirus*, and *Tospovirus*. All of these except the tospoviruses infect vertebrates. Three of these genera, *Orthobunyavirus*, *Phlebovirus*, and *Nairovirus*, are associated with arthropod transmission, while the genus *Hantavirus* is typically spread by rodents. RVFV is the most common virus of the genus *Phlebovirus* [9]. Although RVFV has only one serological and immunological type, strains are generally classified into three genetic lineages: Egyptian, West African, and Central-East African. Still, the RVFV nucleotide and amino acid sequences are highly conserved [10].

RVFV particles are spherical, slightly pleomorphic, measure 80–120 nm in diameter and contain no matrix proteins. The virion consists of an envelope covering the genome; this envelope is composed of a lipid bilayer containing the Gn and Gc glycoproteins forming some 350–375 surface spikes, with spikes 10–18 nm in length and 5 nm in diameter [11,12]. The glycoproteins are organized as protruding hollow cylinders on the surface that cluster into distinct capsomers [13]. The surface is covered with 122 capsomers arranged in icosahedral symmetry with a triangulation number of 12 [14]. RVFV has a tripartite single-stranded ambisense RNA genome; the segments are designated according to their sizes as large (L), medium (M), and small (S) with a total length of approximately 11.9 kilobases (kb) [11]. The L segment (~6.4 kb) encodes for the viral RNA-dependent RNA polymerase. The M segment (~3.8 kb) encodes four proteins, two of which are envelope glycoproteins Gn/Gc and two are nonstructural proteins: the 14 kDa NSm and 78 kDa (NSm and Gn fusion peptides) [15,16]. The S segment (~1.6 kb) encodes the N nucleoprotein and the nonstructural protein (NSs) [17]. The viral ribonucleoproteins (RNPs) are composed of three genomic segments, associated with numerous copies of the N nucleoprotein, and the RNA-dependent RNA polymerase L is packaged into the virion. The RNPs of RVFV have a string-like appearance, while other negative-sense RNA viruses have a structure with helical symmetry [18].

62.1.2 EPIDEMIOLOGY

RVFV was isolated in 1930 after the first outbreak near Lake Naivasha in the Great Rift Valley of Kenya. At that time, RVF was known as a disease of ruminants, particularly of cattle and sheep following sudden death of lambs and ewes [19]. In 1951, severe acute and fatal disease in cattle and sheep caused an estimated death of 100,000 sheep and 500,000 ewes due to abortions that reached 40%–100% among pregnant animals. This was the first epidemic where severe diseases in humans were also reported [20]. Within a short period, RVF was reported in the neighboring countries of Mozambique, Namibia [21], Zambia, and Zimbabwe [22,23].

In 1977, RVFV emerged from the north of the Sahara desert, where it was responsible for a massive epidemic with abortion rates estimated between 80% and 100% in pregnant animals along the Nile River and Delta in Egypt [24]. During the outbreak in Egypt from 1977 to 1978, there were an estimated 200,000 human infections and at least 594 deaths and an antibody prevalence of 29.6% among humans [25]. Common disease symptoms that were associated with RVFV infection in humans included hepatic necrosis with fatal hemorrhagic fever, delayed onset of encephalitis, permanent neurologic deficits, and severe retinitis with permanent blindness [26]. Transmission of the virus to humans was thought to have occurred primarily by direct contact with or exposure to aerosols from RVFV-infected animals or contaminated blood and tissues. The outbreaks in Egypt were possibly the result of windblown-infected mosquitoes, transport of infected animals from the neighboring countries, infected adult mosquitoes emerging from hibernation, or the return of the Egyptian troops from peacekeeping missions in Congo [27].

The introduction of RVFV outside of Africa was reported for the first time in 2000. The outbreaks in Saudi Arabia and Yemen were associated with severe RVFV infections that resulted in fatalities of both humans and livestock. There were an estimated 2000 human infections resulting in at least 245 human deaths and several thousands of sheep, goat, and camel deaths [28]. Phylogenetic analysis of six RVFV strains isolated from *Aedes* mosquitoes during the Saudi Arabia–Yemen outbreak showed no difference to viruses present during the 1997–1998 Kenyan outbreak, and thus the introduction was likely due to imported livestock or infected mosquitoes from Africa [29].

In Madagascar, RVFV was first isolated in 1979 from a pool of mosquitoes from the highlands; however, neither human nor animal cases were reported at that time [30]. The first epizootics occurred in 1990–1991 in humans and livestock in the highlands during two consecutive rainy seasons [31]. In 2008–2009, a large RVFV outbreak occurred,

affecting livestock and humans throughout the country with 59 human cases, 19 of which were fatal [32]. A phylogenetic analysis of RVFV strains isolated in 1979, 1990–1991, and 2008–2009 revealed that strains were genetically different and that the virus was most likely introduced from the African mainland by cattle trade [33]. In West Africa, the first RVF outbreak occurred in Mauritania in 1987 and resulted in 220 human deaths [34]. In 1998, an RVF outbreak occurred in southeastern Mauritania, resulting in 300–400 human cases and 6 deaths [35]. In October 2003, nine human cases of hemorrhagic fever were reported in the three provinces of Mauritania as a result of acute RVFV infection. In this case, RVFV was isolated from the mosquito species *Culex poicilipes* and phylogenetic studies revealed little variation among the isolated strains [36]. In 1997, the largest RVFV outbreak was recorded in East Africa with an estimated 27,500 human infections in the Garissa District of Kenya alone [37]. In December 2006, an RVF outbreak was also identified in the Garissa District; however, virus activity was subsequently identified in other parts of Kenya as well as Somalia and Tanzania [38].

Historically, RVF outbreaks in Africa occur at irregular intervals of 5–15 years in the savannah grasslands and 25–35 years in the semiarid regions. Climate and weather are the factors that influence RVFV transmission patterns. In East Africa, outbreaks are significantly influenced by heavy rainfall, which causes natural depressions called "dambos," where RVFV-infected *Aedes* mosquito eggs are laid and hatched. More recently, many devastating outbreaks affecting both humans and domestic animals have occurred throughout a number of African countries including Kenya (2006) [10], Somalia (2006) [39], Sudan (2007) [40], Tanzania (2007) [41], Madagascar (2008) [42], Botswana (2010) [43], Mauritania (2010) [3], Namibia (2010) [44], and South Africa (2008–2011) [45].

62.1.3 CLINICAL FEATURES AND PATHOGENESIS

The clinical features and pathogenesis of RVFV have been reviewed recently [46–49] and thus will only be summarized here. The review by Davies and Martin [49] provides a thorough description of how to clinically recognize RVF disease. RVFV primarily causes disease in ruminant animals, including sheep, goats, cattle, and camels. The clinical signs of RVF are typically high fever, lymphadenitis, and nasal and ocular discharges in mature animals. Vomiting, abdominal colic, severe prostration, dysgalactia, and jaundice have also been noted. The disease is also associated with profuse hemorrhagic diarrhea. More specific to RVF is the sudden onset of abortion storms, with up to 100% mortality in newborn lambs. Disease outbreaks can affect as few as 15% of susceptible animals to as many as 90% [50]. The disease in other animals such as horses, dogs, cats, and rodents, which can become viremic, is usually subclinical or inapparent. The role of buffaloes, rodents, and other wild animals as potential reservoirs is still unclear [51–53]. Infection in humans can be subclinical to fatal with disease symptoms including high fever, encephalitis with confusion and coma, ocular degeneration, and hemorrhage [54]. More recent case fatality rates as high as 20% to 50% have been reported [38,90]. Clinical features during this particular epidemic included retinal impairment or failure in 27.5% of human patients [56]. Many of the patients with ocular lesions (71%) remained legally blind for up to 5 years after initially acquiring the disease [57].

62.1.4   DIAGNOSIS

Livestock abortions or sudden deaths during the rainy seasons are the first indicators of RVF. Standard diagnostic assays are available for RVF but availability can be limited. The international-approved protocols for these tests are described in the World Organization for Animal Health, Office International des Epizooties Terrestrial Manual [58]. Methods used include virus neutralization test, hemagglutination inhibition tests, immunofluorescence assay, and complement fixation test (CFT). The detection of immunoglobulin G or M (IgG or IgM) antibodies specific for RVFV in serum samples is widely used in laboratories to confirm the disease. A number of viral nucleic acid–based diagnostic tests have also been described [59].

62.1.4.1   Conventional Techniques

Virus neutralization test (VNT): VNTs including microneutralization, plaque reduction neutralization test (PRNT) [60], and neutralization in mice have been used to detect antibodies against RVFV in the serum of a variety of species. VNTs are highly specific and sensitive, but these tests require live viruses; thus, trained personnel can only do them in appropriate biosecurity facilities. Attenuated strains, such as the MP-12 vaccine strain, can be used for VNT assays providing a good estimation of wild-type virus VNT results.

Hemagglutination inhibition (HI): Sucrose/acetone-extracted hamster liver antigen and erythrocytes are used in the HI test, following well-established protocols [58]. The antigen is initially mixed with the RVFV serum sample and incubated overnight at 4°C; erythrocytes are then added to the antigen–serum mixtures. Results are read after 30 min at room temperature. A solid-phase immunosorbent technique can be developed to detect RVFV-specific IgM in human serum samples. Plates coated with goat antihuman IgM are incubated with serum samples, RVFV hemagglutinating antigen, and erythrocytes. The RVFV-specific IgM in the serum samples binds to the RVFV antigen and inhibits hemagglutination of erythrocytes. The advantages of HI assays are that results can be obtained within 5 h, reagents are inexpensive, and several hundred serum samples can be tested daily without the need for specialized equipment. However, there is the potential for false-positive results due to nonspecific agglutinins.

Indirect immunofluorescence assay (IFA): RVFV may be detected by immunofluorescence on impression smears of the liver, spleen, and brain. This is a method to detect viral infection in multiple types of host cells. This procedure can be modified for both high-throughput and low-throughput assays for any virus for which antibodies are available. In comparison with PRNT, IFA is less sensitive in detecting RVF viral antibodies than either HI or enzyme-linked immunosorbent assay (ELISA). Although reagents for IFA are relatively easy to obtain, IFA is very subjective and interpretation of results may vary between observers, thus resulting in false-positives (low specificity). In addition, IFA requires a fluorescence microscope, which may not be readily available to every diagnostic lab.

Complement fixation test (CFT): CFT is generally used for detecting RVF viral antibodies during acute infections. Although CFT is quite specific, the low sensitivity of this method in detecting RVF viral antibodies, even during acute infections, limits its usefulness [58].

62.1.4.2   Molecular Techniques

Molecular diagnostic methods allow the detection and identification of virus-specific nucleic acid, proteins, and antibodies with high specificity and sensitivity. Although they do require specialized reagents and equipment, they are faster than virus isolation or VNTs, which can take up to a week, and they are safer for the laboratory personnel performing the assays, because infectious samples can be inactivated prior to testing.

62.1.4.2.1   Reverse-Transcription Polymerase Chain Reaction

There are a number of sequence-based assays that have been developed with broad to very specific diagnostic capabilities. Reverse-transcription polymerase chain reaction (RT-PCR) assays enable rapid detection with high-throughput and multiplexing capabilities, are specific and sensitive, and can be performed by biosafety level (BSL)-2 laboratories after inactivation of infectious samples. This type of assay can be performed on blood, serum, and homogenized tissues. RT-PCR for the detection of RVFV nucleic acid (RNA) is important for early diagnosis and monitoring acute viremia. A detailed summary of published primers and probes designed for RT-PCR and real-time RT-PCR detection of RVFV from 1997 to 2011 has been reviewed previously [61]. These will be discussed here briefly with the addition of assays published in 2012 and 2013.

Early RT-PCR assays for RVFV were developed as single target assays to detect Gn-specific amplification of the M segment or NSs of the S segment of the RVFV genome [58]. These were effective as diagnostic tools for screening human and animal sera early in infection during the outbreaks of the disease [62]. Advancement to real-time RT-PCR assays allowed for not only detection of RVFV but also quantitation of the virus. Real-time RT-PCR assays were developed to target the L, M, and S segments of the RVFV genome [63]. Highly sensitive and broadly reactive, these assays could detect 21 distinct strains representing all lineages of RVFV with as little as 5 RNA copies per reaction [64] or 37 different RVFV strains with a limit of 100 RNA copies per reaction [65]. To expedite the procedure for a more high-throughput format optimal for diagnostic laboratories, a one-step real-time RT-PCR assay for RVFV was developed [66] based on the primers published by [64]. A more economical real-time RT-PCR assay was published using the SYBR Green for the detection of RVFV with a limit of 300 RNA copies [67]. More recently, a real-time RT-PCR assay able to detect all three RVFV genome segments with differentiation of infected and vaccinated animals (DIVA) compatibility was developed [63]. This assay uses both L and M segments as confirmatory targets and the NSs as the DIVA-compatible marker, which is deleted in the commercially available clone 13 vaccine in South Africa as well as other candidate vaccines in development.

More broad-based and multiplexed assays have been developed as well, which allow detection of multiple viruses per test. A nested RT-PCR assay for the detection of phleboviruses in general was able to detect RVF, Toscana, sandfly fever, Aguacate, and Punta Toro viruses [68]. A multiplexed real-time RT-PCR assay to diagnose 8 different hemorrhagic fever viruses including RVFV was developed with a detection limit of 2.8×10^3 RVFV genome/mL plasma [69]. Another multiplex assay to detect RVFV and other RNA emerging viruses had a limit of 1.5×10^3 RNA copies per reaction for the detection of RVFV [70]. A more sensitive multiplex RT-PCR assay was able to detect and specifically identify RVFV, bluetongue virus, rinderpest virus, and peste des petits ruminants virus, with a limit of $10^{1.1}$ TCID$_{50}$/mL of an RVFV blood sample [71].

62.1.4.2.2   Reverse-Transcription Loop-Mediated Isothermal Amplification

RT loop-mediated isothermal amplification (RT-LAMP) assays that are faster than real-time RT-PCR, are highly sensitive, and have been successful for detecting a wide range of RVFV strains have also been developed. This type of assay uses multiple primer sets and can be performed at a constant temperature without the need for a thermocycler and is therefore advantageous for field diagnostic testing. RT-LAMP assays targeting the L segment are able to detect RVFV within 30 min, with as few as 10 RNA copies per reaction [72,73]. A more recent assay targeting the S segment demonstrated sensitivity of 19 RVFV molecules and was able to detect 20 different RVFV strains [74].

62.1.4.2.3   Enzyme-Linked Immunosorbent Assay

Although rapid and sensitive, RNA-based methods are most effective at an early time during the infection while the animal or individual is viremic, which may only last for a very short period of time. Reactions can also become easily contaminated, compromising the assay. Contamination may come

from the environment where the specimen was collected or from within the lab where samples are prepared. Therefore, confirmatory tests based on the presence of viral antigen or antibodies are desirable. A number of ELISA formats for the detection of RVFV antigens or antibodies have been developed and have been reviewed previously by [48]. The disadvantages of many antigen-based assays include reagents that can be cumbersome and expensive to produce and, in many cases, require handling of live virus. The development of antigen for ELISAs using recombinant proteins has helped address safety concerns since they do not require infectious materials and samples can be inactivated prior to performing the procedure. Recombinant nucleocapsid proteins (rNPs) have been validated extensively and shown to be highly accurate for the detection of anti-RVFV antibody in humans and animals [75–78]. A sandwich ELISA for antigen detection using an RVFV rNP has proven to be effective at detecting a variety of RVFV strains from both human and animal samples at limits as low as $\log_{10}10^{2.2}$ TCID$_{50}$ per reaction and with a little cross-reactivity to other closely related phleboviruses [79]. An indirect ELISA based on RVFV recombinant glycoprotein Gn for the detection of IgG antibodies recently developed demonstrated sensitivity and specificity above 94% compared to viral neutralization [80]. The concurrent detection of IgG and IgM antibodies to RVFV has also been described [81]. Reactivity profiles of several baculovirus-expressed RVFV recombinant proteins against a panel of immune sera from natural hosts were recently evaluated and may be promising candidates for the development of improved diagnostic methods, which incorporate DIVA compatibility [82].

62.1.4.2.4   Antibodies for the Detection of Rift Valley Fever Virus–Specific Antigens

The generation of good monoclonal antibodies is also important for improving the specificity of ELISAs and other rapid antigen detection tests such as dipstick/strip tests or lateral flow devices [83,84]. Anti-RVFV antibodies are also important tools for the detection of RVFV antigen by immunohistochemistry (IHC) in fixed tissues [66]. The generation of monoclonal antibodies with a high specificity against recombinant RVFV proteins is needed for the development of safer diagnostic reagents and methods [84].

It is recommended that RVFV-specific antibody tests be run in parallel with nucleic acid–based and antigen-based detection methods because the short duration of RVFV viremia limits the effective time frame for use of these detection methods. Seroconversion can be determined early after infection, in subclinical or mild infections, and postviremia using RVFV-specific antibody assays [60].

62.1.4.2.5   Fluorescent Microsphere–Based Immunoassays

Fluorescent microsphere–based immunoassays (FMIAs) are promising for rapid, high-throughput, and multiplexed diagnostic tests based on nucleic acid, antigen, or antibody detection. FMIAs are simple to run, are cost-effective, and have the potential to be highly sensitive and specific. Differential assays can be designed for the detection of multiple targets of a single pathogen or the detection of multiple pathogens using different color markers assigned to each target. The development of these types of multiplex assays enables the detection of RVFV and/or other pathogens in a single reaction with a minimal sample requirement. An assay for the simultaneous detection of RVFV antibodies against the virus nucleocapsid and the Gn glycoprotein was recently published [85]. To increase specificity and provide potential DIVA compatibility, additional RVFV antigens were added. The resulting test includes four recombinant antigens, N nucleoprotein, two nonstructural proteins (NSm and NSs), and two glycoproteins (Gn and Gc), which are produced in *E. coli* or baculovirus and affinity purified [86]. The advantage of FMIA is the ability to detect multiple targets in a single assay. Multiplex panels containing domestic pathogen targets in a nonendemic country that include RVFV targets as markers to reroute positive samples to national laboratories could result in early detection and rapid response.

62.2 METHODS

62.2.1 SAMPLE PREPARATION

Handling and preparation of infected samples poses great health risk to personnel and often requires selecting BSL-3 or BSL-4 laboratories for processing suspected RVFV-infected specimens. Furthermore, there are few labs that are capable to handle such samples. Therefore, there is a critical need to safely inactivate infected tissues that would allow safe handling and expedite diagnostic confirmation by BSL-2 laboratories. For assays such as virus isolation or VNTs, inactivation is not an option; however, inactivation is permissible for most molecular diagnostic assays. The inactivation method used is dependent on the type of sample and the type of assay that will be used. Formalin fixation is usually suitable for tissues that are being prepared for histopathology and IHC analysis. For blood and serum samples tested by ELISA or FMIA, thermochemical inactivation is sufficient. For example, 1% Tween-20 added at an equal volume followed by incubation at 56°C for 1 h was demonstrated for the inactivation of RVFV up to $\log_{10} 10^{6.5}$ $TCDI_{50}/mL$ [79]. For RNA-based assays, the virus is inactivated by the addition of guanidinium isothiocyanate and phenol-containing RNA extraction reagents that are available commercially, such as TriPure or TRIzol reagents [87]. The preferred method is to use phenol-containing RNA extraction, as that is the most effective means to inactivate the virus. Magnetic bead–based systems that can be automated are then used to provide cleaner template for amplification [88].

The serum, brain, spleen, and liver are the preferred tissues for diagnostic testing. The serum is ideal for diagnostic screening because sample preparation requires little manipulation and equipment, and it can be collected with minimal stress from the animal or individual. To improve detection by molecular diagnostic assays, one study found out that detection of RVFV was enhanced when whole-blood samples were used versus the serum only, likely due to the association of RVFV with monocytes and macrophages [88]. However, it may not be possible to acquire blood or serum samples especially if the animal is found deceased. Tissues are typically the first samples submitted to diagnostic labs during an outbreak. The condition of the tissues submitted, as well as which diagnostic tools are available to the reference laboratory, determines the type of assay used for diagnostic testing.

62.2.2 DETECTION PROCEDURES

Nucleic acid amplification systems: Given the high sensitivity of nucleic acid amplification assays, the availability of a dedicated "clean," nucleic acid–free PCR hood and workspace is critical in helping reduce contamination risk and increase assay reliability. Gloves should be worn and changed often, and RNase-free reagents and materials should be used. Sample template should be prepared in a physically separated area if possible and added last to the reaction mix just prior to amplification. For RT-PCR, as well as for all other diagnostic tests, standardized positive and negative controls should be prepared and analyzed alongside test samples. Due to issues from potential contamination, closed systems such as real-time RT-PCR are recommended. Real-time RT-PCR samples with amplification occurring at cycle threshold (Ct) < 35.00 should be considered positive. If amplification occurs, but has a Ct value between 35 and 40, and/or the curve is low to the threshold, the real-time RT-PCR should be repeated and/or the RNA should be reextracted and tested again. If amplification does not occur, a Ct value must be obtained for the positive control to ensure the RNA extraction procedure was successful. At a minimum, an external amplification control should be used to monitor for inhibitors. Ideally, an internal or constitutively transcribed gene is also used to monitor for the quality of the RNA in the sample.

The multiplex RVF real-time RT-PCR assay was designed to have internal confirmation and be DIVA compatible with RVFV NSs-deleted vaccines [63]. This assay combines standard real-time RT-PCR assays [64,69] for detection and internal confirmation. The NSs target was specifically designed to be in the NSs coding region, thus DIVA compatible [63]. Finally, an external RNA amplification control was used to monitor amplification inhibitors [89]. The primer and probe designs are listed in Tables 62.1 and 62.2. The reaction mixture is described in Table 62.3. Interpretation of the results is described in Table 62.4. The primers and probe mixtures can be adjusted to meet the individual purposes. For example, if there is no need for a DIVA-compatible diagnostic test, the S primers and probe can be omitted. The amplification conditions consist of 45°C for 10 min and 95°C for 10 min, followed by 40 cycles of 45°C for 15 s and then 60°C for 60 s.

TABLE 62.1

Rift Valley Fever Virus Primers for Real-Time Reverse-Transcription Polymerase Chain Reaction

Primer	Final Conc. (μM)	Nucleotide Sequence 5′–3′
RVFL-2912fwdG[a]	10	TGA-AAA-TTC-CTG-AGA-CAC-ATG-G
RVFL-2981revAC	10	ACT-TCC-TTG-CAT-CAT-CTG-ATG
RVF-MP12–3269F[b]	10	CCT-CAC-TAT-TAC-ACA-CCA-TTC
RVF-MP12–3453R	10	ATC-ATC-AGC-TGG-GAA-GCT
RVFV-M(G2)-F(RVAs)[a]	10	CAC-TTC-TTA-CTA-CCA-TGT-CCT-CCA-AT
RVFV-M(G2)-R (RVS)	10	AAA-GGA-ACA-ATG-GAC-TCT-GGT-CA
RVFV-S(NSs)-F	20	CCC-AAT-CTG-AAA-GAA-GCC-AT
RVFV-S(NSs)-R	20	GAG-TGG-CAA-TCT-GAT-CCC-TT
XIPC F[c]	25	TTC-GGC-GTG-TTA-TGC-TAA-CTT C
XIPC R[c]	25	CCA-CTG-CGC-CCA-ACC-TT

Source: Adapted from Wilson, W.C. et al., *J. Virol. Methods*, 193, 426, 2013.

[a] RVFL and RVFM primers and probes are identical to the previous publications [64,69].

[b] MP12-specific primers and probe substituted for RVFL primer and probe for increase sensitivity detecting RVF MP-12 vaccine strain.

[c] XIPC primers and probes are identical to those described previously [89].

TABLE 62.2

Rift Valley Fever Virus Probes for Real-Time Reverse-Transcription Polymerase Chain Reaction

Probe	Fluorescent Reporter Dye (5′ End)	Quencher (3′ End)	Final Conc. (μM)	Nucleotide Sequence (5′–3′)
RVFL-probe-2950	CAL FLUOR RED 610	BHQ2	10	CAA-TGT-AAG-GGG-CCT-GTG-TGG-ACT-TGT-G
RVF-MP12-3371P	CAL FLUOR RED 610	BHQ2	5	CTG-AGA-TGA-GCA-AGA-GCC-TGG-TTT-GTG-A
RVFV-M(G2)	FAM	BHQ1	1	AAA-GCT-TTG-ATA-TCT-CTC-AGT-GCC-CCA-A
RVFV-S(NSs)	QUASAR 670	BHQ2	10	TCC-TGG-CCT-CTT-GGA-GAA-CCC-TC
XIPC pb	Cy3	BHQ2	15	CTC CGC AGA AAT CCA GGG TCA TCG

Source: Adapted from Wilson, W.C. et al., *J. Virol. Methods*, 193, 426, 2013.

TABLE 62.3

Rift Valley Fever Virus Polymerase Chain Reaction Mix Preparation

Master Mix	1× (μL)
RNase-free water	4.5
25× enzyme mix	1
2× reaction mix	12.5
12.5× primer/probe assay mix	2
Volume of master mix per tube	20
Volume of template	5
Total volume	25

TABLE 62.4

Interpretation of the Rift Valley Fever Virus Multiplex Real-Time Reverse-Transcription Polymerase Chain Reaction Results

L Segment (CFR610) Status	M Segment (FAM) Status	S Segment (QUASAR670) Status	XIPC (Cy3) Internal Control RNA	Sample Interpretation
Negative	Negative	Negative	Positive	The sample is negative for RVFV.
Positive	Negative	Negative	Positive	The sample is positive for L segment and deemed positive for RVFV.
Negative	Positive	Negative	Positive	The sample is positive for M segment and deemed positive for RVFV.
Negative	Negative	Positive	Positive	The sample is positive for S segment and deemed positive for RVFV.
Positive	Positive	Negative	Positive	The RVFV sample is possibly NSs-deleted vaccinate.
Positive	Positive	Positive	Positive	L/M/S segments positive, the sample is positive for RVFV.
Positive/negative	Positive/negative	Positive/negative	Negative	Probable loss of RNA during extraction procedure. There is a new RNA extraction necessary to confirm results of assay.

62.3 CONCLUSION

There are numerous assays available for the detection of antibodies to RVFV, viral antigen, or RNA. The limitation of many of these assays is the requirement for high biosecurity to ensure the safety of diagnostic staff. Although there are ELISAs available for the detection of antibodies to RVFV, often the reagents have limited availability and are expensive to produce. Several research laboratories have developed recombinant protein–based RVFV serological tests that provide safer alternatives to assays that require production of RVFV-infected cell lysates. Two of these assays are commercially available (BDSL, Ayrshire, Scotland, and IdVet, Grabels, France). The molecular-based real-time assays such as the one described here provide rapid inactivation of this zoonotic viral pathogen and high sensitivity for the detection of RVFV RNA. Detection of viral RNA should not be considered as definitive confirmation of infectious virus, but it is a rapid and strong presumptive test. Data from several studies demonstrate the importance of combining assays (e.g., antigen capture, RT-PCR, or virus isolation) for the identification of viruses and diagnosis of the disease, as well as monitoring of virus-specific IgM to ensure that no acutely RVFV-infected animals or patients are missed [56].

REFERENCES

1. Nicholas, D.E., K.H. Jacobsen, and N.M. Waters, risk factors associated with human Rift Valley fever infection: Systematic review and meta-analysis. *Trop. Med. Int. Health*, 19, 1420–1429, 2014.
2. Rich, K.M. and F. Wanyoike, An assessment of the regional and national socio-economic impacts of the 2007 Rift Valley fever outbreak in Kenya. *Am. J. Trop. Med. Hyg.*, 83, 52–57, 2010.
3. Faye, O. et al., Reemergence of Rift Valley fever, Mauritania, 2010. *Emerg. Infect. Dis.*, 20, 300–303, 2014.
4. Anyamba, A. et al., Prediction of a Rift Valley fever outbreak. *Proc. Natl. Acad. Sci. USA*, 106, 955–959, 2009.
5. Britch, S.C. et al., Rift Valley fever risk map model and seroprevalence in selected wild ungulates and camels from Kenya. *PLoS One*, 8, e66626, 2013.
6. Weaver, S.C. and W.K. Reisen, Present and future arboviral threats. *Antiviral Res.*, 85, 328–345, 2010.
7. Chevalier, V., Relevance of Rift Valley fever to public health in the European Union. *Clin. Microbio.l Infect.*, 19, 705–708, 2013.
8. Hartley, D.M. et al., Potential effects of Rift Valley fever in the United States. *Emerg. Infect. Dis.*, 17, e1, 2011.
9. Elliott, R.M., Molecular biology of the Bunyaviridae. *J. Gen. Virol.*, 71, 501–522, 1990.
10. Bird, B.H. et al., Multiple virus lineages sharing recent common ancestry were associated with a large Rift Valley fever outbreak among livestock in Kenya during 2006–2007. *J. Virol.*, 82, 11152–11166, 2008.
11. Bouloy, M. and F. Weber, Molecular biology of Rift Valley fever virus. *Open Virol. J.*, 4, 8–14, 2010.
12. Ellis, D.S. et al., Morphology and development of Rift Valley fever virus in vero cell cultures. *J. Med. Virol.*, 24, 161–174, 1988.
13. Freiberg, A.N. et al., Three-dimensional organization of Rift Valley fever virus revealed by cryoelectron tomography. *J. Virol.*, 82, 10341–10348, 2008.
14. Sherman, M.B. et al., Single-particle cryo-electron microscopy of Rift Valley fever virus. *Virology*, 387, 11–15, 2009.
15. Gerrard, S.R. et al., The NSm proteins of Rift Valley fever virus are dispensable for maturation, replication and infection. *Virology*, 359, 459–465, 2007.
16. Weingartl, H.M. et al., Rift Valley fever virus incorporates the 78 kDa glycoprotein into virions matured in mosquito C6/36 cells. *PLoS One*, 9, e87385, 2014.
17. Brennan, B., S.R. Welch, and R.M. Elliott, The consequences of reconfiguring the ambisense s genome segment of Rift Valley fever virus on viral replication in mammalian and mosquito cells and for genome packaging. *PLoS Pathog.*, 10, e1003922, 2014.
18. Raymond, D.D. et al., Structure of the Rift Valley fever virus nucleocapsid protein reveals another architecture for RNA encapsidation. *Proc. Natl. Acad. Sci. USA*, 107, 11769–11774, 2010.
19. Daubney, R.J., Enzootic hepatitis of Rift Valley fever: An undescribed virus disease of sheep, cattle and man from East Africa. *J. Pathol. Bacteriol.*, 34, 543–579, 1931.

20. Swanepoel, R. and J.A.W. Coetzer, Rift Valley fever, in *Infectious Diseases of Livestock*, Coetzer, J.A.W., Thomson, G.R., and Tustic, R.C., Eds. Oxford University Press, Cape Town, South Africa, pp. 688–717, 1994.

21. Chevalier, V. et al., Epidemiological processes involved in the emergence of vector-borne diseases: West Nile fever, Rift Valley fever, Japanese encephalitis and Crimean-Congo haemorrhagic fever. *Rev. Sci. Tech.*, 23, 535–555, 2004.

22. Dautu, G. et al., Rift Valley fever: Real or perceived threat for Zambia? *Onderstepoort J. Vet. Res.*, 79, 94–99, 2012.

23. Davies, F.G. et al., Patterns of Rift Valley fever activity in Zambia. *Epidemiol. Infect.*, 108, 185–191, 1992.

24. Ghoneim, N.H. and G.T. Woods, Rift Valley fever and its epidemiology in Egypt: A review. *J. Med.*, 14, 55–79, 1983.

25. Meegan, J.M., R.H. Watten, and L.W. Laughlin, Clinical experience with Rift Valley fever in humans during the 1977 Egyptian epizootic. *Contr. Epidem. Biostatist.*, 3, 114–123, 1981.

26. Meegan, J.M., Rift Valley fever in Egypt: An overview of the epizootics in 1977 and 1978. *Contrib. Epidemiol. Biostat.*, 3, 100–113, 1981.

27. Hoogstraal, H. et al., The Rift Valley fever epizootic in Egypt 1977–1978. 2. Ecological and entomological studies. *Trans. R. Soc. Trop. Med. Hyg.*, 73, 624–629, 1979.

28. Ahmad, K., More deaths from Rift Valley fever in Saudi Arabia and Yemen. *Lancet*, 356, 1422, 2000.

29. Abdo-Salem, S. et al., Risk assessment of the introduction of Rift Valley fever from the Horn of Africa to Yemen via legal trade of small ruminants. *Trop. Anim. Health Prod.*, 43, 471–480, 2011.

30. Fontenille, D., C. Mathiot, and P. Coulanges, Hemorrhagic fever viruses in Madagascar. *Arch. Inst. Pasteur Madagascar*, 54, 117–124, 1988.

31. Morvan, J. et al., Rift Valley fever epizootic in the central highlands of Madagascar. *Res. Virol.*, 143, 407–415, 1992.

32. Andriamandimby, S.F. et al., Rift Valley fever during rainy seasons, Madagascar, 2008 and 2009. *Emerg. Infect. Dis.*, 16, 963–970, 2010.

33. Carroll, S.A. et al., Genetic evidence for Rift Valley fever outbreaks in Madagascar resulting from virus introductions from the East African mainland rather than enzootic maintenance. *J. Virol.*, 85, 6162–6267, 2011.

34. Digoutte, J.P. and C.J. Peters, General aspects of the 1987 Rift Valley fever epidemic in Mauritania. *Res. Virol.*, 140, 27–30, 1989.

35. Nabeth, P. et al., Rift Valley fever outbreak, Mauritania, 1998: Seroepidemiologic, virologic, entomologic, and zoologic investigations. *Emerg. Infect. Dis.*, 7, 1052–1054, 2001.

36. Faye, O. et al., Rift Valley fever outbreak with East-Central African virus lineage in Mauritania, 2003. *Emerg. Infect. Dis.*, 13, 1016–1023, 2007.

37. Woods, C.W. et al., An outbreak of Rift Valley fever in Northeastern Kenya, 1997–1998. *Emerg. Infect. Dis.*, 8, 138–144, 2002.

38. Nguku, P.M. et al., An investigation of a major outbreak of Rift Valley fever in Kenya: 2006–2007. *Am. J. Trop. Med. Hyg.*, 83, 05–13, 2010.

39. Nderitu, L. et al., Sequential Rift Valley fever outbreaks in Eastern Africa caused by multiple lineages of the virus. *J. Infect. Dis.*, 203, 655–665, 2011.

40. Hassan, O.A. et al., The 2007 Rift Valley fever outbreak in Sudan. *PLoS Negl. Trop. Dis.*, 5, e1229, 2011.

41. Jost, C.C. et al., Epidemiological assessment of the Rift Valley fever outbreak in Kenya and Tanzania in 2006 and 2007. *Am. J. Trop. Med. Hyg.*, 83, 65–72, 2010.

42. Jeanmaire, E.M. et al., Prevalence of Rift Valley fever infection in ruminants in Madagascar after the 2008 outbreak. *Vector Borne Zoonotic Dis.*, 11, 395–402, 2011.

43. Archer, B.N. et al., Outbreak of Rift Valley fever affecting veterinarians and farmers in South Africa, 2008. *S. Afr. Med. J.*, 101, 263–266, 2011.

44. Monaco, F. et al., Rift Valley fever in Namibia, 2010. *Emerg. Infect. Dis.*, 19, 2025–2027, 2013.

45. Pienaar, N.J. and P.N. Thompson, Temporal and spatial history of Rift Valley fever in South Africa: 1950 to 2011. *Onderstepoort J. Vet. Res.*, 80, 384, 2013.

46. Ikegami, T. and S. Makino, The pathogenesis of Rift Valley fever. *Viruses*, 3, 493–519, 2011.

47. Boshra, H. et al., Rift Valley fever: Recent insights into pathogenesis and prevention. *J. Virol.*, 85, 6098–6105, 2011.

48. Pepin, M. et al., Rift Valley fever virus (Bunyaviridae: Phlebovirus): An update on pathogenesis, molecular epidemiology, vectors, diagnostics and prevention. *Vet. Res.*, 41, 61, 2010.

49. Davies, F.G. and V. Martin, Recognizing Rift Valley fever. *Vet. Ital.*, 42, 31–53, 2006.

50. FAO. Animal Production and Health. Emergency Prevention System for Transboundary Animal and Plant Pests and Diseases (EMPRES). Caracalla, Italy. Available at http://www.fao.org/ag/againfo/programmes/en/empres.html. Accessed on March 31, 2015.

51. Beechler, B.R. et al., Rift Valley fever in Kruger National Park: Do buffalo play a role in the inter-epidemic circulation of virus? *Transbound Emerg. Dis.*, 2013.

52. Olive, M.M., S.M. Goodman, and J.M. Reynes, The role of wild mammals in the maintenance of Rift Valley fever virus. *J. Wildl. Dis.*, 48, 241–266, 2012.

53. LaBeaud, A.D. et al., Rift Valley fever virus infection in African buffalo (*Syncerus caffer*) herds in rural South Africa: Evidence of interepidemic transmission. *Am. J. Trop. Med. Hyg.*, 84, 641–646, 2011.

54. Gerdes, G.H., Rift Valley fever. *Rev. Sci. Tech.*, 23, 613–623, 2004.

55. Flick, R. and M. Bouloy, Rift Valley fever virus. *Curr. Mol. Med.*, 5, 827–834, 2005.

56. Madani, T.A. et al., Rift Valley fever epidemic in Saudi Arabia: Epidemiological, clinical, and laboratory characteristics. *Clin. Infect. Dis.*, 37, 1084–1092, 2003.

57. Al-Hazmi, A. et al., Ocular complications of Rift Valley fever outbreak in Saudi Arabia. *Ophthalmology*, 112, 313–318, 2005.

58. OIE (Office International des Epizooties), *OIE Terrestrial Manual* 2014. Rift valley fever.

59. Wilson, W.C. et al., Diagnostic approaches for Rift Valley fever. *Dev. Biol.* (Basel), 135, 73–78, 2013.

60. Paweska, J.T., F.J. Burt, and R. Swanepoel, Validation of IgG-sandwich and IgM-capture ELISA for the detection of antibody to Rift Valley fever virus in humans. *J. Virol. Methods*, 124, 173–181, 2005.

61. Ikegami, T., Molecular biology and genetic diversity of Rift Valley fever virus. *Antiviral Res.*, 95, 293–310, 2012.

62. Sall, A.A. et al., Use of reverse transcriptase PCR in early diagnosis of Rift Valley fever. *Clin. Diagn. Lab Immunol.*, 9, 713–715, 2002.

63. Wilson, W.C. et al., Development of a Rift Valley fever real-time RT-PCR assay that can detect all three genome segments. *J. Virol. Methods*, 193, 426–431 (Errata *J. Virol Methods* 25, 124, 2014), 2013.

64. Bird, B.H. et al., Highly sensitive and broadly reactive quantitative reverse transcription-PCR assay for high-throughput detection of Rift Valley fever virus. *J. Clin. Microbiol.*, 45, 3506–3513, 2007.

65. Weidmann, M. et al., Rapid detection of important human pathogenic Phleboviruses. *J. Clin. Virol.*, 41, 138–142, 2008.

66. Drolet, B.S. et al., Development and evaluation of one-step rRT-PCR and immunohistochemical methods for detection of Rift Valley fever virus in biosafety level 2 diagnostic laboratories. *J. Virol. Methods*, 179, 373–382, 2012.

67. Naslund, J. et al., Kinetics of Rift Valley fever virus in experimentally infected mice using quantitative real-time RT-PCR. *J. Virol. Methods*, 151, 277–282, 2008.

68. Sanchez-Seco, M.P. et al., Detection and identification of Toscana and other phleboviruses by RT-nested-PCR assays with degenerated primers. *J. Med. Virol.*, 71, 140–149, 2003.

69. Drosten, C. et al., Rapid detection and quantification of RNA of Ebola and Marburg viruses, Lassa virus, Crimean-Congo hemorrhagic fever virus, Rift Valley fever virus, dengue virus, and yellow fever virus by real-time reverse transcription-PCR. *J. Clin. Microbiol.*, 40, 2323–2330, 2002.

70. He, J. et al., Simultaneous detection of CDC Category "A" DNA and RNA bioterrorism agents by use of multiplex PCR & RT-PCR enzyme hybridization assays. *Viruses*, 1, 441–459, 2009.

71. Yeh, J.Y. et al., Simultaneous detection of Rift Valley Fever, bluetongue, rinderpest, and Peste des petits ruminants viruses by a single-tube multiplex reverse transcriptase-PCR assay using a dual-priming oligonucleotide system. *J. Clin. Microbiol.*, 49, 1389–1394, 2011.

72. Peyrefitte, C.N. et al., Real-time reverse-transcription loop-mediated isothermal amplification for rapid detection of rift valley Fever virus. *J. Clin. Microbiol.*, 46, 3653–3659, 2008.

73. Le Roux, C.A. et al., Development and evaluation of a real-time reverse transcription-loop-mediated isothermal amplification assay for rapid detection of Rift Valley fever virus in clinical specimens. *J. Clin. Microbiol.*, 47, 645–651, 2009.

74. Euler, M. et al., Development of a panel of recombinase polymerase amplification assays for the detection of biothreat agents. *J. Clin. Microbiol.*, 2013.

75. Jansen van Vuren, P. et al., Preparation and evaluation of a recombinant Rift Valley fever virus N protein for the detection of IgG and IgM antibodies in humans and animals by indirect ELISA. *J. Virol. Methods*, 140, 106–114, 2007.

76. Paweska, J.T., P. Jansen van Vuren, and R. Swanepoel, Validation of an indirect ELISA based on a recombinant nucleocapsid protein of Rift Valley fever virus for the detection of IgG antibody in humans. *J. Virol. Methods*, 146, 119–124, 2007.

77. Paweska, J.T. et al., Recombinant nucleocapsid-based ELISA for detection of IgG antibody to Rift Valley fever virus in African buffalo. *Vet. Microbiol.*, 127, 21–28, 2008.

78. Williams, R. et al. Validation of an IgM antibody capture ELISA based on a recombinant nucleoprotein for identification of domestic ruminants infected with Rift Valley fever virus. *J. Virol. Methods*, 177, 140–146, 2011.

79. Jansen van Vuren, P. and J.T. Paweska, Laboratory safe detection of nucleocapsid protein of Rift Valley fever virus in human and animal specimens by a sandwich ELISA. *J. Virol. Methods*, 157, 15–24, 2009.

80. Jackel, S. et al., A novel indirect ELISA based on glycoprotein Gn for the detection of IgG antibodies against Rift Valley fever virus in small ruminants. *Res. Vet. Sci.*, 95, 725–730, 2013.

81. Ellis, C.E. et al., Validation of an ELISA for the concurrent detection of total antibodies (IgM and IgG) to Rift Valley fever virus. *Ondersteoort J. Vet. Res.*, 81, #675, 2014.

82. Faburay, B. et al., Rift Valley fever virus structural and nonstructural proteins: Recombinant protein expression and immunoreactivity against antisera from sheep. *Vector Borne Zoonotic Dis.*, 13, 619–629, 2013.

83. Fukushi, S. et al., Antigen-capture ELISA for the detection of Rift Valley fever virus nucleoprotein using new monoclonal antibodies. *J. Virol. Methods*, 180, 68–74, 2012.

84. Fafetine, J.M. et al., Generation and characterization of monoclonal antibodies against Rift Valley fever virus nucleoprotein. *Transbound Emerg. Dis.*, 60(Suppl. 2), 24–30, 2013.

85. van der Wal, F.J. et al., Bead-based suspension array for simultaneous detection of antibodies against the Rift Valley fever virus nucleocapsid and Gn glycoprotein. *J. Virol. Methods*, 183, 99–105, 2012.

86. Hossain, M.M. et al., Multiplex detection of IgG and IgM to Rift Valley fever virus (RVFV) nucleoprotein (N), nonstructural proteins (NSm and NSs), and glycoprotein (Gn) in sheep and cattle. *Vector Borne Zoonotic Dis.*, submitted 2014.

87. Blow, J.A. et al., Virus inactivation by nucleic acid extraction reagents. *J. Virol. Methods*, 119, 195–198, 2004.

88. Wilson, W.C. et al., Evaluation of lamb and calf responses to Rift Valley fever MP-12 vaccination. *Vet. Microbiol.*, 172, 44–50, 2014.

89. Schroeder, M.E. et al., Development and performance evaluation of calf diarrhea pathogen nucleic acid purification and detection workflow. *J. Vet. Diagn. Invest.*, 24, 945–953, 2012.

90. Heald, R. Infectious disease surveillance update: Rift Valley fever in Mauritania. In: *The Lancet Infectious Diseases* 12, 915, 2012.

63 Schmallenberg Virus

Dongyou Liu

CONTENTS

63.1 Introduction 563
    63.1.1 Classification...563
    63.1.2 Morphology and Genome Organization...564
    63.1.3 Biology and Epidemiology 564
    63.1.4 Clinical Features and Pathogenesis 565
    63.1.5 Diagnosis 565
    63.1.6 Treatment and Prevention 566
63.2 Methods 566
    63.2.1 Sample Preparation...566
    63.2.2 Detection Procedures..566
        63.2.2.1 RT-PCR for Schmallenberg Virus Detection 566
        63.2.2.2 RT-PCR for Pan-Simbu Serogroup Detection 567
63.3 Conclusion 567
References..567

63.1 INTRODUCTION

Schmallenberg virus (SBV) is an orthobunyavirus that was first identified using a metagenomic approach from dairy cattle near the town of Schmallenberg in Germany in October 2011. The virus causes fever, diarrhea, and a drop in milk yield in ruminants, as well as congenital malformations and stillbirths in calves, lambs, and goat kids. Since its initial description, SBV has been reported in thousands of cattle, sheep, and goat farms across Europe [1].

63.1.1 CLASSIFICATION

SBV is classified in the Simbu serogroup, genus *Orthobunyavirus*, family *Bunyaviridae*.

The family *Bunyaviridae* covers an expanding group of enveloped, tripartite, negative-sense, single-stranded RNA viruses, which in the latest estimate includes about 330 viruses that form five genera, 97 species, and 81 tentative species, as well as a large number of unassigned isolates. Of the five genera (*Hantavirus*, *Nairovirus*, *Orthobunyavirus*, *Phlebovirus*, and *Tospovirus*) recognized within the family *Bunyaviridae*, the genera *Orthobunyavirus* and *Phlebovirus* involve mosquitoes and mammals in their life cycles, with humans acquiring the viruses via mosquito bites; the genus *Nairovirus* circulates between ticks and mammals, and humans become infected through tick bites; the genus *Hantavirus* is perpetuated in persistently infected small mammals (rodents and occasionally insectivorous shrews and bats), with incidental human infections arising from inhalation of aerosolized rodent excreta; and the genus *Tospovirus* is maintained between trips (commonly known as thunder flies or corn lice) and plants.

Accounting for 170 of the 350 viruses within the family *Bunyaviridae*, the genus *Orthobunyavirus* is separated into 48 species: Acara virus, Akabane virus (AKAV), Alajuela virus, Anopheles A virus, Anopheles B virus, Bakau virus, Batama virus, Benevides virus, Bertioga virus, Bimiti virus, Botambi virus, Bunyamwera virus, Bushbush virus, Bwamba virus, California encephalitis virus, Capim virus, Caraparu virus, Catu virus, Estero Real virus, Gamboa virus, Guajara virus, Guama virus, Guaroa virus, Kaeng Khoi virus, Kairi virus, Koongol virus, M'Poko virus, Madrid virus, Main Drain virus, Manzanilla virus, Marituba virus, Minatitlan virus, Nyando virus, Olifantsvlei virus, Oriboca virus, Oropouche virus, Patois virus, Sathuperi virus (SATV), Shamonda virus (SHAV), Shuni virus, Simbu virus, Tacaiuma virus, Tete virus, Thimiri virus, Timboteua virus, Turlock virus, Wyeomyia virus, and Zegla virus, with Bunyamwera virus being the prototype virus for the genus as well as for the family [2].

Examination of antigenic relationships among members of the genus *Orthobunyavirus* by plaque reduction neutralization, hemagglutination inhibition, complement fixation, and radial immunodiffusion tests led to the differentiation of 19 serogroups: Anopheles, Bakau, Bunyamwera, Bwamba, California, Capim, Gamboa, Group C, Guama, Koongol, Mapputta, Minatitlan, Nyando, Olifantsvlei, Patois, Simbu, Tete, Turlock, and Wyeomyia [3].

As one of the largest serogroups within the genus *Orthobunyavirus*, the Simbu serogroup consists of at least 25 related viruses, including Iquitos virus, Jatobal virus, Leanyer virus, Oropouche virus, Oya virus, Thimiri virus, Akabane serocomplex (AKAV, Sabo virus, and Tinaroo virus), Sathuperi serocomplex (SATV, Douglas virus [DOUV], and SBV), Shuni serocomplex (Aino virus and Shuni virus), Shamonda serocomplex (SHAV, Peaton virus, and Sango virus), and Simbu complex (Simbu virus). Members of the Simbu serogroup are arthropod-borne (or arboviruses), with transmission facilitated by *Culicoides* biting midges and mosquitoes. Some of the viruses in this serogroup produce a syndrome in ruminants referred to as arthrogryposis–hydranencephaly syndrome, presenting with abortions, stillbirths, and congenital defects in neonatal cattle, sheep, and goats, following infection during pregnancy.

First emerged in Europe in 2011, SMV shares 69% identity with AKAV in the L segment, 71% identity with Aino virus in the M segment, and 97% identity with SHAV in the S segment in the initial phylogenetic analysis [4]. Additional sequence comparisons revealed that the SBV shows the highest homology to SATV and DOUV in the M segment and has a close genetic relationship with SHAV in the S and L segments. Recent full-genome and serologic investigations indicate that SBV belongs to the SATV species and represents an ancestor of SHAV, which appears to be a reassortant containing the S and L genomic segments from SBV and the M segment from an unclassified virus [5–7].

The closest relatives to SBV appear to be Sathuperi, Douglas, Shamonda, Akabane, Aino, Peaton, or Sango viruses, but these viruses have not been identified in Europe to date.

Among these, SATV was initially isolated in India and Nigeria in the 1980s and appeared in Japan in 1999.

SHAV was identified from cattle and *Culicoides* biting midges in Nigeria in the 1960s and emerged in Japan in 2002. The fact that the nucleotide sequences of Japanese strains of SHAV isolated in 2005 differ by only 3% from those of strains isolated in Nigeria 30 years before highlights its relative genetic stability [8].

AKAV is a teratogenic virus closely related to SBV, with presence in Australia, Japan, Israel, Turkey, and Cyprus. Spread by *Culicoides imicola*, AKAV infection of dams between the third and sixth month of the 9-month gestation results in congenital defects (primarily in the central nervous system and skeletal muscle) in cattle (see Chapter 59).

63.1.2 MORPHOLOGY AND GENOME ORGANIZATION

Morphologically, SBV is similar to other members of the family *Bunyaviridae* and has an enveloped, spherical virion of 100 nm in diameter, with surface glycoproteins (Gn and Gc) projecting from an outer envelope. The lipid layer of the envelope is acquired during budding at the Golgi apparatus lumen.

Beneath the envelope are three filamentous, helical nucleo-capsids (of 200–3000 nm in length and 2–2.5 nm in width), which are composed of multiple copies of nucleocapsid (N) protein encapsidating each of the three segments of anti-sense, single-stranded RNA, designated S (small, 830 nt), M (medium, 4,415 nt), and L (large, 6,865 nt) segments, with approximately 12,000 nt in total.

The L segment encodes the RNA-dependent RNA polymerase L (or L protein), which interacts with the nucleoprotein

N to form the viral polymerase complex. Inside the virus particle, the viral genome associates with many copies of the nucleoprotein N and a few copies of the polymerase L forming ribonucleoprotein (RNP).

The M segment encodes a precursor polyprotein that is cleaved cotranslationally into the envelope glycoproteins Gn and Gc and the nonstructural protein NSm, which is situated between the glycoproteins Gn and Gc in the precursor polyprotein. Consisting of three hydrophobic and two non-hydrophobic domains, NSm may act as a scaffolding protein during viral assembly.

The S segment encodes the nucleoprotein N and the non-structural protein NSs, which are encoded from the +1 open reading frame (ORF) within the N ORF. The NSs proteins are not essential for virus growth but contribute to viral pathogenesis.

63.1.3 BIOLOGY AND EPIDEMIOLOGY

Similar to other bunyaviruses, SBV utilizes the Gn/Gc het-erodimers on the viral surface to bind to cellular receptors and enters the host cell via endocytosis. This induces conformational changes in the viral glycoproteins and exposes the Gc fusion peptide. This facilitates fusion between viral envelope and endosomal membranes, and release of RNP inside the cytoplasm. Subsequent transcription produces viral mRNA, which is translated into viral proteins by host cell ribosomes. The L protein generates complementary copies of positive-sense antigenomes, from which negative-sense genomes are synthesized. The viral genome interacts with N proteins forming new genome RNP, which accumulates in the Golgi complex. The glycoproteins Gn and Gc form heterodimer complexes in the endoplasmic reticulum and are transported to the Golgi apparatus. The genome RNP interacts with the C-terminal domains of the glycoproteins Gn and Gc, and mature virus particles are then transported to the plasma membrane and then released in the extracellular compartment by exocytosis [9].

After its first description in dairy cattle in Germany in 2011, SBV was identified by reverse transcription polymerase chain reaction (RT-PCR) in sheep with the birth of malformed lambs displaying crooked neck, hydrocephalus, and stiff joints in the Netherlands the following month. By March 2012, the virus was shown to be present in cattle, sheep, or goats in Belgium, the United Kingdom, France, Luxembourg, Italy, and Spain [10,11]. The development of serological assays in early 2012 facilitated detection of anti-SBV antibodies in ruminants in Denmark, Switzerland, Austria, Poland, Sweden, Scotland, Finland, Ireland, Northern Ireland, Norway, Czech Republic, Estonia, Slovenia, Turkey, and Greece within a 2-year period [12–14]. Overall, SBV has been detected in 2062 cattle, 2482 sheep, and 77 goat farms by later 2012. However, relatively small percentages of sheep (up to 6.6%) and cattle (4%) in these farms were positive for SBV, suggesting the limited impact of SBV on animal health [15–17].

Members of the Simbu serogroup are known to be transmitted by *Culicoides* biting midges, and sometimes by mosquitoes of the *Aedes* and *Culex* genera as well as by several tick species. The fact that SBV was detected by RT-PCR in several *Culicoides* species (*C. obsoletus complex, C. scoticus, C. chiopterus, C. dewulfii, C. pulicaris,* or *C. nubeculosus*) in Belgium, Denmark, the Netherlands, Norway, Poland, and Sweden, where animal infections with SBV were diagnosed, confirms the role of biting midges in the transmission of this virus [18–22]. However, the absence of SBV in hibernating mosquitoes during winter indicates a limited role of mosquitoes in the persistence of SBV in colder months [23].

Besides cattle, sheep, and goats, SBV is infective to bison, wild cervids, and llamas, in which the virus does not seem to induce clinical signs or macroscopic abnormalities. Similarly, experimental infection of pigs with SBV induces seroconversion, but virus replication and transmission are ineffective [24]. However, SBV infection in humans has not been reported so far [16].

63.1.4 CLINICAL FEATURES AND PATHOGENESIS

SBV induces various clinical signs in ruminants of different age groups and host species.

In adult cows, clinical symptoms due to SBV infection include loss of appetite, hyperthermia, diarrhea, and reduction (up to 50%) in milk production. Symptoms often abate within a few days, along with transient viremia (lasting for 2–6 days) [25]. In adult sheep and goats, SBV infection causes very mild symptoms for and for only a few days [9,26].

However, SBV infection in pregnant females may facilitate virus transmission to fetuses (ovine, caprine, and bovine), leading to atypical congenital malformations, intrauterine death, death immediately after birth, or premature birth. The aborted and stillborn animals may show a neuromusculoskeletal disorder called arthrogryposis, severe torticollis, ankylosis, kyphosis, lordosis, scoliosis, brachygnathia inferior, and neurological disorders (e.g., amaurosis, ataxia, and/or behavioral abnormalities). In twin gestation, arthrogryposis versus neurological disorders, or malformation versus delayed growth, may be seen in one but not the other [9].

Necropsy of stillborn and malformed newborn reveals hydranencephaly (lack of the brain cerebral hemispheres), hydrocephaly, cerebral and cerebellar hypoplasia, and porencephaly. Calf born at term may display severe central nervous system lesions, dysfunctions of the cerebral cortex, basal ganglia and mesencephalon, porencephaly, or hydranencephaly [9,27].

Histological examination uncovers lymphohistiocytic meningoencephalomyelitis with glial nodules in the mesencephalon and hippocampus in lambs and goats and neuronal degeneration and necrosis in the brain stem of calves. Astrogliosis, microgliosis, and sometimes myofibrillar hypoplasia of skeletal muscles are detected in both calves and lambs. In addition, calf may display evidence of poliomyelitis [9].

SBV RNA is detectable in brain, blood, or spleen samples in malformed aborted or stillborn ruminants [28].

The observation of various neurological signs and associated pathological changes suggests that neurons are the major target for SBV replication in naturally infected newborn lambs and calves. Although the underlying mechanisms of SBV pathogenicity are yet to be elucidated, they may be inferred from findings derived from other bunyaviruses. For example, NSs from Bunyamwera (Bunyamwera serogroup) antagonizes the expression of type I interferon, inhibits the transcription mediated by RNA polymerase II, blocks protein synthesis, and contributes to apoptosis mediated by IRF-3 (interferon regulatory transcription factor) [9,29].

63.1.5 DIAGNOSIS

As SBV infections are usually subclinical and their reporting is not required by the veterinary regulations of most countries, the epizootic situation may be monitored by midge testing. The transovarial transmission of SBV in the midge population has been demonstrated to be possible, which may explain the overwintering of the virus in insects in the absence of acute infections detectable in ruminants. SBV spread into the colder and harsher climates of Europe may be related to the involvement of other vectors, such as that of *Culicoides punctatus* as described here.

Laboratory identification of SBV is possible by virus isolation, serology, and nucleic acid detection [30].

Viral isolation: Vero (African green monkey kidney epithelial), BHK21 (baby hamster kidney fibroblast), or KC (*Culicoides variipennis* larvae) cells may be used for SBV isolation from brain, serum, or blood samples. Typically, the homogenate sample is inoculated on KC cells (cell line L1062). After 5 days of incubation, cytopathic effects are visible. Following initial isolation on KC cells, the virus is passaged 3–5 times in baby hamster kidney (BHK21) cells (cell line L0164) (inoculum I: KC/BHK), alternatively once in BHK21 cells, again in KC cells and again in BHK cells (inoculum II: KC/BHK/KC/BHK) [4].

Serology: When SBV is isolated, virus neutralization tests and a plaque reduction neutralization test may be utilized to detect antibodies present in the serum of infected animals [31]. Indirect ELISA test is useful for serological diagnosis of SBV infection [32]. However, a positive serological result does not indicate that an animal is currently infected or that animal will give birth to an affected fetus [33,34].

Nucleic acid detection: SBV RNA is present in a variety of tissues from affected lambs and calves, particularly the cerebrum, spinal cord, external placental fluid, and umbilical cord. This provides an opportunity to diagnosis of SBV infection based on detection of the viral RNA using RT-PCR or other nucleic acid amplification techniques [35–38]. RT-PCR targeting the conserved regions of the S, M, and L segments of SBV has been developed. A real-time TaqMan assay with primers from the S segment enables detection of 1–10 genome copies of SBV [39,40]. A pan-Simbu real-time RT-PCR system targeting the L segment allows rapid and sensitive detection of members of the Simbu serogroup of SBV.

63.1.6 TREATMENT AND PREVENTION

As no specific treatment is currently available for SBV infection, effective management of outbreaks and strategic deployment of vaccines are important for the control and prevention of SBV [41].

In the event of an SBV outbreak, thorough disinfection with viral disinfectants such as hypochlorite, chlorhexidine, alcohol, and phenols is beneficial. To prevent further SBV outbreak, management priorities should include sentinel monitoring of vectors and cattle, conducting mating in cattle and sheep to outside of the vector season, avoiding transportation of naïve animals to endemic areas for mating or during pregnancy, and application of a sensitive test to detect SBV RNA in semen [42,43].

Vaccination is a vital preventive measure against SBV infection [44]. Inactivated vaccines have been shown to provide complete protection of sheep against SBV infection [45]. Other vaccines for SBV are in the pipeline.

63.2 METHODS

63.2.1 SAMPLE PREPARATION

Serum and fecal samples are collected from affected animals. For malformed lambs or calves, the cerebrum and the spinal cord are collected from areas with malformations. Other suitable samples include external placental fluid and umbilical cord.

For RNA extraction from serum, 200 µL serum is added to 400 µL working solution (High Pure Viral RNA Kit, Roche), vortexed, and incubated at room temperature for 10 min. Subsequently, the mixture is spiked with 5 µL of internal control (IC) RNA (8×10^5 copies/µL). After mixing, the solution is centrifuged through a high pure filter tube. In the next step, 500 µL inhibitor removal buffer is added to the filter tube and again centrifuged. After washing the column twice with the appropriate buffer, RNA is eluted with 50 µL elution buffer and stored at −20°C [39,46].

For RNA extraction from tissues, about 0.5 cm³ of tissue is homogenized in 1 mL phosphate buffered saline (PBS) containing 10–15 1 mm silicon carbide beads (BioSpec Products, Inc.) via high-speed shaking (2 min, 25 Hz) in TissueLyser (Qiagen). The homogenate is then processed by using RNeasy Mini Kit (Qiagen) and RNA is eluted in 50 µL RNase-free water.

For RNA extraction from feces, 1 mL of PBS is added to approximately 100 mg of feces and homogenized in TissueLyser (3 min, 10 Hz). Next, 200 µL of the homogenate is added to 800 µL TRIzol and shaken in TissueLyser (5 min, 10 Hz). Then, 200 µL of chloroform is added, and phase separation is performed (Life Technologies). Total RNA present in the aqueous phase is extracted using Viral RNA Mini Kit (Qiagen) and eluted in 60 µL elution buffer.

For RNA extraction from insect pools, the insect pools are first homogenized in Lysing Matrix D tubes (MP Biomedicals) containing 1.4 mm ceramic beads and 1 mL

of TRI Reagent (Sigma–Aldrich) and in FastPrep 120 ribolyzer (Thermo Electron Corp.) for 90 s at 6.5 m/s speed. Total RNA is extracted using 2-propanol, chloroform, and ethanol. Extracted RNA is suspended in 20 µL of nuclease-free water and stored at −70°C until testing [47].

63.2.2 DETECTION PROCEDURES

63.2.2.1 RT-PCR for Schmallenberg Virus Detection

Bilk et al. [39] described a RT-PCR for rapid identification and quantitation of SBV with primers and probe designed from the S segment of SBV genome (Table 63.1). The inclusion of internal control primers and probe from epidermal growth factor (EGF) gene allows effective monitoring of sample preparation and amplification efficiency.

RT-PCR mixture (25 µL) is composed of 2.5 µL RNase-free water, 12.5 µL 2 × RT-PCR buffer, 1.0 µL 25 × RT-PCR enzyme mix (AgPath-ID One-Step RT-PCR Kit, Applied Biosystems), 2 µL SBV-specific primer–probe mix (10 µM SBV-specific primers +1.875 µM SBV-specific probes) and 2 µL IC-specific primer–probe mix (2.5 µM epidermal growth factor protein (EGFP)-specific primers +1.875 µM EGFP1-hexachlorofluoresceine (HEX) probe), and 5 µL RNA template.

Amplification and detection are conducted with the following temperature profile: 10 min at 45°C (RT), 10 min at 95°C (inactivation reverse transcriptase/activation Taq DNA polymerase), followed by 42 cycles of 15 s at 95°C (denaturation), 20 s at 55°C (annealing), and 30 s at 72°C (elongation).

Note: This real-time RT-PCR is highly sensitive with a detection limit of 1–10 copies of SBV genome per reaction. In addition, the assay is able to detect the highly related SATV species member DOUV as well as the reassortant SHAV [40].

TABLE 63.1

Identity of Schmallenberg Virus and Internal Control Primer and Probes

Primer/Probe	Sequence (5′→3′)	Genome Position	Amount per Reaction (µM)
SBV-S-382F	TCA GAT TGT CAT GCC CCT TGC	382–402	10
SBV-S-469R	TTC GGC CCC AGG TGC AAA TC	450–469	10
SBV-S-408FAM	FAM-TTA AGG GAT GCA CCT GGG CCG ATG GT-BHQ1	408–433	1.875
EGFP1-F	GAC CAC TAC CAG CAG AAC AC	637–656	2.5
EGFP2-R	GAA CTC CAG CAG GAC CAT G	768–750	2.5
EGFP1-HEX	HEX-AGC ACC CAG TCC GCC CTG AGC A-BHQ1	703–724	1.875

63.2.2.2 RT-PCR for Pan-Simbu Serogroup Detection

Fischer et al. [40] developed a pan-Simbu real-time RT-PCR with primers (panOBV-L-2959F: 5′-TTG GAG ART ATG ARG CTA ARA TGT G-3′, and panOBV-L-3274R: 5′-TGA GCA CTC CAT TTN GAC ATR TC-3′) targeting nt 2888–3167 of the L segment.

RT-PCR mixture (25 μL) is made up of 8 μL RNase-free water, 5 μL 5× One-Step RT-PCR Buffer, 1 μL One-Step RT-PCR Enzyme Mix (One-Step RT-PCR Kit Qiagen), 1 μL dNTP Mix (10 mM each), 1 μL ResoLight Dye (Roche), 2 μL of each primer (panOBV-L-2959F and panOBV-L-3274R; 10 μM each), and 5 μL RNA template (or RNase-free water for the no template control).

Amplification and detection is carried out with the following thermal program: 1 cycle of 50°C for 30 min and 95°C for 15 min, followed by 40 cycles of 95°C for 30 s, 55°C for 30 s, 72°C for 30 s, and 78°C for 15 s. The collection of the fluorescence data is performed during the 78°C elongation step.

Note: The pan-Simbu real-time RT-PCR system detects all members of the Simbu serogroup under investigation as well as three Bunyamwera serogroup viruses. Subsequent species classification is possible after sequencing the amplified product.

63.3 CONCLUSION

SBV is a newly emerged arthropod-borne virus of ruminants (e.g., cattle, sheep, and goats) that causes congenital deformities in young born to adults infected in their first trimester (sheep) or second trimester (cattle), and reduces milk production in dairy cows. First identified in Germany in the autumn of 2011, SBV has been since reported in various countries in Europe and Turkey. Given the significant economic impact of SBV on ruminant breeders and dairy farmers, use of rapid, sensitive, and specific laboratory tests such as RT-PCR is essential for its identification and control. The discovery of SBV highlights the power of molecular techniques in the studies of uncharacterized infectious agents that escape the attention of conventional diagnostic procedures. There is no doubt that continuing refinement and application of these state-of-art techniques will help unravel the molecular basis of SBV pathogenicity, leading to improved control and prevention strategies against SBV infection.

REFERENCES

1. Beer M, Conraths FJ, van der Poel WH. 'Schmallenberg virus'—A novel orthobunyavirus emerging in Europe. *Epidemiol Infect.* 2013;141(1):1–8.
2. International Committee on Taxonomy of Viruses (ICTV). http://www.ictvonline.org/virusTaxonomy.asp. Accessed on June 12, 2015.
3. Saeed MF, Li L, Wang H, Weaver SC, Barrett ADT. Phylogeny of the Simbu serogroup of the genus Bunyavirus. *J Gen Virol.* 2001;82(9):2173–2181.
4. Hoffmann B et al. Novel orthobunyavirus in Cattle, Europe, 2011. *Emerg Infect Dis.* 2012;18(3):469–472.
5. Garigliany MM et al. Schmallenberg virus: A new Shamonda/Sathuperi-like virus on the rise in Europe. *Antiviral Res.* 2012;95(2):82–87.
6. Goller KV, Höper D, Schirrmeier H, Mettenleiter TC, Beer M. Schmallenberg virus as possible ancestor of Shamonda virus. *Emerg Infect Dis.* 2012;18(10):1644–1646.
7. Yanase T et al. Genetic reassortment between Sathuperi and Shamonda viruses of the genus Orthobunyavirus in nature: Implications for their genetic relationship to Schmallenberg virus. *Arch Virol.* 2012;157(8):1611–1616.
8. Yanase T et al. The emergence in Japan of Sathuperi virus, a tropical Simbu serogroup virus of the genus Orthobunyavirus. *Arch Virol.* 2004;149(5):1007–1013.
9. Doceul V et al. Epidemiology, molecular virology and diagnostics of Schmallenberg virus, an emerging orthobunyavirus in Europe. *Vet Res.* 2013;44:31.
10. Claine F, Coupeau D, Wiggers L, Muylkens B, Kirschvink N. Schmallenberg virus among female lambs, Belgium, 2012. *Emerg Infect Dis.* 2013;19(7):1115–1117.
11. Monaco F et al. First cases of Schmallenberg virus in Italy: Surveillance strategies. *Vet Ital.* 2013;49(3):269–275.
12. Larska M et al. First report of Schmallenberg virus infection in cattle and midges in Poland. *Transbound Emerg Dis.* 2013;60(2):97–101.
13. Chaintoutis SC et al. Evidence of Schmallenberg virus circulation in ruminants in Greece. *Trop Anim Health Prod.* 2014;46(1):251–255.
14. Yilmaz H et al. Detection and partial sequencing of Schmallenberg virus in cattle and sheep in Turkey. *Vector Borne Zoonotic Dis.* 2014;14(3):223–225.
15. Tarlinton R, Daly J, Dunham S, Kydd J. The challenge of Schmallenberg virus emergence in Europe. *Vet J.* 2012;194(1):10–18.
16. Wernike K, Hoffmann B, Beer M. Schmallenberg virus. *Dev Biol* (Basel). 2013b;135:175–182.
17. Harris KA et al. Impact of Schmallenberg virus on British sheep farms during the 2011/2012 lambing season. *Vet Rec.* 2014;175(7):172. Erratum in: *Vet Rec.* 2014;175(10):246.
18. Lehmann K, Werner D, Hoffmann B, Kampen H. PCR identification of culicoid biting midges (Diptera, Ceratopogonidae) of the Obsoletus complex including putative vectors of bluetongue and Schmallenberg viruses. *Parasit Vectors.* 2012;5:213.
19. Elbers AR et al. Schmallenberg virus in *Culicoides* spp. biting midges, the Netherlands, 2011. *Emerg Infect Dis.* 2013;19(1):106–109.
20. Goffredo M et al. Schmallenberg virus in Italy: A retrospective survey in *Culicoides* stored during the bluetongue Italian surveillance program. *Prev Vet Med.* 2013;111(3–4):230–236.
21. Larska M, Lechowski L, Grochowska M, Żmudziński JF. Detection of the Schmallenberg virus in nulliparous *Culicoides obsoletus/scoticus* complex and C. punctatus—The possibility of transovarial virus transmission in the midge population and of a new vector. *Vet Microbiol.* 2013;166(3–4):467–473.
22. Veronesi E et al. Implicating *Culicoides* biting midges as vectors of Schmallenberg virus using semi-quantitative RT-PCR. *PLoS One.* 2013;8(3):e57747.
23. Wernike K et al. Lack of evidence for the presence of Schmallenberg virus in mosquitoes in Germany, 2011. *Parasit Vectors.* 2014;7:402.

24. Poskin A, Van Campe W, Mostin L, Cay B, De Regge N. Experimental Schmallenberg virus infection of pigs. *Vet Microbiol.* 2014;170(3–4):398–402.

25. Wernike K, Holsteg M, Schirrmeier H, Hoffmann B, Beer M. Natural infection of pregnant cows with Schmallenberg virus—A follow-up study. *PLoS One.* 2014b;9(5):e98223.

26. Wernike K et al. Schmallenberg virus experimental infection of sheep. *Vet Microbiol.* 2013;166(3–4):461–466.

27. van den Brom R et al. Epizootic of ovine congenital malformations associated with Schmallenberg virus infection. *Tijdschr Diergeneeskd.* 2012;137(2):106–111.

28. De Regge N, van den Berg T, Georges L, Cay B. Diagnosis of Schmallenberg virus infection in malformed lambs and calves and first indications for virus clearance in the fetus. *Vet Microbiol.* 2013;162(2–4):595–600.

29. Herder V et al. Immunophenotyping of inflammatory cells associated with Schmallenberg virus infection of the central nervous system of ruminants. *PLoS One.* 2013;8(5):e62939.

30. Bouwstra RJ et al. Schmallenberg virus outbreak in the Netherlands: Routine diagnostics and test results. *Vet Microbiol.* 2013;165(1–2):102–108.

31. Loeffen W et al. Development of a virus neutralisation test to detect antibodies against Schmallenberg virus and serological results in suspect and infected herds. *Acta Vet Scand.* 2012;54:44.

32. Näslund K et al. Development and evaluation of an indirect enzyme-linked immunosorbent assay for serological detection of Schmallenberg virus antibodies in ruminants using whole virus antigen. *Acta Vet Scand.* 2014;56(1):71.

33. Mansfield KL et al. Detection of Schmallenberg virus serum neutralising antibodies. *J Virol Methods.* 2013;188(1–2):139–144.

34. Wernike K et al. A novel panel of monoclonal antibodies against Schmallenberg virus nucleoprotein and glycoprotein Gc allows specific orthobunyavirus detection and reveals antigenic differences. *Vet Res.* 2015;46(1):27.

35. Rosseel T et al. DNase SISPA-next generation sequencing confirms Schmallenberg virus in Belgian field samples and identifies genetic variation in Europe. *PLoS One.* 2012;7(7):e41967.

36. De Regge N et al. Detection of Schmallenberg virus in different *Culicoides* spp. by real-time RT-PCR. *Transbound Emerg Dis.* 2012;59(6):471–475.

37. De Regge N et al. Schmallenberg virus circulation in *Culicoides* in Belgium in 2012: Field validation of a real time RT-PCR approach to assess virus replication and dissemination in midges. *PLoS One.* 2014;9(1):e87005.

38. Aebischer A, Wernike K, Hoffmann B, Beer M. Rapid genome detection of Schmallenberg virus and bovine viral diarrhea virus by use of isothermal amplification methods and high-speed real-time reverse transcriptase PCR. *J Clin Microbiol.* 2014;52(6):1883–1892.

39. Bilk S et al. Organ distribution of Schmallenberg virus RNA in malformed newborns. *Vet Microbiol.* 2012;159(1–2):236–238.

40. Fischer M et al. Development of a pan-Simbu real-time reverse transcriptase PCR for the detection of Simbu serogroup viruses and comparison with SBV diagnostic PCR systems. *Virol J.* 2013;10:327.

41. Conraths FJ, Peters M, Beer M. Schmallenberg virus, a novel orthobunyavirus infection in ruminants in Europe: Potential global impact and preventive measures. *N Z Vet J.* 2013;61(2):63–67.

42. Ponsart C et al. Evidence of excretion of Schmallenberg virus in bull semen. *Vet Res.* 2014;45:37.

43. Van Der Poel WH et al. Schmallenberg virus detection in bovine semen after experimental infection of bulls. *Epidemiol Infect.* 2014;142(7):1495–1500.

44. Rodríguez-Prieto V et al. Natural immunity of sheep and lambs against the Schmallenberg virus infection. *Transbound Emerg Dis.* August 6, 2014. doi: 10.1111/tbed.12256.

45. Hechinger S, Wernike K, Beer M. Single immunization with an inactivated vaccine protects sheep from Schmallenberg virus infection. *Vet Res.* 2014;45:79.

46. Hoffmann B, Schulz C, Beer M. First detection of Schmallenberg virus RNA in bovine semen, Germany, 2012. *Vet Microbiol.* 2013;167(3–4):289–295. Erratum in: *Vet Microbiol.* 2014;170(1–2):179.

47. Rasmussen LD et al. Rapid spread of Schmallenberg virus-infected biting midges (*Culicoides* spp.) across Denmark in 2012. *Transbound Emerg Dis.* 2014;61(1):12–16.

64 Arenaviruses

Randal J. Schoepp and Aileen E. O'Hearn

CONTENTS

64.1 Introduction 569
    64.1.1 Classification 569
    64.1.2 Morphology, Genome Organization, and Biology 571
        64.1.2.1 Lymphocytic Choriomeningitis Virus 572
        64.1.2.2 Lassa Virus 572
        64.1.2.3 Junin Virus 573
        64.1.2.4 Guanarito Virus 573
        64.1.2.5 Machupo Virus 573
    64.1.3 Clinical Features and Pathogenesis 574
        64.1.3.1 Lymphocytic Choriomeningitis Virus 574
        64.1.3.2 Viral Hemorrhagic Fevers 574
    64.1.4 Diagnosis 574
        64.1.4.1 Virus Isolation 575
        64.1.4.2 Immunological Detection 575
        64.1.4.3 Molecular Detection 575
64.2 Methods 576
    64.2.1 Sample Preparation 576
        64.2.1.1 Procedure 576
    64.2.2 Detection Procedures 576
        64.2.2.1 ELISA Detection 576
        64.2.2.2 Real-Time RT-PCR Detection 577
        64.2.2.3 MAGPIX Bead-Based Detection 577
64.3 Conclusion and Future Perspectives 577
Acknowledgments 581
References 581

64.1 INTRODUCTION

The *Arenaviridae* family comprises of a diverse group of single-stranded RNA viruses. There are currently 25 confirmed members, most of which do not cause human disease. They have life cycles intimately associated with rodent reservoirs, a trait that imposes varied geographical limits on each member. Lymphocytic choriomeningitis virus (LCMV), the prototype virus, was first described in 1934 by Armstrong and Lily. It is the most widely distributed of the *Arenaviridae*, being found worldwide in the house mouse (*Mus musculus*). The study of LCMV and its host is to a large extent responsible for our understanding of immunology and virus–host interactions.

With the exception of LCMV, which causes viral encephalitis and meningitis in humans, all other arenaviruses associated with human disease cause viral hemorrhagic fevers. Five arenaviruses are known to consistently cause human diseases: LCMV causes lymphocytic choriomeningitis worldwide; Lassa virus (LASV) causes Lassa fever in West Africa; Junin virus (JUNV) causes Argentine hemorrhagic

fever in Argentina; Machupo virus (MACV) causes Bolivian hemorrhagic fever in Bolivia; and Guanarito virus (GTOV) causes Venezuelan hemorrhagic fever in Venezuela. The viruses causing these diseases will be the subject of this chapter. Several other arenaviruses implicated in human illness are primarily involved in isolated cases of sporadic natural infections or accidental laboratory infections. These are Sabiá virus (SABV) in Brazil [1], Chapare virus (CHAV) in Bolivia [2], Whitewater Arroyo virus (WWAV) [3], Pichinde virus (PICV), Tacaribe virus (TCRV), Flexal virus (FLEV) [4], and Lujo virus (LUJV) [5].

64.1.1 CLASSIFICATION

The family *Arenaviridae* has a single unique genus, *Arenavirus*, which currently has 25 virus species and several tentative members, all with similar morphological, physiochemical, serological, and biological properties (Table 64.1). Traditionally, the arenaviruses are divided into two serocomplexes: the lymphocytic choriomeningitis–Lassa (Old World) complex and

TABLE 64.1

Arenaviruses

Virus Species	Acronym	Lineage	Human Pathogenicity	Geographic Distribution	Primary Natural Host Reservoir
Old World Arenaviruses					
Ippy virus	IPPYV		No evidence	Central African Republic	*Arvicanthis niloticus*
Lassa virus	LASV		HF[a], sore throat, nausea	West Africa (Sierra Leone, Guinea, Liberia, and Nigeria)	*Mastomys* spp.
Lujo virus	LUJV		HF, diarrhea, vomiting, sore throat	Zambia	Unknown
Luna virus	LUNV		No evidence	Zambia	*Mastomys natalensis*
Lymphocytic choriomeningitis virus	LCMV		Meningitis or encephalitis, biphasic fever, nausea, teratogenic	Worldwide	*Mus musculus*
Mobala virus	MOBV		No evidence	Central African Republic	*Praomys* spp.
Mopeia virus	MOPV		No evidence	Mozambique, Zimbabwe	*Mastomys natalensis*
New World Arenaviruses (North and Central America)					
Bear Canyon virus	BCNV	Rec/A	No evidence	United States	*Peromyscus californicus*
Tamiami virus	TAMV	Rec/A	No evidence	Florida, United States	*Sigmodon hispidus*
Whitewater Arroyo virus	WWAV	Rec/A	HF, respiratory distress, liver failure	New Mexico, United States	*Neotoma albigula*
New World Arenaviruses (South America)					
Allpahuayo virus	ALLV	A	No evidence	Peru	*Oecomys bicolor*
Amapari virus	AMAV	B	No evidence	Brazil	*Oryzomys capito*
Chapare virus	CHPV	B	HF, headache, body aches	Bolivia	Unknown
Cupixi virus	CPXV	B	No evidence	Brazil	*Oryzomys capito*
Flexal virus	FLEV	A	Fever	Brazil	*Oryzomys capito*
Guanarito virus	GTOV	B	HF	Venezuela	*Zygodontomys brevicauda*
Junín virus	JUNV	B	HF	Argentina	*Calomys musculinus*
Latino virus	LATV	C	No evidence	Bolivia	*Calomys callosus*
Machupo virus	MACV	B	HF	Bolivia	*Calomys callosus*
Oliveros virus	OLVV	C	No evidence	Argentina	*Bolomys obscurus*
Paraná virus	PARV	A	No evidence	Paraguay	*Oryzomys buccinatus*
Pichinde virus	PICV	A	Subclinical infection	Colombia	*Oryzomys albigularis*
Pirital virus	PARV	A	No evidence	Venezuela	*Sigmodon alstoni*
Sabiá virus	SABV	B	HF	Brazil	Unknown
Tacaribe virus	TCRV	B	Fever, mild neurological signs	Trinidad	*Artibeus* spp.

Sources: Modified from Sauvage, F.F.J. and Gonzalez, J.P., Arenaviruses, in *Manual of Security Sensitive Microbes and Toxins*, Liu, D., Ed., CRC Press, Boca Raton, FL, 2014, 9pp.; Charrel, R.N. and de Lamballerie, X., *Vet. Microbiol.*, 140, 213, 2010; ICTV, Virus taxonomy: 2013 Release, in *International Committee on Taxonomy of Viruses*, Vol. 2014, International Committee on Taxonomy of Viruses, 2014.

[a] Major clinical signs including HF—hemorrhagic fever clinical (petechial, fever, etc.) and biological (thrombocytopenia) signs.

the Tacaribe (New World) complex. Designations of viruses into Old World and New World groups are based on antigenic differences in immunological tests, disease etiology, reservoir host species, geographical range, and protein sequence [7,8]. The Old World viruses are represented by the prototype (LCMV) that is found worldwide and causes viral encephalitis and meningitis in humans. Among the Old World arenaviruses, LCMV causes acute aseptic meningoencephalitis and congenital malformations of the central nervous system (CNS) in humans worldwide. In Africa, two arenaviruses are highly pathogenic for humans and can cause viral hemorrhagic fever.

These include LASV associated with Lassa fever in West Africa and LUJV that has caused disease in Southern Africa. Among the New World arenaviruses, three viruses are highly pathogenic for humans and consistently cause viral hemorrhagic fever. These include MACV causing Bolivian hemorrhagic fever, JUNV causing Argentine hemorrhagic fever, and GTOV causing Venezuelan hemorrhagic fever. Two other viruses, CHAV and SABV, have potential to cause significant human disease. Four other arenaviruses are associated with human disease, but the disease is less severe or occurs infrequently; these are PICV, TCRV, FLEV, and WWAV [4].

64.1.2 Morphology, Genome Organization, and Biology

The arenaviruses (from the Latin word *arenosus*, meaning "sandy") derive their name from the *sandy* appearance in electron micrographs that generally results from host-derived ribosomes (20–25 nm) incorporated into the virion [9]. Arenavirus virions are spherical to pleomorphic in shape with a diameter of 50–300 nm. They are surrounded by a dense host-derived lipid envelope covered with viral glycoproteins, appearing as long (8–10 nm) club-shaped projections.

The arenavirus genome consists of two single-stranded ambisense RNA segments. The large (L) genomic segment (7.2 kb) encodes the L polymerase protein and the Z matrix protein. The small (S) genomic segment (3.5 kb) encodes the nucleoprotein (NP) and the glycoprotein complex (GPC) that is posttranslationally cleaved into a stable signal peptide and two envelope glycoproteins (GP1 and GP2). The proteins are translated by a unique ambisense strategy. Each genome segment encodes two proteins from open reading frames of opposite sense (Figure 64.1). From each segment, one protein is encoded directly from the viral RNA (vRNA) genome sequence, while the second protein is encoded from opposite sense or viral complementary RNA (vcRNA). The genes are separated by noncoding intergenic regions that can form hairpin loops and may function in termination of transcription of mRNA from the vRNA and vcRNA [10]. The 5′ and 3′ untranslated region of each RNA segment have spans of 19 nucleotides of relatively conserved reverse complementarity and form a panhandle structure at the end of the virus genome [11]. The L and NP proteins are translated through an mRNA transcribed by the virus RNA polymerase from the vRNA sequence of the L and S genomic segments, respectively. These translated proteins are thought to take part in the full-length vcRNA synthesis, which then becomes the template for transcription of the Z and GPC mRNA for translation of the respective proteins and synthesis of full-length vRNA [12]. The mRNAs are not polyadenylated but are capped using a unique cap-snatching method where cellular mRNA caps serve as primers for mRNA synthesis [13].

The L protein is an RNA-dependent RNA polymerase and along with the NP is the minimum requirement for RNA replication and transcription [8,14]. The NP encapsidates the genome segments to form the nucleocapsid, making it the most abundant structural protein in the virion. The NP has a variety of functions. It interacts with the L polymerase protein for RNA replication and transcription and with the Z matrix protein for viral assembly. The NP also plays a role in suppression of the innate immune response and exhibits exonuclease and nucleotide binding activity. The GPC and its subsequent GP1 and GP2 forms are responsible for host cell tropism, mediating virus assembly and entry into the host cell and uncoating of the virion. The Z matrix protein is a small zinc-binding protein that interacts with all of the other viral proteins [15]. These interactions play roles in viral transcription and replication with the L polymerase protein, in virus assembly with the glycoproteins, in virus budding with the NP, and in proapoptotic activity.

Rodents are the natural hosts of arenaviruses. TCRV, the prototype of the New World serocomplex, is the only exception, being associated with fruit-eating bats (*Artibeus* species) [16]. Each arenavirus species is closely associated with a

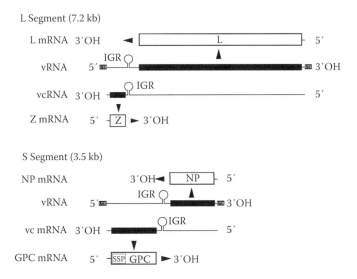

FIGURE 64.1 Organization, transcription, and replication of the arenavirus large and small genomic RNAs. Genomic regions encoding the polymerase (L), matrix (Z), nucleoprotein (NP), and glycoprotein complex (GPC) proteins are shown as solid bars, and the resulting proteins are shown as open boxes with arrowheads indicating the notional direction of translation. GPC is posttranslationally cleaved into a stable signal peptide and two envelope glycoproteins (GP1 and GP2). RNA transcription processes are indicated by vertical arrows. The intergenic regions (IGR) separate the open reading frames. Viral RNA segments have untranslated regions at their 5′ and 3′ termini. (Adapted from Zapata, J.C. and Salvato, M.S., *Viruses*, 5, 241, 2013; Salvato, M.S. et al., Arenaviridae, in *Virus Taxonomy*, *Ninth Report of the International Committee on Taxonomy of Viruses*, King, A.M.Q., Adams, M.J., Carstens, E.B., and Lefkowitz, E.J., Eds., Elsevier, San Diego, CA, 2012, 715pp.)

single rodent species or closely related group of species. This intimate association of rodent host and virus species has led to the speculation that the host and virus have coevolved over millennia. The coevolution is thought to have begun with the split of the rodent family *Muridae* into the subfamily *Murinae* that evolved into the Old World rodents and the subfamily *Sigmodontinae* that evolved into the New World rodents. The ancestral arenaviruses that infected each rodent subfamily codiverged and, through a long, shared evolutionary relationship, cospeciated. In many instances, the phylogeny of the arenavirus species matches the phylogeny of the rodent reservoir [17–19]. The hypothesis for rodent and arenavirus cospeciation is not universally accepted. Contradictions exist that are not explained by the model, such as the isolation of TCRV from bats rather than rodents, or New World viruses of different lineages that have the same rodent host [18,20,21]. Alternatively, a combination of virus transfers between rodent species with subsequent cospeciation would explain not only the virus–host pairings not explained by the coevolution hypothesis but also the bat host of TCRV [18,22]. Regardless, the pattern of geographical distribution of the arenaviruses coincides with their rodent hosts; however, the arenavirus distribution is patchy or discontinuous within the rodent geographical range [23]. Within the arenavirus host ranges, there are areas of greater rodent infection rates that correlate with increased human disease. Seasonal increases in rodent populations and/or greater rodent–human contacts are also associated with increase outbreaks.

Arenavirus infections are generally considered to be asymptomatic in the rodent reservoir host. This asymptomatic infection can lead to a chronic viremia and/or viruria, in which virus is continually shed in urine and feces throughout the life of the rodent host. This significant source of virus in the environment has ramifications for the epidemiology of the diseases they cause. Within the rodent populations, virus is transmitted through horizontal (direct contact), vertical (infected mother to offspring), and sexual transmission. Depending on the age an individual becomes infected, the rodents can be immunocompetent developing immunity and recovering or immunotolerant and not develop immunity and remain chronically infected. Chronic infections can be detrimental to the infected host and reduce their fitness or fertility. Arenaviruses are naturally transmitted to humans through contact with infected rodent hosts either directly, through bites, trapping, or meat preparation, or indirectly by ingesting food or inhalation of fomites contaminated with rodent urine or feces. Nosocomial infections can occur through infected bodily fluids during the care of the sick or preparation of the dead. Transmission has also resulted from tissue transplantation from infected donors.

64.1.2.1 Lymphocytic Choriomeningitis Virus

LCMV is more noted for its contributions as an experimental system for the understanding of viral immunology and pathogenesis than for the disease it causes in humans. The virus was discovered in 1933 by Armstrong and Lillie during an outbreak of St. Louis encephalitis in humans [24]. LCMV has

worldwide geographic distribution by virtue of its natural reservoir host, the common house mouse subspecies *Mus musculus musculus* and *Mus musculus domesticus*. This reservoir host, often living in close association with humans, is persistently infected by intrauterine infection and fails to mount an immune response to the virus. The resulting chronic, lifelong infection results in the mice shedding large amounts of virus in urine and feces, as well as nasal secretions, saliva, milk, and semen [25]. Serological studies of wild mice suggest that the antibody prevalence ranges from 3% to 20% [26].

The incidence of human infection with LCMV is unknown, but considering the worldwide distribution of chronically infected mice, the disease is likely underrecognized and underreported. Humans are often infected through mucosal infection of aerosolized fomites contaminated with mouse excreta or by direct contact through bites. Congenital infections of the fetus can cause a severe form of disease that differs pathologically from its acquired version. In humans, serological studies in the Americas and Europe indicate that prevalence ranges from 1% to 9% [27]. In the United States, overall prevalence is around 5%; however, a 1992 study done in Baltimore, Maryland, found a 9% seroprevalence in mice, with a corresponding 13.4% in humans in inner city locations, the highest recorded in the United States [28].

Infected mice and Syrian hamsters in the pet trade and laboratory research have resulted in large outbreaks of human disease [29,30]. The largest LCMV outbreak occurred in 1973–1974 causing 181 human cases in 12 states. The source of the infections was traced back to infected pet hamsters from a single supplier [31,32]. Person-to-person transmission has occurred through solid-organ transplants and has resulted in a high mortality rate in the immunosuppressed recipients; however, this is not a normal route of infection [33,34]. One transplant-associated LCMV infection was traced to the organ donor infected by a pet hamster [35]. Approximately 3% of the hamsters in the pet supplier population were infected with LCMV and resulted in U.S. CDC issuing guidance to minimize risk of LCMV infection associated with rodents.

64.1.2.2 Lassa Virus

LASV, the cause of Lassa fever, occurs in West Africa and is hyperendemic in Sierra Leone, Guinea, Liberia, and Nigeria. It was first isolated in 1969 from a fatal case in a missionary nurse working in the town of Lassa in northeastern Nigeria [36–38]. Rodent-to-human infection occurs by contact with infected rodent excreta, tissue, or blood. Human-to-human transmission can occur through mucosal or cutaneous contact or nosocomial contamination with blood or other body fluids. Early studies of Lassa fever estimated hundreds of thousands of cases per year throughout its range resulting in about 5000 deaths [39]. LASV is of concern outside the hyperendemic region, being the most prominent exotic hemorrhagic fever imported to Europe, North America, Asia, and the Middle East [4]. The virus causes an acute human disease associated with a range of disease manifestations. In endemic areas, the high prevalence of LASV-specific antibodies in the general population suggests that most infections are mild or asymptomatic [40]. An estimated 20% of infections cause moderate to severe disease that can progress to acute hemorrhagic fever with multi-organ failure [41]. Human infection of Lassa fever tends to peak in the dry season, January to March, but recent evidence suggests that it might persist through the transition from dry to wet season [42].

The primary LASV reservoir host is the multimammate rodent *Mastomys natalensis* that lives in close association with humans; however, an incomplete understanding of the genus taxonomy has led to some uncertainty about which species or subspecies may serve as reservoir hosts [43]. Infected rodent hosts excrete virus in urine and feces that can infect humans via inhalation or ingestion of virus-contaminated materials. Similar to LCMV, the rodent host is thought to be infected early in life and develop persistent infections that are a continuous source of contaminating virus in the environment. In studies in Guinea of natural populations of *M. natalensis*, where it is the only LASV reservoir, antibody data supported a model in which the rodents become horizontally infected, clear the virus, and develop antibodies [44].

64.1.2.3 Junin Virus

JUNV is the etiological agent of Argentine hemorrhagic fever. The virus was isolated in 1957 during an outbreak of hemorrhagic fever in the city of Junin, Buenos Aires Province, from which its name is derived [4,23]. Since its discovery, progressively expanding annual outbreaks of Argentine hemorrhagic fever occur during the agricultural harvest season with peak cases in the month of May [45]. The occupational risk of fieldwork results in adult men becoming disproportionately infected. Humans are usually infected through cuts or skin abrasions or aspiration of particles contaminated with urine, saliva, or blood from infected rodents [46].

The primary rodent reservoir of JUNV is the drylands vesper mouse, *Calomys musculinus* [47–49]. Serological evidence implicates other rodents in the Argentine hemorrhagic fever endemic region as possible reservoirs, but their role in the disease ecology is unclear [25]. Naturally and experimentally infected *C. musculinus* are known to have persistent viremias and shed virus in their saliva. The rodents can be infected horizontally and vertically; however, in nature, evidence suggests that the primary mode of infection is by horizontal transmission [47]. Male *C. musculinus* have a higher prevalence of JUNV infection probably due to greater movement and more aggressive behavior than females. Experimental JUNV virus infection, by vertical or horizontal transmission routes, may have an effect on animal fitness. Animals infected at birth had reduced weight gain that lasted until 60 days of age as compared to uninfected controls [50]. In contrast, when adult *C. musculinus* were infected, the mean body mass was no different than the uninfected controls [51]. Whether similar effects occur in the natural rodent populations is unknown.

64.1.2.4 Guanarito Virus

GTOV is the etiological agent of Venezuelan hemorrhagic fever and was first recognized during an outbreak of hemorrhagic disease in the municipality of Guanarito, Portuguesa

State, Venezuela, in 1989 [52]. The disease was associated with human encroachment into newly forested regions and increased exposure to infected rodents [23,53]. Venezuelan hemorrhagic fever tends to be seasonal with human infections peaking in November to January when the rains end and the dry season begins [54]. The disease coincides with the increased agricultural activity of harvesting of crops and preparations for planting, when humans have greater contact with the rodent reservoir.

The rodent reservoir of GTOV is a grassland rodent, the short-tailed cane mouse, *Zygodontomys brevicauda* [55]. Originally, another rodent species, the Alston's cotton rat, *Sigmodon alstoni*, was also thought to be a reservoir of GTOV; however, the misidentification attributed to serologic cross-reactivity of arenaviruses was corrected with genetic characterization of viruses isolated from each. *S. alstoni* and *Z. brevicauda* coexist in the same geographic locations where GTOV and Pirital virus (PARV), not known to cause human disease, are sympatric arenaviruses. Genetic characterization of virus isolates from each of the rodent populations implicated *Z. brevicauda* as the reservoir for GTOV and *S. alstoni* as the reservoir for PARV [55,56].

Experimental infection of the cane mouse, *Z. brevicauda*, with GTOV has demonstrated that chronic, lifelong infections can be established and virus can be persistently shed in urine and oropharyngeal secretions [57]. In nature, evidence suggests that horizontal transmission in *Z. brevicauda* is the dominant mode of GTOV transmission and is acquired through allogrooming, mating, fighting, and other close physical contact between the rodents [56]. Vertical transmission may occur but is rare in *Z. brevicauda*. Chronic GTOV infection of experimentally infected rodent reservoirs did not cause illness or death; however, other aspects of host fitness were not evaluated [57].

64.1.2.5 Machupo Virus

MACV is the etiological agent of Bolivian hemorrhagic fever. The virus was first recognized in 1962 in samples from a 1959 outbreak of viral hemorrhagic disease in San Joaquin in northeastern Bolivia [58]. Natural human infections are acquired through direct contact with the rodent reservoir or by aerosols of fomites contaminated with infected rodent urine. The incidence of human disease increases from April to June, during the late rainy and early dry season. Epidemiologically, MACV causes sporadic cases and focal outbreaks of disease in Bolivia with years of dormancy intervening [59,60]. Cases of Bolivian hemorrhagic fever were not reported between 1976 and 1992 but reemerged in outbreaks in 1994 and 2007–2008 [61].

The rodent reservoir of MACV is the large vesper mouse, *Calomys callosus*, which is found in grasslands, savannas, and forest fringes throughout southeastern South America, but the disease is limited to northeastern Bolivia [62]. Genetic differences in the *C. callosus* populations in northeastern Bolivia compared to other populations in southeastern South America may explain the incomplete overlap of the virus and the geographic range of the reservoir.

Virus–reservoir dynamics are difficult to study in natural cycles, so the understanding of MACV maintenance in nature is incomplete; however, it is probably similar to other arenavirus reservoirs with most infected individuals becoming viremic, producing neutralizing antibodies, and then clearing the infection. A small group of individuals, usually infected early in life, become persistently infected and excrete virus for life. Experimentally infected *C. callosus*, infected as suckling mice, demonstrate immune tolerance and a persistent infection [63]. In contrast, mice more than 9 days of age can follow either course, clear the infection, or become persistently infected. The virus persistence affects the fitness of the animal causing runting, shorter life spans, and almost total reproductive failure in females as a result of fatal virus infections of the embryos.

64.1.3 CLINICAL FEATURES AND PATHOGENESIS

64.1.3.1 Lymphocytic Choriomeningitis Virus

Acquired LCMV often results in asymptomatic or mild disease. When the virus does cause disease, unlike other arenaviruses that cause hemorrhagic fever, LCMV infections often result in meningitis. Flu-like symptoms including fever, malaise, headache, and nausea develop and usually resolve within a few days. Occasionally in more severe cases, the symptoms retreat temporarily and are soon followed by a bout of aseptic meningitis. Symptoms progress to neck stiffness and CNS involvement with associated arthritis, meningeal inflammation, pancreatitis, and parotitis [64,65]. Mild cases usually resolve within 1–3 weeks with fatalities being rare (<1%); in severe cases, full recovery can take months [66].

LCMV can cause congenital infections, which is associated with severe fetal disease and low survival, particularly if infection occurs during the first trimester. Surviving infants have evidence of extensive neurotropic infection of the developing brain and result in brain dysfunctions that lead to a range of conditions, most commonly chorioretinitis resulting in vision impairment, hydrocephaly, macrocephaly or microcephaly, and varying degrees of mental retardation and motor skill deficiencies [64,66]. At 21 months of age, the mortality rate may reach 35% [67]. Even though LCMV is known to naturally cause meningitis, there have been isolated cases of it causing hemorrhagic disease, similar to Lassa fever. These rare cases were associated with organ donations, where transmission of LCMV to the organ recipients often resulted in fatalities [33,34,66]. Viral hemorrhagic fever symptoms and multi-organ failure were the prominent disease presentations, but evidence of CNS infection also occurred. This severe manifestation may be due to the immunosuppressive drugs and the resulting T-cell depletion, which plays a significant role in the body's defense to LCMV infection.

64.1.3.2 Viral Hemorrhagic Fevers

Arenaviruses are most notably associated with the hemorrhagic fevers they cause; however, the majority of arenaviruses are not known to cause human disease. One of the largest clinical studies of viral hemorrhagic fever–causing arenaviruses collected data from 441 Lassa fever hospitalized patients in the Lassa fever hyperendemic region in Sierra Leone [68]. The clinical course of the disease begins with a range of general symptoms, such as fever, cough, sore throat, and joint, back, and chest pain. Abdominal pain, vomiting, diarrhea, pharyngitis, and conjunctivitis may present midcourse. Late disease symptoms include respiratory difficulty, facial edema, and hearing loss. Neurological symptoms such as deafness, muscle spasms, and seizures have been reported but are rare. In a small subset of hospitalized patients, internal bleeding and mucosal hemorrhaging occurs, but it is a strong indicator of a negative patient outcome [68–70]. The combination of symptoms most diagnostic for Lassa fever was pharyngitis, retrosternal pain, and proteinuria. Fatal outcome was most strongly predicted by a combination of hemorrhage and sore throat. Although initial infection is predominantly in macrophages, it quickly becomes systemic, and virus can be isolated from internal organs, mucous, semen, urine, and blood [40,71]. Viremia is also strongly associated with a fatal outcome, with 91% of patients who succumb to the disease having measurable viremia at some point during infection that persists through the end (compared to 56% of survivors) [71]. Disease is resolved by recovery, or death, within 2 weeks of illness onset.

New World arenaviruses have symptoms and mortality rates similar to Old World arenaviruses. The greatest outbreaks are caused by JUNV (Argentine hemorrhagic fever), GTOV (Venezuelan hemorrhagic fever), and MACV (Bolivian hemorrhagic fever). While all present with the typical viral hemorrhagic fever symptoms as described earlier, each virus has signs and symptoms that are commonly associated with the particular disease. JUNV infections commonly begin with fever, malaise, or chills and may advance to more serious systemic symptoms within a few days, including neurological and gastrointestinal complications [45,65]. GTOV infection most frequently presents with fever, sore throat, abdominal, and joint pain [52,72]. A significant amount of GTOV patients (>18%) present with convulsions, and it is a strong indicator of a fatal outcome [54]. MACV infection case studies report that the most frequent symptoms are bleeding from the gums and gastrointestinal tract, petechiae, and tremors with an accompanying fever [61]. Pathologically, arenavirus viral hemorrhagic fevers produce marked thrombocytopenia and leukopenia. During the later stages of the disease, vascular shock and multi-organ failure is evident [4,73].

Other arenaviruses also cause viral hemorrhagic fever, but clinical data are extremely limited due to their infrequent appearances. These viruses, CHPV, FLEV, PICV, SABV, and LUJV, are limited to isolated laboratory infections or sporadic natural infections but have so far resembled typical viral hemorrhagic fever disease [74,75].

64.1.4 DIAGNOSIS

A clinical diagnosis based on symptoms alone is not reliable. Arenavirus diseases, whether hemorrhagic or meningitis disease forms, have symptoms that resemble a variety of viral infections in the early stages. Laboratory testing is necessary for a definitive diagnosis. The most common laboratory tools for clinical arenavirus diagnosis include isolation techniques and serology, with reverse transcription polymerase chain reaction (RT-PCR) used as a means for discriminating specific species and/or strains.

64.1.4.1 Virus Isolation

Virus isolation is considered the gold standard in the diagnostics of arenaviruses but is often not practical or safe without proper biocontainment controls. Virus isolation attempts to culture virus from patient blood or serum in susceptible cell lines such as Vero (African green monkey kidney) or BHK (baby hamster kidney) cell lines. The infected cultures are monitored for cytopathic effects (CPE); an exception to this method is GTOV, which shows minimal or no CPE [76].

64.1.4.2 Immunological Detection

ELISA is the standard used in many diagnostic laboratories. Detection of virus or virus components by antigen detection ELISA or seroconversion in paired acute and convalescent samples is the most common immunodiagnostics used today. Seroconversion is determined by a rise of IgM antibodies early (acute phase) in the infection and/or a fourfold rise in IgG antibodies in the later stages (convalescent phase) of the infection. A notable exception to the dogmatic rise and fall of IgM antibodies is a Sierra Leone study where convalescent patients were found to have persisting IgM antibody for months to years following LASV infection [77]. Although this would have significant implications on LASV statistics and diagnostic standards, more extensive studies need to be done to confirm this observation. Immunofluorescence assays such as the indirect fluorescent antibody have been used to detect LASV. This antibody-based assay treats slide-mounted, fixed, LASV-infected cell monolayers with patient serum and detects virus with fluorescein isothiocyanate–conjugated antispecies secondary antibodies. Even though the method is less sensitive and specific than current ELISA methods, these assays are still in use in laboratory and clinical settings due to their lower technological requirements and ease of use [78].

Complement fixation was once commonly used for detection of viruses or antibodies. The method was heavily relied upon for identification of LCMV infections, particularly in cerebrospinal fluid [64]. Detection of antibodies or viral antigen was accomplished by adding antigen (for antibody detection) or antibody (for antigen detection) to sheep red blood cells (RBCs) with complement and the patient serum. Absence of the target results in complement lysis of the RBCs, while its presence results in complement utilization and the RBCs stay intact. Modern detection methods have proven to be more efficient and sensitive, and the assay is rarely used anymore.

Today, there are new immunological assays being developed that will have advantages in clinical, laboratory, and field settings. Point-of-care diagnostics bring the diagnostic testing conveniently and immediately to the patient that will improve the patient outcome with quicker results for clinical management by the physician and care team. One example is the recently developed ReLASV™ for LASV, which was

released by Corgenix (Broomfield, Colorado, United States) to the European market [79]. Tulane University, working at Kenema Government Hospital, Sierra Leone, developed the lateral flow immunoassay to rapidly test for LASV. This multiyear study resulted in a 15 min recombinant antigen-based test, which uses a small dipstick to detect LASV in the suspected blood sample. The test continues to be in routine use at the Kenema Government Hospital located in the Lassa fever hyperendemic region in West Africa, although the relative specificity and sensitivity compared to the traditional antigen detection ELISA is still being determined. Such assays are usually of low complexity and can be run by staff with minimal diagnostic training; however, they often suffer from drawbacks, most notably a lack of sensitivity. An understanding of the limitations of the particular assay is important for the correct interpretation of the results.

The ELISA is the workhorse of immunodiagnostics; however, it has changed little since it was first described [80]. Today, the plastic ELISA plate is being replaced with plastic or magnetic beads that act as the solid support. MAGPIX (Luminex Corporation, Austin, Texas, USA) is a particularly versatile system that uses individually addressed bead sets. The addressable microspheres are dyed with a specific ratio of red/infrared dye, which can be distinguished from each other by a charge-coupled device (CCD) camera system when illuminated with LED lights. These unique bead sets allows for the development of multiplex assays that can test for multiple targets in a single sample. Theoretically, the MAGPIX system is capable of detecting up to 50 targets in a single well. For antigen detection, antibodies to viral antigen can be covalently coupled to the microspheres and then mixed with an infected sample. A secondary antibody to the antigen is used for fluorescent detection of target, and beads are counted with the CCD camera system. For antibody detection, the assay is altered by adding an antibody capture step (sandwich or indirect ELISA) or by conjugating purified viral antigen directly to beads. Protocols are also available to conjugate oligonucleotides to the microspheres to detect complementary sequences. The MAGPIX system, while already popular in laboratory settings, is also attractive for field diagnostics because it is portable and does not use lasers requiring critical alignment. The CCD camera system can handle the rigors of travel to and work in austere environments. The system is being tested and evaluated at multiple sites in West Africa to detect and identify LASV, EBOV, and other circulating viruses that occur in the region.

64.1.4.3 Molecular Detection

Immunological cross-reactivity between arenavirus species is not uncommon and can confound diagnostic results. Today, complete genomic sequences are available for all established arenaviruses, making a specific diagnosis easily accomplished with nucleic acid testing. Molecular diagnostics relying on standard (gel-based) or real-time RT-PCR utilizing glycoprotein and/or NP-specific primer sequences are frequently used for clinical and research applications [18]. In real-time RT-PCR, the probe adds additional specificity to the assay to provide even greater confidence in the diagnostic result.

There are a variety of instruments that are used with the real-time RT-PCR assays. For clinical laboratories, there are high-throughput instruments like the Applied Biosystems® 7500 Real-Time PCR System (Applied Biosystems Incorporated, Carlsbad, California, USA) that use 96-well plates. The plate-based systems have replaced the more cumbersome glass capillary systems used in the early LightCycler® instruments (Roche Diagnostics, Indianapolis, Indiana, USA). The U.S. military uses a ruggedized capillary system, specially developed for mobile analytical laboratories and fields hospitals. The Ruggedized Advanced Pathogen Identification Device is a portable real-time PCR system designed to identify biological agents of military concern in challenging environments.

The advantages of PCR-based diagnostics are their extreme sensitivity and specificity; however, these characteristics can also be problematic in certain situations. Lassa fever is often diagnosed in West Africa using RT-PCR and qRT-PCR [81]. LASV strain divergence can hamper clinical molecular diagnostics. Strain sequence variation, when it coincides with the primers and/or probe hybridization regions, results in false negatives. The sequence variation between and within the arenaviruses illustrates the importance of an orthogonal system using both immunological and molecular diagnostics to provide the highest confidence in a diagnostic result.

64.2 METHODS

64.2.1 SAMPLE PREPARATION

Sample preparation for culturing methods and immunological detection methods such as ELISA or MAGPIX is minimal, with serum, plasma, and/or whole blood used directly in the assays. For molecular detection, sample preparation techniques primarily focus on inactivation of viral material and purification of the target nucleic acid for downstream RT-PCR detection. Two different manual sample processing protocols are commonly used for extraction of genomic materials. TRIzol (Invitrogen, Waltham, MA, USA)-based methods utilize a phenol–chloroform phase separation and others, such as the QIAamp Viral RNeasy Kits (Qiagen Valencia, California, USA) utilize silica-membrane binding of nucleic acids in spin columns. These standard methods are used for RNA extractions according to manufacturer recommendations with a variety of matrices. They are relatively user-friendly and efficient at isolation of amplifiable RNA, and many companies have developed benchtop extractors that automate the same protocol.

A more recent advancement in nucleic acid extraction with improved efficiency over traditional methods utilizes binding of nucleic acids to magnetic beads for isolation and subsequent purification. Arenavirus RNA is efficiently isolated from diagnostic samples using kits such as the manual MagMAX AI/ND Viral RNA Isolation Kit (Ambion, Austin, Texas, USA). Viral particles are disrupted in the presence of TRIzol, to inactivate the virus and release the vRNA. Paramagnetic beads with a nucleic acid binding surface are then able to bind

the target nucleic acids. The beads with the RNA bound are captured by magnets, and proteins and other contaminants are washed away. Pure, high-quality RNA is eluted from the beads, relatively free of contaminants and PCR inhibitors.

64.2.1.1 Procedure

1. Incubate aliquots up to 400 µL of TRIzol-treated arenavirus material with 800 µL of Viral Lysis/ Binding Solution and mix by vortex for 30 s.
2. Add 20 µL of resuspended magnetic bead mixture to each sample. To ensure effective release of the RNA and binding to the beads, the samples are mixed by vortex during the 4 min incubation at room temperature.
3. Remove the magnetic beads with the bound RNA from solution by attraction to the magnetic source. The contaminants are removed with the discarded supernatant.
4. Wash the RNA bound beads once with Wash Solution 1, twice with Wash Solution 2, each time discarding the supernatant, and then dry.
5. Elute arenavirus RNA from the beads with 100 µL of water at room temperature, then store at −70°C.

64.2.2 DETECTION PROCEDURES

64.2.2.1 ELISA Detection

ELISA can be used for virus antigen detection or serological detection of IgM and IgG antibodies with minimal changes. The IgM capture ELISA is outlined below. Although there are many variations, standard solutions are a coating buffer of phosphate buffered saline (PBS), wash buffer of PBS with 0.1% Tween (PBS-T), and dilution buffer of PBS-T in 5% skim milk to aid in blocking nonspecific binding. Samples should be tested against a negative antigen as a control. Different combinations of enzyme–substrate detection schemes may be used, although horseradish peroxidase (HRP) and 2,2′-azinobis[3-ethylbenzothiazoline-6-sulphonic acid] (ABTS) are the most common.

64.2.2.1.1 Procedure

1. Coat a 96-well plate with anti-IgM antibody at the recommended concentration diluted in coating buffer. Wrap the plate in plastic and let incubate overnight at 4°C.
2. Wash plate three times with wash buffer and add samples diluted in dilution buffer at desired concentration, usually 1:10 to 1:100. Incubate at 37°C for 1 h.
3. Wash plate three times with wash buffer and add the arenavirus antigen and negative control antigen in dilution buffer to the appropriate wells. Incubate at 37°C for 1 h.
4. Wash plate three times with wash buffer and add antiarenavirus antibody at the recommended concentration diluted in dilution buffer. Incubate at 37°C for 1 h. The arenavirus can be detected directly with

an antibody custom labeled with HRP or detected indirectly with an antiarenavirus antibody followed by an antispecies secondary detector antibody.

5. Wash plate three times with wash buffer and add ABTS substrate. Incubate at 37°C for 30 min.
6. Detect absorbance on a plate reader at 410 nm.

64.2.2.2 Real-Time RT-PCR Detection

Trombley et al. [82] developed virus-specific real-time RT-PCR assays targeting the GP, NP, Z, and/or L polymerase genes for detection of LASV, JUNV, MACV, GTOV, and SABV in diagnostic samples. Strains, primer–probe sets, and their sensitivities are listed in Table 64.2. Although these assays were optimized for the LightCycler (Roche Diagnostics), they can easily be altered with master mix and software adjustments to be used on any cycler platform.

64.2.2.2.1 Procedure

1. The cDNA synthesis and real-time PCR uses the SuperScript™ One-Step RT-PCR Kit (Invitrogen) as described by the manufacturer. The Master Mix (20 μL) contains 0.2 mM each of the dNTPs, the appropriate final concentration of each primer and TaqMan probe, 3 mM MgSO₄, and 1–10 ng/5 μL (or 1–1000 plaque-forming units per 5 μL) of extracted RNA. When using glass capillaries on the LightCycler 2.0 (Roche Diagnostics), bovine serum albumin is added to the reaction mix.
2. Add real-time RT-PCR Master Mix to the reaction tube or plate well followed by 5 μL of genomic RNA sample. Cap the tubes or cover the plates with sealing film.
3. Cycle the real-time RT-PCR tubes or plates as follows: 50°C for 15 min (1 cycle); 95°C for 5 min (1 cycle); 95°C for 1 s; 60°C for 20 s (45 cycles); 40°C for 30 s (1 cycle). A single fluorescence measurement is made at the end of each 60°C step.
4. Analyze the real-time RT-PCR curves with the software packages specific to the instrument utilized.

64.2.2.3 MAGPIX Bead-Based Detection

The MAGPIX (Luminex Corporation, Austin, Texas, USA) can be used for molecular as well as immunological detection. The customizable bead sets are conjugated to the capture antibody, antigen, or oligonucleotide depending on the target and format desired. For multiplexed assays, multiple different bead sets may be used in single wells once the interactions of the singleplex assays are optimized for the multiplex assay. While there are many variations that can be optimized for each protocol, a standard buffer used for both dilutions and washes is PBS with 0.05% Tween (PBS-T). Incubation times may range from 30 min to 1 h. Assays can be arranged with internal negative control bead sets or with separate negative control wells. Fluorescent detection can be observed using biotinylated

detector antibodies and a subsequent fluorescent reporter or a directly labeled detector antibody. An example antigen detection protocol is outlined here. Plates and solutions should be protected from light throughout the procedure.

64.2.2.3.1 Procedure

1. Choose desired sets of capture antibody-coated beads. In an opaque 96-well plate, dilute bead sets to 2000–5000 beads/set/well. Combine bead and test sample at desired concentration, usually 1:100. Incubate at room temperature for 1 h.
2. Attach plate to magnet and wash three times with buffer. Add biotin-labeled, antiarenavirus antibody(s) to wells at the recommended concentration in PBS-T buffer. Incubate at room temperature for 1 h.
3. Attach plate to magnet and wash three times with buffer. Add Streptavidin-R-Phycoerythrin at 4 μg/mL. Incubate at room temperature for 30 min.
4. Attach plate to magnet and wash three times with PBS-T buffer. Insert plate into MAGPIX instrument and read with protocol directed to look for desired bead sets.

64.3 CONCLUSION AND FUTURE PERSPECTIVES

Arenaviruses are a diverse group of viruses commonly associated with rodents and humans; however, TCRV, associated with fruit-eating bats (*Artibeus* species), is a well-known exception. Other arenavirus reservoirs undoubtedly exist, and it is important to remain vigilant and receptive to other virus–host interactions. Recently, highly divergent arenaviruses were isolated from two species of boas suffering from inclusion body disease, an often fatal disease in snakes best characterized in boas and pythons [83–85]. Interestingly, these arenaviruses demonstrated sequence variation that may reflect geographical distances between the evolutionary origins of snake species, similar to what is hypothesized for rodent reservoirs. More in-depth study of the virus and snake association is needed to fully understand this unique group of arenaviruses.

The *Arenaviridae* family will continue to grow in the future as we gain a greater understanding of the viruses and their reservoir hosts. Improvements in high-throughput sequencing (or next-generation sequencing) technologies will provide better access to virus and reservoir host genome sequences, resulting in new species and host relationships that will help guide our understanding of arenavirus evolution. For example, the incomplete overlap of the geographic range of some rodent reservoirs and their associated arenaviruses may result from genetic differences in rodent subpopulations, which may become evident with better genetic characterization of regional populations. Similarly, it is hypothesized that shared structural motifs in the *Arenaviridae*, *Filoviridae*, and *Bunyaviridae* may be derived from common ancestry

TABLE 64.2
Real-Time Reverse Transcription Polymerase Chain Reaction Assays for Arenaviruses

Virus	Target[a]	Genome Accession #	Amplicon Size	Primers/Probe Name and Sequence	Final Conc. (µM)	Sensitivity (PFU/PCR)[b]
Lassa Josiah-MGB	GPC	AY628203	80	F3569 5'-TGC TAG TAC AGA CAG TGC AAT GAG-3' R3648 5'-TAG TGA CAT TCT TCC AGG AAG TGC-3' p3598S 6FAM-TGT TCA TCA CCT CTT C-MGBNFQ	0.9 0.9 0.2	0.1 (268 copies)
Lassa Macenta-MGB	NP	AY628201	73	F1079 5'-CAG GAA GGG CAT GGG AAA-3' R1151 5'-TTG TTG CTC CCA ATT TTT TGT G-3' p1106S 6FAM-TTG ATT TGG AAT CAG GCG AG-MGBNFQ	0.8 0.8 0.2	10–6 (RNA dilution)[d] (257 copies)[c]
Lassa Weller-MGB	NP	AY628206	72	F1695 5'-GCA TTG ATG GAC TGC ATT ATG TTT-3' R1766 5'-CAC AGC TCT TAG GAC CTT TGC AT-3' p1720S 6FAM-ATG CAG CAG TCT CGG GA-MGBNFQ	0.9 0.9 0.2	0.1 (583 copies)[c]
Lassa Pinneo-MGB	GPC	AY628207	78	F3525 5'-CCA ATA ATC CCA CAT GTA GCG ATG-3' R3602 5'-GAA CAT TGT GCT AAT TGC GCT TTC-3' p3559S 6FAM-CCT TCA AGA TTG CCA-MGBNFQ	0.9 0.9 0.2	0.001
Lassa Mozambique (Mopeia)-MGB	NP	DQ328874	78	F1788 5'-TCT GGG GAC CGG CAA TTG TG-3' R1865 5'-ACA CCA CAT TGT GCC TTA CTA GAC-3' p1815S 6FAM-TAT GAC TGC TGC TTC-MGBNFQ	0.9 0.9 0.2	1.0
Lassa Mobala (Acar)-MGB	GPC	AY342390	80	F1858 5'-TAC AGA CCA CAG CTA CAC ACA CC-3' R1937 5'-ACT CAC CGT CAC CTG GTT GG-3' p1915A 6FAM-AGC CGT GCC CAA AG-MGBNFQ	1.0 1.0 0.2	0.001
Lassa Josiah-TM	NP	AY628203	70	F548 5'-GGA ATG AGT GGT GGT AAT CAA GG-3' R617 5'-TTT TCA CAT CCC AAA CTC TCA CC-3' p594A 6FAM-ACT CCA TCT CTC CCA GCC CGA GC-TAMRA-3'	0.6 0.6 0.1	0.1 (11,711 copies)[c]
Lassa Macenta-TM	NP	AY628201	73	F1079 5'-CAG GAA GGG CAT GGG AAA-3' R1151 5'-TTG TTG CTC CCA ATT TTT TGT G-3' p1098S 6FAM-CAC TGT TGT TGA TTT GGA ATC AGG CGA	0.8 0.8 0.1	10–5 (RNA dilution)[d] (257 copies)[c]
Lassa Weller-TM	NP	AY628206	72	F1695 5'-GCA TTG ATG GAC TGC ATT ATG TTT-3' R1766 5'-CAC AGC TCT TAG GAC CTT TGC AT-3' p1720S 6FAM-ATG CAG CAG TCT CGG GAG GGC TC-TAMRA-3'	0.9 0.9 0.1	0.01 (583 copies)[c]
Lassa Pinneo-TM	GP	AY628207	81	F2730 5'-CCC AGT TTC CCT TTC CTG AGT-3' R2810 5'-CCA ACG GAG TGT TGC AAA CA-3' p2757S 6FAM-CAA TGT ATC TTC CAC CCC AGG CCA TTC-TAMRA-3'	0.8 0.8 0.1	0.1 (234 copies)[c]

(Continued)

TABLE 64.2 (Continued)
Real-Time Reverse Transcription Polymerase Chain Reaction Assays for Arenaviruses

Virus	Target[a]	Genome Accession #	Amplicon Size	Primers/Probe Name and Sequence	Final Conc. (µM)	Sensitivity (PFU/PCR)[b]
Lassa Mozambique (Mopeia)-TM	NP	DQ328874	81	F2687 5′-CCT GAT GGT CTC CAG CAT ATT TC-3′	0.9	0.1
				R2768 5′-GCT ACA ATT TCA GCT TGT CTG C-3′	0.9	
				p2712S 6FAM-CCC GTC TAT GAG GCA AGC CCC AGC-TAMRA-3′	0.1	
Lassa Mobala (Acar)-TM	GP	AY342390	85	F1860 5′-CAG ACC ACA GCT ACA CAC ACC-3′	1.0	0.0001
				R1944 5′-AAT TCC AAC TCA CCG TCA CCT G-3′	1.0	
				p1915A 6FAM-AGC CGT GCC CAA AGC CTC ATC GTC TC-TAMARA-3′	0.1	
Machupo Carvallo-MGB	GPC	AY619643	80	F699 5′-ATG ACC CGT GTG AGG AAG GG-3′	0.7	0.001
				R778 5′-GCC ACA GTA GTC AAA GGA ACT GG-3′	0.7	
				p721S 6FAM-AGT GTG CTA CCT GAC CAT-MGBNFQ-3′	0.2	
Machupo Mallele-MGB	GPC	AY619645	80	F406 5′-CAC CTT CCT GAT TCG GGT CTC-3′	0.8	0.001
				R485 5′-CAA GGT CTT CTG GTT CAT AGA CTG-3′	0.8	
				p428S 6FAM-AGT GTA TCT GTC CTC CTC A-MGBNFQ-3′	0.2	
Machupo Carvallo-TM	GPC	AY619643	78	F681 5′-ATC TCT TCA GGG GCT TCC ATG-3′	0.7	0.001
				R758 5′-TGG GGT CAC CAC ACT GAT TG-3′	0.7	
				p728A 6FAM-AGC ACA CTT TCC CTT CCT CAC ACG G-TAMRA-3′	0.1	
Machupo Mallele-TM	Polymerase	AY619642	78	F1009 5′-TTC GAT CAA CAT TGC AGC TAA ATC-3′	0.5	0.001
				R1086 5′-GAT GGT GTA TCG GTT GTT GCA G-3′	0.5	
				p1059A 6FAM-CCA TTG TTC ACC GGG CAG GCC AGT-TAMRA-3′	0.1	
Junin-MGB	NP	AY619641	65	F1933 5′-CAT GGA GGT CAA ACA GCT TCC T-3′	0.8	0.001 (234 copies)[c]
				R1997 5′-GCC TCC AGA CAT GGT TGT GA-3′	0.8	
				p1956S 6FAM-ATG TCA TCG GAT CCT T-MGBNFQ	0.2	
Junin-MGB	Polymerase	AY619640	68	F2427 5′-TTG CCA AAT TGA CCC ATC TGT A-3′	0.8	0.001 (234 copies)[c]
				R2494 5′-GCA ATA TGC CGG GTG TAG TGA-3′	0.8	
				p2452S 6FAM-TAG AGT TCG TCT GAT TTC A-MGBNFQ	0.2	
Junin-TM	NP	AY619641	79	F3032 5′-CAG TTC ATC CCT CCC CAG ATC-3′	1.0	0.0001
				R31110 5′-GGT TGA CAG ACT TAT GTC CAT GAA GA-3′	1.0	
				p3083A 6FAM-TGT TCA ACG AAA CAC AGT TTT CAA GGT GGG-TAMRA-3′	0.1	
Junin-TM	Zinc-binding protein (Z)	AY619640	66	F282 5′-AGG AAT TCG GAC CTC TGC AA-3′	1.0	0.001
				R347 5′-CTC CAC CGG CAC TGT GAT T-3′	1.0	
				p303Sa 6FAM-ATC TGT TGG AAG CCC CTA CCT ACC A-TAMRA-3′	0.1	

(Continued)

TABLE 64.2 (Continued)
Real-Time Reverse Transcription Polymerase Chain Reaction Assays for Arenaviruses

Virus	Target[a]	Genome Accession #	Amplicon Size	Primers/Probe Name and Sequence	Final Conc. (μM)	Sensitivity (PFU/PCR)[b]
Sabiá-MGB	GP	U41071	65	F903 5′-CGT CAG ATC TTA AGT GCT TTG GAA-3′	0.8	0.001
				R967 5′-TTC CGA ATC GTG GTC AAG GT-3′	0.8	
				p928S 6FAM-CAC AGC ACT AGC AAA A-MGBNFQ	0.2	
Sabiá-MGB	Polymerase	AY358026	73	F948 5′-TGT GAT CAT TGG TCG CAG CTA-3′	0.7	0.01
				R1020 5′-CTG AGA GGG AAG AAG GCA GTG A-3′	0.7	
				p970S 6FAM-ATC GTT ATC CAT TAG AAT CT-MGBNFQ	0.2	
Sabiá-TM	GP	U41071	83	F903 5′-CGT CAG ATC TTA AGT GCT TTG GAA-3′	0.8	0.001
				R985 5′-TTT CAA CAT GTC ACA GAA TTC CG-3′	0.8	
				p959A 6FAM-CGT GGT CAA GGT TAC ATT TTG CTA GTG CTG TGT-TAMRA-3′	0.2	
Sabiá-TM	Hairpin intergenic region and RNA polymerase	AY358026	79	F433 5′-CCG GGG TGG TGT GGT TTA-3′	0.7	0.01
				R511A 5′-AAG GCT CAA GGG TGT TAC CTG-3′	0.7	
				p460S 6FAM-TTC AAC GGA CTG CTG CTG TCT AAA TAG TCT-TAMRA-3′	0.1	
Guanarito-MGB	GP1	AY129247	73	F485 5′-TGG ATT CTT GGG TGG ACA ATT AA-3′	0.6	0.01
				R557 5′-TAG GCT CAC AGC AGA TTC TTG GA-3′	0.6	
				p513S 6FAM-TGG GAC ATG ACT TTT T-MGBNFQ-3′	0.2	
Guanarito-MGB	Polymerase	AY358024	80	F521 5′-GCT GCC GGA GCT GTC TGA-3′	0.7	0.1
				R599 5′-ATG GTG CGA GTT TGT GGA CTT-3′	0.7	
				p542S 6FAM-CAC CAA GTC CCT TAA AG-MGBNFQ-3′	0.2	
Guanarito-TM	GP1	AY129247	73	F485 5′-TGG ATT CTT GGG TGG ACA ATT AA-3′	0.8	0.001
				R557 5′-TAG GCT CAC AGC AGA TTC TTG GA-3′	0.8	
				p532A1 6FAM-CCT CAA AAA GTC ATG TCC CAA TCC C-TAMRA-3′	0.1	
Guanarito-TM	Polymerase	AY358024	80	F521 5′-GCT GCC GGA GCT GTC TGA-3′	0.7	0.01
				R599 5′-ATG GTG CGA GTT TGT GGA CTT-3′	0.7	
				p542S 6FAM-CAC CAA GTC CCT TAA AGA TTT GCT ATA GCA GAC AC-TAMRA-3′	0.1	

Source: Adapted from Trombley, A.R. et al., *Am. J. Trop. Med. Hyg.*, 82, 954, 2010.

[a] PCR, polymerase chain reaction; PFU, plaque-forming units; TM, TaqMan; MGB, TaqMan minor groove binder; NP, nucleoprotein; GP, glycoprotein; VP, virus protein; GPC, glycoprotein precursor.

[b] Limit of detection (LOD) for all assays were measured in PFU/PCR unless otherwise stated; all sensitivities were done in triplicate and were positive in three of three replicates.

[c] Synthetic RNA was used to determine LODs in genome copy number.

[d] LODs measured as dilution of stock RNA because PFU information was not available.

Machupo Mallele-TM assay run at 62°C.

or common functions. Genomic sequencing will be a critical tool in advancing our understanding of arenavirus and reservoir host relationships as well as identifying potential new arenavirus candidates. Nevertheless, it is important to remember that experimental work is still critical to demonstrate causality.

ACKNOWLEDGMENTS

The authors were funded in part by the Division of Global Emerging Infections Surveillance and Response System (GEIS) Operations at the Armed Forces Health Surveillance Center (RJS) and by the Department of Defense Cooperative Biological Engagement Program funding a National Research Council Research Associateship Award at the U.S. Army Medical Research Institute of Infectious Diseases (USAMRIID) (AEO).

Opinions, interpretations, conclusions, and recommendations are those of the authors and are not necessarily endorsed by the U.S. Army.

REFERENCES

1. Lisieux, T. et al., New arenavirus isolated in Brazil, *Lancet*, 343, 391 (1994).
2. Delgado, S. et al., Chapare virus, a newly discovered arenavirus isolated from a fatal hemorrhagic fever case in Bolivia, *PLoS Pathog*, 4, e1000047 (2008).
3. Enserink, M., Emerging diseases. New arenavirus blamed for recent deaths in California, *Science*, 289, 842 (2000).
4. Sauvage, F. F. J. and Gonzalez, J. P., Arenaviruses, in *Manual of Security Sensitive Microbes and Toxins*, Liu, D., Ed., CRC Press, Boca Raton, FL, 9pp. (2014).
5. Paweska, J. T. et al., Nosocomial outbreak of novel arenavirus infection, southern Africa, *Emerg Infect Dis*, 15, 1598 (2009).
6. ICTV, Virus taxonomy: 2013 Release, in *International Committee on Taxonomy of Viruses*, Vol. 2014, International Committee on Taxonomy of Viruses (2014).
7. Charrel, R. N., de Lamballerie, X., and Emonet, S., Phylogeny of the genus Arenavirus, *Curr Opin Microbiol*, 11, 362 (2008).
8. Zapata, J. C. and Salvato, M. S., Arenavirus variations due to host-specific adaptation, *Viruses*, 5, 241 (2013).
9. Rowe, W. P. et al., Arenoviruses: Proposed name for a newly defined virus group, *J Virol*, 5, 651 (1970).
10. Pinschewer, D. D., Perez, M., and de la Torre, J. C., Dual role of the lymphocytic choriomeningitis virus intergenic region in transcription termination and virus propagation, *J Virol*, 79, 4519 (2005).
11. Perez, M. and de la Torre, J. C., Characterization of the genomic promoter of the prototypic arenavirus lymphocytic choriomeningitis virus, *J Virol*, 77, 1184 (2003).
12. Salvato, M. S. et al., Arenaviridae, in *Virus Taxonomy, Ninth Report of the International Committee on Taxonomy of Viruses*, King, A. M. Q., Adams, M. J., Carstens, E. B., and Lefkowitz, E. J., Eds., Elsevier, San Diego, CA, 715pp. (2012).
13. Raju, R., Raju, L., Hacker, D., Garcin, D., Compans, R., and Kolakofsky, D., Nontemplated bases at the 5′ ends of Tacaribe virus mRNAs, *Virology*, 174, 53 (1990).
14. McLay, L., Ansari, A., Liang, Y., and Ly, H., Targeting virulence mechanisms for the prevention and therapy of arenaviral hemorrhagic fever, *Antiviral Res*, 97, 81 (2013).
15. Loureiro, M. E., D'Antuono, A., Levingston Macleod, J. M., and Lopez, N., Uncovering viral protein-protein interactions and their role in arenavirus life cycle, *Viruses*, 4, 1651 (2012).
16. Downs, W. G., Anderson, C. R., Spence, L., Aitken, T. H., and Greenhall, A. H., Tacaribe virus, a new agent isolated from Artibeus bats and mosquitoes in Trinidad, West Indies, *Am J Trop Med Hyg*, 12, 640 (1963).
17. Bowen, M. D., Peters, C. J., and Nichol, S. T., The phylogeny of New World (Tacaribe complex) arenaviruses, *Virology*, 219, 285 (1996).
18. Bowen, M. D., Peters, C. J., and Nichol, S. T., Phylogenetic analysis of the Arenaviridae: Patterns of virus evolution and evidence for cospeciation between arenaviruses and their rodent hosts, *Mol Phylogenet Evol*, 8, 301 (1997).
19. Clegg, J. C., Molecular phylogeny of the arenaviruses, *Curr Top Microbiol Immunol*, 262, 1 (2002).
20. Emonet, S. F., de la Torre, J. C., Domingo, E., and Sevilla, N., Arenavirus genetic diversity and its biological implications, *Infect Genet Evol*, 9, 417 (2009).
21. Charrel, R. N. and de Lamballerie, X., Zoonotic aspects of arenavirus infections, *Vet Microbiol*, 140, 213 (2010).
22. Gonzalez, J. P., Emonet, S., de Lamballerie, X., and Charrel, R., Arenaviruses, *Curr Top Microbiol Immunol*, 315, 253 (2007).
23. Jay, M. T., Glaser, C., and Fulhorst, C. F., The arenaviruses, *J Am Vet Med Assoc*, 227, 904 (2005).
24. Armstrong, C. and Lillie, R. D., Experimental lymphocytic choriomeningitis of monkeys and mice produced by a virus encountered in studies of the 1933 St. Louis encephalitis epidemic, *Public Health Rep*, 49, 1019 (1934).
25. Childs, J. E. and Peters, C. J., Ecology and epidemiology of arenaviruses and their hosts, in *The* Arenaviridae, Salvato, M. S., Ed., Plenum Press, New York, 331pp. (1993).
26. Jamieson, D. J., Kourtis, A. P., Bell, M., and Rasmussen, S. A., Lymphocytic choriomeningitis virus: An emerging obstetric pathogen? *Am J Obstet Gynecol*, 194, 1532 (2006).
27. Tomaskova, J., Labudova, M. J. K., Pastorek, J., Petrik, J., and Pastorekova, S., Lymphocytic choriomeningitis virus, in *Manual of Security Sensitive Microbes and Toxins*, Liu, D., Ed., CRC Press, Boca Raton, FL, 735pp. (2014).
28. Childs, J. E., Glass, G. E., Korch, G. W., Ksiazek, T. G., and Leduc, J. W., Lymphocytic choriomeningitis virus infection and house mouse (*Mus musculus*) distribution in urban Baltimore., *Am J Trop Med Hyg*, 47, 27 (1992).
29. Dykewicz, C. A. et al., Lymphocytic choriomeningitis outbreak associated with nude mice in a research institute, *JAMA*, 267, 1349 (1992).
30. Emonet, S., Retornaz, K., Gonzalez, J. P., de Lamballerie, X., and Charrel, R. N., Mouse-to-human transmission of variant lymphocytic choriomeningitis virus, *Emerg Infect Dis*, 13, 472 (2007).
31. Hotchin, J., Sikora, E., Kinch, W., Hinman, A., and Woodall, J., Lymphocytic choriomeningitis in a hamster colony causes infection of hospital personnel, *Science*, 185, 1173 (1974).
32. Gregg, M. B., Recent outbreaks of lymphocytic choriomeningitis in the United States of America, *Bull WHO*, 52, 549 (1975).
33. Fischer, S. A. et al., Transmission of lymphocytic choriomeningitis virus by organ transplantation, *N Engl J Med*, 354, 2235 (2006).
34. Palacios, G. et al., A new arenavirus in a cluster of fatal transplant-associated diseases, *N Engl J Med*, 358, 991 (2008).
35. Centers for Disease Control and Prevention, Lymphocytic Choriomeningitis virus infection in organ transplant recipients—Massachusetts, Rhode Island, 2005, *Morb Mortal Wkly Rep*, 54, 537 (2005).

36. Frame, J. D., Baldwin, J. M., Jr., Gocke, D. J., and Troup, J. M., Lassa fever, a new virus disease of man from West Africa. I. Clinical description and pathological findings, *Am J Trop Med Hyg*, 19, 670 (1970).

37. Buckley, S. M. and Casals, J., Lassa fever, a new virus disease of man from West Africa. III. Isolation and characterization of the virus, *Am J Trop Med Hyg*, 19, 680 (1970).

38. Buckley, S. M., Casals, J., and Downs, W. G., Isolation and antigenic characterization of Lassa virus, *Nature*, 227, 174 (1970).

39. McCormick, J. B., Webb, P. A., Krebs, J. W., Johnson, K. M., and Smith, E. S., A prospective study of the epidemiology and ecology of Lassa fever, *J Infect Dis.*, 155, 437 (1987).

40. Yun, N. E. and Walker, D. H., Pathogenesis of Lassa fever, *Viruses*, 4, 2031 (2012).

41. McCormick, J. B. and Fisher-Hoch, S. P., Lassa fever, *Curr Top Microbiol Immunol*, 262, 75 (2002).

42. Richmond, J. K. and Baglole, D. J., Lassa fever: Epidemiology, clinical features, and social consequences, *BMJ*, 327, 1271 (2003).

43. Salazar-Bravo, J., Ruedas, L. A., and Yates, T. L., Mammalian reservoirs of arenaviruses, *Curr Top Microbiol Immunol*, 262, 25 (2002).

44. Fichet-Calvet, E., Becker-Ziaja, B., Koivogui, L., and Gunther, S., Lassa serology in natural populations of rodents and horizontal transmission, *Vector Borne Zoonotic Dis*, 14, 665 (2014).

45. Grant, A. et al., Junín virus pathogenesis and virus replication, *Viruses* 4, 2317 (2012).

46. Gomez, R. M., Jaquenod de Giusti, C., Sanchez Vallduvi, M. M., Frik, J., Ferrer, M. F., and Schattner, M., Junin virus. A XXI century update, *Microbes Infect*, 13, 303 (2011).

47. Mills, J. N. et al., A longitudinal study of Junin virus activity in the rodent reservoir of Argentine hemorrhagic fever, *Am J Trop Med Hyg*, 47, 749 (1992).

48. Mills, J. N. et al., Junin virus activity in rodents from endemic and nonendemic loci in central Argentina, *Am J Trop Med Hyg*, 44, 589 (1991).

49. Mills, J. N. et al., Prevalence of infection with Junin virus in rodent populations in the epidemic area of Argentine hemorrhagic fever, *Am J Trop Med Hyg*, 51, 554 (1994).

50. Vitullo, A. D., Hodara, V. L., and Merani, M. S., Effect of persistent infection with Junin virus on growth and reproduction of its natural reservoir, *Calomys musculinus*, *Am J Trop Med Hyg*, 37, 663 (1987).

51. Vitullo, A. D. and Merani, M. S., Vertical transmission of Junin virus in experimentally infected adult *Calomys musculinus*, *Intervirology*, 31, 339 (1990).

52. Salas, R. et al., Venezuelan haemorrhagic fever, *Lancet*, 338, 1033 (1991).

53. Doyle, T. J., Bryan, R. T., and Peters, C. J., Viral hemorrhagic fevers and hantavirus infections in the Americas, *Infect Dis Clin North Am*, 12, 95 (1998).

54. de Manzione, N. et al., Venezuelan hemorrhagic fever: Clinical and epidemiological studies of 165 cases, *Clin Infect Dis*, 26, 308 (1998).

55. Fulhorst, C. F. et al., Natural rodent host associations of Guanarito and pirital viruses (Family Arenaviridae) in central Venezuela, *Am J Trop Med Hyg*, 61, 325 (1999).

56. Milazzo, M. L., Cajimat, M. N., Duno, G., Duno, F., Utrera, A., and Fulhorst, C. F., Transmission of Guanarito and Pirital viruses among wild rodents, Venezuela, *Emerg Infect Dis*, 17, 2209 (2011).

57. Fulhorst, C. F., Ksiazek, T. G., Peters, C. J., and Tesh, R. B., Experimental infection of the cane mouse *Zygodontomys brevicauda* (family Muridae) with guanarito virus (Arenaviridae), the etiologic agent of Venezuelan hemorrhagic fever, *J Infect Dis*, 180, 966 (1999).

58. Johnson, K. M. et al., Virus isolations from human cases of hemorrhagic fever in Bolivia, *Proc Soc Exp Biol Med*, 118, 113 (1965).

59. Centers for Disease Control and Prevention, Bolivian hemorrhagic fever—El Beni Department, Bolivia, 1994, *Morb Mortal Wkly Rep*, 43, 943 (1994).

60. World Health Organization, Re-emergence of Bolivian hemorrhagic fever, *Epidemiol Bull*, 15, 4 (1994).

61. Aguilar, P. V. et al., Reemergence of Bolivian hemorrhagic fever, 2007–2008, *Emerg Infect Dis*, 15, 1526 (2009).

62. Salazar-Bravo, J., Dragoo, J. W., Bowen, M. D., Peters, C. J., Ksiazek, T. G., and Yates, T. L., Natural nidality in Bolivian hemorrhagic fever and the systematics of the reservoir species, *Infect Genet Evol*, 1, 191 (2002).

63. Webb, P. A., Justines, G., and Johnson, K. M., Infection of wild and laboratory-animals with Machupo and Latino viruses, *Bull WHO*, 52, 493 (1975).

64. Bonthius, D. J., Lymphocytic choriomeningitis virus: An underrecognized cause of neurologic disease in the fetus, child, and adult, *Semin Pediatr Neurol*, 19, 89 (2012).

65. Iserte, J. A., Lozano, M. E., Enria, D. A., and Levis, S. C., Junin Virus (Argentine Hemorrhagic Fever), in *Molecular Detection of Human Viral Pathogens*, Liu, D., Ed., CRC Press, Boca Raton, FL, 721pp. (2011).

66. Wilson, M. R. and Peters, C. J., Diseases of the central nervous system caused by lymphocytic choriomeningitis virus and other arenaviruses, *Handb Clin Neurol*, 123, 671 (2014).

67. Wright, R. et al., Congenital lymphocytic choriomeningitis virus syndrome: A disease that mimics congenital toxoplasmosis or Cytomegalovirus infection, *Pediatrics*, 100, E9 (1997).

68. McCormick, J. B. et al., A case-control study of the clinical diagnosis and course of Lassa fever, *J Infect Dis*, 155, 445 (1987).

69. White, H. A., Lassa fever. A study of 23 hospital cases, *Trans R Soc Trop Med Hyg*, 66, 390 (1972).

70. Webb, P. A. et al., Lassa fever in children in Sierra Leone, West Africa, *Trans R Soc Trop Med Hyg*, 80, 577 (1986).

71. Johnson, K. M., McCormick, J. B., Webb, P. A., Smith, E. S., Elliott, L. H., and King, I. J., Clinical virology of Lassa fever in hospitalized patients, *J Infect Dis*, 155, 456 (1987).

72. Salas, R. A., de Manzione, N., and Tesh, R., Venezuelan hemorrhagic fever: Eight years of observation, *Acta Cient Venez*, 49(Suppl. 1), 46 (1998).

73. Moraz, M.-L. and Kunz, S., Pathogenesis of arenavirus hemorrhagic fevers., *Exp Rev Anti-infect Ther*, 9, 49 (2011).

74. Bausch, D. G. and Mills, J. N., Arenaviruses: Lassa fever, Lujo hemorrhagic fever, lymphocytic Choriomeningitis, and the South American hemorrhagic fevers, in *Viral Infections of Humans*, Kaslow, R. A., Stanberry, L. R., and Le Duc, J. W., Eds., Springer US, Boston, MA (2014).

75. Centers for Disease Control and Prevention, Old World/New World Arenaviruses|Viral Hemorrhagic Fevers http://www.cdc.gov/vhf/virus-families/arenaviruses.html. Accessed December 9, 2014.

76. Tesh, R. B., Jahrling, P. B., Salas, R., and Shope, R. E., Description of Guanarito virus (Arenaviridae: Arenavirus), the etiologic agent of Venezuelan hemorrhagic fever, *Am J Trop Med Hyg*, 50, 452 (1994).

77. Branco, L. M. et al., Emerging trends in Lassa fever: Redefining the role of immunoglobulin M and inflammation in diagnosing acute infection, *Virol J*, 8, 478 (2011).

78. Bausch, D. G. et al., Diagnosis and clinical virology of Lassa fever as evaluated by enzyme-linked immunosorbent assay, indirect fluorescent-antibody test, and virus isolation, *J Clin Microbiol*, 38, 2670 (2000).

79. Corgenix, Corgenix Gains CE Mark for ReLASV® Antigen Rapid Test for Diagnosis of Lassa Fever (2013). http://www.corgenix.com/news-releases/corgenix-gains-ce-mark-for-relasv-antigen-rapid-test-for-diagnosis-of-lassa-fever/. Accessed on May 7, 2014.

80. Engvall, E. and Perlmann, P., Enzyme-linked immunosorbent assay (ELISA). Quantitative assay of immunoglobulin G, *Immunochemistry*, 8, 871 (1971).

81. Panning, M. et al., Laboratory diagnosis of lassa fever, Liberia, *Emerg Infect Dis*, 16, 1041 (2010).

82. Trombley, A. R. et al., Comprehensive panel of real-time TaqMan polymerase chain reaction assays for detection and absolute quantification of filoviruses, arenaviruses, and New World hantaviruses, *Am J Trop Med Hyg*, 82, 954 (2010).

83. Stenglein, M. D. et al., Identification, characterization, and in vitro culture of highly divergent arenaviruses from boa constrictors and annulated tree boas: Candidate etiological agents for snake inclusion body disease, *MBio*, 3, e00180 (2012).

84. Bodewes, R. et al., Detection of novel divergent arenaviruses in boid snakes with inclusion body disease in The Netherlands, *J Gen Virol*, 94, 1206 (2013).

85. Hetzel, U. et al., Isolation, identification, and characterization of novel arenaviruses, the etiological agents of boid inclusion body disease, *J Virol*, 87, 10918 (2013).

65 Picobirnaviruses

Yashpal S. Malik, Naveen Kumar, Kuldeep Dhama, Krisztián Banyai,
Joana D'Arc Pereira Mascarenhas, and Raj Kumar Singh

CONTENTS

65.1 Introduction 585
    65.1.1 Classification, Morphology, and Genome Organization 586
    65.1.2 Epidemiology and Impact on Health 586
    65.1.3 Transmission, Clinical Features, and Treatment 587
    65.1.4 Diagnosis 588
65.2 Methods 589
    65.2.1 Sample Collection and Preparation 589
        65.2.1.1 Sample Collection 589
        65.2.1.2 Virus Purification 589
        65.2.1.3 PBV Viral RNA Extraction 589
    65.2.2 Detection Procedures...589
        65.2.2.1 Electron Microscopy 589
        65.2.2.2 Polyacrylamide Gel Electrophoresis...589
        65.2.2.3 Multiplex PCR 589
65.3 Conclusion and Future Perspectives 591
References...591

65.1 INTRODUCTION

Enteric infections leading to acute gastroenteritis, together with respiratory tract infections, rank as the most common infectious disease syndromes of humans and animals affecting all age groups. It has been estimated that nearly 5 billion diarrheic episodes occur every year globally, and 15%–30% of severely affected patients succumb to the infection, which is generally higher at the extreme ages of life, particularly in low-income settings of developing countries. The clinical illness caused by enteric pathogens varies but usually involves diarrhea, vomiting, and fever of differing length and severity. Many microbial pathogens are the main causes of infectious gastroenteritis, and amid them, bacterial and parasitic infections have diminished occurrence as a result of progresses in public health practice. However, viral gastroenteritis has not dropped in an analogous manner from these interventions.

Enteric viral pathogens recognized since 1970 include viruses mainly belonging to four families: *Reoviridae* (rotavirus), *Caliciviridae* (calicivirus), *Astroviridae* (astrovirus), and *Adenoviridae* (adenovirus) and are known to cause billions of cases of enteric viral infections every year. During the last decade, the availability of next-generation sequencing as a new research tool to explore unknown viruses of human and veterinary importance has aided in the detection of putative novel pathogens and the documentation of the etiologic agents of numerous infections, cracking the long-standing anonymities of many genetically diverse and rapidly evolving viruses.

This metagenomic approach has been employed in several studies probing fecal samples of livestock and companion animal species, providing evidence on the diversity of the animal fecal virome, helping the elucidation of the etiology of diarrheal disease in animals, and identifying potential zoonotic and emerging viruses in humans. In the past few decades, the molecular characterization of many of these gastroenteritis viruses has led to advances both in our understanding of the pathogens themselves and in the development of state-of-art diagnostics. Although the use of these highly accurate, reliable, and sensitive approaches to identify and characterize individual viral pathogens is in its infancy, it has opened new avenues to review disease burden and contemplate implementation of vaccination and improvements in public health measures and sanitary practices.

In contrast to some bacterial pathogens, there are no national notifiable diseases data on the prevalence of enteric viruses. The possible transmissibility of such viruses between animals and humans should be assessed as a number of known and emerging viruses have been associated with enteric infections in animals. Since the first detection of a small-sized bisegmented double-stranded RNA (dsRNA) virus named *Picobirnavirus* (PBV) in humans and black-footed pygmy rice rats in 1988 [1,2], it has been identified in various domestic and captive animals. The sequencing of partial segment 1 and full-length segment 2 of this virus by Rosen [3] unraveled some of the mysteries regarding its genome. Although accretion of genomic data of PBVs from different mammalian and reptile species across the world resolved some of the ambiguity, quite a few questions and hypotheses regarding pathogenesis, persistence location, and evolution of PBVs remain unreciprocated. PBV, with main emphasis on its biology, epidemiology, viral persistence, and their zoonotic potential, has been reviewed recently [4,5].

65.1.1 Classification, Morphology, and Genome Organization

PBVs are small (35–41 nm in diameter), nonenveloped, double-stranded, bisegmented RNA viruses. The prefix "pico" denotes the small diameter of the viral particle, and "birna" suggests a genome composed of two segments of dsRNA [2]. Due to the bisegmented nature of the genome as revealed in polyacrylamide gel electrophoresis (PAGE), it was initially confused with birnaviruses. PBVs differ from members of *Birnaviridae* in respect of host, virion size, capsid, RNA polymerase, genome size, and organization (Table 65.1).

A complete new taxonomic order *Diplornavirales*, family *Picobirnaviridae*, genus *Picobirnavirus* was assigned to accommodate PBVs [6]. The two species under the genus *Picobirnavirus* are *Human picobirnavirus* (type species) and *Rabbit picobirnavirus* (designated species) [7]. The PBVs have a simple core capsid with a distinctive icosahedral arrangement, displaying 60 twofold symmetric dimers of a coat protein with a new 3D-fold as revealed from VLPs of a rabbit PBV generated in baculovirus [8]. This coat protein undergoes an autoproteolytic cleavage, releasing a posttranslational modified peptide that remains associated with nucleic acid within the capsid. The ability of PBV particles to disrupt biological membranes *in vitro* appears to reflect evolution of animal cell invasion properties of its simple 120 subunits capsid [8]. Also, PBVs are known to carry 4–10 copies of particular sequence motif (ExxRxNxxxE), which encode proteins of various sizes (106–224 residues and without proline and cysteine) [9].

A uniform nomenclature for PBV has been put forward, which reveals the genogroup (genogroup I [GI] or genogroup II [GII]), host, country of origin, strain, and year of isolation for a specific PBV identified in such order [10]. For example, GI/PBV/human/China/1-CHN-97/1997 specifies a PBV with GI specificity and strain name 1-CHN-97 detected in humans from China in 1997.

The segment 1 (2.2–2.7 kb) of PBV encodes the capsid protein, while gene segment 2 (1.2–1.9 kb) codes for the viral RNA-dependent RNA polymerase (RdRp) [11,12]. PBVs are classified into two genogroups, that is, GI (reference strain-1-CHN-97) and GII (reference strain-4-GA-91) based on the RdRp gene (segment 2) of human PBV [3,13]. Remarkably, more than 83% of PBV sequences available in the National Center for Biotechnology Information belong to GI and less than 3% to GII, while the rest remains undefined.

Recently, genogroup III (GIII) PBV has been identified in humans from the Netherlands (PBVIII/Homo sapiens/ VS6600008/2008/NL/KJ206569) [14]. At present, limited full-length PBV genome sequences from humans, turkey, camels, and pigs are available in nucleotide sequences databases, that is, GI/PBV/human/THAI/Hy005102/2002 [15] and GI/ PBV/California sea lion/Hong Kong/HKG-PF080915/2012 [16]. Complete nucleotide sequences of segment 1 of lapine PBV [17] and segment 2 of bovine PBV and monkey RdRp sequences are accessible. The variances in PBV genome from different species are presented in Table 65.2.

65.1.2 Epidemiology and Impact on Health

PBV prevalence reports across the world suggest that PBV has a wide host range. PBVs have been detected in a number of domestic as well as captive animal species [2,18–30]. Species-wise PBVs distribution across the world is presented in Figure 65.1.

Incidence studies carried out across the world in different species by different diagnostic assay indicated the presence of PBVs as 0.4%–27.1% by RNA-PAGE and 18.2%–65% by reverse transcription polymerase reaction (RT-PCR) in swine [22–24,31], 0.7%–3.7% by RNA-PAGE in bovines [27,32], 14.3% in equines by RT-PCR [29], 0.9%–1.8% by RNA-PAGE and 0.6% by RT-PCR in canines [18,33], 3.4%–15.3% by RNA-PAGE and 49.4% by RT-PCR in chickens [19,20], and 2.2%–25% by RNA-PAGE and 2.4%–47.9% by RT-PCR in other animals [18,34].

Reports associating PBV with gastroenteritis gave contradictory results, some with positive correlation [19,20,22], while others with negative correlation [31]. Remarkably, all such association studies in captive animals yielded negative correlation of PBV with diarrhea [18,34]. Concomitant infection having both genogroups (GI and GII) of PBVs in one host has also been detected in humans [35], pigs [36], and recently in bovines [26].

Persistence of PBV infection in asymptomatic carrier, greater rhea (Rhea americana), has elucidated where PBV infection was acquired during the early stages of life and maintained until the beginning of adulthood confirming that the PBV–host interaction is proved as a persistent and asymptomatic infection model [37]. The maintenance and evolution of PBV infection during adulthood was also ascertained in an adult orangutan (*Pongo pygmaeus*) recently, assuming the mechanisms by which persistent infections are maintained involve both modulation of

---

TABLE 65.1

Fundamental Differentiating Features between *Picobirnaviruses* and *Birnaviruses*

Properties	*Picobirnavirus*	*Birnavirus*
Hosts	Mammals	Fish, chicken, and turkey
Virion size	35–41 nm	65–70 nm
Capsid structure	Triangulation of 1, 3, or 4	T = 13 laevo symmetry
RNA polymerase	A–B–C motifs	C–A–B motifs
Genome size	Smaller segment (1.7 kb) and larger segment (2.5 kb)	Smaller segment (2.8 kb) and larger segment (3.3 kb)
Genome organization (segment 1)	Two ORFs	Single ORF

TABLE 65.2
Genomic Features of Different Species Picobirnaviruses

	Human PBV (Hy005102 Strain)	Otarine PBV (PF080915 Strain)	Lapine PBV (Strain 35227/89)	Bovine PBV (RUBV-P Strain)
Segment 1 (encodes for capsid protein)				
Length (bp)	2525	2347	2362	—
ORFs	ORF1 (224aa), ORF2 (552aa)	ORF1 (163 aa), ORF2 (576aa)	ORF1 (591aa), ORF2 (155aa), ORF3 (55aa)	—
Segment 2 (encodes for RNA-dependent RNA polymerase)				
Length (bp)	1745	1688	—	1758
ORFs	ORF1 (534aa)	ORF1 (532aa)	—	ORF1 (554aa)

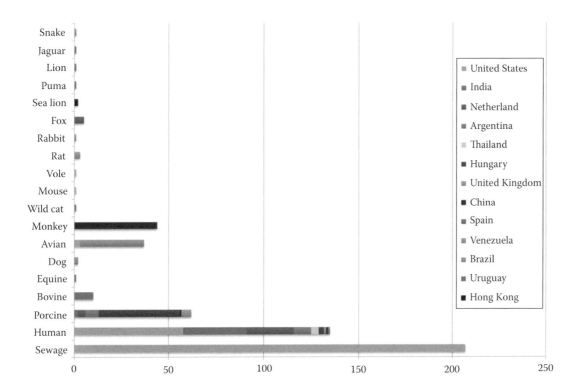

FIGURE 65.1 Global distributions of PBVs in different domestic and captive animal species. (Reprinted from Malik, Y.S. et al., *BioMed Res. Int.*, 2014, 780752, 2014. With permission.)

the virus and cellular gene expression and/or modification of the host immune response [38]. However, there is still a need to understand the molecular mechanisms governing persistent and asymptomatic coexistence with the host and the potential host suitability to maintain this relationship.

Studies on molecular epidemiology and evolutionary aspects of PBVs resulted in some intriguing observations. The sequence variation within any PBV genogroups may extend 50%, although some studies have reported complete or nearly complete sequence identities among PBV strains detected at distant locations and different host species. Prolonged infection and fecal shedding allowed insight into the intrahost evolution of PBVs; these recent studies have indicated significant divergence of genomic sequences over a relatively short period [37,38]. In addition to accumulation of single nucleotide mutations that leads to significant sequence diversity, recent

findings have provided evidence for reassortment within GI porcine PBV genomes [39].

Based on the studies carried out by researchers across the world on association studies of PBVs with gastroenteritis, the authors predicted that PBV may not be the primary cause of diarrhea and may play synergistic effect in association with the primary enteric causative agents [40,41]. Further, captive animals might be acting either as the reservoir or persistent asymptomatic carriers.

65.1.3 TRANSMISSION, CLINICAL FEATURES, AND TREATMENT

PBVs have been associated with gastroenteritis either as sole entity or more commonly with other enteric causative agents. GI PBV is the most common genogroup detected in all the species [5,42]. Transmission is suspected through contaminated water and environment [43] as PBVs have been detected typically in the feces of animals (e.g., contamination of surface waters with runoff from animal). However, PBV has been isolated from the respiratory tract of pigs [36], which may be a secondary location following viremia from the gut. At present, there is no evidence for PBV in airborne droplet. This may suggest the zoonotic potential of these viruses with possible emerging and/or reemerging risk to a number of animals in different geographical locations. Transmission from one species to another is defined in terms of genetic relatedness of one or more segments among the segmented genome viruses like rotavirus or PBVs infecting two different species from the same or different geographical areas. These studies indicated the genetic relatedness among different PBVs based on a short fragment of RdRp gene, for example, between swine and human [24,29], human and equine PBVs [29], and human and rodents PBVs [44].

Animals infected by PBV alone or with other enteric causative agents may have diarrhea (nonbloody and watery), nausea, and fever. Studies suggest that most of the infections caused by PBVs are asymptomatic in captive animals. Since there is no specific antiviral treatment and no vaccine to prevent the infection, prophylactic measures include maintenance of strict hygiene to prevent contamination of food, water, and environment, identification of mode of transmission, and prevention of spread of infection from animals to animals as well as animals to humans. Treatment mainly focuses on supportive care, especially preventing and treating dehydration.

65.1.4 Diagnosis

At present, three diagnostic methods are widely employed for detection of PBV: direct electron microscopy, PAGE, and RT-PCR. Of these, RT-PCR is widely used for detection and genotyping of PBVs.

Conventional techniques: The first report of PBV detection dates back to 1988, where it was detected in the stool samples surveyed for rotavirus, as two migrated segments in PAGE [1,2]. While direct visualization of PBV is possible under electron microscope [2,31], its use for epidemiological studies is impractical because of its requirement of highly skilled personnel and costly equipment. Given the current lack of immunological/serological assays and animal model as well as permissive cell lines for PBVs, molecular approaches play a vital role in the diagnosis and clinicopathological studies of PBV infections.

Molecular techniques: The beginning of the usage of molecular techniques for PBVs was with the first detection of bisegmented dsRNA virus in humans and black-footed pygmy rice rats by using PAGE [1,2]. Since then, PAGE has been used by different countries for detection of PBVs from different species as a reliable diagnostic assay. The most consistent indicator for diagnosis of PBV infection is the presence of PBV RNA in fecal samples. Therefore, the specimen of choice is fecal samples from diarrheal animals. The importance of RNA-PAGE lies in its specificity to resolve the PBV genome into

two discrete bands and also can distinguish genome profiles (small or large genome profiles), which are missed in molecular assays. Electrophoretic patterns with banding of genomic segments in two patterns with larger genome profile and small genome profile, based on migrating distance and size of segment 1 and 2, after PAGE analysis with silver staining have been observed [5,26–28]. In the larger profile, the segments 1 and 2 correspond to 2.7 kb and 1.9 kb respectively, while size of these genome segments remain 2.2 kb and 1.2 kb for the short profile PBVs [5] (Figure 65.2). However, RNA-PAGE is a cumbersome, labor-intensive, and time-consuming technique.

With the successful cloning and sequencing of partial genome of two human PBV strains for the first time [3], RT-PCR assay has been widely employed as a tool for the routine diagnosis of PBV infection. Use of RT-PCR and DNA sequencing techniques for detection and characterization of PBV genetic background has markedly enhanced the understanding of the molecular epidemiology of PBV infection. For genogrouping of PBVs, oligonucleotide primers sequences targeting the RdRp gene are based on two prototype strains, GI/PBV/human/China/1-CHN-97/1997 and GII/PBV/human/USA/4-GA-91/1991, and have been validated across the world in different animals species [10,13,23,26–29,35]. However, to further improve the diagnosis and capture the highly diverse porcine PBVs, diagnostic primers sets (PBV2-19 [+] 5′-CGACGAGGTTGATAAGCGGA-3′ and

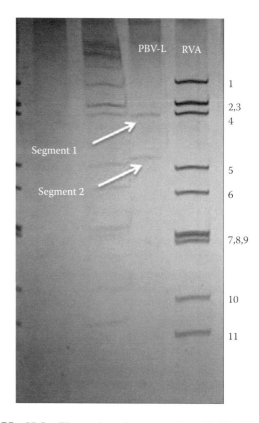

FIGURE 65.2 Electrophoretic separation of *Picobirnavirus* (PBV) segments in silver-stained polyacrylamide gel. PBV-L: large genome profile PBV; RVA: type A rotavirus. (Reprinted from Malik, Y.S. et al., *BioMed Res. Int.*, 2014, 780752, 2014. With permission.)

PBV2-281 [−] 5′-CACAGTTCGGGCCTCCTGA-3′) targeting conserved region of RdRp gene (824–1086 nt) were developed and also allowed detection of porcine-like PBVs in humans [45]. For specific identification of PBV GIII, real-time PCR has been described with primers VS791 (5′-CGATGGATCTTTATGTTCCCG-3′), VS792 (5′-GTAGTTGAAATGTTGATCCATTT-3′), and VS793 (5′-CAAACTTTCCAGCAACCGCTT-3′) [14].

The authors have also developed improved target set of diagnostic oligonucleotide sequences for segment 2–based RT-PCR for bovine PBVs with superior sensitivity and specificity (data not shown), and the same primer sets are also being found useful for detecting PBVs in pigs.

65.2 METHODS

65.2.1 SAMPLE COLLECTION AND PREPARATION

65.2.1.1 Sample Collection
The clinical samples (feces) should be collected from the diarrheic animals, even though samples can also be collected from the nondiarrheic cases, as PBVs have also been isolated from nondiarrheic cases. The 10% fecal suspension is prepared in phosphate-buffered saline (PBS; pH 7.4) and centrifuged at 7500 rpm for 20 min after vortexing for 2 min to remove the particulate matter. The clear supernatant thus obtained is used for RNA extraction.

65.2.1.2 Virus Purification
Virus purification is done by taking clarified fecal suspension (10% w/v) prepared from samples that have been tested positive to PBV by PAGE. Centrifuge the sample at 7500 rpm for 20 min to pellet the unwanted matter. The virus is initially concentrated by passing the sample through a cushion of 40% sucrose in PBS by using ultracentrifugation at $1,00,000 \times g$ for 3 h. Resuspend the pellet thus obtained in 500 µL PBS and finally concentrate the virus preparation by flotation in an isopycnic CsCl gradient. Collect the fractions and measure their refractive index. Pool the fractions with a refractive index between 1.3 and 1.4 with corresponding densities between 1406 and 1435 g/mL and pellet the virus particles by further centrifugation at $160,000 \times g$ for 12 h at 4°C. Suspend the pellet thus obtained in 250 µL PBS and store at −70°C.

65.2.1.3 PBV Viral RNA Extraction
Different methods have been used for RNA extraction from PBVs before subjecting to RNA-PAGE and RT-PCR. If the sample size is large, guanidinium isothiocyanate–phenol–chloroform extraction (Tri reagent LS) method is preferred, considering its low cost and acceptable RNA purity. Briefly, 500 µL of Tri reagent is added to 500 µL of clear supernatant in a 1.5 mL microcentrifuge tube, vortexed thoroughly, and kept at room temperature for 15 min. Further, 100 µL of 1-bromo-2-chloropropane is added to the tube, brief vortexed, and kept at room temperature for 10 min. The tube is then centrifuged at 12,000 rpm for 15 min. The upper aqueous layer, thus obtained after centrifugation, is carefully collected

in a fresh 1.5 mL microcentrifuge tube. To the aqueous layer, 500 µL of isopropanol is added and kept at −20°C for 1 h for precipitation of viral RNA. The PBV RNA is then pelleted by centrifugation at 12,000 rpm for 15 min followed by washing with prechilled 70% ethanol at 12,000 rpm for 10 min. Pellet is air dried. Finally, PBV RNA pellet is dissolved in 20 µL nuclease-free water (NFW) or 1× TE buffer. Concentration and purity of extracted PBV RNA is measured using Nanodrop spectrophotometer. The obtained PBV RNA can be stored at −80°C until further use as template for RT-PCR assay.

If the sample size is small, highly sensitive commercially available RNA extraction kits (QIAamp Viral RNA Mini Kit, Qiagen; Quick-RNA™ Mini Prep Kit, Prolab) are widely used for PBV RNA isolation as per the manufacturer's instructions.

65.2.2 DETECTION PROCEDURES

The most commonly used PBV detection methods in various laboratories include RNA-PAGE and gel-based RT-PCR. Both of these assays should be performed, as PAGE gives indication about the genome profile (large/small), while RT-PCR about genogroups. Electron microscopy is also used for visualization of the complete virion particles.

65.2.2.1 Electron Microscopy
To visualize the PBV particles, purified viral suspensions as prepared earlier are directly observed under the electron microscope. The samples are negative stained with 2% phosphotungstic acid for approximately 1 min. The size of virus particles is measured and identified as PBV as per their morphology. PBV appear as spherical, with approximately 35 nm in diameter, nonenveloped, and with uniform structure and morphology, suggestive of icosahedral symmetry.

65.2.2.2 Polyacrylamide Gel Electrophoresis
Direct visualization of PBV genome (peculiar bisegmented excluding *Birnaviruses*) in PAGE after silver staining [46] has still been used in many parts of the world for reliable diagnosis. The PBV presents at least two genomic profiles in PAGE, that is, large genome profile (segment 1 [2.3–2.6 kb] and segment 2 [1.5–1.9 kb]) (Figure 65.2) and small genome profile (segment 1 [1.75 kb] and segment 2 [1.55 kb]) (Figure 65.3). Notably, a third genome segment has also been documented in chickens [47] and dogs [48]. These viruses with trisegmented dsRNA genome might be due to mixed infection of PBV strains or with other viruses; nevertheless, it needs further investigation to confirm and/or ascertain the identification.

65.2.2.3 Multiplex PCR
Owing to poor sensitivity of PAGE, molecular-based tests like multiplex PCR are the most preferred method for detection as well as genogrouping of PBVs [3]. For genogrouping of PBVs, oligonucleotide primers targeting the RdRp gene are based on two prototype strains (GI/PBV/human/China/1-CHN-97/1997 and GII/PBV/human/USA/4-GA-91/1991) and have been widely employed for genogrouping [13,23,26–28].

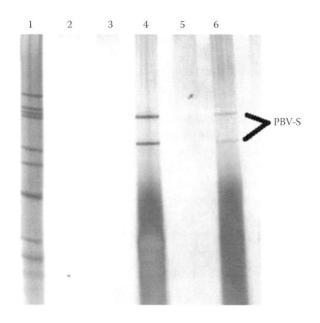

FIGURE 65.3 Electrophoretic separation of *Picobirnavirus* (PBV) segments in silver-stained polyacrylamide gel (small genome profile [PBV-S]). Lane 1 = RVA: type A rotavirus (SA11); lanes 4 and 6 = PBV-S.

TABLE 65.3
Components of Reverse Transcription Mix

Components	Volume (μL)
RNA sample	7.0 (volume can vary depending on the concentration)
Random hexamer (100 ng/μL)	1.0
Dimethyl sulfoxide (DMSO)	1.0
Nuclease-free water (NFW)	4.0
After denaturing at 95°C for 5 min, followed by snap chilling on ice, the following reagents are added to microcentrifuge tube:	
RT buffer (5×)	4.0
Deoxynucleotide triphosphate mix (10 mM)	1.0
Moloney murine leukemia virus reverse transcriptase (M-MLV RT) (200 U/μL)	1.0
RNase inhibitor (Ribolock)	1.0
Total volume	20

65.2.2.3.1   Reverse Transcription
65.2.2.3.1.1   Procedure

1. The RT mix mentioned in Table 65.3 is prepared in a 0.5 mL microcentrifuge tube (volume indicated per sample). It is important to note that four components, that is, sample RNA, random hexamer, dimethyl sulfoxide, and NFW, are mixed first in the mentioned quantity followed by denaturation of dsRNA and snap chilling (Table 65.3).
2. Transfer the remaining reagents to microcentrifuge tube.
3. Spin the tubes briefly for uniform mixing of all the regents.
4. The reaction mixture is kept at 37°C for 1.5 h followed by inactivation of Moloney murine leukemia virus reverse transcriptase at 80°C for 2–3 min.
5. The cDNA thus obtained can be directly used for PCR or stored at –20°C until further use.

65.2.2.3.2   PCR

The two sets of primer pairs mentioned in Table 65.4 are used to amplify the 201 bp fragment of the RdRp gene (segment 2), related to prototype PBV strain 1-CHN-97, representing GI and 369 bp product of the RdRp gene of reference strain 4-GA-91 of GII (Table 65.4).

The following multiplex PCR protocol has been modified in terms of thermal cycling conditions described in earlier methods [3,13]. The presence of a 201 bp product is indicative of PBV GI genogroup, while the presence of a 369 bp product is indicative of PBV GII genogroup.

65.2.2.3.2.1   Procedure

1. Allow all the PCR reagents to thaw on ice and spin briefly.
2. Prepare the multiplex PCR reaction mix in 0.2 mL PCR tube as follows (volume indicated per sample) (Table 65.5).
3. Spin the tubes at 5000 rpm in a microcentrifuge for 5 s at room temperature.
4. Place the sample tubes in a thermal cycler for 30 cycles of 94°C for 2 min 30 s, 48°C for 20 s, and 72°C for 30 s, and final extension at 72°C for 5 min.

TABLE 65.4
Picobirnavirus Genotyping Primers Based on RNA-Dependent RNA Polymerase Gene (Segment 2)

Genogroup Specificity	Primers	Oligonucleotide Sequences (5′–3′)	Nucleotide Positions	Amplicon Size (bp)	Reference Strains
Genogroup I	PicoB25[+]	TGG TGT GGA TGT TTC	(665–679)	201	1-CHN-97
	PicoB43[−]	A(GA)T G(CT)T GGT CGA ACT T	(850–865)		
Genogroup II	PicoB23[+]	CGG TAT GGA TGT TTC	(685–699)	369	4-GA-91
	PicoB24[−]	AAG CGA GCC CAT GTA	(1039–1053)		

Source: Rosen, B.I. et al., *Virology*, 277, 316, 2000.

TABLE 65.5
Components of Multiplex Polymerase Chain Reaction Mix

Components	Volume (μL)
Dream Taq PCR master mix (2×)	12.5
PicoB25 forward primer (10 pmol/μL)	1.0
PicoB43 reverse primer (10 pmol/μL)	1.0
PicoB23 forward primer (10 pmol/μL)	1.0
PicoB24 reverse primer (10 pmol/μL)	1.0
cDNA	2.0
Nuclease-free water (NFW)	6.5
Total volume	25.0

5. Analyze the expected size of specific amplicons in 2% agarose gel electrophoresis and document the images using gel documentation system.

Note: Always include positive and negative controls while performing the multiplex PCR for PBV.

65.3 CONCLUSION AND FUTURE PERSPECTIVES

Among the most recently recognized viruses, PBVs are genetically diverse, small, nonenveloped viruses with a bisegmented dsRNA genome, classified under a new family, *Picobirnaviridae*, imposing possible risk for a number of host species including humans. The availability of genomic data from different PBVs of mammalian species has helped in resolving some of the abstruseness associated with this virus infection. However, there are still some unanswered questions including pathogenesis, persistent location, and evolution of PBVs. The genomic segment 2, encoding "RdRp gene" sequence-based genogrouping (GI, GII, and GIII), has aided in identifying viral genogroups circulating in animals across the different countries. Sequence and phylogenetic analysis studies suggest tremendous sequence diversity and varying evolutionary dynamics of PBVs in different host species. The recent detection of PBVs from respiratory tracts of swine and humans, the proposal for new genogroups [14,49], and the increasing popularity of viral metagenomics using different high-throughput sequencing platforms to detect viral diversity in clinical, pathological, and environmental samples are further expanding our knowledge on this highly diverse virus. Still, not much has been done on exploring replication strategies adopted by the virus and the role of innate and adaptive immunity. The close relatedness of animal PBVs with human along with detection of PBVs from the sewage designate the potential threat in terms of infection acquirement from the sewage and transmission of these viruses across the species. Although PBVs may have an inadequate or ambiguous clinical implication, they do pose a potential public health concern in humans, and the control of PBVs mainly relies on nonvaccinal approach.

REFERENCES

1. Pereira, H. G. et al., Novel viruses in human faeces, *Lancet*, 332, 103, 1988.
2. Pereira, H. G. et al., A virus with a bisegmented double-stranded RNA genome in rat (*Oryzomys nigripes*) intestines, *J. Gen. Virol.*, 69, 2749, 1988.
3. Rosen, B. I. et al., Cloning of human picobirnavirus genomic segments and development of an RT-PCR detection assay, *Virology*, 277, 316, 2000.
4. Nates, S. V., Gatti, M. S. V., and Ludert, J. E., The picobirnavirus: An integrated view on its biology, epidemiology and pathogenic potential, *Future Virol.*, 6, 223, 2011.
5. Ganesh, B. et al., Picobirnavirus infections: Viral persistence and zoonotic potential, *Rev. Med. Virol.*, 22, 245, 2012.
6. International Committee on Taxonomy of Viruses (ICTV). 2014. Viral taxonomy. http://www.ictvonline.org/virusTaxonomy.asp (accessed January 15, 2015).
7. Carstens, E. B. and Ball, L. A., Ratification vote on taxonomic proposals to the International Committee on Taxonomy of Viruses (2008), *Arch. Virol.*, 154, 1181, 2009.
8. Duquerroy, S. et al., The picobirnavirus crystal structure provides functional insights into virion assembly and cell entry, *EMBO J.*, 28, 1655, 2009.
9. Costa, B. D. et al., Picobirnaviruses encode a protein with repeats of the ExxRxNxxxE motif, *Virus Res.*, 158, 251, 2011.
10. Fregolente, M. C. and Gatti, M. S., Nomenclature proposal for Picobirnavirus, *Arch. Virol.*, 154, 1953, 2009.
11. Chandra, R., Picobirnavirus, a novel group of undescribed viruses of mammals and birds: A minireview, *Acta Virol.*, 41, 59, 1997.
12. Rosen, B. I., Molecular characterization and epidemiology of picobirnaviruses. In Desselberger, U. and Gray, J. (eds.), *Viral Gastroenteritis*, Volume 9 of Perspectives in Medical Virology, pp. 633–644, Amsterdam, the Netherlands: Elsevier, 2003.
13. Bányai, K. et al., Sequence heterogeneity among human picobirnaviruses detected in a gastroenteritis outbreak, *Arch. Virol.*, 148, 2281, 2003.
14. Smits S. L. et al., New viruses in idiopathic human diarrhea cases, the Netherlands, *Emerg. Infect. Dis.*, 20, 1218, 2014.
15. Wakuda, M., Pongsuwanna, Y., and Taniguchi, K., Complete nucleotide sequences of two RNA segments of human Picobirnavirus, *J. Virol. Methods*, 126, 165, 2005.
16. Woo, P. C. et al., Complete genome sequence of a novel Picobirnavirus, Otarine Picobirnavirus, discovered in California Sea Lions, *J. Virol.*, 86, 6377, 2012.
17. Green, J. et al., Genomic characterization of the large segment of a rabbit picobirnavirus and comparison with the atypical Picobirnavirus of *Cryptosporidium parvum*, *Arch. Virol.*, 144, 2457, 1999.
18. Fregolente, M. C. et al., Molecular characterization of picobirnaviruses from new hosts, *Virus Res.*, 143, 134, 2009.
19. Alfieri, A. F. et al., A new bi-segmented double stranded RNA virus in avian feces, *Arq. Bras. Med. Vet. Zootec.*, 40, 437, 1988.
20. Silva, R. R. et al., Genogroup I avian picobirnavirus detected in Brazilian broiler chickens: A molecular epidemiology study, *J. Gen. Virol.*, 95, 117, 2014.
21. Pereira, H. G. et al., A virus with bi-segmented double-stranded RNA genome in guinea pig intestines, *Mem. Inst. Oswaldo Cruz.*, 84, 137, 1989.
22. Gatti, M. S. et al., Viruses with bisegmented double-stranded RNA in pig faeces, *Res. Vet. Sci.*, 47, 397, 1989.

23. Bányai, K. et al., Genogroup I picobirnaviruses in pigs: Evidence for genetic diversity and relatedness to human strains, *J. Gen. Virol.*, 89, 534, 2008.

24. Ganesh, B. et al., Detection and molecular characterization of porcine Picobirnavirus in feces of domestic pigs from Kolkata, India, *Indian J. Virol.*, 23, 387, 2012.

25. Vanopdenbosch, E. and Wellemans, G., Birna-type virus in diarrhoeic calf faeces, *Vet. Rec.*, 125, 610, 1989.

26. Malik, Y. S. et al., Identification and characterisation of a novel genogroup II picobirnavirus in a calf in India, *Vet. Rec.*, 174(11), 278, 2014.

27. Malik, Y. S. et al., Picobirnavirus detection in bovine and buffalo calves from foothills of Himalaya and Central India, *Trop. Anim. Health. Prod.*, 43, 1475, 2011.

28. Malik, Y. S. et al., Molecular characterization of a genetically diverse bubaline picobirnavirus strain, India, *Thai J. Vet. Med.*, 43, 609, 2013.

29. Ganesh, B. et al., Genogroup I picobirnavirus in diarrhoeic foals: Can the horse serve as a natural reservoir for human infection? *Vet. Res.*, 42, 52, 2011.

30. Gillman, L., Sánchez, A. M., and Arbiza, J., Picobirnavirus in captive animals from Uruguay: Identification of new hosts, *Intervirology*, 56, 46, 2013.

31. Ludert, J. E. et al., Identification in porcine faeces of a novel virus with a bisegmented double stranded RNA genome, *Arch. Virol.*, 117, 97, 1991.

32. Buzinaro, M. G. et al., Identification of a bi-segmented double-stranded RNA virus (picobirnavirus) in calf faeces, *Vet. J.*, 166, 185, 2003.

33. Costa, A. P. et al., Detection of double-stranded RNA viruses in fecal samples of dogs with gastroenteritis in Rio de Janeiro, *Arq. Bras. Med. Vet. Zootec.*, 56, 554, 2004.

34. Masachessi, G. et al., Picobirnavirus (PBV) natural hosts in captivity and virus excretion pattern in infected animals, *Arch. Virol.*, 152, 989, 2007.

35. Ganesh, B. et al., Detection and molecular characterization of multiple strains of picobirnavirus causing mixed infection in a diarrhoeic child: Emergence of prototype genogroup II-like strain in Kolkata, India, *Int. J. Mol. Epidemiol. Genet.*, 2, 61, 2011.

36. Smits, S. L. et al., Genogroup I and II picobirnaviruses in respiratory tracts of pigs, *Emerg. Infect. Dis.*, 17, 2328, 2011.

37. Masachessi, G. et al., Establishment and maintenance of persistent infection by picobirnavirus in greater rhea (*Rhea americana*), *Arch. Virol.*, 157, 2075, 2012.

38. Masachessi, G. et al., Maintenance of picobirnavirus (PBV) infection in an adult orangutan (*Pongo pygmaeus*) and genetic diversity of excreted viral strains during a three-year period, *Infect. Genet. Evol.*, 29, 196, 2015.

39. Bányai, K. et al., Genome sequencing identifies genetic and antigenic divergence of porcine picobirnaviruses, *J. Gen. Virol.*, 95, 2233, 2014.

40. Alfieri, A. A. et al., Occurrence of *Escherichia coli*, rotavirus, picobirnavirus and *Cryptosporidium parvum* in a post weaning diarrhoea focus in swine, *Cienc. Agrar.*, 15, 5, 1994.

41. Bhattacharya, R. et al., Molecular epidemiology of human astrovirus infections in Kolkata, India, *Infect. Genet. Evol.*, 6, 425, 2006.

42. Malik, Y. S. et al., Epidemiology, phylogeny, and evolution of emerging enteric picobirnaviruses of animal origin and their relationship to human strains, *BioMed. Res. Int.*, 2014, 780752, 2014. doi:10.1155/2014/780752.

43. Symonds, E. M., Griffin, D. W., and Breitbart, M., Eukaryotic viruses in wastewater samples from the United States, *Appl. Environ. Microbiol.*, 75, 1402, 2009.

44. Phan, T. G. et al., The fecal viral flora of wild rodents, *PLoS Pathog.*, 7, e1002218, 2011.

45. Carruyo, G. M. et al., Molecular characterization of porcine picobirnaviruses and development of a specific reverse transcription-PCR assay, *J. Clin. Microbiol.*, 46, 2402, 2008.

46. Herring, A. J. et al., Rapid diagnosis of rotavirus infection by direct detection of viral nucleic acid in silver-stained polyacrylamide gels, *J. Clin. Microbiol.*, 16, 473, 1982.

47. Leite, J. P. et al., A novel avian virus with trisegmented double-stranded RNA and further observations on previously described similar viruses with bi-segmented genome, *Virus Res.*, 16, 119, 1990.

48. Volotão, E. M. et al., First evidence of a trisegmented double-stranded RNA virus in canine faeces, *Vet. J.*, 161, 205, 2001.

49. Woo, P. C. et al., Metagenomic analysis of viromes of dromedary camel fecal samples reveals large number and high diversity of circoviruses and picobirnaviruses, *Virology*, 471, 117, 2014.

66 Infectious Myonecrosis Virus

Maria Erivalda Farias de Aragão, Maria Verônyca Coelho Mello, and Maria de Lourdes Oliveira Otoch

CONTENTS

66.1 Introduction 593
    66.1.1 Classification..593
    66.1.2 Morphology and Genome Organization..593
    66.1.3 Epidemiology..594
    66.1.4 Clinical Features and Pathogenesis...594
    66.1.5 Transmission..595
    66.1.6 Diagnosis 596
66.2 Methods 596
    66.2.1 Sample Preparation..596
    66.2.2 Detection Procedures..596
        66.2.2.1 RT-PCR...596
        66.2.2.2 Real-Time RT-PCR...596
        66.2.2.3 Nested PCR..596
        66.2.2.4 Loop-Mediated Isothermal Amplification Assay 596
        66.2.2.5 Immunoassays..597
66.3 Conclusion and Future Perspectives 597
References...598

66.1 INTRODUCTION

Marine shrimp farming started in Brazil only in the 1970s with Asian species *Marsupenaeus japonicus*. In the 1980s, the practice of shrimp farming emerged as an agribusiness following the introduction of exotic species *Litopenaeus vannamei*. Despite favorable conditions, it is only recently that Brazilian shrimp farming achieved a prominent place in the world [1,2]. Nevertheless, in 2004, the production of *L. vannamei* suffered a drastic drop due to various economic, environmental, and biological factors [3]. Among these, two viral epidemics caused great economic losses to Brazilian producers. The first epidemic, designated as infectious myonecrosis (IMN), occurred in the Northeast region of Brazil in 2002, and the second epidemic, named white spot infection, occurred in the South in 2004 [3–6].

Initially, IMN was recognized as a disease of unknown infectious agent (the so-called idiopathic myonecrosis). Extensive research was conducted into the identification and characterization of the etiological agent as a virus, which was named infectious myonecrosis virus (IMNV) [7]. Further description of IMNV particles was subsequently reported [8,9].

viral sequences allowed its classification into the *Totiviridae* family [8–12].

The *Totiviridae* family encompasses a diverse group of viruses that appear as isometric capsid particles of 40 nm in diameter and possess a nonsegmented, double-stranded RNA (dsRNA) genome. In most cases, the genome encodes only one capsid protein (major capsid protein, or MCP) and RdRp [9,11,13].

According to the International Committee on Taxonomy of Viruses report in 2008, the family *Totiviridae* is made up of three recognized genera: *Totivirus*, *Giardiavirus*, and *Leishmaniavirus*, which together include 18 viral species that infect fungi or protozoa. Viruses infecting yeast and filamentous fungi are placed in the *Totivirus* genus [9,11,13]. Considered as "putative" *Totivirus* by some authors [8,9,11,12], IMNV presents an analogy both in size and in genomic sequence to *Giardia lamblia* virus from the *Giardiavirus* genus. Tang et al. [9] proposed a new monotypic genus in the family *Totiviridae* to accommodate IMNV.

IMNV is the only member of the *Totiviridae* family that infects other hosts than fungi or protozoa and has been implicated in the disease process [8,9,13].

66.1.1 CLASSIFICATION

Analysis of the amino acid sequence of the RNA-dependent RNA polymerase (RdRp) of IMNV and comparison with other

66.1.2 MORPHOLOGY AND GENOME ORGANIZATION

Virion of IMNV measures about 40 nm in diameter, is nonenveloped, contains a capsid with icosahedral symmetry

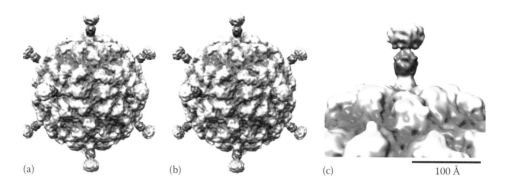

FIGURE 66.1 Fiber-like protrusions at 5f axes of infectious myonecrosis virus (IMNV) virion. (a) Radially color-coded surface view of IMNV virion, in stereo, with complex protrusions evident at the 5f axes. (b) Close-up view of one protrusion and the surrounding capsid. (c) Surface view of IMNV virion in black and white, with overlay of T_1 model and selected major capsid protein (MCP)-A subunits (blue, cyan) and MCP-B subunits (red, pink, orange). (Reprinted with permission from Tang, J., Ochoa, W.F., Sinkovits, R.S. et al., Infectious myonecrosis virus has a totivirus-like, 120-subunit capsid, but with fiber complexes at the fivefold axes, *Proc. Natl. Acad. Sci. USA*, 105, 17526, Copyright 2008 National Academy of Sciences, U.S.A.)

composed of 120 subunits, and shows a buoyant density of 1366 g/mL in cesium chloride (CsCl) [8,9]. Structurally, the 120 subunits consist of the same protein, the MCP. However, this protein shows two forms of arrangement within the capsid, and therefore, it was named MCP-A and MCP-B. Capsids possess 60 copies each of two chemically identical subunits, MCP-A and MCP-B, distinguished by their nonequivalent positions (Figure 66.1a and b). The A subunits form pentameric clusters around each of the 12 5f axes, and the B subunits form trimeric clusters around each of the 20 3f axes. Moreover, five B subunits lie in a staggered arrangement around each A pentamer so that together they form a compact decamer centered at each 5f axis [9].

Unusually for the *Totiviridae* family, IMNV shows fiber-like protrusions along the viral capsid (Figure 66.1a and b) [8,9]. These proteinous protrusions are likely to be important for virus entry through their binding to receptors on target cells or involvement in cellular internalization processes. This may help explain the unique ability of IMNV, among other members of the same family, in being transmitted in an extracellular way between host organisms [9].

IMNV genome harbors a nonsegmented, dsRNA of 7560 bp, localized in the interior to the capsid, and packaged as five or six concentric rings (Figure 66.2a and b). Similar to other members of the family, IMNV genome shows two open reading frames (ORFs). ORF 1 is located at nucleotide 136–4953 and encodes a protein of 179 kDa, which includes the N-terminal sequence of the capsid protein (MCP). The ORF 2 is located between the nucleotides 5241–7451 and encodes a protein of 85 kDa that corresponds to the RdRp [7–9,11,13–15].

66.1.3 Epidemiology

The first records of IMN in cultured *L. vannamei* were documented in Northeast Brazil (Piauí state) in 2002, after a period of heavy rainfall. Since 2003, IMN was reported in several marine shrimp farms, mainly in the states of Ceará, Rio Grande do Norte, Paraíba, and Pernambuco [16,17].

IMNV seems to be limited to the Brazilian Northeast, but shrimps with similar signs have also been found in Indonesia, with IMNV being posteriorly identified.

Species of penaeid shrimp *L. vannamei*, *L. stylirostris*, and *Penaeus monodon* are susceptible to IMNV and among these species, *L. vannamei* is most frequently affected [14].

IMNV is the major cause of mortality in shrimp farming and affects the quality of final product [14,18]. Juvenile and subadult shrimps are most severely affected by the disease, as they are grown in salt or low salinity [5,7].

Besides Brazil, IMNV has spread to other countries, with a confirmed case in Indonesia [15,19]. Senapin et al. [20] further confirmed that IMNV occurred in Indonesia but not in China, India, Malaysia, Thailand, and Vietnam. Thus, IMNV has not yet reached other Asian countries outside Indonesia.

Interestingly, the symptoms similar to those observed in myonecrosis caused by IMNV were observed in *Penaeus aztecuz* in the United States in 1970 and in Mexico in 1978 as well as in *Farfantepenaeus subtilis*, native species in the coastal zone of Piauí, in the 1980s [8,17]. Given these findings, Santana et al. [21] hypothesized that the IMNV may be a recurrence of a disease in a new host, *L. vannamei*. Some researchers believed that *F. subtilis*, infected in the Piauí coast, may be responsible for the contamination of farms of *L. vannamei* and caused the spread of the disease on farms in the Northeast.

66.1.4 Clinical Features and Pathogenesis

The manifestations of the disease are characterized by an extensive area of necrosis in the skeletal muscle, especially in the distal abdominal segment, and loss of transparency in the tail, which becomes opaque with milky areas [5,7,8]. The whitening may begin at the fifth or sixth abdominal segments or on the sides and distal areas of the tail close to pleopods [5]. Relative to the degree of severity, opacity extends progressively toward the telson and uropods. The disease is slowly progressive, with mortality reaching 70% in juvenile and subadult shrimps.

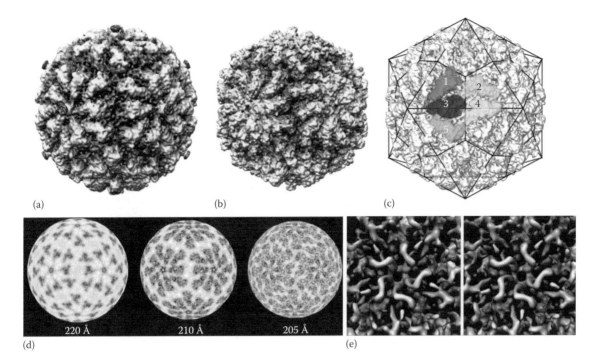

(a)   (b)   (c)

(d)   220 Å   210 Å   205 Å   (e)

FIGURE 66.2 Infectious myonecrosis virus (IMNV) capsid at 8.0 Å resolution and comparison with x-ray model of ScV-L-A capsid computed to 8.0 Å resolution. (a) Radially surface view of IMNV virion, with outer parts of protrusions removed by zeroing densities above radius 225 Å. (b) Comparable surface view of ScV-L-A (14) to which the same radial color table was applied. (c) Surface view of IMNV virion in black and white, with overlay of T = 1 model and selected major capsid protein (MCP)-A subunits (3, 4) and MCP-B subunits (1, 2, e 5), was highlighted for reference in the text. (d) Spherical sections at different radii as indicated below each image, with density values coded in grayscale (darker, denser). (e) Space-filling slab view, in stereo, as seen from the inner surface of the IMNV capsid. Tube-like densities that form two concentric rings around the 5f axis (inner and outer rings from MCP-A and MCP-B subunits, respectively) are interpreted as α-helices. (Reprinted with permission from Tang, J., Ochoa, W.F., Sinkovits, R.S. et al., Infectious myonecrosis virus has a totivirus-like, 120-subunit capsid, but with fiber complexes at the fivefold axes, *Proc. Natl. Acad. Sci. USA*, 105, 17526, Copyright 2008 National Academy of Sciences, U.S.A.)

In the acute phase, it presents coagulative myonecrosis with some swelling between necrotic muscle fibers. Lesions may be present in a focal or multifocal way after the acute phase. In a chronic stage, lesions are accompanied by liquefaction of necrotic muscle fibers. At this stage, the muscles exhibit a reddish color, giving an appearance of cooked shrimp. Due to this, many producers have dubbed the dying individuals "zombie shrimp" or "living dead shrimp" because even with the appearance of being dead (red tail) and debilitated, animals still live for several days [5,17]. In some cases, symptoms may offer evidence of the severity of IMNV grown in population. However, it should not be used alone as a reference because mortality can be accentuated even in mild and moderate stages of the disease [4]. The animal may die at any stage of the disease, but most mortality occurs in the advanced stage of the disease.

The illness causes food restriction, decrease in volume of hepatopancreas (and subsequent reduction or absence of lipid storage), difficulty in hardening carapace, and also erratic swimming. Many animals present deformed intestine, or in the shape of "S" and a rigid bend in the abdominal muscle.

Larger animals die easily due to physical weakness [5,17,22]. Using *in situ* hybridization technique, Tang et al. [14] noted that IMNV infects the skeletal muscle tissue and

also reaches the lymphoid organ and hindgut. On histological analysis, lesions in the muscle tissue and viral inclusion bodies in the cytoplasm of connective tissue cells and also in hemocytes of infected shrimps were observed. Hemocytic infiltration in the hepatopancreas and heart of shrimps also occur during infection [14]. It is also possible to observe the presence of ectopic spheroids in the lymphoid organ in proximity to the heart and antennal gland [8,17]. Tang et al. [14] showed that the presence of lymphoid organ spheroids is a typical histopathological finding of IMNV. The spheroids are a defense mechanism from lymphoid organ to play its role engulfing the invading agent by defense cells, aiming for its expulsion.

66.1.5 Transmission

IMNV transmission usually occurs through horizontal transfer. That is, transmission of an infectious agent from a sick or a bearer individual, either through direct or indirect contact or by means of a vector. In the case of IMNV, there are no specific data on vectors. Transmission occurs through contamination of water and/or cannibalism, for example, healthy shrimp feed on infected shrimp [8]. Vertical transmission is the transfer of an infectious agent from a sick individual to its offspring, particularly from the mother to its offspring either through contamination of eggs or during lay. It is believed that vertical transmission may occur, but this is still not experimentally documented [8,23-25].

In most cases, contamination results from an improperly transboundary movement of broodstock by people or equipment cleaned in an inappropriate way [2]. It is likely that marine birds feeding on contaminated shrimp from farms may play a part in epizootic transmission of IMNV [14,15,23].

Additionally, acquisition of noncertified postlarvae and environmental imbalance represent other determinants of contamination [26].

66.1.6 Diagnosis

Conventional techniques: Histopathological examination of the tissue sections stained with hematoxylin–eosin has been widely employed to the diagnosis of IMN. Positive samples to IMNV show lesions characteristics of the acute phase such as coagulative necrosis of striated skeletal muscles fibers and hemocyte infiltration with edema formation among muscle fibers and fibrosis [7,17,22,27].

Molecular techniques: The diagnosis of IMNV infection based on clinical signs and histological examination is insufficient to rule out diseases caused by other pathogens such as *Penaeus vannamei* nodavirus [28] and *Vibrio* spp. [29].

Therefore, more specific methods such as molecular techniques need to be employed for IMN diagnosis. Reverse transcription polymerase chain reaction (RT-PCR), real-time RT-PCR, nested RT-PCR, and lamp RT-PCR are examples of molecular procedures that are increasingly applied for IMNV identification. In addition, immunoassays with highly specific monoclonal antibodies (MAbs) have also proven valuable for IMNV diagnosis.

66.2   METHODS

66.2.1 Sample Preparation

The first step in sample preparation involves the homogenization of gnathothoraces ("heads") from subadult *P. vannamei* showing signs of disease in a buffer Tris/HCl (pH 7.5) enriched with NaCl. Then, the suspension is cleared through low-speed centrifugation. The use of differential centrifugations further helps separate the viral particles from the cell debris [30]. Poulos et al. [8] used fumed silica and activated charcoal to clarify the suspension. Furthermore, the homogenate may be subjected to a sucrose gradient and an ultracentrifugation to obtain purified viruses.

66.2.2 Detection Procedures

66.2.2.1   RT-PCR

PCR utilizes oligonucleotide primers from regions of interest to amplify a specific DNA sequence from target organism. PCR mixtures containing target DNA, nucleotides, and thermostable DNA polymerase are typically subjected to cycles of temperatures that permit the separation of double strand (denaturation), primers pairing, and elongation of segment.

The entire genome from an organism can be sequenced after amplification of segments of DNA by PCR. Organisms that possess RNA as genome sometimes must be transcribed into DNA before PCR amplification. This is possible by utilizing a reverse transcriptase from retrovirus. The combination of RT and PCR amplification (RT-PCR) forms the basis of molecular diagnosis of RNA viruses such as IMNV.

Poulos et al. [8] determined the complete sequence of genomic RNA from Brazil IMNV, and Senapin et al. [15] cloned and sequenced the full 7.5 kb genome from Indonesia IMNV by using a commercial kit and primers to target regions (Table 66.1). These advances facilitated the development of RT-PCR for IMNV diagnosis. A commercial kit that amplifies a 227 bp fragment is also available, and nested RT-PCR and lamp RT-PCR have been subsequently designed on the basis of this fragment.

66.2.2.2   Real-Time RT-PCR

Real-time RT-PCR represents an improvement of RT-PCR. The introduction of a probe labeled with fluorescent dyes enables direct (real-time) detection of the amplified products.

Andrade et al. [22] described a real-time assay with primers and TaqMan probe derived from the IMNV genome (Genbank accession no. AY570982) [8]. This assay can detect a few IMNV copies (about 10 copies/µL RNA), which is more sensible than the nested RT-PCR (detecting 1000 copies/µL RNA).

Table 66.1 lists the primers utilized by several authors in various RT-PCR protocols. Andrade et al. [22] utilized a segment corresponding to the nucleotides 467–500 as probe (which is labeled with fluorescent dyes 5-carboxyfluoroscein [FAM] on the 5′ end and N,N,N′,N′-tetramethyyl-6-carboxyrhodamine [TAMRA] on the 3′ end).

66.2.2.3   Nested PCR

Nested PCR is a modification of PCR with the goal to reduce the formation of incorrect/undesirable DNA sequences due to binding of the primer to the nonspecific sequences. The technique involves a second set of primers that amplify a secondary target localized within the segment produced by the first set of primers. Nested RT-PCR has been used for the detection of IMNV by several authors including Senapin et al. [20] and Poulos and Lightner [31].

66.2.2.4   Loop-Mediated Isothermal Amplification Assay

Loop-mediated isothermal amplification (LAMP) is a nucleic acid amplification technique that utilizes a set of four specially designed primers and a DNA polymerase with strand displacement activity to amplify DNA under isothermal conditions [32]. It has the advantages of being rapid, efficient, and specific, without relying on elaborate equipment.

TABLE 66.1

Primers and Probes Used in RT-PCR, Nested RT-PCR, Real-Time RT-PCR, and RT-LAMP–AuNP

Method	Primer/Probe	Sequence (5′–3′)	Genome Position	References
RT-PCR	4587 F	CGACGCTGCTAACCATACAA	4587–4607	Mello et al. [30]; Poulos and
	4914 R	ACTCGGCTGTTCGATCAAGT	4914–4934	Lightner [31]; Andrade et al. [22]
Nested RT-PCR	4725NF	GGCACATGCTCAGAGACA	4725–4743	Poulos and Lightner [31]
	4863NR	AGCGCTGAGTCCAGTCTTG	4863–4882	
RT-PCR	IMNV218F	GCTGGACTGTATTGGTTGAG		Andrade et al. [22]
	IMNV682R	AACCAAGTTCTTCTTCTCCAGTT		
Real-time RT-PCR	IMNV412F	GGACCTATCATACATAGCGTTTGCA		
	IMNV545R	AACCCATATCTATTGTCGCTGGAT		
	probe, IMNVp1	6FAMCCACCTTTACTTTCAATACTACATCAT CCCCGGTAMRA	467–500	
RT-PCR	IMNVR	CATATGGGGCAATTACGGTTACAGGG		Borsa et al. [35]
	IMNVF	CGGGATCCGTATACATACCAAATGGCC-		
RT-LAMP–AuNP	F3	GGTTTTAAAGAATACTTTGGAGTGA	5922–5946	Arunrut et al. [33]
	B3	ACTTGACTAACTACAATTTCTCC	6118–6096	
	FIP	TCCAGCTTGTAGTTTCTCAATCAT TTTT CGCTTGACGAAAAACCAG	6014–5990/ TTTT/5947–5964	
	BIP	TGGATTACCATTTGACTATGCATCATTTT ACTCTTCTCACCATTGTCTT	6020–6044/ TTTT/6094–6075	
	LF	ATTTCTCTCTGATGCTGTTG	5986–5967	
	LB	TGATCATCAGCCAACAACTTTCGAAG	6047–6072	
	Thiol probe	(SH-)A10-CAACAGCATCAGAGAGAAATT	A10/5967–5987	

The inclusion of additional primers (termed loop primers) further accelerates the LAMP reaction [32] and reduces the reaction time by half relative to the original protocol [31]. By incorporating a RT step into LAMP protocol together with the use of a ssDNA probe labeled with gold nanoparticle (AuNP), Arunrut et al. [33] demonstrated that IMNV could be detected with decreased testing time and increased sensitivity in comparison with RT-PCR methods.

66.2.2.5 Immunoassays

Immunological methods are widely used to detect and diagnose infectious diseases. Recently, an immunoassay using MAbs against segments of IMNV–MCP has been reported for highly specific detection of IMNV [34]. In another study, a recombinant fragment of the IMNV–MCP gene encoding amino acids 105–297 was heterologously expressed in *Escherichia coli* and the product was used to immunize Balb/c mice to generate Mabs [35]. Of the six MAbs obtained, four recognized IMNV in tissue homogenates from naturally infected shrimps by immunodot blot assay and three detected a 100 kDa protein that corresponds to MCP in western blot assay. Viral inclusion bodies were detected in muscle fibroses by western blot and immunohistochemistry. Two MAbs showed high specificity and sensitivity, as no cross-reaction was observed with healthy shrimp tissues in any assays, indicating their usefulness for IMNV detection.

Mello et al. [30] purified IMNV from infected shrimp and used the virions to immunize mouse. The resulting polyclonal antibodies were screened against crude extracts in western blot assay. Antibodies that did not react with the crude extract from uninfected shrimp appeared to be specific for IMNV and provided an efficient approach to detect IMNV.

66.3 CONCLUSION AND FUTURE PERSPECTIVES

Accurate and rapid diagnosis of IMNV is crucial for the control or eradication of IMN in shrimps. Complementing to classical methods of histopathology and microbiology, molecular and immunological diagnostic methods have played an increasingly important role in safeguarding the health and well-being of farmed shrimp [15,26].

Despite enormous efforts that have been made by researchers, no effective treatment and/or vaccine is currently available for the eradication or at least control of the spread of IMNV among carciniculture farms. Thus, prevention remains the only option against IMNV infection. Recently, a treatment based on interference RNA (iRNA) is being investigated as a way to prevent replication of the virus in shrimps. Bartholomay and Harris [36] showed that after receiving 5 mg of iRNA corresponding to portions of the IMNV genome, shrimps previously exposed to a lethal dose of IMNV did not succumb to the disease.

Other measures to prevent IMNV infection in shrimp consist of the use of specific pathogen-free stocks, selection of IMNV-resistant white shrimp [37], and satisfactory sanitation conditions.

REFERENCES

1. Brazilian Association of Shrimp Farmers (ABCC), Census of shrimp farming, 2004. http://www.abccam.com.br (accessed July 15, 2014).
2. Barbieri, R. C. and Ostrennsky, N. A., *Marine Shrimp Fattening* Viçosa-MG, Brazil: Aprenda Fácil, 370pp., 2002.
3. Rodrigues, J., Marine shrimp performance, *Brazilian Association of Shrimp Farmers (ABCC)* Magazine, 7, 38, 2005.
4. Nunes, A. J. P. and Martins, P. C., Evaluating the state of health of marine shrimp in fattening, *Panorama Aquacult.*, 12, 23, 2002.
5. Nunes, A. J. P., Martins, P. C. C., and Gesteira, T. C. V., Infectious myonecrosis virus (IMNV) threatens shrimp producer, *Panorama Aquacult.*, 83, 37, 2004.
6. Madrid, R. M., The economic crisis of shrimp farming, *Panorama Aquacult.*, 90, 22, 2005.
7. Lightner D.V., Pantoja C.R., Poulos B.T. et al., Infectious myonecrosis: New disease in Pacific white shrimp, *Global Aquaculture Advocat.*, **7**, 85, 2004.
8. Poulos, B. T., Tang, K. F. J., Pantoja, C. R. et al., Purification and characterization of infectious myonecrosis virus of penaeid shrimp, *J. Gen. Virol.*, 87, 987, 2006.
9. Tang, J., Ochoa, W. F., Sinkovits, R. S. et al., Infectious myonecrosis virus has a totivirus-like, 120-subunit capsid, but with fiber complexes at the fivefold axes, *Proc. Natl. Acad. Sci. USA*, 105, 17526, 2008.
10. Lightner, D. V., Virus diseases of farmed shrimp in the Western Hemisphere (the Americas): A review, *J. Invertebr. Pathol.*, 106, 110, 2011.
11. Nibert, M. L., '2A-like' and 'shifty heptamer' motifs in penaeid shrimp infectious myonecrosis virus, a monosegmented double-stranded RNA virus, *J. Gen. Virol.*, 88, 1315, 2007.
12. Fauquet, C. M., Mayo, M. A., Maniloff, J. et al., Totiviridae. In: *Virus Taxonomy: Classification and Nomenclature of Viruses, Eighth Report of the International Committee on the Taxonomy of Viruses*, Elsevier, San Francisco, CA, 571 pp., 2005.
13. Ghabrial, S. A., Totiviruses. In: *Encyclopedia of Virology*, 3rd edn., eds. Mahy, B., van Regenmortel, M. H. V., Elsevier, London, U.K., 163pp., 2008.
14. Tang, K. F. J., Pantoja, C. R., Poulos, B. T. et al., *In situ* hybridization demonstrates that *Litopenaeus vannamei, L. stylirostris* and *Penaeus monodon* are susceptible to experimental infection with infectious myonecrosis virus (IMNV), *Dis. Aquat. Org.*, 63, 261, 2005.
15. Senapin, S., Phewsaiya, K., Briggs M. et al., Outbreaks of infectious myonecrosis virus (IMNV) in Indonesia confirmed by genome sequencing and use of an alternative RT-PCR detection method, *Aquaculture*, 266, 32, 2007.
16. Pinheiro, A. C. A. S., Lima, A. P. S., Souza, M. E. et al., Epidemiological status of Taura syndrome and Infectious myonecrosis viruses in *Penaeus vannamei* reared in Pernambuco (Brazil), *Aquaculture*, 262, 17, 2007.
17. Silva, V. A., Anatomopathological study of infectious myonecrosis viral (IMNV) in the cultured shrimp Litopenaeus vannamei, in Pernambuco, Brazil, *Rev. Acta Sci.: Biol. Sci.*, 15, 2007.
18. Gouveia, R. P., Freitas, M. D. L., and Galli, L., Southern brown shrimp found susceptible to IMNV, *The Advocate. Global Aquaculture Alliance*, February, 2007.
19. Sutanto, Y., IMNV cases in Indonesia, Presented to *Shrimp Club Indonesia Workshop*, Bali, Indonesia, March 2, 2011.
20. Senapin, S., Phiwsalya, K., Gangnonngiw, W. et al., False rumours of disease outbreaks caused by infectious myonecrosis virus (IMNV) in the whiteleg shrimp in Asia, *J. Negat. Results Biomed*, 10, 10, 2011.
21. Santana, W. M. et al., Infectious myonecrosis virus (IMNV): A new disease? Educational Journal of Research and Extension UFRPE, *Anais.* Recife: [s.n], CD-ROM, 2004.
22. Andrade, T. P. D., Srisuvan, T., Tang, K. F. J. et al., Real-time reverse transcription polymerase chain reaction assay using TaqMan probe for detection and quantification of infectious myonecrosis virus (IMNV), *Aquaculture*, 264, 9, 2007.
23. OIE (Office International des Epizooties), *Manual of Diagnostic Tests for Aquatic Animals,* 2006. Available in: http://www.oie.int/eng/en_index.htm (accessed on March 30, 2015).
24. Vanpatten, K. A., Nunan, L. M., and Lightner, D. V., Seabirds as potential vectors of penaeid shrimp viruses and the development of a surrogate laboratory model utilizing domestic chickens, *Aquaculture*, 241, 31, 2004.
25. Abud, M. J. A., Barracco, M. A., Donald, V. L. et al., In: *Technical Guide—Pathology and Immunology of Peneids Diseases,* eds. Morales, V., Cuéllar-Anjel, J. 139, 2008.
26. Lightner, D. V., Redman, R. M., Arce S. et al., Specific pathogen-free shrimp stocks in shrimp farming facilities as a novel method for disease control in Crustaceans. In: *Shellfish Safety and Quality*, eds. Shumway, S. and Rodrick, G. Woodhead Publishers, London, U.K., 384pp., 2009.
27. Bell, T.A. and Lightner, D. V., *A Handbook of Normal Penaeid Shrimp Histology.* World Aquaculture Society, Baton Rouge, LA, 1988.
28. Tang, K. F. J., Pantoja, C. R., Redman, R.M. et al., Development of *in situ* hybridization and RTPCR assay for the detection of a nodavirus (*PvNV*) that causes muscle necrosis in *Penaeus vannamei*, *Dis. Aquat. Org.*, 75, 183, 2007.
29. Longyant, S., Rukpratanporn, S., Chaivisuthangkura, P. et al., Identification of *Vibrio* spp. in vibriosis *Penaeus vannamei* using developed monoclonal antibodies. *J. Invertebr. Pathol.*, 98, 63, 2008.
30. Mello, M. V. C., Aragão, M. E. F., Torres-Franklin, M. L. P. et al., Purification of infectious myonecrosis virus (IMNV) in species of marine shrimp *Litopenaeus vannamei* in the State of Ceará, *J. Virol. Methods*, 177, 10, 2011.
31. Poulos, B.T. and Lightner, D. V., Detection of infectious myonecrosis virus (IMNV) of penaeid shrimp by reverse-transcriptase polymerase chain reaction (RT-PCR), *Dis. Aquat. Org.*, 73, 69, 2006.
32. Nagamine, K., Hase, T., Notomi, T., Accelerated reaction by loop-mediated isothermal amplification using loop primers, *Mol. Cell. Probes*, 16, 223, 2002.
33. Arunrut, N., Kampeera, J., Suebsing, R. et al., Rapid and sensitive detection of shrimp infectious myonecrosis virus using a reverse transcription loop-mediated isothermal amplification and visual colorogenic nanogold hybridization probe assay, *J. Virol. Methods*, 193, 542, 2013.

34. Kunanopparat, A., Chaivisuthangkura, P., Senapin, S. et al., Detection of infectious myonecrosis virus using monoclonal antibody specific to N and C fragments of the capsid protein expressed heterologously, *J. Virol. Methods*, 171, 141, 2011.

35. Borsa, B., Seibert, C. H., Rosa, R. D. et al., Detection of infectious myonecrosis virus in penaeid shrimps using immunoassays: Usefulness of monoclonal antibodies directed to the viral major capsid protein, *Arch. Virol.*, 156, 9, 2011.

36. Bartholomay, L. and Harris, D.L.H., Evaluation of highly targeted dsRNA for the treatment of infectious myonecrosis virus (IMNV) in *Litopenaeus vannamei. Animal Industry Report*, 2012.

37. White-Noble, B. L., Lightner, D. V., Tang, K. F. J. et al., Lab challenge for selection of IMNV-resistant white shrimp, *Global Aquaculture Advocate*, 71pp., July/August 2010.

67 Infectious Bursal Disease Virus

Maja Velhner, Dejan Vidanović, and Ivan Dobrosavljević

CONTENTS

67.1 Introduction 601
    67.1.1 Classification, Morphology, and Genome Organization 601
    67.1.2 Clinical Features and Pathogenesis 602
    67.1.3 Diagnosis 603
        67.1.3.1 Conventional Methods 603
        67.1.3.2 Molecular Techniques 604
67.2 Methods 604
    67.2.1 Sample Preparation 604
    67.2.2 Detection Procedures 604
        67.2.2.1 RNA Extraction 604
        67.2.2.2 Reverse Transcription 604
        67.2.2.3 PCR Amplification 604
        67.2.2.4 Real-Time PCR 605
67.3 Conclusion and Future Perspectives 605
Acknowledgments 605
References 606

67.1 INTRODUCTION

The worldwide spread of infectious bursal disease virus (IBDV) in poultry has been well documented over the past decades [1]. The virus usually affects young chickens, causing moderate to high mortality rate, loss of weight, poor flock performances, and immunodeficiency. In the best-case scenario, birds that recovered from infection may show good productivity later on. However, secondary viral or bacterial disease may hamper the recovery process and contribute to higher mortality rate than IBDV itself. Good management practice and vaccination is currently the most effective way to mitigate against infectious bursal disease (IBD). In the early 1970s, field strains usually caused immunosuppression in affected flocks. Many years of work by poultry specialists highlighted the value of vaccination, given the dramatic changes in pathogenic potential of the IBDV. An approach to vaccinate breeder flocks with inactivated vaccines (after priming with live attenuated viruses), to obtain high levels of maternally derived antibodies in progeny, has held promises for years. Due to the emergence of a new biotype of very virulent IBDV (vvIBDV), attempts have been made to protect broiler and layer flocks as soon as possible by introducing intermediate and intermediate plus vaccines. The concept of subcutaneous injection with immunocomplex vaccines [2] and in ovo vaccination applying automatic injectors is increasingly accepted for early protection purposes [3,4]. Thus, vaccination strategies have essentially evolved around the selection of an appropriate vaccine and the timing of vaccination [5,6]. Paying necessary attention to the epizootiology of IBDV is also important to prevent and combat the infection.

67.1.1 CLASSIFICATION, MORPHOLOGY, AND GENOME ORGANIZATION

Classified in the genus *Avibirnavirus*, family *Birnaviridae*, IBDV is a double-stranded RNA (dsRNA) virus with a bisegmented genome. Of the two distinct serotypes identified, serotype 1 is further divided into three major subtypes that cause disease in poultry, whereas serotype 2 strains are infectious for poultry, turkey, and other free-living birds but do not cause clinical illness and thus are not targeted for immunization.

In the early 1970s, the classical types of IBDV emerged all over the world. The mortality rate in young birds was up to 20% and immunosuppression was an important viral mechanism. In the early 1980s, the variant IBDV were isolated in North America. These viruses were discovered by introducing the sentinel unvaccinated birds in flocks that have received regular IBDV prophylaxis. The bursas of sentinel chickens were atrophic but an inflammatory reaction does not occur [7,8]. In the early 1990s, the very virulent biotype of the virus had emerged in Europe [9,10], and soon after, it was also diagnosed in Southeast Asia, Middle East, Africa, and America.

Analysis of the short linear sequences from the hypervariable domain of the VP2 and subsequent generation of the phylogenetic tree have facilitated improved virus classification. It is important to denote which amino acid exchanges contribute to virulence and virus tropism. Sequence alignment is often performed for all virus proteins, but the hypervariable domain of the VP2 is the most frequently studied.

The next important breakthrough in research was the establishment of the reverse genetic system to obtain infectious cDNA, providing a chance to induce the specific amino acid exchanges, by site-directed mutagenesis technique [11]. When constructing the chimeric viruses, it was discovered that a single amino acid exchange Ala284→Thr at the VP2 was sufficient for partial attenuation and the virus retrieved grew on chicken embryo fibroblast [12]. A mutant virus obtained by the reverse genetic system from a vvIBDV strain UK661 with Ala284→Thr and Gln253→His transitions also became attenuated and induced mild transient lesions in the bursa [13]. It is interesting that such mutations could also occur naturally in the bursa environment [14]. Unexpectedly, after the passage in susceptible chickens, some of the genetically engineered IBDV has resulted in a clone that reverted to a pathogenic phenotype [15]. Since 002/73, F52/70, and UK661 strains derived from the bursa also have Gln253 and Ala284 on the VP2, but induced 0%, 10%, and 80% mortality in specific pathogen-free (SPF) chickens, these amino acids could not be solely contemplated as a marker of virulence. Therefore, the following amino acid composition was suggested for virus classification at molecular level:

Virus Biotype	Amino Acid Composition in the Hypervariable Domain of the VP2				
Very virulent	Ile 242	Gln 253	Ile 256	Ala 284	Ile 294
Classical	Val 242	Gln 253	Val 256	Ala 284	Leu 294
Attenuated	Val 242	His 253	Val 256	Thr 284	Leu 294

In total, four phylogenetic groups were obtained, but cluster II was further divided into IIa and IIb. It was also recorded that Australian virulent IBDV has evolved independently from very virulent viruses in Europe or the Far East countries and presented a different linage even though the same amino acid sequence Ala222 and Ser299 are shared between them (it was previously believed that Ala222 is a marker of virulence) [16]. A more recent phylogenetic work confines nine genetic clusters (from I to IX) in IBDV. Except for a few strains, 70% prove to have the same origin and it was concluded that vvIBDVs are probably clonally related because of the high sequence homology in both segments (A and B), which also suggests their coevolution [17]. Recently, the reverse genetics system and site-directed mutagenesis have been used to develop several recombinant viruses from the parent strain 94432, and it was discovered that amino acid valine at 321 modifies the most exposed part of the VP2 causing reduced pathogenicity. Similar attenuation effect was accomplished with threonine residue at 276 from the finger domain of the VP1 [18]. Thus, single amino acid exchanges at both viral proteins (i.e., VP2 and VP1) are sufficient to shift the virulent form of IBDV. It was also discovered recently that genetic rearrangements between vvIBDV and serotype 2 viruses may occur naturally, and it was postulated that because serotype 2 strains infected

a wide variety of birds, they have a chance to counteract with viruses belonging to serotype 1 group [19].

Virus genome of IBDV is localized within a single-shelled capsid of icosahedral symmetry. With a diameter of 60 nm, the virus particle has a shape of a lettuce ($T = 13$) and is non-enveloped [20,21]. The virus possesses a bisegmented dsRNA genome [22,23]. The larger segment A (3.2 kb) encodes a polyprotein that is autocatalytically cleaved [24] to obtain VP2, VP3, and VP4 proteins. The small open reading frame (ORF) partially overlapping the large ORF encodes a VP5 [25]. This protein has an important role in the apoptosis processes and induces cell lyses [26,27]. The hypervariable region of the capsid protein VP2 encompasses amino acids from 206 to 350 having several important mutational "hot spots" that are affecting, in part, the pathogenic potential of the virus [28]. This region is also carrying important immunogenic determinants [29]. The VP3 is the ribonucleoprotein and scaffolding protein, while the VP4 is the virus protease [18]. The smaller segment B (2.9 kb) encodes the RNA-dependent RNA polymerase (RdRp) [30]. In birnavirus, transcription and replication processes are initiated by protein priming. This process is accomplished by RdRp. Recently, it was discovered that RNA polymerase motifs are arranged in a permuted order of sequence (this has not been found in other dsRNA viruses). The IBDV encodes polyprotein in gene segment A rather than produces single transcripts of the mRNA. The sequence ADN in polymerase motif C on the VP1 was found to be an unusual and new property of birnavirus polymerases that appears to function as a regulator of the growth rate of the virus. These features as well as the crystal structure of the viral capsid suggest the possibility that birnaviruses may have originated from their single-stranded RNA counterparts [31]. The nonstructural protein 17 kDa also has a significant role in IBDV pathogenesis and needs to be demanded when studying the virus replication mechanisms [32].

67.1.2 CLINICAL FEATURES AND PATHOGENESIS

Clinical symptoms appear in chickens infected with virulent IBDV after 24–48 h. The morbidity and mortality are most severe 3–5 days postinfection (dpi), although symptoms could last for another 10 days or longer. The final course of the disease depends excessively on secondary bacterial (e.g., *Escherichia coli*) and viral infections, which albeit influence the health status of the birds after IBDV has diminished from the flock.

The mean symptomatic index (MSI) has been introduced to monitor as close as possible the symptoms of the disease in chickens infected by IBDV [33]. The proposed model comprises the following: no signs of the disease (0); typical IBD signs such as ruffled feathers only in quiet birds; birds stimulated by a sudden change in the environment appear normal; motility is not reduced (score 1); dehydration and reduced motility, conspicuous even when the birds are stimulated (score 2); and prostration and death (score 3). This model enables statistical calculations in evaluating the severity of clinical symptoms of affected birds.

All IBDV strains replicate in dividing IgM-bearing B cells in the bursa of Fabricius. Mild strains (like mild vaccine strains) induce little depletion of B lymphocytes, while the pathogenic viruses cause severe inflammatory response, lymphocyte depletion, and bursal atrophy. The comparison of tissue lesions in birds infected with vvIBDV (strain Ehime/91) and J1 strain (which have been isolated in 1972 and did not induce clinical condition) has shown that vvIBD causes severe depletion of B lymphocytes in the bursa, spleen, cecal tonsils, and severe cortical atrophy in the thymus at 5 dpi. Such pathology is much less prominent in birds infected with the mild J1 virus. In the bone marrow, pathological changes are not detected in chickens infected with mild J1 strain while in those infected with vvIBDV, a reduced number of hematopoietic cells were found from 2 to 7 dpi. Pathological lesions in the bone marrow also include a decrease in myelocyte numbers in the extrasinusoidal space, and an increase in the number of macrophages and a slight decrease of erythrocytes in the sinusoidal space. Lesions in the bone marrow may indicate that chickens have been infected with vvIBDV [34]. Importantly, the recovery process in the bursa takes place after infection with vaccine viruses, while pathogenic viruses may damage the bursa permanently [35,36].

The influx of T lymphocytes in the bursa of the chickens infected with pathogenic IBDV was first discovered in 1997 [37,38]. Until 7 dpi, T cell accumulation in the bursa increases and thereafter declines. The role of the T cell in the bursa of the chickens infected with IBDV is extensively studied. It was discovered that T cells inflowing the bursas are activated and proliferated in vitro when stimulated with the purified IBDV antigen and Concanavalin A (Con A). Con A is a non-antigen-specific plant mitogen known to stimulate proliferative response of T cells [39]. Intrabursal T cells also release IL-2 and IFN-γ, promoting greater inflammatory response compared to T cell–immunodeficient chickens [40], and they actually participate in virus clearance as it was discovered in the model of cyclophosphamide-bursectomized chickens [41].

67.1.3 Diagnosis

Clinical symptoms (high morbidity and high mortality in young chickens, usually after 3 weeks of age) and typical postmortem findings closely indicate the IBDV infection. It is therefore important to dissect as many chickens as possible if an outbreak occurs. The most striking pathological changes are subcutaneous and intramuscular bleedings on tight and pectoral muscles, while the bursa is enlarged in the gelatinous edema. Urate accumulation and swollen kidneys (nephrosonephritis syndrome) also indicate the pathological processes caused by IBDV. Hemorrhages in the proventriculus–gizzard junctions are sometimes seen in severe cases. If hemorrhages in the proventriculus–gizzard junctions are recorded, it is important to exclude Newcastle disease virus infection, especially in endemic regions, even if IBDV infection is suspected with high certainty. For differential diagnosis, the following diseases sometimes

need to be ruled out: coccidiosis, infection with nephrotoxic strains of the infectious bronchitis virus, and hemorrhagic syndrome [42]. If the flock is supervised after the onset of clinical symptoms, the bursa is atrophic but bleeding on the tight and pectoral muscles could be visible for some time. Atrophic bursas may be diagnosed also in Marek's disease and if the birds are infected with adenoviruses [42]. Since the bursa is the primary organ for virus infection in IBDV, it is best to collect bursal specimens during the acute phase of infection. In situations when it is not possible to send bursa samples to a diagnostic laboratory instantly, it is recommended to store the bursas at −20°C (in an appropriate disposable package) and send it to the laboratory at the earliest occasion.

67.1.3.1 Conventional Methods

IBDV is present in the bursa for up to 10 dpi. The virus could be confirmed by applying immunodiffusion assay (also called agar gel immunodiffusion [AGID] test) after obtaining the bursa homogenate [43]. High antibody titer to IBDV in young birds detected by enzyme-linked immunosorbent assay also indicates the recent infection. Approximately 7–10 days after the onset of first symptoms, it is also possible to detect precipitin antibodies conducting the AGID test.

The virus could be isolated in embryonated chicken eggs by applying chorioallantoic membrane inoculation technique. It is preferable to inoculate SPF eggs at 9–11 days of embryonation, but in the case that SPF eggs are not available, the commercial embryonated chicken eggs could be used as well. The chorioallantoic fluid is collected from embryos that have died 3–5 dpi (embryos that have died in the first 48 h are discarded) [43]. Pathological changes in the chicken embryos infected with IBDV could not be used solely for virus confirmation. If possible, a challenge experiment on susceptible chickens or those deprived of IBDV antibodies could be organized, even though it is not the popular method due to welfare reasons. Since the mortality rate in infected chickens accurately implicates the virulence of the strain (see MSI index in the clinical symptoms section), it is still used for research purposes. The challenge experiment has to be done in highly secured experimental rooms (the best is to use Horsfall isolators) under the observation of an experienced technician or veterinarian and after obtaining an agreement from the national regulation on animal welfare. The IBDV could also be confirmed by applying reverse transcription-polymerase chain reaction (RT-PCR), with the specific primers.

Vaccine strains usually grow in a cell culture, but pathogenic viruses require successive passage in the chicken's embryo before an actual adaptation on cell culture [44,45]. However, IBDV that has been adapted to the cell culture can revert to an attenuated phenotype and may restrain the pathogenic form.

Immunohistochemical staining of the bursa section is also commendable because lesions in the bursa caused by IBDV are quite typical and because it is possible to detect the virus in situ.

67.1.3.2 Molecular Techniques

The RT-PCR method is often used to confirm the presence of IBDV in the bursa, allantoic fluid collected from embryonated chicken eggs, or other specimens. This method is solidly replacing conventional techniques of virus detection and provides the possibility of simultaneous detection of vaccinal and pathogenic strains in the same specimen [14,46]. If the sequence of the hypervariable region of the VP2 is available, the alignment experiment is helpful in evaluating the type of the virus in question. Next, the phylogenetic analysis could be done by constructing the tree that places the virus to one of the IX clusters obtained to date [17].

67.2 METHODS

67.2.1 Sample Preparation

It is important to collect only the enlarged bursas in gelatinous edema for virus confirmation. When the bursas are dissected, intrafollicular hemorrhages are often visible or sometimes the whole bursa is hemorrhagic. For antigen preparation, it is recommended to select the enlarged bursas with less bleeding inside the organ. The bursas are homogenized with the tissue blender or manually. If manual homogenization is performed, this is achieved by freezing and thawing the homogenate at −20°C. An alternative approach for sample homogenization is to use sterile sand. When the homogenate is ready (the virus has been released from the cells), it is diluted in a sterile phosphate-buffered saline, pH 7.2–7.4 at a ratio of 1:10 w/v, centrifuged at 3000 × g for 10–15 min, and after which, the supernatant can be collected [43]. The obtained supernatant is enriched with the virus and needs to be handled with care (as for other infectious material, respecting the decontamination procedure before disposing the samples). It could also be used for the challenge experiment, for inoculation of embryonated chicken eggs, for the RT-PCR test, and also to detect the virus in AGID against the monovalent hyperimmune sera. The obtained homogenate could also be used as an antigen to test sera for the presence of IBDV antibodies. Except for the challenge test and for the purpose of inoculation of embryonated chicken eggs, the virus in the homogenate must be chemically inactivated.

Bursas and other lymphoid organs (the spleen, thymus, cecal tonsils, and bone marrow) may be collected for pathohistology examination and/or immunohistochemical staining. For that purpose, pieces of tissue are placed in 10% neutral buffered formalin. For immunostaining procedure, it is important to obtain a tissue section within 72 h. If the hematoxylin–eosin staining is only being performed, the tissue can be held in formalin for a longer period of time.

67.2.2 Detection Procedures

The bursal homogenate is placed in the central well of the agar while positive and negative sera are distributed apart. After 24–48 h of incubation at 37°C (providing moist chamber), the precipitin lines (sometimes three or four appear between wells) are detected under the light in a dark room [47].

Characteristic pathological changes of the embryo are as follows: edematous distention of the abdominal region; cutaneous congestion and petechial hemorrhages particularly along feather tracts; occasional hemorrhages on the toe joint and cerebral region; mottled-appearing necrosis and ecchymotic hemorrhages in the liver; pale and parboiled appearance of the heart; kidney congestion and mottled necrosis; and extreme lung congestion and pale spleen, occasionally with small necrotic foci. Variant strains usually do not induce mortality of embryos but splenomegaly and liver necrosis is visible during the infection [42].

67.2.2.1 RNA Extraction

RNA extraction should be performed using RNase-free gloves, labware, and reagents, which will help to prevent contamination of samples, with ubiquitous RNases and other agents. Double-stranded IBDV RNA can be extracted from infected tissues using commercial kits based on silica membrane or magnetic beads. As an alternative, the standard protocol for phenol/chloroform extraction or the Trizol protocol for RNA extraction can be used, after the digestion of the bursa suspension with lysis buffer and proteinase K. Extracted RNA should be stored at temperatures $\geq-20°C$ or $\geq-70°C$.

67.2.2.2 Reverse Transcription

Extracted IBDV RNA must be denaturized before the RT reaction. This can be done by heating the RNA extract at 92°C for 3 min and then chilling it on ice, or heating the RNA extract at 95°C for 5 min and then incubating it at −80°C or in liquid nitrogen [43]. RT can be done with many commercially available reverse transcriptases by following the manufacturer's instructions and by using specific or random primers. In a routine diagnostic, one-step RT-PCR kits are often used, thus shortening the time of the reaction and avoiding the possibility of contamination.

67.2.2.3 PCR Amplification

Primers used for the detection and molecular characterization of IBDV are mainly designed for amplification of the middle third of the VP2 gene in segment A that contains a variable region (vVP2), or for amplification of the region on the 5′ extremity of the IBDV in segment B. Both regions are suitable for molecular epizootiological studies. Two of such protocols are described in the following [43].

Nucleotide sequence of the U3 and L3 IBDV-specific PCR primers specific for segment A, VP2 gene:

Upper U3: 5′-**TGT-AAA-ACG-ACG-GCC-AGT**-GCA-TGC-*GGT-ATG-TGA-GGC-TTG-GTG-AC*-3′

Lower L3: 5′-**CAG-GAA-ACA-GCT-ATG-ACC**-GAA-TTC-*GAT-CCT-GTT-GCC-ACT-CTT-TC*-3′

Nucleotide sequence of the +226 and −793 IBDV-specific PCR primers specific for segment B, VP1 gene [43]:

Upper +290: 5′-**TGT-AAA-ACG-ACG-GCC-AGT**-GAA-TTC-*AGA-TTC-TGC-AGC-CAC-GGT-CTC-T*-3′

Lower −861: 5′-**CAG-GAA-ACA-GCT-ATG-ACC**-CTG-CAG-*TTG-ATG-ACT-TGA-GGT-TGA-TTT-TG*-3′

The U3/L3 pair generates a 604 base pair (bp) product, 516 bp of which is specific for the amplified IBDV sequence and covers the region encoding the hypervariable region of the VP2 protein. The other pair, +290/−861, generates a 642 bp product, 549 bp of which are specific for the amplified IBDV sequence. Obtained PCR products can be sequenced in both directions and possess the restrictions sites (*Sph*I in primer U3, *Eco*Ri in primer L3 and +290, and *Pst*I in primer −861[underlined]) suitable for cloning experiment [43].

The procedure for both protocols are as follows:

1. Remove all PCR reagents, RNA extracts, and positive and negative controls from the freezer; vortex and spin them down before opening the tubes.
2. Prepare the one-step RT-PCR mix as follows (volumes indicated per specimen).

Component	Volume (µL)
2× reaction mix	25
Template RNA (0.01 pg–1 µg)	5
Lower primer (10 µM)	1
Upper primer (10 µM)	1
SuperScript® III RT/ Platinum® Taq Mix	2
Autoclaved distilled water	to 50

3. Add 45 µL of the prepared RT-PCR mix and 5 µL of extracted RNA to a 0.2 mL PCR tube.
4. Briefly spin the tubes in a microcentrifuge.
5. Program the thermal cycler as follows and put the tubes in the cycler: cDNA synthesis, 1 cycle at 50°C for 30 min; denaturation, 1 cycle at 94°C for 2 min; PCR amplification, 40 cycles at 95°C for 30 s (denature), 64°C for 45 s (anneal), and 68°C for 1 min; and final extension, 1 cycle at 68°C for 5 min.
6. Once the amplification is over, briefly spin the tubes and analyze the PCR products by agarose gel electrophoresis.
7. Electrophoresis of the PCR products and DNA molecular weight markers should be performed in a 1%–2% agarose gel, stained with ethidium bromide. (Caution: Ethidium bromide is toxic and carcinogenic.)
8. Before the sequencing of the amplified product, the unincorporated nucleotides, primers, and nonspecific products must be removed. This step can be done by excision of a specific band from agarose and purification using one of the commercially available spin column kits. The amount of purified DNA can be estimated by electrophoresis in a 1%–2% agar gel in parallel with DNA standards or measured by spectrophotometer or colorimeter.

If no internal amplification control is used, three PCRs should be performed for each cDNA sample (pure, 10-fold, and 100-fold diluted cDNA) to avoid false-negative results due to PCR inhibition in mixes containing high amounts of cDNA preparation. Each PCR should include negative and positive control reactions [43]. Using both segment A–targeted and segment B–targeted RT-PCR protocols increases the likelihood of serotype 1 IBDV detection. This also allows a thorough genetic characterization of the IBDV strains detected.

67.2.2.4 Real-Time PCR

More recently, real-time RT-PCR assays have quickly become one of the most promising methods for improving control and epizootiological surveillance programs [48]. This method is highly sensitive and enables detection of 10^2–10^3 copies of virus template. The advantage of real-time RT-PCR assay over the previously described assays is that it can be fully automated and it is suitable for processing large numbers of samples. The possibility of screening a large number of samples in a rapid, sensitive, and reproducible way makes this assay a suitable tool for routine testing in poultry diagnostic laboratories [49]. Several real-time RT-PCR assays have been developed for IBDV and some of them can simultaneously identify, characterize, and differentiate field strains from vaccinal strains using either fluorogenic probes or intercalating dyes [50–54].

67.3 CONCLUSION AND FUTURE PERSPECTIVES

IBD has a significant economic impact in many countries around the globe. In spite of very intensive research and improving knowledge on IBDV pathogenesis, we do not understand what the exact reason for the mortality in chickens is. We also do not know if the IBDV has reached the limit in its virulent and pathogenic potential. It is possible that some amino acid exchanges in virus proteins are strain specific, or they occur as a random event. For that reason, reverse genetic system is the best way to study their role in viral pathogenesis. The research on the crystallographic structure of virus proteins, the virus–cell synergy in apoptosis, and the replication process of the IBDV will provide a better understanding of virus–cell interactions caused by this family of viruses. The possibility for performing research on the IBDV of diverse virulence provides an opportunity to identify common features in the pathogenic processes they cause and thereafter to identify the differences between them. The goal is also to find the best immunizing strategy for chickens, keeping in mind that if there is no susceptible host, this contagious virus will not be able to spread.

ACKNOWLEDGMENTS

We thank Professor Hermann Müller for critical reading of this chapter. This work was financially supported by a grant from the Ministry of Education, Science and Technological Development, Republic of Serbia, Project number TR 31071.

REFERENCES

1. Lasher H.N., Shane S.M., Infectious bursal disease, *World Poultry Sci. J.*, 50, 133, 1994.
2. Iván J., Velhner M., Ursu K. et al., Delayed vaccine virus replication in chickens vaccinated subcutaneously with immune complex infectious bursal disease vaccines: Quantification of vaccine virus by real-time polymerase chain reaction, *Can. J. Vet. Med.*, 69, 135, 2005.
3. Gagic M., St Hill C.A., Sharma J.M., In ovo vaccination of specific-pathogen-free chickens with vaccines containing multiple agents, *Avian Dis.*, 43, 293, 1999.
4. McCarty J.E., Brown T.P., Giambrone J.J., Delay of Infectious bursal disease virus infection by *in ovo* vaccination of antibody-positive chicken eggs, *J. Appl. Poultry Res.*, 14, 136, 2005.
5. Rautenschlein S., Kraemer Ch., Vanmarcke J., Montiel E., Protective efficacy of intermediate and intermediate plus infectious bursal disease virus (IBDV) vaccines against very virulent IBDV in commercial broilers, *Avian Dis.*, 49, 231, 2005.
6. Müller H., Mundt E., Eterradossi N., Islam M.R., Current status of vaccines against infectious bursal disease, *Avian Pathol.*, 41, 133, 2012.
7. Rosenberger J.K., Cloud S.S., Isolation and characterization of variant infectious bursal disease viruses, *J. Am. Vet. Med. Assoc.*, 189, 357, 1986.
8. Snyder D.B., Changes in the field status of infectious bursal disease virus, *Avian Pathol.*, 19, 419, 1990.
9. Chettle N., Stuart J.C., Wyeth P.J., Outbreak of virulent infectious bursal disease virus in East Anglia, *Vet. Rec.*, 125, 271, 1989.
10. van den Berg T.P., Gonze M., Meulemans G., Acute infectious bursal disease in poultry: Isolation and characterization of a highly virulent strain, *Avian Pathol.*, 20, 133, 1991.
11. Mundt E., Vakharia V.N., Synthetic transcripts of double-stranded Birnavirus genome are infectious, *Proc. Natl. Acad. Sci. USA*, 93, 11131, 1996.
12. Mundt E., Tissue culture infectivity of different strains of infectious bursal disease is determined by distinct amino acids in VP2, *J. Gen. Virol.*, 80, 2067, 1999.
13. van Loon A.A.W.M., de Haas N., Zeyda I., Mundt E., Alteration of amino acids in VP2 of very virulent infectious bursal disease virus results in tissue culture adaptation and attenuation in chickens, *J. Gen. Virol.*, 83, 121, 2002.
14. Velhner M., Mitevski D., Potkonjak D. et al., Biological properties of a naturally attenuated infectious bursal disease virus isolated from a backyard chicken flock, *Acta Vet. Hung.*, 58, 499, 2010.
15. Raue R., Islam M.R., Islam M.N. et al., Reversion of molecularly engineered, partially attenuated very virulent infectious bursal disease virus during infection of commercial chickens, *Avian Pathol.*, 33, 181, 2004.
16. Rudd M.F., Heine H.G., Sapats S.I., Parede L., Ignjatovic J., Characterization of an Indonesian very virulent strain of infectious bursal disease virus, *Arch. Virol.*, 147, 1303, 2002.
17. Le Nouën C., Rivallan G., Toquin D. et al., Very virulent infectious bursal disease virus: Reduced pathogenicity in rare natural segment-B-reassorted isolate, *J. Gen. Virol.*, 87, 209, 2006.
18. Escaffre O., Le Nouën C., Amelot M. et al., Both genome segments contribute to the pathogenicity of very virulent infectious bursal disease virus, *J. Virol.*, 87, 2767, 2013.
19. Jackwood D.J., Sommer-Wagner S.E., Crossley B.M., Stoute S.T., Woolcock P.R., Charlton B.R., Identification and pathogenicity of a natural reassortant between a very virulent serotype 1 infectious bursal disease virus (IBDV) and serotype 2 IBDV, *Virology*, 420, 98, 2011.
20. Nick H., Cursiefen D., Becht H., Structural and growth characteristics of infectious bursal disease virus, *J. Virol.*, 18, 227, 1976.
21. Dobos P., Hill B.J., Hallett R., Kells D.T.C., Becht H., Teninges D., Biophysical and biochemical characterization of five animal viruses with bisegmented double-stranded genome, *J. Virol.*, 32, 593, 1979.
22. Müller H., Scholtissek C., Becht H., The genome of infectious bursal disease virus consists of two segments of double-stranded RNA, *J. Virol.*, 31, 584, 1979.
23. Spies U., Müller H., Becht H., Nucleotide sequence of infectious bursal disease genome segment A delineates two major open reading frames, *Nucleic Acids Res.*, 17, 7982, 1989.
24. Hudson P.J., Mc Kern N.M., Power B.E., Azad A.A., Genomic structure of the large RNA segment of infectious bursal disease virus, *Nucleic Acids Res.*, 14, 5001, 1986.
25. Mundt E., Beyer J., Müller H., Identification of a novel viral protein in infectious bursal disease virus-infected cells, *J. Gen. Virol.*, 76, 437, 1995.
26. Liu M., Vakharia V.N., Nonstructural protein of infectious bursal disease virus inhibits apoptosis at the early stage of virus infection, *J. Virol.*, 80, 3369, 2006.
27. Lombardo E., Maraver A., Espinosa I., Fernández-Arias A., Rodriguez J.F., VP5, the nonstructural polypeptide of infectious bursal disease virus, accumulates within the host plasma membrane and induces cell lysis, *Virology*, 277, 345, 2000.
28. Bayliss C.D., Spies U., Shaw K. et al., A comparison of the sequences of segment A of four infectious bursal disease virus strains and identification of a variable region in VP2, *J. Gen. Virol.*, 71, 1303, 1990.
29. Azad A.A., Jagadish M.N., Brown M.A., Hudson P.J., Deletion mapping and expression in *Escherichia coli* of the large genomic segment of a birnavirus, *Virology*, 161, 145, 1987.
30. von Einem U.I., Gorbalenya A.E., Schirrmeier H., Behrens S.E., Letzel T., Mundt E., VP1 of infectious bursal disease virus is an RNA dependent RNA polymerase, *J. Gen. Virol.*, 85, 2221, 2004.
31. Pan J., Vakharia V.N., Tao Y.J., The structure of a birnavirus polymerase reveals a distinct active site topology, *Proc. Natl. Acad. Sci. USA*, 104, 7385, 2007.
32. Yao K., Goodwin M.A., Vakharia V.N., Generation of a mutant infectious bursal disease virus that does not cause bursal lesions, *J. Virol.*, 72, 2647, 1998.
33. Le Nouën C., Toquin D., Müller H. et al., Different domains of the RNA polymerase of infectious bursal disease virus contribute to virulence, *PLoS One*, 7, e28064, 2012.
34. Tanimura N., Tsukamoto K., Nakamura K., Narita M., Maeda M., Association between pathogenicity of infectious bursal disease virus and viral antigen distribution detected by immunohistochemistry, *Avian Dis.*, 39, 9, 1995.
35. Kim I.J., Gagic M., Sharma J.M., Recovery of antibody producing ability and lymphocyte repopulation of bursal follicles in chickens exposed to infectious bursal disease virus, *Avian Dis.*, 43, 401, 1999.
36. Rautenschlein S., Yeh H.Y., Sharma J.M., Comparative immunopathogenesis of mild, intermediate and virulent strains of classic infectious bursal disease virus, *Avian Dis.*, 47, 66, 2003.
37. Tanimura N., Sharma J.M., Appearance of T cells in the bursa of Fabricius and cecal tonsils during the acute phase of infectious bursal disease virus infection in chickens, *Avian Dis.*, 41, 638, 1997.

38. Vervelde L., Davison T.F., Comparison of the in situ changes in lymphoid cells during infection with infectious bursal disease virus in chickens of different ages, *Avian Pathol.*, 26, 803, 1997.

39. In Jeong K., You S.K., Kim H., Yeh H.Y., Sharma J.M., Characteristics of bursal T lymphocytes induced by infectious bursal disease virus, *J. Virol.*, 74, 8884, 2000.

40. Rautenschelin S., Yeh H.Y., Njenga M.K., Sharma J.M., Role of intrabursal T cells in infectious bursal disease virus (IBDV) infection: T cell promote virus clearance but delay follicular recovery, *Arch. Virol.*, 147, 285, 2002.

41. Yeh H.Y., Rautenschlein S., Sharma J.M., Protective immunity against infectious bursal disease virus in chickens in the absence of virus-specific antibodies, *Vet. Immunol. Immunopathol.*, 89, 149, 2002.

42. Saif Y.M., Luker, P.D., Diseases of poultry, CD-ROM. In *Infectious Bursal Disease*, 11th edn., ed. Y.M. Saif, pp. 161–79. Ames, IA: Iowa State Press/A Blackwell Publishing Company, 2003.

43. OIE (Office International des Epizooties), *OIE Manual of Diagnostic Tests and Vaccines for Terrestrial Animals 2014*, Chapter 2.3.12.

44. Yamaguchi T., Ogawa M., Inoshima Y., Miyoshi M., Fukushi H., Hirai K., Identification of sequence changes responsible for the attenuation of highly virulent infectious bursal disease virus, *Virology*, 223, 219, 1996.

45. Wang X.M., Zeng X.W., Gao H.L., Fu C.Y., Wei P., Changes in VP2 gene during the attenuation of very virulent infectious bursal disease strain Gx isolated in China, *Avian Dis.*, 48, 77, 2004.

46. Dobrosavljević I., Vidanović D., Velhner M., Miljković B., Lako B., Simultaneous detection of vaccinal and field infectious bursal disease virus in layer chickens challenged with a very virulent strain after vaccination, *Acta Vet. Hung.*, 62, 264, 2014.

47. Islam M.T., Islam M.N., Khan M.Z.I., Islam M.A., Comparison of agar gel immunodiffusion test, immunohistochemistry and reverse transcription-polymerase chain reaction for detection of infectious bursal disease virus, *Bangl. J. Vet. Med.*, 9, 121, 2011.

48. Wu C.C., Rubinelli P., Lin T.L., Molecular detection and differentiation of infectious bursal disease virus, *Avian Dis.*, 51, 515, 2007.

49. Peters M.A., Lin T.L., Wu C.C., Real-time RT-PCR differentiation and quantification of infectious bursal disease virus strains using dual-labeled fluorescent probes, *J. Virol. Methods*, 127, 87, 2005.

50. Kong L.L., Omar A.R., Hair Bejo M., Ideris A., Tan S.W., Development of SYBR green I based one-step real-time RT-PCR assay for the detection and differentiation of very virulent and classical strains of infectious bursal disease virus, *J. Virol. Methods*, 161, 271, 2009.

51. Li Y.P., Handberg K.J., Kabell S., Kusk M., Zhang M.F., Jørgensen P.H., Relative quantification and detection of different types of infectious bursal disease virus in bursa of Fabricius and cloacal swabs using real time RT-PCR SYBR green technology, *Res. Vet. Sci.*, 82, 126, 2007.

52. Escaffre O., Queguiner M., Eterradossi N., Development and validation of four Real-Time quantitative RT-PCRs specific for the positive or negative strands of a bisegmented dsRNA viral genome, *J. Virol. Methods*, 170, 1, 2010.

53. Ghorashi S.A., O'Rourke D., Ignjatovic J., Noormohammadi A.H., Differentiation of infectious bursal disease virus strains using real-time RT-PCR and high resolution melt curve analysis, *J. Virol. Methods*, 171, 264, 2011.

54. Wang Y., Kang Z., Gao H. et al., A one-step reverse transcription loop-mediated isothermal amplification for detection and discrimination of infectious bursal disease virus, *Virol. J.*, 8, 108, 2011.

68 African Horse Sickness Virus

Dongyou Liu and Frank W. Austin

CONTENTS

68.1 Introduction 609
    68.1.1 Classification...610
    68.1.2 Morphology and Genome Organization 610
        68.1.2.1 Morphology...610
        68.1.2.2 Viral Proteins 610
    68.1.3 Biology and Epidemiology 611
        68.1.3.1 Biology..611
        68.1.3.2 Epidemiology 612
    68.1.4 Clinical Features and Pathogenesis 612
        68.1.4.1 Clinical Features 612
        68.1.4.2 Pathogenesis..613
    68.1.5 Diagnosis 613
    68.1.6 Treatment and Prevention 613
68.2 Methods 613
    68.2.1 Sample Preparation...613
    68.2.2 Detection Procedures...613
68.3 Conclusion 614
References..615

68.1 INTRODUCTION

African horse sickness virus (AHSV) is the causal agent of African horse sickness (AHS), which is a highly infectious and deadly disease of horses, ponies, European donkeys, mules, camels, goats, and buffalo. The clinical presentations of AHS range from inappetence, severe pyrexia, widespread hemorrhages and edematous exudations, impaired respiratory and vascular functions, to death (with mortality as high as 95% in horses and 80% in mules). However, African donkeys and zebra are largely refractory to the devastating consequences of AHS and rarely show clinical signs. In addition, AHSV may infect carnivores such as dogs that ingest contaminated meat.

Although a disease resembling AHS was first recorded in 1327 in Yemen, with the natural zebra population acting as reservoir host, the disease was reported in imported horses in central and east Africa in 1569 and in southern Africa around 1710s. The first major outbreak of AHS in 1719 claimed the lives of about 1700 animals. Since then, >10 additional outbreaks of AHS were documented on the African continent. Along with the decline in the wild zebra population due to hunting throughout the last century and the introduction of AHS vaccines, the incidence of AHS in southern Africa has steadily declined. Currently, the

endemic zones of AHS are restricted to sub-Saharan Africa, Yemen, and the Arabian Peninsula, but the disease has periodically emerged in other parts of the world, including India, Pakistan, Spain (1987–1990), and Portugal (1989) [1].

While not directly contagious, AHS is known to be transmitted by insect vectors, such as *Culicoides* midges, mosquitoes, black flies, sand flies, or ticks. Given the potential of AHSV to cause significant mortality and debilitating morbidity in naïve equid populations and its capacity for sudden and rapid expansion, AHS is listed as a notifiable disease by the Office International des Epizooties and is included on the list of select agents of the USDA. In order to support effective deployment of vaccination strategies (considering the fact that the current attenuated, inactivated, and subunit vaccines for AHSV are serotype specific), to demonstrate the absence of the virus in individual animals or animal products (for safe movements/export/import), and to declare a "virus free" status, especially after outbreaks in nonendemic countries or zones, it is important to have rapid, sensitive, and specific diagnostic tests available for identification and detection of AHSV type. By first presenting an overview on its classification, morphology, genome organization, epidemiology, clinical features, pathogenesis, diagnosis, treatment, and prevention, this chapter will provide streamlined molecular procedures for rapid detection and typing of AHSV.

68.1.1 CLASSIFICATION

AHSV is a double-stranded RNA (dsRNA) virus classified in the genus *Orbivirus*, subfamily *Sedoreovirinae*, family *Reoviridae*, which is divided into two subfamilies: *Sedoreovirinae* and *Spinareovirinae*. Whereas the subfamily *Sedoreovirinae* comprises six genera: *Cardoreovirus*, *Mimoreovirus*, *Orbivirus*, *Phytoreovirus*, *Rotavirus*, and *Seadornavirus*; the subfamily *Spinareovirinae* includes nine genera: *Aquareovirus*, *Coltivirus*, *Cypovirus*, *Dinovernavirus*, *Fijivirus*, *Idnoreovirus*, *Mycoreovirus*, *Orthoreovirus*, and *Oryzavirus*.

The genus *Orbivirus* is known to contain at least 17 species: AHSV, Bluetongue virus (BTV; the type species of the genus), Changuinola virus, Corriparta virus, epizootic hemorrhagic disease virus (EHDV), equine encephalosis virus (EEV), Eubenangee virus, Great Island virus, Orungo virus, Palyam virus (PALV), Peruvian horse sickness virus (PHSV), St Croix River virus, Umatilla virus, Wallal virus, Warrego virus, Wongorr virus, and Yunnan orbivirus [2]. Of these, AHSV, BTV, EHDV, EEV, PALV, and PHSV are highly pathogenic and economically important. With the exception of PHSV, which is likely spread by mosquitoes, most orbiviruses are transmitted by adult females of hematophagous *Culicoides* midges.

Currently, nine serotypes of AHSV (AHSV 1–9) are recognized, with AHSV serotype 9 being identified in 1960. Among AHSV serotypes, some cross-relatedness appears to exist between serotypes 1 and 2, 3 and 7, 5 and 8, and 6 and 9. Interestingly, while serotypes 1–8 are found only in restricted areas of sub-Saharan Africa, serotype 9 is associated with epidemics outside Africa. However, Spanish–Portuguese outbreaks between 1987 and 1990 were caused by AHSV-4 [3].

68.1.2 MORPHOLOGY AND GENOME ORGANIZATION

68.1.2.1 Morphology

As a nonenveloped particle of about 70 nm in diameter, AHSV virion is made up of an outer two-layered icosahedral capsid and a core that encloses the dsRNA genome [4,5].

The outer capsid is composed of two viral proteins (VP2 and VP5), with VP5 trimers forming globular domains and VP2 trimers in a three entwined leg formation (triskelions) on the virion surface. VP5 is positioned between the peripentonal VP2 molecules and around VP7 on the threefold axis of symmetry.

The core (inner capsid) consists of two major viral proteins (VP3 and VP7) and three minor viral proteins (VP1 [polymerase], VP4 [capping enzyme], and VP6 [helicase]). VP3 and VP7 are conserved among the nine AHSV serotypes and represent the group-specific epitopes. VP3 (a flat, triangular molecule organized in 60 asymmetric dimers) constitutes the innermost protein layer. Attached to the inner surface of the VP3 layer, VP1, VP4, and VP6 form a flowerlike transcriptase complex. VP7 (780 monomers arranged as 260 trimers) forms a second layer on top of VP3.

The AHSV genome (of 19.52 kb) is made up of 10 linear dsRNA segments of different sizes, with segments 1–3 (of 3965, 3203, and 2792 bp) being designated L1–L3 (large), segments 4–6 (of 1978, 1748, and 1566 bp) designated M4–M6 (medium), and segments 7–10 (of 1179, 1166, 1169, and 758 bp) designated S7–S10 (small). While the segments (L3, S7) coding for the core proteins (VP3, VP7) and the segments (M5, S8) coding for NS1 and NS2 proteins are highly conserved between different AHSV serotypes, the segment (S10) encoding NS3 and NS3a and the segments (L2 and M6) encoding the outer capsid proteins (VP2 and VP5) are more variable between AHSV serotypes [6–8].

68.1.2.2 Viral Proteins

The 10 segments of AHSV genome encode 7 structural proteins (VP1–VP7) and 4 nonstructural proteins (NS1, NS2, NS3, and NS3a), with NS3 and NS3a being synthesized from two in-phase overlapping reading frames of segment 10 (S10) [4,9,10].

VP1: Encoded by segment 1 (L1), VP1 is an RNA-dependent RNA polymerase, whose activity is highly dependent upon temperature, with an optimum between 27°C and 42°C and little or no activity below 12°C–15°C. This property permits AHSV undertaking viral RNA synthesis and replication within the insect vectors (ectothermic or ambient temperature) and mammalian hosts (about 37°C). As one of the core structural proteins, VP1 is very highly conserved and is unlikely to play a significant role in the specific adaptation of the virus to vectorial transmission [11].

VP2: Encoded by segment 2 (L2), VP2 (124 kDa) is susceptible to proteolytic cleavages (by equine serum proteases and the saliva of the insect host in vivo), which increase the infectivity of the virus in *Culicoides* insect. This may be attributable to a change in VP2-mediated receptor binding or increased efficiency of uncoating and exposure of VP7 during the early stages of infection and cell entry [12]. Similarly, cleavage of VP2 in the AHSV outer capsid by trypsin or chymotrypsin generates infectious subviral particles that demonstrate unchanged infectivity for mammalian cells but greatly enhanced infectivity for *Culicoides sonorensis* cells. Components (VP2 and VP5) of the outer capsid layer are released from the AHSV core below pH 6, an event that likely occurs in vivo within endosomes during the processes of infection and cell entry. This results in the activation of the transcriptase activity of the virus core, which is released into the host–cell cytoplasm. Therefore, besides its involvement in the initial cell attachment, the activated VP2 also contributes to virus entry in vivo. As a host specificity determinant, VP2 shows remarkable antigenic variability due to selective pressure from the host immune system. The specificity of interactions between components of the outer capsid, particularly protein VP2, and neutralizing antibodies forms the basis of serotype determination. Additionally, as a key immunogen and an important target for virus neutralization, VP2 is capable of inducing a protective immune response. VP2 contains 15 antigenic sites, with a major antigenic domain containing

12 of the 15 sites located in the region between residues 223 and 400 and a second domain contained the three remaining sites between residues 568 and 681. Based on in vitro experiments, 3 of the 15 sites are capable of inducing neutralizing antibodies, and two neutralizing epitopes have been defined, "a" and "b," between residues 321 and 339, and 377 and 400, respectively. A combination of peptides representing both sites induces a more effective neutralizing response [5,13,14].

VP3: Encoded by segment 3 (L3), VP3 is a flat, triangular molecule of about 103 kDa that forms the innermost protein layer (with 60 asymmetric dimers) of AHSV core (inner capsid). Similar to VP1, VP3 is a highly conserved structural protein, showing sequence variations that correlate with the virus species/serogroup and with the geographic origins of the virus isolate.

VP4: Encoded by segment 4 (M4), VP4 is a minor core protein of 642 amino acids (75 kDa) that functions as a capping enzyme.

VP5: Encoded by segment 6 (M6), VP5 (57 kDa) is a major structural protein forming the outer two-layered capsid with VP2. Containing coiled-coil motifs typical of membrane fusion protein, VP5 plays a part in membrane penetration during cell entry as well as introduction of AHSV-specific neutralizing antibodies. The most immunodominant region of VP5 is located in the N-terminal 330 residues, defining two antigenic regions (residues 151–200 and residues 83–120), which include 8 antigenic sites, with neutralizing epitopes situated at positions 85–92 (PDPLSPGE) and 179–185 (EEDLRTR) [15,16].

VP6: Encoded by segment 9 (S9), VP6 is a minor structural protein with putative helicase activity. In vitro experiments showed that VP6 binds double- and single-stranded RNA (ssRNA) and DNA in a NaCl concentration-sensitive reaction. The binding activity appears to vary with the pH of the binding buffer (ranging from pH 6.0 to 8.0).

VP7: Encoded by segment 7 (S7), VP7 is a 38 kDa molecule (with 780 monomers organized as 260 trimers) forming a second (surface) layer on top of VP3 in the AHSV core (inner capsid) [17,18]. Being highly conserved among different serotypes, VP7 may assist cell attachment and virus entry in insect systems, through involvement of a distinct cell surface receptor from that used by VP2 [19].

NS1: Encoded by segment 5 (M5), NS1 forms tubular structures within the cell cytoplasm, which are characteristically produced during virus replication in infected cells. These tubules have an average diameter of 23 ± 2 nm (which is less than half the width of the corresponding BTV tubules) and are more fragile at high salt concentrations or pH. NS1 protein is involved in cell lysis, which is the main mode of virus exit from infected mammalian cells and spreading to other cells.

NS2: Encoded by segment 8 (S8), NS2 is a major component of the granular viral inclusion bodies that form within the cytoplasm of infected cells. Sedimentation analysis and nonreducing sodium dodecyl sulphate-polyacrylamide gel electrophoresis (SDS-PAGE) revealed that NS2 is a 7S multimer with both inter- and intramolecular disulfide bonds, probably consisting of six or more NS2 molecules. The 7S NS2 multimer binds ssRNA nonspecifically.

NS3: Encoded by segment 10 (S10), NS3 is a proline-rich protein with strict sequence conservation and two hydrophobic domains (HD1 and HD2), which are linked to NS3 cytotoxicity. Variations in NS3 are found in HD1 and the adjacent variable region between HD1 and HD2. NS3 is associated with the membrane of infected cells and relatively exposed to the host immune system. In an attempt to avoid host immune recognition, the virus may constantly alter its NS3 through mutation, contributing its high variability. This variability corresponds to genetic differences in the insects that the virus infects. NS3 has the characteristics of lytic viral proteins that modify membrane permeability and contribute to viral pathogenesis. Modification of four amino acids (aa 165–168) from hydrophobic to charged amino residues in a predicted transmembrane region of NS3 alters its ability to interact (anchor) with cellular membrane components and largely abolishes its cytotoxic toxicity [20]. NS3 protein mediates virus budding, which is vital for virus exit from infected insect cells. This is reflected by the fact that NS3 is expressed at higher levels in insect cells than mammalian cells [21–23].

NS3a: Also encoded by segment 10 (S10), but via another in-phase overlapping reading frame (compared to NS3), NS3a is a truncated protein localized in areas of plasma membrane disruption and associated with events of viral release from cells [24].

68.1.3 BIOLOGY AND EPIDEMIOLOGY

68.1.3.1 Biology

The establishment of AHSV infection and subsequent pathogenic changes in susceptible hosts involve a series of events: (1) proteolytic cleavage of VP2 in the serum or in midge saliva, (2) interaction with the host–cell receptor, (3) entry through an endocytic pathway, (4) activation of VP5 at low pH leading to exposure of the fusion peptide and subsequent insertion into the endosomal membrane, (5) release of the double-shelled core into the cytoplasm, (6) multiplication in the regional lymph nodes, (7) spreading to pulmonary microvascular endothelial cells, (8) breaking up the endothelial cell barrier (increased vascular permeability resulting in edema, effusion, and hemorrhage), (9) dissemination via the bloodstream (to the lungs, spleen, other lymphoid tissues, choroid plexus, and pharynx), and (10) replication in these organs producing a secondary viremia. The duration from initial infection to secondary viremia may vary between 2 and 21 days, but it usually takes <9 days.

Being acid sensitive, AHSV is inactivated at pH of <6.0 or ≥12.0, but remains relatively stable at pH 7.0–8.5. The virus is vulnerable to treatment with formalin, β-propiolactone, acetylethyleneimine derivatives, or radiation; however, it is resistant to lipid solvents and heat. In the presence of stabilizers such as serum, sodium oxalate, carbolic acid, and glycerine, its infectivity is remarkably stable at 4°C. The virus shows minimal loss of titer after lyophilization or storage at −70°C in Parker Davis Medium, but it is labile between −20°C and −30°C [25].

68.1.3.2 Epidemiology

The natural reservoir host for AHSV is zebras (*Equus burchelli*), which display a detectable viremia for up to 40 days postinfection without visible clinical symptoms, while virus could be recovered from tissues and cells for up to 48 days. On the other hand, horses are highly susceptible to AHSV infection, with a short period of viremia (4–8 days) and high mortality rates (frequently exceeding 90%). Donkeys typically display a viremia for up to 4 weeks and a mortality rate of around 10% [26]. Other susceptible mammals include mules, asses, elephants, camels, and dogs (after eating infected horsemeat) [27].

The biological vector of AHSV is the *Culicoides* midges (Diptera: Ceratopogonidae), especially *Culicoides imicola*, which measures 1–3 mm, preferring warm, humid conditions, and which is also a transmitting vector for Akabane, bovine ephemeral fever, bluetongue, epizootic hemorrhagic disease, equine encephalosis, Schmallenberg, Oropouche, and the Palyams. *Culicoides variipennis* (*sonorensis*) (the American BTV vector), *Culicoides obsoletus* and *Culicoides pulicaris* (the northern European BTV vectors), and *Culicoides bolitinos* (present at higher altitudes in Africa than *C. imicola*) may also have the capacity to transmit AHSV. In addition, AHSV may be also transmitted by mosquitoes (e.g., *Aedes aegypti*, *Anopheles stephensi*, and *Culex pipiens*) and ticks (e.g., camel tick *Hyalomma dromadarii* and dog tick *Rhipicephalus sanguineus sanguineus*) [28–32].

In South Africa, AHS outbreaks chiefly occur in summer; no cases are reported in winter, from July to September. Apparently, AHSV survives the winter (overwintering) in the adult vector population, via transovarial transmission (vertical transmission of AHSV from adult females to their eggs), or retention in adult *Culicoides* that survive the winter drop in temperatures. Occurrence of AHS is often preceded by seasons of heavy rain that alternate with hot and dry climatic conditions. In particular, the cases of AHS in susceptible animals tend to appear as soon as *Culicoides* populations increase to a critical level.

AHSV serotype 9 (AHSV-9) is endemic in the northern parts of sub-Saharan Africa, while other serotypes are found in eastern and southern Africa (with AHSV-2 detected in Senegal and Nigeria in 2007, Ethiopia in 2008, and Ghana in 2010; AHSV-4 in Kenya in 2007; AHSV-7 in Senegal in 2007; and outbreak involving AHSV serotypes 2, 4, 6, 8, and 9 in Ethiopia in 2010). During 1959–1961, AHSV-9 emerged in the Middle East (Iran, Iraq, Syria, Lebanon, and Jordan), then Cyprus, Afghanistan, Pakistan, India, and Turkey. By 1965, AHSV-9 was identified in Libya, Tunisia, Algeria, Morocco, and Spain [33,34]. AHSV serotype 4 (AHSV-4) was reported in Spain (1988–1990), Portugal (1989), and Morocco (1989–1991) following the importation of a number of subclinically infected zebras from Namibia [35–40]. However, AHS has not been reported in the Americas, eastern Asia, or Australasia.

68.1.4 CLINICAL FEATURES AND PATHOGENESIS

68.1.4.1 Clinical Features

AHSV infections present four forms of clinical diseases of various severities in horses: acute (pulmonary), subacute (cardiac), mixed, or febrile (horse sickness fever) forms [41].

Pulmonary form: This acute form of AHS is characterized by a short incubation period (3–5 days), high fever (39°C–41°C, which usually corresponds with the onset of viremia), depression, respiratory distress, dyspnea, spasmodic coughing, dilated nostrils, and sweating. Standing with its legs apart and head extended, the affected animal experiences breathing difficulty (due to lung congestion), coughs frothy fluid from nostril and mouth, and displays signs of pulmonary edema. As the most serious form of AHS, its mortality rate exceeds 95%, with severely affected horses succumbing to the disease (anoxia) within 24 h of infection. The pulmonary form of AHS is also the form most usually seen in dogs. The virus is mainly localized in the cardiovascular and lymphatic system (which is also the case with the cardiac and mixed forms). At necropsy, the lungs are distended and heavy, with edema being visible in the intralobular space, and frothy fluid is present in the trachea, bronchi, and bronchioles. Also observed are edematous thoracic lymph nodes, congested gastric fundus and abdominal viscera, and petechiae in the pericardium.

Cardiac form: This subacute form of AHS has an incubation period longer than that of the pulmonary form, with clinical signs of disease appearing at days 7–12 after infection. The disease is usually characterized by high fever; progressive dyspnea; edema of the head, neck, chest, and supraorbital fossae; petechial hemorrhages in the eyes; ecchymotic hemorrhages on the tongue; and colic (abdominal pain) leading to notable swelling in the supraorbital fossae, eyelids, facial tissues, neck, thorax, brisket, and shoulders. The mortality rate is between 50% and 70% and surviving animals recover in 7 days. Postmortem examination reveals prominent petechiae and ecchymoses on the epicardium and endocardium; flaccid or slightly edematous lungs; yellow, gelatinous infiltrations of the subcutaneous and intramuscular tissues (especially along the jugular veins and ligamentum nuchae); and other lesions (e.g., hydropericardium, myocarditis, hemorrhagic gastritis, and petechiae on the ventral surface of the tongue and peritoneum).

Mixed form: The mixed form is often the most common form of the disease, with affected horses showing signs of both the pulmonary and cardiac forms of AHS. The mortality rate may exceed 70%, and death often occurs within 3–6 days.

Mild form: The mild (febrile) form (or horse sickness fever) of AHS is a subclinical disease that is seen in zebras and African donkeys as well as mules and partially immune horses. Infected animals may display mild to moderate fever, congested mucous membrane, and some edema of the supraorbital fossae despite having high virus titers in the blood. The survival rate is 100%. The virus is concentrated in the spleen and almost absent from the lungs and heart. Interestingly, the virus isolated from the spleen appears to be less virulent than that from the lungs (as in the case of pulmonary and cardiac forms of the disease).

68.1.4.2 Pathogenesis

In a single infection cycle, AHSV has to pass through the skin of the equid host twice. Besides forming a mechanical barrier, the skin also represents an immunologically active organ that is capable of initiating an immune response against an invading pathogen. To mitigate the damaging aspects of host response in the skin, arthropod insects such as *Culicoides* secrete saliva with components that inhibit the phagocytic activity of host macrophages at the biting site, suppress the host's inflammatory reactions, prevent blood coagulation, and promote viral replication (the so-called saliva-activated transmission) [42].

With a predilection for the vascular endothelial cells throughout the body, AHSV increases vascular permeability, enhances effusions into body cavities and tissues, and causes widespread hemorrhages [43]. The appearance of fever in an infected host may disrupt the barriers to viral replication and dissemination, and invite additional insect feeding due to the elevated temperature. The induction of severe clinical signs in the host by AHSV will incapacitate the host to mount effective defense against vector attack [44].

68.1.5 Diagnosis

Presumptive diagnosis of AHS is often made on the basis of characteristic clinical signs, postmortem lesions, and presence of competent vectors. Differential diagnoses include equine infectious anemia, equine encephalosis, equine viral arthritis, Hendra virus, anthrax, babesiosis (*Babesia equi* and *B. caballi*), trypanosomiasis and purpura hemorrhagica.

For a definitive diagnosis of AHS, isolation and identification of the virus from clinical specimens have been traditionally performed. Clinical samples useful for AHSV isolation and detection include whole blood collected in anticoagulant (preferably EDTA) during the febrile stage of infection and the tissues of spleen, lung, lymph nodes, and salivary glands collected from newly dead animals. Blood may be preserved in OCG solution (50% glycerol, 0.5% potassium oxalate, and 0.5% phenol), and spleen samples may be preserved in 10% buffered glycerol. Usually, whole blood is washed and lysed and tissues ground in a mortar with sterile sand before inoculation in embryonated hen eggs, 2–4-day-old suckling mice, or in vitro cell culture (e.g., baby hamster kidney [BHK], African green monkey [Vero], monkey kidney, and *C. sonorensis* [KC] cells) for virus isolation [45].

Further confirmation of AHS is possible by using immunological techniques such as antigen capture ELISA, complement fixation, agar gel immunodiffusion, immunofluorescence, and virus neutralization for identification and serotyping [46–48]. Due to its reliance on virus isolation from clinical specimens in a suckling mouse brain and adaptation to tissue culture (BHK-21, Vero or KC [*C. sonorensis*] cells), AHSV serotyping using virus neutralization test often takes a minimum of 3–4 weeks. Given their time-consuming nature and their requirement for virus isolation and/or

access to standard reagents (antibodies and antigens) that pose a potential biosecurity risk to lab personnel, immunological tests are increasingly superseded in recent years by molecular techniques such as PCR that facilitate direct detection of viral RNA in whole blood and other tissues with a rapid turnover time [49–58]. These techniques negate the need for virus isolation and reference antisera and thus reduce the use of lab animals. To date, reverse transcription polymerase chain reaction (RT-PCR) targeting conserved AHSV genome segments such as the genes encoding VP3, VP7, NS1, and NS2 have been reported for species and serotype-specific identification of AHSV.

68.1.6 Treatment and Prevention

As no specific treatment is currently available for animals suffering from AHS, rest and good husbandry represent the best approach for their recovery. In view of the fact that AHSV is only spread via the bites of infected vectors, control and prevention of the disease should center on (1) restricting the movement of infected animals, (2) slaughter of viremic animals, (3) husbandry modification (e.g., installing sand fly netting/mesh on windows and doors), (4) vector control (e.g., the use of insecticides such as synthetic pyrethroids, ivermectin, tetrachlorvinphos, and 5% temephos granulated with gypsum), and (5) vaccination of naïve animals against homologous AHSV serotype (including the use of polyvalent, monovalent, and monovalent inactivated vaccines) [59–75].

68.2 METHODS

68.2.1 Sample Preparation

For virus isolation, unclotted whole blood is collected in an appropriate anticoagulant at the early febrile stage. Spleen, lung, and lymph node samples are collected from freshly dead animals and placed in appropriate transport media. The samples are sent at 4°C to the laboratory and kept unfrozen.

For serology, serum samples (preferably paired) are taken 21 days apart and kept frozen at −20°C

AHSV dsRNA is extracted from cell-free tissue culture supernatant using QIAamp Viral RNA Mini Kit (Qiagen) or a Universal BioRobot (Qiagen). Alternatively, AHSV RNA may be prepared from infected cells using Trizol Reagent (Invitrogen).

68.2.2 Detection Procedures

AHSV genome segment 3 (Seg-1) and segment 3 (Seg-3) encode VP1 (viral polymerase) and VP3 (subcore shell protein), respectively, which are highly conserved virus species/orbivirus-serogroup-specific antigens. Primers derived from Seg-1 and Seg-3 provide a useful approach for AHSV species-specific identification [76].

On the other hand, AHSV genome segment 2 (Seg-2) encodes outer capsid protein VP2, which is the most variable of the AHSV proteins and also a primary target for AHSV

TABLE 68.1
Identity of Primers and Probes for African Horse Sickness Virus Species- and Type-Specific Detection

Assay Type Primer/Probe	Sequence (5'–3')
Seg-1 based/virus-species-specific	
AHSV/Seg-1/FP/1310-1329	CCTGAAGATATGTATACATC
AHSV/Seg-1/RP/1444-1461	ACTAACGCTTTAATCCGC
AHSV/Seg-1/P/1400-1423	CCCAACGCRAARGGTGG ACGTGG
Seg-3 based/virus-species-specific	
AHSV/Seg-3/FP/2030-2050	AGATATYGTAAGGTGGAGTCA
AHSV/Seg-3/RP/2115-2136	CTAACATCAARTCTTCAAARTC
AHSV/Seg-3/P/2085-2107	ATCGCCCAAGCTTCCCTA TTCAA#
Seg-2 based/type-specific	
AHSV/SRT-1/FP/1143-1163	GAATGGTAAGCTTTGGATTGA
AHSV/SRT-1/RP/1225-1207	TTAGAGGCGCTCGGTTCTC
AHSV/SRT-1/P/1165-1187	CATAAACAAACGGTGAGTGA GCA
AHSV/SRT-2/FP/1763-1783	AGTGGACTTCGATTATAGATG
AHSV/SRT-2/RP/1897-1878	CTGTCTGAGCGTTAACCTTC
AHSV/SRT-2/P/1863-1842	TTCAACCGTCTCTCCGCCTCTC#
AHSV/SRT-3/FP/809-829	TAGAAAGAATGATGAGCAGTG
AHSV/SRT-3/RP/901-921	TAATGGAATGTCGCCTGTCTT
AHSV/SRT-3/P/845-870	CCGTTATTGAGAGCGTCATA AGATTC
AHSV/SRT-4/FP/1934-1956	GTTTCGTATCATATTGGTATAGA
AHSV/SRT-4/RP/2079-2101	GATATATATATGTGGATGCATGG
AHSV/SRT-4/P/1996-2020	CYGGAYATGAATGAGAAAC AGAAGC
AHSV/SRT-5/FP/1875-1895	GAGACACATCAAGGTTAAAGG
AHSV/SRT-5/RP/2035-2014	CAGGATCAAACTGTGTATACTT
AHSV/SRT-5/P/1964-1985	TTGAAGCAAGGAATCTATT GACTTT
AHSV/SRT-6/FP/1285-1304	GTTGGTCATTGGGTCGATTG
AHSV/SRT-6/RP/1505-1485	GACGAATTTGATCCCTTCTTG
AHSV/SRT-6/P/1385-1408	TGATGTCGGGTATGAACA AACTGG
AHSV/SRT-7/FP/1159-1179	TGGATCGAGCATAAGAAGAAG
AHSV/SRT-7/RP/1282-1302	CCAATCAACCCARTGTGTAAC
AHSV/SRT-7/P/1213-1237	ACCAAAATCGTCCGATGC TAGTGC
AHSV/SRT-8/FP/2122-2141	GGTGAGCGGATTGTTGATAG
AHSV/SRT-8T/RP/2293-2274	CTGAATCCACTCTATACCTC
AHSV/SRT-8/P/2169-2146	CGACAAATCATACCACACG ATCTC#
AHSV/SRT-9/FP/1931-1952	TATCATATTGGTATCGAGTTCG
AHSV/SRT-9/RP/2074-2054	AAGTTGATGCGTGAATACCGA
AHSV/SRT-9/P/1978-2000	ACATCCTCAATCGAYCTC CTCTC#

Source: Bachanek-Bankowska, K. et al., *PLoS One*, 9(4), e93758, 2014.

Notes: Seg-1 and Seg-3 indicate the group-specific assays. SRT-1 to SRT-9 indicate respective type-specific assays. FP, RP, or P stand for "forward primer," "reverse primer," and "probe," respectively. The numbers represent annealing positions on the genome segment. Probes indicated (#) are complementary to the negative strand.

TABLE 68.2
Composition of RT-PCR Mixture (Using SuperScript III/Platinum Taq One-Step qRT-PCR Kit, Invitrogen)

Reagent	Assay Detecting Segment		
	Seg-1	Seg-3	Seg-2
Forward primer (10 µM)(µL)	2	1	2
Reverse primer (10 µM)(µL)	2	1	2
Probe (5 µM)(µL)	1.5	1	0.5
$MgSO_4$ (µL)	0.5	1.5	1
SuperScript III RT/Platinum Taq Mix (µL)	0.5	0.5	0.5
2× reaction mix (µL)	12.5	12.5	10
Nuclease free water (µL)		1.5	
dsRNA (µL)	6	6	4
Total volume (µL)	25	25	20

Source: Bachanek-Bankowska, K. et al., *PLoS One*, 9(4), e93758, 2014.

specific neutralizing antibodies. Primers designed from Seg-2 allow accurate determination of nine AHSV serotypes [55,76].

Bachanek-Bankowska et al. [76] utilized primers with AHSV virus species specificity (targeting AHSV Seg-1 or Seg-3) and type specificity (targeting Seg-2 of the nine AHSV serotypes) in real-time RT-PCR assays for rapid, sensitive, and reliable detection and typing of AHSV RNA in infected blood, tissue samples, homogenized *Culicoides*, or tissue culture supernatant (Table 68.1).

The real-time RT-PCR mixture is prepared using SuperScript III/Platinum Taq One-Step qRT-PCR Kit (Invitrogen) (Table 68.2). dsRNA samples are heat denatured prior to addition to the reaction mix. Amplification for all of the assays is conducted in MX3005p (Stratagene) using the same conditions: 55°C for 30 min, 95°C for 10 min, followed by 45 cycles of 95°C for 30 s and 55°C for 1 min. Fluorescence is measured during the 55°C annealing/extension step. Cycle threshold values are measured at the point at which the sample fluorescence signal crosses a threshold value (the background level).

The real-time RT-PCR assays using AHSV Seg-1 and Seg-3 primers enable species-specific detection of all nine serotypes of AHSV down to 0.029 $TCID_{50}$ (or 60 copies of AHSV genome per reaction). The real-time RT-PCR assays using AHSV Seg-2 primers facilitate identification of AHSV serotypes 1–5, 7–9 down to 1 $TCID_{50}$/mL (0.019 $TCID_{50}$/assay), and AHSV-6 down to 10 $TCID_{50}$/mL (0.19 $TCID_{50}$/assay).

68.3 CONCLUSION

AHS is an economically important, noncontagious but infectious disease. Transmitted by *Culicoides* midges, the disease severely affects horses, ponies, and European donkeys as well as mules, but has little effect on African donkeys and zebra. Although AHSV is largely confined to sub-Saharan Africa, Yemen, and the Arabian Peninsula, it has the potential for sudden and rapid excursions and causes significant mortality in naïve horse populations in other parts of the world [34,77–79].

It is therefore critical that rapid, sensitive, and specific laboratory methods are available for prompt identification and typing of AHSV in case of unexpected AHS outbreaks. In particular, the use of real-time RT-PCR permits highly sensitive and specific detection, typing, and quantitation of AHSV RNA [76]. In addition, the development of effective vaccines against this deadly disease will help prevent the serious consequence of a sudden epidemic due to AHSV.

REFERENCES

1. Mellor PS, Mertens PPC. African horse sickness viruses. *Encycl. Virol.* 2008;1:37–43.
2. Unitprot, http://www.uniprot.org/taxonomy/10892. Accessed on June 12, 2015.
3. Calisher CH, Mertens PP. Taxonomy of African horse sickness viruses. *Arch. Virol. Suppl.* 1998;14:3–11.
4. Roy P, Mertens PP, Casal I. African horsesickness virus structure. *Comp. Immunol. Microbiol. Infect. Dis.* 1994;17:243–273.
5. Manole V et al. Structural insight into African horsesickness virus infection. *J. Virol.* 2012;86(15):7858–7866.
6. Bremer CW, Huismans H, Van Dijk AA. Characterization and cloning of the African horsesickness virus genome. *J. Gen. Virol.* 1990;71:793–799.
7. Williams CF et al. The complete sequence of four major structural proteins of African horse sickness virus serotype 6: Evolutionary relationships within and between the orbiviruses. *Virus Res.* 1998;53:53–73.
8. Matsuo E, Celma CC, Roy P. A reverse genetics system of African horsesickness virus reveals existence of primary replication. *FEBS Lett.* 2010;584:3386–3391.
9. Grubman M, Lewis S. Identification and characterisation of the structural and non-structural proteins of African horse sickness virus and determination of the genome coding assignments. *Virology* 1992;186:444–451.
10. Huismans H et al. A comparison of different orbivirus proteins that could affect virulence and pathogenesis. *Vet. Ital.* 2004;40(4):417–425.
11. Vreede FT, Huismans H. Sequence analysis of the RNA polymerase gene of African horse sickness virus. *Arch. Virol.* 1998;143(2):413–419.
12. Marchi PR et al. Proteolytic cleavage of VP2, an outer capsid protein of African horse sickness virus, by species-specific serum proteases enhances infectivity in *Culicoides. J. Gen. Virol.* 1995;76:2607–2611.
13. Burrage TG et al. Neutralizing epitopes of African horsesickness virus serotype 4 are located on VP2. *Virology* 1993;196:799–803.
14. Martinez-Torrecuadrada JL, Casal JI. Identification of a linear neutralization domain in the protein VP2 of African horsesickness virus. *Virology* 1995;210:391–399.
15. Martinez-Torrecuadrada JL et al. Antigenic profile of African horse sickness virus serotype 4 VP5 and identification of a neutralizing epitope shared with bluetongue virus and epizootic hemorrhagic disease virus. *Virology* 1999;257:449–459.
16. Stassen L, Huismans H, Theron J. Membrane permeabilization of the African horse sickness virus VP5 protein is mediated by two N-terminal amphipathic α-helices. *Arch. Virol.* 2011;156(4):711–715.
17. Basak AK et al. Crystal structure of the top domain of African horse sickness virus VP7: Comparisons with bluetongue virus VP7. *J. Virol.* 1996;70:3797–3806.

18. Stassen L, Huismans H, Theron J. Silencing of African horse sickness virus VP7 protein expression in cultured cells by RNA interference. *Virus Genes* 2007;35(3):777–783.
19. Rutkowska DA et al. The use of soluble African horse sickness viral protein 7 as an antigen delivery and presentation system. *Virus Res.* 2011;156(1–2):35–48.
20. Van Niekerk M et al. Membrane association of African horsesickness virus nonstructural protein NS3 determines its cytotoxicity. *Virology* 2001;279:499–508.
21. Stoltz MA et al. Subcellular localization of the nonstructural protein NS3 of African horsesickness virus. *Onderstepoort J. Vet. Res.* 1996;63:57–61.
22. Quan M et al. Molecular epidemiology of the African horse sickness virus S10 gene. *J. Gen. Virol.* 2008;89:1159–1168.
23. Meiring TL, Huismans H, van Staden V. Genome segment reassortment identifies non-structural protein NS3 as a key protein in African horsesickness virus release and alteration of membrane permeability. *Arch. Virol.* 2009;154(2):263–271.
24. Martin LA et al. Phylogenetic analysis of African horse sickness virus segment 10: Sequence variation, virulence characteristics and cell exit. *Arch. Virol. Suppl.* 1998;14:281–293.
25. Wellby MP et al. Effect of temperature on survival and rate of virogenesis of African horse sickness virus in *Culicoides variipennis sonorensis* (Diptera: Ceratopogonidae) and its significance in relation to the epidemiology of the disease. *Bull. Entomol. Res.* 1996;86:715–720.
26. Hamblin C et al. Donkeys as reservoirs of African horse sickness virus. *Arch. Virol. Suppl.* 1998;14:37–47.
27. Alexander KA et al. African horse sickness and African carnivores. *Vet. Microbiol.* 1995;47:133–140.
28. Venter GJ, Graham S, Hamblin C. African horse sickness epidemiology: Vector competence of South African *Culicoides* species for virus serotypes 3, 5 and 8. *Med. Vet. Entomol.* 2000;14:245–250.
29. Venter GJ, Koekemoer JJ, Paweska JT. Investigations on outbreaks of African horse sickness in the surveillance zone in South Africa. *Rev. Sci. Tech.* 2006;25(3):1097–1109.
30. Venter GJ, Paweska JT. Virus recovery rates for wild-type and live-attenuated vaccine strains of African horse sickness virus serotype 7 in orally infected South African *Culicoides* species. *Med. Vet. Entomol.* 2007;21(4):377–383.
31. Venter GJ et al. The oral susceptibility of South African field populations of *Culicoides* to African horse sickness virus. *Med. Vet. Entomol.* 2009;23:367–378.
32. Meiswinkel R, Paweska JT. Evidence for a new field *Culicoides* vector of African horse sickness in South Africa. *Prev. Vet. Med.* 2003;60:243–253.
33. Zientara S et al. Molecular epidemiology of African horse sickness virus based on analyses and comparisons of genome segments 7 and 10. *Arch. Virol. Suppl.* 1998;14:221–234.
34. Aklilu N et al. African horse sickness outbreaks caused by multiple virus types in Ethiopia. *Transbound Emerg. Dis.* 2014;61(2):185–192.
35. Lubroth J. African horsesickness and the epizootic in Spain 1987. *Equine Pract.* 1988;10:26–33.
36. Rodriguez M, Hooghuis H, Castano M. African horse sickness in Spain. *Vet. Microbiol.* 1992;33:129–142.
37. Mellor PS et al. Isolations of African horse sickness virus from vector insects made during the 1988 epizootic in Spain. *Epidemiol. Infect.* 1990;105:447–454.
38. Scacchia M et al. African horse sickness: A description of outbreaks in Namibia. *Vet. Ital.* 2009;45(2):255–274.

39. Maclachlan NJ, Guthrie AJ. Re-emergence of bluetongue, African horse sickness, and other orbivirus diseases. *Vet. Res.* 2010;41(6):35.

40. Thompson GM, Jess S, Murchie AK. A review of African horse sickness and its implications for Ireland. *Ir. Vet. J.* 2012;65(1):9.

41. Weyer CT et al. African horse sickness in naturally infected, immunised horses. *Equine Vet. J.* 2013;45(1):117–119.

42. Burrage TG, Laegreid WW. African horsesickness: Pathogenesis and immunity. *Comp. Immunol. Microbiol. Infect Dis.* 1994;17:275–285.

43. Laegreid WW et al. Electron microscopic evidence for endothelial infection by African horsesickness virus. *Vet. Pathol.* 1992;29:554–556.

44. Wilson A et al. Adaptive strategies of African horse sickness virus to facilitate vector transmission. *Vet. Res.* 2009;40(2):16.

45. O'Hara RS et al. Development of a mouse model system and identification of individual genome segments of African horse sickness virus, serotypes 3 and 8 involved in determination of virulence. *Arch. Virol.* 1998;S14:259–279.

46. Clift SJ et al. Standardization and validation of an immunoperoxidase assay for the detection of African horse sickness virus in formalin-fixed, paraffin-embedded tissues. *J. Vet. Diagn. Invest.* 2009;21(5):655–667.

47. Clift SJ, Penrith ML. Tissue and cell tropism of African horse sickness virus demonstrated by immunoperoxidase labeling in natural and experimental infection in horses in South Africa. *Vet. Pathol.* 2010;47(4):690–697.

48. Bitew M et al. Serological survey of African horse sickness in selected districts of Jimma zone, Southwestern Ethiopia. *Trop. Anim. Health Prod.* 2011;43(8):1543–1547.

49. Sailleau C et al. African horse sickness in Senegal: Serotype identification and nucleotide sequence determination of segment S10 by RT-PCR. *Vet. Rec.* 2000;146:107–108.

50. Agüero M et al. Real-time fluorogenic reverse transcription polymerase chain reaction assay for detection of African horse sickness virus. *J. Vet. Diagn. Invest.* 2008;20(3):325–328.

51. Rodriguez-Sanchez B et al. Novel gel-based and real-time PCR assays for the improved detection of African horse sickness virus. *J. Virol. Methods* 2008;151(1):87–94.

52. Aradaib IE. PCR detection of African horse sickness virus serogroup based on genome segment three sequence analysis. *J. Virol. Methods* 2009;159(1):1–5.

53. Fernández-Pinero J et al. Rapid and sensitive detection of African horse sickness virus by real-time PCR. *Res. Vet. Sci.* 2009;86(2):353–358.

54. Quan M et al. Development and optimisation of a duplex real-time reverse transcription quantitative PCR assay targeting the VP7 and NS2 genes of African horse sickness virus. *J. Virol. Methods* 2010;167(1):45–52.

55. Maan NS et al. Serotype specific primers and gel-based RT-PCR assays for 'typing' African horse sickness virus: Identification of strains from Africa. *PLoS One* 2011;6(10):e25686.

56. Monaco F et al. A new duplex real-time RT-PCR assay for sensitive and specific detection of African horse sickness virus. *Mol. Cell Probes* 2011;25(2–3):87–93.

57. Scheffer EG et al. Use of real-time quantitative reverse transcription polymerase chain reaction for the detection of African horse sickness virus replication in *Culicoides* imicola. Onderstepoort. *J. Vet. Res.* 2011;78(1):E1–E4.

58. Guthrie AJ et al. Diagnostic accuracy of a duplex real-time reverse transcription quantitative PCR assay for detection of African horse sickness virus. *J. Virol. Methods* 2013;189(1):30–35.

59. Laegreid WW et al. Characterization of virulence variants of African horsesickness virus. *Virology* 1993;195:836–839.

60. Roy P et al. Recombinant baculovirus-synthesized African horsesickness virus (AHSV) outer-capsid protein VP2 provides protection against virulent AHSV challenge. *J. Gen. Virol.* 1996;77:2053–2057.

61. Stone-Marschat MA et al. Immunization with VP2 is sufficient for protection against lethal challenge with African horsesickness virus type 4. *Virology* 1996;220:219–222.

62. Scanlen M et al. The protective efficacy of a recombinant VP2-based African horsesickness subunit vaccine candidate is determined by adjuvant. *Vaccine* 2002;20:1079–1088.

63. Sánchez-Vizcaíno JM. Control and eradication of African horse sickness with vaccine. *Dev. Biol.* (Basel) 2004;119:255–258.

64. Diouf ND et al. Outbreaks of African horse sickness in Senegal, and methods of control of the 2007 epidemic. *Vet. Rec.* 2013;172(6):152.

65. MacLachlan NJ et al. Experiences with new generation vaccines against equine viral arteritis, West Nile disease and African horse sickness. *Vaccine* 2007;25(30):5577–5582.

66. Koekemoer JJ. Serotype-specific detection of African horsesickness virus by real-time PCR and the influence of genetic variations. *J. Virol. Methods* 2008;154(1–2):104–110.

67. von Teichman BF, Smit TK. Evaluation of the pathogenicity of African Horsesickness (AHS) isolates in vaccinated animals. *Vaccine* 2008;26(39):5014–5021.

68. von Teichman BF, Dungu B, Smit TK. In vivo cross-protection to African horse sickness Serotypes 5 and 9 after vaccination with Serotypes 8 and 6. *Vaccine* 2010;28(39):6505–6517.

69. Castillo-Olivares J et al. A modified vaccinia Ankara virus (MVA) vaccine expressing African horse sickness virus (AHSV) VP2 protects against AHSV challenge in an IFNAR -/- mouse model. *PLoS One* 2011;6(1):e16503.

70. de Vos CJ, Hoek CA, Nodelijk G. Risk of introducing African horse sickness virus into the Netherlands by international equine movements. *Prev. Vet. Med.* 2012;106(2):108–122.

71. El Garch H et al. An African horse sickness virus serotype 4 recombinant canarypox virus vaccine elicits specific cell-mediated immune responses in horses. *Vet. Immunol. Immunopathol.* 2012;149(1–2):76–85

72. Oura CA et al. African horse sickness in The Gambia: Circulation of a live-attenuated vaccine-derived strain. *Epidemiol. Infect.* 2012;140(3):462–465.

73. Pretorius A et al. Virus-specific CD8+ T-cells detected in PBMC from horses vaccinated against African horse sickness virus. *Vet. Immunol. Immunopathol.* 2012;146(1):81–86.

74. Scheffer EG et al. Comparison of two trapping methods for *Culicoides* biting midges and determination of African horse sickness virus prevalence in midge populations at Onderstepoort, South Africa. *Vet. Parasitol.* 2012;185(2–4):265–273.

75. Boom R, Oldruitenborgh-Oosterbaan MM. Can Europe learn lessons from African horse sickness in Senegal? *Vet. Rec.* 2013;172(6):150–151.

76. Bachanek-Bankowska K et al. Real time RT-PCR assays for detection and typing of African horse sickness virus. *PLoS One* 2014;9(4):e93758.

77. Venter GJ et al. *Culicoides* species abundance and potential over-wintering of African horse sickness virus in the Onderstepoort area, Gauteng, South Africa. *J. S. Afr. Vet. Assoc.* 2014;85(1):1102.

78. Goffredo M et al. Orbivirus detection from *Culicoides* collected on African horse sickness outbreaks in Namibia. *Vet. Ital.* 2015;51(1):17–23.

79. Ayelet G et al. Outbreak investigation and molecular characterization of African horse sickness virus circulating in selected areas of Ethiopia. *Acta Trop.* 2013;127(2):91–96.

69 Bluetongue Virus

Fan Lee

CONTENTS

69.1 Introduction 619
    69.1.1 Classification and Morphology...619
    69.1.2 Biology and Epidemiology 619
    69.1.3 Clinical Features and Pathogenesis 620
    69.1.4 Diagnosis 620
        69.1.4.1 Conventional Techniques...620
        69.1.4.2 Molecular Techniques..621
69.2 Methods 622
    69.2.1 Sample Preparation...622
        69.2.1.1 Samples for Virus Isolation...622
        69.2.1.2 Samples for BTV-Specific RNA Detection 622
        69.2.1.3 Serum Samples for Antibody Detection 622
    69.2.2 Detection Procedures..622
        69.2.2.1 Virus Isolation Using Embryonated Chicken Eggs 622
        69.2.2.2 Virus Isolation Using Cell Cultures..623
        69.2.2.3 Detection of BTV RNA by RT-PCR..623
        69.2.2.4 Detection of BTV RNA by Real-Time RT-PCR..624
        69.2.2.5 Measurement of BTV-Specific Antibodies by VNT 624
        69.2.2.6 Detection of BTV-Specific Antibodies by cELISA..624
69.3 Conclusion and Future Perspectives 624
References...625

69.1 INTRODUCTION

69.1.1 CLASSIFICATION AND MORPHOLOGY

Bluetongue was first described as "malarial catarrhal fever of sheep" by Hutcheon[1] and Spreull.[2] The causative agent of bluetongue, that is, bluetongue virus (BTV), is the prototype of the genus *Orbivirus* within the family *Reoviridae*. In addition to BTV, the genus *Orbivirus* contains 21 other virus species and some of which, such as African horse sickness virus and epizootic hemorrhagic disease virus, are of veterinary importance.

The number of BTV serotypes had long held at 24 until the finding of Toggenburg Orbivirus, which was later designated the prototype of BTV serotype 25.[3] In 2010, BTV serotype 26 was identified in Kuwait.[4] This makes the total number of BTV serotypes to 26 to date.

The virion of BTV is nonenveloped and is made up of 10 segments of double-stranded RNA and approximately 1000 viral protein molecules that pack the RNA segments. The viral proteins form a double-shelled icosahedral capsid with a diameter of 69.2 nm[5] and a density of 1.337 g/cm.[3,6] The outer layer of the capsid consists of viral proteins VP2 and VP5; the inner layer comprises primarily VP3 and VP7 together with smaller amounts of VP1, VP4, and VP6 proteins.[7] The VP2 and VP5 proteins are in contact with VP7, linking the outer and inner layers.

69.1.2 BIOLOGY AND EPIDEMIOLOGY

Animals susceptible to BTV are primarily ruminants, but BTV infections are not limited to these animals. Domestic cattle (*Bos taurus*), sheep (*Ovis aries*), and goat (*Capra aegagrus hircus*) are the susceptible species of economic concern. Wild ruminants susceptible to BTV include American white-tailed deer (*Odocoileus virginianus*),[8] mules (*Odocoileus hemionus*),[9] pronghorns (*Antilocapra americana*), desert bighorn sheep (*Ovis canadensis*), North American elk (*Cervus canadensis*),[10] roe deer (*Capreolus capreolus*), mouflons (*Ovis aries*), red deer (*Cervus elaphus*),[11] camels,[12] alpacas (*Vicugna pacos*), and llamas (*Lama glama*).[13] Besides ruminants, BTV is able to infect dogs naturally and via the administration of BTV-contaminated vaccines.[14-16] During the European bluetongue epidemic in 2007–2008, BTV-associated deaths in Eurasian lynx (*Lynx lynx*) were also reported.[17] A serological survey conducted in Africa showed that varying levels of seropositivity have been detected in lions (*Panthera leo*) and spotted hyenas (*Crocuta crocuta*).[18] Thus, BTV has the capacity to infect a variety of ruminants and carnivores.

As an arthropod-borne pathogen, BTV is naturally transmitted by blood-sucking, female *Culicoides* biting midges. Among the more than 1000 species of *Culicoides* worldwide, only a limited number of midge species are involved in the transmission of BTV, and the species responsible for transmitting BTV in different geographical areas are varied. For example, *Culicoides imicola* is the primary BTV vector in Europe and Africa.[19,20] *Culicoides sonorensis* (formerly *Culicoides variipennis*) in North America,[21] and *Culicoides insignis* in South America.[22] Although transplacental transmission in sheep has been demonstrated experimentally,[23–25] the primary route of transmission of BTV is via the bite of *Culicoides* biting midges that carry the virus from viremic animals to susceptible animals.

The areas in which BTV infection is prevalent have previously been limited, but BTV-prevalent areas are expanding. The infection is traditionally prevalent in tropical, subtropical, and some temperate areas—roughly contained in a zone between the latitudes of 40°N and 35°S[26]—due to the limited availability of competent *Culicoides* vectors. More recently, however, the European epidemic that has been occurring since 1998, resulting from the introduction of multiple BTV strains from Turkey and northern Africa, has gone beyond these geographical boundaries. The epidemic has even moved northwestward to regions at latitudes higher than 50°N. This development suggests that the existing competent vectors may be extending their habitats northward, or that novel vectors may be involved in BTV transmission, or both, and that the invading BTV strains have developed overwintering mechanisms and adapted to local animal populations.

69.1.3 Clinical Features and Pathogenesis

The BTV infection is usually subclinical; however, it may cause clinical illness in some animals, primarily in some breeds of sheep. Moreover, both natural[27,28] and experimental[29,30] infections with certain BTV strains, such as BTV serotype 8, which caused an epidemic in Europe in the early 2000s, also lead to clinical disease in cattle.

Bluetongue in sheep primarily affects mucosal surfaces and the feet. The clinical signs of bluetongue in sheep include fever; excessive salivation; vomiting; respiratory distress; mucosal hyperemia; edematous and enlarged conical papillae; dilated gingival vessels; petechial hemorrhages of the gingival mucosa; crusting, erosion, and ulceration of the muzzle; necrosis of the oral mucosa; cyanotic tongue; subcutaneous edema of the face, head, ears, and neck; conjunctivitis; hemorrhagic excoriations of the epithelium of the external nares; enlarged lips; coronitis; acute hyperemia of the bulbs of the feet and between digits; stiff gait; and torticollis.[31–34] These symptoms lead to decreased feed consumption, general weakness, reluctance to move, and in some cases, death of the affected sheep.

Bluetongue causes clinical disease in cattle similar to that in sheep. In addition to the clinical signs observed in sheep, the disease may cause debilitation, ulceration of different surfaces of the oral cavity, copious nasal exudate, oral breathing, esophageal paresis, and a decrease in milk production.[35–38]

Bluetongue causes economic losses not only by affecting the health of adult animals but also by causing reproductive failure in pregnant animals and congenital abnormalities in newborns. Abortion and stillbirth have been reported in cows. The infection is also associated with excessive, purplish-red gingival tissue, agnathia, arthrogryposis, domed cranium, prognathism, dwarfism, blindness, and incoordination in newborn calves and weakness, inability to walk, difficulty in sucking, and muscular rigidity in newborn sheep.[29,39–41]

Bluetongue disease is principally a result of BTV-mediated vascular injuries. Although the clinical manifestations of the infection in ruminants can be highly variable between species and individuals, the pathogenesis of BTV infections follows essentially the same pattern in sheep and in cattle: BTV infections are initiated with regional BTV replication, followed by long-lasting viremia, which spreads the virus to other organs and results in damage to the vascular endothelium. When BTV is introduced from *Culicoides* saliva into animals, it initially proliferates in endothelial cells of skin capillaries and in leukocytes in the skin; the skin is considered an important organ for BTV replication in the early stage of infection.[42] Subsequently, BTV replicates in regional lymph nodes and is then disseminated throughout the blood by infecting a variety of blood cells. At this viremic stage, BTV can survive within blood cells, primarily platelets and erythrocytes, for as long as 54 days in sheep and goats[43] and 56 days in cattle.[44,45] In addition, the disseminating BTV particles are distributed unevenly in different viscera and lymph nodes. These differences in tissue predilection are proposed to result from variations in BTV strains and endothelial cell susceptibility.[42]

69.1.4 Diagnosis

Clinical diagnosis of bluetongue is neither easy nor reliable. Bluetongue is difficult to be differentiated clinically from some ruminant diseases such as foot-and-mouth disease, bovine viral diarrhea, and orf virus infection because of the similarities between the symptoms of bluetongue and other diseases. Moreover, BTV infections can be subclinical. Asymptomatic infections allow BTV to transmit silently between susceptible individuals in a population before it is known that BTV is present in the population. Thus, laboratory diagnosis plays an important role in the identification and surveillance of bluetongue.

Methods for the laboratory diagnosis of bluetongue, traditional or contemporary (Table 69.1), have been recommended by the World Organization for Animal Health, or the OIE, to monitor the antibodies against BTV, to estimate the prevalence of BTV infection, and/or to confirm BTV infection. As is true for many other infectious diseases, the laboratory diagnosis of bluetongue can be accomplished in three ways: by isolating the causative agent, by detecting the genome of the agent, and by detecting specific antibodies against the agent.

69.1.4.1 Conventional Techniques

BTV in specimens sampled from infected animals can be isolated by inoculation of the specimen into sheep, embryonated chicken eggs, and cell cultures. Although the isolation of BTV

TABLE 69.1

Methods Available for the Diagnosis of Bluetongue and the Specific Diagnostic Purposes of These Methods, Modified from the OIE Terrestrial Manual

	Purpose					
Method[a]	Population Freedom from Infection	Individual Animal Freedom from Infection	Contribution to Eradication Policies	Confirmation of Clinical Cases	Surveillance of the Prevalence of Infection	Measurement of Immune Status
Virus isolation	—	Recommended	—	Recommended	—	—
RT-PCR	—	Recommended	—	Recommended	Suitable	—
Real-time RT-PCR	—	Recommended	—	Recommended	Suitable	—
VNT	Suitable	Recommended	Suitable	—	Suitable	Suitable
cELISA	Suitable	Recommended	Suitable	—	Suitable	Suitable

[a] RT-PCR, reverse transcription polymerase chain reaction; VNT, virus neutralization test; cELISA, competitive enzyme-linked immunosorbent assay.

provides definitive evidence of the presence of the virus in the specimen, the techniques required for virus isolation have not been changed substantially for decades. Attempting to isolate BTV by inoculating sheep is expensive and time consuming and is therefore not practical for routine diagnosis. Inoculation of embryonated chicken eggs is the most sensitive method for BTV isolation, particularly when the embryonated chicken eggs are inoculated intravenously.[47] Virus isolation is thus technically feasible, but such methods alone are unlikely to meet with the need for a rapid response in emerging outbreaks.

Techniques for detecting specific antibodies against BTV include agar gel immunodiffusion, complement fixation test, virus neutralization test (VNT), and enzyme-linked immunosorbent assay (ELISA). The agar gel immunodiffusion and complement fixation tests are neither efficient nor able to discriminate antibodies against BTV from those against other orbiviruses, particularly epizootic hemorrhagic fever virus; their unsatisfactory sensitivity and high labor intensity have limited their application. On the other hand, the VNT and ELISAs are more reliable and more rapid. The VNT is able to measure relative quantities of anti-BTV antibodies, that is, antibody titers, in serum samples. The assay is a reliable quantitative method for evaluating herd immunity against bluetongue, and it is also a qualitative method that can be used to clarify which BTV serotypes are currently circulating in a population. Nevertheless, the disadvantage of this assay, which limits its usefulness, is its dependence on serotype. The ability to measure antibodies against a given BTV serotype depends entirely on the availability of the corresponding serotype in the testing laboratory. Unfortunately, except for a few reference laboratories, nearly all of the veterinary diagnostic laboratories in the world possess only a very limited number of BTV serotypes. Thus, the diagnostic capacity of most laboratories is typically restricted to the serotypes present in nearby regions. When an unknown serotype is suspected, consultation with and technical support from reference laboratories are absolutely necessary. (The contact details of relevant reference laboratories are available on the World Organisation for Animal Health website: http://www.oie.int/our-scientific-expertise/reference-laboratories/list-of-laboratories/.)

Furthermore, when antibodies against more than one BTV serotype need to be measured, multiple tests must be performed for each serum sample. This conventional method remarkably amplifies the laboratory workload, even when robotic systems are used.

Despite being serotype-specific, the VNT is labor intensive. To enhance the effectiveness and specificity of anti-BTV antibody detection, several competitive ELISAs (cELISAs) have been developed since the 1980s. These cELISAs employ the serogroup-specific capsid protein VP7 as the immobilized antigen and an anti-VP7 monoclonal antibody as the competing antibody.[54–56] These changes improve the specificity of the assay and its ability to detect antibodies against BTV and to exclude cross-activity to other orbiviruses.

69.1.4.2 Molecular Techniques

In contrast, molecular methods, for example, reverse transcription polymerase chain reaction (RT-PCR), may compensate for the difficulties associated with classical viral isolation methods. Following the extraction of RNA or total nucleic acids from a clinical specimen, RT-PCR and real-time RT-PCR can detect a very small amount of BTV genomic RNA in the specimen within a few hours rather than a few days, remarkably shortening the interval between receiving a specimen and obtaining diagnostic results. Although molecular biology methods are often rapid and sensitive, it should be noted that greater diagnostic sensitivity is usually accompanied by a greater risk of false-positive results. To lessen this risk, RT-PCR results, particularly those obtained from interpreting gel electrophoresis data, should be confirmed by determining the nucleotide sequences of the RT-PCR products of interest. Moreover, when using molecular assays to detect BTV, it should be noted that BTV RNA is present in the blood much longer than infectious BTV is.[46]

Thus, detection of viral RNA by RT-PCR does not necessarily indicate that the sampled animal carries infectious and transmissible BTV.

RT-PCR and real-time RT-PCR allow rapid detection and identification of BTV RNA; many potential target genomic regions have been considered for specific amplification in these assays. The selection of regions of the BTV genome to amplify primarily depends on the purpose of the assays being designed. To detect the presence of BTV regardless of the serotype or genotype of the virus present in a sample, the region selected should be conserved among the BTV serotypes and distinct from the other orbiviruses. In this regard, regions in the VP1, VP7, and NS3 genes are preferred. In contrast, for serotype-specific detection, RT-PCR assays typically target the VP2 gene because the VP2 structural protein is the major determinant of serotype. Several RT-PCR and real-time PCR assays for BTV have been described previously.[48–53]

69.2 METHODS

69.2.1 SAMPLE PREPARATION

69.2.1.1 Samples for Virus Isolation

Blood, spleen, liver, kidney, lung, and heart tissues collected at the peak of fever are the samples of choice for BTV isolation. The detailed procedures for preparing these samples have been described by Clavijo et al.[47]

Heparinized blood is centrifuged at $420 \times g$ at 4°C for 10 min, and the supernatant is discarded. The sedimented blood cells are washed with sterile isosmotic buffer several times by repeatedly resuspending and centrifuging the cells and discarding the supernatant. The washed cells are then prepared as a 1/10 dilution in minimum essential medium supplemented with gentamicin (50 µg/mL), after which the blood cells are disrupted by sonication or simply repeated freezing and thawing. Alternatively, the washed cells can be disrupted first, brought to the original volume of the blood sample using the medium, and diluted. Prior to inoculation, dilution to 1:100 or lower concentrations may be necessary because less diluted samples can be lethal to chicken embryos.

To prepare tissue samples, the samples are weighed, phosphate-buffered saline is added, and the samples are emulsified using a tissue grinder to make a 10% (w/v) suspension. The suspension is sonicated and then clarified by centrifuging at $2000 \times g$ at 4°C for 20 min. After centrifugation, the supernatant is collected as the undiluted inoculum. The undiluted inoculum then is diluted to 1:10 or 1:20 concentrations before it is being used to inoculate embryonated chicken eggs.

To use field-collected Culicoides biting midges as specimens for BTV isolation, the collected midges are pooled, with several dozen midges in each pool, and homogenized in 100 µL of cold Dulbecco's modified Eagle's medium (DMEM) using an electrical micro-grinder. The homogenized midges are centrifuged at $14,000 \times g$ for 10 min, and the supernatant is transferred into a 1.5 mL microcentrifuge tube. One hundred microliters of DMEM supplemented with

penicillin (100 IU/mL), streptomycin (100 µg/mL), gentamicin (5 µg/mL), nystatin (50 IU/mL), and fetal calf serum (20%) is then added.[57]

69.2.1.2 Samples for BTV-Specific RNA Detection

For detecting BTV RNA, anticoagulated blood is the specimen most commonly used. The viral RNA can be obtained by manual extraction with phenol/chloroform or extraction reagents (such as TRIzol; Invitrogen, Life Technologies, Carlsbad, CA), or silica-resin-based extraction kit (e.g., RNeasy Mini Kit; Qiagen, Valencia, CA); also, when a number of specimen are needed to be handled, extraction kits incorporating automatic systems can be used. The RNA that has been extracted either manually or automatically is generally pure enough to be used in reactions to detect BTV RNA. However, both the quality and quantity of the extracted RNA should be monitored, preferably through a laboratory accreditation system, because the quality and quantity of the RNA used in detection assays may alter the analytical sensitivity of the assays.

The RNA in collected Culicoides biting midges can be extracted and assayed for the presence of BTV viral RNA. The pooled midges are homogenized in 2 mL of phosphate-buffered saline containing two 4.5 mm copper beads using a TissueLyser (Qiagen). The homogenate is then clarified by centrifugation. After centrifugation, 200 µL of the clarified supernatant is mixed with 600 µL of TRIzol Reagent (Invitrogen) and finally eluted in 50 µL of DNase-free and RNase-free water.[58] Alternatively, a robotic system can be used when a large-scale survey is carried out. Approximately 50 collected biting midges of the same species can be pooled and homogenized in 400 µL of lysis buffer (NucleoSpin 96 Virus kit; Macherey-Nagel, Düren, Germany) containing a 5 mm steel bead using a TissueLyser instrument (Qiagen) for 2 min at 30 Hz. The homogenized midges are centrifuged at $12,000 \times g$ for 1 min, and nucleic acids in 200 µL of the supernatant are extracted using a NucleoSpin 96 Virus kit on a Tecan Freedom EVO automatic platform (Tecan Group Ltd., Männedorf, Switzerland) or using other systems with similar functions.[59]

69.2.1.3 Serum Samples for Antibody Detection

Blood sampled for antibody detection is usually collected without anticoagulants such as EDTA and heparin. After the blood samples are allowed to clot, the serum is separated from the sampled blood by centrifugation.

69.2.2 DETECTION PROCEDURES

69.2.2.1 Virus Isolation Using Embryonated Chicken Eggs

To select an appropriate blood vessel for intravenous injection, embryonated chicken eggs are observed under a strong light source. The triangular area of the egg shell immediately above a straight portion of a blood vessel is marked with pencil, and then the triangular piece of the shell is cut with a small hand drill. The cut piece of the shell is removed, and care should be taken so that the shell membrane underneath remains intact. Before the embryo is inoculated with the clinical specimen, sterile mineral oil is dropped on the exposed shell membrane to render the membrane more transparent. Through the triangular window, a tuberculin syringe with a half-inch 27-gauge needle is used to inoculate 0.1 mL of the sample into the blood vessel. Blood displacement in the penetrated vessel can be observed when the inoculum is successfully introduced into the vessel. After the inoculation, the triangular window of the egg shell is sealed with surgical tape or hot melt adhesive, and the inoculated eggs are incubated at 33.5°C for 7 days.

The inoculated eggs are observed twice a day, and the embryos that die within 24 h are discarded. Those that die between 2 and 7 days are refrigerated, and the embryos' tissues (brain, liver, spleen, heart, and lungs) are used in virus identification assays.

69.2.2.2 Virus Isolation Using Cell Cultures

Baby hamster kidney (BHK-21) cell line is sensitive to BTV infection. Prior to the day of inoculation of clinical sample, BHK-21 cells are seeded in 24-well plates at a concentration such that the cells will have formed a 70%–90% confluent monolayer at the time of inoculation. One 24-well plate should be prepared for each specimen. Immediately before inoculation, the cell culture medium is removed, and 0.3 mL of fresh serum-free medium is added to each well. For each specimen, 0.2 mL of the 10% (w/v) tissue suspension is added to each of the six wells in a row. The same volumes of 10-fold dilutions (1:10 and 1:100) of the tissue suspension are also added to two additional rows of the same plate. Six wells of the cells, the last row, are used as negative controls; 0.2 mL of the serum-free medium alone is added to these wells. The inoculum is allowed to adsorb to the cells by incubating with orbital shaking at 4°C for 60 min in a humidified incubator with 5% CO_2. The inoculum is then removed, and 0.5 mL of the medium supplemented with antibiotics and 4% fetal bovine serum are added to each well. The inoculated cells are incubated at 37°C for 6 days in a humidified incubator with 5% CO_2 and observed daily for cytopathic effect. The cells demonstrating cytopathic effects are stored at 4°C until used for virus identification.

In addition to the BHK-21 cell line, several other cell lines are also susceptible to BTV infection, including the Vero cell line from African green monkey, calf pulmonary artery endothelial cell line, C6/36 cell line from the mosquito *Aedes albopictus*,[60] and HmLu-1 cell line from hamster lung.[61]

69.2.2.3 Detection of BTV RNA by RT-PCR

While several conventional RT-PCR assays targeting a variety of regions of the BTV genome have been described, the effectiveness of these assays varies. All of these assays consist of similar sets of reactions: first, cDNA is generated from RNA via the enzymatic activity of reverse transcriptase; next, samples are heated for a short time period to denature the cDNA template completely; the samples then undergo 30–40 repeats of a denaturing–annealing–extension thermal cycle; and lastly, the samples undergo a final extension period to ensure that all amplified DNA copies are synthesized completely. The major distinctions between the assays lie in the temperature and duration of the RT reaction and in the optimal temperature for primer annealing during thermocycling.

Here, two sets of RT-PCR assays used to detect BTV RNA are described. One is a duplex RT-PCR assay that amplifies full-length segment 7 of the BTV genome, and the other is a set of RT-PCR assays that are designed to identify BTV serotypes.

The following procedures of the duplex RT-PCR were developed by Anthony et al.[48] to target the VP7 gene. Two sets of primer pairs and a commercially available QIAGEN one-step RT-PCR kit (Qiagen) are used. The first primer set is composed of the forward primer BTV/S7/01 (5′-GTTAAAAATCTATAGAGATG-3′) and the reverse primer BTV/S7/02 (5′-GTAAGTGTAATCTAAGAGA-3′). The second set comprises the forward primer BTV/S7/03 (5′-GTTAAAAAATCGTTCAAGATG-3′) and the reverse primer BTV/S7/04 (5′-GTAAGTTTAAATCGCAAGACG-3′). The duplex RT-PCR is carried out in a final volume of 50 μL containing 28 μL of RNase-free water, 10 μL of 5 × QIAGEN one-step RT-PCR buffer, 2 μL of dNTP mix, 2 μL of enzyme mix, 0.5 μL of each of the four primers (0.6 μM), and 6 μL of an extracted RNA sample. The 50 μL reaction is performed as follows: one step of 45°C for 30 min; one step of 94°C for 15 min; 40 cycles of denaturation at 94°C for 1 min, annealing at 45°C for 1 min, and extension at 72°C for 2 min; and the RT-PCR is completed with an extension step at 72°C for 10 min. The products are analyzed on a 0.9% agarose gel, stained with ethidium bromide or other DNA stain such as SYBR Safe (Invitrogen), and then visualized under UV light. The predicted length of the product is 1156 base pairs.

The BTV structural protein VP2, encoded by genomic segment 2, is the outermost protein on the surface of the bluetongue virion. This protein is the main determinant of BTV serotypes.[62] To identify BTV serotypes based on the nucleotide sequences of genomic segment 2, Maan et al. developed a set of RT-PCR assays aiming to identify the 26 known BTV serotypes.[49]

Nucleotide sequences of the serotype-specific RT-PCR primer pairs are described by Maan et al. (Tables S1 and S2).[49] For RT-PCR amplification, two commercially available one-step RT-PCR kits are recommended: QIAGEN one-step RT-PCR Kit (Qiagen) and SuperScript III one-step RT-PCR system with Platinum Taq high fidelity (Invitrogen); both have been proven to be efficient in RT-PCR assays.[63] Before the RT-PCR assays are performed, the RNA samples can be denatured with methyl mercuric hydroxide at room temperature followed by the addition of 2-mercaptoethanol.[64] Alternatively, denaturation may be easier to achieve by incorporating a heat denaturation step right before the RT-PCR steps. The reaction would then be initiated by denaturing the RNA at 95°C for 30 min, after which the samples would be placed at 4°C for 10 min. After the initial denaturation step, RT is performed, and the optimal temperature and duration of the RT step depend on the specific enzyme and kit used; the manufacturer's instructions should be followed. The reverse-transcribed cDNA is then used as a template for 40 cycles of denaturation at 94°C for 15 s, annealing at 55°C for 30 s, and extension at 68°C for 1 min. Finally, an extension step of 68°C for 5 min is performed to complete the RT-PCR assay. Five microliters of the RT-PCR products is analyzed by electrophoresis using a 1% agarose gel stained with ethidium bromide or another DNA stain such as SYBR Safe and then visualized under UV light. The electrophoresis-separated products are interpreted based on the patterns described by Maan et al.[49]

69.2.2.4   Detection of BTV RNA by Real-Time RT-PCR

The following procedures for a real-time RT-PCR assay targeting genomic segment 10 of BTV, which is applicable for most BTV serotypes, have been described previously by Hofmann et al.[65] The forward primer of the real-time RT-PCR is 5′-TGGAYAAAGCRATGTCAAA-3′, and the reverse primer is 5′-ACRTCATCACGAAACGCTTC-3′. The concentration of the working stock of both primers is 20 pmol/μL. The probe used in the reaction is 5′-FAM-ARGCTGCATTCGCATCGTACGC-3′ BHQ1 (FAM: 6-carboxyfluorescein), and the concentration of its working stock is 5 pmol/μL. Two microliters of the RNA sample (or controls) is mixed with 0.5 μL of each primer stock (20 pmol/μL), heated at 95°C for 5 min, and then chilled on ice for 3 min. After the heat-denaturation step, 22 μL of one-step real-time RT-PCR master mix, composed of reverse transcriptase, thermostable DNA polymerase, deoxynucleotides (dNTPs), appropriate salts, and the probe (final concentration of 0.2 pmol/μL), is added to the denatured RNA and primers, and the mixture is placed into a real-time thermocycler programmed as follows: 48°C for 30 min and 95°C for 2 min, followed by 50 cycles of 95°C for 15 s, 56°C for 30 s, and 72°C for 30 s. Samples with Ct-values lower than 50 are interpreted as positive.

69.2.2.5   Measurement of BTV-Specific Antibodies by VNT

Each of the serum sample, positive control serum, and negative control serum is twofold serially diluted with cell culture medium. Fifty microliters of the diluted sera are transferred to 96-well flat-bottomed microtiter plates, after which a virus suspension containing 100 $TCID_{50}$ of BTV of a specific serotype is added to each well. After incubation at 37°C for 1 h, BHK-21 or Vero cells are added to the wells, and the plates are incubated in a humid chamber with 5% CO_2 for 3–5 days. The virus neutralization titer of the tested serum is expressed as the $\log_2$ of the reciprocal of the serum dilution that protects cells from cytopathic effects.

69.2.2.6   Detection of BTV-Specific Antibodies by cELISA

A few cELISAs for the specific detection of antibodies against BTV have been developed and are commercially available. Most of these assays demonstrate satisfactory diagnostic sensitivity and specificity, although some inconsistencies in testing results may be present.[66,67] Although the volume of each serum sample that is needed, the antigen used to coat plates, the competitive labeled antibodies employed, and the incubation time between steps may be different among the kits, their general formats and test procedures are similar.

The following procedures are derived from one of the commercially available kits: Twenty microliters of serum (or controls) and 80 μL of dilution buffer are added to one of the wells in the 96-well plate coated with BTV VP7 protein. The mixture is then homogenized by gentle agitation and incubated at room temperature for 45 min. After the plate is incubated, 100 μL of a diluted, peroxidase-conjugated monoclonal anti-VP7 antibody is added to each well, and the plate is further incubated at room temperature for 45 min. After the second incubation, the plate is emptied and washed three times with washing buffer. After the plate is washed, 100 μL of substrate (3,3′,5,5′-tetramethylbenzidine, or TMB) and hydrogen peroxide solution is added. The plate is then incubated at room temperature for 10 min, avoiding exposure to the light. Finally, 100 μL of stop solution (usually 1 M sulfonic acid) is added. The optical densities of each well at a wavelength of 450 nm (OD_{450}) are read, and the S/N ratio (OD_{450} of the tested serum/OD_{450} of the negative control serum) is calculated. When the S/N ratio of a sample is equal to or lower than 70%, the sample is interpreted as a positive reactor.

Different enzyme substrate can be used in the cELISAs. In addition to the TMB, 2,2′-azino-bis(3-ethylbenzothiazoline-6-sulphonic acid) (ABTS) and o-phenylenediamine dihydrochloride (OPD) are the substrates of peroxidase as well. Notably, when ABTS or OPD are employed as substrate, optic densities of different corresponding wavelengths (410 nm/650 nm and 492 nm, respectively) are measured.

Regardless of the improvements in effectiveness and diagnostic specificity of these tests over those of the conventional VNT, it should be noted that cELISA and VNT assays are inherently different tests and that their results are not parallel. The targets of the two tests, neutralizing antibodies for the VNT and VP7-specific antibodies for the cELISA, share no antigenic determinant, and the results are interpreted and expressed in distinct ways. Therefore, the data obtained from these two assays are not interchangeable. Care should be taken when one intends to use and compare the results of cELISA and VNT assays.

69.3   CONCLUSION AND FUTURE PERSPECTIVES

Bluetongue has continued to threaten the livestock industry in many areas of the world, and the impacts of bluetongue seem to be increasing more than ever before, possibly in association with global climate changes. In particular, the bluetongue epidemic sweeping Europe since 1998 not only reveals the terrestrial expansion of BTV to the historically highest latitudes but also demonstrates the complexity of the disease situation and difficulties in BTV diagnosis and surveillance.

In most attempts to fight arthropod-borne viral diseases, such as dengue fever, yellow fever, and Japanese encephalitis, few of the diseases have been successfully eliminated from an endemic area. Instead, the diseases recur frequently, and a dynamic equilibrium between sporadic outbreaks and severe epidemics is established. People maintain this equilibrium with improvements in hygiene, pest control, and intensive vaccination. Meanwhile, vigorous elimination of arthropod vectors may decrease the incidence of the disease to a certain extent, but such reductions in the vector population may also unbalance the ecosystem simultaneously. In contrast, vaccination may be a better alternative to combat the disease effectively without upsetting the ecological balance.

With regard to the control of bluetongue by vaccination, a few European countries have adopted it as one of their strategies for controlling bluetongue epidemics. When facing multiple BTV serotypes and strains cocirculating in the continent, offering a vaccine with the serotype corresponding to the serotype of BTV that currently exists in the field determines whether the vaccine will be able to deliver protective immunity to vaccinated animals. From this aspect, the abilities to determine the serotype of a BTV isolate and the serotype of BTV against which detected serum antibodies were elicited play a critical role in deciding the direction in which control and preventive measures should be implemented. Currently, although conventional and molecular methods for serotyping are available, adaptations of new techniques continue to be in demand. RT loop-mediated isothermal amplification[68] and next generation sequencing[69] have been tested in this capacity. To detect anti-BTV antibodies, multiplexed immunoassays employing flow cytometry[70] may be a rational approach for simultaneously analyzing antibodies elicited by a variety of serotypes of BTVs in a sample.

There is no doubt that the impact of bluetongue is global in scope. The variety of BTV serotypes and variation in BTV strains will continue to challenge the diagnostic capacities of the available assays, both traditional and molecular. It is essential to continue to develop novel technologies and to evaluate the existing methods.

REFERENCES

1. Hutcheon, D. Malarial catarrhal fever of sheep. *Veterinary Record* 14 (1902): 629–633.
2. Spreull, J. Report from Veterinary surgeon Spreull on the result of his experiments of malarial catarrhal fever of sheep. *Agricultural Journal of Cape of Good Hope* 20 (1902): 469–477.
3. Hofmann, M., Griot, C., Chaignat, V., Perler, L., Thür, B. Bluetongue disease reaches Switzerland. *Schweizer Archiv für Tierheilkunde* 150(2) (2008): 49–56.
4. Maan, S., Maan, N.S., Nomikou, K. et al. Complete genome characterization of a novel 26th bluetongue virus serotype from Kuwait. *PLoS One* 6(10) (2011): e26147.
5. Martin, S.A., Zweerink, H.J. Isolation and characterization of two types of bluetongue virus particles. *Virology* 50(2) (1972): 495–506.
6. Mertens, P.P., Burroughs, J.N., Anderson, J. Purification and properties of virus particle, infectious subviral particle, and core of bluetongue virus serotypes 1 and 4. *Virology* 157(2) (1987): 375–386.
7. Schwartz-Cornil, I., Mertens, P.P.C., Contreras, V. et al. Bluetongue virus: Virology, pathogenesis and immunity. *Veterinary Research* 39(6) (2008): 46.
8. Hoff, G.L., Trainer, D.O. Observation on bluetongue and epizootic hemorrhagic disease viruses in white-tailed deer: (1) distribution of virus in the blood; (2) cross-challenge. *Journal of Wildlife Diseases* 10(1) (1974): 25–31.
9. Hoff, G.L., Issel, C.J., Trainer, D.O. Arbovirus serology in North Dakota mule and white-tailed deer. *Journal of Wildlife Diseases* 9(4) (1973): 291–295.
10. Murray, J.O., Trainer, D.O. Bluetongue virus in North American elk. *Journal of Wildlife Diseases* 6(3) (1970): 144–148.
11. Rodriguez-Sanchez, B., Gortazar, C., Ruiz-Fons, F., Sanchez-Vizcaino, J.M. Bluetongue virus serotypes 1 and 4 in red deer, Spain. *Emerging Infectious Diseases* 16(3) (2010): 518–520.
12. Batten, C.A., Harif, B., Henstock, M.R. et al. Experimental infection of camels with bluetongue virus. *Research in Veterinary Science* 90(3) (2011): 533–535.
13. Schulz, C., Eschbaumer, M., Rudolf, M. et al. Experimental infection of South American camelids with bluetongue virus serotype 8. *Veterinary Microbiology* 154(3–4) (2012): 257–265.
14. Dubovi, E.J., Hawkins, M., Griffin, R.A., Johnson, D.J., Ostlund, E.N. Isolation of bluetongue virus from canine abortions. *Journal Veterinary Diagnostic Investigation* 25(4) (2013): 490–492.
15. Akita, G.Y., Ianconescu, M., MacLachlan, N.J., Osburn, B.I. Bluetongue disease in dogs associated with contaminated vaccine. *Veterinary Record* 134(11) (1994): 283–284.
16. Evermann, J.F., McKeiman, A.J., Wilbur, L.A. et al. Canine fatalities associated with the use of a modified live vaccine administered during late stages of pregnancy. *Journal Veterinary Diagnostic Investigation* 6(3) (1994): 353–357.
17. Jauniaux, T.P., De Clercq, K.E., Cassart, D.E. et al. Bluetongue in Eurasian lynx. *Emerging Infectious Diseases* 14(9) (2008): 1496–1498.
18. Alexander, K.A., MacLachlan, N.J., Kat, P.W. et al. 1994. Evidence of natural bluetongue virus infection among African carnivores. *American Journal of Tropical Medicine and Hygiene* 51(5) (1994): 568–576.
19. Mellor, P.S., Osborne, R., Jennings, D.M. Isolation of bluetongue and related viruses from *Culicoides* spp. in the Sudan. *Journal of Hygiene* 93(3) (1984): 621–628.
20. Mellor, P.S., Jennings, D.M., Wilkinson, P.J., Boorman, J.P. *Culicoides imicola*: A bluetongue virus vector in Spain and Portugal. *Veterinary Record* 116(22) (1985): 589–590.
21. Loomis, E.C., Bushnell, R.B., Osburn, B.I. Epidemiologic study of bluetongue virus on the Tejon Ranch, California: Vector, host, virus relationships. *Progress in Clinical and Biological Research* 178 (1985): 583–588.
22. Tanya, V.N., Greiner, E.C., Gibbs, E.P. Evaluation of *Culicoides insignis* (Diptera: Ceratopogonidae) as a vector of bluetongue virus. *Veterinary Microbiology* 32(1) (1992): 1–14.
23. Belbis, G., Bréard, E., Cordonnier, N. et al. Evidence of transplacental transmission of bluetongue virus serotype 8 in goats. *Veterinary Microbiology* 166(3–4) (2013): 394–404.
24. Rasmussen, L.D., Savini, G., Lorusso, A. et al. Transplacental transmission of field and rescued strains of BTV-2 and BTV-8 in experimentally infected sheep. *Veterinary Research* 44 (2013): 75.

25. Van der Sluijs, M.T.W., Schroer-Joosten, D.P.H., Fid-Fourkour, A. et al. Transplacental transmission of bluetongue virus serotype 1 and serotype 8 in sheep: Virological and pathological findings. *PLoS One* 8(12) (2013): e81429.

26. Purse, B.V., Mellor, P.S., Roger, D.J., Samuel, A.R., Mertens, P.P.C., Baylis, M. Climate change and the recent emergency of bluetongue in Europe. *Nature Reviews Microbiology* 3 (2005): 171–181.

27. Vercauteren, G., Miry, C., Vandenbussche, F. et al. Bluetongue virus serotype 8-associated congenital hydranencephaly in calves. *Transboundary and Emerging Diseases* 55(7) (2008): 293–298.

28. Wouda, W., Roumen, M.P.H.M., Peperkamp, N.H.M.T., Vos, J.H., van Garderen, E., Muskens, J. Hydranencephaly in calves following the bluetongue virus serotype 8 epidemic in the Netherlands. *Veterinary Record* 162(13) (2008): 422–423.

29. Brown, C.C., MacLachlan, N.J. Congenital encephalopathy in a calf. *Veterinary Pathology* 20(6) (1983): 770–773.

30. MacLachlan, N.J., Osburn, B.I., Ghalib, H.W., Stott, J.L. Bluetongue virus-induced encephalopathy in fetal cattle. *Veterinary Pathology* 22(4) (1985): 415–417.

31. Forman, A.J., Hooper, P.T., Le Blanc Smith, P.M. Pathogenicity for sheep of recent Australian bluetongue virus isolates. *Australian Veterinary Journal* 66(8) (1989): 261–262.

32. Hamblin, C., Salt, J.S., Graham, S.D., Hopwood, K., Wade-Evans, A.M. Bluetongue virus serotypes 1 and 3 infection in Poll Dorset sheep. *Australian Veterinary Journal* 76(9) (1998): 622–629.

33. Jeggo, M.J., Corteyn, A.M., Taylor, W.P., Davidson, W.L., Gorman, B.M. Virulence of bluetongue virus for British sheep. *Research in Veterinary Science* 42(1) (1987): 24–28.

34. MacLachlan, N.J., Crafford, J.E., Vernau, W. et al. Experimental reproduction of severe bluetongue in sheep. *Veterinary Pathology* 45(3) (2008): 310–315.

35. Backx, A., Heutink, R., van Rooij, E., van Rijn, P. Transplacental and oral transmission of wild-type bluetongue virus serotype 8 in cattle after experimental infection. *Veterinary Microbiology* 138(3–4) (2009): 235–243.

36. Goltz, J. Bluetongue in cattle: A review. *Canadian Veterinary Journal* 19(4) (1978): 95–98.

37. Pardon, B., Vandenberge, V., Maes, S., De Clercq, K., Ducatelle, R., Deprez, P. Oesophageal paresis associated with bluetongue virus serotype 8 in cattle. *Veterinary Record* 167(15) (2010): 579–581.

38. Santman-Berends, I.M., Hage, J.J., Lam, T.J., Sampimon, O.C., van Schaik, G. The effect of bluetongue virus serotype 8 on milk production and somatic cell count in Dutch dairy cows in 2008. *Journal of Dairy Science* 94(3) (2011): 1347–1354.

39. Luedke, A.J., Jochim, M.M., Jones, R.H. Bluetongue in cattle: Effects of *Culicoides variipennis*-transmitted bluetongue virus on pregnant heifers and their calves. *American Journal of Veterinary Research* 38(11) (1977): 1687–1695.

40. McKercher, D.G., Saito, J.K., Singh, K.V. Serologic evidence of an etiologic role for bluetongue virus in hydranencephaly of calves. *Journal of American Veterinary Medical Association* 156(8) (1970): 1044–1047.

41. Pérez, V., Suárez-Vega, A., Fuertes, M. et al. Hereditary lissencephaly and cerebellar hypoplasia in Churra lambs. *BMC Veterinary Research* 9 (2013): 156.

42. Darpel, K.E., Monaghan, P., Simpson, J. et al. Involvement of the skin during bluetongue virus infection and replication in the ruminant host. *Veterinary Research* 43 (2012): 40.

43. Koumbati, M., Mangana, O., Nomikou, K., Mellor, P.S., Papadopoulos, O. Duration of bluetongue viraemia and serological responses in experimentally infected European breeds of sheep and goats. *Veterinary Microbiology* 64(4) (1999): 277–285.

44. MacLachlan, N.J., Heidner, H.W., Fuller, F.J. Humoral immune response of calves to bluetongue virus. *American Journal Veterinary Research* 48(7) (1987): 1031–1035.

45. Richards, R.G., MacLachlan, N.J., Heidner, H.W., Fuller, F.J. Comparison of virologic and serologic responses of lambs and calves infected with bluetongue virus serotype 10. *Veterinary Microbiology* 18(3–4) (1988): 233–242.

46. MacLachlan, N.J., Nunamaker, R.A., Katz, J.B. et al, Detection of bluetongue virus in the blood of inoculated calves: Comparison of virus isolation, PCR assays, and in vitro feeding of Culicoides variipennis. *Archives of Virology* 136(1–2) (1994): 1–8.

47. Clavijo, A., Hechert, R.A., Dulac, G.C., Afshar, A. Isolation and identification of bluetongue virus. *Journal of Virological Methods* 87(1–2) (2000): 13–23.

48. Anthony, S., Jones, H., Darpel, K.E. et al. A duplex RT-PCR assay for detection of genome segment 7 (VP7 gene) from 24 BTV serotypes. *Journal of Virological Methods* 141(2) (2007): 188–197.

49. Maan, N.S., Maan, S., Belaganahalli, M.N. et al. Identification and differentiation of the twenty six bluetongue virus serotype by RT-PCR amplification of the serotype-specific genome segment 2. *PLoS One* 7(2) (2012): e32601. http://www.plosone.org/article/info%3Adoi%2F10.1371%2Fjournal.pone.0032601.

50. Momtaz, H., Nejat, S., Souod, N., Momeni, M., Safari, S. Comparisons of competitive enzyme-linked immunosorbent assay and one step RT-PCR tests for the detection of bluetongue virus in south west of Iran. *African Journal of Biotechnology* 10(36) (2011): 6857–6862.

51. Shaw, A.E., Monaghan, P., Alpar, H.O. et al. Development and initial evaluation of a real-time RT-PCR assay to detect bluetongue virus genome segment 1. *Journal of Virological Methods* 145(2) (2007): 115–126.

52. Toussaint, J.F., Sailleau, C., Breard, E., Zientara, S., De Clercq, K. Bluetongue virus detection by real-time RT-qPCRs targeting two different genomic segments. *Journal of Virological Methods* 140(1–2) (2007): 115–123.

53. Wilson, W.C., Hindson, B.J., O'Hearn, E.S. et al. A multiplex real-time reverse transcription polymerase chain reaction assay for detection and differentiation of bluetongue virus and epizootic hemorrhagic disease virus serogroups. *Journal of Veterinary Diagnostic Investigation* 21(6) (2009): 760–770.

54. Anderson, J. Use of monoclonal antibody in a blocking ELISA to detect group specific antibodies to bluetongue virus. *Journal of Immunological Methods* 74(1) (1984): 139–149.

55. House, C., House, J.A., Berninger, M.L. Detection of bluetongue group-specific antibody by competitive ELISA. *Journal Veterinary Diagnostic Investigation* 2(2) (1990): 137–139.

56. Lunt, R.A., White, J.R., Blacksell, S.D. Evaluation of a monoclonal antibody blocking ELISA for the detection of group-specific antibodies to bluetongue virus in experimental and field sera. *Journal of General Virology* 69(11) (1988): 2729–2740.

57. Savini, G., Goffredo, M., Monaco, F. et al. The isolation of bluetongue virus from field populations of Obsoletus Complex in central Italy. *Veterinaria Italiana* 40(3) (2004): 286–291.

58. Rizzo, F., Cerutti, F., Ballardini, M. et al. Molecular characterization of flaviviruses from field-collected mosquitoes in northwestern Italy, 2011–2012. *Parasites and Vectors* 7(1) (2014): 395.

59. Hoffmann, B., Bauer, B., Bauer, C. et al. Monitoring of putative vectors of bluetongue virus serotype 8, Germany. *Emerging Infectious Diseases* 15(9) (2009): 1481–1484.

60. Wechsler, S.J., McHolland, L.E. Susceptibilities of 14 cell lines to bluetongue virus infection. *Journal of Clinical Microbiology* 26(11) (1988): 2324–2327.

61. Ianconescu, M., Akita, G.Y., Osburn, B.I. Comparative susceptibility of a canine cell line and bluetongue virus susceptible cell lines to a bluetongue virus isolate pathogenic for dogs. *In Vitro Cellular and Developmental Biology. Animal* 32(4) (1996): 249–254.

62. Kahlon, J., Sugiyama, K., Roy, P. Molecular basis of bluetongue virus neutralization. *Journal of Virology* 48(3) (1983): 627–632.

63. Lee, F., Lin, Y.L., Tsai, H.J. Comparison of primer sets and one-step reverse transcription polymerase chain reaction kits for the detection of bluetongue viral RNA. *Journal of Virological Methods* 200 (2014): 6–9.

64. Mertens, P.P.C., Maan, N.S., Prasad, G. et al. Design of primers and use of RT-PCR assays for typing European bluetongue virus isolates: Differentiation of field and vaccine strains. *Journal of General Virology* 88(10) (2007): 2811–2823.

65. Hofmann, M.A., Renzullo, S., Mader, M., Chaignat, V., Worwa, G., Thuer, B. Genetic characterization of Toggenburg orbivirus, a new bluetongue virus, from goats, Switzerland. *Emerging Infectious Diseases* 14(12) (2008): 1855–1861.

66. Batten, C.A., Bachanek-Bankowska, K., Bin-Tarif, A. et al. Bluetongue virus: European Community inter-laboratory comparison test to evaluate ELISA and RT-PCR detection methods. *Veterinary Microbiology* 129(1–2) (2008): 80–88.

67. Niedbalski, W. Evaluation of commercial ELISA kits for the detection of antibodies against bluetongue virus. *Polish Journal of Veterinary Sciences* 14(4) (2011): 615–619.

68. Mulholland, C., Hoffmann, B., McMenamy, M.J. et al. The development of an accelerated reverse transcription loop mediated isothermal amplification for the serotype specific detection of bluetongue virus 8 in clinical samples. *Journal of Virological Methods* 202 (2014): 95–100.

69. Rao, P.P., Reddy, Y.N., Ganesh, K., Nair, S.G., Niranjan, V., Hegde, N.R. Deep sequencing as a method of typing bluetongue virus isolates. *Journal of Virological Methods* 193(2) (2013): 314–319.

70. Fulton, R.J., McDade, R.L., Smith, P.L., Kienker, L.J., Kettman Jr, J.R. Advanced multiplexed analysis with the FlowMetrix system. *Clinical Chemistry* 43(9) (1997): 1749.

70 Epizootic Hemorrhagic Disease Virus

Srivishnupriya Anbalagan

CONTENTS

70.1 Introduction 629
    70.1.1 Classification, Morphology, and Genome Organization 629
        70.1.1.1 Classification 629
        70.1.1.2 Morphology and Genome Organization 629
    70.1.2 Epidemiology, Transmission, and Pathogenesis 630
    70.1.3 Clinical Features 631
    70.1.4 Diagnosis 631
70.2 Methods 632
    70.2.1 Sample Collection and Preparation 632
        70.2.1.1 RNA Isolation Using QIAamp Viral RNA Isolation Kit (Qiagen, Valencia, CA) 632
        70.2.1.2 RNA Isolation Using Ambion MagMAX-96 Viral RNA Isolation Kit (Life Technologies, Grand Island, NY) 632
    70.2.2 Detection Procedures 633
        70.2.2.1 Multiplex PCR Detection of Eight EHDV Serotypes 633
        70.2.2.2 Whole Genome Sequencing 633
70.3 Conclusion 634
References 634

70.1 INTRODUCTION

Epizootic hemorrhagic disease (EHD) is an insect-transmitted disease of wild and domestic ruminants, resulting from infection with either epizootic hemorrhagic disease virus (EHDV) or bluetongue virus (BTV).[1] First described in white-tailed deer (WTD) in New Jersey in 1955, EHD was associated with severe clinical signs and death in naive population of wild and domestic ruminants including endangered species.[1–4] For this reason, EHD is considered an economically important disease, and a ban has been imposed on trading livestock, ova, and semen from affected countries. This chapter presents an overview on the classification, morphology, genome organization, biology, epidemiology, and diagnosis of EHDV.

70.1.1 CLASSIFICATION, MORPHOLOGY, AND GENOME ORGANIZATION

70.1.1.1 Classification

EHDV is classified in the genus *Orbivirus*, subfamily *Sedoreovirinae*, family *Reoviridae*.[1]

The *Reoviridae* family covers viruses with naked, icosahedral capsids and 10–12 segments of linear double-stranded RNA (dsRNA). It is separated into two subfamilies: *Sedoreovirinae* and *Spinareovirinae*.

The subfamily *Sedoreovirinae* (*sedo* = smooth) contains viruses that are characterized by the absence of a turreted protein on the inner capsid to produce a smooth surface.

Currently, six genera are recognized in this subfamily: *Cardoreovirus*, *Mimoreovirus*, *Orbivirus*, *Phytoreovirus*, *Rotavirus*, and *Seadornavirus*. Apart from EHDV, other notable livestock-infecting viruses in the genus *Orbivirus* include BTV and African horse sickness virus, among others.

The subfamily *Spinareovirinae* consists of viruses that are distinguished by the presence of a turreted protein on the inner capsid. At present, nine genera are recognized in this subfamily: *Aquareovirus*, *Coltivirus*, *Cypovirus*, *Dinovernavirus*, *Fijivirus*, *Idnoreovirus*, *Mycoreovirus*, *Orthoreovirus*, and *Oryzavirus*.

Being transmitted by biting midges in the genus *Culicoides*,[1] EHDV is widespread in WTD and occasionally causes serious epidemics in wild populations and domestic animals.[1–4] Clinically, WTD are often severely affected.[1–4] Mule deer and pronghorn antelope are affected to a lesser extent and show mild disease symptoms.[1,5–7] Black-tailed deer, mule deer, red deer, wapiti, fallow deer, roe deer, and *Gazella subgutturosa subgutturosa* were seropositive for EHDV.[1,6,8,9] Yaks are susceptible to EHDV infection and exhibit histological lesions of hemorrhagic disease as observed in other ruminants.[10] Sheep can be infected experimentally but rarely develop clinical signs and have a negligible role in the epidemiology of EHDV.[11,12] Goats and pigs are not susceptible to EHDV infection.[9,13]

70.1.1.2 Morphology and Genome Organization

EHDV is structurally, antigenically, and genetically related to BTV, the type species of the genus that causes disease in ovine, bovine, and other species.[1,14–16] EHDV contains 10 segmented linear dsRNA genome.[1] The genome encodes a total of seven structural proteins (VP1 to VP7) and four nonstructural proteins (NS1, NS2, NS3, and NS3a).[1,17–19] VP2 and VP5 are the most variable viral proteins and are responsible for viral attachment and entry to the host cell.[1,17,20] Specificity of VP2 interaction with neutralizing antibodies determines the virus serotype.[1,17,20] VP2 and VP5 trimers are located on the outer core layer formed by VP7.[1,17,20] VP3, an RNA-binding protein, forms the inner subcore layer and interacts with both the viral RNA genome and the minor proteins VP1, VP4, and VP6.[1,17,19] Self-assembly of VP3 controls the size and organization of the capsid structure.[19] VP1, a highly conserved protein, functions as a viral RNA-dependent RNA polymerase (RdRp).[1,17,19] VP4 is a capping enzyme, and VP6 is believed to be the viral helicase.[1,17,19]

The outer coat proteins VP2 and VP5 are the most variable proteins in EHDV and they determine the serotype. Although VP2 sequences were quite variable (47%–53% identity) between the serotypes, they have 97%–99% identity within the serotypes.[21] Similarly, VP5 sequences that are quite variable (64%–69% identity) between the serotypes have 97%–99% identity within the serotypes.[21] However, VP7 sequences had 97%–99% nucleotide identity within ST1 and ST2 isolates, and only 78%–80% identity between serotype 1 and 2 viruses.[21]

Whole genome sequence analysis of serotype 1 and 2 viruses demonstrated preferential reassortment of VP2, VP5, and VP7 within the homologous serotype.[21] The results suggest incompatibility between the surface proteins VP2 and VP5 and the outer core protein VP7 from the heterologous serotype.[21] However, VP2 and VP5 from ST6 are compatible with VP7 from serotype 2.[21] Also, high genetic identity (>94%) was observed between ST1 and ST2 reference viruses for segments encoding VP1, VP3, VP4, VP6, NS1, NS2, and NS3, preventing conclusive identification of reassortment events between these segments for serotype 1 and 2 viruses.[21]

Based on the serological reactivity, eight serotypes of EHDV are proposed.[22] EHDV serotypes 3 and 4 are reported in Africa.[4,23–25] EHDV serotypes 5, 6, 7, and 8 are reported in Australia.[4,26] EHDV-6 has been identified in Bahrain, the Sultanate of Oman, The Sudan, Morocco, and Algeria.[27] Historically, in the United States, only two serotypes, EHDV-1 (New Jersey strain) and EHDV-2 (Alberta strain), were identified.[4] However, in 2006, a novel reassortant serotype 6 EHDV (ST6) containing two variable surface antigens (VP2 and VP5) from an exotic EHDV-6 with the remaining segments derived from endemic EHDV-2 was isolated from moribund and dead WTD in Indiana and Illinois.[4] Subsequently, reassortant serotype 6 EHDV was identified in several other states demonstrating that the virus is geographically widespread.[4] This strain is currently endemic in the United States.[4] Reassortment between native and exotic EHDV illustrates the plasticity of EHDV and their ability to become established in new environmental niches.[4]

70.1.2 EPIDEMIOLOGY, TRANSMISSION, AND PATHOGENESIS

Epidemiology: EHDV has been isolated from bovines throughout the world, including North America, Australia, Japan, Turkey, and Israel.[1,8,16,20,28–32] An EHDV serogroup virus, Ibaraki virus, which is now recognized as a strain of EHDV-2, caused serious epidemics in Japan.[29,33] EHDV-7 was associated with outbreaks in bovines from Israel.[16] EHDV-6 was associated with outbreaks in bovines from Morocco and Turkey.[16,20,28,34] In the United States, EHDV infection in bovines is typically subclinical and has only been rarely reported.[35] However, recent reports indicate that EHDV serotype 2 can cause clinical disease in bovines from the United States.[36]

Transmission: EHDV is transmitted by species of *Culicoides* biting midges. In North America, *C. mohave*, *C. lahillei*, *C. sonorensis*, *C. debilipalpis*, and *C. venustus* species are suspected as vectors for EHDV.[37–40] However, *C. sonorensis* is the only species that complies with the criteria provided by the WHO for vector implication.[1] In North America, majority of EHDV cases occur in late summer and autumn when the *C. sonorensis* season is at its peak.[1]

Pathogenesis: EHDV replicates in endothelial cells of lymphatic vessels and lymph nodes present at the site of infection.[1] It then disseminates through the bloodstream to other sites of replication such as the lymph nodes and spleen.[1] In the blood, the virus is associated with lymphocytes and red blood cells, where it is found in high titers for a long period of time.[9,41] Eventually, virus replicates in vascular endothelium causing vascular damage and thrombosis, with disseminated intravascular coagulopathy in some cases.[42–44] The extent of endothelial damage might in turn cause differences in clinical diseases observed with the infected ruminant species.[43,44]

Experimental infection of EHDV-2 in deer identified that the virus could be detected 2 days postinfection (pi).[45–47] All the infected animals were viremic at 4 days pi. In a few deer, EHDV was isolated for more than 50 days pi.[45–47] However, in most of the cases, virus could not be detected after 3 weeks from infection.[45–47] In EHDV-2-infected deer, interferon gamma was identified from day 4 to day 10 pi.[45] In EHDV-1-infected deer, neutralizing antibodies were identified from day 10 to day 14 pi.[9] Although virus and antibodies were identified at the same time, neutralizing antibodies were unable to completely remove the virus from circulation.[9] In serum collected from young deer, maternal antibodies against EHDV were detected up to 17–18 weeks of age.[47] Although passive immunity could not prevent infection or viremia in fawns, it can protect them against several clinical diseases.[47]

In cattle experimentally infected with EHDV-1, EHDV-2, or EHDV-6, virus can be detected anywhere from 9 to 50 days pi.[9,34,48,49] Antibodies against EHDV were first detected 7 days post infection (dpi) and persisted up to 35 dpi.[9,34,48,49] Although cattle remained clinically unaffected after experimental inoculation, they display high levels of viral RNA and virus in their blood, confirming that subclinical infection of cattle likely plays an important role in EHDV transmission.[9,34,48,49]

70.1.3 CLINICAL FEATURES

Clinical signs of EHDV are observed in WTD, cattle, and other wild and domestic ungulates. WTD are most severely affected by EHDV. In cattle, the mortality is usually low. Three clinical forms of EHD were observed in WTD, peracute, acute, and chronic.[1] In the peracute stage, animals display high fever, anorexia, weakness, respiratory distress, and severe and rapid edema of the head and neck.[1] Swelling of tongue and conjunctivae is also common.[1] Hemorrhagic diathesis with bloody diarrhea and/or hematuria and dehydration are commonly found in dead animals.[1] Deer with the peracute form usually die rapidly, within 8–36 h, sometimes without clinical signs.[1] Deer with the acute form of the disease exhibit hemorrhages in many tissues including the skin, heart, and gastrointestinal tract.[1] Hyperemia of the conjunctivae and the mucus membranes of the oral cavity with excessive salivation and nasal discharge that is blood-tinged is common in animals with the acute form of the disease.[1] These animals can also develop ulcers or erosions of the tongue, dental pad, palate, rumen, and abomasums.[1] Mortality rate is high in both the peracute and acute forms. In the chronic form, deer are ill for several weeks but slowly recover.[1] Even after recovery, due to the growth interruptions that occur during the disease, some animals develop breaks or rings in the hooves and become lame.[1] In some cases, due to sloughing of the hoof wall or toe, animals crawl on their knees or chest.[1] In the chronic form, deer might also develop ulcers, scars, or erosions in the rumen.[1] Even when there is no shortage of food, animals might look thin and weak.[1] Notable histopathological lesions include widespread vasculitis with thrombosis, endothelial swelling, hemorrhages, degenerative changes, and necrosis in multiple organs, more specifically in the tongue, salivary glands, walls of forestomachs, aorta, and papillary muscle of the left ventricle of the myocardium.[1] In addition, animals may be hyperthermic and become hypothermic prior to death.[1,50] Testicular lesions and antler abnormalities observed in mule deer are considered a sequel of EHDV infection.[51]

In yaks, clinical signs comparable to the disease in WTD such as intermittent tremors, dysphagia, oral ulcerative lesions, hemorrhagic enteritis, tachypnea, and thrombocytopenia were observed.[52] Recently, head tremors and excessive lip smacking were observed in yaks from Colorado.[52] The dysphagia observed in yaks is comparable to the excess salivation found in WTD and is most likely secondary to decreased swallowing due to pain from oral mucosal erosions and ulcerations.[52]

In bighorn sheep, widespread edema, yellowish fluid accumulation in the thorax and pericardial sac, greenish discharge from the nose, and blood around the anus were also observed.[1] Multifocal hemorrhages in the epicardium and in the papillary muscle of the left ventricle are also reported.[1] Hemorrhages were also reported on the conjunctival membranes of the eye and on the serosal surfaces of the rumen and intestine.[1] Petechial hemorrhages were identified in the central nervous system, myocardium, epicardium, tunica muscularis, and submucosa of the rumen bur.[1] It is hard to determine morbidity and mortality rates due to EHDV infection in

wildlife population.[1] However, the estimated number of affected animals and population in at-risk areas suggests an infection rate of 29% and mortality rate of 20% in WTD.[1,2,53]

The first major outbreak of EHDV in cattle was reported in 1959.[29] During the first epidemic, the disease was characterized by fever, anorexia, and difficulty in swallowing.[29] The second major epidemic occurred in 1997 in Japan, which was characterized by abortions, stillbirths, fetal malformations, and infertility problems.[54] Generally, EHDV infection in cattle characterized by hemorrhages, erosions, and ulcerations were seen in the mouth, on the lips, and around the coronets.[28,33] Cattle can also show stomatitis, swelling of eyelids, respiratory distress, nasal and ocular discharge, redness and scaling of muzzle and lips, lameness, and udder erythema.[28,33] In most cases, animals became stiff and lame, and their skin were thickened and edematous.[28,33] The most noticeable clinical sign of the disease was a sharp reduction in milk production. Animals may also look dehydrated and emaciated, and in some cases, death may occur due to aspiration pneumonia.[28,33]

Until recently, in North America, EHDV infections of bovines are typically considered subclinical. However, recently, EHDV-2 was isolated and characterized from a cow that aborted and showed clinical signs of EHDV disease.[36] Also, the calf born from an EHDV PCR-positive cow in the same herd displayed clinical signs of EHDV at birth.[36] Although the evidence in this case is circumstantial, it is suggestive of vertical EHDV transmission.[36] The clinical signs that the cow exhibited were similar to the EHDV virus that caused an epidemic in western Japan in 1997 suggesting that EHDV infection in bovines has the potential to cause severe epidemic disease in the United States.[36] The bovine EHDV was nearly identical to the WTD EHDV concurrently circulating in a similar geographical area.[36] These results justify further EHDV surveillance and diagnostic investigation given the widespread distribution of these viruses and their ability to cause severe clinical disease, both in wildlife and domestic animals.

70.1.4 DIAGNOSIS

Given that EHD may be caused by either EHDV or BTV,[1] it is important that the causal agent is accurately identified. Methods for identification and detection of EHDV range from virus isolation, serological assays, and molecular techniques.[55–60]

EHDV can be grown on Vero or baby hamster kidney (BHK) cells as well as embryonated chicken eggs. ATM, *Culicoides sonorensis* cell lines (Kc), and cow pulmonary artery endothelial cells are also useful for virus isolation.[3,55] Cytopathic effect (CPE) is observed in mammalian cells anywhere between 2 and 7 days after virus inoculation. Generally, virus isolation takes 2–4 weeks.

Serum neutralization (SN) tests provide a means to identify and quantitate antibodies against EHDV serotypes. SNs require viruses from all the serotypes to be included in the assay making it time consuming and labor intensive. Agar gel immunodiffusion test was used for animal trade to detect antibodies against EHDV.[58–60] It is a very simple and economical test. Since cross-reactions occur among different *Orbivirus* serogroups, this test is not predominantly used to detect EHDV. Complement fixation test detects and quantitates EHDV antibodies[1] and appears to be sensitive and specific for EHDV diagnosis. It detects antibodies for 4–12 months after infection but is less reliable after this period. The test was once relied upon for diagnosis and certification of animals for export.

By contrast, molecular techniques such as reverse transcription polymerase chain reaction (RT-PCR) are less time consuming and much faster than virus isolation and serological assays. It is no surprise that these techniques have been increasingly applied in laboratories worldwide for identification and typing of EHDV. Indeed, several serotype-specific RT-PCR assays have been designed to amplify EHDV VP2.

70.2 METHODS

70.2.1 Sample Collection and Preparation

Sample collection: EHDV has been isolated from tissue samples such as spleen, lung, and lymph nodes of infected animals. Some labs were successful in recovering virus from unclotted blood. If samples have to be collected from dead deer, necropsy has to be conducted as soon as possible after death. Temperature is also a primary factor; necropsy should be done within 6 h at the most and less if the deer is in the direct sun. At least a fist-sized spleen sample should be collected with plenty of ice in an insulated diagnostic shipper box for next-day delivery to the laboratory. Whenever possible, freezing is recommended.

Viral RNA isolation: Several methods have been used for the extraction of EHDV RNA from clinical specimens prior to RT-PCR.

70.2.1.1 RNA Isolation Using QIAamp Viral RNA Isolation Kit (Qiagen, Valencia, CA)

70.2.1.1.1 Procedure

1. Centrifuge samples at $1500 \times g$ for 10 min to remove cell line and debris. Remove supernatant to a 1.5 mL centrifuge tube.
2. Add 140 µL of the supernatant to AVL buffer. Pulse vortex for 15 s to mix.
3. Incubate at room temperature for 10 min.
4. Add 560 µL of ethanol (96%–100%) and mix by pulse vortexing for 15 s.
5. Apply 630 µL of the sample to the spin column and centrifuge at $6000 \times g$ for 1 min.
6. Discard filtrate.
7. Repeat steps 5 and 6 with the remaining sample.
8. Add 500 µL of AW1 buffer to the spin column. Centrifuge at $6000 \times g$ for 1 min. Discard filtrate.
9. Add 500 µL of AW2 buffer and centrifuge at full speed for 3 min. Place spin column in new collection tube and repeat full speed centrifugation step for 1 min.

10. Place spin column in sterile Eppendorf tube.
11. Add 60 µL of AVE buffer directly to the center of the filter in the spin column.
12. Incubate at room temperature for 1 min.
13. Centrifuge at $6000 \times g$ for 1 min.
14. Transfer filtrate containing the RNA to a labeled, sterile microcentrifuge tube.

70.2.1.2 RNA Isolation Using Ambion MagMAX-96 Viral RNA Isolation Kit (Life Technologies, Grand Island, NY)

70.2.1.2.1 Procedure

1. Pipet 130 µL of prepared lysis/binding solution (carrier RNA and isopropanol added) into each well of the processing plate.
2. Transfer 50 µL of sample into the lysis/binding solution in the processing plate and pipet up and down to mix.
3. Shake the plate for 5 min on an orbital shaker.
4. Add 20 µL of bead mix (10 µL of beads and 10 µL of lysis/binding enhancer) to each sample and pipet up and down to mix.
5. Shake the plate for 5 min.
6. Move processing plate to magnetic stand to capture the RNA binding beads. Leave plate on the magnetic stand for at least 5 min.
7. Carefully aspirate and discard the supernatant without disturbing the beads and then remove the plate from the magnetic stand.
8. Add 150 µL of wash solution 1 (isopropanol added) to each sample and shake the plate for 1 min.
9. Capture the RNA binding beads on a magnetic stand for at least 1 min or until the mixture becomes clear, indicating that the capture is complete.
10. Carefully aspirate and discard the supernatant without disturbing the beads and remove the plate from the magnetic stand and repeat steps 9–11 to wash a second time with 150 µL of wash solution 1.
11. Add 150 µL of wash solution 2 (ethanol added) to each sample and shake for 1 min.
12. Capture the RNA binding beads on a magnetic stand for at least 1 min or until the mixture becomes clear, indicating that the capture is complete.
13. Carefully aspirate and discard the supernatant without disturbing the beads and then remove the plate from the magnetic stand and repeat steps 12–13 to wash with a second 150 µL of wash solution 2.
14. Move the processing plate to the shaker (after removing plate lid) and shake for at least 5 min to allow any remaining alcohol from the wash solution 2 to evaporate.
15. Inspect the wells and if there is any residual solution, shake the plate for 1–2 min to let it evaporate.

16. Add 30 μL elution buffer to each sample and pipet up and down to mix. Shake the plate for 5 min.
17. Capture the binding beads as in the previous steps. The supernatant will contain the purified RNA for PCR. The supernatant should be saved in a sterile, nuclease-free microcentrifuge tube.

70.2.2 DETECTION PROCEDURES

70.2.2.1 Multiplex PCR Detection of Eight EHDV Serotypes

70.2.2.1.1 Procedure

1. Denature dsRNA by placing purified RNA at 95°C for 5 min, then immediately place sample on ice and set up PCR mix.
2. Prepare the RT mix as follows:

Component	Volume/Final Concentration	References
2× Reaction Mix with ROX	12.5 μL	
Primer Mix	0.4 μM of each primer	[56]
Probe Mix	0.08 μM of each probe	[56]
SuperScript III Reverse Transcriptase (Life Technologies)	0.5 units	
Molecular Grade Water	Make it up to 22 μL	

3. Aliquot 22 μL/0.2 mL PCR tube on ice.
4. Add 3 μL of extracted RNA.
5. Place the sample tubes in the thermocycler for 55°C for 25 min, 95°C for 2 min, and 95°C for 15 s followed by 40 cycles of 95°C for 10 s and 55°C for 1 min.[56]
6. A sample with Ct value <37 is considered positive.

70.2.2.2 Whole Genome Sequencing

70.2.2.2.1 RNA Isolation

1. BHK cells that showed 100% CPE following EHDV virus infection should be used for RNA extraction.
2. Filter 40 mL of cell culture supernatant using the 0.2 μm bottle top filters (Thermo Fisher Scientific, Lenexa, KS).
3. Centrifuge filtrate at 50,000 × g for 2 h.
4. Discard the supernatant and resuspend the pellet in 1000 μL of water.
5. Concentrate the sample to a final 100 μL volume using Amicon Ultra Centrifugal Filters (0.5 mL; 50 kDa) (Millipore, Tullagreen, Ireland).
6. Remove cellular DNA and RNA removed by incubation with DNase I (25 units) (New England Biolabs [NEB], Ipswich, MA) and RNase A (25 units) (Qiagen) at 37°C for 1 h.
7. Extract RNA using QIAamp viral RNA isolation kit (Qiagen) by standard methods and elute in 35 μL of the supplied elution buffer.

70.2.2.2.2 Genome Amplification

1. Ligate dsRNA to PC3-T7 loop (5′-p–GGATCC CGGGAAT TCGGTAATAC GAC TCACTATA TTTTTATAGTGAGTCGTATTA–OH-3′)[57] as given in the following:
2.

Reagents	Final Concentration
ATP (NEB)	1 mM
RNasin (Promega, Madison, WI)	40 U
PC3-T7 loop (200 ng/μL)	200 ng
DMSO (Sigma-Aldrich)	10%
T4 RNA Ligase buffer (NEB)	1×
T4 RNA Ligase 1 (NEB)	30 U

3. Incubate the reaction mix at 16°C overnight.
4. Purify ligated dsRNA using RNA Clean & Concentrator (Zymo Research, Irvine, CA) according to manufacturer's instructions. Elute in 10 μL of sterile H_2O.
5. Add DMSO (15% [v/v] final concentration) to the purified ligated dsRNA and incubate in a thermal cycler at 95°C for 5 min and 4°C for 5 min.
6. Reverse transcribe RNA to cDNA using GoScript Reverse Transcription System (Promega) as follows:

Reagents	Final Concentration/Volume
GoScript 5× Reaction Buffer	1×
$MgCl_2$	2 mM
PCR Nucleotide Mix	0.5 mM each dNTP
RNasin	20 units
GoScript Reverse Transcriptase	1 U
H_2O	Make it up to 20 μL

7. Incubate the reaction in a thermal cycler at 25°C for 15 min, 42°C for 60 min, and 5°C for 15 min.
8. Remove RNA using RNase H (2.5 units, NEB) by incubating in a thermal cycler at 37°C for 20 min. Anneal cDNA at 65°C for 4 h.
9. Amplify cDNA using primer PC2 (5′-p– CCGAATTCCCGGGATCC-3′) using the PCR mixture given in the following:

Reagents	Final Concentration
10× Ex Taq Buffer (Clontech, Mountain View, CA)	1×
dNTP Mix (NEB)	0.2 mM each dNTP
PC2	80 μM
Takara Ex Taq (Clontech)	2.5 U
H_2O	Make it up to 50 μL

10. Place the samples in a thermocycler and incubate at 72°C for 1 min, 94°C for 2 min followed by 40 cycles of 94°C for 30 s, 67°C for 30 s, and 72°C for 4 min and 1 cycle of 72°C for 5 min.

11. Analyze the PCR products after separation on 1.5% agarose gels (TBE) containing ethidium bromide.

12. Amplified products can be purified and sequenced using Ion Personal Genome Machine™ (PGM™) sequencing platform (Life Technologies) according to manufacturer's instruction.

70.3　CONCLUSION

EHDV was once considered a disease of minor importance and not much attention was given to the virus. However, emergence of exotic serotypes such as EHDV-6 in North America and North African countries, introduction of EHDV-7 in Israel, wide distribution of clinical forms of EHDV in cattle from the United States, and economic losses caused by several outbreaks have put this virus on the spotlight. As a consequence, several diagnostic tests have been developed to distinguish EHDV from other orbiviruses. Nonetheless, there is a considerable lack of information about the susceptibility to different EHDV serotypes, vector capacities, and epidemiology. More work needs to be done to fill these gaps and to guide the future control and prevention campaigns against this highly infectious, economically damaging virus.

REFERENCES

1. Savini, G. et al. Epizootic hemorrhagic disease. *Res Vet Sci* 91, 1–17 (2011).
2. Gaydos, J.K. et al. Epizootiology of an epizootic hemorrhagic disease outbreak in West Virginia. *J Wildl Dis* 40, 383–393 (2004).
3. Murphy, M.D., Hanson, B.A., Howerth, E.W., and Stallknecht, D.E. Molecular characterization of epizootic hemorrhagic disease virus serotype 1 associated with a 1999 epizootic in white-tailed deer in the eastern United States. *J Wildl Dis* 42, 616–624 (2006).
4. Allison, A.B. et al. Detection of a novel reassortant epizootic hemorrhagic disease virus (EHDV) in the USA containing RNA segments derived from both exotic (EHDV-6) and endemic (EHDV-2) serotypes. *J Gen Virol* 91, 430–439 (2010).
5. Dunbar, M.R., Velarde, R., Gregg, M.A., and Bray, M. Health evaluation of a pronghorn antelope population in Oregon. *J Wildl Dis* 35, 496–510 (1999).
6. Dubay, S.A., deVos, J.C., Jr., Noon, T.H., and Boe, S. Epizootiology of hemorrhagic disease in mule deer in central Arizona. *J Wildl Dis* 40, 119–124 (2004).
7. Dubay, S.A., Noon, T.H., deVos, J.C., Jr., and Ockenfels, R.A. Serologic survey for pathogens potentially affecting pronghorn (*Antilocapra americana*) fawn recruitment in Arizona, USA. *J Wildl Dis* 42, 844–848 (2006).
8. Albayrak, H., Ozan, E., and Gur, S. A serologic investigation of epizootic hemorrhagic disease virus (EHDV) in cattle and gazella subgutturosa subgutturosa in Turkey. *Trop Anim Health Prod* 42, 1589–1591 (2010).
9. Gibbs, E.P. and Lawman, M.J. Infection of British deer and farm animals with epizootic haemorrhagic disease of deer virus. *J Comp Pathol* 87, 335–343 (1977).
10. Van Campen, H. et al. Epizootic hemorrhagic disease in yaks (*Bos grunniens*). *J Vet Diagn Invest* 25, 443–446 (2013).
11. Eschbaumer, M. et al. Epizootic hemorrhagic disease virus serotype 7 in European cattle and sheep: Diagnostic considerations and effect of previous BTV exposure. *Vet Microbiol* 159, 298–306 (2012).
12. Kedmi, M. et al. No evidence for involvement of sheep in the epidemiology of cattle virulent epizootic hemorrhagic disease virus. *Vet Microbiol* 148, 408–412 (2011).
13. Nol, P. et al. Epizootic hemorrhagic disease outbreak in a captive facility housing white-tailed deer (*Odocoileus virginianus*), bison (*Bison bison*), elk (*Cervus elaphus*), cattle (*Bos taurus*), and goats (*Capra hircus*) in Colorado, U.S.A. *J Zoo Wildl Med* 41, 510–515 (2010).
14. Jessup, D.A. Epidemiology of two orbiviruses in California's native wild ruminants: Preliminary report. *Prog Clin Biol Res* 178, 53–65 (1985).
15. Stallknecht, D.E., Nettles, V.F., Rollor, E.A., 3rd, and Howerth, E.W. Epizootic hemorrhagic disease virus and bluetongue virus serotype distribution in white-tailed deer in Georgia. *J Wildl Dis* 31, 331–338 (1995).
16. Yadin, H. et al. Epizootic haemorrhagic disease virus type 7 infection in cattle in Israel. *Vet Rec* 162, 53–56 (2008).
17. Mecham, J.O. and Dean, V.C. Protein coding assignment for the genome of epizootic haemorrhagic disease virus. *J Gen Virol* 69 (Pt 6), 1255–1262 (1988).
18. Anthony, S.J. et al. Genetic and phylogenetic analysis of the non-structural proteins NS1, NS2 and NS3 of epizootic haemorrhagic disease virus (EHDV). *Virus Res* 145, 211–219 (2009).
19. Anthony, S.J. et al. Genetic and phylogenetic analysis of the core proteins VP1, VP3, VP4, VP6 and VP7 of epizootic haemorrhagic disease virus (EHDV). *Virus Res* 145, 187–199 (2009).
20. Anthony, S.J. et al. Genetic and phylogenetic analysis of the outer-coat proteins VP2 and VP5 of epizootic haemorrhagic disease virus (EHDV): Comparison of genetic and serological data to characterise the EHDV serogroup. *Virus Res* 145, 200–210 (2009).
21. Anbalagan, S., Cooper, E., Klumper, P., Simonson, R.R., and Hause, B.M. Whole genome analysis of epizootic hemorrhagic disease virus identified limited genome constellations and preferential reassortment. *J Gen Virol* 95, 434–441 (2014).
22. Campbell, C.H. and St George, T.D. A preliminary report of a comparison of epizootic haemorrhagic disease viruses from Australia with others from North America, Japan and Nigeria. *Aust Vet J* 63, 233 (1986).
23. Lee, V.H., Causey, O.R., and Moore, D.L. Bluetongue and related viruses in Ibadan, Nigeria: Isolation and preliminary identification of viruses. *Am J Vet Res* 35, 1105–1108 (1974).
24. Moore, D.L. Bluetongue and related viruses in Ibadan, Nigeria: Serologic comparison of bluetongue, epizootic hemorrhagic disease of deer, and Abadina (Palyam) viral isolates. *Am J Vet Res* 35, 1109–1113 (1974).
25. Moore, D.L. and Kemp, G.E. Bluetongue and related viruses in Ibadan, Nigeria: Serologic studies of domesticated and wild animals. *Am J Vet Res* 35, 1115–1120 (1974).
26. St George, T.D., Cybinski, D.H., Standfast, H.A., Gard, G.P., and Della-Porta, A.J. The isolation of five different viruses of the epizootic haemorrhagic disease of deer serogroup. *Aust Vet J* 60, 216–217 (1983).

27. Mohammed, M.E. et al. Application of molecular biological techniques for detection of epizootic hemorrhagic disease virus (EHDV-318) recovered from a sentinel calf in central Sudan. *Vet Microbiol* 52, 201–208 (1996).

28. Temizel, E.M. et al. Epizootic hemorrhagic disease in cattle, Western Turkey. *Emerg Infect Dis* 15, 317–319 (2009).

29. Omori, T., Inaba, Y., Morimoto, T., Tanaka, Y., and Ishitani, R. Ibaraki virus, an agent of epizootic disease of cattle resembling bluetongue. I. Epidemiologic, clinical and pathologic observations and experimental transmission to calves. *Jpn J Microbiol* 13, 139–157 (1969).

30. Uchinuno, Y. et al. Differences in Ibaraki virus RNA segment 3 sequences from three epidemics. *J Vet Med Sci* 65, 1257–1263 (2003).

31. Weir, R.P. et al. EHDV-1, a new Australian serotype of epizootic haemorrhagic disease virus isolated from sentinel cattle in the Northern Territory. *Vet Microbiol* 58, 135–143 (1997).

32. House, C., Shipman, L.D., and Weybright, G. Serological diagnosis of epizootic hemorrhagic disease in cattle in the USA with lesions suggestive of vesicular disease. *Ann N Y Acad Sci* 849, 497–500 (1998).

33. Inaba, U. Ibaraki disease and its relationship to bluetongue. *Aust Vet J* 51, 178–185 (1975).

34. Batten, C.A., Edwards, L., Bin-Tarif, A., Henstock, M.R., and Oura, C.A. Infection kinetics of Epizootic Haemorrhagic Disease virus serotype 6 in Holstein-Friesian cattle. *Vet Microbiol* 154, 23–28 (2011).

35. Aradaib, I.E., Mederos, R.A., and Osburn, B.I. Evaluation of epizootic haemorrhagic disease virus infection in sentinel calves from the San Joaquin Valley of California. *Vet Res Commun* 29, 447–451 (2005).

36. Anbalagan, S. and Hause, B.M. Characterization of epizootic hemorrhagic disease virus from a bovine with clinical disease with high nucleotide sequence identity to white-tailed deer isolates. *Arch Virol* 159, 2737–2740 (2014).

37. Smith, K.E., Stallknecht, D.E., and Nettles, V.F. Experimental infection of *Culicoides lahillei* (Diptera: Ceratopogonidae) with epizootic hemorrhagic disease virus serotype 2 (Orbivirus: Reoviridae). *J Med Entomol* 33, 117–122 (1996).

38. Mullen, G.R., Hayes, M.E., and Nusbaum, K.E. Potential vectors of bluetongue and epizootic hemorrhagic disease viruses of cattle and white-tailed deer in Alabama. *Prog Clin Biol Res* 178, 201–206 (1985).

39. Rosenstock, S.S., Ramberg, F., Collins, J.K., and Rabe, M.J. *Culicoides mohave* (Diptera: Ceratopogonidae): New occurrence records and potential role in transmission of hemorrhagic disease. *J Med Entomol* 40, 577–579 (2003).

40. Ruder, M.G. et al. Vector competence of *Culicoides sonorensis* (Diptera: Ceratopogonidae) to epizootic hemorrhagic disease virus serotype 7. *Parasit Vectors* 5, 236 (2012).

41. Aradaib, I.E., Brewer, A.W., and Osburn, B.I. Interaction of epizootic hemorrhagic disease virus with bovine erythrocytes in vitro: Electron microscope study. *Comp Immunol Microbiol Infect Dis* 20, 281–283 (1997).

42. Tsai, K. and Karstad, L. The pathogenesis of epizootic hemorrhagic disease of deer: An electron microscopic study. *Am J Pathol* 70, 379–400 (1973).

43. Howerth, E.W., Greene, C.E., and Prestwood, A.K. Experimentally induced bluetongue virus infection in white-tailed deer: Coagulation, clinical pathologic, and gross pathologic changes. *Am J Vet Res* 49, 1906–1913 (1988).

44. Howerth, E.W. and Tyler, D.E. Experimentally induced bluetongue virus infection in white-tailed deer: Ultrastructural findings. *Am J Vet Res* 49, 1914–1922 (1988).

45. Quist, C.F. et al. Host defense responses associated with experimental hemorrhagic disease in white-tailed deer. *J Wildl Dis* 33, 584–599 (1997).

46. Gaydos, J.K., Allison, A.B., Hanson, B.A., and Yellin, A.S. Oral and fecal shedding of epizootic hemorrhagic disease virus, serotype 1 from experimentally infected white-tailed deer. *J Wildl Dis* 38, 166–168 (2002).

47. Gaydos, J.K. et al. Cross-protection between epizootic hemorrhagic disease virus serotypes 1 and 2 in white-tailed deer. *J Wildl Dis* 38, 720–728 (2002).

48. Breard, E. et al. Epizootic hemorrhagic disease virus serotype 6 experimentation on adult cattle. *Res Vet Sci* 95, 794–798 (2013).

49. Aradaib, I.E., Sawyer, M.M., and Osburn, B.I. Experimental epizootic hemorrhagic disease virus infection in calves: Virologic and serologic studies. *J Vet Diagn Invest* 6, 489–492 (1994).

50. Fletch, A.L. and Karstad, L.H. Studies on the pathogenesis of experimental epizootic hemorrhagic disease of white-tailed deer. *Can J Comp Med* 35, 224–229 (1971).

51. Fox, K.A. et al. Testicular lesions and antler abnormalities in Colorado, USA Mule Deer (*Odocoileus hemionus*): A possible role for epizootic hemorrhagic disease virus. *J Wildl Dis* 51, 166–176 (2015).

52. Raabis, S.M., Byers, S.R., Han, S., and Callan, R.J. Epizootic hemorrhagic disease in a yak. *Can Vet J* 55, 369–372 (2014).

53. Drolet, B.S., Mills, K.W., Belden, E.L., and Mecham, J.O. Enzyme-linked immunosorbent assay for efficient detection of antibody to bluetongue virus in pronghorn (*Antilocapra americana*). *J Wildl Dis* 26, 34–40 (1990).

54. Ohashi, S., Yoshida, K., Watanabe, Y., and Tsuda, T. Identification and PCR-restriction fragment length polymorphism analysis of a variant of the Ibaraki virus from naturally infected cattle and aborted fetuses in Japan. *J Clin Microbiol* 37, 3800–3803 (1999).

55. McHolland, L.E. and Mecham, J.O. Characterization of cell lines developed from field populations of *Culicoides sonorensis* (Diptera: Ceratopogonidae). *J Med Entomol* 40, 348–351 (2003).

56. Wilson, W.C. et al. Detection of all eight serotypes of Epizootic hemorrhagic disease virus by real-time reverse transcription polymerase chain reaction. *J Vet Diagn Invest* 21, 220–225 (2009).

57. Potgieter, A.C. et al. Improved strategies for sequence-independent amplification and sequencing of viral double-stranded RNA genomes. *J Gen Virol* 90, 1423–1432 (2009).

58. Stallknecht, D.E. and Davidson, W.R. Antibodies to bluetongue and epizootic hemorrhagic disease viruses from white-tailed deer blood samples dried on paper strips. *J Wildl Dis* 28, 306–310 (1992).

59. Stallknecht, D.E., Kellogg, M.L., Blue, J.L., and Pearson, J.E. Antibodies to bluetongue and epizootic hemorrhagic disease viruses in a barrier island white-tailed deer population. *J Wildl Dis* 27, 668–674 (1991).

60. Stallknecht, D.E. et al. Precipitating antibodies to epizootic hemorrhagic disease and bluetongue viruses in white-tailed deer in the southeastern United States. *J Wildl Dis* 27, 238–247 (1991).

71 Equine Encephalosis Virus

Dongyou Liu

CONTENTS

71.1 Introduction 637
    71.1.1 Classification..637
    71.1.2 Morphology and Genome Organization 638
    71.1.3 Replication and Epidemiology...638
    71.1.4 Clinical Features and Pathogenesis 639
    71.1.5 Diagnosis 639
    71.1.6 Treatment and Prevention 639
71.2 Methods 639
    71.2.1 Sample Preparation..639
    71.2.2 Detection Procedures...639
        71.2.2.1 Procedure 640
71.3 Conclusion 640
References..640

71.1 INTRODUCTION

Equine encephalosis (EE) is an arthropod-borne, febrile, non-contagious disease of equines, initially referred to as equine ephemeral fever by Theiler in 1910 [1]. The causal agent—equine encephalosis virus (EEV)—was first identified in 1967 in South Africa from a thoroughbred mare displaying marked neurological signs, including listlessness, tightening of the muscles of the face, high temperature, and elevated pulse rate about 24 h before death. Transmitted by *Culicoides* biting midges, the virus was also isolated from the blood samples of other horses located nearby, which showed no overt clinical signs apart from a febrile reaction [2]. Although EEV is capable of causing <5% mortality in affected animals, it is noninfective to humans and thus poses no public health risk. The real significance of EEV infections lies in the fact that the mild forms of EE are clinically similar to those of African horse sickness (AHS), which is caused by African horse sickness virus (AHSV) (see Chapter 68). AHSV utilizes identical vectors for transmission and represents one of the most devastating equine pathogens. It is, therefore, important to differentiate EE from AHS, so that tailor-made strategies are implemented for effective control and prevention. In this chapter, we focus on EEV in relation to its classification, morphology, genome organization, epidemiology, clinical features, and diagnosis.

71.1.1 CLASSIFICATION

EEV is a double-stranded RNA (dsRNA) virus classified in the *Orbivirus* genus, *Sedoreovirinae* subfamily, *Reoviridae* family, which contains another subfamily named *Spinareovirinae*. While the subfamily *Sedoreovirinae* is subdivided into six genera (*Cardoreovirus*, *Mimoreovirus*, *Orbivirus*, *Phytoreovirus*, *Rotavirus*, and *Seadornavirus*), the subfamily *Spinareovirinae* is separated into nine genera (*Aquareovirus*, *Coltivirus*, *Cypovirus*, *Dinovernavirus*, *Fijivirus*, *Idnoreovirus*, *Mycoreovirus*, *Orthoreovirus*, and *Oryzavirus*). Currently, the genus *Orbivirus* consists of 22 recognized species and 13 unassigned viruses, involving midges (AHSV, bluetongue virus [BTV], Chuzan virus, epizootic hemorrhagic disease virus [EHDV], EEV, Eubenangee virus, Palyam virus, Wallal virus, and Warrego virus), mosquitoes (Corriparta virus, Peruvian horse sickness virus, Wongorr virus, Umatilla virus, and Yunnan virus), and ticks (Broadhaven virus, Great Island virus, Kemerovo virus, Lipovnik virus, and Tribec viruses) in their transmission. Among the known orbiviruses, BTV (the type species of the genus), AHSV, and EHDV are economically the most important. The significance of EEV lies in its similarity to AHSV in terms of morphology, cytopathology, disease symptoms, host range, and vector usage. In fact, EEV, AHSV, BTV, and EHDV are all transmitted by *Culicoides* biting midges.

Phylogenetically, EEV isolates appear to form a lineage independent from other orbiviruses, which is further distinguished into two clusters (according to the analysis of the NS3 protein) in line with the northern and southern regions of South Africa and in association with the distribution of two transmitting *Culicoides* spp. (*C. imicola* and *C. bolitinos*) [3–5]. Based on serological reactions, seven EEV serotypes (EEV 1–7) have been identified to date: EEV-1 (Bryanston), EEV-2 (Cascara), EEV-3 (Gamil), EEV-4 (Kaalplaas), EEV-5 (Kyalami), EEV-6 (Potchefstroom), and EEV-7 (E21/20), with Kyalami, Bryanston, and Cascara being the best known [6,7]. Although EEV 1–7 have been largely found in Africa, including southern Africa (Kenya, Botswana, and South Africa), eastern Africa (Gambia, Ethiopia, and Ghana), and western Africa, a recent report provided evidence on the emergence of EEV in regions of Israel [8–11]. Given the notable absence of EE in Morocco, it is possible that the Sahara desert may act as a natural barrier to its spread.

71.1.2 Morphology and Genome Organization

Similar to other members of the genus *Orbivirus* (from Latin orbi, meaning "ring" or "circle," referring to the structure at the surface of core particles), EEV possesses a nonenveloped, spherical, icosahedral virion of 73 nm in diameter, with a triple capsid structure [12]. The outer capsid is made up of structural proteins VP2 and VP5 trimers; the intermediate capsid is formed by 780 copies of structural protein VP7 in a T = 13 icosahedral symmetry; the inner capsid is formed by 120 copies of structural protein VP3 in a T = 2 icosahedral symmetry.

Underneath the icosahedral shell, the EEV genome (of 19 kb in total) is located, consisting of 10 dsRNA segments that are packaged along with minor enzymatic proteins VP1, VP4, and VP6. These 10 genomic segments are of 800–3900 bp in length and encode at least 7 structural (VP1–VP7 of 36–120 kDa in molecular weight) and 4 nonstructural (NS1–NS3) proteins [13]. Specifically, segment 1 codes for VP1 Pol (the viral polymerase), segment 2 for VP2 (the outer capsid protein containing serotype-specific antigenic epitopes), segment 3 for VP3 (the inner capsid protein containing serogroup-specific antigens), segment 4 for VP4, segment 5 for NS1, segment 6 for VP5, segment 7 for VP7 (the intermediate capsid protein containing serogroup-specific antigens), segment 8 for NS2, segment 9 for VP6 (the viral helicase), and segment 10 for NS3 and NS3A (via alternative initiation) [14].

Like AHSV and BTV, EEV genes have conserved termini (5′ GUU and UAC 3′). The structural proteins VP2 and VP5 in the outer capsid layer mediate cell attachment and penetration during the initiation of the infection. The VP2 protein is more variable than the core proteins and the most variable among the nonstructural proteins, and the specificity of their reactions with neutralizing antibodies determines the virus serotype. The structural proteins VP3 and VP7 are coexpressed and subsequently assembled into core-like particles with typical orbivirus capsomeres [15]. Phosphorylated by cellular kinases, the NS2 protein is an important matrix component of the granular viral inclusion bodies that form within the cytoplasm of infected cells. The membrane glycoproteins NS3 and NS3a are involved in the release of progeny virus particles from infected cells and also in determination of both vector competence and virulence [5].

71.1.3 Replication and Epidemiology

Replication of orbiviruses in host cells involves a number of steps: (1) virus attachment to host receptors and entry into host cells via clathrin-mediated endocytosis; (2) partial uncoating of virus particle and removal of the outer capsid (preventing cellular activation of antiviral state) in endolysosomes, and then penetration into the cytoplasm; (3) synthesis of capped mRNA from each dsRNA segment by viral polymerase VP1 (reducing potential exposure of dsRNA to the cytoplasm); (4) translocation of the capped mRNA to the cytoplasm; (5) translation of viral proteins and accumulation of genomic RNAs in cytoplasmic viral factories; (6) encapsidation of translated viral proteins and genomic RNAs in the subviral particle, transcription of RNA (+) into RNA (−), and base pairing of RNA (+) and RNA (−) into dsRNA genomes; (7) assembly of capsid on the subviral particle; and (8) release of mature virions after breakdown of host plasma membrane in the event of cell death.

EEV is infective to all species of equids (horse, donkey, and zebra), with zebra and elephants acting as maintenance host [16–18]. Being an arthropod-borne virus, its transmission is facilitated by *Culicoides* biting midges, particularly *C. imicola* (sensu stricto) and *C. bolitinos* during a blood meal. *C. imicola* and *C. bolitinos* are widespread in sub-Saharan Africa, and also present in regions of Northern America, Southern Europe, and South Asia. Regarded as the main vector of EEV, *C. imicola* is closely associated with livestock in the summer rainfall region of southern Africa [3,19]. Other *Culicoides* species such as *C. (Meijerehelea) leucostictus*, *C. magnus*, and *C. (Hoffmania) zuluensis* appear to be susceptible to EEV1 and/or EEV2 infections under experimental conditions. Given the presence of *C. imicola* and *C. bolitinos* as well as other *Culicoides* species in various geographic locations, EEV has the potential to spread to regions of the world besides sub-Saharan Africa and Israel.

EE is endemic in southern, eastern, and western Africa (including Kenya, Botswana, South Africa, Gambia, Ethiopia, and Ghana) [3,7,10]. In South Africa, >75% of horses and 85% of donkeys have been shown to be seropositive for EEV, with serotypes 1 and 6 being the most frequently and extensively identified, while serotypes 2, 3, 5, and 7 being identified as sporadic and localized infections affecting individual horses [20,21]. The prevalence of neutralizing antibodies in thoroughbred horses against one or more serotypes of the EEV was 56.9% [7]. Antibodies to EEV have been also demonstrated in zebra and African elephant [17,22]. The relatively high prevalence of EEV in southern parts of Africa may be attributed to immunization of horses against AHSV or related viruses but not EEV, for which no vaccine is available. In addition, serological evidence suggests that EEV was circulating in Israel as early as 2001, with incidence ranging from 20% to 100% during 2001–2008 [8,9].

Outbreaks of EE in horses often take place in late summer and autumn as these seasons are favorable for the replication of transmitting vectors. Anti-EEV antibodies may be occasionally detected in elephants [17]. Wild animals such as zebra and, to a lesser extent, elephants appear to help in sustenance of the virus in the equine population in the African continent. Therefore, eventual eradication of EE from Africa is reliant on prompt identification and effective control of EEV in zebra and elephants [16].

71.1.4 Clinical Features and Pathogenesis

In the majority of cases, EEV infection in equids is a mild febrile disease without overt clinical signs. However, in some severe cases, EE may produce clinical syndromes that are similar to AHS. These include fever (up to 41°C, typically lasting for 2–5 days), lassitude, unrest, anorexia, cough, edema (of the neck, legs, lips, and eyelids; leading to stiff facial expression, stiffness of the mouth and commissures; in fact, the swelling of the lips and eyelids represents a characteristic of the disease), blood-tinged nasal discharge, petechial hemorrhages of the conjunctivae, abnormally colored mucous membranes (ranging from yellow to a brick red color), accelerated pulse and breathing rates, muscle pain, hindquarter ataxia, convulsions, hyperexcitability, depression, enteritis, encephalitis, and occasional death (<5%) [2,8,23]. In addition, EEV infection may be associated with equine fetus abortion during the first 5–6 months of gestation. Incubation period is short but variable between 3 and 6 days, and clinical signs may appear just 24 h before the death of the animal. Most horses recover from the disease without complications.

Postmortem examinations of the affected horses reveal general venous congestion, fatty liver degeneration, brain edema, localized catarrhal enteritis (particularly in the distal half of the small intestine), degeneration of cardiac myofibers and myocardial fibrosis [20].

The predilection site of EEV attacks is the lining of the blood vessels, and the fluid leaks from damaged blood vessels into the interstitial spaces are responsible for edema and other associated clinical signs. Despite the inclusion of encephalosis in its disease name, it is not primarily a neurological disease.

71.1.5 Diagnosis

As EE presents many nonspecific clinical signs that are indistinguishable from those of AHS and equine viral arteritis, laboratory tests are required for accurate diagnosis.

Traditionally, EEV identification involves virus isolation using baby hamster kidney cells, suckling mice brains, or embryonated hen's eggs, and subsequent serotyping using plaque inhibition neutralization assay. In addition, for group-specific antibody detection, complement fixation, agar gel immunodiffusion, indirect immunofluorescent antibody, and enzyme-linked immunosorbent assay (ELISA) are available [22]. For group-specific detection of EEV antigen, indirect sandwich ELISA may be utilized [24,25].

However, the previous methods suffer from the drawbacks of being time consuming and difficult to interpret, having insufficient sensitivity, and providing only a retrospective diagnosis. Therefore, alternative methods targeting nucleic acids have been developed for improved diagnosis of EEV infection [7]. For example, gene probes derived from the EEV genome segment encoding nonstructural protein NS1 enabled rapid detection for minute quantity of virus-specified RNA in infected cells [26]. EEV was diagnosed using a novel DNA array with subsequent reverse transcription polymerase chain reaction (RT-PCR) and sequence analysis [8]. A novel real-time PCR assay targeting the gene VP7 was reported for specific and sensitive diagnosis of EEV infection [27].

71.1.6 Treatment and Prevention

Currently, no specific treatment is available for EE. Supportive treatment includes administration of anti-inflammatories, vitamins, and appetite stimulants. Horses that are unresponsive to supportive treatment may have to be hospitalized. As 90% of infected horses recover without complications, antibiotics are only provided to prevent secondary infections.

Due to the absence of an effective vaccine, vector control remains the most important way of minimizing exposure of horses to EEV. This may include the following: (1) keeping horses in stable at night (putting in 2 h before sunset and letting out once the dew has dried on the ground; as the period of dawn and dusk is the time when vector is more active); (2) avoiding paddocks with long grass or standing water; (3) turning stable lights off at night; (4) keeping window screen intact and being vigilant with fly repellants and day sheets; (5) using fans in stables (as the midges are weak fliers); and (6) restricting horse movement between stables and farms.

EEV is not a zoonotic pathogen and thus poses no public health risk to human populations.

71.2 METHODS

71.2.1 Sample Preparation

For virus isolation, 1 mL of blood is inoculated onto 1-day-old monolayers of Vero cells (ATCC) at 90% confluency in a 25 cm² flask. The flask is incubated for 1 h at 37°C in 5% CO_2 to allow the virus to adsorb, and then the inoculums are washed off and overlaid with 10 mL of minimum essential medium (MEM) supplemented with 5% fetal bovine serum (Life Technologies). The cultures are examined daily for up to 5 days incubation, before being harvested by freeze/thawing and passaging onto fresh cell cultures [27].

Viral dsRNA is extracted from EEV cell culture isolates. The contents of a flask are agitated and 500 μL is transferred to a 1.5 mL Eppendorf tube. Samples are centrifuged at $11,000\times g$ for 5 min. The supernatant is discarded and the cell pellet is mixed with 50 μL of phosphate-buffered saline. Total nucleic acid extraction from the cell pellet is performed using a MagMax™-96 Total RNA Isolation Kit (Life Technologies) together with a MagMax™ Express Particle Processor (Life Technologies). The extracted RNA is eluted in a 50 μL elution buffer and stored at −20°C until use. Similarly, EEV nucleic acids may be extracted from a 50 μL of heparinized blood using a MagMax-96 Total RNA Isolation Kit (Life Technologies) [27].

71.2.2 Detection Procedures

Rathogwa et al. [27] described a real-time RT-PCR assay using primers from the 5′ end of the S7 (VP7) gene and a TaqMan® minor groove binder (MGB™) hydrolysis probe for the sensitive and specific detection of EEV in samples from horses infected with EEV (Table 71.1).

TABLE 71.1

Primers and Probe from S7 (VP7) Gene for Real-Time RT-PCR Detection of EEV

Identity	Type	Sequence (5′–3′)	Position
EEV_VP7_F0028_0048	Forward primer	GAT AGC GGC TAG AGC CCT TTC	28–48
EEV_VP7_P0054_0072	MGB probe	TAA GAG CAT GTG TTA CTG C	54–72
EEV_VP7_R0085_0106	Reverse primer	AAC TTG AGG AGC CAT R*GT AGC T	85–106

71.2.2.1 Procedure

1. Add 5 μL RNA to a 1 μL 25× primer/probe mixture (400 nM/120 nM final concentrations, with the probe containing NED™ label) (Life Technologies) and a 4 μL nuclease-free water. The probe was labeled with NED (Life Technologies).
2. Denature the mixture by heating at 95°C for 1 min on a heat block, and then cool quickly on ice.
3. Add 12.5 μL 2× RT-PCR buffer, 1.5 μL nuclease-free water, and 1 μL enzyme (VetMAX™-Plus One-Step RT-PCR Kit, Life Technologies) reagents.
4. Conduct the assay on a StepOnePlus™ Real-Time PCR System (Life Technologies) according to the manufacturer's recommendations.
5. Classify samples as positive if the normalized fluorescence for the EEV assay exceeds the 0.1 threshold.

Note: The EEV real-time RT-PCR assay recognizes all 7 serotypes of EEV and is nonreactive with 24 serotypes of BTV and 9 serotypes of AHSV tested. The assay has a sensitivity of $10^{2.9}$ $TCID_{50}$/mL blood (95% confidence interval: $10^{2.7}$–$10^{3.3}$), corresponding to a C_T of 38.42.

71.3 CONCLUSION

As a dsRNA virus in the genus *Orbivirus*, family *Reoviridae*, equine encephalosis virus (EEV) is shown to contain seven antigenically distinct serotypes (EEV1–EEV7). Similar to African horse sickness virus (AHSV) and bluetongue virus (BTV), EEV is transmitted by *Culicoides* midges (particularly *C. imicola* and *C. bolitinos*). EEV is infective to all equine species, with zebra and elephants acting as maintenance hosts. Endemic in southern, eastern, and western Africa, EE has also been detected in Israel.

Clinically, EE often presents as a mild febrile disease, with abortion and death occurring in more severe cases. Other less frequent clinical symptoms include swelling of the eyelids, the supraorbital fossa, and the entire face; ataxia and stiffness; convulsions; respiratory distress and blood-tinged nasal discharge; petechiae in the conjunctiva; and acute heart failure. As these signs are indistinguishable to those caused by AHSV, which is a high impact disease of horses with lethal consequence and requires different control measures, it is essential to discriminate EE from AHS.

Laboratory confirmation of EEV infection has traditionally relied on various serological assays such as complement fixation test, serum neutralization test, and ELISA. Due to

their obvious shortcomings, these phenotypic procedures have been increasingly superseded by new generation molecular techniques, which are much faster, more sensitive, and more specific for identification and typing of EEV. It can be envisaged that molecular methods such as real-time RT-PCR will play a vital role in the tracking and epidemiological investigation of future outbreaks of EE, contributing to effective control and prevention against this disease.

REFERENCES

1. Theiler A. Notes on a fever in horses simulating horsesickness. *Transvaal Agric J.* 1910;8:581–586.
2. Erasmus BJ et al. The isolation and characterization of equine encephalosis virus. *Bull Off Int Epiz.* 1970;74:781–789.
3. Venter GJ, Groenewald DM, Paweska JT, Venter EH, Howell PG. Vector competence of selected South African *Culicoides* species for the Bryanston serotype of equine encephalosis virus. *Med Vet Entomol.* 1999;13(4):393–400.
4. Venter GJ, Groenewald D, Venter E, Hermanides KG, Howell PG. A comparison of the vector competence of the biting midges, *Culicoides* (*Avaritia*) *bolitinos* and *C.* (*A.*) *imicola*, for the Bryanston serotype of equine encephalosis virus. *Med Vet Entomol.* 2002;16(4):372–377.
5. van Niekerk M et al. Variation in the NS3 gene and protein in South African isolates of bluetongue and equine encephalosis viruses. *J Gen Virol.* 2003;84(3):581–590.
6. Gerdes GH, Pieterse LM. The isolation and identification of Potchefstroom virus: A new member of the equine encephalosis group of orbiviruses. *J S Afr Vet Assoc.* 1993;64(3):131–132.
7. Howell PG et al. The classification of seven serotypes of equine encephalosis virus and the prevalence of homologous antibody in horses in South Africa. *Onderstepoort J Vet Res.* 2002;69(1):79–93.
8. Mildenberg Z et al. Equine encephalosis virus in Israel. *Transbound Emerg Dis.* 2009;56(8):291.
9. Aharonson-Raz K et al. Isolation and phylogenetic grouping of equine encephalosis virus in Israel. *Emerg Infect Dis.* 2011;17(10):1883–1886.
10. Oura CA et al. Equine encephalosis virus: Evidence for circulation beyond southern Africa. *Epidemiol Infect.* 2012;140(11):1982–1986.
11. Westcott D et al. Evidence for the circulation of equine encephalosis virus in Israel since 2001. *PLoS One.* 2013;8(8):e70532. eCollection 2013. Erratum in: *PLoS One.* 2013;8(9).
12. Lecatsas G, Erasmus BJ, Els HJ. Electron microscopic studies on equine encephalosis virus. *Onderstepoort J Vet Res.* 1973;40(2):53–57.
13. Viljoen GJ, Huismans H. The characterization of equine encephalosis virus and the development of genomic probes. *J Gen Virol.* 1989;70(8):2007–2015.

14. Dhama K et al. Equine encephalosis virus (EEV): A review. *Asian J Anim Vet Adv.* 2014;9:123–123.

15. Potgieter AC, Steele AD, van Dijk AA. Cloning of complete genome sets of six dsRNA viruses using an improved cloning method for large dsRNA genes. *J Gen Virol.* 2002;83(9):2215–2223.

16. Barnard BJ, Paweska JT. Prevalence of antibodies against some equine viruses in zebra (*Zebra burchelli*) in the Kruger National Park, 1991–1992. *Onderstepoort J Vet Res.* 1993;60:175–179.

17. Barnard BJH. Antibodies against some viruses of domestic animals in southern African wild animals. *Onderstepoort J Vet Res.* 1997;64:95–110.

18. Lord CC, Venter GJ, Mellor PS, Paweska JT, Woolhouse ME. Transmission patterns of African horse sickness and equine encephalosis viruses in South African donkeys. *Epidemiol Infect.* 2002;128(2):265–267.

19. Maclachlan NJ, Guthrie AJ. Re-emergence of bluetongue, African horse sickness, and other orbivirus diseases. *Vet Res.* 2010;41(6):35.

20. Paweska JT, Venter GJ. Vector competence of Culicoides species and the seroprevalence of homologous neutralizing antibody in horses for six serotypes of equine encephalosis virus (EEV) in South Africa. *Med Vet Entomol.* 2004;18(4):398–407.

21. Howell PG, Nurton JP, Nel D, Lourens CW, Guthrie AJ. Prevalence of serotype specific antibody to equine encephalosis virus in Thoroughbred yearlings in South Africa (1999–2004). *Onderstepoort J Vet Res.* 2008;75(2):153–161.

22. Williams R, Du Plessis DH, Van Wyngaardt W. Group-reactive ELISAs for detecting antibodies to African horse sickness and equine encephalosis viruses in horse, donkey, and zebra sera. *J Vet Diagn Invest.* 1993;5(1):3–7.

23. Chaimovitz M. *Equine Encephalosis* Israel. 2009. Online http://www.oie.int/wahis/public.php?page=event_summary&reportid=7957. Accessed on June 12, 2015.

24. Crafford JE et al. A group-specific, indirect sandwich ELISA for the detection of equine encephalosis virus antigen. *J Virol Methods.* 2003;112(1–2):129–135.

25. Crafford JE et al. A competitive ELISA for the detection of group-specific antibody to equine encephalosis virus. *J Virol Methods.* 2011;174(1–2):60–64.

26. Venter EH, Viljoen GJ, Nel LH, Huismans H, van Dijk AA. A comparison of different genomic probes in the detection of virus-specified RNA in Orbivirus-infected cells. *J Gen Virol.* 1991;32(2–3):171–180.

27. Rathogwa NM et al. Development of a real-time polymerase chain reaction assay for equine encephalosis virus. *J Virol Methods.* 2014;195:205–210.

72 Rotavirus

Helen O'Shea, P.J. Collins, Lynda Gunn, Barbara A. Blacklaws, John McKillen, and Miren Iturriza Gómara

CONTENTS

72.1 Introduction 643
    72.1.1 Virus Structure and Classification..643
    72.1.2 Virus Replication...644
    72.1.3 Pathogenesis and Clinical Features 645
    72.1.4 Immune Response 645
    72.1.5 Epidemiology...646
        72.1.5.1 Humans 646
        72.1.5.2 Bovines 646
        72.1.5.3 Porcines 646
        72.1.5.4 Equines 646
        72.1.5.5 Other Animal Species..646
        72.1.5.6 Zoonotic Potential of Rotavirus A 647
    72.1.6 Prevention and Treatment..647
        72.1.6.1 Human Vaccines 647
        72.1.6.2 Animal Vaccines..647
    72.1.7 Diagnosis 648
72.2 Methods 650
    72.2.1 RNA Extraction 650
    72.2.2 Detection Procedures...650
72.3 Conclusion and Future Perspectives 651
References..651

72.1 INTRODUCTION

First identified in 1973 by electron microscopy in epithelial cells of duodenal mucosal samples from children, rotaviruses (RVs) were soon recognized as a major cause of gastroenteritis in the young of many animal species, including humans.[1–3] In humans, RV-associated disease typically occurs in children less than 5 years of age, but RV can infect all ages, including adults[4,5]; in 2008, an estimated 453,000 deaths worldwide were attributed to RV infection, with most deaths occurring in resource-poor countries.[6]

72.1.1 VIRUS STRUCTURE AND CLASSIFICATION

RVs are medium-sized, icosahedral, triple-layered particles (TLPs, ~100 nm in diameter), with a segmented, double-stranded RNA (dsRNA) genome (Figure 72.1). The 11 segments encode six virion proteins (VP1–4, 6–7) and six nonstructural proteins (NSP1–6) (Table 72.1). The genome is enclosed in a triple-layered particle/capsid (consisting of VP2, VP6, and VP7, representing the inner, intermediate, and outer layers, respectively [Figure 72.1]), with the VP4 spikes on the surface giving the virion its distinct "wheel" shape when viewed using the electron microscope ("rota" is Latin for wheel). Proteins VP1 and VP3 form an inner complex, responsible for genome replication and capping of viral mRNAs, respectively. The NSPs are responsible for sequestering host cell components necessary for viral replication (NSP3), evasion of the innate immune response (NSP1), viroplasm formation (NSP2 and NSP5), and formation of the mature virion (NSP4).[7–12]

Taxonomically, RVs are classified in the genus *Rotavirus*, subfamily *Sedoreovirinae*, family *Reoviridae*, which includes an additional subfamily *Spinareovirinae*. While six genera (i.e., *Cardoreovirus*, *Mimoreovirus*, *Orbivirus*, *Phytoreovirus*, *Rotavirus*, and *Seadornavirus*) are recognized in the subfamily *Sedoreovirinae*, nine genera (i.e., *Aquareovirus*, *Coltivirus*, *Cypovirus*, *Dinovernavirus*, *Fijivirus*, *Idnoreovirus*, *Mycoreovirus*, *Orthoreovirus*, and *Oryzavirus*) are known in the subfamily *Spinareovirinae*.

The genus *Rotavirus* is further separated into eight antigenically distinct groups (A–H), using immunological and phylogenetic analysis of the VP6 gene.[3,13,14] RV groups A, B, C, and H have been identified in mammals, including humans, while RV D–G to date have only been identified in nonhuman mammal and avian species. The focus of this chapter is on group A rotaviruses (RVAs). Within the RVA group, the viruses are

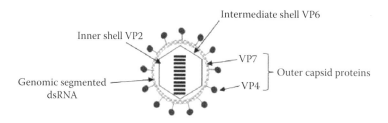

FIGURE 72.1 Rotavirus structure outline.

TABLE 72.1
Rotavirus Genes and Products

(A) Gene Segment (in Order of Size on Gel, Largest to Smallest)	Product	Product Name
1	VP1	**R**NA polymerase (**R**)
2	VP2	**C**ore shell (**C**)
3	VP3	RNA capping **M**ethyl-transferase (**M**)
4	VP4	**P**rotease-activated spike (**P**)
5	NSP1	Interferon **A**ntagonist (**A**)
6	VP6	**I**ntermediate capsid shell (**I**)
7	NSP3	**T**ranslation regulation (**T**)
8	NSP2	Octameric **N**TPase (**N**)
9	VP7	Outer shell **G**lycoprotein (**G**)
10	NSP4	**E**nterotoxin (**E**)
11	NSP5/6	p**H**osphoprotein (**H**)

(B) Order of viral proteins used to format genetic constellations

Protein name	VP7	VP4	VP6	VP1	VP2	VP3	NSP1	NSP2	NSP3	NSP4	NSP5/6
Abbreviated name	G	P	I	R	C	M	A	N	T	E	H
Number identified to date	27	37	16	9	9	8	16	9	12	14	**11**

classified antigenically and genetically, based on the main antigenic determinants, the outer capsid proteins VP7 and VP4, which specify the G and P serotypes/genotypes, respectively. RVAs are a genetically diverse group of viruses with variations in circulating genotypes being observed temporally and geographically.[15] Additionally, similar to influenza virus, RV has a segmented genome (Figure 72.1) and so is capable of genetic reassortment, leading to genetic shift, which can facilitate the emergence of new strains.

A whole genome classification system has been adopted recently for classification of all 11 segments of the RV genome, applying nucleotide homology cutoff values to distinguish genotypes.[16]

72.1.2 VIRUS REPLICATION

RV infects mature enterocytes on the villi of the small intestine, with the outer surface proteins (VP4 and VP7) facilitating virion attachment and entry into host cells. Enzymatic cleavage of the VP4 spikes into VP5* and VP8* is a prerequisite for infection. Different strains of RV utilize different receptor molecules that include sialylated and nonsialylated glycans via interactions with VP8*, and such

interactions can impact on cell tropism and host susceptibility to infection and disease.[17] Different entry mechanisms have also been proposed, such as fusion or Ca^{2+}-dependent endocytosis, with the latter being generally accepted as the more likely model.[18] Once internalized, the outer surface proteins are digested within endosomes, and the release of the double-layered particle (DLP) into the cell cytoplasm[19] is mediated by conformational changes of the VP8* subunit of the viral spike (VP4). The DLP is the transcriptionally active form of the virus; each of the 11 gene segments is packaged with a VP1 (RNA-dependent RNA polymerase) and VP3 (RNA-capping enzyme) complex, which transcribes negative-sense RNA from the double-stranded template, releasing capped mRNAs into the cell cytoplasm.[20,21] Each mRNA acts as a template for translation but also functions as a template for the production of genomic dsRNA. The translation of NSPs 2 and 5 activates viroplasm formation (non-membrane bound perinuclear inclusion bodies surrounded by progeny virions), which is essential for replication and assembly of virus progeny.[22,23] The progeny virions are assembled into the DLP within the viroplasm; here, interactions between dsRNA segments and proteins (VP1, VP3, NSP2, and NSP5) facilitate DLP formation.

Two models have been proposed for assembly and packaging: a concerted model and a core-filling model.[24,25] Newly assembled DLPs then bud into the host endoplasmic reticulum (ER) (facilitated by NSP4, which acts as an intracellular receptor for VP6) and acquire a transient membrane studded with NSP4, VP4, and VP7 proteins.[3,7,8] NSP4 is a multifunctional protein as, in addition to mediating DLP translocation into the ER, it is also associated with the formation of tubelike projections from the cell cytoplasm, which can mediate infection of adjacent cells.[26,27] Within the ER, TLPs are assembled through membrane displacement,[3,28,29] which are then released into the intestinal lumen.

72.1.3 PATHOGENESIS AND CLINICAL FEATURES

Diarrhea is believed to be the result of several factors. First, there is malabsorption of fluids and electrolytes from the intestine due to damaged enterocytes as a result of viral replication. Second, loss of mature enterocytes through viral infection causes the loss of digestive enzymes, especially disaccharidases. This causes lack of digestion of carbohydrates in the lumen and so increases the osmotic pressure of the lumen contents, attracting water into the lumen. Third is the action of NSP4, the only known viral enterotoxin, which stimulates Ca^{2+} secretion, thus altering the osmotic pressure of the cell and causing lysis.[30,31] In animal models, there is onset of diarrhea prior to the detection of histopathological changes, possibly due to the action of the NSP4.[32,33] Clinical signs are further exacerbated by stimulation of the host's enteric nervous system by NSP4.[34] Infection of the enterochromaffin cells within the digestive tract leads to the stimulation of secretion of serotonin and of brain stem structures that control vomiting.[35,36] RVA may spread extraintestinally[37,38]; however, the role of this spread in pathology is unclear.[39,40]

The first infection with RV in infants is characterized by watery dehydrating diarrhea and vomiting, often accompanied by abdominal cramps and low-grade fever, lasting 6–10 days. Subsequent reinfections are associated with mild or subclinical presentations[41,42]; however, such reinfections are important in boosting immunity and maintaining long-term protection from RV disease.

72.1.4 IMMUNE RESPONSE

Although many of the immune responses, both innate and acquired, induced after RV infection or vaccination in relevant species are known, the specific mechanisms mediating protection are less clear.[43–45] This lack of clarity is compounded by the possibility that important mechanisms vary between species.

RV infection stimulates both innate and acquired immune responses. At present, the extent to which the disease is modified or RV is cleared by the action of innate immune responses is not fully clear. Interferon (IFN) production is a common feature after RV infection in mice, pigs, and humans. Mice deficient in IFN signaling ($Stat1^{-/-}$) demonstrate increased

levels of virus replication[46]; however, direct evidence for a protective role for type I and II IFN in adult mice has been more difficult to show,[47–49] although there is evidence for protective effects in young animals.[50,51] It is now thought that type III IFN (IFN-λ) may be the important protective IFN in vivo.[48] Evidence for the importance of the antiviral pathways induced by IFN also comes from how RV actively attempts to evade these responses. RV NSP1 interferes with IFN and NF-κB signaling,[52,53] although the ability of NSP1 to function in this manner can vary with viral strain and host species.[54] VP3 also functions in evasion of the IFN-induced antiviral pathways by interfering with RNase L activation by degrading 2′–5′ adenylate oligomers.[55] In contrast, NSP4 has been associated with the stimulation of proinflammatory cytokines,[56,57] although in the context of infection this signaling may also be inhibited.[58]

Acquired immune responses include both B-cell and T-cell activation. Studies in mice suggest that cellular immunity is important for clearing RV after primary infection, while RV-primed memory B cells are responsible for long-term protection.[59]

The involvement of CD8+ lymphocytes in limiting primary infection by RV has been shown in mice[60] and calves.[61] In humans, the role of CD8+ T cells is less clear, although RV-specific T cells including IFN-γ-secreting CD4+ and CD8+ cells may be detected in the blood of most adults after infection,[62–64] and RV-specific CD4+ T lymphocytes show gut-homing potential.[65]

Humoral responses are primarily directed against VP2 and VP6 in humans,[66] but VP7 and VP4 contain the major neutralizing epitopes of the virus, and many of these antibodies are protective in vivo, as demonstrated by passive transfer in mice and gnotobiotic piglets.[67–69] However, protection from RV infection does not correlate with the concentration or breadth of VP4-specific and VP7-specific neutralizing antibodies in serum. VP6 may contain a protective epitope as VP6-specific antibodies can neutralize RV infection intracellularly in vitro by blocking DLP transcription.[70–73] In vivo, it is thought that mucosal anti-VP6 IgA may inhibit virus replication intracellularly in enterocytes that are transcytosing the antibody.[72,73] However, IgA-deficient mice and humans can still clear RV infection, possibly via compensatory IgM and IgG responses.[74,75]

Similarly in piglets, although non-neutralizing antibodies against VP4 and VP7 are induced after infection, the levels of serum and intestinal nonneutralizing IgA correlate better with protection from reinfection, and these antibodies show immunodominant recognition of VP6.[76–80]

It has been suggested that local mucosal production of antibody (IgA) is important in protection, which correlates with the majority of circulating RV-specific B cells expressing the gut-homing integrin, $\alpha_4\beta_7$, and being IgM or IgA positive.[81,82] In humans, the presence of high serum and fecal total antibody and, more specifically, IgA titers has been associated with protection from reinfection and RV disease,[83–87] and a recent systematic review suggests that a critical level of serum IgA still correlates with protection.[88] However, data from natural infection and vaccination studies in developing countries suggest that IgA seroconversion is not a good correlate of protection in settings with high level of infection.[44,89]

In cattle, RV infection induces virus-specific antibody in feces, nasal secretions (IgM and IgA, probably produced locally), and serum (IgM then IgA, then IgG1 predominates later in the response).[90,91] Anti-RV IgG1 in colostrum fed to calves is associated with reduced incidence of infection and diarrhea when these neonates are infected,[92–94] and it is this information that has led to current vaccine strategies. Similar to other species, VP7 and VP4 neutralizing antibodies do not absolutely correlate with protection, and VP6 vaccines are protective.[95]

72.1.5 Epidemiology

The predominant mode of transmission of RV is fecal–oral. In humans, RVs are ubiquitous, with 95% of children 3–5 years of age being infected worldwide.[96] Prior to the introduction of RV vaccines, RVAs could be detected in up to 50%–60% of all childhood hospitalizations due to acute gastroenteritis each year and were estimated to cause 138 million cases of gastroenteritis annually[97] and 527,000 deaths in children <5 years of age living in developing countries.[6] In other animals, RV disease also occurs in the young of the species and in farm animals, leading to significant economic losses.

72.1.5.1 Humans

Although RVA infection usually occurs in children <5 years old, it can occur in all age groups, with the elderly and immunocompromised being particularly vulnerable.[4,5,98–100] Although, in total, 12 G and 15 P genotypes have been reported to date, the most prevalent human RVAs include G1, G2, G3, G4, and G9, in combination with P types P[8], P[4], or P[6]; strains with these types account for over 80% of the RV-associated gastroenteritis cases worldwide.[15,99,101] The most common genotype combinations are G1P[8], G2P[4], G3P[8], G4P[8], and G9P[8], accounting for 74.7% of all strains detected globally,[15] although in developing countries animallike RV genotypes can also be detected regularly.[101–105] The virulence of different genotypes has been investigated, with some studies indicating that G9 may have been associated with more severe disease in some instances, but these findings have not been observed consistently.[106–109]

72.1.5.2 Bovines

Diarrhea in young calves remains an important problem in cattle herds worldwide, with bovine RV being one of the most common pathogens associated as a causative agent.[110] In the United States, RV was found to be responsible for 46% of scours in dairy herds.[111] As a consequence of RVA infection in cattle, substantial economic losses occur due to increased morbidity and mortality rates, treatment costs, and reduced growth rates of these animals. Most at risk are very young calves between 4 days and 3 weeks old.

Seven G types have been detected in cattle (G1, G3, G6–G8, G10, and G15), with G6, G8, and G10 being the common types reported.[112] Six P types have been detected in bovines (P[1], P[5], P[11], P[14], P[17], and P[21]), with P[1], P[5], and P[11] being the common types reported.[113–115]

72.1.5.3 Porcines

Porcine RVs are one of the most frequently detected viral agents associated with diarrhea, affecting piglets between 1 and 8 weeks of age.[116] RVA infection of pigs has been recognized in both enzootic and epizootic forms of diarrhea, resulting in economic losses in commercial piggeries.[116]

Antigenic and molecular characterization of porcine RVAs has revealed a broad heterogeneity in the VP7 and VP4 genes. The most common G types identified in pigs are G3, G4, G5, and G11, associated with common P types, P[6] and P[7].[117,118] Porcine RVA strains with additional P types, P[13], P[19], P[23], P[26], and P[27], have also been described but more sporadically. Furthermore, porcine strains displaying human-like G types (G1, G2, G9, and G12) and P types (M37-like P[6], P[8]) and bovine-like G types (G6, G8, and G10) and P types (P[1], P[5], and P[11]) have also been described.[117–126]

72.1.5.4 Equines

Control of equine RV is a major concern each foaling season in horse-breeding farms/centers throughout the world.[127–129] It causes diarrheal disease in foals, leading to grave economic loss due to morbidity and mortality attributed to a high prevalence of infection estimated at between 50% and 90% in outbreaks.[129–132] Symptoms of equine RVA infections include diarrhea, dehydration, anorexia, abdominal pain, and depression. The disease is usually self-limiting, but dehydration alone can be fatal in young foals.[133–135] Two genotypes, G3P[12] and G14P[12], are commonly detected, with other genotypes being detected sporadically in individual cases. Infection with genotype G3P[12] is the most common, contributing 48%–72% of RVA cases.[129] Although G3 RVA can be detected in many other animal species, equine G3 RVA appears to be evolutionarily distinct and so is not subclassified into commonly used lineages. Instead, antigenic subtypes have been defined based on cross-neutralization assays, G3A and G3B.[136] Alternatively, G14 RVA is unique to equine RVA and contributes between 10% and 38% of cases.[129] Additionally, other RVA genotypes have been detected in horses, usually in sporadic individual cases: RVA/Horse-tc/GBR/L338/1991/ G13P[18],[137,138] RVA/horse-wt/ARG/E3198/2008/G3P[3],[139] porcine-like RVA strain RVA/Horse-wt/GBR/H-1/1975/ G5P[7],[140,141] or bovine-like RVA strains with the G10P[1] or G8P[1] genotypes.[142,143]

72.1.5.5 Other Animal Species

RVA infection appears to be ubiquitous in nature; infection has been characterized in companion, livestock, and wild (exotic) animals including ungulates, nonhuman primates, bats, and birds. Clinical signs are generally similar to those described previously: diarrhea, depression, dehydration, anorexia, and in some instances stunted growth. RVA infection in certain animal species appears to be associated with particular genotypes, suggesting a host-species barrier to infection.[103]

72.1.5.6 Zoonotic Potential of Rotavirus A

The accumulation of point mutations (genetic drift) and reassortment (genetic shift) is responsible for the huge genetic heterogeneity of RVs. These mechanisms, in combination with the potential for interspecies (zoonotic and anthroponotic) transmission, can also lead to the emergence of "novel" strains in a given species with a potential for epidemic or epizootic spread.[16,103] The zoonotic potential of RVA has been documented on several occasions.[103] Uncommon human RVA G genotypes include G5, G6, G8, G10, and G11, in combination with P types including P[3], P[9], P[10], P[11], and P[14], and are generally considered to be animal-to-human reassortment variants. However, certain animallike genotypes have become established in human circulation, particularly in developing countries. RVA G9 and G12 were formerly considered porcine RVA strains but emerged in human circulation in the 1990s and 2000s, respectively.[15,144] Today, G9 has become the fifth most common genotype detected in humans worldwide, while G12 strains appear to have spread globally.[15] Another animal like genotype, G8, has appeared and become established in certain geographical regions, for example, Africa. G8 RVAs are detected with high frequency in African countries,[15,101] with increasing reports from other countries, for example, Iraq, America, Brazil, and Australia.[15,105,145–147] Other animallike RVA genotypes (i.e., G5 and G10) have been associated with infections in children in Brazil, Thailand, and India.[101,104,105] Conversely, human-to-animal and animal-to-animal reassortments have also been reported.[126,148–150] Although reassortment is a driving force in RV evolution, some genotypes appear to remain stable across species,[103] with others appearing to be restricted to particular hosts, for example, equine RV G14 and porcine P[13].

Studying animal RVs is fundamental to obtaining a better understanding of RV ecology and the mechanisms by which RVs evolve. Along with cattle, pigs are regarded as important reservoirs for the diversification of human RV.[103] Novel human RVA strains are likely to have originated through direct zoonotic transmission from pigs, and porcine–human reassortant strains have spread successfully throughout the human population in Latin America, Southeast Asia, and Europe, demonstrating the impact of animal RVA on human health.[118,124,151–156] However, there are also important animal welfare and economic impact reasons for the study of animal RVs in order to improve the prophylactic tools for use in livestock.

72.1.6 Prevention and Treatment

72.1.6.1 Human Vaccines

Since its discovery, many attempts have been made to produce effective RV vaccines, with an emphasis on those directed to the prevention of human disease. In 1998, the first human vaccine, RotaShield, was licensed. This multivalent vaccine contained the VP7 genes of common human genotypes, G1, G2, and G4, and rhesus RVA G3.[157] However, despite the proven efficacy of RotaShield, it was withdrawn from the market a year later, due to a temporal association with intussusception.[157,158] Today, two oral live-attenuated vaccines are being used in RV immunization programs globally: RotaTeq™, which is a pentavalent human–bovine reassortment containing human RVA-derived G1–G4 and P[8] types within the backbone of the WC3 bovine strain, and Rotarix™, a monovalent human G1P[8] live-attenuated vaccine. Both vaccines are efficacious and endorsed by the WHO, which recommends the implementation of these vaccines in national immunization programs.[159] In 2010, porcine circovirus DNA was detected in Rotarix by researchers at the Blood Systems Research Institute and the University of California, San Francisco, using microarray and high-throughput sequencing. The vaccine was subsequently found to contain infectious virus and was withdrawn temporarily from sale in the United States but not in the European Union. Follow-up research also found the presence of porcine circovirus DNA in RotaTeq.[160]

The introduction of universal mass vaccination (UMV) has resulted in a significant decrease in childhood RV infection morbidity and mortality.[161–166] Despite this, speculation on whether UMV may exert selective pressure on the RVA genotypes in circulation and result in a shift in genotype distribution or lead to changes in age distribution of the disease in the longer term due to differential levels of cross protection still remains. The backbone of Rotarix is a human G1 (Wa-like) virus, while that of Rotateq is a bovine–human reassortant (similar to DS-1).[167] A recent review of genotypes in circulation globally during the prevaccination era shows that fluctuations in genotype distribution from 1996 to 2007 are common.[15] Countries where UMV has been implemented have been monitoring circulating genotypes post-UMV to determine whether vaccination is exerting selective pressure on genotype distribution. Belgium and Brazil have reported a marked increase in G2 RVA from 2006 to 2007[166,168,169]; however, this increase was also noted in other countries where there is no UMV[165]; thus, it is far more likely that this shift was a result of normal fluctuations in RV diversity.[165,166,170] Additionally, a similar study in Finland has not reported any selective pressure effects from UMV.[171]

72.1.6.2 Animal Vaccines

The physiology of the placenta in ruminants and pigs blocks the transfer of antibody and immune cells from the dam to the fetus. These species therefore rely on the uptake of antibody (and immune cells) from colostrum and milk before gut closure to mediate passive transfer of maternal immunity. IgG1 is the predominant antibody in colostrum and milk due to expression of the neonatal Fc receptor (FcRn) by mammary gland epithelial cells that transports IgG1 into the ductal lumen. Neonates in these species also express FcRn on the duodenal and jejunal epithelium that aids the uptake of this IgG1 into the circulation of the neonate. Therefore, neonates that have received sufficient colostrum will have high levels of circulating and gut lumen maternal antibody.[172] Early work demonstrating that feeding RV-immune colostrum to calves limited RV infection and reduced disease severity[92] has been used as the basis for farm animal vaccination strategies. Unlike human vaccination strategies, most veterinary vaccines licensed for use vaccinate dams in late gestation in order to increase antibody titers against the virus in colostrum and milk.

Vaccines in pigs: Currently, there is no vaccine licensed in Europe for the prevention of RVA-associated enteritis in piglets. A porcine RV vaccine is available in the United States and is administered to sows before and after farrowing or to young piglets for protection against RVA enteritis. However, given the broad genetic/antigenic heterogeneity of the RVA strains detected in previous studies, it remains unclear whether this diversity may pose a challenge for future prophylactic programs for the prevention of enteritis in suckling and weaning piglets.

Vaccines in cattle: Currently, a bovine RV vaccine, derived from the G6P[5] UK strain is available for late gestation dams. In a recent study in France,[173] the use of this vaccine did not seem to promote the emergence of RV strains with genotypes or variants different from those of the vaccine, suggesting a significant degree of cross protection given the diversity of circulating bovine RVAs. In addition, RVA vaccination was not associated with increases in other viruses such as caliciviruses, which also cause calf diarrhea, which suggests that RVA vaccination of dams should result in a net decrease of diarrheal illness in calves.

Vaccines in horses: Following rigorous management to try to prevent RV infection, vaccination is the only effective means to prevent RVA outbreaks. In horses, vaccine is also administered to mares in late pregnancy to boost levels of immunity that will be transferred to foals in colostrum. Antigenic/genetic characterization of equine RVAs identified in Japan, Australia, the United Kingdom, and Germany has revealed that most strains are either G3P[12] or G14P[12].[79,130,137,174] Currently, three inactivated vaccines are available for equine RVA immunization; two are monovalent and the third is trivalent. In Ireland, the United Kingdom, and the United States, an inactivated monovalent vaccine is used based on the strain RVA/Horse-wt/UK/H-2/1976/G3AP[12],[175] while in Japan, another inactivated monovalent vaccine based on the strain RVA/Horse-tc/JPN/HO-5/1982/G3BP[12] is in use.[176] In Argentina, a trivalent vaccine incorporating three animal strains is in use: equine H-2, bovine RVA/Cow-tc/JPN/NCDV/200X/G6P[1], and simian RVA/Simian-tc/ZAF/SA11/1958/G3P[2] strains.[177] However, recently, the effectiveness of vaccines administered to horses has been questioned and suggested to be only partially protective, as in most recent studies of RVA outbreaks in horses,

the animals had been vaccinated.[127] This may be caused by a shift in the prevalent strain type to G14, which is not included in any of the vaccines.[131,178,179] Experimental studies on vaccinated animals have also observed the failure of circulating antibodies to neutralize infection in challenged animals.[133,180] An oral passive immunotherapy using equine hyperimmune serum from animals vaccinated with the RVA/Horse-wt/UK/H-2/1976/G3AP[12] vaccine has also been developed for use in clinically affected foals (Veterinary Immunogenics Ltd., U.K.), although no peer-reviewed trials are available for this product.

72.1.7   Diagnosis

RVA infection is diagnosed by collecting stool samples from the suspected cases and investigating the presence RVA particles, antigens, or RNA. A variety of methods can be used, ranging from electron microscopy to molecular methods. Electron microscopy, which led to the discovery of RVs, is currently rarely used for diagnostic purposes. This is a labor-intensive process that requires highly skilled personnel, and it has been superseded by other less expensive and higher-throughput techniques such as antigen detection or molecular methods, with the latter also allowing strain characterization (Table 72.2).

Routine laboratory RVA diagnosis generally relies on the detection of RV antigen in stool samples using a variety of immunological methods such as enzyme-linked immunosorbent assays (ELISAs), latex passive agglutination tests, and immunochromatographic methods. All of these methods are currently available commercially and can be performed in most laboratories as they do not require specific equipment or highly skilled staff (Table 72.2). In many settings, the use of ELISA has diminished in favor of the rapid analyses offered by latex agglutination and immunochromatographic tests. A particular advantage of these tests is that they can be used to test single or small numbers of samples, without reagent waste, and test controls are contained within each test. These methods also offer the advantage that they can be performed by nonlaboratory staff at the bed or pen side, not only reducing diagnosis times very significantly but also reducing sample transport and laboratory costs.

The first ever molecular method for the detection of RVs, and the simplest, is the visualization of the 11 RV genome segments by polyacrylamide gel electrophoresis (PAGE). In addition, this technique can be used to electropherotype isolates, based on banding patterns of the gene segments, which serves to distinguish RV belonging to different groups as well as some genome constellations within RVA. Further characterization can also be achieved through genome–genome hybridization,[181] a method that was central for the detection of mixed infections and interspecies transmission and reassortment, prior to the development and spread of the use of genome amplification methods.

TABLE 72.2

Methods for Rotavirus A Detection, Use for Rotavirus A Diagnosis, Suitability for Rotavirus A Typing, Sensitivity, Specificity, Throughput, and Cost Information

Method	Useful for Diagnosing RVA Gastroenteritis	Suitable for RVA Typing (Reference Work)	Sensitivity	Specificity	Throughput	Cost	Comments
Electron microscopy	Yes	No	++	+++	Low	High	Cannot distinguish RVA from other RV groups unless used in conjunction with antibody capture Few laboratories have equipment and/or skill
ELISA	Yes	No	++	+++	High	Low	Detection of RVA antigen proven to correlate with clinical disease in humans Many commercially available
Immunochromatography	Yes	No	++	++	Medium/high	Low	Cost effective specially for on-demand diagnostic
Passive agglutination	Yes	No	+/++	++	Medium/high	Low	Cost effective specially for on-demand diagnostic
PAGE	Yes	Limited	+	+++	Low	Moderate	Allows group and broad genome constellation discrimination
Conventional RT-PCR	Yes, but some limitations	Yes	+++	+++	Medium/high	Moderate/high	May detect asymptomatic shedding due to high sensitivity
Real time RT-PCR	Yes	Limited	+++	+++	High	High	Allows for the calculation of viral load cutoff that is associated with disease Not widely used for typing as use of multiple probes/dyes is a limiting factor and also increases cost
Sanger sequencing	No	Yes	++	+++	Medium/low	High	Requires amplicon generation and relatively high Different sensitivities for each gene Inherent biases make it difficult to distinguish mixed infections and strain mixtures from reassortant strains
Next-generation sequencing	Only in a metagenomic context	Yes	++/+++	+++	High	High/medium	Highest throughput, allows for high level of multiplexing, reducing cost per base Sensitivity may be compromised unless there is high viral load which is typical in the acute phase Introduces less bias and greater potential for detection of mixed infections but bioinformatic analysis challenging in particular for discrimination of mixtures and reassortment

72.2 METHODS

72.2.1 RNA EXTRACTION

All molecular methods for detection or characterization of RVs require a nucleic acid purification step.

Total nucleic acid can be extracted from the stool samples using a standard phenol–chloroform method for use in molecular diagnostic methods.[182] This method is relatively inexpensive and does not require specialized equipment, yielding good quality RNA for separation by PAGE. If the RNA is to be used in subsequent reverse transcription polymerase chain reaction (RT-PCR) for detection or typing purposes, a further purification step or an alternative nucleic acid method is often recommended in order to minimize the risk of PCR inhibition due to the presence of residual solvents in the purified sample.

Currently, there are many suitable nucleic acid extraction kits commercially available, all based on similar principles; the nucleic acid extraction typically involves the use of detergents and proteases in conjunction with chaotropic agents to inactivate enzymatic activity and prevent nucleic acid degradation of the exposed nucleic acid, and then fractionated silica is used to bind the exposed nucleic acid and allow the removal of all other contaminants in the samples. The silica may be used in solution, coupled to magnetic particles or immobilized onto filters, and the purification requires the use of vacuum or centrifugal force for washing steps and final nucleic acid elution into molecular-grade nuclease-free water or appropriate buffer (e.g., TE buffer).

72.2.2 DETECTION PROCEDURES

After the RNA has been purified from the clinical sample, it must be converted into complementary DNA (cDNA) in order to perform subsequent gene-specific amplification. This may be achieved in a separate RT reaction using either RV gene-specific primers or random primers that will yield cDNA from all RNA present in the extract, which is then followed by a specific PCR. Alternatively, one-step RT-PCRs with specific primers can also be performed in a single reaction mix that contains both a reverse transcriptase and a DNA polymerase. A number of highly sensitive and specific RT-PCR involving the amplification of different RVA genes have been developed and are in use in different laboratories.[183–186] In particular, one-step qRT-PCR (real time RT-PCR) methods are increasingly being used in medical diagnostic laboratories. These methods are amenable to automation and multiplexing with methods for the detection of other enteric pathogens, increasing throughput and reducing costs and turnaround times. However, their high sensitivity of detection can be problematic as this can lead to the detection of low loads of RVA that may not be associated with the disease.

Since the mid-1990s, investigations of RVA diversity have relied upon G and P genotyping. Oligonucleotide primers directed to conserved regions of the VP7[182] or the VP4[187]

genes are effective for consensus amplification of all animal RVA. However, genotyping is performed using nested type-specific primers in multiplex reactions. This means that when deciding on the composition of the multiplex reactions, prior knowledge of the expected types for the species being investigated is necessary. Furthermore, significant differences may occur between sequences of the same viral genotype prevalent in different species, requiring alternative oligonucleotide primers for detection of a given genotype in different species. For example, G3 strains are ubiquitous and prevalent in many species, including humans, monkey, cats, dogs; however, although they are all classified as G3 on the basis of the overall sequence homology of the entire VP7 encoding gene, primers designed for the detection of G3 strains in some species are unable to reliably detect G3 strains in other species.

Since the adoption of RT-PCR for genotyping RVs, many primers have been developed and used worldwide, depending on the particular species or settings involved. For the typing of human RVA, several protocols and a variety of primers are available for use.[188]

For genotyping of RVA in the veterinary field, amplification and typing of the VP7 gene is carried out using nested multiplex PCR in swine,[189] cattle,[190] and horses.[131] For equine RVA, the G type can be determined using primers described by Gouvea et al.,[182] and primers are also available for G3 and G14 genotyping.[174,179] Equine RVA P types can be investigated using primers described by Gentsch et al.[187] and typing primers described by Fukai et al.[191]

Although G and P typing have been used in informing vaccine design and surveillance, there has been a recent drive toward whole genome sequencing to further understand RVA diversity and evolutionary dynamics. To take this approach, primers, which amplify all 11 gene segments, have been described and can be applied to RVA from most animal species.[16] Additionally, an online genotype classification tool can also be used to rapidly analyze sequence data: RotaC, http://rotac.regatools.be/.[192]

A protocol for characterization and strain nomenclature has been detailed by the Rotavirus Classification Working Group, which provides a common baseline to molecular epidemiology studies of RV worldwide.[16,193,194] Phylogenies are studied, based on nucleotide sequences using neighbor-joining analysis, as the nucleotide sequences are deemed more informative than amino acid sequences for phylogenetic and other molecular analyses, including the detection of new genotypes.[16,194] Neighbor-joining phylogenies are based on distance-based models of molecular evolution; however, with advances in maximum likelihood and Bayesian methods (character-based models), distance methods have decreased in popularity for inferring relatedness and evolution.[195] Distance-based models are, however, sufficient to elucidate relationships and clustering among closely related sequences with a low level of divergence. For RVA, these models are capable of distinguishing different genotypes, with the advantages of relative accuracy and computational/time efficiency.

72.3 CONCLUSION AND FUTURE PERSPECTIVES

RV diversity across the animal kingdom is vast. In particular, the segmented nature of the RV genome can lead to gene reassortment during coinfection with more than one RV. This impacts on the diversity associated with the outer capsid proteins VP7 and VP4, targets for neutralizing antibody responses. Numerous reports have described interspecies transmission events leading to sporadic cases of human disease with RVs from different animal species origin,[129,196–199] and the emergence of epidemiologically important strains in the human population such as G9P[8] globally, G10P[11] in India, G8P[4] in Africa, Europe, and the United States is postulated to have resulted from reassortment among human and other animal RVs.[147,200–207] Whole genome sequencing of human RVA strains has revealed that among human RVA strains, with the exception of VP7 and VP4, the remaining gene constellation is relatively well conserved, and most strains can be classified as belonging to a Wa-like G1 or a DS-1 G2 gene constellation. In human infections from suspected direct zoonotic transmission, gene constellations different from those typically associated with human RVAs are seen. However, epidemiologically significant emergent strains, suspected to have originated through zoonosis, are typically associated mostly with a human RVA characteristic gene constellation, likely to have originated through reassortment and to be associated with species adaptation. Despite the increasing number of whole genome data available from human strains, similar data from other animal species are scarce. Structured surveillance in animals that includes whole genome characterization of RV strains from different animal hosts may provide important leads to better understand drivers and constraints to interspecies transmission and reassortment.

REFERENCES

1. Bishop, R., Davidson, G. P., Holmes, I. H., and Ruck, B. J. Virus particles in epithelial cells of duodenal mucosa from children with acute non-bacterial gastroenteritis. *Lancet* 302, 1281–1283 (1973).
2. Flewett, A. S. B. Letter: Virus diarrhoea in foals and other animals. *Vet. Rec.* 96, JMM (1975).
3. Estes, M. K. and Kapikian, A. Z. *Fields Virology*, Vol. 2, pp. 1917–1974. Lippincott, Williams & Wilkins, Philadelphia, PA (2007).
4. Anderson, E. J. et al. Rotavirus in adults requiring hospitalization. *J. Infect.* 64, 89–95 (2011).
5. Gunn, L. et al. Molecular characterization of group A rotavirus found in elderly patients in Ireland; predominance of G1P[8], continued presence of G9P[8], and emergence of G2P[4]. *J. Med. Virol.* 84, 2008–2017 (2012).
6. Tate, J. E. et al. 2008 estimate of worldwide rotavirus-associated mortality in children younger than 5 years before the introduction of universal rotavirus vaccination programmes: A systematic review and meta-analysis. *Lancet Infect. Dis.* 12, 136–141 (2012).
7. Au, K. S., Chan, W. K., Burns, J. W., and Estes, M. K. Receptor activity of rotavirus nonstructural glycoprotein NS28. *J. Virol.* 63, 4553–4562 (1989).
8. Meyer, J. C., Bergmann, C. C., and Bellamy, A. R. Interaction of rotavirus cores with the nonstructural glycoprotein NS28. *Virology* 171, 98–107 (1989).
9. Padilla-Noriega, L., Paniagua, O., and Guzmán-León, S. Rotavirus protein NSP3 shuts off host cell protein synthesis. *Virology* 298, 1–7 (2002).
10. Patton, J. T., Silvestri, L. S., Tortorici, M. A., Carpio, R. V.-D., and Taraporewala, Z. F. Rotavirus genome replication and morphogenesis: role of the viroplasm. In *Reoviruses: Entry, Assembly, and Morphogenesis* (ed. M.Sc, P. R.) pp. 169–187. Springer, Berlin, Germany (2006).
11. Sherry, B. Rotavirus and reovirus modulation of the interferon response. *J. Interferon Cytokine Res.* 29, 559–567 (2009).
12. Taraporewala, Z. F. and Patton, J. T. Nonstructural proteins involved in genome packaging and replication of rotaviruses and other members of the Reoviridae. *Virus Res.* 101, 57–66 (2004).
13. Kindler, E., Trojnar, E., Heckel, G., Otto, P. H., and Johne, R. Analysis of rotavirus species diversity and evolution including the newly determined full-length genome sequences of rotavirus F and G. *Infect. Genet. Evol.* 14, 58–67 (2013).
14. King, A. M. Q., Adams, M. J., Lefkowitz, E. J., and Carstens, E. B. *Virus Taxonomy: Classification and Nomenclature of Viruses: Ninth Report of the International Committee on Taxonomy of Viruses.* Elsevier, Oxford, U.K. (2012).
15. Bányai, K. et al. Systematic review of regional and temporal trends in global rotavirus strain diversity in the pre rotavirus vaccine era: Insights for understanding the impact of rotavirus vaccination programs. *Vaccine* 30(Suppl. 1), A122–A130 (2012).
16. Matthijnssens, J. et al. Full genome-based classification of rotaviruses reveals a common origin between human Wa-like and porcine rotavirus strains and human DS-1-like and bovine rotavirus strains. *J. Virol.* 82, 3204–3219 (2008).
17. Stencel-Baerenwald, J. E., Reiss, K., Reiter, D. M., Stehle, T., and Dermody, T. S. The sweet spot: Defining virus-sialic acid interactions. *Nat. Rev. Microbiol.* 12, 739–749 (2014).
18. Gutierrez, M. et al. Different rotavirus strains enter MA104 cells through different endocytic pathways: The role of clathrin-mediated endocytosis. *J. Virol.* 84, 9161–9169 (2010).
19. Odegard, A. L. et al. Putative autocleavage of outer capsid protein μ1, allowing release of myristoylated peptide μ1N during particle uncoating, is critical for cell entry by Reovirus. *J. Virol.* 78, 8732–8745 (2004).
20. Estrozi, L. F. et al. Location of the dsRNA-dependent polymerase, VP1, in rotavirus particles. *J. Mol. Biol.* 425, 124–132 (2013).
21. Lawton, J. A., Estes, M. K., and Prasad, B. V. V. Three-dimensional visualization of mRNA release from actively transcribing rotavirus particles. *Nat. Struct. Mol. Biol.* 4, 118–121 (1997).
22. Fabbretti, E., Afrikanova, I., Vascotto, F., and Burrone, O. R. Two non-structural rotavirus proteins, NSP2 and NSP5, form viroplasm-like structures *in vivo. J. Gen. Virol.* 80, 333–339 (1999).
23. Petrie, B. L., Greenberg, H. B., Graham, D. Y., and Estes, M. K. Ultrastructural localization of rotavirus antigens using colloidal gold. *Virus Res.* 1, 133–152 (1984).
24. McDonald, S. M. and Patton, J. T. Assortment and packaging of the segmented rotavirus genome. *Trends Microbiol.* 19, 136–144 (2011).

25. Zeng, C. Q.-Y., Estes, M. K., Charpilienne, A., and Cohen, J. The N Terminus of rotavirus VP2 is necessary for encapsidation of VP1 and VP3. *J. Virol.* 72, 201–208 (1998).

26. Boyce, M., Yang, W., and McCrae, M. The multifunctional roles of the rotavirus non-structural protein NSP4. *Virus-Cell and Virus-Gut Interactions: Fifth European Rotavirus Biology-Meeting ERBM*, Valencia, Spain (2013).

27. Yang, W. and McCrae, M. A. The molecular biology of rotaviruses X: Intercellular dissemination of rotavirus NSP4 requires glycosylation and is mediated by direct cell-cell contact through cytoplasmic extrusions. *Arch. Virol.* 157, 305–314 (2012).

28. Chasey, D. Different particle types in tissue culture and intestinal epithelium infected with rotavirus. *J. Gen. Virol.* 37, 443–451 (1977).

29. Silvestri, L. S., Taraporewala, Z. F., and Patton, J. T. Rotavirus replication: Plus-sense templates for double-stranded RNA synthesis are made in viroplasms. *J. Virol.* 78, 7763–7774 (2004).

30. Hyser, J. M., Collinson-Pautz, M. R., Utama, B., and Estes, M. K. Rotavirus disrupts calcium homeostasis by NSP4 viroporin activity. *mBio* 1, e00265-10 (2010).

31. Ramig, R. F. Pathogenesis of intestinal and systemic rotavirus infection. *J. Virol.* 78, 10213–10220 (2004).

32. Ward, L. A., Rosen, B. I., Yuan, L., and Saif, L. J. Pathogenesis of an attenuated and a virulent strain of group A human rotavirus in neonatal gnotobiotic pigs. *J. Gen. Virol.* 77, 1431–1441 (1996).

33. Estes, M. K. and Morris, A. P. A viral enterotoxin. A new mechanism of virus-induced pathogenesis. *Adv. Exp. Med. Biol.* 473, 73–82 (1999).

34. Lundgren, O. et al. Role of the enteric nervous system in the fluid and electrolyte secretion of rotavirus diarrhea. *Science* 287, 491–495 (2000).

35. Hagbom, M. et al. Rotavirus stimulates release of serotonin (5-HT) from human enterochromaffin cells and activates brain structures involved in nausea and vomiting. *PLoS Pathog.* 7, e1002115 (2011).

36. Hagbom, M., Sharma, S., Lundgren, O., and Svensson, L. Towards a human rotavirus disease model. *Curr. Opin. Virol.* 2, 408–418 (2012).

37. Blutt, S.E. and Conner, M.E. The gastrointestinal frontier: IgA and viruses. *Front. Immunol.* 4:402 (2013). doi: 10.3389/fimmu.2013.00402.

38. Blutt, S. E., Miller, A. D., Salmon, S. L., Metzger, D. W., and Conner, M. E. IgA is important for clearance and critical for protection from rotavirus infection. *Mucosal Immunol.* 5, 712–719 (2012).

39. Ramig, R. F. Systemic rotavirus infection. *Expert Rev. Anti Infect. Ther.* 5, 591–612 (2007).

40. Ramani, S. et al. Rotavirus antigenemia in Indian children with rotavirus gastroenteritis and asymptomatic infections. *Clin. Infect. Dis.* 51, 1284–1289 (2010).

41. Velázquez, F. R. et al. Rotavirus infection in infants as protection against subsequent infections. *N. Engl. J. Med.* 335, 1022–1028 (1996).

42. Bishop, R. F., Barnes, G. L., Cipriani, E., and Lund, J. S. Clinical immunity after neonatal rotavirus infection. *N. Engl. J. Med.* 309, 72–76 (1983).

43. Desselberger, U. and Huppertz, H.-I. Immune responses to rotavirus infection and vaccination and associated correlates of protection. *J. Infect. Dis.* 203, 188–195 (2011).

44. Angel, J., Franco, M. A., and Greenberg, H. B. Rotavirus immune responses and correlates of protection. *Curr. Opin. Virol.* 2, 419–425 (2012).

45. Holloway, G. and Coulson, B. S. Innate cellular responses to rotavirus infection. *J. Gen. Virol.* 94, 1151–1160 (2013).

46. Vancott, J. L., McNeal, M. M., Choi, A. H. C., and Ward, R. L. The role of interferons in rotavirus infections and protection. *J. Interferon Cytokine Res.* 23, 163–170 (2003).

47. Angel, J., Franco, M. A., Greenberg, H. B., and Bass, D. Lack of a role for type I and type II interferons in the resolution of rotavirus-induced diarrhea and infection in mice. *J. Interferon Cytokine Res.* 19, 655–659 (1999).

48. Pott, J. et al. IFN-λ determines the intestinal epithelial antiviral host defense. *Proc. Natl. Acad. Sci. USA* 108, 7944–7949 (2011).

49. Franco, M. A. et al. Evidence for CD8+ T-cell immunity to murine rotavirus in the absence of perforin, fas, and gamma interferon. *J. Virol.* 71, 479–486 (1997).

50. Schwers, A. et al. Experimental rotavirus diarrhoea in colostrum-deprived newborn calves: Assay of treatment by administration of bacterially produced human interferon (Hu-IFN alpha 2). *Ann. Rech. Vét. Ann. Vet. Res.* 16, 213–218 (1985).

51. Feng, N. et al. Role of interferon in homologous and heterologous rotavirus infection in the intestines and extraintestinal organs of suckling mice. *J. Virol.* 82, 7578–7590 (2008).

52. Barro, M. and Patton, J. T. Rotavirus nonstructural protein 1 subverts innate immune response by inducing degradation of IFN regulatory factor 3. *Proc. Natl. Acad. Sci. USA* 102, 4114–4119 (2005).

53. Feng, N. et al. Variation in antagonism of the interferon response to rotavirus NSP1 results in differential infectivity in mouse embryonic fibroblasts. *J. Virol.* 83, 6987–6994 (2009).

54. Arnold, M. M. and Patton, J. T. Diversity of interferon antagonist activities mediated by NSP1 proteins of different rotavirus strains. *J. Virol.* 85, 1970–1979 (2011).

55. Zhang, R. et al. Homologous 2′,5′-phosphodiesterases from disparate RNA viruses antagonize antiviral innate immunity. *Proc. Natl. Acad. Sci. USA* 110, 13114–13119 (2013).

56. Ge, Y. et al. Rotavirus NSP4 triggers secretion of proinflammatory cytokines from macrophages via Toll-like receptor 2. *J. Virol.* 87, 11160–11167 (2013).

57. Malik, J., Gupta, S. K., Bhatnagar, S., Bhan, M. K., and Ray, P. Evaluation of IFN-γ response to rotavirus and non-structural protein NSP4 of rotavirus in children following severe rotavirus diarrhea. *J. Clin. Virol.* 43, 202–206 (2008).

58. Holloway, G., Truong, T. T., and Coulson, B. S. Rotavirus antagonizes cellular antiviral responses by inhibiting the nuclear accumulation of STAT1, STAT2, and NF-κB. *J. Virol.* 83, 4942–4951 (2009).

59. Franco, M. A., Angel, J., and Greenberg, H. B. Immunity and correlates of protection for rotavirus vaccines. *Vaccine* 24, 2718–2731 (2006).

60. Offit, P. A. and Dudzik, K. I. Rotavirus-specific cytotoxic T lymphocytes passively protect against gastroenteritis in suckling mice. *J. Virol.* 64, 6325–6328 (1990).

61. Oldham, G., Bridger, J. C., Howard, C. J., and Parsons, K. R. *In vivo* role of lymphocyte subpopulations in the control of virus excretion and mucosal antibody responses of cattle infected with rotavirus. *J. Virol.* 67, 5012–5019 (1993).

62. Jaimes, M. C. et al. Frequencies of virus-specific CD4(+) and CD8(+) T lymphocytes secreting gamma interferon after acute natural rotavirus infection in children and adults. *J. Virol.* 76, 4741–4749 (2002).

63. Mäkelä, M., Marttila, J., Simell, O., and Ilonen, J. Rotavirus-specific T-cell responses in young prospectively followed-up children. *Clin. Exp. Immunol.* 137, 173–178 (2004).

64. Offit, P. A., Hoffenberg, E. J., Santos, N., and Gouvea, V. Rotavirus-specific humoral and cellular immune response after primary, symptomatic infection. *J. Infect. Dis.* 167, 1436–1440 (1993).

65. Rott, L. S. et al. Expression of mucosal homing receptor alpha4beta7 by circulating CD4+ cells with memory for intestinal rotavirus. *J. Clin. Invest.* 100, 1204–1208 (1997).

66. Svensson, L. et al. Serum antibody responses to individual viral polypeptides in human rotavirus infections. *J. Gen. Virol.* 68(Pt 3), 643–651 (1987).

67. Offit, P. A. et al. Enhancement of rotavirus immunogenicity by microencapsulation. *Virology* 203, 134–143 (1994).

68. Franco, M. A. and Greenberg, H. B. Immunity to rotavirus infection in mice. *J. Infect. Dis.* 179(Suppl. 3), S466–S469 (1999).

69. Yuan, L. and Saif, L. J. Induction of mucosal immune responses and protection against enteric viruses: Rotavirus infection of gnotobiotic pigs as a model. *Vet. Immunol. Immunopathol.* 87, 147–160 (2002).

70. Garaicoechea, L. et al. Llama-derived single-chain antibody fragments directed to rotavirus VP6 protein possess broad neutralizing activity *in vitro* and confer protection against diarrhea in mice. *J. Virol.* 82, 9753–9764 (2008).

71. Aladin, F. et al. *In vitro* neutralisation of rotavirus infection by two broadly specific recombinant monovalent llama-derived antibody fragments. *PLoS One* 7, e32949 (2012).

72. Aiyegbo, M. S. et al. Human rotavirus VP6-specific antibodies mediate intracellular neutralization by binding to a quaternary structure in the transcriptional pore. *PLoS One* 8, e61101 (2013).

73. Sapparapu, G. et al. Intracellular neutralization of a virus using a cell-penetrating molecular transporter. *Nanomedicine* 9, 1613–1624 (2013).

74. O'Neal, C. M., Harriman, G. R., and Conner, M. E. Protection of the villus epithelial cells of the small intestine from rotavirus infection does not require immunoglobulin A. *J. Virol.* 74, 4102–4109 (2000).

75. Kuklin, N. A. et al. Protective intestinal anti-rotavirus B cell immunity is dependent on α4β7 integrin expression but does not require IgA antibody production. *J. Immunol.* 166, 1894–1902 (2001).

76. Hoshino, Y., Saif, L. J., Sereno, M. M., Chanock, R. M., and Kapikian, A. Z. Infection immunity of piglets to either VP3 or VP7 outer capsid protein confers resistance to challenge with a virulent rotavirus bearing the corresponding antigen. *J. Virol.* 62, 744–748 (1988).

77. Yuan, L., Ward, L. A., Rosen, B. I., To, T. L., and Saif, L. J. Systematic and intestinal antibody-secreting cell responses and correlates of protective immunity to human rotavirus in a gnotobiotic pig model of disease. *J. Virol.* 70, 3075–3083 (1996).

78. Yuan, L., Kang, S.-Y., Ward, L. A., To, T. L., and Saif, L. J. Antibody-secreting Cell responses and protective immunity assessed in gnotobiotic pigs inoculated orally or intramuscularly with inactivated human rotavirus. *J. Virol.* 72, 330–338 (1998).

79. To, T. L., Ward, L. A., Yuan, L., and Saif, L. J. Serum and intestinal isotype antibody responses and correlates of protective immunity to human rotavirus in a gnotobiotic pig model of disease. *J. Gen. Virol.* 79, 2661–2672 (1998).

80. Chang, K. O., Vandal, O. H., Yuan, L., Hodgins, D. C., and Saif, L. J. Antibody-secreting cell responses to rotavirus proteins in gnotobiotic pigs inoculated with attenuated or virulent human rotavirus. *J. Clin. Microbiol.* 39, 2807–2813 (2001).

81. Gonzalez, A. M. Rotavirus-specific B cells induced by recent infection in adults and children predominantly express the intestinal homing receptor α4β7. *Virology* 305, 93–105 (2003).

82. Jaimes, M. C. et al. Maturation and trafficking markers on rotavirus-specific B cells during acute infection and convalescence in children. *J. Virol.* 78, 10967–10976 (2004).

83. Chiba, S. et al. Protective effect of naturally acquired homotypic and heterotypic rotavirus antibodies. *Lancet* 328, 417–421 (1986).

84. Coulson, B. S., Grimwood, K., Hudson, I. L., Barnes, G. L., and Bishop, R. F. Role of coproantibody in clinical protection of children during reinfection with rotavirus. *J. Clin. Microbiol.* 30, 1678–1684 (1992).

85. Matson, D. O., O'Ryan, M. L., Herrera, I., Pickering, L. K., and Estes, M. K. Fecal antibody responses to symptomatic and asymptomatic rotavirus infections. *J. Infect. Dis.* 167, 577–583 (1993).

86. O'Ryan, M. L., Matson, D. O., Estes, M. K., and Pickering, L. K. Acquisition of serum isotype-specific and G type-specific antirotavirus antibodies among children in day care centers. *Pediatr. Infect. Dis. J.* 13, 890–895 (1994).

87. Hjelt, K., Paerregaard, A., Nielsen, O. H., Krasilnikoff, P. A., and Grauballe, P. C. Protective effect of preexisting rotavirus-specific immunoglobulin A against naturally acquired rotavirus infection in children. *J. Med. Virol.* 21, 39–47 (1987).

88. Patel, M. et al. A systematic review of anti-rotavirus serum IgA antibody titer as a potential correlate of rotavirus vaccine efficacy. *J. Infect. Dis.* 208, 284–294 (2013).

89. Premkumar, P. et al. Association of serum antibodies with protection against rotavirus infection and disease in South Indian children. *Vaccine* 32(Suppl. 1), A55–A61 (2014).

90. Vonderfecht, S. L. and Osburn, B. I. Immunity to rotavirus in conventional neonatal calves. *J. Clin. Microbiol.* 16, 935–942 (1982).

91. Saif, L. J. Development of nasal, fecal and serum isotype-specific antibodies in calves challenged with bovine coronavirus or rotavirus. *Vet. Immunol. Immunopathol.* 17, 425–439 (1987).

92. Snodgrass, D. R., Stewart, J., Taylor, J., Krautil, F. L., and Smith, M. L. Diarrhoea in dairy calves reduced by feeding colostrum from cows vaccinated with rotavirus. *Res. Vet. Sci.* 32, 70–73 (1982).

93. Saif, L. J., Redman, D. R., Smith, K. L., and Theil, K. W. Passive immunity to bovine rotavirus in newborn calves fed colostrum supplements from immunized or nonimmunized cows. *Infect. Immun.* 41, 1118–1131 (1983).

94. Castrucci, G. et al. Field trial evaluation of an inactivated rotavirus vaccine against neonatal diarrhea of calves. *Eur. J. Epidemiol.* 3, 5–9 (1987).

95. Gonzalez, D. D. et al. Evaluation of a bovine rotavirus VP6 vaccine efficacy in the calf model of infection and disease. *Vet. Immunol. Immunopathol.* 137, 155–160 (2010).

96. Parashar, U. D., Bresee, J. S., Gentsch, J. R., and Glass, R. I. Rotavirus. *Emerg. Infect. Dis.* 4, 561–570 (1998).

97. Glass, R. I. et al. Rotavirus vaccines: Current prospects and future challenges. *Lancet* 368, 323–332 (2006).

98. Feeney, S. A. et al. Association of the G4 rotavirus genotype with gastroenteritis in adults. *J. Med. Virol.* 78, 1119–1123 (2006).

99. Iturriza-Gomara, M. et al. Rotavirus genotypes co-circulating in Europe between 2006 and 2009 as determined by EuroRotaNet, a pan-European collaborative strain surveillance network. *Epidemiol. Infect.* 139, 895–909 (2011).

100. Nakajima, H. et al. Winter seasonality and rotavirus diarrhoea in adults. *Lancet* 357, 1950 (2001).

101. Santos, N. and Hoshino, Y. Global distribution of rotavirus serotypes/genotypes and its implication for the development and implementation of an effective rotavirus vaccine. *Rev. Med. Virol.* 15, 29–56 (2005).

102. Armah, G. E. et al. Diversity of rotavirus strains circulating in West Africa from 1996 to 2000. *J. Infect. Dis.* 202, S64–S71 (2010).

103. Martella, V., Bányai, K., Matthijnssens, J., Buonavoglia, C., and Ciarlet, M. Zoonotic aspects of rotaviruses. *Vet. Microbiol.* 140, 246–255 (2010).

104. Miles, M. G., Lewis, K. D. C., Kang, G., Parashar, U. D., and Steele, A. D. A systematic review of rotavirus strain diversity in India, Bangladesh, and Pakistan. *Vaccine* 30(Suppl. 1), A131–A139 (2012).

105. Santos, N., Lima, R. C. C., Pereira, C. F. A., and Gouvea, V. Detection of rotavirus types G8 and G10 among Brazilian children with diarrhea. *J. Clin. Microbiol.* 36, 2727–2729 (1998).

106. Anderson, E. J. Rotavirus G9 severity data revisited. *Clin. Infect. Dis.* 44, 154–155 (2007).

107. Arista, S., Vizzi, E., Ferraro, D., Cascio, A., and Di Stefano, R. Distribution of VP7 serotypes and VP4 genotypes among rotavirus strains recovered from Italian children with diarrhea. *Arch. Virol.* 142, 2065–2071 (1997).

108. Aupiais, C. et al. Severity of acute gastroenteritis in infants infected by G1 or G9 rotaviruses. *J. Clin. Virol.* 46, 282–285 (2009).

109. Linhares, A. C., Verstraeten, T., Bosch, J. W. den, Clemens, R., and Breuer, T. Rotavirus serotype G9 is associated with more-severe disease in Latin America. *Clin. Infect. Dis.* 43, 312–314 (2006).

110. Snodgrass, D. R. et al. Rotavirus serotypes 6 and 10 predominate in cattle. *J. Clin. Microbiol.* 28, 504–507 (1990).

111. Maes, R. K. et al. Evaluation of a human group a rotavirus assay for onsite detection of bovine rotavirus. *J. Clin. Microbiol.* 41, 290–294 (2003).

112. Kapikian, A. Z., Hoshino, Y., Chanock, R. M., and Perez-Schael, I. Jennerian and modified Jennerian approach to vaccination against rotavirus diarrhea using a quadrivalent rhesus rotavirus (RRV) and human-RRV reassortant vaccine. *Arch. Virol. Suppl.* 12, 163–175 (1996).

113. Rao, C. D., Gowda, K., and Reddy, B. S. Sequence analysis of VP4 and VP7 genes of nontypeable strains identifies a new pair of outer capsid proteins representing novel P and G genotypes in bovine rotaviruses. *Virology* 276, 104–113 (2000).

114. Fukai, K., Maeda, Y., Fujimoto, K., Itou, T., and Sakai, T. Changes in the prevalence of rotavirus G and P types in diarrheic calves from the Kagoshima prefecture in Japan. *Vet. Microbiol.* 86, 343–349 (2002).

115. Okada, N. and Matsumoto, Y. Bovine rotavirus G and P types and sequence analysis of the VP7 gene of two G8 bovine rotaviruses from Japan. *Vet. Microbiol.* 84, 297–305 (2002).

116. Saif, L. J. and Fernandez, F. M. Group A rotavirus veterinary vaccines. *J. Infect. Dis.* 174(Suppl. 1), S98–S106 (1996).

117. Gouvea, V., Santos, N., and Timenetsky Mdo, C. Identification of bovine and porcine rotavirus G types by PCR. *J. Clin. Microbiol.* 32, 1338–1340 (1994).

118. Gouvea, V., Santos, N., and Timenetsky Mdo, C. VP4 typing of bovine and porcine group A rotaviruses by PCR. *J. Clin. Microbiol.* 32, 1333–1337 (1994).

119. Chan-it, W. et al. Multiple combinations of P[13]-like genotype with G3, G4, and G5 in porcine rotaviruses. *J. Clin. Microbiol.* 46, 1169–1173 (2008).

120. Ciarlet, M., Ludert, J. E., and Liprandi, F. Comparative amino acid sequence analysis of the major outer capsid protein (VP7) of porcine rotaviruses with G3 and G5 serotype specificities isolated in Venezuela and Argentina. *Arch. Virol.* 140, 437–451 (1995).

121. Ghosh, S. et al. Molecular characterization of a porcine Group A rotavirus strain with G12 genotype specificity. *Arch. Virol.* 151, 1329–1344 (2006).

122. Martella, V. et al. Isolation and genetic characterization of two G3P5A[3] canine rotavirus strains in Italy. *J. Virol. Methods* 96, 43–49 (2001).

123. Martella, V. et al. Sequence analysis of the VP7 and VP4 genes identifies a novel VP7 gene allele of porcine rotaviruses, sharing a common evolutionary origin with human G2 rotaviruses. *Virology* 337, 111–123 (2005).

124. Martella, V. et al. Relationships among porcine and human P[6] rotaviruses: Evidence that the different human P[6] lineages have originated from multiple interspecies transmission events. *Virology* 344, 509–519 (2006).

125. Martella, V. et al. Identification of group A porcine rotavirus strains bearing a novel VP4 (P) Genotype in Italian swine herds. *J. Clin. Microbiol.* 45, 577–580 (2007).

126. Parra, G. I. et al. Phylogenetic analysis of porcine rotavirus in Argentina: Increasing diversity of G4 strains and evidence of interspecies transmission. *Vet. Microbiol.* 126, 243–250 (2008).

127. Bailey, K. E., Gilkerson, J. R., and Browning, G. F. Equine rotaviruses—Current understanding and continuing challenges. *Vet. Microbiol.* 167, 135–144 (2013).

128. Dwyer, R. M. Rotaviral diarrhea. *Vet. Clin. North Am. Equine Pract.* 9, 311–319 (1993).

129. Papp, H., Matthijnssens, J., Martella, V., Ciarlet, M., and Bányai, K. Global distribution of group A rotavirus strains in horses: A systematic review. *Vaccine* 31, 5627–5633 (2013).

130. Browning, G. F. and Begg, A. P. Prevalence of G and P serotypes among equine rotaviruses in the faeces of diarrhoeic foals. *Arch. Virol.* 141, 1077–1089 (1996).

131. Collins, P. J., Cullinane, A., Martella, V., and O'Shea, H. Molecular characterization of equine rotavirus in Ireland. *J. Clin. Microbiol.* 46, 3346–3354 (2008).

132. Imagawa, H. et al. Epidemiology of equine rotavirus infection among foals in the breeding region. *J. Vet. Med. Sci. Jpn. Soc. Vet. Sci.* 53, 1079–1080 (1991).

133. Conner, M. E. and Darlington, R. W. Rotavirus infection in foals. *Am. J. Vet. Res.* 41, 1699–1703 (1980).

134. Strickland, K. L., Lenihan, P., O'Connor, M. G., and Condon, J. C. Diarrhoea in foals associated with rotavirus. *Vet. Rec.* 111, 421 (1982).

135. Studdert, M. J., Mason, R. W., and Patten, B. E. Rotavirus diarrhoea of foals. *Aust. Vet. J.* 54, 363–364 (1978).

136. Browning, G. F., Chalmers, R. M., Fitzgerald, T. A., and Snodgrass, D. R. Evidence for two serotype G3 subtypes among equine rotaviruses. *J. Clin. Microbiol.* 30, 485–491 (1992).

137. Isa, P. and Snodgrass, D. R. Serological and genomic characterization of equine rotavirus VP4 proteins identifies three different P serotypes. *Virology* 201, 364–372 (1994).

138. Matthijnssens, J. et al. Complete molecular genome analyses of equine rotavirus A strains from different continents reveal several novel genotypes and a largely conserved genotype constellation. *J. Gen. Virol.* 93, 866–875 (2011).

139. Miño, S. et al. Equine G3P[3] rotavirus strain E3198 related to simian RRV and feline/canine-like rotaviruses based on complete genome analyses. *Vet. Microbiol.* 161, 239–246 (2013).

140. Ciarlet, M., Isa, P., Conner, M. E., and Liprandi, F. Antigenic and molecular analyses reveal that the equine rotavirus strain H-1 is closely related to porcine, but not equine, rotaviruses: Interspecies transmission from pigs to horses? *Virus Genes* 22, 5–20 (2001).

141. Ghosh, S., Shintani, T., and Kobayashi, N. Evidence for the porcine origin of equine rotavirus strain H-1. *Vet. Microbiol.* 158, 410–414 (2012).

142. Imagawa, H. et al. Genetic analysis of equine rotavirus by RNA-RNA hybridization. *J. Clin. Microbiol.* 32, 2009–2012 (1994).

143. Iša, P. et al. Survey of equine rotaviruses shows conservation of one P genotype in background of two G genotypes. *Arch. Virol.* 141, 1601–1612 (1996).

144. Matthijnssens, J. et al. Phylodynamic analyses of rotavirus genotypes G9 and G12 underscore their potential for swift global spread. *Mol. Biol. Evol.* 27, 2431–2436 (2010).

145. Ahmed, S. et al. Characterization of human rotaviruses circulating in Iraq in 2008: Atypical G8 and high prevalence of P[6] strains. *Infect. Genet. Evol.* 16, 212–217 (2013).

146. Swiatek, D. L. et al. Characterisation of G8 human rotaviruses in Australian children with gastroenteritis. *Virus Res.* 148, 1–7 (2010).

147. Weinberg, G. A. et al. First reports of human rotavirus G8P[4] gastroenteritis in the United States. *J. Clin. Microbiol.* 50, 1118–1121 (2012).

148. Abe, M. et al. Molecular epidemiology of rotaviruses among healthy calves in Japan: Isolation of a novel bovine rotavirus bearing new P and G genotypes. *Virus Res.* 144, 250–257 (2009).

149. Matthijnssens, J. et al. Simian rotaviruses possess divergent gene constellations that originated from interspecies transmission and reassortment. *J. Virol.* 84, 2013–2026 (2010).

150. Palombo, E. A. Genetic analysis of group A rotaviruses: Evidence for interspecies transmission of rotavirus genes. *Virus Genes* 24, 11–20 (2002).

151. Coluchi, N. et al. Detection, subgroup specificity, and genotype diversity of rotavirus strains in children with acute diarrhea in Paraguay. *J. Clin. Microbiol.* 40, 1709–1714 (2002).

152. Ghosh, S. K. and Naik, T. N. Detection of a large number of subgroup 1 human rotaviruses with a 'long' RNA electropherotype. *Arch. Virol.* 105, 119–127 (1989).

153. Krishnan, T., Burke, B., Shen, S., Naik, T. N., and Desselberger, U. Molecular epidemiology of human rotaviruses in Manipur: Genome analysis of rotaviruses of long electropherotype and subgroup I. *Arch. Virol.* 134, 279–292 (1994).

154. Okada, J. I. et al. New P serotype of group A human rotavirus closely related to that of a porcine rotavirus. *J. Med. Virol.* 60, 63–69 (2000).

155. Urasawa, S. et al. Antigenic and genetic analyses of human rotaviruses in Chiang Mai, Thailand: Evidence for a close relationship between human and animal rotaviruses. *J. Infect. Dis.* 166, 227–234 (1992).

156. Varghese, V. et al. Molecular characterization of a human rotavirus reveals porcine characteristics in most of the genes including VP6 and NSP4. *Arch. Virol.* 149, 155–172 (2004).

157. Dennehy, P. H. Rotavirus vaccines: An overview. *Clin. Microbiol. Rev.* 21, 198–208 (2008).

158. Joensuu, J., Koskenniemi, E., Pang, X. L., and Vesikari, T. Randomised placebo-controlled trial of rhesus-human reassortant rotavirus vaccine for prevention of severe rotavirus gastroenteritis. *Lancet* 350, 1205–1209 (1997).

159. WHO. Global use of rotavirus vaccines recommended, 2009. http://www.who.int/mediacentre/news/releases/2009/rotavirus_vaccines_20090605/en/. Accessed on June 5, 2009.

160. Petricciani, J., Sheets, R., Griffiths, E., and Knezevic, I. Adventitious agents in viral vaccines: Lessons learned from 4 case studies. *Biologicals* 42, 223–236 (2014).

161. Curns, A. T. et al. Reduction in acute gastroenteritis hospitalizations among US children after introduction of rotavirus vaccine: Analysis of hospital discharge data from 18 US States. *J. Infect. Dis.* 201, 1617–1624 (2010).

162. Do Carmo, G. M. I. et al. Decline in diarrhea mortality and admissions after routine childhood rotavirus immunization in Brazil: A time-series analysis. *PLoS Med.* 8, e1001024 (2011).

163. Pendleton, A. et al. Impact of rotavirus vaccination in Australian children below 5 years of age: A database study. *Hum. Vaccines Immunother.* 9, 1617–1625 (2013).

164. Tate, J. E. et al. Decline and change in seasonality of US rotavirus activity after the introduction of rotavirus vaccine. *Pediatrics* 124, 465–471 (2009).

165. Usonis, V., Ivaskeviciene, I., Desselberger, U., and Rodrigo, C. The unpredictable diversity of co-circulating rotavirus types in Europe and the possible impact of universal mass vaccination programmes on rotavirus genotype incidence. *Vaccine* 30, 4596–4605 (2012).

166. Zeller, M. et al. Rotavirus incidence and genotype distribution before and after national rotavirus vaccine introduction in Belgium. *Vaccine* 28, 7507–7513 (2010).

167. Matthijnssens, J. and Van Ranst, M. Genotype constellation and evolution of group A rotaviruses infecting humans. *Curr. Opin. Virol.* 2, 426–433 (2012).

168. Gurgel, R. Q. et al. Predominance of rotavirus P[4]G2 in a vaccinated population, Brazil. *Emerg. Infect. Dis.* 13, 1571–1573 (2007).

169. Nakagomi, T. et al. Apparent extinction of non-G2 rotavirus strains from circulation in Recife, Brazil, after the introduction of rotavirus vaccine. *Arch. Virol.* 153, 591–593 (2008).

170. Cilla, G., Montes, M., Gomariz, M., Piñeiro, L., and Pérez-Trallero, E. Rotavirus genotypes in children in the Basque Country (northern Spain) over a 13-year period (July 1996–June 2009). *Eur. J. Clin. Microbiol. Infect. Dis.* 29, 955–960 (2010).

171. Hemming, M. and Vesikari, T. Genetic diversity of G1P[8] rotavirus VP7 and VP8* antigens in Finland over a 20-year period: No evidence for selection pressure by universal mass vaccination with RotaTeq® vaccine. *Infect. Genet. Evol.* 19, 51–58 (2013).

172. Tizard, I. R. *Veterinary Immunology*. Elsevier Health Sciences, St. Louis, MO, (2013).

173. Kaplon, J. et al. Impact of rotavirus vaccine on rotavirus genotypes and caliciviruses circulating in French cattle. *Vaccine* 31, 2433–2440 (2013).

174. Elschner, M. et al. Isolation and molecular characterisation of equine rotaviruses from Germany. *Vet. Microbiol.* 105, 123–129 (2005).

175. Powell, D. G. et al. Field study of the safety, immunogenicity, and efficacy of an inactivated equine rotavirus vaccine. *J. Am. Vet. Med. Assoc.* 211, 193–198 (1997).

176. Imagawa, H. et al. Field study of inactivated equine rotavirus vaccine. *J. Equine Sci.* 16, 35–44 (2005).

177. Barrandeguy, M. et al. Prevention of rotavirus diarrhoea in foals by parenteral vaccination of the mares: Field trial. *Dev. Biol. Stand.* 92, 253–257 (1998).

178. Garaicoechea, L. et al. Molecular characterization of equine rotaviruses circulating in Argentinean foals during a 17-year surveillance period (1992–2008). *Vet. Microbiol.* 148, 150–160 (2011).

179. Tsunemitsu, H. et al. Predominance of G3B and G14 equine group A rotaviruses of a single VP4 serotype in Japan. *Arch. Virol.* 146, 1949–1962 (2001).

180. Higgins, W. P. *Field and Laboratory Studies of Equine Influenza Virus, Equine Rhinopneumonitis Virus, and Equine Rotavirus.* Cornell University, Ithaca, NY (1987).

181. Nakagomi, O., Nakagomi, T., Akatani, K., and Ikegami, N. Identification of rotavirus genogroups by RNA-RNA hybridization. *Mol. Cell. Probes* 3, 251–261 (1989).

182. Gouvea, V. et al. Polymerase chain reaction amplification and typing of rotavirus nucleic acid from stool specimens. *J. Clin. Microbiol.* 28, 276–282 (1990).

183. Freeman, M. M., Kerin, T., Hull, J., McCaustland, K., and Gentsch, J. Enhancement of detection and quantification of rotavirus in stool using a modified real-time RT-PCR assay. *J. Med. Virol.* 80, 1489–1496 (2008).

184. Mijatovic-Rustempasic, S. et al. Sensitive and specific quantitative detection of rotavirus A by one-step real-time reverse transcription-PCR assay without antecedent double-stranded-RNA denaturation. *J. Clin. Microbiol.* 51, 3047–3054 (2013).

185. Iturriza Gomara, M., Wong, C., Blome, S., Desselberger, U., and Gray, J. Molecular characterization of VP6 genes of human rotavirus isolates: Correlation of genogroups with subgroups and evidence of independent segregation. *J. Virol.* 76, 6596–6601 (2002).

186. German, A. C. et al. Molecular epidemiology of rotavirus in UK cats. *J. Clin. Microbiol.* 53, 455–464 (2015).

187. Gentsch, J. R. et al. Identification of group A rotavirus gene 4 types by polymerase chain reaction. *J. Clin. Microbiol.* 30, 1365–1373 (1992).

188. World Health Organization. *Manual of Rotavirus Detection and Characterization.* WHO, Geneva, Switzerland (2009).

189. Collins, P. J., Martella, V., Sleator, R. D., Fanning, S., and O'Shea, H. Detection and characterisation of group A rotavirus in asymptomatic piglets in southern Ireland. *Arch. Virol.* 155, 1247–1259 (2010).

190. Collins, P. J. et al. Detection and characterisation of bovine rotavirus in Ireland from 2006–2008. *Ir. Vet. J.* 67, 13 (2014).

191. Fukai, K. et al. Molecular characterisation of equine group A rotavirus, Nasuno, isolated in Tochigi Prefecture, Japan. *Vet. J.* 172, 369–373 (2006).

192. Maes, P., Matthijnssens, J., Rahman, M., and Van Ranst, M. RotaC: A web-based tool for the complete genome classification of group A rotaviruses. *BMC Microbiol.* 9, 238 (2009).

193. Matthijnssens, J. et al. Uniformity of rotavirus strain nomenclature proposed by the Rotavirus Classification Working Group (RCWG). *Arch. Virol.* 156, 1397–1413 (2011).

194. Matthijnssens, J. et al. Recommendations for the classification of group A rotaviruses using all 11 genomic RNA segments. *Arch. Virol.* 153, 1621–1629 (2008).

195. Yang, Z. and Rannala, B. Molecular phylogenetics: Principles and practice. *Nat. Rev. Genet.* 13, 303–314 (2012).

196. Fredj, M. B. H. et al. Feline origin of rotavirus strain, Tunisia, 2008. *Emerg. Infect. Dis.* 19, 630–634 (2013).

197. Doan, Y. H. et al. Identification by full-genome analysis of a bovine rotavirus transmitted directly to and causing diarrhea in a human child. *J. Clin. Microbiol.* 51, 182–189 (2013).

198. Luchs, A., Cilli, A., Morillo, S. G., Carmona Rde, C., and Timenetsky Mdo, C. S. T. Rare G3P[3] rotavirus strain detected in Brazil: Possible human-canine interspecies transmission. *J. Clin. Virol.* 54, 89–92 (2012).

199. Mukherjee, A., Mullick, S., Deb, A. K., Panda, S., and Chawla-Sarkar, M. First report of human rotavirus G8P[4] gastroenteritis in India: Evidence of ruminants-to-human zoonotic transmission. *J. Med. Virol.* 85, 537–545 (2013).

200. Donno, A. D. et al. Emergence of unusual human rotavirus strains in Salento, Italy, during 2006–2007. *BMC Infect. Dis.* 9, 43 (2009).

201. Iturriza-Gomara, M. et al. Molecular epidemiology of human group A rotavirus infections in the United Kingdom between 1995 and 1998. *J. Clin. Microbiol.* 38, 4394–4401 (2000).

202. Jere, K. C. et al. A first report on the characterization of rotavirus strains in Sierra Leone. *J. Med. Virol.* 83, 540–550 (2011).

203. Leite, J. P. G., Carvalho-Costa, F. A., and Linhares, A. C. Group A rotavirus genotypes and the ongoing Brazilian experience: A review. *Mem. Inst. Oswaldo Cruz* 103, 745–753 (2008).

204. Nyaga, M. M. et al. Sequence analysis of the whole genomes of five African human G9 rotavirus strains. *Infect. Genet. Evol.* 16, 62–77 (2013).

205. Pietsch, C., Petersen, L., Patzer, L., and Liebert, U. G. Molecular characteristics of German G8P[4] rotavirus strain GER1H-09 suggest that a genotyping and subclassification update is required for G8. *J. Clin. Microbiol.* 47, 3569–3576 (2009).

206. Ramani, S. et al. Whole genome characterization of reassortant G10P[11] strain (N155) from a neonate with symptomatic rotavirus infection: Identification of genes of human and animal rotavirus origin. *J. Clin. Virol.* 45, 237–244 (2009).

207. Than, V. T., Kang, H., Lim, I., and Kim, W. Molecular characterization of serotype G9 rotaviruses circulating in South Korea between 2005 and 2010. *J. Med. Virol.* 85, 171–178 (2013).

Section IV

DNA Viruses

73 Beak and Feather Disease Virus

Ana Margarida Henriques, Teresa Fagulha, and Miguel Fevereiro

CONTENTS

73.1 Introduction 659
    73.1.1 Classification 659
    73.1.2 Morphology and Genome Organization 660
    73.1.3 Replication 662
    73.1.4 Clinical Features and Pathogenesis 662
    73.1.5 Diagnosis 663
        73.1.5.1 Conventional Techniques 664
        73.1.5.2 Molecular Techniques 664
73.2 Methods 664
    73.2.1 Sample Preparation 664
    73.2.2 Detection Procedures 665
        73.2.2.1 Polymerase Chain Reaction 665
        73.2.2.2 DNA Sequencing 665
        73.2.2.3 Bayesian Analysis 666
73.3 Conclusion 666
References 666

73.1 INTRODUCTION

Beak and feather disease virus (BFDV) is the causative agent of a disease associated with feather dystrophy and beak deformity in psittacine birds named psittacine beak and feather disease (PBFD).[1] The virus belongs to the genus *Circovirus* in the *Circoviridae* family that includes the smallest known viruses. All members of this genus share similar morphology and genome organization. Circoviruses are nonenveloped viruses with icosahedral nucleocapsids that contain a single-stranded circular DNA genome of about 2000 nucleotides.

There is evidence that BFDV occurs naturally in wild Australian birds for more than one century.[2,3] The first outbreak of PBFD was reported in 1907 by Ashby from wild red-rumped parrots (*Psephotus haematonotus*) in Adelaide hills in 1988.[4] However, BFDV was only officially confirmed in 1975 in Australian cockatoos, including sulfur-crested cockatoos, lovebirds, budgerigars, and galahs.[5] The international trade of exotic parrots, both through legal trade and illegal trafficking, has facilitated the spread of BFDV that is now found in many countries, including Thailand,[6] New Zealand,[7] Germany,[8] South Africa,[9] the United States,[10] Italy,[11] Taiwan,[12] Japan,[13] and Portugal.[14] The prevalence of BFDV is highly variable, with reported viral DNA positive rates ranging from 3.5% to 4.0% in the United States[10] to 41.2% in Taiwan.[12] PBFD is nowadays considered a global disease[15] and is the most important viral infection in psittacine birds in many countries.[16]

All psittacine species are considered susceptible to BFDV since the virus has been reported to affect more than 60 wild and captive species of cockatoos and parrots.[17] The disease is recognized as a key threat to endangered psittacine birds in Australia[18] and other countries.

73.1.1 CLASSIFICATION

BFDV is classified in the family *Circoviridae* that covers the smallest known autonomously replicating animal viruses. The family's name, "circo," is derived from the Greek root "gyro," which means ring or circular and refers to the characteristic circular shape of the viral genome. In the classification of Baltimore,[19] members of this family are included in Class II of viruses with a single-stranded DNA (ssDNA) genome. The family *Circoviridae* contains two accepted genera, *Gyrovirus* and *Circovirus*, and one proposed genus, *Cyclovirus*.

Currently, there is only one recognized member of the *Gyrovirus* genus, chicken anemia virus (CAV), a ubiquitous pathogen of chickens with a worldwide distribution. This virus was discovered in Japan in 1979 and is responsible for anemia and immunosuppression in newly hatched chickens.[20] CAV is distinguishable from members of the *Circovirus* genus both by its size and its genome organization. The recent discovery of viruses with CAV-like genomes in chicken sera (AGV2),[21] human skin swabs (HGyV),[22] and human feces (GyV3 and GyV4)[23,24] indicates that there may be significantly more diversity within this genus. However, these viruses were not yet accepted by the International Committee on Taxonomy of Viruses (ICTV) as members of the genus *Gyrovirus*.

The first member of the *Circovirus* genus reported was porcine circovirus type 1 (PCV1), a nonpathogenic virus detected in 1974 as a contaminant of the porcine kidney cell line PK15.[25,26] BFDV was detected 1 year later in Australian cockatoos. Another type of porcine circovirus, PCV2, was initially identified in a Canadian swine herd in 1991 and has been implicated as the etiological agent of a disease named postweaning multisystemic wasting syndrome.[27] Heretofore, circoviruses have been detected in a large variety of host species. Besides PCV1, PCV2, and BFDV, the best characterized of the circoviruses, there are eight more species within this genus currently accepted by ICTV 2013, all of them are isolated from birds, revealing strong host specificity. These include pigeon circovirus,[28] canary circovirus,[29] goose circovirus,[28] duck circovirus,[30] starling circovirus,[31] finch circovirus,[32] gull circovirus,[32] and mute swan circovirus.[33] Circoviruses were also detected in other hosts, including ostriches,[34] ravens,[35] bats,[36] fishes,[37] dogs,[38] and minks,[39] but to date they were not yet recognized. There are two criteria adopted to differentiate species in the genus *Circovirus*, which are a complete nucleotide sequence identity of less than 75% and capsid protein amino acid sequence identity of less than 70%.[21]

Metagenomics approaches led to the identification of a novel proposed genus, *Cyclovirus*. The identification of viruses similar to circoviruses, but with sequences unknown or sharing low homology with known viral proteins revealed the existence of undescribed species that could eventually be assigned to the *Circoviridae* family. The first cyclovirus was identified from a metagenomic analysis of fecal samples from children, with the finding of a nucleic acid fragment that encodes a protein showing close similarity to the Rep protein of circoviruses.[40] The new genus was proposed by Li et al.,[41] after the identification of circovirus-like, circular DNA viral genomes in human and chimpanzee stool samples. The name *Cyclovirus* was proposed, from the Greek "cyclo" meaning circular. Cycloviruses were also found to be prevalent in the muscle tissue of farm animals, such as chickens, cows, sheep, goats, and camels, which suggests the potential for frequent cross-species exposure and zoonotic transmissions.[41] The identification of cycloviruses in abdomens of dragonflies, together with the evidences found of recombination among different viral isolates, suggests that dragonflies may be natural hosts of cycloviruses.[42] The existence of this proposed genus is pending by the ICTV.

Two different lineages may be considered within BFDV, one including the budgerigar isolates (BG lineage) and the other is composed of viruses from cockatoos and other psittacine species (CT lineage). A high divergence is observed within the CT lineage, and therefore, different classifications have been proposed. Heath et al.[9] suggested the existence of eight genotypes, I–VIII, and a new genotype IX was proposed by Henriques et al.[14] The inclusion of more recent strains in the analysis enabled Varsani et al.[43] to consider 14 genotypes, A to N, within the CT lineage of BFDV. A Bayesian tree showing the 14 genotypes proposed by Varsani et al.[43] and their correspondence to the classification of Henriques et al.[14] is represented in Figure 73.1.

73.1.2 MORPHOLOGY AND GENOME ORGANIZATION

BFDV is a spherical nonenveloped virus of icosahedral morphology. It is among the smallest known viruses, having a diameter of 14–16 nm.[44] BFDV virions consist of 60 identical subunits of capsid protein arranged in 12 pentamer clustered units forming a pentagonal capsomere. Under electron microscope, capsomeres of BFDV exhibit a smooth appearance.[45] The density of this virus cesium chloride is 1.378 g/mL.[46]

The genome of BFDV is monomeric, nonsegmented, and single stranded. It consists of a covalently closed circular DNA molecule of approximately 2 kb, with an ambisense organization, composed by an intergenic region flanked by two head-to-head arranged open reading frames (ORFs).[7] The intergenic region located between the 5′ ends of the two major ORFs includes a potential stem-loop structure formed by a pair of inverted repeats (Figure 73.2). At its apex is present a nonanucleotide motif that constitutes an essential cis-acting element required for DNA replication.[47] This nonanucleotide sequence NANTATTAC is highly conserved among circoviruses, geminiviruses, and nanoviruses, but only part of the conserved sequence (TATTAC) is essential for the cleavage of the viral strand.[48] Indeed, although circoviruses infecting different species have different nonanucleotide motifs, the last six nucleotides of the sequence are conserved among them.[35] The consensus sequence of BFDV is TAGTATTAC.[49] The most common nucleotide numbering used was adopted from Todd et al.,[28] with position 1 corresponding to the "A" residue that occurs at the nucleotide position 8 in the nonanucleotide sequence.

Two direct tandem repeats of eight nucleotides are present in the genome of BFDV. The octamer repeat is located directly downstream the stem-loop structure and probably play a role in virus replication, perhaps by acting as putative binding sites for the Rep protein.[29,31] Although not conserved in all BFDV isolates, the octamer sequence is GGGCACCG in most viruses[50] (Figure 73.2). Also, direct and inverted hexamer repeats may be observed in the genome of BFDV, but between isolates, they are not conserved in number nor in sequence.[14] Indeed, a study performed with PCV1 demonstrated that from the four existing hexamers only two are essential for an efficient replication of the virus.[51]

Two major ORFs are present in the genome of BFDV (Figure 73.2). ORF1 (rep gene) occurs in the viral strand and encodes the replication-associated protein Rep of about 33.5 kDa,[14] essential to initialize viral replication. The Rep protein contains three amino acid motifs, I, II, and III, conserved among circular ssDNA genomes employing the rolling circle replication (RCR) method of DNA replication, indicating that BFDV also replicate via this method.[52] Motifs I, II, and III are located in the N-terminus of the protein and are, respectively, FTLNN, GxxHLQGY, and YxxK.[53] Also, a P-loop for dNTP binding (GPPGCGKS) and three additional conserved motifs (WWDGY, DDFYGWLP, and DRYP) of no known function are present in the amino acid sequence of Rep protein.[14] It has been suggested that this protein is a recombinant between plant and mammalian viruses, with the N-terminal portion showing close similarity to the N-terminal region of the Rep protein

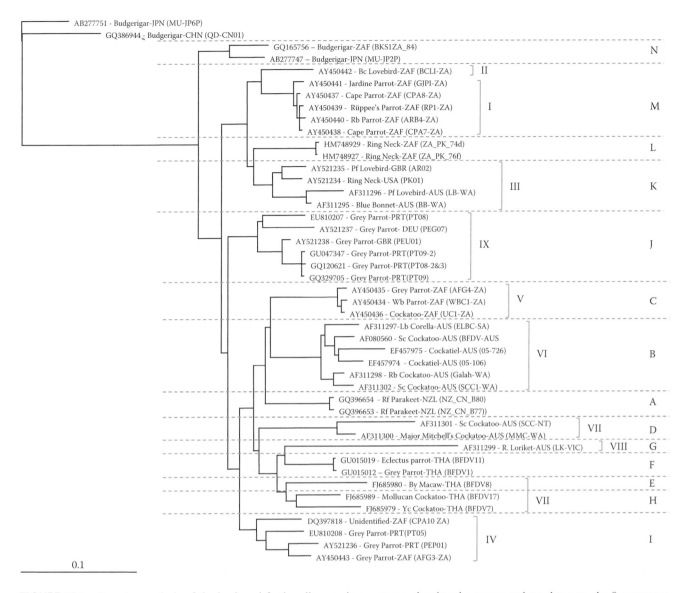

FIGURE 73.1 Bayesian analysis of the beak and feather disease virus genome, showing the correspondence between the 9 genotypes described by Henriques et al. (I–IX) and the 14 strains considered by Varsani et al. (A to N).

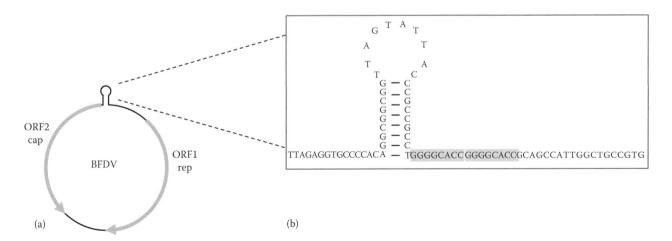

FIGURE 73.2 Schematic representation of the beak and feather disease virus genome. (a) Diagram of the putative replicative form, showing the two principal open reading frames and the stem-loop structure; (b) Nucleotide sequence of the intergenic region. The stem-loop, the nonanucleotide, and the octamer repeats (shaded) are represented.

from nanoviruses and geminiviruses, and the C-terminal portion being related to an RNA-binding protein encoded by calicivirus. This recombination may have occurred through a transition of RNA to DNA made by an endogenous mammalian retrovirus.[54]

ORF2 (cap gene) occurs in the complementary strand of the double-stranded replicative form (RF) and encodes the major structural capsid protein Cap of about 29.3 kDa.[14] Although the start codon for most BFDV isolates is the conventional ATG, a variety of alternative start codons in different positions for this ORF have been described, such as CTG,[50,55] TCT,[55] ATA or AGG,[56] and ATT.[14] The Cap protein contains a putative bipartite nuclear localization signal (NLS) consisting of 41 amino acid residues at N-terminus[57] that directs viral DNA through the nuclear membrane to the nucleus for DNA synthesis.[58] Also present in the Cap protein is a highly basic N-terminal region arginine rich, probably necessary for packaging of the viral genome into the viral capsid.[45]

Five other potential ORFs, ORF3–ORF7, that potentially encode proteins were detected in the putative RF of BFDV,[50] two of which are situated on the virus-sense strand (ORF4 and ORF5) and the other three on the complementary-sense strand (ORF3, ORF6, and ORF7).[14] The protein encoded by ORF3 seems to play a role in virus-induced apoptosis and in viral pathogenesis,[59] although it was shown that the protein was not essential for viral replication.[60] Indeed, there are two isolates from grey parrot from Portugal that lack this ORF.[60] The function of the other four ORFs is unknown and they are not all present in all isolates,[14] raising doubts regarding their requirement and significance.

Two potential TATA boxes are present in the virion strand in most isolates; the first (TATA) is positioned at nt 86–89, upstream Rep protein, while the second (TATAAAA) is located around nt 680. However, the utility of these sequences is unclear since BFDV strains were detected with only one or three TATA boxes.[14]

Three polyadenylation signals, two direct (AATAAA) and one inverted (TTTATT), were also detected in the majority of viruses, but as occurs with TATA boxes, the sequence, number, and location of these signals are not universal in all BFDV isolates.[14]

73.1.3 REPLICATION

Replication of BFDV is dependent on the replication machinery of the host cell since this virus does not have an autonomous DNA polymerase for the *de novo* DNA synthesis.[58] The Rep protein is essential to initialize viral replication, but continuation of the process requires cellular enzymes expressed during S phase of the cell cycle and initiates only after mitosis of the host cell.[61] BFDV is thought to replicate in rapidly dividing tissues, with evidence suggesting that the liver is an important site of replication.[62]

DNA replication occurs in the nucleus of infected cells, and therefore, before establishing a productive infection the viral DNA needs to cross the nuclear membrane through protein-mediated nuclear transport.[63] Proteins are directed in the nucleus

via NLSs such as those existing at the N-terminus of the Cap protein, so is this protein that transports the viral genome across the nuclear membrane before replication can occur.[58]

Once inside the nucleus of the host cell, the circular ssDNA genome replicates using RCR.[64] The initiation of the RCR of the virus has been ascribed to the nonanucleotide motif located at the apex of the stem-loop structure, in the nontranslated intergenic region.[29] The two octanucleotides located downstream of the stem loop are thought to be the binding sites for BFDV replication-associated protein.[64] The ssDNA genome serves as the template for the formation of a complementary strand of DNA, originating a double-stranded DNA (dsDNA) intermediate, known as the replicative form (RF). When one unit of the genome has been replicated, the loop is covalently closed, with the release of a positive circular single-stranded parental DNA molecule and a circular dsDNA molecule composed of the negative parental strand and the newly synthesized positive strand. The ssDNA molecule will then be encapsidated,[65] while the dsDNA RF forms the template for production of mRNA, with ORFs potentially on both the original virion strand and also on the complementary strand of the replicative intermediate.[52]

73.1.4 CLINICAL FEATURES AND PATHOGENESIS

BFDV is the causative agent of a disease typically characterized by symmetric feather dystrophy (Figure 73.3) and beak deformities, therefore named psittacine beak and feather disease (PBFD). As BFDV is dependent on the hosts' machinery for replication, it is predominantly found in rapidly dividing cells such as those of the epithelium and consequently, the skin, feathers, beak, esophagus, and crop, as well as organs of the immune system such as thymus, cloaca, bursa of Fabricius, and bone marrow are affected.[66]

PBFD is a potentially fatal disease that can be present as a peracute, acute, or chronic disease. Peracute and acute forms affect predominantly younger birds, from hatching to 3 years old, in which sudden death occurs with no evidence (peracute) or mild evidence (acute) of feather dystrophy.[67] Chronic disease was not described until now in nestling and fledgling parrots,[15] due to host conditions rather than antigenic or genotypic traits of BFDV.[7] Older birds may overcome the infection to become carriers of the virus, and maternal antibodies have been shown to provide immunity to offspring.[68]

The chronic form of the disease, also known as the "classical form," is the most frequent and affects birds up to 3 years of age. The disease begins with depression, lethargy, and diarrhea followed by a chronic, progressive, bilateral, and symmetrical feather dystrophy occasionally accompanied by beak and occasionally claw deformities. The abnormal appearance of feathers is the result of a dystrophic process that consists of necrosis and hyperplasia of epidermal cells.[69] The developing feathers will have a host of deformities including shortening, retention of a thick outer sheath, constrictions or necrosis of the shaft, stress lines in the vane, curled or clubbed feathers, and hemorrhage into the feather shaft.[44,69,70] Beak deformities are not always present and seem to occur in specific species or

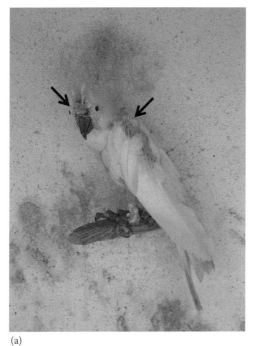

(a)   (b)

FIGURE 73.3 Clinical signs of the beak and feather disease virus. (a) Cockatoo exhibiting feather loss. (b) Grey parrot showing down feathers as a result of cover feathers loss. (Photo courtesy of Dr. Filipe Martinho from Nyctea, Lda, Lisbon, Portugal.)

are dependent on other factors.[46] Indeed, although symptoms are most often associated with the feathers rather than with the beak, species such as sulfur-crested cockatoos (*Cacatua galerita*), galahs (*Eolophus roseicapillus*), little corellas (*Cacatua sanguinea*), and Moluccan cockatoos (*Cacatua moluccensis*) are more susceptible to beak changes.[46] Possible beak deformities include elongation, transverse or longitudinal cracking, and palatine necrosis.[44,69,70] The feather and beak abnormalities associated with BFDV are the result of necrosis resultant by the targeting of feather, follicular, and beak epithelial cells by BFDV, observed in histopathological examinations. In addition to beak and feathers abnormalities, BFDV-infected birds typically display lymphoid depletion associated to an immunosuppression that favors the appearance of opportunistic secondary infections, which are the cause of death in most infected birds.[62] Psittacine species may be susceptible to gram-positive and gram-negative septicemia, localized bacterial infections, and fungal or parasitic infections: chlamydiosis, severe pulmonary aspergillosis, severe enteric cryptosporidiosis,[66] herpes virus, and polyomavirus adenovirus.[67]

Transmission of BFDV is not fully elucidated, but apparently can occur both horizontally and vertically.[71] Horizontal transmission may occur through inhalation or ingestion of virus particles shed in feather dust, crop secretions, or feces, as indicated by the presence of viral antigen in many organs associated with the alimentary tract.[72] Nuclear inclusions were observed within testicular germinal epithelium, showing that BFDV may be gonadotrophic.[66] Horizontal transmission may also occur when adults feed their young. Vertical transmission, from infected hens to embryonated eggs, has

been reported.[8] Also, infection of the ovary or oviduct[66] and the BFDV detection in nestlings of BFDV-positive parents[71] suggest a route for the vertical transmission of the virus. It has been reported that asymptomatic parents may produce offspring showing clinical signs of BFDV, suggesting that a carrier state exists with vertical or horizontal transmission and/or an existence of a genetic predisposition to the disease.[73] Within a captive environment, horizontal spread through environmental contamination is probably the most prevalent route of transmission. It is speculated that humans are also vehicles of disease transmission as the disease can be spread by handling a healthy bird after coming into contact with infected birds.

Currently, there is no cure or effective treatment for the disease caused by BFDV. Moreover, the production of a protective vaccine is hampered by the lack of success in culturing BFDV *in vitro*.[15]

73.1.5 DIAGNOSIS

PBFD is difficult to diagnose based on clinical signs alone as feather abnormalities may be due to other causes such as infection by avian polyomavirus.[46] Histopathological examinations have revealed the presence of basophilic intranuclear and intracytoplasmic inclusion bodies in feather and feather follicle epithelial cells or in macrophages present in the feather epithelium, follicular epithelium, or pulp cavity, with evidence of the presence of BFDV in such inclusion bodies.[66] Basophilic to amphophilic inclusion bodies were found in histological sections of feathers and follicular epithelium using hematoxylin and eosin stains.[66] However, similar basophilic inclusions have been observed with other viruses like avian polyomavirus and adenovirus, hindering diagnosis based solely upon histopathology. For that reason, specific assays for BFDV infection detection have been developed: hemagglutination and hemagglutination inhibition,[74] *in situ* hybridization,[75] electron microscopy,[46] ELISA,[76] polymerase chain reaction (PCR),[77] real-time PCR,[78,79] and RCA.[80]

73.1.5.1 Conventional Techniques

The discovery that BFDV has the ability to agglutinate erythrocytes allowed the development of hemagglutination (HA) test for the detection of BFDV that can be also used to evaluate the amount of virus present in a sample.[81] Virus can be detected in affected feathers.[74] Hemagglutination inhibition (HI) assays can correspondingly be used to detect antibodies against BFDV that can be useful in diagnosing clinically normal birds that have mounted an immune response following exposure to the virus. Antibodies can be detected in blood, serum, plasma, or yolk.[74,82] The HA assay has been shown to be ineffective in identifying carriers of the disease as it does not detect latent or incubating BFDV infection. The use of HA and HI tests as serological tools is limited because persistently infected birds are required and inadequate amounts of virus are usually obtained from feather follicle tracts.[83] The suitability of these techniques is also limited by the possibility of genetic and antigenic diversity of BFDV.[84]

Several ELISA reactions had been developed for the detection of BFDV-specific antibodies. The advantages of these tests include their use as noninvasive diagnostic tests and the ability to screen multiple samples simultaneously. However, as occurs with HA and HI, a drawback of ELISA tests is the inability to produce large amounts of standardized serological diagnostic test antigen, which is safe to use. Also cross-reactivity may occur. For that reason, a competitive ELISA was developed, using a baculovirus-expressed recombinant BFDV capsid protein and a newly developed monoclonal antibody raised against this protein. This competitive ELISA has proven to be a sensitive and specific diagnostic test that should have wide application for the serodiagnosis of BFDV.[76]

As BFDV cannot be cultivated in cell cultures or in embryonated eggs, due to either its high *in vivo* tissue specificity or its specific growth requirements,[62] diagnosis must be confirmed by detection of either viral antigen or viral nucleic acid. The detection of viral nucleic acid may be performed by molecular techniques.

73.1.5.2 Molecular Techniques

Once the BFDV genome was fully sequenced,[50,64] universal primers designed to bind to and amplify a portion of the rep gene were developed[77] and later refined[7] to enable diagnosis of BFDV through the PCR. The oligonucleotide primers hybridize within the ORF1 (Rep) of the BFDV genome and allow for the detection of clinically infected as well as latently infected birds. Compared with HA and HI, PCR has proven to be more specific and have greater sensitivity.[85] The quantitative real-time PCR is even more specific and sensitive, allowing the detection of asymptomatic infections with simultaneous

quantification of the amount of virus present in a sample. The indication of the viral titer may be useful to determine the clinical relevance of the infection. Katoh, Shearer, and colleagues developed real-time PCR based on the SYBR Green[78] or SYTO9[79] assay systems. Eastwood and collaborators[86] developed a real-time PCR based on a TaqMan probe. Real-time PCR assays exhibit low inter- and intra-assay variability and may eventually replace conventional PCR methods as the gold standard for the diagnosis of BFDV infection.[78]

More recently, the bacteriophage phi29 DNA polymerase has been used for the efficient isothermal amplification of circular DNA viral genomes without the need of specific primers by a rolling circle amplification (RCA) mechanism.[80] Major advantages of this technique is that the amplification is performed at isothermal conditions, the single-step amplification of the whole genome is possible and is independent of DNA sequence, making it possible to detect viruses with very low sequence similarities to known viruses.[31] RCA has been successfully applied to BFDV positive samples.[43]

The loop-mediated isothermal amplification has been used with success for the detection of goose,[87] pigeon,[88] and duck[89] circoviruses and, therefore, may be also applied for the diagnosis of BFDV. This molecular method developed by Notomi et al.[90] to amplify DNA or RNA under isothermal conditions is highly specific and sensitive and, advantageously, it can be completed within less than 45 min under isothermal conditions, without the need of a thermal cycler and/or specialized laboratory.

73.2 METHODS

73.2.1 Sample Preparation

The most widely used methods for the diagnosis of BFDV are PCR and real-time PCR since they are cheap, easy to implement, and highly sensitive and specific methods. Before these techniques can be carried out, DNA extraction is required, what may be performed from several matrices.

The presence of BFDV can be detected in live animals in blood, cloacal swabs, or feather samples.[91] After death, several organs may be collected at necropsies and pooled for DNA extraction. Sample preparation depends on the matrix used. When blood samples are used, the sample must be diluted to 5% in phosphate buffered saline (PBS) prior to DNA extraction, since avian erythrocytes are nucleated. Dry cloacal swabs are resuspended in PBS containing 200 units/mL of penicillin, 200 mg/mL of streptomycin, and 100 mg of gentamicin/mL. For DNA extraction from feathers, a 0.5–1 cm section is cut from the terminal portion of the feather quill and placed in a 1.5 mL Eppendorf tube containing 300 μL Buffer ATL (QIAGEN, Valencia, CA) and 20 μL dithiothreitol (DTT) 1 M. Incubation at 56°C must be performed until the sample is completely lysed. When organs collected from a dead animal at necropsy are used, the pool of organs is homogenized and suspended 10% in PBS containing antibodies as described earlier.

Several approaches may be used for DNA extraction, from the classical method of extraction with phenol and chloroform/isomayl alcohol followed by ethanol precipitation to highly sensitive and reproducible DNA extraction kits introduced by various commercial companies and widely applied. For processing of a large number of samples, the nucleic acid extraction workstation BioSprint 96 (QIAGEN, Valencia, CA) may be employed, with the advantage of providing high-quality DNA in a quick way reducing the risk of contamination. This kit uses MagAttract magnetic-particle technology that combines the speed and efficiency of silica-based DNA purification with the convenient handling of magnetic particles. The protocol must be performed according to the manufacturer's instructions. Briefly, 200 µL sample are mixed by pulse vortexing with 200 µL Buffer AL. The mixture is transferred to an s-block followed by the addition of 20 µL QIAGEN Protease. After 10 min incubation at 70°C, 200 µL isopropanol and 30 µL MagAttract Suspension G are added. The BioSprint 96 is loaded with s-block, the other s-blocks already prepared (two with 500 µL Buffer AW1 per well, two with 500 µL Buffer AW2 per well, and one with 500 µL RNase-free water with 0.02% Tween 20 per well) and a 96-well microplate containing 100 µL Buffer AE for elution of the DNA. Start button is pressed to start the protocol run. After the samples are processed, s-blocks are removed and discarded and the elution plate is stored at 4°C. The DNA extracted is ready for use as template for PCR amplification.

73.2.2 DETECTION PROCEDURES

73.2.2.1 Polymerase Chain Reaction

The presence of BFDV can be detected by PCR amplification using degenerate primers 174 and 175 described by Mankertz and collaborators.[56] These primers are directed to regions of the *rep* gene that are conserved in vertebrate circoviruses. Primer 174 (5′-GTGTTTCACBCTYAAYAAYCCTWCMGA-3′) recognizes protein motif FTLNN involved in RCR, while primer 175 (5′-ATCTTCCCCRTSRTAWCCATCCCACCA-3′) is directed against the conserved amino acid motif WWDGY. The reaction is performed with 500 ng of the total DNA and 25 pmol of each primer, using High Fidelity PCR Master (Roche Diagnostics, Indianapolis, IN), according to the manufacturer's protocol.

73.2.2.1.1 Procedure

1. Remove all of PCR reagents from the freezer and allow them to thaw. Vortex and spin down all microcentrifuge tubes of PCR components before use.
2. Prepare the reaction mix as follows (volumes indicated are per sample):

Component	Volume (µL)
High-fidelity PCR master mix (2×)	12.5
Primer 174 (50 pmol/µL)	0.5
Primer 175 (50 pmol/µL)	0.5
H₂O	9.5
Total	23.0

Note: Positive and negative controls are always included along with the test samples in order to monitor the reaction and any possible contamination that might occur during the PCR procedure, respectively.

3. Distribute 23.0 µL of PCR mix into each 0.2 mL PCR tube.
4. Add 2 µL of DNA sample.
5. Spin the sample tubes in a microcentrifuge at 14,000 × *g* for 30 s at room temperature.
6. Place the sample tubes in the thermal cycler for an initial denaturation at 95°C for 10 min, followed by 45 cycles of denaturation at 95°C for 30 s, annealing at 52°C for 30 s, and extension at 72°C for 1 min. A final extension is performed at 72°C for 10 min.
7. Analyze the PCR products by agarose gel electrophoresis through 1% agarose gel in TBE buffer at 100 V for 30 min. Use GelRed (Biotium Inc., Hayward, CA) for staining.
8. Observe under ultraviolet light source.
9. Excise the expected fragments of about 561 bp for DNA sequencing.

For the amplification of the full-length genome of BFDV, four additional PCR may be performed as stated earlier using primers indicated in the following table. The name and the sequence of each primer are shown, as well as their location in the BFDV genome. Also, the annealing temperature at which each reaction must be performed and the length of the amplified fragment are indicated.

Name	Sequence (5′–3′)	Local (nt)	Temp. Annealing (°C)	Fragment Length (Base Pairs)
Pap (F)	CACGTGACTTC AAGACTGAG	588–607	50	509
1096 R	CCTGACGGCCC TGCGAGC	1096–1079		
921 F	TCACTCGTGTT TGGTGGTAC	921–940	55	631
1551 R	CACAAGACCCA GTGGCACC	1551–1533		
1441 F	CGCTGTGGTCAG GTCGTCTATTG	1441–1461	58	530
1970 R	CACCTCTAACTG CGCATGCG	1970–1951		
1889 F	GTATCGCCTGTT GTGACGTC	1889–1908	58	700
Pap (R)	TCACGTGGACT GGAACCAAC	594–575		

73.2.2.2 DNA Sequencing

The nucleotide sequence of the full-length genome of BFDV may be obtained by direct sequencing of the five amplicons obtained in PCR. Prior to sequencing, the amplified fragments must be purified from the agarose gel, for example, using NZYGelpure Kit (NZYTech, Lda., Lisbon, Portugal). DNA sequencing may be performed using a BigDye Terminator Cycle Sequencing Kit (Applied Biosystems, Foster City, CA) according to the manufacturer's instructions and the nucleotide sequences are determined on an automated 3130 Genetic Analyzer System (Applied Biosystems, Foster City, CA).

73.2.2.2.1 Procedure

1. Mix the following reagents in a labeled tube for each sequencing reaction.

Component	Volume (μL)
Terminator ready reaction mix (v1.1)	4.0
Sequencing buffer	2.0
Sequencing primer (5 pmol/μL)	1.0
Purified PCR product	1.0
H_2O	12.0
Total	20.0

2. Spin the sample tubes in a microcentrifuge at $14,000 \times g$ for 30 s at room temperature.
3. Place the sample tubes in a thermocycler for 25 cycles of 96°C for 10 s, 5 s at annealing temperature of the primer, and 60°C for 1 min.
4. Precipitate the DNA sequencing product with 50 μL ethanol 95%, 2 μL EDTA 2 mM, and 2 μL sodium acetate 3 M pH 5.2.
5. Centrifuge for 20 min at $14,000 \times g$ at 4°C.
6. Wash with 200 μL ethanol 70%.
7. Centrifuge for 7 min at $14,000 \times g$ at 4°C.
8. Resuspend the dry pellet in 15 μL formamide.
9. Finally, obtain the nucleotide sequence of the DNA products on an automated 3130 Genetic Analyzer System (Applied Biosystems, Foster City, CA).

73.2.2.3 Bayesian Analysis

In order to evaluate the genotype of the strain in study, it is necessary to perform a phylogenetic analysis with representatives of each genotype retrieved from a database such as GenBank. Multiple alignments of the nucleotide sequences are generated by Clustal W,[92] and the results are converted to the NEXUS format using Mesquite software.[93] The phylogenetic tree is obtained with a Bayesian inference of phylogeny throughout the MrBayes version 3.1.2 software that uses a simulation technique called Markov chain Monte Carlo to approximate the posterior probabilities of trees.[94,95] MrBayes analysis is performed using the general time reversible (GTR) model (nst=6) with gamma-shaped rate variation with a proportion of invariable sites (rates=invgamma). The analysis is run for 10^6 generations (ngen=106) with four chains of temperature (nchains=4), and each chain is sampled every 10th generations (samplefreq=10).

The position of the strain in study in the Bayesian tree obtained will indicate the genotype of the isolate. Genotypic differentiation of BFDV isolates is crucial to establish effective control strategies for the conservation of endangered species and epidemiological investigations of disease outbreaks.

73.3 CONCLUSION

BFDV is an important pathogen responsible for PBFD, characterized mainly by feather dystrophy and beak deformities. The immunosuppression often associated with the presence of the virus favors the appearance of opportunistic secondary infections, which are the cause of death in the majority of infected birds. Nowadays, PBFV is considered a global disease, with prevalence highly variable country by country.

Molecular detection of BFDV genome is of major importance for the diagnosis of the disease, and the application of sensitive molecular techniques all over the world contributes to a great recognition of the overall prevalence and economic impact of this virus. The prompt diagnosis of PBFD is crucial to prevent rapid spread of the virus among birds and also to control international trade of psittacides. This virus remains infectious for long periods in the environment and is highly resistant to disinfection, requiring strict quarantine procedures to prevent contamination and spreading within the facility. Molecular characterization and genotyping of the virus is critical to establish effective control strategies for the conservation of endangered species.

There is currently no cure or treatment for BFDV, with therapeutic intervention being limited to treating secondary infections. The management of the disease lies thus mostly in prevention, and infected birds must be culled to stop viral spread to healthy members of the flock. Candidate vaccines against BFDV infection are being tested, but so far no suitable vaccine is available. Further development and implementation of improved prevention and treatment measures are warranted and indeed essential for the eventual eradication of BFDV infection.

REFERENCES

1. Ritchie, B.W. Circoviridae. In *Avian Viruses: Function and Control*, ed. Ritchie, B.W., pp. 223–252, Wingers Publishing, Lake Worth, FL (1995).
2. Ashby, E. Notes on *Psephotus hematonotus*, the Red-rumped Grass Parrakeet. *The Avicultural Magazine Third Series* XII, 131–133 (1921).
3. Powell, A.M. Nude cockatoos. *Emu* III, 55–56 (1903).
4. Ashby, E. Parakeets moulting. *Emu* 6, 193–194 (1907).
5. Pass, D.A. and Perry, R.A. Psittacine beak and feather disease: An update. *Aust Vet Pract* 15, 55–60 (1985).
6. Kiatipattanasakul-Banlunara, W. et al. Psittacine beak and feather disease in three captive sulphur-crested cockatoos (*Cacatua galerita*) in Thailand. *J Vet Med Sci* 64, 527–529 (2002).
7. Ritchie, P.A., Anderson, I.L., and Lambert, D.M. Evidence for specificity of psittacine beak and feather disease viruses among avian hosts. *Virology* 306, 109–115 (2003).

8. Rahaus, M. and Wolff, M.H. Psittacine beak and feather disease: A first survey of the distribution of beak and feather disease virus inside the population of captive psittacine birds in Germany. *J Vet Med B Infect Dis Vet Public Health* 50, 368–371 (2003).

9. Heath, L. et al. Evidence of unique genotypes of beak and feather disease virus in southern Africa. *J Virol* 78, 9277–9284 (2004).

10. de Kloet, E. and de Kloet, S.R. Analysis of the beak and feather disease viral genome indicates the existence of several genotypes which have a complex psittacine host specificity. *Arch Virol* 149, 2393–2412 (2004).

11. Bert, E., Tomassone, L., Peccati, C., Navarrete, M.G., and Sola, S.C. Detection of beak and feather disease virus (BFDV) and avian polyomavirus (APV) DNA in psittacine birds in Italy. *J Vet Med B Infect Dis Vet Public Health* 52, 64–68 (2005).

12. Hsu, C.M., Ko, C.Y., and Tsaia, H.J. Detection and sequence analysis of avian polyomavirus and psittacine beak and feather disease virus from psittacine birds in Taiwan. *Avian Dis* 50, 348–353 (2006).

13. Sanada, N. and Sanada, Y. Epidemiological survey of psittacine beak and feather disease in Japan. *J Jpn Vet Med Assoc* 60, 61–65 (2007).

14. Henriques, A.M. et al. Phylogenetic analysis of six isolates of beak and feather disease virus from African grey parrots in Portugal. *Avian Dis* 54, 1066–1071 (2010).

15. Robino, P. et al. Molecular analysis and associated pathology of beak and feather disease virus isolated in Italy from young Congo African grey parrots (*Psittacus erithacus*) with an "atypical peracute form" of the disease. *Avian Pathol* 43, 333–344 (2014).

16. Katoh, H., Ogawa, H., Ohya, K., and Fukushi, H. A review of DNA viral infections in psittacine birds. *J Vet Med Sci* 72, 1099–1106 (2010).

17. Katoh, H., Ohya, K., Ise, K., and Fukushi, H. Genetic analysis of beak and feather disease virus derived from a cockatiel (*Nymphicus hollandicus*) in Japan. *J Vet Med Sci* 72, 631–634 (2010).

18. Sarker, S. et al. Phylogeny of beak and feather disease virus in cockatoos demonstrates host generalism and multiple-variant infections within Psittaciformes. *Virology* 460–461, 72–82 (2014).

19. Baltimore, D. Expression of animal virus genomes. *Bacteriol Rev* 35, 235–241 (1971).

20. Schat, K.A. Chicken anemia virus. *Curr Top Microbiol Immunol* 331, 151–183 (2009).

21. Rijsewijk, F.A. et al. Discovery of a genome of a distant relative of chicken anemia virus reveals a new member of the genus *Gyrovirus. Arch Virol* 156, 1097–1100 (2011).

22. Sauvage, V. et al. Identification of the first human gyrovirus, a virus related to chicken anemia virus. *J Virol* 85, 7948–7950 (2011).

23. Phan, T.G. et al. A third gyrovirus species in human faeces. *J Gen Virol* 93, 1356–1361 (2012).

24. Chu, D.K. et al. Characterization of a novel gyrovirus in human stool and chicken meat. *J Clin Virol* 55, 209–213 (2012).

25. Tischer, I., Rasch, R., and Tochtermann, G. Characterization of papovavirus-and picornavirus-like particles in permanent pig kidney cell lines. *Zentralbl Bakteriol Orig A* 226, 153–167 (1974).

26. Dulac, G.C. and Afshar, A. Porcine circovirus antigens in Pk-15 cell-line (Atcc Ccl-33) and evidence of antibodies to circovirus in Canadian Pigs. *Can J Vet Res* 53, 431–433 (1989).

27. Harding, J.C.S. and Clark, E.G. Recognizing and diagnosing postweaning multisystemic wasting syndrome (PMWS). *Swine Health Prod* 5, 201–203 (1997).

28. Todd, D., Weston, J.H., Soike, D., and Smyth, J.A. Genome sequence determinations and analyses of novel circoviruses from goose and pigeon. *Virology* 286, 354–362 (2001).

29. Phenix, K.V. et al. Nucleotide sequence analysis of a novel circovirus of canaries and its relationship to other members of the genus *Circovirus* of the family Circoviridae. *J Gen Virol* 82, 2805–2809 (2001).

30. Hattermann, K., Schmitt, C., Soike, D., and Mankertz, A. Cloning and sequencing of Duck circovirus (DuCV). *Arch Virol* 148, 2471–2480 (2003).

31. Johne, R., Fernandez-de-Luco, D., Hofle, U., and Muller, H. Genome of a novel circovirus of starlings, amplified by multiply primed rolling-circle amplification. *J Gen Virol* 87, 1189–1195 (2006).

32. Todd, D. et al. Molecular characterization of novel circoviruses from finch and gull. *Avian Pathol* 36, 75–81 (2007).

33. Halami, M.Y., Nieper, H., Muller, H., and Johne, R. Detection of a novel circovirus in mute swans (*Cygnus olor*) by using nested broad-spectrum PCR. *Virus Res* 132, 208–212 (2008).

34. Eisenberg, S.W., van Asten, A.J., van Ederen, A.M., and Dorrestein, G.M. Detection of circovirus with a polymerase chain reaction in the ostrich (*Struthio camelus*) on a farm in the Netherlands. *Vet Microbiol* 95, 27–38 (2003).

35. Stewart, M.E., Perry, R., and Raidal, S.R. Identification of a novel circovirus in Australian ravens (*Corvus coronoides*) with feather disease. *Avian Pathol* 35, 86–92 (2006).

36. Ge, X.Y. et al. Genetic diversity of novel circular ssDNA viruses in bats in China. *J Gen Virol* 92, 2646–2653 (2011).

37. Lorincz, M., Csagola, A., Farkas, S.L., Szekely, C., and Tuboly, T. First detection and analysis of a fish circovirus. *J Gen Virol* 92, 1817–1821 (2011).

38. Kapoor, A., Dubovi, E.J., Henriquez-Rivera, J.A., and Lipkin, W.I. Complete genome sequence of the first canine circovirus. *J Virol* 86, 7018 (2012).

39. Lian, H. et al. Novel circovirus from mink, China. *Emerg Infect Dis* 20, 1548–1550 (2014).

40. Victoria, J.G. et al. Metagenomic analyses of viruses in stool samples from children with acute flaccid paralysis. *J Virol* 83, 4642–4651 (2009).

41. Li, L.L. et al. Multiple diverse circoviruses infect farm animals and are commonly found in human and Chimpanzee feces. *J Virol* 84, 1674–1682 (2010).

42. Rosario, K. et al. Dragonfly cyclovirus, a novel single-stranded DNA virus discovered in dragonflies (Odonata: Anisoptera). *J Gen Virol* 92, 1302–1308 (2011).

43. Varsani, A., Regnard, G.L., Bragg, R., Hitzeroth, I.I., and Rybicki, E.P. Global genetic diversity and geographical and host-species distribution of beak and feather disease virus isolates. *J Gen Virol* 92, 752–767 (2011).

44. Ritchie, B.W. et al. Ultrastructural, protein composition, and antigenic comparison of psittacine beak and feather disease virus purified from four genera of psittacine birds. *J Wildl Dis* 26, 196–203 (1990).

45. Crowther, R.A., Berriman, J.A., Curran, W.L., Allan, G.M., and Todd, D. Comparison of the structures of three circoviruses: Chicken anemia virus, porcine circovirus type 2, and beak and feather disease virus. *J Virol* 77, 13036–13041 (2003).

46. Ritchie, B.W., Niagro, F.D., Lukert, P.D., Steffens, W.L., 3rd and Latimer, K.S. Characterization of a new virus from cockatoos with psittacine beak and feather disease. *Virology* 171, 83–88 (1989).

47. Mankertz, A., Persson, F., Mankertz, J., Blaess, G., and Buhk, H.J. Mapping and characterization of the origin of DNA replication of porcine circovirus. *J Virol* 71, 2562–2566 (1997).

48. Steinfeldt, T., Finsterbusch, T., and Mankertz, A. Demonstration of nicking/joining activity at the origin of DNA replication associated with the rep and rep' proteins of porcine circovirus type 1. *J Virol* 80, 6225–6234 (2006).

49. Rosario, K., Duffy, S., and Breitbart, M. A field guide to eukaryotic circular single-stranded DNA viruses: Insights gained from metagenomics. *Arch Virol* 157, 1851–1871 (2012).

50. Bassami, M.R., Berryman, D., Wilcox, G.E., and Raidal, S.R. Psittacine beak and feather disease virus nucleotide sequence analysis and its relationship to porcine circovirus, plant circoviruses, and chicken anaemia virus. *Virology* 249, 453–459 (1998).

51. Steinfeldt, T., Finsterbusch, T., and Mankertz, A. Functional analysis of cis- and trans-acting replication factors of porcine circovirus type 1. *J Virol* 81, 5696–5704 (2007).

52. Ilyina, T.V. and Koonin, E.V. Conserved sequence motifs in the initiator proteins for rolling circle DNA replication encoded by diverse replicons from eubacteria, eucaryotes and archaebacteria. *Nucleic Acids Res* 20, 3279–3285 (1992).

53. Regnard, G.L., Boyes, R.S., Martin, R.O., Hitzeroth, I.I., and Rybicki, E.P. Beak and feather disease viruses circulating in Cape parrots (*Poicepahlus robustus*) in South Africa. *Arch Virol* 160(1), 47–54 (2014).

54. Gibbs, M.J. and Weiller, G.F. Evidence that a plant virus switched hosts to infect a vertebrate and then recombined with a vertebrate-infecting virus. *Proc Natl Acad Sci USA* 96, 8022–8027 (1999).

55. Bassami, M.R., Ypelaar, I., Berryman, D., Wilcox, G.E., and Raidal, S.R. Genetic diversity of beak and feather disease virus detected in psittacine species in Australia. *Virology* 279, 392–400 (2001).

56. Mankertz, A., Hattermann, K., Ehlers, B., and Soike, D. Cloning and sequencing of columbid circovirus (coCV), a new circovirus from pigeons. *Arch Virol* 145, 2469–2479 (2000).

57. Shang, S.B. et al. Development and validation of a recombinant capsid protein-based ELISA for detection of antibody to porcine circovirus type 2. *Res Vet Sci* 84, 150–157 (2008).

58. Heath, L., Williamson, A.L., and Rybicki, E.P. The capsid protein of beak and feather disease virus binds to the viral DNA and is responsible for transporting the replication-associated protein into the nucleus. *J Virol* 80, 7219–7225 (2006).

59. Liu, J. et al. The ORF3 protein of porcine circovirus type 2 interacts with porcine ubiquitin E3 ligase Pirh2 and facilitates p53 expression in viral infection. *J Virol* 81, 9560–9567 (2007).

60. Chaiyakul, M., Hsu, K., Dardari, R., Marshall, F., and Czub, M. Cytotoxicity of ORF3 proteins from a nonpathogenic and a pathogenic porcine circovirus. *J Virol* 84, 11440–11447 (2010).

61. Tischer, I., Peters, D., Rasch, R., and Pociuli, S. Replication of porcine circovirus: Induction by glucosamine and cell cycle dependence. *Arch Virol* 96, 39–57 (1987).

62. Todd, D. Circoviruses: Immunosuppressive threats to avian species: A review. *Avian Pathol* 29, 373–394 (2000).

63. Lin, W.L., Chien, M.S., Du, Y.W., Wu, P.C., and Huang, C. The N-terminus of porcine circovirus type 2 replication protein is required for nuclear localization and ori binding activities. *Biochem Biophys Res Commun* 379, 1066–1071 (2009).

64. Niagro, F.D. et al. Beak and feather disease virus and porcine circovirus genomes: Intermediates between the geminiviruses and plant circoviruses. *Arch Virol* 143, 1723–1744 (1998).

65. Faurez, F., Dory, D., Grasland, B., and Jestin, A. Replication of porcine circoviruses. *Virol J* 6, 60 (2009).

66. Latimer, K.S. et al. A novel DNA virus associated with feather inclusions in psittacine beak and feather disease. *Vet Pathol* 28, 300–304 (1991).

67. Doneley, R.J. Acute beak and feather disease in juvenile African Grey parrots—An uncommon presentation of a common disease. *Aust Vet J* 81, 206–207 (2003).

68. Ritchie, B.W. et al. Antibody response to and maternal immunity from an experimental psittacine beak and feather disease vaccine. *Am J Vet Res* 53, 1512–1518 (1992).

69. Pass, D.A. and Perry, R.A. The pathology of psittacine beak and feather disease. *Aust Vet J* 61, 69–74 (1984).

70. McOrist, S., Black, D.G., Pass, D.A., Scott, P.C., and Marshall, J. Beak and feather dystrophy in wild sulphur-crested cockatoos (*Cacatua galerita*). *J Wildl Dis* 20, 120–124 (1984).

71. Rahaus, M. et al. Detection of beak and feather disease virus DNA in embryonated eggs of psittacine birds. *Vet Med* 53, 53–58 (2008).

72. Ritchie, B.W. et al. Routes and prevalence of shedding of psittacine beak and feather disease virus. *Am J Vet Res* 52, 1804–1809 (1991).

73. Ritchie, B.W. et al. A review of psittacine beak and feather disease characteristics of the PBFD virus. *J Assoc Avian Vet* 3, 143–149 (1989).

74. Ritchie, B.W. et al. Hemagglutination by psittacine beak and feather disease virus and use of hemagglutination inhibition for detection of antibodies against the virus. *Am J Vet Res* 52, 1810–1815 (1991).

75. Ramis, A. et al. Diagnosis of psittacine beak and feather disease (PBFD) viral infection, avian polyomavirus infection, adenovirus infection and herpesvirus infection in psittacine tissues using DNA in situ hybridization. *Avian Pathol* 23, 643–657 (1994).

76. Shearer, P.L., Sharp, M., Bonne, N., Clark, P., and Raidal, S.R. A blocking ELISA for the detection of antibodies to psittacine beak and feather disease virus (BFDV). *J Virol Methods* 158, 136–140 (2009).

77. Ypelaar, I., Bassami, M.R., Wilcox, G.E., and Raidal, S.R. A universal polymerase chain reaction for the detection of psittacine beak and feather disease virus. *Vet Microbiol* 68, 141–148 (1999).

78. Katoh, H., Ohya, K., and Fukushi, H. Development of novel real-time PCR assays for detecting DNA virus infections in psittaciform birds. *J Virol Methods* 154, 92–98 (2008).

79. Shearer, P.L., Sharp, M., Bonne, N., Clark, P., and Raidal, S.R. A quantitative, real-time polymerase chain reaction assay for beak and feather disease virus. *J Virol Methods* 159, 98–104 (2009).

80. Johne, R., Muller, H., Rector, A., van Ranst, M., and Stevens, H. Rolling-circle amplification of viral DNA genomes using phi29 polymerase. *Trends Microbiol* 17, 205–211 (2009).

81. Raidal, S.R. and Cross, G.M. The haemagglutination spectrum of psittacine beak and feather disease virus. *Avian Pathol* 23, 621–630 (1994).

82. Raidal, S.R., Firth, G.A., and Cross, G.M. Vaccination and challenge studies with psittacine beak and feather disease virus. *Aust Vet J* 70, 437–441 (1993).

83. Studdert, M.J. Circoviridae: New viruses of pigs, parrots and chickens. *Aust Vet J* 70, 121–122 (1993).

84. Johne, R., Raue, R., Grund, C., Kaleta, E.F., and Muller, H. Recombinant expression of a truncated capsid protein of beak and feather disease virus and its application in serological tests. *Avian Pathol* 33, 328–336 (2004).

85. Khalesi, B., Bonne, N., Stewart, M., Sharp, M., and Raidal, S. A comparison of haemagglutination, haemagglutination inhibition and PCR for the detection of psittacine beak and feather disease virus infection and a comparison of isolates obtained from loriids. *J Gen Virol* 86, 3039–3046 (2005).

86. Eastwood, J.R. et al. Phylogenetic analysis of beak and feather disease virus across a host ring-species complex. *Proc Natl Acad Sci USA* 111, 14153–14158 (2014).

87. Wozniakowski, G., Kozdrun, W., and Samorek-Salamonowicz, E. Loop-mediated isothermal amplification for the detection of goose circovirus. *Virol J* 9, 110 (2012).

88. Tsai, S.S. et al. Development of a loop-mediated isothermal amplification method for rapid detection of pigeon circovirus. *Arch Virol* 159, 921–926 (2014).

89. Xie, L. et al. A loop-mediated isothermal amplification assay for the visual detection of duck circovirus. *Virol J* 11, 76 (2014).

90. Notomi, T. et al. Loop-mediated isothermal amplification of DNA. *Nucleic Acids Res* 28, E63 (2000).

91. Ledwon, A., Sapierzynski, R., Augustynowicz-Kopec, E., Szeleszczuk, P., and Kozak, M. Experimental inoculation of BFDV-positive budgerigars (*Melopsittacus undulatus*) with two *Mycobacterium avium* subsp. *avium* isolates. *Biomed Res Int* 2014, 418563 (2014).

92. Thompson, J.D., Higgins, D.G., and Gibson, T.J. CLUSTAL W: Improving the sensitivity of progressive multiple sequence alignment through sequence weighting, position-specific gap penalties and weight matrix choice. *Nucleic Acids Res* 22, 4673–4680 (1994).

93. Maddison, W.P. and Maddison, D.R. Mesquite: A modular system for evolutionary analysis. Available from: http://mesquiteproject.org. Version 2.71 edn (2009). Accessed on November 15, 2014.

94. Ronquist, F. and Huelsenbeck, J.P. MrBayes 3: Bayesian phylogenetic inference under mixed models. *Bioinformatics* 19, 1572–1574 (2003).

95. Huelsenbeck, J.P. and Ronquist, F. MRBAYES: Bayesian inference of phylogenetic trees. *Bioinformatics* 17, 754–755 (2001).

Porcine Circovirus

Yu-Liang Huang, Hui-Wen Chang, Victor Fei Pang, and Chian-Ren Jeng

CONTENTS

74.1 Introduction 671
   74.1.1 Classification 671
   74.1.2 Physiochemistry of PCV 672
   74.1.3 Transmission and Epidemiology 672
   74.1.4 Clinical Features and Pathogenesis of PCVAD 672
      74.1.4.1 Subclinical Infection of PCV2 672
      74.1.4.2 Characteristics and Pathogenesis of PMWS 672
      74.1.4.3 Cofactors of PMWS 673
   74.1.5 Diagnosis 673
      74.1.5.1 Viral Isolation 673
      74.1.5.2 Molecular Techniques 674
74.2 Methods 674
   74.2.1 In Situ Hybridization 674
      74.2.1.1 Procedure 674
   74.2.2 Real-Time PCR 674
      74.2.2.1 Procedure 674
74.3 Conclusion 675
References 675

74.1 INTRODUCTION

Porcine circovirus–associated diseases (PCVAD) result from infection with porcine circovirus type 2 (PCV2).[1] PCV2 has been demonstrated to infect the lymphoid, respiratory, nervous, digestive, urinary, reproductive, integumentary, cardiovascular, and skeletal systems.[2] Regardless of the organs affected, all PCV2-induced diseases, including postweaning multisystemic wasting syndrome (PMWS),[3] porcine dermatitis and nephropathy syndrome (PDNS),[4] porcine respiratory disease complex,[5] and reproductive failure,[2] are classified as PCVAD. Commonly occurring in pig industry, PCVAD cause significant economic losses worldwide.[1,6–12]

74.1.1 CLASSIFICATION

PCV2, belonging to the genus *Circovirus* of the family *Circoviridae*, is a small, nonenveloped, single-stranded, circular DNA virus with a diameter of 16–18 nm.[13] Of the three genera (*Circovirus*, *Cyclovirus*, and *Gyrovirus*) within the family *Circoviridae*, the genus *Circovirus* encompasses porcine circovirus (PCV), beak and feather disease virus, canary circovirus, duck circovirus, finch circovirus, goose circovirus, gull circovirus, starling circovirus, and swan circovirus; the genus *Cyclovirus* includes dragonfly cyclovirus and Florida woods cockroach–associated cyclovirus; and the genus *Gyrovirus* comprises chicken anemia virus, avian gyrovirus 2, and human gyrovirus 3 and 4.[14–17]

PCV is composed of two subtypes: PCV type 1 (PCV1) and PCV type 2 (PCV2). First identified in 1974,[18] PCV1 frequently causes contamination in the PK15 cell line (ATCC-CCL31) and is generally considered nonpathogenic in swine. PCV2 was first isolated from PMWS-affected pigs in the 1990s and is pathogenic in swine.[19] PCV2 exhibits 68% and 66% identity to PCV1 in the full genomic and ORF2 sequences, respectively.[20] The various PCV2 strains have 95%–100% sequence identity.[21] According to the sequence diversity, PCV2 is further subdivided into PCV2a (with five subclusters, 2A-2E), PCV2b (with three subclusters, 1A-1C), and PCV2c genotypes,[22,23] which have been retrospectively studied and found as early as 1985, 1988, and 1980, respectively.[23,24] In the past, the genotype PCV2a was the predominant virus causing PCVAD in pigs. The shift from the genotype PCV2a to PCV2b occurred between 2003 and 2006. Since 2006, the genotype PCV2b has been the predominant PCV2 subtype worldwide.[24–28] The genotype PCV2c has only been found retrospectively in Denmark.[23]

The genome of PCV2 is made up of 1767–1768 nucleotides forming a stem-loop structure and 3-hexamer repeats. Of 11 open reading frames (ORF) contained in the genome, the ORF1, ORF2, and ORF3 encode the replication-related proteins, capsid protein, and apoptosis-related protein, respectively.[13] More specifically, the ORF1 encodes two replication-related proteins, Rep and Rep′. The Rep protein is translated from the full-length transcripts consisting of 341 amino acids with a dNTP-binding domain (P-loop) and three motifs of rolling-circle replication (RCR): RCR-I (FTLN), RCR-II (HxQ), and RCR-III.[13] However, the C-terminus of the Rep' protein has entire deletion of the 128 amino acids in the P-loop and consists of only RCR-I, RCR-II, and RCR-III. In the transcription of ORF1, the mRNAs of both Rep and Rep' proteins carry three nuclear location signals (NLS) in their N-terminus. NLS1 and NLS2 mediate the nuclear accumulation, and NLS3 enhances nuclear transport of both Rep and Rep' proteins.[13] The ORF2 encodes the capsid protein. The sequence similarity of ORF2 among various PCV2 strains is 91%–100%.[21] The mRNA of the capsid protein carries an NLS that can carry the capsid protein into the nucleus. The capsid protein, consisting of 233 amino acids, is rich of arginine in the N-terminus, which binds to the viral DNA.[13] The ORF3 encodes an apoptosis-related protein and can induce apoptosis via caspase 8 and 3.[29]

74.1.2 PHYSIOCHEMISTRY OF PCV

PCV has a buoyant density of 1.37 g/mL in CsCl and a sedimentation coefficient of 57S. It does not hemagglutinate erythrocytes and is capable of resisting the acid inactivation of acidity at pH 3, chloroform, and 70°C for 15 min. However, the titer of PCV2 is reduced following exposure to chlorhexidine, formaldehyde, iodine, or alcohol at room temperature for 10 min.[30]

74.1.3 TRANSMISSION AND EPIDEMIOLOGY

Horizontal and vertical transmissions of PCV2 have been demonstrated experimentally.[31] In horizontal transmission, favorable factors for transmission are the ocular, nasal, tonsillar, and bronchial secretions, saliva, feces, urine, milk, and semen.[31–33] PCV2 infection in pigs is usually via direct or indirect contact with these secretions or uncooked tissues of PCV2-infected pigs by the oral–nasal route.[34] Horizontal transmission of PCV2 has also been demonstrated by intermingling of PCV2-infected pigs and naïve pigs.[31] In vertical transmission, PCV2 can pass via the placenta to the fetus in pregnant sows, causing late-term abortions and stillborn piglets.[31,35] PCV2 has also been demonstrated to infect the embryo and fetus with PCV2-infected semen via artificial insemination.[36]

Although PCV2 was initially identified in PMWS-affected pigs from SPF herds in Canada in 1991,[19] several retrospective studies have shown that PCV2 infection can be traced in PCVAD-suspected cases as early as the 1960s.[24,37] The earliest evidence of PCV2 infection in PDNS-associated lesions, including systemic vasculitis, dermal and epidermal necrosis, and necrotizing glomerulonephritis, was in 1985,[37] and PMWS-related lesions, including lymphoid depletion and granulomatous inflammation, were revealed in the 1960s. A retrospective serological survey of PCV2 antibodies also showed that PCV2 is not a new virus; in fact, it may have existed for more than 50 years.[24,37]

74.1.4 CLINICAL FEATURES AND PATHOGENESIS OF PCVAD

74.1.4.1 Subclinical Infection of PCV2

PCV2 infection in pigs can be either subclinical or clinical. Most PCV2-infected pigs are healthy and subclinical. They show no clinical signs and exhibit normal to mild lymphoid depletion, with small amounts of PCV2 antigens present in lymphoid tissues. PCV2-associated lesions and antigens are also restricted in one or two lymphoid tissues.[1,38] Although subclinical PCV2-infected pigs have no significant clinical symptoms, the reduction in the efficacy of porcine reproductive and respiratory syndrome virus (PRRSV) and classic swine fever virus (CSFV) vaccination has been demonstrated in *in vivo* models.[39,40] In addition, subclinical PCV2-infected pigs could develop clinical disease when a cofactor is present.[38,41]

74.1.4.2 Characteristics and Pathogenesis of PMWS

A diagnosis of PMWS in PCV2-infected pigs must meet the following criteria: (1) wasting syndrome with frequent dyspnea and enlargement of inguinal lymph nodes; (2) histopathologically, lymphoid tissues showing moderate to severe lymphoid depletion and formation of basophilic or amphophilic grape-like intracytoplasmic inclusion bodies; and (3) lymphoid tissues containing a moderate to high level of PCV2 load.[1,2,42] None of the three criteria can be excluded in PMWS identification.

PMWS primarily affects pigs at ages between 25 and 120 days, with the highest prevalence between 60 and 80 days. The morbidity and mortality in affected herds range from 4% to 30% and 4% to 20%, respectively.[38,43] The clinical characteristics of PMWS include weight loss, pallor of skin, respiratory distress, diarrhea, and occasionally, jaundice.[38,43] Grossly, the most predominant feature is the enlargement of lymph nodes, mainly in the inguinal, submandibular, mesenteric, and mediastinal lymph nodes. The lungs show extensive to diffuse mottling with consolidation in the anterio-ventral aspect. The thymus is usually atrophied. The liver shows orange-yellow discoloration. In the kidney, multiple variably sized pale foci may be observed.[38,44]

Histopathologically, the characteristic lesions in PMWS-affected pigs are granulomatous inflammation and the presence of basophilic intracytoplasmic inclusion bodies in histiocytic lineage cells. Multifocal to locally extensive coagulative necrosis and variable lymphoid depletion together with histiocytic infiltration and formation of multinucleated giant cells are present in almost all lymphoid tissues, including lymph nodes, tonsils, Peyer's patches, and spleen. In the thymus, the cortex is atrophied. Lesions in the lungs are characterized by the infiltration of lymphocytes, macrophages, and multinucleated giant cells in the alveolar walls. The liver is infiltrated by lymphohistiocytic inflammatory cells in the portal triads, with swelling, vacuolation, and coagulative necrosis of hepatocytes. There is also lymphohistiocytic cell infiltration in the renal interstitium in PMWS-affected pigs.[1,38]

The role of PCV2 in the pathogenesis of PMWS has been demonstrated experimentally.[45,46] PCV2 is transmitted via direct contact between infected and naïve pigs through either the oral or the respiratory route. After exposure to PCV2, the virus first enters the tonsil and lymph nodes of the head. Although PCV2 antigens can be detected within monocytes, macrophages, and follicular dendritic cells, PCV2 usually does not replicate within these cells.[3,38,47] The presence of

PCV2 on the surfaces of dendritic cells does not modulate the MHC class I and II expression.[47] Dendritic cells and monocytes/macrophages can harbor PCV2 without getting killed.[47] These cells function as a vehicle that allows the virus to spread throughout the body and infect lymphoid organs. In secondary lymphoid tissues, PCV2 also infects B and T cells, where viral replication takes place.[38,48] It has been shown that stimulation and/or activation of the immune system in PCV2-infected pigs by other viruses,[49,50] bacteria,[65] or noninfectious agents[51,52] could upregulate PCV2 replication and increase the PCV2 load in the tissue. As PCV2 replication depends on the cell polymerases present in the nuclei of cells during the S phase of the cell cycle, immune stimulation with a mitogen such as concanavalin A may activate lymphocyte proliferation and lead to the upregulation of PCV2 replication.[52,53]

The characteristic pathological changes in lymphoid tissues of PMWS-affected pigs are lymphoid depletion and granulomatous lymphadenitis. Although PCV2 can induce apoptosis of CD79α+ B-lymphocytes and PK-15 cells by ORF3 protein-induced activation of caspase 3 and 8 *in vitro*,[29,54] the apoptotic index of lymphoid tissues in PMWS-affected pigs is either normal or lower than that of normal pigs.[55] In contrast, the proliferation index and apoptosis/proliferation rate in lymphoid tissues of PMWS-affected pigs are lower than those in normal pigs. Therefore, PCV2-induced lymphoid depletion may not be fully due to apoptosis, but rather largely due to reduced lymphocyte proliferation.[56] The downregulation in lymphocyte proliferation caused by PCV2 infection is resulted from the decreased expression of lymphocyte growth factor, IL-2 and IL-4.[57,58] In addition, PCV2 infection may induce the proliferation of blood monocytes and monocyte-derived macrophages; the formation of multinucleated giant cells and expression of monocytic chemokines and proinflammatory cytokines such as monocyte chemoattractant protein-1, macrophage inflammatory protein-1, IL-1β, and IL-8; and the directional movement of blood monocytes and neutrophils. Experimental results confirmed that PCV2 plays a role in the induction of the granulomatous inflammation occurring in PMWS-affected pigs.[59]

74.1.4.3 Cofactors of PMWS

PCV2 is recognized as the essential infectious agent in the development of PMWS, but experimental inoculation of PCV2 alone cannot reproduce all the clinical signs and pathological changes of PMWS in pigs.[1,2,38,49,50] The seroprevalence of PCV2 infection in commercial pig herds is nearly 100%, but only 5%–30% of PCV2-infected pigs show the representative signs of PMWS. Among these PMWS cases, more than 98% of the affected pigs have coinfection with other pathogens. Singular PCV2 infection occurs only in 1.9% of PMWS-affected cases.[48] These observations suggest that the development of PMWS requires the combination of PCV2 and other cofactors, including coinfection,[49,60] immunomodulation, different PCV2 genotypes,[31] and the ages and breeds of pigs.[61–63]

Porcine parvovirus,[49] PRRSV,[64] and *Mycoplasma hyopneumoniae*[65] have been demonstrated to enhance PCV2 replication, increase the severity of PCV2-associated lesions,

and heighten the risk of PMWS development under experimental and field conditions. Nonspecific immune stimulation by various vaccines of *M. hyopneumoniae*, *Actinobacillus pleuropneumoniae*, PRRSV, and CSFV may also promote PCV2 replication and increase the severity of PMWS.[66–68] The oil-in-water adjuvant and keyhole limpet hemocyanin in incomplete Freund's adjuvant have been shown to cause a longer duration of viremia, increased serum and tissue viral loads, and more severe lymphoid depletion as compared to pigs vaccinated with aqueous and aluminum hydroxide products.[69] In addition, cyclosporine- and dexamethasone-induced immunosuppression can enhance the development of PMWS in PCV2-infected pigs.[69,70]

74.1.5 Diagnosis

PCV2 can be diagnosed by several methods: indirect fluorescent antibody (IFA) assay in monolayer,[71] immunoperoxidase monolayer assay,[72] enzyme-linked immunosorbent assay (ELISA),[73] and serum virus neutralization assay[74] for anti-PCV2 antibody detection; PCR,[75,76] multiplex PCR,[77–80] quantitative real-time PCR,[32,81–86] *in situ* PCR,[87] and *in situ* hybridization (ISH)[3,88–90] for PCV2 nucleic acid detection; and virus isolation,[52,91] immunohistochemical (IHC) staining,[92–94] antigen detection capture ELISA,[95,96] and IFA[97] for PCV2 virus or viral antigen detection. Given that most pigs in the field, regardless of their health status, are infected with PCV2,[86] use of conventional PCR, detection of anti-PCV2 antibodies, or direct or indirect demonstration of PCV2 infection in pigs is not considered a replacement for clinical assessment of pigs and microscopic evaluation of tissues for confirmative diagnosis of PCVAD. For PCVAD diagnosis, it is important to demonstrate characteristic lesions in conjunction with the presence of PCV2 antigens or nucleic acids in the affected organs. A confirmative diagnosis of PCVAD cannot be made without evaluating microscopic lesions and demonstrating that PCV2 is associated with the characteristic lesions (lymphoid depletion, histiocytosis, syncytial cells, and/or basophilic cytoplasmic inclusion bodies in lymph nodes, Peyer's patches, or lamina propria of the intestinal villi). Currently, IHC staining and ISH for localizing intralesional PCV2 are considered gold standard methods for the diagnosis of PCVAD.[44,87,98] In addition, quantitative PCR for detecting the amount of PCV2 nucleic acids in serum and tissue has been demonstrated to be predictive for the clinical outcome,[81,86] such as in PCV2-associated reproductive failures.[99]

74.1.5.1 Viral Isolation

Viral isolation is the traditional method for the diagnosis of PCV2 infection. It is performed using the PK-15 cell line treated with D-glucosamine.[52,100] D-glucosamine has been shown to initiate PCV replication by enabling the PCV genome to enter the cell nucleus, which normally can be achieved only by inclusion in the daughter nuclei at the end of mitosis.[100] As PCV2 is a slow-growing virus and its replication in cell culture has no distinct cytopathic effect,[42] it may require a couple of weeks for PCV2 to grow, after which IFA or PCR can be used to confirm the viral isolation. Therefore, even though it has been demonstrated that the quantitative virus isolation using a serial tenfold dilution of clinical specimens could differentiate clinical and subclinical PCV2 infections,[95] virus isolation is laborious.

74.1.5.2 Molecular Techniques

Quantitative PCR can accurately differentiate PCV2 infection from PCVAD by copy numbers of PCV2 DNA present in the serum or tissue. A PCR result of less than 10^6 PCV2 DNA copies is generally not considered a PCVAD case. A threshold of 10^7 or greater PCV2 genomic copies per milliliter of serum has been found to correlate well with severe PCV2-associated lesions and clinical disease, as well as poor prognosis.[84,86]

ISH using a labeled DNA probe corresponding to a specific portion of the PCV2 genome and the IHC staining using a monoclonal or polyclonal antibody targeting PCV2 antigens are able to correlate the distribution of PCV2 with the lesion area using formalin-fixed, paraffin-embedded tissue sections. These two methods are important for the diagnosis of PCVAD. Furthermore, ISH has been found to be more specific than IHC staining, especially when the IHC staining is performed using polyclonal antibodies.[101] Considering the ease of preparing DNA probes and the convenience of using nonradioactive digoxigenin-labeled probes, an ISH using a 5′ end digoxigenin (DIG)-labeled 40 bp oligonucleotide probe specific for PCV2 ORF2 is described herein.

74.2 METHODS

74.2.1 In Situ Hybridization

74.2.1.1 Procedure

The method of ISH for PCV2 nucleic acid detection is modified from previous studies.[51,88,102]

1. Tissue samples are fixed in 10% formalin and paraffin embedded. Sections of the paraffin-embedded tissues are dewaxed in xylene and rehydrated in graded ethanol (100%, 90%, 70%, 50%, and 30%).
2. After being washed twice with phosphate buffered saline (PBS), tissues are digested with 4 µg/mL proteinase K in sterile water for 5 min at room temperature.
3. After being washed twice with PBS, slides are covered and prehybridized with the hybridization buffer (50% formamide, 0.9 M NaCl, 0.02 M Tris HCl, and 0.01% SDS) at 47°C for 1 h.
4. During prehybridization, the 5′ end digoxigenin (DIG)-labeled 40 bp oligonucleotide probe (5′-CAGTAAATACGACCAGGACTACAA TATCCGTGTAACCATG-3′) complementary to nucleotides 1085–1124 of ORF2 from PCV2[88] is diluted with hybridization buffer to a concentration of 5 ng/µL. The probe is denatured at 95°C for 3 min and then quenched on ice.

5. The prehybridization buffer is removed.
6. The probe is added in hybridization buffer for hybridization at 47°C overnight.
7. The probe is removed and the slides are washed with graded (4×, 2×, 0.5×, 0.1×, and 0.01×) saline-sodium citrate (SSC) buffer (1× SSC contains 50 mM NaCl and 15 mM sodium citrate, pH 7.0) at 47°C for 15 min each.
8. After being washed with tris-base solution (TBS) buffer (0.1 M Tris HCl, 0.4 M NaCl, pH 7.5), sections are then treated with the blocking solution (Roche, Nonnenwald, Penzberg, Germany) for 30 min.
9. For detection of hybridized signals, the probe is labeled with 1:200 anti-digoxigenin antibody conjugated with alkaline phosphatase (anti-DIG APase) (Roche, Nonnenwald, Penzberg, Germany) in TBS.
10. After three washes in buffer II (0.1 M Tris HCl, 0.1 M NaCl, 0.05 M $MgCl_2$) for 5 min each, color detection is developed by NBT/5-bromo-4-chloro-3-indolyl phosphate (Roche, Nonnenwald, Penzberg, Germany) and counterstained with 1% methyl green.
11. Positive cells contain bluish-purple intranuclear and/or intracytoplasmic precipitates.

The combination of PCVAD-associated lesions coupled with the presence of abundant viral nucleic acid within the developing granulomas and germinal centers is important in making a diagnosis of PCVAD.

74.2.2 Real-Time PCR

The method of real-time PCR for PCV2 nucleic acid detection is modified from that reported by Huang et al.[40] Tissue DNA extraction follows the protocol of the QIAamp® DNA Mini kit. The real-time PCR is carried out by the LightCycler® 480 system (Roche, Mannheim, Germany).

74.2.2.1 Procedure

1. The tissues are ground to 10% tissue emulsion with minimal essential medium and centrifuged at 3000 × g for 10 min.
2. After tissue preparation, 200 µL of the supernatant is mixed with 20 µL of protease K and 200 µL Buffer AL by pulse vortexing for 15 s.
3. The mixture is incubated at 56°C for 10 min and then mixed with 200 µL of ethanol (96%–100%) by pulse vortexing for 15 s.
4. The mixture from step 3 is transferred to the QIAamp Spin column and centrifuged at 6000 × g for 1 min to allow DNA to bind to the membrane of the spin column.
5. Following DNA binding, the QIAamp Spin column is subsequently washed with 500 µL of Buffer AW1 at 6,000 × g for 1 min and 500 µL of AW2 at 20,000 × g for 3 min.
6. The DNA in the QIAamp Spin column is then eluted by addition of 200 µL of Buffer AE, incubated at

TABLE 74.1
Sequences of Primers and Probes

Primer/Probe	Sequence (5′–3′)
PCV-F	5′-GCTGGCTGAACTTTTGAAA-3′
PCV-R	5′-CCTTTAGTCTCTACAGTCAATGGAT-3′
PCV2-probe	5′-FAM-TGCTAATTTTGCAGACCCGGAAACCAC-BHQ1-3′

TABLE 74.2
Real-Time PCR Mixture for PCV2 Detection

Component	Volume (μL)
DNA	5
PCV2 probe (5 μM)	0.6
PCV-F (10 μM)	1
PCV-R (10 μM)	1
2× Kapa Probe Fast qPCR Master mix	10
Double distilled water	2.4

room temperature for 1 min, and centrifuged at $6000 \times g$ for 1 min.

7. Real-time PCR is carried out by the LightCycler® 480 system (Roche, Mannheim, Germany). The primers and probe are shown in Table 74.1.

8. The reaction mixture of the real-time PCR is prepared as shown in Table 74.2.

9. The reaction conditions consist of initial incubation at 95°C for 10 min, followed by 45 cycles of denaturation at 95°C for 20 s, annealing at 60°C for 30 s, and extension at 72°C for 20 s. A sample is considered PCV2 negative when the threshold cycle is above 40.

10. The quantification of PCV2 load in the tissue is calculated according to the standard curve of PCV2.

Establishment of the standard curve of PCV2

1. The PCR products of PCV2 are cloned using the pGEM-T Easy vector system (Promega, Madison, Wisconsin).

2. The PCV2 plasmid is propagated in *Escherichia coli*, JM109, and purified by the QIAamp plasmid Maxi Kit (Qiagen, Valencia, California).

3. The concentration of PCV2 plasmid is quantified by the measurement of OD_{260} with a spectrophotometer DU®640 (Beckman, Palo Alto, California). The copy number of viral DNA of the extracted plasmid is calculated using the formula

$$[X \text{ (g/μL DNA)} \times 6 \times 10^{23}]/[\text{Plasmid length (bp)} \times 660] = Y \text{ viral copy/μL}.$$

4. The PCV2 plasmid is optimized to 1×10^{10} copy number/mL. Following optimization of the concentration, the standard plasmid is subjected to a 10-fold serial dilution and then detected with real-time PCR for PCV2.

5. The standard curve is drawn using the known concentration of PCV2 plasmid and the corresponding Ct value.

74.3   CONCLUSION

Although PCV2 is the essential causative agent of PCVAD, PCV2 alone rarely leads to the development of clinical disease. The level of PCV2 load and the severity of its associated lesions can be enhanced by concurrent infection with other viruses or bacteria.[11,60,103–107] While the exact mechanisms of how concurrent pathogens or cofactors upregulate PCV2 replication are not completely known, it is important to further identify possible concurrent pathogens and cofactors in PCVAD cases. Such knowledge would be helpful for the development of the most appropriate and effective control strategies against PCVAD.[108]

REFERENCES

1. Opriessnig, T., Meng, X.J., and Halbur, P.G. Porcine circovirus type 2 associated disease: Update on current terminology, clinical manifestations, pathogenesis, diagnosis, and intervention strategies. *J Vet Diagn Invest* 19, 591–615 (2007).

2. Segales, J. Porcine circovirus type 2 (PCV2) infections: Clinical signs, pathology and laboratory diagnosis. *Virus Res* 164, 10–19 (2012).

3. McNeilly, F. et al. A comparison of in situ hybridization and immunohistochemistry for the detection of a new porcine circovirus in formalin-fixed tissues from pigs with postweaning multisystemic wasting syndrome (PMWS). *J Virol Methods* 80, 123–128 (1999).

4. Choi, C. and Chae, C. Colocalization of porcine reproductive and respiratory syndrome virus and porcine circovirus 2 in porcine dermatitis and nephrology syndrome by double-labeling technique. *Vet Pathol* 38, 436–441 (2001).

5. Kim, J., Chung, H.K., and Chae, C. Association of porcine circovirus 2 with porcine respiratory disease complex. *Vet J* 166, 251–256 (2003).

6. Walker, I.W. et al. Development and application of a competitive enzyme-linked immunosorbent assay for the detection of serum antibodies to porcine circovirus type 2. *J Vet Diagn Invest* 12, 400–405 (2000).

7. Rodriguez-Arrioja, G.M. et al. Retrospective study on porcine circovirus type 2 infection in pigs from 1985 to 1997 in Spain. *J Vet Med B Infect Dis Vet Public Health* 50, 99–101 (2003).

8. Vicente, J. et al. Epidemiological study on porcine circovirus type 2 (PCV2) infection in the European wild boar (*Sus scrofa*). *Vet Res* 35, 243–253 (2004).

9. Finlaison, D., Kirkland, P., Luong, R., and Ross, A. Survey of porcine circovirus 2 and postweaning multisystemic wasting syndrome in New South Wales piggeries. *Aust Vet J* 85, 304–310 (2007).

10. Yang, Z.Z., Shuai, J.B., Dai, X.J., and Fang, W.H. A survey on porcine circovirus type 2 infection and phylogenetic analysis of its ORF2 gene in Hangzhou, Zhejiang Province, China. *J Zhejiang Univ Sci B* 9, 148–153 (2008).

11. Wang, L., Ge, C., Wang, D., Li, Y., and Hu, J. The survey of porcine teschoviruses, porcine circovirus and porcine transmissible gastroenteritis virus infecting piglets in clinical specimens in China. *Trop Anim Health Prod* 45, 1087–1091 (2013).

12. Barbosa, C.N., Martins, N.R., Freitas, T.R., and Lobato, Z.I. Serological survey of porcine circovirus-2 in captive wild boars (*Sus scrofa*) from registered farms of South and South-eastregions of Brazil. *Transbound Emerg Dis* (epub ahead of print).

13. Mankertz, A. et al. Molecular biology of Porcine circovirus: Analyses of gene expression and viral replication. *Vet Microbiol* 98, 81–88 (2004).

14. Chu, D.K. et al. Characterization of a novel gyrovirus in human stool and chicken meat. *J Clin Virol* 55, 209–213 (2012).

15. Sauvage, V. et al. Identification of the first human gyrovirus, a virus related to chicken anemia virus. *J Virol* 85, 7948–7950 (2011).

16. Maggi, F. et al. Human gyrovirus DNA in human blood, Italy. *Emerg Infect Dis* 18, 956–959 (2012).

17. Phan, T.G. et al. A third gyrovirus species in human faeces. *J Gen Virol* 93, 1356–1361 (2012).

18. Tischer, I., Rash, R., and Tochtermann, G. Characterization of papovavirus- and picornavirus-like particles in permanent pig kidney cell lines. *Zentralblatt Bakteriol. Parasitenkunde Infektionskrankheiten Hygiene Erste Abteilung Originale Reihe A: Med Mikrobiol Parasitol* 226, 153–174 (1974).

19. Harding, J.C.S. and Clark, E.G. Recognizing and diagnosing postweaning multisystemic wasting syndrome (PMWS). *J Swine Health Prod* 5, 201–203 (1997).

20. Hamel, A.L., Lin, L.L., and Nayar, G.P. Nucleotide sequence of porcine circovirus associated with postweaning multisystemic wasting syndrome in pigs. *J Virol* 72, 5262–5267 (1998).

21. Larochelle, R., Magar, R., and D'Allaire, S. Genetic characterization and phylogenetic analysis of porcine circovirus type 2 (PCV2) strains from cases presenting various clinical conditions. *Virus Res* 90, 101–112 (2002).

22. Cortey, M., Olvera, A., Grau-Roma, L., and Segales, J. Further comments on porcine circovirus type 2 (PCV2) genotype definition and nomenclature. *Vet Microbiol* 149, 522–523 (2011).

23. Dupont, K., Nielsen, E.O., Baekbo, P., and Larsen, L.E. Genomic analysis of PCV2 isolates from Danish archives and a current PMWS case-control study supports a shift in genotypes with time. *Vet Microbiol* 128, 56–64 (2008).

24. Cortey, M. et al. Genotypic shift of porcine circovirus type 2 from PCV-2a to PCV-2b in Spain from 1985 to 2008. *Vet J* 187, 363–368 (2011).

25. Henriques, A.M. et al. Molecular study of porcine circovirus type 2 circulating in Portugal. *Infect Genet Evol* 11, 2162–2172 (2011).

26. Cheung, A.K. et al. Detection of two porcine circovirus type 2 genotypic groups in United States swine herds. *Arch Virol* 152, 1035–1044 (2007).

27. Gagnon, C.A. et al. Development and use of a multiplex real-time quantitative polymerase chain reaction assay for detection and differentiation of Porcine circovirus-2 genotypes 2a and 2b in an epidemiological survey. *J Vet Diagn Invest* 20, 545–558 (2008).

28. Wang, C. et al. Prevalence and genetic variation of porcine circovirus type 2 in Taiwan from 2001 to 2011. *Res Vet Sci* 94, 789–795 (2013).

29. Liu, J., Chen, I., and Kwang, J. Characterization of a previously unidentified viral protein in porcine circovirus type 2-infected cells and its role in virus-induced apoptosis. *J Virol* 79, 8262–8274 (2005).

30. Segales, J., Allan, G.M., and Domingo M. Porcine circovirus diseases. In: *Disease of Swine*, 10th edition. Blackwell Publishing, Oxford, U.K. pp. 405–417 (2012).

31. Grau-Roma, L., Fraile, L., and Segales, J. Recent advances in the epidemiology, diagnosis and control of diseases caused by porcine circovirus type 2. *Vet J* 187, 23–32 (2011).

32. Chung, W.B. et al. Real-time PCR for quantitation of porcine reproductive and respiratory syndrome virus and porcine circovirus type 2 in naturally-infected and challenged pigs. *J Virol Methods* 124, 11–19 (2005).

33. Shibata, I., Okuda, Y., Kitajima, K., and Asai, T. Shedding of porcine circovirus into colostrum of sows. *J Vet Med B Infect Dis Vet Public Health* 53, 278–280 (2006).

34. Segales, J., Allan, G.M., and Domingo, M. Porcine circovirus diseases. *Anim Health Res Rev* 6, 119–142 (2005).

35. Pensaert, M.B., Sanchez, R.E., Jr., Ladekjaer-Mikkelsen, A.S., Allan, G.M., and Nauwynck, H.J. Viremia and effect of fetal infection with porcine viruses with special reference to porcine circovirus 2 infection. *Vet Microbiol* 98, 175–183 (2004).

36. Madson, D.M. et al. Reproductive failure experimentally induced in sows via artificial insemination with semen spiked with porcine circovirus type 2. *Vet Pathol* 46, 707–716 (2009).

37. Jacobsen, B. et al. Retrospective study on the occurrence of porcine circovirus 2 infection and associated entities in Northern Germany. *Vet Microbiol* 138, 27–33 (2009).

38. Darwich, L., Segales, J., and Mateu, E. Pathogenesis of post-weaning multisystemic wasting syndrome caused by Porcine circovirus 2: An immune riddle. *Arch Virol* 149, 857–874 (2004).

39. Opriessnig, T., McKeown, N.E., Harmon, K.L., Meng, X.J., and Halbur, P.G. Porcine circovirus type 2 infection decreases the efficacy of a modified live porcine reproductive and respiratory syndrome virus vaccine. *Clin Vaccine Immunol* 13, 923–929 (2006).

40. Huang, Y.L. et al. Porcine circovirus type 2 (PCV2) infection decreases the efficacy of an attenuated classical swine fever virus (CSFV) vaccine. *Vet Res* 42, 115 (2011).

41. Harding, J.C. The clinical expression and emergence of porcine circovirus 2. *Vet Microbiol* 98, 131–135 (2004).

42. Kim, J. and Chae, C. A comparison of virus isolation, polymerase chain reaction, immunohistochemistry, and in situ hybridization for the detection of porcine circovirus 2 and porcine parvovirus in experimentally and naturally coinfected pigs. *J Vet Diagn Invest* 16, 45–50 (2004).

43. Segales, J., Rosell, C., and Domingo, M. Pathological findings associated with naturally acquired porcine circovirus type 2 associated disease. *Vet Microbiol* 98, 137–149 (2004).

44. Krakowka, S., Ellis, J., McNeilly, F., Waldner, C., and Allan, G. Features of porcine circovirus-2 disease: Correlations between lesions, amount and distribution of virus, and clinical outcome. *J Vet Diagn Invest* 17, 213–222 (2005).

45. Ladekjaer-Mikkelsen, A.S. et al. Reproduction of postweaning multisystemic wasting syndrome (PMWS) in immunostimulated and non-immunostimulated 3-week-old piglets experimentally infected with porcine circovirus type 2 (PCV2). *Vet Microbiol* 89, 97–114 (2002).

46. Magar, R., Larochelle, R., Thibault, S., and Lamontagne, L. Experimental transmission of porcine circovirus type 2 (PCV2) in weaned pigs: A sequential study. *J Comp Pathol* 123, 258–269 (2000).

47. Vincent, I.E. et al. Dendritic cells harbor infectious porcine circovirus type 2 in the absence of apparent cell modulation or replication of the virus. *J Virol* 77, 13288–13300 (2003).

48. Gillespie, J., Opriessnig, T., Meng, X.J., Pelzer, K., and Buechner-Maxwell, V. Porcine circovirus type 2 and porcine circovirus-associated disease. *J Vet Intern Med* 23, 1151–1163 (2009).

49. Allan, G.M. et al. Experimental reproduction of severe wasting disease by co-infection of pigs with porcine circovirus and porcine parvovirus. *J Comp Pathol* 121, 1–11 (1999).

50. Allan, G.M. et al. Experimental infection of colostrum deprived piglets with porcine circovirus 2 (PCV2) and porcine reproductive and respiratory syndrome virus (PRRSV) potentiates PCV2 replication. *Arch Virol* 145, 2421–2129 (2000).

51. Chang, H.W. et al. Bacterial lipopolysaccharide induces porcine circovirus type 2 replication in swine alveolar macrophages. *Vet Microbiol* 115, 311–319 (2006).

52. Yang, X. et al. Comparative analysis of different methods to enhance porcine circovirus 2 replication. *J Virol Methods* 187, 368–371 (2013).

53. Lefebvre, D.J. et al. Increased porcine circovirus type 2 replication in porcine leukocytes in vitro and in vivo by concanavalin A stimulation. *Vet Microbiol* 132, 74–86 (2008).

54. Shibahara, T., Sato, K., Ishikawa, Y., and Kadota, K. Porcine circovirus induces B lymphocyte depletion in pigs with wasting disease syndrome. *J Vet Med Sci* 62, 1125–1131 (2000).

55. Resendes, A.R. et al. Apoptosis in lymphoid organs of pigs naturally infected by porcine circovirus type 2. *J Gen Virol* 85, 2837–2844 (2004).

56. Mandrioli, L., Sarli, G., Panarese, S., Baldoni, S., and Marcato, P.S. Apoptosis and proliferative activity in lymph node reaction in postweaning multisystemic wasting syndrome (PMWS). *Vet Immunol Immunopathol* 97, 25–37 (2004).

57. Shi, K.C., Guo, X., Ge, X.N., Liu, Q., and Yang, H.C. Cytokine mRNA expression profiles in peripheral blood mononuclear cells from piglets experimentally co-infected with porcine reproductive and respiratory syndrome virus and porcine circovirus type 2. *Vet Microbiol* 140, 155–160 (2010).

58. Darwich, L. et al. Cytokine mRNA expression profiles in lymphoid tissues of pigs naturally affected by postweaning multisystemic wasting syndrome. *J Gen Virol* 84, 2117–225 (2003).

59. Tsai, Y.C. et al. Porcine circovirus type 2 (PCV2) induces cell proliferation, fusion, and chemokine expression in swine monocytic cells in vitro. *Vet Res* 41, 60 (2010).

60. Dorr, P.M., Baker, R.B., Almond, G.W., Wayne, S.R., and Gebreyes, W.A. Epidemiologic assessment of porcine circovirus type 2 coinfection with other pathogens in swine. *J Am Vet Med Assoc* 230, 244–250 (2007).

61. Harding, J.C., Ellis, J.A., McIntosh, K.A., and Krakowka, S. Dual heterologous porcine circovirus genogroup 2a/2b infection induces severe disease in germ-free pigs. *Vet Microbiol* 145, 209–219 (2010).

62. Opriessnig, T. et al. Differences in virulence among porcine circovirus type 2 isolates are unrelated to cluster type 2a or 2b and prior infection provides heterologous protection. *J Gen Virol* 89, 2482–2491 (2008).

63. Tomas, A., Fernandes, L.T., Valero, O., and Segales, J. A meta-analysis on experimental infections with porcine circovirus type 2 (PCV2). *Vet Microbiol* 132, 260–273 (2008).

64. Harms, P.A. et al. Experimental reproduction of severe disease in CD/CD pigs concurrently infected with type 2 porcine circovirus and porcine reproductive and respiratory syndrome virus. *Vet Pathol* 38, 528–539 (2001).

65. Opriessnig, T. et al. Experimental reproduction of postweaning multisystemic wasting syndrome in pigs by dual infection with *Mycoplasma hyopneumoniae* and porcine circovirus type 2. *Vet Pathol* 41, 624–640 (2004).

66. Opriessnig, T. et al. Effect of porcine parvovirus vaccination on the development of PMWS in segregated early weaned pigs coinfected with type 2 porcine circovirus and porcine parvovirus. *Vet Microbiol* 98, 209–220 (2004).

67. Ju, C. et al. Immunogenicity of a recombinant pseudorabies virus expressing ORF1-ORF2 fusion protein of porcine circovirus type 2. *Vet Microbiol* 109, 179–190 (2005).

68. Opriessnig, T. et al. Effect of vaccination with selective bacterins on conventional pigs infected with type 2 porcine circovirus. *Vet Pathol* 40, 521–529 (2003).

69. Krakowka, S. et al. Immunologic features of porcine circovirus type 2 infection. *Viral Immunol* 15, 567–582 (2002).

70. Kawashima, K. et al. Effects of dexamethasone on the pathogenesis of porcine circovirus type 2 infection in piglets. *J Comp Pathol* 129, 294–302 (2003).

71. Racine, S., Kheyar, A., Gagnon, C.A., Charbonneau, B., and Dea, S. Eucaryotic expression of the nucleocapsid protein gene of porcine circovirus type 2 and use of the protein in an indirect immunofluorescence assay for serological diagnosis of postweaning multisystemic wasting syndrome in pigs. *Clin Diagn Lab Immunol* 11, 736–741 (2004).

72. Ramirez-Mendoza, H., Castillo-Juarez, H., Hernandez, J., Correa, P., and Segales, J. Retrospective serological survey of Porcine circovirus-2 infection in Mexico. *Can J Vet Res* 73, 21–24 (2009).

73. Pileri, E. et al. Comparison of the immunoperoxidase monolayer assay and three commercial ELISAs for detection of antibodies against porcine circovirus type 2. *Vet J* 201, 429–432 (2014).

74. Meerts, P. et al. Correlation between the presence of neutralizing antibodies against porcine circovirus 2 (PCV2) and protection against replication of the virus and development of PCV2-associated disease. *BMC Vet Res* 2, 6 (2006).

75. Liu, Q., Wang, L., Willson, P., and Babiuk, L.A. Quantitative, competitive PCR analysis of porcine circovirus DNA in serum from pigs with postweaning multisystemic wasting syndrome. *J Clin Microbiol* 38, 3474–3477 (2000).

76. Choi, C., Chae, C., and Clark, E.G. Porcine postweaning multisystemic wasting syndrome in Korean pig: Detection of porcine circovirus 2 infection by immunohistochemistry and polymerase chain reaction. *J Vet Diagn Invest* 12, 151–153 (2000).

77. Liu, J.K., Wei, C.H., Yang, X.Y., Dai, A.L., and Li, X.H. Multiplex PCR for the simultaneous detection of porcine reproductive and respiratory syndrome virus, classical swine fever virus, and porcine circovirus in pigs. *Mol Cell Probes* 27, 149–152 (2013).

78. Liu, X.Q., Xu, K., and Zhang, Z.Y. A touchdown multiplex PCR for porcine circovirus type 2 and pseudorabies virus. *Acta Virol* 56, 163–165 (2012).

79. Yue, F., Cui, S., Zhang, C., and Yoon, K.J. A multiplex PCR for rapid and simultaneous detection of porcine circovirus type 2, porcine parvovirus, porcine pseudorabies virus, and porcine reproductive and respiratory syndrome virus in clinical specimens. *Virus Genes* 38, 392–397 (2009).

80. Kim, J., Han, D.U., Choi, C., and Chae, C. Differentiation of porcine circovirus (PCV)-1 and PCV-2 in boar semen using a multiplex nested polymerase chain reaction. *J Virol Methods* 98, 25–31 (2001).

81. Huang, Y. et al. Preclinical detection of porcine circovirus type 2 infection using an ultrasensitive nanoparticle DNA probe-based PCR assay. *PLoS One* 9, e97869 (2014).

82. Perez, L.J. et al. A multiple SYBR Green I-based real-time PCR system for the simultaneous detection of porcine circovirus type 2, porcine parvovirus, pseudorabies virus and Torque teno sus virus 1 and 2 in pigs. *J Virol Methods* 179, 233–241 (2012).

83. Pal, N. et al. Development and validation of a duplex real-time PCR assay for the simultaneous detection and quantification of porcine circovirus type 2 and an internal control on porcine semen samples. *J Virol Methods* 149, 217–225 (2008).

84. Olvera, A., Sibila, M., Calsamiglia, M., Segales, J., and Domingo, M. Comparison of porcine circovirus type 2 load in serum quantified by a real time PCR in postweaning multisystemic wasting syndrome and porcine dermatitis and nephropathy syndrome naturally affected pigs. *J Virol Methods* 117, 75–80 (2004).

85. Segales, J. et al. Quantification of porcine circovirus type 2 (PCV2) DNA in serum and tonsillar, nasal, tracheo-bronchial, urinary and faecal swabs of pigs with and without postweaning multisystemic wasting syndrome (PMWS). *Vet Microbiol* 111, 223–229 (2005).

86. Brunborg, I.M., Moldal, T., and Jonassen, C.M. Quantitation of porcine circovirus type 2 isolated from serum/plasma and tissue samples of healthy pigs and pigs with postweaning multisystemic wasting syndrome using a TaqMan-based real-time PCR. *J Virol Methods* 122, 171–178 (2004).

87. Lin, C.M. et al. Development and evaluation of an indirect in situ polymerase chain reaction for the detection of porcine circovirus type 2 in formalin-fixed and paraffin-embedded tissue specimens. *Vet Microbiol* 138, 225–234 (2009).

88. Rosell, C., Segales, J., and Domingo, M. Hepatitis and staging of hepatic damage in pigs naturally infected with porcine circovirus type 2. *Vet Pathol* 37, 687–692 (2000).

89. Kim, J. and Chae, C. Differentiation of porcine circovirus 1 and 2 in formalin-fixed, paraffin-wax-embedded tissues from pigs with postweaning multisystemic wasting syndrome by in-situ hybridisation. *Res Vet Sci* 70, 265–269 (2001).

90. Choi, C. and Chae, C. In-situ hybridization for the detection of porcine circovirus in pigs with postweaning multisystemic wasting syndrome. *J Comp Pathol* 121, 265–270 (1999).

91. Wellenberg, G.J. et al. Isolation and characterization of porcine circovirus type 2 from pigs showing signs of postweaning multisystemic wasting syndrome in the Netherlands. *Vet Q* 22, 167–172 (2000).

92. Kim, D. et al. Comparative study of in situ hybridization and immunohistochemistry for the detection of porcine circovirus 2 in formalin-fixed, paraffin-embedded tissues. *J Vet Med Sci* 71, 1001–1004 (2009).

93. Sorden, S.D., Harms, P.A., Nawagitgul, P., Cavanaugh, D., and Paul, P.S. Development of a polyclonal-antibody-based immunohistochemical method for the detection of type 2 porcine circovirus in formalin-fixed, paraffin-embedded tissue. *J Vet Diagn Invest* 11, 528–530 (1999).

94. Ha, Y., Jung, K., Choi, C., Hwang, K.K., and Chae, C. Synthetic peptide-derived antibody-based immunohistochemistry for the detection of porcine circovirus 2 in pigs with postweaning multisystemic wasting syndrome. *J Comp Pathol* 133, 201–204 (2005).

95. McNeilly, F. et al. Evaluation of a porcine circovirus type 2-specific antigen-capture enzyme-linked immunosorbent assay for the diagnosis of postweaning multisystemic wasting syndrome in pigs: Comparison with virus isolation, immunohistochemistry, and the polymerase chain reaction. *J Vet Diagn Invest* 14, 106–112 (2002).

96. Allan, G.M., Mackie, D.P., McNair, J., Adair, B.M., and McNulty, M.S. Production, preliminary characterisation and applications of monoclonal antibodies to porcine circovirus. *Vet Immunol Immunopathol* 43, 357–371 (1994).

97. Gilpin, D.F. et al. In vitro studies on the infection and replication of porcine circovirus type 2 in cells of the porcine immune system. *Vet Immunol Immunopathol* 94, 149–161 (2003).

98. Szczotka, A., Stadejek, T., and Pejsak, Z. A comparison of immunohistochemistry and in situ hybridization for the detection of porcine circovirus type 2 in pigs. *Pol J Vet Sci* 14, 565–571 (2011).

99. Hansen, M.S. et al. Selection of method is crucial for the diagnosis of porcine circovirus type 2 associated reproductive failures. *Vet Microbiol* 144, 203–209 (2010).

100. Tischer, I., Peters, D., Rasch, R., and Pociuli, S. Replication of porcine circovirus: Induction by glucosamine and cell cycle dependence. *Arch Virol* 96, 39–57 (1987).

101. Seo, H.W. et al. Evaluation of commercial polyclonal- and monoclonal-antibody-based immunohistochemical tests for 2 genotypes of Porcine circovirus type 2 and comparison with in-situ hybridization assays. *Can J Vet Res* 78, 233–236 (2014).

102. Lee, Y.Y., Chueh, L.L., Pang, V.F., Chang, H.W., and Jeng, C.R. Detection of type 2 porcine circovirus by in situ hybridization. *Taiwan Vet J* 29, 174–180 (2003).

103. Park, C., Seo, H.W., Park, S.J., Han, K., and Chae, C. Comparison of porcine circovirus type 2 (PCV2)-associated lesions produced by co-infection between two genotypes of PCV2 and two genotypes of porcine reproductive and respiratory syndrome virus. *J Gen Virol* 95, 2486–2494 (2014).

104. Lin, H.X., Ma, Z., Yang, X.Q., Fan, H.J., and Lu, C.P. A novel vaccine against Porcine circovirus type 2 (PCV2) and *Streptococcus equi* ssp. *zooepidemicus* (SEZ) co-infection. *Vet Microbiol* 171, 198–205 (2014).

105. Liu, J. et al. Three new emerging subgroups of Torque teno sus viruses (TTSuVs) and co-infection of TTSuVs with porcine circovirus type 2 in China. *Virol J* 10, 189 (2013).

106. Li, S., Wei, Y., Liu, J., Tang, Q., and Liu, C. Prevalence of porcine hokovirus and its co-infection with porcine circovirus 2 in China. *Arch Virol* 158, 1987–1991 (2013).

107. Takada-Iwao, A. et al. Porcine circovirus type 2 potentiates morbidity of *Salmonella enterica* serovar Choleraesuis in Cesarean-derived, colostrum-deprived pigs. *Vet Microbiol* 154, 104–112 (2011).

108. Opriessnig, T. and Halbur, P.G. Concurrent infections are important for expression of porcine circovirus associated disease. *Virus Res* 164, 20–32 (2012).

75 Aleutian Mink Disease Virus

Trine Hammer Jensen and Åse Uttenthal

CONTENTS

75.1 Introduction 679
    75.1.1 Classification, Morphology, and Genome Organization 679
    75.1.2 Transmission, Clinical Features, and Pathogenesis 680
    75.1.3 Diagnosis 681
        75.1.3.1 Conventional Techniques 681
        75.1.3.2 Molecular Techniques 681
75.2 Methods 682
    75.2.1 Sample Preparation 682
    75.2.2 Detection Procedures 682
        75.2.2.1 PCR Detection 682
        75.2.2.2 Sequencing 682
75.3 Conclusion and Future Perspectives 683
References 683

75.1 INTRODUCTION

Parvoviruses can infect many animals, usually producing acute, systemic disease. Mink are hosting two distinct parvoviruses, mink enteritis virus (MEV) and Aleutian mink disease virus (AMDV). MEV represents the typical parvovirus and is closely related to parvoviruses in cats and dogs; its infection results in highly contagious systemic and enteric disease. Effective vaccines toward MEV are used worldwide to prevent disease. AMDV can cause chronic progressive disease with persistent viremia and extreme antibody production. The disease is related to immune complexes and no vaccines exist. This chapter focuses on the persistent ADMV infection.

Compared to most other domestic animals, commercial mink breeding is rather new. Globally about 50 million mink skin are produced every year. In Denmark, mink farming was initiated around 1930. The rearing of mink for production of fur and the sale of skin generate a turnover of US $2.48 billion per year in Denmark alone [1]. Besides Scandinavia, mink farming also takes place in the Netherlands, Poland, Russia, China, and North America.

Aleutian disease in mink due to infection with the parvovirus AMDV is a significant problem for mink farming worldwide, leading to nonoptimal health status and decreased production. Affecting the purebred Aleutian mink, the disease was originally described by Hartsough and Gorham [2]. The fur of Aleutian mink was bluish-gray, resembling the fur of the Aleutian fox, which also contributed to the name of the disease. The Aleutian mink has a defect in its immune system, the Chediak–Higashi syndrome. Due to this immunodeficiency, the pathology is more severe in the Aleutian genotype of mink, but all mink breeds are susceptible to infection. The disease is chronic and no treatment or prophylactics is currently available. Therefore, the only control method is to detect and eliminate the infected animals.

The mink excrete AMDV DNA 2–8 weeks after experimental infection in feces [3], which results in a very heavy contamination of the premises. For proper eradication, the herd need to be culled and the farm intensively cleaned and disinfected in order to remove the remaining parvovirus. Studies on the closely related parvovirus MEV have shown that virus is infectious in feces after storage under outdoor conditions for up to 10 months [4]. Even after culling and disinfection, ADMV is likely to reoccur in the farm [5,6]. In Norway and Denmark, intensively culling has resulted in AMDV-free areas with improved health of the mink, increased breeding results, and enriched fur quality.

However, when mink kits were born from dams without AMDV antibodies, they were naïve and resultantly suffered from a new type of Aleutian disease characterized by acute pneumonia with high mortality of newborn mink kits [7]. Acute pneumonia caused by AMDV was first detected in mink farms in 1982 [7] and later on reproduced experimentally [8]. The combination of an environmentally extremely stable parvovirus, acute viral pneumonia with high virus excretion and the traditional farming systems with many mink farms in confined areas compromised the eradication program.

75.1.1 CLASSIFICATION, MORPHOLOGY, AND GENOME ORGANIZATION

AMDV is a negative single-stranded DNA virus belonging to the family *Parvoviridae* [9]. The family *Parvoviridae* is divided into two subfamilies: *Densovirinae* and *Parvovirinae*. While the subfamily *Densovirinae* contains five invertebrate-infecting genera: *Ambidensovirus*, *Brevidensovirus*, *Hepandensovirus*, *Iteradensovirus*, and *Penstyldensovirus*, the subfamily *Parvovirinae* comprises eight vertebrate-infecting genera: *Amdoparvovirus*, *Aveparvovirus*, *Bocaparvovirus*, *Copiparvovirus*, *Dependoparvovirus*, *Erythroparvovirus*, *Protoparvovirus*, and *Tetraparvovirus*.

Being the type species of the genus *Amdoparvovirus*, AMDV is a small icosahedral nonenveloped virus with a capsid surface and a genome of about 4748 nt [9]. The virion consists of three nonstructural proteins (NS1, NS2, and NS3) and two structural proteins (VP1 and VP2) [10,11]. All these proteins (NS1, NS2, NS3, VP1, and VP2) are necessary for viral replication [11–16]. The VP genes encode for the majority of the viral capsid [10,12] and together with the NS1 gene, the two VP genes are essential for capsid assembly [12]. The VP2 protein harbors at least one epitope important for pathogenesis of AMDV including antibody binding, virus neutralization, and antibody-dependent enhancement of infection [17]. Genetic variations between different AMDV strains have been described to occur at both the NS1 [18,19] and VP2 genes [10,20].

Parvovirus replication is dependent on actively replicating cells [21]; even with optimal cell lines, wild-type ADMV cannot be grown in cell cultures. Only a few AMDV strains have adapted to cell culture, such as the AMDV-G that is used for the production of antigens for countercurrent immunoelectrophoresis (CIEP) [10].

75.1.2 Transmission, Clinical Features, and Pathogenesis

AMDV can be transmitted by many routes: (1) feces, saliva, and oronasal secretions [3,22]; (2) urine [22,23]; (3) mechanical transmission via fomites [20]; (4) vertical via transplacental transmission [24]; (5) free-ranging mink and other furbearing animals [25–28]; (6) insect vectors [29]; (7) airborne transmission has been found experimentally [30], hypothesized in mink farms [31] but not proven [6].

The outcome of AMDV infection in mink depends on the age and genotype of the mink as well as the virus strain [32–34]. Mink carrying the Aleutian gene (e.g., sapphire and pastel mink) are most susceptible for developing severe and rapidly progressive Aleutian disease [32]. Neonatal kits suffer an AMDV-induced acute pneumonia [35–37]. The interstitial pneumonia is caused by the destruction of the surfactant producing alveolar type II cells [36]. AMDV infection of adult mink can cause either a multisystemic disease [32] or a chronic asymptomatic disease [3,38,39]. Older literature reported potential but not well-documented clearing of the virus infection [40]. The various clinical symptoms experienced by adult mink are related to chronic immune complex-mediated disease with hypergammaglobulinemia, glomerulonephritis [41,42], arteritis, [43] reduced reproductive performance including reduced litter size and abortions [24,35,44], nonsuppurative meningoencephalitis [45,46], and abnormal fur color [47]. These very diverse clinical features are mainly caused by the destruction of the actively dividing cells, which are necessary for the ADMV multiplication. The fetus, the kidneys, and the

hair follicles are all vulnerable due to high cell multiplication rate. In commercial fur production, the single depigmented hairs in brown or black fur severely reduce the fur value.

Antibodies against AMDV develop between 2 and 7 weeks after experimental infection as detected with CIEP [31,32,34,48] and even as short as 1 week after infection using rocket line immunoelectrophoresis [49]. The antibodies persist throughout the course of infection and are therefore used as the primary diagnostic method [3,48]. AMDV DNA has been detected by PCR in serum samples as early as 5 days after sapphire and pastel mink were infected with AMDV-Utah and AMDV-Pullman and 10 days after infection with AMDV-TR [34,48]. AMDV DNA was found in full blood from brown/wild-type mink 1–2 weeks after infection with a Danish field strain of AMDV [3]. In another study, AMDV DNA was found in leucocytes 2 weeks after infection of black mink with an American field AMDV strain [31]. The course of viremia varies in the individual mink; it can be transient with potential relapses or it can persist (Figure 75.1).

The immune complex-mediated disease results from conglomerates of antibodies and virus in the mink. Thus, antibodies in adult mink will have a severe negative effect on the progression of disease, meaning that traditional vaccination by killed vaccines causing antibodies will only aggravate the disease [50]. In spite of much effort, no vaccines have yet resulted in solid protection; only partial protection following DNA vaccination has been observed [51].

American mink (*Neovison vison*) is the primary host for AMDV, while European mink (*Mustela lutreola*) is also susceptible to AMDV [27]. AMDV infection has been described in a number of other carnivores such as raccoons (*Procyon lator*) [25,34], striped skunk (*Mephitis mephitis*) [34,52–54], ferrets (*Mustela putorius furo*) [23,55–57], Eurasian otter (*Lutra lutra*) [27], bobcat (*Lynx rufus*), short-tailed weasels (*Mustela erminea*), and North American river otter (*Lontra Canadensis*) [25].

Experimentally, a number of carnivores including Finn raccoons (*Nyctereutes procyonoides*), ferrets, dogs, mice, and cats have been inoculated with highly virulent ADV-Utah without developing clinical symptoms, but their organs could infect naïve mink [58]. Additionally, Finn raccoon, mink, and dog produced antibodies toward the nonstructural proteins after infection indicating virus replication in these species [58]. Kenyon et al. experimentally infected 14 members of the family *Mustelidae* besides mink, ferret, and striped skunk AMDV inoculation resulted in antibody production in short-tailed weasel, fisher (*Martes pennanti*), and American marten (*Martes americana*) [53].

Serological evidence of AMDV has been found in free-ranging American mink (*Neovison vison*) in Canada, the United States [25,28,59,60], and Europe [26,27,61,62] as well as in a number of other mustelids and small carnivores including free-ranging, European mink (*Mustela lutreola*) [62], polecats (*Mustela putorius*), stone martens (*Martes foina*), pine martens (*Martes martes*), and common genets (*Genetta genetta*) [61].

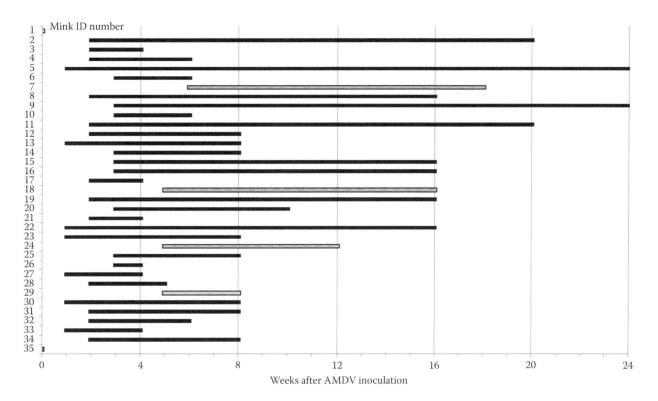

FIGURE 75.1 Viremia in 35 mink following Aleutian mink disease infection. The wild-type mink (black bars) were inoculated with AMDV at week 0, and the sapphire mink (gray bars) were kept as noninoculated sentinels. The bars represent the duration of viremia starting within the first week after challenge inoculation with AMDV. Viremia was detected as AMDV DNA by PCR on full blood. Mink number 1 and number 36 did not develop viremia, and mink number 27 was not included in the experiment. The asterisk indicates that the mink were diseased and euthanized before the others in the group (mink 1–12 were euthanized 24 weeks after AMDV inoculation, mink 13–24 after 16 weeks, and mink 25–36 after 8 weeks). (From Jensen, T.H. et al., *Vet. Microbiol.*, 168, 420, 2014.)

Reports of a possible relationship between AMDV and human infection are rare [63].

75.1.3 DIAGNOSIS

75.1.3.1 Conventional Techniques

Mink infected with ADMV produce massive amounts of antibodies toward ADMV. Serum from healthy mink has approximately 7 mg gammaglobulin/mL serum, whereas the average levels in ADMV-infected mink are 35–56 mg/mL [64]. In spite of the high levels, these antibodies are not able to neutralize the virus neither *in vitro* in cell cultures nor in the persistently infected mink host. The detection of hypergammaglobulinemia [65,66], based on antibody detection, has been the key of the eradication program in Scandinavia. As the nail in carnivores contain blood vessels, it was quite easy for the farmer to obtain a blood sample from mink without needles or special sanitary risks. Blood was collected in capillary tubes for centrifugation on the farm. Initially, the farmers quantified the level of gammaglobulins aspecifically using the iodine test (Mallens test) [67]. As the mink farmers organized, they also build central laboratories for research and improved diagnostic. As part of the international collaboration, an agar gel–based CIEP was developed and has been used worldwide for the diagnosis of ADMV [44,59]. This gel-based test has proven to be a sensitive and

cheap method to analyze mink sera for the presence of specific antibodies toward ADMV [68]. Annually 3.5–4 million sera are tested using this method [69]; a disadvantage is the dependence on manual reading of all samples after gel electrophoresis, and specially trained technicians are necessary. The CIEP are being replaced in Denmark by an automated ELISA based on whole AMDV-G [69]. In Finland, an ELISA based on VP2 recombinant antigen has been used for years [70,71].

75.1.3.2 Molecular Techniques

The pathogenesis of older AMDV strains such as AMDV-Utah, AMDV-Pullman, AMDV-Demark have been extensively studied using immunohistochemistry, southern blots, and *in situ* hybridization together with countercurrent and rocket line immunoelectrophoresis (reviewed by [18,31,35]). Genetic variations between different AMDV strains have been described to occur at both the NS1 [12,13] and VP2 genes [2,14]. More recently, various PCR primers have been used to detect AMDV in spleen and mesenteric lymph nodes, some amplifying capsid gene (VP2) for diagnostic and for molecular biological characterization [19,32] as well as targeting AMDV in the brain [36]. Other primers amplifying the nonstructural protein (NS1) primarily for molecular biological characterization have been used in Finland [37] and Sweden [38] but also for diagnostic purposes in Denmark [39].

75.2 METHODS

75.2.1 SAMPLE PREPARATION

In vivo samples include swabs from oronasal region, blood, and feces, which can be sampled from trapped or anesthetized animals. Swabs are easily processed by direct DNA isolation of the swab media such as phosphate-buffered saline (PBS) using QIAamp® Blood Mini Kit [3]. Blood can be processed either as full blood (e.g., EDTA stabilized [3]), as serum [34,72], as plasma [25], or as blood leukocytes [31], using various protocols such as QIAamp Blood Mini Kit [3], Dynabeads Silane Viral NA Kit [25] or polysucrose–sodium diatrizoate gradients, and phenol–chloroform extraction [31]. Total DNA from feces can be isolated with QIAamp DNA Stool Mini Kit [3].

Environmental samples, such as swabs from various surfaces at the farm and surroundings including soil samples, may be tested by real-time PCR as a way to uncover disease transmission paths [73]. This real-time PCR utilized primers from the NS1 gene and FAM-labeled plus VIC-labeled TaqMan® probes [73].

Postmortem a number of tissues have been investigated for AMDV [3], and the spleen and mesenteric lymph nodes appear to be the most consistent organs for AMDV diagnosis. The assay sensitivity and specificity increase notably if both organs are simultaneously tested [74,75]. Tissues, for example, spleen and mesenteric lymph nodes, can be processed for PCR analysis by homogenizing tissue (0.1 g) either manually or by Qiagen Tissue Lyser (Qiagen, Hilden, Germany). Subsequently, the tissue must be suspended in PBS containing 10% Qiagen protease (Qiagen) and incubated at 56°C for 1 h. Total DNA can be extracted from the tissue by QIAamp Blood Mini Kit according to the manufacturer's instructions (Qiagen, Hilden, Germany).

75.2.2 DETECTION PROCEDURES

The frequent genetic variation of AMDV highlights the importance of using primers targeting a conserved part of either the NS1 or the VP2 genes to enhance the sensitivity of the PCR. PCR amplifying a part of the VP2 gene was used to diagnose AMDV infection in raccoons, mink [27,34,76], ferrets [20,23,34,55–57], and a captive skunk [52]. PCR targeting the NS1 gene has been employed to characterize a high genetic diversity of AMDV strains from mink farms from Sweden [77], Finland [75], and Denmark [5]. An obvious advantage of the PCR as diagnostic tool is the ability to allow subsequent sequencing analysis of the AMDV gene products, which helps reveal the possible introduction and spread of certain AMDV strains. The definition of AMDV strain is not clear. However, there are strong indications for multiple different strains in certain geographical areas [5,75,77,78]. The old AMDV strains seem to vary significantly from the recent strains. There is evidence for wild mustelids to carry similar strains as the farmed mink but also different strains [18,20,27,57].

Primers amplifying a fragment of 374 bp of the NS1 gene have been used for diagnostic purposes and pathogenesis studies in Denmark for years [3,5,63,74]. These primers were blasted against all AMDV sequences available at GenBank and found able to detect them. The PCR procedure employing NS1 primers has also been applied together with the VP2 primers of Saifuddin et al. [57]. This PCR procedure is described in detail as follows:

Primer	Sequence (5′–3′)	Expected Product (bp)	Specificity	Reference
ADV-F-7-H-PN1	CATATTCACTG TTGCTAGGTTA	374	NS1	[3]
ADV-R-7-HPN2	CGTTCTTTGTTA GTTAGGTTGTC	374	NS1	[3]
P1	AGAGCAACCAA ACCAACCC	401	VP2	[57]
P2	TCACCCCAAAA GTGACC	401	VP2	[57]

75.2.2.1 PCR Detection

1. Prepare the PCR amplification reaction mixture in a cool rack as follows:
   a. 0.8 µL dNTPs (25 mM, Stratagene, Edinburgh, Scotland)
   b. 5 µL of each primer AMDV-F-7-H-PN1 and AMDV-R-7-HPN2 (each 2 pmol/µL)
   c. 79.2 µL DNAse-free water
2. Add 10 µL DyNAzyme™II DNA polymerase (Finnzymes, Espoo, Finland) just before use of the amplification mixture.
3. Add the 5 µL sample DNA (extracted as written earlier).
4. Place the samples in a thermocycler programmed to initial preheating (also the lid) at 94°C for 5 min, followed by 45 cycles of denaturation at 94°C for 30 s, annealing at 55°C for 30 s, elongation at 72°C for 30 s, and a final elongation at 72°C for 7 min.
5. Separate the PCR products by 1.5% agarose gel electrophoresis (in TAE buffer) at 75 V for 45 min. Stain the gel with ethidium bromide and visualize under ultraviolet light.

Note: To monitor any possible contamination that could occur during the PCR procedure, negative and positive controls are included along with the DNA extraction. The negative tissue sample is from a Danish AMDV-free farm. The positive control is from a non-Danish AMDV-infected mink to verify potential contamination by sequencing.

75.2.2.2 Sequencing

DNA is purified for sequencing using the QIAquick PCR Purification Kit (Qiagen). Sequencing is performed with the PCR primers either by DNA Technology, Aarhus, Denmark or LGC Genomics, Berlin, Germany. The sequences are assembled and proofread by the use of CLC DNA Workbench (http://www.clcbio.com/). Sequence similarity is detected by BLAST search (http://blast.ncbi.nlm.nih.gov/) and alignment with existing Danish AMDV sequences [5] by ClustalW (24) included in CLC DNA Workbench.

75.3 CONCLUSION AND FUTURE PERSPECTIVES

The optimal diagnostics for AMDV is by PCR because the virus cannot be cultured. Since AMDV is environmentally stable, the samples do not require specific care, and a variety of DNA isolation methods may be used. Organs, especially spleen and mesenteric lymph node, are ideal starting materials for sensitive and specific detection of AMDV at any time during infection, which is often chronic asymptomatic [3,75]. The advantage of PCR diagnostic is that the resulting product can be sequenced and then characterized and compared to other AMDV sequences. This can be valuable for investigation of disease transmission. However, for screening purposes, AMDV serology is found to be superior to PCR detection of AMDV on blood or secretions because AMDV tends to cause a transient unpredictable viremia and virus excretion [3]. ELISA analysis of serum antibodies is amenable to high-throughput automation. Environmental samples have been investigated for screening purposes but their use is limited due to the fact that they often contain very small amount of virus DNA. To eliminate the gel electrophoresis and speed up the diagnosis, adaptation of conventional PCR to a real-time platform is critical, thereby also allowing for a quantification of the virus in target organs. Finally, optimization of a DNA isolation procedure for environmental samples will greatly benefit future screening programs.

REFERENCES

1. Knuuttila, A., Diagnostics and epidemiology of Aleutian mink disease virus. PhD thesis, University of Helsinki, Helsinki, Finland, 2015. Available at: https://helda.helsinki.fi/bitstream/handle/10138/158058/diagnost.pdf?sequence=1; last accessed February 20, 2016.
2. Hartsough, G.R., Gorham, J.R., Aleutian disease in mink. *Nat. Fur News* 28, 10–11, 1956.
3. Jensen, T.H., Hammer, A.S., Chriel, M., Monitoring chronic infection with a field strain of Aleutian mink disease virus. *Vet. Microbiol.* 168, 420, 2014.
4. Uttenthal, A., Lund, E., Hansen, M., Mink enteritis parvovirus. Stability of virus kept under outdoor conditions. *APMIS* 107, 353, 1999.
5. Christensen, L.S., Gram-Hansen, L., Chriel, M., Jensen, T.H., Diversity and stability of Aleutian mink disease virus during bottleneck transitions resulting from eradication in domestic mink in Denmark. *Vet. Microbiol.* 149, 64, 2011.
6. Chriel, M., Impact of outbreaks of acute Aleutian disease in Danish Mink farms. Wageningen Academic Publisher. Proceedings of the Seventh International Scientific Congress in fur animal production. *Scientifur* 24, 16–20, 2000.

7. Larsen, S., Alexandersen, S., Lund, E., Have, P., Hansen, M., Acute interstitial pneumonitis caused by Aleutian disease virus in mink kits. *Acta Pathol. Microbiol. Immunol. Scand.* [A] 92, 391, 1984.
8. Alexandersen, S., Aasted, B., Restricted heterogeneity of the early antibody response to Aleutian disease virus in mink kits. *Acta Pathol. Microbiol. Immunol. Scand.* [C] 94, 137, 1986.
9. King, A., Adams, J.M., Carstens, E., Lefkowitz, LE. Amdovirus. In: King, A., Adams, J.M., Lefkowitz, L.E., Carstens, E. (eds.) *Virus Taxonomy: Classification and Nomenclature of Viruses: Ninth Report of the International Committee on Taxonomy of Viruses.* San Diego, CA: Elsevier Academic Press, 415pp., 2012.
10. Bloom, M.E., Kaaden, O.R., Huggans, E., Cohn, A., Wolfinbarger, J.B., Molecular comparisons of in vivo- and in vitro-derived strains of Aleutian disease of mink parvovirus. *J. Virol.* 62, 132, 1988.
11. Alexandersen, S., Bloom, M.E., Perryman, S., Detailed transcription map of Aleutian mink disease parvovirus. *J. Virol.* 62, 3684, 1988.
12. Qiu, J., Cheng, F., Burger, L.R., Pintel, D., The transcription profile of Aleutian mink disease virus in CRFK cells is generated by alternative processing of pre-mRNAs produced from a single promoter. *J. Virol.* 80, 654, 2006.
13. Christensen, J., Storgaard, T., Bloch, B., Alexandersen, S., Aasted, B., Expression of Aleutian mink disease parvovirus proteins in a baculovirus vector system. *J. Virol.* 67, 229, 1993.
14. Bloom, M.E., Alexandersen, S., Perryman, S., Lechner, D., Wolfinbarger, J.B., Nucleotide sequence and genomic organization of Aleutian mink disease parvovirus (ADV): Sequence comparisons between a nonpathogenic and a pathogenic strain of ADV. *J. Virol.* 62, 2903, 1988.
15. Fox, J.M., Crackin Stevenson, M.A., Bloom, M.E., Replication of Aleutian mink disease parvovirus in vivo is influenced by residues in the VP2 protein. *J. Virol.* 73, 8713, 1999.
16. Best, S.M., Wolfinbarger, J.B., Bloom, M.E., Caspase activation is required for permissive replication of Aleutian mink disease parvovirus in vitro. *Virology* 292, 224, 2002.
17. Best, S.M., Shelton, J.F., Pompey, J.M., Wolfinbarger, J.B., Bloom, M.E., Caspase cleavage of the nonstructural protein NS1 mediates replication of Aleutian mink disease parvovirus. *J. Virol.* 77, 5305, 2003.
18. Gottschalck, E., Alexandersen, S., Cohn, A., Poulsen, L.A., Bloom, M.E., Aasted, B., Nucleotide sequence analysis of Aleutian mink disease parvovirus shows that multiple virus types are present in infected mink. *J. Virol.* 65, 4378, 1991.
19. Gottschalck, E., Alexandersen, S., Storgaard, T., Bloom, M.E., Aasted, B., Sequence comparison of the non-structural genes of four different types of Aleutian mink disease parvovirus indicates an unusual degree of variability. *Arch. Virol.* 138, 213, 1994.
20. Schuierer, S., Bloom, M.E., Kaaden, O.R., Truyen, U., Sequence analysis of the lymphotropic Aleutian disease parvovirus ADV-SL3. *Arch. Virol.* 142, 157, 1997.
21. Cotmore, S.F., Tattersall, P., The autonomously replicating parvoviruses of vertebrates. *Adv. Virus Res.* 33, 91, 1987.
22. Gorham, J.R., Leader, R.W., Henson, J.B., The experimental transmission of a virus causing hypergammaglobulinemia in mink: Sources and modes of infection. *J. Infect. Dis.* 114, 341–345, 1964.
23. Pennick, K.E., Stevenson, M.A., Latimer, K.S., Ritchie, B.W., Gregory, C.R., Persistent viral shedding during asymptomatic Aleutian mink disease parvoviral infection in a ferret. *J. Vet. Diagn. Invest.* 17, 594, 2005.

24. Broll, S., Alexandersen, S., Investigation of the pathogenesis of transplacental transmission of Aleutian mink disease parvovirus in experimentally infected mink. *J. Virol.* 70, 1455, 1996.

25. Farid, A.H., Aleutian mink disease virus in furbearing mammals in Nova Scotia, Canada. *Acta Vet. Scand.* 55, 10, 2013.

26. Jensen, T.H., Christensen, L.S., Chriel, M., Harslund, J., Salomonsen, C.M., Hammer, A.S., High prevalence of Aleutian mink disease virus in free-ranging mink on a remote Danish island. *J. Wildl. Dis.* 48, 497, 2012.

27. Manas, S. et al., Aleutian mink disease parvovirus in wild riparian carnivores in Spain. *J. Wildl. Dis.* 37, 138, 2001.

28. Nituch, L.A., Bowman, J., Beauclerc, K.B., Schulte-Hostedde, A.I., Mink farms predict Aleutian disease exposure in wild American mink. *PLoS One* 6, e21693, 2011.

29. Shen, D.T., Gorham, J.R., Harwood, R.F., Padgett, G.A., The persistence of Aleutian disease virus in the mosquito *Aedes fitchii*. *Arch. Gesamte Virusforsch.* 40, 375, 1973.

30. Geus, B.D., Eck, J.V., van de Louw, A., Wiligen, F.K.V., Bokhout, B., Transmission of Aleutian disease virus by air. *Scientifur* 20, 350–354, 1996.

31. Jackson, M.K., Ellis, L.C., Morrey, J.D., Li, Z.Z., Barnard, D.L., Progression of Aleutian disease in natural and experimentally induced infections of mink. *Am. J. Vet. Res.* 57, 1753, 1996.

32. Hadlow, W.J., Race, R.E., Kennedy, R.C., Comparative pathogenicity of four strains of Aleutian disease virus for pastel and sapphire mink. *Infect. Immun.* 41, 1016, 1983.

33. Alexandersen, S., Pathogenesis of disease caused by Aleutian mink disease parvovirus. *APMIS Suppl.* 14, 1, 1990.

34. Oie, K.L. et al., The relationship between capsid protein (VP2) sequence and pathogenicity of Aleutian mink disease parvovirus (ADV): A possible role for raccoons in the transmission of ADV infections. *J. Virol.* 70, 852, 1996.

35. Alexandersen, S., Acute interstitial pneumonia in mink kits: Experimental reproduction of the disease. *Vet. Pathol.* 23, 579, 1986.

36. Alexandersen, S., Bloom, M.E., Wolfinbarger, J., Race, R.E., In situ molecular hybridization for detection of Aleutian mink disease parvovirus DNA by using strand-specific probes: Identification of target cells for viral replication in cell cultures and in mink kits with virus-induced interstitial pneumonia. *J. Virol.* 61, 2407, 1987.

37. Alexandersen, S., Larsen, S., Aasted, B., Uttenthal, A., Bloom, M.E., Hansen, M., Acute interstitial pneumonia in mink kits inoculated with defined isolates of Aleutian mink disease parvovirus. *Vet. Pathol.* 31, 216, 1994.

38. An, S.H., Ingram, D.G., Detection of inapparent Aleutian disease virus infection in mink. *Am. J. Vet. Res.* 38, 1619, 1977.

39. Bloom, M.E., Race, R.E., Hadlow, W.J., Chesebro, B., Aleutian disease of mink: The antibody response of sapphire and pastel mink to Aleutian disease virus. *J. Immunol.* 115, 1034, 1975.

40. Hadlow, W.J., Race, R.E., Kennedy, R.C., Royal pastel mink respond variously to inoculation with Aleutian disease virus of low virulence. *J. Virol.* 50, 38, 1984.

41. Porter, D.D., Larsen, A.E., Porter, H.G., The pathogenesis of Aleutian disease of mink. I. In vivo viral replication and the host antibody response to viral antigen. *J. Exp. Med.* 130, 575, 1969.

42. Alexandersen, S., Bloom, M.E., Wolfinbarger, J., Evidence of restricted viral replication in adult mink infected with Aleutian disease of mink parvovirus. *J. Virol.* 62, 1495, 1988.

43. Porter, D.D., Larsen, A.E., Porter, H.G., The pathogenesis of Aleutian disease of mink. 3. Immune complex arteritis. *Am. J. Pathol.* 71, 331, 1973.

44. Hansen, M., Lund, E., Pregnancy rate and foetal mortality in Aleutian disease virus infected mink. *Acta Vet. Scand.* 29, 271, 1988.

45. Dyer, N.W., Ching, B., Bloom, M.E., Nonsuppurative meningoencephalitis associated with Aleutian mink disease parvovirus infection in ranch mink. *J. Vet. Diagn. Invest.* 12, 159, 2000.

46. Jahns, H., Daly, P., McElroy, M.C., Sammin, D.J., Bassett, H.F., Callanan, J.J., Neuropathologic features of Aleutian disease in farmed mink in Ireland and molecular characterization of Aleutian mink disease virus detected in brain tissues. *J. Vet. Diagn. Invest.* 22, 101, 2010.

47. Farid, A.H., Ferns, L.E. Aleutian mink disease virus infection may cause hair depigmentation. *Scientifur* 35, 55–59, 2011.

48. Bloom, M.E., Kanno, H., Mori, S., Wolfinbarger, J.B., Aleutian mink disease: Puzzles and paradigms. *Infect. Agents Dis.* 3, 279, 1994.

49. Alexandersen, S., Hau, J., Rocket line immunoelectrophoresis: An improved assay for simultaneous quantification of a mink parvovirus (Aleutian disease virus) antigen and antibody. *J. Virol. Methods* 10, 145, 1985.

50. Porter, D.D., Larsen, A.E., Porter, H.G., The pathogenesis of Aleutian disease of mink. II. Enhancement of tissue lesions following the administration of a killed virus vaccine or passive antibody. *J. Immunol.* 109, 1, 1972.

51. Castelruiz, Y., Blixenkrone-Moller, M., Aasted, B., DNA vaccination with the Aleutian mink disease virus NS1 gene confers partial protection against disease. *Vaccine* 23, 1225, 2005.

52. Allender, M.C. et al., Infection with Aleutian disease virus-like virus in a captive striped skunk. *J. Am. Vet. Med. Assoc.* 232, 742, 2008.

53. Kenyon, A.J., Kenyon, B.J., Hahn, E.C., Protides of the Mustelidae: Immunoresponse of mustelids to Aleutian mink disease virus. *Am. J. Vet. Res.* 39, 1011, 1978.

54. LaDouceur, E.E. et al., Aleutian disease: An emerging disease in free-ranging striped skunks (*Mephitis mephitis*) from California. *Vet. Pathol.* 52, 1250, 2015.

55. Une, Y., Wakimoto, Y., Nakano, Y., Konishi, M., Nomura, Y., Spontaneous Aleutian disease in a ferret. *J. Vet. Med. Sci.* 62, 553, 2000.

56. Murakami, M., Matsuba, C., Une, Y., Nomura, Y., Fujitani, H., Nucleotide sequence and polymerase chain reaction/restriction fragment length polymorphism analyses of Aleutian disease virus in ferrets in Japan. *J. Vet. Diagn. Invest.* 13, 337, 2001.

57. Saifuddin, M., Fox, J.G., Identification of a DNA segment in ferret Aleutian disease virus similar to a hypervariable capsid region of mink Aleutian disease parvovirus. *Arch. Virol.* 141, 1329, 1996.

58. Alexandersen, S., Jensen, A.U., Hansen, M., Aasted, B., Experimental transmission of Aleutian disease virus (ADV) to different animal species. *Acta Pathol. Microbiol. Immunol. Scand. [B]* 93, 195, 1985.

59. Cho, H.J., Greenfield, J., Eradication of Aleutian disease of mink by eliminating positive counterimmunoelectrophoresis test reactors. *J. Clin. Microbiol.* 7, 18, 1978.

60. Farid, A.H., Rupasinghe, P., Mitchell, J.L., Rouvinen-Watt, K., A survey of Aleutian mink disease virus infection of feral American mink in Nova Scotia. *Can. Vet. J.* 51, 75, 2010.

61. Fournier-Chambrillon, C. et al., Antibodies to Aleutian mink disease parvovirus in free-ranging European mink (*Mustela lutreola*) and other small carnivores from southwestern France. *J. Wildl. Dis.* 40, 394, 2004.

62. Yamaguchi, N., Macdonald, D.W., Detection of Aleutian disease antibodies in feral American mink in southern England. *Vet. Rec.* 149, 485, 2001.

63. Jepsen, J.R., d'Amore, F., Baandrup, U., Clausen, M.R., Gottschalck, E., Aasted, B., Aleutian mink disease virus and humans. *Emerg. Infect. Dis.* 15, 2040, 2009.

64. Porter, D.D., Dixon, F.J., Larsen, A.E., Metabolism and function of gammaglobulin in Aleutian disease of mink. *J. Exp. Med.* 121, 889, 1965.

65. Eklund, C.M., Hadlow, W.J., Kennedy, R.C., Boyle, C.C., Jackson, T.A., Aleutian disease of mink: Properties of the etiologic agent and the host responses. *J. Infect. Dis* 118, 510, 1968.

66. Cho, H.J., Ingram, D.G., Antigen and antibody in Aleutian disease in mink. I. Precipitation reaction by agar-gel electrophoresis. *J. Immunol.* 108, 555, 1972.

67. Mallen, M.S., Ugalde, E.L., Bolzar, R.M., Bolivar, J., Meyran, S., Precipitations of abnormal serum by Lugol's solution. *Am. J. Clin. Pathol.* 20, 39, 1950.

68. Uttenthal, A., Screening for antibodies against Aleutian disease virus (ADV) in mink. Elucidation of dubious results by additive counterimmunoelectrophoresis. *Appl. Theor. Electrophor.* 3, 83, 1992.

69. Dam-Tuxen, R., Dahl, J., Jensen, T.H., Dam-Tuxen, T., Struve, T., Bruun, L., Diagnosing Aleutian mink disease infection by a new fully automated ELISA or by counter current immuno-electrophoresis: A comparison of sensitivity and specificity. *J. Virol. Methods* 199, 53, 2014.

70. Knuuttila, A., Aronen, P., Saarinen, A., Vapalahti, O., Development and evaluation of an enzyme-linked immunosorbent assay based on recombinant VP2 capsids for the detection of antibodies to Aleutian mink disease virus. *Clin. Vaccine Immunol.* 16, 1360, 2009.

71. Knuuttila, A., Aronen, P., Eerola, M., Gardner, I.A., Virtala, A.M., Vapalahti, O., Validation of an automated ELISA system for detection of antibodies to Aleutian mink disease virus using blood samples collected in filter paper strips. *Virol. J.* 11, 141, 2014.

72. Bloom, M.E., Oie, K.L., Wolfinbarger, J., Christensen, P., Durrant, G., Evaluation of the polymerase chain reaction (PCR) as a tool for diagnosing infections with the Aleutian mink disease parvovirus (ADV). *Scientifur* 21, 141, 1997.

73. Prieto, A. et al., Application of real-time PCR to detect Aleutian Mink Disease Virus on environmental farm sources. *Vet. Microbiol.* 173, 355, 2014.

74. Jensen, T.H., Christensen, L.S., Chriel, M., Uttenthal, A., Hammer, A.S., Implementation and validation of a sensitive PCR detection method in the eradication campaign against Aleutian mink disease virus. *J. Virol. Methods* 171, 81, 2011.

75. Knuuttila, A., Uzcategui, N., Kankkonen, J., Vapalahti, O., Kinnunen, P., Molecular epidemiology of Aleutian mink disease virus in Finland. *Vet. Microbiol.* 133, 229, 2009.

76. Bloom, M.E. et al., Expression of Aleutian mink disease parvovirus capsid proteins in defined segments: Localization of immunoreactive sites and neutralizing epitopes to specific regions. *J. Virol.* 71, 705, 1997.

77. Olofsson, A., Mittelholzer, C., Treiberg, B.L., Lind, L., Mejerland, T., Belak, S., Unusual, high genetic diversity of Aleutian mink disease virus. *J. Clin. Microbiol.* 37, 4145, 1999.

78. Sang, Y., Ma, J., Hou, Z., Zhang, Y., Phylogenetic analysis of the VP2 gene of Aleutian mink disease parvoviruses isolated from 2009 to 2011 in China. *Virus Genes* 45, 31, 2012.

76 Canine Parvovirus

S. Parthiban

CONTENTS

76.1 Introduction 687
    76.1.1 Classification and Morphology...687
    76.1.2 Biology and Epidemiology 688
    76.1.3 Clinical Features and Pathogenesis 688
    76.1.4 Diagnosis 689
        76.1.4.1 Conventional Techniques...689
        76.1.4.2 Molecular Techniques..690
76.2 Methods 690
    76.2.1 Sample Preparation..690
    76.2.2 Detection Procedures...690
        76.2.2.1 PCR-Based Detection of CPV Using Primer Pair H_{for}/H_{rev} and Subsequent Sequencing...................690
        76.2.2.2 Detection of Canine Parvovirus in Fecal Samples Using Loop-Mediated Isothermal
                Amplification Assay..691
76.3 Conclusion 691
References...692

76.1 INTRODUCTION

Canine parvovirus (CPV), a highly contagious virus causing acute gastroenteritis in domestic dogs, is prevalent all over the world, mainly because this virus can survive in adverse environmental conditions for a long time [1]. Natural CPV-2 infection has been reported in domestic dogs, bush dogs, cats, coyotes, bears, and wolves. The disease is clinically characterized by severe vomiting, pyrexia, anorexia, and diarrhea leading to fatal dehydration as well as myocarditis particularly in young pups. CPV-2 is a nonenveloped, single-stranded DNA virus belonging to the *Parvoviridae* family and genus *Parvovirus*, along with feline panleukopenia virus (FPLV) and other parvoviruses of carnivores [2]. A wide range of species are affected by viruses of the *Parvoviridae* family; however, CPV-2 does not cross species lines [2]. After its emergence in the 1970s, CPV-2 underwent a rapid evolution, and within a few years, new antigenic types emerged at regular intervals, termed CPV-2a in 1978, CPV-2b in 1984, and CPV-2c in 1997 and completely replaced the original CPV-2. Presently, the original CPV-2 type no longer circulates in the dog population, whereas CPV-2a and CPV-2b types are distributed worldwide; now, CPV-2a and CPV-2b types have further undergone several mutations that produced variants of CPV-2a (New CPV-2a) and CPV-2b (New CPV-2b). Mutations affecting the VP2 gene of CPV have been responsible for the evolution of different antigenic variants. Sequence analysis of the VP2 gene of the virus and subsequent characterization is important for molecular epidemiology of CPV [3].

76.1.1 CLASSIFICATION AND MORPHOLOGY

Of the two types of CPV identified to date (i.e., canine minute virus [CPV-1] and CPV-2, CPV-2 causes the most serious disease and affects domesticated dogs and wild canids. CPV-2 was first isolated in 1978 [4,5]; however, further studies on the rate of CPV molecular evolution indicate that the virus likely emerged 10 years earlier [5,6]. The exact origin of CPV-2 is not known, although it is thought to have originated as a host range variant from FPLV or other parvoviruses. Since the original type CVP-2 was discovered, several antigenic variants—CPV-2a, CPV-2b, and CPV-2c—have replaced it entirely and have become distributed throughout the canine population worldwide [2]. Another CPV variant has an amino acid substitution of Asp-426 to Glu, which occurs in a residue of the capsid protein that is considered important for the antigenic properties of CPV-2. This variant (CPV-2/Glu-426 mutant) was named CPV-2c that was first observed in Italy and was broadly distributed worldwide with CPV-2a and CPV-2b types [2]. CPV infections are also caused by New CPV-2a/CPV-2b variants, which have additional amino acid difference in both CPV-2a and CPV-2b at position 297 (Ser to Ala) (Table 76.1). This mutation first appeared in 1993 in German CPV isolates and was designated as New CPV-2a/CPV-2b [3].

The CPV-2 belongs to the family *Parvoviridae*. The genome consists of a single molecule of linear single-stranded DNA, which is 5.2 kb in size. It is a nonenveloped, icosahedral symmetry. The organization of CPV capsid is similar to that of other autonomous parvoviruses, in that it contains two

TABLE 76.1
Amino Acid Variation in the VP2 Protein of Different CPV-2 Types

CPV Strains	Nucleotides in CPV Genome				
	3676	3685	3699	3909	4062
Amino acid residue in VP2	297	300	305	375	426
CPV-2	TCT/Ser	GCT/Ala	GAT/Asp	AAT/Asn	AAT/Asn
CPV-2a	TCT/Ser	GGT/Gly	TAT/Tyr	GAT/Asp	AAT/Asn
CPV-2b	TCT/Ser	GGT/Gly	TAT/Tyr	GAT/Asp	GAT/Asp
New CPV-2a	GCT/Ala	GGT/Gly	TAT/Tyr	GAT/Asp	AAT/Asn
New CPV-2b	GCT/Ala	GGT/Gly	TAT/Tyr	GAT/Asp	GAT/Asp
CPV-2c	GCT/Ala	GGT/Gly	TAT/Tyr	GAT/Asp	GAA/Glu

largely overlapping proteins, VP1 (82 kDa) and VP2 (62 kDa) [7,8]. There is a third protein, VP3 (63 kDa), which is produced by proteolytic processing of VP2. The 22 nm parvovirus capsid comprises of about 10 copies of VP1 and 60–70 copies of VP2. Epitope mapping experiment showed that all of the epitope neutralizing antibodies are within the VP2 region. Amino acid substitutions in the VP2 sequence are known to affect the antigenic properties [9,10].

76.1.2 BIOLOGY AND EPIDEMIOLOGY

All CPV isolates are thought to descend from a related feline parvovirus (FPV) strain. The differences in DNA sequence between CPV and FPV isolates can be as little as 0.5% [11]. Two substitution mutations of two key amino acids—changed 93 from Lys to Asn and 323 from Asp to Asn—on the VP2 residues allowed for CPV to extend its host range to canines [5,11]. Thus, CPV acquired the ability to infect canines as a result of the changes made to the capsid protein surface. CPV is able to infect canine cells by binding to the canine transferrin receptor (TfR). CPV-2 capsids enter cells via "clathrin-mediated endocytosis," colocalize with transferrin (Tf) in perinuclear endosomes, and then slowly leave the endosome and enter the cytoplasm prior to the DNA gaining access to the nucleus for replication [11]. Parvoviruses are small (18–26 nm in diameter), which are made up of single-stranded DNA consisting of 5000–5200 nucleotides. As these viruses contain small DNA molecules with limited genetic information, they must replicate in the cell nuclei and require rapidly dividing cells in the synthesis phase of the cell cycle for optimal replication. This makes tissues with high replication rates—like those of fetuses and neonates, in addition to the hematopoietic and epithelial tissues in adults—targets of parvovirus infection [2].

Prevalence study of CPV revealed that the oldest CPV-2c strain was identified in 1996 in Germany [12], and the results from European epidemiological surveys show that CPV-2c is now predominant in Italy, Germany, and Spain and is also widely codistributed with CPV-2a or CPV-2b in Portugal, France, and Belgium [12]. Outside of Europe, CPV-2a and

CPV-2b isolates are common in the United States [13], whereas CPV-2c is more widespread in Uruguay and Brazil [14]. Surprisingly, variants of CPV-2a and CPV-2b are predominant in Asian countries [3,15]. The adaptation of CPV to dogs led to the natural global replacement of CPV-2 by the CPV-2a variant and later by the CPV-2b strain, through the addition of a point mutation [11]. As CPV is a relatively new virus that continues to evolve and produce new antigenic types, the more predominant mutants are selected for by their ability to bind to the TfR and by extending host range, in newer antigenic types, to include both dogs and cats [5]. Puppies between the age of 4 and 12 weeks are the most susceptible to CPV-2 infection, as this is when the maternally derived antibodies (MDA) typically wanes [2], while unvaccinated young (6 weeks to 6 months old) dogs also remain vulnerable. Some breeds, such as Rottweilers, Doberman pinschers, American pit bull terriers, English springer spaniels, and German shepherd dogs are also thought to be at higher risk to CPV-2 infection [7]. The morbidity may reach up to 100% and mortality may be as high as 10%. CPV-2 is highly contagious and can be spread through direct contact, fecal–oral route, or vectors such as fomites. Most frequently, the virus is spread by indirect contact with infected feces, as they are highly infectious, containing billions of CPV-2 virions per gram [2]. CPV-2 is shed in the feces of infected dogs within 4–5 days prior to the onset of clinical symptoms, throughout the period of illness, and up to 3 weeks after clinical recovery [1]. The clinical manifestation of CPV infection depends on the age and immune status of the animal, the virulence of the virus, the dose of infectious virus, and the preexisting or concurrent parasitic, bacterial, or virus infections [16].

There are three key factors in preventing the spread of CPV-2: (1) sanitation, (2) isolation, and (3) vaccination. Under ideal conditions, CPV-2 can live in an environment for up to a year, and it has been found to be resistant to most common detergents and disinfectants and extreme temperatures. However, the virus can be killed by dilute bleach on hard surfaces. As CPV-2 is highly contagious, infected dogs should be hospitalized in an isolation ward to prevent the spread of the virus. Isolation will also help protect the infected dog against contracting a secondary infection by minimizing their exposure to other sick canines [1]. However, vaccination is the most successful means of preventing the spread of CPV-2 infection in domestic dogs. There are both inactivated and attenuated vaccines available for use against CPV-2. Live vaccines are most commonly used in domestic dogs because the inactivated vaccines only provide short-term immunity. These vaccines are made using either the original type CPV-2 or the CPV-2b variant; however, there are concerns about the effectiveness of CPV-2-based vaccines against variant CPV-2 strains [2].

76.1.3 CLINICAL FEATURES AND PATHOGENESIS

Virus can spread through direct contact, the fecal–oral route, or vectors such as fomites [1]. Virus enters the host by indirect contact with infected feces, as they are highly infectious, containing billions of CPV-2 virions per gram [2]. Clinical features of infected dogs range from asymptomatic infection to fulminant disease and sudden death. Clinical disease is usually seen in young and immunocompromised dogs, as well as predisposed breeds. Clinical signs often start with anorexia, lethargy, and fever. The condition progresses within 1–2 days to include vomiting and diarrhea, which may be yellow, mucoid, or hemorrhagic. More severe GI signs can be expected in puppies with intestinal parasites or other concurrent intestinal disease. Uncommonly, clinical signs of myocarditis may be observed in puppies, typically those younger than 8 weeks of age with failure of passive transfer of maternal immunity. These puppies may die acutely or after exhibiting signs of enteric disease. Those that survive acute myocardial infection may succumb later to congestive heart failure. CPV-2 is shed in the feces of infected dogs within 4–5 days prior to the onset of clinical symptoms, throughout the period of illness, and up to 3 weeks after clinical recovery [1].

After initial infection, the virus replicates in the lymphoid tissues of the oropharynx, mesenteric lymph nodes, and thymus [2] and then spreads through the blood stream [1]. CPV will infect and attack the rapidly dividing tissues or cells of the intestinal crypt, lymphopoietic tissue, and bone marrow [1,7]. The viral attack on the intestinal crypt epithelium results in massive lysis and necrosis of enterocytes, which impairs the absorptive capacity and cell turnover at the villi tips, and results in hemorrhagic diarrhea characteristic of the clinical disease that results from a combination of increased intestinal permeability and malassimilation from an abnormal mucosal function. In addition, the virus infects white blood cells and bone marrow and lymphopoietic tissues, inducing lymphopenia and neutropenia, often resulting in secondary infections. In 2–3-week-old puppies, often lacking MDA, CPV-2 can replicate in cardiac cells, without signs of enteritis, and result in fatal myocarditis [2,7].

The appearance of clinical symptoms of CPV-2 infection generally surfaces within 3–7 days of initial infection. However, subclinical infection is also common and some dogs can be asymptomatic carriers of the virus [1]. Initially, nonspecific symptoms, such as fatigue, fever, dehydration, and inappetence, will manifest. These symptoms will often develop into vomiting and hemorrhagic diarrhea (which has a distinct odor) within 24–48 h [1,7]. Intestinal loops can also become dilated and fluid filled, causing abdominal pain. Animals with severe infections commonly exhibit signs of septic shock, characterized by a decrease in time, tachycardia, and hypothermia. Necropsies reveal CPV-2 infection that can cause thickening of the cell wall, hemorrhagic and watery intestinal contents, and edema of abdominal and thoracic lymph nodes. In cases of CPV-2-induced myocarditis, pale streaks can be observed in the heart muscle tissue [7].

76.1.4 Diagnosis

A rapid diagnosis of CPV infection is especially important in order to isolate infected dogs and prevent secondary infections

of susceptible contact animals. Clinical diagnosis of CPV infection is difficult because the main clinical signs of the disease, which include vomiting and diarrhea, are common to many other enteric diseases like several other viral pathogens such as coronaviruses, adenoviruses, morbilliviruses, rotaviruses, reoviruses, and noroviruses. Several methods have been used to detect CPV, its proteins or nucleic acids, and include electron microscopy (EM), virus isolation (VI), latex agglutination, hemagglutination (HA), in situ hybridization, ELISA, polymerase chain reaction (PCR)-based assays, and loop-mediated isothermal amplification (LAMP) assay. Molecular diagnostic techniques like PCR-based methods are emerging as the most reliable techniques and have a high degree of sensitivity and specificity in detecting CPV from fecal samples [17]. PCR-based molecular typing of CPV also helps to gain new insights into the pathogenesis of CPV-2 infection and is extremely useful to comprehend whether the antigenic differences between CPV type 2a, 2b, or 2c and the original type 2 may account for partial or total failure of the conventional vaccines based on CPV-2 in protecting pups against the antigenic variants [2]. PCR-based assays can only be done in a diagnostic or commercial laboratory with access to specialized equipment not common to veterinary clinics. A new technique called loop-mediated isothermal amplification (LAMP) involves autocycling strand displacement DNA synthesis carried out by the use of a DNA polymerase with high strand displacement activity and a set of specific primers that recognize a total of six distinct sequences in the target DNA. This technique amplifies the target DNA sequence with a high selectivity; the assay can be completed within 1 h, has a high sensitivity, and produces an end product that is visible to the naked eye.

76.1.4.1 Conventional Techniques

EM is a vital diagnostic tool utilized to view the morphological characteristics of extremely small organisms, like parvoviruses, when a standard microscopy will not suffice. Even with all the newest forms of diagnostic methods, EM is still extensively used as a reference technique, especially in cases of specimens expected to contain high concentrations of the virus in diarrheic stool samples [18]. The HA test is simple and rapid for detecting CPV in feces. The HA activity of CPV is utilized diagnostically in estimating amounts of viral hemagglutinin in fecal samples [19] and is the most widely used test for detecting CPV in feces. CPV hemagglutinates with porcine, rhesus monkey, and human "O" RBCs. The HA is considered positive and specific when a HA titer of 32 and above is obtained, and the HA is inhibited by a specific antiserum [19]. But the HA and hemagglutination inhibition tests require continuous supply of quality porcine or human "O" erythrocytes. Moreover, nonspecific HA by fecal material and no HA activity by some strains of CPV [20] may complicate the scenario.

Virus isolation and identification in cell cultures like primary cell cultures of feline kidney and canine kidney cell lines is considered as a very reliable technique for the detection of CPV, but it is too labor intensive and time consuming.

The virus, having low cytopathogenicity, requires 3–5 blind passages for producing minimal cytopathic effect, or one may have to depend on indirect methods of diagnosis like HA and hemagglutination inhibition tests, fluorescent antibody technique, and PCR [21]. Latex agglutination test (LAT) is a rapid test for screening CPV infections, but it fails to detect low concentration of the virus. LAT also gives false-positive results, when hemorrhagic feces are employed for testing [22]. Immunochromatographic test is the most common rapid field diagnostic test used in clinical practice because the test procedure is simple and rapid and can be performed by veterinarians as well as by owners. However, large amounts of viral antigen are required to produce a clearly visible band [23]. ELISA is also used for screening CPV infections and has a high specificity, but it cannot detect samples with low virus concentrations. ELISA needs a minimum of 10^6 PFU/g of feces unlike PCR that can detect very small quantities of virus up to 10^3 PFU/g of feces [24].

76.1.4.2 Molecular Techniques

Diagnosis of CPV infection by traditional methods has been shown to be poorly sensitive, especially in the late stages of infections. New diagnostic approaches based on molecular methods have been developed for sensitive detection of CPV in clinical samples and rapid characterization of the viral type. Parvovirus PCR testing has been shown to be a very effective and sensitive way to detect CPV. The PCR test will also find the presence of fecal parvo DNA even when an antigen test (ELISA) showed a negative result. Real-time PCR is the newest technology using a minor groove binder probe assay to specifically identify CPV-2 vaccine strains and field strain types (2a, 2b, and 2c). It is based on the TaqMan technology, which allows the detection of single nucleotide polymorphisms, and this real-time PCR has been proven to be even more specific, sensitive, and reproducible than other conventional methods such as HA assays, immunochromatographic tests, VI, and even gel-based PCR [15]. Amplification of the VP2 gene fragment of CPV (capsid protein) using primers like H_{for}/H_{rev} and subsequent sequencing of the PCR products covering the informative amino acids would definitely help in detecting genetic variation that exists between CPV-2 and its variants [25].

LAMP is a novel, sensitive, and rapid technique for the detection of genomic DNA. The end product of the technique is a white precipitate of magnesium pyrophosphate that is visible without the use of gel electrophoresis [26]. The LAMP method was applied to the detection of CPV genomic DNA. A set of four primers, two outer and two inner, were designed from the CPV genomic DNA targeting the VP2 gene. The optimal reaction time and temperature for LAMP were determined to be 60 min and 63°C, respectively. This technique amplifies the target DNA sequence with a high selectivity; the assay can be completed within 1 h, has a high sensitivity, and produces an end product that is visible to the naked eye. The detection limit of the LAMP was 0.0001 $TCID_{50}$ mL^{-1}, whereas the detection limit of the PCR was 1000 $TCID_{50}$ mL^{-1} [26].

76.2 METHODS

76.2.1 SAMPLE PREPARATION

The main source of infection is feces from infected dogs. More than 10^9 virus particles per gram of feces were shed during the acute phase in the enteric form. The feces are the suitable clinical material for the detection of virus in the enteric form of the disease. Fecal/rectal swabs samples are collected from CPV-suspected animals as soon as possible after the onset of acute gastroenteritis. The amount of virus shedding begins to decline 6 days postinfection, and after the 12th day, the virus is rarely recovered in feces, which may give false-negative diagnosis. The fecal samples/rectal swabs obtained from the suspected/healthy animals are emulsified in 1 mL of 0.1 M PBS of pH 7.4, stored at 4°C, and transported as early as possible to the laboratory under refrigeration. The emulsions will be centrifuged at 6000 rpm for 15 min at 4°C and the supernatants will be collected and stored at −40°C until further use.

Hundred microliters of the supernatants will be used for template DNA preparation by boiling at 96°C for 10 min and chilling immediately in crushed ice. Then, they will be centrifuged at 10,000 rpm for 10 min and the supernatants will be diluted 1:10 in distilled water to reduce residual inhibitors of the DNA polymerase activity [27,28].

76.2.2 DETECTION PROCEDURES

PCR-based methods had been the most reliable techniques, having high degree of sensitivity and specificity in detecting CPV from fecal samples [28]. Amplification of the VP2 gene fragment (capsid protein) using primers H_{for}/H_{rev} and subsequent sequencing of the PCR products covering the informative amino acids would definitely help in detecting genetic variation that exists between CPV-2 and its variants [29].

Buonavoglia et al. [29] designed a primer pair 555 (F)/555 (R), which also amplified the gene encoding for the VP2 protein located at nucleotide position 4003–4022 and 4561–4585 of the CPV genome yields 583 bp product. Sequencing of the PCR product amplified by this primer pair provides information about amino acid changes in residues 426 and 555 to distinguish the new variants CPV-2a, CPV-2b, and CPV-2c from the original CPV-2.

76.2.2.1 PCR-Based Detection of CPV Using Primer Pair H_{for}/H_{rev} and Subsequent Sequencing

The processed samples were screened by primer pair H_{for} (5′-CAGGTGATGAATTTGCTACA-3′) and H_{rev} (5′-CATTTGGATAAACTGGTGGT-3′) that amplifies 630 bp fragment of the gene encoding the capsid protein. PCR amplification was carried out as per Buonavoglia et al. [29], and the products were analyzed by electrophoresis using a 1.5% agarose gel in Tris-acetate-EDTA buffer (1×). The amplified PCR products were gel extracted and custom sequenced with primer pair H_{for}/H_{rev} using the automated sequencer, Applied Biosystems 3100. The specificity of the sequences obtained and the nucleotide and amino acid variations with respect to the VP2 gene sequence of CPV were determined using the Basic Local Alignment Search Tool (http://www.ebi.ac.uk/) [15]. The nucleotide sequences of the VP2 gene were aligned with the sequences of prototype CPV strains (M38246-FPV; M38245-CPV-2; M24003-CPV-2a; M74849-CPV-2b; AY742953-New CPV-2a; AY742955-New CPV-2b; FJ222821-CPV-2c) using the Clustal W (www.ebi.ac.uk/clustalw) and analyzed for the nucleotide variation of the VP2 gene at positions 3675, 3685, 3699, 3909, and 4064 with the corresponding amino acid residues at 297, 300, 305, 375, and 426, respectively [15]. For the phylogenetic studies, sequences were aligned using the Clustal W 1.8 program, and .aln and .nxs files were generated. The .aln file was converted into .meg file using the Mega 4.1, and the neighbor-joining tree was constructed (bootstrap replicates = 1000; seed = 64,248) using the Kimura 2 method for pairwise deletion at uniform rates [15].

76.2.2.2 Detection of Canine Parvovirus in Fecal Samples Using Loop-Mediated Isothermal Amplification Assay

A new technique called loop-mediated isothermal amplification (LAMP) involves autocycling strand displacement DNA synthesis carried out by the use of a DNA polymerase with a high strand displacement activity, and a set of four primers, two outer and two inner, were designed from CPV genomic DNA targeting the VP2 gene that recognize a total of six distinct sequences in the target DNA (Table 76.2). This technique amplifies the target DNA sequence with a high selectivity; the assay can be completed within 1 h, has a high sensitivity, and produces an end product that is visible to the naked eye [30]. LAMP is a novel, sensitive, and rapid technique for the detection of genomic DNA.

The LAMP technique was performed with a set of four primers (B3, F3, BIP, and FIP) recognizing a total of six distinct sequences (B1–B3 and F1–F3) on the VP2 gene of CPV DNA, which were designed using the Primer Explorer V3 software. The LAMP technique was performed with a 25 µL reaction mixture containing 1 mL (40 pmol) of each of the primers CPV-BIP and CPVFIP, 1 mL (5 pmol) of the primers CPV-F3 and CPV-B3, 12.5 µL of a 2× reaction mixture (40 mM TrisHCl, 20 mM KCl, 16 mM MgSO$_4$, 20 mM [NH$_4$]$_2$ So$_4$, 0.2% Tween 20, 1.6 M betaine, 1.0 µL [8U]

of *Bst* DNA polymerase, and 2.8 mM each dNTP), 1 µL of target DNA, and 7.5 µL of distilled water. To determine the optimal temperature for the LAMP assay, the primer and sample mixtures were incubated at 65°C for 15, 30, 45, and 60 min, respectively, and, at the end of each incubation period, the reaction was terminated by heating at 80°C for 10 min. The reaction temperature was optimized (60°C, 63°C, and 65°C), and the LAMP was carried out for a predetermined period (60 min). The end product of the technique is a white precipitate of magnesium pyrophosphate and by adding propidium iodide (1 mg/mL), a DNA-binding dye and exposure to UV light. A change in color from reddish orange to pink under ambient light and fluorescence under UV light by propidium iodide indicated a positive reaction. The LAMP end product (5 µL) was analyzed by gel electrophoresis with a 1.5% agarose gel. Positive cases yield bands in a ladderlike pattern [31,32].

The LAMP technique offers several advantages over the PCR analysis for the detection of CPV DNA in dog feces; the assay requires only 2 h to complete and is a sensitive method that can amplify a few copies of DNA to about 10^9 molecules in ≤1 h under isothermal condition. The specificity of the LAMP assay can be further confirmed by sequencing the gel-extracted products. The detection limit of the LAMP was 0.0001 TCID$_{50}$/mL, whereas the detection limit of the PCR was 1000 TCID$_{50}$/mL [31,32].

76.3 CONCLUSION

CPV infection, a highly contagious disease of canines, is prevalent in many parts of the world. The disease is characterized by two prominent clinical forms, enteritis with vomiting and diarrhea in dogs of all ages and myocarditis and subsequent heart failure in pups of less than 3 months of age with high morbidity (100%) and frequent mortality up to 10% in adult dogs and 91% in pups. The virus affects dogs' gastrointestinal tracts and is spread by direct dog-to-dog contact and contact with contaminated stools, environments, or people. The virus can also contaminate kennel surfaces, food and water bowls, collars, and leashes, and the hands and clothing of people who handle infected dogs make the disease highly contagious. The disease condition has been complicated further due to the emergence of a number of variants—CPV-2a, CPV-2b, CPV-2c, New CPV-2a, New CPV-2b, and Asp 300 (2a/2b) over the years—and the involvement of domestic and wild canines.

The molecular era of CPV studies begins with successful isolation of CPV from the fecal samples. The availability of molecular techniques to amplify, sequence, and express the genome of CPV stains provides the tools necessary for characterization of CPV stains both genetically and antigenetically, leading to the development of several new diagnostic assays. The application of these sensitive methods for the detection and characterization of CPV infection in different parts of the world contributes a greater recognition of the overall health and economic impact of this virus. Vaccination is the most cost-effective and ideal method to control CPV infections in canines. Both live-attenuated and inactivated vaccines are

TABLE 76.2
Sequence of Each Primer Used in the Loop-Mediated Isothermal Amplification Assay

No.	Primer Identity	Primer Sequence
1	CPV B1P	5′-CATCTCATACTGGAACTAGTGGCA-TGAACATCATCTGGATCTGTAC-3′ (B1c-B2)
2	CPV F1P	5′-TCTCCATGGAGTTGGTATGGT-TATGAGATCTGAGACATTGGGT-3′ (F1c-F2)
3	CPV B3	5′-AGTAAGTGTACTGGCACAG-3′
4	CPV F3	5′-GCCATTTACTCCAGCAGC-3′

available to control the disease in animals. Vaccines used during the late 1970s and early 1980s were of FPLV origin, which was followed by the use of inactivated and live-attenuated vaccines of CPV-2 and CPV-2b of canine origin. High-titer, low-passage CPV vaccines containing a canine origin attenuated virus are currently the vaccines of choice for use in pups of any breed. Live vaccines are most commonly used in domestic dogs because the inactivated vaccines only provide short-term immunity. Continuous evolution of CPV through accumulation of point mutations in the viral genome presents two different, but equally important, implications: (1) the emergence of new antigenic variants displaying different biological and antigenic properties may require an extensive revision of the vaccination programs and an update of the viral strains contained in the commercially available products and (2) the sophisticated molecular methods able to detect and characterize the CPV strains, which are based on perfect matching between viral DNA and assay oligonucleotides, may be affected by the onset of point mutations and should be, therefore, updated as well. Thus, a continuous epidemiological surveillance is needed to detect new CPV variants potentially escaping the host immune system and detection methods.

REFERENCES

1. Johnson, A., Canine parvovirus type 2 (CPV-2). In: *Small Animal Pathology for Veterinary Technicians*. Wiley-Blackwell, CA, pp. 10–13, 2014.
2. Decaro, N. and Buonavoglia, C., Canine parvovirus—A review of epidemiological and diagnostic aspects, with emphasis on type 2c. *Vet. Microbiol.* 155, 1, 2012.
3. Ohshima, T., Hisaka, M., Kawakami, K. et al., Chronological analysis of canine parvovirus type2 isolates in Japan, *J. Vet. Med. Sci.* 70, 769, 2008.
4. Appel, M.J., Scott, F.W., and Carmichael, L.E., Isolation and immunization studies of a canine parvo-like virus from dogs with haemorrhagic enteritis. *Vet. Rec.* 105, 156, 1979.
5. Truyen, U., Evolution of canine parvovirus: A need for new vaccines? *Vet. Microbiol.* 117, 9, 2006.
6. Shackelton, L.A., Parrish, C.R., Truyen, U. et al., High rate of viral evolution associated with the emergence of canine parvovirus. *Proc. Natl. Acad. Sci. USA* 102, 3790, 2005.
7. Aiello, S, E., Moses, M.A., and Steigerwald, M.A., Canine parvovirus. In: *The Merck Veterinary Manual*. 2012.
8. Paradiso, P.R., Rhode, S.L., and Singer, I., Canine parvovirus: A biological and structural characterization. *J. Gen. Virol.* 62, 113, 1982.
9. Wobble, C.R., Mitra, S., and Ramakrishnan, V., Structure of the capsid of kilham rat virus. *Biochemistry* 23, 6565, 1984.
10. Strassheim, L.S., Gruenbreg, A., Veijalainen, P. et al., Two dominant neutralizing antigenic determinants of canine parvovirus are found on the three spike of the virus capsid. *Virology* 198, 175, 1994.
11. Hueffer, K., Parker, J.S.L., Weichert, W.S. et al., The natural host range shift and subsequent evolution of canine parvovirus resulted from virus-specific binding to the canine transferrin receptor. *J. Virol.* 77, 1718, 2003.
12. Decaro, N., Cirone, F., Desario, C. et al., Severe parvovirus in a 12-year-old dog that had been repeatedly vaccinated. *Vet. Rec.* 164, 593, 2009.
13. Macartney, L., McCandlish, I.A.P., Thompson, H. et al., Canine parvovirus enteritis 2: Pathogenesis. *Vet. Rec.* 115, 453, 1984.
14. Decaro, N., Desario, C., Elia, G. et al., Evidence for immunisation failure in vaccinated adult dogs infected with canine parvovirus type 2c. *New Microbiol.* 31, 125, 2008.
15. Mohan Raj, J., Mukhopadhyay, H.K., Thanislass, J. et al., Isolation, molecular characterization and phylogenetic analysis canine parvovirus. *Infect. Genet. Evol.* 10, 1237, 2010.
16. Hagiwara, M.K., Mamizuka, E.M., and Pavan, M.F., Role of intestinal flora in acute haemorrhagic gastroenteritis (parvovirus infection) of dogs. *Braz. J. Vet. Res. Anim. Sci.* 33, 107, 1996.
17. Decaro, N., Elia, G., Martella, V. et al., A real time PCR assay for rapid detection and quantitation of canine parvovirus type 2 DNA in the faeces of dogs. *Vet. Microbiol.* 105, 19, 2005.
18. Kelly, W.R., An enteric disease of dogs resembling feline pan-leucopaenia. *Aust. Vet. J.* 54, 593, 1978.
19. Carmichael, L.E., Haemagglutination (HI) and Haemagglutination Inhibition (HI) tests for canine parvovirus. *Am. J. Vet. Res.* 41, 784, 1980.
20. Cavalli, A., Bozzo, G., Decaro, N. et al., Characterization of a canine parvovirus strain isolated from an adult dog. *New Microbiol.* 24, 239, 2001.
21. Hirayama, K., Kano, R., Kanai, T.H. et al., VP2 gene of a canine parvovirus isolate from stool of a puppy. *J. Vet. Med. Sci.* 67, 139, 2005
22. Veijalainen, M.L., Neuvonen, E., Niskanen, A. et al., Latex agglutination test for detecting feline panleukopenia virus, canine parvovirus, and parvovirus of fur animals. *J. Clin. Microbiol.* 23, 556, 1995.
23. Esfandiari, J. and Klingeborn, B., A comparative study of a new rapid and one-step test for the detection of parvovirus in faeces from dogs, cats and mink. *J. Vet. Med.* 47, 145, 2000.
24. Uwatoko, K., Sunairi, M., Yamamoto, A. et al., Rapid and efficient method to eliminate substances inhibitory to the polymerase chain reaction from animal faecal samples. *Vet. Microbiol.* 52, 73, 1994.
25. Decaro, N., Desario, C., Addie, D.D. et al., Molecular epidemiology of canine parvovirus, Europe. *Emerg. Infect. Dis.* 13, 1222, 2007.
26. Mukhopadhyay, H.K., Amsaveni, S., Matta, S.L. et al., Development and evaluation of loop-mediated isothermal amplification assay for rapid and sensitive detection of canine parvovirus DNA directly in faecal specimens. *Lett. Appl. Microbiol.* 55, 202, 2012.
27. Decaro, N., Martella, V., Desario, C. et al., First detection of canine parvovirus type 2c in pups with haemorrhagic enteritis in Spain, *J. Vet. Med. B: Infect. Dis. Vet. Public Health* 53, 468, 2006.
28. Decaro, N., Gabriella, E., Martella, V. et al., A minor groove binder probe real-time PCR assay for discrimination between type-2 based vaccines and field strains of canine parvovirus. *J. Virol. Methods* 136, 65, 2006.
29. Buonavoglia, C., Martella, V., Pratelli, A. et al., Evidence for evolution of canine parvovirus type 2 in Italy, *J. Gen. Virol.* 82, 3021, 2001.
30. Notomi, T., Okayama, H., Masubuchi, H. et al., Loop-mediated isothermal amplification of DNA. *Nucleic Acids Res.* 28, 63, 2000.
31. Nagamine, K., Hase, T., and Notomi, T., Accelerated reaction by loop-mediated isothermal amplification using loop primers. *Mol. Cell. Probes* 16, 223, 2002.
32. Cho, H.S., Kang, J.I., and Park, N.Y., Detection of canine parvovirus in fecal samples using loop-mediated isothermal amplification. *J. Vet. Diagn. Invest.* 18, 81, 2006.

77 Goose Parvovirus

Yakun Luo and Shangjin Cui

CONTENTS

77.1 Introduction 693
    77.1.1 Classification, Morphology, and Genome Organization 693
    77.1.2 Replication 693
    77.1.3 Clinical Features 694
    77.1.4 Diagnosis 694
77.2 Methods 694
    77.2.1 Sample Collection and Preparation 694
    77.2.2 Detection Procedures...694
        77.2.2.1 Virus Isolation 694
        77.2.2.2 Passive Immunity Test 694
        77.2.2.3 Neutralization Test...694
        77.2.2.4 Cattle Sperm Agglutination Inhibition Test 694
        77.2.2.5 Agar Gel Precipitin Test..695
        77.2.2.6 Enzyme-Linked Immunosorbent Assay 695
        77.2.2.7 Western Blotting..695
        77.2.2.8 PCR..695
        77.2.2.9 Multiplex PCR 695
        77.2.2.10 Fluorescent Quantitative Real-Time PCR 695
77.3 Conclusion 696
References...696

77.1 INTRODUCTION

77.1.1 CLASSIFICATION, MORPHOLOGY, AND GENOME ORGANIZATION

Goose parvovirus (GPV) is a well-known causative agent of gosling plague (GP), which is an acute, contagious, and fatal disease often referred to as Derzsy's disease [1]. Being a member of the genus *Dependoparvovirus* within the family *Parvoviridae* [2–5], GPV was first described as a clinical entity by Fang in 1962 [6]. GPV has attracted much attention owing to its ability to induce high rates of morbidity and mortality in geese, contributing to enormous economic losses in commercial goose production worldwide [1,7,8,9].

The GPV genome is a single-stranded DNA of 5106 nucleotides long and contains two open reading frames (ORFs) that encode regulatory and structural proteins. The left ORF encodes regulatory proteins, while the right ORF encodes three capsid proteins: VP1, VP2, and VP3, which are derived from the same gene by differential splicing. Additionally, VP2 is contained within the carboxyl terminal portion of VP1, and VP3 is contained within that of VP2 [10,11]. The capsids of GPV are nonenveloped, 20–22 nm in diameter, and assembled from 32 capsomers [12–15]. The classical detection of GPV by virus isolation in gosling, cell cultures, and serological assays like enzyme-linked immunosorbent assay (ELISA) is dependent on the availability of specific pathogens free (SPF) gosling embryos and standard positive control sera [16].

77.1.2 REPLICATION

Similar to other parvoviruses, GPV binds to a sialic acid–bearing cell surface receptor before entry into the host cell through clathrin-mediated endocytosis. Making use of the phospholipase A2 activity conferred by the amino-terminal peptide of the capsid VP1 protein, GPV penetrates into the cytoplasm and then permeabilizes the host endosomal membrane to facilitate its microtubular transportation toward the nucleus. Inside the nucleus, GPV begins uncoating, and its ssDNA is converted by cellular proteins into dsDNA, which functions as template for synthesis of mRNAs encoding the nonstructural (NS) proteins, NS1 and NS2, when host cell reaches to S phase. Moving out of the nucleus into the cytoplasm, the viral mRNAs are translated into viral proteins by the host ribosomes. After covalent binding to the 5′ genomic end, the NS1 endonuclease transactivates an internal transcriptional promoter that directs synthesis of the structural VP polypeptides. This initiates viral DNA replication via a unidirectional strand displacement mechanism, with the hairpin structure at the 3′ end acting as a self-primer to synthesize a plus-sense DNA, leading to the formation of double-stranded DNA. The growing strand can replicate back on itself to produce a tetrameric form that is then cleaved to generate two plus-sense and two minus-sense strands of DNA. Individual ssDNA genomes excised from replication concatemers can be further converted to dsDNA and serve as a template for transcription/replication, or encapsidated and packaged in a 3′–5′ direction to form new virions that are released by cell lysis.

77.1.3 Clinical Features

GP was noted among goslings showing anorexia, wheezing, and locomotor dysfunction [1,17]. The specific lesions observed in the affected geese include myopathy of skeletal muscle, hepatitis, myocarditis, sciatic neuritis, and polioencephalomyelitis [3]. The disease affects mainly young goslings between 2 and 4 weeks old. However, infections in older birds also occur and are often asymptomatic except ascites or loss in featheriness. Under natural conditions, the adult geese can be infected without showing clinical symptoms, but can be spread by excreta and eggs of the disease. The disease has been reported in major goose-producing countries in the world.

77.1.4 Diagnosis

For diagnosis of GPV infection, many methods are available, including direct electron microscopy, virus isolation and identification, agar diffusion test, immunoelectron microscopy, latex agglutination test, polymerase chain reaction (PCR), real-time PCR, hemagglutination inhibition (HI), ELISA, and nucleic acid sequence analysis. Agar diffusion test and HA/HI are complex and time consuming, while ELISA, which is used to detect GPV antigen, is prone to false-positive and false-negative results [18]. The advent of molecular diagnostic techniques such as PCR has vastly improved the sensitivity, specificity, and speed of laboratory diagnosis of GPV.

77.2 METHODS

77.2.1 Sample Collection and Preparation

Sample should be collected from affected geese as soon as possible. Collect the liver, spleen, kidney, thymus, and brain and grind them into the suspension. Then clarify the suspension by centrifugation at 3000 rpm for 30 min. Extract DNA by using virus genomic DNA kit [19].

77.2.2 Detection Procedures

77.2.2.1 Virus Isolation

Inoculate susceptible goose embryos (12–14 days of age) via embryo allantoic cavity with 0.2 mL of virus suspension

prepared earlier. Observe daily for a total of 8 days. Embryo's death may start on day 3, and most of the deaths usually occur in 5–7 days. Dead embryos often display thickening in chorioallantoic membrane, congestion in body skin, bleeding (in wing tips, toes, chest pores, neck, and beak), congestion in liver, hemorrhage in the heart and brain, and subcutaneous edema in head skin and ribs. At day 7, the embryo stops developing and appears very small. The virus may be detected in urine.

77.2.2.2 Passive Immunity Test

The LD50 of the virus is first determined in susceptible goslings. Take 10 susceptible goslings, inject subcutaneously 5 goslings with 0.5 mL GPV hyperimmune serum, and the other 5 with 0.5 mL of saline as a control group. After 6–12 h, both groups of goslings are injected subcutaneously or intramuscularly with 100 LD50 dose of virus and observed for 10 days. If serum group has no gosling death, while the control group dies, GPV is present [20].

77.2.2.3 Neutralization Test

The virus to be tested is divided into two equal portions: one is added four times the amount of GPV hyperimmune serum (serum or rehabilitation) as the serum group and the other is added four times the amount of saline as a control group. After mixing, both preparations are kept at 37°C for 60 min.

Virus preparation (0.2 mL) is then inoculated into the chorioallantoic membrane cavity of five susceptible goose embryos (of 12–14 in age), and the embryos are incubated at 37°C–38°C to hatch. Observation is made once daily until the seventh day. If serum group is healthy and alive, while the control group dies, displaying most or all of the typical embryo lesions, GPV is present. Gosling may also be used in place of the susceptible embryos for neutralization test, and the results are determined similarly [21].

In addition, virus neutralization test may be utilized to assess antibody levels in goose serum as well as antigen in dead geese [22]. A neutralizing index of <1.5 indicates the absence of specific parvovirus neutralizing antibodies.

77.2.2.4 Cattle Sperm Agglutination Inhibition Test

GPV can agglutinate cattle sperm, and this feature can be used for virus identification. The virus to be tested is diluted in saline (ratio 1:5) and added to two small test tubes, each containing 0.25 mL. The first tube is added 0.25 mL of GPV-positive serum, while the second tube is added 0.25 mL of physiological saline as control tube. After mixing, the tubes are incubated at 37°C for 50 min, then 0.25 mL cattle semen (one tablet semen added 1 mL 2.9% trisodium citrate solution at 42°C) is added to each tube, mixed, and set at room temperature for 6 h before observation. As the first one agglutinates and the second one does not agglutinate, the virus can be judged as GPV.

77.2.2.5 Agar Gel Precipitin Test

Eight of embryonating Khaki Campbell duck eggs (9 days of age) are each inoculated via the allantoic cavity with 0.2 mL of a 10^{-2} dilution of the 22nd passage of duck embryo adapted parvovirus [22]. After 6–8 days, the dead embryos and amnio-allantoic fluids (AAF) are harvested, the embryos' head and limbs removed, and a 20% (W/V) suspension prepared in the AAF. After homogenization, the suspension is mixed with an equal volume of trichlorotrifluoroethane (Arcton 113, ICI Ltd, London) for 30 min followed by centrifugation at $3000 \times g$ for 15 min. The supernatant is removed and further centrifuged at $46,000 \times g$ for 2 h at $4°C$. The resulting pellet is resuspended in deionized water to 5% of the original volume and treated with 0.1% N-lauroylsarcosine (Sigma Chemical Co., Poole, Dorset). Agar plates are prepared using 1% Ionager No. 2 (Oxoid, London) in 8% sodium chloride solution, buffered to pH 7.8. The gel pattern consists of a central well 5 mm in diameter containing antigen surrounded by six outer wells of the same diameter containing test sera. All the wells are spaced 3 mm apart.

Diffusion of antigen and antibody takes place at room temperature, and the plates are examined for precipitin lines after approximately 24 h. Sera recorded as positive are diluted by twofold series in phosphate buffered saline and further tested to ascertain the agar gel precipitation (AGP) titer. The titers are expressed as the log 2 of the reciprocal of the highest dilution of serum producing a precipitin line.

The concentrated parvovirus antigen used for the AGP tests has an EID 50 titer of between 10^8 and 10^9/mL, calculated by the method of Reed and Muench [23]. Antigen is also prepared from normal duck embryos and used in the AGP test as a control to check for nonspecific precipitin lines.

77.2.2.6 Enzyme-Linked Immunosorbent Assay

ELISA is a common diagnostic test for detection of various infectious pathogens, and different versions are available including direct ELISA, antigen capture ELISA, sandwich ELISA, and dipstick ELISA. In 1996, Kardi and Szegletes used ELISA procedure to detect Derzsy's disease virus of geese and antibodies produced against it. Recently, Wang et al. developed a competitive ELISA using a monoclonal antibody for the detection of GPV virus-like particles and vaccine immunization in goose sera. The procedural details for ELISA and competitive ELISA are available in published reports [24–26].

77.2.2.7 Western Blotting

In 2005, Wang et al. used sera collected from experimentally infected ducks in Western blotting and showed that antibodies against the NS protein appeared first after infection, followed by antibodies against the capsid protein. While a number of methods are available for the detection of antibodies against waterfowl parvoviruses, none is able to differentiate responses against the capsid and NS proteins. Procedures for the Western blotting assay are available in detail elsewhere [21].

77.2.2.8 PCR

The following PCR procedure for GPV detection is based on published literature [18,21,22]:

1. Primers (see Table 77.1).
2. PCR mixture is submitted to 30 cycles of amplification involving denaturation at $95°C$ for 1 min, annealing at $48°C$ for 1 min, and polymerization at $72°C$ for 1 min. PCR products are electrophoresed on 1.2% agarose gels; gels are stained with ethidium bromide and then visualized with ultraviolet (UV) light.

77.2.2.9 Multiplex PCR

Chen et al. [27] developed a multiplex PCR for rapid diagnosis of goose viral infections. The goose viral infections include GPV (GenBank accession number JF333590), goose paramyxovirus (GenBank accession number GQ245793), goose herpesvirus (GenBank accession number EF053033), and goose adenovirus (GenBank accession number EU979376).

1. Primers (see Table 77.2).
2. Prepare PCR premix containing the following components: 1.25 U of Taq DNA polymerase (TaKaRa, Dalian, China), 2.5 μL of 10× reaction buffer (200 mM Tris–HCl, pH 8.55, 160 mM $[NH_4]_2SO_4$, and 20 mM $MgCl_2$), 10 mM of each dNTPs (TaKaRa, Dalian, China), and 50 pmol of each primer (Table 77.1). Mix these components with 4 μL of template cDNA/DNA in a total volume of 25 μL.
3. Perform PCR amplification using the following cycling programs: an initial denaturation step at $95°C$ for 5 min; 35 cycles of denaturation at $94°C$ for 30 s, primer annealing at $58°C$ for 30 s, and elongation at $72°C$ for 30 s; and final elongation step at $72°C$ for 7 min.
4. Separate the PCR products by electrophoresis on 2.5% agarose gels, stain with ethidium bromide and visualize under UV light.

77.2.2.10 Fluorescent Quantitative Real-Time PCR

In 2009, Yang et al. developed a fluorescent quantitative real-time PCR for the detection of GPV in vivo [28].

1. Primers (see Table 77.3).
2. The reaction, data acquisition, and analysis are carried out using iCycler IQ Optical System Software. The **fluorescent quantitative** (FQ)-PCR is performed in a 25 μL reaction mixture containing 1× PCR buffer,

---

TABLE 77.1

Oligonucleotide Sequences of the Primers and Probes Used in the Polymerase Chain Reaction Method

Primer	Fragment Size	Primer Sequence (5′–3′)
AL18F2	800 bp	CCGGGTTGCAGGAGGTAC
AL18R2		AGCTACAACAACCACATC

TABLE 77.2

Oligonucleotide Sequences of the Primers and Probes Used for the Goose Parvovirus Multiplex Polymerase Chain Reaction

Virus[a]	Target Gene	Primer Sequence (5′–3′)[b]	Tm	Target Products (bp)
NDV	F gene	NDV-s:GCTCTTCTATCACAGAACCGACTT	59.1	302
		NDV-as:GCTGACCATCCAGGCACTTT	59.8	
GHV	TK	GHV-s:CAATGGCAGCAACATAACAA	58.4	501
		GHV-as:ATGGGATGGAACGTAACAGC	57.3	
GAV	Hexon	GAV-s:TACGGAGGAACCGCATACAA	59.0	401
		GAV-as:GCCATCACGCCCAGATACTT	59.9	
GPV	VP3	GPV-s:GATTTCAACCGCTTCCACTG	58.1	226
		GPV-as:TAGCCGAGCCCAGCACATAC	59.1	

[a] GPV, goose parvovirus, GenBank accession number JF333590; NDV, goose paramyxovirus, GenBank accession number GQ245793; GHV, goose herpes-virus, GenBank accession number EF053033; GAV, goose adenovirus, GenBank accession number EU979376.

[b] The orientation for each of these sequences is either sense (-s) or antisense (-as).

TABLE 77.3

Oligonucleotide Sequences of the Primers and Probes Used in the Goose Parvovirus Fluorescent Quantitative Polymerase Chain Reaction Method (Oligonucleotide Positions are Located on the Gene Sequence of U25749)

Name	Primer Sequence (5′–3′)	Position	Amplicon Size (bp)
GPV-F	GTGCCGATGGAGTGGGTAAT	3084–3103	60
GPV-R	ACTGTGTTTCCCATCCATTGG	3122–3143	
GPV-FP	6FAM-FTCGCAATGCCAATTTCCCGAGGP—TAMRA	3098–3120	
VP3-1	AAGCTTTGAAATGGCAGAGGGAGGA	3008–3033	1658
VP3-2	GGATCCCGCCAGGAAGTGCTTTATTTGA	4637–4665	

0.3 mmol/L dNTPs, 1.25 U Taq, and 1 μL DNA template according to the manufacturer's instructions. Autoclaved nanopure water is added to make the final volume to 25 μL. Each run comprises an initial activation step of 30 s at 95°C, followed by 40 cycles of denaturation at 94°C for 10 s and annealing at 60°C for 30 s; the fluorescence is measured at the end of the annealing/extension step. The tests are performed using 0.2 mL PCR tubes (ABgene, United Kingdom). FQ-PCR is optimized in triplicate based on the primer, probe, and MgCl₂ concentration selection criteria, which is performed according to 4 × 4 × 4 matrix of primer concentrations (0.10, 0.12, 0.16, and 0.20 μmol/L), probe concentrations (0.10, 0.12, 0.16, and 0.20 μmol/L), and MgCl₂ concentrations (1.0, 5.0, 10.0, and 15.0 mmol/L). Conditions are selected to ensure that both the fluorescence acquisition curves are robust and Ct values are the lowest possible to the known template DNA concentrations.

77.3 CONCLUSION

Goose parvovirus (GPV) infection, also known as gosling plague (GP), is an important disease of geese and Muscovy ducklings under the age of 4–20 days [29]. The genomes of GPV and MDPV contain two major ORFs: the first encoding the NS protein involved in viral replication and regulatory functions and the second the three capsid proteins VP1, VP2, and VP3. The proteins VP1, VP2, and VP3 are derived from the same gene, and the entire amino acid sequences of VP2 and VP3 are contained within the carboxyl terminal portion of VP1 [29]. VP3 appears to be the most abundant of the three capsid proteins in purified virions [30,31]. Sequence comparisons have shown that VP1 of GPV and MDPV have 87.4% amino acid sequence identity and the NS proteins have 90.6% identity [32,33]. Laboratory diagnosis of GP is possible through the use of virus isolation, and various immunological/molecular procedures. The development and application of hyperimmune serum and attenuated vaccine have proven effective for the control of GP in China.

REFERENCES

1. Gough D et al. Isolation and identification of goose parvovirus in the UK. *Vet Rec*. 2005, 13:424.
2. Brown KE, Green SW, Young NS. Goose parvovirus-an autonomous member of the Dependovirus genus. *Virology* 1995, 210:283–291.
3. Glávits R et al. Comparative pathological studies on domestic geese (*Anser anser domestica*) and Muscovy ducks (*Cairina moschata*) experimentally infected with parvovirus strains of goose and Muscovy duck origin. *Acta Vet Hung*. 2005, 53:73–89.

4. Takehara K et al. Isolation, identification, and plaque titration of parvovirus from Muscovy ducks in Japan. *Avian Dis.* 1994, 38:810–815.

5. Woolcock PR et al. Evidence of Muscovy duck parvovirus in Muscovy ducklings in California. *Vet Rec.* 2000, 146:68–72.

6. Fang DY. Recommendation of GPV. *Vet Sci China* 1962, 8:19–20 (in Chinese).

7. Takehara K et al. An outbreak of goose parvovirus infection in Japan. *J Vet Med Sci.* 1995, 4:777–779.

8. Richard E, Gough RE. Goose parvovirus infection. In: Saif YM, Barnes HJ, Fadly AM, Glisson JR, McDougald LR, Swayne DE (Eds.), *Diseases of Poultry.* Iowa State Press, Ames, IA, 2003, vol. 11, pp. 367–374.

9. Holmes JP et al. Goose parvovirus in England and Wales. *Vet Rec.* 2004, 4:127.

10. Tatar-Kis T et al. Phylogenetic analysis of Hungarian goose parvovirus isolates and vaccine strains. *Avian Pathol.* 2004, 33:438–444.

11. Zádori Z, Erdei J, Nagy J. Characteristics of the genome of goose parvovirus. *Avian Pathol.* 1994, 23:359–364.

12. Chu CY, Pan MJ, Cheng JT. Genetic variation of the nucleo-capsid genes of waterfowl parvovirus. *J Vet Med Sci.* 2001, 11:1165–1170.

13. Zádori Z, Erdei J, Nagy J, Kisary J. Characteristics of the genome of goose parvovirus. *Avian Pathol.* 1994, 23:359–364.

14. Zádori Z, Stefancsik R, Rauch T. Analysis of the complete nucleotide sequences of goose and Muscovy duck parvoviruses indicates common ancestral origin with adeno-associated virus 2. *Virology* 1995, 212:562–573.

15. Zhang Y, Geng HW, Guo DC. Sequence analysis of the full length genome of waterfowl parvovirus. *Chin J Prev Vet Med.* 2008, 30:415–419.

16. Gough D, Ceeraz V, Cox B. Isolation and identification of goose parvovirus in the UK. *Vet Rec.* 2005, 13:424.

17. Schettler CH. Isolation of a highly pathogenic virus from geese with hepatitis. *Avian Dis.* 1970, 15:323–325.

18. Limn CK, Yamada T, Masayuyki N. Detection of Goose parvovirus genome by polymerase chain reaction: Distribution of Goose parvovirus in Muscovy ducklings. *Virus Res.* 1996, 42:167–172.

19. Kardi V, Szegletes, E. Use of ELISA procedures for the detection of Derzsy's disease virus of geese and of antibodies produced against it. *Avian Pathol.* 1996, 25:25–34.

20. Wang Q, Ju H, Li Y. Development and evaluation of a competitive ELISA using a monoclonal antibody for antibody detection after goose parvovirus virus-like particles (VLPs) and vaccine immunization in goose sera. *J Virol Methods* 2014, 209:69–75.

21. Wang CY et al. Expression of capsid proteins and non-structural proteins of waterfowl parvoviruses in *Escherichia coli* and their use in serological assays. *Avian Pathol.* 2005, 34:376–382.

22. Gough RE, Spackman D. Studies with a duck embryo adapted goose parvovirus vaccine. *Avian Pathol.* 1982, 11:503–510.

23. Reed LJ, Muench H. A simple method of estimating 50% end points. *Am J Hyg.* 1938, 27:493–497.

24. Baxi M et al. A one-step multiplex real-time RT-PCR for detection and typing of bovine viral diarrhea viruses. *Vet Microbiol.* 2006, 1–3:37–44.

25. Alexandrov M et al. Fluorescent and electronmicroscopy immunoassays employing polyclonal and monoclonal antibodies for detection of goose parvovirus infection. *J Virol Methods* 1999, 79:21–32.

26. Cheng AC et al. Development and application of a reverse transcriptase-polymerase chain reaction detect Chinese isolates of duck hepatitisvirus type 1. *J Microbiol Methods* 2009, 77:332–336.

27. Chen ZY et al. Rapid diagnosis of goose viral infections by multiplex PCR. *J Virol Methods* 2013, 191:101–104.

28. Yang JL, Cheng A, Wang MS. Development of a fluorescent quantitative real-time polymerase chain reaction assay for the detection of Goose parvovirus in vivo. *J Virol.* 2009, 6:142–149.

29. Gough RE. Goose parvovirus infection. In: Saif YM, Barens HJ, Glisson JR, Fadly AM, McDougal LR, Swayne DE (Eds.), *Diseases of Poultry*, 11th edn. Iowa State Press, Ames, IA, 1991, pp. 367–374.

30. Kisary J. Derzsy's disease of geese. In: McFerran JB, McNulty MS (Eds.), *Virus Infection of Birds*. Elsevier Science Publishers, Amsterdam, the Netherlands, 1993, pp. 157–162.

31. Za'dori Z et al. Analysis of the complete nucleotide sequences of goose and muscovy duck parvoviruses indicates common ancestral origin with adeno-associated virus 2. *Virology* 1995, 212:562–573.

32. Le Gall-Recule J V. Biochemical and genomic characterization of muscovy duck parvovirus. *Arch Virol.* 1994, 139:121–131.

33. Chu CY, Pan MJ, Cheng JT. Genetic variation of the nucleo-capsid genes of waterfowl parvovirus. *J Vet Med Sci.* 2001, 63:1165–1170.

78 *Penaeus monodon* Hepandensovirus

Dongyou Liu

CONTENTS

78.1 Introduction 699
    78.1.1 Classification...699
    78.1.2 Morphology and Genome Organization 700
    78.1.3 Biology and Epidemiology 700
    78.1.4 Clinical Features and Pathogenesis 700
    78.1.5 Diagnosis 701
    78.1.6 Treatment and Prevention 702
78.2 Methods 702
    78.2.1 Sample Preparation..702
    78.2.2 Detection Procedures...702
        78.2.2.1 Conventional PCR...702
        78.2.2.2 Real-Time PCR...702
78.3 Conclusion 702
References...703

78.1 INTRODUCTION

Penaeus monodon hepandensovirus (PmoHDV) is a parvovirus that causes high mortality and stunted growth in juvenile shrimp in Asia Pacific regions, Africa, Australia, and North and South America. Along with white spot syndrome virus (WSSV, see Chapter 98), yellow head virus (YHV; see Chapter 33), infectious myonecrosis virus (IMNV, see Chapter 66), Taura syndrome virus (TSV, see Chapter 3), *Penaeus stylirostris* penstyldensovirus (PstDV or infectious hypodermal and hematopoietic necrosis virus [IHHNV]; see Chapter 79), and *Penaeus monodon* nucleopolyhedrovirus (PemoNPV or monodon baculovirus [MBV]), PmoHDV is responsible for approximately 60% of disease losses in shrimp aquaculture. Furthermore, abdominal segment deformity disease due possibly to an undescribed retrovirus-like agent and monodon slow growth syndrome due to Laem-Singh virus create additional problems to *Penaeus vannamei* and *Penaeus monodon* shrimp farming in Asia. Despite their distinct taxonomical and ecological backgrounds, the aforementioned viruses have one thing in common, that is, they all demonstrate a capacity to decimate the cultured and wild shrimp populations throughout the world.

78.1.1 Classification

PmoHDV (formerly hepatopancreatic parvovirus [HPV]) is a single-stranded DNA virus classified in the genus *Hepandensovirus*, subfamily *Densovirinae*, family *Parvoviridae*.

Of the two subfamilies (*Parvovirinae* and *Densovirinae*) within the family *Parvoviridae*, the subfamily *Parvovirinae* consists of viruses that infect vertebrate hosts and is organized into eight genera (*Amdoparvovirus*, *Aveparvovirus*, *Bocaparvovirus*, *Copiparvovirus*, *Dependoparvovirus*, *Erythroparvovirus*, *Protoparvovirus*, and *Tetraparvovirus*), whereas the subfamily *Densovirinae* covers viruses that infect invertebrate (insect) hosts and is divided into five genera (*Ambidensovirus*, *Brevidensovirus*, *Hepandensovirus*, *Iteradensovirus*, and *Penstyldensovirus*) [1].

According to the latest release from the International Committee on Taxonomy of Viruses, the genus *Hepandensovirus* is composed of a single species Decapod hepandensovirus 1, which includes seven isolates: PmoHDV1, PmoHDV2, PmoHDV3, PmoHDV4, *Penaeus merguiensis* hepandensovirus (PmeDV), *Penaeus chinensis* hepandensovirus, and *Fenneropenaeus chinensis* hepandensovirus [1,2]. These viruses are monophyletic and all have relatively large (~6.3 kb), monosense, heterotelomeric genomes that lack discernable PLA2 motifs. Members of the species Decapod hepandensovirus 1 are associated with hepatopancreatic disease in black tiger shrimp (*P. monodon*) and Chinese white shrimp (*Penaeus chinensis*), resulting in stunted growth (with growth stopping at approximately 6 cm in length) but rarely death. For convenience, we will refer to the viruses of the species Decapod hepandensovirus 1 as PmoHDV in this chapter.

Closely relating to the genus *Hepandensovirus* in the subfamily *Densovirinae*, the genus *Penstyldensovirus* comprises a single species Decapod penstyldensovirus 1, which has four isolates: PstDV1, PstDV2, *Penaeus monodon* penstyldensovirus 1 (PmoPDV1), and PmoPDV2 [1,2]. These viruses are monophyletic and all have small (~4 kb), monosense, homotelomeric genomes that lack discernable PLA2 motifs.

Members of the species Decapod penstyldensovirus 1 are responsible for high mortality in juveniles and subadults of American blue shrimp (*Penaeus stylirostris*) and cuticular deformities and growth retardation (collectively called runt-deformity syndrome, or RDS) in pacific whiteleg shrimp (*P. vannamei*, which is currently known as *Litopenaeus vannamei*). For convenience, we will refer to the viruses of the species Decapod penstyldensovirus 1 as PstDV in this chapter (see Chapter 79, Section 79.1.1).

78.1.2 MORPHOLOGY AND GENOME ORGANIZATION

PmoHDV is a nonenveloped, icosahedral virus of 22–24 nm in diameter, with a linear, negative-sense, single-stranded DNA genome of approximately 6000 nucleotides (ranging from 6336, 6321, 6310, 6299, 5936, 5742, to 5740 nt) in PmoHDV isolates from Korea, Thailand, India, Australia, and Madagascar [3–7].

As exemplified by the isolate from infected *P. monodon* in Thailand, the PmoHDV genome of 6321 nucleotides includes 5′- and 3′-noncoding termini (containing hairpin-like structure of 222 and 215 nt in length, respectively) flanking three open reading frames (ORF1 [NS2 gene], ORF2 [NS1 gene], and ORF3 [VP gene]) [3].

The ORF1 (nt 216–1500) encodes putative nonstructural protein-2 (NS-2) of 428 amino acids with a molecular mass of 50 kDa. The NS-2 protein may have a role in the production of other viral replicative forms as well as the enhancement of NS1-associated parvovirus-induced cell killing in some cell lines [3].

The ORF2 (nt 1487–3224) encodes a major nonstructural protein (NS-1) of 579 amino acids with a molecular mass of 68 kDa. The NS-1 protein contains two regions of highly conserved motifs commonly found in all parvoviruses, including replication initiator motifs located between aa 87 and 134 and NTP-binding and helicase domains located between aa 390 and 490. Analysis of the 119 amino acids conserved region of NTP-binding and helicase domain suggests that PmoHDV is closely related to PstDV (or IHHNV) and two mosquito brevidensoviruses *Aedes aegypti* densovirus and *Aedes albopictus* densovirus. The NS-1 protein may be involved in viral capsid assembly and in the generation of viral single-stranded DNA [3].

The ORF3 (nt 3642–6096) encodes a capsid protein (VP) of 818 amino acids with a molecular mass of 92 kDa. This protein may be cleaved at two arginine residues located upstream (aa 271) and downstream (aa 298) of the glycine-rich region to produce a 57 or 54 kDa structural protein, respectively, which display a doublet band on SDS-PAGE. The 57 kDa protein presumably acts as capsid protein and the 54 kDa protein has unknown function. Phylogenetic analysis of the amino acid sequence of the ORF3 allows division of PmoHDV isolates into three genotypes (I, II, III): genotype I (from *P. chinensis* of Korea, and *P. monodon* of Madagascar, Mozambique, and Tanzania), genotype II (from *P. monodon* of Thailand, Indonesia, and India), and genotype III (from *P. merguiensis* of Australia, and New Caledonia), which largely reflect the

geographic distribution of the hosts. Like other parvoviruses, PmoHDV a highly conserved glycine-rich sequence was found in structural protein of HPV, whereas phospholipase A2 (PLA2) signature sequence was found to be absent in this shrimp densovirus [3].

78.1.3 BIOLOGY AND EPIDEMIOLOGY

PmoHDV targets hepatopancreatic epithelial cells of shrimps such as black tiger shrimp (*P. monodon*) and Chinese white shrimp (*P. chinensis*). After uptake by the host, PmoHDV attaches to the microvilli and gains entry into the host cell by pinocytosis. Inside the nucleoplasm, PmoHDV replicates and mature viruses accumulate in the main inclusion bodies. After disrupting the nuclear membrane, the mature virus particles are released into the cytoplasm. Subsequently, the viruses move into the hepatopancreatic tubule lumens and spread to new host cells [8,9].

The natural hosts of PmoHDV include an ever-increasing number of cultured and captured shrimp species, such as *P. monodon*, *P. merguiensis*, *P. semisulcatus*, *P. esculentus*, *P. penicillatus*, *P. japonicus*, *P. schmitti*, *P. (Fenneropenaeus) chinensis*, *P. (Fenneropenaeus) indicus*, *P. (Litopenaeus) vannamei*, and *P. (Litopenaeus) stylirostris*. Other shrimp species (e.g., *Parapenaeopsis stylifera*, *Metapenaeus monoceros*, *M. affinis*, *M. elegans*, *M. dobsoni*, *M. ensis*, *Solenocera choprai*, and *Macrobrachium rosenbergii*), crayfish (*Cherax quadricarinatus*), and mud crab (*Scylla serata*) may be also affected [9,10].

First described in *P. (Fenneropenaeus) chinensis* in Korea, PmoHDV has been shown to occur in wild, cultured, and hatchery reared shrimps throughout the Indo-Pacific region, including Thailand (*P. monodon*), Singapore (*P. merguiensis*), Kuwait (*Penaeus semisulcatus*), the Philippines (*P. monodon*), China, Taiwan, India (*P. semisulcatus*, *P. monodon*, and *P. indicus*), New Caledonia, Indonesia, Malaysia, and Australia (*P. esculentus*, *P. merguiensis*, *P. monodon*, and *P. japonicus*) [11–15]. In addition, PmoHDV has also been identified in Madagascar, Mozambique, Kenya (*P. penicillatus*), and Tanzania in Africa. After introduction to Hawaii through importation of *P. (Fenneropenaeus) chinensis* from Korea and to Israel with the importation of *P. penicillatus* from Kenya, PmoHDV has since spread to wild shrimp in North and South America. The movements of infected *P. (Litopenaeus) vannamei* further facilitated its incursions into the Pacific coast of Western Mexico and along the coast of coastal El Salvador [16].

PmoHDV is capable of both horizontal and vertical transmissions. The horizontal transmission of PmoHDV is made possible by cannibalism that occurs in shrimp farms, and the vertical transmission from parental broodstock to F1 progeny is not uncommon [17].

78.1.4 CLINICAL FEATURES AND PATHOGENESIS

Clinical signs in shrimp due to PmoHDV infection are largely non-specific, and may include anorexia, decreased preening activity, increased surface and gill fouling by epicommensal organisms, sporadic opacity of tail musculature, and atrophy of the hepatopancreas, with mortality rates of 50%–100% in juvenile stages (which is exacerbated by stress and overcrowding conditions) after 4–8 weeks of infection and stunted growth rates in surviving juveniles [18].

More often, PmoHDV infection appears as coinfections with other hepatopancreatic pathogens such as YHV and PemoNPV (formerly MBV) [19,20].

Histopathologically, PmoHDV infection is characterized by the presence of single, large basophilic intranuclear inclusion bodies in hypertrophied nuclei of hepatopancreas tubules and adjacent midgut cells. The fully formed inclusion bodies cause nuclear hypertrophy and compression and displacement of the host cell nucleolus [21]. Transmission electron micrograph of a thin section of hepatopancreatic tissue of *P. monodon* infected with PmoHDV reveals ovoid to spherical inclusion bodies of 5–11 μm in a diameter. These inclusion bodies are composed of electron-dense granular material and virions, which lead to the lateral displacement of the nucleoli and margination of the chromatin [22,23].

78.1.5 DIAGNOSIS

Given the absence of distinctive gross signs, diagnosis of PmoHDV infection in shrimp requires the use of laboratory methods such as histopathology, serology, and molecular techniques [24].

For histopathological examination, smears of hepatopancreatic tissue are prepared on a microscope slide and fixed with a drop of 10% formalin solution containing 2.8% NaCl. After drying and normal H&E staining, intranuclear inclusion bodies are visualized under microscope [22,25].

In vitro culture of PmoHDV may be performed. An insect model, *Acheta domesticus*, allows growth of PmeDV [26].

Serological detection of PmoHDV is feasible with the use of monoclonal antibodies (Mabs) against the 54 kDa protein [27,28].

Molecular techniques detect PmoHDV DNA and permit rapid and nondestructive diagnosis of PmoHDV infection in shrimp larvae, parental broodstock, and carrier host. Early generation molecular techniques are based on using short nucleic acid fragments of PmoHDV as probes through in situ hybridization or dot blot hybridization procedures [25]. With the development of nucleic acid amplification technologies such as polymerase chain reaction (PCR) and loop-mediated isothermal amplification (LAMP), accurate diagnosis of PmoHDV infection is possible. In particular, PCR and its derivatives such as nested PCR, PCR-enzyme-linked immunosorbent assay, and real-time PCR have been widely applied in clinical laboratories for routine diagnosis of PmoHDV infection [5,29–43]. Table 78.1 presents some of the PCR primers reported for PmoHDV identification. In view of their superior sensitivity, specificity, and speed, PCR-based methods are recommended by the World Animal Health Organization (the Office International des Epizooties [OIE]) for the detection of shrimp viruses. LAMP by itself or in combination with lateral flow chromatographic strips offers another useful approach for PmoHDV identification [44–46].

TABLE 78.1
PCR Primers for PmoHDV Detection

Primer	Sequence (5′–3′)	Product (bp)	Reference
121F	GCA CTT ATC ACT GTC TCT AC	156	Sukhumsirichart et al. [29]
276R	GTG AAC TTT GTA AAT ACC TTG		
1120F	GGT GAT GTG GAG GAG AGA	592	Pantoja and Lightner [31]
1120R	GTA ACT ATC GCC GCC AAC		
7490F	TGGAGGTGAGACAGCAGG	350	Phromjai et al. [32]
7852R	CCA ACT GTC CTC GCT CTT		
H441F	GCATTACAAGAGCCAAGCAG	441	Phromjai et al. [33]
H441R	ACACTCAGCCTCTACCTTGT		
HPnF	ATAGAACGCATAGAAAACGCT	265	Umesha et al. [35]
HPnR	GGTGGCGCTGGAATGAATCGCTA		
HPV140F	CTACTCCAATGGAAACTTCTGAGC	140	La Fauce et al. [36]
HPV140R	GTG GCG TTG GAA GGC ACT TC		
HPVF	CATACATACAAGGCTTTCAGACA	659	Khawsak et al. [37]
HPVR	GGCTCATCGTCTTTCTTCTCC		
HPV-2F	GGAAGCCTGTGTTCCTGACT	595	Tang et al. [5]
HPV-2R	CGTCTCCGGATTGCTCTGAT		
RT-HPV-F119	CTA GCA TGG GAG CAG TCG TA	119	Jang et al. [40]
RT-HPV-R119	GAC CTG AAC AGT CTC TGCCA		
RT-2F	ACG ATC AGA GCA ATC CGA AG	129	Jeeva et al. [42]
RT-2R	TGG CTC TCT CCA GGT ATT CG		

78.1.6 Treatment and Prevention

Being endemic in Asia and many other parts of the world, PmoHNV infection causes high mortality and stunted growth in cultured shrimp. In the absence of specific treatment, control and prevention of PmoHNV infection in shrimp farms are mainly achieved by prompt identification and diagnosis, isolation of affected shrimp, disinfection, quarantine and hygiene measures, use of specific pathogen-free stocks, and vaccination [47,48].

78.2 METHODS

78.2.1 Sample Preparation

Virus purification: Hepatopancreata are removed from the gnathothoracies of shrimp and homogenized in TN buffer (0.02 M Tris-HCl, 0.4 M NaCl, pH 7.4). The homogenate is centrifuged at 7,000 and 13,000 × *g*, respectively, for 15 min each at 4°C using Suprafuge 22 12.50 rotor. Liquid supernatant is vacuum filtered through Whatman GF/B, Whatman GF/F filter, and Millipore Nitrocellulose 0.45 μm membrane filter, respectively. The filtered supernatant is subsequently centrifuged at 4°C for 1 h at 142,459 × *g* using Beckman Coulter Optima L-90K Ultracentrifuge (Beckman Coulter) Type 70 Ti rotor. The pellet is resuspended in 500 μL of TN buffer, layered onto the top of a 20%–40% sucrose gradient, and recentrifuged at 113,652 × *g* in a SW 40 Ti rotor for 3 h at 4°C. The pellet is resuspended in 200 μL of TN buffer and stored at −80°C until required [4].

DNA extraction: Pereiopod (about 20 mg) is removed from shrimp and homogenized in 500 μL of buffer containing 0.1 M NaCl, 20 mM Tris-HCl, EDTA, and 0.5% SDS. DNA is extracted from the homogenate using the same volume of phenol/chloroform/isoamylalcohol (25:24:1) mixture. DNA is precipitated with 0.1 volume of 3 M NaOAc (pH 5.3) and 2 volumes of absolute ethanol. The pellet is washed with 70% ethanol, and the air-dried pellet is resuspended in 100 μL of TE buffer. The DNA concentration is determined by measuring the ratio of the optical density at 260 nm to that at 280 nm (OD260/OD280) with an spectrophotometer [42].

Alternatively, nucleic acids from shrimp samples are extracted using CTAB extraction method. Briefly, one swimmeret or uropod of shrimp is homogenized in 500 μL of CTAB buffer (2% [w/v] CTAB, 1.4 M NaCl, 20 mM EDTA, 100 mM Tris-HCl, pH 7.5, 0.25% [v/v] β-mercaptoethanol) and incubated at 25°C for 3 h. The supernatant is collected by centrifugation at 12,000 × *g* for 5 min. Then, 1× phenol/chloroform/isoamyl alcohol (25:24:1, v/v) is added to the supernatant, incubated at 25°C for 5 min, and centrifuged at 12,000 × *g* for 5 min. The aqueous phase is transferred into a new tube and mixed with 1× chloroform/isoamyl alcohol (24:1, v/v). After centrifugation, the upper phase is collected, and nucleic acids are precipitated using 0.9 volumes of cold isopropanol. The pellet is washed with 70% ethanol, air-dried, and resuspended in nuclease free water [41]. Nucleic acids from larvae, gill, hemolymph, and hepatopancreas samples are extracted as described previously [37].

In addition, total DNA may be extracted from 200 μL of the viral suspension using High Pure PCR Template Preparation Kit (Roche Diagnostics) [4] or DNeasy Kit (Qiagen) [40].

78.2.2 Detection Procedures

78.2.2.1 Conventional PCR

Khawsak et al. [37] developed a PCR assay with primers HPV-F (5′-CATACATACAAGGCTTTCAGACA-3′) and HPV-R (5′-GGCTCATCGTCTTTCTTCTCC-3′) for specific detection of PmoHNV.

PCR mixture (10 μL) is made up of 1 μL of 10× PCR buffer, 0.5 μL of 50 mM MgCl$_2$, 1 μL of 10 mM dNTPs, 0.5 μL each of 1 μM of primers HPV-F and HPV-R, 1 U of *Taq* DNA polymerase (Invitrogen), and 1 μL of DNA.

PCR amplification is performed with the following conditions: denaturation at 94°C for 5 min, 30 cycles of denaturation at 94°C for 1 min, annealing at 55°C for 1 min, extension at 72°C for 1 min, and final extension at 72°C for 5 min.

PCR products are mixed with loading dye, separated on agarose gel electrophoresis (1.5% gel), stained with ethidium bromide, and visualized under UV transilluminator.

The use of 100 bp DNA ladder (Fermentas) allows easy observation of a specific 659 bp band, which is indicative of PmoHNV [41].

78.2.2.2 Real-Time PCR

Jeeva et al. [42] developed a real-time PCR kit for identification of PmoHDV with primers RT-2F (5′-ACG ATC AGA GCA ATC CGA AG-3′) and RT-2R (5′-TGG CTC TCT CCA GGT ATT CG-3′) and FAM-conjugated TaqMan probe (5′-CGGAGGGAGCTAACGAAATTAACGA-3′).

PCR mixture (20 μL) is made up of the necessary components for real-time PCR (Bioneer Corporation), 15 pmol of RT-2F and RT-2R primers, 15 pmol of the probe, and 80 ng of total DNA.

Real-time PCR thermal cycling is conducted using Exicycler (Bioneer Corporation) with a program consisting of predenaturation at 95°C for 7 min followed by 40 cycles of denaturation at 95°C for 30 s and primer annealing and DNA synthesis at 55°C for 40 s.

Note: The PCR product (129 bp) may be verified by 2% agarose gel electrophoresis.

78.3 CONCLUSION

PmoHDV is a major cause of slow/stunted growth and mortality in juvenile penaeid shrimp. Considering its high prevalence, widespread distribution, and ferocious pathogenicity, it is vital for us to have the capability to accurately and rapidly identify the culprit agent and develop the preparedness to respond to any future outbreaks of PmoHDV infection. Undoubtedly, the recent advances in molecular detection technologies have adequately prepared us for the diagnostic challenge in dealing with PmoHDV; much has to be done in the areas of antiviral and vaccine research and development against PmoHDV infection in shrimp aquaculture.

REFERENCES

1. Cotmore SF et al. The family Parvoviridae. *Arch Virol.* 2014;159:1239–1247.
2. International Committee on Taxonomy of Viruses (ICTV). http://www.ictvonline.org/virusTaxonomy.asp. Accessed on June 12, 2015.
3. Sukhumsirichart W, Attasart P, Boonsaeng V, Panyim S. Complete nucleotide sequence and genomic organization of hepatopancreatic parvovirus (HPV) of *Penaeus monodon*. *Virology.* 2006;346(2):266–277.
4. La Fauce KA, Elliman J, Owens L. Molecular characterization of hepatopancreatic parvovirus (*Pmerg*DNV) from Australian *Penaeus merguiensis*. *Virology.* 2007;362:397–403.
5. Tang KFJ, Pantoja CR, Lightner DV. Nucleotide sequence of Madagascar hepatopancreatic parvovirus (HPV) and comparison of genetic variation among geographic isolates. *Dis Aquat Org.* 2008;80:105–112.
6. Safeena MP, Tyagi A, Rai P, Karunasagar I, Karunasagar I. Complete nucleic acid sequence of *Penaeus monodon* densovirus (PmDNV) from India. *Virus Res.* 2010;150(1–2):1–11.
7. Jeeva S et al. Complete nucleotide sequence analysis of a Korean strain of hepatopancreatic parvovirus (HPV) from *Fenneropenaeus chinensis*. *Virus Genes.* 2012;44:89–97.
8. Owens L. Bioinformatical analysis of nuclear localisation sequences in penaeid densoviruses. *Mar Genomics.* 2013;12:9–15.
9. Dhar AK, Robles-Sikisaka R, Saksmerprome V, Lakshman DK. Biology, genome organization, and evolution of parvoviruses in marine shrimp. *Adv Virus Res.* 2014;89:85–139.
10. Owens L, Liessmann L, La Fauce K, Nyguyen T, Zeng C. Intranuclear bacilliform virus and hepatopancreatic parvovirus (*Pmerg*DNV) in the mud crab *Scylla serrata* (Forskal) of Australia. *Aquaculture.* 2010;310:47–51.
11. Bonami JR, Mari J, Poulos BT, Lightner DV. Characterization of hepatopancreatic parvo-like virus, a second unusual parvovirus pathogenic for penaeid shrimps. *J Gen Virol.* 1995;76:813–817.
12. Spann KM et al. Hepatopancreatic parvo-like virus (HPV) of *Penaeus japonicus* cultured in Australia. *Dis Aquat Org.* 1997;31:239–241.
13. Uma A, Koteeswaran A, Karunasagar I, Karunasagar I. Prevalence of hepatopancreatic parvovirus (HPV) in *Penaeus monodon* postlarvae from commercial shrimp hatcheries in Tamilnadu, southeast coast of India. *Asian Fish Sci.* 2006;9:113–116.
14. Uma A, Ramanathan N, Saravanabava K. Study by PCR on the prevalence of white spot syndrome virus, monodon baculovirus and hepatopancreatic parvovirus infection in *Penaeus monodon* post larvae in Tamil Nadu, Southeast coast of India. *Asian Fish Sci.* 2007;20:227–239.
15. Flegel TW. Historic emergence, impact and current status of shrimp pathogens in Asia. *J Invertebr Pathol.* 2012;110(2):166–173.
16. Safeena MP, Rai P, Karunasagar I. Molecular biology and epidemiology of hepatopancreatic parvovirus of penaeid shrimp. *Indian J Virol.* 2012;23(2):191–202.
17. Roekring S et al. Comparison of penaeid shrimp and insect parvoviruses suggests that viral transfers may occur between two distantly related arthropod groups. *Virus Res.* 2002;87(1):79–87.
18. Rai P, Pradeep B, Karunasagar I, Karunasagar I. Detection of viruses in *Penaeus monodon* from India showing signs of slow growth syndrome. *Aquaculture.* 2009;289:231–235.

19. Manivannan S, Otta SK, Karunasagar I, Karunasagar I. Multiple viral infections in *Penaeus monodon* prawn post larvae in an Indian hatchery. *Dis Aquat Org.* 2002;48:233–236.
20. Umesha KR, Dass MKB, Manjanaik B, Venugopal MN, Karunasagar I, Karunasagar I. High prevalence of dual and triple viral infections in blacktiger shrimp ponds in India. *Aquaculture.* 2006;258:91–96.
21. Catap ES, Lavilla-Pitogo CR, Maeno Y, Travina RD. Occurrence, histopathology and experimental transmission of hepatopancreatic parvovirus infection in *Penaeus monodon* postlarvae. *Dis Aquat Org.* 2003;57:11–17.
22. Roubal FR, Paynter JL, Lester RJG. Electron microscope observation of hepatopancreatic parvo-like virus (HPV) in the penaeid prawn, *Penaeus merguiensis* de Man, from Australia. *J Fish Dis.* 1989;12:199–201.
23. La Fauce KA, Owens L. Investigation into the pathogenicity of *Penaeus merguiensis* densovirus (*Pmerg*DNV) to juvenile *Cherax quadricarinatus*. *Aquaculture.* 2007;271:31–38.
24. Lightner DV, Redman RM, Moore DW, Park MA. Development and application of a simple and rapid diagnostic method to study hepatopancreatic parvovirus of penaeid shrimp. *Aquaculture.* 1993;116:15–23.
25. Pantoja CR, Lightner DV. Detection of hepatopancreatic parvovirus (HPV) of penaeid shrimp by in situ hybridization at the electron microscope level. *Dis Aquat Org.* 2001;44:87–96.
26. La Fauce KA, Owens L. The use of insects as a bioassay for *Penaeus merguiensis* densovirus (*Pmerg*DNV). *J Invertebr Pathol.* 2008;98:1–6.
27. Srisuk C et al. Improved immunodetection of *Penaeus monodon* densovirus with monoclonal antibodies raised against recombinant capsid protein. *Aquaculture.* 2011;311:19–24.
28. Chaivisuthangkura P, Longyant S, Sithigorngul P. Immunological-based assays for specific detection of shrimp viruses. *World J Virol.* 2014;3(1):1–10.
29. Sukhumsirichart W et al. Characterization and PCR detection of hepatopancreatic parvovirus (HPV) from *Penaeus monodon* in Thailand. *Dis Aquat Org.* 1999;38:1–10.
30. Sukhumsirichart W et al. Detection of hepatopancreatic parvovirus (HPV) *Penaeus monodon* using PCR–ELISA. *Mol Cell Probes.* 2002;16:409–413.
31. Pantoja CR, Lightner DV. A non-destructive method based on the polymerase chain reaction for detection of hepatopancreatic parvovirus (HPV) of penaeid shrimp. *Dis Aquat Org.* 2000;39:177–182.
32. Phromjai J, Sukhumsirichart W, Pantoja C, Lightner DV, Flegel TW. Different reactions obtained using the same DNA detection reagents for Thai and Korean hepatopancreatic parvovirus of penaeid shrimp. *Dis Aquat Org.* 2001;46:153–158.
33. Phromjai J, Boosaeng V, Withyachumnarmkul B, Flegel TW. Detection of hepatopancreatic parvovirus (HPV) in Thai shrimp *Penaeus monodon* by in situ hybridization, dot blot hybridization and PCR amplification. *Dis Aquat Org.* 2002;51:227–232.
34. Manjanaik B, Umesha KR, Karunasagar I, Karunasagar I. Detection of hepatopancreatic parvovirus (HPV) in wild prawn from India by nested polymerase chain reaction (PCR). *Dis Aquat Org.* 2005;63:255–259.
35. Umesha KR, Uma A, Otta SK, Karunasagar I, Karunasagar I. Detection by polymerase chain reaction (PCR) of hepatopancreatic parvovirus (HPV) and other viruses in hatchery-reared postlarvae of *Penaeus monodon*. *Dis Aquat Org.* 2003;57:141–146.
36. La Fauce KA, Layton R, Owens L. TaqMan real-time PCR for detection of hepatopancreatic parvovirus from Australia. *J Virol Methods.* 2007;140:10–16.

37. Khawsak P, Deesukon W, Chaivisuthangkura P, Sukhumsirichart W. Multiplex RT-PCR assay for simultaneous detection of six viruses of penaeid shrimp. *Mol Cell Probes.* 2008;22:177–183.

38. Tang KFJ, Lightner DV. Duplex real-time PCR for detection and quantification of monodon baculovirus (MBV) and hepatopancreatic parvovirus (HPV) in *Penaeus monodon. Dis Aquat Org.* 2011;93:191–198.

39. Yan DC, Tang KF, Lightner DV. A real-time PCR for the detection of hepatopancreatic parvovirus (HPV) of penaeid shrimp. *J Fish Dis.* 2010;33:507–511.

40. Jang IK, Suriakala K, Kim JS, Meng X, Choi TJ. TaqMan real-time PCR assay for quantifying type III hepatopancreatic parvovirus infections in wild broodstocks and hatchery-reared postlarvae of *Fenneropenaeus chinensis* in Korea. *J Microbiol Biotechnol.* 2011;21:1109–1115.

41. Panichareon B, Khawsak P, Deesukon W, Sukhumsirichart W. Multiplex real-time PCR and high-resolution melting analysis for detection of white spot syndrome virus, yellow-head virus, and *Penaeus monodon* densovirus in penaeid shrimp. *J Virol Methods.* 2011;178(1–2):16–21.

42. Jeeva S, Kim NI, Jang IK, Choi TJ. Development of two quantitative real-time PCR diagnostic kits for HPV isolates from Korea. *J Microbiol Biotechnol.* 2012;22(10):1350–1358.

43. Liu T et al. Detection and quantification of hepatopancreatic parvovirus in penaeid shrimp by real-time PCR assay. *J Invertebr Pathol.* 2013;114(3):309–312.

44. Nimitphaka T, Kiatpathomchaia W, Flegel TW. Shrimp hepatopancreatic parvovirus detection by combining loop-mediated isothermal amplification with a lateral flow dipstick. *J Virol Methods.* 2008;154:56–60.

45. Nimitphak T, Meemetta W, Arunrut N, Senapin S, Kiatpathomchai W. Rapid and sensitive detection of *Penaeus monodon* nucleopolyhedrovirus (PemoNPV) by loop-mediated isothermal amplification combined with a lateral-flow dipstick. *Mol Cell Probes.* 2010;24(1):1–5.

46. Chaivisuthangkura P et al. Rapid and sensitive detection of *Penaeus monodon* nucleopolyhedrovirus by loop-mediated isothermal amplification. *J Virol Methods.* 2009;162(1–2):188–193.

47. OIE (Office International des Epizooties). *OIE Aquatic Animal Disease Cards 2007.* Hepatopancreatic parvovirus disease. http://www.oie.int/fileadmin/Home/eng/Internationa_Standard_Setting/docs/pdf/Hepatopancreatic ParvovirusCard2007_Revised92707_.pdf. Accessed on June 12, 2015.

48. Attasart P, Kaewkhaw R, Chimwai C, Kongphom U, Panyim S. Clearance of *Penaeus monodon* densovirus in naturally pre-infected shrimp by combined ns1 and vp dsRNAs. *Virus Res.* 2011;159(1):79–82.

79 *Penaeus stylirostris* Penstyldensovirus

Norma A. Macías-Rodríguez, Erika Camacho-Beltrán, Edgar Rodríguez-Negrete,
Norma E. Leyva-López, and Jesús Méndez-Lozano

CONTENTS

79.1 Introduction 705
    79.1.1 Classification... 705
    79.1.2 Morphology and Biology/Epidemiology 706
    79.1.3 Clinical Features and Pathogenesis 706
    79.1.4 Diagnosis 707
        79.1.4.1 Conventional Techniques.. 707
        79.1.4.2 Molecular Techniques... 707
79.2 Methods 708
    79.2.1 Sample Preparation.. 708
    79.2.2 Detection Procedures... 708
        79.2.2.1 End-Point PCR and Nested PCR 708
        79.2.2.2 Real-Time qPCR 709
79.3 Conclusions and Future Perspectives.. 709
References..710

79.1 INTRODUCTION

With their delicious taste and high nutritional value, shrimps have long been an important food source for human society. In Asia, Indo-Pacific, and the United States, shrimp farming represents an enormously profitable activity [1,2]. Worldwide, the shrimp farming industry has confronted several emergencies due to viral diseases implicated in mass mortalities in cultured shrimps, such as white spot syndrome virus (WSSV), yellow head virus (YHV), Taura syndrome virus (TSV), and infectious hypodermal and hematopoietic necrosis virus (IHHNV), the latter of which was lately classified into the family *Parvoviridae*, as *Penaeus stylirostris* penstyldensovirus (PstDV) [3], which will be referred to as IHHNV or PstDV in this chapter. The social and economic impacts of the pandemics caused by these viral pathogens have been dramatic in countries in which shrimp farming constitutes a significant industry [1].

IHHNV (PstDV) was first reported as a cause of acute epizootics and mass mortalities in juvenile and subadult *Litopenaeus stylirostris* in Hawaii in the early 1980s [4]. It causes a disease condition called runt-deformity syndrome (RDS), characterized by reduced growth, deformities of the cuticle and rostrum in stocks of *L. vannamei* and *L. monodon* [5]. To date, infectious IHHNV (PstDV) is distributed broadly throughout shrimp cultures in several distinct geographical regions, including the United States, Oceania, and Asia [6], and noninfectious IHHNV-related sequences have also been detected in the genome of *L. monodon* from Africa and Australia [7]. According to the World Organization for Animal Health (OIE) report [8], two distinct genotypes of infectious IHHNV

have recently been identified as type 1 (from the United States and East Asia) and type 2 (from Southeast Asia). Furthermore, recent studies have documented the presence of IHHNV in wild crustaceans in an asymptomatic manner [9,10].

79.1.1 CLASSIFICATION

The International Committee on Taxonomy of Viruses has assigned IHHNV to the genus *Penstyldensovirus*, subfamily *Densovirinae,* family *Parvoviridae*, which includes a second subfamily *Parvovirinae* [3].

Currently, the genus *Penstyldensovirus* consists of a single species, *decapod penstyldensovirus 1*, which includes four isolates: PstDV1, PstDV2, *Penaeus monodon penstyldensovirus* (PmoPDV) 1, and PmoPDV2. Being monophyletic, these viruses all have small (~4 kb), monosense, homotelomeric genomes that lack discernible PLA2 motifs.

Members of the species *decapod penstyldensovirus 1* are responsible for the high mortality in juvenile and subadult American blue shrimp (*Penaeus stylirostris*) and cuticular deformities and growth retardation (collectively called runt-deformity syndrome, or RDS) in pacific whiteleg shrimp (*Penaeus vannamei*, which is now known as *Litopenaeus vannamei*). For convenience, we will refer to the viruses of the species *decapod penstyldensovirus 1* as IHHNV or PstDV in this chapter.

Further phylogenetic alignment data suggest that IHHNV is most closely related to the genus *Brevidensovirus*, which includes two mosquito densoviruses (DNVs) isolated from *Aedes aegypti* and *Aedes albopictus* (AaeDNV and AalDNV) [11]. Virus assigned to the subfamily *Densovirinae* infects arthropods and multiplies efficiently in most of the tissues of larvae, nymphs, and adult host species without the involvement of helper viruses [3].

79.1.2 MORPHOLOGY AND BIOLOGY/EPIDEMIOLOGY

IHHNV is the smallest known member of penaeid shrimp viruses. The IHHNV virion is a nonenveloped, icosahedral-shaped particle with a diameter of 22 nm and a density of 1.40 g/mL in CsCl. Its genome is a linear single-stranded DNA molecule of 4.1 kb in length that encodes three open reading frames (ORFs). ORF1 comprises approximately 50% of the genome encoding a polypeptide of 666 amino acids, which is predicted to produce a nonstructural protein 1 (NS1) based upon its degree of homology with 2 mosquito brevidensoviruses. ORF2, which starts 56 nucleotides upstream and overlaps ORF1, theoretically encodes a 343 amino acid putative nonstructural protein 2 (NS2). The 5′ end of ORF3 also overlaps ORF1, and the ORF3 encodes a 329 amino acid polypeptide. At least four structural proteins (74, 47, 39, and 37.5 kDa) are found in the purified virions [12–14]. IHHNV was first described in the early 1980s in Hawaii, mainly in shrimp specimens of *L. stylirostris*. It has been reported from every country or region where either *L. vannamei* or *L. stylirostris* are farmed and documented in both captive *Penaeus monodon* and *Marsupenaeus japonicas* brood stocks collected from the wild and their cultured progeny, which suggests that Asia is within the natural geographic range of the virus. It has been detected in different regions of Australia, Brazil, and Asia and is widely distributed in wild and cultured species in SE Asia, where it apparently does not produce losses [9,15,16]. This virus is also commonly present in North, South, and Central America, the Caribbean, the Indo-Pacific, the Gulf of California, and Mexico, in both shrimp farms and wild populations of *L. Vannamei*, *L. stylirostris*, and *Farfantepenaeus* [17,18].

Molecular studies show considerable variation among Asian isolates of the virus [12,19]. IHHNV-related sequences have also been detected in the genome of *L. monodon* from Africa and Australia exhibiting 86% and 92% identities to the viral genome [20]. The virus-related sequences are classified into two groups: type A, which was reported from Madagascar and Australia, and type B, reported from Tanzania in Africa [19]. Both types of virus-related sequences showed similarity with the IHHNV genome by having the three ORFs in them [21]. This shows a close relationship between the IHHNV and virus-related sequences found in *P. monodon* from Tanzania, Madagascar, and Australia [7]. According to the OIE report [8], two distinct genotypes of infectious IHHNV have recently been identified as type 1 (from the United States and East Asia) and type 2 (from Southeast Asia), although these designations remain somewhat controversial [22]. Different studies have revealed phylogenetic relationships among different geographic IHHNV isolates, which suggest that the Philippines was the most likely source of the original infection in the shrimp farming areas of Latin America [23]. As for the U.S./Philippine genotype of IHHNV, three other genetic variants of the virus have been documented: IHHNV-I; IHHNV-II, the variant

from Southeast Asia; and IHHNV-III, the IHHNV variant from East Africa, Madagascar, Mauritius, and Australia. Both IHHNV-I and IHHNV-II are infectious to the representative penaeids, *L. vannamei* and *P. monodon*, while the genetic variant IHHNV-III has been demonstrated to be not infectious to these but can result in RDS [12,20]. All isolates of IHHNV from the United States are nearly identical to IHHNV from the Philippines. This finding, along with other aspects of its history and epidemiology of IHHNV in the United States, suggests that IHHNV was introduced from the Philippines, perhaps with live *P. monodon* that were imported in the early 1970s as a candidate aquaculture species during the very early development of shrimp farming in the United States [1].

79.1.3 CLINICAL FEATURES AND PATHOGENESIS

The IHHNV is highly contagious to several penaeid species. The initial recognizable signs in *P. stylirostris* are lethargic surface swimming followed by motionless, inverted sinking to the bottom, muscle opacity, lack of mobility, and death; in addition to the high mortalities associated with this virus, IHHNV causes abnormal physical defects, often resulting in permanent slow growth, small size, and deformities of the rostrum, antenna, thoracic, and abdominal areas [18,24]. In contrast, the impact of this virus in other penaeid species is not severe. In *P. vannamei*, it has been associated with RDS, causing significant reduction on the growth rate, and cuticular deformities as bent rostrum, blistered cuticle, bent telson, abdominal anomalies, and curly antennae [18]. In *P. monodon*, no important diminution on growth rate, deformities, or affect fecundity of lightly infected brood stock has been observed [25].

It has been reported that the disease is more severe in groups of small-sized animals (5.0 mg to 3.7 g) than in larger-sized groups (14.5–36.9 g). This virus was recognized as the etiological agent of high mortalities (over 50%) in juvenile and subadult blue shrimp *P. stylirostris* farmed in Hawaii in 1981. Within 14 days of appearance of gross signs of the disease and by the time the disease had run its course, the mortalities increased to 80%–90%; *P. stylirostris* specimens surviving IHHNV may become carriers and transmitter of the virus for life, despite the absence of evident gross signs of disease, and thus, vertically infected larvae and early postlarvae undergo mass mortalities when reaching the juvenile state [26]. Using traditional techniques (tissue histopathology and electron microscopy) of acute and subacute organisms of species *P. stylirostris*, *P. monodon*, and *P. vannamei* have been reported with large Cowdry type A intranuclear inclusions in gills, nerve cord, stomach mucosa, antennal gland, cuticular hypodermis, and hematopoietic and connective tissues. *P. stylirostris* presented a higher degree of tissue damage in the gills and nerve cord ganglia compared with *P. vannamei*. The damage provoked by such histopathology may be the cause for the characteristic clinical signs observed in IHHNV-infected shrimps [27]. Furthermore, in a survey study performed in two stations in the northern part of the Gulf of California, a high prevalence of this virus was found in wild adult females (86% and 89%) and males (56% and 57%) of *P. stylirostris*. On the other hand, it has been shown that

TABLE 79.1

Recommended Primer Sets for End-Point PCR Detection of IHHNV

Primer	Product	Sequence (5′–3′)	GenBank	References
389F	389 bp	CGG-AAC-ACA-ACC-CGA-CTT-TA	AF218266	[26]
389R		GGC-CAA-GAC-CAA-AAT-ACG-AA		
77012F	356 bp	ATC-GGT-GCA-CTA-CTC-GGA	AF218266	[33]
77353R		TCG-TAC-TGG-CTG-TTC-ATC		
392F	392 bp	GGG-CGA-ACC-AGA-ATC-ACT-TA	AF218266	[20,26]
392R		ATC-CGG-AGG-AAT-TCG-ATG-TG		
309F	309 bp	TCC-AAC-ACT-TAG-TCA-AAA-CCA-A	AF218266	[20]
309R		TGT-CTG-CTA-CGA-TGA-TTA-TCC-A		
MG831F	831 bp	TTG-GGG-ATG-CAG-CAA-TAT-CT	DQ228358	[20]
MG831R		GTC-CAT-CCA-CTG-ATC-GGA-CT		

the prevalence of IHHNV in wild shrimp *P. setiferus* (white shrimp) and *P. aztecus* (brown shrimp) from the estuary of La Pesca in La Laguna Madre, Tamaulipas, Mexico, is low (only 4.4%) [28]. However, high prevalence of IHHNV has been also reported in other regions of the world, such as China. Some members of populations of *P. stylirostris* and *P. vannamei* that survive IHHNV infections and/or epizootics may carry the virus for life and pass the virus onto their progeny and other populations by horizontal and vertical transmission [24].

79.1.4 DIAGNOSIS

Due to the importance of penaeid shrimp culture, the availability of easy and rapid methods that allow early diagnosis is essential for the routine monitoring of viral diseases. In early studies, IHHNV detection was achieved mainly by histological examination [29]. Lately, different molecular techniques have been used, such as dot blot and *in situ* hybridization (ISH) assays, end-point polymerase chain reaction (PCR), nested PCR, real-time quantitative PCR (qPCR), duplex qPCR, and loop-mediated isothermal amplification (LAMP) [10,20,29].

79.1.4.1 Conventional Techniques

Traditionally, the diagnosis of IHHNV infection is by examining histological sections where the nuclear Crowdy type A inclusion body, which is the principal histological diagnostic characteristic of this disease, is observed and strongly labeled [29]. The detection and identification of IHHNV with an immunoassay is a simple and low-cost alternative that can be specific and sensitive. An indirect enzyme-linked immunosorbent assay (ELISA) was developed for the detection of IHHNV using 6 IgM murine monoclonal antibodies (MAbs) generated against purified IHHNV [30]. Unfortunately, the detection was not reliable as some of the shrimp samples were negative by ELISA and produced positive results for PstDV by histopathology or DNA hybridization and vice versa. In 2009, Sithigorngul generated MAbs specific to a recombinant capsid fusion protein of PstDV to detect PstDV-infected shrimps by immunodot, Western blot, and immunohistochemical assays [31].

79.1.4.2 Molecular Techniques

79.1.4.2.1 End-Point PCR and Nested PCR

PCR is a highly sensitive and robust technique for the detection of IHHNV infection in shrimps. Besides its specificity, and the sensitivity of this detection system allows diagnosis during early infection. The PCR assay's success is a function of several factors such as primer composition, structure, and homology of the target molecule. PCR methods have been used to develop, maintain, and document the SPF (specific-pathogen-free) status of *L. vannamei* lines developed in the United States for use by the domestic and international shrimp farming industries [32]. A number of commercial PCR kits are available for IHHNV detection. There are multiple geographical variants of IHHNV, some of which are not distinguishable by all of the available methods for IHHNV detection. Different recommended primer sets for end-point PCR detection of IHHNV are enlisted in Table 79.1.

The two primer sets 392F/R and 389F/R are the most suitable for detecting all known genetic variants of IHHNV [19,34], including types 3A and 3B, which are inserted into the genome of certain geographic stocks of *P. monodon* from Western Indo-Pacific, East Africa, Australia, and India [7,20,35]. Primer set 309F/R amplifies only a segment from IHHNV types 1 and 2 (the infectious forms of IHHNV) but not types 3A and 3B, which are noninfectious and part of the *P. monodon* genome [7,20]. Primer set MG831F/R reacts only with types 3A and 3B. The primer set 77012F/77353R binds to a region between the nonstructural and structural (coat protein) protein-coding regions and amplifies a 356 bp product [33]. Different nested PCR–based commercial kits were also developed for improved sensitivity of detection [24]. A nested PCR has also been used to increase sensitivity for IHHNV detection in wild crustaceans and shrimp samples collected in India [21] and Mexico [10].

79.1.4.2.2 Single and Duplex Quantitative PCR

qPCR, which is used to amplify and simultaneously quantify a targeted DNA molecule, enables detection and quantification of the viral pathogen in the tissues of infected shrimps. By using this technique, the viral titer in infected shrimps can be accurately determined, which in turn helps in risk assessment as well as disease monitoring during culture. qPCR assays have been successfully applied for the detection and quantification of IHHNV infection in shrimps. In 2001, Tang and Lightner developed a method using a TaqMan probe in a 50-nuclease assay to detect the virus in penaeid shrimps [36]. A pair of PCR primers that amplified an 81 bp DNA fragment and a fluorogenic probe (TaqMan probe) was selected from a conserved region of ORF1 of the IHHNV genomic sequence (GenBank AF218266). Dhar et al. [37] developed a duplex qPCR for detection and quantification of both IHHNV and WSSV using SYBR Green chemistry. In 2008, a multiplex qPCR for the detection of three viral pathogens WSSV, IHHNV, and TSV of penaeid shrimps was developed, and sensitivity in comparison to routine PCR was 10, 1000, and 10 times higher, respectively, for the three viruses [38]. In a more recent report, Leal et al. [39] developed and standardized a duplex qPCR assay for the simultaneous detection of the WSSV and the IHHNV, showing that this method is more effective than assaying for each virus separately.

79.1.4.2.3 Loop-Mediated Isothermal Amplification

LAMP is a novel DNA amplification method originally developed by Notomi et al. [40]. The LAMP reaction involves an autocycling strand displacement DNA synthesis that is performed by a DNA polymerase with high strand displacement activity and a set of two specially designed inner and outer primers. The amplification products are stem-loop DNAs with several inverted repeats of the target and exhibit cauliflower-like structure with multiple loops. The LAMP reaction can be conducted under isothermal conditions ranging from 60°C to 65°C. Using the LAMP method, a highly specific and sensitive diagnostic system for IHHNV detection was designed [41]. Recently, Arunrut et al. described a rapid and sensitive detection system of IHHNV by LAMP combined with a lateral flow dipstick and it was faster than the usual LAMP reactions as described [41,42]. The results obtained using this method were comparable to nested PCR.

79.1.4.2.4 Dot Blot and In Situ Hybridization

DNA probes are useful for the diagnosis of various viral diseases providing greater diagnostic sensitivity than conventional histopathology procedures. They find application in the nonlethal testing of valuable brood stock. Dot blot and ISH methods for the detection of IHHNV have been developed, and commercial kits are available [8]. The first generation of gene probes was developed in 1993 [29]. The specificity of BS4.5 probe labeled with nonradioactive digoxigenin (DIG)-11-dUTP was investigated by dot blot, ISH, and Southern blotting. Detection of IHHNV DNA was highly specific since cross-reaction with insect parvovirus DNA and HPV-infected tissues was not detected [29,43]. A nonlethal serodiagnostic field kit was used for screening of candidate SPF *P. vannamei* brood stock by using specific gene probe [44]. Lately, a DIG-labeled probe I703 was developed for the detection of IHHNV infection in *P. vannamei* by dot blot and ISH [45].

79.2 METHODS

79.2.1 SAMPLE PREPARATION

Several methods have been used for the DNA extraction from specimens prior to analysis. Tang and Lightner obtained total DNA extracts from either gills or pleopods of shrimps with the High Pure DNA Template Preparation Kit (Roche Bioscience) whereas Kim et al. used gills or shrimp pleopods pooled and homogenized in phosphate-buffered saline (PBS; pH 7.4), and the viral DNA was extracted using DNeasy Blood & Tissue Kit (Qiagen, Valencia, CA) [7,22]. Galván-Álvarez et al. obtained total DNA extracts from pleopods' muscle using glassmilk (GeneClean Spin Kit, MP Biomedicals) according to the manufacturer's instructions. Silva et al. described the sample preparation and DNA extraction of muscular tissue extracted from the third abdominal segment [46,47]. Tissue samples were cut and immediately transferred to a buffer containing proteinase K, according to the specifications of NucleoSpin® Tissue kit. Recently, Macías-Rodríguez et al. published another methodology for wild organisms and shrimps using 50 mg of lamellar, intestinal, or muscle tissue in 400 μL of lysis buffer from DNAzol kit (Invitrogen, Carlsbad, CA) that is centrifuged at $13,000 \times g$ for 10 min, and then, 200 μL of supernatant was transferred to another Eppendorf tube with 150 μL of 100% ethanol, mixed by inversion and incubated for 3 min at room temperature before being centrifuged at $12,000 \times g$ for 5 min. Subsequently, the sample with 150 μL of 75% ethanol is washed and centrifuged at $13,000 \times g$ for 3 min, and DNA pellet is dried at room temperature and finally resuspended in 30 μL of ultrapure water before being stored at −20°C [10].

79.2.2 DETECTION PROCEDURES

A number of molecular methods for IHHNV detection are available: end-point PCR [26], nested PCR [10], and qPCR [37].

79.2.2.1 End-Point PCR and Nested PCR

79.2.2.1.1 Procedure

1. Prepare the PCR mix as follows (volumes indicated per reaction).

Component	End-Point PCR (μL)	Nested PCR (μL)
Buffer 10×	2.5	2.5
MgCl₂ (50 mM)	1.0	1.0
dNTP (10 mM)	0.3	0.5
Primer 392F (10 pmol/μL)	0.5	—
Primer 392R (10 pmol/μL)	0.5	—
Primer NNJ208F (10 pmol/μL)	—	0.5
Primer NNJ208R (10 pmol/μL)	—	0.5
Taq polimerasa (5 U/μL)	0.5	0.2
ddH₂O	Up to 24	Up to 24

2. Add 24.0 μL of the PCR mix and 1 μL of cDNA into a 0.2 mL PCR tube.
3. Turn on a thermocycler and preheat the block to 94°C.
4. Spin the sample tubes in a microcentrifuge at 14,000 × *g* for 30 s at room temperature.
5. Place the sample tubes in the thermal cycler at 95°C for 4 min; 35 cycles of 55°C for 30 s and 72°C for 90 s; and 1 cycle of 72°C for 5 min.
6. Following amplification, pulse spin the reaction tubes to pull down the condensation droplets at the inner wall of the tubes. The samples are now ready for electrophoresis or to be stored frozen at −20°C.
7. Analyze the PCR products by agarose gel electrophoresis through 2.0% agarose gel in 1× TAE buffer at 90 V for 40 min.
8. Stain the gel with ethidium bromide and then visualize under ultraviolet light source.

Note: Negative control is concurrently included along with the test samples in order to monitor any possible contamination during the PCR procedure. The presence of a 392 bp product is indicative of IHHNV by single PCR [26] and a 208 bp fragment is indicative of IHHNV by nested PCR [10].

79.2.2.2 Real-Time qPCR

79.2.2.2.1 Procedure

1. Prepare the real-time qPCR mix as follows.

Component	Volume (μL)	Component	Concentration
SYBR Green Master Mix 2× (PE Applied Biosystems)	7.1	TaqMan Universal PCR Master Mix (Applied Biosystems)	1×
Primer 313F (0.24 μM)	1.0	Primers IHHNV1608F	0.3 μM
Primer 363R (0.24 μM)	1.0	Primers IHHNV1688R	0.3 μM
DNA (1 ng)	1	DNA	5–50 ng
—	—	TaqMan probe	0.15 μM
ddH₂O	Up to 25	ddH₂O	Up to 25 μL

2. The reactions were carried out in a 96-well plate in a 25 μL reaction volume.
3. The thermal profile for all SYBR Green PCRs was 50°C for 2 min and 95°C for 10 min, followed by 40 cycles of 95°C for 10 s and 60°C for 1 min. Melting curves were obtained after a final step of 60 cycles under a ramp rate of 0.5°C every 10 s from 60°C to 90°C. Cycling profile for TaqMan is activation of AmpliTaq Gold for 10 min at 95°C, followed by 40 cycles of denaturation at 95°C for 15 s and annealing/extension at 60°C for 1 min.

4. In each 96-well plate, serial dilutions of plasmids harboring the target sequences were used as standards for the respective virus. Standard curves were run in parallel, with the unknown samples for the corresponding virus and their β-actin amplification for each sample used as reference gene.

The PCR primers and TaqMan probe are selected from a region of the IHHNV genomic sequence (GenBank AF218266) that encodes for nonstructural protein. The primers and TaqMan probe are designed by the Primer Express software (Applied Biosystems). The upstream (IHHNV1608F) and downstream (IHHNV1688R) primer sequences are 5′-TACTCCGGACACCCAACCA-3′ and 5′-GGCTCTGGCAGCAAAGGTAA-3′, respectively. The TaqMan probe (5′-ACCAGACATAGA-GCTACAATCCT CGCCTATTTG-3′), which corresponds to the region from nucleotide 1632–1644, is synthesized and labeled with fluorescent dyes 5-carboxyfluoroscein (FAM) on the 5′ end and *N,N,N′,N′*-tetramethyl-6-carboxyrhodamine (TAMRA) on the 3′ end (Applied Biosystems).

Note: SYBR Green PCR amplifications were performed in a GeneAmp 9600 thermocycler coupled with a GeneAmp 5700 Sequence Detection System [37]. Each sample was replicated two to three times. All reactions were repeated at least three times independently to ensure the reproducibility of the results. The absolute levels of IHHNV in the experimental samples were determined by extrapolating the C_q values from the standard curves of the viruses. TaqMan PCR amplifications were performed in a GeneAmp 5700 Sequence Detection System (Applied Biosystems) or ABI PRISM 7000, 7300, or 7500 [36].

79.3 CONCLUSIONS AND FUTURE PERSPECTIVES

IHHNV (PstDV) is one of the major viral pathogens of penaeid shrimps worldwide causing acute epizootics and mass mortalities in juvenile and subadult *L. stylirostris*. It also causes a disease condition called RDS, deformities of the cuticle and rostrum, bubbleheads along with wide size variations, and reduced growth. Currently, infectious IHHNV (PstDV) occurs throughout shrimp cultures in several distinct geographical regions, including the United States, Oceania, and Asia, and noninfectious IHHNV-related sequences have also been detected in the genome of *L. monodon* from Africa and Australia. Recent studies have shown that IHHNV is prevalent in wild crustacean population and in shrimp brood stock. The available evidence implies that IHHNV has a potential of persisting in wild organisms in the vicinity of shrimp farms. When wild organisms gain entry into the ponds, they may cause disease outbreaks. Further studies are required to determine the role of IHHNV in viral diseases of shrimps in terms of synergistic or antagonistic interaction with other viruses or pathogens. Existing data remain controversial with regard to the ability of IHHNV to modify the susceptibility of a second infection either by virus or bacteria. Therefore, research should be focused on the elucidation of IHHNV host range, IHHNV–pathogens interactions, and epidemiology to better understand the origin of the present and emerging shrimp diseases.

REFERENCES

1. Lightner DV, Status of shrimp diseases and advances in shrimp health management. In: *Diseases in Asian Aquaculture VII*, Bondad-Reantaso MG, Jones JB, Corsin F, and Aoki T, eds. Fish Health Section, Asian Fisheries Society, Selangor, Malaysia, pp. 121–134, 2011.
2. Flegel TW, Historic emergence, impact and current status of shrimp pathogens in Asia, *J Invertebr Pathol*, 110, 166–173, 2012.
3. Andrew M.Q. King, Elliot Lefkowitz, Michael J. Adams and Eric B. Carstens (eds.), Virus Taxonomy, Classification and Nomenclature of Viruses: Ninth Report of the International Committee of Taxonomy of Viruses. Elsevier, Amsterdam, the Netherlands, 2012.
4. Lightner DV and Redman RM, Shrimp diseases and current diagnostic methods, *Aquaculture*, 164, 201–220, 1998.
5. Primavera HJ and Quintio EM, Runt-deformity syndrome in cultured giant tiger prawn *Penaeus monodon*, *J Crusta Biol*, 20, 796–802, 2000.
6. Robles-Sikisaka R et al., Genetic signature of rapid IHHNV (Infectious Hypodermal and Hematopoietic Necrosis Virus) expansion in wild *Penaeus* shrimp populations, *PLos One*, 5, 1–11, 2010.
7. Tang KFJ and Lightner DV, Infectious hypodermal and hematopoietic necrosis virus (IHHNV) in the genome of the black tiger prawn *Penaeus monodon* from Africa and Australia, *Virus Res*, 118, 185–191, 2006.
8. Anon., *OIE Manual of Diagnostic Tests for Aquatic Animals*. Office International des Epizooties, Paris, France, 2009.
9. Souto-Cavalli L et al., Natural occurrence of White spot syndrome virus and Infectious hypodermal and hematopoietic necrosis virus in Neohelice granulata crab, *J Invertebr Pathol*, 114, 86–88, 2013.
10. Macías-Rodríguez N et al., Prevalence of viral pathogens WSSV and IHHNV in wild organism at the Pacific Coast of Mexico, *J Invertebr Pathol*, 116, 8–12, 2014.
11. Shike H et al., Infectious hypodermal and hematopoietic necrosis virus of shrimp is related to mosquito brevidensoviruses, *Virology*, 277, 167–177, 2000.
12. Tang KFJ et al., Geographic variations among infectious hypodermal and hematopoietic necrosis virus (IHHNV) isolates and characteristics of their infection, *Dis Aquat Org*, 53, 91–99, 2003.
13. Zhang C et al., Molecular epidemiological investigation of Infectious hypodermal and hematopoietic necrosis virus and Taura syndrome virus in *Penaeus vannamei* cultured in China, *Virol Sin*, 22, 380–388, 2007.
14. Rai P et al., Genomics, molecular epidemiology and diagnostics of infectious hypodermal and hematopoietic necrosis virus, *Indian J Virol*, 23, 203–214, 2012.
15. Jiménez RR et al., Infection of IHHNV virus in two species of cultured penaeid shrimp *Litopenaeus vannamei* (Boone) and *Litopenaeus stylirostris* (Stimpson) in Ecuador during El Niño 1997–1998, *Aquacult Res*, 30, 695–705, 1999.
16. Flegel TW, Detection of major penaeid shrimp viruses in Asia, a historical perspective with emphasis on Thailand, *Aquaculture*, 258, 1–33, 2006.
17. Nunan L et al., Prevalence of Infectious Hypodermal and Hematopoietic Necrosis Virus (IHHNV) and White Spot Syndrome Virus (WSSV) in *Litopenaeus vannamei* in the Pacific Ocean off the Coast of Panama, *J World Aquat Soc*, 32, 330–334, 2001.
18. Vega-Heredia S, Mendoza-Cano F, and Sánchez-Paz A, The infectious hypodermal and haematopoietic necrosis virus: A brief review of what we do and do not know, *Transbound Emerg Dis*, 1, 11, 2011.
19. Krabsetsve K et al., Rediscovery of the Australian strain of infectious hypodermal and haematopoietic necrosis virus, *Dis Aquat Org*, 61, 153–158, 2004.
20. Tang KFJ, Navarro SA, and Lightner DV, PCR assay for discriminating between infectious hypodermal and hematopoietic necrosis virus (IHHNV) and virus-related sequences in the genome of *Penaeus monodon*, *Dis Aquat Org*, 74, 165–170, 2007.
21. Rai P et al., Detection of viruses in *Penaeus monodon* from India showing signs of slow growth syndrome, *Aquaculture*, 289, 231–235, 2009.
22. Kim JH et al., Detection of infectious hypodermal and hematopoietic necrosis virus (IHHNV) in *Litopenaeus vannamei* shrimp cultured in South Korea, *Aquaculture*, 313, 161–164, 2011.
23. Montgomery-Brock D et al., Reduced replication of infectious hypodermal and hematopoietic necrosis virus (IHHNV) in *Litopenaeus vannamei* held in warm water, *Aquaculture*, 265, 41–48, 2007.
24. Motte E et al., Prevention of IHHNV vertical transmission in the White shrimp *Litopenaeus vannamei*, *Aquaculture*, 219, 57–70, 2003.
25. Withyachumnarnkul B et al., Low impact of infectious hypodermal and hematopoietic necrosis virus (IHHNV) on growth and reproductive performance of *Penaeus monodon*, *Dis Aquat Org*, 69, 129–136, 2006.
26. Tang KFJ et al., Post larvae and juveniles of a selected line of *Penaeus stylirostris* are resistant to infectious hypodermal and hematopoietic necrosis virus infection, *Aquaculture*, 190, 203–210, 2000.
27. Chayaburakul K et al., Different responses to infectious hypodermal and hematopoietic necrosis virus (IHHNV) in *Penaeus monodon* and *P. vannamei*, *Dis Aquat Org*, 67, 191–200, 2005.
28. Guzmán-Sáenz FM et al., Virus de la necrosis hipodérmica y hematopoyética infecciosa (IHHNV) y virus del síndrome de Taura (TSV) en camarón silvestre (*Farfantepenaeus aztecus* Ives, 1891 y *Litopenaeus setiferus* Linnaeus, 1767) de La Laguna Madre, Golfo de México, *Rev Biol Mar Oceanograf*, 44, 663–672, 2009.
29. Mari J, Bonami JR, and Lightner D, Partial cloning of the genome of infectious hypodermal and haematopoietic necrosis virus, an unusual parvovirus pathogenic for penaeid shrimps; Diagnosis of the disease using a specific probe, *J Gen Virol*, 74, 2637–2643, 1993.
30. Poulos BT et al., Monoclonal antibodies to a penaeid shrimp Parvovirus, infectious hypodermal and hematopoietic necrosis virus (IHHNV), *J Aquat Anim Health*, 6, 149–154, 1994.
31. Sithigorngul P et al., Simple immunoblot and immunohistochemical detection of *Penaeus stylirostris* densovirus using monoclonal antibodies to viral capsid protein expressed heterologously, *J Virol Methods*, 162, 126–132, 2009.
32. Wyban JA et al., Development and commercial performance of high health shrimp using specific pathogen free (SPF) brood stock *Penaeus vannamei*. In: *Proceedings*

of the Special Session on Shrimp Farming, Wyban J, ed. World Aquaculture Society, Baton Rouge, LA, pp. 254–259, 1992.

33. Nunan LM, Poulos BT, and Lightner DV, Use of polymerase chain reaction (PCR) for the detection of infectious hypodermal and hematopoietic necrosis virus (IHHNV) in penaeid shrimp, *Mar Biotechnol*, 2, 319–328, 2000.

34. Tang KFJ and Lightner DV, Low sequence variation among isolates of infectious hypodermal and hematopoietic necrosis virus (IHHNV) originating from Hawaii and the Americas, *Dis Aquat Org*, 49, 93–97, 2002.

35. Duda TF Jr. and Palumbi SR, Population structure of the black tiger prawn, *Penaeus monodon*, among western Indian Ocean and western Pacific populations, *Mar Biol*, 134, 705–710, 1999.

36. Tang KFJ and Lightner DV, Detection and quantification of infectious hypodermal and hematopoietic necrosis virus in penaeid shrimp by real-time PCR, *Dis Aquat Org*, 44, 79–85, 2001.

37. Dhar AK, Roux MM, and Klimpel KR, Detection and quantification of infectious hypodermal and hematopoietic necrosis virus and white spot virus in shrimp using real-time quantitative PCR and SYBR green chemistry, *J Clin Microbiol*, 39, 2835–2845, 2001.

38. Xie Z et al., Development of a real-time multiplex PCR assay for detection of viral pathogens of penaeid shrimp, *Arch Virol*, 153, 2245–2251, 2008.

39. Leal AGC et al., Development of duplex real-time PCR for the detection of WSSV and PstDV1 in cultivated shrimp, *Vet Res*, 10, 150, 2014.

40. Notomi T et al., Loop-mediated isothermal amplification of DNA, *Nucleic Acids Res*, 28, 63, 2000.

41. Sun ZF et al., Sensitive and rapid detection of infectious hypodermal and hematopoietic necrosis virus (IHHNV) in shrimps by loop-mediated isothermal amplification, *J Virol Methods*, 131, 41–46, 2006.

42. Arunrut N et al., Rapid and sensitive detection of infectious hypodermal and hematopoietic necrosis virus by loop-mediated isothermal amplification combined with a lateral flow dip stick, *J Virol Methods*, 171, 21–25, 2011.

43. Lightner DV, Epizootiology, distribution and the impact on international trade of two penaeid shrimp viruses in the Americas, *Rev Sci Tech Office Int Epiz*, 15, 579–601, 1996.

44. Carr WH et al., The use of an infectious hypodermal and hematopoietic necrosis virus gene probe serodiagnostic field kit for the screening of candidate specific pathogen-free *Penaeus vannamei* brood stock, *Aquaculture*, 147, 1–8, 1996.

45. Yang B et al., Evidence of existence of infectious hypodermal and hematopoietic necrosis virus in penaeid shrimp cultured in China, *Vet Microbiol*, 120, 63–70, 2007.

46. Galván-Alvarez D et al., Experimental evidence of metabolic disturbance in the white shrimp *Penaeus vannamei* induced by the infectious hypodermal and hematopoietic necrosis virus, *J Invertebr Pathol*, 111, 60–67, 2012.

47. Silva D et al., Infectious hypodermal and hematopoietic necrosis virus from Brazil: Sequencing, comparative analysis, and PCR detection, *Virus Res*, 189, 136–146, 2014.

80 Porcine Parvovirus

Zhengyang Wang, Yu Zhu, Miaomiao Kong, and Shangjin Cui

CONTENTS

80.1 Introduction 713
    80.1.1 Classification... 713
    80.1.2 Morphology and Genome Organization.. 713
    80.1.3 Epidemiology... 714
    80.1.4 Clinical Features and Pathogenesis 714
    80.1.5 Diagnosis 714
        80.1.5.1 Conventional Techniques... 714
        80.1.5.2 Molecular Techniques 715
80.2 Methods 715
    80.2.1 Sample Collection and Preparation 715
    80.2.2 Detection Procedures... 715
        80.2.2.1 Nano-PCR Detection of PPV 715
        80.2.2.2 Nucleic Acid Probe Detection of PPV 716
80.3 Conclusion and Future Perspectives 716
References.. 716

80.1 INTRODUCTION

Porcine parvovirus (PPV) is considered to be one of the major causes of reproductive failure in swine [1,2]. First described from aborted fetal swine in 1967 by Cartwright and Huck [3], PPV is linked to the reoccurrence of estrus, abortion, and the delivery of mummified or stillborn fetuses [4], the so-called SMEDI (stillbirth, mummification, embryonic death, and infertility) syndrome. The virus is endemic in most areas of the world and can be found in all pig herd categories.

80.1.1 CLASSIFICATION

PPV is a single-stranded DNA virus classified in the family *Parvoviridae*. Of the two subfamilies (*Parvovirinae* and *Densovirinae*) within the family *Parvoviridae*, the subfamily *Parvovirinae* covers viruses that infect vertebrate hosts and are separated into eight genera (*Amdoparvovirus*, *Aveparvovirus*, *Bocaparvovirus*, *Copiparvovirus*, *Dependoparvovirus*, *Erythroparvovirus*, *Protoparvovirus*, and *Tetraparvovirus*), whereas the subfamily *Densovirinae* contains viruses that infect insect hosts and are divided into five genera (*Ambidensovirus*, *Brevidensovirus*, *Hepandensovirus*, *Iteradensovirus*, and *Penstyldensovirus*) [1,6].

To date, at least seven genetically divergent parvoviruses that infect pigs have been identified. These include PPV type 1 (PPV1), PPV2, PPV3 (also known as porcine PARV4, hokovirus, or partetravirus), PPV4, PPV5, and PPV6, as well as porcine bocavirus (of the genus *Bocaparvovirus*) [1,6–10].

Recent genetic evidence suggests that PPV2 and PPV3 demonstrate a close relationship to human parvovirus 4 (PARV4) of the genus *Tetraparvovirus*, whereas PPV5 and PPV6 are similar to PPV4 (of the ungulate copiparvovirus 2 group in the genus *Copiparvovirus*) [7–10]. On the other hand, the nonpathogenic PPV strain NADL-2 (PPV-NADL2) (which is currently used as an attenuated vaccine) and the dermatitis-associated PPV strain Kresse belong to the ungulate protoparvovirus 1 group in the genus *Protoparvovirus*. Interestingly, a natural attenuated PPV-N strain, isolated in 1989 from the viscera of a stillborn fetus farrowed by a gilt in Guangxi, southern China, resembles attenuated PPV-NADL-2 strain [11].

Antigenically, PPV appears to be related to canine parvovirus and feline panleukopenia virus. However, it is genetically distinct from parvoviruses of all other species.

80.1.2 MORPHOLOGY AND GENOME ORGANIZATION

PPV is a nonenveloped virus of about 20 nm in diameter, with a round or hexagonal appearance. The capsid of PPV is a spherical shell consisting of 60 identical copies of viral proteins arranged in an icosahedral symmetry [12].

The PPV genome consists of a single-stranded DNA molecule of about 5.2 kb with terminal palindromic sequences [1]. Two large open reading frames (ORFs) are present in the genome, one (located at the 3′ end) coding for the structural (capsid) protein (or VP) and the other (located at the 5′ end) coding for the nonstructural protein (or NSP). It is notable that 60 copies of VP are required to assemble the icosahedral viral capsid. In each of these copies (capsid subunit), a spike on the threefold axis, a depression or canyon around the fivefold axis, and a dimple on the twofold axis can be observed [13,14].

However, in some parvoviruses such as PPV4 and members of the genus *Bocaparvovirus*, an additional ORF3 (encoding accessory protein) has been found in the middle of the viral genome [15,16].

80.1.3 Epidemiology

PPV infection is common in gilts, usually endemic; and occurrence of the disease seems closely related to season, from April to October each year, after farrowing and mating the disease multiples [17]. Once the disease occurs, it can continue for several years. The virus mainly infects newborn piglets, embryo, and embryo pig. When a sow becomes infected in early pregnancy, the mortality of its embryo and embryo pig can be as high as 80%–100%. In addition, PPV infection may induce viremia in the first 6 days, and from days 3 to 7 the infected swine starts to detox to the outside through feces and continue to pollute the environment by detoxing irregularly afterward.

Hemagglutination inhibition antibody can be detected as early as one week after infection. Antibody titers can be up to 1:15,000 within 20 days and can last for several years.

Sources of PPV infection include infected sows and boar, which may sometimes be apparently healthy, asymptomatic carriers. Stillbirths, live births, low earners, and uterine secretions of infected sows contain high titers of the virus, and the environment of pens both inside and outside can be polluted. Sperm cells, spermatic cord, epididymis, and vice gonads of infected boar can contain the virus, so that susceptible sows may acquire infection during mating. If pregnant sows have been infected with the virus for >55 days, the farrow may develop immune tolerance leading to the state of persistent infection. Under this circumstance, the piglets may not have detectable specific antibodies, and will poison and detoxify in their lifetime. The virus is mainly distributed in the lymphoid tissue (such as gut lymphoid tissue, renal interstitial cells, turbinate periosteum) of the infected swine [1].

The disease can be transmitted vertically through the placenta and mating, as well as the respiratory and gastrointestinal routes. Mechanically, rodents can also play a role in the spread of the disease.

80.1.4 Clinical Features and Pathogenesis

Clinical features: Pigs infected with PPV are generally in subclinical form, with reproductive failures in sows being the key clinical sign. In infected sows, the pathological sequela caused by PPV is related mainly to the gestational period in which infection occurs. In the beginning of gestation, the conceptus is protected by the zona pellucida and is therefore resistant to infection. During the embryo stage, infection with PPV results in embryonic death and resorption. From the 35th day onward, ossification starts, and the infection results in death and subsequent mummification. Finally, around the 70th day of gestation, the fetus is already immune competent and may resist viral infection. An infection in this stage is then usually controlled, and the piglet is born with anti-PPV antibodies [18].

Pathogenesis: The primary replication of the PPV is suggested to occur in lymphoid tissues. From there, the virus is distributed systemically via viremia. In the epitheliochorial placenta, six tissue layers completely separate the sow and the fetal blood circulation. Since these cells are closely connected, not allowing the passage of even small molecules such as antibodies, it is still not understood how PPV crosses the placental barrier and reaches the fetus [4]. Probably, as the virus was already detected in lymphoid tissue from pigs and in fetal lymphocytes and is able to remain infectious after phagocytosis by macrophages [1], PPV could use them to infect the fetus. Inside the fetus, PPV replicates due to the high mitotic activities present in the fetal tissues [5].

PPV enters cells through a series of interactions that culminate in the release of viral genetic material into a cell compartment in which replication can occur [1]. The mechanisms of PPV entry are not fully understood but may include clathrin-mediated endocytosis or macropinocytosis followed by transportation through the endosomal pathway. Endosomal trafficking and acidification are essential for PPV to enter in the nucleus [1]. Endosome acidification results in reversible modifications of the virus capsid that allow the virus to escape from the endosome [1]. This motif activity is essential for breaking the vesicular membranes, resulting in the formation of pores. After arriving in the nucleus, PPV replicates using the host cell's own machinery, including cellular DNA polymerase for DNA replication. As virus replication takes place in the replication phase (S) of the cell, cells with a higher replication index are necessary for optimal outcome [1].

80.1.5 Diagnosis

80.1.5.1 Conventional Techniques

Virus isolation represents the primary method for PPV detection. A bacterial suspension mixed with the brain, kidney, testis, lung, liver, and mesenteric lymph nodes of aborted fetuses, stillbirths, mummified and low earners of gilts or placenta, and vaginal secretions of sows is inoculated on cells and cultured. If the specimen contains the virus, the inclusion bodies can be observed in the nucleus after 16–36 h, and the characteristic cytopathic effect can be seen after 5–10 days. Initially, cells appear diffused and granular, and then they turn round, clustered, and condensed. Next, cells disintegrate and their shape is not whole. Finally, the cells detach from the bottle wall.

As the virus isolation is time-consuming and laborious, it is not conducive for routine application. Instead, a range of serological assays are commonly used. These include hemagglutinin assay (HA), hemagglutination inhibition (HI) test, ELISA, serum neutralization test, and indirect immunofluorescence. Of these, HA and HI are the most common methods. Sera for HI test need to be pretreated by heat inactivation followed by adsorption with kaolin [1]. HA and HI are simple, convenient, and fast, but their low sensitivity and low specificity limit their use for secondary diagnosis only [19–23].

80.1.5.2 Molecular Techniques

In comparison with conventional diagnostic techniques, molecular methods such as nucleic acid probes technology [24] and PCR [25] demonstrate relatively high sensitivity and specificity for the detection of PPV.

Not surprisingly, multiplex PCR, real-time PCR, and nano-PCR are increasingly employed as a tool for the diagnosis of PPV infection. Multiplex PCR utilizes a plurality of pairs of primers in the same PCR system so that multiple target genes are amplified at the same time. It enables a sample to be tested simultaneously for multiple pathogens or multiple gene types so various diseases can be diagnosed in a single run, thus shortening the detection time and reducing reagent consumption. Real-time PCR combines the conventional PCR technique and fluorescence detection. By eliminating manual handling after amplification, real-time PCR is not only rapid but also reduces potential cross-contamination [26]. Nano-PCR represents a further advance in PCR technology and enables amplification of DNA with complex structures [27]. Nucleic acid probe technology also has high specificity and sensitivity, but its high cost is inappropriate for use in routine diagnostic laboratories.

80.2 METHODS

80.2.1 SAMPLE COLLECTION AND PREPARATION

Suitable samples for PPV detection include aborted fetuses, stillbirths, mummified and weak earner's brain, kidney, testis, lung, liver, and mesenteric lymph nodes or placenta, and vaginal secretions of sows.

Add 0.5% solution of hydrolyzed milk protein containing penicillin (1000 units/mL), streptomycin (1000 units/mL), and kanamycin (1000 units/mL) to the specimen, with the proportion of specimen and solution being 1:5–10. Grind the specimen, centrifuge it 10 min at 3000 rpm, and store the supernatant at −20°C.

80.2.2 DETECTION PROCEDURES

80.2.2.1 Nano-PCR Detection of PPV

1. *Primers*: Two pairs of primers are designed using Primer Premier 5.0 software from PPV gene sequence of PPV retrieved from GenBank (http://www.ncbi.nlm.nih.gov/nuccore/194277764) [28] (Table 80.1).

2. *Construction of a plasmid containing the NS1 gene*: The complete sequence of the NS1 gene was inserted into the vector pGBKT7 as the standard plasmid, amplified in *E. coli* DH5a, and the recombinant plasmids were purified *using* the related kit. Finally, the constructs were confirmed by digestion with restriction enzymes *EcoRI* and *SalI* and sequencing. The verified products we*re k*ept at −20°C until used [28].

3. *Optimization of PPV nano-PCR assay conditions*: PCR conditions such as temperature and primer volume need to be optimized.

80.2.2.1.1 Procedure

1. Remove all aliquots of PCR reagents (2× nanobuffer, Taq DNA polymerase, ddH$_2$O) from the freezer. Vortex and spin down all reagents before opening the tubes.

2. Prepare multiplex PCR mix as follows (volumes indicated per specimen).

Component	Volume (µL)
2× nanobuffer	6.0
F/R primers (10 µmol/L)	1.0
Taq DNA polymerase (5 U/µL)	0.2
ddH$_2$O	3.3
Total	11.5

3. Add 11.5 µL of PCR mix and 0.5 µL of plasmid into a 0.2 mL PCR tube.

4. Turn on thermocycler and preheat the block to 94°C.

5. Spin the sample tubes in microcentrifuges at $14,000 \times g$ for 30 s at room temperature.

TABLE 80.1

Primers Used to Amplify the NS1 Gene (1989 bp) and the Primers Used to Amplify a Porcine Parvovirus (PPV)-Specific 142-bp Portion of the NS1 Gene

Fragment	Length (bp)	Primer Sequence (5′–3′)	Enzyme Sites
NS1	1989	F: 5′-CCGGAATTCATGGCAGCGGGAAACAC-3′	*EcoRI*
		R: 5′-ACGCGTCGACGTTATTCAAGGTTTGTTG-3′	*SalI*
NS1 (nt 1694–1835)	142	F: 5′-AGCCAAAAATGCAAACCCCAATA-3′	
		R: 5′-CTTTAGCCTTGGAGCCGTGGAG-3′	

Source: Cui, Y. et al., *Lett. Appl. Microbiol.*, 58, 163, 2013.

F, forward primer; R, reverse primer.

6. Place the sample tubes in thermocycler for 30 cycles at 94°C for 30 s, 55°C for 30 s, 72°C for 30 s, and a final elongation of 72°C for 10 min.

7. After amplification, pulse-spin the reaction tubes to pull down the condensation droplets at the inner wall of the tubes. The samples are now ready for the electrophoresis or stored frozen at –20°C.

8. Analyze the PCR products by agarose gel electrophoresis through 1% agarose gel in Tris-acetate-EDTA (TAE) buffer at 120 V for 30 min.

9. Stain the gel with ethidium bromide and then visualize under ultraviolet light source.

80.2.2.2   Nucleic Acid Probe Detection of PPV

DNA probe: Specific gene of PPV is cloned into pUC-plasmid and then digested by endonucleases and labeled with biotin or isotope before use [22].

80.2.2.2.1   Procedure

1. *Spotting*: Take an appropriate size nylon membrane, mark the grid, add DNA samples at the center of membrane, and air dry.

2. *Denaturation*: Put the nylon membrane on the double filter paper saturated with denaturing solution for 10 min.

3. *Neutralization*: Move the nylon membrane onto the double filter paper saturated with neutralized solution and leave it for 5 min; dry the membrane at 37°C for 30 min and then bake it in an oven at 120°C for 30 min to fix the DNA.

4. *Prehybridization*: Transfer the nylon membrane in hybridization tube, add prehybridization solution, and leave it in the hybridization furnace 68°C for 2 h.

5. *Hybridization*: Denature the probes for 10 min in boiling water, immediately quench for 5 min in ice water, remove the hybridization tube, decant prehybridization solution, and then add the denatured DNA probe, and hybridize at 68°C overnight.

6. *Washing the membrane*: Follow instructions of the kit insert.

7. *Chromogenic reaction*: Place the membrane in chromogenic solution; once spotting spots change color, add TE buffer to terminate chromogenic reaction.

80.3   CONCLUSION AND FUTURE PERSPECTIVES

PPV is a small, ssDNA virus that is responsible for causing reproductive failure in swine worldwide. The significant expansion in human populations in recent decades has created a huge demand for large-scale pig farming, which in turn has contributed to a marked rise in the incidence of PPV infections.

For the diagnosis of PPV infection, serological assays (especially hemagglutination and hemagglutination inhibition test) have been widely applied. Nonetheless, virus isolation remains valuable for precise and complete determination of PPV. Due to their high sensitivity, specificity, reliability, and rapid turnover, molecular techniques (e.g., PCR and nucleic acid probes) have been used with increasing frequency in clinical diagnostic laboratories. Further development in automation will help streamline the diagnostic operation for PPV infection.

Currently, no specific treatment is available for PPV infection. The widespread and resistant characteristics of this virus make its control difficult. Regular vaccination of the breeding stock is still the most effective preventive measure for PPV infection [1,5]. Although the first vaccines were developed in the late 1970s [29,30], the constant evolution and mutation of PPV and identification of novel PPVs (e.g., PPV2, PPV3, PPV4, PPV5, and PPV6) [7–10,31–37] have reduced the effectiveness of existing vaccines. Continuing efforts to develop new vaccines, including genetic engineered vaccines, that take account of virus mutations and emergence of novel strains are vital to strengthen control and prevention campaigns against PPV infection.

REFERENCES

1. Streck AF. *Studies on Genetic Properties of Porcine Parvoviruses* [D]. Leipzig, Germany: University of Leipzig, 2013.
2. Wang G, Xie X. Progress of porcine parvovirus. *Shandong Anim Husband Vet*. 1, 78–80, 2012.
3. Cartwright SF, Huck RA. Viruses isolated in association with herd infertility, abortions and stillbirths in pigs. *Vet Rec*. 81, 196–197, 1967.
4. Mengeling WL, Lager KM, Vorwald AC. The effect of porcine parvovirus and porcine reproductive and respiratory syndrome virus on porcine reproductive performance. *Anim Reprod Sci*. 60–61, 199–210, 2000.
5. Truyen U, Streck AF. Porcine parvovirus. In: Zimmerman J, Karriker L, Ramirez A, Schwartz K, Stevenson G (eds.), *Diseases of Swine*, 10th edn. Oxford, U.K.: John Wiley & Sons Inc, pp. 447–455, 2012.
6. Cotmore SF et al. The family Parvoviridae. *Arch Virol*. 159, 1239–1247, 2014.
7. Xiao CT, Giménez-Lirola LG, Jiang YH, Halbur PG, Opriessnig T. Characterization of a novel porcine parvovirus tentatively designated PPV5. *PLoS One*. 8, e65312, 2013.
8. Xiao CT, Halbur PG, Opriessnig T. Complete genome sequence of a novel porcine parvovirus (PPV) provisionally designated PPV5. *Genome Announc*. 1, e00021-12, 2013.
9. Wu R et al. First complete genomic characterization of a porcine parvovirus 5 isolate from China. *Arch Virol*. 159, 1533–1536, 2014.
10. Ni J et al. Identification and genomic characterization of a novel porcine parvovirus (PPV6) in China. *Virol J*. 11, 203, 2014.
11. Su QL et al. Complete genome sequence of porcine parvovirus N strain isolated from Guangxi, China. *Genome Announc*. 3, e01359-14, 2015.
12. Chapman MS, Rossmann MG. Structure, sequence and function correlations among parvoviruses. *Virology*. 194, 491–508, 1993.

13. Hao X et al. Phylogenetic analysis of porcine parvoviruses from swine samples in China. *Virol J.* 8, 320, 2011.

14. Cui J, Wang X, Ren Y, Cui S, Li G, Ren X. Genome sequence of Chinese porcine parvovirus strain PPV2010. *J Virol.* 86, 2379, 2012.

15. Dai XF, Wang QJ, Jiang SJ, Xie ZJ. Complete genome sequence of a novel porcine parvovirus in China. *J Virol.* 86, 13867, 2012.

16. Wang LQ, Wang Y, Chen LB, Fu PF, Chen HY, Cui BA. Complete genome sequence of a porcine parvovirus strain isolated in central China. *Genome Announc.* 2, e01247-13, 2014.

17. Wu Q. *Infectious Diseases* [M]. Beijing, China: China Agricultural University Press, 2002.

18. Joo HS, Donaldson-Wood CD, Johnson RH. Observation on the pathogenesis of the porcine parvovirus infection. *Arch Virol.* 51, 123–129, 1976,

19. Li J, Tang D. Advances in technology of porcine parvovirus disease diagnosis. *Swine Industry Sci.* 7, 89–93, 2012.

20. Chen J, Guo W. Advances in porcine parvovirus detection technology. *Progr Vet Med.* 8, 34–37, 2005.

21. Xie Q, Li C. Advances in molecular biology of porcine parvovirus diagnostic techniques and genetic engineering vaccines. *Anim Husband Vet Med.* 2, 92–95, 2010.

22. Wang G et al. A comparative study of PCR and nucleic acid probe technology diagnosis of MD and RE. *Chin J Prev Vet Med.* 6, 479–482, 2003.

23. Wang K, Pei Z, Liu D. Advances in porcine parvovirus disease diagnostic technology. *China Anim Health Inspect.* 7, 70–71, 2009.

24. Oraveeerakul K, Sooccchoi C, Molitor TW. Detection of porcine parvovirus using nonradioactive nucleic acid hybridization. *J Vet Diagn Invest.* 2, 85–91, 1990.

25. Molitor TW et al. Polymerase chain reaction (PCR) amplification for the detection of porcine parvovirus. *J Virol Methods.* 32, 201–211, 1991.

26. Miao LF, Zhang CF, Chen CM, Cui SJ. Real-time PCR to detect and analyze virulent PPV loads in artificially challenged sows and their fetuses. *Vet Microbiol.* 138, 145–149, 2009.

27. Li H, Rothberg L. Colorimetric detection of DNA sequences based on electrostatic interactions with unmodified gold nanoparticles. *Proc Natl Acad Sci USA.* 101, 14036–14039, 2004.

28. Cui Y et al. A sensitive and specific nanoparticle-assisted PCR assay for rapid detection of porcine parvovirus. *Lett Appl Microbiol.* 58, 163–167, 2013.

29. Zeeuw EJL et al. Study of the virulence and cross-neutralization capability of recent porcine parvovirus field isolates and vaccine viruses in experimentally infected pregnant gilts. *J Gen Virol.* 88, 420–427, 2007.

30. Li B et al. Research progress and application of porcine parvovirus vaccine. *Swine Industry Sci.* 2, 34–37, 2008.

31. Cadar D et al. Emerging novel porcine parvoviruses in Europe: Origin, evolution, phylodynamics and phylogeography. *J Gen Virol.* 94, 2330–2337, 2013.

32. Hijikata M et al. Identification of new parvovirus DNA sequence in swine sera from Myanmar. *Jpn J Infect Dis.* 54, 244–245, 2001.

33. Lau SKP et al. Identification of novel procine and bovine parvoviruses closely related to human parvovirus. *J Gen Virol.* 89, 1840–1848, 2008.

34. Adlhoch C et al. High prevalence of porcine Hokovirus in German wild boar populations. *Virol J.* 7, 171, 2010.

35. Cheung AK et al. Identification and molecular cloning of a novel porcine parvovirus. *Arch Virol.* 155, 801–806, 2010.

36. Huang L et al. Detection of a novel porcine parvovirus, PPV4, in Chinese swine herds. *Virol J.* 7, 333, 2010.

37. Blomström AL et al. Detection of a novel porcine boca-like virus in the background of porcine circovirus type 2 induced post weaning multisystemic wasting syndrome. *Virus Res.* 146, 125–129, 2009.

81 Polyomaviruses

Wojciech Kozdruń

CONTENTS

81.1 Introduction 719
    81.1.1 Classification 719
    81.1.2 Morphology, Genome Organization, and Biology 720
        81.1.2.1 Goose Hemorrhagic Nephritis and Enteritis Virus 720
        81.1.2.2 Budgerigar Fledgling Disease Virus 720
    81.1.3 Clinical Features and Pathogenesis 720
        81.1.3.1 Hemorrhagic Nephritis and Enteritis of Geese Virus 720
        81.1.3.2 Budgerigar Fledgling Disease Virus 721
        81.1.3.3 Polyomaviruses in Other Birds and Mammals 721
    81.1.4 Diagnosis 722
        81.1.4.1 Conventional Techniques 722
        81.1.4.2 Molecular Techniques 722
81.2 Methods 722
    81.2.1 Sample Preparation 722
        81.2.1.1 Hemorrhagic Nephritis and Enteritis of Geese Virus 722
        81.2.1.2 Budgerigar Fledgling Disease Virus 723
    81.2.2 Detection Procedures 723
        81.2.2.1 Isolation of Hemorrhagic Nephritis and Enteritis of Geese Virus 723
        81.2.2.2 Gel-Based PCR 723
        81.2.2.3 Real-Time PCR 724
        81.2.2.4 PCR for Budgerigar Fledgling Disease Virus 724
        81.2.2.5 Histopathological Examination 724
81.3 Conclusion 724
References 725

81.1 INTRODUCTION

Increased production of waterfowl and ornamental birds in recent years has heightened the epidemiological risk of poultry diseases, especially those of viral origins. Viral diseases cause significant losses in birds and reduce the profitability of the poultry industry.

Polyomaviruses represent an important group of pathogens that affect birds and mammals. For example, hemorrhagic nephritis and enteritis of geese (HNEG) viruses are responsible for causing severe disease in waterfowl, and budgerigar fledgling disease virus (BFDV) is associated with health problem in ornamental birds [1,5,28]. Currently, avian polyomaviruses (APyVs) are widely distributed throughout the world. While chronic cases dominate in Europe, acute cases of high mortality are diagnosed in the United States and Canada [5,6,21].

Focusing on polyomaviruses infecting birds, this chapter provides an overview on their characteristics, clinical symptoms, pathological changes, and most aspects of its diagnosis [1,5,6,11,16,21,24,32,33].

81.1.1 CLASSIFICATION

Polyomaviruses are classified in the *Polyomaviridae* family, which consists of three genera: *Avipolyomavirus*, *Orthopolyomavirus*, and *Wukipolyomavirus*, according to the recent recommendation by the International Committee on Taxonomy of Viruses [16]. In addition, there are a number of viruses with similarities to members of the family *Polyomaviridae* that have not been assigned to a polyomavirus species.

The genus *Avipolyomavirus* is divided into five species: avian polyomavirus (formerly known as budgerigar fledgling disease virus), canary polyomavirus, crow polyomavirus, finch polyomavirus, and goose hemorrhagic polyomavirus (GHPV) [15,16]. HNEG syndrome is caused by GHPV. The phylogenetic analysis revealed small differences in the nucleotide sequences of the VP1 protein of GHPV isolates [6].

The genus *Orthopolyomavirus* is known to have at least 16 species, including B-lymphotropic polyomavirus (formerly known as African green monkey polyomavirus), baboon polyomavirus 1 (formerly known as simian agent 12),

BK polyomavirus, bovine polyomavirus (BPyV), John Cuningham (JC) polyomavirus, simian virus 40 (SV40, type species; formerly known as simian vacuolating virus 40), etc.

The genus *Wukipolyomavirus* is composed of four species: human polyomavirus 6, human polyomavirus 7, KI polyomavirus (formerly known as the Karolinska Institutet polyomavirus), and WU polyomavirus (formerly known as the Washington University polyomavirus).

81.1.2 Morphology, Genome Organization, and Biology

81.1.2.1 Goose Hemorrhagic Nephritis and Enteritis Virus

HNEG syndrome was first described in 1969 in Hungary. The disease was initially called the "disease of young geese" or "late Derzsy's disease form [3]." A few years later, it was described in Germany and France. This epizootic has become a serious problem since the late 1980s. So far, it was mostly observed in geese, because mallard and Muscovy ducklings are resistant to the infection. The etiological agent of the disease was described 30 years later by a French researcher, Guerin [11,12].

The etiological agent of the disease is a DNA virus with a virion diameter of 40–50 nm. Its buoyant density is of 1.20 g/cm^3 in sucrose gradient, while in the cesium chloride, it is of 1.35 g/cm^3 [6,11].

GPHV genome consists of a circular double-stranded DNA of approximately 5200 bp (5256 bp). There are five open reading frames (ORFs), which encode the following proteins: large T antigen, small T antigen, and proteins VP1, VP2, and VP3. In addition, one reading frame (ORF-x) is located in the 5′ region of the mRNA and encodes a 169 amino acid sequence, which has a low degree of homology (11.2%) of the VP4 protein of other APyVs. Encoded viral proteins are shown in Table 81.1 [17].

GHPV is very resistant to heating and retains full virulence even after 2 h of heating at 55°C. It is also resistant to freezing and thawing, as well as to fat solvents [17].

81.1.2.2 Budgerigar Fledgling Disease Virus

Budgerigar fledgling disease (BFD) occurs in young, feathering parrots. The first cases of the acute form of the disease were described in North America and Europe in the 1980s. Then, it was given the name "French moult." The disease occurred in the south of France in large aviaries, mainly in budgerigar *Melopsittacus undulatus* genus [4,13].

BFDV has been identified as the first avian virus of the family *Papovaviridae*. Its virions are icosahedral and nonenveloped, and their diameter is 42–49 nm or 50–55 nm [4,7].

By understanding the complete DNA sequence of the virus, its similarity to monkey virus (SV 40) and mouse virus (SE [S.E. polyomavirus]) was demonstrated.

It was found that BFD is caused by BFDV-1, although other strains were isolated from chickens and parrots [19].

Early BFDV regions encode large and small T antigens, and the late region encodes three structural capsid proteins: VP1, VP2 and VP3. Large T antigens play a role in the lytic and neoplastic transformation. The presence of protein agno-2a was also confirmed. The protein appeared to be involved the later phase of the infection in the process of cell lysis. Additionally, BFDV does not agglutinate chicken, turkey, budgerigar, and guinea pig red blood cells [7,19,30].

81.1.3 Clinical Features and Pathogenesis

81.1.3.1 Hemorrhagic Nephritis and Enteritis of Geese Virus

Goslings are susceptible to GHPV infection, although experimental infection was also reported in ducklings. Infected birds excrete virus in the feces to the environment. There was no transmission of the infection by vectors.

So far, the vertical transmission of the virus was not described, but the DNA GHPV already was observed in 4-day-old ducklings, which may indirectly indicate the existence of a vertical route of infection [8].

The incubation period of the disease depends on the age of the birds. After the infection of 1-day-old goslings, the disease symptoms are visible after 68 days, while in the case of 3-week-old goslings after 15 days. In older birds, the infection is usually subclinical.

HNEG syndrome is generally described in growing geese between 4 and 10 weeks of age. The mortality rate of the birds in the flock can range from 10% to 80%, and clinical symptoms are visible only a few hours before death. Birds sit alone or focus on the source of warmth, and listlessness and opisthotonus are visible. In the chronic form, several deaths of birds per day up to the age of 12 weeks were observed. In the feces, partly digested blood can be seen [2,18].

During the necropsy, subcutaneous edema, ascites (liquid with the consistency of gelatin), enlargement and inflammation of the kidneys, and bleeding in the intestinal mucosa are observed (Figure 81.1). Urate deposits in the internal organs and joints are often found.

The only form of prevention of the disease is biosecuration consisting in properly done disinfection of premises, equipment,

TABLE 81.1

Proteins Encoded by the Genome Goose Hemorrhagic Polyomavirus

Name	Genome Localization	Number of Amino Acid	Molecular Weight (kDa)
Large T antigen	5256–4960	636	71,966
	4767–3157		
Small T antigen	5256–4776	160	18,487
VP1	1965–3023	353	38,542
VP2	1085–2062	326	35,109
VP3	1412–2062	217	24,658
ORF-x	349–426	169	16,756
	500–928		

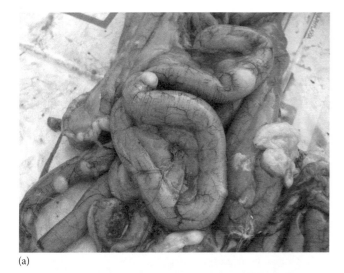

(a)

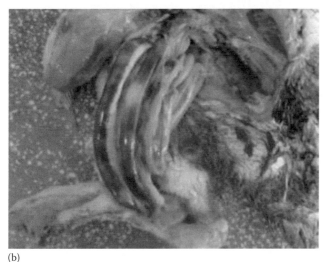

(b)

FIGURE 81.1 Necropsy changes in the intestine of geese with bleeding occurring in the intestinal mucosa (a, external; b, internal).

and catwalks. It is recommended to use disinfectants based on chlorine compounds. In addition, a sound selection of birds during rearing, a good quality feed, and avoiding stress in birds should prevent the occurrence of the disease in geese. HNEG is supposed to have no public health implication [29].

81.1.3.2 Budgerigar Fledgling Disease Virus

All species of parrots show susceptibility to the infection. Young budgies, macaws, great lory, and lovebirds are the most susceptible to the infection.

The route of infection with BFDV is not fully understood. It is believed that the infection may occur in vertical and horizontal route [13].

Asymptomatic carriers may be in the aviary. In the case of a horizontal route, the infection of adult birds should take place during feeding. Moreover, the infection can occur by aerogenic route and by contaminated water.

Budgies aging 1–23 days become frequently affected. However, the occurrence of characteristic clinical symptoms

depends on the age and condition of the birds. Usually, the birds thrive for 10–15 days, despite the existence of BFDV infection in aviary. Then, it comes to a halt in the volumetric content of food and to sudden deaths of birds. On the other hand, abdominal edema, skin discoloration, bruising under the skin, torticollis, and ataxia can be observed in other birds [13].

There is a decline in hatchability by 90%. At the same time, hatched chick mortality rate increases from 10% to 60% [30].

In birds, which survived the infection, plumage disorders, most commonly in the form of dystrophy of feathers including tail feathers, occur [19].

During the necropsy, significant abdominal edema, skin redness, or bruising is observed. In infected birds about 50% less body weight compared to healthy birds is noticed. Additionally, there are hydropericardium, numerous pallor myocardial necrotic foci, hepatomegaly, ascites, and splenomegaly observed. The liver may be enlarged, congested, or with numerous necrotic foci [37].

The best way to prevent the infection is the detection and elimination of sick birds and carriers.

Specific immunoprophylaxis is performed with the use of inactivated vaccine. The vaccination of all birds in the aviary or home breeding is recommended. A suitable biosecuration is also important [29].

81.1.3.3 Polyomaviruses in Other Birds and Mammals

81.1.3.3.1 Polyomaviruses in Other Passeriformes

Polyomaviruses in *Passeriformes* are also morphologically and antigenically similar to each other. However, clinical signs and postmortem lesions vary widely between budgerigar and large birds in the parrot family.

The characteristic feature observed often in large young birds is a peracute form of the disease without any clinical symptoms [13].

The acute form is characterized by mortality among birds occurring within 12–48 h after the infection and clinical symptoms in the form of depression, anorexia, body weight loss, diarrhea, bleeding under the skin, difficulty breathing, and polyuria.

During the postmortem examination, the clear fluid in the abdominal cavity, fibrin deposits in visceral organs, and small pale spleen are observed [13].

81.1.3.3.2 Polyomaviruses in Finches

In young and older birds with acute mortality, disturbances in the shape of the beak (most common in young birds) were demonstrated. In addition, increased mortality, hepatitis, nephritis, and dermatitis on the head were observed [15].

81.1.3.3.3 Bovine Polyomavirus

BPyV was first described by Rang et al. as a contaminant of macaque kidney cell cultures. The source of infection was contaminated calf serum, which was used for the preparation of cell culture. Its occurrence was also described in the sewage and sediment from the slaughterhouse and waters of the rivers polluted by sewage farms. The virus was present in urban sewage. BPyV was detected in about 30% of the bovine urine samples. It has also been shown that 70% of the bovine serum may contain polyomavirus. The virus does not induce any disease in cattle so far; however, it plays an important role in biotechnological techniques in which calf serum is used [14,22,34,35].

BPyV is a circular double-stranded DNA, 4697 bp in size, corresponding with 89.6% of the size of SV40). Its similarity to other polyomaviruses is about 58.6% [14,38].

81.1.3.3.4 Polyomavirus in Ducks

Polyomavirus infection in Muscovy ducks (*Cairina moschata*) and mule ducks (hybrid from a Muscovy duck and Pekin duck) was recently described.

Weight loss, dysfunction in plumage, and increase in the number of secondary bacterial infections and mortality were recorded in the birds [26].

81.1.3.3.5 Polyomavirus in Baboons

Polyomavirus was isolated from the baboon's kidneys and the presence of polyomavirus antibodies in the baboon's sera was observed [10,27].

81.1.4 DIAGNOSIS

81.1.4.1 Conventional Techniques

81.1.4.1.1 Hemorrhagic Nephritis and Enteritis of Geese Virus

In differential diagnosis, Derzsy's disease and mineral–vitamin deficiency should be considered primarily, and then bacterial infections (pasteurellosis) and gout.

Laboratory diagnosis is based on virological examination with regard to virus isolation, molecular biology techniques, and histopathology. Isolation of the virus is performed in cell cultures (kidney cell culture), but especially in the embryos of geese. Serological examination is useless [12].

81.1.4.1.2 Budgerigar Fledgling Disease Virus

Circovirus infection should be considered in the differential diagnosis. In older birds, the circovirus and polyomavirus infection may exist at the same time.

Currently, conventional methods such as serological methods are not widely applied, but they are used in research [13,36].

In budgerigars, specific antibodies can be detected already around 10 days after infection, while in other bird species on 14–18 days of age. The maximum antibody level is found in birds 4–6 weeks of age [31].

Histopathological examination, which allows detecting inclusion bodies in the internal organs, and gel immunodiffusion are among other possible serological methods [37].

Virological methods, which involve the multiplication of the virus in cell cultures prepared from tissues of sick birds,

are also used, but these methods are time consuming. The presence of virus particles in tissues (inclusion bodies) is demonstrated by electron microscopy [13].

81.1.4.2 Molecular Techniques

HNEG virus: PCR and real-time PCR are used to detect DNA of the virus. Total DNA is isolated from the spleen, liver, kidneys, cloacal swabs, and blood. However, it should be remembered that in the case of blood, the virus is present only for about a month after infection. Primers for the amplification reaction are based on the sequence of the VP1 gene [20,21,23,25].

BFDV: Currently, PCR method was applicable for the detection of viral DNA in the blood and cloacal swabs. For molecular examination, the mucosa swabs and feather samples may also be used [9,23,25].

81.2 METHODS

81.2.1 SAMPLE PREPARATION

81.2.1.1 Hemorrhagic Nephritis and Enteritis of Geese Virus

81.2.1.1.1 Preparation of Internal Organ Sample

1. Obtain 5–10 sick live birds from the flock or internal organs (liver, kidneys, intestine) from sick birds.
2. After necropsy, homogenize sections of the internal organs immediately.
3. Suspend 1 mL of the homogenate of internal organs in a liquid PBS containing 500–1000 IU/mL of penicillin and 50–10 mg/mL of streptomycin, or 1 mL/100 mL of a mixture of antibiotics (Antibiotic-Antimycotic, Gibco).
4. Centrifuge at $1600 \times g$ for 20 min at 4°C.
5. Collect supernatants and filter through a Millipore filter with a pore diameter of 45 μm.
6. After confirmation of sterility, inoculate the filtrate into blood agar plates, and/or goose embryo cell cultures.
7. Store remaining filtrate at –20°C.

81.2.1.1.2 Isolation of Total Cellular DNA

Commercial kits are recommended for the isolation of total cellular DNA from 1 mL of organ homogenate.

81.2.1.1.3 Preparation of Goose Embryo Fibroblast Cells (GEF)

1. Disinfect the shells of eggs (at the age of 13–15 days from flocks no evidence of infectious diseases) with 70% alcohol.
2. Remove the embryos to sterile Petri dish and then remove the eyes, wings, legs, and internal organs.
3. Powder the embryos with scissors and rinse four times with PBS at temperature 37°C to remove red blood cells.

4. Trypsinize (0.25% trypsin) the material on a magnetic stirrer two times for 15 min and after each trypsinization filter the material through a layer of gauze. Use about 20 mL of trypsin per embryo.
5. Centrifuge for 5 min at $800 \times g$.
6. Decant supernatant and add 9 mL of Eagle fluid and 1 mL of calf serum to cell pellet.
7. Centrifuge 2–3 min at $1200 \times g$ and remove the supernatant.
8. Suspend the cell pellet at a dilution of 1:400 in growth liquid containing: Eagle's fluid, with 1% of antibiotics (Gibco) and 10% of calf serum. Keep cell density at approx. 1×106 in 1 mL of suspension.

81.2.1.2 Budgerigar Fledgling Disease Virus

1. Blood, cloacal and mucosal swabs, feather follicles, and visceral organs (liver, heart, spleen, kidney, lung) can be used.
2. For the isolation of total cellular DNA from feathers follicles, swabs, and visceral organs, commercial kits are recommended to be used.

81.2.2 Detection Procedures

81.2.2.1 Isolation of Hemorrhagic Nephritis and Enteritis of Geese Virus

1. Inoculation of goose embryos
   a. For the isolation, 10–15-day-old goose embryos are used. The embryos are inoculated with homogenates of internal organs into the allantoic cavity (CAV).
   b. Indicate the air chamber and a point located approximately 1–1.5 cm below the air chamber line on the side of the embryo between the large blood vessels. Marked places on the shell should be disinfected with iodine.
   c. Using a metal drill, make a hole at the site of inoculation and the air chamber.
   d. Insert 0.2 mL of test material into the allantoic cavity.
   e. Introduce 0.2 mL of liquid PBS to control embryos.
   f. Use 3–5 embryos for each material.
   g. The embryos are incubated at 37°C at a relative humidity of about 70%.
   h. The embryos must be examined every day for 8 days.
   i. The dead embryos must be rejected within 24 h p.i., and the fluid of embryonic membranes is taken after 8 days from all dead and cooled (24 h in a refrigerator) embryos and from the liver.
   j. During the collection of the material, necropsy lesions should be evaluated.
   k. In addition to changes characteristic of HNEG in the form of congestion in embryonic membranes,

embryos, and liver, necrotic foci may also be present.
2. Inoculation of the GEF cell culture
   a. The cell cultures are infected with a viral material (homogenate of visceral organs) after the 24 h of incubation period.
   b. The infected cultures are incubated in an incubator at 37°C with 5% CO_2 and 85% humidity. The cell culture should be observed every day under the microscope. The infected cultures create a cytopathic effect (CPE) after 5–7 days of incubation. The cells shrink and become fine and highly refractile. Gradually, the single cell layer becomes disrupted. Syncytia are also visible.
   c. If CPE is not observed in the first passage, two to three other passages are performed.

81.2.2.2 Gel-Based PCR

1. Reagents

Primers (oligonucleotides)

PCR Kit:

   a. PCR buffet
   b. A mixture of deoxynucleotides (dNTPs)
   c. $MgCl_2$
   d. Thermostable DNA polymerase
   e. 1 M betaine solution (N, N, N—trimethylglycine)
   f. Double-distilled water

Reagents for electrophoresis:

   a. 10× TBE (Tris/Borate/EDTA) buffer concentrate: the buffer is used for electrophoresis. To 100 mL of the buffer, add 900 mL of bidistilled water.
   b. 1% or 2% agarose gel: dissolve 1 g (1% agarose gel) or 2 g (2% agarose gel) agarose in 100 mL 1× TBE buffer concentrate. The suspension should be warmed up to complete the dissolution of the agarose.
   c. DNA molecular weight marker (commercial).
   d. Loading buffer for DNA samples (commercial).
   e. Ethidium bromide solution.
2. PCR setup
   a. The isolated DNA is verified by electrophoresis on a 1% agarose gel. Electrophoresis is performed at 120 V for about 1 h. After electrophoresis, the gel is stained with ethidium bromide solution for about 20 min. Visualization and recording of the results of electrophoresis are performed using a gel documentation kit.
   b. The primers should be diluted with redistilled water, in accordance with the manufacturer's instructions, to a concentration of 10 μM. The volume of the reaction mixture used in the PCR

method is 25 μL (for each sample) with the following ingredients:

  i. PCR buffer—2.5 μL
      – dNTPs—1 μL
      – MgCl$_2$—1 μL
      – Primer GPOL1—1 μL
      – Primer GPOL2—1 μL

Primers	Amplified Gen	Primer Sequence
GPOL1	VP1	GAG GTT GTT GGA GTC ACC ACA ATG
GPOL2	VP1	ACA ACC GTG CAA TCC CAA GGG TTC

      – Betaine—4 μL
      – DNA polymerase—0.5 μL
      – Double-distilled water—11 μL
      – 3 μL of the isolated DNA added to the prepared mixture
      – The PCR conditions are outlined as follows:

Reaction	Temperature (°C)	Time (min)	Number of Cycles
Initial denaturation	95	5	1
Denaturation	94	1	35
Annealing	60.2	1	
Elongation	72	1	
Final elongation	72	10	1

3. Electrophoresis of PCR products and result analysis

Electrophoresis is performed in a 2% agarose gel (2 g + 100 mL of electrophoresis buffer 1× TBE). Electrophoresis occurs at a constant voltage of 120 V for approximately 50 min. After the electrophoresis, the gel is stained in ethidium bromide solution for about 20 min and viewed using a gel documentation kit.

A positive PCR result indicates the presence of PCR product of 570 bp.

81.2.2.3 Real-Time PCR

 1. The primers for this reaction are based on the sequence of the VP1 protein GHPV (Leon):

Primers	Sequence (3′–5′)	Position on VP1 Gene
VP1F	GAG GTT GTT GGA GTG ACC ACA ATG	396–419
VP1R	ACA ACC CTG CAA TTC CAA GGG TTC	516–539
VP1-SBF	GAT GGT GCT TAT CCG GTG GA	645–664
VP1-SBR	TTC ATT CCG GGA TGG GTC	678–695

 2. Real-time SYBR Green
 3. 95°C—5 min, 35 cycles (initial denaturation); 94°C—30 s (denaturation); 52°C—30 s (annealing); 72°C—40 s (elongation); 72°C—10 min (final elongation)

81.2.2.4 PCR for Budgerigar Fledgling Disease Virus

 1. Primers (oligonucleotides)

Primers	Sequence (3′–5′)	Position on VP1 Gene
BFDV1	CTT ATG TGG GAG GCT GCA GTG TT	2183–2206
BFDV2	TAC TGA AAT AGC GTG GTA GGC CTC	2709–2732

 2. PCR conditions

Reaction	Temperature (°C)	Time	Number of Cycles
Initial denaturation	95	5 min	1
Denaturation	94	30 s	40
Annealing	56	30 s	
Elongation	72	4 min	
Final elongation	72	10 min	1

Electrophoresis of PCR products is performed in the same way as before. A positive PCR result indicates the presence of PCR product of 550 bp.

81.2.2.5 Histopathological Examination

For both HNEG virus and BFDV, the following apply:

 1. Section of internal organs should be fixed in 10% buffered formalin, and then paraffin sections stained with hematoxylin and eosin are prepared.
 2. During the microscopic examination, the following characteristic lesions for GHPV are observed: necrotic–hemorrhagic foci on the duodenum and jejunum, degenerative epithelial tubular lesions on the kidneys, and marked pyknosis on the medullar area of bursa Fabricius lobules.
 3. During the microscopic examination of the liver, lesions characteristic of BFDV are observed: multifocal to confluent hepatic necrosis, mild accumulation of heterophils, and amphophilic intracellular inclusions in hepatocytes.

81.3 CONCLUSION

Recent findings suggest a significant spread of APyVs in duck populations worldwide [5,8]. Due to the limitations of current diagnostic methods, polyomavirus infections may have been under-detected. Besides occasional isolations from clinically healthy birds, polyomaviruses are often isolated from geese displaying the clinical form of Derzsy's disease. Future study on improving intravital and serological diagnosis is clearly relevant, which will contribute to the elimination of infected birds from flocks.

REFERENCES

1. Bernath, S. et al., Aetiology and epidemiological significance of the haemorrhagic nephritis and enteritis of geese, *Mag. Allatorvosok Lapja*, 123, 522, 2001.
2. Bernath, S., Farsang, A., Kovacs, A., Nagy, E., Dobos-Kovacs, M., Pathology of goose hemorrhagic polyomavirus infection in goose embryos, *Avian Pathol.*, 35, 49, 2006.
3. Bernath, S., Szalai, F., Investigations for clearing the etiology of the disease appeared among goslings in 1969, *Mag. Allatorvosok Lapja*, 25, 531, 1970.
4. Bozeman, L.H. et al., Characterization of a papovirus isolated from fledgling budgerigars, *Avian Dis.*, 25, 972, 1981.
5. Corrand, L., Gelfi, J., Albaric, O., Etievant, M., Pingret, J.L., Guerin, J.L., Pathological and epidemiological significance of goose haemorrhagic polyomavirus infection in ducks, *Avian Pathol.*, 40, 355, 2011.
6. Dobos-Kovacs, M. et al., Haemorrhagic nephritis and enteritis of geese: Pathomorphological investigations and proposed pathogenesis, *Acta Vet. Hung.*, 53, 213, 2005.
7. Dykstra, M.J., Dykstra, C.C., Lukert, P., Bozeman, L.H., Investigations of budgerigar fledgling disease virus, *Am. J. Vet. Res.*, 45, 1183, 1984.
8. Farsang, A., Bernath, S., Dobos-Kovacs, M., Case report of goose haemorrhagic polyomavirus in 4-day-old goslings indicating vertical transmissibility, *Acta Vet. Brno*, 80, 255, 2011.
9. Garcia, A.P. et al., Diagnosis of polyomavirus—Induced hepatic necrosis in psittacine birds using DNA probes, *J. Vet. Diagn. Invest.*, 6, 308, 1994.
10. Gardner, S.D., Knowles, W.A., Hand, J.F., Porter, A.A., Characterization of a new polyomavirus (*polyomavirus papionis-2*) isolated from baboon kidney cell cultures, *Arch. Virol.*, 105, 223, 1989.
11. Guerin, J.L. et al., A novel polyomavirus (goose hemorrhagic polyomavirus) is the agent of hemorrhagic nephritis enteritis of geese, *J. Virol.*, 74, 4523, 2000.
12. Guerin, J.L. et al., Isolation of avian polyomavirus, associated with hemorrhagic nephritis and enteritis of geese, *Proceedings of the First World Waterfowl Conference*, 66pp., 1999.
13. Horton, S., *Avian Polyomavirus in Psittacines and Passerines*, An Avian and Exotic Hospital, Chicago, IL, 2008.
14. Hundesa, A. et al., Development of a quantitative PCR assay for the quantitation of bovine polyomavirus as a microbial source-tracking tool, *J. Virol. Methods*, 163, 385, 2010.
15. Johne, R., Müller, H., Avian polyomaviruses in wild birds: Genome analysis of isolated from falconiformes and psittaciformes, *Arch. Virol.*, 143, 1501, 1998.
16. Johne, R., Müller, H., Polyomaviruses of birds: Etiologic agents of inflammatory diseases in a tumor virus family, *J. Virol.*, 81, 11554, 2007.
17. Johne, R., Müller, H., The genome of goose hemorrhagic polyomavirus, a new member of the proposed subgenus *Avipolyomavirus*, *Virology*, 308, 291, 2003.
18. Lacroux, C., Andreoletti, O., Payre, B., Pingret, J.L., Dissais, A., Guerin, J.L., Pathology of spontaneous and experimental infections by goose haemorrhagic polyomavirus, *Avian Pathol.*, 33, 351, 2004.
19. Ledwoń, A., Szeleszczuk, P., Budgerigar fledgling disease—New data, *Magazyn Wet.*, 16, 74, 2007.
20. Leon, O. et al., Goose hemorrhagic polyomavirus detection in geese using real-time PCR assay, *Avian Dis.*, 57, 797, 2013.
21. Miksch, K., Grossmann, E., Kohler, K., Johne, R., Detection of goose haemorrhagic polyomavirus (GHPV) in flocks with haemorrhagic enteritis and nephritis of geese in southern Germany, *Berl. Munch. Tierarztl.*, 115, 390, 2002.
22. Nairn, C., Lovatt, A., Galbraith, D.N., Detection of infectious bovine polyomavirus, *Biologicals*, 31, 303, 2003.
23. Niagro, F.D. et al., Polymerase chain reaction for detection of PBFD virus and BFD virus in suspect birds, *Proceedings of Association of Avian Veterinarians*, pp. 25–37, 1990.
24. Palya, V., Ivanics, E., Glavits, R., Dan, A., Mato, T., Zarka, P., Epizootic occurrence of haemorrhagic enteritis infections of geese, *Avian Pathol.*, 33, 244, 2004.
25. Phalen, D.N., Wilson, V.G., Graham, D.G., Polymerase chain reaction assay for avian polyomavirus, *J. Clin. Microbiol.*, 29, 1030, 1991.
26. Pingret, J.L., Boucraut-Baralon, C., Guerin, J.L., Goose haemorrhagic polyomavirus infection in ducks, *Vet. Rec.*, 2, 164, 2008.
27. Rangan, S., Lowrie, R., Roberts, J., Johston, P., Warriyck, R., Virus from stump-tailed monkey (*Macaca arctoides*) kidney culture, *Lab. Anim. Sci.*, 24, 211–217, 1974.
28. Ritchie, B.W. et al., Polyomavirus infections in adult psittacine birds, *J. Assoc. Avian Vet.*, 5, 202, 1991.
29. Ritchie, B.W. et al., Prevention of avian polyomavirus infections through vaccination, *Birds n ways*, 21, 34, 1997.
30. Rott, O., Kroger, M., Müller, H., Hobom, G., The genome of budgerigar fledgling disease virus, an avian polyomavirus, *Virology*, 165, 74, 1988.
31. Roy, P., Dhillon, A.S., Lauerman, L., Shivasprasad, H.L., Detection of avian polyomavirus infection by polymerase chain reaction using formalin—fixed, paraffin—embedded tissues, *Avian Dis.*, 48, 400, 2004.
32. Schetler, C.H., Clinical picture and pathology of hemorrhagic and enteritis in geese, *Tieraztliche Prax.*, 8, 313, 1980.
33. Schettler, C.H., Detection en France de la nephrite hemorragique et enterite de l'oie, *Recueil de Medicine Veterinaire D'Alfort*, 153, 353, 1977.
34. Schuurman, R., Sol, C., Van der Noordaa, J., The complete nucleotide sequence of bovine polyomavirus, *J. Gen. Virol.*, 71, 1723, 1990.
35. Schuurman, R., Van Steenis, B., Sol, C., Bovine polyomavirus, a frequent contaminant of calf serum, *Biology*, 19, 265, 1991.
36. Shahidur Rahmann Khan, M., Johne, R., Beck, I., Pawlita, M., Kaleta, E.F., Müller, H., Development of a blocking enzyme-linked immunosorbent assay for the detection of avian polyomavirus–specific antibodies, *J. Virol. Methods*, 89, 39, 2000.
37. Szeleszczuk, P., Ledwoń, A., Budgerigar fledgling disease, *Życie Wet.*, 78, 151, 2003.
38. Wang, J., Horner, G.W., O'Keefe, J.S., Detection and molecular characterization of bovine polyomavirus in bovine sera in New Zealand, *N. Z. Vet. J.*, 53, 26, 2004.

82 Bovine Papillomaviruses

M.A.R. Silva, R.C.P. Lima, M.N. Cordeiro, G. Borzacchiello, and A.C. Freitas

CONTENTS

82.1 Introduction 727
    82.1.1 Classification 728
    82.1.2 Genome Organization 728
    82.1.3 Transmission, Clinical Features, and Pathogenesis 728
    82.1.4 Diagnosis 729
        82.1.4.1 Conventional Techniques 729
        82.1.4.2 Molecular Techniques 729
82.2 Methods 730
    82.2.1 Sample Collection and Preparation 730
    82.2.2 Detection Procedures 731
        82.2.2.1 PCR Using Consensus Primers 731
        82.2.2.2 PCR Using Type-Specific Primers 731
82.3 Conclusion and Future Perspectives 731
References 732

82.1 INTRODUCTION

Papillomatosis is generally described as a group of diseases/disorders in which the most notable clinical symptoms are hyperkeratotic lesions that lead to tissue damage. It has been found in virtually all vertebrate species, including reptiles, birds, and mammals [1]. The correlated lesions are usually characterized as warts or papillomas, and their development may result in different grades of cutaneous and/or mucosal tumors, including cancers. Among mammals, human and bovine papillomatosis have been best studied. It has been shown that bovine papillomavirus (BPV) has the ability to infect other hosts, particularly equines, causing a disease known as sarcoids. Furthermore, malignant tumors are also associated with BPV infection in cattle.

Lesions related to bovine papillomatosis have commonly been found in the form of cutaneous, genital mucosa and orofaringeous papillomas in 2-year-old animals [2]. Bovine papillomas are benign lesions, which generally regress but occasionally persist and undergo a malignant transformation into squamous cell carcinoma, as in the cases of urinary bladder and upper alimentary canal cancers [3]. However, some animals are unable to resolve their lesions and succumb to widespread cutaneous or mucosal involvement. These forms of papillomatosis are troublesome and of economic significance [4], because in addition to causing a depreciation in the value of pelt, teat papillomas reduce the lactation period, and the affected sites may become vulnerable to injuries and infections. Moreover, cattle infected with BPV are susceptible to extensive hyperproliferative lesions in the upper gastrointestinal tract and, as a result, experience difficulty in feeding

and breathing and thus become debilitated [5]. Perhaps, one of the most severe consequences of bovine papillomatosis is enzootic hematuria, in which bladder lesions show malignant progression in cattle that feed on bracken fern (*Pteridium aquilinum*) [6]. Thus, bovine papillomatosis has a serious economic effect on livestock farming.

The widely recognized causative agent associated with bovine papillomatosis is the BPV group [4]. BPVs are members of the *Papillomavirus* (*PV*) genus, a group of epitheliotropic viruses that often cause asymptomatic infections [7]. This group of papillomaviruses was originally described as species specific, but, in recent years, some types have been implicated in cross-species transmission events.

An equine skin tumor, described as equine sarcoid [8], has often been diagnosed in equids [9]. Although its lesions are hyperproliferative, in the same way as those from bovine papillomatosis, sarcoids very rarely regress; in fact, it is more common for them to persist and to be locally aggressive. Six clinical types of sarcoids have been classified so far: occult, verrucous, nodular, fibroblastic, mixed, and malignant [10]. Sarcoid-bearing species include horses, zebras, donkeys, tapirs, and mules [11–15], and its lesions have often been found in the skin of the lower abdomen, head, neck, and limbs [16,17]. In general, these tumors do not metastasize, and although they may have different levels of invasiveness, some sarcoids may cause ulceration as well. When they occur near the eyes or on the eyelids, they may impede vision. Equine sarcoids seem to be a disease confined to young horses, since they rarely affect horses older than 7 years [18].

Despite their differential progression as a disease, bovine papillomatosis and equine sarcoids share a common factor;

both have the same oncogenic virus as an etiological agent: the well-known BPV [4]. While all BPVs are strictly species specific, BPV-1 and BPV-2 have been reported in cross-species infection, which allows infections to occur in a wide range of hosts between bovids and equids. Important financial considerations are involved in both cattle and horse breeding, which means that diagnostic tools are essential to define BPV-related epidemiological factors and therapeutic approaches. In view of this, a good deal of research is being carried out in the hope of making advances in early BPV detection so that new guidelines can be drawn up about how to deal with animal papillomatosis. In this chapter, we describe important aspects of BPV infection, including the most recent and accurate molecular techniques for the diagnosis of these diseases.

82.1.1 Classification

The *Papillomaviridae* family comprises 18 genera (*Alphapapillomavirus* to *Sigmapapillomavirus*), in which each genotype is distributed in accordance with its genomic DNA sequence similarity [19]. The standard method employed to characterize a new BPV type is by sequence determination of the highly conserved L1 region and by phylogenetic analysis [20]. Thirteen BPV types are known, and these have been classified into three genera: *Deltapapillomavirus* (BPV-1 and BPV-2), *Epsilonpapillomavirus* (BPV-5 and BPV-8), and *Xipapillomavirus* (BPV-3, BPV-4, BPV-6, BPV-9, BPV-10, BPV-11, and BPV-12). In the past few years, BPV-7 has been reported and still remains unclassified [21–23]. It has been suggested that BPV-7 may represent a member of a new *PV* genus. BPV-13 is one of the most recent types reported so far [24]. BPV types also cause some characteristic histopathological lesions, which allow a classification to be made. Xipapillomaviruses are strictly epitheliotropic, Deltapapillomaviruses induce fibropapillomas, and Epsilonpapillomaviruses appear to cause both fibropapillomas and true papillomas [22]. Members of the *Deltapapillomavirus* genus have been implicated in the development of equine sarcoids [23]. Until now, only three types of BPV (BPV-1, BPV-2, and BPV-4) have been associated with cancer. BPV-1 and BPV-2 cause urinary bladder tumors, while BPV-4 causes cancer in the upper alimentary tract.

82.1.2 Genome Organization

The BPV viral structure follows a general pattern shared by all virus-like particles. It can be described as a 52–55 nm diameter icosahedral protein capsid composed of two protein types: a major protein structured in 72 oligomeric complexes and 12 copies of minor transcapsid DNA-bound protein, L1 and L2, respectively [25]. While L1 protein plays an essential role in viral particle anchorage [26], L2 minor capsid protein is necessary for viral morphogenesis [25]. This small and nonenveloped capsid has a size of 8 kb, and is a circular, double-stranded DNA genome, which includes a long-control region (LCR) and two gene groups, early (E) and late (L) [27]. The LCR is composed of 500–1000 nucleotides; although it does not comprise genes, it contains transcriptional regulatory and replication

origin sites [28]. The E and L coding regions follow a time-dependent expression pattern during the virus cycle [29].

The E region in the PV genome usually contains six early genes, although the same genes may not be present in all the PVs and their sequences may not be conserved. The E1 protein has helicase activity and plays a role in the viral replication [30]. The E2 gene product is a transcriptional factor and is also involved in recognition and ligation to the replication origin. In addition, it has mitotic chromosome binding activity to ensure an equal distribution of the viral episomes among the daughter cells [31]. The E4 gene produces a small protein within the E2 gene but in a different reading frame. Without a clear function as described in the BPV infection, this small protein is found within profuse cytoplasm of keratinocytes during productive replication [32].

Two early proteins are necessary for the BPV-mediated carcinogenesis: E5 and E6 [3]. BPV E5 is a membrane-associated hydrophobic protein, which plays a role in disrupting cellular growth control. It has been shown that BPV-1 E5-expressing mouse fibroblast cell lines exhibited a transformed phenotype as well as nonimmortalized human foreskin fibroblasts [33–35]. In cultured cells and animals, the transforming activity of BPV-1 E5 can be assumed to be a consequence of its ability to bind and activate the cellular platelet-derived growth factor β receptor [36]. BPV E6 protein is known to have a large number of binding partners and activities in the virus life cycle [37–39]. In *Xipapillomavirus*, the E6 gene is replaced by an E5-like gene, which was initially defined as E8. However, when it was found that most of the functions of this protein are shared with BPV-1 E5, this prompted its redefinition as E5 [4]. Unlike oncogenic human papillomavirus (HPV) E7 protein, which has a great carcinogenic potential, BPV E7 protein just appears to be cooperating with E5- and E6-mediated cell transformation, since its expression only increases the transformation efficiency [40].

82.1.3 Transmission, Clinical Features, and Pathogenesis

The BPV group is responsible for a wide range of diseases, the clinical symptoms of which can be summarized as outcomes of diverse hyperproliferative lesions. However, each subgroup can be specifically associated with some particular lesions in its host, particularly, in bovines.

Deltapapillomaviruses, known as fibropapillomaviruses, cause fibropapillomas, which involve subepithelial fibroblasts, followed by epithelial acanthosis and papillomatosis, in a hyperproliferative process. Although these infected fibroblasts may harbor multiple episomal copies of the BPV genome (commonly, BPV-2), it is not usual to be able to detect a mature virus or its antigens from tumors in the upper alimentary tract [41]. Fibropapilloma persistence, which can be attributed to environmental cofactors, may also allow mesenchymal cancers to develop such as bladder cancer. BPV-1 has been found in teat and penile fibropapillomas, while BPV-2 is linked to cutaneous warts, alimentary fibropapillomas, and urinary bladder tumors; all these are members of the fibropapillomavirus group [42].

Epitheliotropic BPVs (Xipapillomaviruses) induce epithelial papillomas without fibroblast involvement. BPV-3 causes general cutaneous papillomas. BPV-4 has been implicated with epithelial papillomas, which are located in the upper gastrointestinal tract. BPV-6 is often isolated from multiple teat papillomas. BPV-9 and BPV-10 are associated with epithelial squamous papillomas of the udder. Epsilonpapillomaviruses are involved in tumor development and are of both mesenchymal and epithelial origin. BPV-5 induces rice grain fibropapillomas, which are usually found in the udder, while BPV-8 causes cutaneous papillomas. Furthermore, the detection of two or more BPV types in the same animal may be a frequent event that has consequences for diagnosis and treatment. Coinfections by BPV-1 and BPV-2 [43] or even detection of different genera of BPVs in the same animal have already been reported [44,45].

In a similar way to the expression patterns exhibited by several PVs, BPV replication during acute infection is linked to the differentiation status of the infected cell. Papilloma development is preceded by infection of basal keratinocytes, where a set of early genes is expressed in the undifferentiated basal and suprabasal layers. BPV replication has only been detected in keratinocytes undergoing terminal differentiation to squamous epithelium, as well as L gene expression, where it corresponds to structural proteins. In the differentiating spinous and granular layers, new virus particles are encapsidated and released into the environment when the cells die. In fact, a complete viral cycle seems to be strictly performed in the epithelial component of the BPV-associated tumors and only at certain stages of its development. It should be noted that even in BPV-containing fibroblasts, virus replication has never been found in these cells, and the viral genome remains in a nonintegrated episomal form [46]. In addition to environmental cofactors, such as immune suppression by bracken fern ingestion, malignant transformation is an outcome of deregulated expression and is especially linked to the upregulation of the early viral genes (E5 and E6 genes), which results in an uncontrolled proliferation and delay or a lack of cell differentiation [47].

In general, transmission is made possible by the exposition of the basal layers, where viral particles may have access to primary keratinocytes and fibroblasts. Viral-bearing fomites, like contaminated milking equipment, halters, nose leads, grooming and earmarking equipment, rubbing posts, and wire fences, may provide the means of transmission between animals [48]. Since warts are rare in animals that are artificially inseminated, it can be postulated that sexual transmission of papillomatosis can occur among cattle [49]. Moreover, arthropods like flies are probably an aggravating factor in bovine papillomatosis transmission, not only between bovines but also between horses [50,51].

82.1.4 Diagnosis

Traditional methods for virus identification include electron microscopy, cell culture, inoculation studies, and serology [52].

Although these techniques were responsible for the identification of many of the viruses known today, they have some obvious limitations, including time-consumingness and variability. On the other hand, molecular techniques provide a sensitive and rapid means of virus detection and identification [53]. PCR-based methods include conventional PCR, degenerated PCR, sequence-independent PCR, and rolling circle amplification.

82.1.4.1 Conventional Techniques

A commonly employed technique to diagnose whether a cutaneous lesion is induced by PVs is to conduct a histological analysis of the collected tissue. Papillomas induced by BPV in bovine have two classical patterns depending on the type of BPV involved. The true papillomas are composed of multiple papillary fronds of varying width, the surfaces of which are covered with the multiple-layered keratinized epithelium. There are also acanthosis and parakeratosis of the epithelium, and active viral replication takes place during development, resulting in inclusion bodies that are visible in the nuclei of the infected cells.

The BPV-induced fibropapilloma follows different recognized stages. Initially, the tumor develops as a fibroma, either alone or plus acanthosis. Histologically, an active proliferation of anaplastic fibroblasts with large nuclei gives rise to a subcutaneous mass stretching of the overlying epithelium. Subsequently, proliferation of basal and keratinocyte layers of the epithelium spreads downward toward the mesenchymal mass. In the epithelium, the granular cell layer becomes thicker, and clear central areas in the nuclei appear. Viral replication takes place in the nuclei and, once the cells develop into more differentiated layers, the viral particles spread from the cells among the keratinized plates. Finally, the tumor is infiltrated by macrophages and lymphocytes as the regression phenomenon takes place.

82.1.4.2 Molecular Techniques

Since the virus cycle of the PVs requires keratinocyte differentiation to express all the genes, there are no conventional cell lines that allow the growth of these viruses [54]. Moreover, the PVs cannot be transmitted to laboratory animals [55]. Hence, molecular methods have been widely used to discover and characterize the genome of the new PVs as well as to carry out a diagnosis of the PVs that are clinically significant [56].

The most commonly used molecular method to study PVs through their genome is the PCR. This methodology has been widely used to detect a broad range of PV types in human and animals. The PCR-based method has been routinely used on account of its high sensitivity (with the capacity to detect a few copies of PVs, DNA, and viral load samples) and specificity [57,58].

At present, there are several molecular assays that use PCR methods to investigate PVs. These methodologies essentially differ in the extent to which the PV gene segment is amplified and in the number of primers used. For clinical studies of PV prevalence, the most frequently used primer sets are MY9/MY11, FAP59/64, and PGMY9/ PGMY11. These consensus primer sets were designed to amplify a partial sequence of highly conserved PV L1 genes. In addition, these primers made use of degenerate nucleotides that increase the range of PV detection, although they may lower their sensitivity [59]. The MY9/MY11 primer set is the most common method used to detect HPV infection in cervical samples [60]. These primers amplify approximately 450 base pair fragments and allow amplification of 47 HPV DNA types. The PGMY9/PGMY11 primer set is an extended version of the MY9/11 primer set, which allows the sensitivity to be increased to detect HPV types [61]. Alternatively, the association of fixed nucleotide primers, such as GP5+/ GP6+, with degenerate nucleotide primers may be employed to increase their sensitivity to HPV.

Another PCR assay using FAP59/64 primers also amplifies a broad spectrum of HPV types. Similar to the MY09/11 primer set, the FAP59/64 primers have degenerate nucleotides and were designed to amplify a partial sequence of highly conserved PV L1 genes [62]. The FAP59/64 primers allow the detection of 46 HPV types. Apart from HPV, FAP59/64 and MY09/11 can be used to identify a broad range of animal PVs DNA. For instance, Ogawa et al. [63] showed that MY09/11 primers make it possible to detect BPV-1 and BPV-3. Moreover, the FAP59/64 primers can detect 13 BPV types [64,65] as well as putative new BPV types [63]. Furthermore, the use of FAP59/64 primers made it possible to identify PVs in chimpanzees, gorillas, macaques, monkeys, lemurs, cows, and elks [20].

More recently, real-time PCR has been used to detect PVs as well as viral load assessment. This method is based on the release of fluorescent markers at each amplification cycle, which is directly proportional to the amount of amplicon generated. The real-time PCR has been used to detect and quantify the viral load of human and animal PV types, such as BPV-1, BPV-2, BPV-12 [66,67].

BPV DNA is detected by a variety of PCR-based techniques. These techniques often involve the detection of one or two BPV types by using degenerated or type-specific primers. Genotyping is performed either by real-time detection [68] or by sequence analysis [69] or restriction fragment length polymorphism analysis [70] of the generated PCR fragments. The discovery of putative new types, subtypes, and variants are usually made in works that employ consensus primers capable of identifying potentially more than two BPV types [63]. Previous studies that involved consensus primers detected about 31 putative new BPV types [14,20,62,63,65,71]. Thus, it is believed that there may be an underestimation of the extent of the spread of BPV. Only 13 BPV types have been described so far, despite the great diversity found in HPV, and the use of consensus primers is a very important means of increasing our knowledge of BPV diversity. The use of consensus primers is also of great value in the investigation of novel PV types and shows that there is a need for this technique.

PCR that uses a type-specific primer is widely employed in research studies and allows BPV DNA to be amplified from different tissues and cells, such as epithelium, blood,

spermatozoa, and oocytes [64,72–74]. A comparison between these consensus primers and the BPV type-specific primers showed that the BPV type-specific primers are more sensitive than the consensus primers and are able to detect coinfection of BPV in the tissues [74]. Several epidemiological studies employed on the herd in recent years have revealed a large number of coinfections caused by BPV [65,75,76]. This shows the importance of using BPV type-specific primers to reveal the real diversity of BPV that infects the herd.

82.2 METHODS

82.2.1 SAMPLE COLLECTION AND PREPARATION

BPV DNA can be detected in samples of papillomas and the blood of symptomatic and asymptomatic cattle. Papillomas should be collected by parallel incision from different areas of the skin surface of the animal, using a disposable sterile scalpel.

In general, the chosen papillomas should be neither too early nor too chronic. During the procedure, it is important to use local anesthesia subcutaneously with 1–2 mL of 2% lidocaine and restraint. The incision should be made after 5 min of local anesthesia. One fragment of the collected papilloma must be separated and fixed in 10% formalin for later histological analysis, and the another fragment must be cooled for molecular analysis.

The blood should be collected by jugular venipuncture using EDTA-containing tubes. About 200 µL of total blood is used for the DNA extraction.

Several methods have been used for the extraction of BPV DNA from clinical samples prior to PCR amplification. In addition, commercial DNA extraction kits introduced by various companies are widely used for BPV DNA isolation. The DNeasy® Blood & Tissue from Qiagen, Germany, allows the DNA from papillomas and cattle blood to be isolated.

Using this DNA isolation kit, the DNA is extracted from small fragments of papillomas and blood. Briefly, 25 mg of cut papillomas is mixed with buffer ATL and proteinase K and incubated at 56°C, until the tissue is completely lysed. It is important to note that the papillomas are keratinized tissue samples, so vortex should be used during the process. After this, the buffer AL is added to the sample and then the ethanol (96%–100%) is mixed with the sample. The resulting mixture is pipetted to the column and centrifuged for 1 min. The flow-through is discarded and two buffers (AW1 and AW2) are added to the column and centrifuged. Finally, the buffer AE is introduced onto the column and incubated for 1 min. To maximize the DNA yield, the buffer AE may be warmed before use on the column.

In isolating DNA from blood, the first step is to add proteinase K to 100 µL of blood plus 100 µL of PBS. Following this, the buffer AL is added to the mixture, and the sample is incubated at 56°C for 10 min. After this, the ethanol is added to the sample, and the mixture is placed in the column. The final stages are the same as with the previous protocol described for the papilloma sample and involve the addition of buffer AW1 and AW2 and, in the final stage, the addition of buffer AE.

82.2.2 Detection Procedures

Recent molecular studies have been largely based on the amplification of a fragment of L1 BPV ORF with some exceptions (e.g., BPV-4 type-specific primer).

82.2.2.1 PCR Using Consensus Primers

The consensus primers FAP59/64 (Fw 5′-TAACWGTIGG ICAYCCWTATT-3′; Rev 5′-CCWATATCWVHCCATITC ICCATC-3′) and MY 11/09 (Fw 5′-GCM CAG GGW CAT AAY AAT GG-3′; Rev 5′-CGT CCM ARR GGA WAC TGA TC-3′) are very suitable for use in studies that are designed to detect a broad range of BPV types; however, it is necessary to perform the sequencing assay after the PCR technique to distinguish between the BPV types. These primers amplify a conserved fragment of L1 ORF. The conditions of the PCR when combined with these primers are described in the following.

82.2.2.1.1 Procedure

1. Prepare the PCR mix as follows: add 25 units/mL of Taq DNA polymerase, 200 μM dATP, 200 μM dGTP, 200 μM dCTP, 200 μM dTTP, 1.5 mM MgCl$_2$, and 0.25 μM of each oligonucleotide primer (FAP59/64 or MY09/11).
2. Homogenize by pipetting the mixture and then add 23 μL of the mixture to each PCR tube.
3. Add 2 μL of DNA samples at a concentration of 50 ng/μL and pipette the mixture until it is homogenized.
4. Add 2 μL of ultrapure water in a PCR tube to perform the nontemplate control (NTC).
5. Take the samples to a thermocycler to amplify the DNA fragments in accordance with the conditions outlined in Table 82.1.
6. Visualize the amplification products by using 1.5% Tris-acetate buffer (TAE) agarose gel electrophoresis and subsequent ethidium bromide staining.

82.2.2.2 PCR Using Type-Specific Primers

The BPV type-specific primers are very suitable for epidemiological studies because of their sensitivity. However, several rounds of amplifications are necessary to obtain a screening of the major BPV types from a set of samples.

In the following, there is a description of the conditions and the primers required to amplify the 12 first BPV described in the literature (with the exception of BPV-7). Some of them can be used with the same thermal profile as the thermocycler, although since the amplicon has a similar fragment size, they cannot be used as multiplex PCR.

82.2.2.2.1 Procedure

1. Prepare the PCR mix as follows: add 25 units/mL of Taq DNA polymerase, 200 μM dATP, 200 μM dGTP, 200 μM dCTP, 200 μM dTTP, 1.5 mM MgCl$_2$, and 0.25 μM of each oligonucleotide primer. Information about the primers is given in Table 82.2.
2. Homogenize the mixture by pipetting and then add 23 μL to each PCR tube.
3. Add 2 μL of DNA samples at a concentration of 50 ng/μL and pipette the mixture to homogenize.
4. Add 2 μL of ultrapure water in a PCR tube to perform the NTC.
5. Take samples to a thermocycler to amplify the DNA fragments in accordance with the conditions outlined in Table 82.3.
6. Visualize the amplification products by using 2% TAE agarose gel electrophoresis and subsequent ethidium bromide staining.

82.3 CONCLUSION AND FUTURE PERSPECTIVES

Although diseases caused by BPV have been known for a long time, new discoveries about BPV infection such as the classification of new variants, coinfection by multiple genotypes, and the spread of viruses through nonepithelial sites and heterologous infection have given them an increasing prominence. The lack of understanding of BPV biology suggests further need to study this subject and to employ control strategies against papillomaviruses. Hence, a knowledge of BPV diversity and epidemiology is of crucial importance to establish prevention strategies and obtain an understanding of the evolution of this group of viruses. The molecular strategies deployed in recent years have been tremendously helpful for farmers and veterinarians. These strategies have increased our knowledge of BPV biology and epidemiology to a great degree.

BPV infection in its natural host has become an important concern as the disease reduces the value of livestock in financial terms, impairs the quality of the leather, leads to a general decline in animal health, and, in severe cases, causes the death of the animal. Furthermore, the presence of this disease on a farm imposes a threat to neighboring farms, and thus, urgent action is required by the farmers involved or public agencies. In the case of bovine papillomatosis in particular, there is a lack of accurate information about the serious consequences of the disease, and thus, interventionist measures are needed to control its spread and mitigate its effects. Currently, there is a range of therapies with limited

TABLE 82.1
PCR Thermal Profile Using Consensus Primers

Primer	Denaturation	Annealing	Extension	Number of Cycles
FAP59/64	94°C for 1 min	50°C for 1 min	72°C for 1 min	45
MY09/11	95°C for 30 s	48°C for 30 s	72°C for 1 min	35

TABLE 82.2

Polymerase Chain Reaction Primers for Bovine Papillomavirus Detection

Primers	Sequence	Fragment Size (bp)	Region Amplified
BPV-1	BPV-1 F-5′ GGA GCG CCT GCT AAC TAT AGG 3′ BPV-1 R-5′ ATC TGT TGT TTG GGT GGT GAC 3′	301	L1 Gene (5721–6021nt, GeneBank X02346)
BPV-2	BPV-2 F-5′ GTT ATA CCA CCC AAA GAA GAC CCT 3′ BPV-2 R-5′ CTG GTT GCA ACA GCT CTC TTT CTC 3′	164	L1 Gene (6888–7051nt, GeneBank M20219)
BPV-3	BPV-3 F-5′ CAG TCA ATT GCA ACT AGA TGC C 3′ BPV-3 R-5′ GGC TGC TAC TTT CAA AAG TGA 3′	216	L1 Gene (6620–6836nt, GeneBank AF486184)
BPV-4	BPV-4 F-5′ GCT GAC CTT CCA GTC TTA AT 3′ BPV-4 R-5′ CAG TTT CAA TCT CCT CTT CA 3′	170	E7 Gene (642–812nt, GeneBank X05817)
BPV-5	BPV-5 F-5′ GGC ATG TAG AGG AAT ATA AGC 3′ BPV-5 R-5′ TTC TCT GAG ATC AAT ATT CC 3'	262	L1Gene (6646–6908nt, GeneBank AF457465)
BPV-6	BPV-6 F-5′ TTA GAG ACC TGG AAC TTG GG 3′ BPV-6 R-5′ TAC GCT TTG GCG CTT TTT TGC 3′	294	L1 Gene (6829–7123nt GeneBank AJ620208)
BPV-8	BPV-8 F-5′ TAGAGGACACATACCGCTTCCAAAGC 3′ BPV-8 R-5′ TTTGCGAGCACTGCAGGTGATCCC 3′	196	L1Gene (6823–7019nt, GenBank DQ098913)
BPV-9	BPV-9 F-5′ AAAGAGCAAATCGGGAGCACC 3′ BPV-9 R-5′ AACTAATGACCCACTAGGGCTCC 3′	264	L1Gene (6275–6539nt, GenBank AB331650)
BPV-10	BPV-10 F-5′ AAGGCATTTGTGGTCTCGAGG 3′ BPV-10 R-5′ CTAAAGAACCACTTGGAGTGCC 3′	148	L1 Gene (6237–6385nt, GenBank AB331651)
BPV-11	BPV-11 F-5′ TGCAGACACTCAACCAGGAG 3′ BPV-11 R-5′ CCATAAGGGTCGTTGCTCAT 3′	197	L1 Gene (6110–6307nt, GeneBank AB 543507.1)
BPV-12	BPV-12 F-5′ AAAGCTGAACCATGCAAACC 3′ BPV-12 R-5′ TAACAATGTCAAGGGGCACA 3′	159	L1 Gene (5909–6068nt, GeneBank JF 834523.1)

TABLE 82.3

Polymerase Chain Reaction Thermal Profiles Using Various Bovine Papillomavirus Primers

Primer	Denaturation	Annealing	Extension	Number of Cycles
BPV-1	94°C for 40 s	68°C for 40 s	72°C for 1 min	35
BPV-2	94°C for 40 s	55°C for 40 s	72°C for 1 min	35
BPV-3	94°C for 40 s	60°C for 40 s	72°C for 1 min	35
BPV-4	94°C for 40 s	60°C for 40 s	72°C for 1 min	35
BPV-5	94°C for 40 s	60°C for 40 s	72°C for 1 min	35
BPV-6	94°C for 40 s	65°C for 40 s	72°C for 1 min	35
BPV-8	94°C for 40 s	60°C for 40 s	72°C for 1 min	35
BPV-9	94°C for 40 s	60°C for 40 s	72°C for 1 min	35
BPV-10	94°C for 40 s	60°C for 40 s	72°C for 1 min	35
BPV-11	94°C for 40 s	60°C for 40 s	72°C for 1 min	35
BPV-12	94°C for 40 s	54°C for 40 s	72°C for 1 min	35

The ultimate goal should be to develop an effective and affordable vaccine approach, in contrast with the current expensive and relatively ineffective strategies. Thus, there is a good deal of scope in the BPV field to identify and implement improved control, intervention, and treatment strategies, together with the development of highly protective vaccines.

REFERENCES

1. Sundberg JP. Papillomavirus infections in animals. In *Papillomaviruses and Human Disease*, Syrjanen K, Gissmann L, and Koss LG. (eds). Berlin, Germany: Springer Verlag, pp. 40–103 (1987).
2. Correa WM and Corrêa CNM. *Enfermidades infecciosas dos mamíferos domésticos*. Rio de Janeiro, Brazil: Medsi, pp. 23–24 (1992).
3. Nasir L and Campo MS. Bovine papillomaviruses: their role in the aetiology of cutaneous tumours of bovids and equids. *Vet Dermatol.* 19, 243–254 (2008).
4. Campo MS. Bovine papillomavirus: Old system, new lessons? In *Papillomavirus Research from Natural History to Vaccines and Beyond*, Campo MS (ed.). Norfolk, VA: Caister Academic Press, pp. 373–383 (2006).
5. Kumar P et al. Detection of bovine papilloma viruses in wartlike lesions of upper gastrointestinal tract of cattle and buffaloes. *Transbound Emerg Dis.* 62, 264–271 (2015).
6. Resendes AR et al. Association of bovine papillomavirus type 2 (BPV-2) and urinary bladder tumours in cattle from the Azores archipelago. *Res Vet Sci.* 90, 526–529 (2011).
7. Antonsson A et al. Prevalence and type spectrum of human papillomaviruses in healthy skin samples collected in three continents. *J Gen Virol.* 84, 1881–1886 (2003).

efficacy for the treatment of BPV-infected animals. The surgical removal of the lesions is often recommended when the animal shows signs of having papillomas, or alternatively chemical treatment may be used. There has been a great demand for an effective product for the treatment of papillomatosis, due to the difficulties encountered in its control and the increasing incidence of the disease observed in recent years. In the case of equine sarcoid, besides being a locally aggressive tumor due to its infiltrative capacity, it is resistant to different forms of therapy.

8. Jackson C. The mixed tumours. *Onderstepoort J Vet Sci Anim Ind.* 6, 345–385 (1936).

9. Scott DW and Miller WH. Neoplastic and non-neoplastic tumors. In *Equine Dermatology*. Scott DW and Miller WH (eds.), St. Louis, MO: Saunders, pp. 719–731 (2003).

10. Knottenbelt DC. A suggested clinical classification for the equine sarcoid. *Clin Tech Eq Prac.* 4, 278–295 (2005).

11. Reid SW et al. Epidemiological observations on sarcoids in a population of donkeys (*Equus asinus*). *Vet Rec.* 134, 207–211 (1994).

12. Martens A and DeMoor A. Equine sarcoid. 1. Clinical types, prevalence, epidemiology, aetiology and pathogenesis. *Vlaams Diergeneeskundig Tijdschrift.* 65, 10–17 (1996).

13. Chambers G et al. Association of bovine papillomavirus with the equine sarcoid. *Can J Vet Res.* 71, 28–33 (2007).

14. Lohr CV et al. Sarcoids in captive zebras (*Equus burchellii*): Association with bovine papillomavirus type 1 infection. *J Zoo Wildl Med.* 36, 74–81 (2005).

15. Kidney BA and Berrocal A. Sarcoids in two captive tapirs (*Tapirus bairdii*): Clinical, pathological and molecular study. *Vet Dermatol.* 19, 380–384 (2008).

16. Broström H. Equine sarcoids. A clinical and epidemiological study in relation to equine leucocyte antigens (ELA). *Acta Vet Scand.* 36, 223–236 (1995).

17. Postey RC, Appleyard GD, and Kidney BA. Evaluation of equine papillomas, aural plaques, and sarcoids for the presence of Equine papillomavirus DNA and Papillomavirus antigen. *Can J Vet Res.* 71, 28–33 (2007).

18. Foy JM, Rashmir-Raven AM, and Brashier MK. Common equine skin tumors. *Compend Contin Ed Pract Vet.* 24, 242–254 (2002).

19. De Villiers EM et al. Classification of papillomaviruses. *Virology.* 324, 17–27 (2004).

20. Antonsson A and Hansson BG. Healthy skin of many animal species harbors papillomaviruses which are closely related to their human counterparts. *J Virol.* 76, 12537–12542 (2002).

21. Bernard HU et al. Classification of papillomaviruses (pvs) based on 189 pv types and proposal of taxonomic amendments. *Virology.* 401, 70–79 (2010).

22. Hatama S et al. Detection of a novel bovine papillomavirus type 11 (bpv-11) using xipapillomavirus consensus polymerase chain reaction primers. *Arch Virol.* 156, 1281–1285 (2011).

23. Zhu W et al. Characterization of novel bovine papillomavirus type 12 (BPV-12) causing epithelial papilloma. *Arch Virol.* 157, 85–91 (2011).

24. Lunardi M et al. Genetic characterization of a novel bovine papillomavirus member of the Deltapapillomavirus genus. *Vet Microbiol.* 162, 207–213 (2013).

25. Modis Y, Trus BL, and Harrison SC. Atomic model of the papillomavirus capsid. *EMBO J.* 21, 4754–4762 (2002).

26. Joyce JG et al. The L1 major capsid protein of human papillomavirus type 11 recombinant virus-like particles interacts with heparin and cell-surface glycosaminoglycans on human keratinocytes. *J Biol Chem.* 274, 5810–5822 (1999).

27. Doorbar J et al. The biology and life-cycle of human papillomaviruses. *Vaccine.* 20, 30 (2012).

28. Munger K and Howley PM. Human papillomavirus immortalization and transformation functions. *Virus Res.* 89, 213–228 (2002).

29. Johansson C and Schwartz S. Regulation of human papillomavirus gene expression by splicing and polyadenylation. *Nat Rev Microbiol.* 11, 239–251 (2013).

30. Lambert PF. Papillomavirus DNA replication. *J Virol.* 65, 3417–3420 (1991).

31. Baxter MK et al. The mitotic chromosome binding activity of the papillomavirus E2 protein correlates with interaction with the cellular chromosomal protein Brd4. *J Virol.* 79, 4806–4818 (2005).

32. Anderson RA et al. Viral proteins of bovine papillomavirus type 4 during the development of alimentary canal tumours. *Vet J.* 154, 69–78 (1997).

33. Bergman P et al. The E5 gene of bovine papillomavirus type 1 is sufficient for complete oncogenic transformation of mouse fibroblasts. *Oncogene.* 2, 453–459 (1998).

34. DiMaio D, Guralski D, and Schiller JT. Translation of open reading frame E5 of bovine papillomavirus is required for its transforming activity. *Proc Natl Acad Sci USA.* 83, 1797–1801 (1986).

35. Petti LM and Ray FA. Transformation of mortal human fibroblasts and activation of a growth inhibitory pathway by the bovine papillomavirus E5 oncoprotein. *Cell Growth Differ.* 11, 395–408 (2000).

36. Lai CC, Henningson C, and DiMaio D. Bovine papillomavirus E5 protein induces the formation of signal transduction complexes containing dimeric activated platelet-derived growth factor b receptor and associated signaling proteins. *J Biol Chem.* 275, 9832–9840 (2000).

37. Tong X et al. Interaction of the bovine papillomavirus E6 protein with the clathrin adaptor complex AP-1. *J Virol.* 72, 476–482 (1998).

38. Tong X and Howley PM. The bovine papillomavirus E6 oncoprotein interacts with paxillin and disrupts the actin cytoskeleton. *Proc Natl Acad Sci USA.* 94, 4412–4417 (1997).

39. Zimmermann H et al. Interaction with CBP/p300 enables the bovine papillomavirus type 1 E6 oncoprotein to downregulate CBP/p300-mediated transactivation by p53. *J Gen Virol.* 81, 2617–2623 (2000).

40. Bohl J, Hull B, and Vande Pol SB. Cooperative transformation and co-expression of bovine papillomavirus type 1 E5 and E7 proteins. *J Virol.* 75, 513–521 (2001).

41. Jarret WF. Alimentary fibropapilloma in cattle: A spontaneous tumor, nonpermissive for papillomavirus replication. *J Natl Cancer Inst.* 73, 499–504 (1984).

42. Pfister H et al. Partial characterization of a new type of bovine papilloma viruses. *Virology.* 96, 1–8 (1979).

43. Santos EU et al. Detection of different bovine papillomavirus types and co-infection in bloodstream of cattle. *Transbound Emerg Dis.* (2014). doi: 10.1111/tbed.12237.

44. Carvalho CC et al. Detection of bovine papillomavirus types, co-infection and a putative new BPV11 subtype in cattle. *Transbound Emerg Dis.* 59, 441–447 (2012).

45. Claus MP et al. A bovine teat papilloma specimen harboring Deltapapillomavirus (BPV-1) and Xipapillomavirus (BPV-6) representatives. *Braz Arch Biol Technol.* 52, 87–91 (2009).

46. Jarrett WFH. The natural history of bovine papillomavirus infections. In *Advances in Viral Oncology*. Klein G (ed.), New York: Raven Press, pp. 83–101 (1985).

47. Campo MS. Vaccination against papillomavirus in cattle. *Clin Dermatol.* 15, 275–283 (1997).

48. Hama C, Matsumoto T, and Franceschini PH. Papilomatose bovina: Avaliação clínica de diferentes produtos utilizados no controle e tratamento. *Cien Vet.* 22, 14 (1988).

49. Carvalho C et al. Bovine papillomavirus type 2 in reproductive tract and gametes of slaughtered bovine females. *Braz J Microbiol.* 34, 82–84 (2003).

50. Finlay M et al. The detection of Bovine Papillomavirus type 1 DNA in flies. *Virus Res.* 10, 4–15 (2009).

51. Brandt S et al. BPV-1 infection is not confined to the dermis but also involves the epidermis of equine sarcoids. *Vet Microbiol.* 150, 35–40 (2011).

52. Bexfield N and Kellam P. Metagenomics and the molecular identification of novel viruses. *Vet J Lond Engl.* 190, 191–198 (2011).

53. Belák S, Thorén P, LeBlanc N, and Viljoen G. Advances in viral disease diagnostic and molecular epidemiological technologies. *Expert Rev Mol Diagn.* 9, 367–381 (2009).

54. Geimanen J et al. Development of a cellular assay system to study the genome replication of high- and low-risk mucosal and cutaneous human papillomaviruses. *J Virol.* 1, 3315–3329 (2011).

55. De Villiers E-M. Cross-roads in the classification of papillomaviruses. *Virology.* 445, 2–10 (2013).

56. Maeda Y et al. An outbreak of teat papillomatosis in cattle caused by bovine papillomavirus (BPV) type 6 and unclassified BPVs. *Vet Microbiol.* 121, 242–248 (2007).

57. Gravitt PE et al. Reproducibility of HPV 16 and HPV 18 viral load quantitation using TaqMan real-time PCR assays. *J Virol Methods.* 112, 23–33 (2003).

58. Hesselink AT et al. Clinical validation of the Abbott RealTime High Risk (HR) HPV assay according to the guidelines for HPV DNA test requirements for cervical screening. *J Clin Microbiol.* 51, 2409–2410 (2013).

59. Qu W et al. PCR detection of human papillomavirus: Comparison between MY09/MY11 and GP5+/GP6+ primer systems. *J Clin Microbiol.* 35, 1304–1310 (1997).

60. Resnick RM et al. Detection and typing of human papillomavirus in archival cervical cancer specimens by DNA amplification with consensus primers. *J Natl Cancer Inst.* 82, 1477–1484 (1990).

61. Gravitt PE et al. Improved amplification of genital human papillomaviruses. *J Clin Microbiol.* 38, 357–361 (2000).

62. Forslund O. Genetic diversity of cutaneous human papillomaviruses. *J Gen Virol.* 88, 2662–2669 (2007).

63. Ogawa T et al. Broad-spectrum detection of papillomaviruses in bovine teat papillomas and healthy teat skin. *J Gen Virol.* 85, 2191–2197 (2004).

64. Silva MAR, Pontes NE, Da Silva KMG, Guerra MMP, and Freitas AC. Detection of Bovine Papillomavirus type 2 DNA in commercial frozen semen of bulls (*Bos taurus*). *Anim Reprod Sci.* 129, 146–151 (2011).

65. Carvalho CCR, Batista MVA, Silva MAR, Balbino VQ, and Freitas AC. Detection of bovine papillomavirus types, co-infection and a putative new BPV11 subtype in cattle. *Transbound Emerg Dis.* 59, 441–447 (2012).

66. Bogaert L, Martens A, Kast WM, Van Marck E, and De Cock H. Bovine papillomavirus DNA can be detected in keratinocytes of equine sarcoid tumors. *Vet Microbiol.* 146, 269–275 (2010).

67. Pathania S, Dhama K, Saikumar G, Shahi S, and Somvanshi R. Detection and quantification of bovine papillomavirus type 2 (BPV-2) by real-time PCR in urine and urinary bladder lesions in enzootic bovine haematuria (EBH)-affected cows. *Transbound Emerg Dis.* 59, 79–84 (2012).

68. Rai GK, Saxena M, Singh V, Somvanshi R, and Sharma B. Identification of bovine papillomavirus 10 in teat warts of cattle by DNase-SISPA. *Vet Microbiol.* 147, 416–419 (2011).

69. Brandt S, Haralambus R, Schoster A, Kirnbauer R, and Stanek C. Peripheral blood mononuclear cells represent a reservoir of Bovine Papillomavirus DNA in sarcoid-affected equines. *J Gen Virol.* 89, 1390–1395 (2008).

70. Carr EA, Théon AP, Madewell BR, Griffey SM, and Hitchcock ME. Bovine papillomavirus DNA in neoplastic and nonneoplastic tissues obtained from horses with and without sarcoids in the western United States. *Am J Vet Res.* 62, 741–744 (2001).

71. Claus MP, Lunardi M, Alfieri AF, Ferracin LM, Fungaro MHP, and Alfieri AA. Identification of unreported putative new Bovine Papillomavirus types in Brazilian cattle herds. *Vet Microbiol.* 132, 396–401 (2008).

72. Lindsey CL et al. Bovine papillomavirus DNA in milk, blood, urine, semen, and spermatozoa of bovine papillomavirus-infected animals. *Genet Mol Res.* 8, 310–318 (2009).

73. Roperto S et al. Peripheral blood mononuclear cells are additional sites of productive infection of bovine papillomavirus type 2. *J Gen Virol.* 92, 1787–1794 (2011).

74. Silva MAR et al. Comparison of two PCR strategies for the detection of Bovine Papillomavirus. *J Virol Methods.* 192, 55–58 (2013).

75. Schmitt M, Fiedler V, and Muller M. Prevalence of BPV genotypes in a German cowshed determined by a novel multiplex BPV genotyping assay. *J Virol Methods.* 170, 67–72 (2010).

76. Batista MVA et al. Molecular epidemiology of bovine papillomatosis and the identification of a putative new virus type in Brazilian cattle. *Vet J.* 197, 368–373 (2013).

83 Poultry Adenoviruses

Győző L. Kaján

CONTENTS

83.1 Introduction 735
   83.1.1 Classification.. 736
   83.1.2 Morphology and Genome Organization.. 736
   83.1.3 Epidemiology and Clinical Features 736
      83.1.3.1 Aviadenoviruses.. 738
      83.1.3.2 Atadenoviruses 739
      83.1.3.3 Siadenoviruses 739
   83.1.4 Diagnosis 739
      83.1.4.1 Conventional Techniques.. 739
      83.1.4.2 Molecular Techniques... 740
83.2 Methods 740
   83.2.1 Sample Preparation.. 740
      83.2.1.1 Materials Needed 741
      83.2.1.2 Procedure 741
   83.2.2 Detection Procedures..741
      83.2.2.1 Materials Needed 741
      83.2.2.2 Procedure 741
   83.2.3 Phylogenetic Analysis.. 742
83.3 Conclusion and Future Perspectives 742
References.. 743

83.1 INTRODUCTION

Adenoviruses are double-stranded DNA viruses with a nonenveloped icosahedral capsid. With a broad host range, adenoviruses have been isolated from members of every class of vertebrates, ranging from cartilaginous fish, birds, to humans [1–3].

The first avian adenovirus was isolated by accident at the end of the 1940s [4], and a clinical isolation was obtained from a quail bronchitis around the same time [5]. Human adenoviruses were isolated in the 1950s, at an attempt to produce tissue cultures from excised human adenoids. With some cells undergoing a spontaneous degeneration, the causing agent was named "adenoid degeneration agent," abbreviated as "A.D. agent" [2,3].

Adenovirus infection is rather common among birds, including both poultry [6–8] and wild species [9,10]. In most cases, the route of infection is fecal–oral or vertical through the egg. The infection is regularly subclinical; a coinfection or other immunosuppressive effect—for example, laying eggs—is needed for the clinical manifestation of the disease [11,12]. A chronic infection is established in the birds, which results in a continuous virus shedding and the vertical transmission of the infection [13]. But some types among poultry-infecting adenoviruses are primary pathogens, for example, the turkey hemorrhagic enteritis virus [14].

Adenoviruses usually exhibit strong host specificity; the natural host of one specific type is usually only one species or a group of closely related species. *In vitro*, most avian adenoviruses replicate in primary cell cultures originating from the embryonic liver or kidney of the host species. The reproduction of the virus takes place in the nucleus, and intranuclear inclusion bodies are very common both in the propagated cells and during the histopathological examination of animals [15].

Avian adenoviruses were traditionally diagnosed and classified using serological methods. As the number of different serotypes rose, serological approach that relies on the use of reference strains and sera has become increasingly cumbersome and impractical. Molecular methods circumvent this problem by comparing the samples to entries stored in publicly available databases, such as GenBank. Diverse molecular methods have emerged in the recent years. In this chapter a short summary of available methods is given with more details on a generally applicable one.

83.1.1 Classification

The family *Adenoviridae* currently consists of five established genera: while the genera *Mastadenovirus* and *Aviadenovirus* have been recognized long ago [16], the genera *Atadenovirus*, *Siadenovirus* [17], and *Ichtadenovirus* [1] represent more recent additions. A sixth genus is being proposed: *Testadenovirus* [18].

Initially, all bird-infecting adenoviruses were classified into the genus *Aviadenovirus* simply because of the host of the virus. However, it was observed that some virus types differ fundamentally (e.g., antigenically) from a majority of the genus members. This led to the establishment of three distinct groups within the genus: group I, II, and III avian adenoviruses [15]. Phylogenetic analyses accomplished later among members of the family proved that the only member of group II (turkey hemorrhagic enteritis virus) can be classified as a *Siadenovirus*, and the sole member of group III (egg drop syndrome virus) is an *Atadenovirus* [19–21]. Thus, birds can be infected by members of three different genera of the family *Adenoviridae*: *Aviadenovirus*, *Atadenovirus*, and *Siadenovirus*. The current phylogenetic classification, official species and type names, recommended abbreviations, and host species of bird-infecting adenoviruses are summarized in Table 83.1.

The species demarcation criteria are officially set for every genus in the family *Adenoviridae* by the International Committee on Taxonomy of Viruses [1]. The most important and up-to-date criterion for all genera is the phylogenetic distance based on the DNA polymerase gene (>5%–15%), but for some genera other features (e.g., the host range, genome organization, or restriction fragment length polymorphism analysis of the genome) can be used as well.

Unfortunately there are no generally accepted type demarcation criteria for adenoviruses. Although the serotype definition is still valid [1], the neutralization test [22–24] is rarely used nowadays. Novel avian adenovirus isolates are increasingly studied using molecular methods instead, and then classified into the existing serotypes [6,7,25–28]. Thus, the recommended term for strains classified using such methods is "type" rather than "serotype." Marek et al. [29] tried to redefine the classification and typing system of fowl adenoviruses (FAdVs) using only molecular methods, and the usage of "genotype" was proposed. The same topic caused a long and heated debate among human adenovirologists as well [30–33], and a general solution to this problem that can be applied hopefully to all genera of the virus family is really anticipated.

83.1.2 Morphology and Genome Organization

Adenoviruses are medium-sized (70–90 nm in diameter), nonenveloped viruses with an icosahedral capsid. The major capsid proteins are the hexon (numbered as II after protein electrophoreses), the penton base (III), and the fiber proteins (IV). The 20 facets of the capsid are assembled from 240 homotrimers of the hexon protein. The 12 vertex capsomeres are constituted from the basal homopentomer of the penton base proteins and the protruding, elongated homotrimer of the fiber proteins [34]. Most adenoviruses possess one piece of fiber per vertex, but aviadenoviruses possess two [35]. These can be a duplicate of the same fiber or two different ones [36].

The minor capsid proteins or cement proteins are either stabilizing the capsid (IIIa, VI, and VIII) or found within the capsid. Most of the latter ones are bound to the DNA (VII, µ, IVa2, and the terminal protein); the only nonassociated protein is the viral protease [37].

The genome of adenoviruses is a nonsegmented, linear, double-stranded DNA genome coding on both strands. The length of the genome varies between 26 and 45 kb; siadenoviruses tend to have the shortest genomes (~26 kb) among adenoviruses [38] and aviadenoviruses (~43 to 45 kb) the longest [39].

In every adenovirus, the genome ends code for the inverted terminal repeats, which are the longest among aviadenoviruses (duck adenovirus 2: 721 bp) [36]. The central region of the genome contains a gene cassette, conserved in every adenovirus, and codes for basic proteins needed for DNA replication, DNA encapsidation, formation, and structure of the virion. The flanking regions vary in size among the different genera and contain genus-, species-, or type-specific genes [40,41].

83.1.3 Epidemiology and Clinical Features

Adenovirus infections are commonly present in both domestic and wild avian species. As most of the adenovirus strains are opportunistic pathogens, they do not necessarily cause overt disease in the infected birds. Rather, in most cases, secondary immunosuppressive factors are responsible for the onset of an outbreak. The severity of the disease can be affected by the exact strain of the virus; different strains classified even into the same adenovirus type can differ in virulence. Other factors have influence as well: the age of the bird or the infective titer of the virus. Perhaps this can explain the often equivocal results of animal experiments [11].

Avian adenovirus infections can spread both vertically and horizontally. Vertical transmission occurs through the embryonated egg, while horizontal transmission occurs through the fecal–oral route involving the feces or other excretions. The birds often develop a persistent, chronic infection after recovering and can be virus carriers for their whole lives. The virus shedding reoccurs if any immunosuppressive stress effect takes place, for example, egg laying. This ensures maximum transmission of the virus again [11–14].

Avian adenoviruses are relatively resistant to chemical or physical treatment. As they do not possess a viral envelope, they are resistant not only to lipid solvents such as ether and chloroform, but also to variations in pH between 3 and 9. Some strains of aviadenoviruses can withstand 60°C or even 70°C for 30 min, but their stability is reduced in the presence of divalent cations, and they can be inactivated by a 0.1% formaldehyde solution [11]. The infectivity of the egg drop syndrome virus can be decreased by 50% if it is heated for

TABLE 83.1

Current Phylogenetic Classification of Bird-Infecting Adenoviruses

Genus	Species	Type	Reference Strain	Host Species	GenBank Accession No.	Publication
Aviadenovirus	Falcon aviadenovirus A (FaAdV-A)	Falcon adenovirus 1 (FaAdV-1)	N/A	Different falcon species	AY683541	[111,112]
	Fowl aviadenovirus A (FAdV-A)	Fowl adenovirus 1 (FAdV-1)	Phelps=CELO	Chicken, quail	AC_000014	[113]
	Fowl aviadenovirus B (FAdV-B)	Fowl adenovirus 5 (FAdV-5)	340	Chicken	KC493646	[114]
	Fowl aviadenovirus C (FAdV-C)	Fowl adenovirus 4 (FAdV-4)	KR5	Chicken	GU188428	[115]
		Fowl adenovirus 10 (FAdV-10)	C-2B	Chicken	L08450 HM854003	[91,116]
	Fowl aviadenovirus D (FAdV-D)	Fowl adenovirus 2 (FAdV-2)	GAL1	Chicken	DQ208708 HM853995 KC750793 EF458160	[6,91,117,118]
		Fowl adenovirus 3 (FAdV-3)	SR49	Chicken	HM853996 KC750797	[6,91]
		Fowl adenovirus 9 (FAdV-9)	A2	Chicken	AC_000013	[119]
		Fowl adenovirus 11 (FAdV-11)	380	Chicken	HM854004 AF339925	[87,91]
	Fowl aviadenovirus E (FAdV-E)	Fowl adenovirus 6 (FAdV-6)	CR119	Chicken	HM853999 AF508954	[87,91]
		Fowl adenovirus 7 (FAdV-7)	X11	Chicken	HM854000 AF508955	[87,91]
		Fowl adenovirus 8a (FAdV-8a)	TR59	Chicken	HM854001 KC750801 AF155911	[6,91]
		Fowl adenovirus 8b (FAdV-8b)	CFA3	Chicken	GU734104	[120]
	Goose aviadenovirus A (GoAdV-A)	Goose adenovirus 1–5 (GoAdV-1–5)	4: P29 5: D1036/08	Goose	4: JF510462	[65,78]
	Turkey aviadenovirus B (TAdV-B)	Turkey adenovirus 1 (TAdV-1)	D90/2	Turkey	GU936707	[58]
	Turkey aviadenovirus C (TAdV-C)	Turkey adenovirus 4 (TAdV-4)	TNI1	Turkey	KF477312	[8]
	Turkey aviadenovirus D (TAdV-D)	Turkey adenovirus 5 (TAdV-5)	1277BT	Turkey	KF477313	[8]
	Unassigned	Turkey adenovirus 2 (TAdV-2)	TAV-2	Turkey	GU936708	[58]
	Pigeon aviadenovirus A (PiAdV-A)	Pigeon adenovirus 1 (PiAdV-1)	IDA4	Pigeon	FN824512	[36]
	Duck aviadenovirus B	Duck adenovirus 2 (DAdV-2)	GR	Muscovy duck	KJ469653	[36]
	Unassigned	Meyer's parrot adenovirus 1	N/A	Meyer's parrot	AY644731	[101]
	Unassigned	Psittacine adenovirus 1 (PsAdV-1)	GB 818–3	Red-breasted parakeet	EF442329	[121]
	Unassigned	Unassigned (gull adenovirus)	LA010815.1	European herring gull	KC309438 KC309439 KC309440	[122]
Atadenovirus	Duck atadenovirus A (DAdV-A)	Duck adenovirus 1 (DAdV-1, EDSV)	B8/78	Duck, goose, chicken	AC_000004	[123]
	Unassigned	Psittacine adenovirus 3 (PsAdV-3)	HKU/Parrot19	Mealy amazon	KJ675568	[73]

(Continued)

TABLE 83.1 (*Continued*)
Current Phylogenetic Classification of Bird-Infecting Adenoviruses

Genus	Species	Type	Reference Strain	Host Species	GenBank Accession No.	Publication
Siadenovirus	*Raptor siadenovirus A* (RAdV-A)	Raptor adenovirus 1 (RAdV-1)	N/A	Harris' hawk, Indian eagle owl, Verreaux's eagle owl	NC_015455	[124]
	Turkey siadenovirus A (TAdV-A)	Turkey adenovirus 3 (TAdV-3)	N/A	Turkey, chicken, pheasant	AC_000016	[125]
	Great tit siadenovirus A (GTAdV-A)	Great tit adenovirus 1 (GTAdV-1)	5957/SZ	Great tit	FJ849795	[10]
	Skua siadenovirus A	Skua adenovirus 1	T03	South polar skua	HM585353	[76]
	Unassigned	Psittacine adenovirus 2	FL-DAK	Plum headed parakeet, umbrella cockatoo	EU056825	[74]
	Unassigned	Budgerigar adenovirus 1	N/A	Budgerigar	AB485763	[126]
	Unassigned	Gouldian finch adenovirus 1	Hun1	Gouldian finch	KF031569 KF031570	[9]
	Unassigned	N/A	CSPAdV_8	Chinstrap penguin	KC593386	[75]

Italicized species names are officially accepted by the International Committee on Taxonomy of Viruses. Names of species officially not established yet are in
    roman style. Recommended abbreviations are given in parentheses.

30 min at 60°C [42]; a treatment using 0.5% formaldehyde or 0.5% glutaraldehyde is disinfecting [43]. The turkey hemorrhagic enteritis virus can be destroyed only by a 1 h treatment at 70°C or using a 0.0086% sodium hypochlorite solution [44]. Sodium lauryl sulfate, phenolic, and iodine-based disinfectants are also effective against this virus [45].

83.1.3.1 Aviadenoviruses

83.1.3.1.1 In Chickens

Chicken-infecting aviadenoviruses are called fowl adenoviruses. These are the most examined aviadenoviruses because of the economic value of their host and their worldwide distribution; FAdVs are ubiquitous in chicken populations. There are 12 different serotypes described among FAdVs; these serotypes are classified into five species: *Fowl aviadenovirus A–E*. There are several diseases of chickens associated with aviadenoviruses; most significant are the inclusion body hepatitis (IBH), the hydropericardium syndrome, and the gizzard erosion.

IBH occurs mainly in 3–7-week-old broilers. The disease starts with a sudden onset of mortality; sick birds take a crouching position and have ruffled feathers [46]. Pale, friable, swollen livers are found in the dead animals and intranuclear inclusion bodies in the hepatocytes during the histopathological examination [47,48]. Although almost every fowl adenovirus (FAdV) type was isolated already from diseased animals, the disease is mostly associated with species FAdV-D (FAdV-2, -3, -9, -11) and FAdV-E (FAdV-6, -7, -8a, -8b) strains [6,27,29,47,49].

The hydropericardium syndrome causes similar clinical and pathological signs to the IBH but the mortality (20%–80%) and the incidence of the hydropericardium is higher. The name of the disease is of course derived from its pathology: the pericardial sac is often filled with clear straw-colored fluid. Other pathological findings are pulmonary edema, swollen liver, and enlarged kidneys. The etiological agents of the disease are FAdV-4 strains (species FAdV-C) [11,12].

Pathogenic strains of the type FAdV-1 (species FAdV-A) can cause the gizzard erosion in chickens [6,13,50]. The disease has no clinical signs apart from the elevated level of mortality. During necropsy, the most common findings are erosions in the koilin layer and the inflammation of the mucosa. The gizzards might be dilated with bloody fluids. Pathogenicity of the Japanese strains was in connection with their fiber gene sequence [51], but European strains could not been differentiated using this method [52].

FAdV strains are sometimes associated with a fall in egg production, with decreased food consumption and depressed body weights, with respiratory diseases, or with tenosynovitis [53].

83.1.3.1.2 In Turkeys

There are multiple diseases in turkeys associated with adenoviruses. The strains, later classified into the types turkey adenovirus 1 and 2 (TAdV-1, TAdV-2) [54], were isolated first in Northern Ireland. TAdV-1 strains were found in birds suffering from conjunctivitis, nephritis, and airsacculitis [55]; a TAdV-2 strain was recovered from 1-day-old chicks with

IBH [56]. This latter strain caused suboptimal hatchability in experiments [57]. For historical reasons, a new strain of TAdVs was designated later as TAdV-1, originating from the trachea of 10-week-old turkeys showing respiratory signs [58], and the original TAdV-1 strains were reclassified as TAdV-4 [8]. Further strains—now regarded as type TAdV-5—were isolated from diseased fattening turkeys [8]. TAdVs were also isolated from diseased turkeys showing respiratory signs, but these isolates have not been typed [59–61].

83.1.3.1.3 In Geese

The first strains of goose adenoviruses were isolated in Hungary from unhatched eggs and fecal samples [62]. Later, other strains were classified into three types (goose adenovirus 1–3), but these were found to be apathogenic in animal infection studies [63]. Pathogenic strains of goose adenoviruses have caused IBH [64] occasionally combined with hydropericardium syndrome in young goslings [65].

83.1.3.1.4 In Other Birds

Adenoviruses can cause severe respiratory infections in young Muscovy ducks [66]. FAdV-1 can also cause severe respiratory infection, the so-called quail bronchitis in bobwhite quails and Japanese quails with up to 50% mortality [12,67]. Pigeons possess their own aviadenoviruses [36] but can be infected by FAdVs as well [68,69].

83.1.3.2 Atadenoviruses

The most important avian atadenovirus type from an economical point of view is the duck adenovirus 1, which causes the egg drop syndrome in chickens. The natural host of the virus is different species of ducks and geese, hence the official name of the virus type. But the virus is apathogenic in ducks, and the same is true in most of the cases for geese. In young goslings, some strains can cause a severe respiratory disease [70] or the so-called new type gosling viral enteritis [71]. Wild waterfowl species might play an important role in the spread of the disease.

Chickens of all ages are susceptible for the virus, which causes thin-shelled, soft-shelled, or shell-less eggs and a sharp decrease in egg production by up to 40%. Apart from these, the birds seem healthy. The virus replicates in the pouch shell gland; common pathological findings associated with the disease are the inactive ovaries, atrophied oviducts, edema of the uterine folds, and exudate in the pouch shell gland [72].

Atadenoviruses can be found in other bird species as well [73].

83.1.3.3 Siadenoviruses

Economically, the most important siadenovirus type is the TAdV-3. Strains belonging to this type cause the hemorrhagic enteritis in turkeys, the marble spleen disease in pheasants, and the avian adenovirus splenomegaly in chickens. Most important clinical sign of the hemorrhagic enteritis is the bloody droppings; blood-containing feces can be found on the feathers surrounding the cloaca. The average mortality is 10%–15% but may reach more than 60% during the 6–10 days of the disease. Dead poults are usually pale due to the blood loss; the small intestinal tract is usually distended, discolored, and filled with a bloody content; the spleen is enlarged, friable, and mottled [14].

The marble spleen disease in pheasants and the avian adenovirus splenomegaly in chickens are quite similar to each other in their clinical appearance and pathology, but the former one is more severe. In both cases the birds are depressed or might die suddenly. Mortality values may reach 5%–20% for the marble spleen disease but only 8.9% for the avian adenovirus splenomegaly. The spleens of dead or infected animals are enlarged and mottled (marbled); the lungs are edematous and congested [14].

Siadenoviruses are common among wild birds, too [10,74–76]. The raptor adenovirus 1 was diagnosed in three species of birds of prey: Bengal eagle owl, Verreaux's eagle owl, and Harris hawk. In the specimens of these birds, hepatic and splenic necroses were found together with ventricular ulceration and necrosis [77]. A siadenovirus was also found in Gouldian finches with depression and significant respiratory crackles [9].

83.1.4 Diagnosis

83.1.4.1 Conventional Techniques

Conventional techniques in the avian adenovirus research and diagnostics involve the virus isolation, microscopic and histological techniques, and immunological methods.

The sample source for virus isolation is the affected organs, mainly the liver, or the feces from diseased animals. The sample is homogenized in a cell medium containing antibiotics and seeded onto avian cells. The most permissive cell lines for FAdVs are usually primary chicken cells: chicken embryo liver cells, chicken embryo kidney cells, chicken kidney cells, or chicken embryo fibroblasts. Duck adenovirus 1, the causative agent of the egg drop syndrome, can be isolated on embryonated duck or goose eggs; the sample for the isolation attempt might be an altered egg or an affected pouch shell gland. The isolation of TAdV-3, the turkey hemorrhagic enteritis virus, requires either live turkeys or a special lymphoblastoid cell line (MDTC-RP19). Adenoviruses of other birds require usually cell lines originating from the host species; goose aviadenoviruses, for example, are propagated on goose embryonic liver cells [78]. This might be almost impossible for adenoviruses originating from wild, especially from endangered bird species [77]. The success of virus isolation can be confirmed by the occurrence of a cytopathogenic effect. The isolate can be identified as an adenovirus using immunological or molecular methods, an electron microscopic investigation, or a simple hematoxylin–eosin staining by the characteristic intranuclear inclusion bodies. Electron microscope or hematoxylin–eosin staining can be used in the histopathological examination of the animals as well; the same inclusion bodies are also characteristic, for example, in the IBH [47,48,65].

Immunological methods are either detecting the virus (the antigen) or the antibodies produced against adenoviruses. Agar gel precipitation (agar gel precipitin, agar gel immunodiffusion, double immunodiffusion) can be used for both purposes. Several enzyme-linked immunosorbent assays have been developed for adenovirus diagnostics, and immunofluorescence assays [54] are also useful. The immunohistochemistry (immunofluorescent or immunoperoxidase staining) might aid the histological examination. Duck adenovirus 2 agglutinates avian red blood cells, so hemagglutination tests are available for the identification of the virus, and also hemagglutination inhibition tests are available for serological diagnosis.

Serum neutralization tests were widely used for the serotyping of avian adenovirus isolates [22–24,79]. Such tests are rather tedious and time consuming, and for example for the proper serotyping of a FAdV isolate, the 12 standard reference antiserums are also needed. For the establishment of a new, officially accepted serotype, the 12 reference FAdV strains are needed as well, as "a serotype is defined as one that either exhibits no cross-reaction with others or shows a homologous/heterologous titer ratio greater than 16 (in both directions)" [1]. The evaluation of the results might be further complicated by the fact that different diagnostic institutes might use different reference or prototype strains of the same serotype [80] and that chickens are often infected by multiple serotypes of FAdVs [81,82].

83.1.4.2 Molecular Techniques

To aid the detection of avian adenoviruses and simplify the typing of these, numerous molecular methods have been developed. The restriction endonuclease analysis of the viral genomes was one of the first techniques used for such a purpose. Although this technique determines basically only a few nucleotides in the whole genome using the specificity of the restriction endonuclease, it was still sufficient for the genotyping of goose [63], turkey [57], and FAdV [83] strains. This classification of the FAdV types was retained later when the FAdV species were established [17], as phylogenetic analyses based on hexon sequences supported it [29,84,85]. Another molecular method used in the diagnostics of avian adenoviruses is the hybridization of labeled probes to the viral DNA either on the blotting membrane or *in situ*. But as the isolation of the virus strain and the extraction of the viral genomic DNA are still required for both the genomic restriction endonuclease analysis and the DNA hybridization (except the *in situ* hybridization), their usage became rather limited since the development of the polymerase chain reaction (PCR).

Using PCR, a specific fragment is amplified from the targeted genome using custom-made, specific oligonucleotides, the primers. Several PCR methods have been published for the diagnostics of adenoviruses infecting birds, targeting the hexon gene [84–88], the viral DNA polymerase gene [89–91], and other different genes [92,93]. More sophisticated techniques include real-time PCR [92,94] and the high-resolution melting curve analysis of the products

[28,29,95], pyrosequencing [96], and loop-mediated isothermal amplification [97].

Chickens can be infected by FAdVs (genus *Aviadenovirus*), by duck adenovirus 1 (egg drop syndrome virus, genus *Atadenovirus*), and by TAdV-3 (turkey hemorrhagic enteritis virus, genus *Siadenovirus*). These viruses are evolutionarily distant, thus the available conserved gene stretches are scarce, limiting the possible locations for diagnostic primers. So beyond the PCR developed for FAdVs, separate PCR tests were designed for the detection of duck adenovirus 1 [85] and TAdV-3 [98]. More universal PCR systems were also published, but these were tested specifically on duck adenovirus 1, TAdV-3, and FAdV types only [86,99], and it is not certain at all whether these would be able to detect a new adenovirus infecting chickens. These PCRs were not tested either in the diagnostics of other adenoviruses like goose or turkey adenoviruses.

There is one PCR [90] which is regarded the pan-adenovirus PCR, possibly detecting all adenoviruses, as it was tested with multiple samples belonging to all known and proposed genera in the family *Adenoviridae* [10,18,47,74,77,91,100–104]. Thus, it simplifies the diagnostics of avian adenoviruses as it detects at- and siadenoviruses as well, not only aviadenoviruses. It enables the detection of hitherto unknown adenoviruses without the need of virus isolation, and as it is a nested PCR system, it ensures a high sensitivity.

However, every method has its limitations. The nested system also means double workload: two rounds of PCR and an enhanced risk for contamination. And also, it is not a real-time PCR system, so the agarose gel electrophoresis of the PCR products is needed for the detection of the results. But as it is a pan-adenovirus PCR, no other reaction is needed for the detection of any adenovirus. This exceptionally wide target range was achieved by highly degenerate consensus primers targeting the viral gene of the DNA-dependent DNA polymerase. The high level of degeneracy enhances the possibility of nonspecific results as well, therefore, the sequencing of the PCR products is essential in every case, even if they are of the right size. The determined partial polymerase gene sequence is so much conserved that it allows a preliminary, genus, and species classification only. For the intraspecies, type-level identification of avian adenoviruses, a hexon-targeting PCR can be recommended. The materials and methods needed for the DNA polymerase–targeting PCR is discussed in detail in the next section.

83.2 METHODS

83.2.1 Sample Preparation

For molecular diagnostics the adenoviral DNA is needed. This does not mean necessarily that the DNA must be extracted from the sample, for some PCR the diluted viral isolate might be applicable [6]. Obviously an extracted, pure DNA enhances the chance of a successful PCR. For the restriction endonuclease analysis of adenoviruses, the selective extraction of the viral genome is needed [105,106]. But for PCR applications a simpler, total DNA is appropriate. The source of DNA extraction might be the isolated virus; an affected organ, mainly the liver; or, if diarrhea was observed, the feces.

For the DNA extraction, any commercially available DNA extraction kit gives sufficient result. Here a DNA extraction method based on guanidine hydrochloride according to Dán et al. [107] is presented.

83.2.1.1 Materials Needed

1. 1× TE (tris(hydroxymethyl)aminomethane-ethylenediaminetetraacetic acid) buffer
2. 10% sarkosyl (sodium lauroyl sarcosinate)
3. Proteinase K (20 mg/mL)
4. 8 M guanidine hydrochloride
5. 7.5 M ammonium acetate
6. Refrigerated (−20°C) absolute ethanol
7. Refrigerated (−20°C) 70% ethanol
8. Sterile, ultrapure water
9. Equipment: micropipettes; sterile scissors, scalpels; mortar and pestle, or other tissue homogenizer; 1.5 mL tubes; thermoshaker or water bath capable of 55°C; centrifuge (possibly refrigerated)

83.2.1.2 Procedure

1. Collect sample from an affected organ using uncontaminated sterile tools. The desired sample size is the size of a peppercorn.
2. Homogenize the sample in 200 μL TE buffer.
3. Transfer 100 μL of the homogenate to a 1.5 mL tube and add 10 μL sarkosyl and 4 μL Proteinase K.
4. Gently mix the content of the tube and incubate it on 55°C overnight.
5. Add 300 μL guanidine hydrochloride and 20 μL ammonium acetate, mix thoroughly, and incubate the tube on room temperature for 1 h. Shake the tube every 20 min.
6. Add 1 mL absolute ethanol and mix thoroughly.
7. Spin the tube (16,000 × g, 12 min, 4°C).
8. Discard the supernatant and add 1 mL 70% ethanol.
9. Spin the tube (16,000 × g, 5 min, 4°C).
10. Discard the supernatant and let the pellet dry on room temperature.
11. Reconstitute the pellet in 50 μL ultrapure water. If pieces of undigested tissue remain in the tube, transfer the supernatant to a new tube after a short spin.
12. Store the DNA sample on −20°C.

83.2.2 DETECTION PROCEDURES

Here, the nested PCR targeting the adenoviral DNA polymerase gene [90] is described. Slight modifications are applied to the original publication.

83.2.2.1 Materials Needed

1. DNA samples
2. Primers for PCR amplification (see Table 83.2)
3. REDTaq DNA Polymerase (Sigma-Aldrich)
4. 10× reaction buffer (provided with REDTaq)

TABLE 83.2

Polymerase Chain Reaction Primers Targeting Deoxyribonucleic Acid Polymerase Gene

PCR	Primer	Sequence (5′–3′)
First round	polFouter	TNMGNGGNGGNMGNTGYTAYCC
	polRouter	GTDGCRAANSHNCCRTABARNGMRTT
Second round	polFinner	GTNTWYGAYATHTGYGGHATGTAYGC
	polRinner	CCANCCBCDRTTRTGNARNGTRA

Source: Wellehan, J. et al., *J. Virol.*, 78(23), 13366, 2004.
IUPAC nucleotide codes: N = G/A/T/C, M = A/C, R = A/G, W = A/T, S = C/G, Y = C/T, H = A/C/T, D = A/G/T, B = C/G/T.

5. 10 mM deoxynucleotide mix: 10 mM dATP, 10 mM dCTP, 10 mM dGTP, 10 mM dTTP in ultrapure water
6. 25 mM $MgCl_2$
7. Sterile, ultrapure water
8. Adenovirus DNA solution for positive control
9. Agarose
10. TBE (tris(hydroxymethyl)aminomethane-boric acid-ethylenediaminetetraacetic acid) or TAE (tris(hydroxymethyl)aminomethane-acetic acid-ethylenediaminetetraacetic acid) electrophoresis running buffer
11. DNA-staining dye (ethidium bromide, GelRed [Biotium])
12. Equipment: micropipettes, PCR tubes, 1.5 mL tubes, ice, tube racks, gel electrophoresis apparatus, power supply, thermal cycler, and gel imaging system

83.2.2.2 Procedure

1. Thaw all reagents completely, vortex, and spin them briefly. Store them on ice. It is recommended to set up the reaction on ice.
2. Prepare a master mix according to Table 83.3. Do not forget to count the positive and the negative control

TABLE 83.3

Recipe for Polymerase Chain Reaction Targeting Deoxyribonucleic Acid Polymerase Gene

Component	Volume (per Reaction) (μL)
Sterile ultrapure water	37
10× PCR Buffer of the REDTaq DNA Polymerase	5
$MgCl_2$ (25 mM)	1
Deoxynucleotide mix (10 mM)	1.5
polFouter/polFinner primer (50 mM)	1
polRouter/polRinner primer (50 mM)	1
DNA sample	1
REDTaq DNA Polymerase	2.5
Total reaction volume	50

Source: Wellehan, J. et al., *J. Virol.*, 78(23), 13366, 2004.

beyond the samples, and add 10% excess, too, to counterbalance any pipetting inaccuracies. The recipes of the two rounds are almost the same except the primers: use the outer primers for the first round and the inner ones for the second round.

3. Dispense 49 μL from the master mix to each marked PCR tube.
4. Add 1 μL ultrapure water to the negative control, 1 μL DNA sample to each sample tube, and 1 μL adenovirus DNA solution to the positive control.
5. Close the tubes properly, place them into a compatible thermal cycler, and run the program according to Table 83.4.
6. When the thermal cycler has finished, prepare the master mix of the second round, containing the inner primers.
7. Dispense 49 μL from the master mix to each marked PCR tube.
8. Transfer 1 μL from the ready reaction mixture of the first round to the tube of the second round.
9. Close the tubes properly, place them into a compatible thermal cycler, and run the same program again according to Table 83.4.
10. Prepare a 1.5% agarose gel and run 10 μL from the samples of both rounds. The product size of a positive reaction is 318–324 bp, but nonspecific bands of other sizes can be often observed.

Sequencing of the PCR products is essential in every case, even if they are of the right size. The PCR product needs to be purified for this purpose. If nonspecific bands are observable next to the specific product, the leftover 40 μL needs to be loaded onto a gel. The specific band needs to be excised using a sterile scalpel and purified from the agarose gel using a gel extraction kit. If only a specific band is observable, perhaps a commercial sequencing provider might be recommended for the purification and sequencing of the product. It is recommended to determine the nucleotide sequence of the PCR product on both strands.

TABLE 83.4
Thermal Cycling Conditions for Both Rounds of Polymerase Chain Reaction Targeting Deoxyribonucleic Acid Polymerase Gene

Step	Temperature (°C)	Duration
Denature template	96	5 min
45 cycles		
1. Denature template	96	30 s
2. Anneal primers	46	1 min
3. Extension	72	1 min
Final extension	72	3 min

Source: Wellehan, J. et al., *J. Virol.*, 78(23), 13366, 2004.

Birds are often infected by multiple strains of adenoviruses [6,81,82]. In this case the PCR amplifies its product from all the strains, resulting in a mixed product, and at the end also a mixed, undeterminable sequence. This still might provide enough data to prove the presence of adenoviruses in the sample. If exact species determination is needed, the molecular cloning of the PCR product is recommended [6].

83.2.3 PHYLOGENETIC ANALYSIS

For a phylogenetic analysis, molecular typing of the avian adenovirus strains, the resulting forward and reverse sequences of the PCR product need to be assembled to form a consensus sequence. The primer sequences must be omitted before submitting the consensus sequence to a BLAST (basic local alignment search tool) search [108] or to a phylogenetic analysis. A step-by-step description of a basic phylogenetic analysis using adenoviral sequences can be found in Harrach and Benkő [109].

83.3 CONCLUSION AND FUTURE PERSPECTIVES

As adenovirus infection is common in both domestic and wild birds and perfect diagnostic methods are not available yet, a continuous improvement of these methods is needed. An ideal diagnostic method for avian adenoviruses would detect any adenovirus belonging to different genera in a single closed-tube reaction with high sensitivity, specificity, and reproducibility possibly within a few hours from sample collection till the result. It would also enable the typing of the investigated strain, even if it happens to originate from a mixed infection, without the need for the reference strains. Preferably, it should also have an affordable cost. Current diagnostic methods are far from these goals.

Closest solutions to these goals nowadays are real-time PCR methods combined with the high-resolution melting curve analysis of the products [28,29,95]. But even these methods do not fulfill by far the above mentioned aims. High-resolution melting curve analysis methods used in avian adenovirus diagnostics are specific for FAdVs only, which are economically the most important, but there are numerous other avian adenoviruses as well. To use it for molecular typing, reference strains are needed to compare the samples to these. But even with all 12 fowl reference strains, it is questionable whether the high-resolution melting curve analysis can clearly classify FAdV field samples into the nowadays accepted types; the molecular classification of virus strains not necessarily mirrors the classical serotyping [29]. For a clear, safe result new type demarcation criteria are needed, which are based on molecular principles. The outdated type demarcation criteria need to be updated to meet the needs of the molecular age.

The greatest concern about molecular methods might be that they yield a positive result only with the organisms they are designed for. Molecular methods usually operate with a specific oligonucleotide, thus missing any microorganism that they are not specific to. The method of virus isolation enables the discovery of new viruses, but with its tedious and time-consuming nature, it is not used routinely these days. This problem might be solved in the future by the increased usage of deep sequencing. As the cost of sequencing decreases continuously, it will definitely play a bigger role in future diagnostics. The high number of sequencing reads enables the identification of several microorganisms in a clinical sample without any prior cultivation [110].

REFERENCES

1. Harrach, B. et al., Family Adenoviridae, in *Virus Taxonomy: IXth Report of the International Committee on Taxonomy of Viruses*, King, A.M.Q. et al. (eds.), 2011, Elsevier, San Diego, CA, pp. 125–141.

2. Huebner, R.J. et al., Adenoidal-pharyngeal-conjunctival agents: A newly recognized group of common viruses of the respiratory system. *N Engl J Med*, 1954. 251(27): 1077–1086.

3. Rowe, W.P. et al., Isolation of a cytopathogenic agent from human adenoids undergoing spontaneous degeneration in tissue culture. *Proc Soc Exp Biol Med*, 1953. 84(3): 570–573.

4. van den Ende, M., P.A. Don, and A. Kipps, The isolation in eggs of a new filtrable agent which may be the cause of bovine lumpy skin disease. *J Gen Microbiol*, 1949. 3(2): 174–183.

5. Olson, N.O., A respiratory disease (bronchitis) of quail caused by a virus. *Vet Med*, 1951. 46(1): 22.

6. Kaján, G.L. et al., Molecular typing of fowl adenoviruses, isolated in Hungary recently, reveals high diversity. *Vet Microbiol*, 2013. 167(3–4): 357–363.

7. Choi, K.S. et al., Epidemiological investigation of outbreaks of fowl adenovirus infection in commercial chickens in Korea. *Poult Sci*, 2012. 91(10): 2502–2506.

8. Marek, A. et al., Whole-genome sequences of two turkey adenovirus types reveal the existence of two unknown lineages that merit the establishment of novel species within the genus Aviadenovirus. *J Gen Virol*, 2014. 95(1): 156–170.

9. Joseph, H.M. et al., A novel siadenovirus detected in the kidneys and liver of Gouldian finches (*Erythura gouldiae*). *Vet Microbiol*, 2014. 172(1–2): 35–43.

10. Kovács, E.R. et al., Recognition and partial genome characterization by non-specific DNA amplification and PCR of a new siadenovirus species in a sample originating from Parus major, a great tit. *J Virol Methods*, 2010. 163(2): 262–268.

11. Adair, B.M. and S.D. Fitzgerald, Adenovirus Infections. Group I Adenovirus Infections, in *Diseases of Poultry*, Saif, Y.M. et al. (eds.), 2008, Wiley-Blackwell, Ames, IA, pp. 252–266.

12. Smyth, J.A. and M.S. McNulty, Adenoviridae, in *Poultry Diseases*, Pattison, M. et al. (eds.), 2007, Elsevier, Edinburgh, Scotland, pp. 367–381.

13. Grafl, B. et al., Vertical transmission and clinical signs in broiler breeders and broilers experiencing adenoviral gizzard erosion. *Avian Pathol*, 2012. 41(6): 599–604.

14. Pierson, F.W. and S.D. Fitzgerald, Adenovirus Infections. Hemorrhagic enteritis and related infections, in *Diseases of Poultry*, Saif, Y.M. et al. (eds.), 2008, Wiley-Blackwell, Ames, IA, pp. 276–286.

15. McFerran, J.B. and J.A. Smyth, Avian adenoviruses. *Rev Sci Tech*, 2000. 19(2): 589–601.

16. International Committee on Taxonomy of Viruses and F. Fenner, *Classification and Nomenclature of Viruses: Second Report of the International Committee on Taxonomy of Viruses*. Intervirology, Karger, Basel, Switzerland, 1976, 115pp.

17. Benkő, M. et al., Family Adenoviridae, in *Virus Taxonomy: VIIIth Report of the International Committee on Taxonomy of Viruses*, Fauquet, C. et al. (eds.), 2005, Elsevier, New York, pp. 213–228.

18. Doszpoly, A. et al., Partial characterization of a new adenovirus lineage discovered in testudinoid turtles. *Infect Genet Evol*, 2013. 17: 106–112.

19. Benkő, M. and B. Harrach, A proposal for a new (third) genus within the family Adenoviridae. *Arch Virol*, 1998. 143(4): 829–837.

20. Harrach, B., Reptile adenoviruses in cattle? *Acta Vet Hung*, 2000. 48(4): 485–490.

21. Harrach, B. et al., Close phylogenetic relationship between egg drop syndrome virus, bovine adenovirus serotype 7, and ovine adenovirus strain 287. *Virology*, 1997. 229(1): 302–308.

22. Calnek, B.W. and B.S. Cowen, Adenoviruses of chickens: Serologic groups. *Avian Dis*, 1975. 19(1): 91–103.

23. Kefford, B. et al., Serological identification of avian adenoviruses isolated from cases of inclusion body hepatitis in Victoria, Australia. *Avian Dis*, 1980. 24(4): 998–1006.

24. McFerran, J.B., J.K. Clarke, and T.J. Connor, Serological classification of avian adenoviruses. *Arch Gesamte Virusforsch*, 1972. 39(1): 132–139.

25. Lim, T.H. et al., Identification and virulence characterization of fowl adenoviruses in Korea. *Avian Dis*, 2011. 55(4): 554–560.

26. Mase, M., K. Nakamura, and F. Minami, Fowl adenoviruses isolated from chickens with inclusion body hepatitis in Japan, 2009–2010. *J Vet Med Sci*, 2012. 74(8): 1087–1089.

27. Ojkić, D. et al., Genotyping of Canadian isolates of fowl adenoviruses. *Avian Pathol*, 2008. 37(1): 95–100.

28. Steer, P.A. et al., Application of high-resolution melting curve analysis for typing of fowl adenoviruses in field cases of inclusion body hepatitis. *Aust Vet J*, 2011. 89(5): 184–192.

29. Marek, A. et al., Classification of fowl adenoviruses by use of phylogenetic analysis and high-resolution melting-curve analysis of the hexon L1 gene region. *J Virol Methods*, 2010. 170(1–2): 147–154.

30. Seto, D. et al., Using the whole-genome sequence to characterize and name human adenoviruses. *J Virol*, 2011. 85(11): 5701–5702.

31. Aoki, K. et al., Toward an integrated human adenovirus designation system that utilizes molecular and serological data and serves both clinical and fundamental virology. *J Virol*, 2011. 85(11): 5703–5704.

32. Seto, D. et al., Characterizing, typing, and naming human adenovirus type 55 in the era of whole genome data. *J Clin Virol*, 2013. 58(4): 741–742.

33. Kajon, A.E., M. Echavarria, and J.C. de Jong, Designation of human adenovirus types based on sequence data: An unfinished debate. *J Clin Virol*, 2013. 58(4): 743–744.

34. Nemerow, G.R., P.L. Stewart, and V.S. Reddy, Structure of human adenovirus. *Curr Opin Virol*, 2012. 2(2): 115–121.

35. Gelderblom, H. and I. Maichle-Lauppe, The fibers of fowl adenoviruses. *Arch Virol*, 1982. 72(4): 289–298.

36. Marek, A. et al., Complete genome sequences of pigeon adenovirus 1 and duck adenovirus 2 extend the number of species within the genus Aviadenovirus. *Virology*, 2014. 462–463: 107–114.

37. Russell, W.C., Adenoviruses: Update on structure and function. *J Gen Virol*, 2009. 90(1): 1–20.
38. Davison, A.J. and B. Harrach, Siadenovirus, in *The Springer Index of Viruses*, Tidona, C.A. and Darai, G. (eds.) 2011, Springer, New York, pp. 49–56.
39. Harrach, B. and G.L. Kaján, Aviadenovirus, in *The Springer Index of Viruses*, Tidona, C.A. and Darai, G. (eds.) 2011, Springer, New York, pp. 13–28.
40. Davison, A., M. Benkő, and B. Harrach, Genetic content and evolution of adenoviruses. *J Gen Virol*, 2003. 84: 2895–2908.
41. Harrach, B., Adenoviruses: General features, in *Encyclopedia of Virology*, Mahy, B. and van Regenmortel, M. (eds.) 2008, Elsevier, Oxford, U.K., pp. 1–9.
42. Zsák, L. and J. Kisary, Some biological and physico-chemical properties of egg drop syndrome (EDS) avian adenovirus strain B8/78. *Arch Virol*, 1981. 68(3–4): 211–219.
43. Takai, S., M. Higashihara, and M. Matumoto, Purification and hemagglutinating properties of egg drop syndrome 1976 virus. *Arch Virol*, 1984. 80(1): 59–67.
44. Domermuth, C.H. and W.B. Gross, Effect of chlorine on the virus of hemorrhagic enteritis of turkeys. *Avian Dis*, 1972. 16(4): 952–953.
45. Domermuth, C.H. and W.B. Gross, Hemorrhagic enteritis of turkeys. Antiserum—Efficacy, preparation, and use. *Avian Dis*, 1975. 19(4): 657–665.
46. McFerran, J.B. et al., Isolation of viruses from clinical outbreaks of inclusion body hepatitis. *Avian Pathol*, 1976. 5(4): 315–324.
47. Zadravec, M. et al., Inclusion body hepatitis associated with fowl adenovirus type 8b in broiler flock in Slovenia—A case report. *Slov Vet Res*, 2011. 48(3–4): 107–113.
48. Zadravec, M. et al., Inclusion body hepatitis (IBH) outbreak associated with fowl adenovirus type 8b in broilers. *Acta Vet (Beograd)*, 2013. 63(1): 101–110.
49. Ojkić, D. et al., Characterization of fowl adenoviruses isolated in Ontario and Quebec, Canada. *Can J Vet Res*, 2008. 72(3): 236–241.
50. Kecskeméti, S. et al., Observations on gizzard ulcers caused by adenovirus in chickens. *Magyar Állatorvosok Lapja*, 2012. 134(3): 145–149.
51. Okuda, Y. et al., Comparison of the polymerase chain reaction-restriction fragment length polymorphism pattern of the fiber gene and pathogenicity of serotype-1 fowl adenovirus isolates from gizzard erosions and from feces of clinically healthy chickens in Japan. *J Vet Diagn Invest*, 2006. 18(2): 162–167.
52. Marek, A. et al., Comparison of the fibers of Fowl adenovirus A serotype 1 isolates from chickens with gizzard erosions in Europe and apathogenic reference strains. *J Vet Diagn Invest*, 2010. 22(6): 937–941.
53. Kaján, G.L. and S. Kecskeméti, First Hungarian isolation of an adenovirus type belonging to species Fowl adenovirus B. *Magyar Állatorvosok Lapja*, 2011. 133(6): 347–352.
54. Adair, B., J. McFerran, and V. Calvert, Development of a microtitre fluorescent antibody test for serological detection of adenovirus infection in birds. *Avian Pathol*, 1980. 9(3): 291–300.
55. Scott, M. and J. McFerran, Isolation of adenoviruses from turkeys. *Avian Dis*, 1972. 16(2): 413–420.
56. Guy, J.S., J.L. Schaeffer, and H.J. Barnes, Inclusion-body hepatitis in day-old turkeys. *Avian Dis*, 1988. 32(3): 587–590.
57. Guy, J. and H. Barnes, Characterization of an avian adenovirus associated with inclusion body hepatitis in day-old turkeys. *Avian Dis*, 1997. 41(3): 726–731.
58. Kaján, G.L. et al., The first complete genome sequence of a non-chicken aviadenovirus, proposed to be turkey adenovirus 1. *Virus Res*, 2010. 153(2): 226–233.
59. Simmons, D. et al., Isolation and identification of a turkey respiratory adenovirus. *Avian Dis*, 1976. 20(1): 65–74.
60. Sutjipto, S. et al., Physicochemical characterization and pathogenicity studies of two turkey adenovirus isolants. *Avian Dis*, 1977. 21(4): 549–556.
61. Crespo, R. et al., Inclusion body tracheitis associated with avian adenovirus in turkeys. *Avian Dis*, 1998. 42(3): 589–596.
62. Csontos, L., Isolation of adenoviruses from geese. Preliminary report. *Acta Vet Acad Sci Hung*, 1967. 17(2): 217–219.
63. Zsák, L. and J. Kisary, Characterisation of adenoviruses isolated from geese. *Avian Pathol*, 1984. 13(2): 253–264.
64. Riddell, C., Viral hepatitis in domestic geese in Saskatchewan. *Avian Dis*, 1984. 28(3): 774–782.
65. Ivanics, É. et al., Hepatitis and hydropericardium syndrome associated with adenovirus infection in gosling. *Acta Vet Hung*, 2010. 58(1): 47–58.
66. Bergmann, V., R. Heidrich, and E. Kinder, Pathomorphologisch und elektronen-mikroskopische Feststellung einer Adenovirus Tracheitis bei Moschusenten (Cairina moschata). *Monatshefte für Veterinärmedizin*, 1985. 40: 313–315.
67. Reed, W.M. and S.W. Jack, Adenovirus infections. Quail Bronchitis, in *Diseases of Poultry*, Saif, Y.M. et al. (eds.) 2008, Wiley-Blackwell, Ames, IA, pp. 287–290.
68. Hess, M., C. Prusas, and G. Monreal, Growth analysis of adenoviruses isolated from pigeons in chicken cells and serological characterization of the isolates. *Avian Pathol*, 1998. 27(2): 196–199.
69. McFerran, J.B., T.J. Connor, and R.M. McCracken, Isolation of adenoviruses and reoviruses from avian species other than domestic fowl. *Avian Dis*, 1976. 20(3): 519–524.
70. Ivanics, E. et al., The role of egg drop syndrome virus in acute respiratory disease of goslings. *Avian Pathol*, 2001. 30(3): 201–208.
71. Chen, S. et al., Histopathology, immunohistochemistry, in situ apoptosis, and ultrastructure characterization of the digestive and lymphoid organs of new type gosling viral enteritis virus experimentally infected gosling. *Poult Sci*, 2010. 89(4): 668–680.
72. Adair, B.M. and J.A. Smyth, Adenovirus infections. Egg drop syndrome, in *Diseases of Poultry*, Saif, Y.M. et al. (eds.) 2008, Wiley-Blackwell, Ames, IA, pp. 266–276.
73. To, K.K. et al., A novel psittacine adenovirus identified during an outbreak of avian chlamydiosis and human psittacosis: Zoonosis associated with virus-bacterium coinfection in birds. *PLoS Negl Trop Dis*, 2014. 8(12): e3318.
74. Wellehan, J.J. et al., Siadenovirus infection in two psittacine bird species. *Avian Pathol*, 2009. 38(5): 413–417.
75. Lee, S.Y. et al., A novel adenovirus in Chinstrap penguins (*Pygoscelis antarctica*) in Antarctica. *Viruses*, 2014. 6(5): 2052–2061.
76. Park, Y.M. et al., Full genome analysis of a novel adenovirus from the South Polar skua (*Catharacta maccormicki*) in Antarctica. *Virology*, 2012. 422(1): 144–150.
77. Zsivanovits, P. et al., Presumptive identification of a novel adenovirus in a Harris hawk (*Parabuteo unicinctus*), a Bengal eagle owl (*Bubo bengalensis*), and a Verreaux's eagle owl (*Bubo lacteus*). *J Avian Med Surg*, 2006. 20(2): 105–112.
78. Kaján, G.L. et al., Genome sequence of a waterfowl aviadenovirus, goose adenovirus 4. *J Gen Virol*, 2012. 93(11): 2457–2465.

79. McFerran, J.B. and T.J. Connor, Further studies on the classification of fowl adenoviruses. *Avian Dis*, 1977. 21(4): 585–595.

80. Monreal, G., Adenoviruses and adeno-associated viruses of poultry. *Poult Sci Rev*, 1992. 4: 1–27.

81. Mockett, A.P. and J.K. Cook, The use of an enzyme-linked immunosorbent assay to detect IgG antibodies to serotype-specific and group-specific antigens of fowl adenovirus serotypes 2, 3 and 4. *J Virol Methods*, 1983. 7(5–6): 327–335.

82. Erny, K., J. Pallister, and M. Sheppard, Immunological and molecular comparison of fowl adenovirus serotypes 4 and 10. *Arch Virol*, 1995. 140(3): 491–501.

83. Zsák, L. and J. Kisary, Grouping of fowl adenoviruses based upon the restriction patterns of DNA generated by BamHI and HindIII. *Intervirology*, 1984. 22(2): 110–114.

84. Meulemans, G. et al., Phylogenetic analysis of fowl adenoviruses. *Avian Pathol*, 2004. 33(2): 164–170.

85. Raue, R. and M. Hess, Hexon based PCRs combined with restriction enzyme analysis for rapid detection and differentiation of fowl adenoviruses and egg drop syndrome virus. *J Virol Methods*, 1998. 73(2): 211–217.

86. Mase, M. et al., Identification of group I-III avian adenovirus by PCR coupled with direct sequencing of the hexon gene. *J Vet Med Sci*, 2009. 71(9): 1239–1242.

87. Meulemans, G. et al., Polymerase chain reaction combined with restriction enzyme analysis for detection and differentiation of fowl adenoviruses. *Avian Pathol*, 2001. 30(6): 655–660.

88. Raue, R., H. Gerlach, and H. Müller, Phylogenetic analysis of the hexon loop 1 region of an adenovirus from psittacine birds supports the existence of a new psittacine adenovirus (PsAdV). *Arch Virol*, 2005. 150(10): 1933–1943.

89. Hanson, L.A. et al., A broadly applicable method to characterize large DNA viruses and adenoviruses based on the DNA polymerase gene. *Virol J*, 2006. 3: 28.

90. Wellehan, J. et al., Detection and analysis of six lizard adenoviruses by consensus primer PCR provides further evidence of a reptilian origin for the atadenoviruses. *J Virol*, 2004. 78(23): 13366–13369.

91. Kaján, G.L., S. Sameti, and M. Benkő, Partial sequence of the DNA-dependent DNA polymerase gene of fowl adenoviruses: A reference panel for a general diagnostic PCR in poultry. *Acta Vet Hung*, 2011. 59(2): 279–285.

92. Günes, A., A. Marek, and M. Hess, Species determination of fowl adenoviruses based on the 52K gene region. *Avian Dis*, 2013. 57(2): 290–294.

93. Jiang, P. et al., Application of the polymerase chain reaction to detect fowl adenoviruses. *Can J Vet Res*, 1999. 63(2): 124–128.

94. Günes, A. et al., Real-time PCR assay for universal detection and quantitation of all five species of fowl adenoviruses (FAdV-A to FAdV-E). *J Virol Methods*, 2012. 183(2): 147–153.

95. Steer, P. et al., Classification of fowl adenovirus serotypes by use of high-resolution melting-curve analysis of the hexon gene region. *J Clin Microbiol*, 2009. 47(2): 311–321.

96. Pizzuto, M.S. et al., Pyrosequencing analysis for a rapid classification of fowl adenovirus species. *Avian Pathol*, 2010. 39(5): 391–398.

97. Xie, Z. et al., Rapid detection of group I avian adenoviruses by a loop-mediated isothermal amplification. *Avian Dis*, 2011. 55(4): 575–579.

98. Hess, M., R. Raue, and H.M. Hafez, PCR for specific detection of haemorrhagic enteritis virus of turkeys, an avian adenovirus. *J Virol Methods*, 1999. 81(1–2): 199–203.

99. Xie, Z. et al., Detection of avian adenovirus by polymerase chain reaction. *Avian Dis*, 1999. 43(1): 98–105.

100. Rivera, S. et al., Systemic adenovirus infection in Sulawesi tortoises (*Indotestudo forsteni*) caused by a novel siadenovirus. *J Vet Diagn Invest*, 2009. 21(4): 415–426.

101. Wellehan, J. et al., Identification and initial characterization of an adenovirus associated with fatal hepatic and lymphoid necrosis in a Meyer's parrot (*Poicephalus meyeri*). *J Avian Med Surg*, 2005. 19(3): 191–197.

102. Jánoska, M. et al., Novel adenoviruses and herpesviruses detected in bats. *Vet J*, 2011. 189(1): 118–121.

103. Vidovszky, M.Z. and S. Boldogh, Detection of adenoviruses in the Northern Hungarian bat fauna. *Magyar Állatorvosok Lapja*, 2011. 133: 747–753.

104. Papp, T. et al., PCR-sequence characterization of new adenoviruses found in reptiles and the first successful isolation of a lizard adenovirus. *Vet Microbiol*, 2009. 134(3–4): 233–240.

105. Shinagawa, M. et al., A rapid and simple method for preparation of adenovirus DNA from infected cells. *Microbiol Immunol*, 1983. 27(9): 817–822.

106. Kajon, A.E. and D.D. Erdman, Assessment of genetic variability among subspecies B1 human adenoviruses for molecular epidemiology studies. *Methods Mol Med*, 2007. 131: 335–355.

107. Dán, A. et al., Characterisation of Hungarian porcine circovirus 2 genomes associated with PMWS and PDNS cases. *Acta Vet Hung*, 2003. 51(4): 551–562.

108. Altschul, S. et al., Gapped BLAST and PSI-BLAST: A new generation of protein database search programs. *Nucleic Acids Res*, 1997. 25(17): 3389–3402.

109. Harrach, B. and M. Benko, Phylogenetic analysis of adenovirus sequences. *Methods Mol Med*, 2007. 131: 299–334.

110. Phan, T.G. et al., The viruses of wild pigeon droppings. *PLoS One*, 2013. 8(9): e72787.

111. Schrenzel, M. et al., Characterization of a new species of adenovirus in falcons. *J Clin Microbiol*, 2005. 43(7): 3402–3413.

112. Tomaszewski, E. and D. Phalen, Falcon adenovirus in an American kestrel (*Falco sparverius*). *J Avian Med Surg*, 2007. 21(2): 135–139.

113. Chiocca, S. et al., The complete DNA sequence and genomic organization of the avian adenovirus CELO. *J Virol*, 1996. 70(5): 2939–2949.

114. Marek, A. et al., The first whole genome sequence of a Fowl adenovirus B strain enables interspecies comparisons within the genus Aviadenovirus. *Vet Microbiol*, 2013. 166(1–2): 250–256.

115. Griffin, B.D. and E. Nagy, Coding potential and transcript analysis of fowl adenovirus 4: Insight into upstream ORFs as common sequence features in adenoviral transcripts. *J Gen Virol*, 2011. 92(6): 1260–1272.

116. Sheppard, M. and H. Trist, The identification of genes for the major core proteins of fowl adenovirus serotype 10. *Arch Virol*, 1993. 132(3–4): 443–449.

117. Corredor, J., P. Krell, and E. Nagy, Sequence analysis of the left end of fowl adenovirus genomes. *Virus Genes*, 2006. 33(1): 95–106.

118. Corredor, J. et al., Sequence comparison of the right end of fowl adenovirus genomes. *Virus Genes*, 2008. 36(2): 331–344.

119. Ojkić, D. and E. Nagy, The complete nucleotide sequence of fowl adenovirus type 8. *J Gen Virol*, 2000. 81: 1833–1837.

120. Grgić, H., D.H. Yang, and E. Nagy, Pathogenicity and complete genome sequence of a fowl adenovirus serotype 8 isolate. *Virus Res*, 2011. 156(1–2): 91–97.

121. Lüschow, D. et al., Adenovirus of psittacine birds: Investigations on isolation and development of a real-time polymerase chain reaction for specific detection. *Avian Pathol*, 2007. 36(6): 487–494.

122. Bodewes, R. et al., Identification and characterization of a novel adenovirus in the cloacal bursa of gulls. *Virology*, 2013. 440(1): 84–88.

123. Hess, M., H. Blöcker, and P. Brandt, The complete nucleotide sequence of the egg drop syndrome virus: An intermediate between mastadenoviruses and aviadenoviruses. *Virology*, 1997. 238(1): 145–156.

124. Kovács, E.R. and M. Benko, Complete sequence of raptor adenovirus 1 confirms the characteristic genome organization of siadenoviruses. *Infect Genet Evol*, 2011. 11(5): 1058–1065.

125. Pitcovski, J. et al., The complete DNA sequence and genome organization of the avian adenovirus, hemorrhagic enteritis virus. *Virology*, 1998. 249(2): 307–315.

126. Katoh, H. et al., A novel budgerigar-adenovirus belonging to group II avian adenovirus of Siadenovirus. *Virus Res*, 2009. 144(1–2): 294–297.

84 Betaherpesvirinae (*Suid Herpesvirus* 2)

Timothy J. Mahony

CONTENTS

84.1 Introduction 747
    84.1.1 Classification.. 747
    84.1.2 Transmission, Clinical Features, and Pathogenesis.. 748
    84.1.3 Diagnosis 748
        84.1.3.1 Conventional Approaches 748
84.2 Methods 749
    84.2.1 Sample Collection and Preparation 749
    84.2.2 Virus Isolation 749
    84.2.3 Molecular Detection 749
    84.2.4 Sample Extraction... 749
    84.2.5 Loop-Mediated Isothermal Amplification Assay... 749
        84.2.5.1 Procedure 749
    84.2.6 Quantitative Real-Time Polymerase Chain Reaction 750
        84.2.6.1 Procedure 750
84.3 Conclusion 751
References...751

84.1 INTRODUCTION

Members of the subfamily *Betaherpesvirinae* are characterized by host-specific infection and cell type–specific replication in culture. In general, these viruses are cell-associated and slow growing in culture. The site of latency in some members of the subfamily is monocytes. The genomes of viruses belonging to this subfamily are characterized by the presence of homologues of the *Human herpesvirus 5* (HHV-5) gene US22 [1].

The most well studied and characterized of the subfamily *Betaherpesvirinae* is HHV-5, more commonly known as human cytomegalovirus. The descriptive name of viruses belonging to the *Betaherpesvirinae* subfamily, "cytomegalovirus," is derived from the characteristic feature of the infection, which is enlarged cells, or cytomegalo. As with many other herpesviruses, HHV-5 is ubiquitous in the human population and largely asymptomatic. However, HHV-5 can be of concern in immunocompromised patients such as transplant recipients and those with acquired immunodeficiency syndrome. HHV-5 is also of concern to infected pregnant women as it may cross the placenta and have negative effects on the fetus. As a result, considerable effort has been placed on the development of a safe and effective vaccine for preventing transplacental infections in humans.

There are few well-studied betaherpesvirus that infect animals other than some species of primates and rodents. *Bovine herpesvirus 4* and *Equine herpesvirus 2* were initially placed within the betaherpesviruses; however, molecular data resulting in these viruses are being reclassified to the subfamily *Gammaherpesvirus* [2,3]. *Elephantid herpesvirus 1* has also been assigned to the betaherpesviruses and has emerged as an important infection of elephant populations [4].

Currently, the sole member of the subfamily *Betaherpesvirinae* known to infect livestock is the *Suid herpesvirus 2* (SuHV-2), which is also known as the porcine cytomegalovirus. The clinical disease in pigs caused by SuHV-2 is referred to as inclusion body rhinitis, which was first described in 1955 [5]. SuHV-2 has a widespread distribution and generally is only of clinical relevance in young swine. It has gained some interest in recent years as it represents a potential risk in xenotransplantation of porcine cells and tissues to humans [6]. This chapter focuses on the current status, biological properties, and diagnosis/detection of SuHV-2 infection in pigs.

84.1.1 CLASSIFICATION

Forming one of the three subfamilies (*Alphaherpesvirinae*, *Betaherpesvirinae*, and *Gammaherpesvirinae*) in the family *Herpesviridae*, the subfamily *Betaherpesvirinae* is currently divided into four genera (i.e., *Cytomegalovirus*, *Muromegalovirus*, *Proboscivirus*, and *Roseolovirus*). Although these genera share a common lineage within the subfamily, they clearly diverge from other viruses in the family *Herpesviridae*. The characteristic biological feature of the subfamily is the enlargement of virus-infected cells referred to as cytomegalic.

The genus *Cytomegalovirus* includes the type species HHV-5, *Panine herpesvirus 2* (chimpanzee cytomegalovirus), and other viruses that infect primates. The genus *Muromegalovirus* includes the two species that infect rodents, *Murid herpesvirus 1* and *Murid herpesvirus 2*, which infect mice and rats, respectively. The viruses assigned to the *Cytomegalovirus* and *Muromegalovirus* generally have genomes that exceed 200 kb in length.

The genus *Proboscivirus* contains a single virus species, ElHV-1, which is associated with severe and fatal disease in elephants. The genome ElHV-1 is approximately 177 kb in length [7]. The inability to culture strains of ElHV-1 has hampered research efforts due to a complete reliance on material sourced from clinical infected animals and samples obtain at necropsy. The most recent molecular analyses of ElHV-1 genome and strains have led to a proposal that these strains not only represent a number of separate viral species but have also recommended the establishment of a new subfamily, *Deltaherpesvirinae*, within the family *Herpesviridae* [8,9]. This recommendation is yet to be ratified by the International Committee on Taxonomy of Viruses.

The fourth genus of the *Betaherpesvirinae* subfamily is *Roseolovirus*. Viruses assigned to this genus are generally characterized by smaller genomes compared to other genera with genomes less than 200 kb in length. The genus currently includes HHV-6 and HHV-7, which are distinct species although they are serologically related [10]. These viruses are the cause of roseola infantum in infants [11].

Unassigned species within the subfamily are *Caviid herpesvirus 2* and *Tupaiid herpesvirus 1*, which infect guinea pigs and tree shrews, respectively. The most well studied of the betaherpesvirus animal viruses, SuHV-2, has not been assigned to a specific genus. It was recently recommended that SuHV-2 be classified within the genus *Roseolovirus* based on the analyses using the complete viral genome sequence [12]. This recommendation has not been adopted yet by the International Committee on Taxonomy of Viruses.

84.1.2 Transmission, Clinical Features, and Pathogenesis

SuHV-2 is widely distributed and has been reported to be endemic in many swine herds. In countries where the virus has been investigated, the infection rate of herds can vary from 12% to 50% with up to 90% of animals serologically positive in some herds. Where the virus is endemic, it may persist in the absence of clinical signs. Expression of the disease is more likely to occur as a result of introduction, poor management, or coinfections.

SuHV-2 can be transmitted both vertically and horizontally [13]. The clinical presentation can vary depending on the age of the animals in question and the viral status of the herd. A SuHV-2 incursion into a susceptible herd will likely result in transplacental infection and horizontal transmissions, which can result to severe disease. Transplacental infections can result in fetal loss. Animals surviving to birth are likely to exhibit severe respiratory signs along with runting and poor production performance. Infection of young naïve animals, less than 10 weeks of age, can result in generalized clinical signs of rhinitis including ocular and nasal discharge. In very young animals, less than 2 weeks of age, more severe disease

can develop where the nasal discharge becomes mucopurulent resulting in potential blockage of the nasal passages. Piglets with severe disease may be unable to suckle effectively and may die due to rapid loss of body condition.

In endemically infected herds, maternal antibody can prevent transplacental infections and suppress the development of severe disease resulting from horizontal virus transmission [14]. The most likely outcome of infection in these herds is subclinical disease expression.

There are few studies on the pathogenesis of SuHV-2 and the clinical significance of infections in older animals, an area that requires further investigation. An example of this is a comprehensive study on porcine respiratory disease complex (PRDC) reported by Hansen et al. [15]. This study explored the association of five viruses including SuHV-2 and seven bacterial species with PRDC cases by histopathology, pathogen culture, and pathogen detection. Of all the pathogens examined, SuHV-2 was more commonly associated with controls (43%) compared to cases (24%), although this association was not statistically significant. The study did conclude that in acute cases of PRDC, the presence of SuHV-2 was more common compared to subacute and chronic cases. While further research is required, the results suggested that when SuHV-2 was present, more severe diseases were observed.

84.1.3 Diagnosis

The diagnosis of SuHV-2 infection can be confirmed with a range of experimental methodologies.

84.1.3.1 Conventional Approaches

Serological methods have been described for the detection of SuHV-2-specific immunological responses following the infection of swine using virus-derived and recombinant antigens [16,17]. SuHV-2 can be readily grown in cell cultures from a variety of clinical samples. Histological examination of tissue sections or nasal scrapings with hematoxylin and eosin staining, particular tissues of the kidney, and nasal epithelia can be useful in diagnosis. The presence of cells with large, basophilic, intranuclear inclusions bodies exhibiting cytomegalia strongly suggests a SuHV-2 infection. Sekiguchi et al. [18] described *in situ* and immunohistochemistry methodologies for enhanced specificity in the detection of SuHV-2 in tissue sections. Electron microscopy has also been used to confirm the presence of SuHV-2 in glandular epithelial cells [19].

Molecular detection: Various molecular assays have been used to confirm the presence of SuVH-1 in samples and virus isolation experiments. Initially, the herpesviral consensus oligonucleotides designed using conserved motifs of DNA polymerase (*dpol*) provided the first sequence information of SuHV-2 PCR, and these data were used to develop a specific PCR assay [20–22]. After these initial reports, a number of conventional, semiquantitative and quantitative real-time PCR assays have been reported for the detection and quantitation of SuHV-2 [21,23–25]. The impetus for most of these studies was not for the detection of SuHV-2 in clinical samples from pigs but rather in the risk management processes for xenotransplantation studies in primates as models for human transplants.

84.2 METHODS

84.2.1 SAMPLE COLLECTION AND PREPARATION

Due to a commonality of clinical signs, the SuHV-2-associated disease may require a discriminated test. The possible confounding agents will depend on the status of the country in question but could include swine fever virus, porcine circovirus 2, porcine respiratory and reproductive syndrome viruses, porcine parvovirus, and porcine enterovirus. Lee et al. [26] have reported a multiplex conventional PCR assay for the detection of SuHV-1, porcine herpesvirus 1, and porcine circovirus.

The samples of choice from live animals for the isolation and detection of SuHV-2 are nasal swabs, nasal scrapings, or whole blood. Suitable samples collected at necropsy include turbinate mucosa and lung lavage samples for the isolation of pulmonary macrophages and kidney tissues. The virus may be more widely distributed in tissues of aborted fetuses, allowing for a more expansive list of tissues to be utilized for isolation.

84.2.2 VIRUS ISOLATION

Cells: SuHV-2 can be cultured in the laboratory although slow-growing SuHV-2 will reach modest titers ($\leq 1 \times 10^6$ TCID$_{50}$/mL) in primary cells established from swine testis, ovarian, and lung tissues. The virus can also be grown directly from or in porcine pulmonary macrophages recovered from lung lavage samples. The virus will also replicate efficiently in continuous porcine turbinate PT-K75 cells (ATCC CRL-2528) and porcine kidney epithelial PK-15 cells (ATCC CCL-33) to similar tires.

84.2.3 MOLECULAR DETECTION

A number of molecular assays have been reported for the detection of SuHV-2 in clinical materials. The diagnosis of clinical disease to SuHV-2 should take into consideration the clinical presentation as well as the presence of the virus or viral genome. Two potential methodologies are outlined in the succeeding text for the specific detection of the SuHV-2 genome in extracts from various samples such as nasal scrapings, tissues, or cell culture supernatants.

84.2.4 SAMPLE EXTRACTION

Isolation of the SuHV-2 genomic DNA can be readily extracted from a suitable quantity of available tissue (20–30 mg) or serum (200 μL) or whole blood (200 μL) using a suitable commercially available kit such as the Qiagen blood and tissue extraction kit as described by the manufacturer. The integrity of the extracted DNA can be analyzed

using standard gel electrophoresis, and the quantity can be estimated using UV spectroscopy. For most amplification-based detection methodologies, the integrity of the DNA is not critical with successful results being obtained from degraded samples. Despite this flexibility, a negative result in highly degraded samples should be interpreted with caution.

Some operators prefer to standardize the quantity of input DNA for amplification. This is not essential but can aid in determining the comparative viral load of particular samples where this is of interest. While the template concentration is not critical, care should be taken not to exceed 10 ng/μL of template; the standard working range for quantitative real-time PCR (qPCR) is between 2 and 5 ng/μL, which is suitable for most applications. Caution should be used in drawing conclusions based on viral loads with a pathogen qPCR assay alone from tissue samples. Ideally, if comparisons are required between tissues samples, a host gene should be consider as a calibrator/reference gene to account for variation in the starting quantity of the tissue.

84.2.5 LOOP-MEDIATED ISOTHERMAL AMPLIFICATION ASSAY

The presence or absence of SuHV-2 genomic DNA in extracted nucleic acids can be determined using loop-mediated isothermal amplification assay (LAMP) with minimal specialized equipment [27]. The LAMP reaction, targeting the *dpol*, uses the following combination of oligonucleotides.

Oligonucleotide	Sequence (5′–3′)	Nucleotide Position	Amplicon Size (bp)
F3 forward outer	TATCTGGTTCTG GCGGAC	46,356–46,373[a]	211
B3 backward outer	AGCATATTTCTCTT TCTAGTCTC	46,163–46,185	
FIP (forward inner primer)	TAGCAGATGCTTC CATATGGTAATT-GGGCAGATATTGT ATACAGGA	46,266–46,290	
		46,325–465,345[a]	
BIP backward inner	TTCTAAGTTGG CCTACTTGCCC-ATGAAGGATA CACGTGAACAC	46,243–46,264[a]	
LF loop forward	ACAACTCCTTAAC GATCACCGAA	46,186–46,206	
LB loop backward	ATCAGGAAGGTG ATAAATGATGGAC	46,218–46,242[a]	

[a] Complement to viral genome sequence.

84.2.5.1 Procedure

1. Thaw reaction components on ice. Once thawed, gently mix and pulse spin reagents to ensure homogeneity.

2. Prepare required reaction volumes in multiples of the following:

Reaction Component	Volume (µL)
F3 forward outer (20 µM)	0.25
B3 backward outer (20 µM)	0.25
FIP forward inner (20 µM)	1.0
BIP backward inner (20 µM)	1.0
LF loop forward (20 µM)	0.5
LB loop backward (20 µM)	0.5
dATP, dCTP, dGTP, dTTP (25 mM)	1.0
Betaine (25 M)	1.0
Isothermal amplification buffer (×10)[a]	2.5
Bst 2.0 WarmStart® DNA polymerase (8 U/µL)[a]	1.0
DNA/RNA-free water	15
Total	24

[a] Sourced from New England Biolabs.

1. Add 24 µL per reaction tube.
2. Add 1 µL of each nucleic acid extract sample (≤100 ng) to labeled reaction tubes. Include template-free reaction by adding 1 µL of water to one tube.
3. Close/seal tubes, gently mix, and pulse spin.
4. Place reactions at 65°C for 40 min.
5. Three methodologies can be used to determine the results of the LAMP assay:
   a. *Visual inspection*: As a result of the accumulation of the by-product magnesium pyrophosphate in the LAMP reaction, the reaction mixture will appear turbid in those reactions with sufficient quantities of starting SuHV-2 template DNA. Test reactions are compared to the controlled reaction (no template).
   b. *Agarose gel electrophoresis*: Standard agarose gel electrophoresis can be used to assess the amplification results. An aliquot (5 µL) of the reaction is run on 2% agarose gel with appropriate staining. Successful amplification is evident through the observation of multiple amplification products ranging in size from approximately 200 to 1000 bp. No amplification products should be in the reactions in the absence of added templates.
   c. *Addition of fluorescent dye*: The addition of a suitable DNA interchelating fluorescent dye, such as SYBR green, can be used to enhance the detection of LAMP-amplified products. This is performed by adding 1 µL of 1/1000 working solution, which can be added to each reaction. Positive reactions are expected to change to green in color. A 10-fold increase in sensitivity is expected compared to the evaluation of magnesium pyrophosphate accumulation alone.

Note: The LAMP amplification products result from the complex binding reaction of oligonucleotides and strand displacement activity of the *BSt dpol* (see review [28]). Alignment of SuHV-1 *dpol* sequences from GenBank suggests a variation within the primary LAMP amplicon (data not shown). A negative LAMP amplification result, where other observations such as clinical presentation and/or histopathology suggest a SuHV-2 infection, may require additional investigations such as consensus *dpol* PCR and amplicon sequencing [20]. Any subsequence redesign of the LAMP oligonucleotides should be performed using an appropriate software package for optimal performance.

84.2.6 QUANTITATIVE REAL-TIME POLYMERASE CHAIN REACTION

The presence or absence of SuHV-2 genomic DNA in extracted nucleic acids can also be determined using qPCR, which is the oligonucleotide combination of PCMVpolF, PCMVpolR, and PCMV probe dual labeled with 6-carboxyfluorescein (6FAM) and tetramethylrhodamine (TAMRA) [24]. The locations of the oligonucleotides in the SuHV-2 genome sequence are shown, which target *dpol* on the negative strand of the viral genome [12]. The reaction uses the following oligonucleotides:

Oligonucleotide	Sequence (5′–3′)	Nucleotide Position	Amplicon Size (bp)
PCMVpolF	GCTGCCGTGTC TCCCTCTAG	45,699–45,718[a]	82
PCMVpolR	ATTGTTGATAAA GTCACTCGTCTGC	45,637–45,661	
PCMV probe	6FAM-CCATCAC CAGC ATAGGGCGGGAC- TAMRA	45,670–45,692[a]	

[a] Complement to viral genome sequence.

84.2.6.1 Procedure

1. Thaw reaction components on ice. Once thawed, gently mix and pulse spin reagents to ensure homogeneity.
2. Prepare required reaction volumes in multiples of the following:

Reaction Component	Volume (µL)
Qantitect Master Mix (×2)	10.0
Primer PCMVpolF (0.2 µM)	0.2
Primer PCMVpolF (0.2 µM)	0.2
Dual-labeled probe PCMV-P (0.2 µM)	0.2
DNA/RNA-free water	7.4
Total	18.0

3. Add 18 μL per reaction tube.
4. Add 2 μL of each nucleic acid sample (≤100 ng) to labeled reaction tubes. Include template-free reaction by adding 2 μL of water to one tube.
5. Close/seal tubes, gently mix, and pulse spin.
6. Place in qPCR thermocycler and cycle as follows: 1 cycle of 95°C for 5 min; 40 cycles of 95°C for 15 s and 60°C for 75 s.
7. Fluorescent amplification signals are acquired using default instrument settings. Analyze experimental outputs with the proprietary instrument software provided according to the manufacturer's instructions.

Note: The sensitivity of the qPCR makes template control critical, and every effort should be made to prevent reaction contamination. Where possible, each step in the sample analysis process from collection, nucleic acid extraction, qPCR reaction setup, qPCR amplification, and any postamplification manipulations should be physically separated.

The reaction components, setup, and cycling described earlier are generally appropriate for the Master Mix QuantiTec® kit (Qiagen) with cycling conditions and data acquisition using a Rotor Q (Qiagen). The manufacturer has variations of these Master Mix reagents available that are compatible with a range of qPCR cyclers, and it is strongly recommended that the user consults the supporting documentation to ensure that the most suitable Master Mix is selected for the instrument to be used. It should also be noted that the manufacturer recommends reaction volumes of 50 μL. Reduced reaction volumes are used in the authors' laboratory for a wide range of qPCR assays to detect herpesvirus in various samples with no impact on the assay performance. For users new to the technology, it may be more appropriate to initially use the recommended reaction volumes before reducing assay volumes. The user should also ensure that the reduced reaction volume is compatible with the available qPCR instrument.

A comparison of the SuHV-2 *dpol* sequences available from GenBank demonstrated strain-dependent sequence variations within the qPCR 82 bp amplicon. While the primer and probe-binding sites were invariant, a variable nucleotide was identified in approximately half of the 16 sequences examined (data not shown). If any postamplification validations such as high resolution melt analyses are performed, the impact of these variations should be considered.

84.3 CONCLUSION

Given the ubiquity with which herpesviruses are distributed across vertebrate animal species, it is reasonable to conclude that there are many betaherpesviruses that infect livestock, which remain undiscovered. This likely reflects the homeostasis the virus and host have attained through coevolution. It is only when something perturbs this established relationship that the presence of these agents becomes apparent. Further, the lack of in-depth investigation in many veterinary disease episodes may also contribute to the underrepresentation of viral species infecting livestock in this subfamily. Exploratory studies in wild suid and bat species have reported evidence of novel betaherpesviruses [29,30]. The existence of undetected betaherpesviruses in livestock warrants further investigation.

As SuHV-2 is the most characterized of the animal betaherpesviruses, it will likely remain of interest due to the continued research on xenotransplantation between swine and humans. In pig–primate xenotransplant studies, reactivation of SuHV-2 has been demonstrated, and in the immunosuppressed setting of transplant recipients, it is of concern [6]. A benefit of these studies is that unlike many other herpesvirus infections of animals, the sensitivities of SuHV-2 to antiviral drugs have also been reported [23]. However, it is considered unlikely these would ever be used in pig production systems. Similarly, the capacity of SuHV-2 to be transmitted across the placenta and infections in piglets make it an ideal experimental model for testing of vaccination studies to prevent HHV-5 cross-placental infections in human [31].

While SuHV-2 is of importance in swine production, it can be effectively managed with the implementation and maintenance of on-farm biosecurity measures. Studies have been unable to establish links between the presence of SuHV-2 and porcine diseases with complex etiologies such as PRDC and porcine periweaning failure-to-thrive syndrome [15,32]. More recently, molecular investigations suggest novel and complex interactions between SuHV-2 and the porcine immune system resulting in immune suppression [33]. To resolve these issues, the continued surveillance for SuHV-2 using the outlined detection methodologies here will be important to more accurately assess its impact on production.

REFERENCES

1. Pellett, P.E.D. et al., Herpesvirales, in *Virus Taxonomy Classification and Nomenclature of Viruses*, A.M.Q. King, Adams, M.J., Carstens, E.B., and Lefkowitz, E.J., Eds. 2012, Elsevier Academic Press, Amsterdam, the Netherlands, pp. 99–107.
2. Telford, E.A. et al., Equine herpesviruses 2 and 5 are gammaherpesviruses. *Virology*, 1993. 195(2): 492–499.
3. Thiry, E. et al., Molecular biology of bovine herpesvirus type 4. *Vet Microbiol*, 1992. 33(1–4): 79–92.
4. Ehlers, B. et al., Genetic and ultrastructural characterization of a European isolate of the fatal endotheliotropic elephant herpesvirus. *J Gen Virol*, 2001. 82(3): 475–482.
5. Done, J.T., An inclusion body rhinitis of pigs. *Vet Rec*, 1955. 67: 525–527.
6. Mueller, N.J. and J.A. Fishman, Herpesvirus infections in xenotransplantation: Pathogenesis and approaches. *Xenotransplantation*, 2004. 11(6): 486–490.
7. Ling, P.D. et al., Complete genome sequence of Elephant Endotheliotropic Herpesvirus 1A. *Genome Announc*, 2013. 1(2): e0010613.
8. Zong, J.C. et al., Comparative genome analysis of four Elephant Endotheliotropic Herpesviruses, EEHV3, EEHV4, EEHV5, and EEHV6, from cases of hemorrhagic disease or Viremia. *J Virol*, 2014. 88(23): 13547–13569.

9. Richman, L.K. et al., Elephant Endotheliotropic Herpesviruses EEHV1A, EEHV1B, and EEHV2 from cases of hemorrhagic disease are highly diverged from other mammalian herpesviruses and may form a new subfamily. *J Virol*, 2014. 88(23): 13523–13546.

10. Ward, K.N., The natural history and laboratory diagnosis of human herpesviruses-6 and -7 infections in the immunocompetent. *J Clin Virol*, 2005. 32(3): 183–193.

11. Stone, R.C., G.A. Micali, and R.A. Schwartz, Roseola infantum and its causal human herpesviruses. *Int J Dermatol*, 2014. 53(4): 397–403.

12. Gu, W. et al., Genomic organization and molecular characterization of porcine cytomegalovirus. *Virology*, 2014. 460–461: 165–172.

13. Edington, N., R.G. Watt, and W. Plowright, Experimental transplacental transmission of porcine cytomegalovirus. *J Hyg (Lond)*, 1977. 78(2): 243–251.

14. Edington, N., A.E. Wrathall, and J.T. Done, Porcine cytomegalovirus (PCMV) in early gestation. *Vet Microbiol*, 1988. 17(2): 117–128.

15. Hansen, M.S. et al., An investigation of the pathology and pathogens associated with porcine respiratory disease complex in Denmark. *J Comp Pathol*, 2010. 143(2–3): 120–131.

16. Assaf, R., A.M. Bouillant, and E. Di Franco, Enzyme linked immunosorbent assay (ELISA) for the detection of antibodies to porcine cytomegalovirus. *Can J Comp Med*, 1982. 46(2): 183–185.

17. Tajima, T. et al., Application of enzyme-linked immunosorbent assay for the seroepizootiological survey of antibodies against porcine cytomegalovirus. *J Vet Med Sci*, 1993. 55(3): 421–424.

18. Sekiguchi, M. et al., In situ hybridization and immunohistochemistry for the detection of porcine cytomegalovirus. *J Virol Methods*, 2012. 179(1): 272–275.

19. Deim, Z. et al., Inclusion body rhinitis in pigs in Hungary. *Vet Rec*, 2006. 158(24): 832–834.

20. Widen, B.F. et al., Development of a PCR system for porcine cytomegalovirus detection and determination of the putative partial sequence of its DNA polymerase gene. *Epidemiol Infect*, 1999. 123(1): 177–180.

21. Hamel, A.L. et al., PCR assay for detecting porcine cytomegalovirus. *J Clin Microbiol*, 1999. 37(11): 3767–3768.

22. Rupasinghe, V. et al., Analysis of porcine cytomegalovirus DNA polymerase by consensus primer PCR. *J Vet Med Sci*, 1999. 61(11): 1253–1255.

23. Fryer, J.F. et al., Susceptibility of porcine cytomegalovirus to antiviral drugs. *J Antimicrob Chemother*, 2004. 53(6): 975–980.

24. Fryer, J.F. et al., Quantitation of porcine cytomegalovirus in pig tissues by PCR. *J Clin Microbiol*, 2001. 39(3): 1155–1156.

25. Tanaka, N. et al., Quantitative analysis of cytomegalovirus load using a real-time PCR assay. *J Med Virol*, 2000. 60(4): 455–462.

26. Lee, C.S. et al., Multiplex PCR for the simultaneous detection of pseudorabies virus, porcine cytomegalovirus, and porcine circovirus in pigs. *J Virol Methods*, 2007. 139(1): 39–43.

27. Yang, J.L. et al., Development and evaluation of a loop-mediated isothermal amplification assay for the rapid detection of porcine cytomegalovirus under field conditions. *Virol J*, 2012. 9: 321.

28. Parida, M. et al., Loop mediated isothermal amplification (LAMP): A new generation of innovative gene amplification technique; perspectives in clinical diagnosis of infectious diseases. *Rev Med Virol*, 2008. 18(6): 407–421.

29. Wibbelt, G. et al., Discovery of herpesviruses in bats. *J Gen Virol*, 2007. 88(10): 2651–2655.

30. Ehlers, B. and S. Lowden, Novel herpesviruses of Suidae: Indicators for a second genogroup of artiodactyl gammaherpesviruses. *J Gen Virol*, 2004. 85(4): 857–862.

31. Schleiss, M.R., Developing a vaccine against congenital cytomegalovirus (CMV) infection: What have we learned from animal models? Where should we go next? *Future Virol*, 2013. 8(12): 1161–1182.

32. Huang, Y., H. Gauvreau, and J. Harding, Diagnostic investigation of porcine periweaning failure-to-thrive syndrome: Lack of compelling evidence linking to common porcine pathogens. *J Vet Diagn Invest*, 2012. 24(1): 96–106.

33. Liu, X. et al., Transcriptome analysis of porcine thymus following porcine cytomegalovirus infection. *PLoS One*, 2014. 9(11): e113921.

85 Iltovirus (Infectious Laryngotracheitis Virus)

Joseph J. Giambrone and Shan-Chia Ou

CONTENTS

85.1 Introduction 753
    85.1.1 Classification, Morphology, Genome Organization, and Biology 753
        85.1.1.1 Classification 753
        85.1.1.2 Morphology 753
        85.1.1.3 Genome Organization 754
        85.1.1.4 Biology 756
    85.1.2 Transmission and Pathogenesis 756
    85.1.3 Clinical Features 757
    85.1.4 Diagnosis 757
        85.1.4.1 Conventional Techniques 757
        85.1.4.2 Molecular Techniques 757
85.2 Methods 758
    85.2.1 Sample Preparation 758
    85.2.2 Detection Procedures 758
        85.2.2.1 Real-Time PCR 758
        85.2.2.2 LAMP 759
85.3 Conclusion and Future Perspectives 761
References 761

85.1 INTRODUCTION

Infectious laryngotracheitis (ILT), caused by infectious laryngotracheitis virus (ILTV) or gallid herpesvirus 1, is an economically important respiratory disease of chickens worldwide. The natural hosts include chickens, turkeys, and pheasants. Although all ages of chickens are vulnerable to ILTV, the most sensitive age is above 3 weeks [1,46]. In the virulent form of ILT, the clinical signs include gasping, coughing, bloody mucus in trachea, and dyspnea, which lead to suffocation. This form can result in 90%–100% morbidity and 5%–70% mortality. The signs of the mild form include nasal discharge, conjunctivitis, reduced egg production and shell quality, decreased weight gain, and increased condemnation rate. The mortality and morbidity of the mild form is about 5% [2]. Despite the availability of various vaccines (attenuated live, inactivated, recombinant), ILTV still causes significant economic losses.

ILT was first reported in the 1920s in the United States; however, this disease may have existed in chickens much earlier [3,4]. In the 1930s, the name infectious laryngotracheitis was formally used for this disease. The agent of ILT was verified as a herpesvirus in 1963 [4]. From then on, ILT become prevalent in chickens all over the world especially in countries with large poultry industries.

85.1.1 CLASSIFICATION, MORPHOLOGY, GENOME ORGANIZATION, AND BIOLOGY

85.1.1.1 Classification

The *Herpesviridae* family covers a large group of DNA viruses, with more than 200 different species identified in humans and animals. The *Herpesviridae* family can be separated into three subfamilies, *Alphaherpesvirinae*, *Betaherpesvirinae*, and *Gammaherpesvirinae*, which share similar basic biological features. The *Alphaherpesvirinae* consists of four genera, including the genera *Mardivirus* and *Iltovirus*, which are known to infect avian hosts. Specifically, avian Marek's disease viruses and herpesvirus of turkey belong to *Mardivirus*, while ILTV (or gallid herpesvirus 1) and psittacid herpesvirus 1 (PsHV-1) are members of *Iltovirus*. Several alphaherpesviruses cause important diseases in animals, such as pseudorabies virus (PrV), bovine herpesvirus 1 and 5, and equid herpesvirus 1 and 4 [5,6]. The DNA sequence of ILTV is different from those of other alphaherpesviruses [7].

85.1.1.2 Morphology

ILT viral particles, as herpes simplex virus (HSV), have icosahedral symmetry and are 200–350 nm in diameter. The spherical virion comprises core, capsid, tegument, and envelope.

The viral genome is packaged as a single, linear, double-stranded DNA molecule into a capsid. The nucleocapsid of ILTV contains 162 capsomers, which contain 150 hexons and 12 pentons, and the triangulation number (T) is 16. The major capsid proteins are constructed by six copies of hexons and five copies of pentons. The diameter of virion is about 195–250 nm. The envelope surrounding the nucleocapsid is a lipid bilayer, which is intimately associated with the outer surface of the tegument. It contains a number of integral viral glycoproteins [7]. The buoyant density of ILTV DNA in CsCl is 1.704 g/mL, and the molecular weight is 100×10^6 Da. The genome of ILTV is a linear 155 kb double-stranded DNA, which comprises long and short unique regions (U_L, U_S) and inverted repeat sequences (internal repeat, IR; terminal repeat, TR) that flank the U_S regions. The genome forms two isomers, which are different from the orientation of the U_S regions [8]. This genome structure is designated as type D herpesvirus genomes [9].

ILTV is enveloped and thus sensitive to the lipolytic solvents, such as ether and chloroform [10]. ILTV can be inactivated by ether after 24 h [11]. ILTV is also susceptible to heat. At 55°C for 10–15 min or 38°C for 48 h, this virus is readily inactivated. However, different strains of ILTV show varied resistance to heat [10], and the Belgian strain had partial infectivity after 56°C at 1 h. In chicken tracheae and chorioallantoic membranes (CAMs), ILTV can be destroyed in 44 h at 37°C or inactivated in 5 h at 25°C [3]. ILTV can be destroyed in 1 min by treating with 3% cresol or a 1% lye solution [10]. On the chicken farm, 5% hydrogen peroxide mist with fumigating equipment can completely inactivate ILTV. At a lower temperature, ILTV can maintain its infectivity for a long period. The virus can survive for 10–100 days in tracheal exudates and chicken carcasses at 13°C–23°C [2]. When stored at −20°C to −60°C, ILTV can be viable for several months to years. Storage media containing 50% glycerol or sterile skim milk can greatly increase the infectivity in tracheal swabs [16].

85.1.1.3 Genome Organization

ILTV genome is double-stranded linear DNA. By assembling 14 ILTV genome sequence data published by different laboratories, the ILTV genome was confirmed to be 148,687 bp in length, with a G + C content of 48.16% [13] (Figure 85.1). The ILTV genome consists of 77 predicted open reading frames (ORFs), 63 of which are homologous to HSV-1 genes, the prototype of alphaherpesviruses [5]. The U_L region may contain 62 genes, and the U_S region has 9 predicted genes. There are five ORFs (ORF A–E) located in U_L region, which were unique for ILTV and PsHV-1 [14,15]. In the other U_L region, upstream of the conserved UL1 gene, UL0 and UL (-1) were ILTV-specific genes [35]. They were expressed during the late phase of viral replication in cell culture, and both of their mRNAs are spliced with removing introns at the 5′ ends. These two gene products showed 28% homology, which indicated a duplication of ILTV evolution [16]. The ILTV genome contained internal inversion within U_L region from UL22 to UL44 genes, which were similar to the inversion from UL27 to UL44 genes

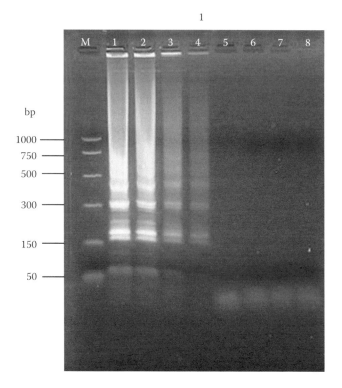

FIGURE 85.1 Sensitivity of the loop-mediated isothermal amplification assay for infectious laryngotracheitis virus detection. Electrophoresis photo of serial 10-fold dilutions of infectious laryngotracheitis virus (ILTV) ICP4. The ILTV-positive products showed many bands with different sizes and a smeared DNA between these bands. The bands smaller than 50 bp were primer dimmers. M: DNA marker (PCR Makers, Promega). Lane 1: 6×10^4 copies/μL; lane 2: 6×10^3 copies/μL; lane 3: 6×10^2 copies/μL; lane 4: 60 copies/μL; lane 5: 30 copies/μL; lane 6: 15 copies/μL; lane 7: 6 copies/μL; lane 8: negative control.

of PrV. The biological functions of these inversions were not clear [15,17]. The UL47 gene of ILTV changed to U_S region, and there was no UL16 homologous gene in ILTV genome [13,18,30]. ILTV had three viral replication regions (Ori), one located in the middle of U_L region (Ori$_L$) and two copies in the U_S region (Ori$_S$). There were 11 structural glycoproteins of ILTV homologous to HSV-1 (gB–gM), but between UL2 and UL3 of ILTV was an ORF, UL3.5, which was conserved in other alphaherpesviruses but not HSV-1 [19].

According to the phylogenetic analysis, comparative DNA, and amino acid sequences, researchers indicated that ILTV and PsHV-1 genes shared limited homology to other alphaherpesviruses and were distinct from the other two avian alphaherpesviruses, Marek's disease virus and herpesvirus of turkeys. Their studies revealed that ILTV was a separate branch in the phylogenetic tree and could be an early type of alphaherpesviruses [6,13].

There are many conserved ORFs in ILTV and other alphaherpesviruses genomes. In recent research, the PsHV-1 was the most similar virus to ILTV in gene arrangement and gene sequences. Therefore, PsHV-1 and ILTV were classified into *Iltovirus* (ILT-like) genus [13]. There are about 40 genes commonly found in all herpesviruses. These gene products include capsid proteins, DNA replication proteins, DNA packing or cleavage proteins, and envelope glycoproteins. They are termed core genes for herpesviruses [20].

Herpesvirus genes can be classified into three groups: immediate early (IE), early (E), and late (L) genes. IE genes are transcribed immediately after infection and are viral genome regulators and activators. E gene products are involved in viral DNA replication and nucleic acid metabolism. Most of the L gene products are virion components, like capsid proteins, tegument proteins, and envelope glycoproteins. In ILTV, many viral proteins have been identified by monoclonal antibodies, monospecific antisera, or direct detected in viral particles. The function and the distributions of ILTV gene products are assigned by [13].

Glycoprotein B (gB): gB is identified in every herpesvirus, and it is the most conserved structural protein. According to amino acid analysis, protein sequence of ILTV gB is similar to other herpesviruses. Therefore, it might have the same function with its homologue [21,22]. The gB is an essential protein for virus attachment and penetration to host cells. It also involves cell-to-cell spread and syncytium formation [23,24]. Furthermore, gB induces humoral and cell-mediated immune response. An ILTV gB subunit vaccine protected chickens from presenting clinic signs and viral replication. Recombinated ILTV gB into fowl pox virus can protect chickens from ILT challenge [25]. gB might be a candidate for an ILTV recombinant or subunit vaccine. In ILTV-infected chicken embryo liver (CEL) cells, gB is synthesized as a 110 kDa monomeric precursor protein in the endoplasmic reticulum and then converted to homodimers, which are composed of two 100 kDa subunits. The mature ILTV gB is composed by two disulfide-linked 58 kDa subunits with N-linked glycohydrate chains and is secreted from the Golgi apparatus.

Glycoprotein C (gC): gC is a structural and nonessential protein for viral replication in cell culture. This glycoprotein involves viral attachment and virulence. In most alphaherpesviruses, gC mediates viral attachment by interacting with the cellular heparan sulfate receptor [26]. However, according to the gC sequence analysis and protein expression research, ILTV gC lacked about 100 amino acids at N-terminal end of the protein, which was the heparin-binding motif, but this did not affect ILTV infectivity. It indicated that ILTV used a different mechanism to attach to the host cells [27,38].

Glycoprotein E and glycoprotein I (gE and gI): gE and gI are nonessential ILTV proteins. They form a heterodimer with noncovalent binding. In HSV-1, gE and gI assists virus cell-to-cell spreading. The gE/gI heterodimer was used to promote cell-to-cell spreading by binding cellular receptor at cell junctions [28]. The gE/gI-deleted mutants of HSV-1, PRV, and bovine herpesvirus 1 reduced their ability to spread cell to cell [29]. In ILTV, gE/gI-deleted mutants showed reduced cell-t-ocell spread, and this gE/gI-deficient ILTV could not form plaques in cell culture. Thus, gE/gI plays a more significant role in cell-to-cell spread of ILTV than other alphaherpesviruses [29].

Glycoprotein G (gG): gG is nonessential and conserved in most of alphaherpesviruses. This glycoprotein is not

assembled into virus particles but secreted from infected cells [9]. The gG of ILTV serves a virulence factor. The gG-deficient ILTV reduced their pathogenicity producing less clinical signs, less effect on weight gain, and less mortality [31]. Workers inoculated this gG-deficient ILTV into birds by eye drop, drinking water, and aerosol routes [31]. Results showed that eye drop and drinking water routes were safe and induced protection from virulent strain challenging. However, aerosol inoculation did not reach the estimated protection level. This gG-deficient ILTV had the potential to become a marker vaccine to quarantine vaccinated and naturally infected birds [31].

Glycoprotein J (gJ): The gJ is the product of ORF 5 gene located in U_S region, a nonessential and structural glycoprotein. This protein was named gp60 because it was identified as a 60 kDa glycoprotein [32]. This ORF 5 gene product was recognized by ILTV gJ monoclonal antibody. Therefore, it was identified as the gJ of ILTV. The ILTV gJ showed different protein forms by separating in different masses [32]. ILTV gJ contains *N*- and *O*-linked carbohydrate chains, and two different translation products are formed by nonspliced and spliced mRNA. The gJ-deficient ILTV has minor effects on cell-to-cell spread, but the virus replication rate was affected. Moreover, the gJ-deficient ILTV reduced their virulence, when compared with parent viral challenge. Therefore, the gJ-deleted ILTV might serve as a marker vaccine for serological differentiation between vaccinated and wild virus-infected birds [33].

Glycoprotein M and glycoprotein N (gM and gN): The gM and gN are gene products of UL10 and UL49.5 in ILTV. They are dispensable for viral replication of most alphaherpesviruses in cell culture [9]. The gM and gN form heterodimeric complex with each other by disulfide link. The gM and gN of many herpesviruses are *O*-glycosylated. However, in ILTV, the gM is not modified by glycosylation [18].

Thymidine kinase (TK): The TK gene is located in the U_L region of viral genome. It is not essential for virus growth in cell culture, but TK is a virulence factor of ILTV. TK-negative alphaherpesviruses reduced the rate of reactivation from latent infection [34]. Chickens, which received TK-deficient ILTV, showed reduced clinical signs and induced protection against virulent ILTV. TK-deficient ILTV may provide a direction for latent infection, reactivation, and as a vaccine candidate.

Other proteins: The UL50 gene of ILTV encodes dUTPase. This protein prevents uracil addition into DNA, when ILTV DNA replication occurs. In ILTV, this protein is nonessential in cell culture. The UL50-deleted ILTV did not attenuate viral pathogenicity. However, when green fluorescent protein (GFP) was inserted into UL50-deleted region, the virulence was readily decreased. The disadvantage was that this GFP-inserted recombinant was unstable [18]. The ILTV-specific UL0 gene can be deleted, and the recombinant virus was almost apathogenic for chickens. Although UL (-1) gene product was structurally homologous to UL0, UL (-1) was indispensable for ILTV [32]. The genes, ORF A–E, were unique in ILTV. They were not essential proteins, but the deficient mutants showed reduced virus titer and plaque size [35].

The gene products of ORF A–E might be important in virus replication [18]. The ILTV genome contains U_L and U_S regions and two inverted repeats (IR, TR). The IR and TR flank the U_S region.

85.1.1.4 Biology

Virus replication of ILTV was similar to that of other alpha-herpesviruses, such as PrV and HSV [36]. At the beginning of infection, the virus attached to the receptors on the cell surface. The envelope fuses with the cell membrane, and virus nucleocapsid was released into the cytoplasm. Viral DNA was transported into the nucleus. Transcription of viral genome, replication of viral DNA, and assembly of new virion occurred in the nucleus [18].

The replication of ILTV DNA was highly regulated in a cascade function. After infecting the cell, viral linear DNA circularized rapidly, and the DNA was replicated by rolling circle mechanism producing concatemers. The DNA concatemers were cleaved into monomers and packaged into newly synthesized nucleocapsids within the nucleus. The viral DNA containing nucleocapsids were transported through the cytoplasm to become enveloped [18]. The enveloped virions were released by cell lysis or budding [18,37].

During ILTV replication, like HSV, protein expression is regulated strictly in a cascade function. The virus turns off host protein synthesis and expresses three groups of viral proteins. When the later viral genes are expressed, the prior genes are downregulated. ILTV uses host RNA polymerase II to transcribe all viral genes. At first stage, the viral α- or IE genes are expressed. The IE gene activation does not need prior protein synthesis. A virion tegument protein, α-transinducing factor, enhances α-genes transcription. The α-gene products stimulate β- or E gene expression. The β-proteins regulate DNA replication, nucleotide metabolism, and γ-gene stimulation. The β-gene expression starts infection and turns off after viral DNA replication. The γ-genes or L genes are divided into two groups based on their timing and requirement for viral DNA synthesis. The γ-genes encode structural components of the virion [18,36].

The capsid proteins are synthesized in the cytoplasm, and then transported into the nucleus for viral capsid assembly. The ILTV progeny DNA concatemers are cleaved into unit length monomers and packaged into capsids by recognizing the specific sequences as signals. After encapsulation of viral DNA, the nucleocapsids are budded into inner nuclear membrane and form primary enveloped virion in the perinuclear space. At primary envelopment steps, UL31 and UL34 proteins form part of primary virion, which is different from mature virus particles, which helps the nucleocapsid to become primary enveloped. The primary enveloped virion then fuses with the outer nuclear membrane and deenvelopes for releasing nucleocapsid into cytoplasm. After transportation into the cytoplasm, the nucleocapsid collects 15 tegument proteins and 10 envelope glycoproteins and is reenveloped by budding into endoplasmic reticulum–Golgi vesicle. This mature virus particle is released by fusing of the vesicle with the cell membrane [44,45].

85.1.2 Transmission and Pathogenesis

The chicken is the primary natural host of ILTV. All ages of chickens are affected, but chickens above 3 weeks old are more susceptible [1]. A report maintained that ILTV can also infect pheasants, pheasant–bantam crosses, and peafowl.

Turkeys can be infected with ILTV experimentally, but they might have an age-dependent resistance, since lesions are restricted in young age birds [39,40]. A natural infection of ILTV in turkeys was reported [40]. The clinical signs in turkeys were similar to that in chickens. Starling, sparrows, crows, pigeons, ducks, and guinea fowls were resisted to ILTV infection. Chicken and turkey embryo eggs were susceptible to ILTV, but guinea fowl and pigeon eggs were not [12].

ILTV can be propagated in chicken embryo eggs and several avian cell cultures. In chicken embryonating eggs, because of necrosis and proliferative tissue, ILTV forms plaques on the CAM. Plaques on the CAM can be observed in 48 h after infection, and embryo death might occur 2–12 days after infection. Different strains of ILTV showed different plaque size and morphology on the CAM [41].

ILTV can be propagated in primary avian cell cultures, such as CEL, chicken embryo kidney (CEK), and chicken kidney (CK) cell cultures [42,43]. Hughes and Jones [52] investigated the sensitivity of ILTV isolation and propagation in different culture methods. They found that CEL was the most sensitive for isolation, and CK was a second choice. The CEK, chicken embryo lung cells, and CAM inoculation were less sensitive [44]. It was reported that ILTV can be propagated in chicken leukocytes from the buffy coat cells [42]. Macrophages from bone marrow or spleen were susceptible to ILTV [45]. Chicken embryo fibroblasts, Vero cells, and quail cells were not satisfactory for primary isolation of ILTV. Other cell types, such as lymphocytes, thymocytes, and activated T cells were not sensitive to ILTV infection [43]. ILTV can also replicate in an avian liver cell line, LHM, which is chicken liver tumor cell induced by chemicals. In order for ILTV to multiply in LHM cells, it must be adapted.

Viral cytopathic effect (CPE) can be observed in cell culture in 4–6 h postinoculation (PI). Cytopathology contains increased swelling of cells, chromatin displacement, and rounding of the nucleoli. Cytoplasmic fusion produces multinucleated giant cells. The intranuclear inclusion bodies can be observed at 12 h PI. Large cytoplasmic vesicles develop in the multinucleated cells and become more basophilic as cells generate [2].

Natural transmission in birds is through the upper respiratory and ocular routes. The sources of ILTV transmission are clinically affected chickens, latent infected carrier chickens, contaminated dust, litter, and fomites that are contaminated with ILTV. Egg transmission is not verified [2,16]. Other sources of mechanical transmission included dog, crows, cats, and beetles [47]. One report investigated that wind-borne transmission as a critical for ILTV spread.

After infection, ILTV replicates in the epithelium of larynx and trachea. Virus particles are present in trachea tissue and secretion for 6–8 days PI. The virus may survive in trachea at least 10 days PI [48]. Like other herpesviruses, ILTV establishes latent infections. The virus can be reisolated by tracheal swabs from 7 weeks after infection or 2 months after infection in tracheal homologized samples [48]. Trigeminal ganglion is the target for ILTV latency. After 4–7 days PI with ILTV in the tracheae, 40% of infected chickens showed that virus migrated to trigeminal ganglion [48]. After vaccination for 15 months, the latent ILTV from trigeminal ganglion could be reactive. Mature laying chickens challenged with wild ILTV DNA were detected in the trigeminal ganglion by polymerase chain reaction (PCR) at 31, 46, and 61 days PI [49]. When birds are stressed, such as onset of lay or rehousing, ILTV can respread from latent infected birds [44].

85.1.3 Clinical Features

Clinical signs appear 6–12 days PI. Experimental challenge with intratracheal route results in an incubation period of 2–4 days [50]. There are two clinical forms for ILT infection. The clinical signs of the severe form include dyspnea and bloody mucus. The severe form of ILT causes 90%–100% morbidity and 5%–70% mortality [2,49].

Clinical signs of the mild form include depression, reduction in egg production and weight gain, conjunctivitis, swelling of infraorbital sinuses (almond shaped eyes), and nasal discharge. The morbidity of the mild form is about 5% and mortality is 0.1%–2%. It takes 10–14 days for recovery, but with some strains, the clinical signs might extend to 1–4 weeks.

Gross lesions are most commonly observed in the larynx and trachea. With the severe form of ILT, the mucosa of the respiratory tract showed inflammation and necrosis with hemorrhage. A characteristic diagnostic feature was intranuclear inclusion bodies in epithelial cells, which might be observed about 3 days PI. These cells have a condensed nucleus surrounded by a halo and margining of chromatin. Inclusion bodies generally were present only for a few days at the early stage of infection before epithelial cells died. Epithelial cell hyperplasia induced multinucleated cells (syncytia), lymphocytes, histiocytes, and plasma cells, which migrated to the lamina propria. Lesions were followed by desquamation of the necrotic epithelium and loss of mucous glands. At that time, bloody mucus in the trachea was observed [51,52].

85.1.4 Diagnosis

85.1.4.1 Conventional Techniques

Laboratory diagnosis is required for ILT certification, because many respiratory diseases cause similar clinical signs and lesions, such as infectious bronchitis, Newcastle disease, avian influenza, mycoplasma, and infectious coryza. ILT infection can be confirmed with several methods. Virus isolation uses susceptible live hosts to propagate pathogens from samples and can be used for investigating the pathologic changes in these hosts. For ILTV isolation, the dropped CAM of 9–12-day-old embryo eggs and cell cultures is most commonly used. Samples from sick birds include trachea,

conjunctiva, larynx, and lung, which are collected and inoculated on the CAM. The CAM route is the most sensitive method for ILTV isolation in embryos [43]. Plaques can be observed in 2 days PI; however, pocks usually develop on the CAM at 5–7 days PI. The plaque size and morphology on the CAM can be used to differentiate the virulence of different strains. The CEL and CEK cells are suitable for ILTV isolation. The typical CPE-multinucleated giant cells might be observed in 24 h PI in cell culture. At least two serial passages of virus isolation are often required to detect CPE [2].

Traditional antigen detection uses ILTV polyclonal or monoclonal antibodies to bind ILTV from clinical samples. Viral antigen can be detected by direct or indirect fluorescent antibodies (FA) in the tracheal smear or tracheal tissues [25]. A more sensitive method, immunoperoxidase (IP)-labeled monoclonal antibodies, was used as immunoprobes to detect ILTV in tracheal smears. This IP method can detect ILTV on the second day PI [54]. Agar gel immunodiffusion (AGID) used ILTV hyperimmune serum to detect antigen in tracheal samples, and it can quickly differentiate ILT from the diphtheritic form of fowl pox. However, the sensitivity of this method was relatively lower than other methods [11]. Antigen capture enzyme-linked immunosorbent assay (AC-ELISA) used ILTV monoclonal antibodies for antigen detection. The AC-ELISA was faster and more accurate than AGIP or FA [25].

There are several methods for detection of ILTV-specific antibodies in chicken serum including AGID, ELISA, virus neutralization (VN), and indirect FA. Among these antibody detection methods, ELISA was more sensitive than VN, and AGID was least sensitive. Both ELISA and IFA were quick and sensitive [55,56] and an ELISA method, which coated *Escherichia coli*–expressed ILTV gE and gp60 (gJ) for ILTV-specific antibody detection. This ELISA test was able to differentiate ILTV infected from noninfected chicken serum [56]. It is difficult to identify different strains of ILTV by serological methods, because the ILTVs have close genomic homology [57].

85.1.4.2 Molecular Techniques

ILT DNA detection methods emerged in recent years. The DNA detection methods can identify ILTV quickly, accurately, and sensitively. The molecular techniques for ILTV detection include cloned DNA probes for dot-blot hybridization [58,59], PCR and nested PCR [59], real-time PCR [60], multiplex PCR [61], *in situ* hybridization [62], and PCR followed by restriction fragment length polymorphism (RFLP) [63–66]. Work [49] compared the sensitivity of different ILTV detection methods and found that PCR was much more sensitive than virus isolation in cell culture and electron microscopy. PCR also detected ILTV samples, which were contaminated with other pathogens [49].

The most common and effective molecular method for ILTV differentiation is PCR-RFLP. PCR-PFLP analysis of single or multiple viral genome regions identified different strains from different geographic areas and vaccine from wild-type strains [8,14,65–68]. Restriction endonuclease analysis of ILTV DNA can also differentiate vaccine strains from wild-type strains [52]. Moreover, PCR-RFLP analysis of partial ICP4 gene, gC gene, and TK gene distinguished wild strains from vaccines in Taiwan. In that investigation, some virulent isolates were from Taiwan poultry houses and could be separated from vaccine strains [69]. Analysis of multiple genes PCR-RFLP combined with DNA sequence analysis of gG gene and TK gene to separate vaccine and nonvaccine strains in Korea [69]. In Austria, researchers demonstrated that multiple gene PCR-RFLP was more reliable to identify the relationship between vaccine and wild strains [64].

In the United States, researchers [53,65] investigated ILTV isolates from commercial poultry samples between 1988 and 2005 using multiple gene PCR-RFLP analysis (ORFB-TK, ICP4, UL47/gG, and gM/UL9). ILTVs in the United States were separated into nine genetic groups. Group I and II were the USDA reference strain and tissue culture origin (TCO) vaccine strain. Group IV isolates were identical to chicken embryo origin (CEO) vaccine strains, whereas group V isolates, which had one PCR-RFLP pattern different from the CEO vaccine strains, were CEO-related isolates. Group VI isolates were different from CEO and TCO vaccine strains. In that report, most of ILTV commercial poultry isolates were closely related to vaccine strains [65]. Researchers [65] investigated 46 ILTV field isolates in the United States from 2006 to 2007. After multiple gene PCR-RFLP analysis, most of the isolates (63%) were closely related to vaccine strains (group III–V) [66]. According to these reports, most ILTV wild isolates in U.S. chicken farms were derived from vaccines.

In Europe, a research investigation collected 104 field isolates during 35 years from eight different countries and analyzed the viruses with PCR-RFLP from the TK gene. The results showed that ILTVs in these countries were separated into three genetic groups. Moreover, 98 field isolates had the same RFLP patterns with vaccine strains [67]. In Australia, PCR-RFLP was used to analyze ILTV gG, TK, ICP4, ICP18.5, and ORFB-TK genes. These isolates could be discriminated into five genetic groups. Some isolates could not be distinguished from vaccine strains [64].

PCR assays such as conventional PCR, nested PCR [67], real-time PCR [60], and multiplex PCR [61] offer a quick, accurate, and highly sensitive approach for ILTV DNA detection. However, PCR samples can show false-negative results, if heavily contaminated pathogens and organisms are present [49].

Due to its speed and sensitivity, reproducibility, and ability to quantify directly from clinical samples, real-time PCR has increasingly become a method of choice in clinical laboratories. The most widely used real-time PCR assays are the SYBR Green PCR melting curve analysis and TaqMan PCR [70,71], the latter of which has a detection limit of <100 genome copies. A sensitive SYBR green PCR assay, with an extended dynamic range of 7 orders of magnitude compared to 7 orders of magnitude reported for the TaqMan, was developed and validated [71].

A novel DNA amplification method, loop-mediated isothermal amplification (LAMP), has been used in disease diagnosis. LAMP method uses *Bst* DNA polymerase, which has strand displacement activity, and a set of two inner and two outer primers to recognize six regions of target sequences. The LAMP method amplifies DNA at isothermal conditions between 60°C and 65°C in 60 min. The LAMP assay provides a rapid, and low-cost, detection method in disease diagnosis. It has been widely used for viral and bacterial nucleic acid detection. In the present study, the LAMP assay and TaqMan® probe-based real-time PCR were developed and compared for ILTV DNA detection.

85.2 METHODS

85.2.1 SAMPLE PREPARATION

Five commercial ILT vaccine strains: four chicken embryo origin (CEO) vaccines—AviPro® LT (Lohmann Animal Health Inc., Winslow, ME), LT Blen® (Merial Select Inc., Gainesville, GA), Laryngo-Vac® (Zoetis Animal Health, Raleigh, NC), Trachivax® (Schering-Plough Animal Health Corp., Kenilworth, NJ), and the TCO vaccine (LT-IVAX®) (Schering-Plough Animal Health Corp., Kenilworth, NJ). *Mycoplasma gallisepticum* (MG) vaccine (MycoVac-L®) (Intervet Inc., Millsboro, DE), fowl pox vaccine (Chicken–N-Pox™ TC) (Zoetis Animal Health), and Marek's disease serotype 3 vaccine (MD-Vac® CFL) (Zoetis Animal Health) were used for test specificity [71].

85.2.2 DETECTION PROCEDURES

85.2.2.1 Real-Time PCR

ILTV DNA is extracted using Qiagen DNeasy® Blood & Tissue Kit (Qiagen, Valencia, CA) with slight modifications. Briefly, 200 µL of each sample is mixed with 20 µL proteinase K and 200 µL of Buffer AL. The mixture is mixed well and incubated at 56°C for 10 min. Then, 200 µL of 100% ethanol is added and vortexed. The mixture is transferred into the DNeasy mini spin column and centrifuged at $6000 \times g$ for 1 min, and then the flow-through was discarded. The column is placed in a 2 mL collection tube, and 500 µL of Buffer AW1 was added. The tube is centrifuged at $6000 \times g$ for 1 min, and the flow-through was discarded. The column is transferred into a 2 mL collection tube and 500 µL of Buffer AW2 was added. The tube is centrifuged at $20,000 \times g$ for 3 min to remove excess reagents from the column membrane. Finally, the DNA is eluted in 100 µL of Buffer AE by centrifuging at $6000 \times g$ for 1 min and stored at −20°C.

A segment of LT Blen ILTV ICP4 gene is amplified by PCR using the following primers: ICP4-F: 5′-CG CAGAGGACCAGCAAAGACCG-3′; ICP4-R: 5′-GAAGCAGA CGCCGCCGTAGGAT-3′. For PCR, 50 µL of reaction is set up as follows: 5 µL of 10× PCR buffer, 5 µL of 25 mM MgCl$_2$, 1 µL of 10 mM dNTP each, 1 µL of 100 µM ICP4-F, 1 µL of 100 µM ICP4-R, 0.25 µL of Taq DNA polymerase (5 U/µL; AmpliTaq® DNA polymerase, Applied BioSystems, Foster City, CA), 28.75 µL of water, and 5 µL of sample DNA. PCR steps were subjected to a 94°C initial denaturation for 2 min and 35 cycles of 94°C denaturation for 30 s, 55°C annealing for 30 s, and 72°C extension for 1 min, followed by 72°C final extension for 5 min.

PCR is conducted in GeneAmp® PCR System 9700 (Applied BioSystems, Foster City, CA). PCR products are detected with 2% agarose gel electrophoresis.

PCR products are purified by Wizard® PCR preps DNA purification system (Promega, Madison, WI). The cDNA is cloned into the pT7Blue-3 vector. A blunt-end cloning kit (Novagen, Darmstadt, Germany) is used according to the manufacturer's instructions. The plasmids are transformed into NovaBlue® Singles™ competent cells (Novagen, Darmstadt, Germany). Several clones are selected and the plasmid DNA extracted using Wizard Plus SV kit (Promega, Madison, WI), according to the manufacturer's directions. Clones containing the proper insert are verified by DNA sequencing. The concentrations of cloned plasmids are measured by the NanoDrop® ND-100UV-Vis Spectrophotometer (Wilmington, DE).

Copy number of ILT ICP4 cloned DNA is calculated with the following formula:

$$ICP4\ DNA\left(copies\ /\ \mu L\right)$$

$$= \frac{6.022 \times 10^{23}\left(\dfrac{molecules}{mol}\right) \times concentration\left(\dfrac{g}{\mu L}\right)}{Weight\left(\dfrac{g}{mol}\right)}$$

$$Weight\ in\ Daltons\ (g/mol) = (bp\ size\ of\ plasmid + insert) \times (330\ Da \times 2\ nucleotide/bp)$$

Real-time PCR amplification is performed in a LightCycler® (Roche, Applied Science, Indianapolis, IN) with a 20 μL reaction volume. For real-time PCR assay, each reaction contains 10 μL of 2× master mix (QuantiTect® Probe PCR kit, Qiagen, Valencia, CA), 0.5 μM of each primer, 0.1 μM of probe, 2.5 μL of water, and 5 μL of DNA template. Real-time PCR mixture is subjected to 95°C initial activation for 15 min, and 40 cycles for 95°C denaturation 0 s, and 60°C combined annealing and extension for 60 s.

Conventional PCR is performed in GeneAmp PCR System 9700 (Applied BioSystems, Foster City, CA) using the same primers with real-time PCR. Totally 50 μL of PCR reagents in a reaction tube contains 5 μL of 10× PCR buffer, 5 μL of 25 mM MgCl₂, 1 μL of 10 mM dNTP each, 1 μL of each primer in 100 μM, 0.25 μL of Taq DNA polymerase (5 U/μL; AmpliTaq DNA polymerase, Applied BioSystems, Foster City, CA), 28.75 μL of water, and 5 μL of sample DNA. The PCR steps are subjected to a 94°C initial denaturation for 2 min and 35 cycles of 94°C denaturation for 30 s, 50°C annealing for 30 s, and 72°C extension for 30 s, followed by 72°C final extension for 5 min. PCR products were detected with 2% agarose gel electrophoresis.

The LAMP assay is performed in 25 μL, which contained 1× ThermolPol buffer (New England Biolabs Inc., Beverly, MA). Each dNTP are used at a final concentration of 1.2 mM, inner primers to a final concentration of 1.6 μM, outer primers to a final concentration of 0.2 μM, loop primers to a final concentration of 0.4 μM, 1.0 M of butaine (Sigma, Saint Louis, MO), 1 μL of 8U *Bst* DNA polymerase (Large Fragment; New England Biolabs Inc., Beverly, MA), 4 mM of MgSO₄, and 5 μL of DNA template. To optimize the reaction, 60°C, 63°C, and 65°C temperatures are tested. Reaction times of 15, 25, 35, 45, 50, and 60 min were tested to optimize the LAMP. The reaction is stopped at 95°C for 3 min to terminate enzyme activity. After LAMP reaction, DNA products are verified by 2% agarose gel electrophoresis, stained with ethidium bromide, and visualized on a UV transilluminator (Fotodyne® Inc., Hartland, WI). In LAMP assay, 1.0 M of betaine is added to the reaction. During DNA amplification (PCR or LAMP), the betaine increased the yield and specificity. It also reduces secondary structure forming in GC-rich regions and eliminates the base pair composition dependence of DNA melting.

Primers and probe for real-time PCR and primers for LAMP are designed in the amplicon of ICP4F/R. The ICP4 gene sequences of four ILTV strains: CEO vaccine (GenBank accession number EU104900), TCO vaccine (accession number EU104908), ILTV-assembled total genome sequence (accession number NC_006623), and 2 ILTV vaccine strains (AviPro LT and LT Blen), are sequenced. They are aligned with AlignX® from vector NTI sequence analysis and data management software v10.3 (Invitrogen Co., Carlsbad, CA). Real-time PCR primers and probe, which generate a 125 bp product, are selected from the conserved regions of ICP4 gene. The target DNA sequence is compared with other sequences using the BLAST service in the GenBank database. Results indicate that the real-time PCR product is located in the ILTV ICP4 gene. Primers and probes are produced by Integrated DNA Technologies®, Inc. (Coralville, CA).

85.2.2.2 LAMP

The LAMP primers are designed using the conserved region of ILTV ICP4 gene in ICP4F/R amplicon using the Primer Explorer V4 software (Eiken Chemical Co., Ltd., Japan). For the end stability, the primer selection followed the rule that the free energy (ΔG) of 3′ end of F2/B2, F3/B3, LF/LB, and the 5′ end of F1c/B1c should be −4 kcal/mol or less. One set of primers is chosen. This primer set comprising two outer, two inner, and two loop primers recognized eight distinct regions in the target sequence. The forward inner primer and the backward inner primer each have two distinct sequences corresponding to the sense and antisense sequence. The DNA strands synthesized from the outer primers (F3 and B3) displace the primary strands. The forward loop primer (LF) recognizes the complementary strand corresponding to the region between F1 and F2, and the reverse loop primer (LB) annealed to the complementary strand corresponding to the region between B1 and B2.

To determine the sensitivity and detection limit for real-time PCR and LAMP, serial 10-fold dilutions of plasmid DNA from 10^9 to 10^0 copies/μL are analyzed. The real-time PCR assay is repeated four times. Standard curves indicating the linear relationships between the threshold crossing points (Cp) and the logarithms of initial ILTV ICP4 gene DNA count are constructed. The serial dilutions are repeated in the same run for evaluating the intraexperiment reproducibility.

To determine specificity, five ILTV strains, nontemplate negative controls, MG, fowl pox, and Marek's disease vaccines are tested. The detection limit and reproducibility of the real-time PCR assays are determined by four independent runs using 10-fold serial dilutions (10^9–10^0) of the ICP4 gene plasmids as template. A standard curve and equation are generated. The ILTV template copy number per amplification reaction is estimated using the standard curve equation. Serial dilution of the ICP4 gene plasmids from 6×10^4, 6×10^3, 6×10^2, 60, 30, 15, and 6 copies/µL is used to determine the sensitivity of the LAMP assay.

To determine viral titer, 50% embryonic infective dosage (EID_{50}) of the LT Blen vaccine is performed as follows. Nine serial 10-fold dilutions of viral stock are made in phosphate-buffered saline. Four 9-day-old specific pathogen-free (SPF) chicken embryonic eggs are used per viral concentration, and each embryonic egg was inoculated with 200 µL of viral dilution via CAM route. The inoculated CAMs are checked for pock formation after 7 days. The viral titer is calculated. The EID_{50} titrations are repeated three times and compared with the ILTV genome copy number with real-time PCR assay.

Both assays are positive for ILTV strains only. No fluorescent signal or positive patterns are detected with nontemplate control and non-ILTV pathogens. The LAMP assay produces stem-loop DNA structure with different length of inverted repeats and cauliflower-like structures of target sequence. Thus, when the LAMP products are analyzed with agarose gel electrophoresis, they show many bands with different sizes and a smeared DNA between these bands at each loading well.

Four repeats of real-time PCR tests generate standard curves, which have an average intercept of 38.28 ± 0.63 and an average slope of -3.14 ± 0.06. The standard curves display a significant correlation between Cp value and copy number with the square of the sample correlation coefficient (R^2) above 0.99 and an average efficiency of 2.063 ± 0.75.

The standard deviation (SD) of Cp value is low; therefore, these data indicate excellent reproducibility. The real-time PCR assay maintains linearity at 10 copies/µL, and the SD and standard error values are stable and low. Although the plasmid concentrations at 1 copy/µL are positive, the machine cannot define the Cp value above 35 cycles. The quantification limits are determined at 10 copies/µL, and the samples with the Cp value ≤35 are considered positive for ILTV. Comparing with real-time PCR, conventional PCR with the same primer set is able to detect 10^3 copies/µL. Therefore, real-time PCR is about 100 times more sensitive than conventional PCR.

Serial dilutions of plasmid are used to determine the sensitivity of the LAMP assay. The LAMP assay can detect ILTV DNA templates at 60 copies/µL. Therefore, the sensitivity of real-time PCR is sixfold higher than LAMP.

Viral titers of ILTV LT Blen are measured using 9-day-old SPF chicken embryonic eggs via CAM route. To check viral concentration in 1 dose of ILTV vaccine, the vaccine is diluted by the manufacture's direction to four doses and the copy number checked with real-time PCR. Results show that 1 EID_{50} was equal to $(2.1 \pm 1.3) \times 10^2$ ILT genomic copies, and 1 dose of ILTV vaccine is equal to $(6.9 \pm 0.85) \times$ 10^5 copies. This viral vaccine titer is compatible with most manufacture results.

The LAMP reaction creates many DNA bands of different size in the target sequence region. The amplification with LAMP assay shows a ladder-like pattern upon agarose electrophoresis. The optimal temperature for *Bst* DNA polymerase is between 60°C and 65°C. The reaction temperature of 60°C, 63°C, and 65°C is performed to check the optimal efficiency. Results show that the LAMP assay products at 65°C are brighter and clearer than at other temperatures. Thus, 65°C is selected as the optimal temperature for this ILTV LAMP assay. Different reaction times also affect LAMP efficiency. Different reaction time intervals are performed to check the optimal reaction timed for the LAMP assay. Results show that the LAMP products can be easily observed with bright bands after the LAMP reaction is performed above 45 min.

The real-time PCR assay is highly sensitive and specific and can detect as few as 10 copies/µL of the ILTV DNA genome and was highly reproducible. Standard curves of multiple repeats are almost identical. According to the statistics analysis, the variation of slop (-3.14 ± 0.06) and intercept (38.28 ± 0.63) of regression equation is small. Thus, this real-time PCR assay is highly reproducible. Viral titers of field samples and vaccines are critical for vaccine production and efficacy experiments. Since the real-time PCR can determine the concentration of nucleic acids in the samples by estimating with standard curve, the viral titer in samples can be accurately predicted. The titers of 1 dose of vaccine determined in 9-day-old chicken embryo eggs (EID_{50}) are compared with the copy number of ILTV genome by real-time PCR. Results indicate that 1 EID_{50} is equivalent to $(2.1 \pm 1.3) \times 10^2$ ILT genomic copies (Table 85.1). One dose of ILTV vaccine is equivalent to $(2.1 \pm 1.3) \times 10^2$ ILT genomic copies. One dose of ILTV vaccine was equivalent to $(6.9 \pm 0.85) \times 10^5$ copies (Table 85.2).

The LAMP assay is highly specific and sensitive. However, the sensitivity of the ILTV LAMP assay (60 copies/µL) is slightly lower than real-time PCR (Figure 85.1). It provides a 10–100-fold higher sensitivity than regular PCR. It is a simple, rapid, and an easy detection method, in which the reaction is performed at 65°C for less than 1 h and does not require a thermal cycler. In addition, the reaction time is faster than real-time PCR. Furthermore, the reagents and equipment of LAMP assay are cheaper than real-time PCR and can be performed in most laboratories. The sensitivity of LAMP assay is less affected by contaminating components, feces, feed, and blood, which are contained in clinical samples, than PCR. Therefore, the DNA purification steps can be omitted.

The detection limit of real-time PCR and LAMP assays is higher than other prior methods. The sensitivity of nested PCR for ILTV gE gene detection reached 31.3–62.5 fg. The sensitivity of traditional PCR reaches 10^3 copies/µL using the same primers with real-time PCR. However, the real-time PCR (10 copies/µL) and LAMP (60 copies/µL) are more sensitive than conventional PCR. Another TaqMan real-time PCR for ILTV gE gene could detect 100 copies/µL [60]. Creelan et al. [60] used real-time PCR to detect ILTV ICP4 gene with SYBR Green I chemistry, and the sensitivity reached 140 copies/µL.

TABLE 85.1

Correlation between Infectious Laryngotracheitis Virus Viral Genomic Copy Number and 50% Embryonic Infective Dosage

	Titration1	Titration 2	Titration 3	Mean	SD
EID_{50}/mL	8.9×10^5	8.0×10^5	5.0×10^5		
Real-time PCR (copies/mL)	3.2×10^8	9.8×10^7	7.6×10^7		
Conversion factor (copies/EID_{50})	3.6×10^2	1.2×10^2	1.5×10^2	2.1×10^2	$\pm 1.3 \times 10^2$

$1\ EID_{50} \approx (2.1 \pm 1.3) \times 10^2$ copies.

TABLE 85.2

Correlation between Vaccine Dosage and Infectious Laryngotracheitis Virus Viral Genomic Copy Number

	4 Doses of ILTV Vaccine				
Real-time PCR (copies/4 doses)	2.4×10^6	3.1×10^6	2.8×10^6	Mean	SD
Conversion factor (copies/dose)	6.0×10^5	7.7×10^5	7.0×10^5	6.9×10^5	$\pm 0.85 \times 10^5$

1 dose $\approx (6.9 \pm 0.85) \times 10^5$ copies to $(6.9 \pm 0.85) \times 10^5$ copies.

The ILTV ICP4 gene real-time PCR is the most sensitive detection method reported so far. As the binding of SYBR Green dye to double-stranded DNA molecules is nonspecific, the SYBR Green–based PCR can be affected by a number of factors, such as the formation of primer–dimmers, and sample concentration [72]. In addition, multiple SYBR Green dye molecules can bind to the same double-stranded DNA fragment. These disadvantages have limited the application of SYBR Green–based real-time PCR in quantitative detections. Although the nested PCR has similar sensitivity as real-time PCR and LAMP assay, it requires two rounds of PCR thermal cycles. Therefore, it is more time consuming, and the possibility of nucleic acid contamination is higher than for real-time PCR and LAMP assays. Since ILT-LAMP method uses six primers, which recognize eight distinct regions of the ILT ICP4 gene, the specificity of LAMP is high. However, the selection region of the LAMP primer designation is limited to the same region of real-time PCR. This could affect the sensitivity of LAMP assay. Designing LAMP primers of different genes or different region of ICP4 gene might increase the sensitivity.

The ILTV real-time PCR and LAMP assays are specific, sensitive, and reliable for ILTV detection (Figure 85.1) [73]. Although the sensitivity of LAMP is slightly lower than that of real-time PCR, it has a value for ILTV detection because of its speed, lower cost, and does not require a temperature cycler or more expensive real-time PCR equipment.

85.3 CONCLUSION AND FUTURE PERSPECTIVES

ILTV molecular detection methods continue to evolve. They provide faster results, increased sensitivity, and reproducibility. Starting with PCR about 15 years ago, they have further developed to nested PCR, multiplex PCR, real-time PCR, and loop-mediated PCR. On the near horizon will be novel automated methods for the rapid extraction of viral DNA. Real-time PCR systems using fast ramping combined with intuitive software packages. The recombinant polymerase amplification assay has been recently developed with a run time of 10 min, and including nucleic acid extraction from samples, results can be obtained in 30 min. The next-generation droplet digital PCR assay employs an emulsion-based endpoint to quantitate the amount of target DNA and is more robust than real-time PCR when analyzing sequence variations. This test could replace PCR and RFLP testing for differentiating vaccine from field viruses. Nanoparticle-assisted PCR is a novel method for the rapid amplification of DNA and has been used for the detection of virus. Rapid detection and characterization of ILTV field isolates will add in epidemiological studies, diagnosis, and development of more efficient vaccines for improved control and prevention of this important pathogen of poultry.

REFERENCES

1. Fahey, K.J., Bagust, T.J., and York, J.J., 1983. Laryngotracheitis herpesvirus infection in the chicken: The role of humoral antibody in immunity to a grade challenge infection. *Avian Pathol.* 12, 505–514.
2. Guy, J.S. and Garcia, M., 2008. Laryngotracheitis. In Saif, Y.M., Barnes, H.J., Glisson, J.R., Fadly, A.M., McDougald, L.K., and Swayne, D.E. (Eds.), *Disease of Poultry*, 11th edn. Iowa State University Press, Ames, IA, pp. 137–152.
3. May, H.G. and Tittsler, R.P., 1925. Tracheo-laryngotracheotis in poultry. *J. Am. Vet. Med. Assoc.* 67, 229–231.
4. Cover, M.S. and Benton, W.J., 1958. The biology variation of infectious laryngotracheitis virus. *Avian Dis.* 2, 375–383.

5. McGeoch, D.J., Dolan, A., and Ralph, A.C., 2000. Toward a comprehensive phylogeny for mammalian and avian herpesviruses. *J. Virol.* 74, 10401–10406.

6. McGeoch, D.J., Rixon, F.J., and Davison, A.J., 2006. Topics in herpesvirus genomics and evolution. *Virus Res.* 117, 90–104.

7. Davison, A.J., Eberle, R., Hayward, G.S. et al., 2005. Herpesviridae. In Pattison, M., McMullin, P.F., Bradbury, J.M., and Alexander, D.J. (Eds.), *Virus Taxonomy: Eighth Report of the International Committee on Taxonomy of Viruses*. Elsevier Academic Press, San Diego, CA, pp. 193–212.

8. Leib, D.A., Bradburt, J.M., Hart, C.A., and McCarthy, K., 1987. Genome isomerism in two alphaherpesviruses: Herpesvirus saimiri-1 (*Herpesvirus tamarinus*) and avian infectious laryngotracheitis virus. *Arch. Virol.* 93, 287–294.

9. Roizman, B. and Knipe, D.M., 2001. Herpes simplex viruses and their replication. In Knipe, D.M. and Howley, P.M. (Eds.), *Fields Virology.* Lippincott Williams & Wilkins, Philadelphia, PA, pp. 2399–2459.

10. Meulemans, G. and Halen, P., 1978. Some physico-chemical and biological properties of a Belgian strain (U76/1035) of infectious laryngotracheitis virus. *Avian Pathol.* 7, 311–315.

11. Fitzgerald, J.E. and Hanson, L.E., 1963. A comparison of some properties of laryngotracheitis and herpes simplex viruses. *Am. J. Vet. Res.* 24, 1297–1303.

12. Jorden, F.T.W., 1966. A review of the literature on infectious laryngotracheitis. *Avian Dis.* 10, 1–26.

13. Thureen, D.R. and Keeler, C.L. Jr., 2006. Psittacid herpesvirus 1 and infectious laryngotracheitis virus: Comparative genome sequence analysis of two avian alphaherpesvirus. *J. Virol.* 80, 7863–7872.

14. Keeler, C.L., Hazel, J.W., Hastings, J.E., and Rosenberger, J.K., 1993. Restriction endonuclease analysis of Delmarva field isolates of infectious laryngotracheitis virus. *Avian Dis.* 37, 418–426.

15. Ziemann, K., Mettenleiter, T.C., and Fuchs, W., 1998. Gene arrangement within the unique long genome region of infectious laryngotracheitis virus is distinct from that of other alphaherpesviruses. *J. Virol.* 72, 847–852.

16. Ziemann, K., Mettenleiter, T.C., and Fuchs, W., 1998. Infectious laryngotracheitis herpesvirus expresses a related pair of unique nuclear proteins which are encoded by spilt genes located at the right end of the UL genome region. *J. Virol.* 72, 6867–6874.

17. Ben-Porat, T., Veach, R.A., and Ihara, S., 1983. Localization of the regions of homology between the genomes of herpes simplex virus type 1 and pseudorabies virus. *Virology* 127, 194–204.

18. Fuchs, W. and Mettenleiter, T.C., 1999. DNA sequence of the UL6 to UL20 of infectious laryngotracheitis virus and characterization of the UL10 gene product as a nonglycosylated and nonessential virion protein. *J. Gen. Virol.* 80, 2173–2182.

19. Fuchs, W., Klupp, B.G., Granzow, H., Rziha, H.-J., and Mettenleiter, T.C., 1996. Identification and characterization of the pseudorabies virus UL3.5 protein, which is involved in virus egress. *J. Virol.* 70, 3517–3527.

20. Nishiyama, Y., 2004. Herpes simplex virus gene products: The accessories reflect her lifestyle well. *Rev. Med. Virol.* 14, 33–46.

21. Griffin, A.M., 1991. The nucleotide sequence of the glycoprotein gB of infectious laryngotracheitis virus: Analysis and evolutionary relationship to the homologous gene from other herpesviruses. *J. Gen. Virol.* 72, 393–398.

22. Kongsuwan, K., Prideaux, C.T., Johnson, M.A., Sheppard, M., and Fahey, K.J., 1991. Nucleotide sequence of the gene encoding infectious laryngotracheitis virus glycoprotein B. *Virology* 184, 404–410.

23. Liang, X.P., Babiuk, L.A., van Den Hurk, S.D., Fitzpatrick, D.R., and Zamb, T.J., 1991. Bovine herpesvirus 1 attachment to permissive cells is mediated by its major glycoproteins gI, gIII, and gIV. *J. Virol.* 65, 1124–1132.

24. Okazaki, O., 2007. Proteolytic cleavage of glycoprotein B is dispensable for in vitro replication, but required for syncytium formation of pseudorabies virus. *J. Gen. Virol.* 88, 1859–1865.

25. York, J.J. and Fahey, K.J., 1988. Diagnosis of infectious laryngotracheitis using a monoclonal antibody ELISA. *Avian Pathol.* 17(1), 173–182.

26. Mettenleiter, T.C., 2004. Budding events in herpesvirus morphogenesis. *Virus Res.* 106, 167–180.

27. Kingsley, D.H. and Keeler, C.L. Jr., 1999. Infectious laryngotracheitis virus, an alphaherpesvirus that does not interact with cell surface heparin sulfate. *Virology* 256, 213–219.

28. Dingwell, K.S., Brunetti, C.R., Hendricks, R.L. et al., 1994. Herpes simplex virus glycoprotein E and I facilitate cell-to-cell spread in vivo and across junction of cultured cells. *J. Virol.* 68, 834–845.

29. Devlin, J.M., Browning, G.F., and Gilkerson, J.R., 2006. A glycoprotein I- and glycoprotein E-deficient mutant of infectious laryngotracheitis virus exhibits impaired cell-to-cell spread in cultured cells. *Arch. Virol.* 151, 1281–1291.

30. Helferich, D., Veits, J., Teifke, J.P., Mettenleiter, T.C., and Fuchs, W., 2007. The UL47 gene of avian infections laryngotracheitis virus is not essential for in vitro replication but is relevant for virulence in chickens. *J. Gen. Virol.* 88, 732–734.

31. Devlin, J.M., Browning, G.F., Hartley, C.A. et al., 2006. Glycoprotein G is a virulence factor in infectious laryngotracheitis virus. *J. Gen. Virol.* 87, 2839–2847.

32. Veits, J., Mettenleiter, T.C., and Fuchs, W., 2003. Five unique open reading frames of infectious laryngotracheitis virus are expressed during infection but are dispensable for virus replication in cell culture. *J. Gen. Virol.* 84, 1415–1425.

33. Fuchs, W., Wiesner, D., Veits, J., Teifke, J.P., and Mettenleiter, T.C., 2005. In vitro and in vivo relevance of infectious laryngotracheitis virus gJ proteins that are expressed from spliced and nonspliced mRNAs. *J. Virol.* 79, 705–716.

34. Efstathiou, S., Kemp, S., Darby, G., and Minson, A.C., 1989. The role of herpes simplex virus type 1 thymidine kinase in pathogenesis. *J. Virol.* 69, 4546–4568.

35. Veits, J., Lüschow, D., Kindermann, K. et al., 2003. Deletion of the non-essential UL0 gene of infectious laryngotracheitis (ILT) virus leads to attenuation in chickens, and UL0 mutants expressing influenza virus haemagglutinin (H7) protect against ILT and fowl plague. *J. Gen. Virol.* 84, 3343–3352.

36. Prideaux, C.T., Kongsuwan K., Johnson, M.A., Sheppard, M., and Fahey, K.J., 1992. Infectious laryngotracheitis virus growth, DNA replication, and protein synthesis. *Arch. Virol.* 123, 181–192

37. Guo, P., Scholz, E., Turek, J., Nodgreen, R., and Maloney, B., 1992. Assembly pathway of avian infectious laryngotracheitis virus. *Am. Vet. Res.* 54, 2031–2039.

38. Mettenleiter, T.C., 2006. Intriguing interplay between viral proteins during herpesvirus assembly or: The herpesvirus assembly puzzle. *Vet. Microbiol.* 113, 163–169.

39. Winterfield, R.W. and So, I.G., 1968. Susceptibility of turkeys to infectious laryngotracheitis. *Avian Dis.* 12, 191–202.

40. Portz, C., Beltrao, N., Furian, T.Q. et al., 2008. Natural infection of turkey by infectious laryngotracheitis virus. *Vet. Microbiol.* 131, 57–64.

41. Burnet, F., 1934. The propagation of the virus of infectious laryngotracheitis on the CAM of developing egg. *Br. J. Ex. Pathol.* 12, 179–198.

42. Chang, P.W., Jasty, V., Fry, D., and Yates, V.J., 1973. Replication of a cell-culture-modified infectious laryngotracheitis virus in experimentally infected chickens. *Avian Dis.* 17, 683–689.

43. Hughes, C.S. and Jones, R.C., 1988. Comparison of cultural methods for primary isolation of infectious laryngotracheitis virus from field materials. *Avian Pathol.* 17, 295–303.

44. Hughes, C.S., Gaskell, R.M., Jones, R.C., Bradbury, J.M., and Jordan, F.T.W., 1989. Effects of certain stress factors on the re-excretion of infectious laryngotracheitis virus from latently infected carrier birds. *Res. Vet. Sci.* 46, 247–276.

45. Bulow, V.V. and Klasen, A., 1983. Effect of avian viruses on cultured chicken bone-marrow-derived macrophages. *Avian Pathol.* 12, 179–198.

46. Bagust, T.J., Jones, R.C., and Guy, J.S., 2000. Avian infectious laryngotracheitis. *Rev. Sci. Tech. Off. Int. Epiz.* 19(2), 483–492.

47. Ou, S.C., Giambrone, J.J., and Macklin, K.S., 2012. Detection of infectious laryngotracheitis virus from darkling beetles and their immature stage (lesser mealworms) by quantitative polymerase chain reaction and virus isolation. *J. Appl. Poult. Res.* 21, 33–38.

48. Bagust, T.J., 1986. Laryngotracheitis (Gallid-1) herpesvirus infection in the chicken. 4 Latency establishment by wild and vaccine strains of ILT virus. *Avian Pathol.* 6(15), 581–595.

49. Williams, R.A, Savage, C.E., and Jones, R.C., 1994. A comparison of direct electron microscopy, virus isolation, and a DNA amplification method for the detection of avian infectious laryngotracheitis virus in field material. *Avian Pathol.* 23, 709–720.

50. Kernohan, G., 1931. Infectious laryngotracheitis in fowls. *J. Am. Vet. Med. Assoc.* 78, 196–202.

51. Nair, V., Jones, R.C., and Gough, R.E., 2008. Herpesviridae. In Fauquet, C.M., Mayo, M.A., Maniloff, J., Desselberger, U., and Ball, L.A. (Eds.), *Poultry Disease*, 6th edn. Saunders Elsevier, San Diego, CA, pp. 267–271.

52. Guy, J.S., Barnes, H.J., Munger, L.L., and Rose, L., 1989. Restriction endonuclease analysis of infectious laryngotracheitis viruses: Comparison of modified-live vaccine viruses and North Carolina field isolates. *Avian Dis.* 33, 316–323.

53. Goodwin, M.A., Smeltzer, M.A., Brown, J., Resurreccion, R.S., and Dickson, T.G., 1991. Comparison of histopathology to the direct immunofluorescent antibody test for the diagnosis of infectious laryngotracheitis in chickens. *Avian Dis.* 35, 389–391.

54. Guy, J.S., Barnes, H.J., and Smith, L.G., 1992. Rapid diagnosis of infectious laryngotracheitis using a monoclonal antibody-based immunoperoxidase procedure. *Avian Pathol.* 21, 77–86.

55. Bauer, B., Lohr, J.E., and Kaleta, E.F., 1999. Comparison of commercial ELISA kits from Australia and the USA with the serum neutralisation test in cell cultures for the detection of antibodies to the infectious laryngotracheitis virus of chickens. *Avian Pathol.* 28, 65–72.

56. Chang, P.C., Chen, K.T., Shien, J.H., and Shieh, H.K., 2002. Expression of infectious laryngotracheitis virus glycoproteins in *Escherichia coli* and their application in enzyme-linked immunosorbent assay. *Avian Dis.* 46, 570–580.

57. Shibley, G.P., Luginbunhl, R.E., and Helmboldt, C.F., 1962. A study of infectious laryngotracheitis virus. I. Comparison of serologic and immunogenic properties. *Avian Dis.* 6, 59–71.

58. Nagy, E., 1992. Detection of infectious laryngotracheitis virus infected cells with cloned DNA probe. *Can. J. Vet. Res.* 56, 34–40.

59. Humberd, J., Garcia, M., Riblet, S.M., Resurreccion, R.S., and Brown, T.P., 2002. Detection of infectious laryngotracheitis virus in formalin-fixed, paraffin-embedded tissues by nested polymerase chain reaction. *Avian Dis.* 46, 64–74.

60. Creelan, J.L., Calvert, V.M., Graham, D.A., and McCullough, S.J., 2006. Rapid detection and characterization from field cases of infectious laryngotracheitis virus by real-time polymerase chain reaction and restriction fragment length polymorphism. *Avian Pathol.* 35, 173–179

61. Pang, Y., Wang, H., Girshick, T., Xie, Z., and Khan, M.I., 2002. Development and application of a multiplex polymerase chain reaction for avian respiratory agents. *Avian Dis.* 46, 691–699.

62. Nielsen, O.L., Handberg, K.J., and Jorgensen, P.H., 1998. In situ hybridization for the detection of infectious laryngotracheitis virus in section of trachea from experimentally infected chickens. *Acta Vet. Scand.* 39, 415–421.

63. Chang, P.C., Lee, Y.L., Shien, J.H., and Shieh, H.K., 1997. Rapid differentiation of vaccine strains and field isolates of infectious laryngotracheitis virus by restriction fragment length polymorphism of PCR products. *J. Virol. Methods* 66, 179–186.

64. Kirkpatrick, N.C., Mahmoudian, C.A., O'Rourke, D., and Noormohammadia, A.H., 2006. Differentiation of infectious laryngotracheitis virus isolates by restriction fragment length polymorphic analysis of polymerase chain reaction products amplified from multiple genes. *Avian Dis.* 50, 28–34.

65. Oldoni, I. and García, M., 2007. Characterization of infectious laryngotracheitis virus isolates from the US by polymerase chain reaction and restriction fragment length polymorphism of multiple genome regions. *Avian Pathol.* 36, 167–176.

66. Oldoni, I., Rodríguez-Avila, A., Riblet, S., and García, M., 2008. Characterization of infectious laryngotracheitis virus (ILTV) isolates from commercial poultry by polymerase chain reaction and restriction fragment length polymorphism (PCR-RFLP). *Avian Dis.* 52, 59–63.

67. Neff, C., Sudler, C., and Hoop, R.K., 2008. Characterization of western European field isolates and vaccine strains of avian infectious laryngotracheitis virus by restriction fragment length polymorphism and sequence analysis. *Avian Dis.* 52, 278–283.

68. Oldoni, I., Rodríguez-Avila, A., Riblet, S.M., Zavala, G., and García, M., 2009. Pathogenicity and growth characteristics of selected infectious laryngotracheitis virus strains from the United States. *Avian Pathol.* 38, 47–53.

69. Han, M.G. and Kim, S.J., 2001. Analysis of Korean strains of infectious laryngotracheitis virus by nucleotide sequences and restriction fragment length polymorphism. *Vet. Microbiol.* 83, 321–313.

70. Callison, S.A., Riblet, S.M., Oldoni, I. et al., 2007. Development and validation of real time PCR assay for the detection and quantification of infectious laryngotracheitis virus in poultry. *J. Virol. Methods* 139, 3–38.

71. Ou, S.C., Giambrone, J.J., and Macklin, K.S., 2012. Comparison of a TaqMan real time PCR with a loop-mediated isothermal amplification assay for detection of infectious laryngotracheitis. *J. Vet. Biotech. Res.* 3, 27–36.

72. Mahmoudian, A., Kilpatrick, N.C., Cooper, M. et al., 2011. Development of a SYBER Green quantitative polymerase chain reaction assay for rapid detection of infectious laryngotracheitis virus. *Avian Pathol.* 40, 37–242.

73. Ou, S.C., 2012. Improved detection and control of infectious laryngotracheitis virus on poultry farms. PhD dissertation, Auburn University, Auburn, AL, 124pp.

86 Macavirus

Richa Sood, Kh. Victoria Chanu, and Sandeep Bhatia

CONTENTS

86.1 Introduction 765
    86.1.1 Classification, Morphology, and Genome Organization 765
    86.1.2 Transmission 766
    86.1.3 Clinical Features and Pathogenesis 766
    86.1.4 Diagnosis 767
        86.1.4.1 Conventional Techniques 767
        86.1.4.2 Molecular Techniques 768
    86.1.5 Treatment and Control 769
86.2 Methods 769
    86.2.1 Sample Collection and Preparation 769
    86.2.2 Detection Procedures 769
        86.2.2.1 Virus Isolation 769
        86.2.2.2 Genomic Detection 769
86.3 Conclusion and Future Perspectives 770
References 770

86.1 INTRODUCTION

Malignant catarrhal fever (MCF) is a clinically noticeable, lethal disease of cloven-hoofed, ruminant mammals that include bison, African buffalo, water buffalo, antelopes, gazelles, sheep, goats, domestic cattle, and deer [1–3]. This disease has also been reported in pigs [4]. The causal agents for MCF are members of the MCF virus (MCFV) group that belongs to the genus *Macavirus,* subfamily *Gammaherpesvirinae,* family *Herpesviridae* [5]. MCF has a worldwide distribution and occurs sporadically and rarely in outbreaks. Clinically, MCF is characterized by high fever, hemorrhagic diarrhea, corneal opacity, and swelling of lymph nodes, usually culminating in death of the susceptible species. Similar to members of the *Herpesviridae* family, MCFV produces lifelong infection with establishment of latency. This chapter focuses on the two most prevalent viruses, alcelaphine herpesvirus 1 (AlHV-1) and ovine herpesvirus 2 (OvHV-2), within the genus *Macavirus.*

86.1.1 CLASSIFICATION, MORPHOLOGY, AND GENOME ORGANIZATION

Gammaherpesvirinae is one of three subfamilies in the family *Herpesviridae* and consists of a large group of viruses that can infect humans as well as animals. After a recent taxonomical reorganization, gammaherpesviruses are now subdivided into four genera, two of which are newly established; *Macavirus* and *Percavirus,* and the other two are well-known *Lymphocryptovirus* and *Rhadinovirus* [6]. The *Lymphocryptovirus* genus contains Epstein–Barr virus and several other lymphocryptoviruses of primates. The *Rhadinovirus* genus contains herpesviruses that infect many mammalian taxa. All the MCF causing viruses belong to the genus *Macavirus* under the subfamily *Gammaherpesvirinae* in the family *Herpesviridae*. To date, 10 closely related but different viruses have been identified within the MCFV group and out of these 6 are associated with clinical diseases [7]. The two most prevalent macaviruses are AlHV-1, which causes wildebeest-associated MCF (WA-MCF), and OvHV-2, which is responsible for sheep-associated MCF (SA-MCF). AlHV-1 is endemic in wildebeest, but does not lead to clinical disease in this host [8]. Domestic and wild sheep are reservoirs for OvHV-2 [2,5]. In both forms of the disease, animals with clinical disease are not a source of infection as the viruses are only excreted by the natural hosts—wildebeest and sheep, respectively. Other MCFVs known to cause disease include caprine herpesvirus-2 (CpHV-2), which is endemic in goats [9], MCFV of unknown origin causing disease in white-tailed deer (MCFV-WTD) [10], ibex MCFV (MCFV-ibex) carried by ibex [11], and an AlHV-2-like virus carried by Jackson hartebeest [12]. The remaining four viruses carried by roan antelope, oryx, muskox, and aoudad have not yet been associated with disease [7].

Both AlHV-1 and OvHV-2 are about 140–220 nm in diameter containing a double-stranded DNA genome. The genomic DNA is enclosed in an icosahedral capsid, which is embedded in a complex amorphous layer composed of several proteins called the tegument. This complex structure is then enclosed by a lipid envelope [13,14]. AlHV-1 has been isolated and sequenced completely. Its DNA genome encompasses 70 open reading frames (ORFs). Among these, 61 ORFs are homologous to other herpesviruses. There are some genes that are shared only by AlHV-1 and other gammaherpesviruses. About 10 specific ORFs (A1–A10) of AlHV-1 have been identified to encode proteins A1–A10 with different functions. Out of these, eight ORFs have clear homologues with OvHV-2 and there are no equivalents of A1 or A4 [5]. Protein A1 is believed to play a role in cell transformation and/or immunoregulatory function [13,15]. A2 and A6 are involved in activation of transcription while A3 is involved in suppression of tumors, immunoregulation, development of bones and heart, arthritis, and the growth of sensitive cones during embryogenesis [13,15–17]. While A4 has no assigned function, A5 is structurally similar to G protein and believed to be involved in cell proliferation [15,18]. A7 and A10 may be related with the virulence of the virus as there is 3′end deletion of A7 gene and modification of A10 ORF in attenuated AlHV-1 [19]. Glycoprotein A8 is responsible for the interaction of the virus and cell receptor. A9 proteins are homologous to Bcl-2 family, which regulates apoptosis [13].

OvHV-2 has never been isolated due to the lack of a compatible tissue-culture system. More recently, an OvHV-2 infected bovine T-lymphoblastoid cell line, BJ1035, that was derived from a clinically affected cow with SA-MCF [20] has been cloned and sequenced [21]. Sequence analysis showed that the unique region of the OvHV-2 genome consists of 130 kb with 73 ORFs. The organization of ORFs in the OvHV-2 genome is quite similar to that of other gammaherpesviruses but matches most closely with that of AlHV-1. Three ORFs are unique to OvHV-2, that is, Ov 2.5, Ov 3.5, and Ov 8.5, whereas 8 ORFs are common in AlHV-1 and OvHV-2, that is, Ov2, Ov3, Ov4.5, Ov6, Ov7, Ov8, Ov9, and Ov10 [21]. Ov2.5 contains five exons and encodes a homologue of cellular IL-10. Ov3.5 encodes a 163 aa long protein, which forms a signal peptide. This protein, therefore, is believed to be secreted from the infected cell. Ov2 contains two exons and encodes a protein with a basic leucine-zipper (bZIP) motif and is believed to be a transcription factor. Ov3 produces a signal peptide to residue 22 and is homologous to proteins of the semaphorin family. Ov3 is predicted to be involved in the modulation of the host response to OvHV-2. The Ov4.5 and Ov9 products promote survival of infected lymphocytes and the establishment of latency in the OvHV-2 infected cells. The Ov5 ORF is located downstream of ORF9/DNA polymerase and overlaps with ORF10. This ORF is predicted to encode a G proteincoupled receptor. In between ORF50 and ORF52 lie three ORFs: Ov6, Ov7, and Ov8. The product of Ov6 contains a leucine-zipper motif in its C-terminal region and plays a role in transcriptional activation. The product of Ov7 contains a predicted signal peptide and N-linked glycosylation motifs and is thus likely to be a viral glycoprotein, involved in receptor binding. The product of Ov8 contains a transmembrane anchor near the C terminus and N-linked glycosylation sites. Ov8.5 is predicted to encode a proline-rich (24%) 42 kDa peptide of unknown function. The predicted Ov10 protein has a potential transmembrane anchor at the C terminus and four consensus nuclear-localization signals, and thus the Ov10 protein may localize to the nucleus of infected cells [21].

86.1.2 Transmission

Ovine herpesvirus 2: All domestic sheep are infected with ovine herpesvirus-2 [22]. Transmission occurs mainly by nasal shedding of the virus from lambs that are above 3 months of age, especially from the adolescent lambs of age 6–9 months [23]. The viral DNA in nasal secretions shows a steep rise and subsequent fall within 24–36 h. In sheep, this occurs multiple times between the ages 6 and 9 months and then declines after 9 months of age. Adult sheep are less frequent shedders than adolescent lambs [24]. Unlike WA-MCF, shedding of the virus might not be associated with lambing and no seasonal trend in viral shedding from adult sheep has been identified [23]. Horizontal transfer between clinically ill cattle has neither been seen in field nor in any experimental observations [3]. Bali cattle, Asian swamp buffalo, the American bison, and some deer species are reported to be more susceptible to disease caused by OvHV-2 than *Bos taurus* and *Bos indicus* cattle species [25,26]. Death in bison occurs within 48 h of infection [27]. Cattle in general are less susceptible to disease caused by OvHV-2 than AlHV-1 [28].

Alcelaphine herpesvirus-1: WA-MCF occurs throughout natural distribution of its carrier host in Africa. It was earlier believed that the virus existed as a cell-associated virus and was later identified that it survived as cell-free virus also and could be transmitted to cattle by nasal secretions of wildebeest [29]. It can also be vertically transmitted to offspring transplacentally [30]. The main source of infection is the wildebeest. The virus has been isolated from the ocular and nasal secretions of young wildebeest less than 3 months of age, and virtually all animals are infected by 4 months of age. Viral shedding in adult wildebeest is quite low and occurs primarily during periods of stress or parturition [31]. AlHV-1 is not transmitted from one clinically susceptible host to another via natural means, as the virus secreted from nonhost animals is cell associated and therefore extremely labile [32,33]. As such, sick animals may be housed with healthy animals without fear of horizontal disease transmission [34]. All ages of cattle are susceptible to the disease though adults are more prone, particularly peri-partuent females [35].

86.1.3 Clinical Features and Pathogenesis

Clinical features: The clinical features have been described in detail [36]. The incubation period in MCF infection is variable. It ranges from 3 to 8 weeks, and in case of experimental infection, it ranges between 14 and 37 days (22 days on an average). MCF is a multisystemic syndrome and is seen in a number of forms: (1) peracute form, (2) alimentary tract form, (3) common "head and eye" form, and (4) mild form; all of these gradations are based on appearance of clinical signs.

In serial transmissions with one strain of the virus, all of these forms may be produced. The most common manifestation is the *head and eye form*, which is characterized by a sudden onset of extreme dejection, anorexia, agalactia, high fever (40°C–43°C), rapid pulse rate (100–120 bpm), profuse mucopurulent nasal discharge, severe dyspnea with stertor due to obstruction of the nasal cavities with exudate, ocular discharge with variable degrees of edema of the eyelids, and congestion of sclera vessels. There is superficial necrosis seen in the anterior nasal mucosa and on the buccal mucosa. Discrete local areas of necrosis may be seen on the hard palate, gums and gingivae, dorsum of the tongue, or commissures of the mouth. At this stage, there is excessive salivation with ropy and bubbly saliva hanging from the lips. Opacity of the cornea is almost always seen, beginning as a gray ring at the corneoscleral junction and spreading centripetally with conjunctival and episcleral hyperemia. Hypopyon is observed in some cases. Lesions may occur at the skin-hoof junction of the feet, especially at the back of the pastern. The skin of the teats, vulva, and scrotum in acute cases may slough off entirely upon touch or become covered with dry, tenacious scabs. Nervous signs, like incoordination, muscle tremor, nystagmus, head-pushing, paralysis, and convulsions, may occur in the final stages. The superficial lymph nodes are visibly and palpably enlarged. In cases of prolonged illness, skin changes, including local papule formation with clumping of the hair into tufts over the loins and withers, may occur. In addition, eczematous weeping may result in crust formation, particularly on the perineum, around the prepuce, in the axillae, and inside the thighs. Infection of the cranial sinuses may occur with pain on percussion over the area. The *peracute and alimentary tract forms*: In the peracute form, the disease runs a short course of 1–3 days and characteristic signs and lesions of the "head and eye" form do not appear. There is usually a high fever, dyspnea, and an acute gastroenteritis. Symptomatically, WA-MCF resembles SA-MCF, except in alimentary tract form where there is marked diarrhea in SA-MCF whereas there is constipation in WA-MCF with only minor eye changes consisting of conjunctivitis rather than ophthalmia as seen in SA-MCF. The *mild form* is generally observed in experimental animals. There is a transient fever and mild erosions on the oral and nasal mucosae. This form may also be followed by complete recovery or recovery with recurrence or chronic MCF. A distinctive clinical feature in chronic MCF is persistent bilateral ocular leukomata. The disease in pigs is similar to the head and eye form in cattle and manifests with fever and tremor, ataxia, hyperesthesia and convulsions, and death [36].

Pathogenesis: Infection of susceptible animal has an incubation period of about 10 days to 2 months, which is followed by febrile condition till death of the animal. Although the pathological changes that occur in MCF infections are marked, the pathogenesis still remains to be fully understood. The production of a bacterial artificial chromosome clone carrying the entire pathogenic AlHV-1 genome has been used in studying the pathogenesis of MCF [37]. This clone grew infectious AlHV-1 virus and produced MCF disease in rabbits, which

was similar to the disease caused by the field virus. In rabbits, the disease was seen as a progressive T-cell hyperplasia involving local proliferation and infiltration of both lymphoid and nonlymphoid organs, associated with extensive vasculitis and the dysregulated cytotoxic lymphocytes that lead to tissue destruction [20,38]. The lesions caused by AlHV-1 were seen in peripheral lymph nodes whereas OvHV-2-associated lesions are more apparent in visceral lymphoid tissue. The AlHV-1-associated lesions contained less areas of necrosis than those of OvHV-2. However, in both viruses, lymphoid cell infiltrations consist of mainly T cells, of which CD8+ T cells predominate, with very few CD4+T cells. The presence of a small number of infected lymphocytes observed in lesions characterized by lymphocyte accumulation also suggests that MCF has an autoimmune-like pathology, caused by the cytotoxic action of uninfected cells under the influence of a small number of infected cells [20,38]. The presence of CD8+ OvHV-2-infected lymphocytes in brain and various tissue sites indicate a direct action of virus-infected dysregulated cytotoxic T cells at sites of lesions [39]. The other lesions like epithelial erosions and keratoconjunctivitis is mainly because of the involvement of the vascular adventitia. The lymph node enlargements occur due to atypical proliferation of sinusoidal cells. The cerebromeningeal changes observed are basically a form of vasculitis. There is synovitis, especially involving tibiotarsal joints and which is also associated with a lymphoid vasculitis. Thus, the pathogenesis of this disease is basically an outcome of virus–cell interactions and immune-mediated responses directed against infected cells.

86.1.4 Diagnosis

Diagnosis of SA-MCF poses significant challenges to veterinarians due to multisystemic involvement of the disease and symptomatic resemblance to many other diseases in the field. Symptomatically, MCF should be differentiated from bovine viral diarrhea caused by *pestivirus*, IBR caused by *bovine herpesvirus* 1, bluetongue caused by *Orbivirus*, and rinderpest caused by *morbillivirus* among viral diseases. It should also be differentiated from salmonellosis, coccidiosis, and besnoitiosis caused by protozoan *Besnoitia besnoiti* [40]. Clinically, only a tentative diagnosis of the disease is possible, and a laboratory confirmation is essential to arrive to a definitive diagnosis. A history of contact with sheep or wildebeest or with pasture recently grazed by either of these species is also a support for the tentative diagnosis. Earlier, histological identification of vasculitis, hyperplasia of lymphoid tissues, and accumulations of lymphoid cells in nonlymphoid organs was considered the gold standard for disease confirmation. However, with the development of molecular tests, an antemortem as well as early confirmatory diagnosis is now possible. Various methods used in diagnosis are enumerated below.

86.1.4.1 Conventional Techniques

Virus isolation: Plowright and associates first isolated AlHV-1 in cell culture and also obtained a cell-free virus of an isolate

WC11 [8,33]. Tissue cultures infected with AIHV-1 exhibit a cytopathic effect (CPE), consisting of multinucleate cellular foci, which is characteristic of AlHV-1. Intranuclear, cytoplasmic, and extracellular herpes-like particles has been demonstrated by electron microscopy in AIHV-1-infected tissue cultures [41]. This virus isolate is still utilized today for the development of diagnostics. In case of OvHV-2, lymphoblastoid cells infected with OvHV-2 have been propagated from cattle, deer, and rabbits with SA-MCF [42,43], but cell-free virus has not yet been cultivated despite numerous attempts till date.

Virus neutralization assay: This assay was developed by Plowright in 1967, utilizing the WC11 virus isolate mentioned previously [44]. The virus neutralization assay in use today still employs the AlHV-1 virus as the target of neutralization, as cell-free virus has never been isolated from other strains. This assay works best for antibody against the alcelaphine group of herpesviruses and is used in studying the extent of natural gammaherpesvirus infections in wildlife. The virus neutralization assay is not used as a diagnostic test in clinically affected animals, as these animals are not able to produce virus neutralizing antibody [45]. Animals with SA-MCF do not produce virus neutralizing antibodies to AlHV-1 [46].

Indirect immunofluorescence assay (IFA): Although the IFA is not as specific as the virus neutralization assay, it is useful in detecting antibodies in AlHV-1-infected cell monolayers. This test is not very specific, as other herpesviruses such as bovine herpesvirus-4 and infectious bovine rhinotracheitis virus cross-react [47]. This assay alone cannot be reliably utilized to detect SA-MCF when this disease is suspected [45].

Enzyme-linked immunosorbent assay (ELISA): Many ELISA assays have been developed to detect antibody to MCFVs [48–50]. The most commonly accepted competitive inhibition CI-ELISA was developed in 1994 to detect antibody to OvHV-2 using a MAb (15-A) targeting an epitope on a complex of glycoproteins that are conserved among all MCFVs [51]. This indirect CI-ELISA utilized enzyme labeled antimouse immunoglobulins for antibody detection and was reformatted as a direct CI-ELISA in 2001 to increase sensitivity [52]. This direct ELISA showed high sensitivity and specificity. However, there were certain issues like uninfected lambs less than 4 months of age give positive results due to the presence of maternal antibodies and low doses of virus infection take longer than usual (almost 4 weeks) for antibody production. Therefore, paired sampling is required for conclusive identification, and moreover, ELISA, though specific, cannot differentiate between different MCFVs [52].

86.1.4.2 Molecular Techniques

Genomic detection of MCF virus: Detection of the viral nucleic acid is currently the accepted method of diagnosis for MCF. Recently developed molecular diagnostic assays have improved the detection and differentiation of MCF causative viruses. A number of different formats of PCR have been developed targeting

various MCFVs since the first PCR for detection of AlHV-1 DNA was reported in 1990 [53]. However, among all the diagnostic tests, the one with the most significant impact is the PCR specific for detection of OvHV-2 DNA developed by Baxter et al. [54]. It is currently the only OIE (World Organisation of Animal Health) approved diagnostic test for detection of OvHV-2 infection. In this assay, the primers target a DNA fragment in the ORF75 of OvHV-2, a gene coding for a viral tegument protein. Two sets of primers are used for a nested OvHV-2 PCR assay: the outer set 556 (5′-AGTCTGGGGTATATGAATCCAGATGGCTCTC-3′) and 755 (5′-AAGATAAGCACCAGTTATGCATCTGATAAA-3′) in the first round and heminested PCR with inner set comprising of 556 and 555 (5′-TTCTGGGGTAGTGGCGAGCGAAGGCTTC-3′). The primers are specific to OvHV-2. The outer set of primers (556 and 755) amplifies a 422 bp fragment of OvHV-2 DNA. The nested primer set (556 and 555) amplifies a 238 bp fragment. The specificity of both of the fragments can be confirmed by restriction enzyme (RE) analysis with RsaI enzyme. The primers are specific and do not amplify DNA from BoHV-1, BoHV-2, BoHV-4, or AlHV-1. This PCR assay has been used in many studies to detect latent OvHV-2 infections in sheep. This nested PCR has high sensitivity and is validated for detection of OvHV-2 DNA in infected sheep as well as in animals with clinical MCF. This assay has been widely used in veterinary diagnostic laboratories and is considered as the gold standard [55]; however, its use as a routine method to detect OvHV-2 DNA for confirmation of clinical SA-MCF in diagnostic laboratories was problematic due to a high potential for carryover amplicon contamination leading to false positive results. In addition, it can only be used as a qualitative method for establishing infection in the *in vivo* model. As repeated attempts to isolate the virus had failed, a need to trace the agent in a quantitative manner along the time axis of a natural infection was felt, and the second important step in MCF diagnosis was the development of quantitative PCR or the real-time PCR for OvHV-2 DNA by Hussy et al. [56]. The OvHV-2-specific primer–probe set for the real-time PCR is based on the same sequence of genomic OvHV-2 DNA that had been used for the nested PCR (ORF75 gene) and is equally sensitive for clinical samples. The development of a multiplex PCR for detection and differentiation of MCF causing viruses was the next significant advancement in MCF molecular diagnostics, especially for mixed-species operations, such as zoos, wildlife parks, and game farms. The multiplex real-time PCR uses one pair of primer in conjunction with fluorescently labeled probes that target a polymorphic region in the viral DNA polymerase gene specific for OvHV-2, CpHV-2, MCFV-WTD, MCFV-ibex, and AlHV-1. This PCR has been optimized and validated for the identification of these pathogenic MCFVs in clinical samples using a single reaction. The assay is able to simultaneously detect and differentiate the viruses in clinical samples with high sensitivity and specificity [57]. Based on the analysis of several conserved genes of MCFV, it has been reported that the use of the glycoprotein region sequence provides an accurate phylogenetic relationship [58]. Sequencing of this region for studying the epidemiology of the virus has also been done indicating high similarity between various herpes viruses [59,60].

86.1.5 Treatment and Control

There is no commercial effective treatment or vaccines available; reduction of stress in subclinical or mildly affected animals is suggested. Nonsteroidal anti-inflammatory drugs may give some symptomatic relief. The prognosis of the disease is usually grave. The segregation of the susceptible hosts from the reservoir hosts is the most important step for control of the disease. If cattle and sheep are reared under natural flock conditions, the grazing areas should be separated. Grazing sheep in the morning hours and cattle in the evening hours on the same land patch may also help as the virus may be destroyed in the day's sun as it is very heat labile. Under natural flock conditions, lambs are not infected until after 2 months of age. If lambs are removed from contact with infected sheep prior to that age, they remain uninfected and can be raised virus-free [23]. This knowledge should be used by sheep producers and zoos to produce sheep free of OvHV-2. If certain flocks continue to be the source of infection, disposal of these flocks for slaughter should be considered. The introduction of carrier species from areas where the disease has occurred to farms with cattle should be avoided. Zoological parks also should introduce seronegative animals only and follow strict quarantine restrictions for newly acquired animals.

86.2 METHODS

86.2.1 Sample Collection and Preparation

Tissue samples intended for AlHV-1 virus isolation must be collected either before or shortly after the death of the animal. Collected tissues should be kept in sterile viral transport media containing antibiotics and antifungal agent and be transported under cold chain conditions to the diagnostic laboratory immediately. The probability of a successful isolation is higher when the time interval between collection and inoculation of the culture is minimized. Transport media containing tissue samples for virus isolation should never be frozen since virus cannot be recovered from frozen material [61].

Makeup of transport medium 199 as recommended by WHO is presented as follows:

1. Tissue culture medium 199 containing 0.5% bovine serum albumin (BSA).
2. To the above medium, add benzylpenicillin (2×10^6 IU/L), streptomycin (200 mg/L), polymyxin B (2×10^6 IU/L), gentamicin (250 mg/L), nystatin (0.5×10^6 IU/L), ofloxacin hydrochloride (60 mg/L), and sulfamethoxazole (0.2 g/L).
3. Sterilize by filtration and distribute in 1.0–2.0 mL volumes in screw-capped tubes.

For molecular diagnosis both in carrier and susceptible animals, 2–3 mL of peripheral blood in EDTA collected in vaccutainers should be couriered to the diagnostic laboratory in ice pack. Widespread pathological changes are observed involving many organs in clinical cases of MCF. These include hemorrhages and erosions throughout gastrointestinal tract, enlarged lymph nodes, formation of diptheritic membrane, and erosions with catarrhal accumulations in respiratory tract. Echymotic hemorrhages are also often seen on the epithelial lining of urinary bladder [45]. Hence, for histopathological examination, different tissue samples including lymph node, brain, liver, adrenal gland, kidney, urinary bladder, salivary gland, and alimentary tract mucosa including pharynx, esophagus, rumen, and peyer's patch can be collected in 10% formalin and should be sent to a diagnostic laboratory.

86.2.2 Detection Procedures

86.2.2.1 Virus Isolation

Isolation of causative agent to confirm MCF could only be performed for AlHV-1 till date. Infectivity of the virus is strictly cell associated and hence can be isolated only from cell suspensions of affected tissues, lymph nodes, and peripheral blood leucocytes. Bovine thyroid cells are commonly used for isolation of AlHV-1 though most primary and low-passage monolayers of ruminant origin may prove to be suitable substrate for virus isolation. Monolayers inoculated with approximately 5×10^6 cells/mL of cell suspensions from affected tissue should be examined microscopically at 40× for CPEs. Characteristic multinucleate foci appear within the monolayers, which retracts progressively forming dense bodies. CPE seldom appears within 1 week and may take as long as 21 days. Since infectivity is cell associated, for further passage it must be ensured that cell viability is maintained [62]. In case of isolation of OvHV-2, several attempts have been made wherein lymphoblastoid cells infected with OvHV-2 have been propagated from cattle, deer, and rabbits [42,43], but cell-free virus has not yet been cultivated.

86.2.2.2 Genomic Detection

86.2.2.2.1 Peripheral Blood Leucocytes Separation

Peripheral blood leucocytes (PBL) are separated from the whole blood. Usually, cells from 1.5 mL blood are sufficient for DNA extraction. The blood is centrifuged at 5000 rpm for 5 min and the supernatant is discarded. About 1 mL of erythrocyte lysis buffer is added to the pellet, mixed with gentle tapping, and kept for incubation for 10 min at room temperature. The centrifugation is at least twice or until clear PBL pellet is visible. The pellet is washed with 1 mL of 1× PBS by vortexing twice and stored at −40°C for further use after resuspending in 250 µL 1× PBS.

Makeup of erythrocyte lysis buffer is presented as follows:

1. Dissolve 1.21 g of Tris base in 800 mL of double distilled water.
2. Adjust pH to 7.4 with conc. HCl.
3. Add 0.584 g of NaCl.
4. Add 8 mL of 1 M (0.76 g) of $MgCl_2$.
5. Make up the volume to 1000 mL with double distilled water and autoclave.

86.2.2.2.2   DNA Extraction

Use 200 µL of PBL suspended in PBS for DNA extraction by conventional method or any commercial kit available.

86.2.2.2.3   PCR

PCR is currently the OIE approved test for confirmatory diagnosis of OvHV-2 detection. Tegument region is unique for all the OvHV-2 so this region is targeted for diagnostic purpose. Amplification of OvHV-2 DNA is carried out by using OIE approved primers developed by Baxter et al. [54]. The Two-step PCR amplification uses the following primer sets: primer set 556 (5′-AGTCTGGGGTATATGAATCCAGATGGC TCTC-3′) and primer set 755 (5′-AAGATAAGCACCAGTTA TGCATCTGATAAA-3′) in the first step and primer sets 556 and 555 (5′-TTCTGGGGTAGTGGCGAGCGAAGGCTTC-3′) in the second step (nested PCR). The amplification reactions are performed in a 25 µL volume containing 10 mM Tris-HCl (pH 8.0); 2 mM MgCl$_2$; 200 mM (each) dATP, dCTP, dGTP, and dTTP; 20 pM (each) primer; and 2 U of Taq DNA polymerase. Thermal cycling conditions for both steps are 5 min at 95°C and then 35 cycles of 94°C for 1 min, 60°C for 1 min, and 72°C for 2 min; this is followed by a final 7 min at 72°C for extension. The expected size of the amplified product for positive samples in the outer PCR and the nested PCR are 422 and 238 bp, respectively. Agarose gel electrophoresis of the amplified product is performed by using 2% agarose gel.

86.3   CONCLUSION AND FUTURE PERSPECTIVES

MCF is widespread to a large part of the world and has become one of the emerging diseases of ruminants of significant economic, trade, and/or food security concerns. The two most prevalent forms are the alcelaphine herpesvirus infection and OvHV-2 infections. Wildebeest for AlHV1 and sheep and goats for OvHV-2 are asymptomatic carrier species for these two infections that transmit this fatal disease to the susceptible ruminant species. The disease is of particular significance in the context where mixed farming is a common practice with these species. The virus is transmitted by aerosol or by contact, and most lambs or wildebeest calves are infected shortly after birth and are capable of then infecting susceptible cattle. The clinically susceptible species either die within 24 h or usually undergo a conspicuous clinical course affecting any or all of these systems: respiratory system, gastrointestinal system, and nervous system. Due to its sporadic nature and overlapping clinical signs with many other diseases, the disease at places has not received the due attention of veterinarians, researchers, and policy makers; thus in most parts of the world, there is no national policy for disease control. There is no treatment and vaccine available for this disease to date. Although the pathogenicity of the disease has become much clearer now, the fact that the OvHV-2 infection agent has not yet been isolated worldwide has hampered the information influx. With development and continuing refinement of PCR-based assays, it has now become easier to recognize the status

of this disease, and a confirmatory diagnosis can be made wherever the facility is available or the samples can be easily sent to the nearest testing lab. Countrywide survey of this disease coupled with molecular characterization of virus genome will provide new insights into epidemiology and economic losses associated with MCF and will help devise an effective control strategy.

REFERENCES

1. Ackermann, M., Pathogenesis of gammaherpesvirus infections. *Vet. Microbiol.* 113, 211, 2006.
2. Sood, R., Hemadri, D., and Bhatia S., Sheep associated malignant catarrhal fever—An emerging disease of bovids in India. *Ind. J. Virol.* 24, 321, 2013.
3. Plowright, W., Malignant catarrhal fever virus. In *Virus Infections of Ruminants*, 3rd edn., Dinter, Z. and Morein, B., Eds., Elsevier Science Publishers, New York, p. 123, 1990.
4. Albini, S., Zimmermann, W., Neff, F. et al., Identification and quantification of ovine gammaherpesvirus 2 DNA in fresh and stored tissues of pigs with symptoms of porcine malignant catarrhal fever. *J. Clin. Microbiol.* 41, 900, 2003.
5. Russell, G.C., Stewart, J.P., and Haig, D.M., Malignant catarrhal fever: A review. *Vet. J.* 179, 324, 2009.
6. Davison, A., Eberle, R., Ehlers, B. et al., The order Herpesvirales. *Arch. Virol.* 154, 171, 2009.
7. Li, H., Cunha, C.W., and Taus, N.S., Malignant catarrhal fever: Understanding molecular diagnostics in context of epidemiology. *Int. J. Mol. Sci.* 12, 6881, 2011.
8. Plowright, W., Ferris, R.D., and Scott, G.R., Blue wildebeest and the aetiological agent of bovine malignant catarrhal fever virus. *Nature* 188, 1167, 1960.
9. Chmielewicz, B., Goltz, M., and Ehlers, B., Detection and multigenic characterization of a novel gammaherpesvirus in goats. *Virus Res.* 75, 87, 2001.
10. Li, H., Dyer, N., Keller, J. et al., Newly recognized herpesvirus causing malignant catarrhal fever in white-tailed deer (*Odocoileus virginianus*). *J. Clin. Microbiol.* 38, 1313, 2000.
11. Li, H., Gailbreath, K., Bender, L.C. et al., Evidence of three new members of malignant catarrhal fever virus group in muskox (*Ovibosmos chatus*), Nubian ibex (*Capra nubiana*), and gemsbok (*Oryx gazella*). *J. Wildl. Dis.* 39, 875, 2003.
12. Klieforth, R., Maalouf, G., Stalis, I. et al., Malignant catarrhal fever-like disease in Barbary red deer (*Cervus elaphus barbarus*) naturally infected with a virus resembling alcelaphine herpesvirus 2. *J. Clin. Microbiol.* 40, 3381, 2002.
13. Ensser, A., Pflanz, R., and Fleckenstein B. Primary structure of the alcelaphine herpesvirus 1 genome. *J. Virol.* 71, 6517, 1997.
14. McGeoch, D.J., Rixon, F.J., and Davison, A.J., Topics in herpesvirus genomics and evolution. *Virus Res.* 117, 90, 2006.
15. Coulter, L.J., Wright, H., and Reid, H.W., Molecular genomic characterization of the viruses of malignant catarrhal fever. *J. Comp. Pathol.* 124, 2, 2001.
16. Ensser, A. and Fleckenstein, B., Alcelaphine herpesvirus type 1 has a semaphorin-like gene. *J. Gen. Virol.* 76, 1063, 1995.
17. Lange, C., Liehr, T., Goen, M. et al., New eukaryotic semaphorins with close homology to semaphorins of DNA viruses. *Genomics* 51, 340, 1998.
18. Arvanitakis, L., Geras-Raaka, E., and Gershengorn, M., Constitutively signaling G-protein-coupled receptors and human disease. *Trends Endocrinol. Metab.* 9, 27, 1998.

19. Handley, J.A., Sargan, D.R., Herring A.J. et al., Identification of a region of the alcelaphine herpesvirus-1 genome associated with virulence for rabbits. *Vet. Microbiol.* 47, 167, 1995.

20. Schock, A. and Reid, H.W., Characterisation of the lymphoproliferation in rabbits experimentally affected with malignant catarrhal fever. *Vet. Microbiol.* 53, 111, 1996.

21. Hart, J., Ackermann, M., Jayawardane, G. et al., Complete sequence and analysis of the ovine herpesvirus 2 genome. *J. Gen. Virol.* 88, 28, 2007.

22. Li, H., Shen, D.T., O'Toole, D. et al., Investigation of sheep associated malignant catarrhal fever virus infection in ruminants by PCR and competitive inhibition enzyme linked immunosorbent assay. *J. Clin. Microbiol.* 33, 2048, 1995.

23. Li, H., Hua, Y., Snowder, G. et al., Levels of ovine herpesvirus 2 DNA in nasal secretions and blood of sheep: Implications for transmission. *Vet. Microbiol.* 79, 301, 2001.

24. Li, H., Taus, N.S., Lewis, G.S. et al., Shedding of ovine herpesvirus 2 in sheep nasal secretions: The predominant mode for transmission. *J. Clin. Microbiol.* 42, 5558, 2004.

25. OToole, D., Li, H., Sourk, C. et al., Malignant catarrhal fever in a bison *Bison bison* feedlot 1993–2000. *J. Vet. Diagn. Invest.* 14, 183, 2002.

26. Schultheiss, P.C., Collins, J.K., Spraker, T.R. et al., Epizootic malignant catarrhal fever in three bison herds: Differences from cattle and association with ovine herpesvirus-2. *J. Vet. Diag. Invest.* 12, 497, 2000.

27. Sood, R., Kumar, M., Bhatia, S. et al., Transboundary diseases: Ovine herpesvirus type 2 infection in captive bison in India. *Vet Rec.* 170, 654, 2012.

28. Loken, T., Bosman, A.M., and van Vuuren, M. Infection with Ovine herpesvirus 2 in Norwegian herds with a history of previous outbreaks of malignant catarrhal fever. *J. Vet. Diagn. Invest.* 21, 257, 2009.

29. Mushi, E.Z., Rossiter, P.B., Jessett, D. et al., Isolation and characterization of a herpesvirus from topi (*Damaliscus korrigum*, Ogilby). *J. Comp. Pathol.* 91, 63, 1981.

30. Plowright W., Kalunda, M., Jessett, D.M. et al., Congenital infection of cattle with herpsevirus causing malignant catarrhal fever. *Res. Vet. Sci.* 13, 37, 1972.

31. Barnard, B.J.H., Bengis, R.G., Griessel, M.D. et al., Excretion of alcelaphine herpes virus 1 by captive and free-living wildebeest (*Connachaetes taurinus*). *Onderstepoort J. Vet. Res.* 56, 131, 1989.

32. Mushi, E.Z. and Rurangirwa, F.R., Epidemiology of bovine malignant catarrhal fevers, a review. *Vet. Res. Commun.* 5, 127, 1981.

33. Plowright, W., Malignant catarrhal fever. *J. Am. Vet. Med. Assoc.,* 152, 795, 1968.

34. Plowright, W., Malignant catharral fever in East Africa. 1. Behaviour of the virus in free living populations of blue wildebeest (*Gorgon taurinus taurinus, Burchell*). *Res. Vet. Sci.* 6, 56, 1965.

35. Barnard, B.J.H., Van der Lugt, J.J., and Mushi, E.Z., Malignant catarrhal fever. In *Infectious Diseases of Livestock*, Coetzer, J.A.W., Thomson, G.R., and Tustin, R.C., Eds., Oxford University Press, New York, 1994.

36. Radostits, O.M., Gay, C.C., Hinchcliff, K.W. et al., *Veterinary Medicine: A Text Book of Diseases of Cattle, Sheep Goat Pigs and Horses*, 10th edn., Saunders, Philadelphia, PA, vol. 21, p. 1245, 2010.

37. Dewals, B., Boudry, C., Gillet, L. et al., Cloning of the genome of Alcelaphine herpesvirus 1 as an infectious and pathogenic bacterial artificial chromosome. *J. Gen. Virol.* 87, 509, 2006.

38. Buxton, D., Reid, H.W., Finlayson, J. et al., Pathogenesis of sheep-associated malignant catarrhal fever in rabbits. *Res. Vet. Sci.* 36, 205, 1984.

39. Simon, S., Li, H., O'Toole, D. et al., The vascular lesions of a cow and 760 bison with sheep-associated malignant catarrhal fever contain ovine herpesvirus 2-infected CD8+ 761 T lymphocytes. *J. Gen. Virol.* 84, 2009, 2003.

40. Schelcher, F., Foucras, G., Meyer, G. et al., Le coryza gangreneux chez les bovins. *Point. Vet.* 32, 30, 2001.

41. Plowright, W. and Macadam, R.F., Growth and characterization of the virus of bovine malignant catarrhal fever in East Africa. *J. Gen. Microbiol.* 39, 253, 1963.

42. Reid, H.W., Buxton, D., Pow, I. et al., Isolation and characterisation of lymphoblastoid cells from cattle and deer affected with "sheep-associated" malignant catarrhal fever. *Res. Vet. Sci.* 47, 90, 1989.

43. Schuller, W., Cemy-Reiterer, S., Cenery, R. et al., Evidence that the sheep associated form of Malignant catarrhal fever is caused by a herepesvirus. *J. Vet. Med. Series B.* 37, 442, 1990.

44. Plowright, W., Malignant catarrhal fever in East Africa III. Neutralizing antibody in free-living wildebeest. *Res. Vet. Sci.* 8, 129, 1967.

45. OIE (Office International des Epizooties), *OIE Terrestrial Manual 2008*. Chapter 2.4.15: Malignant catarrhal fever.

46. Rossiter, P.B., Antibodies to malignant catarrhal fever virus in cattle with non-wildebeest-750 associated malignant catarrhal fever. *J. Comp. Pathol.* 93, 93, 1983.

47. Rossiter, P.B., Mushi, E.Z., and Plowright, W., Development of antibodies in rabbits and cattle infected experimentally with an African strain of malignant catarrhal fever virus. *Vet. Microbiol.* 2, 57, 1977.

48. Fraser, S.J., Nettleton, P.F., Dutia, B.M. et al., Development of an enzyme-linked immunosorbent assay for the detection of antibodies against malignant catarrhal fever viruses in cattle serum. *Vet. Microbiol.* 116, 21, 2006.

49. Frolich, K., Li, H., and Muller-Doblies, U., Sero-survey of antibodies for malignant catarrhal fever associated viruses for free living and captive cervids in Germany. *J. Wildl. Dis.* 34, 777, 1998

50. Wan, S.K., Castro, A.E., and Fulton, R.W., Effect of interferons on the replication of alcelaphine herpesvirus-1 of malignant catarrhal fever. *J. Wildl. Dis.,* 24, 484, 1988.

51. Li, H., Shen, D.T., Knowles, D.P. et al., Competitive inhibition enzyme-linked immunosorbent assay for antibody in sheep and other ruminants to a conserved epitope of malignant catarrhal fever virus. *J. Clin. Microbiol.* 32, 1674, 1994.

52. Li, H., McGuire, T.C., Muller-Doblies, U.U. et al., A simpler, more sensitive competitive-inhibition ELISA for detection of antibody to malignant catarrhal fever viruses. *J. Vet. Diagn. Invest.* 13, 361, 2001.

53. Hsu, D., Shih, L.M., Castro, A.E. et al., A diagnostic method to detect alcelaphine herpesvirus-1 of malignant catarrhal fever using the polymerase chain reaction. *Arch. Virol.* 114, 259, 1990.

54. Baxter, S.I., Pow, I., Bridgen, A. et al., PCR detection of the sheep-associated agent of malignant catarrhal fever. *Arch. Virol.* 132, 145, 1993.

55. Muller-Doblies, U.U., Li, H., Hauser, B. et al., Field validation of laboratory tests for clinical diagnosis of sheep associated malignant catarrhal fever. *J. Clin. Microbiol.* 36, 2970, 1998.

56. Hussy, D., Stauber, N., Leutenegger, C.M. et al., Quantitative fluorogenic PCR assay for measuring ovine herpesvirus 2 replication in sheep. *Clin. Diagn. Lab. Immunol.* 8, 123, 2001.

57. Cunha, C.W., Otto, L., Taus, N.S. et al., Development of a multiplex real-time PCR for detection and differentiation of malignant catarrhal fever viruses in clinical samples. *J. Clin. Microbiol.* 47, 2586, 2009.

58. McGeoch, D.J. and Cook, S., Molecular phylogeny of the alpha herpesvirinae subfamily and a proposed evolutionary timescale. *J. Mol. Biol.* 238, 9, 1994.

59. Goltz, M., Broll, H., Mankertz, A. et al., Glycoprotein B of bovine herpesvirus type 4: Its phylogenetic relationship to gB equivalents of the herpesviruses. *Virus Genes* 9, 53, 1994.

60. Sood, R., Khandia, R., Bhatia, S. et al., Detection and molecular characterization of naturally transmitted sheep associated malignant catarrhal fever in cattle in India. *Trop. Anim. Health Prod.* 46, 1037, 2014.

61. Reid, H.W. and Van Vuuren, M., Bovine malignant catarrhal fever. In *Infectious Diseases of Livestock*, 2nd edn., Coetzer, J.A.W. and Tustin, R.C., Eds., Oxford University Press, Oxford, U.K., 2004.

87 Marek's Disease Virus (*Gallid Herpesvirus* 2) and Herpesvirus of Turkey (*Melagrid Herpesvirus* 1)

Grzegorz Woźniakowski and Jowita Samanta Niczyporuk

CONTENTS

87.1 Introduction 773
    87.1.1 Classification.. 774
    87.1.2 Morphology and Biology 774
    87.1.3 Clinical Features and Pathogenesis 775
    87.1.4 Diagnosis 775
        87.1.4.1 Conventional Techniques.. 775
        87.1.4.2 Molecular Techniques... 776
87.2 Methods 777
    87.2.1 Sample Preparation... 777
    87.2.2 Detection Procedures.. 778
        87.2.2.1 Virus Isolation.. 778
        87.2.2.2 Agar Gel Immunodiffusion 778
        87.2.2.3 Radial Immunodiffusion Assay 778
        87.2.2.4 Polymerase Chain Reaction.. 778
        87.2.2.5 Loop-Mediated Isothermal Amplification 778
87.3 Conclusion 779
References.. 779

87.1 INTRODUCTION

Marek's disease (MD) is known as the most important threat in commercial production of chickens, turkeys, and Japanese quail. The loss due to the disease reaches US$1 billion annually [1], with mortality up to 60% in affected chicken flocks prior to application of protective vaccination [2]. Apart from significant mortality, egg production may also be reduced [3,4].

MD was first described in 1907 by the Hungarian veterinarian József Marek who observed polyneuritis signs along with histopathological lesions in the brain and peripheral nerves of chickens [5,6]. Following this first conference report, the disease was noted in 1914 in the United States, the Netherlands, and Great Britain [1]. For many years, MD was sporadically observed among chickens and turkeys and was manifested by neurological signs and disorders of locomotor system [7]. A notable increase in mortality among commercial chickens was observed between 1920 and 1940. In the 1960s, the observed clinical signs were called by Dr. Biggs [8] as "Marek's disease" but the observed clinical signs and lesions were misdiagnosed as lymphoid leukosis (LL). Therefore, MD was assigned to avian leukosis group. The milestone in MD history was the description of clinical signs by Pappenheimer et al. [7] as lymphoproliferative lesions in the spatial ganglia and peripheral nerves. Subsequently, Churchill and Biggs [9] and Nazerian et al. [10] isolated an etiological agent that was herpesvirus and called it "Marek's disease virus" (MDV). However, early attempt to reisolate the virus was not successful, probably due to the strictly cell-associated nature of MDV and the lack of specific-pathogen-free (SPF) chickens.

MD has been subdivided into acute and classic forms. The classic form of the disease was observed in the 1950s but is seldom detected these days. Following the successful virus isolation, Clanek et al. [11] found that the cell-free virus particles are shed *via* feather follicle epithelium (FFE), which explained MDV contagious features and transmission. MD is also an example of successful disease prevention by prophylactic vaccination of chickens [12]. Additionally, vaccination against MDV was the first application of vaccine against tumoral disease, which paved the way for further research on cancer treatment. However, the attempts to verify the possible role of MD in cancerogenesis in humans failed indicating that the virus is not capable itself to replicate in humans [13].

87.1.1 CLASSIFICATION

The primary classification of MDV was confusing due to the different clinical signs observed during disease process [8]. Some of the virus strains were more or less virulent inducing lymphomas, neurological disorders, or both. MDV shares some lymphotropic features with gammaherpesviruses, but finally, the virus was assigned to *Mardivirus* genus of *Herpesviridae* family [4,14]. According to International Committee on Taxonomy of Viruses, the MDV is represented by three different serotypes (MDV-1, *Gallid herpesvirus 2*; MDV-2, *Gallid herpesvirus 3*; and MDV-3, *Melagrid herpesvirus 1*, or herpesvirus of turkey, HVT). All pathogenic strains fall into the serotype 1 of MDV; however, the most efficient vaccine against MD is based on attenuated CVI988/Rispens strain [2]. The MDV-2 serotype is represented by SB-1, HPRS$_{24}$, and 301B/1 strains or other field isolates that are nonpathogenic for chickens [15,16]. These strains are used in vaccines against MD, which are applied mainly in the United States and Asia. Application of MDV-2-based vaccines in Europe is limited due to the increasing incidence of leukosis (LL) among vaccinated chickens [14,17,18]. The MDV-3 (FC126) strain is also nonpathogenic for chickens. Meanwhile, the vaccine based on the FC126 strain of HVT has been applied since 1970 as a live cell-associated or lyophilized cell-free suspension. On the basis of MDV-1 oncogenicity and virulence, four pathotypes may be distinguished, that is, mildly virulent MDV (mMDV), virulent MDV (vMDV), very virulent MDV (vvMDV), and very virulent plus strains (vv+MDV). The latter one is capable to override postvaccinal protection. The pathotyping of MDV isolates is however laborious and time-consuming as it is based on lesions onset in vaccinated 15 × 7 chickens according to Avian Disease & Oncology Laboratory [19]. An increase of MDV pathogenicity in field may be related to the common vaccination against MDV that leads to the emergence of more virulent strains [14].

87.1.2 MORPHOLOGY AND BIOLOGY

The MDV's particles observed under electron microscopy (EM) are enveloped and have a diameter between 150 and 160 nm. The hexagonal nucleocapsid measures 85–100 nm [4]. The amorphous virus particles can be also visualized in thin sections of feather epithelium cells [20]. The virion particles from different serotypes appear similar. The capsids of FC126 and HVT show a crossed pattern.

MDV genome consists of double-stranded DNA (dsDNA) with a unique long (U$_L$) and unique short (U$_s$) regions flanked by internal (IT$_R$) and terminal (T$_R$) repeats [21,22]. At the terminal ends of IT$_R$ and T$_R$, the alpha (α-like) sequences specific for *Alphaherpesvirinae* subfamily are located [23,24]. These regions have a crucial role in the packaging of viral DNA into the capsids. All three MDV serotypes have genomes of different sizes (from 159 to 180 kb), with encoded homologous genes, as well as genes that are unique for the serotype [25]. As the density of viral DNA equals to approximately 1.70 g/mL, it is difficult to separate the virus from the cellular, chicken DNA.

The MDV genome encodes over 100 specific proteins that are specific for alphaherpesviruses and may be specific only to the particular serotype. The main oncoprotein of MDV is Meq with leucine zipper motif characteristic for all transcriptional regulators [26]. The Meq is present in two copies located within the TR$_L$ region and between the junction of IR$_L$ and IR$_s$. Other MDV genes taking part in replication and oncogenesis are ICP4 (immediately early promoter), pp38 (phosphoprotein from pp38 to pp24 complex), and latency-associated transcripts (LATs), which downregulate MDV replication during latency [27,28]. The complete clones of all three MDV serotypes have been constructed in the bacterial artificial chromosome, which facilitated further genetic manipulation on MDV and determination of a number of MDV gene functions [29]. The MDV DNA is integrated into the telomeric regions of the host chromosomes [30]. This feature may be linked to the sequences located at the terminal ends of virus genome. Therefore, MDV is strictly cell-associated and cell-dependent. However, it was still unclear how the viral DNA is maintained during latency stage [27,31]. The study conducted in tumor cell lines showed the virus may parallely replicate in infected cells while in other cells switch to transformation [31]. Another unique feature of MDV is the possibility of concurrent infection with virulent MDV-1 strain and nonpathogenic MDV-2 and HVT viruses. This may potentially lead to interactions between these three agents but the recombination between different MDV serotypes in the field is rare. The successful recombinant event between MDV-1 and MDV-2 was artificially achieved using MSB1-41C cell line [32,33]. Meanwhile, it has been shown that coinfection with MDV-1 and avian leukosis or reticuloendotheliosis (RE) virus leads to spontaneous insertion of long terminal repeat (LTR) regions into the MDV genome [34,35]. It has been also shown that MDV may interact during coinfections with other immunosuppressive agents especially fowl adenoviruses (FAdV) [36], chicken anemia virus (CIAV) [37], or infectious bursa disease virus. The recent *in vitro* study on the interactions between FAdV-7 and FC126 of HVT in chicken embryo fibroblasts (CEFs) indicated the role of FAdV in HVT replication inhibition. These results were confirmed during *in vivo* studies on SPF chickens [36].

The productive replication of MDV occurs in B lymphocytes. The virus replication initiates by MDV binding to the cellular receptors via glycoprotein B (gB) and other virus proteins [31]. The virus spreads to other cells by direct contact *via* intracellular bridges. Three glycoproteins of MDV called gE, gI, and gM are important in virus transfer from infected to noninfected cell. Replication rate differs between serotypes as well as virus passage level. During productive replication, the number of virus genome may reach 100–1000 copies per cell. In infected cells, the virus leads to the formation of inclusion bodies and cell lysis. Fully productive stage of infection that leads to shedding of the infectious particles undergoes in FFE [5,15]. Latency is an important stage during MDV infection. During this stage, the viral proteins are not expressed but MDV DNA may be detected at low levels. LATs encoded by MDV have been described to play an important role in mapping to antisense sequence of genes thus downregulating protein expression [27]. However, in the case of MDV-1, it is impossible to separate latently infected from the transformed cells. Latency undergoes mainly in CD4+ cells but can be also associated with CD8+ and B cells. The viral copy number may reach below five copies per cell. The latency may then switch to transformation stage with expression of MD tumor-associated surface antigen (MATSA) [38]. Similarly, CD30 antigen is detected during lymphomas onset [31].

MDV is transmitted by direct contact between infected chickens or indirectly by virus particles present in feather debris and dust [20]. Vertical transmission of MDV does not exist. Additionally, MDV may be passively carried by litter mites, darkling beetles (*Alphitobius diaperinus*), or even mosquitoes [39]. The virus present in dust particles remains infectious up to several months in temperatures between 20°C and 25°C. It may persist for a longer time in lower temperatures.

87.1.3　Clinical Features and Pathogenesis

MDV infection may cause distinct clinical signs and anatomopathological lesions. The most common clinical signs include paralysis of legs, torticollis, stunting, and depression. Paralysis or torticollis occurs after 2–12 weeks post infection (PI). In field conditions, the outbreaks of MD occur in chicken aging over 3 weeks. The late MD outbreaks are observed in 8–9-week-old chickens [31]. The early mortality may occur after 8–16 days PI during infection with vMDV and vvMDV strains. Before the prophylactic vaccination was applied, the mortality reached from 20% to 60% but currently was considerably reduced [2].

The neurological signs are related to peripheral nerve dysfunction [14]. Frequently, celiac plexus as well as sciatic and brachial nerves is affected. These nerves lose the cross-striations and have yellowish appearance. The gross lesions may also be found in the ganglia and brain. Visceral lymphomas may occur in different organs including the liver, spleen, lungs, heart, intestines, kidneys, proventriculus, ovaries, testis, skin, and skeletal muscles [40]. The lesions are seen as the considerable enlargement of affected organs and the presence of nodular white or gray tumors. The surface of an organ with tumors is however smooth. The skin lesions are associated with infiltration of feather follicles [11]. The nodular erythematous tumors may be detected especially in feather tracts and are termed "Alabama Red Leg syndrome." Other less specific signs including diarrhea and anorexia may also be observed. Ocular lesions with infiltration of the iris and loss of pigmentation are not frequently observed.

87.1.4　Diagnosis

Diagnosis of MD on the basis of clinical signs or lesions may be difficult or confusing because of some similarities observed during LL or RE. Therefore, different techniques of classic virology and molecular biology are adapted to confirm the presence of MDV as the causative agent. As in the case of any infectious agent, the most important is virus isolation, then the presence of specific antibody or viral DNA in organs of infected birds.

87.1.4.1　Conventional Techniques

Virus isolation: MDV isolation is always conducted in order to confirm the presence and infectivity of the virus [6,9,10], as the presence of DNA alone does not indicate the virus presence. MDV may be isolated from organs of infected chickens as early as 1–2 days PI. Additionally, cell-free virus particles may be retrieved from skin or feather tips of infected chickens [11]. The infected lymphocytes may be separated from the peripheral blood as a "buffy coat." MDV may be isolated in primary cell cultures prepared from SPF chicken embryos or duck embryos. The best cell type for MDV isolation is duck embryo fibroblasts (DEF) prepared from 17- to 18-day-old duck embryos. The strains belonging to serotype 1 as well as 2 and 3 (HVT) show rapid growth in DEF with the presence of evident cytopathic effect (CPE) after 48–96 h PI. Similar good results are achieved when chicken embryo kidney cells (CEK) prepared from 18- to 19-day-old SPF chicken embryos are used. The good results are also obtained when trypsinized chicken kidney cells of MDV-infected chickens are used for virus isolation. In general, CEFs that are prepared from 11-day-old SPF chicken embryos are the good choice for isolation of MDV-2 and MDV-3 strains. These cells are less permissive for inoculation with MDV-1 field strains but after 3 virus passages, some strains are showing specific CPE.

Immunofluorescence assay: The confirmation of isolated virus purity may be examined using indirect immunofluorescence assay (IFA) using monoclonal antibodies targeting different antigens [41,42]. The antigens of MDV may also be detected in infected lymphoid tissues, FFE, and feather tips. The most frequent pp38, gB, and Meq proteins are the targets for monoclonal antibodies in immunofluorescent detection (Figure 87.1).

Electron microscopy: EM allows to confirm the presence of MDV particles in cell cultures and FFE during

meq

FIGURE 87.1 Indirect immunofluorescence assay. Chicken embryo fibroblasts infected with 31/07 Marek's disease virus isolate at third passage 56 h postinfection. Primary anti-*meq*-specific antibody and secondary antimouse Alexa 568 (Life Technologies, Delhi, India) were applied. Magnification 400×. The *meq* conglomerates in cell nuclei were indicated.

productive infection. Also, the sections of affected organs are used to identify the presence of enveloped icosahedral particles with a diameter between 150 and 160 nm. Due to the requirements of sophisticated equipment and experienced staff, this method cannot be considered as a routine technique for MD diagnosis.

Histopathology and immunohistopathology: The classic staining methods of tissue section collected from infected chickens are used to show the cytopathological changes caused by MDV. The tissues are fixed in formalin or frozen and embedded in paraffin to obtain ultra thin sections. The sections are stained with hematoxylin and eosin. The viral and cellular antigens can be detected in the thin sections of tissues using immunocytochemistry. The cells found in MDV tumor sections include infiltrations of large and small lymphocytes, macrophages, and lymphoblasts. The proportion of infected cells may be different, which also reflects the stage and severity of infection. The cells in MD tumors express MATSA and CD30 antigen [38]. Other antigens comprise of MHC class II antigen as well as CD4 and CD8 specific markers.

Specific antibody detection: The humoral immune response against MDV has a marginal role in MD prophylaxis but might be used to indicate the contact of chickens with the virus. In general, enzyme-linked immunosorbent assay (ELISA) and agar gel precipitation assay (AGID) are used to monitor the purity of SPF chickens from the control groups. Also, virus neutralization test may be used to determine the titer of the sera [43,44]. However, the MDV- or HVT-specific antibody is detected as early as 14–21 days PI and their level may decrease after 56 days PI.

Radial immunodiffusion assay: Radial immunodiffusion assay (RID) relies on detection of MDV antigen present in feather tips during the productive stage of infection [45]. The method is conducted in petri dishes covered with agar layer supplemented with sodium chloride and positive serum against MDV. The feather tips picked from the neckal route of infected chickens are put into the agar layer. The prepared dish is then incubated in a humid chamber at 37°C for 24–48 h. In case of the presence of MDV antigen, the whitish ring of precipitates around the feather tips is observed. The method

is useful to monitor the productive stage of MDV infection; however, its application is very limited during latency stage.

87.1.4.2 Molecular Techniques

The classic methods including virus isolation and histopathological examination are still the "gold standard" in MDV identification. However, in spite of their advantages, they are time-consuming and laborious and require experienced laboratory staff or pathologist. The complete sequences of all three MDV serotypes allowed further development of molecular biology techniques including polymerase chain reaction (PCR) and real-time PCR as well as novel techniques of loop-mediated isothermal amplification (LAMP) [46–55]. These techniques facilitated quick and convenient detection of all MDV serotypes. However, molecular biology methods cannot be used as a sole evidence for the virus presence because the DNA detection is not equal to the presence of infectious virus.

87.1.4.2.1 DNA Hybridization

DNA hybridization is useful for confirmation of the presence of sequences specific for MDV. The specific probes targeting genes of MDV are labeled with radioisotope or fluorescent reporter dye. This technique may also be successfully applied for *in situ* hybridization to detect MDV-1 and MDV-3 (HVT) strains. Building on hybridization and use of labeled molecular probes, other molecular biology techniques including PCR and real-time PCR were developed.

87.1.4.2.2 PCR and Real-Time PCR

The primary PCR used 132 bp repeated region of MDV-1 to amplify tandem repeats of this sequence [45,46]. The vMDV has one or two copies of these repeats while the attenuated forms of MDV-1 including CVI988/Rispens have at least several repeats (Figure 87.2).

Similarly, based on a double nucleotide mismatch within the pp38 gene, it was possible to differentiate vMDV from CVI988/Rispens using a mismatch amplification mutation assay (MAMA-PCR) [53]. Other PCR targets include gB, *meq*, ICP4, and more specific MDV-1 genes [47–50].

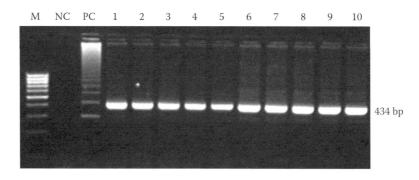

FIGURE 87.2 Polymerase chain reaction. Detection of 132 bp tandem repeats in samples collected from Marek's disease virus (MDV)-infected chickens. Descriptions: M, molecular length marker GeneRuler™ 100 bp DNA Ladder Plus (Thermo Scientific, Waltham, MA); NC, DNA template extracted from noninfected chicken embryo fibroblasts; PC, positive control (DNA of CVI988/Ripens strain isolated from commercial vaccine) (MSD Animal Health, Boxmeer, the Netherlands); 1–10 DNA templates extracted from liver of MDV-infected chickens. The obtained product length –434 bp indicates the presence of virulent MDV-1 DNA.

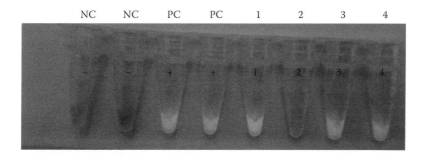

FIGURE 87.3 Loop-mediated isothermal amplification (LAMP): Detection of Marek's disease virus (MDV)-1 strains in chickens sent for Marek's disease (MD) examination. Visual LAMP detection of MDV under UV light after addition of SYBR Green® I dye. NC, DNA template extracted from noninfected chicken embryo fibroblasts; PC, positive control (DNA of HPRS$_{16}$ strain) (Houghton Poultry Research Station, Huntingdon, United Kingdom); 1–4 DNA templates extracted from liver of chickens sent for MD examination.

Given that HVT has nonhomologous genes including HVT068 (SORF1) and HVT070, it is also possible to differentiate this serotype from MDV-1 and MDV-3 strains. Due to the latent nature of MDV, its detection may be difficult because PCR detects approximately up to 100–1000 DNA copies. Taking into account the numbers below 10 copies of MDV DNA in latently infected cells, conventional PCR may produce false-negative results. The more sensitive alternative for PCR is real-time PCR with the use of SYBR Green® dye binding dsDNA products of fluorescently labeled probes (TaqMan®, Scorpion®, etc.). The sensitivity of real-time PCR may reach even a few copies of DNA and may be used to quantify the virus load in different tissues, FFE, or farm dust [54]. Using quantitative PCR (qPCR), it is also possible to detect and differentiate all three MDV serotypes. In the case of qPCR, an additional internal control represented by the "housekeeping" gene ensures the reliability of the obtained results. The DNA templates extracted from the blood or some organs including the spleen or liver may contain PCR inhibitors, thus additional internal control minimizes the risk of false-negative data. Additional advantage of qPCR is the possibility to apply it for the study on replication kinetics, expression patterns of different MDV serotypes and field isolates, or detection of nucleotide point mutations (SNPs).

87.1.4.2.3 *Loop-Mediated Isothermal Amplification*

Alternative methods to standard PCR or real-time PCR include nucleic acid sequence–based amplification, cross-priming amplification, helicase-dependant amplification, and LAMP [55]. LAMP uses two or three primer pairs specific to the chosen target under isothermal conditions. The main advantage of LAMP is its no absolute requirement to use PCR thermocyclers, electrophoresis chambers, or real-time PCR systems. LAMP is carried out in a water bath or dry heating block at different temperatures ranging from 59°C to 68°C. The method uses *Bst* or *Bsm* polymerases isolated from *Bacillus stearothermophilus* and *Bacillus smithi*, respectively. Other more effective and faster polymerases including *Gsp*SSD polymerase have been recently released. These enzymes have DNA strand displacement activity; therefore, LAMP may be conducted in isothermal conditions. The sequence of the target gene is separated into six fragments specific for the

designed primers [55]. Forward inner primer (FIP) and backward inner primer (BIP) are needed to initiate the reaction. The FIP sequence consists of F1c fragment and the F2 fragment complementary to F2c region in target (e.g., *meq*, ICP4, or HVT070) sequence. BIP consists of the B1c fragment complementary to B1 sequence and B2 fragment. The outer primers B3 and F3 mediate displacement of the new DNA strands amplified from FIP and BIP primers. The initial steps of LAMP lead to the formation of a dumbbell-like structure, which is the prerequisite for complexes of stem-loop DNA and autocycling DNA synthesis and subsequent addition of new loops. LAMP is capable of amplifying a few copies of DNA in 30–120 min. The final products are cauliflower-like structures with multiple overhangs and loops; therefore, the method is called "loop-mediated isothermal amplification." LAMP was successfully applied to detect and differentiate all three MDV serotypes [51,52]. Recently, LAMP was used to specifically detect U$_L$49 fragment of attenuated CVI988/Rispens strain. However, due to its high sensitivity, the method is prone to incidental contamination as it is also observed in very sensitive qPCR methods (Figure 87.3).

87.2 METHODS

87.2.1 SAMPLE PREPARATION

The ideal material for MD diagnosis and MDV detection originates from live chickens showing MD clinical signs. It is especially important for successful virus isolation, which is strictly cell-associated. The "buffy coat" layer of infected lymphocytes collected from the peripheral blood of chickens gives the best results for virus isolation or molecular detection. The chicken sera without the signs of hemolysis may also be used to detect the presence of specific antibody in ELISA or AGID assays [43,44]. The good source of the virus is 10% V/w homogenates of the spleen, kidney, ovaries, and liver. The virus may also be isolated from freshly collected feather tips. This material is also recommended for RID [45]. As transportation of whole chickens is sometimes difficult logistically, DNA cards with immobilized "buffy coat" drops, blood, or stamp from the affected organ provide a reliable alternative for PCR diagnosis of MD. However, it should be noted that such kind of material may be used only for the confirmation of DNA of MDV not the infectious virus presence. Both PCR and LAMP facilitate detection of MDV DNA in poultry dust. In the case of LAMP, direct detection of viral DNA in dust particles without DNA extraction is also possible [56].

87.2.2 Detection Procedures

87.2.2.1 Virus Isolation

The optimal cell monolayer for infection should have approximately 1×10^6 cells/mL. After incubation for 24–48 h at 37°C, the virus-free cell monolayer is ready for inoculation. Especially for MDV-1 field strains, the inoculated cell cultures should be incubated for 24 h, and then the inoculum should be washed off. The typical CPE is observed after 72–96 h PI. MDV strains may be propagated up to six passages in CEF or CEK cells but normally three passages are enough to show CPE typical for MDV. Moreover, an excessive passaging of field MDV isolates may lead to loss of its pathogenicity and attenuation. In such case, the back-passaging of MDV isolate in SPF chickens may recover the initial virus features. On the basis of the plaque size, it is also possible to distinguish the three MDV serotypes. The largest plaques are produced by MDV-3 (HVT), while the smallest by MDV-1 strains. The field MDV-1 isolates should be purified from the possible contaminants with vaccine strains using specific antiserum. The CIAV may be eliminated by serial passaging of field isolate. It may also be achieved by specific MDV-1 plaque purification assay. Purification is important to eliminate the vaccine strains that may mask the true pathogenicity of MDV-1 in infected chickens. The purified virus stocks should be gradually frozen then stored in liquid nitrogen at −196°C. An addition of 10%–15% dimethyl sulfoxide protects the cells from the disruption during freezing. The virus stocks collected from liquid nitrogen after long storage should be checked for their titer because of its possible change.

87.2.2.2 Agar Gel Immunodiffusion

AGID is applied to detect specific anti-MDV antibody or the antigen. The test is conducted on glass slides with 2–3 mm layer of 1% agar in phosphate-buffered saline containing 8% sodium chloride. The holes in the gel are cut using a template to obtain six wells in equal distance. The antigen is placed in the center well while the exterior wells with examined sera. The slide is then incubated at 37°C for 24–48 h in a humid chamber to mediate diffusion of sera and antigen. The assay may be used to examine antigens with the standard serum. The MDV antigen may originate from the infected cell cultures or as an extract of feather tips or skin. The results are observed using lamp in a darkened room. The presence of precipitation lines between standard antigen and examined sera is considered as positive.

87.2.2.3 Radial Immunodiffusion Assay

The 1% solution of agar with 8% sodium chloride should be prepared. The whole solution should be heated to 56°C and

then 20% of standard positive serum against MDV with the titer 1:8 is added. Petri dishes are then covered with 3 mm agar layer. The feather tips taken from examined birds are put into the agar layer each in the distance of about 10 mm away from the other wells. Petri dishes should be incubated in a humid chamber at 37.0°C for 24–48 h. The presence of ring precipitates around the feather tips is meant to be the positive result for the viral antigen presence.

87.2.2.4 Polymerase Chain Reaction

PCR for MDV-1 is conducted using a pair of specific primers complementary to one of the specific targets comprising of *meq*, ICP4, pp38, or 132 bp tandem repeated sequence that allows differentiation between virulent and attenuated strains [46–50,53,54]. For MDV-3 (HVT), the primers specific to HVT068 (SORF1) region are applied. Real-time qPCR is known to be a useful tool for quantification of the virus load and measuring dynamics of its replication *in vitro* and *in vivo*. Detection of all serotypes is possible using DNA extracted from internal organs of infected chickens, peripheral blood, feather tips, or dust of poultry houses [49,50,54].

87.2.2.5 Loop-Mediated Isothermal Amplification

The protocol for MDV-1 detection in feather samples along with the differentiation of all MDV serotypes has been previously described [51,52]. The successful attempts to selectively detect CVI988/Rispens strain using U_L49 fragment have been also conducted. Other LAMP primers may be designed using Primer Explorer version 4 (Net Laboratory, Tokyo, Japan, http://primerexplorer.jp/e/). LAMP is conducted with the application of two or three sets of specific primers in 15–25 µL using *Bsm* DNA polymerase kit (Thermo Scientific Fermentas, Vilnius, Lithuania) or Isothermal Mastermix Kit (OptiGene, Horsham, United Kingdom). The latter kit has preoptimized the concentration of *Gsp*SSD isothermal DNA polymerase with partial reverse transcriptase activity. Using the *Bsm* polymerase optimization of $MgSO_4$, *Bsm* and polymerase concentration is recommended. Additional parameters that should be optimized prior to the routine testing of field samples are concentration of inner, outer, and loop primers. The working concentrations of primers in LAMP for MDV are 50 pmol each of inner primers FIP and BIP, 10 pmol each of outer primers F3 and B3, and additionally 25 pmol each of loop primers LF and LB. The reaction mixture is supplemented with 1 µL of 1:10 dilution per reaction of 10,000× concentrated SYBR Green or GelRed™ (Biotum, Delhi, India). The primer sequences used for the detection of three different MDV serotypes and CVI988/Rispens are listed in Table 87.1.

Reaction tubes may be incubated in the broad temperature range from 59.0°C to 68.0°C, but the optimum temperature for most of the isothermal polymerases is 63°C. LAMP reaction time may range from 30 to 90 min and is dependent on not only initial DNA concentration in MDV templates but also the usage of two or three sets of primers. The final results are registered under UV light illumination. In case when too

TABLE 87.1

Sequences of Primers Used for Loop-Mediated Isothermal Amplification and Polymerase Chain Reaction Detection and Differentiation of Marek's Disease Virus Serotype 1, 2, 3, and CVI988/Rispens Strain

Primer	Sequence (5′–3′)
F3 MDV-1	TTCCCTCTTCTGCCCTCC
B3 MDV-1	TCCTGTTCGGGATCCTCG
FIP MDV-1	GTAAACCGTCCCCGGCGATG*TTTT*GGGCATCTTCCCTGCATTG
BIP MDV-1	CTTTGTCCTGTTGGCCAGGCTC*TTTT*GACGAGCATAAAGCCTCTCC
LF MDV-1	TACACGGCTCGGTAACAGGA
LB MDV-1	CCACATCCGGCTCCGGAGCC
F3 MDV-2	GTTCTCCGTCGATGAAGCG
B3-MDV-2	GTCAATTTCGCCAGCACAAC
FIP MDV-2	ATAACGCGAAGACAGCGCGG*TTTTT*ATGCGAGGAACCCAGAAGG
BIP MDV-2	GACTGTGCGATCTCGTTGCCC*TTTT*TGCCACTCGTACACCTAGA
LF MDV-2	TTCTTCTATATAACAGGTCC
LB MDV-2	TATGAAATATGTGCGTTAGA
F3 MDV-3	ATAAATTATATCGCTAGGACAGAC
B3 MDV-3	ACGATGTGCTGTCGTCTA
FIP MDV-3	CCAGGGTATGCATATTCCATAACAG*TTTT*CCAAACGACCTTTATCCCA
BIP MDV-3	CCAGAAATTGCACGCACGAG*TTTT*AGAATTTGTGCATTTAGCCTT
LF MDV-3	TTGAGAAGAGGATCTGACTG
LB MDV-3	GCGTCATTGGTTTTACATTT
F3 Rispens	GATGGGGGAAGCCGAAAC
B3 Rispens	CACTGCCACGATATCCGC
FIP Rispens	GCCGCGTCTTGTACGTTCAGA*TTTT*TTGCCCGCAAACTATTGGAA
BIP Rispens	AATCAAATCGCCAGATCCGGGA*TTTT*GATGGCGACGCGAAGTTG
LF Rispens	CCCCTGGGATAATCCAGACTC
LB Rispens	CTCATCGTACACATAACCCTCGC

The primers are specific for the serotype MDV-1, MDV-2, MDV-3, or attenuated Rispens strain. The nucleotides in FIP and BIP primers written in italics are thymidine linkers.

high background of fluorescence is observed in negative control, an additional glass or plastic mat may be placed between UV light source and the tubes with LAMP mixtures.

virology including virus isolation should not be neglected. Indeed, following molecular detection, confirmation of the virus infectivity using virus isolation is highly valuable.

87.3 CONCLUSION

MD remains a continuous problem in worldwide production of chickens and turkeys. The disease discovered over 100 years ago seems to evolve through the different clinical forms as well as models of pathogenesis. The prophylactic vaccination against the disease onset considerably reduced the loss caused by the mortality of birds or reduced egg production. The unique ability of MDV to avoid host immune system and to transform infected cells complicates the study on this virus. Application of effective vaccines is critical for the control of MD, although vaccination has been implicated in indirect increase of pathogenicity of field MDV isolates. The use of rapid and reliable diagnostic techniques such as PCR or LAMP for MD is equally important to avoid virus transmission to the further flocks of birds. Given the strictly cell-associated character of MDV, the classic methods of

REFERENCES

1. Morrow, C. and Fehler, F., Marek's disease: A worldwide problem, in *Marek's Disease: An Evolving Problem*, eds. F. Davison and V. Nair, Elsevier Academic Press, London, U.K., pp. 49–61, 2004.
2. Witter, R.L., Protective efficacy of Marek's disease vaccines. *Curr. Top. Microbiol. Immunol.*, 255, 57–90, 2001.
3. Adldinger, H.K. and Calnek, B.W., Pathogenesis of Marek's disease: Early distribution of virus and viral antigens in infected chickens. *J. Natl. Cancer Inst.*, 50, 1287–1298, 1973.
4. Biggs, P.M., Marek's disease, in *The Herpesviruses*, ed. A. Kaplan, Academic Press, New York, pp. 557–594, 1973.
5. Calnek, B.W. et al., Comparative pathogenesis studies with oncogenic and nononcogenic Marek's disease viruses and turkey herpesvirus. *Am. J. Vet. Res.*, 40, 541–548, 1979.
6. Schat, K.A., Isolation of Marek's disease virus: Revisited. *Avian Pathol.*, 34, 91–95, 2005.

7. Pappenheimer, A.M. et al., Studies on fowl paralysis (neuro-lymphomatosis gallinarum). *J. Exp. Med.*, 49, 87–102, 1929.

8. Biggs, P.M. and Payne L.N., Relationship of Marek's disease (neural lymphomatosis) to lymphoid leukosis. *Natl. Cancer Inst. Monogr.*, 17, 83–97, 1964.

9. Churchill, A.E. and Biggs, P.M., Agent of Marek's disease in tissue culture. *Nature*, 215, 528–530, 1967.

10. Nazerian, K. et al., Studies on the etiology of Marek's disease. II. Finding of a herpes virus in a cell culture. *Proc. Exp. Biol. Med.*, 127, 177–182, 1968.

11. Calnek, B.W. et al., Feather follicle epithelium: A source of enveloped and infectious cell-free herpesvirus from Marek's disease. *Avian Dis.*, 14, 219–233, 1970.

12. Okazaki, W. et al., Protection against Marek's disease by vaccination with a herpesvirus of turkeys. *Avian Dis.*, 14, 413–429, 1970.

13. Laurent, S. et al., Detection of avian oncogenic Marek's disease herpesvirus DNA in human sera. *J. Gen. Virol.*, 82, 233–240, 2001.

14. Witter, R.L., Increased virulence of Marek's disease virus field isolates. *Avian Dis.*, 41, 149–163, 1997.

15. Witter, R.L. et al., New serotype 2 and attenuated serotype 1 Marek's disease vaccine viruses: Selected biological and molecular characteristics. *Avian Dis.*, 31, 829–840, 1987.

16. Zelník, V., Marek's disease research in the post-sequencing area: New tools for the study of gene function and virus-host interactions. *Avian Pathol.*, 32, 323–333, 2003.

17. von Bulow, V. et al., Characterization of a new serotype of Marek's disease herpesvirus. *IARC Sci. Publ.*, 11, 329–336, 1975.

18. Bacon, L.D. et al., Augmentation of retrovirus-induced lymphoid leukosis by Marek's disease herpesviruses in White Leghorn chickens. *J. Virol.*, 63, 504–512, 1989.

19. Dudnikova, E. et al., Evaluation of Marek's disease field isolates by the "best fit" pathotyping assay. *Avian Pathol.*, 36, 135–143, 2007.

20. Beasley, J.N. et al., Transmission of Marek's disease by poultry house dust and chicken dander. *Am. J. Vet. Res.*, 31, 339–344, 1970.

21. Brunkovskis, P. and Velicer, L.F., The Marek's disease virus (MDV) unique short region: Alphaherpesvirus-homologous, fowlpox virus—Homologous, and MDV-specific genes. *Virology*, 206, 324–338, 1995.

22. Kingham, B.F. et al., The genome of herpesvirus of turkeys: Comparative analysis with Marek's disease viruses. *J. Gen. Virol.*, 82, 1123–1135, 2001.

23. Tulman, E.R. et al., The genome of a very virulent Marek's disease virus. *J. Virol.*, 74, 7980–7988, 2000.

24. Spatz, S.J. and Schat, K.A., Comparative genomic sequence analysis of the Marek's disease vaccine strain SB-1. *Virus Genes*, 42, 331–338, 2011.

25. Silva, R.F. et al., The genomic structure of Marek's disease virus. *Curr. Top. Microbiol. Immunol.*, 255, 143–158, 2001.

26. Brown, A.C. et al., Interaction of MEQ protein and C-terminal-binding protein is critical for induction of lymphomas by Marek's disease virus. *Proc. Natl. Acad. Sci. USA*, 103, 1687–1692, 2006.

27. Cantello, J.L. et al., Identification of latency-associated transcripts that map antisense to the ICP4 homolog gene of Marek's disease virus. *J. Virol.*, 68, 6280–6290, 1994.

28. Xie, Q. et al., Marek's disease virus (MDV) ICP4, pp38, and meq genes are involved in the maintenance of transformation of MDCC-MSB1 MDV-transformed lymphoblastoid cells. *J. Virol.*, 70, 1125–1131, 1996.

29. Schumacher, D. et al., Reconstitution of Marek's disease virus serotype 1 (MDV-1) from DNA cloned as a bacterial artificial chromosome and characterization of a glycoprotein B-negative MDV-1 mutant. *J. Virol.*, 74, 11088–11098, 2000.

30. Kaufer, B.B. et al., Herpesvirus telomerase RNA (vTR)-dependent lymphoma formation does not require interaction of vTR with telomerase reverse transcriptase (TERT). *PLoS Pathog.*, 6, e1001073, 2010.

31. Jarosinski, K.W. et al., Marek's disease virus: Lytic replication, oncogenesis and control. *Expert Rev. Vaccines*, 5, 761–772, 2006.

32. Pratt, W.D. et al., Characterization of reticuloendotheliosis virus-transformed avian T-lymphoblastoid cell lines infected with Marek's disease virus. *J. Virol.*, 66, 7239–7244, 1992.

33. Hirai, K. et al., Replicating Marek's disease virus (MDV) serotype 2 DNA with inserted MDV serotype 1 DNA sequences in a Marek's disease lymphoblastoid cell line MSB1-41C. *Arch. Virol.*, 114, 153–165, 1990.

34. Isfort, R. et al., Retrovirus insertion into herpesvirus in vitro and in vivo. *Proc. Natl. Acad. Sci. USA*, 89, 991–995, 1992.

35. Davidson, I. and Borenshtain, H., In vivo events of retroviral long terminal repeat integration into Marek's disease virus in commercial poultry: Detection of chimeric molecules as a marker. *Avian Dis.*, 45, 102–121, 2001.

36. Niczyporuk, J.S., Molecular characteristic on occurrence of fowl adenovirus field strains and effect of efficiency on prophylactic vaccinations against Marek's disease. Doctoral dissertation, National Veterinary Research Institute, Pulawy, Poland, 2014.

37. Jeurissen, S.H. and de Boer, G.F., Chicken anaemia virus influences the pathogenesis of Marek's disease in experimental infections, depending on the dose of Marek's disease virus. *Vet. Q.*, 15, 81–84, 1993.

38. Murthy, K.K. and Calnek, B.W., Marek's disease tumor-associated surface antigen (MATSA) in resistant versus susceptible chickens. *Avian Dis.*, 23, 831–837, 1979.

39. Chu, T.T. et al., Molecular detection of avian pathogens in poultry red mite (*Dermanyssus gallinae*) collected in chicken farms. *J. Vet. Med. Sci.* 76, 1583–1587, 2014.

40. Baigent, S.J. and Davison, T.F., Development and composition of lymphoid lesions in the spleens of Marek's disease virus-infected chickens: Association with virus spread and the pathogenesis of Marek's disease. *Avian Pathol.*, 28, 287–300, 1999.

41. Spencer, J.L. and Calnek, B.W., Marek's disease: Application of immunofluorescence for detection of antigen and antibody. *Am. J. Vet. Res.*, 31, 345–358, 1970.

42. Purchase, H.G., Immunofluorescence in the study of Marek's disease. *J. Virol.*, 3, 557–565, 1969

43. Cheng, Y.Q. et al., An enzyme-linked immunosorbent assay for the detection of antibodies to Marek's disease virus. *Avian Dis.*, 28, 900–911, 1984.

44. Ikuta, K. et al., Identification with monoclonal antibodies of glycoproteins of Marek's disease virus and herpesvirus of turkeys related to virus neutralization. *J. Virol.*, 49, 1014–1017, 1984.

45. Król, K. et al., Comparison of different methods for detection of Marek's disease virus. *Medycyna Wet.*, 65, 99–101, 2009.

46. Becker, Y. et al., Polymerase chain reaction for differentiation between pathogenic and non-pathogenic serotype 1 Marek's disease viruses (MDV) and vaccine viruses of MDV—Serotypes 2 and 3. *J. Virol. Methods*, 40, 307–322, 1992.

47. Handberg, K.J. et al., The use of serotype 1- and serotype 3-specific polymerase chain reaction for the detection of Marek's disease virus in chickens. *Avian Pathol.*, 30, 243–249, 2001.

48. Abdul-Careem, M.F. et al., Development of a real-time PCR assay using SYBR Green chemistry for monitoring Marek's disease virus genome load in feather tips. *J. Virol. Methods*, 133, 34–40, 2005.

49. Baigent, S.J. et al., Absolute quantitation of Marek's disease virus genome copy number in chicken feathers and lymphocyte samples using real-time PCR. *J. Virol. Methods*, 123, 53–64, 2005.

50. Król, K. et al., Duplex PCR assay for detection and differentiation of pathogenic and vaccine strains of MDV serotype 1. *Bull. Vet. Inst. Pulawy*, 51, 331–335, 2007.

51. Woźniakowski, G. et al., Rapid detection of Marek's disease virus in feather follicles by loop-mediated amplification. *Avian Dis.*, 55, 462–467, 2011.

52. Woźniakowski, G. et al., Comparison of loop-mediated isothermal amplification and PCR for the detection and differentiation of Marek's disease virus serotypes 1, 2, and 3. *Avian Dis.*, 57, 539–543, 2013.

53. Gimeno, I.M. et al., Detection and differentiation of CVI988 (Rispens vaccine) from other serotype 1 Marek's disease viruses. *Avian Dis.*, 58, 232–243, 2014.

54. Islam, A. et al., Absolute quantitation of Marek's disease virus and herpesvirus of turkeys in chicken lymphocytes, feather tips and poultry dust samples using real-time PCR. *J. Virol. Methods*, 132, 127–134, 2006.

55. Notomi, T. et al., Loop-mediated isothermal amplification of DNA. *Nucl. Acid Res.*, 28, 63, 2002.

56. Woźniakowski, G. et al., Direct detection of Marek's disease virus in poultry dust by loop-mediated isothermal amplification. *Arch Virol.*, 159, 3083–3087, 2014.

88 Cypriniviruses

Kenneth A. McColl, Agus Sunarto, and Lijuan Li

CONTENTS

88.1 Introduction 783
    88.1.1 Classification... 783
    88.1.2 Morphology, Biology, and Epidemiology 784
        88.1.2.1 Morphology.. 784
        88.1.2.2 Biology and Epidemiology... 784
    88.1.3 Clinical Features and Pathogenesis 787
        88.1.3.1 Clinical Features 787
        88.1.3.2 Pathogenesis... 787
    88.1.4 Diagnosis 788
        88.1.4.1 Conventional Techniques 788
        88.1.4.2 Molecular Techniques 789
88.2 Methods 789
    88.2.1 Sample Preparation.. 789
    88.2.2 Detection Procedures... 790
        88.2.2.1 Conventional PCR.. 790
        88.2.2.2 Quantitative PCR 791
88.3 Conclusion and Future Perspectives 791
Acknowledgment 792
References.. 792

88.1 INTRODUCTION

Cypriniviruses largely affect common and koi carp (*Cyprinus carpio*), goldfish (*Carassius auratus*), Prussian carp (*Carassius gibelio*), and crucian carp (*Carassius carassius*). In particular, cyprinid herpesviruses 1 (CyHV-1) and 3 (CyHV-3) are important pathogens of *C. carpio*, while CyHV-2 affects *C. auratus*, *C. gibelio*, and *C. carassius*. These host species belong to the family *Cyprinidae*, the largest family of freshwater fishes in the world with over 2000 species in 210 genera.[1] Native species occur in all tropical and temperate regions except Australia, New Zealand, Madagascar, and South America. However, cyprinids have even been introduced to each of these previously free locations. While common carp are an important food source for humans in Asia, Europe, and Israel,[2] koi carp and goldfish are almost purely ornamental fish, albeit sometimes being very expensive. By contrast, in many regions of the world (e.g., Australasia and North America), carp are regarded as an environmental pest species.[3]

Anguillid herpesvirus 1 (AngHV-1) is the fourth known member of the cypriniviruses. Much less work has been done on this virus compared with the other three cypriniviruses, and therefore, only basic comments will be made where appropriate.

88.1.1 CLASSIFICATION

There are four genera in the family *Alloherpesviridae*[4] (herpesviruses of fish and amphibians; see Table 88.1): (1) the *Cyprinivirus* genus containing CyHV-1, CyHV-2, CyHV-3 (type species), and AngHV-1; (2) the *Batrachovirus* genus containing the ranid herpesviruses; (3) the *Ictalurivirus* genus; and (4) the *Salmonivirus* genus.

Within the genus *Cyprinivirus*, genomes of all four viruses have been fully sequenced, and genome maps have been proposed for each.[4] The cyprinid herpesviruses are most closely related to each other, and they share a genome structure—a central unique (U) region flanked by direct terminal repeats. Within the U region of these three viruses, there is a central section (open reading frames [ORFs]28–113) where conserved genes are arranged in an equivalent manner. On the other hand, 18 ORFs are unique to CyHV-1, 13 to CyHV-2, and 21 to CyHV-3.[4] ORF134, which encodes an interleukin (IL)-10-related protein, is one of the ORFs unique to CyHV-3 and AngHV-1,[5] while ORFs encoding the JUNB family of proteins (oncoproteins) are unique to CyHV-1 and have never been reported in herpesviruses previously.

Phylogenetic analysis for viruses in the order *Herpesvirales* can be inferred from full-length deduced amino acid sequences of the concatenated DNA polymerase and

TABLE 88.1

Classification of Cypriniviruses

Taxonomy and Scientific Name	Abbreviation	Common Name	Host	Genome Size (bp)	Accession No.
Order *Herpesvirales*					
Family *Alloherpesviridae*					
Genus *Cyprinivirus*					
Cyprinid herpesvirus 1 (unassigned virus)	CyHV-1	Cyprinid herpesvirus (CHV), Herpesvirus cyprini, Carp pox herpesvirus	*Cyprinus carpio*	291,144	NC_019491
Cyprinid herpesvirus 2 (unassigned virus)	CyHV-2	Goldfish hematopoietic necrosis herpesvirus (GFHNHV)	*Carassius auratus* *C. gibelio* *C. carassius*	290,304	NC_019495
Cyprinid herpesvirus 3	CyHV-3	Koi herpesvirus (KHV)	*Cyprinus carpio*	295,146	NC_009127
Anguillid herpesvirus 1	AngHV-1	Herpesvirus anguillae (HVA)	*Anguilla anguilla* *A. japonica*	248,526	NC_013668

terminase (Figure 88.1). It shows monophyly of the fish and amphibian herpesviruses (*Alloherpesviridae*), separated from invertebrate herpesviruses (*Malacoherpesviridae*) and herpesviruses of mammals, birds, and reptiles (*Herpesviridae*). Within the genus *Cyprinivirus*, CyHV-2 and CyHV-3 are more closely related to each other than either is to CyHV-1 or to AngHV-1.[4,6]

88.1.2 Morphology, Biology, and Epidemiology

88.1.2.1 Morphology

Of the cypriniviruses, the most comprehensive morphological studies have been conducted on CyHV-3.[7–9] It was concluded that the development of virions *in vitro* was the same as for mammalian herpesviruses.[7] A capsid is formed in the cell nucleus and buds into the lumen between the inner and outer nuclear membrane. As it does so, it acquires a primary envelope (derived from the inner nuclear membrane). The enveloped capsid then enters the cytoplasm by fusion of the primary envelope with the outer nuclear membrane. As a result, a naked capsid is released into the cytoplasm. From this point, the capsid acquires a layer of tegument proteins and then acquires a mature envelope by budding into Golgi-derived cytoplasmic vesicles. When the vesicles fuse with the cell membrane, mature infectious virus is released from the cell. *In vivo*, viral particles with typical herpesvirus morphology have been observed in the tissues of CyHV-3-infected fish.[10] The inner capsid was 100–110 nm in diameter, while the envelope increased the size to 170–230 nm.[11]

While the morphology of CyHV-1 and CyHV-2 is similar to CyHV-3 (Figure 88.2), the sizes of the former are slightly different. The diameters of enveloped virions and icosahedral nucleocapsids of CyHV-1 are 157–190 nm and 109–113 nm, respectively,[12,13] whereas the equivalent values for CyHV-2 are 170–200 nm and 95–110 nm, respectively.[14]

88.1.2.2 Biology and Epidemiology

88.1.2.2.1 CyHV-1

CyHV-1, also known as cyprinid herpesvirus, Herpesvirus cyprini, or carp pox herpesvirus, affects *C. carpio* but generally only kills young fish up to 2 months old.[13,15] The disease is seasonal, and clinical signs associated with this acute form of the disease develop when water temperature is around 15°C.[13] CyHV-1 has been reported worldwide including in most European countries, the United States, Russia, Israel, and in some Asian countries.[16]

CyHV-1 is also the cause of "carp pox,"[13] and up to 60% of fry surviving the acute form of the disease may develop epidermal tumors/hyperplastic papillomas, also known as herpesvirus epidermal proliferation in carp (HEPC).[12,15,17] The prevalence of HEPC on carp farms ranges from <1:10,000 to over 70%.[17] It appears with low to moderate water temperatures (from late winter to early summer). HEPC can be transmitted to carp and to carp–goldfish hybrids.

88.1.2.2.2 CyHV-2

While CyHV-1 and CyHV-3 seem to affect *C. carpio* alone, CyHV-2 occurs in *C. auratus*, *C. gibelio*, and *C. carassius*, but not *C. carpio*.[18–21] In *C. auratus*, CyHV-2 is commonly known as goldfish hematopoietic necrosis herpesvirus.

The virus appears to be widespread in goldfish populations throughout the world. It was first reported as a severe pathogen in Japan in 1995.[19] Subsequently, there were a number of outbreaks in other countries, including the United States,[22,23] Taiwan,[24] Australia,[25] the United Kingdom,[26] and China.[20] Disease outbreaks usually occur in spring and autumn when water temperatures are between 15°C and 25°C.[19] Mortality due to the disease varies from 60% to 100%.[19]

88.1.2.2.3 CyHV-3

CyHV-3 causes a severe systemic disease with high mortality in *C. carpio*. It was first described in 1997 in Germany,[27] and from Israel and the United States in 1998.[10] However, there

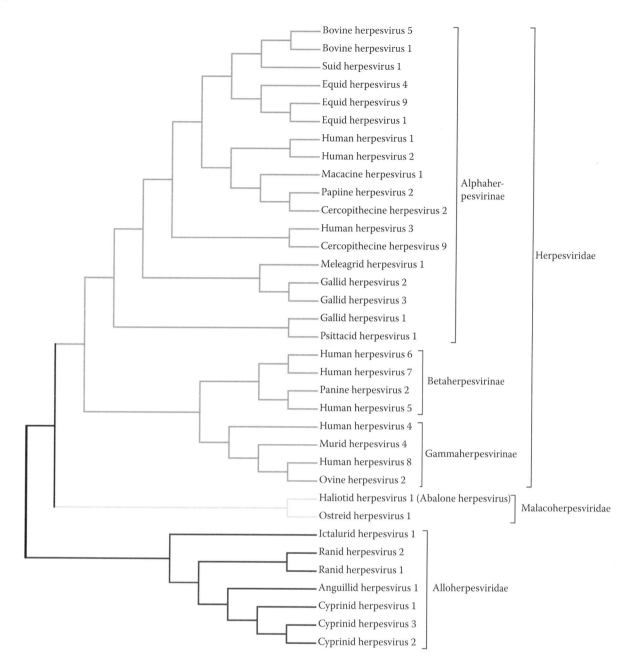

FIGURE 88.1 Molecular phylogenetic analysis of the order *Herpesvirales*. The DNA polymerase and terminase full-length protein sequences were concatenated; then, Mega6 was used to align them with Muscle and infer the evolutionary history using the minimum evolution method (see Figure 94.1 for further details).

is evidence that CyHV-3 may have been present in carp in the United Kingdom in 1996.[28] It has since spread to many other countries around the world and is now considered to represent a significant worldwide threat to the important carp industry.[29]

From sequence analysis, there appear to be two clear genotypes of CyHV-3, Asian and European, the former having two variants and the latter having seven.[30] The variants can be differentiated by a number of methods: a sequence from the C-terminal region of the thymidine kinase gene[30]; an examination of sequences from ORF25 to ORF26, which encode

membrane glycoproteins[31]; a duplex PCR based on differences in three variable domains within two genetic markers[32,33]; or a variable number tandem repeat analysis.[34] However, new isolates, wherever they are found, need to be characterized because Asian isolates may readily be introduced to Europe,[35] and vice versa.[36] The recent PCR discovery of three variants of CyHV-3[37] may portend modifications to the classification of CyHV-3 genotypes, but first they need to be cultured and fully sequenced.

It was hypothesized that CyHV-3 had spread globally as a result of intensive fish culture, koi shows, and local and

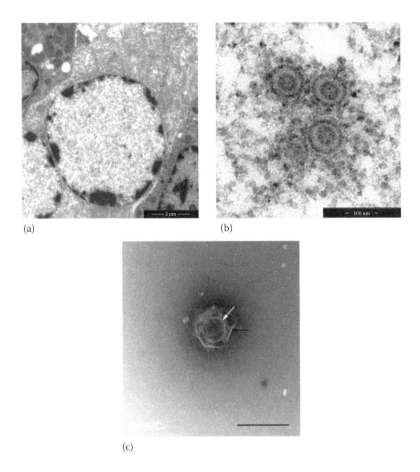

FIGURE 88.2 The morphology of CyHV-2 (a, b)[20] and CyHV-3 (c).[80] (a) Transmission electron micrograph of kidney infected with CyHV-2. Enlarged nucleus with marginalized chromatin. (b) Immature CyHV-2 particles in the nucleus. (c) Negative contrast electron microscopy of semipurified CyHV-3 from tissue culture supernatant. Note an icosahedral capsid (white arrow) and a loose-fitting envelope (black arrow). Bar = 200 nm.

international trade.[2] On the whole, these activities tend to occur without any restrictions or inspections. It was also predicted that movements of live carp were likely to be the main means of transmission of CyHV-3 between locations (in this case, in England and Wales).[38] Although the transmission of the virus by fomites (such as fishing gear) was considered to be of low probability, the high frequency of the use of fishing equipment suggested that there would, nevertheless, be a risk with this activity.[38] Alternatively, the stocking of imported ornamental fish appears to constitute a high-risk activity for the introduction of CyHV-3. Carrier fish represents an important issue in terms of developing control measures. At an International Koi Herpesvirus Workshop in 2004, it was recommended that fish be held in quarantine at 23°C–28°C for 2–4 weeks followed by a PCR check for CyHV-3.[39]

The outcome of the infection is very dependent on environmental temperatures. Mortality occurs when water temperatures range from approximately 16°C to 28°C, with the optimal range being 18°C–25°C.[2,40–42] Survival of the infectious virus in water is up to 3 days within this temperature range,[43] although it may survive longer in fish feces and in mud.[44,45] In different parts of the world, natural disease outbreaks tend to occur in the spring, as water temperatures are warming,[10] or in the spring and autumn, corresponding to the times when water temperatures are in the optimal range.[46]

While many studies have shown that juvenile and adult *C. carpio* are susceptible to CyHV-3.[46–48] Ito et al.[49] demonstrated that while larvae larger than 2 cm were, in general, 100% susceptible, larvae between 1 and 2 cm were of mixed susceptibility (66%), and larvae/fry less than 1 cm were generally insusceptible. The virus could not be detected by PCR in the larvae, suggesting that the larval resistance was to the infection rather than to the disease.

While *C. carpio* are well known for their susceptibility to CyHV-3, the susceptibility of other species of fish is a more contentious issue. The susceptibility of at least 22 species of fish has been tested,[27,46,50,51] and in all cases, there has been no evidence of the disease in these other species. While CyHV-3 DNA was detected in 10 out of 15 species of fish that were exposed to acutely or latently infected carp, only a small proportion of each species was infected, but none showed clinical signs of the disease, and only low copy numbers of CyHV-3 DNA were found.[50] In no case was there an attempt to demonstrate CyHV-3 RNA (as an indicator of virus replication).

Yuasa et al.[52] developed a reverse transcription-polymerase chain reaction (RT-PCR) for CyHV-3 and then commented on the work of various groups that had tried to demonstrate that *C. auratus* (and other species) may carry CyHV-3. Even though one study[53] used an RT-PCR to "demonstrate" that CyHV-3 replication took place in *C. auratus*, their work was slightly flawed in that neither primer in their PCRs was designed over a splice junction nor in separate exons. Therefore, it was possible that they were simply detecting DNA from a contaminant virus. In fact, Yuasa et al.[54] subsequently provided the most compelling evidence that *C. auratus* are not infected by CyHV-3; rather, they simply act as mechanical vectors of the virus.

Due to the high mortality, and specificity, of CyHV-3 for *C. carpio*, the virus is also regarded as a potential biological control agent in those parts of the world where *C. carpio* are regarded as a pest species.[55]

88.1.3 CLINICAL FEATURES AND PATHOGENESIS

88.1.3.1 Clinical Features

Water temperature directly affects the body temperature of fish, which, in turn, affects viral infections of fish (1) directly, by affecting the rate of viral replication, and (2) indirectly, by affecting the host immune response.[2] These are important considerations in understanding the clinical signs and the pathogenesis of the Cypriniviruses.

88.1.3.1.1 CyHV-1

The severity of the acute phase of a CyHV-1 infection is temperature and age dependent.[15,56,57] At 20°C, mortalities over 80% were reported in 2-week-old carp, 20% in 4-week-old carp, and 0% in 8-week-old carp.[15] Other work has shown that mortality is highest at 15°C and decreases with increasing temperature. For example, it was reported that CyHV-1 caused up to 60%, 16%, and 0% mortality in juvenile carp at water temperatures of 15°C, 20°C, and 25°C, respectively.[56] Clinical signs of the disease depend on the age of the fish and may include abnormal behavior such as spiral swimming, occasional immobility on the bottom, loss of appetite, a distended abdomen, exophthalmia, darkened pigmentation, and hemorrhages on the operculum and abdomen.[15]

Fish that survive the acute infection may develop white, mucoid papilloma-like lesions at 5–6 months postinfection.[13] They usually occur in caudal areas, including the caudal fin, but they may also occur on the mandible and, indeed, anywhere on the skin.[58] Eventually, the tumors are sloughed, but they may recur in the same season in subsequent years.[57]

88.1.3.1.2 CyHV-2

There are no consistent clinical signs of the disease, although affected fish may exhibit pale gills, and there may be yellowish discoloration of, and white patches on, the skin.[26] Fish tend to be inappetent and listless, and they sit at the bottom of the pond.[19]

88.1.3.1.3 CyHV-3

Experimental infections of *C. carpio* with CyHV-3 reveal a disease course ranging from 3 to 23 days, the duration being inversely related to both water temperature and virus dose.[2,10] The clinical course of the disease, however, is most frequently 24 h, although it can last up to 3 days.[27] The range of signs is relatively consistent: inappetence; lethargy; gasping in shallow water; excess, followed by decreased, cutaneous, mucus production resulting in pale patches on the skin ("sandpaper-like"); swollen, irregularly colored gills also with excess mucus; and occasionally nervous signs such as uncoordinated movements and erratic swimming.[9,10,29,59]

88.1.3.2 Pathogenesis

88.1.3.2.1 CyHV-1

The acute form of the disease is characterized by necrosis in the liver, kidney, and lamina propria of the intestinal mucosa (reviewed in Fijan[17]). Viral antigen and genome have been demonstrated by indirect fluorescent antibody tests and *in situ* hybridization (ISH), respectively, not only in these tissues but also in gills.[15] Following the acute phase, there is a phase in which virus cannot be isolated nor can viral antigens be detected in any tissues. However, the viral genome may be demonstrated by ISH, and it is presumed that the virus becomes latent in the cranial nerve ganglion, spinal nerves, subcutaneous tissue, and sometimes in gills.[57] These same tissues are also associated with reactivation of the virus, and it is thought that it is this recrudescent virus that results in the induction of the skin tumors (also known as HEPC, see Section 88.1.2.2) at 11–20 weeks postinfection. During this reactivation phase, the viral antigen can once again be detected in developing tumors, and the virus may also be isolated in the early stages of tumor formation, but not in the later stages. Generally, tumors are prevalent in the late fall and winter as ambient temperatures decrease, and they regress in the spring as water temperatures increase above 20°C or 25°C.[12,57,60] As viral antigen can be detected in recurrent tumors, it is speculated that following tumor regression, the virus becomes latent in the cranial nerve ganglion and spinal nerves once again and is reactivated when the temperature becomes permissive in the following year.[15]

Light microscopic examination of the affected skin revealed that epithelial hyperplasia was the primary lesion leading to proliferative papilloma-like tumors.[12,13,17] Within these proliferative lesions, the nuclei of affected cells are enlarged with marginated chromatin. Some cells have Cowdry type A intranuclear inclusion bodies (IBs), and mitoses may be frequent. By electron microscopy, particles with typical herpesvirus morphology may be observed in infected cells. Proliferating cells are not invasive. After the skin tumors slough off, scar tissue may develop, but mortality is rare.[12,15,17]

88.1.3.2.2 CyHV-2

C. auratus of all sizes are susceptible to infection with CyHV-2.[19] While the portal of entry of the virus is unknown, an infected fish can transmit the virus horizontally through the water, and there is also evidence for vertical transmission.[61] The most severe gross changes in the infected fish are swollen kidneys, an empty intestine, and splenomegaly with white nodular lesions in the spleen.[19,26] Histologically, the most severe lesions are necrosis of the hematopoietic tissue of the kidney and spleen.[19] Affected cells may have enlarged nuclei with margination of chromatin and the presence of intranuclear IBs.[26] Viral DNA may be detected for lengthy periods in healthy *C. auratus* brood stock suggesting that latency may be a feature of CyHV-2.[61]

88.1.3.2.3 CyHV-3

Within the optimal temperature range of CyHV-3, *C. carpio* are susceptible to extremely low concentrations of the virus, with most fish dying following bath exposure to 12 or even 1.2 $TCID_{50}$/mL of virus.[2] Pikarsky et al.[9] stated that the early lesions in the gills of carp suggested that these were the portal of entry of the virus, but it was later demonstrated, through the use of bioluminescence studies, that the skin, and later the oral mucosa, may, in fact, be the major portals of entry.[62,63] However, ISH studies[64] and direct *per-gill* infection of carp[8,65] continue to support a role for gills in the infection.

Although the cell receptor for the virus has not yet been identified, it is known that following infection, CyHV-3 spreads rapidly throughout the body of the fish. Viral DNA is found in the blood and kidneys by 1 day postexposure[9] and then progressively spreads to other tissues such as the brain, liver, and spleen.[59] A great deal of work has been conducted on both the innate and adaptive immune responses (IRs) of the host (reviewed by Adamek et al.[66]), including the acute phase response.[67] Both arms of the adaptive IR have been dissected, in particular, the humoral[68,69] and the cell-mediated IRs. For the latter, resistant and susceptible lines of *C. carpio* have been especially useful in understanding the dynamics of the expression of various immune-related genes.[70]

In addition, a number of potentially important immunosuppressive molecules have been identified in the viral genome.[66,71] A viral homologue of the host IL-10 gene (khvIL-10) is likely to be very influential in the pathogenesis of the disease,[72] although this could not be confirmed in one study.[73] Interestingly, a viral IL(vIL)-10 gene is also present in AngHV-1.[5] Although AngHV-1 is closely related to CyHV-3, there is a little sequence homology between the respective vIL-10s. CyHV-1 and CyHV-2 both lack a vIL-10.[4] While the latter observation may indicate an important role for khvIL-10 in the pathogenesis of CyHV-3 infection,[74] clearly much remains to be learned about the function of khvIL-10.[75]

Depending on the water temperature, infected fish may shed virus for days before mortality begins,[41] suggesting the potential for a high risk of transmission from fish that show no clinical signs of the disease. It has been proposed that the loss of osmoregulatory function in the gills, kidney, gut, and skin (all of which contain virus) account for the death of fish.[59]

The histological lesions associated with infection by CyHV-3 are consistent with the pathogenesis of the disease. In the gills, there may be an early mixed inflammatory cell infiltrate, followed by complete effacement of the gill structure characterized by a severe inflammatory response, hyperplasia, and hypertrophy of epithelial cells and fusion of the secondary lamellae.[40] Intranuclear IBs may be present in epithelial cells. Subepithelial inflammation and congestion of the gill rakers, with focal sloughing of the surface epithelium, is often a more consistent finding.[40] Multifocal necrosis of the hematopoietic tissue, and in some fish the tubular epithelium, of the kidney may be prominent.[40] Multifocal necrosis, occasionally with IBs, may also be seen in the oral epithelium, spleen, pancreas, and liver, while a mild to moderate enteritis and a focal meningitis have also been reported.[9,40]

In fish that survive infection with CyHV-3, persistent (possibly latent) infections may develop, and, in this case, B cells have been identified as at least one site of latency.[76]

88.1.4 DIAGNOSIS

A tentative diagnosis of the infection by any of the Cypriniviruses may be based on the use of a number of conventional techniques (clinical signs, gross and histopathology, electron microscopy, virus isolation, immunochemistry, serology, immunochromatography). However, to make a definitive diagnosis, specific molecular techniques (PCR, sequencing) are required. Loop-mediated isothermal amplification (LAMP) and ISH are other molecular techniques. LAMP has some advantages over PCR, but like ISH, it is unable to provide a definitive diagnosis.

88.1.4.1 Conventional Techniques

The clinical, pathological, and electron microscopic diagnoses of infection by CyHV-1, CyHV-2, and CyHV-3 have been discussed earlier (see Sections 88.1.2.2 and 88.1.3.1, respectively).

CyHV-1: Davison et al.,[4] in their sequencing studies on cyprinid herpesviruses, cultured CyHV-1, CyHV-2, and CyHV-3 on koi fin-1[10] (KF-1) cells. CyHV-1 was passaged by using an infected cell culture medium.

CyHV-2: CyHV-2 is difficult to isolate in cell culture. The virus has been isolated from the spleen and kidney of infected fish using FHM, EPC, and TO-2 cyprinid cell lines.[19] The virus induced cytopathic effect (CPE) in infected cells, including vacuolation and pyknosis followed by aggregation of rounded cells. However, the virus was lost after four passages. Therefore, the diagnosis of CyHV-2 was based on the presence of histological lesions and electron microscopic isolation of the virus. Davison et al.[4] cultured CyHV-2 on KF-1 cells by passaging infected cells (rather than cell culture supernatant). Difficulty in detection of the virus may be masking the true distribution and significance of the virus in goldfish populations.

CyHV-3: For isolation of CyHV-3 in cell culture, the earliest, and most commonly used, cell lines are the KF-1 and common carp brain[77] lines. A number of others have subsequently been described (e.g., Dong et al.[78]). The gill, kidney, and spleen are the best tissues for virus isolation, but it is often difficult to isolate the virus from fish that have been dead for a few hours, or fish that have been frozen.[59] It appears that CyHV-3 in the kidney is more stable than in the gills[79] possibly because bacteria in the gills degrade the virus. While Hedrick et al.[10] stated that uninfected cells should be cultured at 25°C, but that virus isolation attempts should be conducted at 20°C, Gilad et al.[2] found that optimal growth of CyHV-3 in KF-1 cells was at temperatures between 15°C and 25°C. Sunarto et al.[80] found that maximum virus titers in the supernatant fluid of a KF-1-infected culture at 25°C were between 4 and 6 dpi. There have been reports of CyHV-3 being cultured on various other cyprinid and salmonid cell lines,[10,27,81–83] but results are often difficult to reproduce with other strains of CyHV-3.

CyHV-1, CyHV-2, and CyHV-3 all share some common antigens,[84] but various polyclonal and monoclonal antibodies have also been reported[29,85] that can be used in immunocytochemical or immunohistochemical assays for the detection of CyHV-3.

Ronen et al.[83] developed an ELISA for the detection of anti-CyHV-3 antibodies. For fish that survived the infection, an elevated titer was first observed at 14 dpi with the peak titer occurring at 21 dpi. The titer remained elevated until at least 51 dpi. Subsequently, another antibody-detection ELISA was developed[84] that could detect specific antibodies to CyHV-3 for up to 1 year after the last known exposure of fish to the virus. While this appears to be the serological test of choice, even this test is said to be more useful in identifying the antibody status of a population of fish rather than any individual within the population.

An immunochromatographic lateral flow device (LFD) for the detection of CyHV-3 antigen has been developed for pondside use as a diagnostic tool.[86] Gill swabs are the preferred site for sampling. The LFD is capable of detecting CyHV-3 up to 2 days prior to mortality, and up to 100% of infected fish can be detected at 7–9 dpi. The test takes about 15 min, and, while being very useful, it certainly does not have the sensitivity of a PCR test.

In general, CyHV-1 and CyHV-3 can be differentiated by (1) differences in cell-line specificity (CyHV-1 grows on EPC and FHM cyprinid cell lines), (2) type of CPE, (3) antigenic properties (Hedrick et al.[10] demonstrated that anti-CyHV-1 antiserum recognized CyHV-1-, but not CyHV-3-infected cells), and (4) timing of lesions (CyHV-1 lesions occur when water temperatures are declining[12]).

88.1.4.2 Molecular Techniques

Conventional PCR, with sequencing of specific products, or quantitative PCR (qPCR) are the most common molecular techniques for the definitive diagnosis of a *Cyprinivirus* infection.

CyHV-1: There is limited information available on conventional PCRs,[87] and none on qPCR, for the diagnosis of CyHV-1.

CyHV-2: A conventional PCR[23] and a quantitative real-time PCR[88] have been developed for the detection of CyHV-2.

CyHV-3: For CyHV-3, PCR is commonly used for diagnosis, and it is much more sensitive than virus isolation for detection of the virus.[29,89] Numerous one-step, and nested,

conventional PCRs have been developed for CyHV-3, two of the early ones being based on noncoding regions of the genome.[89,90] Considering the possibility that conserved regions of the genome might yield a more reliable PCR (both in terms of detection of regional variants of CyHV-3)[29] and the reduced likelihood for mutations that could upset the efficacy of the PCR primers,[91] another conventional PCR was based on the thymidine kinase gene.[91] Although two of these PCRs[90,91] have been adopted by the World Organization for Animal Health,[92] work continues toward a single recommended protocol for the detection of CyHV-3 by a conventional PCR.[93,94] A widely used quantitative TaqMan real-time PCR was also developed for CyHV-3.[59] This test is not only more sensitive than a one-step conventional PCR, but it also dramatically reduces the risk of false-positive results due to inadvertent contamination problems.

Early LAMPs were developed for the detection of CyHV-3.[95,96] It was noted that while the speed and the absence of expensive equipment to run the test were the great advantages of LAMP, the test was no more sensitive than a one-step PCR.[97] However, a later commercial test claimed to be more sensitive.[98] Other LAMP tests for CyHV-3 have also been reported.[99–101]

The use of ISH as a diagnostic tool for CyHV-3 has been reported only infrequently.[94,102,103] Lee et al.[104] used the technique on paraffin-embedded sections to make a retrospective diagnosis of CyHV-3 in common carp in Korea. They showed, in 2012, that an outbreak of the disease in 1998 was, in fact, due to CyHV-3.

88.2 METHODS

88.2.1 Sample Preparation

Given that, of the molecular tests available, only conventional and qPCRs are capable of providing a definitive diagnosis, this section will be devoted to these tests.

Identification of the viral genome by ISH in the gills, kidney, liver, and intestine[105] indicates that these are the tissues of choice for PCR diagnosis of an acute CyHV-1 infection. By contrast, the cranial nerve ganglia, spinal nerve, and subcutaneous tissue are the tissues of choice during latent infections.[57]

The gill, kidney, and spleen are the tissues of choice for PCR diagnosis of CyHV-2.[19]

In acute CyHV-3 infections, the gill, kidney, and spleen are the tissues of choice, regardless of whether samples are collected from clinically normal fish in a suspect population, or from moribund or dead fish. Even decomposed fish may be tested,[90] although, in this case, primers that result in short PCR products are more likely to be useful. For suspected subclinical or persistent infections, the gut and brain are recommended.[92]

For the diagnosis of any *Cyprinivirus* infection, whole fish, or specific tissues, should be sent to the laboratory, either on ice or frozen (assuming that only a PCR diagnosis is being attempted). Alternatively, small tissue samples may be submitted in 80%–100% alcohol (at a tissue:fixative ratio of 1:10), or in RNAlater (Ambion Cat #: AM70), which preserves both RNA and DNA). Viral DNA (and, if necessary, RNA) may be extracted from tissues through the use of a variety of commercially available nucleic acid extraction kits that produce high-quality nucleic acid. The latter can then be held at −20°C or −80°C for short-term or long-term storage, respectively.

88.2.2 Detection Procedures

Ideally, water and equivalent tissues from uninfected *C. carpio* should be used as negative controls, while tissues from known infected fish act as positive controls. Due to the sensitivity of PCR tests, care at every step of sample preparation must be taken to ensure that cross-contamination of diagnostic samples does not occur. Thus, all instruments and sample containers must be clean and uncontaminated. Wherever possible, it is recommended that disposable reagents and plasticware are used.

88.2.2.1 Conventional PCR

1. In a virus-free room, prepare the Master Mix. This consists of the following reagents for a single sample: 9.5 µL of nuclease-free water, 12.5 µL of HotStar Taq Master Mix (QIAGEN Cat #: 203443), 0.5 µL of the forward primer (18 µM), and 0.5 µL of the reverse primer (18 µM). Calculate reagent amounts for the number of samples required, including positive and negative controls plus extras to account for pipette error. Aliquot 23 µL of Master Mix into each PCR tube. Primers for the required PCR are shown in Table 88.2.

2. In a separate extraction room, tissue samples (including blood) and other specimens containing cells are homogenized at approximately 10% w/v in water, and then frozen and thawed. Samples are then microfuged (approximately $13,000 \times g$ for 20–30 s), and the cell-free supernatant fluid is collected. Nucleic acid (including viral DNA) is obtained from cell-free samples using the QIAamp viral RNA extraction kit (QIAGEN Cat #: 52904). The manufacturer also recommends this kit for extraction of DNA viruses.

3. Moving to a separate template room, 2 µL of extracted DNA is then added to each sample of Master Mix, resulting in a total volume of 25 µL.

4. The mixture is then incubated in an automatic thermal cycler that is programmed appropriately for each PCR (Table 88.3).

5. At the completion of conventional PCR, specific PCR fragments of the correct size are identified by agarose gel electrophoresis. The negative control sample must have no evidence of specific amplified products, while a positive control sample must yield a specific CyHV-3 fragment. Amplified fragments of the correct size are then eluted from the gel and used to determine the DNA sequence (by using the PCR primers as sequencing primers).

TABLE 88.2
Primers and Probes for Cyprinivirus Polymerase Chain Reaction Tests

Virus	PCR	Primers/Probe	Sequence	Fragment Size
Conventional PCRs				
CyHV-1	Lievens et al. [87]	F	5′ GGCAGACCCGTTTTGTCTAG 3′	410 bp
		R	5′ CGGTCATCCCCTTGGTCTTT 3′	
CyHV-2	Goodwin et al. [23]	F	5′ CGGAATTCTAGAYTTYGCNWSNYTNTAYCC 3′ (Y=C, T; N=ACGT; W=AT; S=CG; R=AG)	520 bp
		R	5′ CCCGAATTCAGATCTCNGTRTCNCCRTA 3′	
CyHV-3	Bercovier et al. [91]	F	5′ GGGTTACCTGTACGAG 3′	409 bp
		R	5′ CACCCAGTAGATTATGC 3′	
	Yuasa et al. [90]	F	5′ GACACCACATCTGCAAGGAG 3′	292 bp
		R	5′ GACACATGTTACAATGGTCGC 3′	
Quantitative PCRs				
CyHV-1		F	None published	
		R		
		Probe		
CyHV-2	Goodwin et al. [88]	F	5′ TCGGTTGGACTCGGTTTGTG 3′	
		R	5′ CTCGGTCTTGATGCGTTTCTTG 3′	
		Probe	5′ FAM-CCGCTTCCAGTCTGGGCCACTACC-BHQ1 3′	
CyHV-3	Gilad et al. [59]	F	5′ GACGCCGGAGACCTTGTG 3′	
		R	5′ CGGGGTTCTTATTTTTGTCCTTGTT 3′	
		Probe	5′ FAM-CTTCCTCTGCTCGGCGAGCACG-TAMRA 3′	

F, forward; R, reverse.

TABLE 88.3

Cycling Conditions for Cyprinivirus Polymerase Chain Reaction Tests

Virus	PCR	Cycling Conditions
Conventional PCRs		
CyHV-1	Lievens et al. [87]	1 cycle, 94°C for 2 min; 35 cycles, 94°C for 45 s, 59°C for 30 s, and 72°C for 45 s; 1 cycle, 72°C for 10 min
CyHV-2	Goodwin et al. [23]	1 cycle, 94°C for 1 min; 30 cycles, 94°C for 30 s, 45°C for 30 s, and 72°C for 3 min
CyHV-3	Bercovier et al. [91]	1 cycle, 94°C for 15 min; 35 cycles, 94°C for 30 s, 52°C for 30 s, and 72°C for 1 min; 1 cycle, 72°C for 7 min
	Yuasa et al. [90]	1 cycle, 94°C for 15 min; 40 cycles, 94°C for 30 s, 63°C for 30 s, and 72°C for 30 s; 1 cycle, 72°C for 7 min
Quantitative PCRs		
CyHV-1		None published
CyHV-2	Goodwin et al. [88]	1 cycle, 95°C for 2 min; 35 cycles, 95°C for 30 s, 58°C for 45 s, and 72°C for 45 s; 1 cycle, 72°C for 2 min
CyHV-3	Gilad et al. [59]	1 cycle, 50°C for 2 min; 1 cycle, 95°C for 10 min; 40 cycles, 95°C for 15 s and 60°C for 1 min

Note: An apparently specific PCR product only provides a tentative diagnosis. *The PCR product must be sequenced to make a definitive diagnosis.* Sequence identity and genotype are determined by a BLAST search of the GenBank database. An assay is valid only when all controls yield the expected results.

88.2.2.2 Quantitative PCR

1. In a virus-free room, prepare the Master Mix. The PCR mixture for a single sample (1 well only) consists of the following reagents: 3 µL of nuclease-free water, 12.5 µL of TaqMan 2× Universal PCR Master Mix (Applied Biosystems, Cat #: 4304437), 1.25 µL of the forward primer (18 µM), 1.25 µL of the reverse primer (18 µM), 1.25 µL of the TaqMan probe (5 µM), 1.25 µL of the forward 18S primer (5 µM), 1.25 µL of the reverse 18S primer (5 µM), and 1.25 µL of the TaqMan 18S probe (5 µM). For multiple samples (and if more than 1 well per sample is required), the volumes are multiplied appropriately. Aliquot 23 µL of Master Mix into each well of a 96-well plate.

2. The primers and probes for Cyprinivirus qPCR tests, which are designed to be multiplexed with 18S ribosomal RNA (rRNA) qPCR, are shown in Table 88.2. The 18S rRNA gene (Applied Biosystems) is used to validate the nucleic acid presence and integrity, and the absence of PCR inhibitors. qPCR is performed using 96-well plates with samples, and positive and negative controls, tested in duplicate or triplicate.

3. In a separate extraction room, tissue samples (including blood) and other specimens containing cells are homogenized at approximately 10% w/v in water, and then frozen and thawed. Samples are then microfuged (approximately $13,000 \times g$ for 20–30 s), and the cell-free supernatant fluid is collected. Nucleic acid (including viral DNA) is obtained from cell-free samples using the QIAamp viral RNA extraction kit (QIAGEN Cat #: 52904). The manufacturer also recommends this kit for extraction of DNA viruses.

4. Moving to a separate template room, 2 µL of extracted DNA is then added to each well of Master Mix, resulting in a total volume of 25 µL. Transfer the plate to the qPCR thermal cycler that is programmed appropriately for each PCR (Table 88.3).

5. For the assay and test results to be accepted, the following criteria must be fulfilled. The no template control and/or negative sample control must have no evidence of specific amplification curves. Each positive control must yield a specific amplification curve and mean C_T values within the acceptable range according to quality control data. Each sample must have an 18S-specific amplification curve. Test samples with characteristic amplification curves and C_T values are considered test positive.

88.3 CONCLUSION AND FUTURE PERSPECTIVES

Despite their taxonomic similarities, individual cypriniviruses exert vastly different impacts on farmed and wild fish throughout the world. These differences account for the variation in methods, and effort, to control each of the viruses.

As mentioned earlier, CyHV-1 can cause mortality in young cyprinids[13,15] and also causes HEPC[17] in up to 60% of fry surviving the acute infection.[12,15] Control of this disease involves increasing the water temperature, to speed up sloughing of tumors, isolation of the diseased fish, and good husbandry practices to prevent further spread of the disease in a population.[87]

CyHV-2, on the other hand, appears to be restricted to *C. auratus*, *C. gibelio*, and *C. carassius* where it may be associated with low to severe mortality.[25] Control methods, where invoked, would include standard control practices based on isolation of the affected fish, sanitation measures, and good husbandry to prevent spread.[23]

CyHV-3 is considered to represent a significant worldwide threat to the important carp industry.[29] As a consequence, there has been great interest in a variety of potential control methods including the development of vaccines and "natural" vaccination,[65,83,106–110] disinfection of water,[111,112] crossbreeding for resistance,[113–115] RNA interference,[116] and the institution of standard control practices for any infectious disease.[48] Undoubtedly, the importance of the disease will continue to drive advances in many of these approaches. In addition, given that subclinical persistent infections are likely very important in the spread of CyHV-3 both locally and globally, it is critical that we continue to develop a much better understanding of persistent/latent infections for this virus.

Finally, it should be noted that while the specificity for, and high mortality in, *C. carpio* makes CyHV-3 a scourge for farmed carp, these same factors mean that the virus offers great potential as a biological control agent for *C. carpio* in those countries where this fish is regarded as a pest species.[55]

ACKNOWLEDGMENT

We thank Dr. Keith W. Savin (Department of Environment and Primary Industries, AgriBio Centre, Bundoora, Victoria 3083, Australia) for the use of Figure 94.1 in this chapter (as Figure 88.1).

REFERENCES

1. Allen, G., Midgley, S., and Allen, M. *Field Guide to the Freshwater Fishes of Australia*, p. 394. Western Australian Museum, Perth, Western Australian, Australia (2002).
2. Gilad, O., Yun, S., Adkison, M.A. et al. Molecular comparison of isolates of an emerging fish pathogen, the koi herpesvirus, and the effect of water temperature on mortality of experimentally infected koi. *J. Gen. Virol.* 84, 1–8 (2003).
3. Koehn, J.D. Carp (*Cyprinus carpio*) as a powerful invader in Australian waterways. *Freshwater Biol.* 49, 882–894 (2004).
4. Davison, A.J., Kurobe, T., Gatherer, D. et al. Comparative genomics of carp herpesviruses. *J. Virol.* 87, 2908–2922 (2013).
5. Van Beurden, S.J., Forlenza, M., Westphal, A.H., Wiegertjes, G.F., Haenen, O.L.M., and Engelsma, M.Y. The alloherpesviral counterparts of interleukin 10 in European eel and common carp. *Fish. Shellfish Immunol.* 31, 1211–1217 (2011).
6. Waltzek, T.B., Kelley, G.O., Alfaro, M.E., Kurobe, T., Davison, A.J., and Hedrick, R.P. Phylogenetic relationships in the family Alloherpesviridae. *Dis. Aquat. Organ.* 84, 179–194 (2009).
7. Miwa, S., Ito, T., and Sano, M. Morphogenesis of koi herpesvirus observed by electron microscopy. *J. Fish Dis.* 30, 715–722 (2007).
8. Miyazaki, T., Kuzuya, Y., Yasumoto, S., Yasuda, M., and Kobayashi, T. Histopathological and ultrastructural features of Koi herpesvirus (KHV)-infected carp *Cyprinus carpio*, and the morphology and morphogenesis of KHV. *Dis. Aquat. Organ.* 80, 1–11 (2008).
9. Pikarsky, E., Ronen, A., Abramowitz, J. et al. Pathogenesis of acute viral disease induced in fish by carp interstitial nephritis and gill necrosis virus. *J. Virol.* 78, 9544–9551 (2004).
10. Hedrick, R.P., Gilad, O., Yun, S. et al. A herpesvirus associated with mass mortality of juvenile and adult koi, a strain of common carp. *J. Aquat. Animal Health* 12, 44–57 (2000).
11. Hedrick, R.P., Gilad, O., Yun, S.C. et al. Initial isolation and characterization of herpes-like virus (KHV) from koi and common carp. *Bull. Fish. Res. Agency* 2, 1–7 (2005).
12. Hedrick, R.P., Groft, J.M., Okihiro, M.S., and McDowell, T.S. Herpesviruses detected in papillomatous skin growths of koi carp (*Cyprinus carpio*). *J. Wildl. Dis.* 26, 578–581 (1990).
13. Sano, T., Fukuda, H., and Furukawa, M. Herpesvirus cyprini: Biological and oncogenic properties. *Fish Pathol.* 20, 381–388 (1985).
14. Wu, T., Ding, Z., Ren, M. et al. The histo- and ultra-pathological studies on a fatal disease of Prussian carp (*Carassius gibelio*) in mainland China associated with cyprinid herpesvirus 2 (CyHV-2). *Aquaculture* 412–413, 8–13 (2013).
15. Sano, T., Morita, N., Shima, N., and Akimoto, M. *Herpesvirus cyprini*: Lethality and oncogenicity. *J. Fish Dis.* 14, 533–543 (1991).
16. Palmeiro, B. and Scott Weber III, E. Viral pathogens of fish. In *Fundamentals of Ornamental Fish Health* (ed. Robert, H.E.), pp. 113–124. Blackwell Publishing (2010).
17. Fijan, N. Spring viremia of carp and other diseases and agents of warm-water fish. In *Fish Diseases and Disorders*, Vol. 3 (eds. Woo, P.T.K. and Bruno, D.W.), pp. 177–244. CABI Publishing, Oxon, U.K. (1999).
18. Hedrick, R.P., Waltzek, T.B., and McDowell, T.S. Susceptibility of koi carp, common carp, goldfish, and goldfish × common carp hybrids to Cyprinid Herpesvirus 2 and Herpesvirus 3. *J. Aquat. Animal Health* 18, 26–34 (2006).
19. Jung, S.J. and Miyazaki, T. Herpesviral haematopoietic necrosis of goldfish, *Carassius auratus* (L.). *J. Fish Dis.* 18, 211–220 (1995).
20. Luo, Y.Z., Lin, L., Liu, Y. et al. Haematopoietic necrosis of cultured Prussian carp, *Carassius gibelio* (Bloch), associated with Cyprinid herpesvirus 2. *J. Fish Dis.* 36, 1035–1039 (2013).
21. Fichi, G., Cardeti, G., Cocumelli, C. et al. Detection of Cyprinid herpesvirus 2 in association with an *Aeromonas sobria* infection of *Carassius carassius* (L.) in Italy. *J. Fish Dis.* 36, 823–830 (2013).
22. Groff, J.M., LaPatra, S.E., Munn, R.J., and Zinkl, J.G. A viral epizootic in cultured populations of juvenile goldfish due to a putative herpesvirus etiology. *J. Vet. Diag. Invest.* 10, 375–378 (1998).
23. Goodwin, A., Khoo, L., Lapatra, S. et al. Goldfish hematopoietic necrosis herpesvirus (cyprinid herpesvirus 2) in the USA: Molecular confirmation of isolates from diseased fish. *J. Aquat. Anim. Health* 18, 11–18 (2006).
24. Chang, P.H., Lee, S.H., Chiang, H.C., and Jong, M.H. Epizootic of herpes-like virus infection in goldfish, *Carassius auratus* in Taiwan. *Fish Pathol.* 34, 209–210 (1999).
25 Stephens, F., Raidal, S., and Jones, B. Haematopoietic necrosis in a goldfish (*Carassius auratus*) associated with an agent morphologically similar to herpesvirus. *Aust. Vet. J.* 82, 167–169 (2004).
26. Jeffery, K.R., Bateman, K., Bayley, A. et al. Isolation of a cyprinid herpesvirus 2 from goldfish, *Carassius auratus* (L.), in the UK. *J. Fish Dis.* 30, 649–656 (2007).
27. Bretzinger, A., Fischer-Scherl, T., Oumouma, R., Hoffmann, R., and Truyen, U. Mass mortalities in koi, *Cyprinus carpio*, associated with gill and skin disease. *Bull. Eur. Assoc. Fish Pathol.* 19, 182–185 (1999).
28. Aoki, T., Hirono, I., Kurokawa, K. et al. Genome sequences of three koi herpesvirus isolates representing the expanding distribution of an emerging disease threatening koi and common carp worldwide. *J. Virol.* 81, 5058–5065 (2007).

29. Haenen, O., Way, K., Bergmann, S., and Ariel, E. The emergence of koi herpesvirus and its significance to European aquaculture. *Bull. Eur. Assoc. Fish Pathol.* 24, 293–307 (2004).

30. Kurita, J., Yuasa, K., Ito, T. et al. Molecular epidemiology of koi herpesvirus. *Fish Pathol.* 44, 59–66 (2009).

31. Xu, J.R., Bently, J., Beck, L. et al. Analysis of koi herpesvirus latency in wild common carp and ornamental koi in Oregon, USA. *J. Virol. Methods* 187, 372–379 (2013).

32. Bigarre, L., Baud, M., Cabon, J. et al. Differentiation between cyprinid herpesvirus type-3 lineages using duplex PCR. *J. Virol. Methods* 158, 51–57 (2009).

33. Chen, J., Chee, D., Wang, Y. et al. Identification of a novel cyprinid herpesvirus 3 genotype detected in koi from the East Asian and South-East Asian Regions. *J. Fish Dis.* 34, 547–554 (2014).

34. Avarre, J.-C., Madeira, J.-P., Santika, A. et al. Investigation of Cyprinid herpesvirus-3 genetic diversity by a multi-locus variable number of tandem repeats analysis. *J. Virol. Methods* 173, 320–327 (2011).

35. Marek, A., Schachner, O., Bilic, I., and Hess, M. Characterization of Austrian koi herpesvirus samples based on the ORF40 region. *Dis. Aquat. Organ.* 88, 267–270 (2010).

36. Dong, C., Li, X., Weng, S., Xie, S., and He, J. Emergence of fatal European genotype CyHV-3/KHV in mainland China. *Vet. Microbiol.* 162, 239–244 (2013).

37. Engelsma, M.Y., Way, K., Dodge, M.J. et al. Detection of novel strains of cyprinid herpesvirus closely related to koi herpesvirus. *Dis. Aquat. Organ.* 107, 113–120 (2013).

38. Taylor, N.G.H., Norman, R.A., Way, K., and Peeler, E.J. Modelling the koi herpesvirus (KHV) epidemic highlights the importance of active surveillance within a national control policy. *J. Appl. Ecol.* 48, 348–355 (2011).

39. Haenen, O. and Hedrick, R. Koi herpesvirus workshop. *Bull. Eur. Assoc. Fish Pathol.* 26, 26–37 (2006).

40. Ilouze, M., Dishon, A., and Kotler, M. Characterization of a novel virus causing a lethal disease in carp and koi. *Microbiol. Mol. Biol. Rev.* 70, 147–156 (2006).

41. Yuasa, K., Ito, T., and Sano, M. Effect of water temperature on mortality and virus shedding in carp experimentally infected with koi herpesvirus. *Fish Pathol.* 43, 83–85 (2008).

42. Haenen, O.L.M., Way, K., Bergmann, S.M., and Ariel, E. The emergence of koi herpesvirus and its significance to European aquaculture. *Bull. Eur. Assoc. Fish Pathol.* 24, 293–307 (2004).

43. Shimizu, T., Yoshida, N., Kasai, H., and Yoshimizu, M. Survival of koi herpesvirus (KHV) in environmental water. *Fish Pathol.* 41, 153–157 (2006).

44. Dishon, A., Perelberg, A., Bishara-Shieban, J. et al. Detection of carp interstitial nephritis and gill necrosis virus in fish droppings. *Appl. Environ. Microbiol.* 71, 7285–7291 (2005).

45. Honjo, M.N., Minamoto, T., and Kawabata, Z. Reservoirs of Cyprinid herpesvirus 3 (CyHV-3) DNA in sediments of natural lakes and ponds. *Vet. Microbiol.* 155, 183–190 (2012).

46. Perelberg, A., Smirnov, M., Hutoran, M., Diamant, A., Bejerano, Y., and Kotler, M. Epidemiological description of a new viral disease afflicting cultured *Cyprinus carpio* in Israel. *Isr. J. Aquacult.-Bamid.* 55, 5–12 (2003).

47. Ishioka, T., Yoshizumi, M., Izumi, S. et al. Detection and sequence analysis of DNA polymerase and major envelope protein genes in koi herpesviruses derived from *Cyprinus carpio* in Gunma prefecture, Japan. *Vet. Microbiol.* 110, 27–33 (2005).

48. Sunarto, A., Rukyani, A., and Itami, T. Indonesian experience on the outbreak of koi herpesvirus in koi and common carp (*Cyprinus carpio*). *Bull. Fish. Res. Agency Jpn. Suppl.* 2, 15–21 (2005).

49. Ito, T., Sano, M., Kurita, J., Yuasa, K., and Iida, T. Carp larvae are not susceptible to koi herpesvirus. *Fish Pathol.* 42, 107–109 (2007).

50. Fabian, M., Baumer, A., and Steinhagen, D. Do wild fish species contribute to the transmission of koi herpesvirus to carp in hatchery ponds? *J. Fish Dis.* 36, 505–514 (2013).

51. Kempter, J., Kielpinski, M., Panicz, R., Sadowski, J., Myslowski, B., and Bergmann, S.M. Horizontal transmission of koi herpes virus (KHV) from potential vector species to common carp. *Bull. Eur. Assoc. Fish Pathol.* 32, 212–219 (2012).

52. Yuasa, K., Kurita, J., Kawana, M., Kiryu, I., Oseko, N., and Sano, M. Development of mRNA-specific RT-PCR for the detection of koi herpesvirus (KHV) replication stage. *Dis. Aquat. Organ.* 100, 11–18 (2012).

53. El-Matbouli, M. and Soliman, H. Transmission of Cyprinid herpesvirus-3 (CyHV-3) from goldfish to naïve common carp by cohabitation. *Res. Vet. Sci.* 90, 536–539 (2011).

54. Yuasa, K., Sano, M., and Oseko, N. Goldfish is not a susceptible host of koi herpesvirus (KHV) disease. *Fish Pathol.* 48, 52–55 (2013).

55. McColl, K.A., Cooke, B.D., and Sunarto, A. Viral biocontrol of invasive vertebrates: Lessons from the past applied to cyprinid herpesvirus-3 and carp (*Cyprinus carpio*) control in Australia. *Biol. Control* 72, 109–117 (2014).

56. Sano, N., Moriwake, N., and Sano, T. *Herpesvirus cyprini*: Thermal effects on pathogenicity and oncogenicity. *Gyobyo Kenkyu* 28, 171–175 (1993).

57. Sano, N., Moriwake, M., Hondo, R., and Sano, T. *Herpesvirus cyprini*: A search for viral genome in infected fish by in situ hybridization. *J. Fish Dis.* 16, 459–499 (1993).

58. Calle, P.P., McNamara, T., and Kress, Y. Herpesvirus-associated papillomas in koi carp (*Cyprinus carpio*). *J. Zoo Wildl. Med.* 30, 165–169 (1999).

59. Gilad, O., Yun, S., Zagmutt-Vergara, F.J., Leutenegger, C.M., Bercovier, H., and Hedrick, R.P. Concentrations of a herpes-like virus (KHV) in tissues of experimentally-infected *Cyprinus carpio koi* as assessed by real-time TaqMan PCR. *Dis. Aquat. Organ.* 60, 179–187 (2004).

60. Morita, N. and Sano, T. Regression effect of carp, *Cyprinus carpio* L., peripheral blood lymphocytes on CHV-induced carp papilloma. *J. Fish Dis.* 13, 505–511 (1990).

61. Goodwin, A.E., Sadler, J., Merry, G.E., and Marecaux, E.N. Herpesviral haematopoietic necrosis virus (CyHV-2) infection: Case studies from commercial goldfish farms. *J. Fish Dis.* 32, 271–278 (2009).

62. Costes, B., Stalin Raj, V., Michel, B. et al. The major portal of entry of koi herpesvirus in *Cyprinus carpio* is the skin. *J. Virol.* 83, 2819–2830 (2009).

63. Fournier, G., Boutier, M., Stalin Raj, V. et al. Feeding *Cyprinus carpio* with infectious materials mediates cyprinid herpesvirus 3 entry through infection of pharyngeal periodontal mucosa. *Vet. Res.* 43, 6–6 (2012).

64. Monaghan, S.J., Thompson, K.D., Adams, A., Kempter, J., and Bergmann, S.M. Examination of the early infection stages of koi herpesvirus (KHV) in experimentally infected carp, *Cyprinus carpio* L. using *in situ* hybridization. *J. Fish Dis.* 38, 477–489 (2015).

65. Yasumoto, S., Kuzuya, Y., Yasuda, M., Yoshimura, T., and Miyazaki, T. Oral immunization of common carp with a liposome fusing koi herpesvirus antigen. *Fish Pathol.* 41, 141–145 (2006).

66. Adamek, M., Steinhagen, D., Irnazarow, I., Hikima, J., Jung, T.-S., and Aoki, T. Biology and host response to Cyprinid herpesvirus 3 infection in common carp. *Dev. Comp. Immunol.* 43, 151–159 (2014).

67. Pionnier, N., Adamek, M., Miest, J.J. et al. C-reactive protein and complement as acute phase reactants in common carp *Cyprinus carpio* during CyHV-3 infection. *Dis. Aquat. Organ.* 109, 187–199 (2014).

68. Perelberg, A., Ilouze, M., Kotler, M., and Steinitz, M. Antibody response and resistance of *Cyprinus carpio* immunized with cyprinid herpes virus 3 (CyHV-3). *Vaccine* 26, 3750–3756 (2008).

69. St-Hilaire, S., Beevers, N., Joiner, C., Hedrick, R.P., and Way, K. Antibody response of two populations of common carp, *Cyprinus carpio* L., exposed to koi herpesvirus. *J. Fish Dis.* 32, 311–320 (2009).

70. Rakus, K.L., Irnazarow, I., Adamek, M. et al. Gene expression analysis of common carp (*Cyprinus carpio* L.) lines during cyprinid herpesvirus 3 infection yields insights into differential immune responses. *Dev. Comp. Immunol.* 37, 65–76 (2012).

71. Tomé, A.R., Kus, K., Correia, S. et al. Crystal structure of a poxvirus-like zalpha domain from cyprinid herpesvirus 3. *J. Virol.* 87, 3998–4004 (2013).

72. Sunarto, A., Liongue, C., McColl, K.A. et al. Koi herpesvirus encodes and expresses a functional interleukin-10. *J. Virol.* 86, 11512–11520 (2012).

73. Ouyang, P., Rakus, K., Boutier, M. et al. The IL-10 homologue encoded by cyprinid herpesvirus 3 is essential neither for viral replication *in vitro* nor for virulence *in vivo*. *Vet. Res.* 44, 53 (2013).

74. Sunarto, A., McColl, K.A., Crane, M.S. et al. Characteristics of cyprinid herpesvirus 3 in different phases of infection: Implications for disease transmission and control. *Virus Res.* 188, 45–53 (2014).

75. Ouyang, P., Rakus, K., van Beurden, S.J. et al. IL-10 encoded by viruses: A remarkable example of independent acquisition of a cellular gene by viruses and its subsequent evolution in the viral genome. *J. Gen. Virol.* 95, 245–262 (2014).

76. Reed, A.N., Izume, S., Dolan, B.P. et al. Identification of B cells as a major site for cyprinid herpesvirus 3 (CyHV-3) latency. *J. Virol.* 88, 9297–9309 (2014).

77. Neukirch, M. and Kunz, U. Isolation and preliminary characterization of several viruses from koi (*Cyprinus carpio*) suffering gill necrosis and mortality. *Bull. Eur. Assoc. Fish Pathol.* 21, 125–135 (2001).

78. Dong, C., Weng, S., Li, W., Li, X. et al. Characterization of a new cell line from caudal fin of koi, *Cyprinus carpio koi*, and first isolation of cyprinid herpesvirus 3 in China. *Virus Res.* 161, 140–149 (2011).

79. Yuasa, K., Sano, M., and Oseko, N. Effective procedures for culture isolation of koi herpesvirus (KHV). *Fish Pathol.* 47, 97–99 (2012).

80. Sunarto, A., McColl, K.A., Crane, M.S. et al. Isolation and characterization of koi herpesvirus (KHV) from Indonesia: Identification of a new genetic lineage. *J. Fish Dis.* 34, 87–101 (2011).

81. Davidovich, M., Dishon, A., Ilouze, M., and Kotler, M. Susceptibility of cyprinid cultured cells to cyprinid herpesvirus 3. *Arch. Virol.* 152, 1541–1546 (2007).

82. Grimmett, S.G., Warg, J.V., Getchell, R.G., Johnson, D.J., and Bowser, P.R. An unusual koi herpesvirus associated with a mortality event of common carp *Cyprinus carpio* in New York State, USA. *J. Wildl. Dis.* 42, 658–662 (2006).

83. Ronen, A., Perelberg, A., Abramowitz, J. et al. Efficient vaccine against the virus causing a lethal disease in cultured *Cyprinus carpio*. *Vaccine* 21, 4677–4684 (2003).

84. Adkison, M.A., Gilad, O., and Hedrick, R.P. An enzyme linked immunosorbent assay (ELISA) for detection of antibodies to the koi herpesvirus (KHV) in the serum of koi *Cyprinus carpio*. *Fish. Pathol.* 40, 53–62 (2005).

85. Aoki, T., Takano, T., Unajak, S. et al. Generation of monoclonal antibodies specific for ORF68 of koi herpesvirus. *Comp. Immunol. Microbiol. Infect. Dis.* 34, 209–216 (2011).

86. Vrancken, R., Boutier, M., Ronsmans, M. et al. Laboratory validation of a lateral flow device for the detection of CyHV-3 antigens in gill swabs. *J. Virol. Methods* 193, 679–682 (2013).

87. Lievens, B., Frans, I., Heusdens, C. et al. Rapid detection and identification of viral and bacterial fish pathogens using a DNA array-based multiplex assay. *J. Fish Dis.* 34, 861–875 (2011).

88. Goodwin, A.E., Merry, G.E., and Sadler, J. Detection of the herpesviral hematopoietic necrosis disease agent (cyprinid herpesvirus 2) in moribund and healthy goldfish: Validation of a quantitative PCR diagnostic method. *Dis. Aquat. Organ.* 69, 137–143 (2006).

89. Gilad, O., Yun, S., Andree, K.B. et al. Initial characteristics of koi herpesvirus and development of a polymerase chain reaction assay to detect the virus in koi, *Cyprinus carpio koi*. *Dis. Aquat. Organ.* 48, 101–108 (2002).

90. Yuasa, K., Sano, M., Kurita, J., Ito, T., and Iida, T. Improvement of a PCR method with the Sph 1–5 primer set for the detection of koi herpesvirus (KHV). *Fish Pathol.* 40, 37–39 (2005).

91. Bercovier, H., Fishman, Y., Nahary, R. et al. Cloning of the koi herpesvirus (KHV) gene encoding thymidine kinase and its use for a highly sensitive PCR based diagnosis. *BMC Microbiol.* 5, 13 (2005).

92. OIE (Office International des Epizooties). *Manual of Diagnostic Tests for Aquatic Animals.* Koi herpesvirus disease. World Organization for Animal Health (OIE), Paris, France (2014).

93. Meyer, K., Bergmann, S.M., van der Marel, M., and Steinhagen, D. Detection of Koi herpesvirus: Impact of extraction method, primer set and DNA polymerase on the sensitivity of polymerase chain reaction examinations. *Aquacult. Res.* 43, 835–842 (2012).

94. Monaghan, S.J., Thompson, K.D., Adams, A., and Bergmann, S.M. Sensitivity of seven PCRs for early detection of koi herpesvirus in experimentally infected carp, *Cyprinus carpio* L., by lethal and non-lethal sampling methods. *J. Fish Dis.* 38, 303–319 (2015).

95. Gunimaladevi, I., Kono, T., Venugopal, M., and Sakai, M. Detection of koi herpesvirus in common carp, *Cyprinus carpio* L., by loop-mediated isothermal amplification. *J. Fish Dis.* 27, 583–589 (2004).

96. Soliman, H. and El-Matbouli, M. An inexpensive and rapid diagnostic method of Koi Herpesvirus (KHV) infection by loop-mediated isothermal amplification. *Virol. J.* 2, 83 (2005).

97. El-Matbouli, M., Rucker, U., and Soliman, H. Detection of Cyprinid herpesvirus-3 (CyHV-3) DNA in infected fish tissues by nested polymerase chain reaction. *Dis. Aquat. Organ.* 78, 23–28 (2007).

98. Cheng, L., Chen, C.-Y., Tsai, M.-A. et al. Koi herpesvirus epizootic in cultured carp and koi, *Cyprinus carpio* L., in Taiwan. *J. Fish Dis.* 34, 547–554 (2011).

99. Soliman, H. and El-Matbouli, M. Immunocapture and direct binding loop mediated isothermal amplification simplify molecular diagnosis of Cyprinid herpesvirus-3. *J. Virol. Methods* 162, 91–95 (2009).

100. Yoshino, M., Watari, H., Kojima, T., and Ikedo, M. Sensitive and rapid detection of koi herpesvirus by LAMP method. *Fish Pathol.* 41, 19–27 (2006).

101. Yoshino, M., Watari, H., Kojima, T., Ikedo, M., and Kurita, J. Rapid, sensitive and simple detection method for koi herpesvirus using loop-mediated isothermal amplification. *Microbiol. Immunol.* 53, 375–383 (2009).

102. Bergmann, S.M., Kempter, J., Sadowski, J., and Fichtner, D. First detection, confirmation and isolation of koi herpesvirus (KHV) in cultured common carp (*Cyprinus carpio* L.) in Poland. *Bull. Eur. Assoc. Fish Pathol.* 26, 97–104 (2006).

103. Bergmann, S.M., Lutze, P., Schutze, H. et al. Goldfish (*Carassius auratus*) is a susceptible species for koi herpesvirus (KHV) but not for KHV disease (KHVD). *Bull. Eur. Assoc. Fish Pathol.* 30, 74–84 (2010).

104. Lee, N.-S., Jung, S.H., Park, J.W., and Do, J.W. *In situ* hybridization detection of koi herpesvirus in paraffin-embedded tissues of common carp *Cyprinus carpio* collected in 1998 in Korea. *Fish Pathol.* 47, 100–103 (2012).

105. Sano, N., Sano, M., Sano, T., and Hondo, R. *Herpesvirus cyprini*: Detection of the viral genome by in situ hybridization. *J. Fish Dis.* 15, 153–162 (1992).

106. Perelberg, A., Ronen, A., Hutoran, M., Smith, Y., and Kotler, M. Protection of cultured Cyprinus carpio against a lethal viral disease by an attenuated virus vaccine. *Vaccine* 23, 3396–3403 (2005).

107. Fuchs, W., Fichtner, D., Bergmann, S.M., and Mettenleiter, T.C. Generation and characterization of koi herpesvirus recombinants lacking viral enzymes of nucleotide metabolism. *Arch. Virol.* 156, 1059–1063 (2011).

108. Costes, B., Fournier, G., Michel, B. et al. Cloning of the koi herpesvirus genome as an infectious bacterial artificial chromosome demonstrates that disruption of the thymidine kinase locus induces partial attenuation in *Cyprinus carpio koi*. *J. Virol.* 82, 4955–4964 (2008).

109. Zhou, J.-X., Wang, H., Li, X.-W., Zhu, X., Lu, W.-L., and Zhang, D.-M. Construction of KHV-CJ ORF25 DNA vaccine and immune challenge test. *J. Fish Dis.* 37, 319–325 (2014).

110. Zhou, J.X., Xue, J., Wang, Q. et al. Vaccination of plasmid DNA encoding ORF81 gene of CJ strains of KHV provides protection to immunized carp. *In Vitro Cell. Dev. Biol. Anim.* 50, 489–495 (2014).

111. Kasai, H., Muto, Y., and Yoshimizu, M. Virucidal effects of ultraviolet, heat treatment and disinfectants against koi herpesvirus (KHV). *Fish Pathol.* 40, 137–138 (2005).

112. Yoshida, N., Sasaki, R.K., Kasai, H., and Yoshimizu, M. Inactivation of koi-herpesvirus in water using bacteria isolated from carp intestines and carp habitats. *J. Fish Dis.* 36, 997–1005 (2013).

113. Dixon, P.F., Joiner, C.L., Way, K., Reese, R.A., Jeney, G., and Jeney, Z. Comparison of the resistance of selected families of common carp, *Cyprinus carpio* L., to koi herpesvirus: Preliminary study. *J. Fish Dis.* 32, 1035–1039 (2009).

114. Piačková, V., Flajshans, M., Pokorova, D. et al. Sensitivity of common carp, *Cyprinus carpio* L., strains and crossbreeds reared in the Czech Republic to infection by cyprinid herpesvirus 3 (CyHV-3; KHV). *J. Fish Dis.* 36, 75–80 (2013).

115. Shapira, Y., Magen, Y., Zak, T., Kotler, M., Hulata, G., and Levavi-Sivan, B. Differential resistance to koi herpes virus (KHV)/carp interstitial nephritis and gill necrosis virus (CNGV) among common carp (*Cyprinus carpio* L.) strains and crossbreds. *Aquaculture* 245, 1–11 (2005).

116. Gotesman, M., Soliman, H., Besch, R., and El-Matbouli, M. *In vitro* inhibition of Cyprinid herpesvirus-3 replication by RNAi. *J. Virol. Methods* 206, 63–66 (2014).

89 *Ictalurid Herpesvirus 1*

Larry Hanson

CONTENTS

89.1 Introduction 797
    89.1.1 Classification 797
    89.1.2 Morphology, Genome Organization, Biology, and Epidemiology 797
    89.1.3 Clinical Signs and Pathogenesis 799
    89.1.4 Diagnosis 799
89.2 Methods 799
    89.2.1 Sample Preparation 799
    89.2.2 Detection Procedures 800
        89.2.2.1 Cell Culture Isolation 800
        89.2.2.2 Antibody Neutralization 801
        89.2.2.3 PCR 801
89.3 Conclusion and Future Perspectives 804
References 805

89.1 INTRODUCTION

Ictalurid herpesvirus 1 (IcHV1), commonly known as channel catfish virus, is the first identified and characterized herpesvirus of fish [1]. This virus causes channel catfish virus disease (CCVD), a severe hemorrhagic viremia in young channel catfish (*Ictalurus punctatus*). Given that losses to CCVD can exceed 90% of the affected population within a 2-week period, this disease is of concern to aquaculturists that produce channel catfish. This virus is easily identified in acutely infected individuals, but latent IcHV1 infections can be difficult to detect. A related alloherpesvirus designated *Ictalurid herpesvirus* 2 (IcHV2) has been detected in Italy from another North American freshwater catfish, the black bullhead (*Ameiurus melas*) experiencing hemorrhagic viremia similar to CCVD [2]. The distinction between the two virus species is important in diagnostic evaluation because the channel catfish may be extremely sensitive to IcHV2 [3] and it is not known to occur in North America where the largest stocks of wild and aquacultured channel catfish occur.

89.1.1 CLASSIFICATION

IcHV1 is the type species of the *Ictalurivirus* genus, family *Alloherpesviridae*, order *Herpesvirales* [4]. Apart from *Ictalurivirus,* the family *Alloherpesviridae* also includes the genera of *Batrachovirus, Cyprinivirus,* and *Salmonivirus*. Currently, the *Ictalurivirus* genus contains three recognized species: IcHV1, IcHV2, and *Acipenserid herpesvirus* 2 (AciHV2). AciHV2 was isolated from white sturgeon (*Acipenser transmontanus*) in California [5]. Classification within the *Ictalurivirus* genus was based on sequencing

portions of the genomes of IcHV2 and AciHV2 and comparing them to the genome sequence of IcHV1 [6–8].

89.1.2 MORPHOLOGY, GENOME ORGANIZATION, BIOLOGY, AND EPIDEMIOLOGY

Initial electron microscopic characterization of negatively stained IcHV1 particles and infected cells demonstrated that IcHV1 morphology and assembly processes are similar to other herpesviruses [1]. The IcHV1 virion has a buoyant density of 1.15 g/mL [9] and consists of an electron-dense staining core, a capsid, an amorphous tegument, and a lipid bilayer envelope [1]. The enveloped virion is 175–200 nm in diameter, while the capsid is 100 nm in diameter. It has icosahedral symmetry containing 162 capsomeres. The genome replication, capsid assembly, and capsid packaging take place in the nucleus. The complete nucleocapsids are seen budding out of the nucleus and budding into cytoplasmic vacuoles [1]. It is assumed that the nucleocapsids exit the nucleus by budding out of the inner nuclear membrane, and then they lose this nuclear-derived envelope by fusing with the outer nuclear membrane. These cytoplasmic nucleocapsids acquire the tegument components and then acquire the final envelope by budding into cytoplasmic vacuoles and are transported out of the cell by exocytosis [10].

Early electron microscopic analysis, sucrose density gradient sedimentation, reassociation kinetics, and restriction fragment analysis demonstrated that the packaged genome of IcHV1 is a single linear, nonpermuted ~130 kb dsDNA molecule with 18 kb direct repeats [11,12]. There are genetic differences between strains [13] and that the virus loses one copy of the direct repeats and becomes endless in the infected cells (concatemeric or circular) [14]. In 1992, Davison sequenced 134,226 bp linear genome of IcHV1 and predicted that it contained 79 open reading frames (ORFs) encoding 76 genes (the DNA polymerase gene ORFs 57 and 58 and terminase gene ORFs 62, 69, and 71 are spliced) [8]. This was the first fish herpesvirus sequenced, and the lack of gene sequence homology to mammalian herpesviruses indicated a very distant relationship. When compared to other members of *Alloherpesviridae* that have been sequenced, the IcHV1 genome is the smallest. It is less than half of the size and encodes less than half of the genes of *Cyprinid herpesvirus* 3. The purified IcHV1 genome was also shown to be infectious and efficiently undergo homologous recombination allowing for trait and mutation analysis by marker rescue/marker transfer [15]. These studies and, more recently, the ability to efficiently generate recombinants in *Escherichia coli* using overlapping genomic fragments cloned into bacterial artificial chromosomes have opened the way for efficiently evaluating IcHV1 genes by reverse genetics [16].

Proteomic analysis of differentially fractionated preparations from purified virions demonstrated that 11 of the predicted genes encode structural proteins, one on the envelope (gene 59 product—major glycoprotein), two tegument proteins (gene 11 and gene 65 products), four tegument-associated proteins (gene 15, gene 72, gene 73, and gene 74 products), and four capsid proteins (gene 28, gene 27, gene 39, and gene 53 products) [17]. The quantitative data of the capsid proteins was used in conjunction with cryoelectron microscopy and three-dimensional image reconstruction to deduce the detailed morphology and makeup of the capsid [18]. The structure of the IcHV1 capsid is remarkably similar to that of *Human herpesvirus* 1 (HHV1—herpes simplex virus 1). It is composed of 150 hexamers and 12 pentamers. The capsomers have 9 nm (IcHV1) compared to 11 nm (HHV1) chimney-like protrusions with an axial channel through each capsomer, and the gene 39, gene 53, and gene 27 products are structural orthologs of HHV1 VP5, VP19c, and VP23, respectively.

Other characterized gene products include the gene 50 product. It has been shown to be a secreted mucin [19,20]; the product of gene 10 was identified as an envelope protein using antibodies [21] and the thymidine kinase product of gene 5, which can activate nucleotide analogs including acyclovir [22]. The thymidine kinase gene is likely derived from the host deoxycytidine kinase gene [23]. High-throughput mass spectroscopy with probabilistic proteogenomic mapping demonstrated the expression of 37 of the predicted genes and identified the products of 17 novel protein–coding regions [24].

In 1980, Dixon and Farber evaluated the temporal profile of virus protein expression in IcHV1-infected cells and found that the genes are temporally regulated much like the more intensely studied members of *Herpesviridae*. Of 32 identified infected cell proteins, 8 were expressed within 2 h of infection (immediate-early candidates), 8 were expressed 2–4 h after infection (early candidates), and 16 were expressed after 4 h postinfection (late candidates). Furthermore, like other herpesviruses, some of the immediate-early protein candidates were expressed at higher levels following the release from protein synthesis inhibitor, cycloheximide, and the synthesis of late protein candidates was inhibited by ara-C, an inhibitor of DNA synthesis [11]. Similar temporal expression assays were performed on various transcripts coded by the direct repeat regions. However, accurately evaluating of mRNA expression is challenging because some genes utilize the same termination/polyadenylation signals resulting in overlapping transcription [25,26]. Also, when trying to differentiate immediate-early genes from early genes using cycloheximide blocking, some early genes are apparently leaky, and these early gene transcripts can build up to high levels over time [27,28]. The reason for "leaky" early gene transcription is not known, but among the 14 genes in the direct repeats, genes 1 and 3 are clearly immediate-early genes [28].

CCVD-affected populations of catfish may experience high mortality, reduced growth, and a predisposition to bacterial diseases [29]. The severity of a CCVD epizootic is significantly enhanced by environmental stress and crowding. Outbreaks of CCVD are sporadic in heavily stocked channel catfish fingerling ponds. Although older fish can be affected [30], natural outbreaks almost exclusively occur in young-of-the-year channel catfish during the warm summer months. In experimental challenge trials, very young fish are more resistant to the virus and become more sensitive as they age to 60 days posthatch. Furthermore, the resistance was correlated to the maternal parent's IcHV1-neutralizing antibody titer suggesting that the maternal transfer of antibodies may influence early susceptibility [31]. The optimum temperature for disease progression is 27°C or higher [32]. In the catfish-producing region of the United States, these temperatures occur from July through September when fingerlings are less than 4 months of age. Under these conditions, over 90% of the population may die in less than 2 weeks from the first signs of disease, yet most IcHV1 endemic populations will experience no obvious CCVD. Losses due to CCVD on fingerling production operations can vary substantially from year to year without obvious environmental cues, changes in management, or genetics of the stock.

IcHV1 appears to be maintained in a population by vertical transmission. IcHV1-specific PCR on recently hatched fry from five representative fingerling operations demonstrated latent carrier status of 10%–20% in each population [33]. These farms represented approximately 20% of the commercial catfish fingerling production. Therefore, it can be assumed that IcHV1 is endemic in most aquaculture populations of channel catfish in the Southeastern United States. Furthermore, evaluating incidence of latent IcHV1 in endemic fingerling populations over time showed an increase even though no CCVD was detected. This indicates that horizontal transmission occurs in the populations without overt disease [33]. By using PCR to detect and eliminate carrier brooders, it was noted that numbers of carrier offspring were reduced, but vertical transmission to the subsequent generation was not eliminated [33].

89.1.3 CLINICAL SIGNS AND PATHOGENESIS

During a CCVD outbreak, IcHV1 is highly contagious and spreads quickly through the population. In experimental exposures, clinical signs appear 2–3 days postinfection and include erratic swimming, exophthalmia, a distended abdomen, and hemorrhages in the fins (Figure 89.1). Internal gross pathology includes yellow ascites, swollen spleen, and posterior kidney.

The posterior kidney, the most severely affected organ during acute CCVD, typically displays extensive edema, inflammation, and necrosis of hematopoietic tissue and tubules. This is followed by focal necrosis, hemorrhage and edema of the liver and gastrointestinal tract, and necrosis of pancreatic tissue, congested spleen, and focal areas of hemorrhage in musculature [34,35].

Acute CCVD usually occurs in young fingerlings. More chronic CCVD can occur in older fingerlings. This form is associated with less kidney necrosis and inflammation but a higher incidence of secondary bacterial infections, most commonly external infections with *Flavobacterium columnare*. Although rare, IcHV1 can cause disease in yearlings, and IcHV1 has been isolated from adult fish during winter months [36].

A study using radioisotope labeled IcHV1 in immersion challenges indicated that the virus was taken up by the gills and caudal fin. After this, the virus DNA increased in the liver, gallbladder, and gut [37]. Cell culture titrations and quantitative PCR (qPCR) performed on the posterior kidney, gill, blood, skin, intestine, spleen, and liver of immersion-exposed catfish fingerlings demonstrated that the highest levels of virus develop in the posterior kidneys with peak virus production occurring 3–4 days postinfection, and this correlates well with the period of peak mortality [38]. The next highest concentration of virus occurs in the gills. Analysis of water demonstrated detectable shedding on days 2–5 postinfection with day 4 being the peak [38].

FIGURE 89.1 Photograph of a channel catfish fingerling displaying abdominal swelling and exophthalmia due to channel catfish virus disease.

89.1.4 DIAGNOSIS

The identification of IcHV1 infection to diagnose CCVD is easily accomplished and most often done using traditional cell culture techniques. This virus replicates rapidly in cell culture, producing distinctive cytopathic effect (CPE) within 24 h from heavily infected samples. Molecular methods are also effective for detecting the virus in tissues from CCVD-affected fish. However, detecting latent IcHV1 infection is more problematic and requires the use of sensitive DNA-based detection methods.

Conventional techniques: IcHV1 is somewhat host specific in cell culture. It has been shown to replicate in a variety of catfish cells but does not replicate in most cell lines derived from noncatfish species. For diagnostic purposes the virus replicates very well and shows distinctive CPE in the channel catfish ovary (CCO) cell line [39] and the brown bullhead (BB) cell line [1,40]. Also, either cell line can be used for the primary isolation of IcHV2 and causes similar CPE, but IcHV2 can replicate in the bluegill (BF2) cell line, which is refractive to IcHV1. Thus, the use of CPE on the catfish cells but no CPE on BF2 cells has presumptive diagnostic value. These cell lines are available from the American Type Culture Collection. Although conventional cell culture method is effective for diagnosing CCVD, it will not detect latent virus. Thus, it is of little use for inspecting populations of fish for IcHV1 carrier state. The use of IcHV1-specific antibody screening has some value in detecting populations that have CCVD. However, the amounts of antibody can decrease to undetectable levels over time, as antibody levels subside over the winter [41] and many carriers never develop detectable antibody titers.

Molecular techniques: Published nucleic acid–based techniques for detecting IcHV1 include gene-specific PCR [33,42–44] and degenerate PCR followed by sequencing of a fragment of the DNA polymerase gene [45]. A hydrolysis probe (TaqMan)-mediated qPCR method has also been published that is specific for IcHV2 [46], and a sensitive TaqMan qPCR method has been developed in the author's lab for IcHV1 that can distinguish between two observed genomovars and will not detect IcHV2. Conventional PCR and TaqMan PCR are sensitive enough to detect latent IcHV1, but the level of virus is often close to the sensitivity limit. Samples from whole processed sac fry give a relatively strong signal, but most samples from tissues of adult fish are very low, and it is suggested that for highest sensitivity, at least three PCR assays be run per tissue sample evaluated.

89.2 METHODS

89.2.1 SAMPLE PREPARATION

Virus culture: The best samples for disease diagnostic cases are fish that are displaying signs of disease or are freshly dead from the epizootic. If clinically affected fish are available, three of them should be evaluated. Alternatively, if diseased fish cannot be sampled, a minimum of 10 randomly sampled fish from a population of fish that is experiencing an epizootic can be evaluated for culturable virus. Small fish (2 cm or less) are evaluated in entirety or with the yolk sac removed from sac fry. The trunk portion (anterior to the anal fin) is evaluated in 2–3 cm fish, and the viscera can be taken on 3–5 cm long fish. Larger fish are dissected and posterior kidney and spleen sampled. These samples are suspended in the homogenization medium (serum free cell culture medium, pH 7.4–7.6, with 800 IU/mL penicillin, 800 µg/mL streptomycin, and 7.5 µg/mL amphotericin B) at approximately 1:10 tissue to medium and can be stored at 0°C–4°C for up to 48 h. To process the samples, the tissues are simply homogenized using a tissue homogenizer or mortar and pestle. Then the cellular debris is then removed by centrifugation at $1500 \times g$ for 10 min. This tissue supernatant is then filtered through a 0.45 µm syringe filter to remove bacterial and fungal agents. The filtrate is then diluted to a final concentration of 1:100 (tissue to medium) in cell culture medium with serum for use as the cell culture inoculum. As an alternative protocol, the homogenate may be decontaminated by the incubation overnight at 4°C in the high concentration of antibiotics in the homogenization medium, and the supernatant from the centrifugation step (diluted as stated earlier) can then be used directly as the cell culture inoculum. For disease diagnosis, in general, up to 1/3 of the inoculum is used fresh to inoculate cells, 1/3 is stored at 4°C for up to 48 h, and 1/3 is archived at −70°C.

Molecular assays: Samples for PCR analysis can be the tissue pellet from processing described earlier or separate samples. Approximately 100 µg of tissue is processed. DNA can be isolated using a commercial kit such as DNeasy Tissue Kit (Qiagen, Maryland, Delaware). To evaluate latent carriers, a noninvasive biopsy can be taken from the caudal fin using a 3 mm diameter biopsy punch. After extraction, the quality and quantity of the DNA can be assessed using a NanoDrop spectrophotometer.

89.2.2 Detection Procedures

89.2.2.1 Cell Culture Isolation

Case samples should be inoculated on to CCO or BB cells and BF2 cells. The fish cell lines are cultured in Dulbecco's modified minimal essential medium (DMEM) containing γ-irradiated 10% fetal bovine serum and 25 mM HEPES buffer (Table 89.1). The maintenance medium has no antimicrobial or antimycotic agents. These cells are incubated at 20°C–30°C and may be cultured without the use of CO_2. In this case, no bicarbonate is used in the medium. Cell lines are routinely split when cultures reach confluency using 0.25% trypsin/EDTA. All media are prepared using commercially available certified endotoxin-free cell culture water. Each cell line is handled separately during subculturing procedures, and each has separate designated culture solutions. All procedures are performed in laminar flow HEPA filter hoods, using sterile, disposable plastic pipettes, centrifuge tubes, filters, and culture vessels. Countertops in the hoods are swabbed with 70% ethanol between culture

TABLE 89.1

Reagents Used for Cell Culture and Virology

Reagent	Source[a]	Description
DMEM powder	Sigma D1152	DMEM powder (makes 1 L at 1×) with 4500 mg/mL glucose, L-glutamine, and 25 mM HEPES without sodium bicarbonate
Tissue culture water	Sigma W3500	Cell culture water, ultrapure endotoxin-free
L-Glutamine	Gibco 25030	L-Glutamine 200 mM (100×) in water
Penicillin–streptomycin	Gibco 15140-122	10,000 IU/mL penicillin and 10,000 µg/mL streptomycin
Fungizone	Gibco 15290-018	250 µg/mL amphotericin B, 205 µg/mL sodium deoxycholate
Trypsin/EDTA	Gibco 15140-054	(0.5% trypsin, 5.3 mM Na_4EDTA) 10× liquid
FBS	Gibco 10438-026	Fetal bovine serum—qualified and heat inactivated—making 30 mL aliquots to avoid freezing and thawing repeatedly

[a] Company and catalog number, equivalent reagent from another source may be substituted.

procedures with various cell lines. All cell lines in the culture facility are evaluated for mycoplasma contamination every 6 months using a commercially available kit, and any contaminated cultures are discarded.

89.2.2.1.1 Procedure

1. To prepare cells for virus culture, a confluent monolayer is split 1:3 into a 24-well plate or a 12.5 cm^2 flask and incubated at 30°C the day before virus inoculation to produce cells at 80%–90% confluency the next day.
2. To inoculate with the case sample, the medium is removed, and the cells are overlaid with 0.2 mL per well of a 24-well plate or 1.5 mL per flask of inoculum and incubated at 30°C for 30 min.
3. The cultures are then fed 1 mL/well or 6 mL/flask of cell culture medium containing 200 UI/mL penicillin and 200 µg/mL streptomycin. If using a 24-well plate, inoculate at least two wells with each sample. Also a noninfected control flask or set of wells (inoculated with homogenization medium alone) should accompany each case.
4. Observe the cultures daily by microscopic examination at 40–100× total magnification for 10 days. The use of a phase contrast microscope or similar contrast enhancement facilitates the recognition of early cytopathology. The CPE produced by IcHV1 is extensive and rapidly progressing by 48 h postinoculation in cultures from overtly diseased individuals.

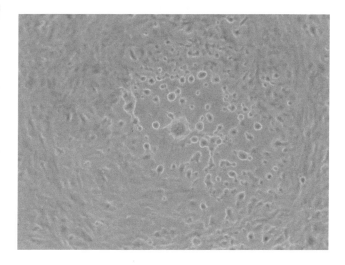

FIGURE 89.2 Photograph of channel catfish ovary cells displaying typical cytopathic effect caused by *Ictalurid herpesvirus* 1 infection (40× total magnification).

This CPE consists of cell fusion (syncytium) formation and contraction leaving cytoplasmic spindles irradiating from the syncytium to points on the flask surface where the cells were originally attached (Figure 89.2). Cultures from chronic or subclinical infections may show CPE later than this.

5. Blind passage is recommended from negative cultures at 7–10 days postinoculation. Some cultures may show signs of stress or unusual growth due to toxicity or growth factors in the cell extract; these samples should be passaged to fresh cells as well. This blind passage consists of inoculating fresh cells with the medium from the original culture wells/flask as described earlier. These passaged cells are observed for at least 7 more days. Cultures that show CPE can be harvested and stored short term (up to 4 days) at 4°C or archived at −70°C.

6. The observation of characteristic CCVD clinical signs and pathology in channel catfish fingerlings and the characteristic CPE in catfish cell lines is a presumptive diagnosis of CCVD. Confirmation requires the use of IcHV1-specific antibody-based analysis or a specific molecular assay.

89.2.2.2 Antibody Neutralization

The use of neutralizing monoclonal antibody (NAb) produced by Arkush et al. [47] provides the specificity to distinguish IcHV1 from IcHV2 [3]. Polyclonal antibodies are also useful, but IcHV1 in general is poor at inducing neutralizing antibodies in immunized rabbits. None of these diagnostic antibodies are commercially available.

89.2.2.2.1 Procedure

1. Seed and incubate at least two 96-well tissue culture plates to provide 80%–90% confluence as described earlier.

2. Collect the culture medium of the cell monolayers exhibiting CPE and centrifuge at $2000 \times g$ for 15 min at 4°C to remove cell debris.

3. Dilute virus-containing medium from 10^{-2} to 10^{-5} in 200 μL aliquots.

4. Mix 100 μL aliquots of each virus dilution with equal volumes of an antibody solution specific for CCV, and similarly treat 100 μL aliquots of each virus dilution with cell culture medium.

5. In parallel, other neutralization tests must be performed against the following:
   a. The type isolate of IcHV1 or other confirmed samples (positive neutralization test)
   b. A heterologous virus (negative neutralization test)

6. Incubate all the mixtures at 25°C for 1 h. Then bring the volume up to 1 mL with cell culture medium.

7. Aspirate the medium off of at least 6 wells per dilution per sample and transfer 100 μL aliquots of each of the aforementioned mixtures on to the cell and then incubate at 30°C.

8. Microscopically observe all cell cultures daily for CPE and note the dilution and time of first detectable CPE and then note positive and negative wells at 96 h postinfection. Then calculate the titers of the virus sample treated and untreated using the proportional distance calculation of Reed and Muench [48] or Karber [49].

9. The tested virus is identified as IcHV1 when the sample virus titer is reduced by treatment with the IcHV1-specific antibody, when compared to the same sample given the medium alone treatment. If the IcHV1 NAb does not reduce the titer, the suspect sample should be evaluated using IcHV1- and IcHV2-specific nucleic acid–based assays. When cell culture supernatant of the monoclonal NAb 95 [47] was used undiluted on IcHV1 case isolates, it demonstrated a 10–100 fold reduction in titer of 90% of the samples.

89.2.2.3 PCR

Latent infections of IcHV1 are not associated with infectious virus, and the genome concentrations in the infected individual are very low. PCR is the only method sensitive enough to detect latent virus reliably. Also these assays are easy to perform and thus are often the method of choice for confirming a presumptive diagnosis of CCVD by IcHV1 isolation in cell culture. There are three types of PCR methods that are of diagnostic value for IcHV1, which are described in the following text: conventional PCR, TaqMan-based qPCR, and degenerate primer PCR. When using PCR assays in a diagnostic laboratory, there must be constant vigil against contamination. The most common source of DNA contamination that causes false-positive results is the product of a previous assay. Given the contamination risk, the use of nested PCR should be avoided in diagnostic assays whenever possible. Other sources of high levels of target DNA are tissues from a CCVD-affected fish and infected cell cultures. Therefore, all labs that use PCR assays must have separate physically isolated areas and pipettors for PCR reagent preparation, and reagents are usually dispersed into inconvenient single assay aliquots. Also assay setup is done in separate areas from where PCR product evaluation is done. To minimize potential for contamination, all sample preparation and reaction setup should be done with aerosol-preventing pipette tips. Each assay must have relevant positive and negative controls to assure that the assay is functioning well and to monitor for contamination problems. The best negative control is a known negative sample that is similar to the test sample that is extracted and prepared using the same reagents in parallel with the test sample. Also, a water blank control is often run. In the positive control, the target sequence should be diluted to the level of sensitivity desired in the assay. A third type of control that helps assure the quality of individual sample results is the use of an internal control. This can be an assay for a host genomic sequence, a spiked modified target [38], or a commercially available exogenous internal control. The internal control will indicate PCR failure or inhibition due to inhibitors in the sample or errors in setup.

89.2.2.3.1   Cell Culture Confirmation–Extraction Procedure

A cell culture sample that displays CPE typical of IcHV1 can be easily tested either by conventional PCR or qPCR. The DNA can be extracted using a commercially available kit or simply proteinase K treated as described here:

1. Collect 1–0.5 mL of cell culture supernatant from the affected wells into a 1.5 mL microcentrifuge tube.
2. Centrifuge at maximum speed in microfuge (18,000–20,000 × g) for 30 min.

3. Discard supernatant and resuspend the pellet in 10 μL proteinase K buffer (50 mM KCl, 15 mM Tris-HCl pH 8.3, and 0.5% Nonidet P-40) containing 0.5 mg/mL proteinase K, and incubate at 55°C for 1 h.
4. Heat-inactivate the proteinase K at 95°C for 10 min. Centrifuge at maximum speed for 10 s. Use 3 μL per PCR.

Other standard DNA extraction methods may be used.

89.2.2.3.2   Conventional PCR

The conventional PCR assay described here is sensitive and does not rely on nested PCR procedures [33]. It uses modified primers (Table 89.2) of the original described PCR method [42] and can use the described internal control [38] to evaluate PCR sensitivity in individual assays.

1. Make the master mix. Prepare at least one extra aliquot for each 10 reactions planned. Include a positive and a negative (water blank or negative cell culture extract) control for every reaction.

Per sample:

Distilled water	33.6 μL
10× buffer	5.0 μL
Upper primer (50 pmol/μL)	0.4 μL
Lower primer (50 pmol/μL)	0.4 μL
dNTPs (2.5 mM each)	2.0 μL
25 mM MgSO$_4$	5.0 μL
Internal control (0.025 pg/μL—optional)	0.3 μL
Taq polymerase (5 U/μL)	0.3 μL

2. Add 47 μL per reaction tube.

TABLE 89.2
Primers, Probes, and Amplification Parameters for *Ictalurid herpesvirus* 1 Quantitative Polymerase Chain Reaction Assays

Assay, Primers, and Probes	Sequence[b]	Cycle Parameters
IcHV1 PCR		30 s at 60°C, 30 s at 72°C, 30 s at 90°C
Upper primer	TCATCCGAATCCGACAACTGA	
Lower primer	CCAAGATCGCGGAGAAAC	
IcHV1 TaqMan		60 s at 64°C, 15 s at 72°C, 15 s at 95°C
Upper primer	CTCCGAGCGATGACACCAC	
Lower primer	TGTGTTCAGAGGAGCGTCG	
IcHV1a Probe[a]	FAM-CCCATCCCTTCCCTCCTCCCTG-BHQ-1#	
IcHV1b Probe[a]	FAM-CCCATCCTTCCCCTCCTCCCTG-BHQ-1	
IcHV2 TaqMan [46]		45 s at 59°C, 45 s at 72°C, 30 s at 95°C
Upper primer	ATACATCGGTCTCACTCAAGAGG	
Lower primer	TAATGGGTATTGGTACAAATCTTCATC	
Probe	FAM-CGC+CTG+AGA+ACC+GAGCA –BHQ-1	

Source: Latorra, D. et al., *Mol. Cell. Probes*, 17, 253, 2003.

[a]   Probe IcHV1a detects genome type A; Probe IcHV1b detects genome type B. Both assays use the same primers and reaction conditions.

[b]   FAM designates 6-FAM fluorescent dye, BHQ-1 indicates black hole quencher, and + indicates use of locked bases to increase its melting point [50].

3. Add 3 μL (up to 1 μg total DNA) of sample or negative control.
4. Run the reaction. Incubate at 95°C for 5 min to activate the polymerase followed by 30 cycles of 90°C for 30 s, 60°C for 30 s, and 72°C for 30 s. After the last cycle, the samples are incubated at 72°C for 5 min.
5. Electrophorese 10 μL of the product on a 10% polyacrylamide gel with a 100 bp size ladder. Stain the gel with ethidium bromide. Observe and photograph under UV transillumination.
6. Evaluate the results. The internal control produces a 149 bp product. CCV produces a 136 bp product. Expect a 149 bp band within the negative control and on negative samples. Expect a 136 bp product and a 149 bp product on CCV-positive samples. If CCV is present at a high level, the 136 bp band will be very strong, and the 149 bp band may be missing. If no bands are present, the PCR did not work indicating a failure of one of the PCR components or an inhibitor in the sample. If the negative control shows a 136 bp product, then there is a contaminant in the PCR setup, and the assay must be redone. To confirm that the PCR product is from CCV, the PCR can be redone using no internal standard, and the product can be directly sequenced. The sequence can then be evaluated using basic local alignment search tool (BLAST) on the National Center for Biological Information internet site (http://blast.ncbi. nlm.nih.gov/Blast.cgi) to identify sequences with high homology. The PCR amplifies the region from 107,827 to 10,7962 of the CCV genome (GenBank accession M75136) representing a portion of ORF 73 [8]. Expect at least 95% identity to this sequence.

A modification to the aforementioned protocol includes the use of 40 cycles to detect low-level genomes such as in latently infected fish [33]. If 40 cycles are used, additional larger non-diagnostic bands may be present in the gel.

89.2.2.3.3 TaqMan PCR

The use of TaqMan qPCR increases the specificity of the PCR assay due to the use of the internal probe. The quantitative analysis assists in evaluating the clinical relevance of a positive result. Also because there are no steps that involve manipulating the products, there is less chance of product contamination of reagents or samples. There are three TaqMan qPCR assays that are of use for *Ictalurid herpesvirus* diagnosis: IcHV2-specific qPCR [46] and two IcHV1-specific qPCR assays that differentiate two identified genome sequences of IcHV1. The two IcHV1 assays use the same primers and reaction conditions so they can be run on the same plate.

89.2.2.3.3.1 Procedure

1. To prepare for the qPCR, dilute the DNA sample to 0.1 μg/μL. For fin tissue extractions using the Gentra Kit, rehydrating using 20 μL/mg of tissue will give you approximately 1 μg/μL.

2. Make a CCV control containing 0.1 μg/μL of CCO DNA and 0.001 μg/μL of positive control virus DNA.
3. Plan an experiment using at least three positive control wells and three negative (water) control wells per plate. Make up a master mix for the plate. Calculate the volume for at least one more well than you will use for every 20 wells.

Reaction makeup for one sample, volumes in μL

Water	8.8
10× PCR buffer	2.5
MgSO₄	1.5
dNTPs	0.5
20 μM primer U	0.5
20 μM primer L	0.5
10 μM probe	0.5
Platinum Taq	0.2
Total	15

4. Add 15 μL of master mix per well of real-time plate that will be used. (Make sure the qPCR thermocycler is warmed up).
5. Add 10 μL of fish DNA or positive or negative control sample.
6. Seal plate using real-time plate sealer.
7. Spin at 1200 rpm for 3 min.
8. Real-time PCR is performed using an iCycler BioRad or similar qPCR thermocycler.

Stage 1—1 cycle at 95°C for 3 min 30 s
Stage 2—45 cycles as indicated in Table 89.2
The real-time reaction takes approximately 2 h to run.

The plate can be saved for polyacrylamide gel electrophoresis to see if the product is generated if you get unusual results. The IcHV2 assay will not detect IcHV1. The IcHV1a assay will not detect IcHV2 and is specific for IcHV1 genomotype A, which represents 72% of the CCVD diagnostic cases in Mississippi. The IcHV1b assays will not detect IcHV2 and is specific for IcHV1 genomotype B, which represents 18% of the CCVD diagnostic cases. Due to the high specificity of these assays, some genomic variants may be overlooked. If so and PCR product is produced by the primers, direct sequencing can be used to characterize these variants. The positive control can be generated using cloned genomic DNA containing the target sequence or purified viral DNA. If the DNA is extracted from the extracellular-packaged IcHV1, each genome will have two copies of the IcHV1 qPCR target, because the target is in the direct repeat region. In the author's lab, the 10 fg control sample has a threshold cycle (CT) of 32. Given the 84 MD molecular mass of the mature genome, this represents 71 genome copies, and the theoretical CT value for one copy would be 38. Using 1 μg of sample DNA, most latently infected fish give a CT value of 33–38, whereas extracts from fish with CCVD and cell culture give CT values of less than 27. Modifications to the aforementioned protocol that are used in the author's lab include the use of a commercially available Exogenous Internal Positive Control Reagents (VIC probe) (Life Technologies) in each sample as described by the manufacturer. This provides an assessment of the individual variation in qPCR efficiency. If a sample has a CT value of over two cycles more than the standards, the sample is reanalyzed.

89.2.2.3.4 Degenerate PCR and Sequencing

The use of degenerate PCR followed by sequencing of a fragment of the DNA polymerase gene [45] is useful for evaluating virus isolates that show unexpected host range, culture characteristics, antigen profile, or PCR results. The concept is based on the use of two regions in the DNA polymerase gene that are conserved. The upstream primer is 5′cggaattctaGAYTTYGCNWSNYTNTAYCC3′ with the upper case being specific for the DNA pol target (Y = C or T, N = ACG or T, W = C or G) and the lower primer being 5′cccgaattcagatcTCNGTRTCNCCRTA3′ (R = A or G). Furthermore, the use of DNase to remove host sequences from virion pellets can be used before DNA purification to increase the effectiveness of the degenerate PCR. This method is most effective on virus from cell cultures but also can be used on productively infected tissues.

89.2.2.3.4.1 Procedure

1. Disrupt the cells/tissue to release the virions. The virus from infected cell cultures in 25 cm^2 flasks is released from the cells by three serial freeze–thaw cycles or by sonication.
2. Pellet the cellular debris by centrifugation at $1000 \times g$ for 5 min.
3. Filter the supernatant through a 0.45 μm syringe filter.
4. Concentrate the virus from the filtrate by centrifugation at $21,000 \times g$ for 30 min.
5. Resuspend the pellet in 80 μL of water, add 20 μL of 10× buffer and 100 μL of RQ1 DNase (buffer and enzyme from Promega), and incubate at 37°C for 2 h. This is followed by the addition of 20 μL of stop buffer (20 mM EGTA, pH 8.0) and a 10 min incubation at 65°C to inactivate the DNase.
6. Extract the DNA using the Puregene genomic DNA isolation system (Qiagen, Maryland, Delaware, United States).
7. Run degenerate PCR: PCR consists of approximately 100 ng of template DNA, 20 pmole of each primer, 4 μL 10 mM dNTP, 5 μL 10× buffer, and 2.5 U Taq polymerase mix (Fisher Scientific) in 50 μL reactions. The reaction conditions are 93°C, 1 min for 1 cycle, followed by 93°C, 30 s; 45°C, 2 min; and 72°C, 3 min for 35 cycles, and then a single cycle at 72°C for 4 min. Product is then evaluated by electrophoresis on 1.5% agarose gels followed by staining with ethidium bromide or a similar stain and UV transillumination. Bands of interest are then excised from the gel, and the DNA was recovered using GenElute Agarose Spin Columns (Supelco, Bellefonte, PA).
8. Clone predominant bands of the appropriate size using the TOPO TA Cloning Kit (Invitrogen) or a similar rapid cloning system. The sizes for large DNA viruses range from 450 to 800 bp; IcHV1 produces a band of 465 bp. Selected candidate clones are evaluated for a DNA insert of the appropriate size using colony PCR. Then plasmid is purified for sequencing from 1 mL cultures using the QIAQUICK Plasmid Purification Kit (Qiagen).
9. Sequence the products and use BLASTx to compare the translated sequence to deduce amino acid sequences in GenBank.

Variations that are helpful are (1) to use larger amounts of cells or tissues in samples that are suspected of having low numbers of virions and (2) to run a negative control of noninfected tissue when multiple weak bands are produced to distinguish cellular products from virus product candidates.

89.3 CONCLUSION AND FUTURE PERSPECTIVES

Ictalurid herpesviruses are important fish pathogens that have impacted aquaculture. As the *Ictalurivirus* genus is smaller and simpler than other alloherpesviruses, it offers a useful model for evaluating the molecular basis of how an alloherpesvirus infects, persists, and causes disease in fish. The alloherpesviruses studied to date are very host specific. The *Ictalurivirus* genus has been detected causing disease in North American catfish and sturgeons, which are genetically distant fish hosts. Therefore, in the future, as more types of fish are cultured at high densities, the recognitions of more members of this genus as pathogens are likely. Detecting new members for this genus using traditional cell culture may be difficult if appropriate cell lines are not available. Therefore, the use of molecular methods such as degenerate primer PCR or next-generation sequencing will likely facilitate the discovery of new *Ictalurid* viruses and allow the establishment of efficient, specific molecular diagnostic assays.

Many regional and federal agencies restrict the importation of catfish that may harbor IcHV1. However, most commercially produced populations of channel catfish carry the virus, and the current diagnostic tests are not sensitive enough to detect all carriers. This virus is efficiently vertically transmitted from parent to offspring. Also the disease outbreaks only occur in very young fish, held at high density at high temperatures. Therefore, many populations of channel catfish are endemic with the pathogen even if they have no history of CCVD. The PCR assays in use are very sensitive, near the one copy threshold per reaction. Therefore, the development of an assay that could evaluate larger amounts of host tissue would be beneficial for screening brood stock. Currently used methods are reliable enough to detect most carrier populations, but higher sensitivity is needed for screening individuals with confidence.

REFERENCES

1. Wolf, K. and Darlington, R. W. 1971. Channel catfish virus: A new herpesvirus of Ictalurid fish. *J. Virol.* 8(4):525–533.

2. Alborali, L., Bovo, G., Lavazza, A., Cappellaro, H., and Guadagnini, P. F. 1996. Isolation of an herpesvirus in breeding catfish (*Ictalurus melas*). *Bull. Eur. Assoc. Fish Pathol.* 16(4):134–137.

3. Hedrick, R. P., McDowell, T. S., Gilad, O., Adkison, M., and Bovo, G. 2003. Systemic herpes-like virus in catfish Ictalurus melas (Italy) differs from IcHV1 (North America). *Dis. Aquat. Organ.* 55(2):85–92.

4. ICTV. 2014. ICTV Master Species List 2013 v2. http://talk.ictvonline.org/files/ictv_documents/m/msl/4911.aspx : International Committee on Taxonomy of Viruses (ICTV). Accessed on June 30, 2014.

5. Watson, L. R., Yun, S. C., Groff, J. M., and Hedrick, R. P. 1995. Characteristics and pathogenicity of a novel herpesvirus isolated from adult and subadult white sturgeon Acipenser transmontanus. *Dis. Aquat. Organ.* 22:199–210.

6. Kurobe, T., Kelley, G. O., Waltzek, T. B., and Hedrick, R. P. 2008. Revised phylogenetic relationships among herpesviruses isolated from sturgeons. *J. Aquat. Anim. Health* 20:96–102.

7. Doszpoly, A., Benko, M., Bovo, G., Lapatra, S. E., and Harrach, B. 2011. Comparative analysis of a conserved gene block from the genome of the members of the genus Ictalurivirus. *Intervirology* 54(5):282–289.

8. Davison, A. J. 1992. Channel catfish virus: A new type of herpesvirus. *Virology* 186:9–14.

9. Robin, J. and Rodrigue, A. 1978. Purification of channel catfish virus, a fish herpesvirus. *Can. J. Microbiol.* 24:1335–1338.

10. Mettenleiter, T. C., Klupp, B. G., and Granzow, H. 2009. Herpesvirus assembly: An update. *Virus Res.* 143(2):222–234.

11. Dixon, R. A. F. and Farber, F. E. 1980. Channel catfish virus: Physicochemical properties of the viral genome and identification of viral polypeptides. *Virology* 103:267–278.

12. Chousterman, S., Lacasa, M., and Sheldrick, P. 1979. Physical map of the channel catfish virus genome: Location of sites for restriction endonucleases *Eco* RI, *Hind* III, *Hpa* I, and *Xba* I. *J. Virol.* 31(1):73–85.

13. Colyer, T. E., Bowser, P. R., Doyle, J., and Boyle, J. A. 1986. Channel catfish virus: Use of nucleic acids in studying viral relationships. *Am. J. Vet. Res.* 47(9):2007–2011.

14. Cebrian, J., Bucchini, D., and Sheldrick, P. 1983. "Endless" Viral DNA in cells infected with channel catfish virus. *J. Virol.* 46(2):405–412.

15. Hanson, L. A., Kousoulas, K. G., and Thune, R. L. 1994. Channel catfish herpesvirus (CCV) encodes a functional thymidine kinase gene: Elucidation of a point mutation that confers resistance to Ara-T. *Virology* 202:659–664.

16. Kunec, D., Hanson, L. A., van Haren, S., Nieuwenhuizen, I. F., and Burgess, S. C. 2008. An over-lapping bacterial artificial chromosome system that generates vectorless progeny for Channel catfish herpesvirus. *J. Virol.* 82(8):3872–3881.

17. Davison, A. J. and Davison, M. D. 1995. Identification of structural proteins of channel catfish virus by mass spectrometry. *Virology* 206(2):1035–1043.

18. Booy, F. P., Trus, B. L., Davison, A. J., and Steven, A. C. 1996. The capsid architecture of channel catfish virus, an evolutionarily distant herpesvirus, is largely conserved in the absence of discernible sequence homology with herpes simplex virus. *Virology* 215:134–141.

19. Vanderheijden, N., Alard, P., Lecomte, C., and Martial, J. A. 1996. The attenuated V60 strain of channel catfish virus possesses a deletion in ORF50 coding for a potentially secreted glycoprotein. *Virology* 218:422–426.

20. Vanderheijden, N., Hanson, L. A., Thiry, E., and Martial, J. A. 1999. Channel catfish virus gene 50 encodes a secreted, mucin-like glycoprotein. *Virology* 257:220–227.

21. Liu, Y. et al. 2012. Identification of envelope protein orf10 of channel catfish herpesvirus. *Can. J. Microbiol.* 58:271–277.

22. Hanson, L. A. and Thune, R. L. 1993. Characterization of thymidine kinase encoded by channel catfish virus. *J. Aquat. Anim. Health* 5:199–204.

23. Harrison, P. T., Thompson, R., and Davison, A. J. 1991. Evolution of herpesvirus thymidine kinases from cellular deoxycytidine kinase. *J. Gen. Virol.* 72:2583–2586.

24. Kunec, D., Nanduri, B., and Burgess, S. C. 2009. Experimental annotation of channel catfish virus by probabilistic proteogenomic mapping. *Proteomics* 9(10):2634–2647.

25. Huang, S. and Hanson, L. A. 1998. Temporal gene regulation of the channel catfish virus (*Ictalurid herpesvirus* 1). *J. Virol.* 72:1910–1917.

26. Silverstein, P. S., Bird, R. C., Santen, V. L. V., and Nusbaum, K. E. 1995. Immediate-early transcription from the channel catfish virus genome: Characterization of two immediate early transcripts. *J. Virol.* 69:3161–3166.

27. Silverstein, P. S., van Santen, V. L., Nusbaum, K. E., and Bird, R. C. 1998. Expression kinetics and mapping of the thymidine kinase transcript and an immediate-early transcript from channel catfish virus. *J. Virol.* 72:3900–3906.

28. Stingley, R. L. and Gray, W. L. 2000. Transcriptional regulation of the channel catfish virus genome direct repeat region. *J. Gen. Virol.* 81:2005–2010.

29. Plumb, J. A. 1978. Epizootiology of channel catfish virus disease. *Mar. Fish. Rev.* 3:26–29.

30. Hedrick, R. P., Groff, J. M., and McDowell, T. 1987. Response of adult channel catfish to waterborne exposures of channel catfish virus. *Prog. Fish Cult.* 49:181–187.

31. Hanson, L. A., Rudis, M. R., and Petrie-Hanson, L. 2004. Susceptibility of channel catfish fry to channel catfish virus (CCV) challenge increases with age. *Dis. Aquat. Organ.* 62:27–34.

32. Plumb, J. A. 1973. Effects of temperature on mortality of fingerling channel catfish (*Ictalurus punctatus*) Experimentally infected with channel catfish virus. *J. Fish. Res. Board Can.* 30(4):568–570.

33. Thompson, D. J., Khoo, L. H., Wise, D. J., and Hanson, L. A. 2005. Evaluation of channel catfish virus latency on fingerling production farms in Mississippi. *J. Aquat. Anim. Health* 17:211–215.

34. Plumb, J. A., Gaines, J. L., Mora, E. C., and Bradley, G. G. 1974. Histopathology and electron microscopy of channel catfish virus in infected channel catfish, *Ictalurus punctatus* (Rafinesque). *J. Fish Biol.* 6:661–664.

35. Wolf, K., Herman, R. L., and Carlson, C. P. 1972. Fish viruses: Histopathologic changes associated with experimental channel catfish virus disease. *J. Fish. Res. Board Can.* 29:149–150.

36. Bowser, P. R., Munson, A. D., Jarboe, H. H., Francis-Floyd, R., and Waterstrat, P. R. 1985. Isolation of channel catfish virus from channel catfish *Ictalurus punctatus* (Rafinesque), broodstock. *J. Fish Dis.* 8:557–561.

37. Nusbaum, K. E. and Grizzle, J. M. 1987. Uptake of channel catfish virus from water by channel catfish and bluegills. *Am. J. Vet. Res.* 48:375–377.

38. Kancharla, S. R. and Hanson, L. A. 1996. Production and shedding of channel catfish virus (CCV) and thymidine kinase negative CCV in immersion exposed channel catfish fingerlings. *Dis. Aquat. Organ.* 27:25–34.

39. Bowser, P. R. and Plumb, J. A. 1980. New cell line. Fish cell lines: Establishment of a line from ovaries of channel catfish. *In Vitro* 16:365–368.

40. Bowser, P. R. and Plumb, J. A. 1980. Channel catfish virus: Comparative replication and sensitivity of cell lines from channel catfish ovary and the brown bullhead. *J. Wildl. Dis.* 16:451–454.

41. Bowser, P. R. and Munson, A. D. 1986. Seasonal variation in channel catfish virus antibody titers in adult channel catfish. *Prog. Fish Cult.* 48:198–199.

42. Boyle, J. and Blackwell, J. 1991. Use of polymerase chain reaction to detect latent channel catfish virus. *Am. J. Vet. Res.* 52:1965–1968.

43. Baek, Y.-S. and Boyle, J. A. 1996. Detection of channel catfish virus in adult channel catfish by use of nested polymerase chain reaction. *J. Aquat. Anim. Health* 8:97–103.

44. Gray, W. L., Williams, R. J., Jordan, R. L., and Griffin, B. R. 1999. Detection of channel catfish virus DNA in latently infected catfish. *J. Gen. Virol.* 80:1817–1822.

45. Hanson, L. A., Rudis, M. R., Vasquez-Lee, M., and Montgomery, R. D. 2006. A broadly applicable method to characterize large DNA viruses and adenoviruses based on the DNA polymerase gene. *Virol. J.* 3:28.

46. Goodwin, A. E. and Marecaux, E. 2010. Validation of a qPCR assay for the detection of Ictalurid herpesvirus-2 (IcHV-2) in fish tissues and cell culture supernatants. *J. Fish Dis.* 33:341–346.

47. Arkush, K. D., McNeill, C., and Hedrick, R. P. 1992. Production and characterization of monoclonal antibodies against channel catfish virus. *J. Aquat. Anim. Health* 4:81–89.

48. Reed, L. J. and Muench, H. 1938. A simple method of estimating fifty per cent endpoints. *Am. J. Epidemiol.* 27:493–497.

49. Karber, G. 1931. Beitrag zur kollektiven Behandlung pharmakologischer Reihenversuche. *Arch. Exp. Path. Pharm.* 162:480–483.

50. Latorra, D., Arar, K., and Hurley, J. M. 2003. Design considerations and effects of LNA in PCR primers. *Mol. Cell. Probes* 17:253–259.

90 Abalone Herpesvirus

Mark St. J. Crane, Kenneth A. McColl, Jeff A. Cowley, Kevin Ellard, Keith W. Savin, Serge Corbeil, Nicholas J.G. Moody, Mark Fegan, and Simone Warner

CONTENTS

90.1 Introduction 807
    90.1.1 Classification..808
    90.1.2 Morphology, Biology, and Epidemiology 809
    90.1.3 Clinical Features and Pathogenesis 809
    90.1.4 Diagnosis 809
        90.1.4.1 Conventional Techniques...809
        90.1.4.2 Molecular Techniques..810
90.2 Methods 810
    90.2.1 Sample Preparation..810
    90.2.2 Detection Procedures...811
        90.2.2.1 TaqMan qPCR..811
        90.2.2.2 Conventional PCR..812
        90.2.2.3 *In Situ* Hybridization..812
90.3 Conclusion and Future Perspectives 813
References...814

90.1 INTRODUCTION

Abalone were first farmed in Japan and China in the 1950s. However, it took until the 1990s before aquaculture of various abalone species began to expand significantly in various countries, including the United States, Mexico, Chile, South Africa, Australia, New Zealand, Japan, China, Taiwan, Korea, Ireland, and Iceland [1]. While the fishery in Tasmania in southern Australia is the predominant world supplier of wild abalone (approx. 2500 tons per annum), China is the largest producer of farmed abalone, with annual production of mostly Taiwan abalone (jiukong; *Haliotis diversicolor supertexta*) in the order of 50,000 tons per annum. As abalone aquaculture has expanded, various parasitic (e.g., perkinsosis [2]), bacterial (e.g., vibriosis [3], withering syndrome [4]), and viral diseases [5–7] have emerged. Of the viral diseases, the most recent to emerge in 2003 in Taiwan was a herpes-like virus associated with mass mortality of farmed *H. diversicolor supertexta* [8]. Diseased abalone displayed necrotic lesions with hemocytic infiltrations in nerve tissues, and the presence of a viral infection was confirmed by electron microscopic examination. Virion ultrastructural and morphogenesis were indicative of a herpesvirus. In 2005, mass mortalities also occurred in greenlip (*Haliotis laevigata*) and blacklip (*Haliotis rubra*) abalone being farmed in Victoria, Australia [9]. As observed in the *H. diversicolor supertexta* [8], diseased abalone exhibited ganglioneuritis pathology, and electron microscopic examination of virions purified from nerve tissues confirmed the presence of a herpes-like virus [10]. Subsequently, in 2008,

abalone herpesvirus (AbHV) was discovered to be the cause of mortality of wild abalone being held in a live-holding system at a processing plant in Tasmania, Australia [11].

Upon DNA genome sequence information being acquired for the Victorian strain of AbHV (AbHV-Vic), various deduced protein sequences were identified that were most similar (amino acid similarity ranging from 19% to 53%) to homologs deduced from the ostreid herpesvirus 1 (OsHV1) genome sequence [12,13], revealing the abalone and oyster herpesviruses to be distantly related [14]. Based on this genetic relatedness, it has been proposed that AbHV be classified alongside OsHV-1 in the family *Malacoherpesviridae*, order *Herpesvirales* [15]. Comparison of three genomic regions of AbHV-Vic and AbHV-Taiwan has confirmed their nucleotide homology of between 92.4% and 96.6%. Complete genome sequences assembled recently indicate that a number of AbHV genotypic variants occur in wild abalone populations inhabiting coastal regions around Tasmania [16].

In Australia, the known host range for AbHV includes *H. laevigata*, *H. rubra*, and the hybrids of these species [9]. Clinical signs consistent with abalone viral ganglioneuritis (AVG), the disease caused by AbHV infection, have not been reported in other molluscan species inhabiting areas where AVG has decimated wild abalone populations. In Taiwan, disease was also not reported among Japanese black abalone (*H. discus*) cohabitating farms at which ganglioneuritis and mortality of *H. diversicolor supertexta* occurred due to acute herpes virus infection [8].

90.1.1   Classification

The International Committee on Taxonomy of Viruses (http://www.ictvonline.org/virusTaxonomy.asp) has formally named AbHV as Haliotid herpesvirus 1 (based on the genus of the host and for consistency with OsHV1), as the type species of a newly created genus, *Aurivirus* (based on the abalone shell shape), which together with the genus *Ostreavirus* (type species OsHV-1) comprises the family *Malacoherpesviridae* within the order *Herpesvirales*. Genome sequence analysis of

Chlamys acute necrobiotic virus, a herpesvirus found recently to infect the Chinese scallop, *Chlamys farreri*, has indicated it to be an OsHV-1 variant [17]. A phylogenetic tree illustrating the evolutionary relatedness of viruses classified within each genus of the *Herpesviridae*, *Malacoherpesviridae*, and *Alloherpesviridae* families comprising the *Herpesvirales* determined for concatenated amino acid sequences of the DNA polymerase and DNA packaging terminase is shown in Figure 90.1.

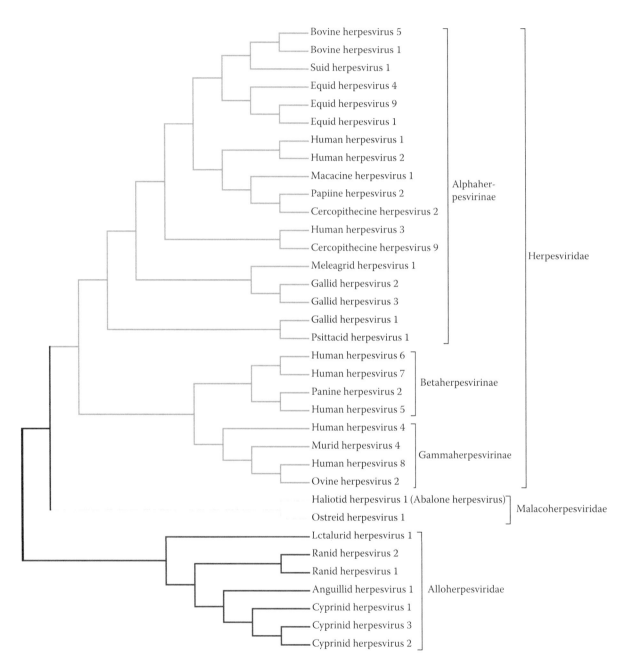

FIGURE 90.1   Molecular phylogenetic analysis of the *Herpesvirales*. The DNA polymerase and terminase full-length protein sequences were concatenated, and then Mega6 [18] was used to align them with muscle and infer the evolutionary history using the minimum evolution method (ME) [19]. The bootstrap consensus tree was inferred from 500 replicates [20]. The evolutionary distances were computed using the number of differences method [21]. The ME tree was searched using the close-neighbor-interchange algorithm [21] at a search level of 1. The neighbor-joining algorithm [22] was used to generate the initial tree.

90.1.2 Morphology, Biology, and Epidemiology

Morphology and physicochemical properties: Transmission electron microscopy of negatively stained AbHV particles purified from AVG-affected abalone demonstrated virions comprising an icosahedral capsid (92–109 nm diameter) surrounded by an envelope (~150 nm diameter) decorated with numerous projecting spikes [10]. Particles similar in morphology have been detected in ultrathin sections of cells within the pleuropedal ganglion of AbHV-infected abalone [23]. When ultracentrifuged to equilibrium in either caesium chloride or isopycnic potassium tartrate density gradients, AbHV particles have a buoyant density of 1.17–1.18 g/mL. The morphology and buoyant density of AbHV particles are typical of those described for other members of the *Herpesviridae* family [24].

Biology and epidemiology: In AVG outbreaks occurring in abalone cultured in Taiwan, in 16°C–19°C water, mortalities began to accumulate 3 days after the appearance of gross signs of disease, and both adults and juveniles suffered 70%–80% cumulative mortality [8]. In Australia, AVG outbreaks in both farmed and wild abalone have similarly been associated with rapid disease onset in all age classes and up to 90% cumulative mortality [25]. In experimental infections undertaken in Taiwan (water temperature 17°C–20°C) and Australia (water temperature 15°C–18°C), gross signs and pathology consistent with AVG have been reproduced typified by a 1–3 day preclinical period and 100% mortality within 5–7 days postexposure [8,23,26].

Horizontal transmission has been demonstrated experimentally either by exposing healthy abalone to water containing diseased abalone using tank systems in which direct contact between the diseased and healthy abalone could not occur or by placing healthy abalone in water that previously contained diseased abalone [8,23,26]. An important epidemiological finding from one study [23] was that AbHV DNA and histological lesions could be demonstrated in infected abalone before the onset of clinical signs of disease. This suggests that preclinical excretion of virus may occur in infected abalone, a finding of great importance in developing control programs for this disease.

90.1.3 Clinical Features and Pathogenesis

In Australia, and similarly in Taiwan, AVG outbreaks have been typified by rapidly increasing mortality and abalone dying within a day of displaying gross signs of disease (e.g., curling of the foot). There have been limited studies on the pathogenesis of the disease. Whereas Hooper et al. [9,27] simply reported lesions in dead and moribund abalone, Chang et al. [8], Corbeil et al. [23], and Crane et al. [28] used bath immersion and time-course studies to gain some preliminary insights into pathogenesis. Based on all of these studies, lesions in infected abalone are largely restricted to neural tissues, possibly occurring first in those around the buccal cavity. In particular, the cerebral, pleuropedal, and circumoesophageal ganglia may be affected in addition to the pedal

nerve cords, the transverse commissures linking these nerve cords, and the associated peripheral nerves. The constant histological lesion is necrosis and/or apoptosis of neurons in the gray matter of the ganglia and nerve cords. This is associated with edema and increased cellularity in affected areas possibly due to proliferation of glial cells or infiltration by hemocytes. Observation of a leucopenia in affected abalone [27] may support sequestration of hemocytes into the neural tissue. Increased cellularity of the white matter of the ganglia and cords is also observed, together with disruption of the perineural sheath investing these sites. As the clinical course continues, the severity of these lesions increases until there is almost complete effacement of normal neural architecture in the ganglia and nerve cords [23]. It has been reported that, occasionally, intranuclear inclusion bodies may be seen in neurons [9].

90.1.4 Diagnosis

AVG caused by AbHV infection can be diagnosed by a combination of methods. Typical disease gross signs associated with rapidly increasing mortality are suggestive, but not pathognomonic, of AVG. Biopsies of nerve tissue can allow AbHV DNA to be detected rapidly by conventional or real-time polymerase chain reaction (PCR) methods [25]. Tissues preserved for histology, *in situ* hybridization (ISH), and electron microscopy can assist in confirming AVG, particularly in circumstances when the PCR tests might not detect a particular AbHV genotype. In the absence of cell lines suitable for virus isolation and amplification, PCR methods provide the only current means of reliably detecting subclinical AbHV infection [25].

90.1.4.1 Conventional Techniques

Gross signs of AVG in abalone resulting from AbHV infection are characterized by irregular elevation of the foot, swollen and protruding mouth parts, eversion of the radula (Figure 90.2), restricted movement of the pedal muscle, excessive mucus production, reduced pedal adhesion to substrates, and an inability to right themselves if turned onto their back. AVG-affected wild blacklip abalone examined from disease episodes at Tasmanian processing plants have also exhibited "hard foot" or tetany, excessive mucus production, abnormal spawning, and "bloating" [11]. Similar gross signs have been reported in herpesvirus-infected abalone in Taiwan [8]. However, compared to outbreaks of AVG in farmed or wild abalone in Victoria, the onset of gross signs and mortality was slower; morbidity and mortality rates were lower and appear to be influenced by environmental conditions under which abalone were held. In some cases, it was noted that blacklip abalone infected with the virus were apparently normal, suggesting that subclinical carriage can occur [11].

Moribund abalone exhibiting loose attachment to the substrate due to pedal muscle weakness or abnormalities should be sampled for AbHV diagnosis. Tissue samples containing neural tissue (see Figure 90.3 for location) should be fixed for histological examination.

(a)

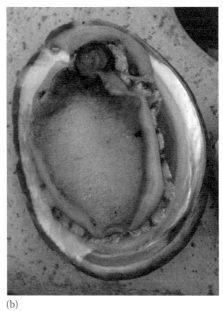
(b)

FIGURE 90.2 Photographs of a healthy greenlip abalone (a) and a greenlip abalone affected by abalone viral ganglioneuritis (b). Note the curling of the foot, the swollen and protruding mouth parts, and the eversion of the radula in the abalone with AVG. (Photographs provided by Kevin Ellard, DPIPWE, and Abalone Farms Australia.)

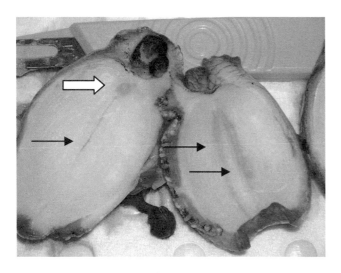

FIGURE 90.3 Photograph of a dissected abalone. A longitudinal cut through the foot muscle reveals the pleuropedal ganglion (white arrow) and nerve cords (black arrows). (Photograph provided by CSIRO AAHL.)

As detected by histopathology, AVG is characterized by increased necrosis/apoptosis of neurons in the gray matter of ganglia and nerve cords with concomitant proliferation of glial cells and infiltration by hemocytes [9,23].

While tissues can also be fixed for electron microscopy, viral particles often occur in relatively low numbers difficult to detect even in cases of acute disease [23]. The intranuclear location of AbHV particles and their ultrastructure [23] are the characteristic of members of the *Herpesviridae* family [24].

90.1.4.2 Molecular Techniques

The complete genome sequence of the AbHV-Vic strain has been reported [14] and used to develop conventional and real-time PCR tests [25] as well as ISH probes [23,29] to detect, quantify, and localize AbHV infection in abalone with either clinical or subclinical infections [30]. Conventional PCR tests targeted to variable genome sequences have also been established to identify AbHV genotypic variants [16]. The ISH test is particularly useful in tracking disease progressing in target organs and in associating AbHV infection with tissue lesions evident by histopathological examination.

90.2 METHODS

90.2.1 SAMPLE PREPARATION

Tissues of live moribund individuals should be sampled when abalone exhibit gross signs of AVG (i.e., abnormal behavior, loose attachment to the substrate, sudden onset of mortality). For screening of healthy farmed abalone, representatives of all class ages should be sampled. For histopathology, dissected tissue pieces (<1 cm^3) containing nerves should be fixed for 24 h in 10% formalin prepared in filtered seawater at a tissue: fixative ratio of at least 1:10, processed for embedding in paraffin wax and tissue sections cut and stained with hematoxylin and eosin using standard methods. For optimal ISH detection of AbHV DNA, tissues should not be fixed for longer than 24 h in formalin before being transferred to alcohol (where they may be held for extended periods).

For electron microscopy, small pieces (<0.1 cm^3) of fresh tissue containing nerves should be fixed in 2.5% glutaraldehyde prepared in filtered seawater at a tissue: fixative ratio of at least 1:10. Following fixation, ultrathin tissue sections can be cut and negatively stained for electron microscopy using standard methods.

For PCR, dissected tissue pieces (<1 cm³) containing nerves should be preserved in 95% ethanol at a tissue: preservative ratio of at least 1:10. As PCR is extremely sensitive, in situations where sample cross contamination could compromise data interpretation, this should be prevented by either using disposable tissue dissection instruments or decontaminating instruments thoroughly between individuals.

If fixatives are not available, either whole abalone or dissected tissue pieces should be kept on ice or refrigerated and processed at a diagnostic laboratory within 24 h of collection. While freezing of diagnostic specimens is acceptable for PCR analysis, frozen tissue is not suitable for processing for histopathology or electron microscopy.

For PCR, DNA is extracted from ethanol-preserved tissue ideally trimmed to primarily contain pleuropedal ganglion and/or pedal nerve cords using the QIAamp® DNA Mini Kit (QIAGEN) or equivalent. Briefly, ~20 mg tissue is digested overnight and DNA bound to a column is eluted in 100 μL Buffer AE (~100 ng/μL DNA concentration) provided in the kit. For larger sample numbers, 50 μL tissue digests can be used to extract total nucleic acid (DNA/RNA) using the MagMAX-96 Viral RNA Isolation Kit and an AB MagMAX Express-96 Magnetic Particle Processor (Applied Biosystems) or equivalent system. DNA should be stored at −80°C until further processing

90.2.2 Detection Procedures

AbHV has been detected in abalone in Taiwan and in Victoria and Tasmania in Australia, with genotypic variants identified in abalone affected by AVG in Tasmania [16]. Of various TaqMan real-time quantitative PCR (qPCR) tests that have been developed, two tests targeted to ORF49 and ORF66 have proved useful for detecting AbHV genomic variants when used in parallel [31]. Plasmid DNA is available for use as a positive control for each test and negative controls should include water and DNA extracted from naïve abalone. Negative controls should not demonstrate any amplification.

90.2.2.1 TaqMan qPCR

Primer and probe sequences used in the AbHV-ORF49 and AbHV-ORF66 TaqMan real-time qPCR tests designed to be multiplexed with an 18S rDNA TaqMan qPCR test are described in Table 90.1. The 18S rDNA PCR test (Applied Biosystems) is used to validate nucleic acid extraction and integrity and the absence of PCR inhibitors. qPCR tests are performed using 96-well PCR plates with diagnostic samples and positive and negative controls being tested in either duplicate or triplicate.

90.2.2.1.1 Procedure

1. PCR primer, probe, and reagent concentrations and volumes used in the PCR tests are detailed in Table 90.2. Master mixes are prepared as for the real-time PCR tests.

TABLE 90.1

Primer and Probe Sequences Used in TaqMan qPCR Tests for AbHV and 18S rDNA

Test	Primers/Probe	Sequence
AbHV-ORF49	ORF49F1	5′-AACCCACACCCAATTTTTGA-3′
	ORF49R1	5′-CCCAAGGCAAGTTTGTTGTT-3′
	49Prb1	5′-6FAM-CCGCTTT CAATCTGATCCGTGG-TAMRA-3′
AbHV-ORF66	ORF66F1	5′-TCCCGGACACCAGTAAGAAC-3′
	ORF66R1	5′-CAAGGCTGCTATGCGTATGA-3′
	66Prb1	5′-6FAM-TGGCCG TCGAGATGTCCATG-TAMRA-3′
18S rDNA	18S Forward	5′-CGGCTACCACATCCAAGGAA-3′
	18S Reverse	5′-GCTGGAATTACCGCGGCT-3′
	18S VIC probe	5′-VIC-TGCTGGCA CCAGACTTGCCCTC-TAMRA-3′

TABLE 90.2

Reagent Volumes Required per Sample in TaqMan Real-Time qPCR Tests Multiplexed with a TaqMan Real-Time 18S rRNA qPCR

Reagent	Stock Conc. (μM)	Volume per Reaction (μL)	Final Conc. (μM)
Deionized water	—	8.46	—
TaqMan Universal PCR Master Mix	—	12.5	—
AbHV F1 primer	20	0.38	0.3
AbHV R1 primer	20	0.38	0.3
AbHV Prb1	5	0.51	0.1
18S Forward primer	20	0.13	0.1
18S Reverse primer	20	0.13	0.1
18S VIC probe	5	0.51	0.1
Total volume		23.0	

2. Briefly vortex and pulse-centrifuge the master mix, dispense 23 μL into each 200 μL PCR tube or plate well, and transfer the tubes/plate to a refrigerator in a dedicated template addition laboratory.

3. Add 2 μL template DNA or water (no template negative control) to each well first and then 2 μL/well of quantified positive control plasmid DNA (20,000 and 200 DNA copies/reaction).

4. Pulse-centrifuge and transfer tubes/plate to an ABI 7500 thermal cycler or equivalent in a dedicated real-time PCR laboratory.

5. Thermal cycling conditions are 95°C for 59 s followed by 50 cycles of 95°C for 3 s and either 62°C for 30 s (AbHV-ORF49 PCR) or 60°C for 30 s (AbHV-ORF66 PCR).

6. At the completion of PCR, fluorescence data are analyzed at threshold value of 0.1.

Note: For samples with characteristic amplification curves and C_T values to be considered positive for AbHV DNA, PCR test results must be validated by

1. There being no evidence of specific amplification curves with the no template and/or negative sample controls
2. Each positive control yielding a specific amplification curve and mean C_T value within the acceptable QC (quality control) range
3. The sample generating specific amplification curve for 18S rDNA

90.2.2.2   Conventional PCR

Real-time PCR results can be confirmed by conventional PCR (AbHV1617 test) utilizing the primers AbHV-16 (5′-GGCTCGTTCGGTCGTAGAATG-3′) and AbHV-17 (5′-TC AGCGTGTACAGATCCATGTC-3′).

90.2.2.2.1   Procedure

PCR primer and reagent concentrations and volumes used in the AbHV1617 PCR test are detailed in Table 90.3.

1. Briefly vortex and pulse-centrifuge the master mix, dispense 23 μL into each 200 μL PCR tube or plate well, and transfer the tubes/plate to a refrigerator in a dedicated template addition laboratory.
2. Add 2 μL template to the appropriate wells. Include the appropriate positive controls and no template control.
3. Pulse-centrifuge briefly and transfer tubes/plate to a thermal cycler in a dedicated PCR laboratory.
4. Thermal cycling conditions are 95°C for 15 min followed by 40 cycles of 94°C for 30s, 52°C for 30 s, and 74°C for 45 s and a final extension at 72°C for 7 min and 4°C hold.
5. Amplified DNA is detected using a blue-light trans-illuminator following electrophoresis in a 1.5% agarose-TAE gel containing SYBR Safe dye or equivalent.
6. DNA sequence analysis is recommended to confirm that AbHV DNA was amplified.

TABLE 90.3
Reagent Volumes Required for per Sample in the AbHV1671 PCR Test

Reagent	Stock Conc. (μM)	Volume per Reaction (μL)	Final Conc. (μM)
Deionized water	—	9.50	—
HotStar Taq Master Mix	—	12.50	—
AbHV Forward primer	18	0.50	0.36
AbHV Reverse primer	18	0.50	0.36
Total volume		23.0	

Note: For samples generating DNA bands to be considered positive for AbHV DNA, PCR test results must be validated by

1. There being no evidence of specific DNA products being amplified with the no template and/or negative sample controls.
2. Each plasmid positive control DNA yielding a DNA product 522–588 bp in length depending on the AbHV genotype used to prepare the plasmid.
3. Test samples generating amplicons in the 522–588 bp size range depending on the AbHV genotypes detected, with DNA authenticity confirmed by sequence analysis of the purified amplicon.

90.2.2.3   *In Situ* Hybridization

The ISH procedure described here uses a digoxigenin (DIG)-labeled DNA probe to detect AbHV DNA in formalin-fixed, paraffin-embedded tissue sections. The following reagents are required:

20 × standard saline citrate (SSC) pH 7.0 (store at room temperature) (made up of 175.32 g/L NaCl, 88.23 g/L sodium citrate)

100 × Denhardt's solution (store at –20°C) (made up of 2% bovine serum albumin (Fraction V), 2% Ficoll 400, 2% polyvinylpyrrolidone)

Hybridization buffer (store at –20°C) (made up of 25 mL formamide, 10 mL 20 × SSC, 2.5 mL 100 × Denhardt's solution, 10 mL 50% dextran sulfate in distilled water, 500 μL 10 mg/mL herring sperm DNA made to 50 mL with Milli-Q water)

10 × Tris-buffered saline (TBS) (store at room temperature) (made up of 23.6 g/L Tris base, 127 g/L Tris-HCl, 87.66 g/L NaCl [pH 7.5])

90.2.2.3.1   Procedure

1. The DIG-labeled probe is prepared by PCR amplification of AbHV plasmid DNA or DNA extracted from AbHV-injected abalone using the modified ORF66 primers AbHVORF66f1 (5′-TCCCGGACACCAGTAAGAAC-3′) and AbH VORF66r2 (5′-GCCGGTCTTTGAAGGATCTA-3′) and reagents in the PCR DIG Probe Synthesis Kit (Roche Cat. No. 11-636-090-910) according to the manufacturer's instructions.
2. Thermal cycling conditions 95°C for 5 min followed by 30 cycles of 95°C for 30 s, 57°C for 30 s, and 72°C for 60 s followed by a final extension at 72°C for 10 min, with 4°C hold.
3. Place tissue sections (3 μm thickness) cut from paraffin block onto Superfrost Plus slides (Menzel Catalogue No. SF41296SP) and allow to dry.
4. Heat sections at 65°C for 30 min and deparaffinize in two changes of xylene.
5. Rehydrate tissue sections by placing slides in absolute ethanol (2 min) followed by 90% ethanol (2 min) and 70% ethanol (2 min) and then into distilled water.

6. Place slides in 0.2 N HCl for 20 min and rinse in distilled water for 5–10 min.

7. Apply 50–100 µL of 100 µg/mL proteinase K in TBS and incubate at 37°C for 30 min.

8. Rinse with 0.2% glycine for 2 min and wash in running water for 10 min.

9. Dehydrate sample in 70% ethanol (2 min), 90% ethanol (2 min), and 100% ethanol (2 min) and allow slides to air-dry.

10. Make 100 µL hybridization solution per tissue section containing ~5 ng/L DIG-labeled DNA probe.

11. Heat the hybridization solution to 95°C–100°C for 5 min to denature the DNA and place on ice until used.

12. Apply sufficient hybridization solution to cover the section (~50 µL) and gently cover with a glass coverslip.

13. Using either a PCR heating block or a specialist hybridization block, heat the slides to 95°C for 5 min to denature the nucleic acid in the tissue specimen.

14. Place the slides in a humidified chamber preheated to 37°C and incubate at 37°C overnight (12–16 h).

15. Remove coverslips by immersing slides in $2 \times$ SSC at room temperature, place slides in a rack, and immerse in $2 \times$ SSC at room temperature.

16. Wash, with gentle rocking/shaking, in $0.5 \times$ SSC at 37°C for 15 min.

17. Wash slides briefly in TBS at room temperature.

18. Incubate slides in blocking solution (0.5% skim milk powder in TBS) for 30 min at room temperature.

19. Cover sections with 100–200 µL alkaline phosphatase conjugated sheep anti-DIG antibody (Roche Cat. No. 1093274) diluted 1:100 in blocking solution and incubate at room temperature for 1 h.

20. Wash (3×3 min) in TBS buffer.

21. Equilibrate in Solution II (either a solution of 0.1 M Tris-HCl (pH 8.0), 0.5 M NaCl, and 0.1 M MgCl$_2$ with final pH 9 or a solution made up of 0.1 M Tris-HCl (pH 8), 0.05 M, and MgSO$_4$·7H$_2$O with final pH 9.5) for 3 min at room temperature.

22. Cover the sections with NBT/BCIP diluted in Solution II (25 µL/mL), cover with a coverslip, and incubate in the dark for 3–4 h in a humidified container, making sure that the slides do not dry out.

23. Periodically monitor for color development in tissue using a light microscope. For additional color development, slides incubation in the dark at room temperature can be continued overnight.

24. To stop color development, remove the coverslip by immersing the slides in distilled water and wash the slides under running water for 5 min.

25. Counter stain the sections for 1 min with 0.5% Bismarck brown or equivalent.

26. Mount the slides with mounting medium (DAKO Cat. No. S3023) and a coverslip.

27. Examine by light microscopy for specific dark blue-black intracellular staining indicating the presence of AbHV DNA.

90.3 CONCLUSION AND FUTURE PERSPECTIVES

AbHV is an emerging viral pathogen of abalone that has caused disease outbreaks associated with high mortality rates in farmed abalone in Taiwan and wild and farmed abalone in Victoria, Australia. It has also caused mortality among wild-captured abalone being held in recirculating live-holding facilities at processing premises in Tasmania, Australia. In this latter case, mortality has been reported to be lower and varying dependent upon where abalone were sourced and the conditions under which they were held. Of interest, genotypic variation has been detected among the Taiwanese, Victorian, and Tasmanian AbHV strains and among AbHV strains identified as the cause of AVG outbreaks occurring at different Tasmanian premises between 2008 and 2011.

As there are no effective treatments for AVG, control methods in Australia currently rely on surveillance, biosecurity systems at farms and processing premises, zoning and restrictions on the movement of live abalone, and rapid response and decontamination systems to contain disease occurrences. Farm disease outbreaks are managed through the destruction and safe disposal of affected abalone, disinfection of water and equipment, and following procedures to effectively prevent recurrences. The use of sentinel abalone at previously affected farms and premises has been proposed as a means of ensuring the absence of infectious AbHV prior to restocking. These practices are consistent with many used to control and eradicate viral diseases of other aquatic species [32,33]. As an example, rigorous surveillance, culling, and decontamination processes have resulted in viral hemorrhagic septicemia (VHS) of rainbow trout being eradicated successfully from farms in the United Kingdom in 2006 [34] and farms in Norway in 2007 [35]. More recently in Denmark, >400 affected farms were declared free from VHS after 45 years of surveillance and control [36].

In cases where abalone might harbor subclinical, possibly persistent, AbHV infections, reduced handling and other potential stressors in farms or processor live-holding facilities (e.g., poor water quality, high stocking density, and poor feed quality) will assist in preventing acute infection from arising that can result in rapid horizontal transmission of disease to cohabiting abalone. The virus has been shown to persist in biofilters of recirculating systems and reemerge when tanks hold infected abalone previously restocked. Strict hygiene measures applied between tanks and between batches of abalone have been shown to provide some degree of effective controls in Tasmania. Another important issue for control of the disease is the early detection of abalone with low-level infections. Current DNA-based PCRs cannot differentiate truly infected abalone from uninfected abalone acting as fomites or cross contamination during laboratory studies [28]. Therefore, a reverse transcription polymerase chain reaction (RT-PCR) capable of detecting AbHV mRNA (similar to that used for CyHV-3, [37]) should be developed in order to identify genuinely active infections.

With aquaculture of species including various molluscs expanding globally and tighter restrictions on the use of chemicals to control disease, there is mounting interest in understanding defense mechanisms that can be stimulated to protect hosts against pathogen challenge. In salmon, carp, and channel catfish, as examples, understanding of their pathogen defense systems [38] has guided the development of experimental and commercial vaccines [39]. Similar investigations are now aiming to better understand the mechanisms underpinning pathogen defense in shrimp [40], oysters [41,42], and abalone [43–45]. In addition to understanding antiviral responses in abalone [46,47], better understanding of persistent infections (which are possibly associated with spread of the disease) and potential protective roles for RNA interference [48] and genetic selection strategies [49,50] to generate disease-resistant abalone may also offer opportunities to minimize environmental and commercial impacts of AVG.

REFERENCES

1. Anon, *2012 Yearbook, Fishery and Aquaculture Statistics*, Food and Agriculture Organisation of the United Nations, Rome, Italy, 2014, ftp://ftp.fao.org/FI/CDrom/CD_yearbook_2012/booklet/i3740t.pdf.
2. Goggin, C. L. and Lester, R. J. G., *Perkinsus*, a protistan parasite of abalone in Australia: A review. *Aust. J. Mar. Freshwater Res.*, 46, 639, 1995.
3. Elston, R. and Lockwood, G. S., Pathogenesis of vibriosis in cultured juvenile red abalone, *Haliotis rufescens* Swainson. *J. Fish Dis.*, 6, 111, 1983.
4. Friedman, C. S. et al., "*Candidatus Xenohaliotis californiensis*" a newly described pathogen of abalone, *Haliotis* spp., along the west coast of North America. *Int. J. Syst. Evol. Microbiol.*, 50, 847, 2000.
5. Nakatsugawa, T., Infectious nature of a disease in cultured juvenile abalone with muscular atrophy. *Fish Pathol.*, 25, 207, 1990.
6. Otsu, R. and Sasaki, K., Virus-like particles detected from juvenile abalones (*Nordotis discus discus*) reared with an epizootic fatal wasting disease. *J. Invert. Pathol.*, 70, 167, 1997.
7. Wang, J. et al., Virus infection in cultured abalone, *Haliotis diversicolor* Reeve in Guangdong Province, China. *J. Shellfish Res.*, 23, 1163, 2004.
8. Chang, P. H. et al., Herpes-like virus infection causing mortality of cultured abalone *Haliotis diversicolor supertexta* in Taiwan. *Dis. Aquat. Organ.*, 65, 23, 2005.
9. Hooper, C. et al., Ganglioneuritis causing high mortalities in farmed Australian abalone (*Haliotis laevigata* and *Haliotis rubra*). *Aus. Vet. J.*, 85, 188, 2007.
10. Tan, J. et al., Purification of a herpes-like virus from abalone (*Haliotis* spp.) with ganglioneuritis and detection by transmission electron microscopy. *J. Virol. Methods*, 149, 338, 2008.
11. Ellard, K. et al., Findings of disease investigations following the recent detection of AVG in Tasmania, *Proceedings of the Fourth National FRDC Aquatic Animal Health Scientific Conference*, Cairns, Queensland, Australia, July 22–24, 2009.
12. Davison, A. J. et al., The order Herpesvirales. *Arch. Virol.*, 154, 171, 2009.
13. Le Deuff, R. M. and Renault, T., Purification and partial genome characterization of a herpes-like virus infecting the Japanese oyster *Crassostrea gigas*. *J. Gen. Virol.*, 80, 1317, 1999.

14. Savin, K. W. et al., A neurotropic herpesvirus infecting the gastropod, abalone, shares ancestry with oyster herpesvirus and a herpesvirus associated with the amphioxus genome. *Virol. J.*, 7, 308, 2010, http://www.virologyj.com/content/7/1/308. Accessed on March 31, 2015.
15. Pellett, P. E. et al., Herpesvirales. In: King, A. M. Q., Lefkowitz, E., Adams, M. J., Carstens, E. B. (eds.), *Virus Taxonomy—Ninth Report of the International Committee on Taxonomy of Viruses*, Elsevier Academic Press, London, U.K., pp. 99–107, 2011.
16. Cowley, J. A. et al., Sequence variations amongst abalone herpes-like virus (AbHV) strains provide insights into its origins in Victoria and Tasmania, *Proceedings of the First FRDC Australasian Scientific Conference on Aquatic Animal Health*, Cairns, Queensland, Australia, July 5–8, 2011.
17. Ren, W. et al., Complete genome sequence of acute viral necrosis virus associated with massive mortality outbreaks in the Chinese scallop, *Chlamys farreri*. *Virol. J.*, 10, 110, 2013. doi: 10.1186/1743-422X-10-110.
18. Tamura, K. et al., MEGA6: Molecular evolutionary genetics analysis version 6.0. *Mol. Biol. Evol.*, 30, 2725, 2013.
19. Rzhetsky, A. and Nei, M., A simple method for estimating and testing minimum evolution trees. *Mol. Biol. Evol.*, 9, 945, 1992.
20. Felsenstein, J., Confidence limits on phylogenies: An approach using the bootstrap. *Evolution*, 39, 783, 1985.
21. Nei, M. and Kumar, S., *Molecular Evolution and Phylogenetics*, Oxford University Press, New York, 2000.
22. Saitou, N. and Nei, M., The neighbor-joining method: A new method for reconstructing phylogenetic trees. *Mol. Biol. Evol.*, 4, 406, 1987.
23. Corbeil, S. et al., Abalone viral ganglioneuritis: Establishment and use of an experimental immersion challenge system for the study of abalone herpes virus infections in Australian abalone. *Virus Res.*, 165, 2017, 2012.
24. Roizman, B. and Pellett, P. E., The family Herpesviridae: A brief introduction. In: Knipe, D. M. and Howley, P. M. (eds.), *Fields Virology*, 4th edn. Lippincott Williams & Wilkins, Philadelphia, PA, pp. 2381–2397, 2001.
25. Corbeil, S. et al., Development and validation of a TaqMan® PCR assay for the Australian abalone herpes-like virus. *Dis. Aquat. Organ.*, 92, 1, 2010.
26. Crane, M. St. J. et al., Aquatic Animal Health Subprogram: Development of molecular diagnostic procedures for the detection and identification of herpes-like virus of abalone (*Haliotis* spp.), Fisheries Research and Development Corporation Project No. 2007/006, 79pp., 2009.
27. Hooper, C. et al., Leucopenia associated with abalone viral ganglioneuritis. *Aus. Vet. J.*, 90, 24, 2012.
28. Crane, M. St. J. et al., Evaluation of abalone viral ganglioneuritis resistance amongst wild abalone populations along the Victorian coast of Australia. *J. Shellfish Res.*, 32, 67, 2013.
29. Mohammad, I. M. et al., Development of an *in situ* hybridisation assay for the detection and identification of the abalone herpes-like virus, *Proceedings of the First FRDC Australasian Scientific Conference on Aquatic Animal Health*, Cairns, Queensland, Australia, July 5–8, 2011.
30. OIE (Office International des Epizooties), *Manual of Diagnostic Tests for Aquatic Animals*. Chapter 2.1.4: Infection with abalone herpesvirus. World Organisation for Animal Health (OIE), Paris, France, 2014, pp. 441–451. http://www.oie.int/fileadmin/Home/eng/Health_standards/aahm/current/2.4.01_INF_ABALONE.pdf. Accessed on March 31, 2015.

31. Corbeil, S. et al., Aquatic Animal Health Subprogram: Characterisation of abalone herpes-like virus infections in abalone. Fisheries Research and Development Corporation Project No. 2009/032, 57pp., 2014.

32. Olesen, N. J., Sanitation of viral haemorrhagic septicaemia (VHS). *J. Appl. Ichthyol.*, 14, 173, 1998.

33. Olesen, N. J. and Korsholm, H., Control measures for viral diseases in aquaculture: Eradication of VHS and IHN. *Bull. Eur. Assoc. Fish Pathol.*, 17, 229, 1997.

34. Stone, D. M. et al., The first report of viral haemorrhagic septicaemia in farmed rainbow trout, *Oncorhynchus mykiss* (Walbaum), in the United Kingdom. *J. Fish Dis.*, 31, 775, 2008.

35. Dale, O. B. et al., Outbreak of viral haemorrhagic septicaemia (VHS) in seawater-farmed rainbow trout in Norway caused by VHS virus Genotype III. *Dis. Aquat. Organ.*, 85, 93, 2009.

36. Jensen, B. B. et al., Spatio-temporal risk factors for viral haemorrhagic septicaemia (VHS) in Danish aquaculture. *Dis. Aquat. Organ.*, 109, 87, 2014.

37. Yuasa, K. et al., Development of mRNA-specific RT-PCR for the detection of koi herpesvirus (KHV) replication stage. *Dis. Aquat. Organ.*, 100, 11, 2012.

38. Uribe, C. et al., Innate and adaptive immunity in teleost fish: A review. *Vet. Med.*, 56, 486, 2011.

39. Gudding, R. and Van Muiswinkel, W. B., A history of fish vaccination: Science-based disease prevention in aquaculture. *Fish Shellfish Immunol.*, 35, 1683, 2013.

40. Rowley, A. F. and Pope, E. C., Vaccines and crustacean aquaculture—A mechanistic exploration. *Aquaculture*, 334–337, 1, 2012.

41. Roch, P., Defense mechanisms and disease prevention in farmed marine invertebrates. *Aquaculture*, 172, 125, 1999.

42. Li, Y. et al., Spawning-dependent stress responses in pacific oysters *Crassostrea gigas*: A simulated bacterial challenge in oysters. *Aquaculture*, 293, 164, 2009.

43. Hooper, C. et al., Stress and immune responses in abalone: Limitations in current knowledge and investigative methods based on other models. *Fish Shellfish Immunol.*, 22, 363, 2007.

44. Wang, B. et al., Molecular cloning and characterization of macrophage migration inhibitory factor from small abalone *Haliotis diversicolor supertexta*. *Fish Shellfish Immunol.*, 27, 57, 2009.

45. Dang, V. T. et al., Immunological changes in response to herpesvirus infection in abalone *Haliotis laevigata* and *Haliotis rubra* hybrids. *Fish Shellfish Immunol.*, 34, 688, 2013.

46. Olicard, C. et al., *In vitro* research of anti-HSV-1 activity in different extracts from Pacific oysters *Crassostrea gigas*. *Dis. Aquat. Organ.*, 67, 141, 2005.

47. Green, T. J. and Montagnani, C., Poly I:C induces a protective antiviral immune response in the Pacific oyster (*Crassostrea gigas*) against subsequent challenge with Ostreid herpesvirus (OsHV-1 mvar). *Fish Shellfish Immunol.*, 35, 382, 2013.

48. Lima, P. C. et al., Exploring RNAi as a therapeutic strategy for controlling disease in aquaculture. *Fish Shellfish Immunol.*, 34, 729, 2013.

49. Nell, J. A. and Perkins, B., Evaluation of the progeny of third-generation Sydney rock oyster *Saccostrea glomerata* (Gould, 1850) breeding lines for resistance to QX disease *Marteilia sydneyi* and winter mortality *Bonamia roughleyi*. *Aquaculture Res.*, 37, 693, 2006.

50. Delaporte, M. et al., Characterisation of physiological and immunological differences between Pacific oysters (*Crassostrea gigas*) genetically selected for high or low survival to summer mortalities and fed different rations under controlled conditions. *J. Exp. Mar. Biol. Ecol.*, 353, 45, 2007.

91 African Swine Fever Virus

Dongyou Liu

CONTENTS

91.1 Introduction 817
    91.1.1 Classification...817
    91.1.2 Morphology and Genome Organization 818
    91.1.3 Biology and Epidemiology 818
    91.1.4 Clinical Features and Pathogenesis 819
    91.1.5 Diagnosis 820
    91.1.6 Treatment and Prevention 820
91.2 Methods 820
    91.2.1 Sample Preparation... 820
    91.2.2 Detection Procedures 820
        91.2.2.1 Conventional PCR.. 820
        91.2.2.2 Real-Time PCR 821
        91.2.2.3 Phylogenetic Analysis 821
91.3 Conclusion 822
References 822

91.1 INTRODUCTION

African swine fever virus (ASFV) is an intracytoplasmically replicating DNA virus that is responsible for African swine fever (ASF). Maintained in warthogs and bushpigs and transmitted by soft ticks of the *Ornithodoros* genus, ASFV infection in domestic pigs leads to high fever, loss of appetite, hemorrhages in the skin and internal organs, and death (with mortality rates as high as 100%) [1].

First described in Kenya in 1921, ASF is endemic in sub-Saharan African countries. The disease spread beyond Africa to Portugal in 1957 and Spain in 1960. In subsequent years, sporadic outbreaks of ASF occurred in Belgium, Italy, Malta, Sardinia, and Holland, and the virus reached the Caribbean (Cuba, Dominican Republic, and Haiti) and Brazil in the 1970s. Apart from Sardinia in Italy, most European countries have managed to eradicate the disease in domestic pigs by the late 1990s. However, with the reintroduction to Georgia in the Caucasus region in 2007, ASF quickly traveled to Armenia, Iran, Azerbaijan, and Chechnya. By 2014, ASF outbreaks have been reported in Ukraine, Belarus, Lithuania, Latvia, and Poland [1].

Being a highly contagious, fatal disease of domestic pigs, ASF contributes to significant losses in pig farming industry and increases economic burdens on both ASF-affected and disease-free countries with trade bans and eradication costs. As such, ASF is listed as a reportable disease by the World Organization for Animal Health (OIE) Terrestrial Animal Health Code.

91.1.1 CLASSIFICATION

ASFV is a double-stranded DNA virus classified in the genus *Asfivirus*, family *Asfarviridae*, which derived its name from ASF and other related viruses.

As the sole member of the family *Asfarviridae*, ASFV demonstrates several distinct features from other viruses. First of all, while ASFV is capable of stimulating host antibody production in natural infection, the resulting antibodies do not seem to consistently neutralize the virus [2]. Second, despite the existence of antigenic variations among ASFV isolates, it has been practically impossible to serotype these viruses. Third, being the only known DNA arbovirus, ASFV is remarkably adaptable. Besides circulating between warthogs and bushpigs (natural host) and *Ornithodoros* soft ticks (vector), ASFV is able to maintain its life cycle in domestic pigs through contaminated feed, without further input from tick vector. Further, free-ranging domestic pigs may acquire the virus by feeding on infected carcasses of warthogs and bushpigs or via tick bites [3].

On the basis of sequence variations in the C-terminal region of the *B646L* gene encoding the major capsid protein VP72, ASFV is distinguished into 22 genotypes (I–XXII) [4]. While genotypes I–XXII are found in eastern and southern Africa, genotype I (the so-called ESAC-WA genotype) is present in Europe, South America, the Caribbean, and western Africa, and genotype II ASFV is identified in Madagascar, the Caucasus region, and Russia [5].

In addition, phylogenetic analysis of the central variable region (CVR) within the *B602L* open reading frame (ORF) detects 12 differently sized products within the ESAC-WA genotype and separates this genotype into 19 subgroups. Moreover, a combined B646L–CVR assessment identifies eight lineages within genotype VIII, and an examination of *E183L* gene encoding the inner envelope transmembrane protein P54 permits finer differentiation within genotype IX [6,7].

Thus, sequencing analysis of the variable 3′ end of the *B646L* gene encoding the major capsid protein p72 provides a valuable approach for genotyping ASFV isolates. Further subtyping of ASFV *B646L* (p72) genotypes into subgroups is facilitated by sequencing the complete *E183L* gene encoding the p54 protein, and analysis of the nucleotide sequence of the CVR as well as the tetramer amino acid repeats within the CVR of the *B602L* gene encoding the J9L protein [8–12].

91.1.2 Morphology and Genome Organization

ASFV is a large, enveloped, icosahedral DNA virus of 175–215 nm in diameter. Extracellular virions are composed of five concentric layers, including the central nucleoid, the core shell, the inner envelope, the icosahedral capsid, and the lipid envelope layer, which is acquired when the virion buds from the plasma membrane.

The genome of ASFV consists of a linear, double-stranded DNA of 170–190 kb in length. Variation in the genome size among ASFV isolates is mainly due to the length heterogeneity (resulting from insertions and deletions of members of different multigene families [MGFs] such as MGFs 100, 110, 300, 360, and 505/530 and family p22) in the regions that flank a central conserved region (CCR) of approximately 125 kb. The left variable region ranges from 38 to 47 kb and the right variable region ranges from 13 to 16 kb. Additionally, minor length variations in the CCR also impact the overall genome size [13–15].

Within the DNA genome of ASFV, between 150 and 167 ORFs are identified. Closely spaced and read from both DNA strands, these ORFs encode 54 structural proteins including the antigenically stable virus protein VP72 and enzymes that are required for transcription and replication of the virus genome such as elements of a base excision repair system. Other ASFV-encoded molecules include factors involved in evading host defense systems and modulating host cell function and proteins that inhibit apoptosis of infected cells to facilitate production of progeny virions [14,15].

91.1.3 Biology and Epidemiology

Similar to other viruses causing viral hemorrhagic fevers, ASFV targets monocyte and macrophage lineage for replication. Following the uptake by the host, ASFV enters the host cell through receptor (dynamin) binding and clathrin-mediated endocytosis. Macropinocytosis also plays a part in the internalization of ASFV into the cells [16]. Inside the cytoplasm, the virus replicates in a highly orchestrated process with at least four stages of transcription—immediate-early, early, intermediate, and late. This activity mostly occurs in discrete, perinuclear regions of the cell called virus factories. The newly assembled progeny virions are then transported to the plasma membrane along the microtubules for a release via budding or actin-propelled cell-to-cell spread [17–22].

ASFV naturally occurs in wild pigs such as warthogs (*Phacochoerus aethiopicus* and *Phacochoerus africanus*), bushpigs (*Potamochoerus porcus*), and giant forest boars (*Hylochoerus meinertzhageni*), in which no clinical signs are apparent, although ASFV-infected wild boar (*Sus scrofa*) and feral pigs often show similar clinical signs and mortality to domestic pigs [23–25].

Soft ticks of the genus *Ornithodoros* (particularly, *O. moubata*, *O. erraticus*, *O. porcinus domesticus*, and *O. porcinus porcinus* and possibly, *O. coriaceus*, *O. turicata*, *O. parkeri*, *O. puertoricensis*, *O. savignyi*, *O. sonrai*, and *O. tholozani*) function as a vector for the virus as they are able to retain the virus for long periods and transmit it to susceptible hosts. In addition, tick-to-tick transmission through transstadial, sexual, and transovarial routes allows the virus to persist in the absence of viremic hosts [26,27].

ASFV is present in all tissues and body fluids of infected pigs with highest concentrations in the blood and the spleen [28]. Blood shedding during necropsies or pig fights and bloody diarrhea from infected pigs are main sources of contamination. The virus remains viable for long periods in the blood, feces, and tissues, with half-lives in the feces or blood ranging from >2 years at temperature up to 12°C to 15 days at temperatures of 30°C and in the tissue samples ranging from 1.7 to 7.4 days at 20°C. In addition, ASFV survives for 30 days in pepperoni and salami sausages and for >100 days in Iberian-cured pork products and Parma hams. ASFV also maintains infectivity to domestic pigs for 3 days in contaminated pigpens in the tropics.

ASFV is transmitted via three cycles: the sylvatic cycle, domestic pig cycle, and wild pig–to–domestic pig cycle. In the *sylvatic cycle*, warthogs become infected and develop a transient viremia lasting 2–3 weeks after biting by resident *O. moubata* ticks harboring the virus. Ticks feeding on viremic warthogs acquire the infection, completing the sylvatic cycle of infection. The sylvatic cycle mainly occurs in eastern and southern Africa [29]. In the *domestic pig cycle*, pigs acquire the virus from other infected/recovered pigs through ingestion of infected pig carcasses or pork products, or exposure to the blood and body fluids during fighting or mating. Ticks may or may not be involved in the transmission. The domestic pig cycle is seen in both west and east Africa [30]. In the *wild pig–to–domestic pig cycle*, free-ranging domestic pigs may acquire the infection through feeding on ASFV-contaminated wild pig carcasses or feces or tick bites during foraging [31].

Endemic in most sub-Saharan Africa including Madagascar, ASFV commonly occurs in feral pigs in Sardinia, Italy. After its reintroduction into the Caucasus in 2007, ASFV has become endemic among wild boars in the region, causing outbreaks among domesticated swine in the Republic of Georgia, Armenia, Azerbaijan, Russia, and other countries in the region [32–37].

91.1.4 Clinical Features and Pathogenesis

With an incubation period of 5–19 days after direct contact with infected pigs or <5 days after exposure to ticks, ASFV infection in domestic pigs may induce peracute, acute, subacute, or chronic disease depending on the virulence of virus strains, age, and immune status of the host [1].

Peracute disease is caused by highly virulent ASFV strains and usually occurs in a naïve pig population. Individual pigs are often found dead without showing apparent clinical symptoms [1].

Acute disease due to a highly virulent ASFV often shows fever (40°C–42°C) by 48 h postinfection. This is followed by loss of appetite, depression, dyspnea, anorexia, lethargy, leg weakness, recumbence (with shivering and coughing), erythema, cyanotic (blue-purple color), skin blotching (on the nose, ears, tail, lower legs, or hams), abdominal pain, constipation or diarrhea (initially mucoid and later bloody), viremia (appearing from 3 days postinfection and rising to a peak of over 10^8 hemadsorption units [HAD]$_{50}$/m), hemorrhages (on the ears and flanks and the internal organs), vomiting, nasal and conjunctival discharges, neurologic signs, abortion, coma, and death (up to 100% within 7–10 days). The infected pigs develop an acute hemorrhagic fever with viremia appearing from 3 days postinfection and rising to a peak of over 10^8 HAD$_{50}$/mL. Mortality of approaching 100% occurs within 8–12 days postinfection. In this form of the disease, the death rate is generally lower in adult swine, but may still be very high in very young animals [1].

Subacute or chronic disease caused by moderately virulent strains produces milder symptoms that may take several weeks to spread through the herd. Clinically, subacute disease is characterized by intermittent low fever, loss of appetite, depression, coughing, diarrhea, vomiting, emaciation, stunting, swollen joints, ulcers, reddened or raised necrotic skin foci over body protrusions, thrombocytopenia, and transient leukopenia. The infected pigs usually die (70%–80% in young pigs but less than 20% in older animals) or recover within 3–4 weeks [1].

Subacute disease due to moderate to highly virulent strains is characterized by intermittent low fever (lasting 30–45 days), anorexia, loss of appetite, coughing, dyspnea, depression, diarrhea, vomiting, emaciation, stunting, swollen joints, ulcers, reddened or raised necrotic skin foci over the body protrusions, abortion, thrombocytopenia, and transient leukopenia. The infected pigs usually die (30%–70% in young pigs but less than 20% in older animals) or recover within 3–4 weeks. This form of disease is also seen in herds with previous exposure [1].

Chronic disease due to strains of low to moderate virulence often occurs in previously exposed herds. Clinical signs include weight loss, unthriftiness, growth retardation, loss of hair, irregular febrile conditions, undulating respiratory peaks, lameness (arthritis), secondary infections, and low-grade perpetual sickness, with mortality (up to 30%). However, conjunctivitis is absent [1].

Asymptomatic infections or mild disease is usually seen in warthogs and bushpigs, while wild boar may develop clinical disease similar to domestic pigs.

At postmortem, pigs with peracute or acute disease may show bluish-purple discoloration and/or hemorrhages in the skin; bloody diarrhea; internal hemorrhages (in the spleen, lymph nodes, kidneys, and heart); large, friable, and dark red to black spleen; swollen and hemorrhagic lymph nodes (especially the gastrohepatic and renal lymph nodes); petechiae on the cortical and cut surfaces of the kidneys and in the renal pelvis; perirenal edema; hydropericardium with hemorrhagic fluid; petechiae and ecchymoses (in the urinary bladder, lungs, heart, stomach, and intestines); congestion or edema (in the liver, gall bladder, or lungs); straw-colored or blood-stained fluid (in the pleura, pericardium, and peritoneum); and congested, edematous, or hemorrhagic brain and meninges. Aborted fetuses are anasarcous, with a mottled liver, petechiae, or ecchymoses in the skin and myocardium as well as in the placenta [38].

Pigs with subacute or chronic disease may have emaciated carcass, enlarged but nonfriable spleen with a normal color, enlarged and hemorrhagic lymph nodes, slight petechiation on the kidneys, focal areas of skin necrosis, skin ulcers, consolidated lobules in the lung, caseous pneumonia, nonseptic fibrinous pericarditis, pleural adhesions, generalized lymphadenopathy, and swollen joints [39].

The pathogenesis of acute ASF reflects the capacity of ASFV to generate a series of molecules that interfere with host signaling pathways controlling the transcription of cytokines and modulate host cell function in relation to the apoptosis, inflammation, and immune responses. For example, ASFV A276R, A528R, and I329L genes are involved in the inhibition of IFN and NF-κB induction via the TLR3, cytosolic, or type I interferon (IFN) signaling pathways. This helps ASFV to evade host immune surveillance and overcome host innate and adapted immunity against the invading virus, leading to its unhindered replication and spread within/between host cells [40,41].

After gaining entry into the host via oronasal route, ASFV migrates to the tonsils or dorsal pharyngeal mucosa, then the mandibular/retropharyngeal lymph nodes, and finally the hematogenous system, leading to viremia as early as 8 h postinfection in newborn piglets and subsequent spread to other body organs within a day of postinfection. ASFV replication in these cells/organs results in pronounced depletion of lymphoid tissues, apoptosis of lymphocyte subsets, impairment of hemostasis and immune functions, extensive necrosis, destruction of endothelial membrane, and hemorrhages (in melena, epistaxis, erythema, renal petechiae, and lymph nodes) [42].

The pathogenesis of subacute/chronic ASF relates to the inability of nonvirulent ASFV strains to completely deplete host CD8(+) lymphocytes and to the increased host natural killer cell activity. This reduces the severity of the disease and protects the host from succumbing to the disease. At the same time, this enhances host autoimmune activity, leading to increased deposition of immune complexes in tissues (the kidneys, lungs, and skin) [42].

Thus, most lesions seen in ASF are attributable to cytokine-mediated interactions triggered by infected and activated monocytes and macrophages, rather than by virus-induced direct cell damage [38,43].

91.1.5 DIAGNOSIS

Pigs displaying fever together with a very large, friable, dark red to black spleen and greatly enlarged and hemorrhagic gastrohepatic and renal lymph nodes at necropsy may be suspected of ASF.

Differential diagnoses include classical swine fever (hog cholera), acute porcine reproductive and respiratory syndrome, porcine dermatitis, nephropathy syndrome, erysipelas, salmonellosis, pasteurellosis, eperythrozoonosis, actinobacillosis, Glasser's disease (*Haemophilus parasuis* infection), Aujeszky's disease (in young animals), thrombocytopenic purpura, warfarin poisoning, heavy metal toxicity, and other generalized septicemic or hemorrhagic conditions.

Laboratory diagnosis of ASF is mainly reliant on virus isolation, detection of viral antigen, antiviral antibody, and viral nucleic acids [44].

Virus isolation: ASFV from the blood (heparin or EDTA) or tissues (the spleen, kidney, tonsils, and lymph nodes) are isolated on the porcine leukocyte, bone marrow, alveolar macrophages, and blood monocyte cultures. ASFV virulent isolates tend to induce hemadsorption of pig erythrocytes to the surface of infected cells (the so-called a hemadsorption "autorosette" test), while avirulent isolates do not. This offers a sensitive approach to isolate ASFV from the blood, spleen, lymph nodes, tonsil, and kidney, but its usefulness is limited by its requirement of regular supply of fresh primary bone marrow cells or pig leukocytes [45,46].

Antigen and antibody detection: ASFV antigens are detected in tissue smears, cryostat sections, or buffy coat by fluorescent antibody test. Anti-ASFV antibodies may be detected from the serum, lymph nodes, or spleen of an infected pig with enzyme-linked immunosorbent assay (ELISA), immunoperoxidase test, immunoblotting, counterimmunoelectrophoresis, etc. [47]. A commercial-blocking ELISA (INGEZIM PPA Compac 11.PPA.K3, INGENASA) targeting VP73 viral protein is recommended by the OIE [48].

Nucleic acid detection: ASFV nucleic acids may be detected by hybridization probes and, more recently, by PCR and other amplification techniques (isothermal amplification assays) [49–65]. Real-time PCR assay targeting a highly conserved structural protein p72 has been shown to be extremely useful for detecting ASFV in samples of unknown origin. A commercial reverse transcription polymerase chain reaction (RT-PCR) Tetracore® assay (Tetracore Inc.) may be also used to detect ASFV DNA. Furthermore, fluorogenic probe hydrolysis (TaqMan) PCR allows specific, one-step, single-tube detection of ASFV in <2 h.

91.1.6 TREATMENT AND PREVENTION

In the absence of specific treatment, control and prevention of ASF center on quarantine, stamping out (slaughter), and eradication [66,67].

In the event of an ASF outbreak, stamping out may be applied to the affected individual farms or the areas within 5 km surrounding the affected farms. The key quarantine and eradication measures include slaughtering of infected and in-contact animals, proper disposal of carcasses (by burying, rendering, or burning), movement control, chemical disinfection (with sodium hypochlorite, 2% sodium hydroxide, 1% formalin, 4% anhydrous sodium carbonate plus 0.1% detergent, 10% crystalline sodium carbonate plus 0.1% detergent, and 1% phosphorylated iodophors, chloroform, and cresol), and control of tick vectors with acaricides. In addition, proper cooking of swill fed to pigs, that is, heating unprocessed meat at 70°C for 30 min and heating serum and bodily fluids at 60°C for 30 min, may help inactivate the virus.

Prevention in countries free of the disease requires stringent import policies to ensure proper disposal of waste food from aircraft, ships, or vehicles coming from infected countries. Application of state-of-the-art diagnostic technologies is also important for early detection and prompt implementation of control measures.

Although vaccination represents the most effective approach to prevent virus diseases, past attempts to develop vaccines against ASF have not had much success. Preliminary data suggest that partial protection of pigs is possible with low-virulence isolates from the field or obtained by passage in tissue culture or by deletion of genes involved in virulence. Further research in this area is warranted [66,67].

91.2 METHODS

91.2.1 SAMPLE PREPARATION

Suitable samples for ASFV diagnosis are the blood, lymph nodes, and spleen. Ticks are collected and placed in small plastic boxes containing a wet piece of filter paper for tick survival. In the laboratory, ticks are stored at −80°C until analysis.

Total DNA is purified from 200 μL of a sample (cell culture supernatant, serum, blood treated with EDTA, or tissue homogenate in 10% phosphate-buffered saline) by using the High Pure PCR Template Preparation Kit (Roche Molecular Biochemicals), DNeasy Blood and Tissue Kit (Qiagen), or QIAamp Viral RNA Mini Kit (Qiagen). The extracted DNA is eluted in a 50 μL volume and stored at −20°C.

91.2.2 DETECTION PROCEDURES

91.2.2.1 Conventional PCR

Aguero et al. [68] developed a PCR for ASFV detection with primers from the *B646L* gene encoding the major capsid protein p72. Use of the primer set PPA1 (5′-AGTTATGGGAAACCCGACCC-3′) and PPA2 (5′-CCCTGAATCGGAGCATCCT-3′) allows amplification of 257 bp.

PCR mixture (25 μL) is made up of 2 μL of sample DNA, 1× PCR buffer II (50 mM KCl, 10 mM Tris-HCl), 2 mM MgCl$_2$, 0.2 mM concentrations of the 4 deoxynucleoside triphosphates (Roche Molecular Biochemicals), and 0.2 μM concentrations of both primers (PPA-1/2 and 0.625 U of Taq Gold polymerase; Applied Biosystems).

Amplification is carried out with (1) incubation at 95°C for 10 min; (2) 40 cycles of 95°C for 15 s, 62°C for 30 s, and 72°C for 30 s; and (3) incubation at 72°C for 7 min.

Amplification products are analyzed by electrophoresis on a 2% agarose gel containing 0.5 µg of ethidium bromide per mL.

Note: The PCR using primer set PPA1 and PPA2 appears to be more sensitive than the OIE-recommended primer set (primer 1: 5′-ATGGATACCGAGGGAATAGC-3′; primer 2: 5′-CTTACCGATGAAAATGATAC-3′), which corresponds to a 278 bp region in the central portion of the p72 gene [48].

91.2.2.2 Real-Time PCR

91.2.2.2.1 Protocol of Haines et al. [69]

Haines et al. [69] described a real-time PCR assay for ASFV identification with primers and probe from the 3′ end of the VP72 gene (Table 91.1). The inclusion of IC-RNA-specific primers EGFP1-F and EGFP2-R and TaqMan probe EGFP1-HEX allows monitoring the efficiency of sample preparation and amplification (Table 91.1).

PCR mixture (25 µL) is composed of 8.4 µL nucleic acid extract, 12.5 µL 2× reaction mix, a final concentration of 5 mM $MgSO_4$, 5 U RNAsin, 50 nM ROX, and 0.5 U Superscript III Reverse Transcriptase/Platinum Taq mix. EGFP1-F and EGFP2-R primers are at a concentration of 200 nM and EGFP-HEX probe is at 100 nM. Primers ASFVp72IVI_L and ASFVp72IVI_R are at a concentration of 900 nM and the ASFV-Cy5 is at 200 nM.

Amplification and detection are conducted in Mx3005P qPCR System (Agilent Technologies) with the following conditions: RT at 50°C for 15 min, followed by incubation at 95°C for 2 min, then 50 cycles of denaturation at 95°C for 15 s, and annealing and extension at 60°C for 1 min. After amplification, a threshold cycle (C_T) value is assigned to each sample.

Note: Samples are considered ASFV positive if the ASFV probe results in a (CY5) C_T value ≤35. Samples are classed as ASFV negative if the IC-RNA probe (HEX) gives C_T values ≤38 but for which no C_T values for FAM or CY5 are obtained. In the event where a sample does not meet the aforementioned criteria, the sample is classified as "inconclusive."

91.2.2.2.2 Protocol of Zsak et al. [70]

Zsak et al. [70] designed a one-step real-time PCR assay for ASFV, with primers and TaqMan probe from ASFV gene encoding a highly conserved structural protein p72. Specifically, the forward primer starts at base position 1466 (5′-CCTCGGCGAGCGCTTTATCAC-3′); reverse primer starts at base position 1528 (5′-GGAAACTCATTCACCAAATC CTT-3′); probe starts at base position 1486 (5′-CGATGC AAGCTTTAT-3′). The TaqMan probe is labeled with a 5′ reporter dye, 6-carboxyfluorescein, and 3′ minor groove binder nonfluorescent quencher (PE Biosystems).

The PCR mixture (25 µL) is made up of 5× buffer solution, 5 mM Mn_2-acetate, 0.3 M concentration of each primer, 0.1 mM dATP, dCTP, or dGTP, 0.2 mM dUTP, 0.1 U of recombinant *Tth* DNA polymerase per µL (EZ-RT-PCR, PE Biosystems), 0.1 µg of bovine serum albumin per µL, and 0.5 M trehalose. The PCR mixture is dried within the reaction tubes (Fisher Scientific) and stored at room temperature for at least 1 year. For sample testing, the dried reagents are rehydrated with 22.5 µL of 1× buffer and 2.5 µL of DNA sample material.

Cycling parameters include 45 amplification cycles (95°C for 2 s and 60°C for 30 s) using the SmartCycler (Cepheid, Inc.).

Note: The real-time PCR assay displays a higher sensitivity with tonsil scrapings (95.4%) and nasal swabs (92.8%) than that with the blood (89.7%).

91.2.2.3 Phylogenetic Analysis

Atuhaire et al. [71] employed primers p72-U/p72-D from the C-terminal region of the p72 gene (478 bp), PPA722/PPA89 from the complete gene encoding the P54 protein (676 bp), and CVR-FL1/CVR-FL2 from the CVR located in the B602L gene for phylogenetic analysis of ASFV isolates (Table 91.2).

TABLE 91.1
Identity of Real-Time Polymerase Chain Reaction Primer and Probes

Target	Primer/Probe	Sequence (5′–3′)
ASFV	ASFVp72IVI_L	GAT GAT GAT TAC CTT YGC TTT GAA
	ASFV-CY5	Cy5-CCA CGG GAG GAA TAC CAA CCC AGT G-DDQ II
	ASFVp72IVI_R	TCT CTT GCT CTR GAT ACR TTA ATA TGA
IC	EGFP1-F	GAC CAC TAC CAG CAG AAC AC
	EGFP1-HEX	HEX-AGC ACC CAG TCC GCC CTG AGC A-BHQ1
	EGFP2-R	GAA CTC CAG CAG GAC CAT G

TABLE 91.2
Identity of Primers for Phylogenetic Analysis of African Swine Fever Virus

Gene	Primer	Sequence (5′–3′)	Product (bp)
p72 gene	p72-U	GGCACAAGTTCGGACATGT	478
	p72-D	GTACTGTAACGCAGCACAG	
P54 gene	PPA722	CGAAGTGCATGTAATAAACGTC	676
	PPA89	TGTAATTTCATTGCGCCACAAC	
CVR (B602L gene)	CVR-FL1	TCGGCCTGAAGCTCATTAG	Variable size
	CVR-FL2	CAGGAAACTAATGATGTTCC	

PCR mixture (100 µL) is made up of 10–50 ng of sample DNA, 1× PCR buffer II (50 mM KCl, 10 mM Tris-HCl), 2.5 mM MgCl$_2$, 0.2 mM concentrations of the 4 deoxynucleoside triphosphates (Roche Molecular Biochemicals), 0.2 µM concentrations of the primers, and 0.625 U of Taq Gold polymerase (Applied Biosystems).

Amplification is carried out in the Perkin–Elmer thermal cycles with (1) denaturation at 95°C for 5 min; (2) 40 cycles of 95°C for 30 s, 50°C for 30 s, and 72°C for 1 min; and (3) incubation at 72°C for 10 min.

PCR products are separated in a 2% agarose gel; bands of the correct size are excised and purified by Gel Extraction Kit (BIORON®) and sequenced.

The resulting sequences are submitted to the Basic Local Alignment Search Tool in order to identify their similarity with sequences obtained from the GenBank. Sequence alignment is performed using the CLUSTAL W with manual adjustments. For phylogenetic analysis, the MEGA 5.0 program is used to construct the neighbor-joining trees with 1000 bootstrap replications. The trees are drawn to scale, with branch lengths in the same units as those of the evolutionary distances used to infer the phylogenetic trees. The evolutionary distances are computed using the p-distance method and are in the units of the number of base differences per site. Codon positions included are 1st + 2nd + 3rd + noncoding. All ambiguous positions are removed for each sequence pair.

91.3 CONCLUSION

ASF is a hemorrhagic, fatal disease of pigs, resulting from infection with ASFV. Due to its remarkable ability to perpetuate in wild pigs, soft ticks, and/or domestic pigs, ASF has spread from sub-Saharan Africa in the 1960s to Europe and subsequently to the Caribbean. Following its reintroduction to the Caucasus in 2007, this disease is associated with a number of outbreaks in eastern Europe and southern Russia, with huge economic impact. In the absence of effective vaccines, control and prevention of ASF are dependent on the application of reliable laboratory diagnostic methods for early detection and monitoring, adoption of stamping out policy, implementation of quarantine and disinfection measures, and prevention of ASF incursion from infected areas to ASF-free areas.

REFERENCES

1. Tulman ER, Delhon GA, Ku BK, Rock DL. African swine fever virus. *Curr Top Microbiol Immunol*. 2009;328:43–87.
2. Escribano JM, Galindo I, Alonso C. Antibody-mediated neutralization of African swine fever virus: Myths and facts. *Virus Res*. 2013;173(1):101–109.
3. Anderson EC, Hutchings GH, Mukarati N, Wilkinson PJ. African swine fever virus infection of the bushpig (*Potamochoerus porcus*) and its significance in the epidemiology of the disease. *Vet Microbiol*. 1998;62(1):1–15.
4. Boshoff CI, Bastos ADS, Gerber LJ, Vosloo W. Genetic characterisation of African swine fever viruses from outbreaks in southern Africa (1973–1999). *Vet Microbiol*. 2007;121:45–55.
5. Costard S, Mur L, Lubroth J, Sanchez-Vizcaino JM, Pfeiffer DU. Epidemiology of African swine fever virus. *Virus Res*. 2013;173(1):191–197.
6. Lubisi BA, Bastos ADS, Dwarka RM, Vosloo W. Molecular epidemiology of African swine fever in East Africa. *Arch Virol*. 2005;150:2439–2452.
7. Leblanc N et al. Development of a suspension microarray for the genotyping of African swine fever virus targeting the SNPs in the C-terminal end of the p72 gene region of the genome. *Transbound Emerg Dis*. 2013;60(4):378–383.
8. Bastos AD et al. Genotyping field strains of African swine fever virus by partial p72 gene characterization. *Arch Virol*. 2003;148:693–706.
9. Bastos AD, Penrith ML, Macome F, Pinto F, Thomson GR. Co-circulation of two genetically distinct viruses in an outbreak of African swine fever in Mozambique: No evidence for individual co-infection. *Vet Microbiol*. 2004;103:169–182.
10. Gallardo C, Mwaengo D, Macharia J. Enhanced discrimination of African swine fever virus isolates through nucleotide sequencing of the p54, p72, and pB602L (CVR) genes. *Virus Genes*. 2009;10:85–95.
11. Gallardo C et al. African swine fever viruses with two different genotypes, both of which occur in domestic pigs, are associated with ticks and adult warthogs, respectively, at a single geographical site. *J Gen Virol*. 2011;92(2):432–444.
12. Owolodun O et al. Molecular characterisation of African swine fever viruses from Nigeria (2003–2006) recovers multiple virus variants and reaffirms CVR epidemiological utility. *Virus Genes*. 2010;41:361–368.
13. Yáñez RJ et al. Analysis of the complete nucleotide sequence of African swine fever virus. *Virology*. 1995;208(1):249–278.
14. Chapman DA, Tcherepanov V, Upton C, Dixon LK. Comparison of the genome sequences of non-pathogenic and pathogenic African swine fever virus isolates. *J Gen Virol*. 2008;89:397–408.
15. de Villiers EP et al. Phylogenomic analysis of 11 complete African swine fever virus genome sequences. *Virology*. 2010;400(1):128–136.
16. Sánchez EG et al. African swine fever virus uses macropinocytosis to enter host cells. *PLoS Pathog*. 2012;8(6):e1002754.
17. Alonso C et al. African swine fever virus-cell interactions: From virus entry to cell survival. *Virus Res*. 2013;173(1):42–57.
18. Dixon LK, Chapman DA, Netherton CL, Upton C. African swine fever virus replication and genomics. *Virus Res*. 2013;173(1):3–14.
19. Netherton CL, Wileman TE. African swine fever virus organelle rearrangements. *Virus Res*. 2013;173(1):76–86.
20. Rodríguez JM, Salas ML. African swine fever virus transcription. *Virus Res*. 2013;173(1):15–28.
21. Salas ML, Andrés G. African swine fever virus morphogenesis. *Virus Res*. 2013;173(1):29–41.
22. Sánchez EG, Quintas A, Nogal M, Castelló A, Revilla Y. African swine fever virus controls the host transcription and cellular machinery of protein synthesis. *Virus Res*. 2013;173(1):58–75.
23. Ravaomanana J et al. First detection of African Swine Fever Virus in Ornithodoros porcinus in Madagascar and new insights into tick distribution and taxonomy. *Parasit Vectors*. 2010;3:115.
24. Boinas FS, Wilson AJ, Hutchings GH, Martins C, Dixon LJ. The persistence of African swine fever virus in field-infected *Ornithodoros erraticus* during the ASF endemic period in Portugal. *PLoS One*. 2011;6(5):20383.

25. Atuhaire DK et al. Prevalence of African swine fever virus in apparently healthy domestic pigs in Uganda. *BMC Vet Res.* 2013;9:263.

26. Bastos ADS, Arnot LF, Jacquier MD, Maree S. A host species-informative internal control for molecular assessment of African swine fever virus infection rates in the African sylvatic cycle *Ornithodoros* vector. *Med Vet Entomol.* 2009;23(4):399–409.

27. Burrage TG. African swine fever virus infection in Ornithodoros ticks. *Virus Res.* 2013;173(1):131–139.

28. Giammarioli M et al. Genetic characterisation of African swine fever viruses from recent and historical outbreaks in Sardinia (1978–2009). *Virus Genes.* 2011;42(3):377–387.

29. Jori F et al. Review of the sylvatic cycle of African swine fever in sub-Saharan Africa and the Indian ocean. *Virus Res.* 2013;173(1):212–227.

30. Misinzo G et al. Molecular characterization of African swine fever virus from domestic pigs in northern Tanzania during an outbreak in 2013. *Trop Anim Health Prod.* 2014;46(7):1199–1207.

31. Sánchez-Vizcaíno JM, Mur L, Martínez-López B. African swine fever: An epidemiological update. *Transbound Emerg Dis.* 2012;59(Suppl. 1):27–35.

32. Rowlands RJ et al. African swine fever virus isolate, Georgia. *Emerg Infect Dis.* 2008;14(12):1870–1874.

33. Rahimi P et al. Emergence of African swine fever virus, northwestern Iran. *Emerg Infect Dis.* 2010;16(12):1946–1948.

34. Oganesyan AS et al. African swine fever in the Russian Federation: Spatio-temporal analysis and epidemiological overview. *Virus Res.* 2013;173(1):204–211.

35. Sánchez-Vizcaíno JM, Mur L, Martínez-López B. African swine fever (ASF): Five years around Europe. *Vet Microbiol.* 2013;165(1–2):45–50.

36. Pejsak Z, Truszczyński M, Niemczuk K, Kozak E, Markowska-Daniel I. Epidemiology of African Swine Fever in Poland since the detection of the first case. *Pol J Vet Sci.* 2014;17(4):665–672.

37. Costard S et al. African swine fever: How can global spread be prevented? *Philos Trans R Soc Lond B Biol Sci.* 2009;364(1530):2683–2696.

38. Galindo-Cardiel I et al. Standardization of pathological investigations in the framework of experimental ASFV infections. *Virus Res.* 2013;173(1):180–190.

39. Muhangi D et al. A longitudinal survey of African swine fever in Uganda reveals high apparent disease incidence rates in domestic pigs, but absence of detectable persistent virus infections in blood and serum. *BMC Vet Res.* 2015;11:106.

40. Blome S, Gabriel C, Beer M. Pathogenesis of African swine fever in domestic pigs and European wild boar. *Virus Res.* 2013;173(1):122–130.

41. Correia S, Ventura S, Parkhouse RM. Identification and utility of innate immune system evasion mechanisms of ASFV. *Virus Res.* 2013;173(1):87–100.

42. Takamatsu HH et al. Cellular immunity in ASFV responses. *Virus Res.* 2013;173(1):110–121.

43. Gómez-Villamandos JC, Bautista MJ, Sánchez-Cordón PJ, Carrasco L. Pathology of African swine fever: The role of monocyte-macrophage. *Virus Res.* 2013;173(1):140–149.

44. de León P, Bustos MJ, Carrascosa AL. Laboratory methods to study African swine fever virus. *Virus Res.* 2013;173(1):168–179.

45. Weingartl HM, Sabara M, Pasick J, van Moorlehem E, Babiuk L. Continuous porcine cell lines developed from alveolar macrophages: Partial characterization and virus susceptibility. *J Virol Methods.* 2002;104(2):203–216.

46. Hurtado C, Bustos MJ, Carrascosa AL. The use of COS-1 cells for studies of field and laboratory African swine fever virus samples. *J Virol Methods.* 2010;164(1–2):131–134.

47. Cubillos C et al. African swine fever virus serodiagnosis: A general review with a focus on the analyses of African serum samples. *Virus Res.* 2013;173(1):159–167.

48. Wilkinson PJ. African swine fever. In *Manual of Standards for Diagnostic Test and Vaccines*, Office International Des Epizooties, Paris, France, 2000, pp. 189–198.

49. Steiger Y, Ackermann M, Mettraux C, Kihm U. Rapid and biologically safe diagnosis of African swine fever virus infection by using polymerase chain reaction. *J Clin Microbiol.* 1992;30(1):1–8.

50. King DP et al. Development of a TaqMan PCR assay with internal amplification control for the detection of African swine fever virus. *J Virol Methods.* 2003;107(1):53–61.

51. Basto AP et al. Development of a nested PCR and its internal control for the detection of African swine fever virus (ASFV) in *Ornithodoros erraticus.* *Arch Virol.* 2006;151(4):819–826.

52. McKillen J et al. Molecular beacon real-time PCR detection of swine viruses. *J Virol Methods.* 2007;140(1–2):155–165.

53. McKillen J et al. Sensitive detection of African swine fever virus using real-time PCR with a 5′ conjugated minor groove binder probe. *J Virol Methods.* 2010;168(1–2):141–146.

54. Giammarioli M, Pellegrini C, Casciari C, De Mia GM. Development of a novel hot-start multiplex PCR for simultaneous detection of classical swine fever virus, African swine fever virus, porcine circovirus type 2, porcine reproductive and respiratory syndrome virus and porcine parvovirus. *Vet Res Commun.* 2008;32(3):255–262.

55. Ballester M et al. Intranuclear detection of African swine fever virus DNA in several cell types from formalin-fixed and paraffin-embedded tissues using a new in situ hybridisation protocol. *J Virol Methods.* 2010;168(1–2):38–43.

56. James HE et al. Detection of African swine fever virus by loop-mediated isothermal amplification. *J Virol Methods.* 2010;164(1–2):68–74.

57. Ronish B et al. Design and verification of a highly reliable Linear-After-The-Exponential PCR (LATE-PCR) assay for the detection of African swine fever virus. *J Virol Methods.* 2011;172(1–2):8–15.

58. Tignon M et al. Development and inter-laboratory validation study of an improved new real-time PCR assay with internal control for detection and laboratory diagnosis of African swine fever virus. *J Virol Methods.* 2011;178(1–2):161–170.

59. de Carvalho Ferreira HC, Weesendorp E, Quak S, Stegeman JA, Loeffen WL. Quantification of airborne African swine fever virus after experimental infection. *Vet Microbiol.* 2013;165(3–4):243–251.

60. de Carvalho Ferreira HC, Weesendorp E, Quak S, Stegeman JA, Loeffen WL. Suitability of faeces and tissue samples as a basis for non-invasive sampling for African swine fever in wild boar. *Vet Microbiol.* 2014;172(3–4):449–454.

61. Oura CA, Edwards L, Batten CA. Virological diagnosis of African swine fever—Comparative study of available tests. *Virus Res.* 2013;173(1):150–158.

62. Fernández-Pinero J et al. Molecular diagnosis of African Swine Fever by a new real-time PCR using universal probe library. *Transbound Emerg Dis.* 2013;60(1):48–58.

63. Wernike K, Hoffmann B, Beer M. Single-tube multiplexed molecular detection of endemic porcine viruses in combination with background screening for transboundary diseases. *J Clin Microbiol*. 2013;51(3):938–944.

64. Braae UC et al. Detection of African swine fever virus DNA in blood samples stored on FTA cards from asymptomatic pigs in Mbeya region, Tanzania. *Transbound Emerg Dis*. 2015;62(1):87–90.

65. Randriamparany T et al. African swine fever diagnosis adapted to tropical conditions by the use of dried-blood filter papers. *Transbound Emerg Dis*. November 27, 2014.

66. Penrith ML, Vosloo W. Review of African swine fever: Transmission, spread and control. *J S Afr Vet Assoc*. 2009;80(2):58–62.

67. Penrith ML, Vosloo W, Jori F, Bastos AD. African swine fever virus eradication in Africa. *Virus Res*. 2013;173(1):228–246.

68. Aguero M et al. Highly sensitive PCR assay for routine diagnosis of African swine fever virus in clinical samples. *J Clin Microbiol*. 2003;41(9):4431–4434.

69. Haines FJ, Hofmann MA, King DP, Drew TW, Crooke HR. Development and validation of a multiplex, real-time RT PCR assay for the simultaneous detection of classical and African swine fever viruses. *PLoS One*. 2013;8(7):e71019.

70. Zsak L et al. Preclinical diagnosis of African swine fever in contact-exposed swine by a real-time PCR assay. *J Clin Microbiol*. 2005;43(1):112–119.

71. Atuhaire DK et al. Molecular characterization and phylogenetic study of African swine fever virus isolates from recent outbreaks in Uganda (2010–2013). *Virol J*. 2013;10:247.

92 Megalocytiviruses

Chun-Shun Wang and Chiu-Ming Wen

CONTENTS

92.1 Introduction 825
   92.1.1 Classification 825
   92.1.2 Morphology, Biology, and Epidemiology 826
   92.1.3 Clinical Features and Pathogenesis 827
   92.1.4 Diagnosis 827
      92.1.4.1 Conventional Techniques 827
      92.1.4.2 Molecular Techniques 829
92.2 Methods 831
   92.2.1 Sample Collection and Preparation 831
      92.2.1.1 DNA Extraction by Using Phenol/Chloroform 831
      92.2.1.2 DNA Extraction by Using Spin Column 832
   92.2.2 Detection Procedures 832
      92.2.2.1 Conventional PCR 832
      92.2.2.2 Nested PCR 832
      92.2.2.3 Real-Time PCR 833
92.3 Conclusion and Future Perspectives 833
References 833

92.1 INTRODUCTION

As a recently identified genus in the family *Iridoviridae*, *Megalocytivirus* contains several double-stranded DNA (dsDNA) viruses that are pathogenic to a wide variety of freshwater and marine fishes, with a significant negative impact on ornamental and food fish aquaculture worldwide [1–5]. The first outbreak of megalocytivirus-induced disease was recorded in cultured red sea bream (*Pagrus major*) in Japan in 1990, and the causative agent was designated as red sea bream iridovirus (RSIV) [6]. Thus far, RSIV disease has been documented in 31 cultured seawater fish in Japan [7,8]. In China, a molecular epidemiology study indicated that over 50 species of cultured and wild seawater fish were infected by megalocytiviruses [9]. More than 10 species of cultured marine fish with iridovirus infection were reported in Taiwan [10–12,68]. In Korea, megalocytiviruses were reported in 10 freshwater ornamental fish species and 7 cultured fish species [13–15]. In addition, the farmed Murray cod (*Maccullochella peelii peelii*) and dwarf gourami (*Colisa lalia*) were observed to be infected in Australia [16]. Since 2007, a systemic megalocytivirus infection in three-spine stickleback (*Gasterosteus aculeatus*) was described in Canada [17]. Given the wide host range and geographic distribution of megalocytivirus, phylogenetic analysis using the major capsid protein (MCP) and ATPase gene has been performed for multiple isolates. This chapter focuses on the classification, morphology, clinical features, pathogenesis, and diagnosis of megalocytivirus. Developments in various vaccines and RNA interference (RNAi) techniques for preventing megalocytivirus infection in fish are also discussed.

92.1.1 CLASSIFICATION

According to the Eighth Report of the International Committee on Taxonomy of Viruses, the family *Iridoviridae* currently consists of five genera: *Iridovirus*, *Lymphocystivirus*, *Chloriridovirus*, *Ranavirus*, and *Megalocytivirus*, with *Megalocytivirus* being the latest addition to the family [4].

Megalocytivirus is composed of several members, including infectious spleen and kidney necrosis virus (ISKNV) and RSIV. ISKNV was first identified in a high-mortality outbreak among cultured mandarin fish (*Siniperca chuatsi*) in China [18]. RSIV, another crucial member of the genus, was listed by the World Organisation for Animal Health, or the Office International des Epizootic (OIE), as a pathogen and isolated from cultured red sea bream (*P. major*) [6]. Since 1992, increasing numbers of megalocytivirus strains genetically similar to these two species have been identified in food and ornamental fish and were assigned with host-derived names such as the sea bass iridovirus [19], rock bream iridovirus (RBIV) [20], Murray cod iridovirus [16], dwarf gourami iridovirus (DGIV), African lampeye iridovirus [21], large yellow croaker iridovirus (LYCIV) [22], orange-spotted grouper iridovirus (OSGIV) [23], Taiwan grouper iridovirus (TGIV) [10], spotted knifejaw iridovirus (SKIV-ZJ07) [56], and turbot red-dish body iridovirus (TRBIV) [24].

Sequence analysis of megalocytiviruses revealed high nucleotide and amino acid sequence identities in many open reading frames (ORFs) [25]. The viral putative MCP, vascular endothelial growth factor, mRNA capping enzyme, tumor necrosis factor receptor-associated protein, and ATPase genes have been used to classify new megalocytivirus isolates [26]. Among these genes, the MCP and ATPase genes are suitable targets for phylogenetic studies because they are highly conserved in the iridovirus [27].

Analyses of the MCP and ATPase genes show that the genus *Megalocytivirus* can be divided into three clusters represented by RSIV, ISKNV, and TRBIV, which could be considered distinct viral species [1,28–30]. The megalocytiviruses on the RSIV cluster can be further divided into two subclusters: genotype I, of which RSIV Ehime-1 is a representative, and genotype II, which includes other RSIV strains found in Japan, Korea, Taiwan, China, and Southeast Asia. The host ranges for RSIV clusters are extremely wide and have been associated with a mass mortality epizootic in the crucial maricultured fish species [29].

The second cluster is represented by ISKNV and related viruses and exhibits less genetic variation than the RSIV cluster does as well as different biological characteristics. ISKNV, originally isolated from the mandarin fish (*S. chuatsi*) raised for food in China, also resulted in an epizootic of more than 10 species of fresh ornamental fishes. A second ISKNV genotype was recently recognized following disease episodes in ornamental and food fish species, including the Banggai cardinalfish iridovirus, isolated from a marine ornamental fish, and marbled sleepy goby iridovirus, isolated from a freshwater species cultured for food in China [31,32]. The host range and geographical distribution of ISKNV are relatively broad, but freshwater and brackish water fish species are the predominant species affected.

The third cluster, TRBIV, has primarily been associated with diseases in a cultured Asian flatfish species, *Scophthalmus maximus* (turbot) [24,33]. The cluster includes flounder iridovirus and other viruses from the bastard halibut (*Paralichthys olivaceus*); the viral hosts belong mainly to the order Heterosomata [34].

Except for slight differences in nucleotide sequence among the three clusters, variations also exist in their biological characteristics. For example, the marine isolates of the ISKNV group of viruses are easier to propagate in cell cultures than those of the RSIV group of viruses. Mortality is observed in experimental infections with ISKNV between 25°C and 34°C, optimal replication of the RSIV group of viruses is observed at 20°C or 25°C, and an outbreak of TRBIV disease in South Korea occurred at 17°C–18°C [1].

Recently, a fourth cluster of TSIV (i.e., three-spine stickleback iridovirus) was proposed by Waltzek et al.; the virus was isolated from an epizootic that occurred in three-spine stickleback fish in Canada [17]. Due to the impact of megalocytiviruses on aquaculture, further collection of new virus strains from affected species is crucial. Continual monitoring

of the iridovirus genotype, with strains being contributed to a worldwide database, may improve the understanding of local viral genotype shifts and their relationship with worldwide epidemiology.

92.1.2 Morphology, Biology, and Epidemiology

Iridoviruses are large cytoplasmic dsDNA viruses with an icosahedral capsid of approximately 120–240 nm in diameter. Complete virions consist of three layers: the outer layer is an icosahedral capsid, the intermediate layer is a thick translucent space, and the inner spheroid core is a homogeneously electron-dense nucleoid [2,3,24]. The viral genomes are both circularly permuted and terminally redundant, which are unique features among the eukaryotic virus genomes [35]. The entire genomes of seven megalocytiviruses have been sequenced: ISKNV, RSIV, RBIV, OSGIV, TRBIV, RBIV-C1, and LYCIV (accession no. AY 779031) [23,36–40]. The genome lengths range from 110104 to 112636 bp, with putative ORFs from 115 to 124. The GC content of megalocytiviruses ranges from 53% to 55%. On the basis of the complete genome sequences, megalocytiviruses were suggested to be more similar to ranaviruses than to viruses in other iridoviral genera [41]. Genes encoding large and small subunits of ribonucleotide reductase (RNRS) are found in all iridoviruses except for megalocytivirus, in which only a small subunit is present [1].

Megalocytiviruses can propagate in cell cultures; a fibroblast cell line (CRF-1) derived from the tail fin of red sea bream (*P. major*) susceptible to RSIV infection was developed. RSIV-infected CRF-1 cells showed typical morphological changes that were associated with apoptosis according to EGFP–annexin V staining. Serial viral passages were successful in CRF-1 cells according to an MTT assay [42]. In Taiwan, a similar cell line derived from the dorsal fin of red sea bream was also established (i.e., RSBF-2 cells). The RSBF-2 cell line was susceptible to RSIV but produced only round and refractory cells and was then subcultured to generate a persistently infected cell line [43]. In addition, a GBC4 cell line susceptible to giant seaperch iridovirus (GSIV) was developed, and the virus titers approached 10^9 TCID$_{50}$/mL. The suitable temperature for GSIV growth was 15°C–30°C [44]. A continuous cell culture derived from the mandarin fish fry (MFF-1) was developed [45]. MFF-1 could produce high titers of ISKNV through continuous viral passages, and these titers were confirmed through indirect immunofluorescence (IF) assay and real-time polymerase chain reaction (quantitative PCR [qPCR]) analysis. In addition, the TK and TF cell lines derived from the kidney and fin of turbot were found to be susceptible to the TRBIV isolate [46,47]. Using cell culture testing, RSIV was determined to be sensitive to acid (pH 3), chloroform, and ether. The virus was unstable when exposed to heat but stable when subjected to ultrasonic treatment, repeated freezing, and thawing *in vitro* [48]. Experimental infection in mandarin fish showed that ISKNV could be inactivated at pH 11 and with UV irradiation and chemicals (e.g., sodium hypochlorite and potassium permanganate) but that iodine was not effective for viral inactivation [49].

Megalocytiviruses have caused mass mortality in various marine and freshwater fish species throughout Asia, indicating that these viruses have a broad host range and geographical distribution [8–11]. The viruses have spread internationally through the translocation of live fish. Recently, a growing number of megalocytiviruses have been identified. As fish are transported worldwide for commercial and ornamental use, one or more species may be the vehicle by which megalocytivirus is introduced to a new host or environment [30,50,51]. In addition, megalocytivirus (RBIV) in Pacific oysters (*Crassostrea gigas*) and blue mussels (*Mytilus edulis*) inhabiting the area around rock bream farms was identified in Korea. Cohabitation of mussels with megalocytivirus-infected rock bream results in the presence of viral particles in the gills, mantle, and digestive gland of mussels. The virus can be retained in the digestive gland for 28 days, regardless of water temperature. Infection with megalocytivirus derived from the digestive gland of cohabiting mussels has led to 100% cumulative mortality in rock bream [52].

92.1.3 CLINICAL FEATURES AND PATHOGENESIS

Megalocytivirus-infected fish may exhibit nonspecific clinical signs including lethargy, anorexia, hyperpigmentation, exophthalmos, skin lesion, unusual swimming behavior, gulping, severe anemia, and white feces [18,22,24,53]. The characteristics of affected fish are systemic formation of enlarged cells and basophilic intracytoplasmic inclusion bodies in multiple organs including the spleen, kidney, liver, heart, brain, gill, intestine, eye, and hematopoietic tissues. Electron microscopy reveals viral particles in an inclusion body within the cytoplasm of infected cells. The heterochromatins of the affected cells are marginated and aggregated to one side of the nuclei during the early stages of infections. The surrounding cytoplasm of infected cells is compressed and contains fragmented organelles and degenerated mitochondria. The viroplasm generally occupies a large proportion of the cytoplasm. Inside the viroplasm, virus particles are assembled around either a ring or disc-shaped virogenic stroma [18,21,22,24,53,56,60].

The pathogenicity of RSIVs isolated from various cultured marine fishes to the red sea bream has been examined through experimental infection. The different megalocytivirus isolates proved to be of high virulence leading to 60%–90% mortality in red sea bream, as the original RSIV isolates [54]. In Australia, an outbreak of high mortality in Murray cod fingerlings caused by megalocytivirus infection was reported. A hypothesis that the outbreak arose through the induction of a virus into an ornamental dwarf gourami (DGIV) imported from Southeast Asia underwent testing. Intraperitoneal injection of Murray cod fingerlings with DGIV resulted in 90% mortality. Mortality was also induced by cohabitating Murray cod fingerlings and dwarf gourami, with researchers speculating that the virus spread between the two species through water [55]. Similar experiments were performed in Korea. The megalocytivirus was isolated from freshwater pearl gourami (*Trichogaster leeri*; PGIV-SP) and the pathogenicity of seawater rock bream (*Oplegnathus fasciatus*; IVS-1) was examined, showing a cross infection between these two fish species. Comparisons of the stability of the two megalocytiviruses showed that both viral genomes decreased much faster in seawater than in freshwater after 4 days of incubation at 25°C. In addition, healthy rock bream that cohabitated with the PGIV-SP-challenged rock bream showed 100% cumulative mortality, implying the potential for megalocytivirus transmission from freshwater ornamental fish to marine fish even in a marine environment [50]. In addition, a new marine megalocytivirus isolated from spotted knifejaw was designated as SKIV-ZJ07. Twenty proteins of SKIV-ZJ07 were found to match proteins derived from RBIV, OSGIV, and ISKNV. Challenge tests showed that SKIV-ZJ07 was highly virulent in freshwater mandarin fish [56].

In another experiment, the megalocytivirus disease progressed more rapidly at higher water temperatures. When the water temperature for fish infected with the megalocytivirus by intramuscular injection was shifted from 13°C to 25°C, the fish experienced rapid onset of the disease, with a cumulative mortality of 100% [57]. In addition, asymptomatic RSIV-infected carriers of fingerling rock bream showed a high cumulative mortality caused by elevated water temperature. However, no morality was observed and no RSIV DNA was detected using PCR in yearling rock bream [58]. Using a two-step PCR method, a high level of asymptomatic iridovirus infection in various marine cultured fish was reported in Korea. The proportion of iridovirus-positive fish varied from month to month, but the overall prevalence was approximately 37.2% and differed depending on the fish species (i.e., rock bream, rockfish, seaperch, and red sea bream) [59]. A zebrafish (*Danio rerio*) model for ISKNV infection was established, and the experimental conditions were characterized. The mortality of adult zebrafish infected with ISKNV through intraperitoneal injection exceeded 60%. ISKNV can be passed stably in zebrafish for over 10 passages. Histological and real-time reverse transcription polymerase chain reaction (RT-PCR) analyses confirmed that ISKNV is replicated in fish. The establishment of a zebrafish model of ISKNV provided a valuable tool for studying the interactions between ISKNV and its host [60].

92.1.4 DIAGNOSIS

Several methods are used for diagnosing megalocytivirus infection in fish, including histopathology, transmission electron microscopy (TEM), cell culture tests, IF tests, immunohistochemistry enzyme-linked immunosorbent assay (ELISA), dot blot assay, *in situ* hybridization (ISH), PCR, nested PCR, multiplex PCR, qPCR, and loop-mediated isothermal amplification (LAMP) (Table 92.1). PCR and DNA sequence analysis are widely used for detecting megalocytivirus infection and identifying the genotype of the virus.

92.1.4.1 Conventional Techniques

Giemsa-stained impression smears of splenic cells have been commonly used for the rapid diagnosis of RSIV in

TABLE 92.1

Different Detection Methods Currently Used to Diagnose Megalocytivirus Infection in Fish

Detection Methods	Viruses Strains	Sensitivity	References
Histopathology	RSIV	—	[6]
	ISKNV	—	[18]
	DGIV, ALIV	—	[16,55]
	TGIV	—	[53]
	TRBIV	—	[24,33]
	TSIV	—	[17]
TEM	RSIV	—	[6]
	ISKNV	—	[18,60]
	DGIV, ALIV	—	[16,55]
	TGIV	—	[53]
	RBIV	—	[64]
	LYCIV	—	[22]
	TRBIV	—	[24]
	TSIV	—	[17]
	SKIV-ZJ07	—	[56]
Cell culture test	ISKNV	—	[45]
	SKIV-ZJ07	—	[56]
IF test	RSIV	—	[62]
	ISKNV	—	[45,60]
Immunohistochemistry	ISKNV	—	[60]
ELISA and dot blot assay	RBIV	0.3 μg of total protein/mL	[63]
ISH	TGIV	—	[53]
	RBIV	—	[64]
PCR	RSIV	—	[1,65–67]
	RSIV	10^{-4} μg of DNA/mL	[66,70]
	TRBIV	—	[33]
	DGIV, ALIV,	—	[16,72]
	TSIV	—	[17]
Nested PCR	RSIV	—	[66]
	TGIV	—	[68]
Multiplex PCR	RSIV	—	[69]
qPCR	RSIV	$10^{3.5}$ genome/mg	[73,74]
	RSIV	10^{-6} μg of DNA/mL	[70]
	DGIV	10–100 fg of DNA	[71]
	DGIV	100 copies viral DNA	[72]
LAMP	RSIV	3 pg of DNA	[75]
	ISKNV	10 copies viral DNA	[76]

The following are the viruses with their corresponding abbreviations used: red sea bream iridovirus (RSIV), rock bream iridovirus (RBIV), infectious spleen and kidney necrosis virus (ISKNV), dwarf gourami iridovirus (DGIV), African lampeye iridovirus (ALIV), large yellow croaker iridovirus (LYCIV), Taiwan grouper iridovirus (TGIV), spotted knifejaw iridovirus (SKIV-ZJ07), three-spine stickleback iridovirus (TSIV), and turbot reddish body iridovirus (TRBIV).

fish in Japan. However, this method provides only a presumptive diagnosis; the sensitivity and accuracy of this method are poor [1]. In addition, isolation of the virus in cell cultures by using susceptible cell lines is time-consuming and requires exquisite manipulation by highly trained technicians; the drawbacks are similar to those of histopathological and TEM observation. Therefore, these three methods for diagnosing megalocytivirus infection are impracticable in field tests. A monoclonal antibody (MAb) M10, which

is a RSIV-specific polypeptide (180–230 kDa), was established and used to develop a method for rapidly diagnosing RSIV by using an indirect IF test [61]. The rate at which infected fish were detected by using the IF test with MAb was much higher than that obtained using Giemsa-stained impression smears [62]. The use of MAb M10 is described in the manual *Diagnostic Methods for Aquatic Animals* published by the World Organisation for Animal Health (OIE) and applicable to both RSIV and ISKNV.

In addition, an RBIV-specific polyclonal antibody was produced using a recombinant RBIV-049L fusion protein with glutathione *S*-transferase in mice. One-step PCR with a dot blot–indirect ELISA combination detection system by using the ORF-49L protein and its antiserum showed high sensitivity and specificity for RBIV [63]. In the zebrafish ISKNV infection model, polyclonal mouse sera against viral protein VP23R can be used to mark the infected cells through immunohistochemistry. Specific staining of the plasma membrane with the anti-VP23R antiserum shows that ISKNV-infected hypertrophic cells are present in different organs of moribund zebrafish. Using a double-stained immunofluorescent assay, VP23R signals were found in an equal number of enlarged cells in zebrafish and mandarin fish, indicating that ISKNV can infect zebrafish as effectively as it infects mandarin fish [60].

92.1.4.2 Molecular Techniques

On the basis of the nucleotide sequences of megalocytiviruses genomes, several specific and sensitive detection methods have been developed for diagnosing iridoviral disease outbreaks. Among these methods, ISH can successfully show the distribution and kinetics of the viral DNA–containing cells of infected fish. Digoxigenin-labeled probes were cloned from parts of the viral genome, and the sections of various tissues were hybridized with specific probes. The results of ISH showed the megalocytivirus-specific DNA located in most of the basophilic but not eosinophilic enlarged cells of various infected organs. Nuclei of infected cells were labeled during an early stage of infection, and the nuclei and cytoplasm were labeled during later stages [53]. Viral DNA was also found in the cytoplasm of leucocytes in blood smears by using the ISH method. The results suggested that virus-infected cells can migrate to the connective tissues of various organs from the blood [64].

Conventional diagnosis of megalocytivirus is based on histopathological, electron microscopic observation, or immunoassays. However, using the increased amount of information available on the genomic regions of megalocytivirus, PCR amplification has been developed to provide a rapid, simple, and sensitive method for detecting virus-specific nucleic acids. Several PCR-based methods have been established for detecting and quantifying megalocytivirus infection. Primer sets employed in PCR have been designed to detect target regions of different viral genes and can detect both ISKNV and RSIV-like megalocytivirus genotypes (Table 92.2). In 1996, a small subunit of the RNRS gene of RSIV was cloned. Subsequently, PCR was used to amplify virus-specific nucleic acids with a primer set based on the RNRS gene [65]. In addition, four oligonucleotide primer sets were synthesized on the basis of the ATPase gene, DNA polymerase (DPOL) gene, and a *Pst* I–restriction fragment of RSIV genomic DNA. A PCR- and nested PCR-based diagnostic system were developed for detecting RSIV infection in various species of affected fish, and none of the PCR products were obtained from frog virus 3 or lymphocystis disease virus [66]. Using the

designed primers from the *Pst* I–restriction fragment of RSIV genomic DNA, four species of cultured fish showing clinical signs of RSIV diseases were identified to have megalocytivirus infection by using PCR in Korea. Nucleotide sequence analysis revealed that these viruses were the same as RSIV or had a slight variation in RSIV DNA sequences [67]. Nested PCR was also developed for detecting grouper iridovirus infection in cultured hybrid grouper, giant seaperch, and largemouth bass in Taiwan. The PCR results and sequence analyses of the amplicons suggested that the etiological agents infecting giant seaperch and largemouth bass might be the same and related to but distinct from TGIV [68]. Considering that variants of RSIV have been found in different species, countries, and years, it is yet to be determined whether PCR primers that have been published in different reports are suitable for their detection. Multiplex PCR with all four primers derived from two variable and two conserved positions in the *Pst* I–restriction fragment of RSIV was developed. Three RSIV strains isolated in Korea could be distinguished according to the sizes of the PCR amplicons [69]. The presence of multiple viral strains among genus *Megalocytivirus* has caused problems in designing primers for diagnostic PCR. By using genomic DNA sequence information for the viral MCP gene, new PCR primers for discriminating among the three viral species– RSIV, ISKNV, and TRBIV, of the genus *Megalocytivirus*– can be developed. The specifics of the primer sets for RSIV and ISKNV have been confirmed; however, those of the TRBIV primer sets remain unknown [1].

qPCR is suitable for use in large research studies because it is a rapid test, enables quantifying a viral load, and is more sensitive compared with conventional PCR. Requiring less time and having a lower risk of contamination, qPCR is an attractive alternative because only one round of amplification is performed and analysis through gel electrophoresis is not necessary. Of the chemicals used to detect PCR products during qPCR, the inexpensive dsDNA-binding dye SYBR Green I is desirable because it eliminates the need for costly probes linked to fluorescent molecules. In addition, many conventional PCRs can be used in a real-time format with few modifications. Thus far, several reports have detailed the development of qPCR assays for megalocytiviruses. In 2003, a novel qPCR detection and quantification method for RSIV was developed using the MCP gene. The detection method is two orders of magnitude more sensitive than conventional PCR. The MCP gene was also shown to be specific to RSIV (i.e., it did not react with hirame rhabdovirus, viral hemorrhagic septicemia virus, or DNA samples from apparently healthy red sea bream). The viral loads of dead juvenile red sea bream experimentally challenged with RSIV were determined using qPCR to have MCP gene copy totals ranging from 2.4×10^5 to 5.4×10^7. The highest viral load was obtained in the heart, whereas the lowest copy total was detected in the blood [70]. Based on the nucleotide sequence of the MCP gene, real-time PCRs that involve using SYBR Green and molecular beacon assays were developed to specifically target gourami iridovirus. The analytical sensitivities

TABLE 92.2
The Primers Used in Different Polymerase Chain Reaction Assay for Megalocytivirus Detection

Assay Type	Virus Strains	Oligonucleotide			Product Size (Base Pairs)	Target	References
		Name	Primer	Sequence(5′–3′)			
Conventional PCR	RSIV	1F	Forward	CTCAAACACTCTGGCTCATC	570	959 bp *Pst* I fragment	[66]
	RSIV	1R	Reverse	GCACCAACACATCTCCTATC			[66]
	RSIV	2F	Forward	TACAACATGCTCCGCCAAGA	564	959 bp *Pst* I fragment	[66]
	RSIV	2R	Reverse	GCGTTAAAGTAGTGAGGGCA			[66]
	RSIV, ISKNV, TRBIV	MCP-uni332-F3	Forward	AGGTGTCGGTGTCATTAACGACCTG	777	MCP	[1]
	RSIV, ISKNV, TRBIV	MCP-uni1108-R8	Reverse	TCTCAGGCATGCTGGGCGCAAAG		MCP	[1]
	RSIV, ISKNV,	MCP-exeT37-F1	Forward	TTCATCGACATCTCCGCGTTT	486	MCP	[1]
	RSIV, ISKNV,	MCP-exeT512-R1	Reverse	AATGGGCAAATTAAGGTAGRCG		MCP	[1]
	TRBIV	MCP-specT37-F1	Forward	TTCATCGACATCTCCGCTTTC	453	MCP	[1]
	TRBIV	MCP-specT490-R1	Reverse	TSTGACCGTTGGTGATACCGGAG		MCP	[1]
	RSIV	MCP-specR697-F4	Forward	CCCGCACTGACCAACGTGTCC	191	MCP	[1]
	RSIV	MCP-spec-R888-R6	Reverse	CACAGGGTGACTGAACCTCAGGTCG		MCP	[1]
	ISKNV	MCP-spec-I465-F3	Forward	GGTGGCCGGCATCACCAACGGC	415	MCP	[1]
	ISKNV	MCP-spec-I879-R6	Reverse	CACGGGGTGACTGAACCTG		MCP	[1]
Nested PCR (first round)	DGIV,RSIV	C1105	Forward	GGGTTCATCGACATCTCCGCG	430	MCP	[72]
	DGIV,RSIV	C1106	Reverse	AGGTCGCTGCGCATGCCAATC		MCP	[72]
qPCR and nested PCR (second round)	DGIV,RSIV	C1073	Forward	AATGCCGTGACCTACTTTGC	167	MCP	[72]
	DGIV,RSIV	C1074	Reverse	GATCTTAACACGCAGCCACA		MCP	[72]

For virus abbreviations see Table 92.1.

of the SYBR Green and molecular beacon assays are 10 and 100 fg, respectively. The analytical specificity of real-time PCR determined that the assays can detect all megalocytiviruses with no cross-reaction, nontarget viruses, and aquatic bacterial species being observed [71]. In addition, a qPCR assay for detecting DGIV and other megalocytiviruses was developed in Australia. The limit of detection for this qPCR assay was 100 copies, and the assay did not cross-react with ranaviruses or other aquatic viral pathogens tested. DNA extraction using the magnetic bead or spin column–based methods had no effect on the sensitivity of the qPCR assay. Compared with the sensitivity of OIE reference PCR protocol, the sensitivity was improved by approximately 3–4 logs. The OIE reference PCR protocol detected only 8% of the samples identified to be positive through qPCR [72]. A qPCR targeting the *Pst* I–fragment gene of RSIV was developed. The quantitative detection limit of the assay was $10^{3.5}$ genomes/mg, corresponding to 100 genomes per reaction. After RSIV inoculation of rock bream, the viral genome was detected in fish at 10 days by using qPCR [73]. The qPCR detection method was employed to investigate the kinetics of RSIV in fish. The results of measurements of the viral DNA of several organs from experimental fish infected using the immersion method showed that the spleen is the target organ of RSIV in Japanese amberjack [74].

PCR-based methods are time-consuming and require expensive equipment and reagents, rendering them unsuitable for use in field tests. LAMP, a novel nucleic acid amplification method, was developed and involved employing a *Bst* DPOL with strand displacement activity and four primers that recognize a total of six distinct sequences on the target DNA. In the LAMP method, target DNA can be amplified from a few copies to 10^9 copies within 1 h under isothermal conditions (60°C–65°C), by using a simple incubator such as a water bath or a thermostat block. At the end of the reaction, the presence or absence of the target DNA is judged visually according to the appearance of a white precipitate magnesium pyrophosphate. The presence of multiple bands of varying sizes during agarose gel electrophoresis of the LAMP reaction products indicates that they are composed of a mixture of stem-loop DNAs with various sizes of stem and cauliflower-like structures with multiple loops induced through annealing between alternatively inverted repeats of the target sequence in the same strand. A LAMP procedure was established to detect the genomic DNA molecule of RSIV targeting the *Pst* I–fragment gene. The method was at least 10 times more sensitive than conventional PCR in detecting the presence of RSIV [75]. A 20 min LAMP amplification of the DPOL gene of ISKNV by using a biotin-labeled primer combined with lateral flow dipstick (LFD) chromatography was developed. The detection limit of LAMP–LFD in detecting ISKNV is 10 copies of the viral genome. The method has higher sensitivity and speed and less dependence on equipment than standard PCR in specifically detecting low levels of ISKNV DNA; this method can be used in the field as a routine diagnostic tool for megalocytiviruses [76].

92.2 METHODS

92.2.1 Sample Collection and Preparation

Moribund or dead fish displaying signs of megalocytivirus infection can be obtained from cultured fish farms or ornamental fish shops. The spleens and kidneys of infected fish are collected and stored at −80°C until used. Several methods have been employed for extracting of megalocytivirus DNA from clinical specimens prior to conventional PCR, nested PCR, or qPCR. These methods include conventional DNA extraction by using phenol/chloroform, extraction by using magnetic beads, and the spin column–based method. Recently, highly sensitive and reproducible DNA extraction kits have been developed by various commercial companies and are prevalently used for genomic DNA extraction in megalocytiviruses. The procedures for extracting DNA by using phenol/chloroform and the spin column–based method (Wizard Genomic DNA Purification System; Promega) are described in the following two sections.

92.2.1.1 DNA Extraction by Using Phenol/Chloroform

92.2.1.1.1 Procedure

1. Homogenize samples of approximately 30–50 mg of tissues from megalocytivirus-infected fish in a 500 μL lysis buffer (50 mM Tris–HCl, pH 8.0; 20 mM NaCl, 3% SDS; 10 mM EDTA, pH 8.0; proteinase K at 100 μg/mL) by using a small homogenizer. Incubate the mixture in a 1.5 mL microcentrifuge tube at 65°C for 1–3 h.

2. Add an equal volume of saturated phenol to the mixture. Mix the phases through vortexing for approximately 10 s. Resolve the phases through microcentrifugation for 2 min. Carefully remove and transfer the upper aqueous phase to a fresh tube by using a micropipettor. If a heavy interface is present, repeat this procedure until the excessive interface is no longer present.

3. Add an equal volume of chloroform. Mix. Resolve the phases through centrifugation for 2 min in a microcentrifuge. Transfer the upper aqueous phase to a fresh tube.

4. Add 1/10 volume of 3 M sodium acetate (pH 7.0) and 2.5 volume of 95% ethanol. Mix through vortexing or repeated inversion. Incubate the mixture on ice for 10 min, or at −20°C for 4 h to overnight (i.e., for small amounts of DNA). Collect the precipitate through microcentrifugation for 15–20 min at 13,000–16,000 × *g*.

5. Decant the supernatant. Add 1 mL 75% ethanol to the tube containing the pellet. Mix. Collect the pellet through centrifugation for 15–20 min at 13,000–16,000 × *g*. Decant the ethanol and dry the pellet in an oven until all residual ethanol is evaporated.

6. Dissolve the pellet in 100 μL of either sterile water or a TE buffer. The DNA can be used directly in the PCR step or stored at −20°C for future use.

92.2.1.2 DNA Extraction by Using Spin Column

92.2.1.2.1 Procedure

1. Add 10–20 mg of fresh or thawed tissues to 600 μL of a chilled nuclei lysis solution and homogenize the mixture for 10 s by using a small homogenizer. Incubate the lysate in a 1.5 mL microcentrifuge tube at 65°C for 15–30 min.
2. Add 3 μL of RNase solution to the nuclear lysate and mix the sample by inverting the tube 2–5 times. Incubate the mixture for 15–30 min at 37°C. Allow the sample to cool to room temperature for 5 min before proceeding.
3. To the room-temperature sample, add 200 μL of protein precipitation solution and vortex vigorously at high speed for 20 s. Chill the sample on ice for 5 min. Centrifuge it for 4 min at 13,000–16,000 × *g*. The precipitate protein will form a tight, white pellet.
4. Carefully remove the supernatant containing the DNA and transfer it to a clean 1.5 mL microcentrifuge tube containing 600 μL of room-temperature isopropanol. Gently mix the solution through inversion until the white threadlike strands of DNA form a visible mass.
5. Centrifuge the mixture for 1 min at 13,000–16,000 × *g* at room temperature. Carefully decant the supernatant. Add 600 μL of room-temperature 70% ethanol, and gently invert the tube several times to wash the DNA. Centrifuge for 1 min at 13,000–16,000 × *g* at room temperature.
6. Carefully aspirate the ethanol by using either a drawn Pasteur pipette or a sequencing pipette tip. Invert the tube on clear absorbent paper, and air dry the pellet for 10–15 min.
7. Add 100 μL of a DNA rehydration solution, and rehydrate the DNA by incubating it at 65°C for 1 h. Alternatively, rehydrate the DNA by incubating the solution overnight at room temperature or at 4°C. The DNA can be used directly in the PCR step or stored at −20°C for future use.

92.2.2 Detection Procedures

The genomic diversity of megalocytiviruses in fish is evidenced by the identification of three genotypes (RSIV, ISKNV, and TRBIV) within the genus *Megalocytivirus*. Several molecular diagnostic methods have been developed and used to detect these three genotypes. The following protocols (conventional PCR, nested PCR, and qPCR) have been modified from previous reports.

92.2.2.1 Conventional PCR

92.2.2.1.1 Procedure

1. Remove all aliquots of PCR reagents (nuclease-free water, 10× PCR buffer, dNTP mix, *Taq* DPOL, 25 mM MgCl₂, primer pairs, and temple DNA) from the freezer and store on ice. Vortex and spin all reagents before opening the tubes.

2. Prepare 25 μL of a PCR mix in a 0.2 mL PCR tube as follows:

Component	Volume (μL)
Nuclease-free water	16.5
10× PCR buffer	2.5
dNTP mix (10 mM of each dNTP)	1.0
Taq DPOL (5 units/μL)	0.5
25 mM MgCl₂	1.5
Forward primer (1F) (20 pmol/μL)	1.0
Reverse primer (1R) (20 pmol/μL)	1.0
Template DNA (100 ng)	1.0
Total	25.0

3. Spin the sample tubes in a microcentrifuge for 30 s at room temperature. Turn on a thermocycler and preheat the block to 94°C. Place the sample tubes in the thermal cycler for 40 cycles at 94°C for 0.5 min, 58°C for 1 min, and 72°C for 1 min, followed by an extended period of 10 min at 72°C after the last cycle.
4. Following the amplification, the samples are ready for electrophoresis or to be stored frozen at −20°C. Analyze the PCR products through agarose electrophoresis using 1% agarose gel in a TAE buffer at 100 V for 30 min.
5. Stain the gel with ethidium bromide and then observe the samples under a UV light source.

Note: Negative control is concurrently included along with the test samples to monitor for possible contamination during the PCR procedure. The presence of a 570 bp product is indicative of RSIV genomes in the samples. However, different primer pairs can be used to detect and discriminate three viral species of the genus *Megalocytivirus*: RSIV, ISKNV, and TRBIV (Table 92.2).

92.2.2.2 Nested PCR

Reaction conditions are the same as those described for conventional PCR amplification. However, the primer set is replaced by 1F and 2R with 848 bp amplicons in the first-round PCR assays. The amplified products are diluted 1:100 in nuclease-free water and subjected to reamplification. The second round of amplification is performed using primers 2F and 1R (Table 92.2), and the 248 bp products are amplified. An alternative nested PCR is also performed using the C1105/C1106 primer set for the first amplification and C1073/C1074 primer set for the second PCR (Table 92.2). The amplified products are 430 bp and 167 bp, respectively. The temperature and conditions for the PCR assays are listed as follows: 1 cycle at 95°C for 3 min and 35 cycles at 95°C for 0.5 min, 55°C for 0.5 min, and 72°C for 1 min, followed by an extended period of 5 min at 72°C after the last cycle.

92.2.2.3 Real-Time PCR

92.2.2.3.1 Procedure

1. Remove all aliquots of qPCR reagents (nuclease-free water, SYBR Green Master Mix, primer pairs, template DNA) from the freezer and store on ice. Vortex and spin all reagents before opening the tubes.
2. Prepare 25 µL of a PCR mix in 0.2 mL of qPCR for 8 tubes (or 12 tubes) as follows:

Component	Volume (µL)
Nuclease-free water	9.75
SYBR Green Master Mix	12.5
Forward primer (C1073) (100 pmol/µL)	0.125
Reverse primer (C1074) (100 pmol/µL)	0.125
Template DNA (100 ng)	2.5
Total	25.0

3. Place the sample (8 tubes or 12 tubes) in the qPCR system for 1 cycle at 95°C for 15 min and 40 cycles at 95°C for 0.5 min, 62°C for 0.5 min, and 72°C for 0.5 min.
4. Standard curves are produced through 10-fold serial dilution of the plasmid-containing target region in Molecular Grade Water to establish template concentrations of 1×10^7 to 1×10^0 copies per reaction. Standard curves, which are generated using the mean data from experiments performed in duplicate, indicate a satisfactory linear relationship between the threshold cycle (Ct) value and the logarithmic plasmid concentration. Positive and no template controls are included in each run.
5. Quantifications of the viral DNA are interpolated from the plasmid DNA standard curve, based on the Ct value determined after 40 cycles of the unknown sample.

92.3 CONCLUSION AND FUTURE PERSPECTIVES

Over the past two decades, epizootic iridoviral disease has caused serious economic losses in aquaculture worldwide. In particular, megalocytivirus infections represent one of the most consequential diseases in cultured fish. The availability of rapid, specific, and sensitive pathogen detection methods is crucial to monitor the disease status, evaluate viral kinetics, reduce virus transmission, and minimize the economic and environmental impact of viral diseases in aquaculture.

To control and prevent megalocytivirus infections, formalin-inactivated RSIV vaccine was developed [77–80]. The vaccine is efficient for protecting red sea bream (*P. major*), yellowtail (*Seriola quinqueradiata*), and other fish species but is not effective for fishes belonging to the genus *Oplegnathus*. This injectable vaccine is now commercially available in

Japan, and a similar megalocytiviral vaccine has been developed in Taiwan for grouper iridoviral disease [1]. A formalin-killed cell (FKC) vaccine was also prepared using ISKNV propagated in the MFF-1 cell line, and more than 90% of fish immunized with the vaccine were protected against ISKNV virulence [81]. An FKC vaccine used to immunize cage-cultured mandarin fish was also shown to be efficient in preventing RSIV infection in field trials [82]. Recombinant subunit vaccines with the RSIV and ISKNV MCP gene have been developed and induced protective immunity against iridovirus disease in fish but have not been commercialized [83,84]. In addition, DNA vaccines based on the RSIV MCP gene and seven genes of RBIV-1 were separately constructed. The results demonstrated that DNA vaccines are able to induce robust protection in fish against RSIV infection and a cellular immune response, as shown by the upregulation of the MHC class I transcript, which may be associated with such protection, after vaccination [85]. The protective effects of RBIV-1 DNA vaccines were examined in a turbot model. The results indicated that pCN 86, which expresses an equivalent of the RSIV 351R gene, is a highly effective DNA vaccine that may be used in controlling megalocytivirus-associated diseases in aquaculture [86].

Given the high level of manual labor and stress to fish caused by injection, oral vaccines based on the MCP gene of megalocytivirus have been developed by using cell-surface display system technology in yeast [87,88]. However, the protective efficiency of the oral vaccines requires further confirmation through field trials. Recently, RNAi has been shown to have activity against a wide range of virus infection and is considered a potential antiviral tool. In aquaculture, relatively few studies have reported on the use of RNAi for protection from megalocytivirus infection. Thus far, the results have only shown that MCP-targeted siRNA and miRNA can effectively and specifically inhibit the expression of the target gene and hinder RSIV replication in a cell culture system [89,90]. The precise mechanism of the RNAi employed in preventing the megalocytivirus infection is not fully understood.

Although commercial vaccines are available in Japan for preventing RSIV infection, the existence of other virus strains in the genus *Megalocytivirus* weakens the protective effect and application of the vaccines. Thus, prevention of exposure, together with the adoption of rapid and accurate diagnostic technologies, remains the most effective method for controlling diseases caused by megalocytivirus.

REFERENCES

1. Kurita, J. and Nakajima, K., Megalocytiviruses. *Viruses*, 4, 521, 2012.
2. Darai, G. et al., Analysis of the genome of fish lymphocystis disease virus isolated directly from epidermal tumours of pleuronectes. *Virology*, 126, 466, 1983.
3. Darai, G. et al., Molecular cloning and physical mapping of the genome of fish lymphocystis disease virus. *Virology*, 146, 292, 1985.

4. Chinchar, V.G. et al., Family Iridoviridae. In: *Virus Taxonomy, Eighth Report of the International Committee on Taxonomy of Viruses*, eds. C.M. Fauquet, M.A. Mayo, J. Maniloff, U. Desselberger, and L.A. Ball, pp. 145–162. Academic Press, San Diego, CA, 2005.

5. Fu, X. et al., Genotype and host range analysis of infectious spleen and kidney necrosis virus (ISKNV). *Virus Genes*, 42, 97, 2011.

6. Inouye, K. et al., Iridovirus infection of cultured red sea bream, *Pagrus major*. *Fish Pathol*, 27, 19, 1992.

7. Matsuoka, S. et al., Cultured fish species affected by red sea bream iridoviral disease from 1991–1995. *Fish Pathol*, 31, 233, 1996.

8. Kawakami, H. et al., Cultured fish species affected by red sea bream iridoviral disease from 1996 to 2000. *Fish Pathol*, 37, 45, 2002.

9. Wang, Y.Q. et al., Molecular epidemiology and phylogenetic analysis of a marine fish infectious spleen and kidney necrosis virus–like (ISKNV–like) virus. *Arch Virol*, 152, 763, 2007.

10. Wang, C.S. et al., Studies on the epizootic of iridovirus infection among red sea bream, *Pagrus major* (Temminck & Schlegel), cultured in Taiwan. *J Fish Dis*, 26, 127, 2003.

11. Wang, C.S. et al., PCR amplification and sequence analysis of the major capsid protein gene of megalocytiviruses isolated in Taiwan. *J Fish Dis*, 32, 543, 2009.

12. Huang, S.M. et al., Genetic analysis of fish iridoviruses isolated in Taiwan during 2001–2009. *Arch Virol*, 156, 1505, 2011.

13. Do, J.W. et al., Sequence variation in the gene encoding the major capsid protein of Korean fish iridoviruses. *Arch Virol*, 150, 351, 2005.

14. Jeong, J.B. et al., Outbreaks and risks of infectious spleen and kidney necrosis virus disease in freshwater ornamental fishes. *Dis Aquat Organ*, 78, 209, 2008.

15. Jeong, J.B. et al., Characterization of the DNA nucleotide sequences in the genome of red sea bream iridoviruses isolated in Korea. *Aquaculture*, 220, 119, 2003.

16. Go, J. et al., The molecular epidemiology of iridovirus in Murray cod (*Maccullochella peelii peelii*) and dwarf gourami (*Colisa lalia*) from distant biogeographical regions suggests a link between trade in ornamental fish and emerging iridoviral diseases. *Mol Cell Probes*, 20, 212, 2006.

17. Waltzek, T.B. et al., Systemic iridovirus from threespine stickleback *Gasterosteus aculeatus* represents a new megalocytivirus species (family *Iridoviridae*). *Dis Aquat Organ*, 98, 41, 2012.

18. He, J.G. et al., Systemic disease caused by an iridovirus–like agent in cultured mandarinfish, *Siniperca chuatsi* (Basilewsky), in China. *J Fish Dis*, 23, 219, 2000.

19. Miyata, M. et al., Genetic similarity of iridoviruses from Japan and Thailand. *J Fish Dis*, 20, 127, 1997.

20. Jung, S.J. and Oh, M.J., Iridovirus–like infection associated with high mortalities of striped beakperch, *Oplegnathus fasciatus* (Temminck et Schlegel), in southern coastal areas of the Korean peninsula. *J Fish Dis*, 23, 223, 2000.

21. Sudthongkong, C. et al., Iridovirus disease in two ornamental tropical freshwater fishes: African lampeye and dwarf gourami. *Dis Aquat Organ*, 48, 163, 2002.

22. Chen, X.H. et al., Outbreaks of an iridovirus disease in maricultured large yellow croaker, *Larimichthys crocea* (Richardson), in China. *J Fish Dis*, 26, 615, 2003.

23. Lu, L. et al., Complete genome sequence analysis of an iridovirus isolated from the orange–spotted grouper, *Epinephelus coioides*. *Virology*, 339, 81, 2005.

24. Shi, C.Y. et al., The first report of an iridovirus–like agent infection in farmed turbot, *Scophthalmus maximus*, in China. *Aquaculture*, 236, 11, 2004.

25. Eaton, H.E. et al., Comparative genomic analysis of the family *Iridoviridae*: Re–annotating and defining the core set of iridovirus genes. *Virol J*, 4, 11, 2007.

26. Sudthongkong, C. et al., Viral DNA sequences of genes encoding the ATPase and the major capsid protein of tropical iridovirus isolates which are pathogenic to fishes in Japan, South China Sea and Southeast Asian countries. *Arch Virol*, 147, 2089, 2002.

27. Hyatt, A.D. et al., Comparative studies of piscine and amphibian iridoviruses. *Arch Virol*, 145, 301, 2000.

28. Song, J.Y. et al., Genetic variation and geographic distribution of megalocytiviruses. *J Microbiol*, 46, 29, 2008.

29. Nakajima, K. and Kurita, J., Red sea bream iridoviral disease. *Uirusu*, 55, 115, 2005.

30. Kim, W.S. et al., Detection of megalocytivirus from imported tropical ornamental fish, paradise fish *Macropodus opercularis*. *Dis Aquat Organ*, 90, 235, 2010.

31. Weber, E.S. et al., Systemic iridovirus infection in the Banggai cardinalfish (*Pterapogon kauderni* Koumans 1933). *J Vet Diagn Invest*, 21, 306, 2009.

32. Wang, Q. et al., Outbreaks of an iridovirus in marbled sleepy goby, *Oxyeleotris marmoratus* (Bleeker), cultured in southern China. *J Fish Dis*, 34, 399, 2011.

33. Kim, W.S. et al., Characterization of an iridovirus detected from cultured turbot *Scophthalmus maximus* in Korea. *Dis Aquat Organ*, 64, 175, 2005.

34. Do, J.W. et al., Phylogenetic analysis of the major capsid protein gene of iridovirus isolates from cultured flounders *Paralichthys olivaceus* in Korea. *Dis Aquat Organ*, 64, 193, 2004.

35. Goorha, R. and Murti, K.G., The genome of frog virus 3, an animal DNA virus, is circularly permuted and terminally redundant. *Proc Natl Acad Sci USA*, 79, 248, 1982.

36. He, J.G. et al., Complete genome analysis of the mandarin fish infectious spleen and kidney necrosis iridovirus. *Virology*, 291, 126, 2001.

37. Kurita, J. et al., Complete genome sequencing of red sea bream iridovirus (RSIV). *Fisheries Sci*, 68, 1113, 2002.

38. Do, J.W. et al., Complete genomic DNA sequence of rock bream iridovirus. *Virology*, 325, 351, 2004.

39. Shi, C.Y. et al., Complete genome sequence of a megalocytivirus (family Iridoviridae) associated with turbot mortality in China. *Virol J*, 7, 159, 2010.

40. Zhang, B.C. et al., Complete genome sequence and transcription profiles of the rock bream iridovirus RBIV–C1. *Dis Aquat Organ*, 104, 203, 2013.

41. Eaton, H.E. et al., The genomic diversity and phylogenetic relationship in the family *Iridoviridae*. *Viruses*, 2, 1458, 2010.

42. Imajoh, M. et al., Characterization of a new fibroblast cell line from a tail fin of red sea bream, *Pagrus major*, and phylogenetic relationships of a recent RSIV isolate in Japan. *Virus Res*, 126, 45, 2007.

43. Ku, C.C. et al., Establishment and characterization of a fibroblast cell line derived from the dorsal fin of red sea bream, *Pagrus major* (Temminck & Schlegel). *J Fish Dis*, 33, 187, 2010.

44. Wen, C.M. et al., Development of two cell lines from *Epinephelus coioides* brain tissue for characterization of betanodavirus and megalocytivirus infectivity and propagation. *Aquaculture*, 278, 14, 2008.

45. Dong, C. et al., Development of a mandarin fish *Siniperca chuatsi* fry cell line suitable for the study of infectious spleen and kidney necrosis virus (ISKNV). *Virus Res*, 135, 273, 2008.

46. Wang, N. et al., Development and characterization of a new marine fish cell line from turbot (*Scophthalmus maximus*). *Fish Physiol Biochem*, 36, 1227, 2010.

47. Fan, T.J. et al., Establishment of a turbot fin cell line and its susceptibility to turbot reddish body iridovirus. *Cytotechnology*, 62, 217, 2010.

48. Nakajima, K. and Sorimachi M., Biological and physico–chemical properties of the iridovirus isolated from cultured red sea bream, *Pagrus major*. *Fish Pathol*, 29, 29, 1994.

49. He, J.G. et al., Experimental transmission, pathologenicity and physical–chemical properties of infectious spleen and kidney necrosis virus (ISKNV). *Aquaculture*, 204, 11, 2004.

50. Jeong, J.B. et al., Transmission of iridovirus from freshwater ornamental fish (pearl gourami) to marine fish (rock bream). *Dis Aquat Organ*, 82, 27, 2008.

51. Sriwanayos, P. et al., Megalocytivirus infection in orbiculate batfish *Platax orbicularis*. *Dis Aquat Organ*, 105, 1, 2013.

52. Jin, J.W. et al., Dynamics of megalocytivirus transmission between bivalve molluscs and rock bream *Oplegnathus fasciatus*. *Aquaculture*, 428, 29, 2014.

53. Chao, C.B. et al., Histological, ultrastructural, and *in situ* hybridization study on enlarged cells in grouper *Epinephelus* hybrids infected by grouper iridovirus in Taiwan (TGIV). *Dis Aquat Organ*, 58, 127, 2004.

54. Nakajima, K. and Maeno, Y., Pathogenicity of red sea bream iridovirus and other fish iridoviruses to red sea bream. *Fish Pathol*, 33, 143, 1998.

55. Go, J. and Whittington, R., Experimental transmission and virulence of a megalocytivirus (Family: *Iridoviridae*) of dwarf gourami (*Colisa lalia*) from Asia in Murray cod (*Maccullochella peelii peelii*) in Australia. *Aquaculture*, 258, 140, 2006.

56. Dong, C. et al., A new marine megalocytivirus from spotted knifejaw, *Oplegnathus punctatus*, and its pathogenicity to freshwater mandarinfish *Siniperca chuatsi*. *Virus Res*, 147, 98, 2010.

57. Jun, L.J. et al., Influence of temperature shifts on the onset and development of red sea bream iridoviral disease in rock bream *Oplegnathus fasciatus*. *Dis Aquat Organ*, 84, 201, 2009.

58. Choi, S.K. et al., Organ distribution of red sea bream iridovirus (RSIV) DNA in asymptomatic yearling and fingerling rock bream (*Oplegnathus fasciatus*) and effects of water temperature on transition of RSIV into acute phase. *Aquaculture*, 256, 23, 2006.

59. Jeong, J.B. et al., Asymptomatic iridovirus infection in various marine fishes detected by a 2–step PCR method. *Aquaculture*, 255, 30, 2006.

60. Xu, X. et al., A zebrafish (*Danio rerio*) model of infectious spleen and kidney necrosis virus (ISKNV) infection. *Virology*, 376, 1, 2008.

61. Nakajima, K. and Sorimachi, M., Production of monoclonal antibodies against red sea bream iridovirus. *Fish Pathol*, 30, 47, 1995.

62. Nakajima, K. et al., Immunofluorescence test for the rapid diagnosis of red sea bream iridovirus infection using monoclonal antibody. *Fish Pathol*, 30, 115, 1995.

63. Kim, Y.I. et al., Production and characterization of polyclonal antibody against recombinant ORF–049L of rock bream (*Oplegnathus fasciatus*) iridovirus. *Proc Biochem*, 42, 134, 2007.

64. Lee, N.S. et al., Characterization of virus distribution in rock bream (*Oplegnathus fasciatus*; Temminck and Schlegel) infected with megalocytivirus. *J Comp Pathol*, 141, 63, 2009.

65. Oshima, S.I. et al., A method for direct DNA amplification of uncharacterized DNA viruses and for development of a viral polymerase chain reaction assay: Application to the red sea bream iridovirus. *Anal Biochem*, 242, 15, 1996.

66. Kurita, J. et al., Polymerase chain reaction (PCR) amplification of DNA of red sea bream iridovirus (RSIV). *Fish Pathol*, 33, 17, 1998.

67. Kim, Y.J. et al., PCR amplification and sequence analysis of irido–like virus infecting fish in Korea. *J Fish Dis*, 25, 121, 2002.

68. Chao, C.B. et al., A nested PCR for the detection of grouper iridovirus in Taiwan (TGIV) in cultured hybrid grouper, giant seaperch, and largemouth bass. *J Aquat Anim Health*, 14, 104, 2002.

69. Jeong, J.B. et al., Multiplex PCR for the diagnosis of red sea bream iridoviruses isolated in Korea. *Aquaculture*, 235, 139, 2004.

70. Caipang, C.M. et al., Development of a real–time PCR assay for the detection and quantification of red seabream iridovirus (RSIV). *Fish Pathol*, 38, 1, 2003.

71. Gias, E. et al., Development of real–time PCR assays for detection of megalocytiviruses in imported ornamental fish. *J Fish Dis*, 34, 609, 2011.

72. Rimmer, A.E. et al., Development of a quantitative polymerase chain reaction (qPCR) assay for the detection of dwarf gourami iridovirus (DGIV) and other megalocytiviruses and comparison with the Office International des Epizooties (OIE) reference PCR protocol. *Aquaculture*, 358, 155, 2012.

73. Oh, S.Y. and Nishizawa, T., Optimizing the quantitative detection of red seabream iridovirus (RSIV) genome from splenic tissues of rock bream *Oplegnathus fasciatus* using a qPCR assay. *Fish Pathol*, 48, 21, 2013.

74. Ito, T. et al., Prevalence of red sea bream iridovirus among organs of Japanese amberjack (*Seriola quinqueradiata*) exposed to cultured red sea bream iridovirus. *J Gen Virol*, 94, 2094, 2013.

75. Caipang, C.M.A. et al., Rapid detection of a fish iridovirus using loop–mediated isothermal amplification (LAMP). *J Virol Methods*, 121,155, 2004.

76. Ding, W.C. et al., Rapid and sensitive detection of infectious spleen and kidney necrosis virus by loop–mediated isothermal amplification combined with a lateral flow dipstick. *Arch Virol*, 155, 385, 2010.

77. Nakajima, K. et al., Vaccination against red sea bream iridoviral disease in red sea bream. *Fish Pathol*, 32, 205, 1997.

78. Nakajima, K. et al., Effectiveness of a vaccine against red sea bream iridoviral disease in various cultured marine fish under laboratory conditions. *Fish Pathol*, 37, 90, 2002.

79. Nakajima, K. et al., Effectiveness of a vaccine against red sea bream iridoviral disease in a field trial test. *Dis Aquat Organ*, 36, 73, 1999.

80. Caipang, C.M.A. et al., Immunogenicity, retention and protective effects of the protein derivatives of formalin–inactivated red seabream iridovirus (RSIV) vaccine in red sea bream, *Pagrus major*. *Fish Shellfish Immunol*, 20, 597, 2006.

81. Dong, C. et al., Efficacy of a formalin–killed cell vaccine against infectious spleen and kidney necrosis virus (ISKNV) and immunoproteomic analysis of its major immunogenic proteins. *Vet Microbiol*, 162, 419, 2013.

82. Dong, Y. et al., Field trial tests of FKC vaccines against RSIV genotype Megalocytivirus in cage–cultured mandarin fish (*Siniperca chuatsi*) in an inland reservoir. *Fish Shellfish Immunol.*, 35, 1598, 2013.

83. Shimmoto, H. et al., Protection of red sea bream *Pagrus major* against red sea bream iridovirus infection by vaccination with a recombinant viral protein. *Microbiol Immunol*, 54, 135, 2010.

84. Fu, X. et al., Protective immunity against iridovirus disease in mandarin fish, induced by recombinant major capsid protein of infectious spleen and kidney necrosis virus. *Fish Shellfish Immunol*, 33, 880, 2012.

85. Caipang, C.M.A. et al., Genetic vaccines protect red seabream, *Pagrus major*, upon challenge with red seabream iridovirus (RSIV). *Fish Shellfish Immunol*, 21, 130, 2006.

86. Zhang, M. et al., Construction and analysis of experimental DNA vaccines against megalocytivirus. *Fish Shellfish Immunol*, 33, 1192, 2012.

87. Tamaru, Y. et al., Application of arming system for the expression of the 380R antigen from red sea bream iridovirus (RSIV) on the surface of yeast cells: A first step for the development of an oral vaccine. *Biotechnol Prog*, 22, 949, 2006.

88. Seo, J.Y. et al., Codon–optimized expression of fish iridovirus capsid protein in yeast and its application as an oral vaccine candidate. *J Fish Dis*, 36, 763, 2013.

89. Dang, L.T. et al., Inhibition of red sea bream iridovirus (RSIV) replication by small interfering RNA (siRNA) in a cell culture system. *Antiviral Res*, 77, 142, 2008.

90. Dang, L.T. et al., Engineered virus–encoded pre–microRNA (pre–miRNA) induces sequence–specific antiviral response in addition to nonspecific immunity in a fish cell line: Convergence of RNAi–related pathways and IFN–related pathways in antiviral response. *Antiviral Res*, 80, 316, 2008.

93 Avipoxvirus

Bimalendu Mondal

CONTENTS

93.1 Introduction 837
    93.1.1 Classification.. 837
    93.1.2 Morphology, Genome Structure, Replication, and Epidemiology 838
    93.1.3 Clinical Features and Pathogenesis 838
    93.1.4 Diagnosis 839
        93.1.4.1 Conventional Techniques.. 840
        93.1.4.2 Molecular Techniques 840
93.2 Methods 841
    93.2.1 Sample Collection and Preparation 841
    93.2.2 Detection Procedures.. 841
        93.2.2.1 Demonstration of Elementary Bodies (Borrel Bodies).. 841
        93.2.2.2 Agar Gel Immunodiffusion 841
        93.2.2.3 Virus Isolation.. 841
        93.2.2.4 Enzyme-Linked Immunosorbent Assay 841
        93.2.2.5 Polymerase Chain Reaction 841
93.3 Future Perspectives... 842
References.. 843

93.1 INTRODUCTION

Avipoxviruses (APVs) are large, complex DNA viruses that belong to the subfamily *Chordopoxvirinae* of the family *Poxviridae* [1]. APVs can naturally infect over 278 species in 23 orders of birds [2] and the effects of APV infection on wild bird species can be severe, which may even cause extinction of species. Undisputed examples come from the Hawaiian Islands, where extinctions of endemic forest birds are attributed to avian pox and avian malaria transmitted from introduced species [3]. The virus is transmitted via biting insects and aerosols and is usually named on the basis of bird species from which the virus was first isolated. The highly visible, wart-like lesions associated with the featherless areas of birds have facilitated recognition of avian pox since ancient times. The disease is characterized by proliferative lesions of the skin and diphtheric membranes of the respiratory tract, mouth, and esophagus of avian species [4]. APV infection in wild birds may produce several negative effects including elevated predation among affected birds [5], trauma, secondary infection, reduced male mating success [6], and death. Avian pox occurs worldwide, but little is known about its prevalence in wild bird populations. The increased frequency of reported cases of this highly visible disease and the involvement of new bird species during recent years suggest that avian pox is an emerging viral disease. Outbreaks are commonly reported at aviaries, rehabilitation centers, and other places where confinement provides close contact among birds. The disease can spread rapidly when avian pox is introduced into such facilities. Avian pox has rarely been reported in wild waterfowl.

Although wild birds can be infected by APV year round, disease outbreaks have been associated with the environmental conditions, the emergence of vector populations, and the habits of the species affected. Environmental factors such as temperature, humidity, moisture, and protective cover all play a role in the occurrence of this disease by affecting virus survival outside the bird host. APVs can withstand considerable dryness, thereby remaining infectious on surfaces or dust particles. Mosquitoes that feed on birds are the most consistent and efficient transmitters of this disease.

Fowlpox is a slow-spreading virus disease of chickens and turkeys caused by the fowlpox virus (FWPV), the prototype species of the genus *Avipoxvirus*. The disease results in significant economic losses to the poultry industry due to decreased egg production, reduced growth rate, blindness, and increased mortality. The mortality rate may go up to 50% particularly in young birds in the diphtheritic form of the disease. Our knowledge of the molecular and biological characteristics of APV is largely restricted to FWPV and canarypox virus (CNPV) for which full genome sequences are available [7,8].

93.1.1 CLASSIFICATION

From the beginning, APVs and other poxviruses have been classified on the basis of original host, growth, and morphological characteristics in the chorioallantoic membrane

(CAM) of embryonated eggs or cell cultures and on clinical manifestations of the disease in different host species [9,10]. These criteria, rather than genetic identity, form the basis for subsequent classification of APVs despite the development of new molecular tools capable of resolving the species identity of APV. All avian poxviruses are assigned to the genus *Avipoxvirus* in the subfamily *Chordopoxvirinae* of the *Poxviridae* family, indicating their shared biological features with other poxviruses. Despite the large number of host species, currently only 10 recognized species exist within the genus *Avipoxvirus*: FWPV, CNPV, *Juncopox virus*, *Mynahpox virus*, *Psittacinepox virus*, *Sparrowpox virus*, *Starlingpox virus*, *Pigeonpox virus*, *Turkeypox virus*, and *Quailpox virus*, according to the International Committee on Taxonomy of Viruses (http://ictvonline.org/). The exact number of avipoxvirus species, strains, and variants is unknown, since new isolates continue to be identified from a wide variety of avian species. These new isolates vary in virulence and host specificity, highlighting the need to further investigate and characterize APVs.

93.1.2 Morphology, Genome Structure, Replication, and Epidemiology

Morphology: The morphology of APVs is like that of other viruses in the *Poxviridae* family. The mature virus (elementary body) is brick shaped and measures about 330 × 280 × 200 nm. The virion consists of an electron-dense centrally located biconcave core or nucleoid with two lateral bodies in each concavity and surrounded by an envelope. The outer coat is composed of random arrangements of surface tubules visible in negative-stained (phosphotungstic acid [PTA]) preparations under electron microscope [11]. APV particles have generally been shown to be resistant to ether, but sensitive to chloroform treatment.

Genome structure: APVs possess a single, linear, double-stranded DNA genome of 260–365 kb in size with low G + C content (30%–40%). The ends of the DNA are covalently closed by hairpin loops. The central region of the genome is flanked by two identical inverted terminal repeats and contains several hundred closely spaced open reading frames [7]. The central region contains about 90–106 homologous genes that are involved in basic replication mechanisms, including viral transcription and RNA modification, viral DNA replication, and proteins involved in structure and assembly of intracellular mature virions and extracellular enveloped virions [8]. In general, genes located in this region have common molecular functions and are relatively conserved among poxviruses. This is in contrast to the more variable, terminally located genes that have been shown to encode a diverse array of proteins involved in host range restriction. Complete genomes of two APVs (a virulent and an attenuated FWPV, and a virulent CNPV) have been sequenced [7,8,12]. Although nucleotide and amino acid sequences for these two viruses are known, the functions of some putative genes and proteins remain to be fully assigned. Virulence genes, which influence the pathological profile of virus in an infected host,

are generally nonconserved in nature and in infected host, and play an important part in viral evolution.

Replication and morphogenesis: APVs are believed to replicate only in avian cells, notably chicken embryo fibroblasts (CEFs). However, recent studies have shown that APV can replicate in mammalian cell cultures, such as embryonic bovine tracheal cells [13] and baby hamster kidney cells [14] that are defined by the presence of infectious viral particles and cytopathic effects (CPE). CEF cells have a good split ratio compared to other cell lines and are thus useful for large-scale propagation of virus for vaccine production or diagnostic tools. APVs have also been shown to replicate in chicken embryo kidney, chicken embryo dermis, and quail cell lines, such as QT-35.

After the virus binds to the cellular membranes, a fusion step, which is generally poorly understood, results in the release of the virion core into the cytoplasm of the cell. The core, which contains the viral RNA polymerase and transcription factors, initiates the transcription of early viral genes. This is followed by the uncoating stage, the release of viral DNA into the cytoplasm where it serves as a precursor for viral DNA replication as well as the source of intermediate and late viral gene transcription. As late viral gene product accumulates, the virus undergoes assembly and morphogenesis of infectious virus particles. During morphogenesis, APVs induce the formation of inclusion bodies in the cytoplasm of infected cells. The inclusions, which may also be termed viral factories, viroplasms, or viral replication complexes, are generally believed to be the sites of active viral replication and particle assembly within infected cells. Assembled virus is released by budding, cytolysis, or exocytosis.

Epidemiology: During avipox outbreaks, mortality can reach 80%–100% in canaries and other finches. This is in contrast to a generally lower mortality seen in chicken and turkey affected by FWPV. The virus is present in large numbers in lesions and usually transmitted by contact through a break in the skin. Aerosols generated from infected birds, or the ingestion of contaminated food or water have also been implicated as a source of transmission. Skin lesions (scabs) shed from recovering birds in poultry houses can become a source of aerosol infection. Mosquitoes and other biting insects may serve as mechanical vectors. Transmission within flocks is rapid when mosquitoes are plentiful. The disease tends to persist for extended periods in multiple-age poultry complexes because of slow spread of the virus and availability of susceptible birds. APVs are resistant and may survive in the environment for extended periods in dried scabs. Photolyase and A-type inclusion body protein genes in the genome of FWPV appear to protect the virus from environmental insults.

93.1.3 Clinical Features and Pathogenesis

Avipox is a slow-spreading disease. In chicken and turkey, it is known as fowlpox and turkeypox, respectively. The disease is most commonly characterized by development of cutaneous proliferative lesions (cutaneous form—dry pox), ranging from small nodules to spherical wart-like masses on the skin of unfeathered areas, such as comb, wattle, legs, feet, eyelids, and the base of the beak (Figure 93.1). Scars are usually visible after recovery and healing of skin lesions. The mortality in wild birds is usually low, depending on the number and size of the proliferative lesions. However, if infection occurs with secondary bacterial infection, mortality may be high. The other and less common form of APV infections is the diphtheritic form (wet pox) where slightly elevated white opaque nodules develop on the mucous membranes. They rapidly increase in size to become a yellowish diphtheritic membrane. Lesions occur on mucous membranes of the upper respiratory tract (larynx or trachea), mouth, and esophagus. Diphtheritic form generally has a higher mortality than the cutaneous form [15]. In some instances, birds display both cutaneous and diphtheritic forms and in those cases, mortality rates are often higher compared to the cutaneous form alone. Despite the variety of hosts and virus strains, associated pathology remains the same in infected domestic birds, although clinical signs vary depending on the virulence of the virus, susceptibility of the host, distribution, and type of lesions.

APV infections are associated with significant levels of morbidity and mortality in domestic and wild bird populations. It is difficult to address the pathogenicity of different APVs in different bird species as most of the investigations and reported cases are based on single APV isolates. Chickens are commonly used to determine the pathogenicity of new isolates, but they may not be the ideal host, since APVs from wild birds may not multiply in chickens. Pathogenicity studies of APVs in turkeys, pigeons, and canaries have been reported. Canaries are highly susceptible to CNPV, but show resistance to turkeypox virus, FWPV, and pigeonpox virus [4]. Pigeonpox virus produces mild infection in chickens and turkeys, but it is more pathogenic for pigeons [16]. Despite the worldwide prevalence of APV infections, experimental infection studies in birds using APVs have centered on few viral isolates especially on FWPV termed the prototype of the genus.

FIGURE 93.1 Fowl pox lesions on the head of a chicken.

FWPV and reticuloendotheliosis virus (REV): In the poultry industry, prophylactic measures against FWPV are achieved primarily by vaccination with live FWPV or antigenically similar pigeonpox virus strains produced in chicken embryo or CEF cells. In the past two decades, numerous outbreaks have been reported in vaccinated flocks, suggesting that vaccines used against the disease were not effective. In the United States, a commercial FWPV vaccine was shown to be contaminated with REV and caused lymphoma among broiler chickens [17]. Reticuloendotheliosis is a tumorigenic and immunosuppressive disease. REV strains have been reported to cause diseases characterized by chronic lymphoma, nonneoplastic lesions, and a runting-stunting syndrome in chickens, turkeys, and quails [18,19]. It has been shown that sequences of REV have been integrated into the DNA of FWPV vaccines as well as in FWPV field isolates [20–24]. The integration site is constant, while the size of the integrated fragments differs between various isolates and strains. Two types of integrated sequences are reported: long terminal repeats (LTRs) with size of approximately 200–600 bp and the near full-length REV provirus of about 8000 bp [21,24,25]. Most vaccine strains carry only an LTR remnant, while most FWPV field isolates carry the near full-length provirus. Singh et al. [20], however, detected REV LTRs of various lengths in the genome of two commercial FWPV vaccine strains and four field isolates, while several studies have shown that the source of REV infection was REV-contaminated FWPV [17,26] and herpesvirus of turkeys vaccines [27]. There exists a relationship between FWPV and REV and it is plausible that the presence of REV provirus in FWPV may exacerbate disease progression. Integration of near-full-length REV provirus in FWPV genome has been seen in a recurring fowlpox infection, which suggests that REV integration may modulate disease progression and contribute to disease slowdown into a chronic form [28]. The mechanisms of disease progression following REV provirus integration into the FWPV genome remain unresolved. The presence of REV in FWPV vaccines and the failure of currently used FWPV vaccines to evoke high level immunological protection against field challenge of FWPVs are of major concern to the poultry industry [29], which emphasizes the need for research into alternative vaccines.

93.1.4 Diagnosis

Clinical diagnosis is based on the presence of multiple skin lesions varying from papules to nodules in infected birds. Gross lesions in both the cutaneous and the diphtheritic forms, seen on birds and during necropsy, are usually sufficient to suspect APV infection. However, these signs are inadequate to provide a definitive diagnosis of APV infection as other agents, such as papilloma virus, scaly leg mites, and mycotoxins, may produce similar lesions in the skin, and conditions like candidiasis, capillariasis, and trichomoniasis may give lesions in the oral cavity similar to the diphtheritic form of APV infection. Laryngeal and tracheal lesions in chickens must be differentiated from those of infectious laryngotracheitis, which is caused by a herpesvirus. It is therefore crucial to secure samples and confirm the viral etiology of the condition in the laboratory.

93.1.4.1 Conventional Techniques

Histopathology and electron microscopy: FWPV multiplies in the cytoplasm of epithelial cells with the formation of large intracytoplasmic inclusion bodies (Bollinger bodies) that contain smaller elementary bodies (Borrel bodies). The inclusions can be demonstrated in sections of cutaneous and diphtheritic lesions by the use of hematoxylin and eosin, acridine orange, or Giemsa stains. Transmission electron microscopy can be used to demonstrate typical APV particles morphology in ultrathin sections of infected tissues or in fluid (cell culture isolate, pustule) by negative staining with 2% PTA.

Virus isolation: Demonstration of infectious virus by inoculation of homogenates of clinical samples of typical APV skin lesions onto the CAMs of embryonated hen's eggs is the gold standard method for the diagnosis of APV, although some strains of APV do not grow readily on chicken embryos [30]. Primary CEFs, chicken embryo kidney cells, chicken embryo dermis cells, or the permanent quail cell line QT-35, can also be used to propagate FWPV.

Serological tests: Although both cell-mediated immunity (CMI) and humoral immunity play an important role in APV infections, routine use of the CMI test is not convenient. Therefore, serological tests, such as virus neutralization (VN), agar gel immunodiffusion (AGID), passive hemagglutination, and fluorescent antibody tests as well as the enzyme-linked immunosorbent assay (ELISA), are used to measure specific humoral antibody responses. The conventional serological techniques of VN and AGID are in continued global use for surveillance and disease control efforts in domestic poultry species despite the availability of modern molecular and immunoassay techniques.

1. *VN*: After virus/serum interaction, the residual virus activity may be assayed in embryonating chicken eggs or in cell cultures. This technically demanding test may not be convenient for routine diagnosis. Only some selected strains of the virus have plaque-forming ability in chicken embryo cells. Neutralizing antibodies develop within 1–2 weeks of infection.

2. *AGID*: Precipitating antibodies can be detected by reacting test sera against viral antigens. The test is less sensitive than the ELISA or the passive hemagglutination test.

3. *Passive hemagglutination*: Passive hemagglutination is more sensitive than the AGID. The test will give cross-reactions among avian poxviruses. Tanned sheep or horse red blood cells are sensitized with a partially purified fowlpox viral antigen.

4. *Fluorescent antibody tests*: Direct or indirect immunofluorescence tests will reveal specific intracytoplasmic fluorescence in infected cells. The latter test is commonly used and involves two steps: the antibody against FWPV is reacted with the antigen in infected cells and followed by a secondary fluorescein isothiocyanate-labeled antibody against chicken gamma globulin (e.g., goat anti-chicken). In this regard, formalin-fixed tissue sections can be used effectively for fluorescent antibody test.

5. *Immunoperoxidase*: Specific staining of cytoplasmic inclusions is achieved when horseradish peroxidase-conjugated specific polyclonal antibody against fowlpoxvirus is reacted with hydrated sections of APV-infected fixed tissues (CAM and skin) or cell culture. Similar results are obtained when either polyclonal or monoclonal antibodies are used in an indirect test. An advantage of the technique is that the sections can be examined with the light microscope and can be stored for an extended period.

6. *Immunoblotting*: Antigenic variations between strains of APV can be evaluated by means of immunoblotting or western blotting. In this method, viral antigens separated by sodium dodecyl sulfate polyacrylamide gel electrophoresis (SDS-PAGE) are reacted either with polyclonal or monoclonal antibodies against the virus. This method is not convenient for routine diagnosis.

7. *ELISA*: ELISA has been described as a non-species-specific test for birds. It is a faster and easier method to detect antibodies against APV, particularly when large numbers of sera are to be screened. The technique is also more sensitive than the neutralization test [31]. ELISA protocols have also been developed to test the efficacy of FWPV vaccines in commercial and wild bird species where AGID is ineffective due to lack of precipitating antibodies [32].

93.1.4.2 Molecular Techniques

APVs are increasingly being detected and characterized by polymerase chain reaction (PCR), restriction fragment length polymorphism (RFLP), Southern blot hybridization, and cycle sequencing, directed at specific genes such as the 4b core protein gene [33,34]. RFLP analysis has been used for comparison of field isolates and vaccine strains of FWPV [35,36]. However, this procedure is not used in routine diagnosis. Cloned genomic fragments of FWPV have been used effectively as nucleic acid probes for diagnosis of fowlpox. Viral DNA isolated from lesions can be detected by hybridization with either radioactively or nonradioactively labeled genomic probes. This method is especially useful for differentiation of fowlpox from infectious laryngotracheitis when tracheal lesions are present [37].

APV or FWPV genomic DNA sequences of various sizes can be amplified by the PCR using specific primers [38]. PCR amplicons could be cloned and sequenced or directly sequenced and amplicon sequence allows a rapid search for homologous sequences in gene databases, to verify and identify the virus in question and to address phylogenetic relationships. This technique is useful when there is only an extremely small amount of viral DNA in the sample. Detection by real-time PCR has been used to identify recombinant APV from individual plaques [39] or to identify mixed genotype in the blood [40].

93.2 METHODS

93.2.1 Sample Collection and Preparation

Samples may be collected from live or dead affected bird. From the live bird, samples should be collected by cutaneous scarification of the scab or lesions on comb or other places in a sterile vial. Blood should also be collected for demonstration of neutralizing or precipitating antibody in serum. Diphtheritic lesions should be collected from the upper respiratory and digestive tract of dead bird. For virus isolation, samples should be collected in phosphate buffered saline (PBS) with no preservative and transported to laboratory on ice. This may be kept preserved in −80°C for a long time. Lesion or tissue should be preserved in formaldehyde for histopathology.

93.2.2 Detection Procedures

93.2.2.1 Demonstration of Elementary Bodies (Borrel Bodies)

APVs multiply in the cell with the formation of large intracytoplasmic inclusion bodies (Bollinger bodies) that contain smaller elementary bodies (Borrel bodies). The elementary bodies can be detected in smears from lesion by a simple staining technique [41]. Briefly, on a clean slide a drop of distilled water and the lesion (cutaneous or diphtheritic) are placed. A thin smear is prepared by pressing the lesion with another clean slide and rotating the upper slide several times. The smear is air died and gently fixed over a flame and stained for 5–10 min with freshly prepared basic fuchsin. After washing, the slide is counterstained with malachite green for 30–60 s. Following a wash and drying, the slide is examined under oil immersion. The elementary bodies appear red and are approximately 0.2–0.3 μm in size.

93.2.2.2 Agar Gel Immunodiffusion

Precipitating antibodies can be detected by reacting test sera against viral antigens. The antigen can be derived from infected skin or CAM lesions as well as infected cell cultures after proper treatment. Gel-diffusion medium is prepared with 1% agar, 8% sodium chloride, and 0.01% thimerosal. The viral antigen is placed in the central well and the test sera are placed in peripheral wells. It is important to include a positive and negative control serum. The plates are incubated at room temperature. Precipitation lines develop in 24–48 h after incubation of the antigen with antibody to homologous or closely related strains.

93.2.2.3 Virus Isolation

APVs and specifically FWPV virus can be isolated by inoculation of suspected material into embryonated chicken eggs. Approximately 0.1 mL of tissue suspension of skin or diphtheritic lesions is inoculated on to the CAMs of 9–12-day-old developing chicken embryos. These are incubated at 37°C for 5–7 days, and then examined for focal white pock lesions or generalized thickening of the CAMs. Histopathological examination of CAM lesions will reveal eosinophilic intracytoplasmic inclusion bodies following staining with hematoxylin and eosin. Primary CEF cells, chicken embryo kidney cells, chicken embryo dermis cells, or the permanent quail cell line QT-35 can be used for isolation of virus directly from the suspected material. A 10% suspension of the suspected material is made in PBS, which is filtered through 0.45 μm membrane filter. Virus in the filtrate is adsorbed onto the preformed layer of the cell for 1 h at 37°C and then washed with media. The infected cells are then incubated in an appropriated medium at 37°C. This results in the formation of CPE within 4–6 days postinoculation, depending on the virus isolate and on the multiplicity of infection. The isolated virus is then confirmed by molecular techniques.

93.2.2.4 Enzyme-Linked Immunosorbent Assay

ELISA can detect humoral antibodies to FWPV 7–10 days after infection [31]. FWPV antigens are prepared from either infected primary or permanent cell (QT-35 cell) monolayers or CAM lesions. Infected cells are pelleted (700 g for 10 min at 4°C), washed with isotonic buffer followed by lysis in hypotonic buffer or sonication. Nuclei and cellular debris are removed by low-speed centrifugation and the resulting supernatant is used as a source of FWPV antigens for ELISA or immunoblotting. To isolate viral antigen from CAM lesions, initial grinding of the lesions with subsequent detergent treatment is required.

Wells of microtiter plates are coated with 1 μg of soluble FWPV antigen in 100 μL of coating buffer (15 mM Na_2CO_3, 35 mM $NaHCO_3$, pH 9.6) and incubated overnight at 4°C. Each well is then rinsed once with wash solution (0.29 M NaCl, 0.05% Tween 20) and then blocked with PBS (pH 7.4) containing 3% bovine serum albumin (BSA) for 1 h at 37°C. After one wash, serial dilutions of the test sera in PBS containing 1% BSA are added to the wells. After rocking for 2 h at 37°C, wells are washed three times prior to the addition of 100 μL/well horseradish peroxidase-conjugated goat anti-chicken IgG antibodies at a recommended dilution in PBS. After 2 h incubation at 37°C and three subsequent washes, 100 μL of the peroxidase substrate tetramethylbenzidine (TMB) is added to each well. Reactions are terminated by the addition of 1 M phosphoric acid and absorbance at 450 nm is recorded using an ELISA plate reader.

93.2.2.5 Polymerase Chain Reaction

APVs are increasingly being detected and characterized by PCR, targeting specific genes such as the 4b core protein gene [38]. PCR allows for sensitive and specific detection of viral nucleic acids when compared to cell culture. From a small amount of suspected tissue material or infected cell or CAM, DNA is isolated by using commercial kits. DNA can also be isolated following the conventional method of treating the tissue with detergent and proteinase K and extraction with phenol–chloroform. Tissue materials (scab and pharyngeal lesions) are triturated and homogenized to make a 10% suspension in PBS. About 0.5 mL of the tissue homogenate is treated with SDS (sodium dodecyl sulfate) (1% final concentration) at 37°C for 30 min and then digested with proteinase K (1 mg/mL final concentration) in a buffer (0.01 M Tris-HCl, 0.1 M NaCl, 0.001 M EDTA, pH 7.6) at 56°C for 2 h. DNA is extracted using equal volume of phenol–chloroform mixture and precipitated by 2.5 volume of ethanol. After washing and drying, DNA is dissolved in 20 μL of TE buffer (10 mM Tris-HCl, 1 mM EDTA, pH 8.0) and used in PCR. PCR is carried out in a total volume of 50 μL, containing 10 mM Tris-HCl, pH 8.8, 50 mM KCl, 0.08% Nonidet P40, 1.5 mM MgCl$_2$, 200 μM each dNTP, 5 μL DNA, 0.2 μM each primer, and 1.5 units of Taq DNA polymerase. Amplification is done in a PCR thermal cycler under the following conditions for 30 cycles: initial denaturing at 94°C for 5 min, subsequent denaturing at 94°C for 1 min, annealing at 60°C for 1 min, and extension at 72°C for 1 min. The final extension is carried out at 72°C for 15 min to ease cloning in TA cloning vector. A volume of 5 μL of the PCR products is separated on 1.2% agarose gel electrophoresis. The size of the amplified DNA fragment using primers designed by Lee and Lee [38] is expected to be 578 bp long (Figure 93.2). For authentication of the amplified fragment, it may be directly sequenced or sequenced after cloning in a plasmid vector. The sequence allows a rapid search for homologous sequences in gene databases, to verify and identify the virus in question and to address phylogenetic relationships.

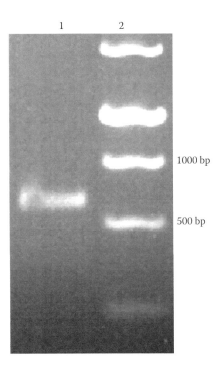

FIGURE 93.2 Polymerase chain reaction amplification of fowlpox virus (FWPV) 4b gene. Total DNA was extracted from scab arterial of a chicken affected with fowlpox and FWPV 4b gene was amplified (lane 1) using the primers described by Lee and Lee [38] with an expected size of 578 bp. Lane 2:100 bp DNA molecular weight marker.

93.3 FUTURE PERSPECTIVES

APVs cause disease in poultry, pet, and wild birds. It is of great economical importance for the poultry industry and thus prophylactic measures, such as vaccination, will always be required. There are very effective vaccines and they have undoubtedly contributed immensely to the prevention of the disease in commercial poultry farming. However, there is a need for more efficient and safe vaccines. It is known that fowlpox vaccines often get contaminated with REV. It has been shown that sequences of REV have been integrated into the DNA of FWPV as well as in field isolates. The reactivation of REV often causes disease in vaccinated birds although the FWPV vaccine virus is attenuated. Therefore, FWPV vaccines should be made from virus strains or isolates free from integrated REV DNA. Field strains of FWPV are known to contain integrated near-full-length REV provirus DNA and to achieve a REV-free FWPV; the integrated sequence has been deleted through the application of modern biotechnological approaches. The recombinant FWPV, thus prepared from a field strain, could be considered a safe candidate for vaccine preparation. Live modified vaccines have been developed mainly to prevent APV infection in chickens, turkeys, pigeons, and quails. There is seldom any vaccine for APV infections in other species of birds. Many APVs have been isolated from a wide range of bird species and only a few isolates have been characterized and so, it has been difficult to develop vaccines for all species. More knowledge of epidemiology, pathology, and molecular biology of these viruses is necessary to develop vaccines that can effectively protect a wide range of bird species.

The traditional vaccine technology based on live viruses has given rise to a wide range of effective vaccines against infectious agents, both in veterinary and human medicine. However, the emergence of new deadly human pathogens and cancers have proven less amenable to the application of traditional vaccine platforms, indicating the need for new approaches. The use of a live virus vector represents novel way to deliver and present vaccine antigens that may provide advantages over traditional platforms. Currently, representatives of a wide range of virus families are under intensive investigation as vaccine vectors for human or veterinary use. Of these, FWPV and CNPV appear to be very promising as vectors, and some veterinary APV-vectored vaccines are already licensed and in commercial use in North America, South America, and Europe. One of the main advantages of APVs as vaccine vectors is its large genome that can accommodate many heterologous genes, such as genes coding for antigens, cytokines, and other immunomodulating factors. The major safety argument for using APVs rather than vaccinia virus or other mammalian viruses as vectors is that APVs are not zoonotic and are unable to conduct a full replication cycle in mammals. However, it was recently shown that FWPV was able to replicate and produce progeny virions in some established mammalian cell lines. This is certainly a matter of concern that requires a good understanding of the molecular properties of APVs for development of safe APV-vectored vaccines.

REFERENCES

1. Fauquet, C.M., Mayo, M.A., Maniloff, J., Desselberger, U., and Ball, L.A., *Virus Taxonomy: VIIIth Report of the International Committee on Taxonomy of Viruses*. Elsevier Academic Press, San Diego, CA, 2005.

2. Van Riper, C. and Forrester, D.J., Avian pox. In *Infectious Diseases of Wild Birds*, eds. N. Thomas, B. Hunter, and C.T. Atkinson. Ames, IA: Blackwell Publishing, 2007, pp. 131–176.

3. Smith, K.F., Sax, D.F., and Lafferty, K.D., Evidence for the role of infectious disease in species extinction and endangerment, *Conserv. Biol.*, 20, 1349, 2006.

4. Tripathy, D.N., Schnitzlein, W.M., Morris, P.J. et al., Characterization of poxviruses from forest birds in Hawaii, *J. Wildl. Dis.*, 36, 225, 2000.

5. Laiolo, P., Serrano, D., Tella, J.L., Carrete, M., López, G., and Navarro, C., Distress calls reflects poxvirus infection in lesser short-toed lark *Calandrella rufescens*, *Behav. Ecol.*, 18, 507, 2007.

6. Kleindorfer, S. and Dudaniec, R.Y., Increasing prevalence of avian poxvirus in Darwin's finches and its effect on male pairing success, *J. Avian Biol.*, 37, 69, 2006.

7. Afonso, C.L., Tulman, E.R., Lu, Z., Zsak, L., Kutish, G.F., and Rock, D.L., The genome of fowlpox virus, *J. Virol.*, 74, 3815, 2000.

8. Tulman, E.R., Afonso, C.L., Lu, Z., Zsak, L., Kutish, G.F., and Rock, D.L., The genome of canarypox virus, *J. Virol.*, 78, 353, 2004.

9. Fenner, F. and Burnet, F.M., A short description of the poxvirus group (vaccinia and related viruses), *Virology*, 4, 305, 1957.

10. Fenner, F., Adventures with poxviruses of vertebrates, *FEMS Microbiol. Rev.*, 24, 123, 2000.

11. Carter, J.K.Y. and Cherville, N.F., Isolation of surface tubules of fowlpoxvirus, *Avian Dis.*, 25, 454, 1981.

12. Laidlaw, S.M. and Skinner, M.A., Comparison of the genome sequence of FP9, an attenuated, tissue culture-adapted European strain of Fowlpox virus, with those of virulent American and European viruses, *J. Gen. Virol.*, 85, 305, 2004.

13. Sainova, I.V., Kril, A.I., Simeonov, K.B., Popova, T.P., and Ivanov, I.G., Investigation of the morphology of cell clones, derived from the mammalian EBTr cell line and their susceptibility to vaccine avian poxvirus strains FK and Dessau, *J. Virol. Methods*, 124, 37, 2005.

14. Weli, S.C., Nilssen, O., and Traavik, T., Avipoxvirus multiplication in a mammalian cell line, *Virus Res.*, 109, 39, 2005.

15. Kim, T. and Tripathy, D.N., Evaluation of pathogenicity of avian poxvirus isolates from endangered Hawaiian wild birds in chickens, *Avian Dis.*, 50, 288, 2006.

16. Boosinger, T.R., Winterfield, R.W., Feldman, D.S., and Dhillon, A.S., Psittacine poxvirus: Virus isolation and identification, transmission and cross-challenge studies in parrots and chickens, *Avian Dis.*, 26, 437, 1982.

17. Fadly, A.M., Witter, R.L., Smith, E.J. et al., An outbreak of lymphomas in commercial broiler breeder chickens vaccinated with a fowlpox vaccine contaminated with reticuloendotheliosis virus, *Avian Pathol.*, 25, 35, 1996.

18. Payne, L.N., Retrovirus-induced disease in poultry, *Poult. Sci.*, 77, 1204, 1998.

19. Witter, R.L. and Fadly, A., Reticuloendothelosis. In *Diseases of Poultry*, 11th edn., eds. Y.M. Saif, H.J. Barnes, J. Glisson, L.R. McDougald, and D. Swayne. Ames, IA: Iowa State University Press, 2003, pp. 517–535.

20. Singh, P., Kim, T.J., and Tripathy, D.N., Re-emerging fowlpox: Evaluation of isolates from vaccinated flocks, *Avian Pathol.*, 2000, 29, 449, 2000.

21. Singh, P., Schnitzlein, W.M., and Tripathy, D.N., Reticuloendotheliosis virus sequences within the genomes of field strains of fowlpox virus display variability, *J. Virol.*, 77, 5855, 2003.

22. García, M., Narang, N., Reed, W.M., and Fadly, A.M., Molecular characterization of reticuloendotheliosis virus insertions in the genome of field and vaccine strains of fowl poxvirus, *Avian Dis.*, 47, 343, 2003.

23. Hertig, C., Coupar, B.E., Gould, A.R., and Boyle, D.B., Field and vaccine strains of fowlpox virus carry integrated sequences from the avian retrovirus, reticuloendotheliosis virus, *Virology*, 235, 367, 1997.

24. Kim, T.J. and Tripathy, D.N., Reticuloendotheliosis virus integration in the fowl poxvirus genome: Not a recent event, *Avian Dis.*, 45, 663, 2001.

25. Diallo, I.S., Mackenzie, M.A., Spradbrow, P.B., and Robinson, W.F., Field isolates of fowlpox virus contaminated with reticuloendotheliosis virus, *Avian Pathol.*, 27, 60, 1998.

26. Bendheim, U., A neoplastic disease in turkey following fowlpox vaccination, *Refu. Vet.*, 30, 35, 1973.

27. Jackson, C.A., Dunn, S.E., Smith, D.I., Gilchrist, P.T., and Macqueen, P.A., Proventriculitis, "nakanuke" and reticuloendotheliosis in chickens following vaccination with herpesvirus of turkeys (HVT), *Aust. Vet. J.*, 53, 457, 1977.

28. Biswas, S.K., Jana, C., Chand, K., Rehman, W., and Mondal, B., Detection of fowlpox virus integrated with reticuloendotheliosis virus sequences from an outbreak in backyard chickens, *Vet. Ital.*, 47, 147, 2011.

29. Fatunmbi, O.O. and Reed, W.M., Evaluation of a commercial quail pox vaccine (Bio-Pox Q) for the control of "variant" fowl poxvirus infections, *Avian Dis.*, 40, 792, 1996.

30. Holt, G. and Krogsrud, J., Pox in wild birds, *Acta Vet. Scand.*, 14, 201, 1973.

31. Buscaglia, C., Bankowski, R.A., and Miers, L., Cell-culture virus-neutralization test and enzyme-linked immunosorbent assay for evaluation of immunity in chickens against fowlpox, *Avian Dis.*, 29, 672, 1985.

32. Wang, J., Meers, J., Spradbrow, P.B., and Robinson, W.F., Evaluation of immune effects of fowlpox vaccine strains and field isolates, *Vet. Microbiol.*, 116, 106, 2006.

33. Luschow, D., Hoffmann, T., and Hafez, H.M., Differentiation of avian poxvirus strains on the basis of nucleotide sequences of 4b gene fragment, *Avian Dis.*, 48, 453, 2004.

34. Weli, S.C., Traavik, T., Tryland, M., Coucheron, D.H., and Nilssen, O., Analysis and comparison of the 4b core protein gene of avipoxviruses from wild birds: Evidence for interspecies spatial phylogenetic variation, *Arch. Virol.*, 49, 2035, 2004.

35. Ghildyal, N., Schnitzlein, W.M., and Tripathy, D.N., Genetic and antigenic differences between fowlpox and quailpox viruses, *Arch. Virol.*, 106, 85, 1989.

36. Schnitzlein, W.M., Ghildyal, N., and Tripathy, D.N., Genomic and antigenic characterisation of avipoxviruses, *Virus Res.*, 10, 65, 1988.

37. Fatunmbi, O.O., Reed, W.M., Schwartz, D.I., and Tripathy, D.N., Dual infection of chickens with pox and infectious laryngotracheitis (ILT) confirmed with specific pox and ILT DNA dot-blot hybridization assays, *Avian Dis.*, 39, 925, 1995.

38. Lee, L.H. and Lee, K.H., Application of the polymerase chain reaction for the diagnosis of fowlpoxvirus infection, *J. Virol. Methods*, 63, 113, 1997.

39. Boyle, D.B., Anderson, M.A., Amos, R., Voysey, R., and Coupar, B.E., Construction of recombinant fowlpox viruses carrying multiple vaccine antigens and immunomodulatory molecules, *Biotechniques*, 37, 104, 2004.

40. Farias, M.E.M., LaPointe, D.A., Atkinson, C.T. et al., TaqMan real-time PCR detects *Avipoxvirus* DNA in blood of Hawaii amakihi (*Hemignathus virens*), *PLoS One*, 5(5), e10745. doi:10.1371/journal.pone.0010745, 2010.

41. Tripathy, D.N. and Hanson, L.E., A smear technique for staining elementary bodies of fowlpox, *Avian Dis.*, 20, 609, 1976.

94 Capripoxviruses (Sheeppox, Goatpox and Lumpy Skin Disease Viruses)

E.H. Venter

CONTENTS

94.1 Introduction 845
    94.1.1 Classification... 845
    94.1.2 Morphology, Genome Organization, Biology, and Epidemiology 846
    94.1.3 Clinical Features and Pathogenesis 847
    94.1.4 Diagnosis 848
        94.1.4.1 Conventional Techniques... 848
        94.1.4.2 Molecular Techniques.. 849
94.2 Methods 849
    94.2.1 Sample Preparation... 849
    94.2.2 Detection Procedures.. 850
        94.2.2.1 Extraction Methods.. 850
        94.2.2.2 Amplification Methods 850
94.3 Conclusion 851
References.. 852

94.1 INTRODUCTION

Capripoxviruses are classified within the family *Poxviridae* and are the etiological agents of sheeppox (SPP), goatpox (GTP), and lumpy skin disease (LSD) of cattle. The causative viruses, sheeppox virus (SPPV), goatpox virus (GTPV), and lumpy skin disease virus (LSDV), are responsible for a generalized pox disease characterized by pyrexia, generalized skin lesions, ulcerative lesions in the mucous membranes of the mouth and nasal cavities, and lymphadenopathy.[1–4] Due to their transboundary capacity, capripoxvirus infections could become emerging diseases; the Office Internationale des Epizooties (OIE), or the World Organization for Animal Health, have to be notified of outbreaks. In endemic areas, they are of great economic significance to farmers and have a major impact on international trade in livestock and livestock products. The morbidity and mortality of SSP and GTP are very high and can reach 100% in naïve animals. In contrast, LSD is an occasionally fatal disease with morbidity averaging 10% and mortality 1% in affected herds, although mortality rates over 75% have been recorded.[5]

Initially, the distribution of capripoxviruses differed geographically from the current distribution. SPPV and GTPV mainly occurred in Africa north of the equator while LSDV was originally identified south of the equator and also in regions free of SPPV and GTPV. These viruses, expanding their distribution and LSD, has now spread to almost the entire Africa and the Middle East with outbreaks reported in 2013 and 2014 from previously disease-free countries such as Turkey, Iraq, Iran, and Azerbaijan. SPP and GTP spread to Asia and as far east as China but inexplicably only recently south of the equator. The spread of capripoxviral diseases is predominantly associated with illegal animal movement through trade.[6,7]

Capripoxviruses are morphologically and serologically indistinguishable, share a group-specific antigen, and have varying degrees of cross-protection. They have a close relationship genomically but differ phylogenetically.[8–11] The availability of sequence data from all three viruses led to the development of molecular laboratory techniques like PCR assays and improves the diagnosis of these diseases.

94.1.1 CLASSIFICATION

The *Poxviridae* family, one of the largest virus families, is divided into two subfamilies: poxviruses affecting insects (*Entomopoxvirinae*) and those affecting vertebrates (*Chordopoxvirinae*). The subfamily *Chordopoxvirinae* consists of at least eight genera: *Avipoxvirus*, *Capripoxvirus*, *Cervidpoxvirus*, *Crocodylipoxvirus*, *Leporipoxvirus*, *Molluscipoxvirus*, *Orthopoxvirus*, *Parapoxvirus*, *Suipoxvirus*, and *Yatapoxvirus*, of which *Orthopoxvirus* and vaccinia virus are the prototypes of the family and the genus, respectively.[12,13] LSDV, SPPV, and GTPV represent the only known members of the genus *Capripoxvirus*. SPPV was first studied during the nineteenth century (cited by Bhanuprakash and coworkers)[14] and SPPV is the type species. The prototype of LSDV is the Neethling strain, which was isolated in South Africa in 1944.[15]

94.1.2 Morphology, Genome Organization, Biology, and Epidemiology

Morphologically, members of the subfamily *Chordopoxvirinae* are very similar and it is difficult to differentiate capripoxviruses from orthopoxviruses on the morphological criteria alone.[16] With the largest virions (measuring 320–260 nm in size) of animal viruses, capripoxviruses are ovoid or brick-shaped.[3,17] They are characterized by the presence of three major structures: an outer coat (the envelope), two lateral bodies, and the inner dumb-bell shaped core, containing the double stranded DNA genome.[16] The virion consists of over 100 polypeptides. The core of the viruses contains proteins that include a transcriptase and several other enzymes.[16] Poxviruses exist with, or without an envelope in the intracellular space and are enveloped in the extracellular space.[16] Both forms are infectious and have the same core and genetic material. The mature virions,[18] also called intracellular mature virions,[16] are surrounded by a single lipid membrane with irregular arrangements of tubular nonglycosylated viral proteins on the surface. The extracellular enveloped virion has an additional outer membrane.[19] All strains of capripoxvirus share a group-specific antigen,[20] and recovery from infection with one strain provides immunity against all other strains.

Comparative DNA sequence analysis of SPPV and GTPV genomes from isolates including pathogenic field isolates and attenuated vaccine viruses revealed that genomes are approximately 150 kb and are strikingly similar to each other, exhibiting 96% nucleotide identity over their entire length. Wild-type genomes share at least 147 putative genes, including conserved poxvirus replicative and structural genes, and genes likely involved in virulence and host range. Not many genomic changes are evident in SPPV and GTPV vaccine viruses compared to their respective field strains. SPPV and GTPV genomes are very similar to LSDV, sharing 97% nucleotide identity, but phylogenetic reconstructions have shown that these viruses can be segregated into three different lineages according to their host origins: the SPPV, GTPV, and LSDV lineages.[9–11,21] In addition, a 21-nucleotide deletion found in all individuals within only the SPPV group allows for diagnostic differentiation of SPPV from GTPV.[21] All genes present in SPPV and GTPV are present in LSDV[11] but nine LSDV genes with possible virulence and host range functions are disrupted, including a unique LSDV gene (LSDV132). The absence of these genes in SPPV and GTPV suggests involvement in the bovine host range. The close genetic relationship among capripoxviruses suggests that SPPV and GTPV are distinct and likely derived from an LSDV-like ancestor.[11]

DNA analysis using restriction endonucleases on both field samples and vaccine strains showed 80% homology between strains of capripoxviruses.[22] Species-specific molecular assays have been developed by sequencing the host-specific G protein-coupled chemokine receptor or the 30 kDa RNA polymerase subunit genes. These genes also enable differentiation and phylogenetic grouping of capripoxviruses.[23–26]

SPPV and GTPV strains are not totally host-specific and some cause disease in both sheep and goats, while others may cause disease in only one species. The severity of SPP is influenced by several factors like host (age, sex, breed, nutritional and immunological status), agent (strain, virulence, pathogenicity), environment (micro and macro), poor management and scarcity of feed. Although the severity varies with age, all age groups of sheep are affected. The disease is more severe in young lambs (4–8 months) than adult sheep and newborn lambs[27] (cited by Bhanuprakash and coworkers[14]).

LSD is, however, specific for cattle of all breeds and sexes. Experimental infection of LSDV in sheep and goats only caused a local erythematous swelling at the site of inoculation.[3] Evidence of infection has been reported in some wildlife species but no clinical disease has been observed in nature.[28–30] These studies do not confirm that wildlife species play a significant role in the spread or maintenance of capripoxviruses and no strong evidence of a wildlife reservoir for these viruses exists. Studies by Hosamani and coworkers[31] indicating that molecular markers for host specificity of capripoxviruses can be determined if sufficient virus strains and genes are analyzed.

The host range also reflects the different geographical distribution of these viral diseases. LSD was known to exist in the central African territories for many years[32] before the disease was reported in Northern Rhodesia (Zambia) in 1929,[2,32,33] and by the 1940s, the disease had spread across southern African countries, affecting large numbers of livestock.[32] The disease then slowly spread northward and is currently present virtually throughout the entire African continent, including Madagascar (World Animal Health Information database, OIE WAHID Interface). The only African countries still disease-free are Libya, Algeria, Morocco, and Tunisia. The disease was first reported in Egypt in 1988, and in August 1989, the disease spread for the first time out of Africa into Israel.[34] In 2006–2007, after an apparent absence of 17 years, LSD reoccurred in Egypt and Israel.[35] LSD is endemic in the Horn of Africa from where many Middle Eastern countries import live cattle and frozen meat.[36] According to the OIE WAHID and Handistatus II Database, LSD has been reported in Kuwait (1991), Yemen (1995), the United Arab Emirates (2000), Bahrain (2009), and Oman (2013). The presence of LSDV in Saudi Arabia (1992) has not been confirmed.[29] Between 2012 and 2013, several outbreaks of LSD occurred in Israel (ProMed 20120728.1218484), Lebanon (ProMed 20130118.1505118), Jordan (ProMed 20130612.1768278), and the West Bank (ProMed 20130311.1581763). No data are available from Syria but due to the LSDV outbreaks in the Golan Heights in Israel and southern Lebanon, which is in close proximity of Syrian borders, it is highly likely that the disease is present in that country.[37] Incursions of the disease to Turkey (ProMed 20130831.1915595) and Iraq (ProMed 20130718.1831781) were reported in the last part of 2013. In July 2014, the disease spread further to Iran (ProMed 20140623.2561202) and Azerbaijan (ProMed 20140719.2621294) (Figure 94.1a).

LSD is most prevalent in the wet season or summer and along water courses[38–40] although outbreaks may also occur

FIGURE 94.1   (a) Occurrence of lumpy skin disease. (b) Occurrence of sheeppox and goatpox.

in the dry season.[1,39] In endemic areas, LSD has a tendency to disappear during the dry and cold months.[4,41]

SPPV infections were studied during the nineteenth century, and during the mid-twentieth century, the disease was reported from African, Asian, and European countries. Many SPP outbreaks occurred during 1847–1866 in Great Britain and eradication was achieved in 1866. The disease persisted in other parts of Europe, however, due to extensive trade between other foreign countries. The disease has been eradicated from many countries, but is still present in Africa (cited by Bhanuprakash and coworkers).[14] SPP and GTP currently mainly occur in Africa north of the equator, but suspected cases were reported in Zambia in 2012 (OIE WAHID and Handistatus II databases 2012). The diseases occur throughout West Asia, (Russia, Kazakhstan, and Kyrgyzstan), India, Turkey, Mongolia, Greece, and as far east as China and Bangladesh (Figure 94.1b) (OIE WAHID and Handistatus II databases, 2012, 2013).[42] Recent cases were reported in 2012 and 2013 in Russia (OIE, RFI 20121010.1335877; OIE, RFI 20131019.2009540 and Promed 2012 20131023.2017272), 2013 in Bulgaria (RFI 20131227.2138858), and 2013 and 2014 in Greece (OIE 20131221.2128611; OIE 201).

SPP occurs throughout the year depending on season, climate, agent, and the different breeds of sheep[43] but outbreaks associated with unusual winters[44] and drought[45] have been reported. In many countries, sheep are exposed to various geoclimatic conditions due to migration; this causes stress, which aids in spreading the disease.[14]

94.1.3   Clinical Features and Pathogenesis

Capripoxviruses cause severe acute diseases that have major impacts on small ruminants (SPP and GTP) and cattle (LSD)

production in countries, where the disease is endemic, due to reduced milk yields, damage to hides, and mortalities.[46] The severity of clinical disease is variable depending on the host species, the breed and age of the susceptible host, and the virus isolate. Mortality rates for SPPV and GTPV reach up to 100%, whereas mortality rates in LSDV infections are usually much lower with morbidity averaging 10% and mortality 1% but 75% in affected herds have been reported (reviewed in reference 42). Different isolates can behave differently in sheep and goats, with most isolates more virulent in either sheep or goats, while others are pathogenic in both species.[47] LSDV, however, will only cause disease in cattle. A number of diseases may show symptoms similar to SPP and some of them warrant differential diagnosis, that is, bluetongue.[14]

Generally, poxviruses are epitheliotropic and cause localized cutaneous (orf and pseudocowpox viruses) or systemic diseases (LSDV and SPPV) involving the skin and many organs and tissues. The incubation period for LSDV under experimental conditions is 4–14 days.[48] Under natural conditions, it may take between 2 and 4 weeks.[3] Early signs include a fever of 40°C–41.5°C, which will last between 4 and 14 days. During this time, the animal will manifest inappetence, salivation, lacrimation, and mucoid to mucopurulent nasal discharge.[49–51] The superficial lymph nodes especially the prescapular and precrural become markedly enlarged.[51–55] Conjunctivitis develops and may be followed by corneal opacity and eventual blindness.[1,15,55] Characteristic circumscribed intradermal, mostly eruptive, skin nodules develop. They are firm, raised, round, about 10–50 mm wide, and are readily seen in animals with short hair unlike in long haired animals. Ulcerative lesions may develop in the conjunctiva, nostrils, muzzle, and mucous membranes of the mouth, larynx, trachea, reproductive, and alimentary tracts, especially the abomasum.[3,48,49,51] The number of lesions may vary from a few in mild cases to multiple lesions covering the entire body in severely infected individuals. Subclinical infections are common in the field. In severe cases, nodules may erupt leaving a raw surface of the muscles. Nodules in the skin of the udder and teats may lead to mastitis. Subcutaneous swelling of the legs may occur, leading to lameness. Abortions or prolonged anoestrous as well as intrauterine infection may occur. Temporal sterility in infected bulls may be observed and secondary complications like pneumonia caused by the virus itself or secondary bacterial infections may develop.[51]

After infection of SPPV and GTPV, fever develops concurrently with the generation of macules in the skin. Rhinitis, conjunctivitis, and excessive salivation occur, and macules enlarge and develop into papules and then scabs. The distribution of pox lesions in the skin can be widespread with over 50% of the skin surface affected. More commonly in endemic areas, however, the lesions are restricted to a few nodules under the tail and are thus only detected on close examination.[42]

New biochemical and microscopic findings led to a better understanding of poxvirus entry and membrane fusion. Four viral proteins associated with the mature virus/virion membrane facilitate attachment by binding to glycosaminoglycans or laminin on the cell surface, whereas enveloped viruses attachment proteins have not yet been identified.[19] Host cell entry of vaccinia virus, the prototypic poxvirus, revealed a membrane fusion event with delivering of the viral core and two proteinaceous lateral bodies into the cytosol[18] where they undergo two stages of uncoating and multiplication. Uncoating of viral cores is poorly characterized, and the composition and function of lateral bodies remain unclear.[56]

After multiplication at the site of entry, the virus enters and multiplies in local draining lymph nodes followed by a primary viremia.[57,58] Draining lymph nodes are often enlarged following infection; however, lymphadenopathy is not associated with high viral replication/load in the nodes.[59] Internal organs such as the lung and stomach also develop characteristic pox-like lesions. In severely affected animals, respiratory distress occurs, followed by death. Like smallpox, capripoxvirus disease pathogenesis is associated with both viral and host factors.[60]

In experimental LSDV-infected animals, the first viremia has been detected as early as day 2–7 postinfection.[48,61,62] Infection also spreads to internal organs especially the lungs, liver, spleen, rumen, and abomasum.[4,42] There is infiltration of epithelial and endothelial cells, pericytes, and fibroblast causing lymphangitis and vasculitis with concomitant thrombosis and infarction, which leads to edema and necrosis.[51,63] The virus has been demonstrated in saliva, semen, nasal secretion, skin lesions, kidneys, abomasum, and rumen but not in urine and feces.[3,51,61]

Viremia of SSP and GTP viruses is cell-associated,[64] starts at the time of lesion occurrence, and clears within 2–3 weeks postinoculation in their respective hosts. This coincides with the time of seroconversion when host antibodies can neutralize the virus. Peak shedding of viral DNA and infectious virus in nasal, conjunctival, and oral secretions occurred between 10 and 14 days postinoculation and persisted at low levels for up to an additional 3–6 weeks. Although gross lesions developed in multiple organ systems, the highest viral titers were detected in skin and in discrete sites within oronasal tissues and the gastrointestinal tract confirming that the greatest viral replication occurs in the skin.[59,65]

Infective mature viruses are released from the host cells by cell eruption.[16] The majority of extracellular enveloped viruses remain attached to the host cell surface and mediate efficient cell-to-cell spread of the virus. The other enveloped viruses are free in extracellular space and responsible of long-range dissemination of the virus within a host.[16] The envelope is fragile and breaks easily outside the cell, particularly during freeze and thaw cycles.[16]

94.1.4 Diagnosis

Although experienced veterinarians can initially diagnose capripoxvirus infections based on animal host origin and clinical signs including overt skin lesions, laboratory confirmation is still necessary. Low virulence strains and the absence of skin lesions present problems for differential diagnosis, and laboratory confirmation is therefore necessary when studying epidemiology, for example.[14]

94.1.4.1 Conventional Techniques

In the past, laboratory confirmation for identification of virus in clinical material relied on classical laboratory techniques including animal transmission, pathology and histopathology, electron microscopy,[66] and virus isolation in cell cultures.[48,67,68]

Histopathology has been used to determine changes in cells using hematoxylin and eosin staining,[51,55,69] while immunohistochemistry confirms infection by demonstrating the viral antigen in the tissues.[63,70] A monoclonal antibody was developed from the LSDV core protein (ORF 057) expressed in *E. coli*.[61] An immunohistochemical method was successfully used in the diagnosis of LSDV during an outbreak in Egypt[70] and to confirm the presence of LSDV in testes of experimentally infected bulls.[63] It was also used to demonstrate the presence of the virus in macrophages and hair follicle epithelium.[61] Immunohistochemistry has also been used in the diagnosis of SPPV.[43,71]

Electron microscopy is technically simple and provides a rapid method for detecting viruses in tissue samples on the basis of their morphology[72] and can therefore not distinguish between capripoxviruses. Negative contrast staining or positive staining of thin sections as well as immune-staining can be used.[72] A disadvantage of this technique is that the limit of detection for negative staining is 10^6 virus particles per mL but it can improve to between 10^4 and 10^5 for ultrathin sections.[73]

Capripoxviruses can be grown in a variety of sheep, goat, and cattle cells[74] and primary lamb kidney, lamb testis cells, or bovine dermis cells are used.[48,67,75,76] These viruses grow slowly in cell cultures and may require several passages for cytopathic effects to be observed.[15] The presence of the virus in cell cultures is demonstrated by cytopathic changes and intracytoplasmic inclusion bodies.[3]

Currently, tests used to distinguish a specific virus are made according to the infected host species from which the virus was isolated, complemented by sequencing and phylogenetic analyses.

Until the late 1990s, the only capripoxvirus-specific antibody tests available were based on immunofluorescence and the virus neutralization test (SNT) using cell cultures. The SNT is considered the gold standard serodiagnostic method for detecting neutralizing antibodies to capripoxviruses.[42] This method has been used as a screening test to confirm infection with LSDV[48,62] and to carry out serological surveys on LSD.[77,78] However, the test cannot differentiate between GTPV, SPPV, and LSDV.[22,64] As immunity to LSDV is mostly cell-mediated, the sensitivity of this test is low and it cannot detect the low level of circulating antibodies in animals.[62] SNT is also difficult to interpret, is time-consuming, and is not available in countries where the viruses are not endemic.

The availability of sequence data led to the study and expression of specific open reading frames, which inspire/stimulate the development of different ELISA tests. Although there is currently no validated ELISA for the detection of capripoxviruses, many attempts have been made to develop different formats of the assay. Antigen trapping and recombinant antigen ELISAs were developed by Carn and coworkers.[79,80] An ELISA was developed based on the envelope protein P32 of LSDV expressed in *Escherichia coli*.[79,81] The 3.7 kb viral DNA fragment of SPPV, which contained open reading frames homologous to the vaccinia virus J6R, H1L, H2R, H3L, and H4L genes, was cloned and sequenced. The vaccinia virus H3L homolog was identified as the capripoxvirus P32 antigen. The P32 proteins of SPPV and LSDV were expressed in *Escherichia coli* as a fusion protein and used as a trapping antigen in an ELISA, but this test had limitations due to the instability of the recombinant protein.[82] Whole inactivated sucrose gradient-purified SPPV as a coating antigen[82] has also been used in an indirect ELISA but has the disadvantage of being too costly to produce in bulk.[83] An indirect ELISA was developed using the virion core proteins ORF095 and ORF103, which were able to detect capripoxvirus antibodies in both acute and convalescent phase sera.[83]

Western blotting assays, although difficult to perform and interpret, can be used and are specific and sensitive.[84]

94.1.4.2 Molecular Techniques

The complexity of capripoxvirus differentiation emphasizes the need to develop more reliable tools for the identification of these viruses. The analysis of genome sequences of capripoxviruses has shown that SPPV, GTPV, and LSDV are genetically distinct.[11] The availability of sequence data initiate the development and validation of several molecular techniques, that is, PCR methods, some of which were refined to detect either vaccine strains or only a specific species of virus. These assays provide a rapid and sensitive diagnostic technique for capripoxvirus genome detection and have the advantage of using internal reaction controls. If used as a diagnostic assay, however, results should be confirmed by at least one additional test.

Conventional PCRs have been used and reported,[48,81,85–89] and real-time PCR has been reported by several authors.[21,25,59,61,69,90–92] A multiplex PCR using species-specific primers was used to detect capripoxvirus strains in Russia and with modification to distinguish vaccine strains.[88] A PCR assay has been developed to identify both capripoxviruses and orf virus (parapoxvirus) in the same procedure.[91,93] A single-tube duplex PCR for the detection of SPPV and GTPV is useful where herds are affected with both viruses.[94]

The development of different platforms for sequencing, that is, next-generation sequencing, although not in general always feasible to use as a diagnostic tool, led to a better understanding of the origin and geographical distribution of strains of capripoxviruses.

94.2 METHODS

94.2.1 Sample Preparation

Samples for virus isolation and antigen-detection ELISA should be taken during the first week of symptoms, before neutralizing antibodies have developed. Samples for PCR, blood (EDTA), and/or skin lesions can be collected after this time. In live animals, biopsy samples of skin nodules or lymph nodes can be used for virus isolation and antigen detection. Scabs, nodular fluid, and skin scrapings may also be collected, specifically also for electron microscopy (EM). The virus can be isolated from blood samples (collected into heparin or EDTA) during the early viremic stage of disease; virus isolation from blood is unlikely to be successful after generalized lesions have been present for more than 4 days. Samples of lesions, including tissues from surrounding areas, should be submitted for histopathology. Acute and convalescent sera are collected for serology. At necropsy, samples for fixation in formalin should be taken of skin lesions and lesions in the respiratory and gastrointestinal tracts, and fresh samples should be taken and treated as described above.[48,51,95]

Tissue or clinical samples are generally homogenized as 10% suspensions in sterile water or phosphate buffer saline and used for extraction of total genomic DNA and stored at −20°C until use.[96] In a method used by Ireland and Binepal,[85] one gram of tissue was ground up with sterile sand and approximately 1 mL of tissue culture medium containing serum and antibiotics in a glass tissue grinder; the mixture was then transferred to a 1.5 mL microcentrifuge tube. The suspension was clarified by centrifugation at full speed for 5 min in a microfuge and stored at 4°C until needed for ELISA or PCR.[85]

94.2.2 Detection Procedures

94.2.2.1 Extraction Methods

Although the use of traditional phenol chloroform extraction methods has recently been reported for scabs and cell culture material,[94] commercially available extraction kits are currently mainly used to obtain nucleic acids from diagnostic samples. DNA was extracted from tissue samples using the DNeasy Blood and Tissue Kit (QIAGEN, USA). After complete lysis of the specimens by ATL buffer and proteinase K, absolute ethanol was added, then the mixture was transferred to a spin column according to the manufacturer's protocol.[97] The Qiagen AllPrep DNA/RNA extraction kit (Qiagen, Hilden, Germany) was used to extract viral DNA by Lamien and coworkers.[25] Homogenates of infected tissues (10% weight/volume in PBS), or infected cell culture supernatants were mixed with the kit lysis buffer, RLT plus (200 µL sample + 800 µL buffer). The DNA was extracted according to the manufacturer's instructions and eluted in 50 µL of elution buffer included in the kit.

Tuppurainen and coworkers[48] used an extraction method containing a lysis buffer with 60% guanidine thiocyanate and digestion by proteinase K at different time periods, following the use of Trizol (Roche products, Mannheim, Germany) for nucleic acid extraction. Briefly, 200 µL of specimen was added to 100 µL of lysis buffer containing 0.378 g KCl, 1 mL Tris (1.0 M, pH 8), 0.5 mL Tween 20%, and 60% guanidine thiocyanate (Roche products, Mannheim, German). To digest proteins, 10 µL of proteinase K (Roche products, Mannheim, Germany) were added to the mixture and incubated at 56°C for 2 h. This was done for blood, saliva, and VI supernatants, and at least 16 h were used for tick homogenates.[48,98,99] The mixture was then heated to 90°C for 10 min to stop enzyme activity. Then, 300 µL of Trizol, a solution of phenol/chloroform/isoamyl alcohol (Roche products, Mannheim, Germany), was added, mixed by vortexing, and incubated at room temperature for 10 min before centrifuging at 13,000 rpm (Centrifuge 5417R, Eppendorph International, Germany) at 4°C for 15 min. The supernatant (200 µL) was collected in a 2 mL tube and DNA precipitated.

94.2.2.2 Amplification Methods

A PCR to detect capripoxviruses in tissue material reported by Ireland and Binepal[85] is widely used. The primers for the attachment gene are

5′ d tccgagctcTTTCCTGATTTTTCTTACTAT 3′
5′ d tatggtacctAAATTATATACGTAAATAAC 3′

And the primers from viral fusion protein encoding gene are

5′ d actagtggatccATGGACAGAGCTTTATCA 3′
5′ d gctgcaggaattcTCATAGTGTTGTACTTCG 3′

The extra sequences shown in lower case letters were added to allow for cloning the PCR products. A 50 mL reaction volume was used that contained 16 mM $(NH_4)_2SO_4$, 67 mM Tris–HCl (pH 8.8 at 25°C), 2.5 mM $MgCl_2$, 0.2 mM dNTPs, 5 units of Taq polymerase, 0.25 mM of each primer for either the attachment gene or fusion gene, 0.1% Tween-20, and 5 mL of cell culture or biopsy supernatant plus 20 mL of GeneReleaser or 20 mL of cell culture, or biopsy supernatant. The PCR had an initial cycle of 94°C for 5 min, 50°C for 30 s, 72°C for 1 min followed by 34 cycles of 94°C for 1 min, 50°C for 30 s, 72°C for 1 min, and a final elongation step of 72°C for 5 min. The reaction products were visualized using a 1.5% agarose gel containing ethidium bromide. To distinguish between capripox-, orthopox-, and parapoxviruses, the PCR products were confirmed by restriction enzyme analysis, using 20 units of *Eco*RI (attachment gene) or *Dra*I (fusion protein encoding gene).

A capripoxvirus-specific PCR assay was developed that differentiated between SPPV and LSDV on the basis of unique restriction sites in the corresponding PCR fragments.[81]

Real-time PCRs based on hydrolysis probe chemistry (TaqMan probes) have been developed and used to study capripoxvirus shedding in sheep, goat, and cattle,[59,61,100] and for a rapid preclinical diagnosis of capripoxvirus.[90] However, these methods were not used for strain discrimination. The method of Balinsky and coworkers[90] used reagents from an EZ-RT PCR kit (Applied Biosystems, Branchburg, NJ) according to the manufacturer's guidelines. The final 25 µL assay mix included an EZ buffer solution, 5 mM manganese acetate, 0.2 mM dNTPs, and 0.1 U of recombinant *rTth* DNA polymerase. A forward primer (5′-GGCGATGTCCATTCCCTG-3′) (200 nM), reverse primer (5′-AGCATTTCATTTCCGTGAGGA-3′) (500 nM), and fluorogenic minor groove-binding TaqMan probe (5′-CAATGGGTAAAAGATTTCTA-3′, labeled with 6-carboxyfluorescein and a nonfluorescent quencher) (200 nM), were included in each reaction mix. Sample template (2.5 µL) was added to the reaction mix in a SmartCycler (Cepheid, Sunnyvale, CA) 25 µL reaction tube. Cycling conditions consisted of an initial denaturation at 95°C for 120 s, followed by 45 amplification cycles (95°C for 2 s and 60°C for 60 s). The assay was run on a Cepheid SmartCycler (Cepheid, Inc., Sunnyvale, CA).

Genotyping can be performed by real-time PCR by using a dual hybridization probe assay (also known as fluorescence resonance energy transfer probe or FRET).[101,102] With this method, PCR chemistry is combined with fluorescent probe detection of the amplified product in the same reaction vial. In this assay, the acceptor probe is designed where differences exist between strains to be able to differentiate between them. The melting point temperature of the probe and its target will therefore differ between different strains. The melting peaks obtained after fluorescence melting curve analysis serve to differentiate the strains. This procedure has been used to discriminate between different species of orthopoxviruses.[103–105]

Recently, a species-specific real-time PCR was reported based on unique molecular markers that were found in the G-protein-coupled chemokine receptor gene sequences of capripoxviruses and that use dual hybridization probes for their simultaneous detection, quantitation, and genotyping. This PCR method described by Lamien and coworkers[26] can be used effectively for differentiation between SPPV and GTPV. Differences in the melting point temperature and fluorescence melting curve analysis were also used to differentiate between capripoxvirus strains.[26] Primers were developed to allow the

TABLE 94.1

Sequences of Real-Time PCR Primers and Probes

Primers and Probes	Sequences (5′–3′)
GpCrf forward	taatgttcgtataaatgtaataaag
GpCrfr reverse	tacaaatccaacaatgagt
CpRt forward	gatagtatcgctaaacaatgg
CpRt reverse	atccaaaccaccatactaag
Cp-LNA-FAM1	acctagcTgtAgttcaCccagtaaa-FAM[a]
Cp-Cy5	Cy5-tcaatttcaataaggacaaaacgatatgga-phosphate

[a] The LNA are in upper characters.

TABLE 94.2

Multiplex Primers for Screening of SPPV and P32 Gene PCR

Target Gene	Primer Name	Sequence (5′–3′)	Amplicon Size (bp)
KL-1 DNA	KS-1.5	GTGTGACTTTCCTGCCGAAT	149
	KS-1.6	TCTATTTTATTTCGTATATC	
Inverted repeats	InS-1.1	AGAAACGAGGTCTCGAAGCA	289
	InS-1.1'	GGAGGTTGCTGGAAATGTGT	
P32 gene	B68	CTAAAATTAGAGAGCTATACT TCTT	390
	B69	CGATTTCCATAAACTAAAGTG	

amplification of a 200 bp target at position 7015–7214 within conserved regions of the G-protein-coupled chemokine receptor gene after alignment of 58 capripoxvirus isolates as target sequences. A pair of dual hybridization probes (donor and acceptor) was designed to bind within the amplicon. The acceptor matched 100% with GTPV sequence but showed 3 and 5 mismatches, respectively, with LSDV and SPPV sequences. The donor (Cp-LNA-FAM1) was labeled at the 3′ end with the blue dye 6-carboxyfluorescein (6-FAM) and further modified by incorporating Locked Nucleic Acid (LNA) into it. The acceptor (Cp-Cy5) was labeled at the 5′ end with a red cyanine dye, Cy5 (KPL, Gaithersburg, MD), and its 3′ end was blocked with a phosphate moiety. All primers and probes were synthesized by VBC-BIOTECH (Vienna, Austria) (Table 94.1).

An iCycler (BioRad, Hercules, CA) thermocycler was used using a volume of 20 µL containing template DNA, 100 nM forward primer, 300 nM reverse primer, 300 nM of each of the dual hybridization probes, and the iQ Supermix (BioRad, Hercules, CA). Reaction cycles proceeded at 95°C for 3 min followed by 40 cycles of 95°C for 10 s, 46°C for 30 s, and 72°C for 30 s. Fluorescence was measured at the end of each annealing step (46°C). Subsequent fluorescence melting curve analysis involved an initial denaturation at 95°C for 1 min and cooling at 40°C for 1 min and then heating to 90°C with a ramping rate of 1°C each 10 s.[25]

A loop-mediated isothermal AMPlification assay for the detection of capripoxvirus DNA was developed using loop primers from the conserved region of the P32 gene. This DNA-dependent amplification method uses a combination of four to six primers targeting six to eight genomic regions, while utilizing the activity of a strand-displacing DNA polymerase. The assay could detect SPPV, GTPV, and LSDV and did not cross-react with other mammalian poxviruses.[106]

A multiplex PCR and phylogenetic analysis were used to detect and characterize the first description of SPPV in a recent outbreak in Saudi Arabia. Primers from the KS-1 DNA fragment and the inverted terminal repeats described by Mangana–Vougiouka and coworkers[87] were used for screening samples and the P32 gene for sequence analysis.[97] The primers used in the PCR assays were KS-1.5/KS-1.6 and InS-1.1/InS-1.1/ for SPPV screening and B68/B69 primers for partial sequencing of the P32 gene are indicated in Table 94.2.

The presence of SPPV was detected by a multiplex PCR using a HotStartTaq® Plus Master Mix Kit (QIAGEN, USA). 2 µL sample of each purified genomic DNAs were amplified in 20 µL of the final volume of a 2× HotStartTaq Plus Master Mix containing 1.5 mM MgCl₂, 200 µM of each dNTP, and 1 unit HotStartTaq Plus DNA polymerase and 10 µM of each KS-1.5/KS-1.6 and InS-1.1/InS-1.1/primers. Thermocycling conditions were enzyme activation and initial denaturation at 95°C for 5 min followed by 35 cycles of 94°C for 30 s, 43°C for 30 s, 72°C for 30 s, and a final extension step at 72°C for 10 min. The amplified PCR products were visualized using a 5% agarose gel stained with ethidium bromide and documented using ultraviolet gel documentation system.

94.3 CONCLUSION

Capripoxviral diseases are economically important diseases of small ruminants and livestock. The spread of the diseases into various geographical areas has been alarming during the past 4 years, with SPP and GTP spreading into South Asia and the Far East and LSD into the Middle East. The threat now is spreading into Europe where naïve animal populations would suffer high morbidity and mortality.

Control of the diseases is possible through vaccination and strict movement control, which is not possible in all countries. An attenuated strain of any capripoxvirus should theoretically protect against all SPPV, GTPV, and LSDV strains. The use of attenuated capripoxvirus virus as a heterologous vaccine has been described but is not used in SPP- and GTP-free areas where a LSDV-specific vaccine is normally used (reviewed by Boshra and coworkers[7]).

The success of control programs needs to be monitored by active serosurveillance. In endemic countries, it is difficult to differentiate between naturally infected and vaccinated animals. Seroconversion to capripoxviruses is also not always detectable using existing laboratory tests. In many countries, socioeconomic and political stability as well as logistics and veterinary support make it difficult to implement surveillance and vaccination programs.

The development of molecular-based assays has made a significant difference to the diagnosis of capripoxviruses enabling rapid genotyping and gene-based classification of viral strains and different isolates. This facilitates rapid responses to outbreaks, improved monitoring of capripoxviruses in endemic regions, and detailed investigation of the epidemiology of the diseases.

REFERENCES

1. Haig, D.A. Lumpy skin disease. *Bull Epizoot Dis Afr* 5, 421–430 (1957).
2. MacOwan, K.D.S. Observations on the epizootiology of lumpy skin disease during the first year of its occurrence in Kenya. *Bull Epiz Disf Afr* 7, 7–20 (1959).
3. Weiss, K.E. Lumpy skin disease virus. *Virology Monographs* 3, 111–131 (1968).
4. Woods, J.A. Lumpy skin disease—A review. *Trop Anim Health Prod* 20, 11–17 (1988).
5. Diesel, A.M. The epizootiology of lumpy skin disease in South Africa. *Report 14th Int Vet Congr* 2, 492–500 (1952).
6. Domenech, J., Lubroth, J., Eddi, C., Martin, V., and Roger, F. Regional and international approaches on prevention and control of animal transboundary and emerging diseases. *Ann N Y Acad Sci* 1081, 90–107 (2006).
7. Boshra, H. et al. Capripoxvirus-vectored vaccines against livestock diseases in Africa. *Antiviral Res* 98, 217–227 (2013).
8. Kitching, R.P. Capripoxviruses. In *Encyclopedia of Virology* (3rd edn.) (ed. Regenmortel, B.W.J.M.H.V.V.), pp. 427–432 (Academic Press, Oxford, U.K., 2008).
9. Tuppurainen, E.S.M. et al. Characterization of sheep pox virus vaccine for cattle against lumpy skin disease virus. *Antiviral Res* 109, 1–6 (2014).
10. Tulman, E.R. et al. Genome of lumpy skin disease virus. *J Virol* 75, 7122–7130 (2001).
11. Tulman, E.R. et al. The genomes of sheeppox and goatpox viruses. *J Virol* 76, 6054–6061 (2002).
12. Moss, B. Poxviridae: The viruses and their replication. In *Fundamental Virology* (eds. Knipe, D.M. and Howley, P.M.), pp. 1249–1283 (Lippincott Williams & Wilkins, Philadelphia, PA, 2001).
13. Lefkowitz, E.J., Wang, C., and Upton, C. Poxviruses: Past, present and future. *Virus Res* 117, 105–118 (2006).
14. Bhanuprakash, V., Indrani, B.K., Hosamani, M., and Singh, R.K. The current status of sheep pox disease. *Comp Immunol Microbiol Infect Dis* 29, 27–60 (2006).
15. Alexander, R.A., Plowright, W., and Haig, D.A. Cytopathogenic agents associated with lumpy skin disease of cattle. *Bull Epiz Disf Afr* 5, 489–492 (1957).
16. Esposito, J.J. and Fenner, F. Poxviruses. In: *Fields Virology*, Fourth Edition (eds. Knipe, D.M. and Howley, P.M.), pp. 2885–2921 (Lippincott Williams & Wilkins, Philadelphia, PA, 2001).
17. Davies, F.G., Krauss, H., Lund, J., and Taylor, M. The laboratory diagnosis of lumpy skin disease. *Res Vet Sci* 12, 123–127 (1971).
18. Moss, B. Poxvirus entry and membrane fusion. *Virology* 344, 48–54 (2006).
19. Moss, B. Poxvirus cell entry: How many proteins does it take? *Viruses* 4, 688–707 (2012).
20. Woodroofe, G.M. and Fenner, F. Serological relationship within the poxvirus group: An antigen common to all members of the group. *Virology* 16, 334–341 (1962).
21. Lamien, C.E. et al. Phylogenetic analysis of the capripoxvirus RPO30 gene and its use in a PCR test for differentiating sheep poxvirus from goat poxvirus. In *Sustainable Improvement of Animal Production and Health* (eds. Odongo, N.E., Garcia, M., and Viljoen, G.J.), pp. 323–326, Food and Agriculture Organization of the United Nations (FAO), Rome, Italy, (2010).
22. Black, D.N., Hammond, J.M., and Kitching, R.P. Genomic relationship between capripoxviruses. *Virus Res* 5, 277–292 (1986).
23. Le Goff, C. et al. Host-range phylogenetic grouping of capripoxviruses. In *Applications of Gene-Based Technologies for Improving Animal Production and Health in Developing Countries* (eds. Makkar, H.S. and Viljoen, G.), pp. 727–733 (Springer, Dordrecht, the Netherlands, 2005).
24. Le Goff, C. et al. Capripoxvirus G-protein-coupled chemokine receptor: A host-range gene suitable for virus animal origin discrimination. *J Gen Virol* 90, 1967–1977 (2009).
25. Lamien, C.E. et al. Real time PCR method for simultaneous detection, quantitation and differentiation of capripoxviruses. *J Virol Methods* 171, 134–140 (2011).
26. Lamien, C.E. et al. Use of the Capripoxvirus homologue of Vaccinia virus 30 kDa RNA polymerase subunit (RPO30) gene as a novel diagnostic and genotyping target: Development of a classical PCR method to differentiate Goat poxvirus from Sheep poxvirus. *Vet Microbiol* 149, 30–39 (2011).
27. Mariner, J.C., House, J.A., Wilson, T.M., Ende, M.V.D., and Diallo, I. Isolation of sheep pox virus from a lamb in Niger. *Trop Anim Health Prod* 23, 27–28 (1991).
28. Ali, A.A., Esmat, M., Attia, H., Selim, A., and Abdel-Hamid, Y.M. Clinical and pathological studies of lumpy skin disease in Egypt. *Vet Rec* 127, 549–550 (1990).
29. Greth, A. et al. Capripoxvirus disease in an Arabian Oryx (*Oryx leucoryx*) from Saudi Arabia. *J Wildl Dis* 28, 295–300 (1992).
30. Barnard, B.J.H. Antibodies against some viruses of domestic animals in southern African wild animals. *Onderstepoort J Vet Res* 64, 95–110 (1997).
31. Hosamani, M. et al. Differentiation of sheep pox and goat poxviruses by sequence analysis and PCR-RFLP of P32 gene. *Virus Genes* 29, 73–80 (2004).
32. Thomas, A.D. and Mare, C.V.E. Knopvelsiekte. *J S Afr Vet Med Assoc* 16, 36–43 (1945).
33. Thomas, A.D. Lumpy skin disease: A new disease of cattle in the Union. *Farm S Afr* 79 (1945).
34. Yeruham, I. et al. Spread of lumpy skin disease in Israeli dairy herds. *Vet Rec* 137, 91–93 (1995).
35. El-Kholy, A.A., Soliman, H.M.T., and Abdelrahman, K.A. Polymerase chain reaction for rapid diagnosis of a recent lumpy skin disease virus incursion to Egypt. *Arab J Biotechnol* 11, 293–302 (2008).
36. Shimshony, A. and Economides, P. Disease prevention and preparedness for animal health emergencies in the Middle East. *Revue Sci Tech Off Int Epiz* 25, 253–269 (2006).
37. Tageldin, M.H. et al. Lumpy skin disease of cattle: An emerging problem in the Sultanate of Oman. *Trop Anim Health Prod* 46, 241–246 (2014).
38. Ali, B.H. and Obeid, H.M. Investigation of the first outbreaks of lumpy skin disease in the Sudan. *Br Vet J* 133, 184–189 (1977).
39. Nawathe, D.R., Asagba, M.O., Abegunde, A., Ajayi, S.A., and Durkwa, L. Some observations on the occurrence of lumpy skin disease in Nigeria. *Zentralblatt fur Veterinarmedizin Reihe B (J Vet Med Series B)* 29, 31–36 (1982).
40. Ahmed, L.A., El-Desawy, O.M., and Mohamed, A.T. Studies on bovine field skin lesions in Fayoum Governorate. *Vet Med J Giza* 53, 73–81 (2005).
41. Hunter, P. and Wallace, D. Lumpy skin disease in southern Africa: A review of the disease and aspects of control. *J S Afr Vet Assoc* 72, 68–71 (2001).
42. Babiuk, S., Bowden, T.R., Boyle, D.B., Wallace, D.B., and Kitching, R.P. Capripoxviruses: An emerging worldwide threat to sheep, goats and cattle. *Transbound Emerg Dis* 55, 263–272 (2008).

43. Gulbahar, M.Y., Cabalar, M., Gul, Y., and Icen, H. Immunohistochemical detection of antigen in lamb tissues naturally infected with sheeppox virus. *J Vet Med Series B* 47, 173–181 (2000).

44. Hailat, N., Al-Rawashdeh, O., Lafi, S., and Al-Bateineh, Z. An outbreak of sheep pox associated with unusual winter conditions in Jordan. *Trop Anim Health Prod* 26, 79–80 (1994).

45. Chamoiseau, G. Poxviral infection in Mauritanian sheep Sheep pox or atypical lumpy skin disease Pox virose chez le mouton mauritanien Clavelee ou maladie nodulaire atypiques? *Revue d'Elevage Med Vet Pays Trop* 38, 119–121 (1985).

46. Yeruham, I. et al. Economic and epidemiological aspects of an outbreak of sheeppox in a dairy sheep flock. *Vet Rec* 160, 236–237 (2007).

47. Kitching, R.P., McGrane, J.J., and Taylor, W.P. Capripox in the Yemen Arab Republic and the sultanate of Oman. *Trop Anim Health Prod* 18, 115–122 (1986).

48. Tuppurainen, E.S., Venter, E.H. and Coetzer, J.A. The detection of lumpy skin disease virus in samples of experimentally infected cattle using different diagnostic techniques. *Onderstepoort J Vet Res* 72, 153–164 (2005).

49. Carn, V.M. and Kitching, R.P. The clinical response of cattle experimentally infected with lumpy skin disease (Neethling) virus. *Arch Virol* 140, 503–513 (1995).

50. Carn, V.M. and Kitching, R.P. An investigation of possible routes of transmission of lumpy skin disease virus (Neethling). *Epidemiol Infect* 114, 219–226 (1995).

51. Prozesky, L. and Barnard, B.J.H. A study of the pathology of lumpy skin disease in cattle. *Onderstepoort J Vet Res* 49, 167–175 (1982).

52. Coetzer, J.A.W. 2004. Lumpy skin disease. In: *Infectious diseases of livestock* (eds. Coetzer, J.A.W. and Tustin, R.C.), pp. 1268–1276 (Oxford University Press Southern Africa, Cape Town, South Africa, 2004).

53. Kumar, S.M. An outbreak of Lumpy Skin Disease in a Holstein dairy herd in Oman: A clinical report. *Asian J Anim Vet Adv* 6, 851–859 (2011).

54. Salib, F.A. and Osman, A.H. Incidence of lumpy skin disease among Egyptian cattle in Giza governorate, Egypt. *Vet World* 4, 162–167 (2011).

55. Body, M. et al. Clinico-histopathological findings and PCR based diagnosis of lumpy skin disease in the Sultanate of Oman. *Pakistan Vet J* 32, 206–210 (2012).

56. Schmidt, F.I. et al. Vaccinia virus entry is followed by core activation and proteasome-mediated release of the immunomodulatory effector VH1 from lateral bodies. *Cell Rep* 4, 464–476 (2013).

57. Woodson, B. Recent progress in poxvirus research. *Bacteriol Rev* 32, 127–137 (1968).

58. Purcell, D.A., Clarke, J.K., McFerran, J.B., and Hughes, D.A. The morphogenesis of pigeonpox virus. *J Gen Virol* 15, 79–83 (1972).

59. Bowden, T.R., Babiuk, S.L., Parkyn, G.R., Copps, J.S., and Boyle, D.B. Capripoxvirus tissue tropism and shedding: A quantitative study in experimentally infected sheep and goats. *Virology* 371, 380–393 (2008).

60. Stanford, M.M., McFadden, G., Karupiah, G., and Chaudhri, G. Immunopathogenesis of poxvirus infections: Forecasting the impending storm. *Immunol Cell Biol* 85, 93–102 (2007).

61. Babiuk, S. et al. Quantification of lumpy skin disease virus following experimental infection in cattle. *Transbound Emerg Dis* 55, 299–307 (2008).

62. Chihota, C.M., Rennie, L.F., Kitching, R.P., and Mellor, P.S. Mechanical transmission of lumpy skin disease virus by *Aedes aegypti* (Diptera: Culicidae). *Epidemiol Infect* 126, 317–321 (2001).

63. Annandale, C.H., Irons, P.C., Bagla, V.P., Osuagwuh, U.I., and Venter, E.H. Sites of persistence of lumpy skin disease virus in the genital tract of experimentally infected bulls. *Reprod Domest Anim* 45, 250–255 (2010).

64. Kitching, R.P. and Taylor, W.P. Clinical and antigenic relationship between isolates of sheep and goat pox viruses. *Trop Anim Health Prod* 17, 64–74 (1985).

65. Plowright, W., MacLeod, W.G., and Ferris, R.D. The pathogenesis of sheep pox in the skin of sheep. *J Comp Pathol* 146, 97–105 (2012).

66. Kitching, R.P. and Smale, C. Comparison of the external dimensions of capripoxvirus isolates. *Res Vet Sci* 41, 425–427 (1986).

67. Bagla, V.P., Osuagwuh, U.I., Annandale, C.H., Irons, P.C., and Venter, E.H. Elimination of toxicity and enhanced detection of lumpy skin disease virus on cell culture from experimentally infected bovine semen samples. *Onderstepoort J Vet Res* 73, 263–268 (2006).

68. Davies, F.G. Characteristics of a virus causing a pox disease in sheep and goats in Kenya, with observations on the epidemiology and control. *J Hyg* 76, 163–171 (1976).

69. Awad, W.S., Ibrahim, A.K., Mahran, K., Fararh, K.M., and Abdel Moniem, M.I. Evaluation of different diagnostic methods for diagnosis of Lumpy skin disease in cows. *Trop Anim Health Prod* 42, 777–783 (2010).

70. Awadin, W., Hussein, H., Elseady, Y., Babiuk, S., and Furuoka, H. Detection of lumpy skin disease virus antigen and genomic DNA in formalin-fixed paraffin-embedded tissues from an Egyptian outbreak in 2006. *Transbound Emerg Dis* 58, 451–457 (2011).

71. Gulbahar, M.Y., Davis, W.C., Yuksel, H., and Cabalar, M. Immunohistochemical evaluation of inflammatory infiltrate in the skin and lung of lambs naturally infected with sheeppox virus. *Vet Patholog* 43, 67–75 (2006).

72. Gardner, P.S. Rapid virus diagnosis (review article). *J Gen Virol* 36, 1–29 (1977).

73. Laue, M. and Bannert, N. Detection limit of negative staining electron microscopy for the diagnosis of bioterrorism-related micro-organisms. *J Appl Microbiol* 109, 1159–1168 (2010).

74. Binepal, Y.S., Ongadi, F.A., and Chepkwony, J.C. Alternative cell lines for the propagation of lumpy skin disease virus. *Onderstepoort J Vet Res* 68, 151–153 (2001).

75. Kalra, S.K. and Sharma, V.K. Adaptation of Jaipur strain of sheeppox virus in primary lamb testicular cell culture. *Ind J Exp Biol* 19, 165–169 (1981).

76. Zhou, J. et al. Culture of ovine testicular cells and observation of pathological changes in the cells innoculated with attenuated sheep pox virus. *Chin J Vet Sci Technol* 34, 71–74 (2004).

77. Gari, G. et al. Lumpy skin disease in Ethiopia: Seroprevalence study across different agro-climate zones. *Acta Trop* 123, 101–106 (2012).

78. Hedger, R.S. and Hamblin, C. Neutralising antibodies to lumpy skin disease virus in African wildlife. *Comp Immunol Microbiol Infect Dis* 6, 209–213 (1983).

79. Carn, V.M., Kitching, R.P., Hammond, J.M., and Chand, P. Use of a recombinant antigen in an indirect ELISA for detecting bovine antibody to capripoxvirus. *J Virol Methods* 49, 285–294 (1994).

80. Carn, V.M. An antigen trapping ELISA for the detection of capripoxvirus in tissue culture supernatant and biopsy samples. *J Virol Methods* 51, 95–102 (1995).

81. Heine, H.G., Stevens, M.P., Foord, A.J., and Boyle, D.B. A capripoxvirus detection PCR and antibody ELISA based on the major antigen P32, the homolog of the vaccinia virus H3L gene. *J Immunol Methods* 227, 187–196 (1999).

82. Babiuk, S. et al. Detection of antibodies against capripoxviruses using an inactivated sheeppox virus ELISA. *Transbound Emerg Dis* 56, 132–141 (2009).

83. Bowden, T.R. et al. Detection of antibodies specific for sheeppox and goatpox viruses using recombinant capripoxvirus antigens in an indirect enzyme-linked immunosorbent assay. *J Virol Methods* 161, 19–29 (2009).

84. Chand, P., Kitching, R.P., and Black, D.N. Western blot analysis of virus-specific antibody responses for capripox and contagious pustular dermatitis viral infections in sheep. *Epidemiol Infect* 113, 377–385 (1994).

85. Ireland, D.C. and Binepal, Y.S. Improved detection of capripoxvirus in biopsy samples by PCR. *J Virol Methods* 74, 1–7 (1998).

86. Mangana-Vougiouka, O. et al. Sheep poxvirus identification by PCR in cell cultures. *J Virol Methods* 77, 75–79 (1999).

87. Markoulatos, P., Mangana-Vougiouka, O., Koptopoulos, G., Nomikou, K., and Papadopoulos, O. Detection of sheep poxvirus in skin biopsy samples by a multiplex polymerase chain reaction. *J Virol Methods* 84, 161–167 (2000).

88. Orlova, E.S., Shcherbakov, A.V., Diev, V.I., and Zakharov, V.M. Differentiation of capripoxvirus species and strains by polymerase chain reaction. *Mol Biol* 40, 139–145 (2006).

89. Stram, Y. et al. The use of lumpy skin disease virus genome termini for detection and phylogenetic analysis. *J Virol Methods* 151, 225–229 (2008).

90. Balinsky, C.A. et al. Rapid preclinical detection of sheeppox virus by a real-time PCR assay. *J Clin Microbiol* 46, 438–442 (2008).

91. Zheng, M. et al. A duplex PCR assay for simultaneous detection and differentiation of Capripoxvirus and Orf virus. *Mol Cell Probes* 21, 276–281 (2007).

92. Kadam, A.S. et al. Detection of sheep and goat pox viruses by polymerase chain reaction. *Ind J Field Vet* 9, 48–51 (2014).

93. Venkatesan, G., Balamurugan, V., and Bhanuprakash, V. TaqMan based real-time duplex PCR for simultaneous detection and quantitation of capripox and orf virus genomes in clinical samples. *J Virol Methods* 201, 44–50 (2014).

94. Sadanand, F. et al. Evaluation of duplex PCR and PCR-RFLP for diagnosis of sheep pox and goat pox. *Int J Trop Med* 1, 66–70 (2006).

95. Irons, P.C., Tuppurainen, E.S., and Venter, E.H. Excretion of lumpy skin disease virus in bull semen. *Theriogenology* 63, 1290–1297 (2005).

96. Venkatesan, G., Balamurugan, V., Yogisharadhya, R., Kumar, A., and Bhanuprakash, V. Differentiation of sheeppox and goatpox viruses by polymerase Chain reaction-restriction fragment length polymorphism. *Virol Sinica* 27, 353–359 (2012).

97. Al-Shabebi, A. et al. Molecular detection and phylogenetic analysis of sheeppox virus in Al-Hassa of Eastern Province of Saudi Arabia. *Adv Anim Vet Sci* 2, 31–34 (2014).

98. Lubinga, J.C. et al. Detection of lumpy skin disease virus in saliva of ticks fed on lumpy skin disease virus-infected cattle. *Exp Appl Acarol* 61, 129–138 (2013).

99. Lubinga, J.C., Tuppurainen, E.S., Coetzer, J.A., Stoltsz, W.H., and Venter, E.H. Evidence of lumpy skin disease virus over-wintering by transstadial persistence in Amblyomma hebraeum and transovarial persistence in Rhipicephalus decoloratus ticks. *Exp Appl Acarol* 62, 77–90 (2014).

100. Babiuk, S. et al. Yemen and Vietnam capripoxviruses demonstrate a distinct host preference for goats compared with sheep. *J Gen Virol* 90, 105–114 (2009).

101. Espy, M.J. et al. Real-time PCR in clinical microbiology: Applications for routine laboratory testing. *Clin Microbiol Rev* 19, 165–256 (2006).

102. Watzinger, F., Ebner, K., and Lion, T. Detection and monitoring of virus infections by real-time PCR. *Mol Aspects Med* 27, 254–298 (2006).

103. Espy, M.J. et al. Detection of smallpox virus DNA by LightCycler PCR. *J Clin Microbiol* 40, 1985–1988 (2002).

104. Nitsche, A., Ellerbrok, H., and Pauli, G. Detection of orthopoxvirus DNA by real-time PCR and identification of variola virus DNA by melting analysis. *J Clin Microbiol* 42, 1207–1213 (2004).

105. Nitsche, A., Steger, B., Ellerbrok, H., and Pauli, G. Detection of vaccinia virus DNA on the LightCycler by fluorescence melting curve analysis. *J Virol Methods* 126, 187–195 (2005).

106. Murray, L. et al. Detection of capripoxvirus DNA using a novel loop-mediated isothermal amplification assay. *BMC Vet Res* 9:90 (May 1, 2013).

95 Myxoma Virus

June Liu and Peter Kerr

CONTENTS

95.1 Introduction 855
    95.1.1 Classification 855
    95.1.2 Morphology, Genome Organization, Biology, and Epidemiology 855
        95.1.2.1 Morphology 855
        95.1.2.2 Genome Organization 856
        95.1.2.3 Natural History 856
        95.1.2.4 Transmission of MYXV 856
        95.1.2.5 Coevolution of MYXV and European Rabbits 857
    95.1.3 Clinical Features and Pathogenesis 857
        95.1.3.1 Clinical Features 857
        95.1.3.2 Pathogenesis 858
    95.1.4 Diagnosis 859
        95.1.4.1 Conventional Techniques 859
        95.1.4.2 Molecular Techniques 860
95.2 Methods 861
    95.2.1 Sample Collection and Processing 861
    95.2.2 Detection Procedures 861
        95.2.2.1 In Vitro Culture of MYXV from Primary Samples 861
        95.2.2.2 Immunohistochemical Detection of MYXV Antigen 861
        95.2.2.3 Serological Determination of Rabbit's Immune Status 862
        95.2.2.4 PCR Amplification of MYXV DNA 863
95.3 Conclusion and Future Perspectives 864
References 864

95.1 INTRODUCTION

95.1.1 CLASSIFICATION

As the type species of the *Leporipoxvirus* genus in the *Chordopoxvirinae* subfamily, *Poxviridae* family, myxoma virus (MYXV) is a large, double-stranded DNA virus that replicates in the cytoplasm of infected cells. MYXV causes the generalized, frequently lethal disease, myxomatosis, in European rabbits (*Oryctolagus cuniculus*). Other leporipoxviruses include hare fibroma virus, rabbit fibroma virus (also called Shope's fibroma virus), and squirrel fibroma virus, which naturally infect various leporid and squirrel species.

95.1.2 MORPHOLOGY, GENOME ORGANIZATION, BIOLOGY, AND EPIDEMIOLOGY

95.1.2.1 Morphology

Electron microscopic images of shadowed MYXV or negatively stained particles show a brick or oval-shaped virus with dimensions of 286 × 230 × 75 nm.[1,2] The MYXV virion is indistinguishable from more intensively studied poxviruses

such as *Vaccinia virus* (VACV).[1,3] More recent structural studies of VACV have suggested that poxviruses resemble a flattened ellipsoid or possibly barrel shape depending on the imaging method (reviewed in reference 4).

As with other poxviruses, there are two infectious forms of MYXV[5]: an intracellular mature form (MV; also called intracellular mature virus [IMV]), released by cell lysis, and an enveloped extracellular form (EV; also called extracellular enveloped virus [EEV]), which is produced by wrapping MV with a double membrane, probably derived from the Golgi or endoplasmic reticulum,[5] forming an intracellular wrapped virion. This contains virus-encoded proteins that are not present in the MV form. The outer of the double membrane layer is lost by fusion with the cell plasma membrane as the virus is exocytosed to produce the final EV.[4] EV may be retained attached to the cell membrane (CEV) or released into the extracellular fluid.

By convention, the outer component of the MV form is referred to as the viral membrane although the structure is much more complex than a typical lipid bilayer; negative-stained electron micrographs show an unstructured array of spicules termed surface tubular elements. This membrane surrounds a peanut-shaped core containing the viral genome complexed with virion structural proteins and proteins needed for early gene expression. By thin section electron microscopy, lateral bodies are seen on either sides of the core concavity, filling the space between the core and the membrane.[4]

No specific cellular receptor has been identified, and it seems likely that MYXV is able to enter cells by binding to common cell membrane molecules followed by uptake in vesicles.[6] VACV can enter cells by direct fusion at the cell membrane or by triggering uptake by macropinocytosis or endocytosis. For EV, rupture of the EV envelope, triggered by binding to molecules such as cell surface glycosaminoglycans, is needed to allow the MV membrane to either fuse with the cell plasma membrane or trigger uptake.[7–9]

Replication of poxviruses occurs entirely within the cytoplasm. Early genes are transcribed from within the virus core by virus-encoded RNA polymerase. Uncoating of the virus core is followed by DNA synthesis and intermediate and late gene expression. Virus assembly occurs in "factories" within the cytoplasm. The replication cycle of MYXV in cell culture takes approximately 16 h.[5]

95.1.2.2   Genome Organization

The genome of the Lausanne reference strain of MYXV (GenBank AF170726.2) consists of 161,777 nucleotides of double-stranded DNA with closed single-stranded hairpin termini and terminal inverted repeats (TIRs) of 11,577 nucleotides.[10,11] There are 158 unique open reading frames (ORFs) annotated, 12 of which are duplicated within the TIRs.[10,11] By convention, genes are numbered from the left-hand end of the genome with the direction of transcription indicated by L (left) or R (right); genes within the TIR are indicated by L/R since the direction of transcription is reversed in the L and R TIR, for example, *M001L/R*, proteins are indicated by omitting the direction of transcription, M001. Like other poxviruses, genes encoding replication and structural proteins tend to be in the center of the genome and conserved with other poxviruses, while genes encoding the large number of proteins involved in immunosuppression or host-range functions tend to be toward the termini and are less conserved.

95.1.2.3   Natural History

MYXV naturally infects the South American tapeti or forest rabbit (*Sylvilagus brasiliensis*), in which it induces a cutaneous fibroma at the inoculation site. Virus adheres to the mouthparts of biting arthropods, such as fleas and mosquitoes, probing through the virus-rich epidermis overlying the fibroma and is mechanically transmitted by these vectors; the virus does not replicate in the vector, and there is no vector specificity.[12] MYXV appears to be present throughout the natural range of the tapeti in South America and into Central America.[13] Although largely innocuous in its natural host, in European rabbits, MYXV causes the lethal, generalized disease termed myxomatosis.

European rabbits probably evolved in the Iberian Peninsula and Southern France but were subsequently spread around Europe as semidomesticated stock, some of which reverted

to the wild.[14–16] The breeds of domestic rabbit farmed for meat and fur, the various pet breeds, the rabbits used in laboratories, and wild rabbits are all *Oryctolagus cuniculus*. European rabbits were introduced to other countries such as Australia and New Zealand following European settlement where they established and became destructive pests.[15] Although European rabbits are not native to the Americas, (native leporids, such as cottontails and jackrabbits, belong to the *Sylvilagus* and *Lepus* genera) domestic breeds were introduced following European settlement and in some countries, such as Chile and Argentina, have established widespread feral populations that have attained serious pest status.[15]

Myxomatosis was first described as a novel disease occurring in laboratory rabbits, at the Institute of Hygiene in Montevideo, Uruguay, in 1896. It was one of the earliest animal diseases ascribed to a filterable virus. MYXV would have had limited significance, except as an occasional cause of disease in domestic rabbits in the Americas, but, in 1950, a South American strain of MYXV was experimentally introduced into the wild European rabbit population of Australia to evaluate its potential as a biological control.[17] The subsequent unprecedented spread and lethality of myxomatosis in Australian rabbits prompted a landholder to introduce a separate strain of South American MYXV into wild rabbits in France in 1952.[16,17] From this introduction, MYXV spread into the wild rabbit populations throughout Europe, Ireland, and Britain and into farmed European rabbits across the continent. MYXV is now endemic in Australia and Europe and has also been introduced and become established in feral European rabbit populations in Chile and Southern Argentina (outside the natural range of the tapeti).

A leporipoxvirus, closely related to MYXV, albeit with significant sequence divergence,[18] occurs in the brush rabbit (*Sylvilagus bachmani*) on the west coast of North America and Mexico. This Californian MYXV (Cal MYXV) also causes lethal myxomatosis in European rabbits but only a cutaneous fibroma in brush rabbits.[19] Another leporipoxvirus, *rabbit fibroma virus* (Shope's fibroma virus), occurs in the eastern cottontail (*S. floridanus*) in the eastern and central parts of the United States.[20] This virus induces an innocuous cutaneous lesion in immunocompetent European rabbits and has been used extensively as a heterologous live vaccine to protect rabbits from myxomatosis. More generalized disease may occur in very young rabbits or in experimentally immunosuppressed rabbits infected with rabbit fibroma virus.

95.1.2.4   Transmission of MYXV

In European rabbits, MYXV can be spread both by biting arthropods and by direct contact. It is mechanically transmitted on the mouthparts of biting arthropods, usually mosquitoes or various species of flea, which pick up the virus by probing through the virus-rich cutaneous lesions and sites such as the base of the ears and eyelids where virus replicates to high titers.[13] Other biting arthropods such as *Culicoides* midges or simuliids (black flies), lice, and mites are also potential vectors. Virus is shed in ocular and respiratory discharges and from the eroded surface of cutaneous lesions and is readily inoculated into the nasal passages, conjunctivae, or epidermis during social interactions or fighting. The virus is not normally infectious orally, and natural aerosol transmission is inefficient.[21] Sexual transmission has been recorded and the virus may be present in semen.[22]

MYXV is only able to infect and transmit from a narrow range of leporids.[21,23] Apart from *S. brasiliensis* and *O. cuniculus*, MYXV can be transmitted by mosquitoes from experimentally infected *S. nuttalli* (mountain cottontail), *S. audubonii* (desert cottontail), and, occasionally, *S. floridanus* (eastern cottontail). The brush rabbit (*S. bachmani*) can be infected, but MYXV does not reach transmissible titers in this species.[24,25] Of these American species, only *S. floridanus* potentially overlaps with the distribution of MYXV. European hares (*Lepus europaeus*) and mountain hares (*L. timidus*) have occasionally been reported with natural myxomatosis, but this is unusual. Cal MYXV also appears to have a very restricted host range.[26,27]

95.1.2.5   Coevolution of MYXV and European Rabbits

The viruses released in Australia (standard laboratory strain, SLS) and Europe (Lausanne strain, Lu) were highly lethal with estimated case fatality rates (CFRs) up to 99.8% in both wild and domestic rabbits.[13,28] The subsequent host–pathogen coevolution has became the classic model for what happens when a novel pathogen shifts into a completely naïve host species.[29–33] Over relatively short time periods, virus strains that were moderately attenuated, albeit, still with CFRs of 75%–90%, became predominant in the field.[31] These slightly attenuated viruses allowed the infected rabbit to survive longer, thus increasing the probability of transmission. At the same time, there was intense selection pressure for rabbits with some degree of resistance to myxomatosis. This may have been aided by the emergence of moderately attenuated viruses and, in Australia, by very hot summer weather, which ameliorates the clinical disease in rabbits infected with attenuated viruses, especially if these rabbits have some genetic resistance.[34] At one study site, CFRs in wild rabbits, tested under controlled laboratory conditions with the same strain of virus, decreased from 90% to 26% within 7 years, representing seven generations of selection.[35,36]

Field isolates of MYXV were classified into five (later extended to six) virulence grades based on CFR and average survival times in small groups of laboratory rabbits infected with low doses of virus (Table 95.1). Grade 1 viruses were of the same virulence as the originally released strains with essentially 100% CFR, whereas grade 5 viruses had CFR of <50% and were never common in the field, probably because the rabbit immune system could quickly reduce titers in the skin lesions below the threshold for efficient mosquito transmission.[28,31] Grade 3 viruses became the most prevalent in both Australia and Europe[31,37]; however, the ongoing selection for resistance in the wild rabbit population may also be selecting for a higher prevalence of more virulent viruses.

It is important to recognize that MYXV strains of different virulence grades and symptomatology are circulating in wild and domestic rabbits and that the clinical signs and disease

TABLE 95.1
Virulence Grades of Myxoma Virus

Virulence Grade	Case Fatality Rate (%)	Average Survival Time (Days)[a]
1	>99	≤13
2	95–99	13–16
3A	90–95	17–22
3B	70–90	23–28
4	50–70	29–50
5	<50	n/a

[a] Survival time (ST) was transformed as $\log_{10}$(ST-8) for each rabbit; average survival times were calculated using the transformed data and then back-transformed.[13,28]

outcomes will depend on both the virulence of the infecting virus and the genetic resistance of the rabbit. In addition, apparent virulence will be influenced by biotic factors such as age of the rabbit, nutritional status, intercurrent disease, and parasite load and abiotic factors such as environmental temperature. The clinical disease in highly resistant wild rabbits may be quite different to that in unselected domestic rabbits (see Section 95.1.3). In addition, infections with attenuated strains of MYXV in farmed rabbits can be confused with bacterial upper respiratory tract disease so that infected rabbits are not recognized and isolated, allowing ongoing transmission.[38,39]

95.1.3   CLINICAL FEATURES AND PATHOGENESIS

95.1.3.1   Clinical Features

Three general forms of clinical disease can be recognized in European rabbits following infection with different strains of MYXV: cutaneous myxomatosis, amyxomatous myxomatosis, and peracute disease.

95.1.3.1.1   Cutaneous Myxomatosis

This is the classic form of the disease. Two to three days after intradermal inoculation, a cutaneous swelling develops at the inoculation site (primary lesion; also termed myxoma or tumor), which, over time, may become as much as 4 cm in diameter and 0.5–2 cm high. This is followed, over the next 2–3 days, by gradually increasing redness and swelling of the eyelid margins and swelling of the head and ears. By around 10–12 days, infected rabbits have a markedly swollen head, sometimes described as leonine; swollen ears, especially the base of the ear; blepharoconjunctivitis with swollen, closed eyelids; and copious mucopurulent discharge, and there may be hair loss from around the head (Figure 95.1). Rhinitis, with profuse mucopurulent discharge obstructing the nasal passages, causes respiratory distress characterized by stertor and gasping with the mouth open and the head and neck extended. Intense swelling of the anogenital region and, in males, of the scrotum and testes is typical. Cutaneous secondary lesions are present over the head, body, and legs (these may merge

FIGURE 95.1 Domestic rabbit with severe myxomatosis. Note the very swollen head with hair loss and swollen eyelids and ears. This rabbit has the amyxomatous form of the disease.

to form a solid mass). Ears may become so swollen that they droop rather than being upright. Erosion, weeping, and scabbing of the surface of the primary cutaneous lesion occur late in infection. Palpable lymph nodes, such as popliteal, can be enlarged. Rectal temperatures are frequently above 40°C from around 4–5 days after infection but may drop below normal prior to death.

Rabbits infected with more attenuated strains of the virus may exhibit milder clinical signs and, depending on the virulence of the virus, have a prolonged disease course often complicated by secondary bacterial infections. These rabbits can lose 15% or more of their body weight and have elevated temperatures for weeks. In survivors, gradual resolution of the clinical disease occurs over a month or so, but rabbits may be left with chronic respiratory problems. The disease in genetically resistant wild rabbits tends to resemble that in domestic rabbits infected with attenuated viruses, but resolution can be surprisingly rapid.[40–42] In some cases, the main clinical finding is multiple gray, hairless, raised fibromas.

95.1.3.1.2 Amyxomatous Myxomatosis

Infected rabbits have the swollen head and ears, blepharoconjunctivitis, rhinitis, nasal obstruction, and discharge together with anogenital swelling that are typical of myxomatosis. However, there are no, or very few, cutaneous lesions apart from the primary inoculation site, but this is absent if the rabbits were infected intranasally or conjunctivally.

Virus strains causing this amyxomatous type of disease were originally described in France and Belgium[43–47] and are reported as common in rabbit production units in Spain and Italy.[22] Some Australian field strains, characterized from the 1990s onward, cause a similar disease in domestic (nonresistant) rabbits. It has been speculated that amyxomatous strains of MYXV are adapted to respiratory transmission because of the absence of cutaneous lesions; however, this has not been demonstrated.[43,46] Whether titers of virus in the swollen base

of the ears and eyelids are compatible with vector transmission has not been reported. As with the cutaneous form of the disease, viruses of very different virulence grades are present in the field, and secondary bacterial infection may be important in the pathogenesis and outcome of the infection.[46,47] Prolonged incubation periods, up to 20 days, have been reported with some strains of amyxomatous viruses. This may allow transmission of virus to uninfected rabbit farms by transfer of apparently healthy rabbits or vaccinated stock that were incubating the disease prior to vaccination.[48,49]

95.1.3.1.3 Peracute Disease

Rabbits are found dead or moribund with minimal clinical signs of myxomatosis 7–12 days after infection. There may be slight to moderate swelling of the head and base of the ears and some swelling of the eyelids and anogenital region, but generally there is little or no ocular or nasal discharge. The lesion at the inoculation site is flat, poorly differentiated from the surrounding skin, and often barely visible; on sectioning it may be only 2–4 mm thick, usually white in color; sometimes the overlying epidermis forms a visible red line. Convulsions at or around death may be seen, and death is often due to pulmonary edema. Rectal temperatures are very variable, sometimes elevated around day 4–6 but only for a short period. Genetically resistant wild rabbits infected with these strains tend to develop more typical myxomatosis.[41] Very young rabbits infected with cutaneous strains of MYXV die with minimal clinical signs of myxomatosis. Cal MYXV strains can cause peracute death in both laboratory and genetically resistant wild rabbits.[28,50]

95.1.3.2 Pathogenesis

Detailed pathogenesis studies of MYXV infection in European rabbits have been done with virulent cutaneous virus such as SLS, the Lu strain, attenuated cutaneous virus, the Cal MYXV MSW strain, and many single-gene knockout viruses derived from Lu[40,50–54]; reviewed in references 32, 33, 55, and 56. There have also been studies on amyxomatous viruses.[43,44,46,47]

Virus inoculated intradermally initially replicates in major histocompatibility complex (MHC)-II-positive cells at the epidermal/dermal interface and then spreads to the overlying epidermis inducing epidermal hyperplasia and hypertrophy with loss of architecture and disruption of the underlying dermis, forming the raised primary lesion. This can be apparent from 2 to 4 days after inoculation depending on the virus strain and the genetic resistance of the rabbit. At the same time, virus spreads to the draining lymph node where it can be demonstrated in the cells of the subcapsular sinus and then in lymphocytes of the cortex.[40,53,54] From here, virus disseminates in lymphocytes to the spleen and lymph nodes, mucocutaneous sites such as eyelids and nasal mucosa, and cutaneous sites, particularly the base of the ears, but secondary cutaneous lesions can be found all over the body. Virus is also found in the liver, lungs, and testes and, at low titers, in the central nervous system.[57,58] Cal MYXV has been reported to reach high titers in brain[28] but this was not observed in another study.[50]

MYXV causes profound immunosuppression of European rabbits with loss of lymphocytes from the lymphoid tissues and anergy in the surviving lymphocytes together with downregulation of macrophage responses.[59–61] Specific viral proteins stimulate cell growth and disrupt intercellular and intracellular inflammatory and antiviral pathways, including type I and type II interferon responses, tumor necrosis factor (TNF), chemokine signaling, NFκB signaling, MHC-1 presentation, inflammasome signaling, and cell death pathways.[32,55,56] Death typically occurs 10–14 days after infection with highly virulent virus. The direct cause of death is unclear as there is no critical pathology in any of the major organs.

95.1.4 DIAGNOSIS

Presumptive diagnosis of myxomatosis can be made on clinical examination, potentially combined with gross autopsy data and supported by histological evaluation. Definitive diagnosis relies on demonstration of virus or viral DNA in samples from clinically affected animals or demonstration of specific IgM or a rising IgG titer (Table 95.2).

TABLE 95.2
Diagnostic Procedures for Myxomatosis

Procedure	Interpretation/Limitations
Clinical examination/autopsy	Indicative in cutaneous form; may be equivocal in amyxomatous or peracute cases.
Histopathology	Epidermal cell hyperplasia, hypertrophy, and degeneration with disruption of underlying dermis in cutaneous form. Eosinophilic inclusions may be present in cytoplasm of epidermal cells.
Immunostaining (immunofluorescence, immunoperoxidase, biotin/streptavidin/peroxidise, etc.)	Specific antiserum required; confirmatory on histological sections or impression smears.
Negative staining electron microscopy	Indicative on extracts from tissues; does not differentiate from other poxviruses.
Agar gel immunodiffusion (AGID)	Confirmatory on tissue extracts; requires specific antiserum.
Polymerase chain reaction (PCR)	Confirmatory on total DNA from tissues or conjunctival swabs.
Virus isolation	Typical cytopathic effect is indicative; confirmation by PCR, immunostaining, or AGID.
Serology (ELISA, immunofluorescence assay [IFA], plaque reduction neutralization assay [PRNA], AGID, complement fixation assay [CFA])	IgM (I-ELISA or C-ELISA) or rising antibody titer by ELISA, IFA, or PRNA together with clinical signs or history. AGID and CFA have low sensitivity.

95.1.4.1 Conventional Techniques

95.1.4.1.1 Clinical Diagnosis
Diagnosis is relatively straightforward in typical cutaneous forms of myxomatosis where the disease has progressed sufficiently for the raised skin lesions, swollen head, and anogenital edema to be apparent, and there is a history of potential exposure either from contact with infected rabbits or by insect transmission.

Differential diagnosis: Amyxomatous forms of the disease are readily confused with bacterial respiratory disease or blepharoconjunctivitis. Peracute forms of the disease need to be differentiated from other causes of acute death such as clostridial infections, bacterial septicemia, and rabbit hemorrhagic disease.[62] Infection with *leporid herpes virus 4* is also a potential differential diagnosis because of the acutely swollen head, conjunctivitis with closed eyelids, skin lesions, anogenital swelling, and respiratory difficulty caused by this virus.[63] *Rabbitpox virus* (an orthopoxvirus derived from VACV) can cause acute death or a more protracted disease with skin rash, blepharoconjunctivitis, and orchitis. Historically, outbreaks of rabbitpox have occurred in laboratory rabbits, but it should now only be seen following experimental infection. Poisoning with sodium fluoroacetate or anticoagulants such as pindone are potential causes of acute death in wild rabbits in countries where lethal control is practiced.

95.1.4.1.2 Autopsy Findings
Swollen head, ears, eyelids, and anogenital swelling are typical. If cutaneous lesions are present, these are firm on sectioning with color ranging from red to white depending on the virus strain and stage of infection. Classic cutaneous forms have very glistening cut sections due to the mucinous nature of the swollen dermis. Lymph nodes can be grossly enlarged 3–5× their normal size and range from firm to flaccid and watery on cut section; the spleen is frequently very enlarged often 2–4× normal size. Testes and epididymides may be swollen or atrophied, depending on the stage of disease, and granulomata may be present. Other internal tissues often appear normal; however, the lungs may show evidence of pneumonia due to secondary bacterial infections, particularly with amyxomatous strains. In chronic cases, there is often complete wastage of the fat depots around the kidneys. Pulmonary edema may cause peracute deaths; massively enlarged, wet lungs with froth and fluid filling the trachea and bronchi are obvious at autopsy.

95.1.4.1.3 Histopathology
Hematoxylin- and eosin-stained sections from the cutaneous lesions or eyelids of rabbits with the cutaneous form of the disease show hyperplasia, hypertrophy, and degeneration of the epidermal cells often with vesicle formation in the cutaneous lesions. There is disruption of the underlying dermal collagen fibers and massive thickening of the dermis and subdermis by the acellular mucinous material.[51,52,54] The small blood vessels are frequently disrupted by large stellate or elongated cells termed "myxoma cells," which may also be seen throughout the dermis. Eosinophilic cytoplasmic inclusions may be

present in degenerating epidermal cells; although these are reported as best seen with fixatives such as Bouin's solution,[44] they can also be seen in formalin-fixed sections. These inclusions are distinct from the eosinophilic A–type inclusion bodies formed by some orthopoxviruses. Large numbers of heterophils (neutrophils) may be present in the underlying tissues, sometimes only in deep tissues, but lymphocytes are normally absent unless the virus is very attenuated.

Cutaneous lesions may not be identifiable in the amyxomatous forms of the disease, but when these are sectioned, the epidermal hyperplasia and degeneration are not as pronounced as in the cutaneous form and the underlying dermis may only be disrupted directly under the epidermis. In amyxomatous cases, typical proliferative and degenerative changes in epidermis and mucosa can often be seen in sections from the eyelids.

Lymph nodes may be completely or partially devoid of lymphocytes with destruction of B cell follicles and loss of cortical lymphocytes. In some cases, proliferation of reticular cells may completely fill the node; in other cases, there may be very large numbers of heterophils present. Similar changes can occur in the spleen. Proliferation of reticular cells in lymphoid follicles has been described in the lungs together with varying degrees of pneumonitis, but frequently the lungs appear histologically normal. Proliferation of connective tissue elements and hepatic necrosis have also been described in the liver.[52]

95.1.4.1.4   Immunohistochemistry

Virus antigen can be demonstrated by immunostaining using impression smears made from primary lesions or conjunctivae or in histological sections using either frozen sections or antigen retrieval on fixed and paraffin-embedded sections[22,50,54,64] (Figure 95.2). Nonspecific autofluorescence, especially around

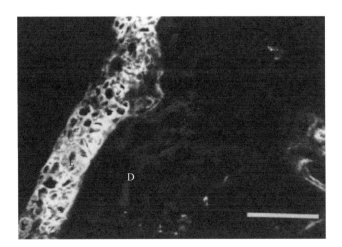

FIGURE 95.2   Immunofluorescent demonstration of MYXV antigen in the epidermis of a primary cutaneous lesion; E: epidermis; D: dermis; scale bar 100 μm. (Reproduced from *Virology*, 277, Best, S.M., Collins, S.V., and Kerr, P.J., Coevolution of host and virus: cellular localization of myxoma virus infection of resistant and susceptible European rabbits, 76–91. Copyright 2000, with permission from Elsevier.)

hair follicles, can be a problem, so it is important to use appropriate controls.

95.1.4.1.5   Electron Microscopy

Poxvirus particles can be demonstrated by negative staining of extracts from cutaneous tissues[22] or in thin sections from infected tissues. These cannot formally be differentiated from rabbitpox virus or rabbit fibroma virus without further immunological or DNA-based procedures.

95.1.4.1.6   Virus Isolation and Culture

MYXV can be readily isolated from tissue samples collected at autopsy or from clinical samples (see Section 95.2.1). Despite its narrow host range, MYXV replicates in cell lines derived from a number of species including: monkey (Vero, CV-1, BGMK), human (Hela), avian (chick embryo fibroblasts), and rabbit (RK-13, SIRC). It can also be grown and titrated on the chorioallantoic membrane of fertile hen eggs. The cytopathic effect (CPE) seen depends somewhat on the cell line and multiplicity of infection (MOI). At low MOI on RK-13 cells, proliferative raised foci develop 24–48 h after infection. At later times, or at high MOI, multinucleate syncytia may be the most prominent feature with progression to cell death and loss of monolayer integrity. The virus is strongly cell associated, and the spread in the culture is mostly to adjoining cells. Some strains may take up to 7 days to show CPE.[22] Confirmation of MYXV can be achieved by immunostaining, polymerase chain reaction (PCR; see Section 95.1.4.2), or purification of viral DNA and restriction endonuclease digest to show the typical MYXV restriction profile.[65,66] Viral antigen can also be demonstrated by the formation of specific precipitin bands using agar gel immunodiffusion (AGID) of cell extracts against specific antiserum. Detailed methods for preparing MYXV stocks, including gradient purification of virus, are given in reference 67. MYXV can be titrated by plaque assay (Vero cell monolayers) or focus assay (RK-13 cell monolayers).

95.1.4.2   Molecular Techniques

PCR and real-time PCR have been used for the molecular diagnosis of MYXV DNA from tissue materials such as lesions from eyelids, ear, internal organs, nasal and conjunctival swabs, or semen.[41,68–70] PCR is highly sensitive and very specific if the primers are designed properly. The high sensitivity can cause false positives unless stringent steps are taken to avoid contamination with MYXV DNA from other sources during sample preparation and setting up the reaction mix. Some live vaccines used in Europe, for example, the Borghi strain, are derived from the MSD strain of Cal MYXV, which means that oligonucleotide primers for MYXV may not work for these viruses. Specific sequences are available.[18,69] The French SG33 vaccine strain has a large deletion, and primers need to be designed to avoid this, plus there has been recombination between this strain and a Cal MYXV strain.[18,71] Specific primers for these strains are available.[22] Real-time PCR assays have also been published.[68,70] DNA sequence analysis can be used to confirm the identity of PCR products.

Restriction endonuclease analysis of purified genomic DNA has been used for epidemiological analysis of MYXV,[66,72] and PCR amplification of strain-specific restriction fragment length polymorphisms has been used for field epidemiological studies.[41,66,73,74] Genomic sequencing and phylogenetic analysis are now becoming readily available for evolutionary studies.[75,76]

95.2 METHODS

95.2.1 Sample Collection and Processing

MYXV can be readily cultured from cutaneous lesions or eyelids collected from clinical cases at autopsy. For field collection, samples should be taken into sterile 5 mL screw-cap plastic vials containing 1:1 (v/v) glycerol/phosphate-buffered saline (PBS; pH 7.2) (antibiotics/antifungals can also be included). The virus is relatively robust, but samples need to be kept cool, preferably frozen, prior to further processing. Tissues from the liver, lung, testis, lymph node, and spleen may also be used for MYXV isolation and have the advantage that they can be collected using sterile technique; virus titers can be quite low in these tissues and secondary bacteria can be present.

To release the virus, tissues are washed with PBS and hair removed using sterile scissors. Impression smears can be taken for immunohistochemistry at this stage, and small subsamples (10–20 mg) of the virus-rich epidermis collected for DNA extraction for PCR diagnosis. Extracts of the virus-rich epidermis can also be used for AGID against specific antiserum to demonstrate virus antigen or for electron microscopy. For virus isolation, the tissues are minced with scissors and can be ground with sterile sand, disrupted using a magnetic bead beater, or simply minced finely and then freeze–thawed two to three times or sonicated to disrupt cells. This is followed by low-speed centrifugation to clarify the supernatant. Passing the clarified supernatant through a 0.45 μm filter can reduce bacterial and fungal contamination; a 0.22 μm filter is effective at removing contaminants but will exclude a large proportion of the virus.

Virus isolation can be attempted from whole blood or, preferably, buffy coat, but virus titers may be very low. Cells need to be freeze–thawed or sonicated to release the virus; there is little or no virus in serum.

For nonlethal clinical sampling, conjunctival or nasal swabs are probably the simplest samples to collect; using a saline-moistened swab causes minimal distress to the animal. It is important to actually collect mucosal cells onto the swab not just the mucopurulent discharge. For processing, cut off the swab tip into a 1.5 mL capped plastic centrifuge tube containing approx 1 mL of minimal essential medium (MEM) or PBS plus 5× antibiotics/fungizone (penicillin 500 U/mL, streptomycin 500 μg/mL, fungizone [amphotericin B] 1.25 μg/mL, 5× P/S/F), vortex vigorously, or squeeze the swab with an applicator stick. Then centrifuge for 5 min in a microcentrifuge at high speed to pellet cells, remove the swab (but retain for further processing if necessary) and either freeze–thaw or sonicate the cell pellet, and then vortex, centrifuge briefly to

clarify, and inoculate the supernatant into cell culture. DNA can also be extracted from the cell pellets for PCR diagnosis.[41] Punch biopsy of cutaneous lesions or needle biopsy of eyelids can be used for virus isolation or PCR (or AGID), and as noted earlier, buffy coat could be used, particularly in the acute phase of the disease, but it is less reliable.

95.2.2 Detection Procedures

95.2.2.1 In Vitro Culture of MYXV from Primary Samples

The following procedure is for RK-13 cells; however, other cell lines (Vero, SIRC, BGMK, CV-1) can be used for the isolation of MYXV using the same method. Seed cells in a T25 flask at approximately 1.4×10^6 cells to allow it reach 95% confluence after 24 h. Cells are cultured in MEM containing 10% fetal calf serum (FCS) at 37°C in an atmosphere of 5% CO_2/95% air.

Remove the medium from the cell monolayer and add 1–2 mL of the sample supernatant in 5× P/S/F and allow to adsorb for 60 min in the incubator. While not essential, removing the inoculum and washing the monolayer twice with PBS after adsorption can help to reduce contamination. Add 5 mL MEM + P/S/F and incubate at 37°C with 5% CO_2 (MEM supplemented with as little as 2% FCS can be used for maintenance once monolayers are confluent; RK-13 cells will tolerate at least 5× P/S/F). An uninfected control monolayer is also useful. On RK-13 cell monolayers, small proliferative, raised foci can be seen as early as 24 h after infection and should be readily seen by 48–72 h. With low-virus concentrations, there may be very few foci. With very high-virus concentrations, large syncytia may be seen instead of foci. Note that normal, uninfected RK-13 cells spontaneously form small syncytia as they age. If no CPE is observed after 5–7 days, a blind passage is usually performed: freeze–thaw the cells twice and centrifuge the suspension at $1500 \times g$ for 5 min and use 1–2 mL of the supernatant. It can be worthwhile reprocessing the original sample, and repeating the infection as the concentration of viable virus in poorly handled samples or samples collected late in the course of the disease may be very low. If bacterial or fungal contamination is a problem in the cell culture, it is often possible to rescue the virus by washing the monolayer several times with PBS + 5× P/S/F, freezing/thawing in PBS + 5×P/S/F to rupture the cells, centrifuging ($3000 \times g$, 10 min, 4°C) to remove debris, and then passing the supernatant through a 0.45 μm filter; use the filtered supernatant to infect cells as mentioned earlier using MEM + 5× P/S/F. Other antibiotic combinations can also be used.[22]

Virus identification should be confirmed by immunohistochemistry, AGID, PCR, or DNA purification.

95.2.2.2 Immunohistochemical Detection of MYXV Antigen

Immunohistochemistry to detect MYXV antigen can be performed on fixed tissue sections, frozen sections, impression smears, or on cells grown in culture. Antisera can be from rabbits immunized against MYXV or, preferably, mouse monoclonal antibody, which has the advantage of high specificity and low background. The following protocols are for immunofluorescence assay (IFA). Protocols for immunoperoxidase staining of cultured cells, antigen retrieval on fixed tissue sections, and streptavidin–biotin–peroxidase staining have been published.[22,50]

95.2.2.2.1 Sample Preparation

Frozen sections: Slices of tissue are embedded in OCT compound (Tissue-Tek) and frozen in liquid N_2 or dry ice and stored at −80°C. Sections are cut at 4–7 µm, using a cryostat, onto poly-L-lysine treated glass slides; fixed in formol calcium (10% v/v formalin, 10% v/v 1 M $CaCl_2$, 80% v/v water), dipped in cold acetone and then chloroform/acetone 1:1 (v/v) for 5 min at −20°C, dipped in cold acetone, washed twice in PBS, and stored frozen at −80°C.[54] At least two sections for each tissue should be prepared: one for specific primary antibody, one for normal serum control; a no antibody control is also useful.

For cell culture samples, cells are grown on sterile glass cover slips in 24-well plates and infected with MYXV for 24–48 h. The cells are fixed in cold methanol or methanol/acetone 1:1 (v/v) for 5 min, washed with cold PBS (pH 7.4), and stored at −20°C prior to staining. Subsequent washing and staining can be done in the 24-well plate.

Impression smears from lesions can be made onto clean glass slides, air-dried, then fixed in methanol/acetone for 2 min at room temperature. These can be stored in the refrigerator or freezer prior to processing.

95.2.2.2.2 Immunofluorescence Procedures[54]

Thaw the slides for 20 min at room temperature and add 100 µL of blocking buffer (3% bovine serum albumin (BSA) w/v in PBS). Incubate for at least 30 min at room temperature or overnight at 4°C to reduce nonspecific staining. Wash with 1% BSA in PBS: (PBS-B).

Add 100 µL of the MYXV-specific antibody diluted in PBS-B (dilutions will need to be determined for each antiserum) onto the slide and incubate at room temperature for at least 2 h in a humidified chamber or at 4°C overnight. Wash slides with 0.01% Tween 20 v/v in PBS three times.

Apply fluorophore-labeled secondary antibody at dilution recommended by the manufacturer and incubate in the dark for 1 h at 37°C.

Wash slides with Tween 20-PBS three times and mount the slide with an antifade mounting medium. Visualize the signal using a fluorescence microscope.

95.2.2.2.3 Result Interpretation

Check the negative control sample first. If the sample with only secondary antibody but not with primary antibody gives low-background fluorescence, the signal from the sample with anti-MYXV antibody indicates the presence of MYXV in the cells or tissues.

95.2.2.3 Serological Determination of Rabbit's Immune Status

Following infection with MYXV, specific IgM antibodies are detectable as early as 6 days and have largely disappeared by 30–40 days; IgG antibodies are present from 10 to 15 days after infection and can be detected for at least 2 years.[40,77] Neutralizing antibodies can be detected by day 10 after infection in laboratory rabbits and as early as 6 days in resistant wild rabbits.[40]

Serological assays to determine the immune status to MYXV include ELISA, IFA, plaque reduction neutralization assays (PRNA), complement fixation test (CFT), and AGID. Due to the low sensitivity, CFT and AGID are not recommended.[22] Two types of ELISA are routinely used to detect MYXV-specific antibodies in rabbits: indirect ELISA (I-ELISA) and competition ELISA (C-ELISA). Alternative ELISA protocols[78] and protocols for IFA, CFT, and AGID are given in reference 22. It should be noted that when using these ELISAs, each laboratory should further optimize them according to the available antigen and the secondary antibody.

95.2.2.3.1 I-ELISA[77]

95.2.2.3.1.1 Preparation of Standard Sera   Standard negative sera: Collect blood from rabbits that have never been exposed to MYXV and have not been vaccinated with rabbit fibroma virus. Confirm that the sera are negative prior to pooling the sera from all the rabbits. Aliquot and store at −20°C.

Standard positive sera: Vaccinate adult rabbits with live MYXV vaccine (this can be the heterologous rabbit fibroma virus or one of the commercial live homologous viruses) and challenge the rabbits with a virulent strain of MYXV 30 days later. Collect a blood sample 20–30 days after challenge to confirm that titers are high before bleeding the rabbits. Pool the sera, measure the specific antibody titer, aliquot, and store at −20°C.

95.2.2.3.1.2 Preparation of Virus Antigen   The international reference MYXV strain (Lu, ATCC code VR-115) or another suitable strain can be used to prepare antigen. This antigen detects antibodies to Cal MYXV as well as MYXV.[27,50]

Infect 6–12 T175 flasks of RK-13 cells at a MOI of 0.2–0.5 and culture the cells in MEM. At this MOI, virus should be ready to harvest at 48–72 h after infection. When uniform CPE is observed, tip off the culture medium and wash the cell monolayers twice with PBS (pH 7.2), and then scrape the cells from the flask and pellet by centrifugation (800 × g for 10 min at 4°C). Resuspend pellets in 5 mL PBS and sonicate the cells for 2 min to release virus. Alternatively, cells can be Dounce homogenized to release the virus. Digest with DNase 1 (25 µg/mL—Sigma DN-25) and RNase A (50 µg/mL—Sigma R-4875) at 37°C for 30 min with frequent agitation. Pellet the virions by centrifugation (250,000 × g, 20 min at 4°C) through a step gradient formed by overlaying 10% of T10 dextran over the same volume of 36% sucrose, both solutions in TE (10 mM Tris, 1 mM EDTA, pH 8.0). Resuspend the pellet in 1 mL PBS. Repeat the gradient centrifugation step. Resuspend the pellet in 0.5–1.0 mL PBS, aliquot, and store frozen at −20°C as antigen stock. Virus titers in the antigen should be around 10^9 pfu/mL.

To establish the dilution at which to use the antigen in the ELISA, the antigen stock is titrated against a positive reference serum to produce an optical density (OD) of 1.0 at a serum dilution of 1/100.

95.2.2.3.1.3 ELISA Protocol

This protocol was optimized with horseradish peroxidise (HRP)–conjugated secondary antibody and using ABTS (2,2-azino bis 3-ethylbenzthiazoline sulfonate) as the chromogen. Other chromogens could be used as per the manufacturer's specifications.

Diluted MYXV at pretitrated dilution in 50 μL of PBS (pH 7.2) is added into each well of 96-well plates and incubated for 2 h at 37°C, followed by three washes with 5% (w/v) skim milk powder in PBS (PBS/M). Block plates with PBS/M overnight at 4°C.

Centrifuge rabbit sera at 13,000 × g in a microcentrifuge for 5 min then serially dilute in PBS/M. Add 50 μL per well at minimum dilution of 1/50. Standard positive and negative sera should be included in each assay. Empirically, the centrifugation step reduced background in the assay.

Incubate plates for 2 h at 37°C, wash three times with PBS/T (PBS with 0.05% v/v Tween 20). Add 50 μL of HRP-conjugated antirabbit IgG or antirabbit IgM (for detection of IgM antibody) diluted in PBS/M (as previously optimized) and incubate for 30 min at 37°C.

Wash plates six times with PBS/T. Add 100 μL of ABTS (2,2-azino bis 3-ethylbenzthiazoline sulfonate, 1 mg/mL) mixed with H_2O_2 (0.06%) in 0.1 M citrate–phosphate buffer (pH 4.0).

Incubate at room temperature for 20 min and read absorbance at 405 nm on a microplate reader. The reading on wells containing only antigen and substrate is set as zero.

95.2.2.3.1.4 Result Interpretation

Results can be reported as either the OD_{405} nm at a dilution of 1:100 or as a titer (the last dilution reciprocal to a positive reading) for each serum. In the original work, the OD_{405} nm cutoff value for negative serum was set at 0.13 based on the results with a large number of negative sera and positive at 0.28 based on the results with known positive sera. Values between 0.13 and 0.28 were regarded as equivocal.[77]

95.2.2.3.2 Competition ELISA[22]

The OIE Reference Laboratory for Myxomatosis uses a C-ELISA. The Reference Laboratory can supply the main reagents for the MYXV C-ELISA in a kit format that includes detailed methods and interpretation.

95.2.2.3.3 Plaque Reduction Neutralization Assay

The PRNA is considered to be the "gold standard" for detecting and measuring antibodies to MYXV.[22]

95.2.2.3.3.1 Procedure[77]

Seed Vero or RK-13 cells in 6-well plates and allow to grow up to 90%–100% confluence as for a standard plaque assay. The same procedure can be used in a 12-well-plate format.

Rabbit serum is heated at 56°C for 30 min to inactivate complement then diluted in twofold intervals from 1/10 to 1/640. Mix 150 μL of each antibody dilution with an equal volume of MYXV diluted to contain approximately 300 pfu. This can be done in a 96-well plate. Incubate at 37°C for 1 h. Controls are incubated with normal serum at 1:10 and 1:100 dilution or no serum.

After removing medium from the 6-well plates, add 100 μL aliquots of each dilution onto duplicate monolayers and adsorb for 60 min at 37°C, 5% CO_2; 2.5 mL of medium is added to each well and the cells are incubated for 4–5 days at 37°C 5% CO_2 (MEM + 2% FCS can be used to overlay the confluent monolayers; 1.5% carboxymethyl cellulose can be added to the overlay medium but is not required). Foci can be readily seen by eye by around 4 days.

Remove medium from the plates and add approximately 1 mL of 0.1% crystal violet in 15%–20% ethanol into each well (stain can be made up using Sigma HT901, 2.3% crystal violet in 20% ethanol: 20 mL crystal violet plus 70 mL ethanol plus 410 mL water). Allow to stain for 15–60 min; gently wash the wells with tap water and air-dry. Count plaques/foci by eye or with a dissecting microscope. Vero cells tend to produce small clear plaques; RK-13 cells produce foci.

95.2.2.3.3.1 Result Interpretation

Titer of each serum is calculated as the reciprocal of the last dilution of serum reducing plaque numbers by >50% compared with controls. Some workers prefer to report 80% neutralization titers.

95.2.2.4 PCR Amplification of MYXV DNA

95.2.2.4.1 Preparation of MYXV DNA

Extraction of total DNA from tissues or virus-infected cells can be readily achieved with commercial DNA purification kits following the manufacturer's instructions or using sodium dodecyl sulfate (SDS)–proteinase K digestion followed by phenol–chloroform extraction and ethanol precipitation.[22,41] Extracted DNA is stored at −20°C.

95.2.2.4.2 PCR Procedure

The *M080L* gene (nucleoside triphosphatase) is highly conserved, with no mutations in MYXV isolates sequenced to date[76]; the following primers work in both conventional and real-time PCR and amplify a 126 bp product: M080F: 5′-TATCAAACAACCTCCGCATACC-3′; M080R: 5′-CTCCCATAACGCTTCCGAC-3′.

PCR mixture: 10× PCR buffer, 2.5 μL; 10 mM dNTPs, 0.5 μL; 10 μM forward primer, 0.5 μL; 10 μM reverse primer, 0.5 μL; 50 mM $MgCl_2$, 0.25 μL; template DNA, 100–250 ng; Taq DNA polymerase, 0.25 μL; add H_2O to 25 μL. Gently mix the reaction. Collect all liquid to the bottom of the tube by brief centrifugation. Use H_2O instead of DNA template in the negative control tube and MYXV DNA as positive control.

PCR cycle: Initial denaturation at 98°C for 30 s; 30 cycles of 98°C for 10 s, 55°C for 30 s, and 72°C for 2 min; final extension at 72°C for 10 min and hold at 4°C.

Gel electrophoresis: Take 10 μL of PCR product and mix with 2 μL of 6× loading dye and load onto a 1% agarose gel containing a suitable DNA stain (e.g., GelRed or SYBR Green). A suitable size ladder should also be run (e.g., 100 bp ladder). Electrophorese for 20 min at 80 V and examine with an appropriate UV source to detect the specific viral band. Gels can be poststained if preferred.

95.2.2.4.3   Result Interpretation

There should be no specific band in the negative control sample. By using a DNA ladder as standard marker, a band of 126 bp in the samples is indicative of MYXV DNA.

95.2.2.4.4   Preparation of Genomic DNA from MYXV

The following protocol is adapted from reference 79; it yields 2–5 μg of clean viral DNA suitable for next-generation sequencing or restriction enzyme analysis and has the advantage of being simple to perform without requiring ultracentrifugation. An alternative protocol is given in reference 65. Detailed protocols for gradient purification of MYXV are provided in reference 67.

Infect 4× T175 flasks of 90%–100% confluent RK-13 cells at an MOI of 0.2. Incubate until confluent CPE is present but before large numbers of cells are lost. This usually takes 2–4 days. [Some field isolates appear to spread poorly in RK-13 cells, for these viruses, using a higher MOI (0.5–1) will often yield sufficient virus.]

Tip off the medium from the flasks into 10% bleach solution (or other suitable disinfectant), and gently wash the monolayers 2–3× with PBS (pH 7.4). Scrape the cells from the plastic and transfer to a sterile 50 mL centrifuge tube, then pellet the cells by centrifuging at 2000× g for 10 min at 4°C. Tip off the supernatant, and drain the tube. Resuspend the cells in 9 mL of 10 mM Tris (pH 7.5). Leave on ice for 20 min and then rupture the cells by pipetting up and down with a 10 mL pipette 50–100 times.

Centrifuge at 2000 × g for 10 min at 4°C. Carefully remove the supernatant into a clean 50 mL centrifuge tube, add 1 mL 10× DNase I buffer (100 mM Tris pH 7.5, 25 mM $MgCl_2$, 5 mM $CaCl_2$), 200 μL DNase I (2.5 mg/mL: Sigma DN-25), and 50 μL RNase A (10 mg/mL; Sigma R-4875). Vortex the tube briefly and incubate at 37°C for 1–2 h with occasional mixing. Total virus levels of >10^9 pfu should be obtained from the preparation.

Add 2 mL 10 mM Tris (pH 7.5) and 3 mL of 40% (w/v) polyethylene glycol in PBS. Vortex to mix thoroughly and leave at 4°C for 1–2 h or overnight. Centrifuge at 3000 × g at 4°C for 60 min to pellet the virus. Carefully tip off the supernatant and drain the tube—it is important to remove as much of the polyethylene glycol (PEG) as possible.

Resuspend the pellet in 150 μL of 1× DNase I buffer (if necessary can use 300 μL but will need to then split into two tubes for subsequent steps) and transfer the suspension to a sterile 1.5 mL tube. Add 20 μL DNase I, vortex briefly to mix and incubate at 37°C for 60 min. Heat the tube at 75°C–80°C for 10–15 min to inactivate the DNase I then cool on ice.

We then extract DNA using the Masterpure DNA kit (Epicorp) but proteinase K/SDS digestion followed by phenol/chloroform extraction and ethanol precipitation can also be used.

95.3   CONCLUSION AND FUTURE PERSPECTIVES

Myxomatosis remains a significant disease of rabbit production and wild rabbits in Europe and of wild rabbits in Australia. It also affects domestic European rabbits and feral rabbits in countries where it is endemic or has been established. MYXV has also become a model for understanding immunology: the virus encodes multiple proteins that manipulate cells and inflammatory and immune responses.[32,55,56] Some of these proteins have been demonstrated to have potential as inflammatory modulators in clinical models.[80,81]

Like other poxviruses, MYXV is readily engineered to express foreign proteins and can induce an immune response to these proteins in rabbits[82] and other species in which it does not replicate.[83] It has been investigated as a means of delivering immunocontraceptive antigens to rabbits as means of population control.[84] Recombinant MYXV expressing the rabbit hemorrhagic disease virus capsid protein have been developed and tested in the field and commercialized as a combined live vaccine against myxomatosis and rabbit hemorrhagic disease.[85,86] A further area of active research is the possible use of MYXV in human cancer therapy.[87]

The coevolution of MYXV with the European rabbit has formed the bedrock of much of the thinking on host-pathogen evolution over the last 50 years.[29,30] MYXV has now been evolving in European rabbits for 65 years. Different forms of the disease have become apparent but whether these have altered transmission mechanisms has not been clearly demonstrated. Genomic sequencing studies indicate that in these large complex DNA viruses there are multiple pathways to virulence and attenuation.[75,76] Further evolutionary studies should be aimed at teasing out the molecular basis of the adaptations that have occurred in the rabbits and the virus.

REFERENCES

1. Farrant, J.L. and Fenner, F. A comparison of the morphology of vaccinia and myxoma viruses. *Aust. J. Exp. Biol. Med. Sci.* 31, 121–126 (1953).
2. Padget, B.L., Wright, M.J., Jayne, A., and Walker, D.L. Electron microscopic structure of myxoma virus and some reactivable derivatives. *J. Bacteriol.* 87, 454 (1964).
3. Chapple, P.J. and Westwood, J.C.N. Electron microscopy of myxoma virus. *Nature* 199, 199–200 (1963).
4. Condit, R.C., Moussatche, N., and Traktman, P. In a nutshell: Structure and assembly of the vaccinia virion. *Adv. Virus Res.* 66, 31–124 (2006).

5. Duteyrat, J.-L., Gelfi, J., and Bertagnoli, S. Ultrastructural study of myxoma virus morphogenesis. *Arch. Virol.* 151, 2160–2180 (2007).

6. Villa, N.Y. et al. Myxoma and vaccinia viruses exploit different mechanisms to enter and infect human cancer cells. *Virology* 401, 266–279 (2010).

7. Roberts, K.L. and Smith, G.L. Vaccinia virus morphogenesis and dissemination. *Trends Microbiol.* 16, 472–479 (2008).

8. Mercer, J. et al. Vaccinia virus strains use distinct forms of macropinocytosis for host-cell entry. *Proc. Natl. Acad. Sci. USA* 107, 9346–9351 (2010).

9. Laliberte, J.P. and Moss, B. Appraising the apoptotic mimicry model and the role of phospholipids for poxvirus entry. *Proc. Natl. Acad. Sci. USA* 106, 17517–17521 (2009).

10. Cameron, C. et al. The complete DNA sequence of myxoma virus. *Virology* 264, 298–318 (1999).

11. Morales, M. et al. Genome comparison of a nonpathogenic myxoma virus field strain with its ancestor, the virulent Lausanne strain. *J. Virol.* 83, 2397–2403 (2009).

12. Day, M.F., Fenner, F., Woodroofe, G.M., and McIntyre, G.A. Further studies on the mechanism of mosquito transmission of myxomatosis in the European rabbit. *J. Hyg. (Camb.)* 54, 258–283 (1956).

13. Fenner, F. and Ratcliffe, F.N. *Myxomatosis.* Cambridge University Press, Cambridge, U.K., 1965.

14. Corbet, G.B. Taxonomy and origins. In *The European Rabbit: The History and Biology of a Successful Colonizer* (eds. Thompson, H.V. and King, C.M.), pp. 1–7 (Oxford University Press, Oxford, U.K., 1994).

15. Flux, J.E.C. World distribution. In *The European Rabbit: The History and Biology of a Successful Colonizer* (eds. Thompson, H.V. and King, C.M.), pp. 8–21 (Oxford University Press, Oxford, U.K., 1994).

16. Rogers, P.M., Arthur, C.P., and Soriguer, R.C. The rabbit in continental Europe. In *The European Rabbit: The History and Biology of a Successful Colonizer* (eds. Thompson, H.V. and King, C.M.), pp. 22–63 (Oxford University Press, Oxford, U.K., 1994).

17. Fenner, F. and Fantini, B. *Biological Control of Vertebrate Pests: The History of Myxomatosis—An Experiment in Evolution.* CAB International, New York, 1999.

18. Kerr, P.J. et al. Comparative analysis of the complete genome sequence of the California MSW strain of myxoma virus reveals potential host adaptations. *J. Virol.* 87, 12080–12089 (2013).

19. Marshall, I.D. and Regnery, D.C. Myxomatosis in a Californian brush rabbit. *Nature* 188, 73–74 (1960).

20. Shope, R.E. A transmissible tumour-like condition in rabbits. *J. Exp. Med.* 56, 793–802 (1932).

21. Martin, C.J. Observations on myxomatosis cuniculi (Sanarelli) made with a view to the use of the virus in the control of rabbit plagues. *CSIR Aust. Res. Bull.* 96, 1–28 (1936).

22. OIE (Office International des Epizooties). *OIE Terrestrial Manual*, Section 2.6. Lagomorpha, Chapter 2.6.1: Myxomatosis (2014).

23. Bull, L.B. and Dickinson, C.G. The specificity of the virus of rabbit myxomatosis. *J. CSIR* 10, 291–294 (1937).

24. Marshall, I.D. and Regnery, D.C. Studies in the epidemiology of myxomatosis in California. III. The response of brush rabbits (*Sylvilagus bachmani*) to infection with exotic and enzootic strains of myxoma virus and the relative infectivity of the tumours for mosquitoes. *Am. J. Hyg.* 77, 213–219 (1963).

25. Regnery, D.C. The epidemic potential of Brazilian myxoma virus (Lausanne strain) for three species of North American cottontails. *Am. J. Epidemiol.* 94, 514–519 (1971).

26. Regnery, D.C. and Marshall, I.D. Studies in the epidemiology of myxomatosis in California. IV. The susceptibility of six leporid species to Californian myxoma virus and the relative infectivity of their tumours for mosquitoes. *Am. J. Epidemiol.* 94, 508–513 (1971).

27. Silvers, L. et al. Host-specificity of myxoma virus: Pathogenesis of South American and North American strains of myxoma virus in two North American lagomorph species. *Vet. Microbiol.* 141, 289–300 (2010).

28. Fenner, F. and Marshall, I.D. A comparison of the virulence for European rabbits (*Oryctolagus cuniculus*) of strains of myxoma virus recovered in the field in Australia, Europe and America. *J. Hyg. (Camb.)* 55, 149–191 (1957).

29. Anderson, R.M. and May, R.M. Coevolution of hosts and parasites. *Parasitology* 85, 411–426 (1982).

30. May, R.M. and Anderson, R.M. Epidemiology and genetics in the coevolution of parasites and hosts. *Proc. R. Soc. Lond. B* 219, 281–313 (1983).

31. Fenner, F. Biological control as exemplified by smallpox eradication and myxomatosis. *Proc. R. Soc. Lond. B* 218, 259–285. (1983).

32. Kerr, P.J. Myxomatosis in Australia and Europe: A model for emerging infectious diseases. *Antiviral Res.* 93, 387–415 (2012).

33. Kerr, P.J. et al. Myxoma virus and the leporipoxviruses: An evolutionary paradigm. *Viruses* 7, 1020–1061 (2015).

34. Marshall, I.D. The influence of ambient temperature on the course of myxomatosis in rabbits. *J. Hyg. (Camb.)* 57, 484–497. (1959).

35. Marshall, I.D. and Fenner, F. Studies in the epidemiology of infectious myxomatosis of rabbits. V. Changes in the innate resistance of wild rabbits exposed to myxomatosis. *J. Hyg. (Camb.)* 56, 288–302 (1958).

36. Marshall, I.D. and Douglas, G.W. Studies in the epidemiology of infectious myxomatosis of rabbits. VIII. Further observations on changes in the innate resistance of Australian wild rabbits exposed to myxomatosis. *J. Hyg. (Camb.)* 59, 117–122 (1961).

37. Fenner, F. and Ross, J. Myxomatosis. In *The European Rabbit: The History and Biology of a Successful Colonizer* (eds. Thompson, H.V. and King, C.M.), pp. 205–240 (Oxford University Press, Oxford, U.K., 1994).

38. Marlier, D., Mainil, J., Linden, A., and Vindevogel, H. Infectious agents associated with rabbit pneumonia: Isolation of amyxomatous myxoma virus strains. *Vet. J.* 159, 171–178 (2000).

39. Marlier, D. et al. Cross-sectional study of the association between pathological conditions and myxoma virus seroprevalence in intensive rabbit farms in Europe. *Prev. Vet. Med.* 48, 55–64 (2001).

40. Best, S.M. and Kerr, P.J. Coevolution of host and virus: The pathogenesis of virulent and attenuated strains of myxoma virus in resistant and susceptible European rabbits. *Virology* 267, 36–48 (2000).

41. Kerr, P.J., Merchant, J.C., Silvers, L., Hood, G., and Robinson, A.J. Monitoring the spread of myxoma virus in rabbit populations in the southern tablelands of New South Wales, Australia. II. Selection of a virus strain that was transmissible and could be monitored by polymerase chain reaction. *Epidemiol. Infect.* 130, 123–133 (2003).

42. Kerr, P.J. et al. Expression of rabbit IL-4 by recombinant myxoma viruses enhances virulence and overcomes genetic resistance to myxomatosis. *Virology* 324, 117–128 (2004).

43. Joubert, L., Duclos, P., and Toaillen, P. La myxomatose des garennes dans le sud-est. La myxomatose amyxomateuse. *Rev. Méd. Vét.* 133, 739–753. (1982).

44. Duclos, P., Tuaillon, P., and Joubert, L. Histopathologie de l'atteinte cutanéo-muqueuse et pulmonaire de la myxomatose. *Bull. Acad. Vet. (France)* 56, 95–104 (1983).

45. Arthur, C.P. and Louzis, C. La myxomatose du lapin en France: Une revue. *Rev. Sci. Tech. Off. Int. Epiz.* 7, 937–957 (1988).

46. Marlier, D., Cassart, D., Boucrat-Baralon, C., Coignoul, F., and Vindevogel, H. Experimental infection of specific pathogen-free New Zealand white rabbits with five strains of amyxomatous myxoma virus. *J. Comp. Pathol.* 121, 369–384 (1999).

47. Marlier, D. et al. Study of the virulence of five strains of amyxomatous myxoma virus in crossbred New Zealand White/Californian conventional rabbits, with evidence of long-term testicular infection in recovered animals. *J. Comp. Pathol.* 122, 101–113 (2000).

48. Brun, A., Godard, A., and Moreau, Y. La vaccination contre la myxomatose: Vaccins heterologues et homologues. *Bull. Soc. Vet. Med. Comp.* 83, 251–254 (1981).

49. Kritas, S.K. et al. A pathogenic myxoma virus in vaccinated and non-vaccinated commercial rabbits. *Res. Vet. Sci.* 85, 622–624 (2008).

50. Silvers, L. et al. Virulence and pathogenesis of the MSW and MSD strains of Californian myxoma virus in European rabbits with genetic resistance to myxoma virus compared to rabbits with no genetic resistance. *Virology* 348, 72–83 (2006).

51. Rivers, T.M. Infectious myxomatosis of rabbits: Observations on the pathological changes induced by virus myxomatosis (Sanarelli). *J. Exp. Med.* 51, 965–975. (1930).

52. Hurst, E.W. Myxoma and the Shope fibroma. I. The histology of myxoma. *Br. J. Exp. Pathol.* 18, 1–15. (1937).

53. Fenner, F. and Woodroofe, G.M. The pathogenesis of infectious myxomatosis: The mechanism of infection and the immunological response in the European rabbit (*Oryctolagus cuniculus*). *Br. J. Exp. Pathol.* 34, 400–410 (1953).

54. Best, S.M., Collins, S.V., and Kerr, P.J. Coevolution of host and virus: Cellular localization of myxoma virus infection of resistant and susceptible European rabbits. *Virology* 277, 76–91 (2000).

55. Stanford, M.M., Werden, S.J., and McFadden, G. Myxoma virus in the European rabbit: Interactions between the virus and its susceptible host. *Vet. Res.* 38, 299–318 (2007).

56. Spiesschaert, B., McFadden, G., Hermans, K., Nauwynck, H., and Van de Walle, G.R. The current status and future directions of myxoma virus, a master in immune evasion. *Vet. Res.* 42, 76 (2011).

57. Hurst, E.W. Myxoma and the Shope fibroma. III. Miscellaneous observations bearing on the relationship between myxoma, neuromyxoma and fibroma viruses. *Br. J. Exp. Pathol.* 18, 23–30 (1937).

58. Swan, C. The route of transmission of the virus of infectious myxomatosis of rabbits to the central nervous system. *Aust. J. Exp. Biol. Med. Sci.* 19, 113–115 (1941).

59. Jeklova, E. et al. Characterisation of immunosuppression in rabbits after infection with myxoma virus. *Vet. Microbiol.* 129, 117–130 (2008).

60. Cameron, C.M., Barrett, J.W., Liu, L., Lucas, A.R., and McFadden, G. Myxoma virus M141R expresses a viral CD200 (vOX-2) that is responsible for down-regulation of macrophage and T-cell activation in vivo. *J. Virol.* 79, 6052–6067 (2005).

61. Cameron, C.M., Barrett, J.W., Mann, M., Lucas, A., and McFadden, G. Myxoma virus M128L is expressed as a cell surface CD47-like virulence factor that contributes to the downregulation of macrophage activation in vivo. *Virology* 337, 55–67 (2005).

62. Kerr, P.J. and Donnelly, T. Viral diseases of rabbits. *Vet. Clin. North Am. Exotic Anim. Pract.* 16, 437–468 (2013).

63. Jin, L. et al. An outbreak of fatal herpesvirus infection in domestic rabbits in Alaska. *Vet. Pathol.* 45, 369–374 (2008).

64. Strayer, D.S. and Sell, S. Immunohistology of malignant rabbit fibroma virus–A comparative study with rabbit myxoma virus. *J. Natl. Cancer Inst.* 71, 106–116 (1983).

65. Russell, R.J. and Robbins, S.J. Cloning and molecular characterization of the myxoma virus genome. *Virology* 170, 147–159 (1989).

66. Saint, K.M., French, N., and Kerr, P. Genetic variation in Australian isolates of myxoma virus: An evolutionary and epidemiological study. *Arch. Virol.* 146, 1105–1123 (2001).

67. Smallwood, S.E., Rahman, M.M., Smith, D.W., and McFadden, G. Myxoma virus: Propagation, purification, quantification and storage. *Curr. Protoc. Microbiol.* Chapter 14, Unit 14A.1 (2010).

68. Albini, S. et al. Development and validation of a myxomoa virus real-time polymerase chain reaction assay. *J. Vet. Diagn. Invest.* 24, 135–137 (2012).

69. Cavadini, P., Botti, G., Barbieri, I., Lavazza, A., and Capucci, L. Molecular characterization of SG33 and Borghi vaccines used against myxomatosis. *Vaccine* 28, 5414–5420 (2010).

70. Duarte, M.D. et al. Development and validation of a real-time PCR for the detection of myxoma virus based on the diploid gene M000.5L/R. *J. Virol. Methods* 196, 219–224 (2014).

71. Camus-Bouclainville, C. et al. Genome sequence of SG33 strain and recombination between wild-type and vaccine myxoma viruses. *Emerg. Infect. Dis.* 17, 633–638 (2011).

72. Kerr, P.J., Hone, J., Perrin, L., French, N., and Williams, C.K. Molecular and serological analysis of the epidemiology of myxoma virus in rabbits. *Vet. Microbiol.* 143, 167–178 (2010).

73. Merchant, J.C. et al. Monitoring the spread of myxoma virus in rabbit (*Oryctolagus cuniculus*) populations on the southern tablelands of New South Wales, Australia. III. Release, persistence and rate of spread of an identifiable strain of myxoma virus. *Epidemiol. Infect.* 130, 135–147 (2003).

74. Berman, D., Kerr, P.J., Stagg, R., van Leeuwen, B.H., and Gonzalez, T. Should the 40-year-old practice of releasing virulent myxoma virus to control rabbits (*Oryctolagus cuniculus*) be continued? *Wildl. Res.* 33, 549–556 (2006).

75. Kerr, P.J. et al. Evolutionary history and attenuation of myxoma virus on two continents. *PLoS Pathog.* 8, e1002950 (2012).

76. Kerr, P.J. et al. Genome scale evolution of myxoma virus reveals host pathogen adaptation and rapid geographic spread. *J. Virol.* 87, 12900–12915 (2013).

77. Kerr, P.J. An ELISA for epidemiological studies of myxomatosis: Persistence of antibodies to myxoma virus in European rabbits (*Oryctolagus cuniculus*). *Wildl. Res.* 24, 53–65 (1997).

78. Gelfi, J., Chantal, J., Phong, T.T., Py, R., and Boucraut-Baralon, C. Development of an ELISA for detection of myxoma virus-specific antibodies; test evaluation for diagnostic applications on vaccinated and wild rabbit sera. *J. Vet. Diagn. Invest.* 11, 240–245 (1999).

79. Dalton, K.P., Ringleb, F., Alonso, J.M.M., and Parra, F. Rapid purification of *myxoma virus* DNA. *J. Virol. Methods* 162, 284–287 (2009).

80. Dai, E. et al. Inhibition of chemokine-glycosaminoglycan interactions in donor tissue reduces mouse allograft vasculopathy and transplant rejection. *PLoS One* 5, e10510 (2010).

81. Brahn, E., Lee, S.Y., Lucas, L., McFadden, G., and Macaulay, C. Suppression of collagen-induced arthritis with a serine proteinase inhibitor (serpin) derived from myxoma virus. *Clin. Immunol.* 153, 254–263 (2014).

82. Kerr, P.J. and Jackson, R.J. Myxoma virus as a vaccine vector for rabbits: Antibody levels to influenza virus haemagglutinin presented by a recombinant myxoma virus. *Vaccine* 13, 1722–1726 (1995).

83. McCabe, V.J., Tarpey, I., and Spibey, N. Vaccination of cats with an attenuated recombinant myxoma virus expressing feline calicivirus capsid protein. *Vaccine* 20, 2454–2462 (2002).

84. van Leeuwen, B.H. and Kerr, P.J. Prospects for fertility control in the European rabbit (*Oryctolagus cuniculus*) using myxoma virus-vectored immunocontraception. *Wildl. Res.* 34, 511–522 (2007).

85. Angulo, E. and Bárcena, J. Towards a unique and transmissible vaccine against myxomatosis and rabbit haemorrhagic disease for rabbit populations. *Wildl. Res.* 34, 567–577 (2007).

86. Spibey, N. et al. Novel bivalent vectored vaccine for control of myxomatosis and rabbit haemorrhagic disease. *Vet. Rec.* 170, 309 (2012).

87. Bell, J. and McFadden, G. Viruses for tumor therapy. *Cell Host Microb.* 15, 260–265 (2014).

96 Orthopoxvirus

Vinayagamurthy Balamurugan, Gnanavel Venkatesan,
and Veerakyathappa Bhanuprakash

CONTENTS

96.1 Introduction 869
    96.1.1 Classification, Morphology, and Physico-Biochemical Properties 870
    96.1.2 Genome, Biology, and Epidemiology 870
        96.1.2.1 Genome 870
        96.1.2.2 Biology 871
        96.1.2.3 Epidemiology 871
    96.1.3 Clinical Features and Pathogenesis 872
    96.1.4 Diagnosis 872
        96.1.4.1 Conventional Techniques 873
        96.1.4.2 Molecular Techniques 873
96.2 Methods 875
    96.2.1 Sample Preparation 875
    96.2.2 Detection Procedures 876
        96.2.2.1 Antigen and Antibody Detection 876
        96.2.2.2 Virus Isolation in Cell Culture System and Its Identification 876
        96.2.2.3 Electron Microscopy 876
        96.2.2.4 Immunohistochemistry 876
        96.2.2.5 Serum Neutralization Test 876
        96.2.2.6 Diagnostic PCR Assay 876
        96.2.2.7 LAMP Assay 877
96.3 Conclusion and Perspectives 877
References 877

96.1 INTRODUCTION

There have been some reports on the emergence of human infections by pox viruses since smallpox was eradicated globally in 1979 [1]. However, different pox viruses such as cowpox virus (CPXV), buffalopox virus (BPXV), and vaccinia virus (VACV) (genus *Orthopoxvirus* [OPV]) and orf virus, bovine papular stomatitis, and pseudocowpox (genus *Parapoxvirus*) can cause very similar lesions in cattle and animal handlers [2]. In general, OPVs carrying genes responsible for host immune evasion pose a threat to livestock and humans as potential biowarfare agents.

As type species of the genus *Orthopoxvirus*, VACV causes only mild complications in humans. Given that VACV and variola virus (VARV), the causative agent of smallpox, are mutually immunogenic, VACV was used as a vaccine against smallpox worldwide by mass immunization strategy promoted by the World Health Organization, which led to the eventual eradication of smallpox.

Camelpox is considered as an emerging public health threat during this decade due to increased cases and outbreaks in animals and humans [3–5]. Camelpox is a contagious, often sporadic, and notifiable skin disease of camelids and is socio-economically significant as it incurs considerable loss in terms of morbidity, mortality, loss of weight, and reduction in milk yield and confined to camel-rearing belts particularly in developing countries [6]. The camelpox virus (CMLV) is genetically closely related to VARV and has gained much attention from researchers due to its recent emergence in humans [3].

Cowpox is a zoonotic OPV infection caused by CPXV, which is closely related to VACV. It is most commonly found in animals other than bovines such as rodents and cats. Clinically, cowpox is similar but milder than smallpox and transferrable to humans causing mild localized skin lesions. CPXV was used initially to control smallpox and remains prevalent in Europe [7,8].

Buffalopox is another emerging and reemerging OPV infection due to BPXV. Being a close relative of VACV, BPXV causes mild to severe skin lesions in buffaloes, cows, and humans [9]. Monkeypox is yet another zoonotic OPV infection caused by monkeypox virus (MPXV). It has been confined to West and Central Africa for several decades where the rodents and squirrels act as reservoirs [10]. However, the disease recently emerged in the United States through imported wild rodents from Africa. Monkeypox is clinically very similar to that of ordinary forms of smallpox, including flu-like symptoms, fever, malaise, back pain, headache, and characteristic rash [11].

Globally, emergence and reemergence of vaccinia-like viruses (VLVs) in Brazil, and BPXV in Egypt, Indonesia, India, Nepal, and Pakistan constitute a public health risk [2,12]. Recent outbreaks of zoonotic camelpox and buffalo-pox in developing countries like India are of particular concern [2–5]. The nature of occurrence, mode of transmission by contact, difficulty in detection, and identification make OPV as potential biological or bioterror weapon like smallpox virus, creating panic and fear among common people [13]. The phylogenetic relationships, ecology, and host range of OPVs such as smallpox (VARV), MPXV, CPXV, VACV, and CMLV as well as the pattern of their evolution suggests that a VARV-like virus could emerge in the course of natural evolution of modern zoonotic OPVs [14].

OPVs have gained much attention from researchers due to close genetic relatedness to VARV and carrying genes responsible for host immune evasion mechanisms. Effective control of any infectious disease warrants diagnostic, epidemiological, and prophylactic studies. Although the disease can be diagnosed based on clinical features, the similar confusing skin lesions necessitate identification, detection, and differentiation of the different OPV infections (CMLV, BPXV, CPXV, MPXV infections, etc.) by molecular techniques. Further, novel highly sensitive and specific techniques would be useful in the identification of emerging and reemerging viruses, thereby therapeutic, prophylactic, preventive measures would be applied on proper time to curtail further spread of OPV infections like other zoonotic diseases. There is an obvious need to coordinate international control efforts over the outbreaks to zoonotic OPV infections in various countries. This will help provide a rapid response and prevent OPV infections from evolving into epidemics.

In this chapter, a comprehensive note on camelpox, cowpox, buffalopox, and other OPVs with particular reference to their classification, biology, epidemiology, pathogenesis, diagnosis, and prevention will be presented.

96.1.1 Classification, Morphology, and Physico-Biochemical Properties

The genus OPV is one of the eight genera in the subfamily *Chordopoxvirinae*, family *Poxviridae* [15]. Members of the genus include several pathogens of veterinary and zoonotic importance—VACV, CPXV, CMLV, VARV, MPXV, BPXV (a variant of VACV), ectromelia virus, rabbitpox virus, taterapox virus, North American OPVs (volepox virus, raccoonpox virus [RCNV], and skunkpox virus)—and an unclassified OPV species, Uasin Gishu disease virus [16,17]. Recently, pox-like infections caused by VACV-like agents affecting cattle and humans, Cantagalo [18] and Aracatuba viruses [19], were also reported from Brazil [12].

Morphologically, OPVs are enveloped brick-shaped large viruses (350 × 270 nm). The average size of the virion estimated by electron microscopy (EM) is 224 × 389 nm [15,20]. The size of BPXV isolates BP Giza 72 (Egyptian) and BP4 (Hisar, India) was reported to be 270 × 700 × 100 Å [9]. There were no differences in appearance between Indian and Egyptian isolates neither of BPXV nor between BPXV and VACV.

An orthopox virion consists of an envelope, outer membrane, two lateral bodies, and a core. It is brick-shaped and the outer membrane is covered with irregularly arranged tubular proteins. The different OPV species cannot be distinguished by means of EM.

OPVs are ether resistant but chloroform sensitive, whereas it is sensitive to both acidic (pH 3–5) and alkaline (pH 8.5–10) conditions [21]. The virus can be destroyed by either autoclaving or boiling for 10 min and ultraviolet rays in a few minutes [22]. In general, it is well recognized that poxvirus virions show high environmental stability (strong tolerance to high temperatures, pH, and chemicals) and can remain contagious over several months. This feature is enhanced by the materials (crusts, serum, blood, and other excretions) in which the virus is released [23].

96.1.2 Genome, Biology, and Epidemiology

96.1.2.1 Genome

The OPV genome contains a single, linear, double-stranded DNA molecule with a size of approximately 200 kb, the ends of which are connected by covalent links (terminated by a hairpin loop) and replicates in the cytoplasm [15]. The genome is AT-rich (66.9%) having cross-links that join the two DNA strands at both ends. The end of each DNA strand has long inverted tandem repeats that form single-stranded loops. The central region of the genome contains genes that are highly conserved among all sequenced OPVs [24]. The OPV genes are tightly packed with little noncoding sequences. The genome of the poxviruses encodes 150–200 different proteins. Most genes at each terminus within 25 kb of the genome are transcribed outward toward the terminus and are variable in nature coding for host range, virulence, and immunomodulation, whereas the genes within the center of the genome transcribed from either DNA strands are of highly conserved and code for proteins involved in viral replication like RNA transcription, DNA replication, and virion assembly.

Molecular details about the genome structure and phylogenetic analyses of some selected genes indicate that CMLV is clearly distinct from VARV and VACV. Among the OPVs, CPXV possesses the largest genome (~224–228 kb) [25] and possibly infects the widest range of host species [26]. The extensive phylogenetic analysis showed that CPXV strains sequenced clearly cluster into several distinct clades, some of which are closely related to VACVs while others represent different clades in a CPXV cluster. Particularly, one CPXV clade is more closely related to CMLV, taterapox virus, and VARV than to any other known OPVs [25]. CPXV is rodent borne with a broad host range and contains the largest and most complete genome of all poxviruses, including parts with high homology to VARV (smallpox).

BPXV is closely related to VACV vaccine strains by sequence and phylogenetic analyses based on structural protein genes including H3L, A27L, D8L, B5R, and nonstructural protein (H4L) homologue genes [9]. It showed that among OPVs, BPXV shared 97.2%–98.9% and 94.6%–98.7% sequence similarity with VACV at the nucleotide and amino acid levels, respectively. Other VLVs reported from Brazil also showed that they are closely related to VACV than other OPVs. Aracatuba virus showed 95.7%–96.4% identity with VACV, in comparison to 97.2%–98.9% sequence identity of BPXV isolates with VACV strains. However, BPXV isolates appeared to be more closely related to VACV than Aracatuba virus to VACV [9].

96.1.2.2 Biology

Unlike other DNA viruses, poxviruses replicate in the cytoplasm of infected cells in so-called virus factories (Guarnieri inclusion bodies). Originally, four different infectious virus particles can be distinguished during OPV replication: intracellular mature enveloped virus particles, intracellular enveloped virus particles, cell-associated enveloped virus particles, and extracellular enveloped virus particles. Both intracellular and extracellular viruses play an important role for pathogenesis. Intracellular and cell-associated viruses are involved in the spreading of the virus from cell to cell, whereas viruses released from the cell enable the dissemination within the infected organism. The growth kinetics of CMLV in human embryonic lung fibroblast cells indicates that CMLV is different from VACV and CPXV [27]. CMLV appears to share biological features with other OPVs, mainly VARV. Both CMLV and VARV are restricted to a single host and induce a similar disease course [20,28]. CMLV was first regarded as an own OPV species in the mid-seventies, sometime before phylogenetic analyses were conducted [29]. Earlier, CMLV was shown to share strong similarities with VARV as both had a narrow host range and were indistinguishable in terms of pock formation on chorioallantoic membrane (CAM) in embryonated eggs, growth in cells, and low or absence of pathogenicity in various animal models [30,31]. Camels have been used as animal models of camelpox infection, mainly for evaluating the efficacy of vaccines [32–34]. Most of CMLV strains produce white pocks, flat in shape, with sizes varying between 0.5 and 1.5 mm diameters at 37°C. Most of the CMLV strains did not induce pock proliferation, necrosis, or hemorrhage, in contrast to what was observed with VACV or CPXV [35–37]. In general, cells derived from camel, lamb, calf, pig, monkey, chicken, hamster, and mouse enable the propagation of CMLV strains [35]. Other than camels, the species that have been infected successfully are monkeys and infant mice [29].

So far, most CPXV cases have occurred individually in unvaccinated animals and humans and were caused by genetically distinguishable virus strains. The ability of CPXV to infect a wide range of mammalian host is likely due to the fact that, among the OPVs, CPXV encodes the most complete set of open reading frames (ORFs) expected to encode immunomodulatory proteins. This renders CPXV particularly interesting for studying poxviral strategies to evade and counteract the host immune responses [26].

Serological studies demonstrated the cross-antigenicity among VACV, VARV, CPXV, and CMLV, but not with parapoxviruses (PPVs) and avipoxvirus infections [21,29]. The serological relationship of BPXV with VACV and CPXV has been studied by different methods and it indicated that a pattern of cross-reaction in agarose gel immunodiffusion (AGID) and immunoelectrophoresis is indicative of a closer relationship of BPXV with VACV than CPXV [9].

96.1.2.3 Epidemiology

Reports on human poxvirus infections have been rare since smallpox eradication. The main OPVs that infect humans include VARV and MPXV infections as seen in Congo and the United States, CPXV in Europe, VLVs in Brazil, and BPXV (Vaccinia variant) and CMLV infections in India. Camelpox is one of the most common contagious OPV diseases of the Old World (both *Camelus dromedarius* and *Camelus bactrianus*) and the New World camelids [17]. CMLV is considered to solely naturally infect Old World camelids [28]. The disease occurs throughout the camel-breeding areas of Africa, north of the equator, the Middle East, and Asia, as the camels are used for nomadic pastoralism, transportation, racing, and production of milk, wool, and meat purposes [38]. Of note, camelpox has never been reported in Australia, even though camel farming is practiced [28], and has not been seen in feral camels in Australia, in wild Bactrian camels in China and Mongolia, or in New World camelids [16]. The disease has been reported initially in Punjab and Rajputana (India) [39] and later from many other countries. It has been reported in the Middle East (Bahrain, Iran, Iraq, Saudi Arabia, UAE, and Yemen), in Asia (India, Afghanistan, and Pakistan), in Africa (Algeria, Egypt, Kenya, Mauretania, Niger, Somalia, Morocco, Ethiopia, Oman, and Sudan), and in the southern parts of former USSR, where the disease is endemic [6,40]. Camelpox has been reported for the first time in two provinces named Hama and Duma in Syria [41]. In general, young camels (calves) under the age of 4 years and pregnant females appear more susceptible to camelpox [41,42]. In the recent past, the first conclusive evidence of zoonotic CMLV infection in humans (unvaccinated smallpox individuals) associated with outbreaks in dromedarian camels in northwest region of India has been reported in three human confirmed cases [3].

CPXV is distributed in Europe, west former USSR, and adjacent areas of Northern and Central Asia, with an increasing number of reports in Europe [7]. Human CPXV infections are commonly described in relation to contact with diseased domestic cats and rarely from rats [43]. Human infections usually remain localized and self-limiting but can become fatal in immunosuppressed and eczematous patients particularly children. CPXV infections in captive exotic animals have been reported to be transmitted by rodents [7]. CPXV has also been reported to infect elephant [43], llama [44], and banded mongooses and jaguarundis [8]. However, the list of animals known to be susceptible to CPXV is still growing. Thus, the likely existence of unknown CPXV hosts and their distribution may present a risk for other exotic animals but also for the general public [8].

Buffalopox was first recorded in Lahore during 1934 in undivided India [45]. Later, regular outbreaks have been reported from different states of India [46]. It affects primarily buffaloes and occasionally cows and humans [9,40]. The disease outbreaks are reported from India, Pakistan, Egypt, Nepal, and Bangladesh. In the recent past, analysis of samples collected from cows suffering with pox-like disease in India revealed that cows suffered mostly from BPXV than CPXV based on sequence and phylogenetic analysis of HA gene [47].

VLVs are reported from Brazil. Their origin in Brazil is hypothesized to account for the usage of VACV strains during the smallpox eradication campaign and may be related to outbreaks of bovine vaccinia, namely, Cantagalo and Aracatuba viruses in cattle and humans [18,19]. Until 2010, these infections were limited to cattle and humans. However, there was an evidence of seropositivity with OPV antibodies in buffaloes, which again hypothesized to account for exposure to VACV, the only OPV known to be circulating in Brazil [48].

Human monkeypox was first identified in humans in 1970 in the Democratic Republic of Congo. Since then, the majority of cases have been reported in rural regions of the Congo Basin and western Africa, particularly in Congo [49]. In Midwest of the United States, the first confirmed report of monkeypox cases outside of the African continent has been documented during 2003. Later, it has been reported in Unity state, Sudan [50]. In Africa, human infections have been documented through the handling of infected monkeys, Gambian rats, or squirrels [51].

96.1.3 CLINICAL FEATURES AND PATHOGENESIS

CMLV and VARV cause illness in a single host species and both viruses are distinguishable. CMLV has rarely caused disease in man. Similarly, VARV is unable to cause disease in camels, although camels immunized with VARV are resistant to subsequent infection with CMLV [31]. The clinical manifestation of camelpox ranges from mild, local, and inapparent infection confined over the skin to less common but severe systemic disease, depending on the CMLV strains involved in the infection [28]. The incubation period of the disease varies from 4 to 15 days with an initial rise in temperature followed by enlarged lymph nodes and skin lesions with development of papules on labia, vesicles, pustules, and scabs [52]. In the generalized form, pox lesions may cover the entire body and even multiple pox-like lesions can be found on the mucous membranes of the mouth, respiratory, and digestive tracts [28]. In contrast to smallpox, in which pustules occur only on the skin and the squamous epithelium of the oropharynx, severely ill camels also develop proliferative poxviral lesions in the bronchi and lungs [53].

The CMLV enters commonly through the skin. However, the oronasal infection is also reported. After local replication and development of a primary skin lesion, the virus spreads to local lymph nodes leading to a leukocyte-associated viremia, which may be associated with pyrexia. Widespread secondary skin lesions appear a few days after the onset of viremia, and new lesions continue to appear for 2–3 days, at that time the viremia subsides. Like other OPVs, CMLV encodes multiple genes that antagonize or affect the antiviral host immune response by interfering with the interferon response, key proinflammatory cytokines (interleukin [IL]-1b, IL-18, and tumor necrosis factors [TNFs]), chemokines, and complements [16].

Cowpox causes skin eruptions with lesions occurring principally on the udder and teats in milking cattle. Human infection may occur while milking an infected animal and also from rats and mice bite [54]. In humans and nonhuman primates, lesion develops facial swelling, gingivitis, and vesicopapules followed by scabs with an incubation of 1 week. In herbivores, it develops pustules and swellings in the skin and gingiva, pox lesions on the vulva, penis, trunk, anal mucosa, gingiva, and tongue, and detached sole horn and stillbirth after an incubation period of 15–22 days [54].

BPXV is associated with severe disease outbreaks among buffaloes [9], cattle [47], and humans in contact with these animals [4,5]. Most human BPXV infections occur in animal attendants and milkers [9]. However, in Pakistan, buffalopox has caused a nosocomial infection in humans [55]. The incubation period in animals is comparatively shorter (2–4 days) than in humans (3–19 days) [56]. The disease may be localized or generalized with mild to severe disease associated invariably with mastitis in almost 50% of the affected animals. The skin lesions are more common on the udder, the base of the ear, and other less hairy areas like the inguinal region. Otitis has also been reported due to secondary bacterial complications [9]. The lesions are mainly confined to the hands, forehead, face, buttocks, and legs [9]. Occasionally, lymphadenopathy has also been recorded. Human-to-human transmission has not been reported so far.

Monkeypox is a rare but potentially serious viral illness, characterized by blister-like rashes that resemble smallpox [49]. However, unlike smallpox, monkeypox causes lymphadenopathy. Humans contract monkeypox by having close contact with animals or other infected people. Primates (monkeys, apes), rodents, and rabbits are known to carry the MPXV including most warm-blooded animals. Monkeypox illness in humans usually starts about 12 days after exposure to MPXV. The illness begins with fever, headache, muscle aches, backache, swollen lymph nodes, a general feeling of discomfort, and exhaustion. Within 1–3 days (sometimes longer) after the appearance of fever, the patient develops a papular rash (i.e., raised bumps), often first on the face [57].

96.1.4 DIAGNOSIS

The diagnosis of OPV infection can be done based on the history of contact with infected animals and clinical manifestations in affected animals and humans in zoonotic cases.

Following the appearance of clinical signs of the disease, tissue samples (skin or organ biopsies) are most useful to identify the infectious agent [6,16]. However, the confounding signs caused by contagious ecthyma (orf [parapox] virus), herpesvirus infections, papillomatosis, and insect bites demand OPV infections to be differentiated from these infections using laboratory-based diagnostic methods. Initially, the diagnosis and differential diagnosis of the OPV infections has been performed by inoculation of suspected clinical material on the CAM and identification of different OPVs based on cultivation at distinct temperature, pock morphology, and histology characteristics [14,58]. It is necessary to apply more than one method for confirmatory diagnosis. Few complementary techniques are required for the diagnosis of OPV infections like transmission electron microscopy (TEM), virus isolation, immunohistochemistry, demonstration of neutralizing antibodies by immunoassays and PCR assays. However, the identity of the causative agent as OPV must be confirmed by TEM, PCR, and/or sequencing.

96.1.4.1 Conventional Techniques

Generally, most of the conventional serological tests are time consuming (less rapid), laborious, less sensitive, and therefore not suitable for primary diagnosis but useful in secondary confirmatory testing and retrospective epidemiological studies.

TEM and restriction enzyme analysis can be used to differentiate CMLV from other infections caused by OPV and PPVs [24,41]. TEM is a rapid primary diagnostic allowing first morphological characterization of pathogens [16]. It is a reliable and rapid method to demonstrate the presence of OPVs in scabs or tissue samples, although a relative high concentration of the virus in the sample is required. This technique enables the morphological differentiation between OPVs, which are brick-shaped, and PPVs, which are ovoid-shaped [20], and also from herpesviruses and *Bacillus anthracis*, which causes comparable type of skin lesions [14].

OPVs including CMLV antigen in infected scabs and pox lesions can be identified by immunohistochemistry [28], antigen-capture ELISA [58], and immunoblot [59]. Similarly, OPV infection can be diagnosed by the presence of specific IgM antibodies in acute infection using ELISA. However, these serological methods are less sensitive than PCR and not useful in the differentiation of OPVs as they are antigenically closely related to each other. The hemagglutination, hemagglutination inhibition, neutralization [60], complement fixation, fluorescent antibody, and indirect ELISA assays/tests are available to detect CMLV antibodies [6,40]. Although various serological assays have been developed for the diagnosis of buffalopox including AGID, counterimmunoelectrophoresis (CIE), serum neutralization test (SNT), ELISA, and immunoperoxidase test, these tests fail in the accurate diagnosis of the disease because of antigenic cross-reactivity [9]. Conventional biological and serological methods have limitations in the identification and differentiation of OPVs.

Most of the OPVs can be cultured in suitable cell cultures. The isolation of the virus using embryonated eggs and various cell lines (HeLa, GMK-AH1, WISH, Vero [29], MA104, and BHK21 cells) as well as primary cell cultures (lamb testis and kidney, camel kidney, calf kidney, and chicken embryo fibroblast [21]) could be of gold standard diagnostic procedure. However, cell lines of monkey and human origin are mostly preferred. Blood, serum, and homogenized scab or tissue samples can be used to infect cell cultures. The infected cells should be monitored for cytopathic effects (CPE) for 10–12 days. CPE includes the formation of multinucleated syncytia, rounding, ballooning, and syncytia with degenerative changes. CAMs can also be used for the growth of CMLV and other OPVs [36], but it is important to consider that pocks produced by VARV and CMLV are indistinguishable [29]. In case of BPXV, it produces characteristic "rosette" or "sunflower" like hemadsorption with 0.4% fowl erythrocytes [46]. The infected cell culture can be identified by known polyclonal or monoclonal OPV antibodies in the immunofluorescence technique [58,59].

96.1.4.2 Molecular Techniques

To overcome the drawbacks associated with conventional diagnostic techniques, molecular techniques like PCR, real-time PCR, and loop-mediated isothermal amplification (LAMP) assays have been used for the rapid and sensitive detection of OPV DNAs from clinical samples. DNA can be extracted from cell cultures, clinical scab samples, or tissue material using numerous commercial kits. Recently, a reliable low-cost two-step extraction method has been developed for isolating CMLV DNA from skin samples [61]. It can be modified for other OPV samples as a cost-effective method.

PCR assays are available for the identification and differentiation of OPV species and for differential diagnosis of OPVs from other viruses such as varicella zoster virus and herpes simplex virus, but also of bacteria such as *Bacillus anthracis* [58]. Human OPV infections should be given careful attention due to their high transmission to humans. Therefore, OPV species are to be differentiated using validated PCR detection methods and also sequencing and phylogenetic analyses of several genome regions in order to achieve reliable classification of these viruses. In case of suspected OPV infection in humans, it can be verified by means of virological or microbiological and molecular methods [14].

PCR-based assays have been described for the diagnosis and differentiation of OPVs like VACV, VARV, MPXV, and CPXV, based on inclusion and HA genes [18,62,63]. PCR using primers CoPV-3 and CoPV-4 has been used for the diagnosis of OPVs including CPXV, CMLV, MPXV, RCNV, VACV [64,65], and BPXV [9]. Amplification of full-length A-type inclusion (ATI) gene using CoPV-1 and CoPV-2 primer set [64] has been used in differentiation of OPVs (CMLV, CPXV, and BPXV). Primer pair ATI-up and ATI-low, which amplifies a fragment of the inclusion gene, is also used for differentiation of OPVs [62,66]. PCR assays in combination with restriction endonuclease (RE) or RE viral DNA analysis enabled differentiation of OPVs [66,67]. But, it needs virus isolation and genomic DNA (gDNA) extraction, which is time consuming and laborious. Further, PCR strategies targeting HA gene [63] and B2L gene [68] have been developed for detection of CMLV and its differentiation from OPV and PPV infections in camels.

Similarly, a PCR assay based on the C18L gene (encoding ankyrin repeat protein) specific for CMLV has been developed [69], which differentiates CMLV from other OPVs, PPVs, and capripoxviruses in both cell culture and clinical samples. Further, a duplex PCR based on the C18L and DNA polymerase (DNA pol) genes for specific and rapid detection and differentiation of CMLV from BPXV has also been developed [69,70]. These assays have the advantage of avoiding an extra step of restriction analysis. As an improvement over conventional PCR approaches, the real-time PCR techniques targeting A36R gene using fluorescence resonance energy transfer (FRET) method [71], A13L, RPO18, and viral early transcription factor (VETF) genes [72,73] using melting curve analysis have been in use for rapid, highly sensitive, and specific detection and quantitation of VARV and CMLV and their differentiation from related OPVs.

Further, C18L gene specific real-time PCR based on SYBR green [69] and TaqMan hydrolysis probe [74] chemistry have been optimized for specific detection of CMLV in clinical samples. A TaqMan-based assay targets the OPV DNA pol gene and detects Eurasian OPVs other than variola. A hybridization assay, utilizing a 3′-minor groove binder (MGB) probe, targeting an envelope protein gene (B6R) specifically detects MPXV in human clinical samples [75]. Further, a quantitative real-time PCR assay targeting ATI gene of MPXV has been described for specific detection and differentiation of MPXV Congo basin and West African strains [76,77] and will be helpful in differential diagnosis of MPXV infections from other rash illnesses. As there is an increase in the number of human infections of cowpox, the real-time PCR for specific detection of CPXV has been reported [78]. A SYBR green- and probe-based real-time PCR assays have been reported for detecting and/or typing DNA of OPV from clinical samples as VARV, MPXV, CPXV, and VACV [79] as alternatives to PCR detection methods for clinical diagnosis of human pathogenic OPV infections in laboratories. Likewise, many real-time PCR assays in either single or multiple format/s for species specific identification of VARV and/or other OPVs and also their differentiation among them had been described [80–85].

LAMP, a simple, rapid, specific, highly sensitive, novel, and pen-side or field disease diagnostic laboratory test, has

TABLE 96.1
Different Primers Used in Molecular Diagnosis of Buffalopox Virus and Camelpox Virus Infections

Assay	Primers	Primer/Probe Sequence (5′–3′)	Target Size/Target Species [Reference]
Diagnostic PCR	DNA pol FP1 and RP1	ATTACCTGTGGTTAAAACCTTTGT[1] AGAGAAATGACGTTTATCGGTGAC	96 bp in all OPVs [70]
	BPXV C18L F and R2	GCGGGTATCACTGTTATGAAACC[2] CATAAATACACTTTTATAGTCCTCG	368 bp in BPXV [70]
	CMLV C18L F and R	GCGTTAACGCGACGTCGTG[3] GATCGGAGATATCATACTTTACTTTAG	243 bp in CMLV [69]
Duplex PCR	CMLV C18L F and R DNA pol FP1 and RP1	Primer sequences[1 and 3]	96 bp and 243 bp in CMLV [69]
	BPXV C18L F and R2 DNA pol FP1 and RP1	Primer sequences[1 and 2]	96 bp and 368 bp in BPXV [70]
SYBR Green quantitative PCR	CMLV C18L F and R	Primer sequence[3]	243 bp in CMLV, with specific melting point at 77.6°C [69]
TaqMan probe real-time PCR	DNA pol FP1 and RP1 DNA Pol probe	Primer sequences[1] **FAM**-5′AATGCAGCGGGTTTGCCTTTCGC3′-**TAMRA**	OPVs detection and quantitation [70]
	CMLV C18L F and R and CMLV C18L Probe	Primer sequence[3] Cy5-CAT CAA TAC AGC CAT CCA GAA AAG TAG-BHQ-2	CMLV detection and quantitation [74]
	BPXV C18L Fwd and Rev and BPXV C18L Probe	CGGTTATTGGAATATGGAGCGAGTG CGTAGTAATCGTCGTAGGGAGAGAC **HEX**-TCACGCTCGATAATCAATACGGCCATCCA-**BHQ 1**	BPXV detection and quantitation [70]
LAMP	CMLV C18L F3 CMLV C18L B3 CMLV C18L FIP CMLV C18L BIP	GCTCGATCATCAATACAGC AAGGAAAATCATCATCTAGGATT AGAGTGGGATGGTAACTCAGTAATA- CATCCAGAAAAGTAGTTACAGAA AGAGAGTATGATTGACGCATTCAA- AAGGCGTATCTGATACAGG	198 bp in CMLV in the form of ladder-like pattern [86]

TABLE 96.2

Molecular Assays for Detection and Differentiation of Orthopoxviruses

Assay	Target Gene	OPV Identification [References]
Genus-specific single PCR	ATIP (full length and partial)/HA/C18L/*DNA Pol*	All OPVs [18,62–66,69]
Species-specific single PCR	TNFR-II/C18L	CMLV/BPXV [69,70,92]
Restriction fragment PCR	A36R	Differentiation of OPVs [93]
Duplex PCR	DNA Pol and C18L	Detection and simultaneous differentiation of BPXV and CMLV from other OPVs [69,70]
SYBR green real-time PCR	C18L	BPXV and CMLV [69,70]
	VETF, RPO18, and A13L genes	All OPVs [94]
	—	OPVs from herpesviruses [72,73]
	HA	All OPVs [85]
TaqMan probe real-time PCR	C18L	BPXV/CMLV [70,74]
	B5R/ATIP	MPXV strains [75–77]
	D11L/D8L	CPXV strains [78,95]
	HA (FRET technology)	All OPVs [71]
	LPMP (late 16 kDa putative membrane protein) and CP (39 kDa core protein)	Pan OPV and also VARV and MPXV [79]
Multiplex real-time format	—	VARV, VACV, MPXV, and CPXV [81–84]
	crmB (OPV), ORF 29, and *DNA Pol* (herpesviruses)	Detection and differentiation of OPVs from herpesviruses [96]
LAMP	C18L	CMLV DNA [86]
	D14L and ATI	MPXV (Congo and West African strains) [88]
	HA	BPXV [89]

been developed to target C18L [86] and HA genes [87] for rapid detection of CMLV and BPXV, respectively, in the clinical samples. A real-time quantitative MPXV genome amplification system targeting D14L and partial ATI genes has been developed using LAMP technology for specific diagnosis and discrimination of Congo basin and West African strains [88], and this tool would be useful for the assessment of MPXV infections. In general, LAMP assays appear to be potential as rapid and sensitive diagnostic tools for their application in less equipped rural disease diagnostic laboratories. Different molecular diagnostics targeting different genes of OPVs along with the primer set for detection and differentiation are summarized in Tables 96.1 and 96.2.

96.2 METHODS

Different techniques for detection of OPVs and diagnosis of OPV infections in terms of antigen, antibody, and nucleic acid in clinical samples are described here particularly for BPXV and CMLV. Suspected clinical samples from OPV-infected cattle, buffaloes, camels, and humans in the form of dry scab or skin lesions in 50% phosphate glycerol saline and blood in case of viremic stage and paired sera collected at 14 days interval should be sent for diagnostic purposes.

96.2.1 SAMPLE PREPARATION

The clinical samples either collected from field outbreaks or submitted to reference laboratory for disease investigation and

confirmation should be processed as described in the following text. A 10% (w/v) suspension of scab material is prepared using phosphate-buffered saline (PBS; pH 7.4) and after freezing and thawing for three times, the homogenate suspension is clarified at $1500 \times g$ for 10 min. The clarified suspension is used for viral antigen identification by CIE. Antibiotics such as penicillin (10,000 IU/mL) and streptomycin (100 μg/mL) as well as antifungal agent (mycostatin 50 IU/mL) is added to the supernatant and filtered through 0.45 μm membrane filters. The filtrates so obtained can be aliquoted and stored at −80°C for further use for either extraction of DNA or isolation of virus in the cell culture system [5,69].

Alternatively, for isolation of the virus, the triturated suspension is repeatedly freeze-thawed thrice at −80°C and then subjected to sonication for 2 min at an amplitude of 30% with a pulse frequency of 0.9 s duration with 9.9 s to extrude virus particles from cells. The suspension is clarified at 5000 $\times$ g for 5 min and the supernatant is filtered using 0.45 μm membrane filters and inoculated onto 48 h old confluent cell monolayers. Infected Vero and primary lamb testes (PLT) cells in 25 cm^2 tissue culture (TC) flasks showing 80%–90% CPE is harvested after decanting the media and the pelleted cells are resuspended in 250 μL of PBS (pH 7.2) for gDNA extraction.

The blood samples collected from infected animals and humans with anticoagulant are used as such or diluted with PBS and used for gDNA extraction or virus isolation. The sera collected at 2 weeks interval (paired sera) from ailing animals are inactivated at 56°C for 30 min before use in SNT and immunohistochemical test or it can be used as such for any other antibody detection procedures like CIE and indirect ELISA.

96.2.2 Detection Procedures

96.2.2.1 Antigen and Antibody Detection

CIE test is being used to identify either unknown soluble antigen in suspected pox lesions or antibody present in the serum samples using standard/reference hyperimmune serum or antigen, respectively, as the case may be [89]. This method can be used for initial screening of OPV suspected clinical samples. The collected samples will be processed as 10% homogeneous suspension in PBS as described earlier and used for the detection of viral antigen. The test is quick as well as more sensitive than the agar gel precipitation test. The positive samples can be further subjected for virus isolation in a suitable cell culture system for definite identification. CIE can also be performed to identify the precipitating antibodies present in serum collected from suspected animals and humans of OPV infections. Identification of specific antibodies is carried out by using specific precipitin antigen prepared from cell culture infected with OPV or skin lesions collected from diseased animals during outbreaks.

96.2.2.2 Virus Isolation in Cell Culture System and Its Identification

Most of OPVs are known to grow in monkey and human cell lines such as Vero and HeLa. For isolation of the virus, the triturated suspension is inoculated onto 48 h old confluent Vero cell monolayers and incubated at 37°C for 1 h with intermittent shaking to allow the virus to adsorb onto cells. After decanting the virus inoculum, infected cells are washed three times with serum-free eagle's minimum essential medium (EMEM) and fed with medium containing 2% bovine calf serum. Infected flasks are incubated further at 37°C and are regularly observed for the appearance of CPE. OPVs (CMLV and BPXV) produce rounding, clumping, increase in refractivity, and detachment of cells in 24–48 h postinfection [5,40]. The identity of the virus isolates is further confirmed by virus neutralization test (VNT), which checks the ability of the virus to neutralize the virus-specific serum antibodies [89]. For VNT, the virus is titrated both in the presence and in the absence of hyperimmune serum and the difference in the virus titer gives the neutralization index.

96.2.2.3 Electron Microscopy

TEM can be used for primary diagnosis of OPV infections and its differentiation from PPV infections based on virion morphology. Here, the methodology of TEM is exemplified using BPXV present in infected Vero cells [90]. Vero cells grown in TC flask until attaining confluency for 48 h is infected with cell culture–isolated BPXV at 0.05 multiplicity of infection. The cells are observed for the appearance of CPE at 24 h interval and, at 80%–90% CPE, cells should be harvested using cell scraper and pelleted by centrifugation at $800 \times g$ for 10 min. Cell pellet should be washed in

chilled PBS (0.1 M) and suspended in a fixative (2.5% glutaraldehyde +, 2% paraformaldehyde, in 0.1 M sodium phosphate buffer, and pH 7.4) for 12 h at 4°C. On fixation, cells are washed once again and resuspended in PBS before processing and viewed under TEM.

96.2.2.4 Immunohistochemistry

Indirect immunoperoxidase technique is performed to identify OPVs in infected cells or specific antibodies in the serum from infected animal using antiserum against OPV or known OPV in cell culture. Briefly, Vero cells showing CPE suspected to any of OPVs in cell culture plates containing cover slips are selected. Cover slips exhibiting characteristic CPE (rounding, ballooning, and syncytia) at 36–48 h postinfection are rinsed gently with PBS and fixed in acetone for 30 min at −20°C. Acetone-fixed cover slips are then washed four times with phosphate-buffered saline containing 0.005% Tween-20 (PBST) and processed as described earlier [3]. The endogenous peroxidase is inactivated by methanol with 3% hydrogen peroxide at 37°C for 20 min followed by blocking with 4% skimmed milk at 37°C for 1 h. Thereafter, cells are incubated at 37°C upon addition of antiserum raised against OPV after diluting in PBST containing 4% skim milk. Subsequently, infected cells are washed four times with PBST and incubated with peroxidase-labeled anti-species IgG diluted (varies from 1:500 to 1000) in 4% skim milk powder. Finally, cells are washed and treated with 3,3-diaminobenzidine tetrahydrochloride for 8–12 min, followed by counter staining with 20% Harris hematoxylin for 2 min. The cover slips are mounted on glass slides and examined microscopically for the presence of antigen antibody reaction, which is exhibited in the form of reddish-brown color in infected cells.

96.2.2.5 Serum Neutralization Test

The serum samples were subjected to SNT for detecting antibodies against BPXV following the method described earlier [91]. Briefly, serum samples (in duplicate rows) are diluted in EMEM in twofold dilution series from 1:2 to 1:64 in flat-bottomed 96-well TC plates (M/s Nunc, Germany). The reference BPXV-BP4 virus suspension (50 μL/well) with a titer of 100 $TCID_{50}$ in 50 μL is added to each serum well and the mixture is incubated for 1 h at 37°C in a humidified CO_2 incubator. An aliquot of 50 μL of Vero cell suspension (≈ 1.5×10^6/mL) is added to each well and incubated. Control wells like serum, virus, and cells are included in the test. The plates should be observed for CPE after 72–96 h. The highest dilution of serum inhibiting the CPE of virus at the 50% endpoint is taken as the neutralizing titer of the serum sample.

96.2.2.6 Diagnostic PCR Assay

The sensitivity and reliability of PCR, real-time PCR, and sequencing for diagnostic and research purposes require efficient unbiased procedures of extraction and purification of nucleic acids. Extraction of DNA is often an early step in many diagnostic processes used to detect OPVs in the environment as well as in clinical samples.

DNA extraction from the suspension of the material could be carried out using commercial gDNA extraction kit following the manufacturer's protocol or conventional nucleic acid isolation method following phenol–chloroform extraction system or simplified nonenzymatic fast DNA extraction protocol are described [61]. These methods can be used to extract total gDNA from the purified virus, virus-infected cells, and clinical samples, and eluted gDNA are stored at −80°C until use.

Clinical scab materials from buffaloes, cows, camels, and humans collected from suspected cases of OPV infections during different outbreaks can be employed for C18L gene–based single PCR and duplex PCR in combination with DNA pol gene and confirmed as camelpox and buffalopox infections with simultaneous differentiation [70]. The primers used for these diagnostic PCRs are shown in Table 96.1. In brief methodology, the total gDNA extracted from clinical samples or infected cells after virus isolation will be used in standard protocol along with positive and negative controls. OPV primers (BPXV C18L F and R2) will produce a 368 bp fragment in BPXV, whereas primer pairs (CMLV C18L F and R) will amplify 243 bp PCR fragment in CMLV suspected samples. For simultaneous detection and differentiation, a duplex PCR (BPXV C18L F and R2 with DNA Pol FP1 and RP1) will produce two bands (368 and 96 bp) in BPXV and only one (96 bp) in CMLV suspected clinical samples [70].

Quantitative PCR (qPCR) or real-time PCR assay targeting DNA Pol gene using TaqMan probe has been reported for rapid and sensitive detection and quantification of CMLV or BPXV as OPV infections. Further, C18L gene–based qPCR has been developed for specific detection and differentiation of these viruses using two different primer pairs with specific probes [70,74].

96.2.2.7 LAMP Assay

Loop-mediated isothermal amplification (LAMP) is quite simple, requiring only a conventional water bath or heat block for incubation under isothermal conditions. The LAMP assay based on C18L gene described earlier is valuable due to its simple plot, rapid reaction, and trouble-free detection of CMLV DNA from infected camels and humans [86]. A negative control containing no template is included in each assay. The positive LAMP reaction produces a ladder-like pattern with a set of bands of different sizes consisting of several inverted repeat structures. LAMP products are separated electrophoretically in 2% agarose gel using standard protocol and visualized under UV light using ethidium bromide. Alternatively, addition of SYBR green I or hydroxyl naphthol blue (HNB) to LAMP reaction will identify positive cases by change in color (orange to apple green in SYBR green I or violet to sky blue in case of HNB). Similar kind of rapid nucleic acid detection method has also been reported for detection of MPXV [88] and BPXV [87] in clinical samples.

96.3 CONCLUSION AND PERSPECTIVES

OPV infections were considered inconsequential till recently and have emerged as an important public health problem during this decade and due to increased cases and outbreaks in camels and buffaloes. In this context, particular attention should be given to camelpox and buffalopox outbreaks, as well as to the identification of any human infections associated with it. Effective prevention and control measures can be achieved through the use of proper diagnostic and prophylactic aids to curtail further spread of these infections as described for most of the zoonotic diseases. In general, several factors related to human activities, environmental changes, and virus could be the determinants of incidence and prevalence of the diseases. To safeguard the public health from pathogens of zoonotic infections, application of skills, knowledge, and resources of veterinary public health is essential. Due to frequent outbreaks of OPV zoonotic infections in different parts of the world, a directed attention is essential on understanding the virus ecology and epidemiology, the identification of potential reservoirs, the transmission pattern, and the immune response upon natural infection in order to control the zoonotic infection onslaughts in the near future. Further, the control measures for emerging and reemerging pathogens are demanding, as there is population explosion. Novel highly sensitive and specific techniques comprising genomics and proteomics along with conventional methods would be useful in the identification of emerging and reemerging pathogens or viruses; thereby, therapeutic/prophylactic/preventive measures would be applied on proper time. Intensified surveillance should be implemented to detect and restrain the spread of these zoonotic infections. Research should focus on molecular biology of these viruses so as to develop diagnostics and prophylactics in a modern way to combat these infections.

REFERENCES

1. Essbauer, S., Pfeffer, M., Meyer, H., Zoonotic poxviruses. *Vet. Microbiol.* 140, 229, 2010.
2. Singh, R.K., Balamurugan, V., Bhanuprakash, V. et al., Emergence and reemergence of vaccinia-like viruses: Global scenario and perspectives. *Ind. J. Virol.* 23, 1, 2012.
3. Bera, B.C., Shanmugasundaram, K., Barua, S. et al., Zoonotic cases of camelpox infection in India. *Vet. Microbiol.* 152, 29, 2011.
4. Gurav, Y.K., Raut, C.G., Yadav, P.D. et al., Buffalopox outbreak in humans and animals in Western Maharashtra, India. *Prev. Vet. Med.* 100, 242, 2011.
5. Venkatesan, G., Balamurugan, V., Prabhu, M. et al., Emerging and re-emerging zoonotic buffalopox infection: A severe outbreak in Kolhapur (Maharashtra), India. *Vet. Ital.* 46, 439, 2010.
6. Balamurugan, V., Venkatesan, G., Bhanuprakash, V. et al., Camelpox, an emerging orthopox viral disease. *Virus Dis.* 24, 295, 2013.
7. Vorou, R.M., Papavassiliou, V.G., Pierroutsakos, I.N., Cowpox virus infection: An emerging health threat. *Curr. Opin. Infect. Dis.* 21, 53, 2008.
8. Kurth, A., Straube, M., Kuczka, A. et al., Cowpox virus outbreak in banded mongooses (Mungos mungo) and jaguarundis (Herpailurus yagouaroundi) with a time-delayed infection to humans. *PLoS One* 4, e6883, 2009.
9. Singh, R.K., Hosamani, M., Balamurugan, V. et al., Buffalopox: An emerging and re-emerging zoonosis. *Anim. Health Res. Rev.* 8, 105, 2007.

10. Orba, Y., Sasaki, M., Yamaguchi, H. et al., Orthopoxvirus infection among wildlife in Zambia. *J. Gen. Virol.* 96(Pt 2), 390–394, 2015.

11. Nalca, A., Rimoin, A.W., Bavari, S. et al., Reemergence of monkeypox: Prevalence, diagnostics, and countermeasures. *Clin. Infect. Dis.* 41, 1765, 2005.

12. Trindade, G.S., Emerson, G.L., Carroll, D.S. et al., Brazilian vaccinia viruses and their origins. *Emerg. Infect. Dis.* 13, 965, 2007.

13. Ferguson, N.M., Keeling, M.J., Edmunds, W.J. et al., Planning for smallpox outbreaks. *Nature* 425, 681, 2003.

14. Pauli, G., Blümel, J., Burger, R. et al., Orthopox viruses: Infections in humans. *Transfus. Med. Hemother.* 37, 351, 2010.

15. Moss, B. Poxviridae: The viruses and their replication. In: Knipe, D.M., Howley, P.M., eds., *Fields Virology*, 5th edn. Philadelphia, PA: Lippincot Williams & Wilkins, pp. 2905, 2007.

16. Duraffour, S., Meyer, H., Andrei, G. et al., Camelpox virus. *Antiviral. Res.* 92, 167, 2011.

17. Office Internationale des epizooties (World Health Organization for Animals), Paris, France, 2008.

18. Damaso, C.R., Esposito, J.J., Condit, R.C. et al., An emergent poxvirus from humans and cattle in Rio de Janeiro state: Cantagalo virus may derive from Brazilian small pox vaccine. *Virology* 277, 439, 2000.

19. de Souza Trindade, G.S., da Fonseca, F.G., Marques, J.T. et al., Araçatuba virus: A vaccinia like virus associated with infection in humans and cattle. *Emerg. Infect. Dis.* 9, 155, 2003.

20. Damon, I. Poxviruses. In: Knipe, D.M., Howley, P.M., eds., *Fields Virology*, 5th edn. Philadelphia, PA: Lippincott Williams & Wilkins, pp. 2947, 2007.

21. Davies, F.G., Mungai, J.N., Shaw, T., Characteristics of a Kenyan camelpox virus. *J. Hyg.* 7, 381, 1975.

22. Coetzer, J.A.W. Poxviridae. In: Coetzer, J.A.W., Tustin, R.C., eds. *Infectious Diseases of Livestock*, vol. 2, 2nd edn. Southern Africa, Africa: Oxford University Press, pp. 1265, 2004.

23. Rheinbaden, F.V., Gebel, J., Exner, M. et al. Environmental resistance, disinfection, and sterilization of poxviruses. In: Schmidt, A.A., Weber, A., Mercer, O., eds., *Poxviruses*. Basel, Switzerland: Birkhauser Verlag, pp. 397, 2007.

24. Fenner, F., Wittek, R., Dumbell, K.R. *The Orthopox Viruses.* New York: Academic Press, Inc., 1989.

25. Dabrowski, P.W., Radonić, A., Kurth, A. et al., Genome-wide comparison of cowpox viruses reveals a new clade related to Variola virus. *PLoS One* 8, e79953, 2013.

26. Alzhanova, D., Fruh, K., Modulation of the host immune response by cowpox virus. *Microbes Infect.* 12, 900, 2010.

27. Duraffour, S., Snoeck, R., Krecmerova, M. et al., Activities of several classes of acyclic nucleoside phosphonates against camelpox virus replication in different cell culture models. *Antimicrob. Agents Chemother.* 51, 4410, 2007.

28. Wernery, U., Kaaden, O.R., *Camelpox: Infectious Diseases in Camelids*, 2nd edn. Berlin, Germany: Blackwell, p. 176, 2002.

29. Baxby, D., Smallpox viruses from camels in Iran. *Lancet* 7786, 1063, 1972.

30. Baxby, D., Differentiation of smallpox and camelpox viruses in cultures of human and monkey cells. *J. Hyg.* 72, 251, 1974.

31. Baxby, D, Ramyar, H., Hessami, M. et al., A comparison of the response of camels to intradermal inoculation with camelpox and smallpox viruses. *Infect. Immun.* 11, 617, 1975.

32. Hafez, S.M., Al-Sukayran, A., dela Cruz, D. et al., Development of a live cell culture camel pox vaccine. *Vaccine* 10, 533, 1992.

33. Nguyen, B.V., Guerre, L., Saint-Martin, G., Preliminary study of the safety and immunogenicity of the attenuated VD47/25 strain of camelpox virus. *Rev. Elev. Med. Vet. Pays Trop.* 49, 189, 1996.

34. Wernery, U., Zachariah, R., Experimental camelpox infection in vaccinated and unvaccinated dromedaries. *Zentralbl Veterin-Armed B* 46, 131, 1999.

35. Renner-Muller, I.C., Meyer, H., Munz, E., Characterization of camelpoxvirus isolates from Africa and Asia. *Vet. Microbiol.* 45, 371, 1995.

36. Sheikh Ali, H.M., Khalafalla, A.I., Nimir, A.H., Detection of Camelpox and vaccinia viruses by polymerase chain reaction. *Trop. Anim. Health Prod.* 41, 1637, 2009.

37. Tantawi, H.H., El Dahaby, H., Fahmy, L.S., Comparative studies on poxvirus strains isolated from camels. *Acta Virol.* 22, 451, 1978.

38. Bett, B., Jost, C., Allport, R. et al., Using participatory epidemiological techniques to estimate the relative incidence and impact on livelihoods of livestock diseases amongst nomadic pastoralists in Turkana South District, Kenya. *Prev. Vet. Med.* 90, 194, 2009.

39. Leese, S., Two diseases of young camels. *J. Trop. Vet. Sci.* 4, 1, 1909.

40. Bhanuprakash, V., Prabhu, M., Venkatesan, G. et al., Camelpox: Epidemiology, diagnosis and control measures. *Expert Rev. Anti Infect. Ther.* 8, 1187, 2010.

41. Al-Ziabi, O., Nishikawa, H., Meyer, H., The first outbreak of camelpox in Syria. *J. Vet. Med. Sci.* 69, 541, 2007.

42. Kriz, B., A study of camelpox in Somalia. *J. Comp. Pathol.* 92, 1, 1982.

43. Kurth, A., Wibbelt, G., Gerber, H.P., Rat-to-elephant-to-human transmission of cowpox virus. *Emerg. Infect. Dis.* 14, 670, 2008.

44. Cardeti, G., Brozzi, A., Eleni, C. et al., Cowpox virus in llama, Italy. *Emerg. Infect. Dis.* 17, 1513, 2011.

45. Sharma, G.K., An interesting outbreak of variola vaccinia in milch cattle of Lahore. The Imperial Council of Agricultural Research. Selected Clinical Articles. *Misc. Bull.* 8, 1, 1934.

46. Sehgal, C.L., Ray, S.N., Ghosh, T.K. et al., Investigation of an outbreak of buffalopox in animals and human beings in Dhulia district, Maharashtra. I. Laboratory studies. *J. Commun. Dis.* 9, 171, 1977.

47. Yadav, S., Hosamani, M., Balamurugan, V. et al., Partial genetic characterization of viruses isolated from pox-like infection in cattle and buffaloes: Evidence of buffalo pox virus circulation in Indian cows. *Arch. Virol.* 155, 255, 2010.

48. deAssis, F.L., Pereira, G., Oliveira, C. et al., Serologic evidence of orthopoxvirus infection in buffaloes, Brazil. *Emerg. Infect. Dis.* 18, 698, 2012.

49. Damon, I.K., Status of human monkeypox: Clinical disease, epidemiology and research. *Vaccine* 29, 54, 2011.

50. Damon, I.K., Roth, C.E., Chowdhary, V., Discovery of monkeypox in Sudan. *N. Engl. J. Med.* 355, 962, 2006.

51. Levine, R.S., Peterson, A.T., Yorita, K.L. et al., Ecological niche and geographic distribution of human monkeypox in Africa. *PLoS One* 2, e176, 2007.

52. Wernery, U., Kaaden, O.R., Ali, M., Orthopox virus infections in dromedary camels in United Arab Emirates (UAE) during winter season. *J. Cam. Pract. Res.* 4, 51, 1997.

53. Kinne, J., Cooper, J.E., Wernery, U., Pathological studies on camelpox lesions of the respiratory system in the United Arab Emirates (UAE). *J. Comp. Pathol.* 118, 257, 1998.

54. Chantrey, J., Meyer, H., Baxby, D. et al., Cowpox: Reservoir hosts and geographic range. *Epidemiol. Infect.* 122, 455, 1999.

55. Zafar, A., Swanepoel, R., Hewson, R. et al., Nosocomial buffalopox virus infection, Karachi, Pakistan. *Emerg. Infect. Dis.* 13, 902, 2007.

56. Ghosh, T.K., Arora, R.R., Sehgal, C.L. et al., An investigation of buffalopox outbreak in animals and human beings in Dhulia district (Maharashtra state). *J. Commun. Dis.* 9, 146, 1977.

57. Huhn, G.D., Bauer, A.M., Yorita, K. et al., Clinical characteristics of human monkeypox, and risk factors for severe disease. *Clin. Infect. Dis.* 41, 1742, 2005.

58. Kurth, A., Nitsche, A., Fast and reliable diagnostic methods for the detection of human poxvirus infections. *Future Virol.* 2, 467, 2007.

59. Kitamoto, N., Kobayashi, T., Kato, Y. et al., Preparation of monoclonal antibodies cross-reactive with orthopoxviruses and their application for direct immunofluorescence test. *Microbiol. Immunol.* 49, 219, 2005.

60. Boulter, E.A., Zwartouw, H.T., Titmuss, D.H. et al., The nature of the immune state produced by inactivated vaccinia virus in rabbits. *Am. J. Epidemiol.* 94, 612, 1971.

61. Yousif, A.A., Al-Naeem, A.A., Al-Ali, M.A., Rapid non-enzymatic extraction method for isolating PCR-quality camelpox virus DNA from skin. *J. Virol. Methods* 169, 138, 2010.

62. Meyer, H., Ropp, S.L., Esposito, J.J., Gene for A-type inclusion body protein is useful for a polymerase chain reaction assay to differentiate orthopoxviruses. *J. Virol. Methods* 64, 217, 1997.

63. Ropp, S.L., Jin, Q., Knight, J.C. et al., Polymerase chain reaction strategy for identification and differentiation of smallpox and other orthopoxviruses. *J. Clin. Microbiol.* 33, 2069, 1995.

64. Funahashi, S., Sato, T., Shida, H., Cloning and characterization of the gene encoding the major protein of the A-type inclusion body of cowpox virus. *J. Gen. Virol.* 69, 35, 1988.

65. Meyer, H., Rziha, H.J., Characterization of the gene encoding the A-type inclusion protein of camelpox virus and sequence comparison with other orthopoxviruses. *J. Gen. Virol.* 74, 1679, 1993.

66. Meyer, H., Pfeffer, M., Rziha, H.J., Sequence alterations within and downstream of the A-type inclusion protein genes allow differentiation of Orthopoxvirus species by polymerase chain reaction. *J. Gen. Virol.* 75, 1975, 1994.

67. Esposito, J.J., Knight, J.C., Orthopoxvirus DNA: A comparison of restriction profiles and maps. *Virology* 143, 230, 1985.

68. Khalafalla, A.I., Buttner, M., Rziha, H.J., Polymerase chain reaction (PCR) for rapid diagnosis and differentiation of para- and orthopox virus infections in camels. FAO/IAEA International Animal Production and Health in Developing Countries, Vienna, Austria, October 6–10, 2003.

69. Balamurugan, V., Bhanuprakash, V., Hosamani, M. et al., A polymerase chain reaction strategy for the diagnosis of camelpox. *J. Vet. Diagn. Invest.* 21, 231, 2009.

70. Singh, R.K., Balamurugan, V., Hosamani, M. et al., Sequence analysis of C18L gene of Buffalopox virus: PCR strategy for specific detection and its differentiation from orthopoxviruses. *J. Virol. Methods* 154, 146, 2008.

71. Panning, M., Asper, M., Kramme, S. et al., Rapid detection and differentiation of human pathogenic orthopox viruses by fluorescence resonance energy transfer real-time PCR assay. *Clin. Chem.* 50, 702, 2004.

72. Nitsche, A., Ellerbrok, H., Pauli, G., Detection of orthopoxvirus DNA by real-time PCR and variola virus DNA by melting analysis. *J. Clin. Microbiol.* 42, 1207, 2004.

73. Olson, V.A., Laue, T., Laker, M.T. et al., Real-time PCR system for detection of orthopoxviruses and simultaneous identification of smallpox virus. *J. Clin. Microbiol.* 42, 1940, 2004.

74. Venkatesan, G., Bhanuprakash, V., Balamurugan, V. et al., TaqMan hydrolysis probe based real time PCR for detection and quantitation of camelpox virus in skin scabs. *J. Virol. Methods* 181, 192, 2012.

75. Li, Y., Olson, V.A., Laue, T. et al., Detection of monkeypox virus with real-time PCR assays. *J. Clin. Virol.* 36, 194, 2006.

76. Saijo, M., Ami, Y., Suzaki, Y. et al., Diagnosis and assessment of monkeypox virus (MPXV) infection by quantitative PCR assay: Differentiation of Congo Basin and West African MPXV strains. *Jpn. J. Infect. Dis.* 61, 140, 2008.

77. Li, Y., Zhao, H., Wilkins, K., Hughes, C. et al., Real-time PCR assays for the specific detection of monkeypox virus West African and Congo Basin strain DNA. *J. Virol. Methods* 169, 223, 2010.

78. Gavrilova, E.V., Shcherbakov, D.N., Maksyutov, R.A., Development of real-time PCR assay for specific detection of cowpox virus. *J. Clin. Virol.* 49, 37, 2010.

79. Kuo, S.C., Wang, Y.M., Identification of pan-Orthopoxvirus, monkeypox-specific and smallpox-specific DNAs by real-time PCR Assay. *J. Med. Sci.* 33, 293, 2013.

80. Kulesh, D.A., Baker, R.O., Loveless, B.M. et al., Smallpox and pan-orthopox virus detection by real-time 3′-minor groove binder TaqMan assays on the roche LightCycler and the Cepheid smart Cycler platforms. *J. Clin. Microbiol.* 42, 601, 2004.

81. Shchelkunov, S.N., Gavrilova, E.V., Babkin, I.V., Multiplex PCR detection and species differentiation of orthopoxviruses pathogenic to humans. *Mol. Cell. Probes* 19, 1, 2005.

82. Shchelkunov, S.N., Shcherbakov, D.N., Maksyutov R.A. et al., Species-specific identification of variola, monkeypox, cowpox, and vaccinia viruses by multiplex real-time PCR assay. *J. Virol. Methods* 175, 163, 2011.

83. Schroeder, K., Nitsche, A., Multicolour, multiplex real-time PCR assay for the detection of human-pathogenic poxviruses. *Mol. Cell. Probes* 24, 110, 2010.

84. Aitichou, M., Saleh, S., Kyusung, P. et al., Dual-probe real-time PCR assay for detection of variola or other orthopoxviruses with dried reagents. *J. Virol. Methods* 153, 190, 2008.

85. Putkuri, N., Piiparinen, H., Vaheri, A. et al., Detection of human orthopoxvirus infections and differentiation of smallpox virus with real-time PCR. *J. Med. Virol.* 81, 146, 2009.

86. Venkatesan, G., Bhanuprakash, V., Balamurugan, V. et al., Development of loop-mediated isothermal amplification assay for specific and rapid detection of camelpox virus in clinical samples. *J. Virol. Methods* 183, 34, 2012.

87. Venkatesan, G., Balamurugan, V., Bhanuprakash, V. et al., Loop-mediated isothermal amplification (LAMP) assay for specific detection of buffalopox virus in clinical samples. Paper presented in *XX Annual Convention of IAAVR and International Conference on Thrust Areas in Veterinary Research, Education, Regulatory Reforms and Governance for Quality Services to Farmers*, KVAFSU, Bangalore, India, April 17–18, 2013.

88. Iizuka, S.M., Shiota, T. et al., Loop-mediated isothermal amplification-based diagnostic assay for monkeypox virus infections. *J. Med. Virol.* 81, 1102, 2009.

89. Sharma, B., Negi, B.S., Pandey, A.B., Detection of goat pox antigen and antibody by the counter immunoelectrophoresis test. *Trop. Anim. Health Prod.* 20, 109, 1988.

90. Bhanuprakash, V., Venkatesan, G., Balamurugan, V. et al., Zoonotic Infections of Buffalopox in India. *Zoonoses Public Health* 57, e149, 2010.

91. Singh, R.K., Hosamani, M., Balamurugan, V. et al., An out-break of buffalopox in buffalo (*Bubalus bubalis*) dairy herds in Aurangabad, India. *Rev. Sci. Tech. OIE* 25, 981, 2006.

92. Marodam, V., Nagendrakumar, S.B., Tanwar, V.K. et al., Isolation and identification of camelpox virus. *Ind. J. Anim. Sci.* 76, 326, 2006.

93. Huemer, H.P., Hönlinger, B., Hopfl, R. et al., A simple restriction fragment PCR approach for discrimination of human-pathogenic Old World animal Orthopoxvirus species. *Can. J. Microbiol.* 54, 159, 2008.

94. Carletti, F., Di Caro, A., Calcaterra, S. et al., Rapid, differential diagnosis of orthopox- and herpesviruses based upon real-time PCR product melting temperature and restriction enzyme analysis of amplicons. *J. Virol. Methods* 129, 97, 2005.

95. Maksyutov, R.A., Gavrilova, E.V., Meyer, H. et al., Real-time PCR assay for specific detection of cowpox virus. *J. Virol. Methods* 211, 8, 2015.

96. Sias, C., Carletti, F., Capobianchi, M.R. et al., A rapid differential diagnosis of Orthopoxviruses and Herpesviruses based upon multiplex real-time PCR. *Infez. Med.* 15, 47, 2007.

97 Parapoxvirus

Graziele Oliveira, Galileu Costa, Felipe Assis, Ana Paula Franco-Luiz, Giliane Trindade, Erna Kroon, Filippo Turrini, and Jonatas Abrahão

CONTENTS

97.1 Introduction 881
    97.1.1 Classification, Morphology, and Genome Organization 881
    97.1.2 Transmission, Clinical Features, and Pathogenesis... 882
        97.1.2.1 Orf Virus.. 882
        97.1.2.2 Bovine Papular Stomatitis Virus 883
        97.1.2.3 Pseudocowpox Virus 884
        97.1.2.4 Parapoxvirus of Red Deer in New Zealand 884
    97.1.3 Diagnosis 884
        97.1.3.1 Conventional Techniques... 884
        97.1.3.2 Molecular Techniques.. 884
97.2 Methods 885
    97.2.1 Sample Collection and Virus Isolation 885
    97.2.2 Detection Procedures.. 885
        97.2.2.1 DNA Extraction 885
        97.2.2.2 qPCR Detection of PPV 885
        97.2.2.3 Nested PCR Detection of PPV 886
    97.2.3 Sequencing and Phylogeny of PPV 886
97.3 Conclusion 888
References.. 888

97.1 INTRODUCTION

The *Parapoxvirus* (PPV) genus of the *Poxviridae* family includes four worldwide-distributed viral species that mainly infect domestic and wild ruminants (e.g., sheep, goat, and cattle). PPV species are transmissible to humans and, for this reason, are considered zoonotic agents of veterinary and public health relevance [1,2]. The fact that PPV infection has been reported in an even broader host range (including camels, seals, and deer species) highlights the emerging importance of this genus [1,3,4]. PPVs are epitheliotropic viruses whose clinical manifestations include exanthematic lesions as painful pustules and vesicles. Clinical diagnosis of PPV infection is often made difficult by both the absence of pathognomonic signs and the great similarity in the symptoms to either different *Poxviridae* infections or unrelated exanthematic diseases, such as cutaneous anthrax, foot-and-mouth disease, blue tongue, and bovine viral diarrhea. In addition, subclinical infections have been recorded. Thus, improvement of molecular analyses and epidemiological studies is considered an urgent priority in order not only to avoid unnecessary treatment due to misdiagnosis but also to lessen the economical impact of PPV diseases and to implement strategies for the limitation of this infection. Clinical evidence and restriction analysis have been traditionally used to diagnose and

characterize PPV diseases. Nevertheless, the historical relevance of the *Poxviridae* family has increased the availability of genomic data in scientific databases. This further facilitates the use of the molecular diagnostic approach for improved diagnosis of PPV infection.

97.1.1 CLASSIFICATION, MORPHOLOGY, AND GENOME ORGANIZATION

The PPV genus belongs to the subfamily *Chordopoxvirinae* (ChPV), family *Poxviridae*. The genus comprises four zoonotic viral species that cause localized skin nodules mainly in domestic and wild ruminants: *Bovine papular stomatitis virus* (BPSV), *Pseudocowpox virus* (PCPV), the genus prototype species *Orf virus* (ORFV), and the newest characterized species *Parapoxvirus of red deer in New Zealand* (PVNZ) [5–7]. PPVs are primarily classified on their natural host range and viral DNA restriction enzyme analyses, while the extensive serological cross-reactivity makes neutralization tests and immunofluorescence useless to differentiate viral species [8].

Viral particles are large, enveloped, and oval shaped, with variable size ranging from 260 nm × 160 nm in ORFV to 300 nm × 190 nm in PCPV. Tubular filiform structures surround virions and can be used to discriminate PPV particle from others poxviruses, such as orthopoxviruses (OPV) by electron microscopy (EM) [9,10]. PPV virions enclose a typical *Poxviridae* genome, consisting of linear double-stranded DNA of approximately 140 kb, composed by one central region and two flanking inverted terminal repeats (ITR) terminally linked by a hairpin structure that allows cross-hybridization between isolates of the same species. High intraspecific variability has been found at either genomic level, by viral DNA restriction pattern analyses, or protein level [8,11–16].

PPV genome contains approximately 130 open reading frames organized in a conserved central core and two external regions. ChPV conserved genes are essential for viral replication and morphogenesis and encode transcription and replication machineries. Nonessential variable genes, conversely located along ITRs, are mainly responsible for pathogenesis and virulence and, in some cases, are genus specific [10,17,18]. *B2L* gene encodes PPV major envelope antigen of approximately 42 kDa. This gene is the homologue of *F13L* gene of *Vaccinia virus* (VACV) and has been widely used in PPV diagnosis and characterization, as well as vascular endothelial growth factor gene (*VEGF*) and virus interferon resistance (*VIR*) genes, which are common targets of phylogenetic analyses. For differential diagnosis, a polymerase chain reaction (PCR) assay targeting the *A32L* gene, which encodes a 34 kDa ATPase protein involved in DNA packaging, has also been developed [19–26]. Like other poxviruses, PPV replicates in the cytoplasm of host cells, exploiting their own machinery for DNA transcription and replication. Notably, PPV shares with the *Molluscipoxvirus* genus the peculiar genome feature of high guanine and cytosine (G+C) content. In fact, conversely to a majority of other poxvirus genomes that enclose approximately 25%–35% of G+C, PPV genome contains about 65% of G+C [12,27,28].

97.1.2 Transmission, Clinical Features, and Pathogenesis

PPV infections occur primarily in ruminants such as sheep, goats, and cattle. Infections in wild species, such as grey seals, serows, cervids, and gazelle, have also been described. Furthermore, PPV infections have been reported in cats expanding the range of the identified hosts [6,29–34]. PPV infections are zoonoses, and their transmission is facilitated by contact with lesions or fomites. Zoonotic transmission to humans represents an occupational hazard that mainly affects either milkers or particular risk category workers like veterinarians, farmers, and butchers, who directly handle infected mammals. Additionally, transmission from wild animals is also possible at low frequency. Often, the presence of painful lesions impairs milking and breastfeeding, and the infected animals become unproductive resulting in economical losses in outbreak-affected farms [35–37]. Finally, the reported cases of human-to-human transmission have highlighted the importance of developing a thorough understanding of the chain of transmissional events [38–44].

The clinical outcome in human manifestations consists of painful lesions on hands and less frequently on the face [9,45,46], while the main clinical signs observed in infected animals are papular lesions around the mouth and in the oral mucosa, skin, and teats. Lesions in the esophagus, stomach, and intestine have been reported with less frequency. Lesion progression is characterized by formation of macules and papules, followed by vesicles and pustules. Disease progression and resolution generally occur over the course of 1–2 months concomitantly with an accumulation of neutrophils, T and B cells, and dendritic cells in proximity to infected epidermal cells. Keratinocytes are the major multiplication site that is characterized by eosinophilic cytoplasmic inclusions and irregularly shaped nuclei.

97.1.2.1 Orf Virus

ORFV is the prototype species of the PPV genus and the causative agent of contagious ecthyma (CE), also known as orf, a zoonosis that primarily affects domestic sheep and goats, although some cases have been also described in wild animals [34,47,48]. CE can decrease the fitness of infected hosts leading to severe economic losses, as well as public health problems, according to the zoonotic potential of this disease [49–52].

Since the first identification of a human ORFV by Newsom and Cross in 1934 [53], several human cases caused by occupational or household exposures have been reported [54–56]. CE is characterized by epidermal proliferative lesions (vesicles) mainly localized in the lips, mouth, nostrils, gums, and tongue. Less frequently, lesions can be observed in the esophagus, stomach, and intestine, while among sheep, the lesions are commonly localized at the teat and udder. Genital lesions have also been described. However, asymptomatic ORFV infections can occur. In addition, when lesions in the mouth, lips, and udders prevent infected animals from feeding, the consequences can be even more severe leading to a rapid weight loss and increased mortality rate [57]. Incubation period in sheep ranges from 3 to 14 days, while in humans is shorter (2–4 days). Based on the clinical symptoms only, CE can be frequently misdiagnosed, leading to unnecessary treatments. Furthermore, cases of coinfection between poxviruses of different genera have been reported, reinforcing the need to apply molecular assay for the differential diagnosis [36,48,52,58]. In human, ORFV causes skin lesions and occasionally malaise and lymphadenopathy, while immunodeficient individuals can develop even severe infection. In such cases, even though the treatment had resulted in successful outcome, no therapeutic options have been yet characterized [59–61]. Studies have shown that the mortality rate in kids can reach 93%; nevertheless, the disease is considered mild by some authors. ORFV transmission is typically associated to animal direct contact (Figure 97.1), and the importance of preventive measure (e.g., wearing gloves and appropriate hand hygiene) is pivotal in order to limit the zoonotic transmission of CE, especially in rural areas where the risk of human-to-human transmission can be very high.

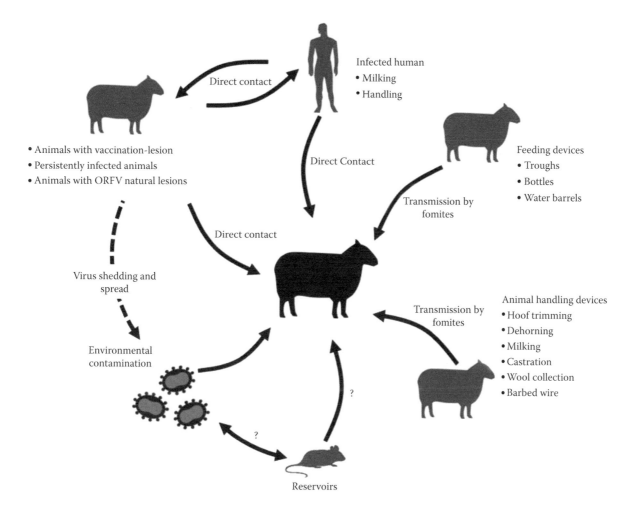

FIGURE 97.1 *Orf virus* hypothetical transmission cycle.

In human, lesions generally appear on the hands and face, but some patients develop ulcerating lesions on oral mucosa, in particular at the level of the upper lip [22,57,62–65]. A case of human-to-human transmission was reported from father to daughter via the common use of tweezers. Despite being uncommon, human-to-human infection reveals new questions concerning the transmissional chain of the ORFV [44]. Like other poxviruses, ORFV contains immune modulatory genes primarily located in terminus of the viral genome. Among them, granulocyte macrophage colony-stimulating factor (GM-CSF) inhibitory factor gene (*GIF*) encodes for an inhibitor of both GM-CSF and viral gene encoding an interferon resistance factor orf virus interferon-resistance gene (OVIFNR). These genes have been used in genetic variation studies of ORFV [66–69].

Immunohistological studies of the skin of infected sheep have shown that there is an accumulation of neutrophils, dendritic cells, and T cells. CD4+ are the predominant T cells in lesions. Reinfection can occur, but smaller lesions are observed, which resolve usually in 3 weeks. For this reason, the disease is most severe in those animals, such as lambs, that have never been exposed to the virus [66,70,71]. In some countries, vaccination with attenuated vaccine is practiced. The virus was attenuated through serial passages on primary chicken embryo fibroblast tissue cultures. A field study was conducted to confirm the efficacy of the vaccine and demonstrated that vaccination was able to block the normal course of the disease and induce rapid recovery. Thus, the ORFV vaccine can limit the severity of the disease [72]. In short, CE is an emerging zoonosis, with an increasing number of outbreaks reported worldwide. It not only has a significant impact on livestock but also poses a public health threat. Further studies and epidemiologic surveys are warranted in order to improve our knowledge on this viral disease [51,63,73–75].

97.1.2.2 Bovine Papular Stomatitis Virus

BPSV infection is characterized by exanthematic lesions mainly localized on the muzzle, around the mouth, oral mucosa, and less frequently in the esophagus and forestomach of calves and udders of cows. In general, lesions begin with papules formation progressing to pustules and vesicles that regress approximately 12–16 days after the onset. Calves usually develop the classical disease outcome, whereas BPSV is the cause of severe esophagitis in adult bull [76–78]. The exanthematic lesions typically resemble those of foot-and-mouth disease, highlighting the importance of differential diagnosis. Commonly, acute forms of disease are described, but chronic forms of bovine papular stomatitis in young cattle have been also demonstrated [79]. Recently, a molecular analysis has identified a new viral strain, revealing the existence of genetic variability in the BPSV envelope gene [80]. Reinfections by BPSV are commonly observed, indicating that this virus does not confer permanent immunity, as it has been shown for other viruses of the family, such as *VACV*, which cause repeated outbreaks of bovine vaccinia mainly in Brazil and ORFV in sheep and goats. BPSV has a worldwide distribution, and infection of calves and cows is of economical importance as it leads to a decrease in milk production [81–84]. Furthermore, since it is a zoonosis, human cases have a significant public health impact. Human transmission usually occurs by direct contact with lesions causing nodules and pustules on the hands and less frequently on the face; occasional cases may develop severe lesions [46]. In this way, milkers and other livestock workers are exposed to infection [41].

97.1.2.3   Pseudocowpox Virus

First isolation of *PCPV*, the causative agent of milker's nodule disease, dates back to 1963, and it was obtained from lesions of cows and calves [85]. Nowadays, PCPV infections are currently worldwide distributed affecting cows and calves and representing a zoonotic threat to human health [86–89]. Moreover, infection in hosts such as camels in the Arabian Peninsula, reindeers, and cats has also been described [90–92]. The virus is usually transmitted by direct contact with infected animals. Once infected, milking cows develop proliferative lesions on teats and udders in a clinical course of approximately 10–15 days. As observed for ORFV and BPSV, cases of reinfection can occur in herds. In addition, chronic lesions may occur. PCPV diagnosis is complicated not only by virion morphology but also by clinical symptoms, which are similar with other poxvirus infections that affect bovine, such as VACV and BPSV. Moreover, cases of PCPV human coinfection with different *Poxviridae* species have been described [93]. Therefore, the correct diagnosis of PCPV infection in cattle mainly relies upon the identification of the viral agent by molecular assays [94].

97.1.2.4   Parapoxvirus of Red Deer in New Zealand

PVNZ has been classified as new PPV species due to its distinct restriction enzyme patterns [5]. PPV in red deer have been firstly reported in New Zealand and, more recently, in Italy among wild ruminants. In red deer, the disease is characterized by lesions mainly located in the muzzle, lips, ears, and neck. Infections are more common in younger animals, with morbidity often reaching 100%. Diagnosis is generally made on clinical signs alone, but histology, EM, and molecular tests have also been used [95,96].

97.1.3   Diagnosis

Diagnostic methods for PPV infection include clinical diagnosis based on the history of contact with animals and the clinical appearance of the lesion. However, PPV infections may be misdiagnosed due to the great similarity in symptoms with other vesicular–pustular diseases. In these cases, the diagnosis can be established by histopathology and EM;

serological methods, such as ELISA, immunofluorescence assay (IFA), agar gel immunodiffusion (AGID), and neutralization test (NT); viral growth in cell cultures and molecular assays as restriction fragment length polymorphism (RFLP); and conventional PCR, multiplex PCR, and real-time PCR (qPCR). Among the available methods, molecular approaches and nucleic acid sequence analysis are widely used for detection and identification of PPV infections.

97.1.3.1   Conventional Techniques

PPV infection can be identified by histopathological analysis that usually shows vascular degeneration of keratinocytes, the primary viral target [97]. PPV infection of keratinocytes is also characterized by eosinophilic cytoplasmic inclusions and irregularly shaped nuclei [34,97]. The diagnosis can be enhanced by negative-stain EM, which identifies the viral particles within infected cells, discriminating PPV genus from OPV [34,97]. However, EM requires morphologically intact clinical samples with high viral load (approximately 10^6 particles per mL), as well as trained personnel [98,99]. High viral load is usually obtained from fresh lesions and crusts of infected individuals and/or animals, whereas in case of low viral load or not correctly processed samples, EM may produce false-negative results with high probability. In this case, virus isolation can also be performed using primary ovine or bovine cell culture. Although once regarded as the gold standard method for detection of poxviruses, virus isolation is not attempted routinely in clinical laboratories [17,100–103].

Determination of serum titers during the acute and convalescent phase of disease is also useful. The presence of specific antibodies can be confirmed by AGID and IFA. In these techniques, Madin–Darby bovine kidney–infected, fetal bovine muscle–infected, and bovine fetal testicle–infected cells with known PPV strains can be used as antigen after lysis or sonication treatments [104–106]. Moreover, ELISA for IgM and IgG determination has also been employed, according to the direct correlation between fast healing of lesions and rise in antibody levels during the infection [58,104,105,107]. Furthermore, NT assays are also applied. Although neutralizing antibodies are associated with protective status, all PPV are immunologically related and cannot be distinguished in NT [108,109]. Western blot analysis is usually confirmatory and has demonstrated that the produced antibodies in response to infection are directed against the viral surface tubule protein [99,102].

97.1.3.2   Molecular Techniques

Traditional diagnostic laboratories rely on conventional culture and serological methods for the confirmation of several viral infections. The access to reliable diagnostic laboratory testing methods is limited, due to the requirement of well-qualified personnel, standard and automated products, and equipments. The identification and characterization of an already isolated viral pathogen is mainly assessed by molecular techniques that represent a rapid tool for the diagnosis of fresh clinical specimens. Indeed, molecular techniques generally in use for PPV diagnostic include RFLP, nucleic acid hybridization, and, especially nowadays, conventional PCR and qPCR. These methods, and in particular PCR, are well recognized to be more rapid, sensitive, and specific than conventional techniques for the identification of target viruses.

In RFLP analysis, viral DNA is digested with one or more restriction enzymes (e.g., EcoRI, HindIII, BamHI), and the resulting restriction fragments are separated according to their size by electrophoresis, generating distinct restriction fragment profiles and allowing species and strains differentiation [5,15,34,80,110,111]. Multiplex PCR can be useful for clinical differential diagnosis as rapid approach for simultaneous detection of coinfections with different *Poxviridae* genera, such as PPV and OPV [112,113], and PPV and capripoxvirus [114].

The PCR assay that has traditionally provided the most valuable approach to identify PPV infections targets the conserved major envelope protein gene (*B2L*), which is the homologue of *VACV* envelope protein antigen-p37K [26]. Besides, *B2L* gene has been widely used in the molecular characterization and phylogenetic analysis, representing the main molecular marker for PPV genetic analysis. Other PPV-specific genes have been described as target of PCR amplification, such as interferon-resistant protein (*VIR*) [101,115], ATPase (*A32L*), dsRNA-binding protein (*E3L*) [116], and GM-CSF/interleukin-2 inhibitory factor (*GIF*) genes [111]. These genes have been extensively used for phylogenetic studies in order to investigate on epidemiology and evolution of PPVs [21,24,75,93,101,111,112,115,116].

Finally, qPCR assays have further improved the diagnosis of PPV infections due to their high efficiency and sensitivity. These features are required for the analysis of human clinical samples, which often have a low viral load and/or poor DNA quality. Different qPCR methods for PPV DNA detection in clinical specimens have been developed [23,116–118]. These methods target ORFV major envelope antigen *B2L* and can be used either in the diagnosis and quantification of ORFV [23,117] or in the diagnosis of several PPV species including PCPV and BPSV [23].

97.2 METHODS

97.2.1 SAMPLE COLLECTION AND VIRUS ISOLATION

Biopsies of lesions are the most frequent and suitable clinical samples collected for the diagnosis of PPV-associated infections. For virus isolation, 1%–10% (weight–volume) of biopsy suspension is prepared in pH 7.4 phosphate-buffered saline (PBS) or nuclease-free water, both supplemented with gentamycin to eliminate bacterial contamination, then clarified by centrifugation at $2000 \times g$ for 3 min to eliminate larger debris. The supernatant is collected and submitted to DNA extraction [119] or, in order to isolate viruses, is directly used to infect cell cultures. There are some permissive cell lines useful for PPV isolation, such as ovine organotypic skin culture, ovine fetal turbinate (OFTu) cells, and Madin–Darby ovine kidney cells [120,121]. Besides, primary bovine lineage esophageal cells (KOP, cell culture collection Friedrich Loeffler Institute,

Riems, Germany), bovine lung, or kidney fibroblasts can be used. OFTu cells, which are the most frequently used for PPV isolation, are cultured in minimal essential medium (Hyclone) with 10% fetal bovine serum (Hyclone), 2 mM of L-glutamine, 100 U of penicillin, 100 µg of streptomycin, 20 µg of nystatin per mL at 37°C, and 5% of CO_2 [121]. In this system, the cytopathic effect (CPE) is usually detected 1–2 days postinfection, according to the starting viral load, and is characterized by cell rounding, pyknosis, and cell detachment. When 50%–75% of CPE is observed, cell cultures are subjected to one cycle of freeze–thawing. The virus-containing supernatants are then clarified by centrifugation, aliquoted, and frozen at −70°C [34,99].

97.2.2 DETECTION PROCEDURES

97.2.2.1 DNA Extraction

DNA from macerated sample supernatant, serum, and virus-infected cells is extracted by commercial kits and phenol–chloroform–isoamyl alcohol, or in some cases, samples can be submitted to PCR with no previous DNA extraction step [112].

97.2.2.2 qPCR Detection of PPV

One of the first diagnostic qPCR for PPV detection was developed by Nitsche and colleagues in 2006. Before that time, the diagnosis of PPV infections had been traditionally performed by EM and/or conventional PCR. The following protocol describes the qPCR currently used for detection of PPV. We have used this protocol in our lab successfully. To date, this assay is capable of specifically amplifying DNA from all PPV species. A conserved region into the major envelope protein gene (*B2L*) was chosen for the design of two primers and one 5-nuclease minor groove-binder (MGB) probe [23].

Primer	Target Gene	Sequence (5′–3′)	Expected Product (bp)	Reference
PPV up	B2L	TCGATGCGGTGCAGCAC	95	[23]
PPV do		GCGGCGTATTCTTCTCGGAC		
PPV TMGB		GTCGTCCACGATGAGCAG		

The reaction is performed into a mix solution containing 1× PCR buffer, 5 mM of $MgCl_2$, 1 mM of deoxynucleotide triphosphate mixture with dUTP, 1 µM of ROX, 1 U of platinum Taq p polymerase (Invitrogen), 7.5 pmol of each primers, 2.5 pmol of the MGB probe, and 5 µL of DNA template [23].

Cycle condition
Incubation at 95°C for 5 min
↓
40 cycles
Denaturation (95°C, 15 s)
Annealing (60°C, 30 s)

This reaction has a linear detection range from 10^6 to 10^1 copies per reaction, with a correlation (R^2) of 0.99 and an amplification efficiency of 97%. Its limit of detection is 4.7 copies per assay (95% confidence interval, 3.7–6.8 copies/reaction). The assay variability is <30% for yield close to the detection limit (10^2 copies) and <15% for higher DNA amount (10^4 and 10^6 copies). The assay specificity has been tested against several viral DNA, such as OPV (*vaccinia*, *cowpox*, *monkeypox*, *ectromelia*, and *camelpox virus*), *Molluscum contagiosum virus*, *Yaba-like disease virus*, human herpesviruses types 1–8, and adenoviruses, and no amplifications were detected [23].

In 2013, Zhao and colleagues have validated a new robust and specific qPCR assay, which amplifies the conserved RNA polymerase gene of all PPV species.

Primer	Target Gene	Sequence (5'–3')	Reference
Forward primer	PAPV J6R	CGCGGTCTGGTCCTTG	[118]
Reverse primer		CAGCATCAACCTCTCCTACATCA	
Probe		CCACGAAGCTGCGCAGCAT	

PCR protocol is as follows.

Component	Concentration	Thermocycle
PPV primers	0.4 μmol/L	Incubation at 95°C for 2 min
FastStart TaqMan Universal PCR Master Mix (Applied Biosystems, Foster City, CA)	1×	45 cycles
TaqMan probe	200 nmol/L	Denaturation (95°C, 5 s)
Template	2 μL	Annealing (62.5°C, 20 s)
Total volume	**25 μL**	

97.2.2.3   Nested PCR Detection of PPV

Abrahão and colleagues in 2009 designed a nested–multiplex PCR to detect orthopox–parapox coinfections. In addition, the reaction was successfully used to specifically detect PPV, with high sensitivity. An adapted nested PCR protocol developed to detect exclusively PPV in scabs is as follows. Clinical sample is prepared by maceration of the scab in PBS (pH 7.4) (0.1 g scab/0.9 mL PBS) and clarification by centrifugation at $2000 \times g$ for 3 min. Supernatant is collected and directly amplified by PCR targeting the gene ORFV011–*B2L* gene (viral envelope antigen). As previously stated, the *B2L* gene is the most common target used for diagnosis and phylogenetic studies of PPV. The technique described in the following has been previously reported [24,26,75,93,112], and here, we proposed an adaptation. It is worth mentioning that all known PPV species are detected by this reaction.

Primer	Target Gene		Sequence (5'–3')	Expected Product (bp)	References
OVB2LF1	B2L	First step	TCCCTGAA GCCCTATTA TTTTTGT	1206	[26]
OVB2LR1			GCTTGCG GGCG TTCGGA CCTTC		
PPP-1		Nested step	GTCGTCCAC GATGA GCAG	592	[24]
PPP-4			TACGTGGGA AGCGCC TCGCT		

Prepare the mix of the first step and perform the reaction in a thermal cycler following the protocol:

Component	Concentration	Thermocycle
PPV primers	0.8 mM	Incubation at 95°C for 9 min
dNTPs	10 mM	↓
MgCl₂	2.0 mM	30 cycles
Bovine serum albumin (BSA)	500 ng	Denaturation (94°C, 1 min)
Taq DNA Polymerase (Promega, Madison, the United States)	2 U	Annealing (45°C, 1 min)
DEPC-treated water q.s	18 μL	Extension (72°C, 1 min)
Template	2 μL	↓
Total volume	**20 μL**	Final extension (72°C, 10 min)

Notes: For the nested step, the same chemical and thermal conditions must be used, but using internal PPV (PPP-1 and PPP-4–0.8 mM) primers. The final PCR products may be electrophoresed on 2% agarose or 8% PAGE gels in parallel with standard DNA, as described elsewhere [119].

97.2.3   Sequencing and Phylogeny of PPV

In order to discriminate the PPV species detected by PCR, the amplified products are usually sequenced and subjected to phylogenetic analysis. Sequencing is usually performed using the primers described earlier and those by Hosamani et al. [26] and Inoshima et al. [24]. In addition, other targets have been described for characterization as shown in the following:

Primer	Target Gene	Sequence (5'–3')	References
PPV011-F	ORF 011 (B2L)	GTG CGC GAA GGT GTC GTC CA	[103]
PPV011-R		ATGTGGCCGTTCTCCTCCATC	
PPV032-F	ORF 032	CGA GCT TTA AAT AGT GGA AAC ACA GC	
PPV032-R		GCA CCA TCA TCC TGT ACT TCC TC	
D27F	VEGF	AAT GTA AAT WMT AAC GCC	[122]

(Continued)

11R		AAC CCA GAA ACG TCC CGC TAC	
D2R		CGT TTG GAT MTG CGG TCC	
VIR1	VIR	TTA GAA GCT GAT GCC GCA G	[34]
VIR2		ACA ATG GCC TGC GAG TG	
FP	A32L	GTG TTG ATC ATC GAA GAC TCG GTG	[123]
RP1		GTC GCC CTT GTC GCC CTT AGT CTC	
RP2		CCG CCG TCA GAG TCG ACG TCG CCC T	

[a] VEGF, vascular endothelial growth factor;

[b] W, A/T. M, A/C;

[c] VIR, virus interferon resistance.

In 2014, Friederichs and colleagues developed a pan-PPV PCR reaction using primers that encompasses all PPV strains, designed on *B2L* gene [103]. They also described a further PCR protocol, targeting a more variable region within the ORF 032. For the ORF 032 gene, the same B2L-PCR protocol must be used, but the annealing temperature is set to 67°C. PCR conditions for B2L and ORF 032 are the following:

Component	Concentration	Thermocycle
PPV primers (B2L/ ORF 032)	0.5 μM	Incubation at 98°C for 30 s
dNTPs	200 μM	↓
Q5 reaction buffer	1×	35 cycles
Q5 high GC enhancer	1×	Denaturation (98°C, 10 s)
Q5 high fidelity DNA polymerase (New England Biolabs, MA)	1 U	Annealing for B2L (65°C, 30 s)
		Annealing for ORF 032 (67°C, 30 s)
DEPC-treated water q.s	49 μL	Extension (72°C, 30 s)
Template	1 μL	↓
Total volume	**50 μL**	Final extension (72°C, 2 min)

Other genes that encode nonstructural proteins—*VEGF* and *VIR*—have been previously described to characterize PPV isolates [122,124]. Worth of note, high variability in *VEGF* gene of PCPV has been reported, suggesting that particular attention is required in choosing this target for PCR diagnosis. PCR conditions for *VEGF* and *VIR* are the following:

Component	Concentration	Thermocycle
PPV primers VEGF/VIR	0.2 μM	Incubation at 95°C for 5 min
DNTPsmix	800 μM	↓
Tris-HCl (pH 8.3)	10 mM	35 cycles
KCl	50 mM	Denaturation (94°C, 30 s)
MgCl₂	1.5 mM	Annealing for VEGF(40°C, 30 s)
AmpliTaq Gold DNA polymerase (Applied Biosystems, CA)	1 U	Annealing for VIR (55°C, 30 s)
DEPC-treated water q.s	49 μL	Extension (72°C, 30 s)

(Continued)

Template	1 μL	↓	
Total volume	**50 μL**	Final extension (72°C, 7 min)	

The PPV strains can be genotyped using a differential PCR for the *A32L* gene, as described previously [123]. This process is based on the polymorphisms among PPV isolates, and the PCR conditions for the *A32L* are described as follows.

Component	Concentration	Thermocycle
FP primer	1 μM	Incubation at 95°C for 9 min
RP1primer	0.4 μM	↓
RP2 primer	0.4 μM	35 cycles
dNTP mix	800 μM	Denaturation (94°C, 30 s)
Tris-HCl (pH 8.3)	10 mM	Annealing (63°C, 30 s)
KCl	50 mM	Extension (72°C, 30 s)
MgCl₂	1.5 mM	↓
AmpliTaq Gold DNA polymerase	1 U	Final extension (72°C, 7 min)
Extracted DNA	3 μL	
Total volume	**50 μL**	

The expected gene fragments are amplified by the protocols mentioned earlier and electrophoresed on 2% agarose gels. The amplicons are purified, cloned, and sequenced. The obtained sequences are compared with the available sequences in the National Center for Biotechnology Information GenBank database, The European Molecular Biology Laboratory, or DNA Data Bank of Japan.

With these results, phylogenetic and molecular evolutionary analyses have been performed. Several studies have traditionally inferred PPV phylogeny by amplification and sequencing of *B2L* gene [15,24,26,96]. However, Friederichs and colleagues recently proposed to include ORF 032 in phylogenetic analyses of PPV. ORFV phylogenetic tree based on *B2L* gene, as well as on ORF 032, shares a similar basic structure and reveals three distinct clusters with high bootstrap support, each representing one of the PPV species: ORFV, PCPV, and BPSV. In particular, ORF 032–derived phylogenetic tree shows a highly diversified branching and consequently provides an improvement in the analysis of PPV phylogeny [103].

As previously reported, *B2L* gene has been extensively used in phylogeny. In 2010, Inoshima and colleagues have reported a highly conservative profile among *B2L* sequences from Japanese's ORFs. Then, they have proposed additional analysis using three nonstructural protein genes, such as *VEGF*, *VIR*, and *A32L*. These additional analyses are suitable to phylogeny allowing the proper clustering of PPV isolates into regional clades [103]. Interestingly, *B2L*, *VIR*, and *VEGF* genes of Brazilian ORFV isolates have been recently analyzed, showing a high variation degree among them, indicating a multiple introduction source of ORFV in Brazil from different sources, such as European countries, Asia, North America, and/or Oceania. The cocirculation of allochthonous-derived isolates may have led to this polytomous phenomenon observed among Brazilian ORFV isolates [21].

Virus isolation, molecular detection, and whole genome sequencing are useful tools for the diagnosis of PPV infections. The genome-wide studies can support the understanding about the origins, evolution, and epidemiology of ORFVs. All of these concerns are of great importance in the veterinary, economy, and public health fields.

97.3 CONCLUSION

The emergence of zoonotic poxviruses has highlighted the importance of molecular techniques in the diagnosis of PPV infections. In recent years, improvement in molecular assays has enhanced the accuracy of pathogen identification and yielded insights into PPV genomic organization and phylogeny. Further refinement of molecular diagnostic techniques in relation to test turnaround time and automation will not only contribute to the discovery of new pathogens but also assist researchers and clinicians in the implementation of effective intervention (e.g., vaccines and drugs) and management strategies for PPV infections.

REFERENCES

1. Buttner, M. and Rziha, H.J., Parapoxviruses: From the lesion to the viral genome, *J. Vet. Med. B Infect. Dis. Vet. Public Health*, 49, 7, 2002.
2. MacNeil, A. et al., Diagnosis of bovine-associated parapoxvirus infections in humans: Molecular and epidemiological evidence, *Zoonoses Public Health*, 57, 161, 2010.
3. Dashtseren, T. et al., Camel contagious ecthyma (pustular dermatitis), *Acta Virol.*, 28, 122, 1984.
4. Sainsbury, A.W. and Gurnell, J., An investigation into the health and welfare of red squirrels, *Sciurus vulgaris*, involved in reintroduction studies, *Vet. Rec.*, 137, 367, 1995.
5. Robinson, A.J. and Mercer, A.A., Parapoxvirus of red deer: Evidence for its inclusion as a new member in the genus *Parapoxvirus*, *Virology*, 208, 812, 1995.
6. Damon, I., Poxviruses. In: Fields, B.N., Knipe, D.M., and Howley, P.M. (eds.), *Fields Virology*, 6th edn. Lippincott, Williams & Wilkins, Philadelphia, PA, p. 2173, 2013.
7. King, A.M.Q. et al. (eds.). *Virus Taxonomy: Classification and Nomenclature of Viruses: Ninth Report of the International Committee on Taxonomy of Viruses.* Elsevier Academic Press, San Diego, CA, 2011.
8. Wittek, R. et al., Genetic and antigenic heterogeneity of different parapoxvirus strains, *Intervirology*, 13, 33, 1980.
9. Robinson, A.J. and Lyttle, D.J., Parapoxviruses: Their biology and potential as recombinant vaccines. In: Binns, M.M. and Smith, G.L. (eds.), *Recombinant Poxviruses*. CRC Press, Inc., Boca Raton, FL, p. 285, 1992.
10. Moss, B., Poxviridae. In: Fields, B.N., Knipe, D.M., and Howley, P.M. (eds.), *Fields Virology*, 6th edn. Lippincott, Williams & Wilkins, Philadelphia, PA, p. 2129, 2013.
11. Buddle, B.M., Dellers, R.W., and Schurig, G.G., Heterogeneity of contagious ecthyma virus isolates, *Am. J. Vet. Res.*, 45, 75, 1984.
12. Gassmann, U., Wyler, R., and Wittek, R., Analysis of parapoxvirus genomes, *Arch. Virol.*, 83, 17, 1985.
13. Robinson, A.J., Ellis, G., and Balassu, T., The genome of orf virus: Restriction endonuclease analysis of viral DNA isolated from lesions of orf in sheep, *Arch. Virol.*, 71, 43, 1982.
14. Robinson, A.J. et al., Conservation and variation in orf virus genomes, *Virology*, 157,13, 1987.
15. Inoshima, Y. et al., Genetic heterogeneity among parapoxviruses isolated from sheep, cattle and Japanese serows (*Capricornis crispus*), *J. Gen. Virol.*, 82, 1215, 2001.
16. Rziha, H.J. et al., Relatedness and heterogeneity at the near-terminal end of the genome of a parapoxvirus bovis 1 strain (B177) compared with parapoxvirus ovis (*Orf virus*), *J. Gen. Virol.*, 84, 1111, 2003.
17. Delhon, G. et al., Genomes of the parapoxviruses orf virus and bovine Papular stomatitis virus, *J. Virol.*, 78, 168, 2004.
18. Mercer, A.A. et al., The establishment of a genetic map of orf virus reveals a pattern of genomic organization that is highly conserved among divergent poxviruses, *Virology*, 212, 698, 1995.
19. Koonin, E.V., Senkevich, T.G., and Chernos, V.I., Gene A32 product of vaccinia virus may be an ATPase involved in viral DNA packaging as indicated by sequence comparisons with other putative viral ATPases, *Virus Genes*, 7, 89, 1993.
20. Sullivan, J.T. et al., Identification and characterization of an orf virus homologue of the vaccinia virus gene encoding the major envelope antigen p37 K, *Virology*, 202, 968, 1994.
21. Abrahão, J.S. et al., Looking back: A genetic retrospective study of Brazilian *Orf virus* isolates, *Vet. Rec.*, 171, 476, 2012.
22. Zhang, K. et al., Comparison and phylogenetic analysis based on the B2L gene of orf virus goats and sheep in China during 2009–2011, *Arch. Virol.*, 159, 1475, 2014.
23. Nitsche, A. et al., Real-time PCR detection of parapoxvirus DNA, *Clin. Chem.*, 52, 316, 2006.
24. Inoshima, Y., Morooka, A., and Sentsui, H., Detection and diagnosis of parapoxvirus by the polymerase chain reaction, *J. Virol. Methods*, 84, 201, 2000.
25. Li, H. et al., Phylogenetic analysis of two Chinese orf virus isolates based on sequences of B2L and VIR genes, *Arch. Virol.*, 158, 1477, 2013.
26. Hosamani, M. et al., Comparative sequence analysis of major envelope protein gene (B2L) of Indian orf viruses isolated from sheep and goats, *Vet. Microbiol.*, 116, 317, 2006.
27. Lefkowitz, E.J., Wang, C., and Upton C., Poxviruses: Past, present and future, *Virus Res.*, 117, 105, 2006.
28. Wittek, R., Kuenzle, C.C., and Wyler, R. High GC content in parapoxvirus DNA, *J. Gen. Virol.*, 43, 231, 1979.
29. Yeruham, I., Nyska, A., and Abraham, A., Parapox infection in a gazelle kid (*Gazella gazella*), *J. Wildl. Dis.*, 30, 260, 1994.
30. Simpson, V.R. et al., Parapox infection in grey seals (*Halichoerus grypus*) in Cornwall, *Vet. Rec.*, 134, 292, 1994.
31. Suzuki, T. et al., Isolation and antibody prevalence of a parapoxvirus in wild Japanese serows (*Capricornis crispus*), *J. Wildl. Dis.*, 29, 384, 1993.
32. Hamblet, C.N., Parapoxvirus in a cat, *Vet. Rec.*, 132, 144, 1993.
33. Fairley, RA. et al., Recurrent localized cutaneous parapoxvirus infection in three cats, *N. Z. Vet J.*, 56, 196, 2008.
34. Guo, J. et al., Genetic characterization of orf viruses isolated from various ruminant species of a zoo, *Vet. Microbiol.*, 99, 81, 2004.
35. Okada, H., Matsukawa, K., and Chihaya, Y., Experimental transmission of contagious pustular dermatitis from a Japanese serow, *Capricornis crispus*, to a calf and goats, *J. Jpn. Vet. Med. Assoc.*, 39, 578, 1986.
36. McKeever, D.J. et al., Studies of the pathogenesis of orf virus infection in sheep, *J. Comp. Pathol.*, 99, 317, 1988.

37. Jenkinson, D.M., Hutchison, G., and Reid, H.W., The B and T cell responses to orf virus infection of ovine skin, *Vet. Dermatol.*, 3, 57, 1992.

38. Almagro, M. et al., Milker's nodes: Transmission by fomites and virological identification, *Enferm. Infecc. Microbiol. Clin.*, 9, 286, 1991.

39. Smith, K.J. et al., Parapoxvirus infections acquired after exposure to wildelife, *Arch. Dermatol.*, 127, 79, 1991.

40. Schuler, G., Hönigsmann, H., and Wolff, K., The syndrome of milker's nodules in burn injury: Evidence for indirect viral transmission, *J. Am. Acad. Dermatol.*, 6, 334, 1982.

41. Carson, C.A. and Kerr, K.M., Bovine papular stomatitis with apparent transmission to man, *J. Am. Vet. Med.*, 151, 183, 1967.

42. Yirrell, D.L. and Vestey, J.P., Human orf infections, *J. Eur. Acad. Dermatol. Venereol.*, 3, 451, 1994.

43. Bayindir, Y. et al., Investigation and analysis of a human orf outbreak among people living on the same farm, *New Microbiol.*, 34, 37, 2011.

44. Ohashi, A. et al., A rare human-to-human transmission of orf, *Int. J. Dermatol.*, 53, 1, 2014.

45. Hosamani, M. et al., Orf: An update on current research and future prospective, *Expert Rev. Anti Infect.*, 7, 879, 2009.

46. Bowman, K.F. et al., Cutaneous form of bovine papular stomatitis in man, *JAMA*, 246, 2813, 1981.

47. Haig, D.M. and Mercer, A.A., Ovine diseases: Orf, *Vet. Res.*, 29, 311, 1998.

48. Nandi, S., De, U.K., and Chowdhury, S., Current status of contagious ecthyma or orf disease in goat and sheep—A global perspective, *Small Ruminant Res.*, 96, 73, 2011.

49. Mazur, C. and Machado, R.D., Detection of contagious pustular dermatitis virus of goats in a severe outbreak, *Vet. Rec.*, 125, 419, 1989.

50. Mazur, C. et al., Molecular characterization of Brazilian isolates of orf virus, *Vet. Microbiol.*, 73, 253, 2000.

51. Zhao, K. et al., Identification and phylogenetic analysis of an orf virus isolated from an outbreak in sheep in the Jilin province of China, *Vet. Microbiol.*, 142, 408, 2010.

52. Billinis, C. et al., Phylogenetic analysis of strains of *Orf virus* isolated from two outbreaks of the disease in sheep in Greece, *Virol. J.*, 9, 24, 2012.

53. Newsom, I.E. and Cross, F., Sore mouth in sheep transmissible to man, *J. Am. Vet. Med. Assoc.*, 84, 799, 1934.

54. Huminer, D., Alcalay, J., and Pitlik, S., Human orf in Israel: Report of three cases, *Isr. J. Med. Sci.*, 24, 54, 1988.

55. Khaled, A. et al., Orf of the hand, *Tunis. Med.*, 87, 352, 2009.

56. Al-Qattan, M.M., Orf infection of the hand, *J. Hand. Surg. Am.*, 36, 1855, 2011.

57. Lederman, E.R. et al., An investigation of a cluster of Parapoxvirus cases in Missouri, Feb–May 2006: Epidemiologic, clinical and molecular aspects, *Animals*, 3, 142, 2013.

58. Sant'Ana, F.J. et al., Coinfection by vaccinia virus and an orf virus-like parapoxvirus in an outbreak of vesicular disease in dairy cows in midwestern Brazil, *J. Vet. Diagn. Invest.*, 25, 267, 2013.

59. Lederman, E.R. et al., Progressive orf virus infection in a patient with lymphoma: Successful treatment using imiquimod, *Clin. Infect. Dis.*, 44, 100, 2007.

60. Geernick, K. et al., A case of human orf in an immunocompromised patient treated successfully with cidofovir cream, *J. Med. Virol.*, 64, 543, 2001.

61. Tan, S.T., Blake, G.B., and Chambers, S., Recurrent orf in an immunocompromised host, *Br. J. Plast. Surg.*, 44, 465, 1991.

62. Darbyshire, J.H., A fatal, ulcerative mucosal condition of sheep associated with the virus of contagious pustular dermatitis, *Br. Vet. J.*, 117, 97, 1961.

63. Scagliarini, A. et al., Orf in South Africa: Endemic but neglected, *Onderstepoort J. Vet. Res.*, 7, 79, 2012

64. Meechan, J.G. and MacLeod, R.I., Human labial orf: A case report, *Br. Dent. J.*, 173, 343, 1992.

65. Nougairede, A. et al., Sheep-to-human transmission of orf virus during Eid al-Adha religious practices, France, *Emerg. Infect. Dis.*, 19, 102, 2013.

66. McInnes, C.J., Wood, A.R., and Mercer, A.A., Orf virus encodes a homolog of the vaccinia virus interferon resistance gene E3L, *Virus Genes*, 17, 107, 1998.

67. Haig, D.M. et al., The orf virus gene OV20.0L gene product is involved in interferon resistance and inhibits an interferon-inducible, double-stranded RNA-dependent kinase, *Immunology*, 93, 335, 1998.

68. Haig, D.M. and Fleming, S., Immunomodulation by virulence proteins of the parapoxvirus orf virus, *Vet. Immunol. Immunopathol.*, 15, 72, 1999.

69. Kottaridi, C. et al., Phylogenetic correlation of Greek and Italian orf virus isolates based on VIR gene, *Vet Microbiol.*, 116, 310, 2006.

70. Lyttle, D.J. et al., Homologs of vascular endothelial growth factor are encoded by the poxvirus orf virus, *J. Virol.*, 68, 84, 1994.

71. Haig, D.M. and McInnes, C.J., Immunity and counter-immunity during infection with the parapoxvirus orf virus, *Virus Res.*, 88, 3, 2002.

72. Mercante, M.T. et al., Production and efficacy of an attenuated live vaccine against contagious ovine ecthyma, *Vet. Ital.*, 44, 537, 2008.

73. Kumar, N. et al., Isolation and phylogenetic analysis of an orf virus from sheep in Makhdoom, India, *Virus Genes*, 48, 3012, 2014.

74. Chan, K.W. et al., Identification and phylogenetic analysis of orf virus from goats in Taiwan, *Virus Genes*, 35, 705, 2007.

75. Abrahão, J.S. et al., Detection and phylogenetic analysis of orf virus from sheep in Brazil: A case report, *Virol. J.*, 4, 47, 2009.

76. Mayr, A. and Buttner, M., Bovine papular stomatitis virus. In: Dinter, Z. and Morein, B. (eds.), *Virus Infections of Ruminants*. Elsevier, Amsterdam, the Netherlands. p. 23, 1990.

77. Leonard, D. et al., Unusual bovine papular stomatitis virus infection in a British dairy cow, *Vet. Rec.*, 164, 311, 2009.

78. Jeckel, S. et al., Severe oesophagitis in an adult bull caused by bovine papular stomatitis virus, *Vet. Rec.*, 169, 317, 2011.

79. Yeruham, I., Abraham, A., and Nyska, A., Clinical and pathological description of a chronic form of bovine popular stomatitis, *J. Comp. Pathol.*, 111, 279, 1994.

80. Yaegashi, G. et al., Molecular analysis of parapoxvirus detected in eight calves in Japan, *J. Vet. Med. Sci.*, 75, 1399, 2013.

81. Aguilar-Setién, A. et al., Bovine papular stomatitis, first report of the disease in Mexico, *Cornell Vet.*, 70, 10, 1980.

82. Inoshima, Y., Nakane, T., and Sentsui, H., Severe dermatitis on cattle teats caused by bovine popular stomatitis virus, *Vet. Rec.*, 164, 311, 2009.

83. Sant'Ana, F.J. et al., Bovine popular stomatitis affecting dairy cows and milkers in Midwestern Brazil, *J. Vet. Diagn. Invest.*, 24, 442, 2012.

84. Oem, J.K. et al., Bovine popular stomatitis virus (BPSV) infections in Korean native cattle, *J. Vet. Med. Sci.*, 75, 675, 2013.

85. Moscovici, C. et al., Isolation of a viral agent from pseudo-cowpox disease, *Science*, 141, 915, 1963.

86. Carter, M.E., Brookbanks, E.O., and Dickson, M.R., Demonstration of a pseudo-cowpox virus in New Zealand, *N. Z. Vet. J.*, 16, 105, 1968.

87. Cargnelutti, J.F. et al., An outbreak of pseudocowpox in fattening calves in southern Brazil, *J. Vet. Diagn. Invest.*, 24, 437, 2012.

88. Asagba, M.O., Recognition of pseudocowpox in Nigeria, *Trop. Anim. Health Prod.*, 14, 184, 1982.

89. Rossi, C.R., Kiesel, G.K., and Jong, M.H., A paravaccinia virus isolated from cattle, *Cornell Vet.*, 67, 72, 1977.

90. Abubakr, M.I. et al., Pseudocowpox virus: The etiological agent of contagious ecthyma (Auzdyk) in camels (*Camelus dromedarius*) in the Arabian Peninsula, *Vector Borne Zoonotic Dis.*, 7, 257, 2007.

91. Hautaniemi, M. et al., The genome of pseudocowpoxvirus: Comparison of a reindeer isolate and a reference strain, *J. Gen. Virol.*, 91, 1560, 2010.

92. Fairley, R.A. et al., Persistent pseudocowpox virus infection of the skin of a foot in a cat, *N. Z. Vet. J.*, 61, 242, 2013.

93. Abrahão, J.S. et al., Human vaccinia virus and pseudocowpox virus co-infection: Clinical description and phylogenetic characterization, *J. Clin. Virol.*, 48, 69, 2010.

94. Munz, E. and Dumbell, K., Pseudocowpox. In: Coetzer, J.A.W., Thomson, G.R., and Tustin, R.C. (eds.), *Infectious Disease of Livestock*. Oxford University Press, Oxford, U.K., vol. 1, p. 625, 1994.

95. Horner, G.W. et al., Parapoxvirus infections in New Zealand farmed red deer (*Cervus elaphus*), *N. Z. Vet. J.*, 35, 41, 1987.

96. Scagliarini, A. et al., Parapoxvirus infections of red deer, Italy, *Emerg. Infect. Dis.*, 17,684, 2011.

97. Dal Pozzo, F. et al., Original findings associated with two cases of bovine papular stomatitis, *J. Clin. Microbiol.*, 49, 4397, 2011.

98. Hazelton, P.R. and Gelderblom, H.R., Electron microscopy for rapid diagnosis of infectious agents in emergent situations, *Emerg. Infect. Dis.*, 9, 294, 2003.

99. Guo, J. et al., Characterization of a North American orf virus isolated from a goat with persistent, proliferative dermatitis, *Virus Res.*, 93, 169, 2003.

100. McInnes, C.J. et al., Genomic comparison of an avirulent strain of orf virus with that of a virulent wild type isolate reveals that the orf virus G2L gene is non-essential for replication, *Virus Genes*, 22, 141, 2001.

101. Kottaridi, C. et al., Laboratory diagnosis of contagious ecthyma: Comparison of different PCR protocols with virus isolation in cell culture, *J. Virol. Methods*, 134, 119, 2006.

102. Li, W. et al., Isolation and phylogenetic analysis of orf virus from the sheep herd outbreak in northeast China, *BMC Vet. Res.*, 23, 229, 2012.

103. Friederichs, S. et al., Comparative and retrospective molecular analysis of Parapoxvirus (PPV) isolates, *Virus Res.*, 6, 11, 2014.

104. Inoshima, Y. et al., Serological survey of parapoxvirus infection in wild ruminants in Japan in 1996–1999, *Epidemiol. Infect.*, 126, 153, 2001.

105. Inoshima, Y. et al., Use of protein AG in an enzyme-linked immunosorbent assay for screening for antibodies against parapoxvirus in wild animals in Japan, *Clin. Diagn. Lab. Immunol.*, 6, 388, 1999.

106. Iketani, Y. et al., Persistent parapoxvirus infection in cattle, *Microbiol. Immunol.*, 46, 285, 2002.

107. Nollens, H.H. et al., Seroepidemiology of parapoxvirus infections in captive and free-ranging California sea lions *Zalophus californianus*, *Dis. Aquat. Organ.*, 6, 153, 2006.

108. Schütze, N. et al., Specific antibodies induced by inactivated parapoxvirus ovis potently enhance oxidative burst in canine blood polymorphonuclear leukocytes and monocytes, *Vet. Microbiol.*, 6, 81, 2010.

109. Zhang, K. et al., Human infection with orf virus from goats in China, 2012, *Vector Borne Zoonotic Dis.*, 14, 365, 2014.

110. Klein, J. and Tryland, M., Characterisation of parapoxviruses isolated from Norwegian semi-domesticated reindeer (*Rangifer tarandus tarandus*), *Virol. J.*, 2, 79, 2005.

111. Hosamani, M. et al., Isolation and characterization of an Indian orf virus from goats, *Zoonoses Public Health*, 54, 204, 2007.

112. Abrahão, J.S. et al., Nested-multiplex PCR detection of Orthopoxvirus and Parapoxvirus directly from exanthematic clinical samples, *Virol. J.*, 6, 140, 2009.

113. Schroeder, K. and Nitsche, A., Multicolour, multiplex real-time PCR assay for the detection of human-pathogenic poxviruses, *Mol. Cell Probes*, 24, 110, 2010.

114. Venkatesan, G., Balamurugan, V., and Bhanuprakash, V., Multiplex PCR for simultaneous detection and differentiation of sheeppox, goatpox and orf viruses from clinical samples of sheep and goats, *J. Virol. Methods*, 195, 1, 2014.

115. de Oliveira, C.H. et al., Multifocal cutaneous orf virus infection in goats in the Amazon region, Brazil, *Vector Borne Zoonotic Dis.*, 12, 336, 2012.

116. Chan, K.W. et al., Phylogenetic analysis of parapoxviruses and the C-terminal heterogeneity of viral ATPase proteins, *Gene*, 432, 44, 2009.

117. Gallina, L. et al., A real time PCR assay for the detection and quantification of orf virus, *J. Virol. Methods*, 134, 140, 2006.

118. Zhao, H. et al., Specific qPCR assays for the detection of orf virus, pseudocowpox virus and bovine papular stomatitis virus, *J. Virol. Methods*, 194, 229, 2013.

119. Sambrook, J., Fritsch, E.F., and Maniatis, T. *Molecular Cloning: A Laboratory Manual*. Cold Spring Harbor University Press, Cold Spring Harbor, NY, 1989.

120. Scagliarini, A. et al., Ovine skin organotypic cultures applied to the ex vivo study of orf virus infection, *Vet. Res. Commun.*, 29, 245, 2005.

121. Gaili, W. et al., In vitro RNA interference targeting the DNA polymerase gene inhibits orf virus replication in primary ovine fetal turbinate cells, *Virology*, 159, 915, 2014.

122. Inoshima, Y., Ishiguro, N., Molecular and biological characterization of vascular endothelial growth factor of parapoxviruses isolated from wild Japanese serows (*Capricornis crispus*), *Vet. Microbiol.*, 140, 63, 2010.

123. Chan, K.W. et al., Differential diagnosis of orf viruses by a single-step PCR, *J. Virol. Methods*, 160, 85, 2009.

124. Inoshima, Y., Ito, M., and Ishiguro, N., Spatial and temporal genetic homogeneity of orf viruses infecting Japanese serows (*Capricornis crispus*), *J. Vet. Med. Sci.*, 72, 701, 2010.

98 White Spot Syndrome Virus

Dongyou Liu

CONTENTS

98.1 Introduction 891
    98.1.1 Classification 891
    98.1.2 Morphology and Genome Organization 892
    98.1.3 Biology and Epidemiology 892
        98.1.3.1 Biology 892
        98.1.3.2 Host Range 892
        98.1.3.3 Transmission 892
        98.1.3.4 Geographic Distribution 893
    98.1.4 Clinical Features 893
    98.1.5 Diagnosis 893
    98.1.6 Treatment and Prevention 894
98.2 Methods 894
    98.2.1 Sample Preparation 894
    98.2.2 Detection Procedures 894
        98.2.2.1 Conventional PCR 894
        98.2.2.2 Real-Time PCR 894
98.3 Conclusion 895
References 895

98.1 INTRODUCTION

White spot syndrome virus (WSSV) is a double-stranded DNA (dsDNA) virus that has emerged as a significant threat to wild and farmed shrimp worldwide. Although WSSV is infective to a wide range of crustaceans (such as salt and brackish water penaeids, crabs, lobsters, freshwater prawns, and crayfish), its impact on penaeid shrimp is nothing short of extraordinary. Besides developing lethargy, emaciation, white spots on the carapace, and reddish discoloration of the body, nearly 100% of shrimp succumb to white spot disease (WSD) within 3–10 days of infection. Clearly, prompt and accurate diagnosis is vital to limit the spread of infection and reduce its impact on aquaculture, which is currently dominated by two shrimp species: the black tiger shrimp (*Penaeus monodon*) and the white Pacific shrimp (*Litopenaeus vannamei*) [1].

98.1.1 CLASSIFICATION

WSSV is a dsDNA virus classified in the genus *Whispovirus*, family *Nimaviridae*. As the sole member of the family *Nimaviridae*, WSSV was regarded as baculovirus prior to 2005, on the basis of its cylindrical morphology and the histological lesions it induces that resemble those of nonoccluded baculoviruses. The other former names for WSSV include hypodermal and hematopoietic necrosis baculovirus, rod-shaped nuclear virus of *Penaeus japonicus*, Chinese baculovirus, systemic ectodermal and mesodermal baculovirus, penaeid rod-shaped DNA virus, and white spot baculovirus (WSBV) [2].

First identified in 1992 from cultured penaeid shrimp in China and Taiwan, WSSV has quickly spread to other parts of Asia; the Middle East; North, Central, and South America; Africa (Mozambique), and other shrimp farming countries of the world. For the penaeid shrimp, WSSV is responsible for a highly lethal disease (WSD) with a mortality of nearly 100% within 7 days. However, in other aquatic species, such as crayfish, crabs, and bivalves, WSSV causes only subclinical infection. These aquatic species act as a reservoir of WSSV [1].

Previous restriction fragment length polymorphism analysis indicated that WSBV isolates from various geographical regions display limited genomic differences and appear to be closely related to each other [3]. With the whole genome sequences of several WSSV isolates becoming available recently, detailed examination of their genetic differences is possible. Indeed, despite sharing an overall sequence identity of >99%, WSSV genomes contain several variable genomic loci that offer useful targets for genotype identification and epidemiologic trait determination. These include the ORF14/15 region (which is prone to recombination), the ORF23/24 region (which is prone to ~13 kb deletions), and the variable number of tandem repeats (VNTRs) within the coding regions of ORF75, ORF94, and ORF125 [4]. In fact, deletions associated with the ORF23/24 and ORF14/15 variable regions are of evolutionary significance, as WSSV genotypes with smaller genomic size (and larger deletion in the ORF23/24 and ORF14/15) are fitter and thrive better than those with larger genomic size [5–10]. Within the ORF94 and ORF125, perfect repeat units (RUs) of 54 and 69 bp, respectively, are identified, and a compound RU of 45 bp interspersed by a 57 bp unit is found in the ORF75 region [11,12]. It appears that for general epidemiological studies of WSSV strains, ORF94 is a better marker than ORF125 and ORF75 due to its maximum variation. In fact, a recent study separated WSSV isolates from India into 31 different genotypes on the basis of ORF94 VNTR [13,14].

98.1.2 Morphology and Genome Organization

Morphologically, WSSV is an enveloped nonoccluded, elliptical to rod-shaped particle of 210–380 nm in length and 70–167 nm in width, with a tail-like appendage at one end. Inside the trilaminar envelope is the nucleocapsid core of 180–420 nm in length, 54–85 nm in width, with a 6 nm thick external wall. Encasing the viral genome, the nucleocapsid core has a crosshatched appearance and is formed by stacks of rings (about 14 in total), which are in turn formed by regular-spaced globular subunits of about 8 nm in diameter, arranged in two parallel rows [15–17].

The genome of WSSV is a monopartite, circular, supercoiled, dsDNA of approximately 300 kb (ranging from 292,967, 295,884, 305,107 to 307,287 bp in isolates originated from Thailand, Korea, China, and Taiwan, respectively) [18,19]. Sequence analysis on the WSSV-TH, WSSV-KR, WSSV-CN, and WSSV-TW genomes reveals a total of 684, 515, 531, and 507 putative open reading frames (ORFs), respectively, of which only 184 (WSSV-TH) and 181 (WSSV-CN) may encode functional proteins [20]. These include at least 39 structural proteins (of which 22 are envelope proteins), as well as enzymes for the nucleotide metabolism, DNA replication, and protein modification [21].

Five major structural proteins (i.e., viral protein [VP] 15, VP19, VP24, VP26, and VP28, with molecular masses of 15, 19, 24, 26, and 28 kDa, respectively) are found in the WSSV particle. Present in the nucleocapsid core, VP15 is a highly basic, histone-like, dsDNA-binding protein with no hydrophobic regions. VP19 and VP28 are associated with the virion envelope. VP26 acts as a tegument protein that links the nucleocapsid associated proteins VP15 and VP24. In addition, about 40 WSSV encoded minor virion proteins are identified from WSSV virions [22,23].

It is notable that VP28 (encoded by ORF 421 or *wsv421*, of 204 amino acids in size) is a major envelope protein of WSSV [24]. VP28 acts as a viral attachment protein to the shrimp cells and interacts with host cellular proteins such as PmRab7, heat shock cognate protein 70, and signal transducer and activator of transcription. The crystal structure of VP28 comprises 12 copies of the protein assembled into four trimers, with each monomer exhibiting a single β-barrel and a α-helix protruding from the β-barrel architecture. The predicted N-terminal transmembrane region of VP28 may anchor on the viral envelope membrane [21].

The nucleocapsid core encasing the viral genome consists mainly of proteins VP15, VP24, VP26, and VP664. Of these, VP664 (with a molecular mass of around 664 kDa) is the major core protein responsible for the striated appearance of the nucleocapsid core [25].

98.1.3 Biology and Epidemiology

98.1.3.1 Biology
WSSV replication initially takes place in the nucleus, and virus morphogenesis begins with the formation of viral envelopes in the nucleoplasm followed by a generation of extended, empty, long tubules, which break up into fragments of 12–14 rings to form empty nucleocapsid shells. The empty capsids are then surrounded by the envelope leaving an open for nucleoproteins and the viral DNA to enter. After the narrowing of the open end, mature virions produce a taillike extension of the envelope and move out of the cell either by budding or by rupture of the nuclear envelope or cell membrane of an infected cell [17].

WSSV persists for a long period in the environment, remains intact in pond soil after 10 months of storage at room temperature, and maintains infectivity in the seawater up to 40 days.

98.1.3.2 Host Range
WSSV is infective to a large variety of aquatic animal species, including penaeids (*P. monodon, P. vannamei, P. indicus, P. japonicus, P. chinensis, P. penicillatus, P. aztecus, P. merguiensis, Farfantepenaeus duorarum, P. stylirostris*), freshwater prawns, crayfish, crabs, marine and brackish water crustaceans, squillas, aquatic arthropods/decapods, copepods, and planktons [26,27]. In farmed shrimp and many other aquatic animals, WSSV is a deadly pathogen, while in some hosts (polychaete worms, microalgae, and rotifer eggs), it causes subclinical infection, which is attributed to viral latency (involving three WSSV ORFs 366, 151, and 427) [28,29]. Animals with asymptomatic infections (including 93 species of arthropods) may act as carriers or reservoirs for WSSV [30,31].

98.1.3.3 Transmission
WSSV is transmitted by both horizontal and vertical modes. Horizontal transmission occurs between individuals of the same host species or between different host species, usually after ingestion of dead infected shrimp or exposure (mainly through the gills and other body surfaces) to water containing infected animals or free virus particles. Various factors may influence the efficiency of horizontal transmission, including biological (host ages, aggression/predation and density, and virulence potential of virus strains) and physiological (water temperature, dissolved oxygen, and salinity) factors [32]. Vertical transmission takes place between mother (shedding virus particles at the time of spawning) and offspring (ingesting virus particles at first feeding). In addition, there is evidence for vertical transmission from infected brood stock to larvae and from adult rotifer (*Brachionus urceus*) to its eggs.

In infected shrimp, WSSV viral particles are present in the connective tissue layer surrounding the seminiferous tubules in testes and in follicle cells, oogonia, oocytes, and connective tissue cells in the ovary [33,34].

98.1.3.4 Geographic Distribution

Since its first emergence in China and Taiwan in 1992, outbreaks of WSD have been recorded across major shrimp farming regions of East Asia, Southeast Asia, Indonesia, and India within a few short years [35–38]. In 1995, WSSV was detected in Texas and South Carolina in the United States. Early in 1999, WSSV spread to Central and South America including Brazil and Argentina [16,39–41]. In 2002, the virus traveled to Europe (France) and the Middle East (Saudi Arabia) and then to Iran and Madagascar. To date, Australia is the only shrimp farming country free of WSSV [42]. Apparently, WSSV transmission between countries is facilitated by the import and export of both live and frozen uncooked shrimp, as well as brood stock [34,42].

98.1.4 Clinical Features

A typical outbreak of WSD in cultured shrimp starts 1 or 2 days after the introduction of the virus or virus-infected shrimp. Clinical signs of WSSV infection include lethargy, anorexia, lack of appetite, swimming near the surface of ponds, formation of white spots (a few mm to 1 cm in diameter) on the exoskeleton especially on the inside surface of the carapace, and finally abdominal segment, loose cuticle, reddish to pink body discoloration, and high mortality (up to 100% within 2–5 days of onset of clinical symptoms [2]).

At the early stages of infection, WSSV is located in the stomach, gills, and cuticular epidermis, and the connective tissue of the hepatopancreas gets infected. At later stages of infection, the virus is found in the lymphoid organ, antennal gland, muscle tissue, hematopoietic tissue, heart, hindgut and parts of the midgut and eventually reaches to the nervous system and the compound eyes. Destruction of ectodermal and mesodermal tissues by WSSV leads to lesions in the infected tissues and organs, with homogeneous hypertrophied/enlarged nuclei containing basophilic central inclusions (called Cowdry A-type inclusions) surrounded by marginated chromatin. The pathognomonic enlarged nuclei containing reddish basophilic inclusions surrounded by unstained cytoplasm are particularly evident in the subcuticular epithelium of the stomach [2,43].

98.1.5 Diagnosis

Although white spots are typically associated with WSD, they may also be caused by bacterial infection (possibly *Bacillus subtilis*) or environmental conditions (such as high alkalinity). Therefore, accurate diagnosis of WSSV infection requires the use of laboratory tests based on histopathological, serological, and molecular principles.

Histopathological examination is useful for the detection of WSSV. In combination with clinical signs, observation of histological changes in affected tissues from moribund prawns provides a presumptive diagnosis of WSD. Typically, live moribund prawns from a suspected WSSV outbreak are fixed by injection with highly acidic Davidson's fixative, which decalcifies the tissues for easy processing and sectioning. For ISH, specimens are transferred from Davidson's fixative after 24–48 h to 70% ethyl alcohol. Then, the specimens are embedded in paraffin wax and processed using standard methods. Observation of inclusion bodies in target tissues is suggestive of WSSV infection.

Further histological details may be obtained by using electron microscopy [44]. Subcuticular tissues, gills, and pleiopods are ideal for this purpose. These specimens may be first screened by rapid-stain tissue squashes or histology to show signs of hypertrophied nuclei and Cowdry A-type inclusions or marginated chromatin surrounding a basophilic inclusion body. Tissues are fixed in a 10:1 fixative to tissue volume of 6% glutaraldehyde at 4°C and buffered with sodium cacodylate or phosphate to pH 7.0 for at least 24 h. For long-term storage, glutaraldehyde concentration is reduced to 1%. The tissues are then fixed in 1% osmium tetroxide and stained with uranyl acetate and lead citrate. WSD virions are rod shaped to elliptical with a trilaminar envelope and measure $80–120 \times 250–380$ nm.

WSSV may be isolated from infected shrimp using primary cultures of lymphoid or ovary cells [45,46]. A recent report showed that mud crab hemocyte culture also supports WSSV replication in vitro [47]. Collected from live animals, hemocytes are grown in double-strength L-15 medium (2× L-15) prepared in crab saline, supplemented with 5% fetal bovine serum and antibiotic–antimycotic solution (penicillin 100 U/mL, streptomycin 100 µg/mL, and amphotericin B 0.25 µg/mL) with osmolality adjusted to 894 mOsm/kg. The hemocytes adhere within 2 h after seeding and show proliferation up to 72 h. After inoculation with WSSV, the cultured hemocytes display cytopathic effects such as detachment, rounding of cells, and clear areas of depleted cells in 48 h [47,48].

Serological assays employing polyclonal or monoclonal antibodies raised against the virus and native or recombinant viral structural proteins (such as VP28 and VP18) have been proven useful for the diagnosis of WSD. The most common serological testing platforms include indirect fluorescent antibody test, immunohistochemistry, enzyme-linked immunosorbent assay, Western blot, and immunodot assay. These techniques are rapid, convenient, and applicable to field use, but their limited specificity can be a drawback [49,50].

Molecular techniques represent recent additions to diagnostic arsenal for microbial infections including WSD. A range of nonamplified and amplified approaches have been explored with excellent outcomes [51–54]. In particular, PCR and its derivatives (e.g., nested, multiplex, and real time) have been widely adopted in clinical laboratories for WSSV identification, due to their high sensitivity, specificity, speed, and amenability for automation [55–72]. Several commercial PCR assays for WSD diagnosis have been marketed (EBTL, DiagXotics, IQ2000, BIOTEC). The nested PCR described by Lo et al. [73] utilizes the primers 146F1 and 146R1 for amplification of a 1447 bp fragment in the first round and the primers 146F2 and 146R2 for amplification of an internal fragment of 941 bp in the second round. This assay has an overall sensitivity of 18–180 target molecules for nondegraded DNA and has been recommended by the OIE for routine detection of WSSV [74].

98.1.6 Treatment and Prevention

Currently, no specific treatment is available for WSSV infection, and control and prevention are dependent on implementing measures that reduce virus exposure and enhance host immune status. These include removal of diseased shrimp, filtering intake water into the pond (with 250 μm or smaller mesh filters), screening stocked animals for the presence of the virus, use of nonspecific immune stimulants and fortified mineral and vitamin diets to increase stress tolerance, and increasing acclimation times before stocking.

Vaccine strategies that have been explored to protect shrimp against WSSV include inactivated WSSV vaccines; recombinant protein vaccines expressing VP19 and/or VP28; DNA vaccines containing VP15, VP28, VP35, and/or VP281 genes; and dsRNA vaccines with specificity for VP26, VP28, and VP281. As envelope structural proteins are crucial for viral entry, assembly, and budding processes, they offer useful targets for WSSV vaccine development. In particular, VP28-based vaccines confer protection or increase survival rate of shrimp against WSSV challenge. Given the lack of diversified molecules or memory cells in invertebrates, the current vaccine preparations generally provide short-term protection against WSSV. Further research is needed to extend the efficacy of the vaccines, including the use of *specific immune priming* protocol [75].

98.2 METHODS

98.2.1 Sample Preparation

Viruses in the water samples are concentrated by microfiltration (MF) followed by ultrafiltration (UF). Water samples are prefiltered using 5 μm cartridge filter to remove particulate matter and then subjected to MF using TFF system (QuixStand™ Benchtop system, GE Healthcare Bio-Sciences Corp) fitted with a 0.2 μm pore size TFF cartridge (CFP-2-E-4MA with a surface area of 420 cm² or CFP-2-E-9A with a surface area of 8400 cm²) to remove bacteria and particulate matter. The permeate is then subjected to UF for concentration of viruses with a 100 kDa MWCO TFF cartridge using either a UFP-100-C-4MA with a surface area of 650 cm² or a UFP-100-C-9A with a surface area of 1.15 m². As a consequence, 20–100 L of water samples is concentrated to 200–1000 fold. After every use, the filter system is sanitized with 200 ppm of free chlorine and 0.5 M sodium hydroxide solution.

Gill tissues from infected shrimp are homogenized in TN buffer (0.2 M Tris–HCl, 0.4 M NaCl, pH 7.4), followed by centrifugation at 5000 × *g* for 20 min. The supernatant is filtered through a 0.45 μm filter and stored at −80°C for later use.

Total DNA is extracted from ~50 μg of gill tissue. The tissue is homogenized using a disposable rod in a 200 μL of 5% (w/v) Chelex X-100 resin (Bio-Rad) solution, 16 μL of 20 mg/ml proteinase K (Promega) in 50 mM Tris–HCl pH 8.0 is then added, and the mixture is incubated at 56°C for at least 6 h to overnight, followed by 95°C for 10 min to inactivate the proteinase K. The mixture is microcentrifuged at 13,000 rpm for 5 min and the supernatant containing DNA is stored at −20°C. Alternatively, total DNA may be extracted by using the Qiagen DNA purification kit.

98.2.2 Detection Procedures

98.2.2.1 Conventional PCR

Dieu et al. [36] described the use of the primers VP26F (5′-ATGGAATTTGGCAACCTAACAAACCTG-3′; nt 228, 835–228,809) and VP26R (5′- GGGCTGTGACGGTAGAGA TGAC-3′; nt 228,532–228,553) for specific amplification of a 304 bp fragment from the WSSV VP26 gene.

PCR mixture (50 μL) consists of 1 μL of gill DNA solution and Taq DNA polymerase from Promega. The cycling parameters include denaturation at 94°C for 3 min, 40 cycles at 94°C for 30 s, 52°C for 30 s, and 72°C for 50 s and an extension at 72°C for 7 min. The resulting product is separated on 1.5% agarose gel containing ethidium bromide and visualized under UV light.

Note: A nested PCR developed by Lo et al. [73] may be performed if a higher sensitivity is desired. While the first-round PCR primers, 146F1 (5′-ACT-ACT-AAC-TTC-AGC-CTA-TCTAG-3′) and 146R1 (5′-TAA-TGC-GGG-TGT-AAT-GTT-CTT-ACG-A-3′), amplify a 1447 bp product; the second-round PCR primers, 146F2 (5′-GTA-ACT-GCC-CCT-TCC-ATC-TCC-A-3′) and 146R2 (5′-TAC-GGC-AGC-TGC-TGC-ACC-TTG-T-3′), produce a 941 bp fragment. This assay is recommended by OIE for WSSV diagnosis [OIE 2012].

98.2.2.2 Real-Time PCR

Mendoza–Cano and Sánchez–Paz [69] reported a SYBR green-based real-time PCR (qPCR) for the detection and quantitation of WSSV, with primers (VP28-140Fw, 5′-AGG-TGT-GGA-ACA-ACA-CAT-CAA-G-3′; VP28-140Rv, 5′-TGC-CAA-CTT-CAT-CCT-CAT-CA-3′) targeting a 141 bp fragment from a highly conserved region of the VP28 gene. This 141 fragment corresponds to the segment of the protrusion emanating outside the viral envelope (VWNNTSRKINITG MQMVPKINPSKAFVGSSNTSSFTPVSIDEDEVG), which plays an essential role in the initial interaction with host receptors.

The qPCR mixture (14 μL) is composed of 1 μL of DNA, 7.5 μL of 2× iQ SYBR® Green Supermix (Bio-Rad), 0.5 μL of each primer (10 pmol), and 4.5 μL of nuclease-free water.

Amplification and detection are carried out in a Rotor-Gene 3000 real-time rotary thermal cycler (Corbett Life Science), with the following parameters: denaturation at 95°C for 5 min; 30 cycles at 94°C for 30 s, 61°C for 30 s, and 72°C for 30 s; and a final extension step at 72°C for 5 min.

Fluorescence measurements (510 nm) are taken at the end the elongation phase for each cycle. Melting curve analysis, to detect the occurrence of primer-dimer formation or the amplification of other nonspecific products, is performed immediately after amplification by slow heating of the samples from 60°C to 99°C with a 0.3°C/s ramping rate and stepwise fluorescence detection at 0.3°C interval. The melting curves are converted to melting peaks by plotting the negative derivatives of fluorescence against temperature ($-dF/dT$). All reactions are done in triplicate, and the quantification cycle (C_q) values are automatically calculated with the Rotor-Gene software version 6.1. The qPCR reaction efficiency and linearity are calculated from the standard curve.

Note: This one-step real-time PCR provides a rapid and sensitive tool for detection and quantitation of WSSV in both penaeid and nonpenaeid crustacean species (e.g., *P. stylirostris*, *P. vannamei*, *Farfantepenaeus paulensis*, *Farfantepenaeus brasiliensis*, *P. monodon*, *P. indicus*, *Lysmata californica*, *Palaemon ritteri*, *Calanus pacificus*, *Macrobrachium rosenbergii*, and *Petrolisthes edwardsii*).

98.3 CONCLUSION

WSSV is an enzootic organism with common occurrence in >98 wild and cultured aquatic animal species throughout the world. Transmitted both vertically and horizontally, WSSV causes a highly contagious, lethal disease known as WSD, most common in cultured shrimp. In the absence of therapeutic treatments and vaccines, control and prevention of WSD rely on good management practices including the use of highly sensitive and specific PCR for disease diagnosis and certification of WSSV-negative shrimp stocks. Further research is warranted in the area of WSSV pathogenesis in order to develop effective, tailor-made control strategies against WSD.

REFERENCES

1. Lightner DV. Virus diseases of farmed shrimp in the Western Hemisphere (the Americas): A review. *J Invertebr Pathol.* 2011;106:110–130.
2. Pradeep B, Rai P, Mohan SA, Shekhar MS, Karunasagar I. Biology, host range, pathogenesis and diagnosis of White spot syndrome virus. *Indian J Virol.* 2012;23(2):161–174.
3. Marks H, Goldbach RW, Vlak JM, van Hulten MCW. Genetic variation among isolates of white spot syndrome virus. *Arch Virol.* 2004;149:673–697.
4. Wongteerasupaya C et al. High variation in repetitive DNA fragment length for white spot syndrome virus (WSSV) isolates in Thailand. *Dis Aquat Org.* 2003;54:253–257.
5. Lan Y, Lu W, Xu X. Genomic instability of prawn white spot bacilliform virus (WSBV) and its association to virus virulence. *Virus Res.* 2002;90:269–274.
6. Marks H, van Duijse JJA, Zuidema D, van Hulten MCW, Vlak JM. Fitness and virulence of an ancestral white spot syndrome virus isolate from shrimp. *Virus Res.* 2005;110:9–20.
7. Pradeep B, Karunasagar I, Karunasagar I. Fitness and virulence of different strains of white spot syndrome virus. *J Fish Dis.* 2009;32:801–805.
8. Dieu BTM, Marks H, Zwart MP, Vlak JM. Evaluation of white spot syndrome virus variable DNA loci as molecular markers of virus spread at intermediate spatiotemporal scales. *J Gen Virol.* 2010;91:1164–1172.
9. Hoa TT et al. Mixed-genotype white spot syndrome virus infections of shrimp are inversely correlated with disease outbreaks in ponds. *J Gen Virol.* 2011;92:675–680.
10. Saravanan P et al. White spot syndrome virus (WSSV) genome stability maintained over six passages through three different penaeid shrimp species. *Dis Aquat Organ.* 2014;111(1):23–29.
11. Kiatpathomchai W et al. Target for standard Thai PCR assay identical in 12 white spot syndrome virus (WSSV) types that differ in DNA multiple repeat length. *J Virol Methods.* 2005;130:79–82.
12. Durán-Avelar MDJ et al. Genotyping WSSV isolates from northwestern Mexican shrimp farms affected by white spot disease outbreaks in 2010–2012. *Dis Aquat Org.* 2015;114:11–20.
13. Pradeep B, Shekar M, Gudkovs N, Karunasagar I, Karunasagar I. Genotyping of white spot syndrome virus prevalent in shrimp farms of India. *Dis Aquat Organ.* 2008;78:189–198.
14. John KR, George MR, Iyappan T, Thangarani AJ, Jeyaseelan MJP. Indian isolates of white spot syndrome virus exhibit variations in their pathogenicity and genomic tandem repeats. *J Fish Dis.* 2010;33:749–758.
15. Durand S, Lightner DV, Redman RM, Bonami JR. Ultrastructure and morphogenesis of white spot syndrome baculovirus (WSSV). *Dis Aquat Organ.* 1997;29:205–211.
16. Rodriguez J et al. White spot syndrome virus infection in cultured *Penaeus vannamei* (Boone) in Ecuador with emphasis on histopathology and ultrastructure. *J Fish Dis.* 2003;26:439–450.
17. Escobedo-Bonilla CM et al. A review on the morphology, molecular characterization, morphogenesis and pathogenesis of white spot syndrome virus. *J Fish Dis.* 2008;31:1–18.
18. Yang F et al. Complete genome sequence of the shrimp white spot bacilliform virus. *J Virol.* 2001;75:11811–11820.
19. Zwart MP, Dieu BTM, Hemerik L, Vlak JM. Evolutionary trajectory of white spot syndrome virus (WSSV) genome shrinkage during spread in Asia. *PLoS One.* 2010;5:13400–13411.
20. Chai CY, Yoon J, Lee YS, Kim YB, Choi TJ. Analysis of the complete nucleotide sequence of a white spot syndrome virus isolated from Pacific white shrimp. *J Microbiol.* 2013;51(5):695–699.
21. van Hulten MCW et al. The white spot syndrome virus DNA genome sequence. *Virology.* 2001;286:7–22.
22. Moon CH et al. Highly conserved sequences of three major virion proteins of a Korean isolate of white spot syndrome virus (WSSV). *Dis Aquat Org.* 2003;53:11–13.
23. Shekhar MS, Ravichandran P. Comparison of white spot syndrome virus structural gene sequences from India with those at GenBank. *Aquac Res.* 2007;38:321–324.
24. Musthaq SS, Sudhakaran R, Ahmed VP, Balasubramanian G, Hameed AS. Variability in the tandem repetitive DNA sequence of white spot syndrome virus (WSSV) genome and stability of VP28 gene to detect different isolates of WSSV from India. *Aquaculture.* 2006;256:34–41.
25. Leu JH et al. The unique stacked rings in the nucleocapsid of the white spot syndrome virus virion are formed by the major structural protein VP664, the largest viral structural protein ever found. *J Virol.* 2005;79(1):140–149.
26. Lotz JM, Soto MA. Model of white spot syndrome virus (WSSV) epidemics in *Litopenaeus vannamei. Dis Aquat Organ.* 2002;50:199–209.

27. Powell JW, Browdy CL, Burge EJ. Blue crabs *Callinectes sapidus* as potential biological reservoirs for white spot syndrome virus (WSSV). *Dis Aquat Organ.* 2015;113(2):163–167.

28. Yan DC, Dong SL, Huang J, Yu XM, Feng MY. White spot syndrome virus (WSSV) detected by PCR in rotifers and rotifer resting eggs from shrimp pond sediments. *Dis Aquat Organ.* 2004;59:69–73.

29. Vijayan KK et al. Polychaete worms—A vector for white spot syndrome virus (WSSV). *Dis Aquat Organ.* 2005;63:107–111.

30. Gopalakrishnan A, Rajkumar M, Sun J, Wang M, Kumar SK. Mud crab, *Scylla tranquebarica* (Decapoda: Portunidae), a new host for the white spot syndrome virus. *Aquat Res.* 2011;42:308–312.

31. Mendoza-Cano F et al. The endemic copepod *Calanus pacificus californicus* as a potential vector of white spot syndrome virus. *J Aquat Anim Health.* 2014;26(2):113–117.

32. Tuyen NX, Verreth J, Vlak JM, de Jong MC. Horizontal transmission dynamics of White spot syndrome virus by cohabitation trials in juvenile *Penaeus monodon* and *P. vannamei*. *Prev Vet Med.* 2014;117(1):286–294.

33. Sanchez-Paz A. White spot syndrome virus: An overview on an emergent concern. *Vet Res.* 2010;41:3.

34. Shekar M, Pradeep B, Karunasagar I. White spot syndrome virus: Genotypes, epidemiology and evolutionary studies. *Indian J Virol.* 2012;23(2):175–183.

35. Peng SE et al. Performance of WSSV-infected and WSSV-negative *Penaeus monodon* postlarvae in culture ponds. *Dis Aquat Organ.* 2001;46:165–172.

36. Dieu BTM et al. Molecular epidemiology of white spot syndrome virus within Vietnam. *J Gen Virol.* 2004;85:3607–3618.

37. Tan Y et al. Molecular detection of three shrimp viruses and genetic variation of white spot syndrome virus in Hainan Province, China, in 2007. *J Fish Dis.* 2009;32:777–784.

38. Otta SK et al. Association of dual viral infection with mortality of Pacific white shrimp (*Litopenaeus vannamei*) in culture ponds in India. *Virus Dis.* 2014;25(1):63–68.

39. Galaviz-Silva L, Molina-Garza ZJ, Alcocer-Gonzalez JM, Rosales-Encinas JL, Ibarra-Gamez C. White spot syndrome virus genetic variants detected in Mexico by a new multiplex PCR method. *Aquaculture.* 2004;242:53–68.

40. Muller IC, Andrade TP, Tang-Nelson KF, Marques MR, Lightner DV. Genotyping of white spot syndrome virus (WSSV) geographical isolates from Brazil and comparison to other isolates from the Americas. *Dis Aquat Organ.* 2010;88:91–98.

41. Ramos-Paredes J, Grijalva-Chon JM, de la Rosa-Vélez J, Enríquez-Paredes LM. New genetic recombination in hypervariable regions of the white spot syndrome virus isolated from *Litopenaeus vannamei* (Boone) in northwest Mexico. *Aqua Res.* 2011;42:1–10.

42. Paz SA. White spot syndrome virus: An overview on an emergent concern. *Vet Res.* 2010;41:43.

43. Shekhar MS, Ponniah AG. Recent insights into host-pathogen interaction in white spot syndrome virus infected penaeid shrimp. *J Fish Dis.* 2015;38:599–612.

44. Inouye K et al. Mass mortality of cultured kuruma shrimp *Penaeus japonicus* in Japan in 1993: Electron microscopic evidence of the causative virus. *Fish Pathol.* 1994;29:149–158.

45. Wang CH et al. Ultrastructure of White spot syndrome virus development in primary lymphoid organ cell cultures. *Dis Aquat Organ.* 2000;41:91–104.

46. Jiravanichpaisal P, Soderha K, Soderha I. Characterization of white spot syndrome virus replication in in vitro-cultured haematopoietic stem cells of freshwater crayfish, *Pacifastacus leniusculus*. *J Gen Virol.* 2006;87:847–854.

47. Deepika A, Makesh M, Rajendran KV. Development of primary cell cultures from mud crab, *Scylla serrata*, and their potential as an in vitro model for the replication of white spot syndrome virus. *In Vitro Cell Dev Biol Anim.* 2014;50(5):406–416.

48. Wu J, Li F, Huang J, Xu L, Yang F. Crayfish hematopoietic tissue cells but not hemocytes are permissive for white spot syndrome virus replication. *Fish Shellfish Immunol.* 2015;43(1):67–74.

49. Chaivisuthangkura P, Longyant S, Sithigorngul P. Immunological-based assays for specific detection of shrimp viruses. *World J Virol.* 2014;3(1):1–10.

50. Li W, Tang X, Xing J, Sheng X, Zhan W. Proteomic analysis of differentially expressed proteins in *Fenneropenaeus chinensis* hemocytes upon white spot syndrome virus infection. *PLoS One.* 2014;9(2):e89962.

51. Khadijah S et al. Identification of white spot syndrome virus latency-related genes in specific-pathogen-free shrimps by use of a microarray. *J Virol.* 2003;77:10162–10167.

52. Kono T, Savan R, Sakai M, Itami T. Detection of white spot syndrome virus in shrimp by loop-mediated isothermal amplification. *J Virol Methods.* 2004;115:59–65.

53. Dutta S et al. Identification of RAPD-SCAR marker linked to white spot syndrome virus resistance in populations of giant black tiger shrimp, *Penaeus monodon* Fabricius. *J Fish Dis.* 2014;37(5):471–480.

54. Kim HJ, Kwon SR. Development of a rapid method for identifying carryover contamination of positive control DNA, using a chimeric positive control and restriction enzyme for the diagnosis of white spot syndrome virus by nested PCR. *Indian J Microbiol.* 2014;54(4):439–443.

55. Kim CK et al. Development of a polymerase chain reaction (PCR) procedure for the detection of baculovirus associated with white spot syndrome (WSBV) in penaeid shrimp. *J Fish Dis.* 1998;21:11–17.

56. Otta SK et al. Polymerase chain reaction (PCR) detection of white spot syndrome virus (WSSV) in cultured and wild crustaceans in India. *Dis Aquat Organ.* 1999;38:67–70.

57. Tang KFJ, Lightner DV. Quantification of white spot syndrome virus DNA through a competitive polymerase chain reaction. *Aquaculture.* 2000;189:11–21.

58. Dhar AK, Roux MM, Klimpel KR. Detection and quantification of infectious hypodermal and hematopoietic necrosis virus and white spot virus in shrimp using real-time quantitative PCR and SYBR green chemistry. *J Clin Microbiol.* 2001;39:2835–2845.

59. Tsai JM, Shiau LJ, Lee HH, Chan PW, Lin CY. Simultaneous detection of white spot syndrome virus (WSSV) and Taura syndrome virus (TSV) by multiplex reverse transcription-polymerase chain reaction (RT-PCR) in pacific white shrimp *Penaeus vannamei*. *Dis Aquat Organ.* 2002;50:9–12.

60. Sritunyalucksana K, Srisala J, Mccoll K, Nielsen L, Flegel TW. Comparison of PCR testing methods for white spot syndrome virus (WSSV) infections in penaeid shrimp. *Aquaculture.* 2006;255:95–104.

61. Khawsak P, Deesukon W, Chaivisuthangkura P, Sukhumsirichart W. Multiplex RT-PCR assay for simultaneous detection of six viruses of penaeid shrimp. *Mol Cell Probes.* 2008;22:177–183.

62. Natividad KDT, Nomura N, Matsumura M. Detection of white spot syndrome virus DNA in pond soil using a 2-step nested PCR. *J Virol Methods.* 2008;149:28–34.

63. Xie Z et al. Development of a real-time multiplex PCR assay for detection of viral pathogens of penaeid shrimp. *Arch Virol.* 2008;153:2245–251.

64. Chou PH, Lin YC, Teng PH, Chen CL, Lee PY. Real-time target-specific detection of loop-mediated isothermal amplification for white spot syndrome virus using fluorescence energy transfer-based probes. *J Virol Methods*. 2011;173:67–74.

65. Jang IK, Kim JS, Kim BR, Meng XH. Comparison of white spot syndrome virus quantification of fleshy shrimp *Fenneropenaeus chinensis* in outdoor ponds between different growing seasons by TaqMan real-time polymerase chain reaction. *Aquac Res*. 2011;42:1869–1877.

66. Nunan LM, Lightner DV. Optimized PCR assay for detection of white spot syndrome virus (WSSV). *J Virol Methods*. 2011;171:318–321.

67. Panichareon B, Khawsak P, Deesukon W, Sukhumsirichart W. Multiplex real-time PCR and high-resolution melting analysis for detection of white spot syndrome virus, yellow-head virus, and *Penaeus monodon densovirus* in penaeid shrimp. *J Virol Methods*. 2011;178(1–2):16–21.

68. Liu T et al. Detection and quantification of hepatopancreatic parvovirus in penaeid shrimp by real-time PCR assay. *J Invertebr Pathol*. 2013;114(3):309–311.

69. Mendoza-Cano F, Sánchez-Paz A. Development and validation of a quantitative real-time polymerase chain assay for universal detection of the White Spot Syndrome Virus in marine crustaceans. *Virol J*. 2013;10:186.

70. Park JY et al. Development of a highly sensitive single-tube nested PCR protocol directed toward the sequence of virion envelope proteins for detection of white spot syndrome virus infection: Improvement of PCR methods for detection of WSSV. *Aquaculture*. 2013; 410–411:225–229.

71. Leal CA, Carvalho AF, Leite RC, Figueiredo HC. Development of duplex real-time PCR for the detection of WSSV and PstDV1 in cultivated shrimp. *BMC Vet Res*. 2014;10:150.

72. Ramesh Kumar D et al. Development of SYBR Green based real time PCR assay for detection of monodon baculovirus in *Penaeus monodon*. *J Virol Methods*. 2014;205C:81–86.

73. Lo CF et al. Detection of baculovirus associated with white spot syndrome (WSBV) in penaeid shrimps using polymerase chain reaction. *Dis Aquat Organ*. 1996;25:133–141.

74. OIE (Office International des Epizooties). *OIE Manual of Diagnostic Tests for Aquatic Animals 2012*. Chapter 2.2.6 White spot disease. http://www.oie.int/fileadmin/Home/eng/Health_standards/aahm/current/2.2.06_WSD.pdf. Accessed on June 12, 2015.

75. Syed Musthaq SK, Kwang J. Evolution of specific immunity in shrimp—A vaccination perspective against white spot syndrome virus. *Dev Comp Immunol*. 2014;46(2):279–290.

Section V

Prions

99 Bovine Spongiform Encephalopathy

Akikazu Sakudo and Takashi Onodera

CONTENTS

99.1 Introduction 901
   99.1.1 Classification 901
   99.1.2 Morphology and Biology/Epidemiology 901
      99.1.2.1 Morphology 901
      99.1.2.2 Biology 902
      99.1.2.3 Epidemiology 902
      99.1.2.4 Atypical BSE 903
   99.1.3 Clinical Features and Pathogenesis 904
   99.1.4 Diagnosis 905
99.2 Methods 908
   99.2.1 Reagents and Equipment 908
      99.2.1.1 Reagents 908
      99.2.1.2 Equipment 908
   99.2.2 Sample Preparation and Detection Procedures 909
   99.2.3 Inactivation Methods 909
99.3 Conclusion 909
Acknowledgments 910
References 910

99.1 INTRODUCTION

Prion diseases, or transmissible spongiform encephalopathies (TSEs), are fatal neurological diseases that include Creutzfeldt–Jakob disease (CJD) in humans, scrapie in sheep and goats, bovine spongiform encephalopathy (BSE) in cattle, chronic wasting disease (CWD) in cervids, transmissible mink encephalopathy in minks, feline spongiform encephalopathy in cats, and exotic ungulate encephalopathy in zoo animals such as the kudu, nyala, gemsbok, eland, and oryx. The prion agent of each disease is named after the disease itself (e.g., CJD agent in CJD, scrapie agent in scrapie). A key event in the development of prion diseases is the conversion of the cellular, host-encoded prion protein (PrP^C) to its abnormal isoform (PrP^{Sc}) predominantly in the central nervous system (CNS) of the infected host [1]. This chapter will introduce the molecular detection of BSE as well as the fundamental aspects of the disease.

99.1.1 CLASSIFICATION

To understand the causative pathogens, information on their chemical composition will provide important clues. Reports published in 1980 indicated that the scrapie agent was inactivated by proteases [2,3] or by treatment with modified or denatured proteins [4,5]. These studies demonstrated that the scrapie agent required a protein but did little to discriminate it structurally from other infectious particles [6]. The knowledge that a protein was involved undoubtedly motivated scientists to improve both the method for purifying the scrapie

agent and techniques for identifying agent-specific proteins. The discovery of PrP^{Sc} and PrP^C initiated a period of remarkable progress in this field by providing physical and diagnostic markers for prion diseases [7]. Numerous compelling observations made over the last three decades support the notion that PrP^{Sc} is the main component of prion agents [8].

99.1.2 MORPHOLOGY AND BIOLOGY/EPIDEMIOLOGY

99.1.2.1 Morphology

The discovery of scrapie-associated fibrils (SAFs) [9] provided the first ultrastructural observation that achieved some degree of consensus as a prion-specific agent. SAFs were verified by several laboratories and found to be similar, though not identical, to prion rods, which were identified shortly afterward [10,11]. A consensus on the issue of SAF/prion rod identity was not achieved because the theoretical perspective of the respective research groups significantly shaped their interpretation of the observations. Some research groups viewed SAFs as filamentous scrapie viruses or later as components of a tubulofilamentous nemavirus structure [12], whereas others viewed SAFs as part of the pathology. Another research group considered SAFs and prion rods to be either related forms of the same phenomenon (i.e., aggregate of PrP^{Sc} [13]) or unrelated structures [14]. However, in both cases, these structures were considered unnecessary for prion biological activity. This is because neither rods nor SAFs are present after sonication or in liposome preparations although high protein titers and PrP^{Sc}

901

remained [15]. In contrast, reports identifying specific virus particles or nucleoprotein complexes [16] have not been reinforced by other reports in terms of a plausible prion agent.

99.1.2.2 Biology

Three biological criteria are used to define strains of prion agents: incubation time, distribution of vacuoles throughout the brain, and distribution of amyloid or PrP^{Sc}. As the existence of more than one prion agent phenotype in a single genetically identical host can be accepted, there should be little doubt that different prion isolates exhibit distinct phenotypes.

Structural differences in PrP^{Sc} conformations can explain the different strains. Specific examples to support this hypothesis have been reported in the literature [17]. Critics of the protein hypothesis cite the existence of at least 20 distinct strains and argue that different conformations of PrP^{Sc} cannot account for all of these variations. These statements could be counteracted by the following two arguments. First, most of the strains cited have not been characterized in a single inbred host, and thus, the actual number of unique strains remains unknown. PrP^{Sc} is actually a population of molecules having at least 400 different glycosylation variants. Therefore, combinations of protein conformation and different patterns of glycosylation could (at least numerically) account for all the strains known at this time.

99.1.2.3 Epidemiology

The BSE agent is believed to have originated from the scrapie agent. Scrapie was considered a disease of sheep that did not infect humans, although the tissues of infected animals were known to contain an infectious agent. As there were no other known natural TSEs, when BSE first appeared, it was immediately considered to be derived from scrapie. Meat originating from sheep had been fed to cattle as part of the food they consumed to increase milk yield. Under the direction of the British Ministry of Agriculture, Fishery, and Foods (MAFF), a change occurred in the way that this meal was prepared in the United Kingdom in the early 1980s. This change in policy coincided with the original infection in cattle.

A small farm in Surrey in the United Kingdom reported the development of a strange neurological disease in multiple cows. The cattle were immediately killed and their brains removed for examination. When it was found that the cattle had developed a previously unrecognized disease, the farmer wanted to publish the data, but he was advised against this by MAFF. It is currently estimated that approximately 100 cattle may have developed BSE symptoms prior to 1987.

In 1987, Wells et al. published data illustrating that a cow developed a spongiform encephalopathy [18], although few extra details were provided. MAFF soon realized that this was no straightforward disease and devised a committee to advise them on what action should be taken to avoid any risk to humans and cattle.

In April 1988, the UK government established the Southwood Committee to investigate the outbreak of BSE. In the same year, the committee published a report stating that there would be minimal risk to humans, because all infected cattle would be slaughtered. The response to the BSE crisis was to prevent all bovine material from entering the meal fed to cattle. This strategy was implemented in July 1988. The feed manufacturers were warned several months in advance of the change in *policy*. The reporting of cases of BSE to MAFF was made obligatory, and farmers were compensated with half the value of a nonsick animal.

Since its first appearance in 1986, BSE has been registered at a clinical level in more than 180,000 bovine cattle in the United Kingdom (Figure 99.1), and it is believed that as many as 1 million undiagnosed infected cattle would have entered the human food chain [19]. This transmissible prion disease was transmitted from bovine tissue through the consumption of contaminated meat-and-bone meal (MBM) [20]. MBM is used as animal feed in various proportions throughout most of the developed world. However, only a few European countries described cases of BSE originating from within their borders following the initial outbreak in the United Kingdom. Epidemiological analysis revealed many years later that almost all European countries were affected by BSE (Table 99.1). Between November 1986 and October 2014, 184,624 cases of BSE were confirmed in the United Kingdom.

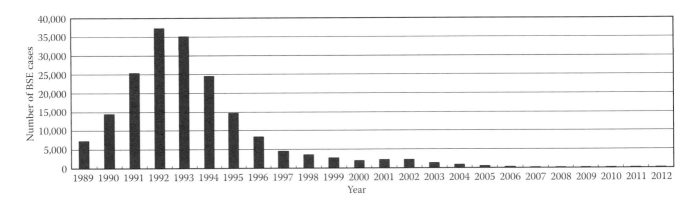

FIGURE 99.1   The annual number of cattle with bovine spongiform encephalopathy (BSE) worldwide. The number of cattle with BSE peaked at 1992 and decreased thereafter. (Cited from Fig. 2 in the report on Food Safety Commission of Japan, *Food Saf.*, 2, 55, 2014. With permission from the Food Safety Commission of Japan.)

TABLE 99.1

Number of Cases of Bovine Spongiform Encephalopathy Worldwide (as of October 12, 2014)

	1992	...	2001	2002	2003	2004	2005	2006	2007	2008	2009	2010	2011	2012	2013	2014[*1]	SUM
Total	37,316	...	2215	2179	1389	878	561	329	179	125	70	45	29	21	7	2	190,652
Europe (excluding the United Kingdom, France, and the Netherlands)	36	...	716	769	616	469	293	189	95	74	46	26	17	15	2	1[*4]	4,855
United Kingdom	37,280	...	1202	1144	611	343	225	114	67	37	12	11	7	3	3	0	184,624
France	0	...	274	239	137	54	31	8	9	8	10	5	3	1	2	—	1,021
Netherlands	0	...	20	24	19	6	3	2	2	1	0	2	1	0	0	0	88
United States	0	...	0	0	0	0	1	1	0	0	0	0	0	1	0	—	3
Canada	0	...	0	0	2[*2]	1	1	5	3	4	1	1	1	0	0	—	20[*3]
Japan	0	...	3	2	4	5	7	10	3	1	1	0	0	0	0	—	36
Israel	0	...	0	1	0	0	0	0	0	0	0	0	0	0	0	—	1
Brazil	0	...	0	0	0	0	0	0	0	0	0	0	0	1	0	1	2

Source: Modified and updated from Fig. 2 in a Food Safety Commission of Japan, *Food Saf.*, 2, 55, 2014. With permission from the Food Safety Commission of Japan.

This table is based on information on the OIE website as of October 12, 2014.

[*1] The data for the United Kingdom (as of October 2, 2014) and the other countries were those reported in 2013 or 2014.

[*2] One case was detected in the United States.

[*3] The total number of cattle tested in Canada included one imported cattle and the first reported case in the United States (December 2003).

[*4] One animal with BSE was found in Romania.

Since 1989, a relatively small number of cases of BSE (6028 in total) have also been reported in native cattle in Austria, Belgium, Brazil, Czech Republic, Denmark, Finland, France, Germany, Greece, Ireland, Israel, Italy, Japan, Liechtenstein, Luxembourg, the Netherlands, Poland, Portugal, Slovakia, Spain, and Switzerland. Since the introduction of monitoring programs to detect BSE in dead and slaughtered cattle, 12 countries have reported their first native case (Austria, Czech Republic, Finland, Germany, Greece, Israel, Italy, Japan, Poland, Slovakia, Slovenia, Spain). A small number of cases have also been reported in Canada, the Falkland Islands (Islas Malvinas), and Oman, but these cases occurred solely in animals imported from the United Kingdom. In the absence of large-scale studies, whether many non-European countries are actually BSE-free remains unknown [21].

In 1996, the British government announced several cases of a new form of CJD in young patients [22] called variant Creutzfeldt–Jakob disease (vCJD) that was caused by a BSE agent entering the body through the oral route. As of June 2014, 229 cases of vCJD have been confirmed (including 177 in the United Kingdom). Many parameters, including the infective dose, susceptibility, and duration of the incubation period, remain unknown. Mathematical models have been developed to analyze the likely spread of the disease, but these remain imprecise. The initial models estimated that the number of cases of vCJD ranged between 100 and more than 100,000. However, more recent models suggest that a total of a few hundred to a few thousand cases of vCJD occurred in the United Kingdom due to the direct contamination of humans through the consumption of beef products [23]. Moreover, the

risk of secondary transmission between humans must be considered. Proteinase K (PK)-resistant prion protein (PrPres), the disease-specific biochemical marker that is believed to be the infectious agent [24], has been found in peripheral lymphoid organs (e.g., tonsils, appendix) of patients with vCJD [25]. The theoretical risk of iatrogenic transmission, *via* blood transfusion, organ transplantation, or contaminated surgical tools, could be higher than that for sporadic CJD [26].

To prevent new cases of vCJD, the first measure has been to stop the consumption of the potent infectious organs of cattle (the CNS and some immune organs) [27], which are termed specified risk materials and which correspond to 99% of the total infectivity. The clinical signs of BSE appear late in the course of the disease, and the clinical diagnosis is complicated by the absence of pathognomonic signatures. Consequently, an infected animal could still enter the human food chain and present a risk to human health. For example, the spinal cord could be incompletely removed, and dorsal root ganglia may be retained in the vertebrae. Moreover, improvements in slaughtering practices did not prevent the contamination of healthy materials by infected tissue. Therefore, the development of sensitive detection and efficient inactivation methods for prion is important.

99.1.2.4 Atypical BSE

The large-scale testing of livestock nervous tissues for the presence of PrPSc has led to the recognition of two molecular signatures that are distinct from BSE. These were termed H-BSE and L-BSE (or BASE) (Table 99.2). The L-type is also referred to as "bovine amyloidotic spongiform encephalopahy" or BASE.

TABLE 99.2
Number of Cases of Atypical Bovine Spongiform
Encephalopathy Worldwide (as of July 2014)

Country	H-BSE	L-BSE	Other	Total Number
Australia	1	2	0	3
Canada	1	1	0	2
Denmark	0	1	0	1
France	15	14	0	29
Germany	2	2	0	4
Ireland	4	0	0	4
Italy	0	5	0	5
Japan	0	2	0	2
Poland	2	13	0	15
Portugal	1	0	0	1
Romania	1	0	0	1
Spain	2	2	0	4
Sweden	1	0	0	1
Switzerland	2	0	2	4
Netherlands	1	3	0	4
United Kingdom	5	4	0	9
United States	2	1	0	3
Total	40	50	2	92

Source: Modified from Table 15 in a Food Safety Commission of Japan, *Food Saf.*, 2, 55, 2014. With permission from the Food Safety Commission of Japan.

Summarized by the Department of Epidemiology of Animal Health Institute of Italy (the 71st Prion Expert Committee Material 3 [May 29, 2012] was partially revised).

Updated from the report of OIE (World Organisation for Animal Health).

Their PrPSc molecular signature differed from that of the classical BSE in terms of the protease-resistant fragment size and glycopattern [28]. The experimental transmission of these cases to different lines of bovine prion protein (PrP) transgenic mice unambiguously demonstrated their infectious nature and confirmed their unique but distinctive strain type as compared to the classical BSE [29,30]. These uncommon cases have been detected in aged asymptomatic cattle during systematic testing at slaughterhouses. In France, a retrospective study of all TSE-positive cattle identified through the compulsory EU surveillance program (2001–2007) was conducted [31]. This study revealed that all cases of H-BSE and L-BSE detected by rapid tests were observed in animals over 8 years of age in either the fallen stock surveillance stream or the abattoir. The frequencies of H-BSE and L-BSE are 0.35 and 0.41 cases per million adult cattle tested, respectively, but it increases to 1.9 and 1.7 cases per million, respectively, in animals more than 8 years of age.

The origin of these atypical BSE cases in cattle is currently unknown, as is the performance of the current active surveillance system for detecting H-BSE- and L-BSE-affected animals. Consequently, there is uncertainty concerning the true prevalence of these conditions within the stock of cattle. All atypical cases of BSE identified in the EU occurred in animals born before the extended or real feed ban that was passed into

law in January 2001. Hence, as observed with classical BSE, exposure of these animals to contaminated feed with low titers of TSE agents cannot be excluded. Equally, additional origins for these TSE forms cannot be ruled out at this stage. In particular, the unusually old age of all animals affected by H-BSE and L-BSE and their apparent low prevalence in the population could suggest that these atypical BSE forms arise spontaneously.

There are some data related to the peripheral distribution of the L-BSE agent in cattle experimentally challenged *via* the intracerebral route. In a Japanese study, infectivity and PrPres were detected in many peripheral nerves tested from midincubation onward. PrPres was not detected in lymphoid tissue [32]. In an Italian study, PrPres was not detected in peripheral nerves [33], but the presence of infectivity in skeletal muscle, presumably linked to nervous structures, of natural cases was described [34].

Both L-BSE and H-BSE agents can propagate in experimentally challenged foreign species such as mice, sheep, voles, primates, and hamsters and in transgenic mice expressing heterologous (i.e., nonbovine PrP) sequences. It is notable that the transmission barrier observed for the L-BSE agent was lower than that for classical BSE. The transmission of an H-BSE isolate originating from France and Poland to bovine PrP transgenic mice has recently been reported [35].

99.1.3 CLINICAL FEATURES AND PATHOGENESIS

BSE, which is also called "mad cow disease," causes progressive and invariably fetal neurodegeneration in cattle. The clinical signs of BSE include tremors; gait abnormalities, particularly of the hind limbs (ataxia); aggressive behavior; apprehension; and hyperreactivity to stimuli [36]. Weight loss and decreased milk production are occasionally observed. The incubation period for BSE is 2–8 years, and most cases of BSE were found in 4–5-year-old dairy cattle [37].

Early studies during the 1980s indicated that the BSE agent can only be found in the brain, spinal cord, and retinal tissue of BSE-diseased cattle [38]. An infectivity assessment in several tissues from orally inoculated cattle, using a bioassay based on RIII mice [39], revealed the presence of BSE infectivity in the CNS, optic nerve, retina, cervical, thoracic, trigeminal ganglia, and facial and sciatic nerves, as well as in the distal ileum. The skeletal muscle, spleen, and other lymphatic tissues were demonstrated to be free from detectable infectivity. More recently, infectivity was discovered on the tonsils of cattle killed 10 months after the oral BSE challenge by intracerebral inoculation in cattle [40]. These results consistently indicate that BSE infectivity after oral uptake only propagates poorly in some intestinal lymphatic tissues (mainly Peyer's patches) and from there spreads to the CNS, probably by an intraneural route *via* the peripheral nervous system [41].

These findings contrast to the manner in which the scrapie agent spreads in infected sheep, mice, and hamsters. In this case, the scrapie agent spreads *via* tissues such as the spleen, other lymphatic tissues, and muscles, even during the preclinical stage [42–45]. However, the BSE agent can be found in the lymphoreticular system and other peripheral tissues when transmitted to sheep or primates [46].

99.1.4 DIAGNOSIS

Several biochemical methods have been used to diagnose BSE. These include enzyme-linked immunosorbent assay (ELISA), western blotting, and immunohistochemistry (IHC) (Figures 99.2 and 99.3). As an index of prion infection, representative changes in the brain were detected in these assays. Using ELISA and western blotting, protease-resistant PrPs, which are accumulated in the brain and form deposits, were biochemically detected after treatment with PK. In the case of ELISA for BSE, a homogenate is usually prepared from the

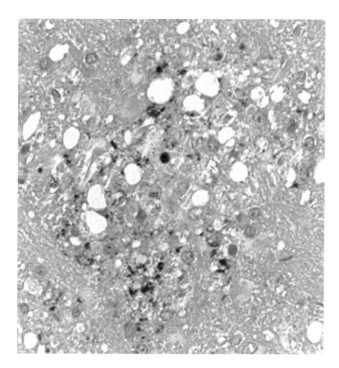

FIGURE 99.3 Immunohistochemistry (IHC) for the diagnosis of prion diseases. IHC is usually used for the definitive diagnosis of prion diseases. In IHC, vacuolation and prion protein (PrP) deposits in brain are usually used as indices of prion diseases. IHC of the brains of terminally prion-diseased mice infected with Obihiro-1 prions is shown. PrP was detected by an anti-PrP antibody (6H4). (Vacuolation and PrP deposits were observed courtesy of Dr. Yuji Inoue, Department of Virology, Center for Infectious Disease Control, Research Institute for Microbial Diseases, Osaka University, Osaka, Japan.)

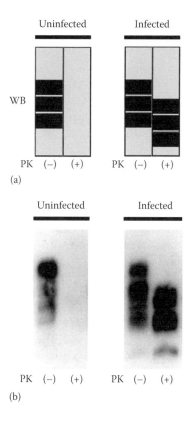

FIGURE 99.2 Western blotting for diagnosing prion diseases. Brain tissue is used for the definitive diagnosis of prion diseases. Therefore, the diagnosis is basically postmortem. Most methods for diagnosing prion diseases are based on the characteristics of the abnormal isoform of a prion protein (PrP^{Sc}), which is resistant to proteinase K (PK). PK completely degrades the cellular isoform of prion protein (PrP^C) but only partially digests PrP^{Sc} because the latter forms protease-resistant aggregates. After the treatment, western blotting with antiprion protein (PrP) antibody detects PK-resistant PrP (PrP^{res}) when PrP^{Sc} is included in the sample. As the N-terminal region of PrP^{Sc} is digested with PK (+), PK-resistant PrP displays a shift to a lower molecular weight compared to untreated PrP (−) in western blotting. (a) Schematic presentation and (b) the data of western blotting for PrP in the brains of prion-infected and uninfected mice before (−) and after (+) PK treatment. (Modified from Fig. 1 in the study by Inoue, Y. et al., *Jpn. J. Infect. Dis.*, 58, 78, 2005. After permission from Japanese Journal of Infectious Diseases.)

brain obex and treated with PK. The treated sample is applied to a microtiter plate for absorption and then reacted with an anti-PrP antibody. This method is commercially exploited in the Bio-Rad TeSE BSE Kit (Bio-Rad, France), Enfer TSE Kit (Abbott Laboratories, United States), and FRELISA BSE Kit (FUJIREBIO Inc., Japan) (Table 99.3). Although the extensively employed ELISA is a sensitive and high-throughput method, the large frequency of false-positives remains a problem. Therefore, if a result is positive, ELISA must be repeated to validate the finding. If a positive result is obtained again, western blotting and IHC are performed. Western blotting uses a membrane to absorb PK-treated proteins separated by sodium dodecyl sulfate (SDS) polyacrylamide gel electrophoresis (PAGE) (Figure 99.2). After the absorption, PK-resistant PrP (PrP^{res}) in the membrane-bound proteins is detected with an anti-PrP antibody. Importantly, western blotting provides information on both prion infections and the mobility of peptides, which is influenced by the host genotype and prion strains [47]. In the case of IHC, the indices for BSE are neuronal cell loss, astrocytosis, and vacuolation in addition to PrP accumulation (amyloid plaques) (Figure 99.3). In IHC-based analyses of brain sections, these changes are examined by light microscopy. Although vacuolation is also used as an index of prion infection, various combinations of prion strains with

TABLE 99.3
Representative BSE Kit

Kit[a]	Company	Methods
Beta Prion BSE EIA Test Kit	Roboscreen	Sandwich ELISA (colorimetric)
Bio-Rad TeSeE BSE	Bio-Rad	Sandwich ELISA (colorimetric)
Bovine Spongiform Encephalopathy Antigen Test Kit, IHC	VMRD	IHC (for BSE, formic acid + heat-treated brain specimen for the detection of accumulated PrP)
BSE PLUS TEST—a presymptomatic test for BSE	Altegen	Detection of antiprion antibody, which is produced at an early stage of prion infection
CediTect BSE	Cedi Diagnostics	Filter assay for the measurement of denatured/ undenatured PrP (chemiluminescence)
Enfer TSE Kit v2.0	Abbott Laboratories	Direct ELISA (automation, chemiluminescence)
Frelisa BSE	Fujirebio	Sandwich ELISA (colorimetric).
IDEXX HerdCheck BSE Antigen Test Kit	IDEXX Laboratories	Direct ELISA (no PK treatment, colorimetric)
Inpro CDI Kit	Beckman Coulter	CDI for the measurement of denatured/undenatured PrP (concentration by PTA precipitation)
NipplBL BSE Test Kit	Nippi	Sandwich ELISA (colorimetric)
PAD-Beads Seprion-PAD	Microsens Biotechnologies	Concentration of abnormal aggregated PrP including PrP^{Sc} (binding of aggregated protein by Seprion ligand)
Platelia BSE Detection Kit	CEA (Bio-Rad)	Sandwich ELISA (colorimetric)
Prion Protein EIA Kit	SPI Bio	Sandwich ELISA (no PK treatment, PrP detection, colorimetric)
Prionics-Check LIA	Prionics	Sandwich ELISA (chemiluminescence)
Prionics-Check PrioSTRIP	Prionics	Immunochromatography (for BSE)
Prionics-Check WB	Prionics	Western blotting (for BSE)
PrionScreen	Roche Applied Science	Sandwich ELISA (colorimetric)
Priontype postmortem	Labor Diagnostik Leipzig	Sandwich ELISA (no PK treatment, colorimetric)
Speed'it BSE	Institut Pourquier	Sandwich ELISA (chemiluminescence)

BSE, bovine spongiform encephalopathy; CDI, conformation dependent immunoassay; ELISA, enzyme-linked immunosorbent assay; IHC, immunohistochemistry; PK, proteinase K; PrP, prion protein; PrP^{Sc}, abnormal prion protein; PTA, phosphotungstic acid; TSE, transmissible spongiform encephalopathy.

[a] Modified and new versions of kits have been developed by each company.

host species result in the accumulation of PrP without vacuolation in brain sections [48]. Many new diagnostic methods for prion diseases have recently been developed (Table 99.4). Most of these methods use PrP^{res} as the index of prion infection. Among them, the protein misfolding cyclic amplification (PMCA) method is the most promising for a sensitive diagnosis [49]. In this method, the ability of PrP^{Sc} to convert PrP^{C} into PrP^{Sc} is used. This enables the *in vitro* amplification of PrP^{res} from a small quantity of PrP^{Sc} as seed *via* sequential cycles of incubation and sonication. The levels of PrP^{res} amplified by this method correlated with the prion infectivity titer [50]. Furthermore, PMCA could detect prions in blood [51]. Thus, PMCA may be used to diagnose both terminally diseased hamsters and prion-infected presymptomatic hamsters [52]. This method has the highest sensitivity when compared with any of the methods used for detecting PrP^{res} reported to date. Recently, PMCA has been modified, leading to recombinant PrP-PMCA [53] and its combination with the quaking-induced conversion reaction [54], as well as the real-time quaking-induced conversion for quantitative analysis [55].

In other diagnostic methods, capillary gel electrophoresis is combined with a competitive antibody-binding assay that uses a fluorescein-labeled synthetic PrP peptide and PrP [56]. The free peptide and antibody-peptide peaks are separated by capillary electrophoresis and used for diagnosis. In fluorescence correlation spectroscopy, single fluorescently labeled molecules in solution are detected [57]. Using this method, PrP^{Sc} can be labeled by the anti-PrP antibody or by conjugation with labeled recombinant PrP for diagnosis. In Fourier transform infrared spectroscopy, the difference of the infrared spectra of samples between prion-infected and uninfected animals is used for diagnosis [58]. A detection tool for PrP^{Sc} was recently developed. Some antibodies and aptamers can distinguish PrP^{Sc} from PrP^{C}. Antibodies having a conformational epitope of PrP^{Sc} were obtained by immunization with a Tyr-Tyr-Arg peptide [59]. The antibodies 15B3, V5B2, mAb132, 3B7, 2C4, 1B12, and 3H6 can also recognize PrP^{Sc}-specific epitopes [60]. PrP^{C}- and/or PrP^{Sc}-specific binding aptamers, which are RNA or DNA molecules that are specifically binding to a target protein, were reported [61]. PrP^{Sc}-specific antibodies and aptamers might both serve as a diagnostic or analytical tool for investigating prion diseases and facilitate the development of useful therapeutic agents, that is, if they are found to inhibit the accumulation of PrP^{Sc}.

There are many genes and proteins with differential abundances in prion diseases [62]. These changes can be used as markers of prion diseases. Examples of deregulated genes associated with prion diseases include adenosine triphosphate (ATP)-binding cassette, amyloid beta (A4) precursor-like protein 1, apolipoprotein D and E, beta-2-microglobulin, CD9 molecule, clusterin, cystatin C, cathepsin S, glial fibrillary acidic protein, osteonectin, bone sialoprotein I, and early T lymphocyte activation 1 [63]. Differentially expressed protein biomarkers for prion diseases include vitronectin, α1-acid glycoprotein, α1-antichymotrypsin, apolipoprotein B100, apolipoprotein E, cathelicidin antimicrobial peptide, clusterin, complement C3 and C4, complement factor

TABLE 99.4

Prion Protein Detection Methods

Method	Principle	Procedural Detail	References
Western blotting	PK-resistant PrP	Detect PK-resistant PrP on the membrane.	Inoue et al. [81]
ELISA	PK-resistant PrP	Detect PK-resistant PrP adsorbed on microtiter plates by anti-PrP antibody.	Grathwohl et al. [82]; Pan et al. [83]
IHC	PK-resistant PrP	Immunostain of tissue sections.	McBride et al. [84]
Bioassay	PK-resistant PrP, incubation time, or infectivity titer	Transmission to mice.	Prusiner et al. [85]
Cell culture assay	PK-resistant PrP or infectivity titer	Transmission to cells.	Klohn et al. [86]
Histoblot	PK-resistant PrP	Cryosection is blotted onto the membrane before the PK treatment and immunolabeling with anti-PrP antibody.	Taraboulos et al. [87]
Cell blot	PK-resistant PrP	Grow the cells on a cover slip, directly transferred to membrane, and detect the PK-resistant PrP using an anti-PrP antibody.	Bosque et al. [88]
Slot blot	PK-resistant PrP	Filter the cell lysate through nitrocellulose membrane, and detect the PK-resistant PrP using an anti-PrP antibody.	Winklhofer et al. [89]
PET blot	PK-resistant PrP	Paraffin-embedded tissue section is collected on membrane, and PK-resistant PrP is immunolabeled with an anti-PrP antibody.	Schulz-Schaeffer et al. [90]; Ritchie et al. [91]
PMCA	PK-resistant PrP	Amplification of misfolding protein by cycles of incubation and sonication.	Castilla et al. [92]
CDI	PrP conformation	Specific antibody binding to denatured and native forms of PrP.	Safar [93]
DELFIA	Insoluble PrP	Measure a percentage of the insoluble PrP extracted by two concentrations of guanidine hydrochloride.	Barnard et al. [94]
Capillary gel electrophoresis (competition assay)	PK-resistant PrP	Competition between fluorescein-labeled synthetic PrP peptide and PrP present in samples is assayed by separation of free and antibody-peptide peaks using capillary electrophoresis.	Schmerr et al. [95]; Jackman et al. [96]
Capillary gel electrophoresis (noncompetition assay)	Protein A–antibody–PrP complex	Complex with PrP is separated using a fluorescein-labeled protein A[a] and an anti-PrP antibody by capillary electrophoresis.	Yang et al. [97]
FCS	Aggregation of PrP	PrP aggregates were labeled by an anti-PrP antibody tagged with fluorescent dyes, resulting in an intensity fluorescent target, which were measured by dual-color fluorescence intensity distribution analysis.	Giese et al. [98]; Bieschke et al. [57]; Miller et al. [99]
Aptamer	PrP conformation	Use RNA aptamers that specifically recognize PrP^C and/or PrP^{Sc} conformation.	Bibby et al. [100]
FT-IR spectroscopy	Alterations of spectral feature	Analyze FT-IR spectra with statistical analysis.	Schmitt et al. [58]
MUFS	Alterations of spectral feature	Analyze spectra of emission excited by ultraviolet radiation.	Rubenstein et al. [101]
Flow microbead immunoassay	PK-resistant PrP	Detect PK-resistant PrP using flow cytometer with an anti-PrP antibody coupled with microbeads.	Murayama et al. [102]
Surrogate marker	Change of expression level of 14-3-3 protein, erythroid-specific marker, or plasminogen	Detect the change by two-dimensional gel electrophoresis, differential display reverse-transcriptase PCR, or western blotting of surrogate marker proteins for prion diseases.	Fischer et al. [73]; Kenney et al. [71]; Miele et al. [72]; Zerr et al. [103]
Mass spectrometry	PrP27-30 (residual PrP after PK treatment)	Detection of MRM signal by mass spectroscopic analysis.	Silva et al. [104]
Mechanical resonance	Frequency sift induced by PrP binding	Sandwich detection of PrP using anti-PrP antibody-bound mechanical resonator and nanoparticle.	Varshney et al. [105]

(Continued)

TABLE 99.4 (*Continued*)
Prion Protein Detection Methods

Method	Principle	Procedural Detail	References
Fluorescent analysis using quantum dot	Increased fluorescence signal of quantum dot induced by sandwich binding of PrP	Sandwich detection of PrP using anti-PrP antibody-bound magnetic microbeads and quantum dots.	Xiao et al. [106]
Filtration blotting	PK-resistant PrP	Detection of filtrated/adsorbed PrP^{res} onto PVDF membrane by anti-PrP antibody after treatment with PK + SDS/heating	Kobayashi et al. [107]
SOFIA	PK-resistant PrP	Fluorescent detection of PrP^{res} using biotinylated anti-PrP antibody and streptavidin labeled with rhodamine red X	Chang et al. [108]
Immuno PCR	PK-resistant PrP	Biotinylated anti-PrP antibody was bound to streptavidin and biotinylated DNA. Then, the DNA was amplified by PCR to detect PrP^{res}.	Barletta et al. [109]
Surface plasmon resonance	Reaction rate of binding DNA aptamer with PrP	Detection of PrP using DNA aptamer coated on the surface of electric conductive polypyrrole.	Miodek et al. [110]
Electrochemical biosensor	Electrochemical signal induced by binding DNA aptamer and PrP	Detection of PrP using a DNA aptamer-bound carbon nanotube.	Miodek et al. [111]

Source: Modified and updated from Table 1 in Sakudo, A. et al., *J. Vet. Med. Sci.*, 69, 329, 2007. With permission from The Japanese Society of Veterinary Science.

CDI, conformation dependent immunoassay; DELFIA, dissociation-enhanced lanthanide fluorescent immunoassay; ELISA, enzyme-linked immunosorbent assay; MRM, multiple reaction monitoring; PCR, polymerase chain reaction; PET, paraffin-embedded tissue; PK, proteinase K; PMCA, protein misfolding cyclic amplification; PrP, prion protein; PrP^C, cellular isoform of PrP; PrP^{res}, PK-resistant PrP; PrP^{Sc}, abnormal isoform of PrP; PVDF, polyvinylidene difluoride; SDS, sodium dodecyl sulfate; SOFIA, surround optical fiber immunoassay.

[a] Protein A binds an antibody.

B and H, haptoglobin α-2 chain, histidine-rich glycoprotein, Ig gamma-2 chain C region, transthyretin, uroguanylin, vitronectin precursor, and α2-macroglobulin [63–67]. Changes in response to prion infection have recently been employed as an index for the diagnosis of prion diseases. For example, the level of urinary α1-antichymotrypsin was drastically increased in the urine of patients with sporadic CJD and other animal models of prion diseases [68]. Moreover, increases of S100β, cystatin C, and heart fatty acid–binding protein levels in the blood of patients with CJD has been reported [67,69]. In particular, 14-3-3 [70], erythroid-specific marker [71], plasminogen [72], tau, phosphor-tau, S100β, and neuron-specific enolase are promising surrogate markers for prion diseases [62].

Taken together, prion diseases can usually be diagnosed by methods such as ELISA, western blotting, and immunohistochemical analysis [73]. Recently, there have been drastic improvements in sensitive diagnostic methods for prion diseases. Nonetheless, further developments of analytical methods to specifically and reliably detect prions are urgently needed.

99.2 METHODS

As an example of a diagnostic method for prion diseases, the materials and methods used for western blotting are presented. In particular, several precautions that should be monitored during the procedures are emphasized.

99.2.1 REAGENTS AND EQUIPMENT

99.2.1.1 Reagents
- Dithiothreitol (Promega, Madison, WI)
- *N*-tetradecyl-*N*,*N*-dimethyl-3-ammonio-1-propanesulfonate and sulfobetaine (Calbiochem or Sigma)
- *N*-lauroyl-sarcosinate (Sigma), protease inhibitors (Roche)
- RNase A (from bovine pancreas, protease-free) and DNase I (Calbiochem or Roche)
- Protein concentration assay kit (Pierce)

99.2.1.2 Equipment
To avoid cross contamination between different scrapie strains, we recommend the use of new glassware and instruments or single-use materials for animal inoculations, the dissection of brains from affected animals, and the purification of PrP^{Sc}. This is particularly important if the isolates are to be used in animal bioassays. Avoid blood contamination during the dissection of brains, rinse excised brains in phosphate-buffered saline (PBS), flash freeze the specimens in liquid nitrogen, and store the dissected brains at −70°C. All solutions must be filtered (0.2 μm) and placed in lint-free vessels previously rinsed with distilled or deionized water. All procedures should be performed in a laminar flow biosafety cabinet in an appropriate laboratory according to governmental regulations. Personal protection equipment (lab coat, gloves, and face mask) should be used, and gloves should be changed frequently.

99.2.2 Sample Preparation and Detection Procedures

PrP^Sc was prepared using a modified version of the procedure described by Sakudo et al. [74]. PrP^Sc from a TSE-affected brain tissue has been successfully purified in several species including humans with CJD and vCJD, sheep with scrapie, cattle with BSE, and cervids with CWD. Numerous experiments have used PrP^Sc isolates purified by this procedure, allowing detailed analyses of several different aspects of TSEs. Typically, the PrP^Sc yield from the hamster 263 K strain is 50–100 μg/g brain. The described method yields PrP^Sc preparations with a total protein content of between 60% and 90%. Non-PK-digested preparations are usually less pure than PK-digested preparations because nearly all other proteins are digested by PK. However, a significant protein contaminant identified in these preparations is ferritin, which survives even with PK treatment [75]. Variations in purity may depend on the TSE strain, the stage of the disease, the manner in which the brain is excised and stored, and the expediency of the purification. In addition to contamination by other proteins, the final PrP^Sc sample most likely contains trace quantities of lipids, nucleic acids, carbohydrates, and ions (e.g., copper, iron, calcium).

Tissues or cells are solubilized in radioimmunoprecipitation assay buffer containing 10 mM Tris–HCl (pH 7.4), 1% deoxycholate, 1% Nonidet P-40, 0.1% sodium dodecyl sulfate (SDS), and 150 mM NaCl and then sonicated at 4°C for 10 min. Then, the protein concentration of the supernatants is measured using a Bio-Rad DC protein assay (Bio-Rad, Hercules, CA). Afterward, the sample (120 μg of protein) is treated with PK at 20 μg/mL for 30 min at 37°C. An equal volume of 2× SDS gel-loading buffer (90 mM Tris–HCl pH 6.8, 10% mercaptoethanol, 2% SDS, 0.02% bromophenol blue, and 20% glycerol) is added and heated at 100°C for 10 min to terminate the reaction prior to western blotting. Samples treated as described previously excluding the digestion by PK should also be included as controls. For protein detection, proteins in the samples are separated by conventional SDS-polyacrylamide gel electrophoresis (PAGE) (15%) as described previously. Then, the proteins are transferred to polyvinylidene difluoride (PVDF) membranes (Amersham Biosciences, Piscataway, NJ) using a semidry blotting system (Bio-Rad). The membranes are blocked with 5% skim milk for 1 h at room temperature. The blocked membranes should then be further incubated for 1 h at room temperature with anti-PrP antibody, which recognizes the C-terminal half of PrP, in PBS-Tween (0.1% Tween 20) containing 0.5% skim milk.

Note: Anti-PrP antibodies recognizing the C-terminal half of PrP such as SAF83 and 6H4 (Prionics AG) should be used for the detection of PrP^Sc in this western blotting system, whereas those recognizing the N-terminal half of PrP such as SAF32 (SPI bio) cannot be used. This is because the C-terminal half of PrP^Sc is resistant to PK, whereas the N-terminal half is susceptible to PK.

After washing the membranes thrice for 10 min in PBS-Tween, they are incubated with horseradish peroxidase-labeled antimouse immunoglobulin secondary antibody in PBS-Tween containing 0.5% skim milk for 1 h at room temperature. The membranes are further developed with an enhanced chemiluminescence (ECL) reagent (Amersham) for 5 min. Blots are exposed to ECL Hypermax film (Amersham). Films are processed in an x-ray film processor.

The results obtained by western blotting using prion-infected and uninfected mouse brain are shown in Figure 99.2. Uninfected brain produces PrP^C but not PrP^Sc, whereas prion-infected brain produces both PrP^C and PrP^Sc. Therefore, PrP in uninfected mouse brain is digestible with PK, whereas that in prion-infected mouse brain is partially resistant to PK. It is obvious that western blotting of PrP using the anti-PrP antibody detects PrP in prion-infected brain but not in uninfected brain after digestion by PK. PK treatment causes a shift in the molecular weight of PrP from 33–35 kDa to 27–30 kDa in prion-infected brain, whereas PK treatment completely degrades PrP in uninfected brain.

99.2.3 Inactivation Methods

Prion agents are highly resistant against various treatments. Since prion has no associated nucleic acid [75], it cannot be inactivated by conventional sterilization procedures such as autoclaving (121°C, 20 min) or exposure to UV or γ-ray irradiation [76]. Treatment with alcohols, such as 70% ethanol, has no effect on prions. Inactivation of prions involves the use of an autoclave under severe conditions (134°C, 18 min), NaOH (1 N, 20°C, 1 h), SDS (30%, 100°C, 10 min), and NaOCl (20,000 ppm, 20°C, 1 min). Fixation using aldehyde slightly decreases the prion titer, but this is insufficient. Therefore, to treat tissue sections from prion-infected animals, formic acid is recommended. The following procedures for prion inactivation are currently recommended by the Society for Healthcare Epidemiology of America [77]: (1) washing with appropriate detergents + SDS treatment (3%, 3–5 min), (2) treating with alkaline detergents (80°C–93°C, 3–10 min) + autoclaving (134°C, 8–10 min), (3) washing with appropriate detergents + autoclaving (134°C, 18 min), and (4) washing with alkaline detergents (at a concentration and temperature according to instructions) + vaporized hydrogen peroxide gas plasma sterilization. Although these methods can be used for apparatuses or instruments, there is no applicable treatment to inactivate prions in food.

99.3 CONCLUSION

BSE was first diagnosed in the United Kingdom in 1986. Since that time, the disease has occurred in many European countries as well as Japan, Brazil, Canada, Israel, and the United States. Most of the reported cases of BSE (95%) have occurred in the United Kingdom.

Most cattle are infected when they orally ingest prion-contaminated "ruminant" MBM. This dietary supplement has been banned from feed since 1999. BSE is a fatal but typically slow developing disease. Infected cattle appear normal for 2–8 years. As the disease develops, the brain is affected. Symptoms such as trembling, stumbling, swaying, and behavioral changes (e.g., nervousness, aggression, or frenzy) are observed. Weight loss and a drop in milk production may be noted.

Humans who eat BSE-contaminated beef products could develop vCJD. The initial signs of vCJD include behavioral changes and abnormal sensations. As the disease progresses, incoordination and dementia develop, followed by coma and death. There is no cure for vCJD. Most people die within a year after the signs occur. Most cases have been observed in people who lived in the United Kingdom during the BSE outbreak in the late 1980s. Recent UK research using appendix biopsy reported 16 carriers of vCJD per 32,441 individuals [78]. This rate implies the presence of 30,000 vCJD carriers with no symptoms in the United Kingdom. Therefore, appropriate global countermeasures are urgently needed with careful monitoring of patient numbers.

Recent experimental evidence demonstrated that a low level of TSE infectivity can be present in blood or blood fractions derived from experimental rodents and sheep models. In addition, three cases of vCJD were observed after the transfusion of contaminated blood in the United Kingdom. Of primary interest is the potential for plasma and biotechnological processes to remove the human form of prion agents.

Prion agents are present in the blood at extremely low levels, thus making them difficult to use for analysis. Thus, most studies attempting to remove prion agents from blood or blood products have been performed using blood spiked with animal brain-derived prion agents as a model system. Moreover, owing to the biochemical similarities between animal and human prion agents, PrP^{Sc} from both species could be expected to partition in the same manner during fractionation. However, there is no evidence that blood PrP^{Sc} has the same or similar properties as brain PrP^{Sc} in addition to species differences between prion agents. Thus, applying sensitive assays for detecting PrP^{Sc} in blood is critical in identifying the plasma and biotechnology manufacturing processes capable of removing animal and human PrP^{Sc} in real systems. Recent developments in diagnostic methods such as PMCA have the potential to facilitate sensitive detection. In addition, further developments of inactivation methods for prion agents are also important in reducing the risks of secondary infection with prion diseases.

ACKNOWLEDGMENTS

This work was supported by Grants-in-Aids for Scientific Research from the Ministry of Education, Science, Culture and Technology of Japan (25450447, 24110717, 23780299, 22110514, 20780219, 17780228), Grants-in-Aid from the Research Committee of Prion Disease and Slow Virus Infection from the Ministry of Health, Labour and Welfare of Japan, and the Practical Research Project for Rare/Intractable Diseases from Japan Agency for Medical Research and Development, AMED.

REFERENCES

1. Aguzzi, A. et al. Mammalian prion biology: One century of evolving concepts. *Cell*, 116, 313, 2004.
2. Cho, H.J. Requirement of a protein component for scrapie infectivity. *Intervirology*, 14, 213, 1980.
3. Prusiner, S.B. et al. Scrapie agent contains a hydrophobic protein. *Proc. Natl. Acad. Sci. USA*, 78, 6675, 1981.
4. Prusiner, S.B. Novel proteinaceous infectious particles cause scrapie. *Science*, 216, 136, 1982.
5. McKinley, M.P. et al. Reversible chemical modification of the scrapie agent. *Science*, 214, 1259, 1981.
6. Chandler, R.L. Encephalopathy in mice produced by inoculation with scrapie brain material. *Lancet*, 1, 1378, 1961.
7. Zlotnik, I. et al. The pathology of the brain of mice inoculated with tissues from scrapie sheep. *J. Comp. Pathol.*, 72, 360, 1962.
8. Somerville, R.A. Do prions exist? PBS/NOVA, http://pbs.org/wgbh/nova/madcow/prions.html, 2001; accessed November 30, 2015.
9. Merz, P.A. et al. Infection-specific particles from the unconventional slow virus disease. *Science*, 225, 437, 1984.
10. Prusiner, S.B. et al. Scrapie prions aggregate to form amyloid-like birefringent rods. *Cell*, 35, 349, 1983.
11. Barry, R.A. et al. Antibodies to the scrapie protein decorate prion rods. *J. Immunol.*, 135, 603, 1985.
12. Narang, H.K. Evidence that single stranded-DNA wrapped around the tubulofilamentous particles termed "nomaviruses" is the genome of the scrapie agent. *Res. Virol.*, 149, 375, 1998.
13. Carp, R.I. et al. The nature of unconventional slow infection agents remains a puzzle. *Alzheimer Dis. Assos. Discord.*, 3, 79, 1989.
14. Prusiner, S.B. Molecular biology of prion disease. *Science*, 252, 1515, 1991.
15. Gabizon, R. et al. Purified prion proteins and scrapie infectivity copartition into liposomes. *Proc. Natl. Acad. Sci. USA*, 84, 4017, 1987.
16. Narang, H.K. Scrapie associated tubulofilamentous particles in scrapie hamsters. *Intervirology*, 34, 105, 1992.
17. Telling, G.C. et al. Evidence for the conformation of the pathologic isoform of prion protein enciphering and propagating prion diversity. *Science*, 274, 2079, 1996.
18. Wells, G.A.H. et al. A novel progressive spongiform encephalopathy in cattle. *Vet. Rec.*, 121, 419, 1987.
19. Kimberlin, R.H. An overview of bovine spongiform encephalopathy. *Dev. Biol. Stand.*, 75, 75, 1991.
20. Nathanson, N. et al. Bovine spongiform encephalopathy (BSE): Causes and consequence of a common source epidemic. *Am. J. Epidemiol.*, 145, 959, 1997.
21. Matthews, D. BSE: A global update. *J. Appl. Microbiol.*, 94(Suppl.), 120S, 2003.
22. Will, R.G. et al. A new variant of Creutzfeldt-Jakob diseases in UK. *Lancet*, 347, 921, 1996.
23. Wroe, S.J. et al. Clinical presentation and pre-mortem diagnosis of variant-Creutzfeldt-Jakob disease associated with blood transfusion: A case report. *Lancet*, 368, 2061, 2006
24. de Marco, M.F. et al. Large-scale immunohistochemical examination for lymphoreticular prion protein in tonsil specimens collected in Britain. *J. Pathol.*, 222, 380, 2010.
25. Ironside, J.W. et al. Retrospective study of prion-protein accumulation in tonsil and appendix tissues. *Lancet*, 355, 1693, 2000.
26. Shlomchik, M. et al. Neuroinvasion by a Creutzfeldt-Jakob disease agent in the absence of B cells and follicular dendritic cells. *Proc. Natl. Acad. Sci. USA*, 98, 9289, 2001.
27. Adkin, A. et al. Estimating the impact on the food chain of changing bovine spongiform encephalopathy (BSE) control measures: The BSE control model. *Prev. Vet. Med.*, 93, 170, 2010.
28. Bushmann, A. et al. Atypical BSE in Germany—Proof on transmissibility and biochemical characterization. *Vet. Microbiol.*, 117, 103, 2006.

29. Casalone, C. et al. Identification of a second bovine amyloidotic spongiform encephalopathy: Molecular similarities with sporadic Creutzfeldt-Jakob disease. *Proc. Natl. Acad. Sci. USA*, 101, 3065, 2004.

30. Capobianco, R. et al. Conversion of BASE prion strain into the BSE strain: The origin of BSE? *PLoS Pathog.*, 3, e31, 2007.

31. Biacabe, A.G. et al. Atypical bovine spongiform encephalopathies, France, 2001–2007. *Emerg. Infect. Dis.*, 14, 298, 2008.

32. Iwamaru, Y. et al. Accumulation of L-type bovine prions in peripheral nerve tissues. *Emerg. Infect. Dis.*, 16, 1151, 2010.

33. Lombardi, G. et al. Intraspecies transmission of BASE induces clinical dullness and amyotrophic changes. *PLoS Pathog.*, 4, 5, 2008.

34. Suardi, S. et al. Infectivity in skeletal muscle of BASE-infected cattle. *Proceeding Prion 2009 Conference*, Porto Carras, Chalkidiki, Greece, p. 47, 2009.

35. Espinosa, J.C. et al. Atypical H-type BSE infection in bovine-PrP transgenic mice let to the emergence of classical BSE strain features. *Prion*, 4, 137, 2010.

36. Imran, M. and Mahmood, S. An overview of animal prion diseases. *Virol. J.*, 8, 493, 2011.

37. Novakofski, J. et al. Prion biology relevant to bovine spongiform encephalopathy. *J. Anim. Sci.*, 83, 1455, 2005.

38. Wilesmith, J.W. et al. Bovine spongiform encephalopathy-epidemiological studies. *Vet. Rec.*, 123, 638, 1988.

39. Wells, G.A.H. et al. Preliminary observations on the pathogenesis of experimental bovine spongiform encephalopathy (BSE): An update. *Vet. Rec.*, 142, 103, 1998.

40. Welles, G.A.H. et al. Pathogenesis of experimental bovine spongiform encephalopathy: Preclinical infectivity in tonsil and observation on the distribution of lingual tonsil in slaughtered cattle. *Vet. Rec.*, 156, 401, 2005

41. Epinosa, J.C. et al. Progression of prion infectivity in asymptomatic cattle after oral bovine spongiform encephalopathy challenge. *J. Gen. Virol.*, 88, 1379, 2007.

42. Bosque, P.J. et al. Prions in skeletal muscle. *Proc. Natl. Acad. Sci. USA*, 99, 3812, 2002.

43. Heggebo, R. et al. Detection of PrP^{Sc} in lymphoid tissues of lambs experimentally exposed to the scrapie agent. *J. Comp. Pathol.*, 128, 172, 2003.

44. Thomzig, A. et al. Widespread PrP^{Sc} accumulation in muscles of hamsters orally infected with scrapie. *EMBO Rep.*, 4, 530, 2003.

45. Thomzig, A. et al. Preclinical disposition of pathological prion protein PrP^{Sc} in muscles of hamsters orally exposed to scrapie. *J. Clin. Invest.*, 113, 1465, 2004.

46. Andreoletti, O. et al. Bovine spongiform encephalopathy agent in spleen from an ARR/ARR orally exposed sheep. *J. Gen. Virol.*, 87, 1043, 2006.

47. Thuring, C.M. et al. Discrimination between scrapie and bovine spongiform encephalopathy in sheep by molecular size, immunoreactivity, and glycoprofile of prion protein. *J. Clin. Microbiol.*, 42, 972, 2004.

48. Iwata, N. et al. Distribution of PrP(Sc) in cattle with bovine spongiform encephalopathy slaughtered at abattoirs in Japan. *Jpn. J. Inf. Dis.*, 59, 100, 2006.

49. Saborio, G.P., Permanne, B., and Soto, C. Sensitive detection of pathological prion protein by cyclic amplification of protein misfolding. *Nature*, 411, 810, 2001.

50. Castilla, J. et al. In vitro generation of infectious scrapie prions. *Cell*, 121, 195, 2005.

51. Thorne, L. and Terry, L.A. In vitro amplification of PrP^{Sc} derived from the brain and blood of sheep infected with scrapie. *J. Gen. Virol.*, 89, 3177, 2008.

52. Saa, P., Castilla, J., and Soto, C. Presymptomatic detection of prions in blood. *Science*, 313, 92, 2006.

53. Atarashi, R. et al. Simplified ultrasensitive prion detection by recombinant PrP conversion with shaking. *Nat. Methods*, 5, 211, 2008.

54. Wiliam, J.M. et al. Rapid end-point quantitation of prion seeding activity with sensitivity comparable to bioassays. *PLoS Pathog.*, 6, e1001217, 2010.

55. Atarashi, R. et al. Real-time quaking-induced conversion: A highly sensitive assay for prion detection. *Prion*, 5, 150, 2011.

56. Jackman, R. and Schmerr, M.J. Analysis of the performance of antibody capture methods using fluorescent peptides with capillary zone electrophoresis with laser-induced fluorescence. *Electrophoresis*, 24, 892, 2003.

57. Bieschke, J. et al. Ultrasensitive detection of pathological prion protein aggregates by dual-color scanning for intensely fluorescent targets. *Proc. Natl. Acad. Sci. USA*, 97, 5468, 2000.

58. Schmitt, J. et al. Identification of scrapie infection from blood serum by Fourier transform infrared spectroscopy. *Anal. Chem.*, 74, 3865, 2002.

59. Paramithiotis, E. et al. A prion protein epitope selective for the pathologically misfolded conformation. *Nat. Med.*, 9, 893, 2003.

60. Yamasaki, T. et al. Characterization of intracellular localization of PrP(Sc) in prion-infected cells using a mAb that recognizes the region consisting of aa 119–127 of mouse PrP. *J. Gen. Virol.*, 93, 668, 2012.

61. Sayer, N.M. et al. Structural determinants of conformationally selective, prion-binding aptamers. *J. Biol. Chem.*, 279, 13102, 2004.

62. Huzarewich, R.L., Siemens, C.G., and Booth, S.A. Application of "omics" to prion biomarker discovery. *J. Biomed. Biotechnol.*, 2010, 613504, 2010.

63. Sorensen, G. et al. Comprehensive transcriptional profiling of prion infection in mouse models reveals networks of responsive genes. *BMC Genomics*, 9, 114, 2008.

64. Miele, G. et al. Urinary α1-antichymotrypsin: A biomarker of prion infection. *PLoS One*, 3, e3870, 2008.

65. Simon, S.L.R. et al. The identification of disease-induced biomarkers in the urine of BSE infected cattle. *Proteome Sci.*, 6, 23, 2008.

66. Sasaki, K. et al. Increased clusterin (apolipoprotein J) expression in human and mouse brains infected with transmissible spongiform encephalopathies. *Acta Neuropathol.*, 103, 199, 2002.

67. Guillaume, E. et al. A potential cerebrospinal fluid and plasmatic marker for the diagnosis of Creutzfeldt-Jakob disease. *Proteomics*, 3, 1495, 2003.

68. Miele, G. et al. Urinary alpha1-antichymotrypsin: A biomarker of prion infection. *PLoS One*, 3, e3870, 2008.

69. Otto, M. et al. Diagnosis of Creutzfeldt-Jakob disease by measurement of S100 protein in serum: Prospective case-control study. *BMJ*, 316, 577, 1998.

70. Kenney, K. et al. An enzyme-linked immunosorbent assay to quantify 14-3-3 proteins in the cerebrospinal fluid of suspected Creutzfeldt-Jakob disease patients. *Ann. Neurol.*, 48, 395, 2000.

71. Miele, G., Manson, J., and Clinton, M. A novel erythroid-specific marker of transmissible spongiform encephalopathies. *Nat. Med.*, 7, 361, 2001.

72. Fischer, M.B. et al. Binding of disease-associated prion protein to plasminogen. *Nature*, 408, 479, 2000.
73. Gavier-Widén, D. et al. Diagnosis of transmissible spongiform encephalopathies in animals: A review. *J. Vet. Diagn. Invest.*, 17, 509, 2005.
74. Sakudo, A. et al. PrP^{Sc} level and incubation time in a transgenic mouse model expressing Borna disease virus phosphoprotein after intracerebral prion infection. *Neurosci. Lett.*, 431, 81, 2008.
75. Bolton, D.C. et al. Isolation and structural studies of the intact scrapie agent protein. *Arch. Biochem. Biophys.*, 258, 579, 1987.
76. Fichet, G. et al. Novel methods for disinfection of prion-contaminated medical devices. *Lancet*, 364, 521, 2004.
77. Siegel, J.D. et al. Guideline for isolation precautions: Preventing transmission of infectious agents in healthcare settings 2007. http://www.cdc.gov/hicpac/pdf/isolation/Isolation 2007.pdf; accessed October 14, 2014.
78. Brandner, S. and Public Health England Press Office. Researchers estimate one in 2,000 people in the UK carry variant CJD proteins. http://www.bmj.com/press-releases/2013/10/14/researchers-estimate-one-2000-people-uk-carry-variant-cjd-proteins; accessed October 14, 2014.
79. Food Safety Commission of Japan. Consideration of risk variations in Japan derived from the proposed revisions of the current countermeasures against BSE (full report). *Food Saf.*, 2, 55, 2014.
80. Sakudo, A. et al. Recent developments in prion disease research: Diagnostic tools and *in vitro* cell culture models. *J. Vet. Med. Sci.*, 69, 329, 2007.
81. Inoue, Y. et al. Infection route-independent accumulation of splenic abnormal prion protein. *Jpn. J. Infect. Dis.*, 58, 78, 2005.
82. Grathwohl, K.U. et al. Sensitive enzyme-linked immunosorbent assay for detection of PrP(Sc) in crude tissue extracts from scrapie-affected mice. *J. Virol. Methods*, 64, 205–216, 1997.
83. Pan, T. et al. An aggregation-specific enzyme-linked immunosorbent assay: detection of conformational differences between recombinant PrP protein dimers and PrP(Sc) aggregates. *J. Virol.*, 79, 12355–12364, 2005.
84. McBride, P.A. et al. Immunostaining of scrapie cerebral amyloid plaques with antisera raised to scrapie-associated fibrils (SAF). *Neuropathol. Appl. Neurobiol.*, 14, 325–336, 1988.
85. Prusiner, S.B. et al. Measurement of the scrapie agent using an incubation time interval assay. *Ann. Neurol.*, 11, 353–358, 1982.
86. Klohn, P.-C. et al. A quantitative, highly sensitive cell-based infectivity assay for mouse scrapie prions. *Proc. Natl. Acad. Sci. USA*, 100, 11666–11671, 2003.
87. Taraboulos, A. et al. Regional mapping of prion proteins in brain. *Proc. Natl. Acad. Sci. USA*, 89, 7620–7624, 1992.
88. Bosque, P.J. et al. Cultured cell sublines highly susceptible to prion infection. *J. Virol.*, 74, 4377–4386, 2000.
89. Winklhofer, K.F. et al. A sensitive filter retention assay for the detection of PrP(Sc) and the screening of anti-prion compounds. FEBS Lett., 503, 41–45, 2001.
90. Schulz-Schaeffer, W.J. et al. The paraffin-embedded tissue blot detects PrP(Sc) early in the incubation time in prion diseases. *Am. J. Pathol.*, 156, 51–56, 2000.
91. Ritchie, D.L. et al. Advances in the detection of prion protein in peripheral tissues of variant Creutzfeldt-Jakob disease patients using paraffin-embedded tissue blotting. *Neuropathol. Appl. Neurobiol.*, 30, 360–368, 2004.
92. Castilla, J. et al. In vitro generation of infectious scrapie prions. *Cell*, 12, 195–206, 2005.
93. Safar, J.G. et al. Eight prion strains have PrP(Sc) molecules with different conformations.*Nat. Med.*, 4, 1157–1165, 1998.
94. Barnard, G. et al. The measurement of prion protein in bovine brain tissue using differential extraction and DELFIA as a diagnostic test for BSE. *Luminescence*, 15, 357–362, 2000.
95. Schmerr, M.J. et al. Use of capillary electrophoresis and fluorescent labeled peptides to detect the abnormal prion protein in the blood of animals that are infected with a transmissible spongiform encephalopathy. *J. Chromatogr.*, 853, 207–214, 1999.
96. Jackman, R. et al. Analysis of the performance of antibody capture methods using fluorescent peptides with capillary zone electrophoresis with laser-induced fluorescence. *Electrophoresis*, 24, 892–896, 2003.
97. Yang, W.C. et al. Capillary electrophoresis-based noncompetitive immunoassay for the prion protein using fluorescein-labeled protein A as a fluorescent probe. *Anal. Chem.*, 7, 4489–4494, 2005.
98. Giese, A. et al. Putting prions into focus: application of single molecule detection to the diagnosis of prion diseases. *Arch. Virol. Suppl.*, 16, 161–171, 2000.
99. Miller, A.E. et al. Fluorescence cross-correlation spectroscopy as a universal method for protein detection with low false positives. *Anal. Chem.*, 81, 5614–5622, 2009.
100. Bibby, D.F. et al. Application of a novel in vitro selection technique to isolate and characterise high affinity DNA aptamers binding mammalian prion proteins. *J. Virol. Methods*, 151, 107–115, 2008.
101. Rubenstein, R. et al. Detection and discrimination of PrPSc by multi-spectral ultraviolet fluorescence. *Biochem. Biophys. Res. Commun.*, 246, 100–106, 1998.
102. Murayama, Y. et al. Specific detection of prion antigenic determinants retained in bovine meat and bone meal by flow microbead immunoassay. *J. Appl. Microbiol.*, 101, 369–376, 2006.
103. Zerr, I. et al. Diagnosis of Creutzfeldt-Jakob disease by two-dimensional gel electrophoresis of cerebrospinal fluid. *Lancet*, 348, 846–849, 1996.
104. Silva, C.J. et al. Utility of mass spectrometry in the diagnosis of prion diseases. *Anal. Chem.*, 83,1609–1615, 2011.
105. Varshney, M. et al. Prion protein detection using nanomechanical resonator arrays and secondary mass labeling. *Anal. Chem.*, 80, 2141–2148, 2008.
106. Xiao, S.I. et al. Sensitive discrimination and detection of prion disease-associated isoform with a dual-aptamer strategy by developing a sandwich structure of magnetic microparticles and quantum dots. *Anal. Chem.*, 82, 9736–9742, 2010.
107. Kobayashi, K. et al. A solid-phase immunoassay of protease-resistant prion protein with filtration blotting involving sodium dodecyl sulfate. *Anal. Biochem.*, 349, 218–228, 2006.
108. Chang, B. et al. Surround optical fiber immunoassay (SOFIA): an ultra-sensitive assay for prion protein detection. *J. Virol. Methods*, 159, 15–22, 2009.
109. Barletta, J.M. et al. Detection of ultra-low levels of pathologic prion protein in scrapie infected hamster brain homogenates using real-time immuno-PCR. *J. Virol. Methods*, 127, 154–164, 2005.
110. Miodek, A. et al. Binding kinetics of human cellular prion detection by DNA aptamers immobilized on a conducting polypyrrole. *Anal. Bioanal. Chem.*, 405, 2505–2514, 2013.
111. Miodek, A. et al. Electrochemical aptasensor of human cellular prion based on multiwalled carbon nanotubes modified with dendrimers: a platform for connecting redox markers and aptamers. *Anal. Chem.*, 85, 7704–7712, 2013.

100 Chronic Wasting Disease

Akikazu Sakudo and Takashi Onodera

CONTENTS

100.1 Introduction 913
   100.1.1 Classification 913
   100.1.2 Biology and Epidemiology 913
   100.1.3 Clinical Features and Pathogenesis 913
   100.1.4 Diagnosis 915
100.2 Methods 918
   100.2.1 Reagents and Equipment 918
      100.2.1.1 Reagents 918
      100.2.1.2 Equipment 918
   100.2.2 Sample Preparation and Detection Procedures 918
100.3 Conclusion and Future Perspectives 919
Acknowledgments 919
References 919

100.1 INTRODUCTION

100.1.1 CLASSIFICATION

Chronic wasting disease (CWD) belongs to a group of fatal neurodegenerative diseases in cervids and is an example of a transmissible spongiform encephalopathy (TSE). An extraordinary feature of CWD is its relatively high incidence in wild animals, despite its capability to affect both captive and free-living cervids. CWD infection in the wild appears to be limited to mule deer (*Odocoileus hemionus*) [1], white-tailed deer (*Odocoileus virginianus*) [2], Rocky Mountain elk (*Cervus elaphus nelson*) [3], and Shiras moose (*Alces alces shirasi*) [4]. Within endemic areas, transmission occurs intraspecifically (among deer of the same species) and interspecifically (among cervids of different species) through direct or indirect contact with infected cervids.

100.1.2 BIOLOGY AND EPIDEMIOLOGY

CWD was first described as a wasting and progressive disorder in mule deer caught for nutritional research near Fort Collins, Colorado, in 1967 [1]. It has since been classified as an example of TSE, because the disease leads to typical vacuolation in the brain [1], prion protein (PrP) accumulation [5], and infectivity [6]. The original epidemic appeared to be confined to northern Colorado and southern Wyoming, but it soon spread in both captive and free-ranging populations across 22 U.S. states and 2 Canadian provinces, as well as the Republic of Korea. Initially, CWD was limited to the captive mule deer and elk but was later identified in captive and free-ranging wapiti, mule, white-tailed deer, and moose. The prevalence of CWD in affected areas between 1996 and 1999 was estimated to be about 5% in mule deer, 2% in white-tailed deer, and <1%

in elk [7]. However, a study published in 2005 reported a high prevalence rate (30%) of CWD in some areas of Colorado [8]. A more recent survey in Colorado covering the period of 2006–2009 indicated that the incidence of CWD varied from <1% to 14.3% among mule deer, from <1% to 2.4% among elk, and <1% among moose [9]. By contrast, the European Food Safety Authority reported no CWD epidemic in the European Union [10]. Currently, CWD testing in other countries is limited, although active surveillance in Japan has not detected the disease [11]. Nevertheless, less than 30 million deer are believed to be infected with CWD in North America [12].

CWD is thought to be horizontally infected at a high rate among cervids [13]. Oral transmission of CWD-infected brain homogenate derived from mule deer to elk has been reported, which is further supported by the finding of CWD-infected free-ranging moose [14]. However, the origins of the disease remain unclear. Since saprovores (such as buzzard and jackal) and other animals (such as mountain lion, fox, raccoon, coyote, and eagle) prey on deer and elk, CWD may potentially spread among a variety of animals. Indeed, experimental infections in the laboratory setting suggest CWD-susceptible animals include not only cervids but also ferret [15], raccoon [16], squirrel monkey [17], cattle [17], sheep [8], goat [18], hamsters [15,19], bank voles [20], mink [21], meadow voles [20], red-backed voles [20], white-footed mice [20], and deer mice [20]. This highlights the potential of CWD to breach the species barrier.

100.1.3 CLINICAL FEATURES AND PATHOGENESIS

An abnormal isoform of PrP (PrP^{Sc}) in cervids, termed PrP^{CWD}, is generated from cervid cellular PrP (PrP^{C}). PrP^{CWD} accumulates in brain tissue causing neurological symptoms and ultimately death. At an early stage of the disease, PrP^{CWD}

localizes in the tonsils, gut-associated lymphoid tissue, and enteric nervous system. For example, orally CWD-infected brain homogenate of mule deer induces PrP^{CWD} accumulation in intestine-associated lymphatic tissues, such as retropharyngeal lymph nodes (RLNs), tonsils, Peyer's patch, and ileocecal lymph nodes, as early as 42 days postinfection [22]. However, it should be noted that PrP^{CWD} depositions in lymphoid tissue of elk are less than those of deer [23]. Thereafter, the distribution of PrP^{CWD} expands to the central nervous system (CNS) [24,25]. Finally, PrP^{CWD} is present in the brain, spinal cord, tonsils, skeletal muscle, lymph nodes, spleen, adrenal glands, peripheral nervous system, antler velvet, and pancreas (Figure 100.1). Significantly, the report using CWD-infected deer showed the presence of PrP^{CWD} in skeletal muscle and its seeding activity even when there were no associated clinical signs [26]. In addition, prion infectivity of TSE-associated PrP has been detected in the heart muscle of CWD-infected cervids [27]. These findings imply that muscle is a source of infection. Furthermore, akin to bovine spongiform encephalopathy (BSE), PrP^{CWD} agent can also be found in the retina of infected deer. Indeed, CWD prions have been detected in the saliva [28], urine [28], feces [29], and blood [30] of infected animals. Therefore, it has been proposed that the CWD agent is transmitted *via* saliva or in the stools of deer [22].

The clinical symptoms of CWD, which appear at least 18 months after infection, include weight loss, emaciation, excessive salivation, teeth grinding, fever, anorexia, polyposia, excessive urination and impaired motor coordination, and respiratory distress [8,31]. However, there is no immune response in CWD-infected deer. As these symptoms are not CWD specific, additional biochemical and pathological analysis is required to verify the diagnosis. After the development of these symptoms, most animals survive for a few weeks up to 3–4 months. Furthermore, most deer displaying symptoms

of CWD are between 3 and 7 years of age, probably developing CWD within 3 years of infection [32]. Experimental transmission to deer showed that CWD has a minimum incubation period of approximately 15 months, and the mean time from oral infection to death is about 23 months (20–25 months). CWD-infected elk developed the disease rapidly 12 months after oral inoculation compared to CWD-infected mule deer (16 months) [13].

It remains unclear how CWD is spread, although potential mechanisms include horizontal and environmental transmission. Unlike BSE, there is evidence that CWD could be transmitted horizontally among cervids [13]. Indeed, experimental model experiments using transgenic mice expressing the normal cervid PrP support the horizontal transmission of CWD [33]. Recently, maternal transmission was also shown to occur in cervids [34]. Experimental maternal transmission of CWD was demonstrated in the Reeves's muntjac deer model (*Muntiacus reevesi*) and PrP^{CWD} was found in the maternal placenta and mammary tissue of the dam [35].

The transfer of infection in saliva and stools are currently considered a likely means of transmission. For example, a carcass put into a field successfully transmitted CWD to experimental mule deer [22], suggesting a contaminated CWD environment can result in the spread of CWD. It has been demonstrated that hamster-derived PrP^{Sc} can adhere to soil minerals [36]. Moreover, some clay components of soil efficiently bind CWD prions [36]. In addition, binding to soil enhances oral transmissibility of hamster-adapted mink prions [37]. Water in areas where CWD is prevalent is known to be contaminated by CWD prions, which can be detected by serial protein misfolding cyclic amplification (sPMCA) [38]. It should be noted that aerosols and nasal exposure can also result in the transmission of CWD in cervid PrP–expressing transgenic mice [39]. A recent study using white-tailed deer confirmed the aerosol transmission of CWD to deer [40].

Some polymorphism in cervid PrP may affect susceptibility to CWD. Polymorphism at codon 95 (glutamine [Q] or histidine [H]) [41], codon 96 (glycine [G] or serine [S]) [41,42], and codon 116 (alanine [A] or glycine [G]) [43] in white-tailed deer has been reported. While all major genotypes were found in deer with CWD, the Q96, G96, A116 allele (QGA) was more frequently found in CWD-affected deer than the QSA allele [41,44]. In addition, some polymorphism of cervid PrP, such as the G96S in white-tailed deer [45] and S225F in mule deer [46], is suggested to reduce susceptibility. The elk PrP also displays polymorphism at codon 132, encoding either methionine (M) or leucine (L) [44,47], corresponding to the polymorphism of human PrP at codon 129. Statistical investigation of wild free-ranging and farm-raised elk [48] and experimental results using elk with all the three genotypes (MM132, ML132, LL132) suggest that the ML132 or LL132 genotype might offer better protection against CWD compared to MM132. Moreover, the inhibitory effect of L132 was supported by a study using transgenic mice [49]. This study showed that transgenic mice expressing cervid PrPL132 failed to develop the disease following challenge with CWD prions, while those expressing cervid PrPM132 were susceptible to infection [49].

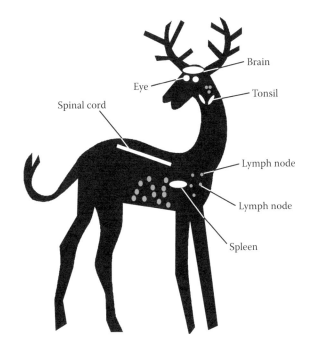

FIGURE 100.1   The distribution of PrP^{CWD}.

Although several TSEs, such as scrapie and BSE, show similar symptoms, there are no relationships between such TSEs and CWD. Transgenic mice overexpressing deer PrP were resistant to BSE from cattle, whereas red deer–adapted BSE prion, which was produced after intracerebral inoculation of European red deer with BSE agent from cattle, could transmit the disease to mice [50].

Similarly, there is no evidence that CWD can transmit to humans, while squirrel monkeys can be infected with CWD prion [17,51]. By contrast, macaques remained healthy after intracerebral inoculation of CWD prion. This resistance of humans against CWD is supported by the fact that transgenic mice expressing human PrP did not show any hallmark symptoms of CWD 600 days after intracerebral inoculation with CWD prions [12]. These findings indicate the presence of a barrier that makes interspecies transmission unlikely, especially between species as diverse as humans and deer [12]. By comparison, the incubation time, which indicates the period required for displaying symptoms after intracerebral infection with CWD prions, using transgenic mice that had been "cervidized" was between 118 and 142 days [12]. Moreover, an *in vitro* assay showed that PrP^{CWD}-mediated conversion of human PrP to a protease-resistant form is very inefficient, suggesting a species barrier at the molecular level that limits the susceptibility of humans to CWD [42].

Recent studies indicate the presence of conformational variants or strains may exist. Based on bioassay studies, including passage of CWD into ferrets, hamsters, and cervidized mice, at least two putative strains of CWD agents have been proposed (e.g., CWD1 and CWD2, SghaCWD^{md-f} and SghaCWD^{md-s}, CWD-CSU and CWD-WI) [19,52–54]. Further evidence comes from the observation that sequential passage of a single mule deer CWD isolate using hamster or hamster PrP–expressing transgenic mice causes two distinct disease phenotypes [19,20]. In addition, LaFauci et al. have reported that elk PrP–expressing transgenic mice developed phenotypically divergent diseases when inoculated with either mule deer or elk CWD, which was suggestive of different strains among cervids [55].

100.1.4 Diagnosis

Western blotting, enzyme-linked immunosorbent assay (ELISA), and immunohistochemistry (IHC) are useful for PrP^{CWD} detection (Figure 100.2). The procedures for detecting PrP^{Sc} in both animal and human prion diseases are similar. While ELISA provides an effective screening approach for CWD, Western blotting and IHC are valuable for confirming the diagnosis [56]. In Chapter 99 of this book, we introduced a diagnostic method for BSE and described an experimental procedure (Western blotting) that is also applicable to CWD.

Although the diagnostic procedures for prion diseases are commonly applicable among animals, it should be noted that the methodology depends on the reactivity of the antibody in each animal. Therefore, it is important to determine if the species of PrP is recognized by the antibody used in the immunoassay (e.g., ELISA, Western blotting, IHC).

To date, reliable diagnoses of CWD can only be performed by postmortem analysis using lymphoid or brain tissues [57]. However, antemortem diagnosis has been attempted using tonsillar or rectal biopsy of live cervids during the incubation stage. The detection of PrP in body fluids (such as blood, cerebrospinal fluid (CSF), and urine) and peripheral tissues (such as lymphatic tissues and peripheral nerves) has also been studied. In this chapter, we firstly describe the detection of PrP by Western blotting, ELISA, and IHC. We then describe the recent development of a CWD diagnosis method using ELISA.

While the amino acid sequence of cervid PrP^C and PrP^{CWD} is identical, these two protein forms are different in terms of the following biochemical characteristics: (1) protease resistance, (2) secondary structure (PrP^C has a high alpha-helix content; PrP^{CWD} is rich in beta-sheet), (3) half-life, and (4) solubility (Table 100.1). PrP^{CWD} is the main component of the CWD agent and can be detected in prion-infected cervids. Therefore, most diagnostic methods of prion diseases rely on an index of the difference between PrP^{CWD} and cervid PrP^C. The most commonly used index relies on protease sensitivity and solubility. These methods depend on the fact that PrP^{CWD} aggregates and is therefore resistant to both denaturing reagents and proteases. In the case of IHC, an autoclaving step is usually incorporated to remove PrP^C. After exposure of the epitope in PrP^C following pretreatment by autoclaving, IHC using anti-PrP antibody can be used to detect PrP^{CWD}. Furthermore, proteinase K (PK) can degrade PrP^C, whereas PrP^{CWD} is resistant to this protease. Thus, Western blotting and ELISA can be used detect PrP^{CWD} after PK treatment. Furthermore, guanidine hydrochloride (GdnHCl) reveals the epitope of PrP^{CWD} because this form of the protein is aggregated, whereas the denaturant does not cause any change in the epitope of PrP^C because PrP^C is solubilized. Thus, the reactivity of antibody against PrP^C is unchanged after GdnHCl treatment. Residual PrP after PK treatment is termed PrP^{res} (PK-resistant PrP). Hence, PrP^{CWD} can be separated from PrP^C on the basis of PK resistance and then subjected to Western blotting and ELISA for detection.

The gold standard for CWD diagnosis is IHC. Fixed and paraffin-embedded tissues are usually subjected to IHC for diagnostic testing. The quantity of accumulated PrP^{CWD} is different among tissues. Generally, medullary obex and lymphatic tissues, such as tonsils and RLNs, are used for the analysis. Since the brain contains the highest quantity of PrP^{CWD}, this tissue is usually used as sample material. However, it should be noted that the quantity of accumulated PrP^{CWD} differs in different regions of the brain. Therefore, it is preferable to use an appropriate tissue region for analysis.

IHC is a method for observing tissue sections stained with anti-PrP antibody using light microscopy. This methodology has merits in terms of its high level of both sensitivity and specificity. In addition, IHC can be used to verify whether the tissue distribution of PrP^{CWD} is as anticipated. Other methods for CWD diagnosis include ELISA and Western blotting. ELISA is used to detect PrP adsorbed on plates after treatment of brain homogenate with PK. Western blotting detects PrP absorbed on a membrane after an electroblotting procedure.

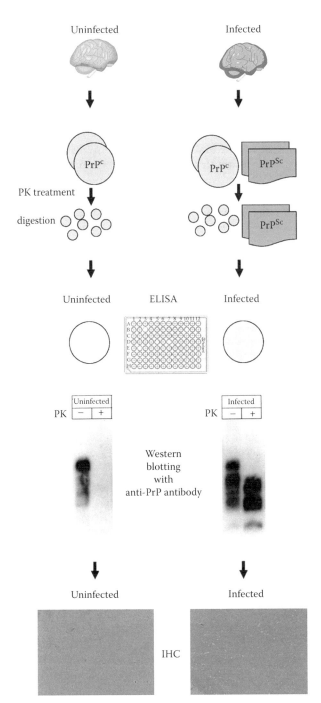

FIGURE 100.2 Diagnostic methods for chronic wasting disease (CWD). Early diagnosis of CWD based on clinical signs is unreliable because clinical symptoms are mild and nonspecific. However, at the late stage of the disease, there is extensive and progressive weight loss followed by hypersalivation, ataxia, polydipsia, and development of various abnormal behaviors. Positive diagnosis of CWD is carried out postmortem using brain tissue (i.e., obex region). Most methods of diagnosing prion diseases are based on the characteristics of PrP^{Sc}, which is resistant to proteinase K (PK). PK completely degrades PrP^{C} but only partially digests PrP^{Sc}, because PrP^{Sc} forms protease-resistant aggregates. After PK treatment, Western blotting and enzyme-linked immunosorbent assay with anti-PrP antibody can be used to detect PK-resistant PrP, when PrP^{Sc} is included in the sample. As the N-terminal region of PrP^{Sc} is digested with PK (+), PK-resistant PrP (27–30 kDa) shows a shift to a lower molecular weight compared to the untreated PrP (−) (33–35 kDa) in Western blotting. Immunohistochemistry (IHC) can be used for definitive diagnosis. In IHC, vacuolation and PrP deposits in brain are usually used as an index of CWD. Given the wide distribution of CWD prions in the animal body, antemortem sampling and testing of lymphatic tissues such as tonsils, retropharyngeal lymph nodes, and rectal biopsy materials have been attempted, and the results gave successful predictions of pre- and subclinical animals in the case of IHC [74]. Unfortunately, however, the sensitivity of this procedure is low compared to tests using brain tissue [74]. Recently, a PrP^{Sc} amplification method, protein misfolding cyclic amplification (PMCA), has been developed for potential antemortem diagnosis of CWD using body fluids such as blood, cerebrospinal fluid (CSF), saliva, and urine. (From Henderson, D.M. et al., *PLoS One* 8, e74377, 2013; Elder, A.M. et al., *PLoS One* 8, e80203, 2013.).

TABLE 100.1

Differences in Biochemical Characteristics between PrP^C and PrP^Sc

	PrP^C	PrP^Sc a
Protease resistance	Low	High
Rate of secondary structure	α-helix 42%, β-sheet 3%	α-helix 30%, β-sheet 43%
Half-life	3–6 h	>24 h
Solubility	High	Low
Cellular distribution	Cell surface	Intracellular
GPI anchor	Yes	Yes
Release from cell surface by PIPLC	Yes	No
Infectivity	No	Yes

PIPLC, phosphatidylinositol-specific phospholipase C; GPI, glycosylphosphatidylinositol.

[a] In the case of CWD, abnormal PrP (PrP^Sc) is sometimes termed PrP^CWD.

TABLE 100.2

Representative Kits for CWD and TSE

Kit[a]	Company	Method
CWD dbELISA	VMRD	Dot blot ELISA
Enfer TSE Kit v2.0	Abbott Laboratories	Direct ELISA
HerdCheck CWD Ag test	IDEXX Laboratories	Antigen capture ELISA
Inpro CDI kit	Beckman Coulter	CDI
PAD-beads Seprion-PAD	Microsens Biotechnologies	Concentration of aggregated proteins including PrP^Sc by Seprion ligand
Prion protein EIA kit	SPI Bio	Sandwich ELISA
Prionics-Check LIA	Prionics	Sandwich ELISA
Prionics-Check PrioSTRIP SR	Prionics	Immunochromatography
PrionScreen	Roche Applied Science	Sandwich ELISA
Priontype postmortem	Labor Diagnostik Leipzig	Sandwich ELISA
TeSeE SAP detection kit	Bio-Rad	Sandwich ELISA

CDI, conformation dependent immunoassay; ELISA, enzyme-linked immunosorbent assay; PrP^Sc, abnormal prion protein; TSE; transmissible spongiform encephalopathy.

[a] Modified new version of kits have been developed in each company.

ELISA is usually preferable for screening, because the technique allows rapid analysis. However, Western blotting facilitates more detailed biochemical analysis, such as the determination of the molecular weight of PrP. Such information may be used to help identify the prion source and strains. Western blot analysis of glycoform patterns of PrP^CWD suggest that CWD in deer and elk must be considered a single disease [58]. However, recent studies suggest that there are at least two strains of CWD agent.

There are currently several tests for CWD, which include recently modified kits. Most kits use lymphoid tissue, preferentially RLN or tissue from the CNS, ideally from the brainstem. In United States, a CWD kit, which uses RLN as test tissue, has been licensed for three different species of cervids, that is, white-tailed deer, mule deer, and elk. In addition, most of these tests were designed for TSE, so it can be applicable to CWD as well as BSE, scrapie, and other TSEs.

Most currently available CWD kits detect PrP^CWD by sandwich ELISA in which PrP is captured on an anti-PrP antibody-coated 96-well microplate and detected using a different anti-PrP antibody (Table 100.2). Alternatively, the Enfer TSE Kit employs a direct ELISA system, which involves the immobilization of PrP^CWD onto microplates. A dot blot ELISA is essentially the same as a direct ELISA except the procedure uses polyvinylidene difluoride membranes instead of microplates. Most of the other tests are based on the use of a sandwich immunoassay. This methodology requires two anti-PrP antibodies recognizing different epitopes. ELISA is preferably used for screening or surveillance of CWD. Positive hits from the ELISA screening should be verified by IHC and Western blotting to confirm a definitive diagnosis of CWD.

More recently, a new method for prion diagnosis has been developed known as protein misfolding cyclic amplification (PMCA). This method can amplify PK-resistant PrP by sequential incubation and sonication using a mixture of PrP^C source and PrP^Sc seed. PMCA can detect PrP^CWD in the urine of presymptomatic deer [59]. Recently, Hoover et al. successfully amplified PrP^CWD using brain homogenate of noncervids (ferret and hamster) [60]. Using this method, Hoover et al. attempted to amplify PrP^CWD using brain homogenate derived from various rodents. The results indicated successful amplification using PrPs derived from bank vole and field mouse but failed using those from prairie dog and coyote [60]. In addition, this method can amplify PK-resistant PrP derived not only from cervids [61] but also from various other species including hamster, mouse, sheep, and human [33]. Moreover, a transgenic mouse expressing cervid PrP [6] and CWD-susceptible cell lines [62,63] have been developed. These powerful research tools will be invaluable to CWD research in helping to reveal the molecular basis of the disease. In particular, these tools will be used to develop bioassays for CWD infectivity in tissues and fluid from CWD-infected cervids [64]. Further development of PMCA enables an estimation of the quantity of prion infectivity in tissues, such as the obex, left ventricle, pancreas, jejunum, and spleen, as well as body fluids, such as saliva and urine, by real-time quaking-induced conversion (RT-QuIC) [65]. Such an assay will contribute to the risk assessment of deer tissues, biological fluids, excreta, or environmental samples obtained from CWD-endemic areas.

Recent developments in detecting CWD have enabled antemortem diagnosis using IHC of tonsil tissue. Generally, sample preparation for IHC is not a high-throughput procedure, and the paraffin-embedded tissues are not readily applicable to further biochemical analysis. Nonetheless, methods have been developed that allow these samples to be examined by Western blotting and ELISA. Continuous warming and freezing separates the tissue from the paraffin, which can then be sonicated and subjected to biochemical analysis (i.e., Western blotting and ELISA).

100.2 METHODS

The detection procedures for CWD are essentially common to those described for BSE. The procedure for the detection of PrP^{Sc} by Western blotting can also be used to detect PrP^{CWD}. Thus, to avoid redundancy, we introduce ELISA as an alternative method for detecting PrP^{CWD} in tissue samples. Fundamentally, the current procedures can be commonly used for all TSE including BSE and scrapie as well as human prion diseases such as Creutzfeldt-Jakob disease (CJD). There are two types of ELISA; direct ELISA and sandwich ELISA. In the direct ELISA method, PrP in the sample adheres to the microplate and is subsequently detected using an anti-PrP antibody. By contrast, in the sandwich ELISA method, anti-PrP antibody is coated onto the microplate to facilitate capture of PrP in the sample, which can then be detected using a different anti-PrP antibody. Sandwich ELISA is recommended because it is both sensitive and reliable. Thus, in this section, we introduce sandwich ELISA for PrP detection.

100.2.1 Reagents and Equipment

100.2.1.1 Reagents

- Anti-PrP antibodies. Anti-PrP antibodies are now available from several companies such as Bertin Pharma (SPI -Bio; Montigny le Bretonneux, France), Prionics (Schlieren, Switzerland), Roboscreen GmbH (Leipzig, Germany), VMRD Inc. (Pullman, Washington), R-Biopharm AG (Darmstadt, Germany) DakoCytomation Denmark (Glostrup, Denmark), and Alicon AG (Zürich-Schlieren, Switzerland).
- Streptavidin-HRP (horseradish peroxidase).
- Lysis buffer (10 mM Tris-HCl [pH 7.4] containing 100 mM NaCl, 10 mM EDTA, 0.5% NP40, 1% sodium deoxycholate).
- Bovine serum albumin (BSA).
- Protein Assay Kit (e.g., DC Protein Assay Kit, Bio-Rad [Hercules, California]).
- EIA buffer (phosphate-buffered saline [PBS] containing 0.1% BSA).
- Washing buffer (PBS containing 0.05% Tween20).
- PK.
- Recombinant PrP.
- Tetramethylbenzidine (TMB) solution containing 3,3′,5,5′-TMB and hydrogen peroxide (Thermo Scientific, Waltham, Massachusetts).
- Stop solution (sulfuric acid or hydrochloric acid).

100.2.1.2 Equipment

- 96-well microplate
- Microplate spectrophotometer

To avoid cross-contamination, we recommend the use of new glassware and instruments or single-use materials. Precautions must be taken to avoid contamination of solutions with highly concentrated reagents. Avoid blood contamination during the dissection of tissues, rinse excised tissues in PBS, flash-freeze the specimens in liquid nitrogen, and store the dissected brains at −70°C. All procedures should be performed in a laminar flow biosafety cabinet in an appropriate laboratory according to governmental regulations. Personal protection equipment (lab coat, gloves, and facemask) should be used, and gloves should be changed frequently. Avoid drying out the microplates. Repeated freeze–thaw cycles of solutions containing recombinant PrP, antibody, or streptavidin-HRP can adversely affect their activity and should be avoided. Thus, the corresponding solutions should be aliquoted and kept at −20°C or −80°C.

100.2.2 Sample Preparation and Detection Procedures

The tonsils, RLN, and brainstem at the obex (with intact dorsal motor nuclei of vagus) were thought to be the best tissues for PrP^{Sc} testing [5,8,56,66,67]. It should be noted that in the case of elk, elk younger than 8.5 years should be used for testing of rectal biopsies, because the number of lymphoid follicles in the recto-anal lymphoid tissues decreases with age [68].

As an example of a diagnostic method for prion diseases, the materials and methods used for ELISA are presented. In particular, several precautions that should be undertaken when carrying out this procedure are emphasized. As described earlier, an index for prion diseases can be generated by analysis of samples for PK-resistant PrP. In addition, as PrP is a glycosylphosphatidylinositol (GPI)-anchored protein on the cell membrane, it must be extracted from tissues using buffer including detergents such as lysis buffer as described earlier. This is also the case for the ELISA detection system.

Before starting ELISA, anti-PrP antibody used in the experiments should be carefully selected by checking information on the species specificity of the antibody recognizing PrP. For example, anti-PrP antibody 3F4 binds human and hamster PrP but not mouse, sheep, deer, and bovine PrP. Anti-PrP antibodies recognizing the C-terminal half of PrP, such as 6H4 (Prionics AG), should be used for the detection of PrP^{Sc} in this Western blotting system, whereas those recognizing the N-terminal half of PrP cannot be used. This is because the C-terminal half of PrP^{Sc} is resistant to PK, whereas the N-terminal half is susceptible to PK. 6H4 can be used for the detection of PrP^{CWD} because it recognizes epitopes of human PrP at amino acid residues 143–153 and cross-reacts with bovine, ovine, and cervine PrP. In addition, epitopes of PrP recognized by capture anti-PrP antibody and detection anti-PrP antibody should not be near the region of PrP. Detection anti-PrP antibody should be labeled with biotin by a standard protocol. Furthermore, the optimum concentration of capture and detection anti-PrP antibodies must be determined before the assay is performed by adjusting nonspecific binding and intensity of PrP-related signals. In addition, optimum protein concentration of the sample for assay should be used.

1. Homogenate tissues or cells (10% w/v) in lysis buffer.
2. Syringe using 18G needles followed by 22G needles to mechanically break tissues or cells combined with sonication or breakage by violent agitation with beads.
3. Transfer samples to 1.5 mL Eppendorf tubes and centrifuge for 5 min at 10,000 rpm.
4. Discard the pellet, and aliquot the supernatant, which is then subjected to a protein assay, such as Bio-Rad DC Protein Assay (Bio-Rad) for measuring protein concentration. Store the samples at −20°C or −80°C until analysis.
5. Add PK solution to a final concentration of 20 μg/mL and incubate for 30 min at 37°C.
6. Mix the solution with 8 M urea to a final concentration of 4 M urea.
7. Heat at 100°C for 10 min to denature proteins.
8. Recombinant PrP can be used to generate a standard curve of PrP by diluting with lysis buffer.
9. Dilute all samples at 10-fold in EIA buffer before performing ELISA.
10. Wash a microplate, which has been coated with an anti-PrP antibody overnight at 4°C, using wash buffer (three times).
11. Dispose samples or standards of diluted recombinant PrP.
12. Incubate for 2 or 3 h at room temperature.
13. Wash the microplate three times with wash buffer.
14. Add biotin-labeled anti-PrP antibody that is different to the anti-PrP antibody coated on the microplate.
15. Seal the microplates with an adhesive cover sheet and incubate for 2 or 3 h at room temperature.
16. Wash the microplate with wash buffer three times.
17. Add streptavidin-HRP.*
18. Incubate for 2 h at room temperature.
19. Wash the microplate three times with wash buffer.
20. Add TMB solution.
21. Allow the enzymatic reaction to proceed at 37°C for 30 min–1 h in darkness.

22. Add stop solution to quench the reaction.
23. Measure absorbance at 450 nm using a microplate reader.
24. Calculate the concentration of PrPSc by comparison of absorbance with a standard curve generated using diluted recombinant PrP.

100.3 CONCLUSION AND FUTURE PERSPECTIVES

CWD is one of the prion diseases that show clinical symptoms such as progressive weight loss, abnormal behavior, and excessive salivation. Incidents have been reported in North America and Korea. Efficient transmission by horizontal infection may occur *via* contact with saliva or feces. Infectivity of the skeletal muscle from diseased deer has been observed, suggesting meat could be a potential source of infection. Recently, possible transmission of CWD to fishes has been reported. However, at the moment, CWD transmission to humans is thought to be very unlikely.

Several modified methods for PrP detection have been developed involving concentration, amplification, and sensitive detection. The most important recent advances in PrPCWD detection have been the development of PMCA and related methods such as RT-QuIC [69,70]. Moreover, enrichment of PrPSc by phosphotungstic acid enhances sensitivity [69]. RT-QuIC enables the detection of PrPSc in the urine and feces of presymptomatic deer [71]. In addition, the production of CWD-susceptible transgenic mice [6,66] and cell lines [63,72] may enable the calculation of CWD infectivity more easily.

Despite these recent advances in diagnostic technologies, currently no treatments or vaccines are available for CWD or any other TSE [73]. In addition, the zoonotic potential of CWD agents, which appear to have various strain types, remains unclear. Thus, further studies of CWD transmission using primate and transgenic mouse models are required to assess the zoonotic potential of CWD agents.

ACKNOWLEDGMENTS

This work was supported by Grants-in-Aids for Scientific Research from the Ministry of Education, Science, Culture, and Technology of Japan (25450447, 24110717, 23780299, 22110514, 20780219, 17780228), Grants-in-Aid from the Research Committee of Prion Disease and Slow Virus Infection, from the Ministry of Health, Labour and Welfare of Japan, and the Practical Research Project for Rare/Intractable Diseases from Japan Agency for Medical Research and Development, AMED.

---

* When using streptavidin-AChE (acetylcholine esterase) as an alternative to streptavidin-HRP, Ellman's reagent (0.75 mM acetylthiocholine and 0.5 mM 5,5′-dithiobis[2-nitrobenzoic acid] in 100 mM phosphate buffer pH 7.4) should be used for the colorimetric substrate AChE. In this case, absorbance at 414 nm should be measured. If HRP-labeled reagents are used, it is essential to remove sodium azide from all buffers, because sodium azide inhibits the activity of HRP. Sodium azide should be replaced by thimerosal.

REFERENCES

1. Williams, E.S. et al., Chronic wasting disease of captive mule deer: A spongiform encephalopathy. *J. Wildl. Dis.* 16, 89, 1980.
2. Willams, E.S. et al., Chronic wasting disease in deer and elk in North America. *Am. Rev. Sci. Tech.* 21, 305, 2002.

3. Williams, E.S. et al., Spongiform encephalopathy of Rocky Mountain elk. *J. Wildl. Dis.* 18, 465, 1982.

4. Kreeger, T.J. et al., Oral transmission of chronic wasting disease in captive Shira's moose. *J. Wildl. Dis.* 42, 640, 2006.

5. Spraker, T.R. et al., Comparison of histological lesions and immunohistochemical staining of proteinase-resistant prion protein in a naturally occurring spongiform encephalopathy of free-ranging mule deer (*Odocoileus hemionus*) with those of chronic wasting disease of captive mule deer. *Vet. Pathol.* 39, 110, 2002.

6. Browning, S.R. et al., Transmission of prions from mule deer and elk with chronic wasting disease to transgenic mice expressing cervid PrP. *J. Virol.* 78, 13345, 2004.

7. Spraker, T.R. et al., Spongiform encephalopathy in free-ranging mule deer (*Odocoileus hemionus*), white-tailed deer (*Odocoileus virginianus*) and Rocky Mountain elk (*Cervus elaphus nelsoni*) in north central Colorado. *J. Wildl. Dis.* 33, 1, 1997.

8. Williams, E.S., Chronic wasting disease. *Vet. Pathol.* 42, 530, 2005.

9. Colorado Parks and Wildlife. CWD info & testing. Available at http://cpw.state.co.us/learn/Pages/ResearchCWD.aspx; accessed on January 25, 2015.

10. EFSA Panel on Biological hazards (BIOHAZ), Scientific opinion on the results of the EU survey for chronic wasting disease (CWD) in cervids. *EFSA J.* 8, 1861, 2010.

11. Kataoka, N. et al., Surveillance of chronic wasting disease in sika deer, Cervus nippon, from Tokachi district in Hokkaido. *J. Vet. Med. Sci.* 67, 349, 2005.

12. Kong, Q. et al., Chronic wasting disease of elk: Transmissibility to humans examined by transgenic mouse models. *J. Neurosci.* 25, 7944, 2005.

13. Miller, M.W. et al., Prion disease: Horizontal prion transmission in mule deer. *Nature* 425, 35, 2003.

14. Baeten, L.A. et al., A natural case of chronic wasting disease in a free-ranging moose (*Alces alces shirasi*). *J. Wildl. Dis.* 43, 309, 2007.

15. Bartz, J.C. et al., The host range of chronic wasting disease is altered on passage in ferrets. *Virology* 251, 297, 1998.

16. Hamir, A.N. et al., Experimental inoculation of scrapie and chronic wasting disease agents in raccoons (*Procyon lotor*). *Vet. Rec.* 153, 121, 2003.

17. Marsh, R.F. et al., Interspecies transmission of chronic wasting disease prions to squirrel monkeys (*Saimiri sciureus*). *J. Virol.* 79, 13794, 2005.

18. Williams, E.S. et al., Spongiform encephalopathies in Cervidae. *Rev. Sci. Tech.* 11, 551, 1992.

19. Raymond, G.J. et al., Transmission and adaptation of chronic wasting disease to hamsters and transgenic mice: Evidence for strains. *J. Virol.* 81, 4305, 2007.

20. Heisey, D.M. et al., Chronic wasting disease (CWD) susceptibility of several North American rodents that are sympatric with cervid CWD epidemics. *J. Virol.* 84, 210, 2010.

21. Harrington, R.D. et al., A species barrier limits transmission of chronic wasting disease to mink (*Mustela vison*). *J. Gen. Virol.* 89, 1086, 2008.

22. Miller, M.W. et al., Environmental sources of prion transmission in mule deer. *Emerg. Infect. Dis.* 10, 1003, 2004.

23. Race, B.L. et al., Levels of abnormal prion protein in deer and elk with chronic wasting disease. *Emerg. Infect. Dis.* 13, 824, 2007.

24. Fox, K.A. et al., Patterns of PrP^CWD accumulation during the course of chronic wasting disease infection in orally inoculated mule deer (*Odocoileus hemionus*). *J. Gen. Virol.* 87, 3451, 2006.

25. Sigurdson, C.J. et al., PrP(CWD) in the myenteric plexus, vagosympathetic trunk and endocrine glands of deer with chronic wasting disease. *J. Gen. Virol.* 82, 2327, 2001.

26. Daus, M.L. et al., Presence and seeding activity of pathological prion protein (PrP(TSE)) in skeletal muscles of white-tailed deer infected with chronic wasting disease. *PLoS One* 6, e18345, 2011.

27. Gilch, S. et al., Chronic wasting disease. *Top. Curr. Chem.* 305, 51, 2011.

28. Haley, N.J. et al., Detection of CWD prions in urine and saliva of deer by transgenic mouse bioassay. *PLoS One* 4, e4848, 2009.

29. Tamgüney, G. et al., Asymptomatic deer excrete infectious prions in faeces. *Nature* 461, 529, 2009.

30. Mathiason, C.K. et al., Infectious prions in the saliva and blood of deer with chronic wasting disease. *Science* 314, 133, 2006.

31. Sohn, H.J. et al., A case of chronic wasting disease in an elk imported to Korea from Canada. *J. Vet. Med. Sci.* 64, 855, 2002.

32. Miller, M.W. et al., Epidemiology of chronic wasting disease in captive Rocky Mountain elk. *J. Wildl. Dis.* 34, 532, 1998.

33. Seelig, D.M. et al., Pathogenesis of chronic wasting disease in cervidized transgenic mice. *Am. J. Pathol.* 176, 2785, 2010.

34. Williams, E.S. et al., Chronic wasting disease of deer and elk: A review. *J. Wildl. Manage* 66, 551, 2002.

35. Mathiason, C.K. et al., Mother to offspring transmission of chronic wasting disease. *Prion* 4, 158, 2010.

36. Johnson, C.J. et al., Prions adhere to soil minerals and remain infectious. *PLoS Pathog.* 2, e32, 2006.

37. Johnson, C.J. et al., Oral transmissibility of prion disease is enhanced by binding to soil particles. *PLoS Pathog.* 3, e93, 2007.

38. Nichols, T.A. et al., Detection of protease-resistant cervid prion protein in water from a CWD-endemic area. *Prion* 3, 171, 2009.

39. Denkers, N.D. et al., Aerosol and nasal transmission of chronic wasting disease in cervidized mice. *J. Gen. Virol.* 91, 1651, 2010.

40. Denkers, N.D. et al., Aerosol transmission of chronic wasting disease in white-tailed deer. *J. Virol.* 87, 1890, 2013.

41. Johnson, C. et al., Prion protein gene heterogeneity in free-ranging white-tailed deer within the chronic wasting disease affected region of Wisconsin. *J. Wildl. Dis.* 39, 576, 2003.

42. Raymond, G.J. et al., Evidence of a molecular barrier limiting susceptibility of humans, cattle and sheep to chronic wasting disease. *EMBO J.* 19, 4425, 2000.

43. Heaton, M.P. et al., Prion gene sequence variation within diverse groups of U.S. sheep, beef cattle, and deer. *Mamm. Genome* 14, 765, 2003.

44. O'rourke, K.I. et al., Polymorphisms in the prion precursor functional gene but not the pseudogene are associated with susceptibility to chronic wasting disease in white-tailed deer. *J. Gen. Virol.* 85, 1339, 2004.

45. Johnson, C. et al., Prion protein polymorphisms in white-tailed deer influence susceptibility to chronic wasting disease. *J. Gen. Virol.* 87, 2109, 2006.

46. Jewell, J.E. et al., Low frequency of PrP genotype 225SF among free-ranging mule deer (*Odocoileus hemionus*) with chronic wasting disease. *J. Gen. Virol.* 86, 2127, 2005.

47. Schatzl, H.M. et al., Is codon 129 of prion protein polymorphic in human beings but not in animals? *Lancet* 349, 1603, 1997.

48. O'rourke, K.I. et al., PrP genotypes of captive and free-ranging Rocky Mountain elk (*Cervus elaphus nelsoni*) with chronic wasting disease. *J. Gen. Virol.* 80, 2765, 1999.

49. Green, K.M. et al., The elk PRNP codon 132 polymorphism controls cervid and scrapie prion propagation. *J. Gen. Virol.* 89, 598, 2008.

50. Vickery, C.M. et al., Assessing the susceptibility of transgenic mice overexpressing deer prion protein to bovine spongiform encephalopathy. *J. Virol.* 88, 1830, 2014.

51. Race, B. et al., Susceptibilities of nonhuman primates to chronic wasting disease. *Emerg. Infect. Dis.* 15, 1366, 2009.

52. Perrott, M.R. et al., Evidence for distinct chronic wasting disease (CWD) strains in experimental CWD in ferrets. *J. Gen. Virol.* 93, 212, 2012.

53. Angers, R.C. et al., Prion strain mutation determined by prion protein conformational compatibility and primary structure. *Science* 328, 1154, 2010.

54. Haley, N.J. et al., Chronic wasting disease of cervids: Current knowledge and future perspectives. *Annu. Rev. Anim. Biosci.* 3, 305, 2015.

55. LaFauci, G. et al., Passage of chronic wasting disease prion into transgenic mice expressing Rocky Mountain elk (*Cervus elaphus nelsoni*) PrP^C. *J. Gen. Virol.* 87, 3773, 2006.

56. EFSA, Opinion on the European Food Safety Authority on a surveillance programme for Chronic Wasting Disease in the European Union. *EFSA J.* 70, 1, 2004.

57. Williams, E.S. et al., Chronic wasting disease in deer and elk in North America. *Revue Sci. Techn. Office Int. Epizoot.* 21, 305, 2002.

58. Race, R.E. et al., Comparison of abnormal prion protein glycoform patterns from transmissible spongiform encephalopathy agent-infected deer, elk, sheep, and cattle. *J. Virol.* 76, 12365, 2002.

59. Rubenstein, R. et al., Prion disease detection, PMCA kinetics, and IgG in urine from sheep naturally/experimentally infected with scrapie and deer with preclinical/clinical chronic wasting disease. *J. Virol.* 85, 9031, 2011.

60. Kurt, T.D. et al., Trans-species amplification of PrP(CWD) and correlation with rigid loop 170N. *Virology* 387, 235, 2009.

61. Kurt, T.D. et al., Efficient in vitro amplification of chronic wasting disease PrP^{RES}. *J. Virol.* 81, 9605, 2007.

62. Raymond, G.J. et al., Inhibition of protease-resistant prion protein formation in a transformed deer cell line infected with chronic wasting disease. *J. Virol.* 80, 596, 2006.

63. Kim, H.J. et al., Establishment of a cell line persistently infected with chronic wasting disease prions. *J. Vet. Med. Sci.* 74, 1377, 2012.

64. Angers, R.C. et al., Prions in skeletal muscles of deer with chronic wasting disease. *Science* 311, 1117, 2006.

65. Henderson, D.M. et al., Quantitative assessment of prion infectivity in tissues and body fluids by real-time quaking-induced conversion. *J. Gen. Virol.* 96, 210, 2015.

66. Angers, R.C. et al., Chronic wasting disease prions in elk antler velvet. *Emerg. Infect. Dis.* 15, 696, 2009.

67. Race, B. et al., Prion infectivity in fat of deer with chronic wasting disease. *J. Virol.* 83, 9608, 2009.

68. Spraker, T.R. et al., Impact of age and sex of Rocky Mountain elk (*Cervus elaphus nelsoni*) on follicle counts from rectal mucosal biopsies for preclinical detection of chronic wasting disease. *J. Vet. Diagn. Invest.* 21, 868, 2009.

69. Henderson, D.M. et al., Rapid antemortem detection of CWD prions in deer saliva. *PLoS One* 8, e74377, 2013.

70. Elder, A.M. et al., *In vitro* detection of prionemia in TSE-infected cervids and hamsters. *PLoS One* 8, e80203, 2013.

71. John, T.R. et al., Early detection of chronic wasting disease prions in urine of pre-symptomatic deer by real-time quaking-induced conversion assay. *Prion* 7, 253, 2013.

72. Bian, J. et al., Cell-based quantification of chronic wasting disease prions. *J. Virol.* 84, 8322, 2010.

73. Williams, E.S. et al., Transmissible spongiform encephalopathies. In: *Infectious Diseases of Wild Mammals*, 3rd edn., eds. Williams, E.S. and Barker, I.K. Iowa State University Press, Ames, IA, p. 292, 2001.

74. Wolfe, L.L. et al., PrP^{CWD} in rectal lymphoid tissue of deer (*Odocoileus* spp.). *J. Gen. Virol.* 88, 2078, 2007.

Index

A

Abalone herpesvirus (AbHV)
  biology, 808–809
  classification, 808
  clinical features, 809
  detection procedures
    conventional PCR, 812
    in situ hybridization, 812–813
    TaqMan qPCR, 811–812
  diagnosis
    conventional techniques, 809–810
    molecular techniques, 810
  discovery, 807
  epidemiology, 808–809
  morphology, 808–809
  pathogenesis, 809
  sample preparation methods, 810–811
ABC, *see* Avidin–biotin complex
Abnormal schooling, 11
ABPV, *see* Acute bee paralysis virus
ABV, *see* Avian bornavirus
ACE, *see* Antigen capture ELISAs
Acute bee paralysis virus (ABPV), 27–29, 33
Acute gastroenteritis, 585
Acute hemorrhagic gastroenteritis, 544
Acute respiratory distress syndrome,
    see Nipah encephalitis
Aerosolization, 278
African horse sickness (AHS), 612–613
African horse sickness virus (AHSV), 637
  biology, 611–612
  classification of, 610
  clinical features, 612–613
  diagnosis, 613
  epidemiology, 612
  methods
    detection procedures, 613–614
    sample preparation, 613
  morphology, 610
  pathogenesis, 613
  prevention, 613
  treatment, 613
  viral proteins, 610–611
African straw-colored fruit bats, 432
African swine fever (ASF), 290, 817
African swine fever virus (ASFV)
  antigen and antibody detection, 820
  biology, 818
  classification, 817–818
  clinical features, 819–820
  detection procedures
    conventional PCR, 820–821
    phylogenetic analysis, 821–822
    real-time PCR, 821
  epidemiology, 818
  genome organization, 818
  morphology, 818
  nucleic acid detection, 820
  pathogenesis, 819–820
  prevention, 820
  sample preparation methods, 820
  treatment, 820
  virus isolation, 820

Agar gel immunodiffusion (AGID), 603
  BTV, 621
  EIAV, 177, 182
  MDV, 778
  NSDV, 550
  PPRV, 467, 471
  RPV, 487, 490
AHS, *see* African horse sickness
AHSV, *see* African horse sickness virus
Aino virus (AINOV), 522
AIV, *see* Avian influenza virus
Akabane virus (AKAV), 563–564
  animal tissue, sample preparation
    methods, 526
  biology, 523–524
  cell cultures, sample preparation
    methods, 526
  classification, 521–523
  clinical features, 524–525
  detection procedures
    cloning and sequencing, 527–528
    nested PCR, 526–527
    quantitative RT-PCR, 528
    RT-PCR, 526–527
  diagnosis
    conventional techniques, 525
    molecular techniques, 526
  epidemiology, 523–524
  morphology, 523–524
  pathogenesis, 524–525
  reveres genetics systems
    plasmid construction, 528–529
    transfection, 529
AKAV, *see* Akabane virus
Aleutian mink disease virus (AMDV)
  classification, 679–680
  clinical features, 680–681
  detection procedures, 682
  diagnosis
    conventional techniques, 681
    molecular techniques, 681
  genome organization, 679–680
  morphology, 679–680
  pathogenesis, 680–681
  sample preparation, 682
  sequencing, 682–683
  transmission, 680–681
Alimentary tract infections, 48
Alphacoronavirus, 331–332; *see also* Porcine
    epidemic diarrhea virus
Alphanodavirus, 9; *see also* Nodamura virus
Alpharetrovirus, 145; *see also* Avian
    leukosis virus
Alphavirus, 255, 269–271; *see also* Eastern and
    Western equine encephalitis viruses;
    Venezuelan equine encephalitis virus
ALSV, *see* Avian leukosis/sarcoma group
ALV, *see* Avian leukosis virus
Amdoparvovirus, 680; *see also* Aleutian mink
    disease virus
AMDV, *see* Aleutian mink disease virus
Ample evidence, 325
aMPV, *see* Avian metapneumovirus
AngHV-1, *see* Anguillid herpesvirus 1

Anguillid herpesvirus 1 (AngHV-1), 783
Antemortem direct immunofluorescence
    testing
    FECVs and FIPVs, 327
Antemortem feline effusions, 327
Anti-BTV antibody detection, 621
Anti-EEV antibodies, 638
Antigen capture ELISAs (ACE), 228
Antigen ELISA, BRSV, 415
Anti-RVFV antibodies, 556
Aparavirus, 17; *see also* Taura syndrome virus
Aphthovirus, 61; *see also* Foot-and-mouth
    disease virus
APVs, *see* Avipoxviruses
APyVs, *see* Avian polyomaviruses
Arboviruses, *see* Arthropod-borne viruses
Arenaviruses
  classification of, 569–570
  clinical features and pathogenesis
    LCMV, 574
    viral hemorrhagic fevers, 574
  diagnosis
    immunological detection, 575
    molecular detection, 575–576
    virus isolation, 575
  methods
    detection procedures, 576–580
    sample preparation, 576
  morphology and genome organization
    biology, 571–572
    GTOV, 573
    JUNV, 573
    LASV, 572–573
    LCMV, 572
    MACV, 573–574
Argentine hemorrhagic fever, 569–570, 573
Arkansas, 310
Arterivirus, 277, 287–288; *see also* Equine
    arteritis virus; Porcine reproductive
    and respiratory syndrome virus
Arthropod-borne viruses, 242, 543
Arthus-type vasculitis, 325
Artiodactyla, 486
Asian swine IAVs, 401
Ataxia, 48, 55
Aujeszky's disease, 290, 820; *see also*
    Varicellovirus
Aviadenovirus, 736; *see also* Poultry
    adenoviruses
Avian astroviruses
  biology, 102–103
  classification, 101–102
  clinical features and pathogenesis
    ANV, 104
    CAstV, 104–105
    DAstV, 105
    TAstV, 105
  detection procedures, 107
  diagnosis
    conventional techniques, 105–106
    molecular techniques, 106
  epidemiology, 102–103
  sample preparation methods, 106–107
  structure, 102–103

Avian bornavirus (ABV)
    classification of, 347
    clinical features, 348–349
    detection procedures
        canary birds, 352–353
        psittacine birds, RT-PCR, 351–352
        real-time RT-PCR, 352–353
    diagnosis
        indirect/direct, conventional techniques,
            349–351
        molecular techniques, 351
    epidemiology, 347–348
    genome structure, 347–348
    morphology, 347–348
    pathogenesis, 348–349
Avian encephalomyelitis virus (AEV)
    chemical composition, 47
    classification, 47
    clinical features, 48
    CNS, 47
    detection procedures
        PCR, 50
        reverse transcription reaction,
            49–50
    diagnosis
        conventional techniques, 48
        molecular techniques, 49
    epidemiology, 47–48
    morphology, 47
    neurologic symptoms, 50
    pathobiology, 47–48
    pathogenesis, 48
    sample collection, 49
Avian infectious bronchitis virus
    biology, 309
    classification, 307–309
    clinical features, 310–311
    clinical signs, 307
    diagnosis
        conventional techniques, 311
        molecular techniques, 311
    epidemiology, 309–310
    flocks, 307
    genome organization, 307–309
    methods
        detection procedures, 312–313
        sample collection, 312
    morphology, 307–309
    pathogenesis, 310–311
    transmission, 310–311
Avian influenza virus (AIV)
    classification of, 377–378
    clinical features, 378
    conventional RT-PCR, 379–380
    diagnosis, 378–379
    epidemiology, 378
    real-time RT-PCR, 380–381
    sample processing methods, 379
Avian leukosis/sarcoma group (ALSV), 145
Avian leukosis virus (ALV), 205
    clinical features, 147–148
    detection procedures
        conventional PCR, 150
        LAMP, 153
        multiplex PCR, 150–151
        quantitative PCR, 152
    diagnosis
        conventional techniques,
            148–149
        molecular techniques,
            149–150

epidemiology, 147
    etiology
        classification, 145–146
        pathogenicity, 147
        structure and replication, 146
        subgroups, 146–147
    pathogenesis, 147–148
    sample preparation methods, 150
    transmission, 147
Avian metapneumovirus (aMPV)
    classification of, 407–408
    clinical features, 408–409
    detection procedures, 410–411
    diagnosis techniques
        conventional, 409
        molecular, 409–410
    epidemiology, 408–409
    genome structure and function, 407–408
    pathogenesis, 408–409
    sample collection and preparation
        methods, 410
Avian nephritis virus (ANV), 104
Avian paramyxovirus serotypes, 447–448
Avian polyomaviruses (APyVs), 719
Avidin–biotin complex (ABC), 280
Avihepatovirus, 54; see also Duck
        hepatitis A virus
Avipolyomavirus, 719; see also Polyomaviruses
Avipoxvirus, 837
Avipoxviruses (APVs)
    classification, 837–838
    clinical features, 838–839
    detection procedures
        agar gel immunodiffusion, 841
        elementary bodies (Borrel bodies), 841
        ELISA, 841
        PCR, 841–842
        virus isolation, 841
    diagnosis
        conventional techniques, 840
        molecular techniques, 840–841
    epidemiology, 838
    genome structure, 838
    morphology, 838
    pathogenesis, 838–839
    replication, 838
    sample collection and preparation
        methods, 841
Avulavirus, 447; see also Newcastle
        disease virus

B

BAL, see Bronchoalveolar lavage
Barfin flounder nervous necrosis virus
        (BFNNV), 9–11
BAstV, see Bovine astrovirus
Bat-borne paramyxovirus, 455
Bayesian analysis, BFDV, 666
BD, see Border disease
BDV, see Border disease virus
Beak and feather disease virus (BFDV)
    classification, 659–660
    clinical features, 662–663
    detection procedures
        Bayesian analysis, 666
        DNA sequencing, 665–666
        PCR, 665
    diagnosis
        conventional techniques, 664
        molecular techniques, 664

genome organization, 660–662
    morphology, 660–662
    pathogenesis, 662–663
    replication of, 662
    sample preparation, 664–665
Beaudette strain, 309
Bee mite parasitic syndrome, 29
Bee paralysis virus
    biology, 29–30
    classification, 28–29
    clinical features, 30
    colonies, 27–28
    detection procedures
        conventional RT-PCR, 32
        real-time PCR, 32–33
    diagnosis, 30–31
    epidemiology, 29–30
    external stresses, 29
    genome organization, 28–29
    morphology, 29–30
    optimization of, 29
    pathogenesis, 30
    sample preparation, 31–32
    types of, 28
Bee viruses, 27
BEF, see Bovine ephemeral fever
BEFV, see Bovine ephemeral fever virus
Beilong virus, 456
Berne virus (BEV),
        see Equine torovirus
Betaherpesvirinae,
        see Suid herpesvirus 2
Betanodavirus
    biology, 11
    classification, 9–10
    clinical features, 11–12
    detection procedures
        conventional RT-PCR, 13
        real-time RT-PCR, 13
    diagnosis, 12
    epidemiology, 11
    genome structure, 10
    genotypic characteristics, 10
    morphology, 10
    pathogenesis, 11–12
    phenotypic characteristics, 10
    prevention, 12
    replication of, 11
    sample preparation, 12–13
    treatment, 12
BFD, see Budgerigar fledgling disease
BFDV, see Beak and feather disease virus;
        Budgerigar fledgling disease virus
BFNNV, see Barfin flounder nervous necrosis
        virus
BHQ, see Black hole quencher
Bile duct hyperplasia, 55
Biotype conversion, 328
Black hole quencher (BHQ), 262–263
Blood–brain barrier (BBB), 227, 424
Blood Systems Research Institute, 647
Blue eye disease virus, see Porcine
        rubulavirus
Bluetongue virus (BTV), 629; see also Epizootic
        hemorrhagic disease virus
    biology, 619–620
    classification of, 619
    clinical features, 620
    diagnosis techniques
        conventional, 620–621
        molecular, 621–622

epidemiology, 619–620
methods
    detection procedures, 622–624
    sample preparation, 622
morphology, 619
pathogenesis, 620
BLV, *see* Bovine leukemia virus
Bootstrap values, 297
Border disease (BD), 211
Border disease virus (BDV)
classification, 211–212
clinical symptoms and
    pathologic changes
      acute infection, 214
      fetal infection, 214–215
diagnosis
    conventional techniques, 215
    molecular techniques, 215–216
epidemiology, 213–214
genome organization, 212
morphology, 213
pathogenesis, 214
RT-PCR, 217–2118
transmission, 214
virus isolation methods, 216–217
Bornavirus, 347; *see also* Avian bornavirus
Bovine astrovirus (BAstV), 101
Bovine coronavirus (BCoV)
biology, 318–319
classification, 317–318
clinical features, 319
diagnosis, 319–320
epidemiology, 318–319
genome organization, 318
methods
    detection procedures, 320–321
    sample preparation, 320
morphology, 318
pathogenesis, 319
phylogenetic examination of, 318
prevention, 320
transmission of, 319
treatment, 320
Bovine ephemeral fever (BEF), 355
Bovine ephemeral fever virus (BEFV)
biology, 355–356
classification of, 355
clinical features, 356–357
diagnosis, 357–358
epidemiology, 355–356
morphology, 355–356
pathogenesis, 356–357
sample preparation methods, 358
Bovine–human reassortant, 647
Bovine leukemia virus (BLV)
biology, 158–159
classification, 157–158
clinical features, 159–160
detection procedures
    nested PCR, 161–162
    real-time PCR, 162–163
diagnosis
    conventional techniques, 160–161
    molecular techniques, 161
epidemiology, 158–159
genome organization, 157–158
morphology, 157–158
pathogenesis, 159–160
sample collection and preparation
    methods, 161
transmission, 158–159

Bovine papillomavirus (BPV)
classification, 728
clinical features, 728–729
diagnosis
    conventional techniques, 729
    molecular techniques, 729–730
genome organization, 728
pathogenesis, 728–729
PCR, detection procedures, 731–732
preparation methods, 730–731
sample collection methods, 730–731
transmission, 728–729
Bovine papular stomatitis virus (BPSV),
    883–884
Bovine polyomavirus (BPyV), 721–722
Bovine respiratory disease (BRD), 413; *see also*
    Bovine respiratory syncytial virus
Bovine respiratory disease complex (BRDC),
    317, 319–320
Bovine respiratory syncytial virus (BRSV)
biology, 413–414
classification of, 413
clinical features, 414
detection procedure, 417
diagnosis techniques
    conventional, 414–416
    molecular, 416
epidemiology, 413–414
genome organization, 413
morphology, 413
pathogenesis, 414
sample collection and preparation methods,
    416–417
Bovine rotavirus A (BRV), 318–320
Bovine RVs, 646
Bovine spongiform encephalopathy (BSE)
atypical BSE, 903–904
biology, 902
classification, 901
clinical features, 904–905
detection procedures, 909
diagnosis, 905–908
epidemiology, 902–903
equipment, 908
inactivation methods, 909
morphology, 901–902
pathogenesis, 904–905
reagents, 908
sample preparation methods, 909
Bovine torovirus (BToV), 339; *see also*
    Toroviruses
Bovine viral diarrhea viruses (BVDV)
    1 and 2
classification, 223–224
clinical features
    persistent infection, 227
    reproductive tract infection,
      226–227
    transient infection, 226
    transmission, 227
conventional techniques
    antigen detection, 228
    serology, 228–229
    virus isolation, 228
detection procedures
    quantitative RT-PCR, 231
    standard RT-PCR, 230
genome organization, 225–226
molecular techniques, 229
morphology, 225–226
phylogenetic analysis, 224

sample preparation
    nasal swab samples, 229
    semen, 230
    serum, 229
    tissue sample, 230
    white blood cells, 229
transmission, 226–227
BPSV, *see* Bovine papular stomatitis virus
BPyV, *see* Bovine polyomavirus
BRDC, *see* Bovine respiratory disease complex
Bronchoalveolar lavage (BAL), 415
BRSV, *see* Bovine respiratory syncytial virus
BRV, *see* Bovine rotavirus A
BSE, *see* Bovine spongiform encephalopathy
BTV, *see* Bluetongue virus
Buckshot, 19, 21
Budgerigar fledgling disease (BFD), 720
Budgerigar fledgling disease virus (BFDV)
biology, 720
clinical features, 721
detection procedures, 723–724
diagnosis, 722
genome organization, 720
morphology, 720
pathogenesis, 721
sample preparation methods, 723
Buffalopox virus (BPXV), 869
Buggy creek virus, 258

C

Cabassou virus (CABV), 270, 272
CAEV, *see* Caprine arthritis–encephalitis virus
Calf diarrhea (CD), 317, 319–320
Calf scours, 319
Caliciviruses
classification, 129–130
clinical features, 131–132
detection procedures, 132
diagnosis, 132
epidemiology, 131
genome organization, 130–131
morphology, 130–131
pathogenesis, 131–132
replication, 131
sample preparation methods, 132
Canarypox virus (CNPV), 837
Canine coronavirus (CCoV), 323–324, 328, 331
Canine distemper encephalitis (CDE), 424
Canine distemper virus (CDV)
biology, 421–422
classification of, 421
clinical features, 422–423
diagnosis techniques
    conventional, 423–425
    molecular, 425
epidemiology, 421–422
genome organization, 421–422
methods
    detection procedures, 426–427
    sample preparation, 425–426
pathogenesis, 422–423
RPV, 486
Canine influenza virus (CIV), 384
Canine parvovirus (CPV)
biology, 688
classification, 687–688
clinical features, 688–689
diagnosis
    conventional techniques, 689–690
    molecular techniques, 690

epidemiology, 688
methods
    LAMP, 691
    PCR-based detection, 690–691
    sample preparation, 690
morphology, 687–688
pathogenesis, 688–689
Cap-independent translation, 234
Caprine arthritis–encephalitis virus (CAEV)
biology, 167–168
classification, 167
clinical forms, 169
diagnosis
    conventional diagnosis, 169–170
    molecular techniques, 170–171
DNA extraction, 172
epidemiology, 167–168
genotype identification, 173
morphology, 167–168
pathogenesis, 168–169
PCR, 172–173
sample collection and preparation methods, 171–172
sequence analysis, 173
SRLV cloning, 173
Caprine RSV, 413
Capripoxviruses
biology, 846–847
classification, 845
clinical features, 847–848
detection procedures
    amplification methods, 850–851
    extraction methods, 850
diagnosis
    conventional techniques, 848–849
    molecular techniques, 849
epidemiology, 846–847
genome organization, 846–847
morphology, 846–847
pathogenesis, 847–848
sample preparation methods, 849–850
Capsid-coding regions, 53
Cardiovirus, 67; see also Encephalomyocarditis virus (EMCV)
Carrier stallions function, 278
Cattle sperm agglutination inhibition test, 694
Cat-to-cat transmission, 324–325
CAV, see Chicken anemia virus
CBPV, see Chronic bee paralysis virus
CCD, see Colony collapse disorder
CCHFV, see Crimean-Congo hemorrhagic fever virus
CCoV, see Canine coronavirus
CCVD, see Channel catfish virus disease
CD, see Calf diarrhea
CDE, see Canine distemper encephalitis
cDNA, see Complementary DNA
cDNA synthesis, chronic bee paralysis virus, 31–32
CDV, see Canine distemper virus
cELISA, see Commercial cELISA
c-ELISA, see Competitive enzyme-linked immunosorbent assays
Centers for Disease Control and Prevention, 364
Central nervous system (CNS)
    AEV, 47
    arenaviruses, 570, 574
    EEEV, 256
    invasion, 259
    LIV, 242
    RABV, 362–363
    VEEV, 271

Cephalothorax, 298
Cerebellar hypoplasia, 227
CFT, see Complement fixation test
Channel catfish virus disease (CCVD), 797–798
Chapare virus (CHAV), 569–570
CHAV, see Chapare virus
Chelex resin method, 31
Chicken anemia virus (CAV), 659
Chicken astrovirus (CAstV), 104–105
Chronic bee paralysis virus (CBPV), 27, 33; see also Bee paralysis virus
Chronic inflammation, 325
Chronic wasting disease (CWD), 901
biology, 913
clinical features, 913–915
diagnosis, 915–917
epidemiology, 913
methods
    detection procedures, 918–919
    equipment, 918
    reagents, 918
    sample preparation, 918–919
pathogenesis, 913–915
Circovirus, 671; see also Porcine circovirus
CIV, see Canine influenza virus
CJD, see Creutzfeldt–Jakob disease
Classical swine fever virus (CSFV)
classification, 233–234
clinical features and pathogenesis
    acute disease, 235
    congenital disease, 235–236
    subacute and chronic disease, 235
detection procedures, 237
    conventional RT-PCR, 237
    real-time RT-PCR, 237
diagnosis, 236
epidemiology, 234–235
genome organization, 234
morphology, 234
phylogenetic analysis, 233
prevention, 236–237
replication, 234–235
RNA arrangement, 234
sample preparation, 237
treatment, 236–237
Classical swine fever (hog cholera) virus (CSFV), 211
Clathrin-mediated endocytosis, 348
    IBV, 309
    PRRSV, 289
    WNV, 248
Climb vegetation, 242
CNPV, see Canarypox virus
CNS, see Central nervous system
Colony collapse disorder (CCD), 27–28
Commercial cELISA, 280, 282
Competitive enzyme-linked immunosorbent assays (c-ELISA)
    BTV, 621, 624
    PPRV, 467, 471
    RPV, 488, 491–492
    VSV, 372
Complementary DNA (cDNA), 312
    RPV, 488
    RVs, 650
Complement fixation test (CFT), 372, 548, 550
Concanavalin A (Con A), 603
Conjunctivitis, 235
Conventional PCR
    AbHV, 812
    ALV, 150

ASFV, 820–821
cypriniviruses, 790–791
megalocytiviruses, 832
PmoHDV, 702
REVs, 208
WSSV, 894
Conventional RT-PCR
    AIV, 378–379
    BCoV, 320–321
    bee paralysis virus, 31–32
    betanodavirus, 13
    CSFV, 237
    ISAV, 396–397
    NDV, 451
    PEDV, 335
    PRRSV, 291
Counterimmunoelectrophoresis (CIEP)
    PPRV, 467, 471
    RPV, 487
Cowpox virus (CPXV), 869
CP, see Cytopathic
CPE, see Cytopathic effects
CPV, see Canine parvovirus
Creutzfeldt–Jakob disease (CJD), 901
Cricket paralysis-like virus, 17; see also Taura syndrome virus
Crimean-Congo hemorrhagic fever virus (CCHFV), 543
Cross protection tests, DHV, 56
Cutting-edge molecular biology techniques, EAV, 282
CWD, see Chronic wasting disease
CyHV-1, see Cyprinid herpesviruses 1
CyHV-2, see Cyprinid herpesviruses 2
CyHV-3, see Cyprinid herpesviruses 3
Cynomolgus macaques, 259
Cyprinid herpesviruses 1 (CyHV-1)
biology, 784
clinical features, 787
conventional techniques, 788
detection procedures, 790–791
epidemiology, 784
molecular techniques, 789
pathogenesis, 787
Cyprinid herpesviruses 2 (CyHV-2)
biology, 784
clinical features, 787
conventional techniques, 788
detection procedures, 790–791
epidemiology, 784
molecular techniques, 789
pathogenesis, 787–788
Cyprinid herpesviruses 3 (CyHV-3)
biology, 784–787
clinical features, 787
conventional techniques, 788–789
detection procedures, 790–791
epidemiology, 784–787
molecular techniques, 789
pathogenesis, 787–788
Cyprinivirus, 788; see also Cypriniviruses
Cypriniviruses
biology, 784–787
classification, 783–784
clinical features, 787
detection procedures
    conventional PCR, 790–791
    quantitative PCR, 791
diagnosis
    conventional techniques, 788–789
    molecular techniques, 789

epidemiology, 784–787
morphology, 784–787
pathogenesis, 787–788
sample preparation methods, 789–790
Cysteine-rich scavenger receptor, 289
Cytomegalovirus
Cytopathic (CP), 224–225, 228
Cytopathic effects (CPEs), 2, 228, 249, 281
   arenaviruses, 575
   EHDV, 631
   Menangle virus, 441
   NDV, 450
   NiV, 458
   NSDV, 548
   PPRV, 468
   toroviruses, 341
Cytoplasmic remnants, 19

D

DAstV, *see* Duck astrovirus
DBH, *see* Dt-blot hybridization
Deformed wing virus (DWV)
   classification, 37–39
   detection procedures
      nested PCR detection, 43–44
      real-time PCR quantification, 44
      reverse transcription, 43
      total RNA isolation, 42–43
   diagnosis
      conventional techniques, 40
      molecular techniques, 40
   epidemiology, 39–40
   members of, 39
   methods
      Iflavirus discovery, 41–42
      sample preparation, 40–41
   oligonucleotide primers for, 44
   organization of, 38
   pathogenesis, 39–40
   phylogram of, 38
   quantification of, 42
   strains, 40
   transmission electron photograph of, 37–38
Deltaretrovirus, 157; *see also* Bovine
      leukemia virus
Dengue virus (DENV), 261
DENV, *see* Dengue virus
Dependoparvovirus, 693; *see also* Goose
      parvovirus
DGIV, *see* Dwarf gourami iridovirus
DHAV, *see* Duck hepatitis A virus
Dhori virus (DHOV), 393
DHOV, *see* Dhori virus
DIC, *see* Disseminated intravascular coagulation
Dicistrovirus, 30
Direct fluorescent antibody test, 320
Disseminated intravascular coagulation (DIC), 500
Distemper virus (DV), 485
Disulfide-linked heterodimers, 234
DLP, *see* Double-layered particle
DNA-based vaccines, 250
DNA extraction
   CAEV, 172
   macavirus, 770
   megalocytiviruses
      using phenol/chloroform, 831
      using spin column, 832
   PmoHDV, 702
   SRLVs, 172
   VMV, 172

DNA viruses
   AbHV
      biology, 808–809
      classification, 808
      clinical features, 809
      conventional PCR, 812
      conventional techniques, 809–810
      discovery, 807
      epidemiology, 808–809
      in situ hybridization, 812–813
      molecular techniques, 810
      morphology, 808–809
      pathogenesis, 809
      sample preparation methods, 810–811
      TaqMan qPCR, 811–812
   AMDV
      classification, 679–680
      clinical features, 680–681
      conventional techniques, 681
      detection procedures, 682
      genome organization, 679–680
      molecular techniques, 681
      morphology, 679–680
      pathogenesis, 680–681
      sample preparation, 682
      sequencing, 682–683
      transmission, 680–681
   APVs
      classification, 837–838
      clinical features, 838–839
      detection procedures, 841–842
      diagnosis, 839–841
      epidemiology, 838
      genome structure, 838
      morphology, 838
      pathogenesis, 838–839
      replication, 838
      sample collection and preparation
         methods, 841
   ASFV
      antigen and antibody detection, 820
      biology, 818
      classification, 817–818
      clinical features, 819–820
      conventional PCR, 820–821
      epidemiology, 818
      genome organization, 818
      morphology, 818
      nucleic acid detection, 820
      pathogenesis, 819–820
      phylogenetic analysis, 821–822
      prevention, 820
      real-time PCR, 821
      sample preparation methods, 820
      treatment, 820
      virus isolation, 820
   BFDV
      Bayesian analysis, 666
      biology, 720
      classification, 659–660
      clinical features, 662–663, 721
      conventional techniques, 664
      detection procedures, 723–724
      diagnosis, 722
      DNA sequencing, 665–666
      genome organization, 660–662, 720
      molecular techniques, 664
      morphology, 660–662, 720
      pathogenesis, 662–663, 721
      PCR, 665
      replication of, 662

      sample preparation, 664–665
      sample preparation methods, 723
   BPV
      classification, 728
      clinical features, 728–729
      conventional techniques, 729
      genome organization, 728
      molecular techniques, 729–730
      pathogenesis, 728–729
      PCR, detection procedures, 731–732
      preparation methods, 730–731
      sample collection methods, 730–731
      transmission, 728–729
   capripoxviruses
      amplification methods, 850–851
      biology, 846–847
      classification, 845
      clinical features, 847–848
      conventional techniques, 848–849
      epidemiology, 846–847
      extraction methods, 850
      genome organization, 846–847
      molecular techniques, 849
      morphology, 846–847
      pathogenesis, 847–848
      sample preparation methods,
         849–850
   CPV
      biology, 688
      classification, 687–688
      clinical features, 688–689
      conventional techniques, 689–690
      epidemiology, 688
      LAMP, 691
      molecular techniques, 690
      morphology, 687–688
      pathogenesis, 688–689
      PCR-based detection, 690–691
      sample preparation, 690
   cypriniviruses
      biology, 784–787
      classification, 783–784
      clinical features, 787
      detection procedures, 790–791
      diagnosis, 788–789
      epidemiology, 784–787
      morphology, 784–787
      pathogenesis, 787–788
      sample preparation methods,
         789–790
   GPV
      agar gel precipitin test, 695
      cattle sperm agglutination inhibition
         test, 694
      classification, 693
      clinical features, 694
      diagnosis, 694
      ELISA, 695
      fluorescent quantitative real-time PCR,
         695–696
      genome organization, 693
      morphology, 693
      multiplex PCR, 695
      neutralization test, 694
      passive immunity test, 694
      PCR, 695
      preparation methods, 694
      replication, 693–694
      sample collection methods, 694
      virus isolation, 694
      Western blotting, 695

IcHV1
    antibody neutralization, 801
    biology, 797–798
    cell culture isolation, 800–801
    classification, 797
    clinical signs, 799
    conventional techniques, 799
    diagnose, 799
    epidemiology, 797–798
    genome organization, 797–798
    molecular assays, 800
    molecular techniques, 799
    morphology, 797–798
    pathogenesis, 799
    PCR, 801–804
    sample preparation methods, 799–800
    virus culture, 799–800
ILTV
    biology, 756
    classification, 753
    clinical features, 757
    conventional techniques, 757
    genome organization, 754–756
    LAMP, 759–761
    molecular techniques, 757–758
    morphology, 753–754
    pathogenesis, 756–757
    real-time PCR, 758–759
    sample preparation methods, 758
    transmission, 756–757
macavirus
    classification, 765–766
    clinical features, 766–767
    detection procedures, 769–770
    diagnosis, 767–769
    genome organization, 765–766
    morphology, 765–766
    pathogenesis, 767
    preparation methods, 769
    sample collection methods, 769
    transmission, 766
MDV
    biology, 774–775
    classification, 774
    clinical features, 775
    detection procedures, 778–779
    diagnosis, 775–777
    morphology, 774–775
    pathogenesis, 775
    sample preparation methods, 777–778
megalocytiviruses
    biology, 826–827
    classification, 825–826
    clinical features, 827
    detection procedures, 832–833
    diagnosis, 827–831
    DNA extraction, 831–832
    epidemiology, 826–827
    morphology, 826–827
    pathogenesis, 827
MYXV
    amyxomatous myxomatosis, 858
    coevolution of, 857
    conventional techniques, 859–860
    cutaneous myxomatosis, 857–858
    genome organization, 856
    immunohistochemical detection,
        861–862
    in vitro culture, 861
    molecular techniques, 860–861
    morphology, 855–856

natural history, 856
pathogenesis, 858–859
PCR amplification, 863–864
peracute disease, 858
sample collection and processing
    methods, 861
serological determination, 862–863
transmission of, 856–857
orthopoxvirus
    biology, 871
    classification, 870
    clinical features, 872
    conventional techniques, 873
    detection procedures, 876–877
    diagnosis, 872–873
    epidemiology, 871–872
    genome, 870–871
    molecular techniques, 873–875
    morphology, 870
    pathogenesis, 872
    physico-biochemical properties, 870
    sample preparation methods, 875–876
PCV
    classification, 671–672
    epidemiology, 672
    physiochemistry of, 672
    transmission, 672
PmoHDV
    biology, 700
    classification, 699–700
    clinical features, 700–701
    conventional PCR, 702
    diagnosis, 701
    DNA extraction methods, 702
    epidemiology, 700
    genome organization, 700
    morphology, 700
    pathogenesis, 700–701
    real-time PCR, 702
    sample preparation methods, 702
    treatment and prevention, 702
    virus purification methods, 702
poultry adenoviruses
    classification, 736
    conventional techniques, 739–740
    detection procedures, 741–742
    epidemiology and clinical features,
        736–739
    genome organization, 736
    molecular techniques, 740
    morphology, 736
    sample preparation methods, 740–741
PPV
    BPSV, 883–884
    classification, 713, 881–882
    clinical features, 714, 882
    conventional techniques, 714–715
    detection procedures, 885–888
    diagnosis, 884–885
    epidemiology, 714
    genome organization, 713–714, 881–882
    molecular techniques, 715
    morphology, 713–714, 881–882
    nano-PCR detection, 715–716
    nucleic acid probe detection, 716
    Orf virus, 882–883
    pathogenesis, 714, 882
    PCPV, 884
    preparation methods, 715
    PVNZ, 884
    sample collection methods, 715, 885

transmission, 882
virus isolation methods, 885
PstDV
    biology/epidemiology, 706
    classification, 705–706
    clinical features, 706–707
    conventional techniques, 707
    dot blot and in situ hybridization, 708
    end-point PCR, 707
    end-point PCR and nested PCR, 708–709
    LAMP, 708
    morphology, 706
    nested PCR, 707
    pathogenesis, 706–707
    real-time qPCR, 709
    sample preparation method, 708
    single and duplex quantitative PCR,
        707–708
SuHV-2
    classification, 747–748
    clinical features, 748
    conventional approaches, 748–749
    LAMP, 749–750
    molecular detection, 749
    pathogenesis, 748
    preparation methods, 749
    quantitative real-time polymerase chain
        reaction, 750–751
    sample collection methods, 749
    sample extraction, 749
    transmission, 748
    virus isolation, 749
WSSV
    biology, 892
    classification, 891–892
    clinical features, 893
    detection procedures, 894–895
    diagnosis, 893–894
    epidemiology, 892–893
    genome organization, 892
    geographic distribution, 893
    host range, 892
    morphology, 892
    sample preparation methods, 894
    transmission, 892–893
    treatment and prevention, 894
Double-layered particle (DLP), 644–645
Double-stranded RNA (dsRNA)
    AHSV, 610
    EEV, 637
    EHDV, 629–630
    IBDV, 601
    IMNV, 593–594
    PBV, 585–586
    RVs, 643
Dry/noneffusive/parenchymatous, 325
dsRNA, see Double-stranded RNA
Dt-blot hybridization (DBH), 20
Duck astrovirus (DAstV), 54, 56–57, 105
Duck hepatitis A virus (DHAV), 54–56
Duck hepatitis virus (DHV)
    classification, 53–55
    clinical features, 55
    diagnosis
        conventional techniques, 56
        molecular techniques, 56–57
    fatal viral disease, 53
    genome organization, 53–55
    methods
        detection procedures, 57–58
        sample collection, 57

morphology, 53–55
pathogenesis, 55
serotypes of, 53
Dugbe virus (DUGV), 544
Dulbecco's modified Eagle's medium (DMEM), 416, 622
Duplex nested polymerase chain reaction detection, 410–411
DV, *see* Distemper virus
Dwarf gourami iridovirus (DGIV), 825
DWV, *see* Deformed wing virus

E

E1, *see* Primary envelope glycoprotein
Eastern and Western equine encephalitis viruses (EEEV) and (WEEV)
    biology, 256–258
    classification, 255–256
    clinical features, 258–259
    diagnosis
        conventional techniques, 260
        molecular techniques, 260–261
    epidemiology, 256–258
    future perspectives, 264–265
    genome organization, 256
    geographic distribution, 257
    methods
        detection procedures, 262–264
        sample preparation, 261–262
    morphology, 256
    pathogenesis, 258–259
    recombination event, 256
Eastern equine encephalitis virus (EEEV), 255–256
Ebola hemorrhagic fever (EHF), 495
Ebola virus
    biology, 498–499
    classification of, 495–496
    clinical features, 499–500
    detection procedures, 502
    diagnosis techniques, 500–501
    epidemiology, 499
    genome organization, 496–498
    list of, 495–496
    morphology, 496–498
    pathogenesis, 499–500
    sample preparation methods, 501
EDTA, *see* Ethylenediaminetetraacetic acid
EEEV, *see* Eastern equine encephalitis virus
EEV, *see* Equine encephalosis virus
EGF, *see* Epidermal growth factor
EHF, *see* Ebola hemorrhagic fever
EIA, *see* Equine infectious anemia
EIAV, *see* Equine infectious anemia virus
18S rRNA, 442–443
EIV, *see* Equine influenza virus
Electron microscopy (EM)
    HeV, 434
    IAV, 402
    ISAV, 396
    Menangle virus, 442
    NiV, 458
    PEDV, 334
    toroviruses, 341
    Vero cells, 334
ELISA, *see* Enzyme-linked immunosorbent assay
EMCV, *see* Encephalomyocarditis virus
Encephalitis, 241, 244–245
Encephalomyocarditis virus (EMCV)

capsid, assembly of, 69
classification, 67
detection procedures
    one-step conventional RT-PCR, 73–74
    real-time RT-PCR, 74–75
diagnosis
    conventional techniques, 72–73
    molecular techniques, 73
epidemiology
    host range, distribution, and transmission, 71
    vaccine strategies, 72
    zoonotic potential, 71–72
genomic organization, 67–69
infectious cycle, 69–71
RNA extraction, 73
rodent, clinical features and pathogenesis, 72
sample preparation methods, 73
swine, clinical features and pathogenesis, 72
three-dimensional structure, 69
viral proteins, 69
Endometrial inflammation, 226
Endoplasmic reticulum (ER), 408, 645
Enteric viral pathogens, 585
Enterovirus, 53, 89; *see also* Duck hepatitis virus; Swine vesicular disease virus (SVDV)
Enzootic transmission cycle, 248
Enzootic viruses, 272
Enzyme-linked immunosorbent assay (ELISA)
    ABV, 349
    AEV, 48
    aMPV, 409
    APVs, 841
    arenaviruses, 576–577
    BCoV, 319–320
    bee paralysis virus, 31
    BRSV, 415
    BTV, 621
    CSFV, 236
    DHV, 56–57
    DWV, 40
    Ebola virus, 500
    EEV, 639
    GPV, 695
    HeV, 434
    IAV, 402
    IBDV, 603
    IBV, 311
    macavirus, 768
    MARV, 515
    NDV, 450
    NiV, 459
    NSDV, 548, 550
    PBV, 480
    PEDV, 333–334
    PRRSV, 290
    RABV, 364
    RVFV, 555
    toroviruses, 341
    VSV, 372
    WNV, 250
Eosinophilic intracytoplasmic inclusion body, 424–425
Epidemic/epizootic viruses, 271–272
Epidemic hemorrhagic fever, 495
Epidermal growth factor (EGF), 566
Epithelial necrosis, 470

Epizootic hemorrhagic disease (EHD), 629
Epizootic hemorrhagic disease virus (EHDV)
    classification of, 629
    clinical features, 631
    diagnosis, 631–632
    epidemiology, 630
    genome organization, 629–630
    methods
        ambion MagMAX-96, 632–633
        detection procedures, 633–634
        QIAamp, 632
        RNA isolation, 632–633
        sample collection and preparation, 632
    morphology, 629–630
    pathogenesis, 630
    transmission, 630
    WTD, 629
Equine arteritis virus (EAV)
    aborted fetus, 277
    biology, 277–279
    clinical features, 279–280
    diagnosis
        conventional techniques, 280
        molecular techniques, 280–281
    epidemiology, 277–279
    future perspectives, 282–283
    methods
        detection procedures, 282
        molecular testing, clinical samples for, 281–282
        sample collection, 281
        virus isolation, clinical samples for, 281
    morphology, 277–279
    multiplication of, 279
    pathogenesis, 279–280
    transmission of, 278–279
    viral proteases, 277
Equine encephalosis virus (EEV)
    classification of, 637–638
    clinical features, 639
    diagnosis, 639
    epidemiology, 638
    equine ephemeral fever, 637
    genome organization, 638
    methods
        detection procedures, 639–640
        sample preparation, 639
    morphology, 638
    pathogenesis, 639
    prevention, 639
Equine infectious anemia (EIA), 177
Equine infectious anemia virus (EIAV)
    AGID, 177, 182
    classification, 178–179
    clinical features, 181–182
    detection procedures, 184–185
    diagnosis
        conventional techniques, 182
        molecular techniques, 183
    epidemiology, 179–180
    genome organization, 178–179
    morphology, 178–179
    pathogenesis, 181–182
    sample collection and preparation methods, 183–184
    transmission, 180–181
Equine influenza (EI), 383
Equine influenza virus (EIV)
    biology, 383–384
    classification of, 383
    clinical features, 384–385

detection procedures, 386–390
diagnosis techniques
conventional, 385–386
molecular, 386
epidemiology, 383–384
morphology, 383–384
pathogenesis, 384–385
sample collection and preparation
methods, 386
Equine morbillivirus, 431
Equine RVs, 646
Equine torovirus (EToV), 339–341
ER, *see* Endoplasmic reticulum
E^{rns} protein, 226
Ethylenediaminetetraacetic acid (EDTA), 281
EToV, *see* Equine torovirus
European swine IAVs, 400–401
Extensive hepatocyte necrosis, 55

F

FA, *see* Fluorescent antibody
FAstV, *see* Feline astrovirus
FAT, *see* Fluorescent antibody test
FBS, *see* Fetal bovine serum
FCV, *see* Feline calicivirus
Fecal–oral route transmission, 324
FECV-to–FIPV mutations, 325
FECV-to–FIPV transformation, 323
FeFV, *see* Feline foamy virus
Feline astrovirus (FAstV), 101
Feline calicivirus (FCV)
biology, 122
classification, 121–122
clinical features, 123
detection procedures
multiplex PCR, 125
PCR, 125
real-time RT-PCR, 125–126
reverse transcription, 125
RNA, extraction of, 125
sequencing, 126
diagnosis
conventional techniques,
123–124
molecular techniques,
124–125
epidemiology, 122
genome organization, 122
morphology, 122
pathogenesis, 123
sample collection methods, 125
Feline enteric coronavirus (FECV) and feline
infectious peritonitis virus (FIPV)
biotypes, 324
classification, 324
clinical features, 324–325
genome organization, 324
internal mutation, 323
methods
detection procedures, 326–328
sample collection and preparation,
325–326
morphology, 324
pathogenesis, 324–325
serotypes, 324
SNP mutation, 324
structural and accessory
genes of, 327
transformation, 323
transmission, 324–325

Feline foamy virus (FeFV), 191
Feline immunodeficiency virus (FIV)
classification, 191
clinical features and pathogenesis, 193
detection procedures, 195
diagnosis, 194
epidemiology, 192–193
genome organization, 192
morphology, 192
prevention, 194
replication, 192–193
sample preparation methods, 194
treatment, 194
Feline infectious peritonitis virus, 331
Feline leukemia virus (FeLV), 191
classification, 197–198
clinical features, 199
detection procedures, real-time PCR,
201–202
diagnosis
nucleic acid detection, 201
serological assays, 200
virus isolation, 200
epidemiology, 198–199
genome organization, 198
morphology, 198
pathogenesis, 199–200
replication, 198–199
sample preparation methods, 201
treatment and prevention, 201
Feline panleukopenia virus (FPLV), 687
FeLV, *see* Feline leukemia virus
Fetal bovine serum (FBS), 373
Filoviridae, 495; *see also* Ebola virus
FIPV antigen, 325
FIV, *see* Feline immunodeficiency virus
Flavivirus, 247, 250; *see also* Louping ill virus;
West nile virus
FLEV, *see* Flexal virus
Flexal virus (FLEV), 569–570, 574
Fluorescent antibody (FA), 53, 55–56
Fluorescent antibody test (FAT)
CSFV, 236
RABV, 363, 365
Fluorescent microsphere-based immunoassays
(FMIAs), 556–557
Fluorescent quantitative real-time PCR,
695–696
FMD, *see* Foot-and-mouth disease
FMDV, *see* Foot-and-mouth disease virus
FMV, *see* Fort Morgan virus
Foot-and-mouth disease (FMD), 61, 89, 369
Foot-and-mouth disease virus (FMDV)
biology, 62–63
classification, 61–62
clinical features, 63
conventional RT-PCR, 65
detection procedures, 65
diagnosis
nucleic acid detection, 64
prevention, 64
serological assays, 63–64
treatment, 64
virus isolation, 63
epidemiology, 62–63
genome organization, 62
morphology, 62
pathogenesis, 63
real-time RT-PCR, 65
sample preparation methods,
64–65

Fort Morgan virus (FMV), 255, 258
Fowlpox virus (FWPV), 837
FPLV, *see* Feline panleukopenia virus
Furin cleavage, 324
FWPV, *see* Fowlpox virus

G

Gallid herpesvirus 2, *see* Marek's disease virus
Gammacoronavirus, 307; *see also* Avian
infectious bronchitis virus
Gammaretrovirus, 197; *see also* Feline leukemia
virus; Reticuloendotheliosis viruses
Ganjam virus (GANV), 544; *see also* Nairobi
sheep disease virus
GAV, *see* Gill-associated virus
Gel-based RT-PCR, 22
Genetic lineage typing methods, 427
Genome-wide association study, 280
GFP, *see* Green fluorescent protein
GHPV, *see* Goose hemorrhagic polyomavirus
Gill-associated virus (GAV), 20
Gill biopsies, 301
Glasser's disease, 820
Global Rinderpest Eradication Program
(GREP), 485
Glycoprotein complex (GPC), 571
Goatpox (GTP), 845
Goatpox virus (GTPV), 845
Goose hemorrhagic polyomavirus (GHPV), 719
Goose parvovirus (GPV)
classification, 693
clinical features, 694
detection procedures
agar gel precipitin test, 695
cattle sperm agglutination inhibition
test, 694
ELISA, 695
fluorescent quantitative real-time PCR,
695–696
multiplex PCR, 695
neutralization test, 694
passive immunity test, 694
PCR, 695
virus isolation, 694
Western blotting, 695
diagnosis, 694
genome organization, 693
morphology, 693
preparation methods, 694
replication, 693–694
sample collection methods, 694
Gosling plague (GP), 693
GPC, *see* Glycoprotein complex
GPV, *see* Goose parvovirus
Granulomatous lesions, 325
Green fluorescent protein (GFP), 262–263, 459
GTOV, *see* Guanarito virus
GTP, *see* Goatpox
GTPV, *see* Goatpox virus
Guanarito virus (GTOV), 569, 573
Guanidine isothiocyanate method, 274
Guanidine isothiocyanate–phenol–chloroform
technique, 31
Guanidinium isothiocyanate–phenol–chloroform
extraction, 589
Guanidinium thiocyanate–phenol–chloroform
technique, 282
Guillain–Barré syndrome, 363
Gyrovirus, 659; *see also* Beak and feather
disease virus

H

HA, *see* Hemagglutination
HAI, *see* Hemagglutination inhibition
H52 and H120 vaccines, 310
Hantavirus, 533; *see also* Hantaviruses
Hantavirus cardiopulmonary syndrome (HCPS), 533, 537
Hantaviruses
    biology, 536
    classification, 533–535
    clinical features, 536–537
    detection procedures
        conventional RT-PCR, 538
        real-time RT-PCR, 538–539
    diagnosis
        genotypic techniques, 537
        phenotypic techniques, 537
    epidemiology, 536
    genome organization, 535–536
    morphology, 535–536
    pathogenesis, 536–537
    prevention, 537–538
    sample preparation methods, 538
    treatment, 537–538
Hantavirus pulmonary syndrome (HPS), 533, 537
HAstV, *see* Human astrovirus
HAV, *see* Hepatitis A virus
Hazara virus (HAZV), 544
HE, *see* Hemagglutinin–esterase
Hemagglutination (HA), 300, 377, 396, 450–451, 548, 664
Hemagglutination inhibition (HI), 270, 273, 320
    LIV, 242–244
    PRV, 480
    RVFV, 555
    toroviruses, 341
Hemagglutinin–esterase (HE), 318, 332, 340, 393
Hemagglutinin–neuraminidase (HN) protein, 448
Heminested PCR and sequencing, 463–464
Hemocytic infiltration, 595
Hemorrhagic fever with renal syndrome (HFRS), 533
Hemorrhagic nephritis and enteritis of geese (HNEG) viruses
    biology, 720
    clinical features, 720–721
    detection procedures, 723–724
    diagnosis, 722
    genome organization, 720
    morphology, 720
    pathogenesis, 720–721
    sample preparation methods, 722–723
Hendra virus (HeV)
    biology, 433
    classification of, 431–432
    clinical features, 433–434
    diagnosis, 434
    epidemiology, 433
    genome organization, 432–433
    methods
        detection procedures, 435
        sample preparation, 435
    morphology, 432–433
    NiV, 456
    pathogenesis, 433–434
    treatment and prevention, 434–435

Hepandensovirus, 699; *see also Penaeus monodon* hepandensovirus
Heparan sulfate (HS), 234
Hepatitis A virus (HAV), 47
Hepatitis E, 111
Hepatitis E virus (HEV)
    classification, 111–112
    clinical features, 112–113
    detection procedures
        conventional RT-PCR, 116
        real-time RT-PCR, 116–117
        sequencing and phylogenetic analyses, 117
    diagnosis
        conventional techniques, 113–114
        molecular techniques, 114–115
    genome organization, 112
    morphology, 112
    pathogenesis, swine, 112–113
    sample preparation methods, 115
Hepatopancreatic parvovirus (HPV), *see Penaeus monodon* hepandensovirus
Hepevirus, 111; *see also* Hepatitis E virus
Herpes simplex virus (HSV), 753
HEV, *see* Hepatitis E virus
HeV, *see* Hendra virus
HHV1, *see Human herpesvirus* 1
HI, *see* Hemagglutination inhibition
Highlands J virus (HJV), 255
High-throughput pyrosequencing, 347
HJV, *see* Highlands J virus
HNEG viruses, *see* Hemorrhagic nephritis and enteritis of geese viruses
Hoof-and-mouth disease, 61
Horizontal transmission
    betanodavirus, 11–12
    DWV, 40
    YHV and GAV, 298–299
Host–cell cytoplasm, 610
Host-specific viral genomes, 29
HRCA, *see* Hyperbranched rolling circle amplification
HRSV, *see* Human RSV
HS, *see* Heparan sulfate
HToV, *see* Human torovirus
Human astrovirus (HAstV), 101
Human herpesvirus 1 (HHV1), 798
Human–mosquito–human transmission, 272
Human RSV (HRSV), 413
Human torovirus (HToV), 339
Hyperbranched rolling circle amplification (HRCA), 20

I

IAPV, *see* Israeli acute paralysis virus
IAV, *see* Influenza A virus
IB, *see* Infectious bronchitis
IBDV, *see* Infectious bursal disease virus
IBV, *see* Infectious bronchitis virus
IC-ELISA, *see* Immunocapture ELISA
IcHV1, *see* Ictalurid herpesvirus 1
IcHV2, *see* Ictalurid herpesvirus 2
Icosahedral capsid, 225
ICPI, *see* Intracerebral pathogenicity index
Ictalurid herpesvirus 1 (IcHV1)
    biology, 797–798
    classification, 797
    clinical signs, 799
    conventional techniques, 799

    detection procedures
        antibody neutralization, 801
        cell culture isolation, 800–801
        PCR, 801–804
    diagnose, 799
    epidemiology, 797–798
    genome organization, 797–798
    molecular assays, 800
    molecular techniques, 799
    morphology, 797–798
    pathogenesis, 799
    sample preparation methods, 799–800
    virus culture, 799–800
Ictalurid herpesvirus 2 (IcHV2), 797
Ictalurivirus, 797; *see also* Ictalurid herpesvirus 1
ICT assay, *see* Immunochromatographic assay
ICTV, *see* International Committee on Taxonomy of Viruses
Idiopathic myonecrosis, *see* Infectious myonecrosis virus
IFA, *see* Immunofluorescence antibody assay
IFAT, *see* Indirect fluorescent antibody tests
Iflaviral polyproteins, 39
Iflavirus, 27, 29; *see also* Deformed wing virus
Iflavirus particles, 41
IFN, *see* Interferon
IgM antibodies, 423–424, 575–576
IGRs, *see* Intergenic regions
IHC, *see* Immunohistochemistry
IHHNV, *see* Infectious hypodermal and hematopoietic necrosis virus
iiRT-PCR, *see* Insulated isothermal RT-PCR
ILT, *see* Infectious laryngotracheitis
Iltovirus, *see* Infectious laryngotracheitis virus
ILTV, *see* Infectious laryngotracheitis virus
Immunoassays, 597
Immunocapture ELISA (IC-ELISA)
    PPRV, 467
    RPV, 487, 490
Immunochromatographic (ICT) assay, 424
    VSV, 372
Immunodiffusion tests, 48
Immunodominant protein, 226
Immunofluorescence antibody assay (IFA)
    CDV, 424
    NiV, 459
    NSDV, 550
    PEDV, 334
    RPV, 487
Immunofluorescent (IF) assay
    EAV, 280
    VEEV, 273
Immunohistochemistry (IHC)
    ABV, 349
    BRSV, 415
    BVDV 1 and 2, 228
    CDV, 424–425
    EEEV and WEEV, 260
    FECVs and FIPVs, 327
    HeV, 434
    IAV, 402
    LIV, 244
    Menangle virus, 442
    MYXV, 861–862
    NiV, 459
    orthopoxvirus, 876
    PEDV, 334
    RVFV, 556
    WNV, 250

Immunological/serological assays
    IMNV, 597
    PBV, 588
Immunoperoxidase monolayer assay, PRRSV, 290
Immunoperoxidase tests, CSFV, 236
IMNV, *see* Infectious myonecrosis virus
Inapparent infection, 30
Indirect fluorescent antibody tests (IFAT)
    ABV, 349
    HeV, 434
    IAV, 402
    NSDV, 548
    PEDV, 334
    RVFV, 555
Infectious bronchitis (IB), 307
Infectious bronchitis virus (IBV), 49; *see also*
    Avian infectious bronchitis virus
Infectious bursal disease (IBD), 601
Infectious bursal disease virus (IBDV), 49
    classification of, 601–602
    clinical features, 602–603
    diagnosis
        conventional methods, 603
        molecular techniques, 604
    genome organization, 601–602
    methods
        detection procedures, 604–605
        sample preparation, 604
    morphology, 601–602
    pathogenesis, 602–603
    in poultry, 601
Infectious hypodermal and hematopoietic
    necrosis virus (IHHNV), 20, 705–707
Infectious laryngotracheitis (ILT), 753
Infectious laryngotracheitis virus (ILTV), 49
    biology, 756
    classification, 753
    clinical features, 757
    detection procedures
        LAMP, 759–761
        real-time PCR, 758–759
    diagnosis
        conventional techniques, 757
        molecular techniques, 757–758
    genome organization, 754–756
    morphology, 753–754
    pathogenesis, 756–757
    sample preparation methods, 758
    transmission, 756–757
Infectious myonecrosis (IMN), 593–594;
    see also Infectious myonecrosis virus
Infectious myonecrosis virus (IMNV)
    classification of, 593
    clinical features, 594–595
    diagnosis, 596
    epidemiology, 594
    genome organization, 593–594
    methods
        detection procedures, 596–597
        sample preparation, 596
    morphology, 593–594
    pathogenesis, 594–595
    transmission, 595–596
Infectious salmon anemia (ISA), 393
Infectious salmon anemia virus (ISAV)
    biology, 394–395
    classification of, 393–394
    clinical features, 395
    detection procedures
        conventional RT-PCR, 396–397
        rRT-PCR, 397

diagnosis, 395–396
epidemiology, 394–395
genome organization, 394
morphology, 394
pathogenesis, 395
sample preparation methods, 396
treatment and prevention, 396
Infectious spleen and kidney necrosis virus
    (ISKNV), 825
Inflamed omentum, 325
Influenza-associated encephalopathy, 385
Influenza A virus (IAV)
    classification, 399–400
    clinical features, 401–402
    detection procedures, 403–404
    diagnosis techniques
        conventional, 402
        molecular, 402–403
    epidemiology and ecology
        Asian swine, 401
        European swine, 400–401
        North American swine, 400
        in South America and Africa, 401
        swine, role of swine, 401
    genome organization, 399–400
    morphology, 399–400
    pathogenesis, 401–402
    sample collection and preparation
        methods, 403
    zoonotic transmission of, 401
Influenzavirus A, 383; *see also* Avian
    influenza virus
Ingestion, 235
In situ hybridization, 301
Insulated isothermal RT-PCR (iiRT-PCR)
    EAV, 282
    EIV, 386, 389
Interferon (IFN), 432, 645
Interferon-induced antiviral protein Mx, 11
Interferon regulatory transcription factor
    (IRF), 565
Intergenic regions (IGRs), 296
Internal ribosome entry site (IRES),
    18, 28, 37
International Committee on Taxonomy of
    Viruses (ICTV), 37, 593, 659
International Office of Epizootics, 223
Intracellular virion assembly, 248
Intracerebral inoculation, 48
Intracerebral pathogenicity index (ICPI), 450
Intranasal (IN) vaccination, 320
In vitro CD3+ T cell–resistant phenotype, 280
IRES, *see* Internal ribosome entry site
IRF, *see* Interferon regulatory transcription
    factor
ISAV, *see* Infectious salmon anemia virus
Isfahan virus (ISFV), 370
ISFV, *see* Isfahan virus
ISGylation pathway, 395
ISKNV, *see* Infectious spleen and kidney
    necrosis virus
Isothermal nucleic acid detection assays,
    AIV, 379
Israeli acute paralysis virus (IAPV), 27–29

J

Japanese encephalitis virus (JEV), 247
JEV, *see* Japanese encephalitis virus
Junin virus (JUNV), 570, 573
JUNV, *see* Junin virus

K

Kakugo virus (KV), 27, 29
Kashmir bee virus (KBV), 27–30
KBV, *see* Kashmir bee virus
Killed virus vaccines (KV), 290, 292
Klenow fragment, 41–42
Koutango virus (KOUV), 247
KOUV, *see* Koutango virus
Kupe virus, 544
KV, *see* Kakugo virus; Killed virus vaccines

L

Lactate dehydrogenase elevating virus
    (LDV), 287
Laminin receptor (LamR), 234
LAMP, *see* Loop-mediated isothermal
    amplification
LAMP-LFD, *see* Loop-mediated isothermal
    amplification-lateral flow dipstick
Lamp RT-PCR, 596
LamR, *see* Laminin receptor
Large yellow croaker iridovirus (LYCIV), 825
Lassa virus (LASV)
    hyperendemic, 572
    immunological detection, 575
    M. natalensis, 573
Lateral flow dipstick (LFD), 20
LCMV, *see* Lymphocytic choriomeningitis virus
LDV, *see* Lactate dehydrogenase elevating virus
Lentivirus, 167, 178, 191; *see also* Equine
    infectious anemia virus; Feline
    immunodeficiency virus; Small
    ruminant lentiviruses
Leporipoxvirus, 855; *see also* Myxoma virus
Lethargy, 55
Leukosis/sarcoma (L/S) group, 145
LFD, *see* Lateral flow dipstick
Live attenuated vaccine, 233, 236–237
Living dead shrimp, *see* Zombie shrimp
Loop-mediated isothermal amplification
    (LAMP)
    ALV, 153
    CPV, 691
    ILTV, 759–761
    MDV, 778–779
    orthopoxvirus, 877
    PstDV, 708
    SuHV-2, 749–750
    TSV, 20, 23
Loop-mediated isothermal amplification-lateral
    flow dipstick (LAMP-LFD), 434
    AuNP, 597
    PPRV, 472
Louping ill virus
    biology, 241–242
    classification, 241
    clinical features, 242
    detection of, 243
    diagnosis, 242–243
    distribution of, 242
    epidemiology, 241–242
    neurological signs, 242–243
    pathogenesis, 242
    PCR methodologies for, 243
    tick life cycle, 241
    tick-transmitted viral disease, 241
    transmission, 241–242
LOV, *see* Lymphoid organ virus
LSDV, *see* Lumpy skin disease virus

Lujo virus (LUJV), 569, 574
LUJV, *see* Lujo virus
Luminex-based multiplexed
            microsphere assays, 460
Lumpy skin disease (LSD), 845
Lumpy skin disease virus (LSDV), 845
LYCIV, *see* Large yellow croaker iridovirus
Lymphocytic choriomeningitis
            virus (LCMV)
    arenaviruses, 569
    causes of, 574
    clinical features, 574
    human infection, 572
    pathogenesis, 574
    prototype virus, 569
    solid-organ transplants, 572
    T-cell depletion, 574
Lymphohistiocytic
            meningoencephalomyelitis, 565
Lymphoid organ virus (LOV), 296

M

MAbs, *see* Monoclonal antibodies
Macavirus
    classification, 765–766
    clinical features, 766–767
    detection procedures
        DNA extraction, 770
        genomic detection, 769–770
        PBL, 769
        PCR, 770
        virus isolation, 769
    diagnosis
        conventional techniques, 767–768
        ELISA, 768
        IFA, 768
        molecular techniques, 768
        treatment and control, 769
        virus isolation, 767–768
        virus neutralization assay, 768
    genome organization, 765–766
    morphology, 765–766
    pathogenesis, 767
    preparation methods, 769
    sample collection methods, 769
    transmission, 766
Machupo virus (MACV), 570, 573
MACV, *see* Machupo virus
Madariaga virus (MADV), 255, 257–258
Madin–Darby canine kidney (MDCK) cells,
            385, 402
MADV, *see* Madariaga virus
Maedi–visna virus (MVV),
            see Visna–maedi virus
Major capsid protein (MCP), 593–595
Malarial catarrhal fever, 619
Malignant catarrhal fever (MCF), 765
Malphigian epithelium layer, 371
MALTs, *see* Mucosaassociated
            lymphoid tissues
Mamastrovirus, 101; *see also* Avian
            Astroviruses
Marburg hemorrhagic fever, 507; *see also*
            MARV disease
Marburgvirus, 507; *see also* Negative-sense
            RNA viruses
Marburg virus (MARV)
    biology, 508, 510
    classification, 507
    clinical features, 509

detection procedures
    antigen detection ELISA, 515
    one-step real-time RT-PCR, 514–515
    RNA extraction, 514
    TaqMan™ real-time RT-PCR, 515
diagnosis
    conventional techniques, 511
    molecular techniques, 511–513
epidemiology, 508–509
morphology, 507–508
pathogenesis, 509
sample collec tion and preparation methods,
            513–514
transmission, 509
Mardivirus, *see* Marek's disease virus
Marek's disease, 205, 603
Marek's disease virus (MDV), 205
    biology, 774–775
    classification, 774
    clinical features, 775
    detection procedures
        AGID, 778
        electron microscopy, 775–776
        histopathology, 776
        immunofluorescence assay, 775
        immunohistopathology, 776
        LAMP, 778–779
        PCR, 778
        radial immunodiffusion assay, 778
        RID, 776
        virus isolation, 775, 778
    diagnosis
        conventional techniques, 775–776
        molecular techniques, 776–777
    morphology, 774–775
    pathogenesis, 775
    sample preparation methods, 777–778
Marine shrimp farming, 593
MARV disease (MVD), 507
Massachusetts (Mass), 309–310
Mastomys natalensis, 573
MAstV, *see* Mink astrovirus
Mayaro virus (MAYV), 270
MAYV, *see* Mayaro virus
MCF, *see* Malignant catarrhal fever
MCP, *see* Major capsid protein
MD, *see* Mucosal disease
Mean symptomatic index (MSI), 602
Measles virus (MV), 485
Megalocytiviruses
    biology, 826–827
    classification, 825–826
    clinical features, 827
    detection procedures
        conventional PCR, 832
        nested PCR, 832
        real-time PCR, 833
    diagnosis
        conventional techniques, 827–829
        molecular techniques, 829–831
    epidemiology, 826–827
    morphology, 826–827
    pathogenesis, 827
    sample collection and preparation methods
        DNA extraction, using phenol/
            chloroform, 831
        DNA extraction, using spin
            column, 832
Melagrid herpesvirus 1, *see* Marek's
            disease virus
MEM, *see* Minimum essential media

Menangle virus
    biology, 438–440
    classification of, 437–438
    clinical features
        fruit bats, 441
        humans, 441
        pigs, 440–441
    diagnosis techniques
        conventional, 441–442
        molecular, 442
    epidemiology, 438–440
    methods
        detection procedures, 443–444
        sample preparation, 442–443
    morphology, 437–438
    pathogenesis, 440–441
    transmission electron micrographs of, 437–438
Metapneumovirus, 407; *see also* Avian
            metapneumovirus
Methionine–to–leucine substitution, 328
MHV, *see* Mouse hepatitis virus
Mimivirus, 1; *see also* Prion
Minimum essential media (MEM), 229, 320
Mink astrovirus (MAstV), 101
Modified live vaccines (MLV), 335
Modified live virus (MLV), 426
Modified live virus vaccines (MLV), 290, 292
Molecular cloning, 53
Monoclonal antibodies (MAbs), 19–21, 300, 396,
            459, 596–597
Morbillivirus, 467; *see also* Peste des petits
            ruminants virus
Morbilliviruses, 422, 471
Mosquito–bird–mosquito transmission, 247
Mosquito vectors, 271–272
Mouse hepatitis virus (MHV), 324
MSI, *see* Mean symptomatic index
Mucambo virus (MUCV), 272
Mucociliary transport system, 290
Mucosaassociated lymphoid tissues (MALTs), 422
Mucosal disease (MD), 227
MUCV, *see* Mucambo virus
Multiplex PCR
    ALV, 150–151
    EHDV, 633
    FCV, 125
    goose parvovirus (GPV), 695
Multiplex RT-PCR, 373
Muscular dystrophy, 48
MV, *see* Measles virus
Myxoma virus (MYXV)
    clinical features
        amyxomatous myxomatosis, 858
        cutaneous myxomatosis, 857–858
        peracute disease, 858
    coevolution of, 857
    detection procedures
        immunohistochemical detection,
            861–862
        in vitro culture, 861
        PCR amplification, 863–864
        serological determination, 862–863
    diagnosis
        conventional techniques, 859–860
        molecular techniques, 860–861
    genome organization, 856
    morphology, 855–856
    natural history, 856
    pathogenesis, 858–859
    sample collection and processing methods, 861
    transmission of, 856–857

N

Nairobi sheep disease virus (NSDV)
  arthropod-borne, 543
  classification and history, 544
  clinical features, 547–548
  diagnosis, 548–549
  epidemiology, 546–547
  genomic organization, 545
  methods
    detection procedures, 551
    sample preparation, 550–551
  morphology, 544
  pathogenesis, 547–548
  phylogenetic analysis, 545–546
  prevention and control, 549–550
  treatment, 550
Nairovirus, 543; *see also* Nairobi sheep
    disease virus
NASBA, *see* Nucleic acid–based amplification
Nasopharyngeal aspirates (NPA), 434
National Animal Health Laboratory Network,
    402–403
National Institute of Allergy and Infectious
    Diseases (NIAID), 434
Natural AEV field strains, 47
Natural recombination, 296
NCLDV, *see* Nucelocytoplasmic large DNA virus
NCP, *see* Noncytopathic
Ncrotic foci, 236
NDV, *see* Newcastle disease virus
Necropsy, 230
Negative-and ambi-sense RNA viruses
  AHSV
    biology, 611–612
    classification of, 610
    clinical features, 612–613
    detection procedures, 613–614
    diagnosis, 613
    epidemiology, 612
    morphology, 610
    pathogenesis, 613
    sample preparation methods, 613
    treatment and prevention, 613
    viral proteins, 610–611
  AKAV
    animal tissue, sample preparation
      methods, 526
    biology, 523–524
    cell cultures, sample preparation
      methods, 526
    classification, 521–523
    clinical features, 524–525
    cloning and sequencing, 527–528
    conventional techniques, 525
    epidemiology, 523–524
    molecular techniques, 526
    morphology, 523–524
    nested PCR, 526–527
    pathogenesis, 524–525
    quantitative RT-PCR, 528
    reveres genetics systems, 528–529
    RT-PCR, 526–527
  arenaviruses
    biology, 571–572
    clinical features, 574
    detection procedures, 576–580
    GTOV, 573
    immunological detection, 575
    JUNV, 573
    LASV, 572–573

    LCMV, 572, 574
    MACV, 573–574
    molecular detection, 575–576
    pathogenesis, 574
    sample preparation, 576
    viral hemorrhagic fevers, 574
    virus isolation, 575
  BTV
    biology, 619–620
    classification of, 619
    clinical features, 620
    detection procedures, 622–624
    diagnosis techniques, 620–622
    epidemiology, 619–620
    morphology, 619
    pathogenesis, 620
    sample preparation methods, 622
    sheep, malarial catarrhal fever of, 619
  EEV
    classification of, 637–638
    clinical features, 639
    detection procedures, 639–640
    diagnosis, 639
    equine ephemeral fever, 637
    genome organization, 638
    morphology, 638
    pathogenesis, 639
    prevention, 639
    replication and epidemiology, 638
    sample preparation methods, 639
    treatment, 639
  EHDV
    ambion MagMAX-96, 632–633
    classification of, 629
    clinical features, 631
    detection procedures, 633–634
    diagnosis, 631–632
    epidemiology, 630
    genome organization, 629–630
    morphology, 629–630
    pathogenesis, 630
    QIAamp, 632
    RNA isolation, 632–633
    sample collection and preparation, 632
    transmission, 630
    WTD, 629
  hantaviruses
    biology, 536
    classification, 533–535
    clinical features, 536–537
    conventional RT-PCR, 538
    epidemiology, 536
    genome organization, 535–536
    genotypic techniques, 537
    morphology, 535–536
    pathogenesis, 536–537
    phenotypic techniques, 537
    prevention, 537–538
    real-time RT-PCR, 538–539
    sample preparation methods, 538
    treatment, 537–538
  IBDV
    classification of, 601–602
    clinical features, 602–603
    detection procedures, 604–605
    diagnosis methods, 603–604
    genome organization, 601–602
    morphology, 601–602
    pathogenesis, 602–603
    in poultry, 601
    sample preparation methods, 604

  IMNV
    classification of, 593
    clinical features, 594–595
    detection procedures, 596–597
    diagnosis, 596
    epidemiology, 594
    genome organization, 593–594
    morphology, 593–594
    pathogenesis, 594–595
    sample preparation, 596
    transmission, 595–596
  NSDV
    arthropod-borne, 543
    classification and history, 544
    clinical features, 547–548
    detection procedures, 551
    diagnosis, 548–549
    epidemiology, 546–547
    genomic organization, 545
    morphology, 544
    pathogenesis, 547–548
    phylogenetic analysis, 545–546
    prevention and control, 549–550
    sample preparation, 550–551
    treatment, 550
  PBV
    classification of, 586
    clinical features, 587–588
    detection procedures, 589–591
    diagnosis, 588–589
    epidemiology, 586–587
    genome organization, 586
    health impact, 586–587
    morphology, 586
    sample collection and preparation
      methods, 589
    transmission, 587–588
    treatment, 587–588
  RVFV
    classification of, 553–554
    clinical features, 554–555
    detection procedures, 557–558
    diagnosis techniques, 555–557
    epidemiology, 554
    genome organization, 553–554
    morphology, 553–554
    pathogenesis, 554–555
    sample preparation, 557
  RVs
    animal species, 646–647
    bovines, 646
    classification, 643–644
    clinical features, 645
    detection procedures, 650
    diagnosis, 648–649
    epidemiology, 646
    equines, 646
    humans, 646
    immune response, 645–646
    pathogenesis, 645
    porcines, 646
    prevention and treatment, 647–648
    RNA extraction, 650
    virus structure, 643–644
    zoonotic potential, 647
  SBV
    biology, 564–565
    classification of, 564
    clinical features, 565
    detection procedures, 566–567
    diagnosis, 565

epidemiology, 564–565
genome organization, 563–564
morphology, 563–564
pathogenesis, 565
sample preparation methods, 566
treatment and prevention, 566
Negative pathogen controls, PCR assay
AEV, 49
Negative-sense RNA viruses
ABV
classification of, 347
clinical features, 348–349
detection procedures, 351–353
diagnosis, 349–351
epidemiology, 347–348
genome structure, 347–348
morphology, 347–348
pathogenesis, 348–349
AIV
classification of, 377–378
clinical features, 378
conventional RT-PCR, 379–380
diagnosis, 378–379
epidemiology, 378
real-time RT-PCR, 380–381
sample processing methods, 379
aMPV
classification of, 407–408
clinical features, 408–409
detection procedures, 410–411
diagnosis techniques, 408–410
epidemiology, 408–409
genome structure and function,
407–408
pathogenesis, 408–409
sample collection and preparation
methods, 410
BEFV
biology, 355–356
classification of, 355
clinical features, 356–357
diagnosis, 357–358
epidemiology, 355–356
morphology, 355–356
pathogenesis, 356–357
sample preparation methods, 358
BRSV
biology and epidemiology, 413–414
classification of, 413
clinical features, 414
detection procedure, 417
diagnosis techniques, 414–416
genome organization, 413
morphology, 413
pathogenesis, 414
sample collection and preparation
methods, 416–417
CDV
biology, 421–422
classification of, 421
clinical features, 422–423
detection procedures, 426–427
diagnosis techniques, 423–425
epidemiology, 421–422
genome organization, 421–422
pathogenesis, 422–423
sample preparation methods,
425–426
Ebola virus
biology, 498–499
classification of, 495–496

clinical features, 499–500
detection procedures, 502
diagnosis techniques, 500–501
epidemiology, 499
genome organization, 496–498
morphology, 496–498
pathogenesis, 499–500
sample preparation methods, 501
EIV
biology, 383–384
classification of, 383
clinical features, 384–385
detection procedures, 386–390
diagnosis techniques, 385–386
epidemiology, 383–384
morphology, 383–384
pathogenesis, 384–385
sample collection and preparation
methods, 386
HeV
biology, 433
classification of, 431–432
clinical features, 433–434
detection procedures, 435
diagnosis, 434
epidemiology, 433
genome organization, 432–433
morphology, 432–433
pathogenesis, 433–434
sample preparation methods, 435
treatment and prevention, 434–435
IAV
classification, 399–400
clinical features, 401–402
detection procedures, 403–404
diagnosis techniques, 402–403
epidemiology and ecology, 400–401
genome organization, 399–400
morphology, 399–400
pathogenesis, 401–402
sample collection and preparation
methods, 403
zoonotic transmission of, 401
ISAV
biology, 394–395
classification of, 393–394
clinical features, 395
detection procedures, 396–397
diagnosis, 395–396
epidemiology, 394–395
genome organization, 394
morphology, 394
pathogenesis, 395
sample preparation methods, 396
treatment and prevention, 396
MARV
antigen detection ELISA, 515
biology, 508, 510
classification, 507
clinical features, 509
conventional techniques, 511
epidemiology, 508–509
molecular techniques, 511–513
morphology, 507–508
one-step real-time RT-PCR, 514–515
pathogenesis, 509
RNA extraction, 514
sample collec tion and preparation
methods, 513–514
TaqMan™ real-time RT-PCR, 515
transmission, 509

Menangle virus
biology, 438–440
classification of, 437–438
clinical features, 440–441
detection procedures, 443–444
diagnosis techniques, 441–442
epidemiology, 438–440
morphology, 437–438
pathogenesis, 440–441
sample preparation, 442–443
transmission electron micrographs of,
437–438
NDV
biology, 449
characteristics of, 447, 449
classification of, 447–448
clinical features, 449–450
detection procedures, 451–452
diagnosis, 450
epidemiology, 449
genome organization, 448–449
morphology, 448–449
pathogenesis, 449–450
sample preparation methods, 451
treatment and prevention, 450–451
NiV
classification of, 456
clinical features, 457–458
detection procedures, 462–464
diagnosis techniques, 458–461
epidemiology outbreak, 456–457
genome organization, 456
morphology, 456
pathogenesis, 457–458
sample collection and preparation, 461–462
PPRV
classification of, 468
clinical features, 470
detection procedure, 472–473
diagnosis techniques, 471–472
epidemiology, 469–470
genome organization, 468–469
geographical distribution, 469
immunity, 471
pathogenesis, 470
postmortem findings, 470
sample preparation methods, 472
virus structure, 468–469
PRV
biology, 478–479
classification of, 477
clinical features, 479
detection procedures, 480–481
diagnosis techniques, 480
epidemiology, 478–479
genome organization, 477–478
morphology, 477–478
pathogenesis, 479–480
sample preparation methods, 480
RABV
biology, 362
classification of, 361–362
clinical features, 363
detection procedures, 365
diagnosis, 363–364
epidemiology, 362
genome structure, 362
morphology, 362
pathogenesis, 363
sample preparation methods, 364–365
treatment and prevention, 364

RPV
    biology, 486–487
    classification of, 486
    clinical features, 487
    detection procedures, 490–492
    diagnosis techniques, 487–489
    epidemiology, 486–487
    morphology, 486–487
    sample preparation methods, 490
VSV
    causes of, 369
    classification of, 369–370
    clinical features, 371
    detection procedures, 373–374
    diagnosis, 372
    genome organization, 370
    morphology, 370
    pathogenesis, 371
    replication and epidemiology, 370–371
    sample preparation methods, 373
    treatment and prevention, 372
NendoU, *see* Uridylate-specific endonuclease
NEP, *see* Nuclear export protein
Nephrosonephritis syndrome, 603
Nested PCR, 462–463
    AKAV, 526–527
    BLV, 161–162
    DWV, 43
    EEEV and WEEV, 264
.   PPV, 886
    PstDV, 707
    VEEV, 274–275
Nested RT-PCR
    IMNV, 596
    YHV and GAV, 302
Neuroinvasiveness, 259
Neurological sequelae, 258
Neutralization assays, 409
Neutrotropic alphaviruses, 271
Neutralization assays, 409
Newcastle disease (ND), 447; *see also* Newcastle
            disease virus
Newcastle disease virus (NDV), 49
    biology, 449
    characteristics of, 447, 449
    classification of, 447–448
    clinical features, 449–450
    diagnosis, 450
    epidemiology, 449
    genome organization, 448–449
    methods
        detection procedures, 451–452
        sample preparation, 451
    morphology, 448–449
    pathogenesis, 449–450
    treatment and prevention, 450–451
NFW, *see* Nuclease-free water
Nipah encephalitis, 456
Nipah virus (NiV)
    classification of, 456
    clinical features and pathogenesis, 457–458
    diagnosis techniques
        conventional, 458–460
        molecular, 460–461
    epidemiology outbreak
        in Bangladesh, 456–457
        infection, bats, 457
        in Malaysia, 456
    methods
        detection procedures, 462–464
        sample collection and preparation, 461–462
    morphology and genome organization, 456

Nodamura virus (NoV), 9
Noncytopathic (NCP), 224, 227–228
Nonstructural accessory protein genes, 308
Nonstructural proteins (NSP), 270, 308–309,
            553, 643–644
Nonsuppurative encephalitis, 259
Nonvaccination stamping out policy, 236
NPA, *see* Nasopharyngeal aspirates
NSDV, *see* Nairobi sheep disease virus
NSP, *see* Nonstructural proteins
Nucelocytoplasmic large DNA virus
            (NCLDV), 1
Nuclear export protein (NEP), 383, 395
Nuclease-free water (NFW), 589
Nucleic acid amplification systems, 557–559
Nucleic acid amplification techniques
    BCoV, 320
Nucleic acid–based amplification (NASBA), 20
Nucleic acid–based molecular diagnostics, 312
Nucleic acid detection
    betanodavirus, 12–13
    HeV, 434
    ISAV, 396
    NDV, 450
    PEDV, 334
    RABV, 364
    SBV, 565
    toroviruses, 341
    WNV, 250
Nucleic acid extraction
    AEV, 49
    arenaviruses, 576
    EEEV and WEEV, 262
Nucleic acid sequence-based amplification
            (NASBA)
    EEEV and WEEV, 261
    TSV, 20
Nucleotide sequence analysis
    DHV, 56

O

OAstV, *see* Ovine astrovirus
Oceanic Institute (OI), 19
Office International des Epizooties (OIE), 20–21,
            23, 450, 845
OI, *see* Oceanic Institute
OIE, *see* Office International des Epizooties
Okavirus, 296; *see also* Gill-associated virus
One-step reverse transcription polymerase chain
            reaction
    BEFV, 358
    NiV, 462
Open reading frames (ORFs), 660, 693
    ABV, 348
    CSFV, 234
    DHAVs, polyadenylated genome of, 54
    EAV, 277
    FCoVs, 324
    HeV, 432
    IBDV, 602
    IBV, 308
    Iflavirus, 27, 29
    IMNV, 594
    IRES, 37
    NDV, 448
    NiV, 456
    SBV, 564
    TSV, 17
    VSV, 370
    WNV, 248

Orange-spotted grouper iridovirus (OSGIV), 825
Orbivirus, 367; *see also* Equine encephalosis
            virus
ORFs, *see* Open reading frames
Ornithodoros, 817; *see also* African swine
            fever virus
Oronasal route, 235
Orthobunyavirus, 563; *see also* Schmallenberg
            virus
Orthobunyavirus, 521, 553; *see also* Akabane
            virus; Rift Valley fever virus
Orthomyxoviridae, 377, 399
Orthopoxvirus
    biology, 871
    classification, 870
    clinical features, 872
    conventional techniques, 873
    detection procedures
        antigen and antibody detection, 876
        cell culture system, virus isolation in, 876
        electron microscopy, 876
        immunohistochemistry, 876
        LAMP, 877
        PCR, 876–877
        SNT, 876
    diagnosis, 872–873
    epidemiology, 871–872
    genome, 870–871
    molecular techniques, 873–875
    morphology, 870
    pathogenesis, 872
    physico-biochemical properties, 870
    sample preparation methods, 875–876
OSGIV, *see* Orange-spotted grouper iridovirus
Ostreavirus, 808; *see also* Abalone herpesvirus
Ovine astrovirus (OAstV), 101
Ovine encephalomyelitis, 241; *see also* Louping
            ill virus

P

PAbs, *see* Polyclonal antibodies
Papain-like protease (PLP), 296
Papain-like proteinase (PL^pro^), 308–309
Papillomavirus (PV), 727; *see also* Bovine
            papillomavirus
Paracrystalline arrays, 296, 298
Paramyxoviridae, 407, 437, 467–468
Parapoxvirus (PPV)
    BPSV, 883–884
    classification, 881–882
    clinical features, 882
    detection procedures
        nested PCR, 886
        qPCR detection, 885–886
        sequencing and phylogeny, 886–888
    diagnosis
        conventional techniques, 884
        molecular techniques, 884–885
    genome organization, 881–882
    morphology, 881–882
    Orf virus, 882–883
    pathogenesis, 882
    PCPV, 884
    PVNZ, 884
    sample collection methods, 885
    transmission, 882
    virus isolation methods, 885
Parapoxvirus of red deer in New Zealand
            (PVNZ), 884
Partial genome sequence analysis, 296

Parvovirus, 687; *see also* Canine parvovirus
Passive hemagglutination
   RPV, 487
PAstV, *see* Porcine astrovirus
PBFD, *see* Psittacine beak and feather disease
PBS, *see* Phosphate-buffered saline
PBV, *see* Picobirnaviruses
PCPV, *see* Pseudocowpox virus
PCR, *see* Polymerase chain reaction
PCV, *see* Porcine circovirus
PCV1, *see* Porcine circovirus type 1
PCV2, *see* Porcine circovirus type 2
PCVAD, *see* Porcine circovirus-associated
    diseases
PdCV, *see* Porcine deltacoronavirus
PDD, *see* Proventricular dilatation disease
PDNS, *see* Porcine dermatitis and nephropathy
    syndrome
PEDV, *see* Porcine epidemic diarrhea virus
Penaeus monodon hepandensovirus
    (PmoHDV)
   biology, 700
   classification, 699–700
   clinical features, 700–701
   conventional PCR, 702
   diagnosis, 701
   DNA extraction methods, 702
   epidemiology, 700
   genome organization, 700
   morphology, 700
   pathogenesis, 700–701
   real-time PCR, 702
   sample preparation methods, 702
   treatment and prevention, 702
   virus purification methods, 702
Penaeus stylirostris penstyldensovirus (PstDV)
   biology/epidemiology, 706
   classification, 705–706
   clinical features, 706–707
   detection procedures
      end-point PCR and nested PCR,
        708–709
      real-time qPCR, 709
   diagnosis
      conventional techniques, 707
      dot blot and in situ hybridization, 708
      end-point PCR, 707
      LAMP, 708
      nested PCR, 707
      single and duplex quantitative PCR,
        707–708
   morphology, 706
   pathogenesis, 706–707
   sample preparation method, 708
Penstyldensovirus, 705; *see also Penaeus*
    stylirostris penstyldensovirus
    (PstDV)
PEP, *see* Postexposure prophylaxis
Peripheral blood leucocytes (PBL)
   separation, 769
Persistent infection (PI), 227
Persistent testicular infection (PTI), 227
Peste des petits ruminants (PPR), 467; *see also*
    Peste des petits ruminants virus
Peste des petits ruminants virus (PPRV)
   classification of, 468
   clinical features, 470
   detection procedure
      RT-PCR amplification, 472
      sequencing analysis and phylogeny,
        472–473

   diagnosis techniques, 471–472
   epidemiology
      disease pattern, 469–470
      host range and pathogenicity, 469
      transmission, 469
   genome organization, 468–469
   geographical distribution, 469
   immunity, 471
   pathogenesis, 470
   postmortem findings, 470
   RPV, 486
   sample preparation methods, 472
   virus structure, 468–469
Pestivirus, 211, 223, 233, 236; *see also* Border
    disease virus; Bovine viral diarrhea
    viruses; Classical swine fever virus
pfu, *see* Plaque forming unit
Phenol–chloroform alkaline extraction
   IBV, 312
   WNV, 250
Phenotypic antiviral susceptibility, 5
PHEV, *see* Porcine hemagglutinating
    encephalomyelitis virus
Phosphate-buffered saline (PBS), 262, 320
PI, *see* Persistent infection
Pichinde virus (PICV), 569
Picobirnaviruses (PBV)
   classification of, 586
   clinical features, 587–588
   diagnosis, 588–589
   epidemiology and impact on health, 586–587
   genome organization, 586
   methods
      detection procedures, 589–591
      sample collection and preparation, 589
   morphology, 586
   transmission, 587–588
   treatment, 587–588
PICV, *see* Pichinde virus
PIRYV, *see* Piry virus
Piry virus (PIRYV), 370
Pixuna virus (PIXV), 272
PIXV, *see* Pixuna virus
PL, *see* Postlarvae
Plaque forming unit (pfu), 261
Plaque reduction neutralization test (PRNT)
   EEEV and WEEV, 260
   VEEV, 273
   WNV, 250
PLP, *see* Papain-like protease
PL^pro, *see* Papain-like proteinase
Plus-stranded genome, 309
PmoHDV, *see Penaeus monodon*
    hepandensovirus
PMWS, *see* Postweaning multisystemic wasting
    syndrome
Polyclonal antibodies (PAbs), 300
Polyclonal antisera, 300
Polymerase chain reaction (PCR)
   ABV, 347, 350–351
   AEV, 49
   APVs, 841–842
   betanodavirus, 13
   BFDV, 665
   BPV, 731–732
   CAEV, 172–173
   CPV, 690–691
   DHAVs and DAstVs, 57–58
   FCV, 125
   GPV, 695
   macavirus, 770

   MDV, 778
   MYXV, 863–864
   orthopoxvirus, 876–877
   PBV
   RPV, 491–492
   YHV and GAV, 301
Polyomaviruses
   in baboons, 722
   biology, 720
   BPyV, 721–722
   classification, 719–720
   clinical features, 720–722
   detection procedures
      gel-based PCR, 723–724
      histopathological examination, 724
      isolation, 723
      real-time PCR, 724
   in ducks, 722
   in finches, 721
   genome organization, 720
   morphology, 720
   in *Passeriformes*, 721
   pathogenesis, 720–722
   sample preparation methods, 722–723
Porcine astrovirus (PAstV), 101
Porcine caliciviruses, *see* Caliciviruses
Porcine circovirus (PCV)
   classification, 671–672
   epidemiology, 672
   physiochemistry of, 672
   transmission, 672
Porcine circovirus-associated diseases
    (PCVAD)
   clinical features and pathogenesis
      PCV2, subclinical infection of, 672
      PMWS, characteristics and pathogenesis
        of, 672–673
      PMWS, cofactors of, 673
   diagnosis
      molecular techniques, 674
      viral isolation, 673–674
   methods
      in situ hybridization, 674
      real-time PCR, 674–675
Porcine circovirus type 1 (PCV1), 660
Porcine circovirus type 2 (PCV2), 290, 671
Porcine deltacoronavirus (PdCV), 332
Porcine dermatitis and nephropathy syndrome
    (PDNS), 672
Porcine encephalomyocarditis virus, *see*
    Encephalomyocarditis virus
Porcine epidemic diarrhea (PED), 331; *see also*
    Porcine epidemic diarrhea virus
Porcine epidemic diarrhea virus (PEDV)
   biology and epidemiology, 332–333
   classification of, 331–332
   clinical features and pathogenesis, 333
   conventional RT-PCR, 335
   detection procedures, 335
   diagnosis
      ELISA and electron microscopy, 334
      immunohistochemistry, 334
      indirect fluorescent antibody test, 334
      nucleic acid detection, 334
      virus isolation, 333–334
   methods, 335
   morphology and genome organization, 332
   pigs, disease of, 331
   real-time RT-PCR, 335
   sample preparation, 335
   treatment and prevention, 334–335

Porcine hemagglutinating encephalomyelitis virus (PHEV), 317–318
Porcine parvovirus (PPV)
classification, 713
clinical features, 714
diagnosis
conventional techniques, 714–715
molecular techniques, 715
epidemiology, 714
genome organization, 713–714
morphology, 713–714
nano-PCR detection, 715–716
nucleic acid probe detection, 716
pathogenesis, 714
preparation methods, 715
sample collection methods, 715
Porcine pulmonary alveolar macrophages, 289
Porcine reproductive and respiratory syndrome virus (PRRSV)
biology, 289
classification, 287–288
clinical features and pathogenesis
boars, 289
piglets, 289
sows, 289
weaners and growers, 289–290
detection procedures
conventional RT-PCR, 291
real-time RT-PCR, 291–292
diagnosis, 290
epidemiology, 289
functions, 288
genome organization, 288–289
genotypes of, 288
internal control primers and probes, 291
morphology, 288–289
prevention, 290–291
proteins, 288
sample preparation, 291
treatment, 290–291
Porcine respiratory coronavirus (PRCV), 331–332
Porcine rubulavirus (PRV), 646
biology, 478–479
classification of, 477
clinical features, 479
detection procedures
HI, 480–481
isolated virus, titration of, 481
RT-PCR, 481
seroneutralization, 481
viral isolation, 481
diagnosis techniques, 480
epidemiology, 478–479
genome organization, 477–478
morphology, 477–478
pathogenesis, 479–480
sample preparation methods, 480
Porcine teschoviruses (PTVs)
biology, 80–82
classification, 79–80
clinical features, 82
detection procedures
conventional RT-PCR, 84
genotyping, 85–86
real-time RT-PCR, 84–85
diagnosis
conventional techniques, 82–83
molecular techniques, 83
epidemiology, 80–82
morphology, 80–82

pathogenesis, 82
sample collection and preparation methods, 84
Porcine torovirus (PToV), 339
Positive-sense RNA viruses
AEV
classification, 47
clinical features, 48
conventional techniques, 48
epidemiology, 47–48
molecular techniques, 49
morphology, 47
pathobiology, 47–48
pathogenesis, 48
PCR, 50
reverse transcription reaction, 49–50
sample collection, 49
ALV
clinical features, 147–148
conventional PCR, 150
diagnosis, 148–150
epidemiology, 147
etiology, 145–147
LAMP, 153
multiplex PCR, 150–151
pathogenesis, 147–148
quantitative PCR, 152
sample preparation methods, 150
transmission, 147
avian astroviruses
biology, 102–103
classification, 101–102
clinical features and pathogenesis, 103–105
detection procedures, 107
diagnosis, 105–106
epidemiology, 102–103
sample preparation methods, 106–107
structure, 102–103
BCoV
biology, 318–319
classification, 317–318
clinical features, 319
detection procedures, 320–321
diagnosis, 319–320
epidemiology, 318–319
genome organization, 318
morphology, 318
pathogenesis, 319
prevention, 320
sample preparation, 320
treatment, 320
BDV
classification, 211–212
clinical symptoms and pathologic changes, 214–215
conventional techniques, 215
epidemiology, 213–214
genome organization, 212
molecular techniques, 215–216
morphology, 213
pathogenesis, 214
RT-PCR, 217–2118
transmission, 214
virus isolation methods, 216–217
bee paralysis virus
biology, 29–30
classification, 28–29
clinical features, 30
conventional RT-PCR, 32
diagnosis, 30–31
epidemiology, 29–30

genome organization, 28–29
morphology, 29–30
pathogenesis, 30
real-time PCR, 32–33
sample preparation, 31–32
betanodavirus
biology, 11
classification, 9–10
clinical features, 11–12
conventional RT-PCR, 13
diagnosis, 12
epidemiology, 11
genome structure, 10
morphology, 10
pathogenesis, 11–12
prevention, 12
real-time RT-PCR, 13
sample preparation, 12–13
treatment, 12
BLV
biology, 158–159
classification, 157–158
clinical features, 159–160
diagnosis, 160–161
epidemiology, 158–159
genome organization, 157–158
morphology, 157–158
nested PCR, 161–162
pathogenesis, 159–160
real-time PCR, 162–163
sample collection and preparation methods, 161
transmission, 158–159
BVDV 1 and 2
classification, 223–224
clinical features, 226–227
conventional techniques, 228–229
genome organization, 225–226
molecular techniques, 229
morphology, 225–226
quantitative RT-PCR, 231
sample preparation, 229–230
standard RT-PCR, 230
transmission, 226–227
CAEV
biology, 167–168
classification, 167
clinical forms, 169
conventional diagnosis, 169–170
DNA extraction, 172
epidemiology, 167–168
genotype identification, 173
molecular techniques, 170–171
morphology, 167–168
pathogenesis, 168–169
PCR, 172–173
sample collection and preparation methods, milk, 171–172
sequence analysis, 173
SRLV cloning, 173
caliciviruses
classification, 129–130
clinical features, 131–132
detection procedures, 132
diagnosis, 132
epidemiology, 131
genome organization, 130–131
morphology, 130–131
pathogenesis, 131–132
replication, 131
sample preparation methods, 132

CSFV
    classification, 233–234
    clinical features, 235–236
    conventional RT-PCR, 237
    diagnosis, 236
    epidemiology, 234–235
    genome organization, 234
    morphology, 234
    pathogenesis, 235–236
    prevention, 236–237
    real-time RT-PCR, 237
    replication, 234–235
    sample preparation, 237
    treatment, 236–237
DHV
    classification, 53–55
    clinical features, 55
    conventional techniques, 56
    detection procedures, 57–58
    genome organization, 53–55
    molecular techniques, 56–57
    morphology, 53–55
    pathogenesis, 55
    sample collection, 57
DWV
    classification, 37–39
    conventional techniques, 40
    discovery, 41–42
    epidemiology, 39–40
    molecular techniques, 40
    nested PCR detection, 43–44
    pathogenesis, 39–40
    real-time PCR quantification, 44
    reverse transcription, 43
    sample preparation, 40–41
    total RNA isolation, 42–43
EAV
    biology, 277–279
    clinical features, 279–280
    conventional techniques, 280
    detection procedures, 282
    epidemiology, 277–279
    future perspectives, 282–283
    molecular techniques, 280–281
    molecular testing, clinical samples for, 281–282
    morphology, 277–279
    pathogenesis, 279–280
    sample collection, 281
    virus isolation, clinical samples for, 281
EEEV and WEEV
    biology, 256–258
    classification, 255–256
    clinical features, 258–259
    conventional techniques, 260
    detection procedures, 262–264
    epidemiology, 256–258
    genome organization, 256
    molecular techniques, 260–261
    morphology, 256
    pathogenesis, 258–259
    sample preparation, 261–262
EIAV
    AGID, 177, 182
    classification, 178–179
    clinical features, 181–182
    conventional techniques, 182
    detection procedures, 184–185
    epidemiology, 179–180
    genome organization, 178–179
    molecular techniques, 183

    morphology, 178–179
    pathogenesis, 181–182
    sample collection and preparation
        methods, 183–184
    transmission, 180–181
EMCV
    capsid, assembly of, 69
    classification, 67
    conventional techniques, 72–73
    genomic organization, 67–69
    host range, distribution, and
        transmission, 71
    infectious cycle, 69–71
    molecular techniques, 73
    one-step conventional RT-PCR, 73–74
    real-time RT-PCR, 74–75
    RNA extraction, 73
    rodent, clinical features and
        pathogenesis, 72
    sample preparation methods, 73
    swine, clinical features and
        pathogenesis, 72
    three-dimensional structure, 69
    vaccine strategies, 72
    viral proteins, 69
    zoonotic potential, 71–72
FCV
    biology, 122
    classification, 121–122
    clinical features, 123
    conventional techniques, 123–124
    epidemiology, 122
    genome organization, 122
    molecular techniques, 124–125
    morphology, 122
    multiplex PCR, 125
    pathogenesis, 123
    PCR, 125
    real-time RT-PCR, 125–126
    reverse transcription, 125
    RNA, extraction of, 125
    sample collection methods, 125
    sequencing, 126
FECV and FIPV
    classification, 324
    clinical features, 324–325
    detection procedures, 326–328
    genome organization, 324
    morphology, 324
    pathogenesis, 324–325
    sample collection, 325–326
    transmission, 324–325
FeLV
    classification, 197–198
    clinical features, 199
    detection procedures, real-time PCR,
        201–202
    epidemiology, 198–199
    genome organization, 198
    morphology, 198
    nucleic acid detection, 201
    pathogenesis, 199–200
    replication, 198–199
    sample preparation methods, 201
    serological assays, 200
    treatment and prevention, 201
    virus isolation, 200
FIV
    classification, 191
    clinical features and pathogenesis, 193
    detection procedures, 195

    diagnosis, 194
    epidemiology, 192–193
    genome organization, 192
    morphology, 192
    prevention, 194
    replication, 192–193
    sample preparation methods, 194
    treatment, 194
FMDV
    biology, 62–63
    classification, 61–62
    clinical features, 63
    conventional RT-PCR, 65
    detection procedures, 65
    epidemiology, 62–63
    genome organization, 62
    morphology, 62
    nucleic acid detection, 64
    pathogenesis, 63
    prevention, 64
    real-time RT-PCR, 65
    sample preparation methods, 64–65
    serological assays, 63–64
    treatment, 64
    virus isolation, 63
HEV
    classification, 111–112
    clinical features, 112–113
    detection procedures, 115–116
    diagnosis, 113–115
    genome organization, 112
    morphology, 112
    pathogenesis, swine, 112–113
    sample preparation methods, 115
IBV
    biology, 309
    classification, 307–309
    clinical features, 310–311
    conventional techniques, 311
    detection procedures, 312–313
    epidemiology, 309–310
    genome organization, 307–309
    molecular techniques, 311
    morphology, 307–309
    pathogenesis, 310–311
    sample collection, 312
LIV
    biology, 241–242
    classification, 241
    clinical features, 242
    diagnosis, 242–243
    epidemiology, 241–242
    pathogenesis, 242
    RT-LAMP, 244
    sample preparation, 243–244
    transmission, 241–242
PEDV
    biology and epidemiology, 332–333
    classification of, 331–332
    clinical features, 333
    conventional RT-PCR, 335
    detection procedures, 335
    diagnosis, 333–334
    genome organization, 332
    methods, 335
    morphology, 332
    pathogenesis, 333
    pigs, disease of, 331
    real-time RT-PCR, 335
    sample preparation, 335
    treatment and prevention, 334–335

PRRSV
  biology, 289
  classification, 287–288
  clinical features, 289–290
  conventional RT-PCR, 291
  diagnosis, 290
  epidemiology, 289
  genome organization, 288–289
  morphology, 288–289
  pathogenesis, 289–290
  prevention, 290–291
  real-time RT-PCR, 291–292
  sample preparation, 291
  treatment, 290–291
PTVs
  biology, 80–82
  classification, 79–80
  clinical features, 82
  detection procedures, 84–86
  diagnosis, 82–83
  epidemiology, 80–82
  morphology, 80–82
  pathogenesis, 82
  sample collection and preparation
      methods, 84
REVs
  biology, 206
  classification, 205–206
  clinical features, 206–207
  conventional PCR, 208
  diagnosis, 207
  epidemiology, 206
  genome organization, 206
  morphology, 206
  pathogenesis, 206–207
  prevention, 207
  real-time PCR, 208–209
  sample preparation methods,
      207–208
  treatment, 207
SVDV
  antigenic structure, 90–91
  classification, 90
  clinical features, 93–94
  conventional RT-PCR, 96–97
  conventional techniques, 94–95
  disease occurrence, geographic range
      of, 92
  disease transmission, 92–93
  genetic and antigenic variation, 92
  host range, 92
  immunity, 93–94
  isolates, molecular characterization of,
      97–98
  molecular techniques, 95
  on-site devices, 95
  pathogenesis, 93–94
  real-time RT-PCR, 97
  sample collection and preparation
      methods, 96
  structural and genome organization, 90
  viral RNA Extraction, 96
  virus-cell interactions, 91
  virus inactivation, 92–93
toroviruses
  biology, 340
  classification of, 339–340
  clinical features, 340–341
  detection procedures, 342–343
  diagnosis of, 341
  epidemiology, 340

  genome organization, 340
  morphology, 340
  pathogenesis, 340–341
  sample preparation methods, 342
  treatment and prevention, 341
TSV
  biology, 17–19
  classification, 17
  clinical features, 19
  conventional techniques, 19
  epidemiology, 17–19
  gel-based RT-PCR, 22
  genome organization, 17–19
  immunological techniques, 20–21
  molecular techniques, 19–20
  pathogenesis, 19
  quantitative RT-PCR, 22–23
  sample preparation, 21–22
VEEV
  classification, 269–270
  clinical features, 272–273
  conventional techniques, 273
  detection procedures, 274–275
  distribution, 271–272
  epidemiology, 271–272
  genome organization, 270
  molecular techniques, 273
  pathogenesis, 270–271
  sample collection, 273–274
  virion structure, 270
VESV
  classification, 136
  clinical features, 137–138
  detection procedures, 140–141
  diagnosis, 139
  genome organization, 136–137
  morphology, 136–137
  pathogenesis, 138
  sample collection and preparation
      methods, 140
  transmission, 137
WNV
  biology, 248–249
  classification, 247–248
  clinical features, 249
  detection procedures, 251–252
  epidemiology, 248–249
  genome structure, 248
  morphology, 248
  nucleic acid detection, 250
  pathogenesis, 249
  prevention, 250
  sample preparation, 250–251
  serological assays, 249–250
  treatment, 250
  virus isolation, 249
YHV and GAV
  biology, 296–298
  classification, 296
  clinical features, 298–299
  conventional techniques, 300
  detection procedures, 301–303
  epidemiology, 296–298
  genome organization, 296
  molecular techniques, 301
  morphology, 296–298
  pathogenesis, 298–299
  sample preparation, 301
Positive single-stranded RNA viruses, 339
Postexposure prophylaxis (PEP), 361, 364
Postlarvae (PL), 299

Postweaning multisystemic wasting syndrome
      (PMWS)
  characteristics and pathogenesis of,
      672–673
  cofactors of, 673
Potential cross-contamination, 262
Poultry adenoviruses
  classification, 736
  detection procedures, 741–742
  diagnosis
    conventional techniques, 739–740
    molecular techniques, 740
  epidemiology and clinical features, 736–738
    atadenoviruses, 739
    aviadenoviruses, 738–739
    siadenoviruses, 739
  genome organization, 736
  morphology, 736
  sample preparation methods, 740–741
PPV, see Parapoxvirus; Porcine parvovirus
PRCV, see Porcine respiratory coronavirus
Preweaning mortality, 290
Primary envelope glycoprotein (E1), 226
Primer–probe energy transfer, 326
Prion, 1–2
Prion diseases, 901; see also Bovine spongiform
      encephalopathy; Chronic wasting
      disease
PRNT, see Plaque reduction neutralization test
Protein-degrading enzymes, 312
Proventricular dilatation disease (PDD), 347–349
Proventriculus–gizzard junctions, 603
PRV, see Porcine rubulavirus
Pseudocowpox virus (PCPV), 884
Psittacine beak and feather disease (PBFD), 659
PstDV, see Penaeus stylirostris
      penstyldensovirus
Pteropus, 455–456; see also Nipah virus
PTI, see Persistent testicular infection
PToV, see Porcine torovirus
PTVs, see Porcine teschoviruses
Purkinje neuron, 48
Putative carrier stallions, 280
PVNZ, see Parapoxvirus of red deer in New
      Zealand
Pyogranulomas, 325
Pyrexia, 242

Q

Qualitative RT-PCR, 303
Quantitative RT-PCR
  AKAV, 528
  BVDV 1 and 2, 231
  SuHV-2, 750–751
  TSV, 22–23
Quasi-species population structure, 28

R

Rabbit hemorrhagic disease virus (RHDV), 130
Rabies virus (RABV)
  biology, 362
  classification of, 361–362
  clinical features, 363
  detection procedures, 365
  diagnosis
    histopathology, 363
    nucleic acid detection, 364
    serological tests, 363–364
    virus isolation, 363

epidemiology, 362
genome structure, 362
morphology, 362
pathogenesis, 363
sample preparation methods, 364–365
treatment and prevention, 364
Radial immunodiffusion assay (RID), 776
RBIV, *see* Rock bream iridovirus
RdRp, *see* RNA-dependent RNA polymerases
Real-time RT-PCR
ABV, 347, 350
AHSV, 614
AIV, 378, 380–381
arenaviruses, 577–580
ASFV, 821
BCoV, 321
betanodavirus, 13
BLV, 162–163
BRSV, 416
BTV, 622
CDV, 426
CSFV, 237
DWV, 40, 44
EAV, 282
Ebola virus, 500
EEEV, 262–263
EIV, 385, 389
FeLV, 201–202
HeV, 434–435
IAV, 402–404
IBV, 312
ILTV, 758–759
IMNV, 596
ISAV, 396–397
LIV, 244
megalocytiviruses, 833
Menangle virus, 442
NDV, 451
NSDV, 549, 551
PCVAD, 674–675
PEDV, 334–335
PmoHDV, 702
polyomaviruses, 724
PPRV, 471–472
PRRSV, 291
REVs, 208–209
vs. RT-PCR, 425
RVs, 650
toroviruses, 342
VSV, 373–374
WNV, 251
WSSV, 894–895
Recombinant nucleocapsid proteins (rNPs), 556
Red sea bream iridovirus (RSIV), 825
Red-spotted grouper nervous necrosis virus
(RGNNV), 9–11, 13
Reproductive tract infection, 226–227
Respiratory tract infections, 319
Restriction fragment length polymorphism
(RFLP), 311–312, 334, 427
Reticuloendotheliosis viruses (REVs)
biology, 206
classification, 205–206
clinical features, 206–207
detection procedures
conventional PCR, 208
real-time PCR, 208–209
diagnosis, 207
epidemiology, 206
genome organization, 206
morphology, 206

pathogenesis, 206–207
prevention, 207
sample preparation methods, 207–208
treatment, 207
Reverse transcription (RT), 334
IBDV, 604
RPV, 488
Reverse-transcription loop-mediated
amplification (RT-LAMP), 55–56
LIV, 243–244
Reverse-transcription polymerase chain reaction
(RT-PCR)
AHSV, 613
AIV, 378
aMPV, 409
arenavirus, 575
bee paralysis virus, 30
BEFV, 357
BRSV, 415
BTV, 621
BVDV 1 and 2, 229–231
CDV
vs. nested RT-PCR, 425
vs. rRT-PCR, 425
DWV, 40
Ebola virus, 500
EEV, 639
EHDV, 632
EIV, 385
FECVs and FIPVs, 326
HeV, 432
IAV, 401–402
IBDV, 603, 605
IBV, 311–312
IMNV, 596
ISAV, 396
LIV, 243–244
NDV, 447
NiV, 457
NSDV, 549, 551
PPRV, 467–468
PRRSV, 290–291
PRV, 480
RABV, 364–365
RVFV, 555–556, 558
RVs, 650
SBV, 564–565
toroviruses, 341–342
VEEV, 273
VSV, 374
YHV and GAV
diseased shrimp, 301–302
sequences primers, 302
Reverse transcription reaction
AEV, 49–50
DHV, 57
DWV, 43
VEEV, 274
REVs, *see* Reticuloendotheliosis viruses
RFLP, *see* Restriction fragment length
polymorphism
RGNNV, *see* Red-spotted grouper nervous
necrosis virus
Rhipicephalus appendiculatus, 544, 546
Ribonucleic acid (RNA), 1, 3
detection, VSV, 372–373
extraction
bee paralysis virus, 57
BVDV 1 and 2, 230
EMCV, 73
FCV, 125

IBDV, 604
MARV, 514
PPRV, 472
RVs, 650
SVDV, 96
isolation, EHDV, 633
replication, CSFV, 234–235
viruses, 495
Ribonucleoproteins (RNPs), 432, 553
ABV, 348
SBV, 564
Ribosomal frame shift, 288, 308
Rift Valley fever (RVF), 553
Rift Valley fever virus (RVFV)
classification of, 553–554
clinical features and pathogenesis, 554–555
diagnosis techniques
conventional, 555
molecular, 555–557
epidemiology, 554
methods
detection procedures, 557–559
sample preparation, 557
morphology and genome organization,
553–554
NSDV, 547
Rinderpest, 485; *see also* Rinderpest virus
Rinderpest virus (RPV)
biology, 486–487
classification of, 486
clinical features, 487
detection procedures
AGID test, 490
competitive ELISA[39], 491, 493
IC-ELISA[33], 490
PCR[43], 491–492
rapid chromatographic strip test[34,35],
490–491
diagnosis techniques, 487–489
epidemiology, 486–487
morphology, 486–487
PPRV, 467, 471
sample preparation methods, 490
Rio Negro virus (RNV), 270, 272
RNA, *see* Ribonucleic acid
RNA-capping enzyme, 644
RNA-dependent DNA polymerases, 43
RNA-dependent RNA polymerases (RdRp), 497,
593–594
AHSV, 610
bee paralysis virus, 28
betanodavirus, 10
HeV, 432–433
IBDV, 602
IBV genome, 308
mRNA replication, 370
ORF, 11
PBV, 586
RVs, 644
RNPs, *see* Ribonucleoproteins
rNPs, *see* Recombinant nucleocapsid proteins
RNV, *see* Rio Negro virus
Rock bream iridovirus (RBIV), 825
Rodent-borne J-virus, 456
Rotavirus Classification Working Group, 650
Rotaviruses (RVs)
animal species, 646–647
bovines, 646
classification, 643–644
clinical features, 645
diagnosis, 648–649

epidemiology, 646
equines, 646
humans, 646
immune response, 645–646
methods
    detection procedures, 650
    RNA extraction, 650
pathogenesis, 645
porcines, 646
prevention and treatment
    animal vaccines, 647–648
    human vaccines, 647
virus replication, 644–645
virus structure, 643–644
zoonotic potential, 647
RPV, *see* Rinderpest virus
RSIV, *see* Red sea bream iridovirus
RT, *see* Reverse transcription
RT loop-mediated isothermal amplification
            (RT-LAMP), 556; *see also* Reverse-
            transcription loop-mediated
            amplification
    BEFV, 358
    CDV, 426–427
    EIV, 389–390
RT-nested PCR (RT-nPCR)
    EAV, 282
    YHC, 302
RT-PCR, *see* Reverse-transcription polymerase
            chain reaction
Rubulavirus, 438
Ruggedized Advanced Pathogen Identification
            Device, 576
Runting, 236
RVFV, *see* Rift Valley fever virus
RVs, *see* Rotaviruses

S

Sabiá virus (SABV), 569, 577
SABV, *see* Sabiá virus
Sacbrood virus (SBV), 27, 29
SAFs, *see* Scrapie-associated fibrils
Saint Louis encephalitis virus (SLEV), 261
Sanger dideoxy method, 42
San Miguel sea lion virus (SMSV), 130
Sathuperi virus (SATV), 563
SATV, *see* Sathuperi virus
SBV, *see* Sacbrood virus; Schmallenberg virus
Schmallenberg virus (SBV)
    biology and epidemiology, 564–565
    classification of, 564
    clinical features and pathogenesis, 565
    detection procedures, 566–567
    diagnosis, 565
    morphology and genome organization,
            563–564
    sample preparation methods, 566
    treatment and prevention, 566
Scrapie-associated fibrils (SAFs), 901
Secondary envelope glycoprotein (E2), 226
Sequence analysis
    DHV, 58
    FECVs and FIPVs, 327–328
Serine protease activity, 226
Serine–to–alanine substitution, 328
Seroconversion, 269, 272–273, 575
Serological assays
    AEV, 48
    betanodavirus, 12–13
    FECVs and FIPVs, 326–327

HeV, 434
NDV, 450
PRV, 481
RABV, 363–364
SBV, 565
WNV, 249–250
YHV, 300
Serotypes/genotypes, 347
Serum neutralization test (SNT), 372, 424,
            459–460
    BVDV 1 and 2, 229
    orthopoxvirus, 876
Seven sample–test combinations, 228
sg mRNAs, *see* Subgenomic mRNAs
Shamonda virus (SHAV), 563
Sheep grazing moors, 242
Sheeppox (SPP), 845
Sheeppox virus (SPPV), 845
SHFV, *see* Simian hemorrhagic fever virus
Shipping fever, 319; *see also* Bovine respiratory
            disease complex
Shrimp disease, 295
SHS, *see* Swollen head syndrome
Signaling lymphocyte activation molecule
            (SLAM), 424
Signal peptide (SP), 308
Simbu serogroup, 563
Simian hemorrhagic fever virus (SHFV), 287
Sindbis virus (SINV), 255–256
Single nucleotide polymorphism (SNP),
            323–324, 328
Single radial hemolysis (SRH), 387
Single-stranded RNA viruses, 361, 569
Sin Nombre virus (SNV), 537
SINV, *see* Sindbis virus (SINV)
SJNNV, *see* Striped jack nervous necrosis virus
SKIV-ZJ07, *see* Spotted knifejaw iridovirus
SLAM, *see* Signaling lymphocyte activation
            molecule
SLEV, *see* Saint Louis encephalitis virus
Slow bee paralysis virus (SBPV), 39
Small ruminant lentiviruses (SRLVs)
    biology, 167–168
    classification, 167
    clinical forms, 169
    diagnosis
        conventional diagnosis, 169–170
        molecular techniques, 170–171
    DNA extraction, 172
    epidemiology, 167–168
    genotype identification, 173
    morphology, 167–168
    pathogenesis, 168–169
    PCR, 172–173
    sample collection and preparation methods,
            milk, 171–172
    sequence analysis, 173
    SRLV cloning, 173
SNP, *see* Single nucleotide polymorphism
SNT, *see* Serum neutralization test
Solar corona, 307
South American genotypes, 422
Spanish sheep encephalitis virus (SSEV),
            241–242, 245
Spawning induction methods, 12
Sperm-rich fraction, 281
Spike protein, 308
Spotted knifejaw iridovirus (SKIV-ZJ07), 826
SPP, *see* Sheeppox
SPPV, *see* Sheeppox virus
SRH, *see* Single radial hemolysis

SSEV, *see* Spanish sheep encephalitis virus
Standard RT-PCR
    BVDV 1 and 2, 230
    EAV, 282
    EEEV and WEEV, 263–264
    EIV, 389
    WNV, 251
Striped jack nervous necrosis virus (SJNNV),
            9–11
Subgenomic mRNAs (sg mRNAs), 277
Suid herpesvirus 2 (SuHV-2)
    classification, 747–748
    clinical features, 748
    conventional approaches, 748–749
    LAMP, 749–750
    molecular detection, 749
    pathogenesis, 748
    preparation methods, 749
    quantitative real-time polymerase chain
            reaction, 750–751
    sample collection methods, 749
    sample extraction, 749
    transmission, 748
    virus isolation, 749
SVD, *see* Swine vesicular disease
Swimming behavior, 11
Swine influenza A virus, *see* Influenza A virus
Swine vesicular disease (SVD), 89, 369
Swine vesicular disease virus (SVDV)
    classification, 90
    clinical features, 93–94
    detection procedures
        conventional RT-PCR, 96–97
        isolates, molecular characterization of,
            97–98
        real-time RT-PCR, 97
    diagnosis
        conventional techniques, 94–95
        molecular techniques, 95
        on-site devices, 95
    epidemiology
        disease occurrence, geographic range
            of, 92
        disease transmission, 92–93
        host range, 92
        virus inactivation, 92–93
    immunity, 93–94
    morphology and biology
        antigenic structure, 90–91
        genetic and antigenic variation, 92
        structural and genome organization, 90
        virus-cell interactions, 91
    pathogenesis, 93–94
    sample collection and preparation
            methods, 96
    viral RNA Extraction, 96
Swollen head syndrome (SHS), 407
SYBR® green assay, bee paralysis virus, 32
Systematic prophylactic vaccination program, 236

T

Tacaribe virus (TCRV), 569, 572
Taiwan grouper iridovirus (TGIV), 825
Taura syndrome virus (TSV), 705
    biology, 17–19
    classification, 17
    clinical features, 19
    detection procedures
        gel-based RT-PCR, 22
        quantitative RT-PCR, 22–23

diagnosis
    conventional techniques, 19
    immunological techniques, 20–21
    molecular techniques, 19–20
epidemiology, 17–19
genome organization, 17–19
horizontal transmission, 17
pathogenesis, 19
PCR, 17
sample preparation, 21–22
vertical transmission, 17
TBEV, *see* Tick-borne encephalitis virus
T-cell depletion, 574
T cell–immunodeficient chickens, 603
TCoV, *see* Turkey coronavirus
TCRV, *see* Tacaribe virus
Teschen–Talfan disease, 79
Teschovirus, 79;
    see also Porcine teschoviruses
Tetraparvovirus, 713; *see also* Porcine
    parvovirus
TGEV, *see* Transmissible gastroenteritis
    coronavirus
TGIV, *see* Taiwan grouper iridovirus
Thogoto virus (THOV), 393
THOV, *see* Thogoto virus
3C-like protease (3CLP), 296
3C-like proteinase (3CL^pro), 308–309
3CLP, *see* 3C-like protease
3-day sickness, *see* Bovine ephemeral fever
Tick-borne encephalitis virus (TBEV),
    241–242, 245
Tiger puffer nervous necrosis virus
    (TPNNV), 9–11
Tissue–cell tropism, 308
Toroviruses
    biology, 340
    classification of, 339–340
    clinical features, 340–341
    detection procedures, 342–343
    diagnosis of, 341
    epidemiology, 340
    genome organization, 340
    morphology, 340
    pathogenesis, 340–341
    sample preparation methods, 342
    treatment and prevention, 341
TPNNV, *see* Tiger puffer nervous
    necrosis virus
Tracheal lesions, 310
Tracheal rings, 311
Transboundary infections, 319
Transcription regulatory sequences
    (TRSs), 309
Transcriptome analysis, DWV, 39–40
Transcriptome sequencing, DWV, 37–39
Transient infection, 226
Transmissible gastroenteritis coronavirus
    (TGEV), 331–332, 335
Transmissible spongiform encephalopathies
    (TSEs), 901, 913
Transplacental infection, 227
TRBIV, *see* Turbot reddish body iridovirus
TRIG, *see* Triple-reassortant internal gene
Triple-reassortant internal gene (TRIG), 400
TRSs, *see* Transcription regulatory sequences
TSEs, *see* Transmissible spongiform
    encephalopathies
TSV-resistant shrimp, 19
Turbot reddish body iridovirus (TRBIV), 826
Turkey astrovirus (TAstV), 105

Turkey coronavirus (TCoV), 309, 311
Turkey rhinotracheitis, 407
Two-way heterologous reactions, 311
Type A influenza, 377

U

Ultraviolet (UV), 58
Universal mass vaccination (UMV), 647
Untranslated region (UTR), 18, 225, 248
Uridylate-specific endonuclease, 287
UTR, *see* Untranslated region
UV, *see* Ultraviolet

V

Vaccine breakdown strains (VBS), 126
Vaccinia-like viruses (VLVs), 870
Vaccinia virus (VACV), 855, 869
Vacuolating encephalopathy and retinopathy
    (VER), 9, 13
Variant Creutzfeldt–Jakob disease (vCJD), 903
Varicellovirus, 290
Varroa destructor virus 1 (VDV-1), 39–40, 43
Varroa parasitism, 30
VBS, *see* Vaccine breakdown strains
VCJD, *see* variant Creutzfeldt–Jakob disease
vcRNA, *see* Viral complementary RNA
VDV-1, *see* Varroa destructor virus 1
VEEV, *see* Venezuelan equine encephalitis virus
Venereal transmission, 279
Venezuelan equine encephalitis virus (VEEV)
    classification, 269–270
    current distribution, 270
    diagnosis
        conventional techniques, 273
        molecular techniques, 273
    distribution, 271–272
    epidemiology, 271–272
    genome organization, 270
    HI test, 270
    methods
        detection procedures, 274–275
        sample collection, 273–274
    origin, 270
    pathogenesis, 270–271
    subtype and strain, 269
    transmission cycles of, 271
    virion structure, 270
Vero cells, 333–334
Vertical transmission
    betanodavirus, 12
    YHV and GAV, 298–299
Very virulent IBDV (vvIBDV)
    tissue lesions, 603
VES, *see* Vesicular exanthema of swine
Vesicular exanthema of swine (VES),
    89, 135, 369
Vesicular exanthema of swine virus
    (VESV), 129
    classification, 136
    clinical features, 137–138
    detection procedures
        real-time RT-PCR, 140
        sequence analysis, 140–141
    diagnosis
        conventional techniques, 139
        molecular techniques, 139
    genome organization, 136–137
    morphology, 136–137

    pathogenesis, 138
    sample collection and preparation
        methods, 140
    transmission, 137
Vesicular stomatitis (VS), 89, 369; *see also*
    Vesicular stomatitis viruses
Vesicular stomatitis viruses (VSV)
    causes of, 369
    classification of, 369–370
    clinical features, 371
    detection procedures, 373–374
    diagnosis, 372
    genome organization, 370
    morphology, 370
    pathogenesis, 371
    replication and epidemiology, 370–371
    sample preparation methods, 373
    treatment and prevention, 372
Vesivirus, 121, 129, 136; *see also* Caliciviruses;
    Feline calicivirus; Vesicular
    exanthema of swine virus
VESV, *see* Vesicular exanthema of swine virus
VHF, *see* Viral hemorrhagic fever
VI, *see* Virus isolation
Viral cDNA, sequence-independent
    amplification, 41–42
Viral complementary RNA (vcRNA), 571
Viral hemorrhagic fever (VHF), 495; *see also*
    Ebola virus
    arenaviruses, 574
    LASV, 574
Viral nervous necrosis (VNN), 9, 11–13
Viral RNA (vRNA)
    arenaviruses, 571, 576
    EHDV, 632–633
    PBV, 589
Virus
    animal-infecting RNA virus families, 3–4
    diseases, 2–4
    filterable agents, 1
    genetic material, 1
    identification, 2
    infections, 2–4
    phenotypic procedures, 2, 5
    taxonomy, 1–2
    typing, 2–4
Virus distribution *vs.* cellular localization, 402
Virus-induced injury *vs.* virus-induced
    proinflammatory mediators, 280
Virus isolation (VI), 363
    ABV, 350
    AIV, 378
    aMPV, 409
    APVs, 841
    arenaviruses, 575
    ASFV, 820
    BDV, 216–217
    betanodavirus, 12–13
    BVDV 1 and 2, 228
    EAV, 280
    Ebola virus, 500
    EEEV and WEEV, 260
    FMDV, 63
    GPV, 694
    HeV, 434
    ISAV, 396
    macavirus, 767–769
    MDV, 775, 778
    Menangle virus, 441
    NDV, 450
    NSDV, 550

orthopoxvirus, 875
PEDV, 333
PPV, 885
PRV, 481
RABV, 363
RPV, 487
SBV, 565
SuHV-2, 749
toroviruses, 341
VEEV, 273
VSV
WNV, 249
Virus neutralization test (VNT)
   BCoV, 318, 320
   BTV, 621, 624
   CSFV, 236
   DHV, 53, 55–56, 58
   EAV, 280, 282
   IBV, 311
   macavirus, 768
   NSDV, 551
   PPRV, 467, 471
Virus–reservoir dynamics, 574
Visna–maedi virus (VMV)
   biology, 167–168
   classification, 167
   clinical forms, 169
   diagnosis
      conventional diagnosis, 169–170
      molecular techniques, 170–171
   DNA extraction, 172
   epidemiology, 167–168
   genotype identification, 173
   morphology, 167–168
   pathogenesis, 168–169
   PCR, 172–173
   sample collection and preparation methods,
      milk, 171–172
   sequence analysis, 173
   SRLV cloning, 173
VN, see Virus neutralization
VNN, see Viral nervous necrosis
VNT, see Virus neutralization test
VP2 gene, 687–688
VS, see Vesicular stomatitis
VSV, see Vesicular stomatitis viruses
vvIBDV, see Very virulent IBDV

W

Wax embedding, 327
WD, see Winter dysentery
WEEV, see Western equine encephalitis virus
Western blotting, 695
Western equine encephalitis virus (WEEV),
   255–256
West Nile virus (WNV), 543
   biology, 248–249
   classification, 247–248
   clinical features, 249
   diagnosis
      nucleic acid detection, 250
      serological assays, 249–250
      virus isolation, 249
   epidemiology, 248–249
   genome structure, 248
   methods
      detection procedures, 251–252
      sample preparation, 250–251
   morphology, 248
   mosquito–bird–mosquito transmission, 247
   mosquito-borne viruses, 247
   pathogenesis, 249
   prevention, 250
   tick-borne viruses, 247
   treatment, 250
   zoonotic pathogen, 247
Wet/effusive/nonparenchymatous, 325
Whataroa virus, 255
Whispovirus, 891; see also White spot
   syndrome virus
White spot baculo virus (WSBV), 891
White spot disease (WSD), 891
White spot syndrome virus (WSSV),
   20, 699, 705
   biology, 892
   classification, 891–892
   clinical features, 893
   detection procedures
      conventional PCR, 894
      real-time PCR, 894–895
   diagnosis, 893–894
   epidemiology, 892–893
   genome organization, 892
   geographic distribution, 893

host range, 892
morphology, 892
sample preparation methods, 894
transmission, 892–893
treatment and prevention, 894
White tailed deer (WTD), 629
Whitewater Arroyo virus (WWAV), 569–570
Winter dysentery (WD), 318–320
Wobbly possum disease virus (WPDV),
   287–288
World Organization for Animal Health, 372,
   377, 620
World Rabies Day, 364
WPDV, see Wobbly possum disease virus
WSSV, see White spot syndrome virus
WTD, see White tailed deer
WWAV, see Whitewater Arroyo virus

Y

Yellow head complex viruses (YHV), including
   gill-associated virus (GAV)
   biology, 296–298
   classification, 296
   clinical features, 298–299
   diagnosis
      conventional techniques, 300
      molecular techniques, 301
   epidemiology, 296–298
   genome organization, 296
   methods
      detection procedures, 301–303
      sample preparation, 301
   morphology, 296–298
   nucleocapsid of, 296, 298
   pathogenesis, 298–299
   penaeid culture, 295
   shrimp disease, 295
   tissue distribution of, 299
Yellow head virus (YHV), 20, 705

Z

Zombie shrimp, 595
Zoonotic potential, 241, 244
Zoonotic viruses, 543
Zygodontomys brevicauda, 573